HARGRAVE'S
COMMUNICATIONS DICTIONARY

HARGRAVE'S
COMMUNICATIONS DICTIONARY

Frank Hargrave

**IEEE
PRESS**

The Institute of Electrical and Electronics Engineers, Inc., New York

This book and other books may be purchased at a discount
from the publisher when ordered in bulk quantities. Contact:

IEEE Press Marketing
Attn: Special Sales
445 Hoes Lane
P.O. Box 1331
Piscataway, NJ 08855-1331
Fax: +1 732 981 9334

For more information about IEEE Press products, visit the
IEEE Online Store & Catalog: http://www.ieee.org/store.

Printed in the United States of America.

10   9   8   7   6   5   4   3   2   1

ISBN 0-7803-6020-6
IEEE Order No. PC5869

Library of Congress Cataloging-in-Publication Data

Hargrave, Frank.
      Hargrave's communications dictionary.
      / Frank Hargrave.
          p.   cm.
      Includes index.
      ISBN 0-7803-6020-6
          1. Telecommunication—Dictionaries. I. Title.
   TK5102.H37 2000
   621.382′03—dc21   00-061416

I would like to dedicate this book to a number of people.
First to my wife, Penny, and to my parents;
and also to all those who took the time
to share their knowledge with me.

Contents

Preface

When I started in electronics, I was told to read everything I could get my hands on and to note on 3″ × 5″ index cards information, tricks and ideas that might be useful for future projects. It turns out that I did read a lot of the technical literature, and I did save much of the information in notebooks. However, the information was never easily accessible; I still had to search through many notebooks to find what I was looking for. (My bookshelves contain over 200 linear feet.) With the advent of the Apple personal computer, I started saving design equations, circuit analyses, and simulations on floppy diskettes. This also made information retrieval difficult as the Apple][I owned did not have a hard disk drive, thus requiring me to store floppy diskettes in notebooks and necessitating a difficult search to find the desired material. Eventually, I acquired a large PC capable of storing essentially all of my collection online to ease the search process.

This book is a collection of many of the terms, acronyms, jargon, charts, equations, and other information I have gathered for more than 30 years as an engineer. The information centers on my exposure to the communications discipline, although many peripheral disciplines that overlap are also included—for example, computer terminology, Internet terms and jargon, fax/modem hardware, optics, and more.

When I started writing this book, it was for my own consumption (it was not to be a book at all, just a personal reference), so I arranged it the way I would like a reference to be arranged. For example, the entries are sorted into the order I expected to find them; the basic order is symbols first (!, $, *, -, ., /, \), numerics next, and then alphabetics last. Finally, spaces are NOT ignored. I sorted numeric entries into strict numerical sequence order; that is, 1 is followed by 2, 3, . . . 9, 10, 11, ... If this sort order is not to your liking, then the index may be—it is sorted in the traditional manner (well, at least the order that Microsoft® has chosen for the world to use).

Frank Hargrave

Symbols

🗀 The 🗀 symbol indicates that the value or condition is frequently one of the values capable of being stored in NVRAM using the *AT&Wn* command if the modem is equipped with NVRAM.

↕ The value preceding this symbol may vary between manufacturers and models. For example, the default value in the S7 register for the SupraFAXModem V.32 bis is 60 seconds, while the default value for DallasFAX 14.4E is 50 seconds and the value for the Intel 144/144e FaxModem is 30 seconds. Therefore, the default value is listed with the ↕ symbol.

<xxx> The "<" and ">" symbols around a name or label indicate either a single key on the keyboard or a single ASCII character. The keyboard key stroke may represent an ASCII character; for example, <CR> is a carriage return or enter (ASCII 013), and <ESC> is the escape key (ASCII 027) or special keyboard control functions such as <F1>, <ALT>, or <CTL>. Some ASCII characters have no equivalent keyboard key and are represented as <name> where "name" is one of the 34 defined control codes; for example, <NUL> is the ASCII 000, <NAK> is an ASCII 021, and <ETB> is the ASCII 023 character. See also *ASCII* for a complete list of the 34 control codes.

[xxx] The "[" and "]" symbols around a name, command, or variable indicate that the function is optional. For example, the Hayes modem command to cause a modem to go on-hook is "*ATH0*"; however, the "0" is optional. Hence the command may be entered as "*ATH*". To denote this optional parameter designation, the command is written as "*ATH[0]*".

! **(1)** See *factorial*. **(2)** See *bang*.

$ See *hexadecimal*.

</> **(1)** In Internet and some operating systems, such as UNIX, the forward *slash* symbol "/" (ASCII 047) is a separator between a directory, subdirectory, and a file name. At the end of a line, it indicates the item is a directory (or subdirectory), not a file. This is the same usage as the backslash symbol <\> (ASCII 092) directory separator in DOS. **(2)** In other operating systems (such as DOS, OS/2, and their derivatives) a *slash* may be used to indicate command line switches or options. For example, the DOS command to list the files in a directory and to stop each time the screen is filled is *DIR /P*. Here the "*/P*" is the option switch indicating that the listing is to pause and await operator action with each displayed screen. See also *Internet, gopher*, and <\>.

<\> In DOS, OS/2, and other operating systems, the *backslash* symbol "\" (ASCII 092) after a name indicates the name is a directory (or subdirectory), not a file. The first "\" indicates the root directory of a particular disk drive.

For example, \dir_1\dir_2\data, indicates the file called "data" is located in a subdirectory named "dir_2" which itself is in a directory called "dir_1" under the root (or top level) directory.

√ See *square root*.

⇒ In mathematics, the symbol meaning *implies*. E.g., A = B and B = C ⇒ A = C.

Σ See *summation*.

μ See *SI* and *mu*.

μ-Law See *mu-law*.

FLOWCHART SYMBOLS

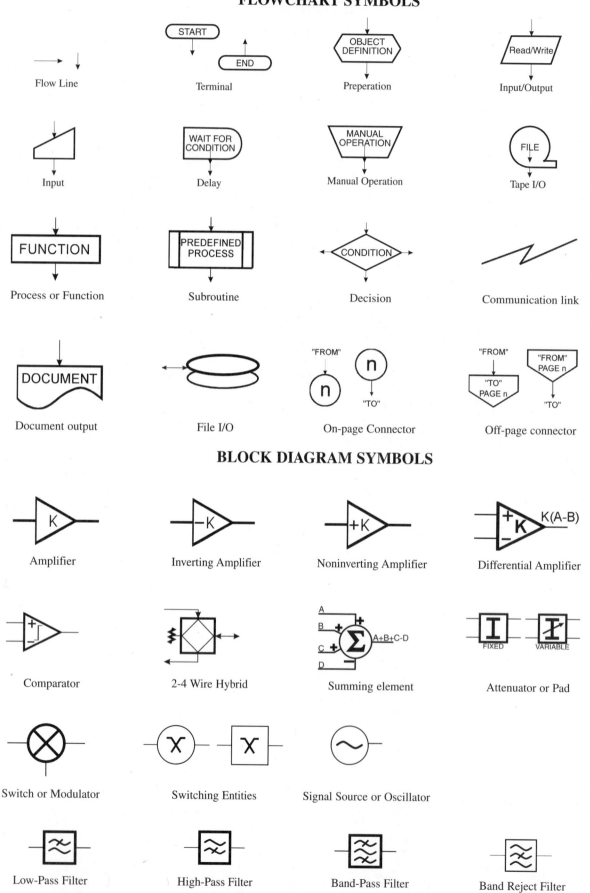

Flow Line

Terminal

Preperation

Input/Output

Input

Delay

Manual Operation

Tape I/O

Process or Function

Subroutine

Decision

Communication link

Document output

File I/O

On-page Connector

Off-page connector

BLOCK DIAGRAM SYMBOLS

Amplifier

Inverting Amplifier

Noninverting Amplifier

Differential Amplifier

Comparator

2-4 Wire Hybrid

Summing element

Attenuator or Pad

Switch or Modulator

Switching Entities

Signal Source or Oscillator

Low-Pass Filter

High-Pass Filter

Band-Pass Filter

Band Reject Filter

BASIC CIRCUIT/SCHEMATIC SYMBOLS

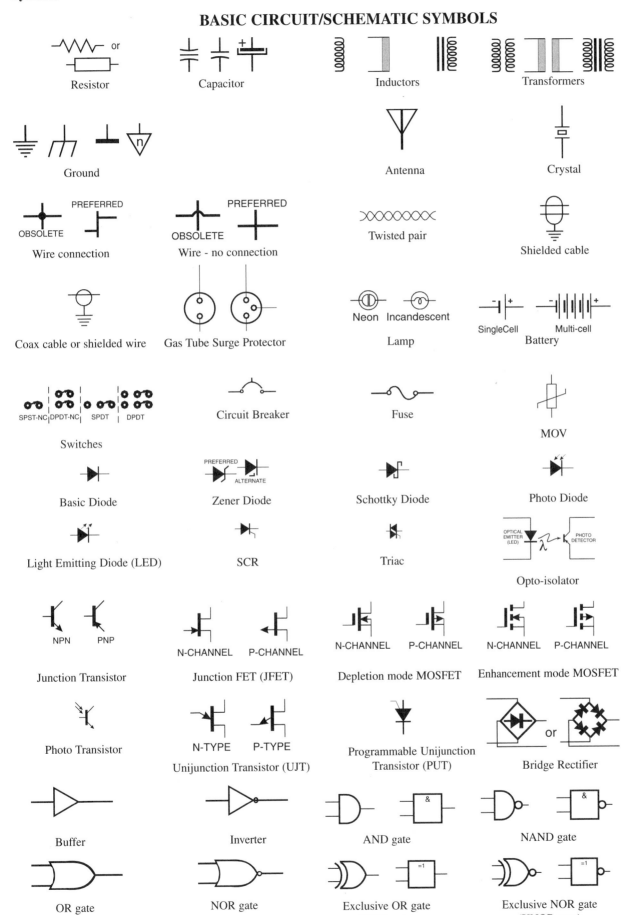

Resistor   Capacitor   Inductors   Transformers

Ground   Antenna   Crystal

Wire connection   Wire - no connection   Twisted pair   Shielded cable

Coax cable or shielded wire   Gas Tube Surge Protector   Neon   Incandescent Lamp   SingleCell   Multi-cell Battery

Switches   Circuit Breaker   Fuse   MOV

Basic Diode   Zener Diode   Schottky Diode   Photo Diode

Light Emitting Diode (LED)   SCR   Triac   Opto-isolator

Junction Transistor   Junction FET (JFET)   Depletion mode MOSFET   Enhancement mode MOSFET

Photo Transistor   Unijunction Transistor (UJT)   Programmable Unijunction Transistor (PUT)   Bridge Rectifier

Buffer   Inverter   AND gate   NAND gate

OR gate   NOR gate   Exclusive OR gate (XOR gate)   Exclusive NOR gate (XNOR gate)

0xnn One way to represent the hexadecimal number "*nn.*" See *hexadecimal*.

1+ call See *one plus call*.

1BASE5 An IEEE 802.3 local area network (LAN) based on a baseband transmission rate of 1 megabit per second over an unshielded twisted pair (UTP). One physical implementation is the AT&T StarLAN standard. The maximum cable segment length is 500 meters. See *IEEE 802.x standard* and *StarLAN*.

2-wire circuit See *two-wire circuit*.

2B+D The ISDN 144-kb/s basic access data rate channel consisting of two 64-kb/s "B" channels for voice or other high-speed information and a 16-kb/s "D" channel for data and control. See also *basic rate access, ISDN,* and *primary rate access*.

2B1Q A *multilevel line coding* technique used for transmission on the *ISDN U interface* wherein 2 bits (2B) are encoded into one quaternary (1Q) symbol.

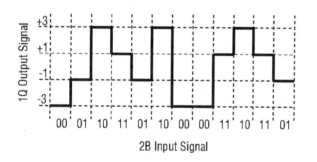

The sample waveform above illustrates the multilevel characteristic of the *2B1Q* encoding technique that follows the encoding algorithm: 00 = –3, 01 = –1, 10 = +3, and 11 = +1.

2-D An acronym from 2 Dimension. The description of points in a system of two variables. Examples include:

- In a physical system, a *2-D* position defines the location in an area (a plane or surface) such as the position of a pixel on a monitor (expressed as an X–Y location) or the location on the surface of the Earth (expressed as a latitude and longitude, azimuth and range, etc.).
- In communications, signals may be described in terms of frequency and amplitude, frequency and phase, and so on.
- In mathematics and engineering, two parts, the real and imaginary parts, frequently describe numbers.

2 dRMS In Global Positioning Systems (GPSs), *2 dRMS* describes distance error within which 95% of the calculated position solutions will fall. See also *Global Positioning System (GPS)*.

2W An abbreviation of 2 Wire.

3+, 3+Open 3COM Corporation's early networking systems, based on parts of the Microsoft/IBM PC LAN and Xerox XNS protocols. Neither the *3+* nor the *3+Open* is supported.

3-D An acronym from 3 Dimension. The description of points or a location in a system of three variables. In a physical system the 3-D position defines a location in a volume. The location may be expressed in either a Cartesian or spherical coordinate system. For example, a location on the Earth may be expressed in terms of latitude, longitude, and altitude or azimuth, range, and elevation. See also *2-D*.

4B3T A multilevel *line coding* method, which encodes four binary bits into one of three ternary states for transmission across the *ISDN U interface*.

4B3T CODE

4B	3T*		4B	3T*	
	+	–		+	–
0000	0 – +	0 – +	1000	0 + –	0 + –
0001	– + 0	– + 0	1001	+ – 0	+ – 0
0010	– 0 +	– 0 +	1010	+ 0 –	+ 0 –
0011	+ – +	– + –	1011	+ 0 0	– 0 0
0100	0 + +	0 – –	1100	+ 0 +	– 0 –
0101	0 + 0	0 – 0	1101	+ + 0	– – 0
0110	0 0 +	0 0 –	1110	+ + –	– – +
0111	– + +	+ – –	1111	+ + +	– – –

*Use the "+" column if the current sum of the pulses transmitted is negative; otherwise use the "–" column.

The 4B3T encoded waveform below demonstrates that only three transitions are required in the same time required for four binary symbols. See also *code* and *waveform codes*.

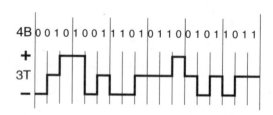

4B/5B encoding A data translation method that prepares the digital data for subsequent waveform encoding in Fiber Distributed Data Interface (FDDI) networks. With *4B/5B encoding*, every group of four data bits is re-coded into a 5-bit symbol. This symbol is then further encoded using standard waveform encoding methods, usually nonreturn to zero inverted (NRZI).

This preprocessing makes the subsequent encoding 80% efficient. That is, with *4B/5B* encoding, a 100-Mbps transmission rate is achieved with a clock speed of only 125-MHz. The Manchester signal-encoding method used in Ethernet (and other types of networks) is only 50% efficient. That is, to achieve a 10-Mbps rate with Manchester encoding, a 20-MHz clock is required. See also *code* and *waveform codes*.

5B/6B encoding A data translation method that prepares the digital data for subsequent waveform encoding in *100BaseVG* networks. In *5B/6B encoding*, each group of 5 data bits is re-coded into a 6-bit symbol. This symbol is then further encoded using standard waveform encoding methods, usually nonreturn to zero inverted (NRZI). See also *code* and *waveform codes*.

8B/10B encoding A data translation method that re-codes 8 data bits into 10-bit symbols. *8B/10B encoding* is used, for example, in IBM's Systems Network Architecture (SNA) networks and in the FC-1 Fiber Channel physical sublayer. *8B/10B* is related to 4B/5B encoding. See also *code* and *waveform codes*.

4-wire circuit See *four-wire circuit*.

5ESS An abbreviation of number 5̲ E̲lectronic S̲witching S̲ystem. An AT&T digital switching system used in number 5 end offices.

7, E, 1 An abbreviation of 7̲ data bits, E̲ven parity, and 1̲ stop bit. Used by CompuServe and some other online services but otherwise an infrequently used communications protocol standard.

8, N, 1 An abbreviation of 8̲ data bits, N̲o parity, and 1̲ stop bit. Used by most bulletin board systems (BBSs) and online services. Modem software refers to the *8, N, 1* setting as a port setting, communication parameter, or sometimes as the data word format.

9.6 kbaud See *communications rate*.

10-n See *10 codes*.

10 codes Numbered codes representing often-used phrases frequently used on mobile voice communications networks to minimize the amount of time a transmitter must be on to convey a message.

10-Net A 1-Mbps baseband CSMA/CD peer-to-peer LAN that runs on a single twisted pair. Fox Research developed *10-Net*.

10Base2 See *IEEE 802.x* standard and *Ethernet*.

10Base5 See *IEEE 802.x* standard and *Ethernet*.

10BaseF See *IEEE 801.x* standard and *Ethernet*.

10BaseT See *IEEE 802.x* standard and *Ethernet*.

10Broad36 See *IEEE 802.x* standard and *Ethernet*.

14.4 kbaud See *communications rate*.

21-type repeater In telephony, a two-wire repeater in which one amplifier serves to amplify the signal in both directions.

22-type repeater In telephony, a two-wire repeater in which two amplifiers are used to amplify the signals, one for the signal in one direction and one in the opposite direction.

23B+D The ISDN primary rate access consisting of 23 64-kbps "B" channels (voice or high-speed data channels) and one 16-kbps data/control channel. The 23B+D structure is designed around the 1.544-Mbps T-1 communication standards.

28.8 kbaud See *communications rate*.

33.6 kbaud See *communications rate*.

100BaseT Also called *Fast Ethernet*. See *IEEE 802.3* standard.

100VG-AnyLAN See *IEEE 802.12* standard.

100BaseX See *IEEE 802.3* standard.

119 The Japanese emergency services number, much like the U.S. 911 service.

193rd Bit A framing bit in a T-1 channel that is attached to every group of 192 bits. These 192 bits represent a single byte from each of the 24 channels multiplexed in a T-1 line.

SOME COMMON "10 CODES"

Code	Definitions	Code	Definitions
10-1	Receiving poorly.	10-33	Emergency traffice, this station.
10-2	Receiving good (and signal strength, if requested).	10-34	Trouble at this station, need help.
10-3	Stop transmitting (off frequency, over modulating, etc.).	10-35	Confidential information.
10-4	OK, afirmative, acknowledge.	10-36	What time is it?
10-5	Relay message.	10-39	Your message is delivered.
10-6	Busy, please stand by.	10-41	Please move to channel _____.
10-7	Station out of service, leaving the air.	10-42	Accident at (location).
10-8	Station in service.	10-43	Traffic tie-up at (location).
10-9	Please repeat, bad reception.	10-44	I have a message for you.
10-10	Transmission completed.	10-54	Accident.
10-11	Talking too rapidly.	10-55	Wrecker or tow truck needed at (location).
10-12	Officials or visitors present.	10-56	Ambulance needed at (location).
10-13	Advise weather and road conditions.	10-60	What is the next message number?
10-16	Make a pickup at _____.	10-62	Unable to copy, use the land line (telephone).
10-17	Urgent business.	10-65	Clear for message.
10-18	Anything for us?	10-68	Repeat message.
10-19	Nothing for you - return to base.	10-70	Fire at (location).
10-20	What is your location?	10-71	Proceed with transmission in sequence.
10-21	Call _____ by telephone.	10-77	Negative contact.
10-22	Report to _____, in person.	10-91	Talk closer to microphone.
10-23	Stand by.	10-92	Transmitter is out of adjustment.
10-24	Completed last assignment.	10-93	Please check my current transmitter frequency.
10-25	Do you have contact with _____?	10-94	Test intermittently with *normal* modulation for (condition).
10-27	Moving to channel _____.	10-95	Test with *no* modulation.
10-28	Please identify your station.	10-99	Unable to receive your signal.
		10-100	Nature calls (toilet stop).

800 number In the United States, a station number in which the called party pays all fees associated with a caller making a connection.

802.x See *IEEE 802.x standards.*

900 number In the United States, a station number in which the *calling station* is charged for information or services rendered by the called party (including the telephone connection fee). The charges may be on a charge-per-call or charge-per-time basis. All charges are collected by the telephone company. The phone company then transfers the collected fee for service to the *called station owner.*

For example, a software company can be called for help on its product where it charges for the assistance.

911 In the United States, an emergency reporting system in which a caller uses the same telephone number, *911*, to report any emergency situation requiring the assistance of fire, police, ambulance, and so on.

All *911* calls are routed to a common location where the caller is asked a series of questions to determine the nature of the emergency. The personnel at this location will then dispatch an appropriate response team to the emergency location. Some *911* locations have an enhanced capability (called *E 911*) wherein the caller's number and location are automatically determined and displayed.

999 Great Britain's emergency services number, similar to the U.S. 911 service.

8250, A, B One of the oldest and most basic universal asynchronous receiver transmitter (UART) integrated circuits. It supports 4, 5, 6, 7, or 8-bit data transfers; 1, 1.5, or 2 stop bits; and odd, even, or no parity bit insertion at data rates to 9600 bps (although not all combinations are possible).

Hardware "bugs" in the original version were fixed in the "-A" release of the design. But because the IBM PC and -XT BIOS ROMs were designed to expect at least one of these bugs, the "-A" version does not work in these machines. The "-B" version fixed bugs of the previous two versions and put the "bug" that the PC and XT expected back in (a spurious interrupt generated at the end of an access). The higher performance 16550 UART replaced the older 8250.

16450 The successor universal asynchronous receiver transmitter (UART) to the 8250 series part. The 16450 is essentially the same as the 8250 except its interface to the processor is faster and there is a one byte general-purpose scratch buffer.

16550 A universal asynchronous receiver transmitter (UART) with all of the capabilities of the 8250 or 16450 devices plus 16-byte input and output first in first out (FIFO) signal buffering (and additional registers to control the FIFO). This buffering allows more efficient data transfers with computer software; hence, a higher end-to-end transfer rate can be achieved.

When the processor transmits, 1, 4, 8, or 16 bytes of data are loaded into the transmit FIFO. The processor is then free to operate on another task while the *16550* serially shifts the FIFO contents to the DCE. When the FIFO is empty, an interrupt is generated by the UART and the processor loads another block of data into the FIFO. Receive data is handled similarly; that is, the receive FIFO accumulates 1, 4, 8, or 16 bytes of information and then generates an interrupt to the processor. The processor unloads the received block of bytes and then proceeds with its internal tasks while the UART accumulates another data block. Because of the buffers, the interrupt rate to the processor is lower with the *16550* than other UARTs *at the same communications rate.* This means either that the processor has more time for other tasks or that a higher data rate is possible.

INTERRUPT RATE (MS)
(AVAILABLE TIME WITHOUT DATA LOSS)

Transfer Rate (bps)	Basic 1 Byte	16550 Buffer Size		
		4 Bytes	8 Bytes	16 Bytes
1200	8.3333	33.333	66.667	133.333
2400	4.1667	16.667	33.333	66.667
9600	1.0417	4.167	8.333	16.667
14400	0.6944	2.778	5.556	11.111
28800	0.3472	1.389	2.778	5.556
56600	0.1736	0.694	1.389	2.778
115200	0.0681	0.347	0.694	1.389

A

Å The symbol for *Angstrom*.

a The SI prefix for *atto* or 10^{-18}.

A The symbol for *ampere*.

A & B signaling A procedure used in most T-1 carrier systems in which one bit is robbed from each of the 24 channels in every sixth frame for control information. The signaling reduces the available bandwidth from 1.544 Mbps to 1.536 Mbps.

A-B switch In telephony, a feature of many cellular telephones which allows the user to select which of two service providers (*wireline* or *non-wireline*) will provide a carrier while *roaming*.

A-B switch box An external switching entity used to select among multiple devices, such as printers, faxes, and modems. The switch may select a specific peripheral (e.g., a fax/modem, a color printer, or a laser printer), or it may select which of several computers are attached to a single shared device (e.g., modem or printer). Also called an *A/B switch box* or *AB box*.

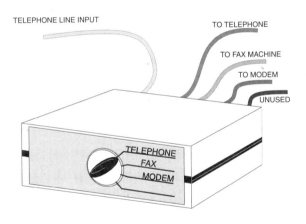

A-condition In start-stop teletypewriter systems, the significant condition of the signal code element immediately preceding a character or block signal. It prepares the receiving equipment for the reception of the code elements. See also *start bit*.

A-D converter (also A/D converter) An abbreviation of <u>A</u>nalog to <u>D</u>igital Converter.

A–law The ITU-T telephony standard for digitizing voice in a nonlinear manner. The analog signal is sampled 8000 times per second and converted to an 8-bit nonlinear word that approximates a logarithmic curve.

The equation describing the actual logarithmic curve is

$$D(a) = Sgn(a) \frac{1 + \log_{10}(A \cdot |a|)}{1 + \log_{10}(A)} \qquad \frac{1}{A} \le |a| \le 1$$

$$D(a) = Sgn(a) \frac{A \cdot |a|}{1 + \log_{10}(A)} \qquad 0 \le |a| \le \frac{1}{A}$$

where

a	is the analog signal,
Sgn(a)	is the sign (polarity) of the signal *a*,
D(a)	is the coded digital equivalent of the signal, and
A	is the factor defining the amount of amplitude compression (87.6 for the ITU-T A-law).

The region between zero and plus full scale (as well as the region between zero and minus full scale) is divided into seven segments (or chords), numbered 1–7. Each segment contains 16 digital values (4 bits) distributed equally (linearly) across the range of the segment. Each segment requires twice as much analog signal change to cause a bit change in the digital word as the next lower segment. Because each segment of the approximation is linear, the approximation is called a *piecewise linear* approximation. The accompanying chart details the binary values of the *A-law* process for each segment.

A LAW ENCODER/DECODER DETAILS

Segment Number	Interval Count & Spacing	Encoder End Points	Binary Code * 1234 5678	Decoder Value
7	16 Δ 128	4096	1111 1111	4032
		2176	1111 0000	2112
6	16 Δ 64	2048	1110 1111	2016
		1086	1110 0000	1056
5	16 Δ 32	1024	1101 1111	1008
		544	1101 0000	528
4	16 Δ 16	512	1100 1111	504
		272	1100 0000	264
3	16 Δ 8	256	1011 1111	252
		136	1011 0000	132
2	16 Δ 4	128	1010 1111	126
		68	1010 0000	66
1	32 Δ 2	64	1001 1111	63
		0	1000 0000	1
1	32 Δ 2	0	0000 0000	−1
		−64	0001 1111	−63
2	16 Δ 4	−68	0010 0000	−66
		−128	0010 1111	−126
3	16 Δ 8	−136	0011 0000	−132
		−256	0011 1111	−252
4	16 Δ 16	−272	0100 0000	−264
		−512	0100 1111	−504
5	16 Δ 32	−544	0101 0000	−528
		−1024	0101 1111	−1008
6	16 Δ 64	−1086	0110 0000	−1056
		−2048	0110 1111	−2016
7	16 Δ 128	−2176	0111 0000	−2112
		−4096	0111 1111	−4032

*Value before inversion of even bit positions.

Bit 1 is the sign bit, bits 2–4 are the segment bits, and bits 5–8 are the weighting bits. Because the positive and negative "ones segments" are collinear, there are only 13 segments (not 14 as expected). In the diagram below, both the analog and digital signals are scaled to 100% of full scale (maximum value); the maximum digital word is +127, which corresponds to a +3.14 dBm0 1020 Hz sine wave.

7

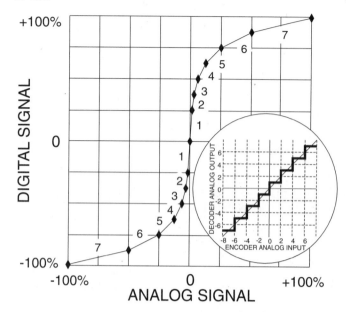

Details of the zero crossing region of segment one are illustrated in the inset. Note that the digital value is NEVER zero for the analog range of −2 to +2. This is in contrast to μ-*law* where the value zero does exist. See also *compander* and *mu-law*.

A.n recommendations An ITU-T series of specifications concerning organization of the work at ITU-T. See Appendix E for a complete list.

A/N An abbreviation of AlphaNumeric.

A/UX A 32-bit Macintosh computer operating system based on UNIX.

A2Mbps™ An Artisoft network adapter that communicates with the network over two pairs at a 2-Mbps rate.

AA **(1)** An abbreviation of Automated Attendant. **(2)** An abbreviation of Auto Answer. **(3)** An abbreviation of Application Association.

AABS An abbreviation of Automated Attendant Billing System. A feature in telephony which allows both *collect* and *bill* to *third-party* calls to be placed by a caller *without* the assistance of an operator.

AAC-n An abbreviation of ATM Access Concentrator for T-1/E-1 (or T-3/E-3). The system from Kentrox uses cell multiplexing to concentrate multiple protocols onto a T-1/E-1 (or T-3/E-3) access facility to the ATM or SMDS network.

AAL An abbreviation of ATM Adaptation Layer.

AALx An abbreviation of ATM Adaptation Layer protocol x (x = 1, 2, 3, 4, or 5).

AAMOF An abbreviation of As A Matter Of Fact.

AAR An abbreviation of Automatic Alternate Routing.

AARNet An acronym from Australian Academic and Research Network.

AARP An abbreviation of AppleTalk Address Resolution Protocol.

AAUI An abbreviation of Apple Attachment Unit Interface. A type of connector used in some Apple Ethernet networks.

AB An abbreviation of ABort session.

AB Box A slang term for a switching device that allows one port on a computer to communicate with any of several peripheral devices or, conversely, any of several computers to communicate with one peripheral (a printer for example). See also *A-B switch box.*

ABAM Western Electric's designation of 22 AWG, 110-Ohm, plastic insulated, twisted pair cable.

abandoned call In telephony, the condition where the calling station returns to the on-hook state after a connection is established but before the called station goes off-hook.

abbreviated dialing In telephony, a feature in a switching system allowing a user to place a call by dialing fewer digits than are required by the number plan. *Abbreviated dialing* may also be implemented within the subscriber's terminating equipment. Here, the *abbreviated number* is translated into a normal network destination number. Also called *speed dialing.*

ABM An abbreviation of Asynchronous Balanced Mode.

ABNT An acronym from Associação Brasileira de Normas Tecnicas. The national standards-setting agency of Brazil.

abort The premature ending of an operation, program, or data transmission. An *abort* may be caused by system error or by operator action. On mainframes the term *Abend* (for Abnormal end) may be used.

ABR **(1)** An abbreviation of Automatic Baud Rate detection. **(2)** An abbreviation of Available Bit Rate.

abr. An abbreviation of ABbReviation.

abrasion The wearing of one surface because of rubbing with another surface.

ABS **(1)** An abbreviation of Alternate Billing Services. **(2)** An abbreviation of Average Busy Season. **(3)** An abbreviation of Acrylonitrile-Butadiene-Styrene. A type of semi-rigid plastic.

ABSBH An abbreviation of Average Busy Season Busy Hour.

absent subscriber service A service offered by the local telephone company to subscribers who will be away from their phone for an extended period of time. When implemented, calls to the subscriber's number are intercepted by an operator or a machine, which delivers a message to the caller. At the termination of the service, the intercept is discontinued and the subscriber's number operates normally.

absolute delay The time interval or phase difference between transmission and reception of a signal.

absolute error The magnitude of the error without regard of the algebraic sign and expressed in the same units as the quantity containing the error.

absolute value The numeral part of a value without regard to the sign. It answers the question "How many?," not "How much greater than or less than." The *absolute value* of a number "x" is written $|x|$. Some examples are:

$$|3|=3; \quad |-4|=4; \quad |-x|=|x|=x$$

absorption loss The loss of transmitted energy due to irreversible conversion of that energy into another energy form (and ultimately into heat). Examples include eddy current losses and optical loss due to impurities in the fiber.

Abstract Syntax Notation One (ASN.1) One element of the *presentation layer* (*layer 6*) of the OSI protocol suite which defines *data type* within a computer database. The real benefit is to shared databases that are accessed by many clients—each with a different program. Here, *ASN.1* would coordinate and identify the *data type* to each program trying to access the same data. See also *data type.*

ABT (1) An abbreviation of Answer Back Tone. **(2)** An abbreviation of ABort Timer.

ac An abbreviation of Alternating Current.

AC (1) An abbreviation of Access Control. **(2)** An abbreviation of Access Charge.

ac adapter The external power supply for devices such as modems and portable computers that converts the high-voltage ac available at the wall outlet (mains ac) into a lower voltage usable by the device. The output voltage supplied by the ac adapter is generally less than 24 V and may be either ac or dc depending on the device to be powered.

ac/dc ringing The most common way of signaling a subscriber telephone. The central office generates two signals toward the subscriber telephone: an alternating current (ac) component and a direct current (dc) component. The ac component (also called the *ring signal*) is used to operate the ringer of the subscriber telephone, while the dc component is used to sense when the subscriber goes off-hook. See also *BORSCHT.*

Academic Computing Research Facility Network (ACRFNET) A network interconnecting U.S. research facilities, such as colleges, research laboratories, and development laboratories.

ACBH An abbreviation of Average Consistent Busy Hour.

ACC An abbreviation of Automated Callback Calling.

acceptable use policy (AUP) A set of rules governing what kind of activities and traffic are allowed on a specified bulletin board system (BBS) or network.

For example, the National Science Foundation Network (NSFNET) policy statement expressly states that the purpose of the network is to provide research communications in and among U.S. research and educational institutions; further, business participation is to be restricted to nonproprietary uses which are in the interest of open scientific collaboration. Another example is CIX (Commercial Internet Exchange), which provides essentially an open, *AUP*-free alternative to the NSFNET backbone. CIX does not restrict commercial use but encourages it.

acceptance angle The angle over which an optical device will accept light (at a specific wavelength). The angle is usually defined as one-half the vertex angle formed by the cone within which useful coupling can occur (coupling into bound modes). The following diagram illustrates the *acceptance angle* for an optical fiber—light entering at an angle greater than the *acceptance angle* is lost in the cladding or reflected out of the device (unbound mode). See also *index of refraction* and *optical fiber.*

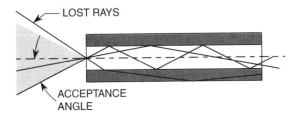

acceptance cone In fiber optics, the three-dimensional analog of an *acceptance angle.* Revolving the *acceptance angle* 360° about the center of the fiber's core as the cone's point generates the cone.

access The means of gaining entry to a system or network so that its resources may be utilized. *Access* (on a computer system) may be divided into two general categories, *direct connection* and *indirect connection.*

- With a *direct connection,* one need only provide appropriate user identification from the user equipment to gain *access* to the system resources.
- With an *indirect connection,* one must establish a communication link from the user equipment to the resource-providing equipment before proceeding with the user identification process and subsequent access to system resources.

Access.bus A high-speed serial input/output communications architecture. The technology, originally developed by Signetics, allows a user to connect more than 100 devices (such as keyboards, mice, trackballs, modems, and LAN interfaces) to a single 100-kbps serial port. Without dip switches or jumpers, individual devices identify themselves to the host and may be connected while the system is running ("hot-plug in"). See also *IEEE P1394.*

access code (1) A unique combination of letters, numbers, and/or symbols used as an identification for gaining admittance to a remote entity (a computer or network for example). On a computer network or an online (dial-up) service, the access code is usually referred to as an *ID code* or a *password.* **(2)** The digits, preceding the called station digits, dialed by a user to select a particular outgoing trunk group or line. On a PABX, for example, one dials a "9" to gain access to the local numbers, an "8" for long-distance numbers, and a two-digit code such as "77" for access to a WATS line. See also *equal access.*

access control (AC) (1) A form of network security used to define the level of access (or restriction) an individual, program, or remote host has to the component parts of a system as a whole. For example, on a computer system, most users are not given access to the payroll database; however, they are all generally given access to the e-mail server. See also *firewall.* **(2)** A feature of most communications systems which allow (or deny) access to certain elements of the communications system. For example, a PABX might block access to all "900" numbers. **(3)** A byte in a token ring network packet that contains the token indicator and frame priority.

Access Control Decision Function (ACDF) A function in open systems that uses various types of information, such as access control information (ACI), and guidelines to decide whether to grant users access to resources in a particular situation.

Access Control Enforcement Function (ACEF) A function in open systems that enforces the decision made by the Access Control Decision Function (ACDF).

Access Control Information (ACI) In ITU's X.500 directory services model, any information used in controlling access to a file or directory.

access control list (ACL) A file that lists the host resources to which a user (an individual, program, or remote host) may gain access. The *ACL* provides a type of computer network security that allows listed access to only certain types of the server's resources. Other users are denied all access. Access for a listed user may be, for example, no access or read only access.

Common access rights may be issued to a group of *Individual Accounts* and/or *Wildcard Accounts* via the *Access Control List Group Account.* See also *individual account* and *wildcard account.*

access customer name abbreviation (ACNA) The three-character abbreviation listed in the Local Exchange Routing Guide (LERG) for each interexchange carrier (IC, IEC, IXC) referenced.

access line **(1)** That portion of a leased line that connects the user with the local central office or wire center. **(2)** The connection from the customer's equipment to the local telephone company for access to the telephone network. The line is assigned a number unique to a regional area. Also called *local loop.*

access method The set of rules by which networks arbitrate (regulate) right of data transmission. Examples include Ethernet's CSMA/CD (listen for availability before transmit) and a token ring's token passing. Also called *access protocol.*

access network A network attached to a backbone network. The connection generally requires either a gateway or a router, depending on the types of networks that comprise the backbone network.

access number **(1)** A number used to establish a connection between nodes of a network. Some examples include:

- The telephone number a person uses to call another person or an online information service.
- The number dialed to gain access to a long distance service provider. *Access code* is usually the preferred term. See also *equal access.*
- The telephone number used by a subscriber to gain access to an online service such as a bulletin board system (BBS) or an Internet service provider (ISP).
- The number used by local data communications equipment (DCE) to gain access to the services of a remote DCE.

(2) A number that allows a user to access specific capabilities of a resource provider. Also called an *access code* or sometimes a *password.*

access privileges See *access rights.*

access protocol The set of procedures specified *between a user and a network* which enable the user to utilize the services and/or facilities of that network. See also *communications protocol.*

access rights A list of privileges that determine what a user can and cannot do with specific network files. *Access rights* form an integral part of network system security; for example, access to the payroll files can be limited to only a few individuals in the accounting department while allowing everyone access to the company newsletter.

Although each network operating system's *access rights* details are different, typical privileges include:

C	User can **create** new files.
D	User can **delete files**.
E	User can **execute** program files.
K	User can **delete subdirectories**.
L	User can **list** file names.
M	User can **make** new subdirectories.
N	User can **rename** files.
R	User can **read** files.
S	User has **supervisory** privileges.
W	User can read, modify, and **write** files.

Access rights can apply to directories or files. They also generally apply to the subdirectories of a specified directory unless a specific exception is made. Also called *access privileges.*

access site A host system that provides connectivity to resources via a telephone connection.

account **(1)** A contract between a user and a service provider wherein the service provider agrees to provide access to certain resources and the user agrees to abide by the acceptable use policy (AUP) (and probably pay a stated fee). **(2)** A bulletin board system (BBS) or network database entry that contains information about a user. Some information stored in the *account database* includes:

- User's name and ID.
- User's password.
- User's security level (defines what information and operations are available).
- Billing rates.
- Time of day access authorization.

Without an *account,* access to the BBS or service provider is generally denied or severely limited.

account policy A set of rules that determines whether a particular user is allowed system access and what resources the user may use.

Accounting Management One of the five Open Systems Interconnection (OSI) network management domains defined by the International Organization for Standardization (ISO) and International Telecommunications Union (ITU). This domain is concerned with the administration of network usage, costs, charges, and access to various resources.

Accunet A family of AT&T-provided data-oriented digital services.

- *Accunet packet service*—a packet switching digital network.
- *Accunet reserved 1.5 service*—a general-purpose digital service using dual 1.544-Mbps land and satellite links.
- *Accunet spectrum of digital services* (*ASDS*)—a 56-kbps digital leased line.

accuracy **(1)** A measure of the *conformity* of a value, measurement, or calculation to its definition or to a standard reference. A measure of how close a value represents or approximates the "true" value.

Accuracy IS NOT the same as *precision,* rather, *accuracy* is *how correct* a value is, while *precision* is an expression of *detail.* For example, a dart in a target 6.315926 cm from the center is more *precise* than a dart at 1.2 cm from the center; however, the latter is more *accurate* (assuming the true target location is zero). See also *precision.* **(2)** The quality of *freedom of error* or mistake. See also *uncertainty.*

ACD An abbreviation of <u>A</u>utomatic <u>C</u>all <u>D</u>istributor (or <u>D</u>istribution) or <u>A</u>daptive <u>C</u>all <u>D</u>istributor.

ACDF An abbreviation of <u>A</u>ccess <u>C</u>ontrol <u>D</u>ecision <u>F</u>unction.

ACE An abbreviation of <u>A</u>dverse <u>C</u>hannel <u>E</u>nhancement. A modem adaptation method that allows the modem to compensate for noisy lines. See *MNP-10*.

ACEF An abbreviation of <u>A</u>ccess <u>C</u>ontrol <u>E</u>nforcement <u>F</u>unction.

ACF An abbreviation of <u>A</u>dvanced <u>C</u>ommunications <u>F</u>unction. A software suite from IBM which enables its computers to communicate.

ACF/NCP An abbreviation of <u>A</u>dvanced <u>C</u>ommunications <u>F</u>unction/<u>N</u>etwork <u>C</u>ontrol <u>P</u>rogram. The control software running on a communications controller that supports the operation of IBM's Systems Network Architecture (SNA) backbone network.

ACF/TCAM An abbreviation of <u>A</u>dvanced <u>C</u>ommunications <u>F</u>unction/<u>Tele</u><u>C</u>ommunications <u>A</u>ccess <u>M</u>ethod.

ACF/VTAM An abbreviation of <u>A</u>dvanced <u>C</u>ommunications <u>F</u>unction/<u>V</u>irtual <u>T</u>erminal <u>A</u>ccess <u>M</u>ethod. The control software on host-based IBM's Systems Network Architecture (SNA) network that allows the host to communicate with terminals on the network.

ACFG An abbreviation of <u>A</u>utomatic <u>C</u>on<u>Fi</u><u>G</u>uration. A feature of the plug-and-play (PnP) BIOS extension.

ACI An abbreviation of <u>A</u>ccess <u>C</u>ontrol <u>I</u>nformation.

ACIS An abbreviation of <u>A</u>utomatic <u>C</u>ustomer <u>I</u>dentification <u>S</u>ervice. A feature of advanced *automatic call distributors* (*ACD*) which captures address information such as the *DID* digits or *caller ID* information for routing or automatic response messaging. See also *ANI, caller ID, calling party ID,* and *DID*.

ACK (1) An acronym from <u>ACK</u>nowledge. (2) A control message sent by the remote DCE to the local DCE to indicate that a data block just received is okay. The ASCII character *<ACK>* (06h) is frequently the total content of the message.

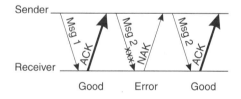

The signal flow diagram above shows two message blocks being sent from a sender to a receiver. Message block 1 from the sender was received without error; therefore, the receiver sent an *ACK* to the sender. The first attempted transmission of the second block is received in error. The receiver sends a *NAK* to the sender which then resends block 2. The second attempt to send message block 2 is received without error, so an *ACK* is the response of the receiver. See also *flow control* and *NAK*.

ACK0 The BiSync acknowledgment for even-numbered messages.

ACK1 The BiSync acknowledgment for odd-numbered messages.

acknowledge An input/output (I/O) signal that indicates that a task has been performed and that the hardware (or software) is ready to perform another task. The I/O signal may be either electronic or message based. The acknowledge signal may indicate that a device has received an information block and is ready for the next block, or it may indicate the completion of a complex process. See also *ACK*.

ACL An abbreviation of <u>A</u>ccess <u>C</u>ontrol <u>L</u>ist.

ACL Group Account See *access control list* (*ACL*).

acoustic coupler A type of *modem* that is designed to use a standard telephone handset as the interface to the telephone line. This is generally accomplished by placing the telephone handset in close proximity to a speaker and microphone pair on the modem. The modem signals are then coupled to the telephone line via audible tones.

ACP An abbreviation of <u>A</u>ccess <u>C</u>ontrol <u>P</u>oint. Sometimes shown as A & CP.

ACR (1) An abbreviation of <u>A</u>bandon <u>C</u>all and <u>R</u>etry. (2) An abbreviation of <u>A</u>utomatic <u>C</u>all <u>R</u>ecording. (3) An abbreviation of <u>A</u>llowed <u>C</u>ell <u>R</u>ate.

ACRFNET An abbreviation of <u>A</u>cademic <u>C</u>omputing <u>R</u>esearch <u>F</u>acility <u>NET</u>work.

acronym A word made up from letters within a series of words. Although not required, the abbreviation should be pronounceable. The letters of the abbreviation are generally the first or most important letters of the phrase; for example, MODEM is taken from <u>MO</u>dulator <u>DEM</u>odulator, and the acronym DOS is derived from <u>D</u>isk <u>O</u>perating <u>S</u>ystem.

Although *acronyms* are useful and serve a legitimate purpose, they are definitely overused. The communications and computer industries probably hold the record for the <u>A</u>cronym <u>O</u>veruse <u>S</u>yndrome (AOS). Some *acronyms* are used so often that they become words in their own right, for example, RADAR (from <u>RA</u>dio <u>D</u>etection <u>A</u>nd <u>R</u>anging) or SPOOL (allegedly from <u>S</u>imultaneous <u>P</u>eripheral <u>O</u>peration <u>O</u>n <u>L</u>ine).

ACS (1) An abbreviation of <u>A</u>synchronous <u>C</u>ommunications <u>S</u>erver. (2) An abbreviation of <u>A</u>ccess <u>C</u>ontrol <u>S</u>tore. (3) An abbreviation of <u>A</u>daptive <u>C</u>ontrol <u>S</u>ystem.

ACSNET An abbreviation of <u>A</u>ustralian <u>C</u>omputer <u>S</u>cience <u>NET</u>work.

active component A *component* or *device* that requires a source of energy in addition to an information input signal(s) for its operation. Generally, an active component delivers more energy to its output than is available at the input. Examples include:

- LEDs, transistors, transmitters, and receivers.
- An amplifier that adds energy to increase the signal level.
- An interface device that injects a current onto a circuit (possibly to power a remote device).

See also *passive component*.

active filter A network of active devices (such as op-amps) and passive components (resistors, capacitors, and inductors) arranged so that it:

- Passes some frequencies and rejects others,
- Performs a phase shift on one signal relative to another signal, or
- Delays one signal relative to another.

See also *all-pass filter, band-pass filter, band-reject filter, high-pass filter,* and *low-pass filter*.

active high A term that designates a digital signal that must go (or be) high to produce an effect. Also called *positive true*.

active laser medium Within a laser, the material that emits radiation or exhibits gain as the result of electron transitions to a lower energy state, from a higher energy state (to which it had been previously stimulated).

active low A term that designates a digital signal that must go (or be) low to produce an effect. Also called *negative true*.

active network A network or circuit that contains at least one source of power. It may contain both active and passive components.

active star See *multiport repeater* and *star coupler.*

ACTLU An abbreviation of <u>ACT</u>ivate <u>L</u>ogical <u>U</u>nit.

ACTPU An abbreviation of <u>ACT</u>ivate <u>P</u>hysical <u>U</u>nit.

ACTS (1) An acronym from <u>A</u>dvanced <u>C</u>ommunications <u>T</u>echnology <u>S</u>atellite. (2) In telephony. An acronym from <u>A</u>utomatic <u>C</u>oin <u>T</u>elephone <u>S</u>ervice.

ACU (1) An abbreviation of <u>A</u>larm <u>C</u>ontrol <u>U</u>nit. (2) An abbreviation of <u>A</u>utomatic <u>C</u>alling <u>U</u>nit.

AD An abbreviation of <u>A</u>ctivity <u>D</u>iscard.

Ada® The official, high-level computer language of the U.S. Department of Defense (DoD) for embedded computer, real-time applications as defined in MIL-STD-1815.

ADA An abbreviation of <u>A</u>ctivity <u>D</u>iscard <u>A</u>cknowledge.

ADAPSO An acronym from <u>A</u>ssociation of <u>DA</u>ta <u>P</u>rocessing <u>S</u>ervice <u>O</u>rganizations.

adapter (1) *Physical.* A device that provides a compatible connection between parts that do not naturally mate. Examples include:

- Computer to serial interface cable connections (i.e., DB-25 to DB-9)
- BNC coaxial cable termination to RJ-11 twisted pair cable termination.

(2) *Electrical.* Any means that enables signals of one device to communicate with another dissimilar device. Examples include:

- The connection of a logic signal (a 0 V or 5 V signal) with an RS-232 signal (a signal defined in two ranges, that is, −15 V <signal<−3 V and +3 V<signal<+15 V).
- The printed circuit board in a computer that allows communication with a network–network interface card (NIC).

adapter card An electronic device designed to be plugged into one of the computer's adapter slots.

Examples of adapter cards include internal modems and network interface devices.

adaptive answer A modem feature in some fax/modems that allows the modem hardware to determine if an incoming call is a facsimile call or a data call. Software that supports this feature will connect the received data stream from the modem to the appropriate computer program. (Data calls are routed to a modem data program, while facsimile calls are routed to the fax-receiving program.) Note that both hardware and software must be supportive.

The accompanying drawing demonstrates the state diagram of the decision process during an *adaptive answer.* The remaining portion of the diagram is the same as that of a dedicated device. See also *silent answer.*

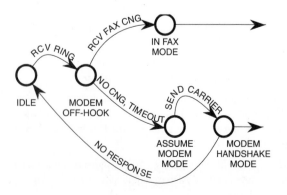

adaptive channel allocation In communications systems, a channel assignment scheme in which the information-handling capacities of active channels are not predetermined but are assigned on demand. This is usually accomplished by means of a multiplexing scheme.

adaptive dialing A feature that allows a local data communications equipment (DCE) to attempt to dial a remote DCE using dual tone multifrequency (DTMF) dialing and then pulse dialing if DTMF fails. The DCE procedure is

1. Wait for dial tone.
2. Send the first digit to the switched network using DTMF signaling.
3. Determine if the dial tone is removed.
4a. If the dial tone is absent, proceed with DTMF, or
4b. If the dial tone remains present, revert to pulse dialing and send the entire dial string (including the first digit).

adaptive differential PCM (ADPCM) A form of pulse code modulation (PCM) in which the generated code word is an incremental value rather than the actual value. The code word is generated by delta sigma modulator means with one additional feature—the magnitude of the code word is adjusted based on the number of immediately preceding corrections to the estimated value with the same sign.

A number of algorithms exist for changing the step size. Some allow the size to be changed multiple times in the same direction; others, such as that shown in the figure, allow only one additional magnitude in each direction. The purpose of changing the step size is to enable the encoding of a higher *slew rate* signal (with some loss of resolution during periods of high slew rate).

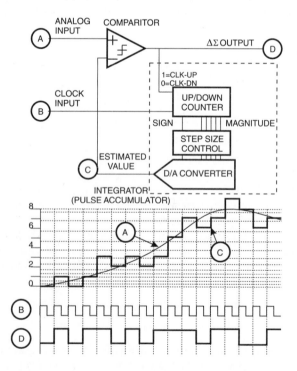

At the receiving station, the bit stream "D" may be used to generate the approximation "C" of the original input signal "A" using an adaptive accumulator analogous to the encoder accumulator. Or a low-pass filter may be used as the D/A converter, which would give the waveform results labeled "E."

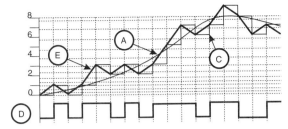

ADPCM is an ITU standard technique (G.726) for voice encoding and compression. It allows analog speech to be carried within a 32-Kbit/s digital channel. See also *delta modulator* and *PCM*.

adaptive equalization The technique for compensating for the effect of different transmission characteristics (generally called distortions) on transported signals. Adaptive equalization has two significant characteristics:

- The process is automatically accomplished while traffic is being transmitted, and
- One or more signal characteristics are dynamically adjusted to compensate for changing transmission path characteristics.

Equalization is usually accomplished by one of two methods, i.e.:

- Inserting a circuit or device in the signal path that has characteristics opposite to the error characteristics introduced by the transmission link.
- Inserting a circuit or device in the signal path that subtracts an artificially created replica of the error signal generated in the real transmission link.

In either case, *adaptation* may occur either on a call-by-call or a semicontinuous basis.

adaptive predictive coding (APC) A narrowband analog-to-digital conversion technique in which the value of the digital signal at each sampling instant is predicted according to a linear function of the past values of the quantized signals. The technique may use either a one-level or multilevel sampling system.

APC like *linear predictive coding* (*LPC*) uses adaptive predictors; however, because *APC* uses fewer coefficients, it requires a higher sampling rate.

adaptive routing Routing that automatically adjusts to network changes, such as changes of traffic patterns, circuit (channel) availability, or element failures. Adaptation is based on historical information from the traffic carried.

adaptive system A system that accommodates or adjusts to changes within its environment in such a way as to optimize its performance. Such a system has:

- A means of monitoring one or more of its own performance characteristics,
- A means of monitoring its operating environment, and
- A means of varying its own parameters in response to this monitoring (called closed-loop control) in order to minimize the effects of changing operating conditions.

Also called *training*.

ADB An abbreviation of <u>A</u>pple <u>D</u>esktop <u>B</u>us. A low-speed serial bus used on Apple Macintosh computers to connect input devices (such as the mouse or keyboard) to the CPU.

ADC (**1**) An abbreviation of <u>A</u>nalog to <u>D</u>igital <u>C</u>onverter (or Conversion). (**2**) A British telecom abbreviation of <u>A</u>ction for <u>D</u>isabled <u>C</u>ustomers. (**3**) An abbreviation of <u>A</u>daptive <u>D</u>ata <u>C</u>ompression. (**4**) An abbreviation of <u>A</u>synchronous <u>D</u>ata <u>C</u>hannel. (**5**) An abbreviation of <u>A</u>synchronous <u>D</u>igital <u>C</u>ombiner.

ADCC An abbreviation of <u>A</u>synchronous <u>D</u>ata <u>C</u>ommunications <u>C</u>hannel.

ADCCP An abbreviation of <u>A</u>dvanced <u>D</u>ata <u>C</u>ommunications <u>C</u>ontrol <u>P</u>rocedure—the ANSI X3.66 specification.

additive white Gaussian noise (AWGN) A model of thermal and shot noise characterized by a flat spectrum (white), a Gaussian amplitude probability, and generally by a power spectral density No/2 in the transmission band. The benefit of this noise model is that the noise contribution in a transmission system can be simply added to the signal. That is:

$$R(t) = S(t) + n(t)$$

where

$R(t)$ is the received signal
$S(t)$ is the transmitted signal
$n(t)$ is the noise

address (**1**) In computers, the coded representation of the location in a memory device where data are stored. (**2**) In communications, a unique identifier assigned to networks, stations, and/or hardware so that each may be individually activated for either control or communication.

Examples include:

- The telephone network: Each subscriber has a unique station number according to a numbering plan. The number includes country code, city or area code, station number, and extension as required.
- The Internet: Each host has a unique IP address.

See also *NIC address*.

address depletion The situation in which more nodes on a network are desired than can be given a unique address.

address resolution The process by which the addresses of nodes or clients on one local area network (LAN) are translated or mapped into clients, physical addresses, or addresses of another network. Translations of incoming datagrams are handled by the Address Resolution Protocol (ARP), while outgoing datagrams are handled by the Reverse Address Resolution Protocol (RARP).

Address Resolution Protocol (ARP) In Internet, a part of the TCP/IP protocol suite that translates *IP addresses* to the physical locations of devices on a network. *ARP* dynamically identifies and maps new node low-level physical hardware addresses (MAC addresses) on the LAN to high-level Internet address (Data Link Control address—DLC), without the need for a network administrator manually updating translation tables.

ARP only operates across a single physical network and is limited to networks supporting hardware broadcast.

ADJ In Internet, the Boolean operator used by WAIS to indicate that the two words on either side of the *ADJ* tag must be <u>*ADJacent*</u> to each other in found documents.

adjacent channel The frequency band immediately above or below a channel of interest.

For example, a frequency division multiplex (FDM) communication path.

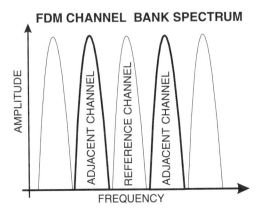

administrative domain   See *autonomous system* (*AS*).

administrator   The person responsible for setting up a network, maintaining its functionality, registering new users, their access rights, and passwords, and so on.

admittance (Y)   The reciprocal of *impedance.* The lack of opposition to the flow of an ac current in a circuit. The units of admittance are *siemens* (formerly called *mhos*) with the symbol of "*S.*"

ADP   An abbreviation of <u>A</u>utomatic <u>D</u>ata <u>P</u>rocessing

ADPCM   An abbreviation of <u>A</u>daptive <u>D</u>ifferential <u>P</u>ulse <u>C</u>ode <u>M</u>odulation.

ADSL   An abbreviation of <u>A</u>symmetric <u>D</u>igital <u>S</u>ubscriber <u>L</u>ine (or <u>L</u>oop).

ADSP   An abbreviation of <u>A</u>ppleTalk <u>D</u>ata <u>S</u>tream <u>P</u>rotocol.

ADU   An abbreviation of <u>A</u>utomatic <u>D</u>ialing <u>U</u>nit.

advertising   A dynamic and automated process used in distributed routing networks for disseminating information about possible routes a packet of data can take to arrive at its desired destination. Routes change because of the addition or deletion of nodes over time.

The process (called *convergence*) involves each node broadcasting (*advertising*) current information from its routing table to all adjacent nodes. These adjacent nodes incorporate appropriate routing information into their routing table and then broadcast the updated information to all nodes directly connected (adjacent nodes). This process continues periodically to develop and maintain an accurate map of the network topology.

advertorial   A contraction of <u>ADVER</u>tisement and edi<u>TORIAL</u>. An advertorial is an advertisement with editorial content.

AE   An abbreviation of <u>A</u>ctivity <u>E</u>nd.

AE-Series adapter   An Artisoft Ethernet adapter that is configured for the network through the use of jumpers. Adapters include the AE-1/T, AE-2, AE-2/T, and AE-3.

AEA   An abbreviation of <u>A</u>ctivity <u>E</u>nd <u>A</u>cknowledge.

AEC   An abbreviation of <u>A</u>daptive <u>E</u>cho <u>C</u>ancellation.

AEP   An abbreviation of <u>A</u>ppleTalk <u>E</u>cho <u>P</u>rotocol.

AESS   An abbreviation of <u>A</u>utomatic <u>E</u>lectronic <u>S</u>witching <u>S</u>ystem.

AF   An abbreviation of <u>A</u>udio <u>F</u>requency.

AFAIK   An abbreviation of <u>A</u>s <u>F</u>ar <u>A</u>s <u>I</u> <u>K</u>now.

AFC   **(1)** An abbreviation of <u>A</u>nalog to <u>F</u>requency <u>C</u>onverter. **(2)** An abbreviation of <u>A</u>utomatic <u>F</u>requency <u>C</u>ontrol. **(3)** An abbreviation of <u>A</u>udio <u>F</u>requency <u>C</u>hoke (inductor). **(4)** An abbreviation of <u>An</u>tenna <u>F</u>or <u>C</u>ommunications.

afc or a.f.c.   An abbreviation of <u>A</u>utomatic <u>F</u>requency <u>C</u>ontrol.

AFK   An abbreviation of <u>A</u>way <u>F</u>rom <u>K</u>eyboard.

AFP   An abbreviation of <u>A</u>ppleTalk <u>F</u>iling <u>P</u>rotocol—the presentation level protocol managing remote file access.

AFRP   An abbreviation of <u>A</u>RCNET <u>FR</u>agmentation <u>P</u>rotocol.

AFS   An abbreviation of <u>A</u>ndrew <u>F</u>ile <u>S</u>ystem.

agc or a.g.c.   An abbreviation of <u>A</u>utomatic <u>G</u>ain <u>C</u>ontrol.

agent   In a client–server network model, that part of a network system which performs a specific task, such as information preparation and exchange, on behalf of a client (or server) application.

A program that accepts complex computer commands, each of which may require the completion of several steps. For example, a user may instruct a *software agent* to search several databases for certain information and format the results in a particular way. Another example is the Simple Network Mail Protocol (SNMP) agent running on a router that provides the ability for the router to exchange information with an SNMP network management system through the use of the SNMP protocol. Contrast with *daemon.*

aggregate bandwidth   The total bandwidth of a channel carrying a multiplexed bit stream.

Ah or Ahr   The symbol for <u>A</u>mp <u>H</u>ou<u>R</u>.

AI   An abbreviation of <u>A</u>rtificial <u>I</u>ntelligence.

AIC   An abbreviation of <u>A</u>utomatic <u>I</u>ntercept <u>C</u>enter.

AIG   **(1)** An abbreviation of <u>A</u>ddress <u>I</u>ndicator <u>G</u>roup. **(2)** An abbreviation of <u>A</u>ddress <u>I</u>ndicating <u>G</u>roup.

AILANBIO.EXE   The LANtastic program that implements NET-BIOS.

AIM   An abbreviation of <u>A</u>mplitude <u>I</u>ntensity <u>M</u>odulation. See *intensity modulation.*

AIOD   An abbreviation of <u>A</u>utomatic <u>I</u>dentified <u>O</u>utward <u>D</u>ialing.

AIOD leads   Terminal equipment leads used solely to transmit *automatic identified outward dialing (AIOD)* data from a PABX to the *public switched telephone network* (*PSTN*) or to switched service networks (e.g., EPSCS). It enables a vendor to provide a detailed monthly bill identifying long-distance usage by individual PABX stations, tie trunks, or the attendant.

aircraft emergency frequency   An international aeronautical emergency frequency for:

- Aircraft stations and stations concerned with safety and regulation of flight along national or international civil air routes and
- Maritime mobile service stations authorized to communicate for safety purposes.

Two such frequencies are 121.5 MHz (civil) and 243.0 MHz (military).

airtime   The actual link connect time (the time available for talking) on a cellular telephone. Most carriers bill customers based on how many minutes of *airtime* they use each month.

AIS   **(1)** An abbreviation of <u>A</u>larm <u>I</u>ndication <u>S</u>ignal. **(2)** An abbreviation of <u>A</u>utomated <u>I</u>nformation <u>S</u>ystem. **(3)** An abbreviation of <u>A</u>utomatic <u>I</u>ntercept <u>S</u>ystem.

AIX®   IBM's version of AT&T's UNIX.

AJ   An abbreviation of <u>A</u>nti-<u>J</u>amming.

AKA   An abbreviation of <u>A</u>lso <u>K</u>nown <u>A</u>s.

ALAP   An acronym from <u>A</u>ppleTalk <u>L</u>ink <u>A</u>ccess <u>P</u>rotocol. The link access layer protocol (data link layer protocol) manages packet transmissions on LocalTalk.

alarm   The *condition* or *state* of a device, circuit, or system indicating a performance degradation or failure. *Alarm,* however, is often used where alert should be used. See also *alert.*

alarm center   A facility that receives local and remote alarms. Usually, the alarm center is in a technical control facility.

Alarm Indication Signal (AIS) A signal to downstream equipment that there is a problem upstream.

For example, T-carrier indicates the condition with all ones (used as a keep-alive signal); and SONET defines four *AIS* categories: DSn AIS, Line AIS, STS Path AIS, and VT Path AIS. Also called a *blue alarm.*

alarm indicator A device that responds to a signal from an *alarm sensor.* Examples of *alarm indicators* include bells, buzzers, gongs, horns, lamps, and messages on a computer screen.

alarm sensor **(1)** In electronic systems, any device that can sense an abnormal condition within the system and provide a signal indicating the presence, magnitude, or nature of the abnormality to an alarm indicator (in either a local or remote location). **(2)** In physical security systems, a device used to indicate a change in the status of the physical environment of a facility.

alarm signal A signal intended to attract the attention of a person in the event of a predefined change of condition or status of equipment or the failure of equipment to correctly respond to a stimulus signal. Alarms may be audible, visual, or both. Also called an *alert signal.*

ALBO An abbreviation of Automatic Line BuildOut.

ALC **(1)** An abbreviation of Advanced Line Circuit. **(2)** An abbreviation of Adaptive Logic Circuit. **(3)** An abbreviation of Antenna Loading Coil. **(4)** An abbreviation of Automatic Load Control.

ALCU An abbreviation of Asynchronous Line Control Unit.

ALD **(1)** An abbreviation of Analog Line Driver. **(2)** An abbreviation of Asynchronous Line Driver.

ALE An abbreviation of Automatic Link Establishment.

alert An indication of a *change* of state from a normal to an alarm condition (or to a normal state from an alarm state). *Alert,* however, is often used where alarm should be used. See also *alarm.*

ALF **(1)** An abbreviation of Automatic Line Feed. **(2)** An abbreviation of Absorption Limiting Frequency.

algorithm A detailed *procedure* for solving a problem that involves a finite number of identified steps. An algorithm can be simple or complex, but MUST yield a solution in a finite number of steps. See also *heuristic.*

ALI An abbreviation of Asynchronous Line Interface.

alias **(1)** An alternate name used in place of the official name. Used to invoke or identify a command, person, network host, a specific group, and so on. **(2)** Frequently used on bulletin board systems as a nickname so as to provide anonymity.

aliasing The "folding" or "mapping" of frequencies higher than a sampling frequency into the frequency band below the sampling frequency.

In the following example, the input signal is shifted from $f_1 = 1/10$ of the sampling clock to $f_2 = 1.1$ times the sampling clock at sampling time 10. Note that the sampled output "looks" the same in both instances. In effect, the higher frequency (f_2) was folded or *aliased* to the lower frequency (f_1). See also *Nyquist interval* and *Shannon limit.*

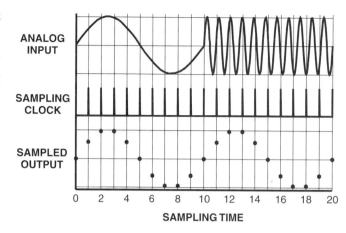

ALICE An acronym from Adaptive LIne Canceller and Enhancer.

aligned bundle A bundle of optical fibers in which the relative position of each fiber is the same at the two ends of the bundle.

Bundles such as these are used for the transmission of images. The top straight-through bundle is direct reading when direct viewed, while the bottom crossed-bundle provides a mirror image. Also called a *coherent bundle.*

alignment The process of adjusting multiple controls of a system for proper interrelationship. For example,

- *Alignment* may refer to the adjustment of tuned circuits in an amplifier so as to achieve a desired frequency response, or
- *Alignment* may refer to the synchronization of subsystems or circuits of a system.

alignment error **(1)** In a network, an error in which a packet has extra bits; that is, the packet does not end on a byte-boundary. Hence, it will have an invalid cyclic redundancy check (CRC) value. An alignment error may be caused by a faulty component, such as a damaged network interface card (NIC), transceiver, or cable. **(2)** In wireless communications systems, an alignment error refers to misdirecting the radiator (RF antenna or optical emitter—LED, LASER, etc.).

ALIT An acronym from Automated Line Insulation Tester.

alkaline cell Strictly, any primary cell that uses any alkaline electrolyte. Examples include the Edison cell, a nickel cadmium cell, or a carbon-zinc cell with potassium hydroxide as the electrolyte. Common usage refers to the carbon-zinc cell. See also *battery* and *cell.*

all call A general broadcast message on a communications channel that does not request or require responses and does not designate any specific receiver addresses. (All receivers capable of message reception on the channel will receive the *all call* message.) Examples of *all call* messages include:

- On radio networks—emergency transmissions (e.g., aircraft mayday transmission on 121.5 MHz),
- Weather bulletins, and
- Network routing messages.

all glass fiber A synonym for *all-silica fiber.*

all-pass filter A circuit or network designed to introduce only phase shift and delay to a signal and not significantly change the amplitude at any frequency in the range of interest. See also *band-pass filter, band reject filter, high-pass filter,* and *low-pass filter.*

all-pass function A circuit or transmission facility that affects only the phase shift and delay characteristics of a signal and does not significantly change the amplitude.

all-silica fiber An optical fiber composed of a silica-based core and cladding. (The presence of additional layers, such as a protective polymer overcoat or a tight buffer does not disqualify a fiber as an *all-silica fiber.*) Also called an *all-glass fiber.*

all trunks busy (ATB) In telephony, a switching equipment condition in which all trunks (paths) in a given trunk group are busy or unavailable. It is indicated by a *fast busy* call progress tone.

allocation unit The basic disk storage unit of DOS-based computers. It is comprised of one or more disk sectors, generally 1, 2, 4, 8, 16, or 32 sectors, depending on the disk's capacity. DOS always allocates disk space in full *allocation units;* that is, no fractional *allocation units* are allowed. (If one byte of information is to be saved on a disk, one full *allocation unit* will be assigned to hold it.)

With early DOS systems (3.0 and earlier), 12 bits were used to address each *allocation unit.* This limited the maximum number of *allocation units* to 4078. Later systems (through DOS 7.×) used 16-bit addressing, limiting this maximum to 65 518. Still later versions use 32-bit addressing. This limit, along with sector size (generally 512 bits) and the size of the disk (or disk partition), determine the number of bytes in an *allocation unit.* Hence, small disks have small *allocation units* and large disks have large *allocation units.* As an example, the maximum disk size supported by an *allocation unit* size of 4K is about 256 M; that is,

$$4096 \frac{\text{bytes}}{\text{cluster}} \times 65518 \text{ cluster} = 268361728 \text{ bytes}$$

$$\approx 255.9 \text{ MB}$$

(Remember 1 MB is 2^{20} B = 1048576 B.)

Also called a *cluster.*

APPROXIMATE ALLOCATION UNIT BOUNDARIES

Disk Size		Allocation Unit Size bytes (sectors)	
Minimum	Maximum	16 bit addr	32 bit addr
0	<32 M	512 (1)	
32 M	<64 M	1024 (2)	
64 M	<128 M	2048 (4)	
128 M	<256 M	4096 (8)	
256 M	<512 M	8192 (16)	
512 M	<1024 M	16384 (32)	4096 (8)
1024 M	<2048 M	32768 (64)	4096 (8)
2048 M	<8 G		4096 (8)
8 G	<16 G		8192 (16)
16 G	<32 G		16384 (32)
32 G	<64 G		32768 (64)

almanac A table of orbital parameters (ephemeris) that allows a Global Positioning System (GPS) receiver to forecast the position and velocity of a GPS satellite. The *almanac* contains information about all satellites in the constellation. A GPS receiver uses the almanac for two purposes:

1. To determine which satellites are visible from the presumed location, and
2. To aid in the acquisition of GPS satellite signals.

The following table illustrates the information available in subframes 4 and 5 of the GPS data bit frame. This information is available from all GPS satellites on a periodic basis and is used to update the GPS receiver tables for acquisition and position calculations. See also *Global Positioning System (GPS).*

TYPICAL ALMANAC INFORMATION

Parameter	Symbol	Units	Value		
ID of satellite	SV	#	1	...	31
Health		#	0	...	0
Reference week of almanac		#	52	...	52
Eccentricity	e		4.942893982E−3	...	9.578227997E−3
Time of Applicability	TOA	s	2.334720000E+5	...	2.334720000E+5
Orbital Inclination	I_0	rad	9.615867138E−1	...	9.488894343E−1
Rate of Right Ascension	$\dot{\Omega}$	rad/s	−7.931759249E−9	...	−8.126052720E−9
Square root of semimajor axis	$\sqrt{a}$	$m^{1/2}$	5.153627441E+3	...	5.153687988E+3
Right Ascension at TOA	Ω_0	rad	2.441418409E+0	...	−7.457513809E−1
Argument of Perigee	Ω	rad	−1.723656654E+0	...	8.336231112E−1
Mean Anomoly	M_0	rad	−5.556287766E−1	...	5.710135102E−1
Af0	Af_0	s	1.411437988E−4	...	8.583068848E−6
Af1	Af_1	s/s	0.000000000E+0	...	0.000000000E+0
Af2	Af_2	s/s/s	0.000000000E+0	...	0.000000000E+0

Aloha Network (ALOHANet) An FM broadband radio LAN developed at the University of Hawaii for interconnecting various islands to the University of Hawaii Computer Center. *ALOHA* was the first LAN to use a form of CSMA/CD based on N. Abramson's 1969 proposed technique.

The *Aloha* system uses a channel in which transmissions occur at random; that is, each station transmits a message packet whenever one is available. Because of the lack of coordination among stations, packets may be transmitted by different stations within the time interval of a packet (an incident called collision). Collided packets of all offending stations are retransmitted after a random delay. (The delay is different for each station.) The maximum usable channel capacity of a *pure ALOHA* protocol is about 18%, as is shown in the figure at the end of this entry. The utilization of the channel can be doubled to about 36% by modifying the channel access algorithm so that packets may be transmitted only within predefined, uniformly spaced time slots. This technique is called *slotted-ALOHA (S-ALOHA).* In both pure and *slotted ALOHA* systems, instabilities arise when large numbers of access attempts are made. Excessive traffic leads to many collisions, and each collision leads to additional access attempts, which increases the total number of collisions further. This runaway process continues until no information throughput is possible. Sometimes also called *random access.* See also *CSMA/CD.*

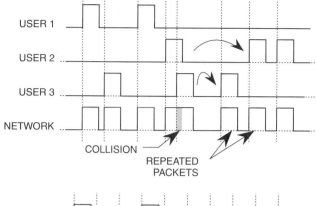

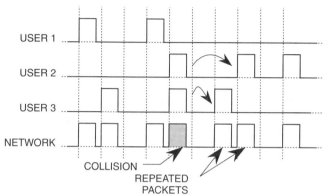

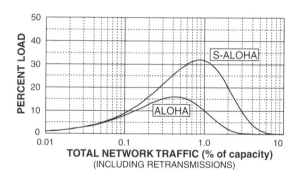

Alphabet transliteration is often necessary in telecommunications systems because of the different alphabets and codes used in different industries (e.g., weather symbols vs. common English) and worldwide language differences. An example of *alphabet transliteration* is the substitution of the Roman letters *a*, *b*, and *u* for the Greek letters α, β, and μ respectively. Formerly called *alphabet translation*.

alphabetic character set A character set that contains letters and may contain control characters, special characters, and the space character but not numeric digits. See also *alphanumeric (A/N)*.

alphabetic code A code according to which data are represented through use of an alphabetic character set.

alphabetic dialing Most telephones have an alphabetic character associated with each numeric digit on the dial. This enables a company to buy a telephone number where the alphabetic information is meaningful and hence, is easier for a customer to remember. For example, in the United States a dentist might try to get the number 866-8437 because it can be made to spell TOOTH DR. This practice is most prevalent in the "800" directory where the *office code* is not restricted to a geographic area.

The numeric digit to alphabetic character mapping is somewhat country-dependent as the following table illustrates.

ALPHABETIC TO NUMERIC MAPPING

Num	UK	USA
1		
2	ABC	ABC
3	DEF	DEF
4	GHI	GHI
5	JKL	JKL
6	MN	MNO
7	PRS	PRS
8	TUV	TUV
9	WXY	WXY
0	OQ	Operator

alpha profile See *power-law index profile*.

alphabet A table of correspondence between an agreed set of symbols and the signal or codes that represent them.

(1) Communications and data processing. The complete set of characters and codes used to represent letters, numbers, punctuation, special symbols, and control codes available. Examples of some alphabets used in communications are the International Alphabet Number 5, the 7-bit *ASCII* code, IBM's *EBCDIC* code, and the Morse code. **(2)** Any set of characters or symbols comprising the letters in a written language. **(3)** An ordered set of all the letters used in a language, including letters with diacritical marks where appropriate but not including punctuation marks.

alphabet translation A deprecated synonym for *alphabet transliteration*.

alphabet transliteration The substitution of the symbols or characters of one alphabet for the corresponding symbols of a different alphabet. Characteristics of the process include:

- The process is executed on a character-by-character substitution basis.
- The process is reversible (lossless).
- No consideration is given to the meaning of the substitute characters or their combinations.

alphabetic string A symbol string consisting solely of letters from the same alphabet. Associated special characters from the same alphabet are also sometimes included.

alphabetic word **(1)** A word consisting solely of letters from the same alphabet. **(2)** A word that consists of letters and associated special characters but not digits.

alphameric See *alphanumeric*.

alphanumeric (A/N) A contraction of ALPHAbetic and NUMERIC pertaining to a character set containing letters, decimal digits, and sometimes special symbols (such as punctuation and math). Also written *alphameric*.

alphanumeric character set A character set that contains letters, numerals, any special characters, punctuation characters, and the space character.

alphanumeric code **(1)** A code derived from an alphanumeric character set. **(2)** A code that, when used, results in a code set that consists of alphanumeric characters.

alphanumeric data Any data represented by symbols from an alphanumeric character set.

ALT One class of *alternative newsgroup hierarchies* on USEnet in which the normal stringent requirements of *world newsgroups* do not apply. *ALT* newsgroups can be established by virtually anyone on the network; consequently, the quality, content, and ethics of each can and does widely vary.

The *ALT newsgroups* cover every possible range of personal interest from archery to zoology, from normal to abnormal to bizarre, and from very serious to ludicrous. A few categories are very interesting, while others are quite distasteful to many viewers. Most have a low *signal-to-noise ratio.* So to paraphrase a famous quote, "let the viewer beware." See also *newsgroup hierarchy.*

altazimuth mount A mounting for directional devices, such as a telescope or directional antenna, in which slewing takes place in either of two directions: i.e., a plane tangent to the surface of the Earth (or other frame of reference) and/or elevation about that plane, i.e., above or below the plane. Also called an *altaz mount* or *x-y mount.*

alternate buffer The second of two buffers used to handle data transfers. The buffers are alternated to achieve continuous throughput. While the input process is filling one, the receiver is reading the other.

alternate mark inversion (AMI) A method of encoding digital data so that it may be injected directly onto a transmission line with a zero average signal-induced dc component. This is accomplished by transmitting successive "marks" ("1s") with equal magnitude but opposite polarities; "spaces" have a transmitted value of 0 volts, so they do not contribute to the dc level.

An *AMI* signal has several consequences:
- The signal will pass through transformers and capacitors with minimal distortion—a benefit.
- A long string of zeros generates no transitions; therefore, clock recovery is not possible—a problem.

To counter this problem, extra pulses are transmitted within the zero string in such a way that they can be detected and removed at the receiving terminal. The process called *bipolar zero suppression* introduces two successive pulses that violate the *AMI* bipolar pulse generation rule (AMI violation). See *BnZS* for more information on zero suppression. This waveform shows a portion of a data stream encoded in an *AMI* format. The average dc level is 0. (There are three marks at +V and three at −V.) Because there are no long zero strings, there are no code violations. *AMI* is the waveform code used on T-carrier lines and ISDN S interfaces. See also *AMI violation, BnZS,* and *waveform codes.*

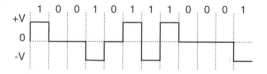

alternate routing A technique enabling communication to continue in the event of failures or congestion in a primary network path. The network design includes multiple paths (alternate routes) which may be used to connect source and destination nodes. Any of the secondary communications paths may be substituted for the primary path if the primary path is unavailable for immediate use.

alternating current (ac or AC) An electric current that reverses its direction of flow on a periodic basis.

The frequency of an *alternating current* is measured in cycles per second and is given the unit name hertz, abbreviated Hz. Some examples of *alternating current* are the ac power in the United States, which has a frequency of 60 Hz (in much of the rest of the world it is 50 Hz); signals that travel on a telephone line which have a frequency range of 300 Hz to 3400 Hz; and radio signals whose frequency can range from tens of kHz to thousands of MHz.

alternative newsgroup hierarchies News categories on USEnet such as *ALT, BIZ,* and *Clari* that bypass the normally strict procedures imposed to set up a new newsgroup.

Because of the lack of control of content on these newsgroups, not all sites carry the alternative newsgroups.

ALU An abbreviation of Arithmetic Logic Unit.

AM (1) An abbreviation of Amplitude Modulation. **(2)** An abbreviation of Accounting Management. **(3)** An abbreviation of Active Monitor.

AMA An abbreviation of Automatic Message Accounting.

AMARC An acronym from Automatic Message Accounting Recording Center.

ambient That which surrounds or encircles a defined region.

ambient light Light present in the environment around a detecting device. The light is considered noise in optical systems.

ambient noise The "all-encompassing" noise associated with a given environment. This noise is generally a composite of several noise sources.

ambient noise level The level of *acoustic noise* existing at a given location, such as in a room, a compartment, or a place out of doors. *Ambient noise level* is measured with a sound level meter and is usually represented in decibels (dB) above the reference pressure level of 20 μPa (in the SI system). In the cgs system, the reference is 0.0002 dyn/cm^2. Sometimes also called *room level noise.*

ambient temperature The temperature of the medium in a designated area, particularly that area surrounding a specified device or system (the temperature of its environment).

ambiguity An uncertainty in the measurement of a parameter in that the measurement is subject to multiple interpretations. For example, the exact number of waves between a transmitter and receiver is generally not known; hence, the exact phase relationship between the transmitted and received waves is unknown. Therefore, a simple phase detector cannot determine if a specific phase is a "1" or a "0."

AME (1) An abbreviation of Asynchronous Modem Eliminator. A null modem. See *null modem.* **(2)** An abbreviation of Amplitude Modulation Equivalent. See *compatible sideband transmission.* **(3)** An abbreviation of Automatic Message Exchange.

America Online (AOL) One of the commercial dial-up information services and an Internet access provider. Some of the services and data available include weather, news, stock quotes, business news, downloadable files and programs, bulletin boards, and conferences.

AMF An abbreviation of Account Metering Function. In the ISO/OSI network management model, the function that keeps track of every user's resource usage.

AMH An abbreviation of Application Message Handling. In the International Standardized Profile (ISP) model, the prefix used to identify Message Handling System (MHS) actions.

AMI An abbreviation of Alternate Mark Inversion.

AMI violation In an alternate mark inversion (AMI) waveform coding scheme, a "mark" that has the same polarity as the previous "mark." Although two consecutive "mark" signal elements of the same polarity generally represent a transmission error, some transmission protocols deliberately insert *AMI violations* to facilitate synchronization, clocking, or signaling. Also called a *bipolar violation.* See also *BnZS* and *waveform codes.*

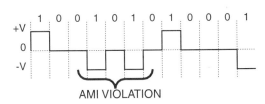

amp An abbreviation of AMPere.

AMP (1) An abbreviation of AMPlifier. (2) An abbreviation of AMPlitude. (3) An acronym from Active Monitor Present. (4) An acronym from Automatic Message Processor. (5) An acronym from Associative Memory Processor.

amp-hour (Ah) A measure of the capacity of a battery (the product of current in amps times the time the battery can deliver the current). The *Ah* rating is approximately a constant for any specific battery type and size. Therefore, a 10-Ah battery could deliver 1 A for 10 hours or 0.1 A for 100 hours. This relationship fails at both very high current delivery rates and very low rates. The shape of the curve depends on the battery chemistry (NiCd, Alkaline, Lithium, NiMH, etc.). One battery's capacity vs. current relationship is shown below.

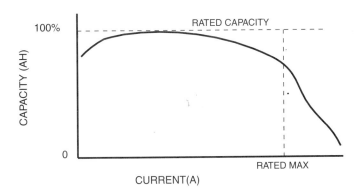

ampere (A) The unit of electric *current,* abbreviated a, A, or amp. One ampere of current is the flow of 1 coulomb of charge past a given point in 1 second. Sometimes abbreviated *amp.*

ampere-turn A unit of magnetomotive force. It is the product of the current (I) flowing in a coil (measured in amps) and the number of turns (N) in the coil. Generally abbreviated *amp turns, AT,* or *NI.*

AMPL An abbreviation of AMPLifier.

amplifier A device used to increase the magnitude or level of an analog signal. The *gain* (the amount the signal is increased) is frequently

expressed with the letter *K. Amplifiers* may be classed in several ways:

- According to the frequency range they are designed to work over; for example,
  - An *audio amplifier* usually has a frequency range of 20 Hz to 20,000 Hz,
  - A *telephonic amplifier* has a frequency range of 300 Hz to 3400 Hz, and
  - An *RF amplifier's* frequency response is generally in the MHz or GHz range.
- According to the type of signal they process; for example,
  - A *single-ended amplifier* amplifies the signal of a single path. A single-ended *amplifier* may invert the signal (180° phase shift) and hence is called an *inverting amplifier.* If no inversion occurs, it is called a *noninverting amplifier.*
  - A *differential amplifier* amplifies the difference between two applied signals.
- According to their basic system function; for example,
  - A *ground isolation amplifier* is designed to provide galvanic isolation between two circuits.
  - A *line driver* is tailored to deliver power to a transmission line.
  - A *signal amplifier* generally cannot deliver more than a few mA to its load.

Note: Not only is the desired signal increased in an amplifier, but also the noise level is increased equally. (That is, the signal-to-noise ratio, SNR, remains essentially constant.) The opposite of *amplification* is *attenuation.* See also *class A amplifier, class B amplifier, class C amplifier, class D amplifier, differential amplifier,* and *single-ended amplifier.*

amplitude A measure of the size of the variation of a signal wave. It may be either the value at an instant of time or a value computed over a period of time. Unless otherwise noted, the *amplitude* is generally taken on the absolute value of the signal.

An example of an instantaneous measurement is the peak or maximum value (the height of the wave). Measurements taken over time could be the average or RMS value.

amplitude distortion One form of distortion occurring in a system, subsystem, or device when the amplitude of signal outputs is not a fixed multiple of the corresponding signal input amplitudes under specified conditions. The deviation is because of a nonlinear relationship between the input and output of the system.

Amplitude distortion is usually measured with the system operating under steady-state conditions and with a sinusoidal input signal. When other frequencies are present, the term *amplitude* refers to the fundamental frequency only. See also *distortion.*

amplitude equalizer A corrective device designed to modify the amplitude characteristics of a circuit or system over a desired frequency range. These devices may be fixed, manually adjusted, or adaptive.

amplitude-frequency distortion Distortion in a transmission system caused by nonuniform attenuation (or gain) in the system with respect to frequency under specified operating conditions. Also called *amplitude-vs.-frequency distortion* and *frequency distortion.*

amplitude frequency response The variation of gain/loss (amplification/attenuation) of a signal traversing a circuit, device, system, or network as a function of frequency in a band of interest. The response is ordinarily expressed in decibels (dB).

The following curve represents the amplitude-frequency response of an entity with a *transfer function* $G(f)$. The response at each frequency across the band is compared to the response at 1000 Hz; consequently, the normalized response at 1000 Hz is 0 dBr. The majority of the displayed response is loss, but in the band 2100–2800 Hz there is gain. At frequencies between 3100 and 3700 Hz, the device loss is greater than 40 dB.

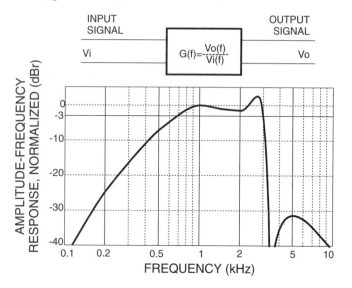

amplitude intensity modulation (AIM) In optical communications (rather than electrical), the modulation method in which the intensity of the light source varies as a function of the signal being transmitted. Analogous to amplitude modulation (AM) in electronic communications.

Although frequently used, the preferred term is *intensity modulation*.

amplitude keying See *amplitude shift key (ASK)*.

amplitude modulation (AM) Modulation in which the amplitude of a carrier wave is varied in accordance with some characteristic of the modulating signal.

Amplitude modulation implies the modulation of a coherent carrier wave with frequency f_0 by mixing it in a nonlinear device with the modulating signal f_m to produce discrete upper and lower sidebands, which are the sum and difference frequencies of the carrier and signal, i.e., $f_0 + f_m$ and $f_0 - f_m$. The envelope of the resultant modulated wave is an analog of the modulating signal. The instantaneous value of the resultant modulated wave is the vector sum of the corresponding instantaneous values of the carrier wave, upper sideband, and lower sideband. In the following figure, a *carrier* is modulated by an analog *information signal* to give an *AM modulated signal*. Recovery of the modulating signal may be by direct detection or by heterodyning.

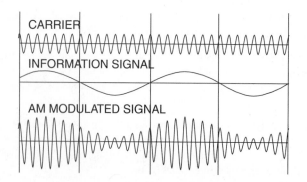

amplitude modulation equivalent (AME) A synonym for *compatible sideband transmission*.

amplitude shift key (ASK) A form of amplitude modulation (AM) in which the modulating signal (data) is a digital signal.

In the following diagram, the information to be transmitted is labeled MODULATING DATA, the unmodulated carrier is labeled CARRIER, and the signal to be presented to the communication channel is labeled ASK MODULATED SIGNAL. The constellation of the two-state ASK MODULATED SIGNAL is shown in the above figure. The amplitude of each of the two states is 0.5 and 1.5, respectively. Both states are at the same relative phase angle, 0°.

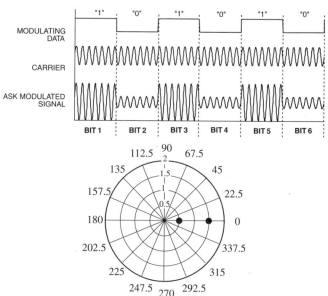

AMPS (1) An acronym from Advanced Mobile Phone Service (or System). An analog-based cellular telephone system in which only one caller occupies a cell carrier at any time. Cell carriers are in the 825–890 MHz frequency band. Contrast this with Time Division Multiple Access (TDMA) in which three callers can utilize the carrier or Code Division Multiple Access (CDMA) that claims an improvement of 10:1 to 20:1 over *AMPS*. **(2)** An acronym from Automatic Message Processing System.

AMSAT An abbreviation of AMateur SATellite corp.

AMTS An abbreviation of Automated Maritime Telecommunications System.

ANAC An acronym from Automatic Number Announcement Circuit.

analog The representation of a variable quantity such as voltage, current, phase angle, temperature, speed, or shaft position that can vary *continuously*. A quantity that can take on *any* arbitrary value between the minimum and maximum as limited by physical constraints. See also *digital*.

analog decoding That portion of the digital-to-analog conversion process that generates an analog signal value from a digital signal. Generally, further processing is required to integrate these samples to obtain a continuous approximation of the analog signal.

analog encoding That portion of the analog-to-digital conversion process that samples an analog signal and creates a digital value representing the value of the analog signal at the instant of sampling. Multiple samples are needed to digitize an analog waveform over a time interval to create a digital signal stream. See also *Shannon limit.*

analog loopback A diagnostic mode of a modem whereby the transmitted analog signal normally connected to the telephone lines is internally routed to the analog receiver. This provides a good test on the functionality of the modem.

analog network A communications network in which transmitted signals are *analog signals.*

analog signal A *continuous* signal that varies in a direct ratio to some stimulus; that is, if the stimulus is changed by some amount, the corresponding analog signal will change by a proportional amount also. Because the analog signal is continuous, *any value* may be represented. See also *digital signal.*

analog switch In telephony, a central office exchange capable of switching either anaog or digital signals without converting them into a set digital format first. They are used to connect circuits between users for real-time transmission of analog signals. Examples include step-by-step exchanges, crossbar exchanges, and most relay-based exchanges.

analog-to-digital converter Abbreviated *A-D converter, A/D converter,* or *ADC.* A device that translates an arbitrary waveform (analog) signal into an equivalent digital form. The converter periodically measures (samples) the analog wave and converts the measured value to a digital word. This process is illustrated in the following drawing.

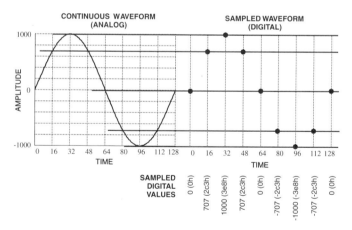

The process that restores the original analog signal is *digital-to-analog conversion (DAC).* See also *adaptive differential PCM (ADPCM).*

analog-to-frequency converter A device that translates an analog input signal into an oscillating output signal whose frequency is proportional to the analog input signal. See also *frequency modulation (FM)* and *voltage-controlled oscillator (VCO).*

ANC An abbreviation of <u>A</u>ll <u>N</u>umber <u>C</u>alling. Calling by means of numerics only, instead of a combination of letters and digits.

AND A logical operation for combining multiple Boolean values (true and false) or two binary (1 and 0) signals. The rule for the *AND* operation is: *The output is true (1) if, and only if, all inputs are true (1); (the output is false (0) otherwise).*

The written symbol for the *AND* operation is a dot (·), ampersand (&), caret (^), or no symbol; therefore, the *AND* of input A and input B may be written A·B as well as "A *AND* B." The *truth table* below shows all possible combinations of input and output for a two-input *AND* operation. A false state is shown as a "0," while a true condition is shown as a "1." See also *AND gate, exclusive OR, exclusive OR gate, NAND, NAND gate, NOT, OR,* and *OR gate.*

AND TRUTH TABLE

A	B	A·B
0	0	0
0	1	0
1	0	0
1	1	1

AND gate A type of digital circuit that performs an *AND* operation on two or more digital signals. The symbol for a two-input *AND Gate* is:

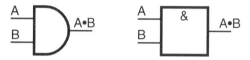

See also *AND, exclusive OR, exclusive OR gate, logic symbol, NAND, NAND gate, NOT, OR,* and *OR gate.*

Andrew File System (AFS) A client program that enables a user to mount a remote directory so that it appears to be part of the host system's structure. *AFS* allows users to access only those files that have been linked to *AFS,* thereby addressing the security problem of granting remote access. See also *Network File System (NFS).*

angle modulation Modulation in which the phase or frequency of a carrier is varied. Phase modulation (PM) and frequency modulation (FM) are both forms of angle modulation.

angle of deviation In optics, the net angular deflection experienced by a light ray after one or more refractions or reflections.

angle of incidence The angle between an incident ray and the normal to a reflecting or refracting surface. See also *critical angle, reflected ray,* and *refracted ray.*

angle of reflection The angle between a reflected ray and the normal of a reflecting surface. See also *critical angle* and *reflected ray.*

angle of refraction The angle between a refracted ray and the normal of a refracting surface. See also *critical angle* and *refracted ray.*

angular alignment loss Signal power loss caused by the deviation from optimum angular alignment of the axes of source to waveguide, waveguide to waveguide, or waveguide to detector. The waveguide may be dielectric, as with an optical fiber or metallic as is used with microwaves. *Angular alignment loss* does not include *gap loss, axial alignment loss, radial alignment loss,* and so on. Also called *angular misalignment loss.* See also *coupling loss.*

Angstrom (Å) A unit of length equal to 10^{-10} meter (in the U.S. numbering system, one ten billionth of a meter). Used to describe the wavelength of light. Though frequently used, Angstrom is not an SI unit; nanometer (nm) should be used instead.

ANF (1) An abbreviation of A̲ppleTalk N̲etworking F̲orum. A consortium of developers and vendors working to encapsulate AppleTalk into other protocols; for example, within the TCP/IP suite. **(2)** An abbreviation of A̲utomatic N̲umber F̲orwarding.

ANI An abbreviation of A̲utomatic N̲umber I̲dentification.

anisochronous From the Greek word meaning not-equal (an-iso) time (chronous). Describes a signal in which the time intervals between consecutive significant instants, events, or entities are not necessarily

- Of equal duration.
- Related by integer multiples of the shortest duration.
- Related to the time interval separating any other two significant instants.

Isochronous and anisochronous are characteristics, while synchronous and asynchronous are relationships. See also *asynchronous, heterochronous, homochronous, isochronous, plesiochronous,* and *synchronous.*

anisotropic Pertaining to a material whose properties (electrical, magnetic, or optical) vary with either the direction of propagation of a traveling wave or with different polarizations of a traveling wave.

In optics, *anisotropy* is exhibited by noncubic crystals, which have different refractive indices for lightwaves propagating in different directions or with different polarizations. *Anisotropy* may be induced in certain materials under mechanical strain.

anisotropic magnet A magnetic material having a preferred orientation so that the magnetic characteristics along one axis are better than any other axis.

Compare with *isotropic magnet.*

ANL (1) An abbreviation of A̲utomatic N̲oise L̲imiter. **(2)** An abbreviation of A̲rgonne N̲ational L̲aboratory.

annex D That part of the ANSI T1.617 standard that specifies a method for indicating permanent virtual circuit (PVC) status in frame relay systems.

anode (1) In energy sinks, the electrode of a device through which the primary stream of electrons *enters* (current *exits*). **(2)** In energy sources, the electrode of a device through which the primary stream of electrons *exits* (current *enters*). See also *cathode.*

anode terminal (1) In a PN junction diode, the terminal from which electrons flow. The terminal is generally not marked; for example,

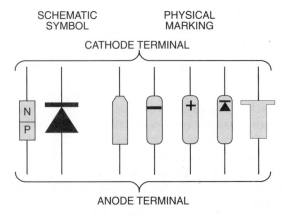

SCHEMATIC SYMBOL   PHYSICAL MARKING

CATHODE TERMINAL

ANODE TERMINAL

(2) In an electrochemical cell or battery, it is the electrode where oxidation takes place. During discharge the *anode* is the negative electrode. During charge, the chemical reaction reverses, and the *anode* is the positive electrode. See also *anode* and *cathode terminal.*

anomalous propagation (AP) Abnormal signal propagation caused by changes in the properties (such as density and refractive index) in the path's medium. *AP* may result in the reception of signals well beyond the distances usually expected or the loss of signals at distances normally achieved.

anonymity In Internet, the creation of an e-mail message or anonymous posting in such a way that the originator's identity is obscured. The link between the originating user and the message is unseen to the extent possible by an *anonymous server.* Note that the link may be difficult to reconstruct but not impossible.

anonymous FTP In Internet, a protocol that allows a user to log in to many remote systems (even nodes from other networks), browse, and retrieve (download) publicly available files without having an account on that system. By using the login name, "anonymous" and a password equal to their Internet address or "guest," users can transfer files without other passwords or login credentials. (*FTP* is an application layer protocol in the Internet's TCP/IP protocol suite.) The remote system may select which files are available for *anonymous FTP.*

anonymous posting A public message on the *USEnet* that has been submitted through an anonymous server. The anonymous server hides the identity of the message originator (but does not destroy it).

anonymous server A program on an Internet host that allows a user to create an *anonymous posting* to USEnet. The program accepts the user's message, strips information that would identify the origin, possibly removes the signature, and sends the file on to its destination. Some caveats:

- Not all *anonymous servers* remove the signature file. So it should be disabled before transmitting the message to the server.
- The *anonymous server* maintains records of its linking of anonymous and real addresses.

ANS (1) An abbreviation of A̲merican N̲ational S̲tandard. **(2)** An abbreviation of A̲NS̲wer.

ANSI An acronym from A̲merican N̲ational S̲tandards I̲nstitute (pronounced "ANN-see"). A nonprofit organization of industry and business that proposes, develops, approves, and maintains telecommunications and data processing standards in the United States. Manufacturing companies voluntarily follow these standards. *ANSI* is a member of International Standards Organization (ISO) and represents the interests of the United States in both ISO and the International Telecommunications Union (ITU).

ANSI T.413 An Asymmetric Digital Subscriber Line (ADSL) specification. An ADSL implementation technique based on discrete multitone (DMT) wherein the entire bandwidth is divided into 255 4-kHz sub-bands. Data transmission is spread over these bands to achieve noise-immune high throughput. Each subcarrier utilizes quadrature amplitude modulation (QAM) and digital signal processing (DSP) techniques to squeeze the maximum information through a twisted pair telephone line.

ANSI BBS A terminal setting used by many bulletin board systems (BBSs) and modem software that allows the display of on-screen graphics.

answer abort timer A timer that determines the maximum time a DCE may remain off-hook in response to a manual answer (or a ring down) during a failed call. Typical delay times are 5.1, 10, 50, or 70 seconds.

This timer is generally in addition to the *call abort timer.*

answer back A signal or message from a receiving device in response to a transmitting device's request message. For example, it may indicate the receiving device is ready to accept data or that it received data correctly.

answer mode The mode the DCE is put into when it is to be the called terminal. In the answer mode, the DCE may automatically respond to a ring signal and may change the operating frequencies.

answer signal A supervisory signal returned from the called telephone to the originating switch when the call station goes off-hook. The *answer signal* stops the ringback signal from being returned to the calling station. The answer signal is returned by means of a closed loop.

answer tone A tone returned by the answering DCE to the calling DCE indicating that a physical connection has been established and that the next step in the handshake process may continue.

answer-only modem A modem that answers incoming calls but cannot originate outgoing calls.

answer/originate modem A modem that can both answer or originate calls. This is the most common type of modem available.

.answers A moderated newsgroup in USEnet that is dedicated to providing information not only about USEnet in general but about specific newsgroups. Within the .answers newsgroups will be found postings of *frequently asked questions* (*FAQs*); for example, "what is USEnet?," how to work with USEnet, and so on.

Check the *.answers* group corresponding to a newsgroup of interest for the FAQs; for example, if the newsgroup is *xxx.yyy.zzz,* try looking in *xxx.answers* for the FAQ posting.

antenna That part of a transmitting (or receiving) system capable of radiating (or receiving) electromagnetic waves. The transmission or reception may be intentional or incidental.

The principle by which an *antenna* operates is based on the 1865 theoretical discovery by James Clerk Maxwell's (1831–1879), that a moving electric charge generates electromagnetic radiation which propagates through space at the speed of light. This was experimentally verified in 1888 by Heinrich Rudolf Hertz (1857–1894).

antenna aperture See *aperture* (1).

antenna array An assembly of antenna elements with dimensions, spacing, and an excitation phase such that the fields of the individual elements of a transmitting antenna combine to produce a maximum radiated power in a particular direction and minimum power in other directions. In a receiving antenna, the array provides a maximum sensitivity in the designated direction and minimum sensitivity in other directions.

anti-sidetone See *hybrid* and *hybrid coil.*

anti-sidetone telephone A telephone subset that includes a two-wire to four-wire hybrid for reducing the amount of *sidetone.* See also *hybrid, speech network,* and *subset.*

anti-spoofing In GPS, the encryption of the P-code by a secret *W-code* to prevent GPS receiver systems from being tricked into delivering false location information by bogus, enemy-generated P-code. The encrypted P-code is called *Y-code.*

anti-virus program A program that detects virus-infected programs, systems, or networks and allows an operator to take appropriate action (delete the infected program, remove the virus, etc.)

AnyNet A networking standard from IBM that uses the Multi-Protocol Transport Networking (MPTN) protocol to provide TCP/IP connectivity over its Systems Network Architecture (SNA).

AOL The abbreviation for America OnLine (pronounced "A," "O," "L").

AOM An abbreviation of Application OSI Management. In the International Standardized Profile (ISP) model, the prefix for functions and services related to network management.

AOS (1) An abbreviation of Acronym Overuse Syndrome. (2) An abbreviation of Alternate Operator Service.

AOW An abbreviation of Asia and Oceania Workshop. One of three regional workshops for implementers of the ISO/OSI Reference Model. The other two are the European Workshop for Open Systems (EWOC) and the OSI Implementers Workshop (OIW).

AP (1) An abbreviation of Application Process. A program that can make use of the ISO/OSI Reference Model application layer services. Application service elements (ASEs) provide the requested services for the *AP.* (2) An abbreviation of Anomalous Propagation.

APC An abbreviation of Adaptive Predictive Coding.

APD An abbreviation of Avalanche PhotoDiode. Also abbreviated *apd* and *a.p.d.*

APDA An abbreviation of Apple Programmers and Developers Association.

APDU An abbreviation of Application Protocol Data Unit. A data packet at the ISO/OSI Reference Model application layer. Also called the application layer PDU. See also *OSI.*

aperiodic Not periodic.

aperiodic antenna An antenna designed to have approximately constant input impedance over a wide range of frequencies. Examples of *aperiodic antennas* include terminated rhombic and wave antennas. Also called a *nonresonant antenna.*

aperture (1) In a directional antenna, that portion of a plane surface very near the antenna and normal to the direction of maximum radiant intensity (the beam) through which the major part of the radiation passes, e.g., the diameter of a parabolic antenna. (2) An opening or a hole in a material opaque to a particular electromagnetic radiation frequency through which that radiation frequency may pass. (3) In an acoustic device that launches an acoustic wave, the passageway, determined by the size of a hole in the inelastic material and the wavelength.

aperture illumination (1) The field distribution, in amplitude and phase, over the antenna physical aperture. (2) The phase and amplitude of the element feed voltages or the distribution of the currents in an array of elements.

API An abbreviation of Application Program Interface.

APIA An abbreviation of Application Program Interface Association. A group that writes APIs for the ITU's X.400 Message Handling System (MHS).

APLT An abbreviation of Advanced Private Line Termination. It provides a PBX user access to all the services of an associated enhanced private switched communications services (EPSCS) network.

apoapsis The orbital point most distant from the center of the primary attracting object. See also *apogee* and *periapsis*.

apochromatic system An optical system that is chromatically corrected for three colors simultaneously.

apogee The orbital point most distant from the gravitational center of the Earth, when the Earth is the center of attraction. See also *apoapsis* and *perigee*.

apparent power In alternating current (ac) power transmission and distribution, the product of the RMS voltage and current. Apparent power is expressed in volt-amps (VA), not watts (W).

When the applied voltage and the current are in phase, the *apparent power* is numerically equal to the *effective power* (the real power—expressed in watts (W)—delivered to or consumed by the load). If the current lags or leads the applied voltage, the *apparent power* is greater than the effective power. See also *effective power*.

apparent solar time A timekeeping scheme based on the Sun's crossing of the local meridian at 12:00 noon. Also called *local apparent time* and *sundial time*. See also *time (3)*.

APPC An abbreviation of <u>A</u>dvanced <u>P</u>rogram to <u>P</u>rogram <u>C</u>ommunications. A collection of protocols executing programs to communicate with each other as peers, that is, without the assistance of the mainframe. *APPC* is used primarily with personal computer communications to the mainframe host in IBM's Systems Network Architecture (SNA) network. *APPC* provides direct peer-to-peer workstation communications, mainframe access, compatible protocols between all types of workstations, and standard application program interfacing to the network. It is defined at a level comparable to the session layer in the ISO/OSI Reference Model. It is supported in various networking environments, including IBM's Systems Network Architecture (SNA), Ethernet, Token Ring, and X.25. Also called *LU6.2*.

APPC/PC A version of APPC developed by IBM to run on PC-based Token Ring networks.

append To attach to an end of something (a file, a data block, etc.). Often used with reference to adding information to a packet of information being transmitted (as in error detection or correction) or adding data to the end of a file.

Apple Datastream Protocol (ADSP) A transport mechanism for interprocess communications between Apple Macintosh and DEC VAX minicomputers.

Apple Desktop Bus (ADB) A high-speed serial interface used on the Macintosh computer. It is where nonperipheral devices such as the keyboard are connected. See also *EIA/RS-232, EIA/RS-422, EIA/RS-449, IEEE P1394,* and *ITU-T V.35*.

Apple Filing Protocol (AFP) A standard means of presenting the filing system of a server to the user with a consistent Apple Macintosh interface.

AppleLink A commercial online information service dedicated to the service of Apple Computer's customers. As with most online information services, a gateway to the Internet is available.

AppleShare Apple Computer's networking system. *Appleshare* runs on a Macintosh network server, providing file access and printer services. It runs on top of the AppleTalk protocols at the uppermost (application) layer and uses the protocol suite to provide services.

applet A small software application performing a specific task.

Traditionally, *applets* have referred to programs like Cardfile and Calculator in Microsoft Windows but now it is used more commonly to describe the small distributed applications created with Sun's Java programming language and are downloaded over the Internet to dress up a web page (e.g., provide motion).

AppleTalk A network protocol suite defined by Apple Computer Inc. for interconnecting Macintosh computers. The original version (Phase 1) allowed only 254 network nodes, however. Phase 2, introduced in 1989, increased this limit to 16 million. Several physical interfaces are defined, i.e.,

- Serial EIA/RS-422 using AppleTalk Remote Access Protocol (ARAP). Apple's basic architecture available on all Macintosh computers. It is a 230.4-kbps architecture that accesses shielded twisted pair (STP) cabling via EIA/RS-422 connections. It allows nodes to be separated by as much as 305 meters (1000 feet).
  The term LocalTalk is sometimes used to refer to an AppleTalk network.
- Apple's network *LocalTalk* utilizing LocalTalk Link Access Protocol (LLAP).
- Ethernet using *EtherTalk* Link Access Protocol (ELAP). Apple's implementation of the 10 megabit per second (Mbps) Ethernet architecture. Two versions of EtherTalk exist. The first, EtherTalk Phase 1, is based on the Blue Book Ethernet 2.0 (not the IEEE 802.3 specification). Phase 1's successor, Phase 2, is modeled on the IEEE 802.3 standard. Because the packets are defined differently in Phase 1 and Phase 2, the two versions cannot communicate directly with each other. *EtherTalk* has replaced *LocalTalk* as the default networking capability in newer Macintosh models.
- Token Ring using *TokenTalk* Link Access Protocol (TLAP). Apple's implementation of the Token Ring architecture. *TokenTalk* supports both the 4-Mbps and 16-Mbps version.
  (*Note:* The Token Ring architecture is supported only in *AppleTalk* Phase 2.)
- FDDI using *FDDITalk* Link Access Protocol (FLAP). Apple's implementation of the 100-Mbps FDDI architecture.

At the network layer, AppleTalk uses the Datagram Delivery Protocol (DDP) regardless of the architecture operating at the data link layer, which makes a best effort at packet delivery, but delivery is not guaranteed. Also at the network layer is the AppleTalk Address Resolution Protocol (AARP) which maps AppleTalk network addresses to Ethernet or Token Ring physical addresses. The relationships of the various members of the protocol suite are outlined in the associated diagram. The diagram also illustrates the relationship of the elements to the ISO/OSI Reference Model.

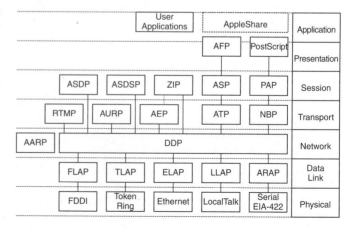

- *AAUI* (Apple Attachment Unit Interface).
- *AARP* (AppleTalk Resolution Protocol). AARP maps AppleTalk network addresses to Ethernet or Token Ring physical addresses.
- *ADSP* (AppleTalk Data Stream Protocol). A session layer protocol that enables two nodes to establish a reliable connection through which data can be transferred.
- *AEP* (AppleTalk Echo Protocol). A transport layer protocol used to determine if two nodes are connected and both are available.
- *AFP* (AppleTalk Filing Protocol). The application/presentation level protocols managing remote file access, i.e., where applications access the *AppleTalk* network.
- *ALAP* (AppleTalk Link Access Protocol). The link access layer protocol (data link layer protocol) manages packet transmissions on LocalTalk. Also called LocalTalk Link Access Protocol (LLAP).
- *ARAP* (AppleTalk Remote Access Protocol). A data link layer protocol that converts packets from higher layers into a form appropriate for transmission on an EIA/RS-422 serial link.
- *ASDSP* (AppleTalk Secure Data Stream Protocol). A session layer protocol similar to ADSP but one that provides additional security against unauthorized use.
- *ASP* (AppleTalk Session Protocol). A session layer protocol used to initiate and terminate sessions, send commands from client to server, and send replies from server to client.
- *ATP* (AppleTalk Transaction Protocol). A transport layer protocol that provides reliable packet transport.
- *AURP* (AppleTalk Update Routing Protocol). A transport layer routing protocol that updates the routing table only when a change has been made to the network.
- *DDP* (Datagram Delivery Protocol). All AppleTalk networks use the *DDP* at the network layer, regardless of the architecture operating at the data link layer. The protocol prepares packets for transmission and, though not guaranteed, makes a best effort at delivery.
- *ELAP* (EtherTalk Link Access Protocol). A data link layer protocol that converts packets from higher layers into a form appropriate for transmission on an Ethernet physical link.
- *FLAP* (FDDITalk Link Access Protocol). A data link layer protocol that converts packets from higher layers into a form appropriate for transmission on an FDDI physical link.
- *LLAP* (LocalTalk Link Access Protocol). A data link layer protocol that converts packets from higher layers into a form appropriate for transmission on a LocalTalk physical link.
- *NBP* (Name Binding Protocol). The transport layer protocol for translating device names into addresses. *NBP* associates easy to remember names (used by users) with the appropriate network addresses.
- *PAP* (Printer Access Protocol). The session layer protocol manages access to *PostScript* printer services. A protocol governing transmissions between workstations and printers (or similar devices).
- *RTMP* (Routing Table Maintenance Protocol). A transport layer routing protocol for transferring packets across networks.
- *TLAP* (TokenTalk Link Access Protocol). A data link layer protocol that converts packets from higher layers into a form appropriate for transmission on a Token Ring physical link.
- *ZIP* (Zone Information Protocol). ZIP is used mainly on larger networks or internetworks, which are more likely to be divided into zones (a zone being a logical grouping of nodes that together make up a subnetwork.)

The concept of a zone was introduced for two reasons: to allow networks with more than 255 nodes and to make addressing and routing tasks easier.

application A computer program that is intended to perform a service or function beneficial to a user; that is, it produces functionally usable results, for example, documents, graphics, navigation, payroll, and spreadsheets. This is in contrast to an *operating system* that provides an interface between the hardware and *applications* or a *utility* that supports computer functionality and maintenance.

Applications may be *stand-alone* or *network based:*

- A *stand-alone application* can execute only one instance of itself at a time and can support only a single user at a time. These applications generally execute on a single machine, which may or may not be connected to a network.
- A *network-based application* executes on a network and uses network resources and devices. The program may be server resident or distributed across several nodes. Multiple users may run the program simultaneously and not worry about others improperly accessing each other's data. Access is controlled by assigned access rights.

Individual *applications* communicate and exchange data using a variety of methods. For example:

- Shared data files are probably the simplest way for programs to communicate.
- Pipes, in which the output of one program is simply "piped" in as input to another program, is one of the simplest ways to share data.
- Dynamic data exchange (DDE), a method for low-level communications that do not need user intervention. When two or more programs that support DDE are running at the same time, they can exchange data and commands by means of a two-way connection (conversation). DDE has been largely replaced by OLE.
- Object linking and embedding (OLE) is a more sophisticated method, which provides much greater flexibility. It makes it possible for updates to be carried over automatically to whatever applications use the updated items.

Examples of applications include word processing programs, spreadsheets, or a Web authoring programs. Also called an *application program.*

application association (AA) Provides the protocols for the cooperation between two application entities (AEs). When an *application association* is established, the AEs involved must agree on an *operation class* and an *association class* for the interaction.

Five operation classes are defined, based on the type of reply the client provides and on whether the interaction is synchronous or asynchronous. The five classes are defined by:

OPERATION CLASS ATTRIBUTES

Operation Class	Synchronous/ Asynchronous	Reply	
		Success	Failure
1	S	X	X
2	A	X	X
3	A	X	
4	A		X
5	A		

Three *association classes* are defined as:

ASSOCIATION CLASS ATTRIBUTES

Association Class	Operations Invoked by	
	Initiator	Responder
1	X	
2		X
3	X	

Application Entity (AE) In the ISO/OSI Reference Model, a process or function (entity) that runs all or part of an application. An *AE* may consist of one or more application service elements (ASEs).

Application File Transfer (AFT) In the International Standardized Profile (ISP) grouping, a prefix that identifies *file transfer, access, and management* (FTAM) profiles. For example, AFT11 represents basic file transfer.

application layer Layer seven, the highest layer, in the ISO/OSI Reference Model for data communications. It defines the protocols for software communications with the network. The protocols typically define:

- Terminals
- Formats for message and e-mail transfers
- File transfer handling

For example, in the Internet, protocols at this level include Telnet, network virtual terminal (NVT), file transfer protocol (FTP), and the simple mail transport protocol (SMTP). See also *OSI.*

Application Process (AP) In the ISO/OSI Reference Model, a program that can make use of application layer services. Application service elements (ASEs) provide the requested services for the AP.

application program interface (API) A means of communication between computer programs so as to give one program access to the other's services or resources. Access is obtained by a published or formalized set of function calls, routines, interrupts, and data formats. Applications, for example, can use these function calls to gain access to the operating system's services. Some *APIs,* such as Net-BIOS, are de facto standards.

Application Program Interface Association (APIA) A group that writes APIs for ITU's X.400 Message Handling System (MHS).

Application Protocol Data Unit (APDU) A data packet at the ISO/OSI application layer. Also called *application layer PDU.*

applique In telephony, circuits or components added to an existing system to provide additional or alternate functions. For example, an *applique* may be used to modify carrier telephone equipment designed for ringdown manual operation to allow for use between points having dial equipment.

APPN An abbreviation of <u>A</u>dvanced <u>P</u>eer to <u>P</u>eer <u>N</u>etworking.

APS An abbreviation of <u>A</u>utomatic <u>P</u>rotection <u>S</u>witching.

AR (1) In telephony, an abbreviation of <u>A</u>lternate <u>R</u>oute. (2) In optics, an abbreviation of <u>A</u>nti-<u>R</u>eflective.

ARA (1) An abbreviation of <u>A</u>ttribute <u>R</u>egistration <u>A</u>uthority. In the ITU X.400 Message Handling System (MHS), the organization that allocates unique attribute values. (2) An abbreviation of <u>A</u>pple <u>R</u>esource <u>A</u>ccess. Hardware and software that provide an asynchronous AppleTalk connection to a Macintosh and its network services through a modem, thus allowing remote file access.

aramid yarn A generic name for the tough synthetic yarn that is often used

- In optical cable construction for the tensile strength member,
- As protective braid, and/or
- As the ripcord for jacket removal.

ARAP An abbreviation of <u>A</u>ppleTalk <u>R</u>emote <u>A</u>ccess <u>P</u>rotocol. A remote access protocol that supports the AppleTalk protocol suite over serial lines.

arbitration The process of monitoring and managing competing demands for a resource by multiple requesters.

arc The flow of current between two electrodes separated by a gas where the gas is in the second of two breakdown states.

The first state is the *glow discharge* state with relatively low currents and with voltages at approximately 65 V or above. The second breakdown state is the *arc discharge* or *arc mode* state that is typified by a low-voltage drop in the immediate vicinity of the cathode (approximately equal to the ionization potential of the gas at the operating temperature and pressure). See also *breakdown.*

ARC One of the older DOS lossless compression/decompression archiving formats. The *ARC* utility was copyrighted by System Enhancement Associates, Inc. (SEA) in 1985 and distributed as Shareware. The extension of files created by the *ARC* utility is ".*ARC*."

The *ARC* utility automatically selects one of several methods of data compression including no compression (none), repeated character compression, dynamic Lempel-Ziv, and Huffman algorithms. The format of the *ARC* command is:

ARC xm arcname [source [source]...]

where:

arcname is the name assigned to the archive.

[*source*] is one or more file names to be compressed. Wild cards (* and ?) are allowed.

x is a command letter selected from:

a	add files to an archive,
c	convert entry to different compression method,
d	delete files from an archive,
f	freshen files in an archive,
l	list file names in an archive,
m	move files to an archive,
p	print, copy files from an archive to the standard output,
r	run files from an archive,
u	update files in an archive,
t	test integrity of archive files,
v	verbose list files names in archive,
x, e	extract files from an archive.

m is an auxiliary command selected from:

b	retain backup copy of archive,
g	encrypt or decrypt with password,
n	suppress notes and comments,
o	overwrite existing files with the same name when extracting,
s	suppress compression (store only),
v	verbose mode,
w	suppress warning messages,
z	include subdirectories in archive.

For a complete list of the commands with examples, see the documentation available with the software. See also *compression (2)*.

arc current The *current* that flows in a circuit after ionization occurs in the gas dielectric separating the electrodes of a spark gap and when the glow to arc transition current has been exceeded. Sometimes called the *arc mode current*. See also *arc* and *breakdown*.

arc voltage (V_{arc}) In a gas tube, for example, a gas surge-protector, the *voltage* across the device during *arc current flow*. Typical values are 7–20 V. Also called the *arc mode voltage*.

Archie In Internet, a service program (developed at the McGill University Department of Computer Science) that attempts to locate files available for *FTP downloading*. *Archie* can automatically search for a file based on file names with a user-specified string (prog) or search file descriptions (whatis).

Detailed descriptions of *Archie* may be downloaded (using FTP) from Internet itself by sending the one-word message "man" to archie@ans.net. An abbreviated description can be obtained by sending the message "help" instead of "man."

architecture The overall structure of a network. It is expressed in terms of its *protocols* (which govern the operation at each layer) and *topology* (which describes the physical structure).

archive Strictly speaking, a collection of information; however, it is frequently used with the additional restriction that the information be *off-line*. Often a file containing a collection of compressed files, records, or other data is referred to as an *archive*.

On the Internet, an *anonymous FTP archive* is a server that is open to callers who log on as "anonymous"; that is, they don't need to set up an account to access files on that machine. An archived file will probably be compressed in order to both save storage space on the server and to reduce the time required to download the file. The file extension indicates the type of compression used and which decompression program must be used to retrieve usable files. See also *filename extensions*.

archive bit An indicator (called a flag) kept for each file that tells whether the file has been modified since it was last copied or backed up.

archive site A host node on a network that has devoted part of its resources to serve as a repository for files. On the Internet, these files can be downloaded via FTP.

ARCnet An acronym from <u>A</u>ttached <u>R</u>esource <u>C</u>omputer <u>NET</u>work. A modified token-passing LAN, developed by Datapoint Corporation in the late 1970s, which supports up to 255 nodes. *ARCnet* became an open standard in 1982. The nodes communicate in the *baseband* at 2.5 Mbps over 93 Ohm RG-62/U or similar coaxial cable (although twisted pair and optical fibers have been used). ARCnetPlus extends the data rate to 20 Mbps, and at least one-third party vendor offers a 100-Mbps version.

Officially, *ARCnet* uses a bus topology, but in practice *ARCnet* networks can use either a star (low-impedance *ARCnet*) or a bus (high-impedance *ARCnet*) wiring scheme. The two types of networks use slightly different components. Most *ARCnet* network interface cards (NICs) have low-impedance transceivers. Unlike most Ethernet NICs, *ARCnet* NICs do not come with *addresses* in a ROM chip. Instead, they have jumpers that must be set to specify an address for the node into which the card is installed. The network administrator must set a unique address (between 1 and 255) for each card in the network. The network administrator also needs to set the interrupt (IRQ) and base I/O addresses on the card. The hardware address is network dependent, while the IRQ and I/O addresses are machine dependent.

ARCNET

Parameter	Value (and Conditions)
Topology	Bus, star, or daisy chain
Data bandwidth	2.5 Mbps baseband (phase 1)
	20 Mbps baseband (phase 2)
	100 Mbps (third party)
Nodes/network	≤ 255 (active hubs count as nodes)
Nodes between hubs	≤ 8 (coaxial)
	≤ 10 (UTP)
Node to Node distance	>1 m (3.25 ft)
Protocol	Token passing
Data encoding	Manchester
Cable	RG-62/U coaxial
	UTP category 2 or better
Terminator	BNC (coaxial) or RJ-11/RJ-45 (UTP)
Characteristic	93 Ω (coaxial)
Impedance	105 Ω ±15% (UTP)
Attenuation	< 11 dB at 5 MHz over entire segment
Propagation delay	< 31 µS between any two nodes
Segment length	≤ 610 m (2000 ft) coaxial
	≤ 122 m (400 ft) UTP.
Network length	≤ 6000 m (20 000 ft)

ARDIS® An acronym from <u>A</u>dvanced <u>R</u>adio <u>D</u>ata <u>I</u>nformation <u>S</u>ervice. A public wireless data network service jointly owned by IBM Corp. and Motorola Inc. It is an outgrowth of IBM's network for its service technicians.

area code A three-digit code that selects the toll center in the United States, Canada, and Mexico to which a call is routed. Also called a *numbering plan area (NPA)*. See also *city code, country code, international access code,* and *NXX*.

area loss In fiber optics, a power loss that is caused by any mismatch in size (*radial alignment loss*), position (*axial alignment loss*), and/or shape (*core ovality loss*) of the cross section of the cores of the mating fibers when optical fibers are joined (by either a splice or a pair of mated connectors).

Either of these conditions may allow light from the core of the "transmitting" fiber to enter the cladding of the "receiving" fiber, where it is quickly lost. *Area loss* may be dependent on the direction of propagation. For example, in coupling a signal from an optical fiber having a smaller core to an otherwise identical one having a larger core, there will be no area loss, but in the opposite direction, there will be *area loss*. See also *coupling loss*.

argument (1) An independent variable. **(2)** Any value of an independent variable. Examples of arguments include search keys, index numbers that identify the location of a data item in a table, and the θ in sin θ.

arithmetic logic unit (ALU) That part of any computer which performs mathematical calculations and comparisons.

ARJ One of the more powerful DOS lossless file compression/decompression archive utilities found on BBSs. The *ARJ* utility was written by Robert K. Jung and is distributed as shareware. Files created by the utility have default extensions of **".ARJ."**

The format of the *ARJ* command is

ARJ x [-sw [-sw...]] arcname [source [source]...]

where

arcname is the name assigned to the archive.

[*source*] is one or more file names to be compressed. Wild cards (* and ?) are allowed.

x is a command letter. Some frequently used commands include:

a	Add files to an archive,
d	Delete files from an archive,
e	Extract files from an archive,
f	Freshen files in an archive,
l	List files in an archive,
m	Move files to an archive,
t	Test integrity of an archive,
u	Update files to an archive,
v	Verbose list of files in an archive,
x	eXtract files with full pathname.

-sw (*or /sw*) is a switch to modify or extend the operation of the command *x*. Some of the more important switches are

-c	skip time-stamp Check,
-e	Exclude paths from names,
-f	Freshen existing files,
-g	Garble with password,
-i	with no progress Indicator,
-m	with Method 0, 1, 2, 3, 4,
-n	only New files (not exist),
-r	Recurs subdirectories,
-s	set archive time-Stamp to newest,
-u	Update files (new and newer),
-v	enable multiple Volumes,
-w	assign Work directory,
-x	eXclude selected files,
-y	assume Yes on all queries.

For a complete list of the commands and switches, either see the documentation available with the program or type the command *ARJ /?*. See also *compression* (2).

ARL An abbreviation of Adjusted Ring Length.

ARM **(1)** An acronym from Asynchronous Response Mode. **(2)** An acronym from Advanced RISC Microprocessor.

armor In communications cables, an outer cable layer intended to protect the cable's internal elements (e.g., electrical conductors, optical fibers, or buffer tubes) from damage by external mechanical attack (e.g., by rodents or abrasion). Armor is usually a steel or aluminum tape wrapped around the inner components. An outer jacket may cover the armor.

ARP An acronym from Address Resolution Protocol.

ARPA An acronym from Advanced Research Projects Agency. The U.S. governmental organization responsible for creating ARPAnet, the beginnings of what is now known as the Internet. The agency was replaced by DARPA, the Defense Advanced Research Projects Agency.

ARPAnet An acronym of Advanced Research Projects Agency NETwork. *ARPAnet*, the predecessor to Internet, is a peer-to-peer wide area data communications network (WAN) developed in the 1960s by the Department of Defense as a nuclear disaster tolerant military command and control system.

Each node in the network was autonomous and communicated with other nodes with packets of data. Each packet of information contained the address of the destination node. The original sending station puts a packet onto the network, and a node receiving the packet forwards it on toward its final destination. (The exact path is unknown and may change over time.) The protocol regulating this packet flow from source to destination was called TCP/IP for Transmission Control Protocol/Internet Protocol. In 1989 the original *ARPAnet* nodes were decommissioned, leaving what is now called Internet. See also *Internet*.

ARQ An abbreviation of Automatic Repeat reQuest.

arrester A device that protects hardware, such as systems, subsystems, circuits, and equipment, from voltage or current surges produced by lightning or electromagnetic pulses (EMP). The device limits the surge voltage across subsequent circuits. Also called a *surge suppressor*. See also *gas-filled protector, MOV,* and *zener diodes*.

arrester discharge capacity The maximum current (with a specified waveform) an arrester can withstand without damage to any of its component parts.

ARRL An abbreviation of Amateur Radio Relay League.

ARS An abbreviation of Automatic Route Selection.

ART In telephony, an acronym from Audible Ringing Tone.

art line A colloquial synonym for artificial transmission line.

articulation index A measure of the intelligibility of voice signals, expressed as a percentage of speech units that are understood by the listener when heard out of context. Level, noise, interference, and distortion affect the index.

artificial antenna A device that has the impedance characteristics and power-handling capability of an antenna but does not radiate or receive radio frequency energy. Energy normally radiated by a transmitting antenna is converted to thermal energy and dissipated as heat. Also called a *dummy antenna* or sometimes a *dummy load*.

artificial ear In telephony, a device for measuring the audible output of a telephone handset while providing a reasonable approximation of the acoustical characteristics of an average human ear.

artificial intelligence (AI) The capability of a device to perform functions that are normally associated with human logic, such as reasoning, learning, and self-improvement (optimization through experience).

AI is the branch of computer science that attempts to approximate the results of human reasoning by organizing and manipulating factual and heuristic knowledge. *AI* activities include expert systems, natural language understanding, speech recognition, visual recognition, and robotics.

artificial load A dissipative but nonradiating device having the impedance characteristics and power-handling capability of the device it is simulating, such as an antenna, a transmission line, and a speaker. Also called a *dummy load*.

artificial mouth In telephony, a device containing an electroacoustical transducer that simulates the sound field produced by an average talker.

artificial transmission line A four-terminal passive electrical network that has the same characteristic impedance, transmission time delay, phase shift, and/or other parameters as a specific real transmission line. One or more of these networks may be cascaded to simulate that specific real transmission line in one or more of these respects. Also commonly called an *art line.*

Artisoft® The company that manufactures LANtastic®.

ARU An abbreviation of <u>A</u>udio <u>R</u>esponse <u>U</u>nit.

AS (1) An abbreviation of <u>A</u>utonomous <u>S</u>ystem. **(2)** An abbreviation of <u>A</u>nalog <u>S</u>ignal. **(3)** An abbreviation of <u>A</u>nswer <u>S</u>upervision. **(4)** An abbreviation of <u>A</u>rticulation <u>S</u>core.

ASAP An acronym from <u>A</u>s <u>S</u>oon <u>A</u>s <u>P</u>ossible.

ASB An abbreviation of <u>A</u>utomatic <u>S</u>peed <u>B</u>uffering.

ASC An abbreviation of <u>A</u>UTODIN <u>S</u>witching <u>C</u>enter.

ascending node That point in an orbit where an orbiting object passes from below to above the equatorial plane. See also *descending node.*

ASCII An acronym from <u>A</u>merican <u>S</u>tandard <u>C</u>ode for <u>I</u>nformation <u>I</u>nterchange (pronounced "AS-key" or "ASK-ee"). A standard code used for information interchange between data processing equipment and over data communications networks.

The code was developed in the 1950s (and revised in 1983) as a means of transmitting written text between computing equipment, such as other computers, terminals, printers, and displays. The International Alphabet Number 5 (IA5), defined by ITU-T T.50 and ISO document 646, is the equivalent international code. These codes are a descendant of the Baudot code, an early 5-bit code used for data transmission between teleprinters. The Baudot code, being only a 5-bit code, is capable of only 32 symbols. But 32 symbols are not enough to provide a unique representation of the upper and lower case alphabet, the numeric digits, special symbols, and any control symbols. The American National Standards Institute (ANSI) developed the ASCII code to solve this symbol limitation problem. Based on 7 bits, the ASCII code is capable of 128 unique symbols, which is sufficient for most requirements. Because computers treat data in 8-bit bytes, an ASCII character will leave 1 bit "unused" in a byte. The eighth bit (sometimes called the high-order bit) is frequently used as a parity bit, which provides a means to detect communication errors. The parity bit can be defined as either even or odd (or no parity). Even parity means that the eighth bit will be set to a logical "0" or "1" in order to force the total number of "1s" in the byte to be 2, 4, 6, or 8. Similarly, odd parity means that the eighth bit will be set to a logical "0" or "1" in order to force the total number of "1s" in the byte to be 1, 3, 5, or 7. When "no parity" is defined for the byte, the eighth bit may be set either always at "0" or at "1." The eighth bit may also be used to define an *extended character set,* giving up to 256 symbols in the communication alphabet. These additional characters are not standardized and are not part of the ASCII character set. If there is a need for special characters such as those used in a foreign language or in a graphics application, the *high code* can be assigned. Within the 128 symbols defined by the ASCII character set there are 33 special control characters. These control symbols are represented as <xx> in documents; for example, the carriage return character (ASCII 013) is shown as <CR>. These control characters are divided into six categories by the ITU-T:

- *Presentation format control characters:*
  - <BS>   *Backspace*—a symbol that moves the active insertion point backward one character position.
  - <HT>   *Horizontal tab*—a symbol that indicates the active position is to be advanced to the next predetermined position on the same line.
  - <LF>   *Line feed*—a symbol used to advance the active position to the same horizontal position of the next line.
  - <VT>   *Vertical tab*—a symbol that causes the active position to be advanced to the next predetermined line and at the same horizontal position.
  - <FF>   *Form feed*—a symbol indicating that the active position is to be advanced to a predetermined position on the next form or page.
  - <CR>   *Carriage return*—a symbol that moves the active insertion point to the beginning of the current line.
- *Device control characters:*
  - <DC1>   *Device control 1*—a symbol intended to turn on or start an ancillary device.
  - <DC2>   *Device control 2*—a symbol intended to turn off or stop an ancillary device.
  - <DC3>   *Device control 3*—a symbol intended to turn on or start an ancillary device.
  - <DC4>   *Device control 4*—a symbol intended to turn off or stop an ancillary device.
- *Information separator characters:*
  - <FS>   *File separator* (also called *information separator 4*)—although application dependent, it is used to split, qualify, and delineate data logically into files.
  - <GS>   *Group separator* (also called *information separator 3*)—although application dependent, it is used to split, qualify, and delineate data files logically into groups.
  - <RS>   *Record separator* (also called *information separator 2*)—although application dependent, it is used to split, qualify, and delineate data groups logically into records.
  - <US>   *Unit separator* (also called *information separator 1*)—although application dependent, it is used to split, qualify, and delineate data records logically into units.
- *Code extension control characters:*
  - <SO>   *Shift out*—a symbol used to indicate an application-dependent alternate meaning of the codes 21h through 7Eh. The alternate definition continues until <SI> is detected.
  - <SI>   *Shift in*—the symbol used to return the definition of the codes 21h through 7Eh to their "normal" meaning.
  - <ESC>   *Escape*—a symbol used to provide additional characters, it alters the meaning of a limited number of subsequent symbols.
- *Miscellaneous control characters:*
  - <NUL>   *Null*—a symbol used as a fill character. It conveys no information.
  - <BEL>   *Bell*—a control character used to set an alarm or as a call for attention.
  - <CAN>   *Cancel*—a character that indicates the previous data is in error and is to be ignored. The exact meaning is application dependent.
  - <EM>   *End of medium*—a symbol used to identify either the end of the **used portion** of a medium, the end of the **wanted portion,** or the **physical end.**
  - <SUB>   *Substitute character*—a symbol used in place of a character that has been found to be invalid or in error.
  - <DEL>   *Delete*—a symbol originally used to obliterate a character on a punched tape. In some applications, it may also be used as a transmission fill character.
- *Transmission control characters:*
  - <SOH>   *Start of header*—a symbol used as the first character of an information message.
  - <STX>   *Start of text*—the symbol used to terminate a header and precede text.
  - <ETX>   *End of text*—a symbol indicating the end of a text.

7 BIT ASCII/ITU-T T.50 CHARACTERS

Dec	Hex	Char	6	5	4	3	2	1
0	00h	<NUL>	X					
1	01h	<SOH>						
2	02h	<STX>						
3	03h	<ETX>						
4	04h	<EOT>						
5	05h	<ENQ>						
6	06h	<ACK>						
7	07h	<BEL>	X					
8	08h	<BS>					X	
9	09h	<HT>					X	
10	0Ah	<LF>					X	
11	0Bh	<VT>					X	
12	0Ch	<FF>					X	
13	0Dh	<CR>					X	
14	0Eh	<SO>				X		
15	0Fh	<SI>				X		
16	10h	<DLE>						
17	11h	<DC1>			X			
18	12h	<DC2>			X			
19	13h	<DC3>			X			
20	14h	<DC4>			X			
21	15h	<NAK>						
22	16h	<SYN>						
23	17h	<ETB>						
24	18h	<CAN>	X					
25	19h	<EM>	X					
26	1Ah	<SUB>	X					
27	1Bh	<ESC>				X		
28	1Ch	<FS>		X				
29	1Dh	<GS>		X				
30	1Eh	<RS>		X				
31	1Fh	<US>		X				
32	20h	<SP>						
33	21h	!						
34	22h	"						
35	23h	❖ # (£)						
36	24h	❖ $ (¤)						
37	25h	%						
38	26h	&						
39	27h	'						
40	28h	(						
41	29h	)						
42	2Ah	*						
43	2Bh	+						
44	2Ch	,						
45	2Dh	-						
46	2Eh	.						
47	2Fh	/						
48	30h	0						
49	31h	1						
50	32	2						
51	33h	3						
52	34h	4						
53	35h	5						
54	36h	6						
55	37h	7						
56	38h	8						
57	39h	9						
58	3Ah	:						
59	3Bh	;						
60	3Ch	<						
61	3Dh	=						
62	3Eh	>						
63	3Fh	?						
64	40h	❖ @						
65	41h	A						
66	42h	B						
67	43h	C						
68	44h	D						
69	45h	E						
70	46h	F						
71	47h	G						
72	48h	H						
73	49h	I						
74	4Ah	J						
75	4Bh	K						
76	4Ch	L						
77	4Dh	M						
78	4Eh	N						
79	4Fh	O						
80	50h	P						
81	51h	Q						
82	52h	R						
83	53h	S						
84	54h	T						
85	55h	U						
86	56h	V						
87	57h	W						
88	58h	X						
89	59h	Y						
90	5Ah	Z						
91	5Bh	❖ [						
92	5Ch	❖ \						
93	5Dh	❖]						
94	5Eh	❖ ^						
95	5Fh	_						
96	60h	❖ `						
97	61h	a						
98	62h	b						
99	63h	c						
100	64h	d						
101	65h	e						
102	66h	f						
103	67h	g						
104	68h	h						
105	69h	i						
106	6Ah	j						
107	6Bh	k						
108	6Ch	l						
109	6Dh	m						
110	6Eh	n						
111	6Fh	o						
112	70h	p						
113	71h	q						
114	72h	r						
115	73h	s						
116	74h	t						
117	75h	u						
118	76h	v						
119	77h	w						
120	78h	x						
121	79h	y						
122	7Ah	z						
123	7Bh	❖ {						
124	7Ch	❖							
125	7Dh	❖ }						
126	7Eh	❖ ~						
127	7Fh	<DEL>	X					

CATEGORY	DEFINITION
Unmarked	Text and punctuation (alternate characters are shown in parenthesis)
	The T.50 character set shown is the preferred *International Reference Version (IRV)* Codes marked ❖ may represent characters other than those shown in some IA5 variants.
1	Transmission control characters
2	Presentation format control
3	Code extension control characters
4	Device control characters
5	Information separator characters
6	Miscellaneous control characters

<EOT> *End of transmission*—a symbol indicating the conclusion of the transmission of one or more texts.

<ENQ> *Inquiry*—A symbol used as a request for response from the remote station. The response may include status and/or identification.

<ACK> *Acknowledge*—a symbol transmitted by the receiving station to the sending station as an **affirmative** response to the sender.

<DLE> *Data link escape*—a symbol that changes the meaning of a limited number of subsequent symbols. It is used exclusively for supplemental transmission control functions.

<NAK> *Negative acknowledge*—a symbol transmitted by the receiving station to the sending station as a negative response to the sender.

<SYN> *Synchronous idle*—a fill character used in synchronous transmission systems to maintain sync during periods of no data transmission (idle time).

<ETB> *End of transmission block*—a symbol used to indicate the end of a data block where data is divided into blocks for transmission.

These characters are entered into a document by one of several methods depending on the application. Two of the possible methods are:

- While holding the control key on the keyboard, type the ASCII character located 64 positions greater than the control character; that is, the <ESC> character (ASCII 27) is entered by holding the control key and typing the "[" an ASCII 91. This is frequently shown as ^[in printed matter.
- While holding the <ALT> key on the keyboard, type the ASCII number of the symbol *using the numeric keypad;* for example the <ESC> character is entered as <ALT>027.

See also *IA5.*

ASCII art A drawing or picture composed exclusively with ASCII symbols. The art ranges from crude glyphs to fairly realistic renditions (when viewed from a distance). The ASCII art below is of whimsical fire-breathing, flying dragon.

```
xxxxxxxxxxxxxxxxxxxxxxxxxxxxxxxxxxxxxxxxxxxxxxxxxxxx
x                                                 x
x                                    /)    (\ /   x
x                    ss.sss. .s'          (  (   ))( x
x                  .ssSSSSSSSSMMs,        ) \  /,/,) x
x                sSSSSSSSSSSSSMS""SSs    /. ( .),  . x
x              "SSSSSSSSSSSSSSSS O SS    , ( .) ( , ) x
x             sSSSSSSSSSSSSSSSSSSSs, ,s )/ / < ./ / x
x            sSSSSSSSSS"SSSSSS"""SSSSSSS"SSSSM,  < ,/ (( x
x           sSSSSSSSSSs""SSSSssss "SSSSSSSS"  '.  ) x
x          sSSSSSSSSSSS'      '"""ss. ^V"    s- x
x         /sSSSSSSSSSSs,              '""""S .SSSs x
x        sssSSSSSSSSSSs,...            xMSS' \ x
x       s:ss..SSSSSSSSSSSSSSSSMMMMs.     xMS"S.  , s- x
x      / 's;;;s.SSSSSSSSSSSSSSSSSMMMMMMMMMSS"  .M.S' x
x     / \;s;;..SSSSSSSSSSSSSSSSSSMMMMa""    xMS$| x
x    s;;"s..SSSSSSSSSSSSSSSSSSSSSSMMs    xMS" S x
x  ,SMMMs.      / \sSS"".SSSSSSSSSSSSSSSXMMMMSMSMSS"  \ x
x -*sssSSSMMM      SS";;"S".$$SSSSSSSSSXM";....;;"M' x
x  "sssSSSMMM       ,"   '  $++wSSSSSSSSXs;../MMs x
x    sssSSSMMs        .X++SSSSSSSSSSSSSSSMMMM" x
x   .sssSSSSSMM     .ssX+++SSSSSSSSSSSSSSMMMM" x
x  *sssssssSSSSMMs .;SX+++++SSSSSSSSSSSSSSMMMMMM" x
x    'ssSSSSMMSs.++++++SSSSSSSSSSSSSSSSSMMMMMM"" x
x       sssSSMMMs.+SSSSSSSSSSSSSSSSSMMMMMMMM" x
x       sssSMMMMnm+SSSSSSSSSSSSSSMMMMMMM""" x
x       SSSMMMMMmsSSSSSSSSMMMMMM""' x
x      .SSSSMMMMMmSSSSMMMMMM"'    ....,SS.... x
x      SSSSMMMM1mSSSSMMMM"'    .sSSSSSSSSSSSSSSSSSsSS x
x      9SSSSSSSMSSSMMMMM"   .sSSSSSSSSSSSSSSSSSSSSSSs. x
x    :SSSSSSSSSSSMMMMM'   SSSSSSSSSSSSSSSSMMMSSSSSSSSS x
x   ((  .SSSSSSSSSSSMMMMM   SSSSSSSSSMMM"  "MMMMSSSSSSSSS x
x  )  (  SSSSSSSSSSSSMMMM.  SSSSSMMM"   "MMMMSSSSSSSSs x
x  (  )  SSSSSSSSSSSSSMMMs. SSSSSMMM"    MMMMSSSSSSSSs x
x  ( (   "SSSSSSSSSSSMMMMM.SSSSSMMM'     .MMMSSSSSSSSS" x
x  )  )  "SSSSSSSSSSSSMMMM.SSMM'        ..MMMSSSSSSSSSS x
x  ( ( \  "SSSSSSSSSSSSSMMMMMM,,,.     ..MMMMSSSSSSSSSS" x
x  )$ ) ) .SSSSSSSSSSSSSMMMMMMMM.   .MMMMMSSSSSSSSSS x
x ( ($$ ( \ "SSSSSSSSSSSSSSSSMMMMMMMMMSSSSSSSSSSSSSS x
x )$$$s )) ,,sSSSSSSMMSSSSSSSSSSSSSSSSSSSSSS"' x
x ( $$$$S/  ,,sSSSSSSMMSSSSSSSSSSSSSSSSSSSSS"" x
x \)_SSSSSSSSSSSSSSSSSSSM""    "";SSSSSSSSSSSS;" x
x   "SSSSSSSSSSSSSSSSSSSM"           """ x
x      "****;;;;"*"""    x
x                                                 x
xxxxxxxxxxxxxxxxxxxxxxxxxxxxxxxxxxxxxxxxxxxxxxxxxxxx
```

On the Internet, the newsgroup *alt.ascii-art* showcases *ASCII art* from around the world.

ASCII file A data file composed of only ASCII characters (000–127). Also called a text file or a text-only file that contains only alphanumeric characters, spaces, punctuation, an end-of-file marker, and sometimes a tab character. An *ASCII file* generally does not contain any formatting information. See also *ASCII* and *binary file*.

ASCII transfer A file transfer protocol involving the transmission of only the letters and symbols found in the ASCII character definition. Almost every communications program supports ASCII file transfer. But it does not offer any error correction or data compression—which means it is slow and somewhat unreliable (because the user may not know that an error occurred in transmission).

ASDS An abbreviation of Accunet Spectrum of Digital Services.

ASDSP An acronym from Application-Specific Digital Signal Processor.

ASE An abbreviation of Applications Service Element.

ASI An abbreviation of Adapter Support Interface.

ASIC An acronym from Application Specific Integrated Circuit (pronounced *A-sick*). Special-purpose chips, with logic designed specifically for a particular application or device. *ASICs* are designed from standard circuit cells from a library and then manufactured at an integrated circuit fabrication facility. *ASIC* advantages include fewer discrete components, lower cost volume manufacturing, lower power consumption, and increased reliability. Also called *gate arrays*.

ASK (1) An abbreviation of Amplitude Shift Key. (2) An abbreviation of American Standard Keyboard. Also called Dvorak.

ASL An abbreviation of Adaptive Speed Leveling. A U.S. Robotics term for automatically adjusting the data transmission rate of a modem when line conditions change. The rate can be adjusted either up (if the line conditions improve) or down (if conditions deteriorate). Training time is in the order of several seconds per adjustment.

ASN.1 An abbreviation of Abstract Syntax Notation One.

ASP (1) An acronym from Association for Shareware Professionals. (2) An acronym from AppleTalk Session Protocol (3) An acronym from Adjunct Service (or Signal) Point.

aspect ratio With respect to a two-dimensional image or object, the ratio of the width to the height.

aspheric In optics, literally meaning not spherical. Any class of usually rotationally symmetrical optical elements with a surface that is not spherical (including spheres with infinite radius, i.e., flat). Although not usually implied by the term, all conic sections are, strictly speaking, *aspheric*. The element may be either refracting or reflecting as with an *aspheric lens or aspheric mirror*.

ASR (1) An abbreviation of Automatic Send and Receive. (2) An abbreviation of Automatic Speech Recognition.

assertion The condition of a signal when it is in the same state as its name; for example, if a signal named "ON" is in the true state, it is said to be *asserted*.

The opposite of *assertion* is *negation*.

assigned number In the Internet, a numerical value that serves to distinguish a particular protocol, application, or organization in some context.

For example, assigned numbers distinguish the different executions of Ethernet protocols used by different implementers. Assigned numbers, which are not addresses, are assigned by the Internet Assigned Numbers Authority (IANA).

assistance call In telephony, a call to an operator for help in establishing a connection.

Association for Shareware Professionals (ASP) A trade group formed in April 1987 to help those actively involved in the shareware community and to strengthen the future of shareware as an alternative to commercial software. The goals of the *ASP* are:

- To inform users and programmers about shareware as a method of marketing and distributing software.
- To inform users about shareware programs.
- To assist members with software marketing.
- To encourage broad distribution of shareware through user groups, BBSs, and disk dealers who agree to identify shareware and educate users about the shareware concept.
- To provide a forum through which *ASP* members can communicate, share ideas, and learn from each other.
- To promote a professional code of ethics among shareware authors by setting standards for programming, marketing, and support requirements. Specifically:
  - *ASP* member evaluation programs must not be limited (so-called crippleware) in any way. In the spirit of the *try-before-buy* concept, users must be able to evaluate all features in a program before paying the registration fee.
  - *ASP* members must respond to every registration; at the minimum, they must send a receipt for payment.
  - *ASP* members must provide support for their product for at least 90 days from the date of registration.

Association of Computer Telephone Integration Users and Suppliers (ACTIUS) A UK forum for users and suppliers to increase awareness of the business benefits of CTI. *ACTIUS* develops education programs and information campaigns on CTI.

associative storage (1) A storage device whose storage locations are identified by their contents, or by a part of their contents, rather than by their names or positions. *Note: Associative storage* can also refer to this process as well as to the device. Also called *content addressable storage* and *content addressable memory* (*CAM*). (2) Storage that supplements another storage.

ASSP An abbreviation of Application Specific Standard Product.

astigmatism A lens or optical system in which orthogonal planes through the optical axis have different focal lengths.

ASVD An abbreviation of Analog Simultaneous Voice and Data.

asymmetric Literally meaning "lacking symmetry," it refers to the condition in which the comparison of similar parameters of two processes or parts of an entity are not equal. Examples include:

- *Asymmetric compression*—a compression algorithm in which the compression process is not the reverse of the decompression process.
- *Asymmetric modem*—a modem that subdivides the communication channel into two unequal channels: one high speed (for data transfer) and one low speed (for control functions).
- *Asymmetric transmission*—a duplex transmission process in which the available bandwidth is greater in one direction than in the other.
- *Asymmetric waveform*—a waveform in which the deviations from a reference in one direction are not the same as the deviations on the other side of that reference.

See also *symmetric*.

asymmetric digital subscriber line (ADSL) A communications system (as defined in ANSI T.413 and other specifications) that uses the existing copper wiring to provide one high-speed data channel to a customer (about 1.5–8 Mbps) and a low-speed reverse channel (about 64–1000 kbps). It is a means of delaying the expense of upgrading/replacing the telephone company's outside plant. *ADSL* operates to distances of about 18 000 ft. See also *xDSL*.

asymmetric transmission A method of high-speed transmission in which the available bandwidth is divided *unequally* into a transmit channel and a receive channel. For example, a telephone channel that has the capacity of 10,000 bps may be divided so that the transmit channel has a 9600 bps data rate while the receive channel is only 300 bps.

asymmetrical modulator A synonym for *unbalanced modulator*.

async. An abbreviation of ASYNChronous.

asynchronous A term describing signals in which corresponding significant events or entities do not necessarily happen at the same average rate. That is, the signals do not have a fixed time relationship. See also *anisochronous, heterochronous, homochronous, isochronous, pleisochronous,* and *synchronous*.

asynchronous balanced mode (ABM) An operating mode of the ISO High Level Data Link Control (HDLC) protocol that gives each node of a point-to-point connection equal opportunity to send. Since each node can be either a primary or a secondary node, either node may communicate without "permission" from any other node. See also *asynchronous response mode* (*ARM*) and *normal response mode* (*NRM*).

asynchronous communication A data communication protocol in which extra signal elements are appended to the data for the purpose of synchronizing individual characters or data blocks with the com-

munication receiver. Each block transmitted is self-sufficient and bears no exact time relationship with either the preceding or following block; that is, the interval between characters or data blocks is not fixed. An asynchronous signal does not have an associated transmitted clocking signal.

In the following diagram, a single character (with start and stop bits appended) is shown, along with its time-independent position in an asynchronous data transfer. For comparison, a synchronous stream is shown with fill characters (<SYN> characters) interspersed. *Asynchronous communication* is the most rudimentary type of communication as the originating and receiving machines do not have to be synchronized. It is cheap, reliable, and common among PCs and minicomputers. Its disadvantage is the large overhead (the number of extra bits needed for data delineation)—typically 25%. Also called a *start-stop system*.

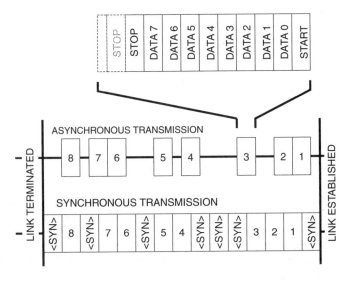

asynchronous network A network in which the clocks of individual members need not be *synchronous* or *mesochronous*. Also called a *nonsynchronous network*.

asynchronous operation (1) A sequence of operations in which the operations occur without a fixed time relationship to any event. (2) An operation that occurs without a regular or predictable time relationship to a specified event.

Asynchronous Response Mode (ARM) A communications mode, in the ISO High Level Data Link Control (HDLC) protocol, in which a secondary (slave) node can initiate communications with a primary (master) node without first getting permission from the primary node. This operation is in contrast to the normal response mode (NRM), in which the primary node must initiate any communication, and to the asynchronous balanced mode (ABM), in which the two nodes are equal.

asynchronous transfer mode (ATM) A high-speed, connection-oriented, packet data transport mechanism designed to operate over multiple physical media—for example, DS1 at 1.544 Mbps, E-1 at 2.048 Mbps, DS3, E-3 or SDH/SONET at 622 Mbps. Generally, digital transmission speeds are from 34 Mbit/s to 622 Mbit/s, although nonstandard rates such as IBM's 25 Mbit/s format are found. It is considered asynchronous because the recurrence of cells does not depend on the bit rate of the transmission system—only the rate of the source. It is intended to support a wide range of services, including voice, data, and multimedia.

The *ATM* standard is an outgrowth of broadband ISDN (B-ISDN) and was established by ITU-T and the *ATM forum* (a group of interested parties that is extending the standard to cover additional applications). The connection in an *ATM* network is a virtual communications circuit; that is, the bandwidth of the physical circuit is allocated only when a non-idle cell is being transferred. As a result, each user can, within limits, use the full network bandwidth when required. All switching is done at the cell rate, not the virtual channel rate. This allows switching cells of different calls, each having a different virtual rate. *ATM* uses a 53-octet (byte) cell (fixed-length packet) containing a 5-octet header and a 48-octet payload (containing the user data). The header contains the information necessary to route the message from point to point. (This eliminates the need for the switching node to interpret a source/destination address into routing information.)

ATM CELL FORMAT

Octet No.	Bit Number							
	8	7	6	5	4	3	2	1
1	GFC			VPI				
2	VPI			VCI				
3	VCI							
4	VCI			PT			CLP	
5	HEC							
6 · · · · 53	Information Payload							

where:

GFC is the *generic flow control*—which allows for various interconnect schemes, that is, ring, star, or bus LAN connections.

VPI is the *virtual path identifier* field—which provides 256 unique path identifiers.

VCI is the *virtual channel identifier field*—which provides the capability of manipulating a set of up to 2^{16} ATM connections as one unique path.

PT is the payload type field—which is used to distinguish cells that carry user information from Connection Associated Layer Management information cells that carry service information that allows the user to monitor the quality of the connection.

PT	Description
000	User data cell, congestion not experienced, SDU-type 0
001	User data cell, congestion not experienced, SDU-type 1
010	User data cell, congestion experienced, SDU-type 0
011	User data cell, congestion experienced, SDU-type 1
100	Segment OAM F5 flow related cell
101	
110	Reserved for future traffic control and resource management
111	Reserved

CLP is the *call loss priority* field (sometimes called the discard eligibility field)—which can be used to identify a level of service and, in the event of network congestion, identify which cells may be discarded.

HEC is the *header error control* field—which detects multiple bit errors and corrects single bit errors.

Information payload is the user-supplied data—none of the ATM routing or switching functions processes any of the contents of the payload, nor do they do any error checking on the field.

ATM, as specified in international standards, is asynchronous in the sense that cells carrying user data need not be periodic. Also known as *fast packet*.

AT (1) From A̲dvanced T̲echnology—the successor to the XT model IBM PC. A 16-bit data bus, Intel 80286-based computer. **(2)** An abbreviation of A̲mpere-T̲urn.

AT command set A *de facto standard* set of commands used to control a modem by a DTE. Also known as the *Hayes Command Set* because it was developed by Hayes Microcomputer Products, Inc. for its modem product line. With few exceptions, all of the commands start with the sequence "AT" (or "at"). The "AT" is a contraction of the word A̲T̲tention. An example of an "AT" command is *ATDT555-0000*. This command directs the modem to go off-hook and dial (D) the number 555-0000 using DTMF signaling (T).

A list of many of the AT commands used is included in the *AT command appendix*. The list is compiled from several manufacturers and from several modem models. Also called the *AT modem control language*.

ATB An abbreviation of A̲ll T̲runks B̲usy. A condition in which all trunks of a routing group are engaged. It is indicated by one or more tones interrupted at a rate of 120-ipm (impulses per minute). See also *call progress tones*.

ATE An abbreviation of A̲utomatic T̲est E̲quipment.

ATM An abbreviation of A̲synchronous T̲ransfer M̲ode.

ATM adaptation layer (AAL) A protocol layer (defined by ITU-T recommendations I.362 and I.363) which resides on top of the *ATM layer* and below the *connectionless convergence protocol layer*. It is a set of standard protocols that translate user information to and from upper layers of the protocol stack into the standard size and format of the payload of an ATM cell. Each of the layers of the *AAL* protocol is subdivided into two sublayers: the segmentation and reassembly (SAR) sublayer and the convergence sublayer.

AAL is implemented by the end-user equipment and is used for a variety of reasons, such as

- Mapping large user data packets into the 48-octet payload field of an ATM cell,
- Solving problems specific to applications, or
- Providing application-dependent switching.

The characteristics of the *AAL* layer services are outlined in the following table.

CLASSIFICATION OF AAL SERVICES

Service Feature	Type			
	1	2	3	4
Timing (between source and destination)	Required		Not Required	
Bit Rate	Fixed		Variable	
Connection Mode	Connection Oriented			Connectionless

- *AAL1*—addresses constant bit rate (CBR) traffic such as voice and video. It is used to address applications that are sensitive to both time delay and cell loss; it is also used to emulate a leased line.

- *AAL2*—addresses time-sensitive, variable bit rate (VBR) traffic, such as packet voice.
- *AAL3/4*—addresses bursty, connection-oriented traffic, for example, LAN file transfers. It is designed for traffic that can tolerate some time delay but not cell loss.
- AAL5 (not shown in the table)—addresses bursty LAN data traffic, as does *AAL3 & 4*, but with less overhead.

In the connectionless transfer mode, *AAL* is compatible with the connectionless transfer mode of IEEE 802.6.

ATM cell delineation A means of determining the cell's boundaries in a continuous bit-flow situation.

ATM cross connect A switch that employs only the *virtual path identifier* control fields to identify cells belonging to a channel; it ignores the *virtual channel identifier* control field.

ATM downlink A 155-Mbps link containing one or more virtual channels.

ATM Forum The organization that defines and maintains the ATM standards.

ATM layer (1) Layer 2 of the ATM architecture. It is responsible for the generation and management of cells (including: routing and error control). Also called the *Cell Layer*. See also *Asynchronous Transfer Mode (ATM)*. (2) The layer in the ATM protocol stack for routing and processing activities. Examples include: building the ATM header, header validation, quality of service specification, cell routing (using virtual path identifiers and virtual channel identifiers), cell multiplexing and demultiplexing, cell reception, payload identification, prioritization, and flow control.

ATM switch A switch that utilizes both the *virtual path identifier* (*VPI*) and the *virtual channel identifier* (*VCI*) control fields. See also *ATM*.

atmospheric absorption The loss of electromagnetic energy traversing the atmosphere due to both true absorption (dissipation) and scattering (by particles in the atmosphere). The dominant effect depends on the wavelength of the propagating electromagnetic energy. See also *rain attenuation*.

atmospheric duct A horizontal layer in the lower atmosphere in which the vertical refractive index gradients affect radio waves of sufficiently high frequency in the following ways:

- They are guided or focused within the duct. (The reduced refractive index at the higher altitudes bends the signals back toward the Earth.)
- They tend to follow the curvature of the Earth. (Signals in the higher refractive index layer tend to remain in that layer because of the refraction above and reflection below.)
- They experience less attenuation in the ducts than they would if the ducts were not present.

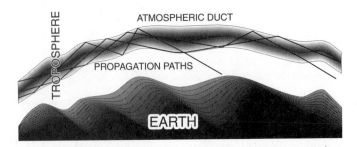

Also called a *tropospheric duct*.

atmospheric noise Radio noise caused by natural atmospheric processes, for example, lightning discharges in thunderstorms.

atmospheric refraction The bending (refraction) of a beam of electromagnetic radiation when it passes through the atmosphere at an oblique angle. The effect is caused by variations in the index of refraction. See also *atmospheric duct* and *ionosphere*.

atmospheric turbulence Perturbations in the atmosphere that cause random spatial and temporal phase and amplitude variations destroying the quality and coherence of laser beams.

ATOB An abbreviation of <u>A</u>SCII <u>TO</u> <u>B</u>inary (pronounced "A" to "B"). A UNIX program that converts ASCII files into binary files. The inverse operator is *BTOA*.

ATOW An abbreviation of <u>A</u>cquisition and <u>T</u>racking <u>O</u>rder<u>W</u>ire.

attached document A document that has been appended to an e-mail message. Attached documents may be text, graphics, programs, and so on.

Attaching text-based documents to Internet e-mail represents little problem; however, binary documents (programs, graphics, etc.) need to be encoded with a program such as BINHEX or UUENCODE in UNIX before attachment and transmission. The received file then must be decoded with the matching decoder, UUDECODE, for example, before it can be used.

attachment unit interface (AUI) (1) The IEEE 802.3 specified cable and connector (local drop) between a network device (data terminal) and the *medium attachment unit* (MAU) on Ethernet 10Base5 communication segments. The cable consists of shielded twisted pairs and is terminated with a 15-pin connector. (2) Refers to the host back panel connector to which an *AUI* transceiver cable attaches.

attack time The time interval between:

- The instant that a signal at the input of a device or circuit exceeds the activation threshold of the device or circuit and
- The instant that the device or circuit reacts in a specified manner to that stimulus.

Attack time occurs in devices such as clippers, peak limiters, compressors, and voice-operated switches (VOXs).

attention signal A signal used by commercial broadcast stations (AM, FM, and TV) to actuate muteable receivers for interstation receipt of emergency cueing announcements and broadcasts involving a range of emergency contingencies posing a threat to the safety of life or property.

attenuation The decrease or weakening in intensity (voltage, current, light level, or power) of a signal, beam, or wave as a result of energy absorption or scattering out of the path to the detector (but not including the reduction due to geometric spreading). A distinction must be made as to whether the *attenuation* is of signal power or signal electric field strength. *Attenuation* is usually expressed in decibels (dB).

Attenuation may be intentional or unintentional. An example of intentional *attenuation* is a volume control, whereas losses in a long-distance telephone connection are unintentional. *Attenuation* is often used as a misnomer for *attenuation coefficient*, which is expressed in decibels per unit length. The opposite of *attenuation* is *amplification*.

attenuation coefficient The rate of reduction of average signal power with respect to distance along a transmission path. It is the sum of the absorption coefficient and the scattering coefficient. *Attenuation coefficient* is expressed in decibels per unit length, for example, decibels per kilometer (dB/km). Also called *attenuation rate*.

attenuation constant **(1)** The real part of the propagation constant in any electromagnetic propagation medium. It is usually expressed as a numerical value per unit length. For each medium, the attenuation constant may be calculated or experimentally determined. **(2)** In fiber optics, the real part of the axial propagation constant for a particular propagation mode in an optical fiber.

attenuation distortion **(1)** The variation in amplitude response of a channel as a *function of frequency.*

The distortion measurement is the comparison of the signal level at a specified frequency in the channel bandwidth to the level at a reference frequency. (In telephone circuits, the reference frequency is 1004 Hz in the United States and 800 Hz or 1020 Hz for ITU-T.) The first drawing below illustrates the allowable (and mandatory) frequency-dependent *attenuation distortion* through a central office switching system. The nonshaded region indicates the acceptable *attenuation distortion* region. In some areas, such as the 45–65 Hz region, attenuation **must** be *greater than* 20 dB. In the region 300–3000 Hz, the distortion must be *less than* about ±1/3 dB. Also called *frequency response.*

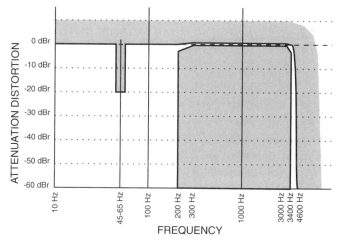

(2) The variation in the response of a channel as a *function of signal level;* also called *gain tracking.* As a simple example, assume that a 1-V input signal level produces a 1-V output signal. Then, reducing the input by exactly 50% should reduce the output by exactly 50%. The actual output may be slightly more or less than 50%, however, and the difference between the expected value and the actual value is the attenuation distortion (or *gain tracking error*). The drawing illustrates one mask defining the maximum allowable *attenuation distortion* for a circuit. The nonshaded area indicates the permissible region, whereas the shaded region is forbidden. The output signal level error must be less than 0.5 dB from the expected level when the input is between −40 dBm and +3 dBm. See also *frequency response* and *gain tracking.*

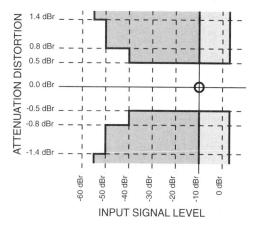

attenuation limited operation A description of the condition that exists when attenuation, rather than bandwidth, limits the performance of a communications link.

attenuation rate A synonym for *attenuation coefficient.*

attenuator In electrical or optical systems, a device that reduces the amplitude of a signal without appreciably distorting its waveform. *Attenuators* are usually passive devices and may be fixed, continuously adjustable, or incrementally adjustable. The input and output impedances of an *electrical attenuator* are usually matched to the impedances of the signal source and load, respectively. A common attenuator symbol is

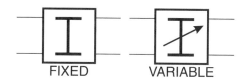

Fixed attenuators are often called *pads,* especially in telephony and audio systems.

attribute **(1)** The characteristic or property of an entity (hardware, programs, or files). The specific attributes defined vary from system to system, but attributes are used in every operating system and networking environment. Certain attributes assume or replace others, and some attributes can override access rights.

For example, the length of a data block, the transmit signal level, and the transmit bit rate are all attributes of a modem's transmission. DOS files have four attributes with an acronym RASH:

R   Read-only
A   Archive required
S   System
H   Hidden

Network operating systems may add others; for example, Novell's NetWare provides 10 attributes in addition to the 4 of DOS:

C	Copy inhibit	Ra	Read audit
D	Delete inhibit	S	Shareable
I	Indexed	T	Transactional used
P	Purge when deleted	Wa	Write audit
R	Rename inhibit	X	Execute only

(2) In database management, a property inherent in an entity or associated with that entity for database purposes. **(3)** In network management, a property of a managed object that has a value. Attributes may be either mandatory or conditional. Mandatory initial values for attributes can be specified as part of the managed object class definition.

audible ringing tone See *ringback tone.*

audio frequency (AF) Any frequency corresponding to a sound wave that can be heard by a human. Normally, frequencies are in the range of 15 Hz to 20,000 Hz.

audio response unit (ARU) A device that provides synthesized voice responses to caller-transmitted dual tone multifrequency (DTMF) signaling. Call processing is based on:

- The call originator's DTMF input.
- Information received from a host database.
- Information in the incoming call, such as the time of day.

ARUs are used to increase the number of information calls handled and to provide consistent quality in information retrieval.

audiotex A communications system that allows a host computer to pass data to a voice mail computer where it is interpreted (text to voice conversion) and delivered over the telephone.

A store-and-forward mechanism for digitized voice.

audit To conduct an independent review and examination of records and activities in order to test the adequacy and effectiveness of some aspect of a system, for example, data integrity, system control, security. The audit is intended to ensure compliance with established policies and procedures, and to recommend any necessary changes in controls, policies, or procedures.

audit review file A log file created and updated by statements included in a computer program for the explicit purpose of providing data for auditing.

audit trail A record of both completed and attempted events (accesses, service, and transactions) that may be reviewed at a time after the event occurred in order to determine cause and effect or accuracy of the event. For example, the audit trail of a LAN may include the time a user logs on and off, the terminal number, the user ID, which files were accessed, and what resources were used.

AUI An abbreviation of <u>A</u>ttachment <u>U</u>nit <u>I</u>nterface.

AUP An abbreviation of <u>A</u>cceptable <u>U</u>se <u>P</u>olicy.

AUP-free Generally refers to the absence of the National Science Foundation's (NSF) Acceptable Use Policies (*AUP*), which restrict commercial activities on the NSFNet backbone.

aurora Sporadic radiant emission (light and radio) from the Earth's upper atmosphere that usually occurs about the north and south magnetic poles (although they may occasionally be observed within 40° or less of the equator). In the northern hemisphere, the *aurora* is called the Aurora Borealis (Northern Lights), and in the southern hemisphere, it is called the Aurora Australis (Southern Lights).

Maximum *auroral* intensity is generally near an altitude of 100 km above the Earth. *Auroras* are most intense at times of intense magnetic storms caused by sunspot activity. They interfere with radio communications and even power distribution networks.

Australian Academic and Research Network (AARNet) An Internet member network with approximately 70 000 hosts. The network services universities, government agencies, companies, and research organizations.

The domain name of *AARNet* is AU.

authentication Any security measure (such as user ID and password, challenge and response, or public key schemes) designed to establish

- That the user of the system or its services is the user of record; that is, the user or device is the one registered in the account.
- An individual's eligibility to receive specific categories of services or information.
- The legitimacy of a transmission attempt.
- The validity of a transmission, message, or originator.
- The integrity of data stored, transmitted, or otherwise exposed to unauthorized modification in an automated information system.

Among the simplest strategies developed is the use of user IDs and passwords. A more powerful scheme, called *digital signatures,* is very effective and almost impossible to fool (unless one has access to the private encryption key). With digital signatures, the message originator uses the receiver's public key to encrypt the transmission and his private key to "sign" it. At the receiving end, the receiver uses the originator's public key to validate the signature and the receiver's private key to decrypt the transmission. The ITU distinguishes two levels of authentication for directory access in its X.509 recommendations:

Simple authentication, which uses just a password and works only for limited directory domains.

Strong authentication, which uses a public key encryption method to ensure the security of a communication.

authoring software A software package that allows a user to create interactive media and multimedia presentations.

authorization **(1)** The rights granted to a user to access, read, modify, insert, or delete certain data, or to execute certain programs. **(2)** Access rights granted to a user, program, or process.

auto answer A mode in which the DCE will answer an incoming call after a preset number of rings *without external intervention.* This feature is not an autonomous behavior of the DCE but a cooperative action of the DCE and DTE software. In contrast, *manual answer* requires a person to initiate the answer sequence. The mode is frequently indicated on external modems by the AA (auto answer) indicator.

auto baud rate (ABR) The automatic resolution and matching of data transmission speed and protocol. The receiving terminal determines the data speed, number of data bits, and number of stop bits based on the reception of the first character—a predesignated sign-on character (generally a <CR>).

auto dial The ability of a DCE to initiate a telephone connection by going off-hook and transmitting the telephone number to the central office switching equipment (either by dial pulse or DTMF signaling). This is in contrast to *manual dialing,* where a person does these operations.

auto magic A slang term stating that an action or process is automatic; the explanation of how it works is too long or complex to explain in the space available. Of course, it may also mean that the person describing the action does not know how it works either! Also spelled *automagic.*

auto partition **(1)** A function of repeaters in which a faulty segment is automatically isolated from the rest of the network. *Auto partitioning* prevents the fault from affecting the entire network. When the fault condition is rectified, the segment is automatically reconnected to the network. **(2)** An error-handling procedure in 10BaseT networks. When 32 consecutive collisions are sensed by a port in a hub or concentrator from its attached workstation or network segment, or when a packet that exceeds the maximum allowable length is received, the port stops forwarding packets. The port continues to monitor traffic and will automatically begin normal forwarding when the first correct packet is received.

auto reliable mode A communication mode in which the modems communicate using the best possible protocol connection (from V.42/MNP error correction to Normal). The modems first attempt to connect using the V.42 LAP-M error control protocol; if that fails, MNP protocol is tried. If a connection is not established with either protocol, a normal connection is established.

AUTODIN An acronym from <u>AUTO</u>matic <u>DI</u>gital <u>N</u>etwork. See also *Defense Data Network (DDN)*.

automated attendant (AA) A feature of a switching system (usually a PABX) which answers incoming calls with a recording (or series of recordings) and allows the caller to route the call to the desired extension by using DTMF signaling.

automated information system (AIS): (1) An assembly of computer hardware, software, and/or firmware configured to accomplish specific information handling operations, such as transmission, reception, dissemination, processing, and storage. **(2)** In information security (INFOSEC), any equipment or interconnected system or subsystem of equipment that is used in the automatic acquisition, storage, manipulation, management, movement, control, display, switching, interchange, transmission, or reception of data. It includes computer software (such as word processing systems), firmware, and hardware (including networks, other electronic information handling systems, and associated equipment).

automated newspaper An application available on many networks which allows a user to select the topics of news to be retrieved and grouped into a "custom newspaper."

For example, on USENet servers that receive *Clari* postings, an *automated newspaper* can be created by subscribing to only those *clari.news.xxxx* categories of interest.

automatic alternate routing (AAR) In networks, the process by which network traffic is automatically routed to maximize throughput, minimize distance, or balance channel usage.

automatic answering A terminal feature in which the called terminal automatically responds to the calling signal and the call may be established whether or not the called terminal is attended by an operator, e.g., an answering machine.

automatic call distributor (ACD) Hardware or software within a telephone switching system that routes calls (based on caller signaling input and the call routing database) to a specific group of terminals. Typically, airlines, hotels, and other telephone marketing organizations use it. It both recognizes and answers incoming calls according to instructions in the database and caller actions, before sending the call to the proper device, operator, or agent.

It also offers management information on the type and volume of calls, and efficiency of the agents

- *ACD basic routing.* If the number of active calls is less than the number of terminals, an incoming call will be routed to the terminal that has been idle the longest. If all terminals are busy, incoming calls are held in a first-in-first-out queue until a terminal becomes available.
- *ACD application bridge.* Pertaining to the link between an ACD and an information database resident on the user's data server. It allows the ACD to communicate with the server's database to assist in call routing.
- *ACD application-based routing.* A means of routing and tracking calls based on the call type (e.g., sales, service).
- *ACD call back messaging.* Enables callers to choose to leave a message for an agent rather than be placed in a hold queue when all agents are busy. Of course, the caller can choose to wait for a live agent if the situation warrants.
- *ACD caller directed call routing.* Enables callers to direct the call to the appropriate agent group without the assistance of an operator. This is accomplished via the infamous "Press 1 for sales, Press 2 for . . ." Also called *auto-attendant capability.*

- *ACD conditional routing.* The ability of the system to monitor various parameters of the system and then route calls based on that information. Some of the parameters monitored include call queue length, average holding time, and agent group loading. This enables the system to refer calls to a secondary group if the primary group is saturated and the secondary group has available personnel to service the calls.
- *ACD data-directed call routing.* The ability of the system to route calls based on user-entered information, such as an account number. The ACD then routes the call appropriately after the information has been verified in the database server.
- *ACD intelligent call processing.* Similar to *data directed call routing,* with the additional capability to monitor system parameters as in *conditional routing.*

automatic callback A service feature (similar to but not the same as camp-on) that permits a user, when encountering a busy condition, to instruct the system to retain the called and calling numbers and to establish the call when a line becomes available. The feature may be implemented in the terminal or in the switching system, or it may be shared between them. Also written *automatic call back.* See also *camp-on.*

automatic calling unit (ACU) A dialing apparatus that permits subscriber devices to automatically originate calls over the communications network.

Generally, a limit is imposed on the maximum number of unsuccessful call attempts to the same address within a specified period. The limit may be by mandate (e.g., FCC regulation) or by network design criteria. Also called an *automatic dialing unit (ADU).*

automatic data handling (ADH) A generalization of automatic data processing (ADP) to include data transfer, that is, the combination of data processing and data transfer.

automatic data processing (ADP) (1) An interacting assembly of procedures, processes, methods, personnel, and equipment to perform, automatically, a series of operations on data. **(2)** Data processing by means of one or more devices that use common storage for all or part of a computer program, and also for all or part of the data necessary for execution of the program; that execute user-written or user-designated programs; that perform user-designated symbol manipulation, such as arithmetic operations, logic operation, or character-string manipulations; and that can execute programs that modify themselves during their execution. *Note:* Automatic data processing may be performed by a stand-alone unit or by several connected units. **(3)** Data processing largely performed by automatic means. **(4)** That branch of computer science and technology concerned with methods and techniques relating to data processing largely performed by automatic means.

automatic dialing See *automatic calling unit.*

Automatic Digital Network (AUTODIN) Formerly, the Defense Communications System worldwide data communications network, superseded by the Defense Switched Network (DSN).

automatic exchange In a telephone system, a switching system in which communication links between a calling user and a called user are established solely by the originating user equipment without human intervention at the central office or branch exchange.

automatic frequency control (AFC): A device or circuit that maintains the frequency of an oscillator within its specified limits.

automatic gain control (agc) A signal processing system or device in which the amplitude of a signal at a specified point is automatically regulated to be approximately the same value regardless of the input excitation level.

automatic identified outward dialing (AIOD) A service feature of some Centrex, switching, or terminal devices that provides the user with an itemized statement of usage on directly dialed calls for all calls initiated by each telephone extension. *AIOD* utilizes the *automatic number identification* (*ANI*) equipment that also provides *automatic message accounting* (*AMA*).

automatic message accounting (AMA) In telephony, a service feature that automatically records data regarding user-dialed calls (such as time of day, connect time, and called number).

automatic message processing system (AMPS) Any set of resources, procedures, and methods used to collect, process, and distribute messages by automatic means.

automatic number identification (ANI) Originally, a term used to describe the automatic identification of a calling station by the telephone company switching equipment for billing (message accounting) purposes. *ANI* provides the originating local telephone number of the calling party. This information is transmitted as part of the digit stream in the signaling protocol and is included in the Call Detail Record for billing purposes.

It is now used (though strictly improperly) to include the *caller ID* feature, that is, to provide the called party the number associated with the telephone station(s) from which switched calls are originated.

automatic operation The supervision and control of systems, devices, or processes in a specified manner and at specified times or conditions by mechanical or electronic means and without human intervention (other than to initially program the controlling algorithm).

automatic partition algorithm A method by which a repeater can automatically disconnect an errant segment from a network—that is, if that segment is not functioning properly. This can happen, for example, when a broken or unterminated cable causes too many collisions. When the fault subsides, the network segment may be reconnected.

automatic protection switching (APS) The ability of a network element to automatically detect a communications link failure and switch to a backup (protection) link. *1:n APS* provides one backup link for a group of *n* main links. If *n* = 1, one backup is provided for each main link, thereby providing 100% redundancy.

automatic redial A service feature that allows the user to dial, by depressing a single key (or at most a few keys), the most recent number dialed at that instrument. Automatic redial is often implemented within the telephone instrument but may be provided by a PABX or the central office. Also called *last number redial*. See also *automatic calling unit*.

automatic repeat request (ARQ) A basic error control technique for data transmission systems in which the receiving DCE detects transmission errors in a message and automatically requests a retransmission from the transmitting DCE. Usually, when the transmitter receives an *ARQ*, it retransmits the message until it is either correctly received or it is transmitted a specified number of times. Also called an *automatic retransmission queue* (*ARQ*), *error-detecting* and *feedback system*, or a *repeat request system*.

automatic ringdown (ARD) A private line providing priority telephone service, typically for key personnel, connecting a station instrument in one location to a station instrument in a distant location with automatic signaling. These circuits provide priority telephone service, typically for key personnel.

The automatic signaling used on these circuits causes the distant station instrument to ring when the local station instrument goes off-hook. Although generally the signaling is two-way, it may also have one-way signaling; that is, station "A" rings Station "B" when Station "A" goes off-hook, but Station "B" cannot ring Station "A." Sometimes called a *hot-line* because urgent communications are typically associated with this service. See also *verified off-hook*.

automatic rollback In NetWare's Transaction Tracking System (TTS), a feature that restores the starting state of a database if a transaction fails before completion.

automatic route selection (ARS) Electronic or mechanical selection and routing of outgoing calls without human intervention. Also called *alternate route selection* or *least cost routing* (*LCR*).

automatic send and receive (ASR) A teleprinter that includes not only a keyboard for data entry and printer for output but a paper tape reader and punch for automatic data sending and receiving. See also *keyboard send and receive* (*KSR*).

automatic signaling service A synonym for *automatic ringdown* or *hot-line service*.

automatic speed buffering (ASB) *ASB* is an attempt to permit computer equipment to transfer data to and from the DCE at a constant rate, regardless of the type or speed of the modem to modem connection.

automatic switching system **(1)** In telephony, a telecommunications exchange in which routine connections between a calling station and a called station are directed by the calling station equipment; that is, all operations required to set up, supervise, and release required connections are automatically performed in response to signals from the calling device. Control (the establishing and disconnecting of the path) is directed by the calling station equipment without the need for operator intervention or direction. **(2)** In data communications, a switching system in which all operations required to execute the three phases of information transfer transactions (access phase, information transfer phase, and disengagement phase) are automatically executed in response to signals from a user DTE. In an *automatic switching system,* the information transfer transaction is performed without human intervention, except for initiation of the access phase and the disengagement phase by a user.

automation **(1)** The implementation of a procedure or process by automatic means. **(2)** The conversion of a procedure, process, or equipment to automatic operation. **(3)** The methods of making procedures, processes, or equipment automatic, self-moving, or self-controlling.

autonomous computer A stand-alone computer. Any computer that is not forcibly controlled, directed, or operated by another computer or system. Examples of *autonomous computers* include both the networked and nonnetworked PCs. An automated teller machine (ATM), however, is not an *autonomous computer* as it is controlled by the *mainframe computer* at a bank data processing center.

autonomous system (AS) A group of one or more interconnected networks under the control of a single administrative organization. Also called an *administrative domain* or a *routing domain* in the Internet.

AUTOSEVOCOM An acronym from <u>AUTO</u>matic <u>SE</u>cure <u>VO</u>ice <u>COM</u>munications network. A worldwide, switched, secure voice network developed specifically for the requirements of the U.S. Department of Defense (DoD) long-haul, secure voice requirements.

AUTOVON An acronym from AUTOmatic VOice Network. Superseded by the Defense Switched Network (DSN). Formerly, the principal long-haul, unsecure voice communications network within the U.S. Defense Communications System.

AUX In personal computers, an abbreviation and logical device name for AUXiliary device. The name is reserved by MS-DOS and usually refers to COM1, the system's first serial port.

auxiliary operation An off-line operation performed by equipment not under control of the processing unit.

auxiliary power Electric power that is provided by an alternate source and that serves as backup for the primary power source at the station main bus or prescribed sub-bus. Classes of power sources include:

- *Class A*—a primary power source, i.e., a source that assures an essentially continuous supply of power.
- *Class B*—an auxiliary (or standby) power plant to cover extended outages, i.e., periods of the order of days.
- *Class C*—a short-term auxiliary power source, i.e., a 10- to 60-second quick-start unit to cover short-term outages of the order of hours.
- *Class D*—an uninterruptible nonbreak unit using stored energy to provide continuous power within specified voltage and frequency tolerances.

Also called an *uninterruptible power supply* (*UPS*).

auxiliary storage (**1**) Storage that is available to a processor only through its input/output channels. (**2**) In a computer, any storage that is not internal memory (main memory), i.e., is not random access memory (RAM), and whose main function is to store large volumes of data and programs not being actively used by the processor.

Examples of *auxiliary storage* media include magnetic disks, optical disks (e.g., CD-ROM), and magnetic tape. Also called *auxiliary memory*. See also *main storage*.

availability (**1**) The fraction of time a component, device, or system is operable and capable of performing its intended function or the ratio of the total time the entity is available for use during a given interval to the length of the interval. The conditions determining operability must be specified. Mathematically, availability is expressed as:

$$Availability = \frac{Uptime}{Uptime + Downtown} = \frac{MTBF}{MTBF + MTTR}$$

$$= 1 - Unavailability$$

Typically, availability objectives are expressed as a decimal fraction, e.g., 0.9995. (**2**) In telephony, the number of output ports of a group that can be reached from a given input port of a switching stage, system, or network.

available bit rate (ABR) An ATM network class of service in which the network makes a "best effort" to meet the traffic bit rate requirements.

available line (**1**) In voice, video, or data communications, a circuit between two points that is ready for service and is in the idle state. Also called a *useful line*. (**2**) In facsimile, that portion of the scanning line that can be specifically used for the image.

available time (**1**) From the point of view of a user, the time during which a functional unit can be used. (*Note:* The unit may be operating in a degraded but functional mode.) (**2**) From the point of view of operating and maintenance personnel, a synonym for uptime, i.e., the time during which a functional unit is fully operational.

avalanche multiplication An effect that occurs in a semiconductor photodiode that is reverse-biased just below its breakdown voltage. That is, photocurrent carriers, (electrons) are swept across the P-N junction with sufficient energy to ionize additional material, creating additional electron-hole pairs in a regenerative action, thereby achieving a signal gain.

avalanche photodiode (APD) A device used to detect light and convert it to electrical signals. When illuminated by even very faint light signals, the *APD's* conductivity dramatically increases due to an "avalanche effect." Avalanche refers to the characteristic in which many electrons are emitted for each detected photon (light particle). *APDs* also have a high signal-to-noise ratio. These properties make *APDs* useful in high bit rate optical receivers where it can regenerate signals that have traveled a long distance over optical fiber.

An *APD* operates with a reverse bias voltage. As the reverse bias voltage increases toward breakdown, electron-hole pairs are created by absorbed photons. Avalanche occurs when these pairs acquire sufficient energy to create additional pairs when incident photons collide with the ions, i.e., the holes and electrons.

avatar An icon used in graphical online chat groups to represent a user. When a user types information at his terminal, that information appears above his *avatar* at all nodes in the group. Similarly, if a user moves an *avatar,* all other users see it move.

AVD circuits An abbreviation of Alternate Voice/Data circuits. Communications circuits that have been conditioned to handle both voice and data traffic.

average See *mean.*

average busy season busy hour (ABSBH) In telephony, an average of individual hourly periods during the season when the items of equipment being measured are the busiest.

average rate of transmission A synonym for *effective transmission rate.*

AVI An abbreviation of Audio-Video Interleaved. The file format Microsoft specifies for Video for Windows. Blocks of video and audio data are interspersed together in this format.

Avogadro's number 6.0221367×10^{23} particles per mole. Named after Amedeo Avogadro (1776–1856), an Italian professor of physics, it is the number of atoms (molecules) in that quantity of a substance that weighs (in grams) the same as its molecular weight (called a mole).

Name	Formula	Weight of 1 mole
Hydrogen	H_2	2.016 g[1]
Helium	He	4.003 g[1]
Carbon	C	12.011 g
Oxygen	O_2	32.000 g[1]
Silicon	Si	28.086 g
Water	H_2O	34.016 g
Sulfuric Acid	H_2SO_4	98.082 g

[1]At 0°C and at 1 atmosphere of pressure.

avoidance routing The assignment of a circuit path to avoid certain known critical or trouble-prone circuit nodes.

AWG (**1**) An abbreviation of <u>A</u>merican <u>W</u>ire <u>G</u>age. A scale of the gage sizes of wire for both solid and stranded conductors. Smaller physical wire sizes have a larger gage number; for example, 26 *AWG* wire is smaller than 20 *AWG* wire. Most telephone wire ranges from 20 to 28 *AWG*. See also *cable, wire,* and *wire gage.* (**2**) An abbreviation of <u>A</u>rbitrary <u>W</u>aveform <u>G</u>enerator.

AWGN An abbreviation of <u>A</u>dditive <u>W</u>hite <u>G</u>aussian <u>N</u>oise. See also *white noise.*

axial propagation constant In fiber optics, the propagation constant evaluated along the optical axis of the fiber in the direction of transmission. It is composed of two parts:

- The real part—called the attenuation constant, and
- The imaginary—called the phase constant.

axial ratio Of an elliptically polarized electromagnetic wave, the ratio of the magnitudes of the major axis and the minor axis of the ellipse described by the electric field vector.

axial ray A light ray that travels along the optical axis of a device.

AZ An abbreviation of <u>AZ</u>imuth.

B

b (1) An abbreviation for bit, binary, or sometimes baud. The meaning is typically clear, based on the context of usage; for example, 11011b indicates the number 11011 is a binary number, whereas 9600 b/s indicates 9600 bits per second. Although usually spelled out or abbreviated *Bd,* baud is sometimes abbreviated as in 2400 b rather than 2400 baud (or 2400 Bd). (2) The symbol for magnetic flux density, measured in Webers per square meter (Wb/m^2), or gauss (G) in the cgs system.

B (1) The symbol for *bel.* (2) An abbreviation for *baud.* (3) A symbol for <u>B</u>ias. (4) A symbol for <u>B</u>yte.

B channel A switchable (and optionally transparent) 64-kbps digital communications channel in an ISDN network. Two *B channels* are included in the ISDN basic rate service along with the *D signaling channel.*

B-CDMA An abbreviation of <u>B</u>roadband <u>C</u>ode <u>D</u>ivision <u>M</u>ultiple <u>Ac</u>cess.

B-ISDN An abbreviation of <u>B</u>roadband <u>ISDN</u>. See *ISDN.*

B.n recommendations An ITU-T series of specifications concerning definitions, symbols, and classifications. See the Appendix E for a complete list.

b/s An abbreviation for *bits per second.*

B3ZS An abbreviation of <u>B</u>ipolar with <u>3</u> <u>Z</u>ero <u>S</u>uppression. See *BnZS.*

B8ZS An abbreviation of <u>B</u>ipolar with <u>8</u> <u>Z</u>ero <u>S</u>uppression. See *BnZS.*

babble Unintelligible crosstalk. The *combined crosstalk,* or interference, in a channel of a multiple channel communications system. Crosstalk induced in a given line by all other lines.

BABT An abbreviation of <u>B</u>ritish <u>A</u>pprovals <u>B</u>oard for <u>T</u>elecommunications. An independent organization that tests telecommunications equipment. *BABT*'s processes are known for being rigorous and labyrinth-like in complexity.

Baby Bell See *regional Bell operating company (RBOC).*

BAC An acronym from <u>B</u>asic <u>A</u>ccess <u>C</u>ontrol.

back end In a client–server architecture, that portion of an application running on the server which does the actual work for the application. It provides services across the network that have been requested by the client. The *front end* runs on the client machine and provides an interface through which the user can send commands to the *back end* and receive the computational results of the application.

For example, a *back end* may be a database server that responds to SQL requests from a workstation running a *front-end* application.

back-to-back connection A direct connection between the signal output of one device and the signal input of an associated second device. The connection is without any intervening long-distance transmission line, modem, and so on. (Generally, a passive device such as an attenuator to accommodate power level constraints is allowed.)

Back-to-back connections are used for equipment measurements or testing purposes because such connections eliminate the effects of the transmission channel or medium.

backbone (1) That part of a communications network onto which multiple or regional networks may be attached. The highest level, or main cable, in a hierarchical network. A high-capacity network tying together several different locations, buildings, or smaller networks of lower capacity.

Backbone sites are usually tied together with high-speed carrier systems such as T-carrier or fiber optic communications links. See also *au-*

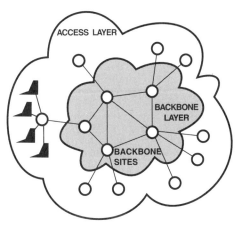

BACKBONE FUNCTIONS

Access Layer	Backbone Layer
Interface to users	Optimal routing
Interface to multiple network services	Network services
Security services	Low blocking
Menu services	High connectivity
Packetizing & protocol termination	Robust network (fault tolerant)

tonomous systems, border gateway protocol (*BGP*), *interior gateway protocol* (*IGP*), and *exterior gateway protocol* (*EGP*). (2) In a 10Base5 Ethernet, the main network cable, to which nodes are attached.

backbone site A computing facility, which serves as a relay and switching node in a wide area network (WAN).

In the Internet, for example, a backbone site serves to carry traffic between major geographical subregions. It may feed all of the *downstream* sites within a country. A USENet message received at one *backbone site* is immediately relayed to other *backbone sites* and on to each of their downstream sites in a matter of a few hours.

background noise Total system noise from all sources due to random or periodic causes in the absence of signal transmissions.

background process The execution of lower priority computer programs when higher priority programs are not using the system resources. Resources may be assigned on a timed basis, in which case the higher priority tasks are given higher percentages of available time than lower priority tasks. Priorities may be assigned by system software, application software, or the operator. More than one *background process* may be running on a multitasking system. Technically, a background process is a task detached from the terminal where it was initiated. Also called a *background task.*

backhaul The terrestrial link between an Earth station and a switching or data center.

backing out In NetWare's Transaction Tracking System (TTS), the process of abandoning an uncompleted database transaction, leaving the database unchanged. This action ensures that the database is not corrupted by information from an incomplete transaction.

backoff algorithm A network load control procedure to prevent *thrashing* (the collapse of throughput in communication equipment). In principle, the backoff algorithm is the heart of load control in multiple access systems such as the IEEE 802.2 and Ethernet networks. It determines when a device has access to the system, not whether it is allowed access.

Each device desiring access to the system has a value associated with it called the *backoff interval*. When the system is jammed with access requests, each machine waits an amount of time equal to the backoff interval before attempting another access request. The backoff interval is adjusted dynamically. That is, if jamming occurs frequently, the interval is increased; if no jamming occurs, the interval is gradually reduced to a minimum. See also *CSMA/CD* and *thrashing*.

backscatter Scattered (dispersed) electromagnetic radiation in a direction generally opposite to that of the incident wave. See also *Rayleigh scattering*.

backup General, a secondary entity to be used in case the primary entity is unavailable or fails.

Examples include:

- A fuse in series with current limiting circuitry or other current-protecting devices (a circuit breaker for example).
- A spark gap in parallel with a primary overvoltage-protecting device.
- A secondary power source capable of operating the computing system in the event of a mains failure. Also called an uninterruptible power source (UPS).
- Copies of either program or data files stored separately from the working files as protection against system or media failure. Generally, the archival copy is stored on an external medium, such as magnetic tape or diskette, and this backup is stored in a different physical location from the original material.

The creation of regular backups is essential in a computing environment. An effective backup system ensures that data stored on the system can be recreated in the event of a virus, crash, or other system failure. Various types of backups are used, including full backups, differential backups, and incremental backups. With full backups, a copy is made of *all of the data*. In differential and incremental backups, only the data that has been added or changed since the previous backup is stored. Differential and incremental backups assume a full backup has been done, and they merely add to this material. Such backups use the Archive flag (attribute), which is supported by DOS and most networking environments. A flag is associated with each file and is set whenever the file is changed and cleared when a file is copied or is backed up. Backups should be done when the system has no other processes executing, i.e., no files open. This generally means that the backups are scheduled for the dead of night or that portions of the system are made unavailable to network users during backup time. File restoration involves restoring the last full backup followed by each incremental backup in the order they were created. Backup procedures should include:

- Full backups on a regular schedule.
- Incremental backups of critical data at frequent intervals.
- Multiple copies of backups; redundancy should be a part of a backup plan.
- Backup storage in a secure, off-site location.
- Media replacement on a scheduled basis.

backup domain controller (BDC) A server in a network domain that keeps a read only copy of the domain's user accounts database and is used to validate logon requests when the *primary domain controller* (*PDC*) fails.

backup file A copy of a file made for purposes of later reconstruction of the file, if necessary (e.g., the original file becomes damaged, the original file is modified for testing). A *backup file* may be used for preserving the integrity of the original file and may be recorded on any suitable medium (removable optical or magnetic diskette, tape, etc.). Also called a *job recovery control file*.

backup server Software or hardware that copies files so that there are always two current copies of each file. Also called a *shadow server*.

backward channel In data transmission, a secondary channel dedicated to carry control information (supervisory, acknowledgment, or error control) in a direction opposite of an associated primary channel. The bandwidth of a *backward channel* is generally much less than that of the primary channel.

backward error correction (BEC) An error-correction scheme in which the receiver detects an error and requests a retransmission. The amount of material that needs to be retransmitted depends on the type of connection, how quickly the error was detected, and the protocols being used. See also *forward error correction* (*FEC*).

backward learning A routing algorithm based on assumed symmetric network paths. That is, a source node assumes that the best path to a given destination is through an adjacent node from which destination information was previously received.

backward recovery The reconstruction of an earlier version of a file by using a newer version of data and incremental process steps recorded in a journal.

backward signal A signal sent in the direction from the called station (remote modem) to the calling station (local modem). *Backward signals* are usually sent via a backward channel and generally consist of supervisory, acknowledgment, or control signals.

backward supervision Supervision signals sent from a secondary station to a primary station.

bacterium One type of computer *virus* that repeatedly replicates itself, eventually filling all available memory. See also *virus,* and *worm*.

bad block re-vectoring In disk data storage management systems, the process by which information written to a defective area of the disk is recovered; that is, it is rewritten to a different, nondefective area of storage. The defective area is identified as such in a *bad block table,* so that future writes will not be made to the area.

Bad block re-vectoring is known as a *Hot Fix* in Novell's NetWare.

bad block table In disk data storage management systems, a table in which all known defective areas of a disk are listed to ensure that nothing will be written to these areas. The process of protecting data in this manner is known as *bad block re-vectoring*.

BADC An abbreviation of Binary Asymmetric Dependent Channel.

BAIC An abbreviation of <u>B</u>inary <u>A</u>symmetric <u>I</u>ndependent <u>C</u>hannel.

bak An acronym from <u>B</u>ack <u>At</u> <u>K</u>eyboard.

bake-off A slang term for a method of testing a router's Internet compatibility. The method involves sending a series of datagrams with all possible combinations of options enabled. (These datagrams are sometimes called *kamikaze packets*.)

balance In electrical circuits and networks, to adjust an element's value to achieve specific objectives.

balance network (**1**) In a two- to four-wire hybrid set, hybrid coil, or resistance hybrid, a circuit composed of discrete elements used to match (balance) the impedance of the associated uniform transmission line. A *balance network* is required to ensure isolation between the two ports of the four-wire side of the hybrid.

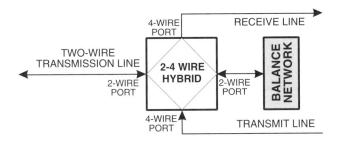

A lumped element circuit (discrete element network) cannot match the characteristics of a distributed element system (transmission line) at all frequencies simultaneously. Therefore, several approximations have come into common usage throughout the world's telephone administrations. Most are based on the circuit model below. Specific component values for several countries or administrations are listed in the table.

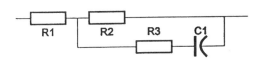

SAMPLE BALANCE NETWORK VALUES

County or Administration	R1 (Ohm)	R2 (Ohm)	R3 (Ohm)	C (nF)
Australia	220	820	0	120
Austria	220	1200	0	150
	220	820	0	115
Belgium	150	830	0	72
	0	1200	0	166.6
	600	open	0	1000
China (Peoples ROC)	200	680	0	100
	200	560	0	100
	0	800	100	50
	0	1650	100	5
	600	0	–	–
	0	1100	0	30
Denmark	500	300	0	200
Finland	390	620	0	100
	270	910	0	120
	600	open	0	1000
	0	910	0	62
France	215	1000	0	137
	600	0	–	–
Germany	100	910	0	86.6
	220	820	0	115
Indonesia	600	0	–	–

SAMPLE BALANCE NETWORK VALUES (*CONTINUED*)

County or Administration	R1 (Ohm)	R2 (Ohm)	R3 (Ohm)	C (nF)
Italy	0	1100	0	33
	180	630	0	60
	400	700	0	200
	600	0	–	–
Japan	150	830	0	72
Korea	600	0	–	–
	0	1650	100	6
	0	800	0	50
	1100	open	0	30
Malaysia	150	510	0	47
Mexico	220	820	0	115
	150	830	0	72
	100	800	0	50
	600	open	0	1000
	900	open	0	1000
Netherlands	383	536	0	215
New Zealand	370	620	0	310
Norway	0	860	0	39
NTT	150	830	0	072
Poland	144	836	0	62.4
	340	402	0	63.7
	240	1299	0	121.7
Portugal	600	0	–	–
Russia	150	510	0	47
Saudia Arabia	360	460	0	200
	600	open	0	1500
	94.3	1556	0	56.2
Spain	144	836	0	62.4
	340	402	0	63.7
	240	1299	0	121.1
Sweden	0	900	0	60
Switzerland	220	820	0	115
Taiwan	0	800	100	50
	0	1650	100	5
	0	open	900	2160
Thailand	0	800	100	50
	0	1650	100	5
	600	0	–	–
	0	1100	0	30
UK–1	310	620	0	310
–2	370	620	0	310
–private	300	1000	0	8.8
USA–loaded	0	1650	100	5
–non-loaded	0	800	100	50
–Test	900	open	0	2160
–Test	600	open	0	2160
–fixed	900	0	–	–
–fixed	600	0	–	–

See also *hybrid, hybrid coil,* and *termination.* (**2**) A device used between a balanced device or line and an unbalanced device or line for the purpose of transforming from balanced to unbalanced or from unbalanced to balanced. Also called a *balun.*

balance return loss (BRL) (**1**) A measure of the effectiveness with which a balance network matches the impedance of a two-wire circuit at a two- to four-wire hybrid. (**2**) A measure of the degree of balance between two impedances connected to two opposite sides of a two- to four-wire hybrid (network, set, coil, or junction). It is mathematically expressed as:

$$BRL = 20 \cdot \log \left| \frac{Z_{2W} + Z_B}{Z_{2W} - Z_B} \right| \text{ dB}$$

In most applications, it is desirable to have *BRL* as high as possible. In a telephone instrument, however, some return signal is advantageous in that the *sidetone* controls a speaker's voice level. The graph illustrates a specification that will satisfy the requirements for *balance return loss* for most of the telephone systems of the world. See also *hybrid, return loss,* and *sidetone.*

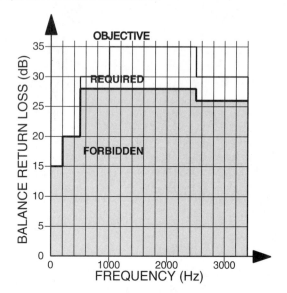

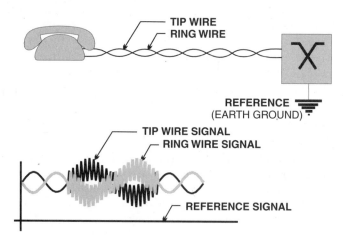

balanced (1) In transmission systems and circuits, refers to the symmetry of *hardware characteristics* (or values) about the direction of transmission, i.e., about the "centerline" of the signal path. The two sides of the path are electrically alike and symmetrical with respect to a third point (usually ground). The opposite of *balanced* is *unbalanced.*

The signal applied to a *balanced* circuit may be either a *differential signal* (or in telephony terms, a *metallic signal*) or *common mode signal* (called *longitudinal* in the telephony industry). Sufficient line balance must be maintained in a long-distance transmission system for two reasons:

- To limit the magnitude of induced common mode currents that are converted into differential signal noise.
- To limit the unbalanced common mode currents (induced differential currents) in cable pairs which may cause crosstalk in adjacent circuits.

See also *balanced line, symmetrical,* and *unbalanced.* (2) Pertaining to an optimum combination of variables or parameters.

balanced code (1) In PCM systems, a code constructed so that the frequency spectrum resulting from the transmission of any code word has no dc component. (2) In PCM, a code that has a finite digital sum variation.

balanced line A transmission line consisting of two conductors in the presence of reference conductor (generally ground), each conductor carrying signals of equal magnitude but opposite in voltage polarity and in the direction of current flow.

The magnitude of the electromagnetic field around the pair is reduced because the individual field associated with each conductor is opposite and tends to cancel each other. This reduces electromagnetic interference (EMI) to (and from) adjacent equipment. A *balanced line* is frequently also a twisted pair, which enhances the field cancellation for the pair—further improving noise immunity. The signal applied to a *balanced line* may be either *differential signal* (or in telephony terms, *metallic signal*) or *common mode signal* (similarly,

longitudinal signal in the telephony industry). In the preceding figure, both differential and common mode signals are depicted. The differential signal is the ac component (speech), while the dc signal offset is the common mode signal. The *tip wire signal* is 180 degrees phase shifted from the *ring wire signal*. The telephone set being connected across the *twisted pair* will be able to receive this differential signal while ignoring the common mode dc signal. The following figure illustrates both an ac longitudinal signal and a metallic signal. Also called a *balanced signal pair.*

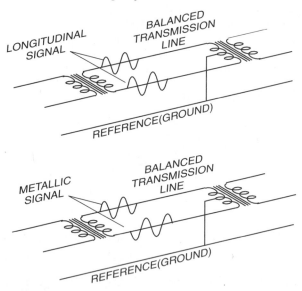

balanced modulator A modulator used in AM transmission systems with the following output characteristics:

- The carrier is suppressed.
- The output contains only upper and lower modulation sidebands.
- Any associated carrier noise is balanced out.

balanced ringing Signaling applied to both the tip and ring sides of the subscriber loop which causes the subset's ringer to operate. The signal voltage applied to both the tip and ring wires is typically one-half the specified ringing source voltage. The tip wire voltage is 180° out of phase from the ring wire voltage. See also *ring, ringer,* and *unbalanced ringing.*

balanced signal pair See *balanced line.*

balun An acronym from <u>BAL</u>anced <u>UN</u>balanced.

A passive device (usually a transformer) used to connect balanced transmission lines to unbalanced lines and to match the impedance of both lines. The usual use is the connection of twisted pair wire (balanced) to coaxial cable (unbalanced). A *balun* controls the electrical signal's passage from one cable type to the other but does not change the signal in any other way. A *balun* may have different connectors at each end to make them compatible with the cable types being connected. For example, a *balun* might have a BNC connector at one end and an RJ-45 connector at the other. In networking, baluns are often used so that IBM 3270 terminals can run off twisted pair or so that coaxial Ethernet can be operated over unshielded twisted pair (UTP). Another use of a *balun* is the connection of a 75-Ohm CATV coaxial cable to the 600-Ohm twin lead input on a television set. One such device is shown in the following drawing.

band (1) In communications, the contiguous frequency spectrum between two defined limits generally dedicated for a specific purpose. (2) In reference to WATS, one of the five specific geographic areas as defined by the carrier. (3) A group of tracks on a magnetic drum or on one side of a magnetic disk. (4) A set of frequencies authorized for use in a geographical area defined for common carriers for purposes of communications system management.

band elimination filter See *band reject filter.*

band reject filter An electronic circuit that *attenuates* (*rejects*) signals within a frequency band (*stopband*) and passes frequencies above and below that band (passband).

The following figure is a *band reject filter* with a center frequency at 1 kHz and band edges at 600 Hz and 1667 Hz. Also called band-stop

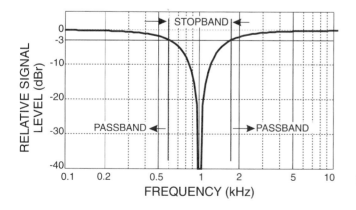

filters, band elimination filters, and band suppression filters. If a *band reject filter* is designed to reject a very narrow range of frequencies, it is also called a *notch filter*. See also *attenuator, all-pass filter, band-pass filter, filter, high-pass filter, low-pass filter, notch filter,* and transition band.

band suppression filter See *band reject filter.*

bandedge That frequency in a transmission system, circuit, or device beyond which a change in performance is observed or defined; that is, the frequencies above or below the respective *band edge* no longer are suitable for the purposes defined within the *band.* Also written *band edge.*

band-pass filter An electronic circuit or device that passes signals within a specified frequency *band* and *attenuates* frequencies both above and below that band. Frequently discussed regions and typical defining points of a band-pass filter include:

- f_1 is the lower 3-dB cutoff frequency.
- f_2 is the upper 3-dB cutoff frequency.
- Passband—all frequencies from f_1 to f_2.
- A *transition band* may be defined either below f_1 and/or above f_2. These transition bands are the region between the maximum allowable *pass-band* attenuation and the minimum specified *stopband* attenuation.
- Stopband—all frequencies above the upper transition band and below the lower transition band (if transition bands are specified).
- The center frequency (f_0) of a band-pass filter is generally the *geometric mean* of the two band edge frequencies (the upper band edge f_2 and the lower band edge f_1); i.e.,

$$f_0 = \sqrt{f_1 \times f_2}$$

- $f_{BW} = f_2 - f_1$ is the bandwidth of the passband.
- Quality or Q of a band-pass filter is the ratio of the center frequency to the bandwidth; i.e.,

$$Q = \frac{f_0}{f_{BW}}$$

Although only loosely adhered to, in the strictest sense, Q only applies to second-order band-pass filters or sections—that is, those that contain two resonant elements (for example, an inductor and a capacitor or an active filter with two capacitors). The figure below is that of a band-pass filter with a center frequency of 1 kHz and band edges at 900 Hz and 1111 Hz. Therefore, the *bandwidth* of this filter is 1111 Hz minus 900 Hz or 211 Hz and the Q is 1000 Hz / 211 Hz or about 5. See also *attenuator, all-pass filter, band reject filter, filter, high-pass filter, low-pass filter, notch filter,* and transition band.

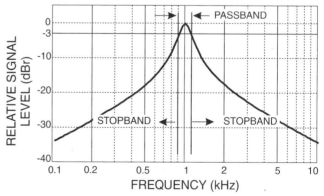

band-pass limiter A device that imposes hard limiting (clipping) on a signal and contains a filter that suppresses the unwanted products (harmonics) of the limiting process.

bandwidth (BW) (1) The difference between the limiting frequencies within which performance of a device, with respect to some characteristic, falls within specified limits. (2) The frequency range or information-carrying capability of a communication channel. The frequency range between the highest and lowest frequencies that are passed through a circuit or system with acceptable levels. Frequently, the highest and lowest frequencies are defined at the points where the signal loss, relative to a specified reference, is 50% (−3dBr). These two frequencies are also called the *band edge.*

Bandwidth is measured in *hertz (Hz)* in analog systems or *bits per second (bps)* in digital systems. As shown in the following figure, the path through a typical public telephone network has a *3-dB bandwidth* of 200 to 3400 Hz or 3200 Hz and a *0.3-dB bandwidth* of 300 to 3000 Hz. For a particular *signal-to-noise ratio (S/N)*, the *digital*

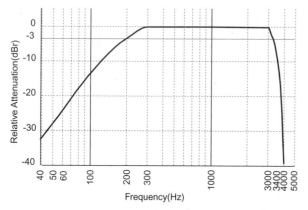

bandwidth is related to the above *frequency bandwidth* by the *Shannon limit.* See also *Shannon limit.* (3) A *relative term* used to describe the data transmission capability of a network. *High-bandwidth* networks are capable of moving large amounts of data quickly. (The data may be from either a large number of users, from a few users transferring large data blocks, or both.) *Low-bandwidth* networks, on the other hand, slow down when too many users try to transfer too much data at once. See also *frequency band.* (4) In memory systems, a function of rated speed and the width of the data path; for example, a 4 MHz transfer clock and a 32-bit bus has 16 times the *bandwidth* of a 1-MHz clock and an 8-bit bus; i.e.,

$$4 \times 32 = 128 = 16 \times (1 \times 8).$$

(5) In optical systems (e.g., fiber optics), the lowest modulation frequency at which the RMS peak-to-valley amplitude (optical power) difference of an intensity-modulated monochromatic source decreases, at the output of the system, to a specified fraction (usually one-half) of the RMS peak-to-valley amplitude difference of an arbitrarily low-modulation frequency. Both modulation frequencies have the same RMS peak-to-valley amplitude difference at the input. (6) In optical fibers, loosely, a synonym for the *bandwidth · distance product.*

bandwidth compression (1) The reduction of the bandwidth needed to transmit a given amount of data in a given time. (2) The reduction of the time needed to transmit a given amount of data in a given bandwidth. *Bandwidth compression* implies a reduction in normal bandwidth of an information-carrying signal without reducing the information content of the signal. See also *compression.*

bandwidth · distance product Of an optical fiber, a figure of merit equal to the product of the fiber's length and the 3-dB bandwidth of the optical signal under specified launching and cabling conditions and at a specified wavelength. It is generally given in megahertz · kilometer (MHz · km) or gigahertz · kilometer (GHz · km).

The normalized *bandwidth · distance product* (i.e., 1 km length) is a useful figure of merit for predicting the effective fiber bandwidth for both contiguous and concatenated fibers of different lengths. Also called *bandwidth · length product.*

bandwidth efficiency (R/W) The ratio of the information bit rate (bps) divided by the signal bandwidth (Hz).

bandwidth hog (1) Slang; descriptive of a large file that slows down a network because it uses a significant amount of the transmission capacity of a network. An example of such a file is a high-resolution photo or graphic. (2) Slang: refers to a person who transfers large amounts of frivolous material across a network.

bandwidth limited operation The condition existing when the system bandwidth limits performance.

Bandwidth limited operation occurs when the system distorts the signal waveform beyond specified tolerances. For a linear system, this is equivalent to *distortion limited operation.*

bandwidth on demand (BOD) The dynamic allocation of transmission channel capacity to only active users or to the users with the highest priority if more capacity is requested than is available.

bang (1) The spoken form of an exclamation point (!); for example, !ecpi!ted is read, "Bang, ecpi, bang, ted." (2) Used as a delimiter in uucp message routing. See also *bang address.*

bang address In Internet, a means of sending messages to users connected to the network using *uucp* store and forward. Using an explicit *uucp* path, a message will be routed to each computer listed in the *bang path.* The address is listed in the order it must travel to reach the final destination; for example, !mach1!mach2!mach3!person directs the message to "person" on machine 3 which is accessible to machine 2 from machine 1.

Bang paths and address originated in the days before automatic routing, because explicit paths were needed when sending to or communicating with another location.

bang path See *bang address.*

barge in In telephony, entering an established communications link without the invitation of the parties on the link. Examples where *barge in* frequently occurs include party line and extension phone connections.

Barlow lens In optics, invented in 1834 by the British optician and professor of mathematics Peter Barlow. It is an optical element placed between the eyepiece and the objective of an optical system that increases the effective focal length of the system. The longer focal length increases the magnification power of the system, i.e.,

$$Magnification = \frac{Objective\ f.l.}{Eyepiece\ f.l}$$

In its simplest form, a *Barlow lens* is a concave (negative) lens (or doublet) which decreases the rate of ray convergence from the objective optics.

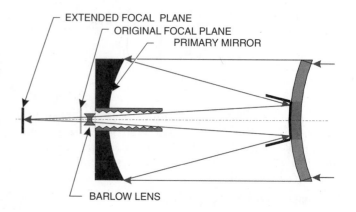

barrel connector A connector designed to connect two like coaxial cable ends. The barrel connector has two terminations, one for each of the coaxial cables being joined, as the BNC *barrel connector* below illustrates. Sometimes also called a *splice connector, inline splice,* or *bullet.*

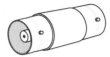

barrel distortion Optical distortion that causes a square grid to appear "barrel shaped"; that is, the sides, top, and bottom are pushed out.

BARRNet An acronym from Bay Area Regional Research Network. A regional branch of the NREN (formerly the NSFNet), a service provider, and an Internet consulting firm in the San Francisco Bay area. The network interconnects more than 200 local sites, including universities, schools, research laboratories, and businesses (including most of Silicon Valley's software and hardware companies).

base (1) In mathematics, the real number that is raised to a power indicated by the exponent; i.e., a number that is multiplied by itself as many times as indicated by the exponent. See also *base x, exponent, mantissa,* and *scientific notation.* (2) A reference value or starting point. (3) The signal input port on a transistor.

base address A base address, in general, defines the starting point or the reference location for a block of contiguous memory. The memory function is unimportant as it may be used for any purpose—for example, general-purpose memory, cache memory, or I/O port memory. Examples of base addresses include:

- An *I/O base address* is the starting location for the memory area allocated to an I/O port. The processor uses this address to communicate with the device connected to the associated port. See *I/O port address* for a list of common PC base addresses.
- A *buffer memory base address* is the starting location for a block of memory reserved as a buffer area or for paged memory (expanded memory).
- A *video memory base address* is the starting location for video output area.

base memory The lower 640 Kbytes of DOS memory (i.e., 00000h-9FFFFh). Also called *conventional memory.* See also *memory map.*

base station (1) In a wireless communications system, any fixed station that communicates with the mobile stations. (2) In personal communication service, the common name for all radio equipment located at a fixed location that is used for serving one or several cells.

base x The number of digits in a numbering system. Typical numbering systems include:

- *Binary* with a base of 2 (2 digits 0 and 1). A base 2 (or *binary* number) may be expressed with either a subscript 2 (indicating the base), a "b" suffix (indicating *binary*), or nothing if the meaning is clear in context; e.g., 0101, 0101b, and 0101_2, all represent the decimal value 5.
- *Octal* with a base of 8 (8 digits 0–7). A base 8 (or *octal* number) is expressed with either a subscript 8 (indicating the base) or no suffix if the meaning is clear.
- *Decimal* with a base of 10 (10 digits 0–9). A base 10 (or *decimal* number) is expressed with either a subscript 10 (indicating the base) or no suffix.
- *Hexadecimal* with a base of 16 (16 digits 0–9 and A, B, C, D, E, F). A base 16 (or *hexadecimal* number) is expressed with either a subscript 16 (indicating the base) or an "h" (indicating *hexadecimal*).

FIRST 16 NUMBERS IN BASE 2, 8, 10, 16

Base 10	Base 2	Base 8	Base 16
0	0000b	00	00h
1	0001b	01	01h
2	0010b	02	02h
3	0011b	03	03h
4	0100b	04	04h
5	0101b	05	05h
6	0110b	06	06h
7	0111b	07	07h
8	1000b	10	08h
9	1001b	11	09h
10	1010b	12	0Ah
11	1011b	13	0Bh
12	1100b	14	0Ch
13	1101b	15	0Dh
14	1110b	16	0Eh
15	1111b	17	0Fh

Also called the *radix.*

baseband (1) A signal in its native frequency band, that is, before modulation onto a carrier. It is the original band of frequencies produced by a transducer, such as a microphone, DTMF tone transmitter, telegraph key, or other signal source **prior to modulation or multiplexing.**

Baseband signals are usually characterized by being much lower in frequency than the frequencies that result when the signal is used to modulate a carrier or subcarrier. *Baseband* transmission generally allows only one signal on the medium at a time. Examples of *basebands* include:

- In a voice communication circuit, the baseband is about 300 to 3400 Hz.
- In a digital carrier system, the *baseband* is the digital signal before modulation and/or multiplexing onto the carrier.
- In a *baseband LAN,* the digital signal is impressed directly onto the network while in *broadband LAN* the digital information is modulated onto a carrier and then placed on the network.

Contrast with *broadband LAN* and *wideband.* (2) In facsimile, the frequency of a signal equal in bandwidth to that between zero frequency and maximum keying frequency.

baseband coding See *waveform codes.*

baseband LAN A LAN where the digital signal is encoded, multiplexed, and transmitted directly onto the network without any carrier. Examples include ARCnet, Ethernet, and Token Ring. See also *baseband network, broadband LAN,* and *LAN.*

baseband modem A modem that does not apply the digital signal to a carrier (it does not modulate) before transmission, but rather applies the digital data directly on the transmission medium. In the strictest sense it, therefore, is not a modem at all. *Baseband modems* cannot be used on the switched telephone network. Also called a *limited distance modem* or *short-haul modem.*

baseband modulation In optical systems, intensity modulation of an optical source, such as a light emitting diode (LED) or injection laser diode (ILD), directly—that is, without first modulating the information signal onto an electrical carrier wave.

baseband network A type of communications network in which the information signal is impressed *directly* onto the transmission medium; that is, there is no carrier onto which data is modulated. Binary data may be encoded so as to be more compatible with the transmission media (e.g., elimination of the normal dc content of a binary signal) or to increase the throughput (e.g., multilevel encoding).

Individual DCEs on a *baseband network* transmit information one at a time and in bursts called *packets*. Each packet contains not only data but information about the source DCE and the destination DCE. When any DCE is ready to transmit, it must wait for an idle window before it can seize the line and transmit its packet. Baseband networks operate over short distances (less than 1.5 to 2 miles) and at low speeds (less than 16 to 20 Mbps). Examples of baseband networks include T-1, ISDN, and Ethernet. See also *broadband network* and *waveform codes*.

baseband signaling Transmission of a digital or analog signal at its original frequencies, i.e., a signal in its original form, not changed by modulation.

baseline The performance of a system, network, or device under normal operating conditions (that is, the system is not being stressed by parameters at their extreme range such as high noise levels or high traffic rates).

Basic Access Control (BAC) In the ITU X.500 directory services model, the more comprehensive of two sets of access control guidelines. The less comprehensive set is called Simplified Access Control (SAC).

basic encoding rules (BER) A set of rules defined by the International Standards Organization (ISO), which are used to encode data so that it is easily accessible to other network users. Using the *BER*, it is possible to specify any ASN.1 element as a byte string. This string includes three components, and the encoding may take any of three forms, depending on the information being encoded. *BER* is most commonly used on systems conforming to the OSI protocol suite.

The components of *BER* are the *Type* (or identifier), *Length,* and *Value* (or contents) fields.

- *Type field*—indicates the class of object as well as the string's form. Examples of ASN.1 types include BOOLEAN, INTEGER, BIT STRING, OCTET STRING, CHOICE, and SEQUENCE OF. Of these, the first two are primitive; the next three may be primitive or constructed types; and the SEQUENCE OF type is always constructed.

  A primitive object consists of a single element of a particular type of information, such as a number or logical value. A constructed type is made up of other simpler elements, such as primitive objects or other constructed types.

- *Length field*—indicates the number of bytes used to encode the value. Values actually may have a definite or an indefinite length. A special value is included in the last byte when *length* indicates an indefinite length.

- *Value field*—represents the information associated with the ASN.1 object as a byte string. For primitive types, this is a single value; for constructed types, there may be several values involved (and possibly with different types).

BER can provide an encoding for any valid ASN.1 object. The three basic forms of encoding are:

- *Primitive/fixed length*—which consists only of a primitive object and which is always a fixed length. For example, an integer variable is of this type.

- *Constructed/fixed length*—which consists of a group of objects and values with a fixed total length. For example, a record with only predefined components, all of which have a fixed and known length.

- *Constructed/variable length*—which consists of a group of objects whose total size may vary from case to case, so that a special value is needed to indicate the end of the value.

Various *BER* variants have been proposed and are being developed. In general, the objective of the variants is to provide faster, simpler, and/or more generic encoding. Several of the alternatives that have been proposed are:

- Canonical encoding rules (CER). The goal of canonical rules is to eliminate any redundant paths, thus speeding up performance.

- Lightweight encoding rules (LWER). The objective of LWER is faster encoding; however, larger transmissions may result.

- Packed encoding rules (PER). These are used to compress the information about an object.

basic group See *group*.

basic information unit (BIU) In SNA network communications, a packet of information created when the transmission control layer adds a request-response header (RH) to a request-response unit (RU). This unit is passed to the path-control layer.

basic link unit (BLU) In IBM's SNA networks, a block, or packet, of information at the data link layer.

basic mode An operating mode in an FDDI II network in which data can be transmitted using packet switching. This is in contrast to hybrid mode, in which both data and voice can be transmitted.

basic mode link control Control of data links by use of the control characters of the 7-bit character set for information processing interchange as given in ISO Standard 646 and ITU-T Recommendation V.3.

basic rate access (BRA) The 144-kb/s ISDN 2B+D basic access channel rate that carries user traffic and signaling information, respectively, to the user via PTSN twisted pair local loop. Each 144-kb/s channel consists of two 64-kb/s voice, data, text, image, etc., channels (called B1 and B2) and a 16-kb/s signaling channel (the D channel) for maintenance, packet information, telemetry, data, and so on.

BRA is defined in the ITU-T Recommendation I.420. Also called *basic rate interface* (BRI). See also *ISDN*.

basic rate interface See *basic rate access*.

basic service A pure transmission capability over a communication path that is ideally transparent in terms of its interaction with customer-supplied information. It is, in reality, limited by the technical parameters of the facility (e.g., distortion, time delay).

Basic Telecommunications Access Method (BTAM) An early access method for communications between IBM mainframes and terminals. *BTAM* is still used but is obsolete because it does not support IBM's Systems Network Architecture (SNA). Advanced Communications Function/Virtual Telecommunications Access Method (ACF/VTAM) has replaced *BTAM* for remote communications with IBM mainframes.

basic transmission unit (BTU) In IBM's SNA communications, an aggregate block of one or more path information units (PIUs) that all have the same destination. Several PIUs can be combined into a single packet, even if they are not all part of the same message. *BTUs* are created at the path-control layer.

batch A method of information retrieval or process control wherein the user submits a set of instruction to be run after the communications link is dissolved. The process sequence is:

- Establish a communications link.
- Submit a sequence of instructions to the host computer.
- Break down the communications link while the computer processes the requests and store the results for later retrieval.
- Wait an appropriate amount of time for processing to occur.
- Reestablish the communications link and retrieve the results of the requests.
- End the communications link and analyze the results.

See also *interactive* and *noninteractive*.

battery (1) Two or more *cells* (a device that converts chemical energy into electrical energy), electrically connected in a series and/or parallel arrangement to either increase the output voltage, current capability, or both. See also *cell* (*3*). (2) Although strictly incorrect, in

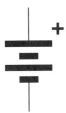

common usage, the term *battery* is often also applied to a single cell. (3) The term used in the telephone industry for the power source that supplies energy to the subscriber loop. The term is derived from the real -48 V battery that supplies this energy (-60 V in Germany and a few other places in the world). This battery is constantly charged from the ac mains. Also called *central office battery*.

battery backup See *uninterruptible power supply*.

battery feed In telephony, the dc supply and injection circuit powering the subscriber loop.

battery reserve The capability of a fully charged battery to carry the working load when commercial power fails. For situations where the load is variable, such as a telephone central office, the load is calculated on some acceptable duty cycle. For a central office, battery reserve is frequently calculated based on the demand during "busy hour."

battery reversal In telephony, normally the *ring* wire is connected to the negative side of the central office battery. Some conditions require that the *tip* lead be made more negative; this is called *battery reversal*.

baud (Bd) A unit of line signaling rate and refers to the number of times the state of a communication signal can change per second. Named after the French engineer and telegrapher Jean-Maurice-Emile Baudot (1845–1903). A *baud* is equal to the maximum number of discrete signal elements, events, or symbols possible per second, i.e., the maximum rate at which changes in the modulated signal can occur. A given *baud* can represent more than one bps depending on the number of symbols in the encoding or modulation technique. A *baud* (the number of symbols per second) and *bit rate* (the number of bits per second) are related by *bps = baud × bits* (and *baud = bps/bits*).

For example, if the modulation method provides only two symbols, then each of the states of a bit will be assigned to one of the symbols and the baud rate will equal the bit rate; i.e.,

$$1\,\frac{symbol}{second} \times 1\,\frac{bit}{symbol} = 1\,\frac{bit}{second}$$

However, if 16 symbols are possible with the encoding method, then each of the possible combinations of four bits can be assigned to a symbol and the bit rate will be four times the baud rate; i.e.,

$$1\,\frac{symbol}{second} \times 4\,\frac{bit}{symbol} = 4\,\frac{bit}{second}$$

The diagram below illustrates one way a binary bit stream may be encoded (using multilevel amplitude modulation) so as to reduce the required number of bauds for a given bit per second rate. The top trace is the unencoded binary bit stream with only two states possible ("0" & "1"); hence, the bit rate is equal to the number of bauds. The second trace encodes each *dibit* into a four-level symbol. Here, the number of bits per second is twice the number of bauds. Finally, the third trace encodes each *tribit* of the original signal into one of eight symbols, yielding a bit rate three times that of the baud. Through the use of a large alphabet, a 28.8-kbps modem transmits at only 3824 Bd!

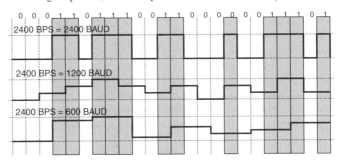

Baudot code An early synchronous 5-bit code (circa 1880) used for communication between teleprinters. Named after Jean-Maurice-Emile Baudot (1845–1903), an officer in the French Telegraph Service who is given credit for devising the first uniform-length code (5 bits) for characters in the alphabet.

The *Baudot code* (also called the International Telegraph Alphabet No. 1, abbreviated ITA1) contains only 32 symbols, not enough to represent the alphabet, numbers, and required punctuation needed for telegraph messages. The solution was to reuse a part of the code much like the "caps lock" key on a typewriter redefines individual keys. Two of the 32 symbols were defined as *letter shift* (LS) and *figures shift* (FS) keys; in this way, 60 symbols are available (62 counting FS and LS). In use, all characters following the LS symbol are interpreted as "letters" and all symbols following the FS symbol are assumed to be "figures." See the following table. In dc teleprinter ap-

THE BAUDOT CODE

Code	Letters	Figures	Code	Letters	Figures
00000	N/A	N/A	10000	A	1
00001	LS	LS	10001	LF	LF
00010	FS	FS	10010	J	6
00011	ER	ER	10011	K	(
00100	Y	3	10100	U	4
00101	S	.	10101	T	N/A
00110	B	8	10110	C	9
00111	R	–	10111	Q	/
01000	E	2	11000	CR	CR
01001	X	,	11001	Z	:
01010	G	7	11010	H	+
01011	M	)	11011	L	=
01100	I	N/A	11100	O	5
01101	W	?	11101	V	'
01110	F	N/A	11110	D	0
01111	N	N/A	11111	P	%

CR = carriage return, ER = error, FS = figures shift, LF = line feed, LS = letters shift, and N/A = not assigned. A space is printed for both LS and FS symbols.

plications, the 5 data bits are preceded by 1 start bit and 1.42 stop bit elements. In electronic systems, the number of stop bits is generally extended to 1.5 for ease of design and manufacture. The Baudot code has been replaced by the International Alphabet No. 2 (IA2). See also *teleprinter codes.*

BB (1) An abbreviation of <u>B</u>road<u>B</u>and. **(2)** An abbreviation of <u>Bul</u>letin <u>B</u>oard. **(3)** An abbreviation of <u>B</u>roadcast <u>B</u>ureau of the FCC.

BBC (1) An abbreviation of <u>B</u>road<u>B</u>and <u>C</u>onducted noise. **(2)** An abbreviation of <u>B</u>ritish <u>B</u>roadcasting <u>C</u>orporation.

BBL An abbreviation of <u>B</u>e <u>B</u>ack <u>L</u>ater.

BBR An abbreviation of <u>B</u>road<u>B</u>and <u>R</u>adiated noise.

BBS An abbreviation of <u>B</u>ulletin <u>B</u>oard <u>S</u>ystem.

BCC (1) An abbreviation of <u>B</u>lind <u>C</u>ourtesy <u>C</u>opy. **(2)** An abbreviation of <u>B</u>lock <u>C</u>heck <u>C</u>haracter.

BCD (1) An abbreviation of <u>B</u>inary <u>C</u>oded <u>D</u>ecimal. The representation of the cardinal numbers 0 through 9 by a *binary code.*

Although many codes are possible, the code usually associated with the *BCD* designation is the code with bit weightings of 8, 4, 2, and 1 in each respective bit position. This is the so-called natural or standard binary code. Numbers greater than 9 are represented by multi-

THE 10 NATURAL BCD DIGITS

Decimal	BCD	Decimal	BCD
0	0000	5	0101
1	0001	6	0110
2	0010	7	0111
3	0011	8	1000
4	0100	9	1001

ple sets of binary codes, one for each decimal digit. For example, the decimal number 13 is represented as 0001 0011, NOT 1101; similarly, 397 decimal is represented in *BCD* as 0011 1001 0111. The six binary values 1010, 1011, 1100, 1101, 1110, and 1111 are forbidden

THE 88 FOUR-BIT WEIGHTED CODES

Positive Weight	Single Negative Weight				Double Negative Weight
5211	531-1	441-2	**543-3**	843-6	63-1-1
4311	**631-1**	541-2	654-3	**753-6**	63-2-1
5311	522-1	**641-2**	621-4	653-7	73-2-1
6311	**622-1**	741-2	721-4		54-2-1
5221	432-1	841-2	821-4		64-2-1
6221	*532-1*	**632-2**	**751-4**		74-2-1
3321	632-1	**443-2**	861-4		**84-2-1**
4321	732-1	543-2	632-4		72-3-1
5321	**442-1**	643-2	**832-4**		75-3-1
6321	542-1	843-2	**652-4**		72-4-1
7321	642-1	621-3	852-4		82-4-1
2421	742-1	721-3	653-4		**86-4-1**
4421	842-1	**651-3**	**842-5**		84-3-2
5421	621-2	751-3	643-5		81-4-2
6421	531-2	542-3	763-5		83-4-2
7421	631-2	**642-3**	841-6		85-4-2
8421	**731-2**	842-3	543-6		**87-4-2**

The code weights are all 9 or less. Highlighted entries are self complementing codes (nines complement).

combinations in *BCD*. As stated above, there are many possible ways to encode the 10 decimal digits with 4-bit binary codes. In fact, there are more than 2.9×10^{10} ways (16 ways to encode any selected digit, 15 ways to encode the next digit, 14 ways for the third, and so on until all 10 symbols are encoded—the total number of combinations possible is then $16 \times 15 \times 14 \times \ldots \times 8 \times 7 = 2.90594 \times 10^{10}$). Of these, only a small number (88) allow the assignment of weights to the bit positions. There are 17 that have all positive weights, 54 that have a single negative weight, and 17 that have two negative weights. See also *code, complement,* and *self-complement.* **(2)** In telephony, an abbreviation of <u>B</u>lock <u>C</u>alls <u>D</u>elayed.

BCH An abbreviation of <u>B</u>locked <u>C</u>alls <u>H</u>eld.

BCH code An abbreviation of <u>B</u>ose-<u>C</u>haudhuri-<u>H</u>ochquenghem, the names of the inventors of a particular forward error correcting (FEC) coding method. It is a multilevel, cyclic, block error correcting, variable-length digital code capable of correcting errors up to about one-fourth of the total number of symbols. In addition to working with binary systems, *BCH codes* may be used with multilevel phase shift keying whenever the number of levels is either a prime number or a power of a prime number, such as 2, 3, 4, 5, 7, 8, 11, and 13.

BCI An abbreviation of <u>B</u>it-<u>C</u>ount <u>I</u>ntegrity.

BCM An abbreviation of <u>B</u>it <u>C</u>ompression <u>M</u>ultiplexer.

BCN An abbreviation of <u>B</u>ea<u>C</u>o<u>N</u>.

BCNU Jargon for Be Seeing (see'n) You. A signoff message frequently seen on USEnet and other message systems.

BCP An abbreviation of <u>B</u>yte <u>C</u>ontrol <u>P</u>rotocols. Protocols that are byte (rather than bit) oriented.

Bd The symbol for *baud.*

beacon (BCN) A token ring frame sent by an NIC indicating that a serious error (such as a broken cable or bad multistation access unit MAU) has been detected in either its node or in that node's nearest addressable upstream neighbor (NAUN). A node sending this frame is said to be *beaconing.* Beaconing is a process that allows a network to self-repair.

Used primarily in IBM's Token Ring and FDDI.

beam A concentrated unidirectional flow of electromagnetic waves. A beam cross section may be circular, elliptical, and so on.

(1) Antenna, the major lobe of the radiation pattern. **(2)** Optics, a collection of rays that may be convergent, parallel, or divergent.

beam diameter (1) The distance between two diametrically opposed points on a perpendicular cross section of an electromagnetic beam where the irradiance is a specified fraction of the maximum (on the beam axis). Common fractions include:

- The one-half power points (3-dB power loss). Generally used in RF antenna systems.

  When the aperture from which the RF beam emerges is comparable to the wavelength of the emission, beam diameter is usually referred to as *beamwidth.*

- In optical systems, the point where the power per unit area (irradiance) is $1/e$ times the maximum power per unit area.

Beam diameter usually, but not necessarily, refers to a beam of circular cross section. A beam may, for example, have an elliptical cross section, in which case the orientation of the *beam diameter* must be specified, e.g., with respect to the major or minor axis of the elliptical cross section. **(2)** The diameter of an aperture that will pass a specified percentage of the beam energy (usually 90%). Also called *beamwidth.*

beam divergence A measure of the increase in an electromagnetic beam diameter with distance from the aperture from which the beam emanates. It is the full angle of beam spread (generally indicated in degrees) defined by the beam diameter and the beam center at the source. For beams not circular in shape, a major and minor *beam divergence* is defined for the two beam diameters corresponding to the maximum and minimum divergence respectively.

Beam divergence is usually used to characterize electromagnetic beams in which the aperture from which the beam emerges is very large with respect to the wavelength, e.g., in optical systems.

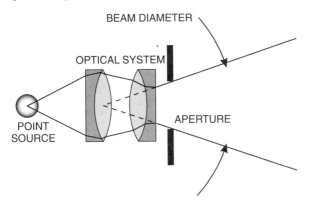

beam expander A combination of elements that will expand the diameter of a beam. Usually pertains to optical beams and devices.

beam splitter A device that divides a beam into two or more separate beams. In optics, this is usually accomplished through the use of partially reflecting mirrors.

The following illustration is of a partially reflecting mirror with an intensity equalization filter in the reflected path. Note that the thickness of the reflector is greatly exaggerated so that the internal ray losses may be more clearly seen.

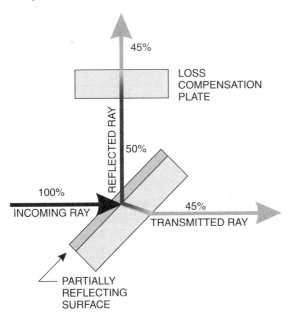

beam steering Changing the direction of the main lobe of a radiation pattern.

In radio systems, *beam steering* may be accomplished by switching antenna elements, by changing the relative phases of the signals applied to the driven elements, or by changing the physical relationship of the feed and reflector. In optical systems, *beam steering* may be accomplished by changing the refractive index of the medium through which the beam is transmitted or by the use of mirrors or lenses. See also *Kerr cell.*

beamwidth (1) With RF antenna patterns, the angle between the half-power (3-dB) points of the main lobe, when referenced to the peak effective radiated power of the main lobe. *Beamwidth* is usually expressed in degrees and for the horizontal plane; however, it may also be expressed for the vertical plane. See also *beam diameter.* **(2)** For optical systems, see *beam diameter* (2) and *beam divergence.*

bearer channel A communications channel that carries information (such as data, voice or images) rather than control signals.

bearer services In ISDN, basic communications services. Includes:
- 64-kbps circuit switched digital services for use with voice telephony and data communications.
- Packet switched services operating to 1.544 Mbps (over T-1 lines in the United States) and 2.048 Mbps on ITU-T compliant lines.

See also *B-channel, ISDN,* and *service access.*

beat frequency oscillator (BFO) An oscillator used to translate a received signal from one frequency band to another through a process called *heterodyning.* See also *heterodyne.*

beat note (1) A wave created by the passage of two (or more) waves through a *nonlinear* device. The frequencies of the created waves are equal to the sums and differences of all combinations of frequencies of the incident waves.

The figure below illustrates this process as it occurs in an *amplitude modulator;* specifically, the top figure shows a *carrier* (the top trace), a *modulating wave* the middle wave, and the resultant *output waveform.* The lower figure is also of the resultant waveform but in the frequency domain rather than the time domain. It shows the carrier as the central peak and the two *sidebands* generated as a result of the nonlinear mixing. The frequency difference between the central peak and each sideband is the same as the modulating waveform. See also *heterodyne.*

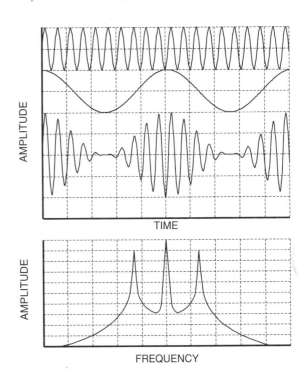

(2) Periodic variations that result from the *linear summation* of periodic signals having different frequencies. The term is sometimes applied to the linear addition of the signals as well as to the more typical definition involving the nonlinear addition. The following figures illustrate the result of the linear summation of two sine wave tones that are close in frequency. The top chart shows the two input tones and the resultant output waveform in a *time vs. amplitude* plot. Note that the peak output amplitude is the sum of the amplitude of both input waveforms and that the *amplitude envelope* changes according to the difference of the two-tone frequencies. The lower chart, also of the resultant output waveform, is displayed in a *frequency vs. amplitude* manner. Each peak is at the frequency of the input signals.

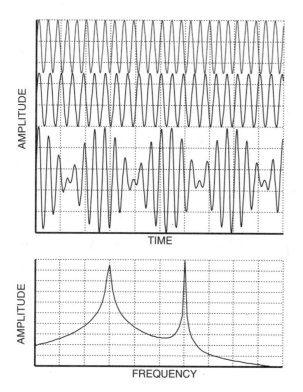

BECN An abbreviation of <u>B</u>ackward <u>E</u>xplicit <u>C</u>ongestion <u>N</u>otification. A flow-control element in ATM and Frame Relay networks.

beeping A synonym for paging or radio paging.

bel (B) The *bel* is named after Alexander Graham Bell (1847–1922) and is the unit for relating two power levels. The number of *bels* (N) denoting this relationship is the logarithm base 10 of the ratio of the two powers (P_1 and P_2); i.e.,

$$N = \log_{10} \frac{P_1}{P_2}$$

Because the *bel* is a large number, the decibel (1/10 *bel* or *dB*) has become the normal way to express this ratio. See also *dB* and *neper.*

bell The ASCII "bell" character (07h).

Bell 103 A 300-baud asynchronous full-duplex FSK data set (modem) with specifications defined by Bell Telephone. The specifications for the line protocol are one of the standard fallback protocols used today. Bell 103 is similar to the ITU-T V.21 specifications.

BELL 103 FREQUENCIES (HZ)

Mode	Transmit		Receive		Answer Tone
	Space	Mark	Space	Mark	
Orig	1070	1270	2025	2225	—
Ans	2025	2225	1070	1270	2225

Bell 201 A 2400-bps synchronous, half-duplex, DPSK data set with a 75-bps FSK reverse channel manufactured to the specifications of Bell Telephone. Performance of the 201B is similar to the ITU-T V.26 specification, whereas the 201C is similar to the V.26 bis specification.

Bell 202 An asynchronous/synchronous 1200/600-bps half-duplex FSK data set (modem) with an asynchronous ASK reverse channel for signaling and error control. The performance of the Bell Telephone data set is similar to the ITU-T V.23 specification.

BELL 202 1200 BAUD FREQUENCIES (HZ)

Channel	Transmit		Receive		Answer Tone
	Space	Mark	Space	Mark	
Half duplex	2200	1200	2200	1200	2025
5 bps Reverse	387	387	387	387	—
150 bps Reverse	487	387	487	387	—

Bell 208 A 4800/2400-bps synchronous, half-duplex, PSK data set with a 75-bps FSK reverse channel. Manufactured to Bell Telephone specifications, the 208A is similar in performance to the ITU-T V.27 ter. Now an obsolete standard because of its incompatibility with the ITU-T standard.

Bell 209 A 9600/7200/4800-bps synchronous, half-duplex, QAM data set manufactured to Bell Telephone specifications. It is similar in performance to the ITU-T V.29 standard. Now an obsolete standard because of its incompatibility with the ITU-T standard.

Bell 212A A 1200-bps asynchronous and synchronous QPSK modem manufactured by Bell Telephone. The specifications for the line protocol are one of the standard fallback protocols used today. The telephone channel is divided into two frequency bands, one centered at 1200 Hz for the originating modem's transmission and one centered at 2400 Hz for the answering modem's transmission. The *Bell 212A* specification is similar to the ITU-T V.22 specification.

BELL 212 MODULATION

Dibit value	Relative Phase
00	90°
01	0°
10	180°
11	270°

Bell Operating Company (BOC) Any of the 22 operating companies that were divested from AT&T by court order in 1984. Two companies, Cincinnati Bell Telephone Co. and Southern New England Telephone Co., were not included. The *BOCs* are now a subsidiary of one of the Regional Bell Operating Companies (RBOC) that provide local telephone service. See also *modified final judgment (MFJ)* and *regional Bell operating company (RBOC).*

Bell protocols The set of modem protocols developed by Bell Laboratories during the 1970s. The protocols known as Bell 103, Bell 201, Bell 212, and so on supported speeds to about 2400 bps.

With the exception of the Bell 103 and 212A standards, these protocols are not used and are, by and large, incompatible with the ITU-T protocols now in use by most high-speed modems today.

BellCore An acronym from <u>Bell</u> <u>CO</u>mmunications <u>RE</u>search. An organization established after the divestiture of AT&T to replace the engineering supplied by Bell Labs (now Lucent Technologies). BellCore represents the interests of, and is funded by, the seven *regional Bell operating companies* (*RBOCs*).

Bellman–Ford algorithm A procedure for finding routes through an internetwork based on distance vectors, as opposed to link states.

The *Bellman–Ford algorithm* is also known as the *old ARPAnet algorithm.*

benchmark A test program used to compare the parameters of one device to another similar device.

Although a *benchmark* does compare fairly, it only gives an *indication* of performance differences. It does not test the true performance for any specific application. Therefore, if it is necessary to compare performance for a specific application, one must run the application on both devices. Also, manufacturers try to optimize a device to perform better with a popular *benchmark* program, further skewing the relationship with reality.

bend loss See *macrobend loss* and *microbend loss.*

BER (**1**) An abbreviation of <u>B</u>it <u>E</u>rror <u>R</u>atio (or <u>R</u>ate). (**2**) An abbreviation of <u>B</u>asic <u>E</u>ncoding <u>R</u>ules.

Berkeley Internet Name Domain (BIND) A domain name system (DNS) server developed and distributed by the University of California at Berkeley.

BERT An acronym from <u>B</u>it <u>E</u>rror <u>R</u>atio (or <u>R</u>ate) <u>T</u>ester. A hardware device for checking a transmission's bit error rate (BER), or the proportion of erroneous bits.

best path In networking, the optimal route between a data source and a data sink (destination) through a wide area network (WAN). The path is determined by protocols such as the Routing Information Protocol (RIP) or Open Shortest Path First (OSPF) protocol. The definition of "optimal" can be based on any of several parameters, e.g., lowest delay or lowest cost.

beta (**1**) The second letter in the Greek alphabet. (**2**) Refers to the final stages of product development and testing, i.e., those tests prior to release to the end customer. *Beta testing* is typically done at cooperative customer sites and by the customer. Most *beta* products have bugs (some serious) when they are first exposed to the "real world."

BETA An acronym from the UK <u>B</u>usiness <u>E</u>quipment <u>T</u>rade <u>A</u>ssociation.

beta test Product testing in a real operating environment generally by real or potential customers. *Beta testing* commences after both engineering and production testing is completed and both believe the product to be ready for sale. Generally, *beta testing* reveals additional problems, quirks, or missing features required by the customer.

BETRS An acronym from <u>B</u>asic <u>E</u>xchange <u>T</u>elecommunications <u>Ra</u>dio <u>S</u>ervice. A service that can extend telephone service to rural areas by replacing the local loop with a radio link.

between-the-lines entry A slang term for unauthorized access to a momentarily inactive terminal, of a legitimate user, assigned to a communications channel.

BEX An acronym from <u>B</u>roadband <u>EX</u>change.

BEZS An abbreviation of <u>B</u>andwidth <u>E</u>fficient <u>Z</u>ero <u>S</u>uppression. A patented T-1 zero suppression technique. It maintains the requisite T-1 pulse density without introducing "errors" into the user data. This is accomplished by using a 32-kbps overhead channel.

BFD An abbreviation of <u>B</u>ig <u>F</u>—ing <u>D</u>eal. Often used in flame mail and posting, indicating a contemptuous attitude.

BFO An abbreviation of <u>B</u>eat <u>F</u>requency <u>O</u>scillator.

BFSK An abbreviation of <u>B</u>inary <u>F</u>requency <u>S</u>hift <u>K</u>ey. See also *FSK* and *modulation.*

BFT The abbreviation of <u>B</u>inary <u>F</u>ile <u>T</u>ransfer.

BGP An abbreviation of <u>B</u>order <u>G</u>ateway <u>P</u>rotocol.

BH In telephony, an abbreviation of <u>B</u>usy <u>H</u>our.

bi The Latin prefix meaning "two."

BIA An abbreviation of <u>B</u>urned <u>I</u>n <u>A</u>ddress.

bias (**1**) A deviation of a value from a reference value in a preferential direction. (**2**) The amount by which the average of a set of values departs from a reference value. (**3**) A force applied to a device to set a reference value or an operating point. The force may be mechanical, electrical (voltage or current), magnetic, mechanical, optical, thermal, and so on. (**4**) In digital signaling systems, the development of any dc voltage at a point on a line that should remain at a specified reference level, such as zero. This bias may be applied or produced by the electrical characteristics of (a) the line, (b) the terminal equipment, or (c) the waveform coding scheme.

bias distortion (**1**) Distortion affecting a data communication system wherein all intervals of either the marking or spacing state are uniformly longer than the theoretical interval.

Using one character of a continuous data stream, the figure below shows the two forms of bias distortion. The top trace is that of a character with no distortion; the second trace has bias distortion with the marking (or idle) state extended beyond the nominal value; and the third trace demonstrates bias distortion with the spacing (or break)

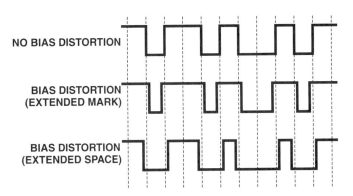

state extended. Mathematically, *bias distortion* (D_B) is generally expressed in one of two ways:

- When the distortion is all assigned to either the mark or space state,

$$D_B = \frac{T_B}{B} \times 100\%$$

which is equivalent to

$$D_B = \frac{Average\ Duty\ Cycle\ (in\ \%) - 50\%}{B} \times 2$$

- When half of the distortion is assigned to the mark state and half to the space state,

$$D_B = \frac{T_B}{2B} \times 100\%$$

Where T_B is the bias distortion time and B is one bit time; i.e., **(2)** Signal distortion resulting from a shift in the bias.

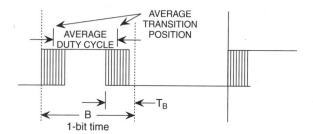

biased ringer A mechanical telephone ringer in which the clapper has been biased (either mechanically or magnetically) with a preference to one of the two gongs. This is one method of reducing bell tapping during pulse dialing.

biased ringing In telephony, the application of the ac ringing current superimposed on a dc loop feed voltage. Typically, the −48 V dc office battery is superimposed as a means of hook switch status detection.

Voltages as high as ±105 V dc are also used to allow the central office to detect *which* station has gone off-hook on a two-station party line loop. The ringers on these loops are isolated by a diode (which acts as a polarity sensitive switch). Also called *AC-DC ringing* (when the feed voltage is −48 V dc) or *superimposed ringing* (when ±105 V dc is added).

biconical antenna An antenna consisting of two conical conductors, having a common axis and vertex, and extending in opposite directions. If one of the cones is reduced to a plane, the antenna is called a *discone*. Excitation is applied at the common vertex.

bi-directional Capable of operation in both a forward and reverse direction.

bifurcated To be divided or separated into two parts; forked.

bifurcated application A term used to describe an application that is divided into two parts, one part running on a server and the other linked by a LAN running on a client machine. See also *client–server network*.

bifurcated routing Routing that may split a message from a source among several paths.

big endian A format for transmitting (or storing) serial binary data in which the most significant bit is the first transferred. This is in contrast to *little endian* in which the least significant bit is first.

The term comes from Jonathan Swift's *Gulliver's Travels,* wherein a war was fought over which end of a hard-boiled egg should be broken first. Similarly, many arguments arise over whether *big endian* or *little endian* data storage is better. The ordering property is also known as the process's byte sex.

big iron A slang term for a mainframe computer.

billboard antenna An array of parallel dipole antennas with flat reflectors usually positioned behind the driven elements in a line or plane. The main lobe of the antenna may be steered to some extent by appropriate phasing of the signals to the individual driven elements of the array. Also called a *broadside antenna.*

billing delay A short delay following the seizure of a line that a DCE must wait before any tones are transmitted. Typically, the delay is 2.5 ± 0.25 s.

billion In the U.S. numbering system (and generally with microcomputers), 1 *billion* is 10^9 (1,000,000,000). The SI prefix for 10^9 is *giga.* In the British system, 1 billion is 10^{12} (1,000,000,000,000). The SI prefix for 10^{12} is *tera.*

bin **(1)** An abbreviation for binary. **(2)** The extension of a binary file. See also *filename extension.* **(3)** On Internet archive sites, often used with directory names to indicate that the directory contains files that can be downloaded with FTP.

binary **(1)** Pertaining to any system choice, condition, or selection that has only two possible states, outcomes, or values. **(2)** A numbering system in which each digit has only two possible values (0 or 1) as opposed to the decimal numbering system wherein each digit may take on any of 10 values (0, 1, 2, . . . 9). The table below shows the first 20 *binary* numbers.

FIRST 20 BINARY NUMBERS

Decimal Number	Binary Coded Number	Decimal Number	Binary Coded Number
0	00000	10	01010
1	00001	11	01011
2	00010	12	01100
3	00011	13	01101
4	00100	14	01110
5	00101	15	01111
6	00110	16	10000
7	00111	17	10001
8	01000	18	10010
9	01001	19	10011

binary digit See *bit.*

binary exponential backoff algorithm The computation each interface card in an Ethernet, CSMA/CA, or CSMA/CD network is required to perform in order to determine its delay time to retransmission after a collision has been detected. Also called *backoff algorithm* or, more formally, the *truncated binary exponential backoff algorithm.* See *CSMA/CD.*

binary file A file that may contain any (or all) of the possible 256 symbols of an 8-bit byte. Generally any *nontextual file.* See also *ASCII file.*

binary file transfer (BFT) ITU-T T.434 standard that allows direct transmission of *actual data files* to a compatible destination facsimile machine instead of first translating the file to fax format.

BFT attachments are not converted to the group 3 fax format but are sent in the *native* format. This saves both processing time (since no conversion is performed) and transmission time. (Binary files are generally shorter than the equivalent fax.) A binary file may be attached to the end of a standard fax transmission. See also *communication application specification (CAS).*

binary filter A software sieve based on bit values (0 or 1) rather on bytes or strings of bytes.

binary modulation The process of varying a parameter of a carrier as a function of two finite, discrete states. Frequently designated as some form of "shift keying," e.g., frequency shift keying (FSK), phase shift keying (PSK), and amplitude shift keying (ASK).

binary notation (1) Any symbolic representation that uses two different symbols; usually, the symbols "0" and "1" are used. Data encoded in *binary notation* need not be in the form of a pure binary numeration system; for example, a transitional code such as Gray, Lippell, or Johnson codes may represent them. See also *code*. **(2)** A scheme for representing numbers, which is characterized by the arrangements of digits in sequence, with the understanding that successive digits are interpreted as coefficients of successive powers of base 2. See also *binary*.

binary number: A number that is expressed in binary notation and is usually characterized by the arrangement of bits in sequence, with the understanding that successive bits are interpreted as coefficients of successive powers of the base 2. See also *base* and *binary*.

binary synchronous communication (BiSync) A synchronous, byte-oriented, serial communication protocol created by IBM. The protocol uses a set of control characters and control character sequences to synchronize the transmission of data between DCEs in a packet-like format called a frame. The format of a BiSync frame is:

SSS HH X mm . . . m x B SS

where:

S   is a synchronizing character,
H   is an optional header,
X   is the STX start of message character,
m   is the message,
x   is the ETX end of message character, and
B   is the BCC error control character.

The *BiSync* protocol is being phased out of most computer communication networks in favor of bit-oriented protocols such as SDLC, HDLC, and ADCCP. Also abbreviated *BSC*.

BIND An acronym from Berkeley Internet Name Domain. An Internet domain name system (DNS) server developed at the University of California, Berkeley.

binder The colored helical wrapping applied to a bundle of wires or fibers in a cable. The binder groups, identifies, and separates the bundled elements of the cable. The *binder* itself may be colored thread, yarn, or plastic ribbon.

binder group That group of elements wrapped by a *binder*. In telephone cables, *binder groups* consist of 25 pairs of wires; the first binder group has a blue binder, the second orange, the third green, and so on.

Bindery The database in Novell's NetWare versions 2.x and 3.x where information about the network clients and resources (such as passwords, access rights, etc.) are stored. The *bindery* is located in the SYS:SYSTEM directory and is maintained by the network operating system (NOS) on each server.

The bindery information determines the activities possible for each user or node. In the bindery, this information is represented as a flat database. The bindery has three types of components:

- *Objects*—users, devices, workgroups, print queues, print servers, and so on. (Most physical and logical entities are regarded as objects.)
- *Properties*—attributes assigned to bindery objects, such as full name, login restrictions, or group membership information.

- *Property data sets*—the values that will be stored in an object's property list.

NetWare version 4.x replaced the *bindery* with the NetWare Directory Services (NDS), in which information is represented hierarchically. Also included is a bindery emulation capability, which makes it possible to integrate such bindery-based objects into a 4.x network.

bindery emulation Allows programs that were written to run under the NetWare versions 2.x/3.x bindery to find the network object information they need in NetWare 4.x's Directory by making the information in the Directory appear as a flat structure.

binding In software, the assigning of a value (or referent) to an identifier; for example,

- The assigning of a value to a parameter.
- The assigning of an absolute address, virtual address, or device identifier to a symbolic address or label.

An example is the process that links a protocol driver and a network adapter.

BINHEX A method of encoding binary files such that they contain only ASCII text. Used with Macintosh and IBM compatible computers to prepare files for transfer over Internet e-mail. Files encoded with *BINHEX* generally have the filename extension .Hqx.

Bionet One of several *USENet* alternative newsgroups. The subject matter includes information and discussions on biological and ecological issues such as agroforestry, the Human Genome Project, molecular biology, and neuroscience. See also *ALT* and *alternative newsgroup hierarchies*.

BIOS An acronym from Basic Input Output System pronounced "BYE-ose." It is software that generally resides in a ROM (or PROM) and supports the transfer of information between hardware elements or subelements of a computer. The *BIOS* provides the primitive control of the devices, such as the keyboard, hard disk drive, and monitor.

BIP-N An acronym from Bit Interleaved Parity-N. A method of monitoring for transmission errors.

biphase modulation A synonym for *phase shift keying* (*PSK*).

BIPM An acronym from Bureau International des Poids et Mesures.

bipolar Literally, having two opposite states. For example, a binary signal with positive and negative voltages representing the "1" and "0" states is said to be *bipolar*. See also *unipolar*.

bipolar signal A binary signal that may assume either of two polarities, neither of which is zero. It is usually symmetrical with respect to zero amplitude; that is, the absolute values of the positive and negative states are nominally equal. Waveform coding of bipolar signals includes two-state nonreturn-to-zero (NRZ) and three-state return-to-zero (RZ) schemes.

bipolar violation (BPV) The presence of two consecutive (though not necessarily adjacent) "one bits" of the same polarity on an alternate mark inversion (AMI) coded signal stream. *Bipolar violations* may indicate a transmission error, or they may be used for signaling. See also *BnZS*.

BiQuinary A numbering system in which each decimal value N is represented by the digit pair XY where $N = 5X + Y$ and $X = 0$ or 1 and $Y = 0, 1, 2, 3,$ or 4. The values of X and Y are represented by several methods as are shown in the following chart.

BIQUINARY REPRESENTATIONS

Decimal	X Y	5 421	50 43210
0	0 0	0 000	01 00001
1	0 1	0 001	01 00010
2	0 2	0 010	01 00100
3	0 3	0 011	01 01000
4	0 4	0 100	01 10000
5	1 0	1 000	10 00001
6	1 1	1 001	10 00010
7	1 2	1 010	10 00100
8	1 3	1 011	10 01000
9	1 4	1 100	10 10000

Birds of a Feather (BOF) A group in which all participants have a common interest and "discuss" a specified topic.

birefringence The term literally means *double fraction* from the Latin bi- (meaning twice) and refringere (meaning to break or fracture). In optics, it is the division of light into two components.

Birefringent materials have different properties as a function of the orientation and polarization angle of an incident light ray (a property called *anisotropism*). These materials have the ability to refract an unpolarized incident light ray into two separate, orthogonally polarized rays, which in the general case take different paths through the material. The path is dependent on orientation of the material with respect to the incident ray and the refractive index for each polarization. The refracted rays are referred to as the "ordinary" (or "O") ray that obeys Snell's Law, and the "extraordinary" (or "E") ray that does not. All crystals, except those of cubic lattice structure, exhibit some degree of anisotropy with regard to their physical properties, including refractive index. Amorphous materials such as glass and liquids are *isotropic* and do not show this behavior. Other materials such as plastics become *birefringent* when they are subjected to a mechanical strain. Hexagonal, tetragonal, and trigonal crystal classes exhibit two indices of refraction, are called *birefringent,* and are uniaxial. Orthorhombic, monoclinic, and triclinic exhibit three indices of refraction, are *trirefringent,* and are biaxial. *Birefringence* (Δn) is defined mathematically by:

$$\Delta n \equiv n_e - n_o$$

where

n_e is the index of refraction for the extraordinary ray, and
n_o is the index of refraction for the ordinary ray.

BIREFRINGENT VALUES FOR SELECTED MATERIALS

Material	Range of Values (Δn)
Synthetic Emerald	.003
Quartz	.009
Tourmaline	.018 − .020+
Zircon	.005 to .059
Calcite	.172
Synthetic Rutile	.287

Also called *double refraction.*

bis Meaning second in Latin. The term is used as a suffix to denote a second version of a standard released by ITU-T.

BISDN An abbreviation of Broadband ISDN.

BIST An acronym from Built In Self Test.

bistable Pertaining to a device capable of assuming either one of two stable states.

BiSync or Bisync An acronym from BInary SYNchronous Communication, pronounced "BYE-sink."

bit An acronym from BInary digiT.

(1) A character used to represent one of the two digits in the numeration system with a base of two, and only two, possible states of a physical entity or system. **(2)** In binary notation, the symbols 0 or 1. **(3)** The smallest unit of digital information equal to one binary decision or the designation of one of two possible and equally likely states of anything used to store or convey information. Bits are combined into working groups called bytes (4 bits), words, fields, records and files, each containing groups of the lower (preceding) collection.

bit bucket The fictitious location to which lost information goes. For example, when scrolling through a long list, information that rolls off the top of the display is said to have fallen into the *bit bucket.*

bit count integrity (BCI) (1) In message communications, the preservation of the exact number of bits that are in the original message. **(2)** In connection-oriented services, preservation of the number of bits per unit time. *Bit count integrity* is not the same as *bit integrity.* The former only requires preservation of the number of bits, while the latter requires that the delivered bits correspond exactly with the original bits.

bit density The number of bits recorded per unit length, area, or volume. (*Bit density* is the reciprocal of *bit pitch.*) Also called *recording density.*

bit error rate (BER) A deprecated term, now replaced by *bit error ratio.*

bit error ratio (BER) The number of erroneous bits divided by the total number of bits transmitted, received, or processed over some stipulated period. Usually, *BER* is expressed as the reciprocal of the average. For example, if 1 error occurs on the average in every 10,000 bits, the *BER* is expressed as 10^{-4}; if 5 errors occur in every 1 million bits, the *BER* is 5×10^{-6}.

A qualifier is frequently used with *BER;* for example,

- *Transmitted BER*—the number of erroneous bits received divided by the total number of bits transmitted.
- *Information BER*—the number of erroneous received bits divided by the total number of received bits.

Formerly called *bit error rate (BER).* See also *burst errors* and *failure in time (FIT).*

bit error ratio tester (BERT) A testing device for checking a system's transmission quality. A *BERT* generally sends a predefined pseudorandom signal through the system under test and compares it with the received signal. The quality is indicated as a bit error ratio (BER). Formerly called a *bit error rate tester.*

bit hierarchy A USENet alternative newsgroup that echoes the most popular LISTSERV mailing lists from BITNET. The topics are widespread, covering medical, technical writing, music, and many others.

bit interval The amount of time a digital signal is left at a particular state (level) to indicate a value. Usually, the level will indicate the value of a single bit, but it is possible to encode more than a single bit in a voltage level, thereby transmitting more than one bit in a single bit interval. In general, the longer the bit interval, the slower the transmission rate. For example, when encoding a single bit at a time, a bit interval of .01 second means a transmission rate of only 100 bps. Also called *bit time,* or *character* or *unit interval.* See also *baud* and *character* or *unit interval.*

bit-mapped image An image wherein each picture element (*pixel*) is described by block of data. The description contains the contribution of each of the primary colors or shades of gray. In the simplest case, only one bit is required to describe the black or white pixel state.

Fax images are bit-mapped images. One square inch of a standard resolution fax contains approximately 40,000 pixels—each pixel being black or white.

bit oriented A communication protocol in which information may be encoded in fields as small as one bit.

bit-oriented protocol (BOP) A protocol in which individual bits are transmitted without regard to their definition. Individual bits are used for both timing (so as to maintain sender and receiver synchronous operation) and for link control. Examples of *bit-oriented protocols* include HDLC, LAPB, and SDLC. See also *byte-oriented protocol.*

bit pairing The practice of establishing, within a code set, a number of subsets that have an identical bit representation except for the state of a specified bit.

For example, in the International Alphabet No. 5 (IA5) and the American Standard Code for Information Interchange (ASCII), where the upper case letters are related to their respective lower case letters by the state of bit six. See also *ASCII.*

bit rate (BR) The speed at which bits in a bit stream are transmitted—from DTE to DTE for example. *Bit rate* is generally expressed in bits per second (bps).

For n-ary operation, the bit rate is equal to $\log_2 n$ times the rate (in bauds), where *n* is the number of significant conditions in the signal. Also called *data rate. Bit rate* and baud rate are not necessarily the same—see *baud.*

bit robbing The use of the least significant bit per channel in every sixth frame of a T-1 transmission for signaling. The *signaling bit* does not materially affect voice transmissions because it is both so infrequent and in a position that contributes the least to the user's message. Furthermore, the bit has a 50% chance of being correct at any time. Also called *speech digit signaling.*

bit serial A term that refers to the transmission of digital data one bit following another over a single wire.

bit slip In digital transmission, the loss of one or more bits, caused by variations in the respective clock rates of the transmitting and receiving devices.

One common cause of bit slippage is the overflow of a receive buffer that occurs when the transmitter's clock rate exceeds the receiver's clock rate. This causes one or more bits to be dropped when the storage capacity is exceeded.

bit stream The sequence of bits representing a flow of digital information through a communication channel.

bit streaming An error on a token ring network where the interface adapter overwrites existing data on the ring.

bit string A sequence of bits. Individual bit strings of a transmission may be separated by data delimiters.

bit stuffing The insertion of extra bits into a transmitted bit stream. *Bit stuffing* is done at the data link level and is used for synchronization and protocol compliance, or to ensure that a specific sequence of bits does not occur.

Examples include:

- Synchronizing bit streams that do not have the same or rationally related bit rates.
- Synchronizing several channels before multiplexing.
- Rate-matching two signal channels to each other.
- Filling buffers or frames when there is insufficient user data to fill a fixed word length packet.
- In HDLC, SDLC, and X.25 communication protocols, six "1s" in a row are allowed only at the beginning or end of a frame. The transmitting DCE inserts or "stuffs" a "0" into the bit stream whenever 5 "one bits" in a row occur.

The receiving DCE removes or "strips" the extra bits from the data stream, restoring the original data. Also called *positive justification.*

BITE An acronym from Built In Test Equipment.

BITNET A large-scale, international computer network connecting primarily academic institutions (research or education). *BITNET* uses the IBM NJE communications protocol, supports e-mail, and file retrieval.

The origin of "BIT" is obscured in folklore and depends on whom you ask. For example, Because Its Time is one frequently cited source of the acronym. In Canada, *BITNET* is called NetNorth, and in Europe it is called EARN (from European Academic Research Network).

Bitronics A specification for a parallel port that allows bidirectional communications on a Centronics cable. Hewlett-Packard pioneered it primarily for use with postscript printers.

bits per inch (b/in) A unit used to express the linear bit density of data in storage. *Note:* The abbreviation "bpi" is not in accordance with international standards; therefore, its use is discouraged.

bits per second (b/s) A unit of measure representing the rate at which individual bits pass a designated point or are transmitted across a communications link or circuit. Also abbreviated *bit/s* and *bps*. See also *baud.*

BIU (1) An abbreviation of Basic Information Unit. **(2)** An abbreviation of Bus Interface Unit. See *network interface device* (*NID*).

BIX An acronym from Byte Information eXchange. A commercial dial-up information service that provides access to the Internet, including e-mail, FTP, and Telnet services.

biz One of several USENet alternative newsgroups. The subject matter includes information and announcements on business products and services (especially as related to the computer industry). See also *ALT* and *alternative newsgroup hierarchies.*

BJT An abbreviation of Bipolar Junction Transistor.

black box A device where the user is only concerned with the functioning of inputs and outputs. The details of the internal operation are not required, nor does the user generally know them.

black noise Noise characterized by a frequency spectrum of predominantly zero power level over all frequencies except for a few narrow bands or spikes.

black recording In facsimile systems, the signal that corresponds to the maximum record density. In AM systems it is the signal with the maximum power, and in FM systems it is the lowest frequency in the passband.

black signal In facsimile, the signal resulting from scanning a maximum-density area of the object.

BLACK signal In cryptology, a signal containing only unclassified or encrypted information.

blackbody A totally nonreflecting body, it may absorb or radiate energy, however. In thermal equilibrium, the absorbed energy is exactly equal to the radiated energy.

blacklist (**1**) In telephony, the prevention of the redial of a number for some period of time after a failed call count has been exceeded. After a predetermined number of call attempts to any one number has failed, further attempts are restricted for a set time. In some countries this time is minutes or hours, and in other cases it requires a manual reset to allow further attempts. (**2**) A feature in some modems or modem software that prevents access to the system by specific calling numbers.

blackout The complete loss of *mains power;* that is, the RMS voltage goes to zero.

blanketing The presence of an AM broadcast signal of sufficient level to cause interference. Generally, the level is defined as 1 V per meter (1 V/m) or greater. These signal strengths are generally confined to an area adjacent to the antenna of the transmitting station. The 1 V/m contour is referred to as the *blanket contour,* and the area within this contour is referred to as the *blanket area.*

BLAST An acronym from <u>BL</u>ocked <u>AS</u>ynchronous <u>T</u>ransmission.

BLER An acronym from <u>Bl</u>ock <u>E</u>rror <u>R</u>atio.

BLERT An acronym from <u>Bl</u>ock <u>E</u>rror <u>R</u>atio <u>T</u>est (or <u>T</u>ester).

BLF An abbreviation of <u>B</u>usy <u>L</u>amp <u>F</u>ield.

blind courtesy copy (BCC) A copy of a message delivered to a receiver without the knowledge of the addressee(s) or copy recipients. In the days preceding xerography and e-mail, *BCC* stood for blind carbon copy (from the carbon paper copy).

blind dialing A feature that allows a modem to dial in the absence of a dial tone. The calling modem, after going off-hook, waits a specified period of time and then proceeds with the dial string. Typical delay values range from 2 to 25 seconds.

Blind dialing is not allowed in most countries.

blind transfer A call transfer from one party to a second party and without the first party informing the second party of the subject or who is being transferred.

blind transmission Transmission without obtaining acknowledgment of reception from the destination receiving station.

Blind transmission may occur or be necessary when security constraints, such as radio silence, are imposed, when technical difficulties with a sender's receiver or receiver's transmitter occur, or when lack of time precludes the delay caused by waiting for acknowledgments.

BLOB An acronym from <u>B</u>inary <u>L</u>arge <u>OB</u>jects. Those items in a database, such as pictures or other entities that consume large amounts of space, that are neither characters, numbers, nor other traditional data base fields.

block (**1**) An information group that is transmitted as a unit and that may be encoded for error control purposes. (**2**) A string of contiguous records, words, or characters, that for technical or logical purposes are treated as a single unit. Blocks have the following distinguishing characteristics:

- They are separated by interblock gaps.
- They are delimited by an end-of-block signal.

- They may append a header and/or trailer for identification and routing.
- They may contain more than one record.

A block is generally subjected to some type of block error control processing, such as multidimensional parity checking and CRC. The *block length* is defined by the communications protocol and determined by physical constraints or characteristics of the transmission medium. For example, the desire to minimize the overhead penalty due to retransmissions in a noisy communications channel indicates short blocks, while the desire to send as quickly as possible in a noiseless channel leads to long blocks. See also *logical record* and *packet.* (**3**) In programming languages, a subdivision of a program that serves to group related statements, delimit routines, specify storage allocation, delineate the applicability of labels, or segment parts of the program for other purposes.

block character See *end of transmission block (ETB).*

block check character (BCC) A character appended to a transmission block to facilitate message error detection.

In systems that utilize *BCC* (such as longitudinal redundancy checking—LRC—and cyclic redundancy checking—CRC), block check characters are computed for, and added to, each transmitted data block. This *BCC* is compared with a second *block check character* (computed by the receiver, over the same data block) to determine whether the transmission is error free.

block diagram A pictorial representation of the significant functional elements of hardware, software, or a system. Appropriately labeled geometric figures are used to represent processes, while arrows are used to show the pathways of information flow or control between the "boxes."

The rudimentaries of an echo-canceling modem are shown in the following block diagram.

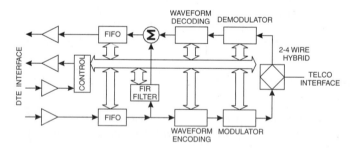

block efficiency In a block, the ratio of the number of user information bits to the total number of bits. For a given block format, *block efficiency* is the maximum possible efficiency over a perfect transmission link as it ignores all errors, overhead, and retransmissions.

block error A discrepancy of information contained in a block as determined by a check code or other means.

block error probability The expected block error ratio. See *block error ratio (BLER).*

block error rate See *block error ratio (BLER).*

block error ratio (BLER) In communications, an error rate based on the proportion of blocks received with errors to the total number transmitted. *BLER* is calculated using empirical measurements. Multiple *block error ratios* may be used to predict *block-error probability.* Formerly called *block error rate (BLER).*

block error ratio tester (BLERT) A hardware device for determining a transmission's block error rate, which is that proportion of blocks with erroneous bits. Also abbreviated as *BKERT* and formerly called a *block error rate tester* (*BLERT*).

block length The number of data elements, such as bits, bytes, characters, or records, in a block.

block loss probability The ratio of the number of lost blocks to the total number of block transfer attempts during a specified period of time.

block mode terminal interface (BMTI) A device used to assemble and disassemble packets being transferred from a block mode device (such as IBM's BiSynch devices) through an ITU X.25 network.

block parity The designation of one (or more) bits within a block as a parity indicator. This parity bit may be either even or odd. It is used to assist in error detection or correction. See also *parity*.

block transfer The process, initiated by a single action, of transferring one or more blocks of data.

block transfer attempt The sequence of activities involved in the effort of transferring a block of data from a source to a destination. A *block transfer attempt* begins when the first bit of the source block crosses the boundary between the station equipment and the transmission system equipment. All *block transfer attempts* end either in successful block transfers or in block transfer failure.

A *block transfer attempt* is successful if both the block is delivered to the intended receiver within the maximum allowable time and the contents of the block are correct.

block transfer efficiency The average ratio of user information bits in a block to the total number of bits transferred successfully.

block transfer failure The failure to successfully deliver a block to its destination. The most common causes of failure are lost, misdelivered, and added blocks.

block transfer rate The number of successful block transfers made in a specified period of time.

block transfer time The average duration of a successful block transfer attempt.

blocked calls delayed (BCD) A variable in queuing theory to describe the effects of holding a caller in a queue because the call is blocked and it cannot be completed instantly.

blocked calls held A variable in queuing theory to describe the effects of a caller redialing a destination the instant a blocked call is detected.

blocked calls released A variable in queuing theory to describe the effects of a caller waiting to redial a destination after a blocked call is detected.

blocking (1) In communications, the inability to connect two idle terminals attached to a communications network because all possible paths or routes are in use.

In the switching system diagrammed below there are "b" fewer links than input or output ports. This leads to a situation where the $m + 1$ call attempt cannot be routed to any destination because all links are in use; the call attempt is said to be blocked. See also *nonblocking*. **(2)** Denying access to, or use of, a facility, system, or component. **(3)** The formatting of data into blocks for purposes of transmission, storage, checking, or other functions.

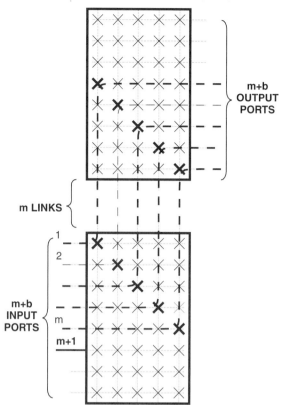

SWITCHING MATRIX WITH BLOCKING

blocking factor The number of records in a block. It is calculated by dividing the block length by the average record length contained in the block. Common *blocking factors* are powers of two, e.g., 128, 256, 512, and 4096 bytes. Also called *grouping factor*.

BLU An abbreviation of Basic Link Unit.

Blue Book Ethernet Ethernet version 2.0. The term is used to distinguish Ethernet 2.0 from the similar, but not identical, Ethernet variant defined by the IEEE 802.3 standard.

blue noise Noise over a specified frequency range in which the spectral density (power per hertz) is proportional to the frequency and is not flat as in white noise.

BMP *.BMP* is the extension of a *graphic bitmap* file format used by Windows and OS/2. The format supports 1, 4, 8, and 24 bit images. The file consists of a 40-bit header (supplying information such as the size and number of bits per pixel) followed by an optional palette and then the image itself. The format supports compression; however, most applications do not implement it.

BMTI An abbreviation of Block Mode Terminal Interface.

BNC A standard locking type of connector used with coaxial cables. The connectors "lock" together when one connector is inserted into the other and rotated 90°. The name may come from Bayonet-Neill-Concelman (its developers), Bayonet Nut Connector (for its attachment mechanism), Bayonet Navy Connector or British Navel Connector (early users), or Baby N-Connector.

The drawing shows the attachment of a coaxial cable to a *BNC* connector. The coax has each layer progressively exposed per manufacturer's instructions. The coax center conductor is soldered to the connector pin, and then the cable is inserted into the connector body. The ferrule is crimped over the sheath holding the assembly together. Also referred to as a *barrel nut connector*.

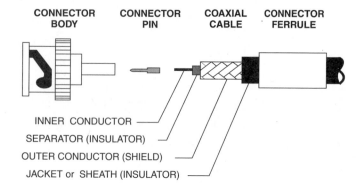

CONNECTOR BODY   **CONNECTOR PIN**   **COAXIAL CABLE**   **CONNECTOR FERRULE**

INNER CONDUCTOR

SEPARATOR (INSULATOR)

OUTER CONDUCTOR (SHIELD)

JACKET or SHEATH (INSULATOR)

BNC T-connector See *T-connector.*

BnZS The abbreviation of <u>B</u>ipolar with "<u>n</u>" <u>Z</u>ero <u>S</u>uppression. *BnZS* codes are normal alternate mark inversion (AMI) codes with strings of "n" zeros replaced by a pattern that is detectable at the receiving terminal. The pattern introduces two successive pulses of the same polarity (a so-called code violation). Values of "n" in use include 3, 4, 6, and 8. The rules governing the symbol substitution are based on both the last nonzero bit transmitted and the last substitution used.

For example, the *B3ZS* code algorithm is:

B3ZS CODING RULES

Last Pulse Transmitted	Last Substitute Sequence Used	
	0 0 + or + 0 +	0 0 − or − 0 −
+	− 0 −	0 0 +
−	0 0 −	+ 0 +

Comparison of a pure AMI and B3ZS coded AMI sequence demonstrates the substitution process outlined in the preceding table. The *B8ZS* rule is even simpler than the B3ZS rule; that is, if the last pulse transmitted was a (+), replace the eight 0s with 000+-0-+; otherwise, replace with 000-+0+-. This always forces a bipolar code violation (BPV) in positions 4 and 7 of the substitution. See also *code, HDBn,* and *waveform codes.*

Binary	1 0 1 1 1 1 0 0 1 0 0 0 0 0 0 1 0 0 0
AMI	+ 0 − + − + 0 0 − 0 0 0 0 0 0 + 0 0 0
B3ZS	+ 0 − + − + 0 0 − 0 0 − + 0 + − 0 0 −

board (1) A specific menu area on a bulletin board system (BBS) for posting/receiving messages. Multiple boards are frequently provided on a single BBS, one for each message topic. Sometimes an entire BBS is referred to as a board. (2) A general term used to describe the device that holds electronic components and has printed or etched conductive paths interconnecting the components. Also called a PCB (from printed circuit board), motherboard (when it is the main board), and card.

BOC An acronym from <u>B</u>ell <u>O</u>perating <u>C</u>ompany.

body The portion of an e-mail communication in which the message is placed, as opposed to the header, address, or signature.

BOF An abbreviation of <u>B</u>irds <u>O</u>f a <u>F</u>eather.

bogus newsgroup A USENet newsgroup which has been deleted from the official list of active newsgroups, even though it still appears in the list of newsgroups to which a user subscribes, as delineated in the file *.newsrc.* The newsgroup may be deleted for several reasons, such as the newsgroup may have had extended periods of inactivity, or it may have been replaced by another newsgroup, or the site manager may have decided not to carry the alternative newsgroup. See also *ALT* and *alternative newsgroup hierarchies.*

Boltzmann's constant (k) Relates the average energy of a molecule to its absolute temperature. The ratio of the universal gas constant (8.314 joules per degree Kelvin) to Avogadro's number (6.022×10^{23}) and is approximately:

$$1.380658 \times 10^{-23} \text{ J/K (joules per Kelvin)}.$$

Named after Ludwig Edward Boltzmann (1844–1906).

bomb See *virus.*

bond Multiple entities joined in an inseparable manner. For example:
- Two physical materials may be "glued," soldered, or welded together.
- Similarly, two ISDN 64-kbps B-channels may be bonded to form a single 128-kbps channel.

BOOK An acronym from <u>B</u>inary <u>O</u>n-<u>O</u>ff <u>K</u>eying.

bookmark An alias and/or tag for the path or address of a user-selected resource (in a network or document).

The resource is usually buried deep within a tree of menus, making its location difficult to find in a future session. The *bookmark* allows the user to return to the marked resource simply by instructing the application software to go to the *bookmark.*

Boolean algebra Proposed in 1847 by the English mathematician George Boole (1815–1864), *Boolean algebra* is a method used to symbolically express the relationships between items. The basic algebraic rules and operators of today's *Boolean algebra* are listed in the associated figure on page 61.

Boolean function (1) A mathematical function that describes *Boolean operations.* (2) A switching function in which the number of possible values of the function and each of its independent variables is two.

Boolean operation (1) Any operation in which both the operands and the operation result take one of two states or values. Typical states are "1 or 0," "on or off," "open or closed," or "present or absent." (2) An operation that follows the rules of *Boolean Algebra.*

boot sector On a hard disk or floppy disk, the first sector of the bootable partition; that is, the partition which contains the system files. On floppy disks, the boot sector is the first sector on the disk. (*Note:* ALL floppy disks have a boot sector, even if they are not bootable.)

BOOLEAN ALGEBRA OPERATORS AND RULES

Operator	Symbol	Rule or Definition	Truth Table		
			P	Q	P · Q
AND (conjunctive)	·, &, Λ, or none	A set of items is true if and only if ALL items in the set are true.	T T F F	T F T F	T F F F
			P		P
NOT (negate)	x̄, −x, x*, ~x, ′	An item is true after the operation if and only if the item was false before the operation.	T F		F T
			P	Q	P+Q
OR (disjunctive)	+, V	The set of items is true if AT LEAST ONE item is true.	T T F F	T F T F	T T T F
			P	Q	P⊕Q
XOR	⊕	A set of items is true only if an odd number of items in the set are true.	T T F F	T F T F	F T T F
Grouping	( )	Groups a set of items so that they may be treated as a single entity.			
			P	Q	P⇒Q
Implication	→, ⇒, ⊃	States the conditional *if the antecedent is true, then the consequent is true*; i.e., the antecedent *implies* the consequent is true. P⇒Q is equivalent to $\overline{(P \cdot \overline{Q})}$.	T T F F	T F T F	T F T T
Commutative		1. The *order of items* in a set joined by the OR connective is not significant; e.g., P+Q+R is the same as Q+P+R. 2. The *order of items* in a set joined by the AND connective is not significant; e.g., P · Q · R is the same as Q · P · R.			
Associative		1. The *order of grouping* items in a set joined by the OR connective is not significant; e.g., P+(Q+R) is the same as (P+Q)+R. 2. The *order of grouping* items in a set joined by the AND connective is not significant; e.g., P · (Q · R) is the same as (P · Q)·R.			
Distributive		1. Given a set of items joined by the OR connective. Another item may be joined to either the entire set by the AND connective or over the set term by term. Thus P · (Q+R+S) is equivalent to (P · Q)+(P · R)+(P · S). 2. Given a set of items joined by the AND connective. Another item may be joined to either the entire set by the OR connective or over the set term by term. Thus P+(Q · R · S) is equivalent to (P+Q) · (P+R) · (P+S).			
			P		P · P
Contradiction		A set of items that is FALSE regardless of the values of the individual items; e.g., the set P · P̄.	T F		F F
			P		P+ P
Tautology		A set of items that is TRUE regardless of the values of the individual items; e.g., the set P+ P̄.	T F		T T
DeMorgan's Law		1. The negation of the AND of two items is logically equivalent to the OR of the negation of each. Or symbolically $\overline{(P \cdot Q)}$ is equivalent to $\overline{P} + \overline{Q}$. 2. The negation of the OR of two items is logically equivalent to the AND of the negation of each. Or symbolically $\overline{(P + Q)}$ is equivalent to $\overline{P} \cdot \overline{Q}$.			

booting Loading a computer memory with the basic information needed to start operation. *Remote booting* refers to loading software over the network.

BOOTP An acronym from <u>Boot</u> <u>P</u>rotocol. A TCP/IP network protocol that allows a network node to request configuration information from a *BOOTP server* node. Sometimes written *BootP*.

bootstrap **(1)** Generally, a technique, procedure, or device designed to bring about a specific state or operating condition by means of its own actions, i.e., to initialize the system. **(2)** A set of computer instructions that cause additional instructions to be loaded until the complete computer program is in main memory. **(3)** The automatic procedure whereby the basic operating system of a processor is reloaded following a complete shutdown, loss of memory, cold start, or warm start. **(4)** That part of a computer program that may be used to establish another version of the computer program.

BOP An acronym from Bit-Oriented Protocol.

border gateway protocol (BGP) An Internet *routing protocol,* defined by RFC 1163, which allows one *autonomous network* to communicate with another. The *border gateway protocol* represents an improvement over the *exterior gateway protocol* (*EGP*) because it allows a datagram to be raised only to the level of network hierarchy required.

The earlier Internet network model required every autonomous network joining the Internet to be attached to the core's backbone. Consequently, every datagram not within the autonomous system had to pass through that core. Later models redefined the Internet as a hierarchical structure of adjacent network domains; hence, a datagram need not necessarily traverse the core's backbone. The *BGP* provides the means of distributing routing decisions among each of the autonomous domains. See also *autonomous system* (*AS*), *exterior gateway protocol* (*EGP*), and *interior gateway protocol* (*IGP*).

border router That router of an *autonomous system* used to connect the system with the Internet. See also *boundary router.*

boresight **(1)** To align a directional device, such as antenna or telescope, using either an optical procedure or a fixed target at a known location. The term was derived from the process of aligning the sights of a gun to a target sighted through the bore of the barrel. **(2)** The physical axis of a directional antenna.

BORSHT In telephony, an acronym from Battery feed, Overvoltage protection, Ringing injection, Supervision or signaling, Hybrid, and Test; the six basic functions of all telephone switching systems. On digital systems, the spelling is sometimes *BORSCHT* where the "C" is for Coding, the analog-to-digital conversion in modern telephone exchanges.

bot A contraction of *robot*. See *knowbot.*

both party control A term that refers to the method of call termination whereby both the originating station and the destination station must cooperatively initiate the call disconnect sequence. For example, if station A originated the connection to station B, then station A *and* station B must initiate the disconnect sequence to end the connection. See also *called party control, calling party control,* and *either party control.*

bounced message An e-mail message returned to its sender when it doesn't go through to the destination because of an error (e.g., incorrect address). The message's header may contain additional information indicating why the message was *bounced.*

bound mode In optical fibers, a transmission mode in which

- The field intensity decays monotonically in the transverse direction everywhere external to the core, and
- Power is not lost due to radiation.

Except for single mode fibers, the power in bound modes is predominantly contained in the core of the fiber. Also called *guided mode* and *trapped mode.*

bound ray A synonym for *guided ray.*

boundary router A router of an *autonomous system* that allows it to connect to other autonomous systems. See also *border router.*

Boundary Routing A 3Com proprietary name for a method of accessing remote networked locations, such as a bank branch office. A form of bridging, it is intended to:

- Reduce the need for technical expertise at the remote site.
- Reduce the cost of equipment at the remote site.
- Manage communications from the head office.

BPDU An abbreviation of Bridge Protocol Data Unit.

bpi An abbreviation of Bits Per Inch.

BPON An acronym from Broadband Passive Optical Network.

bps An abbreviation of Bits Per Second. The number of bits passing through a circuit per second. The information transfer rate. Also sometimes written as b/s.

Note: Frequently bps rate is *improperly* referred to as *baud* rate. They are numerically the same only when the data encoding method has exactly two states (binary). See also *baud* (*Bd*).

BPSK An abbreviation of Binary Phase Shift Key. See *PSK.*

BPV An abbreviation of BiPolar Violation.

BRA An acronym from Basic Rate Access.

braid **(1)** In electrical cables, a cylinder fabricated from woven metallic filaments. The cylinder is generally used as a covering over one or more insulated wires to form a shielded cable. If there is only one center wire and the spacing of that wire is maintained at the center of the *braid,* the assembly is called a coaxial cable. **(2)** In optical fibers, an unwoven parallel bundle of yarn situated inside the jacket and around the buffer of single or multiple fiber bundles. The *braid,* often an aramid yarn, adds tensile strength to the cable and may be anchored to an optical connector or a splice organizer assembly to secure the end of the cable.

In a single-fiber or two-fiber "zip-cord" (loose or tight buffered) fiber optic cable, the *braid* is situated between the buffer tube and jacket. In cables having multiple buffer tubes, the *braid* is usually situated between the inner jacket and outer jacket.

branch **(1)** That portion of a network or graph comprising a section reducible to a two-terminal equivalence and connected between two nodes. **(2)** In a computer program, a conditional jump or departure from the implicit order in which instructions are being executed. **(3)** The verb meaning to select a branch, as in (*2*) above. **(4)** In a power distribution system, a circuit from a distribution device (power panel) of a lower power-handling capability than that of the input circuits to the distribution device.

Branch Systems General License (BSGL) A license that must be obtained by any organization seeking to link its own private network to the Public Switched Telephone Network (PSTN). A separate license must be obtained for each individual site.

branching network A network used for transmission or reception of signals over two or more channels.

branching repeater A repeater with two or more outputs for each input. See also *multiport repeater.*

BRB An abbreviation of Be Right Back.

breadboard **(1)** An assembly of circuits or parts used to prove the feasibility of a device, circuit, system, or principle, with little or no regard to the final configuration or packaging of the parts. **(2)** To prepare a breadboard (*1*).

break (1) In communications, a signaling method used as an "attention command." A *break* occurs when the receiving station interrupts and acquires control of the transmission line.

One method of sending a break is the transmission of a spacing condition for more than the maximum allowable period. This method is derived from a signaling method used in telegraphy where one operator would open the signaling loop (or break the loop) for an extended period of time relative to a normal signaling element. Generally used in *half-duplex* or *simplex* circuits. (2) In dial pulse signaling, that portion of a signal element when the dial contacts are open and no loop current flows. The time the contacts are open is frequently expressed as a percentage break—that is, break time divided by make plus break time.

break in An action an operator or switchboard attendant can take to interrupt a call in progress for an announcement or emergency.

break key A key found on teletypewriters and some modern keyboards used to interrupt the current task running. The key does not generate an ASCII character; it simply generates a string of start bits until the key is released. This is an invalid character symbol, so it can be detected by the host unit that will take appropriate action.

On current loop signaling systems, the break key on the data receiver opens the loop. This signals the sender to stop sending and listen.

break out To separate the individual elements of a cable for the purpose of splicing or installing connectors, i.e.,

- With electrical cables, the individual elements are wire conductors.
- With optical cables, the elements are fibers or buffer tubes.

Also called *fanout* and *furcate*.

breakdown The process of a material changing from an insulating state to a conducting state because of an applied electric field. *Breakdown* occurs in a series of successive steps, namely:

- *Initial state*—no current flow.
- *Initiation*—increasing degree of conductivity.
- *Instability*—runaway current increase due to positive feedback.
- *Voltage collapse*— discharging the electrostatic energy stored in the material through a *breakdown channel*.
- *Final state*—high current flow through the material and low-voltage drop across it.

In a gas, *breakdown* is sometimes referred to as *sparkover* or *ignition*.

breakout box A device that is plugged in between a device and the cable connecting it to another device for the purpose of accessing individual leads of a cable. It allows "cross connects," opening lines, and monitoring signals. Also spelled *break out box*.

Brewster's angle That angle of incidence at which a reflected electromagnetic ray is linearly polarized with the electric vector parallel to the reflective surface. The refracted ray contains the remaining polarization vectors.

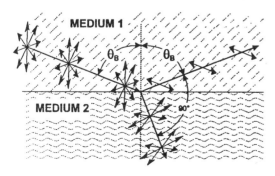

For a ray traveling from medium 1 (whose index of refraction is n_1 and permittivity ϵ_1) to medium 2 (with index n_2 and permittivity ϵ_2), the angle is defined by:

$$\theta_B = \tan^{-1}\left(\frac{n_2}{n_1}\right) = \tan^{-1}\left(\frac{\epsilon_2}{\epsilon_1}\right)$$

BRI An abbreviation of <u>B</u>asic <u>R</u>ate <u>I</u>nterface. See *basic rate access* (*BRA*).

bridge (1) In communications, a device that connects multiple local area networks (LANs) or wide area networks (WANs) at the same *medium access control* (*MAC*) protocol, that is, at the ISO/OSI *data link level* (layer 2) and *physical level* (layer 1) or equivalent. (See the associated figure.) It forwards (or filters) data packets between the LANs based on the packet destination address.

A *bridge* is transparent to OSI layers 3 through 7; therefore, it is up to the individual stations to translate received data. The *bridge* relays packets between LANs by a store-and-forward method. Generally, a *bridge* does not do a protocol conversion; however, if dissimilar LANs are to be connected, protocol conversion on reformatted packets is required. *Bridges* are frequently used to divide large LANs into smaller

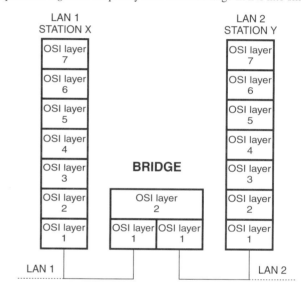

working groups to improve the network's efficiency. Two geographically separated LANs can be joined using a *bridge* at each end of a WAN. See also *brouter, gateway,* and *router.* (2) In electronic circuits, a network with two ports (terminal pairs) having the properties:

- With a fixed voltage polarity at the input port, a voltage of either polarity can be arranged with appropriate selection of element values.
- With an arbitrary voltage at the input port, nonzero valued elements may be found such that the output voltage is zero for all input voltages.

The schematic of a *bridge circuit* is given in the associated figure. (3) The connection of a device or element across another device or element. For example, a DCE *bridged* onto a bus network or a resistor *bridged* across a "T-network." (4) On an electronic assembly, material that spans two conductors not intended to be connected creating a *short circuit*, e.g., a *solder bridge*. See also *hybrid coil.*

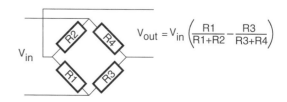

bridge connection A parallel connection to a transmission line where some of the signal energy may be diverted to another destination. Ideally, the connection causes no perceptible effect on the original circuit. In practice, however, the connection will cause reflections and signal attenuation. Bridge connections may be found in LANs, data distribution systems, or voice circuits such as the party line connection depicted.

The following figure shows one instance of bridge taps in the telephone industry—one called a *party line*. A line bridged onto the main

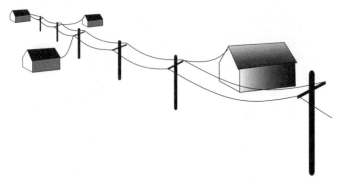

transmission line may not have a terminating device at the far end. (The user simply disconnected the device and left all wiring in place.) This may affect other communications adversely. Also called a *bridged tap.*

bridge lifter See *bridge tap isolator.*

bridge protocol data unit (BPDU) A packet that initiates communications between devices under a *spanning tree* protocol.

bridge tap isolator In telephony, a device (relay, saturable inductor, or semiconductor) inserted in series with a subscriber loop which acts like a switch to voice frequencies. That is, it "opens" the loop when the subscriber is on-hook.

Frequently a current sensitive inductor (*saturable inductor*) is used as a *bridge tap isolator*. With little dc current flowing through the inductor, the inductance is high (about 20 H); therefore, it presents a high impedance to voice frequencies. When dc currents greater than about 20 mA flow through the windings of the inductor, the inductance is low (approximately 20 mH) and the device is a low impedance in the voice band. *Bridge tap isolators* are generally used only when the bridge tap loop length and the primary access line are both long. The ideal location for a *bridge tap* isolator is as close to the

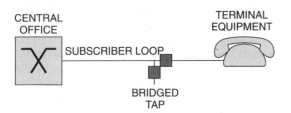

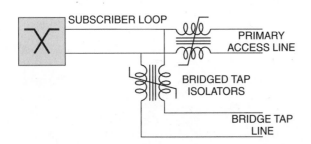

bridge tap as possible. A *bridge tap isolator* adds 20- to 45-Ohm loop resistance and about 0.3- to 0.5-dB insertion loss (at 1000 Hz) to the off-hook loop. Because of these additional losses, *bridge tap isolators* should not be used on long subscriber loops because they exacerbate the problems associated with long loops. Also called *bridged lifters* and *saturable inductors.*

bridge transformer A synonym for *hybrid coil.*

bridged ringing In telephony, the term used to describe the situation where the terminal equipment ringer is connected between the tip and ring wires of the loop. The following diagram illustrates the connection of the ringer in a *bridged ringing* subset. L1, L2, and G are wiring terminals within the subset. See also *coded ringing, divided ringing,* and *isolated ringing.*

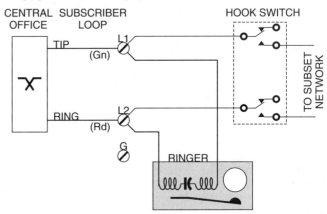

bridged T network A network composed of four branches; three branches are connected as a "T-network," and the fourth is connected across the two series elements of the "T-network" (between the input and output). See also *PAD (3).*

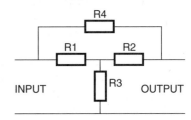

bridging connection A parallel connection used to extract some of the signal energy from a circuit, usually with negligible effect on the normal operation of the circuit.

bridging loss At a given frequency, the loss that results when an impedance is connected across a transmission line. *Bridging loss* is expressed as the before to after bridging ratio of the signal power delivered at a point downstream from the bridging point. The ratio is generally expressed in units of decibels (dB); i.e.,

$$Loss = 20 \log_{10} \left(\frac{Level\_after}{Level\_before} \right) dB$$

Brillouin scattering In optics, the nonlinear scattering of lightwaves in a medium due to thermally driven density fluctuations. The scattering takes place on an atomic level and may cause frequency shifts of several gigahertz at room temperature.

British Standards Institute (BSI) The UK standards body responsible for input into European and international standards setting bodies like International Standards Organization (ISO) and the International Telecommunication Union (ITU-T).

BRL An abbreviation of Balance Return Loss.

broadband (1) Generally, a term meaning *high bandwidth* as compared to the required bandwidth. See *wideband*. (2) A transmission technique that employs some form of channel *multiplexing* (FDM, TDM, or spectral division) to allow signals from more than one source to coexist on the same transmission medium. It allows simultaneous transactions, since the channels are independent. (3) In LAN terminology, it refers to a system in which multiple channels access a common medium, for example, coaxial cable that has a large bandwidth using radio frequency (RF) modems. This may allow the coaxial cable to carry multiple separate LANs whose transmission is being modulated at different frequencies. (4) In cable television (CATV), broadband describes the ability to carry a multiplicity of simultaneous TV channels.

broadband exchange (BEX) A communications switch capable of interconnecting channels having bandwidths greater than voice bandwidth.

broadband ISDN (B-ISDN) An Integrated Services Digital Network (ISDN) offering broadband capabilities to homes, businesses, and schools. It promises universal coverage based on ATM/SDH technologies and optical fiber, supporting data, voice and video traffic.

B-ISDN is an ITU-T proposed ubiquitous service that:

- Includes interfaces operating at data rates from 155 to 622 Mbps.
- Uses asynchronous transfer mode (ATM) to carry all services over a single, integrated, high-speed packet switched network.
- Uses synchronous transfer mode (STM) in situations where ATM is not applicable.
- Provides a service-independent call structure.
- Provides voice/video telephone calls.
- Has LAN interconnection capability.
- Provides teleconferencing (voice-video-data).
- Provides multimedia transmission capability.
- Provides transport for programming services, such as cable TV.
- Provides single-user controlled access to remote video sources (video on demand).
- Provides access to shop at home services.
- Provides access to remote, shared disk servers.
- Provides access to other information services.

The protocol includes code conversion, information compression, multipoint connections, and multiple connection calls and can provide each user with independent control of access features.

broadband LAN A local area network (LAN) where a *carrier* is used to transport the digital data signal between the nodes of the LAN. See also *baseband LAN* and *LAN.*

broadband network A type of local area network (LAN) wherein a carrier is used to transport data. This is in contrast with a *baseband network* in which data are impressed directly onto the transmission line. In a *broadband network,* the transmission line may be a coaxial cable with RF carriers, or the transmission media may be a fiber optic cable with light as the carrier. A *broadband network* is capable of operation at greater than 20 Mbps.

The *broadband network* transmission system is based on the same technology used in cable television. The cable itself may transport many carriers, each with its own independent information such as data, video, or voice. This multichannel capability through multiple carriers is called frequency division multiplexing (FDM). See also *baseband network.*

broadcast A signal or data transmitted without regard to which stations, if any, are receiving it. The broadcast signal may be encrypted so that only selected groups of receivers may recover usable information from the signal. Broadcast messages on a network are generally intended for all nodes in the network.

On the *Internet, broadcast* datagrams, though possible, are generally filtered out by the routers. See also *multicast.*

broadcast address An address indicating that the packet destination is all devices on the network. Broadcast address examples include:

- In Ethernet, 0xFF FF FF FF FF FF.
- In Token Ring, 0xFF FF FF FF FF FF is network wide; that is, it will cross bridges.

  0xC0 00 FF FF FF FF is the *broadcast address* specific to a ring; it will not be forwarded by a bridge.
- On the Internet, the broadcast IP address has all or part of the address set to wild card values (255 or 0) which will go to all hosts within the domain. For example, xxx.yyy.255.255 will go to all hosts on the network numbered xxx.yyy.

broadcast domain The set of all devices to which a particular broadcast transmission is directed. On an Internet, *broadcast domains* are generally bounded by routers.

broadcast list A list of two or more users to whom messages are sent simultaneously. Master broadcast lists are set up by the system administrator, while personal broadcast lists are created and maintained by individual users.

broadcast message A message from a single source to multiple destinations. The destinations may be all users on the network or from a group defined by a *group list.*

broadcast storm A condition in which packets are broadcast, received, and then broadcast again by one or more of the recipients. The effect of a *broadcast storm* is to congest a network with redundant traffic. *Broadcast storms* can arise in bridged networks that contain loops (closed paths).

broadcast storm firewall A mechanism that limits the rate at which either broadcast or multicast packets are propagated through a network.

broadside antenna A synonym for billboard antenna.

brouter A contraction of BRidge rOUTER. A device that can route specific protocols it supports (TCP/IP and IPX, for example) and bridge others, thereby combining the capabilities of both *bridges* and *routers.* Many of them support the IBM source routing protocol (SRP), which allows each LAN station to append routing information to every packet. See also *bridge, gateway,* and *router.*

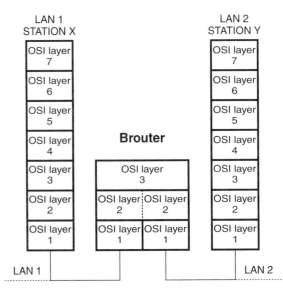

brownout A sustained reduction of mains voltage by 5 to 10% of the nominal value. The reduction may be caused intentionally or by the intrinsic behavior of a power distribution system containing large equipment; i.e.,

- *Intentional undervoltage*—a corrective action imposed by the power utility in response to the situation where more power is being requested by the users than the generating facilities can deliver. If one assumes the load on the power grid is resistive (*R*), then power (*P*) and voltage (*V*) are related by:

$$P = \frac{V^2}{R}.$$

From this, one can see (for a constant load) that a 5% reduction of supply voltage leads to nearly a 10% reduction in supplied power.

- *Intrinsic undervoltage*—the natural response of a power distribution system to high current demands. The response is described by *Ohm's law,* i.e., voltage loss (*V*) is related to supplied current (*I*) by:

$$V = I \cdot R.$$

Here, increasing current (increasing demand) increases the voltage loss in the distribution system's branch with high current. Also called *undervoltage* or *sag*. See also *mains*.

browser In Internet, a "front-end" application program used to navigate (read) through hypertext documents. The *browser* acts on behalf of the user, as his *client*. *Browsers* are resource discovery tools. Two prevalent browsing techniques are tunneling and hypertext.

Tunneling browsers find and go down into the subordinate menus of remote Gopher servers. Hypertext browsers utilize programmed links from within a document to other locations, thus creating a nonsequential method of access information. Examples of *browsers* include:

- Gopher—a tunneling browser.
- Mosaic—a hypertext browser used for viewing the WWW.
- Navigator—Netscape's hypertext browser used for viewing the WWW.
- Explorer—Microsoft's hypertext browser used for viewing the WWW.

browsing The act of searching through an automated information system storage to locate and retrieve information without necessarily knowing of the existence, location, or even the format of the information being sought.

brute force A problem-solving technique wherein a solution to a problem is sought by trial and error. Generally, a computer is programmed to change variables one at a time and to run a test between each change to see if a solution is found. Because of the speed of computers, millions of tests can be run, and it sometimes finds a solution thought to be intractable.

For example, a program can be written to search for a user's password by trying every possible combination of letters, numbers, and spaces of a given word length.

BS5750 A British Standards Institute standard with certification procedures that says an organization is in control of its quality procedures, at least in terms of consistency. Now identical to ISO 9000.

BSA (1) An abbreviation of Basic Switching Arrangement. (2) An abbreviation of Business Software Alliance. (3) An abbreviation of Basic Serving Arrangement.

BSC An abbreviation of Binary Synchronous Communication.

BSD An abbreviation of Berkeley Standard (or Software) Distribution. A version of UNIX developed at the University of California at Berkeley (UCB), which incorporated TCP/IP protocol and the ideal of *sockets.*

UCB no longer supports the software; however, their contributions are part of the standard UNIX features from providers such as AT&T or the Open Software Foundation.

BSD Socket Layer In BSD UNIX, the layer that represents the Application Program Interface (API) between user applications and the networking subsystem in the operating system kernel.

BSD UNIX An abbreviation of Berkeley Software Distribution UNIX. A UNIX version implemented at the University of California, Berkeley. *BSD UNIX* introduced several enhancements to AT&T's original implementation, including virtual memory, networking, and interprocess communication support.

BSE An abbreviation of Basic Service Element.

BSGL An abbreviation of Branch Systems General License. A license that must be obtained by any organization wishing to connect a private network to the British Public Switched Telephone Network (PSTN). Each site must have its own license.

BSI An abbreviation of the British Standards Institute. A source of many standards and reference materials. The *BSI* has English translations of many foreign standards published through a division called the Technical Help for Exporters (THE). They are responsible for input to many international standards organizations such as the ITU and ISO.

BSP (1) An abbreviation of Bell System Practices. A very specific set of rules governing basically everything done in AT&T. For example, the *BSP* defines writing styles for instruction and installation manuals, and defines the standards for climbing poles, answering telephones, collecting debts, installing telephones, and so on. Since divestiture, however, these rules and standards have become much less important. (2) An abbreviation of Bulk Synchronous Parallelism.

BSS An abbreviation of GSM Base Station System.

BSTJ An abbreviation of Bell Systems Technical Journal. A publication of AT&T.

BSTR An abbreviation of Bell Systems Technical Reference. A publication of AT&T.

BT (1) An abbreviation of British Telecom. (2) In telephony, an abbreviation of Busy Tone.

BTAM An acronym from Basic Telecommunications Access Method.

BTI An abbreviation of British Telecom International.

BTL An abbreviation of Bell Telephone Laboratories.

BTLZ An abbreviation of British Telecom Lempel-Zev—a nickname for the ITU-T V.42 bis data compression protocol. It provides up to 4:1 compression on modulation standards V.32, V.32 bis, and V.34. *BTLZ* connections are used with MNP2-4 and LAPM protocols.

BTOA An abbreviation of Binary TO ASCII (pronounced "B" to "A"). A UNIX program that translates binary files into ASCII. The reverse process is accomplished in a program called *ATOB*.

Btrieve Originally introduced in 1983, *Btrieve* was one of the first databases designed for LAN operation. Now owned by Novell, Net-Ware *Btrieve* is a key-indexed record management program that allows a user to access, update, create, delete, or save records from a database on a LAN. The database may be accessed either sequentially or by random access methods. *Btrieve* is a family of programs that can run in either of two versions: client- or server-based.

In addition to record management capabilities, *Btrieve* includes:

- *Communications facilities,* for both local and remote communications between a program and a record base.
- *Requesters* (DOS, OS/2, Windows, and so on) that provide access for applications running on workstations.
- *Utilities* for setting up, monitoring, and maintaining the record base, and so on.
- *Data-protection measures* for dealing with the record base in case of system failure.
- *NetWare Directory Services* (*NDS*) *Support,* which is new with NetWare 4.x. This support is available only beginning with Btrieve version 6.1.

A key-indexed database is one in which keys, or record fields, are used as the basis for creating an index, which is information that guides access to a database. A *Btrieve* record base uses a specially defined data format, which is supported by database programs and other applications from many third-party vendors.

BTRL An abbreviation of British Telecom Research Laboratories.

BTU An abbreviation of Basic Transmission Unit.

BTW Used on most bulletin board systems (BBSs) to save typing, an abbreviation of By The Way.

buffer (1) A routine or storage medium used to compensate for a difference in rate of flow of data, or time of occurrence of events, when transferring data from one device to another.

Buffers are used for many purposes, such as

- Interconnecting two digital circuits operating at different rates. For example, in computer I/O ports (UARTs) the storage of received data until the computer has time to access it results in a communications speed increase.
- Holding data for use at a later time. For example, to temporarily save some number of lines that would have been lost as they scroll off the top of a computer display screen when more than 24 lines are received.
- Allowing timing corrections to be made on a data stream.
- Collecting binary data bits into groups that can then be operated on as a unit. For example, a serial to parallel converter.
- Delaying the transit time of a signal in order to allow other operations to occur. For example, a *store-and-forward* system to store packets from a source while it waits for the link to clear in the direction of the destination.

(2) To use a buffer or buffers. (3) A device inserted between a circuit and its load in order to reduce the effects of the load on the circuit's performance. A typical buffering device is an amplifier (with or without gain). Also called a *buffer amplifier.* (4) In a fiber optic communication cable, a component used to encapsulate one or more optical fibers for the purpose of providing such functions as mechanical isolation, protection from physical damage, and fiber identification. Buffers of this type are available in two forms:

- *Loose buffer*—which may take the form of a miniature conduit, contained within the cable or a *loose buffer tube,* in which one or more fibers may be enclosed, often with a lubricating gel.
- *Tight buffer*—which consists of a polymer coating in intimate contact with the fiber and is applied to the fiber during manufacture.

buffer amplifier See *buffer* (3).

buffered repeater A device inserted in series with a transmission line that can clean and boost signals before sending them on. A *buffered repeater* can hold a message temporarily, for example, when there is already a transmission on the network.

bug (1) A slang description of an error or flaw in the design, concept, or manufacturing implementation of hardware, software, or systems. The term generally does not apply when a formerly working device ceases to function due to a component error.

The origin of the term is not known; however, Thomas Edison (1847–1931), the American inventor, used the term in its current day context in 1878 in a letter to Theodore Puskas 65 years before the famous moth in the Mark I computer story. (Folklore has it that the Mark I computer failed because a moth had gotten into one of the relays, preventing it and the machine from functioning.) (2) In telegraphy, a semiautomatic device (called a key) for generating the correctly spaced dots and dashes of the Morse code. (3) A concealed microphone or listening device, or the installation of the listening device.

building out The process of adding a network (consisting of inductance, capacitance, and resistance) to a cable pair so that its electrical length appears longer. Appropriate use of these networks can control impedance and loss characteristics or make a line appear to be any length the equipment needs to perform correctly. Also called *line build out* (*LBO*).

bulletin board system (BBS) A computer system that automatically receives modem telephone calls and provides its users one or more of the following services:

- A message posting area (called a *board*) where messages may be placed for later retrieval by another caller.
- A means for connecting callers together for the purpose of online telecommunication—much like an interactive electronic letter (called a *conference*).
- A storage area where files may be uploaded or downloaded by the caller. Frequently, an area where *shareware* programs are distributed.
- Access to FidoNet, Internet, and so on.

bundle A group of optical fibers or electrical conductors (such as wires or coaxial cables) within a single cable that share a common color binding and usually in a single jacket. Multiple bundles may be placed in the same cable.

Bureau International des Poids et Mesures (BIPM) The laboratory responsible for the accuracy of Coordinated Universal Time (UTC). See also *time* (*3*).

burn-in A testing process wherein a device is operated for a specified period of time at an elevated temperature so as to accelerate premature failures. Because semiconductors fail in a predictable way, and because each 10-degree rise in temperature halves the time to failure, operating a device at an elevated temperature can "weed out" those devices that would fail early.

The statistics of failure rates show that almost all devices have the same basic failure rate curve, called the "bathtub curve." The failure rate is high early in the life of the equipment owing to manufacturing defects; then it drops rapidly to a quiescent low level. After a long period at the quiescent level, the failure rate again rises owing to various wearout mechanisms. In the case of semiconductors, the wearout time may be several hundred years.

burned in address (BIA) A hardware address for a network interface card (NIC) that is assigned by the manufacturer and is unique for each card.

burning a pole A slang expression to describe an installer accidentally sliding down a telephone pole.

burrow In Internet, a slang term for a system running the *gopher* server program. Also called a *gopher hole.*

burst (1) In communications, a sequence of signals, noise, or interference counted as a unit in accordance with some specific criterion or measure. (2) Separation of perforated continuous form paper into discrete sheets.

burst errors A group of bits in a transmission in which two successive erroneous bits are separated by fewer than *N* correct bits. *N* is an arbitrary number defined by the system designers. An equivalent definition uses time as the criteria, that is, two or more error bits in less than *S* seconds.

Suppose a system has a measured BER of 5 errors in 10^6 bits. The errors could be distributed as 1 error every 200,000 bits, or all 5 bit errors in a string of 5 bits. The latter would be an example of an error burst. The reason for the distinction is both error detecting and error correcting codes have a lower capacity for handling high *burst error* rates than they do for non-bursty error rates. See also *CRC, LRC,* and *parity.*

burst isochronous A deprecated synonym for *isochronous burst transmission.*

burst mode (1) A synchronous data transfer method in which large blocks of data are transferred in an uninterrupted burst rather than in a continual stream of small blocks. (2) A high-speed transmission mode in which the transmitter takes control of the communications channel until its transmission is complete. Frequently used for transfers between subsystems of complex devices, e.g., hard disk drive to memory transfers. (3) A term referring to NetWare's packet burst protocol.

burst mode window Determines the amount of data that can be transferred in a burst mode session before an acknowledgment is required. The window size is dynamic; that is, it is changed in response to bit error rate (BER) changes.

burst speed The maximum speed at which a device or process can operate without interruption (generally only for short periods). This is in contrast to throughput, which indicates the average speed at which a device can operate under ordinary conditions over an entire transaction, such as when transmitting or printing an entire file.

burst transmission Also called a *data burst.* (1) A method of data transmission with high data rates and short transmission times. (2) A method of network operation wherein the flow of data is interrupted at intervals so as to allow other operations to occur. The process enables communications between DTEs and networks operating at different speeds.

burstiness In the ITU recommendations for B-ISDN, a measure of the distribution of data over time. Although not agreed upon, one definition being considered is the ratio of maximum (peak) and mean (average) bit rate.

bursty transmission Information that flows in short intermittent spurts with relatively long quiet intervals between spurts.

bus (1) Generally, one or more conductors used to transfer signals or power among a group of related devices (2) One or more conductors used to transfer signals among multiple devices. Examples include the address and data busses of a computer and the interconnecting network in a 10Base2 or 10Base5 Ethernet.

A signal placed on the bus by one device is seen by all other devices connected to the *bus*. It is up to the individual devices on the *bus* to determine which signals should be responded to and which should be ignored. (3) A conductor used to transfer current from one point to another with little loss due to the conductor's impedance, e.g., a *ground bus.*

bus hog A device connected to a bus which, after gaining access for transmission, refuses to relinquish control (even if other devices are waiting) so that large amounts of data may be transferred.

bus master A device on a bus that can contend for, seize, and control the operations of the bus for the purposes of accessing resources on the bus. No other device (including the CPU) is required to assist in resource utilization after bus control has been seized.

MicroChannel, extended ISA (EISA), VESA local bus (LVB), and peripheral component interconnect (PCI) support bus mastering.

bus network A physical communications *network architecture* wherein all nodes of a network share a common communications means (wire, fiber, RF, etc.). In a bus network, the communications protocol may be token passing, collision detection, or some form of multiplexing such as frequency division (FDM) or code division multiplexing. All terminals are aware of all transmissions on a bus network, but each terminal accepts only those messages addressed to it.

The *bus network* in the following diagram depicts a number of DTEs, each bridged onto the common communications cable. The bridge (depicted as a solid dot) is only an electrical connection; no active devices are involved. Also called *bus topology.* See also *gateway, hierarchical network, hybrid network, mesh network, ring network,* and *star network.*

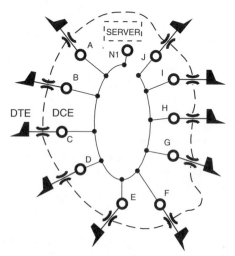

bus slave A device on a bus which can only respond to accesses initiated by a bus master and only to those mapped to its addresses. A slave generally can request an interrupt, but only a bus master can service the interrupt request.

Business Software Alliance (BSA) A coalition of software developers established in 1988 to address both national and international software piracy. The group, based in Washington D.C., represents about 75% of the productivity software market.

busy In telephony, off-hook (or unavailable "busied out") or all circuits busy.

The caller receives an audible tone associated with one of two busy conditions: 1. *Slow busy,* indicating the called line is off-hook (or unavailable), which is a tone pulse 1/2 s on and 1/2 s off, and 2. *Fast busy,* indicating the switching system cannot establish a path to the called line, which is a tone pulse 1/4 s on and 1/4 s off. See also *call progress tone.*

busy back See *busy signal.*

busy call forwarding In telephony, the automatic rerouting of a call to an alternate predetermined destination when the dialed party is off-hook.

busy hour In a communications system, that continuous one-hour period during which the maximum total traffic load occurs in a given 24-hour period. The busy hour is determined by fitting a horizontal line segment equivalent to one hour under the traffic load curve about the peak load point. In cases where more than one busy hour occurs in a 24-hour period, that is, when saturation occurs, the busy hour or hours most applicable to the particular situation are used. Also called the *peak busy hour.*

busy lamp field (BLF) The array of lights, generally on an attendant's station, that tells the operator the status of the individual telephones and lines.

busy out Forcing a line to be unavailable to calling stations (the caller receives a busy tone). An example of *busy out* use would be the period of time a dial-up computer is being maintained.

busy override A feature on some PBX systems that allows the attendant (or other high priority user) to *barge in* on an established connection. Generally, a warning tone is injected into the established connection to inform the user of the *barge in.* Depending on the switching system, the *barge in* party may be heard by both users or only the local user.

busy season An annually recurrent and reasonably predictable three-consecutive-month period of maximum busy hour traffic. Typically in the telephone industry, the *busy season* is the three months preceding Christmas.

busy signal Formerly called *busy back.* **(1)** In telephony, an audible or visual signal that indicates that no transmission path to the called number is available. Also called *busy tone.* **(2)** In telephony, an audible or visual signal that indicates that the called number is occupied or otherwise unavailable. Also called *reorder tone.* See also *call progress tone.*

busy test A method of determining if a channel is carrying traffic (busy), broken (but indicates busy), or available to carry traffic.

busy tone Busy tone is one of the call progress tones generated by the telephone company switching equipment to indicate that a connection to the remote location is not possible because some part of the network path is already in use. Busy tone is generally the sum of several tones in the frequency range of 340 to 650 Hz with a periodic interruption. Also called *slow busy* or *busy signal.* See also *call progress tone.*

busy verification In a public switched telephone network (PSTN), a network-provided service feature that permits an attendant to verify the busy or idle state of station lines and to break into the conversation. To alert both parties on the line that an attendant has bridged onto the line, a 440-Hz tone is applied to the line for 2 seconds, followed by a 0.5-second burst every 10 seconds.

butt set A telephone test handset with electronics designed to test a telephone circuit. There are two explanations as to why it is called a butt set:

- It hangs from the lineman's belt near the wearer's butt, or
- It is used to butt into a line connection to monitor the connection quality.

Also called a *test set.*

button **(1)** A type of actuator on an electronic switch. Another type is a toggle. **(2)** A box or icon used to initiate some operation when it is selected by either keyboard or mouse action. For example, in a hypertext document, the button may activate a link. In this respect, the button is the graphical equivalent of highlighted text.

BW An abbreviation of <u>B</u>and<u>W</u>idth.

BY In telephony, an abbreviation of <u>B</u>us<u>Y</u>.

bypass **(1)** The use of any telecommunications facilities or services that circumvents those of the local exchange common (LEX) carrier. Bypass facilities or services may be either customer provided or vendor supplied.

A connection with an interexchange carrier (IXC) that does not go through a local exchange carrier. For example, a microwave link or satellite link used to directly connect the geographically separated offices of a corporation. **(2)** An alternate circuit that is routed around equipment or system component. Bypasses are often used to allow system operation to continue when the bypassed equipment or a system component is inoperable or unavailable.

bypass mode An operating mode on FDDI and Token Ring networks where an interface has been removed from the ring.

bypass trunk group A trunk group that circumvents one or tandem switching offices in its routing structure.

byte (B) A contiguous set of bits operated on as a single unit. A byte is usually shorter than a word, and usually 8 bits are grouped as a byte.

In pre-1970 literature, *byte* referred to a variable-length bit string. Since that time, the usage has changed so that now it almost always refers to an 8-bit string. This usage predominates in computer and data transmission literature. When used to indicate 8 bits, it is synonymous with an *octet,* especially in UNIX systems and on the Internet.

byte interleaved A type of multiplexing in which individual bytes from multiple sources are concatenated or placed in sequence on a single-output transmission port. For example, bytes from three input channels A, B, and C form the sequence A_1, B_1, C_1, $A_2, B_2, C_2, \ldots, A_n, B_n, C_n, \ldots$; where "n" is the byte number of the respective channel.

byte-oriented protocol A communications protocol in which information is encoded into an 8-bit character set and is transmitted as a string of these characters. Control information is differentiated from message information by the coding algorithm—certain characters in the code alphabet are excluded from the message. Examples of *byte-oriented protocols* are the asynchronous communications protocol used with modems, IBM's Bisync protocol (BSC), and Digital Equipment Corporation's Digital Data Communications Messaging Protocol (DDCMP). See also *bit-oriented protocol (BOP).*

byte sex For a processor, byte sex is a feature that describes the order in which bytes are represented in a word. Processors may be *little endian, big endian, middle endian,* or *bytesexual.*

- *Little endian*—The low-order byte in a word is stored at the lower address, that is, in the order 2–1.
- *Big endian*—The high-order byte is stored in the higher address, that is, in the order 1–2.
- *Middle endian*—This is a representation strategy on 32-bit systems that is neither little endian (2–1 order) nor big-endian (1–2). In a proper 32-bit system, these representations extend to 1–2–3–4 or 4–3–2–1. *Middle endian* systems, on the other hand, use the representation 2–1–4–3 (big endian bytes in little endian words) or 3–4–1–2 (little endian bytes in big endian words).
- *Bytesexual,* a process that is capable of using either *little endian* or *big endian* representations, depending on the value of a flag bit.

byte synchronous A way of mapping payload into virtual tributaries (VT) that

- Synchronize all inputs into the VT.
- Capture framing information.
- Allow access to subrate channels carried in each input.

For example, *byte synchronous* mapping of a channelized DS1 into VT1.5 provides direct access to the DS0 channels carried by the DS1.

Byzantine failure In networking, a situation in which a node fails by performing incorrectly (acting erratically, for example) rather than by failing completely and disappearing from the network.

A network that continues to function in the presence of one or more nodes experiencing *Byzantine failure* has *Byzantine robustness.*

C

c (1) The SI prefix for *centi* or 10^{-2}. **(2)** The symbol for the speed of light.

C (1) The symbol for *coulomb,* the unit of electric charge. **(2)** The symbol for *capacitance* (measured in *farads*). **(3)** The designator for a *capacitor* in an electrical circuit; e.g., C1, C2, C101 refer to three specific capacitors in a circuit. **(4)** A computer programming language. **(5)** The symbol for Celsius, a unit of temperature. **(6)** The symbol for an electrochemical cell or battery capacity, expressed in amp-hours (Ah).

C band That portion of the electromagnetic spectrum in the 4- to 8-GHz range. It is used extensively for satellite communications with downlink frequencies near 4 GHz and uplink frequencies near 6 GHz. See also *frequency band.*

C lead The third wire of the communications trunks between central offices. While two of the three wires are used for voice, the third wire (*C lead*) is used to control such things as seizing, holding, and releasing the trunk. The originating office applies a ground to signal seizure. Also called the *sleeve wire* (the other two are tip and ring).

C-message weighting A method of measuring noise in a telephone communication channel so that the measurement approximates the interference effects noise has on the average telephone listener.

The method is simply the measurement of the signal through a specific band-pass filter shape. The filter shape was derived experimentally by having a large number of telephone listeners judge how loud a tone at a specific frequency must be in order to

1. Be as annoying as a reference 1000-Hz tone with no speech.
2. Interfere with intelligibility and articulation at the same level at the 1000-Hz reference tone.

After many trials with many different individuals, the measurements were combined, averaged, and then plotted, as is shown in the following table and figure. From these results, one can see that an in-

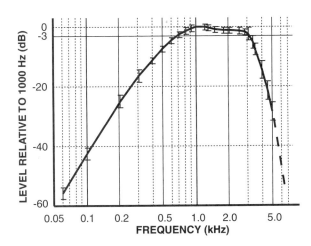

terfering tone at 200 Hz must be 25 dB louder to have the same interference effect as the 1000-Hz reference. See also *psophometric weighting.*

RELATIVE SINGLE TONE INTERFERING EFFECT C-MESSAGE WEIGHTING

Frequency (Hz)	Objective Nominal (dB below ref)	Frequency (Hz)	Objective Nominal (dB below ref)
60	55.7±2	1300	0.5±1
100	42.5±2	1500	1.0±1
200	25.0±2	1800	1.3±1
300	16.5±2	2000	1.3±1
400	11.4±1	2500	1.4±1
500	7.5±1	2800	1.9±1
600	4.7±1	3000	2.5±1
700	2.7±1	3300	5.2±2
800	1.5±1	3500	7.6±2
900	0.6±1	4000	14.5±3
1000	0.0 ref	4500	21.5±3
1200	.02±1	5000	28.5±3

C-notch filter A *C-message filter* with a very narrow band notch filter added at the specified test frequency.

In some transmission systems, certain noise sources are dependent on the signal applied. In order to activate these noise sources, a tone is transmitted through the signal path. At the receiving end, the *C-notch filter* removes the tone and conditions the noise for subsequent measurement.

C.n recommendations An ITU-T series of specifications concerning general telecommunications statistics. See the Appendix E for a complete list.

C/A An abbreviation of Coarse Acquisition code.

C/No An abbreviation of Carrier to Noise Density ratio. The ratio of the carrier signal power level to the power of the noise in a 1-Hz bandwidth. See also *signal-to-noise ratio.*

C7 The European equivalent of the North American Signaling System number 7 (SS7). The two schemes are not compatible; hence, they must be translated in a gateway.

cable (1) *Electrical.* An insulated conductor, transmission line, group of conductors, or transmission lines mechanically bound together to form a flexible interconnecting device. The *cable* may be made up of single or multiple conductors, and these conductors may be electrically shielded. The cable may also contain optical fiber elements. Conductors may be used independently or in groups. A number of conductors in a cable may be bound together in a *binder group;* that is, a helical wrap is applied to the wires in that group.

Cables come in many classifications and grades. Some of the more important types are *unshielded twisted pair* (UTP), *shielded twisted pair* (STP), and *coaxial.* Individual conductors in a cable are generally identified via some form of color coding. The size of an individual conductor is usually either stated in AWG (for American Wire Gauge) or for the SI system in mm (millimeters). The insulation layer on the conductor and the jacket on a cable bundle may be rated for plenum use.

- Plenum cable must be fire resistant and must not give off toxic fumes when heated. Nonflammable fluoroploymers such as Teflon or Kynar are frequently used. Plenum cable should meet the NEC communication plenum cable (CMP) or class 2 plenum (CL2P) cable specifications. They should also meet UL-910 flammability standards.
- Nonplenum cable is less expensive because it uses insulation of polyethylene (PE) or polyvinyl chloride (PVC), which both burns and gives off toxic fumes. These cables must meet the NEC's communications riser cable (CMR) or class 2 riser (CL2R) cable specifications. The cable should also be UL-listed for UL-1666.

See also *AWG, category n, coaxial cable, color coding, level n cable, type n cable,* and *wire.* **(2)** *Optics.* A single optical fiber, multiple fibers (each fiber individually protected), or a fiber bundle (a bundle of unprotected fibers) arranged in a structure such that it satisfies desired optical, mechanical, and environmental specifications. Starting at the center of a fiber cable is the core through which the light travels. Surrounding the core is the cladding (analogous to the individual wire insulation of an electrical cable). Around the core and cladding is a buffer. The buffer is one or more layers of plastic surrounding the cladding. The buffer helps strengthen the cable, thereby decreasing the likelihood of microcracks, which can eventually grow larger into breaks. The buffer can double the diameter of some cables. The cable also has strength members, which are strands of a tough material (such as steel, fiberglass, or Kevlar); these provide extra strength for the cable. Each of the substances has advantages and drawbacks. For example, steel attracts lightning, which will not disrupt an optical signal directly; but it may disrupt the electronic transmission equipment attached to the ends of the cable. Further, enough heat may be generated by the currents to alter the optical fiber characteristics. Finally, the jacket of a fiber optic cable is an outer casing that, as with electrical cable, can be plenum or nonplenum rated. In cable used for networking, the jacket usually houses at least two optical fibers—one for each direction. See also *optical fiber.* **(3)** A message sent by cable, or by any means of telegraphy.

cable assembly An electrical or optical cable that is ready for installation in a specific application and usually terminated with specified connectors.

cable cutoff wavelength (λ_{cc}) For a single mode optical fiber of defined length, bend radius, and operating conditions, the wavelength at which the fiber's second-order mode is attenuated a specified amount when compared to a multimode reference fiber or to a tightly bent single mode fiber.

cable horizontal As defined by the EIA/TIA-568 committee, any cable that goes from a wiring closet, or distribution frame, to the wall outlet in the work area. Distribution frames from a floor or building are connected to other frames using backbone cable.

cable jacket See *cable sheath.*

cable modem A modem, similar to a standard telephone line modem but designed for use on the television cable network. Because of the wide bandwidth available on a TV cable, the modems can be thousands of times faster than telephone line modems.

cable Morse code See *Morse code.*

cable plant A term referring to the physical connection media (e.g., optical fiber, wires, connectors, splices, etc.) of a network. Although more commonly called the outside plant in telephony, *cable plant* is sometimes used to refer to the cables outside the central office.

cable riser The cable running vertically between the floors of a multistory building.

cable run **(1)** The conduit or trays through which cables are routed. **(2)** The route taken by a cable or cables between its source and destination locations.

cable sheath The protective layer covering a cable core. The *sheath* may be one or more layers and may contain a conductive layer. Also called *cable jacket.*

cable shield A metallic component of a cable sheath that has two major functions:

- To prevent outside signals from coupling into the internal wires (reduces interference).
- To drain lightning-induced currents safely to ground rather than allowing them to enter cable-terminating equipment.

cable television (CATV) Originally, a television distribution method—for areas where good broadcast TV direct reception was not possible—in which signals from distant stations are received, amplified, and then distributed by cable (coaxial or fiber) or microwave links to local users.

Now *CATV* refers to a network that consists of a cable distribution system to large metropolitan areas in competition with direct broadcasting. The acronym *CATV* was originally derived from *community antenna television.* Today it is generally understood to mean *cable TV.*

CAC **(1)** An abbreviation of Connection Admission Control. A process in ATM to limit new calls so as to maintain a particular quality of service (QoS). **(2)** In telephony, an abbreviation of Carrier Access Code. See *equal access.*

cache hit The condition that occurs when an application program attempts to fetch information from a cached main storage device (DRAM or hard disk) and it is retrieved from cache memory instead.

The opposite of a *cache hit* is a *cache miss.*

cache memory A high-speed buffer memory that resides logically between a CPU and main memory or between a CPU and disk drive (see the following drawing). *Cache memory* is temporary memory and holds the most recent data accessed by the CPU. *Cache memory* is typically 10 to 100 times faster than DRAM memory and perhaps a million times faster than a disk drive. Because *cache memory* is so much faster than the memory being buffered, re-access of information previously accessed is many times faster than if each read was from the original source. See also *buffer.*

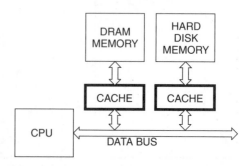

cache miss The condition that occurs when an application program attempts to retrieve information from a cached storage device (DRAM or hard disk) and it is retrieved from the main storage rather than from cache memory.

The opposite of a cache miss is a *cache hit.*

CADS An acronym from <u>C</u>all <u>A</u>buse <u>D</u>etection <u>S</u>ystem.

CAE (1) An abbreviation of <u>C</u>omputer <u>A</u>ided (or <u>A</u>ssisted) Engineering. (2) An abbreviation of <u>C</u>ommon <u>A</u>pplications <u>E</u>nvironment.

CAI (1) An abbreviation of <u>C</u>omputer <u>A</u>ided <u>I</u>nstruction. (2) An abbreviation of <u>C</u>ommon <u>A</u>ir <u>I</u>nterface.

call (1) The action performed by the calling party in establishing a connection to a destination party. (2) The operations necessary to establish a connection. A request from one program (or network node) to begin a communication with another program (node). (3) The occupation of a communications channel. The term used to refer to the resulting communications session. (4) The demand to a switching system to set up a connection. (5) A unit of traffic measurement. (6) In computer software, the action of bringing a computer program, a routine, or a subroutine into use, usually by specifying the entry conditions and the entry point.

call abandons See *abandoned call.*

call abort timer A timer in a DCE that determines how long the device may remain off-hook for a failed call. In the event call progress tone recognition is disabled, this timer provides the means to disconnect when a connection is not properly established. Typical delay times in modems and fax/modems are 5.1, 10, 50, and 70 seconds.

This timer is not the same as the *answer abort timer.*

call accepted signal A control signal generated by the called equipment that indicates to the calling station that it accepts the incoming call.

call accounting system In telephony, a computer attached to (or part of) a local switching system which with appropriate software monitors all aspects of an organization's telephone calling mannerisms. A call accounting system provides the capability to significantly reduce cost by addressing:

- *Telephone abuse*—identifying those who overuse or frivolously use the telephone.
- *Cost allocation*—billing the organization's various departments according to usage. If the telephone is shared by a number of people, the cost accounting system can allocate the costs accordingly.
- *Billing*—billing clients and "projects" for the cost of calling.
- *Network optimization*—allowing the computer to do least cost routing, i.e., selecting the proper carrier and trunk type on a call-by-call basis.
- *Telephone misuse*—checking to verify that automatic least call routing is working. If it isn't present, the information can be used to inform the organization personnel which carrier and which trunk to use and when.
- *Diagnostics*—determining if the telephone system is functioning correctly.
- *Phone company bill verification*—making a crosscheck on the bill from the various suppliers to ensure their billing is correct.
- *Call tracing*—discovering who made which call and to whom.
- *Motivation*—when sales, for example, is a direct function of the number of calls, having a report of who made how many calls and the associated deal closing rate can provide an insight on what works. Giving out awards for improving the ratio could be a real motivator.

call announcer In telephone systems, a device that receives dial pulses (or DTMF tones) and generates audible words corresponding to the dialed digits so that they may be heard by a manual operator.

call associated signaling (CAS) Signaling required for supervision of a bearer service between two end points, including support for call functions, i.e., origination, call delivery, handover, and disconnect.

call attempt In a telecommunications system, a demand by a user for a connection to another user through a switching system. Call attempts are terminated by one of the following: completion, overflow, loss, and abandonment.

call back A procedure for identifying a remote terminal's authenticity. In the procedure, a remote terminal calls a host system and identifies itself; the host system disconnects and then dials the telephone number (from the host database) of the remote terminal to reestablish the connection.

call back modem A type of modem operation where the initiating DCE may only request the destination DCE to *call back* to establish a communication link.

The normal method of doing this requires the calling DCE to supply the called DCE/DTE with an identification number. The called DCE/DTE will then look up the telephone number of the calling DCE (using the identification number as a reference). The called DCE/DTE will then *call back* the DCE that originated the sequence. The identification number may be supplied by either DTMF signaling, a modem message, or the telco's *caller ID service.* There are at least two reasons to use a call back modem in a dial-up server:

- *Security.* The feature ensures that users (unauthorized users?) do not have the ability to log onto a system from unauthorized terminals because data transactions can only access the system from predetermined dial-up locations.
- *Cost.* When a user calls from a remote location to a company, it is generally over standard "high-cost" telephone lines. The company's communications network, however, may have access to low-cost outgoing lines.

Also written *callback modem.*

call barring The ability to prevent all or certain call originations (or terminations) from (to) a telephone.

call block In telephony, a service offered by the telephone company which allows the subscriber's phone to reject certain calling numbers—that is, prevent it from establishing a connection.

call blocking The practice of an alternate service provider (such as a hotel) prohibiting a caller from selecting the long-distance carrier of its choice.

call circuit In telephony, a communications circuit between switching points used for the transmission of switching information.

call clear packet An ITU X.25 packet that terminates the communication session. Equivalent to the on-hook condition in POTS.

call collision (1) A condition that occurs when both ends of a trunk or channel are seized simultaneously. Also called *glare.* (2) Contention that occurs when a terminal transfers an incoming call to a channel and simultaneously a DCE transfers a call request to the same channel.

call completion rate The ratio of the successfully connected calls to the number of call attempts.

The ratio is generally expressed as either a percentage or a decimal fraction.

call concentration The ratio of busy hour calls to total daily calls.

call control (1) Generally, the procedure that sets up the intended connection in a switching system and ultimately disconnects it.

The basic setup procedure follows these steps:

- Scan trunk and line (station) ports for any requests for service.
- Deliver a proceed signal to the requesting device.
- Accept call routing information.
- Check stored programs and call routing tables for appropriate connection information.
- Establish the connection (or deliver an error signal; e.g., busy, out of service).

Also called *call processing*. (2) In telephony, the management of call routing, connection, disconnection, call transfer, call forwarding, and other services such as voice mail. *Call control* does not affect message content (in either voice or data messages).

call control signal Any signal used to establish, maintain, or release a call.

call delay (1) The delay generated when a call reaches a busy switch. In telephone systems, it is considered acceptable to have a delay of up to 3 seconds during busy hour on 1.5% of the call attempts. (2) The time between the instant a system receives a call attempt and the instant of ring initiation at the called instrument.

call detail recording (CDR) The feature of a telephone system that allows the system to collect and record information on outgoing calls (information such as originating station, destination number, time of day, duration, outgoing port; e.g. trunk number). The information is essential for a cost accounting system.

call diverter A device which when connected to a line intercepts calls made to that line and re-routes them to a recorded message, an operator, or another line. The device may intercept the call at the central office or at the subscriber premises. If the device is at the subscriber premises, it will initiate a call out on a second line and then effectively connect the two lines together. Also called a *boomerang box*.

call duration The time during which a link is capable of transferring user information. In telephony, it starts when a call is answered, and it terminates when either party goes on-hook. In data transmission, it is the duration of the information transfer phase of an information transfer transaction.

call failure signal A signal sent to the calling terminal indicating that a call cannot be completed because of a timeout, a fault, or a condition that does not correspond to any other particular signal. See also *call progress tone*.

call for vote (CFV) In USENet, a posting on the newsgroup *news. announce.newsgroup* that invites votes on the creation of a previously proposed new *world newsgroup*.

The voting occurs throughout a 21- to 31-day period. Voters send e-mail messages indicating a yes or no vote. To pass, the proposal must receive not only 100 more yes votes than no votes, but the yes votes must represent two-thirds of the total votes. If the issue does not pass, six months must elapse before the proposed newsgroup may again be proposed. See also *alternative newsgroup hierarchies*.

call forward In telephony, a feature provided by switching equipment where incoming calls to one number are re-routed within the switching equipment to another destination. The request for and control of subsequent re-routing is at the discretion of the person at the original called number.

The feature is available in most PABXs where the call may be redirected from one extension to another. The service is also available in many public switched telephone networks (PSTNs) as a revenue-producing option, where a subscriber may redirect calls to his number to another subscriber's number.

call hold A feature in most, if not all, telephone switching devices which allows a user to temporarily disconnect from the distant party and use the handset for another purpose, such as answer an incoming call, make an announcement on the paging system, or call someone (presumably to get assistance for the party being *held*).

call hour (Ch) A unit of traffic quantity. One *call hour* is the quantity represented by one or more calls having a combined total duration of one hour.

Related units include the call-minute (Cmin) and the call-second (Cs); 1 Ch = 60 Cmin = 3600 Cs.

call identifier A name assigned by the originating network for each established or partially established virtual call. When a *call identifier* is used in conjunction with the calling data terminal equipment (DTE) address, the *call identifier* uniquely identifies the virtual call.

call in absence horn alert A cellular car telephone feature that sounds the car's horn when the telephone is receiving a call.

call in absence indicator A cellular car phone feature that ensures that power to the phone is not lost in the event the car's ignition switch is turned off.

call in progress override A cellular car phone feature that sustains power to the phone during a call, even if the car's ignition switch is turned off.

call indicator In telephony, a device that receives dial pulses (or DTMF tones) and generates a visual indication corresponding to the dialed digits so that they may be viewed at a manual operator switchboard.

call intensity The number of calls in a group of communications channels per unit time. Also called *traffic intensity*.

call minute (Cmin) See *call hour*.

call not accepted signal A control signal generated by the called station equipment that indicates to the calling equipment that it does not accept the incoming call.

call originator See *calling party*.

call packet An ITU X.25 message that carries addressing and other information necessary to establish a *switched virtual circuit* (SVC).

call packing In telephony, a method of selecting paths in a switching network according to a fixed hunting sequence.

call park A feature of many PBX systems that enables a user or attendant to place a call in a special kind of "hold." The held call can be picked up at *any* telephone by dialing a one- or two-digit code. In effect, the call has been transferred from one extension to another via a temporary *parking* location.

call pickup A service feature of many PABX systems wherein a user may answer incoming calls that are directed to another user in a preselected call group by dialing a predetermined code.

call processing (1) The sequence of operations performed by a switching system from the acceptance of an incoming call through the final disposition of the call. (2) The end-to-end sequence of operations performed by a network from the instant a call attempt is initiated until the instant the call release is completed. (3) In data transmission, the operations required to complete all three phases of an information transfer transaction. Also called *call control*.

call progress signal: A call control signal transmitted by the called data circuit-terminating equipment (DCE) to the calling data terminal equipment (DTE) to report either

- The progress of a call by using a positive call progress signal, or
- The reason why a connection cannot be established by using a negative call progress signal.

call progress tone A family of audible signaling tones generated by the communications switching equipment and transmitted to the calling station. The tones may be generated in the local PABX, the local telephone company switching office, or in a remote switching office. The tones indicate the progress and status of the connection being established. Although the tone details are country dependent (as is shown in the table on pages 76–79), they tend to follow the same general format for the basic tones.

Call progress tones include:

Basic tones

- *Busy tone*—A short cadence with a symmetric on/off period tone. Indicates the called subscriber is using the telephone line.
- *Congestion tone (also called fast busy)*—A very short cadence with a symmetric on/off period tone (the cadence is twice as fast as a busy tone). Indicates that a path cannot be established through the network (access to the subscriber line cannot be established).
- *Dial tone*—Generally, a steady uninterrupted tone. Indicates the telephone switching equipment is ready to accept dial information.
- *Ringback tone*—A long cadence, with an asymmetric on/off period tone. Indicates the called subscriber's line has a ring signal applied to it.

Extended function tones

- Coin tone
- Confirmation tone
- Executive override tone
- Function acknowledge tone
- Intercept tone
- Number unobtainable tone
- Pay tone
- Recall dial tone
- Receiver off-hook (ROH) tone (howler tone)
- Recording tone
- Route tone
- Second dial tone
- Special information tone
- Waiting tone
- Warning tone (operator intervening tone)

call rate In telephony, the number of calls per unit of time. See also *call hour.*

call receiver A synonym for *called party.*

call record Recorded data pertaining to a single call.

call release time In communication systems, the time interval from initiation of a clearing signal (call release signal) by a terminal until the link available condition appears on the originating terminal equipment. See also *call request time, call selection time,* and *call setup time.*

call request packet A message that carries information needed to establish an X.25 circuit. The packet is sent by the originating DTE; it contains the requested network terminal number (NTN) and X.29 control information or user data.

call request time In communications systems, the time interval from the initiation of a calling signal (for example, an off-hook signal or call request packet) to the receipt of a proceed to select signal (a dial tone, for example). See also *call release time, call selection time,* and *call setup time.*

call restriction A switching system service feature that prevents selected terminals from exercising one or more service features otherwise available from the switching system or from accessing selected communications links.

call restrictor In telephony, a device (or software option) that prevents a caller from accessing certain numbers or classes of numbers from specific terminal units. For example, a *call restrictor* may prevent access to any long-distance area code except the 800 numbers, or the *restrictor* may prevent access to only the 900 number, or it may only prevent access to the international gateway (011 in the United States).

call return A telephone company optional service that allows a called station to dial the number of the last caller, even if the last call was not answered.

call screening (1) A PBX feature that looks at the calling party's number, compares it to numbers in a database, and then determines if the call should be connected. (2) A receptionist or secretary answers the incoming call and determines if (and/or to whom) the call should be routed.

call second A unit for measuring communications traffic. It is defined as one call with one second of connect time. Two calls of 30 seconds is 60 call seconds of traffic. Other useful units of measure include:

100 call seconds = 1 CCS
3600 call seconds = 36 CCS = 1 call hour
1 call hour per hour = 1 Erlang = 1 traffic unit

call selection time The time from which a call-originating terminal receives a proceed to select signal (e.g., dial tone) until all selection signals have been transmitted (e.g., completion of dialing). See also *call release time, call request time,* and *call setup time.*

call selector A local company's optional service that alerts the subscriber with a distinctive ring that one of several numbers is being called.

call setup time The overall length of time required to establish a circuit switched call between users or terminals, that is, the time between the initiation of the call request signal to the call connected signal. It is the sum of the call request time, call selection time, and the link setup time (the time from the end of the transmission of the selection signals until the delivery of the call connected signal to the originating terminal). See also *call release time, call request time,* and *call selection time.*

call spill over In common channel signaling, the effect on a communications link which an abnormally delayed call control signal of a previous call has on the current call in progress.

call splitting In telephony, the opening of the transmission path between the parties of a call. It allows a switch attendant to talk privately in either direction on an established call.

call tracing: In telephony, a procedure that permits an authorized user to ascertain the routing data for an established connection, identifying the entire route from the origin to the destination. Specifically, it is a means for identifying the equipment number of a calling party.

Country	Dial Frequency	Dial Cadence (on-off)	Ringing Frequency	Ringing Cadence (on-off)	Busy Frequency	Busy Cadence (on-off)	Congestion Frequency	Congestion Cadence (on-off)	Number Unobtainable Frequency	Number Unobtainable Cadence (on-off)
Algeria	450		25, 50	1.5–3.5	400	.5–.5				
Argentina		Continuous	25+450×16.66	1–2	450	.25–.25	400	.375–.375	400	2.5–.5
Austria			450	1–5	450	.25–.25				
Australia	425, 154*	Continuous	400×17	.4–.2–.4–2	400	.375–.375	400	.375–.375	400	Continuous
(PABX)	154	Continuous								
Bahrain	350, 440	Continuous	440 ↕ 450	.4–.2–.4–2	400	.375–.375	400	.4–.35–.225–.525 / .375–.375	400	Continuous
Belgium	450	Continuous	450	1–3	450	.167–.167	450	.5–.5	450	
(1)	450	1–.25								
Botswana	425	Continuous	400 ↕ 450	.4–.2 . .4–2	400	.375–.375	400		400	Continuous
Brazil	425	.975–.05	425	1–4	425	.25–.25			425	.75–.25–.25–.25
(PABX)	50, 25									
British Virgin Isles		Continuous	400 ↕ 450 / 400 ↕ 450	.5–.25–.5–1.75 / .4–.2–.4–2	400	.75–.75	400 / 400	.75–.75 / .4–.6–.2–.4	400	Continuous
Burundi	150		450	1–4	450	.25–.25				
Cameroon	425	Continuous	425	1.66–3.33	425	.5–.5				
Canada	600×120, 350+400 / 350+400	Continuous / Continuous	440+480	2–4	600×120	.5–.5	600×120* / 480+620 / 480+620	.25–.25 / .25–.25 / .25–.25		
(PABX)	350+400	Continuous	440+480	1–3	480+620					
Cayman Isles	425	Continuous	425, 50	2–4	400	.5–.5	400	.75–.75	400	Continuous
(PABX)	50	Continuous	400 ↕ 450	.5–.25–.5–1.75	425	.75–.75	400	.5–.5	425	.2–3
Chile	25	Continuous	400 ↕ 450	.4–.2–.4–2	400	.5–.5	400	.4–.6–.2–.4		
Cuba	400	Continuous	400 / 420+40 / 450	1–3 / 1–2 / 1–5	400 / 600+120 / 450	.5–.5 / .5–.5 / .25–.25	600+120	.25–.25	520	.4–.12–.12–.12–.12–.12
Cyprus	50	Continuous	400×25	.4–.2–.4–2	400	.75–.75	400	.75–.75	400	Continuous
Denmark	425	Continuous	425	.8–7.2	425	.1–.566	425		950 ↕ 1400 ↕ 1800	.33–.03–.33–.03–.33–1
Djibouti	425	Continuous	425	1.5–3.5	425	.5–.5	425		400	Continuous
Dominican Rep.	50	.0166–.0166	400 ↕ 450	.4–.2–.4–2.2	400	.8–.8	400	.8–.8		
El Salvador	33	Continuous	425	1.2–5	425 / 400	.375–.375 / .333–.333	425	.375–.375		
Fiji	425 / 33 / 50	Continuous	133×16.66 / 400×25	.4–.2–.4–2 / .4–.2–.4–2	425 / 400	.75–.75	400		400	2.5–.5
(PABX)	425	Continuous	425	1–4	400	.375–.375	400		400	2.5–.5
Finland	33 1/3	Continuous	400, 450	1–9	425	.3–.3				
Fujeirah	425	.2–.3–.2–.3–.2–.8	400	.5–.25–.5–1.75	400	.75–.75	400	.75–.75	400	Continuous
France	25	Continuous	400	1.66–3.33	440	.5–.5				
French Polynesia			450	1.66–3.33	450	.5–.5				
Germany (Fed Rep)	425, 450	Continuous	425	.96–3.84	425, 450	.5–.5	425	.25–.25		
Germany (Dem. Rep)	400+425 (1) / 425 / 450	Continuous / .75–.75–.25–.25 / .7–.8–.2–.3	450, 425 / 425 / 450	1–4 / 1–5 / 1–9	425 / 425 / 450	.015–.475 / .31–.31 / .25–.25 / .015–.475	425 / 425 / 450	.31–.31 / .25–.25 / .15–.475		

Country										
Ghana			450, 135	1-4	400	.25-.25			400 / 150, 450 / 150, 450	Continuous / .5-1.5 / .5-4.5
Greece	50, 25	Continuous	450	1-4	450	.3-.3				Continuous
Grenadines, The		Continuous	400 ◊ 450	.5-.25-.5-1.75	400	.75-.75	400	.75-.75	150, 450	
Guinea	425	Continuous	450	.4-.2	450	.2-.2			400	Continuous
Hungary	33, 50, 400×25	Continuous	425	1.2-3.7	425	.3-.3				
India		Continuous	133, 400×25	.4-.2-.4-2	400	.75-.75			400 / 400	Continuous / 2.5-.5
Indonesia			400×25	.4-.2-.4-2.2	400	.6-.6				
Iraq			435, 450	1-4	435, 450	.5-.5				
Ireland	33.33, 50	Continuous	425×50	Continuous	400	1-1			400, 425	6-1
Israel	400	Continuous	400	.4-.2-.4-2	400, 425	.4-.333				
Italy	425	.6-1-.2-.2	425	1-3, 1-4	400	.5-.5				
Ivory Coast	425	Continuous	50	1-4	425	.2-.2				
Jamaica	350+400	Continuous	440+480	1.66-3.33	400	.5-.5	480+620 / 480+620	.3-.2 / .2-.3	400	3-.5
Japan	400	Continuous	400, 133	1.5-4.5, 1-4	480+620	.5-.5				
(PABX)	400	.25-.25	400×16	.4-.2-.4-2	400	.5-.5			400	Continuous
Jordan		Continuous	400×20	1-2	400	.5-.5			400	Continuous
Kenya	33, 50	Continuous	400+16.66 / 133+17 / 400+17	.4-.2-.4-2 / .25-.5-.25-2 / .25-.5-.25-2	400	.75-.75 / .75-.75 / .375-.375			400	2.5-.5
Korea, Rep of	350+400	Continuous	440+480	1-2	480+620	.5-.5	480+620	.3-.2	450	.2-.1-.2-1.5
Kuwait	33	Continuous	400	1-3	400	.5-.5			400	Continuous
Lebanon			435	1.2-4.4	435	.4-.2				
Liberia	425	Continuous	425	1-4	425	.5-.5			425	
Luxembourg	425, 450	Continuous	450	1-4, 1-9	450	.5-.5	425	.25-.25		.6-.2-.2-.2-.2-.2-.2-.2
Madagascar			425	.96-3.84	450	.015-.475				
Malawi	425	Continuous	400, 133	1.66-3.33	450	.5-.5			400	2.5-.5
Malaysia			425	.4-.2-.4-2	400	.75-.75				
Maldives	33	Continuous	425 ◊ 50	.4-.2-.4-2	400, 425	.4-.333			425	2.5-.5
Malta		Continuous	133	.4-.2-.4-2	425	.35-.65	400	.4-.35-.225-.525	400	Continuous
Mauritania	425	Continuous	400+450	.4-.2-.4-2	400	.75-.75			400	Continuous
Mexico	425	Continuous	400×25	.4-.2-.4-2	425	.5-.5	425	.25-.25	425	.25-.25
Montserrat	50	Continuous	400×16.66	.4-.2-.4-2	425	.25-.25	425	.8-.8	400	Continuous
Morocco	425	Continuous	425	1.5-3.5	400	.8-.8	400	.4-.6-.2-.4	400	Continuous
Mozambique	400	Continuous	425	1-5	400	.75-.75				
Nauru	400	Continuous	400	1-4	400	.5-.5			400	.2-.2
(PABX)	33	Continuous	425×25	.4-.2-.4-2	425	.5-.5	400	.375-.375	400	.75-.25-.25-.25
Netherlands	150+450	Continuous	425, 450* 50, 450×25(1), 450 (1)	1-4	425, 450(1)	.375-.375 / .5-.5 / .25-.25 / .25-.25	450	.5-.5	400	
(PABX)	425, 450*	Continuous	425, 450*	1-4	425 / 450	.5-.5 / .25-.25	425	.25-.25		

CALL PROGRESS TONES (CONTINUED)

Country	Dial		Ringing		Busy		Congestion		Number Unobtainable	
	Frequency	Cadence (on-off)	Frequency	Cadence (on-off)	Frequency	Cadence (on-off)	Frequency	Cadence (on-off)	Frequency	Cadence (on-off)
New Caledonia			425	1.66–3.33	425	.5–.5				
New Zealand	400	Continuous	400×33	.4–.2–.4–2	400	.5–.5	900 (2) 400 (2)	.5–.5 .25–.25	400	.75–.1–.75–.1– .75–.1–.75–.4
(PABX)	400	Continuous	400×450	.4–.2–.4–2	400	.5–.5	400	.25–.25		
Nigeria	400, 450	Continuous	400, 450	2–4	400	.5–.5	400	.25–.25		
Norway		Continuous	400, 450	1–3 1–4	425	.5–.5	425	.25–.25	450	5.5–.5
Oman	425	Continuous	425	.4–.2–.4–2	400, 450	.2–4	425	.5–.5	425	.2–.3
Pakistan			450 400+16.66	1–4 .4–.2–.4–2	425 450	.5–.5 .4–.675.–.13–.17	450			
Philippines	600×120, 425	Continuous	425+480	1–4	600×120, 425	.5–.5	600×120	.25–.25	600×120	.25–.25
Poland	400 400+425 (1)	.2–2 Continuous	450	1–10	450	.4–2	480+620	.25–.25	400	.5–.5
Portugal	400	Continuous Continuous	400 400	1–4 1–5	400 400	.5–.5 .5–.5	400 400	.25–.25 .5–.5	400	.2–.2
Qatar	350+400	Continuous	400+450 400+450	.4–.2–.4–2 .4–.2–.4–2.2	400, 450 400	.8–8 .375–.375	400	.4–.35–.225–.525	400	Continuous
Ras Al Khaima	25, 50	Continuous	400 400, 400 ‡ 500	.5–.25–.5–1.75 .4–.2–.4–2	400 400 400	.75–75 .375–.375	400 400 400	.75–75 .4–.4–.2–.6 .375–.375	400	Continuous
Romania			16.66, 450-25	1.85–4.15	133, 450	.167–.167			450 400×133	.362–.11– (.092–.11)×6 .125–.075– (.075–.075)×3
St. Lucia	50, 25, 600+120	Continuous	400, 450 400 ‡ 450 400 ‡ 450	2–4 .5–.25–.5–1.75 .4–.2–.4–2	400 400	.75–75 .5–.5	400 600+120 400	.75–75 .5–.5 .4–.6–.2–.4	400 600+120	Continuous .3–2
St. Vincent	50, 25	Continuous	400 ‡ 450 400 ‡ 450	.5–.25–.5–1.75 .4–.2–.4–2	400	.75–75	400	.75–75	400	Continuous
Seychelles	50, 25	Continuous	400 ‡ 450 400 ‡ 450	.5–.25–.5–1.75 .4–.2–.4–2	400 400	.75–75 .375–.375	400 400	.75–75 .375–.375	400	Continuous
Sharjah	50, 150	Continuous	400 400 ‡ 450	1–4 .4–.2–.4–2	400, 450 400 400, 450	.5–.5 .4–.35–.225–.525 .375–.375	400, 450 400	.5–.5 .4–.35–.225–.525	400 400, 450	Continuous .5–.5
Singapore	400×24	Continuous	400×24 400×24	.4–2 .4–.2–.4–2	400	.75–75	400	.25–.25	400	2.5–.5
South Africa		Continuous	400×33.33	.4–.2–.4–2	400	.75–75	400	.25–.25	400	2.5–.5
Spain (International)	425 600	Continuous Continuous	425	1.5–3	425	.2–2	425	.2–.2–.2–.2–.2–.6		

Country	f1	t1	f2	t2	f3	t3	f4	t4	f5	t5
Suriname	600×120	Continuous	450	1-4	600×120	.5-.5				
Swaziland	25+50	Continuous	420×20	1-5	450	.25-.25				
			425, 450	1-4		.5-.5				
			400							
Sweden	425	Continuous	425	.4-.2-.4-.4	425	.25-.25	425	.25-.75	400	4.8-.2
	425	.32-.01								
	500									
Switzerland (PABX)	425	Continuous	425, 500	1-5, 1-9*	425	.5-.5, .25-.25				
	500	Continuous		1-4	500	.5-.5				
Syria	450	Continuous	450 ‡ 50	1-3	450	.5-.5	450	.2-.23-.2-.23-.2-.92	450	.9-.2-.25-.2-.25-.2
			425, 475	1-4	450	.44-.49				
Tanzania	50	Continuous	400	Continuous, 1-2	400	.75-.75			400	2.5-.5
					400	.375-.375				
Thailand			400	1-4	400	.5-.5			400	.3-.1-.1-.1-.1- .1-.1-.1
					450	.333-.333				
Tunisia	425		425	1-5	425	.25-.25				
					400	.75-.75				
Turks & Caicos	350+440	Continuous	400+480	2-4	480+120	.5-.5	400	.75-.75	400	Continuous
			400 ‡ 450	.5-.25-.5-1.75			480+620	.5-.5	480+620	.3-.2
										2.5-.5
Uganda	33, 50	Continuous	133+17	.25-.5-.25-2	400	.375-.375	400		400	
			400+17	.25-.5-.25-2						
			400+450	.25-.5-.25-2						
United Kingdom	50, 350+400	Continuous	400+450	.4-.2-.4-2	400	.375-.375	400	.4-.35-.225-.525	400	Continuous
			400×25	.4-.2-.4-2						
			400×16.66	.4-.2-.4-2						
Uruguay	450	.7-.8-.2-.3	450	1-4	400	.75-.75				
					450	.3-.3				
United States	350+440	Continuous	440+480	2-4	600×120*	.5-.5	620×120*	.25-.25		
	600×120*	Continuous	420×40*	2-4	480+620	.5-.5	480+620	.25-.25		
(PABX)			440+480	1-3						
Yugoslavia	425	.7-.8-.2-.3	425	1-4	425	.5-.5				
			450×25	1-9	425	.2-.4				
Zambia	425, 50	Continuous	425	1-4	400	.75-.75*	400*	.75-.75*	400*	.75-.75
			133, 425	.4-.2-.4-2	425	.5-.5	425	.25-.25	425	.25-.25

‡ Indicates tones follow in sequence; i.e., f1 is applied and removed, then f2 is applied and removed.

. Indicates any of the frequencies or time sequences may be found in service.

+ Indicates tones f1 and f2 are added together (not modulated).

× Indicates tone f1 is modulated by tone f2.

* Old system, being phased out of service.

(1) Special cases or equipment.

(2) Disconnect tone / Overflow tone.

Two categories of *call tracing* are:

- *Permanent call tracing,* which permits tracing of all calls.
- *On demand call tracing,* which permits tracing, upon request, of a specific call, provided that the called party dials a designated code immediately after the call to be traced is disconnected, i.e., before another call is received or placed.

call transfer A PABX (or other switching system) service feature that allows a user to instruct the local switching equipment (or switch attendant) to move a call's existing terminal appearance to another terminal.

call waiting A service offered by the local telephone company to subscribers wherein the subscriber is notified of an incoming call even if a connection to a remote terminal is already established. This is accomplished by sending a 1.5-second tone burst to the subscriber over the voice pair.

Obviously, this will disrupt any digital data transfers in progress. To disable the feature *for the current outgoing call,* send a special dial sequence to the central office before dialing the number of the remote terminal. On many systems this sequence is **"*70"** for DTMF signaling or **"1170"** for rotary dialing. Not all telephone systems use the same sequence; not all telephone systems even have a sequence to disable the feature; and some systems require a short pause after the sequence before proceeding with the dialed number. (This is accomplished in most communications software simply by appending a comma or two to the disable sequence.) For example, to call the number 555-1234 and disable the call waiting feature during the call the Hayes command line

ATDT * 70 , 555 - 1234

would be entered. Another method gives the modem more time to determine that a link failure has occurred. This is accomplished by increasing the "hang up delay from lost carrier" from 1.4 seconds to 4 to 6 seconds. A delay value of 5 seconds is set register S10 (on Hayes compatible modems) by the command ATS10 = 50.

call waiting tone In telephony, the signaling tone generated by the local central office (CO) to a subscriber indicating that a second call is waiting for service.

callback A feature on some private branch exchange (PBX) systems that allows a caller to have the telephone system automatically establish a connection to a called, but busy, number as soon as it becomes available. When the called number becomes available, the system performs a sequence of actions such as:

- Lockout the called number (prevent it from initiating or receiving any calls).
- Ring the callback originator's (caller's) phone.
- When the caller answers the phone, ring the called parties phone and proceed as for a normal call.
- If the callback originator does not go off-hook within a timeout period, abort the process and restore the called number to normal service.

callback modem See *call back modem.*

callback queuing An optional feature on many PBX systems which allows outgoing calls to be put in a holding line (queue) awaiting the availability of an outgoing trunk. When the trunk becomes available, the call process continues. There are actually two ways the call can proceed:

- *Hold on queuing.* The call originator dials a number; the switching system has no outgoing trunks to service the call, so it notifies the caller with a tone. The caller may then elect to wait for a trunk to clear. When a trunk becomes available, the PBX connects to the public network and proceeds as a normal call. Also called *off-hook queuing (OHQ).*
- *Callback queuing.* The call originator dials a number; the switching system has no outgoing trunks to service the call, so it notifies the caller with a tone. The caller then dials several digits to initiate the *callback queuing* feature and hangs up. When the trunk becomes available, the system seizes the trunk and rings the call originator. Upon answering the call, the PBX connects to the public network and proceeds as a normal call. Also called *on-hook queuing.*

called line identification facility A network-provided service feature in which the network notifies a calling terminal of the address to which the call has been connected.

called line identification signal: A sequence of characters transmitted to the calling terminal to permit identification of the called line.

called party An entity, such as a person, equipment, or program to which a call is directed. Also called the *call receiver.*

called party camp-on A switching system service feature that enables the system to automatically complete a call attempt even if the called terminal is unavailable (e.g., the called terminal is busy) when the call attempt was initiated. To provide this feature, the system monitors the called station until the blocking signal ends; it then automatically completes the requested access. This feature permits holding an incoming call until the called party is free. Also called *camp-on* or *camp-on busy.*

called party control A term for the method of call termination where *only* the receiving station may initiate the call disconnect sequence. For example, if station A originated the connection to station B, then only station B could end the connection. Also called *line release.* See also *both party control, calling party control,* and *either party control.*

caller ID (CLID) In some telecommunications environments, a network service feature that includes the sender's identification number in the transmission so that the receiver knows who is calling.

The telephone company service allows a called subscriber device to determine what station number called before the called equipment goes off-hook. The caller's telephone number is transmitted across

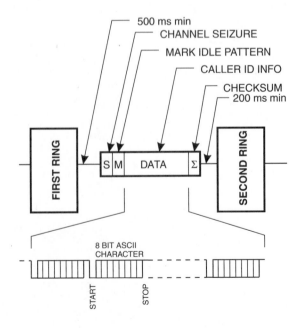

the switched network between the first and second ring signal using the Bell 202 asynchronous 1200 baud FSK specification. The actual message is composed of a number of 8-bit characters with one start bit and one stop bit. The preceding diagram outlines some of the important characteristics of the *caller ID* messaging format. The data portion of the *caller ID* contains the date, time, originating caller telephone number, and possibly a company name. If the modem software supports *caller ID,* the display might look as follows:

RING
DATE = 1231
TIME = 1812
NMBR = 7045551212
NAME = NUMBERHELP
RING

Caller ID may be blocked by the calling party on either a *per call* basis or a *per line* basis. In the per call case, the caller must issue a call block command before entering the number of the remote station. The sequence to temporarily block *caller ID* transmissions is generally "***67**" for DTMF signaling or "**1167**" for rotary dialing. When *caller ID* has been blocked on a per line basis, the calling subscriber must enable the feature on a per call basis if it is desired. The enabling command is frequently "***67**" for DTMF signaling or "**1167**" for rotary dialing. ***82** or **1182** is also used to control the caller ID mode. Also called *automatic number identification (ANI)* and *calling line identification (CLI).*

calling device In telephony, any apparatus that generates the required signals to establish a connection between two stations on the automatic switched network.

calling line identification (CLI) A service available on digital phone networks that informs the apparatus being called which number is calling them. The central office equipment identifies the telephone number of the caller, enabling information about the caller to be sent along with the call itself.

Analogous to Automatic Number Identification (ANI) used on analog lines. See also *caller ID (CLID).*

calling line identification facility A network-provided service feature in which the network notifies a called terminal of the address from which the call has originated. See also *caller ID (CLID).*

calling line identification signal A sequence of characters transmitted to the called terminal to permit identification of the calling line. See also *caller ID (CLID).*

calling party An entity, such as a person, equipment, or program that originates a call. Also called the *call originator.*

calling party camp-on A communication system service feature that enables the system to automatically complete a call attempt even if the transmission or switching facilities required to establish the requested access are temporarily unavailable. To provide this feature, the system monitors the required facilities until they become available; it then automatically completes the requested access. Systems providing called party camp-on may or may not issue a system-blocking signal to the calling party to advise of the access delay. See also *called party camp-on.*

calling party control A term that refers to the method of call termination whereby *only* the originating station may initiate the call disconnect sequence. For example, if station A originated the connec-

tion to station B, then only station A could end the connection. The release may be either immediate or after a predetermined delay, depending on the switching equipment involved. Also called *calling line release.* See also *both party control, called party control,* and *either party control.*

calling rate The *call intensity* per traffic path generally during the *busy hour.* The number of telephone calls originated during a specified time interval such as one hour.

calling sequence A sequence of instructions together with any associated data necessary to perform a call.

calling signal A call control signal transmitted over a circuit that indicates a connection is desired.

calling station identifier (CSID) The information that identifies the calling station to the receiving station. In a fax transmission, the *CSID* information is usually appended to the message as a header on the fax pages.

calling tone (CNG) **(1)** A 1300-Hz (980-Hz) tone generated by the calling DCE to inform the called DCE that a modem (computer) is originating the call. **(2)** A tone generated by the call-originating facsimile machine to indicate that the call is a fax transmission, not a voice transmission. The tone is 1100 Hz with a cadence of 0.5 seconds on and 3.5 seconds off.

calls-barred facility A service feature that permits a terminal either to make outgoing calls or to receive incoming calls but not both.

CALM An acronym from Connection Associated Layer Management.

CAMA An acronym from Centralized Automatic Message Accounting. In telephony, an arrangement at an intermediate office for collecting automatic message accounting information.

Cambridge Ring See *slotted ring.*

camp-on See *automatic callback, called party camp-on, calling party camp-on,* and *queue traffic.*

camp-on busy signal **(1)** A signal that informs a busy telephone user that another call originator is waiting for a connection. **(2)** A teleprinter exchange facility signal that, in response to a destination or circuits busy condition, automatically causes a calling station to retry the called number after a specified time interval. Also called *speed-up tone.*

camp-on with recall Any camp-on in which the calling party terminal is released from the switching system until the called party terminal can be seized.

The calling party can then use the terminal to originate or answer other calls until the *recall signal* is obtained. This prevents the user from the necessity of simply waiting until the called party's line is available. See also *automatic callback, called party camp-on, calling party camp-on,* and *queue traffic.*

Campus Area Network (CAN) A network that connects nodes (and possibly departmental LANs) from multiple locations, which may be separated by a considerable distance. Unlike a wide area network (WAN), however, a campus network does not require remote communications facilities such as modems and telephones.

campus wide information system (CWIS) A computer system developed to make information about a college or university easily accessible to both students and the public in general via the Internet. The information includes things such as campus event calendars, course listings, and job openings. It is a *browser* and should provide a wide variety of information. The *CWIS* developed at the University of Minnesota led to the *Gopher* resource discovery tool.

CAN (1) An acronym from Campus Area Network. (2) An abbreviation for CANcel character.

Canadian Network (CA*NET) The Canadian regional backbone network providing Internet access to both research and educational institutions.

Canadian Radio-Television and Telecommunications Commission (CRTC) The federal agency with exclusive jurisdiction over all aspects of telecommunications in Canada.

Canadian Standards Association (CSA) An independent, non-government, not-for-profit association for the development of Canadian standards and product certification.

cancel character (CAN) (1) A control character used by some conventions to indicate that the data with which it is associated are in error or are to be disregarded. (2) An accuracy control character used to indicate that the data with which it is associated are in error, are to be disregarded, or cannot be represented on a particular device.

CAP (1) An acronym from Carrierless Amplitude and Phase. A modem modulation technique used on ADSLs and 50-Mbps LANs. (2) An acronym from Competitive Access Provider. An alternative method of access to an Interexchange carrier (IXC) rather than the normal local exchange carrier (LEC). (3) A colloquial abbreviation of CAPacitor.

capacitance The property of a system of conductors and dielectrics which permit the storage of an electric charge when a voltage exists between the conductors.

capacitive coupling The transfer of energy from one circuit to another by means of mutual capacitance between the circuits. *Capacitive coupling* favors transfer of the higher frequency components of a signal. The coupling may be deliberate or inadvertent. See also *inductive coupling*.

capacitor A device that stores energy in the form of an electric field, i.e., charge. A capacitor blocks the flow of direct current (dc), and it passes or partially conducts the flow of an alternating current (ac).

The impedance depends on the amount of *capacitance* the device has and the *frequency* of the ac impressed across the device. As either the frequency or the value of the *capacitor* increases, the impedance decreases, an effect called *capacitive reactance* (X_C). Capacitive reactance (X_C, in Ohms) is expressed in Ohms and is related to capacitance (C, in farads) and frequency (f, in hertz) by:

$$X_C = \frac{1}{2\pi f C}$$

Symbols for a *capacitor* include:

capacity (1) In electrochemical cells and batteries, the amount of energy available to do useful work, usually expressed in amp-hours (Ah) or watt-hours (Wh). The amount of energy is dependent on a number of variables, such as the charge on the battery, the discharge rate, and the temperature of the cells of the battery. (2) In data transmission, the number of symbols per second that can be transferred across a communications link. See also *channel capacity* and *traffic capacity*. (3) A measure of the amount of energy a capacitor can store, expressed in basic units of farads (F).

CAPI An acronym from Communications Applications Programming Interface. A programming interface specification used by some ISDN devices.

capture (1) The term used by communications software to describe the storing of information (in a *capture file*) either currently on the terminal screen (i.e., *screen capture*) or incoming information to be routed to the terminal screen. (2) A term used to describe the operation or state of an electronic receiver when it acquires synchronization with a sender's carrier or clock signal. Frequently used to describe the operation of a *phase lock loop* (*PLL*) when the local oscillator is operating synchronously with the sending circuit's station clock. (3) The effect associated with the reception of FM signals wherein two signals on or near the same frequency and differing in strength are presented to the receiver, and only the stronger signal will appear at the output. The suppression of the weaker signal carrier occurs at the receiver's limiter (detector) where it is treated as noise. When both signals are nearly equal in strength or are fading independently, the receiver may switch from one to the other. Also called *capture effect* and *FM capture effect*.

capture effect See *capture* (*3*).

capture range In *phase lock loops* (*PLLs*), the total frequency range over which a loop can lock onto an incoming signal. A related term is the *lock-in range,* which describes how close a signal must be to the center frequency before capture can occur. The lock-in range is therefore usually one-half of the capture range. See also *capture* (*2*) and *lock in range*.

capture ratio The ability of a receiver to select the stronger of two signals (at or near the same frequency) presented to its input. The ratio is expressed in decibels (dB).

carbon block The oldest and most commonly used primary overvoltage protector used on telephone systems. The *carbon block* protector is basically a spark gap device composed of two carbon electrodes (blocks) separated by a small air gap of about 0.003 inches. When a voltage greater than the breakdown (or sparkover) voltage of the gap is applied, an arc between the electrodes dissipates the energy of the applied transient.

Although the *carbon block* protector is a very low-cost device, its breakdown voltage cannot be closely controlled. For a particular type, it is as low as 300 V and as high as 1000 V. The *carbon block* protector suffers a relatively short life. See also *gas tube* and *MOV.*

carbon noise In telephony, the inherent noise voltage generated in the carbon transmitter (microphone) of a telephone handset.

carbon transmitter A type of microphone used in telephone subsets used to directly modulate the current flowing through the subscriber loop. The modulation is based on the variation of resistance of carbon-to-carbon contacts in response to the sound pressure impinging upon it.

CardBus A 32-bit version of the 16-bit PC Card (formerly called a PCMCIA card). *CardBus* is backward compatible with the PC Card. See also *PCMCIA*.

carpet bomb The posting of a message to numerous newsgroups on USENet. Generally, the message is a commercial advertisement or an electronic chain letter, and it contains deliberately offensive material that violates the Internet's acceptable use policy (AUP).

Don't do it! Don't respond to them!

carried load The traffic that occupies a communications system, e.g., the calls on a trunk or the traffic that occupies a group of servers on a LAN.

carried traffic In communications, a measure of the calls served in a specified period of time. That portion of offered traffic that successfully seizes equipment, such as trunks, lines, and LAN servers.

carrier (cxr) (1) A continuous frequency suitable for *modulation* by a signal containing data to be transferred over a communication channel. The *carrier* will travel through the communication path while the information bearing signal alone generally will not. A *carrier* is generally an analog signal (a sine wave) fixed in frequency and amplitude; however, it may be a uniform or predictable series of pulses. Also called a *carrier wave*. **(2)** A company that provides communications service (including telephone and data service) to consumers and is regulated by appropriate federal, state, or local governmental agencies; i.e., *common carrier* or *common communications carrier*. Also called a *common carrier*. **(3)** Sometimes employed as a synonym for *carrier system*.

carrier access code (CAC) The sequence of digits an end user dials to gain access to the switched services of a carrier. *CACs* for Feature Group D are composed of five digits in the form 10XXX, where XXX is the Carrier Identification Code (CIC). See also *equal access*.

carrier band A range of frequencies that can be modulated to carry information on a specific transmission system.

carrier bypass See *bypass*.

carrier detect A signal generated by a fax/modem that indicates when a connection to another fax/modem has been made. This signal is sent to the computer to allow the software to proceed with the connection and data transmission protocol. In a more general sense, a *carrier detect* signal is generated by all communication receivers to indicate that a link between the source and destination exists.

carrier dropout A "short" duration loss of carrier signal.

carrier frequency The frequency of a carrier wave, that is, the frequency of an unmodulated wave capable of being modulated with a second (information bearing) signal. Also called the *center frequency* in FM systems.

carrier identification code (CIC) In telephony, the unique three- or four-digit number dialed by a caller to reach a specific long-distance or interexchange carrier (IXC). The *CIC* is assigned by Bellcore for use with Feature B or D Switched Access Service.

A caller's primary IXC is reached by dialing 1 + area code + number; any other IXC is reached by dialing 10 + XXX + area code + number (alternatively dial 950+0XXX+. . .) where "XXX" is the number of the *CIC*. See also *equal access*.

carrier leak The unwanted carrier that exists in a suppressed carrier modulation system. Sometimes the residual carrier is purposefully maintained and used to provide the reference for an automatic frequency control (afc) system.

carrier loss (1) An output state of a carrier detector. **(2)** In a carrier system that uses zero suppression coding, the reception of more consecutive zeros than is allowed (generally, the threshold is set several times that maximum). For example, in a T-1 system, a carrier loss occurs when 32 consecutive "0s" are detected. The carrier is reacquired when the first "1" is detected.

carrier noise The random variations of an unmodulated carrier. These variations appear as noise to a receiver. Also called *residual modulation*.

carrier sense In a local area network (LAN), that part of a transmission protocol that each network interface card uses to detect network activity in other nodes.

carrier sense multiple access (CSMA) A network control scheme in which a node verifies the absence of other traffic before transmitting.

carrier sense multiple access/collision avoidance (CSMA/CA) See *CSMA/CA*.

carrier sense multiple access/collision detection (CSMA/CD) See *CSMA/CD*.

carrier shift (1) A method of modulation; see *FSK*. **(2)** An error in carrier frequency at a receiver (not present at the transmitter) caused by any of several effects in a transmission system.

carrier signal An electrical signal that is used as the basis for a transmission. This signal has well-defined properties but conveys no information (content). Information is sent by modifying (modulating) some feature of the carrier signal, such as the amplitude, frequency, or timing to represent the values being transmitted.

carrier suppression See *suppressed carrier transmission*.

carrier system A multichannel telecommunications system in which a number of independent circuits are multiplexed together for transmission between nodes of a network on a common medium. Many different forms of multiplexing may be used, e.g., time division multiplexing (TDM), frequency division multiplexing (FDM), or code division multiplexing (CDM).

Multiple layers of multiplexing may be used; that is, the input of one multiplexer may be the output of another multiplexer (e.g., group, supergroup, and master group multiplexing). At any receiving station, specified channels, groups, supergroups, and so on may be demultiplexed without demultiplexing the others.

carrier telephone channel A telephone channel derived from a carrier system.

carrier-to-noise ratio (CNR) The ratio of the carrier signal level to the level of noise in a receiver's bandwidth. The ratio is taken before any nonlinear process, usually in the intermediate frequency (IF) band. Generally, *CNR* is referred to the input of the receiver, unless otherwise stated, and expressed in decibels (dB).

carrier-to-receiver noise density (C/kT) In satellite communications, the ratio of the received carrier power to the receiver noise power density. The carrier-to-receiver noise density is given by C/kT, where:

C is the received carrier power in watts.
k is Boltzmann's constant in joules per kelvin.
T is the receiver system noise temperature in kelvins.

Carrier-to-receiver noise density ratio is usually expressed in decibels (dB). The receiver noise power density (kT), is the receiver noise power per hertz.

carrier wave See *carrier (1)*.

carrier wire A conductive wire (capable of carrying an electrical signal), e.g., the central wire in a coaxial cable, which serves as the medium for the electrical signal.

Carson bandwidth rule (CBR) A rule estimating the approximate bandwidth requirements of communications system components for a carrier signal that is frequency modulated by a continuous or broad spectrum of frequencies rather than a single frequency. Mathematically, the Carson bandwidth rule is expressed as

$$CBR = 2(\Delta f + f_m)$$

where:

CBR   is the bandwidth requirement,
Δf is the carrier peak deviation frequency, and
fm   is the highest modulating frequency.

The rule is often applied to antennas, optical sources, photodetectors, receivers, transmitters, and other communications system components.

Carterphone An acoustically coupled device invented by Thomas Carter for connecting a two-way mobile radio base station to the public telephone system. The Bell System felt the device violated its rule of telephone land, so the courts were asked to decide. In 1968 the courts ruled that Carter could indeed connect his device to the phone lines. In fact, the ruling said that nontelephone company devices could be connected to the phone line if they were "beneficial privately and publicly not harmful." Based on this first decision (and others that have followed), the FCC part 68 Rules were devised. Basically, the rules state that no device shall harm the telephone network and that the phone company has the right to disconnect lines to devices that do harm.

Cartesian coordinates The position of a point on a Cartesian Plot located by its distance from the intersecting axes (*origin*). The system is named after the seventeenth-century French mathematician–philosopher René Descartes (1596–1650) who introduced the method. The position is written in the form (*X,Y*) where *X* is the distance from the origin along the *X*-axis and *Y* is the distance along the *Y*-axis.

In the graph below, three points (3,2), (-1,-3), and (2,-3) are plotted.

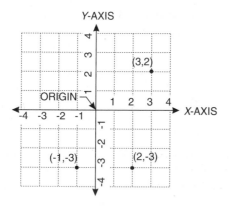

Cartesian plot A method of graphically displaying the relationship of two (or more) variables. One variable is along the horizontal axis, while the other is along the vertical.

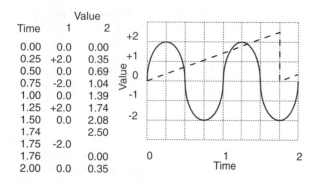

Time	Value 1	Value 2
0.00	0.0	0.00
0.25	+2.0	0.35
0.50	0.0	0.69
0.75	-2.0	1.04
1.00	0.0	1.39
1.25	+2.0	1.74
1.50	0.0	2.08
1.74		2.50
1.75	-2.0	
1.76		0.00
2.00	0.0	0.35

This chart shows a plot or graph of two variables (value 1 and value 2) plotted against a third variable (time). See also *polar plot*.

CAS (1) An acronym from Communicating Applications Specification. (2) An acronym from Column Address Strobe. One of the two addressing signals used to select the desired RAM device within a bank of devices. The other signal is the Row Address Strobe (RAS). (3) An acronym from Call Associated Signaling. (4) An acronym from Centralized Attendant Services.

cascade To connect the output of one device to the input of a second device, often the same as the first. For example, two or more amplifiers may be cascaded to increase gain; two or more networks may be cascaded to increase the number of available nodes and so on.

CASE (1) An acronym from Computer Aided (or Assisted) Software Engineering. (2) An acronym from Computer Aided Systems Engineering. (3) An acronym from Common Application Service Elements—a MAP application protocol.

case sensitive A program or system that distinguishes between upper and lower case characters and will take action only when the individual characters correspond to the expectation of the program.

For example, many password recognition programs are case sensitive and would recognize "My_NamE" but not "my_name" or "My_Name." Another example is the UNIX operating system. Most commands are lower case only. And file names usually must be typed not only with correct "spelling" but with correct "case."

case shift (1) In typesetting or typewriting, the change from (to) lower case letters to (from) upper case letters. (2) In data equipment, the change from (to) letters mode to (from) other character modes.

Cassegrain system An electromagnetic wave focusing system in which external energy is reflected from a large concave main reflector to a secondary convex reflector inside the focus of the main reflector. From this secondary reflector, energy is directed back through a hole in the main reflector to the focal plane. Although an external energy source is described, the source can be placed at the focal plane as is done in a radio wave transmitting antenna.

Cassegrain systems may be found in both RF and optical applications. Benefits include a shorter physical length for a given main reflector focal length, better support for the focal plane elements, and an image with minimal chromatic aberration.

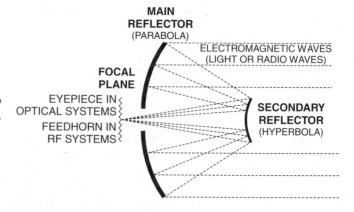

cat A UNIX command to display the contents of a text file located on a remote host.

To read long files, combine the *cat* command with the *more* command, as in:

cat readme.txt | more

This is called *piping* (and the symbol "|" is called a piping symbol).

CAT An acronym from <u>C</u>ommon <u>A</u>uthentication <u>T</u>echnology.

CAT-n An acronym from <u>CAT</u>egory <u>n</u> cable.

catastrophic degradation The rapid reduction of the ability of a system, subsystem, component, equipment, or software to perform its intended function. *Catastrophic degradation* usually results in total failure to perform any function. Also called *catastrophic failure.*

catch up The action of marking all of the articles in a USENet newsgroup as read even if they have not been read. By using the *catch up* command, existing articles do not appear on either the thread selector or article selector. Only new articles appear when the newsgroup is next visited.

category n cable (CAT-n) The EIA/TIA-568 specification on *unshielded twisted pair* (*UTP*) cable for use in LANs. Some manufacturers use "*level n*" instead of *category n* in their product specifications, and IBM specifies wire by "*type n.*" The approximate data rates for each of the cable categories are:

Category 1: Voice grade, UTP telephone cable. The cable that has been used for years in telephone communications. Generally, *category 1* cable is not considered suitable for data grade transmissions (in which every bit must get across correctly). In practice, however, it works fine over short distances, under ordinary working conditions, and at data rates less than 4 Mbps.

Category 2: Data grade UTP, capable of supporting transmission rates of up to 4 megabits per second (Mbps). IBM's Type 3 cable is an example of this category cable.

Category 3: 100 Ω data grade UTP and associated termination hardware, capable of supporting transmission rates up to 16-Mbps. 10BaseT networks, for example, require at least *category 3* cable.

Category 4: 100 Ω data grade UTP with 12 twists per foot and associated termination hardware, capable of supporting transmission rates up to 20 Mbps. IBM's 16 Mbps Token Ring network requires *category 4* cable.

Category 5: 100 Ω data grade UTP and associated termination hardware, capable of supporting transmission rates of up to 100 Mbps (although some applications, ATM for example, allow data rates to 155 Mbps). The proposed Copper Distributed Data Interface (CDDI—a copper-based version of the 100 Mbps FDDI network architecture) network standard and 100BaseX network architecture require *category 5* cable.

An enhanced version of Cat-5 extends the bandwidth to 200 MHz.

Category 6: Data grade SFTP. Four pairs wrapped in foil insulation and twisted around one another. The 600-MHz bandwidth is suitable for gigabit Ethernet and 155 Mbps ATM.

See also *cable, level n cable, type n cable,* and *wire.*

cathode (1) In energy sinks, the electrode of a device through which the primary stream of electrons *exits* (current enters). **(2)** In energy sources, the electrode of a device through which the primary stream of electrons *enters* (current leaves). **(3)** The electrode in an electrochemical cell where reduction takes place. During discharge, the positive electrode of the cell is the cathode. During charge in a rechargeable battery, the negative electrode is the cathode. See also *anode.*

cathode terminal (1) In a PN junction diode, the terminal into which electrons flow. The terminal is generally marked with a chamfer, bar, "+," or other symbol; for example

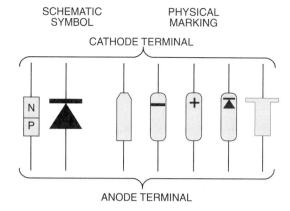

(2) In an electrochemical cell or battery, it is the electrode where reduction takes place. During discharge the *cathode* is the positive electrode. During charge, the chemical reaction reverses and the *cathode* is the negative electrode. See also *anode terminal* and *cathode.*

CATI An acronym from <u>C</u>omputer <u>A</u>ssisted <u>T</u>elephone <u>I</u>nterviewing. A market research term for a call center based on the use of computerized databases.

CATNIP: An acronym from <u>C</u>ommon <u>A</u>rchitecture for <u>N</u>ext generation <u>I</u>nternet <u>P</u>rotocol. Defined in RFC-1707.

CATV (1) An abbreviation of <u>C</u>ommunity <u>A</u>ntenna <u>Tele</u><u>V</u>ision. **(2)** An abbreviation of <u>CA</u>ble <u>Tele</u><u>V</u>ision.

CAU An abbreviation of <u>C</u>ontrolled <u>A</u>ccess <u>U</u>nit.

CAU/LAM An abbreviation of <u>C</u>ontrolled <u>A</u>ccess <u>U</u>nit/<u>L</u>obe <u>A</u>ttachment <u>M</u>odule.

Cauer filter See *elliptic filter.*

CAV An abbreviation of <u>C</u>onstant <u>A</u>ngular <u>V</u>elocity. A term used to describe an interactive LaserDisc where each frame has its own access number and a user can access individual frames randomly and in any sequence desired. Discs such as this can store up to 54,000 individual images or 30 minutes of fully random access motion video. Two methods of accessing images are:

- CAV Level I—The first level of LaserDisc interactivity where you access images and motion sequences from a LaserDisc using a remote control or barcode reader.
- CAV Level III—The third level of interactivity where the LaserDisc is controlled through a computer using special software programs such as Apple's HyperCard or IBM's LinkWay Folders.

cavity A volume defined by a conductor-dielectric and/or dielectric-dielectric reflective boundaries, and having dimensions designed to produce specific resonance or interference effects (constructive or destructive) when excited by an electromagnetic wave.

CB An abbreviation of <u>C</u>itizen's <u>B</u>and radio.

CBC (1) An abbreviation of <u>C</u>ipher <u>B</u>lock <u>C</u>haining. An operating mode of Data Encryption Standard (*DES*). **(2)** An abbreviation of <u>C</u>anadian <u>B</u>roadcasting Corporation.

CBDS An abbreviation of <u>C</u>onnectionless <u>B</u>roadband <u>D</u>ata <u>S</u>ervice.

CBEMA An acronym from <u>C</u>omputer and <u>B</u>usiness <u>E</u>quipment <u>M</u>anufacturers <u>A</u>ssociation.

CBMS An abbreviation of Computer Based Messaging System. An older term for a Message Handling System (MHS) or e-mail.

CBR An abbreviation of Constant Bit Rate or Continuous Bit Rate. A channel or service in an ATM network that provides for voice or synchronous data connections. It provides a steady data flow with low variation in cell relay. See also *variable bit rate* (*VBR*).

CBTA An abbreviation of Canadian Business Telecommunications Alliance.

cc An abbreviation of Courtesy (or Carbon) Copy

CC (**1**) An abbreviation of Clearing Center. (**2**) An abbreviation of Country Code. (**3**) An abbreviation of Connect Confirmed or Connection Confirmed. (**4**) An abbreviation of Cluster Controller. Used in IBM's SNA for a group of dumb terminals. (**5**) An abbreviation of Continuity Check.

CCBS An abbreviation of Completion of Calls to Busy Subscribers. A supplemental service defined for ISDN.

CCC (**1**) An abbreviation of Clear Coded Channel. A 64-kbps channel in which the total capacity is available for data. (**2**) An abbreviation of Clear Channel Capability. The available bandwidth of a communications channel after all control and signaling information is removed.

CCDN An abbreviation of Corporate Consolidated Data Network.

CCEE An abbreviation of China Comm. for Conformity Certification of Electrical Equipment. The Chinese agency responsible for certification of equipment compliance with standards. See also *CSBS*.

cch An abbreviation of Connections per Circuit Hour (in hundreds).

CCIA An abbreviation of Computer and Communications Industry Association. A trade organization that runs seminars, does lobbying, and addresses the common interests of the membership. Members include computer companies, data communications companies, and some common carrier services.

CCIR An abbreviation of Comité Consultatif International des Radiocommunications (International Consultative Committee for Radio Communications). That part of the ITU responsible for developing, issuing, and maintaining worldwide standards in the field of radio.

Merged with the CCITT and called the ITU-TSS in 1993 (see CCITT).

CCIS An abbreviation of Common Channel Interoffice Signaling.

CCITT An abbreviation of Comité Consultatif Internationale Téléphonique et Télégraphique (Consultative Committee for International Telephone and Telegraph). The *CCITT* is part of the United Nations International Telecommunications Union (ITU) and is responsible for developing, issuing, and maintaining worldwide telecommunication standards. In March 1993, *CCITT* merged with the CCIR and changed its name to *ITU-TSS* for ITU Telecommunication Standardization Sector or *ITU-T* for short.

ITU-T (*CCITT*) documents are listed in the format "letter.number"; for example, the *ITU-T* recommendation on the standard character set is in the document T.50. Document revisions are marked "bis" and "ter" for the second and third revision, respectively. Every four years these standards (recommendations) are published in a color-coded series of books, e.g.,

Color of Book	Year
Orange	1980
Red	1984
Blue	1988
White	1992
Green	1996

CCITT X.n A series of recommendations or standards from the *CCITT* (now called the *ITU-T*) covering all aspects of telecommunications. See *ITU-T recommendations* and Appendix E for more information on available standards.

CCS (**1**) An abbreviation of Common Channel Signaling. (**2**) An abbreviation of Common Communications Support. (**3**) An abbreviation of Centum (Hundred) Call Seconds. See *call seconds*.

CCS6 An abbreviation of Common Channel Signaling number 6. The first out of band common channel interoffice signaling (CCIS) system in North America.

CCS7 An abbreviation of Common Channel Signaling number 7.

CCSA An abbreviation of Common Control Switching Arrangement.

CCT (**1**) An abbreviation of CirCuiT or CirCuiT switching. (**2**) An abbreviation of Continuity Check Tone.

CCTA An abbreviation of Central Computer and Telecommunications Agency.

CCTV An abbreviation of Closed-Circuit TV.

CCU An abbreviation of Communications Control Unit.

CCW (**1**) An abbreviation of Cable Cutoff Wavelength. (**2**) An abbreviation of Counter ClockWise.

cd (**1**) Both the DOS and UNIX command to change the current directory.

For example, the UNIX command *cd/modem/standard* causes the host machine to move the current directory pointer to the subdirectory "standard" which resides under the directory "modem." To move the pointer up one level to "modem" the user could type the command *cd ..* where the ".." means to move up one level. Similarly, the DOS command *cd\ modem\ standard* causes exactly the same action as does *cd ..* Note that the delimiters used in UNIX are forward slashes "/." (DOS-based systems use the back slash "\.") (**2**) The symbol for *candela*.

CD (**1**) An abbreviation of Carrier Detect. (**2**) An abbreviation of Collision Detect. (**3**) An abbreviation of laser Compact Disc. (**4**) An abbreviation of Committee Draft. (**5**) An abbreviation of Capability Data. (**6**) An abbreviation of Count Down. A counter that holds the number of queued cells preceding the local message segment in an IEEE 802.6 system.

CD-DA An abbreviation of Compact Disc Digital Audio. A CD that contains digitally encoded musical or audio information. It is based on the standards published by Philips and Sony. This standard allows for up to 74 minutes of digital sound, transferred at 150 kbps (a data rate known as "single-speed").

CD-DA is the standard format used by the music industry and is known as the *Red Book* standard.

CD+G An abbreviation of Compact Disc plus Graphics. It is a format that includes limited video graphics capabilities in a CD-DA format. Mostly used in Karoke (sing-along) devices.

CD-i An abbreviation of Compact Disc-Interactive. It was developed by Philips as a compact disc stand-alone (set top) system that connects to a standard TV. This proprietary system was designed for the home user interested in multimedia entertainment but not willing to invest in a full multimedia computer.

CD-i is capable of storing 19 hours of audio, 7,500 still images, and 72 minutes of full screen/full motion video (in MPEG) in a standard CD format. *CD-i* devices are also capable of playing CD-DA, CD+G, Photo CD, and Video CDs. The standard used for *CD-i* is known as the *Green Book* standard.

CD-R An abbreviation of Compact Disc Recordable. A format that enables one to record data onto compact discs so that regular *yellow book* CD-ROM drives can read it. With a CD-R drive, you can record data onto the recordable disc on different occasions, a process known as *multi-sessions*.

CD-ROM An acronym from Compact Disc-Read Only Memory. It is a laser-read, digitally encoded optical memory storage medium on which digital data is stored. *CD-ROM* is the basis for many existing CD formats, using the same constant linear velocity (CLV) spiral concept as computer audio discs.

CD-ROMs hold about 650 Mb of data, sound, and still and motion video. A *CD-ROM* player will typically play CD-DA discs, but a CD-DA player will not play *CD-ROMs*. The standard used for most *CD-ROM* formats is based on the standard published by Philips and is known as the *Yellow Book* standard.

CD-ROM XA An acronym from CD-ROM Extended Architecture. A hybrid format that combines *CD-ROM* and *CD-i* capabilities to allow interleaved compressed sound and graphics. *CD-ROM XA* is an extension of the *Yellow Book* standard and is the basis for Kodak's Photo CD format.

CDA An abbreviation of Capability Data Acknowledgment.

CDDI An abbreviation of Copper Distributed Data Interface. A copper wire equivalent to the fiber FDDI.

CDF (**1**) An abbreviation of Comma Delimited File. A database file containing one record per line and multiple fields per record, each separated by a comma. (**2**) An abbreviation of Combined (or Cumulative) Distribution Frame.

CDFS An abbreviation of CD File System.

CDMA An abbreviation of Code Division Multiple Access. See also *spread spectrum*.

CDPD An abbreviation of Cellular Digital Packet Data.

CDPSK An abbreviation of Coherent Differential Phase-Shift Keying (PSK).

CDR An abbreviation of Call Detail Recording.

CDU An abbreviation of Control Display Unit.

CDV An abbreviation of Cell Delay Variation. An ATM traffic parameter.

CE An abbreviation of ATM Connection Endpoint.

CEB An abbreviation of Comité Electrotechnique Belge. The Belgium national standards agency.

CEBEC An abbreviation of Comité Electrotechnique BElge Service. The Belgium national certification body.

CECC An abbreviation of CENELEC Electronic Components Committee.

cell (**1**) In packet communications, a fixed-length unit of data. For example, in the Asynchronous Transfer Mode (ATM) network architecture, *cell* refers to a packet. ATM cells are each 53 octets, of which 5 octets are header and 48 are data. (**2**) In cellular telephony, that region served by a single antenna. Each cell contains an antenna and facilities that receive signals from another cell or from a caller and to pass them on to an adjacent cell or to a user within the cell. Cell size can be anywhere from a few kilometers to 32 kilometers (20 miles) in diameter, depending on the expected traffic density. (**3**) In information storage, the basic unit of storage, e.g., binary cell, decimal cell. (**4**) In electrochemical voltage devices, the basic electrochemical unit characterized by an anode, a cathode, and an electrolyte, which is used to receive, store, and deliver electrical energy. An assembly with two or more cells connected in a series and/or parallel arrangement is called a *battery*. Electrical cells are divided into two categories: primary and secondary. Primary cells can only store and deliver a current (nonrechargeable), whereas secondary cells not only store and deliver energy but also receive energy to be stored for future use (rechargeable). Each choice of anode and cathode material, and to a lesser extent the electrolyte, produces a cell with a different output voltage. (**5**) In optics, a single unit whose resistance or generated voltage varies with radiant energy, e.g., a photocell. (**6**) In optics, an entity in an optical system holding one or more lenses.

cell loss priority (CLP) A flag bit in an ATM cell header that specifies whether the cell may be discarded if advisable, e.g., if the network gets too busy. A value of 1 indicates it is expendable, whereas a 0 indicates high priority and should be kept.

cell relay A generic term for a packet transmission technology, based on IEEE 802.6, for metropolitan area networks (MANs) and for ATM wide area networks (WANs). The protocol is based on small fixed-length packets capable of supporting voice, video, and data at very high speeds. The packets (or cells) are 53 octets long.

cell reversal The phenomenon of cell voltage polarity reversal that occurs when a series string of cells is overdischarged.

When several cells are connected in series and discharged, one cell will be completely depleted of energy before the others. If current continues to be delivered when one or more cells is discharged, the current will actually try to charge these cells; however, the current is flowing in the wrong direction for a proper charge so the cell is charged with a reverse polarity.

cell splitting A means of increasing the capacity of the cellular system, in a specific geographic area, by subdividing (splitting) cells into two or more smaller cells.

cell switching Refers to how calls are switched in the cellular system.

cellphone The British term for a cellular telephone (both handheld or automotive based).

cellular data link control (CDLC) A public domain communications protocol used in cellular telephony. A *CDLC* modem allows data terminals to be attached to a cellular phone to send and receive information. *CDLC* features forward error correction (FEC), bit interleaving, selective retransmission, and others to make it a good choice for cellular data transmissions. It is the de facto standard in the United Kingdom for data communications.

cellular digital packet data (CDPD) A system that provides two-way wireless digital communications over existing analog cellular mobile telephone channels.

CDPD exploits the capabilities of the existing EIA Advanced Mobile Phone Services (AMPS) cellular structure by inserting *19.2-kbps data packets* into the idle time of normal voice connections. (Idle time is generally >30%.) In this way it can coexist with existing services. The specifications for *CDPD* were developed by IBM in concert with cellular carriers such as Ameritech, AT&T, McCaw Cellular, Motorola, and Nynex Mobil. Version 1 of the specification was released in July 1993.

cellular modem A modem designed for use with cellular telephones. In order to deal with the poor signal quality found with wireless transmissions, cellular modems differ from basic modems in several ways; specifically:

- They generally come with very advanced error-detection and correction capabilities, such as Microcom's MNP 10 protocol.
- They are more tolerant of timing variations and carrier loss, which can arise, for example, when a transmission is handed off from one cell to another.
- They do not expect to hear a dial tone from a central office.
- They are more expensive.

It is possible to use a "regular" modem with a cellular telephone; however, special adapters may be required.

cellular telephony A radio-telephone system in which a user operates within a grid or matrix of low-powered radio transmitter-receivers operating in the 825- to 890-MHz frequency band. The region served by each transmitter-receiver is called a *cell*. The user is mobile within this grid, and as the user travels from cell to cell, the system automatically switches (hands-off) the routing and operating frequencies of the cells so as to follow the user.

Features of a cellular telephone may include:

- *A/B switch*—permits the user to select the wireline carrier (B) or nonwireline carrier (A) when roaming.
- *Alphanumeric memory*—stores names with phone numbers.
- *Call in absence alert*—user-activated feature that sounds the car horn when the unit receives an incoming call.
- *Call in absence indicator*—a display that shows what incoming calls were attempted while the user was away.
- *Call in progress override*—ensures that power is maintained on the phone if the ignition switch is turned off.
- *Call restriction*—security features that limit a cellular phone's dial-out capability, such as:
  — Dial from memory only.
  — Dial last number only.
  — Dial limited to seven digits only.
  — No memory access, and so on.
- *Call timer*—a display of information about calls; for example,
  — Length of time—current call.
  — Length of time total all calls.
  — Total number of calls.
  — Beep once per minute, and so on.
- *Continuous DTMF*—unit sends DTMF tones continuously while the button is depressed. Necessary for the access of information on some telephone answering devices (TAD) that require a long tone for activation.
- *Dual NAM*—allows two phone numbers with separate carriers to be assigned to one phone.
- *Electronic lock*—security feature that allows the phone to be disabled, thereby preventing unauthorized use.
- *Full Spectrum*—allows access to all 832 analog channels currently available.
- *Hands Free operation*—allows the user to converse via a speaker and microphone similar to the office speakerphone.
- *Memory linkage*—allows linking or chaining phone number registers. Also called *chain dial*.

- *Multi NAM*—allows the cellular phone to have two phone numbers, each assigned to a different cellular carrier if desired. This allows two home areas.
- *Mute*—(1) disables the microphone on the cellular phone for a local private conversation; or (2) disables the car radio for the duration of a call.
- *NAM*—an acronym from Numerical Assignment Module. The device in a cellular phone that actually stores the phone number.
- *On-hook dialing*—the ability to pre-dial a number before removing the handset from the cradle.
- *Roaming*—strictly speaking not a feature of the cellular phone, but allows the user to leave the home service area and still have communications services.
- *Scratch pad*—allows the storage of phone numbers in temporary memory during a call.
- *Signal strength indicator*—a display of the receive carrier strength. It provides an indication of imminent call loss.
- *Speed dialing*—dialing a previously stored number with one or two button presses.
- *Standby time*—a measure of battery life, i.e., the maximum time the device may be left on to receive calls when powered from batteries.
- *Talk time*—a measure of battery life, i.e., the maximum time the device may be left on transmit when powered from batteries.
- *Voice-activated dialing*—a form of hands-free dialing wherein speech recognition is used to equate an utterance with a phone number; then dial it.

Several standards for cellular telephony include AMPS (analog—U.S.), GSM (TDMA—European), IS-54 (TDMA—United States), E-TDMA (U.S.), and IS-95 (CDMA—U.S.). See also *cell* and *hand-off*.

CELP An acronym from Code Excited Linear Predictive coding. A variant of the LPC voice encoding and synthesis algorithm. *CELP* can produce digitized voice output at rates of 4800 bps and lower. See also *linear predictive coding (LPC)*.

Celsius (C) Originally devised by the Swedish astronomer Anders Celsius (1701–1744). It pertains to a temperature scale based on the solid–liquid equilibrium (melting) point and boiling point of water. The melting point is defined as 0.00°, and the boiling point is defined as 100°. Also referred to as the *centigrade scale*. See also *fahrenheit (F)* and *kelvin (K)*.

CEN/CENELEC An abbreviation of the Comité European de Normalisation/Comité European de Normalisation ELECtronic (European Committee for Electrotechnical Standardization). A standards-generating organization whose membership includes Austria, Belgium, Denmark, Finland, France, Germany, Greece, Ireland, Italy, Luxembourg, the Netherlands, Norway, Portugal, Spain, Sweden, Switzerland, and the United Kingdom. The objective of the *CEN/CENELEC* is to remove trade barriers between its members. An item certified to meet *CEN/CENELEC* standards must be accepted for use by all members regardless of the country that actually did the certification.

center frequency (f_c) (1) With filters and transmission systems, generally the geometric mean of the upper and lower band edge frequencies, that is, the square root of the product of these two frequencies. (2) Phase lock loop (PLL). The frequency at which the loops operate when no input signal is present. Also called the *free running frequency*. (3) In frequency modulation systems, the rest or no signal frequency, i.e., the frequency of the unmodulated carrier. Also called the *carrier frequency*. (4) In facsimile systems, the frequency midway between the picture-black and picture-white frequencies.

Central Computer and Telecommunications Agency (CCTA) The U.S. Government Center for Information Systems.

central office (CO) The location of the telephone company switching equipment for one or more office codes and where subscriber loops are connected to the network (the end of the loop in opposition to the customer premises equipment—CPE). It is a *class 5 office* in the U.S. hierarchy of switching systems.

When the loops are connected to a remote concentrator, the term *central office* refers to the combination of the remote concentrator and its host switch. Also called an *end office, exchange, local central office,*

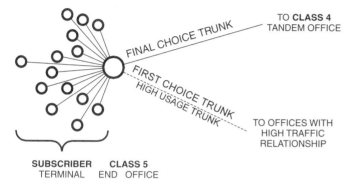

FINAL CHOICE TRUNK — TO **CLASS 4** TANDEM OFFICE

FIRST CHOICE TRUNK
HIGH USAGE TRUNK — TO OFFICES WITH HIGH TRAFFIC RELATIONSHIP

SUBSCRIBER TERMINAL   CLASS 5 END OFFICE

local exchange, local office, switching center, switching exchange, wire center, or in Europe a *telephone exchange.* An older and deprecated term is *switch.* See also *class n office.*

central office battery Generally, a bank of lead acid wet cells arranged so as to provide the −48 V necessary to run the central office switching equipment and provide the subscriber loop current for several hours. The battery is charged from the mains power when it is present or from a secondary generator when there is a mains failure. The battery serves two main functions:

- To provide a constant voltage power source in the event of a mains failure. Generally, the voltage is nominally −48 VDC; however, in Germany, the nominal value is −60 VDC. Depending on the state of charge, the voltage may range between −42 and −56 V (or −56 and −72 V in Germany).
- To act as a filter to remove any mains noise, transients, or surges that may be present.

central office connecting facility A synonym for *central office trunk* when applied to a trunk between public and private switch. See also *trunk.*

central office prefix The first three digits (NXX) of the seven-digit telephone number. See also *office code.*

central office terminal (COT) See *subscriber carrier.*

central office trunk A trunk between a public and private switch. Also called a *central office connecting facility.* See also *trunk.*

centralized attendant services (CAS) A function of a centrally located attendant console that permits the control of multiple communication switches, some of which may be geographically remote.

centralized automatic message accounting (CAMA) An automatic message accounting system that serves more than one communication switch from a central location.

centralized operation The operation of a communications network in which transmission may occur between the control station and any tributary station, but not between tributary stations.

Centrex® A contraction of CENTRal EXchange. A service offered by operating telephone companies which provides functions and features comparable to those in a PABX. Two versions are possible:

- *Centrex CO* (for company)—wherein the service is provided by switching equipment at the service provider, and
- *Centrex CU* (for customer)—wherein the service is provided by switching equipment at the subscriber premises.

In both cases, the attendant's position and station equipment reside at the customer's premises.

Centronics interface A parallel interface with 36 pins that will transmit 8 data bits simultaneously. The interface was developed by the Centronics Company, a printer manufacturer. It has become widely used as the parallel interface standard.

CEP (1) An acronym from Circular Error Probability. **(2)** An acronym from Connection EndPoint.

CEPI An acronym from Connection EndPoint Identifier.

CEPT An acronym from Comité Européen des Administrations des Postes et des Télécommunications (European Committee for the Administration of Post and Telecommunications). Formed in 1958 by the various PTTs with the goal of making standards of the approximately 30-member countries compatible. This responsibility has largely passed on to the European Technical Standards Institute (ETSI).

CEPT format Defines how the bits of a 30-channel E-1 PCM link will be used and in what sequence they will be presented. The E-1 frame consists of 32 eight-bit words, 30 assigned for voice frequency applications, one for signaling, and one for synchronization.

CERFNet An acronym from California Education and Research Federation Network. Established in 1989 as a regional Internet network and service provider that offered connectivity to both educational and business entities throughout southern California. Now expanded to cover the entire U.S. market.

CERN An acronym from Conseil Européen pour la Recherche Nucléaire, the former name of the European Laboratory for Particle Physics Research located in Geneva, Switzerland.

CERN is famous as the birthplace of the World Wide Web (WWW), which was originally conceived as a communications aid for the scientific community. In 1994 *CERN* and *MIT* (Massachusetts Institute of Technology) joined forces to develop international standards addressing Web security, privacy, and commercial use.

CERT An acronym from Computer Emergency Response Team.

Certified Novell® Engineer (CNE) A person who has passed a test given by Novell® about their networking products.

Sometimes called a Certified NetWare Engineer.

cesium clock A clock containing a cesium standard as the main frequency determining element.

cesium standard A primary frequency standard in which electronic transitions between the two hyperfine ground states of cesium-133 atoms are used to control the output frequency. The energy level between these two states corresponds to a frequency of exactly 9 192 631 770 Hz (in the absence of external influences such as the Earth's magnetic field). The *SI second* is defined as the period of time necessary for exactly 9 192 631 770 transitions to occur.

CET (1) An abbreviation of Computer Enhanced Telephony. **(2)** An abbreviation of Certified Electronics Technician.

CF (1) An abbreviation of <u>C</u>onversion <u>F</u>acility. (2) An abbreviation of <u>C</u>arry <u>F</u>orward. (3) An abbreviation of <u>C</u>oin <u>F</u>irst—referring to the operation of a payphone. Other methods of payphone operation include dial tone first (DTF) and post pay (PP).

CFA An abbreviation of <u>C</u>arrier <u>F</u>ailure <u>A</u>larm.

CFB (1) An abbreviation of <u>C</u>ipher <u>F</u>ee<u>B</u>ack. An operating mode of DES. (2) An abbreviation of <u>C</u>all <u>F</u>orward <u>B</u>usy.

CFV An abbreviation of <u>C</u>all <u>F</u>or <u>V</u>ote.

CGA (1) An abbreviation of <u>C</u>arrier <u>G</u>roup <u>A</u>larm. A service alarm generated by a channel bank when an *out of frame* (*OOF*) error persists for a predetermined length of time (0.3 to 2.5 seconds). The condition causes all calls on the trunk to be dropped. (2) An abbreviation of <u>C</u>olor <u>G</u>raphics <u>A</u>dapter.

CGI (1) An abbreviation of <u>C</u>ommon <u>G</u>ateway <u>I</u>nterface. (2) An abbreviation of <u>C</u>omputer <u>G</u>raphics <u>I</u>nterface.

CGM The extension of a CGM file (an abbreviation of <u>C</u>omputer <u>G</u>raphics <u>M</u>etafile). A standard format that allows the interchange of computer graphics (images).

CGMIF An abbreviation of <u>C</u>omputer <u>G</u>raphics <u>M</u>etafile <u>I</u>nterchange <u>F</u>ormat.

cgs An abbreviation of <u>C</u>entimeter-<u>G</u>ram-<u>S</u>econd. A system of measurement in which the centimeter, gram, and second are the fundamental units. The system has generally been replaced by the SI system.

CGSA An abbreviation of <u>C</u>ellular <u>G</u>eographic <u>S</u>ervice <u>A</u>rea. The area in which a cellular company provides cellular service. They may be oblivious to geopolitical boundaries; that is, they may cross city, county, and state lines.

CGSB An abbreviation of <u>C</u>anadian <u>G</u>eneral <u>S</u>trandards <u>B</u>oard.

chad The material removed when a hole or notch is punched in a medium, e.g., the hole in a paper tape or computer punched card.

chain code A cyclic code sequence consisting of some or all of the possible *n*-bit words wherein adjacent words are related. The relationship is such that each word is derivable from the neighbor code by shifting the bits one position left or right, dropping the leading bit, and inserting a bit in the trailing position. The only requirement of the inserted bit is that the code word must not repeat before the end of the cycle.

For example, the sequence 000, 001, 010, 101, 011, 111, 110, 100, 000 is derived by shifting 1 bit left and inserting 1, 0, 1, 1, 1, 0, 0, 0 at each point respectively. See also *code.*

chain dial The concatenation of several dial registers so that station numbers longer than a single register may be automatically dialed. For example, 15-digit registers could not store and dial "99 011 01 704 555-1234 1111"; however, by cascading two registers (for a total of 30 possible digits), the number can be dialed.

channel (1) A combination of equipment and transmission media capable of receiving a signal at one point and delivering it to another remote point. The term *channel* refers to the smallest subdivision of a transmission system; that is, one *channel* is capable of carrying only one information stream (voice or data) from a source node to a destination node. Although in the strictest sense a *channel* signifies a one-way path (providing transmission capability in only one direction), it sometimes also represents a two-way path, providing transmission in both directions.

A *channel* may either be a physical wire link or contain optical or radio links. The *channel* may be a dedicated wire or part of a switched, multiplexed, packet network. A channel may occupy all or only part of the available bandwidth of the transmission medium. Also called a *circuit, facility, link, line,* or *path.* (2) The portion of a storage medium, such as a track or a band, that is accessible to a given reading or writing station or head.

channel bank The part of a carrier multiplex terminal that performs the first step of modulation by multiplexing a group of channels into either a higher bandwidth analog channel or a higher bit rate digital channel (and the corresponding demultiplex back into individual channels). For example, the multiplexing and demultiplexing of twenty-four 64-kbps low-speed channels onto a 1.544-Mbps T-1 or thirty 2.048 Mbps E-1 high-speed link. See also *DS-n* and *T-1.*

channel capacity The maximum speed at which a communication channel can transfer information. Expressed in bits per second, bytes per second, symbols per second, and so on. See also *Shannon's limit.*

channel gate A device for connecting a channel to a highway, or a highway to a channel, at specified times.

channel noise level (1) The ratio of the channel noise at any point in a transmission system to a specified reference level. *Channel noise level* is expressed in several ways:

- dBrn—dB above reference noise.
- dBrnC—dB above reference noise with C-message weighting.
- dBa—adjusted dB.

Each of these expressions reflects a noise reading of a specialized instrument designed to measure different interference effects that occur under specified conditions. (2) The *noise power density spectrum* in the frequency range of interest. (3) The *average noise power* in the frequency range of interest.

channel reliability (ChR) The percentage of time a channel was available for use in a specified period of scheduled availability. *Channel reliability* is given by:

$$\text{ChR} = 100\,(1 - T_o/T_s) = 100\,T_a/T_s$$

where:

T_o is the channel total outage time,
T_a is the channel total available time, and
T_s is the channel total scheduled time ($T_s = T_a + T_o$).

channel service unit (CSU) The device used to connect a digital phone line (Switched 56 or T-1) to either a multiplexer, channel bank, or other device capable of producing a compatible digital signal (including a digital PBX or a data communications device). *CSU* functions include:

- Line conditioning and equalization.
- Loopback (local and remote).
- Digital signal regeneration.
- Diagnostics and signal monitoring.
- Bit stuffing.
- A framing and formatting pattern compatible with the network.

channel spacing The frequency increment between assigned adjacent carrier frequencies in a frequency division multiplex transmission system.

channel time slot That portion of a time division multiplex (TDM) data stream that corresponds to the information of a information source (one channel).

channel utilization index The ratio of the information rate through a channel to the channel capacity. Sometimes called the *channel utilization factor.*

channelization The subdivision of one-wideband (high-speed) channel into several relatively narrowband (lower capacity) independent channels. The total capacity of all subchannels may be less than the original channel owing to necessary guardbands and/or control overhead.

CHAP An abbreviation of Challenge Handshaking Authentication Protocol. A security procedure used in dial-up access.

character (1) An elementary symbol, mark, or event that may be combined with others to represent information, data, or commands. A letter, number, punctuation, or other symbols contained in a message or used in a control function.

The description of the individual symbols, marks, or events is called a *character design.* The set of all character designs within a given system is called the *alphabet.* The use of any character design is called a *character event.* For example, within the word "CHARACTER" there are nine character events and five character designs; within the word "Character" there are again nine character events, but six character designs; and within the word "10110101" there are eight character events and two character designs. Note that a group of characters, in one context, may be considered as a single character in another context. For example, binary coded decimal (BCD) and ASCII. See also *ASCII, Baudot, code, EBCDIC, T.50,* and *teleprinter.* (2) One of the units of an alphabet.

character count integrity (1) In message communications, the preservation of the exact number of characters that are in the original message. (2) In connection-oriented services, preservation of the number of characters per unit time.

Character count integrity is not the same as *character integrity.* The former only requires preservation of the number of characters, while the latter requires that the delivered characters correspond exactly with the original characters.

character density The number of characters that can be stored per unit length, area, or volume on a given medium.

character distortion The normal and predictable malformation of received data bits produced by the characteristics of a given circuit or channel at a specified speed.

character integrity The preservation of a character during processing, storage, and transmission.

character interleaving A form of time division multiplex (TDM) used for asynchronous protocols. This can be used either with extra channels or by carrying RS232-C control signals.

character interval In start-stop systems, the duration of a character expressed as the total number of unit intervals (including not only information but start-stop, control, and error checking) required to transmit any character in the alphabet of a given system.

character-oriented protocol (COP) A communications protocol in which both the beginning and ending of a data block are denoted with unique characters. *Character-oriented protocols* are used in both synchronous and asynchronous systems. IBM's Binary Synchronous Communications (BCS) protocol is one example of a *COP.*

character recognition The identification of characters by automatic means.

character set (1) A finite set of different symbols that a sender and receiver have agreed are complete for some purpose. (2) A set of unique representations called characters. (3) A defined collection of characters.

Examples include the 26 letters of the English alphabet, ASCII, EBCDIC, Cyrillic, Kanji, and Boolean characters 0 and 1. (4) All of the characters that can be represented by a particular coding method or that are recognized by a particular machine. Also called an alphabet. See also *ASCII, EBCDIC,* and *teleprinter codes.*

character stuffing A method of transmitting control information so that the receiver does not misinterpret it as data during character-based transmissions. This is accomplished by having the transmitter insert unique characters into the data stream to delineate various portions of the data stream. The receiver recognizes these characters and strips them from the information stream before passing them to the terminal equipment.

character terminal A terminal that cannot display graphics. A terminal that can display graphical information is called a graphics terminal.

character user interface (CUI) An entirely character-based operating environment. DOS, for example, is a character-based operating system, whereas Windows is a graphical user interface (GUI).

characteristic curve A plot or curve that graphically represents the relationship of two (or more) variable properties of a process or physical phenomenon. For example, the relationship of the current through a device vs. the voltage applied across the device.

characteristic distortion Distortion caused by transients, which as a result of previous modulation, are present in the transmission channel. *Characteristic distortion* effects are not consistent; their effects on a given signal element transition are dependent on transients remaining from previous signal transitions. *Characteristic distortion* is dependent on transmission channel qualities. See also *intersymbol interference.*

characteristic impedance (Z_0) (1) That impedance which when connected to the output of a transmission line of ANY length makes the line appear infinitely long.

A uniform transmission line terminated in its *characteristic impedance* has no standing waves, no reflections from the ends, and a constant ratio of voltage to current at a given frequency at every point on the line. (2) The ratio of the complex voltage between the conductors of a transmission line to the complex current in the conductors in the same transverse plane, and with the sign chosen such that the real part of the ratio is positive. (3) The calculation of the *characteristic impedance* (Z) of transmission lines is dependent on the geometry of the conductors and the environment in which they are immersed. *Note:* Every transmission line geometry is different! Several examples include:

• Coaxial cable

$$Z = \frac{60}{\sqrt{\epsilon}} \ln\left(\frac{D}{d}\right)$$

where:

d is the outside diameter of the inner conductor,
D is the inside diameter of the outer conductor, and
ϵ is the dielectric constant of the material between the inner and outer conductors.

• Two-wire open line—free space

$$Z = 120 \cdot \cosh^{-1}\left(\frac{D}{d}\right) \approx 120 \cdot \ln\left(\frac{2D}{d}\right)$$

where:

    d is the diameter of the conductors and
    D is the center-to-center spacing between conductors.

characters per second (CPS) A measure of information transfer rate between terminal equipment. It is not a measure between communications equipment.

For example, a simple asynchronous connection protocol contains 10 bits per character (8 data, 1 start and 1 stop bit). On a 2400-bit-per-second link, there are approximately 240 characters per second transferred. If a synchronous protocol is used, there are no start and stop bits, so there are only the 8 data bits transferred; hence, that character transfer rate is 300 cps. Both numbers are approximate because signaling overhead is not included.

charge The process of converting electrical energy into chemical energy within an electrochemical cell or a battery.

charge (Q) A property of subatomic particles, which can have either a negative charge or a positive charge. A negative charge results from an excess of electrons, while a positive charge results from a deficiency of electrons. The unit of charge is the *coulomb;* one *coulomb* of charge is a *charge* equal to $6.241\ 460 \times 10^{18}$ electrons. The symbol for *coulomb* is C.

charging current (1) In electrochemical systems, the dc current applied to a cell or battery necessary to convert the chemical composition of the electrodes to a material capable of storing energy for later release in the form of electricity. **(2)** In transmission, the current that flows into the capacitance of a transmission line when a voltage is applied to its terminals.

chat On computer networks and bulletin board systems (BBSs), a real-time "conversation" between two or more online participants. During a *chat,* information typed by a sender will appear on the receiver's monitor. It is not stored in a computer for later retrieval as is done with e-mail.

"Conversations" on BBSs occur in special areas called *rooms.* A room is nothing but a communication program that allows several participants to communicate as if they were all at a large table. A room is entered by going through a *door* (which means starting the communication program). See also *conference, door,* and *room.*

Cheapernet A 10Base2 network. See *Ethernet.*

check A process for determining accuracy.

check bit A bit, such as a parity bit, derived from and appended to a bit string for later use in error detection and possibly error correction. See also *parity bit.*

check character A character, derived from and appended to a data stream, for later use in error detection and possibly error correction.

check digit A digit, derived from and appended to a data stream, for later use in error detection and possibly error correction.

check key A group of characters, derived from and appended to a block of data, used in error control procedures to provide an error detection means.

checksum One of many methods of verifying the accuracy of a received digital transmission by identifying errors at the receiving end of a communication link, in a process called error detection.

The transmitting station calculates what is called a *checksum* by sequentially combining all of the bytes in the data block to be transmitted using either logical or arithmetic addition. This sum is then appended to the block of data. The receiving station then performs the same calculation on the data portion of the packet and compares the received sum with the sum it computed. If the sums match, the received data are presumed to be correct; if they do not match, a transmission error is assumed and retransmission of the block is usually requested by the receiving station. Without significant loss in error-detection capability, only the last 1 or 2 bytes of the sum are transmitted and compared in practice. The *checksum* is the sum of data bytes over some standard data block size, usually between 128 and 2048 bytes. *Checksums* are very fast and easy to implement, and they can detect more than 99% of errors in a packet. This is generally acceptable for most simple communications situations but is less reliable than the more sophisticated *cyclical redundancy check (CRC)* calculations, which have an accuracy of more than 99.9%. See also *cyclic redundancy check (CRC)* and *parity.*

CHI An abbreviation of Computer Human Interface. Also called *man–machine interface (MMI).*

chief ray In optics, the ray that passes through the center of an optical system. Also called the *principle ray.*

CHILL An acronym from the CCITT HIgh Level Language. A programming language developed by the ITU (CCITT) for the standardization telephone industry software applications on telecommunications switching systems. *C* and its variants are more prevalent.

chip (1) A slang term for an integrated circuit or many other devices used in surface mount technology, such as *chip capacitor* or *chip resistor.* **(2)** The length of time to transmit individual elements in a binary string (usually a pseudorandom sequence). A chip, unlike a bit, carries no information, although a sequence of chips may. **(3)** In satellite communications systems, the smallest element in a pseudorandom carrier encoded signal.

chip rate (1) The rate of encoding expressed as a number of chips per second. **(2)** In direct sequence modulation spread spectrum systems, the rate at which the information signal bits are transmitted as a pseudorandom sequence of chips. Generally, the chip rate is usually several times the information bit rate.

chip time In spread spectrum systems, the duration of a chip produced by a frequency-hopping signal generator.

chirp (1) The rapid change of the frequency of an electromagnetic wave. **(2)** A pulse compression technique that uses linear (usually) frequency modulation of the pulse.

CHT An abbreviation of Call Holding Time.

choke Obsolete term; see *inductor.*

choke packet In packet networks, a packet used in flow control. A node detecting congestion generates a *choke packet* and launches it toward the source; the source responds by reducing its transmission rate.

chromatic dispersion In fiber optics, one of the mechanisms that reduces the usable bandwidth of an optical fiber. It is the dispersion of an optical signal because of the different speeds of propagation of light at different wavelengths; also known as *material dispersion.* Those wavelengths around which dispersion is minimal, such as those around 1550, 1300, or 830 nanometers (nm), are commonly used for transmission.

CI (1) An abbreviation of Connect Indication. **(2)** An abbreviation of Connection Identifier. A frame or cell address. **(3)** An abbreviation of Copy Inhibit.

CIB An abbreviation of CRC Indication Bit. A flag bit in SMDS that indicates if the CRC field is used (flag = 1) or not used (flag = 0).

CIC (1) An abbreviation of Carrier Identification Code. (2) An abbreviation of Circuit Identification Code. (3) An abbreviation of CSNet's Coordination and Information Center.

CICS An abbreviation of Customer Information Control System.

CICNet An acronym from Committee on Institutional Cooperation Network.

CIDR An abbreviation of Classless InterDomain Routing.

CIF (1) An abbreviation of Cell Information Field. The 48-octet payload in each ATM cell. (2) An abbreviation of CCITT's Common Intermediate Format. An ISDN videophone standard contained in the ITU H.261 recommendations. It produces a 30 frame per second, 288 line per frame, 352 pixel per line, color image on an ISDN channel. The format uses both B-channels (for a bandwidth of 128 kbps) partitioned into 32 kbps for voice and the remainder for picture.

CIG (1) An abbreviation of Commercial Internet exchanGe. (2) An abbreviation of Call Interconnection Gateway.

CIGOS An acronym from Canadian Interest Group on Open Systems.

CIIG An abbreviation of Canadian ISDN Interest Group.

CIM (1) An abbreviation of Computer-Integrated Manufacturing. (2) An abbreviation of Common Information Model.

cipher (1) Any cryptographic system in which arbitrary symbols, or groups of symbols, represent units of plaintext of regular length, usually single letters, or in which units of plaintext are rearranged, or both, in accordance with specified rules. (2) The resultant message after using a cipher. For example, an enciphered message is a *cipher*.

Cipher Block Chaining (CBC) An operating mode of the Digital Encryption Standard (DES).

Cipher Feedback (CFB) An operating mode of the Digital Encryption Standard (DES).

ciphertext In cryptography, text (plaintext) that has been encrypted (coded) to make it unintelligible to all except the intended recipient; that is, it is unintelligible to anyone who lacks necessary information about the encryption scheme. The required information is generally a specific value, known as a decryption key. Conventional, public, or private key encryption methods are used to create *ciphertext*. See also *plaintext*.

ciphony A contraction of CIpher telePHONY. The process of enciphering audio information.

CIR An abbreviation of Committed Information Rate. The minimum throughput rate guaranteed by a frame relay carrier. See also *explicit congestion notification* (*ECN*).

CIRC An abbreviation of Cross Interleave Reed-Soloman Code, an error-correcting code that tolerates high burst error blocks.

circuit (1) Any path that can carry an electrical current. (2) An interconnection or network of electrical elements; such as resistors, capacitors, and transistors. (3) In communications, a connection between a sending station and a receiving station that allows information to be transferred in either one or both directions.

circuit breaker (CB) A type of overcurrent protection. It is similar to a fuse in that it opens the electrical path when a current exceeds a specified level. It differs, however, in that the fuse must be replaced to reestablish the connection, while a *circuit breaker* may be reset and reused. *Circuit breakers* may be either manually resettable or self-resetting. Self-resetting devices may reset based on time delay, temperature, or removal of the faulty condition.

Circuit breakers in general are used to protect power sources. Whereas fuses are used to protect the load. See also *fuse*.

Circuit Identification Code (CIC) That part of Signaling System Number 7 (SS7) used to identify the circuit being established between two exchanges.

circuit noise level At any point in a transmission system, the ratio of the *circuit noise* at that point to a specified reference level. *Circuit noise level* is usually expressed in either dBrn0 or dBa0.

circuit reliability (CiR) The percentage of time a circuit was available for use in a specified period of scheduled availability. *Circuit reliability* is given by

$$CiR = 100 \left(1 - T_o/T_s\right) = 100\, T_a/T_s$$

where:

T_o is the circuit total outage time,
T_a is the circuit total available time, and
T_s is the circuit total scheduled time $T_s = T_a + T_o$.

Also called *time availability*.

circuit switched cellular (CSC) A cellular telephone link established for the duration of a transaction, then released for use by another caller. A caller is charged only for the time the connection is established.

circuit switched network A network that establishes a physical circuit temporarily between two users upon demand and releases the physical connection when a disconnect signal is received.

Contrast with *packet switched network*.

circuit switching A method of establishing a connection from a sending station to a specific but arbitrary receiving station utilizing spatially discrete paths. The connection between the sender and receiver is accomplished in switching centers where a fixed bandwidth is dedicated to the call for as long as both parties require it. At the end of the requirement, the switching equipment releases the circuits and makes them available to other stations. The PSTN is an example of a *circuit switching* system. See also *connection oriented, dedicated circuit, message switching*, and *packet switching*.

circuit transfer mode In ISDN applications, a transfer mode by means of permanent allocation of channels or bandwidth between connections.

circular error probability (CEP) A measurement of position accuracy in navigation. Any circle with a sufficiently large radius such that there is a 50% probability that the true horizontal location is located within the circle.

For example, a particular survey of a location might produce a plot such as is shown in the following figure. In this plot, 1000 measurements were made and a circle was drawn so that the center was at the true location and the radius was just large enough that the circle encompassed 500 measurements. *Note:* Many other circles could be chosen to satisfy the *CEP* criterion.

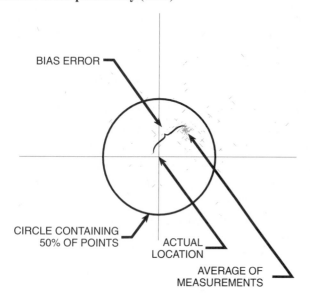

From the plot, it can be observed that the *average value* need not coincide with the *actual value*. The difference between the actual value and some value representing the set of measured values is called a *bias error*.

circular hunting In telephony, a particular method of searching a *hunt group* for an available line number. Specifically, the search starts just after the last line number used and sequentially searches to the end of the list and then continues to search starting at the list beginning. When an available number is found, it becomes the predecessor of the new starting point for the next search.

circular mil (cmil) A measure of the cross-sectional area of wire. Mathematically, it is the square of the diameter of the wire, where the diameter is expressed in mils (thousandths of an inch).

$$1 \; cmil = 0.7854 \; mil^2 = 785.4 \times 10^{-3} \; in^2$$
$$1974 \; cmil = 1 \; mm^2$$

See also *AWG, cable,* and *wire.*

circular polarization Electromagnetic radiation in which the electric vectors can be broken into two perpendicular elements that have equal amplitude and differ in phase by 1/4 wavelength (90°). Properties of circular polarization include:

- The tip of the electric field vector describes a helix as the wave propagates through space.
- The magnitude of the electric field vector is constant.
- The projection of the tip of the electric field vector upon any fixed plane intersecting, and normal to, the direction of propagation describes a circle.
- Circular polarization may be referred to as "right-hand" or "left-hand," depending on whether the helix describes the thread of a right-hand or left-hand twist.

circulator A passive device with three or more branches (ports) designed so that power applied to one branch is transferred exclusively to the next branch in sequence. It is the first branch following the last branch in order so as to form a circular appearance.

Circulators are commonly used in microwave systems. For example, in radar systems a circulator is used to switch the antenna alternately between the transmitter and receiver.

CIS An abbreviation of Compuserve Information Service.

CISC An acronym from Complex Instruction Set Computer. A processor design strategy that provides the processor with a large number of basic instructions, many of which are complex and very powerful. These complex instructions generally require several clock cycles to complete, which can slow overall processing.

citizen's band (CB) Radio frequencies set aside by the FCC for use by the general public. It allows private, two-way, short-distance voice communications by individuals or businesses on any of the available 40 carrier frequencies between 26.965 MHz and 27.405 MHz and additional channels in the 462.5375- to 467.7375-MHz band.

CB FREQUENCY ASSIGNMENTS

			(27 MHz band)				
CH	Freq	CH	Freq	CH	Freq	CH	Freq
1	26.965	11	27.085	21	27.215	31	27.315
2	26.975	12	27.105	22	27.225	32	27.325
3	26.985	13	27.115	23	27.235	33	27.335
4	27.005	14	27.125	24	27.245	34	27.345
5	27.015	15	27.135	25	27.255	35	27.355
6	27.025	16	27.155	26	27.265	36	27.365
7	27.035	17	27.165	27	27.275	37	27.375
8	27.055	18	27.175	28	27.285	38	27.385
9	27.065	19	27.185	29	27.295	39	27.395
10	27.075	20	27.205	30	27.305	40	27.405

Channel 9 (27.065 MHz) may be used only for emergency or traveler assistance transmissions. The transmit power is limited to 4 watts for an AM carrier and 12 watts for single sideband (SSB). Restrictions apply to antenna location and height. The complete requirements are specified in the *U.S. Code of Federal Regulations* 47 Part 95.

city code (CC) In telephony, that part of the telephone routing code that informs the public switch telephone network's (PSTN's) high-level switches to which group of end offices the caller should be connected. See also *area code, country code, international access code, local access code,* and *local number.*

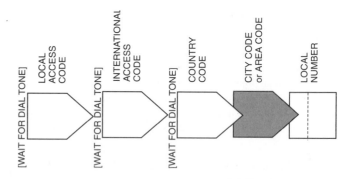

CIU An abbreviation of Communications Interface Unit.

civision A contraction of CIpher teleVISION. (1) The application of cryptography to television signals. (2) A television signal that has been enciphered to preserve the confidentiality of the transmitted information.

CIX An abbreviation of Commercial Internet eXchange (pronounced *kicks*). An agreement among Internet service providers which allows them to account for commercial network traffic.

CKT An abbreviation of circuit or circuit switching.

CL An abbreviation of ConnectionLess.

CL-TK An abbreviation of <u>CL</u>aim <u>To</u><u>K</u>en

cladding **(1)** In fiber optics, the dielectric material surrounding the core of an optical waveguide. The cladding has a lower index of refraction than the core, which means that light hitting the cladding at angles greater than a critical *angle of incidence* will be reflected back into the core to continue its path along the cable. **(2)** In electrical cables, a process of covering the cable with metal.

cladding diameter In fiber optics, the average of the diameters of the smallest circle that can be circumscribed about the cladding, and the largest circle that can be inscribed within the cladding. The cross section of a realizable optical fiber is generally not circular but can be assumed to be approximately elliptical as is illustrated below.

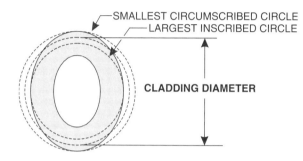

cladding eccentricity See *ovality.*

cladding glass In fiber optics, the glass (or plastic) that surrounds the core of an optical waveguide fiber. The cladding has a lower index of refraction than the core.

cladding mode In fiber optics, an undesired mode in which a light ray is confined to the cladding of an optical fiber because the cladding has a higher refractive index than its surrounding medium (air or the primary polymer overcoat). To alleviate this problem, a primary polymer overcoat with a refractive index that is slightly higher than that of the cladding is used. This *strips off* cladding modes after a short distance of propagation. See also *cladding mode stripper* and *optical fiber.*

cladding mode stripper A device for converting optical fiber cladding modes to radiation modes; as a result, the cladding modes are removed from the fiber. One implementation is the application of a fiber coating in intimate contact with the cladding and with a refractive index greater than the cladding. This generally strips the unwanted mode within several centimeters of propagation.

cladding ray See *cladding mode.*

clamp An electronic circuit or device that limits the amplitude excursions of a signal. *Clamps* may be either hard limiting or soft limiting.

- Hard limiting *clamps* simply "chop" off the top of any waveform's excursion beyond that which the clamp allows (see also *clipping*). Hard limiting clamps are frequently used as transient protection devices. When used as a protector, the majority of the energy is absorbed within the device and then dissipated over time.

- Soft limiters reduce the gain of the transmission device as the signal amplitude increases beyond the *clamp* threshold. Soft limiting *clamps* tend to produce less harmonic distortion than do hard limiting *clamps*. For this reason they are used in linear transmission systems.

Also called a *clipper* or *limiter.*

clamping time In transient protection, the amount of time required for a surge protector to turn on and limit a voltage spike or surge to acceptable levels.

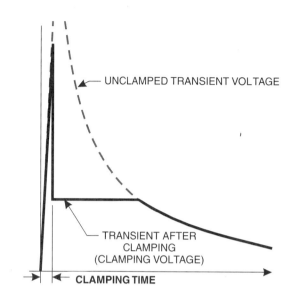

clari One of several USENet alternative newsgroups. Clari newsgroups carry the moderated postings of ClariNet (a for-profit organization). All of the newsgroups, with the exception of *clari.news.talk,* accept postings only from "qualified news organizations" such as United Press International (UPI) or other national and international news services. Some of the *clari* newsgroups include:

clari.biz.economy	business economy
clari.biz.top	top business news
clari.biz.urgent	very hot business news
clari.nb.news	news about computers
clari.news.top	top news of the day.

Any site wishing to carry the *clari* newsgroups must pay a fee and sign a licensing agreement. Because the material is copyrighted, it cannot be redistributed. See also *ALT* and *alternative newsgroup hierarchies.*

CLASS An acronym from <u>C</u>ustom <u>L</u>ocal <u>A</u>rea <u>S</u>ignaling <u>S</u>ervices. For example, ANI or tracing.

class 1 A fax/modem command set standard in which most of the protocol and image processing is done by the computer. The fax/modem handles only the basic communication protocols (modulation and data conversion). The *class 1* standard was developed jointly by the Electronics Industry Association (EIA) and the Telecommunications Industry Association (TIA) as the EIA/TIA-578 standard. See also *class 2, class 2.0,* and *Group 1-4.*

class 2 A fax/modem interim command set standard in which most of the ITU-T T.30 protocol processing is done by the fax/modem. The computer only provides the image in the proper format and manages the session.

Built on the assumption that a personal computer did not have enough signal processing power to support *class 1,* the EIA and TIA jointly developed the *class 2* fax/modem standard. The idea was that the modem could provide the needed auxiliary power to digitize and compress the images. The problem is, the interim standards were implemented by several manufacturers and they were somewhat different. Not only were these standards slightly different, but the manufacturers inserted some proprietary differences as well. The official version (version 2.0) of the standard was released in 1992, and it was again somewhat different from that of its predecessors. As a result, not all *class 2* modems perform the same. See also *class 1, class 2.0,* and *Group 1-4.*

class 2.0 The official class 2 fax/modem command set standard in which most of the ITU-T T.30 protocol processing is done by the fax/modem. The standard was developed jointly by the EIA and TIA and released in "final form" in 1992 as the EIA/TIA-592 standard. See also *class 1, class 2,* and *Group 1-4.*

class 3 A proposed fax/modem class that will turn over the entire task of creating and transmitting a fax to the modem. That is, it will handle both the ITU-T T.30 and T.4 tasks.

class 4 Fax/modems that have buffers, which allow the CPU to go for a short time without responding to the fax data stream.

class A amplifier A type of analog amplifier in which the output device conducts current at all points on the waveform as shown in the figure below. The amplifier is biased midway between cutoff and saturation and is said to have a *conduction angle* of 360°. Although the instantaneous current in the output device does vary as the signal amplitude changes, the average current is the same as the "no signal" current, i.e., near 50%. *Class A amplifiers* generally introduce very little distortion into the output signal. However, because they conduct large currents even when no signal is present, they are very inefficient (50% maximum theoretical efficiency). See also *class AB, class B, class C,* and *class D amplifiers.*

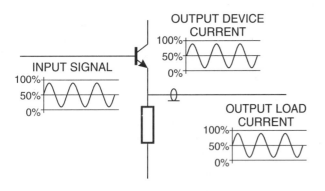

class A certification See *FCC.*

class A network The highest level in the Internet network hierarchy. The current *IP address* specifications allow 128 *class A networks,* each of which can have 16 777 215 hosts in lower levels of the hierarchy. The *IP address* of a *class A network* is in the range of 0.XXX.XXX.XXX and 127.XXX.XXX.XXX. *Class A networks* are reserved for very large institutions such as government agencies, universities, and regional servers. See also *IP address.*

class AB amplifier A type of analog amplifier that attempts to solve the high-power dissipation of *class A amplifiers* and the high *crossover distortion* of *class B amplifiers.* To solve these problems, the output devices of a *class AB amplifier* each conduct a small current at the zero signal point. The conduction angle of each output device is therefore between 180° and 360°. This has the effect of smoothing the signal in the region of zero crossing; hence, it reduces distortion. The region of overlap is shown highlighted on the output load current waveform in the following figure.

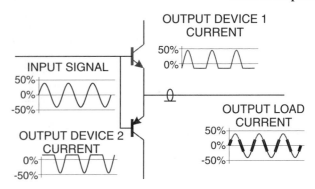

There are various subclasses within *class AB* designated class AB1 and AB2. Although originally defined in terms of the bias on the control grid of a vacuum tube, it can be thought of as a percentage of overlap of the conduction; that is, AB2 has a greater overlap than does AB1. See also *class A, class B, class C,* and *class D amplifiers.*

class B amplifier A type of analog amplifier in which the output devices conduct current only during excursion above a reference level. Generally, *class B amplifiers* are arranged with two output devices; one conducts when the signal is above 50% of a reference and the other when the signal is below 50%. Each output device is biased at the cutoff point and is said to have a conduction angle of 180°. When no signal is present, both device currents are 0 amp. As the signal amplitude increases, the excursions from 0 increase. *Class B amplifiers* have a theoretical maximum efficiency of 78.5% and generally introduce very little distortion except at the point where one output device turns off and the other device turns on (called *crossover distortion*). See also *class A, class AB, class C,* and *class D amplifiers.*

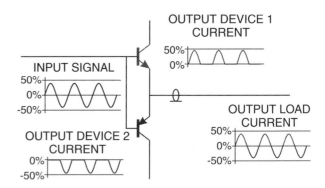

class B certification See *FCC.*

class B network In the Internet network hierarchy, the current *IP address* specifications allow 16 384 *class B networks,* each of which can have 65 536 hosts in lower levels of the hierarchy. The *IP address* of a *class B network* is in the range of 128.XXX.XXX.XXX and 191.XXX.XXX.XXX. *Class B networks* are reserved for organizations and businesses with the potential for significant growth. See also *IP address.*

class C amplifier A type of amplifier in which the amplifier's output device conducts current for periods less than 50% of the input signal cycle. When no signal is present, the output device currents are 0 amps. As the signal amplitude increases above the threshold, the excursions from 0 increase. *Class C amplifiers* produce a significant amount of waveform distortion even when arranged as a push-pull output. However, their efficiency in carrier devices is very high, and the quality of the waveform is generally of little significance.

If given a sinewave input, the output device will conduct current over less than 50% of the waveform. In the figure below, the threshold is set to about 65% of the input; therefore, the output will conduct only 35% of the time. See also *class A, class AB, class B,* and *class D amplifiers.*

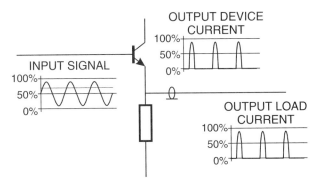

class C network In the Internet network hierarchy, the current *IP address* specifications allow 2 097 152 *class C networks,* each of which can have 256 hosts. The *IP address* of a *class C network* is in the range of 192.XXX.XXX.XXX and 221.XXX.XXX.XXX. *Class C networks* are reserved for small businesses, schools, bulletin board systems (BBSs), and others with direct Internet connections. See also *IP address.*

class D amplifier A type of amplifier in which the output devices are used as switches; that is, they are either turned on or off. The input to these switches is a pulse width modulated (PWM) representation of the analog signal to be "amplified." Following the amplifier is a low-pass filter that removes the PWM carrier "switching noise." To be effective, the PWM carrier frequency must be much higher than the highest frequency to be amplified.

The following figure shows the waveforms at various points of a *class D* system. Also called a *class S amplifier, pulse width modulation amplifier,* or *sampling amplifier.* See also *class A, class AB, class B,* and *class C amplifiers.*

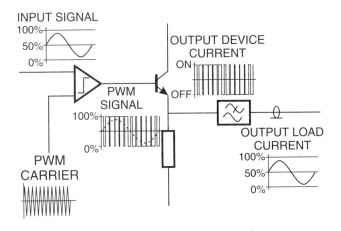

class n office The designation of various levels of the hierarchy of the PSTN in the United States. Each level allows connection to adjacent class offices along routes of heaviest traffic flow while providing redundant paths and auxiliary paths for overflow and infrequent connections.

Class 5 office, the *local central office,* or the *end office* (*EO*) is the switching center to which subscriber lines are connected and access to the hierarchical switch network begins.

Class 4 office or *toll center* (*TC*) interconnects class 5 offices in the local area, other class 4 offices, and extends the hierarchical access to class 3 offices. If operators are present, the center is called a toll center; otherwise, it is a *toll point* (*TP*).

Class 3 office or the *primary center* (*PC*) interconnects class 3 and 4 offices while further extending the hierarchical network to the class 2 office.

Class 2 office or the *sectional center* (*SC*), interconnects class 1, 2, and 3 offices.

Class 1 office or the *regional center* (*RC*) interconnects class 1 and 2 offices.

Any selected center handles traffic from one level to two or more lower levels in the hierarchy. In specific cases, special high-usage trunks may be installed between certain centers. Since divestiture, however, these designations have become somewhat blurred. The telephone network

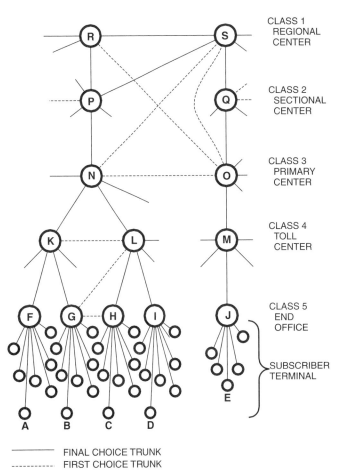

switching hierarchy is shown in the drawing. The basic structure is shown in solid paths (*final choice trunks* or paths) and *high-usage trunk groups* or auxiliary paths shown dashed. High-usage paths are installed between selected offices based on the amount of traffic between the two offices (e.g., between offices G and H, K and L, N and O, etc.). The switching or routing algorithm is based on making the connection from the calling to called subscriber station via the shortest path and through the lowest hierarchy level possible, i.e., going up through higher class offices only when no other alternative is available. Utilization of a high-usage trunk at the same hierarchical level is the first choice path if the called party is not within the calling party office. The second choice path is a high-usage trunk to the next higher class office of the called party hierarchy. And the final choice path is to the next higher office level of the calling party.

Example 1: A caller at station A dials the number of subscriber B; the switch F cannot connect directly, so the routing is passed on to the next higher office (K). Office K then routes the call on to office G, which then establishes the final link to subscriber B.

Example 2: Subscriber B dials subscriber C. End office G first attempts a connection to end office H; if the attempt fails because the trunk is busy, a routing attempt via the high-usage trunk to toll center L is made. If this also fails, the final choice path to toll center K is selected. From toll center K the first choice path is again a high-usage trunk to toll center L, with the final choice path to the primary center N then to L. From toll center L the path is routed to local office H and finally to subscriber C.

Example 3: Subscriber A dials subscriber E. The first choice path is A-F-K-N-O-M-J-E. Failing this at the N-O high-usage trunk, a second choice path is attempted, i.e., A-F-K-N-S-O-M-J-E. (A fallback from regional center S is via sectional center Q.) The theoretical final route (TFR) is A-F-K-N-P-S-Q-O-M-J-E.

class of emission A means of classifying transmitted signals by a set of three (or more) characters. The first character identifies the *type of modulation* used, the second character describes the *nature of the modulating signal* or signals, and the third character denotes the *type of information* to be transmitted. Optional additional characters may be used to refine the description.

For example, A3E describes an amplitude-modulated carrier with a single analog information channel of telephone sound quality. The following table lists the emission class codes defined by the ITU-T.

EMISSION CLASS CODE DEFINITIONS

Modulation on Main Carrier	Symbol
Unmodulated	N
Amplitude modulated (AM) double sideband	A
Amplitude modulated — SSB full carrier	H
Amplitude modulated — SSB reduced carrier	R
Amplitude modulated — SSB suppressed carrier	J
Amplitude modulated — SSB independent sidebands	B
Amplitude modulated — SSB vestigial sideband	C
Frequency modulation (AM)	F
Phase modulation (PM)	G
AM and PM either together or sequential	D
Unmodulated pulses	P
Amplitude modulated pulse train	K
Pulse width (duration) modulated pulse train	L
Pulse position modulated pulse train	M
Combination of K, L, and/or M or other pulse means	Q
Other pulse modulation methods	V
Combination of amplitude, angle, and/or pulse not above	W
Any other modulation method	X

Nature of Modulating Signal(s)

No modulating signal	0
Single digital channel w/o modulating subcarrier	1
Single digital channel with modulating subcarrier	2
Single analog channel	3
Multiple digital channels	7
Multiple analog channels	8
Composite analog & digital, single or multiple channels	9
Any other signal	X

EMISSION CLASS CODE DEFINITIONS (*CONTINUED*)

Type of Information Transmitted	Symbol
No information	N
Telegraphy for aural reception	A
Telegraphy for automatic reception	B
Facsimile	C
Data transmission, telemetry, . . .	D
Telephony and sound broadcasting	E
Television video	F
Combination of above	W
Any other	X

Signal Details (optional)

2 condion code with differing numbers and/or duration	A
2 condion code with same numbers and/or duration no EC	B
2 condion code with same numbers and/or duration with EC	C
4 condition code, each representing a symbol of 1 or more bits	D
Multicondition code, each representing a symbol of 1 or more	E
Multicondition code, each representing a character	F
Mono sound, broadcast quality	G
Stereo or quadraphonic, sound broadcast quality	H
Other commercial quality sound	J
Commercial quality sound with band splitting or frequency inversion	K
Commercial quality sound with separate FM signal to control demodulation levels	L
Monochrome	M
Color	N
Combination of above	W
Any other	X

Multiplexing Nature (optional)

None	N
Code division and spread spectrum	C
FDM	F
TDM	T
Combination of FDM and TDM	
Other multiplex methods	X

class of service (**1**) The privileges and features of a telephone or group of telephones in a system. Examples include:

- Local calls only.
- WATS accessibility allowed.
- National call only.
- International call allowed.
- All calls except 900 numbers and so on.

Also called *user class of service.* (**2**) A category of data transmission defined in the ITU X.1 recommendations. It provides a standardized structure for data signaling rates, terminal operating modes, and so forth, in a public data network (PDN). (**3**) A designation assigned to describe the access rights and privileges given to a particular terminal. (**4**) The categories of ATM traffic types, e.g., real vs. non-real time and CBR vs. VBR.

class of service mark See *classmark.*

Classless InterDomain Routing (CIDR) In Internet, a method of using the existing 32-bit Internet address word more efficiently. A proposal adopted by the Internet Engineering Task Force (IETF) that provides for the creation of large networks on the user side of the gateway router. See also *subnet* and *subnet address.* Also known as *subnetting.*

classmark A designator used to describe the service feature privileges, restrictions, and circuit characteristics for the lines and/or trunks that access a communications switch.

Examples of classmarks include conference privilege, priority level, security level, and zone restriction. Also called *class of service mark* and *toll restriction.*

clear To cause one or more storage locations to be forced to a prescribed state or value, usually that value corresponding to a zero, null, or the space character. Also called *reset.*

clear channel (1) In radio broadcasting, a frequency assigned to one station for its near exclusive use. For example, in the United States, an AM radio station that has exclusive rights of its frequency and a transmitting power of 50 kW does not have to reduce its power at night and is protected from interference by other stations for distances of 750 miles or greater. (2) In networking, a signal path that provides its full bandwidth for a user's service. No control or signaling is performed on this path.

Examples include:

- A T-1 circuit in which all 192 payload bits of each frame are available; that is, a 1.536-Mbps channel is available.
- An E-1 circuit in which either 1.920 Mbps or 1.984 Mbps is available.
- A SONET frame in which the synchronous payload envelope (SPE) is assigned to a single channel yielding 50-Mbps capacity.

clear collision In packet networks, contention that occurs when simultaneously:

- A data terminal equipment (DTE) transmits a *clear request packet.*
- A data circuit terminating equipment (DCE) transmits a *clear indication packet.*
- Both specify the same logical channel.

The DCE will consider that the clearing is completed and will not transmit a *DCE clear confirmation packet.*

clear confirmation signal A call control signal used to acknowledge either

- The receipt of the data terminal equipment (DTE) *clear request* by the data circuit terminating equipment (DCE), or
- The reception of the DCE *clear indication* by a DTE.

clear text In cryptography, intelligible data. An unencrypted message. Also called *clear message* and *plaintext.*

clearing (1) A sequence of events used to disconnect a call and return equipment and transmission links to the ready or idle state. (2) Erasing data from an automated information system (AIS), its storage devices, and other storage devices associated with its peripherals, in such a way that the data may not be reconstructed using "normal" system capabilities.

Clearing does not necessarily remove all traces of the data; hence, it does not allow the system to be moved from a classified to a non-classified environment. It does allow its reuse within the secure facility.

clearing center (CC) In electronic data interchange (EDI), a message switching element through which documents are passed on the way to their destinations.

cleave (1) In fiber optics, a deliberate, controlled break, ideally with a perfectly flat endface, perpendicular to the longitudinal axis of the optical fiber.

A good cleave is required when:

- Splicing by mechanical or fusion means (or coupling losses will occur).
- Terminating with some connector types. Some types of fiber optic connectors do not employ abrasives and polishers. Instead, they use some type of cleaving technique to trim the fiber to its proper length and produce the required smooth, flat perpendicular endface.

The procedure to cleave a fiber is basically a two-step process:

- First, a microscopic fracture ("nick") on the surface of the fiber is created with a special tool, called a *cleaving tool.* The tool has a sharp blade of hard material, such as diamond, sapphire, or tungsten carbide.
- Finally, tension is applied to the fiber (either as the nick is made or immediately afterward) to force the fracture to propagate in the desired fashion. Some tools do this automatically; others require it be done manually.

See also *coupling loss.* (2) To break a fiber in such a controlled fashion as described above.

CLI An abbreviation of Calling Line Identification.

CLID (1) An acronym from CaLler ID. (2) An acronym from Calling Line IDentification.

client (1) A program (or computer) that is dependent on another program (or computer) for information, resources, or services. The program (or computer) on which it is dependent is called the *host* or *server.* A *client device* does not share any of its resources with the other users of the network.

For example, workstations are network clients because they use services and/or resources of the server. As another example, a client application is an application that makes requests of other applications, on the same or on different machines, for its services, information, or resources. (2) In ISO/OSI specifications, a *client* is simply a service requestor.

CLIENT.EXE The NetWare Lite program, which allows users to access the network as a client.

client–server computing A distributed computing network system in which transactions are divided into two parts—a front end (the client part), and a back end (the server part). A *client* is defined as a requester of services, and a *server* is defined as the provider of services. This definition does not preclude one machine from being both a server and a client.

Functions of these two parts include:

- Clients are defined as a requester of services. They may be personal computers (PCs) or workstations on which user applications are run. They rely on servers for resources such as files, devices, and even computing power.
- Servers are defined as the provider of services. They are generally more powerful computers than client machines. They provide services such as managing disk drives (file servers), printers (print servers), network traffic (network servers), and application resources.

The term was first used in the early 1980s in reference to PCs on a network. By the late 1980s the model started gaining acceptance. The *client–server computing* architecture is message-based and generally improves performance, usability, flexibility, interoperability, and scalability as compared to mainframe–monitor or file server computing. The *client–server computing* model introduced a database server to replace the file server of file-sharing architectures. Using a relational database management system (DBMS), one can answer user queries directly. This reduces network traffic by providing a query response rather than total file transfer.

client–server network A network of computers in which at least one computer requires the services, resources, or information of another in order to function. A *client-server network* may be any physical structure (star, ring, bus, etc.).

To be a true client–server application, real work (information processing) must be done in the server. For example, a database query is *processed* by the server, and only the results are returned to the client program. See also *peer-to-peer network*.

CLIP An acronym from <u>C</u>alling <u>L</u>ine <u>I</u>dentification <u>P</u>resentation.

clipper A circuit or device that limits the instantaneous output signal amplitude to a predetermined maximum value, regardless of the amplitude of the input signal. See also *clipping*.

Clipper chip Officially known as the MYK-78, it is a low-cost data encryption device proposed by the U.S. government for public use. It would allow businesses and private parties to send and receive confidential messages. However, it is a *key-escrow system* in which the police and government have access to a set of keys that allow them to decrypt *any message sent by anyone!* See also *pretty good privacy (PGP)*.

clipping In communications circuits, the loss of either amplitude- or time-related information due to imprecise circuit operation or excessive signal level (overload). **(1)** *Amplitude related.* A type of signal *distortion* where the peaks of a signal are cut off. Generally, this is caused by equipment attempting to transmit a signal with an amplitude larger than the signal-handling device or circuit is designed to handle. Positive peaks, negative peaks, or both may be clipped in any given situation.

The figure below displays a two-unit peak-to-peak amplitude sine wave with clipping on the positive peak at 0.75 unit. **(2)** *Time re-*

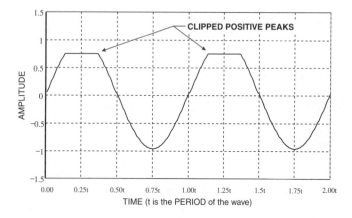

lated. The loss of the initial or final parts of a word or phrase due to, for example, the nonideal operation of voice-operated switches (VOX). **(3)** *In a display device.* The removal of those parts of display elements that lie outside of a given boundary.

CLIR An acronym from <u>C</u>alling <u>L</u>ine <u>I</u>dentification <u>R</u>estriction.

CLLM An abbreviation of <u>C</u>onsolidated <u>L</u>ink-<u>L</u>ayer <u>M</u>anagement.

CLNP An abbreviation of <u>C</u>onnection<u>L</u>ess mode <u>N</u>etwork <u>P</u>rotocol.

CLNS An abbreviation of <u>C</u>onnection<u>L</u>ess-Mode <u>N</u>etwork <u>S</u>ervice.

clock (1) A device used to maintain (and possibly display) date and/or time. **(2)** *System clock.* A periodic *signal* used to synchronize digital data transfers or circuit actions. An independent timekeeping *device* or *circuit* within an electronic apparatus used to synchronize digital actions.

In the following diagram, the clock signal can be used to transfer the data signal on the *rising edge* (also called the low-to-high transition). The *falling edge* cannot be used (in this example) because it is uncertain what the value of the data signal is at that point.

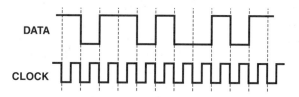

clock bias The difference between a clock's indicated time and a specific time reference (for example, Universal Time). Subtracting the clock difference from the indicated time yields the reference time. Also called *clock error.*

clock phase slew The rate of relative phase change between a given clock signal and a stable reference signal. The two signals are generally at the same frequency or have an integral multiple frequency relationship.

clock rate A synonym for *clock speed.* The rate at which the system clock in an electronic device operates (the oscillator rate).

clock recovery The process of extracting a clocking signal from a data stream for use in a synchronous communications system.

Clock and data are combined at the sending terminal using one of many coding methods. This combined signal is sent across the network where the receiving terminal separates them. See also *clock* and *waveform code.*

clock speed The number of time slices per second into which a process is divided. *Clock speed* is generally expressed in millions of cycles per second (megahertz, or MHz) or pulses per second (pps). For example, the CPU of the original IBM PC had a clock speed of 4.77 MHz. This, however, is slow when compared with today's processors, with clock speeds that exceed 500 MHz. Also called *clock rate.*

clockwise polarization A circular (or elliptically) polarized electromagnetic wave in which the direction of rotation of the electric vector is clockwise (as observed looking in the direction of propagation). Also called *right-hand polarization.* See *circular polarization.*

close talking microphone An acoustic transducer (microphone) designed specifically to operate close to a speaker's mouth, generally less than 4 inches away. The purpose of the design is to reduce the pickup of extraneous background noise.

closed circuit (1) In radio and television, pertaining to an arrangement in which information or programming is directly transmitted to specific terminals or users and not broadcast to the general public. **(2)** In telecommunications, a circuit dedicated to specific users. **(3)** In electrical circuits, a continuous unbroken path in which current can flow, i.e., a completed electrical circuit.

closed user group In a network, a group of users permitted to communicate with each other but not with users outside the group. A user data terminal equipment (DTE) may belong to more than one closed user group.

closed user group with outgoing access A closed user group in which at least one member of the group has a facility that permits communication with one or more users external to the closed user group.

cloud The phone company name for a digital connect service which more closely resembles a LAN than a switched system. To send a message, the user puts the address of the destination on the header of the packet and launches it. The service reads the address and routes it to the proper destination. As with a LAN, user terminals are always connected and "set up," thus, eliminating the dial and switching delays associated with the POTS system. Services such as frame relay use this *cloud* designation. Also called *connectionless networks.*

cloud attenuation In electromagnetic signal transmission, attenuation caused by absorption and scattering by water or ice particles in clouds. The amount of cloud attenuation depends on factors such as

- The transmission frequency.
- The density, particle size, and turbulence of the clouds.
- The transmission path length in the cloud.

CLP An abbreviation of <u>C</u>all <u>L</u>oss <u>P</u>riority.

CLR An abbreviation of <u>C</u>ell <u>L</u>oss <u>R</u>atio.

CLS (1) An abbreviation of <u>C</u>onnection<u>L</u>ess <u>S</u>erver. (2) An abbreviation of <u>CL</u>ear <u>S</u>creen.

CLSF An abbreviation of <u>C</u>onnection<u>L</u>ess <u>S</u>ervice <u>F</u>unctions.

CLTP An abbreviation of <u>C</u>onnection<u>L</u>ess mode <u>T</u>ransport <u>P</u>rotocol.

CLTS An abbreviation of <u>C</u>onnection<u>L</u>ess <u>T</u>ransport <u>S</u>ervice.

CLU An abbreviation of <u>C</u>ommand <u>L</u>ine <u>U</u>tility.

cluster (1) In networks, a group of input/output (I/O) devices, such as terminals, computers, or printers, that share a common communication path to a host machine. Communications between the devices in a *cluster* and the host are generally managed by a *cluster controller.* (2) A group of separate computer servers that work together to process data and handle transactions on a network. For cluster processing to take place, the network operating system (NOS) must know how to distribute the tasks. *Clusters* need high-capacity network connections between individual server nodes to smooth data flow. (3) See *allocation unit.*

cluster controller A device that allows multiple 3270 terminals to be linked directly to a host computer, or into a SNA network through a communications controller. Also called a *control unit* in IBM circles.

CLV An abbreviation of <u>C</u>onstant <u>L</u>inear <u>V</u>elocity. Describes a LaserDisc with information stored in a linear or straight-line fashion. In a typical operation, the disc plays from beginning to end with minimal interaction. Using software or other control, limited random access can be accomplished. See also *CRV.*

CM An abbreviation of <u>C</u>onfiguration <u>M</u>anagement.

CMI (1) An abbreviation of <u>C</u>oded <u>M</u>ark <u>I</u>nversion. A line signal for STS-3. (2) An abbreviation of <u>C</u>onstant <u>M</u>ark, <u>I</u>nverted. A line coding method for a T-1 local loop in Japan.

CMIP An abbreviation of <u>C</u>ommon network <u>M</u>anagement Information <u>P</u>rotocol.

CMIP/CMIS An abbreviation of <u>C</u>ommon <u>M</u>anagement <u>I</u>nformation <u>P</u>rotocol/<u>C</u>ommon <u>M</u>anagement <u>I</u>nformation <u>S</u>ervices.

CMIP/CMIS over TCP (CMOT) The use of ISO CMIP/CMIS network management protocols to manage gateways in a TCP/IP Internet. *CMOT* is a co-recommended standard with SNMP.

CMIPDU An abbreviation of <u>C</u>ommon <u>M</u>anagement <u>I</u>nformation <u>P</u>rotocol <u>D</u>ata <u>U</u>nit.

CMIPM An abbreviation of <u>C</u>ommon <u>M</u>anagement <u>I</u>nformation <u>P</u>rotocol <u>M</u>achine.

CMIS An abbreviation of <u>C</u>ommon network <u>M</u>anagement Information <u>S</u>ervice.

CMISE An abbreviation of <u>C</u>ommon <u>M</u>anagement <u>I</u>nformation <u>Se</u>rvice <u>E</u>lement.

CML An abbreviation of <u>C</u>urrent-<u>M</u>ode <u>L</u>ogic.

CMMU An abbreviation of <u>C</u>ache <u>M</u>emory <u>M</u>anagement <u>U</u>nit.

CMOS An acronym from <u>C</u>omplementary <u>M</u>etal <u>O</u>xide <u>S</u>emiconductor (pronounced "see-moss"). A type of transistor and circuit that provides medium speed and very low power dissipation. On the negative side, CMOS tends to be damaged by static discharge. See *transistor.*

CMOT An acronym from <u>CM</u>IP <u>O</u>ver <u>T</u>CP/IP. A proposed Internet standard defining the use of CMIP for managing TCP/IP LANs.

CMR (1) An abbreviation of <u>C</u>ommon <u>M</u>ode <u>R</u>ejection. (2) An abbreviation of <u>C</u>ellular <u>M</u>obile <u>R</u>adio.

CMRR An abbreviation of <u>C</u>ommon <u>M</u>ode <u>R</u>ejection <u>R</u>atio.

CMS An abbreviation of <u>C</u>onversational <u>M</u>onitor <u>S</u>ystem. A subsystem in IBM's SNA.

CMT An abbreviation of <u>C</u>onnection <u>M</u>anagemen<u>T</u>. That part of station management (SMT) in an FDDI network that establishes physical links between adjacent stations.

CMTS An abbreviation of <u>C</u>ellular <u>M</u>obil <u>T</u>elephone <u>S</u>ystem.

CN (1) An abbreviation of <u>C</u>ommon <u>N</u>ame. In the NetWare Directory Services (NDS) for Novell's NetWare 4.x, a name associated with a leaf object in the NDS Directory tree. For a user object, this would be the user's login name. (2) An abbreviation of <u>C</u>ountry <u>N</u>ame. (3) An abbreviation of <u>C</u>ustomer <u>N</u>etwork. (4) An abbreviation of <u>C</u>o<u>N</u>nect.

CNA An abbreviation of <u>C</u>ertified <u>N</u>ovell <u>A</u>dministrator or <u>C</u>ertified <u>N</u>et<u>W</u>are <u>A</u>dministrator. The title given to people who successfully complete Novell authorized courses on administering a NetWare network and pass their comprehensive administrator exam.

CNC An abbreviation of <u>C</u>omputerized <u>N</u>umerical <u>C</u>ontrol.

CND An abbreviation of <u>C</u>alling <u>N</u>umber <u>D</u>elivery. A synonym for ANI.

CNE (1) An abbreviation of <u>C</u>ertified <u>N</u>ovell <u>E</u>ngineer or <u>C</u>ertified <u>N</u>et<u>W</u>are <u>E</u>ngineer. The title given to people who successfully complete a series of Novell authorized courses on becoming technicians or consultants for NetWare networks and pass their comprehensive exam. (2) An abbreviation of <u>C</u>ertified <u>N</u>etwork <u>E</u>ngineer.

CNET An abbreviation of <u>C</u>entre <u>N</u>ational d'<u>É</u>tudes de <u>T</u>élécommunication (National Center for the Study of Telecommunications). The French organization that approves telecommunications products sold in France.

CNG An abbreviation of Calli<u>NG</u> tone.

CNI (1) An abbreviation of <u>C</u>ertified <u>N</u>ovell <u>I</u>nstructor or <u>C</u>ertified <u>N</u>et<u>W</u>are <u>I</u>nstructor. A title given to people successfully completing a series of Novell authorized courses on becoming instructors for Novell training courses and pass their comprehensive instructor exam. **(2)** An abbreviation of <u>C</u>oalition for <u>N</u>etwork <u>I</u>nformation. **(3)** An abbreviation of <u>C</u>alling <u>N</u>umber <u>I</u>dentification.

CNIS An abbreviation of <u>C</u>alling <u>N</u>umber <u>I</u>dentification <u>S</u>ervice. A service that provides, screens, or delivers the calling party's number in ISDN.

CNM (1) An abbreviation of <u>C</u>ustomer <u>N</u>etwork <u>M</u>anagement. All activities customers perform to manage their communications network. In Switched Megabit Data Services (SMDS), the service enables customers to directly manage many aspects of the SMDS service provided by the carrier. **(2)** An abbreviation of IBM's Systems Network Architecture (SNA) <u>C</u>ommunication <u>N</u>etwork <u>M</u>anagement.

CNR (1) In telephony, an abbreviation of <u>C</u>ustomer <u>N</u>ot <u>R</u>eady. This means the installation must be rescheduled. **(2)** An abbreviation of <u>C</u>arrier to <u>N</u>oise <u>R</u>atio.

CNRI An abbreviation of the <u>C</u>orporation for <u>N</u>ational <u>R</u>esearch <u>I</u>ncentives.

CNRS An abbreviation of <u>C</u>entre <u>N</u>ationale de <u>R</u>écherche <u>S</u>cientifique (National Center for Scientific Research).

CNS (1) An abbreviation of <u>C</u>omplementary <u>N</u>etwork <u>S</u>ervices. **(2)** An abbreviation of <u>C</u>hinese <u>N</u>ational <u>S</u>tandards (ROC).

CO (1) An abbreviation of <u>C</u>entral <u>O</u>ffice (pronounced "CEE-oh"). Also written C.O. **(2)** An abbreviation of <u>C</u>onnection <u>O</u>riented. **(3)** An abbreviation of <u>C</u>ustomer <u>O</u>wned.

co-channel interference Interference resulting from two or more simultaneous transmissions on the same channel.

Coalition for Network Information (CNI) A nonprofit organization consisting of approximately 200 member organizations dedicated to the development of computer networking. They are interested in questions concerning "networking and intellectual property, navigation tools for research, the improvement of network bibliographic standards, equity of access to networked resources, and educational applications of networking technology."

COAM An acronym from <u>C</u>ustomer <u>O</u>wned <u>A</u>nd <u>M</u>aintained (equipment). A deprecated term. See *customer premises equipment (CPE)*.

coarse acquisition (C/A) code A family of direct sequence spread spectrum codes used in the Global Positioning System (GPS) to determine the range between a receiver and the transmitting satellite. Each satellite is assigned one of 32 unique Gold codes. Each code is a 1023-bit pseudorandom sequence bi-phase modulated onto the GPS L1-carrier at a 1.023-Mbps chipping rate (hence, it repeats every millisecond). The code is subject to an accuracy degradation introduced by the U.S. Department of Defense called *selective availability (SA)*. The *C/A* code is used primarily by commercial receivers and as an aid in P-code acquisition. Also called the *civilian code*. See also *Global Positioning System (GPS)*.

coasting mode In timing-dependent systems, a free-running operational timing mode in which the normal continuous or periodic measurement of clock error and the subsequent automatic frequency or phase correction of the clock to minimize this error is not made.

Although closed-loop corrections, as described above, are not made, open-loop corrections may be made when they are based on a prior knowledge of the clock behavior, e.g., temperature or aging drift.

coated optics Optical elements (either reflecting or refracting) in which one or more optical surfaces have been overlaid (coated) with one or more layers of dielectric or metallic material. These layers are designed to enhance either the transmission or the reflection of the surface. Some coatings are applied simply as a protection of the underlying material; that is, the coating is harder than the optical material.

coating (1) An epoxy or plastic layer applied to electronic circuit assemblies to make them more moisture and fungus resistant. **(2)** A protective material applied to an optical fiber during manufacture to strengthen it and to protect the fiber from forces that could lead to microbending losses. **(3)** A material applied to an optical device to change its characteristics—for example, to increase its transmission or reflection. See *coated optics.*

coax A contraction of *coaxial* cable.

coax booster An amplifier or repeater designed to be inserted into a coaxial cable run so as to increase the maximum cable run length possible.

coaxial cable A cable consisting of two concentric conductors: a central wire surrounded by a cylindrical shield and separated by a dielectric. The shield minimizes electrical and radio frequency interference (EMI and RFI). To be effective, the *characteristic impedance* of a coaxial cable must match the electronic circuit to which they are connected. To this end, coaxial cables are rated by their *characteristic impedance* (expressed in Ohms).

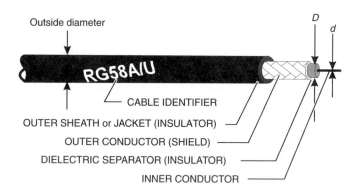

Some of the many criteria (and characteristics) used to define coaxial cables include:

- *Physical size.* The physical outside diameters of coaxial cables may range from less than a millimeter to perhaps 10 centimeters. Generally, the reason behind the larger sizes is the need to carry more power.
- *Flexibility.* Depending on the application, a coaxial cable may be constructed with an ability to bend ranging between extremely flexible (may easily be bent 180°1000 times to a radius five times the outside diameter of the cable) to what amounts to a rigid pipe (which may only be bent one time and only to a radius of 20 times the outside diameter).
- *Characteristic impedance.* The average characteristic impedance (Z) of a coaxial cable is determined by the ratio of the diameters of the inner conductor (d) and the outer conductor (D) and by the dielectric constant (ϵ) of the insulation between them. This im-

pedance may vary from a few Ohms to several hundred Ohms. It is described mathematically by:

$$Z = \sqrt{\frac{L}{C}} = \frac{138.16}{\sqrt{\epsilon}} \log_{10}\left(\frac{D}{d}\right) \Omega.$$

- *Shield type.* The coaxial cable outer conductor (or shield) may be made of a wire braid, a solid extrusion, or a metalized plastic film. Shield materials include bare copper, silver-plated copper, tinned copper, copper-plated steel, high-resistance wire, and silver plated alloys.

- *Shield coverage.* The shield coverage is given in percent and may be as little as 50% on low-performance cable (some single shield braids) to near 100% on an extruded shield. Some braided systems use two layers to improve the percent coverage. For most LANs, 95% coverage is considered good.

- *Attenuation.* Attenuation is the amount of loss introduced by the insertion of the cable between a transmitter and receiver. It is typically expressed as loss in dB per unit length, e.g., dB/km, dB/100 ft.

Attenuation (α) in dB/100 ft at a frequency (f) can be *estimated* by:

$$\alpha = \frac{.00317\left(\frac{D}{d} + 1\right)\sqrt{\epsilon \cdot f}}{D \cdot \log_{10}\left(\frac{D}{d}\right)} + 2.78 \cdot \sqrt{\epsilon} \cdot PF \cdot f$$

where ϵ is the dielectric constant, PF is the power factor of the dielectric, D is the inside diameter of the shield, and d is the outside diameter of the inner conductor.

- *Capacitance.* Capacitance is the natural consequence of the physical relationship of the inner conductor and outer conductor separated by a dielectric. It is measured in picofarads per unit length; e.g., pF/ft. For most coaxial cables the capacitance falls in the range of 5 pF/ft to 50 pF/ft. Mathematically, capacitance is expressed as

$$C = \frac{7.36 \cdot \epsilon}{\log_{10}\left(\frac{D}{d}\right)} \text{ pF/ft.}$$

- *Inductance.* Inductance is the natural result of physical conductors that carry current. Mathematically, inductance is expressed:

$$L = 140 \log_{10}\left(\frac{D}{d}\right) \mu\text{H/ft.}$$

- *Maximum power-handling capability.* The maximum power capability is directly related to the current handling capability of the inner conductor and the frequency-dependent losses of the dielectric. The power-handling capacity of a coax may be only a few watts or hundreds of thousands of watts. When determining the maximum power-handling capability of a coaxial cable, environmental conditions as well as circuit conditions must be taken into account, specifically, altitude, temperature, humidity, VSWR, and the like. This effective power (P_e) can be estimated by including a correction factor for temperature (T_C), altitude (H_C), and VSWR to the average power (P_A); i.e.:

$$P_e = \frac{P_A\left[\left(VSWR + \frac{1}{VSWR}\right) + \text{K} \cdot \left(VSWR - \frac{1}{VSWR}\right)\right]}{2 \cdot T_C \cdot H_C}$$

where some typical values of T_C, H_C, and K are given in the following charts.

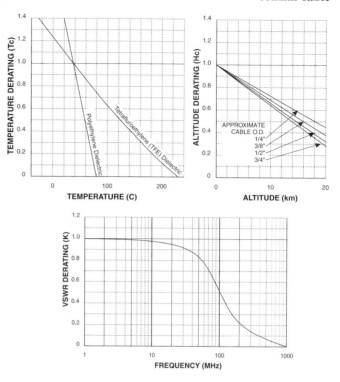

- *Maximum operating voltage.* The maximum operating voltage on a cable must be less than the corona level (extinction voltage) of the cable. (Operation at or above this voltage produces excessive noise and permanent dielectric damage.) As with maximum power, the effects of VSWR, altitude, and temperature must be considered when determining the maximum permissible operating voltage. The effective maximum voltage (V_e) can be calculated from the actual voltage (V_A) and the *VSWR;* i.e.:

$$V_e = V_A \cdot VSWR$$

- *Cutoff frequency.* The cutoff frequency is defined as the frequency above which unwanted modes of propagation may be generated at impedance discontinuities. The cutoff frequency of a coaxial cable may be estimated by:

$$f_c = \frac{7.5}{(D + d)\sqrt{\epsilon}} \text{ GHz}$$

- *Propagation velocity.* Propagation velocity is the speed of transmission of electrical energy through the cable compared with its speed in a vacuum (the speed of light). It is therefore generally expressed as a percent of the speed of light, i.e.,

$$V = \frac{100}{\sqrt{\epsilon}} \%.$$

- *Propagation time.* Propagation time or time delay is the time it takes a signal to transit a unit length of cable.

$$t_d = 1.017\sqrt{\epsilon} \text{ ns/ft.}$$

- *Dielectric material.* Many materials are used as the dielectric separating the inner conductor from the outer conductor. The dielectric may be a solid, liquid, gas, or a combination. Dielectric materials used include air, sulfur hexafluoride, petroleum and silicon oils, polyethylene, polystyrene, polypropylene, nylon, fluorinated ethylene-propylene (FEP Teflon®), and so on.

- *Environmental stability.* Exposure to sunlight (ultraviolet radiation), high humidity, corrosive vapors, underground burial, and high temperatures all affect the life and performance of the materials used to construct a coaxial cable.

The following chart lists a few of the hundreds of coaxial cables with a few of their basic characteristics. Also referred to as *coax, coax cable,* or *coaxial transmission line.*

COAXIAL CABLE CHARACTERISTICS

ID	Characteristic Impedance (Ohms)	Capacitance (pF/ft)	Loss dB/100ft Freq. in MHz			OD (in.)
			10	50	100	
RG57A/U	95	17.0	0.65	1.60	2.40	0.625
RG58A/U	52	28.5	1.40	3.30	4.90	0.195
RG58B/U	53.5	28.5	1.20	3.10	4.60	0.195
RG58C/U	50	30.8	1.40	3.30	4.90	0.195
RG59/U	73	21.0	1.10	2.30	3.30	0.242
RG59A/U	75	20.6	1.10	2.30	3.30	0.242
RG59B/U	75	20.6	1.10	2.30	3.30	0.242
RG62A/U	93	13.5	0.90	1.90	2.80	0.242
RG62B/U	93	13.5	0.90	2.10	3.00	0.242
RG63B/U	125	10.0	0.50	1.10	1.50	0.405
RG72/U	150	7.8	0.50	1.10	2.10	0.630

COB (**1**) An acronym from Close Of Business. (**2**) An acronym from Chip On Board.

cobalt glass A type of glass that transmits near-ultraviolet radiation but is opaque in the visible light region. Also called *Woods Glass.*

COC An abbreviation of Central Office Connections.

COCF An abbreviation of Connection Oriented Convergence Function.

COCOT An abbreviation of Customer Owned Coin Operated Telephone.

code (**1**) A set of rules that maps the elements of one set, the coded set, onto the elements of another set, the code element set. Also called a *coding scheme.* (**2**) A system in which the cryptographic equivalents (usually called a *code group*), typically consisting of letters and/or digits in otherwise meaningless combinations, are substituted for plaintext elements which are primarily words, phrases, or sentences. (**3**) A set of items, such as abbreviations, that represents corresponding members of another set. (**4**) A way to write a computer routine or program. (**5**) The representation of data, computer program, or parts of computer programs in a symbolic form that can be accepted by a computer. (**6**) Any system of communication in which arbitrary groups of letters, numbers, or symbols represent units of plaintext of varying length. Codes may or may not provide security. Common uses include:

- Converting information into a form suitable for communications or encryption.
- Reducing the length of time required to transmit information.
- Describing the instructions that control the operation of a computer.
- Converting plaintext to meaningless combinations of letters or numbers and vice versa.

(**7**) A set of unambiguous rules specifying: (**a**) The way all characters (or symbols) in a given character set (or alphabet) may be represented, (**b**) The encryption of data for security or error-detection or correction purposes, (**c**) The conversion of a binary data stream into a format more suitable for transmission or modulation, (**d**) A set of items (such as abbreviations) representing members of another set.

From definition (a) it may seem that any representation of a symbol would be acceptable; however, in practice additional constraints such as:

- The order relationship of the original symbol set should be maintained in the encoded form.
- Operations applied to the original symbol set and comparable functions applied to the binary code should produce corresponding results.

With these "rules" in mind, one can construct a multitude of binary codes to satisfy specific purposes. Specifically, various codes can be created to represent *decimal numbers* or *alphanumeric symbols. Alphanumeric codes* represent a complete communications symbol set (alphabetic symbols, numeric symbols, punctuation and control symbols). Alphanumeric codes include *ASCII, Baudot, EBCDIC, Hollerith,* and so on. *Numeric codes* provide a representation of only numeric values; for example, *decimal codes* represent only the 10 decimal symbols 0 through 9. *Numeric codes* generally fall into one of two categories: *weighted* and *transitional* codes:

- *Weighted.* A weighted code assigns a different mathematical value (weight) to each bit position in the code. For example, the rightmost bit in the normal BCD code has a weight of one, the next bit to the left has a weight of two , the next bit is weighted four, and finally the rightmost bit is weighted eight. (The name of this code is therefore the 8421 code.)

The code can be described with an equation; i.e.,

$$D = W_3b_2 + W_2b_2 + W_1b_1 + W_0b_0$$

where D is the decimal digit to be coded, W_n is the weighting of the bit position, and b_n is the binary value 0 or 1. The coded value is expressed by the binary string $b_3b_2b_1b_0$. (The weighting factors Wn are implied and not written.)

Restrictions in defining the weighting factors W_n include:

- Once the weighting factors W_n are selected, there must be at least one combination of bits b_n when taken with the corresponding weights sum to every value in the symbol set.
- When more than one combination of bits b_n yields a specific value, an auxiliary arbitration rule (such as use the code with the fewest number of "1s") must be imposed.

The table on page 105 provides a small sampling of the many different weighted codes possible.

- *Transitional.* Rules may be developed to indicate how an individual code element is encoded based on predecessor elements. The most common of these codes are the *Gray codes* (used in mechanical position encoding devices). Other specialized codes have been implemented to solve specific problems; for example, the *Lippell codes* allow the encoding of a decimal number with only two code tracks. (There are still four readout sensors, however.)

See also *ASCII, Baudot, BCD, DES, EBCDIC, encode, encryption, Gold code, Gray code, Hamming code, Lippell code, self-checking code, teleprinter codes,* and *waveform codes.*

code bit The smallest signaling element used on the ISO/OSI Reference Model Physical Layer (layer 1) for transmission.

code blocking In telephony, the ability of a switching system to prevent access to certain types of calls or to certain geographic areas.

WEIGHTED CODE EXAMPLES

Code Name Bit Weight	Hex 8421	BCD 8421	XS3 8421+3	MBQ 5421	Aiken 2421	Berkeley 4221	2 of 5 74210	BiQuinary 50 43210	1 of 10 9876543210	Hamming AB8C421	7421	742-1
Digit												
0	0000	0000	0011	0000	0000	0000	11000	01 00001	0000000001	0000000	0000	0000
1	0001	0001	0100	0001	0001	0001	00011	01 00010	0000000010	1101001	0001	0011
2	0010	0010	0101	0010	0010 (1000)	0010 (0100)	00101	01 00100	0000000100	0101010	0010	0010
3	0011	0011	0110	0011	0011 (1001)	0011 (0101)	00110	01 01000	0000001000	1000011	0011	0101
4	0100	0100	0111	0100	0100 (1010)	0110 (1000)	01001	01 10000	0000010000	1001100	0100	0100
5	0101	0101	1000	1000 (0101)	1011 (0101)	0111 (1001)	01010	01 00001	0000100000	0100101	0101	0111
6	0110	0110	1001	1001 (0110)	1100 (0110)	1100 (1010)	01100	10 00010	0001000000	1100110	0110	1001 (0110)
7	0111	0111	1010	1010 (0111)	1101 (0111)	1101 (1011)	10001	10 00100	0010000000	0001111	1000	1000
8	1000	1000	1011	1011	1110	1110	10010	10 01000	0100000000	1110000	1001	1011
9	1001	1001	1100	1100	1111	1111	10100	10 10000	1000000000	0011001	1010	1010
A	1010	(1010)	(1101)	(1101)	—	—	—	—	—	1011010	(1011)	(1101)
B	1011	(1011)	(1110)	(1110)	—	—	—	—	—	0110011	(1100)	(1100)
C	1100	(1100)	(1111)	(1111)	—	—	—	—	—	0111100	(1101)	(1111)
D	1101	(1101)	(0000)	—	—	—	—	—	—	1010101	(1110)	(1110)
E	1110	(1110)	(0001)	—	—	—	—	—	—	0010110	(1111)	—
F	1111	(1111)	(0010)	—	—	—	—	—	—	1111111	—	—

Codes in parentheses indicate an error condition; e.g., in BCD only the digits 0–9 are defined, so the six codes from 1010 through 1111 have no meaning.

TRANSITION RULE BASED CODE EXAMPLES

Digit Name	Gray Natural	Gray GXS-3	Gray 10-16	Gray Watts	Lippell 1	Lippell 2	Lippell 3	Johnson	Chain L	Chain R
0	0000	0010	0000	0000	0000	0010	0000	00000	0000	0000
1	0001	0110	0001	0001	0001	0011	0010	00001	0001	1000
2	0011	0111	1001	0011	0010	0110	0011	00011	0011	0100
3	0010	0101	1101	0010	0011	0111	0111	00111	0110	1010
4	0110	0100	0101	0110	0110	1110	0110	01111	1101	1101
5	0111	1100	0111	1110	1111	1111	1110	11111	1010	0110
6	0101	1101	1111	1010	1110	1100	1100	11110	0100	0011
7	0100	1111	1011	1011	1101	1101	1101	11100	1001	1001
8	1100	1110	0011	1001	1100	1000	1001	11000	0010	1100
9	1101	1010	0010	1000	1001	1001	1000	10000	0101	1110
A	1111	1011	1010	—	—	—	—	—	1011	1111
B	1110	1001	1110	—	—	—	—	—	0111	0111
C	1010	1000	0110	—	—	—	—	—	1111	1011
D	1011	0000	0100	—	—	—	—	—	1110	0101
E	1001	0001	1100	—	—	—	—	—	1100	0010
F	1000	0011	1000	—	—	—	—	—	1000	0001

The "L" chain code shifts each preceding field left and inserts one symbol from the sequence 1, 1, 0, 1, 0, 0, 1, 0, 1, 1, 1, 1, 0, 0, 0, 0 as the least significant bit (LSB). The "R" chain code shifts each preceding field right and inserts one symbol from the sequence 1, 0, 1, 1, 0, 0, 1, 1, 1, 1, 0, 1, 0, 0, 0, 0 as the most significant bit (MSB).

For example, the switch could block access to certain area codes, central office codes, or specific telephone numbers.

code call access A feature on many PBX systems wherein an attendant or other user may dial a special number followed by the extension of the party being called. The PBX will then operate audible and/or visual signaling devices located throughout the facility to alert the called party. Generally, these devices are coded so that only a specific called party need respond. The called party then dials the *meet me* extension to complete the connection.

code character A character with the following properties:

- It is derived in accordance with a code.
- It is used to represent a discrete value or symbol.

See *code* (7).

code conversion (**1**) Conversion of signals, or groups of signals, of one code into corresponding signals, or groups of signals, of another code. (**2**) A process for converting a code of some predetermined bit structure, such as 5, 7, 8, or 16 bits per character, to another code with the same or a different number of bits per character interval.

This is frequently necessary when information is sent from one network to another network based on a different scheme, e.g., the conversion of Baudot codes on the telex network to ASCII codes on the TWX network, or the conversion of ASCII codes on most small computer systems to IBM's EBCDIC code used on some mainframe systems. See also *code* (*7*).

code division multiple access (CDMA) (**1**) A method of creating more than one information channel *in the same frequency bandwidth* through the use of *spread spectrum* modulation techniques.

Characteristics of *CDMA* include:

- Can be used as an access method that permits carriers from different stations to use the same transmission equipment.
- Permits simultaneous access of the available bandwidth by multiple stations. Each transmitter's carrier is modulated with a unique code.
- On reception, each channel can be distinguished and isolated from the others by means of the unique modulation code, thereby enabling the reception of signals that were originally overlapping in space, time, and frequency.
- The mutual interference is reduced by the degree of *orthogonality* of the unique codes used in each transmission.
- A more uniform distribution of energy in the allocated bandwidth.

See also *spread spectrum*. (**2**) In cellular telephony, the transmission method claims to fit up to 20 times as much information into a channel as time division multiple access (TDMA). Each user's transmit signal is "spread" over the frequency band using a different code. When the receiver decodes the received signals, only the signal with a matching code will be meaningful; the other signals will be received as noise. *CDMA* uses something called soft hand-off when switching a transmission from one cell to another to ensure that no bits are lost in the transmission. In this type of hand-off, both cells transmit the transitional bits at the same time. This way, one of the transmissions will be within range of the receiver. *CDMA* is not compatible with the 1989 TDMA standard. Some digital cellular telephone transmission systems use this technology.

code element One of the discrete conditions or events of a code, e.g., the presence or absence of a pulse. See also *character*.

code excited linear predictive coding (CELP) An analog-to-digital conversion and voice coding scheme. It is a variant of the linear predictive coding (LPC) voice encoding algorithm. *CELP* can produce digitized voice output at rates of 4800 bps and lower.

code group A group of letters, numbers, and/or symbols in a code system used to represent a plaintext (words, phrases, and sentences).

code independent data communication See *code transparent data communication*.

code level The number of bits used to represent a character. For example, Baudot code is a five-level code, ASCII is a seven-level code, and EBCDIC is an eight-level code.

Code of Federal Regulations (CFR) A listing of the general and permanent rules published in the Federal Register. It is divided into 50 titles, each of which represents a broad area subject to federal regulation. Title 47 pertains to telecommunications and contains all of the rules, requirements, and standards that must be met with any device that has the potential to communicate across state boundaries or to interfere with other devices that do. Within Title 47 are several subparts that are widely referenced, i.e.,

Part 15—addresses electromagnetic emanations.

Part 22—deals with common carriers.

Part 68—addresses the protection of the PSTN from attached subscriber devices.

Part 90—deals with private carriers.

Part 95—addresses Citizens Band equipment manufacture and use.

code restriction A condition in which specific terminals are prevented from accessing certain features of the network.

code set The complete set of representations defined by a particular code and language. See also *code*.

code tracking loop That portion of a receiver that aligns the receiver internal pseudorandom code generator to the received sequence. Alignment is accomplished by shifting the locally generated sequence in time so that a particular chip in the sequence is generated at the same instant as the corresponding chip of the received sequence arrives at the receiver. Also called a *delay lock loop*.

code transparent data communication A mode of data communication that uses protocols that do not depend for their correct functioning on the data character set or data code used. Also called *code independent data communication*.

code violation A breach of a primary coding rule. For example, the AMI code states that successive mark conditions must be of alternate polarity; that is, the code sequence $+V$, $+V$, 0, $-V$ is illegal (i.e., a code violation). These code violations, however, are sometimes used as a means of out-of-band signaling (a secondary coding rule) between equipment since they cannot represent data from the user.

code word In a code, a word that consists of a sequence of symbols assembled in accordance with the specific rules of the code and assigned a unique meaning.

- The meaning may be public knowledge and used to shorten frequently used transmissions, e.g., SOS, MAYDAY, ROGER, OVER, OUT, TEN-FOUR.
- The meaning may be known only to a few people and is used to identify classified material so as to safeguard intentions and information regarding a classified plan, operation, or other entity.
- The meaning may be known to only one person and is used as a mnemonic for a password, i.e., the code word is the password.
- A secret code word (cryptonym) is used to identify sensitive intelligence data.

CODEC An acronym from <u>CO</u>der/<u>DEC</u>oder. The devices that transform analog data into a digital bit stream (coder) and a digital bit stream into analog signals (decoder). Used to convert analog signals to (and from) digital bit streams for storage or transmission on digital networks.

coded character set A character set established in accordance with unambiguous rules that define the character set and the one-to-one relationships between the characters of the set and their coded representations.

coded ringing In telephony, a subscriber ringing method wherein the number and/or the duration of rings indicate which party is being called.

It is a signaling method used on party lines where all primary telephones on a particular loop ring when the central office ring signal is applied. Each subscriber has a particular ring cadence that is aurally identified by the subscriber. Some possible *coded ringing* cadences are follows:

Coded Ringing Examples						
One long						
One short						
Two short						
Three short						
Long + short						
Short, long, short						

Although many combinations are possible, it is not practical to rely on the subscriber to differentiate among a large population of codes. Therefore, usually only two or three codes are utilized on any one line. See also *bridged ringing, distinctive ringing, divided ringing, isolated ringing, selective ring,* and *semiselective ring.*

coded set   A set of elements onto which another set of elements has been mapped according to a prescribed algorithm (code). Examples of coded sets include:

- The names of the months of the year mapped onto the set of decimal numbers 01–12.
- The list of country names mapped onto a set of corresponding country codes.
- The list of airport names mapped onto a set of corresponding three-letter representations.

coding   **(1)** The process of transforming messages or signals from one form or format to another in accordance with a strict set of rules to make the signal more suitable for an intended application.

The purpose of coding may be:

- Security—as in *encryption.*
- Transmission improvement—as with *waveform encoding.*
- Bandwidth conservation—as with any of the many *data compression* schemes.
- *Error control* (detection and correction).
- The representation of a human readable symbol in a machine readable format. For example, ASCII and EBCDIC are two widely used coding methods.

See also *data compression, encryption,* and *waveform codes.* **(2)** In communications and computer systems, implementing rules that are used to map the elements of one set onto the elements of another set, usually on a one-to-one basis. **(3)** The digital *encoding* of an analog signal. The inverse is *decoding* to an analog signal.

coding scheme   A synonym for *code (1).*

coding theory   The mathematical theory describing how to encode data streams into digital symbol streams at a transmitter and how to decode the symbol streams into data streams at a receiver such that maximum accuracy and speed are obtained.

codress message   In military communications systems, a message in which the entire address is encrypted with the message text.

coercive force (H_c)   The demagnetizing force that must be applied to a magnetic material to reduce the induction to zero. The units of coercive force are amps/meter (SI system) or oersteds (cgs system).

COFA   An abbreviation of <u>C</u>hange <u>O</u>f <u>F</u>rame <u>A</u>lignment. Movement of the Synchronous Payload Envelope (SPE) within an STS frame.

coherence   Pertaining to electromagnetic waves in which the phase relationship between any two points in the radiation field is constant throughout the duration of the radiation. There is a fixed relationship of like points on the electromagnetic wave separated in space or time.

An ideal *coherent wave* would be coherent at all points in space. In practice, however, the region of high coherence extends over only a finite distance.

coherence area   With respect to an electromagnetic wave, the area in a plane perpendicular to direction of propagation that the wave may be considered coherent. That is, the area over which the wave maintains a specified *degree of coherence.* (The specified degree of coherence is usually taken to be 0.88 or greater with light.) See also *degree of coherence.*

coherence degree   See *degree of coherence.*

coherence length   The propagation distance from a coherent source over which an electromagnetic wave maintains a specified *degree of coherence.* Although the definition applies to all electromagnetic waves, it is usually utilized with optical systems.

In optical communications, the coherence length (L) is given approximately by:

$$L = \frac{\lambda^2}{n \cdot \Delta\lambda}$$

where:

λ is the central wavelength of the source,
n is the refractive index of the medium, and
$\Delta\lambda$ is the spectral width of the source.

In long-distance transmission systems, the *coherence length* may be reduced by factors such as dispersion, scattering, and diffraction.

coherence time   For an electromagnetic wave, the time over which a propagating wave may be considered coherent. Although the definition applies to all electromagnetic waves, it is usually utilized with optical systems.

In long-distance transmission systems, the *coherence time* may be reduced by factors such as dispersion, scattering, and diffraction. In optical communications, *coherence time* (τ) is calculated by dividing the *coherence length* by the phase velocity of light in a medium, which is approximately given by

$$\tau = \frac{\lambda^2}{c \cdot \Delta\lambda}$$

where:

λ is the central wavelength of the source,
$\Delta\lambda$ is the spectral width of the source, and
c is the velocity of light.

In long-distance transmission systems, the *coherence time* may be reduced by factors such as dispersion, scattering, and diffraction.

coherent bundle   See *aligned bundle.*

coherent differential phase shift keying (CDPSK)   A phase shift keying system in which the transmitted carrier phase is modulated in discrete steps according to the data to be transmitted and its relation to a reference signal. Ideally, the amplitude and average frequency remain constant.

Information is recovered at the receiver by comparing the phase transitions between the carrier and successive pulses rather than by the absolute phases of the pulses.

coherent noise The presence of electromagnetic waves from scatterers outside the plane of the object in a coherent system.

coherent radiation See *coherence.*

coherent source An electromagnetic wave source that is capable of producing radiation in which the propagating waves have a fixed phase relationship. A laser is an example of a coherent light source.

coil (1) An assembly of multiple turns of wire arranged in such a way as to generate a magnetic field when current is passed through the wire. The magnetic field may then operate a mechanical element, e.g., electric motors and relays, solenoids. (2) See *inductor.*

coil Q See *Q.*

coin box In telephony, a telephone set containing a device for collecting coins in payment for connection time. Commonly called a *coin telephone* or *pay telephone.*

coin call In telephony, a call in which a coin collection device is used.

coin control signal In telephony, on a coin call, a family of signals used to collect or return the coin(s) in the collecting mechanism. One such signal is a ± 150 V pulse—one to collect the coins and the other to return the coins to the caller.

coin phone A synonym for *pay telephone.* The pay telephone was originally developed by the American inventor William Gray.

coin tone In telephony, a class of service tone that indicates to an operator that a call has originated from a coin telephone.

cold flow The deformation of a material due solely to constant mechanical force or pressure—the deformation is not due to heat softening. The deformation (or flow displacement) does not occur immediately with the application of force but continuously over the entire time period a force is applied.

cold standby Pertaining to spare electronic equipment that is wired in and available for substitute use but is not powered up and ready for use.

collaborative computing The use of computer tools that make it easier for groups of people (possibly geographically separated) to work together as teams. For example, software suites that provide multi-user communication (e-mail, BBS services, and teleconferencing) and document services (groups of people working on parts of a document). Also called *groupware.*

collapsed backbone A backbone network in which all subordinate networks are connected at a single hub. This provides the possibility for much faster operation than the conventional backbone architecture.

collect call In telephony, a connection in which the called party agrees to pay for the connection time.

collective routing Routing in which a switching center automatically delivers messages to a specified list of destinations. It avoids the need to list each destination address in the message header.

collimation (1) The process by which a convergent or divergent beam of electromagnetic radiation is aligned so as to produce a beam with the minimum divergence or convergence possible for that system, i.e., to make parallel. Ideally, the generation of a bundle of parallel rays. (2) The process of adjusting two (or more) lens (and/or mirror) axes with respect to each other or to a common reference.

collimator A device that renders divergent or convergent rays more nearly parallel. The degree of collimation (parallelism) should be stated.

collinear A term in mathematics meaning *in a straight line.* For example, three points are said to be collinear if and only if they can be connected with a straight line.

collision (1) In local area networks (LANs), multiple concurrent transmissions on a shared communications channel resulting in garbled data at one or more receivers on the channel. For example, two nodes on a LAN attempting to transmit on the LAN at nearly the same time. Also called *head-on collision.* (2) In a data transmission system, the situation that occurs when two or more demands are simultaneously made on equipment that can handle only one at any given instant. See also *Aloha network, collision avoidance, collision detection,* and *CSMA/CD.*

collision avoidance A procedure executed by any node attempting to access a local area network (LAN) whereby it monitors the LAN for activity before transmitting anything.

For example, the CSMA/CD access method uses Request To Send (RTS) and Clear To Send (CTS) signals before sending a frame onto the network. A node transmits only after the node has requested access to the line and been granted access. Other nodes are aware of the RTS/CTS transmission and will not try to transmit at the same time. See also *CSMA/CA* and *CSMA/CD.*

collision detection The process by which a node on a local area network (LAN) detects a collision between itself and another node. To detect a collision, nodes check the dc voltage level on the line. A voltage level two or more times as high as the expected level indicates a collision (since this means there are multiple signals traveling along the wires at the same time). *Collision detection* in broadband networks involves a separate bandwidth for collision detection and is somewhat more complex, since there may not be any dc voltage to test.

When a *collision* occurs, both nodes wait a random amount of time before attempting to access the LAN again. See also *contention* and *CSMA/CD.*

collision presence In network interface devices, a signal, provided by the physical layer to the media access sublayer (which is part of the data link layer), which indicates multiple stations are contending for access to the physical transmission medium. See also *CSMA/CD.*

collision window The maximum time period it takes a signal to travel from any node on a network to any other node. On simple networks, the maximum time is obtained when the cable length between nodes is maximum; however, on complex networks routers, bridges, and even repeaters add to the signal's propagation delay between nodes.

color code A system of standard colors adopted for identifying components that are too small to be marked with a part number (such as surface mounted devices), component values, polarities of wires, or as an aid in cabling.

Component identification generally requires a conversion chart from the manufacturer, while component values are based on the EIA color code standard.

EIA color code. The EIA color code is used to identify the value of many electronic components such as resistors, capacitors, and inductors as well as some active components. The general color code format used for passive components is:

Color Band 1—the most significant digit.

Color Band 2—the next significant digit.

Color Band 3—the least significant digit (used only on high accuracy components, 2% or better).

Color Band 3 or 4—the number of zeros (the multiplier).

Color Band 4 or 5—the tolerance of the device.

Color Band 5 or 6—the failure rate for 100 hours of operation—expressed in percent (optional).

The following table lists the EIA definitions of the various colors as applied to component values and tolerances.

EIA COLOR CODE

	Digit	Color
	0	Black
	1	Brown
	2	Red
	3	Orange
Digits 1, 2, 3, and	4	Yellow
Multiplier	5	Green
	6	Blue
	7	Violet
	8	Gray
	9	White
	5%	Gold
Tolerance	10%	Silver
	20%	No color
Failure (%) for 1000 Hr	1.0%	Brown
Operation (Optional)	0.1%	Red
	0.01%	Orange
	0.001%	Yellow

Telephone cable color code. The telephone cable color code is based on pairs of colors taken from two sets of five colors. Since most telephone cabling is based on pairs of wires, each wire pair uses the same color pair where the tip wire is predominantly the tip color with only a dot or spiral band of the ring color. The ring wire uses just the opposite combination. From the table, pair number 1 is a white/blue tip wire and a blue/white ring wire; pair 13 is black/green and green/black and so on to cable pair number 25. Multiple sets of 25 conductor pairs are grouped into a single cable with a colored wrap,

TELEPHONE CABLE COLOR CODE

Tip Color	Ring Color				
	Blue	Orange	Green	Brown	Slate
White	1	2	3	4	5
Red	6	7	8	9	10
Black	11	12	13	14	15
Yellow	16	17	18	19	20
Violet (purple)	21	22	23	24	25

again based on the color code. For example, a 100-pair cable would have 4 sets of 25-pair bundles, the first 25-pair bundle would be wrapped with a white/blue binder, the second bundle (pairs 26–50) would have a white orange wrapper and the third bundle (pairs 51–75) would have a white/green binder and so on.

Telephone quad. Pair number 1 is the primary voice pair, and pair number 2 is used for anything else, e.g., carrying power to a specialized subset, a second voice pair, or as a ground lead (the yellow conductor alone—or both yellow and black) for divided ringing.

TELEPHONE QUAD COLOR CODE

	Pair 1		Pair 2	
Country	Tip(+)	Ring(−)		
Germany	Rd-Bk	Rd		
Netherlands	Red	Blue		
United Kingdom	White	Blue		
USA/Canada (normal)	Green	Red	Yellow	Black
USA/Canada (special)	Bl-Wh	Wh-Bl	Or-Wh	Wh-Or

.COM (1) A type of file executable on a MS-DOS/PC-DOS computer in which calculation of relocation addresses is not necessary. See also *filename extensions.* (2) In Internet, one of the top-level *domain names.* See also *domain.*

COM (1) An acronym from Continuation Of Message. A type of ATM or SMDS segment between the beginning of message (BOM) and end of message (EOM). (2) An acronym from Component Object Model. (3) An acronym from Computer Output Microfilm (or Microfiche).

COM n The logical name for the serial communications port reserved by the MS-DOS operating system. Up to four serial ports may be installed and are identified as *COM1, COM2, COM3,* and *COM4.* Although not defined by a standard, the following table lists the settings most manufacturers use.

COMMON ADDRESS AND IRQ SETTINGS

PORT	ADDRESS	IRQ
COM 1	3F8	4
COM 2	2F8	3
COM 3	3E8	4
COM 4	2E8	3

Generally, a PC cannot share the interrupt (IRQ) between two high-activity devices without problems. (For example, a modem on port 3 and a mouse on port 1 share IRQ 4.) Internal modems sometimes get around this problem by allowing an interrupt on IRQ 2, 5, or 9. IRQ 5 is usually a good choice since most systems do not have a second parallel port (LPT2) installed.

COM port An abbreviation for COMmunication Port. See *COM n.*

comb filter A filter whose insertion loss forms a series of narrow-pass bands (or narrow stopbands) centered at fixed multiples of a specified frequency. See also *all-pass filter, band-pass filter, band reject filter, filter, high-pass filter,* and *low-pass filter.*

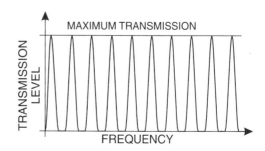

combined distribution frame (CDF) A distribution frame that combines the functions of main and intermediate distribution frames (MDF and IDF) and contains both vertical and horizontal terminating blocks.

- The vertical blocks are used to terminate the permanent outside lines entering the station.
- The horizontal blocks are used to terminate inside plant equipment.
- Horizontal and vertical blocks are interconnected with semipermanent connections (called *jumpers*). This arrangement permits the association of any outside line with any desired terminal equipment. The jumpers may be twisted pair, coaxial cable, or optical fiber as required.
- In technical control facilities, the vertical side may be used to terminate equipment as well as outside lines. The horizontal side is used for jackfields and battery terminations.

See also *intermediate distribution field* (*IDF*) and *main distribution frame* (*MDF*).

combined symbol matching (CSM) A method of facsimile data bandwidth compression using optical character recognition. It is based on the detection of recurrent patterns (such as alphanumeric symbols) in a document to be transmitted.

combiner A coupler that combines multiple incoming signals into a single outgoing signal.

For example, a combiner in a wavelength division multiplexer (WDM), in which signals from multiple input channels are all transmitting at different wavelengths, combines the signals to a single output path. A *combiner* is sometimes known as a *combiner coupler*. See also *diversity*.

COMBS An acronym from <u>C</u>ustomer-<u>O</u>riented <u>M</u>essage <u>B</u>uffer <u>S</u>ystem.

comm An abbreviation of <u>COMM</u>unication(s).

comma free code A code constructed such that the bit length of all code words is an integer multiple of a basic block length; that is, a code word may not terminate in the middle of a block. This property allows proper code word framing if two conditions are met:

- Initial framing is established by an external means; that is, the start of the first code word is determined by some method.
- There are no uncorrected errors in transmission. Huffman codes are examples of variable-length *comma-free codes*.

Also called a *prefix-free code*.

command (**1**) Generally, an instruction for an action to take place. Commands may be issued by an operator via a keyboard (or mouse), a hardware device, or from a computer program. For example, sending *ATH1* to a Hayes compatible modem will cause it to go off-hook. (**2**) In data transmission, an instruction sent by the primary station instructing a secondary station to perform some specific function. (**3**) In signaling systems, a control signal. (**4**) In computer programming, that part of a computer instruction word that specifies the operation to be performed. (**5**) Loosely, a mathematical or logic operator.

command line interface In a computer system, the text line on which a user enters direction to an operating system. Specifically, it allows the user to run programs by typing a specific character sequence rather than by using a mouse, or other user interface, to select programs.

Examples of *command line interface* programs include MS-DOS, many Internet utilities (archie, FTP, ping, and telnet for example), as well as many UNIX utilities. Compare with *graphical user interface* (*GUI*).

command menu A list (menu) of all the available instructions (commands) that may be given to a computer or communications system in its current state by an operator.

Assuming appropriate hardware is installed, the operator may select commands on a command menu by:

- Typing the letter or number associated with the displayed command,
- Positioning a cursor or reverse video bar with the "arrow or tab keys," mouse, or joystick over the displayed command,
- Touching the displayed command on the screen with a finger,
- Using an electromechanical pointer, such as a light pen, or
- Speaking to a voice-recognition system.

command state The condition or state of the local modem when it will accept instructions sent to it over the data path from the DTE. In this state, signals from the DTE are not passed on to the remote modem but are taken to be commands that the modem must act upon. Other states the modem can be placed in include: *online, signaling, training,* and *data*. Also called *command mode, escape state, local mode,* or *terminal mode*.

committed information rate (CIR) In frame relay networks, the bandwidth, or information rate, that represents the average level for a user. If the user's network activity exceeds this rate, the frame relay controller will mark the user's packets to indicate that they can be discarded if necessary.

Committee on Institutional Cooperation Network (CICNet) A regional Internet network and service provider specializing in the U.S. Midwest (Illinois, Indiana, Iowa, Michigan, Minnesota, Ohio, and Wisconsin). The *CICNet* offers interconnectivity among colleges, universities, public schools, and businesses.

common air interface (CAI) The CT2 international mobile communications standard that allows any compliant equipment to be used on any network of the same type. *CAI*-compliant telepoint handsets from different vendors may therefore be used on a telepoint network.

common applications environment (CAE) A computer environment in which applications can be paired across various manufacturers' X/Open system. The *CAE* contains standards for the operating system, networking protocol, and data management.

Common Authentication Technology (CAT) In the Internet community, a specification for a distributed authentication under development. *CAT* supports authentication measures based on either public or private key encryption strategies. With *CAT,* both client and server programs must use the services of a common interface, which will provide the authentication services. This interface will connect to either Distributed Authentication Security Service (DASS), which uses public key encryption, or Kerberos, which uses private key encryption.

common battery signaling Signaling in which the signaling power of a telephone is supplied by the serving switch. Talk power, however, may be supplied by either a common or local battery.

common carrier In telecommunications, a company or organization that provides communications services to the public, such as telephony or data communications.

In the United States, a common carrier is a public utility company licensed to furnish communication services to the general public and is regulated by state and federal agencies. In much of the rest of the world, common carriers are agencies of the countries government. Also called *commercial carrier,* especially by the Department of Defense (DoD), *communications common carrier,* or just *carrier*.

common channel interoffice signaling (CCIS) In telephony, signaling between switching offices which uses a separate channel from the user information-bearing channel. The control signals are sent by a separate fast, packet switch network. Three major benefits of *CCIS* are:

- Much faster call setup and teardown.
- Extra information above the normal call setup and teardown (such as caller ID and billing information) can be sent.
- The out-of-band signaling completely thwarts "phone phreaks'" ability to steal services. Since the control channel is isolated from the voice channel, there is no way for erroneous information to be injected into the signaling channel from a subset appearance.

common channel signaling (CCS) A method of signaling in which signaling information related to a specified group of channels is conveyed over a dedicated channel (i.e., it does not carry user information). Specific channel information is inserted and extracted based on addresses or timing. Channel 24 is used in T-1 systems, while TS-16 is used in E-1. See also *inband* and *out-of-band signaling.*

common control An automatic switching arrangement in which the control equipment necessary for establishing connections is shared; that is, it is associated with a given call only during the period required to accomplish the control function for the given call.

Common control channels are are not used for message traffic.

common equipment Equipment used by more than one system, subsystem, component, or other equipment, such as a channel or switch.

common gateway interface (CGI) In Internet, a standard designed to enable programmers to extend the functionality of servers. Originally, the method the CERN and NCSA web servers on UNIX used to allow interaction between servers and programs. Also called a *gateway script.*

Common Management Information Protocol (CMIP) An ISO standard for network resource management introduced in 1990. It is used to convey CMIS operations over an ISO network. *CMIP* is functionally equivalent to the older, and probably more widely used, Simple Network Management Protocol (SNMP). See also *SNMP.*

Common Management Information Protocol Data Unit (CMIPDU) A packet, in the OSI network management model, that conforms to the *CMIP*. The packet's contents depend on the request from a Common Management Information Service Element (CMISE), which relies on the *CMIP* to deliver the user's requests and to return with answers from the appropriate application or agent.

Common Management Information Protocol Machine (CMIPM) In the OSI network management model, software that accepts operations from a CMISE user and initiates the actions needed to respond and sends valid CMIP packets to a CMISE user.

Common Management Information Service (CMIS) The ITU X.710 and ISO 9595 standard for network monitoring and control services interface to the Common Management Information Protocol (CMIP) in the OSI network management model.

Common Management Information Service Element (CMISE) An entity that provides network management and control services in the OSI network management model. Seven services are specified, i.e.:

Action

Cancel get

Create

Delete

Event report

Get

Set

The services provided are used by the system management functions (SMFs). The SMFs are in turn used to carry out the tasks specified for the five system management functional areas (SMFAs) defined in the OSI network management model.

Common Management Information Services and Protocol over TCP/IP (CMOT) A protocol intended to automate the management of Internet network devices such as routers. These routers implement a version of the *OSI Common Management Information Services/ Common Management Information Protocol* (CMIS/CMIP).

The protocol is intended to replace the *Simple Network Management Protocol* (SNMP), so that Internet-linked networks conform to the OSI network management model.

common mode interference (1) Interference that appears between signal leads, or the terminals of a measuring circuit, and ground. **(2)** A form of coherent interference that affects two or more elements of a network in a similar manner (i.e., highly coupled) as distinct from locally generated noise or interference that is statistically independent between pairs of network elements.

common mode rejection ratio (CMRR) The ratio of the effect a differential mode signal $V_{IN(DM)}$ and a common mode signal $V_{IN(CM)}$ have on the output V_{OUT} of a system. It is therefore the ratio of the differential mode gain ($A_{V(DM)} = V_{OUT}/V_{IN(DM)}$) to the common mode gain ($A_{V(CM)} = V_{OUT}/V_{IN(CM)}$) of a differential signal input system. This ratio is generally expressed in decibels (dB); i.e.:

$$CMRR = 20 \cdot \log_{10} \left(\frac{A_{V(DM)}}{A_{V(CM)}} \right) dB$$

common mode signal The instantaneous algebraic average of two signals applied to the two conductors of a *balanced transmission line.* Both signals are relative to a common reference.

For example, if *a* +1 volt signal is impressed on one of the two conductors, and *a* +2 volt signal is impressed on the other conductor; then, the *common mode signal* is +1.5 V. Also called the *longitudinal signal* in telephony. See also *differential mode signal.*

common mode voltage (1) The voltage common to all terminals of a multiterminal device with respect to a common reference. **(2)** In a differential amplifier, the unwanted part of the voltage between each input and ground that is added to the desired signal voltage.

In telephony, also called a *longitudinal voltage.*

Common Name (CN) In the NetWare Directory Services (NDS) for Novell's NetWare 4.x, a *CN* is a name associated with a leaf object in the NDS Directory tree. For example, for a user object, this would be the user's login name.

Common Object Model (COM) An object-oriented, open architecture that is intended to allow client–server applications to communicate with each other in a transparent manner, even if these applications are running on different platforms. The *COM* model was developed jointly by Microsoft and Digital Equipment Corporation (DEC).

COM's goal is to enable networks or machines that use Microsoft's *Object Linking and Embedding* (OLE) technology to communicate transparently with networks or machines that use DEC's *Object-Broker* technology. To provide these cross-platform capabilities, *COM* uses a protocol based on the Distributed Computing Environment/ Remote Procedure Call (DCE/RPC) protocol. *COM* will allow machines running Microsoft Windows, Windows NT, and Macintosh environments to communicate in a transparent manner with machines running DEC's OpenVMS operating system or any of several UNIX implementations.

Common Object Request Broker Architecture (CORBA) A multiplatform architecture (developed by the Object Management Group) designed to allow applications operating in object-oriented environments to communicate and exchange information, even when they are running on different hardware.

CORBA has been implemented in products such as DEC's Object-Broker.

common return An electrical path that is shared by two or more circuits and serves to complete the current flow to their source or to ground.

common return offset In a line or circuit, the dc potential difference between ground and the common return.

common spectrum multiple access (CSMA) A method of providing multiple access to a common communications medium (such as a satellite radio link) wherein all transmitting nodes use a common time/frequency domain. The common technique for accomplishing this is one of the *forms of spread spectrum* transmission.

common trunk In telephony, a trunk accessible to all grading groups in a system that has a grading arrangement.

Common User Access (CUA) Specifications for user interfaces, in IBM's Systems Applications Architecture (SAA), that are intended to provide a consistent look across applications and platforms.

communicating application specification (CAS) A fax/modem application programming interface (API) standard developed by Intel and DCA and introduced in 1988. It enables the integration of fax capability and other communications functions into application software.

This standard competes with the Class *n* standards hierarchy developed by Electronics Industries Association (EIA).

communication The transmission of information from one point to another separate and isolated point—frequently by electromagnetic wave propagation means (i.e., telecommunications), although, smoke signals, drums, and whistling have been successfully used. The transmission must not alter the sequence, structure, or content of the transferred information.

communication buffer Memory set aside on a file server or in a communications device (a modem, for example) for temporarily holding packets until they can be processed or sent on to the network. The memory in a server is typically allocated as a number of buffers, each with a predetermined size. Also called a *routing buffer* or *packet receive buffer.*

communication medium The physical environment over which a communications signal travels. Examples include:

Wire cable (coaxial and twisted pair)

Optical fiber

The "ether" as in wireless (infrared or radio)

communication port Refers to any data channel on a DTE where a communication device might be attached. Examples are the serial ports, the parallel port, and a LAN port on a PC.

communications (1) Information transfer, among users or processes, according to agreed conventions. (2) The branch of technology concerned with the representation, transfer, interpretation, and processing of data among persons, places, and machines. *Note:* The meaning assigned to the data must be preserved during these operations.

communications blackout (1) A loss of communications or communications capability. For example, because of a lack of power to facility or equipment or because of a judicial decree. (2) A total lack of communications capability caused by propagation anomalies, e.g., those present during strong auroral activity or during the reentry of a spacecraft into the Earth's atmosphere.

communications controller A switching unit central to the implementation of host-based IBM SNA networks. Typically, the network is built around a backbone of interlinked *communications controllers* to which host computers and Control Units (CUs) (cluster controllers) are attached.

communications link The connection between sending and receiving equipment upon which data can be transferred. The link may embody any physical means such as microwave, wire, or power line carrier.

communications program Communications software that provides the means for a person to direct the connection of one DTE to another and to transfer information between them.

Communications program capabilities include:

- Control of hardware-related tasks such as setting up the DTE bit rate, parity, and DCE baud rate.
- Network functions, such as call progress tone monitoring and automatic number dialing.
- The ability to emulate several different types of terminals, such as VT-100 and ANSI.
- Storing and executing logon procedures.
- Information flow control including saving incoming messages and keeping a log of activities.
- Data transfer integrity control—packetizing data and messages, data compression, error control encoding and decoding, and coordinating transmissions to and from the remote DCE.

Some popular communication programs include *BitCom, Crosstalk, MTEZ, Procomm and Procomm Plus, Qmodem, Telink, Telix,* and *Unicom.* Communication programs are sometimes also called MODEM programs.

communications protocol A hierarchical set of conventions or procedures for communication between like processes at an interface. The protocol provides the means to control the orderly flow of information between the processes with minimum errors. The rules (or procedures) can include the modulation technique, handshaking method, error control approach, data compression procedure, and line control functions.

The ITU-T *X.200* series recommendations (the OSI model) outlines the protocol generally accepted for standardizing overall data communications. An earlier protocol is IBM's System Network Architecture (SNA). Internet uses the TCP/IP protocol. The hierarchical protocol structure affects different aspects of the communication process and hardware. Each aspect has its own standard. The RS-232 or ITU-T V.24 standards address the DTE to DCE interface, for example. Other protocol standards affect data flow control and handshaking methods such as XON/XOFF in asynchronous communications or modulation and encoding techniques such as ITU-T V.32 bis and V.42 bis. There are also protocols such as KERMIT, XMODEM, YMODEM, and ZMODEM that govern file transfers. On LANs, protocols such as CSMA/CD and ITU-T X.25 define the method by which information packets are passed from node to node. Collectively, the various standards are an attempt to facilitate the interconnection of different makes and models of hardware and different vendors' communication software.

communications rate The speed at which information is transferred across a communications link. Communications rate may refer to the speed across any transmission link, such as a telephone line or a cable between a computer and a modem. Typically, the rate is expressed in bits per second (bps or b/s).

Common telephone modem *communications rates* include:

Speed (bps)	Common Reference	Speed (bps)	Common Reference
300	Bell 103, V.21	19200	ATT
1200	Bell 202, V.22	28800	V.34
2400	Bell 212A, V.22bis	33600	V.34
9600	Bell 209, V.29	56000	V.90
14400	V.32bis		

See also *baud*.

communications reliability The probability that information transmitted from a sending station will arrive at the intended destination in a timely manner without loss of content.

communications satellite An Earth-orbiting satellite, generally in a stationary position relative to a point on the Earth, which functions as a repeater or transponder for communications circuits.

communications server A type of gateway that translates packets on a local area network (LAN) into a communication protocol suitable for use by asynchronous devices, e.g., RS-232 modems. The server provides a means for all nodes on the LAN to gain access to this device.

communications software See *communications program*.

communications system A collection of individual communications networks, transmission systems, relay stations, tributary stations, and data terminal equipment (DTE) usually capable of interconnection and interoperation to form an integrated entity. The components of a *communications system* serve a common purpose, are technically compatible, use common procedures, respond to controls, and operate in unison.

communications theory The mathematical explanation of the capabilities and limitations of the communication of messages from one point to another separate and isolated point. It is devoted to the probabilistic characteristics of the transmission of data in the presence of noise.

Communications Toolbox A feature on Macintosh System 6 (and later) that eases the process of getting modems to communicate with each other. It provides protocol conversions and drivers necessary for communications tasks.

comp hierarchy One of the *world newsgroups* in USENet dedicated to the reporting of news about computers. The hierarchy includes special interest newsgroups such as for specific applications, specific manufacturers, and developing technology. See also *newsgroup hierarchy*.

compact disc (CD) A generic term used to represent all optical laser *Compact Disc* audio and digital formats including but, not limited to, CD-DA, CD-i, CD-R, CD-ROM, Video CD, CD+G, and so on. *CDs* are generally considered to be 4.72 inches in diameter or smaller while LaserDiscs can be 12 inches. See also *LaserDisc*.

compander A contraction of the words <u>COMP</u>ressor and ex-p<u>ANDER</u>. A *compressor* is a device that reduces signal volume range; that is, it may reduce the level of large signals and/or increase the level of small signals according to a known procedure. The *expander* performs the reverse process, that is, making large signals larger and/or small signals smaller.

The purpose of the *compander* is twofold:

- To reduce the *dynamic range* of the information-carrying signal for either storage or transmission with minimal loss of information.
- To improve the *signal-to-noise ratio* (*SNR*) and *crosstalk* levels at the receiver.

For example, the dynamic range of signals from a subscriber telephone line are too large to send through a 64-kb digital channel, so they are reduced using either the μ-law or ITU-T A-law coding algorithm. These algorithms do not affect the smallest signals but attenuate the largest signals by a factor of about 64. Also spelled *compandor*. See also *A-law, mu-law,* and *dynamic range*.

comparator (**1**) Generally, a circuit, device, or software routine that indicates differences of two variables (or a variable and a constant). (**2**) In analog computing, a device or functional unit that compares two analog variables and indicates the result of that comparison. (**3**) A device that compares two items of data and indicates the result of that comparison. (**4**) A device for determining the dissimilarity of two items such as two pulse patterns or words.

compartmentation The segregation of programs, components, and/or information to provide isolation. *Compartmentation* provides some protection against compromise, contamination, or unauthorized access of the overall entity.

compass pin technology A connector pin design that provides four contact points on a single pin position. Each pin position is an insulator post with independent electrical connections on four sides; each connection runs the length of the post.

compatibility (**1**) Generally, the ability of two or more entities (devices, components of equipment, or material, or software) to exist or function in the same system or environment without mutual interference. *Compatibility* is sometimes built into the entity; if not in the entity, *compatibility* must be achieved through the use of adapters, drivers, or filters. (**2**) In computing, the ability to execute a given program on different types of computers without modification of the program or the computers. (**3**) In computing, the ability of a program function to call the resources of another program or subsystem to achieve desired results. (**4**) The capability that allows the substitution of one subsystem (storage facility), or of one functional unit (e.g., hardware, software), for the originally designated system or functional unit in a relatively transparent manner, without loss of information and without the introduction of errors. (**5**) The ability of the output of one device to communicate with the input of another device in a transparent manner, without loss of information and without the introduction of errors.

compatible sideband transmission Sideband suppressed carrier modulation in which a carrier is deliberately reinserted at a lower level after its normal suppression to permit reception by conventional AM receivers.

Compatible sideband transmission usually contains the upper sideband plus the carrier. Also called *amplitude modulation equivalent* (*AME*).

compelled signaling A signaling method in which the transmission of each signaling element to a destination is inhibited until the satisfactory reception of the previous element is acknowledged.

complement The complement of a given number is equivalent to the negative of that number *for a specific fixed word length*. (Therefore, the complement of a value in an 8-bit register will be different from the complement in a 16-bit register.)

The complement of a given value is the *diminished complement* of the given value plus 1. The *diminished complement* of the value is derived by subtracting each digit of the given value from the largest digit in the number base.

Example 1. The 6-bit, two's complement of 100110 is 011010; i.e.,

Original number	1	0	0	1	1	0
Diminished complement	0	1	1	0	0	1
(Each digit subtracted from 1)						
One	0	0	0	0	0	1
Two's complement	0	1	1	0	1	0

The *diminished complement* in the *binary system* is also called the *one's complement*.

Example 1A. The 8-bit, two's complement of 00100110 is 11011010; i.e.,

Original number	0	0	1	0	0	1	1	0
Diminished complement	1	1	0	1	1	0	0	1
(Each digit subtracted from 1)								
One			0	0	0	0	0	1
Two's complement	1	1	0	1	1	0	1	0

Example 2. The 3-digit, ten's complement of 132 is 868; i.e.,

Original number	1	3	2
Diminished complement	8	6	7
(Each digit subtracted from 9)			
One			1
Ten's complement	8	6	8

The *diminished complement* of a value in the *decimal system* is also called the *nine's complement*.

Example 3. The 5-digit, ten's complement of 00132 is 99868; i.e.,

Original number	0	0	1	3	2
Diminished complement	9	9	8	6	7
(Each digit subtracted from 9)					
One					1
Ten's complement	9	9	8	6	8

completed call In telephony, a call answered by the called party and subsequently released.

completion ratio In telephony, that fraction of call attempts that end in established communication links.

complex tone A sound containing two or more sinusoidal frequencies (more than one pitch).

complex variable In mathematics, a single variable having two parts: a *real part* and an *imaginary part*.

An example of a complex variable is the frequency-dependent impedance $Z(f)$ of a circuit or device. The real part corresponds to pure resistance (R), while the imaginary part refers to the capacitive or inductive reactance $X(f)$. The impedance that would be expressed in equation form as:

$$Z(f) = R + j \cdot X(f)$$

composite cable A cable composed of different types of information-carrying media combined within one sheath. For example, a cable composed of:

- Different wire gages.
- Different mechanical structures (such as single conductor and twisted pair).
- Different wire types (such as single or twisted pair conductor and coaxial).
- Different technologies (such as optical fiber and metallic conductors).

Cable containing only one information medium and a metallic strength member or metallic armor does not qualify as a composite cable.

composite signaling (CX) In telephony, a direct current signaling scheme that separates the signals from the voice band by filters. Also called *CX signaling*. Contrast with *direct current signaling* (*DX*).

composite two-tone test signal A test signal composed of two different frequencies and used for intermodulation distortion measurements.

composite waveform A waveform that either is (or for illustrative or analytical purposes is treated as) the algebraic sum of two or more waveforms. For example, the following figure shows a composite analog waveform composed of a pure tone, periodic switching noise, and random noise.

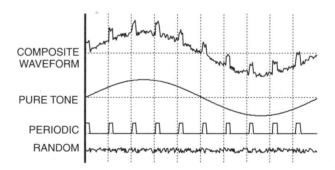

composited circuit A frequency division multiplex (FDM) circuit that can be used simultaneously either for telephony and low-speed data transmission or for telephony and signaling. Also called *voice plus circuit*.

compound signal In ac signaling, a signal consisting of the simultaneous transmission of more than one frequency.

Examples of compound signaling include:

- Dual tone multifrequency (DTMF) signaling, and
- Multifrequency (MF) trunk signaling.

compress A UNIX utility for data compression and decompression. The file extension of the output of the compression program is generally .Z. See also *compression* and *filename extension*.

compressed file A file whose contents have been reduced in size by a special utility program. Size reductions vary depending on the original file type, but the range is about 1.5:1 to 10:1 for lossless compression and up to 200:1 (or more) for *lossy compression*.

The utility program "looks at" the contents of the file and removes redundant information (duplicate information) in such a way that it can be restored when the compressed file is expanded. Some utility programs that are available to do compression are: *ARJ, LHARC, PKZIP/PKUNZIP,* and *STUFFIT.*

compression (1) In signal transmission, the process of reducing the effective gain applied to a large signal as compared to a small signal. The gain change associated with *compression* may be applied to the signal in an instantaneous manner as with a *compander,* or it may be slowly applied as with an *automatic gain control.* **(2)** In software, a program that identifies the redundancies in a file and removes them. The process may be lossless or lossy; that is, some information is lost in a lossy compression method. However, all information is preserved with a lossless compression algorithm. Lossy compression is generally used only either to compress picture information where the loss of information results in a slight "fuzzing" of the picture, or it is used only to compress an audio file where a slight increase in noise level can be tolerated. Most lossless methods create a temporary "dictionary of symbols" found in the file and then transmit a pointer to the dictionary entry rather than retransmit the symbol. The differences between methods involve the size of the dictionary, the number of characters comprising a symbol, and time of "temporary." Also called *compaction.* See also *ARC, ARJ, compressed file, data compression, Huffman coding, JPEG, Lempel-Ziv compression, LHA, LZW, PAK, repeated character compression, RLE, RLL, SDN, STUFFIT, TAR,* and *ZIP.*

compression algorithm The mathematical procedure by which a signal's bandwidth is reduced. The bandwidth reduction may be accomplished by removing redundancy (lossless compression) or by removing portions of real information that do not materially influence the outcome of the transferred information (lossy compression). The results of lossy compression generally look like increased noise to the destination receiver.

compression ratio (1) In signal compression, the ratio of the dynamic range of compressor input signals to the dynamic range of the compressor output signals. The compression ratio is usually expressed in decibels (dB). For example, a 50-dB input dynamic range compressed to a 30-dB output dynamic range would be 20 dB of compression. **(2)** In data compression, the ratio of the compressed message to the uncompressed message. The compression ratio is generally expressed as a percentage. **(3)** In digital facsimile, the ratio of the total pels scanned for an image to the total encoded bits sent for picture information. **(4)** The ratio of the gain of a device at a low power level to the gain at some higher level. The compression ratio is usually expressed in decibels (dB).

compressor A nonlinear signal device that has a lower gain with higher input levels than with lower input levels. The input signal's dynamic range is therefore reduced according to the prescribed nonlinear algorithm.

Compressors are used in conjunction with *expanders* to allow signals with a large dynamic range to be sent through devices or circuits with a smaller dynamic range and to reduce the effects of noise introduced in the communications medium. See also *compander* and *compression (1).*

CompuServe Information Service (CIS) One of the oldest and largest online information networks accessible via a modem.

computer A machine with at least the following characteristics:

- The ability to store and retrieve numbers. (Letters are represented by numbers—see ASCII, for example.)
- The ability to add and subtract numbers.
- The ability to perform a predetermined sequence of instructions.
- The ability to compare numbers.
- The ability to alter the sequence of instructions executed, based on the comparison of numbers.
- The ability to accept information from a user interface (keyboard, card reader, tape reader, etc.).
- The ability to write information to a user interface (monitor, card punch, printer, tape punch, etc.).

computer aided instruction (CAI) A teaching technique that uses computers to help students practice lesson materials, e.g., mathematics problem solving, spelling drills, grammar, and geography. Because the learning is interactive, students who are not as quick to pick up a concept may be given extra assistance or may repeat presented material as many times as is desired.

Computer and Business Equipment Manufacturers Association (CBEMA) An association of manufacturers that lobbies Congress and the FCC to promote the interests of terminal, computer, and peripheral equipment industries.

Computer and Communications Industry Association (CCIA) An organization of communications and data processing companies that promotes their interests before Congress and the FCC.

Computer Emergency Response Team (CERT) In Internet, a group formed in 1988 in response to security problems on the network. *CERT* works with network users to:

- Facilitate responses to security problems;
- Conduct research on existing and emerging Internet security systems;
- Provide a variety of security services and products, such as an anonymous FTP server where security-related documents and tools are archived; and
- Raise the awareness of the Internet users to security issues and problems.

computer network (1) A collection of autonomous computers interconnected in such a way as to allow sharing of resources and/or provide data communications (e-mail, for example). **(2)** A communications network in which the end instruments are computers.

Computer + Science Network (CSNET) Formerly an independent network; merged with BITNET in 1989.

computer rate The rate information is transferred from the DTE (computer) to the DCE (modem). Also called *communications rate, DTE-DCE rate, modem-to-computer speed,* and *serial port speed.*

computer science The discipline that is concerned with methods and techniques relating to data processing performed by automatic means.

Computer Supported Telephony Application (CSTA) A European Computer Manufacturer's Association (ECMA) standard for linking computers to telephone systems.

computer system A functional unit consisting of one or more computers and associated software. It may be a stand-alone system, or it may consist of several interconnected systems. The basic attributes of a *computer system* include:

- The use of common storage for all or part of a program and for all or part of the data necessary for the execution of the program.
- The execution of user-written or user-installed programs.
- User-designated manipulations (including arithmetic and logical operations) of user-designated data.

Also called an *automatic data processing (ADP) system* or a *computing system.* See also *computer.*

computer system fault tolerance The ability of a *computer system* to continue to operate correctly, even though one or more of its components are malfunctioning. The system's performance (in aspects such as speed or processing throughput) may be reduced (called graceful degradation) until the faults are corrected. Also called *computer system resilience.*

computer telephone integration (CTI) A generic term for the convergence of voice and data into one platform. The technology relates computers and PABXs via applications such as Automatic Call Distribution (ACD), power dialing, Interactive Voice Response (IVR), and other customer-facing or agent-facing services. Access to the features is via keyboard or mouse action on the computer.

- *Caller ID.* Lists the phone number and other pertinent data such as name of the calling party and other data from the local computer database. Reports who called whether or not the phone was answered. The Caller ID data is from a telephone company service.
- *Contact management/organizer.* Stores names, phone numbers, addresses, background information, birthdays, notes, followup data, and so on, of friends, family members, and business acquaintances in the local computer database.
- *Telephone dialer.* Dials a selected phone number from the local computer database. Automatically adds the appropriate access codes, area codes, and billing codes as required.
- *Message recording.* Provides answering-machinelike services but with the potential for delivering unique messages to a specific caller. Based on the Caller ID number, for example.
- *Cost accounting.* Generates an estimate of the monthly bill based on the destination and duration of each outgoing call.
- *Least cost routing* (*LCR*). When multiple outgoing phone numbers are available with different office codes, the software will select the phone number with the lowest billing rate.
- *Conferencing, call transfer,* and *call waiting.*
- *Voice command.* Uses voice recognition to access the computer database and to issue commands such as "dial Mr. Jones."

Also known by older, proprietary names such as *Computer Integrated Telephony* (*CIT*) and *Computer Supported Telephony Applications* (*CSTA*).

computer word See *word.*

COMSAT An acronym from <u>COM</u>munications <u>SAT</u>ellite Corporation. A private corporation authorized by the United States Satellite Communications Act of 1962 to represent the United States in international satellite communications and to operate domestic and international satellites.

CON (**1**) An acronym from <u>C</u>onnection <u>O</u>riented <u>N</u>etwork. A network that defines one path per logical connection. (**2**) An abbreviation of <u>con</u>sole. Originally, the term referred to the switches and lights on a computer used for control and monitoring. Later it was extended to include teminals and teletypewriters. Still later it incorporated keyboards and monitors, and now it includes PCs used to communicate with and control other systems.

concatenate To join together sequentially. For example, if given the two strings "eat" and "fish," one could concatenate "fish" to "eat" and get "eat fish." If "eat" were to be concatenated to "fish," the result would be "fish eat."

concentration The function associated with switching networks having fewer output ports than input ports. The inverse function is *expansion.*

concentrator (**1**) Generally, any communications device that allows a shared transmission medium to accommodate more data sources than there are allocated channels in the medium. (**2**) In telephony, a switching entity for connecting a number of input ports to a smaller number of outlet ports. (**3**) A device that combines data from several low-speed data streams into a single high-speed data stream before sending it to the communication link. At the receiving end of the communication link a *distributor* separates the combined signal stream into its constituent components and then forwards these data on to the appropriate terminating DTE.

Both a statistical multiplexer and concentrator are used to focus selected multiple-channel inputs onto a composite link. A concentrator, however, interleaves the selected channel inputs in rotation, whereas a statistical multiplexer removes those portions of the transmission carrying no data. See also *distributor.* (**4**) In Ethernet, a hub with multiple 10BaseT ports, a 10Base2 BNC port, and an AUI 10Base5 port. See also *hub.*

concentric The characterization of two or more entities having the same center. A circle and a square with the same center are said to be *concentric.*

concentricity In coaxial systems (wire or optical fiber), the measure of location of the central member's axis with respect to the axis of the outer elements.

concentricity error The amount by which the axis of a coaxial system is offset from the axis of the outer member's axis, that is, the distance between the centers of two circles representing the inner and outer members of the structure. The members of the coaxial system may be the inner and outer conductors of an electrical coaxial cable or the core and cladding of an optical fiber.

In fiber optics, the *concentricity error* is frequently specified with tolerance fields that characterize the optical fiber core and cladding geometry. In fiber optic systems, concentricity error is also called *core eccentricity, core-to-cladding concentricity, core-to-cladding eccentricity,* and *core-to-cladding offset.* See also *coupling loss* and *tolerance field.*

concurrency control In computer networks, rules that regulate data file access by multiple users on a network server.

concurrent operation In data link operations, the operation in which two or more data links are used during the same, usually short, time interval, while adhering to the protocols of each link without providing data forwarding among the links. Also called *multitasking.*

concurrent processing A term describing the situation where two or more programs effectively execute in parallel on either multiple processors or asynchronously on a single processor. In the latter case, the programs appear to be executing *concurrently* because the microprocessor is much faster than people can directly observe. *Concurrent processes* may or may not be related to each other. Each program is free to suspend processing and wait for an external event before proceeding. Also called *multitasking.*

conditioned baseband representation A synonym for *nonreturn to zero mark.*

conditioned circuit A communications circuit that has been "tuned" (i.e., one or more parameters are optimized) to obtain a desired characteristic by the addition of *conditioning equipment.* Usually used for voice or data transmission improvement. Also called a *conditioned analog line.* See also *conditioning.*

conditioned loop A loop that has been "tuned" for a desired characteristic by the addition of *conditioning* equipment.

conditioning The use of special equipment, inserted in the communications path, to improve the characteristics of a dedicated communication link in order that the transfer of information is at an acceptable error rate. Conditioning equipment (also called an *equalizer*) controls or compensates for attenuation, noise, and various forms of distortion.

Initially, *line conditioning* was specified by AT&T as C and D conditioning. C conditioning controls amplitude distortion and envelope delay distortion, whereas D conditioning controls SNR and nonlinear distortions. C-conditioned noise limits are dependent on circuit length (mileage); that is, for 0 to 50 mi the maximum noise is 31 dBrnC0. These specifications are applied to end-to-end connection, including the subscriber's loop and interoffice trunks. Following the divestiture of the AT&T system, BellCore developed standardized Voice Grade (VG) circuit definitions as outlined in the Voice Grade Type Acceptance Limits chart. These specifications are being adopted by many non-Bell companies. The VG types define performance for end links only and do not cover interoffice trunking. As with C conditioning, noise is dependent on circuit length (mileage). That is, 0 to 50 mi the maximum noise is 30 dBrnC0. Also following divestiture, AT&T developed standards that apply to the interoffice trunk circuits (lines that make up the long-distance portion of the network). These standards are called Service Type specifications, some of which are outlined in the AT&T Service Type Acceptance chart on page 118.

C CONDITIONING SPECIFICATION

Conditioning Specification	Frequency Range (Hz)	Attenuation Distortion (+ = loss) (dBr)	Envelope Delay Distortion (μs)
None	300–500	−3 to +12	
	500–800	−2 to +8	
	800–2500	−2 to +8	1750
	2500–2600	−3 to +12	1750
	2500–3000	−3 to +12	
C1	300–800	−2 to +6	
	800–1000	−2 to +6	1000
	1000–2400	−1 to +3	1750
	2400–2700	−2 to +6	1000
	2700–3000	−3 to +12	
C2	300–500	−2 to +6	
	500–600	−1 to +3	3000
	600–1000	−1 to +3	1500
	1000–2600	−1 to +3	500
	2600–2800	−1 to +3	3000
	2800–3200	−2 to +6	
C3	300–500	−0.8 to +3	
	500–600	−0.5 to +1.5	650
	600–1000	−0.5 to +1.5	300
	1000–2600	−0.5 to +1.5	110
	2600–2800	−0.5 to +1.5	650
	2800–3000	−0.8 to +3	
C4	300–500	−2 to +6	
	500–600	−2 to +3	3000
	600–800	−2 to +3	1500
	800–1000	−2 to +3	500
	1000–2600	−2 to +3	300
	2600–2800	−2 to +3	500
	2800–3000	−2 to +3	3000
	3000–3200	−2 to +6	
C5	300–500	−3.0 to +3.0	
	500–600	−0.5 to +1.5	600
	600–1000	−0.5 to +1.5	300
	1000–2600	−0.5 to +1.5	100
	2600–2800	−0.5 to +1.5	600
	2800–3000	−3.0 to +3.0	

Attenuation distortion is relative to the response at 1004 Hz; envelope delay distortion numbers are relative to the minimum envelope delay in the band.

BELLCORE VOICE GRADE TYPE ACCEPTANCE LIMITS

Conditioning Specification	Frequency Range (Hz)	Attenuation Distortion (+ = loss) (dBr)	Envelope Delay Distortion (μs)
VG-1	304–404	−2.5 to +7.5	
	404–504	−1.5 to +9.5	Not
	504–2504	−1.5 to +7.5	specified
	2504–2804	−1.5 to +9.5	
	2804–3004	−2.5 to +7.5	
VG-2	304–404	−0.5 to +4.5	Not
	404–2804	−0.5 to +3.5	specified
	2804–3004	−0.5 to +4.5	
VG-3	304–404	−0.5 to +4.5	Not
	404–2804	−0.5 to +2.5	specified
	2804–3004	−0.5 to +4.5	
VG-4	304–504	−0.5 to +3.0	
	504–2504	−0.5 to +1.5	Not
	2504–2804	−0.5 to +2.5	specified
	2804–3004	−0.5 to +3.5	
VG-5	404–2804	−0.5 to +4.5	Not specified
VG-6	404–504	−0.5 to +3.5	
	504–804	−0.5 to +2.5	650
	804–2504	−0.5 to +2.5	650
	2504–2604	−0.5 to +2.5	650
	2604–2804	−0.5 to +3.5	
	2804–3004	−0.5 to +4.5	
VG-7	304–404	−0.5 to +4.5	
	404–804	−0.5 to +1.5	
	804–2604	−0.5 to +1.5	650
	2604–2804	−0.5 to +1.5	
	2804–3004	−0.5 to +4.5	
VG-8	304–404	−0.5 to +4.5	
	404–804	−0.5 to +1.5	
	804–2604	−0.5 to +1.5	650
	2604–2804	−0.5 to +1.5	
	2804–3004	−0.5 to +4.5	
VG-9	304–404	−0.5 to +4.5	
	404–804	−0.5 to +1.5	
	804–2604	−0.5 to +1.5	650
	2604–2804	−0.5 to +1.5	
	2804–3004	−0.5 to +4.5	
VG-10	304–404	−2.5 to +11.5	
	404–504	−1.5 to +9.5	
	504–804	−1.5 to +7.5	
	804–2504	−1.5 to +7.5	1700
	2504–2604	−1.5 to +9.5	1700
	2604–2804	−1.5 to +9.5	
	2804–3004	−2.5 to +11.5	
VG-11	304–804	−0.5 to +4.5	
	804–1204	−0.5 to +4.5	650
	1204–2604	−0.5 to +0.5	650
	2604–3004	−0.5 to +4.5	
VG-12	304–504	−0.5 to +2.0	
	504–804	−0.5 to +0.5	
	804–2604	−0.5 to +0.5	650
	2604–2804	−0.5 to +0.5	
	2804–3004	−0.5 to +2.0	

Attenuation distortion is relative to the response at 1004 Hz; envelope delay distortion numbers are relative to the minimum envelope delay in the band.

conditioning equipment (**1**) Equipment installed at the junctions of circuits, used to obtain desired interface characteristics, such as matched transmission levels, matched impedances, and equalization between facilities. (**2**) Corrective networks installed on transmission lines used to improve data transmission performance. These net-

AT&T SERVICE TYPE ACCEPTANCE LIMITS

Conditioning Specification (Service Type)	Frequency Range (Hz)	Attenuation Distortion (+ = loss) (dBr)	Envelope Delay Distortion (µs)
1	304–404	−3 to +12	Not
	404–2804	−2 to +9	specified
	2804–3004	−3 to +12	
2	304–404	−3 to +12	Not
	404–2804	−2 to +6	specified
	2804–3004	−3 to +12	
3	304–504	−2 to +9	
	504–2504	−2 to +6	Not
	2504–2804	−2 to +8	specified
	2804–3004	−2 to +11	
4	404–2804	−4 to +12	Not specified
5	304–404	−3 to +12	
	404–504	−2 to +10	
	504–804	−2 to +8	1750
	804–2504	−2 to +8	1750
	2504–2604	−2 to +10	
	2604–2804	−2 to +10	
	2804–3004	−3 to +12	
6	304–404	−3 to +12	
	404–804	−2 to +6	
	804–2604	−2 to +6	1250
	2604–2804	−2 to +6	
	2804–3004	−3 to +12	
7	304–404	−3 to +12	
	404–804	−2 to +5	
	804–2604	−2 to +5	550
	2604–2804	−2 to +5	
	2804–3004	−3 to +12	
8	304–404	−3 to +12	
	404–804	−2 to +4	
	804–2604	−2 to +4	400
	2604–2804	−2 to +4	
	2804–3004	−3 to +12	
9	304–804	−3 to +12	
	804–1204	−3 to +12	600
	1204–2604	−3 to +3	600
	2604–3004	−3 to +12	
10	304–504	−2 to +6	
	504–804	−1 to +3	
	804–2604	−1 to +3	2000
	2604–2804	−1 to +3	
	2804–3004	−2 to +6	

Attenuation distortion is relative to the response at 1004 Hz; envelope delay distortion numbers are relative to the minimum envelope delay in the band.

works optimize characteristics, such as equalization of the insertion loss vs. frequency characteristic and/or the envelope delay distortion over a desired frequency range. See also *conditioning*.

conductance The "opposite" of resistance, that is, a measure of the ability of a conductor to carry a current. The units of conductance are siemens (formerly Mhos). **(1)** The real part of *admittance*. **(2)** That property of an element, device, or system which when multiplied by the mean-square voltage ($2/RNS$) yields the value of the power lost due to radiation of heat, light, or other electromagnetic energy.

conducted noise Noise produced by one device or system that enters a second device or system *by propagation along a conductor,* i.e., any physical path. The conductor may be either the mains power cable, a communications line, or any other circuit conductor.

conductive coupling Energy transfer by means of physical contact, i.e., coupling other than inductive, capacitive, or optical coupling. *Conductive coupling* may be by wire, resistor, or other conductive (generally metallic) element. *Conductive coupling* passes the full spectrum of frequencies, including dc, and is limited only by stray and parasitic capacitance and inductance. Also called *direct coupling*.

conductivity A term used to describe the ability of a material to carry an electrical current.

conductor A material with the ability to pass an electric current with little loss. Most metals are good conductors. See also *insulator* and *semiconductor.*

conductor loop resistance In telephony, the series resistance of both conductors of a trunk or subscriber line, excluding any terminal equipment. Loop resistance is used as a measure rather than distance because different gauge wires are used to construct loops. For example, AWG 26 wire might be used on a short loop, whereas AWG 19 wire might be used on a very long loop and both loops might have the same loop resistance.

cone of protection Lightning protection whose volume is defined by the highest point on the lightning rod or protecting tower as the apex of a cone. The cone diameter is one to two times the apex height (depending on the apex height and the cloud height). When overhead ground wires are also used, the volume is called the *protected zone* or *zone of protection.*

conference (1) In bulletin board systems (BBS), an activity on a BBS where multiple people can communicate with one another as if they were all sitting at the same table. A *conference* is a special case of a chat in which a single topic is discussed and frequently a "guest speaker" or moderator is present. As with a chat, all typed messages are broadcast to all participants of the *conference*. See also *chat, door,* and *room.* **(2)** In telephony, a shortened form of *conference call.*

conference bridge In telephony, a device that allows multiple callers to be connected together for a *conference call*. The *conference bridge* contains electronics that amplify and balance each caller's signal level so that all parties can hear each other. Generally, bridges continuously monitor call progress and signal conditions in order to compensate for callers entering and leaving the conference.

conference call In telephony, a call in which more than two main stations are interconnected for concurrent communications.

configuration (1) Generally, a term used to refer to the way a device is set up. **(2)** The specific hardware and software settings that allow a device to operate in a system. One *configures* hardware or software by manipulating switches or jumpers, or by setting appropriate values in registers. **(3)** The total combination of hardware components that make a system. For example, the *configuration* of a computer system includes not only the CPU but the keyboard, monitor, modem, printer, tape, disk drives, and any other peripheral attached.

Configuration Management One of five OSI network management domains specified by the ISO and ITU. *Configuration Management* is concerned with:

- Identifying objects on the network and their attributes. Objects include entities such as stations, bridges, routers, and even circuits. Attributes of objects can include information such as interface settings (e.g., speed, parity, jumper settings), model and vendor information (e.g., serial number, operating system, memory, hardware address), other details (including installed drivers, peripherals, maintenance and testing schedules), and so on.

- Determining states, settings, and other information about these objects. The OSI model defines four operational states for an object:
  - *Active*: The object is available and in use but has the capacity to accept services or requests.
  - *Busy:* The object is available and in use, but currently is not able to deal with any more requests.
  - *Disabled:* The object is not available.
  - *Enabled:* The object is operational and available but not currently in use.
- Storing this information for later retrieval or modification.
- Reporting this information if requested by an appropriate and authorized process or user.
- Modifying the settings for objects, if and when necessary.
- Topology management, which involves managing the connections and relationships among the objects.
- Initiating and terminating network operations.

configuration profile A DCE's configurable operating characteristics. Modems have a set of registers (called S-registers) that determine how a modem will operate. The values in the registers (and the related operating characteristics) are called the *configuration profile*.

congestion The condition that occurs in a communications system (or switching system) when the number of requests for service exceeds the capacity of a selected path (or device). This can result in performance degradation or loss of data, or simply, excessive network traffic.

congestion control In packet switching systems, the process in which packets are discarded to clear buffer overflow (congestion). Discarded packets are discarded in a priority order when possible.

congestion tone In telephony, a call progress tone that indicates a path cannot be established through the switching network. There may be no interoffice trunks available, or the switching system may be blocked. The state of the user line is unknown. Generally, congestion conditions are short-lived unless there is an emergency in the area. Also called *fast busy*. See also *blocking, call progress tones,* and *trunk*.

conjunction A synonym for the Boolean AND operation.

connect time (1) The amount of time one system is connected to another system whether or not data are being transferred. *Connect time* is generally measured in hours and minutes, but sometimes it is in fractional hours or in just minutes. *Connect time* is important because on many systems the user is billed for connect time, whether or not information is being transferred. (2) The time interval during which a request for a connection is being processed and completed.

connected (1) In communications, the state a communication path is in when a path is established and it is ready for information flow. (2) In speech recognition, natural speech patterns in which there is less than approximately 50 ms between utterances.

connecting arrangement In public switched telephone networks, the equipment provided by a common carrier to accomplish electrical interconnection between customer-provided equipment (CPE) and that of the common carrier.

connecting block A wire terminating device that can be used to join two or more wire ends. The terminations are generally an insulation displacement (IDC) type; that is, the insulation on the wire needs to be stripped off before the connection is made. The terminator automatically pierces the wire's jacket and makes a gas tight connection to the conductor. Also called a *cross connect block, punchdown block, quick connect block,* and *terminal block*.

connection (1) An electrical continuity between two conductors or devices. (2) A provision (physical or virtual path) for a signal to propagate from one communicating device to another, such as from one circuit, line, subassembly, or component to another. Examples of communicating devices include telephones, computers, and network nodes. (3) An association established between functional units for conveying information. (4) The point of attachment between devices or systems.

connection admission control (CAC) Actions performed by a network to enforce network admission policies. These actions may be performed during call setup or during call renegotiation, and they determine whether or not to grant a Virtual Channel Connection or a Virtual Path Connection to a user. Information from the Traffic Contract is used to make this determination.

connection number The unique number assigned to each node that attaches to a file server. The network operating system (NOS) on the file server uses the *connection number* to control how nodes communicate with each other. A node will not necessarily be assigned the same connection number each time it attaches to the network.

connection oriented A service in which a semipermanent connection (a path) must be established between the source and destination before any information transfer takes place. With this service, packets are guaranteed to reach their destination in the order sent because all packets travel along the same, "no-passing" path. The communications protocol has three distinct phases:

- *Establishing* the connection between a calling station and a called station.
- *Transferring* information, either on a continuous or a packet basis, but always on the path previously established.
- *Releasing* the connection at the end of the call (returning the communications equipment to the idle state so that other stations may use the physical circuits).

Examples include dial-up telephone service, FTP, ITU X.25, and NetWare's Sequenced Packet Exchange (SPX) protocol. See also *circuit switching, connectionless mode transfer, dedicated circuit, message switching,* and *packet switching*.

connection-oriented transmission Information transference across a path established prior to the transfer and maintained throughout the entire transmission. The path is dissolved only at the end of the transaction.

Connection-Oriented Convergence Function (COCF) In the Distributed Queue Dual Bus (DQDB) network architecture, a function that prepares data coming from or going to a connection-oriented service. The service first establishes a fixed, but temporary, connection, then transmits the data, and finally breaks the connection. See also *IEEE 802.x*.

connection rate The rate information is transferred from DCE (local modem) to DCE (remote modem) over the telephone lines. Also called *DCE-to-DCE speed, line speed,* and *modem-to-modem speed*. See also *baud* (Bd).

connection-related function (CRF) The term used in the ATM Forum UNI 3.0 for switch, concentrator, or other network equipment that handles virtual channels CRF(VC) or virtual paths CRF(VP).

connectionless A data transmission method that does not need to wait for a path to be established between end points before transmission can commence. Individual packets are routed independently from the source to their destinations, so that two packets from the same message or transmission might take two different paths. Because packets travel independently, they may not arrive in the transmitted order. Consequently, the original sequence needs to be reconstructed at the destination end. This is generally accomplished at the transport layer in the ISO/OSI Reference Model.

Examples of connectionless protocols include the ISO/OSI Connectionless mode Network Protocol (CLNP) and Connectionless mode Transport Protocol (CLTP); the TCP/IP User Datagram Protocol (UDP); and the NetWare Internetwork Packet Exchange (IPX) and Message Handling Service (MHS) protocols. *Connectionless* is sometimes abbreviated *CL*. See also *connection oriented*.

connectionless broadband data service (CBDS) A European Telecommunications Standards Institute high-speed switched data network specification submitted to ITU-T. Basically similar to and compatible with BellCore's Switched Multimegabit Data Service (SMDS).

connectionless convergence protocol A set of ITU-T recommendations that govern the protocol layer above the ATM adaptation layer (AAL). It corresponds to layer 3 of Switched Multimegabit Data Service (SMDS). With this correspondence, a network can offer SMDS regardless of the network fabric. That is, the network can be based on either ATM or IEEE 802.6.

connectionless mode transfer In packed switched networks, the individual transfer of each data packet, independent of those that precede or follow it. Each packet is routed (transferred) based on information contained within its header. The concept of an end-to-end connection is meaningless.

In *connectionless mode transfer,* a path is established between an originating station and a destination station only for the duration of each packet. The physical route may vary from one packet to the next based on many factors, including traffic loading. Because each packet is numbered at the time of transmission, the receiving station can reassemble the original message sequence even if the packets arrive out of order at the final destination. ITU-T recommendation X.25 describes a set of standards for packet switching on networks. See also *connection oriented*.

connectionless network A packet switching network that uses connectionless mode transfers.

connectionless transmission A data packet transmission mode in which no predetermined path is established prior to the launch of the data packet. The data packet contains the address of the destination embedded in the header. Switching equipment reads the header and routes the packet toward the desired destination.

Connectionless Transport Service (CLTS) In the ISO/OSI Reference Model, a transport layer service that does not guarantee delivery but makes a best effort, does error checking, and uses end-to-end addressing.

connectivity The ability to make hardware and/or software work together as desired. The property that allows dissimilar devices to communicate with each other.

connector A coupling device used to join the conductors (electrical or optical) of one circuit or transmission element with the conductors of another circuit or transmission element on a semipermanent basis. The conductors may be electrical wires, optical fibers, or waveguides.

Most connectors are available as male or female types. The male connector, or plug, has one or more exposed pins, while the female, or jack, connector has one or more sockets designed to accept the pins

of the male connector. A third connector type is genderless and is called a hermaphrodite. There are literally thousands of connector families, each with its own designation. The following table lists just a few. A *connector* is distinguished from a splice, which is a permanent joint.

CONNECTOR TYPES WITH BRIEF DESCRIPTION

Type	Description
BNC	Bayonet locking coaxial cable connector.
D-4	Threaded coupling 2 mm ferrule optical fiber connector. Life is about 1000 matings.
DB	Multi-contact plug; 9, 15, 25, & 37 pin popular. Used on RS-232, RS-422, printer, and PC video cables.
DIN	Electrical multi-conductor connector.
ESCON	Similar to the MIC fiber connector except it has a retractable cover. 500 mating, 0.5 dB insertion loss, and > 35dB reflection loss.
F	Electrical. Applications in 10Broad36, CATV, and IEEE 802.4 token bus.
FC	Threaded coupling, 2.5 mm ferrule fiber optic connector. 1000 matings, 0.3 dB insertion loss, 40 dB reflection loss. Older Japanese type, being superseded by SC and MIC
MMJ	Modified version of the RJ series.
MIC	Dual fiber FDDI connector. Designed by ANSI. Attaches to either an MIC or two Sts. 500 matings, 0.3 to 0.5 dB insertion loss, and >35dB reflection loss.
N	Threaded coupling coaxial cable connector.
RJ	Standard telephone twisted pair connector.
SC	Locking fiber optic connector. 1000 matings, 0.3 dB insertion loss, and 40 dB reflection loss.
SMA	(1) Miniature threaded coaxial cable connector. (2) Threaded coupling fiber optic connector. 200 matings and 1.5 dB insertion loss.
ST	AT&T's straight tip, BNC, 2.5 mm ferrule fiber optic connector. 250 matings, 0.3–0.8 dB insertion loss, >40 dB reflection loss.
TNC	Threaded version of BNC.
V.32	Multi-contact similar to the DB series.

connector loss Energy loss encountered at the union of two transmission entities. Losses may be due to mechanical, material, or electronic causes. See also *characteristic impedance* (Z_o) and *coupling loss*.

connections per circuit hour (cch) A unit of traffic measurement expressed as the number of connections established at a switching point per hour.

CONP An abbreviation of Connection Oriented Network Protocol.

CONS (**1**) An acronym from Connection Oriented Network Service. An upper layer protocol of IBM's System Network Architecture (SNA). (**2**) An acronym from COnnection Mode Network Service.

conservation of radiance A basic principle of optics stating that no passive optical system can increase the quantity

$$L/n^2$$

where:

L is the radiance of a beam and
n is the local refractive index.

Conservation of radiance was formerly called *conservation of brightness* or the *brightness theorem.*

constant bit rate (CBR) A term that refers to processes such as voice which require constant, repetitive, or uniform information. These processes tend to be sensitive to variable delays and hence, must be transported without any interruption in the flow. See also *variable bit rate* (*VBR*).

constellation (1) The pattern obtained by plotting the location of the carrier phase and amplitude for each possible bit combination of a modulator on a Cartesian or polar plot.

It is a pictorial or graphical way of viewing all of the possible states (the alphabet) of a carrier for the modulation method. A constellation is generally displayed as a polar plot depicting signal amplitude as the distance from the origin (center) and phase as an angle from the zero angle-axis. An equivalent Cartesian plot can also be used if the information is more easily understood in that format. The distance between points gives an indication of the relative sensitivity for each symbol. The constellations in these two polar plots depict a two-state

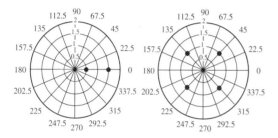

amplitude shift key (ASK) modulated carrier (with constant phase) and a four-phase phase shift key (PSK) modulated carrier (with constant amplitude). (2) The arrangement of a specified set of objects in space, e.g., stars or satellites.

contact bounce In mechanical switches, the undesired, unpredictable, and intermittent opening of a contact during the closure of that open contact. Conversely, the closing of a contact during the opening action. The result in either case is an irregular rise or fall of the signal transition associated with the switch action. The diagram illustrates the signal waveform of an ideal switch contact and the waveform of a contact with *bounce*.

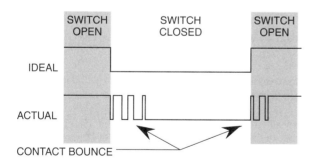

contention (1) The facility of a network or selector that allows multiple devices to compete on a first-come-first-serve basis. No device on the channel is designated a master or has a higher priority; all devices have equal access to the channel. Each device on the channel competes for access by sending signals at will after determining that the channel is idle. *Note:* Several terminals may determine that the channel is idle, at the same time creating a condition known as collision. (2) The condition on a multipoint communication channel when two or more devices attempt to seize the channel at the same time.

context That portion of the Directory Information Tree (DIT), as defined in the ITU X.500 Directory Services (DS) model, which contains information about all directory objects. In Novell's NetWare 4.x NDS, the location in the Directory tree.

contiguous Having a shared or joined boundary. No breaks, spaces, or gaps between adjacent segments.

continuity check (1) Generally, a test to verify a connection between two specified points. (2) In common channel signaling, a test performed to check that a path exists for speech or data transmission.

continuous In speech recognition, generally refers to natural speech patterns in which there is less than approximately 50 ms between utterances.

continuous phase frequency shift key (CPFSK) Frequency shift modulation wherein the phase of the output does not change at the transition between adjacent bits. See also *frequency shift key (FSK)*.

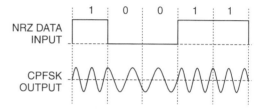

continuous wave (CW) (1) In transmission, a succession of waves traversing a medium such that, in the steady state, each wave is substantially the same as the preceding wave; that is, it has a constant amplitude and constant frequency over time. Also abbreviated *cw*. (2) In lasers, the laser's output may be considered continuous rather than pulsed if there is output for periods greater than 0.25 s.

continuously variable slope delta (CVSD) modulation A type of delta modulation in which the size of the steps of the approximated signal is progressively increased or decreased as required to make the approximated signal more closely match the input analog wave. The step size is increased as the slew rate of the input signal exceeds the slew rate of the current step size and is decreased otherwise. See also *delta sigma modulation* ($\Delta\epsilon M$).

control character A character that initiates, modifies, or stops a function, event, operation, or control operation. Control characters may or may not have a graphic representation. Generally, one of the first 32 characters in the *ASCII* character set (0 through 1Fh) plus the <space> and <nul>. See *ASCII* for its complete list and *EBCDIC*.

control code A character or sequence of characters that produce some action in a device other than the normal passage of data through the device. In some cases a "normal" character (or character string) may be treated as a control character if preceded by the proper control element.

Examples:
- *Control character alone.* Many data transmission devices will start or stop sending information if they receive a <DC1> or <DC3> control character. There are 32 control codes defined in the ASCII character set that are used primarily to control hardware.
- *Control character plus alphanumeric.* The <ESC> character followed by one or more characters is used to control a printer's operation and output.
- *Control sequence plus alphanumeric.* The AT command string is composed of "normal" characters and is treated as a "normal" character string by a Hayes compatible modem *unless* it is preceded by the control code string <guard> +++ <guard>.

control field An area (field) in a frame (packet) allocated for the purpose of control. For example, in a protocol data unit (PDU), the field that contains data interpreted by the destination's logical link controller (LLC). (The field may immediately follow the destination service access point [DSAP] and source service access point [SSAP] address fields of the PDU.)

control operation Any action that affects the transmission, recording, processing, or interpretation of data by a device or system.

control point A program that manages an Advanced Peer to Peer Network (APPN) node and its resources, enabling communications to other control points in the network.

control segment A network of Global Positioning System (GPS) monitor and control stations that ensure the accuracy of the satellite positions and their clocks. The *control segment* updates the almanac for use by the ground-based GPS receivers.

control station In a communications network, the station that selects the master station and supervises operational procedures, such as polling, selecting, and recovery. The *control station* has overall responsibility for the orderly operation of the entire network.

control unit terminal (CUT) A mode that allows an IBM 3270 terminal to have a single session with a mainframe computer.

controlled access unit (CAU) The term for an intelligent hub in IBM Token Ring networks. *CAUs* can determine whether nodes are operating, connect and disconnect nodes, monitor node activity, and pass data to the LAN Network Manager program.

controlled access unit/lobe attachment module (CAU/LAM) In IBM Token Ring networks, a hub (the *CAU*) containing one or more devices (the *LAM*) with multiple ports to which new nodes can be attached.

conventional memory See *base memory*.

convergence An automatic process by which routers on a packet switching LAN build and maintain an accurate description of a dynamic network topology. Specifically, the point at which a substantial majority of the routing information is in agreement. See also *advertising*.

converter (1) A device for changing energy from one form to another form. The converter may change mechanical energy to electrical, optical to electrical, ac to dc, analog to digital, and so on. (2) A device for changing one code to another, e.g., changing a simple binary sequence to an alternate mark invert (AMI) code for transmission to telephone T-carrier. See also *waveform codes*. (3) A repeater that also converts from one medium type to another, e.g., from fiber to copper. Also called a *media adapter*.

convolutional code An error-correction code for continuous data streams in which each *m*-bit information symbol is transformed into an *n*-bit symbol ($n>m$). The transformation is based on the previous *k* information symbols, and *k* is the *constraint length* of the code.

cookies Slang for a software technique used to obtain information about a user's activities while accessing a WWW site. For example, it can observe what products are purchased for future mailing list use. *Cookies* cannot read data from the user's system; further, most Web browsers can be set to disable *cookies*.

The *cookie* itself is a text file saved in the browser's directory or folder. Most of the information in a *cookie* is benign, but some Web sites use *cookies* to store personal preferences. To view the information being stored, view the contents of the file called cookie.txt or MagicCookie located in the browser's folder or directory. Also called a *magic cookie*.

coordinate The location of an element relative to some other element in the same system. For example, the coordinates (or location) of an arbitrary pixel on a display may be referenced to the bottom left corner pixel; or the coordinates of a point in a graph are generally relative to the origin (the zero point for all axes). See also *Cartesian coordinate* and *polar plot*.

coordinated clock One of a set of clocks distributed over a spatial region, producing time scales that are synchronized to the time scale of a reference clock at a specified location. See also *time* (3).

coordinated time scale A time scale synchronized within specified tolerances to a reference time scale.

Coordinated Universal Time (UTC) A time scale, based on the atomic second (as defined by the CCIR recommendations). For many, if not most, applications UTC is equivalent to mean solar time at the prime meridian (0° longitude) in Greenwich, England. Formerly expressed in *GMT*. See also *time* (3).

COP An acronym from <u>C</u>haracter <u>O</u>riented <u>P</u>rotocol. A protocol in which a specific set of communication control character sequences configure and manage the data link.

COPP An acronym from <u>C</u>onnection-<u>O</u>riented <u>P</u>resentation <u>P</u>rotocol.

COPS An acronym from <u>C</u>onnection-<u>O</u>riented <u>P</u>resentation <u>S</u>ervice.

coprocessor An auxiliary processing device (specializing in one particular kind of computation) that executes instructions concurrent with the primary processor. For example, the 80 × 87 math coprocessor executes floating point math instructions for the 80 × 86 CPU.

copy (1) A code word meaning to receive a message. (2) To understand a transmitted message. (3) A recorded or duplicate message or file. (4) To read data from a source, leaving the source data unchanged at the source, and to write the same data elsewhere, though they may be in a physical form that differs from that of the source.

copy protection Pertaining to the use of one or more methods designed to prevent a user from duplicating information on a data medium such as a diskette. The stated intent is to prevent copying in violation of copyright law or security considerations.

copyleft A *copyright* that permits the free distribution and copying of a work for NONCOMMERCIAL purposes. Further, a *copyleft* generally states that no changes to the material can be made without the expressed permission of the author.

copyright The exclusive right granted by law (to an author or owner) to make, copy, distribute, and otherwise control copies of a written work (literary, musical, or software). The law gives the copyright owner broad rights to restrict how their work is distributed. It also provides penalties for those who violate these restrictions. When a program is copyrighted, it must be used in accordance with the copyright owner's restrictions regarding distribution and payment. Usually, these are clearly stated in the program documentation.

Copyrights in the United States are automatically created when a work is created. They may be registered with the U.S. Copyright Office for a small fee. The term of a copyright is generally the life of the author plus 50 years. A *copyrighted program* is one in which the author has asserted his or her legal right to control the program's use and distribution by placing the legally required copyright notices in the program and documentation. *Copyright* is the opposite of *public domain*.

COR An acronym from <u>C</u>onfirmation <u>O</u>f <u>R</u>eceipt.

CORA An acronym from <u>C</u>anadian <u>OSI</u> <u>R</u>egistration <u>A</u>uthority.

CORBA An acronym from <u>C</u>ommon <u>O</u>bject <u>R</u>equest <u>B</u>roker <u>A</u>rchitecture.

cord circuit A communications switchboard in which a plug-terminated cord is used to establish connections manually between user lines or between trunks and user lines.

A number of cord circuits are furnished as part of the switchboard position equipment. The cords may be referred to as *front cord* and *rear cord* (or *trunk cord* and *station cord*). In cordless switchboards, the cord circuit function is replaced by a switch that may be manually operated or may be programmable.

cordless switchboard A communications switchboard in which manually operated keys are used to make connections.

core (1) A concept of many networks including ARPAnet and the early Internet model where a backbone carries all Internet traffic. Autonomous networks associated with this backbone are connected via *core gateways.*

The core model of the Internet was superseded by the philosophy of autonomous domains, each with its own borders and gateways. **(2)** The centermost element of an *optical fiber.* **(3)** The ferromagnetic magnetic material on which one or more coils of wire are wound to create "iron core" or "ferrite core" transformers. The core material significantly increases the *inductance* per unit turn of the coils. The magnetic material becomes the dominant element in the magnetic circuit. **(4)** The ferromagnetic material at the center of an electromechanical relay or solenoid, about which the coil is wound. **(5)** See *core memory.*

core area The central region of an optical fiber in which the refractive index is everywhere greater than the innermost homogeneous cladding by a specified amount. The specified amount is a fraction of the difference between the maximum refractive index of the core and the refractive index of the innermost cladding. Mathematically, the core area is the cross-sectional area with a refractive index greater than n_3 where n_3 is given by:

$$n_3 = [n_2 + m(n_1 - n_2)]$$

where:

n_1 is the maximum refractive index of the core,
n_2 is the refractive index of the homogeneous cladding adjacent to the core,
n_3 is the defining refractive index, and
m is a fraction, usually not greater than 0.05.

When computing *core area,* manufacturing process artifacts, such as refractive index dip, are ignored in computing the demarcation points defined by n_3. See also *optical fiber.*

core cladding offset See *concentricity error.*

core diameter In fiber optics, the average of the diameters of the smallest circle that can be circumscribed about the core and the largest circle that can be inscribed within the cladding. The cross section of a realizeable optical fiber is generally not circular but can be assumed to be approximately elliptical as is illustrated below.

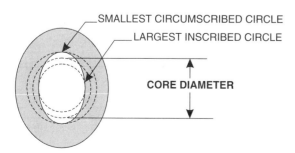

core dump A printout, usually in hexadecimal characters, of the contents of memory (originally from a core memory, now from any memory). A core dump is a diagnostic technique useful for analyzing an abnormally terminated computer program, i.e., for finding bugs in a computer program. Also called a *memory dump.*

core gateway On the Internet, any one of several key routers to which all Internet networks must provide a path.

core loss The power dissipated in a magnetic core when subjected to an alternating magnetic field. Core losses include those from hysteresis and eddy currents.

core memory An electronic memory system based on the bistable magnetic properties of ferrites and other materials.

The system uses one magnetic element in the shape of a toroid for each bit of stored information. Through the center of this toroid are three wires: (1) A horizontal address line, (2) a vertical address line, and (3) a data sense line. Multiple toroids are arranged in a plane as shown in the following figure. Multiple planes are used to store byte- or word-

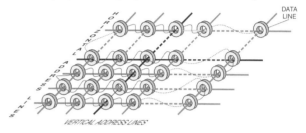

wide fields. To write information into a specific bit, one-half of the minimum current required to change the magnetic orientation of the toroid is applied to the vertical strobe line of the desired bit. At the same time, a similar current is applied to the horizontal strobe line. Together, these currents force the selected core into the desired state. In the above drawing, the bit in the second row and third column is selected. (No other bit is activated by both currents.) In reality, two operations are used: the first clears the bit at the selected address to zero in all planes, and the second sets the desired bits to one. To read the information from a core memory, one must know that any time a core changes state a pulse is generated on the data sense line. The actual read operation, like the write operation, is really two steps: in the first step, the bits being read are first cleared. (This generates a pulse on the data lines that are set to one; all others have no pulse.) The second step restores the original value in memory by rewriting the one cleared in the clear operation.

core model An old model of the Internet in which the network is composed of a single high-speed backbone to which smaller autonomous networks are attached.

corner frequency That frequency indicated by the intersection of two asymptotes on a frequency response plot. The following figure illustrates a corner frequency at 1 kHz for a gain vs. frequency chart.

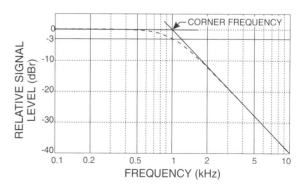

The corner frequency is the point where 3 dB of attenuation occurs.

corner reflector (1) A reflector consisting of three mutually perpendicular intersecting flat, smooth surfaces, which return a reflected electromagnetic wave to its point of origin regardless of the angle of incidence. For reflection in the RF region, the surfaces are conductive, while the only requirement in the optical domain is that the surfaces be optically reflective. Such reflectors are often used as optical, laser, or radar targets. (2) A directional antenna using two mutually intersecting conducting flat surfaces.

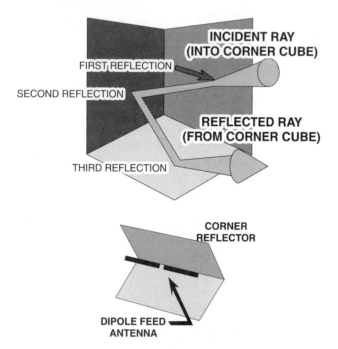

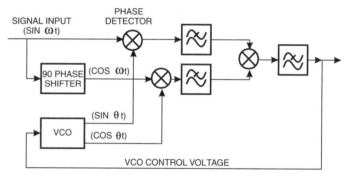

corner reflector antenna See *corner reflector (2)*.

corona A luminous discharge due to ionization of the gas surrounding an electrode caused by a voltage gradient exceeding a critical value. The gas may be air, and the electrodes may be any lines, conductors, or insulators. The critical voltage gradient is dependent on the constituents of the gas, the temperature and pressure of the gas, and the shape of the electrode. The corona is limited to the vicinity surrounding the electrode.

corona modes Depending on the geometry of the electric field, the characteristics of the atmosphere in which the partial discharge occurs, and the polarity of the electrode, two principal modes are displayed: i.e., glow discharge and streamer discharge.

Corporation for Regional and Enterprise Computing (CoREN) An Internet service provider and backbone formed in 1992 by MCI Communications Corporation and regional service providers. *CoREN* provides an alternative backbone through which commercial traffic can be routed.

Corporation for Research and Educational Networking (CREN) A group created in 1989 when CSNET and BITNET merged.

COS (1) The abbreviation for the mathematical cosine function. The ratio of the side adjacent to a given angle of a right triangle to the hypotenuse. Also written *cos*. (2) An abbreviation of Corporation for Open Systems, an organization of computer and communications equipment suppliers and users. (3) An abbreviation of Call Originate Status. (4) An abbreviation of Class Of Service.

COSAC An acronym from Canadian Open Systems Applications Criteria.

COSE (1) An abbreviation of Common Open Software Environment. (2) An abbreviation of Common Operating System Environment.

COSINE An acronym from Cooperation for Open Systems Interconnection Networking-Europe.

cosine emission law A synonym for Lambert's cosine law.

COSMIC (1) An acronym from COmputer Software Management and Information Center. (2) An acronym from COmmon System Main InterConnecting frame.

cosmic noise Random noise that originates outside the Earth's atmosphere.

Cosmic noise characteristics are similar to those of thermal noise. Cosmic noise is experienced at frequencies above about 15 MHz when highly directional antennas are pointed toward the Sun or to certain other regions of the sky such as the center of the Milky Way Galaxy or other radio galaxies. Also called *galactic radio noise* and *Jansky noise*.

COSMOS An acronym from COmputer System for Mainframe OperationS.

COSN An acronym from COnsortium for School Networking.

COSSS An abbreviation of Committee on Open Systems Support Services.

Costas loop A phase locked loop (PLL) used for carrier phase recovery from suppressed carrier modulation signals, such as from double-sideband suppressed carrier signals. One implementation is illustrated in the following diagram:

Also called an *IQ loop* (in-phase and quadrature loop).

COT An acronym from Central Office Terminal. See *subscriber carrier*.

COTP An abbreviation of Connection Oriented Transport Protocol.

COTS (1) An acronym from Connection-Oriented Transport Service. (2) An acronym from Commercial Off-The-Shelf. Indicates the purchase of commercially available hardware and/or software rather than the design of a special low-volume, high-priced entity.

coulomb (C) The SI unit of electric charge, named after Charles-Augustin Coulomb (1736–1806)—a French scientist. One *coulomb* is the quantity of electric charge that passes any cross section of a conductor in one second when one ampere of current is flowing (equivalent to approximately $6.241\ 460 \times 10^{18}$ electrons). The sign of the charge is determined by the number of electrons relative to the number of protons. An excess of electrons results in a net negative charge, whereas a deficit of electrons results in a positive charge.

Coulomb's law The universal law of attraction and repulsion of electric charges; that is, like charges repel and opposite charges attract.

counterclockwise polarized A circular (or elliptically) polarized electromagnetic wave in which the direction of rotation of the electric vector is counterclockwise (as observed looking in the direction of propagation).

counterpoise A conductor or system of conductors used as a substitute for Earth or ground in an antenna system.

country code In telephony, that part of the telephone routing code which informs a national public switch telephone network (PSTN) that the call being placed is to be routed through an international gateway switching office.

A *country code* is a one-, two-, or three-digit number that identifies a country or integrated numbering plan area in the world. The initial digit is the world zone number; subsequent digits further define the geographic area. The original nine zones were defined roughly as

APPROXIMATE COUNTRY CODE ZONE LOCATIONS

1	U.S.A. & Canada	6	Southwestern Pacific
2	Africa	7	Former Soviet Union
3	Europe, Iceland,	8	Northwestern Pacific
4	Malta, and Cyprus	9	East
5	South America		

In addition to the *country code*, it may be necessary to use a *city code*. For example, to call Berlin, Germany, the country code 49 is followed by the city code 30, and Munich is 49 89. Two *country code* charts are shown, one sorted by country and the other by code. See also *area code, city code, domain, international access code, local access code,* and *local number.*

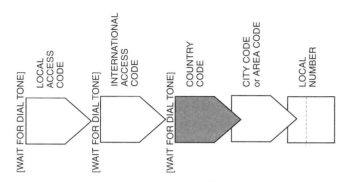

COUNTRY CODES SORTED BY COUNTRY

Albania	355	Cook Islands	682	Hungary	36	Montenegro	38	Singapore	65
Algeria	213	Costa Rica	506	Iceland	354	Montserrat	809	Slovakia	42
American Samoa	684	Croatia	38	India	91	Morocco	212	Slovenia	38
Andorra	628	Cuba		Indonesia	62	Mozambique	258	Solomon Islands	677
Angola	244	Cyprus	357	Iran	98	Namibia	264	South Africa	27
Antarctica	672	Czech Republic	42	Iraq	964	Nauru	674	Spain	34
Antigua	809	Denmark	45	Ireland	353	Nepal	977	Sri Lanka	94
Argentina	54	Diego Garcia	246	Israel	972	Netherlands	31	Suriname	597
Armenia	7	Djibouti	253	Italy	39	Netherlands Antilles	599	Swaziland	268
Aruba	297	Dominica	809	Ivory Coast	225	New Caledonia	687	Sweden	46
Ascension Isle	247	Dominican Republic	809	Jamaica	809	New Zealand	64	Switzerland	41
Australia	61	Ecuador	593	Japan	81	Nicaragua	505	Syrian Arab Republic	963
Austria	43	Egypt	20	Jordan	962	Niger Republic	227	Taiwan	886
Azerbaijan	7	El Salvador	503	Kazakhstan	7	Nigeria	234	Tajikistan	7
Bahamas	809	Equatorial Guinea	240	Kenya	254	Niue	683	Tanzania	255
Bahrain	973	Estonia	372	Kiribati	686	Norfolk Island	672	Thailand	66
Bangladesh	880	Ethiopia	251	Korea, Rep. of	82	Norway	47	Togo	228
Barbados	809	Faeroe Islands	298	Kuwait	965	Oman	968	Tonga Islands	676
Belarus	7	Falkland Islands	500	Kyrgyzstan	7	Pakistan	92	Trinidad & Tobago	809
Belgium	32	Fiji Islands	679	Laos	856	Palau	680	Tunisia	216
Belize	501	Finland	358	Latvia	371	Panama	507	Turkey	90
Benin	229	France	33	Lebanon	961	Papua New Guinea	675	Turks & Calcos Isles	809
Bermuda	809	French Antilles	596	Lesotho	266	Paraguay	595	Tukmenistan	7
Bhutan	975	French Guiana	594	Liberia	231	Peru	51	Tuvalu	688
Bolivia	591	French Polynesia	689	Libya	218	Philippines	63	Uganda	256
Bosnia-Herzegovina	38	Gabon	241	Liechtenstein	41	Poland	48	Ukraine	7
Botswana	267	Gambia	220	Lithuania	370	Portugal	351	United Arab Emirates	971
Brazil	55	Georgia	7	Luxembourg	352	Qatar	974	United Kingdom	44
British Virgin Isles	809	Germany	49	Macao	853	Reunion Island	262	U.S.A.	1
Brunei	673	Ghana	233	Macedonia	38	Romania	40	Universal 800 service	800
Bulgaria	359	Gibraltar	350	Madagascar	261	Russia	7	Uruguay	598
Burkina Faso	266	Greece	30	Malawi	265	Rwanda	250	Uzbekistan	7
Burundi	257	Greenland	299	Malaysia	60	St. Helena	290	Vanuatu	678
Cambodia	855	Grenada	809	Maldives	960	St. Kitts	809	Vatican City	39
Cameroon	237	Grenadines	809	Mali Republic	223	St. Lucia	809	Venezuela	58
Canada	1	Guadeloupe	590	Malta	356	St. Pierre & Miquelon	508	Vietnam	84
Cape Verde Islands	238	Guam	671	Marshall Islands	692	St. Vincent	809	Wallis & Futuna Isles	681
Cayman Islands	809	Guantanamo Bay	5399	Martinique	596	Saipan	670	Western Samoa	685
Central African Rep.	236	Guatemala	502	Mauritius	230	San Marino	39	Yemen, Republic of	967
Chad Republic	235	Guinea	224	Mayotte Island	269	São Tomé	239	Zaire	243
Chile	56	Guinea-Bissau	245	Mexico	52	Saudi Arabia	966	Zambia	260
China	86	Guyana	592	Micronesia	691	Senegal Republic	221	Zimbabwe	263
Columbia	57	Haiti	509	Moldova	373	Serbia	38		
Comoros	269	Honduras	504	Monaco	33	Seychelles Island	248		
Congo	242	Hong Kong	852	Mongolian Peo. Rep.	976	Sierra Leone	232		

COUNTRY CODES SORTED BY CODE

Code	Country	Code	Country	Code	Country	Code	Country	Code	Country
1	Canada	53	Cuba	241	Gabon	373	Moldova	692	Marshall Islands
1	U.S.A.	54	Argentina	242	Congo	500	Falkland Islands	800	Universal 800 service
7	Armenia	55	Brazil	243	Zaire	501	Belize	809	Antigua
7	Azerbaijan	56	Chile	244	Angola	502	Guatemala	809	Bahamas
7	Belarus	57	Columbia	245	Guinea-Bissau	503	El Salvador	809	Barbados
7	Georgia	58	Venezuela	246	Diego Garcia	504	Honduras	809	Bermuda
7	Kazakhstan	60	Malaysia	247	Ascension Isle	505	Nicaragua	809	British Virgin Isles
7	Kyrgyzstan	61	Australia	248	Seychelles Island	506	Costa Rica	809	Cayman Islands
7	Russia	62	Indonesia	250	Rwanda	507	Panama	809	Dominica
7	Tajikistan	63	Philippines	251	Ethiopia	508	St. Pierre & Miquelon	809	Dominican Rep.
7	Tukmenistan	64	New Zealand	253	Djibouti	509	Haiti	809	Grenada
7	Ukraine	65	Singapore	254	Kenya	590	Guadeloupe	809	Grenadines
7	Uzbekistan	66	Thailand	255	Tanzania	591	Bolivia	809	Jamaica
20	Egypt	81	Japan	256	Uganda	592	Guyana	809	Montserrat
27	South Africa	82	Korea, Rep. of	257	Burundi	593	Ecuador	809	St. Kitts
30	Greece	84	Vietnam	258	Mozambique	594	French Guiana	809	St. Lucia
31	Netherlands	86	China	260	Zambia	595	Paraguay	809	St. Vincent
32	Belgium	90	Turkey	261	Madagascar	596	French Antilles	809	Trinidad & Tobago
33	France	91	India	262	Reunion Island	596	Martinique	809	Turks & Calcos Isles
33	Monaco	92	Pakistan	263	Zimbabwe	597	Suriname	852	Hong Kong
34	Spain	94	Sri Lanka	264	Namibia	598	Uruguay	853	Macao
36	Hungary	98	Iran	265	Malawi	599	Netherlands Antilles	855	Cambodia
38	Bosnia-Herzegovina	212	Morocco	266	Burkina Faso	628	Andorra	856	Laos
38	Croatia	213	Algeria	266	Lesotho	670	Saipan	880	Bangladesh
38	Macedonia	216	Tunisia	267	Botswana	671	Guam	886	Taiwan
38	Montenegro	218	Libya	268	Swaziland	672	Antarctica	960	Maldives
38	Serbia	220	Gambia	269	Comoros	672	Norfolk Island	961	Lebanon
38	Slovenia	221	Senegal Republic	269	Mayotte Island	673	Brunei	962	Jordan
39	Italy	223	Mali Republic	290	St. Helena	674	Nauru	963	Syrian Arab Rep.
39	San Marino	224	Guinea	297	Aruba	675	Papua New Guinea	964	Iraq
39	Vatican City	225	Ivory Coast	298	Faeroe Islands	676	Tonga Islands	965	Kuwait
40	Romania	227	Niger Republic	299	Greenland	677	Solomon Islands	966	Saudi Arabia
41	Liechtenstein	228	Togo	350	Gibraltar	678	Vanuatu	967	Yemen, Republic of
41	Switzerland	229	Benin	351	Portugal	679	Fiji Islands	968	Oman
42	Czech Republic	230	Mauritius	352	Luxembourg	680	Palau	971	United Arab Emirates
42	Slovakia	231	Liberia	353	Ireland	681	Wallis & Futuna Isles	972	Israel
43	Austria	232	Sierra Leone	354	Iceland	682	Cook Islands	973	Bahrain
44	United Kingdom	233	Ghana	355	Albania	683	Niue	974	Qatar
45	Denmark	234	Nigeria	356	Malta	684	American Samoa	975	Bhutan
46	Sweden	235	Chad Republic	357	Cyprus	685	Western Samoa	976	Mongolian Peo Rep.
47	Norway	236	Central African Rep.	358	Finland	686	Kiribati	977	Nepal
48	Poland	237	Cameroon	359	Bulgaria	687	New Caledonia	5399	Guantanamo Bay
49	Germany	238	Cape Verde Isles	370	Lithuania	688	Tuvalu		
51	Peru	239	São Tomé	371	Latvia	689	French Polynesia		
52	Mexico	240	Equatorial Guinea	372	Estonia	691	Micronesia		

coupled modes (1) In fiber optics, a mode that shares energy among one or more other modes, all of which propagate together. The distribution of energy among *coupled modes* changes with propagation distance. (2) In microwave transmission, a condition where energy is transferred from the fundamental mode to higher order modes. Generally, energy transferred to these higher modes is undesirable.

coupler A device that enables power or signal information to be transferred from one circuit or system to another. *Couplers* may be classified in several ways: specifically, *close-coupling* vs. *loose-coupling* and *directional-coupling* vs. *bidirectional-coupling*. Close-coupled systems have a large mutual effect and a small phase shift of coupled variables. Directional-couplers transfer power or information predominately in only one direction. Unless the coupler is active, that is, it has amplifiers associated with it, the input power is divided among the output ports. For example, if a passive coupler splits a signal into two equal signals, each of those derived signals is half as strong; it loses 3 decibels (dB) relative to the original signal. Couplers can be designed to split a signal equally or unequally. A coupler may be used to split one path into several output paths (a splitter coupler or simply a splitter); or, the coupler may be designed to combine several input paths to a single output (called a combiner coupler or simply a combiner); or the coupler may be designed to both split and combine. Hence, couplers are often described in terms of the number of input and output ports. For example, a 3×2 coupler has three input and two output ports. If the coupler is bidirectional, it may also be described as a 2×3 coupler.

coupling A term describing both the desirable or undesirable transfer of energy from one circuit, device, or system to another circuit, device, or system. Coupling may be intentional or fortuitous.

Coupling mechanisms include:

- Electrostatic (capacitive).
- Magnetic (inductive).
- Conductive (resistive or hard wire).
- Electromagnetic (RF).
- Optical.

coupling coefficient A number that expresses the degree of electrical coupling between two circuits. The coupling coefficient (γ) is calculated as the ratio of the mutual impedance (Zm) to the square root of the product of the self-impedances of the coupled circuits (Z_1 and Z_2), all impedances being expressed in the same units; i.e.,

$$\gamma = \frac{Z_m}{\sqrt{Z_1 Z_2}}$$

coupling efficiency In fiber optics, the efficiency of optical power transfer between two optical components. It is usually calculated as the ratio of available power of a source component to the power transferred to the second component and expressed as a percentage.

The transfer may take place between:

- An active component and a passive component, e.g., an laser, LED, or detector to an optical fiber, or
- Passive components, e.g., two optical fibers.

coupling loss (1) In fiber optics, the losses from all causes that arise within a splice between two optical fibers. There are multiple sources of *coupling loss* including:

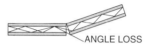
ANGLE LOSS

- *Angular alignment loss*—the optical power loss caused by the angular error of the fiber axis of otherwise aligned optic fibers. Angular misalignment not only can cause signal loss, but it can also cause light to enter the second fiber at an angle different from its original path, which causes signal distortion. Also called *angular misalignment loss.*
- *Area loss*—the combination of radial loss and core ovality loss.

AXIAL LOSS

- *Axial alignment loss*—the optical power loss caused by the non-collinearity of the fiber axis in an otherwise aligned optic fiber junction. Cores that are not both centered in the cladding can lead to spillage from the transmitter's core into the receiver's cladding. Also called *lateral placement.*
- *Core ovality*—connecting cores wherein one or both are elliptical rather than perfectly round. Again, this results in losses from the sending core.
- *Contaminants*—allowing contaminants in the connector can interfere with the connection between the fibers, block light between cores, or refract light at unwanted angles.

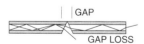
GAP
GAP LOSS

- *Fiber end cuts*—the optical power loss due to angular cut fiber ends in an otherwise aligned optic fiber. Connecting fibers that are not cut cleanly and straight at the ends. The more angular the cut, the greater the signal loss.

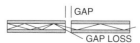
GAP
GAP LOSS

- *Gap loss*—the optical power loss caused by the space between otherwise aligned optic fibers. This light is not only lost for the signal, but some of it can also be reflected back into the sender's

fiber. This reflected light can interfere with the signals traveling in the proper direction. Also called *longitudinal offset loss.*

RADIAL LOSS

- *Radial alignment loss*—the optical power loss caused by differences in the core radii of otherwise aligned optic fibers. Depending on the degree of mismatch, losses can range to more than 10 dB. (*Note:* There is no loss of this type if the sender's smaller core is connected to a larger core at the receiving end.) This loss source is particularly bothersome for single-mode fiber because the cores are so small. Also called *core concentricity.*

LOSS

- *Rough surface*—if the surface of either or both connector end(s) is rough, there will not be a complete union, which will leave space for light to escape.
- *NA mismatch*—connecting a core with a given NA (numerical aperture) to a core with a smaller NA. NA is related to the acceptance angle; hence, all rays of a large transmitted angle will not be captured if the receiving acceptance angle is smaller.

(2) In electric circuits, the loss that occurs when energy is transferred from one circuit, circuit element, or medium to another. Coupling loss is usually expressed in decibels (dB), watts, or sometimes percentage.

cp A UNIX command that copies a file from one location to another. The format of the command is:

cp source_name destination_name

CP (1) An abbreviation of <u>C</u>ustomer <u>P</u>remises. (2) An abbreviation of <u>C</u>ontrol <u>P</u>oint. A function in an IBM Advanced Peer-to-Peer Networking architecture (APPN) node for routing, configuration and directory services. (3) An abbreviation of <u>C</u>onnect <u>P</u>resentation.

CPA An abbreviation of <u>C</u>onnect <u>P</u>resentation <u>A</u>ccept.

CPE (1) An abbreviation of <u>C</u>ustomer <u>P</u>remises <u>E</u>quipment. Refers to both owned and leased equipment, including phones, modems, fax machines, PABXs, and wiring. (2) An abbreviation of <u>C</u>ustomer <u>P</u>rovided <u>E</u>quipment. (3) An abbreviation of <u>C</u>onvergence <u>P</u>rotocol <u>E</u>ntity.

CPFM An abbreviation of <u>C</u>ontinuous <u>P</u>hase <u>F</u>requency <u>M</u>odulation.

CPFSK An abbreviation of <u>C</u>ontinuous <u>P</u>hase <u>F</u>requency <u>S</u>hift <u>K</u>ey.

CPI (1) An abbreviation of <u>C</u>omputer to <u>P</u>BX (or PABX) <u>I</u>nterface. An interface through which a computer can communicate with a private branch exchange (PBX). A proprietary hardware/software modified T-carrier interface developed by Northern Telecom, Inc. and supported by Digital Equipment Corp. (DEC). (2) An abbreviation of <u>C</u>ommon <u>P</u>rogramming <u>I</u>nterface. (3) An abbreviation of <u>C</u>haracters <u>P</u>er <u>I</u>nch.

CPIC An acronym from <u>C</u>ommon <u>P</u>rogramming <u>I</u>nterface for <u>C</u>ommunications. Application Program Interfaces (APIs) for program-to-program communications in IBM's Systems Application Architecture (SAA) environment, e.g., between System Network Architecture (SNA) and OSI environments. The APIs are designed for LU 6.2 protocols, that is, for interactions in which the programs are peers. Also written *CPI-C.*

CPIW An abbreviation of <u>C</u>ustomer <u>P</u>rovided <u>I</u>nside <u>W</u>iring.

CPM (**1**) An abbreviation of <u>C</u>ontinuous <u>P</u>hase <u>M</u>odulation. (**2**) An abbreviation of <u>C</u>ustomer <u>P</u>remises <u>M</u>anagement. (**3**) An abbreviation of <u>C</u>ritical <u>P</u>ath <u>M</u>ethod. A project management method. (**4**) An abbreviation of <u>C</u>ost <u>P</u>er <u>M</u>inute. (**5**) An abbreviation of <u>C</u>ounts <u>P</u>er <u>M</u>inute. (**6**) An abbreviation of <u>C</u>ontrol <u>P</u>rogram <u>M</u>onitor. A family of operating systems developed by Digital Research, Inc. Also written *CP/M.*

CPN (**1**) An abbreviation of <u>C</u>ustomer <u>P</u>remises <u>N</u>etwork. (**2**) An abbreviation of <u>C</u>alling <u>P</u>arty <u>N</u>umber. The directory number of an ISDN calling party. (**3**) An abbreviation of <u>C</u>ustomer <u>P</u>remises <u>N</u>ode.

CPNI An abbreviation of <u>C</u>ustomer <u>P</u>roprietary <u>N</u>etwork <u>I</u>nformation. Customer data held by the local telephone company.

CPODA An acronym from <u>C</u>ontention <u>P</u>riority <u>O</u>riented <u>D</u>emand <u>As</u>signment.

CPR An abbreviation of <u>C</u>onnect <u>P</u>resentation <u>R</u>eject.

CPS or cps (**1**) Obsolete abbreviation of <u>C</u>ycles <u>P</u>er <u>S</u>econd. Replaced by the SI unit hertz (Hz). (**2**) An abbreviation of <u>C</u>haracters <u>P</u>er <u>S</u>econd.

CPSR An abbreviation of <u>C</u>omputer <u>P</u>rofessionals for <u>S</u>ocial <u>R</u>esponsibility.

CPU (**1**) An abbreviation of <u>C</u>entral <u>P</u>rocessing <u>U</u>nit. A term formally used to describe a *mainframe computer;* now frequently applied to most computing devices. (**2**) An abbreviation of <u>C</u>entral <u>P</u>rocessing <u>U</u>nit. It includes those portions of a computer or personal computer (PC) that interpret and execute programmed instructions, perform arithmetic and logical operations on data, and control input/output (I/O) functions. Also called just the *central processor.* (**3**) An abbreviation of <u>C</u>ommunications <u>P</u>rocessor <u>U</u>nit. A computer embedded in a communications system. For example, the portion of a *digital communications switch* that executes programmed instructions, performs arithmetic and logical operations on signals, and controls input/output (I/O) functions.

CR (**1**) An abbreviation of carriage return. Generally written <CR>. (**2**) An abbreviation of <u>C</u>onnection <u>R</u>equest. (**3**) An abbreviation of <u>C</u>hannel <u>R</u>eliability. (**4**) An abbreviation of <u>C</u>ircuit <u>R</u>eliability.

cracker A person who gains unauthorized (illegal) access to another person's computer system. Although many *crackers* are motivated by the challenge of defeating the safeguards, others are motivated by malicious intents, i.e., stealing, modifying, or destroying data. See also *hacker.*

crash Generally, an operation that brings a system to an unexpected halt. The system becomes unresponsive to any operator input or normal watchdog timer "timeouts." The only way to recover is to reboot the system.

A fatal hardware or software error.

CRC An abbreviation of <u>C</u>yclic <u>R</u>edundancy <u>C</u>heck.

CREN The acronym from <u>C</u>orporation for <u>R</u>esearch and <u>E</u>ducational <u>N</u>etworking.

crest See *peak.*

crest factor The ratio of the *peak* (or maximum) value of a waveform to the *average* value. The *crest factor* of a sine wave is 1.414.

CRF (**1**) An abbreviation of <u>C</u>ommunication <u>R</u>elated <u>F</u>unction. (**2**) An abbreviation of <u>C</u>able <u>R</u>etransmission <u>F</u>acility.

crimping tool A special tool used to attach connectors to cables. Each connector type has a unique tool.

crippleware A shareware program whose nonregistered distribution version has some of the features removed that are included in the registered version.

critical angle In optics, the smallest angle of incidence at which total internal reflection will occur in the medium of higher refractive index. In the following figure, medium 1 has a lower index of refraction than medium 2; therefore, a source (*S*) will be both reflected and refracted at all angles less than the critical angle (θ_e). At angles greater than θ_e, the rays are totally reflected.

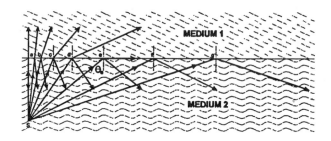

The angle can be calculated by Snell's law:

$$\sin\theta_c = \frac{n_1}{n_2}$$

where θ_c is the *critical angle,* n_1 is the index of medium 1, and n_2 is the index of medium 2. If the two media are air and glass, the value becomes:

$$\sin\theta_c = \frac{1.00}{1.50} \Rightarrow \theta_c = 41.81°$$

For an optical fiber with a core index of 1.48 and a cladding index of 1.46, the *critical angle* is 80.57°.

critical frequency In ionospheric radio propagation, the limiting frequency at (or below) which a wave component is reflected by, and above which it penetrates through, an ionospheric layer. The existence of a *critical frequency* is the result of the limited number of free electrons existing to support reflection at higher frequencies.

cross connect (**1**) In telephony, an easily moved or removed wire that is run loosely between terminal blocks on the two sides of a distribution frame or between terminals on the same terminal block to establish an electrical association. An example of the *cross connect* is its use to associate a specific subscriber pair with a specific switch port address on the switching equipment in a central office. Connections between terminals on the same block are also called *straps.* Also called a *jumper.* (**2**) An ATM switch usually comprising three functional areas:

- *System control*—the central control unit, which also provides the management interface of the system;
- *ATM "fabric block"*—the area providing the system switching capacity;
- *Termination groups*—the area providing the external interface and functions of the ATM layer of the network node.

Each of these functional system areas is configured according to the specified needs of the respective network node. Each functional area usually has its own monitoring and control units for safeguarding the high availability of the complete system.

cross coupling The undesirable transfer of power or signal from one channel to another. Also called *crosstalk.*

cross extension cable A cable wired such that specified pairs terminate at complementary pins on opposite ends of the cable. For example, in a null modem cable the pair terminated at pins 2 and 3 on one cable end, and at pins 3 and 2 at the other end. Similarly, the cable connecting one RJ-11 to another is wired as:

cross modulation *Intermodulation* caused by the modulation of the carrier of a desired signal by an undesired signal.

cross pinned cable A synonym for *null modem cable.*

cross polarized operation A method of multiplexing wherein two transmitters are operating on the same frequency, with one transmitter-receiver pair being vertically polarized and the other pair horizontally polarized. That is, the two carriers are *orthogonally polarized.*

cross post Posting a message in more than one location on a network. For example, on USENet one may post a message to several *related* newsgroups. Cross posting to unrelated newsgroups is called *carpet bombing* and is a violation of USENet netiquette.

cross wye A cable used at the host system or network equipment interface that changes the pin/signal assignments of the cable to conform to a given wiring standard, e.g., from USOC wiring to EIA-568B.

cross-bar An automatic telephone switching system, generally with the following features:

- The switch fabric is composed of cross-bar switches.
- Common electronics select, test, and control the switch fabric.

cross-bar switch In telephony, a multipath switch having a number of vertical paths, a number of horizontal paths, and an electromechanical means of interconnecting any vertical path with any of the horizontal paths such that a signal introduced on the vertical path will be available on the selected horizontal path (and vice versa).

crossover cable A synonym for *null modem cable.*

crosspoint Another name for a controlled switch. A two-state switching device having a low transmission impedance in one state and a very high impedance in the other. A crosspoint is generally considered part of a switching matrix in a communications switching system rather than a stand-alone switch.

SINGLE CROSSPOINT SWITCH

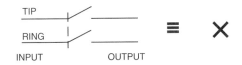

n x m CROSSPOINT SWITCHES IN A MATRIX

crosstalk (XT) One type of noise. The undesirable phenomenon in which a signal on one circuit or transmission line (the disturbing circuit) is coupled into another circuit or transmission line (the disturbed circuit).

In the accompanying drawing, a signal source is connected through a switching network to the signal destination. A second path is established through the same network. Any signal from the signal source that is received in the second path is *crosstalk. Crosstalk* received at the same end as the signal source is called *near end crosstalk* (*NEXT*), and *crosstalk* received at the remote end is called *far end crosstalk* (*FEXT*). See also *far end crosstalk* (*FEXT*) and *near end crosstalk* (*NEXT*).

LOCAL END   REMOTE END

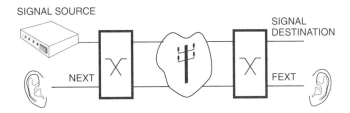

(2) Any phenomenon by which a signal transmitted on one or more circuits or channels of a transmission system create an undesired effect in one or more other circuits or channels.

crosstalk attenuation A measure of the degree to which a system resists crosstalk. It is the ratio of the measured crosstalk level to the exciting signal level; generally, it is expressed in decibels.

crosstalk coupling The ratio of the power in a disturbing circuit to the induced power in the disturbed circuit observed at specified points of the circuits under specified terminal conditions. *Crosstalk coupling* is generally expressed in decibels (dB). Also called *crosstalk coupling loss.*

CRQ (1) An abbreviation of Call ReQuest. **(2)** An abbreviation of Command Response Queue.

CRS An abbreviation of Configuration Report Server.

CRSO An abbreviation of Cellular Radio Switching Office.

CRT An abbreviation of Cathode Ray Tube. The element in a computer monitor, television, or radar display in which video information is displayed. Though strictly incorrect, *CRT* is often used to refer to the entire monitor or display.

CRTC An abbreviation of the Canadian Radio-television and Telecommunications Commission.

CRV An abbreviation of Coding Rule Violation. A unique bit signal for F bit in frame one of *Coded Mark Inversion* (*CMI*).

cryptography The science of coding messages so that they can be read only by the intended recipient.

A related field is *cryptoanalysis,* the science devoted to the breaking of coded messages (where the code key is unknown). See also *Clipper chip, decryption,* and *encryption.*

cryptonym A secret *code word.* See also *code word.*

crystal oscillator (XO) Any oscillator in which the primary frequency determining element is a piezoelectric crystal—typically made of quartz. Additional controls may be added to the oscillator to improve its stability or change its operating frequency; examples include:

- *Temperature-compensated crystal oscillator* (*TCXO*)—includes components to correct for the natural frequency changes of the piezoelectric element due to environmental temperature changes.

- *Oven-controlled crystal oscillator (OCXO)*—placement of the piezoelectric electric element in a constant temperature oven to prevent frequency changes due to environmental temperature changes.
- *Temperature-compensated voltage-controlled crystal oscillator (TCVCXO)*—an oscillator whose frequency is precisely controlled by both temperature control and compensation.
- *Voltage-controlled crystal oscillator (VCXO)*—a piezoelectric-controlled oscillator in which small frequency adjustments may be achieved by applying an input control voltage.
- *Oven-controlled voltage-controlled crystal oscillator (OCVCXO)*—a crystal oscillator in which temperature control is used to stabilize the frequency over environmental temperature changes, but the precise frequency can be set by an externally applied control voltage.
- *Microcomputer-compensated crystal oscillator (MCXO)*—a piezoelectric oscillator in which deviations from a desired frequency are minimized by a computer whose control output is algorithmically related to the oscillator's open-loop characteristics and sense inputs.

CS (1) An abbreviation of Circuit Switched. (2) An abbreviation of Convergence Sublayer. In ATM, where header and trailer information is added; before segmentation. (3) An abbreviation of Coordinated Single-Layer. (4) An abbreviation from Carrier System.

CS-MUX An acronym from Circuit Switched MUltipleX.

CSA (1) An abbreviation of the Canadian Standards Association. The Canadian national standards-setting and certification agency (equivalent to the Underwriters Laboratories). (2) An abbreviation of Carrier Service Area. The region defined by the local loop length out of a central office (CO) or remote terminal.

CSA T-528 The Canadian equivalent of the EIA-606 standard.

CSA T-529 The Canadian equivalent of the EIA-568 standard.

CSA T-530 The Canadian equivalent of the EIA-569 standard.

CSBS An abbreviation of China State Bureau for Standardization. The Chinese national standards-setting agency. See also *CCEE*.

CSC (1) An abbreviation of Circuit Switched Channel (or Connection). (2) An abbreviation of Circuit Switched Cellular. (3) An abbreviation of Common Signaling Channel.

CSDC An abbreviation of Circuit Switched Digital Capability.

CSDN An abbreviation of Circuit Switched Digital Network.

CSE An abbreviation of Coordinated Single-Layer Embedded.

CSELT An acronym of Centro Studi E Laboratori Telecommunicazioni (Telecommunications Study Center and Laboratory).

CSFS An abbreviation of Cable Signal Fault Signature. The unique signal reflected back from a transmission line when using time domain reflectometry (TDR) to test the soundness of the line.

CSID An abbreviation of Calling Station IDentifier.

CSLIP An acronym from Compressed Serial Line Internet Protocol. Like SLIP, a protocol that can provide a serial modem connection to a network; however, it is faster than SLIP.

CSM An abbreviation of Combined Symbol Matching.

CSMA (1) An abbreviation of Common Spectrum Multiple Access. (2) An abbreviation of Carrier Sense Multiple Access. A local area network medium access technique in which multiple stations connected to the same channel are able to sense the transmission activity of other nodes on that channel and to defer transmission while the channel is active. Also called *collision sense multiple access*.

CSMA/CA An abbreviation of Carrier Sense Multiple Access with Collision Avoidance. A network protocol for addressing the problem of two or more nodes attempting to access the LAN at the same time. With *CSMA/CA*, a node wishing to send information first monitors the line for a activity; if none is heard, it sends a request to send (*RTS*) to the designated receiving station. If the sending node receives the receiving station clear to send (*CTS*) message within a predefined time period, transmission begins. If no *CTS* is received, the sending node assumes there is a collision and waits to try later (called the *deferral time*).

In Apple's LocalTalk network architecture, the minimum interframe gap (IFG)—the time between successive frames (such as RTS and CTS or between CTS and data transmission)—is 200 µs.

CSMA/CD An abbreviation of Carrier Sense Multiple Access with Collision Detection. A network protocol, defined by IEEE 802.3, for addressing the problem of two or more nodes attempting to access a LAN at the same time.

- *CS—Carrier sense* means that any node wishing to transmit, monitors the LAN first. If the LAN is idle, the node proceeds with transmission. If the LAN is busy (that is, a carrier is detected), the node waits at least 9.6 µs (the minimum interpacket gap time) after the LAN is idle before transmitting. (The delay is called the *deferral time.*)
- *MA—Multiple access* means that any node with pending traffic may gain admittance to the LAN essentially through autonomous behavior. No central station or master node is needed to decide which node is able to transmit and when.
- *CD—Collision detection* means that the circumstance of two or more nodes attempting to access an idle network at the same time (collision) is detectable and that an appropriate retry procedure is instituted.

In the event of a collision, any node detecting the collision will continue to transmit for a fixed time in order to ensure that all other interfering nodes also detect the collision (a process known as *jamming*). After jamming, the node stops transmitting and waits a random period of time before retrying. In an attempt to reduce traffic offerings, the magnitude of the maximum random delay time is increased each time a consecutive collision is detected. The load-shedding algorithm is called the *truncated binary exponential backoff algorithm* (*binary exponential backoff* or *backoff algorithm*). The maximum random delay value is given by: $Max = (2^n - 1) \cdot 51.2$ µs, where n is the number of consecutive collisions detected. The table on page 131 lists the maximum value for the random delay at a node detecting consecutive collisions. After 16 consecutive collisions are detected, an error is reported. Collision detection is generally accomplished by analog signal level detection at each transceiver. Each transmitter, when active, applies a modulated signal of approximately 2 V peak to peak onto the transmission line. When a signal greater than approximately 2.2 V is detected, a collision is assumed to have occurred. Because impedance discontinuities (and therefore reflections) are minimized at each node, a signal greater than 2 V must be due to the sum of two (or more) transmitters sending at the same time; hence, collision. See also *Aloha network*.

CSN An abbreviation of Carrier Service Node.

BINARY EXPONENTIAL BACKOFF ALGORITHM VALUES

Number of Collisions Detected	Maximum Random Delay (μs)	Comments
0	9.6	Interpacket gap delay
1	51.2	1×51.2
2	153.6	3×51.2
3	358.4	7×51.2
4	768.0	15×51.2
5	1587.2	31×51.2
6	3225.6	63×51.2
7	6502.4	127×51.2
8	13056.0	255×51.2
9	26163.2	511×51.2
10	52377.6	1023×51.2
11	52377.6	
12	52377.6	
13	52377.6	
14	52377.6	
15	52377.6	
16	52377.6	Error is reported on 17

CSNET The acronym from Computer + Science NETwork. Merged with BITnet to form *CREN*.

CSP An abbreviation of Control Switching Point.

CSPDN An abbreviation of Circuit Switched Public Data Network.

CSR An abbreviation of Control (or Command) and Status Register.

CSTA An abbreviation of Computer Supported Telecommunications Application.

CSU (1) An abbreviation of Channel Service Unit. (2) An abbreviation of Central Switching Unit. (3) An abbreviation of Circuit Switching Unit. (4) An abbreviation of Customer Service Unit.

CSUnet An acronym from California State University Network. A network originally conceived to interconnect the campuses of the California state universities. It now is expanded to include Internet connectivity to all schools, community colleges, and libraries.

CT1 The "first generation" analog Cordless Telephone standard used in Europe (noncellular).

CT2 An abbreviation of Cordless Telephone 2nd generation. An interim ETSI cordless telephone standard using the 864-868 MHz band and FDMA/TDD modulation. Superseded by *DECT*. See also *DECT* and *IEEE 802.11*.

CT3 Ericsson's proprietary cordless telecommunications system.

CTAK An acronym from Cipher Text Auto Key.

CTD An abbreviation of Cumulative Transit Delay.

CTERM An acronym from Command TERMinal Protocol. Digital Equipment Corp's (DEC's) command terminal protocol that provides terminal sessions over DECnet.

CTI An abbreviation of Computer Telephony Integration.

CTIA (1) An abbreviation of the Cellular Telecommunications Industry Association. (2) An abbreviation sometimes used for the Computer Technology Industry Association (formerly the Microcomputer Industry Association).

CTIP An acronym from Commission on Computing, Telecommunications, and Information Policies.

CTL An abbreviation of control. Sometimes abbreviated CTRL. Generally shown as <CTL> (or <CTRL>) indicating a single key.

CTNE An abbreviation of Compañia Telecommunicion Nacional de España (National Telephone Company of Spain).

CTR An abbreviation of Common Technical Requirements.

CTRG An abbreviation of Collaboration Technology Research Group.

CTS (1) An abbreviation of Clear To Send (RS-232 signal CB, RS-449 signal CS, and ITU-T signal 108). It is a signal generated by the DCE to the DTE indicating that the DCE is ready to receive data from the DTE. (2) An abbreviation of Communications Technology Satellite. (3) An abbreviation of Conformance Testing Service.

CTX (1) An abbreviation of CenTreX®. (2) An abbreviation of Clear to transmit (TX).

CU Shorthand for See You.

CUA (1) An abbreviation of Common User Access. In IBM's SAA environment, specifications for user interfaces that are intended to provide a consistent look across applications and platforms. That is, it sets guidelines for the appearance and actions of menu and dialog boxes, buttons, and help windows in a GUI environment. (2) An abbreviation of Commonly Used Acronyms.

CUI An abbreviation of Common User Interface.

CUL Shorthand for See You Later.

current The amount of charge (electrons) flowing past a point in a conductor per second. Current is measured in units of *amperes* (abbreviated a, A, or amp). 1 A of current is the flow of 1 *coulomb* of charge past a given point in a conductor in 1 second.

current loop A serial baseband transmission technique in which current rather than voltage is used to carry information. Currents in the order of a few milliamps to tens of milliamps are typically used in the loop. Current loop signaling is traditionally used in teletypewriter communications, a current flow indicating a marking condition (logical one) and no current indicating a spacing condition (logical zero).

curvature loss A synonym for *macrobend loss*.

curve fitting compaction A data compaction technique in which an analytical expression is substituted for data to be stored or transmitted.

Examples of *curve-fitting compaction* include:

* The breaking up of a continuous function or curve into a series of straight-line segments, each approximating the arbitrary curve between its end points. The compacted information consists of each line segment's slope, intercept, and range.
* Use of a mathematical expression, such as a polynomial or a trigonometric function, and a single point on the corresponding curve instead of storing or transmitting the entire graphic curve or a series of points on it.
* The use of an expression (representing an a priori mathematical equation or procedure) and critical parameters to define the corresponding curve instead of storing or transmitting the entire graphic curve or a series of points on it. For example, a circle might be represented as n, x, y, r, f where n is the number of the mathematical expression, x, y represents the center, r is the radius, and f indicates the figure fill type (solid empty cross-hatch, etc.)

custom local area signaling service (CLASS) A group of enhanced service features provided by a telephone service provider.

A *CLASS* group for a given network usually includes several enhanced service offerings, such as automatic return of the most recent incoming call, call blocking, call redial, call trace, distinctive ringing, incoming call identification, and selective forwarding.

Custom Ringing See *distinctive ring.*

customer The person who generally (though not always) decides on product or service requirements and pays for them.

Customer Information Control System (CICS) A terminal that provides transaction processing capabilities for IBM mainframes. *CICS* supports the Systems Network Architecture (SNA).

customer-owned and -maintained equipment (COAM) A deprecated term. See *customer premises equipment* (*CPE*).

customer premises equipment (CPE) A term that refers to both owned and leased terminal equipment, including phones, PABXs, modems, and associated equipment at a subscriber's premises as well as the wiring connecting to a carrier's communication channel(s) at the demarcation point.

CPE exclusions include overvoltage protection equipment and some pay telephones.

customer-provided equipment A deprecated term. See *customer premises equipment* (*CPE*).

customer service unit (CSU) A device that provides access at a user location to either switched or point-to-point, data-conditioned circuits (at a specifically established data signaling rate). A *CSU* provides local loop equalization, transient protection, isolation, and *central office* loopback testing capability.

CUT (1) An acronym from <u>C</u>ontrol <u>U</u>nit <u>T</u>erminal. A terminal operating mode that allows only one session, such as running an application, per terminal. (2) To transfer a service from one facility to another.

cut through (1) An Ethernet switching technique whereby only the first few bytes of a packet are examined before forwarding (or filtering) the entire packet. Although the technique is faster then examining the entire packet, it does have the potential of allowing some packets onto the wrong switched path. See also *store and forward* (*S-F*). (2) In telephony, the act of connecting one circuit to another, or a telephone to a circuit.

cut through resistance A measure of a material's ability to withstand penetration by sharp edges. See also *cold flow.*

cutoff attenuator A waveguide, of adjustable length, which varies the attenuation of signals passing through it.

cutoff frequency (1) The frequency either above which or below which the output of a line, circuit, device, or system is reduced to a specified level. In most systems, the *cutoff frequency* is specified at the half-power point (the point where attenuation is 3 dB). However, some systems use the 1/e attenuation level to define cutoff. (2) Frequencies above or below which signals in a transmission path are attenuated to the point of being unusable for communications purposes.

Examples include:

- The frequency below and above which a radio wave penetrates the ionosphere at the incidence angle required for transmission between two specified points by reflection from the layer.

- The frequency above which attenuation in a transmission line is sufficient to make reception at a receiver unreliable.

Also written *cut-off frequency.*

cutoff mode The highest order mode that will propagate in a given waveguide at a given frequency.

cutoff wavelength (1) The wavelength (λ_c) corresponding to the *cutoff frequency* (f_c).

$$\lambda_c = 1 / f_c$$

(2) The shortest wavelength at which a signal will take a single path through the core of a single mode optical fiber. (3) In an uncabled single mode optical fiber, the wavelength greater than which a particular waveguide mode ceases to be a bound mode. That wavelength is usually taken to be the wavelength at which the normalized frequency is equal to 2.405. The *cabled cutoff wavelength* is usually considered to be a better measure because it includes the effects of cabling the fiber. Also written *cut off wavelength.*

cutover The physical changing of the lines of one phone system to another, e.g., changing from an old switching system to a new one. There are two kinds of *cutover: flash cutover* and *parallel cutover.*

With parallel cutover, both systems are in place. Both the old and the new systems are functional for some period of time. This type of cutover allows the new system to be debugged and stressed while maintaining the safety net of the old system. Sometimes there is a gradual transition from the old system to the new system; that is, the communications responsibilities are transferred from the old system to the new system "a few at a time." A flash cutover occurs all in one transaction. The old system is disconnected (and probably removed), and the new system is installed. There is no safety net; that is, if the new system does not work, one cannot just switch to the old system and fix the new system in a leisurely manner. Flash cutovers are usually scheduled to occur at times of minimal traffic, e.g., on a weekend to change a PBX switching system.

CV (1) An abbreviation of <u>C</u>ode <u>V</u>iolation. (2) An abbreviation of <u>C</u>onstant <u>V</u>oltage. (3) An abbreviation of <u>C</u>onstant <u>V</u>elocity.

CVCP An abbreviation of <u>C</u>ode <u>V</u>iolation, <u>CP</u>-Bit Parity.

CVCRC An abbreviation of <u>C</u>ode <u>V</u>iolation, <u>C</u>yclical <u>R</u>edundancy <u>C</u>heck.

CVFE An abbreviation of <u>C</u>ode <u>V</u>iolation, <u>F</u>ar <u>E</u>nd.

CVP An abbreviation of <u>C</u>ode <u>V</u>iolation, "<u>P</u>" Bit.

CVSD An abbreviation of <u>C</u>ontinuously <u>V</u>ariable <u>S</u>lope <u>D</u>elta modulation.

CVT An abbreviation of <u>C</u>onstant <u>V</u>oltage <u>T</u>ransformer. Regulation may be accomplished by means of a ferroresonant transformer.

CW (1) An abbreviation of <u>C</u>ontinuous <u>W</u>ave. (2) An abbreviation of <u>C</u>arrier <u>W</u>ave. (3) An abbreviation of <u>C</u>omposite <u>W</u>ave. (4) An abbreviation of <u>C</u>all <u>W</u>aiting. (5) An abbreviation of <u>C</u>lock<u>W</u>ise.

CWARC An acronym from <u>C</u>anadian <u>W</u>orkplace <u>A</u>utomation <u>Re</u>search <u>C</u>enter.

CWC An abbreviation of <u>C</u>enter <u>W</u>eighted <u>C</u>ontrol. A network arrangement in which a central processor addresses networkwide functions, while the other nodes address local tasks.

CWI An abbreviation of <u>C</u>entrum voor <u>W</u>iskunde en <u>I</u>nformatica (Center for Mathematics and Informatics).

CWIS An acronym from <u>C</u>ampus <u>W</u>ide <u>I</u>nformation <u>S</u>ystem.

CX signaling See *composite signaling* (*CX*).

CXI An acronym from <u>C</u>ommon <u>X</u>-windows <u>I</u>nterface.

cxr An abbreviation for carrier.

CYA An abbreviation of <u>C</u>over <u>Y</u>our <u>A</u>ctions (ass, anatomy, . . .). Any behavior designed to protect one's self, not to address the actual problem or situation.

cybernaut An experienced and knowledgeable Internet user. One who is capable of traversing *cyberspace*'s myriad network paths, interconnections, and resources.

cybersex A form of human sexuality made possible by large multiregional networks in which users engage in real-time sexually explicit message exchanges.

cyberspace The name given to electronic nonphysical places (virtual reality) such as bulletin boards, teleconferences, and particularly the Internet. *Cyberspace* includes the computers, programs, data, and the community of users found on these networks. The term *cyberspace* is intended to contrast the encounters one finds within networks and the "real world." *Cyberspace* was coined by William Gibson in his science fiction fantasy *Neuromancer.*

Cyberspace also refers to the artificial world created by computer-based virtual reality games.

cycle (1) See *FDDI-II* and *period.* (2) With electrochemical cells or batteries. The process of one complete discharge and recharge for a secondary (rechargeable) system.

cycle life The number of cycles, under specified conditions, that a device can perform its function before it fails to meet specified performance criteria, e.g., the number of times a secondary battery can be charged and discharged.

cycle slip A discontinuity in a detected periodic waveform due to accumulation of more phase error than can be corrected for in a continuous manner or due to the temporary loss of carrier in some systems (such as phase lock loops).

cyclic binary code See *Gray code.*

cyclic distortion Distortion that is neither bias, characteristic, nor fortuitous in nature and has a general periodic property.

In mechanical systems (such as a teletypewriter), the causes may be due to irregularities in the duration of contact time of the brushes of a transmitter distributor and interference by disturbing alternating currents.

cyclic redundancy check (CRC) The *cyclic redundancy check* is one of many methods of verifying the accuracy of a received digital transmission by identifying errors at the receiving end of a communication channel. *CRC* error detection is better than the simple checksum method. For example, a sixteenth-order *CRC* is capable of detecting:

- 100% of all odd numbers of errors.
- 100% of all single and double bit errors.
- 100% of all error bursts less than 17 bits.
- 99.9969% of single error bursts 17 or more bits in length.
- 99.9984% of all errors longer than 16 bits.

The process involves dividing the data block by a predefined polynomial at both the transmitting and receiving stations. The transmitting station appends the remainder of this division to the end of the transmitted data block. The remainder of the division computed by the receiving station is compared to the remainder computed by the transmitting station. If the two remainders are not equal, a transmission error has occurred. The dividing polynomials most frequently used (though not the only ones) are the ANSI CRC-16 polynomial $X^{16}+X^{15}+X^5+1$ and the ITU-T CRC-16 polynomial $X^{16}+X^{12}+X^5+1$. Because the polynomials are 16-bit words, the remainder will also be a 16-bit word; hence, the *CRC* code word appended to the data block is only a 16-bit word. 16-bit CRCs are usually limited to blocks of less than 4 kbytes. 32-bit CRCs are used for blocks up to 64 kbytes in length. 32-bit CRCs can detect 99.999999977% of all errors. A 32-bit CRC generator polynomial used for both Ethernet and Token Ring is

$$X^{32} + X^{26} + X^{23} + X^{22} + X^{16} + X^{12} + X^{11}$$
$$+ X^{10} + X^8 + X^7 + X^5 + X^4 + X^2 + X + 1$$

Protocols such as KERMIT, XMODEM, YMODEM, and ZMODEM use *CRC* error-detection and *automatic repeat request* (*ARQ*) control methods. See also *checksum, longitudinal redundancy check* (*LRC*), *parity,* and *vertical redundancy check* (*VRC*).

D

d The SI prefix for *deci* or 10^{-1}.

D (1) A symbol for delta, although the Greek letters Δ or δ are more common. (2) A symbol for data; for example, in the ISDN $2B + D$, the *D* channel is a 16-kbps data or signaling channel.

D bit In an X.25 packet switching network (PSN), a control element used to request end-to-end acknowledgment. Also called a *delivery confirmation bit*.

D channel An ISDN communications channel whose primary purpose is to convey signaling information between the user terminal and the network switch. It primarily controls circuit switched calls on associated B channels at the user interface. The channel (usually in HDLC format) operates at 16 kbps for basic rate access (BRI) and at 64 kbps for primary rate access (PRI). Any data bandwidth above the requirements for out-of-band signaling may be utilized for user packet data. (Generally, up to 9600 bps are available.) Sometimes also called the *Delta channel*. See also *B channel* and *ISDN*.

D conditioning A type of wire conditioning intended to control harmonic distortion and signal-to-noise ratio (SNR) to within specified limits. See also *conditioning*.

D layer See *D region* and *ionosphere*.

D region That portion of the ionosphere existing approximately 50 to 95 km above the surface of the Earth. Free electrons generated by solar radiation ($\lambda \approx 121.6$ nm) attenuate (absorbs) radio waves of certain wavelengths during the daylight hours. At night, however, ionization ceases, and so attenuation of radio waves also ceases. The region is sufficiently reflective to provide a bound on VLF and LF transmissions and irregular enough to scatter VHF transmissions. Also called the *D layer*. See also *ionosphere*.

D* In fiber optics, *D** (pronounced "D-star") is a figure of merit used to characterize a detector's performance. See *specific detectivity (D*)*.

D-AMPS An acronym from **D**igital **A**dvanced **M**obil **P**hone **S**ervice.

D-bank A synonym for channel bank.

D-Star(D*) See *specific detectivity (D*)*.

D.n recommendations An ITU-T series of specifications concerning general tariff principles. See Appendix E for a complete list.

D/A (1) An abbreviation of **D**igital to **A**nalog. (2) An abbreviation of **D**rop and **A**dd.

D/L An abbreviation of **D**own **L**ink.

D1, D1D, D2, D3, D4, D5 Various T-carrier multiplex and framing formats. All formats contain an additional 193rd bit position.

For example, *D4* is a framing standard that describes user channels multiplexed onto a trunk that has been segmented (framed) into 24 bytes of 8 bits each. The multiplexing function is performed in the *D4* framing structure by interleaving bits of consecutive bytes as they are presented from individual circuits into each *D4* frame. See also *channel bank, DS-n,* and *T1*.

D4 channelization A term that infers compliance with AT&T Technical Reference TR 62411 with respect to the sequential assignment of channels and time slots within the frame layout.

D4 framing The *bit robbing* signaling method on a *D4* superframe that identifies individual channels and provides interoffice signaling. A *D4* superframe consists of twelve 193-bit frames (twenty-four 8-bit data channels plus 1 signaling bit).

D66 See *loaded line*.

da The SI symbol for deka 10^1.

DA (1) An abbreviation of **D**estination **A**ddress. (2) An abbreviation of **D**ata **A**vailability.

DAA An abbreviation of **D**ata **A**ccess **A**rrangement. In 1968, a legal decision was handed down to allow non-AT&T data sets (modems) to be attached to phone lines. However, they must be connected via a *DAA*—a protective device rented from the telephone company. In 1976, Part 68 of the FCC rules was changed to allow manufacturers to include the protective devices and to connect directly via the now familiar RJ connectors eliminating the *DAA* device. (The *DAA* is still required if the equipment being attached to the public telephone network does not meet FCC standards.)

Data access arrangements are an integral part of all modems built for the U.S. public telephone network.

DAC (1) An acronym from **D**igital to **A**nalog **C**onverter. (2) An acronym from **D**ual **A**ttachment **C**oncentrator.

DACC An acronym from **D**igital **A**ccess **C**ross **C**onnect.

DACCS An acronym from **D**igital **A**ccess **C**ross **C**onnect **S**ystem.

DACD An abbreviation of **D**irectory **A**ccess **C**ontrol **D**omain.

DACS An acronym from **D**igital **A**ccess and **C**ross connect **S**ystem.

DAD An acronym from **D**raft **AD**dendum.

daemon A small UNIX program that runs frequently to test for the occurrence of a particular event. If the event occurred, the program carries out a predetermined sequence of steps; if the event did not occur, the program goes back to sleep.

DAF (1) An abbreviation of **D**estination **A**ddress **F**ield. (2) An abbreviation of **D**irectory **A**uthentication **F**ramework. (3) An abbreviation of **D**istributed **A**pplication **F**ramework.

daisy chain A method of interconnecting electrically adjacent devices or modules (nodes). Each node has two ports: an "in" and an "out"; the out port of the first device is connected to the in port of the second device. Similarly, the out of device two is connected to the in of device three, and so on. If the out of the last device is connected to the in of the first device, a structure called a *ring* is generated.

DAL (1) An abbreviation of **D**ata **A**ccess **L**anguage. (2) An abbreviation of **D**edicated **A**ccess **L**ine. (3) An abbreviation of **D**ata **A**ccess **L**ine.

DAM (1) An acronym from **D**ata **A**ccess **M**anager. (2) An acronym from **D**raft **AM**endment.

DAMA (1) An acronym from <u>D</u>emand <u>A</u>ssigned <u>M</u>ultiple <u>A</u>ccess. (2) An acronym from <u>D</u>ata <u>A</u>ssigned <u>M</u>ultiple <u>A</u>ccess.

damped response In linear systems including circuits, networks, and mechanical devices, a term describing the system output (response) to an abrupt input stimulus (a step or impulse for example). There are three classifications of response—*critically damped, underdamped,* and *overdamped.*

- *Critically damped*—In response to an abrupt input stimulus, the output of a critically damped system will change in the minimum time possible *without* overshoot or ringing. It is the boundary between an underdamped and overdamped response.
- *Underdamped*—In response to an abrupt input stimulus, the output of an underdamped system has an overshoot and possible ringing. Overshoot is the condition of a signal to exceed the final value when changing from one state to another.
- *Overdamped*—As with a critically damped system, an overdamped system will not have either overshoot or ringing. However, the system output response to an abrupt input signal change will be slower than that of the critically damped system.

A comparison of the three damping conditions is shown in the accompanying figure.

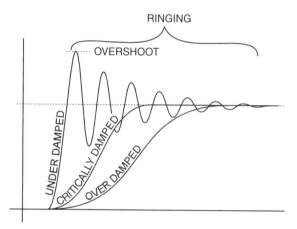

damping The progressive diminishing of specified values characteristic of a phenomenon with respect to time, e.g., the progressive decay in the amplitude of the free oscillations of a circuit with time. See also *damped response.*

damping factor (ζ) A number expressing the ratio of the actual damping to the damping of a critically damped second-order linear system (or subsystem). *Damping factors* less than one are underdamped, while those greater than one are overdamped. See also *damped response.*

DAN An acronym from <u>D</u>epartmental <u>A</u>rea <u>N</u>etwork.

DAP (1) An acronym from <u>D</u>ata <u>A</u>ccess <u>P</u>rotocol. (2) An acronym from <u>D</u>irectory <u>A</u>ccess <u>P</u>rotocol. (3) An acronym from <u>D</u>ocument <u>A</u>pplication <u>P</u>rofile.

dark current In electro-optics, the quiescent current that flows in a photosensitive device when there is no incident radiation.

dark fiber An unused optic fiber. A fiber optic cable not carrying a signal; that is, there is no light energy in the fiber.

DARPA An acronym from <u>D</u>efense <u>A</u>dvanced <u>R</u>esearch <u>P</u>rojects <u>A</u>gency.

DAS (1) An acronym from <u>D</u>ual <u>A</u>ttachment <u>S</u>tation. A device on an FDDI fiber optic ring to which the two rings are attached. (2) An acronym from <u>D</u>ynamically <u>A</u>ssigned <u>S</u>ocket. (3) An acronym from <u>D</u>isk <u>A</u>rray <u>S</u>tation.

DASD An acronym from <u>D</u>irect <u>A</u>ccess <u>S</u>torage <u>D</u>evice (pronounced "DAZ-dee").

DASS (1) An acronym from <u>D</u>istributed <u>A</u>uthentication <u>S</u>ecurity <u>S</u>ervice. (2) An acronym from <u>D</u>irect <u>A</u>ccess <u>S</u>econdary <u>S</u>torage. Storage facilities that are online but slower than the mainstay hard disk drive.

DASS n A British acronym from <u>D</u>igital <u>A</u>ccess <u>S</u>ignaling <u>S</u>ystem. DASS 1 was the original British Telecom ISDN signaling scheme developed for both single line and multiline Integrated Digital Access (IDA) applications (although it was used only with single lines). DASS 2, also developed by British Telecom, is a message-based signaling system that follows the ISO model and provides multiline access to the British Telecom network.

DassII A message-based signaling system, following the ISO model, developed by British Telecom (BT), to provide multiline integrated digital access (IDA) interconnection to the BT network.

DAT (1) An acronym from <u>D</u>igital <u>A</u>udio <u>T</u>ape. (2) An acronym from <u>D</u>uplicate <u>A</u>ddress <u>T</u>est.

data A representation of a collection of facts, concepts, instructions, or information to which meaning has been assigned. The representation may be analog, digital, or any symbolic form suitable for storage, communication, interpretation, or processing by human or automatic means.

"Data" is the plural of the Latin *datum,* meaning one item of information. To be correct, a single item should be called a datum and more than one should be called *data,* i.e., "one datum is . . ." and "two *data are . . .*"

data above voice (DAV) See *data over voice.*

data access arrangement (DAA) See *DAA.*

Data Access Language (DAL) In Macintosh-based client–server environments, an extension to the Structured Query Language (SQL) database language intended to provide uniform access to any database that supports SQL.

data access manager (DAM) In the Apple Computer's System 7 operating system software for Macintoshes, *DAM* is a built-in capability for accessing databases on a network. The *DAM* mediates between an application and the database being accessed. It uses database extensions to communicate with the database. These are database-specific system files that contain the commands necessary to interact with a particular database.

data attribute A characteristic of a data element such as length (number of bits or bytes), method of representation (fixed point, floating point, alphanumeric), or value.

data bank A set of data related to a given subject and organized in such a way that it can be accessed and retrieved by local and/or remote users. The databank's characteristics and attributes may include:

- Information on a single subject or multiple subjects.
- Any rational organization (random, sequential, and so on).
- More than one database.
- More than one data bank in order to achieve a complete database.
- May be geographically distributed.

data base See *database.*

data bits In asynchronous communications, the group of 5, 6, 7, or 8 bits following the start bit (7 or 8 bits are most commonly used). These bits represent a single character or symbol for transmission. Following the *data bits* is an optional parity bit and 1, 1.5, or 2 stop bits.

data bus One of several communication paths in a computer, in particular, the bus used to transfer data among subsystems of the computer.

Other busses include the *address bus, I/O bus, video bus,* and so on.

data circuit connection The interconnection of lines and trunks, on a tandem basis, by means of switching equipment. Once established the connection enables information to be exchanged between end terminals.

data circuit transparency The ability of a data transmission link to transfer data without altering its content or structure.

data communications The transfer of data between separated locations, generally by electronic or optical means and over any transport medium. It includes all of the necessary operations, both manual and machine, necessary to effect the information transfer. For there to be communication at all, there must be at least one sender, at least one receiver, and one or more communications links.

The sender encodes and transmits the data, and the receiver receives and decodes the data. Data encoding includes compression (to eliminate redundancy), encryption (to minimize interception by third parties), and waveform encoding (to prepare the data for efficient transport). Data communications may be found on all types of networks, including broadcast, circuit switched, multicast, packet switched, point-to-point, and store-and-forward.

data communications channel In SONET, a three-byte, 192-kbps field that contains alarm, surveillance, and performance information. The information may be used for internally and externally generated messages or for manufacturer-specific messages.

data communications equipment (DCE) Any device capable of communicating with specified data terminal equipment (DTE) and of providing access to an appropriate type of communications line. For example, a modem can communicate with a computer and provide access to an analog telephone line. Similarly, in digital telecommunications, a data service unit (DSU) and a communications service unit (CSU) together make up a DCE and provide access to the digital lines. Also called *data circuit terminating equipment.*

data compression A method or procedure used to reduce the number of symbols in a message without loss of information. The purpose of *data compression* is to reduce either the size of a data file for storage or the time required for transmission. A compressed file must be restored (or reconstructed) before it can be used. Although both *lossy* and *lossless* compression techniques exist, *data compression* generally refers to lossless methods.

The basis for lossless compression can be any of the following:

- Patterns in bit sequences, as in run-length limited (RLL) encoding.
- Patterns of occurrences of particular byte values, as in Huffman or LZW encoding.
- Commonly occurring words or phrases, as in the use of abbreviations or acronyms.

In a transmission system, the receiving station will expand the received signal and reconstruct the original message. Several data compression techniques are MNP 5—which provides up to 2:1 compression; MNP 7—which is capable of 3:1 compression; and ITU-T V.42 bis—which provides up to 4:1 compression. However, compressed data streams never achieve the maximum possible compression ratios. If a previously compressed file is compressed again, the result might actually be an increase in file size. Therefore, when using a modem to transfer compressed files from a data source, it may be advisable to disable the compression feature of the modem. See also *compression* (2) and *error correction.*

data conferencing repeater A device that enables any user or group of users to transmit a message to all other members of that group. Also called a *technical control hubbing repeater.*

data contamination A synonym for *data corruption.*

data corruption The violation of data integrity. Also called *data contamination.*

data country code (DCC) A 3-digit numerical country identifier that is part of the 14-digit international data numbering plan described in the ITU X.121 recommendation. The *DCC* is of the form ZXX, where Z is the zone and is any digit 2–7 and X is any digit 0–9.

With the exception of codes 1111, 1112, and 1113, zones 0XX and 1XX are reserved for future use. These three codes are used to define nonzoned systems; specifically, they are used for the Maritime satellite packet switched data network. 1111 addresses the Atlantic Ocean, 1112 addresses the Pacific Ocean, and 1113 addresses the Indian Ocean. Zone 8 is used with international telex services, and zone 9 is used with the international telephone numbering plan. The format of the complete international numbering scheme is outlined in the following.

$$P + DCC + N + n\ nnn\ nnn\ nnn$$

where:

P	is the international prefix,
DCC	is the Data Country Code,
N	is an optional national network code, and
n nnn nnn nnn	is the network terminal number in the destination country. It must be between 5 and 10 digits in length.

See also *Data Network Identification Code (DNIC).*

data element A basic unit of information that has a unique meaning.

data encryption algorithm (DEA) See *data encryption standard (DES).*

data encryption key (DEK) A code value used for message encryption and electronic signatures on documents. The key is used by the encryption algorithm to encode the message and may be used by a decryption algorithm to decode the message. More sophisticated encryption strategies use different keys for encrypting and for decrypting.

data encryption standard (DES) A nonproprietary, private-key cryptographic algorithm (data encryption method) developed by the U.S. National Bureau of Standards (now called the National Institute of Standards and Technology—NIST) and IBM. The algorithm is defined in ANSI standard X3.92, while the Federal Information Processing Standard 46 (FIPS-46-1) defines how to implement the algorithm.

The original encryption key was to be 128 bits long; however, the final version used a 64-bit key (8 for error control and 56 for the encryption/decryption process). The standard was certified in 1977 and decertified in 1988; however, it is still widely used. Because *DES* is relatively weak, it is not approved for protection of national security classified information. With only 72 quadrillion keys possible (that is 72 057 594 037 927 936) and the proliferation of high-speed computers, some concern for the security of the *DES* code has prompted *Skipjack*—a code-based 80-bit keys. The Data Encryption Algorithm (DEA) divides a message into 64-bit blocks, each of which is encrypted separately, one character at a time. During the encryption of a block, the computer scrambles the characters in the block 16 times,

COUNTRY OR GEOGRAPHIC CODES OF THE ITU INTERNATIONAL DATA NUMBERING PLAN

Afghanistan	412	Czech Republic	230	Iraq	418	New Caledonia	546	Sweden	240
Albania	276	Denmark	238	Ireland	272	New Hebrides	541	Switzerland	228
Algeria	603	Djibouti, Rep of	638	Israel	425	New Zealand	530	Syrian Arab Republic	417
American Samoa	544	Dominica	366	Italy	222	Nicaragua	710	Tanzania	640
Angola	631	Dominican Republic	370	Ivory Coast	612	Niger Republic	614	Thailand	520
Antigua & Barbuda	344	Ecuador	740	Jamaica	338	Nigeria	621	Togolese Rep	615
Argentina	722	Egypt, Arab Rep of	602	Japan	440	Norway	242	Tonga Islands	539
Australia	505	El Salvador	706	Jordan	416	Oman	422	Trinidad & Tobago	374
Austria	232	Equatorial Guinea	627	Kampuchea	456	Pakistan	410	Tunisia	605
Bahamas	364	Ethiopia	636	Kenya	639	Panama	714	Turkey	286
Bahrain	426	Faeroe Islands	288	Korea, (D.P.R.)	467	Papua New Guinea	537	Turks & Calcos Islands	376
Bangladesh	470	Fiji Islands	542	Korea, Rep. of	450	Paraguay	744	U.S.A.	310
Barbados	342	Finland	244	Kuwait	419	Peru	716	U.S.A.	311
Belgium	206	France	208	Laos	457	Philippines, Rep of	515	U.S.A.	312
Belize	702	French Antilles	340	Lebanon	415	Poland	260	U.S.A.	313
Benin	616	French Polynesia	547	Lesotho	651	Portugal	268	U.S.A.	314
Bermuda	350	Gabon	628	Liberia	618	Puerto Rico	330	U.S.A.	315
Bolivia	736	Gambia	607	Libya	606	Qatar	427	U.S.A.	316
Botswana	652	Germany	262	Luxembourg	270	Reunion Island	647	Uganda	641
Brazil	724	Germany	218	Macao	455	Romania	226	United Arab Emirates	424
British Virgin Islands	348	Ghana	620	Madagascar	646	Russia	250	United Arab Emirates	430
Brunei	528	Gibraltar	266	Malawi	650	Rwanda	635	(AbuDhabi)	
Bulgaria	284	Gilbert & Ellice Isls	545	Malaysia	502	São Tomé	626	United Arab Emirates	431
Burkina Faso	613	Greece	202	Maldives	472	Saudi Arabia	420	(Dubai)	
Burma	414	Greenland	290	Mali Republic	610	Senegal Republic	608	United Kingdom	234
Burundi	642	Grenada	352	Malta	278	Seychelles Island	633	United Kingdom	235
Cameroon	624	Grenadines	360	Mauritania	609	Sierra Leone	619	Uruguay	748
Canada	302	Guam	535	Mauritius	617	Singapore, Rep of	525	Venezuela	734
Cape Verde Islands	625	Guatemala	704	Mexico	334	Solomon Islands	540	Vietnam	452
Cayman Islands	346	Guiana, French	742	Monaco	212	Somali Democratic	637	Virgin Islands, U.S.	332
Central African Rep.	623	Guinea, Rep of	611	Mongolian People's		Rep		Wallis & Futuna Isls.	543
Chad Republic	622	Guinea-Bissau	632	Rep.	428	South Africa	655	Western Samoa	549
Chile	730	Guyana	738	Montserrat	354	Spain	214	Yemen, Republic of	421
China	460	Haiti	372	Morocco,	604	Sri Lanka	413	Yugoslavia	220
Columbia	732	Honduras	708	Kingdom of		St. Kitts	356	Zaire	630
Comoros	654	Hong Kong	454	Mozambique	643	St. Pierre & Miquelon	308	Zambia	645
Congo	629	Hungary	216	Namibia	649	St. Vincent	360	Zimbabwe	648
Cook Islands	548	Iceland	274	Nauru, Rep of	536	St. Lucia	358		
Costa Rica	712	India	404	Nepal	429	Sudan, Rep of	634		
Cuba	368	Indonesia, Rep. of	510	Netherlands	204	Suriname	746		
Cyprus	280	Iran	432	Netherlands Antilles	362	Swaziland	653		

and the encryption method changes after each scrambling. The key determines the details of the scrambling and the character encryption. In short, each 64-bit block goes through multiple transformations during encryption. The encryption algorithm involves three basic steps:

- Transforming (switching the order of) the bits in the block.
- Repeating a computation that uses the data encryption key and that involves substitution and transposition operations.
- Transforming the bits in the block to restore the original order.

DES can operate in any of four modes, each more secure than the previous:

- *Electronic Cookbook* (*ECB*)—the simplest encryption method wherein the encryption process is the same for each block, and it is based on the encryption algorithm and a key. Repeated character patterns, such as names, are always encoded in the same way.
- *Cipher Block Chaining* (*CBC*)—a more involved encryption method in which the encryption for each block depends on the encryption for the preceding block, as well as on the algorithm and key. Here, the same pattern is encoded differently in each block.
- *Cipher Feedback* (*CFB*)—a still more involved method in which ciphertext is used to generate pseudorandom values. These values

are combined with plaintext, and the results are then encrypted. CFB may encrypt an individual character differently each time it is encountered.

- *Output Feedback* (*OFB*)—similar to CFB, except that actual DES output is used to generate the pseudorandom values that are combined with plaintext. This mode is used to encrypt communications via satellite.

See also *Clipper chip.*

data flow control The fifth layer in IBM's Systems Network Architecture (SNA).

data flow diagram A graphical representation of a system showing data sources, data sinks, storage, and the processing performed on the data. The processing steps are shown as nodes and are arranged in the logical order of the steps. The nodes are connected by links indicating the flow of data.

Data Fork The data portion of a Macintosh file. It is the part of a Macintosh file that is transferred to non-Macintosh environments, such as DOS or UNIX.

data format The structure an application program expects data to have so that it can be properly interpreted.

data integrity The condition that exists when data are unaltered after a process as compared to data before the process. That is, there is no alteration or destruction either accidentally or maliciously. Processes include transfer, storage, and retrieval.

data interchange format (DIF) A standardized data file format designed to allow interchange of information among various programs on personal computers.

data link (1) The means of connecting one location to another for the purpose of transmitting and receiving data. (2) A logical connection between two nodes or stations. The connection is medium independent. (3) A collection of entities, consisting of two data terminal equipments (DTEs) and an interconnecting communications circuit. The circuit is controlled by a link protocol that enables data to be transferred from a data source (one DTE) to a data sink (the other DTE).

Data Link Connection Identifier (DLCI) In frame relay communications, a field in the frame relay header that identifies a logical connection. The *DLCI* represents the virtual circuit number associated with a particular destination.

Data Link Control (DLC) A term that refers to the functions provided at the data link layer of the ISO/OSI reference model. These functions are generally provided by a logical link control (LLC) sublayer.

data link escape character <DLE> A transmission control character used in data communications that functions to:

- Identify the beginning of a sequence of a limited number of characters or coded representations.
- Change the meaning of the sequence of immediately following characters.

See also *ASCII.*

data link layer The second layer of the ISO/OSI seven-layer protocol reference model. The *data link layer* defines protocols for frames or packets and how they are transmitted to and from each network device and ensures the error-free delivery of the information through use of error detection, error recovery, and flow control. It is a medium-independent "link-level" interchange facility residing immediately above the *physical layer* (layer 1). The *data link layer* is divided into two sublayers: the medium access control (MAC) and the logical link control (LLC).

Examples of *data link layer protocols* include XMODEM, YMODEM, ZMODEM, KERMIT, HDLC, SLIP, PPP, and contention access methods such as CSMA/CD and Token passing, to mention just a few. See also *OSI.*

data link switching (DLS) An enhancement to source routing that transports source route packets over a resilient Internet protocol/open shortest path first (IP/OSPF) network and provides local termination of LLC2 sessions to avoid LLC timeouts in large or busy networks. It is a good mechanism for mixed LAN-to-LAN and interactive SNA traffic because it can recover from network problems quickly using OSPF.

data management The supervision and control of data-handling operations (including acquisition, analysis, coding, translation, storage, retrieval, and distribution) but not necessarily the generation or use of data.

data medium The material on or in information is stored either on a permanent or temporary basis. Any changeable characteristic of the material may be used to represent the information. Examples of data

media include compact optical disks, magnetic disks, magnetic drums, magnetic tape, optical film, optical tape, paper, and punch cards.

data mode In a communications network, the state of data communications equipment (DCE) when it is connected to a communications channel and is ready to transmit or receive data to or from another DCE. When in the *data mode,* the DCE is not in a talk, dial mode, or command mode. Also called *communications mode.*

data network An information network designed to transfer digital signals as opposed to a voice network that is optimized for analog transmissions.

Data Network Identification Code (DNIC) In the ITU X.121 international data number standard, generally, a four-digit address consisting of the three-digit data country code (DCC) followed by a one-digit national network code. However, provisions are made for countries not wishing to receive the DCC portion of the *DNIC.* See also *data country code (DCC).*

data over voice (DOV) A method of transmitting data and voice simultaneously on the same communication media using *frequency division multiplex (FDM).*

The following figure shows the spectrum of a voice channel and five 50-baud above voice data channels used in a satellite orderwire circuit. Also called *data above voice (DAV).*

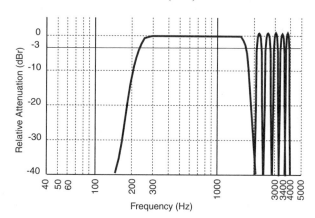

data packet A well-defined block that generally contains user or application data. When transmitted, a data packet will also include some amount of administrative information (overhead, not data) in the packet header and trailer (or footer).

A data packet structure is defined for each particular protocol and message type within the protocol. For example, an X.25 data packet can contain up to 1024 bytes of user data, whereas an XMODEM packet is limited to 128 bytes in its most basic form.

data PBX A private switching system for data traffic that allows terminals and workstations to selectively link to one or more host computers over asynchronous circuits through the use of contention. Terminals and workstations are connected by individual cables to the *data PBX.*

data phase That portion of a data call during which data signals may be transferred between data terminal equipment (DTE) connected to the network. The *data phase* of a data call corresponds to the information transfer phase of an information transfer transaction.

data processing The systematic performance of mathematical, statistical, or selection operations on data. Although the processed data may be significantly modified, the original data should not be changed. Also called *information processing.*

data rate The aggregate rate at which information passes a point in the transmission path of a data transmission system. The rate is generally expressed in bits per second (b/s). Over a multichannel link, the rate is given by:

$$\sum_{i=1}^{m} \frac{\log_2 n_i}{T_i} \text{ b/s}$$

where:

m is the number of parallel channels in the transmission system,
i is a channel number,
n_i is the number of significant conditions of the modulation in the *ith* channel, and
T_i is the signal element unit interval for the *ith* channel (in seconds).

For a single channel ($i = 1$) serial transmission ($m = 1$) system, *DSR* reduces to

$$\frac{\log_2 n}{T} \text{ b/s.}$$

If the transmission system is also a binary system ($n = 2$), this further reduces to

$$1/T \text{ b/s.}$$

For a parallel transmission system with equal unit intervals ($T_1 = T_2 = \ldots T_m$) and equal numbers of significant conditions ($n_1 = n_2 = \ldots n_m$) on each channel, *DSR* reduces to

$$\frac{m \cdot \log_2 n}{T} \text{ b/s}$$

and when the system is binary ($n = 2$) *DSR* becomes

$$m/T \text{ b/s.}$$

Also called *data signaling rate* (*DSR*). See also *baud* (*Bd*) and *Shannon limit*.

data security Pertaining to the protection of data from unauthorized (accidental or intentional) modification, destruction, or disclosure.

data service unit (DSU) **(1)** A device used for interfacing *data terminal equipment* (*DTE*) to the public telephone network. **(2)** One of two components that make up the *data communications equipment* (*DCE*) that provides access to synchronous lines such as T-1 at a customer site. (The other element is a *channel service unit* or *CSU*.) The *DSU* is a type of short-haul, synchronous data line driver that interfaces a servicing central office to the user's equipment over a four-wire circuit at a preset transmission rate. The *DSU* performs the following tasks:

- Connects to the data terminal equipment (DTE) (usually a router or remote bridge) through a synchronous serial interface (generally a RS-422, RS-449, or a V.35, connection). RS-232 connections are also possible for subrate services, i.e., low-speed services. *DSUs* can convert data to or from a native port on a router to an E-1, E-2, or E-3 leased line, primary rate ISDN, or SMDS. A DSU with an HSSI interface will deliver E-2 or E-3 bandwidth from the WAN to an HSSI router on a LAN.
- Formats data for transmission over the digital lines, e.g., converts the customer's data stream, such as X.21 to E-1 or T-1 for transmission through the CSU.
- Controls data flow between the network and CSU.

A *DSU* is frequently combined physically with the CSU.

data set **(1)** A deprecated term from Bell Telephone for a device used to send digital data over a telephone line.

Now called a *modem* or a *line adapter.* **(2)** A collection of data records, each with a logical relation of one to another.

data set ready (DSR) A signal generated by the DCE to the DTE indicating that the local modem is off-hook, is not in test mode, and has completed any call setup requirements. Specific signals are: EIA/RS-232 signal CC, EIA/RS-449 signal DM, and ITU-T signal 107.

data signaling rate See *data rate.*

data sink That portion of a DTE that receives data.

data source The portion of a DTE that originates or sends data.

data state The mode the local modem is in when information presented to it by the DTE is passed through to the remote modem and, conversely, information from the remote modem is passed to the DTE. In data mode, all information is assumed to be data by the modem except the *Escape sequence.* The data state is also called *on-line mode* or *on-line state.*

Other states are *command state, signaling state,* and *training state.*

data stream The uninterrupted flow of digitally encoded signals (data) at a particular granularity used to represent information in transmission.

For example, a continuous flow of bytes across a transmission circuit could be a *data stream.* The flow of data blocks or packets would not be a *data stream* at this level of granularity; however, at a higher level of granularity the continuous flow of data blocks could be considered a *data stream.*

data switch A communications device that accepts incoming digital data and routes it to any selected data sink.

data switching exchange (DSE) The equipment installed at a single location that performs communications switching functions, i.e., circuit switching, message switching, or packet switching.

data terminal equipment See *DTE* (*1*).

data transfer rate The average number of useful bits, characters, or blocks per unit time passing between corresponding equipment in a data transmission system. See also *transfer rate of information bits* (*TRIB*).

data transfer request signal A call control signal originated by a distant data terminal equipment (DTE) to a local DTE (via included data communications equipment—DCE) indicating a request to exchange data.

data transfer time The time between the instant a user information unit (byte, word, block, packet, or message) is made available for transmission over a communications link by a DTE and the instant the complete information unit is accepted by a second receiving DTE.

data type A definition of the kind of data that can be inserted into a specified field of records within a database. For example, within a database of employee records, the "name" field may contain alphanumeric characters, the "date" field must contain a date, and the "salary" field must contain a decimal number.

Some examples of *data types* include:

- *Alphanumeric*—any character (possibly some limitations on punctuation and/or control characters).
- *Integer*—numeric values with no fractional part.

- *Floating point*—numeric values with fractional parts.
- *Date*—an expression of the date in one of many formats (e.g., m/d/y, d/m/y, all numeric, month expressed alphabetically, etc.).
- *Double precision*—a numeric value with more digits of precision than a number expressed in the native machine format. *Double precision* may be applied to either integer or floating point numbers, or both.

data under voice (DUV) A method of transmitting low-speed data and voice simultaneously on the same communication media using frequency division multiplex (FDM).

The diagram shows a 10-Hz data channel below the standard 3100-Hz voice channel. Schemes like this have been used to signal the collection of coins in pay telephones. See also *data over voice* (*DOV*).

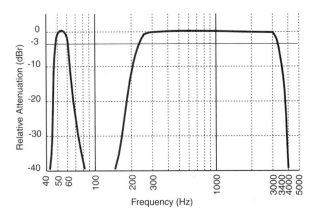

data volatility Pertaining to the rate of change in the values stored in a database over a period of time. For example, the values of particular company stock may be extremely volatile during trading hours but stable during nontrading hours.

data warehouse (1) A computing model in which information is loaded onto a network file server rather than within multiple individual nodes. This makes the data more accessible to users throughout an organization. **(2)** A strategy in which data from a single large transactional database are stored in smaller specialized databases. The information is formatted in a way to make analysis easier, which in turn leads to quicker and better decisions by the directors and other leaders of the organization.

database (1) A set of data that is required for a specific purpose or is fundamental to an operation, system, or enterprise. **(2)** A formally structured collection of data, i.e., a collection of information on a single topic or group of related topics that has been organized in a manner enabling easy retrieval. Usually, it is information that can be stored in categories, such as names, addresses, phone numbers, and salaries. A *database* can be viewed as a table of rows and columns. The rows are called *records,* and the column entries are *fields.*

1st Field Name	Address		
	Street	Town	State
1st record				
Adams, R	123 Xray	Roentgen	NC	•••
Baker, C	1 Modem	Atlanta	GA	•••
Duffus, J	3 Nowhere	Sticks	ND	•••
4th record				
Edwards, X	13 Lucky	Washington	DC	•••
•	•	•	•	•
•	•	•	•	•

A fraction of a *database* table is shown above. Each *record* extends further to the right and contains fields of anything such as salary, place of birth, and marital status. The number of *records* extends downward and may be several hundred or thousand in length.

A database may be organized in many ways; including:

- *Distributed database*—any database type whose contents are stored in multiple locations. Each location contains only a fraction of the total database, and no one location contains the total database.
- *Flat file database*—an organization in which all information is contained in a single file. It consists of individual records that are not necessarily ordered in any particular way. (Access to specific records is by way of lookup tables.) Each record is composed of one or more fields, and each field may contain a particular item of information.
- *Hierarchical database*—an organization in the structure of one or more trees. Each record in a tree has exactly one parent and may have children. Any two records in a hierarchical database are related in exactly one way. Relationships of a hierarchical database include "is a submember of" and "in the same branch" as . . . For example, a DOS directory and file system is arranged as a hierarchical database. Files, therefore, can be described as belonging to a subdirectory of a parent directory or as being in the same directory as . . .
- *Inverted list database*—an organization in which information is organized in tables, but the tables are more content constrained (less abstract) and therefore are less easily manipulated or modified. In addition to tables, an inverted list database also has records whose contents help simplify certain searches. Indexes are used to keep track of records and to speed access.
- *Object oriented*—an organization where the information is arranged into objects, which consist of properties and allowable operations involving the objects. Objects can be defined in terms of other objects (for example, as special cases or as variants of a specific object) and can inherit properties from such "ancestor" objects.
- *Network database*—a term that does not refer to a computer or communication network, but to a mathematical concept in which elements are connected by links. It is a structure in which there are links between records, but there may be no parent, one parent, or multiple parents.
- *Relational database*—an organization wherein the contents are organized as a set of tables in which rows represent records and columns represent fields. Certain fields may be found in multiple tables, and the values of these fields are used to guide searches. Database access and manipulation are a matter of combining information from various tables into new combinations.

In automated information systems, the database is manipulated using a *database management system (DBMS)*.

database administrator The person or group of people responsible for maintaining a database and ensuring the accuracy and integrity of the information contained within the database. Any user with an administrator account security level clearance can oversee the database.

database engineering The discipline involving

- The conception, modeling, and creation of a database.
- Analysis and administration of the database.
- Documentation of the structures and methodologies used.

database management system (DBMS) A computer program or family of programs that

- Enables the creation and maintenance of one or more databases, and
- Assists users or other computer programs using these databases.

database object Any one of the components of a database, e.g., a table, index, procedure, column, rule, and so on.

database server A networked computer node (called a backend or server) dedicated to storing and providing access to the required portions of databases requested by multiple other network nodes (called frontends or clients). Typically, the database is accessed by clients via a query language such as SQL.

The server program is responsible for updating records, ensuring that multiple access is available to authorized users, protecting the data, and communicating with other servers holding relevant data. The client program requests records, modifies them, and adds new ones. A *database server* differs from a file server in that a file server transfers the entire database to the requesting node, while a database server "picks out" the information the requesting node needs and transfers only that portion of the database.

datagram In packet switching, a self-contained, single-packet message (or item of data) that can traverse a network at ISO/OSI Reference Model level 3 (the network layer) and has the following characteristics:

- It is independent of other packets preceding or following, i.e., connectionless mode.
- It contains information sufficient for routing from the originating data terminal equipment (DTE) to the destination DTE without relying on prior exchanges between the equipment and the network. That is, it contains a destination address, a source address, and a sequence number in addition to the user data.
- The packets may arrive at the destination address in the same order as they were sent from the source address.
- The recipient machine must reassemble the packets into the correct sequence.

The term *datagram* refers specifically to the Internet Protocol (IP) rules for organizing information in a packet. The IPv4 header, for example, is six 32-bit words containing the source address, destination address, routing information, and other information. Unlike virtual call services, datagram services provide no call establishing or clearing procedures. Therefore, the network cannot protect against loss, duplication, or misdelivery. *Datagrams* exceeding the *maximum transmission unit* (*MTU*) of a particular network are broken into smaller datagrams (a process called *fragmentation*). When the datagrams arrive at the destination, they are reassembled into the transmitted message. Note that *datagram* and *packet* are not synonyms. *Packet* refers to the physical unit of data, while *datagram* refers to the rules for mapping information into the packet.

Datapak A packet switch network linking Denmark, Finland, Norway, and Sweden. It is run by their respective governments.

date The identification of an instant in the passage of time, described with the desired precision by an appropriate clock and calendar. Examples include:

- *Circa 1890*—the event occurred approximately in the year 1890.
- *January 1, 2000*—anytime during the first day of the first month of the year 2000.
- *18:32:26.05 26 July 1999*—26.05 seconds after 18:32 (6:32 PM) on July 26, 1999. Another representation of this same date is 1999JULY183226.05.

ISO 8601:1988, the international date standard, specifies the numeric format to be yyyy-mm-dd. See also *time*.

date and time stamp The automatically appended date and time of a message. Many programs apply the stamp when the message is created; others apply the stamp when the message is received. These are frequently found on e-mail and facsimile messages.

date format The format used to express the instant of an event.

The *date format* of an event on the UTC time scale is given in the sequence:

24hour:min:sec UT, day month year

For example, 1326:47:58.9 UT, 15 May 1997. See also *date*.

Datex P A packet switched network linking Austria and Germany. It is operated by their respective governments.

datum (**1**) The singular form of *data*. (**2**) A reference point from which measurements or dimensions are taken. (**3**) In geodetic positioning, *datum* refers to the particular mathematical model used to describe the shape of the Earth or a particular local region on the Earth.

In the Global Positioning System (GPS) the WGS-84 model is used to describe the entire Earth. Local models such as NAD-27 datum (for North America), Alaska/Canada datum, Hawaiian datum, and Tokyo datum, are used in specific regions.

DAV An abbreviation of Data Above Voice. See *data over voice* (*DOV*).

daylight savings time (DST) A time standard based on *standard time* but shifted one hour ahead during some seasons. *Daylight savings time* is not universally used throughout the countries of the world. Neither is the starting date and ending date uniform. In fact, it may not be uniformly applied in a given country. (In the United States, for example, neither Hawaii nor Arizona uses *DST*, although the northeast part of Arizona does recognize it.) See also *time* (3).

dB An abbreviation for decibel, literally one-tenth of a bel (after Alexander Graham Bell). A *dB* is an expression relating the ratio of two power levels, *P*—the measured power and P_{ref} the reference power. Mathematically speaking, a decibel is defined by:

$$dB = 10 \cdot \log_{10}\left(\frac{P}{P_{ref}}\right)$$

One of the most frequently expressed values is the 3 *dB* point or the half power point; that is, the power *P* is one-half of the power P_{ref}. Note that the actual numeric value under these conditions is -3.0103 dB, but it is almost always stated -3 dB.

A COMPARISON OF dB AND P:P_REF RATIOS

Ratio	dB	Ratio	dB	Ratio	dB
0.1:1	−10.000	0.9:1	−0.458	2.0:1	3.010
0.25:1	−6.020	1.0:1	0.000	3.0:1	4.771
0.5:1	−3.010	1.1:1	0.413	4.0:1	6.020
0.6:1	−2.218	1.2:1	0.792	5.0:1	6.989
0.7:1	−1.549	1.5:1	1.761	8.0:1	9.030
0.8:1	−0.969	1.8:1	2.553	10.0:1	10.000

Power ratios may also be expressed in terms of voltage (*E*) and impedance (*Z*), or current (*I*) and impedance (*Z*), since

$$P = I^2 Z = \frac{E^2}{Z}$$

a dB can be evaluated by

$$dB = 10 \cdot \log_{10}\left(\frac{E_1^2/Z_1}{E_2^2/Z_2}\right) = 10 \cdot \log_{10}\left(\frac{I_1^2/Z_1}{I_2^2/Z_2}\right)$$

If $Z_1 = Z_2$, these reduce to

$$dB = 20 \cdot \log_{10}\left(\frac{E_1}{E_2}\right) = 20 \cdot \log_{10}\left(\frac{I_1}{I_2}\right)$$

The *dB* is used rather than arithmetic ratios or percentages because the total power gain or loss of a system of cascaded elements is simply the sum of the gains and losses of the individual elements. A positive number of *dB*s indicates gain while a negative number indicates loss. For example, if given a system that consists of a transmission line with 10 dB of loss (−10 dB) followed by a repeater with 20 dB of gain (+20 dB) which in turn is followed by 15 dB of transmission loss (−15 dB), the aggregate system gain is

$$\text{gain} = (-10 \text{ dB}) + (+20 \text{ dB}) + (-15 \text{ dB}) = -5 \text{ dB}$$

or 5 dB of loss. The reference power (P_{ref}) is frequently taken to be 1 watt yielding expressions of dBW, 1 milliwatt yielding dBm, or 1 microwatt yielding dBμ; the following chart gives the relationship of these to absolute power.

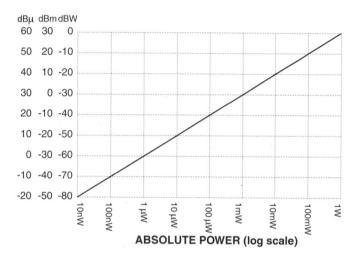

DB-9 The DB-9S or DB-9P is a common 9-pin connector used for making connections to token rings and serial I/O ports on many computers. The S denotes the socket (female) version of the connector, while the P denotes the pin (male) version.

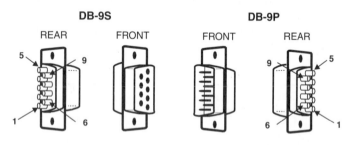

DB-15 The common name of the 15-pin AUI connector used to connect to Ethernet transceivers and game ports for joystick connection on PCs.

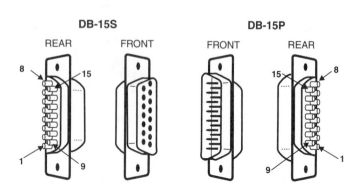

DB-25 The name of the 25-position connector used for making connections to the serial or parallel ports on many computers. Usually, the DB-25S is used for the parallel I/O port on the computer, while the DB-25P is used for the serial I/O port. The S denotes the socket (female) version of the connector, while the P denotes the pin (male) version.

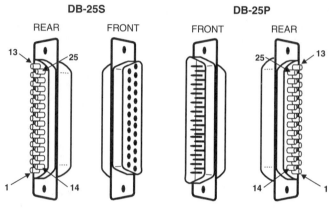

dBa An abbreviation of <u>dB</u> <u>a</u>djusted. A weighted absolute noise power referenced to 3.16 picowatts (−85 dBm). The weighting is specified as F1A-line or HA1-receiver weighting and is indicated in parentheses following the dBa symbol, i.e.,

- *dBa(F1A)* is the weighted absolute noise power in dBa, measured by a noise-measuring set or equivalent to F1A-line weighting.
- *dBa(HA1)* is the weighted absolute noise power in dBa, measured across the receiver of a 302-type (or similar) subset, by a noise-measuring set with HA1-receiver weighting.

A 1-mW 1000-Hz test tone will read +85 *dBa*; however, 1 mW of white noise in the bandwidth of 300–3300 Hz will read only +82 *dBa* due to frequency weighting. dBa(F1A) and dBa(HA1) should be converted to dBa0 to eliminate the necessity of stating the relative transmission level at the point of actual measurement. Also called *dBrn adjusted.*

dBa0 Noise power in dBa referenced to the zero transmission level point (0TLP); also called a point of zero relative transmission level (0 dBr).

DBA (**1**) An abbreviation of <u>Data</u><u>B</u>ase <u>A</u>dministrator. (**2**) An abbreviation of <u>D</u>ynamic <u>B</u>andwidth <u>A</u>llocation.

dBc An abbreviation of <u>dB</u> relative <u>c</u>arrier power.

dBi An expression of antenna gain. The number of decibels of gain of an antenna referenced to the hypothetical 0-dB gain free-space isotropic radiator.

dBm (**1**) An abbreviation of <u>deci</u>bel referred to 1 <u>milli</u>watt (i.e., $P_{ref} = 1$ mW). The dBm in the communication industry is the comparison of a power P to the reference 1 mW at 1004 Hz. 0 dBm = 1 mW, + dBm is power above 1 mW, and −dBm is power below 1 mW. See also *dB*. (**2**) In European practice, the usage context of dBm **may** imply psophometric weighting. It is equivalent to *dBmp* or *dBm0p,* the preferred abbreviations.

dBm(psoph) Noise power in dBm, measured with psophometric weighting where

$$\text{dBm(psoph)} = 10 \log_{10} (pWp) - 90$$
$$= \text{dBa} - 84$$

where:

pWp is power in picowatts psophometrically weighted, and
dBa is the weighted noise power in dB referenced to 3.16 picowatts.

Also written *dBm0p*.

dBm0 (1) An expression of power in dBm referred to or measured at a *zero transmission level point* (*0TLP*). A 0TLP is also called a point of *zero relative transmission level* (*0 dBr0*). That is, a signal designated as X *dBm0* represents a signal X dBm different from the level of a *previously defined reference* signal.

For example, in the following figure, the reference point is defined as the input to the line driver, i.e., the 0 dBm0 point. A reference test signal (which, in this case is used to define the 0 dBm0 level) is given as +3 dBm at the line driver, −16 dBm at the output of the driver, and −19 dBm at the input to the line receiver. A signal defined as −10 dBm0 would then *imply* a level of −7 dBm at the line driver, −6 dBm at the output of the driver, and −9 dBm at the input to the line receiver. (2) Some international documents use dBm0 to mean noise power in dBm0p (psophometrically weighted dBm0).

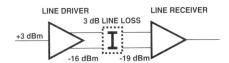

dBm0p Noise power in dBm0, measured by a psophometer (a noise-measuring set having psophometric weighting). See *dBm(psoph)*.

dBmC A *dBmC* is the same as a dBm except the measured signal is band limited to better approximate the responses through a telephone. The band-limiting device is based on measurements taken on a Western Electric®-type 500 telephone. The name given to this filter is a *C-message filter*, hence, the designation *dBmC*. See also *C-message weighting*.

dBmp Similar to the *dBmC* except a slightly different filter is defined by ITU-T. The filter is called a *psophometric filter*; consequently, the units of the measurement are *dBmp*.

0 dBm psophometrically weighted noise is equivalent to:

- An 800-Hz tone at 0 dBm,
- 3-kHz band-limited white noise at +2.5 dBm, or
- A 1000-Hz tone at −1 dBm.

See also *psophometric weighting*.

DBMS An abbreviation of Data Base Management System.

dBmV An abbreviation of dB referenced to 1 millivolt across 75 Ohms.

$$1\ dBmV = 1.33 \times 10^{-5}\ mW$$

dBr An abbreviation of a dB relative. The dBr is the power *ratio* (expressed in dB) between the measured point and a reference point. The two points could be the same physical point but under different operating conditions. Because the dBr is a difference, it does not express an absolute power.

For example, if the original transmit signal level from a particular modem is −13 dBm and the power is reduced to −16 dBm, the relative power is −3 dBr; i.e.,

$$-16dBm - (-13dBm) = -3dBr.$$

dBrn An abbreviation of dB above a reference noise. The reference noise (or 0 dBrn) is defined as 1 picowatt (−90 dBm). The bandwidth of the noise measurement must be specified and may be indicated in parentheses following the symbol. Examples include dBrn(flat), dBrn(f_1-f_2), dBrn(144-line), dBrn(144-receive), and dBrn(C-message) or simply dBrnC.

dBrn adjusted See *dBa*.

dBrn(144) Weighted noise power in dBrn, measured by a noise-measuring set with 144 weightings.

A 1-mW 1000-Hz test tone will read +90 dBrn(144); however, 1 mW of white noise in the bandwidth of 300–3300 Hz will read only +82 dBrn(144) due to frequency weighting. See also *dBrn*.

dBrnC Weighted noise power in dBrn, measured by a noise-measuring set with C-message weighting.

A 1-mW 1000-Hz test tone will read +90 dBrnC; however, 1 mW of white noise in the bandwidth of 300–3300 Hz will read only +88.5 dBrnC due to frequency weighting. See also *C-message weighting*.

dBrnC0 An acronym from dB C-message weighted noise relative to the zero transmission level point (0TLP) (pronounced "*de-brink-o*"). Noise power in dBrnC is referred to or measured at a zero transmission level point (0TLP).

dBrn(f_1-f_2) Flat noise power expressed in dBrn, measured over the frequency band between frequencies f_1 and f_2. See also *C-message weighting*.

dBrnp The same as a dBrnC except the filter inserted into the meter is defined by ITU-T standards rather than based on the Western Electric type 500 telephone. The "p" stands for psophometric.

DBS An abbreviation of Direct Broadcast via Satellite.

DBT An abbreviation of Deutsche Bundespost Telecom.

dBu or dBμ An abbreviation of decibel referred to 1 microwatt (μW) (i.e., $P_{ref} = 10^{-6}$ W); therefore, $dBu = 60log_{10}(P)$.

dBv An expression of the ratio of the peak signal voltage to 1 volt peak regardless of the impedance; i.e.,

$$dBv = 20 \cdot \log_{10}\left(\frac{\text{Peak Signal Voltage}}{\text{1 Volt Peak}}\right)$$

The *dBv* is usually used to express television video signal levels.

dBW An abbreviation of decibel referred to as 1 watt (i.e., $P_{ref} = 1$ W); therefore, $dBW = 10\ log_{10}(P)$.

dBx An abbreviation of dB above reference coupling. It is used to express the amount of crosstalk coupling in telephone circuits and is measured with a noise-measuring set.

dc or DC An abbreviation of Direct Current.

DCA (1) An abbreviation of the Defense Communications Agency. Now called the Defense Information Systems Agency (DISA). (2) An abbreviation of Document Content Architecture.

DCC (1) An abbreviation of Data Communications Channel. (2) An abbreviation of Digital Cross Connect.

DCD (1) An abbreviation of Data Carrier Detect (RS-232 signal CF, RS-449 signal RR, and ITU-T signal 109). It is a signal generated by the DCE to the DTE, indicating that the local modem has detected the carrier from the remote modem. (2) An abbreviation of Dynamically Configurable Device. Another name for a Plug-and-Play device in which the system need not be rebooted after the device is installed hot (that is, with power on).

DCE (1) An abbreviation of Data Communication Equipment (or Data Circuit terminating Equipment). The interface equipment required to couple information from the DTE into a transmission facility (telephone line, LAN, etc.). A modem is an example of a *DCE*.

The *DCE* establishes, maintains, and terminates a data communication session and provides appropriate modulation, encoding, or signal conversion. Deprecated synonyms include *data communications equipment* and *data set.* (2) An abbreviation of Distributed Computing Environment.

DCS (1) An abbreviation of Defined Context Set. ITU's X.216 recommendation of an agreed-upon context for the delivery and use of presentation-level services. (2) An abbreviation of Digital Cross connect System. (3) An abbreviation of Distributed Communication System. (4) An abbreviation of Digital Cellular System. (5) An abbreviation of Defense Communications System.

DCT An abbreviation of Discrete Cosine Transform.

DCT⁻¹ An abbreviation of Inverse Discrete Cosine Transform.

DDA An abbreviation of Domain Defined Attribute. Additional information added to an e-mail address in order to identify people with the same name.

DDB An abbreviation of Distributed DataBase.

DDBMS An abbreviation of Distributed DataBase Management System.

DDC An abbreviation of Digital Data Converter.

DDCMP An abbreviation of Digital Data Communication Message Protocol. The Digital Equipment Corporation (DEC) data link layer communications protocol used to transmit messages over synchronous, asynchronous, or parallel full- or half-duplex communications lines.

DDD An abbreviation of Direct Distance Dialing.

DDE An abbreviation of Dynamic Data Exchange.

DDI An abbreviation of the British Direct Dial Inward.

DDL An abbreviation of Data Definition Language.

DDM An abbreviation of Distributed Data Management.

DDN An abbreviation of Defense Data Network.

DDN NIC An abbreviation of Defense Data Network Network Information Center.

DDP An abbreviation of Distributed Data Processing. A network of interconnected but separate data processing equipment that share common resources and in which some (or all) of the processing and/or input/output (I/O) functions are distributed over multiple machines.

DDR An abbreviation of Dial on Demand Routing.

DDS (1) An abbreviation of Dataphone Digital Service. A 2.4- to 56-kbps common carrier service offering. Also called *switched 56.* (2) An abbreviation of Direct Digital Synthesis. (3) An abbreviation of Digital Data Service.

DDSD An abbreviation of Delay Dial Start Dial. A start-stop signaling protocol for dialing into a telephone switching system.

DE An abbreviation of Discard Eligibility.

de facto standard A standard that develops naturally through widespread acceptance and use, not because of a preissued document.

There are many examples of *de facto standards,* some of which have become official standards, i.e., the Hayes modem command set, the TCP/IP protocols, and so on.

de Forest, Lee An American inventor (1873–1961) who invented the vacuum tube (the Audion) that was patented in 1907. Though not the first vacuum tube per se, it was the first with a control element that made it possible to transmit voice and other analog signals.

Vacuum tubes are called *valves* in some countries.

de jure standard A standard that has been imposed by a nationally or internationally recognized standardization organization, e.g., the ITU, ISO, or the EIA. One example of a *de jure standard* is the open systems interconnection (OSI) model.

de-emphasis In FM transmission, the process of restoring the amplitude vs. frequency characteristics of a received signal that has undergone pre-emphasis in the transmitter. See also *pre-emphasis.*

de-ossification The conversion of definitions that conform to OSI network management model to definitions that conform to the IP network management model. The term is used in TCP/IP environments that use Simple Network Management Protocol (SNMP).

de-stuffing The removal of stuffing bits from a stuffed digital signal, to recover the original signal. Also called *negative justification* and *negative pulse stuffing.*

DEA An abbreviation of Data Encryption Algorithm. See *data encryption standard* (*DES*).

dead band The range through which the input signal to a system can be varied without causing a response on the output.

dead spot In radio communications, a location from which reliable communications cannot be established—generally, areas surrounding the *dead spot* exhibit no such difficulties. *Dead spots* have many causes ranging from hills to structures, electronic interference, and even excessive foliage.

deadlock (1) The undecided contention for the use of a system or component. (2) In computer and data processing systems, an error condition in which each of two components or processes is continuously waiting for an action or response from the other. The result is that neither proceeds. (3) A permanent condition in which a device or system cannot continue to function unless some manual corrective action is taken.

debounce A means of removing the make-break bounce of mechanical switches so as to generate a clean transition from one signal level to another. See also *contact bounce.*

DEC An acronym from Digital Equipment Corporation (purchased by Compaq Computer in 1998).

decade Pertaining to a power of ten. For example, the range of 1–10, 10–100, or 10^6–10^7 represent powers of ten.

decapsulation The stripping off of a protocol header/trailer from a packet of another protocol.

The opposite of *encapsulation.*

decibel See *dB.*

DECmcc An acronym from DEC Management Control Center. Network management software for Digital's DECnet networks.

DECnet™ An acronym from Digital Equipment Corporation NETwork—a proprietary peer-to-peer LAN network architecture that has undergone several revisions; the latest, phase V, was released in 1987.

Phase V was designed to be in compliance with the ISO/OSI Reference Model. It has only seven layers, which correspond to the seven OSI layers (as opposed to the eight layers of its predecessor). DECnet Phase V not only generally supports OSI-compliant protocols at each level, but it also supports DEC's proprietary protocols (such as DDCMP and DAP) for backward compatibility with Phase IV networks. With up to 20 bytes of address information, very large networks can be handled; the network can also be divided into domains for routing or administrative purposes. The address field includes an Initial Domain Part (IDP) value, which is unique for every network. It is the implementation of the Digital Network Architecture (DNA).

decode **(1)** To convert data from one representation to another by reversing the effect of previous encoding. **(2)** To interpret a code. **(3)** To convert encoded text into equivalent plaintext by means of a code. Decoding does not include deriving plaintext by cryptanalysis.

decoder A device or software that converts coded data back to its original form. Decryption, decompression, and code converters are all forms of decoding See also *encryption.*

decollimation In a collimated electromagnetic beam (i.e., a beam with near parallel rays), any mechanism by which the rays are caused to diverge or converge from parallelism.

Decollimation may be deliberate for systems reasons or incidental due to environmental factors; e.g., deflection, diffraction, inhomogeneities, occlusions, reflection, refraction, refractive index, and scattering.

decoupling capacitor A capacitor that provides a low-impedance path to ground to prevent common coupling between circuits.

decrypt **(1)** A generic term encompassing the terms *decode* and *decypher.* **(2)** To convert encrypted text into its equivalent plaintext by means of a cryptosystem. (This does not include solution by cryptanalysis.)

decryption In cryptology, the process of restoring encrypted data to its original form. The *decryption* process requires not only knowledge of the decoding algorithm but knowledge of a specific decoding key.

Decryption is the reciprocal process to encryption. See also *encryption.*

DECT An abbreviation of D̲igital E̲uropean C̲ordless T̲elecommunications (or T̲elephone).

dedicated access A method of connecting a large computer system to the Internet by means of a router and a high-speed communications link. The connected machine becomes a host with a unique IP address and is capable of supporting tens to hundreds of simultaneous users (depending on the bandwidth of the high-speed interconnect).

dedicated channel A channel leased from a service provider for the exclusive use of a specified end user. See also *leased line.*

dedicated circuit A path from a user location directly to a telephone company point of presence (POP)—that is, the path that includes the subscriber's leased or long-distance lines and extends to the point where it connects to the telephone company's lines. A *dedicated circuit* is designated for the exclusive use of specified end users. See also *leased line.*

dedicated connection A permanent connection between two users. See also *leased line.*

dedicated LAN A network segment allocated to a single device. Used in LAN switched network topologies.

dedicated line A transmissions circuit installed between two physical separate locations and available to carry traffic at all times. See also *leased line.*

dedicated machine A machine designed to perform only one task or family of tasks; for example, the computer in a telephone switching system is generally capable of only controlling the switching system and could not do payroll.

General-purpose computers may be programmed to do a single specific task; however, a dedicated machine cannot be easily reprogrammed to anything except those things the original designers intended.

deeply depressed cladding fiber A doubly clad optical fiber construction, usually a single mode fiber, that has an outer cladding of approximately the same refractive index as the core and an inner cladding of very low (depressed) refractive index material between them. See also *doubly clad fiber, optical fiber,* and *refractive index profile.*

default The predefined value, action, or settings that a device, system, or software assumes, unless explicit user action is taken to alter them.

default path In packet networks, a path used by a router to forward a packet when the packet itself contains no explicit routing instructions and the router has no predefined path to the packet's ultimate destination. The default path is generally one to a router that is likely to have more detailed routing information.

default route The communications path to which a router will direct any frame for which there is no definitive listing for the next hop.

default setting **(1)** Initial register values or switch settings defined by the original equipment manufacturer (OEM) at the time of manufacture. **(2)** Initial values to be used every time a program or piece of equipment is started.

Defense Communications Agency (DCA) An agency within the United States Department of Defense. In 1975, the *DCA* acquired control of the ARPAnet (the foundation for the Defense Data Network-DDN). The *DCA* is now called the Defense Information Systems Agency (DISA).

Defense Data Network (DDN) The wide area packet network formed in 1990 to integrate the ARPAnet research network and the MILNET defense network into a single network. It is used by the U.S. Department of Defense (DoD) to interconnect military installations. Parts of the *DDN* are accessible from the Internet, and parts are classified.

Defense Data Network Network Information Center (DDN NIC) A Network Information Center (NIC) maintained by the Defense Data Network (DDN), sometimes called "The NIC."

The DDN NIC is a control center that provides information and services through the Internet. The DDN NIC responsibilities include the following:

- Serves as the chief and authoritative repository for network information, including Requests for Comments (RFCs), that are used to define standards, report results, and suggest planning directions for the Internet community; For Your Information (FYI), and Internet Drafts.
- Assigns IP network addresses.
- Assigns numbers to domains (or autonomous systems, as they are called in the Internet jargon).

Defense Information Systems Agency (DISA) An agency within the U.S. Department of Defense responsible for managing MILNET, the DDN portion of Internet, and other classified networks. *DISA* was formerly called the Defense Communications Agency (DCA).

deferral time The time delay between an attempt to transmit on a collision sense multiple access (CSMA) network when activity is detected and the next attempt. The delay depends on the following:

- The activity level of the network. The deferral time is longer if the network is very active, and shorter when there is little activity.
- A random value added to the base deferral time. This ensures that two nodes that defer at the same instant do not try to retransmit at the same time.

See also *CSMA/CA* and *CSMA/CD*.

degradation The deterioration in quality, level, or standard of performance of a functional unit. That is, a condition in which one or more parameters lie outside specified tolerances (generally resulting in deterioration of performance). *Degradation* is usually categorized as either of two ways:

- *Graceful degradation,* in which the equipment continues to perform but at a reduced capability, or
- *Catastrophic degradation,* in which the equipment is no longer capable of performing its required task(s).

Causes of *degradation* can be equipment failure or external conditions, such as transmission line noise. See also *degraded service state.*

degraded service state The condition that exists when one or more required performance parameters fall outside specified limits for a sustained period of time, resulting in a lower quality of service.

A *degraded service state* is considered to exist when a specified level of degradation persists for a specified period of time.

degauss To demagnetize. It is generally accomplished by passing an alternating magnetic field across the object to be demagnetized. In most cases, it is more practical to gradually reduce the field strength than to move the field across the object. For example, on a CRT (a picture tube) the field is turned on full strength at power on and then reduced to zero over several seconds.

degree of coherence With respect to electromagnetic waves, a dimensionless unit used to indicate the extent of coherence of an electromagnetic wave.

For light, the magnitude of the *degree of coherence* is the visibility (V), of the fringes of a two-beam interference test, as given by:

$$V = \frac{I_{MAX} - I_{min}}{I_{MAX} + I_{min}}$$

where:

I_{MAX} is the intensity at a maximum of the interference pattern, and
I_{min} is the intensity at a minimum.

Light is considered highly coherent when the *degree of coherence* exceeds 0.88, incoherent for values near (or at) zero, and partially coherent otherwise. Also called *coherence degree.*

degree of distortion A measure of the shift of the actual transitions between signal states and the position of the transition. The *degree of distortion* is usually represented as a percentage of the unit interval. See also *bias distortion* and *jitter.*

degree of isochronous distortion The ratio of the absolute value of the maximum difference between the actual and theoretical intervals separating two transmission signaling elements to the theoretical interval. Generally, the maximum is taken over some specified measurement period. The *degree of isochronous distortion* is usually represented as a percentage of the unit interval. See also *isochronous distortion.*

degree of longitudinal balance The ratio of a disturbing longitudinal signal (V_L) to the induced metallic signal (V_M) expressed in decibels; i.e.,

$$longitudinal\ balance = 20 \cdot log\left(\frac{V_L}{V_M}\right) dB.$$

DEK An abbreviation of <u>D</u>ata <u>E</u>ncryption <u>K</u>ey.

delay The time difference between a measured event and a reference event. Delay is typically measured in time (*delay time*) or as an angle (*delay angle* or *phase angle*) in the case of periodic waves.

delay dispersion The differences in signal delay over a specified band of frequencies.

delay distortion The difference in travel time for different frequencies of a transmitted signal. Distortion occurring when the *envelope delay* of a circuit, device, or system is not constant over the frequency range required for transmission. Numerically, *delay distortion* is the difference between the envelope delay at any frequency and the minimum envelope delay within the frequency band of interest. *Delay distortion* is the result of a nonlinear phase frequency response. Also called *envelope delay distortion, group delay distortion,* and *phase distortion.*

delay encoding A method of encoding a digital waveform so that the information is contained in the transitions (not the logical levels). Specifically:

- A "1" always causes a transition from one level to the other level in the middle of the bit period.
- A "0" causes a transition from one level to the other level at the beginning of a bit period unless it follows a "1."

Delay encoding is used primarily for encoding radio signals because the frequency spectrum of the encoded signal is shaped to contain less low-frequency energy than a conventional nonreturn to zero (NRZ) signal and less high-frequency energy than a biphase signal. Also called *Miller encoding* or *modulation* See also *waveform encodes.*

delay equalizer A corrective network designed to make the *phase delay* or *envelope delay* of a channel substantially constant over a specific frequency range. All delay equalizers add to the total path delay however. The delay slope of the equalizer is opposite that of the uncorrected channel. Also called a *phase equalizer* or *delay equalizer* See also *equalization.*

delay lock loop See *code tracking loop.*

delay line A device that introduces a specified time delay into a signal path. Ideally, the delay is constant for all frequencies within the specified passband.

delay pulsing In telephony, a method of pulse signaling wherein the sender delays transmission of address pulses until it receives an off-hook signal flash (a short off-hook pulse followed by a steady on-hook condition) from the far end. The method provides a positive trunk integrity check.

delay slope The ratio of delay dispersion to the bandwidth over which the dispersion is measured.

delay transmission A feature of facsimile machines and fax/modem systems that enables them to store a message in memory until a specified time. This is usually done to allow the sender to take advantage of the lower telephone rates in the evening or nighttime. Also called *delayed sending.*

delayed file backup A LAN feature that lets the user specify the time a file is sent to the backup server; generally used to transfer large files after hours.

delayed printing A LAN feature that allows a user to schedule print jobs (usually large jobs) to a more convenient time (after hours).

delimiter A marker that delineates or separates data strings or fields. *Delimiters* are application dependent; that is, the *delimiter* may be one ASCII character in one application and a different character in another application. Typical *delimiters* include <SPACE>, <COMMA>, <">, and <CR>.

In the Hayes command set, the guard time before and after the "+++" string could be thought of as a *delimiter*. It separates the transmitted data string from command strings.

delivered block A block of information successfully transferred from a source across a communications link to an information sink (receiver or data user).

delivered overhead block A block of information successfully transferred from a source across a communications link to an information sink; however, the block contains no user information.

delivery confirmation A message returned to a source that indicates that a specific information block was successfully delivered to the intended destination.

DELPHI An online service, based in Cambridge, Massachusetts, that can provide access to the Internet, including e-mail, File Transfer Protocol (FTP), World Wide Web (WWW), and Telnet services.

delta (1) The fourth letter in the Greek alphabet, i.e., "Δ" in upper case and "δ" in lower case. (2) In mathematics and engineering, generally indicates a difference between two values or a change from a reference value.

delta modulation (DM or ΔM) An analog-to-digital converter where the generated digital words represent *differences* (deltas) between successive sampled points on the input waveform rather than the actual value at each point. These differences are transmitted to a remote receiver where they are used to reconstruct an approximation of the original waveform. The reconstruction is simply a running sum of all values received.

For the following waveform, values at each of the 16 sampling times as well as the difference between each sample value and the previous sample value are shown. Simply summing these difference values

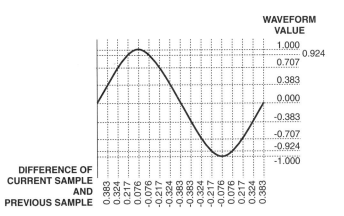

and holding the value until the next sample is received yields the following plot. Examples of delta modulation include *continuously variable slope delta modulation, delta sigma modulation,* and *differential modulation.*

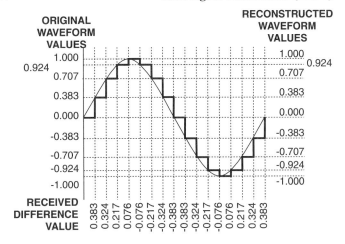

delta sigma modulation (ΔΣM) An analog-to-digital converter where the generated digital words represent *differences* (or delta) between sampled points on the input waveform and an estimate of the value based on previous samples. These differences are transmitted to a remote receiver where they are used to reconstruct an approximation of the original waveform. The reconstruction is simply a running sum of all values received.

The associated figure illustrates the block diagram of one implementation of a *one-bit ΔΣ-modulator* and the waveforms one would expect to see, i.e.,

- The analog input is designated A.
- The sampling clock is waveform B.
- The waveform estimating the analog input is labeled C.
- The ΔΣ-*output* is waveform D.

The single bit output (D) is the sign of the difference (A–C); if it is positive (a "1"), the estimate is too small; hence the counter must increase its value. If the output is negative (a "0"), the counter must decrease its value.

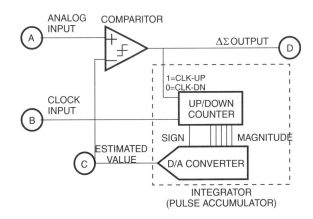

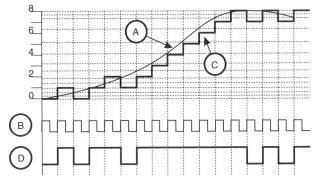

At the receiving station waveform, "C" could be reconstructed by counting the number of "1s" and "0s" as was done in the modulator. However, a low-pass filter could also be used, and a waveform similar to "E" of the next figure would be achieved.

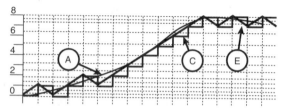

Because the digital word at the output of the up/down counter tracks the original signal at each clock cycle, it is common to use this estimated value in data acquisition A/D converters. Also called a *sigma delta modulator* and *tracking A/D converter.* See also *adaptive differential PCM (ADPCM)*.

demagnetization　The partial or complete reduction of magnetic flux.

demand assigned multiple access (DAMA)　A method of allocating access to communications channels when there are more users than channels.

All idle channels are kept in a pool and are not assigned to any user. When a channel is requested by a user, an arbitrary idle channel is selected from the pool and assigned to the requesting user. DAMA may be implemented with time division multiplex (TDM), frequency division multiplex (FDM), code division multiplex (CDM), or space division multiplex as is the case in a telephone network when there are fewer interconnecting trunks than users. See also *blocking*.

demand assignment　An operation in which multiple users share access to a common communications channel or group of channels on a real-time basis. In operation, any user needing to communicate with another user on the same network requests a channel, uses it, and, when finished, releases it (making it available to other users). See also *blocking*.

demand priority　A media access method used in 100BaseVG, a 100-Mbps Ethernet implementation proposed by Hewlett Packard (HP), and AT&T Microelectronics that uses a star topology.

In operation, a node with traffic to send indicates this to the hub and requests high or regular priority service for its transmission. After it obtains permission, the node begins transmitting to the hub. The hub then passes the transmission on to the destination node. The hub will pass high-priority transmissions through immediately and regular priority transmissions through as the opportunity arises. By letting the hub manage access, the architecture is better able to grant required bandwidths and requested service priority to particular applications or nodes as well as improve security. It also guarantees that the network can be scaled up (enlarged) without loss of bandwidth. *Demand priority* accomplishes this in the following ways:

- A node does not need to keep checking for an idle network before transmitting.

  In current Ethernet implementations, a wire pair is dedicated to this task. By making network checking unnecessary, demand priority frees a wire pair that makes the pair available for information transfer. 100BaseVG specifications use quartet signaling, which uses the four available wire pairs.
- Heavy traffic can effectively bring standard Ethernet networks to a standstill because nodes spend most of their time trying to access the network (see the chart accompanying *Aloha network*). With *demand priority*, the hub needs to pass a transmission on

only to its destination, so that overall network traffic is decreased; hence, more bandwidth is available for heavy network traffic.
- By giving the hub control of transmission, the message can be passed to only its intended destination node or nodes. This makes eavesdropping more difficult.

demarc　An abbreviation of DEMARCation point.

demarcation point (demarc)　That point at which operational control or ownership of communications facilities changes from one organizational entity to another.

In telephony, that point at which telephone company-owned and -maintained equipment ends and the subscriber's responsibilities begin. The *demarcation point* is the interface point between customer premises equipment (CPE) and external network service provider equipment. It is usually at the customer side of a terminal block with built-in transient protection (carbon blocks, for example) or at a distribution frame. Also called the *network terminating interface* or the *network interface (NI)*.

demarcation strip　In telephony, the terminals at which the local telephone company's service ends and to which the customer's equipment is connected. See also *demarcation point (demarc)*.

DEMKO　An acronym from Denmark Elektriske MaterielKOntrol. The Danish testing laboratory.

democratically synchronized network　A method of network clock synchronization in which all node clocks have equal status and equal control on each other. The network operating clock rate (frequency) is the mean of the natural (uncontrolled) clock rates of the total population of clocks. See also *despotically synchronized network* and *oligarchically synchronized network*.

demodulation　The process of separating or recovering the information-bearing signal from a modulated carrier in such a manner that the extracted signal has essentially the same characteristics as the original carrier modulating signal.

demodulator　A device for extracting and reconstructing the information signal from a modulated carrier. Also called a *detector.*

DeMorgan's law　Rules established by Agustus DeMorgan (1806–1871) that equate two logical statements; namely, that

$$\text{The denial of the} \left\{ \begin{matrix} \text{conjunction} \\ \text{disjunction} \end{matrix} \right\} \text{of two}$$

statements P and Q is logically equivalent

$$\text{to the} \left\{ \begin{matrix} \text{conjunction} \\ \text{disjunction} \end{matrix} \right\} \text{of their denials.}$$

This can be written in the Boolean algebraic form as:

$$\overline{P \cdot Q} \equiv \overline{P} + \overline{Q} \quad \text{and} \quad \overline{P + Q} \equiv \overline{P} \cdot \overline{Q}$$

See also *Boolean algebra* and *truth table*.

demultiplex　The reverse process of multiplex, that is, the process of separating an incoming data stream into its constituent parts. These individual parts are directed each to its own designated destination.

demux　An acronym from DEMUltipleX, DEMUltipleXer or DEMUltipleXing.

denial of service　To turn down a request for use or access by a network user. Usually, this is a response for either misuse of the service or nonpayment of bills.

dense binary code　A binary code in which all possible bit combinations of a specified number of bits are used to encode user information. It excludes overhead information such as error detection. Examples of *dense binary codes* include:

- The pure binary representation of hexadecimal digits; i.e., all 16 possible combinations of the four bits are used.
- The pure binary representation of octal digits; i.e., all eight possible combinations of the three bits are used.

Binary coded decimal (BCD) is not a dense binary code as only 10 of the 16 possible combinations are defined. If a binary code is not dense, the unused bit combinations can be used to detect errors since they occur only if there is an error.

density **(1)** The number of bits (or bytes) that can be stored per unit length, area, or volume of a storage medium. For example, a magnetic tape may have a density of 1600 bpi, whereas an optical storage device may have a capacity of 106 bits per square inch and a hard disk drive may have a density of 100 MB per cubic inch. **(2)** In facsimile, a measure of the reflection (or transmission) properties of the source information. It is generally expressed as the logarithm, base 10, of the ratio of the incident to reflected (transmitted) light flux. **(3)** In optics, a measure of *transmittance* (the ratio of the radiant power transmitted by an object to the incident radiant power). *Density* is the logarithm, base 10, of the inverse of the transmittance.

There are many types of *density*, each of which will usually have different numerical values for different materials; e.g.:

- *Diffuse density*—a measurement resulting when the sample is diffusely illuminated.
- *Double diffuse density*—a measurement derived from calculation when the incident flux of a photographic negative is entirely diffuse and all transmissions are included.

The relevant type of density depends on the type of optical system, the component materials of the object, and the surface characteristics of the object.

departure angle The angle between the axis of the main lobe of an antenna pattern and the horizontal plane at the transmitting antenna. Also called *takeoff angle*.

dependent node In network analysis, a node having one or more incoming branches.

depolarization **(1)** With electromagnetic waves, the condition wherein a polarized wave passes through a nonhomogeneous medium and has its degree of polarization reduced or randomized. **(2)** The prevention of polarization in an electrochemical cell or battery.

depopulate The removal of some elements of a system.

depressed cladding fiber A synonym for *doubly clad fiber*.

depressed inner cladding fiber A synonym for *doubly clad fiber*.

dequeue An acronym from **D**ouble-**E**nded **que**ue.

derivation equipment Equipment that provides a number of reduced bandwidth channels from a single wider bandwidth channel.

DES An abbreviation of **D**ata **E**ncryption **S**tandard.

descending node That point in an orbit where an orbiting object passes from above to below the equatorial plane. See also *ascending node*.

descrambler A device to extract a pure signal from an encrypted signal stream, provided no transmission errors have occurred. For example, the signals received from a satellite TV system are frequently distorted (scrambled) in order to prevent unauthorized viewing. The *descrambler* removes the distortions.

design margin The additional performance capability of a device, circuit, or system above specified system parameters. They are included by the design engineer to compensate for uncertainties such as those found in manufacturing processes, component values, and component degradation.

designated router In a multiaccess network with at least two attached routers and using the open shortest path first (OSPF) protocol, one router is assigned special protocol duties, such as generating a link state advertisement for the multi-access network.

The *designated router* concept helps reduce the number of agencies required on a multiaccess network. This in turn cuts routing protocol traffic, as well as the size of the topological database.

designation strip The small strip of paper that slides into a telephone instrument and identifies the function of each button. Also called *desi strip*.

desktop management interface (DMI) A protocol independent API architecture (developed by the Desktop Management Task Force (DMTF) (a consortium of 300 vendors including DEC, IBM, Intel, HP, Microsoft, and Novell) that allows users to access and manage all of the components attached to a computer (including add-in boards, application s/w, peripherals, etc.). The specification is independent of hardware platform and operating system (e.g., stand-alone or networked PCs, Apple Macintosh, MS-DOS, NetWare, OS/2, UNIX, Windows). *DMI* consists of three components:

- The service layer.
- The component interface.
- The management interface.

DMI identifies the manufacturer, component name, version, serial number (if appropriate), installation time, and date for any component installed in a system. See also *Plug and Play* (*PnP*).

despotically synchronized network A method of network clock synchronization in which all node clocks are controlled by a unique master clock. The network operating clock rate (frequency) is therefore the same as the master clock regardless of the natural (uncontrolled) clock rates of the individual node's clock. See also *democratically synchronized network* and *oligarichically synchronized network*.

destination The location to which the result of an operation is directed, such as the transmission of a message from an originator to a receiver (the receiver being the destination).

destination address (DA) That portion of a packet that uniquely defines the location to which the packet is to be delivered. In the Internet Protocol (IP), the header contains the *destination address*.

destination address filtering A feature of bridges that allows only messages intended for the external network to be forwarded.

destination code In telephony, the combination of digits that provide a unique termination address in a communications network.

destination code routing In telephony, the method of using country code, city code or area code, and office codes to direct a call to a particular destination regardless of its point of origin within the network.

destination field The field in a packet header that contains the network address of the station to which the message is directed.

DET An abbreviation of **D**irectory **E**ntry **T**able.

detailed-billed call In telephony, a call for which there is a record of not only the calling party number but the called party number, and the call duration appears in the customer billing statement.

detailed-record call In telephony, a call for which a record of the calling party number, called party number, bill to party number, and call duration are kept for billing and analysis purposes.

detectivity The reciprocal of *noise equivalent power* (*NEP*).

detector (**1**) A synonym for *demodulator.*

- In AM reception—a device or circuit that recovers the information-bearing signal from the modulated wave.
- In FM reception—the detector called a *discriminator* and used to convert the frequency variations of the carrier to amplitude variations corresponding to the information signal.
- In optics—a device that converts a received optical signal to another form, such as electrical or another optical form.

(**2**) Any device or component that determines the presence or magnitude of an entity (physical or electromagnetic) and allows the observation and/or measurement of the indication. Examples include:

- Photo *detector* (electromagnetic).
- Smoke *detector* (physical).
- Heat *detector* (electromagnetic).
- Nuclear radiation *detector* (physical).
- Motion *detector* (either).
- Voltage or current level *detector.*

deterministic routing Routing in which the paths between given node pairs are determined in advance of any transmission.

In a switched network, switching in which the possible paths (both primary paths and alternates) between given node pairs are predetermined and stored in routing tables maintained in each switch database. The tables assign the trunks that are to be used to reach each switch code, area code, and International Access Prefix (IAP).

deterministic transfer mode An asynchronous transfer mode in which the maximum information transfer capacity of a telecommunication service is made available for the duration of a call.

deviation distortion Distortion in an FM receiver due to either inadequate bandwidth and inadequate amplitude rejection or excessive discriminator nonlinearity.

deviation ratio The ratio of the maximum frequency shift of an FM modulating system to the maximum frequency of the modulating signal.

device A general term referring to any element (component, assembly, or subsystem) in a system used to perform a specific function. Some examples include relays, resistors, capacitors, transformers, modems, keyboards, monitors, disk drives, and printers.

Within the context of a computer a device is also called a *peripheral* if it is generally external to the base equipment, e.g., printers, tape drives, and monitors.

device driver A software routine that directly controls the operation and data throughput of a piece of hardware (a device). The *device driver* enables other programs to use the device via standardized commands.

device number A method of identifying a device on a system, e.g., network nodes, hard disk drives, I/O ports, scanners, or floppy drives. Three numbers are used to define each device:

- *Hardware address:* The address associated with the board or controller for the device. This value is set either through software or by setting jumpers on the board itself.
- *Device code:* A value determined by the location of the device's board, the device itself, and possibly by auxiliary components (such as controllers) associated with the board. For example, a device code for a hard disk includes values for disk type, controller, board, and disk numbers.
- *Logical number:* A value based on the boards to which the devices are attached, on the controller, and on the order in which devices are loaded.

DF (**1**) An abbreviation of **D**irection **F**inder. (**2**) An abbreviation of **D**issipation **F**actor.

DFA An abbreviation of **D**oped **F**iber **A**mplifier.

DFE An abbreviation of **D**ecision **F**eedback **E**qualization.

DFS An abbreviation of **D**istributed **F**ile **S**ystem.

DFSK An abbreviation of **D**ouble **F**requency **S**hift **K**eying.

DFT (**1**) An abbreviation of **D**istributed **F**unction **T**erminal. (**2**) An abbreviation of **D**irect **F**acility **T**ermination. A telephone company trunk that terminates directly on one or more telephones. (**3**) An abbreviation of **D**esign **F**or **T**estability. (**4**) An abbreviation of **D**iscrete **F**ourier **T**ransform.

DGPS An abbreviation of **D**ifferential **G**lobal **P**ositioning **S**ystem (GPS).

DGT An abbreviation of **D**ireccion **G**eneral de **T**elecommunicaciones. The Spanish General Directorate of Telecommunications.

DHCP An abbreviation of **D**ynamic **H**ost **C**onfiguration **P**rotocol. A specification for allocating IP addresses (as well as other configuration information) for networked systems based on network adapter addresses.

DIA An abbreviation of **D**ocument **I**nterchange **A**rchitecture.

DIA/DCA An abbreviation of **D**ocument **I**nterchange **A**rchitecture/**D**ocument **C**ontent **A**rchitecture.

diad A synonym for *dibit.*

diagnostic (**1**) A procedure for determining that a piece of hardware or software is functioning according to a predetermined set of rules. A *diagnostic* may also isolate a faulty item or module when improper functionality is detected.

An example of a *diagnostic* is a modem loopback test in which a DTE sends a data stream to the modem and compares the *loopback* response to the original data. If the data are different, then something is likely wrong with either the modem, the transmission path, or the DTE to DCE cabling. (**2**) A message generated by a computer program that indicates a possible fault in another system component.

diagnostic routine A computer program designed to locate either a computer malfunction, a peripheral malfunction, or an error in another programs coding.

diagnostic test A quantitative or qualitative examination of a device with the purpose of:

- Assuring the device's compliance with a standard, or
- Isolating a fault to a lower level of assembly.

dial back In network operations, a security measure to prevent unauthorized dial-up access to a network. The networking software maintains a list of users and the numbers from which they are allowed dial access. When a user dials into the network, the server accepts the call, obtains the user's login information, and then ends the call. The server software then looks up the user login ID in the dial-up table and calls back the number listed for that user.

As an access control and security measure, dial back works reasonably well; however, it can fail when either (1) the authorized user needs to dial in from a nonlisted location, or (2) when an unauthorized person has gained access to the location from which an authorized user generally dials in. (The network calls a number, not a person.) The latter is still somewhat protected by the normal password at login. Also called *call back*. See also *call back modem*.

dial back modem See *call back modem*.

dial call pickup A feature of PBX systems in which a user can answer calls ringing on another telephone by dialing a special code. The feature may be arranged so as to allow access to all phones or access to only phones within a predefined pickup group. Also called *dial pickup*.

dial-in tie trunk A trunk that may be seized by dialing an access code. The trunk provides access to a PBX that in turn gives the dial-in user access to the feature set of the PBX (depending, of course, on the class of service and restrictions assigned to the trunk).

dial mode An operation condition of data communications equipment (DCE) in which equipment associated with call origination is directly connected to the communications channel.

dial prefix A series of numbers that precede the actual called station number.

There are several reasons a *dial prefix* might be used, e.g.,

- The prefix may be required by some PBXs to gain access to a public line.
- It may be required to gain access to a specific common carrier (AT&T, MCI, etc.).
- It may be used for billing purposes (a credit card number).

dial pulse In telephony, a signaling method wherein the dc current in a communications loop is interrupted. Interruptions are produced when a digit is dialed on a telephone subset. There is one interruption for each unit of value of the dialed digit; for example, the number six produces six interruptions. Also called *pulsing* and *dial pulsing*. See also *pulse dialing*.

dial speed The rate at which a dialed number can be accepted by the receiving equipment. Typically, the rate is 10 pulses per second for pulse equipment and 10 characters per second for DTMF equipment. Some equipment standards allow speeds as fast as 20 pulses per second and 20 characters per second, however.

dial tone A signal generated by the communication switching equipment that indicates the switch is ready to accept dialing information. The dial tone is removed after the first digit is recognized. Dial tone is generally the sum of several tones in the frequency range of 340 to 650 Hz. See also *call progress tone*.

SOME DIAL TONE EXAMPLES

Country	f_1, f_2 (Hz)	Level (dBm0)	On Time (ms)	Off Time (ms)
Belgium	450		1000	250
ITU-T	425	-10 ± 5	∞	0
Germany	425 or 450		∞	0
Germany	400+425		∞	0
Germany	450		700,200	800,300
Mexico	425	-10	∞	0
USA	350+440	-10	∞	0
USA	600×120	-13	∞	0
UK	350+440		∞	0

Note: $f_1 + f_2$ indicates that the tones are added, while $f_1 \times f_2$ indicates that tone 1 is modulated by tone 2.

dial tone delay The elapsed time between the instant a user goes off-hook to the instant a dial tone is detected.

dial-up access A method of connecting to a bulletin board or network with a *modem,* a *communications program,* and the *public switched telephone network* (*PSTN*). The user calls the telephone number of the equipment to which a connection is desired (called a *host*). After the connection is established and the appropriate logon procedure has been completed, the user's equipment becomes a *remote terminal* to the host.

dial-up IP Similar to dial-up access except the communications software is designed to make the station look like a *node* of the network to which the connection is being established. In Internet two protocols are available for dial-up IP connections, *SLIP* and *PPP*.

dial-up line Any communication circuit that is established by the switching system of the *public switch telephone network* (*PSTN*). Also called a *dial line* and a *switched network line dial-in line*.

dial-up modem A modem designed to operate with the *public switch telephone network* (*PSTN*). The modem may have the capability to originate a call (dial) or answer a call (go off-hook in response to an incoming ring signal). It must conform to certain regulations so that it will not damage the PSTN or cause interference with other users of the PSTN.

dial-up site A computer or network that accesses another network, such as the Internet, by *dial-up access* means.

dialing In telephony, the act of using the signaling device (rotary or DTMF) on a telephone.

dialing parity The ability of a company that is not an affiliate of a local phone company to provide phone services to their subscribers in such a manner that the subscribers may route their calls automatically without the use of any access code.

DIB An acronym from Directory Information Base.

DIBI An acronym from Device Independent Backup Interface. An interface proposed by Novell intended to make it easier to move information between different environments on the network.

dibit *Dibit,* pronounced "DYE-bit," literally means two bits. The four possible states for a *dibit* are 00, 01, 10, and 11. It generally refers to any modulation technique that encodes 2 bits at a time to generate one of four symbols for transmission (1 baud). For example, in a phase modulation scheme the 00 *dibit* could be transmitted as a single frequency carrier; the *dibit* 01 could be the same frequency but phase shifted by 90° from *dibit* 00; *dibit* 10 could be transmitted at a phase shift of 180°; and 11 at a phase shift of 270° ($-90°$). *Dibit* encoding is a specific form of *multilevel encoding*. Also called a *diad*. See also *quadbit* and *tribit*.

dichroic filter In optics, a filter that reflects one or more wavelengths (colors) and selectively transmits others, and has a nearly zero coefficient of absorption for all wavelengths of interest. A dichroic filter may be high-pass, low-pass, band-pass, or band rejection in nature.

dichroic mirror In optics, a mirror designed to reflect light selectively according to wavelength. See also *dichroic filter*.

dictionary flame A message that avoids the issue of a debate and focuses instead on the definition of a term in a meaningless way.

DID (**1**) An acronym from Direct Inward Dialing. A telephone company service wherein received calls are routed directly to specific terminal equipment (telephone, modem, fax, etc.) without switchboard assistance. (**2**) An acronym from Destination ID. The destination node address in an ARCnet packet.

dielectric Basically, another name for an insulator. Any material (or a vacuum) in which electrostatic energy (an electric field) can be temporarily stored.

dielectric constant (ε) The ratio of the amount of electrostatic energy that can be stored per unit volume of material to the amount that can be stored in the same volume without any material present (a vacuum).

DIELECTRIC CONSTANT OF SELECTED MATERIALS

Material	ε	Material	ε
Vacuum	1.00000	Polyethylene	2.26
Air	1.00059	Polyimide	3–3.5
Aluminum oxide	8.8	Polystyrene	2.56
FEP (Teflon®)	2.10	Polyvinyl-chloride	2.5–8
Foamed Polystyrene	1.03	Silicon Rubber	2.08–3
Nylon	2.7–4.6	TFE (Teflon®)	2.1
Polycarbonate	2.96–3		

Dielectric constant may be somewhat of a misnomer as some materials do vary with operating frequency; for example, barium titanate varies from 1250 at 60 Hz to 100 at 2.5×10^{10} Hz. Polyethylene, on the other hand, remains at a constant 2.26 over the same frequency range.

dielectric filter See *interference filter.*

dielectric lens A lens made of a dielectric material and used for the refraction of radio waves, just as an optical lens refracts light.

dielectric loss Energy dissipated by the high but finite resistance of a dielectric.

dielectric strength (1) The maximum electric field strength that an insulating material can withstand intrinsically without experiencing failure of its insulating properties (breaking down). The electric field strength is usually given in volts per meter (V/m).

The theoretical *dielectric strength* of a material is an intrinsic property of the bulk material. It is not dependent on the shape of either the material or the electrodes with which the field is applied. **(2)** The minimum electric field strength that produces breakdown in an insulator. At breakdown, the electric field frees electrons bound in the material's atoms or molecules, thereby turning the material into a conductor. The actual field strength at which breakdown occurs in a given case is dependent on the respective geometry of the dielectric (insulator) and the electrodes with which the electric field is applied, as well as the rate that the electric field is increased.

dielectric waveguide A waveguide that consists of a dielectric material surrounded by another dielectric material with a lower refractive index, e.g., an optical fiber. A metallic waveguide filled with a dielectric material, however, is not a *dielectric waveguide.*

DIF An acronym from Data Interchange Format.

differential amplifier An analog device that amplifies the algebraic *difference* between the signals applied to the two inputs by an amount "K." The output response to the "+" input is in phase, while the output response to the "−" input is inverted. The symbol for a *differential amplifier* is:

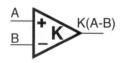

Also called a *difference amplifier.* See also *amplifier.*

differential coding Coding in which signal significant conditions represent changes to informational values, not the actual value.

For example, with differential phase shift keying (DPSK) information is not conveyed by the absolute phase of the signal but by the difference between phases of successive symbols (thus eliminating the requirement for an absolute phase reference at the receiver).

differential GPS (DGPS) A technique to improve the accuracy of a GPS position determination. The method employs an auxiliary receiver at a known location to record the pseudorange errors to the satellites. This information is then offered to GPS receivers in the same general geographic area to be used as a correction factor and to remove the effects of selective availability (SA).

Differences among GPS measurements can be taken between receivers, satellites, and epochs. By convention, the differences are taken between receivers (single difference), then between satellites (double difference), and finally between measurement epochs (triple difference). *Differential GPS* provides position accuracy of 5 meters or better, and when averaged over time the accuracy is further improved. For a three-minute averaging period, the accuracy is typically better than 2 meters.

differential Manchester coding One of many waveform coding methods used to encode both clock and data into a single serial bit stream. One of the two bits, "0" or "1," is represented by no state transition at the beginning of the pulse period. (Both have a transition at the pulse midpoint.) The algorithm that generates a double transition for a "0" states:

(1) Invert the signal at the leading edge of all "0" bits. Do not invert for "1" bits.

(2) Always invert the signal at mid-bit regardless of the "1" or "0" input condition. Mid-bit transitions contain the clock synchronizing information.

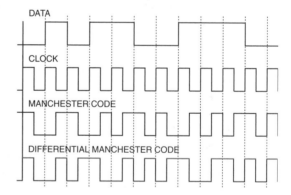

Differential Manchester coding is polarity insensitive at the receiver because only transitions contain information. See also *waveform codes.*

differential mode attenuation In fiber optics, the variation in attenuation among the propagating modes of an optical fiber.

differential mode delay In an optical fiber, the variation in propagation delay that occurs because of the different group velocities of different modes. Also called *multimode group delay.*

differential mode interference Interference that causes a change in potential on one side of a two-wire signal transmission path relative to the other side.

differential mode signal A signal impressed on a transmission line such that voltage *between* the lines changes in direct response to the impressed signal change. This is the normal data signal mode.

In the telephone industry it is called the *metallic signal mode.* See also *common mode signal.*

differential modulation Modulation in which the symbol choice for any signaling element is dependent on not only the current data value but also the previous signal element.

differential phase shift keying (DPSK) A method of phase modulating a carrier with a digital signal (*phase shift keying*) in which the reference phase for a given signal element is the phase of the preceding signal element. See also *modulation* and *PSK*.

differential positioning In the Global Positioning System (GPS), a method of removing certain types of errors that lead to an inaccurate determination of a GPS receiver location. It is based on two receivers, one at a known location and the other at an unknown location. See also *differential GPS* (*DGPS*) and *Global Positioning System* (*GPS*).

differential pulse code modulation (DPCM) Pulse code modulation (PCM) in which an analog signal is sampled and the difference between the actual value of each sample and its predicted value is quantized and converted to a digital signal. The predicted value is derived from the previous sample or samples. Several variations of differential pulse code modulation are used.

differentially coherent phase shift keying See *coherent differential phase shift keying* (*CDPSK*).

differentiating network A circuit that produces an output waveform that is the time derivative of an input waveform. For example, a differentiating network is used for producing short timing pulses from square waves; i.e.,

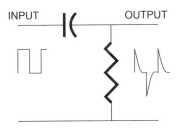

Diffie-Hellman key The first encryption technique that allows exchanging decryption keys in a public manner. The technique was described in a paper by W. Diffie and M. Hellman, published in the 1976 *IEEE Transactions on Information Theory*. See also *encryption* and *public key*.

diffraction The deviation of a wavefront from the geometrically predicted path when the wave passes through an aperture in (or by the edge of) an opaque object. The effects are most noticeable when the aperture's size is of the same order of magnitude as the wavelength of the electromagnetic radiation. The deviations do not include those caused by reflections or refractions.

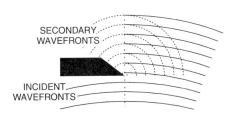

diffraction grating An array of fine, parallel, equally spaced discontinuities in a medium that mutually enhances the effects of diffraction to concentrate the diffracted energy in only a few directions called orders or spectral orders. The discontinuity dimensions and spacing are on the order of the wavelength in question and may be either reflective or transmissive. The discontinuity's cross section may be shaped to force most of the energy into a single spectral order (a grating called a *blazed grating*).

With a normal angle of incidence (θ), the deviation of the diffracted ray from the direction predicted by geometric optics is given by

$$\theta = \pm \sin^{-1}\left(\frac{n\lambda}{d}\right)$$

where:

n is the spectral order,
λ is the wavelength, and
d is the spacing between corresponding parts of adjacent grooves.

Other characteristics of a diffraction grating are as follows:

- The angular dispersion is inversely proportional to line spacing.
- A diffraction grating is wavelength dispersive; that is, it separates the incident beam spatially into its constituent wavelength components (a spectrum).
- Order zero corresponds to direct transmission or specular reflection.
- The spectral orders produced may overlap, depending on the spectral content of the incident beam and the number of grooves per unit distance on the grating. The higher the spectral order, the greater the overlap into the adjacent order.

diffuse reflection A reflection from a rough or irregular surface in which the integrity of the incident wavefront is not maintained by the reflected wavefronts.

Contrast with *specular reflection*.

digest A moderated newsgroup in which postings consist of *summaries* of articles, contributions, and comments. The digest generally points to the location of the full text.

digestified The process by which individual postings to a mailing list or newsgroup are concatenated into a single file.

digit A symbol used to represent one of the integers less than the radix, e.g., in the decimal system (radix 10) any of the numerals 0 to 9 and in binary either of the symbols 0 or 1.

digit absorption In telephony, the interpretation and rejection of those digits received but not required for the control of an automatic switching system.

digital Characterized by discrete states.

digital access and cross connect system (DACS) See *digital cross connect system* (*DCS*).

digital alphabet A coded character set in which the characters have a one-to-one relationship with their digitally coded representations.

digital audio tape (DAT) A magnetic recording tape onto which digital information is directly stored.

digital cash **(1)** A token residing on a PC that allows a consumer to purchase something via a network transaction. A third party debits the consumer's bank account and provides the electronic tokens for spending, e.g., Digicash and Cybercash. **(2)** The value stored in a portable device (cash card) similar in appearance to a credit card but with embedded active electronics that allows the secure storage of information representing a cash value.

digital cellular A cellular telephone technology in which voice is transmitted in digital form rather than analog. Several standards can be found around the world, e.g., digital advanced mobile phone system (D-AMPS) in the United States, Groupe Speciale Mobile (GSM) in Europe, and PDC in Japan.

digital check An encryption and processing method proposed by the Financial Services Technology Consortium for fund transfers over the Internet.

digital circuit (1) An electronic device that processes information presented in digital form, that is, as a sequence of discrete states rather than as a smooth continuum of values. (2) In communications, transmission lines that transmit data as unmodulated square waves, which represent 0 or 1 values. The waveform may, however, be encoded in order to improve transmission characteristics.

digital communications A method of information transmission that uses digital signals (generally binary values) to represent the information, even if the original signal was obtained in analog form. To transmit voice or other analog information, the signal must first pass through an analog-to-digital converter called a *codec*. In the case of telephony, the codec samples the voice signal 8000 times per second, yielding a usable bandwidth of less than 4000 Hz. See also *codec* and *digital-to-analog conversion* (*DAC*).

digital cross connect system (**DCS**) In digital telephony, a digital switching device for routing individual DS-0 time slots among multiple T-1 or E-1 ports. That is, the *DCS* cross connects one digital channel from one piece of equipment to another. With a *DCS*, this cross connection can take place as fast as the rate supported by the slower of the two lines. The mechanism enables switching a 64-kbps DS0 channel from one T-1 line to another. The method was originally developed for use in telephony switching, but it is proving useful in networking contexts.

Although developed for T-carrier systems, modern systems may accommodate high data rates such as those of SONET. Also called *digital access and cross connect system* (*DACS*).

digital data (1) Pertaining to data represented by discrete or quantized values or conditions, as opposed to the continuous values of analog data. See also *analog signal* and *digital signal*. (2) Discrete representations of quantized values of variables or intervals, e.g., the representation of numbers by digits.

Digital Data Communication Message Protocol (**DDCMP**) The DECnet-specific Link Level protocol that operates at Layer 2 of the Digital Network Architecture.

digital device A device that operates on the basis of discrete numerical techniques in which the variables are generally represented by digital data; that is, they are quantized values, represented by coded pulses, words, or states.

digital error A discrepancy between the digital signal actually received and the digital signal that should have been received.

Digital European Cordless Telecommunications (**DECT**) A TDMA/TDD cordless data and telephone standard operating in the 1.880–1.900 GHz frequency band with 250-mW maximum power output. The standard, fully approved by the European Telecommunications Standards Institute (ETSI), provides 12 channels per base station spaced at 1.728 MHz. The digital bit rate is 1.152 Mbps.

DECT covers cordless PBXs, telepoint, and residential cordless telephony. Other wireless data/telephone communications methods include *IEEE 802.11* and *ISM*.

digital filter A signal processing filter that operates in the discrete time domain (in opposition to the continuous time domain of analog filters). Because digital filters are normally implemented with digital computers or processors, the parameters are generally more stable than the parameters of commonly used analog (continuous) filters.

Commonly used topologies include the *finite impulse response* (*FIR*) and *infinite impulse response* (*IIR*) forms. See also *digital signal processor* (*DSP*).

digital group See *digroup*.

digital ID An authenticated credential that certifies the sender's identification to the receiver (similar to a *password* or *digital signature standard*). Used, for example, on the connection between a person and a distant e-mail address or between a company and its computer server. See also *digital signature*.

digital loopback A diagnostic feature of a digital communications device (such as a modem or other digital transmission equipment) that allows a signal source to command that device to loop back (echo) information it receives to the source. A comparison of the transmitted and received information by the source can determine whether the loop is serviceable. See also *analog loopback*.

digital loop carrier (**DLC**) A system that increases the number of subscribers a cable pair can carry and extends the range of an existing central office.

A *DLC* system is a digital *pair gain system* generally employing T-1 carrier systems. It uses time division multiplex (TDM) and pulse code modulation (PCM) to provide the increased capacity. A 24-channel T-1 digital line uses two cable pairs, one for the transmit signal and one for the receive signal. Hence, there is a channel gain of 12 times. In addition to these gains, it is also possible to add circuit multiplexing before the carrier portion of *DLC*. This will further increase the effective number of subscribers a pair will support. However, it does introduce the possibility of blocking when a large number of subscribe connections are attempted. As shown in the figure on the next page, a *DLC* system consists of two terminals (the central office terminal and the remote terminal) connected by a T-1 digital line (that may contain repeaters). Each terminal is essentially the mirror image of the other, with the possible exception of individual channel unit interface specifics. For example, if the central office switching entity is digital, the channel unit at the remote end may be a two-wire analog interface, while the channel unit at the central office end is a four-wire digital interface (called a *direct digital interface*). The channel units provide the appropriate interface to both the remote terminals and the central office facilities. That is, for the basic subscriber it must provide regular two-wire analog service (POTS) including all BORSCHT functions (battery feed, overvoltage protection, ringing, signaling, codec, hybrid, and test). The channel units of the central office terminal provide either the analog or digital equivalent of the normal subscriber to network interface. Time slot interchange switching allows any physical channel to be placed on any DS-1 time slot. In the case of a system with more than one T-1 line, the time slot interchanger generally has access to all time slots of all T-1 lines. This provides a better grade of service when the system is used in the concentrating mode (because each subscriber's line has a larger pool of communication channels from which a connection may be established). See also *SLC96* and *subscriber carrier*.

digital milliwatt In telephony, a digital test signal consisting of eight 8-bit (A-law or mu-law encoded) words corresponding to one cycle of a sinusoidal signal approximately 1 kHz in frequency and 1 milliwatt rms in power, at an analog zero test level point (0TLP).

Since one full cycle of the test waveform is defined, a continuous signal of arbitrary length, i.e., an indefinite number of cycles, may be generated simply by continuously repeating the eight words into the data stream to be converted into analog form. Because the tone is

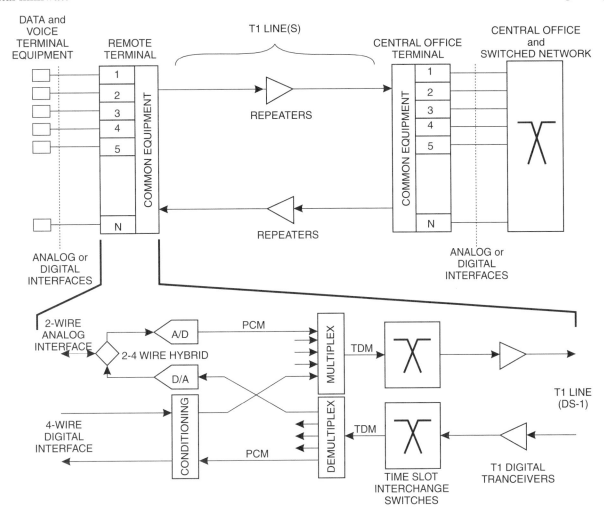

usually generated in the communications equipment itself, several benefits are realized, e.g.,

- No separate test equipment is required to generate the test tone.
- The frequency accuracy and stability are tied to the system's clock.

digital multiplex hierarchy An ordered scheme of combining digital data streams through the repeated use of multiplexing. Because there are many standards (between countries, administrations, and even manufacturers), extreme care must be exercised when selecting equipment for a specific application in order to ensure interoperability.

Digital multiplexing hierarchies are implemented in different configurations depending on parameters such as:

- The number of channels desired,
- The signaling system to be used, and
- The bit rate allowed by the link media.

Currently available *digital multiplexers* include Dl-, DS-, or M-series, all of which operate at T-carrier rates. See also *DS-n*.

digital multiplexed interface (DMI) In digital telecommunications, a T-1 interface between a private branch exchange (PBX) and a computer.

digital network A communications network that transfers digital data directly; that is, there is no need for an analog-to-digital conversion process. This is in contrast to an analog network (such as the traditional telephone system) where a modem is required to change digital data into an analog signal before the network will accept the signal.

Digital Network Architecture (DNA) Digital Equipment Corporation's (DEC) eight-layered data communications protocol used in various implementations of DECnet. The DNA is similar in structure to the ISO/OSI Reference Model at lower levels; however, the top three layers of the DNA correspond to the top two layers in the OSI model. See also *DECnet™*.

digital phase locked loop A phase locked loop (PLL) in which one or more of the reference signal, the controlled signal, or the controlling signal is in digital form.

digital reference signal (DRS) In telephony, one of several specific sequences of 1s and 0s (as defined in the ITU-T standards, Bell standards, etc.) that represent a particular frequency and amplitude sine wave. Generally, the amplitude is 0 dBm, and the frequency is 1004 Hz or 1020 Hz. It is never exactly 1000 Hz as that is an exact submultiple of the 8-kHz sampling frequency in both A-law and μ-law encoding methods.

digital signal A *discontinuous* electronic signal that varies in correlation to some stimulus in *discrete steps;* that is, if the stimulus is changed by some amount, the corresponding digital signal will change by some *finite* amount. If the stimulus change is smaller than the value corresponding to the minimum step size at the output, no change in output will occur. The digital signal has a finite number of steps; therefore, only a finite number of values can be represented.

Contrast with *analog signal*.

digital signal "n" See *DS-n*.

digital signal processor (DSP) A device that can extract and process information directly from a digital signal stream. Advantages of digital signal processing over analog processing include:

- *Stability*—Digital systems are not subject to drift.
- *Precision*—A digital system can be designed to manipulate a data stream with any degree of precision desired. The original analog samples generally limit the precision and accuracy of a digital processing system.
- *Repeatability*—A digital system can be mass-produced, and each device will perform identically. Analog systems are subject to individual device tolerances and manufacturing process differences.
- *Reprogramability*—A digital signal processor is just a computer (albeit, a specialized computer at times); hence, changing the program can change the performance or function. For example, a device could at one moment be a modem, in the next instant be programmed to be a fax, and still later be programmed to be a voice recognition controller.

digital signature In digital transmission systems, a unique and identifiable value associated with a transaction. The signature is used to verify the identity of the sender and the origin of the message. Today, most digital signature systems use a *public key* system. See also *public-key encryption.*

digital signature standard (DSS) A standard approved by the National Institute of Standards and Technology (NIST) that provides a means of verifying the authenticity of an electronic document. The technique is basically a public key cryptography system (an encoding system utilizing two mathematically related keys). One key (a secret key) encodes a digital message (the signature), and the second key (the public and complementary key) decodes and verifies the encoded signature. See also *public-key encryption.*

digital simultaneous voice and data (DSVD) A modem protocol for transmitting both digitized voice and high-speed digital data signals over a single analog telephone line simultaneously.

digital slip In the reception of a digital data stream, the loss of a bit, or the insertion by the receiver of a bit that was not transmitted, because of a difference in the bit rates of the incoming data stream and the local clock. See also *slip.*

digital speech interpolation (DSI) In digital telecommunications, a type of multiplexing. It is a strategy for improving the efficiency of a communications channel by transmitting from a secondary source during the "quiet" periods that occur in normal conversation. On the average, *DSI* can double the number of voice signals that can be carried on a group of lines. (Generally, only one party is talking at a time, so only half of the transmission capacity is in use.)

Although very good for voice circuits, *DSI* can cause problems with data transmission because of its tendency to clip the beginning of any transmitted information block. See also *statistical multiplex* and *TASI.*

digital subscriber line (DSL) Another name for an ISDN basic rate interface (BRI) link; that is, a 144-kbps link that supports two 64-kbps bearer channels (or B-channels) and one 16-kbps data channel (D-channel).

The physical termination of the *DSL* at the network end is the *line termination;* the physical termination at the customer end is the *network termination.* See also *xDSL.*

digital switch In telephony, a communications switching system in which only digital signals are switched (routed). This necessitates the conversion of the analog subscriber signal into a digital signal in a device called a CODEC (from coder decoder). A *digital switch* is inherently a four-wire device; hence, it requires a two-wire to four-wire hybrid at the channel interface when connected to a two-wire analog transmission line.

The actual strategies or implementation of switching may take several forms, e.g.,

- *Space division switching*—in which a digital signal is routed across selected spatially distinct paths between input and output ports.
- *Time division switching*—in which the signal is routed between input and output ports via the selection of appropriate time slots on a PCM multiplex bus.

See also *blocking, n-stage switching, nonblocking,* and *switch (3) & (4).*

digital switching Switching in which digitized signals are routed between input and output ports without converting them to or from analog signals.

digital switching exchange (DSE) A node in a digital telecommunications network.

Digital Termination Service (DTS) In telecommunications, a service by which private networks may gain access to carrier networks using digital microwave equipment within a frequency band allocated by the FCC (Federal Communications Commission) for this purpose.

digital-to-analog conversion (DAC) A device that converts a digital signal to an analog signal.

The figures on the next page illustrate the conversion process from true digital data (numbers as would appear in a computer file) to a sampled electronic waveform (the dotted line in the first drawing) to a true continuous analog signal (the second drawing). Conversion from the sampled waveform to the continuous waveform is generally accomplished with a low-pass filter. In these drawings, the low-pass filter was not selected properly. Consequently, the output waveform contains excessive residual switching noise.

digital transmission The transmission of either analog information encoded in digital form or digital data across either an analog or digital communications path, but in a manner that allows the digital signal to be extracted, regenerated, and retransmitted at intermediate points in the communications channel in order to reduce the effects of noise, distortion, and attenuation.

digital transmission group A group of digitized voice and/or data channels multiplexed into a single digital bit stream for transmission over communications link. Digital transmission groups are usually categorized by their maximum capacity, not by a specific number of channels.

digital transmission system An information transmission system in which all links carry only digital signals and the signals include all framing and supervisory signals.

A-D/D-A conversion, if required, is accomplished external to the system.

Digital Video Interactive (DVI) A set of proprietary hardware and software products by Intel for video digitizing and playback.

digitize To convert information represented in the analog domain to an equivalent representation in the digital domain. Examples include the conversion of an analog signal to a digital signal, the scanning of a picture into a form suitable for storage or transmission, or the conversion of a point on a surface or map to its coordinates.

DIGITAL VALUES

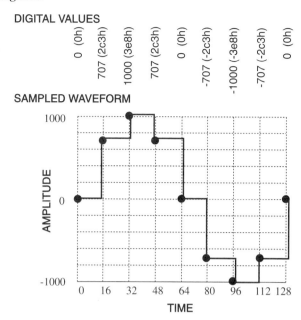

SAMPLED WAVEFORM

CONTINUOUS (ANALOG) WAVEFORM

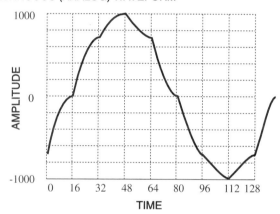

digitizer **(1)** A device that converts an analog signal into an equivalent digital representation. In general, a *digitizer* samples an analog signal's value at constant time intervals and encodes each sample into a numeric representation. See also *digitize* and *Shannon limit*. **(2)** A device that converts the position of a point on a surface into digital coordinate data. For example, the conversion of an aerial photographic map into latitude, longitude, and altitude values.

digroup an acronym from **DI**gital **group**. In telephony, a basic digital multiplexing group. A DS-1 is composed of two *digroups* plus overhead bits, i.e.,

- In the North American and Japanese T-carrier digital hierarchies, each *digroup* supports 12 PCM voice channels (or the equivalent in other services). The DS-1 line rate (*2 digroups* plus overhead bits) is 1.544 Mb/s.
- In the European digital hierarchy, each *digroup* supports 15 PCM channels (or the equivalent in other services). The DS-1 line rate (2 digroups plus overhead bits) is 2.048 Mb/s.

See also *DS-n*.

Dijkstra's algorithm A procedure for selecting a path through a network of nodes. It is essentially the selection of the lowest weighted branch at each node traversed. For example, given the network in the following diagram with links weighted as indicated, determine the path through the network.

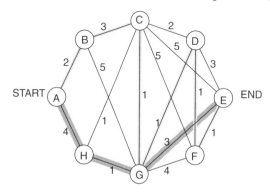

Starting at node A, the path with the lowest weighting factor is to node B ($2 < 4$), and the path from node B is to node C ($3 < 5$); similarly, the path to the end is found to be A-B-C-G-D-F-E. Although each internode link is chosen to have the lowest weighting factor, the resultant path does not necessarily have the lowest composite weighting factor. In the above example, the path chosen has a weight of 9, while the path A-H-G-E has a weight of 8.

DIIK An abbreviation of **D**amned **I**f **I** **K**now.

dilution of precision (DOP) In the Global Positioning System (GPS), a measure of the determined position uncertainty due exclusively to geometrical contributions. The *dilution of precision* is time varying because both the selection of GPS satellites and the position of the selected satellites are continuously changing with respect to the receiver; hence, the geometry is continuously changing. *DOP* has a best case value of 1, with higher numbers being worse, and has a multiplicative effect on the user range error (URE). That is, a URE of 30 m has a best accuracy of 30 m with a *DOP* of 1, and a DOP of 2 yields an accuracy of 60 m.

The following figure illustrates the *dilution of precision* on a two-dimensional system of measurements. Each of the range values has the same degree of uncertainty; yet, when these range values are used to determine a physical location, the positional uncertainty is much greater in one dimension than in the other. This increased uncertainty (*dilution of precision—DOP*) is caused by the shallow angle of intersection when the satellites are close together.

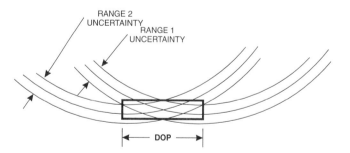

Dilution of precision may occur in any dimension of space or time or combination of dimensions. Standard terms for the contributors include:

- *Geometric dilution of precision (GDOP)*—includes the three-dimensional coordinates and the clock offset in the solution.
- *Horizontal dilution of precision (HDOP)*—the two horizontal dimensional coordinate components.
- *Position dilution of precision (PDOP)*—the three-dimensional coordinates.
- *Relative dilution of precision (RDOP)*—normalized to 60 seconds.
- *Time dilution of precision (TDOP)*—the time offset.
- *Vertical dilution of precision (VDOP)*—the altitude contribution.

See also *Global Positioning System (GPS)*.

diminished radix complement The *diminished radix complement* (sometimes shortened to *diminished complement*) of a number is the number derived by subtracting each digit of the number from the largest digit in the numbering system. See also *complement*.

EXAMPLES IN VARIOUS NUMBER BASES

	n	Diminished Complement of "n"
Base 2 (Largest digit = 1)	0101	1010
	1111	0000
Base 8 (Largest digit = 7)	124	653
	703	074
Base 10 (Largest digit = 9)	956	043
	1380	8619
Base 16 (Largest digit = F)	A03C	5FC3
	14F6	EB09

DIN An acronym from the German national standards organization <u>D</u>eutsches <u>I</u>nstitut für <u>N</u>ormung, e.V. (German Standardization Institute). Most standards are published in German; however, translations are usually available in English from several sources (EMACO and Global Engineering Documents, for example).

diode A two-terminal electronic device that conducts current better in one direction than in the other.

Other characteristics of the diode may be optimized to give additional personalities. For example, a *light emitting diode* (*LED*) will radiate light when a current is passed through it, a *zener diode* will maintain a constant voltage drop for a wide range of currents, and a *noise diode* will generate white noise within a particular current range. Other diode types are optimized to produce a voltage or modulate the flowing current when exposed to light (*photo diode*), produce minimum voltage drop when a current is flowing (*Schottky*), turn on and off very fast (*switching diodes*), and provide a variable capacitance when a variable voltage is impressed across the diode terminals (*varicap*). Some of the diode schematic symbols are shown in the following diagram.

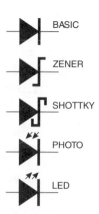

BASIC

ZENER

SHOTTKY

PHOTO

LED

DIP An acronym from <u>D</u>ual <u>I</u>nline <u>P</u>ackage.

DIP switch A series of small toggle or slide switches built into the body of a <u>D</u>ual <u>I</u>nline <u>P</u>ackage (DIP). The *DIP switch body* is approximately 0.4 inches wide and 0.3, 0.4, 0.5, . . . long, depending on the number of switches. (One switch is 0.3 inches long, and each additional switch adds 0.1 inch.)

diplex operation The simultaneous transmission (or reception) of two independent signals on two different frequencies, using a specified common feature such as a common antenna or channel.

For example, the use of one antenna for two radio transmitters on different frequencies.

diplexer A three-port frequency-dependent device that may be used to separate or combine signals. Full-duplex transmission through a *diplexer* is not possible.

dipole antenna A basic antenna, usually a straight, center-fed, 1/2 wavelength antenna. The length may be made other than 1/2 wavelength through the addition of loading coils.

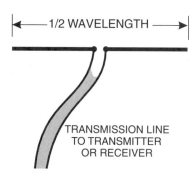

1/2 WAVELENGTH

TRANSMISSION LINE TO TRANSMITTER OR RECEIVER

dipulse coding A waveform coding scheme in which two pulses of equal and opposite polarity are generated for binary symbol and no pulse is generated for the other. A generating algorithm in which a "1" generates pulses is:

- A "1" is represented by a positive pulse excursion followed by a negative excursion of the same amplitude and in the same bit period.
- A "0" is represented by no pulse.

See also *waveform codes*.

Dirac delta function A mathematical expression or tool consisting of an impulse with infinite amplitude, zero time duration, and with an area of unity under the curve. As illustrated in the figure, the impulse starts rising at time t_0^- and is complete at time t_0^+. See also *unit impulse*, a synonym for the *Dirac delta function*.

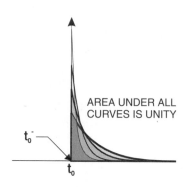

AREA UNDER ALL CURVES IS UNITY

t_0^-

t_0

direct access storage device (DASD) Pronounced "*DAZ-dee,*" it is a data storage system in which a user may access and retrieve any specific piece of information directly—that is, without scanning all preceding information as is required with a magnetic tape system.

direct attachment The IBM term for linking a device or LAN directly to a host computer through an appropriate Control Unit (CU), such as a cluster controller.

direct broadcast satellite (DBS) A satellite messages distribution system in which the effective transmit power of the satellite is sufficiently high that inexpensive Earth stations can be used directly by a user. The user antenna is generally less than 0.5 meter in diameter.

direct connect modem A modem designed to connect directly to the telephone network using appropriate country standards. In the United States, compliance with FCC Part 68 regulations is required. The modem is physically wired to the telephone network as opposed to the acoustically coupled modem that uses air-transmitted sounds to link into the telephone line. Compare with *acoustic coupler.*

direct connection (1) In networking, an unmediated connection to the network. For example, a direct connection might be through a network cable attached to the network interface card (NIC). **(2)** In telecommunications and wide area networks (WAN), a connection to long-distance equipment that does not go through a local carrier. The connection could be via least pair, fiber optic cable, or private microwave link. This type of connection is in contrast to a switched access method, in which the connection is routed through local carrier equipment. Sometimes also called *direct connect* or *by-pass.* **(3)** A connection of user equipment to the telephone equipment by wires, not acoustically or inductively.

direct connection mode A DTE-DCE-DCE-DTE connection without error compression, flow control, or data compression within the DCE hardware.

The purpose of a *direct connection mode* is to make the modem "dumb" for compatibility with older modems or when the DTE's software assumes error control responsibility. Another condition for which the *direct connection mode* is applicable is during a diagnostic test.

direct control switching The path through the network is established by signals from the signal source without the use of central control. For example, dial pulses that control step-by-step relays in a telephone office.

direct coupling A synonym for *conductive coupling.*

direct current (DC) An electric current whose direction of flow and magnitude are constant. The direction does not reverse, and the amplitude is constant. Current from a battery is *direct current,* for example. Compare with *alternating current.*

direct current signaling (DX signaling) In telephony, a signaling method wherein the signaling circuit E & M leads use the same cable pair(s) as the voice circuit and no filter is required to separate the control signals from the voice transmission.

direct dialing service In telephony, a feature that permits a user to place information concerning credit card calls, collect calls, and special billing calls into the public telephone network without operator assistance.

direct distance dialing (DDD) A telephone exchange service wherein one subscriber can call another subscriber outside of the local area without assistance from an operator. The term was introduced by the Bell System to encourage the public to dial their own long-distance call directly rather than use a long-distance operator.

In the United Kingdom and other countries, this service is called *subscriber trunk dialing (STD).*

direct inward dialing (DID) A PABX feature that allows an outside caller to call a specific extension without the assistance of an operator or attendant. The central office passes the necessary dialed digits to the company PABX, which then completes the connection to the desired internal extension. *DID* is also a feature of Centrex systems. Also called *network inward dialing (NID).*

direct inward system access (DISA) A feature of PBX systems wherein an outside caller is given full rights to the PBX feature set, including access to tie lines, outgoing lines, trunks, and WATS, after dialing a password number.

direct link A connection, or circuit, that connects two stations directly, without any intermediate stations.

direct memory access (DMA) A technique for high-speed data transfer between a device such as LAN network adapter card and the computer memory that does not require Central Processing Unit (CPU) software assistance. *DMA* bypasses the CPU of the computer, PC, or workstation, allowing the device to transfer a block of information directly across the bus into system memory. Therefore, a device that has *DMA* capability can transfer data to and from memory without interrupting ongoing CPU software operation. The *DMA* process is managed by a specialized controller that is generally faster than the CPU. When the data transfer is finished, the controller informs the CPU, which can then proceed as if the CPU had managed the transfer.

direct mode The *direct mode* on a modem is a nonbuffered mode of operation. Because of this, the DTE to modem transmission rate MUST be the same as the connection rate.

direct orbit With satellites, a satellite orbit with an inclination between zero degrees (0°) and ninety degrees (90°). That is, it revolves about the Earth in the same direction as the rotation of the Earth. See also *retrograde orbit.*

direct outward dialing (DOD) In telephony, a feature of a private automatic exchange (PABX) or a Centrex® that permits stations to dial outside numbers without the assistance of an attendant. Also called *network outward dialing (NOD).*

direct ray A ray of electromagnetic radiation that follows the path of lowest possible propagation time between transmitting and receiving entities (e.g., antennas in RF and emitters and detectors in optics). The path of least propagation time is not always the shortest distance path.

direct sequence modulation In spread spectrum systems, modulation in which a pseudorandom sequence of binary pulses is used directly to modulate a carrier. Also called *direct spread modulation.*

direct wire circuit In telephony, a supervisory circuit consisting of one metallic conductor and a ground return. Generally, equipment at the receiving station is designed to respond to an increase (or decrease) of current supplied by the sending terminal.

directed broadcast address A special Internet Protocol address that specifies the message destination is "all hosts" on a specified network. A single copy of a directed broadcast is routed to the specified network, where it is then broadcast to all terminals on that network.

directed information Information aimed at a specific user.

directed transmission (1) In AppleTalk's LocalTalk network architecture—and its LocalTalk Link Access Protocol (LLAP)—a transmission intended for a specific node. It is in contrast to a broadcast transmission, which is intended for all nodes. **(2)** In radiated electromagnetic communications, a method in which a signal is aimed at a central reflective target and read by receiving nodes as the signal bounces off the target. This is in contrast to a *diffuse transmission* in which the transmission travels in multiple directions (and therefore is much weaker in any direction).

directional antenna An antenna in which the radiation pattern is not uniform in all directions. Also called a *nonisotropic antenna.*

directional coupler A multiport device inserted into a transmission line for either **(1)** Separately sampling either the forward (incident) wave and the backward (reflected) wave in transmission lines, or **(2)** It may be used to inject a signal into the transmission line in either the forward or backward direction.

directional phase shifter A passive phase shifting device in which the phase shift for signal transmission in one direction is different from the phase shift in the other direction. Also called a directional *phase changer* or a *nonreciprocal phase shifter.*

directory The name of a storage area on the medium where files are stored.

On a large storage device, it is possible to have 10,000 to 100,000 (or more) files available. It would be difficult or impossible to locate a desired file if they were randomly arranged in a single list (a *flat directory*). To make it easier to find files, the computer operator breaks the list of files into smaller lists where each item is related in some manner. The list, called a directory, may be broken further into sublists, called *subdirectories*. A *directory* may contain both *subdirectories* and *data files*. When drawn, this structure resembles an inverted tree; hence, it is called a directory tree. The top directory (list) is called the root directory. This arrangement of files and directories is called a *hierarchical directory* system. The diagram depicts one possible relationship of 15 directories in a typical computer.

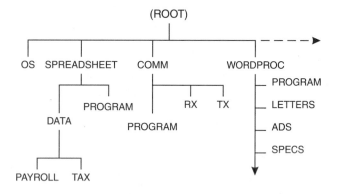

Frequently, directories contain additional information about the file such as file size, creation date, date of the last modification, and number of attributes (read only, system, archived, hidden, etc.) See also *access rights.*

directory access protocol (DAP) An X.500 protocol used to communicate between a Directory User Agent and a Directory System Agent.

directory assistance call In telephony, a call placed to request the station number (directory number) of a customer.

directory caching A method of decreasing the time to find the location of a file on the disk drive.

To speed up the process, it uses a fast memory area (from which information can be quickly retrieved) to store the file allocation table (FAT) and directory entry table (DET) information about the most

commonly used directory entries. As the *directory cache* fills up, the least used directory entries are eliminated from the cache. In addition to directory caching, *file caching* may be used to reduce information access time further. *Directory caching* is used in Novell's NetWare.

directory entry table (DET) In Novell's NetWare, one of two tables used to keep track of directory information. The other table is the file allocation table (FAT). The *DET,* stored on a hard disk, contains information about a volume's file, directory names, and properties. For example, an entry might contain the following:

- File name
- File owner
- Date and time of last update
- Trustee assignments (or user rights)
- Location of the file's first block on the network hard disk

The contents of the *DET* are stored in special storage allocation units called directory entry blocks (DEBs), each of which is 4 kilobytes. NetWare can support up to 65,536 of these blocks. To improve performance, NetWare can use directory caching or hashing. Directory caching keeps currently used directory blocks and the FAT in a reserved area of RAM. Frequently used directory entries will be loaded into a cache memory. Directory hashing is the indexing of the directory entries, which speeds access to directory information.

directory hashing A method for organizing/indexing directory entries to minimize the search time for an entry. The hashing provides guided access to the desired entry, so that fewer entries need to be checked along the way.

directory ID In an AppleTalk network, a unique value associated with a directory when the directory is created.

directory information base (DIB) The body of directory-related information in the ITU-T X.500 Directory Services model. Directory system agents (DSAs) access the DIB on behalf of directory user agents (DUAs).

directory information tree (DIT) Information about a directory information base (DIB) in the ITU-T X.500 Directory Services (DS) model. Since information in a DIT can get quite large, it is generally distributed. This provides faster access to the information at the distributed locations and helps keep down the size of the *DIT* materials at any single location.

A *DIT* does not contain actual objects—just information about them and a pointer to the actual body of information. Therefore, an object that appears at multiple locations in the DIT will have only one body of information associated with it. Each location in the tree has predefined attributes associated with it; the attributes will depend on the object class to which the entry belongs. An object class such as country or organization determines which attributes are mandatory and which are optional for objects belonging to that class. Objects in the tree will have specific values associated with these attributes. Two general classes of operations are possible in a DIT: retrieval (reading) and modification (creating and writing). Furthermore, a given DIT operation may apply to a single entry or to a group of entries; hence, four operation classes are possible. The X.500 model supports three of the four possible operation classes:

- Retrieve a single entry.
- Retrieve a group of entries.
- Modify a single entry.

The fourth operation class, modify a group of entries, is not supported in X.500. Both end users and processes can access the information in the *DIT* by either of two procedures:

- A *directory user agent* (*DUA*) that provides access to the *DIT* through an access point. A particular access point may support one or more of the operation classes.
- A *directory system agent* (*DSA*) that provides the requested services for the DUA (and for other DSAs). Since the *DIT* can be large and may be distributed, more than one DSA may be involved. One particular DSA is generally responsible for a portion of the *DIT*. This portion is known as a context.

directory number In telephony, all of the digits required to identify a customer uniquely in a directory.

directory service (**1**) In telephony, a directory number information service provided by the telephone companies. In the United States it is accessed by dialing 411 for local assistance or XXX-555-1212 for long distance assistance. (XXX is the area code of the region of the desired phone number.) (**2**) The tools within networking software that provide information on resources available on both the local network and any extended networks, including files, users, printers, data sources, and applications.

directory tree See *directory*.

dirty power A term that refers to mains power in which the frequency, waveshape, voltage, or noise content is greater than a specified amount. For example, the power may have excessive noise spikes due to the operation of nearby heavy machinery, or the voltage may surge to greater than 110% of nominal or droop to less than 90% of nominal.

In some parts of the world, mains power is so unreliable and dirty that communications and computing equipment must be operated from an uninterruptible power supply (UPS).

DIS An abbreviation of Draft International Standard. An ISO standards document that has been registered and numbered but has not been finally approved.

DISA (**1**) An acronym from the Defense Information Systems Agency, formerly called the Defense Communications Agency (DCA). (**2**) An acronym from Direct Inward System Access. (**3**) An acronym from Data Interchange Standards Association. An organization with the responsibility of developing Electronic Data Interchange (EDI) standards.

disable To stop or prevent the operation of a device, circuit, or system.

disabling tone A specific tone transmitted over a communication circuit to control some aspect of the communication channel. It is normally used to place an echo suppresser in a nonoperate condition during data transmission on a telephone line.

DISC (**1**) An acronym from Digital International Switching Center. (**2**) An abbreviation of DISConnect.

discard eligibility (DE) bit A bit in a frame relay packet header that can be set to indicate that the packet can be discarded if network traffic warrants it. If network traffic gets too heavy at a particular node, any packet with the *DE* bit set can be discarded by that node. See also *ATM*.

discharge (**1**) Batteries, the conversion of chemical energy into electrical energy. (**2**) Protectors, the conduction of a current through the protecting device, e.g., through the gas in a gas tube protector, through the metal oxide in an MOV, or across the gap in a spark gap protector. (**3**) The act of removing the stored energy in a capacitor (electrostatic energy) or inductor (electromagnetic energy) by allowing the current to flow through a dissipative element such as a resistor.

discharge block A protective device through which unwanted currents are routed to ground. See also *transient* and *transient suppresser*.

discharge current The current that flows during discharge.

discharge voltage The voltage that appears across the terminals of the device conducting a discharge current.

discone antenna See *biconical antenna*.

disconnect In telephony, the release of the connection between two stations on a switched circuit and placing the apparatus into an idle state. See also *on-hook*.

disconnect command (DISC) In Link Layer protocols, such as high-level data link control (HDLC), synchronous data link control (SDLC), and advanced data communication control procedure (ADCCP), an unnumbered command used to terminate a previously set operational mode.

disconnect frame The final transmission of a facsimile call. The sending machine transmits the disconnect frame and then goes on-hook; it does not wait for a response from the destination.

disconnect signal In telephony, a supervisory signal transmitted from one end of a subscriber line or trunk to indicate that the party has released the circuit and that the circuit and all associated equipment should be released (made idle).

discontinuity An abrupt change in the characteristics of an otherwise uniform entity, e.g., at the junction of two different cable types.

discrete cosine transform (DCT) A mathematical algorithm for converting time-domain data into frequency-domain data. It is a form of coding used by many image compression systems to reduce the transmission bit rate, e.g., the ITU Px64 videoconferencing compression standard and both JPEG and MPEG image compression.

discrete multitone (DMT) A modulation technique wherein the entire communications link bandwidth is divided into a large number of sub-bands. One high-speed data channel is distributed across all sub-bands. To achieve maximum throughput, each sub-band is processed separately to compensate for its own anomalous behavior and distortions. Also called *orthogonal frequency division multiplex* (*OFDM*) because it uses code division as well as frequency division.

discriminator A circuit in which the output is dependent on how the input signal differs in some aspect from a standard, reference, or other signal. Examples include:

- *Pulse height discrimination*—The output becomes active only if the pulse exceeds a specified amplitude.
- *FM discriminator*—It demodulates an FM signal; i.e., it extracts the desired signal from an incoming FM wave by changing frequency variations into amplitude variations.

Also called a *detector*.

disengagement attempt An attempt to terminate a telecommunications system access. It may be initiated by either a user or the telecommunications system itself.

disengagement denial After a *disengagement attempt*, a failure to terminate the telecommunications system access. *Denial* is usually caused by excessive delay in the telecommunications system.

disengagement failure The failure of a disengagement attempt to return a communication system to the idle state, for a given user, within a specified time.

disengagement originator The user (or system functional entity) that initiates a disengagement attempt. A *disengagement originator* may be the originating user, the destination user, or the communications system. The communications system may, for example, deliberately originate the disengagement because of preemption.

disengagement phase The third stage of an information transfer transaction, the state during which successful disengagement occurs. See also *access phase* and *information transfer phase*.

disengagement request A supervisory control signal issued by a disengagement originator for initiating a disengagement attempt.

disengagement time The elapsed time between the start of a disengagement attempt for a particular source or destination user and the successful disengagement of that user; i.e., the elapsed time between the start of a disengagement attempt and a successful disengagement.

dish A parabolic reflecting microwave antenna.

disk An abbreviation from DISKette.

disk cache A high-speed memory interposed between a hard disk drive and a CPU. See also *cache memory*.

disk duplexing A method of improving a network file server reliability by using multiple disk drives and possibly multiple disk controllers. See also *RAID*.

disk mirroring A method of improving a network file server reliability by using multiple disk drives. See also *RAID*.

disk server A server equipped with large disk storage facilities and programs permitting users to create and store files on those disks. Each user may have access to his or her own section of disk on the disk server and/or to globally available storage. The aim is to give users access to disk space that they would not normally have on their client machine. The *disk server* is linked to the client via a LAN. See also *communications server, database server,* and *file server*.

diskette A thin, flexible plastic disc coated with a magnetizable medium (usually ferric oxide) and permanently enclosed in a semi-rigid protective jacket. The magnetic material stores the digital data. Several types of diskettes are available, varying in both physical size and magnetic storage capacity. The capacity of the diskettes is typically 360K, 1.2M, 1.44M, or 2.88M bytes. Also called a *disk, flexible disk, floppy,* and *floppy disk*.

diskless workstation A PC or workstation attached to a LAN that has neither floppy nor hard disks but relies on disk storage provided by a server attached to the same LAN. When the diskless workstation is first initialized, it uses a remote boot program stored in a remote boot PROM/EPROM on its network interface card (NIC) to initialize a session with the server. The workstation then loads its operating system, such as MS-DOS, from the server and executes the normal server login procedure.

DISNET An acronym from Defense Integrated Secure NETwork.

DISOSS An acronym from DIStributed Office Support Systems.

disparity In pulse code modulation (PCM), the algebraic sum of a set of signal elements. *Disparity* will be zero (and there will be no cumulative or drifting polarization) if there are as many positive elements (those that represent 1) as there are negative elements (those that represent 0).

dispersion (1) A form of signal distortion caused by the differences in propagation speed of the various component frequencies of the signal. More properly identified as *time dispersion*. (2) Scattering that takes place along a beam of electromagnetic radiation. More properly identified as *spatial dispersion*. (3) In communications, any phenomenon in which the velocity of propagation of an electromagnetic wave is wavelength or frequency dependent. *Dispersion* is used to describe any process by which an electromagnetic signal propagating in a physical medium is degraded because the various wave components (i.e., frequencies) of the signal have different propagation velocities within the physical medium.

In an optical fiber, there are several significant dispersion effects. Three types of dispersion relating to optical fibers (and to communications therein) are defined as:

- *Material dispersion*—the effect in which propagation velocity is dependent on both the wavelength of the optical signal and the properties of the bulk material from which the fiber core is constructed.

  Because an optical signal has a nonzero spectral width (bandwidth due to modulation), *material dispersion* results in *time dispersion* (spreading) of the signal.

  Pure silica, the material from which most telecommunication grade fibers are fabricated, has a minimum *material dispersion* at wavelengths near 1.27 µm, or slightly longer in practical fibers.

  The term *chromatic dispersion* is sometimes used; however, its use is discouraged.

- *Profile dispersion*—dispersion attributable to the variation of refractive index contrast with wavelength. Profile dispersion is a function of the profile dispersion parameter.
- *Waveguide dispersion*—dispersion caused by the dependence of the phase and group velocities on core radius, numerical aperture, and wavelength. For circular waveguides, the dependence is on the ratio of the core radius (r) and the core radius (λ), i.e., r/λ. Practical single mode fibers are designed so that *material dispersion* and *waveguide dispersion* then cancel each other at the transmission wavelength.

Neither the term *multimode dispersion* nor *intermodal dispersion* should be used as a synonym for the term *multimode distortion*. (4) In classical optics, i.e., as used by optical lens designers, *dispersion* is used to mean the *wavelength dependence of refractive index* in matter caused by interaction between the matter and light.

dispersion coefficient (D(λ)) See *material dispersion coefficient (M(λ))*.

dispersion limited operation Operation of a communications link in which the link's dispersive effects dominate signal degradation and therefore limit link performance.

Dispersion limited operation is not the same as *distortion limited operation*, although they are often confused.

dispersion shifted fiber A single mode optical fiber with dopants added to the core material that shifts the *minimum dispersion* wavelength toward the *minimum loss* wavelength. Also called an *EIA Class IVb fiber*. Contrast with *dispersion unshifted fiber*.

dispersion unshifted fiber A single mode optical fiber that has a nominal zero dispersion wavelength in the 1.3-µm transmission window. Also called a *dispersion unmodified fiber, EIA Class IVa fiber,* and *nonshifted fiber*. Contrast with *dispersion shifted fiber*.

dissipation The loss of electrical or mechanical energy as radiated heat.

dissipation factor The ratio of the energy lost as dissipation to the energy stored in an element over one cycle.

distance vector A type of packet routing algorithm in which each router computes the distance between itself and all possible destinations. This is achieved by computing the distance between a router and all of its immediate router neighbors, and adding each neighboring router's computations for the distances between that neighbor and all of its immediate neighbors. Two commonly used implementations are the Bellman-Ford algorithm and the ISO Interdomain Routing Protocol (IDRP).

Distance Vector Multicast Routing Protocol (DVMRP) See *IP multicast.*

distance vector routing algorithm A routing algorithm that computes the best route through a network for a given *packet* based on the number of routers (or hops).

Each router using the *distance vector* technique builds a routing table (network map) based on information it receives from adjacent routers. Every router periodically transmits (a process called *advertising*) the information it possesses about adjacent routers and passes on information about more distant devices. Over time, all routers will accumulate a complete picture of the network and all possible routes—a process called *convergence.* An Internet *interior gateway protocol* (*IGP*) that uses the *distance vector routing algorithm* is called the *routing information protocol* (*RIP*).

distinctive dial tone A PBX feature that presents a different dial tone to the user based on the network to which the user is connected. For example, one dial tone may indicate connections to internal extensions, whereas another dial tone may indicate that the call will be routed by the public telephone network.

distinctive ring (1) A service provided by the local telephone operating company that allows an individual subscriber line to have several telephone numbers, each with a unique sounding ring signal. Each number, when called, will ring the same subscriber's telephone instrument but with a unique ring cadence.

For example, a subscriber with three numbers associated with a single telephone line may have the following unique ring signals:

- The first number will use the standard 1-second ring 4-second silent ring cadence.
- The second number ring cadence has two short ring bursts followed by a long silent period.
- The third number ring cadence is a burst of three rings followed by a silent period; the first and third rings in the burst are short, while the second ring is long.

Also called *CustomRinging* (by USWest), *IdentaRing* (by Bell Atlantic), *Multi Ring* (by Ameritech), *Personalized Ringing* (by Southwestern Bell), *RingMaster* (by Bell South), *Ringmate* (by NYNEX), *Smart Ring* (by GTE), *RingPlus* (by Alltel), and so on. See also *coded ringing, selective ringing,* and *semiselective ringing.* (2) A PBX feature that allows users to distinguish between an inside call and an outside call.

distortion (1) In signal transmissions, any undesirable change in the waveform of a signal. *Distortion* can occur at any point or at multiple points in a transmission path. The results of a *distorted* signal range from an increase in the error rate to the complete loss of the signal. All aspects of the waveform are subject to *distortion.* Some common distortions are listed below.

Amplitude distortion—an unwanted change in the signal amplitude (independent of frequency) wherein the output signal envelope is not proportional to the input signal envelope.

Asymmetrical distortion—see *bias distortion.*

Attenuation distortion—can refer to either of two types of distortion:

1. A variation in the response of a channel as a function of frequency. (That is, changes in output amplitude relative to an input signal with constant amplitude but changing frequency.) The distortion measurement is a comparison of the output levels at specified frequencies in the channel bandwidth with the level at a reference frequency. In telephone circuits, the reference frequency is 1004 Hz in the United States and 800 Hz or 1020 Hz for compliance with ITU-T recommendations.

2. Changes in the output amplitude due to device gain changes at different input levels. Also called *amplitude distortion* and *nonlinear distortion.* See *attenuation distortion* for a typical graph.

Bias distortion—distortion affecting a binary modulation system wherein all time durations of one state are uniformly longer (or shorter) than the theoretical duration.

Cyclic distortion—distortion which, in general, has a periodic character.

Delay distortion—distortion resulting from the nonuniform speed of propagation of different frequency components of the signal. Also called *group delay.*

Deviation distortion—distortion in an FM receiver resulting from either inadequate bandwidth and inadequate amplitude rejection or excessive discriminator non-linearity.

Envelope delay distortion (EDD)—the difference between the envelope delay at one frequency and the envelope delay at a reference frequency.

Fortuitous distortion—distortions resulting from all random causes.

Frequency distortion—distortion in which the relative magnitude of the different frequency components of a complex wave are changed during transmission.

Harmonic distortion—the development of harmonics of the input signal in the output signal due to nonlinearities of the communication channel.

Intermodulation distortion—the nonlinear distortion of a device or system characterized by the appearance of frequencies in the output of the device that are not present in the input when two or more frequencies are presented at the input. The unwanted output frequencies are at the sum and difference of the input frequencies.

Linear distortion—any distortion that is not correlated with signal amplitude. For example, small signal frequency response.

Nonlinear distortion—any distortion that is correlated with signal amplitude. For example, harmonic distortion.

Phase delay distortion—the difference between phase delay at a specified frequency and phase delay at a reference frequency.

Phase distortion—deviations from a constant slope of the output phase vs. frequency response of a device or transmission system. Also called *phase frequency distortion.*

Distortion may result from many mechanisms; however, most are caused by

- A nonlinear relationship of the input signal to the output signal at a given frequency, e.g., the nonlinearities in the transfer function of an active device, such as a vacuum tube, transistor, or operational amplifier.
- Nonuniform amplitude transmission at different frequencies within the band of interest such as occurs with a coaxial cable or optical fiber.
- Nonproportional phase shift with respect to frequency.
- Inhomogeneities, reflections, and so on, in the propagation path.

(2) In start-stop teletypewriter signaling, the shifting of the actual signaling pulses from their expected position relative to the beginning of the start pulse. The magnitude of the distortion is generally expressed as a percentage of an ideal unit pulse length.

distortion factor (DF) The ratio of the root-mean-square (RMS) value of the residue of a wave after elimination of the fundamental frequencies (the total distortion) to the RMS value of the original signal (fundamental plus distortion).

$$DF = \sqrt{\frac{\text{sum of squares of amplitudes of all harmonics}}{\text{sum of squares of amplitudes of all harmonics and fundamentals}}}$$

In some definitions, the *distortion factor* is defined as the ratio of the total RMS distortion to the RMS fundamental only. For large signal to distortion ratios there is little numeric difference; that is,

$$DF = \sqrt{\frac{\text{sum of squares of amplitudes of all harmonics}}{\text{sum of squares of amplitudes of all fundamentals}}}$$

distortion limited operation The condition that exists when the distortion of a received signal limits system performance under stated operational conditions and limits. Other limiting parameters include *amplitude, bandwidth, dispersion,* and *power.*

Distortion limited operation is reached when the received waveform shape is distorted beyond specified limits. For a linear system, *distortion limited operation* is equivalent to *bandwidth limited operation.*

distributed Spread out over an electrically significant geographic area.

distributed authentication security service (DASS) A system for authenticating users logging onto a network from unattended workstations. These workstations must be considered suspect, or untrusted, because their physical security cannot be guaranteed. *DASS* uses a public key encryption method, which supports the more stringent, strong authentication methods defined in the ITU's X.509 specifications. Another system, Kerberos, uses a distributed authentication system that employs a private key encryption method.

distributed computing The trend away from having big, centralized computers (such as minicomputers and mainframes) to bring processing power to the desktop. Contrast with *distributed processing.*

Distributed Computing Environment (DCE) A suite of software utilities and operating system extensions that will, in theory, create applications on networks of heterogeneous hardware, e.g., workstations, minicomputers, PCs, and mainframes. This open networking architecture is promoted by the Open Software Foundation (OSF), a consortium of vendors that includes Digital Equipment Corporation (DEC), Hewlett-Packard (HP), and IBM. When implemented, the entire network appears to a user as one giant, very fast, and powerful computer.

The architecture provides the elements, tools, and services needed to distribute applications and their operation across networks in a transparent fashion; that is, *DCE* protects the user from any implementation details. *DCE* sits on top of whatever network operating system is running, so that a user interacts with the *DCE* environment. It is designed to simplify the building of mixed client–server applications by providing seven general services:

- *Distributed File System.*
- *Naming (directory).*
- *PC Integration.*
- *Remote Procedure Call (RPC)*—enables a local machine to call an application or function on a different machine, just as if the resource were local or even part of the application.

- *Security*—automatically applied to the entire network. This means that a user on any machine is protected automatically from a virus or an unauthorized user on another machine.
- *Threads*—an independently executable program segment. This can be distributed across different machines and executed simultaneously, thereby reducing processing time considerably.
- *Time.*

By making the entire network's resources available in a completely transparent manner, *DCEs* attempt to make the fullest use of available resources and to make it more likely that a resource will be available when it is needed.

distributed control A system or network control strategy in which control is administered from multiple entities or points. Each entity controls a portion of the network, using either local information or information transmitted over the network from distant points.

Distributed Data Management architecture (DDM) An IBM Systems Network Architecture (SNA) Logical Unit (LU) 6.2 transaction providing remote users with proper facilities to locate and access data in the network. It involves two structures:

- *DDM source,* which works with a transaction application to retrieve distributed data and transmits commands to the *target program* on another system where the data that has been requested is stored.
- *DDM target,* which interprets DDM commands, retrieves the data, and sends it back to the requesting source.

distributed database (DDB) A database whose components are dispersed among different locations. The locations may be different disk drives on the same machine, or they may be geographically separated and managed by different machines, as is the case with the Internet domain naming system (DNS). The database is split up across these locations and is not necessarily replicated.

distributed element A circuit element that exists continuously over the length of a transmission line, e.g., the capacitance of a coaxial cable. See also *lumped element.*

distributed file system (DFS) A file system with files located on multiple machines, but accessible to an end user or a process as if all the files were in a single location. See also *distributed database (DDB).*

distributed frame alignment signal A frame alignment signal in which adjacent signal elements occupy noncontiguous positions in the transmitted frame.

distributed function terminal (DFT) A terminal mode in IBM's System Network Architecture (SNA) wherein a terminal may support up to five different simultaneous sessions. A user can then access up to five applications through the same terminal. See also *control unit terminal (CUT).*

Distributed Name Service (DNS) A technique for storing network node names throughout the network. The information can be requested from and supplied by any node.

distributed network A network in which resources, such as switching equipment, file servers, communications servers, and processors, are distributed throughout the geographical area being served by the network. Such a network may be either central or distributed control in nature.

Distributed Office Support Systems (DISOSS) IBM software typically forming part of an IBM Office System Node.

distributed processing (1) A system in which each node performs its own data processing and management and any interconnecting network is provided only to facilitate communications between nodes.

(2) Data processing in which an integrated set of functions is performed within multiple, physically separated devices. **(3)** An approach that allows one application program to execute on multiple computers linked together by a network. The networked computers share the work between them. Contrast with *distributed computing.*

distributed queue dual bus (DQDB) A 150-Mbps distributed multiaccess, packet switched protocol used on Metropolitan Area Networks (MANs). It operates as a dual bus, each typically carrying data in both directions. A queuing system (an implementation of a reservation strategy) maintains transmission order. It is part of the medium access control (MAC) sublayer of the data link layer (DLL). It supports connectionless data transfer, connection-oriented data transfer, and isochronous communications (such as voice communications). See also *IEEE 802.6.*

distributed routing In a packet switching network, the partitioning of routing responsibilities to multiple devices throughout the network. Each router runs a routing algorithm, which allows it to make routing decisions and maintain an accurate network map (routing table) over time. As the network is expanded, more routers are added to handle the additional traffic.

In contrast, a centralized router must make all decisions. This simplifies the design but limits the overall size to which the network can expand. See also *distance vector routing algorithm.*

distributed switching Switching in which many switching units are distributed, usually close to concentrations of users, and operated in conjunction with a host switch.

In telephony, the technique is used for reducing the amount of outside plant wiring. It involves the installation of small switching systems located very close to the neighborhoods containing the local loops. Because the switching equipment is located close to subscriber locations, the length of the subscriber loop is reduced. Traffic considerations allow the number of loops connecting the remote switch to the central office to be significantly less than if no switch were present. Also called *concentrators, remote switches,* or *slave switches.*

distributed system A strategy in which significant elements of a system are spread throughout the geographic area comprising the system. No single element performs all requisite functions for all other elements.

distributing frame In telephony, a device for terminating permanent wires of a switching system (central office, private branch exchange, or private exchange), building, or outside plant wiring, and for permitting the easy change of connection between them by means of cross-connection wires.

A distributing frame allows the association of any equipment port with any subscriber loop on a semipermanent basis. The association may be changed or deleted at any time by moving or removing the cross-connection wires.

distribution list (DL) A tool for reaching multiple destinations with a single transmission in the ITU X.400 Message Handling System (MHS). The *DL* includes all addresses to which a message is to be sent.

distribution voltage drop The reduction in mains voltage at defined points of the power distribution system. The voltage reductions are due to power dissipation losses associated with the resistance of the conductors. Given the current in the distribution line (I) and the resistance of the line (R), and Ohm's law, one can calculate the voltage drop, i.e.,

$$V_{loss} = I \times R.$$

distributor A device at the receiving end of a communication link that separates a composite signal into its constituent message components and then forwards each message on to the appropriate terminating DTE. The composite signal was created by a *concentrator* at the transmitting end of the communication link. See also *concentrator.*

DISU An abbreviation of <u>D</u>igital <u>I</u>nternational <u>S</u>witching <u>U</u>nit.

DIT (1) An acronym from <u>D</u>irectory <u>I</u>nformation <u>T</u>ree. **(2)** An acronym from <u>D</u>ata <u>I</u>dentification <u>T</u>able.

dithering (1) The process of introducing rapid changes (as compared to the signal) between the two states of a binary signal in order to approximate intermediate values. For example, a pattern of black and white dots may be used to approximate various shades of gray. In another example, the A-law digital representation of an analog signal, there is no zero value, only $+1$ and -1; by oscillating between these two values, the average tends toward zero. **(2)** In the Global Positioning System (GPS), the intentional injection of digital noise to reduce measurement accuracies.

DITS An acronym from <u>D</u>igital <u>I</u>nformation <u>T</u>ransfer <u>S</u>et.

DITU An abbreviation of <u>D</u>igital <u>I</u>nterface <u>T</u>est <u>U</u>nit.

DIU An abbreviation of <u>D</u>igital <u>I</u>nterface <u>U</u>nit.

diurnal changes Alterations of a signal characteristic in response to naturally occurring daily stimuli. Examples include:
- Day-night temperature changes that affect phase shifts in cable transmission systems.
- Day-night variations in solar radiation that affect the ionosphere, which in turn affects radio wave propagation and reflection.
- Lunar tidal changes that affect reflection along a transmission path.

DIV An abbreviation of <u>D</u>ata <u>In</u> <u>V</u>oice.

DIVA An acronym from <u>D</u>igital <u>I</u>nquiry-<u>V</u>oice <u>A</u>nswerback.

diversity In radio, the use of multiple frequencies and/or paths to improve the probability of establishing and maintaining a path between a sender and receiver to a desired level of performance. It safeguards against both equipment failure and naturally occurring communications impediments
- *Frequency diversity*—A separate frequency band is allocated for use in case the main band cannot be used (for example, because of noise or other interference).
- *Space diversity*—Two receiving antennas are set up in the same general vicinity. If the primary antenna malfunctions or cannot receive for some reason (rain fade, for example), the auxiliary antenna can be used to receive the signals.

diversity combiner A device that derives a single signal output from multiple input signals. All input signals carry the same information and are combined in such a way that the resultant output signal has superior quality (e.g., lower error rate, lower S/N, lower distortion) to any single contributing input signal. See also *combiner, equal gain combiner, linear combiner,* and *maximal ratio combiner.*

diversity gain In radio communications, the ratio of the signal field strength obtained by diversity combining to the signal strength obtained by a single path. *Diversity gain* is usually expressed in decibels (dB).

diversity reception Radio reception in which a resultant signal is obtained by either combining or selecting signals, from two or more independent transmission paths, each path transporting a modulated signal containing identical information-bearing signals. Each path may vary in its fading or interference characteristics at any given instant.

Diversity reception is used to minimize the effects of fading or interference. The amount that the received signal is improved when using *diversity reception* is a direct function of the independence of the fading or interfering characteristics.

diversity transmission A radio communication technique in which a resultant receiver output signal is obtained by combining or selecting from multiple received signals. These received signals have a common information source but arrive at the receiver over multiple independent paths. Each path is subject to different fading, interference, and failure times and possible mechanisms.

Diversity transmission techniques are used to improve reliability by overcoming the effects of fading, outages, and circuit failures through redundancy. When using the *diversity transmission* technique, the amount of received signal reliability improvement depends on the time independence of the fading, failure, and interference characteristics of the signal.

divestiture The 1984 court-ordered breakup of AT&T into the seven regional Bell operating companies (BOCs) and AT&T long lines. The decree basically stated:

- BOCs were not allowed into long-distance service, equipment manufacturing, or information processing.
- AT&T was not allowed into local communications.

divided ringing In telephony, a method of ringing only one telephone of a two-party line circuit. The central office applies the ring voltage between either the tip and ground for one station (called tip party ringing) or between the ring lead and ground for the other station (called ring party ringing).

Divided ringing has the advantage of being able to ring more ringers over longer distances than bridged ringing. However, there is a noticeable reduction in transmission quality due to the *longitudinal im-*

DIVIDED RINGING (Tip Party)

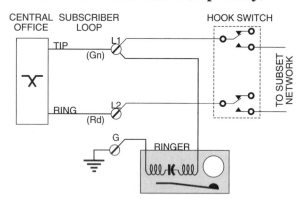

DIVIDED RINGING (Ring Party)

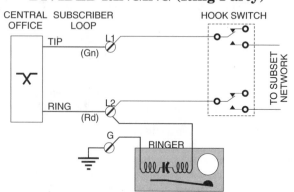

balance. The degradation can be as much as 10 dB unless measures are taken to prevent the imbalance. Several methods of preventing longitudinal imbalances include:

- Installation of the same number of ringers on both tip part and ring party services—even if the addition of dummy ringers (standard ringers with the gong disabled) are used.
- The use of *ringer isolators*. A ringer isolator is simply a device that connects the ringer to the line only while ringing is present.
- The use of the hook switch to disconnect the ringer circuit when the subscriber is off-hook, i.e.,

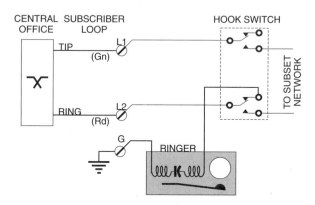

See also *bridged ringing* and *coded ringing*.

DIX Ethernet A local area network (LAN) whose specification was published by Digital Equipment Corporation, Intel Corporation, and Xerox Corporation in 1982. Revised in 1984 and forming the basis of the IEEE 802 network protocols, DIX Ethernet is a 10-Mbps, baseband coaxial network.

Although the IEEE version is somewhat different from the DIX version, both are generally called Ethernet. It is more correct, however, to refer to them as the *DIX Ethernet* and the *IEEE 802 network*. Also referred to as the *Blue Book Ethernet* standard.

DIY An abbreviation of <u>D</u>o <u>I</u>t <u>Y</u>ourself.

DJ An abbreviation of <u>D</u>igital <u>J</u>unction.

DK An abbreviation of <u>D</u>isplay and <u>K</u>eyboard.

DKE An abbreviation of <u>D</u>eutsche <u>E</u>lektrotechnische <u>K</u>ommission im DIN und VD<u>E</u> (German Electrotechnical Commission of DIN and VDE). The German national standards-setting agency. See also *DIN* and *VDE*.

DL (1) An abbreviation of <u>D</u>istribution <u>L</u>ist. (2) An abbreviation of <u>D</u>ata <u>L</u>ink. (3) An abbreviation of <u>D</u>own<u>L</u>ink.

DLA (1) An abbreviation of <u>D</u>ata <u>L</u>ink <u>A</u>dapter. (2) An abbreviation of <u>D</u>ata <u>L</u>ink <u>A</u>ddress.

DLC (1) An abbreviation of <u>D</u>igital <u>L</u>oop <u>C</u>arrier. (2) An abbreviation of <u>D</u>ata <u>L</u>ink <u>C</u>ontrol. (3) An abbreviation of <u>D</u>igital <u>L</u>ogic <u>C</u>ircuit.

DLCF An abbreviation of <u>D</u>ata <u>L</u>ink <u>C</u>ontrol <u>F</u>ield.

DLCI An abbreviation of <u>D</u>ata <u>L</u>ink <u>C</u>onnection <u>I</u>dentifier. A field in the frame relay header that represents the virtual circuit number associated with a particular destination.

DLE An abbreviation of <u>D</u>ata <u>L</u>ink <u>E</u>scape. A control character used exclusively to provide supplementary line signals, control character sequences, or DLE sequences.

For example, in packet switched networks, *DLE* is the name of the <CTL-P> character used to change the packet assembler disassembler (PAD) from data mode to command mode.

DLL (1) An abbreviation of Dynamic Link Library. (2) An abbreviation of Data Link Layer. (3) An abbreviation of Dial Long Line.

DLM An abbreviation of Data Line Monitor.

DLRP An abbreviation of Data Link Reference Point.

DLS (1) An abbreviation of Data Link Services. The services provided at the data link layer in the ISO/OSI Reference Model. (2) An abbreviation of Data Link Switching. IBM's method of transporting SNA and NetBIOS over TCP/IP. (3) An abbreviation of Digital Line System.

DLSw An abbreviation of Data Link Switching. A standard that provides a means for integrating and interoperating SNA and NetBIOS over TCP/IP protocol.

DLT (1) An abbreviation of Digital Loop Termination. (2) An abbreviation of Data Link Terminal. (3) An abbreviation of Data Link Translator. (4) An abbreviation of Data Line Transceiver.

DLTM An abbreviation of Data Link Test Message.

DLTR An abbreviation of Data Link Transmission Repeater.

DLU An abbreviation of Digital Line Unit.

DLY An abbreviation of DeLaY.

DM (1) An abbreviation of Delta Modulation. (2) An abbreviation of Data Manager (or Management). (3) An abbreviation of Debug Mode. (4) An abbreviation of Disconnect Mode. (5) An abbreviation of Data Memory. (6) An abbreviation of Digital Module. (7) An abbreviation of Distribution Module.

DMA (1) An abbreviation of Direct Memory Access. (2) An abbreviation of Direct Memory Address. (3) An abbreviation of Digital Major Alarm.

DMB An abbreviation of Disconnect and Make Busy.

DMC (1) An abbreviation of Discrete Memoryless Channel. (2) An abbreviation of Direct Multiplexer Channel. (3) An abbreviation of Direct Memory Channel. (4) An abbreviation of Data Management Channel. (5) An abbreviation of Data Management Center.

DMD An abbreviation of Directory Management Domain. In the ITU X.500 Directory Management Services, a collection of one or more directory system agents (DSAs), and possibly of some directory user agents (DUAs), all managed by a single organization.

DME (1) An abbreviation of Distance Measuring Equipment. (2) An abbreviation of Digital Multiplex Equipment.

DMF (1) An abbreviation of Data Management Facility. (2) An abbreviation of Digital Multiplexing and Formatting.

DMI (1) An abbreviation of Desktop Management Interface. (2) An abbreviation of Digital Multiplexed Interface. (3) An abbreviation of Direct Memory Interface.

DMI-BOS An abbreviation of Digital Multiplexed Interface Bit Oriented Signaling. A form of signaling that uses the 24th channel of each DS1 to carry signaling information. This makes available the full 64-kbps rate in each of the remaining 23 channels.

DML An abbreviation of Digital Equipment Corporation's (DEC's) Data Management Language.

DMM (1) An abbreviation of Digital MultiMeter. (2) An abbreviation of Direct Mail Manager.

DMOS (1) An acronym from Diffusion Metal Oxide Semiconductor. (2) An acronym from Depletion Metal Oxide Semiconductor.

DMP (1) An abbreviation of DuMP. (2) An abbreviation of Direct Memory Processor.

DMS (1) An abbreviation of Digital Multiplex System. (2) An abbreviation of Data Management Service. (3) An abbreviation of Data Monitor System.

DMSS An abbreviation of Digital Multiplex SubSystem.

DMT (1) An abbreviation of Direct Memory Transfer. (2) An abbreviation of Discrete MultiTone.

DMU (1) An abbreviation of Digital Message Unit. (2) An abbreviation of Data Management Unit.

DMUX An abbreviation of DeMUltipleX.

DmW The symbol for Digital MilliWatt. The digital signal stream that represents a 1-milliwatt analog signal at a specified point in a mixed analog/digital communications system.

DN (1) An abbreviation of Directory Number. (2) An abbreviation of Date Number. (3) An abbreviation of Data Name.

DNA (1) An abbreviation of Digital Network Architecture. A layered architecture from Digital Equipment Corporation (DEC). See *DECnet*. (2) An abbreviation of Distributed Network Architecture. A network in which processing capabilities and services are distributed across the network, as opposed to being centralized in a single host or server. (3) An abbreviation of DeoxyriboNucleic Acid. The strands of information in living cells that determine the hereditary characteristics of individual living entities.

DNAM An acronym from Data Network Access Method.

DNHR An abbreviation of Dynamic NonHierarchical Routing.

DNIC (1) An acronym from Data Network Identification Code. (2) An acronym from Digital Network Interface Circuit.

DNIS An abbreviation of Dialed Number Identification Service. A telephone company service available on 800 and 900 numbers that allows a single-user termination to be reached from several called numbers. Furthermore, it reports the number the caller used to reach the termination. This allows automatic routing of incoming calls based on the called number.

DNR (1) An abbreviation of Digital Noise Reduction. (2) An abbreviation of Dynamic Noise Reduction.

DNS An abbreviation of Domain Name Service. A TCP/IP protocol for collecting and maintaining network resource information that is distributed among various servers on the network. It includes a method of translating network host names into network addresses; that is, it can provide a machine's IP address if given the machine's domain names. *DNS* is described in Internet RFCs 1034 and 1035. The Berkeley Internet Name Domain (BIND) is one product developed to provide DNS.

do not disturb A feature on PBX systems that makes the telephone look busy to incoming calls even if the phone is idle. The feature may apply only to extensions or outside calls, or to both.

DOA (1) An abbreviation of Dead On Arrival. (2) An abbreviation of Direction of Arrival.

DoC An abbreviation of Department of Communications. The Canadian agency responsible for regulating the various aspects of communication. *DoC* was formed in 1969 and was replaced by the Canadian Radio-Television and Telecommunications Commission (CRTC).

DOC (1) Generally, a file that a user could expect to be a document. (2) An acronym from Data Output Channel. (3) An abbreviation of Dynamic Overload Control. (4) An abbreviation of Decimal to Octal Conversion.

Document Content Architecture (DCA) An IBM-defined method for storing and using text documents in various computer environments. Three standard formats are specified:

- *Revisable Form Text (RFT):* the primary format, in which text can still be edited.
- *Final Form Text (FFT):* a format in which text has been formatted for a particular output device and cannot be edited.
- *Mixed Form Text (MFT):* a format that contains more than just text, e.g., a document that also includes graphics.

Document Interchange Architecture (DIA) An IBM term defining the sets of functions needed for document handling in an IBM environment, including storage and distribution.

DoD An acronym for U.S. Department of Defense.

DOD An acronym for Direct Outward Dialing.

Doherty amplifier A high-efficiency class-C RF power amplifier system in which two amplifiers are connected in parallel by 90° phase shift networks (quarter wave networks). The operating points of the two sections are so adjusted that one stage delivers all power to the load up to 50% of the maximum. For input levels greater than 50%, input power is diverted to section 2. The output of section 2 is added to the output of section 1, thereby increasing the total power delivered to the load.

The advantage of such a system is the overall efficiency at power delivery points less than 100%; i.e., at the half power point, section 1 is operating at 100% efficiency, and section 2 is off. When the system is delivering full power to the load, both sections 1 and 2 are operating at 100% efficiency. The overall efficiency of the Doherty circuit is 60–65% essentially independent of the modulation depth.

DOM An acronym from Data Output Multiplexer.

domain (1) In mathematics, the independent variable used to express a function. Examples of domains include time, frequency, and voltage. (2) A group of similar or related entities. (3) In distributed networks, all the hardware and software under the control of a specified set of one or more host processors. (4) A group of nodes (workstations, servers, peer-to-peer stations, etc.) on a network that form an administrative entity. Every computer on the LAN belongs to at least one domain. Being logged in on one domain, however, does not limit resources in other domains to which the user has access permissions. (5) In the Open System Interconnection (OSI), a domain is a division created for administrative purposes and is based on functional differences. The five management domains defined in the OSI model are accounting, configuration, fault, performance, and security. (6) In IBM's Systems Network Architecture (SNA), a domain represents all the terminals and other resources controlled by a single processor or processor group. (7) In Novell's NetWare Name Service, the collection of servers that share bindery information. (8) In Novell's NetWare 4.x, they are special areas in which NetWare Loadable Modules (NLM) can run; i.e., the *OS_PROTECTED domain,* where untested NLMs cannot corrupt the operating system memory and the *OS domain* where proven NLMs can run more efficiently. (9) In the Internet, *domains* are the top levels of the *domain naming system (DNS). A domain* is a single word or abbreviation describing the most general level of responsibility for naming subordinate groups or hosts. There are currently only seven top-level organizational domains but many geographic domains (see the chart on the next page).

TOP LEVEL ORGANIZATIONAL DOMAINS IN THE UNITED STATES

com	Commercial	mil	Military
edu	Educational & academic	net	Network
gov	U.S. governmental	org	Organization
int	International		(other)

domain name On the Internet, a name that uniquely identifies a host computer on the Internet. The name is translated into dotted quads (the IP address) by the server. See also *domain naming system (DNS)* and *fully qualified domain name (FQDN).*

Domain Name Service (DNS) In Internet, a program that keeps track of the alphabetic names of other machines and their corresponding numeric IP addresses. The *Domain Name Service* translates a request to a named machine into the numeric IP address necessary to make the connection.

Within every *domain* of Internet, there is an *authoritative server* that maintains up-to-date information about its domain. When any router needs to deliver information to an unknown host within a domain, the router requests information from the authoritative server of that domain. The router updates its IP address—domain name table—for future use. Over time, the relationship of domain name to IP address diffuses throughout the network. See also *distributed routing.*

domain naming system (DNS) In Internet, a method of choosing host names so that they are unique within the entire network. The system assigns names in a hierarchical manner with the most significant portion of the name on the right, much like an address

1011 Main Street, Town, State, Country, Planet.

There may be several "Main Streets" in several identical "Towns" but not within the same "State." Similarly, the address of a host is assigned a name in the form

name.machine.domain.top_domain.

Additional levels can be added if necessary.

Each alphabetic name corresponds to an IP address (*dotted quad*), and the *Domain Name Service (DNS)* maintains a current translation table of relationship. *DNS* also supports separate mappings between mail destinations and IP addresses. See also *Domain Name Service (DNS).*

domain specific part (DSP) In the OSI Reference Model, part of the address for the network layer service access point (NSAP). That part of the domain address that refers to the part of the network under the control of a particular authority or organization.

dominant mode In a waveguide (microwave or optical) that can support more than one propagation mode, the propagating mode with the minimum degradation, i.e., the mode with the lowest cutoff frequency. Designations for the dominant mode are TE_{10} for rectangular waveguides and TE_{11} for circular waveguides.

dominant wave The guided wave having the lowest cutoff frequency.

DOMSAT An acronym for DOMestic communications SATellite.

dongle See *hardware key.*

door An area on many bulletin board systems (BBSs) where a user can exit the main board and run a stand-alone program. One common program run under a *door* is a conference program that allows a number of people to chat with each other. See also *room.*

DOP An abbreviation of Dilution Of Precision.

TOP LEVEL GEOGRAPHIC DOMAINS

ad	Andorra	ec	Ecuador	kz	Kazakhstan	sa	Saudi Arabia
ae	United Arab Em	ee	Estonia	lb	Lebanon	sb	Solomon Isls
af	Afghanistan	eg	Egypt	lc	St Lucia	sc	Seychelles
ag	Antigua & Barbuda	eh	Western Sahara	li	Liechtenstein	sd	Sudan
ai	Anguilla	er	Eritea	lk	Sri Lanka	se	Sweden
al	Albania	es	Spain	lr	Liberia	sg	Singapore
am	Armenia	et	Ethiopia	ls	Lesotho	sh	St Helena
an	Antilles, Nether	fi	Finland	lt	Lithuania	si	Slovenian
ao	Angola	fj	Fiji Isls	lu	Luxembourg	sj	Svalbard
aq	Antarctica	fk	Falkland Isls	lv	Latvia	sk	Slovakia
ar	Argentina	fm	Micronesia	ma	Morocco	sl	Sierra Leone
as	American Samoa	fo	Faroe Isls	mc	Monaco	sm	San Marino
at	Austria	fr	France	md	Moldova, Rep of	sn	Senegal
au	Australia	fx	France, Metrop	mg	Madascar	so	Somalia
aw	Aruba	ga	Gabon	mh	Marshall Isls	sr	Suriname
ax	Antartica	gb	Great Britain	mk	Macedonia	st	Sao Tome
az	Azerbaijan	gd	Grenada	ml	Mali	sv	El Salvador
ba	Bosnia	ge	Georgia	mm	Myanmar	sy	Syria
bb	Barbados	gf	French Guiana	mn	Mongolia	sz	Swaziland
bd	Bangladesh	gg	Guernsey	mo	Macau	tc	Turks/Caicos Isls
be	Belgium	gh	Ghana	mp	N Mariana Isls	td	Chad
bf	Burkina Faso	gi	Gibraltar	mq	Martinique	tf	French S Terr
bg	Bulgaria	gl	Greenland	mr	Mauritania	tg	Togo
bh	Baharain	gm	Gambia	ms	Monserrat	th	Thailand
bi	Burundi	gn	Guinea	mt	Malta	tj	Tajikistan
bj	Benin	gp	Guadeloupe	mu	Mauritius	tk	Tokelau
bm	Bermuda	gq	Equatorial Guinea	mv	Maldives	tm	Turkmenistan
bn	Brunei Darussalam	gr	Greece	mw	Malawi	tn	Tunisia
bo	Bolivia	gs	S Georgia	mx	Mexico	to	Tonga
br	Brazil	gt	Guatemala	mz	Mozambique	tr	Turkey
bs	Bahamas	gu	Guam	na	Namibia	tt	Trinidad/Tobago
bt	Bhutan	gy	Guyana	nc	New Caledonia	tv	Tuvalu
bv	Bouvet Island	hk	Hong Kong	ne	Niger	tw	Taiwan
bw	Botswana	hr	Croatia	nf	Norfolk Island	tz	Tanzania
by	Belarus	ht	Haiti	ni	Nicaragua	ua	Ukraine
bz	Belize	hn	Honduras	nl	Netherlands	ug	Uganda
ca	Canada	hu	Hungary	no	Norway	uk	United Kingdom
cf	Cent African Rep	id	Indonesia	np	Nepal	um	US outlying isls
cg	Congo	ie	Ireland	nr	Nauru	us	United States
ch	Switzerland	il	Israel	nu	Niue	uy	Uruguay
ci	Cote D'Ivoire	in	India	nz	New Zealand	uz	Uzbekistan
ck	Cook Island	iq	Iraq	om	Oman	va	Vatican City
cl	Chile	ir	Iran	pe	Peru	vc	St Vincent
cm	Cameroon	is	Iceland	pf	French Polynesia	ve	Venezuela
cn	China	it	Italy	pg	Paua New Guinea	vg	British Virgin Isls
co	Columbia	je	Jersey	ph	Philippines	vi	US Virgin Isls
cr	Costa Rica	jm	Jamaica	pk	Pakistan	vn	Viet Nam
cs	Slovak Republic	jo	Jordan	pl	Poland	vu	Vanuatu
cu	Cuba	jp	Japan	pm	St Pierre/Miquelon	ws	Samoa
cv	Cape Verde	ke	Kenya	pn	Pitcairn	ye	Yemen
cx	Christmas Isls	kg	Kygryzstan	pr	Puerto Rico	yt	Mayotte
cy	Cypress	kh	Cambodia	pt	Portugal	yu	Yugoslavia
cz	Czech Republic	ki	Kiribati	pw	Palau	za	South Africa
de	Germany	km	Comoros	py	Paraguay	zm	Zambia
dj	Djibouti	kn	St Kitts & Nevis	qa	Qatar	zw	Zimbabwe
dk	Denmark	kp	Korea, North	re	Reunion	zr	Zaire
dm	Dominica	kr	Korea, South	ro	Rumenia		
do	Dominican Rep	kw	Kuwait	ru	Russia		
dz	Algeria	ky	Cayman Isls	rw	Rwanda		

dopant A material added to a medium to change its properties and to produce a desired effect or property in the substance.

- In fiber optics, *dopants* are used to control the refractive index profile and other refractive properties of the fiber.
- In lasers, *dopants* are used to make nonlasing bulk material into a lasing composite material.
- In semiconductors, *dopants* are used to control the electrical characteristics of the material. Two types of *dopants* may be used: those that induce hole conduction (an electron acceptor *dopant*) and those that induce electron conduction (an electron donor *dopant*).

Also called an *impurity*.

doppler aiding A technique for improving the accuracy of Global Positioning System (GPS) measurements that use the measured Doppler shift to help the receiver more smoothly track the GPS signal.

doppler effect The apparent change in frequency of a wave at a receiver due to the rate at which the distance between the transmitter and receiver is changing. When the transmitter is approaching the receiver, the frequency is shifted higher; when the transmitter is receding, the frequency is shifted below the "at rest" frequency. This *effect* was first stated in 1842 by the Austrian physicist Christian Johann Doppler (1803–1853). Also referred to as the *doppler shift* and *doppler principle.*

doppler shift The degree of observed change in frequency (or wavelength) of a wave due to the doppler effect.

DOS (1) An acronym from Disk Operating System, comprising one or a suite of programs managing a disk-based computer system. *DOS* schedules and supervises work, allocating computer resources and the operation of peripherals. Versions of *DOS* exist for many machines and different vendors supply versions for the same machine. **(2)** Often used as a synonym for Microsoft's MS-DOS® and other compatible PC operating systems, e.g., PC-DOS®, DR-DOS®. **(3)** An abbreviation of Department of State.

DOS LAN Manager A DOS version of Microsoft's network operating system LAN Manager. It gives Named Pipes (an applications interface) support to DOS machines, enabling them to use the client–server environment.

dot address See *IP address.*

dot zero A term that refers to the revision of newly released software, e.g., MS-DOS version 4.0, 5.0, 6.0, or 7.0. The significance is new releases tend to have more bugs than subsequent releases; for example, MS-DOS version 4.01 was a bug fix release. The theory does not always work, however.

dotted quad See *IP address.*

double buffering The use of two buffers (rather than a single buffer) to compensate for speed differences between connected devices. Because each buffer is filled while the other is being emptied, an improvement in data throughput is achieved.

double connection In telephony, a fault condition in which two separate call paths are connected together.

double frequency shift keying (DFSK) Frequency shift keying (FSK) in which two digital signals are multiplexed and transmitted simultaneously by frequency shifting among four frequencies.

double modulation Modulation in which a first carrier (subcarrier) is modulated with an information-bearing signal, and the resulting modulated subcarrier is then used to modulate a second carrier of higher frequency for transmission. For example, the commercial FM stereo system.

double refraction A synonym for *birefringence.*

double sideband (DSB) transmission Amplitude Modulation (AM) transmission in which both sidebands generated by the modulation process and the carrier are transmitted.

double sideband reduced carrier (DSB-RC) transmission Transmission in which both sidebands produced by amplitude modulation and a fixed fraction carrier provided to the modulator are transmitted.

In *DSB-RC* transmission, the carrier is frequently transmitted at a level sufficient for a receiver to use to recover (demodulate) the information.

double sideband-suppressed carrier (DSB-SC) transmission Transmission in which only the sidebands produced by amplitude modulation are transmitted. The carrier level is reduced to the lowest practical level, ideally completely suppressed.

double-tuned amplifier An amplifier system with two or more stages in which each stage utilizes coupled circuits having two frequencies of resonance for the purpose of obtaining wider bandwidths than are obtainable with single-tuned systems.

doubly clad fiber A particular construction of a single mode optical fiber that has two claddings. The inner cladding's refractive index is lower than that of the outer cladding, and both claddings' refractive indices are lower than that of the core. The purpose of the second cladding is to trap and dissipate cladding mode rays.

Advantages of a *doubly clad fiber* include:

- Very low macrobending losses of guided mode rays,
- Two zero-dispersion points, and
- Low dispersion over a much wider wavelength range than a singly clad fiber.

Also called *depressed cladding fiber, depressed inner cladding fiber,* and *W-profile fiber* (from the fact that a symmetrical plot of its refractive index profile superficially resembles the letter W). See also *refractive index profile (n, 7).*

DOV An abbreviation of Data Over Voice.

down A term describing the situation when a host is unavailable for any reason.

down converter A device that performs a frequency translation such that the output frequency is lower than the input frequency.

The opposite is an *up converter.*

downlink (D/L) The link from a communication satellite to an Earth station receiver.

download The process of transferring information (usually in the form of a data file) from a remote location (customarily the host) to the local DTE via a modem or network, as in *downloading* a file from a bulletin board system (BBS). *Downloading* generally is from a "big" node such as a mainframe host computer to a "small" node such as a PC.

To download a file, a *communications program* must be used. The communications program provides the necessary flow control and error control protocols to allow a successful transfer. Many communications programs are available; however, they all offer one or more protocols such as KERMIT, XMODEM, YMODEM, or ZMODEM. See also *FTP* and *upload.*

downstream (1) In communications, the direction of transmission flow from the source toward the sink. **(2)** Network nodes that receive information from another node (or nodes). (All nodes that pass information serially from the source node to a specified node are said to be *upstream* nodes relative to that specified node.)

downstream site A host system with only one connection to the Internet.

A *downstream site* may have limited USENet selections because an upstream site has decided not to carry certain newsgroups.

downtime The interval during which a functional unit is inoperable or unavailable.

downward modulation Modulation in which the amplitude of the modulated carrier is never greater than the amplitude of the unmodulated carrier.

DOY An abbreviation of Day Of Year.

DP (1) An abbreviation of Data Processing. **(2)** An abbreviation of Draft Proposal. A preliminary version of specifications or standards from standards committees in the Internet. The *DP* is circulated for a limited time, during which comments and critiques are collected by the standards committee. (**3**) An abbreviation of Differential Phase. (**4**) An abbreviation of Dial Pulse. (**5**) An abbreviation of Distribution Point. (**6**) An abbreviation of Data Packet.

DPA (**1**) An abbreviation of Demand Protocol Architecture. (**2**) An abbreviation of Document Printing Application. (**3**) An abbreviation of Dial Pulse Access.

DPC (**1**) An abbreviation of Data Processing Center. (**2**) An abbreviation of Deferred Procedure Call. (**3**) An abbreviation of Destination Point Code.

DPCM An abbreviation of Differential Pulse Code Modulation.

DPE An abbreviation of Dynamic Phase Error.

DPI An abbreviation of Detected Pulse Interference.

dpi An abbreviation of Dots Per Inch.

DPLL An abbreviation of Digital Phase Locked Loop.

DPLM An abbreviation of Domestic Public Land Mobile communications.

DPMI An abbreviation of DOS Protected Mode Interface.

DPMS An abbreviation of DOS Protected Mode Services.

DPNSS An abbreviation of Digital Private Networks Signaling System. A signaling standard for digital private networks within the United Kingdom formulated jointly by British Telecom (BT) and PABX manufacturers.

DPO (**1**) An abbreviation of Dial Pulse Originating. (**2**) An abbreviation of Double Pulse Operation.

DPOC An abbreviation of Dynamic Processor Overload Control.

DPS (**1**) An abbreviation of Digital Power Supply. (**2**) An abbreviation of Digital Phase Shift(er). (**3**) An abbreviation of Data Processing System.

DPSK An abbreviation of Differential Phase Shift Key.

DPST An abbreviation of Double Pole Single Throw switch.

DPT An abbreviation of Dial Pulse Terminating.

DQDB An abbreviation of Distributed Queue Dual Bus.

DQPSK An abbreviation of Differential Quadrature Phase Shift Key modulation.

drain wire An uninsulated wire laid over the cable components (under the sheath) and used as a ground connection. Generally in contact with a shield.

DRAM An acronym from Dynamic Random Access Memory (pronounced "DEE-ram"). *DRAM* is a type of RAM memory with a relatively short hold time. Therefore, to prevent information from being lost, it must be refreshed (or rewritten) periodically.

dribble packet A packet that ends on an odd byte.

drift The slow change of an operational parameter of a device, circuit, or system over a specified period of time, generally very long as compared to the information rate. Drift is usually undesirable, but it may be of such long-term duration or of such a low excursion as to be negligible. Drift is generally unidirectional; however, it may be bidirectional, cyclic, or random.

driver (**1**) A program that has the ability to operate a device connected to a computer and provides an interface that can be controlled by an application program running on the computer. The application program generally cannot operate the device directly but requires the

assistance of a *device driver* to interpret messages and to control the device. (**2**) An electronic device that supplies the required signal levels to another circuit.

DRN (**1**) An abbreviation of Data Routing Network. (**2**) An abbreviation of Data Record Number. (**3**) An abbreviation of Data Reference Number.

drop (**1**) In telephony, the line from a telephone cable to a subscriber's building.

More correctly called a *subscriber's drop.* (**2**) In networks, an attachment to a horizontal cabling system (for example, through a wallplate). This is generally the point through which a computer or other device is connected to the transmission medium on a network. Also called a *drop line* or *drop cable.* (**3**) In a communications network, the portion of a device directly connected to the internal station facilities, such as toward a switchboard or toward a switching center. (**4**) The central office side of test jacks. (**5**) To delete, either intentionally or unintentionally, or to lose part of a signal, such as dropping bits from a bit stream.

drop and insert A switching scheme in a multichannel transmission system in which selected signals at an intermediate point are diverted (dropped), discarded, or a new signal may be introduced (inserted). The system may be a time division multiplex (TDM), frequency division multiplex (FDM), or other method.

- *No action.* The signal passes through the drop and insert point without incident.
- *Drop (diversion).* The signal may be demodulated and passed to the final user or reinserted into another transmission path in the same or another time slot or frequency band.
- *Insert.* A different signal for subsequent transmission along the path may be introduced in the position of a dropped signal (that is, in the time slot or frequency band previously occupied by the diverted signal).

The time slot or frequency band vacated by the diverted signal need not be reoccupied by another signal (i.e., it may be idle). Likewise, a previously unoccupied time slot or frequency band may be occupied by a signal inserted at the *drop* and *insert* point. Also called an *add-drop point.*

drop cable A cable that links a network adapter to an external transceiver attached to the main LAN coaxial cable. Also called an *Attachment Unit Interface (AUI) cable, transceiver cable,* or just a *drop.*

drop set All the components needed to connect a machine or other component to the horizontal cabling. At a minimum, this includes cable and an adapter or connector.

drop side All the components needed to connect a device or other component to the patch panel or punch down block that connects to the distribution frame.

dropout (**1**) In a communications system, a momentary loss of signal. *Dropouts* can be caused by noise, propagation anomalies, or system malfunctions. This loss of signal will produce a data transmission error that may be stored in the received file if proper error detection and correction measures are not observed by the communications software. (**2**) A failure to properly read a binary character from data storage. The dropout may be caused either by a defect in the storage medium or by a malfunction of the read mechanism. (**3**) In magnetic recordable media (tape, disk, card, or drum systems), a recorded signal level whose amplitude is less than a specified percentage of a reference signal.

dropped call In cellular telephony, a condition in which neither the calling party nor the called party initiates a disconnect; however, the call becomes disconnected. This may be due to excessive noise or fading in the signal path or an equipment malfunction.

DRP **(1)** An abbreviation of <u>D</u>ECnet <u>R</u>outing <u>P</u>rotocol. **(2)** An abbreviation of <u>D</u>istinctive <u>R</u>ing <u>P</u>attern. See also *coded ringing* and *distinctive ring*. **(3)** An abbreviation of <u>D</u>irectional <u>R</u>adiated <u>P</u>ower.

dry loop In telephony, a circuit that carries only a low-level signal and no dc loop current or ringing current. The lack of high currents (greater than 10 to 30 mA) does not prevent or remove any oxides that may form on wire junctions. Therefore, it is necessary that all terminations be "gas tight" such as soldering, wirewrap, or insulation displacement connectors. See also *wet loop*.

DS **(1)** An abbreviation of <u>D</u>igital <u>S</u>ervice. **(2)** An abbreviation of <u>D</u>igital <u>S</u>ignal. **(3)** An abbreviation of the Danish Board for Standardization; that is, the <u>D</u>ansk <u>S</u>tandardiseringsrad. **(4)** An abbreviation of <u>D</u>raft <u>S</u>tandard.

DS-n An abbreviation of <u>D</u>igital <u>S</u>ignal level "<u>*n*</u>." Refers to the PCM digital encoding and multiplexing of voice and data signals for transmission over digital transmission systems using T-carrier systems (e.g., T-1 and T-3).

Originally, the telephone operating companies used these digital communications methods as long-distance interconnects between switching offices. End users, however, are now using the circuits as the basis of private networks.

DIGITAL SIGNAL CHANNEL HIERARCHIES

Level	North American Line Rate (b/s)	Description	Japanese Line Rate (b/s)	Description	ITU (CEPT) Line Rate (b/s)	Description
DS-0	64 000	The basic one way digitally encoded voice channel in telecommunications networks. It consists of an 8-bit PCM stream at a sampling rate of 8 kHz. When used for voice transmission, the 8 bits are encoded using μ-255 coding.	64 000	The basic one-way digitally encoded voice channel in telecommunications networks. It consists of an 8-bit PCM stream at a sampling rate of 8 kHz. When used for voice transmission, the 8 bits are encoded using μ-255 coding.	64 000	The basic one-way digitally encoded voice channel in telecommunications networks. It consists of an 8-bit PCM stream at a sampling rate of 8 kHz. When used for voice transmission, the 8 bits are encoded using A-law coding.
DS-0A DS-0B	64 000	20 2.4 kbps, 10 4.8 kbps, 5 9.67 kbps, or a 56 kbps channels multiplexed into one DS-0 circuit.				
DS-1	1 544 000	24 DS-0 channels plus one framing bit multiplexed together to form a single 193 bit frame. (T-1 carrier)	1 544 000	24 DS-0 channels plus one framing bit multiplexed together to form a single 193 bitframe. (T-1 carrier)	2 048 000	30 DS-0 channels plus synchronization and framing combined to form a single frame. (E-1 carrier)
DS-1C	3 152 000	44 or 48 voice channels on a T-1C carrier.				
DS-2	6 312 000	96 DS-0, 4 DS-1 streams, or equivalent, plus synchronization and overhead bits multiplexed into a single serial stream. (T-2 carrier)	6 312 000	96 DS-0, 4 DS-1 streams, or equivalent, plus synchronization and overhead bits multiplexed into a single serial stream. (T-2 carrier)	8 448 000	120 DS-0, 4 DS-1 streams, or equivalent, plus synchronization and overhead bits multiplexed into a single serial stream. (E-2 carrier)
DS-3	44 736 000	672 DS-0, 7 DS-2 streams, or equivalent, plus synchronization and overhead bits multiplexed into a single serial stream. The rate is frequently called 45 Mbps. (T-3 carrier)	32 064 000	480 DS-0, 5 DS-2 streams, or equivalent, plus synchronization and overhead bits multiplexed into a single serial stream.	34 368 000	480 DS-0, 4 DS-2 streams, or equivalent, plus synchronization and overhead bits multiplexed into a single serial stream. (E-3 carrier)
DS-4	274 176 000	4032 DS-0 channels or equivalent with appropriate overhead.	97 728 000	1440 DS-0, 3 DS-3 streams, or equivalent, plus synchronization and overhead bits multiplexed into a single serial stream.	139 268 000	1920 DS-0, 6 DS-3 streams, or equivalent, plus synchronization and overhead bits multiplexed into a single serial stream.
DS-5	400 352 000	5760 DS-0 channels or equivalent with appropriate overhead.	400 352 000	5760 DS-0, 4 DS-4 streams, or equivalent, plus synchronization and overhead bits multiplexed into a single serial stream.	565 148 000	7680 DS-0, 4 DS-4 streams, or equivalent, plus synchronization and overhead bits multiplexed into a single serial stream.

The following diagrams depict the multiplexing of up to 672 64-kbps independent voice channels into a single 45-Mbps data stream. See also *E-n, OC-n, SDH, SONET, SDH, STM-n, STS-n,* and *T-n*.

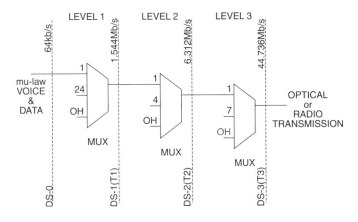

DSA (1) An abbreviation of <u>D</u>irectory <u>S</u>ervice <u>A</u>rea. In telephony, a term used to describe the calling area covered by a directory service. **(2)** An abbreviation of <u>D</u>irectory <u>S</u>ystem <u>A</u>gent. In the ITU X.500 Directory Services model, software that provides services for accessing, using, and, possibly, for updating a directory information base (DIB) or tree (DIT), generally for a single organization. **(3)** An abbreviation of <u>D</u>istributed <u>S</u>ystems <u>A</u>rchitecture. An OSI-compliant architecture from Honeywell. **(4)** An abbreviation of <u>D</u>ial <u>S</u>ervice <u>A</u>ssistance.

DSA board A local dial office switchboard at which the following call types are handled:

- Assistance calls
- Intercepted calls
- Calls from miscellaneous lines and trunks
- Calls from certain toll trunks

DSB An abbreviation of <u>D</u>ouble <u>S</u>ide<u>B</u>and.

DSB-SC An abbreviation of <u>D</u>ouble <u>S</u>ide<u>B</u>and-<u>S</u>uppressed <u>C</u>arrier.

DSB board A switchboard of a dial system for completing incoming calls received from manual offices.

DSC (1) An abbreviation of <u>D</u>ata <u>S</u>tream <u>C</u>ompatibility. In IBM's Systems Network Architecture (SNA), a basic, barebones printing mode. **(2)** An abbreviation of <u>D</u>igital <u>S</u>elective <u>C</u>alling.

DSE An abbreviation of <u>D</u>igital <u>S</u>witching <u>E</u>xchange.

DSI An abbreviation of <u>D</u>igital <u>S</u>peech <u>I</u>nterpolation.

DSN An abbreviation of <u>D</u>efense <u>S</u>witched <u>N</u>etwork.

DSP (1) An abbreviation of <u>D</u>igital <u>S</u>ignal <u>P</u>rocessor. **(2)** An abbreviation of <u>D</u>omain <u>S</u>pecific <u>P</u>art. **(3)** An abbreviation of <u>D</u>isplay <u>S</u>ystems <u>P</u>rotocol. An IBM protocol that allows 3270 control units, printers, terminals, and so on to interface with a Public Data Network (PDN).

DSR (1) An abbreviation of <u>D</u>ata <u>S</u>et <u>R</u>eady. **(2)** An abbreviation of <u>D</u>ata <u>S</u>ignaling <u>R</u>ate. See *data rate*.

DSS An abbreviation of <u>D</u>igital <u>S</u>ignature <u>S</u>tandard.

DSSS An abbreviation of <u>D</u>irect <u>S</u>equence <u>S</u>pread <u>S</u>pectrum.

DSTN An abbreviation of <u>D</u>ual scan <u>S</u>uper <u>T</u>wist <u>N</u>ematic. *DSTN* is the dual scan, passive matrix, color LCD display technology used in many laptop computers. See also *LCD* (*liquid crystal display*) and *TFT* (*thin film transistor*).

DSU An abbreviation of <u>D</u>ata <u>S</u>ervice <u>U</u>nit.

DSVD An abbreviation of <u>D</u>igital <u>S</u>imultaneous <u>V</u>oice and <u>D</u>ata.

DSX1/3 An abbreviation of <u>D</u>igital <u>S</u>ignal <u>C</u>ross-Connect between levels <u>1</u> and <u>3</u>. In digital communications, it specifies the interfaces for connecting DS-1 and DS-3 signals (i.e., connecting T-1 and T-3 lines).

DTE (1) An abbreviation of <u>D</u>ata <u>T</u>erminal <u>E</u>quipment. Any entity that generates information to be transferred over a transmission facility, i.e.,

- An end instrument that encodes user information into signals for transmission or decodes the received signals into user-readable information.
- The functional unit of a data station that serves as a data source and/or a data sink and attends to the data communication control function in accordance with link protocols.

The *data terminal equipment* (*DTE*) may be a single piece of equipment (including PCs, dumb terminals, or mainframe computers) or an interconnected subsystem of multiple pieces of equipment that perform all the required functions necessary for users to communicate. A user interacts with the DTE, and the DTE interacts with the data circuit communications equipment (DCE). **(2)** An abbreviation of <u>D</u>efense <u>T</u>echnology <u>E</u>nterprise.

DTE clear signal A call control signal sent by data terminal equipment (DTE) to initiate clearing, i.e., the restoration of the link to the idle state.

DTE waiting signal A call control signal that indicates that the data terminal equipment (DTE) is waiting for a call control signal from the data circuit terminating equipment (DCE).

DTF An abbreviation of <u>D</u>ial <u>T</u>one <u>F</u>irst payphone. Other methods of payphone operation include *coin first* (*CF*) and *post pay* (*PP*).

DTMF An abbreviation of <u>D</u>ual <u>T</u>one <u>M</u>ulti<u>F</u>requency. A telephone signaling method (developed at Bell Labs) from the subscriber equipment to the CO employing the transmission of two frequencies simultaneously. The frequency pairs are chosen from two sets of four frequencies: one from a high band set, and the other from a low-band set. This pair of frequencies can represent any of 16 symbols, i.e., the digits 0–9, A, B, C, D, #, and *. Also called *Digitone* (*Northern Telecom*), *multifrequency pulsing, multifrequency signaling, pushbutton dialing, Tel-Touch®, Touch Calling* (*GTE*), *and Touch-Tone* (*AT&T*).

Low Group	High Group			
	1209 Hz	1336 Hz	1477 Hz	1633 Hz
697 Hz	1	2	3	A
770 Hz	4	5	6	B
852 Hz	7	8	9	C
941 Hz	*	0	#	D

DTMF dialing A method of signaling to the telephone company switching equipment using pairs of tones. These tones can be presented to the line for as little as 30 ms on most networks (although most specifications recommend a minimum of 50 ms). This represents a significant reduction in signaling time when compared to pulse dialing (which requires hundreds of ms per digit).

DTR (1) An abbreviation of <u>D</u>ata <u>T</u>erminal <u>R</u>eady (RS-232 signal CD, RS-449 signal TR, and ITU-T signal 108.2). The DTR signal is generated by the DTE and is used to prepare the DCE to be connected to the communication channel (and maintains an established connection). **(2)** An abbreviation of <u>D</u>ata <u>T</u>ransfer <u>R</u>ate.

DTS (1) An abbreviation of Digital Termination Service. (2) An abbreviation of Digital Tandem Switch.

DTSY An abbreviation of Digital Transmission SYstem.

DTU (1) An abbreviation of Data Terminal Unit. (2) An abbreviation of Data Terminating Unit. (3) An abbreviation of Data Transfer Unit. (4) An abbreviation of Display Terminal Unit.

DU (1) An abbreviation of Dimensioning Unit. (2) An abbreviation of Display Unit.

DUA An abbreviation of Directory User Agent. In the ITU X.500 Directory Services model, a program that provides access to the directory services. It mediates between an end-user or client program and a directory system agent (DSA), which provides the requested services.

dual access (1) In telephony, the connection of a user to two switching centers by separate access lines using a single message routing indicator or telephone number. (2) In satellite communications, the transmission of two (or more) carriers simultaneously through a single repeater.

dual attachment concentrator (DAC) In a Fiber Distributed Data Interface (FDDI) network architecture, a concentrator used to attach single attachment stations or station clusters to both FDDI rings.

dual attachment station (DAS) In a Fiber Distributed Data Interface (FDDI) network architecture, a station, or node, that is connected physically to *both* the primary and secondary rings. A station can be connected directly to the ring through a port on the *DAS*. See also *single attachment station (SAS)*, which must be attached to a concentrator.

dual diversity The simultaneous combining of (or selection from) two (or more) independently routed signals, so that the resultant signal can be detected through the use of space, frequency, angle, time, or polarization characteristics.

dual homing In networking, a configuration in which a node is connected to the network through more than one physical link. In telephony, the connection is served by either of two switching centers. If one link fails, the station can still communicate via the other link. A single directory number or a single routing indicator is used.

dual in line package (DIP) An electronic package with a rectangular housing and a row of pins along each of two opposite sides. The package is typically 0.3, 0.4, or 0.6 inches wide and between 1 and 2 inches in length. *DIP* packages are used for integrated circuits (ICs) and for discrete components, such as resistors or toggle switches.

dual in line package switch See *DIP switch*.

dual line registration The ability of a single cellular telephone to have two registered telephone numbers. This allows the device to have two home areas, perhaps San Diego and New York. Also called *dual nam*.

dual mode In cellular telephones, the capability of the device to operate in either analog and digital environments. However, there are several digital standards, so the digital capabilities of one specific device may not work in the area desired.

dual tone multifrequency See *DTMF*.

dual use access line A user line normally used for analog voice communication but conditioned for use as a digital transmission circuit.

duct (1) A synonym for a cable conduit that may be buried directly in the Earth or encased in concrete. It is used to enclose power or communications cables.

When used to deter rodent attack, it should have an outside diameter of at least 6 cm (2.25"). (2) See *atmospheric duct*.

ducting The confinement and propagation of radio waves within an atmospheric duct.

dumb terminal A terminal that does not have the ability to process data, does not use a transmission control protocol, and sends and/or receives data sequentially (one character at a time). Typically, a *dumb terminal* displays only alphanumeric information and responds to only the most basic screen control codes. It generally does not have any graphic capability.

dummy load A dissipative but nonradiating substitute device having impedance characteristics and power-handling capability simulating those of the substituted device. It is designed to absorb all incident energy, generally converting it to thermal energy.

Examples include:

- A *dummy load* is used while testing a transmitter when no RF radiation is desired; in this case the device is also called a *dummy antenna*.
- A *dummy load* is used to test a power source when the actual load is unavailable or when the overload conditions are to be tested.

DUN An acronym from Dial-Up Network(ing).

duobinary A waveform encoding technique in which the logical 0 state is represented by a 0 signaling element (0 volts or amps). Logical 1 is represented in one of two ways, (1) If the number of previous "1s" is even, the waveform is positive and (2) if the number of previous "1s" is odd, the waveform is negative.

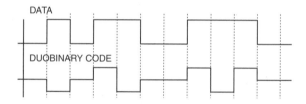

Duobinary signals require less bandwidth than nonreturn to zero (NRZ) encoded signals. They also permit the detection of some errors without the addition of error-checking bits. See also *waveform codes*.

duplex *Simultaneous*, independent transmission capability in both the forward and reverse directions on a communication circuit. *Duplex* operation may apply to the primary information signal or to the control channel. A normal voice telephone connection is an example of *duplex* communication. Also called *full duplex*. See also *half-duplex* and *simplex*.

duplex cable A cable that contains two isolated signal conductors. The conductors may be either electrical (e.g., two coaxial cables) or optical (two optical fibers).

duplex circuit A circuit that permits transmission in both directions.

For simultaneous two-way transmission, see *full-duplex* circuit.

duplexer In radar systems, a device that isolates the receiver from the transmitter while permitting them to share a common antenna. To function correctly, a duplexer must:

- Be designed for operation in both the frequency band used by the receiver and transmitter.
- Be capable of handling the output power of the transmitter.
- Provide adequate rejection of transmitter "noise" occurring in the receive frequency band.
- Provide sufficient isolation to prevent receiver desensitization.

See also *hybrid*.

DUT An abbreviation of <u>D</u>evice <u>U</u>nder <u>T</u>est.

duty cycle (**1**) In a periodic phenomenon, the ratio of on time (event duration) to total time (period). *Duty cycle* is generally expressed as a percent or a decimal fraction. For example, the duty cycle of a pulse train in which the pulse duration is 1 ms and the pulse period is 10 ms is 0.1 or 10%. (**2**) The ratio of the sum of all pulse durations during a specified period of continuous operation to the total specified period of operation.

DUV (**1**) An abbreviation of <u>D</u>ata <u>U</u>nder <u>V</u>oice. (**2**) An abbreviation of <u>D</u>eep <u>U</u>ltra<u>V</u>iolet.

DVC An abbreviation of <u>D</u>igital <u>V</u>oice <u>C</u>ommunications.

DVD An abbreviation of <u>D</u>igital <u>V</u>ersatile <u>D</u>isk. A standard that allows the storage of up to 135 minutes of movie-playing time on a standard 5.25" CD.

DVI An abbreviation of <u>D</u>igital <u>V</u>ideo <u>I</u>nteractive.

DVM An abbreviation of <u>D</u>igital <u>V</u>olt-<u>M</u>eter.

DVMRP An abbreviation of <u>D</u>istance <u>V</u>ector <u>M</u>ulticast <u>R</u>outing <u>P</u>rotocol.

DVP An abbreviation of <u>D</u>igital <u>V</u>oice <u>P</u>rivacy.

DVX An abbreviation of <u>D</u>igital <u>V</u>oice e<u>X</u>change.

dwell time The period during which a dynamic process remains stable or is halted so that another process may occur.

DWIM An acronym from <u>D</u>o <u>W</u>hat <u>I</u> <u>M</u>ean, i.e., guess at my poorly worded instructions.

DX (**1**) In ham radio, indicates long-distance communications. (**2**) An abbreviation of <u>D</u>uple<u>X</u>.

DX signaling An acronym from *direct current signaling.*

DX signaling unit A duplex signaling unit that repeats "E" and "M" lead signals into a cable pair via "A" and "B" leads. These signals are transmitted on the same cable pair that transmit the message.

DXF (**1**) An abbreviation of <u>D</u>rawing e<u>X</u>change <u>F</u>ormat. A format used by many computer aided design (CAD) programs including AutoCAD. The usual extension for files in this format is .DXF. (**2**) An abbreviation of <u>D</u>ata e<u>X</u>change <u>F</u>ormat.

dynamic addressing In an AppleTalk network, a patented strategy by which nodes automatically pick unique addresses each time it is initialized. A new node keeps trying addresses until it finds one that is not already claimed by another node. That is,

- The new node selects a valid address at random and sends an inquire control packet to that address.
- If the address belongs to an existing node, the new node responds with an acknowledge control packet. The new node then selects another address at random and repeats the process.
- If the address does not belong to an old node, the new node uses it as its address.

Also called *dynamic node addressing.*

dynamic bandwidth allocation The capability of a multiplex system to subdivide the capacity of a high-speed link into several lower speed links, each with only enough capacity to supply the needs of the signal source. As sources terminate their connection, their portion of the overall capacity is returned to the "capacity pool." When a new source requests a link, capacity is taken from the pool and assigned to the new link. Also called *dynamic capacity allocation.*

dynamic configuration In networking, a system capability in which the file server can allocate memory as needed, subject to availability, while the network is running. Dynamic reconfiguration enables the server to allocate more resources (such as buffers, tables, and so on) as necessary in order to avoid congestion or overload on the network.

Dynamic Data Exchange (DDE) One of several techniques that enables application-to-application communications. It is implemented in several operating systems, including Microsoft Windows, Macintosh System 7, and IBM's OS/2. When two or more programs that support *DDE* are running at the same time, they can exchange data and commands by means of conversations. (A *DDE* conversation is a two-way connection between applications.) *DDE* is used for low-level communications that do not need user intervention.

DDE has been superseded by a more complex but more capable mechanism known as *Object Linking and Embedding* (*OLE*).

Dynamic Link Library (DLL) A precompiled collection of executable functions that are called directly from application programs. Instead of linking the code for called *DLL* functions into an application program when it is compiled, the application program is given a pointer to the *DLL* at runtime. Because of this, the same *DLL* can be accessed by many application programs. *DLLs* are used extensively in Microsoft Windows, Windows NT, and in IBM's OS/2. *DLLs* have file name extensions of .DLL, .DRV, or .FON.

dynamic memory Generally abbreviated DRAM. A semiconductor memory device in which information is stored as a charge on a capacitor. The charge dissipates within several milliseconds; hence, the memory must be refreshed (recharged) every few milliseconds to prevent data loss. Contrast with *static memory.*

dynamic range (**1**) The ratio of the maximum and minimum values of a specified parameter for acceptable performance of a device or system expressed in decibels (dB). The specified parameter may be any variable related to the system in question. Typical parameters include power, amplitude, voltage, and current.

For example, the *dynamic range* of the amplitude response of an amplifier is

$$20 \cdot \log_{10}\left(\frac{A_{Max}}{A_{min}}\right) \text{dB}$$

where A_{Max} is the maximum amplitude before the amplifier begins to *saturate* and A_{min} is the minimum amplitude before system noise reduces the signal-to-noise ratio (SNR) to unacceptable levels. (**2**) In a transmission system, the ratio of the overload level (the maximum signal power that the system can tolerate without signal distortion) to the noise level of the system. The *dynamic range* of a transmission system is usually expressed in decibels (dB).

dynamic routing A process for selecting (and automatically adjusting for changes in network topology or traffic loading) the most appropriate path for a packet or datagram to traverse a network. At each interim node of a packet's journey across the network, the router decides on the most appropriate path for that packet or datagram to follow if multiple routes are available. Routing decisions are based on network status information gathered from around the Internet and passed from router to router through the use of routing information protocols. Two routing methods are the *distance vector algorithm* and the *link-state algorithm.*

In telephony, *dynamic routing* maximizes throughput and/or balances traffic on transmission channels and automatically routes traffic around congested, damaged, or destroyed switches and trunks. Also called *dynamic adaptive routing.*

dynamic variation A short-term departure (as opposed to long-term drift) from steady-state conditions or values of a specified characteristic of a device or system.

dynamically assigned socket (DAS) A unique socket value assigned, upon request, to a particular client in an AppleTalk Internetwork. (A socket is an entity through which a program or process, known as a socket client, communicates with a network or with another process.) Each AppleTalk socket is associated with an 8-bit value.

A process running on a node can request a *DAS*. Values between 128 and 254, inclusive, are dynamically allocated as requested. While this process is running, the assigned value cannot be reassigned for another socket. Values between 1 and 127, inclusive, are used for statically assigned sockets (SASs). (SASs are allocated for use by various low-level protocols, such as Name Binding Protocol [NBP] and Routing Table Maintenance Protocol [RTMP] in the AppleTalk protocol suite.) Values between 1 and 63 are used exclusively by Apple, while values between 64 and 127 can be used by whatever processes request the values.

dynamicizer A synonym for parallel-to-serial conversion.

DYP An abbreviation of <u>D</u>irectory <u>Y</u>ellow <u>P</u>ages.

E

e A mathematical constant related to many natural physical phenomena, its value is approximately 2.718 281 828 459 045. . . . The number is a transcendental value; that is, it cannot be expressed as the ratio of two whole numbers—the decimal value can be carried on an infinite number of places. The value of *e* can be calculated to any desired precision by evaluating the following expression with a sufficiently large *x*.

$$e \cong \left(1 + \frac{1}{x}\right)^x$$

The following graph shows the value of the preceding expression for successively larger values of *x*. Notice that as *x* gets larger, the expression gets closer to the *asymptote* (the value when *x* is infinitely large).

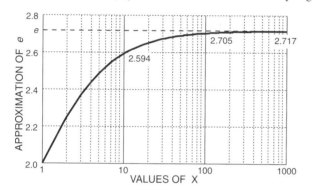

E (1) The SI prefix for *exa* or 10^{18}. (2) The symbol for the unit of traffic intensity erlang.

E & M signaling In telephony, a type of *out-of-band* signaling used on four-wire interoffice trunk circuits for signaling and supervision. It indicates the "hook-switch" status of each end of the circuit. The control signals are in separate paths from the voice paths; the *M-lead* transmits signaling, while the *E-lead* receives the signals. (The *M-lead* of each end of the communications loop is logically connected to the *E-lead* of the opposite end.)

On physical connections, the *E-lead* and the *M-lead* are separate wires. On carrier systems, the E and M signals are out-of-band tones (usually 3825 Hz).

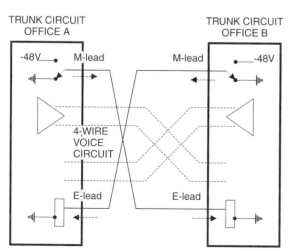

A helpful mnemonic for remembering which signal is the sending and receiving lead is to associate the control with Ear and Mouth. The Ear (and *E-lead*) rEceive, while the Mouth (and *M-lead*) transMit.

E bend A smooth turn in an RF waveguide in which the axis remains in a plane parallel to the direction of the electric E-field (transverse) polarization. Also called an *E plane bend*. See also *H bend*.

E layer See *E region* and *ionosphere*.

E lead A signaling lead associated with the receive communication path on telephone company interoffice trunks. See also *E & M signaling*.

e-mail An abbreviation of electronic mail.

E region That portion of the ionosphere existing between approximately 90 and 140 km above the surface of the Earth. As expected, the *E region* lies between the D and F regions.

Because the mean free path of ions at these relatively low altitudes is very short, continuous solar ultraviolet and X-ray radiation is required to maintain its existence. In addition to the regular E region, irregular cloudlike layers of ionization called sporadic E (or Es) produce partial reflections and scattering of frequencies to about 150 MHz. The causes of these ionized regions range from wind shear effects of the E region to disturbed magnetic conditions and the aurora, to the Earth's plasma stream instabilities. Also called the *Heaviside layer* and the *Kennelly-Heaviside layer*. See also *ionosphere*.

E-TDMA An abbreviation of Extended Time Division Multiple Access. A proposed standard for cellular communications developed by Hughes Network Systems. It squeezes as many as 15 times as many concurrent cellular connections as does the standard analog system.

E-0 through E-4 An ITU-T defined multichannel PCM digital transmission system. *E-1, E-2, E-3,* and *E-4* are used by the European CEPT (Conference of European Postal and Telecommunications Administration) carriers and in fact most of the world, the major exceptions being the United States, Canada, and Japan where T-carrier (T-1, T-2, and T-3) is used. It is often referred to as *MegaStream*, a reference to British Telecom's (BT's) name for its 2Mbit/s leased circuits.

ITU-T CARRIER DESCRIPTION

Rule	Line Rate (b/s)	Description
E-0	64 000	A basic one-way digitally encoded voice channel in telecommunications networks. The channel consists of an 8-bit PCM stream at a sampling rate of 8 kHz. When used for voice transmission, the 8 bits are encoded using A-law coding.
E-1	2 048 000	32 PCM channels are multiplexed together to form a single serial stream.
E-2	8 448 000	4 E-1 streams, or equivalent, plus synchronization and overhead bits multiplexed into a single serial stream.
E-3	34 368 000	4 E-2 streams, or equivalent, plus synchronization and overhead bits multiplexed into a single serial stream.
E-4	139 264 000	4 E-3 streams, or equivalent, plus synchronization and overhead bits multiplexed into a single serial stream.

The basic E-1 format consists of 30 64-kbps A-law encoded voice channels, a 64-kbps signaling channel, plus a 64-kbps channel for framing (synchronization) and maintenance. Multiple E-1 channels are combined in a rigid hierarchical structure as shown in the preceding chart. The line coding is HDB3 for all levels of the hierarchy. The following block diagram shows the hierarichal relationship of each of the multiplexers. The G.7xx specifications reference ITU-T standards, and OH represents all of the overhead bits required. See also *DS-n*, *OC-n*, *SDH*, *SONET*, *STM-n*, and *STS-n*.

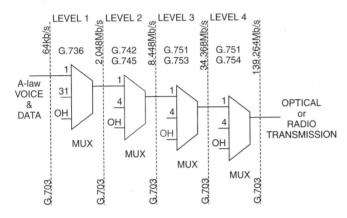

E.n recommendations An ITU-T series of specifications concerning overall network operation. See Appendix E for a complete list.

E/N ratio In the transmission of a pulse of an electromagnetic wave representing a bit, the ratio of the energy in each bit (E) to the noise energy density per hertz (N).

E is usually expressed in joules per bit (J/b), and N is usually expressed in watts per hertz (W/Hz). Hence, the units of the E/N ratio are

$$\frac{E}{N} = \frac{J/b}{W/Hz} = \frac{W \cdot s/b}{W/c/s} = \frac{c}{b}$$

Thus, the units of the *E/N ratio* are actually cycles per bit. If a cycle is a bit, however, then the *E/N ratio* is dimensionless.

EARN An acronym from the European Academic Research Network.

EAROM An acronym from Electrically Alterable Read Only Memory. Equivalent term to *EEPROM* but less frequently used.

earphone An electroacoustic transducer intended to be used in very close proximity of the ear.

Sometimes also called a *receiver*, especially in telephony; however, this term should generally be avoided.

Earth Centered Earth Fixed (ECEF) A Cartesian coordinate system used to describe three-dimensional locations on the Earth. The origin of these coordinate systems is at the Earth's center, and each axis is located at a fixed position relative to fixed locations on the Earth.

One such coordinate system is defined by the World Geodetic System 1984 (WGS 84) for use in the Global Positioning System (GPS). It defines the *ECEF* coordinates such that the *Z*-axis is collinear with the Earth's spin axis, the *X*-axis is through the intersection of the prime meridian and the equator, and the *Y*-axis is rotated 90 degrees east of the *X*-axis about the *Z*-axis. See also *Global Positioning System (GPS)* and *World Geodetic System 1984 (WGS 84)*.

Earth coverage (EC) In satellite communications, the coverage that occurs when the satellite to Earth (downlink) beam is sufficiently wide to cover all of the surface of the Earth visible to the satellite. That is, the footprint is nearly one-half of the surface of the Earth.

Earth electrode subsystem A network of electrically interconnected rods, plates, mats, grids, or other structures in intimate contact with the Earth, installed and connected, for the purpose of establishing a low-resistance electrical contact with Earth.

Earth ground The connection of an electrical system to Earth. This connection provides a path for lightning-induced transients and a zero voltage reference for the system. Sometimes called simply *Earth*. See also *ground*.

Earth station In satellite communications, a station located either on the Earth's surface or within the major portion of the Earth's atmosphere and intended for communication:

- With one or more stations located in space (satellites), or
- With one or more similar stations (Earth stations) by means of one or more reflecting satellites or other objects in space.

Also called *ground station*.

Earth terminal In a satellite link, one of the nonorbiting communications stations that receives, processes, and transmits signals between itself and a satellite. Earth terminals may be mobile, fixed, airborne, and waterborne. Also called a *satellite Earth terminal*.

EAS An abbreviation of Extended Area Service.

EASINET An acronym from the European Academic Supercomputer Initiative NETwork.

Easter eggs Slang for the surprise screen display, feature, or option a programmer has hidden within an application. In a Windows program, the function is sometimes accessed by clicking the "help-about" in the application's main menu with some combination of <shift>, <ctl>, and <alt>. Although it is fun to hunt for them, it is far more likely that a user will find them by accident or by being told about them. The function usually credits the team that developed the application. Sometimes also called a *gang screen*.

EAX An abbreviation of Electronic Automatic eXchange. An electronic central office. The term is used mainly by non-AT&T companies as AT&T uses ESS (from electronic switching system).

E_b/N_0 An expression of signal energy per bit per hertz of thermal noise.

EBCDIC An acronym from Extended Binary Coded Decimal Interchange Code (pronounced "EB-suh-dik" or "eb-see-dik"). *EBCDIC* is an IBM-developed 8-bit code rather than the 7-bit code defined by ASCII, and therefore has 256 characters in its alphabet. Because some characters in the ASCII code are not in the *EBCDIC* code (and vice versa), an exact mapping or translation between the codes is not possible.

EBCDIC is used primarily by IBM mainframe computers rather than PCs. A table of **EBCDIC** codes is provided on page 179.

EBONE An acronym from European BackbOne NEtwork.

EBS An abbreviation of Emergency Broadcast System.

EBU An abbreviation of European Broadcasting Union.

EC (1) An abbreviation of European Community. An organization of nations in Western Europe whose goals are to promote trade and reduce barriers between its members. The 12 members include Belgium, Denmark, France, Germany, Greece, Ireland, Italy, Luxembourg, the Netherlands, Portugal, Spain, and the United Kingdom. Also called the *common market* and the European Union. **(2)** An abbreviation of Exchange Carrier. **(3)** An abbreviation of Earth Coverage. **(4)** An abbreviation of Earth Curvature.

EBCDIC CODE COMBINATIONS

Least Significant Byte	Most Significant Byte																
	0	1	2	3	4	5	6	7	8	9	A	B	C	D	E	F	
0	NUL	DLE	DS		SP	&	-										0
1	SOH	DC1	SOS			/			a	j			A	J			1
2	STX	DC2	FS	SYN					b	k	s		B	K	S		2
3	ETX	TM							c	l	t		C	L	T		3
4	PF	RES	BYP	PN					d	m	u		D	M	U		4
5	HT	NL	LF	RS					e	n	v		E	N	V		5
6	LC	BS	ETB	UC					f	o	w		F	O	W		6
7	DEL	IL	ESC	EOT					g	p	x		G	P	X		7
8		CAN							h	q	y		H	Q	Y		8
9		EM							i	r	z		I	R	Z		9
A	SMM	CC	SM		¢	!		:									
B	VT	CU1	CU2	CU3	.	$	,	#									
C	FF	IFS		DC4	<	*	%	@									
D	CR	IGS	ENQ	NAK	(	)	-	'									
E	SO	IRS	ACK		+	;	>	=									
F	SI	IUS	BEL	SUB	|	¬	?	"									

ECB An abbreviation of <u>E</u>lectronic <u>C</u>ook<u>B</u>ook. An operating mode of the Data Encryption Standard (DES).

ECC An abbreviation of <u>E</u>rror <u>C</u>orrecting <u>C</u>ode. In digital communications, a term applied to any of several types of codes used to detect and correct errors that occur during transmission. Also called *Error Checking and Correction.*

eccentricity **(1)** A measure of the displacement of the center of the central element of a coaxial system to the center of the outer element. **(2)** The ratio of the difference between the average and minimum thickness to the average thickness of an annular element, generally expressed as a percent.

ECEF An abbreviation of <u>E</u>arth <u>C</u>entered <u>E</u>arth <u>F</u>ixed.

ECH An abbreviation of <u>E</u>nhanced <u>C</u>all <u>H</u>andling.

echo **(1)** A wave that has been reflected by a discontinuity in the propagation medium. **(2)** Any signal in a transmission circuit which has been reflected or otherwise returned to a point with sufficient energy and delay to be distinguished in some manner as a wave distinct from the transmitted wave. Because this signal is unwanted energy, it is a source of noise in communication circuits. *Echo* is frequently expressed in decibels (dB) relative to the incident wave.

The *echo* may be desirable (as in sonar, radar, or time domain reflectometry—TDR) or undesireable (as in multipath reception, communications transmission lines, or two- to four-wire hybrids). **(3)** In computing, to print or display characters under any of several conditions, that is,

- As they are entered from an input device, e.g., characters returned to the local data terminal equipment (DTE) by the remote host computer for the purpose of error checking. (remote *loopback testing* and *echoplex*),
- As instructions are executed, or
- As retransmitted characters are received from a remote terminal.

echo attenuation In a communications circuit (two- or four-wire) in which the forward and reverse directions of transmission can be separated, the attenuation of echo signals that return to the input of the circuit under consideration. *Echo attenuation* is expressed as the ratio of the transmitted power to the received echo power in decibels (dB).

echo canceller A device used in bidirectional communications devices such as hands-free telephones or modems to remove the reflected signals (echoes) in the transmission path.

During the initial setup of the communication path, the local station determines what the echo will be by transmitting a test signal. The variable transmission line simulator is then adjusted to approximate the echo. This is then subtracted from the real echo to cancel it. For example, the transmit signal from a local modem is reflected from any nonuniformity in the telephone circuit back to the local modem. This echo, being at the same carrier frequency as the remote modem, will interfere with reception. The *echo canceller* subtracts an appropriately modified portion of the transmit signal (approximately equal to the reflected signal) from the received signal before it is passed on to the demodulator as is shown in the following diagram. Also spelled *echo cancellor*. See also *echo suppresser*.

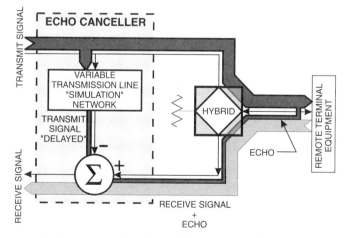

echo check An error-detection technique wherein data received are returned to the sending station for comparison with the original data. If differences are detected, an error in transmission is assumed, and appropriate corrective action is taken. Also called *loop check*. See also *echoplex*.

echo control The reduction of reflected signals in a transmission path. The transmission path may be constrained as in a wire or optical waveguide, or it may be unconstrained as would be found in the environment around a handsfree telephone.

echo return loss (ERL) In telephony, a measure of the relative annoyance to a user of an echo at various frequencies. It is the weighted return loss over the band of 500 to 2500 Hz. The weighting filter has a bandwidth from 560 Hz to 1965 Hz. Generally, *ERL* is expressed in decibels; for example, a typical central office has a return loss of 11 dB or better.

echo suppresser A device in a two-way communications circuit that allows signal transmission in only one direction at a time. An *echo suppresser* is undesirable in data transmission circuits.

An *echo suppresser* is a *switched attenuator* controlled by a device that compares both send and receive signal levels (*compactor*). When the send level is sufficiently higher than the receive level, it attenuates any received signal from the far end of a telephone line. Attenuation in this direction can be as much as 35 dB. Although designed to reduce the unwanted voice echo on the Public Switched Telephone Network (PSTN), it will attenuate a modem signal equally well. If allowed, this reduction in signal level would increase the number of errors in data transmission. The *echo suppresser* can be temporarily disabled (for the duration of the call) by injecting a single frequency tone of 2010 to 2240 Hz onto the line immediately after the connection is established. This is accomplished by the *tone disabler* portion of an echo suppresser. The *echo suppresser* in the figure below shows the transmit signal greater than the receive level. Hence, the attenuator is on, thereby reducing the total signal level to the receiving equipment. Also spelled *echo suppressor*. See also *echo canceller*.

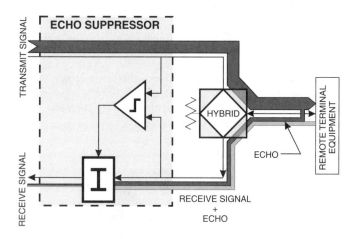

echo/echo reply In networking environments, a test packet that can be used to determine whether a target node is able to receive and acknowledge transmissions. In operation, a node transmits an *echo packet* signal to a target destination; then the sender waits for an acknowledgment. If the destination is connected and able to communicate, it responds with an *echo reply packet*. The method provides a simple mechanism for checking network connections.

echomail A system by which public and private data are transmitted via electronic bulletin board systems (BBSs). See also *FidoNet*.

echoplex A method of error detection whereby characters typed at a sending data terminal equipment (DTE) are retransmitted at the receiving DTE back to the sending DTE. There each character is displayed and visually checked by the operator. This provides the operator with positive feedback that the information entered into the communication path was received at the receiving DTE correctly.

For example if the operator types "download" on the sending DTE and sees "@own=oad" on the screen, it will be known that an error occurred and corrective measures should be taken. If on the other hand the word "download" appears correctly on the screen, the operator can be fairly certain that the information was transferred to the remote DTE correctly. (It is unlikely that two errors would cancel—one on the local to remote transmission and one on the echo from the remote back to the local.)

ECM (1) An abbreviation of Error Correction Mode. **(2)** An abbreviation of Electronic Counter Measures. **(3)** An abbreviation of European Common Market.

ECMA An abbreviation of European Computer Manufacturers Association. A western European trade organization that is a member of the International Standards Organization (ISO) but issues its own standards.

ECN An abbreviation of Explicit Congestion Notification.

ECP An abbreviation of Extended (or Enhanced) Capabilities Port.

ECPA An abbreviation of Electronic Computer Privacy Act.

ED An abbreviation of End Delimiter.

EDAC An acronym for Error Detection And Correction. See also *Hamming code*.

EDD An abbreviation of Envelope Delay Distortion.

EDGAR An acronym for Electronic Data Gathering, Analysis, and Retrieval.

edge emitting LED A light emitting diode (LED) that has a physical structure superficially resembling that of an injection laser diode (ILD) but is operated below the lasing threshold, consequently emitting incoherent light. *Edge emitting LEDs* have a smaller beam divergence than conventional surface-emitting LEDs and are therefore capable of launching more optical power into an optical fiber.

EDI An abbreviation of Electronic Data (or Document) Interchange.

EDIF An acronym for Electronic Design Interchange Format (pronounced "EE-diff").

EDO RAM An acronym for Extended Data Out Random Access Memory. A variant of the basic dynamic RAM (DRAM) that reduces memory access time by assuming each successive access will be at the next contiguous address. Based on this assumption, the device pre-fetches the data. If the next address to be read is not the expected address, the data are discarded and the correct location is read. The assumption reduces access times by up to 10% over standard DRAM.

EDP An abbreviation of Electronic Data Processing.

edutainment From education and entertainment. Software that combines game playing and learning activities. The software is typically multimedia in nature.

EEC An abbreviation of European Economic Community.

EEMA An acronym from European Electronic Mail Association.

EEPROM An acronym from Electrically Erasable Programmable Read Only Memory (pronounced "EE-EE prom"). An *EEPROM* is a memory device that functions primarily as a read only memory (ROM). Unlike a true ROM, an *EEPROM* can be programmed with new information any number of times. Unlike an EPROM, however, it can be erased electrically. Because it can be erased while it is the circuit, it can be used as NVRAM (nonvolatile RAM). Although *EEPROMs* will not lose stored information when operating power is removed, they do suffer from several drawbacks; that is, they are slow to erase and write, and they have a finite number of times they can be written to before they wear out (about 10 000 to 100 000 times). Also written E^2PROM. See also *EPROM* and *NVRAM*.

EES An abbreviation of Escrow Encryption Standard.

EF An abbreviation of Extended superFrame.

EFF An abbreviation of Electronic Frontier Foundation.

effective bandwidth The central part of the total bandwidth in a communications channel where the signal is strongest and cleanest. Generally, it is defined as that band of frequencies within which the total attenuation is less than 3 dB. (The 3-dB frequencies are the point where the signal power is reduced by half.)

effective Earth radius The radius the Earth would have to be if a propagated wave followed a straight line to the horizon rather than the actual curved path (caused by the vertical gradient of the atmospheric refractive index). For a standard atmosphere (with a uniform vertical gradient), the effective Earth radius is 4/3 that of the actual Earth radius (about 5300 mi or 8500 km).

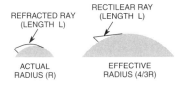

effective isotropic radiated power (EIRP or e.i.r.p.) Antenna systems and transmitters. The term *effective radiated isotropic power* is simply the real transmitter **power** times the **gain** of the transmitting antenna referred to an *isotropic radiator* (a radiator that transmits equally in all directions). *EIRP* and effective radiated power (*ERP*) are related by:

$$EIRP = ERP \times 1.64.$$

See also *effective radiated power* (*ERP* or *e.r.p.*).

effective mode volume In fiber optics, a measure of "optical volume" equal to the product of area and solid angle. It is proportional to the extent of the relative distribution of power among the modes in a multimode fiber. Mathematically, it is given by:

$$D_{nf} \cdot \sin(\phi_{ff})$$

where:

D_{nf} is the diameter of the near-field pattern (defined as the full width at half maximum and the radiation angle at half maximum intensity), and

ϕ_{ff} is radiation angle of the far-field pattern.

effective monopole radiated power (e.m.r.p.) The product of the power supplied to an antenna and its gain, relative to a short vertical antenna, in a given direction.

effective power In alternating current (ac) power circuits, the product of the root-mean-square (RMS) voltage and current (apparent power

multiplied by the power factor—the cosine of the phase angle between the voltage and current). The units of *effective power* are expressed in watts (W).

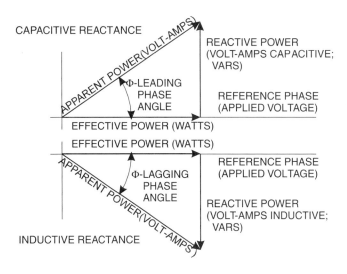

Apparent power is expressed in volt-amps (VA), not watts (W). *Effective power* is also called *true power*.

effective radiated power (ERP or e.r.p.) Antenna systems and transmitters. The term *effective radiated power* is simply the real transmitter **power** times the **gain** of the transmitting antenna referred to a reference antenna (generally a *standard dipole antenna* unless otherwise specified). If the direction is not specified, the direction of maximum gain is assumed. EIRP and ERP are related by:

$$EIRP = ERP \times 1.64.$$

Also called *equivalent radiated power*. See also *effective isotropic radiated power* (EIRP or *e.i.r.p.*).

effective rights In Novell's NetWare environment, the rights a user can exercise in a particular directory, file, or Directory tree created by the NetWare Directory Services (NDS). Effective rights are defined in various categories by:

- Directory rights in the file system—Directory effective rights are determined by any trustee assignments; if no such assignments exist, they are determined by the user's effective rights in the parent directory and the directory's Inherited Rights Mask (NetWare 3.x) or Maximum Rights Mask (NetWare 2.x).
- File rights in the file system—File effective rights are determined by any trustee assignments for the file. Otherwise, the user's effective rights in the directory apply.
- Object rights in the NDS—Object effective rights (NetWare 4.x only) define what a user is allowed to do with an object entry in the NDS Directory tree. These rights apply to the object as a single structure in the tree, not to the properties associated with the object or to the object itself.
- Property rights in the NDS—Property effective rights (NetWare 4.x only) define what kind of access a user has to the information associated with an object.

Effective rights for NDS objects and properties are determined by:

- Inherited rights associated with the object or property, taking into account any Inherited Rights Filters (IRFs) that apply.
- Trustee assignments associated with a user or group.
- Applicable security restrictions.

effective sound pressure The root-mean-square (RMS) value of instantaneous sound pressure over a time period sufficiently long that small changes in the time period do not affect the RMS value. See also *sound pressure*.

effective speed of transmission See *effective transmission rate*.

effective throughput The number of data bits transmitted within a specified time. This is in contrast to ordinary, or simple, throughput, in which the total number of bits transmitted (both data and administrative) are considered.

effective transmission rate The rate at which information is processed by a transmission facility. It is calculated as the measured number of information units (e.g., bits, bytes, characters, blocks, or frames) transmitted during a specified time interval divided by the specified time interval. The *effective transmission rate* is usually normalized and expressed as a number of information units per unit time, e.g., bits per second (b/s), bytes per second (B/s) or characters per second (cps). Also called *average rate of transmission* and *effective speed of transmission*. See also *transfer rate of information bits (TRIB)*.

effective value A synonym for *root-mean-square (RMS) value*.

effective wavelength The wavelength of a monochromatic beam that has the same depth of penetration in a particular medium as an ordinary light beam.

efficiency factor In data communications, the ratio of the time to transfer a message at a specified transmission rate across a perfect (errorless) channel to the actual time required to receive the same message at a specified maximum error rate.

EFM An abbreviation of Eight to Fourteen Modulation. A coding system used in optical compact disk systems in which 8 data bits along with error-correction information are mapped into a 14-bit block code. To this 14-bit word are added 3 merging bits and synchronization.

The error-correction scheme used is the cross-interleave Reed-Solomon code (CIRC).

EFNet An acronym from Eris-free Net. Eris is from eris.berkeley.edu. It is a major Internet Relay Chat (IRC) network.

EFOC An acronym from European Fiber Optic Communications/Local Area Network.

EFS (1) An abbreviation of Error-Free Second. One second of transmission without errors. Either the total or average number of error-free seconds can be used as an index of transmission quality. (2) An abbreviation of End Frame Sequence. The last field in a token ring data packet.

EFT (1) An abbreviation of Electronic Funds Transfer. (2) An abbreviation of Electrical Fast Transient.

EFTA An abbreviation of European Free Trade Association.

EGP An abbreviation of Exterior Gateway Protocol.

EHF An abbreviation of Extremely High Frequency. See *frequency band*.

EHP An abbreviation of Electron Hole Pair. The charge carriers in semiconductors that are responsible for the flow of current when the material is placed in an electric field. Electron-hole pairs may be generated by the application of energy to the material, e.g., thermal or optical.

EHz The SI notation for 10^{18} Hz.

EIA An abbreviation of Electronics Industry Association. Formerly called the *Radio Electronics Television Manufacturers Association (RETMA)*. In 1988 a new organization, the Telecommunications Industry Association (TIA), was formed by a merger of the U.S. Telecommunications Suppliers Association (USTSA) and the EIA's Telecommunications and Information Group (EIA/TIG). Following the merger, created standards have been referred to as TIA/EIA-nnn standards (nnn is the numeric reference ID value).

EIA Class IVa fiber A synonym for *dispersion-unshifted fiber*.

EIA Class IVb fiber A synonym for *dispersion-shifted fiber*.

EIA interface Any of a number of equipment interfaces compliant with voluntary industry standards developed by the Electronic Industries Association (EIA) to define interface parameters. The standards are designated EIA-nnn. Some of the EIA interface standards have been adopted by the federal government as Federal Standards or Federal Information Processing Standards (FIPS-nnn). See also *EIA* and specific *EIA-nnn* entries.

EIA-232 An ANSI/EIA specification (standard) describing the physical, electrical, and control characteristics of a serial communication interface, specifically between data terminal equipment (DTE) and data communications equipment (DCE). Other similar/related standards include the ITU-T V.24, V.28 and ISO 2110, ISO 4902 recommendations.

As is shown in the figure, the EIA-232 signal is defined only for the range −3 to −15 V for a marking condition (logical 1) and +3 to +5 V for the spacing condition (logical 0). (Early versions of the RS-232 standard allowed the voltage excursion to extend to +25 V and −25 V.) The range between −3 to +3 V is the transition region and does not uniquely define either state. The *EIA-232* interface uses the DB25 P connector as the DTE interface connector and the DB25 S connector as the DCE interface connector. An alternative 9-pin connector

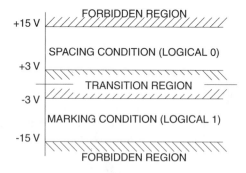

(DE9 commonly called the DB9) is provided for in the EIA-574 specification. The signal definitions, abbreviation, and pin assignments are listed in the following table. *Note:* the transmit direction is from the DTE to the DCE. Also called *EIA/TIA-232* and *RS-232* standard. See also *DB-9, DB-25*.

EIA-366 A specification for automatic dialing equipment. For example, a business might use such equipment to dial remote terminals to allow them to transmit accumulated daily information.

Automatic dialing equipment must not only be able to go off-hook and dial, but must also be able to recognize and properly respond to call progress tones (dial tone, busy tone, etc.).

EIA-232 PIN ASSIGNMENTS

Signal Description	Signal Name	Circuit Name	Connector DB25	Connector DE9
Shield	SHLD	AA	1	
Transmit data	TXD	BA >	2	3
Receive data	RXD	< BB	3	2
Request to send	RTS	CA >	4	7
Clear to send	CTS	< CB	5	8
DCE ready	DSR	< CC	6	6
Signal ground	SGND	AB	7	5
Data carrier detect	DCD	< CF	8	1
Reserved			9	
Reserved			10	
Unassigned/spare			11	
Secondary carrier detect	DCD2	< SCF	12	
Secondary signal rate selector		< CI		
Secondary clear to send	CTS2	< SCB	13	
Secondary transmit data	TXD2	SBA >	14	
Transmit clock (DCE source)	TCLK_c	< DB	15	
Secondary receive data	RXD2	< SBB	16	
Receive clock	RCLK	< DD	17	
Local loopback	LL	LL	18	
Secondary request to send	RTS2	SCA >	19	
Data terminal ready	DTR	CD >	20	4
Remote loopback or	RL	RL	21	
Signal quality detector	SQ	< CG		
Ring indicator	RI	< CE	22	9
Data signal rate selector (DCE)		CH >	23	
Data signal rate selector (DTE)		< CI		
Transmit clock and Country dependent	TCLK_t	DA >	24	
Test mode	TM	TM >	25	

< Indicates flow *from* DCE to DTE.
> Indicates flow *to* DCE from DTE.

EIA-422 A standard that defines a balanced transmission line for high-speed data transfers. The technique minimizes the effects of ground potential differences. Hence, the undefined transition region can be significantly reduced from that specified in the EIA-232 standard. That is, it is reduced from ±3V to ±0.2V.

BASIC EIA-422 SPECIFICATIONS

Parameter	Specification
Number of drivers & receivers allowed on a line	1 Driver 10 Receivers
Maximum cable length	4000 ft (1.2 km)
Maximum data rate	10 Mbps
Driver output voltage	±2 V to ±6V
Driver load	100 Ω
Driver short circuit current limit	150 mA to gnd
Driver output resistance—power off	60 kΩ
Maximum common mode voltage	$+6/-0.25$ V
Receiver input resistance	>4 kΩ
Receiver input sensitivity	±200 mV

The maximum cable length is determined by a number of factors including common mode noise, ground potential differences, shunt cable capacitance, and cable balance. For example, a 24-AWG twisted pair cable with 16 pF/ft shunt capacitance should perform, as is shown in the following figure (or better). In all cases, the cable must be terminated in the characteristic impedance (generally 100 Ohms).

The EIA-422 standard is fully compatible with the ITU V.11 / X.27 recommendations.

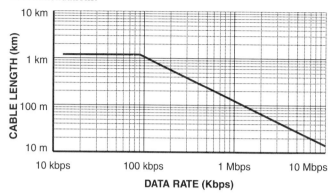

EIA-423 A data transmission standard that can operate in both the unbalanced EIA-232 and balanced EIA-422 environments. The standard specifies an unbalanced line driver; however, a common return path is provided for each direction—that is, the return wire is connected to ground at only the transmitter.

The transmitter drive level is specified at ±4V, which will allow it to transmit to EIA-232 receivers. The receiver transition band is ±0.2V, which means it will operate with an EIA-422 transmitter. The following chart relates cable length, data rate, and symbol rise time for

BASIC EIA-423 SPECIFICATIONS

Parameter	Specification
Number of drivers & receivers allowed on a line	1 Driver 10 Receivers
Maximum cable length	4000 ft (1.2 km)
Maximum data rate	100 kbps
Driver output voltage	±4 V to ±6V
Driver load	100 Ω
Driver short circuit current limit	150 mA to gnd
Driver output resistance—power off	60 kΩ
Maximum common mode voltage	$+6/-0.25$ V
Receiver input resistance	>4 kΩ
Receiver input sensitivity	±200 mV

24 AWG, 16 pF/ft shunt capacitance, twisted pair telephone cable. It does not include effects of common mode noise, ground potential differences, or other signal degradations.

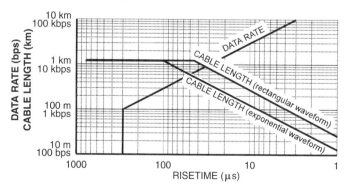

Examples: (1) A 64-kbps data rate requires a rise time of less than 5μs. This rise time can be supported on a 60-m (196 ft) cable if the rise time is exponential or a 105-m (343-ft) cable with rectangular rise times. (2) Given a cable length of 300 m (981 ft), the fastest data rate is about 30 kbps for rectangular waveforms, or only about 10 kbps with the more realistic exponential waveforms. The *EIA-423* standard is fully compatible with the ITU V.10 / X.26 recommendations.

EIA-449 Intended to be the successor to the EIA-232 standard, it provides improved speed, distance, and features over EIA-232. All signals use the EIA-423 transmission standard except those that may optionally use EIA-422 for higher speed links. The primary signals are specified on a 37-pin connector, and the secondary signals utilize a 9-pin connector. Because of the very large installed base of EIA-232 compatible equipment, it is unlikely that this standard will become the success its designers hoped for.

EIA-449 is approximately the same as the ITU-T V.36 recommendation. See also *EIA-530*.

EIA-470 A specification for telephone instruments in the United States. The specification defines ringer characteristics, dial pulse and tone generation, microphone and earphone sensitivity, cords, and so on.

EIA-485 A serial interface standard for multipoint lines utilizing balanced transmission lines, and unlike its predecessors (RS-232, RS-422, etc.), it uses tristate drivers. The third state (an off or high-impedance state) allows two active devices to communicate over shared wiring while other devices are idle.

BASIC EIA-485 SPECIFICATIONS

Parameter	Specification
Number of drivers & receivers allowed on line	30 Driver/Receivers pairs
Maximum cable length	4000 ft (1220 m)
Maximum data rate	10 Mbps
Driver output	± 1.5 V min.
Driver load	60 Ω min.
Driver short circuit current limit	150 mA to gnd 250mA to $-8/+12$ V
Driver output resistance—power on	120 kΩ
Driver output resistance—power off	120 kΩ
Maximum common mode voltage	$+12/-7$ V
Receiver input resistance	12 kΩ
Receiver input sensitivity	± 200 mV

EIA-530 Intended to replace the EIA-232 and EIA-449 standard. It is a balanced signal, synchronous or asynchronous 2-Mbps, 610-meter, serial interface specification that designates a 25-pin DB series connector.

EIA-530 PIN ASSIGNMENTS

Signal Description	Signal Name	Circuit Name	Connector (DB-25)
Shield	SHLD		1
Transmit data	TXD	BA	2
Receive data	RXD	BB	3
Request to send	RTS	CA	4
Clear to send	CTS	CB	5
DCE ready	DSR	CC	6
Signal ground	SGND	AB	7
Data carrier detect	DCD	CF	8
Receiver clock (DCE source)	RCLK	DD	9
Data carrier detect	DCD	CF	10
Transmit clock (DTE source)	TCLK$_T$	DA	11
Transmit clock (DCE source)	TCLK$_C$	DB	12
Clear to send	CTS	CB	13
Transmit data	TXD	BA	14
Transmit clock (DCE source)	TCLK$_C$	DB	15
Receive data	RXD	BB	16
Receive clock	RCLK	DD	17

EIA-530 PIN ASSIGNMENTS (*CONTINUED*)

Signal Description	Signal Name	Circuit Name	Connector (DB-25)
Local loopback	LL	LL	18
Request to send	RTS	CA	19
Data terminal ready	DTR	CD	20
Remote loopback	RL	RL	21
DCE ready	DSR	CC	22
Data terminal ready	DTR	CD	23
Transmit clock (DTE source)	TCLK$_T$	DA	24
Test mode	TM	TM	25

EIA-561 The EIA-232 specification, but defined for a DIN-8 connector.

EIA-568 A commercial building telecom cabling specification. It defines generic building wiring specifications that will support a variety of products and vendors. Its topics include:

- *Media*—a family of cable specifications, including unshielded twisted pair (UTP), shielded twisted pair (STP), coaxial cable, and optical fiber.

  See also *category n cable*.

- *Cable lengths, topology, and performance*—e.g., a maximum horizontal cable run of 90 meters, allowing 10 meters for hub and device attachment, resulting in a total cable length of 1000 meters.

- Interface (termination) standards; for example,

RJ-45	EIA/TIA568A	EIA/TIA568B
Pair 1 (BL/W - W/BL)	Pins 4-5	Pins 4-5
Pair 2 (W/O - O/W)	Pins 3-6	Pins 1-2
Pair 3 (W/G - G/W)	Pins 1-2	Pins 3-6
Pair 4 (W/BR - BR/W)	Pins 7-8	Pins 7-8

- Wiring practices.
- Hardware practices.
- Administration.

The standard was adopted as *FIPS Pub 174*.

EIA-569 A commercial building standard for pathways and spaces. It provides guidelines for the design/layout of telecommunications closets, equipment rooms, conduits, cable trays, and so on.

EIA-574 The EIA-232 specification but defined for a DE-9 connector (commonly called a DB-9 connector).

EIA-578 See *class 1*.

EIA-592 See *class 2.0*.

EIA-606 The telecommunications administration standard for commercial buildings that provides guidelines to the design (and identification) of two-level backbone cabling for both individual buildings and campuses.

EIB An abbreviation of Enterprise Information Base.

EIRP An abbreviation of Effective (or Equivalent) Isotropic Radiated Power, the combination of transmitted power and antenna gain.

EISA An acronym from Extended (or Enhanced) Industry Standard Architecture.

either party control A method of call termination whereby the originating station or the destination station may initiate the call disconnect sequence. For example, if station A originated the connection to station B, then either station A or station B could end the connection. See also *both party control, called party control,* and *calling party control.*

EIU An abbreviation of <u>E</u>thernet <u>I</u>nterface <u>U</u>nit.

elastic buffer (**1**) A buffer that has an adjustable capacity for data. Data are accepted into the buffer under one clock but delivered under another; hence, short-term differences (jitter) in either clock are tolerable. See also *FIFO.* (**2**) A buffer that introduces an adjustable delay of signals.

elbow connector A connector with a right angle in it, for use in corners or in other locations where the cabling needs to abruptly change direction.

electric field The effect produced by the existence of an electric charge (electrons, ions, or protons) in the volume of space or medium that surrounds it.

Only one *electric field* exists, even with a multiplicity of electric charges. Each of the charges contributes to any specified point in the field on the basis of superposition, i.e., the direction and magnitude of the charges as algebraically additive. An electric charge placed in an *electric field* has a force exerted on it.

electric wave filter See *filter.*

electrical distance The distance between two points expressed in terms of the *time* it takes electromagnetic radiation to traverse the distance in a vacuum. Frequently, the units are expressed in microseconds (μs); hence, 1 μs is equivalent to approximately 300 m (981 ft).

electrical length (**1**) In transmission media, the physical transmission path length multiplied by the ratio of the propagation time of a signal through the medium to the propagation time of an electromagnetic wave in free space over a distance equal to the physical length of the medium. Because the velocity of propagation is always slower in a physical medium than in free space, *electrical length* will always be longer than actual length. The velocity of propagation in both coaxial cables and optical fibers is approximately two-thirds that of free space. Consequently, the electrical length is approximately 1.5 times the physical length. (**2**) In transmission media, length expressed as a multiple of the wavelength of a periodic electromagnetic or electrical signal propagating within a medium. Using the example above, a two thirds velocity of propagation implies the wavelength is also two thirds that of free space and that the electrical length is 1.5 times the physical length. (**3**) In antennas, the effective length of an element, usually expressed in wavelengths. The electrical length is in general different from the physical length because of accidental or purposeful additions of reactive components. The addition of an appropriate reactive element (capacitive or inductive) will make the *electrical length* shorter or longer than the physical length. (**4**) Length expressed in angular units (i.e., radians or degrees). Radians are found by multiplying wavelengths by 2π and degrees by multiplying wavelengths by 360.

electrical noise Any unwanted electrical energy, other than crosstalk, in a transmission system.

electrically powered telephone A telephone in which the operating power is obtained either from an external power source, such as a battery located at the telephone (a local battery) or from a telephone central office (a common battery). See also *battery* (*2*).

electro-optic effect Any of the effects that occur when a wave in the optical portion of the electromagnetic spectrum interacts with matter (e.g., changes the refractive index of the material) under the influence of an electric field or the electric field itself. Two of the most important *electro-optic effects* are:

- *Kerr effect,* in which birefringence is induced or modified in a liquid, and the
- *Pockels effect,* in which birefringence is induced or modified in a solid.

The term *electro-optic* is often erroneously used as a synonym for *optoelectronic* (a device that converts optical to electric or electric to optical energy).

electro-optic modulator An optical device in which a signal-controlled device is used to modulate a beam of light. The control may modulate the phase, frequency, amplitude, or direction of the modulated beam. See also *electro-optic effect.*

electro-optics The field of study and the technology associated with those components, devices, and systems that utilize the interaction between electromagnetic (optical) and electric (electronic) states. The operation of electro-optic devices depends on modification of the refractive index of a material by electric fields such as is found in the Kerr and Pockels cells.

The term *electro-optic* is often erroneously used as a synonym for *optoelectronic* (a device that converts optical to electric or electric to optical energy).

electrode (**1**) An electrical conductor that permits the emitting, collection, or control of the movement of electrons or ions. (**2**) In an electrochemical cell, the site at which electrochemical processes occur.

electrode current The current passing to or from an electrode through an interelectrode space. The names cathode current, anode current, grid current, and so on, are used to identify the currents for specific electrodes.

electrode dissipation The power radiated in the form of heat by an electrode due to electron flow (current), ion bombardment, and thermal transfers from other electrodes or elements.

electrode voltage The voltage between the terminals of an electrode and the cathode (or other designated electrode).

electroluminescence The conversion of electrical energy into light by nonthermal mechanisms. For example, in a light-emitting diode (LED) the photon emission resulting from electron-hole recombination in the pn junction. See also *incandescence.*

electromagnet A device that exhibits magnetism only when a current flows through it. Most commonly, an *electromagnet* is just a coil of wire wrapped around an iron core.

electromagnetic compatibility (EMC) (**1**) The condition that prevails when a device or system operates as intended in a specified environment without causing or suffering unacceptable degradation due to unintentional *electromagnetic interference (EMI)* to or from other equipment in the same environment. (**2**) The ability of systems, equipment, and devices that utilize the electromagnetic spectrum to operate in their intended environments without suffering unacceptable degradation or causing unintentional degradation to other systems because of electromagnetic radiation. **(3)** A 1995 European Community (EC) directive-standard for control of radio emissions and tolerance of electrostatic discharge (ESD) and transients.

electromagnetic environment (EME) For a telecommunications system, the distribution of electromagnetic fields surrounding a specified site. The *electromagnetic environment* may be expressed in terms of the physical distribution of electric field strength (volts/meter), irradiance (watts/meter^2), or energy density (joules/meter^3).

electromagnetic interference (EMI) Any electromagnetic radiation (random or periodic) emanating from external sources that interrupts, obstructs, or otherwise degrades or limits the effective performance of electronic or electrical equipment. These sources include:

- *Artifacts*—such as other electronic equipment, e.g., spurious emissions or intermodulation products, motors, or lighting (particularly fluorescent lighting and some types of light dimmers),
- *Natural phenomena*—such as atmospheric, solar activity, or even astronomical radio sources, or
- *Intentional*—as in some forms of electronic warfare.

Compare *radio frequency interference (RFI)*.

electromagnetic noise An electromagnetic disturbance that is not sinusoidal in nature.

electromagnetic pulse (EMP) A broadband, high-intensity, short-duration burst of electromagnetic radiation from a nuclear explosion (or other means). It is caused by Compton-recoil electrons and photoelectrons from photons scattered in the materials of the nuclear device or in a surrounding medium. The resulting electric and magnetic fields may couple with electrical and/or electronic systems to produce damaging current and voltage surges.

In the case of a nuclear detonation, the *electromagnetic pulse* consists of a continuous frequency spectrum in which most of the energy is distributed throughout the lower frequencies, i.e., between 3 Hz and 30 kHz.

electromagnetic radiation (EMR) Radiation energy made up of oscillating electric and magnetic fields propagating at the speed of light (in the medium of propagation). *EMR* includes all entries in the electromagnetic spetrum chart from ULF radio through microwave and optical (infrared, visible, and ultraviolet) to X-rays and gamma radiation. See also *frequency spectrum*.

electromagnetic spectrum The range of all frequencies of electromagnetic radiation from zero to infinity, including what is commonly called radio waves, microwaves, infrared, visible light, ultraviolet light, X-rays, and gamma rays. The radio portion of the *electromagnetic spectrum* was, by custom and practice, formerly divided into alphabetically designated bands. This usage still prevails to some degree; however, the ITU formally recognizes 12 numbered bands, from 30 Hz to 3 THz. See also *frequency band* and *frequency spectrum*.

electromagnetic vulnerability (EMV) Those characteristics of a device or system that causes it to suffer a degradation sufficient to render it incapable of performing its designated function as a result of being exposed to a level of electromagnetic energy greater than a certain level. See also *electromagnetic interference (EMI)* and *electromagnetic pulse (EMP)*.

electromagnetic wave (EMW) A wave produced by the interaction of time-varying electric and magnetic fields.

electromechanical ringer The traditional bell or buzzer type ringer found in a telephone that indicates an incoming call. See also *electronic ringer*.

electromechanical switching In telephony, an automatic switching system in which the control functions are performed principally by electromechanical devices.

electromotive force (EMF) See *voltage*.

electronic bulletin board See *bulletin board system (BBS)*.

electronic commerce The sale of goods and services over network connections, e.g., the Internet, Prodigy, AOL, or private networks between individual companies. It includes the use of facsimile and electronic mail (e-mail) as well as direct computer-to-computer communication.

Electronic Computer Privacy Act (ECPA) A federal law that criminalizes unauthorized access to any computer system in which private communications are transmitted. Some provisions of the act include:

- Data in transit or temporary storage (less than 180 days) require a court order for access by federal agencies. Permanently stored or archived data still require only a subpoena for access, effectively bypassing the law.
- The unauthorized access of a computer system with the intent of reading private communications is prohibited.

Note: Within a business, your boss has the right to peruse your computer files with complete immunity.

electronic cookbook (ECB) One of the operating modes of the data encryption standard (DES).

electronic data interchange (EDI) The connection of an organization's computers to vendors', customers', and other partners' computers for the purpose of exchanging documents. This interconnection *can* reduce human error, paperwork, and transaction time. For example, purchase orders are not written and transferred by mail or fax; they are directly entered into one company's computer by another company.

EDI allows a business and its suppliers to electronically exchange purchase orders, invoices, and other documents. The goal is to allow one business to electronically:

- Accept purchase orders from its distribution businesses (which triggers shipping of requested materials from inventory).
- Update inventories.
- Exchange invoice and other documents.
- Issue purchase orders to suppliers for materials to replace sold inventories.

EDI is not without problems, not only legal, but also physical and political. For example,

- There is no signature on an electronic purchase order; does the exchange constitute a binding contract?
- International communication lines are not always as reliable as desired.
- Data formats are likely to be different between company computers.
- Billing cycles between companies differ.
- Foreign governments may not want direct computer-to-computer communications without monitoring. (Until recently, Japan did not allow direct connections with U.S. computers; they required that all data go through a government-controlled service bureau.)
- Country social customs differ.

Various *EDI* standards have been proposed. Some standards are general purpose, while others are concerned with only certain types of transfers or data. Some of the special purpose standards include:

- EDIFACT: EDI for Finance, Administration, Commerce, and Transport.
- EDIM[E]: EDI Message [Environment].
- EDIMS: EDI Messaging System, or EDI Message Store.
- EDIN: EDI Notification.
- EDIUA: EDI User Agent.
- EDIUser: a directory object in X.500.

Electronic Frontier Foundation (EFF) Founded in 1990, the *EFF* is a public advocacy organization dedicated to the extension of the traditions of privacy, civil liberties, and free information flow into the electronic media world.

In essence, the organization is attempting to extend the *common carrier principles* into all digital communications systems. These principles dictate that a carrier transport all speech (regardless of its potentially controversial content) and provide equitable, nondiscriminatory access.

electronic jamming The deliberate radiation, reradiation, or reflection of electromagnetic energy for the purpose of disrupting use by other (enemy) electronic equipment or systems.

electronic line of sight The path traversed by electromagnetic waves that is not subject to reflection or refraction.

electronic mail The generation, transmission, and display of documents and correspondence by electronic means.

The transmission of messages over a communications network for purposes analogous to mail delivered by the postal service. Used on local area networks (LANs), wide area networks (WANs), bulletin board systems, and dial-up services' electronic mail, it allows users to send and receive text messages and in some systems graphics, data files, or voice messages. A message may be sent to a single recipient, to several recipients, or broadcast to all users of a network. A delivered message (that is, one received by the intended receiver) may be viewed, saved, forwarded to one or more users, deleted, or responded to with an additional message. On the Internet, electronic mail client programs such as *Pine* or *Eudora* are used to receive, send, print, and delete mail. Although the details of an electronic mail system can vary, they all need to provide the following types of services:

- Terminal and/or node handling—interprets user requests and responds to those requests.
- File handling—stores messages as files in the appropriate mailbox(es).
- Local mail services—enables mail reception from and mail delivery to local users.
- Communications handling—enables message exchanges with other systems at remote locations.
- Mail transfer—makes it possible for a mail server to deliver electronic messages to another server and to receive electronic messages from it.
- Encryption (optional but, . . .)—provides a means to transfer private messages on public communications networks. It also provides a means to authenticate the message source (i.e., a digital signature).
- Multicast/broadcast (optional)—provides the capability to send the same message to multiple destinations with only one origination.

Also called *e-mail* or *email.*

electronic mail address The location to which electronic mail is sent. For the end user in Internet, an *e-mail address* is written as a sequence of names, separated by periods or other special characters, e.g., *name@place.domain.*

Electronic Mail Association (EMA) An association of developers and vendors of e-mail (electronic-mail) products.

electronic mail protocols In the past, most systems used proprietary protocols (i.e., each company used its own rules); hence, few could talk to each other. Now, most e-mail products support one or both of two standards:

- The Simple Mail Transfer Protocol (SMTP) from the TCP/IP protocol suite, or
- The protocols specified in ITU X.400.

electronic mail terrorism See *mail bomb.*

electronic mailbox In an e-mail system, a directory provided to store messages for a single user. Each e-mail user has a unique ID and a unique mailbox.

electronic ringer A device in a telephone that audibly announces incoming calls; however, instead of the device generating sounds by electromechanical means, it generates them by electronic means (an oscillator or music synthesizer). The sound is then "launched" by means of a speaker. Typical sounds include chirps, warbles, beeps, chimes, and emulations of electromechanical bells. Novelty ringers include cows mooing, ducks quacking, laughing, dogs barking, and so on. See also *electromechanical ringer.*

Electronic Security Number (ESN) A unique identification number associated with each cellular telephone manufactured. The *ESN* is "burned" into the device's programmable read only memory (PROM); therefore, it is not programmable as is the telephone number. Each time the cellular telephone is used, the *ESN* is transmitted to the Mobil Telephone Switching Office (MTSO) by means of dual-tone multifrequency (DTMF) signaling. The MTSO can determine if an *ESN* is valid or invalid; thus, specific numbers may be denied service.

electronic switching system (ESS) (1) An automatic telephone switching system based on the principles of time and/or space division multiplexing of digitized analog signals. An *ESS* digitizes analog signals from subscribers' loops and interconnects them by assigning the digitized signals to appropriate time slots or physical ports. Some switching systems use a combination of both time and space switching to interconnect the subscribers' signals. **(2)** A switching system with major elements constructed of semiconductor components. A semi-electronic switching system that has reed relays or crossbar matrices, as well as semiconductor components, is also considered to be an *ESS,* i.e., any switching system except an electromechanical relay or stepper-based system.

electronic tandem network (1) In telephony, a switching system used to interconnect class 5 offices in the same geographic area. **(2)** Two or more switching systems operating in parallel but as part of a single network.

electronic text A computer or network-accessible version of literary or reference work. The term *electronic text* (or *e-text*) is generally applied only to computerized *reproductions* of works of significance; e.g., Shakespeare's plays, *Compton's Online Encyclopedia,* the Bible, or federal regulations.

Although many *electronic text* documents are available as pure ASCII documents, many others are designed to be used with a search engine (a program called a parser) which allows a user to locate specific items: words, phrases, titles, and so on. One common format is defined by ISO standard 8879, the Standard Generalized Markup Language (SGML). Many *e-text* documents are copyrighted and cannot be accessed via a network unless prior arrangements with the host archive are made and an appropriate fee is paid. Others are in the public domain and are freely available. Project Gutenberg is attempting to put 10 000 ASCII readable *e-texts* onto the network by the year 2001.

electronvolt (eV) A measure of energy. The amount of kinetic energy an electron acquires (or loosens) when it moves through a field with a potential difference of 1 V in a vacuum. The value must be obtained experimentally and hence is not known exactly.

$$1 \ eV = 1.6021917 \times 10^{-19} \ J$$

An *electronvolt* is also associated with the wavelength of a photon by way of Planck's constant; that is,

$$1 \ eV = 2.179659 \times 10^{14} \ Hz = 1.2398541 \ \mu m$$

electrostatic charge A term applied to electric charges that are *not flowing* along a conductive path; that is, they are static.

electrostatic discharge (ESD) A potentially damaging pulse of current (discharge) into a conducting circuit or device from an electrostatic source. One such source is a person's body. The electrostatic charge may be the result of walking across a carpet or even sliding across a chair. When the conducting circuit happens to be a piece of electronics, the very real potential for damage arises.

elegant An entity that is efficient, small, and frequently innovative. An elegant computer program is one that is efficiently written, occupies the smallest amount of memory, and uses the smallest number of instructions.

element (1) Any electronic device with terminals to which other devices may be connected. Devices in this sense include fundamental devices such as resistors, capacitors, inductors, and transmission lines or complex subsystems, including a combination of fundamental devices or other subsystems. **(2)** The minimum subdivision within a code group representing a character of the code alphabet. **(3)** A representation of all or part of a logical function or subfunction within a single drawing outline. The drawing outline may contain subelements or may itself be a subelement of a larger logical function.

Element Management System (EMS) That part of AT&T's network management system concerned with collecting network management information from and setting parameters on the network elements. Network elements consist of data communications and telecommunications equipment.

elevated duct An *atmospheric duct* consisting of a high-density air layer that starts at high altitudes and continues upward or remains at high altitudes. It affects primarily very-high-frequency (VHF) transmission.

elevation angle The angle between the horizontal plane and a line from a point on the plane to a point in space above or below the plane. Points above the plane form positive angles from 0 to +90°, while points below the plane range from 0 to −90°.

ELF An abbreviation of Extremely Low Frequency. See *frequency band* and *frequency spectrum*.

Eligible Telecommunications Carrier A telecommunications carrier is eligible to receive universal service support if:

- It offers phone service to all customers throughout a service area without preference, and
- It advertises the available supported services through the mass media.

ELINT An acronym for ELectronics (or ELectromagnetics) INTelligence.

ELIU An abbreviation of Electrical Line Interface Unit.

elliptic filter A filter with an equiripple pass band and equi-minima stop band. In the following figure, the allowable range of passband ripple is bounded by 0 dB and A_{max}dB. The stop band minimum ripple is bounded only by A_{min}dB. Both regions are shown shaded on the figure.

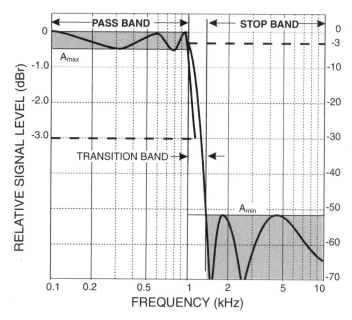

The mathematics describing these filter designs involve elliptic integrals, hence the designation *elliptic filter*. Also called a *Cauer filter*.

elliptical polarization In electromagnetic wave propagation, polarization such that the tip of the electric field vector describes an ellipse in any fixed plane intersecting, and normal to, the direction of propagation. An *elliptically polarized* wave may be resolved into two linearly polarized waves in phase quadrature, with their polarization planes at right angles to each other.

Circular and linear polarization are special cases of *elliptical polarization*.

ELOT An acronym from the Greek Hellenic Organization for Standardization.

ELSEC An acronym for ELectronics SECurity.

EM An abbreviation of End of Medium.

EMA (1) An abbreviation of Digital Equipment Corporation's Enterprise Management Architecture. **(2)** An abbreviation of Electronic Mail Association. An association of developers and vendors of electronic mail products.

EMACS A UNIX text editor. Older versions, although powerful, were not very friendly. Newer versions are mouse and menu oriented.

email A contraction of *electronic mail*.

EMC An abbreviation of ElectroMagnetic Compatibility.

EMD An abbreviation of Equilibrium Mode Distribution.

EME An abbreviation of ElectroMagnetic Environment.

emergency dialing A predesignated button on electronic telephones that calls a specified emergency service. Essentially, *emergency dialing* is a version of speed dialing to call police, fire, ambulance, and so on.

emergency hold A feature of the 911 service that allows the caller's line to be held open, even if the caller attempts to go on-hook. This gives the Public Service Answering Position (PSAP) full control of the connection; that is, the call cannot be released until the agent allows it to be released. Also called *Emergi-Hold*.

emergency ringback A feature of the 911 service that allows the Public Service Answering Position (PSAP) agent to signal a caller that has gone on-hook or left the phone. It enables the phone to be directly rung or to have a howler tone inserted on the line to attract the attention of anyone in the vicinity of the off-hook but abandoned phone.

emergency telephone A telephone on a private branch exchange (PBX) that is functional even when all power to the PBX/PABX (private automatic branch exchange) is lost. It is generally a single-line phone connected to the central office (either directly or by way of a fail-safe relay in the PBX/PABX).

emf **(1)** An abbreviation of ElectroMotive Force. See also *voltage*. **(2)** An abbreviation of ElectroMagnetic Force.

EMI An abbreviation of ElectroMagnetic Interference.

EMI/RFI filter A circuit or device that blocks interfering signals and passes desired signals. Generally, it is a passive low-pass filter, i.e., a filter composed of inductors and capacitors.

emission **(1)** Electromagnetic energy emanating from a source by either radiated or conducted means. The energy may be intentional or incidental and desirable or undesirable, and it may occur at any frequency of the entire electromagnetic spectrum. **(2)** Radiation produced, or the production of radiation, by a radio transmitting station. For example, the radiated energy produced by the local oscillator of a radio receiver would not be an *emission* but a radiation.

emissivity The ratio of power radiated by a substance to the power radiated by a blackbody at the same temperature.

emoticon A contraction of EMOTional ICON. A pictorial representation (using ASCII characters) which attempts to convey the emotional content of the preceding typed message. (Some are obvious when the page is turned to the right.) Used in e-mail and messages on bulletin boards to reduce the amount of typing required of the sender.

Having a hard time seeing the faces? Turn the page 90° clockwise <G>. The "<G>" symbol is frequently used to indicate a grin. Also called *smileys*.

EMP An abbreviation of ElectroMagnetic Pulse.

emphasis In FM transmission, the intentional alteration of the amplitude frequency characteristics of the signal for the purpose of reducing the adverse effects of noise in a communication system. The emphasis network boosts high-frequency signal components relative to low-frequency components so as to produce a more equal modulation index for the transmitted frequency spectrum (hence, a better signal-to-noise ratio for the entire spectral range). The original amplitude frequency characteristics are restored by the receiver's de-emphasis network.

SOME EMOTICON EXAMPLES	
:) or :-)	Smile or humor
;) or ;-)	Wink, raised eyebrow, irony, joking, sarcasm
:(or :-(	Frown or sad
(:(	Very sad
;-r	Displeasure (sticking out one's tongue)
:'(	Cry
:D	Laughter
O:)	An angel
<:)	A dunce or clown
:-O	Yell or surprise
:-o	Bored (yawn)
B)	Wearing sun glasses
:P	Sticking out ones' tongue
:Q	Smoking
:-# or :X	My lips are sealed
:-&	Tongue tied
%(or :-l	Confused or puzzled
:-J	Tongue in cheek
8-)	Wearing glasses and smiling
:-8	Talking out of both sides of one's mouth
>:-<	Anger
:*)	Drunk
:-...	Drooling
#-)	Partied all night
[:{>	Strange guy: crew cut, handlebar mustache, and goatee
= l:-)	Uncle Sam
@>- - -	A rose, romantic message
{}	A hug
{*}	A hug and a kiss

EMR An abbreviation of ElectroMagnetic Radiation.

EMS **(1)** An abbreviation of Expanded Memory Specification. The Lotus, Intel, and Microsoft (LIM) memory management systems. **(2)** An abbreviation of Element Management System. **(3)** An abbreviation of Electronic Message System.

emulator One hardware device or software designed or programmed to imitate the behavior of another device.

For example, a terminal emulator is a communication program that enables a microprocessor to mimic the behavior of a mainframe terminal by using the protocols expected by the mainframe computer. As a second example, a computer based on the 68xxx microprocessor can be programmed to mimic the behavior of a computer based on the 80xxx microprocessor. Although the functions will be correct, the *emulation* will not generally run as fast as the device being emulated. See also *simulate*.

en-bloc signaling Signaling in which address digits are transmitted in one or more blocks, each block containing sufficient address information to enable switching centers to carry out progressive onward routing.

enable To allow the operation of a device, circuit, or system. The enabling signal may or may not also initiate that operation.

enabling signal A signal that permits the occurrence of an event.

encapsulation **(1)** In a layered networking model, the process by which each layer incorporates the protocol data unit (PDU) from the preceding layer into a larger PDU. This is accomplished by adding a header and possibly a trailer to the higher layer PDU. (A PDU is a packet, originating at a particular layer, and used for communicating with a program at the same layer on a different machine.) The inverse process (that is, removing the lower layer headers at the receiving end) is called *decapsulation.* **(2)** The process of sending data encoded in one protocol format across a network operating a different protocol, where it is not possible or desirable to convert between the two protocols. For example, where an Ethernet local area network (LAN) attaches to a fiber distributed data interface (FDDI) backbone, it is not possible to convert between the different packet formats, so the Ethernet packet is *encapsulated* in its entirety inside an FDDI packet as it crosses the bridge on to the FDDI network. When the *encapsulated* Ethernet packet reaches the bridge connecting the destination Ethernet LAN to the FDDI network, the Ethernet packet is removed from the FDDI packet (decapsulated) and put, unchanged, on to the destination Ethernet LAN. Also known as *protocol tunneling.* **(3)** The process of enclosing an electronic assembly in a protective material. The protection may be to prevent damage to the enclosed device, or it may be to prevent access by users (or competitors). Encapsulating materials can be any of a number of materials, including tar, epoxy, plastic, and silicon elastomers.

encapsulation bridging A method of bridging through dissimilar networks where the entire frame from the source network is simply enclosed in the data portion of the link layer frame of the bridging network. See also *encapsulation* (2).

encipher To convert plaintext into an unintelligible form by means of a cipher.

encode Generally, to apply a set of coding rules. **(1)** To *encrypt* data for security. The conversion of plaintext to equivalent cipher text by means of a code. **(2)** To change one data format into a different format or code, e.g., changing a decimal number into binary. Or the mapping of all the symbols of an alphabet into a digital representation, e.g., ASCII. **(3)** To convert a binary data stream into a format more suitable for transmission or modulation in a manner such that the original data may be recovered at the receiver. See also *code* and *waveform codes.* **(4)** To append redundant check symbols to a message for the purpose of generating an error-detection and correction code.

encoding law A rule or specification defining the relative values of the quantum steps used in quantizing and encoding signals. See also *A-law* and *mu-law.*

encryption **(1)** A generic term encompassing both enciphering and encoding. **(2)** The process of making readable information (*plaintext*) indecipherable (*ciphertext*) through normal means in order to protect it from unauthorized viewing or use. *Encryption* is usually based on a coding and decoding key without which the encrypted information cannot be decoded (decrypted). One of the most frequently used *encryption* schemes is one developed by the U.S. National Bureau of Standards, called *DES* (for Data Encryption Standard). Information is encrypted when the information is deemed "sensitive" by the originator and the information exposure cannot be limited, e.g., when a file is transmitted to a remote data terminal equipment (DTE) via public facilities.

Two forms of key *encryption* are possible: *private key* (or secret key) and *public key* (which is really a pair of keys, one public and the other private). See also *Clipper chip, cryptography,* and *public key.*

encryption algorithm The technique for coding a message so that only the intended recipient can read it. See also *encryption.*

end delimiter (ED) A field in a Token Ring or data frame that indicates the end of a token or data frame.

end distortion Asynchronous communication, a specific form of *bias distortion* where the end of all marking pulses (except the stop pulse) is shifted from the theoretical position (relative to the beginning of the *start pulse*). That is, the mark pulses are either shorter or longer than desired.

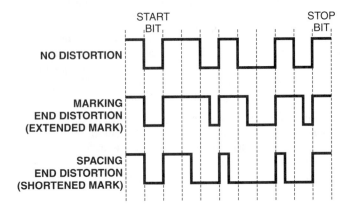

The magnitude of *end distortion* is generally expressed as a percentage of the ideal pulse length.

end exchange A synonym for *end office.*

end finish In fiber optics, the surface condition at the termination of an optical fiber. Refers to the optical quality.

end instrument The device that is connected to the terminus of a communications circuit.

end node A network station (node) that only transmits and receives information for its own use. It can neither route nor forward information to any other station on the network.

end of block (EOB) A transmission control character used to signify the end of a block of data.

end of content (EOC) In telecommunications, a specific symbol used to indicate the end of a message or page.

end of file (EOF) An internal label, immediately following the last record of a file, signaling the end of that file. It may include control totals for error control processing.

It is a unique code placed after the last byte in a file and informs the operating program that no additional information follows. The *EOF* marker is required because it is possible, even likely that the last byte of a file falls in the middle of a data block allocation (either in disk storage or in a transmission packet). In ASCII the *<EOF>* character is *<^Z>* (decimal 26, or hexadecimal 1Ah).

end of medium (EM) A control code that indicates the end of the used (or useful) portion of a serial storage medium. In ASCII, the <EM> character is represented by decimal 25 (or hexadecimal 019h).

end of message (EOM) A control symbol used in data communications to indicate the end of a message.

end of selection character A control code character that indicates the end of the selection signal.

end of text (ETX) A unique code placed after the last character in a text file. The character signifies only the end of user information, not necessarily the end of the transmission as error control and packet control characters may follow the user text. In ASCII the *<ETX>* character is represented by decimal 03 (or hexadecimal 03h).

end of transmission (EOT) A unique transmission control code used to indicate the conclusion of a transmission. The transmission may have included one or more message blocks and associated message headers. It may also initiate other functions, such as releasing circuits, disconnecting terminals, or placing receive terminals in a standby condition. In ASCII the <*EOT*> character is decimal 04 (or hexadecimal 04h).

end of transmission block (ETB) A transmission control character used to indicate the end of a transmission block of data when data are divided into such blocks for transmission purposes. In ASCII, the <ETB> character is represented by decimal 23 (or hexadecimal 017h).

end office (EO) Another name for the *class 5 office (central office switch)* to which subscriber lines are directly attached. The switching system connects subscriber loops to subscriber loops and subscriber loops to interoffice trunks. Also called *end exchange*. See also *central office* and *class n office*.

end system (ES) Any network entity, such as a node, that uses or provides network services or resources in the ISO/OSI Reference Model. It is a system containing the application processes that are the ultimate source and sink of user traffic. Architecturally, an *end system* uses all seven layers of the Reference Model. This is in contrast to an *intermediate system (IS)*, or router, which uses only the bottom three layers (the subnet layers) of the model. Also called a *host*.

end-to-end digital transmission All circuit elements are digital. That is, no modems are used to convert digital signals to or from analog at any point.

end-to-end encryption The encryption of information at its origin and decryption at its intended destination without any intermediate decryption.

end-to-end loss The sum of all transmission losses and gains of a transmission path. The losses are from all sources and include cable, fiber, connector, and splice losses.

end-to-end routing A routing strategy in which the entire path is determined before the message is sent. This is in contrast to node-to-node routing in which the route is built step-by-step.

end-to-end service Interexchange service that extends from one customer premises to another customer premises. It usually consists of the local loops on each end and an IEC leg in the middle.

end-to-end signaling A signaling system in which an originating station may generate and transmit signaling information to a destination station after a connection is established and without the assistance of any intervening entity. For example, in dual tone multifrequency (DTMF) signaling a user may send signals directly to the destination equipment.

end user The ultimate user of a system or service. An entity that uses products and services (but does not necessarily pay for said services, e.g., a person called by a paying customer). Users are usually people but can also be computers, objects, switches, or other types of computer systems or communication equipment.

endpoint node In network topology, any node connected to one and only one branch. Also called a *peripheral node*.

engineering channel A synonym for an *orderwire* circuit.

engineering orderwire (EOW) A synonym for an *orderwire* circuit.

enhanced parallel port (EPP) Developed by Intel, Zenith, and Xircom, it is an improved performance parallel port that boosts the transfer rate from 150 kbps to 16 Mbps. See also *extended capabilities port (ECP)* and *IEEE 1284*.

enhanced serial port (ESP) An extension of the standard serial interface proposed and placed into the public domain by Hayes Microcomputer Products, Inc. It includes the definition of high-speed I/O, control registers, buffer control, direct memory access (DMA), and the interactions with modems. It defines an interface between the DTE and DCE so as to allow greater than 38-kbps reliable transfers over a telephone line. Also *enhanced serial interface (ESI)*.

ENIAC An acronym from Electronic Numerical Integrator And Calculator. One of the first electronic computers; circa 1946.

ENQ An abbreviation of ENQuiry character.

enquiry character (ENQ) A transmission control character used as a request for a response from the station with which a connection has been established. The response may include station identification, type of equipment in service, and status of the remote station. The ASCII code is 05.

ENR An abbreviation of Excess Noise Ratio.

ENSB An abbreviation of Equivalent Noise Sideband.

ENT An abbreviation of Equivalent Noise Temperature.

enterprise information base (EIB) In enterprisewide networks, the information base containing management, performance, and other related information about the network. The information in this type of database is used by network management or monitoring software.

Enterprise Management Architecture (EMA) A network management model, developed by Digital Equipment Corporation (DEC), which strives to provide the tools needed to manage enterprise networks, regardless of the network configuration. The architecture is designed to conform to ISO's Common Management Information Protocol (CMIP).

enterprise network A geographically dispersed network under the auspices of one organization.

Enterprise Network Services (ENS) An extension to Banyan's VINES network operating system (NOS). It enables StreetTalk to keep track of servers using NOSs other than VINES, such as any version of Novell's NetWare or Apple's AppleTalk.

enterprisewide computing Basically, a combination of several local area networks (LANs) joined with a wide area network (WAN) and all for the service of one organization. Each LAN serves the needs of its local region, while the WAN interconnects the regions to serve organizationwide requirements. The joined networks encompass most or all of a company's computing resources.

In most cases, an enterprisewide computing network will include LANs and computers which may be running different operating systems and belong to different types of networks. Therefore, one of the biggest challenges for *enterprisewide computing* is to achieve interoperability for all its components.

entity (1) A term used to refer the abstraction of a device, i.e., to any device on a system to which an action, service, or subject might apply. **(2)** In layered networking models, a reference to an abstract device, such as a program, function, or protocol, that implements the services for a particular layer on a single machine. An entity provides services for entities at the layer above it and requests services of the entities at the layers below it.

entrance facilities In a premises distribution system (PDS), the location at which the building's internal wiring and the external wiring meet.

entropy coding A coding technique that encodes frequently occurring patterns with fewer bits than it does for infrequently occurring patterns.

entry point (1) In networking, the point at which a node (hardware) is connected to the network or the point at which a program, module, or function begins executing (software). (2) An IBM network management term. An entry point provides management functions for itself and the devices attached to it. It has to be a Systems Network Architecture (SNA)-addressable unit, allowing it to participate in network management by monitoring its own environment and exchanging information and messages with a focal point.

entry state In an AppleTalk network routing table, a value that indicates the status of a path. Such an entry may have the value good, suspect, or bad, depending on how recently the path was verified as being valid.

entry switch The *first point of switching* in a switched network.

enumeration list A finite collection that identifies all possible (allowable) values for a variable, field, data attribute, object type, and so on.

envelope (1) In communications, a unit of information consisting of transmitted block to which extra bits have been added for routing, control, and error checking. (2) In communications, the boundary of a wave shape when one or more parameters of the wave are varied. The heavy line in the following figure denotes the *envelope* of an amplitude-modulated carrier.

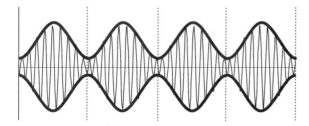

envelope delay The time required for a point on the envelope of a modulated wave to pass between two specified points on the transmission path.

Variation of signal delay with respect to any frequency in the communications channel bandwidth. (The difference in travel time for different frequencies in a transmitted signal.) Mathematically, *envelope delay (D(ω))* is the slope of phase ϕ(ω) with respect to frequency (ω), that is,

$$D(\omega) = \frac{d\phi(\omega)}{d\omega} \approx \frac{\Delta\phi}{\Delta\omega}$$

Phase Frequency Curve

$$\text{Envelope Delay} = \frac{d\phi_1}{d\omega_1} \approx \frac{\Delta\phi_1}{\Delta\omega_1}$$

$$\text{Phase Delay} = \frac{\phi_1}{\omega_1}$$

PHASE (ϕ rad)

ϕ₁

FREQUENCY (ω rad/sec)   ω₁

Also called *group delay*. See also *phase delay*.

envelope delay distortion (EDD) Signal distortion that results when the rate of change of phase shift (Δϕ) with frequency (Δf) over the necessary signal bandwidth is not constant (i.e., *envelope delay* does not vary linearly with frequency). It is usually expressed as one-half the difference between the delays of the two extremes of the necessary bandwidth. *Phase distortion* always accompanies *EDD*.

EO An abbreviation of <u>E</u>nd <u>O</u>ffice.

EOA An abbreviation of <u>E</u>nd <u>O</u>f <u>A</u>ddress. A code in a packet header.

EOB An abbreviation of <u>E</u>nd <u>O</u>f <u>B</u>lock.

EOC (1) An abbreviation of <u>E</u>nd <u>O</u>f <u>C</u>ontent. (2) An abbreviation of <u>E</u>mbedded <u>O</u>perations <u>C</u>hannel.

EOD An abbreviation of <u>E</u>nd <u>O</u>f <u>D</u>ata.

EOF An abbreviation of <u>E</u>nd <u>O</u>f <u>F</u>ile.

EOM An abbreviation of <u>E</u>nd <u>O</u>f <u>M</u>essage.

EOT (1) An abbreviation of <u>E</u>nd <u>O</u>f <u>T</u>ape. (2) An abbreviation of <u>E</u>nd <u>O</u>f <u>T</u>ransmission. (3) An abbreviation of <u>E</u>nd <u>O</u>f <u>T</u>hread.

EOW An abbreviation of <u>E</u>ngineering <u>O</u>rder <u>W</u>ire.

ephemeris A table predicting the position (coordinates) of a celestial object with respect to time.

For example, in the Global Positioning System (GPS), the *ephemeris* provides satellite orbital parameters that the GPS receivers use to calculate the precise position of the GPS satellites. See also *Global Positioning System (GPS)*.

epoch The time interval between measurements (the data frequency).

epoch date Any date in history, chosen as the reference date from which time is measured. Examples include:

• The beginning instant (0000 UT, i.e., midnight) of January 1, 1900, Universal Time, for Transmission Control Protocol/Internet Protocol (TCP/IP). TCP/IP programs exchange date and time of day information with time expressed as the number of seconds past this epoch date.

• The beginning instant (0000 UT, i.e., midnight) of January 6, 1980, Universal Time, for the Global Positioning System (GPS) clock. GPS time is not adjusted and therefore is offset from Coordinated Universal Time (UTC) by an integer number of seconds, due to the insertion of leap seconds. The number remains constant until the next leap second occurs.

EPP An abbreviation of <u>E</u>nhanced <u>P</u>arallel <u>P</u>ort.

EPROM An acronym from <u>E</u>rasable <u>P</u>rogrammable <u>R</u>ead <u>O</u>nly <u>M</u>emory (pronounced "EE-prom"). An EPROM is a nonvolatile memory device that functions primarily as a ROM. However, unlike a true ROM, an EPROM can be programmed with new information many times. As a nonvolatile device, an EPROM will not lose stored information when operating power is removed.

Because an EPROM must be exposed to ultraviolet (UV) light to erase previously stored information, it is not practical to store new information while the device is installed and in normal operation. Also because the device must be erased with UV light, an EPROM is sometimes called a UVPROM.

EPS (1) An abbreviation of <u>E</u>ncapsulated <u>P</u>ost<u>S</u>cript. Files created for postscript printing usually have a name with an ".EPS" extension. See also *filename extension*. (2) An abbreviation of <u>E</u>lectronic <u>Pu</u>blishing <u>S</u>ystem.

equal access (1) A provision of the 1982 Modified Final Judgment (MFJ) to the AT&T divestiture decree that gives subscribers the right to select the interLATA carrier of their choice. All subscribers are given a choice of their primary long-distance carrier which is accessed via the 1+ scheme. All other carriers can be accessed by dialing a *carrier access code (CAC)* followed by the long-distance number. Until July 1, 1998, the access code was five digits with the form 10+XXX. After this date, the FCC extended it to seven digits (101+XXXX) due to the large number of requests for numbers.

CARRIER ACCESS CODES EXAMPLES

	Access Code		For Information
Service Provider	10+	101+	
Allnet	444	0444	800-783-2020
American Telecom	813	0813	800-945-3344
AMNEX	370	0370	800-366-2850
AT&T	288	0288	800-222-0300
BIZ-TEL-CORP.	606	0606	800-231-0989
Business Telecom, Inc. (BTI)	833	0838	800-742-1663
CCSI	329	0329	800-888-9711
Dial&Save	457	0457	800-787-3333
Excel Telecommunications, Inc.	752	0752	800-875-9235
LCI International	432	0432	800-860-1020
Long Distance Savers	036	0036	800-789-7000
Long Distance Wholesale Club	297	0297	800-787-7887
MCI	222	0222	800-888-0800
Metromedia	488	0488	800-275-2273
National Tel	657	0657	800-881-9300
Phone One	393	0393	800-393-9000
Sprint	333	0333	800-767-7759
TelecomUSA	321	0321	
Thrift Call, Inc	923	0923	800-554-3057
Touch One	797	0797	800-286-8241
TransAmerica Communications			407-241-6515
TTI Long Distance	848	0848	800-226-4884
U.S. Long Distance			800-460-USLD
VarTec Telecom, Inc.	811	0811	800-583-8811
World Pass Communications	840	0840	

Sometimes called a *dial-around code.* See also *carrier identification code (CIC).* (2) Equal Access also refers to a more generic concept under which the Bell Operating Companies (BOCs) must provide access services to AT&T's competitors that are equivalent to those provided to AT&T.

equal charge rule A rule contained in the 1982 Modified Final Judgment (MFJ) which required Bell Operating Companies (BOCs) to charge access rates that do not vary with the volume of traffic.

equal gain combiner A *diversity combiner* in which the signals on each input channel are added. Automatic gain control (AGC) on the channels ensures that the resultant output signal remains approximately constant. See also *combiner* and *diversity combiner.*

equalization (1) In general, the maintenance of system transfer function characteristics within specified limits by modifying circuit parameters. (2) In transmission systems, signal conditioning used to compensate for undesirable communication channel characteristics and distortions such as frequency, amplitude, gain, delay, and/or phase distortion.

The following figure demonstrates the *equalization* process. The curve labeled "original response" is the transmission response of the system without any compensation. The "equalizer response" is de-

signed such that the shape of the curve in a band of interest is the mirror image of the uncorrected response. When a signal passes through both the transmission system and the equalizer, the two curves are added together, resulting in the composite response shown. The response magnitude could represent amplitude, phase, delay, and so on.

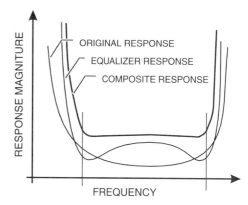

Note that the response magnitude is much flatter in the range of interest; however, both the overall bandwidth is narrower and that the magnitude of the response is greater across the band. See also *subset equalizer.*

equalizer A device inserted into a signal transmission path to improve its transfer characteristics. See also *equalization.*

equatorial orbit A satellite orbit with a zero degree inclination angle; that is, the orbital plane and the Earth's equatorial plane are coincident.

equilibrium length In fiber optics, the length of multimode optical fiber necessary to attain equilibrium mode distribution given a specific excitation condition. It is sometimes used to refer to the longest such length, as would result from a worst case, but undefined, excitation. Also called *equilibrium coupling length* and *equilibrium mode distribution length.*

equilibrium mode distribution (EMD) That condition in a multimode fiber wherein after propagation has taken place for a distance, called the *equilibrium length,* the relative power distribution among propagating modes becomes statistically constant and remains so for any further propagation down the fiber. In practice, this length may vary from a fraction of a kilometer to more than a kilometer.

At distances greater than the equilibrium length, the numerical aperture of the fiber's output is independent of the numerical aperture of the optical source that drives the fiber. (This is because of mode coupling and stripping, primarily by small perturbations in the fiber's geometry which result from the manufacturing and cabling processes.) In the ray optics analogy, the *equilibrium mode distribution* may be loosely thought of as a condition in which the "outermost rays" in the fiber core are stripped off by such phenomena as microbends, and only the "innermost rays" continue to propagate. In a 50-μm core multimode graded-index fiber, light propagating under equilibrium conditions occupies approximately only the middle 70% of the core and has a numerical aperture approximately 70% that of the full numerical aperture of the fiber. Also called the *equilibrium mode power distribution* and the *steady-state condition.*

equivalent circuit A collection of circuit elements that have essentially the same characteristics as a reference entity over a specified operating range. The circuit may not be practical to build; however, it may be far simpler to analyze. For example, see *Norton's* and *Thevenin's theorems.*

equivalent four-wire In communications, the use of frequency division multiplex (FDM) techniques to provide an independent forward and reverse channel on a single two-wire transmission line. The two transmission bands created are frequently called the "high band" and the "low band."

equivalent network **(1)** A network that, under appropriate circumstances, may replace another network without substantial effect on system performance. **(2)** A network with external characteristics that are identical to those of another network. **(3)** A mathematical model or theoretical representation of an actual network.

equivalent noise bandwidth (B_N) The frequency interval, determined by the frequency-amplitude response of a system, that defines the system output noise power given an input noise source of specified characteristic.

equivalent noise resistance A representation, in resistance units, of a noise voltage generator at a specified frequency. Two such relationships are:

$$R_N = \frac{\text{mean}(e^2)}{4kT_0\Delta f} \text{ or } R_N = \frac{\pi W_n}{kT_0}$$

where:

R_N	is the equivalent noise resistance.
k	is Boltzmann's constant.
T_0	is the standard noise-temperature of 290 kelvins.
Δf	is the specified bandwidth.
$\text{mean}(e^2)$	is the mean-square noise generator.
W_n	is the spectral density.

See also *Johnson noise.*

equivalent noise temperature The temperature, in kelvins, of a hypothetical matched resistance at the input of a device or system (assumed to be noiseless) that would account for the measured output noise of the device or system. See also *Johnson noise.*

equivalent PCM noise The amount of thermal noise needed on an analog circuit to be judged as having the same effect on speech quality as the quantizing noise generated in the analog-to-digital conversion of a CODEC. Generally, 33.5 dBrnC ± 2.5 dB is considered the approximate *equivalent PCM noise* of a 7-bit PCM system.

equivocation Information theory, the conditional information content of an input symbol, given an output symbol, averaged over all input-output pairs.

erect position In frequency division multiplexing (FDM), a position of a translated channel in which an increased signal frequency in the untranslated channel causes an increased signal frequency in the translated channel. Also called an *upright position.*

ERL An abbreviation of <u>E</u>cho <u>R</u>eturn <u>L</u>oss.

erlang (E) The international unit of traffic intensity (occupancy) with a range between 0 and 1. The average number of simultaneous connections observed in a measurement period (generally, the busy hour). Named after A. K. Erlang (1878–1929), one *erlang* is the intensity in a communications channel continuously occupied for one hour, or in several channels where the aggregate time is one hour. That is, the number of *erlangs* is the ratio of the time the facility is occupied (continuously or cumulatively) to the time the facility is available. One *erlang* therefore can be expressed as one call hour per hour, one call minute per minute, or one call second per second. Also called a *traffic unit.*

erlang B formula An equation that relates the probability of lost calls to the number of circuits and the traffic offered. It is used to estimate the number of communications circuits needed to carry a given amount of traffic. It assumes:

1. Offered traffic arrives at random times.
2. Holding times of connected calls are distributed exponentially, and calls with longer holding times occur less frequently than do short holding time calls.
3. Offered traffic not immediately served is discarded.
4. Discarded traffic is not queued and does not reattempt a connection (it is lost).

$$E_{1,n}(y) = p = \frac{\dfrac{y_n}{n!}}{1 + \dfrac{y}{1} + \dfrac{y^2}{2!} + \cdot \cdot + \dfrac{y^n}{n!}}$$

where:

p	= the loss probability,
y	= the traffic offered (in erlangs), and
n	= the number of circuits.

Also called the *erlang loss formula* or the *lost calls cleared formula.*

erlang C formula An equation that relates the call queue holding time to the number of circuits and the traffic offered. It is used to estimate the number of communications circuits needed to carry a given amount of traffic. It assumes:

1. Offered traffic arrives at random times, and
2. Offered traffic not immediately served is queued indefinitely.

As a consequence, the offered traffic cannot exceed the number of circuits, or, holding time becomes infinite!

erlang loss formula See also *erlang B and C formula.*

ERP An abbreviation of <u>E</u>ffective (or Equivalent) <u>R</u>adiated <u>P</u>ower. Also written *e.r.p.*

erroneous block A block in which there are one or more erroneous bits.

error The difference between what an observed or measured value and what the true or theoretical value should be.

The *error* (*E*) is the algebraic difference between an indicated value (*I*) and the true value (*T*); that is,

$$E = I - T$$

In communication systems, the indicated value is the received data, while the actual value is taken to be the transmitted data; that is, an error occurs when a bit transmitted as a one is received as a zero (or vice versa).

error burst A group of bits in which two successive bits are separated by fewer than "m" correct bits. The integer parameter "m" is called the guard band of the error burst. When describing an error burst, the parameter "m" should be specified.

error checking A process employed during data transmissions to *detect* discrepancies between transmitted and received data. Parity, check sums, and cyclic redundancy checks are all methods of *error checking.*

error control A method or set of methods used by communication equipment to *identify* the occurrence of an error in a received message and to *correct* the error either by using error-correcting message coding or by requesting that the message in error be retransmitted. ITU-T V.42 uses LAP-M as the primary protocol and MNP class 1 through class 4 for backward compatibility. LAP-B is a protocol used in switched packet networks.

error-correcting code (ECC) A code in which each transmitted signal conforms to specific rules of construction so that departures from this construction can be detected at the receiving end of the communication channel. The construction of the transmitted signal contains enough redundant information so that the *original data may be extracted* even if part of the received signal is lost. The two main classes of error-correcting codes are block codes and convolutional codes. Several *error-correcting codes* are Hamming code, CRC codes, and the Reed-Muller code.

error correction A technique to restore data integrity in received data that has been corrupted during transmission. *Error-correction* techniques involve sending extra data along with the original data being sent. This extra data at least provides the receiver with the ability to detect errors and then to respond in one of several ways, that is,

- Request the data packet in error be retransmitted, or
- Reconstruct the original data using the extra (redundant) data. *Error-correction* schemes that automatically reconstruct the original data are termed *forward error correction* (FEC) methods.

See also *error-correcting code* (*ECC*) and *error-detecting code.*

error-correction mode (ECM) An enhancement to Group III fax capability. Encapsulated data within high-level data link control (HDLC) frames provides the receiving station with error detection and retransmission capability.

error-detecting code A code in which each transmitted signal conforms to specific rules of construction so that departures from this construction can be *detected* at the receiving end of the communication channel.

Error-detection methods include Hamming codes, cyclic redundancy checks (CRC) or longitudinal redundancy checks (LRC), and the use of parity bits. Parity bits, CRC, and LRC values are sometimes referred to as error-correction codes (ECC), even though they can only detect errors. Hamming codes, on the other hand, are true ECCs because they provide enough information to determine the nature of the error and to replace it with a correct value.

error-free seconds A measure of the quality of a signal being transmitted. It is the percentage representing the total amount of time over a 24-hour period that the signal contained bit errors, and it is calculated using a test pattern defined in the ITU-T 0.151 recommendation.

error handling The process of dealing with errors as they occur. *Error-handling* procedures range from ignoring the error and proceeding as if no exception occurred to complex procedures that attempt to correct or reacquire proper information.

error messages A message from the software to the operator advising that some abnormal condition has arisen and that some action is required.

error rate A deprecated term that has been replaced by *error ratio.*

error ratio The ratio of received bits (characters, words, or blocks) incorrectly received to the total number of bits (characters, words, or blocks) transmitted in a specified amount of time. Many error detectors measure the ratio of incorrectly received elements to the total number of received elements.

For a given communication system, the *error ratio* will be affected by both the data transmission rate and the signal power margin. Formerly called *error rate.* See also *Nyquist Theorem.*

ERT An abbreviation of Estimated Repair Time.

ES (1) An abbreviation of Echo Suppressor. **(2)** An abbreviation of Electronic Switching. **(3)** An abbreviation of End System. **(4)** An abbreviation of Expert System.

ESC (1) An abbreviation of ESCape character or code (ASCII 01Bh). **(2)** An abbreviation of Echo Suppressor Control. **(3)** An abbreviation of Electronic Switching Center.

escape character See *escape code.*

escape code (1) The ASCII character <ESC> decimal 027 (or hexadecimal 1Bh) followed by a string of characters used to control a terminal or printer.

In Telnet the escape character is entered as <^[>. It interrupts communications with the host and allows the user to communicate directly with Telnet. In this mode, the user can obtain status, cancel processes, or determine if the host is still responding. **(2)** Sometimes the three repetitions of the ASCII character defined in a Hayes compatible modem's S2-register are referred to as an *escape code.*

escape sequence By the strictest definition, a sequence of ASCII characters beginning with the ASCII <ESC> character (decimal 27, hex 01Bh). However, the definition is generally expanded to include any sequence that a device might interpret as a command or prelude to a command.

The *escape sequence,* sent by the DTE to the DCE, is how a modem "escapes" or changes from the *data mode* of operation to the *command mode* of operation. The general form of the escape sequence is <guard time> <escape code> <guard time>. Typically, the guard time is 1 second, and the escape code is "+ + +". Both guard time and escape code are usually software setable.

ESD An abbreviation of ElectroStatic Discharge.

ESF An abbreviation of Extended Super Frame or Extended Superframe Format.

ESMR An abbreviation of Enhanced Specialized Mobile Radio.

ESN An abbreviation of Electronic Security Number.

Esprit An acronym from European Strategic Program for Research and Development in Information Technology.

ESS In telephony, an abbreviation of Electronic Switching System. The Bell System's term for a computer-controlled telephone exchange.

ETB An abbreviation of End of Transmission Block.

ETC An abbreviation of Enhanced Throughput Cellular. An AT&T asynchronous modem protocol based on the ITU-T V.42 and V.32 bis protocols but modified to provide higher average throughput and more reliable connections on cellular networks.

AT&T's testing indicates *ETC* is superior to Microcom's MNP-10 protocol in three important performance areas:

- File completion percentage (placing a call and successfully completing a file transfer).
- Minimizing total online time for a call (i.e. lowest cost).
- Minimizing start-up time (allows *ETC* to complete short to medium-length file transfers normally in under one minute).

ETFD An abbreviation of Electronic Toll Fraud Device.

Ethernet A 10-Mbps, baseband, carrier sense multiple access with collision detector (CSMA/CD) local area packet network (LAN) architecture invented by Xerox and developed by DEC, Intel, and Xerox in 1976 from which the IEEE 802.3 standard was developed. Later, both ANSI and the ISO standards groups adopted the IEEE standard with only slight modifications. The protocol uses Manchester waveform encoding.

Strictly speaking, the original *Ethernet* should be referred to as a *DIX Ethernet,* while the IEEE version is called an *IEEE 802.3 network.* This distinction has not generally been observed except by LAN experts. The packet structure has the form:

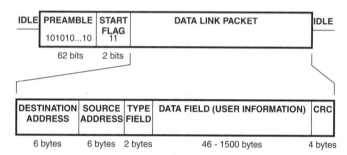

The term *Ethernet* is derived from "luminiferous ether," the substance nineteenth-century scientists thought was necessary for transporting lightwaves and radio waves. (They were wrong!) Several wiring standards are defined: the 1Base5 twisted pair system, the 10Base2 coaxial system, the 10Base5 coaxial system, and the 10BaseT twisted pair system. The number before "Base" indicates the data transfer rate; that is, 10 indicates a 10-Mbps data rate, "Base" stands for baseband transmission, and the final digit 2, 5, or T stands for 200 meters (185 meters rounded off to 1 digit), 500 meters, and twisted pair.

1Base5 (StarLAN)—the IEEE 802.3 designation for a 1-Mbps Ethernet LAN that uses unshielded twisted pair (UTP) cable. The physical topology is a bus with nodes attached to the common cable. AT&T's StarLAN is an example of a *1Base5* network.

10Base2 (Thin Ethernet)—the IEEE 802.3 designation for a 10-Mbps Ethernet in which a wiring segment is a single-shielded 50-Ohm coaxial cable (RG58) with BNC connectors and can be up to 185 meters in length. The maximum number of nodes per segment is 30, separated by at least 0.5 meters.

10Base5 (Thick Ethernet)—a 10-Mbps Ethernet in which the backbone wiring segment can be up to 500 meters and is a 50-Ohm twinaxial cable. Each node is connected to the Thick Ethernet cable via a transceiver and transceiver cable. Up to 100 transceivers may be installed on the backbone cable. The maximum length of the transceiver cable is 50 meters.

10BaseT (Twisted Pair Ethernet)—the IEEE 802.3 10-Mbps Ethernet in which each twisted pair wiring segment can be up to 100 meters. The cable may be either unshielded twisted pair (UTP) or shielded twisted pair (STP). The physical network is a star topology; that is, each node is connected to a central hub. The network may be expanded to five sections through four repeaters.

10BaseF—an IEEE 802.3 10-Mbps Ethernet that uses fiber optic cable for transmission.

Switched Ethernet. A method of improving the performance of traditional Ethernet networks. The technique splits the network into a number of subnets interconnected by a hub. Each subnet can use the full data transmission capacity simultaneously unless packets are di-

rected through the hub to adjacent subnets. If the subnets are arranged into natural working groups, the amount of trans-hub traffic is minimal.

Fast Ethernet. Any "extension" of the Ethernet technology which increases the data rate from 10 Mbps to 100 Mbps. See also *IEEE 802.x.*

Ethernet switch A data switch that allows users on different network segments to exchange data. When users on different segments exchange data, an Ethernet switch dynamically connects the two separate Ethernet channels without interfering with other network segments. The switch can create multiple independent connections between separate segments, allowing multiple parallel data exchanges. This multiplies network bandwidth without modification to Ethernet end station hardware or software.

EtherTalk Apple Computer's protocol for Ethernet transmissions. It is the driver used to communicate between the Macintosh and an Ethernet network interface card. Two versions of EtherTalk have been developed:

- EtherTalk Phase 1 is based on the DIX Ethernet 2 version, also known as Blue Book Ethernet.
- EtherTalk Phase 2 is based on the IEEE 802.3 Ethernet.

See also *AppleTalk.*

Ethertype A two-byte code indicating the protocol type in an Ethernet packet.

ETL An abbreviation of <u>E</u>ffective <u>T</u>esting <u>L</u>oss.

ETSI An abbreviation of the <u>E</u>uropean <u>T</u>elecommunications <u>S</u>tandards <u>I</u>nstitute.

ETX An abbreviation of <u>E</u>nd of <u>TeXt</u>.

EU An abbreviation of <u>E</u>uropean <u>U</u>nion.

Eudora A client program for Internet mail users. It not only provides support for mail sending, receiving, printing, and deleting but also provides multimedia support (voice and video) and encryption for mail privacy. The program runs on UNIX, IBM PC, and Macintosh-based systems.

Eudora was originally developed by Steve Dorner at the University of Illinois circa 1989. The program's name is from the author of the short story "Why I Live at the P. O."—Eudora Welty. The original versions are *freeware* and are still available on the Internet; however, the newest version is a commercial product.

EUnet A major network in Europe, interconnecting several thousand sites across 25 countries. It is funded by subscriptions from the participating organizations. The network emphasis is on education and research.

European Academic and Research Network (EARN) Essentially, the European branch of BITNET, it links the universities and research facilities of more than 27 countries throughout Europe, the Middle East, and Africa. As with BITNET, *EARN* uses IBM's network job entry (NJE) protocol. Access to the Internet is via a gateway.

European Backbone Network (EBONE) A multi-protocol backbone service available for networking throughout Europe. It is intended to encourage the development of research, educational, and commercial networking in Europe. EBONE is the regional backbone for many European autonomous networks.

European Computer Manufacturer's Association (ECMA) A trade organization dedicated to the cooperative development of standards applicable to computer technology. It works closely with certain ITU-TS study groups and ISO subcommittees.

European Telecommunications Standards Institute (ETSI) A European standards body established in 1988 by a decision of the Conférence Européenne des Administrations des Postes et Télécommunications (CEPT). It has taken over the work of the CEPT in the area of developing the Net-Normes Européene de Telecommunication, Net standards.

European Workshop for Open Systems (EWOS) One of three regional workshops for implementers of the ISO/OSI Reference Model. The other two are the Asia and Oceania Workshop (AOW) and the OSI Implementers Workshop (OIW). The forum is aimed at promoting ISO/OSI standards and undertaking the development of functional profiles. Its work includes OSI Reference Model Layers 1–4, FTAM, MHS, ODA, directory services, and the VT protocol.

eV The symbol for electronvolt.

evanescent field In a waveguide, a time-varying field having an amplitude that decreases monotonically as a function of radial distance from a waveguide but without an accompanying phase shift. It is a surface wave and is coupled (bound) to an electromagnetic wave or mode propagating inside the waveguide.

In fiber optics, the *evanescent field* may be used to provide coupling to another fiber.

even parity See *parity*.

event A milestone, a signal, or the completion of something that is of interest to an object, a process, or a system; a change in the state or operation of a system. It is usually significant to the performance of a function operation or task.

event reporting A data-gathering method in which agents report on the status of the objects under the agents' range. The agent generates a report containing the relevant information and initiates the transmission of the report to the management package.

This is in contrast to polling, in which the management program periodically requests such reports from agents.

EWOS An acronym from <u>E</u>uropean <u>W</u>orkshop for <u>O</u>pen <u>S</u>ystems.

excess 3 code See *code*.

excess noise ratio (ENR) That noise produced by a device which is at a higher level than the thermal noise generated by a resistor at 290 K; i.e.,

$$ENR = 10 \log_{10}\left(\frac{T_x - 290}{290}\right)$$

where T_x is the noise temperature of the device. See also *Johnson noise*.

EXCH An abbreviation of <u>EXCH</u>ange.

exchange (1) A room or building equipped so that telephone lines terminating there may be interconnected as required. The equipment may include manual or automatic switching equipment. Also called a *telephone switching center*. **(2)** In the telephone industry, a geographic area (such as one or more villages, towns, or cities and their surrounding region) established by a regulated telephone company for the provision of local telephone services. It consists of one or more switching offices together with the associated equipment. Also called an *exchange area*. **(3)** In the Modified Final Judgment (MFJ), a local access and transport area (LATA).

exchange access (1) In telephone networks, access in which exchange services are provided for originating or terminating interexchange telecommunications within the exchange area. **(2)** The offering of telephone exchange services or facilities for the purpose of the origination or termination of telephone toll services.

exchange access signaling A signaling system that is used by equal access offices to: 1. Transmit originating information (including address digits) to the customer's premises and 2. Provide the means of verifying the receipt of these digits.

Features of this system include:

- Overlap outpulsing.
- Identification of the type of call.
- Calling party number identification.
- Acknowledgment wink supervisory signals.

exchange area A geographic area within which there is a single uniform set of charges for telephone service. A call between any two terminals in an *exchange area* is a local call.

exchange facilities The facilities included within a local access and transport area.

exclusive hold A feature of a private branch exchange (PBX) wherein only the party putting the line on hold can take it off of hold.

exclusive OR (XOR) The *exclusive or* (abbreviated *XOR* or sometimes *EOR*) is a logical operation for combining multiple Boolean values (true and false) or multiple binary (1 and 0) signals. The rule for the *XOR* operation is: The output is *true* if and only if *an odd number of inputs are true;* the output is false (0) if all inputs are false (0) or an even number of inputs are true (1).

The written symbol for the *XOR* function is a plus in a circle ("$\oplus$"); therefore, the *XOR* of input A and input B is written A$\oplus$B. The *truth table* shows all possible combinations of input and output for a two-input *XOR* operation. A true condition is shown as a "1," while a false condition is shown as a "0."

XOR TRUTH TABLE

A	B	A⊕B
0	0	0
0	1	1
1	0	1
1	1	0

See also *AND, AND gate, exclusive OR gate, NAND, NAND gate, NOT, OR,* and *OR gate.*

exclusive OR gate A type of digital circuit that reforms an exclusive OR operation on two or more digital signals. The symbol for an exclusive OR gate is

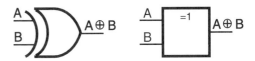

See also *AND, AND gate, exclusive OR, gate, logic symbol, NAND, NAND gate, NOT, OR,* and *OR gate.*

.EXE A PC-DOS/MS-DOS-type *executable program* or *file* in which the calculation of relocation addresses is necessary when the file is loaded into memory for execution. *.EXE* files have addresses or data that change based on the actual location in memory into which the program is placed.

executable program A binary file that is ready to run on the computer system for which it was designed (the *target system*).

For example, on a PC, DOS executable files have the extension .BAT, .COM, or .EXE.

executive camp on A feature of a PBX system in which only certain people are allowed to camp-on to another's extension.

executive override A feature of a private automatic branch exchange (PABX) system in which certain people are allowed to barge into another's established connection. Generally, only the PABX extensions can communicate; outside lines are either totally excluded or can only hear the original party.

exempted addressee An organization, activity, or person included in the group address of a specific message but deemed by the message originator as having no need to receive the message and is therefore explicitly excluded from the list of destinations.

expander (1) A device that increases the dynamic range of an output signal. This is accomplished by a nonlinear amplifier having the characteristic of increasing the magnitude of high-volume signals by a greater percentage than low-volume signals. It is used to restore the dynamic range of a compressed signal to its original dynamic range. See also *compander* and *compressor*. (2) In telephony, a switching device with more output ports than input ports.

expansion (1) The function associated with a network having more output ports than input ports. The inverse function is *concentration*. (2) A process in which the effective transmission gain is a function of signal amplitude; specifically, the gain for low-level signals is more than it is for high-level signals. See also *compressor* and *expander*.

expert system (ES) A computer system that helps solve problems in a given field or application by drawing inference from a knowledge base developed from human expertise. Armed with a set of rules, similar to those that a human expert in the field uses to make judgments, the program can answer questions about some problem area or make decisions based on data inputs. Some *expert systems* are able to expand their knowledge base and develop new inference rules based on their experience with previous problems.

The term *expert system* is sometimes used as a synonym for *knowledge-based system;* however, the expert system is usually taken to emphasize expert knowledge.

expire A term used to describe the automatic deletion of a file after a predetermined amount of time. This is most frequently applied to e-mail messages so that they don't accumulate and waste storage space. *Note:* Unread messages will be deleted just as read messages.

explicit congestion notification (ECN) In frame relay transmissions, a mechanism for indicating that there is traffic congestion on the network. Congestion can be indicated by either or both signaling bit values in a packet header, i.e.,

- *Backward Explicit Congestion Notification (BECN)*—a bit set in headers moving in the direction *opposite* the congestion. It serves to warn source nodes that congestion is occurring "down line."
- *Forward Explicit Congestion Notification (FECN)*—a bit set in the header to warn a destination node that there is congestion.

explorer frame In networks that use source routing, such as IBM's Token Ring network, an *explorer frame* is used to determine a route from the source node to a destination. There are two types of explorer frames:

- An *all routes explorer frame,* which explores all possible routes between source and destination, and

- A *spanning tree explorer frame,* which follows only routes on the spanning tree for the network.

Also known as a *discovery packet,* particularly in the Internet community.

explorer superframe (ESF) A frame sent out by a networked device in a source route bridge environment to determine the optimal route to another networked device.

exponent In mathematics, a number that shows how many times another number is multiplied by itself. For example, if an exponent is 3 and the second number is 4, then the result of the *exponent's* operation is $4 \times 4 \times 4 = 64$. The shorthand way of writing this is 4^3 and is read 4 to the power 3.

If the exponent of any number is 0, the resulting operation yields a value of 1 by *definition,* e.g., $1^0 = 1$, $10^0 = 1$, $8192^0 = 1$, $x^0 = 1$. The only exception occurs when x itself is zero; in this case, the result is said to be *indeterminate.*

In computer technology, the most frequent number raised to a power is 2. Some significant powers of two are listed in the table in the three equivalent forms.

POWERS OF TWO

Notation	Meaning	Value
2^0	1	1
2^1	2	2
2^2	2×2	4
2^3	2×2×2	8
2^4	2×2×2×2	16
2^5	2×2×2×2×2	32
2^6	2×2×2×2×2×2	64
2^7	2×2×2×2×2×2×2	128
2^8	2×2×2×2×2×2×2×2	256
2^9	2×2×2×2×2×2×2×2×2	512
2^{10} (1K)	2×2×2×2×2×2×2×2×2×2	1024
2^{11} (2K)	2×2×2×2×2×2×2×2×2×2×2	2048
2^{16}	2×2×2×2×2×2×2×2×2×2×2×2×2×2×2×2	65 536
2^{20} (1Meg)	2×2×2×2×2×2×2×2×2×2×2×2×2×2×2× 2×2×2×2×2	1 048 576

In scientific notation, the exponent is the power of ten. For example, the exponent in 1.234×10^5 is 5. 1.234 is the mantissa. The number shown is:

$$1.234 \times 10 \times 10 \times 10 \times 10 \times 10 \text{ or } 123\,400.$$

express orderwire A permanently connected voice circuit between selected stations for technical control purposes.

extended area service (EAS) (1) In telephony, the use of load coils to extend the range of both toll and nontoll office to office trunks as well as subscriber lines. (2) A network-provided service feature in which a user pays a higher flat rate to obtain wider geographical coverage without paying per call charges for calls within the wider area.

Extended Binary Coded Decimal Interchange Code (EBCDIC) See *EBCDIC*.

extended capabilities port (ECP) An improved parallel port that increases the data transfer rate from 150 Kbps to 2–5 Mbps and provides bidirectional capability. A technology originally developed by Hewlett-Packard called Zippy was renamed *ECP* when Microsoft joined with HP in implementing the development. See also *enhanced parallel port (EPP)* and *IEEE 1284*.

extended character set An extension of the basic 128 characters of the 7-bit ASCII character set to 256 characters. Unfortunately, there is no universal standard for the extended set; for example, the IBM PC extended character set is different from the Macintosh character set, which in turn is different from the extended set in countries around the world.

Extended Industry Standard Architecture (EISA) A 32-bit backward compatible adaptation of the 8/16-bit ISA bus originally developed by IBM. It was used in PCs that employ the 80X86 processor chips, some UNIX workstations, and servers. *EISA* was a joint development from Compaq and other PC manufacturers. Newer PCs use the PCI bus. See also *ISA, MicroChannel, Peripheral Component Interconnect (PCI),* and *VESA.*

extended LAN A network consisting of multiple local area networks (LANs) interconnected by bridges. An Ethernet *extended LAN* may contain up to seven bridges.

extended superframe format (EF or ESF) A T-carrier framing technique requiring less frequent synchronization than the original T-carrier superframe format. Less frequent synchronization frees overhead bits for use in testing and monitoring. The industry standard enhanced version of D4 formatting is composed of 24 frames of DS1 channels with 192 bits, each preceded by one added signaling bit (*F-bit*) to provide synchronization, supervisory control, and maintenance capabilities.

Specifically:

- Framing, as nonextended version (frames 4, 8, 12, . . . , 24) at 2 kbps.
- A 4-kbps data link between end points for maintenance communications (frames 1, 3, 5, . . . , 23).
- A 6-bit cyclic redundancy check (CRC-6) value (frames 2, 6, 10, . . . , 22) at 2 kbps.
- The eighth bit in every channel of frames 6, 12, 18, and 24 is used for signaling between central offices.

Formerly called F_e.

Exterior Gateway Protocol (EGP) A TCP/IP protocol that routes Internet datagrams from an *autonomous system* into the Internet. The protocol advertises the IP addresses of networks in that system to a gateway in another autonomous system. The *EGP* enables a *boundary router* to determine which networks can be reached through other boundary routers on the Internet. Hence, it is able to direct traffic across the network to the desired destination.

The *EGP* protocol is based on a previous Internet model in which there was a single backbone with many autonomous networks attached. Because of the philosophical change in the Internet structure to a hierarchical model, this protocol has largely been replaced by the *Border Gateway Protocol (BGP)*. Also called *routers* or just *gateways*.

external data representation (XDR) A standard concerning data format representation. Documented in RFC 1014.

external modem Stand-alone modem (DCE) that is connected to a DTE via only a signal cable. External modems have their own power source so they do not need a power connection to the host DTE. They also generally have front panel indicators and controls not available on their counterpart, the *internal modem*.

extinction coefficient The sum of the absorption coefficient and the scattering coefficient.

extinction ratio (r_e) The ratio of two optical power levels as a result of digital modulation of an optical source (e.g., a laser diode), i.e.,

$$r_e = P_1/P_2$$

where:

P_1 is the optical power level generated when the source is on, and
P_2 is the power level generated when the light is off.

Extinction ratio may be expressed as a fraction or in decibels (dB).

extinction voltage The voltage across a gas tube at which the discharge ceases.

extinguishing voltage The voltage across a gas tube at which an abrupt decrease in current flow occurs and the glow around the electrode ceases. This current is dependent on the series impedance external to the gas tube. Also called *extinction voltage*. See also *gasfilled protector.*

extrinsic joint loss In fiber optics, all losses in a connection that are not intrinsic to the fiber, e.g., end separation, angular misalignment, lateral misalignment, and radial misalignment. See also *coupling loss.*

eye pattern The superposition of all analog waveshapes of a demodulator resulting from all possible code sequences.

The following diagram is an *eye pattern* for a ternary coded signal. Hence, the test pattern transmitted must include all pulse sequences involving the three states -1, 0, and $+1$. The trace of the *ideal eye* pattern is shown in heavy lines. The fine lines represent the extremes of amplitude jitter (vertical offsets) and time jitter (horizontal offsets).

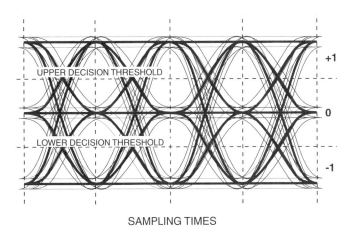

SAMPLING TIMES

As the amplitude jitter and time jitter increase, the "eyes" will close, making the location of the decision crosshairs more critical. Also, as the eyes close, random noise will cause random errors to occur, thereby deteriorating the system bit error rate (BER) performance.

eye relief The distance between the eye and the eyepiece of an optical instrument. Proper distance is determined by the eyepiece design. Typically, *eye relief* distance is 5 to 20 mm. Short *eye relief* (less than about 10 mm) necessitates a closeness of the eye to the lens; eyelashes may be felt touching the lens when a user blinks. Longer eye relief allows a user to wear corrective glasses during viewing.

eyepiece The lens system used between the final *real image* of an optical system and the eye. It acts as an image magnifier. Several common *eyepiece* designs are shown here.

- *Simple—*

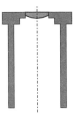

- *Huygen*—Although fields of view to 50° or more are possible, the simple construction is not well corrected for optical defects; hence, their use is limited to low magnification.

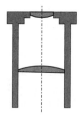

- *Ramsden*—An improvement over the Huygen eyepiece in that it gives a somewhat flatter field. However, the color correction is poor with the formation of color fringes around the object. The field of view is a usable 25° out of a theoretical 40°, and the short eye relief only allows the use of medium to long focal lengths.

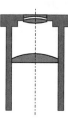

- *Kellner*—An improvement over the Ramsden eyepiece due to the addition of an achromatic eye lens. It provides reasonable performance at low to medium powers. At higher powers, the eye relief may not be adequate.

- *Orthoscopic*—provides
  - reasonably good eye relief to moderate powers,
  - chromatic aberration correction,
  - spherical aberration correction, and
  - moderately wide field of view (~40°).

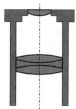

- *Plössl*—Like the orthoscopic eyepiece, the Plössl provides:
  - good eye relief,
  - chromatic aberration correction,
  - spherical aberration correction, but
  - a wider field of view (~50°).

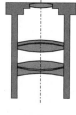

- *Erfle*—A 6-element eyepiece optimized for extra wide viewing. The wide field of view (60–70°) is, however, at the expense of sharpness at the image edges.

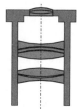

- *Ultra-wide*—A multielement (6-8) design that provides extremely wide fields of view (~85°) and chromatic and spherical aberration correction. Because of the larger number of elements, greater light loss is encountered (due to absorption).

Also called an *ocular*.

F

f The SI prefix for *femto* or 10^{-15}.

F (1) The symbol for *farad*. (2) The symbol for the units of temperature in degrees *fahrenheit*.

F connector A connector used in 10Broad36 (broadband Ethernet) networks, in the broadband versions of the IEEE 802.4 token-bus architecture, and in CATV distribution.

The drawing illustrates the female panel mounted connector on the right and a male cable termination on the left.

F layer See *F region* and *ionosphere*.

F region Reflects normal incident frequencies at or below the critical frequency (approximately 10 MHz) and partially absorbs waves of higher frequency.

The F_1 layer exists from about 160 to 250 km above the surface of the Earth and only during daylight hours. Though fairly regular in its characteristics, it is not observable everywhere nor on all days. The principal reflecting layer during the summer for paths of 2000 to 3500 km is the F_1 layer. The F_1 layer has approximately 5×10^5 e-/cm^3 (free electrons per cubic centimeter) at noontime and minimum sunspot activity, and increases to roughly 2×10^6 e-/cm^3 during maximum sunspot activity. The F_2 layer exists from about 200 to 400 km above the surface of the Earth. The F_2 layer is the principal reflecting layer for high-frequency communications during both day and night. The horizon-limited distance for one-hop F_2 propagation is usually around 4000 km. The F_2 layer has about 10^6 e-/cm^3. However, variations are usually large, irregular, and particularly pronounced during periods of high solar activity and magnetic storms. The F_1 layer merges into the F_2 layer at night and exists about 280 km above the surface of the Earth. The free electron density falls to below 10^4 e-/cm^3 at night. See also *ionosphere*.

F.n recommendations An ITU-T series of specifications concerning telephone services other than telephone. See Appendix E for a complete list.

F1 layer The lower of the two ionized layers normally existing in the F region of the daytime half of the ionosphere. See also *F region* and *ionosphere*.

F1A line weighting In telephony, a noise-measuring set weighting factor used to approximate the noise performance of a type 302 (or similar) receiver.

F2 layer (1) The ionized layer normally existing in the F region of the nighttime half of the ionosphere. (2) The upper of the two ionized layers normally existing in the F region of the daytime half of the ionosphere. See also *F region* and *ionosphere*.

F2F An abbreviation of Face to Face. A meeting in *real life (RL)* as opposed to cyberspace.

fabric (1) In telephony switching, a term that refers to the switching means itself. The capability of connection of every input to every output. (2) In Internet, a nonblocking switch that interconnects all nodes to which it is attached.

FACD An abbreviation of Foreign Area Customer Dialing.

facility (1) In telecommunications, a generic term for a logical component of a system, such as a channel or link between two locations or stations, cross connect, switch, computer, control center, and building. (2) A feature or capability offered by a system, item of hardware, or software. (3) A network-provided service to users or the network operating administration. (4) In a protocol applicable to a data unit, such as a block or frame, an additional item of information or a constraint encoded within the protocol to provide the required control. (5) In an X.25 packet, a field through which users can request special services from the network. (6) A fixed, mobile, or transportable structure, including all installed electrical and electronic wiring, cabling, and equipment and all supporting structures, such as utility, ground network, and electrical supporting structures.

facility bypass In telecommunications, a communication strategy that bypasses the telephone company's central office. For example, wireless transmissions might use *facility bypass*.

facility data link (FDL) In an extended super frame (ESF) digital transmission format, a 4-kbps communications link between the sender's station and the telephone company's monitors. This 4-kbps band is created by taking half of the 24 framing bits in the ESF and using them for the link.

facility grounding system The composite of all electrically interconnected conductors that provides multiple current paths to the Earth, including:

- The wire and cable joining equipment and other subsystems.
- The Earth electrode subsystem.
- The lightning protection subsystem.
- The fault protection subsystem.

facsimile Abbreviated *fax*. A system for transmitting fixed images (text or graphics) over a communications circuit with the end goal of presenting a likeness of the original image in permanent form (paper copy). Resolution and encoding schemes are described in ITU-T Group 1–4 recommendations.

This definition does not exclude the use of computers either to generate or receive images in a paperless manner. See also *group 1, 2, 3, 3 bis, and 4 facsimile*.

facsimile converter (1) In a facsimile receiver, a device that changes signal modulation from frequency shift keying (FSK) to amplitude modulation (AM). (2) In a facsimile transmitter, a device that changes signal modulation from amplitude modulation (AM) to frequency shift keying (FSK).

facsimile picture signal In facsimile systems, the information signal (baseband signal) generated by the scanning process.

factorial (!) In mathematics, the product of all integers from 1 to the specified integer. For example, 6 factorial is written:

$$6! = 6 \times 5 \times 4 \times 3 \times 2 \times 1 = 720$$

fade margin A term that refers to the amount of signal that can be lost due to fading before the signal becomes unusable. It may be stated in either of two ways, i.e.,

- The amount by which a receiving system's gain or sensitivity is in excess of the value necessary to provide the desired quality of service.
- The amount by which a received signal level may be attenuated or reduced without causing the system's operation to fall below a specified performance threshold.

Fade margin is usually expressed in decibels (dB). Also called *fading margin.*

fading In signal receivers, the variations (loss) over time of received amplitude, relative phase, or both, of one or more frequencies in the signal. *Fading* is caused by changes in the characteristics of the transmission path with time. It may be flat; that is, all frequencies change at the same time and rate, or it may be selective, affecting only a limited frequency range.

A few causes of fading effects include:

Blocking—attenuation due to the presence of an object between a transmitter and a receiver. Generally occurs with mobile communications systems, though smoke or fog may affect fixed stations.

Diffraction—attenuation of signal due to loss of ground clearance (because of effects such as atmospheric refraction leading to beam bending).

Distance—attenuation caused by an increased distance between the transmission source and receiver.

Ionospheric movement—changes in the altitude or shape of the ionosphere or changes in the density of electrons in the ionosphere which change the location and strength of reflected waves. This leads to attenuation and/or multipath *fading.*

Multipath—attenuation or boosting of signal strength due to the cancellation or reinforcement of reflected waves at a receiver. Multiple paths may be created by atmospheric effects as well as reflections from physical objects.

Rain—attenuation of signal due to absorption of certain radio frequencies by water (rain, ice, snow, etc.) in the transmission path.

Rayleigh—attenuation caused by the scattering of signal energy due to inhomogenites of the transmission medium composition, which are small compared to the wavelength of the signal carrier. For example, in optics the small variations in the index of refraction on a scale less than the wavelength of transmitted light.

Sometimes referred to as just *fade.*

fading margin See *fade margin.*

FADU An abbreviation of File Access Data Unit.

Fahrenheit (F) A temperature scale developed in 1717 by Gabriel Fahrenheit which assigned a temperature of 32° to the freezing point of water and 96° to the normal temperature of the human body. This put the boiling point of water at 212°, the now used upper reference temperature. See also *Celsius (C)* and *kelvin (K).*

fail safe (1) Of a device or system, a design property that prevents any failure from being critical or leading to a critical condition for another device, system, or person. (2) Pertaining to the automatic protection of programs and/or processing systems to maintain safety when a hardware or software failure is detected in a system. (3) Pertaining to the structuring of a system such that either it cannot fail to accomplish its assigned mission regardless of environmental factors or the probability of such failure is extremely low. (4) A backup mechanism to a primary protective device. For example, in telephony, a device on a subscriber loop to short the loop to ground if a power cross condition exists with sufficient energy to damage the exchange.

fail-safe operation (1) An operation that ensures that a failure of equipment, a process, or system does not spread beyond the immediate vicinity of the failing entity. (2) A control means that prevents improper system operation or catastrophic degradation in the event of circuit malfunction or operator error.

failure (1) The point at which a device or system is not capable of performing its required function within a designated specification (the *failure criteria*). (2) The temporary or permanent loss of an entity's ability to perform its designated function.

failure in time (FIT) An expression estimating the field failure rate as number of failures expected per 10^9 hours of *powered operation.* (100 *FITs* is a failure rate of 0.01% per 1000 hours of operation.)

failure mode Failures are classified in many ways. One way relates to how the device or system itself performs after the failure event without regard to the cause of the failure or its consequences. Three modes of device or system failures are:

- *Catastrophic*—a failure that is both sudden and complete in one or more of the fundamental functions.
- *Degraded*—a failure that is gradual and/or partial in any function. With a failure of this type, the device or system continues to function, though with compromises in the device performance. That is, the device is functioning outside its specified limits.
- *Incipient*—an imperfection in an element of a system or device, which left uncorrected may lead to degraded performance or catastrophic failure. A device or system with an *incipient failure* is performing all of its functions within specified limits. However, there exists some indication that continued operation will lead to a degradation or catastrophic failure. For example, an engine indicating over-temperature may be performing satisfactorily, but without attention its performance will degrade and ultimately fail.

fair queuing In networking, a gateway congestion-controlling scheme in which every host is restricted to an equal share of the gateway's bandwidth. *Fair queuing* does not distinguish between small and large hosts or between hosts with few or many active connections.

fake root In Novell's NetWare version 2.2 and above, a drive mapping to a subdirectory that makes the subdirectory appear to be the root directory. A *fake root* allows you to install programs into subdirectories, even though they insist on executing in the root directory. With the programs in a subdirectory, administrators can be more specific about where they allow users to have rights, and can avoid granting rights at the true root of the volume.

FAL An abbreviation of File Access Listener.

fall back (1) The ability to default to an alternate device or parametric value should the primary fail to perform satisfactorily. (2) The reduction in transmission speed of a modem. This reduction may be due to several causes for example,

- The modem at the remote end does not support the speeds of the local modem.
- Communications channel noise has increased the *bit error rate (BER)* to unacceptable levels. Therefore, the modems renegotiate at a lower speed where the noise is tolerable.

See also *bit error rate* and *Shannon limit.*

fall time Typically, the time interval on the trailing edge of a pulse waveform between the first occurrence of the 90% of final value and the last occurrence of the 10% of final value instances. These values are exclusive of any overshoot, undershoot, or ringing. Also called the *last transition duration* and *pulse decay time*. See also *pulse*.

false lock A condition in which a phase locked loop (PLL) locks to either a frequency other than the correct one, or to an improper phase.

false ringing A recording or synthesis of the ringback signal that is played when a call is being transferred or while a switching system is listening for a facsimile calling (CNG) tone.

falsing In telephony, the condition whereby dual-tone multifrequency (DTMF) tones are falsely detected. Although the tone detectors are designed to minimize the effect, detection can occur with certain speakers, noise, and music.

fanout **(1)** A circuit configuration in which a device output is delivered to more than one receiving device. **(2)** In communications and signaling, a configuration in which there are more output lines than input lines. See also *breakout*.

FAQ An abbreviation of <u>F</u>requently <u>A</u>sked <u>Q</u>uestions. Lists of answers to questions regularly asked on a given bulletin board system (BBS) or network. The lists are frequently posted in news groups or new user forums on the Internet.

far end block error (FEBE) In broadband ISDN (BISDN) networks, an error reported to the sender by the receiver when the receiver's computed checksum result does not match the sender's checksum.

far end crosstalk (FEXT) A particular form of crosstalk in which the disturbing signal is coupled from the local end of one communication circuit to the remote end of another circuit. In the figure, three possible paths of *far end crosstalk (FEXT)* are diagrammed: in the local end office (or its outside plant cabling), in the interoffice network, or in the remote office (or its outside plant cabling).

The remote office for the signal's specified destination need not be the same as the office with far end crosstalk. See also *Near End Crosstalk (NEXT)* and *crosstalk*.

LOCAL END   REMOTE END

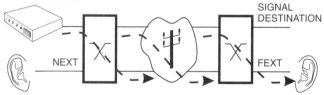

far end receive failure (FERF) In broadband ISDN (BISDN) networks, a signal sent upstream to indicate that an error has been detected downstream.

far zone A synonym for *far-field region*.

far-field A synonym for *far-field region*.

far-field diffraction pattern The diffraction pattern of a monochromatic point source (such as a light emitting diode—LED—an injection laser diode—ILD—or the output of an optical waveguide) observed at an infinite distance from the source. A *far-field pattern* exists at distances that are large compared with

$$s^2/\lambda$$

where:

  s Is a characteristic dimension of the source. For example, if the source is a uniformly radiating circle, then *s* is the radius of the circle, and

  λ Is the wavelength.

Also called a *Fraunhofer diffraction pattern*. Contrast with *near-field diffraction pattern*. See also *far-field region*.

far-field radiation pattern A radiation pattern measured in the *far-field region* of an emitter (RF antenna, LED, sound transducer, etc.)

far-field region The region where the angular field distribution is essentially independent of distance from the source. If the source has a maximum overall dimension *s* that is large compared to the wavelength λ, the *far-field region* is commonly taken to exist at distances greater than $2s^2/\lambda$ from the source.

Also called *far field, far zone*, and *radiation field*. For a beam focused at infinity, the *far-field region* is also referred to as the *Fraunhofer region*.

farad (F) The unit of capacitance (the ability to hold a charge between two conductive plates separated by a dielectric). A 1 *farad* capacitor holds a 1 *coulomb* charge with a potential difference of 1 V across its terminals. The term is named for the nineteenth-century English physicist Michael Faraday (1791–1867).

Faraday effect A magneto-optic effect in which the polarization plane of an electromagnetic wave is rotated under the influence of a magnetic field parallel to the direction of propagation. This effect can be used to modulate a lightwave.

fast busy A call progress tone that indicates all trunks are busy. The tone frequencies are the same as normal busy, but the interruptions are at 120 per minute. See also *call progress tone*.

fast connect circuit switching The use of fast, electronic switching to establish a path (circuit) between two stations.

fast Ethernet See *Ethernet* and *IEEE 802.x*.

fast Fourier transform (FFT) A set of algorithms used to compute the spectral content of an arbitrary periodic waveform.

fast operate See *relay*.

fast packet A general term for various streamlined packet technologies, including frame relay, BISDN, and ATM. Compared to X.25 packet switching, *fast packet* contains a much reduced functionality and lower overhead. Therefore, *fast packet* systems can operate at higher rates for the same processing cost.

fast packet switch (FPS) A packet switching strategy that achieves high throughput by simplifying the switching process. Steps to accomplish this include the following:

- Letting higher level protocols do error checking and acknowledgments.
- Using fixed-size packets.
- Using simplified addresses.
- Switching packets on-the-fly, i.e., as they come in, rather than buffering the entire packet before sending it on.
- Utilization of statistical multiplexing.

Examples of such switches include *ATM, cell relay*, and *frame relay*.

fast switching channel In a Global Positioning System (GPS), a single channel that rapidly samples information to determine ranges from the visible satellites. The switching rate is 2 to 5 ms.

FastPath A high-speed gateway between AppleTalk and Ethernet networks.

FAT An acronym from File Allocation Table. It is the disk file system used by MS-DOS, PC-DOS, and compatible operating systems.

Incompatible file systems include High Performance File System (HPFS) and New Technology File System (NTFS).

fault (1) An accidental condition that prevents an entity from performing its specified function. (2) Any defect that causes a reproducible degradation or catastrophic malfunction. (A malfunction is reproducible if it occurs consistently under the same circumstances.) (3) In power systems, an unintentional current path (short circuit or partial short circuit) between an energized conductor and another energized conductor, a return conductor, or Earth ground. (4) The point of failure in a malfunctioning or inoperative device.

fault management One of five basic OSI network management tasks specified by the ISO and ITU. *Fault management* is used to:

- Detect, isolate, and correct malfunctions in a telecommunications network. It includes maintaining and examining error logs, accepting and acting on error detection notifications, tracing and identifying faults, carrying out diagnostics tests, correcting faults, reporting error conditions, and localizing and tracing faults by examining and manipulating database information.
- Compensate for environmental changes.

fault tolerance The ability of a system to automatically recover from (or operate in the presence of) an error or failure without the loss of data.

In communication systems, forward error correction methods or receive error detectors with retransmit are frequently used to make the transmission of data *fault tolerant*. Multiple devices operating in parallel are frequently used as a means of implementing *fault tolerant* systems, e.g., RAID memory systems, mirror servers, and redundant transmission paths.

FAX An acronym for *facsimile*.

FAX back An interactive automated response system that sends specified facsimile documents to a calling party in response to DTMF signaling. The calling party selects desired data from the available database in response to voice messages sent by the called terminal.

The system, when called, answers and delivers a menu of choices to the caller. The caller, using DTMF signaling, selects an option/document from the menu. The caller then enters the phone number of the FAX machine to which the document is to be delivered and goes on-hook. The *FAX back* system then calls the specified number and transmits the document. Also called *FAX on demand (FOD)*.

FAX class 1 Electronics Industry Association (EIA) computer to modem interface standard based on the AT command set wherein the DTE software controls the protocols.

FAX class 2 A computer to modem interface standard based on the AT command set wherein the fax/modem hardware controls the protocols. The working proposal, TR29.2, was adopted by industry as a standard before it was officially released as a standard.

FAX class 2.0 The official ITU-T release of the computer to modem interface standard based on the AT command set in which hardware protocol is defined.

FAX mailbox A facility that stores received facsimile messages for later retrieval. For example, a business traveler might check the service to see if any fax messages were received; if any messages were received, they can be re-sent to a convenient location.

FAX modem A modem designed to send and receive facsimile messages as well as the digital data normally associated with a modem. *Fax modems* communicate with a computer using class 1 or class 2 signaling, and they communicate with remote machines using Group III protocols. (Relevant ITU-T standards include V.17, V.27 ter, V.29, and V.34.) *FAX modems* do not provide the facilities for scanning a document for transmission, nor do they generate a paper copy of the document. Even without hardware to enter or print paper copy of the FAX message *FAX modems* have several advantages:

- They are much less expensive than a full facsimile machine.
- The transmitted copy is much better quality because it is not scanned in. It is simply translated from one digital form to another.
- Transmission time can be faster because the computer can handle all of the dialing and retries should the first attempt fail (busy line, bad telephone line noise, . . .).
- The transmission can be scheduled for a time when telephone rates are the lowest.

fax on demand (FOD) See *FAX back*.

fax server A specialized Interactive Voice Response (IVR) system that sends requested facsimile messages to a fax machine designated by DTMF tones.

Essentially, a database of fax images resides in the server. A user calls the server, and the server delivers a series of voice messages in response to user signaling. At the end of the messages, the user selects one or more fax messages to be transmitted and the fax machine to which the messages are to be delivered. User signaling is generally by DTMF tones; however, voice recognition systems are sometimes found.

FB An abbreviation of Framing Bit.

FBE An abbreviation of Free Buffer Enquiry.

FC (1) An abbreviation of Frame Control. (2) An abbreviation of Flat Cable—see *ribbon cable*. (3) An abbreviation of Functional Component. (4) An abbreviation of Fiber Channel.

FC connector A connector used for fiber optic cable. It uses a threaded coupling nut for the attachment and 2.5-millimeter ceramic ferrules to hold the fiber.

FCC An abbreviation of Federal Communications Commission. An agency of the federal government created in 1934 to regulate all nonfederal government interstate and U.S. terminated international wire and wireless transmissions, including telecommunications (radio, telephone, television, and telegraph). Five presidential appointees make up the board.

The agency develops and publishes guidelines that govern the operation of communications and other electrical equipment in the United States in the Code of Federal Regulations (CFR). They also allocate (in concert with the International Telecommunications Union—ITU) portions of the electromagnetic spectrum for designated applications and grant licenses to users to operate equipment in the assigned frequency range. Within Title 47 of the Code of Federal Regulations, several parts are frequently cited, namely:

FCC Part 15—regulations under which intentional, unintentional, and incidental radiators may be operated without an individual license. These regulations include **any** device that radiates electromagnetic power between 9 kHz and approximately 300 GHz.

Class A Certification—a certification for computer or other electronic equipment intended for industrial, commercial, or office use, rather than for personal use at home. The Class A commercial certification is less restrictive than the Class B certification.

Class B Certification—a certification for electronic equipment, including PCs, laptops, and portables intended for use in the home rather than in a commercial setting. Class B certification is more stringent than the commercial Class A certification.

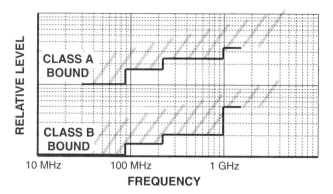

FCC Part 22—regulations concerning the operation and specifics of the cellular telephone network.

FCC Part 68—regulations that protect the telephone network from damage by subscriber equipment. It does not guarantee that a device with Part 68 certification is functional or useful (only harmless to the network).

FCC Part 90—regulations concerning the specifics of Citizen's Band radio.

A similar authority for regulation of federal government telecommunications is vested in the National Telecommunications and Information Administration (NTIA).

FCC registration program The Federal Communications Commission (FCC) program and associated directives intended to assure that all connected terminal equipment and protective circuitry will not harm the public switched telephone network or certain private line services.

The program requires that any terminal equipment or protective device connected to the public telephone network be in compliance with the Code of Federal Regulations (CFR), Title 47, part 68 (including the appropriate testing of the equipment and the assignment of identification numbers). The *FCC registration program,* however, contains no requirement that registered equipment be compatible, or even function, with the network.

FCFS An abbreviation of First Come, First Serve. A description of the order of packet transmission, as opposed to a round robin method. Applies in multiple access to linear bus local area networks (LANs) where stations may place reservations in a separate logical channel.

FCS (**1**) An abbreviation of Fiber Channel Standard. The specifications for optical fiber in the Fiber Distributed Data Interface (FDDI) network architecture. (**2**) An abbreviation of Frame Check Sequence. In packet transmission systems, a computed value that is used to check for errors in the transmitted message. An *FCS* value is determined before sending the message, and it is stored in the packet's *FCS field.* A second *FCS* value is computed from the received packet. If the second value does not match the original, a transmission error has occurred. See also *CRC, error detection,* and *parity.*

FCVC An abbreviation of Flow Control Virtual Circuit.

FDD (**1**) An abbreviation of Frequency Division Duplex. (**2**) An abbreviation of Floppy Disk Drive.

FDDI An abbreviation of Fiber Distributed Data Interface.

FDDI documents The FDDI standard consists of four documents: physical medium dependent (PMD), physical (PHY), media access control (MAC), and station management (SMT).

FDDI-II A variant of FDDI that supports isochronous traffic and both packet and circuit switched traffic. It also expands the types of acceptable cable types and excitation sources.

Officially named *hybrid ring control (HRC)* FDDI.

FDDI-II Cycle A 12,500-bit protocol data unit (PDU), or packet, that provides the basic framing for the Fiber Distributed Data Interface (FDDI) transmission in a FDDI II network operating in hybrid mode. The cycle is repeated 8000 times per second, which yields 100 Mbps of bandwidth for the network. The cycle contains the following fields:

Preamble—establishes and maintains synchronization. The preamble is nominally 20 bits but may be either 16 or 24 bits in a particular cycle to adjust for jitter, or loss of phase alignment, in the transmission.

Cycle header—specifies how the cycle is to be used. One part of the information specified in the 12 bytes in the header is whether each of the wideband channels is being used for packet switched or isochronous data.

Dedicated packet group (DPG)—used for packet-transfer control. The DPG consists of 12 bytes.

Wideband channel (WBC)—the payload, used for actual data transmission. There are 16 WBCs in each cycle. Each WBC consists of 96 bytes, or octets, and may be subdivided into subchannels. Depending on the number of bits allocated each cycle, subchannels may have bandwidths ranging from 8 kilobits per second (kbps) to 6.144 Mbps. For example, an 8-bit-per-cycle subchannel yields a 64-kbps data rate, corresponding to a B channel in the ISDN telecommunications model; using 193 bits per cycle yields a 1.544-Mbps T-1 line. The default FDDI II WBC uses all 768 bits for a single channel.

FDDITalk Apple's implementation of Fiber Distributed Data Interface (FDDI) protocols and drivers for use in an AppleTalk network.

FDL An abbreviation of Facility Data Link.

FDM An abbreviation of Frequency Division Multiplex.

FDMA An abbreviation of Frequency Division Multiple Access.

FDX An abbreviation of Full DupleX.

F$_e$ (**1**) From Frame Extended (read "F sub e"), the old name for extended super frame (ESF). (**2**) The atomic symbol for iron (from ferrous).

FE-1 An abbreviation of Fractional E-1.

FEBE An abbreviation of Far End Block Error.

FEC An acronym for Forward Error Correction.

FECN An abbreviation of Forward Explicit Congestion Notification. See *explicit congestion notification (ECN).*

FED An acronym from Field Emission Display.

FED-STD A set of standards from the Federal Telecommunications Standards Committee (FTSC) which define all aspects of data modulation and transmission.

Federal Information Exchange (FIX) A connection point between the Internet and any of the federal government's internets.

Federal Networking Council (FNC) A committee of representatives from those federal agencies involved in the development and use of federal networking, especially those networks using TCP/IP and the connected Internet. The *FNC* coordinates research and engineering. Members include representatives from the Department of Defense, Department of Energy, Defense Advanced Research Projects Agency (DARPA), National Science Foundation, National Aeronautics and Space Administration (NASA), and Health and Human Services (HHS). Formerly called the *FRICC*.

Federal Radionavigation Plan (FRP) The U.S. government document that defines the official plan for commercial use of GPS. See also *Global Positioning System (GPS)*.

Federal Telecommunications Standards Committee (FTSC) A U.S. government agency established in 1973 to establish and promote communications standards and interfaces. The organization's charter states that its functions are:

- "To develop, coordinate, and promulgate the technical and procedural standards required to achieve inter-operability among functionally similar telecommunications networks of the Federal components of the National Communications System.
- In concert with the National Bureau of Standards (NBS), to develop and coordinate technical and procedural standards for data transmission and the computer-telecommunications interface.
- Increase cohesiveness and effectiveness of the Federal telecommunications community's participation in national/international standards programs and in the Federal Information Processing Standards (FIPS) program."

Federal Telecommunications System (FTS) A switched long-distance telecommunications service formerly provided for official federal government use. The *FTS* has been replaced by *Federal Telecommunications System 2000 (FTS2000)*.

Federal Telecommunications System 2000 See *FTS2000*.

federated database A database that consists of multiple parts, each part residing on a separate host, and is linked by a network so that it appears as a single database to the user. Databases of this type occur when there are collaborative developments by groups geographically separated.

feed (1) In general, to supply data, energy, or materials to an entity. For example, to send data to a DCE or paper to a printer. (2) Pertaining to the function of inserting one thing into another, such as in a feed horn, paper feed, card feed, and line feed. (3) In telephony, the supply of current to a subscriber's terminal equipment (e.g., telephone or modem). (4) A transmission facility between a signal source, such as is generated in a radio or television studio, and the head-end of a distribution facility, such as a broadcasting station or a cable distribution network. (5) With antennas, a coupling device between the antenna and its transmission line. A *feed* may consist of a distribution network or a primary radiator. (6) To supply a signal to the input of a system, subsystem, equipment, or component, such as a transmission line or antenna. Also called *feed current*.

feed circuit In telephony, the device used to apply dc current to a loop to power a telephone subset while passing the voice band frequencies on to switching equipment. See also *BORSCHT, Hayes bridge,* and *Stone bridge.*

feed current In telephony, the supply of current to a subscriber's terminal equipment (e.g., telephone or modem). The current is provided by the central office (CO) and ranges from about 20 mA to 150 mA, depending on the loop length. The current serves two functions:

- *Signaling*—When no current is flowing, the CO assumes the terminal equipment is *on-hook* and when currents greater than about 10 mA are flowing, the CO puts the line in *off-hook* status.
- *Power*—The current flowing in the loop is used to power many devices such as a telephone or pocket modem.

feed line The transmission line connecting a transmitter to its antenna.

feedback (1) The returning of a fraction of the output or processed signal to the input of a device or system. *Feedback* may be classified in one of two ways:

- *Negative feedback*—That is, *feedback* in which the signal returned to the input is substantially out of phase with the input signal reduces the circuit gain and distortion, and increases linearity and stability. Also called *degenerative feedback* (subtractive).
- *Positive feedback*—That is, *feedback* in which the signal returned to the input is substantially in phase with the input signal increases the circuit gain and distortion, and decreases linearity and stability. Positive feedback may be used to speed up a systems response; however, it more often leads to system oscillation. Also called *regenerative feedback* (additive).

Feedback may occur inadvertently and is then usually detrimental. (2) Information returned as a response to an originating source.

FEFO An acronym for First Ended, First Out. A queuing procedure.

FEMF An abbreviation of Foreign EMF. Also called *battery cross.*

FEP An abbreviation of Front End Processor.

FERF An abbreviation of Far End Receive Failure.

Fermat's principle In optics, a principle stating that an electromagnetic ray (e.g., light) follows the path that requires the least time to travel from one point to another, including any reflections and refractions that may occur. Also called the *least-time principle.*

ferreed relay A type of reed relay with bistable (latching) operating characteristics. The name was coined by Bell Telephone Laboratories for a reed relay constructed with a magnetic material that retains a magnetic field after the operating current is removed.

ferrule (1) In coaxial connectors, a metal cylinder used both for holding the cable onto the connector and for providing electrical continuity from the shield to the connector body. The *ferrule* is crimped to hold it in the desired position. (2) In fiber optic connectors, a cylinder that serves to keep the optical core and cladding aligned and immobile. The fiber cladding may be glued to the *ferrule* with epoxy. *Ferrules* may be made of ceramic (the most reliable), plastic, or stainless steel.

FET An acronym for Field Effect Transistor. See *transistor.*

FEXT An acronym for Far End CrossTalk.

FF An abbreviation of Form Feed. The ASCII control character is 12 decimal (0Ch).

FFT An abbreviation of Fast Fourier Transform.

FHMA An abbreviation of Frequency Hopping Multiple Access.

FHSS An abbreviation of Frequency Hopping Spread Spectrum.

fiber A dielectric waveguide that "conducts" light. See also *fiber optics (FO)* and *optical fiber.*

fiber amplifier A device that amplifies an optical signal directly, that is, without the need to first convert it to an electrical signal, then amplify it electrically, and finally reconvert it to an optical signal.

One type of fiber amplifier uses an erbium doped optical fiber segment, which both carries the communication signal and is optically pumped with a laser. The laser has a high-powered continuous output at a wavelength slightly shorter than that of the communication signal. The signal is intensified by *Raman amplification.* Because the amplifier does not require rigorous wavelength control of the pumping laser, it is relatively simple and low cost. Also called a *Raman amplifier.*

fiber axis That line flowing lengthwise through an optical fiber and positioned at the center of successive core cross sections.

fiber buffer The protective material surrounding and adjacent to the optical fiber. It provides mechanical isolation and protection; however, it is softer than the outer jacket.

fiber distributed data interface (FDDI) The IEEE 802.x and ANSI standard for a high-speed Token Ring local area network (LAN) architecture using fiber optic cable. Some of the characteristics include:

- Supports nodes up to 2 kilometers (1.25 miles) apart when using multimode cable and up to 40 kilometers (25 miles) when using single mode cable.
- Supports up to a 1000-node network.
- Supports a network span of up to 100 kilometers (62 miles).
- Uses either multimode or single mode fiber optic cable.
- Supports transmission speeds of up to 100 Mbps. Operation to 200 Mbps is possible with the loss of redundancy.
- Uses a dual ring topology in which information travels in opposite directions.
- Uses token passing as the media access method. However, in order to support a high transmission rate, FDDI can have multiple frames circulating the ring at a time, just as with early token release (ETR) in an ordinary Token Ring network.
- Uses optical rather than electrical means to transport signals. The light source can be either a light emitting diode (LED) or a laser operating at 1300 nm.
- Uses a 4B/5B signal encoding. This scheme transmits 5 bits for every 4 bits of information. (This means that an FDDI network needs a clock speed of 125 Mbps to support a 100-Mbps transmission rate.) The actual bits are encoded using NRZ-I waveform encoding.
- Tolerates an allowable power loss of 11 dB between nodes. This value means that about 92% of the signal's power can be lost between two nodes. (The signal is at least partially regenerated by the transceiver at each node.)
- Can handle packets either from the logical link control (LLC) sublayer of the data link layer or from the network layer.
- Supports hybrid networks, which can be created by attaching a subnetwork to the ring through a concentrator. For example, a collection of stations arranged in a star or a tree is possible.

The four ANSI standards are:

- ANSI X3T9.5, containing Physical Media Dependent (PMD) specifications.
- ANSI X3T9.5, containing the Physical (PHY) specifications.
- ANSI X3.139, containing Media Access Control (MAC) specifications.
- ANSI X39.5, containing the Station Management (SMT) specifications.

A variant of FDDI that supports *isochronous* traffic is called FDDI-II.

fiber loss The attenuation of the light signal propagating down an optical waveguide.

fiber optic cable A transmission medium that uses glass or plastic fibers, rather than copper wire, to carry signals. The signal is imposed on the fiber via modulated light (from either a laser or light emitting diode, LED). Because of the high bandwidth and lack of susceptibility to interference, fiber optic cables are frequently used in long-haul or noisy applications.

The actual optical fibers are surrounded by buffers, strength members, and jackets for protection, stiffness, and strength. A *fiber optic cable* may be an all-fiber cable or contain both optical fibers and metallic conductors. Also called *optical cable* and *optical fiber cable.* See also *optical fiber.*

fiber optic inter-repeater link (FIRL or FOIRL) A link segment that uses fiber optic cable to connect two repeaters in an Ethernet network. It is defined in IEEE 802.3 and implemented with two fibers, one transmit and one receive. The link can be up to 1 kilometer long and cannot have any intervening nodes. The standard connector for such a link is the SMA (submineature type-A) connector (IEC 874-2). See also *IRL.*

fiber optics (FO) That branch of optical technology concerned with the transmission of modulated or unmodulated light through fibers made of transparent materials such as glass and plastic. See also *optical fiber.*

fidelity The degree to which a circuit, device, or system (or specified portion thereof) accurately reproduces, at its output, designated characteristics of the signal applied to its input.

FidoNet A noncommercial store-and-forward network consisting of thousands of bulletin board system (BBS) computers operated by private individuals. They operate their BBSs on their own time and with their own money.

FidoNet uses *echomail* for mail and file transfer. It works a little like the Pony Express; that is, all the BBSs on the network schedule a time in the middle of the night to "shut down" and forward mail and other network traffic through predefined pathways. Mail may take overnight, or it may take several days, or it may not arrive at all. There are many thousands of *FidoNet* boards all over the world, and they were usually there before commercial boards and networks.

field (1) The column designation of a database.

A *field* is a group of bits or bytes designated to hold a specific data item within a record or collection of data; for example, a record might consist of the three *fields* called name, address, and phone number. See also *database.* **(2)** That region of space in which the effects of a physical property such as gravity, pressure, or electromagnetic radiation can be measured. For example, the volume around a magnet in which the magnetic force can be measured. A *field* reduces in strength as the distance from the source increases; however, it never reduces to exactly zero. Because a *field* extends to infinity in a mathematical sense, the boundaries are frequently defined as the point at which the *field* strength is reduced to one-half. The region need not be spherical in nature; in fact, with the exception of a *gravitational field,* it rarely is. A common example is the *field* generated by a *dipole antenna,* where the shape is approximately a toroid as shown in the following figure. (The shape of a field is generally illustrated by a plane created by plotting all equipotential field strength values.)

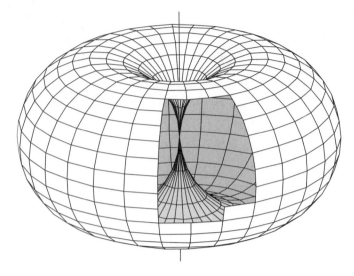

field emission display (FED) A type of flat screen display in which a cathode ejects electrons into a controlled electric field gradient which directs them to a phosphorescent target (anode).

field intensity The irradiance of an electromagnetic wave under specified conditions. *Field intensity* is generally expressed in watts per square meter (W/m^2).

field of view A measure of the angular diameter, expressed in degrees, of the circle of light that is seen by a detector in an optical instrument. Two expressions are used to describe *field of view;* that is,

- *Apparent field of view*—the angular diameter of the virtual region viewed by the detector.
- *True field of view*—the actual angular diameter from the detector to the observed region disregarding the magnification of the optics.

Apparent and true field of views in a telescope are related by the approximation:

$$Apparent\ Field \cong True\ Field \times magnification.$$

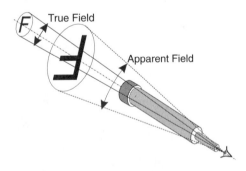

field strength The magnitude of an electric, magnetic, or electromagnetic field at a given point. The units are dependent on the type of field being expressed; that is,

- For an electromagnetic wave, *field strength* is usually expressed as the rms value of the electric field, in volts per meter (V/m).
- For a magnetic field, *field strength* is usually expressed in ampere-turns per meter (AT/m) or oersteds (Oe).

Also called *radio field intensity.*

field wire A flexible insulated wire used in field telephone and telegraph systems. In addition to the normal signal conductors, these wires commonly have a high tensile-strength strand that serves as a strength member. See also *cable* and *wire.*

FIFO An acronym for First In First Out (pronounced "FYE-foe"). A multibyte buffer or memory where the first data element entered (written) is the first element removed (read). See also *first-in first-out (FIFO).*

file (1) A *file* is a named collection of related records. The records may be data, program instructions, or both.

When a *file* is text or graphics material, it is frequently also called a *document.* **(2)** The largest unit of storage structure that consists of a named collection of all occurrences in a database of records of a particular record type. **(3)** For UNIX-based systems, the definition is even more general. It refers to any array of bytes; even devices are treated as *files.*

file access data unit (FADU) In the OSI's File Transfer, Access, and Management (FTAM) service, a packet that contains information about accessing a directory tree in the file system.

file access listener (FAL) In Digital Equipment Corporation's (DEC) DECnet environment, a program that implements the data access protocol (DAP) and that can accept remote requests from processes that use DAP.

file compression A method of reducing the size of a data file. This has the effect of reducing the storage requirements on a host computer as well as reducing the transmission time when the file is transferred across a communication link. See also *compression (2)* and *filename extension.*

file extension See *filename extension.*

file extraction The decoding of a previously encoded file into a binary file. In UNIX-based systems *uuencode* translates a binary file into ASCII, while *uudecode* converts the ASCII file back into binary.

file gap That area on a storage medium, such as a tape, used to indicate the end of file.

file ownership The set of access restrictions associated with a file in a multi-user system. Access restrictions include the ability to *read* the file, *copy* the file, *write* to the file, and *delete* the file. Access restrictions also may be applied to an individual user or to a group.

A UNIX file will have directory entries marked somewhat like the following

- r w x r - - modem_installation_help

The first dash (-) indicates the object is a file. The following three symbols indicate that the owner has read, write, and delete capability. The final three symbols indicate that the public can only read the file. (The dashes indicate no write and no delete rights.) Finally, *modem_installation_help* is the name of the file.

file server (1) A node on a network used to provide special services to all users of the network, such as information storage, sharing, or electronic mail.

A *file server* not only stores massive amounts of data but arbitrates or manages multiple requests for files and enforces *access rights* for file creation, modification, and deletion. Generally, the files are stored on disk; however, they may also be stored on tape or other removable media. A *file server* may be dedicated (that is, no other processes other than network management are allowed while the network is active), or nondedicated (normal user applications can run concurrent with the network). Often a *file server* is simply called a *server.* **(2)** A program, running on a computer, that allows other programs, running on other computers, to access files of that computer.

file sharing An arrangement by which multiple users can access the same file(s) simultaneously. File access has restrictions, however, and it is generally controlled by both application and networking software. For example, certain parts of the file may be locked (made inaccessible) if a user is already accessing that file. Two people cannot change the same area of a file simultaneously.

file site In Internet, a server on which files are stored and anyone can retrieve. Also called *archive site* or *FTP site*.

file system In an operating system, the structure used to name, organize, store, and access file entries on a mass storage device. The file system organizes information about files, such as their names, attributes, and locations. Examples of common file systems include:

- CDFS—the CD-ROM File System used to store information about files on a compact disk.
- FAT—the File Allocation Table used by various versions of DOS.
- HPFS—the High Performance File System used in OS/2.
- NTFS—the New Technology File System used by Windows NT and the NT Advanced File Server.
- HFS—the Hierarchical File System used by the System 7 (the Apple Macintosh operating system).
- NFS—the Novell NetWare 4.x term used in preference to directory structure. (This usage is to avoid confusion between the file system information and the contents of Novell's Directory—the information tree created by the global naming service that replaced the NetWare bindery from earlier versions.) *NFS* has three major levels:
  - *Volume,* the highest level, which refers to a partition created by the NetWare installation program. A volume may encompass any amount of storage from as little as a fraction of one hard disk to as much as multiple disks.
  - *Directory,* an intermediate level that contains other directories or files.
  - *File,* the most specific level (the level at which users or processes generally operate).

file transfer A general phrase used to describe the movement or copying of a file from one computer to another. *File transfers* include uploads, downloads, e-mail transfers, and so on.

When a file is transferred across a network, the file must first be divided into packets for transmission. The details of this "packetization" depend on the transfer protocol (communications and packaging rules) being used. In digital packet networks, File Transfer Protocol (FTP) and File Transfer, Access, and Management (FTAM) are two frequently used protocols. With switched network transfers, that is, transfers using modems, KERMIT, XMODEM, YMODEM, and ZMODEM are commonly available protocols. If the file is being transferred between different operating system environments, the file may also be reformatted during the transfer. For example, in transferring text files between UNIX and DOS environments, the ends of lines must be changed. In transferring from a Macintosh to a DOS environment, the Macintosh file's resource fork will be discarded, and the data fork may also need to be reformatted. These translations may or may not be automatic in the file transfer protocol.

File Transfer Access and Management (FTAM) An ISO 8671 standard involved with integrated message handling as the vehicle for interchanges of EDI information between applications. *FTAM* controls the transfer of whole files or parts of files between end systems. *FTAM* allows remote access to various levels in a file structure and provides a comprehensive set of file management capabilities.

file transfer protocol (FTP) (1) Generally, a protocol that lets a user on one computer system access and copy files to and from another computer system using a network connection. Examples include XMODEM, YMODEM, ZMODEM, KERMIT, and FTP. **(2)** The Transmission Control Protocol/ Internet Protocol (TCP/IP) protocol that is

- A standard high-level protocol for transferring files between computers,
- Usually implemented as an application level program, and
- Uses the Telnet and TCP protocols.

In conjunction with the proper local software, FTP allows computers connected to the Internet to exchange files, regardless of the computer platform. The FTP protocol is described in RFC959. **(3)** In Internet, *FTP* is a TCP/IP application protocol controlling file transfer between hosts. *FTP* is capable of transferring ASCII, binary, or EBCDIC files; however, it assumes ASCII transfers unless otherwise instructed. One logs onto an *FTP* host with either a valid account or with "anonymous" as the name and the e-mail address as the password. Hence, *FTP* is also referred to as *anonymous FTP*. A complete description of the *FTP* command may be obtained from the Internet itself by sending a message to *ftpmail@decwrl.dec.com.* The message field is simply "help." On the Internet, the act of transferring a file is said to be FTP'd, not downloaded.

file transfer system (FTS) A term that refers to any of a multitude of application layer services for handling files and moving them from one location to another. For example,

- Computer Graphics Metafile (CGM).
- Document Filing and Retrieval (DFR).
- Document Printing Application (DPA).
- Electronic Data Interchange (EDI).
- Message Handling System (MHS).
- Remote Database Access (RDA).
- Open Document Architecture/Open Document Interchange Format (ODA/ODIF).
- Virtual Terminal (VT), and many more.

filename The name under which a file is stored, operated upon, and retrieved. The name is generally unique within a given subdirectory.

filename extension A multiletter code at the end of a file name that indicates the type of file. Some of the many file types (and extensions) found on bulletin board systems (BBSs), the Internet, and in an operating PC are shown on the next page.

fill (1) A bit sequence (0s, 1s, or a pattern) used to occupy unused space in a fixed length packet. **(2)** A bit sequence as above used to maintain synchronization between a transmitter and receiver in the absence of message information. Also called *bit stuffing*.

fill character A character that may be inserted into or removed from a data stream without affecting the information content of the stream; it may, however, affect the information layout and/or the control of equipment. Fill characters may be used to occupy either unused transmission time or media space.

filled cable A cable that has a nonhygroscopic material (usually a gel) inside its jacket or sheath to fill the spaces between the interior parts of the cable. This filler prevents moisture from entering minor leaks in the outer sheath and from migrating inside the cable when moisture does enter.

FILO An acronym for First In Last Out. See also *first-in last-out (FILO)*.

filter (1) A computer program that selectively accepts or rejects portions of a data stream based on some defined procedure or criteria. For example, the command, "print all records that have a name start-

COMMON FILE TYPES & EXTENSIONS

Extension	Description	Extension	Description
FONTS and GRAPHICS		**COMPRESSED or ENCODED**	
AI	Encapsulated PostScript	ARC	System Enhancement Associate's archiver.
ATT	AT&T Group 4	ARJ	ARJ compressed files
ART	PFS First Publisher Clip Art	gz	A "gzip" compressed file. Use gunzip or gzip -d.
AVI	Windows Media Player movie files	Hqx	BINHEX binary to hex encoded file
BMF	CorelGallery	sh, shar	SHAR archive without compression (UNIX)
BMP	Windows & OS/2 bitmap	SIT	StuffIt compressed file extension (Macintosh)
CDR	Corel Draw	TAR	TAR grouped file
CGM	Computer Graphics Metafile (ISO 8632)	UU, UUD, UUE	UU-encoded files (UNIX)
CUT	Dr. Halo	z	UNIX based compression/decompression
DCX	Intel & SpectraFAX FAX format	ZIP	PKZIP/PKUNZIP/WINZIP compression
DWG	Micrograpx Designer	ZOO	ZOO compressed file
DXF	Autodesk AutoCAD	**SOUND**	
EPS, WPD	Encapsulated PostScript printer information	AU	Sun & NEXT
FAX	FAX format of many companies	PCM	Pulse code modulation
FFF	JetFAX FAX format	SND	Macintosh
FLC/FLI	Autodesk (animation)	SVX	Amiga
GIF	CompuServe graphics interchange file	WAV, MIDI	Windows sound file and MIDI files
ICO	Windows and OS/2 icon files	VOC	Creative Labs
IFF	Amiga	**EXECUTABLE**	
IGF	Inset Systems Internal Graphics Format	BAS	Executable only from BASIC program
IMG	GEM Paint, IBM image support facility	BAT	Batch files
JPG, JIF, JFI	JPEG format	COM	PC program command file
LBM	Delux Paint	EXE	PC program file
MAC	Apple Macintosh Macpaint	DLL	Windows & OS/2 Dynamic Link Librarys
MCS	MathCAD	**DOCUMENT or TEXT**	
MET	PM Metafile	TXT, ASC, DOC, DAT, HLP	Basic ASCII (usually)
MPG	MPEG video & audio compression		
MSP	Microsoft Windows Paint	ERR, LOG	Log files
PCD	Photo CD	DOC	Microsoft WORD, WordPerfect, and others
PCX	PC Paintbrush	WP*	Word Perfect
PGL	HP 7475A Plotter Language	SAM	AMI Pro
PIC	Lotus, Mouse Systems PC Paint	EPS, PS, WPD	PostScript text and graphics
PIC	Macintosh (on MS-DOS machines)	PDF	Adobe's Portable Document File (read with Acrobat)
PICS	Macintosh (animation)	**SPREADSHEET**	
PICT	Macintosh	XL*	Microsoft EXCEL
PIX	Inset Systems	WK*	Lotus 1-2-3
PRN	Any print image file	WQ1	Quattro Pro
TGA	TargaTruevision™	**DATABASE**	
TIF, TIFF	Tagged Information File Format	DAT	Data files, general
FON, FOT, TTF	Fonts. FOT and TTF are True Type fonts	CRD	Windows Cardfile program
WMF	Windows meta file	INI	Windows program initialization data files
WPG	Word Perfect Graphics	**MISC**	
FIF	Fractal Image Format	BAK	Backup files created by many programs
		CHK	Recovered files created by CHKDSK

ing with 'M,' " is a *filter* that rejects all except names starting with "M." **(2)** In optics, a device that transmits only a part of the incident spectral energy, thereby changing the spectral distribution of energy at the output. The filter may be classed according to the band of transmission; that is,

- High-pass filter—a filter that transmits energy below a certain wavelength.
- Low-pass filter—a filter that transmits energy above a certain wavelength.
- Band-pass filter—a filter that transmits energy in a specified range of wavelengths.
- Band-stop filter—a filter transmits energy outside a specified range of wavelengths.
- All-pass—a filter that transmits all wavelengths. However, it introduces a specified phase shift or delay for specified wavelengths.

(3) In electronics, a frequency-selective circuit designed to pass a desired band of frequencies (*passband*) with less than some maximum attenuation (A_{max}) and reject all other frequencies (*stopband*) with a minimum attenuation (A_{min}). At times, a third band called a *transition band* is defined between the passband and stopband. The transition region is an acknowledgment that filters in the real world cannot have an abrupt change at the *band edge* (the last frequency in the passband). Filtering may be accomplished in either the analog or digital domain.

The filter may be classed according to the band of transmission, that is,

- High-pass filter—a filter that transmits energy above a certain frequency.
- Low-pass filter—a filter that transmits energy below a certain frequency.

- Band-pass filter—a filter that transmits energy in a specified range of frequencies.
- Band-stop filter—a filter that transmits energy outside a specified range of frequencies.
- All-pass—a filter that transmits all frequencies. However, it introduces a specified phaseshift or delay for specified frequencies.

The low-pass filter in the figure below has the following specifications:

- *Passband (BW):* dc to 1 kHz.
- *Cutoff frequency (band edge):* 1 kHz.
- *Maximum passband attenuation (A_{max}):* 0.5 dBr.
- *Minimum stopband attenuation (A_{min}):* 50 dBr.
- *Stopband:* starts at 1.3 kHz. The transition band is therefore from 1 kHz to 1.3 kHz.

Note that the −3 dBr frequency is in the transition region by the preceding definition. See also *all-pass filter, band-pass filter, band reject*

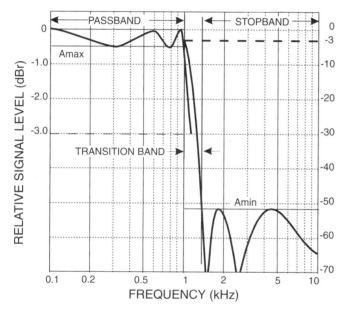

filter, comb filter, high-pass filter, and *low-pass filter.* (**4**) A process whereby a network switch, router, or bridge reads the contents of a packet header, determines that it need not be forwarded, and then discards it in order to control access to a network or to resources, such as files and devices. For example, bridges filter network traffic so that local packets stay on their networks rather than being passed to another network. Similarly, packets from external networks are filtered as a security measure. The basis for the filtering can be addresses or protocols. Packets not filtered are forwarded to an intermediate or final destination. The rate at which packets are checked and filtered is called the *filtering rate.* For a bridge, this is generally a better indicator of the bridge's performance than simple throughput. See also *forwarding.* (**5**) *Filters* may also be used to screen incoming mail or newsgroups to just those sources the user wants to receive.

filtered symmetric differential phase shift keying (FSDPSK) A method of encoding information for digital transmission in which a binary 0 is encoded as a +90° phase change on the carrier and a binary 1 is encoded as a −90° change. Abrupt phase transitions are smoothed by filtering (or other functionally equivalent pulse shaping techniques) so as to limit the spectral content of the modulated signal.

filtering rate The measure of how fast a network bridge or router can check, filter, and forward packet streams. See also *filter (3).*

final mile The communications elements required to transport information from the wideband distribution part of a system to the point where the information is to be used.

Examples include the delivery of program information from a satellite Earth station receiver to a local broadcasting station; or the connection of a home computer terminal to the Internet. Also called the *last mile.*

finder (**1**) In telephony, an automatic switch for finding a calling subscriber line (*line finder*) or trunk (*trunk finder*) and connecting it to the switching apparatus. (**2**) An older Apple Macintosh program that provides access to applications and documents. It has been replaced by MultiFinder.

finger In Internet, a program that helps a user access information about another online user. *Finger* results show information such as full user name, last logon time, and accumulated time on a specific system. It will also give any information the "fingered user" may want revealed by the *finger* program.

finished call (**1**) In an information transaction, a call that is terminated, that is, a call in which the calling party (originator) or called party (receiver) goes on-hook (hangs up). (**2**) In an information transfer transaction, the termination of the information transfer phase. See also *access phase, disengagement phase,* and *information transfer phase.*

finite impulse response (FIR) A system whose reaction to an impulse type signal input is mathematically finite in time; that is, after some period of time, no signal will be present at the output. *FIR* systems do not have any *recursive* paths. See also *infinite impulse response (IIR).*

finite sampling theorem A version of Shannon's sampling theorem that states that a class of functions can be reconstructed exactly by a sufficient number of spectral samples. The reconstructed function is an explicit function of these samples. See also *Shannon limit.*

finite state machine (FSM) A machine (real or abstract) consisting of a set of states including an initial state, a constellation of possible input events, a set of output events, and a state transition function. In general, the transition function accepts both an input event and the current state and then returns a new output state and output event. An *FSM* is deterministic (or can be translated into one); that is, the next state is uniquely determined by the current state and a single input event.

FIOC An abbreviation of Frame Input/Output Controller.

FIP (**1**) An abbreviation of Federal Information Processing. (**2**) An abbreviation of Facility Interface Processor.

FIP equipment In the federal government, any equipment, interconnected system, or subsystem of equipment (as defined in Title 41 of the Code of Federal Regulations—41CFR) used in the automatic acquisition, control, display, interchange, management, manipulation, movement, reception, storage, switching, or transmission of data or information.

FIP system In the federal government, any organized association of *FIP* equipment, software, services, support services, or related supplies.

FIPS An acronym for Federal Information Processing Standard. The government's computer communications standardization program. It covers several major catagories, including:

- Data programs and components.
- Data communications.
- Computer performance.

- Applications and data.
- Personnel and environment.
- Acquisition and reassignment of ADP products.

FEDERAL INFORMATION PROCESSING STANDARDS

FIPS	Description	Specifications
1-2	Code for information interchange, its representations, subsets and extensions	ANSI X3.4, X3.32, X3.41
3-1	Recorded magnetic tape for information interchange (800 cpi, NRZI)	ANSI X3.22
4-1	Representation for calendar date and ordinal date for information interchange	ANSI X3.30
16-1	Bit sequencing of code for information interchange in serial-by-bit data transmission	ANSI X3.15
17-1	Character structure and character parity sense for serial-by-bit data communication in the code for information interchange	ANSI X3.16
18-1	Character structure and character parity sense for parallel-by-bit data communication in the code for information interchange	ANSI X3.25
21-3	COBOL	ANSI X3.23, X3.23A
22-1	Synchronous signaling rates between data terminal and DCE	ANSI X3.1
25	Recorded magnetic tape for information interchange (1600 bpi, phase encoded)	ANSI X3.39
32-1	Character sets for optical character recognition (OCR)	ANSI X3.17, X3.2, X3.49
33-1	Character set for handprinting	ANSI X3.45
46-1	Data encryption standard (DES)	ANSI X3.92
50	Recorded magnetic tape for information interchange (6250 cpi [246 c/mm], group coded recording)	ANSI X3.54
51	Magnetic tape cassettes for information interchange (3.810 mm [0.150 in.], tape at 32 b/mm [800 bpi], phase encoded)	ANSI X3.48
52	Recorded magnetic tape cartridge for information interchange, 4 track, 6.30 mm (1/4 in.) 63 b/mm (1600 bpi) phase encoded	ANSI X3.56
53	Transmittal form for describing computer magnetic tape file properties	
54-1	Computer output microform (COM) formats and reduction ratios, 16mm and 105mm	ANSI/ AIIM MS5, MS14
58-1	Representations of local time of the day for information interchange	ANSI X3.43 (Withdrawn 1998)
59	Representations of universal time, local time differentials and united states time zone references for information interchange	ANSI X3.51 (Withdrawn 1998)
66	Standard industrial classification (SIC) codes	
68-2	BASIC	ANSI X3.113
69-1	FORTRAN	ANSI X3.9
71	Advanced data communication control procedures (ADCCP)	ANSI X3.66
79	Magnetic tape labels and file structure for information interchange	ANSI X3.27
81	DES modes of operation	ANSI X3.106

FEDERAL INFORMATION PROCESSING STANDARDS (*CONTINUED*)

FIPS	Description	Specifications
85	Optical character recognition (OCR) inks	ANSI X3.86
86	Additional controls for use with ASCII	ANSI X3.64
89	Optical character recognition (OCR) character positioning	ANSI X3.93M
93	Parallel recorded magnetic tape cartridge for information interchange, 4-track 6.30 mm (1/4 in.), 63 b/mm (1600 bpi), phase encoded	ANSI X3.72
100-1	Interface between data terminal equipment (DTE) and data circuit terminating equipment (DCE) for operation with packet-switched data communications networks (PSDN) or between two DTEs by dedicated circuits	ANSI X3.100 ITU-T X.25
104-1	American national standard codes for the representation of names of countries, dependencies, and areas of special sovereignty for information interchange	ANSI Z39.27 ISO 3166
107	Local area networks: baseband carrier sense multiple access with collision detection access method and physical layer specifications and link layer protocol	ANSI/IEEE 802.2, 802.3
109	Pascal	ANSI/IEEE 770X3.97
112	Password usage	
113	Computer data authentication	ANSI X3.92, X9.9
114	200 mm (8 in.) flexible disk cartridge track format using two-frequency recording at 6631 b/rad on one side and 1.9 t/mm (48 tpi) for information interchange	ISO 5654/2
115	200 mm (8 in.) flexible disk cartridge track format using modified frequency modulation (MFM) recording at 13,262 b/rad on two sides and 1.9 t/mm (48 tpi) for information interchange	ISO 7065/2
116	130 mm (5.25 in.) flexible disk cartridge track format using two-frequency recording at 3979 b/rad on one side and 1.9 t/mm (48 tpi) for information interchange	ISO 6596/2
117	130 mm (5.25 in.) flexible disk cartridge track format using modified frequency modulation (MFM) recording at 7958 b/rad on two sides and 1.9 t/mm (48 tpi) for information interchange	ISO 7487/3
118	Flexible disk cartridge labeling and file structure for information interchange	ISO 7665
119	ADA	ANSI/ MIL-STD 1815A
120-1	Graphical Kernel System (GKS)	ANSI X3.124, X3.124.1, X3.124.2, X3.124.3
121	Videotex/teletext presentation level protocol systax (North American PLPS)	ANSI X3.110/ CST 500
123	Specification for a data descriptive file for information interchange (DDF)	ANSI/ISO 8211

FIPS	Description	Specifications
125	MUMPS	ANSI/MDC X11.1
126	Database language NDL	ANSI X3.133
127-1	Database language SQL	ANSI X3.135, X3.168
128	Computer graphics metafile (CGM)	ANSI X3.122
129	Optical character recognition (OCR) dot matrix for OCR-MA character sets	ANSI X3.111
133	Coding and modulation requirements for 2400 bps modems	
134-1	Coding and modulation requirements for 4800 bps modems	EIA RS-449
135	Coding and modulation requirements for duplex 9600 bps modems	
136	Coding and modulation requirements for 600 and 1200 bps modems	
137	Analog to digital conversion of voice by 2400 bps linear predictive coding (LPC)	
138	Electrical characteristics of balanced voltage digital interface circuits	EIA RS-422-A
139	Interoperability and security requirements for use of the data encryption standard in the physical layer of data communications	
140	General security requirements for equipment using the data encryption standard (DES)	
141	Interoperability and security requirements for use of the data encryption standard with ITU group 3 facsimile equipment	
142	Electrical characteristics of unbalanced voltage digital interface circuits	EIA RS-423-A
143	General-purpose 37-position and 9-position interface between data terminal equipment (DTE) and data circuit-terminating equipment (DCE)	EIA RS-449
144	Data communication systems and services user-oriented performance parameters	ANSI X3.102
146	GOSIP: government open systems interconnection profile	
147	Group 3 facsimile apparatus for document transmission	EIA RS-465
148	Procedures for document facsimile transmission	EIA RS-466
149	General aspects of group 4 facsimile apparatus	EIA 536
150	Facsimile coding schemes and coding control functions for group 4 facsimile apparatus	EIA 538
151-1	POSIX: portable operating system interface for computer environments	IEEE 1003.1/POSIX
152	Standard generalized markup language (SGML)	ISO 8879
153	Programer's hierarchical graphics system (PHIGS)	ANSI X3.144, X3.144.1
154	High speed 25 position interface for data terminal equipment and data circuit-terminating equipment	EIA 530
155	Data communication systems and services user-oriented performance measurement methods	ANSI X3.141
156	Information resources dictionary system (IRDS)	ANSI X3.138

FIPS	Description	Specifications
159	Detail specifications for 62.5; micron core diameter/125-micron cladding diameter class IA multimode graded-index optical waveguide fibers	ANSI/EIA/TIA-492.AAAA
160	C (the programming language)	ANSI X3.159
174	Generic twisted pair telecommunications cabling system	ANSI/EIA/TIA 568A & B
186	Digital Signature Standard	

FIR (**1**) An abbreviation of Finite Impulse Response. (**2**) An abbreviation of Far InfraRed. (**3**) An abbreviation of Fast InfraRed.

firewall A device and/or software between two networks which restricts the access of one network's users into the other network. For example, between a public network (such as the Internet) and a private network a *firewall* would restrict Internet users from logging onto the private network, but the private network users could access Internet.

Unlike a router that forwards packets to destination hosts, *firewalls* block the transmission of Internet Protocol (IP) packets and store them for later retrieval by users on the "protected" side of the *firewall*. This process is inconvenient at best for the insulated users. Two additional implementation tools reduce this inconvenience.

- First is a *proxy server*—a program running on a *firewall* machine which handles **known** risk-free data exchanges with machines on the Internet.
- Second is the partitioning of the network environment at the protected site. That part of the site with sensitive information is put behind a firewall while the other part with nonsensitive material is given full access.

FIRL An acronym for Fiber Optic Inter-Repeater Link.

FIRMR An abbreviation of Federal Information Resources Management Regulations.

firmware Software resident in either semipermanent or permanent memory, such as flash memory, electrically erasable read only memory (EEROM), or true read only memory (ROM). Device startup routines, such as a PC's BIOS, are frequently stored in ROM and are therefore firmware. Firmware is not modifiable by a standard application program.

first-in first-out (FIFO) A queuing (buffering) technique in which entities in a queue are removed from the queue in the same order in which they arrive.

first-in last-out (FILO) A queuing (buffering) technique in which entities in a queue are removed from the queue in the reverse order in which they arrived.

first level interrupt handler (FLIH) An interrupt handler whose job is to determine which device or channel generated the interrupt and then to invoke a second-level interrupt handler to actually process the request behind the interrupt.

first line release In telephony, release of a communications circuit by the first terminal to go on-hook, i.e., to return to idle state. Also called *either party control.*

first-party call control A call control methodology that allows either the calling or called party to control the communications path *while they are connected to the path.* If the path is transferred to a third party, control is also transferred. See also *third-party call control.*

first point of switching The first telephone company location at which routing occurs.

first window Of silica-based optical fibers, the transmission window at approximately 830 to 850 nm. See also *window*.

FIT An acronym for Failure In Time.

FITL An abbreviation of Fiber In The Loop.

FIX An acronym for Federal Information eXchange.

fixed access In personal communications service (PCS), access to a network in which there is a set relationship between a terminal and the access interface. A single *identifier* serves both the access interface and the terminal. If the terminal moves to another access interface, it assumes the identity of the new interface.

fixed priority-oriented demand assignment (FPODA) In networking, a medium access protocol in which stations must reserve slots on the network from a master station. These slots are allocated according to the stations' priority levels.

fixed routing A routing strategy in which packets or messages are transmitted between the source and destination over a well-defined and constant path.

fixed tolerance band compression A data compression method in which data values are stored or transmitted only when they fall outside a specified limit. The recipient of the data must assume that values are in the specified range unless an alert signal indicating otherwise occurs. As an example, assume the temperature at a particular point in a chemical plant is to be kept between 150°C and 180°C, a *fixed tolerance band compression* telemetry system would transmit temperature values only above 180°C or below 150°C.

flag (1) A two-state indicator/variable such as yes/no, true/false, on/off, error/no-error, and ready/not-ready. It is a variable set at one point in a program for use later in the same or different programs. It informs these programs that a condition has been met or not met. **(2)** In synchronous transmission systems, the *flag* is a specific bit pattern that marks the beginning and end of a packet (frame). See also *flag character.* **(3)** A deprecated term for a mark (stop) signaling element.

flag character In X.25 packet switching technology, a special character (0111 1110) that is included at the beginning and end of every link access protocol-balanced (LAP-B) frame to indicate a frame boundary. The protocol uses *bit stuffing* to ensure that this sequence never occurs anywhere else in the packet. *Flag characters* are used in bit-oriented protocols, such as Advanced Data Communication Control Procedures (ADCCP), Synchronous Data Link Control (SDLC), and High Level Data Link Control (HDLC). Also called a *flag sequence.*

flame On bulletin board systems, a vitriolic criticism or insult of a *person* for his or her ideas, beliefs, postings; just because of a disagreement with those postings, ideas, and so on. Some network users tend to state their opinion in an inflammatory manner that they would never use were it not for the anonymity of a network connection. *Flaming* is not a respected pastime. DON'T do it.

flame bait Any network posting that is virtually guaranteed to draw a plethora of flame responses; e.g., any type of advertisement posted to a newsgroup.

flame war A sustained exchange of flames between two (or more) network users. Generally, they start as disputes as to the "proper" resolution of some issue and degrade into messages bent on undermining the opponent (ignoring the original issue).

flash A quick depression and release of the hook switch on a telephone instrument. The *flash* is a signal to the PBX or Centrex that a system control command will follow, e.g., a command to transfer the call to a different extension. Also called a *hook-flash* or *hook-switch flash.*

flash cut The "instantaneous" transfer from one switching system to another. For example, when installing a new PBX, the old system is removed and the new one is installed; the user calls out using the old system on Friday, and Monday the new system must be used.

flash memory A type of nonvolatile memory that contains both RAM and EEPROM memory. During normal operation, data are written to and read from the RAM portion; during power-off times, data are stored in the EEPROM. Before power is removed from the RAM portion of the chip, a signal tells the chip to copy relevant portions of the RAM to the EEPROM. When power is restored, these data are restored to the proper locations in RAM.

Flash memory is an attempt to solve two shortcomings of EEPROM memory, primarily the inherent slow writing speed and secondarily the relatively limited number of times to which EEPROM memory can be written. Also called *flash ROM.*

flat fading Fading in which all frequency components of a received radio signal vary in the same proportion simultaneously.

flat rate service Telephone service in which a single payment permits an unlimited number of local calls to be made without further charge for a specified period of time.

flat structure A strategy in which each entity is unique and in which there is no logical, physical, or other relationship between them. This strategy may be used for files names or network node names. Accessibility to such a structure is through a lookup table. See also *hierarchical name structure.*

flat weighting In a noise-measuring set, a measurement based on an amplitude-frequency characteristic that is flat over a stated frequency range. Noise power is expressed in $dBrn(f_1 - f_2)$ or $dBm(f_1 - f_2)$.

In telephony, the first frequency (f_1) is frequently omitted and assumed to be 30 Hz. For example: a "3 kHz flat weighting" assumes a flat frequency response between 30 Hz and 3 kHz, and a "15 kHz flat weighting" assumes a flat frequency response between 30 Hz and 15 kHz.

flavor Slang for a different version of software or hardware. For example, Berkeley Software Distribution (BSD) and system V are two *flavors* of UNIX.

Fleming's rule A mnemonic rule stating that if the thumb of the right hand points in the direction of current flow, then the curled fingers point in the direction of the magnetic field that encircles the current. Further, if the curled fingers of the right hand point in the direction of current flow in a solenoid, then the thumb points in the direction of the magnetic field inside the solenoid. Also called the *right-hand rule.*

flexible release The ability of a switching system to release an established circuit when either station goes on-hook. Also called *either party control.*

flexible routing The ability to choose different physical paths through a network for different calls to the same station as the circumstances warrant.

FLIH An abbreviation of First Level Interrupt Handler.

flip-flop A circuit or device that may assume one of two alternative stable states. During the transition from one stable state to the other, the output state is unstable, i.e., for the very short period during which the transition takes place, both outputs may assume the same state. (The state is implementation dependent and may be unpredictable to the user.)

One use is as a basic register memory element in computer and communications systems. Each flip-flop can store one bit of information. Also called a *bistable circuit, bistable multivibrator,* or *bistable trigger circuit.*

FLL In telephony, an abbreviation of Fixed Loss Loop.

floating (1) *Batteries.* A method of keeping a secondary battery at full operating capacity by continuously charging at a low rate (*trickle charging*). **(2)** *Circuits.* A network or device having no terminal directly connected to ground potential.

flood search routing A packet switch routing methodology that determines the optimum route for traffic within a specific network; it avoids failed or congested links/nodes.

flooding (1) The uncontrolled propagation of packets in a network. **(2)** A packet switch routing methodology of transmitting numerous identical packets in various directions to ensure they reach the intended destination. It is neither a very efficient nor a desirable method.

flow control Any of several methods used to start and stop the flow of data between devices when the receiving device cannot handle any new inputs. The function performed by a receiving device to limit the rate of data delivery by the transmitting device. Both hardware and software *flow control* are used between DTE and DCE, that is, *request to send* and *clear to send* (RTS/CTS) in hardware or the specific ASCII characters XON/XOFF in software. *Flow control* between DCE and DCE is always via messages embedded in the data steam, e.g., XON/XOFF characters.

Flow control is required whenever the data flow rate is not the same in all parts of the transmission path. Also called *pacing.* See also *ACK* and *NAK.*

flow diagram See *flowchart.*

flowchart A graphical representation of a system that shows the significant logical *paths* of control and data flow through a program or device and the *processes* or *operations* performed on the data at significant points (nodes) along the path. Flowcharts use geometric symbols to indicate data sources, data sinks, storage, processes performed on data at nodes, as well as written commentary to elaborate.

Some flowchart symbols with their respective meanings are:

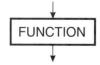

Process—a symbol indicating a *process* or *operation* (or a group of operations) performed by software or hardware. The name "FUNCTION" is replaced by a description of the process or action taking place on a data stream or device.

Decision—a flowchart symbol indicating a change of information processing based on a logical or arithmetic test. Typically, the test results in either the two-state *decision* yes/no (true/false) or the three-state *decision* greater than/less than/equal to ($>/</=$). Although many decision levels *can* be indicated, it is rare to see more than three. The name "CONDITION" is replaced by the test or tests on which the decision is based.

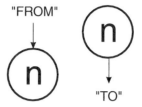

Connector—a pair of flowchart symbols indicating a connection to/from another part of the flowchart. (The location of the destination is generally on the same page but may be on a different page.) When the point is off page, the "*off-page connector*" may also be used. The symbol "n" is typically replaced by a unique number or letter indicating the points to be connected.

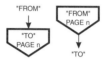

Off-page connector—a pair of flowchart symbols used to indicate a continuation or connection to another page in the flowchart. The symbol may be used on any side of the chart.

Document—a flowchart symbol used to represent a printed output or hard copy. Indicates a person-readable document as opposed to a machine-readable file.

File—a flowchart symbol used to indicate information is to be written to (or read from) a file. Files may be on a floppy disk, hard disk, or a network server.

Communication link—a flowchart symbol used in LAN or WAN diagrams to indicate phone lines, modems, cables, or other telecom links.

Subroutine—a flowchart symbol for a *predefined process* or sequence of operations (subroutine). The subroutine may be referenced several times in the overall flowchart but is stored in only one place in memory.

A sample flowchart diagramming the process of establishing a telephone connection is given in the next drawing. Also called *flow diagrams* and *flow graphs*.

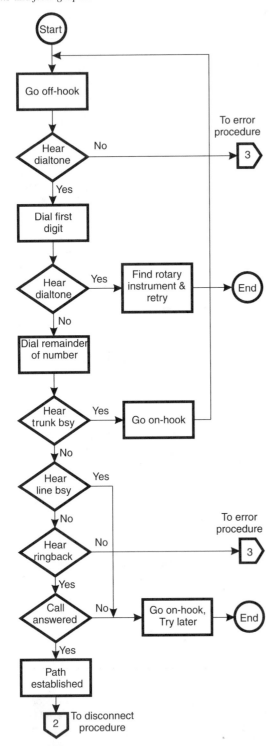

flux **(1)** The rate of flow of energy at a surface. **(2)** The rate of flow of particles at a surface. **(3)** A solid or liquid material applied to a material to be soldered. When heated, the *flux* cleans and removes oxidation from the surfaces so that the solder will "wet" the surface. **(4)** The lines of force of a magnetic field. **(5)** An obsolete synonym for radiant power.

flux budget In fiber optic transmission, the amount of light that can be lost between adjacent nodes or repeaters without having the transmission error rate exceed a specified threshold.

flywheel effect In an electronic circuit or device, the continuation of oscillations after removal of the stimulus (ringing).

The effect may be desirable, as with clock recovery circuits used in synchronous systems, or undesirable, as with a voltage controlled oscillator (VCO). Also called *flywheeling.*

FM **(1)** An abbreviation of F̲requency M̲odulation. **(2)** An abbreviation of the slang F̲'n M̲agic. Indicating the speaker does not care to explain (or does not know).

FM 0 encoding Frequency modulation 0 (FM 0) is a signal encoding method used for LocalTalk networks in Macintosh environment. It uses $+V$ and $-V$ voltage levels to represent bit values. The encoding rules are:

- Logical 1 bits are encoded alternately as $+V$ and $-V$, depending on the previous voltage level. The voltage level remains constant for an entire bit interval (bit cell).
- Logical 0 bits are encoded as $+V$ or $-V$, depending on the immediately preceding voltage level. The voltage changes to the other value at midbit.

FM 0 encoding is self-clocking code because there is always a transition at the bit cell boundary for the receiver's clock to lock onto. The waveform is identical to differential Manchester coding except for a 1/2-bit shift.

FM improvement factor A measure of an FM receiver's performance. It is expressed as the quotient of the signal-to-noise ratio (SNR) at the output of the receiver and the carrier-to-noise ratio (CNR) at its input, i.e., SNR_{out}/CNR_{in}.

FM subcarrier A method of broadcasting a second message channel on a single FM carrier. In principle, the FM carrier is modulated with two signals, one at baseband and the other a high-frequency tone (outside the bandwidth of the baseband signal). The high-frequency tone is itself modulated by an information-bearing signal. In essence it is a form of frequency division multiplex with one channel at baseband.

FMR An abbreviation of F̲ollow M̲e R̲oaming.

FNC An abbreviation of F̲ederal N̲etworking C̲ouncil.

FNPA An abbreviation of F̲oreign N̲umber P̲lan A̲rea.

FO An abbreviation of F̲iber O̲ptics.

focal point A term, in IBM's Network Management Architecture (NMA), for the node on which the network management software is running. This is generally a mainframe host in NMA. Other nodes and devices communicate with the *focal point* through either entry points (in the case of System Network Architecture [SNA]-compliant devices) or service points (in the case of non-IBM devices or networks). NetView is IBM's key implementation of the focal point.

FOD An abbreviation of F̲ax O̲n D̲emand.

FOG An acronym for F̲iber O̲ptic G̲yroscope.

FOIA An abbreviation of F̲reedom O̲f I̲nformation A̲ct.

foil shield A construction of coaxial cables in which the outer conductor (shield) is made from a metalized tape—generally aluminum bonded to mylar or like plastic. The technique provides a very thin shield and a percentage of coverage greater than a simple braid construction.

FOIRL An abbreviation of F̲iber O̲ptic I̲nter-R̲epeater L̲ink.

folded network A switching network in which each port is both an input and an output and a path may be established between any two ports. Also called a *triangular* switching network. The figure below is that of a *full availability, non-blocking, folded* switch. See also *availability, blocking network, crosspoint, nonblocking network,* and *nonfolded network.*

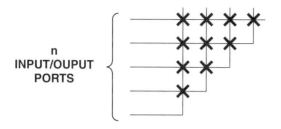

foo A word commonly used by an author to stand for just about anything (a textual "wild card"?). It is frequently used with another "wild card word" *bar,* for example, if giving an example of a persons address at an arbitrary domain the author might use

johnsmith@foo.bar.net

Together they may imply a connection with the acronym FUBAR (fouled up beyond all recognition).

footprint (**1**) The area defined by the intersection of that portion of a radiated beam (radio or optic) from an antenna at or above a specified minimum level on a specified surface (the Earth, for example). That portion of a surface illuminated by a radiated beam which is above a minimum threshold, e.g., the area on the Earth capable of receiving a transmission from a satellite. (The figure shows a footprint

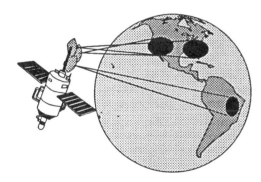

with three target areas). (**2**) In computers, *footprint* is used to refer to the amount of RAM (random access memory) an application uses during execution. (**3**) The amount of space a device occupies.

forbidden code A combination of bits that is defined to be nonpermissible in a particular application and whose occurrence indicates a mistake, error, or malfunction.

An example of a *forbidden code* is the set of binary values 1010, 1011, 1100, 1101, 1110, and 1111 in the binary coded decimal (BCD) system. Only values 0 through 9 are possible digits in the decimal system; therefore, only representations of those values have meaning in a translation. Another example is a code where 10 represents male and 01 represents female. Here, 00 and 11 are forbidden codes.

forced release In telephony, the release of a circuit by a source other than the calling or called station.

FORD (**1**) An acronym for <u>F</u>ix <u>O</u>r <u>R</u>epair <u>D</u>aily. Generally intended to be humor, though sometimes just sarcastic. (**2**) An acronym for <u>F</u>ound <u>O</u>n <u>R</u>oad <u>D</u>ead. Equivalent to dead on arrival (DOA).

foreign area In telephony, a numbering plan area other than the one in which a calling subscriber is located.

foreign exchange (FX) line In telephony, a subscriber telephone line or loop served by a central office different from the office from which service would normally be delivered.

The line will have an office (exchange) code other than that of the central office normally serving the subscriber.

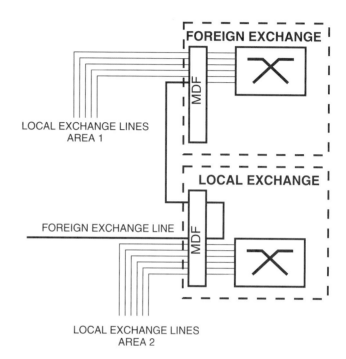

foreign number plan area (FNPA) Any number plan area (NPA) outside the geographic area of the customer's NPA.

fork (**1**) In the Apple Macintosh file system, the two components of a file: the data fork, which contains the actual information in the file, or the resource fork, which contains application specific data. (**2**) In telephony, a synonym for *hybrid.*

forklift upgrade A slang term for an upgrade that requires the replacement of most or all of the installed equipment. Essentially, it is the equivalent of buying completely new equipment and discarding the old system.

form A, B, C, . . . See *switch contact.*

form factor The dimensionless ratio of the root-mean-square value of a periodic function to its average absolute value (taken over a full period of the function). Mathematically, it is,

$$Form\ factor = \frac{r.m.s}{|average|}$$

format (**1**) In data transmission, the arrangement of code elements within a group, block, or subgroup. (**2**) The general order in which information is placed on a storage medium. (**3**) The shape, size, and general makeup of a document. (**4**) The writing of a basic bit pattern on a recording medium which enables the medium to be written to and read by a particular operating system. In DOS, for example, the FORMAT command, among other things, sets up the *boot sector,* at least one copy of the *file allocation table (FAT),* and an empty *root directory.*

fortuitous distortion Signal distortion caused by random events, that is, events that cannot be classified as bias or characteristic distortion.

forum An area within an online service, such as CompuServe, Prodigy, America Online, and Microsoft Network, where a user can obtain information on specific subjects. *Forums* on almost any subject can be found at one time or another. Many manufacturers provide a *forum* as a means of delivering information to and answering questions from customers. A *forum* may have an associated library from which a caller may download an updated driver or a useful utility.

forward busying In a telecommunications system, a control mechanism in which supervisory signals are sent in advance of address signals in order to seize assets of the network before attempting to establish a connection.

forward channel The channel of a communications circuit that transmits information from the originating user to the destination user. The forward channel carries message traffic and some control information. Contrast with *reverse channel.*

forward direction In the direction of propagation between ports of the primary signal channel.

forward echo In a transmission line, an echo propagating in the same direction as the original wave. It consists of energy reflected back by one discontinuity and then forward again by another discontinuity.

In metallic transmission lines (twisted pair, coaxial, or open lines), discontinuities may be caused by impedance mismatches between the source (or load) and the characteristic impedance of the transmission medium, by defects in the medium itself, and so on. In optical transmission lines, reflections are caused by splices or other discontinuities in the transmission medium.

forward error correction (FEC) In communications, an error-correction method involving the addition of redundant bits into the transmitted data stream. The receiver uses the redundant data to correct errors (where possible) introduced in the communication channel. Examples of forward error correction include the Hamming code, BCH (Bose-Chandhuri-Hochquenglen) codes, and trellis coding. Another type of error-correction scheme is the detect and retransmit method: The receiver detects an error and then requests the transmitter to resend the information block which contained the error.

forward explicit congestion notification (FECN) See *explicit congestion notification (ECN).*

forward scatter The deflection, in a medium, of a portion of an incident electromagnetic wave in such a manner that the deflected energy propagates in a direction that is within 90° of the direction of propagation of the incident wave. The deflection may be caused by one or more of the following effects: diffraction, nonhomogeneous refraction, or nonspecular reflection by particulate matter of dimensions that are large with respect to the wavelength in question but small with respect to the beam diameter. The scattering process may be polarization sensitive.

forward signal A signal sent in the direction from the calling to the called station. The forward signal is transmitted on the forward channel. See also *reverse channel.*

forwarding A process whereby a network switch, router, bridge, or gateway reads the contents of a packet header, determines where the packet is to be sent, and passes the packet on to the appropriate attached segment. This is in contrast to *filtering,* in which a packet is discarded. The basis for both filtering and forwarding can be either addresses or protocols.

Generally, a forwarding device performs three functions:
- Reads and buffers the entire packet.
- Checks the address or protocol.
- Filters or forwards the packet, depending on the value found and on the filtering criteria.

In a variant called on-the-fly forwarding, a device begins forwarding the packet as soon as the device determines that this is the appropriate action. This means that the packet can be on its way to a new destination while still being read by the bridge; hence, it is a faster operation. See also *filter (3).*

FOT (1) An acronym for Fiber Optic Transceiver. **(2)** An acronym for Frequency of Optimum Transmission. In radio transmission systems utilizing ionospheric reflection, the highest effective (working) frequency that is predicted to be usable for a specified path and time for 90% of the days of the month. Generally, the *FOT* is taken just below the value of the *maximum usable frequency (MUF),* i.e., 15% below the monthly median MUF. This is usually the most effective frequency for ionospheric reflection of radio waves between two specified points on Earth. Also called the *frequency of optimum traffic, optimum traffic frequency (FOT), optimum transmission frequency,* and the *optimum working frequency.*

FOTCL An abbreviation of Falling Off The Chair Laughing.

FOTS An acronym from Fiber Optic Transmission Systems.

four-wire circuit A transmission facility in which signals transmitted to a remote terminal are routed over one transmission path, whereas signals received from the remote terminal are on a second path. A facility such as this provides full-duplex operation by default.

The *four-wire circuit* gets its name from historical telephone circuits in which two conductors were used in each of two directions for full-duplex operation. Note that the transmission paths may or may not employ four physical wires. For example, the term is applied to a communications link supported by optical fibers, even though only one fiber is required for transmission in each direction. Contrast with *two-wire circuit.*

four-wire repeater In telephony, a device used to separately amplify (or regenerate) the signals flowing in both channels of a four-wire circuit.

four-wire switching In telephony, switch wherein a separate communications channel is used for the forward and reverse channels of a four-wire circuit.

four-wire terminating set In telephony, a transformer two- to four-wire hybrid set used for interconnecting a four-wire circuit and a two-wire circuit.

Four-wire terminating sets have been largely supplanted by resistance hybrids.

Fourier analysis The definition or description of a given periodic waveform of arbitrary shape as a summation of sine waves having specific amplitudes and phases, and having frequencies corresponding to the harmonics of the waveform being defined. See also *wavelet.*

Fourier theorem A proof, by the French mathematician Jean-Baptiste Joseph Fourier (1768–1830), that any arbitrary periodic (repeating) waveform may be represented by a series of sine waves of appropriate amplitude, frequency, and phase. From this proof, two conclusions may be drawn:
- Transmitting a complex waveform requires a bandwidth much larger than the fundamental frequency.

- Any complex periodic waveform may be approximated with a sufficient number of sine waves.

See also *Gibb's phenomenon.*

fox message A standard test message that includes all the alphanumerics plus certain functions keys (space, figures shift, letters shift) on a teletypewriter. The message is:

THE QUICK BROWN FOX JUMPED OVER THE LAZY DOG'S BACK 1234567890.

FPGA An abbreviation of Field Programmable Gate Array. See also *gate array.*

FPIS An abbreviation of Forward Propagation Ionospheric Scatter. See also *ionospheric scatter.*

FPLA An abbreviation of Field Programmable Logic Array. See also *gate array.*

FPODA An acronym for Fixed Priority Oriented Demand Assignment.

FPS An abbreviation of Fast Packet Switching.

FQDN An abbreviation of Fully Qualified Domain Name.

fractional E-1 (FE-1) In the European (or other countries using the ITU-T standards) digital telephony hierarchies, a portion of a 30-channel 2.048-Mbps E-1 carrier, or line. *Fractional E-1* lines have bandwidths in 64-kbps increments from 128 kbps to the full 2.048-Mbps rate.

fractional E-3 (FE-3) In the European (or other countries using the ITU-T standards) digital telephony hierarchies, a portion of a 480-channel 34.368-Mbps E-3 carrier, or line.

fractional frequency fluctuation The deviation of an oscillator's nominal frequency, normalized to the nominal frequency.

fractional T-1 (FT-1) In the North American or Japanese digital telephony hierarchies, a portion of a 24-channel 1.544-Mbps T-1 carrier, or line. *Fractional T-1* lines are available from interexchange carriers (IXCs) and can have bandwidths in 64-kbps increments from 128 kbps to the full 1.544-Mbps rate. Typical rates include 256, 512, 768, and 1024 kbps. See also *DS-n.*

fractional T-3 (FT-3) In digital telephony, a portion of a 672-channel 44.7364-Mbps T-3 carrier, or line.

FRAD An acronym for Frame Relay Assembler Disassembler (or Frame Relay Access Device). Used to interface a customer's local area network (LAN) with the frame relay wide area network (WAN). This device interfaces the Local Management Interface (LMI) with an Internetwork Packet Exchange (IPX) switch port.

fragmentation (1) A condition that develops on hard disk drives (HDD) when files are erased and new files are written.

As the drive information ages, gaps develop between areas of information (gaps where deleted files previously resided). New files are written into these gaps; however, the gap may not be large enough to hold the entire file and so, the file is *fragmented* (broken into pieces). That is, part is placed in one gap and other parts are placed in other gaps. The result is slower disk access and file retrieval. To repair fragmented disk drives, a defragmentation program is run. The program rearranges file locations and concatenates the file fragments. **(2)** A *fragment* is part of a packet, which may be created by accident. For example, in an Ethernet network, packet fragments may be created unintentionally by the collision between two packets transmitted at the same time. When such a collision is detected, jam packets are transmitted. This not only en-

sures that none of the nodes attempts to use the fragments but also allows the network to become resynchronized. **(3)** *Intentional fragmentation* also occurs because of the inevitable division of a packet traversing the Internet or any other network in which different protocols exist. *Fragmentation* occurs when a transmitting system with a *maximum transmission unit (MTU)* is longer than the MTU of the receiving system. For example, Ethernet-based networks allow packets 1500 octets in length, whereas ARPAnet derivative networks only allow 1000 octets and ATM cells are 53 octets long. Fragmentation produces new datagrams, each with its own header. Each datagram is launched independently. All received datagrams are reassembled when they reach their destination. The reverse of fragmentation (removing redundant headers and recombining several fragments into the original packet) is known as reassembly. Also called *segmentation.*

FRAM An acronym for Ferro-electric Random Access Memory (pronounced "EFF-ram"). The ferro-electric effect is the tendency of dipoles within some crystalline materials, such as a thin film of ceramic lead zirconate titanate (PZT), to align (polarize) when subjected to an electric field. The alignment remains even after the field is removed. When the polarity of the electric field is reversed, the polarity of the crystalline alignment is also reversed. With these two stable states a nonvolatile memory can be designed.

frame (1) In communications, a sequence of time slots repeated periodically, for example, in T-carrier all channels are repeated every 125 µs. **(2)** In asynchronous serial communications, a measure of time from the beginning of the start bit to the end of the last stop bit. The *frame* therefore includes the start bit, a character, an optional parity bit, and one or more stop bits. **(3)** In synchronous communications, a packet of information transmitted as a single unit. Every frame has the same basic structure, including some or all of the following:

- Start of frame sequence.
- Control information (e.g., packet number).
- Calling station address.
- Called station address.
- Data.
- Error control.
- End of frame sequence.

The information preceding the data is often called the *header,* and the information following the data is the *trailer.* Additional information may be added to accomplish tasks such as synchronization or overhead not directly associated with the data. Examples of frame protocols include High Level Data Link Control (HDLC), Synchronous Data Link Control (SDLC), and Systems Network Architecture (SNA). **(4)** In ISDN, a block of variable length, labeled as the Data Link Layer of the ISO/OSI Reference Model. **(5)** A device used to hold electronic equipment. Also called a rack or sometimes a cabinet (although cabinets generally have side covering panels whereas *frames* do not).

frame alignment The extent to which receiving equipment is correctly synchronized with the signal frame being received.

frame alignment signal A unique bit pattern inserted once (and only once) in each frame. It is always inserted in the same position relative to the beginning of the frame. It is used to establish and maintain frame alignment (synchronization). Also called the *framing signal.*

frame check sequence (FCS) A value that is used to check for errors in a transmitted message block. An *FCS* value is computed before sending the message and is appended to the transmitted message block. When the block is received, a new FCS is calculated and compared to the old value. If the new *FCS* value does not match the original, a transmission error has occurred. See also *CRC, error-detecting code,* and *parity.*

frame control (FC) A field in a Token Ring data packet or frame. The *FC* data provides the frame type information. The value designates the frame is a MAC-layer management packet or is carrying logical-link control (LLC) data.

frame duration The time or number of bits, characters, or octets between the beginning of a frame and the end of that frame.

frame pitch The time or number of bits, characters, or octets between corresponding points in two consecutive frames.

frame rate The number of frames transmitted or received per unit time, usually expressed as frames per second. In television transmission, *frame rate* is distinguished from *field rate*. (The field rate in the National Television Standards Committee [NTSC] and other systems is twice the frame rate.) Also called *frame frequency*.

frame reject response (FRMR) In a connection using the SDLC (Synchronous Data Link Control) protocol, a signal from the receiving station indicating that an invalid frame or packet has been received.

frame relay A data communications interface for statistically multiplexed, data-oriented, packet-switched communications. It derives its name from using the data link or "frame" ISO/OSI layer 2 to route or "relay" a packet directly to its destination instead of terminating the packet at each switching node.

Frame relay was defined in 1991 and is a variation of X.25. It is based on the ITU-T Lap-D standard, uses variable-length packets, and is applicable to T-3/E-3 or lower data transmission. It does not, however, have as tight an error-control or flow-control strategy as X.25. (X.25 was designed for data transmission over noisy analog lines, not digital lines.) Because of the reduced *overhead*, greater data bandwidth is available. It has become a very popular *data link protocol* not only because of its capabilities, but because the software can be installed directly on existing X.25 hardware.

Characteristics and features of *frame relay* include:

- Correspondence to the ISO/OSI Reference Model Layers 1 and 2.

  Because it operates at the physical layer and the lower part of the ISO/OSI Reference Model data link layer, *frame relay* is protocol independent, and it can transmit packets from TCP/IP, IPX/SPX, SNA, or other protocol families.

- Variable-sized packets (frames) are used that completely enclose the variable-sized user packets they transport. One *frame relay* packet transports one user packet. Frame lengths are from 7 to 1024 bytes.

  The following figure illustrates the contents of a *frame relay packet*. Dotted regions are optional, and the data field is variable length.

- Suitability for LAN-to-LAN sporadic, bursty traffic.

- Minimal flow-control capability. *Frame relay* discards any eligible packets that cannot be delivered, either because their destination cannot be found or because there are too many packets coming in at once. (Discarding is an acceptable flow-control strategy because higher protocol layers have their own error-detection mechanisms which will request retransmissions if packets are lost or discarded.) This means that *frame relay* should be used over "clean" lines, so that there are not too many errors for the higher level protocols to contend with.

  Frame relay can notify sources and/or destinations if there is heavy traffic (congestion) on the network via the *forward explicit congestion notification (FECN)* bit and *backward explicit congestion notification (BECN)* bit, respectively. Notified nodes are expected (but not required) to adjust their transmissions in order to reduce the congestion.

- Like Ethernet, or Token Ring, *frame relay* is an assumption that connections are reliable; hence, it does not have error control (which helps to speed up the protocol). When errors occur, *frame relay* relies on higher level protocols for error detection and correction.

- Implementation of fast-packet technology for connection-oriented frame relay services.

- The capability to handle time-delay insensitive traffic, such as LAN interworking and image transfer.

- Bandwidths from 64 kbps to 1.544 Mbps.

- Voice switching capability.

- Capability as an access method for ATM based WANs.

Frame relay is referred to as the local management interface (LMI) standard and is specified in ANSI T1.617. See also *ATM* and *cell relay*.

frame slip (1) In the reception of framed data, the loss of synchronization between received data and the receiver clock, causing frame misalignment and resulting in the loss of the data contained in the received frame: (A *frame slip* should not be confused with a dropped frame where synchronization is not lost; see definition 2.) **(2)** More properly called a dropped or repeated frame. The dropping or repeating of exactly one frame by a digital transmission facility (or switching system) without the loss of frame synchronization due to buffer overflow or underflow. Also called just *slip*.

frame switch A device that forwards frames based on the frame's layer 2 address. Frame switches are found in two basic forms:

- Cut-through switches (also called on the fly switching).
- Store-and-forward switches.

frame synchronization (1) The process whereby a receiver on a communications channel is aligned to the signal emanating from the transmitter. Alignment is a time-related function, although it may be expressed as phase, frequency, and so on.

The *frame synchronization* process generally involves the detection of a unique bit sequence (the *frame synchronization pattern*) known to reside in a particular location of a frame structure, followed by the forcing of receive timing circuits to the same frame phase. **(2)** The adjustment of an image to a desired position on a device such as a facsimile machine or a video monitor. *Frame synchronization* is sometimes called simply *framing*.

frame synchronization pattern In digital communications, a prescribed pattern of bits periodically transmitted to enable a receiver to achieve *frame synchronization*. The pattern of bits may be as simple as a single bit, a long word with high auto-correlation properties, or a pattern designed to violate information coding rules.

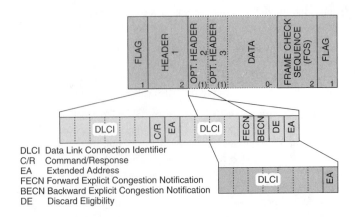

DLCI   Data Link Connection Identifier
C/R   Command/Response
EA   Extended Address
FECN Forward Explicit Congestion Notification
BECN Backward Explicit Congestion Notification
DE   Discard Eligibility

framing The division of a large set of data bits into smaller groups of bits and the appending of a header, check sequence, and trailer to the group for packet transmission. See also *frame* and *frame synchronization.*

framing bits The general term for the bits that surround a character. For example, the start and stop bits used in asynchronous communication are *framing bits.*

Framing bits are noninformational bits used in determining the beginning and/or end of a frame. In synchronous systems, the *framing bits* are usually repetitive and occur at a specific positions in the frame. See also *frame alignment signal.*

framing error An error that occurs when a receiver sees (misinterprets) a set of bits in a frame as the framing flag. Generally due to data transmission errors.

framing signal See *frame alignment signal* and *framing bit.*

Fraunhofer diffraction pattern A synonym for *far-field diffraction pattern.*

Fraunhofer region A synonym for *far-field region.*

free buffer enquiry (FBE) A field in an ARCnet frame.

free code call In telephony, a call made to a *service* or *office code* for which there is no charge to the subscriber.

free line call In telephony, a call made to a *directory number* for which there is no charge to the subscriber.

free routing A message routing technique in which messages are forwarded toward their destination or addressee over any available channel without dependence on predetermined routing.

free software Software (both the source code and executable files) distributed freely to users who are in turn given the right to freely use, modify, and redistribute it (sometimes called a *copyleft*). The only restrictions are:

- All modifications are clearly marked.
- The original author and copyright notice are neither deleted nor modified.
- No commercial use of the software is made of the underlying source code.

The *free software* concept was originated by the *Free Software Foundation,* a nonprofit organization in Cambridge, Massachusetts. See also *freeware, public domain software,* and *shareware.*

free space A theoretical concept of space devoid of (or at least remote from) all matter that could influence the propagation of electromagnetic waves.

free space communications Any form of communications that does not use conducted signals (wire, coaxial, or fiber optics). Any form of radiated communications (infrared or radio).

free space coupling Coupling of magnetic, electric, or electromagnetic fields that is not aided by a device such as a capacitor or inductor.

free space loss Signal attenuation (*path loss*) between two isotropic radiators (antennas) that would result if all absorbing, diffracting, obstructing, reflecting, refracting, and scattering influences had no effect on propagation, i.e., due only to distance. Free space loss is caused primarily by beam divergence, i.e., the spreading of signal energy over larger areas at distances increase from the signal source. It is generally expressed in decibels and is given by

$$\text{Loss} = 20 \cdot \log_{10}\left(\frac{4\pi D}{\lambda}\right) \text{ dB}$$

where   D is the separating distance and
       λ is the wavelength.

Freenet A grass-roots organization whose goal it is to provide free access to the Internet within a community (frequently by working with local schools or libraries). Typically, *Freenets* are designed as model "towns" wherein the user can visit places such as the library, the mayor's office, an e-mail post office, a bulletin board system, or the Internet itself. The first *Freenet* was the *Cleveland Freenet.*

freeware Copyrighted software for which the author or owner does not request a fee for its use. The fact that a program is free does not mean it is public domain (though this is a common misconception). The owner retains all rights not explicitly given to the user. In addition to the copyright, the author (or owner) often retains the source code and all rights to distribute. *Freeware* is often available on bulletin board systems (BBSs). See also *free software, public domain software,* and *shareware.*

freeze-out In telephony, a short time interval during which a subscriber connection is interrupted by a speech interpolation system (e.g., TASI).

FREQ (1) An abbreviation of frequency. **(2)** An acronym for File REQuest. In FidoNet, the ability to transfer files among bulletin board systems (BBSs) automatically.

frequency A measure of how often a periodic event occurs; the number of times per second that event repeats. The number of waves passing a given point per second. The units of frequency are Hertz (abbreviated Hz), named for the German inventor Heinrich Hertz (1857–1894).

frequency agile device In a frequency division multiplex (FDM) system, a device that can switch frequencies in order to allow communications over different available frequency bands at different times.

For example, a *frequency agile modem,* in a broadband system, can switch frequencies in order to allow communications over different channels (different frequency bands) when new links are established.

frequency agility The ability of a device (such as a cellular telephone) to automatically change operating frequencies as conditions justify.

frequency averaging (1) The process by which the relative phases of precision clocks are compared in order to define a single time standard. **(2)** A process in which overall network clock synchronization is achieved. In the process, the frequency of each node's local oscillator is adjusted to the average frequency of the received digital bit stream from all connected nodes. In this scheme, all oscillators are assigned equal weight in determining the ultimate network frequency.

frequency band A contiguous range of frequencies extending between two specified limiting frequencies.

For convenience of speaking about ranges of frequencies, the spectrum is broken up into several arbitrary bands with assigned names. The following chart shows two such naming conventions: the ITU naming system and a common usage system. Letter designators of radio frequency bands are imprecise, deprecated, and legally obsolete. The ITU frequency range is defined as 0.3×10^N through 3.0×10^N where N is the *band number.* In speaking about one of these frequency ranges, it is common to take the first letter of the adjectival description as an acronym. For example, band 7 is the high-frequency band or the HF band; similarly, band 11 is the EHF band. See also *frequency spectrum.*

BAND DESIGNATIONS

Common		ITU		
Band letter	Frequency Range	Band No.	Frequency Range	Adjectival Description
ULF		1	<30 Hz	Ultra low frequency
ELF	30-300 Hz	2	30-300 Hz	Extremely low freq
VF	0.3-3 kHz	3	0.3-3 kHz	Voice frequency
VLF	3-30 kHz	4	3-30 kHz	Very low frequency
LF	30-300 kHz	5	30-300 kHz	Low frequency
MF	0.3-3 MHz	6	0.3-3 MHz	Medium frequency
HF	3-30 MHz	7	3-3 0MHz	High frequency
VHF	30-300 MHz	8	30-300 MHz	Very high frequency
UHF	0.3-1 GHz			
L	1-2 GHz	9	0.3-3 GHz	Ultra high frequency
S	2-4 GHz			
C	4-8 GHz			
X	8-12.5 GHz	10	3-30 GHz	Super
K_u	12.5-18 GHz			high frequency
K	18-26.5 GHz			
K_a	26.5-40 GHz			
V	40-75 GHz	11	30-300 GHz	Extremely
W	75-110 GHz			high frequency
mm	110-300 GHz			
		12	0.3-3 THz	

frequency converter A device that can be used to translate between the sender's and the receiver's frequency bands in a broadband system.

For example, in a broadband network, such as a cable TV system, the headend (main transmitter) may need to convert (relocate) the incoming signal channel before sending it on to network nodes (or cable subscribers). Also called a *frequency translator*.

frequency delay In transmission systems, a delay caused by the fact that signals of different frequencies travel at slightly different speeds through a given medium and, therefore, reach the destination at slightly different times. This delay can result in signal distortion.

frequency departure The unintentional deviation from the expected nominal frequency value.

frequency deviation (1) Generally, the amount by which the actual frequency differs from a prescribed or theoretical value, e.g., the amount an oscillator frequency drifts from its nominal frequency. **(2)** In frequency modulation (FM), the absolute difference between the peak (maximum or minimum) permissible instantaneous frequency of the modulated wave and the unmodulated carrier frequency. **(3)** In frequency modulation (FM), the maximum absolute difference, during a specified period, between the instantaneous frequency of the modulated wave and the unmodulated carrier frequency.

frequency displacement The end-to-end shift in frequency that results from independent frequency translation errors in a communications link.

frequency distortion Distortion in which the *relative magnitude* (amplitude) of different frequency components of a complex wave are changed while propagating through a transmission circuit. Also called *amplitude-vs.-frequency distortion*. See also *distortion*.

frequency diversity A way of providing a redundant signal path between stations such that a failure of either path will not disrupt communications. The redundancy is provided by the use of two or more separate and independently fading frequency bands.

frequency division multiple access (FDMA) (1) The use of frequency division multiplexing to provide multiple and simultaneous transmis-

sions to a single transponder. **(2)** A method of providing multiple Earth station access to a single communications satellite transponder. Several methods are possible, each providing a portion of the total available transponder bandwidth to each Earth station; for example,

- Each Earth station may be assigned one transmit carrier onto which the data to all destination stations are multiplexed.
- In a second alternative, a separate carrier is transmitted by each station for each destination being addressed.
- A third alternative assigns a separate channel for each active voice channel (called a single channel per carrier system—SCPC).

frequency division multiplex (FDM) The deriving of two or more simultaneous, continuous channels in a communication link by splitting the total available spectral bandwidth into smaller sub-bands. Each of the sub-bands is then available as a separate communications channel.

For example, the total bandwidth of a normal telephone line is about 3100 Hz (ranging from 300 Hz to 3400 Hz). This band can be divided into two 1400-Hz bands separated by a 300-Hz *guard band*.

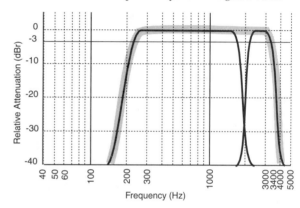

frequency division switching A method of switching in which a common physical path is provided with separate frequency bands into which each simultaneous call may be placed.

frequency doubler A device whose output is a signal at twice the frequency of its input.

frequency drift The gradual, undesired change of frequency over a period of time which is long when compared to the information rate. *Drift* may be in either direction, and the rate of frequency change over time may change. The change may be due to component aging, temperature changes, and the like.

frequency droop The change in frequency between the no load and the full load steady-state condition of a power source.

frequency exchange signaling A form of *frequency shift keying (FSK)* in which the change from one significant condition to another is accompanied by the decay in amplitude of one or more frequencies and by buildup in amplitude of one or more other frequencies. Also called *two-source frequency keying*.

frequency fluctuation In oscillators, the short-term variation of frequency with respect to time. *Frequency fluctuation, f(t)*, is given by:

$$f(t) = \frac{1}{2\pi} \cdot \frac{d^2\theta(t)}{dt^2}$$

where $\theta(t)$ is the phase angle of the sinusoidal wave with respect to time (*t*).

frequency frogging See *frogging*.

frequency hopping A modulation technique in which the carrier frequency is switched among multiple frequencies (according to a specified algorithm) in a given band at a rate less than or equal to the sampling rate of the information being transmitted. Selection of a particular frequency to be transmitted can be either a fixed sequence or a pseudorandom sequence. Although the technique minimizes the possibilities of jamming and unauthorized reception, it does use a much wider bandwidth than that required to transmit the same information on a single carrier. See also *spread spectrum*.

frequency hopping multiple access (FHMA) A multiple access technique in which the transmit frequency is switched at or below the data sampling rate. Selection of the specific transmit frequencies may be in either a fixed sequence or pseudorandom manner. The target receiver must frequency hop in the exact same manner as the transmitter in order to retrieve the information. See also *spread spectrum*.

frequency lock The condition in which an automatic frequency correcting system maintains control of a variable frequency oscillator (VFO) or source within the limits of one cycle.

Frequency lock does not imply *phase lock*, but *phase lock* does imply *frequency lock*. See also *phase lock loop (PLL)*.

frequency modulation (FM) A form of *angle modulation* in which the *instantaneous frequency* of the carrier wave is varied by an amount proportional to the instantaneous value of the information-bearing modulating signal. The amplitude does not change with pure *frequency modulation (FM)*. The unmodulated carrier is called the center frequency. The figure below illustrates the relationship of the modulating signal, the carrier, and the FM modulated signal.

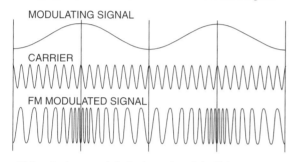

In an *FM optical system*, it is the intensity of the lightwave that is varied (modulated) by an electrical FM carrier. That is, the lightwave's intensity is varied at an instantaneous rate corresponding to the instantaneous frequency of the electrical FM carrier.

frequency modulation noise The frequency modulation of a signal resulting in distortion, which is perceived as noise added to the signal. The noise is not present in the absence of the signal.

frequency of optimum transmission (FOT) In the transmission of radio waves via ionospheric reflection, the highest frequency that is predicted to be usable for a specified path and time for 90% of the days of the month. Generally, it is the most effective frequency for ionospheric reflection of radio waves between two specified points on Earth.

FOT is normally just below the value of the maximum usable frequency (MUF). Frequently, it is taken to be 15% below the monthly median value of the MUF for the specified time and path. Also called *frequency of optimum traffic, optimum traffic frequency FOT, optimum transmission frequency*, and *optimum working frequency*.

frequency offset The difference between the frequency of a source and a reference frequency.

frequency prediction A prediction of:
- The maximum usable frequency (MUF),
- The optimum traffic frequency, and
- The lowest usable frequency (LUF)

for RF transmissions between two specified locations or geographical areas during various times throughout a 24-hour period. The prediction is usually presented as a graph for each frequency as a function of time. See also *FOT*.

frequency range A continuous range or band of frequencies that extends from one limiting frequency to another. Also called *frequency band*.

frequency response (1) The amplitude change of a circuit or system output signal in response to a frequency change of the circuit or system input signal. Also called *insertion loss vs. frequency characteristics*. **(2)** The range of frequencies a device can accept as input and still meet its rated performance. See also *attenuation distortion (1)*.

frequency reuse The ability to use the same frequency within the same system and for a different signal.

For example, in the cellular telephone system, the same carrier frequency may be reused by any cell not adjacent to a cell currently using it.

frequency shift Any change in frequency whether intentional (as in modulation) or not (as in drift) and without regard to information content. Frequency shift may be slow or fast. Sources of frequency shift include FM modulation, frequency shift modulation, and oscillator drift. Also called *rf shift* in the RF regime.

frequency shift key (FSK) A form of *frequency modulation* in which the modulating signal shifts the nominal carrier frequency between predetermined values.

The modulating signal is generally a digital signal with only two states (termed *mark* and *space*). Hence, the carrier will be shifted between two frequencies. The figure is an example of a carrier modulated by a digital data stream to produce an *FSK* signal.

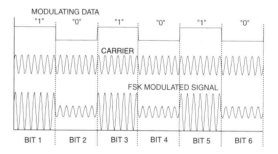

The rate at which the carrier is shifted can be gradual or essentially instantaneous. In systems where the change is gradual, the phase also changes gradually. Hence, it is frequently called *phase continuous FSK*. If the carrier frequency is abruptly shifted between the two binary states, the phase may or may not be continuous. When the system is designed to force phase continuity at signaling element changes, the modulation scheme is frequently called a *phase coherent FSK* system. Phase coherence is important, for it provides a better bit error rate (BER) performance than noncoherent modulation. If the carrier is shifted between more than two frequencies, the modulation scheme is called *M-ary* (or *multilevel*) *FSK* or simply *MFSK*. Two-state *FSK* schemes are also called *binary FSK (BFSK)*, *frequency shift modulation, frequency shift signaling*, or sometimes *frequency shift pulsing*. See also *minimum shift key (MSK)* and *modulation*.

frequency shift pulsing See *frequency shift key (FSK)*.

frequency spectrum A term describing the range of electromagnetic waves. The range technically starts just above zero and has no upper limit; however, the spectrum is usually divided into bands. Specifically, radio energy is generally defined as 3 Hz to 3 THz, followed by light with frequencies from 3×10^{12} Hz to 3×10^{16} Hz (or in terms of wavelength 100 μm to 100 Å), then x-rays with wavelengths from 100 Å to 1 Å (or in terms of energy 100 eV to 10 keV), and finally gamma rays with energies above 10 keV.

Frequency, wavelength, and energy are different ways of expressing the same phenomenon and are related by:

f *frequency*, in Hz, is the number of repetitions of an event per second.

λ *wavelength,* in cm, is the propagation velocity (generally the speed of light—*c*) divided by frequency (*f*); that is,

$$\lambda = \frac{c}{f} \cong \frac{2.99792458 \times 10^{10}}{f} \ cm, \ and$$

E photon energy is the energy, in electronvolts (*eV*), of the radiation and is given by the product of Planck's constant (*h*) and frequency (*f*); that is,

$$E = h \cdot f \cong 4.1357 \times 10^{-15} \cdot f \, eV.$$

Alternately

$$E = \frac{h \cdot c}{\lambda} = \frac{1.2398541}{\lambda(\mu m)} \ eV.$$

See also *frequency band.* Also called the *electromagnetic spectrum.*

frequency standard A very stable oscillator used for frequency calibration or as a reference. It generates its fundamental frequency with a high degree of both accuracy and precision.

frequency swing In frequency modulation (FM) systems, *frequency swing* is the peak difference between the maximum and minimum values of instantaneous frequency.

The specific conditions of maximum and minimum need to be specified, e.g., maximum permissible input level, at nominal operating levels.

frequency tolerance The maximum permissible departure by the center frequency of an electromagnetic emission from the assigned frequency (or reference frequency) for all causes. The departure is expressed directly in Hz or as a percentage.

frequency translation The shifting of signals occupying a specified band of frequencies, such as a channel or group of channels, from one portion of the frequency spectrum to another, and in such a way that the arithmetic difference of frequency between signals within both bands are the same. The translation may be intentional as in the case of a transponder, or it may represent an error in the signal processing path.

frequency translator See *frequency converter* and *frequency translation.*

Fresnel diffraction pattern A synonym for *near-field diffraction pattern.*

Fresnel reflection In optics, the *reflection* of a fraction of incident light at an interface between two media having different refractive indices.

For example, a *Fresnel reflection* occurs at the air–glass interfaces at both the entrance and exit ends of an optical fiber, resulting in transmission losses of about 4% per interface. (The reflection, and corresponding loss, can be significantly reduced by the use of index matching materials at the interface.) The coefficient of reflection depends not only on the refractive index difference, but on the angle of incidence and the polarization of the incident ray. For a normal ray, the fraction of reflected incident power is given by:

$$R = \frac{(n_1 - n_2)^2}{(n_1 + n_2)^2}$$

where:

R is the reflection coefficient,
n_1 is the refractive index of material 1, and
n_2 is the refractive indices of material 2.

In general, the greater the angle of incidence with respect to the normal, the greater the *Fresnel reflection* coefficient, but for radiation

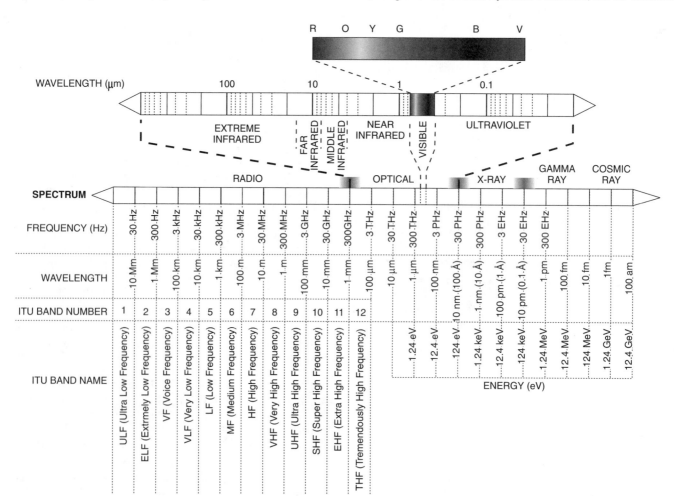

that is linearly polarized in the plane of incidence, there is zero reflection at *Brewster's angle*. Another method of reducing reflection is through the use of antireflection coatings. These coatings consist of one or more thin film layers dielectric having specific refractive indices and thicknesses. These antireflection coatings reduce overall *Fresnel reflection* by mutual interference of individual Fresnel reflections at the boundaries of the individual layers.

frogging (1) In telephony, a process used during the measurement of longitudinal balance wherein the tip and ring leads of the unit under test are reversed relative to the source or termination or both. **(2)** In line-of-sight microwave systems, the interchanging of frequency bands between adjacent repeater sites to reduce crosstalk, prevent singing (oscillation), and correct for frequency response slope anomalies (see definition 3). **(3)** In broadband communications, shifting of the signal carrier frequencies in order to equalize the distortion and loss across the transmission's bandwidth. The incoming channel with the highest frequency will go out as the lowest frequency band, the second highest in will be the second lowest out, and so on. Also called *frequency frogging.*

front end (1) That part of a receiver intended to receive very low level signals from the transmission medium. **(2)** The client part of a client–server application that requests services across a network from a server, or back end. It typically provides an interactive interface to the user, for example, a data entry front end, allowing data to be entered into a server through the use of Structural Query Language (SQL). **(3)** The beginning of a process, e.g., a development. **(4)** A device or software that preprocesses information before transferring it on, e.g., equipment interposed between a computer and the communications line whose purpose it is to organize data being transmitted and received.

front-end noise temperature A measure of the thermal noise generated by the first stage of a receiver.

front-end processor (FEP) (1) A stored program (or programmed logic) device that interfaces data communication equipment (DCE) with an input/output bus or memory of a data processing computer. **(2)** In IBM's Systems Network Architecture (SNA) network, a component that controls access to the host computer (generally a mainframe). It is generally attached to the host by a fast, direct connection (often a fiber optic link) and is controlled by the host through a network control program (NCP) loaded and executed on the *FEP.*

The *FEP* is responsible for:

- Aiding the host of tasks such as establishing connections and monitoring links.
- Any data compression or translation as the data moves between host and remote device.

Also called a *communication controller.*

front to back ratio (1) Of an antenna. The gain in a specified direction, usually the direction of maximum gain, compared to the gain in a direction 180° from that specified direction. *Front to back ratio is* usually expressed in decibels (dB). **(2)** The ratio of a specific parameter (current, resistance, signal strength, etc.) in the forward direction to that in the reverse direction. It is used to characterize rectifiers or other devices. For example, in a rectifier the front to back current ratio may be 10 000:1.

FS (1) An abbreviation of Frame Status. A field in a token-ring data packet. **(2)** An abbreviation of Field Service. **(3)** An abbreviation of File System. **(4)** An abbreviation of Full Standard.

FSDPSK An abbreviation of Filtered Symmetric Differential Phase Shift Keying.

FSF An abbreviation of Free Software Foundation.

FSK An abbreviation of Frequency Shift Key.

FT-1 An abbreviation of Fractional T-1.

FT-3 An abbreviation of Fractional T-3.

FTAM An acronym for File Transfer, Access, and Management.

FTP An abbreviation of File Transfer Protocol.

FTS An abbreviation of File Transfer Service.

FTS2000 An abbreviation of Federal Telecommunications System 2000. A long-distance service whose use is usually mandatory by government agencies except in instances related to maximum security.

The system provides full services, including:

- Switched analog service for voice or data up to 4.8 kbps.
- Switched data service at 56 and 64 kbps.
- Switched digital integrated service for voice, data, image, and video up to 1.544 Mbps.
- Switched packet service for data in packet form.
- Video transmission for both compressed and wideband video.
- Dedicated point-to-point private line for voice and data.

FTTB An abbreviation of Fiber To The Building.

FTTC An abbreviation of Fiber To The Curb.

FTTH An abbreviation of Fiber To The Home.

FUBAR An acronym for Fouled Up Beyond All Recognition (Repair). (The original acronym was from the military: consequently, a slightly different choice of words was probably used.) See also *foo.*

full availability A communications switch or switching system capable of providing a communications channel from every input port to every output port in the absence of traffic. A group of traffic-carrying circuits in which every circuit is accessible to all sources. Also called *nonblocking.*

full carrier A carrier that is transmitted without reduction in power, i.e., a carrier that is of sufficient level to demodulate the sideband(s).

full carrier single sideband A single sideband emission without reduction of the carrier.

full direct trunk group A full trunk group between end offices.

full duplex See *duplex* and *half duplex* (*HDX*).

full-duplex (FDX) circuit A communications circuit that permits simultaneous transmission in both directions.

full load The maximum load a device or system is designed to handle.

full screen editor A text editor that allows the cursor to be moved to any location within the active text entry window. A *full screen editor* will allow a user to go back to previously written material and make additions, corrections, or deletions.

Most *full screen editors* have the editing features of advanced word processors. However, they do not as a rule have text formatting capability. See also *line editor.*

full trunk group In telephony, a trunk group, other than a final trunk group, that does not overflow calls to another trunk group.

fully connected network A network in which each node is directly connected to every other node. Also called a *fully connected mesh network.*

fully qualified domain name (FQDN) In Internet, the entire address following the "@" symbol in a network address, i.e., the *somewhere.domain* part of:

someone@somewhere.domain.

It represents the unambiguous name of a specific system.

fully restricted stations Telephone stations in a PBX environment that cannot make outside calls directly. (They must go through the operator.) They can, however, receive outside calls and make intercom calls.

function management layer The topmost layer in IBM's Systems Network Architecture. The layer an end user deals with directly. This layer in turn, deals with the data-flow control layer.

functional profile A document that characterizes the requirements of a standard or group of standards and specifies how the options and ambiguities in the standard(s) should be interpreted or implemented to:

- Provide a particular function.
- Promote interoperability among different network elements and terminal equipment that implement a specific profile.
- Provide for the development of uniform tests.

Functional profiles were developed in order to ensure that, when defined, ISO/OSI stacks could interoperate. Because of the different protocol elements at each OSI layer, it was possible to define stacks that were syntactically correct but would not be able to exchange information due to differences at particular layers. A *functional profile* that has been defined as a standard is a standardized profile. Similarly, an International Standard Profile is an OSI *functional profile.*

fundamental The sinusoidal component of a periodic wave having the lowest frequency. According to the Fourier theorem, every periodic waveform may be expressed as the summation of the fundamental and its harmonics. See also *Fourier theorem* and *harmonic.*

fundamental mode The lowest order mode of a waveguide. In optical fibers, the fundamental mode is designated HE_{11}.

FUNI An acronym for <u>F</u>rame <u>U</u>ser-to-<u>N</u>etwork <u>I</u>nterface.

fuse (1) In electronics, a protective device that interrupts the flow of current when that current exceeds a specified value. The critical element in a fuse is a conductive wire or strip that will melt when heated by a designated current. When the *fuse* element melts, the fuse is said to have "blown."

There are two major characterizations of fuses:

- Fast-blow—fuses that open nearly instantaneously when exposed to an overcurrent condition.
- Slow-blow—fuses that can tolerate a transient overcurrent condition but will open if the overcurrent condition is sustained. Also called *delayed-blow.*

Some of the more common rated parameters of a fuse are:

- Rated breaking current—the current that the fuse, under specified conditions (rated voltage, ac/dc, temperature, humidity, etc.), can interrupt without destruction or arcing being maintained. Also called *interrupting current.*
- Rated current (I_{rated})—the maximum current the fuse can conduct under specified adverse conditions and not open the circuit. The value is derived by the manufacturer using test data derived from both non-fusing and fusing currents at elevated ambient temperatures.
- Rated voltage (V_{rated})—the maximum voltage that safe fault current interruption can be achieved. Any fault current up to the maximum specified value must be safely interrupted.
- Power dissipation—the maximum possible self-heating of the fuse in its overload region (determined using the non-fusing test current). This parameter is important when selecting the fuse holder.
- Time-current characteristic (I/t-curve) describes the circuit opening time of a fuse as a function of the fault current it is conducting. Generally the characteristic is presented as a curve and at ambient temperature. This characteristic is the major parameter that determines if the fuse is quick acting, normal or slow-blow. The following chart indicates the ranges of possible fuse characteristics.

A related parameter is the *melting integral.*

Other parameters sometimes included are:

- *Cold resistance*—the internal resistance of the fuse in a non-loaded condition. Usually it determined at currents less than 10% of the rated current.
- *Voltage drop*—the internal resistance of the fuse under operating conditions. When specified, it is generally designated at rated current.
- *Melting integral*—the energy required to melt the fuse link when

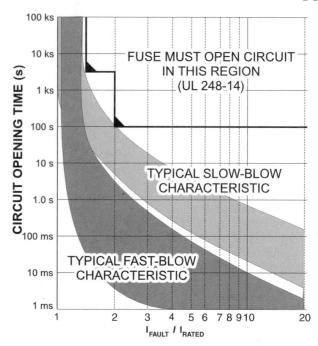

there is no time for heat dissipation. It is a pulse rating and is generally specified for times less than 5 ms. It may be derived from the square of 10 times the rated current and the corresponding circuit opening time.

Fuses, in contrast with circuit breakers, cannot be reset or reused. (2) The process of partially melting the surface of an object in order to join it with another object to form a single unit when cooled. Note, both objects may be partially melted in the process. An example is the joining of the endfaces of a pair of optical fibers by melting (welding) the endfaces together.

fused silica A synonym for *vitreous silica.*

fusion splice In fiber optics, a permanent junction of two optical fibers (splice) created by localized heating of the ends of the two fibers. A fusion splice yields a continuous length of fiber with minimal discontinuities at the junction.

fuzzy logic A method of programming devices to deal with imprecise inputs. The technique in essence expands the YES/NO world of conventional computers to one of multiple degrees, for example,

YES/mostly yes/maybe/mostly no/NO.

From these inexact inputs, the machine assigns relative values representing certain characteristics, calculates a weighted average of the inputs, and finally decides what action to take.

FVC An abbreviation of <u>F</u>requency to <u>V</u>oltage <u>C</u>onverter.

FWHM An abbreviation of <u>F</u>ull <u>W</u>idth at <u>H</u>alf <u>M</u>aximum.

FWIW An abbreviation of <u>F</u>or <u>W</u>hat <u>I</u>ts <u>W</u>orth.

FX An abbreviation of <u>F</u>ixed <u>S</u>ervice.

FX line An abbreviation of <u>F</u>oreign e<u>X</u>change <u>line</u>.

FYA An abbreviation of <u>F</u>or <u>Y</u>our <u>A</u>musement.

FYI An abbreviation of <u>F</u>or <u>Y</u>our <u>I</u>nformation.

FYI papers In the Internet, short documents about various topics of relevance to Internet users. The papers are similar to Requests for Comments (RFCs), except that the FYIs are generally written as background material and are somewhat less technical.

FYn An abbreviation of <u>F</u>iscal <u>Y</u>ear where "n" is the year being discussed. Unlike a calendar year, a fiscal year may begin on any day of the year and will end 365(6) days later.

G

g The SI symbol for gram.

G (**1**) The SI prefix for *giga* or 10^9. It is also taken to be 2^{30} (1 073 741 824) in the binary system. (**2**) The symbol for magnetic flux in the cgs system, i.e., gauss. (**3**) A symbol for <u>G</u>ain. (**4**) A symbol for <u>G</u>round. (**5**) A symbol for one gravitational unit.

G.n Recommendations A series of International Telecommunication Union (ITU) standards covering transmission systems and media, digital systems and networks. See Appendix E for complete list.

G/T ratio The ratio of the gain to the noise temperature of an antenna. Also called a *figure of merit.*

GA (**1**) An abbreviation of <u>Go</u> <u>A</u>head. (**2**) An abbreviation of <u>G</u>lobal <u>A</u>ddress. (**3**) An abbreviation of <u>G</u>ain of <u>A</u>ntenna.

GaAs The atomic formula for <u>Ga</u>llium <u>As</u>enide. A semiconductor used to make very high-speed transistors and integrated circuits.

GAB An acronym for <u>G</u>roup <u>A</u>udio <u>B</u>ridging.

GAIC An abbreviation of <u>G</u>allium <u>A</u>rsenide <u>I</u>ntegrated <u>C</u>ircuit.

gain (**1**) In general, the increase of a parameter of a signal or process. The *gain* is the ratio of the increased value to the original value. The gain may be expressed as a numeric multiplier (such as a gain of 4) or in decibels, for example, an input of 1 V and an output of 2 V yields a gain of 2 (the ratio 2/1) or as +6 dB ($20 \log_{10} 2$). If the input is larger than the output, the gain is fractional, for example, an input of 2 A and an output of 1 A yield a gain of 0.5 (the ratio 1/2) or as −6 dB ($20 \log_{10}(1/2)$). For power ratio gain, i.e., P_1 W in and P_2 W out, the gain is $10 \log_{10}(P_2/P_1)$. (**2**) In antennas, the ratio of the field strength in a given direction as compared to the field strength that would be obtained if the same power were radiated by a reference antenna.

gain hit A data transmission error caused by a momentary increase in circuit gain. Gain surges greater than 3 dB typically last less than 4 ms.

gain tracking A measure of the error in gain for different input signal levels. In a perfect system the output is related to the input by the constant gain multiplier. If, however, the multiplier is not constant for all signal levels, a signal error called *gain tracking error* will occur. See also *attenuation distortion* (2).

GAN An acronym for <u>G</u>lobal <u>A</u>rea <u>N</u>etwork.

gang screen See *Easter eggs.*

gap The space between two electrodes.

gap loss In fiber optics, one of several sources of *coupling loss;* it is the optical power loss caused by the space between axially aligned optic fibers or optical emitters and fibers. The gap allows light from the transmitting fiber to spread out as it leaves the fiber endface. When it enters the receiving fiber, some of the light will enter the cladding, where it is quickly lost.

Gap loss is not usually significant at the optical detector because the sensitive area of the detector is normally larger than the cross section of the fiber core. Also called *longitudinal offset loss.* See also *coupling loss.*

gap loss attenuator An optical attenuator that exploits the principle of *gap loss* to reduce the optical power level when inserted in line with an optical fiber.

To be effective, *gap loss attenuators* should be used near the optical transmitter because at distances greater than the equilibrium length the effective numerical aperture is reduced. For example, a 50-μm core multimode graded-index fiber has a numerical aperture approximately 70% that of the full numerical aperture of the fiber, and light propagating under these conditions occupies approximately the middle 70% of the core. The attenuator, therefore, should be inserted near the optical transmitter, where the core is fully filled. See also *equilibrium length, equilibrium mode distribution (EMD),* and *gap loss.*

garble (**1**) An error in transmission, reception, encryption, or decryption that changes any portion of a message in such a manner that it is incorrect or undecryptable. (**2**) In a telephone circuit or channel, one or more interfering speech signals from other channels with sufficient amplitude to be easily heard but nonetheless unintelligible.

Garble may, for example, take place in a frequency division multiplier (FDM) telephone carrier system in which an interfering signal from another channel or system is demodulated in such a fashion that it has an objectionable audio power level but remains unintelligible.

gas-filled protector An overvoltage protection device used to safeguard sensitive electronics (such as communication circuits) from transient voltages that exceed safe operating values. The device is essentially one or more spark gaps in a sealed enclosure filled with an inert gas mixture at some known pressure. The choice of gasses and operating pressure helps determine the size and breakdown voltage of the spark gap.

When more than one spark gap is contained within the enclosure, the elements are arranged so that any spark gap activated will trigger all of the other spark gaps. This reduces the differential voltages in the circuits being protected. One terminal of the protector is connected to Earth ground, and the others are connected to the circuits to be protected. Symbols for *gas-filled protectors* include:

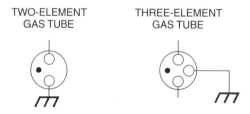

A typical voltage/current over time curve is presented in the following figure. A transient voltage curve is shown as a solid line, while the resulting current is illustrated as a dashed line. The initial breakdown voltage is substantially higher than the glow discharge voltage. Standards such as the ITU-T K.12, Bellcore GR-1089, or IEC 1000-4-2 specify minimum peak voltage, current, and waveshape to be handled. Also called a *gas discharge tube (GDT), gas tube protector,* or *gas tube surge arrester.* See also *carbon block* and *spark gap.*

227

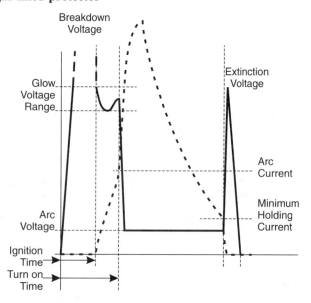

gas tube A gas-filled electron device in which the pressure of the gas or vapor is such that it substantially influences the electrical characteristics of the device.

gas tube surge arrester See *gas-filled protector.*

gaseous discharge The emission of light from excited gas atoms. Excitation may be from an electric current flowing through the gas, a strong electric field, or fast-moving atomic particles.

gate (1) An electronic device or switch that performs one or more of the Boolean operations AND, OR, or NOT. See also *AND gate, exclusive OR gate, inverter, NAND gate, NOR gate,* and *OR gate.* (2) That element in certain semiconductor devices (for example, thyristors and field effect transistors) which controls the flow of current through the device. (3) The octathorp symbol "#."

gate array (GA) A class of integrated circuit (IC) that contains a number of predefined functions that can be interconnected for an original equipment manufacturer (OEM) after the majority of the manufacturing processes are complete. They are classified by the means by which customization is implemented. Two common means are *mask programmed gate arrays* and *field programmable gate arrays (FPGAs).*

Mask programmed gate array is a type of application-specific integrated circuit (ASIC) that is configured for the original equipment manufacturer (OEM) only at the metal interconnect level(s). That is, only the metal layers are custom, and all the other layers that make the device are standard. The standard layers are prefabricated on the wafer, called a *master slice,* and contain an array of some number of gates that are not interconnected. The OEM specifies how this field of gates is to be interconnected, and the *gate array* manufacturer adds the final metal layer(s) to complete the device. Architectures include *channeled array, gridded array,* and *sea of gates.*

- *Channeled array*—an array of transistors or gates arranged with dedicated channels between rows of sets of transistors. The channels provide for the metal layer interconnections.
- *Gridded array*—an array of transistors or gates arranged as a grid with spacing for metal interconnections on all four sides of each "core" cell.
- *Sea of gates*—an array of transistors or gates with routing spacing around each device but no defined routing areas or channels.

Also called a *channel-less gate array.*

Field programmable gate array (FPGA) is a class of ASIC for which both the logic and a number of possible interconnects are defined

during the manufacturing process. The OEM activates the desired interconnects after the IC has been manufactured by means of software and a programming device.

Gate arrays are usually more expensive than standard cell ASICs because on average they have a bigger die area. This is because unused gates are still on the die. For example, a 100k gate array contains 100k gates worth of transistors. The OEM's ASIC may require only 75k gates, but the remaining unconnected or unused 25k gates are still on the device.

gateway (1) An entrance/exit to a communications network. Gateways exist between any two communications service providers, e.g., between AT&T and the local telephone company that provides a connection to a subscriber or between AT&T and an international provider. (2) A network node (hardware and software) that permits devices on one network (LAN) to communicate with or gain access to the facilities of another possibly dissimilar network. It can translate the protocols up to the application layer (e.g., ISO/OSI layer 7) of one network type into the corresponding protocols of a different network. In addition to protocol translation, it may contain capabilities such as fault isolation, rate conversion, or signal conversion necessary to provide system interoperability. It also requires that mutually acceptable administrative procedures be established among the interconnected networks.

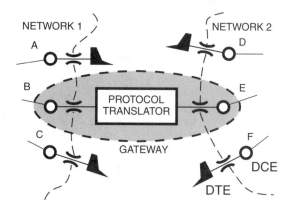

Gateways encompass all layers of the used protocol's model, and hence allow translation of protocols between LANs, for example, between OSI, DECnet, SNA, TCP/IP, and XNS. In the following diagram, a LAN based on the ISO/OSI layer 7 model is connected to a LAN utilizing TCP/IP protocols. See also *bridge, bus network, hierarchical network, hybrid network, mesh network, ring network, router,* and *star network.*

LAN 1 STATION X	**GATEWAY**		LAN 2 STATION Y
OSI layer 7	OSI layer 7	TCP/IP layer 4	TCP/IP layer 4
OSI layer 6	OSI layer 6		
OSI layer 5	OSI layer 5	TCP/IP layer 3	TCP/IP layer 3
OSI layer 4	OSI layer 4		
OSI layer 3	OSI layer 3	TCP/IP layer 2	TCP/IP layer 2
OSI layer 2	OSI layer 2		
OSI layer 1	OSI layer 1	TCP/IP layer 1	TCP/IP layer 1

LAN 1   LAN 2

gateway city   In telephony, the city through which all international calls must be routed. In the United States, Miami, New Orleans, New York, San Francisco, and Washington DC are gateway cities.

gateway server   A particular type of communications server that provides access between networks that use different protocols. See also *gateway*.

Gateway-to-Gateway Protocol (GGP)   The protocol that core gateways use to exchange routing information. It implements a distributed shortest path routing computation.

gauge   A measure of electrical wire diameter. Under the American Wire Gauge (AWG) standards, higher gauge numbers indicate a thinner cable. When measuring wire gauges, the insulation thickness is ignored. See also *AWG* and *wire gauge.*

gauss (G)   A measure of magnetic flux (or field) density in the centimeter-gram-second (cgs) measurement system expressed as lines per square centimeter. One gauss is equal to 10^{-4} webers per square meter or 1 maxwell per square centimeter.

gaussian beam   A beam of electromagnetic radiation having an approximately spherical wave front at any point along the beam and having transverse field intensities over any wavefront which is a gaussian function of distance from the beam-axis. For a beam width circular cross section, it is given by:

$$E(r) = E(0)e^{-\left(\frac{r}{w}\right)^2}$$

where:

r is the distance from beam axis,
$E(r)$ is the field intensity at a distance r,
$E(0)$ is the field intensity at the beam axis, and
W is the beamwidth (beam diameter). The value of r at which the beam intensity is *1/e* of its values on the axis, i.e., $W = 1/e \times E(0)$.

The following figure illustrates this intensity along one wavefront.

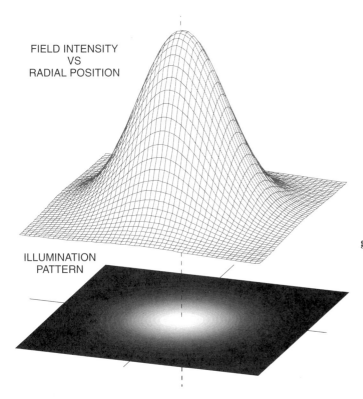

FIELD INTENSITY
VS
RADIAL POSITION

ILLUMINATION
PATTERN

Gaussian distribution   The relative frequency (probability) of occurrence of an event occurring over a large number of trials. It occurs whenever:

- A large number of random causes or events produce additive effects, and
- An appreciable fraction of the causes or events produce effects of nearly maximum variance.

The normal probability density function $\varphi(x)$ is given by:

$$\varphi(x) = \frac{1}{\sigma\sqrt{2\pi}}\ e^{-\left(\frac{(x-m)^2}{2\sigma^2}\right)}$$

where:

x is the random variable,
m is the mean, and
σ is the standard deviation.

The following graph illustrates three related curves:

- The *probability function* $\varphi(x)$ with a *median* (m) equal 0 and the standard deviation (σ) equal 1.
- The *cumulative distribution (C).*
- The probability that the error (absolute deviation-E) exceeds the value indicated.

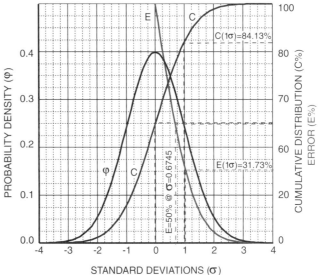

Some examples include:

- The probability of finding x between $-\sigma$ and $+\sigma$ is 68% (i.e., σ yields $C = 84.13\%$, and $-\sigma$ yields $C = 16\%$; and 84-16 = 68).
- The probability of finding x between $-\sigma$ and $+2\sigma$ is 81% (i.e., 2σ yields $C = 97\%$, and $-\sigma$ yields $C = 16\%$; and 97-16 = 81).
- The probability of finding |x|>1σ is 31.73% (i.e., E at 1σ is 31.73%).
- The probability of finding |x|>2 σ is 4.5% (i.e., E at 2σ is 4.5%).
- The probability of finding |x|>3σ is 0.3% (i.e., E at 3σ is 0.3%).

Also called the *normal distribution.*

gaussian filter   A filter whose magnitude frequency response approximates that of an ideal gaussian response. Because of the good transient response of a *gaussian filter* (small overshoot and ringing), these filters are frequently used in communications and pulse systems.

The ideal *gaussian* response is given by:

$$|H(j\omega)| = e^{-\frac{\ln2}{2}\left(\frac{\omega}{\omega_c}\right)^2}$$

where $H(j\omega)$ is the response at "frequency" ω with a 3-dB cutoff frequency of ω_c. With sufficient circuit complexity, the filter response can be made as close to the ideal response as desired.

gaussian minimum shift key (GMSK) A form of minimum shift key (MSK) modulation in which the bandwidth of the modulated carrier is reduced from that of a pure MSK modulator. The bandwidth reduction is accomplished through the use of a gaussian characteristic premodulation low-pass filter, as shown in the diagram. The bandwidth reduction is related to the ratio of the filter bandwidth to the bit rate. (The smaller the filter bandwidth, the smaller the carrier bandwidth.)

Although the reduced carrier bandwidth allows tighter channel spacing, it also reduces the tolerance to noise and allows more *intersymbol interference*. See also *minimum shift key (MSK)* and *modulation*.

gaussian noise Background noise that is produced in all conductors by the normal random motion of electrons. The noise occurs at all frequencies of the electromagnetic spectrum, and it increases with increasing temperature.

gaussian pulse A pulse that has the waveform of a Gaussian distribution. In the time domain, the pulse is of the form:

$$V(t) = V_{peak}e^{-\left(\frac{t}{t_0}\right)^2}$$

where:

V_{peak} is the maximum pulse value,
t is time, and
t_0 is the pulse half duration, i.e., the $1/e$ points.

Gb The symbol for gigabit (2^{30} bits).

GB The symbol for gigabyte (2^{30} bytes).

GBH An abbreviation of Group Busy Hour.

GBP An abbreviation of Gain-Bandwidth Product.

GCT An abbreviation of Greenwich Civil Time.

GD&R An abbreviation of Grinning, Ducking, & Running (generally, after a snide remark).

GDF An abbreviation of Group Distribution Frame.

GDMO An abbreviation of Guidelines for the Definition of Managed Objects. An ISO specification that provides notation for describing managed objects and actions involving such objects.

gel (1) A substance that surrounds one or more fibers that are enclosed in a loose buffer tube. The gel serves to lubricate and support the fibers in the buffer tube as well as to prevent water intrusion in the event the buffer tube is breached. (2) An index-matching material in the form of a gel used to reduce losses at a fiber connection. Also called *index-matching gel*. See also *index-matching material*.

gender changer A device for joining like connector types (either both pin or both socket). The *gender changer* has two like connectors (but opposite the target connectors) wired back to back. (That is, pin one of one connector is wired to pin one of the other connector.) A gender changer has like connectors on both sides, i.e., two socket connectors or two pin connectors. Also called a *gender bender*.

Generalized Data Stream (GDS) The format for mapped data in the Advanced Program to Program Communications (APPC) extension of IBM's Systems Network Architecture (SNA). The data from high-level applications are converted to the intermediate *GDS* format before transmission. This tends to isolate the applications from format differences, e.g., when one application uses the ASCII character set and the other uses EBCDIC.

General Public Virus A slang term for the *General Public License (GPL)* included with the *Free Software Foundation*'s products. The license states that their product may be redistributed or altered, as long as the resulting product itself carries the GPL. In effect, the *GPL* infects everything made from code containing the GPL—just like a virus.

general-purpose computer A computer designed to be programmed to solve many different classes of problems or with appropriate peripheral devices control external equipment. A general-purpose computer's function may be changed simply by loading a new program into its memory.

generalized property The characterization of an entity's physical attributes in a qualitative manner but not in quantitative magnitudes.

generic flow control (GFC) A 4-bit field in an ATM cell header with only local significance (i.e., its value is not carried end-to-end, and it is overwritten by ATM switches to 0000). The field is intended to provide standardized local functions such as flow control and to make sure all nodes get access to the transmission medium. See also *ATM*.

genetic algorithm A goal-seeking procedure that learns or adapts to its environment. The algorithm is modeled on the way a strand of DNA mutates; that is, the algorithm "splices" together various alternatives, projecting each alternative to its ultimate outcome. The algorithm then selects the best of the tried alternatives as the solution. The algorithm discards bad alternatives and mutates good alternatives in its quest for a best alternative. Many algorithms include a random mutation factor that forces completely new lines of trial in their search. Genetic algorithms were pioneered by computer scientist John Holland in the 1960s.

geocentric latitude The acute angle between the two lines described by:

1. A line joining a point (on the latitude to be described) and the center of the Earth, and

2. A line from the center of the Earth to the Earth's equatorial plane.

geodesic The shortest distance between two points on any mathematically derived surface.

geodetic datum A mathematical model designed to approximate the shape of part or all of the *geoid*. It is defined by an ellipsoid and the relationship of the ellipsoid to a point on the topographic surface defined as the origin of datum. Typically, eight parameters are used to define the datum:

- Two define the dimensions of the ellipsoid (major and minor axis).
- Three define the location of the center of the ellipsoid with respect to the Earth's center of mass.
- Three define the orientation of the ellipsoid orientation with respect to the average spin axis of the Earth and the Greenwich reference meridian (the *horizontal datum*).

Many different models have been developed. Some optimize the error of the approximation to a local region, and others attempt to generate a best fit for the entire planet. An example of the latter is WGS-84, which is used by the Global Positioning System (GPS).

geodetic latitude The acute angle, on the ellipsoid approximation to the Earth, between:

1. The normal to the spheroid, and
2. The equatorial plane of the ellipsoid.

geoid That smooth surface about the Earth with constant gravity which is coincident with the mean sea level and extends *through* the continents. The surface is perpendicular to gravity at every point. As one might imagine, the actual mathematical model is very difficult to describe because of local variations (land masses). Hence, an ellipsoid is used to approximate the shape. See also *geodetic datum.*

geometric dilution of precision (GDOP) In the satellite Global Positioning System (GPS), *GDOP* is an expression of the uncertainties of determined location due to both the geometric position of the satellites and the geometric effects on the estimates of time. It provides a measure of the quality of a satellite geometry. (The lower the value of *GDOP,* the smaller the error of the calculated position.) *GDOP* is determined by:

$$GDOP = \sqrt{(PDOP)^2 + (TDOP)^2}$$

where:

PDOP   is the three-dimensional position dilution of precision, which reflects geometric effects on three-dimensional position estimates, and

TDOP   is the time dilution of precision, which reflects geometric effects on time estimates.

See also *dilution of precision (DOP)* and *Global Positioning System (GPS).*

geometric optics The branch of optics that describes light propagation in terms of rays.

Rays are bent at the interface between two media with different refractive indices and may be curved in a medium in which the refractive index is a function of position. The ray in geometric optics is perpendicular to the wavefront in physical optics. Also called *ray optics.*

geosynchronous An Earth-orbiting satellite directly over the equator at an altitude of approximately 35 787 km above mean sea level (42 164 km from the Earth's center). A satellite in this orbit will maintain the same relative position over the surface of the Earth because its period and the rotation of the Earth are equal. It is therefore said to be *geostationary.* Also called *geostationary.*

geosynchronous orbit A specific circular orbit around the Earth in which objects appear to be stationary of a point on the Earth, e.g., the orbit of a communications satellite. The satellites are about 35 787 km (22,366 miles) above mean sea level of the Earth. Their orbit is "synchronous" because the satellite makes a revolution in 24 hours and in the same direction as the Earth rotates. Hence, they appear to be stationary over a location. A satellite in *geosynchronous orbit* is known as a *geosynchronous* or *geostationary* satellite. Also called a *geostationary orbit.*

germanium photodiode A photodiode (a PN or PIN junction diode) based on the element germanium (Ge) rather than silicon (Si).

Germanium photodiodes are used for direct detection of optical wavelengths from approximately 1 μm to several tens of μm. Silicon-based photo detectors are usually preferred for wavelengths shorter than 1 μm because they are less noisy than germanium-based detectors.

getter A material placed in a vacuum electron-emitting device, such as a vacuum tube (e.g., CRT, magnetron, or other radio tube), field emission display (FED), or vacuum deposition chamber, to absorb gasses released by materials in the vacuum. Without the *getter,* the released gasses would eventually degrade the quality of the vacuum, hence, the life of the vacuum device.

GF An abbreviation of <u>G</u>ain <u>F</u>actor.

GFC An abbreviation of <u>G</u>eneric <u>F</u>low <u>C</u>ontrol.

GFE An abbreviation of <u>G</u>overnment <u>F</u>urnished <u>E</u>quipment.

GFI (1) An abbreviation of <u>G</u>roup <u>F</u>ormat <u>I</u>dentifier. In an X.25 packet, a 4-bit field in the header that indicates packet format and other features. It contains the Q-bit, the D-bit, and the modulus value. **(2)** An abbreviation of <u>G</u>round <u>F</u>ault <u>I</u>nterrupter.

GGP An abbreviation of <u>G</u>ateway to <u>G</u>ateway <u>P</u>rotocol.

ghost A secondary signal or image resulting from echo, envelope delay distortion, or multipath reception (reflection).

Gibb's phenomenon Overshoot and ringing at or near discontinuity points of a signal caused by the abrupt truncation of the spectrum of the signal. The sequence in the following figure demonstrates the effects of eliminating high-frequency terms from the spectrum of a square wave. The first waveform includes all spectral components; the second includes only the first 50 components, and so on.

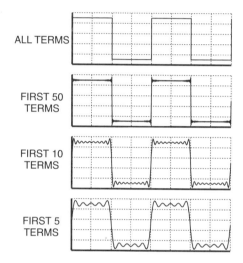

ALL TERMS

FIRST 50 TERMS

FIRST 10 TERMS

FIRST 5 TERMS

GID An abbreviation of <u>G</u>roup <u>ID</u>.

GIF An acronym from <u>G</u>raphics <u>I</u>nterchange <u>F</u>ormat (generally pronounced "jiff"). A platform-independent file format for color raster-based picture distribution and storage, developed by CompuServe.

The basic structure of a *GIF* file is shown in the following chart. For details, download a complete description from CompuServ. The extension of a *GIF* file is generally .GIF. See also *filename extension* and *JPEG.*

BASIC GIF FILE FORMAT

Name	Description
GIF Signature	Identifies the data as a GIF image and its version, e.g., GIF87A.
Screen Descriptor	Describes parameters such as screen width, height, number of colors, etc.
Global Color Map	Equates the red, green, and blue intensity to a hardware related index.
Image Descriptor Local Color Map	Defines placement and extents of following image.
Raster Data	Defines actual image as a series of pixel color index values.
• • •	Additional images
GIF Terminator	Character ";" (3Bh). Pauses decoding S/W and awaits user input.

giga The SI prefix indicating 10^9, that is, 1 000 000 000. In the computer and binary data world, the prefix indicates 2^{30} (1 073 741 824). Some examples include:

Example	Meaning
gigabit (Gb)	2^{30} bits,
gigabyte (GB)	2^{30} bytes,
gigaFLOP	1 000 000 000 floating point instructions (FLOPs)
gigahertz (GHz)	1 000 000 000 Hz

GIGO An acronym for <u>G</u>arbage <u>I</u>n, <u>G</u>arbage <u>O</u>ut (pronounced "GUY-go"). A computer axiom intended to indicate that a computer is not a thinking device and that the quality of the results of a process are no better than either the data presented or the design of the process. Specifically, if erroneous or incorrect data are supplied, the output will be just as invalid as the input.

glare In telephony, a deprecated term indicating the condition of an incoming call colliding with an outgoing call. This occurs frequently with PBXs but can also occur with any terminal equipment. A signaling system called *ground start* is used to eliminate such conditions. Also called *call collision*.

glass (1) In the strict sense, a state of matter. (2) In fiber optics, any of a number of noncrystalline, amorphous inorganic substances fused from metallic or semiconductor oxides or halides. The most common glasses are based on silicon dioxide (SiO_2) and have small amounts of other materials added to control color, dispersion, and refractive index.

GLG An abbreviation of <u>G</u>oofy <u>L</u>ittle <u>G</u>rin.

glitch A jargon term indicating

- A transient condition, e.g., a momentary interruption in mains power,
- An extraneous bit in a data stream (noise),
- An unexpected delay in a process or procedure, or
- An unexpected problem.

global (1) Pertaining to, or involving, the entire planet. (2) Pertaining to an item that is defined in one subsection of an entity and is used in at least one other subsection of the same entity. (3) In computers, data processing, and communications systems, pertaining to that which is applicable to a region beyond the immediate area of consideration. Examples include:

- In computer programming, an entity (global variable) that is defined in one subdivision of a computer program and used in at least one other subdivision of that program.
- In PCs and their software packages, a setting, definition, or condition that applies to the entire software system, e.g., the screen colors of a window in a Windows or Macintosh operating system.

global address In a communications network, the predefined address that is used as an address for all users of that network and that may not be the address of an individual user, or subgroup of users, of the network. See also *global name*.

global kill file A *kill file* that contains the *kill* specifications (delete keys such as topics, sites, or authors) for all of the newsgroups to which a user has subscribed. The kill file enables a newsgroup reader to exclude unwanted postings from ALL newsgroups. See also *kill* and *newsgroup kill file*.

global name In a network (or internetwork) a name known to all nodes and servers. This is in contrast to a local name (a name associated with a particular server). A *global name* is fully qualified; that is, it includes all the intermediate levels of membership associated with the name. See also *global address*.

global naming service A service that provides the mechanisms for naming resources attached to any of several file servers in a network. First developed in the StreetTalk service in Banyan's VINES software, these capabilities have been added to other network operating systems, such as Novell's NetWare version 4.0. Names in a *global naming service* have a predefined format, which reflects the different levels of operations in the network. For example, StreetTalk names have the format:

Item@Group@Organization

Global Positioning System (GPS) A space base radio position and navigation system, developed by the U.S. government, which provides information necessary to determine the latitude, longitude, altitude, speed, and direction of a receiver.

The system consists of a minimum of 24 satellites (21 + 3 operational spares) circling the Earth at an altitude of 20 183 km (a 12-hr orbital period). The *GPS* constellation consists of six different orbital planes inclined at 55°, each with four equally spaced satellites. Each satellite transmits two signals:

- L1 at 1.57542 GHz, which carries both a precision code (P-code) and a coarse/acquisition code (C/A code), and
- L2 at 1.22760 GHz, which carries P-code only.

At any instant, at least four satellites are visible to a receiver on the ground. The receiver can calculate its position on the surface by measuring the time interval between signals received from at least three of those satellites and their orbital mechanics information. Using the fourth satellite allows the determination of altitude as well as latitude and longitude. The *Precision Positioning Service (PPS)* P-code provides a predictable positioning accuracy of at least 22 m horizontally and 27.7 m vertically and time transfer accuracy to Coordinated Universal Time (UTC) within 200 ns, all with a confidence of 95%. Nonmilitary *GPS* navigation receivers are allowed access to the Standard Positioning Service (SPS) L1 C/A-code only. The C/A code may also be subject to a Department of Defense injected degradation known as *selective availability (SA)*. SA degrades the attainable position accuracy to a *circular error probability (CEP)* of 40 m. Therefore, the directly attainable position accuracy is limited to 40 m, with a confidence of 50%, or to 100 m latitude or longitude and 156 m elevation and 340 ns time accuracy, with a confidence of 95%. The illustration shows actual measurements taken over a one-hour period. Each concentric circle represents 25 m from the true position. In addition to the intentional *SA* degradation, other error sources also degrade the Global Positioning System from attaining perfection. Some of these errors are:

- Multipath (reflected) signals—which degrades position typically 1 to 1.5 meters (but can be tens of meters in severe cases).

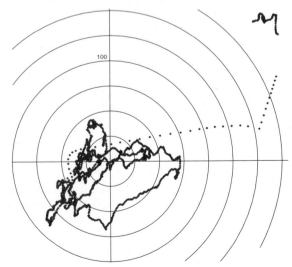

- Ionospheric (atmospheric) irregularities and time-varying propagation delays—which degrade accuracy by 5 to 6 meters.
- Satellite clock errors—1- to 2-meter degradation.
- Satellite orbital positional (ephemeris) errors—2- to 3-meter degradation.
- GPS receiver noise—typically less than 1-meter degradation.
- Rain fade and other water-related losses, such as water in and on an intervening tree canopy.
- Signal loss due to obstruction.

These and other accuracy degradations can be corrected by using time-averaged measurements and sophisticated processing techniques. Using these techniques, one can achieve position accuracies of 1 cm or better. See also *differential GPS, GPS time,* and *GPS week.*

GLOC An acronym for Ground Line Of Communications.

GLOMR An acronym for Global Low Orbiting Message Relay (pronounced "glow-mer").

GLONASS An acronym for GLObal'naya NAvigatsionnaya Sputnikovaya Sistema. The Russian global navigation system similar to the U.S. Global Positioning System.

glow current In gas tubes, the current that flows after breakdown when circuit impedances limit the current to values below the glow to arc transition current. Sometimes called the *glow mode current.* See also *arc* and *breakdown.*

glow discharge A discharge of current through a gas characterized by:
- A change of space potential in the immediate vicinity of the cathode that is much higher than the ionization potential of the gas.
- A low, approximately constant, current density at the cathode, and a low cathode temperature.
- The presence of a cathode glow.

glow to arc current The current threshold that must be exceeded for a spark gap immersed in a gas to pass from glow mode to arc mode.

glow voltage The approximately constant voltage across a spark gap immersed in a gas during glow current. Values range from 60 to 150 V depending on the gas, geometry, and so on. Sometimes called *glow mode voltage.*

GMSK An abbreviation of Gaussian Minimum Shift Key.

GMT An abbreviation of Greenwich Mean Time—a deprecated term. *GMT* is the time at the Greenwich England observatory. This is generally used as the reference time when a standardized value is needed. *GMT* (Zulu Time) is always the same worldwide. It is five hours later than Eastern Standard Time (EST) and four hours later than Eastern Daylight Savings Time (EDT). This official name has been changed to UTC (a permuted abbreviation of Coordinated Universal Time). See also *time (3).*

GND An abbreviation of GrouND.

GNE An abbreviation of Gateway Network Element. A SONET network element.

GNU A "recursive" acronym from GNU's Not UNIX. Developed by Richard Stallman and the Free Software Foundation, *GNU* is an operating system like UNIX. However unlike UNIX, it is free of charge and freely modifiable by its users. It is distributed along with the source code at no cost.

gnu hierarchy One of several USENet alternative newsgroups. The newsgroups offer discussion of the various products and services of the Free Software Foundation. See also *ALT* and *alternative newsgroup hierarchies.*

gold code **(1)** A coding scheme, developed by Robert Gold used in multiple access spread spectrum systems. Each code is essentially a pseudorandom (PN) sequence that serves as the "carrier" in a code division multiplex system; that is, each information data stream is modulated onto a unique code bit stream. Characteristics of *gold codes* include:

- *Gold codes* are composite codes, constructed by modulo-2 addition of two equal length, maximal PN sequences, but are themselves not maximal.
- The length of the *gold code* is the same as code length of the two generating codes.
- Over the set of codes available from a given generator, the cross-correlation between the two codes is uniform and bounded (unlike that for maximal PN sequences).
- Phase shifting either of the maximal length PN generators produces a new code output.

Gold codes have two features that are very beneficial to signal reception and recovery.

- Auto-correlation—the correlation of the code with a delayed version of itself will deliver a low output for any offset (phase shift) except zero. That is, the code is not self-similar for any offsets other than zero. This allows a receiver to synchronize to a transmitter and recover the signal if it is using the same code that a transmitter is using.
- Cross-correlation—the correlation of one code with other codes delivers low output. This characteristic allows a receiver to reject (treat as noise) any signal using a different code.

With these two attributes, a spread spectrum receiver can recover the desired signal while rejecting other signals in the same frequency band. An example of a length 7 *gold code* set is:

```
1001011
1010110
0100010
1001110
1110001
0011000
0000101
0111111
1101100
```

The Global Positioning System (GPS) uses a *gold code* of length 1023 for the C/A code and approximately 2.355×10^{14} for the P-code. **(2)** In software, an expression indicating that the version of a program (code) is the released version ready for production. Note, this definition does not imply a bug-free implementation only that the number of bugs is tolerable by the manufacturer.

Gopher In Internet, an information retrieval program written at the University of Minnesota. *Gopher* resides on many Internet systems and provides a menu-driven way of accessing the plethora of information and resources available.

After the user selects the desired subject, Gopher will locate the Internet system(s) which contain the selected subjects. Gopher then connects the user to allow browsing the available information. It may be named for either (or both) the University's Golden Gophers or the "go for" items in the menu. The systems that run *Gopher* are frequently called *burrows* or *gopher holes.* Information accessible by *Gopher* is sometimes called *gopherspace.*

GoS An acronym for Grade of Service.

GOSIP An acronym for Government Open Systems Interconnection Profile. A country-specific variant of the ISO/OSI functional protocols that have been defined as part of national procurement policies. The standard is intended to allow equipment from different manufacturers to communicate over a common set of protocols. *GOSIP* is a subset of the OSI protocols and is based on agreements reached by vendors and users of computer networks.

The United States has US GOSIP, which is defined as a Federal Information Processing Standard (FIPS), i.e., FIPS PUB 146. The United Kingdom has the UK GOSIP, which is defined by the Central

Computer and Telecommunications Agency. Governments will not buy equipment unless it supports OSI as specified in *GOSIP*. Unfortunately, the various *GOSIPs,* as published by the UK, U.S., and Japanese governments, for example, are all slightly different. Also called *Government OSI Protocol.*

GOTFIA An abbreviation of <u>G</u>roaning <u>O</u>n <u>T</u>he <u>F</u>loor <u>I</u>n <u>A</u>gony.

GPD An abbreviation of <u>G</u>lobal <u>P</u>ositioning System with <u>D</u>ifferential correction applied.

GPIB An abbreviation of <u>G</u>eneral <u>P</u>urpose <u>I</u>nterface <u>B</u>us. A parallel interface used for connecting scientific apparatus to computers. This interface was developed at Hewlett-Packard (HP) for in-house use and is still sometimes known as the Hewlett-Packard interface bus (HPIB). *GPIB* is now an IEEE standard (IEEE-488). See also *IEEE-488.*

GPS An abbreviation of <u>G</u>lobal <u>P</u>ositioning <u>S</u>ervice.

GPS time The time standard to which GPS signals are referenced. The number of seconds that have elapsed since the Saturday to Sunday universal time (UTC) midnight. The indicated time is adjusted to maintain an accuracy of 1 μs. Zero defines midnight itself.

Unlike UTC, *GPS time* does not use leap seconds for time corrections. Therefore, UTC and *GPS time* differ by some integral number of seconds. See also *time.*

GPS week The number of weeks, modulo 1024, that have elapsed since January 6, 1980. Week zero is defined as the week of January 6 itself. Week numbers increment at Saturday/Sunday midnight. The completion of GPS week 1023 and the *rollover* to week 0 occurred on midnight of the evening of 21 August 1999/morning of 22 August 1999.

grace login A system login in which the user logs in with an expired password. In many networks, passwords expire after a limited time. After that time has elapsed, a user must change the password or risk being locked out. Most of these networks, however, allow a limited number of *grace logins* before the user must change the password.

graceful degradation A term indicating the ability of equipment, systems, or communications channels to tolerate some degree of deterioration, albeit with a reduced level of service rather than failing completely. Generally, the error rate will increase or the system will slow down; however, a user will still be able to obtain results.

For example, the loss of a single node in the Internet does not materially affect the performance of the overall network. However, a few users connected directly to the node will be lost, and other users may experience lower throughput rate.

grade of service (GoS) **(1)** In telephony, the probability of a call's being blocked or delayed more than a specified interval, expressed as a decimal fraction. For example, if 1 call in 1000 gets a busy signal, the *GoS* probability will be 0.001.

Grade of service may be applied to the busy hour or to some other specified period and/or set of traffic conditions. Grade of service may be identified independently for incoming vs. outgoing calls and is not necessarily equal in each direction. **(2)** In telephony, the quality of service for which a circuit is designed or conditioned to provide, e.g., voice grade or program grade. The criteria for different grades of service may include equalization for amplitude over a specified band of frequencies (and phase in the case of digital data transported via analog circuits).

gradient index fiber In fiber optics, a fiber having a core with a refractive index that decreases with increasing radial distance from the fiber axis. (This is in contrast to a *stepped index profile.*) The most common *gradient index fiber* profile is approximately parabolic. This profile results in the continual redirection of rays in the core toward the center of the core. Because the refractive index is lower near the outer regions of the core than near its center, rays entering that region travel faster than rays near the center, which compensates for and reduces multimode distortion. Also called *graded index.* See also *optical fiber.*

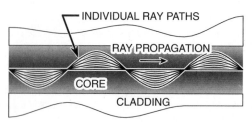

gradient-index lens A lens made from a disc whose index of refraction varies with radial position. A converging (positive) or diverging (negative) lens can be manufactured from a plano parallel disc depending on the direction of the index gradient.

GRAN An acronym for <u>G</u>eneric <u>R</u>adio <u>A</u>ccess <u>N</u>etwork. A wireless telecommunications network capable of handling both GSM and W-CDMA traffic.

grandfathered systems Following a mandated change in system or equipment specification or procedure, the allowance of installed equipment to remain in place and continue to operate. The mandate may be initiated by a government, court, or company (concerning their own systems).

For example, the divestiture decree allowed certain systems connected to the public switched telephone network (PSTN) before a prescribed date to remain permanently connected without registration unless subsequently modified, that is,

- PABX and key telephone systems on or before June 1, 1978.
- PABX and key telephone systems of the same type as those installed before June 1, 1978 and added before January 1, 1980.
- Terminal equipment and protective circuitry connected before July 1, 1978.

graphical user interface (GUI) A symbolic interface in which operating system information and commands are presented to a user through <u>w</u>indows, <u>i</u>cons (pictures), <u>m</u>enus, and <u>p</u>ointers (WIMPs) rather than by textual means. Users enter commands by pointing at icons with a mouse (or other pointing device) and clicking a button on the pointing device. Although invented at the Xerox Palo Alto Research Center in the 1970s, *GUIs* (pronounced "goo-ee") first became popular on the Apple Macintosh and later in Microsoft's Windows operating systems. Other *GUIs* include IBM's OS/2, Motif, and X-Windows.

GUIs are in contrast to character-based interfaces, such as the default interfaces for MS-DOS or UNIX.

Gray code A *monostrophic* code; that is, a code where only one code element changes between adjacent code values. The code was invented and patented by Frank Gray.

This type of code is desirable in applications where mechanical position is to be electronically or optically encoded. If the standard 8421 binary code is used, several numeric transitions could generate grossly false values. For example, going from seven to eight requires that all four bits change state simultaneously (i.e., 0111 → 1000). If the mechanical readout is not absolutely perfect (an impossible situation), one bit will change before the other three bits change. This could lead to the false value 1111 (15) being generated instead of the desired 1000 (8) value. If a code that only changes one bit between adjacent positions were used, the false value could not occur. Two of the possible *unit distance* or *Gray* codes are listed in the following chart: the *natural Gray code* and a code that exhibits a wraparound at either length 10 or length 16, depending on the designer's need. The *natural Gray code* (also called a *reflected code*) is generated by the following relationships:

$$G_3 = b_3$$
$$G_2 = b_3 \oplus b_2$$
$$G_1 = b_3 \oplus b_2 \oplus b_1$$
$$G_0 = b_3 \oplus b_2 \oplus b_1 \oplus b_0$$

Decimal Number (Hex)	GRAY Codes											
	Natural				10-16				8421 Binary Code			
	G_3	G_2	G_1	G_0	L_3	L_2	L_1	L_0	b_3	b_2	b_1	b_0
0	0	0	0	0	0	0	0	0	0	0	0	0
1	0	0	0	1	0	0	0	1	0	0	0	1
2	0	0	1	1	1	0	0	1	0	0	1	0
3	0	0	1	0	1	1	0	1	0	0	1	1
4	0	1	1	0	0	1	0	1	0	1	0	0
5	0	1	1	1	0	1	1	1	0	1	0	1
6	0	1	0	1	1	1	1	1	0	1	1	0
7	0	1	0	0	1	0	1	1	0	1	1	1
8	1	1	0	0	0	0	1	1	1	0	0	0
9	1	1	0	1	0	0	1	0	1	0	0	1
(A)	1	1	1	1	1	0	1	0	1	0	1	0
(B)	1	1	1	0	1	1	1	0	1	0	1	1
(C)	1	0	1	0	0	1	1	0	1	1	0	0
(D)	1	0	1	1	0	1	0	0	1	1	0	1
(E)	1	0	0	1	1	1	0	0	1	1	1	0
(F)	1	0	0	0	1	0	0	0	1	1	1	1

The 10-16 *Gray code* has a benefit over the natural *Gray code* in that a single bit change will transition from 9 to 0; that is, 9 = 0010 and 0 = 0000. (The natural *Gray code* must change three bits 9 = 1101 and 0 = 0000.) This is useful when decimal shaft position encoding is desired. Although messy, the 10–16 code is related to the 8421 binary code with the relationships:

$$L_0 = \bar{b}_3 \cdot (b_2 + b_1 + b_0) + b_3 \cdot \bar{b}_2 \cdot \bar{b}_1 \cdot \bar{b}_0$$
$$L_1 = \bar{b}_3 \cdot b_2 \cdot (b_1 + b_0) + b_3 \cdot \bar{b}_2 \cdot b_3 \cdot b_2 \cdot \bar{b}_1 \cdot \bar{b}_0$$
$$L_2 = b_2 \cdot (b_1 + b_0) + \bar{b}_2 \cdot b_1 \cdot b_0$$
$$L_3 = b_2$$

An encoding disc for both the 8421 binary code and the natural *Gray code* are compared in the drawing. The outer track is the least significant bit (LSB), the shaded sectors represent 1, and the segments are labeled according to position and not according to the encoded value. If the discs are rotated clockwise, it can be seen that the most significant bit (MSB) of the binary coder will engage the readout before any other bit. The code will then indicate the 0, 8, C, E, and finally F as the wheel position where the *Gray code* disc would yield the proper transition 0, then 15. Also called a *reflected code*. See also *code*.

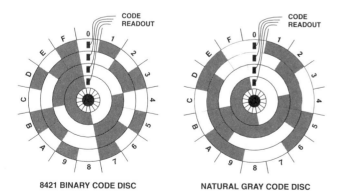

8421 BINARY CODE DISC   **NATURAL GRAY CODE DISC**

gray scale An optical pattern consisting of discrete steps or shades of gray between black and white.

great circle A circle defined by the intersection of the surface of a sphere (such as the Earth) and any plane that passes through the sphere's center. The shortest distance between any two points on the sphere lies along a *great circle*.

Green Book The compact disc specification created by Sony and Philips to work on CD-i players. Currently, no other CD-playing platform makes use of Green Book CDs. *Green Book* was superseded by the White Book CD-ROM in March 1994. See also *CD-i*.

Greenwich Civil Time (GCT) A synonym for Greenwich Mean Time (GMT), an obsolete term replaced by Coordinated Universal Time (UTC). See *time* (3).

Greenwich Mean Time (GMT) The mean solar time at the 0° meridian at the Old Royal Observatory at Greenwich, England, formerly used as a basis for Standard Time throughout the world.

Now an obsolete term replaced by *Coordinated Universal Time (UTC)*. See also *time* (3).

grommet A part used in a hole to protect an entity that must pass through the hole. For example, a rubber or plastic washer-like part is inserted into a hole that a cable or wire must pass through, so that the sharp edges of the hole do not abrade or cut the wire.

ground (1) Frequently abbreviated G, GND, or GRND. A common reference conductor that may or may not be connected to the Earth. *Ground* is at a zero volt potential and serves as either

1. A reference for other voltages and signals in the circuit, or
2. A safety connection in the event the normal current return path is broken.

There are several symbols for a circuit *ground,* i.e.:

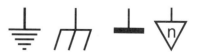

The first symbol is general and may be applied to either circuit or Earth return. The second symbol represents a connection to the chassis. The last two generally represent circuit common, and the fourth symbol is used to distinguish several different classes of return. For example, the analog return might have an "A" designation, while the digital return would have a "D." Also called *circuit common* and *return.* **(2)** An electrical connection to Earth through an Earth–electrode subsystem.

ground constants The electrical parameters of Earth, such as conductivity, permittivity, and magnetic permeability. The specific values of these parameters vary with the local chemical composition, density of the Earth, and water content. They also vary with frequency and the direction of wave propagation.

ground current The current that flows in the protective ground wire of a power distribution system under the condition of an electrical fault. Contrast with *ground loop*.

ground fault With respect to the delivery of ac power, a ground fault is any circuit path between the ac supply and the ground wire (neutral), that is, the current supplied that does not return via the ac return wire.

ground fault interrupter (GFI) A device that senses the current difference between the two ac mains conductors and opens the circuit if the differential current exceeds a prescribed threshold (usually 4 to 6 mA). Also called *ground fault circuit interrupter (GFCI)*.

ground lead (**1**) The conductor connecting a circuit to the signal or power return wire. (**2**) The wire connecting a device to Earth.

ground loop In an electrical or electronic system, a potentially problematic condition that occurs when a circuit is grounded at two or more points. The ground need not be metallic in nature, only conductive at the frequencies of interest. The multiple ground points may result in an unwanted current that flows in a conductor connecting two points that are nominally at the same potential, i.e., ground, but are actually at different potentials. Such a condition can occur for several reasons:

- Solar wind can generate hundreds of volts in potential difference between two points on the Earth. Such a condition can be hazardous when working on long grounded conductors, such as telecommunications cables.
- If the local voltage reference system (ground) has a relatively high impedance at the frequency of operation, any current flowing through the ground will generate a voltage drop (due to Ohm's law $V = IR$).

Contrast with *ground current*.

ground plane An electrically conductive surface that serves as the near-field reflection point for an antenna. It may consist of a natural surface (Earth), an artificial surface of opportunity (e.g., the roof of a motor vehicle), or a specially designed artificial surface (e.g., the disc of a discone antenna).

ground potential The zero reference level used to apply and measure voltages in a system. The ground potential may be nonzero when compared to the potential of the Earth. This potential may also vary with geographic location, soil conditions, and other phenomena.

ground return circuit A circuit using a common return path that is at a system reference (ground) potential. Although not necessary, Earth may serve as a portion of the ground return circuit.

ground start In telephony, a signaling method frequently used between the central office and terminal equipment, such as a PABX, to eliminate a condition called *glare*.

With *ground start signaling*, the originating end applies a ground to one of the two conductors of the loop to seize the line.

- When the central office seizes a line, it places a ground on the tip conductor and then applies the ringing signal.

- When the end terminal seizes a line, it places a ground on the ring conductor. The central office detects the seizure, prevents other calls from attempting to seize the line, and then places a ground on the tip conductor.

Some advantages of *ground start signaling* include:

- A dial start indication to the terminal equipment (which eliminates the need for the terminal equipment to have a dial tone detector).
- A positive indication of line seizure whether or not ringing voltage is present.
- A reduced likelihood of glare by virtue of the positive indication of line seizure, even during the silent period of the ringing signal.

See also *loop start*.

ground state The lowest energy level of an atom. In this state, the atom cannot emit electromagnetic radiation. All other states are called the excited state. Also called the *ground level*.

ground wave In wireless communications, a surface wave that propagates close to the surface of the Earth. The Earth has one refractive index and the atmosphere has another, thus constituting an interface that supports surface wave transmission. These refractive indices are both time varying and different in different geographic areas. Ground waves do not include waves reflected by the ionospheric and tropospheric layers (sky waves).

grounding The connection, by either intentional or accidental means, of a section of a circuit to a common reference point called *ground*. There are two general categories of intentional grounding; namely,

- Earth grounding, which provides:
  - Protection against lightning,
  - A reduction in noise pickup, and
  - A zero volt reference for system devices.
- Equipment grounding, which provides:
  - A zero volt potential for all equipment cases and enclosures, and
  - A high current path for fault currents.

group (**1**) In the context of network security, a *group* is a set of users who share common permissions for one or more resources. Individually assigned user permissions take precedence over those assigned through *groups*. (**2**) A set of characters forming a unit for transmission or cryptographic treatment. (**3**) In telephony, in a frequency division multiplexing (FDM) transmission system, usually an assembly of 12 associated 4-kHz voice channels, each using a different carrier frequency, either within a supergroup or as an independent entity. Two groups are defined, i.e.,

- ITU Group B—Channel banks form groups by modulating carriers of frequency 112-4n kHz and multiplexing the lower sidebands to form a group occupying the frequency band from 60 kHz to 108 kHz.

  See the associated diagram for a pictorial representation.
- ITU Group A—This group occupies upper sidebands in the 12-kHz to 60-kHz band and is used in carrier systems.

Groups are combined into a *supergroup* which usually consists of 60 voice channels (5 groups of 12 voice channels each) and occupies the frequency band from 312 kHz to 552 kHz. These in turn are combined into a mastergroup. The ITU standard *mastergroup* consists of five supergroups. The U.S. commercial carrier standard mastergroup consists of 10 supergroups (or 600 voice channels). The terms *supermaster group* or *jumbo group* are sometimes used to refer to six mastergroups. A group is also called *group channel* or *primary group*. See also *multiplex hierarchy*.

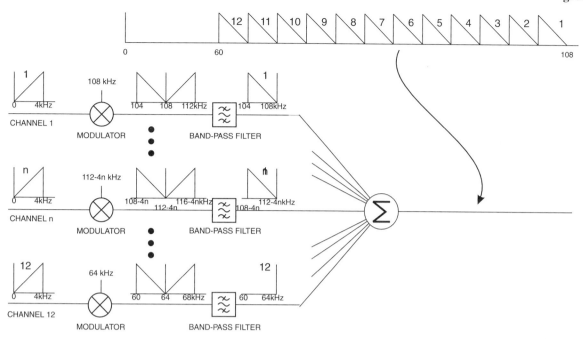

group 1, 2, 3, 3 bis, and 4 facsimile Four ITU-T standards covering the encoding and transmission of images over a communication link using facsimile machines.

Group 1 and Group 2 standards are concerned with analog devices and are, for the most part, not used any longer.

Group 1 uses frequency modulation of analog signals, supports only slow transmission speeds (6 minutes per page), and only offers low (100 dpi) resolution. The specifications are outlined in the ITU-T T.2 recommendation. The ITU frequencies used are 1300 Hz for white and 2300 Hz for black. The North American standard is 1500 Hz for white and either 2300 or 2400 Hz for black.

Group 2 uses both frequency and amplitude modulation to achieve higher speeds (between 2 and 3 minutes per page) and also offers only low (100 dpi) resolution. The specifications are outlined in the ITU-T T.3 recommendation.

Group 3 standard, the most prevalent machine standard in use today, uses quadrature amplitude modulation (QAM) and data compression to increase transmission speeds to about one page per minute. With the ITU-T V.17 protocol, Group 3 faxes are capable of transmission at speeds to 14,400 bps. The ITU-T T.30 protocol specifies the method by which Group 3 faxes manage fax sessions and negotiate the capabilities supported by each fax in the connection. The T.4 protocol controls page size, resolution, transmission time, and coding schemes for Group 3 faxes.

Group 3 specifies:

• Three image resolutions, that is,

IMAGE RESOLUTION

Horizontal (dpi)	Vertical (dpi)
203	98
203	198
203	391

• Two methods of data compression. One, based on Huffman coding, reduces the image file size to 10 to 20% of the original file size. The other (Relative Element Address Designate or READ) reduces the image size to 6 to 12% of the original image file size.

• Password protection.
• Polling, so that a receiving machine can request transmission as appropriate.
• Group 3 bis increases the line data rate from 9600 bps to 14 400 bps.

Group 3C standard is the Group 3 digital mode of facsimile operation defined in ITU-T T.30. Also called *Group 3 facsimile, Group 3 Option C,* or *Group 3-64 kb/s.*

Group 4 standard, defined in ITU-T T.6 and T.563, supports higher-speed digital transmissions and utilizes bandwidth compression, so that a page can be transmitted in less than 20 seconds. The standard, though not widely implemented, provides:

• Resolution up to 400 dpi;
• A data compression algorithm based on image changes from the previous line. The algorithm yields a compressed file 3 to 10% of the original file size;
• No error correction in transmission; and
• An ISDN line rather than a standard dial-up line.

See also *Class 1, 2, 2.0, 3, 4.*

group address In a communications network, a predefined address that is used to address specified set of users or that refers to multiple network devices. Also called a *collective address* or a *multicast address.*

group alert In telephony, a central office feature for simultaneously signaling a group of customers from a control station providing an oral or recorded announcement.

group busy hour (GBH) The busy hour for a given trunk group.

group busy tone In telephony, a tone that indicates to an operator that all trunks in a group are busy.

group delay (1) See *envelope delay.* **(2)** In fiber optics, the transit time required for an optical signal, traveling at a given mode's group velocity, to travel a given distance.

For optical fiber dispersion measurement purposes, the quantity of interest is group delay per unit length, which is the reciprocal of the group velocity of a particular mode. The measured group delay of a signal through an optical fiber is wavelength dependent due to the various dispersion mechanisms present in the fiber.

group delay time In a group of waves that have slightly different individual frequencies, the time required for any defined point on the envelope (i.e., the envelope determined by the additive resultant of the group of waves) to travel through a device or transmission facility.

group distribution frame (GDF) In frequency division multiplexing (FDM) systems, an entity that provides terminating and interconnecting facilities at the group level, i.e., at the group modulator output and group demodulator input circuits of the FDM carrier equipment (the 60-kHz to 108-kHz band). See also *group*.

group index (N) In fiber optics, for a given propagation mode in a medium of refractive index η, the velocity of light in vacuum (c), divided by the group velocity of the mode. Mathematically, for a plane wave of wavelength λ, the group index (N) may be expressed,

$$N = n\,\lambda\,dn/d\lambda$$

where n is the phase index of wavelength λ.

group object In Novell's NetWare 4.x Directory tree, a type of leaf object that has several user objects associated with it. Group objects allow administrators to grant several users rights at the same time, in the same way rights can be granted to network groups.

group velocity (1) The velocity of propagation of an envelope produced when an electromagnetic wave is modulated by, or mixed with, other waves of different frequencies. **(2)** The *group velocity* is the velocity of information propagation and, loosely, of energy propagation. **(3)** In optical fiber transmission, for a particular mode, the reciprocal of the rate of change of the phase constant with respect to angular frequency. The group velocity equals the phase velocity if the phase constant is a linear function of the angular frequency, $\omega = 2\pi f$, where f is the frequency. **(4)** In optical fiber transmission, the velocity of the modulated optical power.

grouping factor A synonym for *blocking factor*.

groupware Software designed to be run on a local area network (LAN) and serve multiple simultaneous users collaborating on a project. It may provide any or all of the following: scheduling, messaging, version managers (that is, a method to track project changes—both documentation and design changes), collaborative computing, and document preparation. *Groupware* users may work cooperatively on projects even if the users are separated geographically.

GSM An abbreviation of Groupe Speciale Mobile or Global Standard (or System) for Mobile communications in the United States. Originally defined by the European Telecommunications Standards Institute (ETSI) as the internationally accepted two-way, digital cellular telephone standard.

It is a Time Division Multiple Access (TDMA) system providing user speech bit rates of 13 kbps. Its specification is in line with ISDN and ITU-T System 7 signaling. Mobile cellular receivers operate between 935.2 and 959.8 MHz and the transmitter operates between 890.2 and 914.8 MHz. *GSM* services include current digital subscriber services and the unique Short Message Service—a form of paging offering up to 160 alphanumeric characters with guaranteed delivery. See also *AMPS*, *CDMA*, *IS-n*, and *TDMA*.

GSMP An abbreviation of General Switch Management Protocol.

GSTN An abbreviation of General Switched Telephone Network. A synonym for public switched telephone network (PSTN).

GTP An abbreviation of Government Telecommunications Program.

GTS An abbreviation of Government Telecommunications System.

GUI An acronym for Graphical User Interface.

guaranteed bandwidth In networking and telecommunications, the capability to transmit continuously and reliably at a specified transmission speed. The guarantee makes it possible to send time-dependent data (such as voice, video, or multimedia) over the line.

guard band A frequency or time interval between channels or events which provides a margin of safety against mutual interference. Also written *guardband*. See also *guard time*.

guard conductor An electrical conductor placed between a noise source and a signal path in such a way as to divert any noise currents into a return path without interfering with the signal currents. See also *guard shield*.

guard shield A conducting enclosure that surrounds all or part of a sensitive signal path and is connected to the return circuit. The *guard shield* shunts not only conducted noise currents to the return but capacitively coupled noise currents as well. See also *guard conductor*.

guard time (1) As part of an escape sequence, the period of time that a modem must not receive characters. The escape sequence used to tell a modem to change from data mode to command mode is generally <guard> +++ <guard> where the *guard time* is 1 second. The *guard time* (indicated by <guard>) isolates the escape code "+++" so that there is no chance of a "+++" embedded in a data stream from being recognized improperly. See also *escape sequence*. **(2)** In a burst transmission (such as a TDM or packet transmission), the *guard time* is the time between bursts. The *guard time* is necessary because of individual differences in transmitter delays. These differences could cause two signals to overlap in time.

guard tone A tone generated by a high-speed dial-up modem to ensure that there is sufficient bandwidth available on the Public Switched Telephone Network (PSTN) circuit for transmission.

guarded release In telephony, the temporary retention of a busy condition during the restoration of a circuit to an idle state.

guest A special account or user name on many networks which allows access by anyone who needs to log in to the network for public information. The access time is generally limited, and the account is given only minimal access rights.

GUI An acronym for Graphical User Interface (pronounced "goo-ee").

guided mode A synonym for *bound mode*.

guided ray In fiber optics, a ray that is confined primarily to the core. A guided ray satisfies the relation given by:

$$0 \leq \sin\theta_r \leq \mathrm{sqrt}\,(n_r^2 - n_a^2)$$

where:

θ_r is the angle the ray makes with the fiber axis,

r is the radial position, i.e., radial distance, of the ray from the fiber axis,

n_r is the refractive index at the radial distance r from the fiber axis, and

n_a is the refractive index at the core radius a, i.e., at the core-cladding interface.

Guided rays correspond to bound modes, i.e., guided modes, in terms of modes rather than rays. Also called *bound ray* and *trapped ray*. See also *optical fiber*.

guided wave An electromagnetic wave whose energy is concentrated near a boundary (or between substantially parallel boundaries) separating materials of different properties (e.g., different refractive indices) and a propagation direction effectively parallel to these boundaries.

.gz The extension used by the GZIP compression program (GNU's version of the ZIP compression program).

GW In networking. An abbreviation of Gate Way.

H

h **(1)** A suffix on a number indicating that the number is expressed in *hexadecimal* (base 16); for example, 0Ah is 10 decimal and 012h is 18 decimal. "H" is sometimes also used as the suffix. **(2)** The SI prefix for *hecto* or 10^2, a term not recommended for use in scientific applications.

H **(1)** The symbol for *henry*. **(2)** The suffix indicating *hexadecimal*. See also *h*.

H bend A smooth turn in an RF waveguide in which the axis remains in a plane parallel to the direction of the magnetic H-field (transverse) polarization. Also called an *H-plane bend*. See also *E bend*.

H channel (Hx) In an Integrated Services Digital Network (ISDN) system, an *H channel* is a high bit rate channel that can be used for transmitting user data. It can be leased as a single unit and can then be subdivided into lower bandwidth channels. These high-speed channels are defined for situations where such high bandwidth is required, such as when transmitting video or other graphics information. Defined H_x channels include:

H0—A 384-kbps channel, which is equivalent to six 64-kbps B channels (bearer channels).

H10—A 1.472-Mbps channel, which represents the 23 B channels of a Primary Rate Interface (PRI) line. This H channel is used primarily in the United States.

H11—A 1.536-Mbps channel, which is equivalent to the PRI in the United States, Canada, and Japan. It consists of 23 64-kbps B channels and one 64-kbps D channel. The D channel is generally used for signaling.

H12—A 1.92-Mbps channel, which is equivalent to the 30 B channels of the European PRI.

H network A network composed of five branches—two connected in series between an input terminal and output terminal, another two connected in series between the other input terminal and output terminal, and the final branch connected between the junctions of the two series branches.

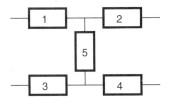

H.n Recommendations A series of International Telecommunication Union (ITU) standards covering "line transmission of nontelephone signals." See Appendix E for complete list.

HACD An abbreviation of Home Area Customer Dialing.

hacker Originally defined as a person totally engrossed in all aspects of computer programming and technology. One who enjoys taking apart a system or software to find out "what makes it tick." The person possesses considerable knowledge and expertise in the field of computers and computing, and uses this skill for constructive purposes.

With the advent of dial-up networks and easy availability of modems, the term has been perverted, generally by the press, to include someone who secretly breaks into other people's computer systems, inspecting programs or data. The person with malicious intent; that is, a *hacker* who does break into another person's system and steals, modifies, or destroys data is called a *cracker.*

HAD An acronym for Half Amplitude Duration. See *T pulse.*

Hagelbarger code A convolutional code, described by D. W. Hagelbarger in 1959, which enables error bursts to be corrected provided there are relatively long error-free intervals between error bursts. Parity check bits are dispersed in time so that an error burst is not likely to affect more than one of the groups in which parity is checked.

HAL An acronym for Hardware Abstraction Layer.

half bridge In wide area networks, a pair of bridges between two widely separated networks. Each bridge connects to a telecommunications link rather than directly to the opposite network as a full bridge would, i.e.,

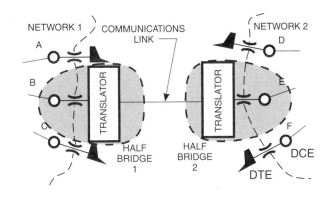

Half bridge was originally an Apple Computer term.

half-channel Those parts of a communications channel which carry information from the calling party to the called party (or from the called party to the calling party).

half duplex (HDX) A mode of communication in which the circuit allows transmission in *either* the forward *or* backward direction but *only one direction at a time.* An example is a normal conversation between two people, i.e., where one person talks while the other listens. With a half-duplex operation, the entire bandwidth can be used for the transmission. In contrast, a full-duplex operation must split the bandwidth between the two directions. Also called *one-way reversible operation* and *two-way alternate operation.* See also *duplex* and *simplex operation.*

half-duplex repeater A duplex digital repeater equipped with an interlock that restricts the transmission of information of signals to one direction at a time.

half section The bisected section of a symmetric circuit such as "pi," "T," or "bridged T" network. *Half sections* may be either unbalanced or balanced depending on the nature of the prototype network. For example, a "T" network prototype gives rise to an unbalanced *half section,* while an "H" network prototype yields a balanced *half section.*

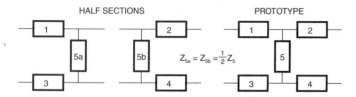

HALF SECTIONS   PROTOTYPE

$$Z_{5a} = Z_{5b} = \tfrac{1}{2} Z_5$$

ham A nickname for people who pursue the hobby of using a personal radio station to communicate, purely for noncommercial purposes, with other radio hobbyists. The origin of this nickname is for all practical purposes lost.

In the United States, ham radio operators are licensed by the U.S. government. There are about 600,000 hams in the United States. With few exceptions, ham radio operators are allowed to do essentially anything a commercial or government station can do. Among the exceptions are:

- They are not allowed to do anything with their radios that make them money in any way.
- They cannot "broadcast" to the public. This means that ham radio transmissions are meant to be received by other ham radio operators. (The public can use short-wave radios or scanners to listen to the ham radio bands. What will be heard is hams talking to other hams, not programs of general interest.)
- Other restrictions include certain power limitations, frequencies of use, and so on.

Also called *Amateur Radio operators* and *ham radio operators*.

Hamming code Any of several error-detecting and forward error-correcting binary codes, used in data transmission. The *Hamming code* is named after its inventor Richard Wesley Hamming. A *Hamming code* satisfies the relation

$$2^m \geq n + 1$$

where

n is the total number of bits in the block, and

m is the number of check bits in the block. $m = n - k$ where k is the number of information bits in the block.

The code in which $k = 4$, $m = 3$, and $n = 7$ can detect all single- and double-bit errors and can correct all single-bit errors.

The *Hamming code* adds three check bits (A, B, and C in the following chart) to the normal 8421 binary code. The check bits, ABC, are mathematically related to the value of the 8421 code in such a way that the receiver can calculate an analysis value ZYX based on the received AB8C421 value. If ZYX is nonzero, a transmission error occurred or, more importantly, the erroneous bit can be identified and corrected based on the ZYX value. The equations relating the 8421 binary value to the *transmitted ABC* code are:

$$A = b_3 \oplus b_2 \oplus b_0$$
$$B = b_3 \oplus b_1 \oplus b_0$$
$$C = b_2 \oplus b_1 \oplus b_0$$

Where b_3 corresponds to the value in the "8s" position, b_2 corresponds to the value in the "4s" position, b_1 corresponds to the "2s" position and finally b_0 is the "1s" value.

HAMMING CODED NUMBERS

Number	A	B	8	C	4	2	1
0	0	0	0	0	0	0	0
1	1	1	0	1	0	0	1
2	0	1	0	1	0	1	0
3	1	0	0	0	0	1	1
4	1	0	0	1	1	0	0
5	0	1	0	0	1	0	1
6	1	1	0	0	1	1	0
7	0	0	0	1	1	1	1
8	1	1	1	0	0	0	0
9	0	0	1	1	0	0	1
A	1	0	1	1	0	1	0
B	0	1	1	0	0	1	1
C	0	1	1	1	1	0	0
D	1	0	1	0	1	0	1
E	0	0	1	0	1	1	0
F	1	1	1	1	1	1	1

At the *receiver,* the AB8C421 value is processed to determine if all seven bits were received correctly and if not, which bit is wrong (knowing which bit is wrong enables it to be corrected). The analysis equations are:

$$X = A \oplus b_3 \oplus b_2 \oplus b_0$$
$$Y = B \oplus b_3 \oplus b_1 \oplus b_0$$
$$Z = C \oplus b_2 \oplus b_1 \oplus b_0$$

The value of the binary number ZYX is the bit position number in error (000 implies errors), i.e.:

HAMMING ERROR LOCATION DECODING

Z	Y	X	Error Location
0	0	0	No errors
0	0	1	A
0	1	0	B
0	1	1	8
1	0	0	C
1	0	1	4
1	1	0	2
1	1	1	1

Hamming distance The number of digit positions by which two n-digit binary words of the same length differ. For example, given the two numbers 1010101 and 1011011, the *Hamming distance* is 3.

The concept can be extended to other number bases and even to alphanumeric sequences; for example, "peddle" and "meddle" have a Hamming distance of 1. Also called the *signal distance.*

Hamming weight The number of nonzero symbols in a symbol sequence. For binary signaling, *Hamming weight* is therefore the number of "1" bits in the binary sequence.

hand-off In cellular communications, *hand-off* (or *handoff*) refers to the transfer of a connection from one cell to another. *Hand-off* time is generally between 200 and 1200 ms, which accounts for the delay sometimes heard when listening to someone on a cellular telephone. This delay can cause problems for devices that require frequent connectivity acknowledgments and for data transmission. For example, some modems will disconnect if a long delay occurs in a connection and during the delay. Not only can no data be transferred, but the packets transmitted at the time of signal loss will be severely corrupted. See also *cellular telephony*.

handover word (HOW) The Global Positioning System (GPS) message word that contains the synchronization information to allow the transfer of tracking the C/A code to P-code.

It is the second word in each subframe of the navigation message and contains the Z-count at the leading edge of the next subframe. This count tells the receiver where in the generated P-code sequence to start the acquisition process (correlation).

handset See *subset*.

handshaking (1) The sequence of signals exchanged when a connection is being established between DTE and DCE or local DCE and remote DCE. The *handshake* involves an exchange of predetermined tokens, characters, or signals between two devices in order to establish synchronism, a common protocol, or a common compression method. (2) An exchange of signaling information between two communications systems or nodes. It establishes when the two systems will transmit data. *Handshaking* may be directed by either hardware or software.

- *Hardware handshaking* uses the request to send (RTS) and clear to send (CTS) pins to control transmissions.
- *Software handshaking* uses the XON and XOFF characters to signal when to stop and start the transmission.

More properly called *flow control*.

hang (1) A slang term meaning to wait for an event that will or can never occur. For example, a system may *hang* if the keyboard is redirected to another device and the computer's program is waiting for keyboard-entered data. (2) To wait for an event to occur that can occur but is slow to arrive (to *hang* around until something happens). For example, a computer program displays a selection menu and then waits (hangs) for operator input. (3) To attach a peripheral device to another device. Generally implies the device is attached with cables and is outside the host device (rather than something that is strictly inside the chassis of the machine). For example, one can *hang* a modem on the computer.

hang-up signal In telephony, a signal transmitted over a line or trunk to indicate that a party has released the circuit. A subscriber's *hang-up signal* is generated upon going on-hook and must be present for more than 800 ms.

hard error An error in hardware or software processing that can be repeatedly demonstrated by a specific test or procedure. The error is not caused by random events but can be attributed to one of several categories:

- *Hardware:* component failure; wearout or external cause (such as lightning);
- *Software:* coding error in the failing program OR any system support program that is accessed;
- *Algorithmic:* an error in the procedure that is to solve the problem (the error may only occur for a particular set of data, or conditions);
- *Operator:* the user enters a wrong command or a parameter outside the acceptable range (an error in the documentation?); or
- *Bad data:* incorrectly entered data.

See also *soft error*.

hard line A slang term for rigid coaxial cable.

hardware (1) Generally, physical equipment as opposed to programs, procedures, rules, and associated documentation. (2) The generic term dealing with "physical items" as distinguished from its capability or function, such as equipment, tools, implements, instruments, devices,

fittings, trimmings, assemblies, subassemblies, components, and parts. (3) Often used in regard to a stage of product development, as in the passage of a device or component from the design stage or conceptual stage into the *hardware* stage as the finished object. (4) In data automation, the physical equipment or devices forming a computer and peripheral components. *Hardware* may refer to any device such as a printer, modem, or the computer itself. See also *firmware* and *software*.

hardware abstraction layer (HAL) A component of an operating system (such as WindowsNT) that functions similar to an application programming interface (API). It, however, resides at the device level (a layer below the standard API level). It is the layer that mediates between the operating system kernel and specific hardware. By implementing functions for interfaces, caches, interrupts, and so on, it can make every piece of hardware look the same to the higher layers. *HAL* allows programmers to write applications with the device-independent advantages of writing to an API but without the large processing overhead that an API normally demands. This helps make programs more transportable to other machines.

hardware address A data link layer address associated with a specific device. Also called the *MAC address* or *physical address*.

hardware flow control A method that data communication equipment (DCE) uses to temporarily stop the flow of data on a communications channel while the data terminal equipment (DTE) "catches up" on information already sent. This is accomplished through the use of the RTS/CTS (or similar) signals.

hardware independent A computer program or communications protocol designed to function correctly on or with several different *computing machines*.

hardware key A physical device used to secure a computer system or specific program from unauthorized use.

The key may be a physical key as on the front panel of most computers or a device that plugs into one of the bus positions on the computer itself. Midway between these is a device that plugs into a port on the computer (called a *dongle*). Each dongle is unique and contains the built-in equivalent to a *password* that only the protected program knows and can access. The program will not run unless it finds the correct dongle. The dongle therefore allows the user to back up (copy) the original disks but does not allow a program to be run on multiple machines simultaneously.

hardwire (1) In general, to connect or attach equipment or components permanently in contrast to using switches, plugs, or connectors or other temporary means. (2) In computers, to wire in fixed logic or read only storage that cannot be altered by program changes. Also called *card-coded*.

harmonic A sinusoidal component of a periodic wave having a frequency that is an integer multiple of the fundamental frequency of the periodic wave.

For example, a component frequency of a waveform that is twice the fundamental frequency is called the *second harmonic*. Also called an *overtone*.

harmonic distortion Nonlinear distortion of a system characterized by the appearance of harmonics in the output signal that do not appear in the input signal. Note that it is possible to have *subharmonic* components as well as harmonics above the desired signal.

harmonic telephone ringer A telephone ringer that responds to an ac current within a very narrow frequency range. See also *ring*.

harmonica In cabling, a device that can convert a 25-pair cable into multiple 2-, 3-, or 4-pair cables.

harmonica block In cabling, a wiring block that can be used to connect a limited number (up to a dozen) of RJ-11 plugs, each coming from different nodes, into a common wiring center via a 25-pair cable.

hartley In communications, a unit of information content equal to one decadal decision, i.e., equal to one of ten possible and equally likely values or states used to store or convey information. One *hartley* is $\log_2 10$ bits (or approximately 3.3219 bits).

hash bucket One of a set of notional stacks (buckets or receptacles) used to hold data items for sorting or lookup purposes.

For example, when looking up an entry in a phone book, hashing typically is by the first letter; the hash buckets are the alphabetically ordered letter sections.

hash coding See *hashing*.

hash collision When two different keys hash to the same value. Also called *hash clash*. See also *hash table*.

hash function A function that assigns a data item distinguished by some *key* into one of a number of possible *hash buckets*. The *hash function* is usually combined with another function that is more precise. Ideally, a *hash function* should distribute items evenly between the buckets to reduce the number of hash collisions.

hash table An array of pointers indexed by a hash function, used in a hashing scheme to provide rapid access to data items which are distinguished by some key.

hashing A process by which access to files or other information can be accelerated. It is accomplished with an indexing function (key) that decreases the number of elements that need to be searched. *Hashing* is commonly used for improving access to lists such as dictionaries and directory lists.

Each data item to be stored is associated with a key, e.g., the name of a person. A hash function is applied to the item's key, and the resulting hash value is used as an index into a hash table. The table contains pointers to the original items. If the hash table already has an entry at the indicated location, then that entry's key must be compared with the given key to see if it is the same. If two items' keys hash to the same value (a hash collision), then an alternative location is used (e.g., the next free location following the indicated one). For best performance, the hash table size and hash function must be tailored to the range of keys to be used. Also called *hash coding*.

Hayes bridge One of many circuits that provide dc feed to a telephone loop. The *Hayes bridge* injects the dc feed current in series with a split winding on the voice coupling transformer as is shown in the figure. Series resistors limit the current to the subscriber when the loop is short.

CONVENTIONAL SERIES FEED (HAYES BRIDGE)

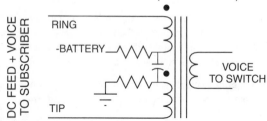

Some general characteristics of the circuit include:

- The circuit is simple and cost effective.
- The "RC" filters any battery noise.
- The two- to four-wire hybrid function (not shown) is nearly free as it only requires the addition of another winding and a line matching network.
- A large transformer core is required due to the high dc feed current flowing in the windings.
- With short loops, there is high power dissipation.
- Longitudinal balance is determined by both the feed resistors and the split transformer winding.
- Ringing must be done from an external source which is switched onto the subscriber line during ringing.

See also *BORSCHT, line circuit,* and *Stone bridge.*

Hayes-compatible A term used to describe a modem that responds to the same basic command set as a modem manufactured by Hayes Microcomputer Products; the originators of the de facto standard "AT" command set.

HBA An abbreviation of Host Bus Adapter.

HCF Humorous. An abbreviation of Halt and Catch Fire. Any of several undocumented and mythical (?) machine code instructions supposedly included by the manufacturer for test purposes but with destructive side effects. Reports of *HCF* instructions are rumored to exist on many architectures and go as far back as the IBM 360.

HCS (1) An abbreviation of Hundred Call Seconds. **(2)** An abbreviation of Heterogeneous Computer System.

HD An abbreviation of Hard Disk.

HDBn An abbreviation of High Density Bipolar with "n-zero" substitutions. *HDBn* encoding is simply normal *AMI* with successive zeros replaced with a code (containing + or − bits) identifiable by the receiving terminal.

On European systems, *HDB3* is used to suppress strings of four zeros. When four successive zeros are encountered, a substitute string containing a bipolar violation in the fourth position is transmitted. The string is dependent on the polarity of the last transmitted pulse and the number of pulses transmitted since the last substitution was made. The following chart tabulates the rule:

HDB3 CODE SUBSTITUTION RULE

Last Pulse Transmitted	ODD Number Transmitted	EVEN Number Transmitted
+	000+	−00−
−	000−	+00+

See also *BnZS* and *waveform codes.*

HDC (1) An abbreviation of Heterogeneous Distributed Computing. **(2)** An abbreviation of Hard Disk Controller.

HDD An abbreviation of Hard Disk Drive.

HDLC An abbreviation of High-level Data Link Control. An ITU-T standard bit oriented synchronous data communication protocol detailed in the ISO standards 3309 and 4435. The *HDLC* protocol applies to the data link layer of the ISO/OSI model for computer-to-computer communications. Messages are transmitted in variable-length packets (called frames) that are organized as shown in the following figure. The three frame types—I-frame or information frame, the U-frame or unnumbered frame, and the S-frame or supervisory frame—are shown.

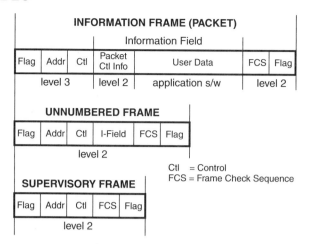

INFORMATION FRAME (PACKET)

			Information Field			
Flag	Addr	Ctl	Packet Ctl Info	User Data	FCS	Flag
	level 3		level 2	application s/w	level 2	

UNNUMBERED FRAME

Flag	Addr	Ctl	I-Field	FCS	Flag
		level 2			

Ctl = Control
FCS = Frame Check Sequence

SUPERVISORY FRAME

Flag	Addr	Ctl	FCS	Flag
		level 2		

HDOP An acronym for <u>H</u>orizontal <u>D</u>ilution <u>O</u>f <u>P</u>recision. A measure of how much the geometry of satellite positions affects the horizontal position estimate of a GPS receiver. The estimate is computed from satellite range measurements. See also *DOP (dilution of precision)* and *Global Positioning System (GPS)*.

HDSL An abbreviation of <u>H</u>igh <u>S</u>peed <u>D</u>igital <u>S</u>ubscriber <u>L</u>ine.

HDTV An abbreviation of <u>H</u>igh <u>D</u>efinition <u>TeleV</u>ision.

HDX An abbreviation of <u>H</u>alf-<u>D</u>uple<u>X</u> transmission.

HE$_{11}$ mode Designation for the fundamental hybrid mode of an optical fiber.

head A device that writes, reads, or erases data on a storage medium. Examples include:

- A *print head,* which writes data onto paper.
- A *magnetic read head,* which reads information from a magnetic disk or magnetic tape.

head end **(1)** In local area networks (LANs) and wide area networks (WANs), a term used to refer to the base, or root, node in a tree topology, or a node on either of the buses in a Distributed Queue Dual Bus (DQDB) architecture. It may provide such centralized functions as retiming, message accountability, contention control, diagnostic control, remodulation, and access to a gateway. **(2)** In broadcast networks, the starting point for every transmission to end users. It is a central point or hub in broadband networks that receives signals on one set frequency band and retransmits them on another. Every transmission goes through the head-end in a broadband network.

For example, a cable network's (CATV) broadcast station and control center is a *head-end.* End-user stations may be allowed to transmit control and error information, but no data, toward the *head-end.*

head on collision A conflict that occurs on a communications channel when two or more users begin to transmit on the channel at approximately the same instant. Also called *collision.* See also *CSMA/CA* and *CSMA/CD.*

header Generally, a unit of information that precedes and identifies the body of whatever follows.

1. In communications, a header is a block of bits (or bytes) that precedes the body of a packet of data and contains control information such as the packet number, sending station identifier, destination station identifier, precedence level, and routing instructions.

2. In e-mail or other network postings, the leading part that contains information about the message, such as, who sent it and when it was sent.

header error control (HEC) See *ATM.*

heartbeat **(1)** In networking, the signal emitted by a Level 2 Ethernet transceiver at the end of every packet to show that the collision-detection circuit is still connected. In the older Ethernet designs (that is pre-IEEE 802.3), it is the Signal Quality Error (SQE) signal quality test function. **(2)** A signal issued at regular intervals by software to demonstrate that it is still alive. Hardware monitors the signal and restarts the machine if it stops seeing it. See also *watchdog timer (WDT).* **(3)** A periodic signal used by software or hardware to synchronize multiple devices. **(4)** The oscillation frequency of a computer's clock crystal, before frequency division down to the machine's clock rate.

heat coil In telephony, a type of fuse used in the main distributing frame (MDF) to protect the switching equipment from prolonged overvoltage conditions.

heat sink A device that absorbs heat from one part of a system and conducts it to another part where it is dissipated.

Heaviside layer See *E region* and *ionosphere.*

HEC An abbreviation of <u>H</u>eader <u>E</u>rror <u>C</u>ontrol. See *ATM.*

helical antenna An antenna in which the radiating element has the form of a helix. When the circumference is much smaller than one wavelength, the antenna radiates at right angles to the helix. When the helix circumference is one wavelength, maximum radiation is in the direction of the helix axis.

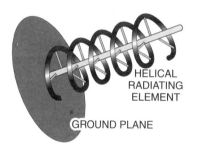

HELICAL RADIATING ELEMENT

GROUND PLANE

heisenbug Jargon. A corruption from Heisenberg's Uncertainty Principle in quantum physics. It is a bug that disappears or alters its behavior when an attempt is made to probe or isolate it. This is based on the unfortunate fact that the use of a debugger sometimes alters a program's operating environment significantly enough that buggy code (such as that which relies on the values of uninitialized memory, or uses software timing loops) behaves quite differently or even correctly. See also *Bohr bug, mandelbug, schrödinbug,* and *software rot.*

help desk A client–server application that lets users track support requests or problems and their respective resolutions for future reuse. That is, it is a database of information structured so that it can be rapidly retrieved to assist with the resolution of the same or similar requests.

henry (H) The unit of inductance. A current changing at a rate of 1 amp per second will generate 1 V across an inductance of 1 henry. Named after Joseph Henry (1797–1878), the American physicist who discovered the relationship between an inductor and a moving magnetic field (1831).

HEPnet An acronym for <u>H</u>igh <u>E</u>nergy <u>P</u>hysics <u>net</u>work. The network links together many nuclear and physics research facilities. The USENet *HEPnet* alternative newsgroup carries postings relating to physics conferences, workshops, job opportunities, and announcements of general interest. See also *ALT* and *alternative newsgroup hierarchies.*

hertz (Hz) *Hertz* is the unit of measure for frequency or bandwidth. A measurement of how often a periodic event (such as a wave) occurs, 1 Hz is equal to 1 cycle per second and 10 Hz is 10 cycles per second. Named in honor of Heinrich Rudolph Hertz (1857–1894), a German physicist who produced the electromagnetic waves in 1888, which were predicted by James Clerk Maxwell (1831–1879).

Hertzian wave   A synonym for *radio wave.*

hetero   From the Greek word meaning difference.

heterochronous   From Greek meaning different (hetero) time (chronous). Describes signals in which corresponding significant events or entities occur at different nominal rates.

For example, two signals having different nominal bit rates and not derived from the same clock (or from homochronous clocks) are usually heterochronous. See also *anisochronous, asynchronous, homochronous, isochronous, plesiochronous,* and *synchronous.*

heterodyne   The process of translating a signal carrier from one frequency band to another by mixing it with a local oscillator (LO) output (sometimes called the *beat frequency oscillator*—BFO). The output of the mixing process is the sum and difference of the carrier and local oscillator frequencies. A *filter* is used to select either the sum frequency or the difference frequency for further processing.

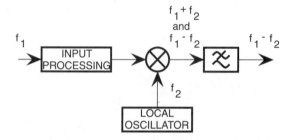

A major advantage of translating the received signal to another band is the ability to select one of many receive channel frequencies and shift it into one preselected frequency. For example, in an AM radio, the input signal is in the range of 550 to 1500 kHz, and the mixer output is at 455 kHz; in an FM radio, the input signal is in the range of 88 to 108 MHz, and the mixer output is at 10.7 MHz. This translation to a common frequency is accomplished by tuning the LO so that the desired frequency is always the same number of Hz from the signal frequency. Also called *beating.*

Heterogeneous LAN Management Specification (HLMS)   A network management specification developed by IBM and 3Com. It provides an underlying structure for the development of network management products which can function with a variety of network operating systems (NOS) and adapter cards.

heterogeneous multiplexing   Multiplexing in which not all the information bearer channels operate at the same data signaling rate.

heterogeneous network   A network that uses multiple protocols at the network layer. In contrast, a homogeneous network uses a single protocol at the network layer. Heterogeneous networks tend to be implemented by different vendors.

heuristic   Pertaining to a trial and error method of problem solving. It may involve a rule of thumb, simplification, or educated guess that reduces or limits the search for solutions in domains that are difficult and poorly understood. An approach that may lead to a correct (or usable) solution by nonrigorous, possibly "self-learning" means. Solutions are discovered by evaluation of test results. Unlike algorithms, *heuristics* do not guarantee solutions. See also *algorithm.*

heuristic routing   A message routing strategy in which data (such as time delay) are extracted from incoming messages, during specified periods and over different routes, and are used to determine the optimum routing for transmitting data back to the sources. Heuristic routing provides route optimization based on recent empirical knowledge of the state of the network.

hex (1)   An abbreviation of *hexadecimal.* **(2)** A quantity of six of anything. For example, a particular integrated circuit with six identical inverters is known as a *hex* inverter.

hexadecimal (hex)   Pertaining to the number system with a base (radix) of 16. Each digit in a number may take on any of the 16 values 0, 1, 2, 3, 4, 5, 6, 7, 8, 9, A, B, C, D, E, and F. The decimal value 247 is represented in *hexadecimal* by 0 x F7, F7h, or $F7. The "h or H" following the "F7" and the "0x" or "$" preceding the "F7" are common methods of denoting that a number is expressed as a base 16 number.

EXAMPLES OF HEXADECIMAL NUMBERS

Decimal	Binary		Hexadecimal	
0	0000 0000	0×00	00h	$00
1	0000 0001	0×01	01h	$01
15	0000 1111	0×0F	0Fh	$0F
123	0111 1011	0×7B	7Bh	$7B
255	1111 1111	0×FF	FFh	$FF

The term was coined in the early 1960s to replace the earlier term *sexadecimal* (which was too racy for stuffy IBM). Later it was adopted by the rest of the industry. Also called *sexadecimal.*

HF   An abbreviation of <u>H</u>igh <u>F</u>requency.

HFC   An abbreviation of <u>H</u>ybrid <u>F</u>iber <u>C</u>oaxial.

HFDF   An abbreviation of <u>H</u>igh <u>F</u>requency <u>D</u>istribution <u>F</u>rame.

HFS   An abbreviation of <u>H</u>ierarchical <u>F</u>ile <u>S</u>ystem. The file system for the Macintosh operating system.

HHOJ   An abbreviation of <u>Ha</u> <u>Ha</u> <u>O</u>nly <u>J</u>oking.

HHOK   An abbreviation of <u>Ha</u> <u>Ha</u> <u>O</u>nly <u>K</u>idding.

hidden Markov method (HMM)   A common algorithm used in voice recognition systems. It uses probabilistic techniques for recognizing speech patterns, both continuous and isolated.

hierarchical computer network   A computer network in which processing and control functions are performed at several levels. Frequently found in industrial process control, inventory control, database control, or hospital automation.

hierarchical decomposition   A system design methodology in which the system is broken down into its components through a series of "top-down" refinements, much like a book is broken down into an outline.

hierarchical file system   A file system in which the files are organized into a hierarchy. The nodes of the hierarchy are called directories, subdirectories, or folders.

hierarchical name structure   A naming strategy that relies on the relationship levels between two entities. This naming method is used, for example, for files or network entities. In a network context, a node's name is based on the name of the parent node, which sits immediately above the node in a hierarchy.

hierarchical network (1) A physical network architecture in which a multiplicity of nonredundant switching nodes share the responsibility of establishing a path from an originating node to a destination node. To each switching node, some of the user nodes (but not all) are attached—called a *switching group*. The switching node will either connect an originating node to a destination node (if the destination node is also connected to that switching node) or pass the responsibility of completing the path to another switching node (the specific node is determined by routing tables in each node).

Using the diagram as the model, assume node E attempts to establish a path to node D then only the switching node N2 is involved. If, however, node E attempts to establish a path to node H, the switching node N2 cannot complete the connection, so N2 enlists the help of node N4. Node N4 then directs the appropriate node (node N3 in this case) to establish the connection. When a large number of connections occur between two switching groups, a high traffic path may be installed between these groups (bypassing a portion of the hierarchy). An example of this is shown between nodes N1 and N2. The

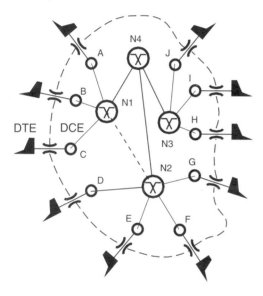

best example of this type of network is the public telephone network. Each of the DTEs (telephones) are connected to a local class 5 switching office (switch group), and each is interconnected via higher level switching offices (class 1–4 offices). See also *bus network, class n office, gateway, hybrid network, mesh network, ring network,* and *star network.* (2) In the Internet USENet, *newsgroup hierarchy* is the organization of *world* and *alternative newsgroups* to which users may subscribe. See *newsgroup hierarchy.*

hierarchical routing A method of routing information flow in large networks in which the network is broken into multiple levels of smaller interconnected networks, and each level is responsible for its own routing. Routing is then accomplished by means of hierarchical addressing such as the Transmission Control Protocol/Internet Protocol (TCP/IP).

Most TCP/IP routing is based on a two-level *hierarchical routing* in which an IP address is divided into a network portion and a host portion. Successive gateways use only the network portion until an IP datagram reaches a gateway that can deliver it directly. Additional levels of hierarchical routing are introduced by the addition of subnetworks. The Internet, for example, has basically three levels: backbones, midlevels, and subnetworks. The backbones route between the midlevels, the midlevels route between subnets, and each subnet (being an autonomous system) knows how to route internally. See also *Exterior Gateway Protocol* and *Interior Gateway Protocol.*

hierarchically synchronized network A mutually synchronized network in which some clocks exert more control than others, the network operating frequency being a weighted mean of the natural frequencies of the population of clocks.

hierarchy Any organization of entities with only one entity (or at most a few entities) at the top of the structure and the remaining entities below the topmost item; a pyramid or inverted tree structure.

high and dry The condition in which a telephone user initiates a long-distance call and no connection is established, i.e., no ringing, no call progress tones, "no nothing."

high-capacity storage system (HCSS) In Novell's NetWare 4.x, a storage system that includes optical disks as part of the file system. These provide slower, but much higher capacity, storage for files. The *HCSS* oversees the use of these media, so that access to the files on these media is transparent to the user.

high definition television (HDTV) A system for transmitting television signals with far greater resolution than existing methods (NTSC, PAL, etc.). Both the vertical and horizontal resolution is roughly twice that of these existing standards.

high frequency distribution frame (HFDF) A distribution frame that provides terminating and interconnecting facilities for those combined supergroup modulator output circuits and combined supergroup demodulator input circuits that contain signals occupying the baseband spectrum.

high impedance A condition in which one circuit's impedance is sufficiently greater than a second circuit's impedance that the connection of the first circuit to the second circuit does not materially affect the performance or characteristics of the second circuit.

high impedance state A condition of a circuit (or device) in which it is effectively isolated from another circuit.

high level language application program interface (HLLAPI) In the IBM PC environment, software used for creating interfaces between mainframes and PC applications. *HLLAPI* is designed for use with high-level programming languages such as C, Pascal, or BASIC.

high memory area (HMA) The first 64 Kbytes of memory (less 16 bytes) above 1 Mbyte on a PC. From version 5.0 on, parts of MS-DOS or PC-DOS, the operating system, can be loaded into this area by adding the line DOS=HIGH to the CONFIG.SYS file. See also *extended memory* and *upper memory block (UMB).*

high order position The leftmost position in a string of characters. See also *big endian, little endian,* and *most significant digit (MSD).*

high-pass filter An electronic circuit or device that passes all frequencies above a specified frequency (greater than zero) and attenuates those frequencies below the specified frequency.

The *high-pass filter* in the following figure has its band edge defined at 1 kHz. Signals above this frequency are allowed to pass, while those below are attenuated. See also *band-pass filter, band reject filter,* and *low-pass filter.*

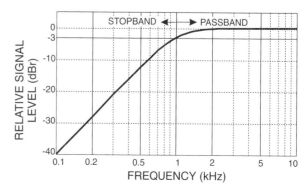

High-Performance File System (HPFS) An OS/2 file system that has faster input/output than the File Allocation Table (FAT) file system. It does not restrict file naming to eight characters with a three-character extension. However, it is backward compatible with the FAT file system.

Other file systems include the File Allocation Table (FAT) and the New Technology File System (NTFS).

High-Performance Parallel Interface (HiPPI) A computer bus for use over fairly short distances at speeds of 800 and 1600 Mbps. (When clocked at 25 Mbps and 64 parallel paths are used, the transfer rate is 1.6 Gbps—including overhead, of course.) It was developed at Los Alamos National Laboratory and is now ANSI standard *X3T9.3/88-127*. *HIPPI* is often used to connect a computer to routers, frame buffers, mass-storage peripherals, and other computers.

High-Performance Serial Bus See *IEEE 1394*.

high-speed circuit In telecommunications, a circuit capable of faster transmission rates than are needed for voice communication. *High-speed circuits* generally support speeds of 20 kbps or more.

high-speed LAN (HSLAN) A term used to describe local area network (LAN) architectures with transmission speeds of 100 Mbps or more. Most of the architectures proposed are designed for larger networks, such as metropolitan area networks (MANs) or wide area networks (WANs).

Some *HSLAN* architectures include:

- Asynchronous Transfer Mode (ATM), which is a broadband extension of the Integrated Services Digital Network (ISDN) architecture that has been poised for great things for many years now. ATM is most suitable for WANs.
- Fiber Distributed Data Interface (FDDI), which uses optical signals and media to achieve its high speeds. FDDI is already widely used for special-purpose networks, such as those connecting mainframes to controllers or connecting high-end workstations to each other.
- 100 Mbps Ethernets, which include Hewlett-Packard's 100BaseVG, Grand Junction's 100BaseX, and MicroAccess's fastEthernet.

high-speed serial interface (HSSI) See *High-speed Synchronous Serial Interface*.

High-speed Synchronous Serial Interface (HSSI) An interface for transferring data to or from a wide area network (WAN) leased line, or to and from a local area network (LAN) via an *HSSI*-capable data service unit (DSU) and *HSSI* router. It is a serial port that supports serial transmission speeds of up to 52 Mbps and is typically used for leased lines such as DS3 (44.736 Mbps) and E-3 (34 Mbps). Also called a *high-speed serial interface*.

high-usage trunk group In telephony, a group of trunks for which an alternate route has been provided to absorb the relatively high rate of overflow traffic. It is a trunk group engineered for efficiencies and economic considerations, but which will overflow traffic. It is intended as the primary path between two switching stations and is designed to transport the majority of the traffic between the two stations.

highway (1) A digital, serial-coded, bit stream with time slots allotted to each call on a sequential basis. (2) A common path or a set of parallel paths over which signals from more than one channel pass with separation achieved by time division.

HILI An abbreviation of HIgher Level Interface.

HiPPI An acronym for High Performance Parallel Interface.

hiss Noise in the audio frequency range, having subjective characteristics analogous to prolonged sibilant sounds (Sssssssss). Noise in which there are no pronounced low-frequency components is considered hiss.

histogram A graphical display plotting the density of occurrences against a reference. For example, the following plot from a Global Positioning System (GPS) measurement system shows the distribution of likely measurements over several hours.

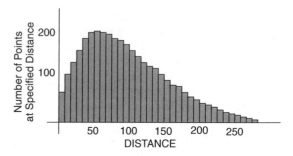

hit (1) Generally, the occurrence of a defined event. (2) A transient disturbance to a device or communications network, e.g.,

- A transient change of a parameter of a signal, i.e., the phase (timing) or amplitude (strength). This gives rise to signal distortion and can increase the error rate.
- The occurrence of a large-voltage transient at a port of a device; for example, a lightning-induced transient on a telephone line or power line is frequently called a *hit*.

(3) A match of data to a prescribed criterion, e.g.,

- A match in a data search. For example, in the Internet, a search for "ITU" will give few hits, while a search for "pictures" will be massive.
- The number of times an Internet Web site is accessed or the number of times a graphic or text file from a site is downloaded. *Hit rate* is frequently used as a measure of popularity. It can be misleading, however, because it does not quantify either activity or holding time.

HLLAPI An abbreviation of High Level Language Application Program Interface.

HLMS An abbreviation of Heterogeneous LAN Management Specification.

HMA An abbreviation of High Memory Area.

HMI (1) An abbreviation of Human Machine Interface. See also *man–machine interface (MMI)*. (2) An abbreviation of Hub Management Interface.

HMUX An acronym for Hybrid Multiplexer.

hobbit An acronym for High Order (b)bit. The most significant bit (of a byte). Also called the *meta bit* or *high bit*.

HOBIS An acronym for HOtel Billing Information System.

hog Indicates that a device, system, or person is using excessive system resources, usually which noticeably degrade interactive response. Usually, it is used with a qualifier, such as *memory hog* (a program that uses extreme amounts of working memory), *disk hog* (a program or person that uses extreme amounts of storage memory), or *bus hog* (a controller that never gives up the I/O bus and is released only after the bus hog timer expires).

The term is not used to describe programs that are simply extremely large or complex or that run painfully slow themselves. (They are said to *run like a pig*.)

hold The condition in which a station is disconnected from a communications link, but the link remains established. A user may reconnect the station to the held circuit.

hold range The total frequency span by which the input signal can deviate at the input of a phase lock loop (PLL) and still maintain lock. Also called *hold-in range, lock range,* and *tracking range.* See also *phase lock loop (PLL).*

hold recall The telephonic feature that allows a user to reconnect to a link previously placed on hold.

holding time In telephony, the length of time a communications channel is in use for each call. The holding time includes the call setup time, the message time, any idle time, and the time to clear the circuit. It is usually expressed in seconds.

holding tone A tone transmitted over a switched circuit for performing noise tests on systems that use commanders or sampling. *Holding tones* are also used during the measurement of jitter and transients. The tone is injected at a specified level and notch filtered out within the measuring set. In the United States, the tone is generally 1004 Hz.

hole **(1)** A region in memory which has been reserved. For example, the memory-mapped I/O region in an IBM PC allows for many devices but they may not actually be present, thereby creating an *I/O hole.* **(2)** The result of some file systems wherein they can store large files with holes so that unused regions of the file are never actually stored on disk. Technically, these are called *sparse files.* **(3)** In semiconductor electronics, the absence of an electron in a semiconductor material. In the electron model, a hole can be thought of as an incomplete outer electron shell in a doping substance. Holes can also be thought of as positive charge carriers. While this is in a sense fiction, it is a useful abstraction.

hole model In semiconductor electronics, a model of semiconductor behavior in which *donors* contribute a positive charge equal in magnitude to the charge of an electron, and *acceptors* contribute space for such a charge within the crystal lattice.

Hollerith code Named after its inventor Herman Hollerith (1860–1929), it is a method of storing information in an 80-column punched card.

hollised In the Internet and USENet in particular, to be ordered by one's employer not to post any even remotely job-related material to USENet (or other Internet resource).

The term comes from an event at Lockheed in 1994 in which one of its employees (Ken Hollis) posted publicly available material on access to Space Shuttle launches to the news group *sci.space.* At the direction of NASA public relations officers, he was gagged under threat of being fired. Of course, NASA got a publicity black eye, but others have since suffered for the same activities. Use of the term strongly implies that those doing the gagging are blinded to their own best interests by territorial reflexes.

holy war Within a large network community, of which the Internet is an example, a *holy war* is a prolonged and unresolvable argument concerning almost any aspect of computing in which two sides of an issue are essentially equal. The argument takes on almost religious attributes in that the controversy boils down to irreducible propositions about the correct way to resolve or implement the issue. It is a situation in which neither side can offer proof nor will either side back down.

The characteristic that distinguishes *holy wars* from a normal technical dispute is the instance of participants to use personal value choices and cultural attachments as objective technical evaluations. Examples of *holy wars* include:

- The big-endian vs. little-endian data format. That is, should the most significant element be first or last?
- The Mac vs. PC hardware architecture.
- EISA vs. MCI bus architectures.
- OS/2 vs. Windows (vs. *DOS*) battle.
- Internet vs. OSI protocol suites.
- C vs. Pascal vs. Fortran vs. assembly programming languages.
- Java vs. ActiveX, and on, and on, . . . !

Many times these arguments degenerate from technical *holy wars* to *flame wars.* Also call a *jihad.*

home area In telephony, the numbering plan area in which the calling subscriber is located.

home box A user's personal machine, especially one he or she owns. Also called a *home machine.*

home page In the World Wide Web (WWW), the top-level hypertext page of a hierarchical WWW client. Typically, the *home page* lists general information about the area and provides *hot links* to frequently accessed pages. The address or Universal Resource Locator (URL) of the home page is frequently just the host name, e.g., http://www.nasa.gov.

homing **(1)** A process in which a mobile station is directed (or directs itself) toward a source of radiant energy (electromagnetic, thermal, sonic, etc.) whether a primary or reflected source. The process may follow a vector force field or a gradient of a scalar force field. **(2)** In telephony, the resetting of a sequential switching device to its fixed starting point (its home).

homo The Greek prefix meaning "the same."

homochronous From the Greek meaning same (homo) time (chronous). Describes signals in which corresponding significant events or entities have a constant but uncontrolled phase relationship with each other. See also *anisochronous, asynchronous, heterochronous, isochronous, plesiochronous,* and *synchronous.*

homogeneous cladding In fiber optics, that part of an optic fiber which has a constant refractive index as a function of distance along the radius. (Some optical fibers may have several homogeneous claddings, each with a different refractive index.) See also *graded index fiber* and *optic fiber.*

homogeneous multiplexing Multiplexing in which all of the information bearer channels operate at the same data signaling rate.

homogeneous network A network that uses a single protocol at the network layer. This is in contrast to a heterogeneous network that uses multiple protocols at the network layer.

hook A software or hardware feature included in the original design in order to simplify later additions or changes.

For example, most Web browsers can properly display text and some graphics, but they may fail to recognize a sound file, a motion graphic (video clip), or an as yet "uninvented" application. To solve this limitation, a hook is provided in the browser to call a helper program that knows what to do with the incoming information. All that need be provided to the browser is a key by which the appropriate application is selected. (Generally the key is simply a file extension.) Also called a *user exit.*

hook flash A signal to the local switching equipment caused by going on-hook and then back off-hook in a "short" period of time. Although the hook flash period is different between various equipment, it is generally between 50 ms and 500 ms but must be less than 800 ms, or some switching systems may interpret it as a call termination signal. A *hook flash* is a request for attention from the switching equipment so that some service may be performed (such as transferring a call in a PABX). Also called *hookflash* and *register recall*.

hook switch The switch that is activated in a telephone instrument when the handset is removed from the cradle. More generally, the switch that electrically connects a DCE to the telephone network. When the hook switch has connected the DCE to the network, the DCE is said to be *off-hook*. Also called *hookswitch, switch hook* and *switchhook*.

Hoot-n-Holler A voice only full-time circuit that connects a speakerphone in one location to a speakerphone in a distant location.

This type of circuit is normally open at all times to allow two-way communications without having to pick up the receiver or dial the phone. That is, to use the system, a user need only talk into the microphone; there is no need to dial or select the destination because it is always connected to one (and only one) destination. The speakerphones used in this type of circuit are full-duplex, transmit, and receive units. Also known as a *shout down*.

hop (1) In radio transmission, the excursion of a radio wave from the Earth to the ionosphere and back to the Earth (*sky wave*). The number of hops is the same as the number of reflections from the ionosphere. **(2)** *In networks,* the transmission of an information unit (message, packet, file, etc.) across one link on a store-and-forward network composed of one or more series links. (The connection between a subscriber and a host, or between two hosts, may involve more than one link or *hop*.) Each link may be either wire, radio, or optical, and each may have a different *modulation* or *transmission protocol*.

On such networks (e.g., UUCPNET and FidoNet), an important intermachine parameter is the number of *hops* in the shortest path between them, which is frequently more significant than their geographical separation. **(3)** A waveform transmitted for the duration of each relocation of the carrier frequency of a frequency hopped system such as a frequency hopping spread spectrum system. **(4)** To modify a modulated waveform with constant center frequency so that its frequency *hops* (3). **(5)** In jargon, to log in to a remote machine, especially via rlogin or telnet. For example, *hop* over to ftp://foo.files to FTP magic.file.

hop count (1) Generally, the number of legs traversed by a signal between its source and destination (e.g., repeaters or switches). **(2)** In network routing, the number of nodes through which a packet passes between the source and destination. Some protocols or services keep track of the number of *hops* a packet has traversed and will discard the packet and display an error message if the *hop count* exceeds a predefined value.

In the Internet TCP/IP protocol, the *hop count* in the IP header is decremented at each node. When the *hop count* equals zero, the packet is discarded. Therefore, a packet must reach its destination before it passes through the specified number of nodes (usually 20) such as bridges, gateways, and routers. This count therefore determines the *time to live (TTL)* for some packets.

horizontal cross connect A cross connect in the telecommunications equipment room to the horizontal distribution cabling.

horizontal parity check See *longitudinal redundancy check*.

horizontal redundancy check A synonym for *longitudinal redundancy check (LRC)*.

horizontal wiring That portion of a wiring system that extends from a workstation to the backbone horizontal crossconnect (BHC) in the telecommunications equipment room.

horn (1) In radio transmission, an electrically open-ended waveguide, of increasing cross-sectional area, which radiates either directly in a desired direction or feeds a reflector that forms a desired beam.

The curves describing the increasing cross-sectional area are quite numerous (e.g., elliptical, conical, hyperbolic, or parabolic curves) and not necessarily the same in the E-plane and H-plane. A wide range of beam patterns may be formed by controlling horn dimensions and shapes, placement of the reflector, and reflector shape and dimensions. **(2)** A portion of a waveguide in which the cross section is smoothly increased along the axial direction. **(3)** In audio systems, a tube, many having a rectangular transverse cross section and a linearly or exponentially increasing cross-sectional area, used for radiating or receiving acoustic waves.

hose (1) To make nonfunctional or greatly degraded in performance. For example, a high-priority application program that uses all of a computer's computational capabilities for an extended time will leave other users with very poor performance. The system is then said to be "hosed." **(2)** A term denoting data paths that represent performance bottlenecks. **(3)** Cabling, especially thick Ethernet cable (which is sometimes also called a *bit hose* or *hosery*, a play on "hosiery," or an *etherhose*).

host (1) Any computer, system, or network that supplies information, services, or resources to a subordinate computer, terminal, or system (*client*).

The *host* system may be connected to a number of other computers or terminals by communication links. The system may contain only one computer with a number of *dumb terminals,* or it may be a computer facility accessed via the dial-up network by other computers. Though generally the *host* is a *node* on a network that can be used interactively (e.g., it can be logged onto), it need not be the most "powerful" computer in the network; it need only supply information or services required by the *client*. In the PC environment, the host may be a computer to which a device is connected. For example, a PC can be the host for a network interface card (NIC), or a printer, or both. On the Internet, a host is a machine through which users can communicate with other machines. For example, a minicomputer at a university may serve as a host for access to the Internet. Also called a *host processor* or just a *host*. **(2)** A computer on which is developed software intended to be used on another computer. **(3)** In packet- and message-switching communications networks, the collection of hardware and software that makes use of packet or message switching to support user-to-user (end-to-end), interprocess communications and distributed data processing.

host bus adapter (HBA) A "smart" controller board designed to do data storage and retrieval tasks, thereby saving the processing time in the computer. A disk channel consists of an *HBA* and the hard disk(s) associated with it. For example, Novell's Disk Coprocessor board is a SCSI HBA adapter.

host table A list of TCP/IP host names on the network paired with their IP addresses.

hot cut The transfer from an old system to a new system, without the benefit of retaining the old system in case the new one does not perform as desired.

hot fixing The ability to detect and mark bad sectors of a disk and then assign alternate disk sectors during routine LAN operation. This automatically updates the original defect map.

hot-line service A private, point-to-point telephone connection. With such a connection, there is no need to dial; one telephone rings as soon as the other is picked up.

hot link (1) In the World Wide Web (WWW), a "button" on a Web page which takes the user immediately to another network location. The new location may be as close as a different point within the same document, or it may be on a different network on the other side of the world. (2) A mechanism for sharing data between two application programs where changes to the data made by one application appear instantly in the other's copy.

hot potato algorithm A packet-routing algorithm in which a node routes a packet or message to the output path with the shortest queue—even if it is not the optimum path in other respects.

hot spot (1) Generally, any place in a system design that turns into a performance bottleneck due to resource contention. (2) In computer programs, an extension of the 80/20 rule as applied to computer programs; that is, in most programs, less than 20% of the code uses 80% of the execution time.

A graph of frequency of visits vs. code addresses would typically yield a chart with a few huge spikes amidst a low-level reference. Such spikes are called *hot spots* and are good areas for optimization. The term applies especially to tight loops and recursions in the code's central algorithm. (3) The active location of a cursor on a bit-map display. "Put the mouse's *hot spot* on the address and click the left button." (4) A screen region that is sensitive to mouse gestures; that is, they trigger some action. For example, WWW browsers present hypertext links as *hot spots* which, when clicked with the left mouse button, point the browser at another document. (These are specifically called *hot links*.)

hot standby A strategy to increase a system's overall reliability wherein two or more subelements are operated in parallel, all of which are fully powered and processing input signal. However, at a given time, only one is delivering output signal. If the subunit that is delivering signal malfunctions, the *hot standby* immediately replaces it and takes over the transmission duties.

hot swap The ability of a system to have parts or subsystems removed and inserted without removing power. When combined with redundant subsystems, removal may be accomplished without affecting the system's operation.

For example, a failed modem card or fan may be replaced in a system without powering down the overall system.

hotlink A hot spot on a World Wide Web page; that is, an area, which, when clicked or selected, fetches the information to which the link points. Also spelled *hot link.*

house wizard A person occupying a technical-specialist, R&D, or systems position at a commercial enterprise. A really effective *house wizard* can have influence out of all proportion to the position's corporate rank and still not have to wear a suit. Term used especially of UNIX wizards. Also called a *house guru.*

housekeeping Equipment operations or routines that do not directly contribute to the solution at task (communication, problem solving, etc.), but they do contribute directly to the operation of the equipment itself.

howler tone A synonym for the receiver off-hook (ROH) tone.

HPCC An abbreviation of High Performance Computing and Communications.

HPFS An abbreviation of IBM's OS/2 High Performance File System.

HPGL An abbreviation of Hewlett-Packard Graphics Language. A set of commands designed to convey vector graphic image information to compatible devices (such as the HP LaserJet III and IV). Also written *HP-GL.*

HPIB An abbreviation of Hewlett-Packard Interface Bus. See *IEEE-488.*

HRC An abbreviation of Hybrid Ring Control FDDI. See *FDDI-II.*

HSLAN An acronym for High Speed LAN.

HSSI An abbreviation of High Speed Synchronous Serial Interface.

HST An abbreviation of U.S. Robotics' High Speed Technology. A proprietary modulation technique allowing speeds up to 14,400 bps in the forward channel and 450 bps in the reverse channel. The high-speed forward channel is dynamically assigned to the modem with the most data to be transferred.

HTML An abbreviation of HyperText Markup Language. The standard format used to code text files with *hot links* for use with World Wide Web (WWW) documents. The standard allows the documents to be read by Web browsers that support the format.

HTTP An abbreviation of HyperText Transfer Protocol. An application-level protocol designed for distributed, collaborative, hypermedia information systems. It is a generic, stateless, object-oriented protocol that can be used for many tasks, such as name servers and distributed object management systems, through extension of its request methods (commands). A feature of *HTTP* is the typing and negotiation of data representation, allowing systems to be built independently of the data being transferred.

Used in the World Wide Web (WWW), it enables the transfer of hypertext-based files between local and remote systems.

HTTP server A server on the World Wide Web (WWW) which can communicate using HTTP (Hypertext Transfer Protocol) and provide *hypertext* documents to WWW clients.

hub A network component that:

- Serves as a common termination and distribution point for multiple nodes (a physical star topology) and
- Accepts a signal from one node and redistributes it to other nodes.

Hubs accommodate not only a number of nodes, but many include connectors for linking to other *hubs. Hubs* usually connect nodes of a common architecture, such as Ethernet, ARCnet, FDDI, or Token Ring. This is in contrast to a bridge, router, or gateway, which may support multiple architectures and protocols. Although the boundary may not always be clear, *hubs* are generally simpler and less expensive. *Hub*-to-node connections for a particular network all use the same type of cable (which may be coaxial, twisted pair, or fiber optic). *Hubs* are frequently located in a wiring closet, and they may be connected to a higher-level wiring center, known as an *intermediate distribution frame (IDF)* or *main distribution frame (MDF)*. All *hubs* provide connectivity; that is, they pass on signals that enter the input. The simplest hub simply rebroadcasts incoming signals to all connected nodes, while more intelligent hubs will selectively transmit signals. In addition, many *hubs* provide supplemental features, for example,

- *Active hubs* may regenerate the signal before passing it on; that is, they clean up the signal before forwarding it. *Passive hubs* simply pass the signal through.

- Token Ring hubs do some internal routing of the node connections to create a logical ring.
- Some intelligent hubs can monitor network activity and report these data to a management program on the network.
- Other intelligent hubs may be able to automatically isolate defective nodes so as to maintain network functionality.

Some intelligent hubs have the ability to be configured and controlled remotely.

hub and spoke A term describing an arrangement with a central device and multiple peripheral, or outlying, devices. For example, a central office with connections to smaller branch offices or remote concentrators would have a hub and spoke arrangement. Also called a *star*.

huff To compress data using a Huffman code. Decompression is obviously referred to as *puff*.

Huffman coding A method of compressing a set of data based on the relative frequency of occurrence of individual elements comprising the data set. That is, the more often an element occurs within the data set, the shorter its encoded bit pattern is (a coding method called *entropy coding*).

The *Huffman coding* technique was one of the earliest methods of data compression and, with modifications, remains in widespread use. See also *compression* (2) and *data compression*.

human–machine interface The boundary at which people communicate with and make use of machines. Typical interface devices are keyboards, monitors, and mice (although any functional switch or indicator on any device is an interface). This boundary is also known as the *user interface* or *man–machine interface*.

hundred call seconds (CCS) In telephony, a measure of circuit activity. One *CCS* is 100 seconds of conversation on a line, so that an hour of line usage is 36 *CCS*; 36 *CCS* is equal to one erlang and indicates continuous use of the line.

hung A truncation of *hung up*. The state in which a program or process is stuck or incapable of proceeding without help (e.g., it is waiting for input), but the system is otherwise functioning. A *hung* state is distinguished from crashed or down, where the program or system is also unusable but because it is not running rather than because it is waiting for something. However, the recovery from both situations is often the same—reboot. Also called *hosed, locked up*, and *wedged*.

hunt group In telephony, a group of lines that are successively tried (hunted), until an idle one is found. That idle line is then connected to the subscriber desiring service. Also called an incoming *service group (ISG)*.

hunting (1) In telephony, pertaining to the operation of a selector or other similar device to find and establish a connection with an idle circuit of a chosen group. See also *hunt group*. **(2)** Pertaining to the failure of a device to achieve a state of equilibrium, usually by alternately overshooting and undershooting the point of equilibrium. Sometimes referred to as *ringing* or *oscillations*.

hybrid (1) A functional unit in which two or more different technologies are combined to satisfy a given requirement. Examples include:

- An electronic circuit having both vacuum tubes and transistors.
- A mixture of thin-film and discrete integrated circuits.
- A computer that has both analog and digital capability.

(2) In telephony, a circuit that converts a two-wire full-duplex circuit (the communications line) into two two-wire *simplex circuits*, one transmit circuit and one receive. A *hybrid* is fundamentally a balanced bridge circuit (as is illustrated in the following circuit) with the following characteristics:

- A signal received from the remote device (port 4) is routed to the receive output (port 2).
- The transmit signal (injected at port 1) is *split* between the communications line (port 4) and a *matching network* (port 3). The signal to the communications line is routed to both a difference circuit and the remote device. The signal on the matching network is also routed to the difference circuit where it cancels with the signal on the communications line. Since the receive signal (at port 2) is the *difference* between the signals on the communications line (port 4) and the matching (port 3) network there is no transmit signal present. It is said to be rejected.

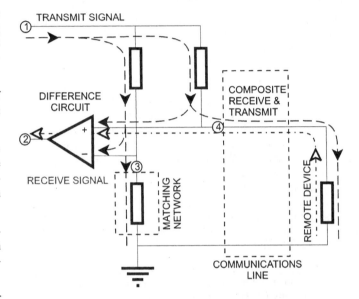

Because it is not possible to exactly balance the *matching network* to the transmission line and remote device, the difference will not be exactly zero; that is, there will be some amount of transmitted signal that leaks into the receive path. The measure of the balance between the matching network and the transmission line is called the *balance return loss (BRL)*. The total loss, called the *transhybrid loss (THL)*, from input port 1 to the output port 2 is the sum of all losses; that is

$$THL = BRL + 6\text{ dB} + \text{insertion loss}$$

The 6-dB loss is due to the splitting of the signal among each of the branches of the bridge. The *insertion loss* is the inherent loss of the components. Also called a *two- to four-wire hybrid, anti-sidetone circuit*, a *fork*, and a *duplexer*. See also *balance return loss (BRL), bridge* (2), *hybrid coil, insertion loss, sidetone*, and *transhybrid loss (THL)*.

hybrid balance A measure of the degree of electrical symmetry between two impedances connected to the two conjugate sides of a hybrid set. Hybrid balance is generally expressed in decibels (dB). See also *balance return loss (BRL)* and *hybrid coil*.

hybrid bridge/router See *brouter*.

hybrid cable An optical communications cable having two or more different types of optical fibers, e.g., single mode and multimode fibers. Contrast with *composite cable*.

hybrid coil An *anti-sidetone* circuit involving one or two transformers arranged in such a way as to create four ports in *conjugate pairs*. For example, a signal delivered to port 1 will appear on ports 3 and 4 but not its conjugate port 2. Similarly, a signal on port 4 will appear on ports 1 and 2 but not port 3 (the conjugate of port 4). The conjugate pairs occur only when the circuit is properly balanced, that is, when the impedance on port 1 is the same as the impedance on port 2 and the impedance on port 3 equals that on port 4 and the transformer turns ratio is correct. Absolutely perfect balance is not possible, so a small portion of the injected signal does appear on the conjugate port (called *transhybrid loss*).

A *two-transformer hybrid* is shown in the following circuit diagram. Port 4 is typically connected to the two-wire telephone line, while ports 1 and 2 are connected to a four-wire circuit. Port 3 is connected to a circuit that attempts to match the *impedance* of the two-wire telephone line.

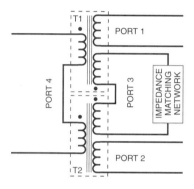

A *single transformer,* properly interconnected, can also create four ports and in conjugate pairs. The next diagram illustrates one such circuit. Also called a *bridge transformer.* See also *hybrid.*

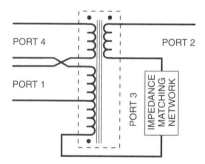

hybrid connector A connector that contains connections for more than one type of service or media. Examples of hybrid connectors include those that have contacts for both optical fibers and twisted pairs, electric power and twisted pairs, or shielded and unshielded twisted pairs or any other combination.

hybrid coupler In an antenna system, a hybrid junction used as a directional coupler. The loss through a *hybrid coupler* is approximately 3 dB.

hybrid fiber coaxial (HFC) A high-bandwidth interconnection and distribution scheme intended to provide telephone subscribers with a multitude of services spanning the range from basic telephony through interactive video and information services. The technique is also used by the CATV industry.

The scheme combines both optical fiber and coaxial cable distribution lines. Optical fiber feeders run from the head-end to distribution points serving neighborhoods of several hundred subscribers. From the distribution points, coaxial cable completes the distribution to each subscriber. *HFC* networks provide many of the reliability and bandwidth benefits of fiber at a lower cost than pure fiber distributions.

hybrid junction A waveguide or transmission line terminating device such that:

- There are four ports,
- Each port is terminated in its characteristic impedance, and
- Energy entering any one port is transferred, usually equally, to two of the three remaining ports.

Hybrid junctions are used as mixing or dividing devices.

hybrid mode (**1**) A mode consisting of components of both electrical and magnetic field vectors in the direction of propagation. In fiber optics, such modes correspond to skew (nonmeridional) rays. (**2**) In an FDDI-II network, an operating mode that makes both packet switched and circuit switched links available. Therefore, both time-independent data and time-dependent data (voice, video, etc.) can be transmitted on the network. This mode is in contrast to the basic mode, wherein only packet switching is done. Hence, only time-independent data transmission is effective (no voice).

hybrid multiplexer (HMUX) In the FDDI-II network architecture, a component at the media access control (MAC) layer that multiplexes network data from the MAC layer and isochronous (time-dependent) data (such as voice or video) from the isochronous MAC (IMAC) layer.

hybrid network (**1**) A communications network that uses a combination of line facilities (i.e., trunks, loops, or links), some of which use only analog or quasi-analog signals and some only digital signals. Also called a *hybrid system* or *hybrid communications network.* (**2**) A network (LAN) whose physical implementation does not fit into any of the basic architectures (bus, hierarchical, mesh, ring, or star) but rather has attributes of several.

For example, the network as shown in the drawing contains a switching group from each of the basic network types (star, ring, hierarchical, and bus). Another way to view this structure is as three networks joined with gateways. Other hybrid networks are not as clearly defined. Also called *hybrid communications network* and *hybrid topology.* See also *bus network, class n office, gateway, hierarchical network, mesh network, ring network,* and *star network.*

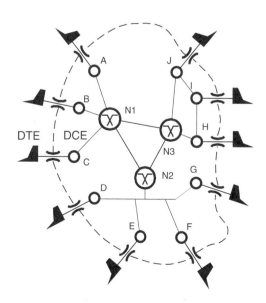

hybrid ring control (HRC) FDDI See *FDDI-II*.

hybrid set Two or more transformers interconnected in such a way as to create a four-port network. When the ports are terminated in appropriate impedances, the ports are made into conjugate pairs.

hydrogen loss Attenuation of an optical signal in a fiber optic cable due to the light absorption characteristics of diffused hydrogen in the glass itself.

hydroxyl ion absorption In fiber optics, the absorption of electromagnetic waves, including the near infrared, due to the presence of trapped hydroxyl ions remaining from water as a contaminant. The hydroxyl (OH-) ion can penetrate glass during or after fabrication, resulting in significant attenuation of discrete optical wavelengths, i.e., approximately 1.3 μm, used for communications via optical fibers.

hygroscopic A term that refers to the characteristic of a material in which it absorbs or attracts moisture from its environment.

hypermedia A type of document that contains pointers for linking to multimedia information sources, such as text, graphics, video, or sound files, which may be in the same or other documents or even in another computer. Operator use of *hypertext* links is known as navigating.

hypertext A term created by Ted Nelson, circa 1965, to describe the nonlinear nature of textual documents when the path followed is not serial but jumps around, following lines of associated information. That is, a user can jump to topics by selecting highlighted words or phrases (called an *anchor*) within the document being read. Selection activates a link (*hot link*) to another location either within the same or different document. The matrix created by the links within and among various documents is called a *web*.

It allows the user to select certain words or phrases and immediately display related information for the selected item. *Hypertext* requires a "tag" language (such as, Hypertext Markup Language—HTML) to specify links to within document or to another document.

HyperText Markup Language (HTML) The standard markup language for documents made available on the World Wide Web (WWW). When a hypertext document is viewed with a browser such as Navigator or Mosaic, the *hot link* will be highlighted. When the links are activated, the user will be taken to a new location as directed by the link.

HTML is a simplified version of the *Standard Generalized Markup Language (SGML)*—a language frequently used to mark up electronic texts for display.

HyperText Transport Protocol (HTTP) The name of both the server software and the protocol for *hypertext* browsing on the World Wide Web (WWW). The protocol allows a user to jump to another location indicated by the links embedded in a hypertext document. *HTTP* is that form of *Universal Resource Locator (URL)* statement which allows an author of hypertext documents to embed commands to jump to the resources of another, possibly remote, system.

hysteresis The tendency of a device, circuit, or system to behave differently when responding to an increasing input parameter change than to a decreasing change.

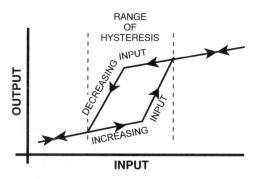

Hytelnet A resource discovery tool, developed by Peter Scott of the University of Saskatchewan, that allows a user to browse through hypertext documents listing Telnet-accessible sites on the Internet as well as their resources. It is a program for finding out what resources are available and how to access them.

Hz The SI abbreviation for <u>Hertz</u>.

I

i Primarily mathematics, although sometimes engineering. The symbol for the imaginary number $\sqrt{-1}$. See also *j*.

I frame An X.25 Information frame.

I.n recommendations An ITU-T series of specifications concerning the Integrated Services Digital Networks (ISDN).

I/F An abbreviation of Inter<u>F</u>ace.

I/G bit A bit in the IEEE 802 MAC address field that distinguishes between Individual and Group addresses.

I/O An abbreviation of Input Output.

I/O bound A description of the operation of a system when it spends most of its time waiting for peripherals transactions to complete. Such devices include modems, disk drives, keyboard, and video. Systems that are I/O bound will not benefit from a more powerful or faster processing element (CPU).

I/O port address The unique "name" of each input or output device in a computer. The port address in hardware is generally selected with DIP switches, jumpers, or in some cases software. (Some manufacturers may not provide a selectable address when *they* feel there will never be a conflict with other devices.)

The port address is the starting location for a block of memory dedicated to an I/O port. Also called an *I/O base address*. See also *base address*.

IA An abbreviation of International Alphabet.

IA5 The abbreviation of International Alphabet number 5. A 7-bit code that is defined in the ITU-T recommendation T.50 and ISO document 646 that characterizes a character set used for message transfers. In its default coding, *IA5* is nearly identical to the ASCII code. However, several character encodings can be changed. In these cases, *IA5* can take on a non-ASCII form. Specifically, the following encodings may be redefined:

- Two possible representations can be used for each of the characters corresponding to codes 35 and 36 (decimal). The ASCII encoding uses # and $, respectively.
- Ten characters (codes 64, 91–94, 96, and 123–126 decimal) may be redefined according to national needs. For example, characters may be redefined to represent characters with diacritical marks (umlauts, accents, or tildes, depending on the country).

IA5 code variants may be created and registered, provided that the variant is defined according to the constraints above. Various national alphabets have been registered with the European Computer Manufacturers Association (ECMA). See also *ASCII*.

IAB An abbreviation of Internet Architecture Board originally called Internet Activities Board.

IAC (1) An abbreviation of Inter-Application Communication. **(2)** An abbreviation of International Access Code.

IAE An abbreviation of In Any Event.

IAM An acronym for Initial Address Message.

IANA An acronym for Internet Assigned Number Authority.

IANAL An abbreviation of I Am Not A Lawyer (but, . . .).

IBG An abbreviation of InterBlock Gap. See *inter-record gap*.

IBM Cable System (IBMCS or ICS) A cabling system introduced by IBM in 1984 based on shielded twisted pair (STP) cabling.

IBMNM An abbreviation of IBM Network Management. A protocol used for network management in an IBM Token Ring network.

IC (1) An abbreviation of Interexchange Carrier (also IEC and IXC). **(2)** An abbreviation of Integrated Circuit. **(3)** An abbreviation of InterCom. **(4)** An abbreviation of Intermediate Cross-connect.

ICA An abbreviation of International Communications Association. Originally founded in 1949 as the National Committee of Communications Supervisors, the *ICA* is a trade association of more than 700 corporations (those that spend the most in the telecommunications arena). The organization is based in Dallas, Texas. The *ICA* holds an annual convention every May.

ICCM An abbreviation of Inter-client Communications Convention Manual. Standards that govern communications between X-client applications.

ICEA An abbreviation of Insulated Cable Engineers Association.

ICI (1) An abbreviation of Interexchange Carrier Interface. The interface between carriers that support Switched Multimegabit Data Service (SMDS). **(2)** An abbreviation of Incoming Call Identification.

ICMP An abbreviation of Internet Control Message Protocol. That part of the TCP/IP protocol suite that allows nodes to share IP status and reports datagram delivery errors. The packet Internet gopher (ping) application is based on *ICMP*.

icon In computer systems, a small, pictorial representation of an application software package, idea, or concept used in a window or a menu to represent commands, files, or options.

ICP An abbreviation of Instituto das Communicacoes de Portugal. The Portuguese Institute of Telecommunications.

ICR An abbreviation of Intelligent Character Recognition.

ICTA An abbreviation of International Computer Telephony Association. Based at the University of Colorado, Boulder, Colorado.

ICV An abbreviation of Integrity Check Value. A mathematically derived check based on a message content. When a message is received with a valid check value, there is a high probability that the message has not been tampered with. Also called *Message Authentication Code*.

ICW An abbreviation of Interrupted Continuous Wave.

IDA An abbreviation of Integrated Digital Access.

IDAPI An acronym for Integrated Database Application Programming Interface.

IDC An abbreviation of Insulation Displacement Connector. A type of wire termination in which the connector cuts through the cable's insulating jacket when the connector is attached. *IDCs* are used on most ribbon cables and telephonic twisted pair terminations.

IDDD An abbreviation of International Direct Distance Dialing.

IDE An abbreviation of Integrated Drive Electronics or Intelligent Drive Electronics.

A standard for connecting hard disks and other drives with an embedded controller to the AT bus. Introduced in 1989 as a low-cost alternative to Enhanced Small Disk Interface (ESDI) and Small Computer Systems Interface (SCSI). IDE devices are limited to 528 Mbytes of storage.

IDE bus A 40-pin cable with a maximum length of 24 inches. Defined in the ANSI X3T9.2 specification. Also called the ATA-2 interface (from AT Attachment)

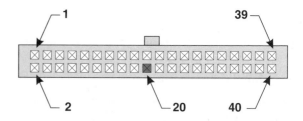

EIDE (ATA-2) PIN ASSIGNMENT

Pin	Signal	Description		Description
1	RESET	Drive Reset	»	Host asserts for >25 s after power stable. May be re-asserted if necessary
2	Ground	Ground		
3	DD7	Data Bit 7	«»	
4	DD8	Data Bit 8	«»	
5	DD6	Data Bit 6	«»	
6	DD9	Data Bit 9	«»	
7	DD5	Data Bit 5	«»	
8	DD10	Data Bit 10	«»	
9	DD4	Data Bit 4	«»	8 or 16 bit bidirectional
10	DD11	Data Bit 11	«»	data bus between host and
11	DD3	Data Bit 3	«»	drives. Bits 0-7 are used for
12	DD12	Data Bit 12	«»	8 bit transfers; e.g. registers
13	DD2	Data Bit 2	«»	and ECC bytes.
14	DD13	Data Bit 13	«»	
15	DD1	Data Bit 1	«»	
16	DD14	Data Bit 14	«»	
17	DD0	Data Bit 0	«»	
18	DD15	Data Bit 15	«»	
19	Ground	Ground		
20	KEY	Unused (keying)		Not electrically used.
21	DMARQ	DMA Request	«	Asserted when drive ready to transfer data either direction.
22	Ground	Ground		
23	DIOW-	I/O Write Data	»	Rising edge strobes data into register or data port.
24	Ground	Ground		
25	DIOR-	I/O Read Data	»	Falling edge enables register or data port and rising edge strobes data into host.
26	Ground	Ground		
27	IORDY-	I/O Channel Ready	«	Negated to indicate drive not ready for transfers.
28	SPSYNC/ CSEL	Spindle Sync/ Cable Select	»*	1. SPSYNC generated by master (host or drive), slave(s) syncronize spindle speed to the master's signal. 2. CSEL selects drive number, ground = drive 0 open = 1.
29	DMACK-	DMA Acknowledge	»	Host acknowledge of DMARQ, i.e., data ready or data received.
30	Ground	Ground		
31	INTRQ-	Interrupt Request	«	Asserted by the drive when an interrupt to the host is pending.
32	IOCS16-	Host 16 Bit I/O	«	Indicates drive is ready for 16 bit transfers.
33	DA1	Host Addr Bus Bit 1	»	3-bit address selects drive's internal register.
34	PDIAG-	Passed diagnostics	«»*	Asserted by drive 1 to drive 0 indicating diagnostic tests complete.
35	DA0	Host Addr Bus Bit 0	»	3-bit address selects drive's internal register.
36	DA2	Host Addr Bus Bit 2	»	3-bit address selects drive's internal register.
37	CS1FX-	Host Chip Select 0	»	Selects Command Block Register.
38	CS3FX-	Host Chip Select 1	»	Selects Command Block Register.
39	DASP-	Drive active/Drive 1 Present	«*	1. Asserted by drive 1 during RESET- to indicate its presence. 2. After RESET- either drive asserts DASP- to indicate drive active.
40	Ground	Ground		

« Drive is signal source.
» Host is signal source.
«» Signal is bidirectional.
* Interdrive communications signal.

ideal filter A hypothetical filter that has the following traits:

- In the frequency domain, passes all frequencies within a specified passband without attenuation and completely rejects all other frequencies. Or,
- In the time domain, has an output response identical to the input excitation except for a constant delay.

An *ideal filter* cannot truly exist because of *Gibb's phenomenon*. However, with sufficient circuit complexity it can be approximated as closely as desired.

IdentaRing See *distinctive ringing.*

identity daemon A UNIX program that provides a means for other programs to determine a remote user's logon name. Used, for example, on the Internet to determine what user ID is addressing the host after a connection is established.

IDF An abbreviation of Intermediate Distribution Frame.

IDG An abbreviation of Inter Dialog Gap.

IDI An abbreviation of Initial Domain Identifier. In the ISO/OSI Reference Model, that part of a network address which represents the domain.

IDK An abbreviation of I Don't Know.

idle The state of a device, process, or system when it is enabled and active but not in use and it is awaiting a command. Also called the *idle state.* The time (*idle time*) during which the device, process, or system is operational but not active.

idle cell In ATM, a cell that is transmitted by a node when there is not enough offered network traffic to keep the transmission rate at a specified level. *Idle cells* can be discarded by any node in the transmission path when necessary, e.g., when the offered network traffic increases to a level at which the idle cell is no longer needed.

idle channel code A specific code pattern transmitted by a station that identifies the channel as idle. It may also provide transmitter/receiver synchronization. Also called an *idle signal.*

idle channel noise Noise that exists in a communications channel when no signals are present. To be valid, the channel conditions and terminations must be stated.

idle character In communications, the character or symbol transmitted when no other information is being sent. See also *synchronization code.*

idle circuit In communications, the condition of a transmission channel *in the talk state* when no signal is present. Also called an *idle line.*

idle line termination A line termination network that is actively controlled by the circuit switching equipment. When the circuit is seized, the destination equipment presents an impedance to the line. However, an idle circuit has no load impedance. The *idle line termination* is switched to the idle circuit so as to maintain a constant impedance.

idle signal A specific code pattern transmitted by a station that identifies the channel as idle. Also called an *idle channel code.*

idle time The time during which a device, process, or system is enabled and operational but not processing commands or data.

idling signaling Any signal placed on a communications circuit during periods of inactivity that indicates that no data are being transferred.

IDN An abbreviation of Integrated Digital Network. A network that uses digital signaling and circuitry.

IDRP An abbreviation of InterDomain Routing Protocol.

IDU An abbreviation of Interface Data Unit.

IEC (1) An abbreviation of International Electrotechnical Commission. An international standards organization for electrical and electronic engineering. Standards are created by committees made up of worldwide experts in the area of concern. Many national standards are based on *IEC* publications. The French name is *Commission Electotechnique Internationale* (*CEI*). (2) An abbreviation of InterExchange Carrier.

IEC Miles Interexchange Carrier (Long Distance) Miles. A method used to calculate the cost of mileage-dependent line charges. It is based on the coordinates (latitude and longitude) of the two stations for which the pricing is being calculated.

IEEE An abbreviation of the Institute of Electrical and Electronics Engineers (pronounced "eye-triple-ee"). An international professional society that produces technical publications, standards, and conferences to promote the development and application of technology.

IEEE-488 The IEEE-488 standard describes a general-purpose interface bus (GPIB) which was originally developed by Hewlett-Packard as a proprietary interface called the *Hewlett-Packard interface bus* (*HPIB*).

IEEE 802.x standards A set of protocols developed by the IEEE which defines methods of access to and control of local area networks (LANs). The protocols correspond to the ISO/OSI model physical and data link layers (layers 1 and 2).

The *IEEE 802.x standards,* however, divide layer 2 into two sublayers; the *logical link control* (*LLC*) adjacent to OSI layer 3 (data link layer), and the *media access control* (*MAC*) adjacent to OSI layer 1 (physical layer). The following figure illustrates the relationship of the various committees, working groups, and technical advisory groups (TAGs), as well as the standards that bear the same numerical name.

The 802.1 through 802.11 standards have been adopted and superseded by the International Standards Organization (ISO). These internationally accepted standards are issued as ISO/IEC 8802-1 through 8802-11.

IEEE 802.1—details the *relationship* of the *IEEE 802.x standards* and the ISO/OSI reference model. It specifies standards for systems management, internetworking, and network management at the hardware level, including the spanning tree algorithm. (The spanning algorithm ensures that only one path is selected when bridges or routers pass messages between networks and it finds a secondary path if the selected path fails.)

IEEE 802.1b—the standard for *network management.*

IEEE 802.1d—the standard of inter-LAN *bridges* between 802.3, 802.4, and 802.5 networks at the MAC level. For example, there should be no more than seven bridges between any two nodes on a network.

The algorithm is used to prevent bridging loops by creating a spanning tree. The algorithm originally developed by Digital Equipment Corporation (DEC) is not the same as, nor is it compatible with, the version documented in 802.1d.

IEEE 802.1j—the standard for both passive and active fiber optic *Star-based* Ethernet segments.

IEEE 802.2—based on the 1982 *DIX Ethernet* specifications (developed by Digital Equipment, Intel, and Xerox corporations), *IEEE 802.2* specifies, in essence, a multipoint peer communication link protocol (the *LLC* sublayer of the ISO/OSI model's data link layer). The LLC provides an interface between media access methods and the network layer. The functions provided by the LLC, which are transparent to upper layers, include framing, addressing, and error control. The *LLC* sublayer is the general portion and applies to all IEEE 802.x standards (but not to DIX Ethernets—Ethernet 2). It addresses station-to-station connections, message frame generation, and error control.

IEEE 802.3—defines the *MAC* sublayer, which is the application-dependent layer; that is, it varies from standard to standard. It deals with network access and collision detection protocols, i.e., the *MAC* protocol for *bus* networks that use CSMA/CD (Carrier Detect Multiple Access with Collision Detection). Data are transmitted least significant bit first (so-called little endian). Networks using CSMA/CD are limited to about 2.5 km.

The standard includes both baseband networks (such as the Ethernet network developed by Xerox or AT&T's StarLAN) and broadband networks. Common 802.3 Ethernet networks include:

1Base5 (StarLAN): The designation for a 1-Mbps Ethernet LAN that uses unshielded twisted pair (UTP) cable. The physical topology is a bus with nodes attached to the common cable. AT&T's StarLAN is one example of a *1Base5* network.

1BASE5

Parameter	Value (and Conditions)
Topology	Physical star or daisy chain
Nodes per hub	10 to 12 (vendor dependent)
Nodes/network	<1210
Data bandwidth	1 Mbps baseband
Protocol	CSMA/CD
Data encoding	Manchester
Cable	UTP category 2 or better
Terminator	RJ-45 connector
Characteristic impedance	100 Ω ±15%, averaged over 5.0–10 MHz
Segment length	≤120 m (400 ft) daisy chain
	≤ 488 m (1600 ft) star.

10BaseT (twisted pair Ethernet): A 10-Mbps IEEE 802.3 Ethernet in which each wiring segment can be up to 100 m. The cable may be either unshielded twisted pair (UTP) or shielded twisted pair (STP). The cable is coupled to the terminal's network interface card (NIC) via an RJ-45 connector wired as shown in the following chart. The physical network topology is a star; that is, each node is connected to a central hub. The network may be expanded to five sections through four repeaters. Sometimes referred to as *10BT, twisted Ethernet, UTP Ethernet,* or *twisted sister.*

RJ-45 CONNECTOR PIN ASSIGNMENT

Pin	Function (at hub)	Function (at NIC)
1	TD+ (Transmit to pair 1)	RD+ (Receive from pair 1)
2	TD− (Transmit to pair 1)	RD− (Receive from pair 1)
3	RD+ (Receive from pair 2)	TD+ (Transmit to pair 2)
4		
5		
6	RD− (Receive from pair 2)	TD− (Transmit to pair 2)
7		
8		

10BASET

Parameter	Value (and Conditions)
Topology	Physical star or tree
Data bandwidth	10 Mbps baseband
Protocol	CSMA/CD, CSMA/CA
Cable	Dual UTP category 3 or better 22, 24, or 26 AWG with 1 or more twists per foot
Connector	RJ-45 connector transmit (+) pin 1, (−) pin 2 receive (+) pin 3, (−) pin 6
Characteristic impedance	100 Ω ±15%, averaged over 5.0–10 MHz; 85–115 at 10 MHz
Segment length	≤ 100 m (328 ft)
Segment loss	< 11.5 dB for all cable & connectors at 10 MHz
Segment delay	< 1 ms
Segment NEXT	< 23 dB at 10 MHz
Network span	≤ 500 m (1640 ft)
Nodes	1 / segment; 1024 / network max.
Signal level	2.2 to 2.8 V peak
Impulse noise	≤ 2 impulses over 264 mV per 10 s

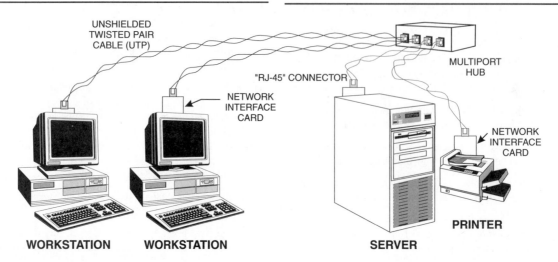

WORKSTATION   WORKSTATION   SERVER   PRINTER

10Base2 (*Thin Ethernet*): The designation of a 10-Mbps IEEE 802.3 Ethernet in which a wiring segment is a single-shielded 50-ohm coaxial cable (RG58) with BNC connectors and can be up to 185 meters in length. The maximum number of nodes per segment is 30, separated by at least 0.5 meters. A 50-ohm termination must be placed on the two "end" nodes. Also referred to as *Thinnet* or *CheaperNet* because the thin RG-58 coax cable is considerably less expensive than the thick coaxial cable used in 10Base5.

AUI CONNECTOR (DB-15) PIN ASSIGNMENT

Pin	Function	Pin	Function
1	Control in (shld)	9	Control in (ret) (CI-B)
2	Control in (CI-A)	10	Tx data (ret) (DO-B)
3	Tx data (DO-A)	11	Tx data (shld)
4	Rx data (shld)	12	Rx data (ret) (DI-B)
5	Rx data (DI-A)	13	Voltage
6	Voltage	14	Voltage (shld)
7	Control out (CO-A)	15	Control out (CO-B)
8	Control out (shld)		

10BASE2

Parameter	Value (and Conditions)
Topology	Bus
Data bandwidth	10 Mbps baseband
Protocol	CSMA/CD, CSMA/CA
Data encoding	Manchester
Cable	RG-58 A/U or RG-58 C/U coaxial #20 AWG, 83.3pF/m (25.4pF/ft)
Terminator	BNC connector (UG-274)
Characteristic impedance	50 ±2 ohms
Segment length	≤ 185 m (604 ft)
Node spacing	≥ 0.5 m (1.5 ft)
Nodes/segment	≤ 30
Nodes/network	≤ 90
Network span	≤ 925m (3035 ft)
Repeaters	≤ 2 between nodes
	≤ 4 per network
Attenuation	< 8.5 dB @ 10 MHz & 185 m; or, <6.0 dB @ 5 MHz
DC loop resistance	< 50 mΩ / meter @ 20°C. Total of center conductor + shield.
Center conductor	Stranded tinned copper. 0.89 ±0.05 mm diameter.
Propagation velocity	>65% of C at 20°C (80% nominal)
Edge jitter	< 8.0 ns at 185 m with a pseudorandom Manchester coded pattern.

10BASE5

Parameter	Value (and Conditions)
Topology	Bus
Data bandwidth	10 Mbps baseband
Protocol	CSMA/CD, CSMA/CA
Data encoding	Manchester
Trunk cable	RG-8, Belden 9880 twin-axial. 10 AWG; 85pF/m (26pF/ft)
Terminator	N-series connector
Characteristic impedance	50 ±2 ohms
Segment length	≤ 500 m (1640 ft)
Node spacing	≥ 2.5 m (8.2 ft)
Total bus length	≤ 2500 m (8200 ft)
Nodes/segment	≤ 100
Nodes/network	≤ 1024
AUI cable	Four STP with solid core; 20 AWG; 84.8pF/m (16.7pF/ft) between conductors; 96.8pF/m (29.5pF/ft) to shield
AUI cable length	≤ 50 m (164 ft)
AUI connector	DB-15
Repeaters	≤ 2 between nodes
	≤ 4 per network

10Base5 (*Thick Ethernet*): A 10-Mbps Ethernet in which the backbone wiring segment can be up to 500 meters and is a 50-ohm twinaxial cable. Each node is connected to the Thick Ethernet cable via a transceiver and transceiver cable. Up to 100 transceivers may be installed on the backbone cable. The maximum length of the transceiver cable is 50 meters. Connectors (AUI) on the transceiver cable are wired as illustrated in the following chart. Also referred to as *standard Ethernet, thicknet,* or *yellow cable.*

10BaseF: An IEEE 802.3 10-Mbps Ethernet that uses fiber optic cable for transmission. Several specific variants include:

10BaseFB The fiber backbone (FB) for facility lines between buildings.
10BaseFL Fiber link (FL) for intermediate hubs and workgroups.
10BaseFP Fiber passive (FP) for connection to desktop PCs.

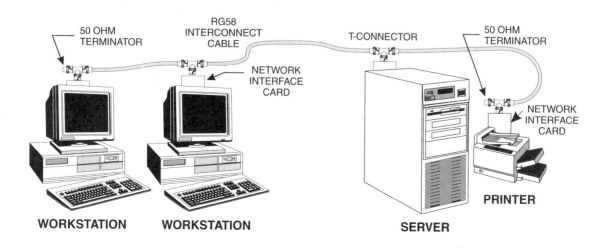

WORKSTATION   **WORKSTATION**   **SERVER**

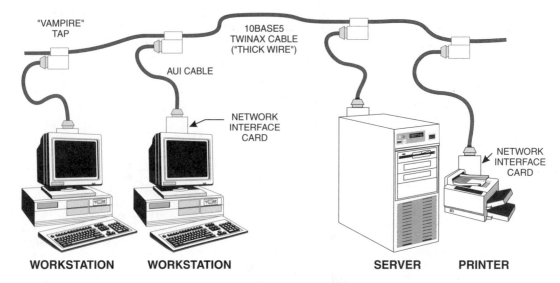

WORKSTATION   WORKSTATION   SERVER   PRINTER

10Broad36 (CATV): A 36-MHz broadband Ethernet LAN specified to operate on 75-ohm CATV coaxial cable. It supports 10-Mbps bidirectional data rates over cable segments up to 3600 meters. The physical topology may be either a bus or a tree structure.

The transmit and receive link must each have an 18-MHz bandwidth. This may be supplied by a single 36-MHz bandwidth coaxial cable or by dual 18-MHz cables. Each 18-MHz channel is further partitioned into 14-MHz to encode the 10-Mbps data stream and 4-MHz for overhead (collision detection and reporting). *10Broad36* networks use differential phase shift keying (DPSK) to encode data for transmission.

10BROAD36

Parameter	Value (and Conditions)
Topology	Bus or tree
Data bandwidth	10 Mbps broadband
Protocol	CSMA/CD
Data encoding	Manchester
Cable	RG-59 coaxial
Terminator	F-series connector
Characteristic impedance	75 ±2 ohms
Segment length	≤ 3600 m (11 800 ft)
Nodes/channel	≤ 1024 to # 64 000 (vendor dependent)
Channels	1 to 40

10BaseF (fiber)

100BaseT: An IEEE standard from proposals by the Fast Ethernet Alliance (including 3Com and SynOptics). It will support Category 3, 4, and 5 UTP cabling.

100BASET

Parameter	Value (and Conditions)
Topology	Physical star or tree
Data bandwidth	100 Mbps
Protocol	
Data encoding	
Cable	2 pair for 100BaseTX. 4 pair for 100BaseT4. Category 3, 4, & 5 UTP; and optical fiber
Terminator	RJ-45 connector
Characteristic impedance	
Segment length	<100 m; <1000 m optical fiber
Nodes/network	
Hubs	2 maximum with <10 m separation

100BaseX: A *fastEthernet* proposed by Grand Junction, which combines the CSMA/CD media access method defined in 802.3 with the physical medium sublayer defined for FDDI networks by ANSI X3T9.5 to obtain 100-Mbps data transfers over category 5 unshielded twisted pair (UTP) cable. The maximum segment length is reduced to about 250 meters. See also *IEEE 802.12.*

IEEE 802.4—defines the physical layer and MAC protocol for 1, 5, 10, and 20-Mbps baseband and broadband *bus* networks using *token passing* to regulate access and traffic control. In essence, the token is passed around the network in a logical ring. Data are transmitted least significant bit first (so-called little endian).

It is typically used with General Motors' 10-Mbps manufacturing automation protocol (MAP) LANs.

MAP

Parameter	Value (and Conditions)
Topology	Bus
Data bandwidth	1, 5, 10, 20 Mbps/channel broadband
Protocol	Token passing
Data encoding	Manchester
Cable	RG-59 coaxial
Terminator	F-series connector
Characteristic impedance	75 ±2 ohms
Segment length	(vendor dependent)
Nodes/channel	(vendor dependent)
Channels	(vendor dependent)

IEEE 802.5—defines the physical layer and MAC protocol for 1, 4, and 16-Mbps physical *ring* networks using *token passing* to regulate access and traffic control. Data are transmitted most significant bit first (so-called big endian).

IEEE 802.6—the technical advisory group that defines standards for the Metropolitan Area Network (MAN). A high-speed data communications network formerly known as QPSX (from Queued Packet and Synchronous eXchange), it is now called DQDB (from Distributed Queue Dual Bus). It operates over distances to about 80 kilometers. Some of the characteristics include:

- Performance which is independent of the number of nodes and of the distances involved.
- Operation at the physical layer and at the media-access control (MAC) sublayer.

TOKEN RING

Parameter	Value (and Conditions)
Data bandwidth	4 or 16 Mbps baseband
Data encoding	Manchester
Cable	IBM type 1, 2, 6, or 9 STP; or IBM type 3 UTP
Terminator	IBM universal data connector (STP) RJ-45 (UTP)
Characteristic impedance	150 Ω ±16% (STP), 100 Ω ±15% (UTP)
Nodes/main ring	≤ 260 IBM cabling ≤ 72 telephone UTP
Main ring length	≤ 366 m (1200 ft) total length
MSAU's/ring	≤ 33 (type 1, 2, 9 STP) ≤ 9 (type 3 UTP) (type 6 cable is for patching only.)
Node to MSAU distance	≤ 300 m (980 ft) STP ≤ 100 m (328 ft) UTP
MSAU to MSAU distance	≤ 200 m (650 ft) STP ≤ 120 m (394 ft) UTP
IBM 8228 equiv. MSAU distance	4.9 m (16 ft)

MSAU = Multi-Station Access Unit.

- Use of two buses for the network access. Each bus operates in a single direction, and the buses operate in opposite directions. A node on the network may transmit and receive on one or both buses, depending on where the node is located in relation to the bus ends.
- Use of fiber optic cable as the main physical bus medium. Copper cable is generally not used because it has difficulty supporting both the distances and the bandwidth required. Copper cable is used, however, in many MANs as the access cable to connect individual nodes or subnetworks to the MAN bus.
- Support for circuit switched voice, data, and video.
- Provisions for both synchronous or asynchronous transmissions.
- Provisions for both connection-oriented, connectionless, and isochronous communications services.
- Dynamic bandwidth allocation, using time slots.
- 53-octet time slots for transmissions.
- Support for transmission speeds to 150 Mbps now and ultimately to approximately 600 Mbps.

See also *Metropolitan Area Network* (*MAN*).

IEEE 802.7—a technical advisory group (TAG) report that defines broadband LAN standards. The document specifies the minimal physical, electrical, and mechanical features of broadband cables, and discusses issues related to their installation and maintenance.

IEEE 802.8—a technical advisory group (TAG) report that defines standards for multimode fiber optic networks. The document discusses the use of optical fiber in networks defined in 802.3–802.6 and provides recommendations concerning the installation of fiber optic cables.

IEEE 802.9—a technical advisory group (TAG) specification addressing the integration of voice and data. It specifies architectures and interfaces for devices that can transmit both voice and data over the same lines (integrated voice and data—IVD). The standard, which was accepted in 1993, is compatible with ISDN, uses the LLC sublayer specified in 802.2, and supports unshielded twisted pair cable.

IEEE 802.10—a technical advisory group (TAG) report of a working group addressing LAN security issues, including data exchange, encryption, network management, and security in architectures that are compatible with the OSI Reference Model.

IEEE 802.11—the technical advisory group (TAG) that defines standards for wireless LANs, including data security, interference, and spread spectrum modulation methods (frequency hopping and direct sequence) at 2.4–2.5 GHz.

IEEE 802.12—an extension of 10BaseT networks to higher speeds, specifically to 100 Mbps. In addition, the upgrade ideally does not require cable replacement. The upgrade path from a 10-Mbps Ethernet to a 100-Mbps 100VG AnyLAN network requires:

- New 100VG AnyLAN network interface card (NIC) at each node (replacing the old 10-Mbps NICs).
- A new 100VG AnyLAN hub replacing the old 10BaseT hub.
- Possibly replacing the twisted pair cabling if it does not have four pairs or does not meet other transmission requirements.

100VG AnyLAN (100VG): A fastEthernet LAN standard with 100-Mbps performance over standard unshielded, voice grade (category 3), or better twisted pairs. It was developed by Hewlett-Packard (HP), IBM, Proteon, and AT&T Microelectronics. The standard differs from the standard 10-Mbps Ethernet in several ways. For example:

- While Ethernet uses a CSMA/CD protocol, *100VG AnyLAN* uses a *demand access protocol.* In theory, the protocol provides a wider available bandwidth by eliminating collisions.
- *100VG AnyLAN* uses 5B/6B NRZI waveform encoding rather than the less efficient Manchester waveform coding of standard Ethernet.
- *100VG AnyLAN* utilizes all four twisted pairs for both transmitting and receiving data in a half-duplex manner. Each of the four pairs operates at 25 Mbps yielding the total network bandwidth of 100 Mbps. Standard Ethernet uses one pair for data transmission and the other for data reception.
- The entire cable path from the hub to each device must be *unbundled;* that is, it must not go through any 25-pair (or similar) cables. Otherwise, severe crosstalk is likely to result in poor or unreliable performance.
- *100VG AnyLAN* can be used with either Ethernet or Token Ring cards, but not both at the same time or in the same network.

100VG ANYLAN

Parameter	Value (and Conditions)
Topology	Physical star or tree
Data bandwidth	100 Mbps
Protocol	Demand access
Data encoding	5B/6B NRZI
Cable	4 category 3 and 5 UTP; and optical fiber
Terminator	
Characteristic impedance	
Segment length	<100 m category 3 cable <200 m category 5 cable <2000 m optical fiber
Node-to-node	<2500 m
Nodes/Network	<1024
Hubs	

IEEE 802.14—the Cable-TV Protocol Working Group. Chartered to create data transport and Internet access standards for use on traditional cable TV networks.

Note that although the 802.3, 802.4 and 802.5 networks are all frequently referred to as Ethernets, they are NOT compatible. See also *bus network, CSMA/CD, Ethernet, ISO/OSI Model, ring network,* and *token passing.*

IEEE 1284—an enhanced parallel port standard based on both the enhanced parallel port (EPP) and the extended capabilities port (ECP) strategies. See also *enhanced parallel port (EPP)* and *extended capabilities port (ECP)*.

IEEE 1394—a 1995 serial bus interface standard offering high-speed communications and isochronous real-time data services for Macintosh and IBM PCs. It is meant to be a replacement of the ADB, RS-232, RS-422, SCSI, and even the parallel port. Features include the following:

- There are data transfer rates between devices of 100, 200, or 400 Mbps, with a planned increase to 1.2 Gbps.
- Cable is 6 conductors with lengths to 5 m.
- It can supply up to 60 watts of power, allowing low-consumption devices to operate without a separate power cord.
- It can *daisy-chain* up to 63 devices in a treelike structure (as opposed to SCSI's linear structure).
- No bus terminator is required.
- It allows peer-to-peer device communication, such as communication between a scanner and a printer, to take place without using system memory or the computer itself.
- Each device identifies itself dynamically (that is, it will not require dip switch or jumper settings).
- It supports plug-and-play (PnP) and allows connection and disconnection of devices without interrupting system operation.

Also called *Firewire, High Performance Serial Bus,* and *i.Link.* See also *Access bus* and *Universal Serial Bus (USB).*

IEN An abbreviation of Internet Experimental Notes.

IEPG An abbreviation of Internet Engineering and Planning Group.

IES (1) An abbreviation of Inter-Enterprise System. An Electronic Data Interchange (EDI) network connection that allows the intercompany exchange of e-mail, fax, electronic fund transfers, CAD/CAM, etc. (2) An abbreviation of Information Exchange Services. (3) An abbreviation of Incoming Echo Suppressor.

IESG An abbreviation of Internet Engineering Steering Group.

IETF An abbreviation of Internet Engineering Task Force.

IF An abbreviation of Intermediate Frequency.

IFAX An acronym for International facsimile (FAX) service.

IFG An abbreviation of InterFrame Gap.

IFM An abbreviation of Instantaneous Frequency Measurement.

IFRB An abbreviation of International Frequency Registration Board. The ITU group responsible for allocating frequency bands in the electromagnetic spectrum.

IFS An abbreviation of Ionospheric Forward Scatter. See *ionospheric scatter.*

IGES An acronym for Initial Graphics Exchange Standard.

IGFET An acronym for Insulated Gate Field Effect Transistor. See *transistor.*

IGMP An abbreviation of Internet Group Membership Protocol.

IGP An abbreviation of Interior Gateway Protocol.

IGRP An abbreviation of Interior Gateway Routing Protocol. An Internetworking protocol for routers developed by Cisco Systems Inc.

IHL An abbreviation of Internet Header Length. A 4-bit field in an Internet Protocol (IP) datagram (packet) that specifies the length of the datagram's header in 32-bit words.

IIR An abbreviation of Infinite Impulse Response.

IIR Filter An abbreviation of Infinite Impulse Response Filter.

IISP An abbreviation of Interim Inter-switch Signaling Protocol.

ILD An abbreviation of Injection Laser Diode.

illegal In computer terminology, a term that signifies an element or procedure that is not allowed because it is impossible or would produce invalid results. For example, an 8-bit character is *illegal* in a system that is designed for 7-bit ASCII. (It's not possible to put an 8-bit character into a 7-bit long field.)

illegal character A combination of bits (character), that is not valid in a particular system according to its specification (such as with respect to a specified alphabet, a particular pattern of bits, a rule of formation, or a check code), e.g., any 8-bit ASCII character with a most significant bit of one. (It may be a valid extended ASCII character, however.) Also called *false character, forbidden character, improper character, unallowable character,* and *unused character.*

IM (1) An abbreviation of Intensity Modulation. (2) An abbreviation of InterModulation.

IMAC An acronym for Isochronous Media Access Control.

image (1) A fully processed unit of operational information that is ready for transmission, execution, or use. Examples include:

- An *exact* copy of a computer's working RAM is transferred to a secondary storage device such as a diskette.
- A file, when loaded into the control storage of a unit, determines the operations of that unit.

(2) A graphic, which may contain pictures and/or text. It may be represented as a *bit map (raster image)* or a *vector image.* (3) In the field of image processing, a two-dimensional representation of a scene. Also called a *picture.*

image antenna A hypothetical, mirror-image antenna considered to be located as far below ground (ground plane) as the actual antenna is above ground. It is in the same position and orientation as the actual antenna; however, it is reversed.

image frequency In heterodyne receivers, the frequency of an undesired input signal to a *heterodyne* receiver capable of producing the same intermediate frequency (IF) as the desired input frequency.

For example, given a local oscillator frequency of 655 kHz (f) and an IF frequency of 455 kHz (Δf), the two input frequencies are 1110 kHz ($f+\Delta f$) and 200 kHz ($f-\Delta f$), one of which is the desired frequency and the other is the *image frequency.*

image rejection ratio In heterodyne receivers, the ratio of the intermediate frequency (IF) signal level produced by the desired input frequency to that produced by the image frequency of equal power. The image rejection ratio is usually expressed in decibels (dB). Also called *image frequency rejection ratio.*

image resolution The degree of clarity of an image due to the number of pixels per unit area. Facsimile images typically are 203 by 98 or 203 by 198 dots per inch. Laser printers, on the other hand, may be 1200 by 1200 dots per inch or better.

imaginary number A number that multiplied by itself will yield a *negative number. Imaginary numbers* are represented by a numeric value and the letter i or j where i or j represents $\sqrt{-1}$. (i is used primarily in the field of mathematics, while j is usually used in engineering.)

Since the product of two positive numbers or two negative numbers is always a positive number, one cannot find the square root of a negative number with only real numbers. To get around this problem, it is assumed that a square root exists and it is called an *imaginary number*. Examples:

$$\sqrt{-4} = 2i = j2$$
$$\sqrt{-64} = 8i = j8$$
$$3i * 4i = -12$$

imaginary part That part of a complex quantity represented by an imaginary number. For example, given the complex number $A + jB$, A is called the real part and B is called the imaginary part (j is $\sqrt{-1}$).

IMAP (1) An acronym for <u>I</u>nternet <u>M</u>essage <u>A</u>ccess <u>P</u>rotocol. (2) An acronym for <u>I</u>nternational <u>M</u>icroelectronics <u>a</u>nd <u>P</u>ackaging Society.

IMCO An abbreviation of <u>I</u>n <u>M</u>y <u>C</u>onsidered <u>O</u>pinion.

IMD An abbreviation of <u>I</u>nter<u>M</u>odulation <u>D</u>istortion.

IMEI An abbreviation of <u>I</u>nternational <u>M</u>obile <u>E</u>quipment <u>I</u>dentity.

IMHO An abbreviation of <u>I</u>n <u>M</u>y <u>H</u>umble <u>O</u>pinion. Used in bulletin board system (BBS) conferences as a preface to a comment or opinion—generally though, not humble.

immediate ringing A feature of telephone switching systems in which the called telephone instrument begins ringing within fractions of a second of being selected by the switch.

This is in contrast to older switching systems where the connection may be established at any point in the ring generator's cycle. Hence, the called subset may not ring for several seconds after connections. The delay is due to the sharing of a single ring generator by many line circuits.

immittance A general term for both *impedance* and *admittance*, it is used where the distinction is irrelevant.

IMNSHO An abbreviation of <u>I</u>n <u>M</u>y <u>N</u>ot <u>S</u>o <u>H</u>umble <u>O</u>pinion. (Generally worth just what you think it is worth.)

IMO An abbreviation of <u>I</u>n <u>M</u>y <u>O</u>pinion.

IMP An abbreviation of <u>I</u>nterface <u>M</u>essage <u>P</u>rocessor. See *router*.

impairment Any type of signal degradation. Impairment may be caused by attenuation, noise, distortion, and other conditions.

impedance (Z) (1) Opposition to the flow of current. Impedance has two components: *resistance* (R), which impedes the flow of both direct current (dc) and alternating current (ac), and *reactance* (jX) which impedes only the flow of alternating current. *Impedance* is therefore expressed as

$$Z = R + jX$$

Reactance is frequency dependent; that is, it changes as the exciting frequency changes. This variation of reactance is not the same for all devices. It will increase as frequency increases for some devices (*inductive reactance,* expressed as $2j\pi fL$), and it will decrease for increasing frequency for other devices (*capacitive reactance,* expressed as $-j/2\pi fC$). Hence, the impedance in terms resistors, inductors, and capacitors is

$$Z = R + j\left(2\pi fL - \frac{1}{2\pi fC}\right)$$

(2) The ratio of *complex voltage* between two conductors to the *complex current* in either of the two conductors.

impedance irregularity A term that denotes an impedance nonuniformity in a transmission line. Usually, the source of the irregularity is a connector or a splice, but it can arise from distortions in the cable due to overly tight cable clamps or tie wraps or from pressure generated as the cable is pulled over a sharp corner.

impedance matching The connection of a special circuit to a transmission line in order to improve the performance of the transmission path. The matching network generally is designed to be the same as the *characteristic impedance* of the transmission line. The matching network minimizes reflections that may otherwise occur at the ends of transmission lines.

For example, the matching network on a 10Base2 Ethernet transmission cable is a 50-ohm resistor. 10BaseT networks are terminated with a 100-ohm resistor, CATV a 75-ohm, and ARCnet uses a 93-ohm resistive termination. The matching network (balance network) for a two- to four-wire hybrid is a circuit composed of several resistors and capacitors. See also *balance network.*

implication A Boolean operator relating statements P and Q such that P implies Q is true in all cases except when P is true and Q is false.

The written symbol for *implication* is a $\Rightarrow$, $\rightarrow$, or $\supset$; therefore, P implies Q may be written P $\Rightarrow$ Q. The *truth table* shown here presents all possible combinations of input and output for a two-variable *implication* operation. A false state is shown as a "0" while a true condition is shown as a "1." See also *AND, Boolean algebra, NOT, OR,* and *XOR.*

IMPLICATION TRUTH TABLE

P	Q	P⇒Q
0	0	1
0	1	1
1	0	0
1	1	1

implicit congestion notification A means of determining that congestion exists on a network. Some transport protocols, such as TCP from the Internet TCP/IP protocol suite, can infer when network congestion is occurring. This notification is in contrast to explicit notification methods, such as the *explicit congestion notification* (*ECN*) method, used in frame relay networks.

impulse A surge (pulse) of unidirectional polarity. The beginning and ending of the surge occur within a short period of time as compared to other system parameters.

impulse hits Data transmission errors caused by voltage surges (impulses). On the telephone line, the impulses are generally 0.3 to 4 ms in duration and within 6 dB of the normal signal level.

impulse noise One of the many types of noise found on communications circuits. It is characterized by random occurrences of high amplitude, random spectral content, and short duration energy. Common causes include lightning, electric sparks, starting and stopping electric motors, or digital circuits.

impulse response The time domain response of a circuit, network, or system to a unit impulse excitation.

Mathematically, the response is the inverse Laplace transform of the network function in the frequency domain.

impulse strength The area under the curve defining the pulse shape. The area is related to the product of amplitude and time duration.

IMR An abbreviation of <u>I</u>nternet <u>M</u>onthly <u>R</u>eport.

IMS An abbreviation of <u>I</u>nformation <u>M</u>anagement <u>S</u>ystems.

IMT An abbreviation of <u>I</u>nter-<u>M</u>achine <u>T</u>runk. A high-capacity trunk connection between telephone switching systems.

IMTS An abbreviation of <u>I</u>mproved <u>M</u>obile <u>T</u>elephone <u>S</u>ervice. In mobile telephony, a service that allows direct dialing between a mobile telephone and an ordinary (wired) phone.

IN An abbreviation of <u>I</u>ntelligent <u>N</u>etwork.

in-band signaling Signaling and control that uses the same communications path and frequencies that the subscriber uses, i.e., the same frequencies as are used for voice and data transmissions. Sometimes written *in-band signaling*. See also *common signaling channel* and *out-of-band signaling.*

incandescence The conversion of electrical energy into light by a thermal process. See also *electroluminescence.*

incident angle In optics, the angle between an incident ray (an incoming ray) and a line perpendicular to an optical surface.

incidental radiation Radio frequency energy radiated during the course of a device's operation, although the device is not intentionally designed to produce this energy.

inclined orbit Any nonequatorial satellite orbit. The orbits may be circular or elliptical, synchronous or nonsynchronous, retrograde or direct.

inclusion The presence of extraneous or foreign material in an otherwise homogeneous substance.

inclusive OR See *OR.*

incoherent light **(1)** Light in which the phase relation of any wavefront is random with respect to any other wavefront. *Incoherent light* may be monochromatic (one color) as with a light emitting diode (LED). **(2)** Characterized by a degree of coherence significantly less than 0.88.

incoherent scattering A term that describes the propagation of electromagnetic waves such that random variations of phase, amplitude, polarization, and direction are induced when an incident wave encounters matter.

incoming call identification (ICI) A switching system feature that allows an attendant to identify visually the type of service or trunk group associated with a call directed to the attendant's position.

incoming calls barred A switching system feature in which all call delivery attempts to a specific number are blocked. Outgoing calls are not affected.

incoming trunk See *one-way trunk.*

incorrect block An information block successfully delivered to the intended destination but containing one or more incorrect bits, additions, or deletions.

increment A small change in the value of an entity or quantity. These changes are frequently indicated by the Greek letter delta (in lower case "δ" or upper case "Δ").

incremental compression Data compression accomplished by specifying only the initial value and all subsequent changes. Two different schemes are used to create the increments: the changes may be referenced to the last value transmitted or to the initial value.

For example, given the sequence 1000, 1035, 1001, 999, 1000, 1000, . . . , then the sequence generated by incremental compression using the last value transmitted would be 1000 (initial value), $+35$, -34, -2, $+1$, 0, . . . ; the sequence generated by using the initial value is 1000 (initial value), $+35$, $+1$, -1, 0, 0, ...

Indeo Intel's compression algorithm that works with the proprietary DVI system or as a stand-alone CODEC for use with QuickTime and Video for Windows.

independent clocks In communication network timing subsystems, free-running precision clocks used at each node for synchronization purposes. Due to the inevitable differences between clock frequencies and the associated difference in data rate, a variable storage buffer is used at each node to accommodate these variations in transmission. Traffic may occasionally be interrupted to allow the buffers to be emptied of some or all of their stored data.

independent ground A ground electrode or system whose voltage to a reference is not materially affected by currents flowing to ground in other electrodes or systems.

independent sideband transmission (ISB) A method of double-sideband transmission in which two separate information-bearing channels exist (one on each sideband). The carrier may or may not be suppressed.

independent telephone company (ITC) A local exchange carrier (LEC) that is not a Bell Operating Company (BOC). There are more than 1500 *ITCs* in the United States. They serve more than 50% of the land area but only about 15% of the telephones.

index dip In fiber optics, the undesired decrease in the index of refraction at the center of an fiber optic core caused by some fabrication techniques. Also called a *profile dip.*

index-matching material In fiber optics, a material interposed between the ends of two optical fibers. The material may be a liquid, gel, or cement (adhesive). However, the refractive index of the material is as close to the index of the fiber's core as practical. Use of the material reduces Fresnel reflections at the fiber endfaces.

index of refraction (*n* in the equation) The ratio of the speed of light in a vacuum (*c*) to the speed of light in the material the light is traversing (*v*). Also called *refractive index.*

$$n = \frac{c}{v}$$

index profile In fiber optics, the index of refraction as a function of radial distance from the core center.

indicator **(1)** A light, meter, or dial that displays information about or the status of a device or system. For example, the "OH" light on the front panel of an external modem indicates that the modem is off-hook when the light is illuminated. **(2)** An item of data that may be interrogated to determine whether a particular condition has occurred or has been satisfied in the execution of a program, e.g., the overflow indicator bit.

individual account A computer network account that provides selected access privileges to a single user. See also *access control list* and *wildcard account.*

individual line In telephony, a subscriber line arranged to serve only one main station. Although additional stations may be connected to the line as extensions, no discriminatory ringing is provided by the switching equipment.

individual trunk A trunk, link, or junctor that serves only one input group of a grading.

induced voltage A voltage produced on a closed path or circuit by a change in magnetic flux linking in the path.

For example, lightning strikes by virtue of the large current flow generating a pulsed magnetic field. When this changing field is linked into a communications loop, an *induced voltage* pulse is generated.

inductance A property of an electric circuit wherein a changing current in one part of the circuit induces a voltage in the same (*self-inductance*) or neighboring part of the circuit (*mutual inductance*). The effect is due to the magnetic field that arises when a current flows through a conductor. The units of inductance are henries (H).

induction The transfer of energy from one coil to another by way of an electromagnetic field.

inductive connection The connection of a device to a circuit through inductive coupling. For example, a modem's receiver may be coupled to the telephone line by means of the electromagnetic field generated in the earphone of a telephone subset. A hearing aid is another device that may use inductive coupling to establish a telephonic connection.

inductive coupling The association of two or more circuits among one another by means of mutual inductance.

inductive pickup A coil of wire used to tap into a magnetic field. It may be used to establish an *inductive connection* to a telephone circuit, for example.

inductor A device that stores energy in the form of a magnetic field. Although an inductor allows the flow of direct current, it impedes the flow of an alternating current (ac). The impedance depends on the amount of *inductance* the device has and the *frequency* of the ac impressed across the device. As frequency increases, the impedance increases; this effect is called *inductive reactance* (X_L). Inductive reactance (X_L, in ohms) is expressed in ohms and is related to inductance (L, in henries) and frequency (f, in hertz) by:

$$X_L = 2 \pi f L$$

Symbols for an *inductor* include:

The first two symbols are "air core" inductors; that is, there is no magnetic material associated with the device. The bars in the third example indicate the *inductor* is an "iron core" *inductor*. The iron (or other magnetic material) increases the amount of inductance per turn of the coil. *Inductors* are used in a variety of applications including transmission line conditioners surge protectors, and noise reduction circuits. They can help remove noise caused by electromagnetic and radio frequency interference (EMI and RFI).

infinite impulse response (IIR) A theoretical description of a system's response to an impulse. Specifically, the system (in a mathematical sense) will generate an output signal forever in response to an input impulse.

In reality, this rarely occurs because the response generally decreases in amplitude with elapsed time and at some point the signal level becomes less than the noise level. In digital systems, an additional limitation occurs because of the finite word lengths; that is, when the signal drops below the value of the least significant bit level, it simply ceases to exist. Contrast with *finite impulse response* (*FIR*).

info (1) One of several USENet alternative newsgroups. The newsgroups originated from mailing lists at the University of Illinois. The topics have a definite bias toward the technical. See also *ALT* and *alternative newsgroup hierarchies*. (2) An abbreviation of <u>info</u>rmation.

information The meaning assigned to data by a known rule or set of rules.

information agent A computer program that can search numerous databases for specific information without the user needing to know where the database resides. Examples include Archie and Veronica in the Internet.

information bearer channel A channel capable of transmitting all the information required for communication, such as user data, synchronizing sequences, and control signals. The channel data rate may be higher than that required for user data alone.

information element The name of the data fields within an ISDN layer 3 message.

information field In packet transmission, a field assigned to contain user information. The contents of the *information field* are not used at the link level.

information frame A frame in a protocol, such as High Level Data Link Control (HDLC), that contains user data.

Information Management Systems (IMS) IBM's hierarchical database management and communications package for use in its Systems Network Architecture (SNA).

information security The protection of information against unauthorized disclosure, transfer, modification, or destruction, whether accidental or intentional.

information service A service offering of an organization to provide the capability for generating, acquiring, storing, processing, retrieving, utilizing, or making available information via telecommunications, e.g., electronic publishing.

It does not include the management, control, or operation of a telecommunications system or telecommunications service through any use of such capability.

information superhighway See *NII*.

information system A system, whether automated or manual, that comprises people, machines, software, and/or methods organized to collect, process, store, display, transmit, and/or disseminate data that represent user information.

Information Systems Network (ISN) An AT&T high-speed switching network that can handle both voice and data transmissions. It may be connected to other networks, Ethernet or SNA, for example.

information theory Study of the characteristics and transmission of information. Emphasis in the area of communication includes channel capacity, maximum transmission rate, and accuracy (or error rate). The field of information theory was ushered in by Claude Elwood Shannon in 1948 with the publication of a paper, "The Mathematical Theory of Communication" in the *Bell System Technical Journal*.

information transfer The final result of data transfer from a data source to a data sink. The information transfer rate is not necessarily the same as the transmission modulation rate (line rate). See also *baud*.

information transfer phase In an information transfer transaction, the phase during which user information blocks are transferred from the source node to a destination node. See also *access phase* and *disengagement phase*.

InfoSource A storehouse of information and tutorials concerning the Internet. It is maintained by General Atomics as part of the InterNIC project and is administered by the NSF (National Science Foundation).

infrared (IR) That portion of the electromagnetic spectrum between visible red light (longer wavelength and lower frequency) and the shortest microwaves, i.e., between about 0.7 μm and 0.1 mm. The infrared band is usually divided into four arbitrary sub-bands or categories based on wavelength, i.e.,

Near infrared	750–1500 nm
Middle infrared	1500–6000 nm
Far infrared	6000–40,000 nm
Far-far infrared	40,000 nm–1 mm

(nm is the abbreviation for nanometer, and mm is the abbreviation for millimeter.)

Infrared light provides a short-range communications technology suitable for data transfer between computers and peripheral devices such as a printer, mouse, and keyboard. *Infrared* can also be used on fiber optic links as the information carrier. See also *frequency spectrum.*

infrared communications A communications technology that uses infrared (IR) light beamed between devices to provide a wireless communications link. These links are generally intended for close-range line-of-sight applications. Devices using infrared links include computer mices, keyboards, printers, telephones, and wireless earphones.

An infrared communications link can be focused (point-to-point) or diffuse (broadcast). When focused, the transmitted signal is aimed directly at the receiver (target or cell) or reflected off a surface to the receiver. A focused link can travel over a greater range than a diffuse link but only to a specific target. In contrast, the signal in a diffuse link travels in multiple directions but is much weaker in each direction. As a result, the range is much smaller for a diffuse link than for a focused link.

infrared network A specific type of wireless network that utilizes infrared light as the information carrier.

inheritance In object-oriented programming, the transfer of object features (data attributes and operations) from a "class" that defines the common features of similar objects. A class might be a member of a superclass and may itself have subclasses.

inhibit (1) Pertaining to the state of a computing device in which some or all of the interrupt inputs are ignored. **(2)** Pertaining to the condition of a communications circuit in which it will not accept incoming requests for service.

inhibiting signal A signal that prevents the occurrence of an event in a device circuit or system.

INIC An acronym for ISDN Network Identification Code.

INIT The Macintosh System 7 equivalent of a DOS terminate and stay resident (TSR) program.

initial address message (IAM) An Integrated Services Digital Network User Part (ISUP) message sent in the forward direction which initiates the seizure of an outgoing circuit. It contains address, routing, and other information relating to the handling of the call. See also *Integrated Services Digital Network User Part* (*ISUP*).

initial domain identifier (IDI) In the ISO/OSI Reference Model, that part of a network address that represents the domain.

initial sequence number (ISN) A number generated at each end of a TCP connection to help uniquely identify that connection.

initialization A process that sets all hardware and software conditions of a device to known starting values.

initialization string A set of commands sent to a modem by communications software which puts the modem in a particular operating mode. For example, the modem may be set up to automatically answer incoming calls, or it may be set up to receive facsimile messages rather than modem messages. For a Hayes compatible mode, the string starts with "AT" and is followed by a series of letter commands.

injection fiber A synonym for *launching fiber.*

injection laser diode (ILD) A laser that uses a forward-biased semiconductor junction as the active medium. Stimulated emission of coherent light occurs at the p-n junction where electrons and holes are driven into the junction. Also called *diode laser, laser diode,* and *semiconductor laser.*

INL An abbreviation of InterNode Link.

INMARSAT An acronym from INternational MARitime SATellite service. A service run by Comsat that provides mobile communications to ships, aircraft, vehicles, and fixed location. The services include dial-up telephone, telex, facsimile, data connections, and e-mail. INMARSAT-3 provides 4.8-kb/s telephony, 2.4-kb/s facsimile, and 2.4-kb/s real-time facilities.

INMS An abbreviation of Integrated Network Management System. Part of AT&T's unified network management architecture (UNMA).

INN An abbreviation of InterNode Network.

INPA An abbreviation of Interchangeable Number Plan Area.

input (1) Data presented to a device for processing or transmission. **(2)** The point on a device or system where information is presented for processing or storage.

input/output (I/O) device A device that inserts data into or extracts data from a system.

input/output port (I/O Port) The channels or paths used for transferring data into and out of a computer. I/O channels include the serial ports, parallel ports, keyboards, monitors, and other communication devices. Each I/O port is accessed via a unique address. For example, on a PC, COM1 is at address 3F8h, while the first address of COM2 is 2F8h. (The addresses are expressed in *hexadecimal* notation as indicated by the "h" following the number.)

input buffer limiting A buffering strategy that divides buffers at a node into two classes. Both classes are available for use by transit packets, but only one class may be used by new packets entering the system at that node.

input data (1) Data received or to be received by a physical device or a computer program. **(2)** Data to be processed.

input impedance The impedance presented to a signal source by a signal processing or transmission entity. *Input impedance* is independent of the signal source impedance.

input output channel For a computer, a device that handles the transfer of data between internal memory and peripheral equipment.

input output controller (IOC) A functional unit that controls one or more input-output channels. Also called an *I/O controller.*

input protection For analog input channels, protection against overvoltages that may be applied between any two input connectors or between any input connector and ground. See also *MOV* and *transient protection.*

insertion gain/loss Given the insertion of a signal processing element into a signal path, *insertion gain/loss* is the ratio of signal power before and after the element has been inserted, measured at an arbitrary point in the system following the new elements location. *Insertion gain/loss* is generally expressed in decibels (dB).

Insertion gain is the ratio of the signal power delivered to a point in a system following a signal processing element *after insertion of the signal processing element* to the signal power delivered to the same point in the system *before insertion of the signal processing element,* that is,

$$G_i = 20 \cdot \log_{10}\left(\frac{V_{1a}}{V_{1b}}\right) = 20 \cdot \log_{10}\left(\frac{V_{2a}}{V_{1b}}\right)$$

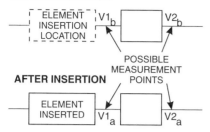

BEFORE INSERTION

AFTER INSERTION

Insertion loss is the ratio of the signal power delivered to a point in a system following a signal processing element *before insertion of the signal processing element* to the signal power delivered to the same point in the system *after insertion of the signal processing element,* that is,

$$L_i = 20 \cdot \log_{10}\left(\frac{V_{1b}}{V_{1a}}\right) = 20 \cdot \log_{10}\left(\frac{V_{2b}}{V_{2a}}\right)$$

inside dial tone A dial tone that users hear when they are connected to an intercom switch. Generally, the dial tone sounds different, either in frequency or cadence, from the public telephone network dial tone.

inside plant **(1)** That portion of a cable system that resides within buildings. **(2)** In telephony, everything inside the telephone company central office building.

inside wiring Wiring within a customer's premises that is used for internal communications or to connect customer premises equipment (CPE) to the network interface (the demarcation point).

in-situ Latin, meaning in place.

instantaneous frequency The rate of change of the angle of an angle-modulated wave with respect to time.

Institute of Electrical and Electronics Engineers (IEEE) See *IEEE*.

insulated conductor A conductor covered with any dielectric other than air having a rated insulating strength greater than the maximum voltage expected on the conductor.

insulating material class A system of rating insulating materials according to their temperature-withstanding capability. The class number is the minimum rated operating temperature. For example, Class 90 is rated for operation to 90°C, while Class 220 is rated for operation to 220°C.

MATERIAL CLASSES

Number	①	②	Typical Materials
90	A		Cotton, silk, or paper without impregnation.
105	B	A	Cotton, silk, paper, or enamel with coatings or impregnation.
130	C	B	Mica, silicon elastomer, or glass fiber with appropriate bonding materials.
155	D	F	Mica, silicon elastomer, or glass fiber with appropriate bonding materials.
180	E	H	Mica, silicon elastomer, or glass fiber with appropriate bonding materials.
200		N	
220	F	R	
>220	G		Mica, porcelain, glass, quartz, etc.
240		S	
>240		C	Mica, porcelain, glass, quartz, etc.

① Electric machine windings, electric cables.

② Used by random-wound ac electric machinery.

The letter designation is different depending on the application and agency specification.

insulator Any material that is a very poor conductor of electricity, i.e., effectively does not conduct electricity in a specified application. Also called a *nonconductor.*

INT 14h The PC software interrupt used to re-route messages from the serial port to a modem or network interface card. This interrupt is used by communications programs.

integer A positive or negative whole number, i.e., a number without a fractional part. Examples include 1, 2, 100, and $-32\,766$.

integral modem A modem that is built into a computer, as opposed to an internal modem that is installed after the computer is built. See also *external modem* and *internal modem.*

Integrated Database Application Programming Interface (IDAPI) A standard from Borland and others providing a standard interface to a wide range of databases. It is claimed to be a superset of Open Database Connectivity (ODBC).

integrated circuit (IC) An electronic circuit that consists of many individual interconnected circuit elements, such as transistors, diodes, resistors, capacitors, inductors, and other active and passive semiconductor devices, formed on a single piece of semiconductor material (called a die). This die is generally mounted on a substrate with leads for attachment to electronic assemblies. Also called a *chip* or *microcircuit.*

integrated digital access (IDA) The means of providing digital access for subscribers to the British Telecom (BT) ISDN service. Two versions are available: *single line IDA* and *multiline IDA.* Services available include data, video, and voice.

integrated digital network (IDN) A communication network that combines digital transmission and digital switching.

integrated numbering plan In telephony, within the world numbering plan, the means for identifying telephone stations inside a geographical area identified by a world-zone-number/country-code-number. See also *country code.*

integrated optical circuit (IOC) A circuit, or group of interconnected circuits, consisting of miniature solid-state optical components on either a semiconductor or dielectric substrate. *IOC* components include light sources (LED and ILD), optical filters, photodetectors, and thin-film optical waveguides.

integrated services digital network See *ISDN.*

Integrated Services Digital Network User Part (ISUP) The call control portion of the Signaling System 7 (SS7) protocol described in the ITU-T Q.761 and Q.764 recommendations. It determines the procedures for setting up, coordinating, and breaking down trunk calls on a SS7 network. Functions in *ISUP* include:

- Calling party number information (including privacy indicator),
- Call status checking (maintains the same trunk state and both trunk ends),
- Trunk management, and
- Application of appropriate tones and/or messages in the event of errors, blocking, or busy conditions.

Seven messages are associated with *ISUP*. They are listed here in the natural order of occurrence:

- Initial Address Message (IAM)
- Continuity Check Message (COT)
- Address Complete Message (ACM)
- Answer Message (ANM)
- Release Message (REL)
- Release Complete Message (RLC)
- Exit Message (EXM)

integrated system A telecommunications system that transfers analog and digital traffic over the same switched network.

integrator A circuit element whose output is essentially the running sum (over time) of the input signal. The diagram illustrates the action of an integrator on several waveforms.

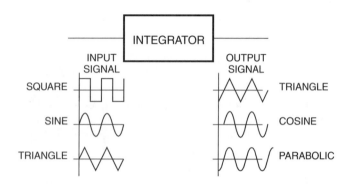

intelligent character recognition (ICR) Software similar to OCR, but it learns to recognize different fonts and character styles and so is more efficient. Useful to make paper documents accessible via groupware.

intelligent hub A network hub that functions as both a bridge and a router.

intelligent manageability A term created by Compaq Computer to define the set of hardware and software tools used to manage networked computers. The tools fall into three categories:

- *Asset control tools,* which inventory and monitor network resources.
- *Fault management tools,* which minimize the impact of errors and hardware failures.
- *Security management tools,* which control user access and improve data security.

Intelligent Printer Data Stream (IPDS) A printing mode, in a Systems Network Architecture (SNA) environment, that provides access to advanced function printer (AFP) capabilities, such as the ability to output text, graphics, and color (if supported) simultaneously on a printer.

intelligibility For voice communications, the capability of being understood. (Intelligibility does not imply recognition of the speaker.)

intelligible crosstalk **(1)** A crosstalk signal from which information can be extracted. **(2)** Crosstalk in a telephone system in which understandable signals may be heard.

INTELSAT An acronym for INternational TELecommunications SATellite. Founded in 1964 to develop a global satellite communications system, *Intelsat* has approximately 120 members. The Earlybird (Intelsat 1) satellite was launched in 1965 as the first in a series.

intensity In optics, the square of the electric field amplitude of a light wave. It is proportional to irradiance and may be used in place of the term *irradiance* when only relative values are important.

intensity distribution curve Generally a polar curve that represents the variation of luminous intensity of a radiating source in a plane through the center of the source. The plane is ordinarily taken on either the horizontal or vertical axis. The drawing shows the *intensity distribution curve* of a particular light emitting diode (LED).

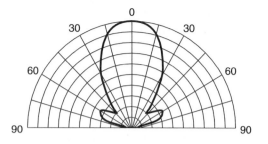

intensity modulation (IM) In optical communications, a form of modulation in which the optical power output of a source is varied in accordance with some characteristic of the modulating signal. Essentially the optical equivalent of electronic amplitude modulation (AM). In intensity modulation, however, there are no discrete upper and lower sidebands in the usually understood sense of these terms (because current optical sources lack sufficient coherence to produce them). Recovery of the modulating signal is by direct detection. Also called *analog intensity modulation (AIM).*

inter- A prefix meaning *between* two or more entities.

inter-application communication (IAC) In the Macintosh System 7 operating system, a process by which applications can communicate and exchange data among each other. Several forms may occur, depending on what is being communicated and which resources are involved in the communication.

- *Copy and paste* provides the most perfunctory form of *IAC.* This type of communication uses the Clipboard as the communication point (a commonly accessible storage area). This form of *IAC* exchanges information that does not change frequently, such as protocol definitions.
- *Publish and subscribe* is used to convey information that may be revised and updated, such as spreadsheets, databases, and text files.

  The most recent version of the information is stored in a file known as the *edition.* Any application that needs this information subscribes to the edition; in this way, the application is always notified when the edition is updated. (This makes it possible to create a document from various source materials, even while these sources are being created.)
- *Events* are used to drive program execution and to control the flow of data in a communications (or other) program. Macintosh processes and servers use Apple events (low-level events that adhere to the Apple Event Interprocess Messaging Protocol) to initiate other processes. Higher level events are requests from an application to the operating system (or another application). Both Apple and higher level events can be used to enable one program to control another program.
- *Program to Program Communications (PPC) Toolbox* provides low-level, powerful, routines to enable applications to communicate with each other.

inter-axial spacing The spacing between the centers of two entities, such as two wires.

inter-block gap See *inter-record gap (IRG).* Also spelled *interblock gap.*

inter-dialog gap (IDG) In the LocalTalk variant of AppleTalk, *IDG* is the minimum gap between dialogs. For LocalTalk Link Access Protocol (LLAP), the gap is approximately 400 μS.

inter-process communications (IPC) **(1)** Communications between several programs either within one computer or across a number of machines. **(2)** Communication across a network between different processes of the same program between different computers running parts of a single program, or between two programs working together. The most common approaches to *IPC* in networking circles are probably Application Programming Interfaces (APIs) such as APPC and NetBios.

inter-record gap (IRG) A deliberately unused space between data on a physical device (tape or diskettes for example) or in a data stream. In either case the gap allows imprecise positioning of the data block due to tape start-stop operations or timing jitter of the sending device or the communication medium. Also called an *inter-block gap* or *interblock gap.*

inter-repeater link (IRL) In an IEEE 802.3 Ethernet network, a cable segment between two repeaters. An *IRL* cannot have any nodes attached.

If the cable is optical fiber, it is known as a fiber optic inter repeater link (FOIRL).

interaction crosstalk Crosstalk caused by coupling between carrier and noncarrier circuits. If the *interaction crosstalk* is, in turn, coupled to another carrier circuit, that crosstalk is called *indexing.*

interactive Time-dependent (real-time) operations, actions, data processing, or transmission in which pacing is determined by the actions of both the user and the destination processes. For example, a user may list a document one screen at a time until the information being searched for is found. At that point, the transfer may be aborted and a new action is started. See also *batch* and *noninteractive.*

interactive voice response (IVR) A telephone answering system that allows a caller to select and control a desired service by audio command (voice or DTMF signaling for example). Responses from an *IVR* system range from simply connecting the caller to a human operator, to speech synthesized or prerecorded message feedback, to faxback documents, to e-mailed messages, and even to paper feedback via the postal service. Examples of IVR services are as follows:

- Bank account transactions may be accomplished any time simply by calling the *IVR* server, supplying the user's ID number, and specifying the action (report balance, transfer funds, etc.).
- A stolen credit card may be reported at any time of day by calling the credit card company and requesting that service.
- Assistance on installing modems, hard disk drives, or a computer may be obtained from an *IVR* system.
- Other capabilities include the ability to retrieve a selected product's technical specification or other documents.

intercept (1) In telephony, to stop a call directed to an unassigned, improper, disconnected, or restricted number, and to redirect that call to an operator, recording, or vacant-code tone. Also called *intercept call.* **(2)** To gain possession of information intended for others without their consent and, ordinarily, without delaying or preventing the transmission. An *intercept* may be an authorized or unauthorized action (wire tap). **(3)** The acquiring of a signal with the intent of delaying or eliminating receipt of that signal by the intended destination user.

intercept trunk In telephony, a central office termination that may be reached by a call to an out-of-service line, unassigned code, or changed number.

interchangeable number plan area (INPA) Code in the NXX format used as a central office code (NNX format) but can also be used as an NPA code.

interchannel interference In a given channel, the noise (interference) resulting from signals in other channels.

intercharacter interval In asynchronous transmission, the time interval between the end of the stop signal of one character and the beginning of the start signal of the next character. The interval may be any duration; it is not limited to an integral number of bit times. The signal sense is always the same as the sense of the stop element, i.e., "1" or "mark."

intercom (1) A privately owned two-way communication system without a central switchboard, usually limited to a single building or area. Stations may not have the ability to initiate a call but can answer any call. There are three basic classifications of intercoms:

- *Dial:* A user at a station may dial another station directly.
- *Automatic:* The destination phone rings as soon as the origination phone goes off-hook. Sometimes called *ringdown.*
- *Manual:* The user must manually initiate the calling signal at the destination phone, e.g., the boss-secretary buzzer arrangement.

Sometimes called an *interphone.* **(2)** A dedicated voice service within a specified user environment.

interconnect company A company that supplies telecommunications equipment to connect to telephone lines. *Note:* This equipment must be registered with the telephone company before it can be connected to the telephone company's lines.

interconnection (1) The linking together of interoperable systems. **(2)** The linkage used to join two or more communications systems, networks, links, nodes, equipment, circuits, or devices.

interdigit time In telephony, in dial-pulse signaling, the extended "make" interval used to separate and delineate dialed address digits. Most central offices require that the interdigital time be greater than 400 ms (300 ms in Mexico). This assumes 10 pulse per second dialers are being used. Central offices with 20 pulse per second dialing reduces the minimum interdigital time to 200 ms.

With systems using tone signaling, the *interdigital pause* is the period between signaling tone bursts.

Interdomain Routing Protocol (IDRP) A protocol that routes packets between domains (subnetworks under the control of a single organization) in an Internetwork. It uses a distance vector algorithm and is based on the Border Gateway Protocol (BGP), which is used in the TCP/IP suite.

IDRP is also the name of a specific interdomain routing protocol. It is the ISO/OSI standard equivalent to the Internet Exterior Gateway Protocol (EGP).

interexchange carrier (IEC) A company that provides connectivity among local access and transport areas (LATAs), i.e., local exchange carriers (LECs). These companies are allowed to provide long-distance service; they include AT&T, MCI, Sprint, and WilTel.

The acronyms IC and IXC are also sometimes used for *interexchange carrier;* however, use of IXC is discouraged because of its confusion with *interexchange channel* (*IXC*).

interexchange channel (IXC) A communications path among several telephone exchanges.

interexchange customer service center The telephone company's primary point of contact for handling the service needs of all long-distance carriers.

interface (I/F) (1) The shared boundary between systems, devices, or software defined by common physical, electrical, or functional attributes, common signal characteristics, and/or common protocols. The region through which information can be exchanged. Interfaces are frequently categorized as either hardware interfaces or software interfaces.

- *Hardware Interface:* The *physical interface* specifies details such as shape, number of pins/sockets, and arrangement of pins/sockets.

  The *electrical interface* specifies the magnitude and sign of electrical signals, signal duration, and timing. For example, it specifies the voltage level and duration for logical 0 and 1 values. Three types of electrical interface are commonly used: voltage, current loop, and contact closure.

  The *functional interface* specifies the meaning of the signals on each wire. For example, pins 2 and 3 are for transmitting and receiving data on an EIA/RS-232 serial interface.

- *Software Interface:* The *format* of the parameters or variables passed between two programs or program elements, including number of bytes and bit arrangement within the individual bytes.

  Parameter or variable *definition*—the classification of the parameter as to its type (pre-interpreted or raw bit streams).

  Evaluation order—the determination of the order of processing (least significant to most significant or the reverse).

  Housekeeping responsibilities—the determination of which routine, the calling or the called, is to remove the used parameters from system stacks.

(2) The means by which multiple devices or a device and a person interact with each other. That is:

- The hardware (cards, plugs, or other devices) used to move data or control information from one location to another.
- The user interface, where people communicate with software via commands or other devices.

(3) To interconnect two or more entities at a common point or shared boundary.

interface data unit (IDU) A data structure, in the ISO/OSI Reference Model, that is passed between layers, specifically, when an entity at one level provides a service for an entity at a higher level.

interface message processor (IMP) See *router.*

interface standard A standard that describes certain functional characteristics (such as code conversion, line assignments, or protocol compliance) or physical characteristics (such as electrical, mechanical, or optical characteristics) necessary to allow the exchange of information between two or more (usually different) systems or equipment. It may also include operational characteristics and acceptable performance levels.

interference In data transmission, any extraneous energy introduced into a signal transmission path that tends to hinder the reception of the desired signal. The unwanted energy may be from either natural or synthetic sources.

The sources of *interference* depend on the type of signals and the media involved. For example, electrical signals are susceptible to other electrical signals, magnetic fields, jamming, and atmospheric conditions. In contrast, optical signals are relatively impervious to these types of *interference.* The order of interference immunity is optical fiber, coax, shielded twisted pair, twisted pair, balanced unshielded pair, unbalanced pair.

interference filter In optics, a filter that reflects one or more spectral bands or lines and transmits others, while maintaining a nearly zero coefficient of absorption for all wavelengths of interest. *Interference filters* consist of multiple thin layers of dielectric material (and possibly additional metallic layers) having different refractive indices. The filters are inherently wavelength selective due to the interference effects that take place between the incident and reflected waves at the thin-film boundaries.

All four forms of filtering may be implemented with *interference filters:* low-pass, high-pass, band-pass, and band reject.

interframe coding Data compression techniques that consider changes between adjacent frames as well as changes within individual frames. This results in higher compression than intraframe compression alone. Both lossless and lossy interframe coding can be implemented.

Interframe Gap (IFG) The amount of time between successive frames, or packets, in a network transmission, e.g., the time between a request to send (RTS) frame and clear to send (CTS) frame, or between the CTS and data transmission.

For example, on an IEEE 802.3 Ethernet LAN, the minimum *IGF* is 9.6 µs. In the LocalTalk variant of AppleTalk, an *IFG* of 200 µs is normal.

interframe time fill In digital data transmission, a sequence of bits transmitted between consecutive frames. (It does not include bits stuffed within a frame.)

Interior Gateway Protocol (IGP) A routing protocol that governs packet routing within an autonomous network. Two examples of *IGP* are the *Routing Information Protocol* (*RIP*) and the *Open Shortest Path First* (*OSPF*) protocols. The term *gateway* is historical; the current preferred term is *router.*

inter-LATA In telephony, services, circuits, revenues, and functions associated with telecommunications that originate in one LATA and terminate in destinations outside of that LATA (e.g., another LATA). The *inter-LATA* services are provided by an interexchange carrier (IC, IEC).

The 1982 modified final judgment (MFJ) decreed that all local exchange carriers (LECs) use an interexchange carrier (IEC) for *inter-LATA* services. Also written *interLATA.*

interLATA call A call that originates in one LATA and terminates in another. Calls carried by the long-distance carrier.

interleave The arrangement of one sequence of items with another sequence of similar items in an alternating manner. Examples of interleaving include:

- A time division multiplex (TDM) data stream. Several data channels are interleaved onto a single bit stream.
- An error-detection technique wherein individual bits or bytes of the transmit messages are reordered (interleaved) at the transmitter. The receiver then restores the original message sequence. If any burst errors occurred in transmission, they would be redistributed so that the error-detecting algorithm might better be able to detect all errors.
- The ordering of individual sectors on a hard disk drive so that maximum throughput is achieved.
- The alternating of odd and even banks of dynamic random access memory (DRAM) in order to improve speed.

INTERLNK A program in MS DOS 6.x that makes it possible to interconnect two computers (through either the serial or parallel ports) and to share drives and printer ports on the computers. The client computer can access the drives and printers on the server. The connection requires:

- For *serial connections:* three-wire serial cable or seven-wire null-modem cable and a free serial port on each computer.

- For *parallel connections:* a bidirectional parallel cable and free parallel ports on each computer.
- *DOS 6.x* on the server computer and *DOS 3.3* or later on the client.
- The *INTERLNK.EXE* program on the client computer, and an entry in the client's CONFIG.SYS file to load this driver.
- *16 kilobytes* of free memory on the client and *130 kilobytes* on the server.

To start the server, the INTERSVR program is started.

intermediate cross connect (IC) An interconnection point within backbone wiring, e.g., the interconnection between the building entrance facility and the main cross connect.

intermediate distribution frame (IDF) A transitional location for wire routing in a building. The *IDF* is connected to a main distribution frame (MDF) at one end and to end users at the other end. In a multifloor building, each floor is likely to have an *IDF,* partly to ease the difficulty in running large wire bundles vertically in buildings, partly to conserve the total number of wires required, and partly to facilitate the administration of the cable plant. Also called a *local distribution frame (LDF).* See also *main distribution frame (MDF).*

intermediate field region Antenna. The transition region (lying between the near-field region and the far-field region) in which the field strength of an electromagnetic wave is dependent on the inverse distance, inverse square of the distance, and inverse cube of the distance from the antenna. For an antenna that is small compared to the wavelength in question, the intermediate-field region is considered to exist at all distances between 0.1 wavelength and 1.0 wavelength from the antenna. Also called *intermediate field, intermediate zone,* and *transition zone.*

intermediate frequency (IF or if) (1) A frequency to which a carrier frequency is shifted as an intermediate step in transmission or reception. **(2)** Generally, within a particular signal receiver design, that specific frequency to which all received signals are converted for further processing. Because the *intermediate frequency* represents only one channel (rather than the multitude of channels available at the receiver's input), the performance of the signal processing can be optimized. Examples of specific *IF* frequencies include 455 kHz for AM radios, 10.7 MHz for FM radios, and 70 MHz for many communications systems.

As is illustrated in the figure, the *intermediate frequency* is generated by mixing a local oscillator frequency (reference frequency) with the input frequencies. For each input frequency, there are two mixer output frequencies ($|f_{in}+f_{LO}|$ and $|f_{in}-f_{LO}|$). Hence, a *filter* is placed at the mixer output to select the desired input channel frequency.

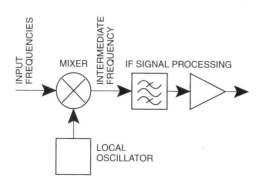

To keep the *intermediate frequency* the same for all input frequencies, the local oscillator frequency must be changed, for example,

FREQUENCY RELATIONSHIPS (MHZ)

Input	Local Oscillator	Intermediate
88.9	78.2	10.7
89.9	88.2	10.7
96.1	85.4	10.7
107.9	97.2	10.7

intermediate office In telephony, a switching system in which trunks are terminated for the purpose of interconnecting other offices.

intermediate system (IS) In the ISO/OSI Reference Model, a network entity that serves as a relay between two or more subnetworks. For example, repeaters, bridges, routers, and X.25 circuits are all *intermediate systems* at the physical, data link, network, and network layers, respectively.

Architecturally, an *intermediate system* uses at most the bottom three layers of the ISO/OSI Reference Model (that is, the network, data link, and physical layers). These are the so-called subnet layers. This is in contrast to an end system (ES), which uses all seven layers of the model. A node is an end system. Also called a *relay open system* in the ISO/OSI Reference Model or *internetworking unit (IWU).*

intermodulation (IM) The production of frequencies not in the original complex input wave to a system, in a nonlinear element of that system. The frequencies correspond to the sum and difference frequencies of all fundamentals and harmonics thereof of the signal passing through the nonlinear element.

intermodulation distortion The generation of spurious frequencies in a communication device or channel output when multiple signals are present at the input. The cause of the generation of the extra frequencies is a nonlinearity in the signal path. The unwanted output frequencies are at the sum and difference of the input frequencies.

For example, if the input frequencies are 1000 Hz and 1400 Hz, the generated frequencies will be 400 Hz and 2400 Hz. It is possible for the generated frequencies to mix with themselves or other input frequencies to generate a plethora of unwanted interference.

intermodulation noise In a transmission path or device, noise generated during modulation and demodulation, which results from nonlinear characteristics in the path or device.

internal bias In a start-stop asynchronous communications system, bias generated locally by the receiver, which has the same effect on the operating margin as bias external to the receiver (applied bias). Bias may be marking or spacing in nature.

internal blocking In switching systems, the unavailability of *paths* in a switching network between a given input port and an available output port.

internal modem A modem capable of being installed *into* a computer after the computer has been built. The modem is fabricated on an expansion board and installed in an available option connector of the computer (usually on the motherboard).

internal traffic In switching systems, traffic originating and terminating within a specified network or switch, e.g., a call originating in a PABX and terminating in the same PABX and not using the public switched network (PSTN). Also called *inside call* and *internal call.*

international access code (IAC) In telephony, that part of the call routing code which informs the local switching system that the call be-

ing placed is an *international direct distance call*. Also called an *international prefix* or *international direct dial (IDD) prefix*. See also *area code, city code, country code, local access code,* and *local number*.

international alphabet number 5 See *IA5*.

International Atomic Time (TAI) The time scale established by the *International Time Bureau* (*BIH*) on the basis of atomic clock data supplied by cooperating institutions. The acronyms *TAI* and *BIH* are derived from literal translation of the official international French names. As of July 1, 1997, TAI time is ahead of Coordinated Universal Time (UTC) by 31 seconds. See also *time*.

international call Any call to a destination outside the national boundaries of the calling party.

H5011101.CDR

INTERNATIONAL ACCESS CODES BY COUNTRY

Country	IAC	Country	IAC	Country	IAC
Albania	00	Guam	011	Oman	00
Algeria	00	Guatemala	0	Pakistan	00
Anguilla	011*	Guinea Rep.	00	Panama	00
Argentina	00	Guyana	001	Peru	00
Australia (Telstra)	0011	Haiti	00	Philippines	00
Australia (OPTUS)	1 0011	Honduras	001	Poland	0.0**
Austria	00	Hong Kong	001	Portugal	07
Bahamas	011*	Hungary	00	Puerto Rico	011*
Bangladesh	00	Iceland	90	Qatar	00
Barbados	011*	India	900	Rwanda Rep.	00
Belgium	00**	Indonesia	00	Saudi Arabia	00
Benin	00	Iran	00	Senegal	00
Bermuda	011*	Iraq	00	Serbia	99
Bosnia	99	Ireland	00	Seychelles	00
Brazil	00	Israel	00	Singapore	005
British Virgin Isls	011*	Italy	00	Slovakia	00
Bulgaria	00	Ivory Coast	00**	Slovenia	99
Burkina-Faso	00	Jamaica	011*	South Africa	09
C.I.S. (old U.S.S.R.)	8** 10	Japan (KDD)	001	Spain	07**
Cameroun	00	Japan (IDC)	0061	Sri Lanka	00
Canada	011*	Jordan	00	St-Kitts	011*
Chile	00	Kenya	00	St-Lucia	011*
China	00	Kuwait	00	St-Vincent & Grenadines	011*
Colombia	90	Korea (South)	001	Sudan	00
Congo Rep.	00	Lebanon	00	Suriname	002
Cook Islands	00	Lesotho	00	Sweden	009**
Costa Rica	00	Liberia	00	Switzerland	00
Croatia Rep.	99	Libya	00	Syria	00
Cyprus	00	Luxembourg	00	Taiwan	002
Czech Republic	00	Macau	00	Thailand	001
Denmark	00	Madagascar	16	Togo	00
Dominica	011*	Malawi	101	Tonga	09
Dominican Republic	011*	Malaysia	007	Trinidad & Tobago	011*
Djibouti	00	Mali	00	Tunisia	00
Egypt	00	Malta	00	Turkey	99
El Salvador	0	Mauritania	00	Turks & Caicos	011*
Ethiopia	00	Mexico	98	United Arab Emirates	00
Faeroe Island	009	Montserrat	011*	UK	00
Falkland Isl	01	Morocco	00**	U.S.A.	011*
Finland	00	Namibia	00	Uruguay	00
France	00	Netherlands	00	U.S. Virgin Isls	011*
Gabon Rep.	00**	Nevis Island	011*	Vatican City	00
Gambia	00	New Guinea (Papua)	31	Venezuela	00
Germany	00	New Zealand	00	Yugoslavia (ex-)	99
Gibraltar	00	Nicaragua	00	Zaire	00
Greece	00	Niger Rep.	00	Zambia	00
Greenland	009	Nigeria	009	Zimbabwe	110
Grenada	011*	Norway	095		

* Part of North American Numbering Plan (NANP).
** Wait for dial tone after code is dialed before proceeding.

International Data Encryption Algorithm (IDEA) An encryption algorithm written by Xuejia Lai and James Massey in 1992. It is a block cipher, which is considered to be the best and most secure available. It operates on 64-bit blocks with a 128-bit key. It is used by the Pretty Good Privacy (PGP) encryption program.

international direct distance dialing (IDDD) The automatic establishing of an international call by means of signaling from the device at the calling party premises.

international gateway A switching office through which all international calls of a region are routed. In the United States carriers (MCI, AT&T, etc.) interface their networks with international telecommunications networks. All U.S. international calls are routed to a gateway switch when the "011" access code is dialed.

international interzone call **(1)** A call to a destination outside of a national or integrated numbering plan area. **(2)** Calls to destinations within the boundaries of an integrated numbering plan area but outside the national boundaries of the caller.

International Mobile Equipment Identity (IMEI) The Electronic Serial Number used in Digital Phones (the Digital equivalent to the ESN—Electronic Serial Number—for analog phones). It is up to the network provider to use the *IMEIs;* some do, some don't. According to the GSM specification, a user "should" be able to find out the *IMEI* on any GSM phone by entering *#06# on the keypad.

international Morse code A system of long and short (dash and dot) signals used in international radio and telegraphy. The international code varies slightly from the American Morse code. See also *Morse code.*

international number In telephony, the combination of digits representing a country code plus the national number. This number consists of an area code, office code, and a line code.

international numbering plan A strategy developed by the ITU-T for assigning telephone numbers around the world. There are subplans, each for different regions of the world.

international operating center In telephony, in World Zone 1, a switching center where telephone operators handle originating and terminating international interzone and intrazone calls.

international originating center In telephony, in World Zone 1, a toll center where telephone operators handle originating interzone calls.

international prefix See *international access code (IAC).*

international record carrier (IRC) A common carrier engaged in providing overseas (international) telecommunications services, e.g., telex, private line service, and alternate voice-data services.

International Reference Version (IRV) That specific variant of the International Alphabet No. 5 (IA5) character-encoding scheme which is identical to the ASCII encoding scheme.

International Standards Organization (ISO) An international standards setting organization. Examples include:

- CIE—International Commission on Illumination
- IEC—International Electrotechnical Commission
- IEEE—Institute of Electrical and Electronics Engineers
- IETF—Internet Engineering Task Force and the Internet Society
- ISO—International Organization for Standardization
- ITU—International Telecommunication Union WWW and gopher
- W3C—World Wide Web Consortium

international switching office A toll office that serves as the gateway for international interzone calls.

International System of Units (SI) See *SI.*

International Telecommunications Union (ITU) An organization originally established in 1865 as the *Union Telegraphique.* In 1947 it became the telecommunications agency of the United Nations and was renamed the *ITU.* Its stated charter is:

1. To develop and promote worldwide standards, procedures, and practices for telecommunication.
2. To promote optimum use of scarce frequency resources, including frequency allocation and radio regulations worldwide.
3. To encourage telecom growth in developing countries.

The CCITT and CCIR are committees under the *ITU.* The CCITT is now called the *ITU-T* (for ITU-Telecommunications standardization sector), and the CCIR is now known as the *ITU-R* (ITU-Radiocommunication sector).

internet In communications, a contraction of <u>INTER</u>connected <u>NET</u>works. *Internet* is a set of communications networks interconnected or joined together by means of *gateways.*

Internet When capitalized, *Internet* refers to a specific collection of networks and gateways that evolved from the ARPAnet. It is referred to as "the Internet" rather than "an internet," which refers to any collection of computer networks.

Internet is now a huge worldwide network and interconnects innumerable smaller networks and their resources via *gateways* and the TCP/IP protocol suite. *Internet* includes many government agencies, colleges and universities, businesses, as well as many commercial networks. Although composed of a multitude of heterogeneous networks internally, the network appears externally to be a single network owing to the nature of the TCP/IP protocol suite. See also *ARPAnet* and *TCP/IP.*

internet address **(1)** A node address in which the address consists of a network number and a host number that are unique for the network. Also called an Internet Protocol (IP) address. IP addresses are normally given in dotted quad form, i.e., in the form 159.3.50.25. **(2)** The AppleTalk address and network number of a socket.

Internet Architecture Board (IAB) A group of invited volunteers that make decisions about Internet standards and manage aspects of the Internet such as address allocation and the development of protocols. Within *IAB* are two task forces: the Internet Engineering Task Force (IETF) and the Internet Research Task Force (IRTF). The IETF is the committee responsible for formulating the Remote Network Monitoring Management Information Base (RMON MIB), which is expected to become the standard for monitoring and reporting network activity in the Internet environment.

The Internet Architecture Board was formerly called the Internet Activities Board (IAB).

Internet Assigned Number Authority (IANA) The group responsible for maintaining standard addressing on the Internet and ensuring that the same identifier values are not assigned to two different entities. Within *IANA* are two subgroups: the Internet Registry (IR) and the Internet Network Information Centers (InterNICs). The IR maintains the database of IP addresses and registered domains.

InterNICs provide documentation, guidelines, assistance, and advice to users. They also provide users with domain-name and IP-address registration forms, maintain File Transfer Protocol (FTP), e-mail accessible archives of Requests for Comments (RFC), and other related documents. This service was operated by the University of Southern California Information Sciences Institute (USC-ISI).

Internet Control Message Protocol (ICMP) A protocol that deals with the problem of router congestion and other causes which prevent it from sending datagrams toward their destination. *ICMP* provides flow control by signaling the preceding hosts to stop sending datagrams until the congested node is again able to accept traffic. Documented in RFC792.

ICMP delivers error and control messages from hosts to message requesters. An *ICMP* test may determine whether a destination is reachable. An *ICMP* echo is also called a *ping*.

Internet draft The current working version of documents of the Internet Engineering Task Force (IETF).

Internet Engineering Steering Group (IESG) In Internet, the executive committee for the Internet Engineering Task Force (IETF).

Internet Engineering Task Force (IETF) A subgroup of the Internet Activities Board (IAB) that concerns itself with solving technical problems on the Internet. The *IETF* is an all-volunteer group that identifies network problems and ultimately proposes solutions to the Internet Architecture Board (IAB) for approval.

For example, the *IETF* is responsible for developing the Remote Network Monitoring Management Information Base (RMON MIB), which will become the standard for monitoring and reporting network activity in the Internet environment.

Internet Group Membership Protocol (IGMP) A protocol used by an IP host to report multicast group memberships to adjacent multicast routers.

Internet Message Access Protocol (IMAP) An e-mail management protocol that improves the administration of messages and allows access to mailboxes on remote mail servers as if they were on the local host. Its features include the ability to download only message headers, create multi-user mailboxes, and build server-based storage folders.

Internet Monthly Report (IMR) A summary of the current state of the Internet, including problems, usage, and reports of the technical committees. It is an online report created for the Internet Architecture Board (IAB) and can be found at:

nis.nsf.net/Internet/newsletter.monthly.report

Individual reports are text files with names such as *imr95-03.txt* (Internet Monthly Report for March of 1995). Reports may be obtained automatically by joining the mailing list or simply by downloading individual reports.

Internet number The dotted quad Internet address.

Internet packet exchange (IPX) Novell's NetWare native LAN communications protocol, which enables communications among servers and clients running on different network nodes.

Internet phone A telephonic link established over an Internet connection. The computers at both ends of the connection must be fast machines, have modems with data rates of 14 400 bps or greater, and have full-duplex sound cards (if both users want to be able to speak simultaneously).

Internet Protocol (IP) At the network layer (equivalent to OSI layer 3) in the Internet architecture, a connectionless protocol that defines the *datagram,* the *addressing* scheme, and *routing* of datagrams to remote hosts. Documented in RFC791 and RFC1577.

Unlike connection-oriented protocols, *IP* does not attempt to establish a connection with the final destination before sending datagrams. Furthermore, it is said to be "unreliable" because it does not provide

any error-correction means. This, it turns out, is not a problem since both are handled by another layer of the protocol hierarchy, specifically, the *transport layer* (equivalent OSI layer 4). Within the *transport layer,* the *Transport Control Protocol* (*TCP*) and the *User Datagram Protocol* (*UDP*) accept data from an application, operate on it, and pass it on the IP network layer. TCP provides a connection-oriented reliable protocol (at the expense of high overhead), while UDP is a faster but connectionless unreliable protocol. *FTP* (*file transfer protocol*) and *SMTP* (*simple mail transport protocol*) use the *TCP protocol,* whereas *SNMP* (*simple network management protocol*) uses the UDP protocol. See also *IPv6* and *TCP/IP*.

Internet protocol (IP) spoofing (**1**) The creation of IP packets with counterfeit (spoofed) IP source addresses. (**2**) A method of attack used by network intruders (crackers) to defeat network security measures, such as authentication based on IP addresses.

- An attack using IP spoofing may lead to unauthorized user access on the targeted system.
- IP spoofing is possible even if no reply packets can reach the attacker.
- A packet-filtering-router firewall may not provide adequate protection against IP spoofing attacks. It is possible to route packets through this type of firewall if the router is not configured to filter incoming packets having source addresses on the local domain.
- A method for preventing IP spoofing problems is to install a filtering router that does not allow incoming packets to have a source address different from the local domain. In addition, outgoing packets should not be allowed to contain a source address different from the local domain, in order to prevent an IP spoofing attack from originating from the local network.

Internet rating system (IRS) A standard formula for rating Internet offerings (such as Web sites and news groups) for violence, language, and sexual content. Software at the user's location may be used to control access based on these ratings.

Internet Registry (IR) A central database that contains the network addresses of machines and ID numbers of autonomous systems (domains) on the Internet. The task of maintaining the *IR* is delegated by the Internet Assigned Numbers Authority (IANA) and is being carried out by the Defense Data Network Network Information Center (DDN NIC).

Internet Relay Chat (IRC) A real-time, multi-user client program that allows users to "chat" with other Internet users *interactively*. The users choose from any of the many channels available at the time. Channels are created and destroyed by the users themselves; consequently, the subjects are varied and fleeting so that an interesting channel may evaporate overnight.

To use the *IRC,* one must join a channel. After joining, the user will see what others are typing. When the user types a line of text and <return>, everyone else on the channel will see that line. Some basic commands include:

/who <#channelname>	Lists who is on the specified channel; including nickname, user name, host, and real name.
/join <#channelname>	Joins a channel
/part	Leaves a channel
/list	Lists ALL channels unless limited by:
[string]	Lists only channels containing "string"
[-min n]	Lists only channels with n or more users
[-max m]	Lists only channels with n or fewer users
/help <topic>	Gets help on commands

/nick	Changes nickname
/whois <nick>	Gives "true identity" of user "nick"
/msg <nick> <ms>	Sends private message to "nick"
/quit	Exits IRC

Online IRC help may be obtained in the #irchelp channel. The original *IRC* was written by Jarkko Oikarinen of Finland in 1988 as a replacement for the UNIX TALK command.

Internet Research Steering Group (IRSG) In the Internet, the group that oversees the Internet Research Task Force (IRTF).

Internet Research Task Force (IRTF) The subgroup of the Internet Architecture Board (IAB) that deals with the long-range problems and aspects of the Internet. It handles problems arising from additional demands on the network that existing protocols, for example, cannot handle. The *IRTF* is the research arm of the Internet. The *IRTF* chairperson is a member of the IAB.

Internet Resource Guide An electronic directory published by the NSF Network Service Center (NNSC) focusing on the research and educational resources available on the Internet. Some of the resources listed include archives, computational resources, library catalogs, and *white pages* (information about users).

Internet router (IR) In an AppleTalk Internetwork, a device that uses network numbering to filter and route packets.

Internet service provider (ISP) A company that sells access to the Internet for a monthly fee (and sometimes, additionally for an hourly connect time charge). The service provider is directly connected to the Internet via high-speed communications links (T-1, E-1, or higher) and generally provides dial-up modem access to subscribers. *ISPs* range in size from small one- or two-person operations to the giants, including MCI, Sprint, and AT&T. Also called *Internet Access Provider* (*IAP*).

Internet shopping mall A text list of Internet-accessible businesses that is periodically posted to USENet *news.answers*. To be included on the list, a business must make information and ordering available by means of the Internet.

Internet Society (ISOC) An organization formed in 1992 that supports the Internet and is the governing body to which IAB reports. *ISOC* activities include:

- Assigning and tracking IP addresses and domain names.
- Generating and promulgating Internet technical standards.
- Publishing a quarterly paper, the *Internet Society News*.
- Resolving problems by examining evolving technology and the changing needs of users.

The *ISOC* is open to anyone willing to pay the nominal dues.

Internet Standard (IS) A specification that has withstood the formal evaluation and testing processes of the IETF; it has proven to be stable and viable, and has been widely implemented. The three stages that lead to an Internet Standard are:

- *Proposed Standard* (*PS*)—a specification that appears robust and is submitted for testing. The specification must be sufficiently detailed and stable to warrant implementation.
- *Draft Standard* (*DS*)—a specification that has been a *Proposed Standard* for at least six months and has been tested in at least two implementations that have interacted with each other.
- *Internet Standard* (*IS*)—a specification that has been a *Draft Standard* for at least four months and has general acceptance as worthy of implementation and use.

See also the list of Internet standards in Appendix G.

Internet Talk Radio A nonprofit service that distributes a "radio show" over the Internet in the form of sound files. The distribution of files is in the form of multicast packets that are delivered in real time via a *multicast backbone* (*MBONE*) protocol. To obtain more information, send an e-mail message to *info@radio.com*. The service was founded by Carl Malamud. See also *IP multicast.*

Internet worm The 1988 incident in which a particular type of virus called a *worm* infected the Internet causing thousands of hosts to crash.

The program was written by Robert Morris, Jr., a graduate student in computer science at Cornell University. The program was written to test certain problems with protocols used to transfer mail. However, as one story goes, "because of a bug, the program got out of hand." Morris was convicted under the Computer Fraud and Abuse Act of 1986 and sentenced to three years probation, 400 hours of community service, and a $10,000 fine.

internetwork A network that consists of two or more smaller (and possibly dissimilar) networks that can communicate with each other, by means of a bridge, router, or gateway. The goal of *internetworking* is to transport data from one user (the source) to another (the destination)—that is, to provide end-to-end connectivity, without the users needing to know anything about intermediate devices.

internetwork connection See *gateway.*

internetworking The process of interconnecting two or more individual networks so as to allow communications among the nodes of all networks. Each network is distinct, with its own addresses, internal protocols, access methods, and administration.

InterNIC An acronym for <u>Inter</u>net <u>N</u>etwork <u>I</u>nformation <u>C</u>enters. A consortium of three organizations (General Atomics, AT&T, and Networking Solutions, Inc.) which provides networking information to the NSFNET community. The areas of service include networking information (General Atomics), directory databases (AT&T), and networking registration (Networking Solutions, Inc.).

internode A communications path originating at one node and terminating at another.

interoffice trunk In telephony, a direct transmission channel connecting two local offices in the same exchange.

interoperability A term that refers to the ability of different systems to exchange information so that it can be processed meaningfully by all systems involved, e.g., between different software products or subsystem elements, hardware devices, or multiple networks.

Interoperability is generally assumed when the networks are homogeneous, that is, when they use the same architecture. When networks are heterogeneous, problems may occur. However, some degree of *interoperability* is almost always possible with data converters or human intervention (although the cost of performance degradation or required equipment may be unacceptably high).

interoperability standard A document that delineates the engineering and technical requirements necessary in the design of devices or systems that enable them to operate effectively together.

interphone See *intercom.*

Interphone A contraction of <u>Inter</u>net <u>phone</u>.

interpolation The mathematical process of constructing a best guess value between two (or more) known values.

The technique is used in motion picture compression where the current video frame is reconstructed (uncompressed) using the differences between it and past and future frames. Also called *forward and backward prediction.*

inter-repeater link (IRL) See *IRL*.

interrupt **(1)** The suspension of a process, such as the operation of a computer program or routine, caused by an event external to that process and in such a way that the original process may later be resumed.

The *interrupt* is a *request for service signal* sent by either hardware or software to the computer. If the computer acts on the *interrupt*, it will suspend operation on the current task and transfer control to another task. It is generally up to the interrupting task to return control to the original task. See also *IRQn*. **(2)** In communications, an action taken by a receiving station that causes the transmitting station to cease the current transmission.

interrupt flag **(1)** One or more particular bits of information in a CPU that a software routine can interrogate to find out which interrupt input line is active. **(2)** One or more bits in a CPU that control the ability of an external interrupt signal to actually interrupt the CPU. During some time-critical operations, it may be desirable to have the CPU complete a process before it is allowed to be interrupted.

interrupt handler A program in a CPU that responds to interrupt requests by executing the appropriate software routine corresponding to the specific interrupt request.

Interrupt handler locations are defined in DOS's interrupt descriptor table (IDT). When an interrupt occurs, DOS looks up the matching handler in the IDT, masks further interrupts, and transfers control to the routine. When an interrupt return call (IRET) is received, the task is complete and the CPU is allowed to continue with its pre-interrupt tasks.

interrupt latency The delay between the application of an external interrupt signal and its servicing by the interrupt routine.

interrupt overhead The total percentage of CPU processing power (time) required to service interrupt requests.

interrupt request (IRQ) A communication channel from computer peripheral devices to the CPU that in effect say, "I need attention." The *IRQ* is issued by the peripheral and is serviced by software in the CPU. See also *IRQn*.

interrupted continuous wave (ICW) Modulation in which there is on-off keying of a continuous wave.

interswitch trunk A single direct transmission channel between switching nodes.

intersymbol interference Energy from within the signal itself emanating from one or more timing intervals that tends to interfere with the signal in another timing interval.

When a square pulse is transferred across a band limited transmission line, the energy of the pulse is spread out in time. This is because the propagation delays for each constituent frequency comprising the pulse are different. If the energy spreading is significant relative to the width of the pulses in the data stream, the energy of one pulse will overflow into another pulse position at the receiver and corrupt the data. See the drawing.

intertoll trunk A transmission channel between two toll offices.

intra- A prefix meaning within or inside an entity.

intra-LATA Within the boundaries of a local access and transport area (LATA).

intra-office call A call between lines of the same central office.

intra-office trunk A single direct transmission channel within a given switching center.

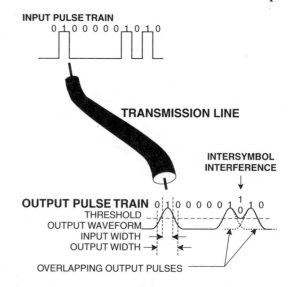

intramodal distortion In fiber optics, distortion resulting from the dispersion of group velocity of a propagating mode. It is the only distortion occurring in a single mode fiber.

intranet A private network that uses Internet technologies, such as the TCP/IP protocol, browser programs, and other World Wide Web (WWW) standards. Intranets are isolated/protected from outside networks via firewalls.

intrinsic joint loss In fiber optics, those losses caused by intrinsic parameter mismatches (e.g., core dimensions, index profiles, numerical aperture, and mode field diameter) when two dissimilar fibers are joined.

intrinsic noise In a transmission path or device, that noise inherent to the path or device and not conditional upon modulation.

INTUG An acronym for **IN**ternational **T**elecommunications **U**sers **G**roup.

invalid Erroneous or unrecognizable because of an error in input.

invalid frame A protocol data unit (PDU) that

- Does not contain an integral number of octets,
- Does not contain at least two octets and a control octet, or
- Is identified by the physical layer or medium access control (MAC) sublayer as containing data bit errors.

inverse discrete cosine transform (DCT $^{-1}$) The converse of the discrete cosine transform, i.e., a mathematical algorithm for converting frequency domain into time domain data.

inverse square law A physical law stating that irradiance, i.e., the power per unit area in the direction of propagation, of a spherical wavefront varies inversely as the square of the distance from the source, assuming there are no losses caused by absorption or scattering. Diffuse and incoherent radiation are similarly affected.

For example, the power radiated from a point source (or from any source at very large distances from the source compared to the size of the source) must spread itself over larger and larger spherical surfaces as the distance from the source increases. That is, each time the distance from an isotropic source is doubled, the power per unit area is reduced by a factor of four.

inverted position In frequency division multiplexing (FDM), a condition of a translated channel in which an increasing signal frequency in the untranslated channel causes a decreasing signal frequency in the translated channel.

inverter In electronics, **(1)** a device that performs the Boolean NOT operation. A logic circuit that reverses (inverts) the input signal, i.e., converts a logic low (or high) at the input to a logic high (or low) at the output. The obsolete synonym *negation circuit* is sometimes seen. **(2)** A device that converts dc into ac.

inverting amplifier An amplifier that produces an output signal equal to the negative of the input signal.

invitation to send (ITS) A character or short message that calls for a station to begin transmission. This process is generally part of a polling scheme. Also called Invitation to Transmit (ITT).

inward trunk A trunk that can be used only for incoming calls, such as an "800" line.

inward WATS service In telephony, a reverse-charge, flat-rate, or measured time direct distance dialing service to a specific directory number. Also called *INWATS*. See also *WATS*.

IOC An abbreviation of InterOffice Channel.

IOCHK A nonmaskable interrupt (NMI) generated by a PC adapter card when it receives a hardware signal or reaches a point where it cannot operate. The card generates the *IOCHK* to the BIOS (Basic Input Output System), the BIOS in turn generates an NMI and an error message, e.g.,

Message	Cause
I/O card parity interrupt at address xxxxh. Type (S)hut off NMI, (R)eboot, other key to continue.	Peripheral card memory failure.
Memory parity error at address xxxxh. Type (S)hut off NMI, (R)eboot, other key to continue.	A chip in main memory failed.
Unexpected HW interrupt at address xxxxh. Type (R)eboot, other key to continue.	Hardware problem. Error not displayed if extended interrupt handler not enabled.
Unexpected SW interrupt at address xxxxh. Type (R)eboot, other key to continue.	Program error(s). Error not displayed if extended interrupt handler not enabled.
Unexpected type 02 interrupt at xxxxh. Type (S)hut off NMI, (R)eboot, other key to continue.	Parity error reported from unknown source.

IODE An abbreviation of Issue Of Data, Ephemeris.

IOF An abbreviation of InterOffice Facility.

ion The generic term for any electrically charged atom or radical, whether it is positively or negatively charged.

ionization The process of stripping electrons off of atoms or molecules. This process creates both negatively charged particles (the electrons) and positively charged particles (everything else). Both charged particles are called *ions*.

ionogram A plot showing paths of ionospherically returned echoes as a function of frequency. The plot has some general tendencies based on time of year, location on the Earth, and time of day. However, due to the fluctuations of the ionosphere, one cannot predict exactly what an *ionogram* will look like at any given instant.

ionosonde A device for generating ionograms. It is a radarlike device that transmits radio signals toward the ionosphere and plots the reflected signal level as a function of frequency.

ionosphere A region in the upper atmosphere (from about 50 to 500 km) where free ions and electrons exist at a sufficient density to affect radio waves, i.e., to reflect or refract them. This ionization is caused by a number of effects, including solar radiation, particle radiation, magnetic and plasma stream interaction, meteors, and cosmic rays. The *ionosphere* is not a single region but is divided into several bands at different altitudes. Each band consists of a central region of relatively dense ionization which tapers off in density both above and below.

The lowest layer is the D region, which, at about 50 to 90 km, is present in the daytime due to solar radiation ($\lambda = 121.6$ nm). The D region generally attenuates (absorbs) radio waves passing through it. However, it is sufficiently reflective to provide an upper bound on VLF and LF transmissions and irregular enough to scatter VHF transmissions. The next layer, at about 90 to 140 km is the regular E region. Because the mean free path of ions at these relatively low altitudes is very short, continuous solar ultraviolet and X-ray radiation are required to maintain its existence. In addition to the regular E region, irregular cloudlike layers of ionization called sporadic E (or ES) produce partial reflections and scattering of frequencies to about 150 MHz. The causes of these ionized regions range from wind shear effects of the E region to disturbed magnetic conditions and the aurora, to the Earth's plasma stream instabilities. The most useful high-frequency reflecting layer (the F region) is at about 280 km at night. During the day, the F region splits into two regions, the F1 and F2 regions, at about 175–220 km and 200–400 km, respectively. At these altitudes, the ionization generally decreases during the nighttime, reaching a minimum just before sunrise. Unlike the D, E, and F1 regions, the F2 region does not directly follow the sun's zenith radiation but rather is more closely related to the average sunspot activity. See also *D, E*, and *F region*.

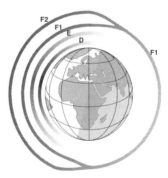

ionospheric absorption The attenuation of certain frequencies of the electromagnetic spectrum as a result of interaction between an electromagnetic wave and free electrons in the ionosphere or water molecules.

ionospheric cross modulation See *Luxemburg effect*.

ionospheric disturbance An increase in the ionization of the D region of the ionosphere, resulting in an increase in absorption of some radio waves. This is caused by an increase in solar activity (sunspots).

ionospheric focusing Variations in the shape of those ionospheric layers that reflect radio waves in such a way that the reflected energy is either focused or defocused at the receiving antenna. The effect is observed as an decrease or increase in path loss.

ionospheric forward scatter (IFS) See *ionospheric scatter*.

ionospheric reflection Of electromagnetic waves propagating in the ionosphere, a redirection of the waves back toward the Earth. (The processing is complex and involves reflection and refraction.) The amount of bending depends on many factors, including:

- The extent of penetration (a function of frequency),
- The angle of incidence, polarization of the wave, and
- The ionospheric conditions (such as the ionization density).

ionospheric scatter The propagation of radio waves by scattering as a result of irregularities or discontinuities in the ionization of the ionosphere. Also called *forward propagation ionospheric scatter* and *ionospheric forward scatter (IFS)*.

ionospheric turbulence Ongoing physical disturbances of the ionosphere that scatter incident electromagnetic waves. The perturbations result in irregularities in the composition and "surface" of the ionosphere that change with time. This causes changes in reflection properties, which in turn, cause changes in skip distance, fading, local intensification, and distortion of incident waves.

IOW An abbreviation of In Other Words.

IP (1) An abbreviation of Internet Protocol. That part of the TCP/IP protocol that governs how packets of information are addressed for delivery throughout the Internet network. **(2)** An abbreviation of Intelligent Peripheral. **(3)** An abbreviation of Information Provider. **(4)** An abbreviation of Intermediate Point. **(5)** An abbreviation of Intellectual Property. **(6)** An abbreviation of Interface Processor.

IP address In Internet, the numeric address created from a *Fully Qualified Domain Name (FQDN)*. The format of the *IP address* (written in *dotted decimal notation*) is xxx.xxx.xxx.xxx where x can be any digit and xxx is less than 256. (At the machine level, the address is actually a 32-bit number unique to each system on the network.)

Users address host machines by names (as defined by the Domain Name Server *DNS*) which is not only easier for users to remember, but easier for administrators to move the physical locations of a host. To move a host to a different machine, the network administrator need only change the database that links (or maps) host names to IP addresses. Also called the *dot address, dotted decimal, dotted digit,* or *dotted quad.*

IP Address Class The five classifications of network addresses that have been defined for use on the Internet are:

Class A: Reserved for very large networks (i.e., networks with a large number of nodes). The 32-bit address word always starts with "0b," and uses 7 bits for Networks and 24 bits for Nodes.

Class B: Allocated for medium-size networks, such as networks that span a large college campus. The 32-bit address word always starts with "10b," and uses 14 bits for Networks and 16 bits for Nodes. This address class is most frequently used for local area networks (LANs), particularly if they use subnetting.

Class C: Used for small networks (those with no more than 255 nodes). The 32-bit address word starts with "110b," and allocates 21 bits for Networks and only 8 bits for Nodes.

Class D: Defined for special multicast purposes, the 32-bit address word starts with "1110b" and allocates the remaining 28 bits for multicast addressing, an address in which a group of targets are specified.

Class E: Reserved "for experimental use" and cannot be guaranteed to be unique. The first 4 bits of this type of address are always "1111b."

Two particular network addresses, "0b" and "1111b," are reserved. Network address "0b" is reserved for the originating entity (network or host), and address 255 ("1111b") is used for broadcasts.

IP datagram The basic unit of information passed across a TCP/IP Internet.

IP hijacking A method of attacking a system in which an active session is intercepted and taken over by the attacker.

ADDRESS CLASS	ADDRESS BIT ALLOCATION		
A	0 NETWORKS	HOST	
	7 BITS / 128 Networks	24 BITS / 16 777 216 Nodes / Network	
B	1 0 NETWORKS	HOST	
	14 BITS / 16 384 Networks	16 BITS / 65 536 Nodes / Network	
C	1 1 0 NETWORKS		HOST
	21 BITS / 2 097 152 Networks		8 BITS / 256 Nodes / Network
D	1 1 1 0 MULTICAST ADDRESS		
	28 BITS / 268 435 456 Possible Address Combinations		
E	1 1 1 1 EXPERIMENTAL		
	28 BITS / 268 435 456 Possible Address Combinations		

IP multicast A routing technique that provides an application on one network node with the ability to "simultaneously" communicate with many other nodes. When multiple nodes on the network can communicate with various other nodes, applications such as collaborative computing and desktop conferencing become possible.

With *IP multicast,* individual nodes dynamically register for receipt of different types of *multicast traffic.* Each node registers itself as a member of selected multicast groups through the use of the Internet Group Membership Protocol (IGMP). Routers keep track of these group registrations dynamically and maintain routing tables (trees) which chart the interconnection paths of each sender to all receivers. When a router receives traffic for a particular group, it uses the tree to distribute the message to appropriate receivers. Protocols that provide this multicast tree-building capability include *Distance Vector Multicast Routing Protocol (DVMRP), Multicast Open Shortest Path First (MOSPF), Protocol Independent Multicast (PIM),* and Apple Computer's *Simple Multicast Routing Protocol (SMRP).*

IP router A router designed to work with the Internet Protocol (IP) and to send and receive datagrams. The datagrams may be either other routers or Internet host computers.

IP spoofing See *Internet protocol (IP) spoofing.*

IP subnet address A subnet is a portion of a network or an internetwork that can be viewed, from the outside, as a single element. An IP address with subnetting contains three types of information: network address, subnet address, and host address. The subnet address can be identified by ANDing the IP address with a mask which has a bit pattern that cancels unwanted bits and passes the remaining bits of interest. RFC791 defines subnet addressing for the Internet.

IPC (1) An abbreviation of InterProcess Communications. **(2)** An abbreviation of Information Processing Center.

IPDS An abbreviation of Intelligent Printer Data Stream.

Iphone An acronym for Internet phone.

IPM (1) An abbreviation of Interruptions (or Impulses) Per Minute. **(2)** An abbreviation of Internal Polarization Modulation. **(3)** An abbreviation of Interference Prediction Model.

IPng An acronym for Internet Protocol Next Generation. A protocol intended to supersede the current IPv4 protocol. RFC1550 is a white paper discussing IPng. Also written *IP(ng).* See also *IPv6.*

IPNS An abbreviation of International Private Network Service. Essentially an international private line with graded data rates from 9.6 kbps to T-1 or E-1 rates.

IPS An abbreviation of Internet Protocol Suite.

IPv4 An abbreviation of Internet Protocol version 4.

IPv6 An abbreviation of Internet Protocol version 6. The Internet Engineering Task Force (IETF) specification on the Internet Protocol succeeding IPv4. Information is available in RFCs 1752, 1809, 1881–1888, 1924, 1933, 1970–1972, 2019, 2030. Formerly called *IP next generation* (*IPng*).

COMPARISON OF IPV4 AND IPV6

	IPv4	IPv6
Source address	32 bit	128 bit
Destination address	32 bit	128 bit
Routing	unicast	anycast
Header	20 octets	40 octets
Fields in header	12	8

40 OCTET HEADER STRUCTURE

Ver	Priority	Flow Lable		
Payload Length			Next Header	Hop Limit
Source Address				
Destination Address				

IPX An abbreviation of Internetwork Packet eXchange. The Novell NetWare transmission protocol similar to the Internet Protocol (IP). It provides addressing that allows multiple LANs to communicate with each other. In *IPX,* both external and internal network numbers are assigned.

The external network number is a unique hexadecimal value associated with a network or network cable segment. The value may be from 1 to 8 hexadecimal digits (up to 4 bytes) and is assigned arbitrarily. The *IPX* internal network number is also a hexadecimal number and uniquely identifies an individual file server. This value can also be from 1 to 8 hexadecimal digits, and it is assigned arbitrarily to the server during the installation of the networking software. An Internetwork address is a 12-byte address broken into three subparts. The first part (4 bytes) is the external network number; the middle part (6 bytes) is the node number; and the third part (2 bytes) is the socket number, with which a particular device or process is associated. *IPX* Internetwork addresses are generally represented as hexadecimal values, so a complete address may have as many as 24 digits associated with it.

IPX/SPX An abbreviation from Internetwork Packet eXchange/Sequenced Packet eXchange. Transport protocols used in Novell's NetWare networks.

IQ An abbreviation of In-phase-Quadrature. See *Costas loop.*

IR (1) An abbreviation of InfraRed. (2) An abbreviation of Internet Registry. (3) An abbreviation of Internet Router.

IRC (1) An abbreviation of Internet Relay Chat. (2) An abbreviation of International Record Carrier.

IREQ An abbreviation of Interrupt REQuest. The interrupt request line from a PC Card defined by the Personal Computer Memory Card International Association (PCMCIA).

IRG An abbreviation of Inter-Record Gap.

iridium An element; number 77 in the periodic table.

Iridium Motorola's personal communications system that used a swarm of 77 low Earth-orbiting satellites (LEOS) until 1994 when the number was reduced to 66. In the system, users carry a small telephone similar to a cellular telephone which communicates with one of the satellites rather than a cellular transceiver. The call is then routed from the satellite to a groundstation, into the switched network, and on to the destination user (possible through another satellite link). The system's advantage is worldwide connectivity: The system can contact the portable telephone anywhere in the world.

IRL An abbreviation of Inter-Repeater Link. An Ethernet segment joining two repeaters and not containing network stations.

IRQn An abbreviation of Interrupt ReQuest line number "n." The IRQ lines are hardware connections between a device and the CPU. The device can interrupt the CPU processes by sending a signal on the IRQ line. There are 8 IRQ lines on an XT (numbered 0–7) and 16 on AT-type machines (0–15). On some devices the IRQ selection is fixed, and on other devices it is assignable by either a hardware jumper or software selection. See also *interrupt.*

COMMON IRQ ASSIGNMENTS

IRQ	Hardware Use
0	Timer
1	Keyboard
2	Slave interrupt (I/O channel)
3	Serial port COM2 or COM4
4	Serial port COM1 or COM3
5	Parallel port LPT2, Available for: Network . . .
6	Floppy diskette controller
7	Parallel port LPT1
8	Real time clock
9	Software redirected to IRQ2
10	Available for: Network, SCSI controller
11	Available for: Network . . .
12	PS/2 mouse, Available for: Network . . .
13	Math coprocessor
14	Hard disk controller
15	Available for: Network . . .

irradiance In fiber optics, radiant power incident per unit area upon a surface, expressed in watts per square meter (or joules per square meter). Formerly called *power density.*

irradiation The product of irradiance and time, i.e., the time integral of irradiance.

For example, an *irradiation* of 125 J/m^2 (joules per square meter) is obtained when an irradiance of 25 W/m^2 (watts per square meter) is continuously incident for 5 seconds.

irrecoverable error An error that makes recovery impossible without the use of error-recovery methods external to the process. See also *recoverable error.*

IRS An abbreviation of Internet Rating System.

IRSG An abbreviation of Internet Research Steering Group.

IRT An abbreviation of Interrupted Ring Tone.

IRTF An abbreviation of Internet Research Task Force.

IRV An abbreviation of International Reference Version.

IS (1) An abbreviation of Information Systems. (2) An abbreviation of Internet Standard. (3) An abbreviation of Intermediate Standard. (3) An abbreviation of Information Separator. A control character used to separate and qualify data logically. The specific definition is determined by the application in which it is being used.

IS-IS An abbreviation of Intermediate System to Intermediate System. An ISO/OSI link state hierarchical routing protocol based on the DECnet Phase V routing protocol.

IS-n An abbreviation for TIA/EIA Interim Standard number n.

- **IS-3**—the original analog cellular standard, now replaced by the ANSI EIA/TIA-553 standard and interim standard IS-91.
- **IS-41**—an intersystem protocol for *roaming* within the USA. It describes how services should handover between operators.
- **IS-54**—a dual mode analog/digital (AMPS/TDMA) cellular service used in North America. In the digital mode, it places three conversations into one cellular channel using Time Division Multiple Access (TDMA) technology.
- **IS-88**—a narrowband analog cellular system developed by Motorola that squeezes three conversations into one 30 kHz cellular channel using analog frequency division multiplexing (FDM). Now incorporated into IS-91. Also called NAMPS.
- **IS-91**—a version of the analog cellular standard, incorporating the functionality of IS-88 (narrowband analog) and IS-94 as well as authentication and PCS band operation.
- **IS-94**—a standard for in-building operation of analog cellular systems using extremely low power. IS-94 is now incorporated in IS-91.
- **IS-95**—a dual-mode CDMA/AMPS digital cellular standard based on the CDMA system pioneered by Qualcomm Inc. IS-95 has a capacity 10–20 times greater that a purely analog system, such as AMPS. This is accomplished by combining 30 kHz cellular channels into a single 1.25 MHz channel and using code division multiplexing to combine and recover the individual conversations.
- **IS-136**—an enhancement to IS-54 TDMA that includes a more advanced control channel (known as the *digital control channel* (DCCH), to distinguish it from the *analog control channel,* which although less sophisticated, is still digital!). Also called Digital AMPS (DAMPS) and North American Digital Cellular (NADC).
- **IS-634**—a standard for 800 MHz cellular base-station to switch interface. Supports CDMA.
- **IS-651**—a standard for an open interface between the PCS switching center and the radio base-station subsystem in a PCS network. Supports both GSM and CDMA.
- **IS-652**—a PCN-to-PCN intersystem operations standard based on DCS 1900.
- **IS-661**—a patented GSM-based composite CDMA/TDMA technology in which adjacent cells used different frequency division multiplex (FDM) channels in a minimum frequency reuse architecture. It uses a unique method of encryption to thwart eavesdropping and cloning. Developed by Omnipoint Corporation.

See also *AMPS, CDMA, E-TDMA,* and *GSM.*

ISA An abbreviation of Industry Standard Architecture. The 8/16-bit option card communication bus architecture originally developed by IBM and now standard in almost all PCs that use Intel's 80X86 chips. See also *EISA, MicroChannel, Peripheral Component Interconnect (PCI),* and *VESA.*

ISAKMP An abbreviation of Internet Security Association Key Management Protocol.

ISAM An acronym for Indexed Sequential Access Method.

isarithmic flow control A token flow-control method in which a token is circulated around an idle network. A station wishing to transmit captures and destroys the token. After the transmission is completed, the station recreates and launches a new token.

ISB An abbreviation of Independent Sideband.

ISC An abbreviation of International Switching Center.

ISDN An abbreviation of Integrated Services Digital Network. *ISDN,* conceived in 1972 by the CCITT (now the ITU-T), is a completely digital communications Public Switched Telephone Network (PSTN) network that uses the existing subscriber telephone lines, a digital switching and communications network, and special terminal equipment at the subscriber premises. There are two versions of *ISDN:* the original now called narrow band ISDN (*N-ISDN*) and broadband ISDN (*B-ISDN*).

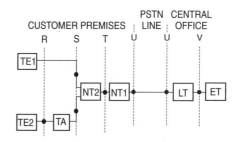

ISDN allows a subscriber to connect a digital device (TE1 or TE2 in the figure) directly to the network (PSTN line) without the need for an analog modem. If the equipment is not *ISDN* compatible, a terminal adapter (TA) is required. NT1 and NT2 are network termination modules; they interface the physical line and address the network protocols. At the exchange, the line terminator (LT) and exchange terminator (ET) perform similar functions. *N-ISDN* uses the integrated digital network (IDN) to provide switching and routing. It provides direct switching and transmission capability for all forms of information transfer, such as voice, computer data, and facsimile. All of the interfaces (R, S, T, U and V) must be able to operate at 192 kbps since they carry two 64-kbps voice and data bearer (or "B") channels, one 16-kbps data (or "D") channel, and 48 kbps of framing and maintenance information in a time division multiplex (TDM) arrangement. *Basic Rate Interface* (*BRI*). A 144-kbps subscriber link divided into two 64-kbps bearer channels (voice or data channels) and one 16-kbps delta channel (data and control channel). Also called 2B+D. *Primary Rate Interface* (*PRI*). Either a 1.544-Mbps link (T-1), in North America, divided into 23 64-kbps "B" channels (voice/data channels) and one "D" channel (data/control channel) or 23B+D. Or a 2048-Mbps link (E-1) divided into 30 64-kbps "B" channels and a "D" channel (30B+D), the ITU-T standard. Also called *ISDN-30.*

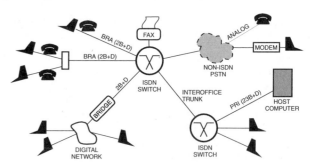

B-ISDN, with a transmission capacity of up to 2.488 Gbps, is designed around the asynchronous transfer mode (ATM) network and the fiber optic synchronous digital hierarchy (SDH). It can offer end users data rates of 155.52 Mbps, 622.08 Mbps (and 1.244 Gbps or 2.488 Gbps).

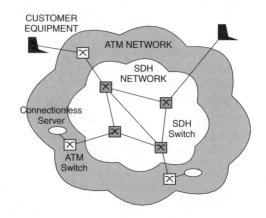

The differences between *N-ISDN* and *B-ISDN* are summarized in the following chart:

N-ISDN / B-ISDN DIFFERENCES

Property	N-ISDN	B-ISDN
Transmission	Uses existing copper distribution pairs	Uses optical fiber
Switching	Circuit switched on B channels packet switched on D channel	Packet switch only
Data rate	Pre-assigned channels with data rates of 2 × 65 kbps + 16 kbps	Virtual channels with unspecified rates (limited to user to network rates of 155.52 Mbps or 622.08 Mbps)

See also *E-1, NT1, NT2, T-1, TA, TE1,* and *TE2.*

ISDN-30 ISDN on an E-1 line, that is, the 30B+D form of ISDN. See also *ISDN.*

ISDN adapter A device attached to a computer that allows users to communicate over ISDN networks. The device is similar to a modem in that it converts the digital data of the computer to a form suitable for transmission over a particular type of network. The device may be an interal PC card or an external device.

ISDN standards In addition to the ITU-T I.n recommendations, many others may come into play. For example, Bellcore's standards include:

TR268—establishes the protocol between caller and network.

TR317—defines the protocols for interfacing among local exchange carriers (LECs) intraLATA switches.

TR394—defines the protocols for operating interLATA switching and networking.

TR444/TR448—defines protocols that allow ISDN on SS7 networks.

Another standard includes the ISO 8877 and the Electronics Industry Association's **T-568,** which define the pin assignments for the 8-termination modular connector.

ISG An abbreviation of Incoming Service Group. A synonym for a *hunt group.*

ISI An abbreviation of InterSymbol Interference.

ISM (1) An abbreviation of Industrial, Scientific, and Medical. Prior to 1985, a set of three frequencies in the United States allocated for industrial, scientific, and medical uses. Now allocated for unlicensed spread spectrum wireless "radio LAN." (2) An abbreviation of ISDN Subscriber Module.

ISM FREQUENCY ALLOCATIONS

Specification	Frequency Allocated
ISM	U.S. 902–928 MHz
ISM (FCC 15.247)	U.S. 2.4000–2.4835 GHz
ISM (FCC 15.247)	U.S. 5.725–5.850 GHz
ISM (CEPT T/R10-01)	Europe 2.400–2.500 GHz

ISN (1) An abbreviation of Information Systems Network. A high-speed switching network from AT&T. It can handle both voice and data transmission, and it can connect to many popular networks, including Ethernet and SNA-based mainframes. (2) An abbreviation of Internet Society News. The official newsletter of the Internet Society (ISOC).

ISO (1) Frequently listed as an acronym from the International Standards Organization, it is in reality not an acronym for the organization in any language. It is a wordplay based on the English initials and the Greek-derived prefix iso- meaning same.

The organization is properly known as the *International Organization for Standards. ISO,* founded in 1946, is an international association of standards-setting organizations dealing mainly with information technology. It is associated with the United Nations. Members (about 90 nations strong) are primarily national standard-making organizations, such as the American National Standards Institute (ANSI), the Association Française de Normalisation (AFNOR), the British Standards Institution (BSI), the Deutsche Institut für Normung (DIN), and so on. (2) An abbreviation of In Search Of.

ISO 7776 The international standard that describes X.25 High Level Data Link Control Procedures.

ISO 8208 The international standard that describes X.25 packet level protocol for DTE.

ISO 8877 The international standard that describes the Q interface connector and its pin arrangement for the ISDN Basic Access Interface at reference points S and T. It also describes the eight-pin modular connector used in the application.

The inner four contacts are used for transmit and receive signals, and the outer four contacts are for power. The assignments are the same as the EIA T-568 standard.

ISO 8879 The international standard that describes the Standard Generalized Markup Language (SGML).

ISO 9000 A set of generic standards that provide quality assurance requirements and quality management guidelines. It has become a mandatory requirement for products (such as telecommunications and medical equipment) sold in Europe and elsewhere.

ISO 9001 A rigorous international standard addressing a company's design, development, production, installation, documentation, and service procedures. Compliance with the standard does not guarantee acceptance in Europe where ISO 9000 registration is widely required. But without it chances of a sale are significantly reduced.

ISO 9660 A format adopted by the International Standards Organization (ISO) for organizing and placing data onto a CD-ROM. Most CD-ROM drives now come with drivers that allow one to read ISO 9660 discs. See also *Yellow Book.*

ISO/OSI reference model See *OSI.*

ISOC An acronym for the Internet SOCiety.

isochronous From Greek meaning equal (iso) time (chronous). *Isochronous* refers to a characteristic, whereas synchronous indicates a relationship. (1) Describes a signal in which the time intervals of consecutive significant events or entities are of equal duration or are integer multiples of the shortest duration. (2) Pertaining to data transmission in which corresponding significant instants of two or more sequential signals have a constant phase relationship. See also *anisochronous, asynchronous, heterochronous, homochronous, plesiochronous,* and *synchronous.*

isochronous demodulation Demodulation in which the time interval separating any two significant instants is equal to the unit interval or a multiple of the unit interval.

isochronous distortion (1) A measure of the total distortion on a pseudorandom serial bit stream—including bias distortion and jitter. *Isochronous distortion* (D_I) is defined mathematically by:

$$D_I = \frac{T_I}{B} \times 100\%$$

Where B is the ideal bit time and T_I is the total spread of data transitions, that is,

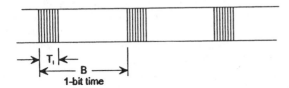

(2) The difference between the measured modulation rate and the theoretical modulation rate in a digital system.

isochronous media access control (IMAC) In the FDDI II architecture, an element of the architecture's *media access control* (*MAC*) layer that can handle time-dependent data (such as voice or video) received through a **circuit switched multiplexer.** *IMAC* is in contrast to the ordinary MAC component, which operates as a packet switched multiplexer.

isochronous modulation Modulation in which the time interval separating any two significant instants is equal to the unit interval or a multiple of the unit interval.

ISODE An acronym for Internationl Standards Organization Development Environment (pronounced "I sew dee ee"). An implementation of the higher layers of the ISO/OSI Reference Model, which enables them to operate in a TCP/IP network.

IsoENET A protocol developed by National Semiconductor and IBM, which adds isochronous services to an established Ethernet without degrading normal Ethernet traffic. It allocates a high-speed isochronous service channel (approximately 6 Mbps) that can be subdivided into 96 channels of 64 kbps each. These subchannels have the capacity to handle constant bit rate, isochronous data with low delays. Also called *ISOEthernet.*

isolated ringing In telephony, a method of ringing a particular telephone of a party line.

For example, the dc bias on the telephone line can be used to select one of two stations. That is, a diode is placed in series with the equipment at each station: forward bias at one station and reverse biased at the other. When ringing is applied, only the station with the forward-biased diode will ring. To ring the other station, the dc polarity to the line is reversed, which reverses the sense of the diode switches. (The forward-biased diode becomes reverse biased, and the diode with reverse bias becomes forward biased.) Again, when ringing is applied, only the forward-biased diode will allow the ringing signal to pass.

isotropic (1) Pertaining to an entity with homogeneous properties, i.e., properties such as density, electrical conductivity, electric permitivity, magnetic permeability, or refractive index that do not vary with distance or direction. (2) Pertaining to an entity with magnetic, electrical, or electromagnetic properties that do not vary with the direction of static or propagating magnetic, electrical, or electromagnetic fields within the material.

isotropic gain See *absolute gain* (1).

isotropic magnet A magnetic material having no preferred axis of magnetic characteristic. Compare with *ansiotropic magnet.*

isotropic radiator A hypothetical radiator or antenna that radiates (transmits) or receives equally well in all directions (nondirectional). *Isotropic radiators* do not exist physically but represent convenient reference antennas for expressing the directional properties of physical antennas. Also called *isotropic antenna.*

ISP An abbreviation of Internet Service Provider.

ISPBX An abbreviation of ISDN PBX.

ISPT An abbreviation of Istituto Superiore delle Poste e delle Telecomunicazioni. Italy's Superior Institute for Post and Telecommunications.

issue of data, ephemeris (IODE) In the Global Positioning System (GPS), it is the serial number of the ephemeris update information. New information is issued periodically (generally hourly) and is used to determine the satellite location. In differential GPS, the information is vital to maintaining the best possible accuracy.

ISTO An abbreviation of Information Systems Techniques Office, the group within Defense Advanced Research Projects Agency (DARPA) responsible for the ARPAnet. *ISTO* was formerly the Information Processing Techniques Office (IPTO).

ISU An abbreviation of Integrated Service Unit. In digital telephone services, a device that consists of a channel service unit (CSU) and a digital service unit (DSU), and replaces a modem on a digital data service (DDS) line.

ISUP An abbreviation of Integrated Services Digital Network User Part.

ISV An abbreviation of Independent Software Vendor.

IT An abbreviation of Information Technology.

ITA1 An abbreviation of International Telegraph Alphabet number 1. The *Baudot code.* See also *ASCII, Baudot code,* and *teleprinter code.*

ITA2 An abbreviation of International Telegraph Alphabet number 2, a five-level code. See also *ASCII, Baudot code,* and *teleprinter code.*

ITA5 An abbreviation of International Telegraph Alphabet number 5. An alphabet based on 7 bit strings, which yields 128 unique symbols to encode upper- and lower-case letters, 10 decimal numerals, symbols, diacritical marks, data delimiters, transmission control characters, and 12 unassigned symbols. The unassigned symbols are open for use in a given country to satisfy their unique requirements (e.g., monetary symbols and diacritical marks). An eighth bit is also defined for use as a parity bit. The standard is published in ITU-T V.3, as ISO 646, and in the United States as the ANSI ASCII code. See also *ASCII.*

ITAR An acronym from International Traffic in Arms Regulations.

ITB (1) An abbreviation of InTermediate Block character. A transmission control character that terminates an intermediate block. Generally, a block check character (BCC) follows the *ITB.* (2) An abbreviation of Intermediate Text Block. (3) An abbreviation of Intermediate Transmission Block.

ITC (1) An abbreviation of Independent Telephone Company. (2) An abbreviation of International Telegraph Congress. (3) An abbreviation of Intermediate Toll Center. (4) Japan's Telecommunications Technology Committee. (5) An abbreviation of InTerCept. (6) An abbreviation of Inter-Task Communication.

ITS An abbreviation of Institute for Telecommunications Science.

ITSO An abbreviation of the International Telecommunications Satellite Organization.

ITT (1) An abbreviation of <u>I</u>nvitation <u>T</u>o <u>T</u>ransmit. In an ARCnet network architecture, the token frame. (2) An abbreviation of <u>I</u>nternational <u>T</u>elephone and <u>T</u>elegraph Company. Formerly the largest telephone company outside the United States. After selling all of their telecommunications interests to the French company Alcatel, the company name was legally changed to ITT with no meaning.

ITU An abbreviation of <u>I</u>nternational <u>T</u>elecommunication <u>U</u>nion.

ITU-R An abbreviation of <u>I</u>nternational <u>T</u>elecommunications <u>U</u>nion <u>R</u>adiocommunications Sector. It has responsiblities for studying technical issues related to radiocommunications, and it has some regulatory powers. Previously called the *CCIR.*

ITU-T An abbreviation of <u>I</u>nternational <u>T</u>elecommunication <u>U</u>nion <u>T</u>elecommunication Standardization Sector. It is responsible for studying technical, operating, and tariff questions and issuing recommendations on them, with the goal of standardizing telecommunications worldwide.

It combines the standards-setting activities of the predecessor organizations formerly called the *International Telegraph and Telephone Consultative Committee (CCITT)* and the *International Radio Consultative Committee (CCIR).*

ITU-T Recommendations A family of specifications covering a significant portion of telecommunications activities, requirements, operations, and functions. The following is the list of categories available. Within each category are the actual specifications. The number in parentheses indicates the number of specifications in each category (at the time of this writing), and the spelling is as presented by the ITU itself.

Series A Recommendations—(12)

Organization of the work of the ITU-T

Series B Recommendations—(13)

Means of expression (definitions, symbols, classification)

Series C Recommendations—(4)

General telecommunication statistics

Series D Recommendations—(125)

General tariff principles

Series E Recommendations—(169)

Overall network operation (numbering, routing, network management, operational performance and traffic engineering); telephone service, service operation and human factors

Series F Recommendations—(123)

Telecommunication services other than telephone (operations, quality of service, service definitions, and human factors)

Series G Recommendations—(214)

Transmission systems and media, digital systems, and networks

Series H Recommendations—(19)

Line transmission of nontelephone signals

Series I Recommendations—(122)

Integrated Services Digital Networks (ISDN)

Series J Recommendations—(35)

Transmission of sound program and television signals

Series K Recommendations—(35)

Protection against interference

Series L Recommendations—(18)

Construction, installation, and protection of cable and other elements of outside plant

Series M Recommendations—(64)

Maintenance: international transmission systems, telephone circuits, telegraphy, facsimile, and leased circuits

Series N Recommendations—(9)

Maintenance: international sound program and television transmission circuits

Series O Recommendations—(13)

Specifications of measuring equipment

Series P Recommendations—(44)

Telephone transmission quality, telephone installations, local line networks

Series Q Recommendations—(450)

Switching and Signalling

Series R Recommendations—(70)

Telegraph transmission

Series S Recommendations—(29)

Telegraph services terminal equipment

Series T Recommendations—(86)

Terminal characteristics and higher layer protocols for telematic services, document architecture

Series U Recommendations—(47)

Telegraph switching

Series V Recommendations—(58)

Data communication over the telephone network

Series X Recommendations—(215)

Data networks and open system communication

Series Z Recommendations—(29)

Programming languages

ITU-TSS An abbreviation of the <u>I</u>nternational <u>T</u>elecommunication <u>U</u>nion—<u>T</u>elecommunication <u>S</u>tandardization <u>S</u>ector or ITU-T for short. See also *ITU-T.*

IVD An abbreviation of <u>I</u>ntegrated <u>V</u>oice and <u>D</u>ata. See also *IEEE 802.9.*

IVDT An abbreviation of <u>I</u>ntegrated <u>V</u>oice <u>D</u>ata <u>T</u>erminal.

IVR An abbreviation of <u>I</u>nteractive <u>V</u>oice <u>R</u>esponse system.

IWBNI An abbreviation of <u>I</u>t <u>W</u>ould <u>B</u>e <u>N</u>ice <u>I</u>f . . .

IWU An abbreviation of <u>I</u>nternet<u>W</u>orking <u>U</u>nit. An intermediate system (IS). See also *intermediate system (IS)* and *IS-IS.*

IX An abbreviation of <u>I</u>nter<u>X</u>change. It refers to any service that crosses an exchange boundary.

IXC (1) A depreciated abbreviation of <u>I</u>nter<u>X</u>change <u>C</u>arrier. The preferred abbreviation is *IEC.* (2) An abbreviation of <u>I</u>nter<u>X</u>change <u>C</u>hannel.

IYFEG An abbreviation of <u>I</u>nsert <u>Y</u>our <u>F</u>avorite <u>E</u>thnic <u>G</u>roup.

IYKWIM An abbreviation of <u>I</u>f <u>Y</u>ou <u>K</u>now <u>W</u>hat <u>I</u> <u>M</u>ean.

IYKWIMAITYD An abbreviation of <u>I</u>f <u>Y</u>ou <u>K</u>now <u>W</u>hat <u>I</u> <u>M</u>ean <u>A</u>nd <u>I</u> <u>T</u>hink <u>Y</u>ou <u>D</u>o.

J

j Primarily engineering. The symbol for the imaginary number $\sqrt{-1}$. See also *i*.

J The symbol for *joule,* the SI unit energy.

J.n recommendations An ITU-T series of specifications concerning the transmission of sound and television. See Appendix E for a complete list.

jabber A network error condition whereby a meaningless transmission (garbage) is injected continually by a node because of a network failure (a faulty transceiver for example). A *jabber packet* is larger than the maximum size (1518 bytes for Ethernet) and contains a bad cyclic redundancy check (CRC) value.

Although long frames may exceed the maximum packet length, they can be distinguished from *jabber packets* because long frames have a valid CRC value.

jabber detector An electronic circuit, in a network that uses the carrier sense multiple access with collision detection (CSMA/CD) media access method, which helps prevent a node from transmitting constantly (e.g., if the node is malfunctioning).

jabber packet In an Ethernet network, a meaningless transmission generated by a network node because of a network malfunction (such as a faulty transceiver). A *jabber packet* is larger than the maximum size (1518 bytes for Ethernet) and contains a bad CRC value. Long frames, on the other hand, exceed the maximum frame length but have a valid CRC value.

jabod An acronym from Just A Bunch Of Disks. A low-end data storage system that uses multiple disks. However, they are not arranged to provide the data protection that higher systems such as RAID (redundant array of inexpensive disks) systems provide. Further, the system does not appear as a single system to the user.

jack A female electrical connecting device designed as a receptacle for a mating plug. Although generally panel mounted, a *jack* may also be found as a cable termination.

jacket See *sheath.*

jam signal A signal sent by a data station to inform the other stations on a network that they must not transmit. Two different uses of the *jam signal* are:

- In carrier sense multiple access with collision detection (CSMA/CD) networks, the *jam signal* indicates that a *collision* has occurred. The signal is produced by the node(s) detecting a collision in an network. It is generated to ensure that all nodes on the network detect a collision if any node detects a collision.

  In an Ethernet/IEEE 802.3 network, the signal consists of a 32- or 48-bit transmission whose contents are unspecified except that it cannot be identical to the cyclical redundancy check (CRC) value of the partial packet sent prior to the collision.

  Each node involved sends a *jam signal* and then waits a random amount of time (called the *backoff interval*) before trying to access the network again.

- In CSMA/CA networks, the *jam signal* indicates that the station sending the *jam signal* intends to transmit a message.

A *jam signal* is not the same as *jamming.*

jamming Intentional (active jamming) or accidental (passive jamming) interference with a radio transmission such that the intended receiver cannot extract information from the source. *Jamming* is caused by an interfering carrier at or near the frequency of the desired channel.

JANET An acronym for Joint Academic NETwork, Great Britain's national network. Within the network, the addressing structure is backwards from that of Internet; that is, the largest domain is first, and the smallest is last. Fortunately, most gateways into JANET perform the translation automatically.

jansky (Jy) The unit used for flux density (or strength) of extraterrestrial radio sources. It is expressed as

$$1Jy = 10^{-26} W/m^2/Hz$$

Typical strong radio sources have strengths of 10 to 100 janskys (10 Jy to 100 Jy), while weaker ones may be measured in millijanskys (mJy) or microjanskys (μJy).

Jansky noise A reference to any noise of extraterrestrial origin. It is named for the engineer who discovered the phenomenon.

JAPATIC An acronym for JApan PATent Information Center.

jargon The shorthand language that develops in a specialized field, generally made up of acronyms and coined words. *Jargon* is usually incomprehensible to those not in the field.

JASE An acronym for Just Another System Error.

JATE An acronym for Japan Approvals Institute for Telecommunications Equipment. The Japanese equivalent of the U.S. FCC part 68 certification for equipment to be attached to the telephone network in Japan. Obtaining a *JATE* certification is expensive, complex, and very time consuming.

JAVA™ A programming language, developed by Sun Microsystems, Inc., that allows small programs (called applets) to be executed as they are being transferred from one machine to another (downloaded). This is in opposition to the previous procedure wherein the file was downloaded and then executed or viewed off-line. The program HOT JAVA, for example, allows users to view live, animated graphics. Although generally thought of as being for the Internet World Wide Web (WWW), *JAVA* works well with many operating systems and computers.

JCL An abbreviation of Job Control Language.

JCT An abbreviation of JunCTion.

JEIDA An abbreviation of Japan Electronic Industry Development Association.

jellyware Slang for "people." Also liveware and wetware.

JF (1) An abbreviation of Junction Frequency. **(2)** An abbreviation of Junctor Frame.

JFET An acronym for Junction Field Effect Transistor.

jitter The random, abrupt variation of a measured parameter over time with respect to its expected value.

Since *jitter* may be related to frequency, amplitude, phase, position, and so on, the name of the category must be included, e.g., delay jitter, pulse position jitter. The magnitude of *jitter* must also be expressed as average, peak, peak-to-peak, percent, RMS, dB, and so on. For example, *jitter* (D_J) expressed as a percentage of a bit time is:

$$D_J = \frac{T_J}{B} \times 100\%$$

The following diagram is of a recovered clock signal as it relates to the original transmitted signal. The jitter category is duration or position; without further measurements, one cannot tell the difference.

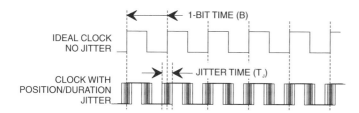

At least three sources of *jitter* may be identified:

- *Data-dependent jitter*—caused by limited bandwidth and baseline wander. Also called *intersymbol interference*.
- *Duty cycle distortion*—caused by imprecise decision thresholds in receivers. Also called *bias distortion*.
- *Random jitter*—caused by random noise.

job control language (JCL) A command language that provides the instructions for an operating system to run an application program.

job transfer and manipulation (JTM) In the ISO/OSI Reference Model, one of the file transfer services (FTSs) defined at the application layer. *JTM* enables an application to do data processing on a remote machine.

joe account A system in which the password is the user's first name (or some other equally obvious word). Although this is easy to remember, it is also easy for a *cracker* to discover. Certainly it does not contribute to system security.

Johnson noise Noise caused by the thermal agitation of electrons in a dissipative material. The available *Johnson noise* power (N) from a resistor is calculated by

$$N = kT\Delta f$$

where:

k is Boltzmann's constant,
T is temperature in kelvins, and
Δf is the measurement bandwidth.

The generated noise is white noise, that is, its distribution is the same at all frequencies throughout the radio frequency spectrum. Noise power is equal in all equal spectral bandwidths. Also called *thermal noise*.

Josephson junction A superconducting device, named after Brian David Josephson (1940–), in which electron pairs tunnel through a thin insulating barrier placed between two superconducting materials (i.e., current flows).

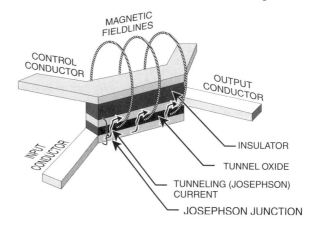

Below a critical current, the voltage drop at the junction is zero; above the critical current, the voltage is given by $2eV = hf$ where e is the charge on an electron, h is Planck's constant, and f is the frequency of the radiation given off by the electron pair as it tunnels across the barrier. The critical current is a function of many variables, including the surrounding magnetic field strength (as shown in the following figure, adapted from J. M. Rowell, *Phys. Rev Let.* 1963).

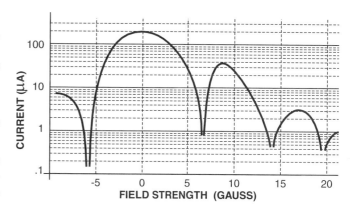

Josephson junctions are used in magnetic sensing devices (called superconducting quantum interference devices or SQUID) and potentially as electronic switches for computer logic elements.

joule (J) A unit of energy in the SI system.

In electrical systems, one *joule* is the amount of energy used to dissipate one watt for one second. In mechanical systems, one *joule* is the amount of energy expended when one newton of force moves an object one meter.

Joule's law The rate at which heat is produced in a constant resistance circuit is equal to the product of the square of the current and the resistance, that is,

$$P = I^2R$$

journal (1) In general, a record of events, expressed in chronological order. Also called a *log*. (2) In computer processing, a time-ordered record of data processing commands and operations that may be used to reconstruct a previous version of a file.

journaling A failure recovery/debug technique in transaction processing. It is a strategy in which every transaction is recorded, so that the original database or file can be re-created in the event of a failure, bug, or malfunction.

joystick A pointing device used in graphical user interface (GUI) systems to manipulate a pointer on the computer screen. It has at least the ability to move the screen pointer vertically and horizontally.

JPD An abbreviation of Japan Publishers Directory.

JPEG An acronym for Joint Photographic Experts Group (pronounced "jay peg"). A group that defined a lossy data compression algorithm for image files. The compression ratio can be 100:1 or higher, but at the expense of some loss in image quality. Files containing *JPEG* compressed images usually have the extension .JPG or .JPEG.

The compression takes advantage of the limitations of the human eye, namely, that the perception of slight color differences is far less sensitive than its perception of contrast and brightness differences. Some loss of sharpness can generally be tolerated. (Sometimes it is even perceived as an improvement.) See also *compression* (2).

JPL An abbreviation of Jet Propulsion Laboratory.

JRSC An abbreviation of Jam Resistant Secure Communications.

JRSVC An abbreviation of Jam Resistant Secure Voice Communications.

JSA An abbreviation of Japan Standards Association. It is the Japanese counterpart of the American National Standards Institute (ANSI) in the United States or the Canadian Standards Association (CSA) in Canada.

JSC An abbreviation of Joint Steering Committee (now called the Joint Telecommunications Standards Steering Group or JTSSG).

JSF An abbreviation of Junctor Switch Frame.

JT-2 A 6.312 Mbps communications link defined by ITU-T recommendation G.704. It is essentially the same as T-2.

JTC An abbreviation of Joint Technical Committee. Any of several technical committees formed by the International Standards Committee (ISO) and the International Electrotechnical Commission (IEC). For example, the JTC1 is the committee that was largely responsible for the ISO/OSI Reference Model.

JTEC An acronym for Japanese Technology Evaluation Center.

JTIDS An acronym for Joint Tactical Information Distribution System.

JTM An abbreviation of Job Transfer and Manipulation.

JTSCC An abbreviation of Joint Telecommunications Standards Coordinating Committee.

JTSSG An abbreviation of Joint Telecommunications Standards Steering Group (formerly called the JSC).

JUG An abbreviation of Joint Users Group.

Jughead A search client that scans an automatically compiled index of directory names that can be retrieved in *gopherspace*. The output includes only directory names; it does not include file names that appear in Gopher submenus. The advantage of *Jughead* over *Veronica* is a shorter output list.

jukebox A data storage device that can automatically switch between multiple removable disks, tapes, or CD-ROMs.

Julian Day (1) A date-numbering system developed by Joseph Justus Scaliger (1540–1609), a French scholar, as a means of correcting ancient calculations of historical time and establishing positive years for all recorded history. He in essence set forth the scientific method of chronology that became the basis for the modern study of ancient history.

The dates were expressed in the form (S,G,I) where S represents a number from the 28-year solar cycle, G (or Golden number) is from the 19-year lunar cycle, and I is from the 15-year "induction cycle" (a Roman tax cycle of unknown origin). The product of these cycles yields a 7980-year cycle. Initializing the system by using the date of Christ's birth at (9,1,3) implies an initial date (1,1,1) of 1 January 4712 BC of the modern calendar. The zero point (midnight) of January 1, 1979 was *Julian Day* 2 450 448.5. (2) The day count using the January 1 of the current year as a reference. This is a corruption of the traditional meaning; therefore, to avoid ambiguity, "day of year" rather than "Julian day" should be used.

jump (1) Changes of the sequential flow of a program to a new location within the program. Several kinds of jump instruction are possible; for example, there are *conditional jumps* that change the program address only when a specified condition is satisfied (one number equal another, less than, greater than, positive, negative, etc.). There are *unconditional jumps* that change the flow to a new location under all circumstances, jumps caused by hardware interrupts, jumps to subroutines, and more. (2) In a *hypertext document,* the movement from one position in a document to another position in the same or different document by selecting a highlighted word or phrase (called an *anchor*). Selecting the *anchor* word activates a *link* to the new location from which sequential display continues.

jump hunting See *nonconsecutive hunting.*

jumper (1) A small installable (and/or removable) wire or plug used to select a specific aspect of a hardware configuration. This selection is accomplished by connecting different points in an electronic circuit. (2) In telephone switching systems, a *cross connect,* i.e., a wire used to interconnect equipment appearances and line appearances on a distributing frame. (3) A wire permanently attached to a circuit board by the manufacturer to correct an error or to effect an upgrade or improvement. (4) A patch cord or wire used to establish a circuit (often temporarily). The *jumper* may be installed for testing, diagnostics, or establishing a communications path.

junction (1) The point at which two or more circuit branches are joined. (2) The region of transition between semiconductor sections of different electrical properties, e.g., P and N, N and N^+, or P and P^- materials.

junction circuit See *toll circuit.*

junctor A circuit extending between frames of a switching unit and terminating in a switching device on each frame.

Within a switching system, a connection or circuit between input ports and output ports of the same or different switching networks.

JUNET A Japanese UNIX-based research network for noncommercial institutions and organizations interconnected with the NSFNet, JANET, and others.

junk The marking of a news article in USENet as read before it is actually completely read. The purpose of *junking* an article is to ensure that the newsreader software will remove it from both the subject selector list and the thread selector list.

K

k **(1)** The SI symbol for the prefix *kilo*. A multiplier of 1000. See also *K*. **(2)** The symbol for Boltzmann's constant.

K **(1)** The SI symbol for the units of *temperature* in kelvins. See *thermodynamic temperature*. **(2)** A symbol for *absorption coefficient*. **(3)** When referring to binary systems, *K* indicates a multiplier of 1024 (2^{10}).

k-factor **(1)** In tropospheric radio propagation, the ratio of the effective Earth radius to the actual Earth radius—approximately 4/3. **(2)** In ionospheric radio propagation, a correction factor that is applied in calculations related to curved layers. It is a function of both distance and the real height of ionospheric reflection.

K-type patch bay A facility designed for patching and monitoring of balanced digital data circuits that support data rates up to 1 Mb/s.

K.n recommendations An ITU-T series of specifications concerning protection from interference. See Appendix E for a complete list.

K12 One of several USENet alternative newsgroups. The *K12* newsgroups are from a gateway connection to the *K12Net* system (a store-and-forward system based on FidoNet). The newsgroups focus on all aspects of education from kindergarten to grade 12. See also *ALT, alternative newsgroup hierarchies,* and *K12Net*.

K12Net Founded in 1990, *K12Net* is a loose, decentralized alliance of school-based bulletin board systems (BBSs) distributed throughout Australia, Europe, North America, and nations of the former Soviet Union. Communication between the various nodes is via FidoNet (a store-and-forward network that uses late night long-distance dial-up connection to propagate messages).

Although focusing on educational matters, the subject matter available on the *K12Net* is as varied as the personalities of the participants. It includes subject matter for both students and teachers.

K56FLEX Rockwell, Lucent Technology, and Motorola's name for a technique for achieving a 56-kbps data rate from a service provider to a user over standard telephone lines. Formerly called *K56PLUS* by Rockwell. See also *V.90*.

K band That portion of the electromagnetic spectrum with frequencies between 18 and 26.5 GHz. See also *frequency band*.

K$_a$ band That portion of the electromagnetic spectrum with frequencies between 26.5 and 40 GHz. See also *frequency band*.

KA9Q An implementation of the TCP/IP protocol suite for packet radio systems. KA9Q is discussed in RFC 1208.

Kalman filter A numerical method used to track time-varying signals in the presence of noise. If the signal can be characterized by a finite number of parameters that change slowly in time, then *Kalman filtering* can be used to tell how the received signal should be processed to best estimate the parameters as a function of time.

kamikaze packets Term for the test datagrams with all possible combinations of options enable user to test the compatibility and interoperability of a gateway with the next hierarchical network level. Described in RFC1025. Also called a *Christmas tree packet*.

Karnaugh map A method of simplifying Boolean expressions published in 1953 by Maurice Karnaugh. It is a rectangular diagram of a logical expression drawn with overlapping regions representing all combinations of the variables of the expression. The rows and columns represent the individual variables of the expression and are equivalent to the *truth table* representation of an expression.

The map is used in the design, simplification, and optimization of logic circuits. For example, the truth table and Karnaugh map of a function of the four variables A, B, C, and D are shown in the diagram.

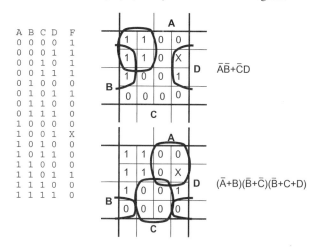

The rules for applying *Karnaugh maps* in minimizing Boolean expressions and logic are:

- Plot the values on the map such that only one bit changes in adjacent columns or rows. True conditions are marked "1," false conditions "0," and "don't care" conditions "X." "Don't care" conditions may be treated as either 1s or 0s, whichever is more beneficial.
- The left edge and right edges are adjacent, as are the top and bottom edges.
- Circle groups of 1s (or 0s) and write the equation for each group.
  Minimization is possible only when there is a symmetrical grouping of 2, 4, 8, 16, . . . squares with 1s and Xs (or 0s and Xs). The groups must be either square or rectangles with length equal twice the width.
- Combine the individual groups to yield the complete expression. Join groups of 1s with the OR and groups of 0s with the AND.

An empty map of eight variables can be drawn as shown in the diagram:

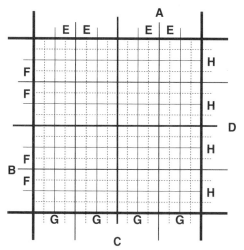

285

KDC An abbreviation of Key Distribution Center.

KDD An abbreviation of Kokusai Denshin Denwa.

keep alive signal A T-1 signal generated when a DTE detects the loss of input (from the customer equipment) for 150 ms. Also called *AIS* or *blue alarm.*

kelvin (K) The SI unit of thermodynamic temperature. The scale range starts at *absolute zero (0 K)*—the temperature at which no atomic motion exists. On this scale, the triple point of water is 273.16 K, and the boiling point is 373.16 K.

No degree symbol is written with K, the symbol for kelvin(s). The kelvin was formerly called "degree Kelvin," a term that is now obsolete. The Celsius temperature scale is defined by setting 0°C equal to 273.16 K. The intervals between each degree Celsius is equal to 1 kelvin.

kelvin temperature scale See *thermodynamic temperature.*

Kennelly-Heaviside layer A synonym for the ionospheric *E region.*

Keplerian elements A set of six parameters used to describe the position and velocity of a satellite in an elliptical orbit (Keplerian orbit). The parameters are:

- The semimajor axis of the ellipse,
- The eccentricity of the ellipse,
- The inclination of the orbital plane with respect to the celestial equator,
- The right ascension of the ascending node of the orbit,
- The argument of the perigee, and
- The time the satellite passes through perigee.

These parameters only provide an approximation to an orbit in which the satellite position/motion is perturbed by external forces or encounters a nonuniform gravitational field about which it is orbiting.

Kerberos A network security system developed for Project Athena in the 1980s at the Massachusetts Institute of Technology (MIT). It is named for the three-headed dog, Cerberus, who guarded Hades in Greek mythology because it has three main parts: a database, an authentication server, and a ticket-granting server.

Kerberos is a distributed authentication system that uses special keys, called tickets, to encrypt transmissions between *Kerberos* and a user. It uses a private key encryption scheme. The system verifies that a user is legitimate when a login occurs, as well as every time a user requests services. The system is designed to provide authentication for users who log onto the network from an unattended workstation. (Such stations must be regarded as suspect or entrusted because their physical security cannot be guaranteed.)

Kermit A file transfer protocol (named after Jim Henson's famous frog—though some say it is an acronym from KL-10 Error-free Reciprocal Micro Interconnect over TTY lines) designed at Columbia University for maximum universality. The original protocol transfers variable-length packets of up to 96 bytes (later extended to 1024 bytes). Each packet is checked for transmission errors. Enhancements of the original protocol include data compression. The DTE converts control characters into standard printable ASCII characters (decimal codes 32 through 126) for transmission by the DCE. As a result, *Kermit* will allow communication on transmission circuits limited to 7 bits.

Kermit has capabilities that allow it to be used between almost any two computers regardless of system differences. But because of the built-in flexibility, it is relatively slow at file transfers and should be used only when no other transfer protocol is available. Other file transfer protocols include *FTP, XMODEM, YMODEM,* and *ZMODEM.*

kernel The essential part of an operating system. It is responsible for resource allocation, low-level hardware interfaces, security, and so on.

kerning In documents and text, the process of reducing the spacing between certain pairs of letters to improve their appearance. When a font is designed, each character is given a width that includes some space on both sides. (This is so letters don't run into each other when displayed or printed.) Some pairs of characters, such as A and V, look better, however, if the space is reduced or even overlaps slightly (but not to the point of touching). See also *leading* and *tracking.*

Kerr cell An electro-optic device in which a material's refractive index changes in response to an applied electric field. The refractive index change is proportional to the square of the electric field, and the material is usually a liquid. See also *Pockels cell.*

Kerr effect (1) Named for its discoverer, the Scottish physicist John Kerr, it is the development of birefringence in an isotropic transparent material when the material is immersed in an electric field. That is, the refractive index to light polarization is different with the electric field applied than with the field off.

The effect is nonlinear, and the degree of birefringence is directly proportional to the square of the applied electric field strength, that is,

$$\Delta n = \lambda_0 K E^2$$

where:

Δn is the difference between the refractive index of the ordinary ray (n_o) and extraordinary ray (n_e),
K is the Kerr constant of the material,
λ_0 is the wavelength of the light, and
E is the electric field strength.

This birefringence effect, produced by the applied electric field, can be used in conjunction with polarizers to modulate light. Devices that use this effect are called Kerr cells. **(2)** The effect in which the polarization plane of a reflected laser beam will rotate, either clockwise or counterclockwise, depending on the magnetic polarity of the reflecting surface. Also called *magneto-optic Kerr effect (MOKE).* **(3)** In fiber optics, an effect that occurs when the refractive index of an optical fiber varies with the intensity of transmitted light. The process is nonlinear and occurs only when the product of the illuminating power and system length becomes a significant fraction of the nonlinearity constant. In systems with span lengths of hundreds of kilometers, the effect can be significant even if the source is only a few mW. The major distortion caused by the effect is self-phase modulation of the signal.

key (1) In telephony, a hand-operated mechanical switch. **(2)** In encryption, the character string used to encrypt a plaintext message. The character string used to decrypt an encrypted message. Generally, the decryption key is different from the encryption key. The key is usually a sequence of random or pseudorandom bits.

Some older terms include *variable, key(ing) variable,* and *cryptovariable.* **(3)** One or more characters that identify data and possibly control its use or the user's access rights to it.

key distribution center (KDC) In data encryption terminology, a center for storing, managing, and distributing encryption electronic keys.

key management The process by which key information is generated, stored, protected, transferred, loaded, used, and destroyed.

key management protocol (KMP) A protocol, in a secure network, used for checking security keys.

key pad The dialing buttons on a pushbutton telephone.

key pulse (KP) In telephony, a signaling system in which digits are transmitted using pushbuttons. Each pushbutton corresponds to a digit and generates a unique tone pair. Also called *pulsing*. See also *DTMF*.

key pulsing signal In multifrequency and key pulsing signaling systems, a signal used to prepare the distant equipment for receiving digits.

key service unit (KSU) The switching portion of a *key telephone system*. It resides at the subscriber's premises between the telephone company subscriber lines from the central office and subscriber's *key sets*.

key set A telephone subset equipped with switches (keys) for line selection, intercom, line holding, and other features. Also called a *key telephone set*.

key stream A sequence of symbols, produced in a cryptosystem, produce a key (2), control transmission security processes, or when combined with plaintext, produce cipher text.

key telephone system (KTS) A small business telephone system in which each telephone instrument (*key set*) has direct access to incoming telephone and intercom lines. The user is therefore permitted to select incoming and outgoing central office lines directly. That is, there is no need to dial "9" for an outside line as must be done with a PABX. The switching and control functions are accomplished in a shared device called a *key service unit* (*KSU*). Features of a KTS include:

- Calling or answering on a selected line,
- Contacting a party over an intercom,
- Putting a caller on hold, and
- Transferring a call to another line.

key variable A deprecated synonym for *key* (2).

keyboard An input device used to enter data by manual depression of keys (3), which causes the generation of the selected code element or symbol.

keyboard send and receive (KSR) A communications device that is a combination of teleprinter transmitter and receiver (printer) with transmission capability from the keyboard only. That is, there is no paper tape punch or reader. Because the device has no storage, messages are printed as they are received and are transmitted as they are typed at the keyboard. See also *automatic send and receive (ASR)*.

keying (1) A modulation system in which the modulator output selections are from a finite set of discrete states. Examples include *frequency shift keying (FSK)*, *phase shift keying (PSK)*, and *minimum shift keying (MSK)*. (2) The formation of signaling elements by an abrupt change in modulation state, such as the interruption of current, the reversal of voltage polarity, or the interruption of a signaling tone. (3) A physical device that prevents a connector from being plugged into a similar looking but incorrect connector. *Keying* is usually accomplished by one of two means:

- By filling one (or more) of the socket positions with an obstruction and removing the pins in the corresponding position of the mating connector. The pins of an improper (but otherwise mating) connector cannot be inserted because of the socket obstruction.
- By shaping the shell or body of connector pairs so that only they fit together. For example, a keyed RJ-45 plug has a small bump that fits into a corresponding notch on the female RJ-45 (an MMJ). The keyed RJ-45 will not plug into a nonkeyed RJ-45 receptacle.

keying interval The interval of time starting at the beginning of the shortest signaling element to the end of the same element. The *keying interval* is therefore equal to the symbol duration.

keying variable A deprecated synonym for *key* (2).

keypunch A keyboard-actuated punch that punches holes in a data medium. Also called a *keyboard punch*.

kg The SI symbol for kilogram.

kHz The abbreviation for kilohertz (1000 Hz).

KIBO An acronym from Knowledge In, Bullshit Out. What happens whenever valid data is passed through an organization (or person) that deliberately or accidentally disregards or ignores its significance.

kill A capability in USENet which allows a user to exclude selected subjects, words, or even other individuals from the list of articles displayed. The *kill* may be applied to specific newsgroups or globally (to all newsgroups).

To *kill* a subject, the subject key word is entered into a *kill file*. When the newsreader program lists articles, it excludes those items entered into the *kill file*. The *kill file* can exclude words, phrases, subjects, specific sites, specific authors, or threads. Most newsreaders allow exclusion on two levels, specific newsgroups and global (all newsgroups). See also *global kill file* and *newsgroup kill file*.

killer channel In digital telecommunications, a transmission channel whose timing is off, and therefore overlaps and interferes with other channels.

kilo (k) A prefix. (1) In most usages and sometimes even in computer usage, "kilo" is equal to 1000; hence, 1 kilogram is 1000 grams, etc. (2) In computer systems "kilo" is equal to 1024 (or 2^{10}); therefore, 1 Kbyte is 1024 bytes and 2 Kbytes is 2048 bytes.

Although not strictly adhered to, the symbol "k" is applied to the multiplier 1000, while the symbol "K" is applied to 1024.

kilocharacter 1000 characters. Used as a measure of billing for data communications by some overseas telephone administrations. See also *kilosegment*.

kilohertz (kHz) A frequency of 1000 Hz.

kilosegment 64 000 characters. Used as a measure of billing for data communications by some overseas telephone administrations. See also *kilocharacter*.

kiosk A Web browser mode that eliminates button bars, menus, borders, the URL window, etc., from the display to leave more room for actual web page information. Also called *presentation mode*.

Kirchhoff's laws (1) *Kirchhoff's current law*. The algebraic sum of all currents into any point of a network is zero. (2) *Kirchhoff's voltage law*. The sum of all voltages (drops and sources) around a closed loop is zero.

KIS An acronym for Knowbot Information Service.

KISS An acronym for Keep It Simple, Stupid.

kludge This word appears to have been derived from the Scots "kludge" or "kludgie" for a common toilet, via British military slang. An improvised solution to a hardware or software problem, generally composed of makeshift parts or haphazard coding. A *kludge* is not elegant and is generally intended to be only a temporary fix.

kluge From the German *klug*, meaning clever; possibly related to the Polish *klucza*, meaning a trick or hook. (1) A clever programming trick intended to solve a particularly nasty problem in an expedient, if not clear, manner. Often used to repair bugs. (2) Something that works for the wrong reason. (3) A feature that is implemented in a coarse manner.

km The SI symbol for kilometer.

KMP An abbreviation of <u>K</u>ey <u>M</u>anagement <u>P</u>rotocol.

knife-edge effect In wave propagation (optical, radio, or mechanical waves), a redirection by *diffraction* of a portion of the incident radiation that strikes a well-defined boundary of opaque and transparent material. The effect is explained by *Huygens' principle,* which states that a well-defined obstruction to a wave acts as a secondary source and creates a new wavefront. This new wavefront propagates into the geometric shadow area of the obstruction.

This effect is sometimes used with line-of-sight communications systems to allow communications with locations that would otherwise be out of range.

knowbot A contraction of <u>KNOW</u>ledge ro<u>BOT</u>. An independent, self-acting computer program, known as a software agent, which automates the search for and collection of information from distributed databases such as are found on the Internet. The *knowbot* may replicate itself on several host systems in its quest to complete the assigned task. After completing its designated task, it sends reports back to the initiating user and then self-destructs. The development of *knowbot* technology is part of a research project at the Corporation for National Research Initiatives. Sometimes also called a *bot.*

Knowbot Information Service (KIS) On the Internet, an experimental service that can query directory services in order to retrieve requested information. *KIS* uses knowbot programs to search the directory services for the information.

Kokusai Denshin Denwa (KDD) A Japanese long-distance telephone service provider.

KP An abbreviation of <u>K</u>ey <u>P</u>ulse.

KSR An abbreviation of <u>K</u>eyboard <u>S</u>end and <u>R</u>eceive.

KSU In telephony, an abbreviation of <u>K</u>ey <u>S</u>ervice <u>U</u>nit.

kT See *noise-power density.*

KTILA The Greek Development Center for Telecommunications.

KTS An abbreviation of <u>K</u>ey <u>T</u>elephone <u>S</u>ystem.

KTU An abbreviation of <u>K</u>ey <u>T</u>elephone <u>U</u>nit.

K_u band That portion of the electromagnetic spectrum with frequencies between 12.5 and 18 GHz. See also *frequency bands.*

kVA The symbol for kilovolt-amp. A rating of high-reactance transformers. The value is determined with the secondary of the transformer shorted. See also *effective power.*

kWh or kWhr The abbreviation for kilowatt-hour. A measure of energy—the product of power (watts) and time (hours), e.g., 1000 watts consumed for one hour, 1 watt consumed for 1000 hours. One *kWhr* is equal to 3 600 000 joules of energy.

L

L The circuit designator for an *inductor* or a symbol for *inductance.*

L band That portion of the electromagnetic spectrum with frequencies between 1.0 and 2.0 GHz. See also *frequency band.*

L network A particular network topology with two branches connected in series. The two free ends are connected to one terminal and the junction of the branches, along with either end connected to the other terminal.

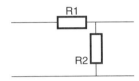

L-UNI An acronym for <u>L</u>AN emulation <u>U</u>ser <u>N</u>etwork <u>I</u>nterface.

L.n recommendations An ITU-T series of specifications concerning the outside plant. See Appendix E for a complete list.

L2TP An abbreviation of <u>L</u>evel <u>2</u> <u>T</u>unneling <u>P</u>rotocol.

LAA An abbreviation of <u>L</u>ocally <u>A</u>dministered <u>A</u>ddress.

label (1) An identifier within or attached to a set of data elements. (2) One or more characters that are both within or attached to a set of data elements and represent information about the set, including its identification.

labeling algorithm An algorithm for shortest path routing in which individual nodes are assigned labels and are updated as the algorithm progresses toward a solution.

LAD An acronym for <u>L</u>ATA <u>A</u>rchitecture <u>D</u>atabase.

LADT An abbreviation of <u>L</u>ocal <u>A</u>rea <u>D</u>ata <u>T</u>ransport.

lag A condition of the phase or time relationship of two or more waveforms where one waveform is used as a reference and the others are behind in phase or time.

The time delay between a stimulus and a response. In communications, the lag refers to the delay between the input of a circuit, device, or system and the output of the element. See also *lead.*

LAM An acronym for <u>L</u>obe <u>A</u>ttachment <u>M</u>odule.

LAMA An acronym for <u>L</u>ocal <u>A</u>utomatic <u>M</u>essage <u>A</u>ccounting.

lambda (Λ and λ) The eleventh letter in the Greek alphabet. The lower case λ is the symbol for wavelength.

Lambertian source An optical source that obeys Lambert's cosine law. Conventional (surface-emitting) light emitting diodes (LEDs) approximate a Lambertian source. That is, they have a large beam divergence and a radiation pattern that approximates a sphere. Thus, most of their optical output does not couple into communications fibers.

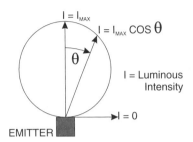

Lambert's cosine law The mathematical statement that the radiance of certain idealized optical sources is directly proportional to the cosine of the angle (with respect to the direction of maximum radiance) from which the source is viewed. The law may also apply to certain idealized diffuse reflectors. Also called *cosine emission law* and *Lambert's emission law.* See also *Lambertian source.*

lamp The proper term for a *light bulb.*

LAN An acronym for <u>L</u>ocal <u>A</u>rea <u>N</u>etwork (rhymes with *can*).

LAN application (software) An application software package specifically designed to operate in a local area network environment.

LAN aware Application programs that have file and record-locking capability for use on networks.

LAN bridge server (LBS) In an IBM Token Ring network, a server whose job is to keep track of and provide access to any bridges connected to the network.

LAN emulation configuration server (LECS) An ATM Forum-defined specification in support of LAN-to-LAN connectivity called LAN emulation. It defines that set of functions implemented in an ATM network that provide LAN DTEs with information regarding the location of other LAN emulation services.

LAN emulation network to network interface (LNNI) A specification that enables each vendor's implementation of LAN emulation to interoperate. A very useful feature of multivendor ATM networks.

LAN emulation server (LES) An ATM Forum-defined specification in support of LAN-to-LAN connectivity called LAN emulation. It defines that set of functions implemented in an ATM network in support of LAN-to-LAN establishing connections.

LAN emulation user network interface (L-UNI) The definition of how legacy LAN protocols and applications coexist within an ATM network.

LAN emulator Software on an ATM that lets users on the ATM network communicate with legacy networks based on other protocols, such as Ethernet or Token Ring.

LAN ignorant Application programs written for use on a single-user system. They should not be used on network systems.

LAN intrinsic Applications written specifically for use on networks.

LAN manager (1) The person who manages a LAN. Duties vary but generally include: adding new users, deleting specified users, setting up and enforcing network security measures, installing new hardware and software, performing routine backups, and diagnosing network problems. (2) A multi-user network operating system jointly developed by Microsoft and 3Com. The system has largely been replaced by WindowsNT.

LAN Manager for UNIX Systems (LM/X) An implementation of LAN Manager for UNIX.

LAN Network Manager IBM's network management software for Token Ring networks.

LAN operating system See *network operating system (NOS)*.

LAN segment A part of a LAN that is separated from the rest by one or more bridges.

LAN segmentation Dividing a single LAN into multiple connected LANs in order to improve performance. Nodes within each divided portion are selected to be more likely to communicate with each other than with nodes in other groups. Individual LAN segments are connected with bridges to allow these infrequent communications.

land line That portion of a communications link that travels over terrestrial circuits. These terrestrial circuits may be wire, fiber optic, or even microwave.

LANE An acronym for Local Area Network Emulator. Found in ATMs.

language code In telephony, on an international call, an address digit that allows the originating operator to obtain assistance in a specified language.

LANPUP A terminate-and-stay resident version of Artisoft's LANtastic NET program. It allows access to many NET functions while the user remains in a DOS program.

LANtastic® Artisoft's peer-to-peer network operating system (NOS). Several versions are available, e.g., *LANtastic* for DOS, *LANtastic* Windows Version, *LANtastic* for Macintosh, *LANtastic* for NetWare, *LANtastic* for TCP/IP, and so on.

LAP-B Acronym for Link Access Protocol-Balanced. The ITU-T standard error control data link level protocol used by switched packet networks, such as X.25. It is a full-duplex, point-to-point, bit-synchronous protocol and a compatible subset of High-Level Data Link Control (HDLC). All public data networks support LAP-B.

LAP-D Acronym for Link Access Protocol D-channel. The ITU-T X.25 bit oriented standard data link protocol used by switched packet networks. A compatible subset of High-Level Data Link Control (HDLC). Used on the ISDN D-channel.

LAP-D+ An LAP-D protocol for use on information channels other than ISDN D-channels, e.g., ISDN B-channels.

LAP-M Acronym for Link Access Protocol for Modems. A ITU-T standard bit-oriented error control protocol. LAP-M provides error control only if both the sending and receiving modems support the protocol.

large squaring capability A key system feature that allows all lines to appear on all telephone instruments.

laser An acronym for Light Amplification by Stimulated Emission of Radiation that has been elevated to the status of a word. It is a device that produces a coherent beam of monochromatic optical radiation (single color light) by stimulating electronic, ionic, or molecular transitions to higher energy levels so that when they return to lower energy levels they emit energy. *Laser* radiation may be either temporally coherent, spatially coherent, or both, and the *degree of coherence* exceeds 0.88.

laser chirp An abrupt change of the center wavelength of a laser, caused by laser instability.

laser diode A synonym for *injection laser diode.*

LaserDisc Generally, a 12-in. diameter medium in which digital information (video, sound, or data) is stored. There are two types of *LaserDiscs:* CAV (or constant angular velocity) and CLV (constant linear velocity). They are differentiated by the way information is stored and accessed on the disc.

LaserDisc players use these 12-in. discs to display high-quality video on a standard television or monitor. See also *compact disc.*

LaserStacks Macintosh software which, when used with HyperCard and a corresponding LaserDisc, allows rapid access of frames and sophisticated branching.

last-in-first-out (LIFO) Pronounced "LYE-foe." A queuing methodology in which the last item, action, or entity inserted into the queue is the first item extracted. (The first entity stored will be the last removed—called a FILO for first-in-last-out.) See also *first-in-first-out (FIFO)*.

last line released In telephony, release of a communications circuit by the last terminal to go on-hook, i.e., to return to idle state.

last mile problem The problem of delivering high-speed (high bandwidth) information to a user from an existing high-speed network. There are two "last mile delivery" capabilities in the United States: the telephone twisted pair and the television coaxial cable (CATV). Each of them provides an aspect of the solution, but neither provides a total solution.

The twisted pair provides a switched, bidirectional communications link but at low speed. The CATV provides wide bandwidth but neither switch capability nor, in general, bidirectional communications. Both networks will undergo changes. The analog twisted pair will go digital (probably the route of N-ISDN, which provides two 64-kbps channels). But, this does not provide even close to the bandwidth required for contemplated entertainment services. CATV is currently a broadcast type of network with very limited, if any, capability for reverse channel communications. Multicast capability can be added relatively easily; however, full user reverse channel capability will require the addition of some form of switching, which is a very expensive proposition. Two other solutions have been suggested: mergers or joint ventures between cable television companies and regional Bell Operating Companies (RBOCs). This solution provides the RBOC with a broadband delivery system, and it provides the CATV company with a switched reverse channel. Another solution proposed is the augmentation of the existing telephone system with high-speed fiber optic connections to the user.

LAT (1) An acronym for Local Area Transport. A DECnet proprietary protocol developed by Digital Equipment Corporation (DEC), which oversees communications between terminals and hosts. The protocol works on the principle of a relatively small, known number of hosts on a local network sending short packets at regular intervals. *LAT*™ will not work on a wide area network. (2) An abbreviation of latitude. (3) An acronym for Local Apparent Time. See *time (3)*.

LATA An acronym for <u>L</u>ocal <u>A</u>ccess and <u>T</u>ransport <u>A</u>rea. In the United States, a telephone geographic service region incorporating local exchanges. Typically, a *LATA* is serviced by a single telephone company, *a local exchange company* (*LEC*), for subscriber connections and multiple *interexchange carriers* (*IXC*) for some *intra-LATA* connections and all *inter-LATA* connections.

latch **(1)** In logic circuits, a positive feedback loop in a digital circuit used to maintain a specified state. It is used as a 1-bit memory element. A latch is the fundamental element of all flip-flops. Also called an *R-S flip-flop.* **(2)** A condition of a device in which,

* It has locked into one state, and
* The only method of recovery is reset or power removal.

latency **(1)** In communications, the amount of time between a request for a channel and the time at which it is made available for a information transfer, i.e., waiting time. **(2)** The time it takes for a packet to cross a network connection from sender to receiver. **(3)** The period of time that a frame is held by a network device before it is forwarded.

lateral offset loss In fiber optics, a loss of optical power caused when a transmitting entity is not radially alligned with the receiving fiber's core axis. The transmitting entity may be another fiber (which occurs at a splice or a connector) or an optical source. The effect of a given amount of *lateral offset* will depend on other parameters such as the relative diameters of the respective cores (in the case of fiber-to-fiber coupling). For example, if, because of manufacturing tolerances, the transmitting fiber core is smaller than the receiving core, the effect will be less than if both cores were the same size. Also called *transverse offset loss.* See also *coupling loss.*

Latin 1 The character set defined by ISO 8859-1:1987.

lattice network A particular network topology with four branches connected in series: two nonadjacent junctions serve as the input, while the remaining two junctions are the output.

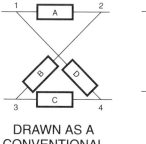 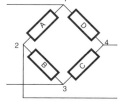

DRAWN AS A CONVENTIONAL MESH   **DRAWN AS A BRIDGE**

launch angle **(1)** The angle at which a light ray emerges from a surface measure with respect to the normal of that surface. **(2)** The beam divergence at an emitting surface, such as that of a light emitting diode (LED), laser, lens, prism, optical fiber endface, or other optical device. **(3)** At an endface of an optical fiber, the angle between an input ray and the fiber axis. If the endface of the fiber is perpendicular to the fiber axis, the launch angle is equal to the incidence angle when the ray is external to the fiber and the refraction angle when initially inside the fiber.

launch numerical aperture (LNA) The *numerical aperture* (*NA*) of an optical system used to couple (launch) power into an optical fiber. *LNA* may differ from the stated NA of a final focusing element if, for example, that element is underfilled or the focus is other than that for which the element is specified. *LNA* is one of the parameters that determine the initial distribution of power among the modes of an optical fiber.

launching fiber An optical fiber used in conjunction with a source to excite the modes of another fiber in a particular fashion. They are most often used in test systems to improve the precision of measurements. Also called an *injection fiber.*

LAWN An acronym for <u>L</u>ocal <u>A</u>rea <u>W</u>ireless <u>N</u>etwork.

Lawrence Livermore National Laboratory (LLNL) A research organization operated by the University of California founded on September 2, 1952 and under contract with the U.S. Department of Energy (DOE).

lay The length of the helical arrangement formed by wrapping the individual wire pairs, fibers, and the like, of a cable around a common center. The term may apply to individual elements or to bundles of elements within the cable. Also called *lay length, pitch,* and *strand lay.* See also *strand lay* and *twisted pair* (*TNP*).

lay length **(1)** In general, in cables (including: electronic, mechanical support, and fiber optic cables) having the media wrapped helically around a central axis, the longitudinal distance along the cable required for one complete helical wrap, i.e., the total cable length divided by the total number of wraps. **(2)** In fiber optics, the longitudinal distance along the cable required for one complete helical wrap of the optical fiber around a central strength member. In many fiber optic cable designs, the *lay length* is shorter than in metallic communications cables of similar diameter, so as to avoid overstressing the fibers during the pulling associated with the installation operation. Wraps in fiber optic cables are for mechanical reasons, not electronic reasons as is the case with metallic twisted pairs. Wire pairs are twisted to reduce electromagnetic coupling; fiber optic pairs are not given such twists. Also called *lay, pitch,* and *strand lay.* See also *strand lay* and *twisted pair.*

layer **(1)** In communications networks, the hardware and software protocols comprising the hierarchical architecture of a specific network type. Each *layer* (or level) performs specific functions for the next higher level while shielding the higher level from the functions of lower level. An example is the seven-layer ISO/OSI network Reference Model. More formally called a *protocol stack.* See also *OSI.* **(2)** See *ionosphere.*

layer management entity (LME) A mechanism, in the ISO/OSI network management model, by which layers can communicate with each other to (1) exchange information and (2) to access management elements at different layers. Also called *hooks.*

layer models In communications and networking, layers are used to partition communications responsibilities and to distinguish the types of network and application-based activities that are carried out.

Examples of layered communications models include:

* The ISO/OSI seven-layer Reference Model for describing network activities. Its layers range from the physical layer (where details of cable connections and electrical signaling are specified) to the application layer (where details of the current application and network services are defined),
* IBM's Systems Network Architecture (SNA),
* Digital's DECnet, and
* The Internet TCP/IP model.

LBA An abbreviation of <u>L</u>ogical <u>B</u>lock <u>A</u>ddress.

LBO An abbreviation of <u>L</u>ine <u>B</u>uild <u>O</u>ut.

LBRV An abbreviation of <u>L</u>ow <u>B</u>it <u>R</u>ate <u>V</u>oice.

LBS An abbreviation of <u>L</u>AN <u>B</u>ridge <u>S</u>erver.

LBT An abbreviation of <u>L</u>isten <u>B</u>efore <u>T</u>alk.

LC An abbreviation of <u>L</u>ocal <u>C</u>hannel.

LCC An abbreviation of <u>L</u>ost <u>C</u>alls <u>C</u>leared.

LCD **(1)** An abbreviation of <u>L</u>iquid <u>C</u>rystal <u>D</u>isplay. A type of display made of a material whose light transmission characteristics change when a voltage is applied.

The material (a liquid crystal) is sandwiched between two layers of glass. The patterns to be displayed are printed on the glass (next to the liquid crystal material) with a transparent electrical conductor (e.g., tin oxide). When a voltage is applied to the conductors, the liquid crystal next to the electrode becomes visible when viewed in polarized light. Below the rear polarizer is either a light source (for a

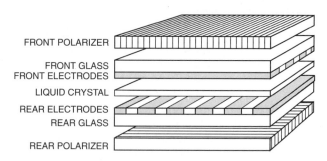

FRONT POLARIZER
FRONT GLASS
FRONT ELECTRODES
LIQUID CRYSTAL
REAR ELECTRODES
REAR GLASS
REAR POLARIZER

transmissive LCD) or a reflector (for a *reflective LCD*). **(2)** An abbreviation of <u>L</u>ost <u>C</u>alls <u>D</u>elayed. **(3)** An abbreviation of <u>L</u>oss of <u>C</u>ell <u>D</u>elineation. An ATM receiver cannot identify any cells in the receive channel.

LCI An abbreviation of <u>L</u>ogical <u>C</u>hannel <u>I</u>dentifier.

LCM An abbreviation of <u>L</u>east (or <u>L</u>owest) <u>C</u>ommon <u>M</u>ultiple.

LCR **(1)** An abbreviation of <u>L</u>east <u>C</u>ost <u>R</u>outing. **(2)** An abbreviation of <u>L</u>ine <u>C</u>ontrol <u>R</u>egister.

LD **(1)** In telephony. An abbreviation of <u>L</u>ong <u>D</u>istance. **(2)** An abbreviation of <u>L</u>aser <u>D</u>iode.

LD-CELP An abbreviation of <u>L</u>ow <u>D</u>elay <u>CELP</u>. A voice compression scheme with little processing delay described in the ITU-T G.728 recommendation.

LDAP An acronym for <u>L</u>ightweight <u>D</u>irectory <u>A</u>ccess <u>P</u>rotocol. A specification for directory structures that allow directory-independent applications to be developed. *LDAP* version 2.0 is described in Internet RFC1777 and RFC1778.

LDC In telephony, an abbreviation of <u>L</u>ong <u>D</u>istance <u>C</u>arrier.

LDDS An abbreviation of <u>L</u>imited <u>D</u>istance <u>D</u>ata <u>S</u>ervice.

LDF An abbreviation of <u>L</u>ocal <u>D</u>istribution <u>F</u>rame.

LDM An abbreviation of <u>L</u>imited <u>D</u>istance <u>M</u>odem.

lead **(1)** A condition of the phase or time relationship of two or more waveforms where one waveform is used as a reference and the others are ahead in phase or time. See also *lag*. **(2)** A wire or cable to a device or instrument.

lead in The wire from an antenna to a receiver.

leader That section of a tape used to secure the tape to its take up reel. The *leader* generally does not have any information on it and may not even have the capacity to store information. *Leader* is used on film and magnetic tape.

leading The spacing between lines of text. This is defined when a font is designed but can often be altered in order to change the appearance of the text or for special effects. It is measured in points and is normally 120% of the height of the text. See also *kerning* and *tracking*.

leading edge The first step or transition in a pulse.

LEAF An acronym for <u>L</u>aw <u>E</u>nforcement <u>A</u>ccess <u>F</u>ield. See also *Clipper chip.*

leaf object In Novell's NetWare Directory Services (NDS), an object that represents an actual network entity, such as users, devices, and lists. Five basic types of leaf objects are defined: user-related, server-related, printer-related, informational, and miscellaneous. A list of these leaf objects is shown in the following table. Each of these types includes several subobject types. See also *container object* and *NetWare Directory Service (NDS)*.

NDS Leaf Object	Description
AFP Server	A NetWare node that supports the AppleTalk Filing Protocol and that is probably functioning as a server in an AppleTalk network.
Alias	Refers, or points, to a different location. An alias can be used to help simplify access to a particular object (for example, by using a local object to point to the object entry in a different part of the Directory).
Bindery	Included for backward-compatibility with earlier NetWare versions. Bindery objects are placed in the Directory by the migration (network upgrade) utilities, so the binderies from version 3.x servers have something to access in the Directory.
Bindery Queue	Included for backward-compatibility with earlier NetWare versions.
Computer	Represents a particular node on the network.
Directory Map	Contains information about the network's file system, which is not encompassed by the NDS Directory. The information in a Directory Map provides path information, rather than actually showing the structure of the file system's directory. This information is useful for login scripts.
Group	Represents a list of User objects. The network supervisor can assign rights to all the users on this list simply by assigning the rights to the group.
NetWare Server	Represents any server running any version of NetWare.
Organizational Role	Represents a function or position within an organization, such as Leader, Consultant, or Moderator.
Printer	Represents a network printer.
Print Queue	Represents a network print queue.
Print Server	Represents a network print server.
Profile	Represents a shared login script. The script might be shared, for example, by users who need to do similar things during the login process but who are located in different containers.
User	Represents an individual who can log into the network and use resources. Properties associated with User objects include those concerned with the actual person as an individual (name, telephone number, address, and so on) and as a network entity (password and account information, access rights, and so on).
Unknown	Used for an object that cannot be identified as belonging to any other object type, possibly because the object has become corrupted in some way.
Volume	Represents a physical volume on the network.

leaf site A node on USENet that talks to only one other node instead of passing information on to others, i.e., it has no downstream nodes.

leakage (1) Undesired currents flowing between parts of a signal transmission system or between a part of the transmission system and points outside the system. **(2)** Currents flowing through or around an insulator. **(3)** Currents flowing through a semiconductor device when it is in the "off" state.

leakage flux Any magnetic flux lines produced by a current in a coil that are not coupled into all turns of all windings of the device.

leakage inductance That part of an inductance which is not related to all parts of the exciting winding.

leaky bridge A term used to describe a LAN bridge that forwards packets from one LAN to another, even if the packet should not be forwarded.

leaky bucket algorithm A form of *flow control* that checks an arriving data stream against the traffic-shaping parameters specified by the sending node. Cells arriving at an ATM network switch are transferred to a buffer memory (the bucket). The buffer is allowed to reach capacity but not overflow. Cells transferred out (the leak) of the buffer are routed toward their destinations.

leaky bucket counter An event counter that is incremented each time an event occurs and is periodically decremented by a fixed value.

leaky mode In fiber optics, a mode in which the field decays monotonically for a finite distance in the transverse direction but becomes oscillatory everywhere beyond that finite distance. *Leaky modes* experience attenuation, even if the waveguide is perfect in every respect. Also called *tunneling mode* and *leaky ray* in the terminology of geometric optics.

leaky PBX A PBX that allows outgoing calls from incoming outside lines.

leaky ray In fiber optics, a ray for which geometric optics would predict total internal reflection at the boundary between the core and the cladding but which suffers loss by virtue of the curved core boundary. Also called a *tunneling mode* or a *leaky mode* in the terminology of mode descriptors.

leap second An occasional one-second adjustment, added to or subtracted from *Coordinated Universal Time* (*UTC*) to bring it into approximate synchronism with *UT-1* (the time scale based on the rotation of the Earth). Adjustments are made such that UTC never deviates from UT-1 by more than 0.9 s. Adjustments, when required, are made primarily in the last minute of December 31; however they can be made on June 30. The last minute of the day on which an adjustment is made has 61 or 59 seconds. See also *time.*

leased channel A point-to-point communication channel reserved for the sole use of a single leasing customer.

leased line A communication line separate from the switched telephone network which has a discrete and permanent path between its end points. A *leased line* is a private telephone line reserved exclusively for the customer's use. Several different bandwidths are available from the location telephone administration ranging from a single 3000-Hz voice channel to a 45-Mbps T-3 line or even higher.

Some of the services include:

- *Dataphone Digital Services* (*DDS*): DDS provides synchronous transmission of digital signals at up to 56 kbps. Subrate (lower-speed) services are also available, at 2400 to 19,200 bps.

- *56/64 kbps lines:* In countries conforming to the ITU-T recommendations (Mexico and Europe, for example), the full 64 kbps is available. However, in the United States and Japan, an 8-kbps bandwidth is used for administrative and control overhead, leaving only 56 kbps for the subscriber.

  Similar lines are also available on a dial-up (nondedicated lines) basis—Switched 56 services, for example.
- *Fractional T-1 lines:* Lines are built up in increments of 64 kbps.
- *T-1/E-1 lines:* These lines provide 1.544 Mbps for T-1 (available in the United States and Japan) and 2.048 Mbps for E-1 (available in countries conforming to the ITU-T recommendations).
- *T-3/E-3:* This is similar to T-1/E-1, but the rates are 44.736 Mbps and 34.368 Mbps, respectively.

Also called a *dedicated channel, dedicated circuit, dedicated connection, dedicated line,* and *private line.*

least cost routing (LCR) In private branch exchange (PBX) telephone systems, a feature that selects the most economical path to a destination from the available resources. The algorithm can be implemented in a number of ways, e.g.,

- It may look at only available lines and exclude the dial-up public switched network (PSTN) completely.
- It may include caller holding time or number of call attempts to allow access to the PSTN.
- Or it may simply try the lowest cost line and fall back to the PSTN if the first choice is unavailable.

least significant bit (LSB) In a sequence of bits (either alone or as a group of bytes), the low-order bit of the group (usually the rightmost bit).

least significant character (LSC) In a sequence or string of characters, the low-order character of the group (usually the rightmost character).

least significant digit (LSD) In a sequence of one or more digits, the low-order digit of the group (usually the rightmost digit).

LEC (1) An acronym for Local Exchange Carrier. **(2)** An acronym from Light Energy Converter. **(3)** An acronym from LAN Emulation Client.

LECS An acronym for LAN Emulation Configuration Server.

LED An acronym for Light Emitting Diode.

leeched line A slang expression for a connection to the Internet wherein PC users on a LAN (already connected to the Internet) become full-fledged Internet hosts simply by installing software on the PC. The required programs (such as Eudora or Mosaic) provide the necessary TCP/IP protocol support.

left-hand polarized wave An elliptically or circularly polarized wave in which the electric field vector, observed on a fixed plane, normal to the direction of propagation, and in the direction of propagation, rotates with time in a left-hand or anticlockwise direction. Also called *anticlockwise polarized wave.*

leg One segment of a multisegment route, path, or link that traverses one or more networks. Examples of *legs* include:

- Sequential microwave links between switching centers.
- A transoceanic cable between shore transmission facilities.
- The connection between nodes in a hierarchical network.

Sometimes also called a *hop.*

legacy system An old installed system perceived as too costly to replace or upgrade. For example, the mainframe computer of a large company may contain the company's payroll, inventory, and billing, and the cost of transferring all of the information to a new system may be prohibitive.

legacy wiring Wiring that is already installed in a business or residence, generally from a previous application. Legacy wiring may or may not be suitable for the new application.

LEIN An abbreviation of Law Enforcement Information Network.

LEM (1) An acronym for Logic Enhanced Memory. (2) An acronym from antenna effective LEngth for Magnetic-field antennas. (3) An acronym from Logical End of Media.

Lempel-Ziv compression (LZ) A substitutional data compression scheme, proposed by Jakob Ziv and Abraham Lempel in 1977 and 1978, that uses a sliding dictionary. There are two main schemes, LZ77 and LZ78. It is the basis of the data compression algorithm used by ITU-T V.42 bis. See also *compression* (2).

Lempel-Ziv-Welch (LZW) A patented data compression algorithm designed by Terry Welch in 1984 for hardware for high-performance disk controllers. It is a variant of LZ78, one of the two Lempel-Ziv compression schemes. *LZW* compression and decompression are licensed under Unisys Corporation's 1984 U.S. Patent 4,558,302 and equivalent foreign patents. (A patent can't describe algorithms or mathematical methods in most countries.) *LSW* is used by many data compression programs, including CompuServe's GIF files and Mosaic.

The *LZW* algorithm relies on the recurrence of byte sequences (strings) in its input to achieve data compression. The algorithm maintains a table, mapping input strings to their associated output codes (compressed data). Functionally, the table is initialized to contain mappings for all possible strings of length one (one byte). Input is then accepted one byte at a time. A table search is performed to find the longest initial string present in the table. The code for that string is output, and the string is extended with one more input byte. A new entry is added to the table mapping the extended string to the next unused code. This process repeats on subsequent bytes. The number of bits in an output code, and hence the maximum number of entries in the table, is fixed, and once this limit is reached, no more entries may be added. See also *compression* (2).

LEO An acronym for Low Earth Orbit.

LEOS An acronym for Low Earth Orbiting Satellite system.

LES An abbreviation of LAN Emulation Server.

letterbomb A piece of e-mail containing live data intended to do wicked things to the recipient's machine or terminal. It is possible, for example, to send letterbombs that will lock up some specific kinds of terminals when they are viewed, so thoroughly that the user must reset the machine by turning the power off and on again. Under UNIX, a letterbomb can also try to get part of its contents interpreted as a shell command to the mailer.

Letterbombs may under some circumstances be classed as Trojan horses. For example, under UNIX, a letterbomb can try to get part of its content interpreted as a shell command to the mailer (which could lead to disastrous results). Also called a *mail bomb*.

level (1) The absolute or relative signal *magnitude* (voltage, current, or power) at a specified point in a circuit or system. (2) A tier or layer of a hierarchical system, e.g., the seven layers of the ISO/OSI Reference Model.

level alignment The transmission level adjustment of individual legs so as to prevent signal overloading problems in the subsystems.

level n cable A five-level system to categorize UTP cable similar to the EIA/TIA-568 "category n cable" system.

- *Level 1:* Voice grade cable, which is suitable for use in the "plain old telephone system" (or POTS). *Level 1* cable can handle data rates up to 1 Mbps.
- *Level 2:* Data grade cable that is capable of transmission speeds to 4 Mbps. This level corresponds roughly to IBM's Type 3 cable described in IBM's Cabling System. *Level 2* cable meets the requirements for the 1Base5 (StarLAN) Ethernet network developed by AT&T and Artisoft's LANtastic.
- *Level 3:* Data grade cable that is capable of transmission speeds to 16 Mbps. *Level 3* cable corresponds to the EIA/TIA Category 3 cable specification. *Level 3* cable is used in IBM's 4-Mbps or 16-Mbps Token Ring networks and in 10BaseT networks.
- *Level 4:* Data grade cable that is capable of transmission speeds to 20 Mbps. *Level 4* cable corresponds to the EIA/TIA Category 4 cable specification. *Level 4* cable is in ARCnet Plus networks, a 20-Mbps version of the ARCnet network architecture.
- *Level 5:* Data grade cable that is capable of transmission speeds to 100 Mbps. *Level 5* cable corresponds to the EIA/TIA Category 5 cable specification. *Level 5* cable is used for Copper Distributed Data Interface (CDDI—a copper-based implementation of the 100-Mbps FDDI network architecture) and for the proposed 100-Mbps 100BaseX version of Ethernet.

See also *cable, category n cable, type n cable, UL certified cable levels,* and *wire.*

level n relay An interconnection device between two entities at the nth level of the ISO/OSI Reference Model. For example:

- *Level 1 relay*—a synonym for a repeater. A device that operates at the physical layer (layer 1).
- *Level 2 relay*—a synonym for a bridge. A device that operates at the data link layer (layer 2).
- *Level 3 relay*—a synonym for a bridge-router. A device that operates at the network layer (layer 3).
- *Level 7 relay*—a synonym for a full router or a gateway. A device that operates at the application layer (layer 7).

LEX An acronym for Line EXchange.

LF (1) An abbreviation of Low Frequency. (2) An abbreviation of Line Feed. (3) An abbreviation of Linear Filter. (4) An abbreviation of Line Finder. (5) An abbreviation of Logical File.

LFB (1) An abbreviation of Limited Frequency Band. (2) An abbreviation of Low Frequency Beacon. (3) An abbreviation of Look ahead For Busy.

LFC An abbreviation of Logic Flow Chart.

LFD (1) An abbreviation of Low Frequency Disturbance. (2) An abbreviation of telephone Line Fault Detector. (3) An abbreviation of Local Frequency Distributioin.

LFN An abbreviation of Logical File Name.

LFPS An abbreviation of Low Frequency Phase Shifter.

LFSA An abbreviation of List of Frequently Seen Acronyms.

LFSR An abbreviation of Linear Feedback Shift Register. A method of generating a pseudorandom sequence.

LFU An abbreviation of Least Frequently Used.

LGA An abbreviation of Low Gain Antenna.

LGN An abbreviation of Logical Group Number.

LH An abbreviation of <u>L</u>inear <u>H</u>ybrid.

LHA A successor to LHARC, hence LHA is likely an abbreviation from <u>L</u>empel <u>H</u>uffman <u>A</u>rchiver. It is a shareware data compression and archive program developed by Haruyasu Yoshizaki circa 1988. Files generated by the program usually have the extension .LZH. The *LHA* command format is:

LHA <C> [/*x*[*n*|WDIR]] arcfile [source_files. . .]

where:

<C> is the command, that is,

COMMANDS

<C>	Description
a	Add files
d	Delete files
e	Extract files
f	Freshen files
l	List of files
m	Move files
p	disPlay files
s	make a Self-extracting archive
t	Test the integrity of an archive
u	Update files
v	View listing of files with pathnames
x	eXtract files with pathnames

/*x*[*n*] or -*x*[*n*] is the option, that is,

OPTIONS

/x	n	Description
a	0(−)	ignore hidden & system files
	1(+)	include all files and preserve Attributes
c	0(−)	
	1(+)	skip time-stamp Check
h	0(−)	
	1(+)	select Header level (default = 1)
i	0(−)	Ignore character case
	1(+)	do not Ignore character case
l	0	dispLay file name only
	1	full pathname stored in two line display
	2	full pathname accessed in two line display
m	0	Message on duplicate file
	1	no Message, overwrite duplicate filespec
	2	no Message, duplicate file extension = .nnn
n	0	display progress, name, and compression
	1	Disable progress indicator
	2	Disable file and compression indicator
o	0(−)	
	1(+)	use Old lh113 method, set -h0
p	0(−)	
	1(+)	distinguish full Path names
r	0	non-Recursive, specified path names only
	1	Recursively collect files
	2	tree structure Recursed and archived
s	0(−)	
	1(+)	Skips for later date/time are not reported
t	0(−)	archive's Time-stamp is creation date
	1(+)	archive's Time-stamp set to latest source_file
w	0(−)	use directory where "outfile" resides
	1(+)	assign Work directory
x	0(−)	
	1(+)	allow eXtended file names (full filespec)

OPTIONS (*CONTINUED*)

/x	n	Description
z	0	compress all files
	1	Zero compression (store only)
	2	compress all files except *.ARC, *.LZH, *.LZS, *.PAK, *. ZIP, *.ZOO
	<ext>	compress all files except *.ext
-	1	allow "@" in filenames
	2	allow "@" and/or "-" in filenames

[WDIR] is the working directory,

"arcfile" is the path and filename of the program outout (the compacted file of a compression command is specified or the recovered files if extraction is commanded), and

"source_file. . ." is an optional list of path and filenames to be used as the input to LHA. See also *compression* (*2*).

LHARC A shareware file compression utility, developed and introduced by Haruyasu Yoshizaki in 1988. See also *LHA*.

LIC An acronym for <u>L</u>inear <u>I</u>ntegrated <u>C</u>ircuit.

LIFO An acronym for <u>L</u>ast-<u>I</u>n-<u>F</u>irst-<u>O</u>ut.

light (1) In the strictest sense, that portion of the electromagnetic spectrum which can be perceived by a human eye—generally about 0.38 μm to 0.77 μm. **(2)** In a more general definition, that portion of the electromagnetic spectrum that can be handled with normal optical techniques. It ranges from ultraviolet to infrared (about 0.3 μm to 30 μm). See also *frequency spectrum*.

light emitting diode (LED) A two-terminal semiconductor PN junction diode that radiates monochromatic incoherent light when a current is flowing in the forward direction. Available colors include infrared, red, orange, yellow, green, and blue. The symbol for a *LED* is:

light pipe An optical transmission element that utilizes unfocused transmission and reflections to "conduct" light from point to point. See also *optic fiber*.

light ray In geometric optics, the path of a point on a wavefront as the wavefront travels in time. The path is generally perpendicular to the wavefront.

light valve A synonym for *optical switch*.

lightguide See *optical fiber*.

lightning An electrical discharge that occurs in the atmosphere between clouds or between a cloud and ground (or a structure on the ground).

lightning protection A circuit in a DCE that protects the DCE from damage if there is an induced transient (caused by lightning) on the communication line. The circuit is a high-speed device that clamps the voltage on the telephone line to a voltage less than the failure voltage of the DCE line interface circuits. Devices such as metal oxide varistors (MOVs) and zener diodes are typically used.

Lightweight Directory Access Protocol (LDAP) An Internet standard intended to serve as a universal mechanism for locating and managing network resources and users across diversified directories.

LIM (1) An acronym for Lotus Intel Microsoft, members of the consortium that developed the expanded memory specification (EMS). The expanded memory specification was developed in order to make more memory available to 8086 microprocessors (which cannot operate in protected mode, as is needed to access memory addresses above 1 MB).

Expanded memory is composed of a number of pages of 16 kilobytes each. Each page is able to be switched into a common address area in the upper memory area (between 640 KB and 1MB)—one at a time. (2) An acronym for Line Interface Module.

limited availability In telephony, availability that is less than the number of output ports. See also *blocking.*

limited distance data service (LDDS) In telecommunications, a class of service offered by some carriers which provides digital transmission capabilities over short distances. It uses line drivers (limited distance modems or short-haul modems) instead of true modems.

limited distance modem (LDM) A short-haul modem, which is generally designed for very high-speed transmissions over short distances. They are not modems in the strictest sense because they operate at baseband rather than by modulating a carrier. (That is, the signal remains in digital form and is never converted to an analog signal.) Also called *baseband modems, line driver,* and *short-haul modems.*

limiter A device or circuit that controls a signal passing through it in such a manner that the signal does not exceed a specified value. The signal may be limited in analog amplitude or digital flow rate or in several other ways. The device may limit voltage, current, or some other characteristic of the signal so as to automatically prevent it from exceeding the specified value. See also *clipping.*

limiting Any process by which a specified characteristic (usually amplitude) of the output waveform of a device is prevented from exceeding a predetermined value. Two classifications of limiting include:
- *Hard limiting* (clipping)—a limiting action in which:
    - The limiter exerts negligible effect on the outout signal for input signals within the permitted dynamic range, and
    - The limiter maintains the output at the maximum permitted level for input signals outside the permitted input dynamic range. See also *clipping.*
- *Soft limiting*—a limiting action in which:
    The output level is attenuated by either a fixed amount or an increasing amount as the instantaneous input level exceeds a specified threshold. The output waveform is therefore distorted but not clipped. Compare with *automatic gain control (AGC).*

line (1) Any wire or cable (electrical or optical) that is used to transmit power or signals. (2) In communications, a connection between a sending and receiving terminal—also called a *link.* (3) In telephony, the circuit between a central office and a customer terminal.

line barrier The device that protects the PTT (Post, Telephone, and Telegraph) for any hazardous voltage generated by equipment at the subscriber premises and also protects the subscriber from any hazardous voltage appearing on the line. Devices used for this protection include *carbon blocks, gas tubes,* and Metal Oxide Varistors (MOVs).

line build out (LBO) network A network installed in series with a transmission line which makes it appear longer than its physical length to other system components.

For example, *LBOs* are used on loaded telephone lines where the last section is too short to satisfy the load coil placement rules. They are also used on T-1 lines to set the required transmission loss values. (Typical T-1 *LBOs* provide 0.0, 7.5, and 15 dB of loss.)

line busy tone In telephony, a tone generated by the switching office connected to the called party which indicates that the called party line is off-hook and not available. See also *call progress tones.*

line card In telephony, a plug-in printed circuit board which serves as the interface between a line and communications or switching equipment. *Line cards* frequently operate lamps, ringing, holding, and other features associated with one or more telephone lines in a switching system. See also *BORSCHT.*

line circuit In telephony, that portion of a switching system which interfaces the subscriber line to the switch fabric. It generally detects the subscriber on/off hook status, injects ringing, and provides the two- to four-wire hybrid function. See also *BORSCHT.*

line coding See *waveform codes.*

line conditioner A device for keeping the voltage supply to a device within an "acceptable" range. Line conditioners are used where there is a likelihood of brownouts (power sags), excessive voltage, or voltage spikes (transients). Also called *line stabilizer/line conditioners (LS/LC), power conditioners,* and *voltage regulators.* See also *conditioning.*

line control register (LCR) In a Universal Asynchronus Receiver Transmitter, a register that is used to specify a parity type.

line current In telephony, the current flowing in a subscriber loop. In an off-hook circuit, it may range from over 100 mA to about 20 mA, depending on the loop length, gage of wire used in the loop, and type of equipment at the subscriber premises. Although telephone instruments will operate over a wide current range, 40 to 50 mA is desired for optimal operation.

line driver (1) A baseband transmission device. (2) A component used for matching the impedance of a transmission line and for delivering more power to the line than is available from the signal source. *Line drivers* generally include both a transmitter and a receiver and are used to extend the transmission range between devices that are connected directly to each other. In some situations, *line drivers* can be used instead of modems but only for short distances (a few miles). Also called a *limited distance modem (LDM)* or a *short-haul modem (SHM).*

line editor A basic text editor that allows a user to modify only one line at a time. See also *full screen editor.*

line equipment Equipment in a central office that connects to a subscriber line and serves the communications needs of that line.

line finder In telephony, the switching element in a step-by-step office that recognizes that a calling party is awaiting a dial tone. It identifies the line and connects it to the switching system. A *line finder* is shared by 100 to 200 lines.

line group In telephony, multiple lines that can be activated or deactivated as a group.

line hit In transmission systems, a brief burst of interference on a line that causes a loss of bit integrity in the data stream.

line hold A function in telephone switching systems (a PBX, for example) that releases the telephone instrument from a line but provides the necessary conditions on the line to prevent the remote switching equipment from terminating the connection. The local switching equipment (PBX) generally provides a winking lamp indication at every telephone which has that line appearing on it.

The telephone instrument may be used for a variety of functions while a line is held—e.g., to establish a connection to another line, broadcast a message on a paging system, transfer the call, or reestablish a connection to the held line.

line insulation test (LIT) In telephony, a test performed from a central office that automatically checks the lines for shorts, grounds, and foreign voltages.

line level In transmission systems, the power of a signal at a specified point in the transmission path. The value is generally expressed in decibels (dB).

line load In telephony, a representation of the amount of usage a line is getting at a particular time. *Line load* is usually expressed as a percentage of capacity.

line load control In telephony, a means of selectively restricting call attempts during emergencies so that essential traffic may be handled.

line lockout A means of handling permanent line signals to prevent further recognition as a call attempt.

line noise Any of several forms of electrical interference that may be present on a communications line. Noise sources include both natural causes (e.g., lightning or Johnson) and man-made (crosstalk or switching equipment on and off for example).

line of sight In wireless communications, the condition in which a signal is transmitted from a source to a destination without reflections or refractions. There must be no obstructions in the path between the transmitter and receiver.

Examples of *line-of-sight* transmissions include those of laser and microwave communications links.

line powered In telephony, equipment powered exclusively by the energy delivered through the communications transmission line from the central office talk battery.

line printer daemon (LPD) In UNIX, a daemon program in Berkeley UNIX spooler implementations that provides printer support.

line side The connections to a communications circuit between two DTEs. The connection of a transmission path to outside plant facilities such as the subscriber side of a switching office or trunks and channels.

line signal standards See *DS-n* and *OC-n*.

line speed The maximum rate (in bits per second—bps) at which data can reliably be transferred across a communications link using a specific hardware arrangement. See also *baud code* and *connection rate*.

line switching A synonym for circuit switching.

line termination equipment In telecommunications, any equipment that can be used to send signals, such as line cards, modems, telephones, multiplexers, hubs, concentrators, and so on.

line transformer A transformer for interconnecting a transmission line and DTE for such purposes as isolation, impedance matching, or line balance matching. See also *balun*.

line turnaround The reversing of transmission direction on a half-duplex circuit.

line turnaround time The time interval required to accomplish *line turnaround*.

linear A term that describes how the output signal of a circuit or system is related to the input signal. Specifically, the output is a *constant* multiple of the input at all signal levels. In a nonlinear device, circuit, or system, the multiplier is not constant for all signal levels.

linear combiner A *diversity combiner* in which the the signals on each input channel are simply added together. See also *combiner* and *diversity combiner*.

linear device A device for which the output is, within a given dynamic range, linearly proportional to the input. Mathematically,

$$\text{Output} = \text{Gain} \cdot \text{Input} + \text{Offset}$$

linear detection An amplitude demodulation method in which the output level is a linear function of the envelope of the input wave.

linear distortion Any distortion that is independent of the input signal amplitude, e.g., small signal frequency response. See also *distortion*.

linear map In mathematics, a function transforming a value from one space to another space that obeys the additive and multiplicative rules of algebra. For example, given two vectors (U and V) in the source vector space and any scalar (k) in the field over which it is a vector space, then a *linear map* (f) satisfies:

$$f(U + kV) = f(U) + kf(V)$$

Also called a *linear transform*.

linear polarization Of an electromagnetic wave, confinement of the E-field vector or H-field vector to a given plane. Historically, the orientation of a polarized electromagnetic wave has been defined in the optical regime by the orientation of the electric vector, and in the radio regime, by the orientation of the magnetic vector. Also called *plane polarization*. See also *elliptical polarization*.

linear predictive coding (LPC) A method of digitally encoding analog signals, using a single-level or multilevel sampling system in which the value of the signal at each sample time is predicted to be a linear function of the past values of the quantized signal.

LPC is related to *adaptive predictive coding* (APC) in that both use adaptive predictors. However, *LPC* uses more prediction coefficients to permit use of a lower information bit rate than APC. *LPC* can produce a digitized voice signal at low bit rates (2400 bps). For example, *LPC* is used in secure telephone units (STU-III), which were developed by the National Security Agency.

linear programming (LP) A procedure for locating the maximum or minimum of a linear function of variables that are subject to linear constraints. Also called *linear optimization*.

linear system Any system, circuit, or device whose output signal is represented by a constant multiplier times the input signal (for any input level). That is, the output is proportional to the input. See also *linearity*.

linearity The property of a device or system in which its output is related to the input by a constant multiplier for all defined input values. For example, if input signals I_1 and I_2 result in output signals O_1 and O_2, then input signal $a \cdot I_1 + b \cdot I_2$ will result in the output signal $a \cdot O_1 + b \cdot O_2$ where a and b are arbitrary but fixed values.

linearly polarized (LP) mode A transmission mode for which the field components in the direction of propagation are small compared to components perpendicular to that direction. The *LP* description is a valid approximation for a weakly guiding optical fiber.

link **(1)** In communications, a general term used to indicate the existence of communications facilities between two points. **(2)** The circuit (or logical circuit) between a sender and receiver. The circuit includes all equipment in the transmission path including transmission media, repeaters, and switches (even when different physical paths are used). In all cases, the type of link, such as data link, downlink, duplex link, fiber optic link, line-of-sight link, point-to-point link, radio link, and satellite link, should be identified. A link may be simplex, half-duplex, or duplex. **(3)** The communications facilities between adjacent nodes of a network. **(4)** A portion of a circuit connected in tandem (series) with other portions. **(5)** A connection between switching stages of a switching system. **(6)** The over-current protecting element in a fuse, i.e., a *fusible link.* **(7)** The tie between two hypertext documents or elements with the document.

link control ITU-T X.25 level 2 control for linking DTE and DCE. It includes link initialization, link establishment, link disconnect, packet flow control, and reset.

link state algorithm A routing procedure in which each router knows the parameters to each adjacent router and can broadcast this information to all other routers in a *link state packet* (*LSP*). If a router updates its link state database, a new LSP is broadcast which replaces the older information at all other routers. Rather than storing actual routes, *link state algorithms* store the information needed to generate paths (e.g., network traffic, connection speed, cost, etc.). For example, the ISO *open shortest path first* (*OSPF*) algorithm is a *link state algorithm.* See also *distance vector algorithm.*

link state packet (LSP) A packet, in a link state routing protocol, that contains information about all the connections for a router, including information about all the neighbors for that packet and the "cost" (in money, time, error rate, or other currency) of the link to each neighbor. This packet is broadcast to all other routers of the internetwork.

Lippell code Any of several codes that result from reducing the number of encoding tracks on a mechanical position encoding disc to less than the number of bits in the encoded digital word. The codes may be designed to be either *polystrophic* or *monostrophic codes.*

The following figure demonstrates two polystrophic implementations (Lippell 1 and 2) and a unit distance (Lippell 3) coding discs. The discs are all in position 1, and the shaded area represents a 1 in the code output. See also *codes.*

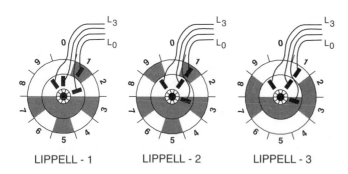

LIPPELL - 1   LIPPELL - 2   LIPPELL - 3

CODE OUTPUT AT EACH POSITION

Position	Lippell - 1				Lippell - 2				Lippell - 3			
0	0	0	0	0	0	0	0	0	0	0	0	0
1	0	0	0	1	0	0	1	1	0	0	1	0
2	0	0	1	0	0	1	1	0	0	0	1	1
3	0	0	1	1	0	1	1	1	0	1	1	1
4	0	1	1	0	1	1	1	0	0	1	1	0
5	1	1	1	1	1	1	1	1	1	1	1	0
6	1	1	1	0	1	1	0	0	1	1	0	0
7	1	1	0	1	1	1	0	1	1	1	0	1
8	1	1	0	0	1	0	0	0	1	0	0	1
9	1	0	0	1	1	0	0	1	1	0	0	0

listen before talk (LBT) A fundamental rule of a carrier sense multiple access, collision detect (CSMA/CD) media access method. The rule states: A node wishing to send a packet onto the network first listens for signals that indicate that the network is in use. If no signal is heard, the node begins transmitting.

listen while talk (LWT) A fundamental rule of a carrier sense multiple access, collision detect (CSMA/CD) media access method. The rule states: A node should keep listening for a collision signal even while transmitting.

listserv A mailing list utility program found on network job entry (NJE) systems (such as BITNET and EARN). E-mail based forums, on a variety of subjects, are based on *listserv;* that is, the *listserv* program automatically distributes all e-mail received to all subscribing users.

LIT An acronym for Line Insulation Test.

little endian A format for transmitting (or storing) serial binary data in which the least significant bit is the first transferred. This is in contrast to *big endian* in which the most significant bit is first.

The term comes from Jonathan Swift's *Gulliver's Travels,* wherein wars were fought over which end of a hard-boiled egg should be broken first. Similarly, many arguments focus on whether *big endian* or *little endian* data storage is better.

LIU An abbreviation of Line Interface Unit.

liveware Slang for "people." Also called *jellyware* and *wetware.*

litz A term derived from the German word *litzendraht* meaning woven wire. It is constructed of multiple insulated strands woven and twisted in a pattern to minimize losses due to *skin effect.* See also *lay, strand lay,* and *twisted pair.*

LL **(1)** An abbreviation of Local Loopback (RS-232 signal LL, RS-449 signal LL, and ITU-T signal 141). The signal generated by the local DTE causes the local DCE to internally loop its transmitted analog signal back onto the receive path (and back to the local DTE). **(2)** An abbreviation of Land Line. A communications link whose circuits are primarily cable based and fixed as opposed to being wireless (radio) and potentially mobile. **(3)** An abbreviation of Leased Line. A private telephone or telegraph line. **(4)** An abbreviation of Local Line. **(5)** An abbreviation of Long Line. **(6)** An abbreviation of Line Leg.

LLA An abbreviation of Leased Line Adapter.

LLB An abbreviation of Line LoopBack.

LLC An abbreviation of Logical Link Control. See also *IEEE 802.x Standard* and *protocol data unit* (*PDU*).

LLC2 An abbreviation of Logical Link Control Type 2.

LLE An abbreviation of Long Line Equipment.

LLF An abbreviation of Line Link Frame.

LLG An abbreviation of Logical Line Group.

LLI An abbreviation of Low Level Interface.

LLN An abbreviation of Line Link Network.

LLP An abbreviation of Line Link Pulsing.

LM/X An abbreviation of LAN Manager for UNIX systems.

LMAO An abbreviation of Laughed (or Laughing) My Ass Off.

LMDS An abbreviation of Local Multipoint Distribution Service.

LME (**1**) An abbreviation of <u>L</u>ayer <u>M</u>anagement <u>E</u>ntity. (**2**) An abbreviation of <u>L</u>ongest <u>M</u>atch search <u>E</u>ngine.

LMI An abbreviation of <u>L</u>ocal <u>M</u>anagement <u>I</u>nterface.

LMS An abbreviation of <u>L</u>ocal <u>M</u>easured <u>S</u>ervice. A method of telephone rate calculation that is based on usage. An alternative to flat rate billing.

LMT An abbreviation of <u>L</u>ocal <u>M</u>ean <u>T</u>ime. See *time* (*3*).

LMU An abbreviation of <u>L</u>AN <u>M</u>anager for <u>UNIX</u>. An implementation of LAN Manager for UNIX servers. LAN Manager is Microsoft's server-based network operating system.

LMX An abbreviation of <u>L</u> <u>M</u>ultiple<u>X</u>.

In analog communications, a hierarchy of channel groupings (group, supergroup, mastergroup, and jumbo group). See also *multiplex hierarchy*.

LN An abbreviation of <u>L</u>oad <u>N</u>umber. See also *ringer equivalence number* (*REN*).

LNA (**1**) An abbreviation of <u>L</u>ow <u>N</u>oise <u>A</u>mplifier. (**2**) An abbreviation of <u>L</u>aunch <u>N</u>umerical <u>A</u>perture.

LNI An abbreviation of <u>L</u>ocal <u>N</u>etwork <u>I</u>nterconnect.

LNNI An abbreviation of <u>L</u>AN emulation <u>N</u>etwork to <u>N</u>etwork <u>I</u>nterface.

LNR An abbreviation of <u>L</u>ow <u>N</u>oise <u>R</u>eceiver.

LNRU An abbreviation of <u>L</u>ike <u>N</u>ew, <u>R</u>epaired and <u>U</u>pdated.

A term indicating the equipment has not only been repaired to make function but has been updated to make it compliant with current specifications. Further, exposed surfaces are replaced to make the equipment appear new as well as function like new. See also *repair and quick clean* (*RQC*), *repair only* (*RO*), and *repair, update, and refurbish* (*RUR*).

LO (**1**) An abbreviation of <u>L</u>ocal <u>O</u>scillator. (**2**) An abbreviation of <u>L</u>ine <u>O</u>ccupancy. (**3**) An abbreviation of <u>L</u>ock-<u>O</u>ut.

load (**1**) In electronics, the amount of current or power required or consumed by a device. (**2**) In electronics, a power-consuming device connected to a circuit or system. (**3**) In telephony, the amount of traffic on a line (measured in *erlangs*). (**4**) In computer networks, the amount of user activity requiring network resources. (**5**) In programming, to install a program or data into storage (either main storage, working storage, or individual registers). (**6**) To insert new data values into a database. (**7**) To place a magnetic tape reel on a tape drive, a CD-ROM in a drive, or cards into the card hopper of a card reader or punch.

load balancing The distribution of a specified quantity over a number of entities. For example:

- In computer networks, the shifting of jobs from heavily loaded resource to less loaded resource.
- In switching systems, a strategy in which messages are distributed across all available channels or routes. *Load balancing* attempts to make the traffic on available channels as evenly distributed as possible (*traffic equalization*).
- In power distribution systems, in multiple source power supplies, each source delivers an equal share of the total required power.

load capacity In pulse code modulation (PCM), the level of a sinusoidal signal, usually expressed as dBm0, that has positive and negative peaks that coincide with the positive and negative maximum values of the encoder. Also called *overload point*.

load coil In telephony, the *inductor* used in a *loaded line*. *Load coils* are constructed using *bifilar* winding on a *toroidal* core in such a way that a coupled pair is created. Also called *loading coil*. See also *loaded line*.

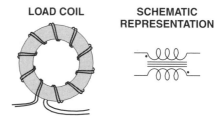

LOAD COIL   SCHEMATIC REPRESENTATION

load factor The ratio of the average load over a designated period of time to the peak load occurring during that period.

load impedance The *impedance* presented by a load to a device, circuit, or system.

load leveling In communications, the distribution of traffic over more than one route in order to relieve congestion. See also *load balancing*.

load losses Those losses in a circuit, device, or system that are due to the carrying of current. They include I^2R losses, eddy current losses in magnetic devices, and losses due to leakage fluxes in the windings of an electromagnetic device.

load number (LN) A method of determining the burden on the central office switching system line circuit (in particular, the ring current generator) in Canada. Typical telephone equipment is rated around 10 to 20 *LNs*. To determine the total burden, one simply adds up all of the *LNs* on a line. The maximum number of *LNs* on a Canadian line is 100. See also *ringer equivalent number* (*REN*).

load sharing A technique in which multiple computers are used to share the processing requirements of the system. See also *load balancing*.

loaded line In telephony, a method of compensating for *amplitude vs. frequency distortion* on a telephone line (usually the line from the central office to the subscriber premises). *Loaded lines* minimize the distortion in a range of frequencies but at the expense of reduced available bandwidth The compensation method places inductors (called *load coils*) in series with the line at regular intervals to compensate for the distributed capacitance of the cable. Load coils are used only on 19, 22, and 24 AWG loops greater than 18 000 feet (5.5 km). For 26 AWG, the loop length must be greater than 15 000 feet (4.6 km).

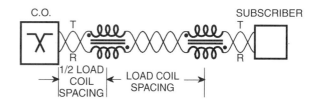

C.O.   SUBSCRIBER

COMMON LOAD COIL CHARACTERISTICS

Coil Type	Coil Value	Coil Spacing
B44	44 mH	3000 ft
D66	66 mH	4500 ft
H88	88 mH	6000 ft

Because load coils severely attenuate the high-frequency response of a transmission line, they can be used only on baseband audio circuits, that is, they cannot be used on any T-carrier (E-carrier) ISDN, or xDSL systems. The following graph plots both frequency vs. attenuation and frequency vs. delay distortion for nonloaded, B44, D66, and H88 loaded loops. All loops are 22 AWG 83 nF per mile cable. Also called *loaded loops.*

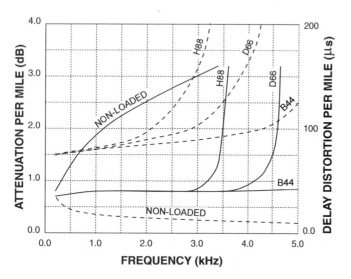

lobe (1) The cable between a Token Ring station and the Trunk Coupling Unit (TCU—lobe attaching unit) to which it is connected. *Lobe* length comprises a patch cable from the TCU to the main wiring panel, the length of the main wiring to the user station's location, then a patch cable from a floor/desk socket to the station. Called a *network node* in most other systems. (2) A pair of channels between a data station and a lobe attaching unit, one channel for sending and one for receiving. (3) Those portions of a directional radiation pattern which are bounded by areas of minimum radiation.

The diagram is of a hypothetical radiator with a *main lobe* pointing in the direction of desired transmission, while the *side lobes* represent undesired transmissions. The side lobes represent not only wasted power but areas of potential interference with other systems.

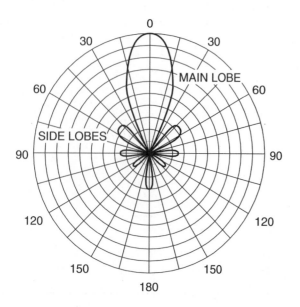

lobe attachment module (LAM) A device in a Token Ring network with multiple interfaces to which new nodes (called lobes) for the network can be attached. A *LAM* may have interfaces for up to 20 lobes. Functionally, a *LAM* is a multistation access unit (MAU), but with a capacity of 20 nodes (instead of 8 with an MAU). *LAM* interfaces may use either IBM connectors or RJ-45 plugs. They can be daisy chained and connected to a hub (a controlled access unit [CAU] in Token Ring terminology). Each CAU can handle up to four LAMs, for a total of 80 lobes. Also called a *lobe attaching unit.*

local A description of entities closest to a designated user. The term is not related to absolute geography but rather to relative geography and specific characteristics of entities (e.g., addressing).

For example, a local area network (LAN) of a large corporation may have nodes covering the world and a connection to laboratories network (which is confined to one building). All nodes of the large corporation are considered local while the laboratories nodes are *remote.*

local access and transport area See *LATA.*

local access code In telephony, that part of the call routing code which informs a local switching system (e.g., PABX) to connect the caller to the public switch telephone network (PSTN). See also *area code, city code, country code, international access code,* and *local number.*

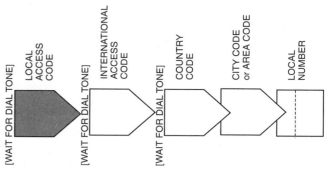

local apparent time (LAT) A timekeeping scheme based on the Sun's crossing of the local meridian at 12:00 noon. Also called *apparent solar time* and *sundial time.* See also *time (3).*

local area data transport (LADT) A service of the local telephone company which provides users a synchronous data transport means.

local area DGPS (LADGPS) A form of differential Global Positioning Systems (*DGPS*) in which the user's receiver receives real-time satellite pseudorange and carrier phase corrections from a nearby reference station. These corrections account for the combined effects of satellite, atmospheric, and SA-induced errors.

local area network (LAN) Two or more computing stations interconnected such that they can share the resources of one or more other stations. The interconnection is over a communication path limited to a "small" area such as a group of offices in a building (or buildings on a campus) and does not extend across the public telephone switched network. Computing stations are connected to the *LAN* via an access card within each computing station. The *LAN* allows computing stations to have access to common peripherals (such as printers, modems, plotters, etc.), common data on file servers, and computing resources. Devices on a network are called *nodes* and are generally connected by a cable (usually twisted pair, coax, or fiber optic), although wireless connections (radio or infrared) connections, called local area wireless networks (LAWNs), are becoming popular.

Nodes on a LAN are interconnected in one of four basic topologies called *bus, daisy chain, ring,* and *star.* Nodes are basically connected in parallel on a bus. A daisy chain is fundamentally the linear connection of the nodes, while in a ring all nodes are connected linearly with the first node connected to the last node. Finally, in the star topology, all nodes are connected to a central hub with wiring radiating outward. To avoid collisions when two or more nodes attempt to access the LAN at the same time, two strategies are used: *contention/collision detection* and *token passing.* See also *baseband network, broadband network, bus network, collision detection, contention, CSMA/CD, network, protocol, ring network, star network,* and *token passing.*

local area signaling services (LASS) In telephony, a set of central office revenue-enhancing features based on the delivery of the calling party number (caller ID—CLID) via the local signaling network. The features can be activated, modified, or suspended by the user station by using appropriate DTMF signaling to his local central office. The features include:

- *Automatic callback*—enables a customer to automatically return a call to the last station number received when both stations become idle, (a form of camp-on).
- *Caller ID* (*CLID*)—a two-faceted feature concerning the delivery of the calling station number to the called station, i.e., *delivery* and *privacy.* The calling party is ultimately in control of the delivery of the identification number.
  - *Calling number delivery*—which transmits the calling party's station number through the network to the called station equipment.
  - *Calling number privacy*—which inhibits the delivery of the calling station number to the called party.
- *Customer originated trace*—allows the called party to request an automatic trace of the pervious incoming call.
- *Selective call acceptance*—enables a customer to preselect which calling station numbers will be recognized. All other calling station numbers will be routed to a prerecorded announcement or to an alternate station number.
- *Selective call forwarding*—enables a subscriber to preselect which calling numbers will be forwarded to an alternate destination.
- *Selective call rejection*—enables a subscriber to preselect which calling numbers will be rejected and re-routed to a prerecorded announcement.
- *Selective distinctive alert*—enables a subscriber to preselect which calling numbers will generate a distinctive ring pattern at the called station.

local area transport (LAT) A Digital Equipment Corporation (DEC) nonroutable protocol for high-speed asynchronous communication between hosts and terminal servers over Ethernet.

local automatic message accounting (LAMA) The process by which the local telephone company handles automatic billing for local and toll calls. The method requires automatic number identification (ANI), a capability that has been adapted to provide caller ID services. An alternative accounting strategy, centralized automatic message accounting (CAMA), accomplishes the same thing but at a central office.

local battery The battery used to power local communications equipment as opposed to the central office battery. The term originated with telegraphy. It was used to distinguish between the power required to operate the station and the power required to transmit the signal down the telegraph line.

local bridge A bridge between multiple networks at a local site (e.g., within a building).

local bus A bus connecting a processor to memory and other high-throughput devices (e.g., video display boards, hard disk controllers, and network interface cards). It is usually on the same board as opposed to a backplane and therefore is faster. Examples of local busses on PCs include VESA (Video Electronic Standards Association) Local Bus (VLB or VL bus) and PCI (Peripheral Component Interconnect).

local bypass A telephone connection that links multiple facilities but bypasses the telephone company.

local call In telephony, any call within that geographic service area of a calling station for which a flat rate is charged (generally on a monthly basis).

local carrier In telephony, the company that provides local telephone service, that is, within an exchange, or calling area. *Local carriers* are connected by interexchange carriers (IXCs). Also called a *local exchange carrier* (*LEC*).

local channel (LC) A digital communications link between a subscriber's premises and a local central office (CO).

local distribution frame (LDF) A synonym for *intermediate distribution frame.* See also *main distribution frame* (*MDF*).

local drive The disk drive that physically exists on the machine the user is operating from. The *remote drive* is on the network server.

local echo A modem feature that reflects digital data transmitted from a DTE back toward the DTE. For example, keyboard data may be echoed back to the screen of a terminal or computer. See also the *ATEn* and *ATFn* commands in Appendix A.

local exchange In telephony, the exchange onto which subscriber lines are terminated. Also called a *class 5 office* and an *end office.*

local exchange carrier (LEC) The telephone company that provides local access and local switching, that is, it supports local calling capability. Typically, it is a regulated monopoly, certified to operate in one or more areas called LATAs (local access and transport areas).

local loop In telephony, the pair of twisted wires that connects the signal switching equipment at the local PSTN central office (CO) to the terminals of a communication device such as a telephone or modem at the subscriber's premises. The local loop carries dc power for the subscriber subset (*loop current*), signaling information (hook switch status, dial information, *call progress tones,* and ringing voltage) as well as voice. Also called a *local line, loop,* or *subscriber loop.*

local loopback address A special Internet address (127.0.0.1) defined by the Internet Protocol. A host can use the *local loopback address* to send messages to itself.

local management interface (LMI) An ITU-T specification for the use of frame relay products that exchange management-related information (e.g., status) between a network and various hardware devices (such as printers, storage devices, and routers).

local mean time (LMT) A timekeeping scheme based on a hypothetical average (mean) Sun crossing the local meridian at 12:00 noon. See also *time* (*3*).

local mode See *command mode.*

local modem Your modem. Generally, the modem that initiates the connection.

local multipoint distribution service (LMDS) A K$_a$ band (28 GHz) wideband (~1 GHz) wireless, digital, duplex communications technology. Services using the technology include corporate LANs, wide area internetworking, and interactive video.

local name In a network or an internetwork, a local name is known only to a single server or domain in a network. In contrast, *global name* is known by all nodes.

local network interconnect (LNI) A concentrator supporting multiple active devices (or communications controllers). It may be used as a stand-alone device, or it may be attached to a standard Ethernet cable.

local number In telephony, that part of the telephone call routing code which determines the specific end subscriber to which the switching network will connect. In the United States, the *local number* is generally composed of a three-digit office code and a four-digit equipment code (or line code). In some countries, the local number may not be fixed in length. Also called a *subscriber number.* See also *area code, city code, country code, international access code,* and *local access code.*

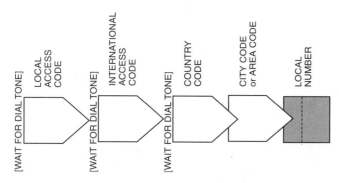

local orderwire A technician's communication link between a control center and a selected terminal equipment locations, such as a repeater or remote concentrator.

local security A security method available for servers running IBM's High Performance File System (HPFS386). This method extends the LAN Manager security measures to protect the files on a server by restricting access of the users working at the server. With *local security,* a user must be assigned permissions to access any file or directory in an HPFS386 partition.

local service area In telephony, that geographic area in which a caller may place calls of any duration for a flat fee.

local switch In telephony, the communications switching system to which a user device is directly connected.

local tandem In telephony, a switching system that interconnects class 5 exchanges (end offices) serving noncontiguous local service areas. Interconnections are via end office trunks. See also *class n office.*

locally administered address (LAA) A parameter used by a 3174 controller to determine whether the node of a Token Ring network can access the mainframe.

LocalTalk® Apple Computer, Inc's proprietary 230.4-kbps, baseband, carrier sense multiple access with collision avoidance (CSMA/CA) network protocol for use in networks that run Macintosh AppleTalk networking software. *LocalTalk* is characterized by:

- Use of unshielded twisted pair cable.
- Use of an RS-422 interface implemented on either a DB-9 or a DIN-8 connector and two DIN-3 connectors. (The DIN-3 connectors are designed so that a node can easily drop out of a network without disrupting the remaining network.)

- Support to 255 nodes per network.
- Internode spacing to 1000 feet.
- Use of operating system services so that all Macintosh computers have built-in networking capabilities.
- Use of the LocalTalk Link Access Procedure (LLAP) at the ISO/OSI data link layer to access the network.
- Use of (CSMA/CA) to access the network.

See also *AppleTalk.*

lock A software flag placed on a database file, record, or field that prevents access or modification by unauthorized users.

lock code In cellular telephony, a code programmed into the Numerical Assignment Module (NAM) that locks the telephone instrument to prevent unauthorized use.

lock in The synchronizing of at least one frequency generator to another generator such that their frequencies are related by the ratio of two integers. Also called *lock on.* See also *phase lock loop (PLL).*

lock-in frequency A frequency at which a closed-loop system, such as a phase lock loop (PLL), can acquire and track a signal. See also *lock-in range.*

lock-in range (1) The span of frequencies over which a phase lock loop (PLL) will acquire lock without cycle slip, that is, the span within which the system can acquire and track a signal. **(2)** The dynamic range over which a phase lock loop (PLL) will acquire lock without cycle slip, that is, the span within which the system can acquire and track a signal. See also *phase lock loop (PLL).*

locking (1) A mechanism for ensuring that two network users or programs do not try to access the same data simultaneously. Access may prevent both reading and writing, or it may prevent only writing.

A *lock* may be *advisory* or *physical.* An advisory lock is a warning to the user that the file is already in use. Advisory locks can be overridden. A physical lock permanently bars access, and they cannot be overridden. The following types of locks are common:

- *File locking*—a scheme in which a file server prevents a user from accessing any part of a file, while another user is already accessing that file. This is the simplest and least efficient method of locking.
- *Record locking*—a scheme in which a file server prevents users from accessing a record in a file while another user is already accessing that same record. *Record locking* is more efficient and less restrictive than file locking.
- *Logical locking*—a scheme in which logical units (e.g., records or strings) in a file are made inaccessible as required.
- *Physical*—a scheme in which actual sectors or allocation units on a hard disk are made inaccessible as required. This is the standard locking scheme used by MS-DOS.

(2) *Locking* is also used as a security measure to prevent or restrict a user's access to files and resources. See also *lockout (3).*

lockout Also spelled *lock out.* **(1)** The condition when either one or both end terminals of a communication path cannot receive the transmissions of the opposite end of the path. The condition occurs on circuits equipped with voice-operated switches in conjunction with high noise conditions.

Lockout occurs when two conditions are met:

- There is excessive noise (or continuous high-level speech) at the near end, which "tells" the voice-operated switch to transmit into the network and to attenuate the receive signal.
- The signal level from the far end is too low to cause the voice-operated switch to attenuate transmissions and allow passage of the receive signals.

Lockout can occur in long-distance circuits with echo suppressors (used in conjunction with two- to four-wire hybrids), or it can occur with hands-free telephones. **(2)** A feature of PBX systems that prevents an attendant from reentering an established connection, unless recalled by the user. **(3)** An action or condition in a network in which a potential user or application is denied access to particular services or resources of the network. *Lockout* may even prevent a potential user from accessing the network itself. See also *locking*.

LOF **(1)** An abbreviation of <u>L</u>oss <u>O</u>f <u>F</u>rame. A condition where a multiplexer cannot find framing for 2.5 s or more, i.e., out of frame (OOF). **(2)** An abbreviation of <u>L</u>owest <u>O</u>perating <u>F</u>requency.

LOFC An abbreviation of <u>L</u>oss <u>O</u>f <u>F</u>rame <u>C</u>ount. The number of LOFs.

log **(1)** A synonym for journal. A record of actions associated with an entity or user. **(2)** An abbreviation of <u>log</u>arithm. The power (x) to which a given base (b) number must be raised in order to be a specified value (n).

$$n = b^x$$

Common base numbers are 10, 2, and e (e is approximately 2.718281828). The *log* of a number n with a base b is generally written as $log_b(n) = x$; there are, however, two exceptions. The *log* base 10 of n is abbreviated $log(n)$, and the *log* base e is abbreviated $ln(n)$.

EXAMPLES OF X = LOG$_b$(N)

Example	Base b	Number n	Value x
log 2	10	2	0.301029995 . . .
log 10	10	10	1
log 100	10	100	2
ln 2	e	2	0.69314718 . . .
ln e	e	e	1
ln e²	e	e²	2
log₂ 2	2	2	1
log₂ 3	2	3	1.5849625 . . .
log₃ 2	3	2	0.630929753 . . .
log₃ 3	3	3	1

To calculate the *log* of a number x to a base b not available on a calculator, use the following relationship:

$$log_b(x) = \frac{log(x)}{log(b)}$$

where the $log(x)$ and $log(b)$ are to whatever base is available, generally base 10 or base e.

log-in See *logon*.

log-off See *logoff*.

log-on See *logon*.

log-out See *logoff*.

logic bomb Code covertly inserted into software (an application or operating system, OS) that causes it to perform some unauthorized activity whenever specified conditions are met.

logic diagram A symbolic representation of the logical elements, the interconnections, and supplemental notations showing the details of a digital device, circuit, or system. The diagram details signal flow, control, and possibly critical timing but does not give any physical information.

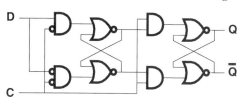

The preceding *logic diagram* is of a rising edge-triggered data latch. Data (D) is transferred from the input to the first-stage latch, while the clock (C) is in the low state. On the rising edge (the *low* to *high* transition) of C, the input to the first stage is disabled and its contents are transferred to the second stage latch and its outputs Q and $\overline{Q}$.

logic level Any value within one of two defined ranges of a physical quantity used to represent logic states. A logic variable may be assigned to any physical quantity for which at least two distinct ranges may be defined, e.g., the flow of current vs. no current, a voltage above $+3$ V vs. below -3 V, a positive pulse vs. a negative pulse, frequency 1 vs. frequency 2, and so on. Each of the two states is referred to as a *logic level* and is assigned a designation L (from *logic low*) and H (from *logic high*), or alternatively 0 and 1, corresponding to the false and true logical states, respectively.

logic symbol A graphical representation of a process that performs one or more logical operations on a set of input signals and generates one or more output signals based on these operations.

There are two major symbol sets in use: a commonly used set, and the official International Electrotechnical Commission (IEC) set as published in Publication 617-12. This standard is also published by IEEE and ANSI. In the commonly used set, many functions are defined by a distinctive shape, while the IEC symbol set relies on one basic symbol (a box) with numerous qualifiers on signal lines and input/output (I/O) terminals. The commonly used symbols include:

The BUFFER, which provides high currents or power to a load.

The INVERTER, which inverts the sense of the input signal; that is, a logical 1 at the input becomes a logical 0 at the output.

The AND gate, which performs the logical AND function on positive true data streams.

The NAND gate, which performs the logical NAND function on positive true data streams.

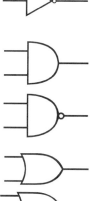

The OR gate, which performs the logical OR function on positive true data streams.

The NOR gate, which performs the logical NOR function on positive true data streams.

The EXCLUSIVE-OR gate, which performs the logical XOR function on positive true data streams.

The EXCLUSIVE-NOR gate, which performs the logical XOR function on positive true data streams and then inverts the signal output.

The DATA storage element (D flip-flop). Data present at the "D" input are transferred to the "Q" output on the rising edge of the clock "C."

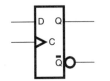

The SET/RESET storage element (RS flip-flop). A logic "0" at the set input "S" causes the "Q" output to be set to "1." A "0" at the reset input "R" forces "Q" low.

The JK storage element (JK flip-flop). Named for its inventor, John Kardash. The Set input "S" forces the "Q" output high, the reset input "R" forces the "Q" output low, and the falling edge of the clock input "C" transfers J and K data to the outputs according to the truth table.

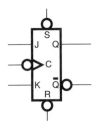

J	K	Q
0	0	no change
1	0	1
0	1	0
1	1	toggle

More complex logic elements are created with rectangular symbols and text describing the functions performed. The IEC/IEEE/ANSI symbol set (illustrated in the figure on pages 305–6) provides for the creation of complex logic symbols. The basic symbol is a box containing a general qualifier that identifies what general functions are performed within the specific element. Each input and output terminal may be qualified by a standard symbol indicating the characteristics of the signal being impressed on the terminal (active low, analog, digital, etc.). Each input and output terminal has a function qualifier to denote the way the element treats the signals (e.g., edge sensitive input, open collector output, tristate output, buffered output, grouped inputs). Finally, the relationship between various inputs and outputs can be expressed through a *dependency notation;* that is, it provides the means of denoting the relationship between inputs, outputs, and/or inputs, and outputs without actually showing all of the elements and interconnections involved. Application of dependency notation is accomplished by executing the following rules:

- Label the input (or output) terminal *affecting* other terminal signals with the letter indicating the relationship involved (see the table below), followed by an identifying number.
  If the label denoting the functions of *affected* terminals must be numeric (e.g., the outputs of a coder), the identifier may be replaced with another suitable symbol (e.g., Greek letters).
  If the complement of the internal signal *affecting* another terminal is required, a bar is placed over the *affected* terminal identifier.
- Label each *affected* terminal with the identifying number selected in the previous rule.
  If the *affected* terminal requires a label to denote its function, the label is prefixed by the identifying number.

If any terminal is *affected* by more than one *affecting* signal, the identifying numbers of each *affecting* signal appear in the label of the *affected* terminal separated by commas. The order of these numbers is the same as the sequence of the *affecting* relationships.

Dependency notation types, symbols, and descriptions include:

Symbol	Definition
A	Address—The purpose of an address is to allow the symbolic representation of an array of elements.
	An affecting address input is labeled "Am" where "m" represents the identifying number of the array affected by the input. Affected terminals are labeled "A," which represents the identifying numbers, i.e., the addresses of the section.
C	Control—When the internal state of a "C" terminal is a "1," terminals affected by the "C" terminal have the normally defined effect on the function of the element. When the "C" terminal is at "0," affected terminals have no effect on the element function.
EN	Enable—When an EN input is at a "1," affected terminals are enabled to perform their function in the element. When EN=0, terminals affected by EN have no effect on the element function.
G	AND—When a "G" terminal is at a "1" state, all affected terminals are at their normally defined internal state. When the "G" terminal is at a "0" state, the affected terminals are also at a "0" state.
M	Mode—Used to indicate that the state of terminals affected by the "M" terminal are dependent upon the mode in which the element is operating.
N	Negate (exclusive-OR)—When an "N" terminal is at a "1" state, all affected terminals are at the complement of their normally defined internal state. When the "N" terminal is at a "0" state, the affected terminals are at their normally defined internal state.
R	Reset—When a "R" input is at a "1," terminals affected by "R" react, regardless of the state of "S," as if the conditions $S = 0, R = 1$ were asserted. When $R = 0$, terminals affected by "R" are free to behave as defined by the element function. The condition $R = 1, S = 1$ generally produces unknown results.
S	Set—When a "S" input is at a "1," terminals affected by "S" react, regardless of the state of "R," as if the conditions $S = 1, R = 0$ were asserted. When $S = 0$, terminals affected by "S" are free to behave as defined by the element function. The condition $S = 1, R = 1$, generally produces unknown results.
V	OR—When a "V" terminal is at a "1" state, all affected terminals are also at a "1" state. When "V" is at the "0" state, all affected terminals are at their normally defined internal state.
X	Transmission—Used to indicate a controlled bidirectional connections between terminals.
	When an "X" terminal is a "1," all terminals affected by this terminal are bidirectionally connected together. When the "X" terminal is a "0," no connection is established.
Z	Interconnection—Used to indicate the existence of internal logic connections between terminals.
	The internal logic state of a terminal affected by a "Z" terminal is the same as the state of the "Z" terminal.

See also *logical operator.*

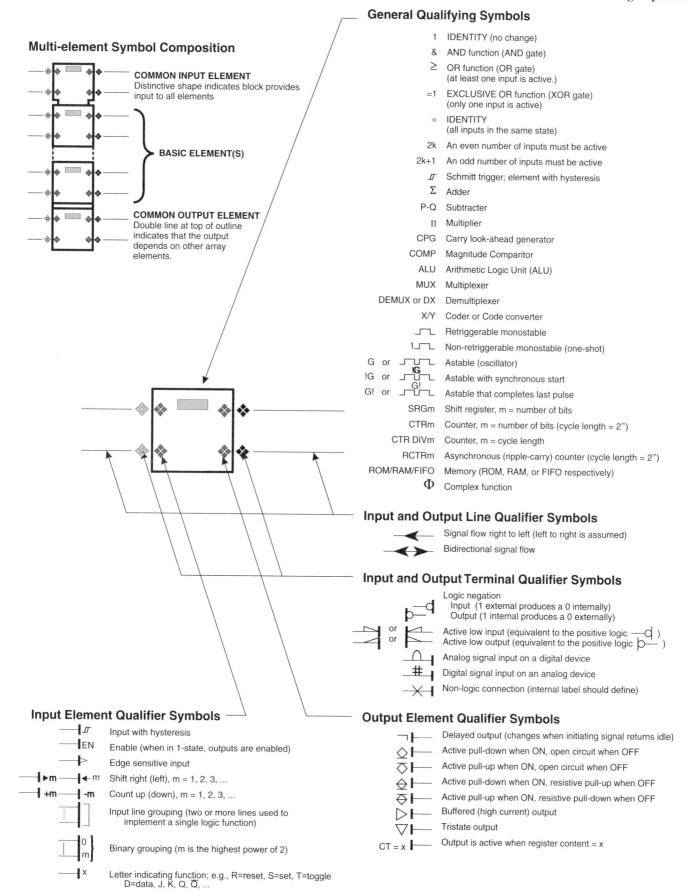

Multi-element Symbol Composition

COMMON INPUT ELEMENT
Distinctive shape indicates block provides input to all elements

BASIC ELEMENT(S)

COMMON OUTPUT ELEMENT
Double line at top of outline indicates that the output depends on other array elements.

General Qualifying Symbols

1	IDENTITY (no change)
&	AND function (AND gate)
≥	OR function (OR gate) (at least one input is active.)
=1	EXCLUSIVE OR function (XOR gate) (only one input is active)
=	IDENTITY (all inputs in the same state)
2k	An even number of inputs must be active
2k+1	An odd number of inputs must be active
⊓	Schmitt trigger; element with hysteresis
Σ	Adder
P-Q	Subtracter
Π	Multiplier
CPG	Carry look-ahead generator
COMP	Magnitude Comparitor
ALU	Arithmetic Logic Unit (ALU)
MUX	Multiplexer
DEMUX or DX	Demultiplexer
X/Y	Coder or Code converter
⌐⌐	Retriggerable monostable
1⌐⌐	Non-retriggerable monostable (one-shot)
G or ⌐⌐⌐	Astable (oscillator)
!G or ⌐⌐ !G	Astable with synchronous start
G! or ⌐⌐ G!	Astable that completes last pulse
SRGm	Shift register, m = number of bits
CTRm	Counter, m = number of bits (cycle length = 2^m)
CTR DIVm	Counter, m = cycle length
RCTRm	Asynchronous (ripple-carry) counter (cycle length = 2^m)
ROM/RAM/FIFO	Memory (ROM, RAM, or FIFO respectively)
Φ	Complex function

Input and Output Line Qualifier Symbols

Signal flow right to left (left to right is assumed)

Bidirectional signal flow

Input and Output Terminal Qualifier Symbols

Logic negation
Input (1 external produces a 0 internally)
Output (1 internal produces a 0 externally)

Active low input (equivalent to the positive logic —◁)
Active low output (equivalent to the positive logic ▷—)

Analog signal input on a digital device

Digital signal input on an analog device

Non-logic connection (internal label should define)

Input Element Qualifier Symbols

⊓	Input with hysteresis
EN	Enable (when in 1-state, outputs are enabled)
▷	Edge sensitive input
▶m ◀ m	Shift right (left), m = 1, 2, 3, ...
+m -m	Count up (down), m = 1, 2, 3, ...
	Input line grouping (two or more lines used to implement a single logic function)
0 m	Binary grouping (m is the highest power of 2)
x	Letter indicating function; e.g., R=reset, S=set, T=toggle D=data, J, K, Q, $\overline{Q}$, ...

Output Element Qualifier Symbols

¬	Delayed output (changes when initiating signal returns idle)
◇	Active pull-down when ON, open circuit when OFF
◇	Active pull-up when ON, open circuit when OFF
◇	Active pull-down when ON, resistive pull-up when OFF
◇	Active pull-up when ON, resistive pull-down when OFF
▷	Buffered (high current) output
▽	Tristate output
CT = x	Output is active when register content = x

IEC LOGIC SYMBOL COMPOSITION

306

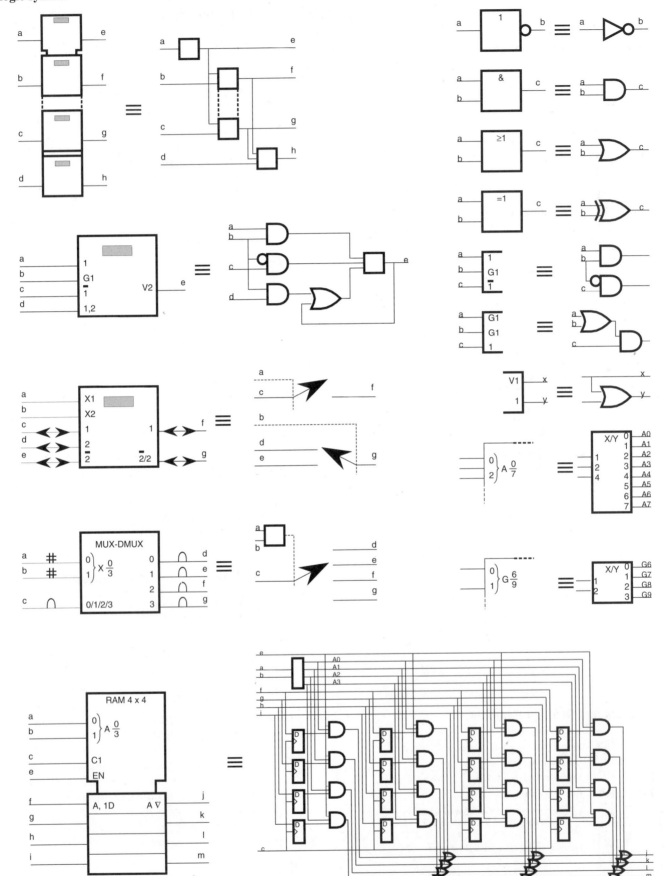

logical address In a network, a value assigned by software during network installation, configuration, or when new a node is added to a network. In contrast, hardware addresses are assigned at the time of manufacture.

Logical Block Addressing (LBA) A method used on IBM PCs (and compatibles) to increase the size limit for a single hard disk drive (HDD) from 528 MB to 8.4 GB. An addressing conversion is performed on the IDE (Integrated Drive Electronics) disk controller card. Most modern PCs will select this mode automatically; some older PCs allow selection of LBA mode manually in the BIOS; and even older versions of BIOS ignore the capability.

logical bus A LAN topology that shares a common communications channel. See also *bus* and *topology*.

logical channel See *logical link*.

logical channel identifier (LCI) A field in a switching packet (e.g., X.25) that indicates the virtual circuit (logical channel) being used for the packet.

logical channel number In packet-switched networks, a circuit identifying number, at the packet level of a virtual circuit, assigned when a virtual call is placed.

logical circuit See *logical link*.

logical complement In Boolean algebra, the logical negation of value; for example, the *logical complement* of TRUE is FALSE. See also *negation*.

logical link A virtual connection or path over a physical connection that is capable of supporting several simultaneous virtual connections. The connection is generally temporary. Also called a *logical channel, logical connection, logical route,* or *virtual circuit*.

logical link control (LLC) One of two sublayers in the ISO/OSI Data Link Layer defined by the IEEE 802 LAN standards. The *LLC* generates protocol data units (PDUs) and response PDUs for transmission, and it interprets command PDUs and response PDUs. The Tasks of the LLC can include:

- Initialization of the signal interchange.
- Organization of data flow.
- Interpretation of command PDUs.
- Generation of command response PDUs.
- Error control and recovery.

The LLC can provide three types of deliver services, that is,

- Type 1 is a connectionless service without acknowledgment. This is the fastest but least reliable service offered by the *LLC*. In a connectionless service, there is neither a predefined path nor a permanent circuit between sender and receiver. Without an acknowledgment, there is no way to determine whether a transmitted packet reached its destination.
  Type 1, however, is the most popular service because most higher level protocols include both error-checking and delivery acknowledgments. Hence, there is no need for the redundant services. Examples of connectionless protocols include the network layer IP protocol (of TCP/IP) and NetWare's IPX protocol.
- Type 2 is a connection-oriented service. In a connection-oriented service, a circuit is established before data transmission begins. Examples of connection-oriented services include the transport layer TCP protocol used on the Internet (and many other systems), the X.25 network layer protocol, and NetWare's SPX protocol.

With a connection (even with virtual connections) the service can provide:

- *Sequence control.* Enables the receiver to correctly reassemble message elements in the correct order.
- *Error control.*
- *Flow control.* Two common methods are: *Stop and wait,* in which each LLC frame must be acknowledged before the next one is sent. *Sliding window,* in which "n" LLC frames can be sent before an acknowledgment is required. The value of "n" represents the window size.
- Type 3 is also a connectionless service but one with acknowledgment.

logical link control type 2 (LLC2) A protocol and packet format for use in SNA-based networks. It is more versatile and widely supported than the SDLC protocol also common in SNA environments.

logical network The division or concatenation of networks, for the convenience of the user, such that it appears the user is on a single network (possibly connected to an external network).

Within a large organization, it may be advantageous to create a network for each of several working groups. This, however, is very expensive, so a single large network may be logically partitioned so that each group operates as if it has its own network (and may or may not be given access to the other *logical networks*). The Internet, on the other hand, appears to be a single network but is in fact a collection of thousands of individual autonomous networks.

logical number A value assigned in a device addressing scheme for a specific hardware device. The *logical number* is based on parameters such as what other devices are attached and the order in which these were installed.

logical operators The logical (Boolean) relationship between two entities. The entities may be single variables, a string, or entire fields in a data base. See also *Boolean algebra*.

SOME LOGICAL OPERATORS

Operator	Definition
=	True if A equal B
>	True if A greater than B
<	True if A less than B
<>	True if A not equal B
< =	True if A less than or equal B
= >	True if A equal to or greater than B
AND	True if A and B true
OR	True if A or B true
NOT	True if A is not true
XOR	True if A equal B
STARTS_WITH	True if A starts with "x"
CONTAINS	True if A contains "x"
ENDS_WITH	True if A ends with "x"
ADJACENT	True if A is adjacent to "x"

logical record A record independent of its physical environment.

Portions of a *logical record* may be stored in different physical records, or multiple *logical records* (or parts of multiple *logical records*) may be stored in one physical record.

logical ring A network that operates as if it were a ring topology, even if it is physically cabled as a star or bus topology.

logical states One of two possible conditions that may be taken on by a binary variable, i.e., true and false.

login See *logon.*

logoff The process of terminating a working session with a computer accessed through a communications network from a remote terminal. The act of *logging off* does not necessarily terminate the communication link connection; it only signifies that the current user is finished with the session.

Although one can simply go on-hook (hang up) to terminate the connection, it is considered at least bad form and may even cause problems for the user. (Billing, for example, may not be turned off.) Also called *log-off, logout,* and *log-out.* Although not a rule, logoff is more likely used when referring to a mainframe, whereas logout is used when discussing networks.

logon The process of establishing a connection and identifying oneself to a host computer as an authorized user. A normal *log-on* procedure asks for the user's name and password or other account identifier. The *log-on* procedure provides the host a means:

- To determine the resources it must provide (or is allowed to provide).
- To provide a security means by identifying authorized users.
- For accounting, to keep track of used (billable?) time.

Also called *log-in, login,* and *log-on.* Although not a rule, logon is more likely used when referring to a mainframe, whereas login is used when discussing networks.

logon script A sequence of operating system commands, automatically executed when a user logs onto a network or system. These commands initialize the communications hardware and the node's and user's operating environments, map directories, allocate resources for the user, and perform other startup tasks for the user.

Examples of three classes of logon scripts include:

- *System script*—which is created by the network administrator and is used to set general parameters and mappings and to execute commands that are appropriate for all users.
- *User script*—which belongs to an individual user and does whatever remains to be done to initialize the hardware and environment for that particular user.
- *Profile script*—which includes commands that initialize the environment for all users of a group in the NetWare Name Service (NNS) and in NetWare 4.x. If defined, the script executes between the *system script* and *user script.*

logon server (1) For a domain, a logon server is the primary domain controller and the backup domain controller. (2) For a user, the server that processes the user's logon request.

logout See *logoff.*

LOL An abbreviation of <u>L</u>aughing <u>O</u>ut <u>L</u>oud.

LON An acronym for <u>L</u>AN <u>O</u>uter <u>N</u>etwork. A virtual network (mobile terminals) or remote network that is connected to an organization or regional LAN, as required, where organization resources reside.

LONAL An acronym for <u>L</u>ocal <u>O</u>ff-<u>N</u>et <u>A</u>ccess <u>L</u>ine.

long distance call In telephony, any call or connection terminating outside the local exchange service area of the calling station, whether inter-LATA or intra-LATA, and for which there is a charge beyond that for basic service. Also called a *toll call.*

long-haul carrier A carrier system designed for long-distance communications, i.e., one whose distance can range from hundreds of miles to transcontinental or international distances. The term encompasses both medium (generally coaxial cable, microwave, or satellite links) and signaling (including modulation) specifications. In the U.S. public switched networks, it pertains to circuits that span inter-LATA, interstate, and international boundaries.

long-haul modem Any modem designed to operate on the public switched telephone network (PSTN) or leased telephone lines.

A *long-haul modem* uses a modulated carrier to convey the digital information through the communications link, while a *short-haul modem* impresses the digital information directly on the link. Most modems fall into the *long-haul* category. See also *short-haul modem.*

long line adapter In telephony, equipment inserted in series with the loop between a line circuit and its associated station(s) to allow the conductor a resistance greater than the maximum for which the system was designed. Also called a *loop extender.*

long wavelength In fiber optics, communications pertaining to optical wavelengths greater than 1 μm. See also *frequency spectrum.*

longitudinal balance (1) The electrical symmetry of a two-wire circuit with respect to a common reference (generally ground). In telephony, for example, the difference between tip-to-ground and ring-to-ground voltages.

The measurement is the ratio of an applied longitudinal signal to any resulting metallic signal and is generally expressed in decibels (dB). **(2)** An expression of the difference in impedance of the two sides of a circuit.

longitudinal circuit In telephony, a circuit formed by two communication conductors and a return in a third conductor or ground. See also *metallic circuit.*

longitudinal current A current acting in series with a *longitudinal circuit.* Mathematically, it is one-half the algebraic sum of the current in each conductor.

longitudinal impedance In telephony, the impedance presented by a longitudinal circuit to a signal source. In general, the *longitudinal impedance* is significantly different from the metallic impedance.

longitudinal offset loss See *gap loss.*

longitudinal redundancy check (LRC) (1) An error-detection method based on the accumulation of the numerical values of characters transmitted in a block.

The transmitting device sums the values of each transmitted character in a block of a specified length. The sum is divided by an integer value (usually 256), and the remainder is appended to the transmitted block (a block check character—BCC). The receiver performs the same operation on the data block and compares its result with the appended block check character. If the two sums are equal, the data block is presumed to be error free. However, if they are different, a transmission error probably occurred. Also called a *block check.* **(2)** An error-detection method in which a running parity of each bit is appended to the end of a stored or transmitted information block. Upon recalling (from a memory device) or receiving the information block, the check word is again calculated and compared to the retrieved value. If the two values match, the data are assumed to be correct; if the values are different, an error has occurred. The *LRC* is sometimes also called *serial parity* or *run-length parity.* When combined with a *vertical redundancy check* (*VRC*), single bit error correction can be obtained from the "*x-y*" location of the VRC and *LRC,* respectively. In the example on page 309, the bit at location H5 was transmitted as a "1" but was received as a "0." By comparing the received *LRC*-VRC data with the regenerated *LRC*-VRC, the location H5 is identified and the bit can be "flipped" to the correct state. Also called *horizontal parity check.* See also *parity* and *vertical redundancy check* (*VRC*).

Bit Number	Received Information Packet													LRC¹	LRC²
	A	B	C	D	E	F	G	H	I	J	K	L	M		
7	1	0	0	1	1	1	0	1	0	1	0	0	1	0	0
6	0	1	0	0	1	0	1	0	0	1	0	0	1	0	0
5	0	0	1	0	1	0	0	0	0	1	1	0	1	1	0
4	1	1	0	1	0	1	0	1	0	1	0	1	0	0	0
3	0	0	1	0	1	0	1	0	0	1	0	1	0	0	0
2	0	1	0	1	0	1	1	0	1	0	1	1	1	1	1
1	1	1	1	1	0	0	1	0	1	1	0	0	1	1	1
0	1	1	0	0	1	1	0	0	1	1	0	0	0	1	1
VRC¹	1	0	0	1	0	1	1	0	0	0	1	0	0		
VRC²	1	0	0	1	0	1	1	1	0	0	1	0	0		

1. Bits as received at the destination DTE.
2. Bits are regenerated by the destination DTE.

longitudinal signal In telephony, the same as *common mode* signal. The *longitudinal voltage* is one-half the algebraic sum of the voltages on each conductor with respect to a reference (usually ground). The *longitudinal current* is one-half the algebraic sum of the current in each conductor. See also *metallic signal.*

longitudinal transmission check (LTC) A parity check (odd or even) inserted at uniform intervals during data transmission.

longitudinal voltage A voltage acting in series with a *longitudinal circuit.* Mathematically, it is one-half the algebraic sum of the voltages on each conductor with respect to a reference (usually ground). Also called *common mode voltage.*

look ahead queuing In telephony, an automatic call distribution feature in which the secondary queue is checked for congestion *before* switching traffic to that queue.

look back queuing In telephony, an automatic call distribution feature in which the secondary queue can check whether congestion on the primary queue has cleared up and, if so, return calls to that queue.

loop (1) In telephony, the pair of wires between an individual subscriber's premises and the local PSTN central office (CO) that carries dc power (loop current), signaling (dial and ringing), and voice. Also called a *local loop, subscriber loop,* and *user line.* See also *local loop.* **(2)** In networks, a communications channel from a switching center or an individual message distribution point to the user terminal. Also called a *subscriber line.* **(3)** In networks, the path from a signal source into a transmission system, through a reflecting device (such as a transmission line mismatch), and back to the receiver colocated with the signal source. See also *loop gain.* **(4)** In electronics, a set of series branch circuits forming a closed current path. Also called a *closed circuit.* **(5)** In antennas, a type of antenna, physically in the form of a closed geometric figure such as a circle or rectangle. It is generally used in direction-finding equipment and in UHF reception. **(6)** In computers, a sequence of instructions that may be executed iteratively while a specified condition prevails.

loop back See *loopback.*

loop check An error-detection method wherein the transmitted data are returned to the originating station by the terminating station. The originating station compares the transmitted data with the returned data; if they are not the same, it is presumed that the transmission was received incorrectly and that it should be retransmitted. Also called *echo check.*

loop current (1) The current provided by a central office (CO) to a customer premises device (DCE or telephone) for both supervision and operating the telephone. The current is proportional to the length of the line between the CO and the customer equipment. The minimum current should be greater than 23 mA in order to guarantee proper operation. The maximum loop current will be less than 140 mA in most countries. **(2)** In a telegraph or teletypewriter circuit, the current is interrupted to convey the digital signals. The current is generally fixed at either 20 mA or 40 mA and is independent of loop length.

loop extender See *long line adapter.*

loop filter In a phase locked loop (PLL), a filter located between the phase detector (or time discriminator) and the voltage-controlled oscillator (or phase shifter). See also *phase lock loop (PLL).*

loop gain The product of all gains and losses encountered by a signal traversing a closed loop. The loop may encompass only a portion of a system or the system as a whole. All loops must have a gain less than unit for all frequencies or they are likely to sing (oscillate).

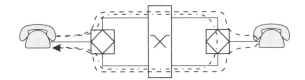

The preceding figure shows two possible loops in a telephone transmission system. The first loop is formed by the transhybrid loss imperfection in the two- to four-wire hybrid on the right. The second loop is formed by imperfect impedance matching at the right-hand station itself. Two additional loops are formed by the circuit on the left. Each of the four has a separate *loop gain* value. Because the actual gains of system elements tend to be frequency dependent (and complex, i.e., they have real and imaginary components), the product of the gains is generally expressed in terms of the S-plane and therefore has units of s^{-1}.

loop noise In telephony, the noise contributed by one or both loops of a telephone circuit to the total circuit noise. In a given case, it should be stated whether the loop noise is for one or both loops.

loop noise bandwidth A phase lock loop property related to the damping factor and natural frequency that describe the effective bandwidth of the received signal.

It is equivalent to the bandwidth of an ideal low-pass filter (square amplitude response), which admits the same noise from a flat noise source as is observed in a phase lock loop (PLL).

loop pulsing In telephony, dial pulsing using loop signaling.

loop signaling In telephony, a method of signaling over direct current paths (that is, that use metallic conductors and terminating bridges). A number of signaling methods are available, including:

- Application of a terminating impedance which starts current flow,
- Interruption of current flow, and
- Current flow reversal.

loop start In telephony, a method of signaling the central office (CO) that a subscriber is requesting service.

Electrically, the subscriber equipment places a resistance (the subset itself) across the tip and ring leads, allowing current to flow in the loop. The CO detects this current and applies dial tone for outgoing calls or removes ringing for incoming calls. Power for the loop is supplied by the CO; specifically, the ring wire is connected to -48 V, and the tip wire is at 0 V. See also *ground start.*

loop timing In digital communications, a synchronization method in which the clock signal (timing information) is extracted from the incoming pulse stream.

loop test A test that uses a closed circuit (a loop) to detect and locate faults.

loop transmission In networks, multipoint transmission in which:

- All stations in a network are serially connected in one closed-loop (ring).
- There are no cross-connections.
- Each station serves as a regenerative repeater, forwarding messages around the loop until they arrive at their destination stations.
- Any station can introduce a message (packet) into the loop by interleaving it with other messages.

See also *Token Ring* and *topology*.

loopback A method of performing transmission and equipment diagnostic tests from only one end of a communication path. For example, a test signal is injected at the local data communications equipment (DCE); this signal is passed on to the remote DCE over an established communication path. The remote DCE then loops back (or echoes) this signal to the originating local DCE for error analysis.

loopback mode An operating mode found in many devices, such as modems. The mode is generally used for testing. In this mode, signals are returned to their origin (looped back) rather than being sent on.

loopback plug A device plugged into a computer communications port for testing the port hardware. When test software is run, it will send a signal out to the transmit pin(s) of the port and compare the response on the receive pin(s). If they are the same, the hardware is probably good. The *loopback plug* connects the transmit pin(s) directly to the receive pin(s) of the port being tested. *Loopback plugs* are designed for specific port types and connectors, i.e., serial or parallel ports and DB, DIN, or other connectors.

The wiring for several *loopback plug* types are shown in the following table.

LOOPBACK PLUG EXAMPLES

Description	Wiring example
Serial port loopback plug (DB-25S).	**DB25S** SHLD 1, TXD 2, RXD 3, RTS 4, CTS 5, DSR 6, S-GND 7, CD 8, DTR 20
Serial port loopback plug (DB-9S).	**DB9S** DCD 1, RXD 2, TXD 3, DTR 4, S-GND 5, DSR 6, RTS 7, CTS 8, RI 20

LOOPBACK PLUG EXAMPLES (*CONTINUED*)

Description	Wiring example
Parallel port loopback plug (DB-25P)	**DB25P** -STROBE 1, D0 2, D1 3, D2 4, D3 5, D4 6, D5 7, D6 8, D7 9, -ACK 10, BUSY 11, PE 12, SLCT 13, -AUTO FD 14, -ERROR 15, -INIT 16, -SLCT IN 17, GND 18, GND 19, GND 20, GND 21, GND 22, GND 23, GND 24, GND 25
	DB25P -STROBE 1, D0 2, D1 3, D2 4, D3 5, D4 6, D5 7, D6 8, D7 9, -ACK 10, BUSY 11, PE 12, SLCT 13, -AUTO FD 14, -ERROR 15, -INIT 16, -SLCT IN 17, GND 18, GND 19, GND 20, GND 21, GND 22, GND 23, GND 24, GND 25

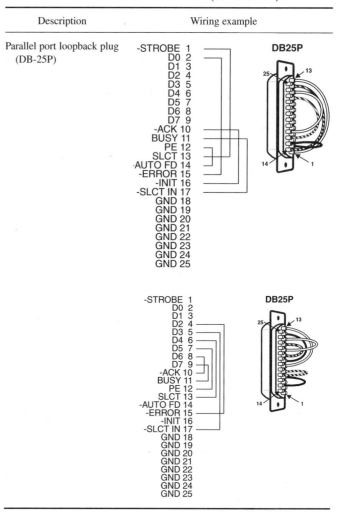

looped dual bus A distributed-queue dual-bus (DQDB) scheme in which the head-of-bus functions for both buses are at the same location.

loose source and record route (LSRR) In Internet transmissions, an Internet Protocol (IP) option that enables a datagram source to specify routing information and to record the route taken by the datagram. This option helps ensure that datagrams take only routes that have a level of security appropriate to the datagram's security classification.

LOP (1) An acronym for Loss Of Pointer. A SONET error condition. **(2)** An acronym for Line-Oriented Protocol.

LOREC An acronym for LOng Range Earth current Communications.

LOS (1) An acronym for Loss Of Signal. **(2)** An acronym for Line Of Sight.

loss The reduction of a parameter between two points in a circuit.

1. *Power.* Power expended without doing useful work. Generally expressed in watts.
2. *Communications.* The ratio of the power that could be delivered to a load under some specified reference condition to the power delivered under actual operating conditions. *Loss* is generally expressed in dB.
3. *Switching (packet or circuit).* The disappearance of a packet or a call, which can occur if a packet is discarded because of heavy traffic, because of an addressing error, or because of a switching system error.

loss budget In electrical or optical signaling, the combination of all contributors that cause signal loss (or gain) between the source and destination.

When designing a system, the maximum allowable system loss is apportioned to the various contributors in a "reasonable manner"; that is, high-loss items such as transmission lines are allowed more loss than switching systems.

lossless compression Data compression in which all information of the original data is preserved and can be reconstituted with decompression. See also *compression* (2) and *data compression.*

lossy compression A method of data compression used on picture and sound data in which some loss of information is tolerated. See also *compression* (2).

lossy material A material that absorbs or scatters a significant amount of energy applied to it. Energy absorbed is eventually converted from one part of the spectrum to heat, while scattered energy is deflected out of the transmission path. In either case, the transmitted signal level is attenuated.

lost call In telephony, a call that cannot be completed because of blocking, i.e., for any reason other than cases where the call receiver (termination) is busy.

lost call attempt A call attempt that cannot be processed beyond some point in switching owing to either blocking or equipment failure.

lost call cleared (LCC) In switching systems, a call-handling strategy in which blocked calls are discarded (lost). An engineering assumption used in erlang C traffic calculations.

lost call delayed (LCD) In switching systems, a call-handling strategy in which blocked calls are queued for later, or delayed, processing. An engineering assumption used in Poisson traffic calculations.

low-battery cutoff point That point in the operation of a system from rechargeable batteries where further discharge could result in damage to the battery. In many systems, automatic means are provided to turn off the system when it reaches the *low-battery cutoff point,* thus preventing the battery from being taken into deep discharge.

For example, in uninterruptible power supplies (UPSs), automatic cutoff switches disconnect the batteries before they exceed safe limits of operation.

low bit rate voice (LBRV) Any digitized voice signal that is transmitted at speeds lower than the 64-kbps channel capacity, generally either at 2400 bps or 4800 bps. Voice data is either compressed or uses sophisticated encoding methods.

low frequency (LF) Frequencies between 30 kHz and 300 kHz. See also *frequency band* and *frequency spectrum.*

low-noise amplifier (LNA) A term that generally refers to a parametric amplifier in a satellite receiver (both space or ground based).

low-order position The right-most position in a string of characters. See also *big endian, least significant digit,* and *little endian.*

low-pass filter An electronic circuit or device that passes all frequencies *below* a specified frequency (the cutoff frequency) and attenuates those frequencies *above* the specified frequency. Generally, the cutoff frequency is considered the −3 dB (or half-power frequency).

The *low-pass filter* curve in the chart is a second order, maximally flat filter with a −3 dBr *cutoff frequency* of 1 kHz. See also *bandpass filter, band reject filter,* and *high-pass filter.*

lowest usable high frequency (LUF) The lowest frequency in the HF band at which the field intensity of a received sky wave is sufficient to provide the required signal-to-noise ratio for a specified time period on 90% of the undisturbed days of the month.

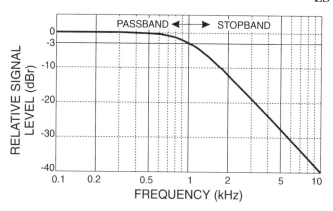

LP (1) An abbreviation of <u>L</u>ine <u>P</u>rinter. **(2)** An abbreviation of <u>Lin</u>ear <u>P</u>hase. **(3)** An abbreviation of <u>L</u>ongitudinal <u>P</u>arity. **(4)** An abbreviation of <u>L</u>inking <u>P</u>rotection.

LP mode An abbreviation of <u>L</u>inearly <u>P</u>olarized <u>mode</u>.

LP$_{01}$ mode The designation of the fundamental LP mode.

LPA (1) An abbreviation of <u>L</u>ow <u>P</u>ower <u>A</u>mplifier. **(2)** An abbreviation of <u>L</u>inear <u>P</u>ower <u>A</u>mplifier. **(3)** An abbreviation of <u>L</u>og <u>P</u>eriodic <u>A</u>ntenna.

LPC An abbreviation of <u>L</u>inear <u>P</u>redictive <u>C</u>oding.

LPCM An abbreviation of <u>L</u>inear <u>P</u>hase <u>C</u>ode <u>M</u>odulation.

lpd An abbreviation of <u>L</u>ine <u>P</u>rinter <u>D</u>aemon.

LPF An abbreviation of <u>L</u>ow <u>P</u>ass <u>F</u>ilter.

LPI An abbreviation of <u>L</u>ines <u>P</u>er <u>I</u>nch.

LPID An acronym for <u>L</u>ocal <u>P</u>age <u>ID</u>entifier.

LPM (1) An abbreviation of <u>L</u>ines <u>P</u>er <u>M</u>inute. **(2)** An abbreviation of <u>L</u>inearly <u>P</u>olarized <u>M</u>ode.

LPN (1) An abbreviation of <u>L</u>ocal <u>P</u>acket <u>N</u>etwork. **(2)** An abbreviation of <u>L</u>ow <u>P</u>ass <u>N</u>etwork.

LPR (1) An abbreviation of <u>L</u>ine <u>PR</u>inter. **(2)** A command used to queue print jobs on Berkeley queuing systems.

LPSN An abbreviation of <u>L</u>ocal <u>P</u>acket <u>S</u>witched <u>N</u>etwork.

LPTn From <u>L</u>ine <u>P</u>rin<u>T</u>er port <u>n</u>. The logical names of parallel ports on a PC. "n" ranges from 1 to 3, and LPT1, the primary parallel port, is also known as PRN.

LQA An abbreviation of <u>L</u>ine <u>Q</u>uality <u>A</u>nalysis.

LRC An abbreviation of <u>L</u>ongitudinal <u>R</u>edundancy <u>C</u>heck.

LRCC An abbreviation of <u>L</u>ongitudinal <u>R</u>edundancy <u>C</u>heck <u>C</u>haracter.

LRL An abbreviation of <u>L</u>ogical <u>R</u>ecord <u>L</u>ength.

LRR An abbreviation of <u>L</u>oop <u>R</u>egenerative <u>R</u>epeater.

LRU (1) An abbreviation of <u>L</u>east <u>R</u>ecently <u>U</u>sed. **(2)** An abbreviation of <u>L</u>ine <u>R</u>eplaceable (or <u>R</u>emoveable) <u>U</u>nit.

ls The UNIX command to list the contents of a directory. For long directory lists, it is advantageous to *pipe* (|) the output of the list command (*ls*) through the *more* command. (The *more* command displays one page of information at a time and then waits for user input before displaying the next page.)

LS (1) An abbreviation of <u>L</u>east <u>S</u>quares. **(2)** An abbreviation of <u>L</u>ine <u>S</u>witch. **(3)** An abbreviation of <u>L</u>oading <u>S</u>plice. **(4)** An abbreviation of <u>L</u>ow <u>S</u>peed.

LS trunk An abbreviation of <u>L</u>oop <u>S</u>tart <u>trunk</u>. A trunk in which the seized (off-hook) condition is initiated by a closure between the tip and ring leads. A high-voltage ringing signals an incoming call.

LSAP An acronym for <u>L</u>ink <u>S</u>ervice <u>A</u>ccess <u>P</u>oint.

LSB (1) An abbreviation of <u>L</u>east <u>S</u>ignificant <u>B</u>it. **(2)** An abbreviation of <u>L</u>ower <u>S</u>ide<u>B</u>and.

LSC An abbreviation of <u>L</u>east <u>S</u>ignificant <u>C</u>haracter.

LSCIE An abbreviation of <u>L</u>ightguide <u>S</u>tranded <u>C</u>able <u>I</u>nterconnect <u>E</u>quipment.

LSCIM An abbreviation of <u>L</u>ightguide <u>S</u>tranded <u>C</u>able <u>I</u>nterconnect <u>M</u>odule.

LSCIT An abbreviation of <u>L</u>ightguide <u>S</u>tranded <u>C</u>able <u>I</u>nterconnect <u>T</u>erminal.

LSD (1) An abbreviation of <u>L</u>east <u>S</u>ignificant <u>D</u>igit. **(2)** An abbreviation of <u>L</u>ine <u>S</u>ignal <u>D</u>etector.

LSF An abbreviation of <u>L</u>ine <u>S</u>witch <u>F</u>rame.

LSI (1) An abbreviation of <u>L</u>arge <u>S</u>cale <u>I</u>ntegrated circuits. An integrated circuit containing tens to hundreds of thousands of transistors. The definition is somewhat nebulous because the number of devices put on a chip grows yearly. Currently, millions of transistors are incorporated in commercial microprocessor *LSIs*. **(2)** An abbreviation of <u>L</u>ine <u>S</u>tatus <u>I</u>ndication.

LSID An abbreviation of <u>L</u>ocal <u>S</u>ession <u>ID</u>entifier.

LSL An abbreviation of <u>L</u>ink <u>S</u>upport <u>L</u>ayer.

LSP (1) An abbreviation of <u>L</u>ink <u>S</u>tate <u>P</u>acket. **(2)** An abbreviation of <u>L</u>east <u>S</u>ignificant <u>P</u>osition.

LSRR An abbreviation of <u>L</u>oose <u>S</u>ource and <u>R</u>ecord <u>R</u>oute.

LSS An abbreviation of <u>L</u>oop <u>S</u>witching <u>S</u>ystem.

LSSGR An abbreviation of <u>L</u>ATA (sometimes identified as <u>L</u>ocal) <u>S</u>witching <u>S</u>ystem <u>G</u>eneral <u>R</u>equirements. A specification published by Bell Communications Research.

LSSU An abbreviation of <u>L</u>ink <u>S</u>tatus <u>S</u>ignaling <u>U</u>nit.

LTA An abbreviation of <u>L</u>ine <u>T</u>urn <u>A</u>round. In half-duplex communications, the amount of time it takes to set up the circuit so that transmissions flow in the opposite direction.

LTC An abbreviation of <u>L</u>ongitudinal <u>T</u>ransmission <u>C</u>heck.

LTE An abbreviation of <u>L</u>ine <u>T</u>erminating <u>E</u>quipment.

LU An abbreviation of <u>L</u>ogical <u>U</u>nit. An *LU* defines a set of protocols for communication between two users on IBM's Systems Network Architecture (SNA). *LUs* are one of three types of addressable units in an SNA network. (The other two units are physical units, PUs, and system services protocol units, SSCPs.)

Several types of *LUs* are possible, depending on the type of communications and protocols used, that is, see table.

LUA An abbreviation of <u>L</u>ogical <u>U</u>nit <u>A</u>pplications programming interface. A standard applications programming interface (API) for writing applications for IBM's Systems Network Architecture (SNA).

LUF An abbreviation of <u>L</u>owest <u>U</u>sable high <u>F</u>requency.

lumen The SI unit of luminous flux. It is defined as the luminous flux emitted within a solid angle of one steradian by a point source having an intensity of one candela.

LOGICAL UNIT TYPES

Type	Description
LU 0	Communication from program to device.
LU 1	Communication from a master program and slave device. Used for mainframe batch systems and printers that use the SNA character string (SCS) data format.
LU 2	Communication from a master program and slave device, e.g., Host to 3270 terminals.
LU 3	Communication from a master program and slave device. Used for 3270 Data Stream terminals.
LU 4	Communication from program to program or from program to device, with a master/slave or a peer-to-peer relationship between the elements. Used for printers using the SCS data format.
LU 6.0	Communication from program to program, with a peer-to-peer relationship between programs. Used for host-to-host communications using either CICS or IMS subsystems.
LU 6.1	Same as LU 6.0.
LU 6.2	Communications from program to program, with a peer-to-peer relationship between programs. Used for dialog-oriented connections that use the General Data Stream (GDS) format. Also known as APPC.
LU 7	Data Stream terminals used on AS/400, System 36, System 38, and so on.

lumped element A circuit element that exists essentially at a point. For example, a resistor, capacitor, or inductor. See also *distributed element*.

LUN An acronym for <u>L</u>ogical <u>U</u>nit <u>N</u>umber.

lurk On bulletin board system (BBS) conferences, someone who reads the online messages of a BBS conference but does not participate (contribute) in the conference. It is generally not a derogatory term; it is just an expression to describe the presence of a nonparticipant. Actually, *lurking* is a good way for a beginner to learn about a BBS, and if something is "heard" that is not understood, most BBS participants will be happy to answer questions.

In Internet's USENet, lurking is the regular reading of a newsgroup without posting anything.

LUT An acronym for <u>L</u>ook <u>U</u>p <u>T</u>able.

lux The SI unit of illuminance. One lux is one lumen per square meter.

Luxemburg effect A nonlinear effect in the ionosphere which results in the modulation of a weak signal by another strong signal in the same region. Also called *ionospheric cross modulation*.

LV An abbreviation of <u>L</u>ow <u>V</u>oltage.

LVDS An abbreviation of <u>L</u>ow <u>V</u>oltage <u>D</u>ifferential <u>S</u>ignaling.

LW An abbreviation of <u>L</u>ong <u>W</u>ave.

LWT An abbreviation of <u>L</u>isten <u>W</u>hile <u>T</u>alk.

lynx A text-based World Wide Web (WWW) browser. *Lynx* is a program that allows users without full Internet connections to examine and download resources on the Web, but without graphics and sound. Users must have an Internet access account; at the account main prompt, type *lynx,* and use the keystroke instructions at the bottom of the screen.

LZ An abbreviation of <u>L</u>empel-<u>Z</u>iv.

LZH An abbreviation of <u>L</u>empel-<u>Z</u>iv-<u>H</u>uffman.

LZW An abbreviation of <u>L</u>empel-<u>Z</u>iv-<u>W</u>elch.

M

m **(1)** The SI prefix for *milli* meaning 10^{-3}. For example, 1/4 W is written 250 mW. **(2)** The SI symbol for the unit of length in *meters*. For example, 3 mm represents a length of 3 millimeters.

M **(1)** The SI prefix for *mega* meaning 10^6. For example, 2.7×10^6 W is written 2.7 MW. **(2)** When M is used in binary systems such as with hard disk drive capacities, it is interpreted to mean 2^{20} or 1 048 576.

M lead A signaling lead associated with the transmit communication path on telco interoffice trunks. See also *E & M signaling*.

m-ary code See *N-ary.*

m-ary phase shift key (MPSK) See *MPSK.*

M.n recommendations A series of International Telecommunications Union (ITU) standards covering "Maintenance." See Appendix E for complete list.

M12 The designation of an interface multiplexer that interconnects four DS1s and one DS2 signal streams. See also *DS-n.*

M13 The designation of an interface multiplexer that interconnects 28 DS1s and one DS3 signal streams. See also *DS-n.*

M24 The designation of an interface multiplexer that interconnects 24 DS0s and one DS1 signal streams. *M24* compatibility infers compliance with coding and channelization techniques of AT&T's TR62411 specification. See also *DS-n.*

M34 The designation of an interface multiplexer that interconnects six DS3s and one DS4 signal streams. See also *DS-n.*

M44 The designation of an interface multiplexer that interconnects 44 voice channels and one T-1 rate signal stream using adaptive differential pulse code modulation (ADPCM).

If no separate signaling is included, 48 channels may be inserted onto the T-1 link. See also *DS-n.*

M48 A multiplexer that puts 48 ADPCM voice channels onto one T-1 and with signaling in each voice channel. See also *DS-n.*

M55 A multiplexer that puts 55 ADPCM voice channels, in five bundles, onto one E-1. See also *DS-n.*

mA A symbol for milliamp or milliampere.

MAC An acronym for <u>M</u>edium <u>A</u>ccess <u>C</u>ontrol. A sublayer residing between the ISO/OSI physical layer and the data link layer. It describes the protocols for broadcast networks, such as LANs and WANs. See also *IEEE 802.x standards* and *medium access control (MAC).*

MAC address A 48-bit number unique to each network interface card (NIC). Generally, the number is programmed into the NIC at the time of manufacture; hence, it is LAN and location independent. Source and destination MAC addresses are included in LAN packets and are used by bridges to filter and forward packets. Also called a *hardware address, MAC name, physical address,* or *universal address.*

MAC convergence function (MCF) A function, in a Distributed Queue Dual Buss (DQDB) network architecture that is responsible for preparing data from a connectionless service.

MAC name See *MAC address.*

Mach An operating system created at Carnegie-Mellon University. Although written from scratch, it is based on UC Berkeley's BSD 4.3 version of UNIX. *Mach* includes innovations such as:

- Multiprocessing (support for multiple processors, or CPUs).
- Multitasking (the ability to work on more than one task at a time).
- A microkernel as an alternative to the traditional operating system kernel.

The NeXTSTEP operating system is a version of *Mach* originally implemented on NeXT computers and now ported to Intel processors.

machine name The network name for a specific computer node. (It is much easier for people to use than the numeric node address, which may be 10 or 15 digits long.)

machine ringing In telephony, subscriber line ringing that once initiated continues both automatically and rhythmically until the call is answered by the called station, abandoned by the calling station, or terminated by the communications network (after an excessive number of rings without an answer).

machine type The designation of a machine as a workstation or a server, or both.

Macintosh® A family of personal computers introduced by Apple Computer, Inc. in January 1984. The computers were the first widely used machines to feature a graphical user interface (GUI), icon, and mouse operation.

The older computers were based on Motorola's 68000 series processor chip (i.e., 68030 and 68040). Later machines used the 6100/66, 7100/66, and 8100/80 RISC processors developed by Motorola, IBM, and Apple. These newer machines include emulation software that enables DOS and Microsoft Windows programs to execute. All but the earliest *Macintosh* computers include built-in networking capabilities, so they require no special NICs or adapters. By default, *Macintoshes* use AppleTalk as the network operating system (NOS) and LocalTalk at the data link layer. They can use a different NOS; however, they will need a NIC in these cases.

Macintosh file system (MFS) An older file system used in early Macintosh systems. The *MFS* uses a flat file structure rather than the more common hierarchical structure. Newer Macintoshes are backward compatible in that they can read disks that use the *MFS.*

Macintosh files Files used on Apple Computer Inc.'s *Macintosh* computers. A *Macintosh file* contains two parts, the data fork and the resource fork.

- The *data fork* contains user-specified information, i.e., the actual file information such as a document's text or a program's code. When a PC reads a *Macintosh file,* only the data fork is read.
- The *resource fork* contains file resource information, including *Macintosh* specific information such as the windows, icons, drivers, and programs used with the file.

macro A predefined sequence of instructions that is given a unique identifier (name). The sequence may be used within another program simply by referring to the identifier. (The sequence is inserted into the program at the point where the macro name is invoked.) The macro name may be followed by an argument list, i.e., a list of parameters/variables used by the routine to perform its function. See also *script.*

macro- A prefix meaning large.

macrobend In fiber optics, a relatively large-radius bend in an optical waveguide (optical fiber). A macrobend will result in no significant radiation loss if it is of sufficiently large radius.

The meaning of sufficiently large depends on the fiber type and numerical aperature (NA), for example,

- Single mode fibers with low NA (typically less than 0.15) are more susceptible to bend losses than other fiber types. Normally, they will not tolerate a minimum bend radius of less than 6.5 to 7.5 cm (2.5 to 3 inches). Certain specialized types of single mode fibers, however, can tolerate a far shorter minimum bend radius without appreciable loss.
- Graded-index multimode fiber having a core diameter of 50 μm (with a NA of 0.20) will typically tolerate a minimum bend radius of about 3.8 cm (1.5 in.).
- Common 62.5 μm core multimode fibers used in customer premises applications have a higher NA (approximately 0.27) and can tolerate a bend radius of less than 2.5 cm (an inch).

See also *microbending.*

macrobend loss In fiber optics, losses attributable to macrobending. Also called *curvature loss.*

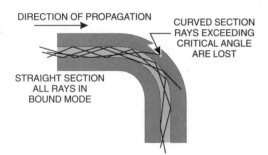

MADN An acronym for Multiple Appearance Directory Number.

Magic Packet A method of remotely awakening a network node that has shut down because of inactivity. Network administrators use it when performing file backups and other administrative duties. The technology was developed jointly by Hewlett-Packard and Advanced Micro Devices.

magnet wire A single strand wire with a thin flexible insulation (such as enamel, polymide, polyurethane, and many others). Generally used for winding coils and transformers. See also *cable* and *wire.*

magnetic amplifier An amplifying device based on one or more saturable reactors.

magnetic card A card with a magnetizable surface on which data can be stored and retrieved.

magnetic circuit (1) The complete closed path taken by magnetic flux. (2) A region of ferromagnetic material, such as the core of a transformer or solenoid, that contains essentially all of the magnetic flux.

magnetic closed circuit A magnetic circuit in which the magnetic flux is guided continually around a closed path through ferromagnetic materials.

magnetic core storage In computer memory technology, a storage device that uses ferromagnetic materials (e.g., iron, iron oxide, or ferrite) and in such shapes as wires, toroids, and rods. See also *core memory.*

magnetic delay line A delay line whose operation is based on the propagation time of magnetic waves.

magnetic field A condition in a region of space, established by the presence of a permanent magnet or a flowing electrical current, characterized by the existence of a measurable magnetic force at every point in the region. The force on a moving charged particle is proportional not only to the charge and field strength but also to the velocity of the moving particle.

magnetic field strength (H) The units of *magnetic field strength* are amperes/meter (A/m) in the SI system (or oersted, Oe, in the cgs system). Also called *magnetizing force.* See also *magnetic flux* (φ) (2).

magnetic flux (φ) (1) A mathematical quantity for which there is no associated physical entity. The unit of *magnetic flux,* called a weber (Wb) in the SI system, is defined as that amount of flux which, changing uniformly in 1 second, induces 1 volt in one turn of a conductor. Therefore, the weber's basic SI units are volts/second (V/s). (2) In the early days of magnetics, *magnetic flux* was defined (in the cgs system) in terms of a unit pole (a pole that exerts 1 dyne of force on a second like pole located 1 cm away and in empty space). Coulomb's inverse square law relates force (F) between these poles (m and m') at a distance r as: $F = kmm'/r^2$ (where k is a constant that includes the medium permeability, $k = 1$).

When a pole is immersed in a magnetic field, a force is exerted on that pole. *Magnetic field intensity (H)* was defined as the field that produces 1 dyne of force on a unit pole (an oersted [Oe]). In order to produce a unit field (1 flux line/cm^2 or 1 gauss) on the surface of a sphere of unit radius, enclosing the unit pole, there must be 4π flux lines (each line is called a maxwell, Mx) emerging from the unit pole. Lines and webers are related by:

$$1 \text{ weber} = 10^8 \text{ lines (webers)}$$

Also called *magnetizing field intensity.*

magnetic flux density (B) The amount of *magnetic flux* (φ) per unit area in a section normal to the direction of flux. Typically expressed in teslas (T) or webers per square meter (Wb/m^2) in the SI system (or gauss [G] or lines/cm^2 in the cgs system). Teslas and gauss are related by:

$$1 \text{ tesla} = 10^4 \text{ gauss}$$

Also called *magnetic induction.* See also *magnetic flux* (φ) (2).

magnetic flux leakage That portion of total magnetic flux in a circuit that does not intercept or couple into the desired element(s), i.e., that part of the total flux that is not used.

magnetic gap (1) That portion of a magnetic circuit that does not contain ferromagnetic material. Also called an *air gap.* (2) The distance between magnetic poles filled with nonmagnetic material.

magnetic hysteresis The property of a magnetic material in which the magnetic induction for a given magnetic force depends on the previous levels of magnetization. See also *magnetic field strength (H), magnetic flux* (φ), and *magnetic flux density (B).*

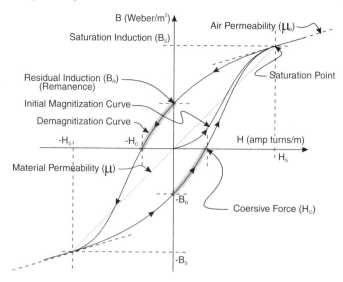

magnetic hysteresis loss (**1**) The power expended as a result of magnetic hysteresis when the magnetic induction is periodic. (**2**) The energy loss per cycle in a magnetic material as a result of magnetic hysteresis when the induction is cyclic (not necessarily periodic).

Although both definitions are in common usage, they are not equivalent.

magnetic medium Any data storage medium in which different patterns of magnetization are used to represent the information being stored. Examples of magnetic media include magnetic tapes, magnetic diskettes, magnetic card stripes, and magneto optics.

magnetic open circuit A magnetic circuit in which the flux path is interrupted by one or more air gaps.

magnetic storm A disturbance of the Earth's magnetic field, caused by abnormal solar activity, usually lasting for a brief period (several days) and characterized by large deviations from the usual value of at least one component of the field. Magnetic storms can affect radio propagation (because of associated ionospheric disturbances), wire communications, and power distribution systems (because of energy coupled into long electric transmission lines).

magnetic tape A ribbon of any material coated (or impregnated) with a magnetic medium, such as ferric oxide, onto which information may be written and retrieved. The information is in the form of magnetically polarized spots.

magneto An ac generator that uses a permanent magnet to supply the field flux.

magneto ionic double refraction The combined effect of the Earth's magnetic field and atmospheric ionization, whereby a linearly polarized wave entering the ionosphere is split into two components called the ordinary wave and the extraordinary wave. These two component waves, in general, follow different paths, experience different attenuations, have different phase velocities, and are elliptically polarized in opposite directions.

magneto optic effect Any of a number of phenomena (such as refractive index or polarization) in which an electromagnetic wave interacts with a magnetic field or with matter under the influence of a magnetic field.

An important *magneto optic effect* having application to optical communication is the *Faraday effect,* in which the plane of polarization is rotated under the influence of a magnetic field parallel to the direction of propagation. This effect is one method used to modulate a lightwave.

magneto telephone set A telephone subset in which microphone current is supplied by a local battery and signaling is generated by a local hand-cranked ac generator (magneto).

mAhr The symbol for milliamp hour, 1/1000 of an amp hour. See also *amp hour (Ahr).*

mail body The content of a mail message, as opposed to the header that indicates the source, destination, subject, and possible other information.

mail bomb A form of e-mail terrorism in which one user inundates another user's mailbox with copious quantities of lengthy, irrelevant, random subjects.

Sending a *mail bomb* is sufficient grounds on most systems to have the sender's account suspended or revoked (*squelched*).

mail bridge A *mail gateway* that connects similar mail services.

mail delivery system In networking, the collection of elements necessary to transport electronic mail (e-mail) from one location to another. A mail delivery system may include some or all of the elements: a *mail directory, mail exploder, mail server, mail switch,* and *mailbox.*

mail directory The file directory for a network in which each user on a network has a unique electronic mailbox. This mailbox is usually a subdirectory of the *mail directory.* It is used to store messages until the mailbox owner is ready to read them.

mail exploder A program that forwards received mail to every person on a mailing list. With a *mail exploder,* a user sends a message to a single address, and the *mail exploder* delivers the message to all listed addresses. Also called a *mail reflector.*

mail gateway A node that exists on two or more networks and can transfer mail among them. Mail gateways generally use a store-and-forward scheme to transfer mail between services. See also *mail bridge.*

mail manager A program that allows a user to write, send, receive, edit, delete and save e-mail.

mail path The series of machine names used to direct e-mail from one user to another.

mail reflector In Internet, a special mail address that will automatically forward any mail received to a number of other addresses. It is typically used to implement a mail discussion group. Also called a *mail exploder.*

mail server (**1**) A program that sends information to a user in response to an e-mail request. It manages delivery of e-mail and other related information. *Mail servers* are generally implemented in the applications layer of the ISO/OSI Reference Model. (**2**) The network computer that executes the mail server program.

mail switch Similar to a gateway in that it routes input e-mail messages to their appropriate output destinations. However, in addition to making the connection and forwarding the information, a *mail switch* can also translate the messages from one e-mail protocol to another. In many cases, a standardized intermediate protocol, such as MHS or the X.400, is used. The intermediate protocol is subsequently translated into the desired output protocol.

mailbox A place (generally a unique directory) where electronic mail is stored in a local area network (LAN) or bulletin board system (BBS) until the addressee has processed the message. Each e-mail user has a unique ID and a unique electronic mailbox address (directory).

On the Internet, mail can be read using the programs PINE or ELM. Most browsers on the WWW also provide a means of accessing e-mail.

mailbox name The name of the person or group to which the message is directed. On the Internet, the complete address consists of a *mailbox name* and a *host name* written in the form:

mailbox_name@host_name.

The "@" symbol is read as "at."

mailer A program that allows bulletin board systems (BBSs) using different software to communicate with each other. Also called a *front end.*

Mailer Daemon A program on an Internet machine which automatically tells the sender of an e-mail message why it could not be delivered and sends a confirmation of delivery if it is requested.

mailing list A discussion group that has been organized by means of e-mail. Contributions to the group are sent to a list manager by e-mail. The list manager then forwards a copy of the message to all users that have subscribed to the group. The list manager places a user on the mailing list whenever a request to subscribe is received. Similarly, a user may request to be unsubscribed.

main beam See *main lobe.*

main distribution frame (MDF) In telephony, a wire termination and interconnection device, one part of which terminates external cables entering a facility and another part of which terminates internal lines and cabling. *MDFs* are associated with both public and private switching systems. The functions of the *MDF* include:

- Semipermanent cross connection of any outside line with any internal line (associated with specific switching equipment port) or other outside line.
- Provision for primary overvoltage protector.
- Provision for test and monitoring between a line and the office.

Also called a *main frame.*

main frame **(1)** In telephony, a synonym for *main distribution frame (MDF).***(2)** In computers, a term originally referring to the cabinet (main frame) containing the central processor unit (CPU) of a room-filling to which peripherals such as tape drives and card reader/punches are attached. See also *mainframe.* **(3)** In computers, the large central computer and associated equipment to which workstations are attached. Usually spelled as a single word. See also *mainframe.*

main lobe That portion of a radiation pattern containing the maximum power (greatest field strength) and bounded by nulls.

Both horizontal and vertical patterns may be plotted. For the horizontal pattern, generally, signal power is plotted as a function of azimuth on a polar coordinate system. The vertical pattern is similarly plotted, but elevation is substituted for azimuth. The width of the main lobe is usually specified as the angle encompassed between the points where the power has fallen 3 dB below the maximum value. Also called the *main beam* and *major lobe.* See also *lobe* and *minor lobe.*

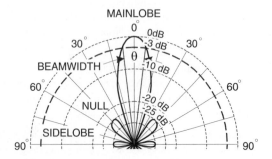

main station In telephony, user equipment (telephone set or terminal) connected to a local loop with a distinct call number designation. This equipment is used for both call origination and for answering incoming calls from the exchange.

main station code The digits, generally following the office code, which designate a main station. See also *area code, city code, country code, international access code, local access code,* and *local number.*

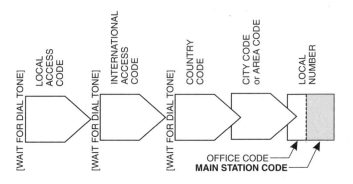

main storage In a computer, program-addressable storage from which instructions or data can be loaded directly into a processor's registers for subsequent execution or processing. It includes the total program-addressable execution space, which may include one or more storage devices.

A distinction is sometimes made between *main storage* (referring to large and intermediate computers or mainframe computers), and *memory* (referring to microcomputers, minicomputers, and calculators). See also *auxiliary storage.*

mainframe When used as a single word, it designates a large to very large, expensive, high-powered computer that is traditionally used in centralized computer processes. *Mainframes* are capable of processing data at very high speeds (many millions of instructions per second [MIPS]) and may have access to terabytes of storage. It is capable of supporting huge databases and hundreds, or even thousands, of users simultaneously.

In the hierarchy of computers that starts with a simple microprocessor (in watches, for example) at the bottom and moves to supercomputers at the top, *mainframes* are just below supercomputers. (Some argue that *mainframes* are more powerful than supercomputers because they support more simultaneous programs or users, but supercomputers can execute a single program faster than a mainframe.) The distinction between a small *mainframe* and a minicomputer is vague. It probably depends on how the manufacturer wants to market its computers more than any technical differentiation. A desktop PC has more computing power and online storage capacity than many mainframes had five years ago. In today's business environment the tendency is away from the traditional mainframe computers and toward client/server local area networks (LANs). Mainframes are sometimes called *big iron.* Contrast with *microcomputer, microprocessor, minicomputer, personal computer,* and *workststion.* See also *main frame (2).*

mains The electrical power distribution system.

Typical *mains* disturbances (distortions) include:

- *Undervoltage (sag, brownouts)*—generally sustained voltages less than nominal (approximately 10% below nominal). Frequently imposed by a power utility as a corrective action taken when the power demand exceeds the generating capacity.

- *Outage (blackout)*—the sustained loss of voltage.
- *Overvoltage (surge)*—generally sustained voltages greater than nominal (approximately 10% above nominal).
- *Transients (spikes)*—short-term overvoltages (less than one or two cycles in duration). The magnitude of a *transient* generally exceeds the *overvoltage* upper limit by many times.
- *Dropouts*—the loss of only several cycles of *mains* current. The number of cycles lost before it ceases to be a dropout and becomes an outage is generally defined in the vicinity of 100 ms.
- *Frequency error*—errors in the number of cycles over a weekly or monthly time period.
- *Noise*—electrical activity (EMI or RFI) that disrupts or distorts the sine wave pattern on which power is delivered. Noise harms signals and information, not physical components.

TYPICAL MAINS CHARACTERISTICS

Country	Voltage (V)	Frequency (Hz)
Australia	240	50
Austria	220–230	50
Bahamas	120	60
Bangladesh	220	50
Brazil	110–220	60
Canada	120	60
Denmark	220–230	50
Dominican Republic	110	60
England	240	50
France	220–230	50
Germany	220–230	50
Greenland	230	50
Hong Kong	200	50
India	220–250	50
Israel	230	50
Japan	100	50/60
Mexico	127	60
Netherlands	220–230	50
New Zealand	230	50
Poland	220	50
Russia	220	50
Saudi Arabia	127/220	50/60
Taiwan	110	60
United States	120/240	60
Venezuela	120	60
Yugoslavia	220	50
Zaire	220	50

major alarm In telephony, an alarm indicating a problem or the presence of a condition that requires immediate attention in order to return or maintain system integrity.

major lobe That radiation lobe containing the maximum radiation. Also called the *main lobe*. See also *lobe* and *minor lobe*.

major trading area (MTA) The 51 geographic territories of the United States that the Federal Communications Commission licenses for broadband wireless services.

make break operation A means of signaling or data transmission involving closing and opening a circuit to produce a series of current pulses.

make busy signal In telephony, a signal transmitted from the terminating end of a trunk to prevent the trunk from being seized by the originating end.

male connector A type of connector whose electrical contacts are pins. The mating connector, a female, has receptacles into which the pins fit.

malfunction An error resulting from a hardware failure.

malicious call Any call that is harassing, threatening, abusive, obscene, and the like.

man Both the abbreviation for manual and a UNIX command for providing pages of the system manual online. In usage, one types *man* followed by the desired subject.

MAN An acronym for Metropolitan Area Network. See *IEEE 802.6.*

man–machine interface (MMI) A phrase used to designate the ease (or effort) of a person working with a computer. Also called *computer–human interface.*

man-made noise Noise generated by machines and other technical devices.

managed object In a network management model, any element in the network that can be used or monitored.

In addition to objects such as nodes, hubs, and so on, less tangible elements such as services, protocols, files, programs, and even algorithms and connections are also considered managed objects.

Management Domain (MD) In the ITU-T X.400 Message Handling System (MHS), a limited (but not necessarily contiguous) area whose messaging capabilities operate under the control of a single management authority. This authority can be a university, an organization, or other group. Two types of management domains are defined: Administrative Management Domain (ADMD)—always run by the ITU/national PTT; and Private Management Domain (PRMD)—run by a local organization.

management information base (MIB) The set of variables or database that a gateway running simple network management protocol (SNMP), CMIP/CMIS over TCP (CMOT), or Common Management Information Protocol (CMIP) network management protocols maintains. It defines variables the SNMP protocol needs to monitor and control components in a network (e.g., change network device settings). It also provides a logical naming for all resources on the network that are pertinent to the management of the network.

MIB-II refers to an extended SNMP management database that contains variables not shared by both CMOT and SNMP. The CMIP and SNMP MIB formats differ in structure and complexity. One MIB specification is RFC1513.

management information system (MIS) An automated method of providing information and information processing to support a company and its organizational activities, such as financial planning, manufacturing planning, decision making, competitive strategy planning, and other planning and analysis models.

Because management applications require data that are best supplied by a database, databases and database management software are generally part of an *MIS.*

management services Network services performed between a host *system services control point* (*SSCP*) and remote *physical units* (*PU*) in IBM's *System Network Architecture* (*SNA*). Services include the request and retrieval of network statistics.

Manchester code A method encoding clock and data signals into a single bit stream for transmission. The receiving station can then decode the bit stream into its clock and data components for use in synchronous protocols.

The encoding rule states:

- For a logical 1 input, the output transitions from a −V to a +V at midbit.

- For a logical 0 input, the output transitions from a $-V$ to a $+V$ at the leading edge of the bit cell and from a $+V$ to a $-V$ at mid-bit.

Each encoded bit contains a transition at the midpoint of a bit period, the direction of transition determines whether the bit is a "0" or a "1," and the first half of the bit period is the true bit value while the second half is the complement of the true bit value. Also called *Manchester II* and *biphase level*. See also *waveform codes*.

mandelbug Jargon for a programming bug whose underlying causes are so complex and obscure as to make its behavior appear chaotic or even nondeterministic (hence its derivation from the Mandelbrot set). This term implies that the bug is a deterministic Bohr bug rather than a heisenbug (whose characteristics change during testing). See also *Bohr bug, heisenbug, schrödinbug,* and *software rot*.

mandrel wrapping The technique of filtering out excessive high-order modes from an optical waveguide by wrapping several turns of the fiber around a small diameter mandrel (1 cm or so). The actual number of turns and mandrel size depend on the fiber characteristics and the desired final modal distribution. See also *macrobend loss*.

mantissa **(1)** The significant digits of a number expressed in scientific notation. The mantissa of 1.234×10^5 is 1.234. Five (5) is the exponent. The number shown is 123 400. **(2)** The fractional part of a logarithm.

manual dialing A person must take the telephone off-hook and dial the number of the remote data communications equipment (DCE). This is in contrast to *automatic dialing,* in which the DCE handles these operations automatically in response to a command from the DTE.

manual exchange A telephone exchange wherein connections between stations are accomplished with plugs and jacks or by means of manual key switches.

manual ringing Ringing that is initiated by the manual operation of a switch (key) and that continues only while the switch is held operated. Contrast with *machine ringing*.

manual telecommunications system A telecommunications system in which connections between stations are established manually by operators in accordance with orders given orally by the calling party.

manufacturing automation protocol (MAP) An ISO/OSI specification model indicating how to automate tasks in computer-integrated manufacturing (CIM) and other factory contexts. The original version of the model was developed by General Motors for its own use. The 1988 version (version 3.0) differs considerably from the original specifications and by agreement was left unchanged for a six-year period (which ended in 1994).

The manufacturing hierarchy is divided into three levels. The upper level corresponds to offices and shops. The intermediate level corresponds to work cells and stations (machines). The lowest level corresponds to individual devices (components). To address these three levels, three types of networks are defined in the MAP model:

- *Type 1 networks*—which connect mainframes, minicomputers, and PCs operating at the highest levels in the automation hierarchy. Their main tasks are non-time-critical activities including information management, task scheduling, and resource allocation, e.g., the exchange of e-mail and files, and various database operations.
- *Type 2 networks*—which connect work cells and workstations. The devices serve as process or machine controllers in which at least some tasks are time-critical. Examples include the exchange of programs, alarms, and synchronization signals.

- *Type 3 networks*—which connect machines (and their components including individual sensors and actuators) with the machine's controllers for these components. The components must operate in real time and must be able to operate in full-duplex mode. Information and command exchanges are constantly occurring due to the real-time operational requirement.

Two protocol suites are defined for MAP systems: the full protocol suite and a "barebones" protocol suite. The full system provides all of the services of the ISO/OSI Reference Model, while the barebones suite is optimized to handle time-critical tasks.

MAP PROTOCOL SUITES

ISO/OSI Layer	Protocol Function	FullMAP	MiniMAP
7	ACSE (Association Control Service Element)	X	
	DS (Directory Service)	X	X
	FTAM (File Transfer & Management)		
	MAP Network Management	X	X
	MMS (Manufacturing Message Specification)	X	X
6	COPP (Connection Oriented Presentation Protocol)	X	
5	COSP (Connection Oriented Session Protocol)	X	
4	COTP (Connection Oriented Transport Protocol)	X	
3	CLNP (Connectionless Network layer Protocol)	X	
	CONP (Connection Oriented Network layer Protocol)	X	
	SNA (Subnet Access Protocol)	X	
	SNDC (Subnet convergence Dependent Protocol)	X	
	SNIC (Subnet convergence Independent Protocol)	X	
2	LLC Class 1	X	
	LLC Class 3	X	X
	Token Bus MAC	X	X
1	Token Bus Broadband	X	
	Token Bus Carrier	X	X

To accommodate these two protocol stacks, three end systems are defined: FullMAP, MiniMAP, and Enhanced Performance Architecture (EPA).

FullMAP stations are used in Type 1 networks, where time is not critical. These stations use a full protocol suite for their activities.

MiniMAP stations are used in both Type 2 and Type 3 networks, which may handle time-critical traffic. To handle time-critical tasks, these nodes communicate with a barebones protocol suite. Because the middle three layers of the ISO/OSI Reference Model are not used in the barebones protocol, several consequences result, that is,

- No routing is possible. (Packets must stay within the MiniMAP network segment.)
- Packet fragmentation is not possible. (There is no transport layer to fragment or reassemble packets.)
- Full-duplex communication is required. (There is no session control layer.)

The *Enhanced Performance Architecture (EPA)* is provided as a mediator among incompatible FullMAP and MiniMAP networks. EPA objects use a protocol suite that supports both the FullMAP and MiniMAP suites. See also *IEEE 802.4*.

manufacturing message service (MMS) In the ISO/OSI Reference Model, a service that enables an application on a control computer to communicate with an application on a slave machine. It defines the framework for distributing manufacturing messages. The specification is used in MAP version 3.x.

MAP An acronym for <u>M</u>anufacturing <u>A</u>utomation <u>P</u>rotocol.

MAP/TOP An acronym for <u>M</u>anufacturing <u>A</u>utomation <u>P</u>rotocol/ <u>T</u>echnical <u>O</u>ffice <u>P</u>rotocol.

MAPI An acronym for <u>M</u>essaging (or <u>M</u>ail) <u>A</u>pplication <u>P</u>rogram <u>I</u>nterface.

mapping (1) The logical association of an arbitrary address to a physical device, e.g., the assignment of a letter to a particular logical disk drive or the assignment of a name to a network. **(2)** The graphical display of two or more sets of numerical data. **(3)** The process of associating each bit of one transmitted data stream into the payload portion of another protocol. For example, one can map a DS-1 service into a SONET VT1.5.

Marcuse loss theory An approximation formula for evaluating the loss of the dominant mode (the HE_{11} mode) in a step index optical fiber.

margin (1) The excess of a specified parameter (gain, phase, distortion, voltage, . . .) over that which is required for proper operation of a device or system. *Margin* is generally expressed in the units of the comparison or as a percentage. Some examples include the following:

- Given a certain system that requires at least a 2-V input signal for proper operation and the actual input level is 3 V, then there is a 1-V *signal margin.*
- Given that the gain in a loop must be less than unity (0 dB) for a system to be stable and that the actual gain around the loop is −3 dB (that is, a loss of 3 dB), the *gain margin* is 3 dB.
- Given a particular operation takes 10 ms to complete, a 100% margin implies 20 ms are allocated for the task.

(2) The allowable error rate, deviation from normal, or degradation of the performance of a system or device.

marginal ray Light rays that pass through an optical system near the edge of the aperture.

mark (1) In binary communications, one of the two significant conditions of encoding, specifically, the idle condition of a signal (i.e., when no information is being transmitted). In current loop signaling, the mark is the presence of current flow (e.g., no dial pulses). The *mark* is the state of the "1" bit and the signaling stop bit for asynchronous transmission. Also called *marking pulse, marking signal, stop polarity,* and *1 bit polarity.*

The complementary signal is called a *space.* **(2)** A symbol or collection of symbols that indicate the beginning or the end of a field, a word, or a data item in a file, record, or block.

Mark 1 A first-generation computer conceived by Howard Hathaway Aiken of Harvard University 1937 and developed at the IBM Research Laboratory at Endicott, New York (by Howard Aiken, Claire D. Lake, Francis E. Hamilton, and Benjamin Durfee). Put into operation in 1944, the Mark 1 is considered the first full-sized digital computer. It weighed 5 tons, had 500 miles of wiring, was used only for numeric calculations, and had a 300-ms clock cycle time. (One multiplication and division took three seconds or more.)

mark hold The normal "no traffic" condition of a serial communications line wherein a steady mark condition (binary 1) is transmitted. See also *space hold.*

marker thread A colored string installed in a cable along with and generally parallel to the individual insulated wires of the cable. The thread is used for identification purposes. It may identify manufacturer, temperature rating, or other specifications.

marking bias The uniform lengthening of all marking signal pulse widths at the expense of the pulse widths of all spacing pulses. See also *bias distortion.*

Markov chain A sequence of symbols (or events) in which the next symbol is influenced only by the current symbol and not by any previous symbols in the sequence. The current state of the system provides all historical information relevant to the future behavior of a Markov process. Named after the Russian mathematician Andrei Andreyevich Markov (1856–1922).

If a symbol is influenced by "m" of its previous symbols, the chain is called an *m^{th}-order Markov* chain.

markup language A standardized set of marking characters (called *tags*) that describe parts of a document and formats (including fonts, paragraph marks, indents, and justification). The purpose of a *markup language* is to allow a document with a complex format to be viewed or printed on virtually any hardware. Because the goal is universality, the tags (as well as the body of the document) are generally reduced to the standard ASCII character set.

Markup languages are divided into two categories: *declarative markup languages* (such as hypertext markup language—HTML— and standard generalized markup language—*SGML*) and *procedural markup languages* (such as PostScript). The declarative *markup language* simply marks the parts of the document, while the procedural *markup language* defines how a document is to be displayed. In the Internet, HTML is the markup language used on hypertext documents for the WWW (World Wide Web). For electronic texts, SGML is the language of preference.

martian A slang term referring to packets that appear on the wrong network. These packets can arrive on the wrong network for reasons such as faulty routing information or, in the case of the Internet, an unregistered address.

maser An acronym for <u>M</u>icrowave <u>A</u>mplification by <u>S</u>timulated <u>E</u>mission of <u>R</u>adiation. A member of the general class of microwave oscillators based on molecular interaction with electromagnetic radiation.

mask A device used to hide all or part of another entity. For example,

- In computer programs, a *mask* is used to expose desired bits in word for testing or other manipulation. The *mask* is the same length as the field to be operated upon and contains logical "1s" in the positions of interest (other positions are "0s"). The mask is logically "ANDed" with the target field, for example,

$$xxxSxxxx \text{ AND } 00010000 \rightarrow 000S000$$
$$\text{field} \qquad \text{mask} \qquad \text{result.}$$

- In communications, to obscure, hide, or otherwise prevent information from being derived from a signal. Masking may be the result of interaction with another signal, such as noise, static, jamming, or other forms of interference.
- In optics, an opaque screen that prevents selected rays from reaching a target.

Masking is not synonymous with erasing or deleting information.

masked A synonym for *disabled.* See also *mask.*

masked threshold The level at which the signal-to-noise ratio (S/N) becomes high enough to allow a signal of interest to be distinguishable from other signals or noise. This level is usually expressed in decibels (dB).

master control center (MCC) Part of Digital Equipment Corporation's (DEC) umbrella network management system Enterprise Management Architecture (EMA).

master slave timing Timing in which one station (or network node) supplies the timing reference for all other interconnected stations (or nodes).

master station **(1)** The unit or terminal having control of all other nodes on a multistation network for purposes of polling and/or selection. On a token passing LAN, the device that enables recovery from error conditions, such as lost, busy, or duplicate tokens, usually by generating a new token.

Servers are sometimes called master stations. **(2)** In a data network, the station that is designated by the control station to ensure data transfer to one or more slave stations. A master station controls one or more data links of the data communications network at any given instant. The assignment of master status to a given station may be temporary and is controlled by the control station according to the procedures set forth in the operational protocol. Master status is normally conferred upon a station so that it may transmit a message, but a station need not have a message to send to be designated the master station. **(3)** In systems using precise time dissemination (such as navigation systems), a station that has the clock used to synchronize the clocks of subordinate stations. **(4)** In basic mode link control, the data station that has accepted an invitation to ensure a data transfer to one or more slave stations. At a given instant, there can be only one *master station* on a data link.

matched The condition of identity with respect to a selected parameter(s).

matched filter A filter that maximizes signal-to-noise ratio so that a waveform of known shape can be separated from random noise.

matched transmission line A transmission line is said to be matched at any transverse section if there is no reflection at that section.

matching network A network interposed between two circuits having unequal attributes for the purpose of transparently joining the circuits.

For example, a matching network might be placed between two transmission circuits with different impedances so as to eliminate signal reflections. See also *balance network*.

matching transformer A transformer designed to simultaneously match the impedance characteristics of both the load and the source it is connected between.

material dispersion coefficient [M(λ)] In fiber optics, pulse broadening per unit length of fiber and per unit of spectral width, usually expressed in picoseconds per nanometer·kilometer (ps/nm·km). For many optical fiber materials, $M(\lambda)$ approaches zero at a specific wavelength (λ_0) between 1.3 μm and 1.5 μm. At wavelengths shorter than λ_0, $M(\lambda)$ is negative and increases with wavelength; at wavelengths longer than λ_0, $M(\lambda)$ is positive and decreases with wavelength.

Pulse broadening caused by material dispersion in a unit length of optical fiber is given by the product of $M(\lambda)$ and spectral width ($\Delta\lambda$),

$$M(\lambda) = \frac{1}{c}\frac{dN}{d\lambda} = -\frac{1}{c}\frac{d^2 n}{d\lambda^2}$$

where:

n is the refractive index of the material,
N is the group index expressed as

$$N = n - \lambda\frac{dn}{d\lambda}$$

λ is the wavelength of interest, and
c is the velocity of light in vacuum.

material scattering That part of total scattering attributable to the intrinsic properties of the materials through which the wave is propagating. Examples of *material scattering* include ionospheric scattering and Rayleigh scattering.

In an optical fiber, *material scattering* is caused by inhomogeneities (on a microscopic scale) in the refractive indices of the materials used to fabricate the fiber, including the dopants used to modify the refractive index profile.

matrix **(1)** The "switching fabric" of a telephone exchange. The switches may be any of several different structures (step-by-step, cross-bar, or a semiconductor time-space-time system). All switching matrices fall into one of two categories, blocking and nonblocking. See also *blocking* and *nonblocking*. **(2)** In mathematics, a regular array of numeric or algebraic quantities that may be treated as a single entity. **(3)** The surrounding/binding material of a composite substance, such as the plastic surrounding the fibers in a filled composite or the cement in concrete. **(4)** What the Opus BBS software and sysops call FidoNet. **(5)** The totality of present-day computer networks.

matrix switching A form of data switching at the heart of ATM, enabling the appropriate bandwidth to be available end to end for the duration of the session without contention.

MAU **(1)** An abbreviation of <u>M</u>ultistation <u>A</u>ccess <u>U</u>nit (also abbreviated MSAU). **(2)** An abbreviation of <u>M</u>edium <u>A</u>ttachment (or <u>A</u>ccess) <u>U</u>nit.

maximal ratio combiner: A *diversity combiner* with the following conditions:

- The gain of each input channel is made proportional to the rms signal level and inversely proportional to the mean square noise level in that channel.
- The same proportionality constant is used for all channels.
- The signals from each input channel are added together.

Also called a *ratio-squared combiner, post-detection combiner, pre-detection combiner,* or *selective combiner*. See also *combiner* and *diversity combiner*.

maximum access time The greatest amount of time allowed between the initiation of an access attempt and the successful completion of the access.

maximum burst size (MBS) In ATM, the maximum number of cells that may be sent at peak cell rate (PCR) without exceeding the sustainable cell rate (SCR).

maximum modulating frequency The highest information frequency a communications system will support.

maximum transmission unit (MTU) The largest size packet that can be transferred across a specified network link. If the incoming packet is larger than the network can handle (greater than the *MTU*), it must be *fragmented* (segmented) and repackaged in lower protocol layer. At the destination, the individual packets are reassembled into the original transmitted message. In Ethernet, for example, the *MTU* is 1500 bytes.

maximum usable frequency (MUF) In radio transmission systems that utilize reflection from the regular ionized layers of the ionosphere, the upper frequency limit that can be used for transmission between two points at a specified time. *MUF* is a median frequency applicable to 50% of the days of a month, as opposed to 90% cited for the *lowest usable high frequency (LUF)* and the *optimum traffic frequency (FOT)*.

maximum user signaling rate The maximum rate (in bits per second—bps) at which information can be transferred in a given direction between users over a link dedicated to a particular information transfer transaction, under conditions of continuous transmission and no overhead information. For communications links consisting of parallel end-to-end communications channels, the parallel channel signaling rate (PCSR) is given by:

$$PCSR = \sum_{i=1}^{m} \frac{\log_2 n_i}{T_i} \text{ bps}$$

where:

i is the index to the Ith channel,
m is the number of channels,
T_i is the minimum interval between significant instants for the Ith channel, and
n_i is the number of significant conditions of modulation for the Ith channel.

For a single end-to-end serial channel (i.e., $m = 1$), the signaling rate (SCSR) reduces to:

$$SCSR = \frac{\log_2 n}{T} \text{ bps}$$

In the case where an end-to-end telecommunications service is provided by nonhomogeneous tandem channels, the end-to-end signaling rate is the lowest signaling rate among the individual channel links. See also *Shannon limit.*

Maxwell's equations A set of partial differential equations that describe and predict the behavior of electromagnetic waves in free space, in dielectrics, and at conductor to dielectric boundaries. *Maxwell's equations* expand upon and unify the laws of Ampere, Faraday, and Gauss; in fact, they form the foundation of modern electromagnetic theory.

Mb An abbreviation of M̲ega b̲its.

MB An abbreviation of M̲ega B̲ytes.

MBONE An acronym for M̲ulticast B̲ackb̲ONE̲.

Mbps An abbreviation of M̲ega̲b̲it p̲er s̲econd. Also written Mb/s.

MBps An abbreviation of M̲ega̲B̲yte p̲er S̲econd. Also written MB/s.

MBS An abbreviation of M̲aximum B̲urst S̲ize.

MCA An obsolete abbreviation of M̲icroC̲hannel A̲rchitecture.

MCC (**1**) An abbreviation of M̲aster C̲ontrol C̲enter. (**2**) An abbreviation of M̲aintenance C̲ontrol C̲ircuit.

MCF (**1**) An abbreviation of M̲AC C̲onvergence F̲unction. (**2**) An abbreviation of M̲essage C̲onfirmation F̲rame.

MCM An abbreviation of M̲ultiC̲arrier M̲odulation.

MCVFT An abbreviation of M̲ultiC̲hannel V̲oice F̲requency T̲elegraph.

MCW An abbreviation of M̲odulated C̲ontinuous W̲ave.

MD (**1**) An abbreviation of M̲anagement D̲omain. (**2**) An abbreviation of M̲ediation D̲evice.

MD5 algorithm An abbreviation of M̲essage D̲igest 5̲ algorithm. A proposed encryption strategy for the Internet Simple Network Management Protocol (SNMP). The algorithm uses the message, an authentication key, and time information to compute a checksum value (the digest). Defined in RFC1321.

MDF An abbreviation of M̲ain D̲istributing F̲rame.

MDI An abbreviation of M̲edium D̲ependent I̲nterface.

MDRAM An acronym for M̲ultibank D̲ynamic R̲andom A̲ccess M̲emory.

MDT An abbreviation of M̲ean D̲own T̲ime.

MDW An abbreviation of M̲ultiple D̲rop W̲ire.

mean In statistics, one of several representations of a group or set of data by means of a single number that is, in some sense, representative of the entire set.

- *Arithmetic mean:* The *arithmetic mean,* written $\bar{x}$, is the sum of all values in the set (x_1 through x_n) divided by the number of items in the set (n), that is,

$$\bar{x} = \frac{x_1 + x_2 + \cdots + x_n}{n} = \frac{\sum_{i=1}^{n} x_i}{n}$$

Also called simply the *mean* or the *average.*
- *Geometric mean.* The *geometric mean* (G) is the n̲th root of the product of all terms (x_1 through x_n) in the set, that is,

$$G = n\sqrt{x_1 \bullet x_2 \bullet \cdots \bullet x_n}$$

The *geometric mean* is used to determine the center frequency of a range of frequencies defined by the two band edges.
- *Harmonic mean:* The *harmonic mean* is defined as n divided by the sum of the reciprocals of the values in the set (x_1 through x_n), that is,

$$H = \frac{n}{\sum_{i=1}^{n} \frac{1}{x_i}}$$

See also *median.*

mean busy hour In telephony, for a telephone line, trunk group of circuits, or switching system, the 60-minute period of time in which the traffic is greatest.

mean time between failures (MTBF) An indicator of expected system reliability. It is calculated on a statistical basis from the known failure rates of various components of the system over a known time period. *MTBF* is usually expressed in hours.

MTBF is generally calculated in one of two ways:

- *For one item*—when operating over a **long** measurement period, the measurement period divided by the total number of failures that have occurred during that measurement period.
- *For a population of items*—during a measurement period, the total functioning life of the population (i.e., the summation of the operating life of every item in the population over the measurement period) of items divided by the total number of failures within the population during that measurement period.

When computing *MTBF,* any measure of operating life may be used, e.g., on-time, cycles, kilometers, or any other event. For example, if a total of 1000 events, such as data transfers, power on/off cycles, system boots, or door openings occurs in a population of items during a measurement period of 500 hours and there are a total of 25 failures among the entire population, then the MTBF for each item is

$$(1000)(500)/25 = 2 \times 10^4 \text{ hours.}$$

Also called *mean time to failure* (*MTTF*).

mean time between outages (MTBO) The average time between failures or events of sufficient severity that unacceptable degradation or total functionality is lost. For example, the average time between transmission failures that renders communications useless. *MTBO* is calculated by the equation:

$$MTBO = \frac{MTBF}{1 - FFAS}$$

where:

MTBF is the mean time between failures and
FFAS is the fraction of failures for which the failed equipment is automatically bypassed.

When there is no redundancy, *FFAS* = 0 and *MTBO* = *MTBF*.

mean time to repair (MTTR) The average time it takes to repair a failed device. The average does not assume any particular failure but is gathered from all reasonable failures possible. It is the total corrective maintenance time divided by the total number of corrective maintenance actions during a given period of time.

mean time to service restoral (MTSR) The average time it takes to restore service following a failure that results in an outage. The time to restore is measured from the time of failure occurrence (not the time of reporting) to the time of service restoration.

MTSR is not the same as mean time to repair (MTTR). For example, a particular communications circuit may be disabled because of transmission line damage, and the damage may take several days to repair. However, service restoration may be accomplished in a matter of seconds by re-routing traffic through redundant circuits.

measured rate service Telephone service for which charges are made in accordance with the total connection time of the line.

measured service Telephone service in which charges are assessed in terms of the number of message units accumulated during the billing period.

mechanical splice In fiber optics, the permanent joining of two optical fibers by mechanical means only, e.g., by means of a fixture, crimping, or adhesive, but not by fusing or melting together.

When the fibers are secured by mechanical means alone (e.g., crimping), the gap between them is usually filled with an index matching gel to reduce Fresnel reflection. Similarly, the optical adhesives have a refractive index close to that of the fiber core and also serve to reduce Fresnel reflection.

MED (1) An abbreviation of MEDia. **(2)** An acronym for Message Entry Device. **(3)** An acronym for Microwave Emission Detector.

media access control (MAC) driver A local area network (LAN) device driver that works directly with the network adapter cards, acting as an intermediary between the transport driver and the hardware.

media filter A device that enables transition from one type of cable transmission scheme to another.

Generally refers to a device for converting the output signal of a Token Ring network interface card to work with unshielded twisted pair (UTP) cable. It is a passive device that can convert between UTP and shielded twisted pair (STP) cables and is designed to eliminate undesirable high-frequency emanations.

media server A synonym for *file server;* however, it implies the existence of and support for multimedia information.

median In statistics, an expression of the "middle" value of a set of numbers arranged in an increasing or decreasing order. Specifically, given a set of *n* items arranged by either increasing or decreasing value, the *median* is either the value of the middle item if *n* is odd, or the mean of the middle two items if *n* is even. See also *mean*.

mediation device (MD) A SONET device that performs mediation functions between network elements (NEs) and operating systems (OSs), such as protocol conversion, language conversion, and message processing.

MEDINET An acronym for MEDical Information NETwork.

medium (1) The material on which data are recorded, e.g., paper, magnetic tape, diskettes, CD-ROM, and punch cards. **(2)** Any material substance that is, or can be, used for the propagation of signals, usually in the form of modulated radio, light, or acoustic waves, from one point to another, such as optical fiber, cable, wire, dielectric slab, water, air, or free space.

medium access control (MAC) (1) In a communications network, the lower part of the ISO/OSI Reference Model *data link layer* (the upper part being the logical link control—LLC) that supports topology-dependent functions and uses the services of the *physical layer* to provide services to the LLC.

For example, it is a layer in the IEEE 802.x protocol standards that defines frame formats and the medium access method used by a broadcast network, such as local area networks (LANs) and wide area networks (WANs). In a broadcast network, signals transmitted by one node are present at the input of all other nodes on the network. The node to which the data are addressed receives the packet, while all other nodes simply ignore the data. A problem can occur, however, when a node attempts to transmit: another node may already be transmitting (called *contention*). The *MAC* protocol defines a procedure for resolving the problem, called carrier sense multiple access with collision detection (CSMA/CD). **(2)** One of the four documents defining Fiber Distributed Data Interface (FDDI). The *MAC* and physical (PHY) layers are implemented directly in an FDDI chip set. Other media access methods include CSMA/CA, polling, and token passing. See also *IEEE 802.x standards, physical (PHY), physical-layer medium dependent (PMD)*, and *station management (SMT)*.

medium access method The strategy used by a node, or station, on a network to access a network's transmission medium. Access methods are defined at the data link layer in the ISO/OSI Reference Model. More specifically, they are determined at the medium access control (MAC) sublayer (as defined in IEEE 802.x). Two main classes of access methods are defined: *probabilistic* and *deterministic*.

- *Probabilistic:* A node checks the line when the node wants to transmit. If the line is busy, or if the node's transmission collides with another transmission, the transmission is canceled. The node then waits a random amount of time before trying again.
Probabilistic access methods can be used only in networks in which transmissions are broadcast, so that each node gets a transmission at just about the same time. These methods are most suitable in smaller networks and in those with light traffic.
Ethernet's CSMA/CD (carrier sense multiple access/collision detect) is an example of a probabilistic access method.
- *Deterministic:* A node gets exclusive access to the network in a predetermined sequence. The sequence is determined either by the server or by the arrangement of the nodes themselves. Common protocols used in deterministic access methods include polling (used in mainframe environments) and token passing (used in several LAN environments).

Deterministic methods are most suitable for large networks and networks with heavy traffic. ARCnet and Token Ring networks are examples of token passing networks.

Medium Attachment Unit (MAU) A transceiver (transmitter/receiver) consisting of hardware that provides the correct electrical or optical connection between the computer and the local area network (LAN) medium. Since *MAUs* typically support only one type of network medium, a choice of *MAUs* is available to support different media. A *MAU* detects carrier and collision activity, passing the information to the computer. The MAU can be a stand-alone unit or incorporated in a circuit board inside the computer. Examples of MAUs include:

- The device used to convert signals from one Ethernet medium to another, e.g., from the 10Base5 thickwire backbone to the access unit interface cable (AUI), or
- An optical transceiver, which interfaces an electrical signal at one port and converts it to an optical signal at its other port.

medium frequency (MF) Frequencies between 300 kHz and 3 MHz. See also *frequency band* and *frequency spectrum*.

meet me conference A feature of some telephone switching systems in which any telephone on the system may dial a special access code to join a conference with other users who have dialed the same code. Participants may arrange to call in a preset time (i.e., "meet me at 9:00") or through the direction of an attendant.

As with all conference connections, there is generally a maximum number of simultaneous users allowed.

mega (M) (1) The SI prefix indicating 1 000 000. Some examples of usage include:

Mbps   megabit per second—one million bits per second.

MHz   megahertz—one million cycles per second.

Mflop   megaflop—one million floating point operations.

(2) In computer terminology, *mega* generally indicates 2^{20} or 1 048 576.

megaflops An acronym for Million (mega) FLoating point Operations Per Second. A measure of computing power (bigger is better).

memory (1) A circuit or device into which information can be stored (or written) and held for some period of time. At a future time (less than the holding time of the device), the original stored information can be recovered (or read).

The holding time of memory devices is dependent on the technology of the device. For example, static random access memory (SRAM) will retain information as long as power is applied. Dynamic random access memory (DRAM) will retain information only a few milliseconds. It therefore must be refreshed frequently. A read only memory (ROM) will retain information with or without power until the part electrically fails. See *DRAM, EPROM, EEPROM, NVRAM, PROM, RAM, ROM, SRAM,* and *UVPROM* for details of various types of memory. **(2)** In secondary electrochemical cells (rechargeable battery systems). The effect wherein a cell that is not cycled through the full range of fully charged to discharged will lose the ability to deliver all of the theoretical potential energy at full charge. This effect is especially pronounced in some cell types, i.e., nickel cadmium (NiCd).

memory caching A hardware technique for increasing the performance of a computing system by storing current sequences in a very high-speed *cache memory* separate from the computer's main memory. See also *cache memory*.

memory dump A displayed, saved, or printed copy of a specified area of computer memory, which reveals the current values of information stored in that area. A memory dump provides information about the "state-of-the-machine" (at that instant) and is used for debug or diagnostics.

memory map An indication of what type of data can be found in each area of memory. The following table illustrates a memory map of a DOS-based PC.

Address (hex)	Usual Data Type	Memory Name
: 110000	User application programs	Extended [2]
10FFFF : 100000	DOS	HMA
0FFFFF 0F0000	ROM BIOS	
:	EMS page frame[1]	Upper
0C8000	HDD BIOS	
0C7FFF : 0A0000	EGA, VGA, SVGA Monochrome	Video
09FFFF :	User application programs	Conventional
000000	DOS, IRQ vectors, TSRs, etc.	

[1]EMS memory is mapped into pages of the upper memory block (UMB).

[2]Extended memory includes all memory above 1MB, including the high memory area (HMA).

menu A listing of commands, options, or subprograms available and generally selectable by one or two keystrokes or mouse click.

meridional ray A light ray that passes through the optical axis of a device. See also *skew ray*.

mesh connectivity A network wherein multiple paths exist between nodes.

mesh network The name of the network topology wherein every node in the network is *directly* connected to every other node in the network via a dedicated communications link. This type of structure is rarely, if ever, used as the primary method of interconnecting more than a few nodes because the number of interconnects becomes prohibitively expensive. That is, the number of links required is $n(n-1)/2$ where n is the number of nodes and a network interface device (modem or similar device) is needed at each end of the link. The *mesh network* structure is used, however, as a subnetwork in *hierarchical networks*.

COMMUNICATION LINKS REQUIRED FOR AN "n" NODE *MESH NETWORK*

n	Links	n	Links
2	1	11	55
3	3	12	66
4	6	13	78
5	10	14	91
6	15	15	105
7	21	20	190
8	28	25	300
9	36	50	1230
10	45	100	4950

The following diagram depicts the totally unwieldy interconnect required for a network with only 10 DTEs. Each DCE must have ports to the nine other DCEs (and one to a server if implemented), as well as 45 interconnecting cables (55 if a server is provided).

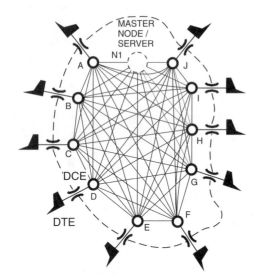

meso- A prefix from the Greek meaning middle.

mesochronous The relationship between two signals when corresponding significant instants occur at the same *average* time.

message (1) In communications, any complete unit of information transmitted from one device to another. A device in this context includes not just hardware but software and even people. **(2)** The information content of a received call. The sequence of symbols used to convey information or data. **(3)** In packet communications, a predefined assemblage of symbols consisting of a header and user information. The header contains at least the address of the destination to which the user information is to be delivered.

message alert A light or other indicator announcing that a phone call was received and that there may be a message waiting. Also called *call in absence.*

message alignment indicator In a signal message, data transmitted between the user part and the message transfer part to identify the boundaries of the signal message.

message feedback An error control method in which a message receiving station transmits the data back to the sender for comparison with the original data.

message format The predefined physical and/or time sequential arrangement of the parts of a message that are recorded in or on a data storage or transmission medium.

Message Handling Service (MHS) A utility in Novell NetWare LANs that provides a common format for exchanging information among applications. It is most commonly used for e-mail transfers.

message handling system (MHS) An application-level service element, in ITU-T's X.400, that enables applications to exchange messages. For example, an electronic mail (e-mail) facility with store-and-forward capabilities is an *MHS*. In the X.400 recommendations, an *MHS* can transfer messages among end users, or between end users and a variety of ITU-T defined services, such as fax, videotext, and so on.

The X.400 *MHS* includes:

- *User agents (UAs),* which provide interfaces for the end users at one end.

- *Message Transfer System (MTS),* which provides the interface at the other end.

  An MTS, along with its message transfer agents (MTAs), performs the actual transfer of the message from one end to the other. The MTAs are responsible for storing and/or forwarding messages to another MTA, to a user agent (UA), or to another authorized recipient. The *MTS* is a connectionless but reliable transfer capability. Connectionless means that parts of the message are transported independently of each other and may take different paths. Reliable means that a message part will be delivered correctly or the sender will be informed that this was not possible.

- *Access units (AUs),* which provide interfaces for the ITU-T services at one end and the MTS at the other end.

- A *message store (MS),* which provides temporary storage for messages before they are forwarded to their destination.

  The MS is a general archive in which mail can be held until an appropriate user retrieves it through a UA or until the allowable storage time for the message is exceeded. The MS is distinct from the mailboxes associated with individual users. UAs and other services use the *Message Store Access Protocol (MSAP)* to access the message store.

message header The sequence of bits or bytes at the beginning of a message that provides one or more of the following: timing, packet number, calling station ID, called station ID, routing information, and packet length.

message ID (MID) In electronic mail or message handling, a unique value associated with a particular message.

message-oriented text interchange system (MOTIS) A set of ISO/OSI message handling standards, adopted in 1988, which expand the capabilities of SMTP (Simple Mail Transfer Protocol) e-mail capabilities. *MOTIS* features include the expected capabilities (message composition, sending, receiving, forwarding, deleting, and saving) as well as enhancements and new abilities (a receipt indicating the transmitted message was received and read, the ability to include graphics and sounds).

From a structural point of view, the addressing information has been separated from the body of the message. This allows a much cleaner way for a system to forward only the information portion. The basic elements of this system are compatible with the model in the ITU-T's X.400 specifications.

message packet A packet containing information originating or terminating in a user application program, i.e., in the ISO/OSI level 7 protocol layer.

message part (1) In radio communications, one of the three major subdivisions of a message (the heading, the text, and the ending). **(2)** In cryptosystems, text that results from the division of a long message into several shorter messages of different lengths as a transmission security measure. *Message parts* are usually designed to appear unrelated externally. Elements that identify the subparts for assembly at the receiving station are encrypted in the texts.

message source That part of a communications system from which a message appears to emanate.

message store (MS) In X.400 e-mail, a staging point (similar to a post office) in which messages are temporarily held for subsequent transmission to one or more destinations.

message switch A switching device in which individual messages are routed to their intended receiver by way of embedded address codes rather than by switching the communications channel itself. See also *packet switching.*

message switching A method of connecting one terminal to another terminal through a communication network on a message-by-message (or packet-by-packet) basis. The connection is not necessarily routed through the same paths for each packet as they would be on a *circuit switched connection*. At each switching point in the fabric of the network, the switch makes a decision as to the best route to send a packet to its final destination. If a path toward the destination is not immediately available, the message packet is temporarily stored until an appropriate outgoing line is available. See also *circuit switching* and *packet switching*.

message timed release A system by which a channel is released after a measured interval of communication.

message transfer agent (MTA) A program residing in a local area network (LAN) server or host that relays e-mail messages created by electronic mail clients. See also *message handling system (MHS)* and *sendmail*.

message transfer part (MTP) That part of a common channel signaling system, such as Signaling System number 7 (SS7), that is used to place formatted signaling messages into packets, send those signaling packets, receive signaling packets, and strip formatted signaling messages from the received packets. It also performs functions such as error control and signaling link security.

message unit **(1)** In telephony, a basic billable amount for local telephone service. The number of *message units* applicable to any given call is a function of the time of day, the distance between calling and called stations, the duration of the connect time, and of course, a monetary multiplier. **(2)** In IBM's Systems Network Architecture (SNA), that portion of a message block that is passed to a specific software layer for processing.

message unit call Any call for which billing is based on message units.

message waiting A feature of some PBX systems that informs the user that there is a message waiting. The message may be held in a voice mailbox or a special message center. Frequently, the indicator is a light on the telephone, or a few characters on the telephone display.

Messaging Application Program Interface (MAPI) An interface for messaging and mail services. Microsoft's *MAPI* provides functions for using Microsoft Mail within a Microsoft Windows application. Simple *MAPI* consists of 12 functions, including Mapi-DeleteMail(), MapiReadMail(), and MapiSendMail(). Appropriate utilization of these functions allows a Windows application to address, send, and receive mail messages while running. See also *Vendor Independent Messaging (VIM)*.

meta bit The top bit of an 8-bit character. When in the off state, character values range from 0 to 127; when in the on state, values are between 128 and 255. Also called *alt bit, high bit,* or *hobbit*.

metal oxide varistor (MOV) A voltage-dependent, nonlinear device with electrical behavior similar to back-to-back zener diodes. A *MOV* is composed primarily of zinc oxide (ZnO), with small amounts of bismuth, cobalt, manganese and other metal oxides added. *MOVs* are used in *transient* protection circuits such as the power mains protectors or telephone line interface circuit protection.

MOVs have a threshold voltage where they change from an open circuit to a near 0 Ω resistance. For a short period of time, the *MOV* can absorb the energy resulting from several thousand amps of current. Therefore, the ratings on the device are not only *voltage* and *current* but *energy* (given in joules-J or Watt-seconds). The amount of energy a *MOV* can withstand is based on its physical size; for example, a 5-mm 22-V MOV is rated at 0.25J, while a 20-mm 22V device is rated at 100J. See also *carbon block, gas tube protector,* and *zener diode*.

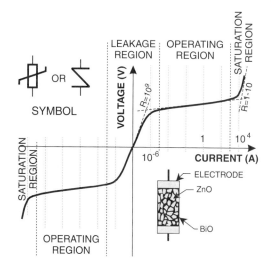

metallic circuit In telephony. **(1)** A complete circuit in which ground or Earth connections are not made, e.g., a telephone subscriber loop. **(2)** A communications path composed entirely of electrical conductors; that is, there are no radio link, no optical fiber link, and no carrier system.

metallic signal In telephony,

- The metallic voltage is equal to the algebraic difference of the voltages on each wire of a loop to a reference (generally ground).
- The metallic current is one-half the algebraic difference of the currents in the two conductors.

Also called a *differential signal*. See also *longitudinal signal*.

meteor burst communications Communications by the propagation of radio signals reflected by ionized meteor trails in much the same manner as a sky wave is reflected from the ionosphere.

meter (m) The basic unit of length in the SI system, equal to 39.37 inches or 3.28 feet.

metering pulse Periodic pulses generated in the telephone network which determine the cost of a call. The cost associated with each pulse is the same (for a given administration); however, the rate at which pulses are delivered is related to the distance called (the further the calling distance, the faster the pulses are counted). Also called *periodic pulse metering (PPM)*. See also *metering tone*.

metering tone A signaling tone pulse used mainly in Europe to control coin collection on coin-operated pay telephones. The tone is usually 12 kHz or 16 kHz and is transmitted at a very high signal level (typically 2.2 V into 200 ohms). Coins are "collected" each time a tone pulse is received at the pay phone. The cadence of the *metering*

TYPICAL METERING TONE VALUES

Country	Frequency (kHz±Hz)	Level (dBm)	On Time (ms)	Off Time (ms)
Belgium	16±160	+17	80	220
Denmark	12	+2.2		
France	12±120	+13	125±25	160
Germany	16±80	+22	120	220
Greece	16±500	+18	50	90
Ireland	12±100	+9	120±20	120±20
Italy	12±120	+9	80	20
Norway	16±160	+6	120	220
Portugal	12±120	+15	120	250
Spain	12±120	+9	100	500
Sweden	12±60	+22	200-500	90
Switzerland	12±300	+22	50-150	90-10,000

tone is dependent on the cost per minute of the telephonic connection and can vary at various stages of a call (e.g., off-hook, dialing, ringing, or connected). See also *metering pulse.*

metric system A decimal system of weights and measures based on the meter (m) as a unit of length, the kilogram (kg) as a unit of mass, and the second (s) as the unit of time. Hence it is frequently referred to as the MKS (Meter Kilogram Second) system. The modern form of the *metric system* is the International System of Units (SI). See also *SI.*

metropolitan area network (MAN) A high-speed network with a maximum range of about 75 km (45 miles) or so. Most *MANs* include some type of telecommunications components and activity to handle long-distance transmissions.

Although *MANs* have many similarities with local area networks (LANs) and wide area networks (WANs), there are some differences, such as the following:

- *MANs* generally provide greater distances than LANs or campus area networks (CANs), but less than WANs; such as an entire metropolitan area (e.g., a large city and its suburbs).
- *MANs* generally include provisions for both voice and data transmissions.
- *MANs* generally involve higher speeds than either LANs or WANs, such as T3 (44+ Mbps) or fiber (600+ Mbps).
- *MANs* interconnect two or more LANs.

MANs often include several LANs connected to each other via telephone lines. Most *MAN* networks use one of two network architectures:

- Fiber Distributed Data Interface (FDDI), which supports transmission speeds of 100+ Mbps, uses a dual ring topology, and has optical fiber as the medium.
- Distributed Queue Dual Bus (DQDB), which is specified in IEEE 802.6. DQDB supports transmission speeds ranging to 600 Mbps over distances as large as 50 kilometers (30 miles). DQDB also uses a two-bus topology.

See also *IEEE 802.6.*

MF (1) An abbreviation of MultiFrequency. A signaling method used to set up connections between central offices. *MF* signals are frequency pairs similar to DTMF signals; the frequencies are different, however. See also *trunk multifrequency (MF) signaling.* (2) An abbreviation of Medium Frequency.

MFD (1) An abbreviation of Mode Field Diameter. (2) The units of capacitance MicroFaraD, although it is usually shown in lower case (mfd) or in the SI system as μF.

MFJ An abbreviation of Modified Final Judgment.

MFLOPS An abbreviation of Million (10^6) Floating point Operations Per Second.

MFM An abbreviation of Modified Frequency Modulation.

MFP An abbreviation of MultiFunction Peripheral.

MFS An abbreviation of Macintosh File System.

MFSK An abbreviation of M-ary Frequency Shift Key, Multi-level Frequency Shift Key, or Multiple Frequency Shift Key. See also *frequency shift key (FSK)* and *modulation.*

mho An obsolete term for the units of measure of *admittance* or *conductance.* Replaced by the SI unit *siemens (S).*

MHS (1) An abbreviation of ITU-T's Message Handling System. (2) An abbreviation of Novell NetWare's Message Handling Service.

MHz An abbreviation of MegaHertz. Meaning 10^6 cycles per second.

MIB An abbreviation of Management Information Base.

micro (μ) (1) The SI prefix indicating one one-millionth (10^{-6}). (2) An abbreviation for MICROprocessor or MICROcomputer.

microbend loss In an optical fiber, optical power loss caused by a microbend.

microbending Deviations of the axis (bending) which are small compared to the wavelength of the transmitted energy. Microbends can result from waveguide coating, cabling, packaging, and installation. Generally applied to optical waveguides. *Microbending* can cause high losses of transmitted energy (radiative loss) and mode coupling. See also *macrobend.*

MicroChannel® IBM's proprietary 32-bit computer communication bus architecture introduced in 1988 that offers improved performance over the Industry Standard Architecture (ISA) bus. It is used primarily by some of the PS/2 computers.

A *MicroChannel* environment allows software to set addresses and interrupts for hardware devices. This means that manual jumper or dipswitch on the boards is not necessary. Hence, it helps reduce the number of address and interrupt conflicts. *MicroChannel* was formerly known as MCA, but this name was dropped after the Music Corporation of America filed a lawsuit. See also *EISA, ISA, Peripheral Component Interconnect (PCI),* and *VESA.*

microcircuit A synonym for *integrated circuit (IC).*

microcode Very low-level software that is resident in ROM and embedded within a computer or microcomputer. It performs the sequence of operations necessary to accomplish the higher level machine language instructions, such as MOVE, ADD, and so on. Generally, several microcode instructions (*microinstruction*) are executed for each machine-level instruction in a non-RTSC (reduced instruction set computing) processor.

Microcom Networking Protocol (MNP) One of the de facto error-correction and data compression protocol suites used in data communications. Portions of the standard are incorporated into the ITU-T recommendations.

Since the original definition of the *MNP* protocol, nine enhancements (called classes) have been incorporated. Classes 1 through 4 have been placed in the public domain. Classes 2 through 4 (as well as the link access protocol for modems (LAP-M) protocol) are required elements of the ITU-T V.42 recommendation for modem error control. See also *MNP Class n.*

microcomputer (1) A computer built onto a single-silicon integrated circuit (chip). The microcomputer contains not only the central processing unit (CPU), but memory for program and data as well. (2) A computer (such as a PC or Macintosh) based on a single-chip microprocessor. It is less powerful than either a minicomputer or a mainframe computer.

microcontroller A microcomputer (or microprocessor) used to control some aspect(s) of a device or system. These mechanisms are found in toys, kitchen appliances, automobile ignition, pollution or airbag control systems, modems, or any of thousands of items in today's electronically controlled environment.

A *microcontroller* generally has a separate port for each sensory input (environmental or user) and control output, and has a program dedicated to a single function or set of functions. This program may not be modified by the user (although parameters of the program may sometimes be selectable). An example of a *microcontroller* is a "smart thermostat" to control temperature. The inputs may include:

- Temperature (perhaps both inside and outside temperature).
- Time of day and day of the week.
- Desired temperature profile.
- Availability of energy sources (e.g., solar or fossil).

The corresponding outputs can then include switches to control:

- Indicator lights that inform a user if heating or cooling is occurring.
- Refrigeration or heating units.
- Blowers, water pumps, and valves.

microfarad (μF and mfd) One millionth of a farad. The most common unit of capacitance. See also *farad (F)*.

micron A unit of measurement corresponding to one millionth of a meter (roughly 1/25,000 inch), also called a micrometer. *Microns* are frequently used to specify the diameter of optical fibers, as in 62.5 or 100 micron fibers.

microphonics Electronic noise introduced into an electronic system through mechanical shock or vibration of elements within the system.

microprocessor A central processing unit (CPU) built on a single-silicon integrated circuit (chip). With the addition of memory, power, and interfaces to peripherals, a *microprocessor* becomes a *microcomputer*.

microprogram A sequence of *microinstructions* that are in special storage where they can be dynamically accessed to perform various functions.

microsecond (μs) One one-millionth of a second (10^{-6} s or 1/1 000 000 s), frequently abbreviated μs.

microwave Loosely, an electromagnetic wave having a wavelength less than 300 mm (a frequency greater than 1 GHz). *Microwaves* exhibit many of the properties usually associated with optical waves. For example, they are easily concentrated into a beam, and they tend to be useful only in line-of-sight applications. See also *millimeter wave*.

MID An abbreviation of Message ID.

middle endian A representation strategy on 32-bit systems that is neither little endian nor big endian.

Given four bytes (numbered 1, 2, 3, and 4) which make up a 32-bit word, there are many ways to order the bytes to form a word. In a proper 32-bit system, the representation sequence is 1-2-3-4 or 4-3-2-1. *Middle endian* systems, on the other hand, use the representation 2-1-4-3 (big endian bytes in little endian words) or 3-4-1-2 (little endian bytes in big endian words). Another way of expressing the ambiguity problem is frequently called the NUXI problem (i.e., how to represent the letters of the word "UNIX" in a 32-bit word). The two proper-endian solutions are UNIX and XINU. The *middle endian* representations are NUXI and IXUN. See also *big endian, byte sex,* and *little endian*.

middleware A program that essentially translates information from a variety of databases, each with a different format and/or operating system, to a common format. This allows an internetwork user to access information from a wider variety of sources.

MIDI An abbreviation of Musical Instrument Digital Interface. *MIDI* is an international industry specification for control of digital audio devices and musical instruments.

mil One one-thousandth of an inch.

Miller code Also called delay encoding or modulation. Information is contained in the transitions, not the logical levels. See also *waveform codes*.

milliampere (mA) One one-thousandth of an ampere (1/1000 ampere).

millimeter wave Loosely, an electromagnetic wave having a wavelength between 1 mm and 0.1 mm (300 GHz to 3000 GHz). *Millimeter waves* exhibit many of the properties usually associated with optical waves. For example, they are easily concentrated into a beam, and they tend to be useful only in line-of-sight applications. See also *microwave*.

million One *million* is 10^6. Mega is the SI prefix.

millisecond (ms) One one-thousandth of a second (10^{-3} s or 1/1000 s).

millivolt (mV) One one-thousandth of a volt (1/1000 volt). Abbreviated mV.

milliwatt (mW) One one-thousandth of a watt (1/1000 Watt). Abbreviated mW.

MILNET An acronym for MILitary NETwork. Originally part of ARPAnet, *MILNET* was separated in 1984 to make it possible for military installations to have a reliable network service while the ARPAnet continued to be used for research. Under normal circumstances, *MILNET* is part of the Internet.

MIMD An abbreviation of Multiple Instruction Multiple Data. A type of parallel processing computer. It is, in effect, composed of a multiplicity of individual computers, each capable of simultaneously processing different parts of a problem. See also *single instruction multiple data (SIMD)*.

MIME An acronym for Multipurpose Internet Mail Extension, an Internet standard for transferring nontextual information as attachments to a message. Attached files can be image or sound data.

Originally *MIME* was developed for the Internet as an open and extensible standard. It is independent of the computing platform and can (in principle) be used to send multimedia messages across different platforms and operating environments. Support for *MIME* has been built into several e-mail packages. Because of its flexibility and extensibility, *MIME* enables more powerful messaging services. For example, a message might contain a program that can execute as part of the message in order to do a demonstration or a calculation. However, these capabilities also raise unresolved issues relating to security and compatibility. For example:

- The message must not do damage to the recipient's files or system.
- The receiver must be able to select which parts of a message to read so that the user need not wait for "nonuseful" transfers, e.g., a multimedia animation on a slow laptop.

There are several Internet documents on the subject: RFCs 1341, 1342, 1343, and 1344. Approved as draft standards in RFCs 1521 and 1522.

minicomputer A digital computer whose performance lies between that of a mainframe and a personal computer or workstation. A *minicomputer* is often used to provide time-sharing services for many users or as a stand-alone real-time processor.

MiniMAP See manufacturing automation protocol (MAP).

minimum bend radius The radius below which a coaxial cable, optical fiber, or fiber optic cable should not be bent. Although the *minimum bend radius* should be specified by the cable manufacturer, it isn't always. If no *minimum bend radius* is specified, it is usually safe to assume a radius at least 15 times the cable diameter.

minimum dispersion wavelength A synonym for *zero dispersion wavelength*.

minimum dispersion window (1) The window of an optical fiber at which *material dispersion* is very small.

In silica-based fibers, the *minimum dispersion window* occurs at an approximate wavelength of 1.3 μm. This may be shifted toward the *minimum loss window*, i.e., 1.55 μm, by the addition of dopants during manufacturing of the fiber. **(2)** In a single mode fiber, the window (or windows, in the case of doubly or quadruply clad fibers) at which *material* and *waveguide dispersion* cancel one another, resulting in extremely wide bandwidth, i.e., extremely low dispersion, over a narrow range of wavelengths. Also called *zero dispersion window*.

minimum distance code A binary error detection code in which the signal distance does not fall below a specified minimum value. See *Hamming distance*.

minimum loss window Of an optical fiber, the transmission window at which the attenuation coefficient is at or near the theoretical (quantum-limited) minimum.

If the losses from various mechanisms are plotted on a single graph as a function of wavelength, the *minimum loss window* occurs near the wavelength at which the Rayleigh scattering attenuation curve and the infrared-phonon-absorption curve intersect. For silica-based fibers, this occurs at approximately 1.55 μm.

minimum shift key (MSK) A special case of frequency shift key (FSK) modulation in which there is phase coherency at bit boundaries. *MSK* characteristics include the following:

- There are no phase discontinuities in the modulated carrier.
- The frequency changes occur at the carrier zero crossings.
- The frequency difference between a logical zero and a logical one is always equal to one-half the data rate (i.e., the modulation index is 0.5).

See also *frequency shift key (FSK), modulation,* and *modulation index.*

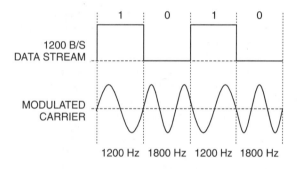

MINn Cellular telephone communications. An abbreviation of Mobile Identification Number n.

- *MIN1* is the seven-digit telephone number assigned to a cellular telephone by the local carrier. The number is used both for receiving calls and for billing.
- *MIN2* is the area code of the cellular telephone number.

minor alarm In telephony, an alarm indicating a problem that does not seriously affect operation or safety to the equipment or personnel.

minor lobe Any radiation lobe other than the major or main lobe. See also *lobe* and *major lobe.*

MIPS (1) An acronym for Million Instructions Per Second. A measure of a computer's speed based on the number of machine-language instructions performed in one second. (2) An acronym for Meaningless Information Per Second [a joke, probably a transmogrification from (1)].

mirror site In Internet, a file transfer protocol (FTP) node that contains the same information as another node. *Mirror sites* are used to distribute the load on a single frequently accessed node.

mirroring A method of constructing a fault-tolerant data storage device through the use of two identical disk storage systems. See also *RAID.*

MIS (1) An abbreviation of Management oriented Information Systems. (2) An abbreviation of Management Initiated Separation. Translation, "you're fired." The term reportedly originated in IBM.

misalignment loss In fiber optics, any of several mechanical offsets (such as gap, angular offset, lateral offset, etc.) that lead to signal level loss. See also *coupling loss.*

misc An abbreviation of miscellaneous.

misc hierarchy One of the *world newsgroups* in USENet. The categories of these newsgroups include anything that does not fit into any of the other six world hierarchy categories. See also *newsgroup hierarchy.*

mismatch The condition in which the impedance of a source is not the same as the impedance of the load to which it is connected.

mistake A human action that produces unintended results, e.g., software programming errors or the incorrect manual operation of a device.

MITI An acronym for the Japanese Ministry of International Trade and Industry.

mixed radix A system in which more than one number base is used. For example, the bi-quinary system uses base 2 and base 5 glyphs to represent a value.

mixer (1) In radio systems, a nonlinear device in a heterodyne receiver which *modulates* the received signal with the output of a local oscillator to create an intermediate frequency.

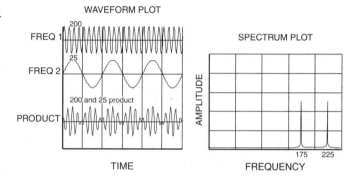

The *mixer* output contains signals equal in frequency to the sum of the frequencies of the input signals, to the difference between the frequencies of the input signals, and, if they are not removed by filtering, to the original input frequencies. See also *heterodyne.* (2) In audio systems, a device that *linearly adds* two or more input signals to create a single composite output signal. The output of the *mixer* contains only signals equal in frequency to the input frequencies.

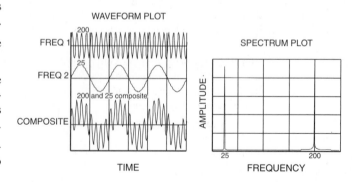

MJ An abbreviation of Modular Jack. A device used to connect voice devices to their respective network interface connector; for example, an RJ-11 is used to connect a telephone to the wall outlet. See also *RJ-nn.*

MLHG An abbreviation of MultiLine Hunt Group.

MLI An abbreviation of Multiple Link Interface.

MLID An abbreviation of Multiple Link Interface Driver.

MLT An abbreviation of Multiple Logical Terminals.

MLT-3 Code A data-encoding method utilizing three voltage levels. Transition from one level to an adjacent level represents a logic 1, while no transition is a logic 0. One advantage of the *MLT-3 code* is the absence of a dc bias (as is present in the normal NRZ digital code). However, it is not a self-clocking code.

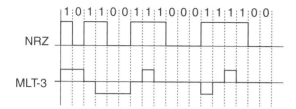

The code is ANSI approved for data transmission on an Fiber Distributed Data Interface (FDDI) network using both shielded and unshielded twisted pair cable. See also *waveform codes*.

MMDS An abbreviation of Multchannel, Multipoint Distribution Service. A TV distribution system in which a microwave signal is used to distribute cable TV channels to users with special receiving equipment. Current systems distribute 33 analog channels; however, new digital systems can distribute up to 200 channels. Also called *wireless cable*.

MMF An abbreviation of MultiMode Fiber.

MMI An abbreviation of Man–Machine Interface.

MMJ An abbreviation of Modified Modular Jack. A six-pin connector similar to the familiar RJ-11 telephone jack frequently used to connect a serial terminal line to the terminal device. The device is easily identified by the locking tab on the top-right rather than on the top-center as is the case with the RJ series connectors.

MMS An abbreviation of Manufacturing Message Service.

MMT An abbreviation of Multimedia Multiparty Teleconferencing.

MMU An abbreviation of Memory Management Unit.

MMX According to Intel, NOT an abbreviation of MultiMedia eXtension, but their brand name of a Pentium processor with 57 extra instructions for single instruction multiple destination (SIMD) operations on multimedia and communications data types.

MNOS An acronym for Metal-Nitride-Oxide Semiconductor.

mnemonic A phrase, or "ditty" that aids the human memory, e.g., the first letters in the saying, "Bashful boys respect our young girls but Violet gives willingly," have the same first letters as are in the colors of the Electronics Industry Association (EIA) color code, i.e., "black, brown, red, orange, yellow, green, blue, violet, gray, white."

MNP® Class n An abbreviation of Microcom Networking Protocol. *MNP* is a family of data link protocols that use error-detection algorithms to ensure data link integrity and may incorporate data compression to increase data throughput. The various *MNP* specifications are numbered *MNP Class 1* through *MNP Class 10*. A brief description of each *MNP* Class standard is listed in the following table.

MNP Class (n)	DESCRIPTION	bps	Protocol Efficiency (η)
1	Basic error detection & control, asynchronous, byte-oriented, half-duplex exchange. Also called *Block Mode*.	1690	70%
2	Basic error detection & control, asynchronous, byte-oriented, full-duplex exchange. Also called *Stream Mode*.	2000	84%
3	Basic error detection & control, synchronous, bit-oriented, full duplex exchange. DCE to DCE transfers are synchronous; however, DTE to DCE transfers are asynchronous.	2600	108%
4	MNP Class 3 plus *Adaptive Packet Assembly* and *Data Phase Optimization*. *Adaptive packet assembly* adjusts the size of packets transmitted based on the communication channel SNR; the more noise, the smaller the packet size (which reduces the penalty for retransmits); higher SNR have longer packets to reduce packet overhead. *Data Phase Optimization* removes repetitive control information from the data stream.	2900	120%
5	MNP Class 4 plus real-time *adaptive data compression* (up to 2:1). The amount of data compression depends on the type of data currently being transmitted and can vary from 1:1 to 2:1 with a typical compression ratio near 1.6:1.	4800	200%
6	MNP Class 5 plus *universal link negotiation, statistical duplexing,* and includes 9600 bps V.29 modulation. *Universal link negotiation* allows two dissimilar modems to find the highest common operating rate in the range of 300 to 9600 bps. Negotiation begins at 300 bps, then automatically increases the rate to the maximum rate both modems support. *Statistical duplexing* is a technique for simulating full-duplex service over half-duplex high-speed carriers. The modem monitors the data stream and assigns the faster forward channel to the modem with the most data to be sent.	9600 to 19200	
7	MNP Class 4 plus *enhanced data compression* (a modified Huffman encoding method). On the average compression ratios of 2.4:1 are achieved with 3:1 realizable.		300%
8	MNP Class 7 plus ITU-T V.29 "*Fast Train.*" This allows emulation of full-duplex operation with half-duplex systems.		
9	MNP Class 7 plus *enhanced universal link negotiation* and ITU-T V.32 modulation. The basic throughput of 9600 bps can be increased by as much as 300% due to the MNP Class 7 data encoding.	9600 to 28800	1200%
10	Also called *Adverse Channel Enhancement* (ACE). A protocol designed to maintain maximum data throughput in a channel that *changes during* the connection. After a connection is established at the best rate (at the time of negotiation), the error-correcting protocol will dynamically adapt data speed and packet size (8 to 256 bytes) to accommodate fluctuating channel conditions. This protocol is most useful on communications channels such as are encountered on cellular or "overseas" calls. The enhancements can be grouped into three categories; i.e., • Multiple aggressive attempts at link setup. • Adaptive pack size to accommodate changing levels of interference. • Renegotiation of transmission speed to accommodate changes in transmission line quality.		
10EC	A variant of Microcom's MNP-10 developed by Rockwell International Corp. to improve performance.		

mobile station A station in the mobile service intended to be used while in motion or during stops at unspecified points.

Mobile Telephone Switching Office (MTSO) In an automatic cellular mobile communications system, an *MTSO* is a central computer that monitors all cellular telephone transmissions and interfaces the radio portion of the system to the public switched telephone network (PSTN). It performs all signaling functions that are necessary to establish and maintain calls to and from mobile stations. If a connection is too noisy, the *MTSO* searches for a less noisy channel or cell site, and does a hand-off by transferring the connection to another channel in the next cell. Also called the *mobile services switching center (MSC)*. See also *hand-off*.

modal dispersion Regarded by some as an incorrect synonym for *multimode distortion*.

modal distortion A synonym for *multimode distortion*.

modal distribution In an optical waveguide operating at a given wavelength, the number of modes supported and their unit propagation time differences.

modal loss In an open waveguide, a loss of energy on the part of an electromagnetic wave due to obstacles outside the waveguide, abrupt changes in direction of the waveguide, or other anomalies that cause changes in the propagation mode of the wave in the waveguide.

modal noise Noise generated in an optical fiber system by the combination of mode-dependent optical losses and fluctuation in the distribution of optical energy among the guided modes or in the relative phases of the guided modes. Also called *speckle noise*.

mode **(1)** The operational state of a device or program. For example, a modem in the *command mode* will accept programming information from the DTE and will not transmit it to the remote modem, whereas in the communications mode the modem will transmit the information to the remote modem and will not process it as programming data. **(2)** In statistics, the value or values that occur most often in a given set of items. For example, if given a set of numbers {1, 2, 3, 4, 5, 6, 7, 8, 9, 9}, the *mode* is 9—that is, there are more 9s than any other number. For the set {1, 2, 2, 3, 5, 7, 9, 9} there are two *modes* 2 and 9; this set is said to be *bimodal*. **(3)** In transmission lines, waveguides, or cavities, one of the various field patterns or patterns of propagation dependent on physical geometric boundaries, frequency of operation, and the refractive index or medium dielectric constant.

Each mode is characterized by frequency, polarization, electric field strength, and magnetic field strength. **(4)** Any electromagnetic field distribution that satisfies Maxwell's equations and the applicable boundary conditions.

mode coupling In an electromagnetic waveguide, the exchange of power among modes.

In a multimode optical fiber, *mode coupling* reaches statistical equilibrium (the equilibrium mode distribution) after the equilibrium length has been traversed.

mode field diameter (MFD) An expression of distribution of the irradiance (optical power) across the endface of a single mode fiber.

For a gaussian power distribution in a single mode optical fiber, the mode field diameter is that at which the electric and magnetic field strengths are reduced to 1/e of their maximum values. (Because power is proportional to the square of the field strength, it is also the diameter at which power is reduced to $1/e^2$ of the maximum power.)

mode filter **(1)** In fiber optics, a device used to select, reject, or attenuate one or more transmission modes. **(2)** In waveguide components, a device designed to pass energy along a waveguide in one or more selected propagation modes and to reject or substantially attenuate energy in other modes.

mode partition noise In an optical communications link, phase jitter of the signal caused by the combined effects of mode hopping in the optical source and intramodal distortion in the fiber.

Mode hopping causes random wavelength changes that in turn affect the group velocity (propagation time). Over a long length of fiber, the cumulative effect is to create jitter called *mode partition noise*.

mode scrambler A device for inducing mode coupling in an optical fiber. Mode scramblers are used to provide a modal distribution that is independent of the optical source, for purposes of laboratory or field measurements or tests. Also called a *mode mixer*.

mode stripper See *cladding mode stripper*.

mode volume The number of bound modes at a particular wavelength that an optical fiber is capable of supporting.

The mode volume is approximately given by

$$M_S = \frac{V^2}{2} \text{ and}$$

$$M_P = \frac{V^2}{2} \cdot \frac{g}{g+2}$$

where:

M_S is the mode volume for a step index fiber,
M_P is the mode volume for a power-law profile fiber,
V is the normalized frequency, and
g is the profile parameter.

See also *effective mode volume*.

modem An acronym for <u>MO</u>dulator–<u>DEM</u>odulator.

(1) A type of DCE which, at the sending station, converts digital data from the DTE to an analog signal suitable for transmission over a communication channel and at the receiving station converts the analog signal back into digital data. Generally, the modem incorporates both the modulating and demodulating functions. The block diagram shows the major blocks of a modem. The computer interface (digital interface) is on the left, and the telephone interface (analog interface) is on the right. To communicate with each other, *modems* must use the same protocol suite. That is, the *modulation* method, the *data*

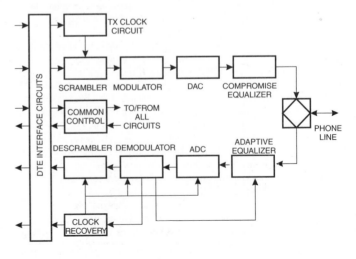

rate, and *data format* must all be the same. These protocols are generally defined by ITU-T (formerly the CCITT) "V.x standards." For example, V.32 bis provides speeds to 14 400 bps, and V.34 defines the protocol for connections to 33 600 bps. **(2)** In FDM carrier systems, a device that converts voice band to, and recovers it from, the first level of frequency translation.

Modem Approvals Group Established in January 1993 to raise awareness of the UK law against connecting unapproved devices to the PSTN, this group demands a level playing field in the modem market with better law enforcement or more open standards.

modem eliminator A device that can replace a modem, in some instances, when the distance to be covered is short. It takes the power it needs to operate from the transmission line. See also *null modem.*

modem patch A method of electrically interconnecting circuits by using back-to-back modems.

modem program See *communications program.*

modem server A networked computer with one or more modems available as resources for workstations to initiate outgoing calls.

moderated newsgroup A newsgroup in which all postings are screened by a human coordinator to:

- Ensure that the postings conform to the stated subject matter of the newsgroup.
- Reduce the high volume of postings in the newsgroup reports summaries.
- Eliminate e-mail terrorism (such as mail bombs, flame wars, trolling, etc.).

moderator A person who reads all submissions to a mailing list or newsgroup to make sure they are appropriate before posting them.

modified AMI code A T-carrier (or E-carrier) alternate mark inversion (AMI) line code in which bipolar violations may be deliberately inserted to maintain system clock synchronization.

The clock rate of a received carrier signal is extracted from its bipolar line code. Although this is acceptable for digitally encoded voice signals, it may fail when a true digital message is transmitted. This is because a true AMI line code may fail to have sufficient marks, i.e., "1s," to permit recovery of the incoming clock, and clock synchronization is lost. This happens when there are too many consecutive zeros in the user data being transported. To prevent this loss of clock synchronization when a long string of zeros is present in the user data, additional marks are inserted by the transmitting equipment so as to maintain synchronization at the receiver. The marks are inserted in such a way as to create a bipolar signal violation. The receiving terminal equipment recognizes the bipolar violations and removes them from the data stream, thus restoring the user data. The exact pattern of bipolar violations transmitted in any given case depends on the line rate and the polarity of the last valid mark in the user data prior to the unacceptably long string of zeros. The number of consecutive zeros that can be tolerated in user data depends on the level of the line code in the carrier hierarchy. See also *bipolar violation* and *BnZS.*

ZERO SUPPRESSION FOR SELECTED CARRIER SYSTEMS

	North American		ITU-T
T-1 (1.544Mb/s)	AMI	E1 (2.048Mb/s)	B8ZS
T-2 (6.312Mb/s)	B6ZS		
T-3 (44.736Mb/s)	B3ZS		

B6ZS means bipolar with six-zero substitution.

modified final judgment (MFJ) A legal document, approved by the United States District Court for the District of Columbia in 1984, which required the divestiture of the Bell Operating Companies from American Telephone and Telegraph Company (AT&T). It delineates the territories of the seven newly formed Regional Bell Operating Companies (RBOCs) and AT&T, and defines the business activities in which each is allowed to participate.

modified frequency modulation (MFM) A signal-encoding process in which both the amplitude and frequency of the write signal to a magnetic storage medium are varied. The number of bytes that can be stored per unit area is twice that of single-density modulation methods. The method of encoding makes the code *self-clocking.* See also *waveform codes.*

modified Huffman compression A one-dimensional data compression scheme with no transmission redundancy.

modified Read compression A two-dimensional data compression scheme used with facsimile equipment.

modular jack A device that conforms to the Code of Federal Regulations, Title 47, part 68, which defines the size and configuration of all units that are permitted for connection to the public exchange facilities. See also *RJ-nn.*

modulation *Modulation* is the process of varying some characteristic of one signal (the carrier) in accordance with another signal (the message signal). Some of the parameters of a carrier that may be modulated are:

- Amplitude (yielding AM or ASK).
- Frequency (yielding FM, FSK, MFSK, MSK, etc.).
- Phase (giving rise to PM, PSK, DPSK, QPSK, 4PSK, 8PSK, etc.).
- Both amplitude and phase may be changed independently, giving rise to QAM (Quadrature Amplitude Modulation).

AM (amplitude modulation), FM (frequency modulation), and PM (phase modulation) are the generic terms for the modulation method; that is, the information signal may be analog or digital in form. ASK (amplitude shift keying), FSK (frequency shift keying), and PSK (phase shift keying) apply only when the information signal is a digital signal. DPSK (differential phase shift key), 4PSK and 8PSK describe the number of phases possible in a PSK modulator. Three basic digitally modulated carrier waveforms are displayed in the diagram. The top trace is the information signal, and the second trace is

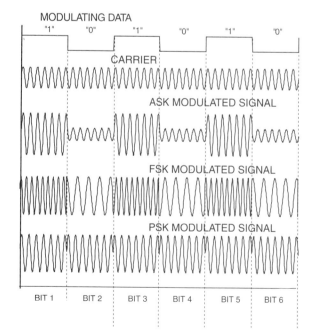

MODULATING DATA
"1" "0" "1" "0" "1" "0"
CARRIER
ASK MODULATED SIGNAL
FSK MODULATED SIGNAL
PSK MODULATED SIGNAL
BIT 1 | BIT 2 | BIT 3 | BIT 4 | BIT 5 | BIT 6

the carrier. The three remaining traces show what happens to a carrier when it is modulated with a digital data stream using ASK, FSK, and PSK modulation techniques. Several forms of pulse modulation are illustrated in the following figure. The top trace is the modulating wave having values of 5, 8, 10, 8, 5, 2, 0, 2, 5 at the nine indicated sampling times. The modulation methods shown are *pulse amplitude modulation (PAM), pulse width modulation (PWM)* sometimes called *pulse density modulation (PDM), pulse position modulation (PPM),* and three forms of *pulse code modulation (PCM)*. The three PCM waveforms are pulse count, direct code word, and difference code word (called delta modulation—DM). See also *amplitude modulation (AM), frequency modulation (FM), gaussian minimum shift key (GMSK), minimum shift key (MSK), phase modulation (PM), pulse amplitude modulation (PAM), pulse code modulation (PCM), pulse position modulation (PPM),* and *pulse width modulation (PWM)*.

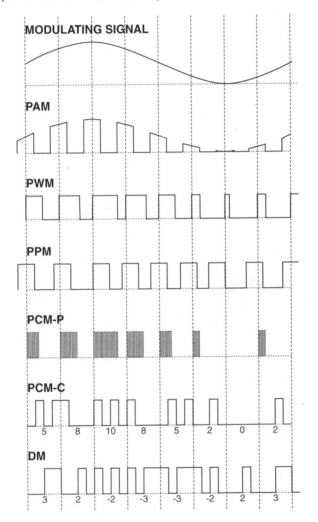

modulation index In frequency or phase modulation systems, the *modulation index* is the ratio of the carrier frequency deviation to the modulating frequency; that is,

$$h = \frac{\textit{carrier frequency deviation}}{\textit{modulation frequency}}$$

Numerically, it is equal to the phase deviation in radians. For example, assume the maximum deviation from the center frequency of a transmitter is 3000 Hz and the modulating frequency is 1000 Hz; then, the modulation index (h) is 3. For a modulating frequency of 6000 Hz, the *modulation index* is 0.5.

modulator A device that impresses a signal (generally an information bearing signal) on a carrier. See also *modulation*.

modulo A mathematical operation whose result is the remainder of a division by a specified base number. For example, 13 modulo 3 is 1—because 13 divided by 3 (the base number) is 4 with a remainder of 1. *Modulo* arithmetic is used in error-detecting and correcting schemes.

modulo N checking A form of error detection in which the check word appended to the transmitted message is the modulo N sum of the preceding data. At the receiver, the data are again used to calculate a modulo N check word. The newly calculated check word is compared to the received check word. If they are different, it is assumed that a transmission error has occurred (in either the data or the check word).

For example, given the transmit data stream 2,7,4,5,6,7,5,2,4,0,3,7 and modulo 8 error checking, the modulo 8 check word is 4; that is,

$$\frac{2+7+4+5+6+7+5+2+4+0+3+7}{8} \rightarrow Q = 6,$$
$$R = 4$$

If the receiver "sees" the data stream 2,7,4,5,6,7,5,2,4,1,3,7, the check word is then 5. When the received check word is compared to the receiver calculated check word, an error condition is discovered.

moiré pattern A virtual pattern that develops when two nearly repetitive patterns are superimposed. Named after French watered silk, which it resembles.

monitor (MON) (1) Software or hardware that observes, supervises, controls, and/or verifies the operation of a circuit, device, or system. **(2)** An interface to a computer that allows a human to observe real-time, but temporary, intermediate and final results of computer processes. The *monitor* may display text or graphical information.

Synonyms include *video display terminal (VDT), video display unit (VDU),* and *visual display unit*.

monitor jack A test point (connector) used to access communications circuits to observe signal conditions without interrupting the services.

monochromatic Electromagnetic radiation of a single frequency or wavelength. Although, in practice, radiation is never truly *monochromatic* (due to noise modulation, for example), it can have a very narrow bandwidth and for most purposes be considered *monochromatic*.

monochromator In optics, an instrument for isolating narrow portions of a spectrum.

monolithic driver A network device driver that acts as network adapter card driver and transport protocol driver combined.

monomode optical fiber A synonym for a single mode optical fiber.

monostrophic code A binary code in which only one bit changes between any two adjacent code positions. *Monostrophic codes* are also called *unit-distance* codes. An example of a *monostrophic code* is the *Gray* code. See also *polystrophic code*.

monotonic A variable, function, or set of values characterized by consecutive values exclusively increasing or decreasing. That is, if any point is greater in value than a previous point, then any point value must be greater than or equal to all previous point values.

Moore's Law An observation, made in 1964 by semiconductor engineer Gordon Moore (who, four years later, co-founded Intel), that the amount of information storable on a given amount of silicon has roughly doubled every year since the technology was invented in 1962. In the late 1970s, the rate slowed to doubling every 18 months. Unfortunately, the laws of physics guarantee this cannot continue indefinitely.

MOP An acronym for Maintenance Operation Protocol. A Digital Equipment Corporation (DEC) protocol used for remote communications between a host and servers.

more A command/program that allows a user to display only enough information to fill one screen at a time. If it were not for the *more* program, information would flow past the reader's eyes faster than it could be read. The program can be used by *piping* (|) the output of another command into the *more* command, for example,

ls | more

It can also be used by itself to list the contents of a text file, as in

more information.txt

To advance to the next page, the user taps the space bar <space>. To quit, the user types "q."

MORF An acronym for Male OR Female.

Morse code Named after its inventor Samuel F.B. Morse (1791–1872), the code is a method of representing the letters of the alphabet, punctuation, and certain phrases using a series of short and long pulses. The ratio of the transmitted pulses is generally about 1:3. The code when spoken uses *dit* for the short pulse and *dah* for the long pulse. When concatenated to form a symbol, the *dit* is shortened to *di*, e.g., the letter A is *didah*. The original code has been adapted and expanded to include more punctuation and phrases, as is shown in the International Morse Code.

International Morse Code

A	• —	N	— •	1	• — — — —	.	• — • — • —
B	— • • •	O	— — —	2	• • — — —	;	— • — • — •
C	— • — •	P	• — — •	3	• • • — —	,	— — • • — —
D	— • •	Q	— — • —	4	• • • • —	:	— — — • • •
E	•	R	• — •	5	• • • • •	?	• • — — • •
F	• • — •	S	• • •	6	— • • • •	'	• — — — — •
G	— — •	T	—	7	— — • • •	-	— • • • • —
H	• • • •	U	• • —	8	— — — • •	/	— • • — •
I	• •	V	• • • —	9	— — — — •	"	• — • • — •
J	• — — —	W	• — —	0	— — — — —	—	• • — — • •
K	— • —	X	— • • —	'	• — — — — •	=	— • • • —
L	• — • •	Y	— • — —	+	• — • — •	( )	— • — — • —
M	— —	Z	— — • •	X	• — • • —	SOS	• • • — — — • • •

Attention	— • — • —	Go Ahead	— • —
Break	— • • • — • —	Error	• • • • • • • •
Understand	• • • — •	Beginning of Msg	— • — • —
OK	• — •	End of Msg	• — • — •
Invitation to Tx	— • —	End of Work	• • • — • —
Wait	• — • • •	Closing station	— • — • • — •

A variation of the code called *cable Morse code,* used mainly in submarine cables, uses a three-element code to represent the dots dash signaling. Specifically, dots and dashes are represented by current pulses of equal magnitude and duration but of opposite polarity. The spaces between symbols are represented by the absence of current.

MOS An acronym for Metal Oxide Semiconductor.

MOSA An acronym for Metal Oxide Surge Arrestor. See *metal oxide varistor (MOV).*

Mosaic Internet, free software developed by the National Center for Supercomputing Applications (NCSA) at the University of Illinois, which allows the exploration and use of the World Wide Web (WWW). *Mosaic* is an easy-to-use graphical interface (GUI) that allows the utilization of most of the features on the WWW including *hypertext links.*

To use *Mosaic,* a direct or dial-up IP connection (such as SLIP or PPP) is required. It is highly recommended that a high-speed modem (minimum 14 400 bps), a fast computer, and plenty of RAM be used to obtain satisfactory performance.

MOSFET An acronym for Metal Oxide Semiconductor Field Effect Transistor. See *transistor.*

MOSPF An abbreviation of Multicast Open Shortest Path First.

most significant bit (MSB) In a sequence of bits (either alone or as a group of bytes), the high-order bit of the group (usually the leftmost bit NOT including the sign bit).

most significant character (MSC) In a sequence or string of characters, the high-order character of the group (usually the leftmost character).

most significant digit (MSD) In a sequence of one or more digits, the high-order digit of the group (usually the leftmost digit).

MOTD An abbreviation of Message Of The Day. A short informational message or quip displayed when logging on to many bulletin board systems.

Motif The name given to the Open Software Foundation's (OSF) Application Programming Interface toolkit "look and feel."

MOTAS An acronym for Member Of The Appropriate Sex. Jargon from the Internet.

motherboard The main interconnection means or circuit board in a device such as a computer. It has connectors (slots) into which secondary boards (*daughterboards* or *option boards*) are inserted to customize or complete the system.

In a PC, the *motherboard* generally contains vital system components such as the microprocessor and support chips. The daughterboards generally contain optional functions such as modems, network interface cards, or additional memory. Functions such as the video controller or disk controller may be on either the *motherboard* or daughterboard.

MOTIS (1) An acronym for Message Oriented Text Interchange System. (2) An acronym for Member Of The Inappropriate Sex. Jargon from the Internet.

MOTOS An acronym for Member Of The Opposite Sex. Jargon from the Internet.

MOTSS An abbreviation of Member Of The Same Sex. Jargon from the Internet. The gay-issues newsgroup on Usenet is called *soc.motss.*

MOU An abbreviation of Minute Of Usage. A usage measure employed in the telephone industry to calculate billing information in certain measured rate services.

MOV An acronym for Metal Oxide Varistor.

MPC (1) An abbreviation of Marker Pulse Conversion. (2) An abbreviation of Message Processing Center. (3) An abbreviation of Multimedia PC. Formed by computer industry leaders, the Multimedia PC (MPC) Marketing Council is a subsidiary of the Software Publishers Association. This council determines the standards for multimedia hardware platforms, primarily IBM PCs and compatibles. Two such standards are:

- *MPC Level 1 (MPC1)*—a specification that requires a computer with a minimum of 16 MHz 386 SX cpu, 2 MB RAM, 3.5" floppy, 30 Mbyte hard disk drive, single speed (150 kB/s) CD-ROM drive, 8-bit audio, VGA display Windows 3.1 (or Windows 3.0 with Multimedia Extensions), and an 8-note synthesizer with MIDI playback. Sample rates of 22.05 and 11.025 kHz must be supported by no more than 10% of CPU bandwidth, preferably 44.1 kHz at no more than 15% of CPU bandwidth.
- *MPC Level 2 (MPC2)*—a specification that requires a computer with a minimum of 25 MHz 486 SX cpu, 4 MB RAM, 3.5" floppy, 160 Mbyte hard disk drive, double speed (300 kB/s) CD-ROM drive (multisession and XA ready), 16-bit audio, 16-bit display (640 x 480), Windows 3.1, and an 8-note synthesizer, and MIDI playback. A sample rate of 44.1 kHz must be available on stereo channels at no more than 15% of CPU bandwidth.

MPE An abbreviation of Memory Parity Error.

MPEG An abbreviation of Motion Picture Experts Group, part of the International Standards Organization (ISO) that defined digital video format. A *lossy data compression* method used to reduce the size of picture files. Compression ratios of 40:1 to 200:1 are possible with this scheme. Because the compression method eliminates some information, it cannot be used on text, spreadsheet, database, or other data files.

Lossy compression of an image eliminates the "high-frequency" components that translate into edge sharpness and some subtle color differences. A reconstructed picture may therefore have a softer look (e.g., fuzzy edges). It can also introduce "blockiness," color bleed, and a shimmering effect in addition to the lack of detail. Standards include:

- *MPEG 1*—a form of video compression optimized for playback from CD-ROM and T-1 communications links at near VHS quality. This compression uses Huffman coding to remove spatial redundancy within a frame and block-based motion compensated prediction (MCP) to remove data that is temporally redundant between frames to allow for data rates to 1.5 Mbps (150 kBps—the same rate as a single-speed CD-ROM drive). Images can be 325 × 240 pixels.
- *MPEG 2*—an International Standard (IS-13818) that increases data rates to 3–10 Mbps, images to 720 × 486 pixels, and support of variable bit rates. This compression method offers broadcast quality video and support for high-definition television. A Video CD with *MPEG 2* video would contain over 2 hours of video.
- *MPEG 3*—a standard supplanted by MPEG 2, which proved more useful at high-definition TV.
- *MPEG 4*—a standard that provides compression of 176 × 144 pixel images and transmission rates from 2 to 64 kbps.

MPEG data files generally have the filename extension .MPG.

MPI An abbreviation of Multiple Protocol Interface. The top part of the link-support layer (LSL) in the generic Open Data-link Interface (ODI) for LAN drivers.

MPL An abbreviation of Maximum Packet Lifetime. The number of hops allowable before an internetwork packet is discarded.

MPM (1) An abbreviation of Magnetic Phase Modulator. (2) An abbreviation of Message Processing Module. (3) An abbreviation of MicroProgram Memory.

MPP (1) An abbreviation of Multiple Parallel Processing. (2) An abbreviation of Massively Parallel Processing. (3) An abbreviation of Message Processing Program.

MPPP An abbreviation of Multilink Point-to-Point Protocol. Sometimes written MPP.

MPR An abbreviation of MultiPort Repeater. A repeater in an Ethernet, usually a thin Ethernet, network.

MPR II A "green" standard published by the Swedish Board for Technical Accreditation (SWEDAC) that limits the maximum amount of electromagnetic radiation (ELF and VLF) a computer monitor may emit. Most personal computer monitors comply with this standard or the more stringent TCO (The Swedish Confederation of Professional Employees) requirement.

MPS (1) An abbreviation of Message Processing System. (2) An abbreviation of Microwave Phase Shifter. (3) An abbreviation of Modular Power Supply.

MPSK An abbreviation of M-ary Phase Shift Key or Multilevel Phase Shift Key. A phase shift modulation method in which a multitude of carrier phases are transmitted in response to the data stream. There is a one-to-one correspondence between a data symbol and a transmitted phase angle. See also *Modulation* and *phase shift key (PSK)*.

MPT An abbreviation of the Japanese Ministry of Posts and Telecommunications.

MPTN An abbreviation of Multi-Protocol Transport Networking.

MPU An abbreviation of MicroProcessor Unit.

MR (1) An abbreviation of Modem Ready. (2) An abbreviation of Message Register. (3) An abbreviation of Message Repeat. (4) An abbreviation of MagnetoResistive.

MROM An acronym for Masked Read Only Memory.

MRP (1) An abbreviation of Message Routing Protocol. (2) An abbreviation of Message Routing Process.

ms An abbreviation of millisecond (10^{-3} s).

MS-DOS An acronym for MicroSoft Disk Operating System.

MS-Net Microsoft's DOS-based networking system software product.

MSA An abbreviation of Metropolitan Service Area.

MSAU An abbreviation of MultiStation Access Unit.

MSB An abbreviation of Most Significant Bit.

MSC (1) An abbreviation of Mobile Switching Center. (2) An abbreviation of Message Switching Center. (3) An abbreviation of Message Switching Computer. (4) An abbreviation of Most Significant Character.

MSD An abbreviation of Most Significant Digit.

MSK An abbreviation of Minimum Shift Key.

MSL An abbreviation of Mean Sea Level.

MSN (1) An abbreviation of Manhattan Street Network. A mesh architecture utilizing wavelength division multiplexing. (2) An abbreviation of MicroSoft Network. Microsoft's answer to AOL, CompuServe, Prodigy, and the like.

MSRP An abbreviation of Manufacturer's Suggested Retail Price.

MSS (1) An abbreviation of Mass Storage System. (2) An abbreviation of TCP/IP Maximum Segment Size. See also *MTU*.

MST An abbreviation of Minimum Spanning Tree.

MTA (1) An abbreviation of Major Trading Area. (2) An abbreviation of Message Transfer Agent. (3) An abbreviation of Mail Transfer Agent.

MTBF An abbreviation of Mean Time Between Failure.

MTBO An abbreviation of Mean Time Between Outages.

MTSO An abbreviation of Mobile Telephone Switching Office.

MTSR An abbreviation of Mean Time to Service Restoral.

MTTR (1) An abbreviation of Mean Time To Repair. (2) An abbreviation of Mean Time To Recovery.

MTU An abbreviation of Maximum Transmission Unit.

Mu (1) The Greek letter (μ). (2) In the SI system, the symbol meaning 10^{-6}, sometimes indicated with the letter *u* (the closest ASCII lookalike).

mu-law Often written μ *law,* μ-*255 law,* or *u-law.* The North American telephony standard for digitizing voice in a nonlinear manner. The analog signal is sampled 8000 times per second and converted to an 8-bit nonlinear word that approximates a logarithmic curve.

The logarithmic curve is described by the equation:

$$D(a) = Sgn(a) \frac{\ln(1 + \mu|a|)}{\ln(1 + \mu)} \quad -1 \leq a \leq 1$$

where:

a is the analog signal,

$D(a)$ is the coded digital equivalent of the signal, and

μ   is the factor defining the amount of amplitude compression, 0 is no compression, and 255 is the value used in North America.

The approximation is called a piecewise linear approximation, that is, the logarithmic curve is divided into eight segments (or chords) between zero and full scale, and each of these segments is approximated by a straight line. Because the positive and negative "ones segments" are collinear (that is, in a line), the μ *law* is said to have 15 segments (not 16 as expected). Each higher segment requires twice as much analog signal change to affect a bit change in the digital word as the next lower segment. In the μ *law* diagram shown here, both the analog and digital signals are scaled to 100% of full scale (maximum value). The maximum digital word is +127 which corresponds to a +3.17 dBm0 1004 Hz sine wave. Details of the zero crossing region

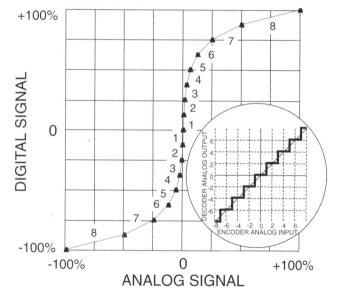

of chord or segment one are illustrated in the inset. Note that the digital value is zero for the analog range of −1 to +1. This is in contrast to A-law where there is no digital zero at all. Bit 1 is the sign bit which is transmitted first. Bits 2 through 4 are the segment bits, and bits 5 through 8 are the weighting bits. See also *A-law* and *compander.*

MUA An abbreviation of Mail User Agent.

MUD (1) An acronym for Multi-User Dungeons. A text-oriented fantasy role-playing adventure game that allows multiple users (perhaps thousands) to simultaneously explore and interact in a mazelike virtual world. Each user takes control of a specific virtual character which can solve puzzles and interact with other characters, probably also controlled by another online user.

MUD has its origins in a combat-oriented game written by Richard Bartle and Roy Trubishaw of Aberstywyth University around 1980. There are now many versions of the original *MUD* on the Internet, each with its own unique style. **(2)** An acronym for Multi-User Domain.

μ LAW ENCODER/DECODER DETAILS

Segment Number	Interval Count & Spacing	Encoder End Points	Binary Code 1234 5678	Decoder Value
8	16 Δ 256	8159	1000 0000	8031
		4319	1000 1111	4191
7	16 Δ 128	4063	1001 0000	3935
		2143	1001 1111	2079
6	16 Δ 64	2015	1010 0000	1951
		1055	1010 1111	1023
5	16 Δ 32	991	1011 0000	959
		511	1011 1111	495
4	16 Δ 16	479	1100 0000	463
		239	1100 1111	231
3	16 Δ 8	223	1101 0000	215
		108	1101 1111	99
2	16 Δ 4	95	1110 0000	91
		35	1110 1111	33
1	15 Δ 2	31	1111 0000	29
		1	1111 1110	2
	1 Δ 1	0	1111 1111	0
	1 Δ 1	0	0111 1111	0
1	15 Δ 2	−1	0111 1110	−2
		−31	0111 0000	−29
2	16 Δ 4	−35	0110 1111	−33
		−95	0110 0000	−91
3	16 Δ 8	−108	0101 1111	−99
		−223	0101 0000	−215
4	16 Δ 16	−239	0100 1111	−231
		−479	0100 0000	−463
5	16 Δ 32	−511	0011 1111	−495
		−991	0011 0000	−959
6	16 Δ 64	−1055	0010 1111	−1023
		−2015	0010 0000	−1951
7	16 Δ 128	−2143	0001 1111	−2079
		−4063	0001 0000	−3935
8	16 Δ 256	−4319	0000 1111	−4191
		−8159	0000 0000	−8031

MUF An abbreviation of Maximum Usable Frequency. The maximum frequency usable during sunspot activity.

muldem An acronym for multiplexer/demultiplexer.

multi-protocol transport networking (MPTN) An open architecture from IBM that enables the integration of a multivendor network environment. It allows communications using Advanced Program-to-Program Communications (APPC) over TCP/IP networks and TCP/IP over Systems Network Architecture (SNA). See also *AnyLan.*

Multi-Session CD-ROM A CD technology that allows additional data to be appended onto an ISO 9660 CD-ROM. Additions can be made to the disc as many times as desired—up to the full capacity of the disc or until the disc is fixated. A *multi-session CD-ROM drive* is required to read CDs that are not fixated. Any drive that is PhotoCD capable is also a multi-session drive. See also *single-session CD-ROM drive.*

Multi-Station Access Unit (MSAU) A wiring concentrator on a Token Ring network that allows devices, 8 to 12 Token Ring stations, to be connected to the ring. Relays in the *MSAU* ensure the integrity of the network when devices are attached or removed. A Managed Multi-Port/Multi-Station Access Unit has built-in network management support.

multiaccess A term that describes the ability of several users or processes to communicate with a computer at the same time.

multiaddress calling A service feature that permits a user to designate more than one addressee for the same data. Functionally, multiaddress calling may be performed sequentially or simultaneously. See also *broadcast.*

multibus Intel's central path (channel) for transmitting electrical signals and data; it was developed for use in 8- and 16-bit computer systems.

multibyte character In encoding, a character represented by two or more bytes. These characters arise in languages whose alphabet contains more than 256 characters, as is the case with Chinese and Japanese.

multicarrier modulation (MCM) A method of transmitting data by dividing the data into several bit streams, each of which is used to modulate a separate carrier. *MCM* is a form of frequency division multiplexing (FDM).

multicast A message sent from one source to a subset of all possible destination hosts that could receive the message. The subset is called a *multicast group,* and hosts that desire to participate must subscribe.

Multicasting is applicable to a number of applications, such as teleconferencing and the Internet Talk Radio. Protocols that support multicast include Apple Computer's Simple Multicast Routing Protocol (SMRP), and *IP multicast.* See also *IP multicast.*

multicast address A routing address that is used to address simultaneously all computers in a specified group and usually identifies a group of computers that share a common protocol, as opposed to a group of computers that share a common network. In the Internet protocol, it is called a *class d address.*

Multicast Backbone (MBONE) One implementation of *real-time* video and audio services for the Internet.

The basic premise of data delivery on the Internet is to make it error free, even at the expense of delay. Both audio and video, on the other hand, can tolerate errors but cannot tolerate delay. Therefore, new protocols had to be developed, the *multicast protocol* being one of them. To support the *multicast protocol* with its time-sensitive transmission requirements, a new backbone is required. This alternative backbone consists of computers equipped with special hardware interfaces, new software, and high-speed transmission facilities. Even with all of this equipment in place, the message throughput is not capable of full-motion multimedia transmission. Full motion requires approximately 30 picture frames per second, and the *multicast backbone* can support only about one-tenth of that. Even that data rate is useful for audiovisual teleconferencing and collaborative worksheets (a kind of electronic chalkboard in which all participants can view and change the contents).

multicast bit A bit in the Ethernet addressing structure used to indicate a broadcast message, i.e., a message to be sent to all stations.

multicast group The list of IP addresses to which multicast messages are sent. See also *IP multicast.*

multicast message A message that is intended for a set of stations on a network.

Multicast Open Shortest Path First (MOSPF) See *IP multicast.*

multidrop line A transmission configuration wherein a single transmission line is shared by several end stations.

For example, an Ethernet bus topology provides a *multidrop* connection, as does a telephone party line. Also called a *multi-drop line* or *multipoint line.*

multifiber cable A fiber optic cable having two or more separate fibers, each of which is capable of serving as an independent optical transmission channel.

multihomed host A host that is connected to more than one network.

Each network may operate with a different protocol. The host will have an IP address on each network, and that address may be different. The host may or may not allow transmissions to cross from one network onto another (as a bridge).

multilayer filter See *interference filter.*

multileaving The transmission of a variable number of data streams between user devices and a computer, usually via BiSync facilities and using BiSync protocols.

multilevel code A code where a single signal variable may take on more than two values. Examples include:

- *DPSK,* which has four possible phase angles, 0°, 90°, 180°, and 270°, per transmitted symbol.
- *AMI,* which has three possible voltage levels per transmitted symbol (+V, 0, and −V). Both +V and −V are usually encoded as logical one.
- *2B1Q,* which has four voltage levels, representing the combined value of two bits, in each symbol transmitted.

See also *waveform codes.*

multilevel phase shift key (MPSK) See *MPSK.*

multilink point-to-point protocol (MPPP) A standard communications protocol used to bond separate ISDN data-carrying B-channels together to transfer data effectively through a larger "pipe." At the basic rate, it allows both B-channels to be used for either voice or data transmissions and supports dynamic bandwidth allocation. That is, one of the two channels can be reallocated automatically for an incoming phone call. Upon call completion, the channel can be reconnected to continue data transfer over *MPPP.*

multimedia **(1)** A generic description of the generation, presentation, or simultaneous transfer of information in more than one way. Media types include text, graphic (drawings), still images (photographs), motion video, and sound. Multimedia therefore involves two or more simultaneous media types to communicate information. Note that multimedia presentations tend to consume huge amounts of resources, computer processing capability, disk memory, and transmission bandwidth. See also *MPC.* **(2)** In local area networks (LAN) applications, the use of mixed types of transmission media such as coax, UTP, and fiber optics.

Multimedia and Hypermedia information coding Expert Group (MHEG) An ISO standard encoding for multimedia and hypermedia information. It is designed to simplify the use and interchange of information such as with games, electronic publishing, and medical applications.

multimedia mail An e-mail system in which users are allowed to include graphics, sounds, and video in addition to text. Also called *MIME* (Multipurpose Internet Mail Extension).

multimode distortion In multimode optical fibers, the gradual spreading of an optical pulse with increasing distance, i.e., the rounding of the digital light pulse as it traverses the fiber. Two methods can be used to explain the effect:

- In wave optics, the signal is spread in time because the propagation velocity of the optical signal is not the same for all modes.

- In ray optics, each ray takes a separate path through the fiber core. The direct path is the shortest and the quickest path. All other paths are represented by reflections at the core-cladding boundary of a step index fiber and must travel a longer distance and hence must take longer to travel the length of the fiber. The direct signal is therefore distorted by these late-arriving reflected signals.

Multimode distortion limits the bandwidth of multimode fibers. For example, a typical step index fiber with a 50-μm core is limited to approximately 20 MHz for a 1 kilometer length (a bandwidth of 20 MHz·km). *Multimode distortion* may be significantly reduced, but not completely eliminated, by using graded index core optical fibers. The bandwidth of a typical graded index multimode fiber, with a 50-μm core, may be 1 GHz·km or more. (Bandwidths near 3 GHz·km have been produced.) Because of its similarity to dispersion in its ef-

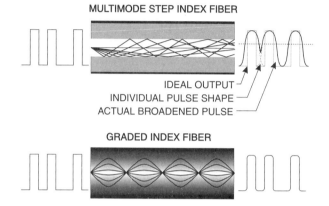

MULTIMODE STEP INDEX FIBER

IDEAL OUTPUT
INDIVIDUAL PULSE SHAPE
ACTUAL BROADENED PULSE

GRADED INDEX FIBER

fect on the optical signal, *multimode distortion* is frequently, and **incorrectly,** referred to as *intermodal dispersion, modal dispersion,* or *multimode dispersion.* (Such usage is incorrect because *multimode distortion* is not a true dispersive effect. Dispersion is a wavelength-dependent phenomenon, whereas multimode distortion may occur at a single wavelength.) True synonyms are *intermodal delay distortion, intermodal distortion,* and *modal distortion.* See also *intersymbol interference* and *optical fiber.*

multimode fiber (MMF) In fiber optics, an optical fiber that supports propagation of multiple transmission modes (paths) simultaneously. A *multimode fiber* may be either a *graded index (GI)* fiber or a *step index (SI)* fiber. *Multimode fibers* are less expensive to make than single mode fibers, but they can be noisier and can introduce errors due to multimode distortion. See also *optical fiber.*

multimode waveguide A waveguide (either optical or radio wave) that allows energy in more than one mode at a frequency of interest to propagate.

In fiber optics, a fiber with a core thick enough for light to take several paths (known as modes) through the core. This is in contrast to a single mode fiber, whose core is thin enough so that light can take only a single path through the core. See also *optical fiber.*

multioffice exchange A telecommunications exchange served by more than one central office.

multiparty line A synonym for *party line.*

multiparty ringing In telephony, an arrangement that allows a central office to selectively ring a selected main station on a multiparty line. Several techniques are available to accomplish this goal. For example:

- *Frequency selective ringers* have been used. The central office transmits a ringing signal at the frequency of the ringer at the called party premises. See also *ring (3).*

- *Bridged and divided ringing* can be used directly to provide three distinct ring parties; that is, a station's ringer may be placed between the ring lead and ground, between the tip and ground, or between the tip and ring. See also *bridged* and *divided ringing.*
- *Coded ringing* is a system in which the central office generates a distinctive ring pattern that a subscriber must recognize as his own (all telephones on the loop ring). See *coded ringing.*
- Diode *isolated ringing* utilizes the switching characteristic of diodes when subjected to a polarity reversal. In this method of multiparty ringing, the central office either applies the normal battery voltage to the loop (positive at ground and negative to the conductor) or the reverse polarity. The ringer at the subscriber premises is placed in series with a diode. This diode will allow the ring signal to pass only when the bias voltage is of the correct polarity. See also *isolated ringing.*
- *Active signaling* allows not only selective multiparty ringing, but affords calling privacy at any active multiparty station. One method involves the use of DTMF or MF signals generated at the central office and transmitted to the subscriber premises. At the subscriber premises an active electronic device recognizes the signaling tones and either connects the selected subscriber to the loop (allowing ringing signals to pass) or blocks all nonspecified users from gaining access to the loop, thus preventing ringing.

multipath The propagation of a wave (acoustic or electromagnetic) from one point to another by more than one course. *Multipaths* usually consist of the direct path and one or more indirect paths created from reflections off objects such as the Earth and its oceans, large man-made structures, and the ionosphere. The longer reflected paths cause interference at the receiver's antenna because of the phase difference between the direct path and reflected paths.

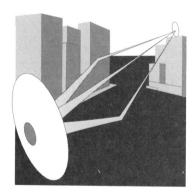

multipath error Receive errors (or noise) caused by the interference of a desired signal by later arriving reflected signals.

multiple In telephony, a term meaning to connect in parallel, or to make a circuit, line, or group of lines accessible at a number of points. Also called *multipoint.*

multiple access (1) The ability of a number of stations to have simultaneous access to one communications transponder. Frequently applied to orbital satellite systems. Three types of multiple access are code division (CDMA), frequency division (FDMA), and time division multiple access (TDMA). **(2)** In computer networking, a scheme that allows temporary access to the network by individual users, on a demand basis, for the purpose of transmitting information. Examples of network *multiple access* are carrier sense multiple access with collision avoidance (CSMA/CA) and carrier sense multiple access with collision detection (CSMA/CD). **(3)** Simultaneous access to the same file for multiple users. *Multiple access* is generally allowed only for reading files. If users are allowed to make changes to a file, some sort of locking mechanism is required to prevent users from interfering with each other's work. **(4)** The connection of a user to two or more switching centers by separate access lines using a single message routing indicator or telephone number.

multiple call A synonym for a *conference call.*

multiple circuit Two or more circuits connected in parallel.

multiple frame transmission The transmission of more than one frame when the token is captured by a node in a Token Ring access system.

multiple frequency shift keying (MFSK) Frequency shift keying (FSK) in which multiple frequencies are used in the transmission of digital signals. Several schemes are used to establish these multiple code levels. For example,

- The sequential transmission of any one of a family of frequencies, each of which represents a particular code state. Two frequencies allow the transmission of two states (normal binary FSK), four frequencies allow the transmission of 2 bits, eight frequencies enable 3 bits, and so on.
- The concurrent transmission of a number of frequencies within a specified family. For example, a family of four discrete frequencies taken two at a time allows six discrete transmission states, five frequencies allows 10 states, six frequencies allows 15 states, and so on.

  DTMF is an example of this type of signaling. It uses eight frequencies arranged in two groups. Two frequencies (one from each group) are transmitted for each of the 16 possible symbols.

multiple link interface (MLI) Part of the *Open Data-link Interface (ODI, ODLI)* generic network driver interface. It sits under the *link-support layer (LSL)* which deals with the protocol stacks, while the *MLI driver (MLID)* deals with the various network interface cards, or adapters, that support ODI.

multiple logical terminals (MLT) A feature of an IBM 3174 establishment controller, in a System Network Architecture (SNA) environment that allows control user terminal (CUT) components to support multiple sessions simultaneously.

multiplex (MUX) To interleave or simultaneously transmit two or more messages on a single communications channel.

multiplex aggregate bit rate In a time division multiplexer, the bit rate that is equal to the sum of all selected input channel data signaling rates available to the user plus the rate of all overhead bits required.

multiplex baseband **(1)** In frequency division multiplexing, the frequency band occupied by the aggregate of the signals in the line interconnecting the multiplexer and the next device of the system. **(2)** In frequency division multiplexed carrier systems, the frequency band presented to the input to any stage of frequency translation.

For example, in telephony's analog multiplex hierarchy each of the 12 4-kHz voice frequency input channels is the baseband to the group multiplexer. The aggregate output consists of a band of frequencies from 60 to 108 kHz. This is the group level baseband that results from combining the 12 voice frequency input channels. Five group-level baseband signals are multiplexed into a supergroup having a baseband of 312 to 552 kHz. Ten supergroups are in turn multiplexed into one master group, the output of which is a baseband that may be used to modulate a microwave frequency carrier or may be further multiplexed. See also *multiplex hierarchy.*

multiplex hierarchy The structure of the frequency division multiplex method used in analog carrier systems. That is,

MULTIPLEX HIERARCHY

	·U. S.	ITU
1 Jumbo group	6 Mastergroups (3600 channels)	6 Mastergroups (1800 channels)
1 Mastergroup	10 Supergroups (600 channels)	5 Supergroups (300 channels)
1 Supergroup	5 groups (60 channels)	5 groups (60 channels)
1 group	12 channels	12 channels
1 channel	300-3400 Hz	300-3400 Hz

See also *group.*

multiplexer (MUX) A device that allows several users to share a single resource (such as a communications circuit). It combines (or funnels) multiple input data streams into an aggregate stream for transport over a single communication channel. At the receiving end, another *multiplexer* separates the single data stream into its constituent parts and directs each on to its destination.

- A device that interleaves two or more signals on a single path.
- A device for selecting one of a number of signal sources and switching it to an output.

multiplexing The process of combining several signals from separate sources into a single signal suitable for delivery on a transmission system and subsequent recovery of the original signals at their respective destinations. The transmission facility is divided into two or more channels such that several independent signals may be transported essentially simultaneously. There are several methods of accomplishing the *multiplexing* task. For example,

- *Code division multiplexing* is accomplished by using two stages of modulation. That is, the data of each channel are first impressed onto a unique *pseudorandom sequence.* Then the carrier is modulated by this signal. A receiver can recover the original data stream only if it uses the same pseudorandom code in the detection process. A different code will cause the signal to look like noise. See also *spread spectrum.*
- *Frequency division multiplexing* (*FDM*) is accomplished by splitting the channel bandwidth into some number of smaller channel bandwidths.
- *Wavelength division multiplexing* (*WDM*) is essentially the same as frequency diffusion multiplexing except light of different wavelengths is specified rather than different frequencies.
- *Statistical multiplexing* is accomplished by providing bandwidth only to those data lines having *activity* or information ready to transmit. No bandwidth is wasted on channels that are idle. A form of *time division multiplexing.*
- *Time division multiplexing* (*TDM*) is accomplished by allocating each subchannel its own time slot where transmission occurs. There are several variants of TDM. That is,
  - *Asynchronous time division multiplexing* (*ATDM*)—multiplexing in which the data are transmitted asynchronously.
  - *Statistical time division multiplexing* (*STDM*)—a multiplexing method that polls nodes and immediately skips any nodes that have nothing to send.
  - *Synchronous transfer mode* (*STM*)—designed for use in broadband ISDN (BISDN) and also supported in the Synchronous Optical Network (SONET) architecture.

multipoint access Access in which more than one terminal is supported by a single network termination.

multipoint circuit A circuit that interconnects three or more separate points.

multipoint connection A condition in which a fraction T-carrier (FT-1) or E-carrier (FE-1) network routes communications channels to two or more locations through Digital Cross-connect Switch (DCS) nodes. (Framing, CRC, and data link channels are not propagated through the DCS nodes.)

multipoint link A data communications link that interconnects three or more terminals.

multipoint network A term that describes a network configuration in which several transmission facilities connect several end stations to a master station. This usually means using either polling techniques with each terminal or terminal address or using a collision detection scheme. See also *multiple access* (2) and *multidrop*.

multipoint service In telecommunications, the distribution of telecommunications services to two or more destinations.

multiport repeater (1) In networking, an active device, having multiple input/output (I/O) ports in which a signal presented to the input of any port generally appears at the output of every port.

Multiport repeaters usually perform signal regenerative functions, that is, they amplify and reshape the digital signals. Depending on the application, a *multiport repeater* may be designed to not repeat a signal back to the port from which it originated. See also *multistation access unit (MAU)*. (2) In Ethernet, a device for interconnecting a number of 10Base2 (thinwire) Ethernet segments. Typically, up to eight segments can be interconnected.

multiprocessing Simultaneous processing by two or more processors acting in concert.

A computing strategy in which multiple processors work on the same task or multiple parts of the same task simultaneously. Sometimes the term is incorrectly used to indicate the concurrent execution of two or more computer programs or sequences of instructions (tasks) by a single processor, i.e., *multitasking*. (In multiprocessing one task is operated on by multiple processors, while in multitasking one processor operates on multiple tasks.) Contrast with *multitask*.

multiprocessor A computer that has two or more processors that have common access to a main storage and may or may not be directed by a common control. See also *MIMD* and *SIMD*.

multiprogramming Pertaining to the concurrent execution of multiple programs or tasks by a computer. The mode of computer operation that provides for the interleaved execution of two or more programs by a single processor. See also *multitasking*.

MultiRing See *distinctive ringing*.

multiserver network A single network that has two or more servers operating. Workstations can access files on any of the servers to which they have been granted access rights.

A *multiserver network* is not the same as interconnected networks, i.e., where two or more networks are connected by a router.

multistation access unit (MAU or MSAU) IBM's term for a wiring hub in its Token Ring architecture. Basic hub functions include:

- A termination point for multiple nodes. Each *MAU* can have up to eight nodes (or lobes as IBM calls them) connected.

  Lobes are connected to *MAUs* using IBM Type 1, 2, or 3 cable. Because Type 3 cable is unshielded, a media filter is needed between the cable and the *MAU* to clean up noise from the signals before they reach the *MAU*.

- A connection to other *MAUs* via two connectors, called ring in (RI) and ring out (RO). When *MAUs* are connected, it is possible to create a main and a secondary redundant ring path.

  Use of IBM Type 1 or 2 cable allows more than 30 *MAUs* to be interconnected, supporting a total of 260 nodes. When using IBM Type 3 cable, a maximum of 9 *MAUs* can be connected, supporting up to 72 nodes. Type 6 cable is sometimes used to interconnect *MAUs*, when the distance between them is just a few meters.

- A repeater, which cleans and boosts (regenerates) a signal as it passes packets around the ring.

An *MAU* organizes the nodes connected to it into an internal ring, and uses the Ring IN (RI) and ring out (RO) connectors to extend the ring across *MAUs*. The most widely supported standard is that of IBM's model 8228 *MAU;* hence, one often see references to "8228-compliant MAUs." This standard is the minimal set of capabilities that most *MAUs* support. In fact, most *MAUs* have capabilities and features beyond those of the 8228. These additional capabilities can make a network more efficient but can also increase the likelihood of compatibility problems. Some additional features include:

- LEDs (light emitting diodes) that indicate the status of each *MAU* port (lobe).
- Automatic disconnect, which disconnects faulty lobes, without affecting the other lobes or disrupting the network.

Also called a *wiring center* in IEEE 802.5.

multisystem network A network in which two or more host computers can be accessed by the network users.

multitasking The ability of a computer operating system (or program) to concurrently operate on or process more than one task. Actually, only one of these tasks is being processed at any given instant. However, each task gets only a few milliseconds of computer time before another task is allowed to operate. Hence, the concurrency is only apparent.

There are two methods of transferring control among various tasks in a multitask system, *preemptive* and *nonpreemptive*.

- In *preemptive multitasking,* the multitasking control system manages the switching between tasks and gives every task its turn in a predictable fashion. OS/2, Windows NT, Windows95, and UNIX support preemptive multitasking.
- In *nonpreemptive multitasking,* an application or process is allowed to execute until it stops itself. The application cannot be interrupted and must be trusted to give up control. System 7, Windows 3.x, and Novell's NetWare use nonpreemptive multitasking.

Additional protocols modify the way basic preemptive/nonpreemptive task switching is accomplished. For example,

- *Context switching:* The simplest form of multitasking loads two or more processes (or tasks), each with its own data and execution environment (context). The operating system switches between tasks, usually when it wants to run another program.

  Task managers, which can be part of an operating system or of a shell, provide context-switching capabilities. In addition to task switching, they must provide and manage storage for each of the loaded tasks.

- *Cooperative switching:* Background processes are allowed to vie for processor time during periods when the foreground process is idle.

  For example, a lengthy data sort program may be running in the background while the user is doing text editing. When the user is not actively typing, the operating system will let the sort program execute.

The Macintosh System 7 operating system, for example, uses this strategy.

- *Time slice switching:* Each process is given a segment of the available processor time. All tasks may receive equal time slices, or each will get a time slice whose length is proportional to the task's priority. The operating system runs each of the tasks in succession, for the duration of the task's time slice.

OS/2 and many mainframe operating systems support preemptive time slice multitasking.

Multitasking differs from *multiprocessing* in that multiprocessing uses multiple processors to work on the same task. Also called *concurrent operation.*

multithreading A special form of multitasking in which all the tasks are part of the same program. Hence, multiple processes from a single program execute, concurrently. This concurrency is only apparent, however, because, as with multitasking in general, the processor is actually switching its resources very rapidly among all of the threads.

A thread is an executable object (with its own stacks, registers, and instruction counter), that belongs to a single process or program. See also *multitasking.*

multiuser system A computer system or network that can be used by more than one user *simultaneously.* UNIX is an example of a multiuser operating system, while DOS and OS/2 are single-user systems.

Murray code The International Telegraph Code No. 2. A five-level code invented circa 1900 replaced the Baudot code. Also called the Murray/Baudot code and sometimes improperly referenced as the Baudot code. See also *Baudot code* and *teleprinter codes.*

Musical Instrument Digital Interface (MIDI) A method of representing sounds digitally for storage and use on a PC. *MIDI* data files can be synthesized, edited, or played (through an appropriate interface card and speakers) and with software running on the PC.

mutual synchronization Synchronization in which the frequency of the clock at a particular node is controlled by a weighted average of the timing on all signals received from neighboring nodes.

mutually synchronized network A network that has a synchronizing arrangement in which each clock in the network exerts a degree of control on all others.

Also called a *democratically synchronized network.*

mux A contraction of *multiplex* or *multiplexer.*

mv A UNIX command to rename a file. For example, to rename the file foo.txt to bar.doc, one enters:

$$mv \quad foo.txt \quad bar.doc$$

mV An abbreviation of millivolt.

MVIP An abbreviation of Multi-Vendor Integration Protocol.

mW An abbreviation of millwatt.

MWI An abbreviation of Message Waiting Indicator.

MWV An abbreviation of Maximum Working Voltage.

MX record An abbreviation of Mail eXchange record.

MYK-78 See *Clipper chip.*

MYOB An abbreviation of Mind Your Own Business.

N

n **(1)** The *SI* prefix for *nano* or 10^{-9} or one-billionth (in the U.S. counting system).

In the binary system, *n* is generally 2^{-30}. **(2)** A commonly used symbol to indicate the index of refraction. Frequently, a subscript is added to indicate the wavelength of light for which the index is measured. The subscript may also be used to identify which of several materials is being referenced. **(3)** A symbol for *refractive index*.

N The SI symbol for the unit of *force* measured in *newtons*.

N-ary A code whose alphabet consists of *N* symbols, significant conditions, or quantizing states, where *N* is a positive integer greater than 1. *N* can also be a prefix that represents an integer, e.g., "bi" (as in binary or 2-ary). Examples include:

- A 2-ary code which is a *binary* code and has an alphabet of two symbols (0 and 1).
- A 3-ary code which is a *ternary* code and has an alphabet of three symbols.
- A 4-ary code which is a *quaternary* code with four symbols in the alphabet and can convey 2 bits per code symbol.
- An 8-ary code which has eight significant conditions and can convey 3 bits per code symbol.
- A 32-ary code which is specified in the ITU-T V.32 modem specification and has a 32-symbol alphabet and can encode 4 data bits and 1 redundancy bit per code symbol.

n-entity An active element in the *n*th layer of a layered open system, such as the seven-layer International Standards Organization (ISO) Open Systems Interconnection Reference Model (ISO/OSI-RM). It is defined by a unique set of rules (both syntax and format), performs a defined set of functions, and interacts with entities in layers immediately above (the *N* + 1 layer) and below (the *N* − 1 layer).

n-function A defined action performed by an *n-entity*. It may be either a single action (a primitive function) or a set of actions.

N-ISDN An abbreviation of <u>N</u>arrow band <u>ISDN</u>. See also *ISDN*.

n-sequence A pseudorandom binary sequence of *n* bits which:

- Is the output of a linear shift register in which the output is logically combined with at least one intermediate output on the shift register and then presented to the shift register input.
- Has the property that, if the shift register (with *m* stages) is set to any nonzero value and then cycled, a pseudorandom binary sequence of maximum length $n = 2^m - 1$ bit will be generated repetitively.

The sequence is a fundamental building block of both transmission test equipment (e.g., bit error ratio test equipment) and spread spectrum transmission systems.

N-series connector A large-diameter threaded coaxial connector; used, for example, with Thick Ethernet coaxial cable. The figure illustrates a male cable end and a female chassis mounted *n*-series connector pair. Standard 50 Ω connectors do not mate with standard 75 Ω connectors. Also called *type N connectors*.

n-stage switching A switching system composed of a number of interconnected subswitching matrices.

In the diagram, a three-stage switching system is depicted. "m" inputs are available at the input to the overall switching system, "n" ports interconnect stages 2 and 3, "o" ports are available to connect the final stage to stage 2, and finally "p" output ports are available. There is no requirement that the number of ports n, o, or p be smaller, larger, or the same as each other or as the number of input ports. Usually, however, m > n > o > p results in a *blocking network*. If the p ports are connected back on themselves, the switching system becomes a *folded network*. See also *switching stage*.

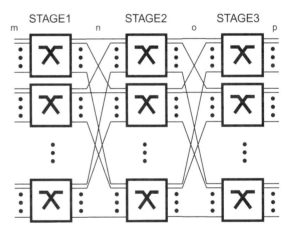

n-user In the International Standards Organization (ISO) Open Systems Interconnection Reference Model (ISO/OSI-RM), an *n* + 1 entity that uses the services of the *n*-layer, and below, to communicate with another *n* + 1 entity. See also *n-entity*.

N.n recommendations An ITU-T series of specifications addressing maintenance. See Appendix E for a complete list.

NA **(1)** An abbreviation of <u>N</u>umerical <u>A</u>perture. **(2)** An abbreviation of <u>N</u>ight <u>A</u>nswer.

NAC An abbreviation of <u>N</u>etwork <u>A</u>ccess <u>C</u>ontroller.

nagware The variety of shareware that displays a large, annoying opening or closing screen reminding the user to register. Generally, it requires some sort of keystroke to continue operation so that one can't use the software in batch mode or pipe data to it from other programs. See also *crippleware*.

nailed connection An obsolete term indicating a permanent (or semipermanent) circuit established through a circuit switching system for point-to-point connectivity. Also obsolete is *nailed-up connection*. See *dedicated circuit* and *permanent virtual circuit* (*PVC*).

341

NAK **(1)** An acronym for <u>N</u>egative Ac<u>K</u>nowledge. A message sent by the remote DCE to the local DCE to indicate that a data block just received was received incorrectly (NOT OK). In asynchronous systems, the ASCII character 21 (15h) is the <NAK> character.

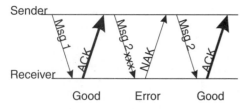

The signal flow diagram shows two message blocks being sent from a sender to a receiver. The first attempted transmission of the second block is received in error. The receiver sends a *NAK* to the sender, which then re-sends block 2. ACKs indicate the message block was received correctly. **(2)** In multipoint systems, the not ready reply to a poll. **(3)** An online answer to a request for chat that roughly translates, "I'm not available."

NAK attack In communications security systems, an infiltration technique that makes use of the negative acknowledge transmission control character (the NAK) and capitalizes on a potential weakness in a system that handles asynchronous transmission interruption poorly.

NAM **(1)** An acronym for <u>Na</u>me and <u>A</u>ddress <u>M</u>odule. A cellular telephone changeable ROM. **(2)** An acronym for <u>N</u>umber <u>A</u>ssignment <u>M</u>odule. A cellular telephone changeable ROM. **(3)** An acronym for <u>N</u>ational <u>A</u>ccount <u>M</u>anager.

name resolution In a network or internetwork, the process of mapping the name of a device or node to an address.

named pipe A connection used to exchange data between two or more separate processes on the same or different computers. The pipe can be referred to by name, and the storage allocated for the pipe can be accessed and used for reading and writing, much like a file, except that the storage and the pipe disappear when the programs involved finish executing.

naming protocol A protocol used by *AppleTalk* to associate a name with the physical address of a network service.

naming service A mechanism that makes it possible to name resources on the network and to access them through the use of those names. That is, the service associates a mnemonic with a network entity, which can then be used instead of the resource's network address. *Naming services* are available in most network operating systems. Two types of naming services are:

- A *local naming service,* which is associated with a single server.
- A *global naming service,* which is associated with a network or an internetwork. With a global naming service, each object on an internetwork has a unique name. Therefore, the server name need not be known to find an object associated with that server.

Examples include:

- Novell's NetWare versions prior to 4.0 use a local naming service; information about the resources associated with a server is stored in a resource database known as the bindery.
- The NetWare Directory Services (NDS) used in NetWare 4.x uses a global naming service.
- Banyan's VINES StreetTalk also is a global naming service.

NAMPS An acronym for <u>N</u>arrowband <u>A</u>dvanced <u>M</u>obile <u>P</u>hone Service. See also *IS-n.*

NAND A contraction of <u>N</u>ot <u>AND</u>. A Boolean operator composed of the logical function AND and the negation (NOT).

The *truth table* shows all possible combinations of input and output for a two-input *NAND* operation. A false state is shown as an "0," while a true condition is shown as a "1." See also *AND, AND gate, exclusive OR, exclusive OR gate, NOT, OR,* and *OR gate.*

NAND TRUTH TABLE

A	B	$\overline{A \cdot B}$
0	0	1
0	1	1
1	0	1
1	1	0

NAND gate A type of digital circuit that performs a *NAND* operation on two or more digital signals. The symbol for a two-input *NAND gate* is:

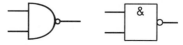

See also *AND, AND gate, exclusive OR, logic symbol, NAND, NOR, NOR gate, NOT,* and *OR.*

nano (n) The *SI* prefix representing the value 10^{-9}. One-billionth in the American numbering system and one thousand millionth in the British system.

nanoacre A unit (about 2 mm square) of real estate on a VLSI (very large scale integration) chip. The term has significance from the fact that VLSI nanoacres have costs that are about the same as real acres if figures include design and fabrication-setup costs.

nanosecond (ns) 10^{-9} second. See *SI* units.

NANP An abbreviation of <u>N</u>orth <u>A</u>merican <u>N</u>umbering <u>P</u>lan.

NAP An acronym for <u>N</u>etwork <u>A</u>ccess <u>P</u>oint.

NAPLPS An abbreviation of <u>N</u>orth <u>A</u>merican <u>P</u>resentation <u>L</u>evel <u>P</u>rotocol <u>S</u>tandard. The ANSI videotext presentation layer protocol for encoding graphic information in a standard and compact manner.

narrow correlator A correlator in a code tracking loop in which the spacing between the early and late versions of the receiver-generated pseudorandom reference code is less than one chip. The use of a *narrow correlator* can result in a lower signal-to-noise ratio (SNR).

narrowband **(1)** A term used to indicate low relative bandwidth. That is, the bandwidth of the channel is only slightly larger than the bandwidth of the signal being transported. Also spelled *narrow band.* **(2)** A transmission technique that does not employ multiplexing; that is, the transmission medium carries only one signal at a time. Compare with *broadband.*

narrowband ISDN (N-ISDN) The original ITU-T ISDN standard based on twisted pair transmission. *N-ISDN* supports two 64-kbps bearer channels (B-channels) and one 16-kbps data or control channel (D-channel). Each 64-kbps channel may be used for either voice or digital data transmission. A key feature of *N-ISDN* is its ability to use existing telephone twisted pair transmission lines. See also *broadband ISDN (B-ISDN)* and *ISDN.*

narrowband modem A modem whose modulated output signal has an essential frequency spectrum (bandwidth) that is limited to that which can be wholly contained within, and faithfully transmitted through, a communications voice channel.

narrowband network A network that does not use multiplexing to combine multiple signals for simultaneous transmission. Generally, these networks have data rates of less than about 20 Mbps. Compare with *broadband network*.

narrowband noise Noise having one or more distinct spectral peaks with a bandwidth less than that of the communications channel in which it is being measured.

narrowband signal **(1)** Any analog signal (or analog representation of a digital signal) whose essential spectral content (bandwidth) is limited to that which can be contained within a systems voice channel bandwidth (typically, a 3-kHz bandwidth). **(2)** Any transmitted signal whose essential spectral content is less than the overall bandwidth of the transmission channel.

narrowcasting A CATV term distinguishing cable from broadcasting, it describes the function of distributing a range of TV channels or programs designed for minority interests rather than mass appeal.

NAS An abbreviation of <u>N</u>etwork <u>A</u>pplications <u>S</u>upport.

NASA Science Internet (NSI) A computer networking project started by NASA's Office of Space Science and Applications. It contains two major networks, Space Physics Analysis Network (SPAN) and NASA Science Network (NSN). See also *NSN* and *SPAN*.

NASA Science Network (NSN) Using TCP/IP, NSN is part of the connected Internet.

NASC An abbreviation of <u>N</u>umber <u>A</u>dministration <u>S</u>ervice <u>C</u>enter.

NASI An acronym for <u>N</u>et<u>W</u>are <u>A</u>synchronous <u>S</u>ervices <u>I</u>nterface.

NATA An acronym for <u>N</u>orth <u>A</u>merican <u>T</u>elecommunications <u>A</u>ssociation.

national call In telephony, a call to a destination outside of the local service area of the calling customer, but within the national boundaries of the country in which the call is being originated.

National Center for Supercomputing Applications (NCSA) A group that has written a great deal of public domain software for the scientific community, including NCSA Telnet and NCSA Mosaic for Windows, X-Windows, and Macintosh.

National Electrical Code® (NEC®) A nationally recognized set of safety standards covering the design, construction, and maintenance of electronic and electrical equipment (including not only power circuits but telecommunications, computer network, and fiber optic cabling). The code itself does not have the force of law; however, many communities and states have passed laws requiring conformance. The code is developed by the NEC Committee of the American National Standards Institute (ANSI) and is sponsored by the National Fire Protection Association (NFPA). It is identified by ANSI/NFPA 70-YYYY, where YYYY is the revision year.

National Information Infrastructure (NII) A proposed, advanced, seamless web of public and private communications networks, interactive services, interoperable hardware and software, computers, databases, and consumer electronics to make vast amounts of information available to essentially anyone. Also called the *information superhighway* and sometimes the *Internet*. See also *NII*.

National Institute of Standards and Technology (NIST) Formerly called the National Bureau of Standards (NBS), it is part of the U.S. Department of Commerce.

It produces Federal Information Processing Standards (FIPS) for all governmental agencies except the Department of Defense.

national number The combination of digits representing an area code (or city code) and a station code (directory number) that, for the purposes of distance dialing, uniquely identifies each main station within the national numbering plan.

national numbering plan In telephony, any scheme for identifying telephone stations within a geographic area identified by a unique *country code*.

National Research and Education Network (NREN) The successor to the NSFNet.

NREN is a broadband network under development that will serve research and educational facilities. It was envisioned in the High Performance Computing Act of 1991. Although it will develop from the Internet, it will expand the capabilities and permit more advanced applications. It is also commissioned to, as stated by NREN, "serve as a catalyst for the development of the National Information Infrastructure (NII)." Although the original goal was to have a gigabit network by 1996, the network will likely top out at 622 Mbps as supplied by ATM technology—at least in the near term. Other issues include:

- Does the network need to serve K-12 schools as well as colleges, universities, and research institutions?
- How do the commercial possibilities of *NREN* resolve?
- Who pays for the *NREN* backbone? How is the cost passed on to the consumer?

National Science Foundation (NSF) An independent agency of the federal government established in 1950 to promote science and engineering through sponsorship research and education. As part of the methodology to achieve these goals, the *NSF* supports a high-speed communications network (the NSFNet—to be superseded by NREN).

National Television Standards Committee (NTSC) The part of the Electronic Industries Association (EIA) that prepared the standard specification for commercial television broadcasting, which was approved by the Federal Communications Commission (FCC) in 1953. An *NTSC* signal is a composite video signal used by televisions and VCRs in the United States. In Europe, the PAL or SECAM standard is used.

natural frequency (ω) The lowest frequency at which an electrical or physical device resonates (freely oscillates).

natural log (ln) A logarithm based on the transcendental number e (approximately 2.7182818 . . .). See also *log*.

natural noise Any noise having its source in natural phenomena such as lightning, electron movement in a conductor, or even the stars (*cosmic noise*). Noise generated in machines and other technical devices is not *natural noise*.

NAU An abbreviation of <u>N</u>etwork <u>A</u>ddressable <u>U</u>nit.

NAUN An abbreviation of <u>N</u>earest <u>A</u>ctive (<u>A</u>ddressable) <u>U</u>pstream <u>N</u>eighbor.

nautical mile (nmi) A unit of distance used in navigation and based on the length of one minute of arc taken along a great circle. Several values are possible because the Earth is not a perfect sphere; internationally, the value 1852 m (6076.104 ft) has been adopted.

Another related unit of distance is the *geographical mile* which is equal to 1 min of arc on the great circle called the equator 1997.096 m (6087.15 ft).

navigating the net Finding one's way around a large network. Locating and retrieving desired information on the network. Often used of the Internet, particularly the World Wide Web (WWW). Navigation is often accomplished with the help of a browser, a tool for navigating hypertext documents.

navigation message In the Global Positioning System (GPS), a 37 500-bit message embedded in the GPS signal. It contains information such as the satellite ephemeris, clock data, and almanac. The data are transmitted at 50 bps so it will take about 12.5 minutes to receive the complete message.

NAVSTAR An acronym for NAVigation Satellite (or System) with Time And Ranging. The name given to the Global Positioning System (GPS) satellites built by Rockwell International.

NBFM An abbreviation of NarrowBand Frequency Modulation.

NBH An abbreviation of Network Busy Hour.

NBP An abbreviation of Name Binding Protocol. See also *AppleTalk*.

NBRVF An abbreviation of NarrowBand Radio Voice Frequency.

NBS An abbreviation of the United States National Bureau of Standards, now called the *National Institute of Standards and Technology* (*NIST*).

NC (1) An abbreviation of Network Computer. (2) An abbreviation of No Charge. (3) An abbreviation of No Connection. (4) An abbreviation of Normally Closed.

NC Codes An abbreviation of Network Channel Codes. In telephony, industry standard codes that define the type of service being provided at each end of a circuit.

NCC (1) An abbreviation of National Coordinating Center for Telecommunications. (2) An abbreviation of Network Control Center.

NCCF An abbreviation of Network Communications Control Facility.

NCD An abbreviation of Network Computing Devices.

NCoP An abbreviation of Network Code of Practice.

NCP (1) An abbreviation of Network Control Point. AT&T's routing, billing, and call control database system. (2) An abbreviation of Network Control Program. IBM's Systems Network Architecture (SNA) program for control of communications controllers. (3) An abbreviation of NetWare Core Protocol. Novell NetWare's rules for requesting and responding to requests for network services. It is used to encode requests to the server and responses to the workstation.

NCP Packet Signature A security feature in Novell's NetWare 4.x that helps prevent a workstation from forging an NCP (NetWare Core Protocol) request packet and using it to gain supervisory rights on the network. Each NCP packet must be signed by the server or workstation sending the packet. The signature is different for each packet. If an invalid NCP packet is received, an alert is entered into the error log and sent to both server and workstation. This alert specifies the workstation and its address.

Four packet signature levels are possible for both the server and workstation, or client. That is,

NCP PACKET SIGNATURE LEVELS

Level	Server	Client
0	Server does not sign packets.	Client does not sign packets.
1	Server signs packets only if client requests it and if client level is 2 or 3.	Client signs packets only if server requests it and if server level is 2 or 3. Default.
2	Server signs packets if client can sign (i.e., if client level is 1 or higher). Default.	Client signs packets if server can sign (i.e., if server level is 1 or higher).
3	Server signs packets and requires clients to sign (or login fail).	Client signs packets and requires server to sign (or login fail).

The four levels for each node yield 16 possible packet signature combinations, only some of which actually are useful. Some levels can slow down performance considerably, and others make it impossible to log in to the network. (For example, if either the server or workstation is set to level 3 and the other's level is set to 0, login is not possible.) Packet signatures occur only if both server and client are set to 2 or higher or if either is set to 1 and the other to 2.

NCS (1) An abbreviation of Network Control System. (2) An abbreviation of Net Control Station. (3) An abbreviation of National Communications System.

NCSA An abbreviation of the National Center for Supercomputing Applications.

NCTA An abbreviation of the National Cable Television Association.

NDD An abbreviation of NetWare Directory Database. See also *NetWare Directory Service* (*NDS*).

NDIS An acronym for Network Driver Interface Specification. A Microsoft/Intel specification for a generic device driver for network interface cards. (That is, it is independent of hardware and protocol.)

NDS An abbreviation of NetWare Directory Services.

near end crosstalk (NEXT) A particular form of crosstalk in which the disturbing signal is coupled from the local end of one communication circuit to the local end of another circuit. In the figure, three possible paths of *near end crosstalk* (*NEXT*) are diagrammed: in the local end office (or it's outside plant cabling), in the interoffice network, or in the remote office (or it's outside plant cabling).

LOCAL END   REMOTE END

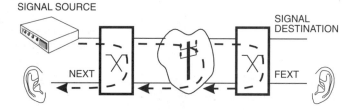

Neither the interfering signal source nor its destination need be in the local or remote office of the channel experiencing crosstalk reception. See also *crosstalk* and *far end crosstalk* (*FEXT*).

near-field diffraction pattern The diffraction pattern of an electromagnetic wave, which is observed close to a source or aperture. The pattern in the output plane is called the *near-field radiation pattern*. The diffraction pattern varies with distance from the source, and the near-field pattern typically differs significantly from that observed at infinity. Also called *Fresnel diffraction pattern*. Contrast with *far-field diffraction pattern*.

near-field region (1) In antennas, the close-in region of an antenna wherein the angular field distribution is dependent on distance from the antenna. Also called *near field* and *near zone*. (2) In optical fiber communications, the region close to a source or aperture.

Nearest Addressable Upstream Neighbor (NAUN) That node, in a Token Ring network, from which a specified node receives packets and the token. Each node in a Token Ring network receives transmissions only from its *NAUN*.

NEC® The acronym for National Electric Code.

necessary bandwidth The width of the frequency band that is just sufficient to ensure the transmission of information at a designated rate and with the desired quality under specified conditions.

negate To perform the logic operation *NOT*.

negotiation The process whereby the local and remote devices determine what type of connection to establish (including transmission speed, error correction, and flow control). The usual objective is to find the fastest, mutually compatible method of transferring data. Also called the *detection phase* or *handshaking*.

neighbor notification (NN) A process in Token Ring networks that tells each lobe about its upstream and downstream neighbors. The process uses the Frame Status and Source Address fields in certain types of MAC frames.

NEMA An acronym for National Electrical Manufacturers Association. A standards-setting association in the United States.

NEP An abbreviation of Noise Equivalent Power.

neper (Np) A dimensionless unit used to express ratios (such as voltage, current, power, gain, loss, etc.). The *neper* is similar to the decibel (dB) except it uses the *natural logarithm* of the ratio, i.e., a logarithm with the *Naperian base e* (2.718281828 . . .). If V is the measured value and V_{ref} is the reference value, then the equation for a *neper* is:

$$Np = ln_e(V/V_{ref})$$

The *neper* is often used to express voltage and current ratios, while the decibel is usually used to express power ratios. To convert a voltage ratio expressed in *Np* to *dB* multiply *Np* by $20/ln(10)$ dB/Np $\approx$ 8.68588963 dB/Np. See also *dB*.

NES An abbreviation of Noise Equivalent Signal.

NET **(1)** An abbreviation of NETwork. **(2)** Frequently a reference to the Internet. **(3)** The Artisoft LANtastic program which allows a user access to servers, printers, e-mail, etc., on a local area network (LAN). **(4)** In the Internet, the top-level domain that is assigned to an Internet administrative organization. See also *domain*. **(5)** An acronym from Normes Europenne de Telecommunication. The *Net* program will produce a range of mandatory standards for type approval of telecommunications equipment in all European Community (EC) states. Once a piece of equipment has passed its *NET* compliance tests, which could be undertaken in any member state, that piece of equipment can be sold in all EC countries for connection to the public network. Key *Nets* include:

NET 1—X.21 Access.

NET 2—X.25 Access.

NET 3—ISDN Basic Access with subdivisions: Part 1, which corresponds to the ITU-T Q.921 standard, and Part 2, which is functionally equivalent to ITU-T Q.931.

NET 4—Public Switched Telephone Network Access.

NET 5—Primary Rate ISDN Access.

Net Address The location on a network where a user's mail is held until it is retrieved. The *Net Address,* in the header of the message, gives the message routing switch information required to deliver the message to the addressee.

net gain The overall gain of a transmission circuit, i.e., the sum of all device gains minus the sum of device losses. *Net gain* may be measured by applying a test signal (at an acceptable power level) at the input port of a circuit and measuring the power delivered at the output port.

Net gain is expressed either as the ratio of the output power to the input power or in decibels (dB), wherein the number of dBs is 10 times the logarithm of this ratio. This gain (expressed in dB) may be positive or negative. If the level at the output is greater than the input level, the gain is positive; if the output is less than the input, the gain is negative. If the *net gain* is negative, it is also called the *net loss*.

net heavy A term describing Internet system administrators.

net information content A measure of the essential information contained in a message. It is expressed as the minimum number of bits required to transmit the message with a specified accuracy over a noiseless channel.

net lag Message delay introduced by network *latency* and *congestion*. Total *net lag* can range from seconds to ten minutes or more.

Latency is the intrinsic delay of a communications path. It includes the *propagation delay* of the transmission medium (which, except for a satellite link, is generally small) and the delay introduced by the store-and-forward nature of routers when they relay packets on to the next *hop* of their journey. If the packet traverses 15 or 20 hops, considerable delay may be encountered. Congestion occurs when a host in the signal path is unable to receive packets. In cases like this, the only remedy the network has is rerouting or waiting.

net loss The sum of all transmission losses occurring between channel ends minus the sum of transmission gains. See also *net gain*.

net police A USENet participant who views himself as a protector of the Internet core values (such as the avoidance of commercial gain and the conservation of network bandwidth) and flames posters of those who violate these values. Also written *net.police*.

net surfing Exploring the Internet in search of interesting information, graphics, and people.

NET_MGR The Artisoft LANtastic program that assists in the management of servers, user accounts, and resources.

NetBEUI An acronym for NetBIOS Extended User Interface (pronounced "net-boo-ee"), an extension of IBM's NetBIOS transport layer protocol. It is a fast protocol used in Microsoft's LAN Manager and LAN Server and communicates with a network through Microsoft's Network Driver Interface Specification (NDIS) interface for the network interface card. It can bind with as many as eight media access control drivers. The protocol is not routable.

NetBEUI is often confused with NetBIOS. NetBIOS is the applications programming interface (API), and *NetBEUI* is the transport protocol.

NetBIOS An acronym for Network Basic Input/Output System. Software developed by IBM that provides the interface between the PC operating system, the I/O bus, and the network. Now a de facto standard protocol governing user network resource access and data exchange. The basic application that allows user applications to communicate on the network, that is, it provides a Session layer interface between network applications running on a PC and the underlying protocol software of the Transport and Network layers.

NETBLT An acronym for NETwork BLock Transfer.

netdead Slang on the Internet Relay Chat (IRC), indicating the state of a user who has signed off IRC; that is, he is "dead to the net."

netILLINOIS An Illinois state network that provides high-speed Internet connectivity for member institutions (universities, community colleges, schools, nonprofit organizations, government agencies, and business).

netiquette A contraction of NETwork etIQUETTE. The manners one should use while conversing on a network. It is an informal, open-ended set of guidelines for use on the Internet in general and USENet in particular.

Some do's and don'ts on networks include:

- **Do** keep postings line lengths less than 80 characters, so that users of 80-column terminals can read them.

- **Do** conserve bandwidth. Read the FAQ (frequently asked questions) before submitting a question to a group that may already have posted answers.
- **Do** use a descriptive subject line so that users can better guess the contents of your posting.
- **Don't** type in all upper case. It is the online equivalent of shouting.
- **Don't** "stand-up" anyone. On the network, just as in the real world, if you make an appointment to meet someone at a certain time—be there!
- **Don't** post messages in many places on the net. A practice known as cross posting or *spamming*.
- **Don't** post things that the average person wouldn't want to see in his hometown newspaper.

netlogon service A local area network (LAN) manager service that implements logon security. The service verifies the username and password supplied by each user logging on to the local area network. See also *LAN manager*.

netnews A synonym for *USENet*.

NetNorth A Canadian branch of BITNET. Like BITNET, *NetNorth* uses IBM's NJE protocol. *NetNorth* accesses the Internet via a gateway.

Netscape Communications Corporation The company, formerly Mosaic Communications Corporation, that produces several Internet software packages, including the browser called Navigator.

NetView® An IBM mainframe network management product used for monitoring Systems Network Architecture (SNA) compliant networks. *NetView* runs as a Virtual Telecommunications Access Method (VTAM) application on the mainframe that is the network manager.

It integrates the three separate Communications Network Management (CNM) programs and features of several other programs:

- *Network Communications Control Facility* (*NCCF*) from CNM.
- *Network Logical Data Manager* (*NLDM*) from CNM—which helps locate problems along the logical connection of an SNA session.
- *Network Problem Determination Application* (*NPDA*) from CNM—which allows diagnostic information and alerts to be displayed.
- *Virtual Telecommunications Access Method/Node Control Application* (*VNCA*)—which monitors the status and activity of all resources in the domain.
- *Network Management Productivity Facility* (*NMPF*)—which helps the network operator install, learn, and use many network management facilities.

NetView/PC An implementation of IBM's NetView running on an OS/2 PC-based platform that allows non-IBM networks and equipment to be managed via an SNA network management station. NetView/PC is a service point in IBM's Open Network Management Architecture.

NetWare® Novell's network operating system (NOS) which uses internetwork packet exchange (IPX) and sequential packet exchange (SPX) protocols. It provides file and printer sharing among at least one file server and a number of networked PCs. The file server provides access control including logins, access to client nodes, printers, and fax/modems.

NetWare® Asynchronous Services Interface (NASI) Specifications for accessing communications servers across a Novell NetWare network.

NetWare® Directory Services (NDS) A global naming service used in NetWare 4.x. The service provides a global directory containing information about all the objects in a network, regardless of which server or servers on which the objects may reside.

The *NetWare Directory Database* (*NDD*) is organized as a tree and contains information about objects of several types:

- *Physical objects*—such as users, nodes, and devices.
- *Logical objects*—such as groups, queues, and partitions.
- *Container objects*—intermediate elements in the Directory tree which help provide a logical organization for other objects in the Directory tree. A container can include other containers, leaf objects, or both. Container objects include:
  - *Organization* (*O*) *object*—which represent the first level of grouping on most networks. Depending on the network, this level could represent a company, division, or department. At least one Organization object is required in each *NDS* Directory tree. An Organization object can contain Organizational Unit or leaf objects.
  - *Organizational Unit* (*OU*) *object*—which can be used as a secondary grouping level. These objects are optional but must be below an Organization or another Organizational Unit object if used. An Organizational Unit object can contain Organizational Unit or leaf objects.
  - *Country* (*C*) *and Locality* (*L*)—objects defined for compatibility with ITU's X.500 Directory Services, but rarely used and, are not required for compliance with the X.500 specifications.
- *Leaf objects*—which represent information about actual network entities, such as users, devices, and lists. See also *leaf object*.

NetWare® Loadable Module (NLM) A driver that runs on a network server (running Novell's NetWare operating system) that can be loaded and unloaded on-the-fly as is needed. *NLMs* make network operation more efficient because services can be selectively loaded. With the availability of *NLMs,* servers only need to load the core of the network operating system (NOS). The core capabilities can be extended by adding the modules that are needed.

Examples of Novell's *NLM* include:

- *NetWare Express*—a fee-based private electronic information service from Novell. Subscribers can access the Novell Support Encyclopedia and the NetWare Buyer's Guide, and they can get product information and technical support. The service is delivered over the GE Information Services Network; users need the appropriate software and an asynchronous modem.
- *NetWare TCP/IP*—NLMs that provide support for the Internet Protocol (IP) as a routing protocol for version 3.x and 4.x servers. With NetWare/IP, a NetWare server can function as a gateway between NetWare and TCP/IP networks.
- *NetWare for Macintosh*—a collection of *NLMs* that provide various NetWare services to Macintosh clients on a NetWare network. Services include file handling, printing, network administration, and AppleTalk routing. With NetWare for Macintosh, Macintosh users can access network resources, files, and applications, send print jobs to network printers, and utilize NetWare features, such as network security.
- *NetWare Mail Handling System* (*MHS*)—provides support for modules that connect to ITU's X.400, IBM's Systems Network Architecture (SNA), and the SMTP protocols. The MHS software provides store-and-forward capability for fax and e-mail services.
- *NetWare Management Agents*—a collection of *NLMs* in NetWare 4.x that enable communication between a server and external management software. The *NLMs* are used to identify and register entities (software, hardware, and data) on the server as manageable objects.

- *NetWare NFS*—a collection of *NLMs* that provide file handling and printing services for UNIX clients in a NetWare network. It uses the Network File System (NFS) application-layer protocol from Sun Microsystems.
- *NetWare for SAA*—a gateway package for connecting NetWare networks to various machines that support IBM's Systems Network Architecture (SNA), including AS/400s, 3090s, and 370s. NetWare for SAA is a series of *NLMs* available in NetWare 3.x or 4.x, and it supports up to several hundred sessions for each gateway.
- *NetWare for UNIX* (formerly called Portable NetWare)—a program that provides NetWare support on machines running the UNIX operating systems, or its derivatives. The software enables the host to provide file handling, printing, and backup services to clients, regardless of whether clients are running DOS, Microsoft Windows, or the Macintosh operating system.

NetWare® Multiprotocol Router (MPR) Software that can route any of several types of protocols over various local area network (LAN) types.

- Intermediate System to Intermediate System (IS-IS) protocols from an ISO/OSI-compliant protocol stack.
- Point-to-Point Protocol (PPP) on the Internet.
- Router Information Protocol (RIP) from Novell's Internetwork Packet Exchange (IPX) protocol stack. Also called IPX RIP.
- Routing Information Protocol (RIP) from the Internet Protocol (IP) stack.
- Routing Table Maintenance Protocol (RTMP) from AppleTalk's protocol stack.
- Simple Network Management Protocol (SNMP) for managing routing activity on the Internet.
- Systems Networking Architecture (SNA).
- X.25 protocols for wide area networks (WANs).

NetWare® NFS Gateway That software installed on a Novell NetWare server which allows NetWare clients (using DOS or Microsoft Windows) to access files on a Network File System (NFS) server. To the client, the files on the NFS server appear to be on the NetWare server.

NetWare® Protocol Suite Some of the features/capabilities found in Novell's NetWare operating system.

Protocol	Description
Burst mode	Used when large amounts of data need to be transmitted (rather than NCP). Burst mode can be used to make NCP more efficient when transmitting large blocks of data (such as entire files) over slower WAN links.
IPX (Internetwork Packet Exchange)	NetWare's standard network-layer protocol. IPX is used to route data packets from the transport layer across a network.
NCP (NetWare Core Protocol)	The protocol NetWare uses to generate and respond to workstation requests. It includes procedures for handling any service a workstation might request (including file or directory handling, printing, and so on).
RIP (Routing Information Protocol)	Used by routers and servers to exchange routing information on an internetwork. RIP packets use NetWare's IPX protocol to move between stations. It is generally known as IPX RIP to distinguish Novell's version from the RIP protocol in the Internet TCP/IP protocol suite.

Protocol	Description
SAP (Service Advertising Protocol)	Used by NetWare services to broadcast their availability across the network. The protocol supports broadcast, query, and response packets.
SPX (Sequenced Packet Exchange)	NetWare's standard transport-layer protocol. It is used to ensure that data packets have been delivered successfully by the IPX services. SPX requests and receives acknowledgments from its counterpart on the receiving node and keeps track of fragmented messages (those consisting of multiple packets).
Watchdog	Used for maintenance purposes. It can determine whether the NetWare shell is still running on workstations that have been idle for a long time.

NetWare® Tools A collection of basic *end-user utilities* for NetWare version 4.x. They can be used to accomplish various tasks on the network, such as mapping drives, sending messages, and setting up printing.

Although NetWare Tools are designed for end-user tasks, administrative tasks are performed using utilities such as the NetWare Administrator.

network (1) A collection of generally passive, electronic components (e.g., resistors, capacitors, and inductors) interconnected in some way that performs a specific function; usually limited in scope (e.g., simulation of a transmission line or pulse shaping). **(2)** A collection of two or more autonomous information sources and sinks interconnected by one or more communication links. The components of a network include:

- Nodes (computers, printers, network interface cards[—NICs], etc.).
- Connection elements (cabling, wiring centers, optical fibers, switching systems, etc.).
  The interconnecting link(s) may either be temporary (as with the dial-up telephone network) or permanent, such as with cables. The data passing through the interconnecting link is examined for errors, in contrast with a *multiprocessor system* wherein the data is accepted "at face value."
- Topology (physical and logical):
  - Physical topology describes how nodes are wired or interconnected. (Various topologies include the bus, ring, and star networks.)
  - Logical topology describes how network packets are treated. For example, a logical ring may be created on a physical star network by addressing a token packet sequentially to each node.
- Auxiliary components (peripheral devices, safety devices, and tools).
- Network operating system (NOS) and workstation software.

Networks are often classified according to their geographic extent or according to the transmission protocol used. Some examples of voice and/or data networks include the public switched telephone network (PSTN), integrated services digital network (ISDN), Ethernet (local area network), and the Internet (a world wide computer network). See also *network classifications.*

network access controller (NAC) A device that provides access to a network, for remote callers or for another network.

network access point (NAP) The switching point through which networks are connected.

network adapter A device used to connect a computer to a local area network (LAN). The device may be an expansion card which is plugged directly into the internal computer bus, or it may be an external "black box" which communicates with the computer via one of its ports. Also called a *network interface card* (*NIC*).

network address (1) A pointer to a particular node on a network. It is a network layer address referring to a logical rather than a physical network device.

Every network node has at least one network address, including a fixed hardware address (assigned by the device's manufacturer), and most have a protocol-specific address (the *protocol address*) assigned by the system administrator. (2) In telephony, the numeric character string used to specify the location of a called station.

Network Addressable Unit (NAU) Any location, in IBM's System Network Architecture (SNA) networks, with one or more ports for communicating over the network. The three types of *NAUs* are *physical units* (*PUs*), *logical units* (*LUs*), and *system service control points* (*SSCPs*).

network administration A group of network management functions that

- Provide support services,
- Ensure that the network is used efficiently, and
- Ensure prescribed service quality objectives are met.

Network administration may include activities such as network address assignment, assignment of routing protocols and routing table configuration, and directory service configuration.

network administrator The person in charge of maintaining the operation, security, and integrity of the network, i.e., network administration.

Network Application Support (NAS) Part of Digital Equipment Corporation's (DECs) Enterprise Management Architecture (EMA) to provide a uniform environment for software running on different platforms (such as VAXes and PCs), so that applications can be integrated with each other, regardless of the platforms involved.

NAS is designed to use international standards to support the multiple platforms. This is in contrast to IBM's strategy used with Systems Application Architecture (SAA), which relies on proprietary protocols to provide support for multiple platforms.

network architecture (1) The design principles, physical configuration, functional organization, operational procedures, and data formats used as the bases for the construction, modification, and operation of a communications network. See also *topology*. (2) The structure of an existing communications network, including the physical configuration, facilities, operational structure, operational procedures, and the data formats in use. It includes the communication equipment, protocols, and transmission links that constitute a network, and the methods by which they are arranged.

network block transfer (NETBLT) A Transport Level, flow-controlled, bulk data transfer protocol used with TCP/IP internets. *NETBLT* controls the rate at which data are sent to allow a steady, high-speed flow.

network board An expansion board that makes a computer network capable. Also called a *LAN card, network adapter,* and *network interface card* (*NIC*).

network busy hour (NBH) See *busy hour*.

network classifications Communications networks can be classified by any of a multitude of parameters. These classifications are neither exclusive nor exhaustive, but they do yield a fruitful way of discussing the aspects of the network. Some of the more common classifications are:

- *Message capacity*—whether the network can transmit one or more messages at a time. Networks are either baseband, carrierband, or broadband.
  - A *baseband network* can transmit exactly one message (packet) at a time.
  - A *carrierband network* is a special case of a baseband network, that is, the channel's entire bandwidth is used for a single transmission, and the signal is modulated before being transmitted.
  - A *broadband network* can transmit more than one message at a time by using a different carrier frequency for each message and then multiplexing these multiple channels (frequency division multiplex—FDM).

  Most LANs are baseband networks.
- *Physical topology*—defines how individual network nodes are interconnected. The basic structures are bus, mesh, ring, and star.
  - *Bus topology:* A central cable forms the backbone of the network, and individual nodes are attached to this bus, either directly or by means of a shorter piece of cable. Signals travel along the bus, and each node hears all messages, but reading only those addressed to the node. Ethernet and some versions of ARCnet use a bus topology.

  Variants of the bus topology include tree and branching tree topologies.
  - *Mesh topology:* A node may be connected to a large number of other nodes. (In the extreme case, every network node is connected directly to every other node.) Although very redundant and fast, the advantage of direct access to each node is more than offset by the number of interconnecting wires and the number of ports required to terminate each interconnection.
  - *Ring topology:* The nodes are arranged in a loop; that is, each node input is connected to the output of the node immediately before it. Messages are passed around the ring in sequence. A node receives the message if the node is the recipient, or passes the message on otherwise. FDDI and IBM's Token Ring networks use a ring topology.
  - *Star topology:* All nodes are connected to a central node or to a wiring center (such as a hub). Messages can be sent directly to their destinations from the center. Some versions of ARCnet and AT&T's StarLAN use a star topology.

  A variant of the star topology is the distributed star in which several hubs, each of which forms a star, are connected to each other.
- *Logical topology*—includes two main logical topologies, bus and ring.
  - *Logical bus:* Information packets are broadcast across the network in a manner that all nodes attached to the network can hear the information at roughly the same time. Only nodes for whom the information is intended actually receive and process the transmitted packets.
  - *Logical ring:* Information is passed sequentially from node to node in a ring. Each node receives information from exactly one node and transmits it to exactly one node.
- *Range*—the geographical or administrative range over which the nodes are distributed. Networks can be categorized as:
  - *Campus area network* (*CAN*)—connects nodes (and/or departmental LANs) from multiple locations, which may be separated by a large distance. Unlike a wide area network (WAN), however, a campus network does not require remote communications facilities, such as modems and telephones.

- *Departmental area network* (*DAN*)—a small network of 20 to 30 nodes, connected so that users can share common resources. DANs are typically used in governmental agencies.
- *Enterprise network*—a network that connects machines for an entire corporate operation. The network may connect very diverse machines from different parts of the company. These machines may be geographically separated by rooms, buildings, cities, or even countries. Enterprise networks are increasingly likely to cross national boundaries in this age of multinational corporations and may be global area networks (GANs) also.
- *Global area network* (*GAN*)—a wide area network or internetwork that extends across national boundaries and may connect nodes on opposite sides of the world. As with very widely distributed WANs, most GANs are likely to be internetworks in disguise.
- *Local area network* (*LAN*)—a network in which the nodes are connected within a relatively small geographical region (e.g., within an office, floor, or a building) and by a particular type of medium. Functionally, a LAN consists of a group of computers interconnected so that users can share files, printers, and other resources.
  A *local area wireless network* (*LAWN*) is a special type of LAN that uses radio or infrared transmissions instead of cabling.
- *Metropolitan area network* (*MAN*)—a network that covers a radius of up to 50 or 75 miles. These types of networks use fast data transmission rates (over 100 Mbps) and are generally capable of handling voice transmission as well as data transmissions.
- *Wide area network* (*WAN*)—a network that consists of machines that are spread out over larger areas; such as across a city, state, or country. WANs usually include some type of remote bridges or routers, which are used to connect groups of nodes by telephone or other dedicated lines. Because of this, the bandwidth for WANs tends to be considerably smaller than for LANs.
- *Node types*—may be PCs, minicomputers, mainframes, or other networks.
  - PC-based networks are generally used for general-purpose computing and operations.
  - Minicomputers or mainframes are most likely found in MIS departments and universities.
  - Backbone networks are networks whose "nodes" are actually smaller networks, known as access networks.
- *Node relationships*—the relationship among the nodes that make up the network. Networks categorized by relationship are known as distributed, peer-to-peer, server-based, and client–server.
  - *Client–server*—a more complex version of a server-based network. While workstations in server-based networks can gain access to many resources through the server, the workstation must do most of the work. That is, the server makes the resources available (downloads files and, possibly, applications to the workstation) and then lets the workstation run programs and operate on the downloaded data.
    In a client–server network, the client issues a query or request for service; the server does the necessary work and then returns a response or the results.
  - *Distributed*—a network with no master station, i.e., one in which any node can talk to any other. In a distributed network, servers are just that! Machines, devices, or programs that provide services, as opposed to controlling network activity. An example of a distributed network is Internet's USENet.

- *Peer-to-peer*—a network where every node can be both client and server, that is, all nodes are equal. These networks are useful if the number of machines that need to be connected is small (generally, fewer than 10) and if no user will be running programs that push available resources to the limit.
- *Server based*—a network in which the server runs the network, granting other nodes access to resources. Most middle to large-sized networks are server based. The most popular PC-based server network operating systems include Novell's NetWare, Microsoft's LAN Manager, IBM's LAN Server, and Banyan's VINES.
- *Packet protocol*—the rules by which packets are launched onto the network, e.g., CSMA/CD, token passing, polling, out-of-band signaling, and so on.
- *Network Architecture*—the physical, electrical, and packet protocol taken as a group, e.g., DIX Ethernet, IEEE 802.3 Ethernet, IBM Token Ring, IBM Systems Network Architecture (SNA), ARCnet, FDDI, and DECnet.
- *Network access possibilities*—the means by which a node gains access to the network, i.e., *shared media*, or *switched*.
  - In a shared media network, only one node may transmit at any instant; otherwise collision will occur. Access to the medium is controlled by a protocol such as CSMA/CD, token passing, or polling.
  - Switched access, on the other hand, allows multiple nodes to transmit simultaneously because an exclusive but temporary connection is established between the sender and the receiver. Switched access networks may be circuit switched (as with the telephone network) or packet switched (as with the Internet).
- *Network composition*—a reference to the makeup and consistency of the network. That is,
  - *Heterogeneous networks*—networks that use multiple protocols at the network layer.
  - *Homogeneous networks*—networks that use a single protocol at the network layer.
  - *Hybrid networks*—networks that include a mixture of topologies, such as both bus and star.

Network Code of Practice (NCoP) A voluntary code of practice for the design of networks covering transmission quality, safety, and technical standards.

Network Communications Control Facility (NCCF) A component of IBM's NetView network management software that can be used to monitor and control the operation of a network.

network computer (NC) A computer designed for operation on the Internet (or other network). An *NC* is similar to a personal computer but has minimal memory and a fast modem. It runs programs loaded from a network host while the network connection is intact. When the communications link is severed, the program is lost. Also called a *network appliance* or a *thin client*.

network computing A term analogous to client–server computing.

network connectivity The topological description of a network that specifies, in terms of circuit termination locations and quantities, the interconnection of the transmission nodes. See also *network classifications* and *topology*.

Network Control Center (NCC) A designated node in a computer network that runs the network management tasks. These tasks receive reports from the agent processes running on workstations.

Network Control Point In IBM Systems Network Architecture (SNA) networks, a host program that directs the operation of a communications controller (such as an IBM 3705 or 3725).

network control program (NCP) In a switch or network node, software designed to store and forward frames between nodes. An NCP may be used in local area networks (LANs) or larger networks.

network control signaling The transmission of signals used in the telecommunications system which perform functions such as:

- Supervision (control, status, and charge signals), including audible call progress tones (ringing, busy, coin collect, coin return, reorder, etc.).
- Addressing (dialing).
- Number identification (calling and called).
- Flow control.
- Error control.

network control system (NCS) A software tool used to monitor and modify network activity. It is generally used to refer to older systems, which were run in a low-speed, secondary data channel created using time division multiplexing. These components have been replaced by the network management systems (NMSs).

network demarcation point In telephony, the point in the public telephone network where the local exchange carrier's equipment and responsibilities end and the subscriber's begin. Generally, the point is at the subscriber's premises on the subscriber's side of the telephone company's protector. Also called the *demarcation point* or the *network interface* (*NI*).

network device A computer, peripheral, or other related communications equipment attached to a network.

network device driver A program that enables the operating system software to communicate the network adapter cards. Also, a software module running on a host or workstation that is responsible for the communications between the computer and the network or a device attached to the network.

Network Driver Interface Specification (NDIS) A de facto standard interface for *network interface card* (*NIC*) drivers. The standard was developed by Microsoft, IBM, and 3Com, and is supported by many NIC manufacturers.

It allows multiple transport protocols to use the same NIC; hence, it helps ensure the NIC's compatibility with multiple network operating systems. It supports multiple concurrent stacks as *NDIS* matches a packet from the NIC's driver with the proper protocol stack by polling each stack until one claims the packet. This is in contrast to the competing Open Data Link Interface (ODI) standard from Novell and Apple in which the link-support layer (LSL) matches the packet with the appropriate protocol.

network element (1) A facility or piece of network equipment used to provide a telecommunications service that can be managed through an element manager as part of a network management system. The term includes subscriber numbers, databases, signaling systems, and information sufficient for billing and collection or used in the transmission, routing, or other provision of a telecommunications service. (2) In integrated services digital networks, a piece of telecommunications equipment that provides support or services to the user.

network equalizer A device connected to a transmission path in order to alter or optimize a specific transmission parameter (such as frequency response, phase shift, or group delay).

network fax server A node on a network that is equipped with one or more fax/modems connected to the public telephone network. The server sends and receives facsimile messages for all network users.

Network File System (NFS) A network file utility that enables a user to access directories and files on a remote system as if they resided on the local system. With *NFS* a user *mounts* the remote directory on his computer. Thereafter, the files of the remote host are accessed with the same commands used on the local computer.

The system was originally developed in 1986 by Sun Microsystems, Inc. and was subsequently published for use on the Internet in RFC 1094 (NFS2) and RFC 1813 (NFS3). Later, it was incorporated into most versions of UNIX. For all of the convenience *NFS* offers, it represents a security risk for system administrators because it allows direct access to the files of remote hosts. *Firewalls* are frequently installed to prevent the malicious use of *NFS*. See also *Alex, Andrew File System* (*AFS*), and *Prospero*.

network information center (NIC) An organization that provides information for any of the subnetworks in the Internet, e.g., RFCs, Internet drafts, and FYIs. The chief and authoritative NIC is the Defense Data Network Network Information Center (DDN NIC).

Network Information Service (NIS) A set of protocols developed by Sun Microsystems that identifies hosts and addresses across networks. It is now an Internet utility program that automatically develops and maintains accurate maps of the Internet data pathways and makes them available to all hosts and routers. Each host and router on the Internet then maintains its internal routing table from this information.

network interface (NI) (1) The point of interconnection between a user terminal and a private or public network. (2) The point of interconnection between one network and another network. (3) The point is defined in the Code of Federal Regulations (CFR) Title 47, Part 68.3 as "that point at which telephone company communication facilities and subscriber terminal equipment interconnect." The point shall be located on the subscriber's side of the telephone company's protector or its equivalent where a protector is not supplied. Also called the *demarcation point*.

network interface controller (NIC) An adapter card (a specific type of network interface device, NID) that is inserted into a computer, which allows the computer to gain physical and electrical access to a specific network. The card contains the necessary software and electronics to support the communications protocol of a particular network type such as Ethernet, Token Ring, DECnet, and FDDI ring interface cards. Also called a *network interface card* (*NIC*), *network interface module* (*NIM*), or *network interface unit* (*NIU*).

network interface device (NID) (1) In telephony, a device interposed between the telephone company's protector and the premise's inside wiring and is used to isolate the customer's equipment from the network. (2) A device inserted between a network and communications equipment that performs functions such as buffering, code, and/or protocol conversion. (3) A device used within a local area network (LAN) to allow a number of independent nodes, each with a different protocol, to communicate among themselves. The task is accomplished by translating each independent protocol to a common intermediate transmission protocol. See also *network interface controller* (*NIC*).

network interface module (NIM) (1) The device that allows connection of customer premises equipment to the telephone network. See also *DCE*. (2) A synonym of network interface card or controller (NIC).

network interface unit (NIU) A synonym for *network interface controller* (*NIC*).

network inward dialing (NID) A synonym for *direct inward dialing* (*DID*).

Network Job Entry (NJE) IBM's proprietary protocol suite used with its mainframe computers. *NJE* is used with BITNET and several other wide area networks attached to the Internet; however, it is incompatible with the Internet protocol suite. Exchanges between *NJE*-based networks and the Internet must be through gateways.

network layer See *OSI.*

network layer address (NLA) An address appended to a LAN packet that indicates where a node is located within an internetwork. TCP/IP, DECnet, and IPX support *NLA,* and, as expected, each has its own unique format. Protocol-dependent routers use *NLA* to make routing decisions.

network management That set of software, hardware, procedures, and operations necessary to keep a network operating in a satisfactory manner. The International Standard Organization's network management model is divided into five categories:

- *Configuration management*—the process of identifying, tracking, and modifying the setup of individual devices on the network. It addresses initial installation and ongoing reconfiguration, as well as the tracking of hardware and software configuration parameters.
- *Performance management*—the process of measuring the performance of various network components. It addresses the real-time operating statistics of the system, e.g., the number of users logged on, the packet transfer rates, and other network performances.
- *Security management*—the process of controlling (granting, limiting, restricting, or denying) access to the network and its resources. It addresses the restrictions and accesses to network resources by individuals and groups and cryptographic key distribution authorization.
  For example, password management allows users onto the network, while access right management allows users to use selected resources.
- *Fault/repair management*—the process of identifying and locating faults in the network. It addresses the inevitable problem of equipment failure.
  For example, redundant resources or transmission path rerouting may be enabled in the event of a component failure.
- *Accounting management*—the process of identifying individual and group access to various resources to either ensure proper access capabilities (bandwidth and security) or to properly charge the various individuals and department for such access. That is, it addresses the problem of allocating the cost of resources among the users.

Network Management Architecture (NMA) IBM's centralized mainframe-oriented network management model. It is used in IBM's NetView network management package and in its SystemView. The model defines four categories:

- *Configuration management,* which is concerned with identifying the network elements and the relationships among them.
- *Change management,* which is concerned with changes in hardware, software, or microcode.
- *Problem management,* which addresses identification, diagnosing, tracking, and resolving problems that arise.
- *Performance and accounting management,* which monitors the availability and use of the network's resources, and manages the billing for use of these resources.

All tasks are carried out by the central host/network manager called the focal point. Devices may be connected to the network manager in either of two ways:

- Devices that support IBM's Systems Network Architecture (SNA) can connect through entry points. Such entry points serve as agents in reporting to the network manager at the focal point.
- Non-SNA compliant devices must be connected through service points (nodes running special software, e.g., NetView/PC, that can communicate with the *NMA* package). Or they may be connected if they can function as logical unit 6.2 (LU 6.2) devices. The NetView/PC software can be used to connect a wide range of devices and networks to an *NMA* network. For example, it can connect one or more Ethernet or Token Ring LANs, a PBX, or a single machine to a network. NetView itself runs as a Virtual Telecommunication Access Method (VTAM) application on the host machine.

network management control center (NMCC) The central place from which a network is maintained and changed, and from where operational statistical information is collected.

network management entity (NME) Software and/or hardware in the OSI network management model that gives a network node the ability to collect, store, and report data about the node's activities.

network management protocol (NMP) The protocol used to transfer network management NetView, and Hewlett-Packard's HP Open-View.

network management vector transport (NMVT) In IBM's Network Management Architecture (NMA), the protocol used to exchange management data. The protocol uses management service request units to request/return information about the status or performance of elements on the network.

network manager (1) Manages and maintains a network to make sure all programs are up-to-date, all hardware is functioning properly, and all authorized users are able to access and work on the network. A network manager or administrator must address tasks such as:

- Set up new accounts.
- Assign user privileges, permissions.
- Perform accounting chores, e.g., billing.
- Install and test new software or hardware.
- Troubleshoot existing hardware/software.
- Handle backup and file management.

(2) In network management, the entity that initiates requests for management information from managed systems or receives spontaneous management-related notifications from managed systems.

network modem A modem that is a separate node on a network. This modem has its own network interface card and network address, and it is connected directly to the network. A remote caller accesses the network through this node. A network modem can work as an *access server.*

Network News Transport Protocol (NNTP) The protocol in USENet that controls the distribution of USENet postings throughout the sites that have subscribed. It provides the mechanism for posting transport, expiration date control, group access control, and which domains have access to it. It is described in RFC977.

network number A unique number assigned to each network in an interconnection of multiple networks. The *network number* is assigned by a seed router.

In the Novell NetWare environment, it is also called the *IPX external network number.*

network operating system (NOS) The fundamental software that allows the implementation and maintenance of a network and the sharing of resources and services over that network. A *NOS* provides some or all of the facilities.

- The network communications protocol, that is, enabling nodes on the network to communicate with each other controlling message traffic and queues.
- Interprocess Communications (IPC), that is, enabling processes on the network to communicate with each other.
- Mapping requests and paths to the appropriate places on the network.
- Response to requests from applications or users on the network.
- Access to shared files and resources (for example, printers) on the network.
- Messaging and/or electronic mail (e-mail) services.
- System access/resource security.
- Administrative tools and functions.

A *NOS* is generally used with local area networks (LANs) and wide area networks (WANs) but could have application to larger networks. Examples include Artisoft's LANtastic; Banyan's Vines; IBM's LAN Server; Microsoft's LAN Manager, WindowsNT, Windows for Workgroups; and Novell's NetWare.

Network Operations Center (NOC) An organization responsible for the day-to-day operation of the subnetworks that make up the Internet.

network outward dialing (NOD) A feature of PABX systems that allows its users to dial any telephone network number directly, i.e., without the assistance of an operator.

network printer A printer shared by multiple nodes/users of a network.

network protection A nebulous term used to describe the strategies used to keep a network in operation in the event a component element fails.

network relay A device that allows the interconnection of dissimilar networks.

network service **(1)** A running application program that is available to perform a function for the users of the network. **(2)** The services within network-addressable units (NAUs) of IBM's Systems Network Architecture (SNA) that control network operations via sessions to and from the Systems Services Control Point (SSCP).

network service access point (NSAP) In the ISO/OSI Reference Model, the location through which a transport layer entity can gain access to network layer services. Each *NSAP* has a unique OSI network address. It is a generic standard for a network address consisting of 20 octets.

ATM has specified E.164 for public network addressing and the *NSAP* address structure for private network addressing.

network station A computer (workstation or server) attached to a network.

Network Support Encyclopedia (NSE) An electronic database of information about networking technology available on CD-ROM or through NetWare Express (Novell's private online information service). *NSE* is available in two versions:

- The Standard Volume includes Technotes, Novell Lab Bulletins, and product information.
- The Professional Volume includes information from the Standard Volume plus application notes, product manuals, and troubleshooting decision trees.

network terminal number (NTN) In the ITU-T X.121 Data Network Identification Code (DNIC) format, the complete number (address) assigned to a data terminal end point.

When the *NTN* is part of a national integrated numbering plan, the *NTN* is the 11 digits of the ITU-T X.25 14-digit address that follow the DNIC. When the *NTN* is not part of a national integrated numbering plan, the *NTN* is the 10 digits of the ITU-T X.25 14-digit address that follow the DNIC.

network terminating interface (NTI) **(1)** The point where the network service provider's responsibilities end and the customer's begin. See also *demarcation point* and *network interface*. **(2)** The interface between data communications equipment (DCE) and its connected data terminal equipment (DTE).

network termination Network equipment that provides functions necessary for ISDN access protocols. See also *ISDN, NT1,* and *NT2*.

network time protocol (NTP) A protocol above the transmission control protocol (TCP) that assures accurate local timekeeping by referencing radio and atomic clocks located on the Internet. The protocol can synchronize distributed clocks to within milliseconds over long time periods. It is defined in the Internet STD 12 (RFC1119).

network-to-network interface (NNI) The interface between two public network equipments; generally refers to ATM equipment. Defined in the ATM Forum's users network interface (UNI). Also called a *network node interface*.

network topology The organization of nodes (real or logical) and interconnecting communication media commonly forming a star, mesh, ring, bus, or some combination of these. See also *bus network, mesh network, network, ring network, star network,* and *topology*.

network transparency A feature of an operating system or other service which lets a user access remote resources through a network without having to know if the resource is remote or local.

network virtual terminal (NVT) A generic terminal protocol that allows programmers to create applications without regard to what hardware the application user may have. Two such terminals are teletypewriter (TTY) and VT100. The TTY is essentially the lowest level machine; it recognizes only the ASCII characters. The VT100 is a higher level machine, allowing direct cursor movement.

neural network (NN) Generally used to mean *artificial neural network (ANN)*. An ANN is a network of many very simple processors (units or neurons), each possibly having a (small amount of) local memory. The neurons are connected by unidirectional communication channels (connections), which carry numeric (as opposed to symbolic) data. The neurons operate only on their local data and on the inputs they receive via their connections.

A *neural network* is a processing device (either an algorithm or actual hardware) whose design was modeled after the design and functioning of animal brains and nerves and components thereof. Most neural networks have some sort of training rule whereby the weights of connections are adjusted on the basis of presented patterns. In other words, neural networks learn from examples, just as animals learn from example stimuli. *ANNs* differ from Von Neumann architectures in that:

- They are composed of large numbers of neurons, which are often very elementary nonlinear signal processors. (In the extreme case, they are simple threshold discriminators.)
- There is a high degree of interneuron interconnection, which allows a high degree of parallelism.
- There is no idle memory containing data and programs, but rather each neuron is preprogrammed and continuously active.

neutral **(1)** The conductor in the ac power distribution system that is intentionally Earth grounded on the supply side of the service disconnect and, under normal circumstances, carries load current. It is the low potential side of the power delivery system. The neutral normally provides the current path, while the safety ground carries current only when there is a fault condition. **(2)** In three-phase ac wye ("Y") power distribution, the low potential fourth wire that conducts only that current required to achieve electrical balance, i.e., to provide a return path for any current imbalance among the three phases.

neutral direct current telegraph system A telegraph transmission system in which line current flows during idle and signal element marking intervals and no current flows during signal element spacing intervals. The direction of current flow is immaterial. Also called *neutral current loop, single current system, single current transmission system,* and *single Morse system.*

neutral ground An intentional ground connection applied to the neutral conductor or neutral point of a circuit, transformer, system, or device. The ground may be applied directly, i.e., via a conductor, or through a component (resistor, capacitor, or inductor) or a combination of components.

neutral operation A method of teletypewriter operation in which marking signals are formed by current pulses (of either polarity) and spacing signals are formed by reducing the current to zero or nearly zero.

neutral transmission A method of sending serial digital data wherein a mark is represented by the flow of current and a space is the absence of current. Also called *unipolar signaling.* See also *polar operation.*

neutralization A method of nullifying an unwanted feedback signal in certain types of amplifiers so as to prevent oscillation. In effect, additional feedback is generated by an external circuit in such a manner as to be the same voltage but opposite phase as the unwanted feedback signal.

NevadaNet A Nevada state network that links universities, research institutions, and community colleges to each other and to the Internet. The network is funded by participating institutions, the National Science Foundation, and the State of Nevada.

new customer premises equipment All customer premises equipment not in service or in the inventory of a regulated telephone utility as of December 31, 1982. See also *modified final judgment* (*MFJ*).

New England Academic and Research Network (NEARnet) A New England regional Internet network and service provider that interconnects universities, laboratories, and businesses. Originally founded by Harvard University, Boston University, and Massachusetts Institute of Technology (MIT), it is currently operated by Bolt, Beranak, and Newman (BBN).

NEARnet is a member of *CIX* and offers unrestricted commercial use.

New York State Education and Research Network (NYSERNet) A New York state nonprofit organization that provides interconnectivity for universities, research centers, K-12 schools, and public libraries as well as Internet access.

newbie In bulletin board systems (BBSs) and especially the Internet, a new user of a system who has not yet learned how to access information without violating the specific network culture, ethics, or personality. Someone who disrupts network operations, not by malicious intent but through ignorance.

Criteria for being considered a *newbie* vary wildly; a person can be called a *newbie* in one newsgroup, while remaining a respected regular in another.

news hierarchy One of the seven *world newsgroups* in USENet. The categories of these newsgroups include anything about USENet itself, e.g., policies, technical aspects, conferences, announcements, information resources, and software. See also *newsgroup hierarchy.*

newsgroup A system for conducting discussions on the Internet's USENet emphasizing discussions about a specific topic. *Newsgroups* are similar to worldwide bulletin boards, that is, messages are stored in a central location (server), and those users interested in the topic of the discussion group may read and post messages without actually receiving a copy of each posting. (This is in contrast to *listservs,* which automatically send messages to all subscribers' mailboxes.)

On Internet, more than 20 000 such groups exist. These newsgroups are organized into a defined *newsgroup hierarchy.* Participation in USENet newsgroups is passive and requires special software (a *newsreader*) to be able to read and post.

newsgroup feed The delivery of USENet newsgroups to a remote computer from another.

newsgroup hierarchy On the Internet a collection of newsgroups in *USENet* which have been grouped under a common name. There are two major categories of newsgroups: the *world newsgroups* and the *alternative newsgroups.* Each of these is further organized into subgroups, which in turn are divided again into specific topics.

The *world newsgroups* are by default distributed to every USENet site (with the possible exception of the *talk* newsgroups). They are organized into the seven subcategories:

comp	Computers and computer applications.
misc	Newsgroups that don't fit into any of the other six categories.
news	Newsgroups of news of USENet itself.
rec	Newsgroups of sports, recreational, and entertainment activities.
sci	Newsgroups of the many sciences.
soc	A newsgroup dedicated to social issues and interactions.
talk	A newsgroup for discussions.

The *alternative newsgroup* hierarchies are not carried by every USENet site; only those that request these newsgroups receive them. A short list of these newsgroups includes:

alt	Contains many newsgroup subcategories covering every conceivable topic.
bionet	Biological and ecological newsgroups.
bit	Echoes of LISTSERV's most popular BITNET mailing lists.
biz	Business-related newsgroups.
clari	Newsgroups containing moderated postings of "official" news sources.
gnu	Newsgroups with discussion about gnu.
HEPnet	Nuclear physics newsgroups.
info	Mostly newsgroups with a technical side.
K12	Educational material and information.
Microsoft	One of many companies.
Relcom	Newsgroups presented in Cyrillic script. Presented mainly in Russia and Europe.
test	A newsgroup for test postings.
ca, de, it, japan, . .	Some of the many country-specific newsgroups.

There are also local newsgroups with distribution limited to a particular geographic area or domain, e.g., within a university, company, or Florida. Most newsgroups are organized as threads, that is, the initial posting followed by responses or follow-on news items. To read these newsgroup postings or to reply, a *newsreader* is required. Newsreaders such as *nn, tin,* and *trn* allow the user to group like postings so that they may be read together. See also the individual *newsgroup hierarchies.*

newsgroup kill file A *kill file* that contains the *kill* specifications (delete keys such as topics, sites, or authors) for a specific newsgroup to which a user has subscribed. The *newsgroup kill file* enables a newsgroup reader to exclude unwanted postings from a selected newsgroup.

Multiple *newsgroup kill files* may be created, one for each newsgroup to which the user has subscribed. See also *global kill file* and *kill.*

newsgroup selector The level within a newsreader program where the user chooses which newgroups are to be read. The user can subscribe to new newsgroups, add or delete newsgroups from the subscription list, and change the presentation order of subscribed newsgroups.

.newsrc The file that lists the names of the USENet newsgroups to which a user has subscribed. The file is generally found in the user's home directory.

newsreader A program that enables a user to read news and to follow and delete threads. Newsreaders are classified into one of two groups: *threaded* and *unthreaded.* An unthreaded newsreader forces the presentation to be in chronological order, while threaded newsreaders allow the presentation to be grouped by subject.

Newsreaders are available as stand-alone programs *rn, nn, tin, trn,* and *Agent,* or are built into the browser, e.g., Navigator or Mosaic. See also *nn, rn* (unthreaded) *tin,* and *trn* (threaded).

NEXT An acronym for Near End CrossTalk.

Next ID (NID) In an ARCnet frame, the address of the next node to receive the token.

NF An abbreviation of Noise Figure.

NFPA An abbreviation of the National Fire Protection Agency. A standards-setting group in the United States.

NFS An abbreviation of Network File System. A distributed file-sharing protocol suite developed by Sun Microsystems that allows computers running different operating systems to share files transparently. It employs the user datagram protocol (UDP) for data transfer.

NFT An abbreviation of Network File Transfer. It copies files between any two nodes on a network, either interactively or programmatically. It allows a user to:

- Copy a remote file.
- Translate file attributes.
- Access remote accounts.

NHOH An abbreviation of Never Heard Of Him (or Her).

NI An abbreviation of Network Interface.

nibble Obsolete. A group of 4 bits (half a byte) that are treated as a single unit. It is frequently represented by a single hex character.

NIC (1) An acronym for Internet's Network Information Center. **(2)** An acronym for Network Interface Controller (or Card).

NIC address A unique network address composed of several parts: a physical part (in hardware), network part, and a node (or station) part.

- The *hardware address* is "wired" into or set on the board. For most NICs (e.g., Ethernet and Token Ring NICs), the hardware address is assigned by the manufacturer and is "burned" into the cards ROM. The hardware address is completely independent of the network in which the board is ultimately used. The hardware address simply identifies the card's manufacturer and includes a unique serial number for that manufacturer's products. Part of this address is assigned according to guidelines specified by the IEEE.
  Some NICs addresses, ARCnet cards, for example, are set by jumpers or DIP switch settings on the board. On these cards the system administrator must set this address manually and ensure that each card in his network has a unique address.
- Each physical network in an internetwork must have a unique address; further, it can have only one *network address.* This address must be unique if the network is connected to other networks via a router, bridge, or gateway. The network address of each physical network is assigned an eight hexadecimal digit (4-byte) value between 01h and 0FFFFFFFFh.
- Each node of a network also is assigned a *node* (or *station*) *address.* A node address uniquely identifies the node within a specific network. Hence, a node that is a member of two networks will have two node addresses.
  For example, a file server that is attached to two different networks will have two different network addresses and two, possible different, node addresses. Bridges, gateways, and routers also have addresses in all connected networks.

NiCd An alkaline secondary storage battery based on a chemistry with nickel (Ni) oxide as the positive electrode, cadmium (Cd) as the negative electrode, and an alkaline electrolyte. As with all secondary batteries, *NiCd* batteries are rechargeable. In some literature, *NiCd* systems are called *nicad.* The nominal voltage per cell is 1.25 V.

nickname (1) A type of alias that allows e-mail to be addressed to a second name (in addition to the mailbox name). Obviously, some use the alias as a nickname such as "the Troll." However, it is far more useful in circumstances where spelling of the official name is difficult. For example, Lebedevitch Kszywienski might have the nickname LK. **(2)** The local area network (LAN) protocol specified in RFC-812. It requests information about a user or hostname from the Network Information Center (NIC) name database service.

NID (1) An abbreviation of Network Inward Dialing. A feature of a PABX in which an outside caller may dial an inside extension directly, without the aid of an attendant. **(2)** An abbreviation of Network Interface Device. **(3)** An abbreviation of Next ID. **(4)** An abbreviation of Network Interface Demarcation.

NIH (1) An abbreviation of Not Invented Here. Expresses the tendency for individuals, organizations, and countries to reject ideas and inventions that they did not think of. It is a major problem of working with another similar and potentially competing group. **(2)** An abbreviation of National Information Highway. A term coined by Al Gore while vice president of the United States. See also *NII.*

NII An abbreviation of the National Information Infrastructure. A futuristic public information system envisioned by the U.S. government's Advisory Council on the *NII.* It will combine the features of telephony, television, fax, electronic libraries, and public accessible databases on a seamless composite of multiple computer networks. With the system one could:

- "Attend" any school without limitations imposed by geographic distances or disabilities. Access to the best teachers would be available to all.
- "Consult" with professional specialists who could bring the latest expertise to you, again without limitations of distance.
- "Visit" museums, "attend" plays and musical productions, or "see" national landmarks, all in high-resolution video hi-fi sound.
- Live anywhere and "work" anywhere else.

The nine stated goals of NII are:

1. Promote private-sector investment.
2. Extend and expand the "universal service" concept.
3. Spur technological innovations and new applications.
4. Impel seamless, interactive, user-driven NII operations.
5. Safeguard information privacy, security, and NII network reliability.
6. Reform spectrum management.
7. Protect intellectual property.
8. Coordinate with other domestic and national governmental bodies.
9. Ensure and expand access to governmental information and procedures.

We now return you to the real world. Although technically possible, many experts disagree as to whether such a system will ever (or should ever) happen. Current information providers (CATV, RBOCs, broadcast TV, and local telephone companies) have a great deal of money invested in the current technology—and much of the investment may have to be scrapped. They view the system as a high-bandwidth broadcast medium with only control signaling in the upstream direction, not what was stated above. Political action committees are placing tremendous pressure on Congress to stress the broadcast aspects of the system. Consequently, the low and middle-income neighborhoods may be left out of the high-bandwidth delivery to the home. Also called the *information superhighway.*

NiMH An abbreviation of Nickel Metal Hydride. The designation of a particular rechargeable battery chemistry. The cells have about 1¼ times as much power as a Nickel Cadmium (NiCd) cell, and they don't have the "memory" problems of NiCds. NiMHs are not without drawbacks: they are more expensive, they do not hold a charge as long as other chemistries, and they cannot deliver current at the same rates as NiCds.

nine's complement The diminished complement of a decimal number. See also *complement.*

NIS An abbreviation of Network Information Services.

NIST An abbreviation of National Institute of Standards and Technology.

NIU An abbreviation of Network Interface Unit. See also *network interface device* (*NID*).

NJE An abbreviation of IBM's Network Job Entry communications protocol suite.

NKT An abbreviation of Nederlands Keuringsinstituut voor Telecommunicatie-apparatuur. A private laboratory for regulatory testing in Denmark.

NLA An abbreviation of Network Layer Address.

NLM An abbreviation of NetWare Loadable Module.

NLP (**1**) An abbreviation of Network Layer Protocol. (**2**) An abbreviation of Natural Language Processing.

nm The SI symbol for nanometer (10^{-9} m).

NMA An abbreviation of IBM's Network Management Architecture.

NME An abbreviation of Network Management Entity.

NMEA An abbreviation of National Marine Electronics Association. An association that defines marine electronic interface standards.

One such standard is the *NMEA 0183,* which defines the message format used to communicate among marine devices or components using an ASCII sentence library. One talker and one or more listeners communicate using a twisted pair and the EIA/RS-422 serial protocol.

nmi (**1**) An abbreviation of Non-Maskable Interrupt. (**2**) An abbreviation of Nautical MIle.

NMT An abbreviation of Nordic Mobile Telephone. A cellular telephone technology used in Eastern Europe. One of two frequency bands is used, either 450 MHz (designated *NMT 450*) or 900 MHz (*NMT 900*).

NMVT An abbreviation of Network Management Vector Transport.

nn A popular UNIX threaded newsreader. Originally written by Kim F. Storm of Texas Instruments in Denmark and released in 1989. In addition to the grouping capabilities of a threaded newsreader, *nn* provides a menu-driven *uudecode* facility that provides decryption and graphics assembly. It also provides word processing capability for replying to postings. See also *newsreader.*

NN (**1**) An abbreviation of Neighbor Notification. (**2**) An abbreviation of Neural Network.

NNI An abbreviation of Network Node Interface.

NNTP An abbreviation of Network News Transport Protocol.

NNX The old three-digit number system used to identify a specific telephone office in a particular area code in North America. *N* represents the digits 2 through 9, while *X* represents any digit 0 through 9. Because the need to distinguish between area codes and office code disappeared with the requirement that a caller always dial "1" before a long-distance call, the system has been expanded to the form NXX. See also *NXX.*

NOC An acronym for Network Operations Center.

NOD An acronym for Network Outward Dialing.

node (**1**) In networks, any device (such as a computer) connected to the network which is capable of communicating with other network devices. Each *node* has a unique address that identifies that device to all others on the network. On the Internet the term *host* is preferentially used however. (**2**) In circuits, the connection of any two or more branches. Also called a *junction point* or *nodal point.* (**3**) The termination of two or more communication links. (**4**) A technical control facility (TCF). (**5**) A point in a standing or stationary wave at which the amplitude is a minimum. Sometimes also called the *null.*

node address A unique numerical value associated with a specific node in a particular network. In general, this value is assigned to the network interface card (NIC) installed in the node.

A complete address for a node will include a *network address* that is common to all nodes in the same physical network, as well as the *node address* that is unique to the node within its physical network. Also called a *NIC address, node number, physical node address,* or a *station address.* See also *NIC address.*

node name The alphabetic name of a computer (node) on a network. In Internet *node names* are that part of an address after the "@" symbol.

node number The numerical address that distinguishes one node from all others on a particular network.

Node-to-Node Routing A routing method used to deliver a packet from its source node to its destination, as opposed to simply direct a packet to the router nearest the destination node.

NodeRunner™ An Artisoft software configurable Ethernet adapter card.

noise An unwanted disturbance superimposed on, and in the band of interest of, a useful signal which tends to obscure its information content.

Any undesired signal or disturbance on a communication path or device that interferes with proper operation. These disturbances convey no information and can be of any origin, natural or man-made. The disturbances may interfere with the flow of information or meaningful signals, commonly called *transmission errors*. Noise is classified in many ways; a few of the many possible examples include:

Its Source
- *Crosstalk*—interference on one wire from another circuit.
- *Electrical noise*—any unwanted electrical energy, other than crosstalk.
- *Intermodulation noise*—spurious or false frequencies that are products of the input signal transmitted through a nonlinear circuit. Typically, the spurious frequencies are the sum and difference of input frequencies.
- *Intrinsic noise*—noise inherent in a device, medium, or communications path. It is not dependent on modulation or coupling to external noise sources.
- *Natural noise*—noise from natural causes such as lightning, the cosmos, or thermal effects at the atomic level (Johnson noise). It does not include noise from man-made equipment.
- *Man-made noise*—noise from electric motors, neon or florescent lights, and power lines. The tendency is a decrease in amplitude as frequency increases.

Its Band of Energy
- *Microwave noise*—electromagnetic energy in the microwave spectral range.
- *Optical noise*—electromagnetic disturbances in the spectral range from infrared to ultraviolet. The noise interferes directly with optical processing or optical to electrical conversion process. See also *modal noise* and *mode partition noise*.
- *Acoustic noise*—an undesired *acoustic wave* or pressure wave disturbance.

Its Character
- *Random noise*—that part of *noise* that is unpredictable except in a statistical sense. Also called *gaussian, Johnson, thermal,* or *white noise.*
- *Pink noise*—noise with constant power per octave.
- *White noise*—noise with constant power spectral density over frequency.
- *Black noise*—noise characterized by a frequency spectrum of predominately zero power level over all frequencies except for a few narrow bands or spikes.
- *Blue noise*—noise over a specified frequency range in which the spectral density (power per hertz) is proportional to the frequency, not flat as in white noise.
- *Impulse noise*—intermittent, short-duration bursts that contain energy over a very large portion of the spectrum. In the time domain, it is characterized by short duration and high-amplitude pulses.

- *Quantization noise*—noise generated in an analog-to-digital conversion process. It is due to the mapping of the arbitrary values possible in the analog domain into the fixed amplitude values of the digital domain.
- *Phase noise*—generally short-term frequency variations of a signal.

noise bandwidth The bandwidth of an equivalent ideal rectangular filter that transmits the same noise power as the system being measured.

noise-equivalent power (NEP) (1) In fiber optics, the radiant power (at a given wavelength, modulation frequency, and effective noise bandwidth) that produces a signal-to-noise ratio (SNR) of 1 at the output of a given detector i.e., the amount of optical power needed to produce an electronic signal as strong as the receiver's base noise level.

(2) Some manufacturers/authors define *NEP* as the minimum detectable power per square root bandwidth. When defined this way, *NEP* has the units of *watts per (hertz)$^{1/2}$*. This, however, is dimensionally incorrect since the units of power are watts. **(3)** Other manufacturers define *NEP* as the radiant power that produces a signal-to-dark current noise ratio of unity. This measurement is valid only if the dark current noise dominates the noise level.

noise figure (NF) At a selected frequency, the ratio of the total power delivered into a load by a system to the apparent noise introduced at the input (corrected for the gain of the system). The *noise figure* is thus the ratio of actual output noise (from all sources) to that which would remain if the device itself did not introduce noise. In equation form, *noise figure* is expressed as:

$$\overline{F}(f) = \frac{P_{out}}{GP_{in}}$$

where:

f is the frequency at which the *noise figure* is to be taken.
$\overline{F}(f)$ is the noise figure at frequency f.
P_{out} is the power to the load at frequency f and 290 kelvins.
P_{in} is the power to the system input at frequency f and 290 kelvins.
G is the system gain at frequency f.

Also called *noise factor*.

noise filter A passive or active, linear or nonlinear circuit or device intended to reduce either the noise level generated by a transmitter or the amount of noise detected in a receiver.

noise floor The minimum noise level with a specified set of measurement parameters. The parameters may specify no signal, a tone at a specified frequency and level, or some other operating condition.

noise level The noise power. It may be expressed as an absolute power (generally in picowatts, pW) or as a ratio to a specified reference. When stated as a ratio, it is usually expressed in decibels (dB).

A suffix is added to denote a particular reference base, bandwidth, or specific qualities of the measurement. Examples of *noise level* measurement units are:
- Absolute power wideband pW.
- Absolute power band limited pWp and pWp0.
- As a ratio wideband dBa, dBa0, dBm, dBrn, and dBm0.
- As a band limited ratio dBa(F1A), dBa(HA1), dBm(psoph), dBm0P, dBrnC, dBrn(f_1-f_2), dBrn(144-line).

noise power (1) The power generated by a random electromagnetic process. **(2)** Interfering and unwanted power in an electrical device or system.

noise-power density (NPD) Thermal noise power expressed in a 1-Hz bandwidth, i.e., the noise power per hertz at a specified frequency in a noise spectrum.

The *noise power density* of the internal noise that is contributed by a receiving system to an incoming signal is expressed as the product of Boltzmann's constant (k) and the equivalent noise temperature (T). Thus, the noise power density is often expressed simply as kT. See also *Johnson noise* and *thermal noise.*

noise suppresser Filtering in a system that is intended to automatically reduce the amount of noise generated or received.

noise suppression **(1)** Reduction of the noise power level in electrical circuits. **(2)** The process of automatically reducing the noise output of a receiver during periods when no carrier is being received. See also *squelch.*

noise temperature At a pair of terminals and at a specific frequency, the temperature of a passive system (e.g., a resistor) having an available noise power per unit bandwidth equal to that of the actual system or device. Therefore, the *noise temperature* of a resistor is the actual temperature of the resistor. The *noise temperature* of a diode, however, may be many times the observed actual temperature.

The standard reference temperature for noise measurements is 290°K.

noise transmission impairment (NTI) The necessary increase in distortionless transmission loss that impairs a noiseless telephonic transmission by an amount equal to the impairment caused by the noise.

noise weighting In telephony, a method of assigning a specific value to the amount of signal impairment due to the noise encountered in transmission. The measure is an approximation to the effect noise will have on an "average" listener using a particular class telephone instrument.

The noise-measuring set has a specific amplitude vs. frequency response characteristic (filter) that permits the set to give the numerical readings. The most widely used noise weightings were established by agencies concerned with public telephone service. They are based on characteristics of specific commercial telephone instruments. The two most common weighting methods used are *C-message* and *psophometric.* See also *C-message weighting* and *psophometric weighting.*

noise window A notch or dip in the noise frequency spectrum characteristic of a device, such as a transmitter, receiver, channel, or amplifier, from external sources or internal sources. The *noise window* is usually represented as a band of lower amplitude noise embedded in a wider band of higher amplitude noise.

For example, by worldwide agreement certain radio frequencies are not used by anybody on the Earth for transmissions. Instead, they are reserved strictly for astronomical receivers. This lack of use creates a *noise window* through which astronomers may observe the stars in that portion of the electromagnetic spectrum.

nom de ligne The name a user employs on an online computer service. It is typically not the user's real name; hence, it provides a degree of anonymity. Also called a *handle.*

nominal bandwidth The widest band of frequencies, including any guard bands, assigned to a channel.

Contrast *nominal bandwidth* with the terms necessary bandwidth and occupied bandwidth.

nominal bit stuffing rate The rate at which synchronization bits are inserted and/or deleted in a transmission system when both the transmitter and receiver are operating at their nominal speeds.

nominal velocity of propagation (NVP) In a transmission system, the value indicating the signal speed, as a proportion of the maximum speed theoretically possible (the speed of light). This value varies with the medium through which the electromagnetic wave is traveling. Values for electrically based local area network cables range from about 60 to 85% of the speed of light.

On optical fibers, the *NVP* is reciprocal of the index of refraction (n), that is,

$$NVP = 100/n\%$$

Also called *propagation velocity* or *velocity of propagation (VOP).*

nonblocking In communications, a switching network in which there exists at least one path over which two idle lines may be connected, regardless of the number of lines already in service.

In the switching system diagrammed here, there are more output ports than input ports (b ≥ 0). Since every input request can be connected to an output, the switching system is said to be *nonblocking.* Also referred to as *full availability.* See also *blocking network.*

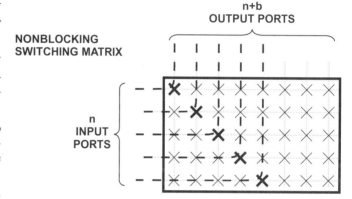

nonconsecutive hunting In telephony, a hunt group in which the incoming telephone numbers are not in sequence. The switching equipment is programmed to search a specific set of line or trunk numbers in a specified order in order to find an available (nonbusy) line or trunk. Also called *jump hunting.*

nondata bit A bit in a data stream with a format encoding violation that is used for special control purposes. For example, zero suppression waveform encoding schemes use violations to prevent long periods with no transitions (only logical ones have transitions).

nondedicated server A network node that can process user applications concurrent with network maintenance applications.

nondeterministic A term that refers to the inability to predict the performance or outcome of a system's response to stimuli, except perhaps in a statistical sense. For example, on a busy carrier sense multiple access network, the message delay cannot be calculated for an arbitrary message.

nondisruptive test A diagnostic test or performance test that can be run in the background and that has little or no effect on ordinary network activity.

nonerasable memory A synonym for *read only memory (ROM).*

nonfolded network A switching network in which both input and output ports are provided and a path may be established between any input port and any output port.

The following figure is that of a *full availability* rectangular *nonfolded* switch network (if $m \geq n$ the network is also *nonblocking*). See also *availability, blocking network, crosspoint, folded network,* and *nonblocking network.*

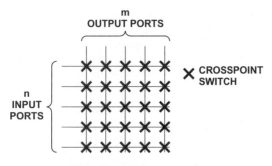

nonintelligible crosstalk A subjective description of crosstalk in which the level is sufficiently low that a listener cannot understand it; however, the listener feels it could be. This type of crosstalk is generally more objectionable than higher levels of crosstalk that approximate white noise.

noninteractive Operations or actions of a network access are controlled by a program at the user location rather than by human intervention. For example, a long text file may be transferred to a computer file much faster than can be read interactively. The user may disconnect and scrutinize the file without concern of elapsed time. See also *batch* and *interactive*.

nonlinear (1) In mathematics. System or device behavior in which output changes are not proportional to input changes; that is, the changes do not follow the first-order polynomial $f(x) = a x + b$. (2) Slang. Hardware or software behaving in an erratic and unpredictable fashion; unstable. It may suggest that the entity is being forced to run far outside of design specifications. (3) Slang. When describing the behavior of a person, suggests a reaction far beyond the behavior of an average person (e.g., a tantrum or tends to "blow-up").

nonlinear distortion Signal distortions which are dependent on the transmitted signal level, that is, the output signal does not have a linear relationship with the input signal, e.g., *harmonic distortion*. See also *distortion*.

nonlinear scattering Direct conversion of a photon from one wavelength to one or more other wavelengths. In an optical fiber, nonlinear scattering is usually not important below the threshold irradiance for stimulated nonlinear scattering. Examples of nonlinear scattering include Raman and Brillouin scattering.

nonpersistent In local area networking, a carrier sense multiple access (CSMA) LAN in which the nodes involved in a collision do not immediately attempt to retransmit but wait some "random" period before retransmission. See also *CSMA/CD*.

nonrepudiation A network security measure that makes it impossible for a sender to deny having sent a message (*origin nonrepudiation*) and for a recipient to deny having received the message (*destination nonrepudiation*).

nonresonant antenna A synonym for *aperiodic antenna*.

nonreturn to zero (NRZ) A code where the value of the bit is held constant over the entire bit period. The upper trace is the "normal" *NRZ* format of a serial bit stream, whereas the second trace is that of a bipolar *NRZ* stream.

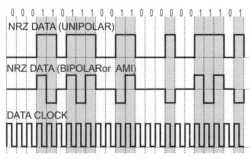

NRZ coding may code logical ones (marks) as the active high state (NRZ-M), or it may code logical zeros (spaces) as the active state (NRZ-S). NRZ-S is the inverse of NRZ-M. See also *return to zero (RZ)* and *waveform codes.*

nonreturn to zero invert (NRZ-I) See *nonreturn to zero mark (NRZ-M)*.

nonreturn to zero mark (NRZ-M) A binary waveform coding method in which a logical one is a transition in the line signaling element and a logical zero is without transition. Also called *nonreturn to zero invert (NRZ-I), nonreturn to zero one (NRZ-1), nonreturn to zero mast, conditioned baseband representation, differentially encoded baseband,* and *NRZ-B*. See also *nonreturn to zero (NRZ), nonreturn to zero space (NRZ-S),* and *waveform codes.*

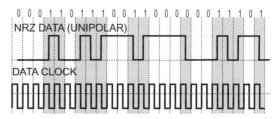

nonreturn to zero on ones (NRZ-1) See *nonreturn to zero mark (NRZ-M)*.

nonreturn to zero space (NRZ-S) A binary waveform coding method in which a logical zero is a transition in the line signaling element and a logical one is without transition. See also *nonreturn to zero (NRZ), nonreturn to zero invert (NRZ-M),* and *waveform codes.*

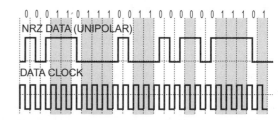

nonroutable protocol Any local area network (LAN) protocol, such as IBM's NetBIOS LAN server and System Network Architecture (SNA), that use names rather than network layer addresses to identify devices and therefore supply no intrinsic routing information. Internetworking devices must therefore find other means to route traffic in networks that use these protocols.

nonshareable A file, device, process, or other resource that is available to only one user at a time.

nonshifted fiber A synonym for *dispersion unshifted fiber.*

nonsynchronous Meaning not synchronous; see *asynchronous.*

nontransparent mode A transmission method in which control characters and sequences are found through the examination of all characters passing through the device. All such control characters are interpreted and used by the system, rather than simply as data to be passed on. The scheme is found mainly in Bi-Sync transmissions. See also *transparent mode.*

nonvolatile memory A storage device or system that does not lose information when power is removed from it. See also *EPROM, flash memory, NVRAM,* and *ROM.*

NOR A contraction of <u>N</u>ot <u>OR</u>. A Boolean operator composed of the logical function OR and the negation (NOT).

The truth table shows all possible combinations of inputs and output for a two-input *NOR* operation. A false input or output is tabulated as a "0," while the true is shown as a "1." See also *AND, AND gate, exclusive OR gate, NOT, OR, OR gate,* and *exclusive OR.*

NOR TRUTH TABLE

A	B	$\overline{A + B}$
0	0	1
0	1	0
1	0	0
1	1	0

NOR gate A type of digital circuit that performs a *NOR* operation on two or more digital signals. The symbol for a two-input *NOR gate* is:

See also *AND, exclusive OR, logic symbol, NOR, NOR gate, NOT,* and *OR*.

NORDUnet The Scandinavian countries' regional high-speed backbone. Founded in 1987, NORDUnet emphasizes research and education. National networks linked to *NORDUnet* include:

DENet	Denmark
FUNET	Finland
SUNET	Sweden
SURIS	Iceland
UNINETT	Norway

normal **(1)** In optics, an axis that forms a right angle with a surface. The *normal* is used as the reference for determining the angles of incident, reflected and refracted rays. **(2)** The state of a material when it is not subjected to a changing stimulus. For example:

- In superconductive materials (such as lead, mercury, . . .) the state the material is in when it does not exhibit superconductivity.
- In birefringent materials (such as are used in Kerr cells), the state the material is in when it does not exibit birefringence.

(3) The state of being within specified acceptable values or limits.

normal connection mode A communication mode where the modems communicate using no hardware error correcting protocol. Error control is the responsibility of the DTE software.

normal distribution See *Gaussian distribution*.

normal glow discharge In gas tubes (e.g., surge protectors), the glow discharge characterized by the fact that the voltage that remains across the device remains constant or decreases as current increases.

normal mode A transmission mode using two signal conductors in which the signal is applied differentially to the conductors. In telephony, also called the *metallic mode*. See also *common mode* and *differential*.

normal mode noise Noise that appears differentially between two signal wires and that acts on the signal receiver in the same manner as the desired signal. It may be caused in several ways:

- Electrostatic induction and differences in the distributed capacitance between each signal wire and its environment.
- Electromagnetic induction and differences in the coupling of magnetic fields to each conductor.
- Intrinsic noise in junctions.
- Thermal potentials generated by the use of dissimilar materials in the connective structure.
- Common mode to normal mode conversion.

normal response mode (NRM) A high-level data link control (HDLC) mode for use on links with one primary and one or more secondaries. In such a situation, a secondary can transmit only when it receives a poll addressed to it from the primary. It may then transmit a number of responses. Upon completion, it sets the F-bit in a response and waits for another poll.

normalize **(1)** In circuits and systems, the ratio of a parameter or quantity to a like reference parameter or quantity, thereby making the result dimensionless. **(2)** In mathematics, to adjust the characteristic and fraction of a number so as to eliminate leading zeros. **(3)** In mathematics, to adjust the fixed point part (mantissa) and exponent of a number so that the mantissa lies within a specified range. **(4)** In instrumentation, to adjust a measured parameter so that its value falls within an acceptable range of a measuring device.

normalized average transfer delay At a particular signaling rate, the ratio of the average transfer delay to the packet transmission time.

normalized frequency (V) **(1)** In fiber optics, a dimensionless quantity (denoted by *V*) and given by

$$V = \frac{2\pi a}{\lambda} \sqrt{n_1^2 - n_2^2}$$

where:

- a is the core radius,
- λ is the wavelength in vacuum,
- n_1 is the maximum refractive index of the core, and
- n_2 is the refractive index of the homogeneous cladding.

For single-mode operation, $V < 2.405$. Also called *V number*. **(2)** The ratio between an actual frequency and a reference value (or nominal frequency).

normalized impedance The ratio of a measured impedance to a reference impedance.

For transmission lines, the reference is generally the characteristic cable impedance.

normalized network throughput At a particular signaling rate, the ratio of the throughput in packets per second to the maximum throughput possible.

Note: The ratio may exceed one.

normalized offered traffic At a particular signaling rate, the ratio of the average number of attempted packet transmissions to the average number of packet transmissions per second possible.

normally closed contact A current-carrying element of a switching device that is engaged and carrying current when the device is in the "normal" mode or position. Such devices include toggle switches and relays.

normally open contact A current-carrying element of a switching device that is not engaged and is not carrying current when the device is in the "normal" mode or position. Such devices include toggle switches and relays.

North American Numbering Plan (NANP) Assigns area codes and set rules for calls to be routed in the World Numbering Zone 1 (essentially North America). It is administered by BellCore. The plan has three parts: a three-digit area code, a three-digit exchange (or office) number, and a four-digit line (subscriber) number.

- *Area code:* Prior to January 1, 1995, the area code was of the form NPX, where N is 2–9, P is 0 or 1, and X is 0–9. After this date, the form became NXX. This increased the number of available area codes from 152 to 792. Several codes are reserved for special purposes:

Code	Function
200	Reserved for special services
211	Assigned to local operators
300	Assigned to special services
311	Reserved for special local services
400	Reserved for special services
500	Reserved for special services
511	Reserved for special local services
600	Reserved for special services
700	Assigned special access code for interLATA carriers/resellers
711	Reserved for special local services

- *Office code*—a three-digit number of the form NXX which identifies a specific central office.
- *Line code (station code)*—a four-digit number of the form XXXX which identifies a specific subscriber line number.

The current numbering plan supports more than 6 000 000 000 terminations.

NorthWestNet A network operated by the nonprofit Northwest Academic Computing Consortium, Inc. It is a member of Commercial Internet Exchange (CIX) and a regional component of the NSFNet network. Its members include universities, colleges, K-12 grade schools, hospitals, businesses, libraries, and state agencies from the northwestern region of the United States (including Alaska, Idaho, North Dakota, Oregon, and Washington).

Norton's theorem A circuit principle that states that any linear, time-invariant, complex circuit containing only resistive elements, voltage sources, and current sources can be reduced to a single equivalent current source and an equivalent shunt resistance. See also *Thevenin's theorem.*

NOS An acronym for Network Operating System.

NOT A logical operator that performs Boolean negate. The rule for the *NOT* operation is: *If the input is true, the output is NOT true (false)*; that is, NOT true = false and NOT false = true. The diagrammatic symbol for the *NOT* function is "**O**."

The written symbol for the *NOT* function is a minus ("−") or a bar over the letter. Therefore, the *NOT* of input A is written A−, ~A, A*, A′, or $\overline{A}$. The symbol "**O**" is never used alone but is combined with other logic symbols to form other common *gates*. Some examples are shown in the figures. See also *AND, AND gate, exclusive OR, exclusive OR gate, logic symbol, NAND, NAND gate, OR,* and *OR gate.*

NOTA An acronym for Novell Open Technology Association. A group of companies that have adopted a software platform suggested by Novell that will lead to the development of a new range of computer telephone interface (CTI) applications.

notarization In network security, the use of a trusted third party, called a notary, to verify that a communication between two entities is legitimate. The "notary" has information that is used to verify the identity of the sender and receiver and also of the time and origin of a message.

notch filter A *band reject filter* designed to reject a very narrow range of frequencies or even a single frequency. See also *band-pass filter, band reject filter, high-pass filter,* and *low-pass filter.*

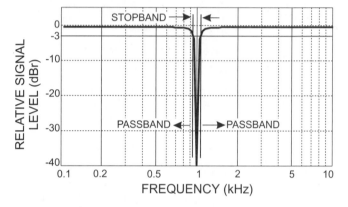

notched noise Broadband noise from which a narrow band of frequencies has been removed. *Notched noise* is generally used for testing devices or circuits.

notwork A slang term used to describe a network that is operating unreliably or not at all.

NoXQS An acronym for No eXQuseS (excuses).

Np An abbreviation of *neper.*

NPA An abbreviation of Number Plan Area.

NPAP An abbreviation of Network Printing Alliance Protocol. A bidirectional protocol to be used for communication among network printers. It allows the exchange of configuration and other data (independent of the printer control or page descriptor language being used).

NPD (1) An abbreviation of Network Protection Device. (2) An abbreviation of Noise Power Density.

NPDU An abbreviation of Network Protocol Data Unit.

NPI An abbreviation of Number Plan Indicator. A field in a message directory number (DN) to specify local, national, or international ISDN calls.

NPSI An abbreviation of Network Packet Switch Interface. An interface used in IBM's Systems Network Architecture (SNA).

NQS An abbreviation of Network Queuing System.

NR An abbreviation of Number Received. A control field sequence that informs the sender the number sent (NS) that the receiver expects in the next frame.

NRAM An acronym for Nonvolatile RAM (pronounced "EN-ram"). See *NVRAM.*

NREN An acronym for the National Research and Education Network.

NRM An abbreviation of Normal Response Mode.

NRN An abbreviation of <u>N</u>o <u>R</u>eply <u>N</u>ecessary. An e-mail convention intended to prevent endless message replies and counter replies.

NRZ An abbreviation of <u>N</u>on<u>R</u>eturn to <u>Z</u>ero. See also *waveform codes.*

NRZ-I An abbreviation of <u>N</u>on<u>R</u>eturn to <u>Z</u>ero Invert.

NRZ-L An abbreviation of <u>N</u>on<u>R</u>eturn to <u>Z</u>ero <u>L</u>evel. The same as *NRZ.*

NRZ-M An abbreviation of <u>N</u>on<u>R</u>eturn to <u>Z</u>ero <u>M</u>ark. See also *waveform codes.*

NRZ-S An abbreviation of <u>N</u>on<u>R</u>eturn to <u>Z</u>ero <u>S</u>pace.

ns A symbol for *nanosecond.* A measure of time (10^{-9} seconds).

NS An abbreviation of <u>N</u>umber <u>S</u>ent. The sequence number of the frame.

NSA (1) An abbreviation of the <u>N</u>ational <u>S</u>ecurity <u>A</u>gency. (2) An abbreviation of <u>N</u>ext <u>S</u>tation <u>A</u>ddressing. In FDDI an addressing mode by which a station can send a packet, or frame, to the next station in the ring, without knowing that station's address. (3) An abbreviation of <u>N</u>on-<u>S</u>ervice <u>A</u>ffecting. A fault that does not interrupt transmission.

NSA bait Any word, phrase, or article that is suggestive of terrorist activities.

A term that refers to the false premise that the U.S. National Security Agency (NSA) uses a program (*NSA line eater*) to scan all USENet postings for keywords (such as, AK-47, plastic explosive, overthrow government, etc.) indicating seditious activities.

NSA line eater The infamous, but probably nonexistent, National Security Agency (NSA) trawling program that is assumed to be watching the net for the U.S. government's spooks. Most users describe it as a mythical beast, a few believe it actually exists, more aren't sure, and many believe in acting as though it exists—just in case.

NSAP An acronym for <u>N</u>etwork <u>S</u>ervice <u>A</u>ccess <u>P</u>oint. In the ISO/OSI Reference Model, the location through which a transport layer entity can gain access to network layer services. Each *NSAP* has a unique OSI network address.

NSDU An abbreviation of <u>N</u>etwork <u>S</u>ervice <u>D</u>ata <u>U</u>nit.

NSF (1) An abbreviation of the <u>N</u>ational <u>S</u>cience <u>F</u>oundation. (2) An abbreviation of <u>N</u>orges <u>S</u>tandardiserings<u>f</u>orbund, the national standards organization of Norway.

NSFNet An acronym for the <u>N</u>ational <u>S</u>cience <u>F</u>oundation <u>Net</u>-work. A high-speed network providing one of the backbones of the Internet and linking users with supercomputer sites around the country. The *NSFNet* is to be superseded by the National Research and Education Network (NREN).

NSI An abbreviation of <u>N</u>ASA <u>S</u>cience <u>I</u>nternet.

NSN An abbreviation of <u>N</u>ASA <u>S</u>cience <u>N</u>etwork.

NT1 An abbreviation of <u>N</u>etwork <u>T</u>ermination <u>1</u>. An ISDN device (located at the customer premises) which provides the OSI layer 1 services. These include the physical and electrical termination of the transmission loop (U interface), line code conversion (2B1Q or 4B3T into the S interface format—ITU-T I.430), line monitoring, timing, and power extraction. See also *ISDN.*

NT2 An abbreviation of <u>N</u>etwork <u>T</u>ermination <u>2</u>. An ISDN device similar to the NT1 except it can accept several S interface lines and it does the switching and concentration functions equivalent to layers 1–3 of the OSI reference model. *NT2* equipment can include PABXs, LANs, multiplexes, concentrators, etc. See also *ISDN.*

NTFS An abbreviation of Windows <u>NT</u> <u>F</u>ile <u>S</u>ystem.

NTI (1) An abbreviation of <u>N</u>oise <u>T</u>ransmission <u>I</u>mpairment. (2) An abbreviation of <u>N</u>etwork <u>T</u>erminating <u>I</u>nterface.

NTIA An abbreviation of the <u>N</u>ational <u>T</u>elecommunications and <u>In</u>formation <u>A</u>dministration (U.S. Department of Commerce).

NTM An abbreviation of <u>NT</u> <u>T</u>est <u>M</u>ode. A control bit in the ISDN basic rate interface (BRI).

NTN An abbreviation of <u>N</u>etwork <u>T</u>erminal <u>N</u>umber.

NTP An abbreviation of <u>N</u>etwork <u>T</u>ime <u>P</u>rotocol.

NTSC An abbreviation of <u>N</u>ational <u>T</u>elevision <u>S</u>tandards <u>C</u>ommittee.

NTSC standard An abbreviation of <u>N</u>ational <u>T</u>elevision <u>S</u>tandards <u>C</u>ommittee <u>standard</u>. The North American (Central America, a few South American countries, and some Asian countries, including Japan) 525-line interlaced raster-scanned video standard for the generation, transmission, and reception of television signals. The standard specifies:

- Picture information is transmitted in vestigial-sideband AM and
- Sound information is transmitted in FM.

See also *PAL, PAL-M, and SECAM.*

NTT The abbreviation of the <u>N</u>ippon <u>T</u>elephone & <u>T</u>elegraph. The Japanese domestic telephone company.

NTU An abbreviation of <u>N</u>etwork <u>T</u>erminating <u>U</u>nit.

NuBus A high-speed bus (pronounced "new-bus") used in Apple Computer's Macintosh-II family of computers. *NuBus* is designed such that a card may be placed into any slot without priority conflicts between cards.

nude The state of a computer which is delivered without an operating system.

NUI (1) An acronym for <u>N</u>etwork <u>U</u>ser <u>I</u>nterface. (2) An acronym for <u>N</u>etwork <u>U</u>ser <u>I</u>dentifier. A unique alphanumeric key provided to dial-up users to identify them to packet switch networks worldwide. The number allows the user to gain network access and provides a means for billing. *NUI* replaces the Network Terminal Number (NTN) in newer networks.

nul A character code <NUL> with a *null* value, i.e., a character meaning "nothing." Although the character is real (an ASCII 00h for example) and can be transmitted and received, it conveys no data. It is used in synchronous systems when nothing else is being transmitted so that the receiving data communications equipment (DCE) can maintain synchronism with the transmitting DCE's clock. A *nul* is also used to fill out the unused portion of a packet of information so that a minimum or fixed length can be achieved.

nul character See *nul.*

null (1) In an antenna, a zone in a radiation pattern in which the effective radiated power is at a minimum relative to the maximum effective radiated power of the main beam.

The null generally has a very narrow directivity angle compared to that of the main beam. Thus, the null is useful for radio navigation. Because there is reciprocity between the transmitting and receiving characteristics of an antenna, there will be corresponding nulls for both the transmitting and receiving functions. These nulls are useful in the suppression of interfering signals in a given direction. See also *lobe* and *main beam.* (2) In database management systems, a special value assigned to a row or a column indicating either unknown values or inappropriate usage. (3) In signal measurement systems, the condition of zero error-signal achieved by equality at a summing junction between a signal input and an automatically or manually adjusted balancing signal in which the phase is opposite to the input signal. **(4)** A dummy letter, letter symbol, or code group inserted in an encrypted message to delay or prevent its solution, or to complete encrypted groups for transmission security purposes.

null modem A cable or device that allows serial data communication between two DTEs using standard communication software, but, without a modem. The *null modem* accomplishes this by attempting to look like a modem to each of the DTEs. The cable (or device) is wired so that the transmit of each DTE is the receive of the other DTE.

Because the Electronics Industries Association (EIA) has not defined the interface between a DTE and another DTE, several variations of interconnect cables have developed. All of the variations are serial cables with socket-type connectors on both ends and with the serial data pins *cross coupled*. (That is, the transmit pin of one connector is crossed to the receive pin of the opposite connector). The difference between the variants is in the way the hardware handshaking signals

(RTS, CTS, DSR, DTR, and CD) are utilized. All variants attempt to "fake" the DTE into thinking a DCE is connected to it. Each of the approaches will work with some file transfer software and not others. In general it is sufficient to loop the "request" signals to the "clear" signals at each DTE independently. However, several forms of cross connected hardware handshake are shown. A null modem cable is also called a *cross pinned cable* or *crossover cable.*

Number Administration and Service Center (NASC) An agency that provides centralized reservation, registration, and administration of the Service Management System (SMS) database of all North American 800 numbers for all carriers. They keep track of the ownership and availability of all 800 numbers.

number plan area (NPA) The official name for area code. See also *area code* and *North American Numbering Plan* (*NANP*).

number portability In telephony, a scheme that allows customers to retain their existing telephone numbers when switching from one carrier to another.

SOME NULL MODEM VARIANTS

BASIC NULL MODEM
Only the serial data lines, ground, and possibly the shield are connected from end to end. All hardware control signals are ignored.
Null modems of this type will work correctly only when neither DTE uses hardware flow control. Software flow control (XON/XOFF protocol) or packet-based file transfer protocols such as Kermit, Xmodem, or Ymodem will still function correctly.

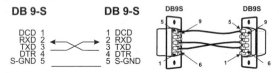

Variant 1
As in the basic *null modem*, the serial data lines are "cross coupled." In addition, each DTE provides its own CTS and CD signals from the RTS line. Similarly, the DTR is provided by the DSR.
As with the basic null modem cable, all combinations of terminating connectors are possible. Pin numbers in parentheses are for a DB9-S connector.

Variant 2
As in the basic *null modem*, the serial data lines are "cross coupled." Each DTE provides its own CTS signal from the RTS line while the DTR line provides the DSR and CD signals.
As with the basic null modem cable, all combinations of terminating connectors are possible. Pin numbers in parentheses are for a DB9-S connector.

Variant 3
As in the basic *null modem,* the serial data lines are "cross coupled." Further, RTS is "cross coupled" to CTS of the opposite end and DTR is "cross coupled" to DSR.
As with the basic null modem cable, all combinations of terminating connectors are possible. Pin numbers in parentheses are for a DB9-S connector.

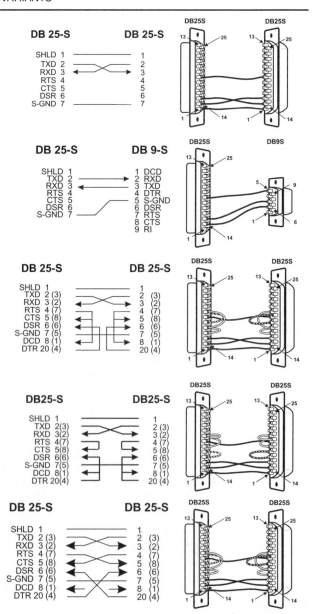

SOME NULL MODEM VARIANTS (*CONTINUED*)

Variant 4

In this variant, not only are the serial data lines "cross coupled," but the DTR line is cross coupled to the CTS line.

As with the basic null modem cable, all combinations of terminating connectors are possible. Pin numbers in parentheses are for a DB9-S connector.

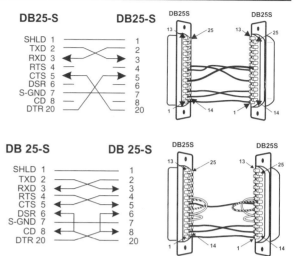

Variant 5

As with all null modems, the serial data lines are "cross coupled." So too are the CTS and RTS crossed as are the DTR to CD/DSR.

As with the basic null modem cable, all combinations of terminating connectors are possible. Pin numbers in parentheses are for a DB9-S connector.

---

numerical aperture (NA) **(1)** The sine of the vertex angle of the largest cone of *meridional rays* that can enter or leave an optical system, device, or element, multiplied by the index of refraction of the medium in which the vertex is located. Generally, this is measured with respect to an object or image point as *NA* varies as that point moves. **(2)** For an optical fiber in which the index decreases monotonically from the core axis to the cladding, *NA* is given by

$$NA = \sqrt{n_1^2 - n_2^2}$$

(3) A frequently encountered but imprecise approximation in fiber optic systems: NA is the sine of the acceptance angle times the index of the material in contact with the fiber core's face.

numerical aperture loss A loss of optical power that occurs at a splice or a pair of mated connectors when the numerical aperture of the *transmitting* fiber exceeds that of the *receiving* fiber, even if the cores are precisely the same diameter and are perfectly aligned. That is, the higher numerical aperture of the transmitting fiber means that it emits a larger cone of light than the receiving fiber is capable of accepting, resulting in a coupling loss.

In cases where the transmitting fiber has a lower numerical aperture, no *numerical aperture loss* occurs because the receiving fiber is capable of accepting light from any bound mode of the transmitting fiber.

numerical assignment module (NAM) The device in a cellular telephone that contains the telephone number(s).

NUT An acronym for Number Unattainable Tone.

NVE An abbreviation of Network Visible Entity. A resource that can be addressed through an AppleTalk network. It is identified by name, type, and zone. The entity type specifies the generic class (such as LaserWriter or AFPServer) to which the resource belongs. Apple maintains a registry of entity types.

NVIS An abbreviation of Near Vertical Incidence Skywave.

NVP An abbreviation of Nominal Velocity of Propagation.

NVRAM An acronym for NonVolatile RAM (Pronounced "EN-VEE-ram"). A memory device or system in which stored data remain undisturbed by the removal of operating power. An *NVRAM* may be made by using either a special memory semiconductor (called an EEPROM) that will retain information after operating power is removed or by providing a battery backup on a standard low-power RAM memory. The advantage of EEPROMs is their ability to store information without any power. The disadvantage is their finite life.

(An EEPROM can only be written to about 50 000 to 100 000 times before it is likely to fail). The advantage of battery-backed RAM if the very long life of the semiconductor and the disadvantage is the relatively short life of the battery.

NVSM An abbreviation of NonVolatile Semiconductor Memory.

NVT An abbreviation of Network Virtual Terminal.

NWD An abbreviation of Network Wide Directory.

NXX The three-digit number used to identify a specific telephone office in a particular area code in North America. *N* represents the digits 2 through 9, while *X* represents any digit 0 through 9. See also *North American Numbering Plan* (NANP).

nybble An alternate but less frequently used spelling of *nibble*.

nyetwork Slang for a nonfunctioning network. (From the Russian word *nyet*).

NYSERNet An abbreviation of New York State Educational and Research Network.

Nyquist interval The maximum time interval between equally spaced samples of an analog signal that will enable the signal waveform to be completely determined.

The *Nyquist interval* is equal to the reciprocal of twice the highest frequency component of the sampled signal. In practice, when analog signals are sampled for the purpose of digital transmission or other processing, the sampling rate must be more frequent than that defined by Nyquist's theorem because of quantization error introduced by the digitizing process. The actual required sampling rate is therefore determined not only by the highest frequency in the signal but also by the accuracy of the digitizing process. Also called the *Nyquist sampling interval*.

Nyquist rate The reciprocal of the interval defined by the *Nyquist interval*. The minimum sampling frequency of an analog waveform is twice the frequency of the highest component in the waveform. In practice, the rate must be somewhat higher due to quantization errors introduced by the sampling process.

Nyquist theorem Named after Harry Nyquist (1889–1976), the theorem defines the maximum time interval (called the Nyquist or sampling interval) which can be given to regularly spaced, instantaneous samples of a complex analog wave of bandwidth B_w for complete determination of the waveform as $1/(2\,B_w)$ seconds. Stated another way, the sampling frequency must be more than twice the highest frequency component of the signal. Also called the *sampling theorem*.

O

O&M An abbreviation of <u>O</u>perations <u>and</u> <u>M</u>aintenance.

O.n recommendations An ITU-T series of specifications concerning "Specifications of measuring equipment."

OA An abbreviation of <u>O</u>ptical <u>A</u>mplifier.

OAI An abbreviation of <u>O</u>pen <u>A</u>pplication <u>I</u>nterface.

OAM An abbreviation of <u>O</u>perations, <u>A</u>dministration, and <u>M</u>aintenance.

OAS An abbreviation of <u>O</u>ffice <u>A</u>utomation <u>S</u>ystem.

object (1) In optics, an entity seen through (or imaged by) an optical system. It may be the real or virtual image formed by another optical system. An image may contain one or more objects. **(2)** As a computing buzzword, any type of entity that can have properties and actions (or methods) associated with it. Each property represents a "container" into which a specific value for the property can be entered. A particular combination of properties defines an object or object type, and a particular combination of values for the properties defines a specific instance of that object type.

Anything (real or imagined) with clear boundaries and meanings within a particular context, view, or domain, e.g., business concepts, contracts, customers, databases records, equipment, goals, locations, parts, products, reports, resources, services, software modules (routines), systems, vendors. **(3)** In networking, an entity in some type of grouping, listing, or definition. The entity may be real (e.g., users, machines, devices, and servers) or abstract (e.g., groups, queues, and functions). **(4)** In facsimile systems, the image, the likeness of which is to be transmitted.

object code Computer instructions in machine language as generated by a compiler or assembler. The code is essentially numeric in nature and does not have any associated alphabetic information to make human reading practical. The original human written code is called source code.

object ID The name that uniquely identifies one object from all others within its environment. A short form may be unique on a local machine (or LAN), while a longer form may be required to uniquely identify it on a wide area network (WAN). If the local machine name (or LAN name) is part of the *Object ID*, special consideration is required to support Object Mobility.

Object Mobility The ability to move an *object* from one machine (or LAN) to another without disrupting operations or modifying *source code*. Also called *location transparency*.

object model A computer representation of an object that encapsulates data attributes and behavioral processes (operations) for that object. An object model is a graphical representation of the structure of objects in a system including their attributes, identity, operations, and associations with other objects.

object modeling technique (OMT) An application life cycle development methodology and graphical notation scheme that spans object models, dynamic models, and functional models from analysis, through design, and implementation.

object oriented (OO) A computer analysis, design, and system development method where "real-world" concepts (like customers, orders, products, etc.) are modeled as *encapsulated objects* with attributes and operations. (This is in contrast to conventional, or procedual, methods that separate database design from program design). With *OO* methods, similar objects are grouped together in *classes* with common data attributes and operations. Objects can inherit attributes from other objects.

Objects communicate with other objects by sending *messages*. An object may be composed of other objects; hence, proper design of the base objects allows them to be reused in a number of higher level objects. This has a number of benefits, notably,

- Combining reusable subcomponent objects in various ways can define a wide variety of business object models.
- Reduction of "reinventing the wheel."
- More compatible applications (better reliability).
- Parallel development.
- Shorter development times.

OO methods are used to create a wide variety of computer applications and communications systems.

OC (1) An abbreviation of <u>O</u>ptical <u>C</u>arrier. A SONET optical signal. **(2)** An abbreviation of <u>O</u>perations <u>C</u>enter.

OC-n An abbreviation of <u>O</u>ptical <u>C</u>ommunication level "<u>n</u>." A set of full-duplex Optical Communication standards where n is related to the communications rate, that is, the communications rate is $n \times 51.840$ Mb/s. It is the optical signal resulting from the conversion of the electrical STS-n signal. See also *DS-n, SONET, SDH, STM-n, & STS-n*.

TRANSPORT LEVEL COMPARISON

SONET		ITU-T (SDH)	Line Rate (Mb/s)	DS-n Equivalents		
Electrical	Optical			DS3	DS1	DS0
STS-1	OC-1		51.84	1	28	672
STS-3	OC-3	STM-1	155.52	3	84	2016
STS-9	OC-9	STM-3	466.56	9	252	6048
STS-12	OC-12	STM-4	622.08	12	336	8064
STS-18	OC-18	STM-6	933.12	18	504	12096
STS-24	OC-24	STM-8	1244.16	24	672	16128
STS-36	OC-36	STM-12	1866.24	36	1008	24192
STS-48	OC-48	STM-16	2488.32	48	1344	32256
STS-96	OC-96	STM-32	4976.64	96	2688	64512
STS-192	OC-192	STM-64	9953.28	192	5376	129024

OCC An abbreviation of <u>O</u>ther <u>C</u>ommon <u>C</u>arrier.

Occam's Razor A concept stating that the simplest of multiple explanations of a phenomenon should be chosen. It is named for the English philosopher, William of Occam (1300–1349), who advanced the idea and stated, "Entia non sunt multiplicanda praeter necessitatem" ("Entities should not be multiplied more than necessary"). That is, the fewer assumptions an explanation of a phenomenon depends on, the better it is.

occupancy For equipment, the ratio of the actual time that a circuit, switch, or communications link is in use (occupied) to the total available time. It is normally expressed as a percentage of the maximum possible available time during a specified period of time (usually one hour). Also called *usage*.

OCD An abbreviation of <u>O</u>ut of <u>C</u>ell <u>D</u>elineation. An ATM receiver is searching for cell alignment.

OCR (1) An abbreviation of <u>O</u>ptical <u>C</u>haracter <u>R</u>eader (or <u>R</u>ecognition). A device or software that converts a scanned (bit mapped) image of a textual document into a character-based file. The "recognition" of characters is accomplished by comparing the bit-mapped shapes with a table of known shapes and letter correlations.

A typical application for *OCR* would be the conversion of a received fax (a bit-mapped image) into a text file that a word processor would be able to edit. (2) An abbreviation of <u>O</u>ffice <u>C</u>hannel <u>R</u>epeater. Also called *Office Channel Unit* (*OCU*).

octad A group of 3 bits used to represent one octal digit.

octal (1) A numbering system with a base (or radix) of eight. (2) A digital system with eight states, 0–7.

octathorpe (#) Also called the number, pound, crosshatch, gate symbol, and tic-tac-toe sign. Sometimes spelled *octothorp*.

octave The interval between any two frequencies having a ratio of 2:1, e.g., 100 Hz and 50 Hz, 5000 Hz, and 2500 Hz. The *center frequency* is defined as the *geometric mean* of the *bandedge* frequencies. For example, bandedges of 5000 Hz and 2500 Hz yield a center frequency of 3535.5 Hz. In equation form, these relationships are:

$$f_c = \sqrt{f_l \cdot f_h}, \quad f_h = f_c \cdot \sqrt{2}, \quad f_l = \frac{f_c}{\sqrt{2}}$$

where:

f_c is the center frequency,
f_l is the lower bandedge frequency, and
f_h is the upper bandedge frequency.

See also *third octave*.

octet A sequence of 8 bits treated as a single item. The term is frequently used on the Internet because of the differences in data word lengths of some systems attached. Also called a *byte* in the world of PCs.

octet alignment The configuration of a field composed of an integral number of octets. If the field is not divisible by eight exactly, bits (usually zeros) are added to either the first octet (*left justification*) or the last octet (*right justification*).

OD An abbreviation of <u>O</u>ptical <u>D</u>ensity.

ODA An abbreviation of <u>O</u>pen <u>D</u>ocument <u>A</u>rchitecture.

ODBC An abbreviation of <u>O</u>pen <u>D</u>ata<u>B</u>ase <u>C</u>onnectivity.

odd even check A synonym for *parity check*.

ODI An abbreviation of <u>O</u>pen <u>D</u>ata link <u>I</u>nterface.

ODLI An abbreviation of <u>O</u>pen <u>D</u>ata <u>L</u>ink <u>I</u>nterface.

ODS An abbreviation of Microsoft's <u>O</u>pen <u>D</u>ata <u>S</u>ervices.

ODSI An abbreviation of <u>O</u>pen <u>D</u>irectory <u>S</u>ervices <u>I</u>nterface.

OED An abbreviation of <u>O</u>xford <u>E</u>nglish <u>D</u>ictionary.

OEM An abbreviation of <u>O</u>riginal <u>E</u>quipment <u>M</u>anufacturer.

oersted (Oe) The cgs unit of magnetic field strength (H) (magnitizing force). A field of 1 Oe is represented by one magnetic line per square centimeter. In the SI system of units, magnetic field strength (H) is represented by amperes per meter (A/m). The SI and cgs units are related by:

$$1 \text{ Oe} = 79.57747 \text{ A/m}$$

OF An abbreviation of <u>O</u>ptical <u>F</u>iber.

OFB An abbreviation of <u>O</u>utput <u>F</u>eed<u>B</u>ack. An operating mode for the data encryption standard (DES).

OFC An abbreviation of <u>O</u>ptical <u>F</u>iber, <u>C</u>onductive. The Underwriters Laboratory (UL) and National Fire Protection Association (NFPA) designation given to interior-use fiber optic cables that contain at least one electrically conductive, noncurrent-carrying component (e.g., a metallic strength member or vapor barrier) and that are not certified for use in plenum or riser applications.

OFCP An abbreviation of <u>O</u>ptical <u>F</u>iber, <u>C</u>onductive, <u>P</u>lenum. The Underwriters Laboratory (UL) and National Fire Protection Association (NFPA) designation given to interior-use fiber optic cables that contain at least one electrically conductive, noncurrent-carrying component (e.g., a metallic strength member or vapor barrier) and that are certified for use in plenum applications.

OFCR An abbreviation of <u>O</u>ptical <u>F</u>iber, <u>C</u>onductive, <u>R</u>iser. The Underwriters Laboratory (UL) and National Fire Protection Association (NFPA) designation given to interior-use fiber optic cables that contain at least one electrically conductive, noncurrent-carrying component (e.g., a metallic strength member or vapor barrier) and that are certified for use in riser applications.

off-axis optical system An optical system in which the aperture's optical axis is not coincident with the aperture's mechanical center.

off-hook (OH) (1) In telephony, the condition that exists when an operational telephone instrument or other user equipment (DCE) is in use, i.e., is electrically connected to the local loop (either a telco or PBX loop) and current is flowing (the active state). The condition exists both during dialing and communication. (It is the mark state during pulse dialing.)

Off-hook originally referred to the condition that exists when an instrument's separate earpiece (the receiver) was removed from its resting hook (the hookswitch). Removal of the receiver from the hook actuated a spring-operated switch (the hookswitch) that connected the telephone instrument to the telephone line. The opposite state is *on-hook*. (2) One of two possible signaling states, such as tone or no tone, current or no current, and ground connection vs. battery connection. Off-hook may indicate either state while *on-hook* pertains to the other. (3) The active state of a subscriber or PBX user loop (closed loop). (4) The operating state of a communications link in which data transmission is enabled either for voice (or data communications) or for network signaling.

off-hook queuing (OHQ) See *callback queuing*.

off-hook service A synonym for a *hot-line*.

off-hook signal In telephony, a signal indicating seizure, request for service, or a busy condition.

off-line (1) The state in which a *device* or *circuit* cannot communicate with or be controlled by another device. (2) Information not automatically available to a computer. It must be manually loaded, e.g., mounting a magnetic tape, inserting a CD-ROM disk into a reader. (3) The status of a device that is disconnected from service.

off-line reader A USENet newsreader designed to run on an autonomous computer which attaches to the Internet via a dial-up connection. The purpose of the program is to allow reading and replying to e-mail and newsgroup postings without tying up the telephone line during periods when the connection is not required.

In operation, the program goes through the following steps:

- The user dials up a service provider using a modem and establishes a telephone line connection.
- The program downloads all postings (to which the user has subscribed) into the users computer.
- The telephone connection is terminated.
- The user reads the postings.
- Optionally, the user creates a response to selected postings and the process continues. If no response is required, the process is terminated at this point.
- A connection to the service provider is again established.
- The program uploads the responses to the appropriate newsgroups.
- The telephone connection is again terminated.

off-line storage Information stored in a manner that is not under control of the processing computer.

off-net call A call from a switched service network to a destination outside that network.

off-peak In communications, those periods of time during which little business activity occurs. The actual time periods vary among carriers and services and may be multitiered.

For example, in telephony, the peak time may be defined as 8:00 A.M. to 5:00 P.M. where the highest calling cost occurs. The first tier of rate reduction may be from 5:00 P.M. to 11:00 P.M. and the second tier between 11:00 P.M. and 8:00 A.M. On a bulletin board system (BBS) the off-peak may be between 2:00 A.M. and 3:00 P.M.

off-premises extension (OPX) A station located in a physically separate location from the main station and with all of the rights of the main station. For example, a telephone line that has appearances in both an office and at the owner's home is an off-premises extension. Also called *off premises station (OPS)*.

offered call In telephony, a call that is offered to a trunk or group of trunks.

offered load In telephony, the total traffic offered, including re-attempts, to a group of servers.

offered traffic In telephony, a measure of the calls requesting service during a given period of time. The total number of attempts to seize a group of servers.

office automation systems (OAS) Programs to assist a user in every-day office activities, such as word processors, spreadsheets, contact managers, and calendars.

office class See *class n office*.

office code Those digits of a dialed number that designate a block of main station codes within a numbering plan area; the NXX portion of the NPA-NXX-XXXX number in North America. Also called *central office prefix* and *local office prefix*.

office drop The network cable that connects to a node.

Office Systems Node (OSN) An IBM concept describing a set of functions and services provided to connected nodes in an IBM office system. Office systems with more limited IBM function sets are termed Source or Recipient Nodes (SRN). A typical *OSN* is a host computer running software as the IBM Distributed Office Support System (DIOSS), while an SRN is typically a text processing PC.

OSN has a central role in a Document Interchange Architecture (DIA) defined office system. A user at a workstation, using commands defined in DIA, can request an *OSN* to supply document library, document distribution, fiber transfer and applications processing services. DIA enables an SNA network to build an office systems network with multiple remote locations sharing an *OSN*.

offset (**1**) In memory systems, the difference between a base address and the address to be operated upon. (**2**) In analog systems, a dc voltage error between the desired signal and the actual signal. Generally equal to the measured voltage with no input signal present.

OFN An abbreviation of Optical Fiber, Nonconductive. The Underwriters Laboratory (UL) and National Fire Protection Association (NFPA) designation given to interior-use fiber optic cables that contain no electrically conductive components and that are not certified for use in plenum or riser applications.

OFNP An abbreviation of Optical Fiber, Nonconductive, Plenum. The Underwriters Laboratory (UL) and National Fire Protection Association (NFPA) designation given to interior-use fiber optic cables that contain no electrically conductive components and that are certified for use in plenum applications.

OFNR An abbreviation of Optical Fiber, Nonconductive, Riser. The Underwriters Laboratory (UL) and National Fire Protection Association (NFPA) designation given to interior-use fiber optic cables that contain no electrically conductive components and that are certified for use in riser applications.

OGM An abbreviation of Out Going Message. The message generated by an automatic telephone-answering device (TAD) to the calling party. For example, "Sorry I'm not here, after the 'beep' please leave your name and number and I will get back to you."

OH An abbreviation of Off-Hook.

Ohio Regional Network (OARnet) An Ohio statewide network and service provider that links cities with a high-speed backbone. It provides Internet connectivity to individuals and organizations throughout the state. It also provides Gopher services and assistance with Internet installations.

ohm (Ω) The unit of electrical resistance, named for Georg Simon Ohm (1787–1854), a German teacher who discoverered the relationship of resistance, current, and voltage. Specifically, a 1-ohm resistor will allow 1 ampere of current to flow in a circuit if 1 volt is applied.

Ohm's law The relationship between voltage, current, and resistance in a circuit. The law states that the current in an electrical circuit (in amps—I) is inversely proportional to the resistance (in ohms—R) of the circuit and directly proportional to the voltage (in volts—V) across the circuit. Or mathematically,

$$I = \frac{V}{R}$$

OHQ An abbreviation of Off-Hook Queuing.

OIC An abbreviation of Oh, I see.

OIW An abbreviation of OSI Implementers Workshop. One of three regional workshops for implementers of the ISO/OSI Reference Model. The other two are Asia and Oceania Workshop (AOW) and European Workshop for Open Systems (EWOC).

OJT An abbreviation of On the Job Training.

OLE An abbreviation of <u>O</u>bject <u>L</u>inking and <u>E</u>mbedding (pronounced "O-lay"). Microsoft's standard which permits programs to "communicate" by allowing a user to cut and paste information between compliant applications.

Linking and *embedding* are two alternate approaches to the communications:

- *Link* inserts a <u>reference</u> to the actual object into the document file. Before the object can be accessed in the document, it needs to be loaded from its source. By retrieving the object only when needed, two benefits are achieved: (1) Assurance that the latest version of the object will be retrieved, and (2) a generally smaller document file.
- *Embedding* inserts a <u>copy</u> of the object into the document file at the desired location. Changes to the original object after embedding do not affect the object in the document file. However, the program used to create the object may be invoked from the document.

OLI An abbreviation of <u>O</u>ptical <u>L</u>ine <u>I</u>nterface.

oligarchically synchronized network A synchronized network in which the timing of all clocks is controlled by a selected few clocks. See also *democratically synchronized network* and *despotically synchronized network*

OLIU An abbreviation of <u>O</u>ptical <u>L</u>ine <u>I</u>nterface <u>U</u>nit.

OLTP An abbreviation of <u>OnL</u>ine <u>T</u>ransaction <u>P</u>rocessing.

omnidirectional The radiation pattern of a signal source or receiving pattern of an input device that is uniform in all defined directions. Examples might include antennas, lamps, and microphones.

omnidirectional antenna An antenna that has a radiation pattern nondirectional in azimuth. However, the vertical radiation pattern may be of any shape.

Omninet A popular 4-Mbps network in Europe, Japan, and other countries developed by Corvus Systems primarily for PCs.

OMT An abbreviation of <u>O</u>bject <u>M</u>odeling <u>T</u>echnique.

ON An abbreviation of <u>O</u>sterrisches <u>N</u>ormungistut. The Austrian standards organization.

on-hook **(1)** In telephony, the condition that exists when neither an operational telephone instrument nor other user equipment (DCE) is in use, i.e., is not electrically connected to the local loop (either a telco or PBX loop) and no current is flowing (the idle state). The space state during pulse dialing.

On-hook originally referred to the condition that existed when an instrument's separate earpiece (the receiver) was in place on its resting hook (the hookswitch). Removal of the receiver from the hook actuated a spring-operated switch (the hookswitch), which connected the telephone instrument to the telephone line. The opposite state is *off-hook*. **(2)** One of two possible signaling states, such as tone or no tone, current or no current, and ground connection vs. battery connection. Off-hook may indicate either state while *on-hook* pertains to the other. **(3)** The idle state of a subscriber or PBX user loop (i.e., an open loop). **(4)** An operating state of a communications link in which information transmission is disabled and a high impedance (i.e., open circuit) is presented to the link by all end instruments. (In telephony, during the *on-hook* condition, the link is responsive to ringing signals.)

on-hook dialing The process of dialing a destination station number while the calling station is on-hook. The number is temporarily stored in the calling station equipment. When the equipment goes off-hook, it then forwards the signaling information to the PSTN switching equipment. This technique is frequently used by equipment that must redial a destination until a connection is established and the required information is sent, e.g., a facsimile machine. Also called *predial*.

on-hook queuing See *callback queuing.*

on-hook signal In telephony, a signal indicating a disconnect, unanswered call, or an idle condition on a circuit.

on-net call A call that originates and terminates within a single switched service network.

on-premises equipment Customer-owned and maintained equipment and cables. It encompasses all communications facilities to transport current or future data, voice, local area network (LAN), and image information. Examples include:

- An extension telephone, PBX station, or key system station located on property that is contiguous with that on which the demarcation point is located.
- Customer-owned metallic or optical fiber communications transmission cables, installed within or between buildings. In network wiring, it may consist of horizontal wiring, vertical wiring, and backbone wiring, and it may extend from the external network interface to the user workstation areas.

on-ramp A metaphor of access to the Electronic Superhighway where all services available can be put to use. An *on-ramp* is usually a *gateway* node, although to some vendors it may refer to a telephone line, a two-way cable connection, or ISDN.

ONA An abbreviation of <u>O</u>pen <u>N</u>etwork <u>A</u>rchitecture.

ONC An abbreviation of <u>O</u>pen <u>N</u>etwork <u>C</u>omputing.

one-dimensional coding A data compression method for facsimile transmissions that considers each scan line independently, that is, it does not use information from the preceding or following lines in the compression process. It operates only in the horizontal direction.

one plus call (1+) Within the United States, a type of station-to-station call in which the digit "1" is dialed before the end station access code. The "1" informs the local telephone company switching system that the call is to a destination outside the local area of the caller. Typically, the "1" is followed by an area code and station code.

one shot A circuit that generates or delivers one output pulse of a specified duration for each input signal (trigger). See also *complement.*

one-way communication Communication in which information is always transferred in only one preassigned direction. Examples of one-way communications systems include radio and television broadcast stations, one-way intercom systems, and wireline news and stock market quote reporting services.

one-way reversible operation A synonym for *half-duplex operation.*

one-way trunk A trunk between two switching systems which is accessible by calls from one end only (though traffic may be two-way). At the call-originating end, the trunk is known as an *outgoing trunk;* at the terminating end, it is known as an *incoming trunk.*

one's complement The *diminished complement* of a binary number. The *one's complement* of a binary number can be calculated simply by inverting each bit in the number (zeros are replaced with ones and vice versa), for example, the *one's complement* of 0110 is 1001.

Few devices use this system to represent negative numbers because there are two representations for zero and simplistic arithmetics can lead to strange results [e.g., $+1 + (-0) = +0$.] Values near zero are:

$$00 \ldots 0011 = +3$$
$$00 \ldots 0010 = +2$$
$$00 \ldots 0001 = +1$$
$$00 \ldots 0000 = +0$$
$$11 \ldots 1111 = -0$$
$$11 \ldots 1110 = -1$$
$$11 \ldots 1101 = -2$$
$$11 \ldots 1100 = -3$$

See also *two's complement.*

ones density The number of electronic "1's" in a block of specified length. The requirement for digital transmission systems to periodically receive a "1" in order to maintain synchronization. In many systems, no more than seven "0s" are allowed before synchronization is lost. See also *BnZS.*

ONI (1) An abbreviation of O̲perator N̲umber I̲dentification. (2) An abbreviation of O̲ptical N̲etwork I̲nterface.

online (1) The state of a *device* or *circuit* when it is activated and ready for normal operation.

For example, the local modem state when it is capable of communicating with a remote modem. All data, except the escape sequence, sent to the local modem from the DTE are transmitted to the remote modem. The escape sequence will put the modem into the *command mode* without causing a disconnect. The modem is also said to be *connected.* (2) Information available for a computer to directly access, no manual intervention is required. For example, the information in a computer's hard disk drive is always *online,* while information in a CD-ROM is *online* only when the CD-ROM is mounted in the reader. Also spelled *on-line.*

online information service A for-profit, dial-up service that provides information such as encyclopedia access, stock quotes, airline and hotel reservation services, e-mail, and much more. In addition to the internal services these companies offer, they offer access to the Internet. Some only offer e-mail connections, while others offer full access in the form of a proprietary Serial Line Interface Protocol/Point-to-Point Protocol (SLIP/PPP)-like connection. The leading services include Prodigy, America Online, CompuServe, GEnie, and Delphi.

ONMA An abbreviation of O̲pen N̲etwork MA̲nagement.

OO An abbreviation of O̲bject O̲riented.

OOF An acronym from O̲ut O̲f F̲rame.

OOK An abbreviation of O̲n-O̲ff K̲eying.

OOLR An abbreviation of O̲verall O̲bjective L̲oudness R̲ating.

OOP An acronym for O̲bject O̲riented P̲rogramming.

OOPART An acronym for O̲ut O̲f P̲lace ARTifact.

OOPS An acronym for O̲bject-O̲riented P̲rogramming S̲ystem.

OOS (1) An abbreviation of O̲ut O̲f S̲ynchronization. (2) An abbreviation of O̲ut O̲f S̲ervice.

opacity The reciprocal of transmission. For example, a material that does not transmit light is opaque to light, a material that does not transmit sound is opaque to sound, and a material that does not allow radio wave to pass is opaque to radio waves.

open (1) In a wire or cable, a gap or separation in the conductive material somewhere along the cable's path, such as in one wire in a pair. (2) In an optical fiber, a gap in the inner core such that the light signal cannot propagate beyond the gap. (3) In networks and computer-related devices, an adjective that refers to elements or interfaces whose specifications have been made public so that they can be used by third parties to create compatible products. This is in contrast to closed, or proprietary, environments. See also *open systems.* (4) A term that refers to the condition of a communications link that is available for use.

open air transmission A term that refers to a transmission system that uses no physical medium to transport the signal, e.g., radio or infrared transmission systems. Also called *through-the-air* transmission.

open application interface (OAI) In telephony, an interface that can be used to program features and change the operation of a private automatic branch exchange (PABX).

open circuit (1) An electrical system in which the current path is not complete; that is, the loop contains an infinite impedance.

An open circuit may be intentional (as with a switch) or accidental (e.g., a cable is severed). (2) In communications, a circuit that is available for use.

open circuit voltage The voltage at the terminals of a device without any external load attached, i.e., the voltage when the device under test does not deliver any current to any external device. Also called *open circuit potential.*

open data link interface (ODI™ or ODLI) An architecture developed jointly by Novell and Apple to provide a standard interface for network interface cards (NICs) or device drivers. It makes possible the use of multiple protocols and multiple local area network (LAN) drivers with a single NIC. For example, *ODI* enables a single workstation to access a Novell NetWare network, a UNIX-based network, or an AppleTalk network, each through its own protocol.

ODI sits between LAN drivers (which talk to the NIC) and the protocol stacks. By providing separate interfaces to the protocols and the NICs, *ODI* allows these two levels to be mixed and matched in a transparent manner. In effect, *ODI* can make communications (at least partially) independent of both protocols and media.

Open Data Services (ODS) The part of Windows Open Service Architecture (WOSA) that supports access from Microsoft's Structural Query Language (SQL) Server to a wide range of data sources and formats, including information from major mainframe databases.

open database connectivity (ODBC) An Application Program Interface (API) developed by Microsoft for accessing both relational and nonrelational databases under Windows. An alternative to *ODBC* is the Borland, IBM, Novell, and WordPerfect standard Integrated Database Application Programming Interface (IDAPI).

open document architecture (ODA) The ISO 8613-1/8 standard for the interchange of compound documents which may contain fonts and graphics in addition to text. The standard seeks to maintain layout and graphics in the document. It defines three levels of document representation:

- Level 1—text-only data.
- Level 2—text and graphical data from a word processing environment.
- Level 3—text and graphical data from a desktop publishing environment.

The standard/interchange format has been adopted by MAP/TOP 3.0, by GOSIP, and by ECMA as ECMA-101.

Open Document Interchange Format (ODIF) Part of the open document architecture (ODA) standard.

open dual bus A dual bus in which the head-of-bus functions for each bus is at a different location.

Open Graphics Library (OpenGL) A multi-platform software interface for graphics hardware. The *OpenGL* interface (based on Silicon Graphics' proprietary IRIS GL) was developed by Silicon Graphics, which licenses it to other vendors. The interface consists of several hundred functions operating on 2D and 3D objects, supporting basic techniques, such as rendering, imaging operations, smooth shading, and advanced operations, such as texture mapping and motion blur. *OpenGL* is network-transparent (and a common extension to the X Window System) which allows an *OpenGL* client to communicate across a network with a different vendor's *OpenGL* server.

open loop system A control system that does not use feedback to determine the output.

For example, the throttle in a car determines how much fuel is delivered to the engine, not the vehicle speed (a fixed fuel setting will make the vehicle go faster downhill than it will uphill). To control speed, the driver observes the speedometer and adjusts the throttle position up or down as conditions change (a process called feedback).

open network architecture (ONA) A set of standards (protocols) available to the public, which allow any telecommunications vendor to design equipment and software to interconnect with any network using the standards.

In the context of the Federal Communications Commission's (FCC's) Computer Inquiry III, the overall design of a communication carrier's basic network facilities and services to permit all users of the basic network to interconnect to specific basic network functions and interfaces on an unbundled, equal-access basis. The ONA concept consists of three integral components:

- Basic serving arrangements (BSAs).
- Basic service elements (BSEs).
- Complementary network services.

open network computing (ONC) A model for distributed computing, originally developed by Sun Microsystems and now supported in most UNIX implementations. The *ONC* model uses Sun's Network File System (NFS) for handling files distributed over remote locations. Communication with remote servers and devices is through remote procedure calls (RPCs). The *ONC* model supports the TCP/IP protocol stack.

Open Network Management Architecture (ONMA) IBM's network management architecture comprised of Entry Points, Service Points, and Focal Points—implemented as NetView.

open numbering plan In telephony, a numbering plan in which the number of digits dialed varies according to the requirements of the destination's network.

open path In network analysis, a path along which no node appears more than once.

open pipe A term that indicates the data path between a sender and receiver is circuit switched or a leased line. The intent is to indicate that the data flows directly between the two locations (through an open pipe) rather than needing to be broken into packets and routed through various paths.

Open Software Foundation (OSF) A consortium of vendors (including Apollo, Digital Equipment Corporation, IBM, Hewlett-Packard, Philips, Siemens, Nixdorf, and Hitachi) formed to develop and support *open computing*. Common operating systems based on the X Window system and AT&T's UNIX are planned. Their version of the UNIX operating system was released on October 23, 1990 and is called *OSF/1*.

open system A nonproprietary system defined by a set of standards available to all vendors, suppliers, and users; without a profit motive and in the hope of promoting standardization. An *open system* will:

- Provide a standard set of relationships.
- Be available to all who request information.
- Be relevant for all hardware types.
- Be portable.

The opposite of an *open system* is a *closed* or *proprietary system*.

Open Systems Interconnection—Reference Model (OSI-RM) In networking and telecommunications, the concept of the International Standards Organization's (ISO's) seven-layer communications reference model.

The *OSI* model is described in a set of ITU-T standards (X.200 series) intended to define the workings of a digital packet switching communication network in seven independent layers of communication protocol. The model defines how dissimilar network devices such as network interface controllers (NICs), routers, and bridges exchange data over a network. Each layer enhances the capabilities of the communication layer below it and shields the layer above it from the implementation details of lower layers. Each layer responds to service requests from the layer above and issues service requests to the layer below. In effect, each layer on a given machine communicates with the corresponding layer on another machine. If there is not a real physical connection between machines at the layer of communication, the connection is said to be "virtual." The seven protocol layers defined in the *OSI* publication are:

Layer 1 The *physical layer* (lowest layer)—hardware connections. Defines the link medium (twisted pair, coaxial cable, optical fiber, etc.), voltages, connectors, pin assignments, communication bit rate, line discipline (full duplex, half duplex, CSMA/CD), flow control (DTE to DTE), and so on. The major services include:

- Establishment and termination of a connection to a communications medium.
- Participation in the process in which the communication is shared among multiple users, e.g., contention resolution and flow control.
- Waveform coding of equipment digital signals to transmission line waveforms.

Equivalent Internet protocols: none.

Layer 2 The *data link layer*—concerned with how data are transferred across a given point to point link. Layer 2 includes coding, addressing, packetizing, and transmitting information. It provides framing, timeouts sequencing, acknowledgments, error detection, and point-to-point flow control. HDLC (High-level data link control) is an example of such a protocol.

Equivalent Internet protocols: *address resolution protocol* (*ARP*) and *SLIP/PPP*.

Layer 3 The *network layer*—provides the functional and procedural means of transferring variable-length data sequences from a source to a destination via one or more networks. It deals with network routing, message handling, and transfer and provides segmentation/desegmentation, message sequencing, flow control, congestion control, and accounting.

Equivalent Internet protocol: IP protocol.

Layer 4 The *transport layer*—concerned with managing accurate point A to Point B delivery. Provides message packetizing, multiplexing, establishes host-to-host connection, flow control, and error control.

Equivalent Internet protocols: the *Internet Control Message Protocol* (*ICMP*), *transport control protocol* (*TCP*), and *User Datagram Protocol* (*UDP*).

Layer 5 The *session layer*—coordination of communications. Establishes and terminates connections between processes, and provides restart procedures, virtual circuit service if the transport layer does not do so, and half-duplex (HDX) or full-duplex (FDX) operation. Prevents two machines from trying to access the local area network (LAN) at the same time.

Equivalent Internet protocols: none.

Layer 6 The *presentation layer*—formatting, code conversion, and display. Performs text compression, converts file formats (e.g., EBCDIC to ASCII), encrypts data, and so on.

Equivalent Internet protocols: data compression and encryption, privacy enhanced mail (PEM).

Layer 7 The *application layer* (highest layer)—program-to-program transfers, login, password checking, e-mail protocols, and file transfer protocol. Provides the services specific to a particular application.

Equivalent Internet protocols: network virtual terminal (NVT), file transfer protocol (FTP), simple mail transfer protocol (SMTP), domain naming system (DNS), and others.

The associated diagram displays the relationship of two nodes on a communications network, each with a user and the seven-layer protocol stack. In the diagram, user X is transmitting a message to user Y. The application layer is not the only layer to generate (or terminate) packets. Some examples include:

- Control messages to set up, maintain, and disconnect a node-to-node connection, i.e., connection request, connection confirmation, disconnect request, and acknowledge.
- Error control messages. With error checking, part or all of a packet's contents may be qualified with a check word. If the check fails, the software at that layer may issue a request to retransmit the associated layer in the originating node.
- Flow-control messages. When a layer on a receiving node is accumulating data in its buffer faster than it can be processed, a message may be generated by the receiving node and sent to the originating node to stop transmitting until further notice.

open waveguide An all-dielectric waveguide in which electromagnetic waves are guided by refractive index differentials or gradients so that the waves are confined to the waveguide by refraction within or reflection from the surface of the waveguide.

In an open waveguide, the electromagnetic waves propagate (without radiation) within the waveguide, although evanescent waves coupled to internal waves may travel in the space immediately outside the waveguide. An optical fiber is an example of an *open waveguide*.

open wire Conductors that are individually supported with insulators on poles or towers above the surface of the Earth. Although each conductor may be insulated or uninsulated, they are not bundled into a common insulated group.

Open wires are used in both communication and power applications. Open wire feed lines are sometimes used between a transmitter and an antenna.

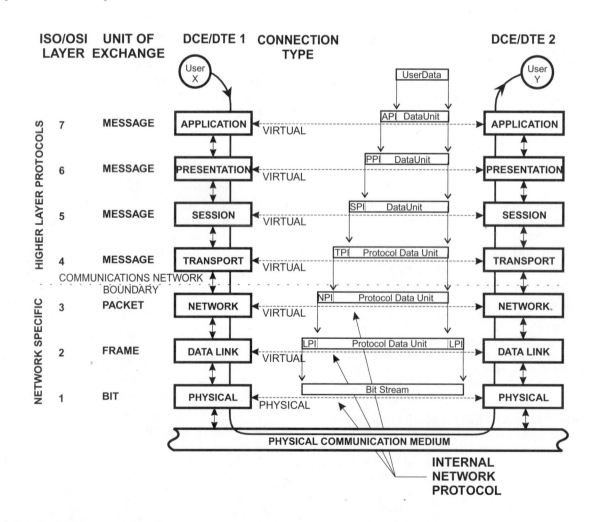

operand An entity to which an operation is applied. The operation may be mathematical (add, subtract, multiply, etc.) or part of a computer instruction (move, test, clear, etc.).

operating noise temperature The temperature in degrees kelvin defined by:

$$T_{op} = \frac{N_o}{kG_s}$$

where:

T_{op} is the effective operating noise temperature,

N_o is the output noise power per unit bandwidth at a specified frequency,

k is Boltzmann's constant (approximately 1.38×10^{-23} joules/kelvin), and

G_s is the ratio of the signal power delivered toward the output circuit to the signal power available at the system input terminals. All signals are under normal operating conditions and at the specified frequency.

See also *noise temperature.*

operating system (OS) An integrated collection of routines that direct or control the sequencing and processing of programs by a computer, that is, that software (or firmware) which controls the operations of a computer, including:

- Controlling the execution of other programs.
- Providing an environment and interface for users.
- Providing services such as resource allocation, e.g., memory management, I/O control; e.g., UART flow control, scheduling, or data management.
- Providing a standardized interface to hardware implementations.

Operating systems are divided into two segments: the *kernel* and the *shell*. The kernel is the underlying portion of the code that actually interacts or controls the computer, whereas the shell provides the interface to the user. Examples of operating systems include CPM, DOS, OS/2, NetWare, System 7, UNIX, VMS, VRTX, Windows NT, and Windows95. (Windows 3.x is a graphics-oriented shell.) A computer's "native" (built-in or default) operating system may be supplemented or replaced by a different operating system, e.g., a network operating system (NOS) and Windows 3.x. Although *operating systems* are predominantly software, partial or complete hardware implementations may be made in the form of firmware, i.e., nonerasable instruction in read only memory (ROM). See also *network operating system (NOS)*.

operating time (1) The time interval between the instant of occurrence of a specified stimulus to a device or system and the instant of completion of a specified operation. (2) In communications, computer, and information processing systems. The time interval between the instant a request for service is received from a user and the instant of final release of all facilities by the controlling user. See also *called, calling,* and *either party control.* (3) In communications systems conference calls. The time interval between the instant a request for service is received and the instant all but one of the users have released all facilities.

operations, administration, and maintenance (OAM) functions A set of functions defined by the ITU-T for managing the lower layers in an Asynchronous Transfer Mode (ATM) network, or a broadband ISDN (BISDN) network. The functions fall into the following categories:

- *Performance monitoring*—functions that determine if the network is functioning at the specified level. They also generate information useful for maintenance.

- *Defect detection*—functions that identify defects or malfunctions in the network.
- *System protection*—functions responsible for isolating a malfunctioning element and switching over to redundant elements.
- *Failure reporting*—functions that inform other management entities (such as network management software or the other party) of a malfunction.
- *Fault localization*—functions that determine where a detected malfunction occurred in order to enable the system to take the appropriate protection and failure reporting measures.

operator A symbol used as a function on one or more operands. Examples include: "+," "−," "×," "÷," or the Boolean operators "AND," and "OR" which operate on two operands while "sin," "cos," "$\sqrt{}$," or the Boolean operator "NOT" have only one operand.

ophthalmic Pertaining to the eyes.

opm An abbreviation of Operations per Minute.

OPR An abbreviation of Optical Power Received.

OPS An abbreviation of Off-Premises Station. See also *off-premises extension (OPX)*.

optic axis A direction through a double refracting crystal along which double refraction does not take place.

A uniaxial crystal has but one optic axis while a biaxial crystal has two. Contrast with *optical axis.*

optical amplifier A device used to amplify an optical signal without converting it to and from an electrical signal. Two examples are erbium doped fiber amplifiers (EDFAs) and semiconductor laser amplifiers. See also *fiber amplifier* and *optical repeater.*

optical attenuator A device used to reduce the intensity (power level) of an optical signal. Optical attenuators may be constructed so as to provide a known fixed loss, a range of stepwise variable loss, or a continuously variable range of loss.

Attenuators use a variety of principles to introduce loss, such as gap loss, reflective loss or refractive loss, and absorption loss. Attenuators using the gap loss principle, however, are sensitive to modal distribution and should be used at or near the transmitting end, or they may introduce less loss than intended. (Absorptive or reflective techniques avoid this problem.)

optical axis (1) Of a refractive or reflective element, the straight line that is coincident with the axis of symmetry of the surfaces. The optical axis is generally coincident with the mechanical axis of the element, however, it need not be. (2) In a lens, the straight line that passes through the centers of curvature of the lens surfaces. (3) In an optical system, the line that passes through the principal axis (centers of curvatures of individual element surfaces) of the series of optical elements. (4) A line connecting the centers of circles circumscribing the core of an optical fiber. Also called the *fiber axis* (the preferred term). Contrast with *optic axis.*

optical cavity The region bounded by two (or more) optically reflecting surfaces (called mirrors or cavity mirrors) which are aligned to provide multiple reflections within the cavity. An example of an optical cavity is the resonator in a laser. Also called a *resonant cavity.*

optical cement A permanent and transparent adhesive used to join optical elements. The classic optical cement is Canada balsam, however, methacrylates, caprinates, and epoxies are now generally used.

optical center The point on the axis of a lens that is the image of the nodal points. Rays passing through the optical center emerge parallel to the rays entering.

optical character reader (OCR) A device that optically scans the markings of a document and converts them into specific symbols usable in a computer-based editor (a word processor, for example). When all goes well, the symbols accurately represent the information on the original document. However, the conversion accuracy is less than 100%, so manual cleanup is almost always required.

optical character recognition (OCR) The automatic detection and identification of graphic characters by means of photosensitive devices. The graphic characters are generally embedded in a single bit mapped document, while the detected characters are generally ASCII characters.

optical combiner A passive optical device in which power from several inputs is distributed among one or more outputs.

optical computer A computer that uses photons rather than electrons to process information.

optical conductor A deprecated synonym for *optical fiber.*

optical connector A device designed to temporarily or semipermanently join two optical fibers (or an optical fiber and a device such as an emitter or detector) so that, to an optical signal, it appears as a contiguous fiber. A connector is distinguished by the fact that it may be disconnected and reconnected, as opposed to a splice, which permanently joins two fibers.

Optical connectors are sometimes erroneously referred to as "couplers." Such usage is incorrect and is to be avoided.

optical contact The adhering of two sufficiently clean and close-fitting surfaces without the use of cement.

optical coupler (1) See *optical isolator.* (2) See *directional coupler, star coupler,* and *T-coupler.*

optical density (OD) For a given wavelength, an expression related to the transmittance of an optical element (or developed photographic image). Higher *optical densities* correspond to lower transmittance.

Mathematically, *optical density* is expressed by $log_{10}(1/T)$ where T is the transmittance. *Optical density* times 10 is equal to transmission loss expressed in decibels for example, an optical density of 0.3 corresponds to a transmission loss of 3 dB.

optical detector A transducer that generates an output signal when irradiated with light.

optical disk A flat, circular, plastic disk coated with material (aluminum or gold, for example) on which bits may be stored in the form of highly reflective areas and significantly less reflective areas. Data may be retrieved from the disk when illuminated with a narrow-beam source, such as a laser diode. Unlike a magnetic computer disk, the bits in an optical disk are stored sequentially on a continuous spiral track.

Optical disks types include video disks, audio and data CD-ROMS, and WORMS. See also *CD-DA, CD+G, CD-i, CD-R, CD-ROM, CD-ROM XA,* and *WORM.*

optical distance The physical path length of a ray times the refractive index of the medium through which it passes. Also called *equivalent air path* or *optical path.* See also *optical path length.*

optical element Any optical part constructed of a single piece of optical material, e.g., a lens, mirror, prism, filter, grating, or fiber.

optical fiber (1) Generally, any filament or fiber, made of a dielectric material, that guides light, whether or not it is used to transmit information. (2) In communications, a thin strand of transparent material, such as glass, used to carry optical signals much as a wire carries an electrical signal. The fiber is specially constructed to keep a light pulse contained within the walls of the fiber.

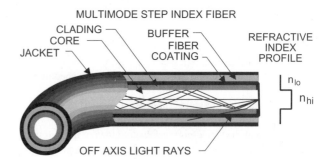

MULTIMODE STEP INDEX FIBER

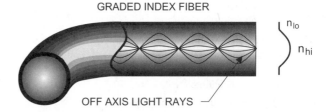

GRADED INDEX FIBER

In the diagram, two fibers are shown. The first is known as a *step index* fiber because the index of refraction makes an abrupt transition from a low index material to a high index material at the core to cladding boundary. The second fiber is a *graded index* because the index of refraction varies continuously from the center outward. (See the index profile next to each fiber.) A fiber is said to be *multimode* if the diameter of the core is large enough for multiple reflection paths to exist. Because the paths with large numbers of reflections are physically longer than paths with few reflections, it takes longer for the light to traverse the fiber (*"modal dispersion"*). This leads to a phenomenon called *pulse spreading.* If the percentage of pulse spreading relative to the input pulse width is high enough (due to either cable length, core diameter, or the pulse repetition rate is high enough), fractions of consecutive pulses will arrive at the receiver at the same time causing interference. The number of propagating modes (N) in a given fiber cable can be *estimated* by

$$N \cong \frac{1}{2} \cdot \left(\frac{\pi \cdot d}{\lambda}\right)^2 \cdot (n_1^2 - n_2^2)$$

where:

d is the diameter of the core,
λ is the wavelength of the optical source,
n_1 is the index of refraction of the core material, and
n_2 is the index of refraction of the cladding material.

One solution to the pulse spreading problem is to manufacture a stepped index cable that is capable of transmitting only one mode (called, logically, a *single mode* cable). This can be accomplished by making N appropriately small (< 2.89). And this can only be accomplished by increasing λ (which may increase cable attenuation), decreasing d (values in the μm are not only hard to manufacture but mechanically fragile), or reducing the difference between n_1 and n_2 perilously close to zero. Another, and better, solution is the graded index fiber. It is designed such that the refractive index decreases continuously with radial distance from the center of the fiber. Light propagation occurs through refraction (rather than reflection as in a step index fiber), i.e., through a continuous bending of a light ray toward the center line of the fiber. Although light rays travel further as they diverge from the center line, they travel faster in the lower index medium. If the rate of change of refractive index is chosen properly, the modal dispersion can be minimized. For minimum dispersion, the optimum profile is nearly parabolic, that is, the refractive index n_r tapers from the maximum n_1 at the core center to n_2 at the outer radius. The following graph charts the differential refractive index Δn against the radial distance r. Where Δn is $n_r - n_1$. Fiber optic cables

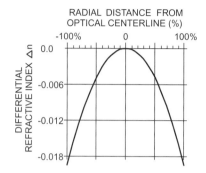

are specified in terms of their core and cladding diameters. For example, a 62.5/125 cable has a core with a 62.5-micron diameter and cladding with a 125-micron diameter. Core diameters for single mode fibers are generally 8–12 microns and 50–100 microns for multimode. The following are some commonly used fiber optic cable configurations:

8/125—a single mode cable with an 8-micron core and a 125-micron cladding. This type of cable is expensive and used only where extremely large bandwidths are needed (such as in some real-time applications) or where long distances are involved. An 8/125 cable is likely to broadcast at a light wavelength of 1300 or 1550 nm.

62.5/125—the most popular fiber-optic cable configuration, used in most network applications. Both 850- and 1300-nm wavelengths can be used with this cable.

100/140—the configuration that IBM first specified for its fiber optic Token Ring network. Because of the tremendous popularity of the 62.5/125 configuration, IBM now supports both configurations.

The fiber core and cladding may be made of plastic or glass. The following list summarizes the composition combinations, going from highest quality to lowest:

Single mode glass—has a narrow core, so only one signal can travel through.

Graded index glass—not tight enough to be single mode, but the gradual change in refractive index helps give more control over the light signal.

Step index glass—the abrupt change from the refractive index of the core to that of the cladding means the signal is less controllable.

Plastic coated silica (PCS)—has a relatively wide core (200 microns) and a relatively low-bandwidth (20 MHz).

Plastic—should be used only for *very short* distances.

Outside the fiber's core and cladding is frequently a 125-micron polymer coating, and finally this is surrounded by the buffer and jacket, bringing the structure to 2.5 mm. Also called a *lightguide*. See also *graded index* and *step index*.

optical fiber, conductive   See *OFC*.

optical fiber, conductive, plenum   See *OFCP*.

optical fiber, conductive, riser   See *OFCR*.

optical fiber, nonconductive   See *OFN*.

optical fiber, nonconductive, plenum   See *OFNP*.

optical fiber, nonconductive, riser   See *OFNR*.

optical fibre   British spelling of optical fiber.

optical filter   An optical element that selectively transmits or blocks specified wavelengths of light.

optical heterodyning   See *optical mixing*.

optical interface   In a fiber optic communications link, a point at which an optical signal is passed from one device or medium to another and without conversion to an electrical signal.

optical isolator   A device that provides electrical isolation between two circuits by converting the electrical signal to an optical signal, transferring it across an optical link and then converting it back into an electrical signal. The optical link may be air or a dielectric waveguide.

The transmitting and receiving elements of an optical isolator may be contained within a single module or separated into multiple cooperative entities (transmitter, lightguide, and receiver). Also called a *optical coupler, optoisolator,* and *photon coupler*. See also *optical waveguide coupler*.

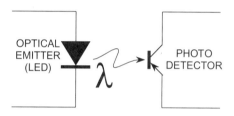

optical junction   Any physical interface in an optical system such as a fiber optic system. Examples include source to fiber, fiber to fiber, fiber to detector, fiber to lens, and lens to lens.

optical link   An optical transmission channel (including repeaters, regenerators, and optical fiber) designed to connect two electronic or optoelectronic communications terminals or in series with other communications channels. Some definitions also include the terminal optical transmitters and receivers, especially in the case of a communications link utilizing separate electronic terminals originally designed for metallic transmission and retrofitted for optical transmission.

optical mixing   Optical beating (heterodyning) of two lightwaves (the received signal and a local reference) in a nonlinear device to produce a beat frequency low enough to be further processed by conventional electronic circuitry. Optical mixing is the optical analog of heterodyne reception of radio signals. Also called *optical heterodyning*.

optical multiplexing   See *wavelength-division multiplexing*.

optical path length   The product of a ray's geometric distance (s) and the refractive index (n) of the material along the path.

- In a medium of constant refractive index,

$$l = n \cdot s$$

- In a medium of varying refractive index, it is the integral of $n \cdot \delta s$ (where δs is an element of length along the path and n is the local refractive index). The length can be approximated by

$$l \cong \sum_{i=1} n_i \cdot \Delta s_i$$

where i is a specific element of length Δs_i and refractive index n_i.

Optical path length is proportional to the phase shift per unit length that a lightwave undergoes along a path. Also called *equivalent air path, optical distance,* and *optical thickness.*

optical power margin In an optical communications link, the difference between the optical power (P_{in}) that is launched by a given transmitter into the fiber, less transmission losses (P_{loss}), and the minimum optical power that is required by the receiver (P_{min}) for a specified level of performance, i.e.

$$P_{margin} = P_{in} - P_{loss} - P_{min}$$

P_{in} is the power launched into the fiber (not the power emitted by the transmitter). It depends on the nature of the active optical source (LED or laser diode), the type of fiber, and parameters such as core diameter and numerical aperture. P_{loss} includes losses from all causes, e.g., coupling, mode coupling, bend, and scattering.

Optical power margin is usually expressed in decibels (dB) and is generally greater than 2 or 3 dB.

optical receiver (**1**) A device that detects an optical signal and converts it to an electrical signal for further processing. (**2**) A device as in (1) above, but including subsequent signal processing.

optical regenerator See *optical repeater.*

optical repeater In lightwave communications systems. A device that receives a signal, amplifies it, and retransmits it. In the case of digital systems, the repeater may also reshape, re-time, or otherwise reconstruct the signal. Also called an *optical regenerator* (especially with digital signals).

optical source (**1**) In optical communications. A device that converts a signal in the electronic domain into a signal in the optical domain. Two commonly used optical sources are light emitting diodes (LEDs) and laser diodes. (**2**) Test equipment that generates a stable optical signal for the purpose of making optical transmission loss measurements.

optical spectrum Generally considered that portion of the electromagnetic spectrum extending from the vacuum ultraviolet (~1 nm) to the far infrared (~0.1 mm). Originally, the term applied to only that portion visible to the human eye. However, it is now considered to extend between the shortest radio waves to the longest wavelength x-rays. See also *frequency band* and *frequency spectrum.*

optical splitter See *directional coupler.*

optical switch (**1**) A switch that enables signals in optical fibers or *integrated optical circuits* (IOCs) to be selectively routed from one circuit to another. An *optical switch* may operate by mechanical means (such as physically moving an output optical fiber among alternative input fibers), electro-optic effects, magneto-optic effects, or other methods.

Slow *optical switches,* such as those using moving fibers, are used for alternate path routing (e.g., routing around a fault). Fast *optical switches* may be used in packet switching systems or to perform logic operations. (**2**) In networking, a switching system that routes optical signals directly in the optical domain, i.e., without conversion to electrical signals, then routes and finally converts back to optical signals.

optical thickness (**1**) The product of the physical thickness of an isotropic (uniform) optical material and its refractive index. (**2**) Of an optical system, the total *optical path length* through all elements. Also called *equivalent air path, optical distance,* or *optical path.* See also *optical path length.*

optical time domain reflectometer (OTDR) In fiber optics, a tool for testing fiber optic cables for integrity and functionality. An *OTDR* analyzes a cable by sending out a pulsed light signal and then checking the amount, timing, and type of light scattered and reflected back from the various discontinuities in the cable. Reflections can occur at any connector, break, crack, sharp bend (kink), or any other flaw (ansiotropic feature) in the medium.

OTDR can aid in:

- The estimation of the attenuation coefficient as a function of distance.
- The identification and location of defects.
- The identification and location of localized losses.

OTDR operates much like RADAR; that is, a pulse of light is launched down the optical fiber, and the time (t) between the transmitted pulse and each reflection is measured. If one knows the index of refraction (n), the exact distance (d) to each anomaly can be determined. That is,

$$d = \frac{c \cdot t}{2 \cdot n}$$

where c is the speed of light.

optical transmitter A device that accepts an electronic input signal, processes the signal, and uses it to modulate an opto-electronic device, such as a light emitting diode (LED) or an injection laser diode (ILD), to produce an optical output signal suitable for transmission on an optical medium.

optical waveguide Any structure capable of guiding optical power. Examples include:

- A glass or plastic sheet that is edge lit.
- An optical fiber.
- A thin film structure used in integrated optical circuits (IOCs).

optical waveguide coupler (**1**) A device that is intended to distribute an optical signal among a number of output ports. (**2**) A device intended to transfer an optical signal between a source and optical waveguide or between an optical waveguide and an optical detector.

optical waveguide splice The permanent joining of two optical fibers in a manner that allows the transfer of an optical signal from one fiber to the other.

optical waveguide termination A device or material placed at the end of an optical fiber to prevent reflection.

optically active material A material that can rotate the polarization of light passing through it. It exhibits different refractive indices for left- and right-circularly polarized light.

optimum traffic frequency See *FOT.*

optimum transmission frequency See *FOT.*

optimum working frequency See *FOT.*

optoelectronic A term that refers to any of a wide range of devices that operate with both electrons (electricity) and photons (light). A device that responds to optical power, emits or modifies optical radiation, or utilizes optical radiation for its internal operation.

Examples of optoelectronic devices include:

- Light emitting diodes (LED)—electricity in and light out.
- Injection laser diode (ILD)—electricity in and light out.
- Kerr cell—electric field controls the transmission of light through the device. (The optical material in a Kerr cell, e.g., nitrobenzine, is an electro-optic material.)

- Photodiodes and transistors—light in controls electrical current flow.
- Photo cell—light in electric current generated.
- Optic coupler—electric current in control electric current out; but the two current paths are isolated through the use of an optical link.
- Integrated optical circuits (IOCs).

See also *electro-optic* (which is often erroneously used as a synonym).

optoisolator A synonym for optical isolator.

OPX An abbreviation of O̲ff-P̲remises eX̲tension.

OR (1) A logical operation for combining multiple Boolean values (true and false) or multiple binary (1 and 0) signals. The rule for the OR operation is: *If any input is true (1), the output is true (1); the output is false (0) only if all inputs are false (0).*

The written symbol for the *OR* function is a plus ("+"); therefore, the *OR* of input A and input B is written A + B. The truth table shows all possible combinations of input and output for a two-input OR operation. A false input or output is tabulated as a "0," while the true is shown as a "1." See also *AND, AND gate, exclusive OR, exclusive OR gate, NAND, NAND gate, NOT,* and *OR gate.*

OR TRUTH TABLE

A	B	A + B
0	0	0
0	1	1
1	0	1
1	1	1

(2) An abbreviation of O̲ff-R̲oute service.

OR gate A type of digital circuit that performs an *OR* operation on two or more digital signals. The symbol for a two-input *OR Gate* is:

See also *AND, AND gate, exclusive OR, exclusive OR gate, logic symbol, NAND, NAND gate, NOR, NOR gate, NOT, OR.*

Orange Book (1) The 1980 compilation of the CCITT/ITU-T standards for telecommunications. **(2)** Specifications relating to CD-R and WORM drive standards. It defines the format that enables CD-R drives to record discs that regular CD-ROM players can read.

The major difference between the *Orange Book* specification and other formats (such as Red and Yellow Book specifications) is that *Orange Book* defines how CD-R devices can append index data to an existing disc's directory in multiple sessions. See also *CD-R* and *WORM.* **(3)** The U.S. government's standards document *Trusted Computer System Evaluation Criteria,* DOD standard 5200.28-STD, December 1985, which characterizes secure computing architectures and defines criteria for trusted computer products. There are four basic levels—D, C, B, and A—each of which adds more features and requirements. Levels B and A provide mandatory control access to which is based on standard *Department of Defense (DOD)* clearances. Within each level may be several sublevels:

- D is a nonsecure system.
- C1 requires user logon but allows group ID.

- C2 requires individual logon with password and an audit mechanism. (Most UNIX implementations are roughly C1 and can be upgraded to about C2 without excessive difficulty.)
- B1 requires DoD clearance levels.
- B2 guarantees the path between the user and the security system and provides assurances that the system can be tested and clearances cannot be downgraded.
- B3 requires that the system is characterized by a mathematical model that must be viable.
- A1 requires a system characterized by a mathematical model that can be proven.

orbit The path of an object (subjected primarily to natural forces, mainly the force of gravity) relative to a celestial object around which it revolves.

orbital inclination The angle between the orbital plane of an orbiting object and the plane of the equator around which it is orbiting, measured at the ascending node.

orbital plane The plane of an orbiting object that contains the radius vector and the velocity vector (in a specified reference system). In the ideal case, the orbital plane remains fixed with respect to the equatorial plane of the primary body.

order of diversity The number of independently fading propagation paths and/or frequencies used in diversity reception.

order of magnitude A way of expressing the *approximate ratio* of two numbers. Generally, the order of magnitude is determined by the base of the number system in use. Hence, in decimal systems, changes magnitudes are a power of 10, while in the binary system, magnitudes are powers of 2.

For example, the two numbers 11 and 90 are said to differ by one order of magnitude, i.e.,

$$\frac{90}{11} = 8.18 \approx 10$$

Similarly, the numbers 3 and 407 differ by two orders of magnitude. See also *SI.*

order tone In telephony, a tone that indicates to an operator that verbal information can be transferred to another operator.

orderwire (OW) An auxiliary communication circuit (voice or data) used in the technical control, coordination, and maintenance of communication facilities. The *orderwire* circuits do not provide any communication paths for customers. Also called *engineering channel, engineering orderwire,* and *service channel.*

ordinate In mathematics, the *y*-coordinate on an (*x,y*) graph. Generally, the output of a function plotted against its input (the *x*-axis or abcissa).

org In the Internet, the top-level domain that is assigned to a nonprofit organization, such as a scientific or scholarly organization. The IEEE (Institute of Electrical and Electronic Engineers) and the EFF (Electronic Frontier Foundation) are examples of organizations using the *.org* domain name. See also *domain.*

Original Equipment Manufacturer (OEM) The maker of equipment marketed by another vendor, usually under the name of the reseller. The *OEM* may make only certain components or complete devices, which can then be configured with software and/or hardware by the reseller.

originate mode In communications, the mode of the device that initiates the call and that waits for the remote device to respond.

originating traffic In telephony, traffic received by a central office switching system from subscriber lines. See also *one-way trunk.*

ORL An abbreviation of Optical Return Loss.

OROM An acronym for Optical Read Only Memory.

ORT An abbreviation of Overload Recovery Time.

orthogonal (1) Geometric, pertaining to objects at right angles (90°) to each other. (2) In mathematics, mutually independent or well separated. (3) A generalization of the mathematical meaning, describing sets of primitives or capabilities that span the entire "capability space" of the system and are in some sense nonoverlapping or mutually independent. (4) In coding, codes that have no correlation among themselves.

orthogonal frequency division multiplex (OFDM) See *discrete multitone (DMT).*

orthometric height The height of a point above the geoid.

OS (1) An abbreviation of Operating System. (2) An abbreviation of Outage Seconds.

OS/2 An operating system originally devised by Microsoft and IBM for PCs based on Intel's 80 × 86 microprocessors. Microsoft and IBM parted ways, with IBM continuing to support *OS/2* while Microsoft developed its Windows products. *OS/2* supports multitasking and programs needing more than 640 kbytes of memory, as well as program-to-program communications. It is a building block on which to base distributed processing.

OSCAR An acronym for Orbiting Satellite Carrying Amateur Radio. The first satellite was launched 12 December 1961.

oscillator An electronic circuit that produces a periodic ac signal (ideally with a perfectly stable fundamental frequency). The signal is frequently used as a stable clock in synchronous systems. See *clock.*

OSD An abbreviation of On Screen Display.

OSF An abbreviation of Open Software Foundation.

OSI An abbreviation of Open Systems Interconnection.

OSI network address An address associated with an entity at the ISO/OSI Reference Model transport layer. This address may be up to 20 octets long and contains two components: a standardized initial domain part (which is the responsibility of the addressing authority for that domain) and a domain specific part (which is under the control of the network administrator).

OSI Network Management Model A network management model that provides a set of concepts and guidelines (not specifications or standards) for various aspects of network management. Although the model does not provide standards or specifications, it furnishes the basis for such specifications. Also called the *ISO network management model,* for the International Standardization Organization (ISO), which developed the model.

OSI presentation address An address associated with an entity at the ISO/OSI Reference Model application layer. This address consists of an *OSI network address* and up to three selectors that identify *service access points (SAPs)* for the presentation, session, and transport layers. The selector values provide layer-specific addresses.

OSI reference model See *Open Systems Interconnection—Reference Model (OSI-RM).*

OSI-RM An abbreviation of Open Systems Interconnection—Reference Model. See *Open Systems Interconnection—Reference Model (OSI-RM).*

OSInet A test network designed to provide a place products based on the OSI Reference Model may be tested for interoperability. (Sponsored by the National Bureau of Standards—NBS.)

OSN An abbreviation of Office Systems Node.

OSPF An abbreviation of Open Shortest Path First. A packet routing algorithm similar to the RIP algorithm; however, the *OSPF* algorithm is expanded and improved.

- It computes the distance between a router and a destination node in "number of hop" units, including attributes such as the data rate of each link, the physical length of each link, and congestion effects.
- Routing tables are built from information broadcast by all *OSPF* routers on the network. Each router broadcasts the information it possesses about routes known to it. Each receiving router uses the information to build an accurate map of possible routes.
- Routing tables are exchanged only when there is a change in link status.
- In case of congestion, the algorithm allows messages to be sent to the data source requesting that it reduce the transmission rate.

See also *RIP.*

OSPFIGP An abbreviation of Open Shortest-Path First Internal Gateway Protocol.

OSME An abbreviation of Open Systems Message Exchange.

OSU An abbreviation of Officially Sanctioned User; a user who is recognized as such by the computer authorities and is allowed to use the computer over the objections of the security monitor.

OTC An abbreviation of Operating Telephone Company. A local exchange carrier (LEC).

OTDR An abbreviation of Optical Time Domain Reflectrometer.

other common carrier (OCC) A communications common carrier (usually an interexchange carrier—IXC) that offers communications services in competition with AT&T and/or the established U.S. telephone local exchange carriers.

OTOH An abbreviation of On The Other Hand.

OTP An abbreviation of One Time Programmable. Refers to a memory device that can be written to only one time. After that, it effectively becomes a read only device.

OTQ An abbreviation of Outgoing Trunk Queue.

OTT An abbreviation of Over The Top. Meaning excessive or uncalled for.

out-of-band signaling (1) In communications, the transmission of some signals (typically control signals) in a channel outside the channel assigned for data. In telephony, for example, the data band for a subscriber is restricted to 300 to 3400 Hz. Therefore, the telephone company could (and does) inject control signals at frequencies greater than 3400 Hz. To keep the subscriber from interfering with out-of-band control signals, filters are used on the subscriber loops that reject out of band signals.

Another *out-of-band signaling* method is to separate the signaling path from the information path. This may be done on a per channel basis as with E & M signaling or on a group of channels as with common channel signaling. See also *common signaling channel* and *in-band signaling.* (2) In personal communication, using methods other than e-mail, e.g., telephone or snail-mail.

out of frame (OOF) The condition whereby a network or its transmission equipment is searching for the framing bit. An *OOF* error condition is declared when either 2 of 4 or 2 of 5 framing bits are missed. A local *red alarm* is generated when the *OOF* error persists for more than 2.5 seconds.

out of synchronization (OOS) The condition whereby a network or its transmission equipment is searching for proper clock phasing or other time alignment.

outage (1) The state of a component or device when it is not available to properly perform its intended function. **(2)** A service interruption, for any cause. **(3)** A telecommunications system service condition in which a user is completely deprived of acceptable service by the system. For a particular system or a given situation, an *outage* may be a service condition that is below a defined system operational threshold, i.e., below a threshold of acceptable performance. See also *outage threshold*.

outage duration The time interval between the onset of an *outage* and the restoration of service.

outage probability The probability that an *outage* will occur within a specified time interval.

outage ratio The ratio of the sum of all *outage durations* to the stated reference time interval.

outage threshold For a particular performance level of a system, the value of a specified parameter below which the minimum acceptable level of performance is not achieved and the system is considered nonoperational.

outgoing access The capability of one network user to initiate communication with a user of another network.

outgoing calls barred A switching system configuration that blocks call origination attempts; only incoming calls are allowed.

outgoing line restriction The ability of a communications system to selectively restrict any line to call reception only.

outgoing station restriction The ability of a switching system to restrict any selected phone (station) from originating calls outside the switching system, that is, only incoming calls are allowed.

outgoing traffic In telephony, traffic delivered directly to trunks from a switching system.

outgoing trunk See *one-way trunk*.

outgoing trunk queuing (OTQ) A PABX feature that allows a call originator to gain access to a busy trunk as soon as it becomes available. In operation, the user dials a busy trunk group, is placed in a waiting queue (after entering an activation code), and is called back by the system when a trunk in the group is available.

OTQ allows a more efficient use of special service lines such as foreign exchange (FX) or wide area telephone service (WATS) lines. See also *callback queuing*.

outpulse dialing A synonym for *rotary dialing*.

outpulsing In telephony, dial pulses from a sender. The process of transmitting address information over a trunk from one switching center or switchboard to another.

output (1) Data that have been processed. **(2)** The state (or sequence of states) that occurs on a specified channel. **(3)** A channel used for expressing the state (or sequence of states) of a device. **(4)** The process of transferring information from an internal device to an external device.

output angle A synonym for *radiation angle*.

output current The total root-mean-square current delivered to a load by a device.

output feedback (OFB) An operating mode for the data encryption standard (DES).

output impedance The impedance presented by the output of a device, circuit, or system to a load, i.e., the impedance seen when looking into the device's output.

output power The power delivered by a device, transducer, or system to its load.

output rating (1) The maximum specified power available at the output terminals of a transmitter when connected to its nominal load. **(2)** Under specified conditions (ambient temperature, supply voltage, etc.), the power that can be continuously delivered by a device without overheating or other deleterious effect.

output span The algebraic difference between the maximum and minimum values of the output range.

outside awareness port (OAP) A humorous IBM term for a window, the glass kind rather than the GUI kind.

outside plant (1) In telephony, that part of the telecommunications network extending between a demarcation point in a switching facility and a demarcation point in another switching facility or customer premises. The demarcation point may be at a distribution frame (generally the main distribution frame, MDF), cable head, subscriber premises protector (or connecting block), or microwave transmitter.

All cables, conduits, ducts, poles, towers, repeaters, repeater huts, and other equipment located between demarcation points are included. However, microwave towers, antennas, and associated repeaters may not be considered part of the outside plant in some administrations. **(2)** In Department of Defense (DoD) communications, the portion of intrabase communications equipment between the main distribution frame (MDF) and a user end instrument or the terminal connection for a user instrument.

ovality In fiber optics, the degree of deviation, from perfect circularity, of the cross section of the core or cladding. Two methods are used to express *ovality;* that is,

- As a first approximation, the cross sections of the core and cladding are assumed to be elliptical. Ovality is then expressed in terms of the lengths of the major axis (D_1) and minor axis (D_2) by

$$\frac{2 \cdot (D_1 - D_2)}{(D_1 + D_2)}$$

- Alternatively, *ovality* may be expressed by a *tolerance field* consisting of two concentric circles, within which the fiber's cross-sectional boundaries must lie. The core or cladding diameter is defined as the average of the diameters of these two circles, that is, $(D_1 + D_2) / 2$.

Also called *noncircularity*. See also *tolerance field*.

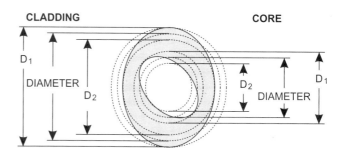

overall objective loudness rating (OOLR) In telephony, a loudness rating of telephone instruments. It is the ratio of the sound pressure level at the mouth reference point (S_M) to the sound pressure level at the ear reference point (S_E) expressed in decibels. (Sound pressure levels are generally expressed in pascals.) Mathematically, *OOLR* is given by:

$$OOLR = -20 \cdot \log_{10}\left(\frac{S_E}{S_M}\right)$$

overcharge The condition of an electrochemical cell or battery when a charging current is applied after all of the active material has been converted to the charged state, that is, charging is continued after the device reaches 100% of its state-of-charge. Generally, the practice is to be avoided because it may reduce the life of the cells or even damage the cells, e.g., because of gassing.

overcurrent Any current in excess of the specified rated value.

overdischarge The process of discharging an electrochemical cell or battery beyond its cutoff voltage. The practice should be avoided because one or more cells may be forced into voltage reversal—damaging the device.

overfill (1) The condition that exists when the numerical aperture (NA) of an optical source, such as a laser, light emitting diode, or optical fiber, exceeds the NA of the driven element, e.g., optical fiber core. (2) The condition that exists when the beam diameter of an optical source, such as a laser, light emitting diode, or optical fiber, exceeds the acceptance angle of the driven element, e.g., optical fiber core.

overflow (1) In telephony, additional potential traffic beyond the capacity of a specified communications element (trunk group, switch, system, or subsystem). The traffic may be lost or handled by designated overflow equipment. (2) In telephony, a count of call attempts made on groups of busy trunks or lines. (3) In telephony, any traffic handled by overflow equipment. (4) In telephony, traffic that exceeds the capacity of the switching equipment and is lost. (5) In telephony, on a particular route, excess traffic that is offered to an alternate route. (6) In computers, the condition that occurs when a mathematical operation generates a result that exceeds the capacity of the number representation system in use. Also called *arithmetic overflow*. (7) In digital computers, the flag bit that indicates the condition of definition (6). (8) In digital communications, the condition that exists when the incoming data rate exceeds that which can be accommodated by a buffer, resulting in the loss of information.

overflow traffic In telephony, that part of the offered traffic that cannot be carried by a server, group of servers, or a carrier. Also called *overflow load*. See also *overflow*.

overhead (1) Generally, an expenditure that does not directly contribute to the desired end result. The expenditure may be power, bits, bandwidth, money, and so on. (2) In communications, that part of a transmission that does not contain user data. These bits are necessary but do not directly contribute to the payload of data, they are therefore *overhead*.

Examples of overhead include:

- Information about the data that follows (e.g., packet size).
- Information to control the channel (e.g., destination address).
- Information to reduce error rate (e.g., parity, CRC, or checksum).
- Retransmissions data received in error.
- Guard time between transmission packets.
- Information concerning network status and operational command.
- Latency time associated with setting up and retransmitting a packet through a routing node.

Stated another way, *overhead* is the amount of bandwidth consumed by the protocols required to transport data across the network. Overhead information originated by a user is not considered to be system overhead information. Overhead information generated within the communications system and not delivered to the user is system overhead information. The user throughput is therefore reduced by both overheads, while system throughput is reduced only by system overhead.

overlap The distance the control of one signal extends into the territory that is governed by another signal or signals. Distance may be measured by any convenient means, e.g., time, angle, frequency, and so on.

overlay In computers, the repeated use of a storage area by different segments of the same program during execution. The area may be reused by subroutines called as needed by the main program, or by different stages of execution of the main program. In all cases, only one segment occupies the storage area at a given time.

overlay network A separate network for a particular service covering essentially the same geographic area as a primary network but operating independently.

overload (1) Generally, loading in excess of the rated maximum capacity of a circuit, device, or system. (2) In communications networks, excessive activity on a network that causes calls to be blocked or lost. See also *overflow*. (3) In signaling, the presentation of an input signal to a device with amplitude sufficiently large that small changes in the input are not reflected in device output. For example, if given an amplifier with a gain of 100 and a maximum output signal capability of 10 V, then, any input signal greater than 100 mV will *overload* the amplifier (100 mV × 100 = 10 V). (4) In electrical circuits, an electrical current that will result in damage from overheating.

overload compression The reduction of the gain of a system or device as the signal level increases.

As an example, most telephone systems specify overload compression with a curve similar to the one below.

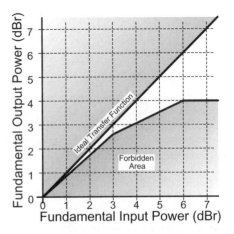

overload point (1) Generally, that point in a linear system when the system behavior becomes nonlinear, that is, the output is not a linear function of the input signal. (2) The signal level in a circuit or transmission system when an increase in input level does not cause a proportional increase in the output. Any attempt to send signals at levels higher than the *overload point* will cause the output signal to be *clipped* or distorted. Generally, the overload point is defined as the signal level that causes the distortion to exceed a specified value.

To a listener a great deal of distortion or noise may be heard. In a digital system, the error rate may become excessive. **(3)** The level of carried traffic when any attempt to increase carried traffic results in a specified degradation of service, e.g., long delays or lost calls. Also called *load capacity*.

overload recovery time The time it takes for the output of a device to return to the active region after momentarily being driven into hard overload.

overmodulation (1) The condition that exists when the *instantaneous level* of the modulating signal exceeds the value necessary to produce 100% modulation of the carrier.

This *overmodulation* results in spurious emissions by the modulated carrier and distortion of the recovered modulating signal and is therefore generally considered a fault condition. **(2)** The condition that exists when the *mean level* of the modulating signal is such that peaks in the modulating signal exceed the value necessary to produce 100% modulation of the carrier. Overmodulation in the sense of this definition may be considered permissible.

overrun (1) The loss of data at a receiving element because it could not accept data at the speed at which it was transmitted. **(2)** A condition that occurs when data in a buffer or register is not read before new data is loaded into the buffer or an attempt to load new data is made.

oversampling The sampling of a signal faster than is required per the Nyquist sampling theorem.

overshoot (1) Of a pulse or step change. The amount by which the first maximum of a step signal change exceeds the final value of the step change (the leading edge). A similar phenomenon may occur at the last transition of a pulse (the falling transition) and is called *undershoot.*

Overshoot and undershoot are generally expressed either as a percentage of their respective final value or as the difference between the maximum value and the final value. See also *pulse.* **(2)** In the transition of a parameter from a low value to a higher value, the transitory value of the parameter that exceeds the final value of the high value. (When the transition is from a higher value to a lower value, and the parameter takes a transitory value that is less than the final value, the phenomenon is called undershoot.) **(3)** The result of an unusual atmospheric condition (e.g., ionospheric skip or ducts) that causes radio signals to be received at ranges beyond the expected distance, i.e., where they are not intended.

oversubscription In frame relay, the condition when the sum of the committed information rates (CIRs) for all permanent virtual circuits (PVCs) on a port exceed the port connection speed. Subscription levels of 200% are typically possible because of dynamic capacity allocation in modern data networks.

overtone An integral multiple of the frequency of a sinusoidal wave (the fundamental). The first overtone is twice the frequency of the fundamental, which corresponds to the second harmonic; the second overtone is three times the fundamental's frequency, which corresponds to the third harmonic, and so forth.

Use of the term is generally confined to:

- Acoustic waves, especially in applications related to music, and
- Piezoelectric crystals such as those used in high-frequency oscillators.

See also *fundamental* and *harmonic.*

overvoltage Any abnormal voltage between two points in a circuit or system that is higher than the highest voltage under normal operating conditions.

overvoltage protector A device that operates when an excessive voltage (such as from lightning or a power cross) is detected. When operated, the voltage or power to the protected circuit is diverted or reduced. The device may be used to attenuate continuous or transient overvoltages. See also *surge protector.*

overwrite To replace the contents of an entity, e.g., a file or a record, with new data.

OVP An abbreviation of <u>O</u>ver <u>V</u>oltage <u>P</u>rotection.

OW An abbreviation of <u>O</u>rder<u>W</u>ire.

P

p The SI prefix for *pico* or 10^{-12}.

P (1) The SI prefix for *peta* or 10^{15}. (2) The symbol for power.

P* An abbreviation for the Prodigy network, especially by members of the various boards on the network itself.

P/AR An abbreviation of Peak to Average Ratio.

P-code The Precise or Protected code used by the Global Positioning System (GPS). It is a very long pseudorandom sequence (about 2.35 × 10^{14} bits long) transmitted on each GPS satellite's carriers at about 10.23 Mbps. The sequence, if allowed to run its course, would repeat about 267 days; however, the code in each satellite is reset weekly to a unique segment. Reset occurs at Saturday/Sunday midnight. The *P-code* is transmitted on both the L1 and L2 carriers. See also *coarse acquisition (C/A) code.*

P.n recommendations An ITU-T series of specifications concerning, "Maintenance: international sound programme and television transmission circuits."

P.nn The grade of service for a telephone system. The digits "nn" indicate the number of calls per 100 that can be blocked by the system, generally during the peak busy hour (the peak hour being the busiest hour of the busiest day of the year). Also written *Pnn.*

PA (1) An abbreviation of Public Address. (2) An abbreviation of PreAmble. (3) An abbreviation of Pre-Arbitrated. (4) An abbreviation of Power Amplifier. (5) An abbreviation of Permanently Associated.

PAA (1) An abbreviation of Phased Array Antenna. (2) An abbreviation of Planar Array Antenna.

PABX An abbreviation of Private Automatic Branch eXchange. An automatic, private telephone-switching device linked to the local central office via analog or digital lines or trunks. The *PABX* may have additional features or services, such as conference, automatic busy, call back, abbreviated dialing, and voice messaging.

The automatic switching differentiates a *PABX* from a private branch exchange (PBX); however, since almost all modern exchanges are automatic, the term PBX is frequently used to mean *PABX.*

pacing In communications, the temporary use of a lower transmission speed. For example, *pacing* may be used to give the receiver time to catch process data previously sent. *Pacing* is the *flow-control* method used in IBM's Systems Network Architecture (SNA). See also *flow control* and *overrun.*

pacing algorithm The mathematical rules established to control the rate at which calls are placed by an automatic dialing machine. Also called a *predictive dialer.*

pacing control A Systems Network Architecture (SNA) term for flow control.

pacing group The number of path information units (PIUs) that can be sent to a destination in IBM's Systems Network Architecture (SNA) before a response is received. A PIU is a data unit.

A *pacing group* is called a Window in other systems.

pack (1) To reduce the size of a block by eliminating unnecessary characters and through the use of data compression. See also *data compression.* (2) In database applications, compression through the elimination of material previously marked for deletion and the reorganization of records.

packet A group of data, control, error control, and sequence information arranged in a specific format suitable for transmission as a single unit across a network. In a *packet* switch network, it is likely that successive packets passing a given point are both from and to different stations.

The largest unit of data is defined by the network's maximum transmission unit (MTU). If the message is longer than the MTU allows, it will be divided into a number of fragments. Each fragment will be placed into the payload of a packet, transported across the network, and recombined to form the original message. Also called a *frame.* See also *packet format.*

packet assembler/disassembler (PAD) An entity that enables data terminal equipment (DTE) not equipped for packet switching to access a packet-switched network. A device that accepts serial data from a DTE and composes packets of information suitable for transmission on the packet network and decomposes packet information from the network and delivers it to the DTE.

One specification is contained in ITU-T recommendations X.3, X.28, and X.29.

packet buffer Memory reserved for temporarily storing packets until they can be processed by the server or sent onto the network. The memory may be either on the network interface controller (NIC) or in the computer to which the controller is attached. Also called a *communication buffer* or a *routing buffer.*

packet burst protocol The ability of a network node to send a number of packets (burst) to a destination before it requires an acknowledgment. A Novell NetWare protocol built on top of the internetwork packet exchange (IPX) protocol that speeds the transfer of data between a workstation and server.

packet filter A process that searches a received packet to determine its destination and then to route it accordingly.

packet format The structure of data, address, and control information in a packet. The size and content of the various fields in a packet are defined by the rules of each network protocol. The format will look something like the following figure.

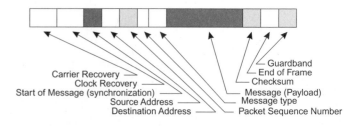

packet forwarding Copying a packet to another node *without* looking at the destination address.

packet header Information about the packet, its destination, and point of origin, e.g., the sequence number, the hop count, the source address, the destination address, and so on.

Sometimes the preamble is included in the definition of packet header. See also *preamble*.

Packet Internet Groper (ping) A system diagnostic that indicates whether a remote host is actually connected to the Internet. To use the diagnostic, simply type *ping* and the host name (URL), for example,

ping remote_host_name

If the remote host is connected, *ping* will display some connection statistics or the message "alive." If it fails, an error message telling why it failed will be displayed.

packet level Level 3 of the ITU-T X.25 recommendation which defines how a user's messages are broken into packets, how data flows across the packet network, and how to handle missing and duplicate packets.

packet level procedure (PLP) A full-duplex protocol that defines the packet transfer strategy between an X.25 data communications equipment (DCE) and an X.25 data terminal equipment (DTE). It supports packet sequencing, flow control, transmission speed maintenance, error detection, and error recovery.

packet mode A method of communications network operation in which packet switching is used rather than message switching.

packet mode terminal Data terminal equipment (DTE) that can control, format, transmit, and receive packets, that is, it has packet assembly/disassembly (PAD) capabilities built in.

packet overhead (1) Those bits in a packet needed for routing and error control. **(2)** A measure of the ratio of the bits needed for non-payload functions (routing, error control, and network control) to the number of data bits. The ratio is generally expressed as a percentage.

packet radio network (PRN) A packet network that uses radio waves as the transmission medium.

packet size The length of a packet; expressed in bits, bytes, or octets.

packet switched data transmission service A communications service that:

- Provides for the transmission of data in the form of packets,
- Switches (routes) data at the packet level, and
- Optionally provides for the assembly and disassembly of data packets.

In addition to the public telephone companies, several commercial companies offer packet switching capabilities to subscribers. CompuServe, SprintNet, and Tymnet are a few of the available services.

packet switching A transmission method in which data to be transmitted are divided into blocks (packets), each containing a destination address. The process routes these addressed packets so that a channel is occupied only during the transmission of the packet. Upon completion of each packet transmission, the channel is made available for the transfer of other packet traffic.

These packets are routed from a calling station to the called station along a path based on the destination address contained in the packet header. The network may operate in a *connection-oriented mode* (where the path is maintained for the exclusive use of the two stations for the duration of the call) or in a *connectionless mode transfer* (where the individual transfer of each data packet is independent of those that precede or follow it). In the illustration, packets are labeled source node, destination node, and packet number, e.g., AF1 indicates that the source node is "A" the destination node is "F" and that it is packet 1 of the message. Both connection-oriented transfers and connectionless transfers are illustrated. The message from node "C" to node "H" is connection-oriented (the circuit is not released during

the idle time between packet 1 and 2) while all other transfers are connectionless. For example, the message from node "B" to node "G" uses the three paths B-N1a-N2a-G, B-N1b-N2b-G, and B-N1-N3-N2-G because, at the time of each packet transmission, the router N1 determines which path is fastest. Note all station to switch links are connection-oriented while the interswitch links may be either.

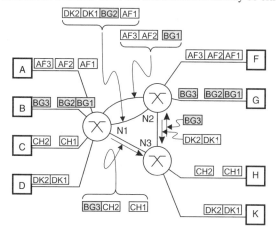

The path that an individual packet takes through the network is determined by the address portion of the packet and the *routing algorithm* within each switch (*router*). See also *circuit switching, connection oriented, connectionless mode transfer, OSPF, RIP,* and *router.*

packet switching network A switched network designed to transport data in the form of packets as opposed to continuous data streams. Each packet in a multipacket message may take a different route from its predecessor, the route being determined by the best path at the time of transmission.

packet switching node (PSN) In a packet switching network (e.g., ARPAnet or MILnet), a dedicated machine (node) that contains data switches and equipment for receiving, formatting, controlling, routing, and transmitting data packets. The term replaced Interface Message Processor (IMP), which in turn was replaced by the term *router.*

In the Defense Data Network (DDN), a packet switching node is usually configured to support up to thirty-two X.25 56 kb/s host connections, as many as six 56 kb/s interswitch trunk (IST) lines to other packet switching nodes, and at least one Terminal Access Controller (TAC).

packet transfer mode A method of information transfer that permits dynamic sharing of network resources among many connections. It is accomplished by means of *packet transmission* and *packet switching.*

packet type identifier Information in a packet header that identifies the packet's function and sequence number (if applicable).

packing density The number of elements per unit space (e.g., length, area, or volume) of storage media.

Examples of the expression of *packing density* include:

- The number of bits or characters stored per unit length of magnetic tape, expressed as bpi or cpi, respectively.
- The number of characters per unit area on a printed page.
- The number of bits stored per unit volume in a hard disk drive.

packing fraction In fiber optics, the ratio of the active area of a fiber bundle to the total area at its emitting or receiving end.

PAD (1) An acronym for Packet Assembler Disassembler. A device that converts a serial data stream into discrete packets in the transmit direction and converts the received packets back into a serial data stream. It adds header information in the transmit packet to allow it to be routed to the proper destination. It is used, for example, for connecting a terminal or computer to an X.25 network. Also called a *Packet Access Device (PAD).* **(2)** A character used to fill out a message block so that the block meets a prescribed length requirement. (Frequently, the .ASCII <SYN> character is used as the fill character.) Also called *fill characters*. **(3)** A device or circuit inserted into an analog signal path for the purpose of reducing (attenuating) signal level (voltage or power), matching the impedance of the source and sink, and without causing any distortions. If the input impedance Z_i is not equal to the output impedance Z_o, the pad may be called a *taper pad*.

Depending on the circuit topology and frequency of use in industry, a pad may be given a unique name, such as a bridged T-pad, an H-pad, L-pad, O-pad, T-pad, or a Π-pad (pi-pad). The attenuation and impedance characteristics of some of the possible networks are listed in the following tables. The following variable definitions are used in all figures:

A Is the power attenuation in dB,

K Is the voltage attenuation ratio and is unitless,

Z_i Is the input impedance in ohms,

Z_o Is the output impedance in ohms, and

R_1, R_2, and R_3 are the individual circuit element in ohms.

Only two of the three parameters Z_i, Z_o, or A may be *independently* set for the L-pad (also called a *minimum loss* or *impedance matching* pad). The third value is *dependent* on those two.

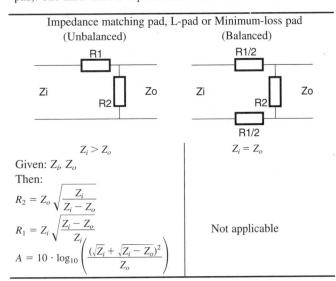

Impedance matching pad, L-pad or Minimum-loss pad

(Unbalanced) (Balanced)

$Z_i > Z_o$ $Z_i = Z_o$

Given: Z_i, Z_o

Then:

$$R_2 = Z_o \sqrt{\frac{Z_i}{Z_i - Z_o}}$$

$$R_1 = Z_i \sqrt{\frac{Z_i - Z_o}{Z_i}}$$

$$A = 10 \cdot \log_{10}\left(\frac{(\sqrt{Z_i} + \sqrt{Z_i - Z_o})^2}{Z_o}\right)$$

Not applicable

The T-pad (and its balanced equivalent, the H-pad) may have all three of the variables Z_i, Z_o, and A chosen *independently* (as long as A > A_{min}).

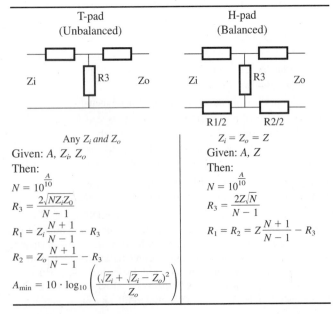

T-pad H-pad
(Unbalanced) (Balanced)

Any Z_i and Z_o $Z_i = Z_o = Z$

Given: A, Z_i, Z_o Given: A, Z

Then: Then:

$$N = 10^{\frac{A}{10}}$$

$$R_3 = \frac{2\sqrt{NZ_iZ_o}}{N - 1}$$

$$R_1 = Z_i \frac{N + 1}{N - 1} - R_3$$

$$R_2 = Z_o \frac{N + 1}{N - 1} - R_3$$

$$A_{min} = 10 \cdot \log_{10}\left(\frac{(\sqrt{Z_i} + \sqrt{Z_i - Z_o})^2}{Z_o}\right)$$

$$N = 10^{\frac{A}{10}}$$

$$R_3 = \frac{2Z\sqrt{N}}{N - 1}$$

$$R_1 = R_2 = Z \frac{N + 1}{N - 1} - R_3$$

The Π-pad (and its balanced equivalent, the O-pad) are like the T-pad / H-pad pair in that all three of the variables Z_i, Z_o, and A may be chosen *independently* (as long as A > A_{min}).

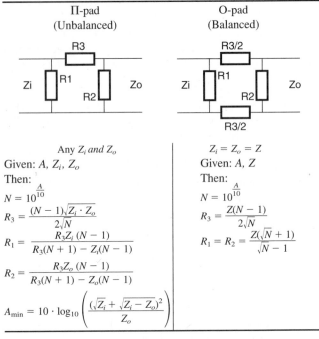

Π-pad O-pad
(Unbalanced) (Balanced)

Any Z_i and Z_o $Z_i = Z_o = Z$

Given: A, Z_i, Z_o Given: A, Z

Then: Then:

$$N = 10^{\frac{A}{10}}$$

$$R_3 = \frac{(N - 1)\sqrt{Z_i \cdot Z_o}}{2\sqrt{N}}$$

$$R_1 = \frac{R_3 Z_i (N - 1)}{R_3(N + 1) - Z_i(N - 1)}$$

$$R_2 = \frac{R_3 Z_o (N - 1)}{R_3(N + 1) - Z_o(N - 1)}$$

$$A_{min} = 10 \cdot \log_{10}\left(\frac{(\sqrt{Z_i} + \sqrt{Z_i - Z_o})^2}{Z_o}\right)$$

$$N = 10^{\frac{A}{10}}$$

$$R_3 = \frac{Z(N - 1)}{2\sqrt{N}}$$

$$R_1 = R_2 = \frac{Z(\sqrt{N} + 1)}{\sqrt{N} - 1}$$

The bridge (or mesh) are balanced pads and have two independent parameters, typically impedance (Z) and attenuation (A).

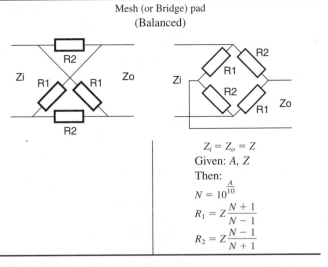

Mesh (or Bridge) pad
(Balanced)

$Z_i = Z_o = Z$

Given: A, Z

Then:

$$N = 10^{\frac{A}{10}}$$

$$R_1 = Z \frac{N + 1}{N - 1}$$

$$R_2 = Z \frac{N - 1}{N + 1}$$

pad switching A technique of automatically or manually maintaining approximately constant transmission levels for different network operating conditions.

pager A mobile receiver for paging communications, also known as a beeper. See also *paging* (2).

paging (1) In computers, a memory allocation strategy that effectively increases memory or allows more flexible use of available memory. A page is a contiguous block of memory in random access memory (RAM) of predefined size. The specifics of paging strategies differ quite drastically among operating systems.

For example,

• In DOS, a virtual memory is created on disk. When portions of working memory (random access memory—RAM) need to be removed temporarily, those portions can be stored on disk to make room.

- In Novell's NetWare, 4-KB memory pages are assigned as needed for process use. *Page tables* map between the physical memory associated with the page and the logical address space (for the process) provided by the pages.

(2) A means of alerting someone that there is a message for them even when that person's exact location is unknown. Two methods of *paging* include:

- A public address (PA) system in which an audible broadcast announcement is made. If the person being paged is in range of the system, he will respond as directed by the page.
- A radio broadcast system that transmits a signal to a portable receiver (pager) carried by the person to be notified. When a message is waiting for the person with a pager, the paging system transmits a signal that only the designated pager receives. The pager, upon acceptance of the signal responds with an alert (such as a tone, optical readout, or vibration) that the user hears (or feels) and appropriately responds.

pair In communications, two equal conductors employed to form an electric circuit (loop); one conductor carries current to a load, while the second conductor forms the return path for the current. Generally, the pair is also twisted to reduce its susceptibility to external noise and to reduce the radiated electromagnetic field strength. See also *twisted pair* (*TWP*).

pair gain In telephony, the multiplexing of a number of phone conversations of a lesser number of physical circuits. Generally, *pair gain* refers to facilities with concentrators outside the central office itself, i.e., near the group of subscribers for which the concentration is to be applied.

The pair *gain* is defined as the number of simultaneous conversations possible with the specific system minus the number of cable pairs required to transport them. See also *digital loop carrier* and *SLC96*.

paired cable A cable in which all conductors are arranged as *twisted pairs* and with each pair not associated with any other pair (i.e., no quads are formed). Also called *nonquaded cable.*

paired disparity code A code in which some or all of the symbols are represented by two sets of digits of opposite connotation that are used in sequence so as to minimize the total inequality of a longer sequence of digits. The digits may be represented by any of several physical quantities, such as two different frequencies, phases, voltage levels, magnetic polarities, or electrical polarities, each one of the pair representing a 0 or a 1.

An alternate mark inversion (AMI) signal is an implementation of a *paired disparity code* in which opposite voltage polarities are used to minimize the long-term voltage offset.

PAK A data compression utility for DOS-based machines. The file extension is generally ".PAK." See also *compression* (2).

PAL (1) An acronym for Programmable Array Logic. **(2)** An acronym for Proprietary ALgorithm. A privately designed and owned method of coding, encrypting, or otherwise operating on a digital data stream. **(3)** An acronym for Phase Alternating by Line system. A 625-line, 25-frame-per second, color television system used in many parts of the world, e.g., the United Kingdom and much of the rest of western Europe, several South American countries, some Middle East and Asian countries, several African countries, Australia, New Zealand, and other Pacific island countries. The system avoids the color distortion that appears in the National Television Standards Committee (NTSC) systems.

Other systems include NTSC, the U.S. 525-line, 30-frame-per-second interlaced color system; and the Système Electronique Couleur Arec Memoire (SECAM) (Sequential and Memory) 819-line system used in France.

PAL-M A modified version of the phase-alternation-by-line (PAL) television signal standard (525 lines, 50 Hz, 220 V primary power), used in Brazil.

PAM An acronym for Pulse Amplitude Modulation.

panel system In telephony, an early automatic telephone switching system with the following features and characteristics:

- Contacts of the multiple switch banks are mounted vertically in flat panels.
- The wiper of the switches is raised and lowered by a motor that is common to a number of selecting mechanisms.
- The dialing information is received and stored by controlling equipment that governs subsequent operations necessary to establish a communications channel.

PANS An acronym for Pretty Amazing New Stuff.

PAP (1) An acronym for Printer Access Protocol. See also *AppleTalk.* **(2)** An acronym for Password Authentication Protocol.

paper-net Found on the Internet, in particular USEnet. A way of referring to the standard postal service, i.e., comparing it to a very slow, low-reliability network. Also called *papernet* and *P-Net.*

PAR (1) An acronym for Positive Acknowledge with Retransmission. **(2)** An acronym for Peak to Average Ratio. **(3)** An acronym for Performance Analysis and Review.

parabola A geometric shape that reflects energy from a distant source to a single point or in the reverse, reflects energy from a point source to a narrow beam.

The general equation describing a parabola is:

$$y = ax^2 + b$$

an example of the shape generated by this equation (with a-/and b-0) is:

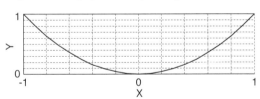

parabolic antenna An antenna consisting of a *parabolic reflector* (the cross section is a parabola) and a radiating (or receiving) element at or near its primary focus. Commonly called a *dish* or *dish antenna.*

parabolic profile In fiber optics, a graded index optical fiber with a *power law index profile* in which the profile parameter $g = 2$. That is, the refractive index of the fiber's core varies as a parabolic function of radial distance. Also called *quadratic profile.* See also *graded index* and *power law index profile.*

parallel (**1**) Two or more devices arranged so that an action or operation on one device is also taken on the other device.

- In electronics, an electronic device is parallel with another when they are connected between the same circuit nodes. In this arrangement, a voltage applied to one device is also applied to the other device.
- In systems, when two or more systems receive the same input information, perform essentially the same operations on the information, and deliver like responses to input stimuli, the devices are said to be in parallel.

(**2**) Pertaining to the concurrent or simultaneous operation of two or more devices. (**3**) Pertaining to the concurrent or simultaneous activities in multiple devices or channels. (**4**) Pertaining to the simultaneous processing of individual parts of a whole, such as the simultaneous processing of bits within a character, using separate facilities for each part.

parallel communications The transfer of information between devices by means of multiple circuits, each circuit handling one signaling element.

The most common example of a parallel communication circuit is the link between a computer and a printer. On a personal computer, for example, the link transfers 8 bits at a time. Compare with *serial communications*.

parallel computer A computer that has multiple arithmetic units or logic units that are used to accomplish parallel operations or parallel processing.

parallel interface An interface that can transfer multiple data bits simultaneously by sending each bit over a separate wire or communications channel. Data transmission timing is controlled by a strobe pulse.

Examples in a computer include a parallel printer port, the communications link to a disk drive (IDE, EIDE, and SCSI), and the internal busses (memory, ISA, VESA, PCI . . .).

parallel port One of the I/O ports on data terminal equipment (DTE) where information is transferred byte by byte (8 bits at a time) as opposed to the *serial port* where information is transferred bit by bit. A printer is an example of the kind of device connected to a *parallel port*.

Probably the best known *parallel port* is the one developed by Centronics, a printer manufacturer, that has 36 pins and can transfer an 8-bit byte of data at a time. Also called a *parallel interface*. See also *serial port*.

parallel processing The simultaneous use of more than one computer to solve a problem. The basic premise can be illustrated by the following reasoning:

> If it takes one person one hour to dig a post-hole, then sixty people can dig it in one minute.

The reasoning also illustrates that not all problems lend themselves to parallel processing because of the difficulty of cooperation. *Parallel processing* computers may either communicate in order to cooperate, run independently, or operate under the control of a master computer that distributes activities and collects results (sometimes called a *processing farm*). Communication among individual processors may be by bus, a network (e.g., a local area network—LAN), shared memory, or a combination. See also *multiple instruction, multiple data* (*MIMD*) and *single instruction, multiple data* (*SIMD*).

parallel resonance The sinusoidal steady-state condition that exists in a circuit composed of an inductor in parallel with a capacitor when the applied frequency is such that:

- The driving point impedance is maximum.
- The magnitude of the impedance of the inductor (X_L) is equal to the impedance of the capacitor (X_C).
- The phase angle of the driving point impedance is zero.

parallel T network See *twin T network*.

parallel transmission (**1**) The simultaneous transmission of multiple bits of an information packet (usually a character) over multiple channels or on different carrier frequencies of the same channel. For example, dual-tone multifrequency (DTMF) signaling requires the simultaneous transmission of two of eight frequencies for each symbol. (**2**) The simultaneous transmission of multiple related signal elements over an equal number of communications channels or paths. See also *parallel interface.*

parallax The optical illusion that causes a foreground object's position to seemingly change when the observer's position changes.

parameter (**1**) A variable that is given a constant value for a specified application. (**2**) A value that is supplied as a specification for the configuration or performance of a device or to provide a control instruction to a computer program. In most instances, a parameter will have or get a default value if neither the user nor the application specifies such a value.

Parameters may be determined at the time of hardware manufacture and then be reported in the produce specification. Alternatively, they may be assigned to a register before the beginning of an operation as a means of customizing the operation. For example, most modems have a number of registers that can be altered to modify its behavior. In software, a *parameter* may be used to modify the behavior of a program. For example, software switches, such as /P in the DOS command DIR /P, tells DOS to list only one page of the directory rather than the entire directory. (**3**) A name in a procedure that is used to refer to an argument passed to that procedure.

parameter RAM (PRAM) A small battery backed-up memory area in a Macintosh computer that stores important system configuration information (parameters); such as the node's network address, desktop pattern, and selectable memory configuration.

parametric amplifier (paramp) (**1**) A very *low noise amplifer* (*LNA*) using a nonlinear or time-varying reactance (generally capacitive but could be inductive) to achieve gain. Types of *parametric amplifiers* include:

- One-port
- Two-port (traveling wave tube)
- Degenerate (pump frequency twice the signal frequency)
- Nondegenerate
- Multiple pumps
- Multiple idlers

The most common *paramp* is the nondegenerate one-port configuration with a circulator because of its low noise figure and low circuit complexity; conceptually shown as

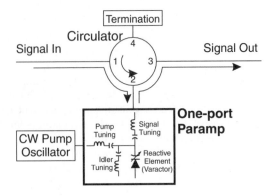

The noise temperature of a *paramp* cooled to the temperature of liquid helium (about 5 K) can be about 20 K with gains of 20–40 dB. **(2)** In optics, a means of amplifying optical waves whereby an intense coherent pump wave (usually from a laser) is made to interact with a nonlinear optical crystal to produce amplification at two other wavelengths.

parametric device A device whose operation depends on the time variation of a characteristic parameter, generally a reactance.

parametric oscillator In optics, a device using a parametric amplified inside a resonant optical cavity to generate a wavelength tunable coherent light beam from an intense, fixed wavelength laser. The device is tuned by changing the phase matching properties of the nonlinear optical material.

parasitic element **(1)** Generally, any unwanted element that is an unavoidable consequence of using real, wanted devices. For example, a simple piece of wire wanted as a conductor also has parasitic characteristics as an inductor and as a capacitor to neighboring elements. **(2)** In an antenna, a directive element that is not connected to a radio transmitter or receiver either directly or via a feeder but is coupled to the driven element only by the fields. Also called a *passive element.*

parasitic oscillation Any unintended self-sustaining oscillation in a device.

paraxial Characteristics of an optical analyses that are limited to infinitesimally small apertures. Also called *first order* or *gaussian.*

paraxial ray In fiber optics, a light ray that is close to and nearly parallel with the optical axis of an optical device.

PARC Universal Packet (PUP) In the Internet system developed by Xerox Corporation, a *PUP* is the fundamental unit of transfer, just as the Internet Protocol (IP) datagram in a Transmission Control Protocol/Internet Protocol (TCP/IP) internet. The name was derived from the name of the laboratory at which the Xerox internet was developed, the Palo Alto Research Center (PARC).

parity A form of error checking. In a binary system, the condition where the total number of bits in the "1" state is always odd (or even). An auxiliary bit, called the *parity bit,* is appended to the message bits and is controlled in such a way that the sum of all bits (including the parity bit) will be even for even parity and odd for odd parity.

Parity checking is not a very strong method of error detection. For example, if two bits are in error, the parity is unchanged and therefore assumed to be correct. However, for short sequences, such as over a single 7- or 8-bit character, it is often sufficient. If parity is checked on a character-by-character basis, it is also called *vertical redundancy checking* (*VRC*). If the parity is checked on a block-by-block basis, it is called *block parity* or *longitudinal redundancy checking* (*LRC*). The communication protocol must be the same at both sending and receiving data communication equipment (DCE); therefore, the DCEs must agree on the parity parameter before data transmission starts. The possible settings of the parity bit parameter in communication programs are:

PARITY BIT DEFINITION

Designation	Description
No parity	No parity bit is used.
Odd parity	The parity bit is set so that the *total number* of "1s" in each block is an *odd* number.
Even parity	The parity bit is set so that the *total number* of "1s" in each block is an *even* number.
Mark parity	The parity bit is used and is always set to 1.
Space parity	The parity bit is used and is always set to 0.

parity bit An extra bit used in checking for data transmission errors. It is computed by the sending terminal and appended to each data block prior to transmission; the receiving terminal checks the bit for validity. If the bit is not correct, an error is assumed to have occurred and appropriate action is taken by the receiving terminal. See also *parity.*

parity check See *parity.*

park **(1)** A private automatic branch exchange (PABX) feature that allows a user to transfer a call from the active station to another station without ringing that station. It is useful if the user needs to go to a different phone to get information the other party wants. **(2)** A PABX feature that allows a user to put the phone on a type of hold that any user of that pickup group can pick up. **(3)** The "home" position of the heads on a hard disk drive. In this position, the heads are mechanically lifted off of the magnetic media so that physical damage cannot occur during moving.

park timeout The period of time before an unanswered call on *park* is redirected to the prime phone for that call.

parser A program that enables a user to read a document that has been prepared with a markup language, such as SGML (Standard Generalized Markup Language) or HTML (HyperText Markup Language). The parser replaces each tag with an appropriately formatted text display so that it is easily read.

Part 15 The section of Title 47 of the U.S. Code of Federal Regulations (CFR) governing radiated electromagnetic waves, i.e., RF radiation. See also *FCC.*

Part 68 The section of Title 47 of the U.S. Code of Federal Regulations (CFR) governing the direct connection of telecommunications equipment and customer premises wiring with the public switched telephone network. See also *FCC.*

party The person or device at either end of a communications path. The *originating party* is also called the *calling party,* while the *destination party* is called the *called party.*

party line A *communications line* that serves more than one main station. In the case of telephony, it is a subscriber line arranged to serve more than one main station, with discriminatory ringing for each station. (Each main station may have extension stations, and discriminatory ringing may be accomplished by distinctive ringing, frequency selective ringing, tip/ring/bridge ringing, or other methods.)

Party lines are used primarily in rural areas where loops are long. Privacy is generally limited (or nonexistent), and congestion often occurs. Also called *multiparty line.*

passband The range of frequencies (or wavelengths in optical devices) which will traverse a device, communication channel, circuit, or system with minimum attenuation.

The limiting frequencies denoting the edge of the *passband* are defined as those at which the relative power (or intensity) decreases to a specified fraction of the maximum level. This decrease in power is often specified to be the half-power points, i.e., 3 dB below the maximum power. The difference between the limiting frequencies is called the bandwidth and is expressed in hertz (in the optical portion of the spectrum, in nanometers or micrometers). Also called the *passband* or *transmission band.* See also *band-pass filter, band reject filter, high-pass filter, low-pass filter, stopband,* and *transition band.*

passband ripple The difference between the maxima and minima loss in a filter passband. If the differences are of constant amplitude, then the filter is called an equiripple filter.

passive component A *device* in which energy may only be lost. Examples include an *attenuator* that *dissipates* energy to decrease signal level or an impedance-matching transformer that interfaces a 50-

Ohm transmission circuit to a 100-Ohm digital driver. See also *active component.*

passive coupler In fiber optic communications, a coupler that splits a signal among a number of output ports. There is always signal loss with a passive coupler as the input power is divided among the outputs and there is no signal regeneration. See also *coupler.*

passive element Any component, device, or system that receives no power other than that from the signal source itself. Examples include wire, resistors, capacitors, inductors, diodes, optical fibers, lenses, and optical filters.

passive filter A filter network containing only passive components, e.g., inductors, capacitors, resistors, and transformers.

passive hub A passive hub serves as a wiring and relay center and passes the signal on, without changing it in any way. *Passive hubs* do not require a power supply. They are used in low-impedance ARCnet networks.

passive network A network that requires no source of energy for operation.

passive optical network A technique for the public network which allows the multiplexing of many individual local loops onto a single fiber cable.

passive reflector A simple reflector (such as a mirror) used to change the direction of a radiated beam (microwave or optic).

passive repeater A passive system used to redirect a microwave beam. It consists of a receiving antenna (horn or parabolic antenna) directly coupled to the transmitting antenna via a short waveguide link. There are no active electronics in the system.

passive satellite A communications satellite that is a reflector, that is, it does not process the signal in any way. Although the satellite acts passively by reflecting signals, it may contain active devices for station keeping.

passive splice An optical cable splice made without monitoring the splice losses.

passive star A network configuration in which the central node of a star topology passes a signal on but does not process the signal in any way. This is in contrast to an active star configuration in which signals are regenerated before being passed on.

passive star coupler A fiber optic coupler (optical signal redirector) created by fusing multiple optical fibers together at their meeting point, creating the center of a star configuration. This type of coupler is used in an IEEE 802.4 optical Token Bus network that uses a passive star topology. Also called a *retransmissive star.*

passive station All stations on a network except those actively engaging in the transfer of information; generally, the master and one or more slaves. The passive stations only monitor the line for supervisory messages.

password A method employed by computer systems of identifying a user in a manner more secure than simply the user's name. A *password* is a unique string of alphanumeric characters and symbols chosen by the user. It is to be kept secret by the user (to be known only by the user, i.e., don't write it down!). The *password* is given to the computer at the time of logon; the computer matches the user name and *password* to its internal record. If there is a match, the computer grants the user access to the services that have been approved by the system administrator. Systems that allow access only to users with valid passwords are said to be *password protected.*

A good password is a random mix of numbers, symbols and alphabetic characters, such as $3Fxq!M. Some examples of poor passwords include:

- The system default!
- Your name, part of your name, or a transformation of your name, e.g., reversed order or reversed initials.
- Someone else's name—wife, girlfriend, boss, etc.
- An identification associated with the user, e.g., social security number, address, birthday, or phone number.
- Common acronyms or jargon.
- Easy-to-type sequences such as QWERTY or !@#JKL.
- Words found in a common dictionary.

Passwords should be changed regularly and should change whenever someone leaves the organization. Some systems use dynamic password protection. Dynamic password systems provide a special type of password scheme in which a user's password is changed every time the user logs into a network. In these systems, the user uses a device, called a *remote password generator* (*RPG*), to generate the password, that is, when the user requests a login, the network host responds with a number, which the user must enter into the RPG, together with the user's own personal identification number (PIN). The RPG then generates the session password. See also *access codes* and *joe account.*

password aging A system security scheme utilized by some hosts wherein the user is asked to change his or her password after a period of time (one to six months).

password authentication protocol (PAP) Internet RFC1334.

password length equation An equation that determines an appropriate password length (M), which provides an acceptable probability (P) that a password will be guessed in its lifetime.

The *password length* is given by

$$M = (\log S)/(\log N)$$

where:
S is the size of the password space, and
N is the number of characters available.

The *password space* is given by

$$S = LR/P$$

where:
L is the maximum lifetime of a password, and
R is the number of guesses per unit of time.

password length parameters Basic parameters affecting the *password length* needed to provide a given degree of security.

Password length parameters are related by

$$P = LR/S$$

where:
P is the probability that a password can be guessed in its lifetime,
L is the maximum lifetime a password can be used to log in to a system,
R is the number of guesses per unit of time, and
S is the number of unique algorithm generated passwords (the password space).

The degree of password security is determined by the probability that a password can be guessed in its lifetime. See also *password length equation.*

patch (1) A temporary connection of circuits, generally with a *patch cord.* **(2)** Instructions added to a software routine (generally, at the last minute or after it is delivered to the customer) to remedy a problem or oversight not discovered in qualification testing. A patch is not part of the original planning or logic of the program. Patches tend to accumulate between revisions or new releases. In fact, they sometimes motivate the next revision as too many patches are difficult to manage over a long term.

patch and test facility (PTF) A facility in which supporting functions, including

- Quality control checking and testing of equipment, links, and circuits,
- Troubleshooting,
- Activating, changing, and deactivating of circuits, and
- Technical coordinating and reporting,

are performed.

patch bay An equipment rack so arranged that a number of circuits, usually of the same or similar type, appear on jacks for monitoring, interconnecting, and testing purposes. They are used in locations such as telephone exchanges and technical control facilities.

patch cable (1) A cable used in a patch panel. **(2)** A cable used to connect two hubs (or medium attachment unit—MAUs). IBM Type 1 or Type 6 patch cables can be used for Token Ring networks. These cables will have special IBM data connectors at each end.

patch cord A short length of wire (or optical) cable with connectors on each end used to join circuits on a temporary basis.

patch panel One segment of a patch bay. A device in which temporary connections can be made between circuits by using a patch cord. They are used, for example, to connect test equipment to specific circuits or to modify or reconfigure a communications system while such testing is being done.

path (1) The list of directories through which DOS searches to find a program. **(2)** In communications, the string of links across which a message traverses when it moves from a source to a destination. The links may be copper wire, or optical fiber, or radio (wireless). Also called a *circuit, link,* or *connection route.* **(3)** In a computer program, the logical sequence of instructions executed by a computer. **(4)** In database management systems, a series of physical or logical connections between records or segments, usually requiring the use of pointers. **(5)** In radio communications, the route (which may consist of multiple concatenated links) that lies between a transmitter and a receiver. Such paths may be line-of-sight and/or ionospheric.

path control layer The network processing layer in IBM's Systems Network Architecture (SNA) that **(1)** handles the routing of data units as they travel through the network and **(2)** manages shared link resources.

path information unit (PIU) A packet created in IBM's Systems Network Architecture (SNA) network communications when the path control layer adds a transmission header to a basic information unit from the transmission control layer above.

path loss In a communication system, the attenuation experienced by an electromagnetic wave in transit between a transmitter and a receiver. It may be due to many effects such as free-space loss, refraction, reflection, aperture-medium coupling loss, and absorption. *Path loss* is usually expressed in decibels (dB). Also called *path attenuation.*

pathname The complete description of a filename on a computer or network. It may include the computer name, disk drive name, and all of the subdirectory names as well as the name assigned to the file.

Pawsey stub A device for connecting an unbalanced coaxial feeder to a balanced antenna.

PAX In telephony, an abbreviation of Private Automatic eXchange. See also *PABX, PBX,* and *private exchange (PX).*

pay telephone A telephone station designed to generally operate only when monetary tokens are inserted. In some instances, the telephone is enabled to call emergency numbers and/or the operator without depositing any tokens. The pay telephone was invented by William Gray, an American inventor.

Pay telephones may require the call to be paid for before the connection is established, called *payphone-prepay,* or the call may be paid for after the call is completed, called *payphone-postpay,* which typically is a credit card type of transaction. Also called *payphone, paystation,* or *public phone.*

pay tone See *metering tone.*

payload That part of a digital packet, cell, or datagram that contains user data, i.e., exclusive of the header, trailer, and any other overhead information. In this case the error control is considered part of the trailer.

For example, the payload in an asynchronous transfer mode (ATM) cell is the data portion of that cell. The cell consists of a 5-octet header and a 48-octet payload. Also called *mission bit stream.*

payload type (PT) The indicator in the header of an asynchronous transfer mode (ATM) cell that makes it possible to identify the classification of information held in the payload of a cell. See *ATM.*

PBA (1) An abbreviation of Printed Board Assembly. **(2)** An abbreviation of Pencil Beam Antenna. **(3)** An abbreviation of Pill Box Antenna.

PBD An abbreviation of Programmer Brain Damaged. A commentary about poorly written software. Applied to bug reports revealing places where the program was obviously broken by an incompetent or short-sighted programmer.

PBEM An abbreviation of Play By Electronic Mail.

PBER An abbreviation of Pseudo Bit Error Ratio. In adaptive high-frequency (HF) radio, a bit error ratio derived by a majority decoder that processes redundant transmissions. It is determined by the extent of error correction.

PBM An acronym for Play By Mail.

PBN An abbreviation of Physical Block Number.

PBX In telephony, an abbreviation of Private Branch eXchange. A *PBX* is a manually operated telephone exchange serving a single organization and having connections to a public telephone exchange. The term *PBX* is often used to indicate an automatic switching system, i.e., private automatic branch exchange (PABX). See also *PABX.*

PBX tie line A tie line connecting two private automatic brand exchanges (PABXs), thereby permitting stations in one PABX to call stations in the other PABX without using the public switched network.

PBX trunk A circuit connecting a private branch exchange/private automatic branch exchange (PBX/PABX) to the local telephone company's central office switching equipment.

PC (1) An abbreviation of Personal Computer. **(2)** The name of a particular IBM computer. **(3)** An abbreviation of Physical Contact. **(4)** An abbreviation of Peg Count. **(5)** An abbreviation of Printed Circuit. **(6)** An abbreviation of carrier power (of a radio transmitter). **(7)** An abbreviation of Primary Center.

PC Card™ The peripheral device standard, promulgated by Personal Computer Memory Card International Association (PCMCIA), which deals with certain electrical and mechanical parameters. See also *PCMCIA.*

PC-DOS An acronym for IBM's Personal Computer Disk Operating System.

PCA (1) An abbreviation of Protective Connecting Arrangement. **(2)** An abbreviation of Printer Communications Adapter. **(3)** An abbreviation of Printed Circuit Assembly.

PCB (1) An abbreviation of Printed Circuit Board. **(2)** An abbreviation of Power Circuit Breaker. **(3)** An abbreviation of Power Control Box. **(4)** An abbreviation of Process Control Block. **(5)** An abbreviation of Product Configuration Baseline. **(6)** An abbreviation of Program Control Block. **(7)** An abbreviation of Protocol Control Block.

PCF An abbreviation of Physical Control Fields. The *PCF* consists of the Access Control (AC) and Frame Control (FC) bytes in a Token Ring network header.

PCI (1) An abbreviation of Peripheral Control Interface. **(2)** An abbreviation of Peripheral Component Interconnect (local bus). **(3)** An abbreviation of Protocol Control Information.

PCL® An abbreviation of Hewlett-Packard's Printer Control Language. A Document description language used by Hewlett-Packard (HP) Laserjet printers. It is a superset of HP-GL/2.

PCM An abbreviation of Pulse Code Modulation.

PCMCIA (1) An abbreviation of People Can't Memorize Computer Industry Acronyms. **(2)** An abbreviation of Personal Computer Memory Card International Association, a group of manufacturers and vendors who developed and are promoting a standard for *PC Card™* based peripherals (and the card slot designed to hold them). Designed primarily for portable notebook and laptop computers, the *PC Card™* itself is a removable assembly containing devices such as memory, modems, disk drives, network access, or even a global positioning satellite receiver.

The release 2.01 *PCMCIA PC Card™* slot will accept a card that is 54 mm wide and 85.6 mm long (about the size of a credit card) and conforms to the *Type I, II, III* thickness specifications. The connector in the *PCMCIA PC Card™* slot is a 68-pin female connector. The *PCMCIA PC Card™* slot may be one of the following three official types and a fourth unofficial extension:

- *Type I:* Release 1.0 (introduced September 1990) defines a slot that will accommodate a device up to 3.3 mm thick. *Type I* cards are expected to be used primarily for memory (RAM, ROM, EEP-ROM, etc.).
- *Type II:* Release 2.0 (introduced September 1991) defines a slot that will accommodate a device up to 5 mm thick. *Type II* cards include devices such as modem, facsimile, and network cards.
- *Type III:* Release 2.01 fixed some problems in release 2.0 and added the *Type III* slot, which will accommodate one device up to 10.5 mm thick or two *Type I* or *Type II* cards. *Type III* cards include disk drives and other devices that require more height.
- *Type IV (unofficial)*: A 16-mm slot, used by several manufacturers, may become a de facto standard because of its ability to accept an RJ-11 jack and large-capacity hard drives. It has not been adopted as a *PCMCIA* standard and is opposed by many because it is too large for subnotebooks and personal digital assistants (PDAs).

Socket Services software provides a standard interface to PC Card hardware, and Card Services software coordinates access to the actual cards. Up to 4080 cards can be supported on a single computer.

PCMCIA modem A modem on a PCMCIA PC Card.

PCN An abbreviation of Personal Communication Network. A network composed of wireless networks and metropolitan area network (MANs).

It is a low-power, short-range, digital radio link for voice and data terminals that can be accessed by a personal user number. It allows portability only in local areas of coverage; however, it can be programmed to "travel" with a user to different local areas.

PCR An abbreviation of Peak Cell Rate.

PCS (1) An abbreviation of Personal Communication Services. Technology that utilizes the *PCS* capabilities includes the personal digital assistant (PDA). A portable electronic data entry device consisting of a combination of a computer, modem, fax, and cellular telephone in one device. A typical *PCS* furnishes some or all of the following services:

SOME PCS SERVICES

Appointment scheduling	Calendar	E-mail
Pen-based notepad	Calculator	Facsimile
Word processing	Spreadsheets	Paging
Simple data entry	File transfer	Printing
Cellular telephone	Phone book	

See also *DECT, ISM, LEOS,* and *SMR*. **(2)** An abbreviation of Plastic Clad Silica. A type of optical fiber, with a glass core and plastic cladding. The performance of this fiber type is inferior to that of an all-glass fiber.

PCSA An abbreviation of Personal Computer Systems Architecture.

.PCX See *filename extension*.

PD (1) An abbreviation of Public Domain. **(2)** An abbreviation of Physical Delivery.

PDA An abbreviation for Personal Digital Assistant, a handheld device combining computing, telephony/fax, and networking capabilities. Most *PDAs* use a stylus and touch screen rather than a keyboard for input. The first PDA was Apple Computer Inc's Newton MessagePad, introduced in 1993.

PDAD An abbreviation of Proposed Draft ADdendum.

PDOP An acronym for Position Dilution of Precision.

PDAM An acronym for Proposed Draft AMendment.

PDAU An abbreviation of Physical Delivery Access Unit.

PDL An abbreviation of Page Description Language.

PDM An abbreviation of Pulse Duration Modulation.

PDN An abbreviation of Public Data Network. A communications network designed specifically for the transmission of data in digital form, such as from a computer, and accessed by many individual public subscribers—just as is the public telephone network (PTN). Most *PDNs* are packet switched networks rather than circuit switched as are PTNs.

PDP An abbreviation of Digital Equipment Corporation's (DEC's) Programmed Data Processor computer family. Members of the family include the PDP-8, PDP-10, and PDP-11/xx.

PDQ An abbreviation of Pretty Damned Quick.

PDS (1) An abbreviation of Premises Distribution System. **(2)** An abbreviation of Protected Distribution System.

PDU An abbreviation of Protocol Data Unit. In the ISO/OSI Reference model, a packet. Specifically, it is a packet originating at a specified layer in one network entity and terminating in the same layer of another entity. The *PDU* is used to communicate between the same layers on different machines.

PDU lifetime A value that indicates the number of routers a protocol data unit (PDU) can encounter before it must reach its destination or be discarded. This discard measure is necessary to keep "lost" packets from traveling around from node to node forever and as more "lost" packets enter the network, eventually using all available bandwidth.

PE (1) An abbreviation of Processing Element. (2) An abbreviation of Phase Encoding. (3) An abbreviation of Professional Engineer.

peak (1) In general, a synonym for maximum. (2) In electronics, the highest point or maximum value on a waveform. Also called the *crest*.

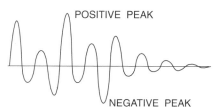

peak cell rate (PCR) An asynchronous transfer mode (ATM) traffic management parameter defined by the ATM Forum.

peak emission wavelength In optics, the spectral line of an optical emitter having the greatest power. Also called *peak wavelength* and *peak emission wavelength*.

peak envelope power (PEP, pX, or PX) The average power delivered to an antenna transmission line by a radio transmitter during one radio frequency cycle at the crest of the modulation envelope under normal operating conditions.

peak hour See *busy hour.*

peak limiting A process by which the absolute instantaneous value of a signal parameter is prevented from exceeding a specified value. See also *clipping.*

peak load (1) For a network, the maximum load that can be (or is) placed on a network. This value may be expressed in any of several performance measures, including transactions, packets, or bits per second. (2) In telephony, the peak load is generally expressed for the busiest one-hour period during the busiest day of the year.

peak power output Output power averaged over that cycle of an electromagnetic wave having the maximum peak value that can occur during transmission.

peak signal level (1) In a transmission link, the maximum *instantaneous* signal power, voltage, or current at a specified point. (2) In a transmission link, the maximum signal power, voltage, or current at a specified point that occurs over a specified period.

peak to average ratio (P/AR or PAR) (1) The ratio of the instantaneous peak value (maximum magnitude) of a signal parameter (voltage, current, power, phase, or frequency) to its time-averaged value. Also called *crest factor.* (2) The ratio of a standard analog transmission line test signal of varying frequencies and amplitudes as compared to the received version of the signal. The resulting composite number has a value of 1 to 100. PAR is a quick way to test the quality of a transmission channel and is essentially a measure of envelope delay. In the Bell standard, a value of 48 is the minimum allowable for a medium-speed data channel.

peak-to-peak value The absolute value of the difference between the maximum and the minimum magnitudes of a varying quantity.

peak wavelength (1) The wavelength of an optical band-pass filter that has the lowest loss. (2) A synonym for *peak emission wavelength.*

peaked load The load resulting from *peaked traffic.*

peaked traffic Random traffic in which the ratio of the variance to the mean is greater than one.

peer Any of the devices on a layered communications network that operate at the same protocol level or the equivalent layer in another protocol.

peer hub A hub that is implemented on a card that plugs into an expansion slot of a PC.

peer layers In a layered network architecture, corresponding layers on two stations.

peer protocol The sequence of message exchanges between two entities at the same layer which lead to the transfer of data, control information, or both from one location to the other.

peer-to-peer network A network in which all nodes have equal opportunity to access the network and the ability to share resources with other nodes. Therefore, a dedicated server is not necessary but can be implemented.

peg count In telephony, the notation of the number of occurrences of an event or condition.

The term originated in the early days of telephony (before fancy electronics and computers) when wooden boards with pegs (arranged in units of 1, 10, 100, and 1000) were used to record the number of events. Each time a specified event occurred, a person would move the pegs to reflect the new value.

PEK An abbreviation of Phase Exchange Keying.

pel An acronym for Picture ELement. The smallest area on a display that is controlled by software. A *pel* may consist of multiple pixels. See also *pixel.*

PEM An abbreviation of Privacy Enhanced Mail.

penetration (1) The passage through a partition or wall of an equipment or enclosure by a wire, cable, or other object. (2) The unauthorized act of bypassing the security mechanisms of a cryptographic system. (3) The passage of a radio frequency through a physical barrier, such as a partition, wall, or building.

PEP (1) An acronym for Telebit's Packetized Ensemble Protocol. A proprietary modulation protocol that analyzes the transmission characteristics of the telephone line at the time of connection and dynamically compensates for telco line impairments. The analysis is done at 511 frequencies. (2) A deprecated abbreviation for *peak envelope power.* Either PX or pX is now the preferred symbol. See also *peak envelope power* (*PEP, pX, or PX*) and *power* (*P*).

percentage modulation (1) In angle modulation, the fraction of a specified reference modulation, expressed in a percentage. (2) In amplitude modulation (AM):

- The modulation factor expressed as a percentage.
- The number of decibels (dB) below 100% modulation.

performance management One of five Open Systems Interconnection (OSI) network management domains specified by the International Standard Organization (ISO) and International Telecommunications Union (ITU). This domain is concerned with the following:

- Monitoring the day-to-day network activity. Addresses data collection.
- Gathering and logging data based on this activity, such as utilization, throughput, and delay values. Addresses data collection.
- Storing performance data as historical archives to serve as a database for planning network optimization and expansion. Addresses data analysis.
- Analyzing performance data to identify actual and potential bottlenecks. Addresses data analysis.
- Changing configuration settings in order to help optimize network performance. Addresses control and optimization.

periapsis The orbital point of a satellite at its closest point to the gravitational center of the primary attracting body. Synonymous with:

perigee when the primary body is the Earth,

perilune when it is the Moon, and

perihelion when it is the Sun.

See also *apoapsis* and *perigee.*

perigee The point at which an Earth-orbiting satellite is closest to the center of the Earth. See also *apogee* and *periapsis*.

period (1) In general, the time required for any uniformly repeating process to complete one iteration or cycle. (2) In electronics, the entire description of a wave from one unique point on a wave to the same corresponding point on the next wave (measured in units of time). Any point on the wave may be used as the starting point as long as the same point on the next cycle is used as the ending point. Two common points are the peaks of two successive waves and the zero crossings of two successive rising (or falling) waveforms. *Frequency* (f) in hertz and *period* (t) in seconds are related to each other by a reciprocal. That is, $f = \dfrac{1}{t}$ and $t = \dfrac{1}{f}$. See also *wave*.

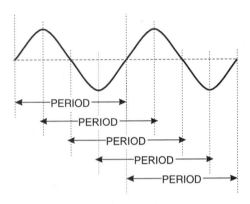

periodic antenna An antenna that has an approximately constant input impedance over a narrow range of frequencies. A dipole array antenna is an example of a periodic antenna. Also called a *resonant antenna*.

periodic function Any function whose behavior over time can be completely described by repeating the behavior over a finite time interval, the *period*. Mathematically, this relationship is expressed $f(x) = f(x + nk)$ where n is any integer and k is the period, e.g., $\sin(x) = \sin(x + n \cdot 2\pi)$.

periodic wave Any wave in which the displacement at each point is a *periodic function* of time.

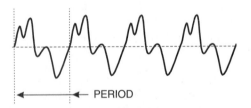

Peripheral Component Interconnect (PCI®) The PC local bus design from Intel. A local bus is a bus that is connected directly to the central processing unit (CPU).

The PCI bus specification includes:

- 64-bit data paths.
- Arbitrated bus mastering (interrupt handling based on priority levels).
- Secondary caching to speed up operations.
- Accommodations for increases in processor speeds.

See also *EISA, ISA, MicroChannel,* and *VESA.*

peripheral equipment Any equipment or device not an integral part of a system but working with that system in a cooperative way. The peripheral device may operate independently of the host device after it has been given appropriate commands.

Some examples of peripheral equipment include:

- In a computer system, a printer, fax/modem, tape drive, CD-ROM, optical scanner, disk drives, joysticks, keyboards, mice, and monitors. (Because keyboards, mice, and monitors are nearly essential parts of a computing system, they are sometimes considered extensions of the processor rather than *peripherals.*)
- In telephony, a printer, CRT display, voice mail system, and so on.

peripheral router A router that primarily connects a network to a larger Internetwork. This is in contrast to a central router, which is a transfer point for multiple networks.

periscope antenna An antenna configuration in which the initial transmitting antenna is arranged to produce a vertical beam, and a second flat or off-axis parabolic reflecting antenna, mounted above the transmitting antenna, is used to direct the beam in a horizontal path toward the receiving antenna. This arrangement has two benefits:

- It enables increased terrain clearance without long transmission lines.
- It permits active equipment to be located at or near ground level for easier maintenance.

permahold That infinitely long maze and waiting queue one encounters when calling many businesses today. You know the one, the one answered by a computer, not the one answered by a person (usually).

permanent connection A connection that will retain its mechanical and electrical integrity for the design life of the conductor.

permanent signal (PS) In telephony, a sustained off-hook supervisory signal originating outside a switching system not related to a call in progress, i.e., not followed by dialing, and that persists beyond a designated timeout period.

permanent virtual circuit (PVC) A type of connection between nodes in a packet switching network such as *frame relay*. The connection is a fixed logical path (a virtual circuit) established between two locations. Since the path is fixed, a *PVC* is the equivalent of a dedicated line but over a packet switched network. It is used to guarantee a minimum committed information rate (CIR).

Once a *PVC* is defined, it requires no call setup procedure before data are sent and no disconnect following data transmission. Previously called a *nailed-up circuit.*

permeability (μ) The ratio of magnetic induction in a given medium to that which would be produced in free space. For all practical purposes, air can be used as the free-space reference.

permission Privileges granted to each network user on an individual and group basis that determines the user's access rights to such things as system access, file or folder access, and commands available. Permissions are under the control of the system administrator.

person call A call wherein (1) the switching equipment attempts to locate the person at known locations, e.g., office, home, or car, or (2) the call is a cellular or wireless call to a mobile type telephone. See also *place call.*

person-to-person call An operator-assisted call wherein the communications path is established only if the designated person is reached by the operator at the destination.

personal communicator A hand-held computer (with fax/modem and cellular communications capability) that combines computing, communications, and personal organizer functions. The device provides some or all of the functions or services in the following table. Also called a *Personal Digital Assistant* (*PDA*) by Apple Computer, Inc. among others.

SOME PDA FUNCTIONS & SERVICES

Appointment scheduling	Calendar	E-mail
Pen-based notepad	Calculator	Facsimile
Word processing	Spreadsheets	Paging
Simple data entry	File transfer	Printing
Cellular telephone	Phone book	

personal computer (PC) Strictly speaking, a single-user microcomputer or minicomputer designed for applications wherein the user has complete control over the data and software accessed. It should be noted, however, that PCs are frequently networked and used as servers so that users may more easily share information. See also *mainframe, microcomputer, microprocessor, minicomputer,* and *workstation.*

Personal Computer Systems Architecture (PCSA) Digital Equipment Corporation's (DEC's) systems and networking architecture that merges DOS, OS/2, and VMS operating system environments. It is based on a client–server model and is part of DEC's Pathworks.

personal digital assistant (PDA) See *personal communicator.*

personal identification number (PIN) A unique code, assigned to an individual for use in transactions on certain types of networks, e.g., to do banking transactions through an automatic teller machine (ATM) or to log onto networks that use dynamic passwords.

personal mobility In *universal personal telecommunications* (*UPT*), a term describing:

- The ability of a user to access telecommunication services (both originate and receive) at any *UPT* terminal on the basis of a personal identifier.
- The capability of the network to provide those services in accord with the user's service profile.
- The network's ability to locate the terminal associated with a user for the purposes of addressing, routing, and charging the user for calls.

The personal mobility aspects of personal communications are based on the *UPT* number. Management of the service profile by the user, however, is not part of personal mobility.

personal registration In *universal personal telecommunications* (*UPT*), the process of associating a UPT user with a specific terminal.

personal terminal In *personal communication services* (*PCS*), a small, lightweight, portable terminal that provides the capability for a user to be either stationary or in motion while using telecommunication services.

Personalized Ringing See *distinctive ringing.*

PF (1) An abbreviation of <u>P</u>ower <u>F</u>actor. (2) An abbreviation of <u>P</u>ower <u>F</u>rame.

PFC An abbreviation of <u>P</u>ower <u>F</u>actor <u>C</u>orrection.

PFD An abbreviation of <u>P</u>ower <u>F</u>lux <u>D</u>ensity.

PFN An abbreviation of <u>P</u>ulse <u>F</u>orming <u>N</u>etwork.

PGA (1) An abbreviation of <u>P</u>rogrammable <u>G</u>ate <u>A</u>rray. See also *gate array.* (2) An abbreviation of <u>P</u>in <u>G</u>rid <u>A</u>rray. (3) An abbreviation of <u>P</u>rofessional <u>G</u>raphics <u>A</u>dapter. An IBM PC monitor resolution. (4) An abbreviation of <u>P</u>ower <u>G</u>ain <u>A</u>ntenna.

PGC An abbreviation of <u>P</u>ort <u>G</u>roup <u>C</u>ontroller.

PGP An abbreviation of <u>P</u>retty <u>G</u>ood <u>P</u>rivacy—a data encryption program.

PH (1) An abbreviation of <u>P</u>acket <u>H</u>eader. (2) An abbreviation of <u>PH</u>antom circuit.

phantom circuit In telephony, a third circuit derived from two physical two-wire circuits or a single four-wire circuit. The physical circuits are called *side circuits* to the *phantom*. One implementation is shown in the following figure. The phantom circuit may be used for delivering power to remote circuits, or it may be used as a third voice channel. The transformers are usually called *repeat coils*.

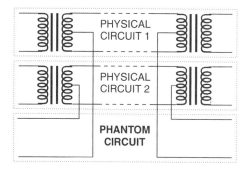

Current from the *phantom circuit* is split evenly at the transformer center taps. This cancels crosstalk from the phantom circuit to the side circuits. Also called a *simplex circuit*.

phantom group The three circuits that are derived from simplexing two physical circuits to form a phantom circuit. See also *phantom circuit*.

phase (1) The position of one wave relative to another wave (the reference wave) of the same frequency. *Phase* is measured in degrees (or in radians—there are 2π radians in $360°$). One wave by definition is $360°$ in length, so another wave that starts at 50% of the reference wave has a *phase angle* of $180°$.

The highlighted wave below has been shifted 25%, or $45°$, from the reference wave (dotted wave). Signal relationship may be represented in polar coordinates by $M < \theta$, where M is the magnitude and θ is the

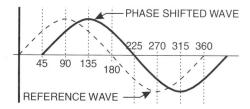

phase angle or in Cartesian coordinates, i.e., an *Argand diagram,* as $(a + jb)$, where a is a real component and b is an imaginary component such that $\tan \theta = (b/a)$, where θ is the phase angle, and the magnitude, $M = (a^2 + b^2)^{1/2}$. See also *lag, lead,* and *phase shift.* (2) A distinguishable state of a phenomenon. (3) A method of delineating various points in a time sequence or sequential list of functions. For example, in the reception of a fax there are five *phases:*

Phase A—the establishing of the call.

Phase B—premessage procedure in which capabilities are exchanged.

Phase C—data transmission and line monitoring, synchronization, and problem detection.

Phase D—end of page procedure that loops back to Phase C if more pages are to be transferred or to Phase E if done.

Phase E—the call release procedure.

phase angle The measure of the difference of phases of a signal and a reference signal. See also *phase*.

phase bandwidth The width of the continuous frequency range over which the phase-vs.-frequency characteristic of a network or device does not depart from linearity by more than a stated amount.

phase coherent A condition in which two signals maintain a fixed phase relationship with each other. If two signals are *phase coherent* with a third (reference) signal, then they are *phase coherent* with each other.

phase constant In fiber optics, the imaginary part of the axial propagation constant for a particular transmission mode. Generally, it is expressed in radians per unit length.

phase corrector A network inserted into a signal path that is designed to correct *phase distortion* within a specified range of frequencies.

phase crossover frequency The frequency at which the phase shift of a network is 180°.

phase delay In the transmission of a single frequency wave from one point to another, the delay of an arbitrary point in the wave that identifies its phase. Phase delay may be expressed in any convenient unit, such as seconds, degrees, radians, or wavelengths.

Phase delay is the phase shift at a specific frequency divided by that frequency. See also *envelope delay*.

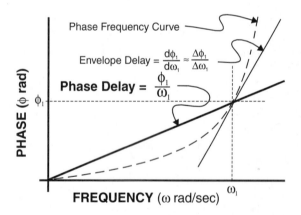

phase delay distortion *Phase delay distortion* occurs when the *phase delay* does not vary linearly with frequency. It is possible to have *phase delay distortion* and not *envelope delay distortion*.

phase departure (1) A phase deviation from a specified value. (2) An unintentional deviation from the nominal phase value.

phase detector A circuit that compares the phase relationship of two signals (without regard to other parameters such as amplitude or waveshape) and generates an output voltage that is proportional to the phase difference between the two signals.

Phase detectors are used in a variety of systems extending from phase lock loops (PLL) to phase modulation receivers. See also *phase lock loop (PLL)* and *phase shift*.

phase detector gain signal The factor relating a phase detector output voltage to the phase difference between the two input signals.

phase deviation In phase modulation (PM), the maximum difference between the instantaneous phase angle of the modulated wave and the phase angle of the unmodulated carrier. For a sinusoidal modulating wave, the phase deviation, expressed in radians, is equal to the *modulation index*.

phase difference The difference in *phase* between two waves of the same frequency.

phase distortion Distortion that occurs when:

- The phase frequency characteristic is not linear over the frequency range of interest. That is, the phase shift introduced by a circuit or device is not directly proportional to frequency, or
- The zero-frequency intercept of the phase frequency characteristic is not 0 or an integral multiple of 2π radians.

Also called *phase frequency distortion*. See also *delay distortion*.

phase encoding See *Manchester coding, modulation*, and *waveform codes*.

phase equalizer See *delay equalizer*.

phase frequency characteristic A Cartesian coordinate plot of phase shift as the dependent variable vs. frequency as the independent variable. The *phase frequency characteristic* is linear if the introduced phase shift is the same for all frequencies of interest.

phase hit A short term and unwanted significant phase shift in an analog signal. Phase hits are more severe than phase jitter as they often cause errors in digital transmission systems.

phase instability A fluctuation of the phase of a wave, relative to a reference.

phase interference fading Variations of signal amplitude produced by the interaction of two or more signal elements with different relative phases, e.g., elements due to time-varying reflections.

phase jitter Usually random fluctuations of the phase angle of a signal relative to a reference. Phase jitter may be expressed in degrees, radians, or seconds and, if periodic, it may be expressed in hertz also. Typically, *phase jitter* is distinguished from other forms of phase perturbation (e.g., drift) by being smaller in magnitude but more rapid in change (approximately the same order of magnitude as the information rate).

Phase jitter may make it difficult to synchronize a local clock to the signal.

phase jump A sudden phase change in a signal.

phase linearity Direct proportionality of phase shift to frequency over the frequency range of interest.

phase lock loop (PLL) A circuit used for synchronizing a locally generated signal (clock) with the phase of a received signal. The circuit generates two outputs, though it generally is not used at the same time:

- The "average" frequency of the input signal with a fixed phase difference relative to the input signal, e.g., the data clock if the input signal is a data stream.
- The phase error signal between the input signal and the local voltage controlled oscillator (VCO). This error signal may represent a modulating signal at the generator of the input signal, i.e., data.

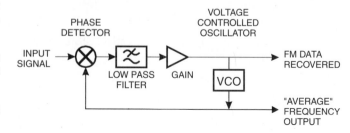

There are several bands of frequencies over which the performance of a *PLL* changes depending on the previous state of the device:

- *Hold range*—the frequency band ($-\Delta\omega_H$ to $+\Delta\omega_H$) over which a *PLL* will remain locked if the input frequency is slowly changed. Also called the *tracking range* or *lock range*.
- *Lock-in range*—the frequency band ($-\Delta\omega_L$ to $+\Delta\omega_L$) in which a *PLL* will acquire lock without cycle slipping.
- *Pull-in range*—the maximum frequency band ($-\Delta\omega_{PI}$ to $+\Delta\omega_{PI}$) over which a *PLL* will acquire lock. Also called the *capture range*.
- *Pull-out range*—the frequency range ($-\Delta\omega_{PI}$ to $+\Delta\omega_{PI}$) over which a stepped input frequency change will not cause the *PLL* to lose lock. For frequencies greater than the pull-out frequency but less than the pull-in range, the *PLL* will reacquire lock.

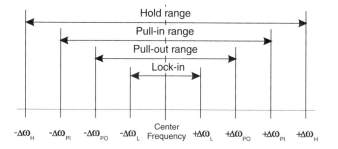

phase margin In feedback systems, 180° minus the value of the phase angle at a frequency where the loop gain is unity.

phase modulation (PM) A form of angle *modulation* in which the *phase* of the carrier wave is varied in accordance with some information-bearing signal. When the modulating signal is a digital signal, the modulation is referred to as *phase shift keying* (*PSK*). The diagram shows a digital bit stream modulating a carrier such that four phases are generated, that is, dibits equal to 00 produce a modulator output of 0°. Similarly, 01 yields +90°, 10 yields −90°, and 11 yields 180° relative to the input carrier. Also called *angle modulation*. See *modulation, phase shift key,* and *waveform codes.*

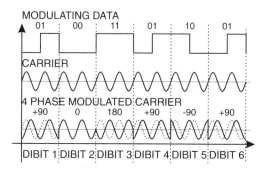

phase noise **(1)** A term generally used to describe short-term random frequency fluctuations of a signal. **(2)** Rapid, short-term, random fluctuations in the phase of a wave. Phase noise, $\phi(f)$ in decibels relative to carrier power (dBc) on a 1-Hz bandwidth, is given by

$$\phi(f) = 10\log_{10}\left(\frac{S \cdot \Phi(f)}{2}\right)$$

where $S\,\Phi(f)$ is the spectral density of phase fluctuations. In an oscillator, *phase noise* is caused by time domain instabilities.

phase nonlinearity Lack of direct proportionality of phase shift to frequency over the frequency range of interest.

phase offset A synonym for *phase difference.*

phase perturbation Any shifting, random or periodic, from any cause, and at any rate, in the phase of a signal with respect to a reference. Examples of *phase perturbation* include:

- *Phase drift*—the relatively slow change of signal phase angle.
- *Phase jitter*—the relatively fast change of signal phase angle.
- *Phase jump*—a sudden change of signal phase angle.

Phase perturbations may be expressed in degrees, radians, or seconds. If there is a periodic component, it may be expressed in hertz as well.

phase quadrature See *quadrature.*

phase shift The lead or lag of one signal parameter with respect to another signal parameter. The signals and parameters may be the same or different, for example, the phase shift between the voltage on signal A and current on signal A could be of interest, or in other cases the phase shift between signal A and signal B could be with respect to either voltage or current. To determine *phase shift,* the waveshapes need not be the same shape; they need only be periodic and the same frequency. See also *lag* and *lead.*

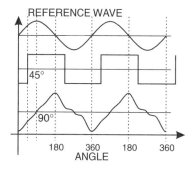

phase shift key (PSK) Phase modulation of a carrier with a digital information signal. The figure diagrams the result of modulating a carrier with a binary digital signal, that is, each of the two states of the data produces a unique phase at the output of the modulator.

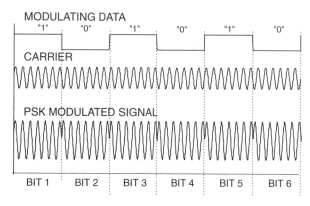

The phase shift may be in relation either to a reference phase or to the phase of the immediately preceding signal element, in accordance with data being transmitted. Because the modulating signal is a digital signal, the phase changes are discrete points on a constellation plot, as is shown in the diagrams that follow. The two preceding *PSK* constellations are of a two-phase (biphase) and an eight-phase *PSK*-modulated carrier. The amplitude of all signals is a constant 1.0; only the phase is changed during modulation. Some of the common *phase shift keying* schemes include:

  BPSK—binary PSK or biphase PSK

  QPSK—quaternary PSK

MPSK—multilevel PSK

DPSK—differential PSK

Also called *phase shift signaling*. See also *modulation* and *phase modulation*.

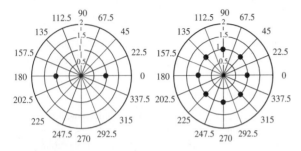

phase splitter A device that produces, from a single input signal, two or more output signals that differ by a constant phase angle from one another.

phase term In the propagation of an electromagnetic wave in a uniform waveguide (an optical fiber or metal waveguide), the parameter that indicates the phase change per unit distance of the wave at any point along the waveguide.

phase velocity (1) Of a traveling wave at a given frequency, the velocity of an equiphase surface in the direction of propagation. (2) In fiber optics, for a particular mode, the ratio of the angular frequency to the phase constant.

phased array A group of cooperative antennas in which the relative phases of the respective signals feeding the antennas are controlled in such a way that the effective radiation pattern of the array is reinforced in a desired direction and suppressed in undesired directions. That is, the beam shape and direction are controlled electronically rather than mechanically.

The relative amplitudes, physical characteristics, and physical location of the individual antennas in addition to the driving signals determine the constructive and destructive interference effects, and therefore, the effective radiation pattern of the array as a whole.

PHF An abbreviation of Packet Handling Function. The switching capability that processes and routes X.25 virtual calls.

PHM An abbreviation of PHase Modulation.

phon In acoustics, a unit of subjective loudness level equal to the sound pressure level in decibels (dB) compared to that of an equally loud reference standard sound. The generally accepted standard is a 1-kHz pure sine wave tone or narrowband noise centered at 1 kHz.

phone (1) An abbreviation of telephone or telephony. (2) Colloquially, the voice operation mode in radio communications.

phone phreak A communications "hobbyist" who likes to figure out how the telephone system operates and how to bypass its billing aspects.

phonetic alphabet A list of standard words used to identify letters in a message transmitted by radio or telephone. The table lists the authorized words for each letter in the alphabet:

PHONETIC ALPHABET

A lpha	H otel	O scar	V ictor
B ravo	I ndia	P apa	W hiskey
C harlie	J uliet	Q uebec	X -ray
D elta	K ilo	R omeo	Y ankee
E cho	L ima	S ierra	Z ulu
F oxtrot	M ike	T ango	
G olf	N ovember	U niform	

phonon A quantum of acoustic energy, the level of which is a function of the frequency of the acoustic wave. Phonons in acoustics are analogous to photons in electromagnetics. The energy of a phonon is usually less than 0.1 eV (electron-volt) and thus is one or two orders of magnitude less than that of a photon.

Photo CD A standard developed by Kodak and Philips that allows the recording and viewing of photographic images on a CD-ROM. Up to 100 high-resolution images can be stored on a Photo CD. Any computer with appropriate software and a Photo CD compatible CD-ROM drive, a Photo CD Player, or a CD-i Player is required to view images on a Photo CD.

photocell A contraction of photoelectric cell. A light-sensitive device in which use is made of the current/voltage variations as a function of incident radiation.

photoconductivity The conductivity increase of some nonmetallic materials due to the absorption of photons (light energy). The increase results from increases in the number of free carriers generated when photons with sufficient quantum energy to overcome the bandgap are absorbed.

photocurrent The current that flows through a photosensitive device, such as a photodiode, as the result of exposure to radiant power.

The *photocurrent* may occur as a result of photoelectric, photoemissive, or photovoltaic effects. The photocurrent may be enhanced by internal gain caused by interaction among ions and photons under the influence of applied fields, such as occurs in an avalanche photodiode (APD).

photodetector A device that converts received optical signals into electrical signals. Two main types of semiconductor photodetectors are the photodiode (PD) and the avalanche photodiode (APD).

photodiode A semiconductor diode that, as a result of the absorption of photons (i.e., is exposed to light), produces

- A photovoltage or
- Free carriers that support the conduction of photocurrent.

Photodiodes are used in the detection of optical signals and for the conversion of optical power to electrical power. Sometimes also spelled *photo diode*.

photoelectric effect In certain materials, the changes in the electrical characteristics caused by photon absorption. Two such effects are:

- The emission of electrons from a material's surface when optically irradiated (external effect).
- The decrease in resistivity due to the exposure of radiant power (internal effect).

See also *photoconductivity*.

photon A discrete packet (basic unit) of light, analogous to the electron. It is a quantum of electromagnetic energy whose energy (w) is given by Einstein's quantum formula:

$$w = \frac{h \cdot c}{\lambda} = h \cdot f$$

where

h is Planck's constant,

c is the speed of light,

λ is the wavelength of the radiation, and

f is the electromagnetic wave frequency.

See also *frequency spectrum*.

photon noise In an optical communication link, noise attributable to the statistical nature of optical quanta. See also *quantum noise*.

photonic layer The lowest of four layers of Synchronous Optical Network (SONET). It specifies the type of optical fiber, emitter, and detector to be used.

phototransistor A transistor that is sensitive to light, and hence, is capable of amplifying small changes of incident light.

photovoltaic effect A term that describes a material that generates a voltage when illuminated with light, e.g., that voltage generated across a PN junction as a result of absorbed photons. The voltage is a result of the internal drift of holes and electrons. See also *PN (3)*.

PHP An abbreviation of P̲ersonal H̲andyP̲hone, a Japanese MPT cordless telephony standard. The scheme uses TDMA/TDD technology in the 1.895–1.918 GHz frequency band. Also called *Japanese Digital Communications (JDC)*, *Personal Digital Communications (PDC)*, and *Personal Handyphone System (PHS)*. See also *DECT, IEEE 802.11,* and *ISM*.

PHR An abbreviation of P̲H̲ysical R̲ecord.

phreaking From *phone phreak.* **(1)** The "art and science" of cracking the telephone network (so as, for example, to make free long-distance calls). **(2)** By extension of **(1)**, security-cracking in any other context especially, but not exclusively, on communications networks. See also *cracker.*

PHS An abbreviation of P̲ersonal H̲andyphone S̲ystem. See also *Personal HandyPhone (PHP)*.

physical (PHY) One of the four documents defining Fiber Distributed Digital Interface (FDDI). The layer that mediates between the media access control (MAC) layer above and the physical medium dependent (PMD) layer below it. Unlike the PMD layer, this is an electronic layer. Signal encoding and decoding schemes are defined at the *PHY* layer. Functionally, *PHY* corresponds to the upper parts of the ISO/OSI Reference Model physical layer. See also *media access control (MAC)*, *physical-layer medium dependent (PMD)*, and *station management (SMT)*.

physical address A binary address that refers to the *actual location* of hardware registers or information in a real device such as a hard disk drive.

physical circuit A two-wire metallic circuit not arranged for phantom use. See also *phantom circuit.*

physical connection The full-duplex association between adjacent physical (PHY) in a Fiber Distributed Digital Interface (FDDI) ring.

physical contact (PC) A term used mainly in connection with optical fiber to indicate that the optical fiber elements involved in a connection are actually touching.

physical delivery access unit (PDAU) In ITU-T's X.400 Message Handling System (MHS), an application process that provides a letter mail service with access to a Message Transfer System (MTS). The MTS can deliver an image of the letter to any location accessible through the MHS.

physical layer The lowest layer (layer 1) in the ISO/OSI Reference Model. It is the layer responsible for interfacing with the communication medium (twisted pair, coaxial cable, optical fiber, etc.), generating and detecting signal on the medium, and interchanging signals with the access control layer. Layer 1 defines the protocols that govern transmission and signaling on the medium. See also *OSI.*

physical-layer medium dependent (PMD) One of the four documents defining Fiber Distributed Digital Interface (FDDI). The lowest sublayer supported by FDDI and the only optic (as opposed to electrical) level. It corresponds roughly to the lower parts of the

physical layer in the ISO/OSI Reference Model. It specifies the requirements for the optical power sources (at least 25 µW output), photodetectors (must respond to 2 µW input), transceivers, the medium interface connector (MIC), and cabling. The MIC for FDDI connections serves as the interface between the electrical and optical components of the architecture. This connector was specially designed by American National Standards Institutes (ANSI) for FDDI and is also known as the FDDI connector.

The cabling specified at this sublayer calls for two rings running in opposite directions. The primary ring is the main transmission medium. A secondary ring provides redundancy by making it possible to transmit the data in the opposite direction if necessary. When the primary ring is working properly, the secondary ring is generally idle. See also *media access control (MAC)*, *physical (PHY)*, and *station management (SMT)*.

physical layer signaling (PLS) The topmost component of the physical layer in the ISO/OSI and IEEE 802.x layer models. This element serves as the interface between the physical layer and the media access control (MAC) sublayer above it. It is the portion of the *physical layer* that

- Interfaces with the MAC sublayer;
- Performs character encoding, transmission, reception, and decoding; and
- Performs optional isolation functions.

physical media In the ISO/OSI Reference Model, any physical means for transmitting data, e.g., twisted pair (basic phone lines), coaxial cable, Earth-orbiting satellite, optical fiber, microwave relay, T-1, and T-3. The bottom of the model's physical layer provides an interface to such media. Specifications for the physical media themselves are not part of the OSI model. Sometimes referred to as *layer 0.*

physical optics That branch of optics that treats light as a wave phenomenon rather than a ray phenomenon (geometric optics).

physical protocols The actual electronic signals used to communicate between devices. Some examples include those described by RS-232, ITU-T V.21/V.22/V.32/V.32 bis/V.34, Bell 103/212A, etc.

physical system A part of the real physical world that is directly or indirectly observable, controllable, or employed by humankind.

physical topology The actual (physical) arrangement of cables and hardware that comprise a network interconnection. See also *bus, ring,* and *star network.*

physical unit (PU) In IBM's Systems Network Architecture (SNA) networks, a term for a physical device (such as a cluster or communications controller or a host computer) and its resources on a network. It manages and monitors the resources of a node.

pi (π) The ratio of the length of the circumference of a circle to the diameter. The ratio is a constant and is approximately 3.141 592 653 589 793.

pi network A network composed of three branches in series with each other. The junction of two branches is designated the common node (1), the input is across a second node (2) and the common model and the output is across the third node (3) and the common node.

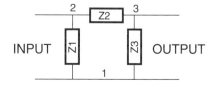

PI An abbreviation of <u>P</u>rimary <u>I</u>nput on Fiber Distributed Digital Interface (FDDI).

PIA An abbreviation of <u>P</u>eripheral <u>I</u>nterface <u>A</u>dapter.

PIC (1) An acronym for <u>P</u>ersonal <u>I</u>dentification <u>C</u>ode. **(2)** An acronym for <u>P</u>rimary <u>I</u>nterexchange (or <u>I</u>nterlata) <u>C</u>arrier. **(3)** In telephony, an acronym for <u>P</u>lastic (or <u>P</u>olyethelene) <u>I</u>nsulated <u>C</u>able. **(4)** An acronym for <u>P</u>rogrammable <u>I</u>nterrupt <u>C</u>ontroller. **(5)** An acronym for <u>P</u>osition <u>I</u>ndependant <u>C</u>ode. **(6)** An acronym for <u>P</u>riority <u>I</u>nterrupt <u>C</u>ontroller.

pickup (1) To answer an incoming call. The call may be ringing on any phone that is the same *pickup group* as the answering phone. **(2)** The action of a relay when energized. **(3)** A transducer that converts sound to electrical signals (e.g., microphone, phonograph pickup), scenes to electrical signals (e.g., video camera), and so on.

pickup value The minimum input signal that will cause a device (such as a relay) to operate.

pico The *SI* prefix meaning 10^{-12} or one-trillionth or one-millionth of a millionth. The SI symbol for *pico-* is *p*. See also *SI*.

picofarad (pF) A very small unit of capacitance—about the amount of capacitance between a twisted pair of wires one inch long.

picosecond (ps) 10^{-12} second.

picowatt See *pW*.

PICS An acronym for <u>P</u>latform for <u>I</u>nternet <u>C</u>ontent <u>S</u>election.

PID (1) An acronym for <u>P</u>rotocol <u>ID</u>entifier. **(2)** An acronym from <u>P</u>roportional <u>I</u>ntegral <u>D</u>erivative.

piecewise linear The approximation of a complex nonlinear relationship between variables with a series of straight lines as is illustrated in the diagram.

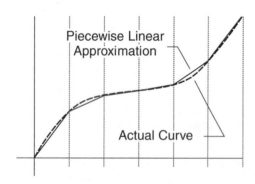

For example, both the A-law and μ-law analog to digital conversion algorithms are piecewise linear approximations to a logarithmic curve.

piezoelectric effect An electromechanical effect some materials possess wherein an electric potential applied to a piezoelectric material will cause the material to change shape (deformation), and conversely an applied pressure or deformation (stress) on the material will cause a potential difference between surfaces of the material.

piggybacking A transmission method in which acknowledgments for packets received are included in (piggybacked on) an ordinary data packet.

pigtail One or more short pieces of wire, cable, or optical fiber permanently affixed to a device for the purpose of attaching the device to other system elements. The *pigtail* generally does not have an attached connector, or if it does, it is a single circuit type. Examples include:

- A short length of optical fiber that is permanently affixed to an active device, e.g, light emitting diode (LED) or laser diode, and is used to splice the device to a longer fiber.
- A short length of wire (electrical conductor and insulation) permanently attached to an electrical component and used to connect the component to another conductor or termination point.

pilot A signal transmitted over a communication channel for control, supervisory, equalization, continuity, synchronization, or reference purposes. The signal is frequently a single frequency giving rise to the term *pilot tone*.

pilot channel A channel over which a pilot signal is transmitted.

pilot lamp A light that indicates the condition of an associated circuit. Also called a *pilot light*.

pilot subcarrier A 19 000 ± 2 Hz subcarrier serving as a control signal for use in reception of FM stereophonic broadcasts.

PIM An acronym for <u>P</u>rotocol <u>I</u>ndependent <u>M</u>ulticast.

pin (1) In general, a long slender projecting part. **(2)** A rodlike structure used to mechanically align two parts by sliding into a socket or hole in the corresponding position of the mating part. **(3)** The electrical contact on a male connector designed to mechanically and electrically interface with the contacts of the mating connector. **(4)** The

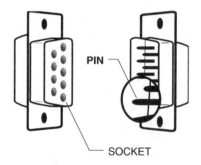

electrical contact on many electronic components, e.g., the connections on a dual inline or quad integrated circuits. See also *socket*.

PIN (1) An acronym for <u>P</u>ersonal <u>I</u>dentification <u>N</u>umber. A unique code number assigned to a user for the purpose of identification and security. *PINs* are frequently used in association with credit card numbers for gaining access to long-distance telephone networks. (Actually the *PIN* is required only for the billing aspects of the connection.) They are also used to log into networks that use dynamic passwords. **(2)** An acronym from <u>P</u>rocedure <u>I</u>nterrupt <u>N</u>umber. A term used in facsimile control procedures.

PIN diode An acronym for <u>P</u>ositive <u>I</u>ntrinsic <u>N</u>egative <u>d</u>iode. A semiconductor constructed with a large intrinsic region (nondoped region) between a p-type (positively doped) semiconductor region and an n-type (negatively doped) region. Applications of the PIN diode include:

- Fiber optic photodetector (PIN photodiode)—photons absorbed in the intrinsic region create electron-hole pairs that are separated by an electric field, thereby generating an electric current in response to the impinging light.

- Attenuator—provides a known amount of transmission line attenuation in response to a known bias.
- Limiter—a passive microwave power limiter based on the nonlinear characteristics of the PIN device.
- Charged particle detectors, radiation detectors, x-ray detectors, etc.

Also written *p-i-n diode*.

pine An acronym for Program for Internet News and E-mail. (Some say it stands for Pine Is No-longer ELM.) Now found on most UNIX systems, *pine* is the University of Washington's e-mail client program. It not only allows a user to view mail, but it includes an easy-to-use word processor (complete with a spell checker) for easy message generation. It supports several Internet protocols including:

IMAP—Internet Message Access Protocol

MIME—Multipurpose Internet Mail Extension

NNTP—Network News Transport Protocol

SMTP—Simple Mail Transport Protocol

ping (**1**) An acronym for Packet INternet Groper that has been elevated to the status of a word. Generally, it is used as a verb, as in "*ping* station x to see if its up." (**2**) A Transmission Control Protocol/Internet Protocol (TCP/IP) protocol function that tests the ability of a computer to communicate with a remote computer by sending a query and receiving a confirmation response. Message types include:

- Type 0—Echo reply
- Type 3—Destination unreachable
- Type 4—Source quench
- Type 5—Redirect
- Type 8—Echo request
- Type 11—Time exceeded
- Type 12—Parameter problem
- Type 13/14—Timestamp request & reply
- Type 17/18—Mark request & reply

ping-pong A technique in communications for achieving a pseudo full-duplex operation on a *half-duplex* channel. A data communication equipment (DCE) alternates between being a sending device and a receiving device on every packet (transmit one, receive one, etc.). The DCEs at each end of the channel are always opposite in nature, that is, when either one is a transmitter, the other is a receiver.

ping-ponging A routing that causes a packet to bounce back and forth between two nodes.

pink noise Random noise with a frequency spectrum that decreases with increasing frequency with a 10-dB-per decade slope. The energy is distributed logarithmically so that there is equal energy in each octave. Also called *1/f noise*.

PIO (**1**) An abbreviation of Parallel Input Output. (**2**) An abbreviation of Private Input Output.

PIP (**1**) An acronym for Peripheral Interchange Program. (**2**) An acronym for Path Independent Protocol.

pipe (|) A form of redirection in many operating system environments that allows the output of one program to be fed into the input of another program.

For example, in DOS the command DIR can write hundreds of lines to the display, but only the last 24 or so will be visible. (The rest scroll off the top of the monitor into oblivion.) By piping the output of DIR to a program that displays only one screen at a time, all lines can be read. The form of the command is DIR | MORE.

PIR An abbreviation of Protocol Independent Routing. Packet routing that is handled independently of the packet format and protocol being used. It is an alternative to tunneling (in which a packet is wrapped in another format in order to allow routing through a foreign domain).

piracy Theft of proprietary software by unauthorized copying, use, or distribution of copyrighted programs. See *freeware, public domain,* and *shareware.*

PITA An acronym for Pain In The Ass.

pitch (**1**) A synonym for *lay* or *lay length.* (**2**) A subjective quality of sound dependent on frequency and loudness. (**3**) The relative position of a tone in a scale as determined by its frequency. (**4**) A tarlike substance used to encapsulate some electrical or electronic equipment.

PIU An abbreviation of Path Information Unit.

pixel An acronym for picture (PIXture) ELement. The smallest area of a display that can be controlled. The actual size is screen and controller dependent. See also *pel.*

PKE An abbreviation of Public Key Encryption.

PKT An abbreviation of PacKeT or packet switching.

PKUNZIP A *shareware* utility program used to uncompress files that were compressed with the *PKZIP* utility. The PKUNZIP command format is:

PKUNZIP [-x] zipfile [@list] [files. . .]

Where "zipfile" is the path and file name of the compressed file, "@list" is the optional file containing a list of commands to be used in the decompression process, and "files" is the optional list of destination file specifications (if omitted, all files are placed in the current directory), and "-x" is one of the following commands:

-c[m]	**Extract** files to Console [with More]
-d	Restore / create the **Directory** structure stored in .ZIP file
-e[c	d	e	n	p	r	s]	**Extract** files. Sort by [CRC	Date	Extension	Name	Percentage	Reverse	Size]
-f	**Freshen** files in destination directory
-j	J<h,r,s>	**Mask**	don't mask <Hidden/System/Readonly> files (default=jhrs)
-n	Extract only **Newer** files
-o	**Overwrite** previously existing files
-p[a/b][c][#]	Extract to **Printer** [Asc mode, Bin mode, Com port] [port #]
-q	Enable ANSI **comments**
-s[pwd]	**Decrypt** with password [If no pwd is given, prompt for pwd]
-t	**Test** .ZIP file integrity
-v[b][r][m][t] [c,d,e,n,o,p,s]	**View** .ZIP [Brief] [Reverse] [More] [Technical] sort by [CRC	Date	Extension	Name	natural Order(default)	Percentage	Size]
-x<filespec>	**eXclude** file(s) from extraction
-$	Restore **volume label** on destination drive
-@listfile	Generate **list file**

See also *compressed files, data compression,* and *PKZIP.*

PKZIP A shareware utility program used to compress data files. Developed by PKWARE in 1989, it is one of the most popular file compression utilities available. *PKZIP* (and the companion uncompressing program *PKUNZIP)* are available on most bulletin board systems. The *PKZIP* command format is:

PKZIP [-x] zipfile [@list] [source_files. . .]

Where "zipfile" is the path and file name into which the compressed files are to be placed, "@list" is the optional file containing a list of commands to be used in the compression process, "source_files" is the optional list of files to be compressed (if omitted, all files in the current directory is assumed), and "-x" is one of the following commands:

-a [+]	**Add** file(s) to zipfile [turn off archive attribute]
-b D:path	Create **temporary** file on drive D:path
-c I C	Create/edit **comments** c=all files I C=new files
-d	**Delete** files
-e[xInIfIsI0]	**Compression** [eXtraINormalIFastI SuperfastInone]
-f	**Freshen** files
-i[-]	**Add** files with archive set [don't turn attribute off]
-jIJ<h,r,s>	**Mask**Idon't mask <hidden,read only,system>
-k[-]	Keep original .zip **date**
-l	Display software **License**
-m[flu]	**Move** files with [FreshenIUpdate]
-o	Set .zip **date** to latest file date
-pIP	Store recurse **path** names p=intoIP=specified & into
-q	Enable **ANSI codes** in comments
-r	**Recurse** subdirectories
-s [pwd]	**Scramble** with pwd, if none prompt
-t [date]	Take files equal to or newer than **date**
-T [date]	Take files older than **date**
-u	**Update** files
-v[b][r][m][t] [c][d,e,n,o,p,s]	**View** .zip [Brief] [Reverse] [More] [Technical] [Comment] sort by [DateIExtensionINameInoneI PercentageISize]
-wIW<h,s>	**include**I**exclude** <hidden,system>
-x filename	**eXclude** filename
-x @file.lst	**eXclude** filenames in file.lst
-z	Create or modify .zip **comment**
-!	add **authenticity verification**
-@ file.lst	**generate** file.lst file
-&[flIluIuI] wIv][s[drive]]	**Span** disks [FormatIformat Low] densityIUnconditional formatIUnconditional [Low densityIWipe diskIVerify][backup entire [drive] with subdirectories]
-=	**compatibility mode** (bypass share)
-?	Display **help**

See also *compressed files* and *data compression*.

PLA An abbreviation of Programmable Logic Array. See also *gate array*.

place call One of two types of call types. One in which the number dialed reaches a specific physical location. Not necessarily the location of the person to which the caller wishes to speak. See also *person call*.

plaintext In cryptography, intelligible data. An unencrypted, unencoded message. Also called *cleartext*. See also *ciphertext*.

planar array An antenna in which all of the elements, both active and parasitic, are in one plane. A planar array provides a large aperture. Uses of planar arrays include:

- Directional beam control by varying the relative phase of each element.
- A reflecting screen behind the active plane.

planar network A network or circuit that can be drawn on a flat sheet without any branches crossing.

Planck's constant (h) A universal constant that relates the energy of a quantum of radiation to the frequency of the source that radiated it. It has units of energy times time (J ×s, for example). *Planck's constant (h)* is related to energy (*E*) in joules and frequency (*f*) in hertz by

$$h = \frac{E}{f} \cong (6.62606891 \pm 0.00000087) \times 10^{-34} \text{ J s}$$

Also called the *elementary quantum of action.*

plane wave (1) A wave whose surfaces of constant phase are infinite parallel planes normal to the direction of propagation. **(2)** An electromagnetic wave that predominates in the far-field region of an antenna and has a wavefront that is essentially in a plane.

In free space, the *characteristic impedance* of a plane wave is 377 Ω.

plant All facilities and equipment used to provide telecommunications services. It is usually characterized as *outside plant* and *inside plant* where:

- *Outside plant* includes all poles, repeaters, and unoccupied buildings housing them, ducts, and cables—including the "inside" portion of interfacility cables outward from the main distributing frame (MDF) in a central office or switching center.
- *Inside plant* includes the MDF itself and all equipment and facilities within a central office or switching center.

plant test number (PTN) A seven-digit local exchange number that corresponds to the local switching part of an 800 in-WATS number. Although the number exists for testing purposes, its use should be minimized because its cost of use is the same as any incoming call on the 800 number itself.

PLAR An acronym for Private Line Automatic Ringdown.

plastic clad silica (PCS) fiber An optical fiber that has a silica-based core and a plastic cladding. *PCS fibers* in general have significantly lower performance characteristics, i.e., higher transmission losses and lower bandwidths, than all glass (silica) fibers.

The plastic cladding of a *PCS fiber* should not be confused with the polymer *overcoat* of a conventional all-glass fiber. Also called *polymer clad silica fiber.*

platform for Internet content selection (PICS) A specification for a universal content-filtering mechanism. It allows groups or individuals to set their own standards as to what material to allow on their portion of the network. When released, it will allow all online services and software companies to put filtering features into their product.

PLC An abbreviation of Programmable Logic Controller.

PLCP An abbreviation of Physical Layer Convergence Procedure.

plenum A large air chamber or cavity with two or more openings through which air is distributed. At least one opening is for air entry, while the remaining is for air distribution.

plenum cable A cable that has a fire-resistant sheath (jacket), which will neither support combustion nor give off toxic fumes when exposed to high heat. In order to comply with building fire safety codes, plenum cable is used in plenums, conduits, or shafts running inside ceilings, walls, or floors of office buildings.

plesio A Greek word meaning almost or near.

plesiochronous Meaning almost synchronous. **(1)** A term that describes signals in which corresponding significant events or entities occur at essentially the same rate but with no constraint on the phase relationship between events or entities. Signals that derive their timing from different sources but at frequencies arbitrarily close. See also *anisochronous, asynchronous, heterochronous, homochronous, isochronous,* and *synchronous.* **(2)** In multiplexing and networking, elements that derive their timing from different sources. The elements adjust for timing differences by inserting or deleting extra bits (bit stuffing).

PLL An abbreviation of <u>P</u>hase <u>L</u>ock <u>L</u>oop.

plokta An acronym for <u>P</u>ress <u>L</u>ots <u>O</u>f <u>K</u>eys <u>T</u>o <u>A</u>bort. To press random keys in an attempt to obtain some response from a system. Generally used:

- When the abort procedure for a program is not known,
- When trying to figure out if the system is just sluggish or really hung up, or
- When trying to figure out any unknown key sequence for a particular operation.

The generalized form is not individual key strokes but complete commands.

plonk To create (or add to) a kill file so that unwanted material does not appear in a user's newsgroup display. *Plonk* is supposed to evoke the picture of an item hitting the bottom of a trash can.

PLP An abbreviation of <u>P</u>acket <u>L</u>evel <u>P</u>rocedure.

PLS **(1)** An abbreviation of <u>P</u>hysical <u>L</u>ayer <u>S</u>ignaling. **(2)** An abbreviation of <u>P</u>rimary <u>L</u>ink <u>S</u>tation. **(3)** An abbreviation of <u>P</u>remises <u>L</u>ightwave <u>S</u>ystem.

plug A male connector. A device designed to be inserted into a jack or receptacle (female connector). Generally, the *plug* is attached to a cable, and the insertion of the *plug* into the jack extends the electrical circuit at the jack through the *plug* and cable.

Plug and Play (PnP) A concept wherein a new device is simply inserted into a computer system and the computer Basic Input/Output System (BIOS) and/or system software configure the system to operate with the new device. It is supposed to do away with the DIP switch programming and device conflicts. Also spelled *plug 'n play.*

plug and pray A humorous commentary on the fact that Plug and Play doesn't always work as advertised.

plug compatible Devices made by different manufacturers that are totally interchangeable. The term is derived from the fact that one can unplug one device and plug in the other and the system performance does not degrade.

plumbing A slang expression for coaxial and especially waveguide transmission lines.

PM **(1)** An abbreviation of <u>P</u>hase <u>M</u>odulation. Also abbreviated *pm.* **(2)** An abbreviation of <u>P</u>ulse <u>M</u>odulation. **(3)** An abbreviation of <u>P</u>ermanent <u>M</u>agnet. **(4)** An abbreviation of <u>P</u>erformance <u>M</u>onitoring. An asynchronous transfer mode (ATM) function. **(5)** An abbreviation of <u>P</u>reventative <u>M</u>aintenance.

PM/X An abbreviation of <u>P</u>resentation <u>M</u>anager for <u>X</u> windows.

PMD An abbreviation of <u>P</u>hysical <u>M</u>edium <u>D</u>ependent.

PMFJI An abbreviation of <u>P</u>ardon <u>M</u>e <u>F</u>or <u>J</u>umping <u>I</u>n. A polite way to get into a conversation.

PMJI An abbreviation of <u>P</u>ardon <u>M</u>y <u>J</u>umping <u>I</u>n. A polite way to get into a conversation.

PN **(1)** An abbreviation of <u>P</u>seudo-<u>N</u>oise. **(2)** An abbreviation of <u>P</u>ositive <u>N</u>otification. **(3)** In semiconductors, a reference to the junction of P-type material and N-type material. The majority carrier in N-type material is the electron, which has a negative charge (hence the N-type designation). In P-type material, the majority current carrier is a hole and is treated as a positive charge. (Therefore, it is called P-type material.)

PNNI An abbreviation of <u>P</u>rivate <u>N</u>etwork to <u>N</u>etwork <u>I</u>nterface. An asynchronous transfer mode (ATM) protocol that finds a connection that delivers the required Quality of Service (QoS) requested, if one is available.

PnP An abbreviation of <u>P</u>lug <u>a</u>nd <u>P</u>lay.

PnP BIOS A Basic Input Output System (BIOS) with the capability of configuring Plug and Play (PnP) devices at system power-up and of providing runtime configuration services after power-up.

PNS An abbreviation of <u>P</u>PTP (point-to-point tunneling protocol) <u>N</u>etwork <u>S</u>erver.

PO An abbreviation of <u>P</u>rimary <u>O</u>utput on FDDI.

POC An abbreviation of <u>P</u>oint <u>O</u>f <u>C</u>ontact. An individual associated with a particular Internet entity (IP network, domain, . . .).

Pockels cell An electro-optic device in which birefringence (refractive index) is modified under the influence of an applied electric field. The refractive index change varies linearly with the electric field, and the material is a crystal. A Pockels cell may be used as an intensity modulator at optical wavelengths. See also *Kerr cell.*

POH An abbreviation of <u>P</u>ath <u>O</u>ver<u>H</u>ead.

POI An abbreviation of <u>P</u>oint <u>O</u>f <u>I</u>nterface.

point code The identification numbering scheme for a Signaling System number 7 (SS7) network element. The numbering scheme is made up of the *network* field, *network cluster* field, and *network cluster member* field.

point of interface (POI) The physical interface (a demarcation point) between the local exchange carrier (LEC) LATA Local Access and Transport Area access and the interLATA service provider. The point establishes the technical interface, test points, and the point for division of operational responsibility between the organizations. Also called the *interface point.*

point of presence (POP) In a *public network* a geographic area in which it is possible to directly connect to the network by means of leased or dialup access. Also abbreviated *PoP.*

point of termination (POT) The demarcation point at which the local service provider responsibility ends and the customer's begins. Also called *demarcation point, network demarcation point,* or *network interface.*

point source An electromagnetic radiation source such that it appears so distant from a receiver (point of observation or measurement) that the wavefronts of the radiation are planar rather than a curved surface, regardless of the shape of the source. That is, the dimensions of the source are small enough in comparison to the distance between the source and receiver to be ignored in mathematical calculations. The source need not necessarily radiate isotropically.

point to multipoint A signal distribution system in which a signal emanates from an origination point and travels to many destinations. For example, signals from a satellite are generally distributed to multiple receiving stations. See also *broadcast* and *multicast.*

point to point A generic term describing a dedicated connection between any two nodes on a network.

point-to-point channel A network configuration in which two stations are connected by a single nonswitched communications channel that is not shared by any other stations. See also *leased line.*

point-to-point connection (1) In a network, a direct connection between two nodes, i.e., a connection without any intervening nodes or switches. (2) In an internetwork, a direct connection between two networks.

point-to-point infrared A short-range wireless communications method that uses a focused narrow beam of infrared light as the carrier between two points.

point-to-point link A dedicated communications link that connects only two stations.

point-to-point network A network in which a message originates at one location and travels to one or more destinations but not to every possible destination on the network. Wide area networks (WANs) are point-to-point networks, while most local area networks (LANs) are *broadcast networks.*

Point-to-Point Protocol (PPP) The protocol defined in RFC1171 which provides the Internet standard method for transmitting Internet Protocol (IP) packets over serial point-to-point links. RFC1220 describes how *PPP* can be used with remote bridging.

PPP has advantages over the Serial Line Interface Protocol (SLIP), for example, it is designed to operate both over asynchronous connections and bit-oriented synchronous systems, it can configure connections to a remote network dynamically, and it can test that the link is usable.

point-to-point topology A network wherein one node directly connects to another node. See also *mesh topology.*

point-to-point transmission Communications between two designated stations only, e.g., a microwave connection between a television studio and its transmitter.

pointer (1) A function indicator that is under the direct control of a computer operator and is used to indicate displayed information, to highlight data, to identify areas of interest, to serve as a graphic display cursor, and/or to select icons, e.g., a mouse pointer. (2) In computer programming, an identifier that indicates the location of an address or a data element.

pointing accuracy With respect to satellites, the directional accuracy of the aiming of a directional antenna toward the satellite, i.e., the angular difference between the actual aiming direction and the required aiming direction.

Poisson distribution A mathematical curve used in statistics to approximate the distribution and probability of occurrence for events. Named for the French mathematician Simeon D. Poisson (1781–1840).

It describes the probability that a random event will occur in some interval (of time, space, etc.) under the conditions that the probability of the event occurring is very small and the interval is very large (so that the event actually does occur), that is,

$$P = \frac{e^{-m} \cdot m^k}{k!}$$

Where P is the probability that the count k will be achieved and m is the *average* number of occurrences of the event in the interval. *Poisson distributions* are used in communications when traffic flows and waiting times are significant. It assumes that blocked calls are held (unlike the Erlang B formula that assumes blocked calls are cleared). Hence, when Poisson is used to estimate the number of circuits required for a particular situation, the predictions tend to be more pessimistic than the Erlang B estimates.

Poisson shot noise A stationary noise that occurs for visible light photodetection when a steady light source dominates the signal. Examples include high background light or a heterodyne reference beam.

polar direct current transmission A form of transmission in which positive and negative direct currents denote the significant conditions. Also called *double current transmission.*

polar operation A system in which mark and space signaling states are represented by opposite polarities of current flow (or voltages). See also *neutral transmission.*

polar plot A method of graphically displaying the relationship of two (or more) variables. One variable is an *angle,* while the other is a *magnitude.*

Note that the angle and magnitude may be called by different names depending on the particular discipline, for example,

- In radar systems they are called bearing and range, respectively.
- In modulation systems, phase and amplitude are used.

Angle	Magnitude
0°	1.5
45°	2.0
135°	1.5
225°	0.5
180°	1.5
337.5°	1.0
202.5°	2.0
45°	1.0

The *polar plot* above reflects the angle and magnitude data presented in the table. See also *Cartesian plot.*

polar relay A relay (electromechanical switch) in which the direction of movement of the armature depends on the direction of the current flow.

polarential telegraph system A direct current telegraph system employing polar transmission in one direction and a form of differential duplex transmission in the other. Two types of *polarential* systems are in use.

- In type A, the direct-current balance is independent of line resistance.
- In type B, the direct current tends to be independent of the line leakage.

Type A is better for cable loops where leakage is negligible, but resistance varies with temperature. Type B is better for open wire where variable line leakage is frequent.

polarity Any condition in which there are at least two opposing states: voltage levels (positive and negative), connectors (key aligned or reversed), and so on.

polarization (1) For connectors, the shape or form of the connector body or the position of a mechanical "key." *Polarization* both maintains path continuity and prevents mating incompatible devices. For example, the RJ11, RJ45, and MMJ connectors terminate twisted pair wire, but each has a different *polarization*. Another example is the ac mains plug which locates the "hot" wire in a specific position. See also *keying*. **(2)** Of an electromagnetic wave. The property that describes the time-varying direction and amplitude of the electric field vector.

States of polarization are described in terms of the figures traced onto a fixed plane in space (the plane being perpendicular to the direction of wave propagation) by a representation of the electric vector. The figures are viewed in the direction of propagation. *Polarization* states include:

- Unpolarized (random figure traced).
- Clockwise (right-hand) elliptical polarization.
- Counterclockwise (left-hand) elliptical polarization.
- Clockwise (right-hand) circular polarization.
- Counterclockwise (left-hand) circular polarization.
- Linear (plane) polarization.

Mathematically, an elliptically polarized wave may be described as the vector sum of two waves of equal wavelength but unequal amplitude, and in quadrature (i.e., having their respective electric vectors at right angles and $\pi/2$ radians out of phase). See also *Brewster's angle, clockwise polarized,* and *counterclockwise polarized.* **(3)** In batteries, the lowering of a cell's terminal voltage when a current is flowing. Numerically, it is the difference between the loaded and open circuit voltage values.

polarization diversity Concurrent signal transmission and reception wherein the same information is transmitted and received simultaneously on orthogonally polarized waves with fade independent propagation characteristics.

polarization maintaining (PM) optical fiber An optical fiber (generally elliptical or rectangular in cross section) in which the polarization planes of lightwaves launched into the fiber are maintained during propagation with little or no cross coupling of optical power between the polarization modes. Also called *polarization preserving (PP) optical fiber.*

polling A deterministic communication access method wherein a control node systematically "asks" each system element if it has a task to be serviced. (A task is any activity that requires the services of the control mode, e.g. data transfer.) If there is a task, a block of time is allocated to that task; if not, the controller goes on to the next element. Two forms of polling are used: *roll-call polling,* wherein the polling sequence is based on a list of elements available to the controller, and *hub-polling,* in which each element simply polls the next element in the sequence. *Polling* is an alternative to *contention* and *token passing.* See also *CSMA/CA, CSMA/CD,* and *interrupt.*

polymer clad silica fiber A synonym for *plastic clad silica fiber.*

polymorphic A term used to describe variable, encrypted viruses. That is, each separate infection by a polymorphic virus generates a different signature. Hence, a simple scan for a specific pattern will not detect the virus.

polynomial An algebraic expression using literal numbers (i.e., letters to represent values) and with more than one term.

A monomial has only one term, a binomial has two terms, a trinomial has three terms, etc. Generally, one writes the terms of a polynomial in descending order for one of the literals, for example,

$$x^4 + 3x^2 - 12x$$

$$x^3y + x^2y^2 + 5x$$

polystrophic code A binary code in which two or more bits may change between any two adjacent code positions. For example the basic binary code is a *polystrophic code*—the change from three (0011b) to four (0100b) requires three bits to change. See also *monostrophic code.*

POM An abbreviation of Phase Of the Moon. Something that depends on the phase of the moon is flaky, unreliable, or produces unexpected results.

PON An abbreviation of Passive Optical Network.

pool A collection of entities available for sharing.

POP **(1)** An acronym for Post Office Protocol. Used on the Internet. **(2)** An acronym for Point of Presence.

port A somewhat ambiguous term with multiple meanings, for example:

- An interface on a computer capable of being connected to a peripheral device such as a modem or printer. Two basic kinds of *ports* are generally available: *serial ports* and *parallel ports.* Serial ports transfer information one bit at a time, while parallel ports transfer information multiple bits simultaneously.
- In a device or network, a point of access where signals may be inserted or extracted, or where the device or network variables may be observed or measured.
- A communications network entry or exit point.
- A measure of equipment capacity, generally based on the sum of the number of station lines and trunks, (e.g., a private automatic branch exchange [PABX] with 128 station lines and 16 trunks has 144 ports).
- A connection point on a multiplexer.
- In computer software, the process of transporting something (like an application program) from one environment to another.

port address **(1)** The location where information is transferred to or from a device. On Intel 80xxx microprocessors, port addresses range from 0 to 65535 (0h–FFFFh) and are separate from the 0 page memory addresses. The chart on page 402 illustrates common PC *port addresses.* Ports are accessed by special IN and OUT instructions. **(2)** A number that identifies the location of an application program that is running on an Internet host. The number appearing on the header of datagrams sent to the host tells the host to which application the payload is to be delivered. Some port numbers, called *well-known ports,* are standardized because they provide access to common services such as file transfer protocol (FTP). Others are dynamically allocated on an as-required basis.

port concentrator A device that allows several data terminals to share a single computer port.

port selector The hardware or software that selects a particular port for a communications session. The selection algorithm may be random, first available, next available, or some other selection criterion.

portability **(1)** In programs, the ability of software or a system to run on more than one type or size of computer and with more than one type of rating system (OS). See also *Portable Operating System Interface (POSIX).* **(2)** In equipment, the quality of being able to function normally while being relocated. **(3)** In data, the ability to transfer data from one system to another without being required to recreate or reenter data descriptions or to modify the applications on either system.

Portable Operating System Interface (POSIX) IEEE standards that define the interface between applications and operating systems. These standards include IEEE1003.1-1988 which defines a UNIX-like operating system interface, 1003.2 which defines the shell and utilities, and 1003.4 which defines real-time extensions.

SOME COMMON PC
PORT ADDRESS ASSIGNMENTS

Port Address (hex)	Function
000-01F	DMA controller 1
020-03F	Interrupt controller 1
040-05F	Timer
060-06F	Keyboard
070-07F	Nonmaskable interrupt (NMI) mask
080-09F	DMA page register
0A0-0BF	Interrupt controller 2
0C0-0DF	DMA controller 2
0F0-0FF	Math coprocessor
100-1EF	
1F0-1F8	Hard disk controller
1F9-1FF	
200-207	Game controller (joystick)
208-277	
278-27F	Parallel port 2
280-2F7	
2F8-2FF	Serial port 2
300-31F	
320-377	
378-37F	Parallel port 1
380-3AF	
3B0-3BF	Monochrome display (MDA) / Printer
3C0-3CF	
3D0-3DF	Color graphics adapter (CGA)
3E0-3EF	
3F0-3F7	Floppy diskette controller
3F8-3FF	Serial port 1
400-78F	
790-793	Cluster adapter 1
794-88F	
890-893	Cluster adapter 2
894-138F	
1390-1393	Cluster adapter 3
1394-238F	
2390-2393	Cluster adapter 4
2394-FFFF	

Originally developed to provide a common interface for UNIX (and UNIX-like) implementations (thus explaining the acronym Portable Operating System for UNIX), *POSIX* has become more widely adopted, and many support various parts of the standard. The standard has been adopted and published as a Federal Information Processing Standard (FIPS PUB 151-1) for a vendor-independent interface between an operating system and an application program, including operating system interfaces and source code functions.

portmapper In networking, a server that converts Transmission Control Protocol/Internet Protocol (TCP/IP) protocol port numbers into *remote procedure calls (RPC)* program numbers.

When an RPC server starts, it tells *portmap* the port number it is listening on and which RPC program numbers it serves. Before a client can call a given RPC program number, it must contact *portmap* on the server machine to determine the port number to which RPC packets should be sent.

POS (1) An abbreviation of Professional Operating System, Digital Equipment Corp's (DEC's). **(2)** An abbreviation of POSitive.

position dilution of precision (PDOP) In the satellite Global Positioning System (GPS), an expression of the uncertainties of deter-

mined location due to the geometric position the satellites. *PDOP* can be determined by

$$PDOP = \sqrt{(HDOP)^2 + (VDOP)^2}$$

where:

   HDOP   is the horizontal dilution of precision (latitude and longitude), and

   VDOP   is the vertical dilution of precision (altitude).

See also *dilution of precision (DOP)* and *Global Positioning System (GPS)*.

positioned channel In *integrated services digital networks (ISDN)*, a channel that occupies dedicated bit positions in the framed data stream. Examples of positioned channels are the B, H, and D channels.

positioned interface structure Within a framed interface, a structure in which positioned channels provide all services and signaling.

positive acknowledge with retransmission (PAR) A method of ensuring the reliable delivery of data across a communications link. In the basic operation of the *PAR* protocol, the sending terminal keeps sending a packet to the destination until an acknowledgment from the destination is received by the sending terminal.

TCP (transport control protocol), one of the Transmission Control Protocol/Internet Protocol (TCP/IP) protocols, uses *PAR*.

positive feedback A process in which a fraction of the output signal of a device is fed back to the input of that device and in phase with the original input signal. The process can increase the apparent gain of the device but it can also lead to unwanted oscillation if carried to extremes. Also called *regeneration*.

positive justification A synonym for *bit stuffing*.

positive logic The representation of the logical true (1 state) and logical false (0 state) by the high (H) and low (L) levels, respectively.

positive modulation Modulation in which an increase in signal level results in an increase in transmitted power.

positive true A synonym for *active high*.

POSIX An acronym for Portable Operating System for UNIX.

post (1) On a bulletin board system (BBS) or the Internet, to send a message to a *discussion group, mailing list,* or *newsgroup* (in USENet). *Post* implies that the message is sent indiscriminately to multiple (and unknown) recipients; that is, the message is public in nature, and anyone wishing to view the message is allowed to. In contrast, *mail* implies one or more specifically selected (known) individual recipients.

A message should be *posted* only if it will be of interest to a significant proportion of the readers of the group or list; otherwise private *e-mail* should be used. Also called *posting*. **(2)** A cylindrical rod placed in a transverse plane of a waveguide and behaving substantially as a shunt susceptance.

POST An acronym for Power On Self Test. A set of diagnostic routines the device executes every time it is powered on. The *POST* ensures that all of the components in the device are functional.

post-detection combiner A synonym for *maximal ratio combiner*.

post-dialing delay In telephony, the delay in switching systems between the reception of the last dialed address digit and the application of ringing to the called station.

post office Any part of an e-mail system that directs or delivers mail. To refer to a specific entity within the mail system, one refers to a local post office, host post office, domain post office, etc. It may exist only as a directory hierarchy and not as active components.

post office protocol (POP) In Internet, an e-mail protocol that functions as a *message transfer agent (MTA)* for users who are accessing a host from an autonomous machine (e.g., a PC).

The POP protocol

- Provides the means to store incoming messages, at the host, until the user logs onto that host.
- Downloads the message to the user's machine.
- Uploads the replies after reviewing the messages with a client such as Eudora and optionally creating replies to selected messages.

post office server A network server that has been designated as the machine to receive and store all network mail.

postalize In communications, to structure rates or prices so that they depend on call duration, type of service, and time of day, not on the distance between stations.

postamble In a packet or message, a sequence of bits or fields that follows the actual data or contents. It generally contains a frame check sequence (FCS) or another error-checking field and may include one or more flags or a predefined bit sequence to indicate the end of a packet. Also called a *trailer.* See also *preamble.*

postmaster (1) The person responsible for ensuring that e-mail is delivered and answering user questions. The Internet standard for e-mail (RFC822) requires that each machine have a *postmaster* address; usually it is aliased to this person. **(2)** The e-mail contact and maintenance person at a site connected to the Internet or UUCPNET. Often, but not always, the same as the administrator.

postmortem (1) Analysis of an operation after it is completed. **(2)** Slang for failure analysis.

postpay (PP) More properly called semi-postpay; it is a type of pay telephone coin/token collection scheme in which the caller is given a dial tone immediately upon going off-hook and is allowed to establish a connection to the called party. However, if the call is a chargeable call, the pay telephone talking circuit is disabled until appropriate coinage is collected.

A postpay telephone will always collect inserted tokens, even if the call is a "free call." Other methods of payphone operation include *coin first (CF)* and *dial tone first (DTF).* See also *pay telephone* and *prepay.*

postproduction processing The processing of audio and video information after creation or contribution but prior to final use.

PostScript® A printer/display protocol and programming language developed by Adobe Corporation. Unlike ASCII printers, which print verbatim on a symbol by symbol basis, a *PostScript* printer accepts and interprets an entire page of information before printing text and images.

pot An abbreviation/slang for *potentiometer.* An electronic circuit element that can be adjusted to provide any selected resistance between some minimum and maximum value.

POT An acronym for Point Of Termination.

potential difference A synonym for voltage difference.

POTS An acronym for Plain Old Telephone Service. The basic private line telephone service with either rotary or dual tone multifrequency (DTMF) dial capability. *POTS* provides only the most basic functions, i.e., battery feed, ringing, and supervision.

POV An abbreviation of Peak Operating Voltage.

power (P) (1) In general, the rate of energy generation, transfer, absorption, or consumption per unit time. *Power* is measured in watts (W) and is given by:

$$P = I^2 \cdot R \text{ or } P = \frac{V^2}{R}.$$

If both the voltage and current are sinusoidal functions of time, the product of the root-mean-square (RMS) values of voltage and current (V_{RMS} and I_{RMS}) is called the *apparent power;* the product of V_{RMS} and the in-phase component of I_{RMS} is called the *active power;* and the product of V_{RMS} and the in-phase component of I_{RMS} is called the *reactive power.* **(2)** In a radio transmitter, it is expressed in one of the following forms, according to the class of emission, using the arbitrary symbols indicated:

- Peak envelope power (PX or pX);
- Mean power (PY or pY);
- Carrier power (PZ or pZ).

For use in formulas, the symbol p denotes power expressed in watts (W) and the symbol P denotes power expressed in decibels (dB) relative to a reference level. **(3)** In optics, the reciprocal of the focal length of a lens. **(4)** In a telescope or microscope, an abbreviation of *magnifying power.*

power budget In transmission systems, the difference between the transmitter's effective output signal power and the receiver's effective input sensitivity, generally expressed in decibels (dBs). This difference is the maximum amount of signal loss that can be tolerated without signal amplification and/or regeneration.

The *power budget* is generally divided up and allocated to various sources of lossey elements. For example, a particular transmitter that can deliver +10 dB of signal to the transmission line and a receiver that requires −40 dB of input signal level is said to have a 50-dB *power budget.* The transmission losses in this example may be 30 dB and the switching losses another 3 dB; this yields a total *power margin* of 17 dB. Also called the *system budget.*

power circuit breaker (PCB) (1) The primary switch used to apply and remove power from equipment. **(2)** A circuit breaker used on ac circuits rated in excess of 1500 V.

power conditioning A generic term describing the process of protecting equipment from various forms of mains distortions. A pure mains signal is a sine wave of a specified amplitude and frequency. Distortions such as undervoltage, overvoltage, spikes, surges, frequency errors, waveform purity, and noise can corrupt data or cause equipment damage. Conditioning is divided into several classes, that is,

- *Isolation*—which tries to contain the disturbance before it reaches the protected device. Isolation protects against noise or interference, and also against over/undervoltage fluctuations.
- *Regulation*—which tries to maintain a constant power supply through brownouts, surges, and blackouts. An uninterruptible power supply (UPS) is generally used to accomplish this.
- *Suppression*—which tries to stop unexpected or massive power surges and spikes. Surge protectors (MOVs [metal oxide varisters], breakdown diodes, gas tubes, etc.) are most commonly used.

See also *mains.*

power cross In telephony, the description of a power line accidentally (and generally momentarily) connecting with a telephone line. Often caused by the wind blowing a tree branch into the distribution systems. Also called *power contact.*

power detection That form of detection in which the power output of the detecting device is used to supply a substantial amount of power to the subsequent devices.

power density In optics, a deprecated synonym for *irradiance.*

power dissipation The amount of energy converted to heat in a device. Mathematically, it is the square of the current through the device times its resistance ($P = I^2 R$).

power disturbance Any of several forms of mains waveform distortions and deviations from the nominal value—e.g., blackout, brownout, spike, surge, frequency stability, waveform purity, and noise. See also *mains.*

power factor (**1**) The ratio of active power (watts) to apparent power (voltamps).

In ac power transmission and distribution, the cosine of the phase angle between the voltage and current. When the load is inductive (e.g., an induction motor), the current lags the applied voltage, and the power factor is said to be a lagging power factor. When the load is capacitive (e.g., a synchronous motor or a capacitive network), the current leads the applied voltage, and the power factor is said to be a leading power factor. *Power factors* other than unity have deleterious effects on power transmission systems—i.e., excessive transmission losses and reduced system capacity. See also *power (P).* (**2**) In switchmode power supplies, the existence of significant distortion in the current waveform due to both the high-input capacitance of the converter and the effect of charging the capacitor only at the peak input voltage. This leads to a current waveform such as is shown in the drawing along with high associated harmonic content. By regulation,

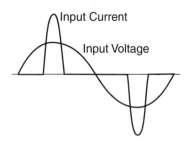

these harmonics must be reduced to levels below those illustrated in the table.

MAXIMUM PERMISSIBLE HARMONIC CURRENT

Harmonic Order (n)	EN60555 Maximum Harmonic Current (A)	IEC 1000-3-2 Maximum Harmonic Current (mA/W)	IEC 1000-3-2 Maximum Harmonic Current (A)
2	1.08		
3	2.30	3.4	2.30
4	0.43		
5	1.14	1.9	1.14
6	0.30		
7	0.77	1.0	0.77
9	0.40	0.5	0.40
11	0.33	0.35	0.33
13	0.21	0.30	0.21
$8 \leq n_{EVEN} \leq 40$	1.84 / n		
$15 \leq n_{ODD} \leq 39$	2.25 / n	3.85 / n	2.25 / n

power failure transfer (**1**) In power, the switching of a load from a primary source to a secondary (backup) source when a measured parameter of the primary source is outside specified limits. Examples include:

- *Power distribution.* The switching of primary utility mains to a secondary (backup) whenever the primary source operates outside specified parameters.
- *Telephony.* The switching from commercial power sources at the central office (CO) to locally generated (backup) power in the event commercial power falls outside desired limits.

(**2**) In telephony, the transfer of certain analog telephone instruments from private automatic branch exchange (PABX) station lines to predesigned central office analog trunks in the event of commercial power failure and a low PABX battery condition. *Power failure transfer* is an emergency mode of operation in which only one instrument at the subscriber location may be powered from each central office trunk. Also called *power fail bypass.*

power gain (**1**) In general, the ratio of the power a device delivers to its load to the power the device receives at its input. Power gain is usually expressed in decibels (dB). (**2**) With an antenna, the ratio of the power flux (at a point in the far field) per unit area of the antenna to an isotropic radiator.

The Friis transmission formula is usually used to define far-field distance (r), that is,

$$r \geq 2 \cdot D^2 \cdot \lambda$$

where D is the maximum aperture dimension of the antenna and λ is the wavelength of the radiation. The isotropic radiator must be contained in the smallest sphere containing the antenna; common points are the antenna terminals and points of symmetry.

power law index profile In optical fibers, a class of *graded index profiles* characterized by:

$$n(r) = \begin{cases} n_1 \sqrt{1 - 2\Delta\left(\dfrac{r}{r_c}\right)^g} & , r \leq r_c \\ n_1 \sqrt{1 - 2\Delta} & , r \geq r_c \end{cases}$$

where $\Delta = \dfrac{n_1^2 - n_2^2}{2n_1^2}$ and

r is the distance from the fiber axis,
$n(r)$ is the nominal refractive index as a function of distance from the fiber axis,
n_1 is the nominal refractive index on axis,
n_2 is the refractive index of the homogeneous cladding ($n(r) = n_2$ for $r \geq a$),
r_c is the core radius, and
g is the profile parameter, i.e., defines the shape of the index profile. (α is often used in place of g. Hence, this is sometimes called an *alpha profile.*)

Multimode distortion is minimum when g takes a particular value depending on the material used. For most materials, this optimum value is approximately 2. When g increases without limit, the profile tends to a *step index profile.*

power level The magnitude of power averaged over a specified time interval. Expressed either in watts (W) or as a ratio to a reference power in dBm or dBW.

power line carrier A communications method employing a low-frequency RF carrier (usually less than 600 kHz) to transmit information over the ac mains.

power margin The difference between *available signal power* and the *minimum signal power* required to satisfy the minimum input requirements of a receiver for a given performance level.

Power margin reflects the excess signal level, present at the input of the receiver, that is available to compensate for:

- The effects of component aging in the transmitter, receiver, or physical transmission medium, and
- The effects of deterioration of propagation conditions.

Also called *system power margin*. Compare with *power budget*.

power rating The maximum power that may be applied to a device (under specified conditions) which will either:

1. Permit satisfactory operation of the device with adequate safety margin for both personnel and operating life of the device, or
2. Not produce a permanent change that causes any performance characteristic to be outside specified limits.

power supply That part of an electronic system that converts the raw electrical energy (from 120/240 volt mains, lead-acid batteries, solar cells, or other source) into the voltages necessary to operate the various subsystems and components. Common power supply output voltages are $\pm$ 5 V, $\pm$ 12 V, and +24 for computers and telecom switching equipment and -48 V (or -60 V) for telephone loop power. However, the number of power supply voltages, currents, and frequencies in use is limitless.

PP (**1**) An abbreviation of Post**P**ay, referring to the operation of a payphone. (**2**) An abbreviation of **P**olarization-**P**reserving optical fiber.

PPDN An abbreviation of **P**ublic **P**acket **D**ata **N**etwork.

ppm (**1**) An abbreviation of **P**arts **P**er **M**illion. (**2**) An abbreviation of **P**ulses **P**er **M**inute.

PPM (**1**) An abbreviation of **P**ulse **P**osition **M**odulation. (**2**) An abbreviation of **P**arts **P**er **M**illion (although it is generally written *ppm*). (**3**) An abbreviation of **P**eriodic **P**ulse **M**etering. (**4**) An abbreviation of **P**ulses **P**er **M**inute.

PPN An abbreviation of **P**rivate **P**acket **N**etwork.

PPP An abbreviation of **P**oint-to-**P**oint Protocol. In Internet, a protocol that allows multiple terminal types to connect to a network over a serial link (e.g., a standard telephone line modem connection). *PPP*, a successor to Serial Line Interface Protocol (SLIP), provides router-to-router and host-to-network connections over both synchronous and asynchronous circuits.

- *PPP* defines and controls the establishment and configuration of remote lines.
- *PPP* is a serial line protocol capable of transmitting multiple protocols simultaneously over the serial link.
- *PPP* provides for data compression.
- *PPP* provides standard messages for link quality monitoring and error correction.
- *PPP* provides a password authentication method.

PPP allows a computer to emulate a full Internet machine with a modem and standard dial-up telephone connection. Defined in RFC1661. See also *shell account* and *SLIP*.

PPRAM An acronym for **P**arallel **P**rocessing **R**andom **A**ccess **M**emory. A memory architecture devised by Professor Kazuaki Murakami at Japan's Kyushu University in which reduced instruction set computer (RISC) processors and dynamic random access memory (DRAM) are combined on one die.

PPS (**1**) An abbreviation of **P**ulses **P**er **S**econd. Frequently written *pps*. (**2**) In the Global Positioning System (GPS), an abbreviation of **P**recision **P**ositioning **S**ervice. (**3**) An abbreviation of **P**ackets **P**er **S**econd. (**4**) An abbreviation of **P**ath **P**rotection **S**witched ring. From Bellcore's TA-496 specification, a duplicated Synchronous Optical Network (SONET) ring in which the signal travels over diverse physical paths. When one path fails, the redundant path can take the load, thus preventing service outages. (**5**) An abbreviation of **P**rogrammable **P**ower **S**upply.

PPSN An abbreviation of **P**ublic **P**acket **S**witched **N**etwork.

PPSS An abbreviation of **P**ublic **P**acket **S**witched **S**ervice.

PPT (**1**) An abbreviation of **P**rocess **P**age **T**able. (**2**) An abbreviation of **P**unched **P**aper **T**ape. (**3**) An abbreviation of **P**ulse **P**air **T**iming.

PPTP An abbreviation of **P**oint to **P**oint **T**unneling **P**rotocol. One of several virtual private network (VPM) protocols. Supporters include Microsoft, Ascend, 3Com, and Shiva. See also *L2TP* (*Level 2 Tunneling Protocol*).

PPTR An abbreviation of **P**unched **P**aper **T**ape **R**eader.

PPX (**1**) An abbreviation of **P**acket **P**rotocol e**X**tension. (**2**) An abbreviation of **P**rivate **P**acket e**X**change.

PR An abbreviation of **P**ulse **R**ate.

PRA An abbreviation of **P**rimary **R**ate **A**ccess.

practices (**1**) In software, requirements used to define a disciplined, uniform approach to the software development process. (**2**) In telephony, especially in the Bell System, a synonym for requirements, procedures, and standards. The technical and installation manuals used by the Bell operating companies to define how things are done and how things work.

PRAM An acronym for **P**arameter **RAM**.

pre-arbitrated (**PA**) A type of traffic on a distributed queue data bus (DQDB) MAN, usually associated with isochronous connections. It is a slot dedicated by the head-of-bus function for transferring isochronous service octets.

pre-emphasis The process of intentionally shaping the amplitude vs. frequency characteristic in one part of a system in order to improve the overall system signal-to-noise ratio (by minimizing the adverse effects of such phenomena as attenuation differences, or saturation of recording media, in subsequent parts of the system). Pre-emphasis exaggerates the amplitude of some frequencies in the band of interest with respect to frequencies in the same band. *Pre-emphasis* is generally used before FM or phase modulation transmission or before audio recording.

In FM or phase modulation transmitters, pre-emphasis is accomplished by passing the signal through a high-pass network (called the *pre-emphasis network*) which increases the magnitude of high frequencies with respect to the low frequencies. This equalizes the modulating signal drive power in terms of deviation ratio. The receiver demodulation process or audio playback circuit includes a reciprocal network (called a *de-emphasis network*) that restores the original signal power distribution, i.e., the original amplitude-frequency response. Also called *pre-distortion* and *pre-equalization*.

pre-emphasis improvement In FM broadcasting, the improvement in the overall system's signal-to-noise ratio (SNR) of the high-frequency portion of the modulating signal, resulting from the passing of the modulating signal through a high-frequency *pre-emphasis* network. At the FM discriminator output in the receiver, a de-emphasis network restores the signal to its original frequency distribution.

preamble The first portion of a packet transmission that allows a receiver to obtain proper timing and information about the packet itself. A preamble contains one or more of the following parts:

- An unmodulated carrier—allows the receiver to determine the carrier frequency.
- A carrier modulated by the transmitter's clock—allows the receiver to obtain phase synchronization with the transmitter clock.
- A known symbol sequence—allows the receiver to remove any time ambiguities, that is, it enables the receiver to determine the beginning of the packet payload.
- Administrative information—includes the source address, destination address, packet type (PT), and packet size.

See also *postamble* and *trailer.*

preamplifier An amplifier specially designed to be connected to a low-level signal source and with the following characteristics:

- Provides appropriate input transmission line and transducer matching.
- Provides low-noise signal amplification.
- Provides sufficient gain that further signal processing and distribution may be accomplished without further degradation of the signal-to-noise ratio.

preassignment access plan In satellite communications system operations, a bandwidth allocation plan in which fixed channel assignments are made to each station. This is in contrast to a demand assignment plan in which a station is allocated a channel from a single pool of channels (available to all stations) only when it is required.

precedence A priority designation assigned to the handling of a call and specified by the caller. It indicates to the communications personnel the degree of urgency of the call; hence, the order in which it is handled and to the called person it indicates the order in which it should be addressed. The descending order of precedence for military messages is FLASH, IMMEDIATE, PRIORITY, and ROUTINE.

precedence call In telephony, a call on which the call originator has opted to use one of several available levels of priority.

precedence prosign A preamble character (or characters) to a message that indicates to the receiving unit how the message is to be handled.

precipitation attenuation The loss of electromagnetic wave energy due to scattering, refraction, and/or absorption as it passes through a volume of atmosphere containing precipitation such as rain, snow, hail, or sleet. Also called *rain attenuation* or *rain fade.*

precipitation static (p-static) Radio interference caused by the impact of charged particles against an antenna. *P-static* may occur in a receiver during certain weather conditions that generate these charged particles, e.g., during windy snowstorms, hail storms, rain storms, or dust storms.

precise frequency A frequency that is maintained to a known accuracy of an accepted reference frequency standard (e.g., 1 part in 10^6 or 1 part in 10^{14}).

precise time A timing mark that is accurately known with respect to an accepted reference time standard.

precision (1) The amount of *detail* used in expressing a number. A measure of how fine grained the value is expressed.

Precision and *accuracy* are not the same; *precision* expresses detail, whereas *accuracy* expresses correctness. For example, given 6.2831852 is a *precise* and *accurate* approximation of the ratio of the circumference of any circle to its diameter, then 6.28 is more *accurate* than 6.3000, even though 6.3000 is more *precise.* See also *accuracy.* (2) A synonym for *reproducibility* in the following sense:

- The ability of a set of *independent devices* of the same design to produce the same value or result, given the same input conditions and operating in the same environment.
- The ability of a *single device* to repeatedly produce the same value or result on independent trials, without adjustments, and given the same input conditions and operating in the same environment.

(3) In computers, a measure of the ability to distinguish between nearly equal values. An 8-bit word can distinguish between only 256 values (a precision of about 0.4%), while a 16-bit word distinguishes 32 768 values (0.003%) and a 32-bit word yields $2.3 \times 10^{-8}\%$. (4) The degree of discrimination with which a quantity is stated, for example, a three-digit numeral in the base 10 discriminates among 1000 possibilities.

precision code (P-code) See *P-code.*

Precision Positioning Service (PPS) The full accuracy, single-receiver Global Positioning Service (GPS) provided by the U.S. government for its military and other select users. It includes access to un-encoded P-code, and it provides a predictable positioning accuracy of at least 22 m (2dRMS) horizontally and 27.7 m (2 sigma) vertically and timing/time interval accuracy within 90 ns (95% probability). This accuracy results from two characteristics:

- Receivers are not subjected to Selective Availability (SA), i.e., purposeful system degradation.
- Receivers have access to the L2 channel which enables a correction for atmospheric propagation errors.

Sometimes called the *Precise Positioning Service.* See also *anti-spoofing (A-S), Standard Positioning Service (SPS),* and *static positioning.*

precision sleeve splicing In fiber optics, an optical fiber splicing technique that uses a capillary tube, of suitable material, to align the mating fibers. The capillary tube has an inside diameter slightly larger than the outside diameter of the cladding of the fibers to be spliced. The capillary tube may contain an index-matching gel or an adhesive having a refractive index that approximates that of the fibers. The fibers are inserted, one from either end, to form a butt joint.

predial See *on-hook dialing.*

predictive code A code method based on the difference between the current value of the signal and the mathematically predicted value of the signal (based on passed values).

predictive dialing An automated dialing feature in which the system predicts from preset algorithms when an agent/operator will become free and makes the call in advance. See also *preview dialing.*

preemptive In telephony, the disconnection and subsequent reuse of a part of an established connection of lower priority by a *precedence call.* The disconnect and reuse only occur when all circuits by the precedence call are busy. See also *precedence.*

prefix code In telephony, one or more digits preceding the national or international number to implement direct distance dialing, e.g., the digit "1" preceding the area code or the digits "011" preceding international calls from the United States.

preform A glass rod with the same cross-sectional structure (except magnified many times) as the optical fiber into which it will be drawn.

premise A thesis. A proposition assumed or proven and used as the basis of an argument or inference.

It is not the same as premises, although it is frequently misused in this way.

premises A building together with its grounds and adjunct structures.

premises distribution system (PDS) (1) A cabling system that covers an entire building or campus. (2) The name of a premises wiring system from AT&T.

premises network A network confined to a single building.

premises wire In communications, the quad, twisted pair, coax, or other cable installed at a user's location used to provide both inter-building and intrabuilding wiring.

premises wiring In telephony, all interior and exterior wiring (permanent and temporary) together with associated fixtures and hardware that extend the local exchange carrier's (LEC) service loop from the demarcation point to the communications device or connector (RJ11, etc.).

prepay One of several types of pay telephones in which the user must insert a coin or token before a connection to a chargeable number. In operation two types exist—those that present dial tone immediately when the user goes off-hook (designated tone-first), and those that require the insertion of a coin before dial tone is delivered (designated coin-first). The chart compares prepay and postpay characteristics. See also *pay telephone* and *postpay*.

Type	Characteristics	Coin handling
Semi-postpay	No coin required for dial tone or dialing	Always collected
Coin first-prepay	Must deposit coin before dialing	Collected or returned as required
Tone first-prepay	No coin required for dial tone or dialing	Collected or returned as required

PREPnet An acronym for Pennsylvania Research & Economic Partnership Network.

preselector A band-pass filter preceding a frequency converter or other signal processing entity that passes some frequencies and attenuates others.

presentation layer Layer 6 of the ISO/OSI communications Reference Model. The *presentation layer* is responsible for the representation of the data to be transmitted, including encryption and data compression. See also *OSI*.

preshoot See *pulse*.

press-to-talk A type of duplex transmission system in which the talker must press a switch to turn on the transmitter. When the button is released, the system is in the listen mode. Also called *push to talk*.

pretrip A central office line circuit malfunction in which the subscriber's telephone rings only once. The line circuit sees the ring signal as an off-hook condition, so it removes the ring signal (expecting a connection to be established). But the subscriber's line is still on-hook, and so the switching system terminates the call. Also spelled *pre-trip*.

pretty good privacy (PGP) The name of an encryption program developed by Phil Zimmermann in 1992.

preventive maintenance (PM) (1) The care and servicing of a device or equipment for the purpose of maintaining its satisfactory operation by providing for systematic inspection, detection, and correction of incipient failures either before they occur or before they develop into major defects. (2) Maintenance, including periodic or scheduled tests, measurements, adjustments, and parts replacement, performed specifically to prevent faults from occurring.

preview dialing An automated dialing feature in which the system determines the next call to be made but allows the agent/operator as a supervisor to check and activate the call. See also *predictive dialing*.

prf An abbreviation of Pulse Repetition Frequency.

PRI An abbreviation of *Primary Rate Interface*. See *primary rate access* (*PRA*).

primary cell A cell that produces an electric current by electrochemical reactions without regard to the reversibility of the reaction. The cells generally cannot be recharged; hence, they are discarded after the reaction materials have been used up. See also *secondary cell*.

primary center (PC) In telephony, a Class 3 switching office, i.e., a toll office to which toll centers and toll points may be connected. See also *class n office*.

primary channel (1) The data transmission channel in a DCE (data communications equipment). See also *secondary channel*. (2) In a communications network, the channel that has the highest data rate of all channels sharing a common interface. A primary channel may be simplex, half duplex, or full duplex. (3) The channel that is designated as a prime transmission channel and is used as the first choice in restoring priority circuits.

primary circuit The feeding circuit on the input side of a power supply, voltage regulator, or transformer.

primary coating The plastic overcoat in intimate contact with the cladding of an optical fiber, applied during the manufacturing process. The primary coating is typically many layers and has an outside diameter of approximately 250 to 750 μm. It serves the following functions:

- Protects the integrity of the fiber surface from both mechanical and chemical damage
- Strips off cladding modes.
- Suppresses cross-coupling of optical signals from one fiber to another in the case where multiple fibers are contained inside a single buffer tube.
- Distinguishes fibers from one another by virtue of color coding.

The primary coating should not be confused with a *tight buffer* or the *plastic cladding* of a plastic-clad-silica (PCS) fiber. Also called *primary polymer coating* and *primary polymer overcoat*.

primary domain controller (PDC) The server at which the master copy of a domain's user accounts database is maintained. The primary domain controller also validates logon requests. A LAN (local area network) Manager term.

primary frequency (1) A frequency that is assigned for usual use on a particular circuit. (2) The first-choice frequency that is assigned to a fixed or mobile station for radiotelephone communications.

primary frequency standard A frequency source that inherently meets national and/or international accuracy standards and operates without the need for calibration against an external standard. Examples of *primary frequency standards* are hydrogen masers and cesium beam frequency standards.

primary group The lowest level of the multiplex hierarchy. See also *group*.

primary interexchange carrier (PIC) The long-distance company (interexchange carrier—IEC) that is automatically accessed when a customer dials 1+.

primary link station (PLS) In environments that use IBM's Synchronous Data Link Control (SDLC) protocol, a primary link station is a node that initiates communications, either with another primary or with a secondary link station (SLS). Also called simply a *primary*.

primary path The preferred route from one switched node to another. See also *primary route*.

primary power The source of electrical power that usually supplies the station main bus. The primary power source may be a government-owned generating plant, a privately owned generator, or the public utility power system.

primary radiation Radiation that is incident upon a material and produces secondary emission from the material, e.g., in photo multipliers.

primary rate access (PRA) The multiplexed communications link to an Integrated Services Digital Network (ISDN) where the link has a capacity of 23 B channels and one D channel (23+D) in the United States (PCM24 using 1.544 Mbps T-1 links) or 30 B channels and a D channel (30 + D) for ITU-T compliant networks (PCM30 using 2.048 Mbps E1 links). The "B" is from the bearer channel, while the "D" is from the data channel. The bandwidth of all channels is 64 kbps.

The interface is a full digital connection, typically connecting a private branch exchange (PBX) and a central office (CO) or a CO with another CO. Also called *primary rate interface (PRI)*.

primary route The predetermined or preferential path of a message from its source (sending or originating station) to a message sink (receiving, addressee, or destination station).

In telephone switchboard operations, the primary route is the route that is attempted first by the operators or equipment when completing a call. Subsequent attempts use *alternate routing*, which is based on network traffic conditions and supervisory policy.

primary service area The service area of a broadcast station in which the groundwave is not subject to objectionable interference or objectionable fading.

primary substation Power distribution. Equipment that switches or modifies voltage, frequency, and/or other characteristics of primary power.

primary time standard A time standard that does not require calibration against another time standard.

Examples of primary time (i.e., frequency standards) include cesium standards and hydrogen masers. The international second is defined as 9 192 631 770 transitions of the microwave frequency associated with the atomic resonance of the hyperfine ground-state levels of the cesium-133 atom in a magnetically neutral environment. See also *second* and *time*.

principle city office In telephony, an intermediate switching office that has the screening and routing capabilities to accept traffic to all end offices (Class 5) within one or more numbering plan area. See also *class n office*.

print server A computer and/or software to provide users on a network with access to a central printer. It queues print requests (allowing job printing in the order they arrive or giving priority to particular users), manages printers, and spools print jobs (holds the information to be printed in memory until the printer is free). Print servers allow the cost of expensive printers to be shared by all users of the network.

print spooler A small program that manages printing from one or more users and/or applications. Print jobs are transferred from the user's application to a print queue (managed by the *print spooler*) until the printer is available. This frees up the user's application and hardware to do other tasks while the spooler is printing the information in the queue. Several users on a network may request printing simultaneously, and the *print spooler* will store all information in the queue until previous jobs are completed. Often called just a *spooler*.

printed circuit A pattern of electrical conductors applied to one or more layers of a substrate (called a *printed circuit board* or *PCB*). The conductors electrically interconnect devices physically mounted to the surface(s) of the substrate. (A printed circuit board with components installed is called a *printed circuit assembly*.)

printer driver A program that controls printing and sets options such as print quality and paper size for a particular printer.

priority (1) In computers, the ranking in receiving the attention and use of a processor or system resources. Within a computer, various devices (such as keyboards, modems, timers, and memory) are vying for the attention of the processor at unscheduled times. If two of these devices request service at the same time, the different priority levels are used as an arbitrator to grant service—thus preventing clashes. **(2)** On communications networks, nodes assigned privileges that determine when and how often they have access to the network resources. **(3)** In radio transmissions, the right to occupy a specific frequency for authorized uses, free of harmful interference from stations of other agencies. **(4)** In Department of Defense (DOD) record communications systems, one of the four levels of precedence used to establish the time frame for handling a given message. **(5)** In Department of Defense (DOD) voice communications systems, one of the levels of precedence assigned to a subscriber telephone for the purpose of preemption of telephone services. **(6)** A synonym for *priority level*.

priority call (1) Emergency calls to an attendant which bypass the normal queue and alert the attendant with a special signal. **(2)** A service offered by some local exchange carriers (LECs) in which a unique ringing signal is generated when any of a few specific caller numbers is received.

priority level In the Telecommunications Service Priority system, the level that may be assigned to a National Security or Emergency Preparedness (NS/EP) telecommunications service, which specifies the order in which provisioning or restoration of the service is to occur relative to other NS/EP or non-NS/EP telecommunication services. (NS/EP services are used to manage local, national, or international crises that could cause harm to the population, damage property, or threaten national security of the United States.) Also called *priority*.

PRIORITY LEVELS AUTHORIZED

	Provisioning	Restoration
Highest	E	
	1	1
	2	2
	3	3
	4	4
Lowest	5	5

priority ringing   The ringing signal indicating a *priority call* (2).

priority transport   The capability of certain network traffic to have priority over other traffic, thereby having lower delay and better performance.

privacy   **(1)** A private automatic branch exchange (PABX) feature in which a line that is occupied by a user cannot be seized by any other user—even if the line appears on the key set. The privacy option may be automatic or manually initiated. **(2)** In a communications system or network, the protection given to information to conceal it from persons having access to the system or network. Also called *segregation*. **(3)** In a communications system, protection given to unclassified information that requires safeguarding from unauthorized persons. **(4)** In a communications system, the protection given to prevent unauthorized disclosure of the information in the system.

privacy enhanced mail (PEM)   A public key encryption e-mail scheme developed for the Internet and bulletin board system (BBS) use.

privacy override   The ability of a privileged user to override the private automatic branch exchange (PABX) privacy feature.

privacy release   The release of the private automatic branch exchange (PABX) privacy feature so that others may join the conversation.

privacy system   A communication system designed to make unauthorized message reception and interpretation difficult. See also *encryption*.

private automatic branch exchange (PABX)   See *PABX* and *PBX*.

private branch exchange (PBX)   See *PABX* and *PBX*.

private exchange (PX)   A telephone switching system serving a single organization and generally having no means for connection with the public telephone network. A *private exchange* may be a manual exchange or contain an automated switching means (i.e., private automatic exchange—*PAX*).

private key   In cryptography, a key used for both encoding and decoding a message. The security of the system depends on the security of the key. The data encryption standard (DES) developed by the U.S. National Bureau of Standards is an example of a private key system.

In the following example, the message "word" is encoded with the key "Asdf." The resulting ciphertext "6<FS><SYN><STX>" is transmitted to the recipient where the key is again used to decode the string into the original message "word." Both the encoding and decoding in this example are simply the "exclusive or" of the string with the key.

Text	ASCII	Binary
Clear	word	01110111 01101111 01110010 01100100
Key	Asdf	01000001 01110011 01100100 01100110
Cipher	6<FS><SYN><STX>	00110110 00011100 00010110 00000010
Key	Asdf	01000001 01110011 01100100 01100110
Clear	word	01110111 01101111 01110010 01100100

private line   **(1)** In telephony, a service that involves dedicated circuits, private switching arrangements, and/or predefined transmission paths (whether virtual or physical), which provide communications between specific locations. A synonym for *leased line*. **(2)** Common usage among users of a public switched telephone network (PSTN). The term *private line* is often used to mean a one-party switched access line.

private line automatic ringdown (PLAR)   A communications system in which two telephone instruments are connected by a leased line and are arranged so that either phone going off-hook causes the other phone to ring.

private management domain (PRMD)   In ITU-T's X.400 model, a Message Handling System (MHS) or an electronic-mail system operated by a private organization, such as a corporation or university.

private network   A network established and operated by a private organization or corporation.

privilege level   For user-level security, one of three settings (user, administrator, or guest) assigned for each user account.

privileges   The access rights to a program, file, directory, or network. *Privileges* may be extended to or denied on any or all of the available attributes, e.g., read, write, delete, create, and run.

PRMA   An abbreviation of P̲acket R̲eservation M̲ultiple A̲ccess.

PRMD   An abbreviation of P̲R̲ivate M̲anagement D̲omain.

PRN   **(1)** An abbreviation of *P̲seudo-R̲andom N̲oise*. See also *pseudo-noise* (*PN*). **(2)** An abbreviation of P̲seudo R̲andom N̲umber.

probability   The likelihood that an event will occur. In computing and communications, *probability* is used to determine the likelihood of a failure or an error in a system, device, or message.

probe   **(1)** A sensing device, e.g., a pencil-shaped device on the end of an instrument lead wire. **(2)** A packet sent to the remote end of the network in an AppleTalk network. The probe requests an acknowledgment from the node at the end, which serves to indicate the end of the network and to acknowledge that the node is functioning.

procedure   **(1)** In general, the steps or actions taken during the process of solving a problem or completing a task. **(2)** In software, a portion of an overall program that is named and performs a specific task.

proceed to select signal   In communications, a signal that indicates that the system is ready to receive a selection signal, e.g., the dial tone in telephony.

process   A program or subprogram that is executing on a host computer.

process gain   In a spread spectrum communications system, the effective signal gain, signal-to-noise ratio (SNR), signal shape, or other signal improvement over a nonspread spectrum system, obtained by coherent band spreading, remapping, and reconstitution of the desired signal. Processing gain is usually measured in decibels (dB). See also *spread spectrum*.

process shrink   A cost-reduction technique used by the semiconductor industry. It is the practice of applying current design techniques to older chip technologies in order to reduce *feature size* and overall die size. A smaller part allows more parts to be fabricated from each silicon wafer, thereby reducing cost.

processing delay   In digital communications networks, the time required by a network node to receive a packet, determine its destination, and modify the packet for retransmission to its final destination. *Processing delays* include:

- Routing information lookup (typically from tables).
- Modification of the packet header.
- Switching overhead (e.g., queuing priorities).
- Editing the packet hop count.
- Re-encapsulating the packet.

processor (1) In a computer, the functional unit that interprets and executes instructions. A processor consists of at least an instruction control unit and an arithmetic unit. (2) A deprecated term for a *processing program.*

processing unit A functional unit that consists of one or more processors and their associated internal storage.

Prodigy® One of the commercial dial-up information services and an Internet access provider. Some of the services and data available include weather, news, stock quotes, business news, downloadable files and programs, bulletin boards, and conferences.

profile (1) The set of parameters that define the way a device or system performs. In a local area network (LAN), a *profile* is often used by one or more workstations to determine the connection they will have with other devices. *Profiles* work like batch files, executing a number of commands to save a user time and effort. (2) In optical fibers, the way the index of refraction changes as a function of distance from the center of the fiber, e.g., step index, graded index, parabolic index, or power law index. See also *power law index profile.*

profile dip A synonym for *index dip.*

profile dispersion In optical fibers, the dispersion attributable to the variation of refractive index profile with respect to wavelength. The profile variations have two sources:

- Variation in index contrast, and
- Variation in the profile parameter.

See also *dispersion.*

profile parameter (g) In fiber optics, the refractive index profile shape-defining variable for a power law index optical fiber. For most silicas, the optimum value of *g* for minimum dispersion is approximately 2. See *power law index profile.*

program (1) In computing, a sequence of instructions that can be carried out (executed) by a computing device. A *program* can refer to either the original source code or the final executable form. Also called *code* and *software.* (2) A sequence of signals transmitted for information or entertainment, e.g., a television program. (3) A collection of processes working together to accomplish a specific task. (4) The process of designing, writing, and testing the sequence of instructions of (1).

program circuit A communications line used for the transmission of radio program material. It is a conditioned line that has been equalized to provide a wider available bandwidth than a normal voice telephone line.

program level The magnitude of a signal in an audio system; expressed in volume units (VU).

program tracking A technique of aiming at or following a moving object by using knowledge about the mathematics of the motion of the object. For example, *program tracking* may be used to point an antenna toward a satellite by having a computer calculate the current position based on its orbital parameters.

programmable An entity that has the ability to change a feature or operating characteristic without rewiring. That is, it is capable of accepting a set of instructions (program) for performing a task or operation.

programmable logic array (PLA) An array of gates having interconnections that can be programmed to perform a specific logical function.

programmer (1) The part of digital equipment that controls the timing and sequencing of operations. (2) A person who prepares computer *programs* (1), i.e., writes sequences of instructions for execution by a computer.

programming language A set of codes and rules (grammar and syntax) used to create instructions for a computer to follow. The *programming language* allows a programmer to write in a form that a person can more easily read. Most programs are written by using a text editor (a simple word processor) to create what is called *source code.*

Generally, the computer can neither directly read nor execute source code. It must first be translated into *object code* (a sequence of machine readable instructions) by a process called assembly, compilation, or interpretation.

- *Assembly*—the process of translating source code into object code where each source code instruction generally translates to one machine instruction. Assembly language programming is the most tedious to write; however, it is the fastest when run by the computer. Assembly programs are machine specific, that is, a program written for an 80xxx processor chip will not run on a 68xxx based computer.
- *Compilation*—the process of translating the source code into object code where each source code instruction translates into many machine instructions (perhaps even to a complete program). Programs that are compiler based are transportable, that is, with the proper compiler the source code may be translated to object code that will run on any computer. Programs such as these are more easily written because the form is more closely related to human form than machine form. The resultant programs, however, are generally slower and larger than assembly language based programs.
- *Interpretation*—the process of translating source code into object code at the same time the program is run. The interpreter reads one source code command, translates it into machine instructions, executes the instructions, and then reads the next source code command. These programs are the slowest of programming languages because the code is compiled as it is needed and then discarded. (If it is needed again later, it must be re-compiled.)

Project 2000 The overall effort to solve the Y2K (year 2000) problem. That is, the effort to resolve the effects of software that uses only the last two digits of the year (effects like year 2001 coming before 1998, i.e., $01 < 98$).

PROM An acronym for Programmable Read Only Memory (pronounced "prom"). A *PROM* is a read only memory (ROM) that can be programmed by the user but only once. After a *PROM* is programmed, it functions as a ROM. See also *EEPROM, EPROM, ROM,* and *UVPROM.*

promiscuous mode A network interface card (NIC) operating mode in which the NIC passes all packets that arrive to higher layers, regardless of whether the packet is addressed to the node. This operating mode allows a remote server to deliver packets to remote users. It also makes it possible to pass everything that happens at a NIC on to a network analyzer program.

propagation Waves traveling through or along a medium, e.g., a transmission line or optical fiber. For electromagnetic waves, propagation occurs in a vacuum as well as in tangible media.

propagation constant For an electromagnetic field mode varying sinusoidally with time at a given frequency, the logarithmic rate of change, with respect to distance in a given direction, of the complex amplitude of any field component. The propagation constant (λ), is a complex quantity given by

$$\lambda = \alpha + j\beta,$$

where:

α is the attenuation constant (the real part), and
β is the phase constant (the imaginary part).

propagation delay The time required for a signal to pass through an entity (e.g., a device, transmission line, or an entire network). (It is the time between a significant instant at the input of an entity and the corresponding significant instant at its output.)

Total propagation delay is important because it may determine the maximum cable lengths of a particular network configuration.

propagation mode (1) The arrangement or configuration of the electric and magnetic fields associated with an electromagnetic wave traveling along a transmission line. Each configuration is called a mode. An infinite number of modes are possible as long as there is no upper bound on the frequency. There is, however, only one possible mode for the lowest frequency that can be transmitted (called the *dominant mode*).

All modes can be divided into one of two general types: the transverse magnetic (TM) and the transverse electric (TE) modes. The TM mode has the entire magnetic field (and a component of the electric field) transverse to the direction of propagation, while the TE mode has the entire electric field (and a component of the magnetic field) transverse to the direction of propagation. A specific *propagation mode* is identified by the mode group identifier (TE or TM) and a two-digit subscript indicating the specific mode. For example, the dominant mode is indicated by TM_{00} and TE_{00}. (2) The manner in which radio signals travel from a transmitting antenna to a receiving antenna, such as ground wave, sky wave, direct wave, ground reflection, refraction, ducted wave, or scatter.

propagation path obstruction In radio propagation systems, an object (man-made or natural) that lies near enough to a radio path to cause a measurable effect on path loss, exclusive of reflection effects. Hills, bridges, cliffs, buildings, and trees are examples of obstructions.

If the clearance from the nearest anticipated path position, over the expected range of Earth radius k-factor, exceeds 0.6 of the first Fresnel Zone radius, the feature is not normally considered an obstruction.

propagation time The time required for an electromagnetic wave to travel between two specified points (in free space or on a transmission line). See also *propagation delay*.

propagation velocity The speed of an electrical or optical signal down a transmission line compared to the speed of light in a vacuum. It is generally expressed as a percentage of the speed of light.

For a coaxial cable, the *propagation velocity* is the reciprocal of the square root of the dielectric constant (i.e., $\lambda / \sqrt{\epsilon}$) of the material between the inner and outer conductors.

proprietary A formula, design, or procedure owned by a company and with secret details known only to that company. *Proprietary* designs are generally incompatible with any similar product on the market. For example, the AT&T *proprietary* 19 200 V.turbo modem protocol is incompatible with U.S. Robotic's *proprietary* 19 200 protocol. In another example, the Rockwell V.fc 28.8 protocol is close to, but not compatible with, the ITU-T V.34 28.8kbps protocol.

The motivation of a proprietary design may be either to exclude competition, to lock users into exclusively purchasing that company's product, or to retain the benefits of the innovation and expense involved in the development.

proprietary network A network developed by a vendor that is not based on any open or public protocol standard. Hence, it will run only on that vendor's equipment.

proprietary software A program owned or copyrighted by an individual or company and available only through the permission of the owner. See also *public domain software.*

proprietary standard A standard for the design of software, hardware, protocols, or other practices or operations that is developed by a single commercial entity or consortium of manufacturers.

If the design is very useful and becomes popular throughout the industry, it may become a *de facto standard,* i.e., a standard all users expect whether or not it is requested by name. For example, Hayes Microcomputer's "AT" modem command structure is so pervasive that it is now a standard to be expected on all modems. See also *standard.*

proration (1) The proportional distribution or allocation of parameters (such as noise, power, or losses) among a number of entities (such as antennas, transmitters, receivers, cables, telephone lines or trunks, etc.) in order to balance the performance of communications circuits. Also called *budgeting.* (2) In shared equipment switching systems, the distribution or allocation of equipment or devices proportionally across a number of entities, to provide a desired grade of service.

Prospero On the Internet, a tool for accessing, organizing, and using files that may be located in diverse remote locations. It is a client program that enables a user to mount a remote host directory so that it appears to be part of the local host machine's filing system. Unlike other systems, however, *Prospero* allows the user to manipulate the view for his convenience.

Information about *Prospero* is available via file transfer protocol (FTP) from prospero.isi.edu. Other similar programs include Alex, Andrew File System (AFS), and Network File System (NFS).

protected distribution system (PDS) A telecommunications system with sufficient acoustic, electrical, electromagnetic, and physical safeguards to permit the transport of nonencrypted classified information. The safeguards must include the subscriber terminals as well as the transport (e.g., wire or optical fiber) and switching system. Formerly called an *approved circuit.*

protective coupling arrangement (PCA) A device placed between the local exchange company's (LEC's) lines and subscriber's equipment if the subscriber's equipment does not have FCC Part 68 approval. The PCA is not needed if the equipment has a Part 68 registration number.

protective ratio (1) The ratio of the withstand voltage of the protected equipment to the protector value. (2) The minimum value of the wanted to unwanted signal ratio, usually expressed in decibels, at the receiver input determined under specified conditions, such that a specified reception quality of the wanted signal is achieved at the receiver output.

protector In telecommunications systems, a device for absorbing and/or shunting to ground excess energy that may be introduced onto the communication medium from a fault condition or natural phenomenon. It protects equipment, facilities, and personnel from abnormally high voltages or currents.

Protectors may be designed to operate on short-duration phenomena (transient), or long-duration phenomena (power cross).

- *Transient protectors* may use spark gap, carbon blocks, metal oxide varistors (MOVs), or gas tubes technologies.
- *Long-duration protectors* also include a type of slow-acting fuse called a *heat coil*. These are used to protect the equipment from a fault condition where the ac mains might come in contact with the communications medium.

In telephone systems, *protectors* are frequently located at the main distribution frame (MDF) in a switching office and at the demarcation point at the subscriber's premises.

protocol **(1)** In general, a strict set of rules (procedures) governing how one entity initiates, maintains, and controls communication with another entity. Various protocols may exist on many levels within a network. **(2)** A formal set of conventions governing the format, timing, and structure of message exchange between two or more devices. The *protocol* may facilitate communication between elements of the same machine or different machines. A *protocol* can range from a simple XON/XOFF flow control to an error-correcting packet transfer protocol such as is used in the *ZMODEM* or *X.25 packet* protocol.

Protocols can be distinguished by any of several types of properties:

- The level, or layer, at which the protocol operates.
- The network architecture (ring, bus, star, etc.) for which the protocol is designed.
- Whether the protocol is synchronous or asynchronous.
- Whether the protocol is connection-oriented or connectionless.
- Whether the protocol is character- or bit-oriented.

See also *communications protocol, MNP, OSI,* and the various *ITU-T protocols.*

protocol analyzer A troubleshooting device that monitors network activity and produces displays indicating network performance.

protocol control information (PCI) In the ISO/OSI Reference Model, protocol-dependent information added to a data packet before the packet is passed to a lower layer for further processing.

protocol converter Any device that translates a data stream from one format (protocol) to another format according to a fixed algorithm. For example, software in a *bridge* connecting two dissimilar networks converts the protocols of one network into the protocols of the other network.

protocol data unit (PDU) The sequence of contiguous octets delivered as a unit to or from the media access control (MAC) sublayer. Information delivered as a block between peer entities (e.g., on a Token Ring) which contain at least control information and possibly data.

A valid logical link control, *PDU* is at least three octets long and contains two address fields and a control field. The *PDU* may (or may not) contain an information field. A *PDU* has four major components:

- Destination service access point (DSAP)—an 8-bit value that identifies the higher-level protocol using the logical link control (LLC) services.
- Source service access point (SSAP)—an 8-bit value that indicates the local user of the LLC service.
- Control—a 1- or 2-byte field that indicates the *PDU* type. The contents of the field depend on whether the *PDU* is an information (I), supervisory (S), or unnumbered (U) frame.
  I frames are used for transmitting data.

S frames are used to oversee the transfer of I frames and are found only in type 2 (connection-oriented) services.

U frames are used to set up and break down the logical link between network nodes in either type 1 or type 2 services. They are also used to transmit data in connectionless (type 1 or type 3) services.

- Data or information—a variable-length field that contains the packet received from the network level protocol. The allowable length for this field depends on the type of access method being used (CSMA/CD or token passing). S frames do not have a data field.

protocol-dependent routing Any routing method in which routing decisions are made on the basis of information provided by the specific local area network (LAN) protocol used by the communicating devices.

For example, Transmission Control Protocol/Internet Protocol (TCP/IP) and multi-protocol routers are protocol dependent.

protocol filtering A programmable feature of some network bridges that allows the bridge to always reject (or forward) transmission associated with specified protocols.

protocol hierarchy In open systems architecture, the distribution of network protocol among the various layers of the network.

protocol-independent multicast (PIM) A multicast routing technique that enables Internet Protocol (IP) multicast routing on existing IP networks. Several algorithms are used to establish routing trees, e.g., *Dense-Mode PIM* and *Sparse-Mode PIM*.

- *Dense-Mode PIM*—a technique designed for applications which send traffic to high concentrations of local area networks (LANs) (e.g., LAN TV and financial broadcasts). *Dense-Mode PIM* works by flooding *all* ports with a new incoming multicast transmission. Routers receiving the flood broadcast but with no need for the traffic will reply with a *prune* message, which causes the router flooding the network to prune the entry from its routing table (also called a flooding list or tree). Also known as *reverse-path forwarding*.
- *Sparse-Mode PIM*—a routing algorithm that assumes there is a sparse population of receivers in the network (i.e., there are relatively few nodes in the network that want to receive multicast data-

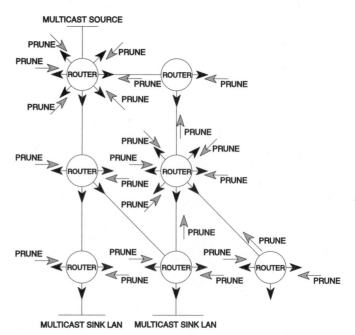

MULTICAST SINK LAN   MULTICAST SINK LAN

grams). One PIM router is configured as the Rendezvous Point (RP) for each multicast group that uses the Sparse Mode PIM protocol. Host systems must request to join a multicast group before they can receive multicast datagrams. With *sparse-mode,* network traffic is reduced as multicast datagrams are only sent along paths to group member nodes. However, the protocol overhead is higher, so the benefits are achieved only if there is a low population of receivers.

protocol-independent routing (PIR) **(1)** A routing method in which routing decisions are made without reference to the protocol being used by the communicating devices. These routers can route non-routable protocols.

Shortest path first (SPF) and the OSI (Open Systems Interconnection) routing standard IS-IS (Intermediate System to Intermediate System) are examples of technologies that support protocol-independent routing. **(2)** A proprietary routing method developed by General Data-Comm for IBM networks is an improvement on Source Routing. The router checks all local area network (LAN) packets for their source address contained in each frame. Adopting the proprietary Discover Shortest Path First (DSPF) router-to-router protocol, a *PIR* router finds the best path between pairs of LANs rather than simply between pairs of routers as does the Open Shortest Path First (OSPF) method. Because a router can have several LANs connected to it, OSPF does not necessarily produce the best result, whereas *PIR* load-balances across each available LAN link for improved throughput and performance. In a Token Ring LAN, *PIR* looks like a Source Routing bridge and in an Ethernet, *PIR* appears as a Transparent bridge.

protocol stack A hierarchical set of related protocols in a layered network model, where each layer's operation is governed by a protocol designed to perform a specific task. Also called a *protocol suite.* See also *OSI.*

protocol transfer A data transfer that uses:

- Error checking to ensure that the receive data are correct and
- Flow-control techniques to regulate the rate at which data blocks are transferred.

protocol translator In a communications system, the collection of hardware, software, and/or firmware that is required or used to convert the protocols used in one network to those used in another network. Also called a *protocol converter.*

prototype **(1)** A functional, preproduction example that is the first of its type. It is typically used to evaluate design performance and/or production processes. **(2)** An engineering model suitable for evaluation of certain design and/or performance parameters.

proxy An element that responds on behalf of another element to a request using a particular protocol. Also called a *proxy agent.*

proxy agent A network management agent that sits in front of a device attached to a network that needs to be managed but has no intrinsic network management capabilities. The *proxy agent* provides network management "by proxy" for that device.

proxy ARP A proxy arrangement in which one device (usually a router) answers address resolution requests on behalf of another device. The proxy agent (the router) is responsible for making sure that packets get to their real destination.

PRR An abbreviation of Pulse Repetition Rate.

PS **(1)** An abbreviation of Power Supply. **(2)** An abbreviation of PostScript. **(3)** An abbreviation of Presentation Service. Level 6 of IBM's Systems Network Architecture (SNA). **(4)** An abbreviation of Print Server. **(5)** An abbreviation of Proposed Standard. **(6)** An abbreviation of Power Status. A 2-bit control field at the basic rate interface (BRI). **(7)** An abbreviation of Permanent Signal.

PS/2® An abbreviation for Personal System/2®. A family of IBM personal computers that succeed the (AT) model computer.

PSAP **(1)** An acronym for Presentation Service Access Point. **(2)** An acronym for Public Safety Answering Point. The point in the U.S. public telephone system where 911 calls are answered and initially handled.

PSD An abbreviation of Power Spectral Density.

PSDN **(1)** An abbreviation of Packet Switched Data Network. **(2)** An abbreviation of Public Switched Data Network.

PSDS An abbreviation of Public Switched Digital (or Data) Service.

PSDU An abbreviation of Presentation Service Data Unit.

PSE **(1)** An abbreviation of Packet Switching Exchange. A device that performs packet switching in a network. **(2)** An abbreviation of Power Series Expansion.

PSEL An acronym for Presentation SELector.

pseudo-noise (PN) A binary *pseudorandom number sequence* that satisfies one or more of the standard tests for statistical randomness. Although the sequence seems to lack any definite pattern, it is a repeating sequence of pulses, albeit a long sequence of pulses. These characteristics allow the sequence to be detected within true noise by a suitable receiver (a receiver with a priori knowledge of the *PN sequence* generation algorithm details).

For example, in spread spectrum systems, a transmitter's signal appears as noise to a receiver not locked on the transmitter frequencies or that are incapable of correlating a locally generated pseudorandom code with the received signal. Also called *pseudorandom noise (PRN)* or *pseudorandom noise code.*

pseudo-range In the Global Positioning System (GPS), the calculated range between the GPS receiver and the satellite antenna phase centers. The range is determined by multiplying the speed of light by the time difference between signal transmission and signal reception. The range is called *pseudo-range* because of time offset errors in the satellite and receiver, atmospheric propagation delay uncertainties, and other errors. Also written *pseudorange.*

pseudo top-level domain In the Internet, a domain name that indicates to a router that the host with this name resides in a network that lacks full Internet connectivity. For example, in the host name

asdf.qwerty.bitnet

bitnet appears to be a top level domain because it is the last name in the sequence.

pseudorandom sequence A sequence of mathematically generated numbers, determined by some arithmetic process to be acceptably random for a given purpose. The sequence may be designed to approximate a uniform distribution, a normal Gaussian distribution, or some other shape as required. Also written *pseudo random sequence.* Also called a *pseudo random code.*

PSG **(1)** An abbreviation of Programmable Sound Generator. **(2)** An abbreviation of Programmable Signal Generator.

PSI **(1)** An abbreviation of Primary Subnet Identifier. Part of the Manufacturing Automation Protocol (MAP) network address in the network level header. **(2)** An abbreviation of Packet Switching Interface. **(3)** An abbreviation of Performance Summary Interval. **(4)** An abbreviation of Process to Support Interoperability.

PSK An abbreviation of Phase Shift Key.

PSM An abbreviation of Phase Shift Modulation.

PSN (1) An abbreviation of P̲acket S̲witched N̲etwork. (2) An abbreviation of P̲acket S̲witching N̲ode. (3) An abbreviation of P̲rivate S̲witching N̲etwork. (4) An abbreviation of P̲ublic S̲witched N̲etwork.

PSNR An abbreviation of P̲ower S̲ignal to N̲oise R̲atio.

psophometer Essentially a voltmeter or powermeter that provides a numerical indication of the audible effects of disturbing voltages of frequencies as measured in a specified band of interest. The instrument incorporates a frequency weighting network (a filter) whose characteristics depend on the type of circuit under investigation (e.g., telephone circuits, high-fidelity music).

psophometric weighting In telephony, a noise weighting established by the International Consultative Committee for Telephony (CCIF, which became CCITT and, more recently, ITU-T), designated as CCIF-1951 weighting, for use in a noise-measuring set or psophometer. Although the shape of the CCIF-1951 characteristic is essentially identical to that of F1A weighting, it is calibrated with an 800-Hz, 0-dBm tone. Therefore, the corresponding voltage across 600 ohms produces a reading of 0.775 V, which introduces a 1-dBm adjustment in the formulas for conversion with dBa.

PSP (1) An abbreviation of P̲resentation S̲ervices P̲rocess. (2) An abbreviation of P̲acket S̲witching P̲rocessor. (3) An abbreviation of P̲eak S̲ideband P̲ower. (4) An abbreviation of P̲ayphone S̲ervice P̲roviders.

PSPDN An abbreviation of P̲acket S̲witched P̲ublic D̲ata N̲etwork.

PSRR An abbreviation of P̲ower S̲upply R̲ejection R̲atio. An expression of the attenuation of a signal (or noise) present at a device's power supply input terminals to the corresponding level at the device's signal output terminals. *PSRR* is generally expressed in decibels (dB), that is,

$$PSRR = 10 \cdot \log_{10} \left(\frac{Pwr_{output}}{Pwr_{input}} \right)$$

PSS (1) An abbreviation of P̲acket S̲witching S̲ervice. (2) An abbreviation of P̲acket S̲witchS̲tream.

PSTC An abbreviation of P̲ublic S̲witched T̲elephone C̲ircuits.

PSTN An abbreviation of P̲ublic S̲witched T̲elephone N̲etwork.

PSU (1) An abbreviation of P̲acket S̲witching U̲nit. (2) An abbreviation in telephony from P̲ath S̲etU̲p. (3) An abbreviation of P̲ower S̲upply U̲nit.

PT (1) An abbreviation of P̲ayload T̲ype. See *ATM*. (2) An abbreviation of P̲ass T̲hrough. (3) An abbreviation of P̲ay T̲one.

PTC (1) An abbreviation of P̲ublic T̲elephone C̲ompany. (2) An abbreviation of P̲ositive T̲emperature C̲oefficient. Refers to the charartTeristic of a device that increases in value as temperature is increased.

PTE An abbreviation of P̲ath T̲erminating E̲ntity.

PTF An abbreviation of P̲atch and T̲est F̲acility.

PTG In telephony, an abbreviation of P̲recise T̲one G̲enerator.

PTI An abbreviation of P̲ayload T̲ype I̲dentifier. A control field in an asynchronous transfer mode (ATM) header.

PTM (1) An abbreviation of P̲ulse T̲ime M̲odulation. (2) An abbreviation of P̲hase T̲ime M̲odulation. (3) An abbreviation of P̲ulse T̲ime M̲ultiplex. (4) An abbreviation of P̲ulse T̲ransmission M̲ode. (5) An abbreviation of P̲ortable T̲raffic M̲onitor.

PTN (1) An abbreviation of P̲ersonal T̲elecommunications N̲umber. (2) An abbreviation of P̲ublic T̲elephone N̲etwork. (3) An abbreviation of P̲lant T̲est N̲umber.

PTNX An abbreviation of P̲rivate T̲elecommunications N̲etwork eX̲change.

PTP An abbreviation of P̲oint T̲o P̲oint.

PTR An abbreviation of P̲oinTeR̲.

PTS (1) An abbreviation of P̲rocede T̲o S̲end. (2)An abbreviation of P̲rocede T̲o S̲elect. (3) An abbreviation of P̲ublic T̲elephone S̲ystem (or S̲ervice).

PTSE An abbreviation of P̲NNI (Private Network to Network Interface) T̲opology S̲tate E̲lement. PTSEs contain node information, topology state, and reachability information. Found in asynchronous transfer mode (ATM) networks.

PTSP An abbreviation of P̲NNI (Private Network to Network Interface) T̲opology S̲tate P̲acket. Found in asynchronous transfer mode (ATM) networks.

PTT (1) An abbreviation of P̲ost, T̲elephone, and T̲elegraph. The generic term for the authority and monopoly in each country governing postal, telephony, and telegraphy activities (equivalent to Regional Bell Operating Company [RBOC] when describing the seven baby-bells.) (2) An abbreviation of P̲ush T̲o T̲alk.

PTTC An abbreviation of P̲aper T̲ape T̲ransmission C̲ode.

PTTI An abbreviation of P̲recise T̲ime and T̲ime I̲nterval.

PTTXAU An abbreviation of P̲ublic T̲eleT̲eX̲ A̲ccess U̲nit.

PTXAU An abbreviation of P̲ublic T̲eleX̲ A̲ccess U̲nit.

PU (1) An abbreviation of P̲hysical U̲nit. A *PU* provides network and resource control services for its associated logical units (LU) on IBM's Systems Network Architecture (SNA). (2) An abbreviation of P̲ower U̲nit.

public access provider An organization that provides (often for a fee) access to a network for individuals or other organizations.

public data network (PDN) A network operated and maintained by a telecommunications administration, Post, Telephone, and Telegraph (PTT), or a recognized private operating agency for the specific purpose of providing data transmission services for the public. It may include circuit switched, packet switched, and/or leased circuit data transmission capabilities.

public data transmission service A data transmission service operated by a telecommunications administration or a recognized private operating agency, and using a public data network (PDN).

public domain software The term *public domain* has a very specific legal meaning. It refers to the status of the work (software) of a creator, who had legal ownership of that work, has given up ownership of that work, and has dedicated that work to the public at large, e.g., a program donated for anyone's use by its owner or a work without copyrights (or with expired copyrights).

Once a program is in the *public domain,* anyone can use it in any way they choose. It is freely available for copying, modifying, and distribution. The author has no control over its use and cannot demand payment for it. Because of the frequent misuse (or misunderstanding) of the phrase, programs are often described by their authors as being *public domain* when, in fact, they are *shareware* (copyrighted commercial software in which the owner allows evaluation before paying for it) or *freeware* (free but copyrighted software). To be sure a program is *public domain,* look for an explicit statement from the owner to that effect. See also *freeware, proprietary software,* and *shareware.*

public exchange A synonym for central office (CO), the preferred term in most of the world.

public key In cryptography, a method of encrypting a cleartext message with one key and decrypting the ciphertext with a second key. The encoding key is made public so that anyone may encrypt a message to the owner of the key pair. Upon receiving the message, the second, private, key is used to decrypt the ciphertext into cleartext.

A feature of the *public key* system is the ability to "ensure" that not only is the message read by only the intended recipient but that the recipient knows that the message is from the claimed sender. This is accomplished by having the sender encrypt the message with the recipient's *public key* AND the sender's private key. The recipient then decrypts with his private key AND the sender's public key.

public key encryption A method of encrypting a message so that the sender does not have to divulge a private key to the receiver for the receiver to be able to decrypt the message.

In *public key encryption* schemes, the encryption and decryption keys are different. The encryption key is made public, while the decryption key is kept private. Therefore, anyone may use the receiver's encryption key to encode a message and then transmit it on a nonsecure channel to the receiver. No one can read the encrypted file except the receiver because only the receiver has the decryption key. A further twist is the ability of the receiver to know for certain who the sender is. This procedure involves the sender encrypting the message twice: once as described above, and the second time the sender uses *his decryption* key to encrypt the message and the receiver uses the *sender's public encryption* key to decrypt the message. Of course, these methods only work if (1) the decryption keys are kept private and (2) the private decryption key cannot be calculated from the public encryption key. Also called *nonsecret encryption*. See also *public key*.

public network A network operated by a common carrier or telecommunications organization for the purpose of providing transmission services to the public. A switched communications network with unrestricted access.

public switched network (PSN) Any common carrier network that provides circuit switching among public users. Although commonly applied to the public switched telephone network (PSTN), it also includes other switched networks, e.g., packet switched public data networks.

public switched telephone network (PSTN) A domestic telecommunications network accessed by telephones, key telephone systems, private branch exchange (PABX) trunks, and data arrangements. Also called a *Switched Services Network (SSN)*.

Public Utilities Commission (PUC) In the United States, a state agency that sets the telephone rate structure (called *tariffs*) for intrastate communications. In some states this regulatory function is performed by public service commissions or state corporation commissions.

PUC An abbreviation of Public Utilities (or Utility) Commission.

PUCP An abbreviation of Physical Unit Control Point.

puff (1) Slang for picofarad. For example, 33 pF is spoken 33 *puff*. **(2)** To decompress data previously compressed using a Huffman code. Compression is obviously referred to as *huff*.

pull-in frequency range The maximum frequency difference between the local oscillator or clock and the reference frequency of a phase locked loop (PLL) over which the local oscillator can be locked. See also *phase lock loop (PLL)*.

pull-in time The time required for a phase lock loop (PLL) to achieve synchronism, i.e., the time for the phase of the local oscillator or clock to be shifted into alignment with an input signal. See also *phase lock loop (PLL)*.

pulsating direct current A current whose value changes with time but always flows in the same direction when it is present. The change in value with respect to time may be at regular or irregular intervals and may be interrupted (i.e., have a zero value), but, it cannot change direction (sign).

In many applications, pulsating dc may be modeled as if it has two components: a fixed dc value and a superimposed ac value.

pulse A transient signal, usually brief and with a finite beginning and ending time. The signal deviates from a nominal or *base* value for only a brief period. The deviation may be either in a positive or negative direction. The names of some of the features of a *pulse* are shown in the diagram.

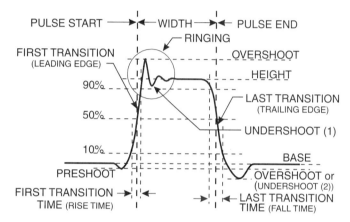

Some parameters used to describe a pulse include:
- *Preshoot*—a distortion that occurs before a major transition.
- *First transition* (*leading edge*)—the major abrupt change of a pulse waveform extending between the base and the top.
- *First transition time* (*rise time*)—the amount of time needed for the first transition to change from the first occurrence of the 10% of final value to the last occurrence of the 90% value and before any overshoot.
- *Overshoot*—the amount by which the first maximum following the first transition exceeds the pulse top. Overshoot is expressed as a percentage of pulse height or in absolute amplitude units. Sometimes called *afterkick*.
- *Ringing*—the amount by which the instantaneous waveform values of a pulse deviates from a straight-line approximation to the pulse top.
- *Droop*—a distortion of an otherwise flat pulse top characterized by an amplitude decline (a height drift toward the baseline) from the first transition toward the last transition. Also called *tilt*.
- *Start*—the 50% point of the first transition.
- *Width* (*duration, length*)—the time duration between the first and last instants of a specified fraction of the pulse amplitude. Generally, the specified fraction is 50%, 90%, or 1/e; however, other values may be used in special cases.
- *Height* (*amplitude*)—the algebraic difference between the top magnitude and the base magnitude. Pulse height may be measured with respect to a specified reference and therefore should be modified by qualifiers, such as *average, instantaneous, peak*, or *root-mean-square*.

- *Final value*—the steady-state value following either major transition (first or last).
- *End*—the 50% point of the last transition.
- *Last transition* (*trailing edge*)—the major abrupt change of a pulse waveform extending between the top and the base.
- *Last transition time* (*decay time, fall time*)—the amount of time needed for the last transition to change from the first occurrence of the 90% of final value to the last occurrence of the 10% value and before any undershoot.
- *Undershoot*—**(1)** That portion of a pulse waveform following any overshoot that is less than the nominal or final value. **(2)** That portion of the waveform that follows the last transition, is of opposite sign of the last transition, and is before the final base value.
- *Base*—that major segment of the waveform with lowest amplitude displacement from a reference baseline.
- *Settling time*—the time required for a pulse waveform to enter and remain within a specified bound following a major transition.
- *Spacing*—the interval between corresponding points of one pulse and the next pulse. Formerly called *pulse interval;* however, the term is deprecated because it may be taken to mean the duration of a single pulse rather than the interval from one pulse to the next.

pulse-address multiple access (PAMA) The ability of a communication satellite to receive signals from several Earth terminals simultaneously and to amplify, translate, and relay the signals back to Earth, based on the addressing of each station by an assignment of a unique combination of time and frequency slots. This ability may be restricted by allowing only some of the terminals access to the satellite at any given time.

pulse amplifier An amplifier specifically designed for the amplification of pulses, e.g., fast rise/fall times and slew rate.

pulse amplitude modulation (PAM) A method of modulating a carrier (consisting of a train of pulses) with another signal. The amplitude of each pulse in the carrier is varied in accordance with the amplitude of the modulating signal, the width of the pulses remaining constant. Two forms of *PAM* are found. In the first, all of the pulses go in the same direction (*unipolar PAM*), and in the second, negative excursions of the signal are represented by negative going pulses (*bipolar PAM*).

The top of the pulse (in either form) may follow the shape of the modulating signal—called *natural* or *top* sampling; or the pulse tops may be flat—called *square topped* or *instantaneous* sampling. See also *modulation.*

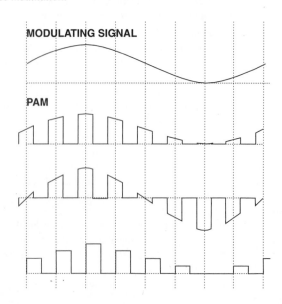

pulse bandwidth The continuous range of frequencies outside of which the amplitudes of *all* frequencies are less than a specified fraction of the amplitude of a reference frequency (generally, the frequency with the largest amplitude). Strictly speaking, the term should be pulse spectrum bandwidth (because the spectrum, not the pulse itself, has a bandwidth). However, common usage of the contraction has forced the term's acceptance.

In the example, a portion of the spectral content of a pulse train is shown. The *pulse bandwidth* is shown for a band edge of −11 dB.

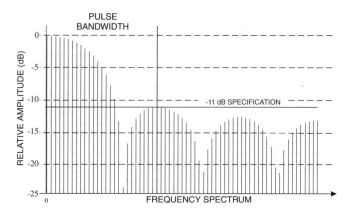

pulse broadening In optical fibers, an increase in pulse duration due to the different path lengths traveled by light rays in the various propagating modes. *Pulse broadening* constraints may be specified by the impulse response, the root-mean-square pulse broadening, or the full duration at half-maximum pulse broadening. Also called *pulse dispersion* and *pulse spreading.* See also *intersymbol interference, multimode distortion,* and *optical fiber.*

pulse burst A finite sequence of pulse waveforms sent as a group followed by a quiet period.

pulse capacity The number of pulses per unit time interval that a pulse receiver can accept and register without loss.

pulse carrier An electromagnetic wave that consists of a series of constant rapid pulses (i.e., constant length, amplitude, and repetition rate) when not modulated.

pulse code (1) A pulse train modulated so as to represent information. **(2)** A code consisting of pulses such as Morse, binary, or Baudot. See also *code.*

pulse code modulation (PCM) A modulation technique in which a digital pulse train is created in accordance with some coding rule. Examples of possible coding rules include:

- The number of pulses in the sampling interval is proportional to the value of the modulation signal (PCM-P for pulse code modulation-pulse).
- Within each sampling interval, a digital word whose value represents the value of the modulating signal is created (PCM-C for pulse code modulation code).
- Within each sampling interval, a digital word whose values represent the difference between the previous sampled value and the current sampled value. This process is frequently called *delta modulation* (*DM*).
- Within each sampling interval, a digital word whose value represents the value of the modulating signal is created. However, the word is not expressed linearly as is done in PCM-C but is compressed according to an approximately logarithmically law. The expression of this compression rule is either A-law for those parts

of the world that implement the *ITU-T* recommendations or the mu-law in the United States, Canada, and several other countries. See also *A-law,* μ-*law,* and *modulation.*

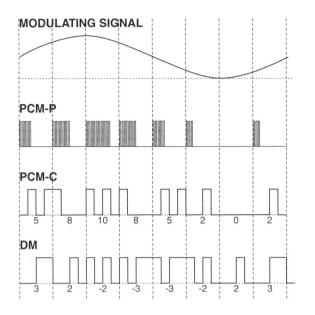

pulse coder A device for varying one or more parameters or a pulse or pulse train so as to convey information.

pulse corrector In telephony, a pulse regenerator used to repeat and restore the make/break ratio of a pulse train.

pulse decay time A synonym for *fall time.*

pulse decoder A device for extracting information from a pulse coded signal.

pulse density A term in certain coding methods, such as those used in T-1. In these coding methods, a waveform transition must occur periodically. *Pulse density* refers to the number of "no transition" signaling elements allowed before an element with a transition is inserted.

In T-1 systems, logical zeros (0) do not generate transitions while logical ones (1) generate alternate positive and negative pulses. In T-1, the maximum number of consecutive 0s is generally 15.

pulse density violation The occurrence of more than the prescribed maximum number of "no transition" signaling elements. See also *BnZS.*

pulse dialing A signaling method consisting of a sequence of momentary openings of the subscriber line. In the figure, an outgoing call is detailed starting from an idle circuit through connection and back to idle.

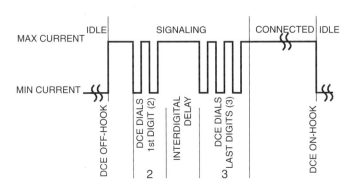

The numerical value of the digit being dialed is typically identical to the number of breaks or line openings in the pulse train (with the exception of the digit "0" which consists of 10 pulses or breaks). The pulse rate is generally 10 pulses per second (pps), and the break time is 60 to 70 ms. Both the pulse rate and the Make/Break ratio are country dependent and can be set by one of the AT commands. The time between digits (the interdigit time) is greater than 700 ms. Some common *make/break* ratios (expressed as a percentage of

Digit Dialed	Pulses Generated		
	Typical	Sweden	Norway New Zealand
0	10	1	10
1	1	2	9
2	2	3	8
3	3	4	7
4	4	5	6
5	5	6	5
6	6	7	4
7	7	8	3
8	8	9	2
9	9	10	1

break) are 50, 56, 60, 61.5, 66, and 67%. Minimum *interdigital* delay times are specified at 750, 800, 850, and 900 ms, depending on the country. *Pulse dialing* is sometimes called *rotary dialing.*

pulse dispersion See *pulse broadening.*

pulse distortion See *pulse.*

pulse duration modulation (PDM) A method of modulating a carrier consisting of a train of pulses with another signal. The only parameter of the carrier to be affected by the modulation process is the duration (width) of the individual pulse.

Pulse duration modulation is sometimes called *pulse length modulation* or *pulse width modulation,* both of which are deprecated terms. See also *modulation.*

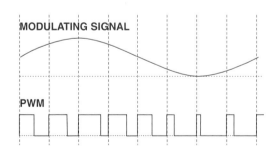

pulse duty factor In a periodic pulse train, the ratio of the pulse duration to pulse period.

pulse frequency modulation (PFM) Modulation in which the pulse repetition rate is varied in accordance with some characteristic of the modulating signal. *Pulse frequency modulation* is analogous to frequency modulation of a carrier wave, in which the instantaneous frequency is a continuous function of the modulating signal.

pulse interval modulation See *pulse position modulation.*

pulse jitter A relatively small variation of pulse spacing in a pulse train. The jitter may be random or systematic, but is generally not coherent with any modulation applied to the pulse train.

pulse length A deprecated synonym for *pulse duration.*

pulse length modulation A deprecated synonym for pulse duration modulation. See *pulse duration modulation.*

pulse period The reciprocal of the pulse repetition rate.

pulse position modulation (PPM) A method of modulating a carrier consisting of a train of pulses with another signal. The only parameter of the carrier to be affected by the modulation process is the temporal position of the individual pulses as compared to a reference clock. That is, individual pulses are advanced or retarded according to the amplitude of the modulating carrier. See also *modulation.*

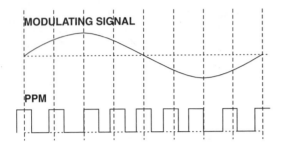

pulse repetition rate The number of pulses per unit time.

pulse shaper Any device used to change one or more characteristics of a pulse or pulse train, e.g., a repeater or regenerator.

pulse spectrum The frequency distribution of the sinusoidal components that comprise a pulse train in relative phase and amplitude.

pulse speed In telephony, the number of dial pulses transferred to/from a device per second.

pulse spreading See *pulse broadening.*

pulse string A synonym for *pulse train.*

pulse stuffing The insertion or deletion of signaling elements in a synchronous data stream for the purpose of maintaining frame synchronization when the transmitting and receiving clocks are slightly different frequencies. See also *bit stuffing.*

pulse time modulation (PTM) Any of several digital modulation methods in which a time-dependent parameter of a pulse is varied to encode an analog signal that is being converted to digital form. For example, pulse width (PWM), pulse duration (PDM), or pulse position (PPM) schemes are all *pulse time modulation* techniques.

pulse train A periodic sequence of pulses with constant amplitude, frequency, and pulse duration. Also called a *pulse string.*

pulse width A deprecated synonym for *pulse duration.*

pulse width modulation (PWM) A deprecated synonym for *pulse duration modulation.* See also *pulse duration modulation.*

pulsing In telephony, the transmission of address information to a switching office by means of pulses, i.e., signals that originate from subscriber (user) equipment.

Examples of pulsing methods include

- *Dual tone multifrequency (DTMF)* signaling, in which a unique pair of audio frequencies represents each of the respective numerals or other characters on a keypad.
- *Rotary dialing,* in which pulses in the DC feed current are generated by a rotary dial.

Also called *key pulsing* (when using a keypad), and *dial pulsing* (when using a rotary dial).

pump A source of ac power to a nonlinear device that causes the device to behave as a time-varying reactance.

pump efficiency The ratio of the power (or energy) absorbed from a pump to the power (or energy) available from the pump source.

pump frequency The frequency of an oscillator used to provide sustaining power to a device (e.g., a laser or parametric amplifier) that requires radio frequency or optical power.

pumping The action of an oscillator that provides cyclic inputs to an oscillating reaction device.

Examples include the action that results in amplification of a signal by a *parametric amplifier* and the action that provides a laser or maser with an input signal at the appropriate frequency to sustain stimulated emission.

punchdown block In telephony, a connecting device for terminating unshielded twisted pair (UTP) or shielded twisted pair (STP) wires of a cable for further distribution. Generally, the device is designed for insulation displacement connections (IDC), although other terminations, such as wire wrap and solder types are in use.

PUP An acronym for Parc Universal Packet.

PURL An acronym for Persistent Uniform Resource Locator.

Functionally, a *PURL* is a Uniform Resource Locator (URL). However, instead of pointing directly to the location of the desired Internet resource, it points to an intermediate resolution service. The resolution service associates the *PURL* with the actual URL and returns that URL to the client. The client can then complete the URL transaction in the normal fashion. Also called a *standard HTTP redirect* in Web language. See also Uniform Resource Locator (URL).

push-pull operation The operation of two like elements which are driven in 180° phase opposition to produce additive output components of a desired wave and cancellation of unwanted components.

pushbutton dialing Dialing in which keys (switches or pushbuttons) are used to select a unique addressing symbol to be transmitted to the telephone switching center. Each key represents a unique digit or symbol.

Pushbutton dialing is usually associated with dual tone multifrequency (DTMF) signaling; however, it may also be used for pulse dialing.

- In DTMF signaling, each button actuates and connects a corresponding audible pair of tone oscillators to the line.
- In pulse signaling, each button actuates a pulse burst generator which transmits a specific number of dial pulses corresponding to the button depressed.

PVC (1) An abbreviation of PolyVinyl Chloride, a common plastic insulation used on wire. **(2)** An abbreviation of Permanent Virtual Circuit.

pW An abbreviation of picowatt. A unit of power equal to 10^{-12} W.

PWG An abbreviation of Permanent Working Group.

PWM An abbreviation of Pulse Width Modulation. A deprecated synonym for *pulse duration modulation.*

pWp The abbreviation of pico Watts of psophometrically weighted noise power. The relationship between phophometrically weighted noise measurements and wide band measurements is:

$$1\text{pWp} = \frac{1 \text{ pW}}{1.78} = 0.56 \text{ pW}$$

1 pWp is equivalent to:

- An 800-Hz tone at a power of −90 dBm.
- A 1000-Hz tone at a power of −91 dBm.
- 3-kHz band limited white noise at −88 dBm.

pWp0 An abbreviation of picoWatts, psophometrically weighted, measured at a 0 dBm transmission level point. See also *dBm(psoph)* and *psophometer.*

PWR An abbreviation of PoWeR.

pX An abbreviation of *peak envelope power* of a radio transmitter (generally in dB). See also *power.*

PX **(1)** An abbreviation of Private eXchange. See also *PABX* and *PBX.* **(2)** An abbreviation of *peak envelope power* (generally in watts). See also *power.*

pY A symbol for mean power (in dB).

PY A symbol for mean power (in watts).

pZ A symbol for carrier power (in dB).

PZ A symbol for carrier power (in watts).

Q

Q (1) The symbol for *electric charge*. **(2)** In general, two pi times the ratio of the maximum stored energy to the energy dissipated per cycle at a given frequency. **(3)** With respect to reactive circuits, the magnitude of the ratio of the reactance (X) in ohms divided by the effective series resistance in ohms.

- With inductors (coils), it is the ratio of the coil reactance (X_L) to its effective series resistance (R).

$$Q = \frac{X_L}{R}$$

- With capacitors, it is the ratio of the capacitor's susceptance (Y_C) to its effective shunt conductance (G).

Also called *quality factor*. **(4)** A measure of the sharpness of the frequency response of band-pass elements and circuits. The numeric value for Q is expressed by the ratio of the center frequency f_c to the bandwidth (BW), that is,

$$Q = \frac{f_c}{BW}$$

In the example, four band-pass characteristics are shown, one with a bandwidth of 2.5 kHz, another with a BW=500 Hz, the third a BW=200 Hz, and the last a BW=50 Hz. This yields the four Qs 0.4, 2.0, 5.0, and 20, respectively. **(5)** The symbol for the Quiet line state

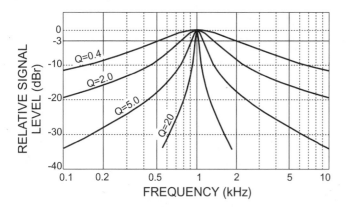

in Fiber Distributed Digital Interface (FDDI). **(6)** The symbol/abbreviation for Quadrature.

Q.n recommendations The ITU-T series of specifications concerning "Switching and Signaling." See Appendix E for a complete list of document titles.

QA (1) An abbreviation of Queued Arbitrated. That portion of packet traffic on an IEEE 802.6 distributed queue dual bus (DQDB) that contends for bandwidth. **(2)** An abbreviation of Quality Assurance.

QAM An abbreviation of Quadrature Amplitude Modulation.

QC An abbreviation of Quality Control.

QDU An abbreviation of Quantizing Distortion Units. The amount of degradation introduced into a signal path by the conversion from analog to pulse code modulation (PCM) and back to analog.

QIC An acronym for Quarter Inch Cartridge tape drive standard. It is a set of tape standards defined by the Quarter-Inch Cartridge Drive Standards Organization, a trade association established in 1987. Several *QIC* standards in common usage are shown in the table.

QIC STANDARDS IN COMMON USE

Designation	Structure	Cartridge	Capacity
QIC-40	10 kbpi/20 tracks 205', 6.35 mm	DC-2000	120 MB
QIC-80	14.7 kbps/28 tracks 307.5', 6.35 mm	DC-2120	250 MB
QIC-80		DC-2120XL	350 MB
QIC-wide	400', 8mm, 900 Oe	DC-3010XL	850 MB
QIC-1350			1.25 GB
QIC-2100			2.1 GB

QIC 80 can read QIC 40 tapes; however, the reverse is not true. The tape capacity is assuming that a 2:1 data compression algorithm is used. The QIC-117 standard specifies the tape drive interface protocol.

QLLC (1) An abbreviation of Qualified Logical Link Control. A frame format of IBM's Systems Network Architecture (SNA). **(2)** An abbreviation of Qualified Link Level Control. Routines to provide Synchronous Data Link Control.

QoS An abbreviation of Quality of Service. **(1)** The Internet performance goals. **(2)** A set of parameters in asynchronous transfer mode (ATM) networks for describing a transmission. They apply to both virtual channel connections (VCC) and virtual path connections (VPC), which specify paths between two entities.

The parameters include values such as allowable delay variation in cell transmission and allowable cell loss (in relation to total cells transmitted).

QOSPF An abbreviation of Quality of service extension for Open Shortest Path First.

QPR An abbreviation of Quadrature Partial Response.

QPSK (1) An abbreviation of Quadrature Phase Shift Key (or Keying). **(2)** An abbreviation of Quaternary Phase Shift Key.

QPSX An abbreviation of Queued Packet and Synchronous circuit eXchange.

QRC An abbreviation of Quick Reaction Capability.

QSAM An acronym for Quadrature Sideband Amplitude Modulation.

QTAM An acronym for Queued Telecommunications Access Method.

quad (1) A group of four entities. **(2)** A particular cable structure consisting of two separate pairs of insulated conductors twisted together. A cable with at least one set of conductors arranged as a *quad* is said to be a *quadded cable*. See also *star quad* and *color code* for the typical telephone quad wiring designations.

quadbit Literally means 4 bits. It generally refers to any modulation technique that encodes 4 bits at a time to generate 1 of 16 possible symbols for transmission.

For example, in a phase modulation scheme the 0000 *quadbit* symbol could be transmitted as a single frequency carrier, the *quadbit* symbol 0001 could be the same frequency but phase shifted by 22.5° from that of *quadbit* 0000, *quadbit* 0010 could be transmitted at a phase shift of 45°, *quadbit* 0011 at a phase shift of 67.5° and so on. *Quadbit* encoding is a specific form of *multilevel encoding*. See also *dibit, quadrature modulation,* and *tribit.*

quadded cable A cable in which some of the conductors are formed of *quads,* grouped and separately bound or insulated, and contained under a common jacket. See also *quad.*

quadratic phase terms Conceptual formulas that characterize both the transmittance functions of lenses and propagation in the Fresnel zone.

quadratic profile A synonym for *parabolic profile.*

quadrature (1) The relationship between two periodic functions with the same period (frequency) and a phase difference of ¼ period (the same as 90° or π/2 radians). **(2)** The state of being separated in phase by 90° (π/2 radians).

quadrature amplitude modulation (QAM) A method of modulating a carrier with both amplitude modulation and phase modulation techniques to create a constellation of signal points, each representing one of the data points defined by a multibit data sample.

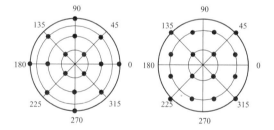

The two *QAM* constellations in the figure represent two ITU-T 9600 bps modem specifications. Each point of the constellation represents one of the 16 states defined by 4 bits of data. The signal amplitude increases further from the center of the *XY*-axis and the modulation angle is the counterclockwise angle from the *X*-axis. Practical implementations amplitude modulate two carriers which are in quadrature. See also *amplitude modulation, phase shift keying, quadbit,* and *trellis coding.*

quadrature hybrid A transmission line or waveguide hybrid device that has the property that waves leaving one port are in phase quadrature with waves leaving a second port.

quadrature modulation Modulation using two carrier components out of phase by 90° (in quadrature) and modulated by separate signals. In digital systems, each signal is generally a portion of a single bit stream. For example,

• In quadrature phase shift keying (QPSK) each bit of the dibit is used to modulate one carrier.

• In quadrature amplitude modulation (QAM), 2 bits of each quadbit are used to amplitude modulate each carrier.

See also *dibit, quadbit, quadrature, quadrature amplitude modulation (QAM),* and *quadrature phase shift key (QPSK).*

quadrature phase shift key (QPSK) A modulation technique whereby 2 bits of signaling data are taken together to cause the carrier's phase to be one of four states. For example, the 00 data bit combination might be represented by a 90° carrier phase angle. Similarly, 01 is represented by 0°, 10 by 180°, and 11 by 270° (or −90°). In effect, each line signaling state carries 2 data bits.

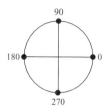

A *QPSK* constellation graphically showing that the four phases of the carrier are points on a circle at 0°, 90°, 180°, and 270°. The distance from the center of the circle to each point is the amplitude of the carrier wave at that phase angle. Also called *quaternary phase shift keying.* See also *PSK.*

quadrifilar helix A type of circularly polarized antenna. It consists of two orthogonal, fractional-turn, bifilar helixes connected in phase quadrature. Also called a *volute antenna.*

quadruple clad fiber In fiber optics, a single mode optical fiber that has four claddings. All claddings have a refractive index lower than that of the core, and with respect to one another the indices alternate, as is shown in the figure.

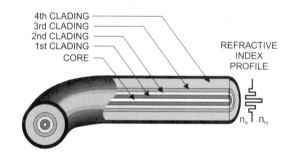

A *quadruply clad* fiber's advantages include:

• Very low macrobending losses,
• Two zero-dispersion points, and
• Moderately low dispersion over a wider range of wavelengths than with a singly or doubly clad fiber.

See also *refractive index profile.*

quality factor See *Q.*

quality of service (QoS) (1) The performance specification of a communications channel or system. *QOS* may be quantitatively indicated by channel or system performance parameters, such as *signal-to-noise ratio (S/N), bit error ratio (BER),* message throughput rate, or call blocking probability. **(2)** A subjective rating of telephone communications quality in which listeners judge transmissions by qualifiers, such as excellent, good, fair, poor, or unsatisfactory.

quantization distortion The inherent distortion introduced in the process of quantization.

quantization error The difference between a continuous value (analog signal) and its quantized value (quantized signal). Also called *quantizing distortion.* See also *analog-to-digital converter.*

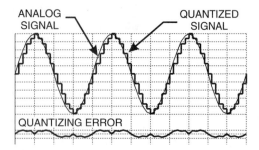

ANALOG SIGNAL — QUANTIZED SIGNAL

QUANTIZING ERROR

quantization level The discrete value assigned to a particular subrange of the analog signal being quantized in the quantization process. See also *analog-to-digital converter.*

quantize To divide the continuous range of values of a variable or signal into a finite number of nonoverlapping subranges or intervals. Each subinterval is assigned a value from within the subrange.

An example is the representation of a person's age by whole numbers only. See also *analog-to-digital converter.*

quantizing distortion Distortion that results from the *quantization process.* A synonym for *quantization error.*

quantizing levels In digital transmission, the number of discrete signal levels transmitted as the result of signal digitization. See also *quantization level.*

quantizing noise The noise introduced during the process of digitally encoding (quantizing) an analog signal. This type of noise, due to *quantizing error,* can be made as small as desired by choosing sufficiently small *quantizing levels.* It cannot, however, be eliminated. *Quantizing noise* is dependent on the particular quantization process used, the statistical characteristics of the quantized signal, and the level of the analog signal. For an N bit linear signal representation, the signal-to-noise ratio (SNR) can be approximated by:

$$SNR = N \cdot 20 \cdot \log_{10} 2 \approx 6.02 \cdot N$$

Also called *quantization noise.* See also *quantization error.*

quantum efficiency In either an optical source or detector, the ratio of the output quanta to the input quanta, i.e., the ratio of the number of photoelectrically emitted electrons or photons by a material to the number of absorbed particles.

quantum limited operation A synonym for *quantum noise limited operation.*

quantum noise Noise attributable to the discrete and probabilistic nature of physical phenomena and their interactions, e.g., noise caused by the motion of individual photons of light in optical fibers (photon noise) or *shot noise* in an electrical conductor or semiconductor.

Quantum noise represents the fundamental limit of the achievable signal-to-noise (SNR) ratio of an optical communication system. This limit, however, is never achieved in practice.

quantum noise limited operation Operation in a system or medium (such as an optical fiber, conductor, or semiconductor) wherein the minimum detectable signal is limited by *quantum noise.* Also called *quantum limited operation.*

quartet A group of four entities operated on as a single unit.

quartet signaling A strategy used in the 100BaseVG Ethernet implementation developed by Hewlett-Packard (HP) and AT&T Microelectronics. The scheme uses four-wire pairs simultaneously and relies on the fact that the wire pairs need not be used for sending and receiving simultaneously.

Wire availability is guaranteed because demand priority (the media access method used in 100BaseVG) enables hubs to handle network access for the nodes. Thus, quartet signaling provides four times as many channels as ordinary (10 Mbps) Ethernet. *Quartet signaling* also uses a more efficient encoding scheme, 5B/6B encoding, as opposed to the Manchester encoding used by ordinary Ethernet. This more efficient encoding, together with the four channels and a slightly higher signal frequency, make it possible to increase the bandwidth for such Ethernet networks to 100 Mbps.

quartz clock A clock containing a *quartz oscillator* that determines the accuracy and precision of the clock.

quartz oscillator An oscillator in which a *quartz crystal* is used to determine and stabilize the frequency of oscillation.

The piezoelectric property of the *quartz crystal* is utilized in the oscillator circuit to produce a nearly constant frequency dependent only on the mechanical properties of the crystal (size and shape) and excitation. Frequency accuracies and stabilites of one part in a billion (10^9) or better are achievable.

quasi-analog signal A digital signal that has been converted to a form suitable for transmission over a specified analog channel. The specification of the analog channel should include frequency range, bandwidth, signal-to-noise ratio (SNR), and envelope delay distortion. For example, a modem may be used for this conversion.

When quasi-analog signaling is used to convey message traffic over dial-up telephone systems, it is often referred to as voice data.

quaternary code A code whose alphabet consists of four symbols. See also *binary, code,* and *ternary.*

quaternary phase shift keying A synonym for quadrature phase shift keying.

quaternary signal A digital signal having four significant conditions.

query (1) In a computer database search system, a *query* is a search "question" that has been written in a manner that the search system can understand and use. In essence, a user is requesting that a particular part of the database be singled out. The *query* is constructed by variable names and Boolean operators. For example, to find database entries where salary is greater than $100 000, the *query* might look like:

salary > 100000

(2) The operation of initiating an information transfer from a remote device to a local device. For example, the action of a master station, in a polled system, when it asks for the identity or status of a slave station.

query language A programming language in a database management system that allows a user to extract and display specific information from the database. For example, structured query language (SQL) is an international database query language that allows the user to create or modify data or the database structure.

queue Literally, meaning a line. In communications and computers, an ordered, temporary holding structure of items waiting to be processed, generally as a first-in-first-out (FIFO) list.

For example, an input *queue* is a buffer containing information waiting to be processed by receive circuits or software, while an output *queue* is a buffer containing data to be transmitted.

queue traffic (1) A series of outgoing or incoming calls waiting for service. (2) In a store-and-forward switching center, the outgoing messages awaiting transmission at the outgoing line position.

queuing The process of entering elements into or removing elements from a *queue.*

queuing delay **(1)** In a switched network, the time between the completion of signaling by the call originator and the arrival of a ringing signal at the call receiver. Delays may arise at any or all of the originating, intermediate, or call termination switches. **(2)** In a data network, the sum of all delays between the request for service and the establishment of a circuit to the called data terminal equipment (DTE). **(3)** In a packet switched network, the sum of the delays encountered by a packet between the time of insertion into the network and the time of delivery to the addressee.

queuing theory A subset of traffic theory that deals with handling entities waiting for service. When an entity requests service and the service facilities are busy, two basic actions are possible: 1. The request is discarded and the entity must initiate the request again, or 2. the request is placed on a waiting list and remains within the system until services are available. *Queuing theory* deals with the latter case, the characteristics of the waiting list, and the criteria for removing entities from the list. Several methods of addressing the entities on the queue are possible. Among them are:

- *First-in first-out* (*FIFO*) queue: When services become available, the entity with the longest wait time is the next serviced.
- *Last in first out* (*LIFO*) queue: When services become available, the entity with the shortest wait time is the next serviced. Commonly used in computer controlled systems.

  Also called a *push down stack.*

- *Random queue:* When services become available, the next entity to be serviced is chosen at random from the pool of waiting entities.
- *Nearest neighbor service queue:* When services become available, the next entity to receive service is the closest entity to the last one serviced. An example is an elevator that goes to the closest requesting floor to its current position without regard to the waiting time of other requests.
- *Priority queue:* When services become available, the entity with the highest priority receives service. For example, the baby bird who shouts the loudest generally is served before the smallest/quietest one.

QuickTime A standard established by Apple Computer for the integration of digital video, animation, and sound. The *MooV* movie file format is used by *QuickTime.* System extensions once were required to view MooV files on the Macintosh. They are now incorporated into the system software. Special drivers are required to view *Quicktime* in Windows.

quiescent The state of no activity or the idle condition.

quieting In an FM receiver, an effect that results in less noise when an unmodulated carrier is present than when no carrier is present. Quieting is generally expressed in decibels (dB).

QWERTY The first six keys on the left top alphabetic row of standard keyboard. Used to identify the type of keyboard. Another layout is the DVORAK keyboard.

R

R **(1)** The circuit designator for a resistor and the symbol for resistance. **(2)** The designation of the ring wire in a standard twisted pair, two-wire local loop. **(3)** An interface reference point in the Integrated Services Digital Network (ISDN) model to non-ISDN terminal equipment.

R interface point The reference point in an Integrated Services Digital Network (ISDN) customer premises terminal equipment set located between a TE2 device (a non-ISDN device) and the TA (terminal adapter). See also *ISDN*.

race condition A condition that occurs in systems in which the simultaneous occurrence of two or more signals are required at a point. In digital electronic circuits, for example, the output of a logic gate depends on the exact timing of multiple input signals.

For example, in the following logic circuit the Boolean equation $(A \cdot \overline{A} = 0)$ indicates that the output never changes state; however, because of the propagation delay in the inverter a *glitch* is created. This pulse may be long enough to be used or to do logical harm to following circuits.

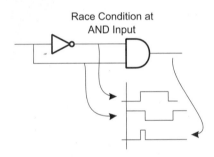

Race Condition at AND Input

raceway A metal or plastic channel or enclosure used in buildings and equipment rooms to hold and protect loose bundles of cabling. They are generally located in the floor and sometimes in the ceiling. Raceways perform the same functions as a conduit, except they provide space for larger cable bundles.

rack A frame into or on which one or more units of equipment are mounted.

rad **(1)** An abbreviation of <u>Rad</u>iation <u>A</u>bsorbed <u>D</u>ose. A basic unit of absorbed radiation per unit mass, equivalent to 0.01 J/kg (100 ergs/g) of absorber. Absorbed radiant energy may heat, ionize, and/or destroy the material on which it is incident. **(2)** An abbreviation of <u>rad</u>ian.

radian (rad) The angle formed by two radii of a circle separated by an arc on the circumference of the circle equal in length to the radius. Numerically, a radian is:

$$rad = \frac{180°}{\pi} \cong 57° \ 17' \ 44.6''$$

radian frequency (ω) The number of radians per unit time interval (generally one second). Hence,

$$\omega = 2 \pi f$$

where:

  f is the frequency in hertz (Hz).

radiance Radiant power per unit solid angle per unit of projected area of the source, in a given direction, as viewed from the given direction. *Radiance* is usually expressed in watts per steradian per square meter (W/sr/m²).

radiant Pertaining to electromagnetic radiation, with the contributions at all wavelengths of interest and all equally weighted.

radiant density Radiant energy per unit volume. Generally expressed in J/m³. Also called *radiant energy density.*

radiant efficiency The ratio of the radiant flux emitted by a source to the total power supplied to the source.

radiant emittance Radiant power emitted into a full sphere (4π steradians) by a unit area of a source. Generally expressed in watts per square meter (W/m²). Also called *radiant exitance.*

radiant energy Energy in the form of electromagnetic waves, such as heat, light, or radio waves. *Radiant energy* is usually expressed in joules (J) or kilowatt-hours (kWhr).

Radiant energy may be calculated by integrating *radiant power* with respect to time.

radiant exitance See *radiant emittance.*

radiant flux The time rate of flow of radiant energy. *Radiant power* emitted, transferred, or received. Generally expressed in watts (W).

Radiant flux is a deprecated synonym for *radiant power.*

radiant heat Infrared radiation emitted from a source not hot enough to emit visible light.

radiant intensity The radiant power (energy emitted within a time period) per unit solid angle. Generally expressed in watts/steradian (W/sr).

radiant power The time rate of flow of electromagnetic energy (radiant energy), such as heat, light, radio waves, or x-rays. *Radiant power* is usually expressed in watts (W) or joules per second (J.s).

When the meaning is clear, the modifier radiant is often dropped and "power" is used to mean "radiant power." Also called *optical power. Flux* and *radiant flux* are deprecated synonyms for *radiant power.*

radiating element A basic subdivision of a radiator (such as an antenna) that is capable of effectively radiating or receiving electromagnetic radiation. Examples of antenna-radiating elements include slots, horns, and dipoles.

radiation **(1)** The outward flow (emission) of radiant electromagnetic energy (such as, radio waves, heat, light, x-rays, and cosmic rays) from any source (e.g., lamps, antennas, light emitting diodes, lasers, etc.). **(2)** In communication, the emission of energy in the form of electromagnetic waves.

radiation angle In optical fibers, one-half the vertex angle of the cone of optical energy emitted at the exit face of an optical fiber. The cone boundary is usually defined by:

- The angle at which the far-field irradiance has decreased to a specified fraction of its maximum value (usually half power or 1/e), or
- The cone within which there is a specified fraction of the total radiated power at any point in the far field.

Also called the *output angle.* See also *acceptance angle.*

radiation detector Any device that can be used to ascertain the presence of radiation in a specified portion of the electromagnetic spectrum, that is, it produces some physical effect suitable for observation and/or measurement.

radiation efficiency At a given frequency, the ratio of the total power radiated by a device to the total power delivered to that device.

radiation field A synonym for *far-field region*.

radiation intensity The power broadcast in a given direction per unit solid angle in that direction.

radiation loss Power or signal loss due to the radiation of electromagnetic energy from a network or device.

radiation mode In an optical waveguide, a mode having fields that are transversely (crosswise) oscillatory everywhere external to the waveguide and that exists even at the limit of zero wavelength. Specifically, a radiation mode is one for which:

$$\beta = \sqrt{n^2(a)k^2 - (l/a)^2}$$

where:

β is the imaginary part (phase term) of the axial propagation constant,

l is the azimuthal index of the mode,

a is the core radius,

$n(a)$ is the refractive index, and

k is the free space wave number given by $k = 2/\pi\lambda$, where λ is the wavelength.

Radiation modes correspond to refracted rays in the terminology of geometric optics. Also called *unbound mode* and *leaky mode*.

radiation pattern (1) A graphical depiction of the points of equal radiation properties emanating from a radiator. Radiational properties include power, flux density, phase, field strength, and polarization. Generally, the representation is in the form of a polar plot where the radius represents the distance from the radiator and the angle is the direction of the measurement relative to the axis of the radiator. Also called an *amplitude pattern, antenna pattern, field pattern, power pattern*, and *voltage pattern*. See also *intensity distribution curve* and *lobe*. (2) In fiber optics, the relative power distribution at the output of a fiber or active device as a function of position or angle.

- The near-field *radiation pattern* describes the *radiant emittance* (in W/m^2) as a function of position in the plane of the exit face of an optical fiber.
- The far-field *radiation pattern* describes the *irradiance* as a function of angle in the far-field region of the exit face of an optical fiber.

The radiation pattern may be a function of the length of the optical waveguide, the wavelength, and the manner in which the excitation is applied. Also called *directivity pattern*.

radiation resistance The resistance that would consume the same amount of power that is radiated by an antenna if inserted in place of the antenna. Also called a *dummy load*.

radiation scattering The diversion of radiation (thermal, electromagnetic, or nuclear) from its original path as a result of interaction or collisions with atoms, molecules, or larger particles in the medium in which it is propagating or traversing.

radiator Any device that emits electromagnetic radiation, e.g., an antenna, light emitting diode (LED), or laser.

radio (1) A general term applied to the use of radio waves. (2) Telecommunication by modulation and radiation of electromagnetic waves. (3) A transmitter, receiver, or transceiver used for communication via electromagnetic waves.

radio and wire integration (RWI) The combining of wire circuits with radio facilities.

radio channel A band of radiated frequencies sufficiently wide to permit its use for radio communications. The bandwidth of the channel is dependent on the type of transmission, the transmission fidelity required, and frequency tolerances.

radio common carrier (RCC) A common carrier that is engaged in the provision of Public Mobile Service and is not also in the business of providing landline local exchange telephone service. Formerly called *miscellaneous common carriers*.

radio fadeout A phenomenon in radio propagation in which radio waves that are normally reflected by the E region of the ionosphere are partially or totally absorbed. Also called the *Dellinger effect*.

radio field intensity A synonym for *field strength*.

radio frequency (RF) That portion of the electromagnetic spectrum between audio and infrared frequencies. Typically defined as frequencies between about 10 kHz and 3000 GHz or wavelengths between 30 km and 0.1 mm, respectively. See also *frequency bands* and *frequency spectrum*.

radio frequency interference (RFI) Noise from any source that generates radio waves in the same frequency range and along the same path as a desired signal and that interferes with information transmissions. Also called *electromagnetic interference*.

radio horizon Those points at which direct rays from an antenna are tangential to the surface of the Earth.

If the Earth were a perfect sphere and there were no atmospheric anomalies, the radio horizon would be a circle. In practice, however, the distance to the radio horizon is affected by the height of the transmitting antenna, the height of the receiving antenna, atmospheric conditions, and the presence of obstructions, e.g., mountains and man-made objects.

radio horizon range (RHR) The distance at which a direct radio wave can reach a receiving antenna of given height from a transmitting antenna of given height.

RHR, is given by the relation

$$RHR = k \cdot (h_t^{1/2} + h_r^{1/2}),$$

where:

k is 1.23 when R, h_t and h_r are in feet and 2.23 when h_t and h_r are in meters,

h_t is the height of the transmitting antenna, and

h_r is the height of the receiving antennas.

The *effective Earth radius* (4/3 times the actual Earth radius) is used in deriving the formulas. Second-order effects are neglected as they are of the order of 0.1%.

radio paging The use of a miniature radio receiver capable of alerting its wearer that there is a phone call waiting, either by displaying a phone number or a mutually agreed upon number indicating a predetermined message between the calling and called parties. Also called *beeping* and *paging*.

Radio Regulations Board A permanent subgroup of the International Telecommunication Union (ITU) that implements frequency assignment policy and maintains the Master International Frequency Register (MIFR). Formerly called the *International Frequency Registration Board* (*IFRB*).

radio spectrum See *frequency spectrum.*

radio wave An electromagnetic wave whose frequency is arbitrarily defined as less than 3 THz. Also called a *Hertzian wave.* See also *electromagnetic spectrum* and *frequency spectrum.*

radiocommunication Telecommunication by means of radio waves.

radiometry The science of radiation measurement.

TABLE OF RADIOMETRIC TERMS

Term	Symbol	Quantity	Unit
Radiant	Q	energy	joule(J)
Radiant power (optical power)	ϕ	power	watt (W)
Irradiance	E	power incident per unit area (regardless of angle)	$W \cdot m^2$
Spectral irradiance	E_λ	irradiance per unit wavelength interval at a given wavelength	$W \cdot m^{-2} \cdot nm^{-1}$
Radiant emittance (radiant exitance)	W	power emitted (into a full sphere) per unit area	W/m^2
Radiant intensity	I	power per solid angle	W/sr
Radiance	L	power per unit angle per projected area	$W/sr \cdot m^2$
Spectral radiance	L_λ	radiance per unit wavelength interval at a given wavelength	$W/sr \cdot m^2 \cdot nm$

radix The base of a numbering system. For example, in the binary system the radix is 2, and in the decimal numbering system it is 10. See also *base n.*

RAI An abbreviation of Remote Alarm Indication.

RAID An acronym for Redundant Array of Inexpensive (or Independent) Disks. A family of techniques for managing a group of disk drives in an attempt to improve reliability, data availability, and performance. These techniques were formally defined in 1987 by professors Garth Gibson, Randy Katz, and David Patterson at the University of California at Berkeley (UCB). The technique illustrates how to use an array of small, inexpensive disks to replace a large expensive storage system. One or more controllers interlink the disks of the array so that they appear to be a single device to the system software. The controller may also provide error detection or correction. There are several different architectures or levels defined, that is,

- *RAID level 0:* The most basic method distributes data block by block across multiple drives but does not provide any *parity.* Level 0 is, strictly speaking, not a RAID because fault tolerance is not provided. Also called *disk spanning* or *disk striping.*
- *RAID level 1:* Two identical controllers and drives that duplicate each other's contents (*mirroring*). If either drive fails, the system operates as a normal single drive system. Data throughput can potentially be doubled by the controller if alternate sectors are accessed on each drive. Also called *disk mirroring.*
- *RAID level 2:* Data bits are spread over multiple drives. Each data bit is written to a separate disk drive (a technique called *bit inter-*

leaving). Further, additional error-correcting information is written to additional drives. For example, Hamming error correction would add 3 bits to an 8-bit field and would require three additional disk drives to store the information. Errors are corrected based on one read of information from the drive.

- *RAID level 3:* Data bytes are distributed across multiple drives (called *byte interleaving*) and a single drive is used for parity. Upon error detection, a re-read is initiated to resolve the error. Level 3 is more reliable than level 2 because there is only one parity drive that can fail.
- *RAID level 4:* Files are split between drives at the sector level rather than the bit or byte level (called *block interleaving*). Error-detection data (parity) is written to an additional drive.
- *RAID level 5:* This level utilizes *block interleaving,* as does RAID level 4, but error-detection data is spread across multiple drives. It is written to another sector on the same drive to which data is written. Further, level 5 can do simultaneous data reads and writes. Level 5 is faster and more reliable than the other levels.
- *RAID level 6:* This is essentially RAID level 5 with a second level of parity and the addition of redundant controllers, power supplies, etc. The same group at UCB that developed the initial 6 RAID levels developed RAID level 6.

The chief benefit of RAID level 6 is that any two drives in the array can fail without the loss of data. This added drive enables the array to remain in active service while any individual physical drive is being replaced and still remain fault tolerant. (In effect, a RAID level 6 array with a single failed physical disk becomes a RAID level 5 array.) The major drawback is its slow write performance. (It must write two parity blocks during every write operation.)

Other RAID levels proposed or manufactured are in general combinations of the previous six levels.

- *RAID level 10:* The layering of RAID levels 0 and 1. To improve input/output performance, it employs data striping, splitting data blocks between multiple drives (as in RAID level 0). To improve reliability, it uses mirroring (as in RAID level 1) so that the striped arrays are exactly duplicated.

One benefit is the ability of the array management software to speed up read operations by filling multiple requests simultaneously from the two mirrored arrays when they are both available. The drawback is chiefly cost. As with any mirroring system, the number of devices is doubled, which in turn increases the cost. Also called *RAID level 0&1.*

- *RAID level 53:* The layering of RAID levels 0 and 3. Data are striped between two RAID level 3 arrays.

Because the simple striping of the top RAID level 0 layer adds no redundancy, the reliability of RAID level 53 is reduced somewhat. The RAID level 3 arrays, however, are sufficiently fault tolerant that the overall reliability of the RAID level 53 array far exceeds that of an individual hard disk drive.

RAID storage systems provide an alternative to a *single large expensive drive* (*SLED*). However, it has been debated whether the amount of increased reliability provided by RAID technology is worth the additional cost.

rain attenuation Signal loss due to radio wave absorption in water or water vapor. Generally, the higher the frequency, the greater the loss for a given path. The following chart illustrates relative atmospheric losses due to water and oxygen molecules. Low angles are likely with microwave links, while higher angles are encountered with satellite communications links. Also called *rain fade.*

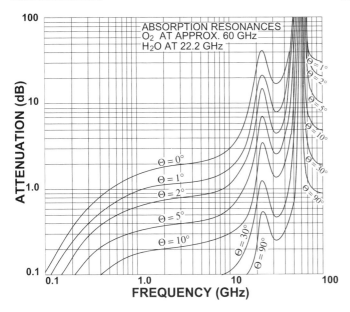

rain barrel effect A signal distortion caused by insufficient attenuation of echoes on the return path. To a listener, the distortion sounds like someone talking in a large rain barrel. Also called a *drain pipe effect*.

RAM An acronym for <u>R</u>andom <u>A</u>ccess <u>M</u>emory (pronounced "ram"). *RAM* is memory that has information stored in such a way that any data element may be retrieved in the same amount of time.

A *RAM* device may be fabricated with several different technologies; one of the most common is based on transistors. These memory systems are fast, inexpensive, and reliable but they do have one drawback: when power is removed, all stored information is lost. Technically most memory systems are random access, for example, a read only memory (ROM) is a random access device. The term *RAM*, however, is generally applied only to semiconductor read/write memory. Various types of RAM are distinguished in the literature:

- *DRAM* (*dynamic RAM*)—which must be refreshed periodically in order to retain its information. Refresh periods are on the order of every few milliseconds.
- *SRAM* (*static RAM*)—which retains its contents as long as power is supplied.
- *VRAM* (*video RAM*)—which is used to provide memory for graphics processing or temporary image storage.

See also *cache memory, DRAM, memory, SRAM,* and *VRAM.*

Raman amplifier A type of *fiber optic amplifier.*

Raman effect The generation of scattered light with wavelengths different from the monochromatic incident light source. The effect is due to the vibration of the scattering molecules in the transmission medium. This scattering is generated:

- By means of lasing action and interaction with molecules, thereby creating many different excited molecular energy levels that will produce photons of various energy levels (wavelengths) when transitions to lower excited states occur and
- By the beating together of two frequencies inducing dipole moments in molecules at the difference frequencies, which causes modulation of laser molecule interaction, which, in turn, produces light at wavelengths near, but not coincident with, the nominal incident wavelength.

If the incident energy is sufficiently high, a threshold can be reached beyond which light at the *Raman* frequencies is amplified, builds up strongly, and exhibits characteristics of stimulated emission (an effect called *stimulated Raman effect or coherent Raman effect*). Also called *Raman scattering.*

ramp A linear feature of a waveform.

random A condition, event, or value not predictable in position, time, or magnitude except in a statistical sense. That is, the future value can only be estimated by a nondeterministic means.

random error Errors that cannot be predicted except on a statistical basis.

random noise A "signal" in which there is no predictability or relationship of future amplitude, frequency, or time with past events. The pattern and magnitude of any signal parameter occurs without pattern or predictability, except in a statistical sense. Thermal noise is an example of *random noise.* See also *pseudo-noise (PN).*

random number (1) A number selected from a known set of numbers in such a way that each number in the set has the same probability of occurrence. See also *pseudo-noise (PN)* and *pseudorandom sequence.* (2) A number obtained by chance. (3) One of a sequence of numbers considered appropriate for satisfying certain statistical tests or believed to be free from conditions that might bias the result of a calculation.

range (1) The region between a lower limit or value and an upper limit. (2) A distance.

range extender In telephony, a device inserted into a subscriber loop so as to counter the effect of excessive loop resistance. The device therefore allows either:

- The subscriber to be located a greater distance from the switching equipment, or
- The telephone company to use a larger gauge wire (smaller diameter), hence, saving the cost of copper.

The device frequently is simply a boost battery.

RARP An acronym for <u>R</u>everse <u>A</u>ddress <u>R</u>esolution <u>P</u>rotocol.

RAS (1) An acronym for <u>R</u>ow <u>A</u>ddress <u>S</u>trobe. One of the two addressing signals used to select the desired random access memory (RAM) device within a bank of devices. The other signal is the Column Address Strobe (CAS). (2) An acronym for <u>R</u>emote <u>A</u>ccess <u>S</u>erver (or <u>S</u>ervices).

RASH An acronym for <u>R</u>ead-only, <u>A</u>rchive, <u>S</u>ystem, <u>H</u>idden—the four file attributes in the PC-DOS/MS-DOS operating system. The individual attribute functions are:

R	*Read only.* If set (1 or on), the file can only be read.
A	*Archive.* If set, the file has not been archived; if clear (0 or off), the file has been archived.
S	*System.* If set, the file is a system file.
H	*Hidden.* If set, neither the file name nor any of its attributes is displayed with the DIR command.

rate (1) The number of changes possible per unit time on a channel. For example, bits per second, characters per second, or symbols per second. A synonym for *speed.* (2) In telephony, a synonym for cost.

rate table A database that contains the cost of calls with respect to area code, calling time, call duration, etc.

RATS An acronym for <u>R</u>adio <u>A</u>mateur <u>T</u>elecommunications <u>S</u>ociety.

RATT An acronym for <u>RA</u>dio <u>T</u>ele<u>T</u>ypewriter.

ray A geometric representation of a light path through an optical device or system. It is a line normal to the electromagnetic wavefronts and indicates the direction of radiant energy flow, i.e., the direction of propagation.

ray intercept plot A graph of the intersections of a fan of rays with the final image plane, plotted as a function of the positions of the rays in the pupil of the system.

ray optics A synonym for *geometric optics.*

ray tracing The mathematical calculation of the path traveled by a ray through an optical component or system.

Rayleigh criterion of resolving power When a lens is free of aberrations, the images of point objects appear as diffraction patterns. When the principal maximum of one pattern strikes the first minimum of another, the images are described as being resolved.

When the optics are circular, this occurs when the angular separation (ϕ) of the point objects, viewed from the objective lens is:

$$\phi = \sin^{-1}\left(\frac{1.22\lambda}{d}\right) \cong \frac{1.22\lambda}{d} \text{ for } d \text{ large.}$$

where λ is the wavelength and d is the distance from the objective lens.

Rayleigh distribution A mathematical statement for the case in which two orthogonal variables are independent and normally distributed with *unit variance.* It is usually applied to frequency distributions of random variables.

Rayleigh fading In electromagnetic wave propagation, phase interference fading that is caused by multipath and may be approximated by the Rayleigh distribution.

Rayleigh limit In optics, the restriction of wavefront errors to less than one-fourth wavelength (λ) of a true spherical surface to assure a near perfect image. See also *Rayleigh criterion of resolving power.*

Rayleigh scattering Electromagnetic wave scattering in a medium caused by refractive index inhomogeneities (variations or particles), which are small compared to the wavelength of the incident wave. *Rayleigh scattered* flux is inversely proportional to the fourth power of the wavelength.

Ionospheric scattering is caused partly by *Rayleigh scattering.* Blue light is scattered more strongly by molecules in the atmosphere than are longer wavelengths; hence, the sky appears blue.

RB (1) An abbreviation of **R**everse **B**attery. (2) An abbreviation of **R**equest **B**lock.

RBHC An abbreviation of **R**egional **B**ell **H**olding **C**ompany.

RBOC An acronym for **R**egional **B**ell **O**perating **C**ompany (pronounced "ARE-bock").

RBS An abbreviation of **R**obbed **B**it **S**ignaling.

RBT An abbreviation of **R**ing**B**ack **T**one.

RC (1) An abbreviation of **R**adio **C**ontrolled. (2) An abbreviation of **R**esistance/**C**apacitance. (3) An abbreviation of **R**eflection **C**oefficient. (4) An abbreviation of **R**egional **C**enter.

RC filter An electronic wave filter composed of only the passive components resistors (R) and capacitors (C). The components are frequently arranged so as to yield *high-pass, low-pass,* and *notch* filter characteristics. Although *band-pass* and *band reject* filters are possible, their performance is not adequate for most applications. Normalized curves are shown in the diagrams:

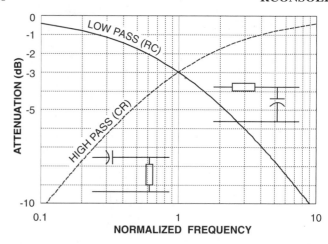

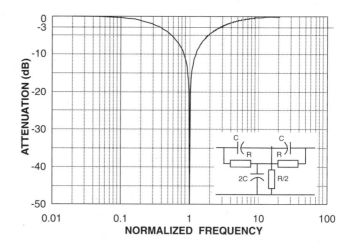

RCA (1) An abbreviation of **R**egional **C**alling **A**rea. The geographical area covered by a telephone company. (2) Before their name change, it stood for **R**adio **C**orporation of **A**merica.

RCA connector A low-cost type of connector used for wire pairs and shielded cable. (However, it is generally not suitable for terminating coaxial cable because of the poor control of impedances.) *RCA connectors* are used predominately for audio and low-frequency communications (hundreds of kHz or so). The figure shows a female jack (left) and male plug (right) *RCA connector.* Also called *phono plugs* and *jacks.*

RCC An abbreviation of **R**adio **C**ommon **C**arrier.

RCLK An acronym for **R**eceive **CL**oc**K** (RS-232 signal DD, RS-449 signal RT, and ITU-T signal 115). A clock signal generated by the data communications equipment (DCE) and supplied to the DTE in synchronism with the receive data (RXD) signal. Provides the timing information for synchronous signal reception.

RCONSOLE A Novell NetWare utility program that enables a network supervisor to manage a server from a workstation. That is, the supervisor can issue commands and accomplish tasks, just as if the commands were entered directly at the server. In NetWare 4.x, *RCONSOLE* was expanded to include asynchronous communication capabilities, thereby allowing remote access via modem.

RCP An abbreviation of the University of California at Berkeley's UNIX Remote Copy Program. A utility, similar to file transfer protocol (FTP), for copying files over Ethernet.

RD (1) An abbreviation of Received Data. A signal on the EIA-232 interface. (2) An abbreviation of Request Disconnect. A secondary station unnumbered frame asking a primary station for disconnect (layer 2—DISC).

RDA An abbreviation of Remote Database Access.

RDF An abbreviation of Radio Direction Finding.

RDL An abbreviation of Resistor Diode Logic.

RDRAM An acronym for Rambus Dynamic Random Access Memory. A high-bandwidth DRAM designed by Rambus Inc. of Mountain View, California.

Currently used mainly for video accelerators and high performance game machines, it offers sustained transfer rates of about 1000 Mbps, compared to 200 Mbps for DRAM and 100 Mbps for synchronous dynamic random access memory (SDRAM). Therefore, it is likely that it will replace DRAM and SDRAM as the main memory system in personal computers as the bus speeds required by these machines increase.

RDS An abbreviation of ReaD Strobe.

RDT An abbreviation of Recall Dial Tone.

RDTL An abbreviation of Resistor Diode Transistor Logic.

RDY An abbreviation of ReaDY.

RE (1) An abbreviation of REference. (2) An abbreviation of Reset. (3) An abbreviation of Reference Equivalent. (4) An abbreviation of Routing Element.

re-ring In telephony, a signal generated by the operator at the calling end of an established communication path to recall either the operator at the called end, the calling party, or the called party.

REA An abbreviation of Rural Electrification Administration, Department of Agriculture.

reactance The opposition to the flow of alternating current (ac) due to the presence of inductance or capacitance. Reactance is analogous to resistance; however, it is existent only in ac systems. The imaginary part of impedance. See also *capacitive* and *inductive reactance*.

reactor A synonym for *inductor*.

read (1) In computers, to acquire information, usually from some form of storage. (2) The interpretation of data.

read after write verification A data verification measure in which the information written to disk is compared with the information in memory. If the two match, the memory containing the information is released; if there is no match, an error-correction or notification procedure is initiated.

READ code An acronym for Relative Element Address Designate code.

read error An error encountered while a computer is in the process of obtaining information from another device (typically, a storage device or another node on a network).

read head A magnetic head designated for or capable of reading only.

read mostly memory A metal-nitride-oxide semiconductor (MNOS) memory that will retain stored values in excess of one year without applied power. With many systems, the write rate is many times slower than the read rate.

read only An attribute that allows an entity to be read but not changed. Examples include:
- A read only file may be opened, read, or copied; however, it cannot be modified or deleted. Also called *fixed storage* and *read only storage*.
- A read only memory (ROM) may, under normal conditions, be only read. Also called *nonerasable storage*.

READ.ME See *README.xxx*.

README.xxx The name of a data file traditionally used for a document containing information that a user either needs or will find informative and might not be included in the formal or released documentation. README.xxx files are placed on the diskette in plaintext form (ASCII without any formatting) so that they can be read essentially by any text display program.

read/write memory A memory whose contents can be continuously, quickly, and easily changed or retrieved during normal system operation.

read/write slot An opening in the jacket of a diskette that allows access by the read write heads. Also called *read/write opening*.

ready-for-data signal (1) A call control signal transmitted by the *data circuit-terminating equipment* (*DCE*) to the *data terminal equipment* (*DTE*), which indicates that the connection is available for data transfer between both DTEs.

Equivalent to the *data set ready* (*DSR*) signal generated by a modem. (2) A signal sent by the user terminal indicating that a through connection is established. Within an interexchange data channel, it may be:
- Sent in the backward direction to indicate that all the succeeding exchanges involved in the connection have been through-connected, or
- Sent in the forward direction to indicate that all the preceding exchanges involved in the connection have been through connected.

Real audio A program for playing audio over the Internet. The system is implemented as a client–server architecture. The *Real audio* server incorporates an encoder which compresses sound into *Real audio files*. The client side is a web browser plug-in or add-on that allows the data stream sent from the server to be uncompressed and output using the normal sound facilities of the computer, such as a sound card.

A 14.4-kbps or better modem is generally required, and a 28.8-kbps connection is recommended for music quality sound. Further, the system requires that the network have the available bandwidth to support continuous transmission at these rates.

real mode The operating mode for memory usage of an 8086/8088 Intel microprocessor and one of the two modes for other members of the 80x86 microprocessor family. *Real mode* can address up to 1 megabyte of memory and can have only one process executing at a time. This is in contrast to the *protected mode* available in the 80286 and later microprocessors. In protected mode, several gigabytes of memory can be addressed, multiple processes can run at the same time, and each process can have its own (protected) memory area.

real power See *effective power*.

real time Pertaining to:

- The actual time during which a physical process occurs,
- The computation of control information during the time of an actual process so that the results may be used to direct the process,
- The processing of data by a computer in relation to an external process and in accordance with the time requirements of the external process, or
- A time-keeping circuit arranged so that its output is synchronous with the time used in every day life. This is also called a *real-time clock*.

real-time kinematics (RTK) A differential Global Positioning System (DGPS) procedure in which carrier phase correction information is transmitted from the reference receiver to the user receiver as it is received (in real time).

real-time transport protocol (RTP) An Internet protocol developed by the Internet Engineering Task Force (IETF) to address the problems encountered with the transmission of time-sensitive data (such as speech or video) across a local area network (LAN). The basic action of *RTP* is to give these time-critical data packets higher priority than normal connectionless data. The *RTP protocol* layer resides above the Internet protocol (IP) and datagram protocol layers.

reasonableness check A test on the data received by a real-time system to verify that the data falls into an expected range of values or that it conforms to specified criteria for the particular operating state in which the system currently resides. It can be used in error detection or to eliminate questionable data points from subsequent processing. Also called *wild point detection*.

reassembly In a data packet system, the process a receiving station performs to reconstitute the original message. The process is basically the concatenation (joining end to end) of the data portion of received packets. See also *fragmentation (3)*.

REC (1) An abbreviation of <u>REC</u>eive. **(2)** An abbreviation of <u>REC</u>over.

rec hierarchy One of the seven *world newsgroups* in USENet. The *rec* newsgroups report on recreational, entertainment, and sports activities such as aviation, backpacking, cars, comics, fishing, ham radio, hiking, hunting, kites, movies, music, pets, scuba diving, skiing, and on, and on. See also *newsgroup hierarchy*.

recall A feature of private branch exchange (PBX) systems that allows a user to obtain a second dial tone without going on- and off-hook. The feature is frequently activated by a *recall button* on the subset.

receive after transmit time delay In half-duplex operation, the time interval between the instant a local transmitter is keyed-off and the instant the local receiver output has increased to 90% of its steady-state value in response to an existing radio frequency signal from a distant transmitter.

receive characteristic In telephony, the acoustic output level of a telephone set as a function of the electrical input power. The input signal is from a test generator delivering known power at a specified impedance, and the output is measured in an artificial ear.

receive only (RO) A term that describes a device that is capable of receiving signals but is not arranged to send or transmit signals.

received noise power (1) The calculated or measured noise power, within the bandwidth being used, at the receive end of a transmission circuit, channel, link, or system. **(2)** The absolute power of the noise measured or calculated at a receive point. The related bandwidth and the noise weighting must also be specified. **(3)** In telephony, the value of noise power, from all sources, measured at the line terminals of the telephone set's receiver. The specific amplitude-frequency characteristic or noise-weighting characteristic (flat, C-message, psophometric) must be designated with the measurement.

received signal level (RSL) The signal level, in a specified bandwidth, at a receiver's input terminal. *RSL* is frequently expressed in decibels (dB) relative to 1 mW, i.e., above or below 0 dBm.

receiver (1) One of the three essential components of a communications system. (The other two are a transmitter and a communications channel.) It is a device employed to translate an incoming signal from the communications channel to a format suitable for a user's recording device, action, or for further transmission. **(2)** In telephony, the earpiece section of a telephone handset. **(3)** In telephone switching systems, that part of an automatic switching system that receives signals from a calling device (or other source) for interpretation or other action.

receiver attack time delay The time interval from the instant a step RF signal, at a level equal to the receiver threshold of sensitivity, is applied to the receiver's input terminals to the instant the receiver's output amplitude reaches 90% of its steady-state value, including the time required to break squelch if a squelch circuit is present.

receiver off-hook (ROH) A tone sent by the central office (CO) to the subscriber's equipment in an abnormal off-hook condition. *ROH* is sent if the subscriber's line has been in the off-hook state for greater than some predetermined time without any dialing. The predetermined time varies but is typically about 1 minute.

The *ROH* tone is intended to inform the subscriber that the telephone needs to be "hung-up," even if the subscriber is in another room. Hence, the *ROH* tone is very loud. *ROH* is also called a *howler tone*.

receiver release time delay The time interval from the instant of removal of radio frequency energy at the receiver's input until the instant the receiver's output is squelched.

receiving objective loudness rating (ROLR) A rating of telephone receivers (earpieces). The units of the numeric value of *ROLR* have no physical meaning but are expressed in terms of dB. A typical receiver will have *ROLRs* in the range of 40 to 55 dB.

$$ROLR = -20 \cdot \log\left(\frac{S_E}{0.5 \cdot V_W}\right)$$

where:

S_E is the sound pressure level at the artificial ear reference point (in pascals), and

V_W is the open circuit voltage of the signal source (in mV).

reciprocal (1) In mathematics, the new value created by dividing 1 by the old value. For example, 2 is the reciprocal of 0.5 and ⅔ is the reciprocal of ¾. **(2)** Indicates interchangeable, mutual, or complementary entities (real or imagined). Items exchanged or owed mutually.

reciprocal agreement An agreement between two parties to exchange or mutually share resources.

recognized private operation agency (RPOA) The ITU-T term for a packet interexchange carrier.

reconfiguration The process of physically altering the location or functionality of network or system elements. Automatic configuration describes the way sophisticated networks can readjust themselves in the event of a link or device failing, enabling the network to continue operation.

reconfiguration burst A unique bit pattern that is repeatedly transmitted in ARCnet networks whenever a node wants to force the creation of a new token or when a new node joins a network. In essence, a *reconfiguration burst* resets the network.

reconstructed sample An analog sample generated at the output of a decoder (digital-to-analog converter) when a digital signal symbol is applied at its input. The amplitude of the reconstructed sample is proportional to the value of the corresponding encoded sample. See also *analog-to-digital conversion* and *digital-to-analog conversion*.

record (1) A set of data treated as a unit. (2) The row designation of a database. The data structure that is a collection of related data fields (each with its own name and type) treated as a single unit and saved in a position dependent fashion in a file. See also *database*. (3) To write data on a medium, such as magnetic tape, magnetic disk, or optical disk.

record gap The space between records on serial device or communication link used to indicate the end of the record.

record layout The organization of data fields within a record.

record length The amount of space required to store a record, typically given in bytes.

record locking A method of managing shared data on a network by preventing more than one user from modifying a data segment in the same session. (Some *record locking procedures* go further and prevent more than one user from any concurrent access of the same data segment.)

record number A unique number assigned to a record in a database. The record number can be assigned in one of two ways, that is,

- It can be assigned as an absolute number from the beginning of the entire group of records, e.g., the sixth record from the beginning, or
- It can be assigned at the time of creation and inclusion into the database.

record structure An ordered list of the field names and types that compose a record along with the definition of each field domain (range of acceptable values).

recorded announcement In telephony, a prerecorded voice message switched to a subscriber when an abnormal condition is encountered, e.g., number disconnected, area code missing, number changed.

recorder warning tone A half-second 1400-Hz tone burst applied to a telephone line every 15 seconds to indicate to the called party that the calling party is recording the conversation. The tone is recorded together with the conversation.

recording density A synonym for *bit* or *byte density*.

recording trunk In telephony, a trunk connection between a local central office or private branch exchange (PBX) to a toll office used only for communicating with the operator and not used for completing a toll connection.

recoverable error An error condition that allows continued operation of a process. See also *irrecoverable error*.

rectifier A device that converts an alternating current (ac) into a direct current (dc). See also *semiconductor*.

recursion (1) An operation in which the output of a functional block or process is used by that same functional block or process as input in a later calculation.

An example of a process that uses *recursion* is the way a computer calculates the square root of a number. Also called an *iterative* process. (2) A computer program or routine that has the ability to call itself to perform repeated processing tasks. These routines are also called *re-entrant* routines.

recursive definition A definition that uses itself to define itself, for example, see *recursive definition*.

recursive filter One type of digital filter in which a fraction of the output data is combined with the input data stream for reprocessing. Also called an *infinite impulse response* (*IIR*) filter.

RED/BLACK theory The separation of circuits, components, equipment, and systems that handle unclassified or encrypted classified information (*BLACK*) from those circuits, components, equipment, and systems that handle nonencrypted classified information (plaintext or *RED*). At no point, other than the encrypting or decrypting engine, are the two signals (red and black) allowed to coexist. See also *BLACK signal* and *RED signal*.

red alarm A T-1 alarm generated for locally detected failures that interrupt significant portions of a communications link.

Red Book (1) The 1984 compilation of the Comité Consultatif Internationale Téléphonique et Télégraphique/International Telecommunications Union (CCITT/ITU-TS's) standards for telecommunications. (2) Another name for the specifications of the CD-DA audio CD format introduced by Sony and Philips. The standard defines the number of tracks on the disc that contain digital audio data and the error-correction methods that save sound from minor data loss. The format allows for a total of 74 minutes of digital sound to be transferred at a rate of 150 kilobytes per second (kBps). See also *CD-DA*. (3) Informal name for one of the four standard references on PostScript®. The other three guides are known as the Blue Book, the Green Book, and the White Book. (4) The National Security Agency (NSA) *Trusted Network Interpretation* companion to the *Orange Book*. (5) The new version of IEEE 1003.1-1990 (also known as ISO 9945-1 or the Green Book). Also called, "the Ugly Red Book that won't fit on the shelf," in the United States because it is printed on A4 paper. In Europe it is called, "the Ugly Red Book that's a sensible size."

RED signal (1) A telecommunications or automated information system signal that would yield classified information if recovered and analyzed. (2) In cryptographic systems, a signal containing classified information that has not been encrypted. See also *RED/BLACK theory*.

redirection The diversion of data (or signals) from a default or intended destination to an alternate destination.

- In packet switching, a function that routes a call to an alternative network address if the link to the original network is not available. It is carried out by the end point switches.
- In networking contexts, it is usually transparent to the user. For example, a print request may be redirected from the printer port to a spooler, or a workstation's request for access to a (supposedly) local drive is redirected to the server's disk.
- In other contexts, the redirection may be explicit. For example, redirection can be accomplished by using the DOS redirection operators >, <, and >>, or the pipe (|) operator.

redirector A system program (a networking component) that intercepts file or print input-output (I/O) requests and translates them into network requests. It gives a client computer access to the network and operates at the ISO/OSI Presentation Layer.

reduced carrier single sideband emission A single sideband (SSB) emission in which the degree of carrier suppression enables the carrier to be reconstituted and to be used for demodulation.

reduced carrier transmission A form of amplitude modulation (AM) in which the carrier is transmitted at a controlled level below that which is required for demodulation but at a level sufficient to serve as a frequency reference.

redundancy That portion of the total (system, device, message, etc.) that can be eliminated without loss of essential function or information when all elements are performing without error. *Redundancy* may be applied to hardware, software, or information.

- In hardware, redundancy is provided by means of more than one device to perform the same function. For example, the use of multiple hard disks in disk mirroring, disk duplexing, and other redundant array to inexpensive disks (RAID) configurations increase a system's fault tolerance.

- With information/communications, data is included in a transmission for detecting an error or automatically correcting a detected error at the receiving station. For example, parity, Hamming coding, and cyclic redundancy check (CRC) coding are methods of improving information reliability.

- With software, multiple routines each using a different algorithm may be used to detect output errors. Or extra copies of critical code segments may be created when the code is needed by different processes to help prevent programs from accidentally corrupting the code.

redundancy check A check that uses extra digits to help detect the absence of errors in a block of information. The number of extra digits is generally only a small fraction of the number in the information block. See also *CRC, LRC,* and *parity.*

Reed-Soloman code A forward error-correcting method.

ref An abbreviation of RE**F**erence.

reference clock (**1**) A clock to which another clock is compared. (**2**) A clock with a stability sufficiently high to allow mutual synchronization of multiple circuit or network elements.

reference level A measured value from which further measurements are compared. In telephony, a common reference level is 0 dBm (I mW).

reference loudness In telephony, (**1**) a calibratable, comparative reference system that establishes a loudness rating of telephone connections (to the extent that an artificial mouth and ear simulate their "average" human counterparts). (**2**) A system that provides 0-dB acoustic gain between a mouth at a point 25 mm in front of the speakers lips and an ear reference point at the entrance to the ear canal of a listener. The frequency response of such a system is ideally flat with 0-dB attenuation from 300 to 3300 Hz and infinite attenuation outside the band.

In practical systems, however, the band edge is to be attenuated 3 dB relative to the midband response, and the system gain is to be adjusted to compensate for the filter transition bands (as well as the deviation from passband flatness). The amount of adjustment can be determined by first calculating the *objective loudness rating* (OLR) over a frequency range that includes the 50-dB attenuation points of the real response, and next calculating the OLR of an ideal response across the same frequency range. Finally, the difference between these OLR values is the required gain adjustment.

reference noise (RN) The amount of circuit noise that will produce the same reading on a noise meter as a reference tone. See also *dBa, dBa(F1A), dBa(HA1), dBa0, dBm, dBm(psoph), dBm0, dBrn, dBrnC, dBrnC0, dBrn(f₁-f₂), dBrn(144-line),* and *dBx.*

reference plane A plane perpendicular to the direction of wave propagation in a transmission line or waveguide to which measurements such as electrical length, reflection coefficients, scattering coefficients, etc., may be referred.

reference point (**1**) A logical point between two nonoverlapping functional groups. When equipment is placed at a reference point, that reference point is designated an interface. (**2**) The virtual or real point in a system to which transmission-level measurements are referenced. See also *dBr, relative transmission level,* and *transmission level point* (*TLP*).

reference surface In optical fibers, that surface of an optical fiber that is used as the primary transversal alignment means with a component such as a connector or mechanical splice. For example, it may be the fiber core, cladding, buffer layer, or an entity not an integral part of the fiber. For telecommunications grade fibers, the reference surface is the outer surface of the cladding. For plastic-clad silica (PCS) fibers, which have a strippable polymer cladding (not the polymer overcoat of an all-glass fiber), the reference surface may be the core.

reflectance The ratio of reflected light power to incident light power. Generally expressed in decibels (dB) or percent.

The optical equivalent of *return loss.*

reflectance factor The ratio of the flux actually reflected by a sample surface to that which would be reflected into the same beam geometry by an ideal diffuse standard surface when illuminated by the same conditions as the same. Also called *reflection factor.*

reflected code A synonym for *Gray code.*

reflected ray The light ray leaving a reflecting surface, indicating the path of the light after reflection. See also *critical angle* and *reflection.*

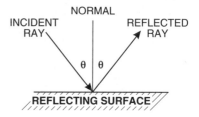

reflected wave A "new" wave generated when a reference wave encounters a discontinuity in a medium or a different medium. The *reflected wave* is contained in the first medium along with the reference wave. See also *critical angle.*

reflecting layer A layer in the ionosphere that has a free electron density sufficiently high to reflect radio waves. The principal reflecting layers, in the daylight hemisphere, are the E, F_1, and F_2 layers. Each layer has an associated critical frequency. See also *ionosphere.*

reflection The abrupt change in the direction of propagation of a wavefront at an interface between two dissimilar media so that the wavefront returns into the medium from which it originated.

Reflection may be specular (i.e., image retaining as with a mirror), diffuse (i.e., not retaining the image, only the energy, as with a rough surface), or a combination of the two, according to the nature of the interface. Depending on the nature of the interface (dielectric-conductor or dielectric-dielectric), the phase of the reflected wave may or may not be inverted. See also *reflected ray.*

reflection coefficient (ρ or RC) (**1**) In optical fibers, the ratio of the forward and reverse flow of power through an optical fiber or cable. (**2**) At a given frequency, geometric point, and propagation mode, the ratio of some quantity associated with the reflected wave to the corresponding quantity in the incident wave (e.g., amplitude or power). The *reflection coefficient* (ρ) may be distinct for different choices of associated quantity; hence, the chosen quantity should be specified. **(3)** At a discontinuity in a transmission line, the complex ratio of the electric field strength of the reflected wave to that of the incident wave. The reflection coefficient is given by

$$\rho = \left| \frac{Z_1 - Z_2}{Z_1 + Z_2} \right| = \frac{\text{SWR} - 1}{\text{SWR} + 1}$$

where:

Z_1 is the impedance toward the source,
Z_2 is the impedance toward the load,
SWR   is the standing wave ratio, and
The vertical bars indicate absolute magnitude.

reflection factor The ratio of the power delivered to a load (whose impedance Z_L is not matched to the source impedance Z_S) to the power that would be delivered if the impedances were matched. Mathematically,

$$R_f = \frac{2 \cdot \sqrt{Z_S \cdot Z_L}}{Z_S + Z_L}$$

reflection loss **(1)** The ratio (generally expressed in dB) of the power delivered to a load of the same impedance as the source (P_{ZS}), to the power delivered into the real load (P_{ZL}).

$$Loss = 10 \cdot \log_{10}\left(\frac{P_{ZS}}{P_{ZL}}\right) \text{dB}$$

An equivalent expression relating source and load impedances is:

$$Loss = 20 \cdot \log_{10}\left(\frac{Z_S - Z_L}{2 \cdot \sqrt{Z_S Z_L}}\right) \text{dB}$$

where:

Z_S is the source impedance and
Z_L is the load impedance.

(2) At a discontinuity or impedance mismatch, e.g., in a transmission line, the ratio of the incident power to the reflected power. The *reflection loss* (*Loss*) is given by

$$Loss = 20 \log_{10}\left| \frac{Z_S - Z_L}{Z_S + Z_L} \right| = 10 \log_{10}\frac{(Z_S - Z_L)^2}{(Z_S + Z_L)^2}$$

where Z_S and Z_L are the source and load impedances, respectively, and the vertical bars indicate absolute magnitude. **(3)** In an optical fiber, the loss that takes place at any discontinuity of refractive index at which a fraction of the optical signal is reflected back toward the source (e.g., at an air–glass interface such as a fiber endface). At *normal incidence,* the fraction of reflected power is expressed by

$$L_f = 10 \log_{10}\frac{(n_1 - n_2)^2}{(n_1 + n_2)^2}$$

where n_1 and n_2 are the respective refractive indices. Also called *Fresnel reflection loss* or *Fresnel loss* in optics.

reflection reduction coating A transparent, single, or multilayer film of specific material(s) applied to a reflecting surface for the purpose of reducing the amount of reflection. The effectiveness of the coating is wavelength dependent as is illustrated in the drawing. Also called *antireflection coating.*

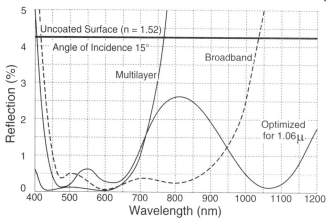

reflectivity The ratio of the intensity of the total radiation reflected from a surface to the total incident radiation on that surface. It is the *reflectance* at the surface of a material so thick that the reflectance does not change with increasing thickness, i.e., the intrinsic reflectance of the surface, regardless of other parameters such as the reflectance of the rear surface. *Reflectivity* is a deprecated term. Reflectance is the preferred term.

reflector **(1)** In an antenna, one or more conducting elements or surfaces that reflect incident radiant energy. **(2)** In a network, a node that retransmits received information to multiple users. Each of the users desiring to receive the information must *subscribe* (sign up) to the reflector service.

reflex circuit A circuit through which a signal passes multiple times. For example, an amplifier through which the signal passes once in the intermediate frequency (IF) band and once in the audio band.

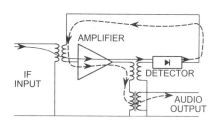

refracted ray **(1)** That part of an incident wave that travels from one medium into a second. Also called the *transmitted ray* or *transmitted*

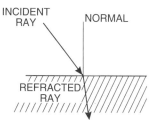

wave in *wave optics.* See also *critical angle* and *reflected ray.* **(2)** A ray that undergoes a change of velocity, or in the general case, both velocity and direction, as a result of interaction with the media in which it travels. **(3)** In an optical fiber, a ray that is refracted from the core into the cladding. Specifically, a ray having direction such that

$$\frac{n^2(r) - n^2(a)}{1 - (r/a)^2 \cos^2 \phi(r)} \leq \sin^2 \theta(r)$$

where:

r is the radial distance from the fiber axis,

$\phi(r)$ is the azimuthal angle of projection of the ray at r on the transverse plane,

$\theta(r)$ is the angle the ray makes with the fiber axis,

$n(r)$ is the refractive index at r,

$n(a)$ is the refractive index at the core radius, a.

Refracted rays correspond to radiation modes in the terminology of mode descriptors. See also *optical fiber.*

refraction Retardation and, in general, the bending of a beam of electromagnetic radiation as it traverses the interface of two mediums with different refractive index or a medium with a continuously varying index as a function of position. For two media of different refractive indices, the angle of refraction is closely approximated by *Snell's Law.* See also *critical angle, refracted ray,* and *Snell's Law.*

refraction loss In fiber optics, that part of a transmission lost due to the refraction resulting from nonuniformity of the medium.

refraction profile See *refractive index profile.*

refractive index (*n*, η) The ratio of the velocity of propagation of an electromagnetic wave in a vacuum to its velocity in the dielectric medium. The value can also be found by the following equation:

$$n = \frac{\sin \theta_i}{\sin \theta_r} = \sqrt{\epsilon'}$$

where:

n is the refractive index,

θ_i is the angle of incidence,

θ_r is the angle of refraction, and

ϵ' is the real dielectric constant.

Both the refractive index (n) and dielectric constant (ϵ') vary with frequency or wavelength. Hence, the equation is valid only if all quantities are measured at the same frequency. The *refractive index* varies from just under 1.5 to a little over 2 for the various types of glasses available and at visible wavelengths. Also called the *index of refraction.* See also *Snell's law.*

refractive index contrast (Δ) In fiber optics, a measure of the relative difference in a refractive index of the core (n_{core}) and cladding ($n_{cladding}$) of an optical fiber. Denoted by Δ, the *refractive index contrast* is given by:

$$\Delta = \frac{n_{core}^2 - n_{cladding}^2}{2 \cdot n_{core}^2}$$

where:

n_{core} is the maximum refractive index in the core and

$n_{cladding}$ is the cladding's index.

refractive index profile In optical fibers, the description or plot of refractive index values as a function of position across the diameter of the fiber. Also called *index profile* and *refraction profile.* See also *optical fiber.*

refractivity The amount by which the refractive index exceeds unity. Often measured in parts per million, called N-units.

reframing time The time interval between the instant at which a valid frame alignment signal is available at the receiving data terminal equipment and the instant at which frame alignment is established. The *reframing time* includes the time required for replicated verification of the validity of the frame alignment signal. Also called *frame alignment recovery time.*

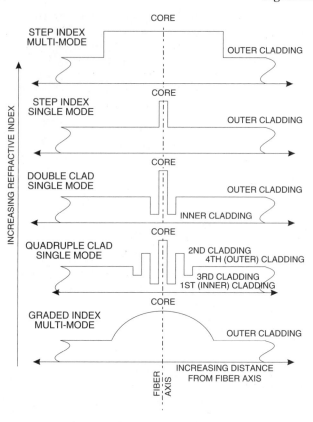

refresh To repeatedly reproduce a display image on a display surface, so that the image remains visible. The display surface generally has a short presistence (i.e., maintains the image only for a short period of time), so that motion can be exhibited.

refresh rate In a display, the maximum number of frames that can be displayed (painted) on a monitor per second, generally expressed in Hertz. The rate is set by the vertical sync signal generated by the video controller. It is limited by the monitor's maximum horizontal scan rate and resolution. (Higher resolution means more scan lines.) Increasing the *refresh rate* decreases flickering, thus reducing eye strain; however, few people perceive changes above 60-72 Hz. Also called the *scan rate, vertical refresh rate,* or *vertical scan rate.*

reg An abbreviation of REGister.

regeneration (1) In feedback systems, the introduction of a fraction of the in-phase component of a device's output signal into the input of the same device. The amount of in-phase or positive feedback varies with the application. For example:

- Small amounts of positive feedback can increase the apparent gain of a device.
- Larger amounts of positive feedback are used to cause oscillation.
- And still larger amounts of feedback can cause a device to lock up in one of two states, such as in a flip-flop.

Also called *positive feedback.* (2) In a *regenerative repeater,* the process or method by which digital signals are amplified, reshaped, retimed, and retransmitted. See also *regenerative repeater.* (3) In a storage or display device, the restoration of stored or displayed data that have deteriorated or faded. Examples include:

- Conventional cathode-ray tube (CRT) displays that must be regenerated many times per second for the data to remain displayed (more than about 25 times per second to avoid flicker).
- In a computer's random access memory (RAM), information "leaks" out of the memory element and must be regenerated to avoid loss.

regenerative feedback Feedback in which a portion of the output signal is returned to the input with a component that is in phase with the input signal. See *regeneration (1)*.

regenerative repeater A device that receives digital data from one segment of a transmission link, restores the pulse integrity (waveform restoration and retiming), and retransmits the reconstituted pulse on the next segment of the transmission link.

The diagram displays the major blocks of a unidirectional *regenerative repeater,* that is, the signal is regenerated in only one direction. Two of these circuits would be used in a four-wire communications path. Also called simply a *repeater* or a *regenerator.* See also *repeater (4).*

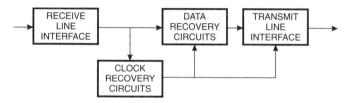

regional Bell operating company (RBOC) Fixed by the *modified final judgment (MFJ)* divestiture decree, the *RBOCs* are responsible for local services in the *Local Access and Transport Areas (LATAs)* as defined by the Federal Communications Commission (FCC). Seven *RBOCs* were created from the AT&T Bell operating companies (BOC) that existed prior to divestiture; that is,

ORIGINAL SEVEN RBOCs

RBOC	Area	BOCs
Ameritech Corp	IL, IN, MI, OH, WI	IL Bell, IN Bell, MI Bell, OH Bell, WI Bell
Bell Atlantic Corp	DE, MD, NJ, PA, WV, VA	Bell of PA, Chesapeake and Potomac of MD, Chesapeake and Potomac of VA, Chesapeake and Potomac of Washington, DC, Chesapeake and Potomac of WV, Diamond State Telephone, NJ Bell
BellSouth Corp	AL, FL, GA, KY, LA MS, NC, SC, TN	South Central Bell Southern Bell
Nynex Corp	MA, ME, NH, NY, RI, VT, CT	New England Telephone, New York Telephone
Pacific Telesis Group	CA, NV	NV Bell, Pacific Bell
SBC Communications Inc. (formerly Southwestern Bell)	AR, KS, MO, OK, TX	Southwestern Bell
US West	AZ, CO, ID, MN MT, NB, NM, ND SD, UT, WA, WY	Mountain Bell Northwestern Bell Pacific NW Bell

Since the creation of RBOCs in December 1983, many have merged and renamed themselves. As of July 2000 the list is: *SBC Ameritech* (Ameritech combined with SBC Communications which itself was the combination of Southwestern Bell, Pacific Telesis and SNET—and independent), *Verizon* (Bell Atlantic combined with NYNEX and GTE and independent), *Bell South,* and *US West.*

RBOCs are also called *Baby Bells* and *regional Bell holding companies (RBHCs).*

regional center (RC) In telephony, a toll office to which a number of *sectional centers* are connected. Also called a *class 1 office.* See also *class n office.*

regional network A wide area network (WAN) with a high-speed backbone that provides Internet access to a defined region, such as a state, country, or contiguous group of states or countries.

register **(1)** A type of memory used to store active data, control information, or status information. Various bits in the register may be read-write, read only, or write only. (Not all bits in the register need be the same type of memory).

Registers are very fast compared to conventional random access memory (RAM) in a computer; typically, two registers can be read and a third written (all in a single machine cycle). RAM memory, in contrast, may require several machine cycles for a single access. Typically, only a few registers are associated with a central processing unit (CPU); in contrast, there are millions of words of main memory (RAM). Computers typically contain a variety of registers. General-purpose registers may perform many functions, such as holding constants or accumulating arithmetic results. Special-purpose registers, on the other hand, perform special functions, such as holding the instruction being executed, the address of a storage location, or data being retrieved from or sent to storage. **(2)** An addressable location in a memory-mapped peripheral device. For example, the transmit and receive data registers in a universal asynchronous receiver transmitter (UART). **(3)** The process of informing a host system or program of the existence and characteristics of a client. Once registered, the services of the host are made available. **(4)** That part of the telephone switching system that receives and stores signaling information from another source (the subscriber, an operator, or another switching system).

register insertion A media access method used in some older (1970s and 1980s) ring topologies. With register insertion, a node that has information to transmit simply inserts a register (a buffer) into the ring's data stream at an appropriate point in the data stream. The inserted register contains the packet to be transmitted (including data, addressing, and error-handling information). Depending on system specifics, the node may be able to insert its register before passing a received packet on to the next node or only during a break in the data stream.

An advantage of an insertion strategy is that multiple nodes can be transmitting at the same time. This is in contrast to simple token passing, in which only the node with the token gets access to the network. Disadvantages of this access method include:

- Inserting a register lengthens the effective logical ring, which means the transfer times increase slightly for each concurrent transmission.
- The ring can become overloaded if many nodes want to transmit at the same time. There is no way to control this, since the protocol does not have any provisions for preventing a node from trying to access the ring.

Register insertion media access methods have been superseded by token passing as the preferable access method for ring networks.

register recall See *hook flash.*

registered equipment In telephony, any apparatus that has passed a series of tests and is certified to be in compliance with the Federal Code of Regulations part 68 subpart C. Properly registered equipment will have a *registration number* permanently attached to it.

registered host An Internet host that uses an Internet Protocol (IP) address assigned by the Network Information Center (NIC).

registered jack (RJ) Any of the *RJ* series of jacks described in title 47, part 68 of the Federal Code of Regulations. See also *RJ-nn.*

registration (1) The accurate positioning of an entity (mechanical or optical) relative to a reference. (2) The recording of an entity's existence or parameters with another entity.

registration number A number assigned to equipment that has passed a series of tests outlined in the Federal Code of Regulations, part 68, subpart C. This number certifies that the equipment will not cause harm to the public telephone network. It *does not* state anything about the functionality, commercial value, or quality of the device.

regression In mathematics, a method where an empirical function is derived from a set of experimental data.

regression analysis A method of predicting future events by plotting past events and extending the trend line. The method assumes a constant cause and effect relationship of the predicted events and past events.

regression testing Part of the test phase of software development where, as new functional modules are integrated into the system and tested, previously tested functionality is re-tested to assure that no new module has corrupted the system.

REJ An abbreviation of RE**J**ect. A supervisory frame that acknowledges received data units while requesting retransmission of a specified error frame.

rejection band See *stopband.*

REL An abbreviation of RELease. The fifth of the Integrated Services Digital network user Part (ISUP) call setup messages. It indicates that the identified circuit is being released for a specified reason and will be returned to the idle state upon receipt of the RELease Complete message (RELC). The REL message may be sent by either end of the transmission link.

relative aperture The ratio of the diameter of the entrance pupil in an optical system to the equivalent focal length of the system. The value is denoted as a fraction in which the numerator is the letter f and the denominator is the decimal value of the ratio. For example, if a given optical system has a focal length of 3 units and an entrance pupil of 2 units, then the *relative aperture* is f/1.5.

On many systems, an adjustable iris or aperture provides a variable pupil. A series of *f/numbers* printed on the device indicate the position of the diaphragm, i.e., f/44, f/32, f/22, f/16, f/11, f/8, f/5.6, f/4, f/2.8, f/2, f/1.4. The values are selected and arranged so that each position admits twice as much light to the system as the previous position allows; for example, an f/8 aperture passes twice as much light as an f/11. Also called an *f/number* or *f/stop.*

relative error The ratio of an absolute error to the true, specified, or theoretically correct value of the quantity that is in error.

relative index The relative propagation velocities of light in two media, neither of which is air.

relative refractive index A value equal to the ratio of the refractive index of one medium divided by the relative index of a second medium.

relative transmission level The ratio of the signal power, at a given point in a transmission system, to a reference signal power.

The ratio is usually determined by applying a standard test tone at a zero transmission level point, 0TLP (or applying an adjusted test tone power at any other point) and measuring the gain or loss to the given point. Note that the standard test tone power is generally not the same as the actual signal power in the circuit during normal operation.

relay (1) Any switch operated by an electrical control signal (either a voltage or current).

Relays may be optimized for the operate time (the time between application of a control signal and switch closure) and/or release time (the time between removal of the control signal and switch opening). Some of the combinations include:

- *Fast-operate, fast-release*—a high-speed relay designed for both short operate and short release times.
- *Fast operate*—a high-speed relay designed for short operate time, but not necessarily short release times, e.g., time delay relays.
- *Slow operate*—a relay designed to operate only after the control signal persists for a predetermined period of time.
- *Slow release*—a relay designed to ignore short period release signals, such as dial pulses.
- *Etc.*

Relays are available in many forms, e.g., reed, mercury, wire, power, coaxial, hermetic, stepping, latching, and open frame, and may be electromechanical or solid state. See also *switch contacts.* (2) To retransmit a received message from one station to another.

relay configuration A network arrangement in which a circuit is established between two stations via an intermediate *relay station.*

relay point In a switching network or system, that point at which packets or messages are switched to other circuits or channels.

relay station An intermediate station that passes information between terminals, nodes, or other relay stations.

relaying The process of actually moving data along the path determined by a routing process. The data is *relayed* from a source and a destination.

Relaying is one of the two major functions of the network layer in the ISO/OSI Reference Model; the other is routing.

RELC An abbreviation of RELease Complete. The sixth of the Integrated Services Digital Network User Part (ISUP) call setup messages. It is a packet to acknowledge release (REL) or disconnect. Sometimes written *RLC.*

Relcom One of several USENet alternative newsgroups. Newsgroups in which the information is presented mainly in Cyrillic script. These newsgroups are found mainly in Russia, Europe, and a few other systems capable of displaying Cyrillic. See also *ALT* and *alternative newsgroup hierarchies.*

release (1) In general, the return of a system, device, or circuit to its idle state. (2) In telephony, the disconnection of a communications link and the restoration of all associated equipment to the idle state. (3) In software, a synonym for version.

release time The time interval between the instant that a function-enabling signal is discontinued and the instant at which the function ceases. Examples include:

- The time to disable an echo suppressor following the removal of the enabling signal power.
- The time interval between the instant a relay coil is deenergized and the instant that contact closure ceases (or, depending on the nature of the relay, is established).

release with howler The release of line equipment and the application of howler tone to a line by a central office (CO) when that line is off-hook for an extended period without dialing.

released loop A synonym for *switched loop.*

reliability A measure of the ability of an entity to perform its required function under stated conditions for a specified time period. See also *availability, MTBF,* and *MTTR.*

reliable connection A connection between local and remote data communication equipment (DCE) that utilizes error-correction and/or data compression protocols.

The local modem and the remote modem negotiate the highest common protocol with V.42/MNP error correction after a connection is established. If no common protocol can be negotiated, the modem will go back on-hook. The modems enable data buffers in the transmit and receive path; therefore, flow control must also be enabled to avoid buffer overflow and underflow.

reliable mode A communications mode found in many modems and other communications equipment that enables them to tolerate noise and other faults and to still function.

One example can be found in the modem's error-correction protocol *V.42* or the various *Microcom Network Protocols* (*MNP*). Another example is the Transmission Control Protocol/Internet Protocol (TCP/IP) protocol which assumes that some part of the Internet will be unreliable (noisy) or unavailable, and assures delivery through retransmission and finding alternate routing (*space diversity transmissions*).

reliable transfer A transfer mode, in the ISO/OSI Reference Model, that guarantees that either the message will be transmitted without error or the sender will be informed if the message cannot be transmitted without error.

reliable transfer service element (RTSE) An application layer service element (ASE) in the ISO/OSI Reference Model that helps ensure that protocol data units (PDUs), or packets, are transferred reliably between applications. *RTSE* services can sometimes survive an equipment failure because they use transport layer services. In the United States, sources other than *RTSEs* are usually used to provide these services.

remanence The magnetic induction that remains in a magnetic circuit after the removal of an applied magnetomotive force. See also *magnetic hysteresis*.

remote Meaning "at a distance," it generally refers to the situation where a device is unable to be connected with local wiring to a host device but requires communications devices and transmission systems to be connected to the host.

remote access The ability to gain entry to a network or switching system from a long distance, using telephone lines or other communication channels. For example:

- A private automatic branch exchange (PABX) service feature that allows a user at a remote location to access by telephone PABX features, such as access to wide area telephone service (WATS) lines. For remote access, individual authorization codes are usually required.
- The ability of a user to gain access to and control of his office computer or network by means of a dial-up telephone connection.

remote bridge A LAN-to-LAN (local area network) bridge that links physically separated network segments via a wide area network (WAN) link.

remote call forwarding A switching system service feature that allows calls directed to a remote call forwarding number to be automatically forwarded to any answering location designated by the call receiver. A remote forwarding number may exist in a telephone central switching office without the user having any other local telephone service in that office.

remote connection A long-distance connection between a workstation and a network; a connection that generally involves telephone lines and that may require modems. Remote connections often require special timing considerations because many network transactions must take place within very limited time periods.

remote console A networking utility that enables a network supervisor to manage a server from a workstation or from a remote location using a modem. The supervisor can issue commands and accomplish tasks just as if all the commands were being issued directly at the server console or terminal.

remote control A remote access method in which one computer controls a second computer operation or parameters through a modem link.

remote control equipment Devices used to perform monitoring, controlling, and/or supervisory functions from a distance.

remote database access (RDA) An ISO/OSI specification to allow remote access to databases across a network.

remote diagnostics The ability of a local device to access a distant system and perform some form of testing on the distant system. When the distant system is not totally disabled, the local device may be able to determine what is wrong and sometimes fix the problem; other times it dispatches a technician to repair the problem.

remote digital loopback A diagnostic mode of modems that connect the received digital output back to the digital transmit input. This allows a distant device to compare a transmitted message with the looped back message. Differences between the two indicate transmission errors or modem malfunctions.

remote host On the Internet, a host system that is accessed by way of Telnet, FTP (file transfer protocol), Mosaic, or other tool.

remote job entry (RJE) The submission of a task from a remote site to a centralized host computer (mainframe) via a terminal that gains access to the computer only through a communications link.

remote line concentrator A multiplexer that concentrates a number of users onto a smaller number of communication circuits. Methods include time division multiplex, frequency division multiplex, and statistical multiplexing (e.g., TASI).

remote login On the Internet, the procedure used to establish a connection with a *remote host*.

remote mode A terminal operating mode in which data are transmitted to and received from a host computer.

Remote Operations Service Element (ROSE) In the ISO/OSI Reference Model, a general-purpose application layer service element (ASE) that supports interactive cooperation between two applications.

The application requesting the association is known as the *initiator*. The application responding to it is the *responder*. The application that requests an operation is known as the *invoker*. The other application, called the *performer*, carries out the requested operation. An application association provides the context for the cooperation between the two application entities (AEs). When an application association is established, the AEs involved must agree on an operation class and an association class for the interaction. Five operation classes are defined, based on the type of reply the performer provided and on whether the interaction is synchronous or asynchronous, that is,

- Class 1 (synchronous) reports both success and failure.
- Class 2 (asynchronous) reports both success and failure.
- Class 3 (asynchronous) reports only in case of failure.
- Class 4 (asynchronous) reports only in case of success.
- Class 5 (asynchronous) reports neither success nor failure.

The three association classes are:

- Association class 1: Only the initiator can invoke operations.
- Association class 2: Only the responder can invoke operations.
- Association class 3: Either the initiator or the responder can invoke operations.

ROSE provides a mechanism for enabling applications to cooperate; however, ROSE does not know how to carry out the actual operations. The details of the operations must be agreed upon by the applications independently of ROSE. Similarly, the processes necessary to carry out the operation must be available once the association is established.

remote orderwire An extension of a local orderwire to a point convenient for personnel to perform required operational and maintenance functions.

remote password generator (RPG) A device that can generate a new unique password every time a user wants to log in to a network. The device uses a special number, which is generated by the network, and the user's personal identification number (PIN) to generate the password.

remote procedure call (RPC) A mechanism by which a procedure on one computer can be used in a transparent manner by a program running on another machine. This mechanism provides an easy way to implement a client–server relationship. That is, an *RPC* allows a client to cause a program stored on a remote host to start executing on that remote host.

The procedure was developed by Sun Microsystems and is documented in RFC 1057.

remote program load (RPL) Starting a computer and loading the operating system into its memory when the operating system software is provided by a server on the network.

remote server In a client–server network model, a server that is located on a computer system other than the local network.

An example is the Internet system in which any user may access any of the thousands of Gopher sites around the world.

remote switching entity A device or entity for switching signals from inlet ports to outlet ports located at a distance from the serving system control.

remote system Any system at the "other end" of a communication link away from the user. The *remote computer* could be in another city, another building, or on the same table as the *local system*.

remote terminal A terminal that is located some distance from a host computer or network. *Remote terminals* are generally connected by telephone lines to the host or network.

remote trunk arrangement (RTA) An arrangement that permits the extension of traffic service position system (TSPS) functions to remote locations.

Remote Write Protocol (RWP) A proposed Internet protocol (detailed in RFC 1756) for exchanging short messages between terminals.

REMS An acronym for Remote Electronic Mail System.

REN An acronym for Ringer Equivalence Number.

reorder tone A call progress tone indicating the call is blocked within the network, that is, all paths in the switching equipment are busy, there is no available toll trunk, or there is some other equipment blockage. The tone frequencies are generally the same as a standard busy tone. However, the interruption rate is twice as fast, i.e., 120 interruptions per minute. Also called *channel busy* and *fast busy*. See also call *progress tones*.

repair and quick clean (RQC) The repair, burn-in, operational testing, and minor "cosmetic" cleaning of equipment. See also *LNRU, repair only (RO)*, and *repair, update, and refurbish (RUR)*.

repair only (RO) The equipment is repaired to the original working condition. No refurbishing or updating is performed except as is necessary to repair. See also *LNRU, repair, update, and refurbish (RUR)*, and *repair and quick clean (RQC)*.

repair, update, and refurbish (RUR) Cosmetic cleaning, repair, and updating of equipment, including the installation of changes necessary to bring the equipment up to current company specifications. See also *LNRU, repair only (RO)*, and *repair and quick clean (RQC)*.

repeat In ring networks, the action of a node receiving a packet or token from an upstream node and retransmitting it toward the downstream node. The repeating station may copy the packet to a buffer, examine it (e.g., to determine to which node it is addressed) or even modify the control bits prior to retransmission.

repeat dialing A telephone company service that automatically examines a previously dialed busy number for availability. When the line is free, the system rings the caller and completes the connection.

Called *repeat call* or *return call* by some telcos and *camp-on* in the terminology of private branch exchanges (PBXs).

repeat-request (RQ) A synonym for *automatic repeat request (ARQ)*.

repeated character compression A data compression scheme which, in its simplest form, indicates what symbol is present and how many consecutive symbols are to be represented. For example, given the data string AAABBBCCCC, the compressed version might look like A3B3C4. Compression methods like this are frequently used on data representing images such as facsimiles. See also *compression (2)*.

repeater (1) A term that originated with telegraphy and referred to an electromechanical device used to *regenerate* telegraph signals. Use of the term has continued in both telephony and data communications. **(2)** A receiver-transmitter combination arranged so that all signals received by the receiver are retransmitted by the transmitter. Such systems are frequently used in satellites and ground-based microwave communications links. Sometimes also called a *transponder*. **(3)** An analog device inserted into a communications channel which amplifies and signals (regardless of its nature, i.e., analog or digital) to compensate for losses and distortions in the communication channel. The device functions at the physical layer of the ISO/OSI Reference Model as in the figure. A *repeater* is used to extend the length, topology, or interconnectivity of the transmission medium beyond the limits imposed by a single transmission segment. It is used to connect two segments of the same network. See also

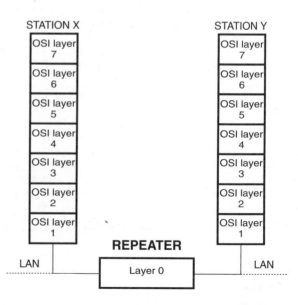

regenerative repeater. (**4**) A digital device that amplifies, reshapes, and/or retimes a digital input signal for retransmission, that is, it regenerates the original signal as closely as possible.

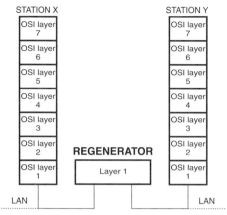

There are several caveats associated with the use of repeaters and regenerators:

- A repeater can increase segment length only to overcome electrical restrictions (e.g., attenuation and loading); it cannot be used to increase the propagation time limitations inherent in the network's layout. For example, a repeater cannot stretch the network so that a transmission could take more than the allowable slot time to reach all the nodes in an Ethernet network.
- A particular repeater, in general, works with only one specific type of network architecture. This has to do with the fact that different architectures use different cabling (e.g., coaxial versus twisted pair) or use cabling with different electrical characteristics (e.g., 50-ohm vs. 93-ohm impedance).
- Only a limited number of repeaters may be used in any branch of a network. For example, in the IEEE 802.3 Ethernet no more than four repeaters may be interposed between nodes.
- A repeater may not allow some features of the network to be operational. For example, the IEEE 802.3 specifications stipulate that repeaters may not be used with transceivers that generate a signal quality error (SQE) test signal.

Also called a *regenerator* or *regenerative repeater.* See also *regenerative repeater.*

repeating coil In telephony, a transformer that connects the voice frequency portion of one telephone circuit to another circuit while blocking the passage of dc. The transformer may also match different circuit impedances or couple balanced and unbalanced circuits.

repertory dialing A feature of many telephone instruments, private automatic branch exchanges (PABX) systems, and the local exchange carriers (LECs) that allows a user to select and dial from a group of stored numbers using only two or three key strokes. Typically, there are between 10 and 100 numbers available to the user for programming. The length of the programmed number may be up to 40 digits in some devices. Also called *memory dialing* and *speed dialing.*

repudiation In network transmissions, the denial by a sending node that the message was sent (origin repudiation) or by the recipient that the message was received (destination repudiation). One security measure that may be used in a network is nonrepudiation, which makes it impossible for a sender or receiver to make such denials.

REQ An abbreviation of RE**Q**uest.

REQD An abbreviation of RE**Q**uire**D**.

Request for Comment (RFC) In Internet, a series of numbered documents containing the standards, proposed standards, and other relevant details regarding the operation of the network. Among them are found the Transmission Control Protocol/Internet Protocol (TCP/IP) protocols, e-mail information, statements regarding the Internet, minutes of technical committees, and notes for new users.

The document series was begun in 1969. It is created by and circulated among Internet participants. See Appendix G for a list of titles.

Request for Discussion That part of a newsgroup creation process wherein a group is proposed and discussion begins. Such postings are found on the *new.announce.newsgroups* newsgroup. See also *call for votes* (*CFV*) and *newsgroups hierarchies.*

request to send (RTS) (**1**) A receiver-generated signal that conditions a remote transmitter to commence transmission. (**2**) One of the hardware control signals in the EIA-232/V.24 specification. When asserted, it places the data communication equipment (DCE) in the originate mode and prepares it for transmission. The DCE responds to the *RTS* with a clear to send (CTS) signal when it is ready.

Request/Response Header (RH) In IBM's Systems Network Architecture (SNA) network communications, a 3-byte field added to a request/response unit (RU) at the transmission control layer to create a basic information unit (BIU).

Request/Response Unit (RU) In IBM's Systems Network Architecture (SNA) network, the type of packet exchanged by network addressable units (NAUs), which are network elements with associated ports (or addresses).

reradiation (**1**) Radiation, from a device, at the same or different frequency (wavelength) received from an incident wave. (**2**) Undesirable radiation of signals locally generated in a radio receiver. Such radiation might cause interference or reveal the location of the device.

rering signal See *re-ring.*

rerouting The temporary change in routing of communications traffic. It may be planned (e.g., when preventative maintenance is scheduled) or a reaction to changing network conditions (e.g., a node failure).

reservation protocol A protocol that allows a node to take exclusive control of a communications channel for a limited period. Such control is needed in some types of communications, such as communications between a satellite and a receiving station.

reserve capacity Installed capacity of a device or system that is not normally utilized but can be made available when required (e.g., memory, power).

Reserve not in operation and requiring switching is frequently called *standby.*

reserved character A keyboard character that has special meaning to a currently running program. It therefore cannot be used within commands, file names, or macros. Common reserved characters include *, /, |, \, ?, >, <.

reset (RES or RS) (**1**) To change the value in a digital register to zero. (**2**) To restore the quantity in a storage device to a specified value. (**3**) To place a system or device in a default or startup state. See also *reset mode.*

reset mode The parameters programmed into a device or system for its basic operation. The parameters adopted by a device upon receipt of a reset signal.

reset packet An X.25 packet that identifies an error condition on the communications link. The packet notifies the communicating data terminal equipment (DTE) of error conditions at a known point in the data packet transfer sequence. It is up to the DTE to initiate corrective action.

reset signal See *reset (3)*.

resident Pertaining to computer programs that remain on a particular storage device.

residual error (1) In general, the sum of random errors and uncorrected systematic errors. (2) In communications, an error that survives the system's error-detection and correction mechanisms.

For example, all occurrences of even numbers or bit errors will survive in a data communications system that relies on just parity for protection. Most, but not all, of these errors will be detected with checksum or cyclic redundancy check (CRC) error-detection methods.

residual error rate (1) The ratio of the number of transmission units (bits, bytes, blocks, characters, or packets) received and not identified as incorrectly received to the total number of transmission units received. This is a measure of the effectiveness of the error control equipment/algorithm. (2) The error ratio that remains after attempts at correction are made. Also called *residual error ratio* and *undetected error rate* or *undetected error ratio*.

residual flux density The magnetic flux density at which the magnetizing force is zero when the material is in a symmetrically cyclically magnetized condition. See also *magnetic hysteresis*.

residual magnetism A property by which ferromagnetic entities retain a certain magnetization after the magnetizing force is removed. See also *magnetic hysteresis*.

residual modulation A synonym for *carrier noise level*.

residue check A data integrity checking method in which each datum is transmitted along with a remainder (from the division by a value *n*). A difference from the receiving node's remainder recalculation indicates an error.

resistance (1) A physical property of materials that impedes the flow of current in both ac or dc circuits. Resistance is measured in units of ohms. All materials (with the exception of superconductors) have some *resistance*. Those materials with a low *resistance* are said to be conductors while those with a high *resistance* are called insulators. (2) The real part of impedance. (3) That property of a circuit that dissipates power (*P*) when current (*I*) is flowing. The power dissipated is given by:

$$P = I^2 R$$

See also *impedance* and *Ohm*.

resistance hybrid A network of resistors to which four branches of a circuit may be connected to make them conjugate in pairs. See also *hybrid*.

resistive coupling The association of multiple circuits by means of resistance mutual to all of the circuits.

resistor A device used in electronic circuits designed to provide a specific amount of *resistance* to current flow. The symbol for a resistor is:

See also *resistance*.

resolution (1) In general, the minimum difference between discrete entities or values that can be discerned by an appropriate measuring device. A measurement of the smallest detail that can be distinguished or measured by a sensor under specific conditions.

Note that high resolution does not necessarily mean high accuracy. See also *accuracy* and *precision*. (2) In optics, the ability of an optical system to reproduce the points, lines, and surfaces in an objective as separate entities in the image. (3) In transmission, the degree to which nearly equal values of a quantity can be discriminated. (4) In data acquisition, the smallest value of a measured quantity that can be distinguished by a sensor. (5) In a spectrum analyzer, the ability to display adjacent frequency responses with a 3-dB notch between them. The optimum resolution (R_o) is defined theoretically by:

$$R_o = K \sqrt{\frac{\text{Frequency Span}}{\text{Sweep Time}}}$$

where *K* is unity unless otherwise specified.

resolution error The inability of a device to express or exhibit variances between two signals whose difference is smaller than a given increment.

resolving power The ability of an optical system to produce separate and distinct images of two closely spaced objects.

Optical systems have limited resolution due to the diffraction pattern produced by the aperture. Because of the diffraction pattern, a point object is imaged as a small disk (called the *Airy disk*) instead of a point. When two point objects are at a critical separation, the first dark ring of the diffraction pattern of one image falls on the central disk of the second image. Under these conditions, the points are resolved as separate entities and are said to be at the *limit of resolution*. See also *Rayleigh criterion of resolving power*.

resonance (1) A condition where the excitation frequency (or vibration) is equal to the natural frequency of a system, structure, or device. (2) The condition in an electric circuit where the inductive reactance is equal to the capacitive reactance (causing electrical energy to oscillate between the magnetic field of the inductor and the electric field of the capacitor). This point exists at only one frequency for a given inductor and capacitor; the circuit is therefore said to be *tuned* to a resonant frequency. The impedance of a series resonant circuit is minimal at resonance, while the impedance of a parallel resonant circuit is maximum.

Resonance occurs because of the cyclic process wherein:

- The collapsing magnetic field of the inductor generates an electric current in its windings that charges the capacitor, and
- The discharging capacitor provides an electric current that builds the magnetic field in the inductor.

At resonance the inductive reactance (X_L) and the capacitive reactance (X_C) are of equal magnitude. Therefore, $2\pi fL = 1/2\pi fC$ and

$$f = \frac{\pi}{2\sqrt{LC}}$$

where:

f   is the resonant frequency in hertz (Hz),
L   is the inductance in henries (H), and
C   is the capacitance in farads (F).

The above equation is valid for series RLC circuits and parallel RLC circuits with a Q greater than about 10. If the Q of the inductor is less than 10, the wire resistance (R_w) of the inductor should be included and the equation becomes

$$f = \frac{\sqrt{1 - \left(R_W^2\, C / L\right)}}{2\pi\sqrt{LC}}$$

resonant antenna A synonym for *periodic antenna*.

resource Any entity that is shared or can be shared on a local area network (LAN), including but not limited to: text files, binary files, phone books, encyclopedia, disk drives, CD-ROMs, printers, plotters, memory, modems, facsimile, and tape storage units.

resource discovery tool Programs developed to assist Internet users find and retrieve information (such as files, programs, and graphics). Some such tools include Archie, WAIS (wide area information server), Gopher, and the hypertext browsers of the World Wide Web (WWW) such as Mosaic.

Resource Fork That portion of a Macintosh file containing information about the resources (windows, applications, drivers, and so on) used by the file. This information is environment-specific and is generally meaningless in non-Macintosh operating systems (such as DOS or UNIX).

resource reservation protocol (RSVP) A flow-control protocol developed by the Internet Engineering Task Force (IETF), which gives user's applications a committed level of end-to-end service from the network. It is designed to work with both normal unicast (one source, one destination) and multicast (one source, multiple destinations). The reservation consists of a traffic specification (TSPEC) announced by the traffic source and a reservation made by each recipient.

RESP ORG An abbreviation of RESPonsible ORGanization.

response (1) In general, a reply to a query. **(2)** The effect of an active or passive device on an input signal. **(3)** In data transmission, the content of the control field of a response frame advising the primary station concerning the processing by the secondary station of one or more command frames. **(4)** In IBM's Systems Network Architecture (SNA), the Synchronous Data Link Control (SDLC) control information sent from a secondary station to a primary station, i.e., an answer to an inquiry.

response frame In data transmission, all frames that may be transmitted by a secondary station.

response mode In communications, the operating mode of a device in which a response must be generated when a call is received.

response PDU A *protocol data unit* (*PDU*) transmitted by a *logical link control* (*LLC*) sublayer in which the PDU *command/response* (*C/R*) bit is set to "1."

response time The time between the issuance of a command, request, or stimulus and an answer, reply, or completion of the requested action (such as supplying of data).

In networking contexts, response time is the time required for a request at a workstation to reach the server and for the server's reply to return to the workstation. Response time is inversely proportional to the transmission speed of the specific network architecture in use. The minimum return time value is a function of several parameters, including:

- Delays introduced by the network interface cards (NICs) in both the workstation and the server.
- Delays in network transmission through either slow access or multiple retransmissions (due to excessive activity for example).
- Delays in the server's reply (e.g., the server is otherwise occupied when the request is received).
- Delays in accessing the server's hard disk and writing or reading any required data.

responsivity The ratio of an optical detector's electrical output to its optical input power. Generally expressed in amps per watt (A/W) or volts per watt (V/W) of incident radiant power. *Responsivity* is a function of

- The wavelength of the incident radiation,
- The bandgap of the material from which the photodetector is fabricated, and
- The construction of the photodetector.

Formerly called *sensitivity*.

restart (1) A synonym for hardware reboot. **(2)** In software, to begin the execution of a routine using data recorded at an intermediate calculation point.

restart packet An X.25 packet that notifies data terminal equipment (DTEs) that an irrecoverable error exists within the network. The restart packet clears all existing switched virtual circuits (SVCs) and resynchronizes all permanent virtual circuits (PVCs).

restore To install data and software that have previously been backed up. The restoration process uses the backup media. When doing a total restoration, first the most recent complete backup, and then each of the incremental or differential backups that followed it are installed.

restricted access A class of service in which certain users are denied access to one or more of the system features or operating levels.

restricted channel In digital communications systems, a channel that has a useful capacity of only 56 kbps instead of 64 kbps. The restricted channel, currently common in North America, was originally developed to satisfy a ones-density limitation in T-1 circuits.

restricted service tone In telephony, a class of service tone that indicates to an operator that certain services are denied to the caller.

result code The response or progress of a modem to an AT command. Result codes may be either *verbose* (expressed in words) or *terse* (expressed by a numeric value), depending on the settings of register S14 in Hayes compatible modems. See Appendix A for a list of AT commands and Appendix C for a list of result codes.

RETMA An acronym for Radio Electronics Television Manufacturers Association. See also *EIA*.

retractile cord A coiled cable that springs back to its original length when a stretching force is removed. Telephone handset cords are a common example of *retractile cords*.

retrain Many high-speed modems automatically re-optimize transmission parameters during an active connection. The is done in response to changes in signal strength or noise on the transmission path. The parameters adjusted during the retraining are internal filters, echo cancelers, and baud rate.

retransmission An error control method in which the receiving station's response to a detected error is either the transmission of a negative acknowledge (NAK) or no acknowledge at all. The sending station responds by re-sending the faulty packet.

retrograde orbit In satellite communications, a satellite orbit with an inclination between ninety degrees (90°) and one hundred eighty degrees (180°), i.e., an orbit in which the satellite's position (as projected onto the Earth's equatorial plane) revolves in the direction opposite that of the rotation of the Earth. See also *direct orbit*.

retry In the bisynchronous protocol, the process of retransmitting the current data block a specified number of times or until it is accepted by the destination.

return band With communications using frequency division multiplexing (FDM), a one-directional (simplex) channel over which remote devices respond to a central controller. Also called a *backward channel*.

return call See *repeat dialing*.

return loss A measure of the dissimilarity between two impedances in metallic transmission lines and loads or between refractive indices in dielectric media. The measure of *return loss* (L_R, generally given in dB) can be expressed as:

- The ratio of the power incident on the discontinuity to the power reflected from the discontinuity, that is,

$$L_R = 10 \cdot \log_{10}\left(\frac{P_{incident}}{P_{reflected}}\right) \text{ dB, or}$$

- The ratio of the sum and difference of the two impedances (Z_1 and Z_2), that is,

$$L_R = 20 \cdot \log_{10}\left|\frac{Z_1 + Z_2}{Z_1 - Z_2}\right| \text{ dB.}$$

- A function of *reflection coefficient* (ρ)

$$L_R = 20 \cdot \log_{10}(1/(1 - \rho^2)) \text{ dB.}$$

- A function of *standing wave ratio* (*SWR*)

$$L_R = 20 \cdot \log_{10}\left(\frac{SWR + 1}{SWR - 1}\right) \text{ dB.}$$

Generally, large *return loss* values are better, that is, they indicate a better match between impedances. See also *table*. Also called *reflection loss* in dielectric media, such as optical fibers. See also *balance return loss* and *standing wave ratio* (*SWR*).

RETURN LOSS AND POWER/IMPEDANCE RATIOS

Return Loss (dB)	P_i/P_r	Z_1/Z_2 or SWR	% Reflected Power
3.01	2.0000	5.8284	50.00%
3.15	2.0634	5.5825	48.46%
3.29	2.1317	5.3474	46.91%
3.44	2.2056	5.1227	45.34%
3.59	2.2855	4.9078	43.75%
3.75	2.3721	4.7025	42.16%
3.92	2.4661	4.5063	40.55%
4.10	2.5684	4.3188	38.93%
4.28	2.6798	4.1397	37.32%
4.47	2.8013	3.9686	35.70%
4.67	2.9342	3.8052	34.08%
4.89	3.0798	3.6492	32.47%
5.11	3.2397	3.5002	30.87%
5.33	3.4157	3.3580	29.28%
5.57	3.6099	3.2223	27.70%
5.83	3.8245	3.0928	26.15%
6.09	4.0625	2.9693	24.62%
6.36	4.3270	2.8516	23.11%
6.65	4.6219	2.7393	21.64%
6.95	4.9515	2.6324	20.20%
7.26	5.3211	2.5305	18.79%
7.59	5.7368	2.4335	17.43%
7.93	6.2059	2.3412	16.11%
8.28	6.7373	2.2534	14.84%
8.66	7.3412	2.1700	13.62%
9.05	8.0302	2.0907	12.45%
9.45	8.8194	2.0154	11.34%
9.88	9.7271	1.9439	10.28%
10.32	10.7756	1.8762	9.28%
10.79	11.9923	1.8120	8.34%
11.27	13.4108	1.7513	7.46%
11.78	15.0726	1.6939	6.63%
12.31	17.0297	1.6397	5.87%
12.87	19.3469	1.5885	5.17%
13.45	22.1059	1.5403	4.52%
14.05	25.4104	1.4949	3.94%
14.68	29.3925	1.4523	3.40%
15.34	34.2222	1.4124	2.92%
16.03	40.1192	1.3750	2.49%
16.76	47.3699	1.3400	2.11%

RETURN LOSS AND POWER/IMPEDANCE RATIOS (*CONTINUED*)

Return Loss (dB)	P_i/P_r	Z_1/Z_2 or SWR	% Reflected Power
17.51	56.3508	1.3074	1.77%
18.30	67.5602	1.2770	1.48%
19.12	81.6633	1.2489	1.22%
19.98	99.5561	1.2228	1.00%
20.88	122.4563	1.1987	0.82%
21.82	152.0338	1.1765	0.66%
22.80	190.6020	1.1562	0.52%
23.83	241.3978	1.1376	0.41%
24.90	308.9984	1.1206	0.32%
26.02	399.9487	1.1053	0.25%
27.19	523.7144	1.0914	0.19%
28.41	694.1507	1.0789	0.14%
29.69	931.7924	1.0677	0.11%
31.03	1267.4725	1.0578	0.08%
32.43	1748.1181	1.0490	0.06%
33.88	2446.1689	1.0413	0.04%
35.41	3475.1083	1.0345	0.03%
37.00	5015.4732	1.0286	0.02%
38.67	7359.1211	1.0236	0.01%
40.41	10985.8362	1.0193	0.01%
42.23	16698.2365	1.0156	0.01%

return to zero (RZ) A code pattern in which each bit returns to the zero state any time during a reference cycle, i.e., in less than 1 bit time. The code is self-clocking, if bipolar signal elements are employed. See also *nonreturn to zero* (*NRZ*) and *waveform codes*.

REV An abbreviation of <u>REV</u>erse.

reversal In secondary cells, the interchange of the normal electrode polarity.

reverse address resolution protocol (RARP) A Transmission Control Protocol/Internet Protocol (TCP/IP) governing the translation of an Ethernet data link control (DCL) address to an Internet IP address.

The translation is required because most networks attached to the Internet do not use the Internet method of addressing nodes. The action of *RARP* makes the network seem as if it were homogeneous. (That is, it appears to be a single logical network to all users.) See also *address resolution protocol (ARP)*.

reverse battery signaling In telephony, a type of loop signaling in which the battery and return are reversed (on tip and ring) to signal an off-hook condition when the called station answers. Depending on the system, the condition may be momentary or sustained until the circuit is returned to idle.

reverse battery supervision A means of signaling the originating central office that the called station has gone off-hook. This information is used, among other things, to initiate toll billing.

reverse channel A technique for providing a means of simultaneous communication from the communication receiver to the communication transmitter on a two-wire transmission facility. The signal on the reverse channel is used for data flow-control and error-correction procedures. Generally, the reverse channel is the lower speed channel if the forward and reverse channel rates are different.

reverting call In telephony, a call between two stations on the same party line.

reverting call tone In telephony, a tone (generated in the central office switching equipment) to the calling station that indicates the call is a reverting call, that is, the call is to a party on the same line.

revertive pulse In telephony, ground pulses sent back to the originating office from the various selector frames to control the selection process.

revertive pulsing In telephony, a means of controlling distance selections by revertive pulses; that is, the near end receives pulse signals from the distant end.

REW An abbreviation of REWind.

RF (1) An abbreviation of Radio Frequency. (2) An abbreviation of ReFerence. (3) An abbreviation of Register File.

RFA (1) An abbreviation of Recurrent Fault Analysis. (2) An abbreviation of Remote File Access.

RFC (1) An abbreviation of Request For Comment. (2) An abbreviation of Radio Frequency Choke.

RFD (1) An abbreviation of Request For Discussion. (2) An abbreviation of Ready For Data.

RFH An abbreviation of Remote Frame Handler. Frame relay switch or network accessed over circuit switched links.

RFI (1) An abbreviation of Radio Frequency Interference. (2) An abbreviation of Request For Information.

RFN An abbreviation of Radio Frequency Noise.

RFP An abbreviation of Request For Proposal.

RFQ An abbreviation of Request For Quote.

RFS (1) An abbreviation of Ready For Service. (2) An abbreviation of Remote File System.

RG (1) An abbreviation of Release Guard. (2) An abbreviation of Ring Generator.

RG-nn An abbreviation of Radio Government cable nn specification, the World War II military specification/designation for coaxial cables. This has been superseded by the MIL C-17 specification. See also *coaxial cable.*

COMMON RG CABLES & NETWORKS

ID	Z	Typical Applications
RG-6A	75 ohm	CATV, broadband
RG-8	52 ohm	Thick Ethernet
RG-11	75 ohm	CATV trunk distribution
RG-58A	52 ohm	10Base2 Thin Ethernet
RG-58B	53.5 ohm	
RG-58C	50 ohm	
RG-59A,B	75 ohm	ARCnet, CATV in the house
RG-62	93 ohm	ARCnet

RGA An abbreviation of Return Goods Authorization.

RH (1) An abbreviation of Request/Response Header. (2) An abbreviation of Receive Hub.

rhombic antenna A directional antenna constructed of long wire radiators that form the sides of a rhombus. At one apex each of the sides is driven equally in amplitude but opposite in phase. If the apex opposite the driven apex is terminated, the antenna is unidirectional; if unterminated, it is bidirectional.

RHR An abbreviation of Radio Horizon Range.

RI (1) An abbreviation of Ring Indicator (EIA-232 signal CE, EIA-449 signal IC, and ITU-T signal 125). A signal generated by the data communications equipment (DCE) in response to an incoming ring signal. (2) An abbreviation of Routing Indicator.

ribbon cable A flat cable with individually insulated conductors and/or optical fibers that lie parallel and are bonded together. The conductors may be individual wires. twisted pairs, coaxial conductors, or a mix. Generally, the conductor to be attached to pin one is marked in some unique way (color, a painted stripe, etc.). See also *cable* and *wire.*

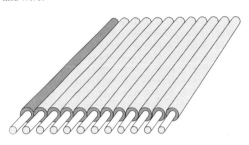

RICI An abbreviation of Remote Intercomputer Communications Interface.

RIFF (1) An acronym for Raster Image File Format. (2) An acronym for Resource Interchange File Format. A specification developed by Microsoft and others that allows audio, image, animation, and other multimedia elements to be stored in a platform-independent format.

right-hand polarization An elliptically or circularly polarized electromagnetic wave in which the electric field vector rotates clockwise as observed in the direction of propagation. Also called *clockwise polarization.*

right-hand rule A mnemonic for determining the direction of magnetic lines of flux around a conductor when a current is flowing in the conductor. That is, grasp the conductor with the right hand and with the thumb pointing in the direction of current flow—the fingers will point in the direction of magnetic flux. Also called *Fleming's rule.*

rights In network environments, values, or settings, assigned to an object, which determine what the object (such as other objects or a user) can do with files, directories, and other resources. See also *access rights.*

RIM (1) An acronym for Request Initialization Mode. (2) An acronym for Receiver InterModulation.

RINEX An acronym for Receiver INdependent EXchange format. A compressed format used for Global Positioning System (GPS) data transmissions. It contains information such as pseudorange, carrier-phase, and Doppler observations.

ring (1) One of the two wires in the telephone line from the central office (CO) to the subscriber. The *ring* wire is the more negative voltage of the two wires. The second wire is called the *tip.* (2) One of the terminations on the plug used in manual switchboards to connect the subscriber lines and interoffice trunks, for example,

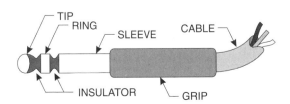

(3) The telephone signal generated by the local central office switching equipment that causes a subscriber's telephone instrument to generate an audible calling signal. Although country dependent, the *ring* signal is typically a 20-Hz sine wave with amplitude greater than 60 Vrms. Several specific *ring* signal characteristics are shown in the table. Three series of frequencies are used to ring a subscriber line: the decimonic series, the harmonic series, and the sychromonic se-

PRIMARY RING SIGNAL AT THE CENTRAL OFFICE GENERATOR

Country	Frequency (Hz)	Voltage (Vrms)	Max. REN (unitless)
Belgium	25	65-90	3
Canada	20		100 LN*
Denmark	50	65-90	3
Germany	25	60	4
Italy	25	60-100	
Mexico	25	90	4
Spain	25	75	5
UK	20 & 26	35-100	
USA	20	85-160	5

* Rather than using the total Ringer Equivalence Number (REN), the Canadian telephone administration uses a rating called a Load Number (LN). In either case, the sum total of all devices connected to the telephone line may not exceed the value given.

ries. Various frequencies are chosen from these pools for services such as party line. (If the ringer is frequency selective, it will ring only when the frequency of resonance is presented). See also *balanced ringing* and *unbalanced ringing*. **(4)** A particular network topology wherein nodes are connected in a closed loop and data packets are transmitted from node to node around the loop in only one direction. See also *ring network*.

Series	Frequencies (Hz)
Decimonic	20, 30, 40, 50, 60
Harmonic	16 ⅔, 25, 33 ⅓, 50, 66 ⅔
Synchromonic	16, 30, 42, 54, 66

ring back tone See *ringback tone.*

ring cycle See *ringing cycle.*

ring down See *ringdown circuit.*

ring generator A component of nearly all telephone systems that supplies the ringing signal power that is used to ring the called telephone instrument. The generated signal is typically 90 Vac at 20 or 25 Hz; however, other voltages and frequencies are used throughout the world. See also *ring (3).*

Ring In and Ring Out Token Ring connectors on the multistation access unit (MSAU) connecting the unit to trunk cabling. The Wrap feature is implemented at these interfaces.

ring isolator A noise-reduction device placed on a telephone line to disconnect the ringer when there is no ringing voltage present.

ring latency (RL) In local area networks (LANs), the total time required for a token to circulate once around a LAN whose network architecture is a ring. *Ring latency* time includes not only the delay of the network medium but all connected devices through which the token must pass. See also *ring network.*

ring modulator A type of amplitude modulator in which the carrier is used to rapidly switch the information-bearing signal between a normal polarity path and an inverted polarity path.

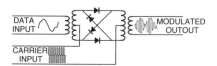

ring network A physical implementation (topology) of a local area network wherein each node or computing station is connected in series in a closed loop. (That is, the output port of each node is connected to the input port of the next node, and the output of the last node is connected to the input of the first node.) The ring may transfer data unidirectionaly or bidirectionaly. Although any of several protocols may be used to transport information (CSMA/CD, token passing, etc.), token passing is generally the preferred protocol.

Each node, in turn, intercepts messages traveling on the ring and examines the destination address of the message. If the message is for that node, the message is received and passed on to the data terminal equipment (DTE). If the message is not for that node, it is regenerated and retransmitted to the next physical node. Response time is determined by the number of nodes on the ring. (The more there are, the slower the response). If one node fails, the loop is broken. Most rings, however, have a self-healing capacity to reconfigure and continue operation. For example, IBM's Token passing ring ensures that the failed station is removed and its neighbors are then directly connected. The diagram of a ring network illustrates that the nodes need not be physically in the same sequence as the logical sequence. A message from node B to node A must pass through nodes C, D, E, F, G, H, I, J, and N1 before it arrives at node A. The response from A need only travel from A to B. Also called *ring topology*. See also *bus network, gateway, hierarchical network, hybrid network, mesh network,* and *star network.*

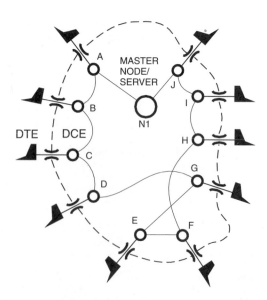

ring trip The process of stopping the ringing signal at the central office from being delivered to the subscriber's line when the subscriber goes off-hook.

ringaround A fault condition in telephony. The improper routing of a call back through a switching center already engaged in attempting to complete the same call.

ringback signal **(1)** In telephony, a signal initiated by the operator at the called end of an established connection to recall the originating operator. **(2)** In telephony, a signal, usually consisting of an audio tone interrupted at a slow rate (generally once or twice per second), provided to a caller which indicates that the called party instrument is receiving the ringing signal. See also *call progress tones, ring (3),* and *ringback tone.* **(3)** A ringing signal returned to a caller to indicate that one of the types of delayed automatic calling (e.g., camp-on) is now ringing the called party.

ringback tone In telephony, a *call progress tone* that indicates to the calling party that a connection to the remote location is complete and is being sent a ringing signal.

Ringback tone may be generated by the called-party servicing switch or by the calling-party servicing switch, but it is not generated by the called telephone instrument. The *ringback* tone frequency is generally the sum of several tones in the frequency range of 340 to 650 Hz and usually the same as the dial tone frequencies. *Ringback* tone cadence, both the period and the on to off ratio, is country dependent but is usually the same as the ring cadence. Also called *audible tone* or *audible ringing tone.* See also *call progress tones* and *ring (3).*

ringdown In telephony, a method of signaling an operator in which telephone ringing current is sent over the line to operate a lamp and cause the drop of a latching relay.

The term *ringdown* originated in magneto telephone signaling in which cranking the magneto in a telephone set would not only *ring* its bell but also cause a lever to fall *down* at the central office switchboard. *Ringdown* attributes include:

- Use in manually operated circuits, as distinguished from dialing,
- Use of continuous or pulsed ac signal transmitted over the line, and
- Use with or without a switchboard.

ringdown circuit **(1)** A communications path between two stations wherein signaling commences when the *calling station* goes off-hook and terminates when the *called station* goes off-hook or the called station returns on-hook. **(2)** A method of signaling subscribers and operators using a 20-Hz or 135-Hz continuous ac signal or a 100-Hz signal interrupted 20 times per second. **(3)** A circuit in which manually generated signaling power is used to perform ringdown.

ringdown interface A private line two-wire interface. Also called a *loop start interface.*

ringdown signaling In telephony, the application of a signal to a line to:

- Operate a line signal lamp or a supervisory signal lamp at a switchboard or
- To ring a called receiver instrument.

ringer In telephony, any device used to convert the electrical ringing signal from a central office into an audible signal indicating an incoming call. The ringer need not be a bell in the strictest sense.

ringer equivalence number (REN) A number specified in the Code of Federal Regulations (CFR), Title 47, part 68, which represents the ringer loading effect on a line. One *REN* represents the load presented to the telephone line by one ringer in a Western Electric type 500 telephone instrument. For testing purposes, the *REN* is defined in the CFR at 20 Hz as 7 kΩ in parallel with 8 μF. Per Federal Communications Commission (FCC) regulations, the maximum total *REN* allowed in the United States is usually five.

Modern telephone devices may have a *REN* less than 1; hence, the actual number of devices across the loop may be greater than 5. All devices attached to a telephone line must have a *REN* rating so that this total may be computed. A related measure of line loading, used in Canada, is the *load number (LN).* Here the maximum number of LNs allowed on a line is 100. See also *ring (3).*

ringing **(1)** In telephony, the production of an audible and/or visual signal at a called station to indicate the presence of an incoming call. **(2)** A pulse waveform distortion in the form of a superimposed damped sinusoidal oscillation. The oscillations, when present, follow the step transitions of the primary waveform. See also *pulse.*

ringing cycle In telephony, the period of the recurring sequence of ringing signals and the silent intervals between them. See also *ring (3).*

ringing signal Any ac or dc signal transmitted over a line or trunk for the purpose of informing the person or equipment at the called station of an incoming call.

ringing tone See *ringback tone.*

ringing voltage The 60–160 volt, 16–50 Hz ac signal generated by telephone switching equipment for use as the *ring signal.* See also *ring (3).*

RingMaster See *distinctive ringing.*

Ringmate See *distinctive ringing.*

RIP **(1)** An acronym for Remote Imaging Protocol. A graphics protocol used on some bulletin board systems (BBSs) to provide a graphical user interface (GUI) rather than a textual interface to the user. **(2)** An acronym for Raster Image Processor. **(3)** An acronym for Retired In Place. **(4)** An acronym for Routing Information Protocol.

rip-cord A string or cord placed immediately beneath the jacket of a cable for the purpose of aiding in the removal of the jacket. The *rip cord* is first exposed by carefully removing or severing a portion of the jacket near the end of the cable. It is then grasped, usually with a pair of pliers, and pulled to split the jacket for the remainder of the desired distance.

RIPE An acronym for Réseaux IP Européens. A group formed to coordinate Transmission Control Protocol/Internet Protocol (TCP/IP) networks in Europe. It, like the Internet Engineering Task Force (IETF), holds periodic conferences to coordinate technical issues and runs a Network Control Center (NCC) to handle operational issues (such as the administration of domain names and routing tables).

RIPEM An acronym for Riordan's Internet Privacy Enhanced Mail. An implementation of Privacy Enhanced Mail (PEM). *RIPEM* allows electronic mail to have the four security facilities provided by PEM, that is,

- Disclosure protection (optional),
- Originator authenticity,
- Message integrity measures, and
- Nonrepudiation of origin (always).

RIPEM was written primarily by Mark Riordan. Most of the code is in the public domain, except for the RSA routines, which are a library called RSAREF licensed from RSA Data Security Inc.

ripple Generally, a periodic alternating current (ac) superimposed on a direct current (dc) supply. *Ripple* is expressed as a peak, a peak-to-peak, a root-mean-square (RMS) or a percent RMS (the ratio of the RMS component to the total value). *Ripple* may be expressed in terms of either voltage or current.

ripple factor The ratio of the ripple magnitude to the arithmetic mean value.

ripple filter A low-pass filter designed to pass the dc component and attenuate the ac components.

ripple voltage In a dc voltage, an alternating component from one of two sources, that is,

- The residual from the rectification of the mains ac power or from generation and commutation (such as with a dc-dc converter).
- The component that is coupled into the circuit from a source of interference.

RIS (1) An acronym for Recorded Information Service. (2) An acronym for Retransmission Identity Signal.

RISC An acronym for Reduced Instruction Set Computing. A microprocessor design strategy with a relatively simple instruction set with the following characteristics:

- The processor executes at least one instruction per clock cycle. (Some execute five or more per clock cycle).
- Each instruction is processed directly by hardware, that is, there is no microcode program executed to implement any instruction in the user instruction set.
- The number of instructions referencing external random access memory (RAM) are minimized, while the register referencing instructions are maximized. (Accessing external RAM is much slower than accessing internal registers). Generally, the number of internal registers is huge compared to *complex instruction set* (*CISC*) architectures.

rise time (1) Typically, the time interval on the leading edge of a pulse waveform between the first occurrence of the 10% of final value and the 90% of final value instances. See also *pulse*. (2) In the Global Positioning System (GPS), the time at which a satellite is above a predetermined elevation angle. (It is said to be visible above the angle). The time when the satellite descends below the mask angle is called the *set time*.

riser The vertical conduit (path) between the floors of a building. It is the duct into which telephone, local area network (LAN), power, and other utility cables are routed to bring services between and to individual floors.

riser cable A cable that runs vertically between floors in a building, often through shafts (such as for the elevator). In many cases, the *riser cable* runs through areas of high electrical interference. Consequently, optical fiber (which is impervious to electrical interference) is frequently used for *riser cable*.

riser closet The location where riser cables are terminated and cross connected to either other riser cables or the horizontal distribution cables.

RIT An abbreviation of Rate of Information Throughput.

RJ-nnC or RJ-nnW A series of standardized connectors for communication adopted by the Federal Communications Commission (FCC) and listed in the Code of Federal Regulations (CFR) Part 68 Subpart F. *RJ* is an abbreviation of Registered Jack, and *nn* identifies a specific configuration.

Various configurations are available and are distinguished by the following:

- Number of wire pairs used: 2, 3, 4, 8, or 25.
- Which wire pairs are used (known as the wiring sequence).
- Keying or other modifications to the plug and jack, designed to make correct connections easier and incorrect connections less likely.

The series provides an access arrangement for most telephonic connection requirements including single line subscriber loops, key system loops, standard "quad cable" termination, 50-pair termination, weather-proof connectors, and various adapters. The RJ-nnC is the surface or flush-mounted version, while the RJ-nnW is the wall-mounted version. The most common connector is the RJ-11C—a 6-position single-pair connector normally associated with single-line devices such as telephones and modems. Also called a *modular connector*.

Connector Type	Description
RJ-1CX	8 position, single tie trunk, E&M interface
RJ-1DC	6 position, single 4 wire line
RJ-2MB	A 50 position 12 pair bridging connector with make busy provision. The typical wiring assignment is:

	T	R	MB	MB1
Line 1	26	1	27	2
Line 2	28	3	29	4
:	:	:	:	:

Connector Type	Description
RJ-4MB	8 position 1 pair bridging with make busy and programmed loss
RJ-11C,W	6 position 1 bridging pair modular connector used for telephone line to equipment termination. The two central positions (pair 4/3) terminate the green and red wires (tip and ring lines, respectively).
RJ-14C,W	6 position 2 pair bridging connector. Terminals 4/3 are bridged on tip/ring of pair one and 2/5 are bridged on tip/ring of pair two.
RJ-14X	Same as RJ-14C,W but with sliding cover.
RJ-15C	3 position 1 bridging pair weatherproof connector.
RJ-17C	6 position 1 bridging pair used in critical care areas.
RJ-18C	6 position 1 bridging pair with a make busy arrangement. Terminals 4/3 are tip/ring pair while terminals 1/6 are used with the make busy circuit.
RJ-21X	50 position 25 pair bridging connector. Typically line one tip/ring is on pair 26/1, line two on pair 25/2, etc.
RJ-22	4 position 2 pair. Used between handset and instrument.
RJ-25C	6 position 3 pair bridging.
RJ-26X	A 50 position 8 bridging pair connector with fixed and programmable loss provision. Typical wiring is:

	Fixed		Pgm'd		Resistor	
	T	R	T	R	PR	PC
Line 1	26	1	27	2	28	3
Line 2	29	4	30	5	31	6
:	:	:	:	:	:	:

Connector Type	Description
RJ-27X	A 50 position 8 bridging pair connector with programmable loss provision. Typical wiring is:

			Pgm'd		Resistor	
	N/C	N/C	T	R	PR	PC
Line 1	26	1	27	2	28	3
Line 2	29	4	30	5	31	6
:	:	:	:	:	:	:

Connector Type	Description
RJ-31M	Up to 8 RJ-31X connectors as an assembly.
RJ-31X	An 8 position 1 pair series connection. Pair 5/4 is the tip/ring in while pair 8/1 is tip/ring out.
RJ-38X	8 position 1 pair series connection with continuity circuit. Pair 5/4 is the tip/ring in while pair 8/1 is tip/ring out and terminals 7/2 provide the continuity loop.
RJ-41M	Up to 8 RJ-41S connectors as an assembly.
RJ-41S	8 position 1 bridged pair keyed connector with fixed and programmable loss provisions. Terminals 5/4 bridged tip/ring pair, 2/1 fixed loss tip/ring pair, and 7/8 programming resistor.
RJ-45M	Up to 8 RJ-45S connectors as an assembly.

Connector Type	Description
RJ-45S	8 position 1 bridge pair keyed connector with programmable loss provisions. Pair 5/4 is the tip/ring and terminals 7/8 are the programming resistor. Used by IEEE 802.3 Ethernet 10Base-T and StarLAN networks.
RJ-48C	8 position 2 pair keyed connector. 1.544 Mbps T-1
RJ-61X	8 position 4 pair
ISO 8877	A variant of the RJ-45 which is compatible with international standards.

RJE An abbreviation of Remote Job Entry.

RJO An abbreviation of Remote Job Output.

RJP An abbreviation of Remote Job Processing.

RJS An abbreviation of Remote Job System.

RJT (1) An abbreviation of REjecT. (2) An abbreviation of ReJecT message.

RL (1) An abbreviation of Real Life. (2) An abbreviation of Random Logic. (3) An abbreviation of Record Length. (4) An abbreviation of Reflection Loss. (5) An abbreviation of Random Logic. (6) An abbreviation of Ring Latency.

RLA (1) An abbreviation of Remote Line Adapter. (2) An abbreviation of Remote Loop Adapter.

RLC (1) An abbreviation of resistor (R), inductor (L), and capacitor (C). (2) An abbreviation of Run Length Coding. (3) An abbreviation of ReLease Complete. See also *RELC*.

RLCM An abbreviation of Remote Line Concentrating Module.

RLCU An abbreviation of Reference Link Control Unit.

RLE An abbreviation of Run Length Encoding, one of several lossless data compression methods.

RLE basically takes a string of repeating symbols and encodes it into two parts; the first part is the symbol, and the second part is the number of times the symbol is to be repeated. The method works well on data with long strings of repeating symbols. For example, a fax file might have a line with 100 consecutive all-white symbols, and an *RLE* encoding method would reduce this 100-byte string to 1 byte indicating "white" and another byte (or two) indicating the number of times to repeat the preceding symbol. (One byte could indicate up to 255, and two bytes could indicate 65535 repeats). The *RLE* encoding method is a good compression algorithm for line artwork but a poor choice for data with frequent changes, such as gray scale images or text files. See also *compression, encoding,* and *Lempel-Ziv*.

RLG An abbreviation of ReLease Guard.

RLL An abbreviation of Run Length Limited.

RLL encoding An abbreviation of Run Length Limited encoding.

RLM An abbreviation of Remote Line Module.

rlogin A remote login service provided as part of the Berkeley 4.2BSD UNIX environment. It allows users of one machine to connect to other UNIX systems across an internet and to interact as if their terminals were connected to the host directly. It is an application layer service that provides a terminal interface between UNIX hosts using the Transmission Control Protocol/Internet Protocol (TCP/IP) network protocol. It is comparable to the Internet's Telnet service. However, many feel it is superior because the software passes information about the user's environment to the remote machine. It requires the remote host to be (or behave like) a UNIX machine.

RLR An abbreviation of Receive Loudness Rating.

RLSD An abbreviation of Received Line Signal Detector. Also called *DCD*.

RLT An abbreviation of Release Link Trunk.

RLU An abbreviation of Remote Line Unit.

rm The UNIX command to delete (remove) a file.

RM An abbreviation of Reference Model.

RMA An abbreviation of Return Material Authorization.

RMATS (1) An acronym for Remote Maintenance And Testing System. (2) An acronym for Remote Maintenance Administration and Traffic System.

RMDM An abbreviation of Reference Model of Data Management.

RMN An abbreviation of Remote Multiplexing Node.

RMON An acronym for Remote MONitoring. A standard Management Information Base (MIB) defined in RFC 1271 to allow remote monitoring of networked devices. It is a capability added to remote electronic hardware such as network hubs and switching equipment that allows a technician to perform real-time diagnostics on the hardware or analysis of other management data from a location away from the equipment under test.

RMS An abbreviation of Root Mean Square. *RMS* is a way of expressing the value of a signal in such a way that it is *independent of the waveform*. For example, a 1 Vrms sine wave will have the same heating effect on a resistance as a 1 Vdc signal.

RMS is sometimes called the effective value. Mathematically, *RMS* is defined as the square root of the mean value (average) of the square of the signal over time. For periodic functions, such as sine waves, the time period is 1 cycle; for a pulse the time is the pulse duration. The equation describing the *RMS* value (Y_{RMS}) of an arbitrary waveform sampled at times $\frac{1}{n}T, \frac{2}{n}T, \cdots, T$ is approximately:

$$Y_{RMS} = \sqrt{\frac{Y_{\frac{1}{n}T}^2 + Y_{\frac{2}{n}T}^2 + Y_{\frac{3}{n}T}^2 + \cdots + Y_T^2}{}}$$

rn One of the most common newsreaders in the world, it is available on all UNIX systems with a newsfeed. Originally written by Larry Wall in the mid-1980s, *rn* is an *unthreaded newreader* and therefore has a serious limitation on today's USENet postings. That is, the presentation cannot be organized by threads. (*rn* presents in date-time order only). See also *newsreader*.

RN (1) An abbreviation of Redirecting Number. The directory number of the party that forwarded a call via the Integrated Services Digital Network (ISDN). (2) An abbreviation of Receipt Notification. (3) An abbreviation of Reference Noise.

RNR An abbreviation of Receiver Not Ready.

RO (1) An abbreviation of Receive Only. (2) An abbreviation of Repair Only. (3) An abbreviation of Remote Operations. (4) An abbreviation of Ring Out.

RO terminal An abbreviation of Receive Only terminal. A terminal that can receive data but cannot generate or send data. Most printers are *RO terminals*.

roamer In telephony, a cellular telephone user who uses services in multiple calling areas.

roaming The ability of a mobile communications device to move freely from one part of a network operator's system to another.

robbed bit signaling The periodic use of the least significant bit in an information-bearing signal on T-1 circuits as a control signal. The effect of bit stealing a fraction of the time is a slight increase in noise.

robot A program used on the Internet World Wide Web (WWW) that automatically explores the Web by retrieving a document and recursively retrieving some or all of the documents that are referenced in it. (This is in contrast with normal Web browsers that are operated by a human and don't automatically follow links other than inline images and redirections.)

The algorithm used by the *robot* to pick which references to follow depends on the purpose of the *robot*. For example:

- Index building *robots* usually retrieve a significant part of the references.
- The other extreme are robots that try to validate the references within a document. (These usually do not retrieve any of the links apart from redirections).

Also called a *crawler* or *spider*.

robust Able to continue functioning despite some communications channel or equipment degradation or failure. See also *graceful degradation*.

ROCC An acronym for Regional Operations Control Center.

ROFL An abbreviation of Rolling On the Floor Laughing.

ROH An abbreviation of Receiver Off Hook.

ROI An abbreviation of Return On Investment.

roll-call polling A sampling technique in which every station is interrogated sequentially for service requests.

ROLR An abbreviation of Receiving Objective Loudness Rating.

ROM An acronym for Read Only Memory (pronunciation rhymes with "bomb"). A type of semiconductor memory device in which stored information is permanent, that is, it cannot be changed after manufacture. The information in a *ROM* is entered during the manufacturing of the basic device and cannot be entered by a user. See also *EEPROM, EPROM, PROM,* and *UVPROM*.

roofing filter A low-pass filter.

room A term describing a program on a bulletin board system (BBS) which allows multiple users to carry on a real-time "conversation." See also *door*.

room cutoff: A hotel/motel private branch exchange (PBX) feature that makes phones in unoccupied rooms receive only.

room noise level A synonym for *ambient noise level*.

root In computer science, the highest level of a hierarchy.

root directory In a hierarchical file system, the highest level directory. All other directories are called subdirectories of the root directory.

root mean square See *RMS*.

root object In the Novell NetWare Directory Service (NDS) global tree, the highest-level object. All country and organization objects are contained in the root object. Granting a user access rights to the root object effectively grants the user rights to the entire Directory tree.

ROSE An acronym for Remote Operation Service Element protocol.

rot13 An algorithm from rotate 13 characters. A method of encrypting a potentially offensive message on USENet so that it cannot be accidentally read. The algorithm basically rotates each letter of the alphabet 13 characters forward (or backward), that is, "a" becomes "n," "b" becomes "o," etc.

A major advantage of rot13 over rot(N) for other N is that it is self-inverse, so the same code can be used for encoding and decoding. That is, a shift of 13 characters twice is equivalent to a shift of 26 characters or a shift of 0 characters.

rotary dialing The same as *pulse dialing*.

rotary hunt See *hunt group*.

rotary switching In telephony, an electromechanical switching method whereby the selecting mechanism consists of a rotating element using several groups of wipers, brushes, and contacts, e.g., a Strowger switch.

ROTF An abbreviation of Rolling On The Floor.

ROTFL An abbreviation of Rolling On The Floor Laughing.

ROTL An abbreviation of Remote Office Test Line.

ROTM An abbreviation of Right On The Money.

round off To discard the least significant digits of a value after applying some correction to the part of the value retained. One of the most common rules increases the magnitude of the value retained by one least significant figure if the most significant digit of the discard is 5, 6, 7, 8, or 9. For example,

123	rounded to two significant places is	120
127	rounded to two significant places is	130
1.236	rounded to three significant places is	1.24

See also *truncate*.

round robin A method of allocating common resources in a sequential, cyclical manner to a number of calling processes or devices. A specific case of polling where the devices are polled *sequentially*.

round trip delay time In communications, the elapsed time for a signal to travel across a closed circuit.

Round trip delay time is significant in systems where the round trip time directly affects the throughput rate. For example, voice telephony (excessive delay makes it difficult for people to communicate), or ACK/NAK (acknowledge/nonacknowledge) flow-control data systems where the round trip time directly affects the throughput rate. *Round trip delay time* may range from a few microseconds for a short line-of-sight (LOS) radio system to many seconds for a multiple-link circuit with one or more satellite links involved. *Round trip delay time* includes all equipment in the loop (such as the node delays) as well as the media transit time.

rounding error (1) In general, the error resulting from the discarding of less significant digits of a value after applying some rule of correction. The most common rule is *round off*. (2) In sampled systems, an error that occurs because of the finite digital word length. For example, given the analog value is exactly one-third, then the three-place digital sampled value is 0.333 and the rounding error is 0.000333. . . .

route **(1)** As a noun, the physical path defined by the selection of a particular set of switches (within one or more central office switching systems or *nodes* on a data network) through which communication may take place. Also called a *path.* **(2)** As a verb, to determine the path that a message or call is to take in a communications network.

In a Transmission Control Protocol/Internet Protocol (TCP/IP) internet, each IP datagram is routed separately. The route a datagram follows may include many gateways and many physical networks. **(3)** As a verb, to construct the path that a call or message is to take in a communications network in going from one station to another or from a source user end instrument to a destination user end instrument.

route caching A strategy to speed the transfer of multiple packets to a destination. The forwarding information obtained during the processing of the first packet is temporarily stored and used on subsequent packets to the same destination.

route discovery In network architectures that use source routing, such as token ring networks, the process of determining possible routes from a source to a destination node.

route diversity The allocation of circuits between two points over more than one geographic or physical route with no geographic points in common. Also called *space diversity.*

router A networking device that:

- Forwards data from one network to another and
- Decides which of several alternative paths a data packet will take across a network.

It is a device that interconnects networks at the OSI/ISO layers 3, as shown in the figure, using protocols such as Transmission Control Protocol/Internet Protocol (TCP/IP), DecNet, Xerox Network Services (XNS), System Network Architecture (SNA), Open Systems Interconnect Internet Protocol (OSI IP), or Internetwork Packet Exchange (IPX). A *router* can manage traffic congestion through its use of flow control and its ability to redirect traffic to alternative, less congested paths. Traffic is assigned on the most efficient path from a sending station to a receiving station on a packet-by-packet basis. A router can be used to link local area networks (LANs) together locally or remotely as part of a wide area network (WAN). A network built using *routers* is often termed an *internetwork.* For simple network configurations, a fixed or static routing table may suffice; how-

ever, in more complex configurations each *router* maintains a dynamic routing table. The dynamic table is updated through the exchange of *address resolution packets* between *routers.* The algorithm used by many routers to pick the shortest path is the *routing information protocol* (*RIP*) algorithm. Another algorithm is the *open shortest path first* (*OSPF*) algorithm. As network complexity increases, other more complex algorithms are required. *Routers* have been called *interface message processors* or *internet message processors* (*IMP*). See also *bridge, gateway, OSPF,* and *RIP.*

routine A sequence of steps or instructions that has a general or frequent use. For example,

- A software program that performs a specific task.
- A program that is called by another program.
- Testing carried out by a manufacturer during the manufacturing process.

routing The process of determining which of several alternative routers to use in sending a data packet toward its destination. The packet may travel through several networks, each with its own routers, before reaching the final destination.

If a host has only one gateway into the Internet, there is no routing decisions. However, with a multihomed host, routing decisions are necessary. These decisions are made by a routing program executing one of several *routing algorithms,* such as *open shortest path first* (*OSPF*) or *routing information protocol* (*RIP*). The algorithms in general attempt to locate the shortest available path.

routing algorithm The procedure used by a program (running in a router) which decides how to direct a packet to the final destination. The program dynamically determines which of several available paths to send the packet. *Distance vector routing algorithms* are most frequently used to determine the best path. Examples include the *open shortest path first* (*OSPF*) and the *routing information protocol* (*RIP*).

routing code A digit or combination of digits used to direct a call towards its destination.

In packet networks the *routing code* is called a *destination address* or *routing indicator.*

routing domain Any collection of networks under the control of a single administrative organization. Within this collection, the administrative organization has the right to determine network topology, subnets, and domain names and addresses without direction of any other organization. Called an *autonomous system* (*AS*) in the Internet.

routing flexibility The ability of a switching network to send information over otherwise underutilized paths to avoid congested links. Also called *load leveling.*

routing indicator (RI) **(1)** A bit in a local area network (LAN) packet header to distinguish transparent routed packets from source-routed packets. **(2)** In a message header, an address, that is, a group of characters, that specify routing instructions for the transmission of the message to its final destination. Routing indicators may also include addresses of intermediate points. Also called a *destination address.* **(3)** A group of letters assigned to indicate:

- The geographic location of a station;
- A fixed headquarters of a command, activity, or unit at a geographic location; and
- The general location of a tape relay or tributary station to facilitate the routing of traffic over the tape relay networks.

LAN 1 STATION X / LAN 2 STATION Y / ROUTER — OSI layer 7 through OSI layer 1, LAN 1, LAN 2

routing information protocol (RIP) A router algorithm that computes the distance between a router and a destination node in "number of hop" units but ignores other attributes such as the data rate of each link and the physical length of each link. Moreover, the *RIP* algorithm requires each router to send a copy of its routing table to adjacent routers every 30 seconds—a significant traffic burden on some networks.

In an attempt to relieve the burden caused by the frequent transfer of routing tables, two alternate updating schemes are used. The first simply uses less frequent updates. This unfortunately has the undesirable side effect of slowing the response to congestion or to link failures in the network. A second method sends copies of the routing table only when there are link *changes* to be reported. Originally developed at Xerox and later adapted for Berkeley's 4BSD UNIX system. See also *Open Shortest Path First (OSPF)*.

routing label That part of a signaling message that identifies its destination.

routing metric The method by which a routing algorithm determines that one path is better than another. The result is stored in the *routing table*. A few possible *metrics* include:

- Reliability,
- Delay,
- Bandwidth,
- Current traffic load,
- Communications cost, and
- Hop count.

routing pattern The implementation of that part of a routing plan that applies to a specific switching office.

routing plan In telephony, a strategy for directing calls through a group of central office switching machines.

routing protocol The routing rules that govern the process of deciding how to best send a data packet toward its destination. The *routing protocol* defines a *routing algorithm* and a means for building a *routing table*.

Within the Internet, there are two levels of routing protocols.

- Inside an autonomous network, routing is determined by an *Interior Gateway Protocol (IGP)*. Some of the protocols are proprietary, that is, they are unique to a specific router manufacturer. More popularly, open protocols such as the *Routing Information Protocol (RIP)* or the *Open Shortest Path First (OSPF)* are used.
- To interconnect these autonomous networks (creating the Internet), a protocol called the *Exterior Gateway Protocol (EGP)* or the newer *Border Gateway Protocol (BGP)* is used.

See also *Border Gateway Protocol (BGP)*, *Exterior Gateway Protocol (EGP)*, *Interior Gateway Protocol (IGP)*, *Open Shortest Path First (OSPF)*, *routing algorithm*, and *Routing Information Protocol (RIP)*.

routing table **(1)** A map of possible internetwork paths contained in each router which is used by the routing program's algorithm in determining the best path to send a data packet. The *routing table* only lists the possible paths; it does not address congestion. That is the job of the *routing protocol*.

The table contains paths and distances between routers on the internetwork. Distances are generally measured in hops, and they may change. As a result, routing tables may be updated frequently. See also *routing algorithm* and *routing protocol*. **(2)** In a private automated branch exchange (PABX) a list of preferred outgoing circuits associated with destination numbers (area codes, country codes, or office codes).

RPC An abbreviation of <u>R</u>emote <u>P</u>rocedure <u>C</u>all.

RPG An abbreviation of <u>R</u>emote <u>P</u>assword <u>G</u>enerator.

RPL An abbreviation of <u>R</u>emote <u>P</u>rogram <u>L</u>oad.

RPM An abbreviation of <u>R</u>evolutions <u>P</u>er <u>M</u>inute. Also frequently abbreviated *rpm*.

RPOA An abbreviation of <u>R</u>ecognized <u>P</u>rivate <u>O</u>perating <u>A</u>gency. An ITU-T term for nongovernment-run Past Telephone and Telegraph (PTT) administrations.

RQ An abbreviation of <u>R</u>epeat-re<u>Q</u>uest. See also *ARQ*.

RQC An abbreviation of <u>R</u>epair and <u>Q</u>uick <u>C</u>lean.

RR **(1)** An abbreviation of <u>R</u>eceive <u>R</u>eady. A supervisory logical link control (LLC) frame that acknowledges the receipt of data and indicates the ability to accept the next frame. **(2)** An abbreviation of <u>R</u>epetition <u>R</u>ate.

RS **(1)** An abbreviation of <u>R</u>ecommended <u>S</u>tandard; standards from the Electronic Industries Association (EIA). **(2)** An abbreviation of <u>R</u>ecord <u>S</u>eparator.

RS-232 An ANSI/EIA specification (standard) describing the physical, electrical, and control characteristics of a serial communication interface, specifically between a data terminal equipment (DTE) and data communications equipment (DCE). See also *EIA-232*.

RS-366 See *EIA-366*.

RS-422 See *EIA-422*.

RS-423 See *EIA-423*.

RS-449 See *EIA-449*.

RS-485 See *EIA 485*.

RS-530 See *EIA-530*.

RS-562 See *EIA-562*.

RSA **(1)** A patented public key encryption algorithm named after its authors Ron <u>R</u>ivesi, Adi <u>S</u>hamir, and Leonard <u>A</u>dleman of the Massachusetts Institute of Technology. **(2)** An abbreviation of <u>R</u>ural <u>S</u>ervice <u>A</u>rea.

RSA encryption A public key cryptosystem for both encryption and authentication. It was invented in 1977 by Ron Rivest, Adi Shamir, and Leonard Adleman of the Massachusetts Institute of Technology.

The scheme is based on the assumption that factoring large numbers is very difficult and time consuming. The basic method is:

- Select two large prime numbers, p and q, keep their values secret, or discard after calculating the public and private keys.
- Calculate the modulus n, that is, $n = p \cdot q$.
- Choose a number e, less than n and relatively prime to $(p - 1)(q - 1)$.
- Find its inverse d, mod $(p - 1)(q - 1)$. [This implies $e \cdot d = 1$, mod $(p - 1)(q - 1)$.]
- Publish n, e as the *public key*.
- Retain d as the *private key*.

RSET An abbreviation of <u>RESeT</u>. A supervisory frame to zero counters.

RSL **(1)** An abbreviation of <u>R</u>equest and <u>S</u>tatus <u>L</u>ink. **(2)** An abbreviation of <u>R</u>eceived <u>S</u>ignal <u>L</u>evel.

RSN An abbreviation of <u>R</u>eal <u>S</u>oon <u>N</u>ow; however, it is intended to be somewhat sarcastic and really means, "maybe never!"

RSP An abbreviation of ReSPonse.

RSU An abbreviation of Remote Service Unit.

RSVP An abbreviation of ReSource reserVation Protocol.

RTA An abbreviation of Remote Trunk Arrangement.

RTCM An abbreviation of Radio Technical Commission for Maritime Services. The commission that recommends standards for differential Global Positioning System (GPS) services.

For example, the *RTCM SC-104* is the *RTCM* special committee no. 104 specification that defines a communications protocol for sending differential corrections from a differential reference station to remote GPS stations.

RTF An abbreviation of Rich Text Format. A declarative *markup language* used to describe a documents format (such as a specific font and size).

RTFM An abbreviation of Read The Fascinating Manual (well, its been cleaned up from the usual definition). A frequently used "flame" on many networks in response to a "stupid" question or question answered in the FAQ (frequently asked question) files.

RTK An abbreviation of Real Time Kinematics.

RTOS An acronym for Real Time Operating System.

RTR An abbreviation of RARE Technical Report.

RTS (1) An abbreviation of Request To Send (EIA-232 signal CA, EIA-449 signal RS, and ITU-T signal 105). The signal is generated by the data terminal equipment (DTE) and used by the data communications equipment (DCE) to connect the transmit data stream to the higher speed communication channel or in half-duplex channels to control the direction of transmission. (2) An abbreviation of Residual Time Stamp. Asynchronous transfer mode (ATM) control information used to support constant bit rate (CBR) services.

RTSE An abbreviation of Reliable Transfer Service Element.

RTTY An abbreviation of Radio TeleTYpewriter.

RTU An abbreviation of Remote Terminal Unit.

RU (1) An abbreviation of Request/response Unit. In IBM's Systems Network Architecture (SNA), an unframed block of used data (up to 256 bytes). (2) An abbreviation of aRe yoU.

rubidium clock A clock containing a quartz oscillator stabilized by a rubidium standard. See also *cesium clock.*

rubidium standard A frequency standard in which a specified hyperfine transition of electrons in rubidium-87 atoms is used to control an oscillator output frequency.

A rubidium standard consists of a gas cell, which has an inherent long-term instability. This instability relegates the rubidium standard to its status as a secondary standard. See also *cesium standard.*

RUMORF An acronym from aRe yoU Male OR Female?

run (1) An electrical or optical circuit connection between two points. (2) A synonym for *install* when dealing with cable installation, for example, one can *run* the cable. (3) The completion of a single execution of a computer program, routine, or job. (A job may require the execution of one or more programs.)

run length limited (RLL) encoding A method of encoding data so that the length of the data stream is reduced without loss of information. It uses codes based on the runs of 0 and 1 values rather than on the individual bit values. See also *compression* (2).

run-length parity See *longitudinal redundancy check.*

runt packet A packet with too few bits.

RUR An abbreviation of Repair, Update, and Refurbish.

rural radio service A public radio service, described in the Code of Federal Regulations, Title 47, performed by fixed stations on frequencies below 1000 MHz used to provide:

- Basic Exchange Telecommunications Radio Service, which is public message communication service between a central office and subscribers located in rural areas,
- Public message communication service between landline central offices and different exchange areas which it is impracticable to interconnect by any other means, or
- Private line telephone, telegraph, or facsimile service between two or more points to which it is impracticable to extend service via landline.

RWI An abbreviation of Radio and Wire Integration.

RX A symbol for receive or receiver.

RXD An abbreviation of Receive Data (RS-232 signal BB, RS-449 signal RD, and ITU-T signal 104). The data recovered from the carrier by the local data communications equipment (DCE) and sent to the local data terminal equipment (DTE).

RZ An abbreviation of Return to Zero.

RZI An abbreviation of Return to Zero Inverted.

S

s The SI symbol for seconds. Used for both time and angle measurement.

S (1) The SI symbol for the units of *conductance* measured in *siemens*. **(2)** In telephony, the designation of the sleeve or control lead. See also *ring, sleeve,* and *tip.*

S band 2 to 4 GHz. See also *frequency bands.*

S interface point The reference point in an Integrated Services Digital Network (ISDN) customer premises terminal equipment set between either a TE1 device and the NT2 (network terminator 2) or between a TA (terminal adapter) and the NT2. The *S interface* is a four-wire basic access rate interface capable of supporting 1000-meter cable separation. See also *ISDN.*

S-ALOHA See *ALOHA.*

S-register Memory in a modem used to store the *configuration profile.* Information in the *S-registers* is changed via the ATSr=n command. Information is read from the register by the ATSr? command. See Appendix B for a detail list.

S.n recommendations An ITU-T series of specifications concerning telegraph services terminal equipment. See Appendix E for a detail list.

S/D An abbreviation of Signal to Distortion ratio.

S/F An abbreviation of Store and Forward. Also abbreviated S-F.

S/N An abbreviation of Signal-to-Noise ratio.

S/W An abbreviation of *software.*

S0 The European designation for basic rate interface (BRI).

S2 The European designation for primary rate interface (PRI), i.e., 30B+D.

SA (1) An abbreviation of Selective Availability. **(2)** An abbreviation of Source Address. **(3)** An abbreviation of Synchronous Allocation. In the IEEE 802.6, time allocated to a Fiber Distributed Data Interface (FDDI) station for sending sync frames. **(4)** The symbol for Source Address. The media access control (MAC, the lower part of the ISO/OSI layer two) address of the IEEE 802.10 security committee. **(5)** An abbreviation of Sex Appeal (found on the Internet). **(6)** An abbreviation of Subject to Approval (found on the Internet).

SAA (1) An abbreviation of the Standards Association of Australia. The national standards-setting and certification agency. **(2)** An abbreviation of Systems Applications Architecture. In IBM's Systems Network Architecture (SNA), the compatibility scheme for application communications among different computers. It allows development of consistent applications across six software environments (TSO/F, CICS/MVS, IMS/ESA/TM, VM/CMS, OS400, and OS/2 extended edition), running on three hardware computing platforms (systems 370 ESA, OS/400, and PS/2).

SAAL An abbreviation of Signaling ATM Adaptation Layer.

SABM An abbreviation of Set Asynchronous Balanced Mode. Connection request between high level data link control (HDLC) controllers or logical link control (LLC) entities.

SABME An abbreviation of SABM Extended. Uses the optional 16-bit control field.

SAC (1) An acronym for Single-Attachment Concentrator. **(2)** An acronym for Simplified Access Control. **(3)** An acronym for Service Access Code. **(4)** An acronym for Special Area Code. In telephony, a representation of services, not places.

SACF An abbreviation of Single Association Control Function.

SACK An acronym for Selective ACKnowledgment.

SADL An abbreviation of Synchronous Auto Dial Language.

SAF An abbreviation of Subnetwork Access Facility.

SAFENET An acronym for Survivable Adaptable Fiber-optic Embedded NETwork.

sag (1) In power distribution systems, a short-duration decrease in mains voltage level. Specifically, a sag occurs when the voltage is 20% or more below the nominal root-mean-square (RMS) voltage. It may last for a few seconds or longer. **(2)** In wire and cable systems, the downward curvature of (droop) of a wire or cable due to its own weight. The curve is closely approximated by the mathematical shape called a catenary. *Sag* is measured from the midpoint of a span and is measured vertically. The perpendicular distance from a straight line between the two points of support to the wire or cable is called the *apparent sag.*

SAG An acronym for Structured Query Language Access Group.

sagan Slang for a large quantity of anything. From the Carl Sagan TV series *Cosmos* and his purported phrase, "billions and billions of . . ."

SAI An abbreviation of S/T Activity Indicator. An Integrated Services Digital Network Basic Rate Interface (ISDN BRI) control bit.

SAK An abbreviation of Selective ACKnowledgment.

SALMON An acronym for SNA AppLication MONitor.

sample The instantaneous measure of a signal waveform or set of values.

SALT An acronym for Subscriber Apparatus Line Testing.

sample and hold A device used with analog-to-digital converters to measure a time-varying waveform (sample) and then maintain the value at a constant level (hold) while the analog-to-digital conversion process takes place.

The following circuit illustrates the principle of a sample and hold circuit. The sampling is accomplished by closing a switch for a short period of time, and hold is accomplished by the charge placed on a capacitor during sampling.

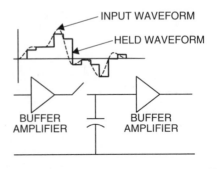

sampled data Data in which the information content is determined only at discrete time intervals.

sampled signal The sequence of values of a signal taken at discrete time intervals.

sampling The process of acquiring the representation of data from a source and storing it. In communications, the conversion of an analog signal to a digital format. Samples of the analog signal are taken at regular intervals (called the *sampling period*). These samples are then converted to binary and stored. The binary value may be any degree of *precision* and *accuracy* required. (8 bits of precision gives a maximum accuracy of 1 part in 256, while 16-bit precision yields a maximum accuracy of 1 part in 65536.) The sequence of digital samples is a *quantized* form of the original signal. The faster the *sampling rate* and the greater the precision of the binary sample, the closer the digitized sample stream will approximate the original waveform. See also *accuracy, analog-to-digital converter, precision,* and *quantize.*

sampling interval The reciprocal of the sampling rate, i.e., the interval between corresponding points on two successive sampling pulses of the sampling signal.

sampling period The *period of time* between samples of the signal. Mathematically, the inverse of the sampling rate. For example, a *sampling rate* of 1000 samples per second is 1 ms. See also *sampling* and *sampling rate.*

sampling rate The *frequency* with which samples of a signal are taken, i.e., the rate at which signals are sampled for subsequent use, such as for modulation, coding, and quantization. The higher the *sampling rate* (i.e., the more samples per second taken), the closer the sampled signal approximates the original waveform.

In order to represent all frequencies in a source signal, the *sampling rate* must be at least twice the highest frequency in the source signal. For example, if the highest signal frequency is 3 kHz, the *sampling rate* must be at least 6000 samples per second. Formerly called *sampling frequency.* See also *Nyquist theorem.*

sampling theorem See *Nyquist theorem.*

sanity check (1) A test to determine the "reasonableness" of a result (from either a transmission, operation, or calculation). For example, in a database, one might compare values in specified fields against a range of expected values. If the field is outside the range, then the record is flagged for further investigation. For example, if a person's age indicates 315 years, it is probably wrong. (2) A check to confirm the functionality of a piece of equipment.

SANZ An acronym for the Standards Association of New Zealand.

SAP (1) An acronym for Service Access Point. (2) An acronym for Service Advertising Protocol.

SAPI An acronym for Service Access Point Identifier.

SAR (1) An acronym for Segmentation And Reassembly. The basic function of an asynchronous transfer mode (ATM) access device protocol that apportions information into cells. (2) An acronym for Successive Approximation Register. (3) An acronym for Segment Address Register. (4) An acronym for Street Address Record. (5) An acronym for Source Address Register.

SARF An acronym for Security Alarm Reporting Function.

SARM An acronym for Set Asynchronous Response Mode. A high level data link control (HDLC) layer 2 unnumbered frame connection request.

SARME An acronym for SARM Extended. Uses the optional 16-bit control field.

SARTS An acronym for Switched Access Remote Test System. The way local exchange carriers (LECs) test leased lines.

SAS (1) An abbreviation of Single Attached Station. A Fiber Distributed Data Interface (FDDI) node attached to only a single ring of the network, i.e., by only two optical fibers. See also *Dual Attached Station* (*DAS*). (2) An abbreviation of Statically Assigned Sockets. (3) An abbreviation of SWITCH Access System.

SASE (1) An abbreviation of Special Application Service Element. (2) An abbreviation of Self Addressed Stamped Envelope.

SASFE An abbreviation of SEF/AIS Alarm Signal, Far End.

SASG An abbreviation of Special Autonomous Study Group. An ITU-T study group set up to produce technical and administrative handbooks on basic telecommunications subjects for developing countries.

SASI An acronym for Shugart Associates Systems Interface. A small computer systems interface specification defined by Shugart, a disk drive manufacturer. Later, the specification was revised and renamed the Small Computer Systems Interface (SCSI).

SASO An acronym for Saudi Arabian Standards Organization.

satellite (1) An entity that revolves around another primary body and whose motion is determined by forces of attraction between the bodies and the speed at which the secondary body revolves around the primary body. A parent body and its satellite revolve about their common center of gravity. (2) A secondary entity at a distance from a primary entity. For example, a telephone line concentrator separated from a central office switching system is a *satellite* piece of equipment to the central office.

satellite communications A telecommunications service provided via one or more satellite relays and the associated uplinks and downlinks with associated Earth terminals.

satellite link A radio link between a transmitting Earth station and a receiving Earth station through a satellite. A satellite link comprises one uplink and one downlink (and links between satellites if direct satellite to satellite transmission is used).

satellite network A satellite system or a part of a satellite system, consisting of only one satellite and the cooperating Earth stations.

satellite PABX A private automatic branch exchange (PABX) system that is not equipped with an attendant position and is associated with an attended main PABX system. The main PABX attendant provides attendant functions for the satellite system.

satellite system A space system using one or more artificial Earth satellites.

SATF (1) An abbreviation of Security Audit Trail Function. (2) An abbreviation of Shortest Access Time First.

SATS An acronym for Selected Abstract Test Suite.

SATT An abbreviation of Strowger Automatic Toll Ticketing.

saturation (1) The condition of a signal processing circuit, system, or device when an increase of the input signal value no longer produces a corresponding increase in output value. (2) The point at which the output of a linear device (e.g., linear amplifier) deviates significantly from being a linear function of the input when the input signal is increased. (3) The condition in which a component of a communications system has reached its maximum traffic-handling capacity.

saturation testing A particular type of test utilizing an artificial traffic generator to flood a communications switch with messages. The test is intended to expose errors that are rare or that occur only during times of peak traffic.

SAW An acronym for <u>S</u>urface <u>A</u>coustic <u>W</u>ave.

SAWO An acronym for <u>S</u>urface <u>A</u>coustic <u>W</u>ave <u>O</u>scillator.

SAX An acronym for <u>S</u>mall <u>A</u>utomatic e<u>X</u>change.

SB (1) An abbreviation of <u>S</u>ignal <u>B</u>attery. (2) An abbreviation of <u>S</u>ide<u>B</u>and. (3) An abbreviation of <u>S</u>tand<u>B</u>y. (4) An abbreviation of <u>S</u>imultaneous <u>B</u>roadcast.

SBED An acronym for <u>S</u>erial <u>BIT</u> <u>E</u>rror <u>D</u>etector.

SBH An abbreviation of <u>S</u>witch <u>B</u>usy <u>H</u>our.

SBI An abbreviation of <u>S</u>ingle <u>B</u>yte <u>I</u>nterleaved.

SBT (1) An abbreviation of <u>S</u>ix <u>B</u>it <u>T</u>ranscode. (2) An abbreviation of <u>S</u>ystem <u>B</u>ackup <u>T</u>ape drive.

SC (1) An abbreviation of <u>S</u>ub<u>C</u>ommittee. (2) An abbreviation of <u>S</u>ubscriber <u>C</u>onnector. (3) An abbreviation of <u>S</u>econdary <u>C</u>hannel. (4) An abbreviation of <u>S</u>ending <u>C</u>omplete. (5) An abbreviation of <u>S</u>ervice <u>C</u>ode. (6) An abbreviation of <u>S</u>uppressed <u>C</u>arrier.

SCA An abbreviation of <u>S</u>hort <u>C</u>ode <u>A</u>ddress. (2) An abbreviation of <u>S</u>ubsidiary <u>C</u>ommunication <u>A</u>uthorization. (3) An abbreviation of <u>S</u>ynchronous <u>C</u>ommunications <u>A</u>dapter.

scada An acronym for <u>S</u>upervisory <u>C</u>ontrol <u>A</u>nd <u>D</u>ata <u>A</u>cquisition. A communications system between a master and remote station wherein supervisory control, data acquisition, and automatic control information are exchanged. Frequently used in oil and gas production networks.

An example is shown in the figure.

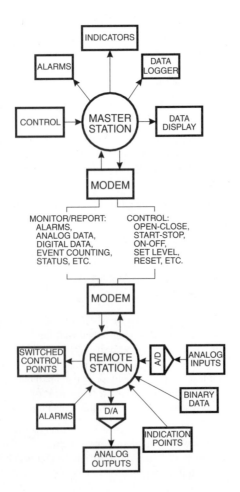

SCAI An abbreviation of <u>S</u>witch to <u>C</u>omputer <u>A</u>pplication <u>I</u>nterface. A link between a host computer and a voice switching system that allows application integration.

scalar A quantity that is completely defined by a single number. Among examples of *scalar quantities* are 27 MHz, a dozen eggs, 6 volts, and pi (3.14159 . . .). See also *vector.*

scaling The linear expansion of a network or system by the addition of more nodes or modules.

scan (1) In general, the process of examining sequentially, step by step or part by part. (2) In software, to examine every record, reference, or field of a file as part of a retrieval scheme. (3) In software, to examine every entry of a data input area for specific information or delimiters. (4) In antennas, the motion of a beam while a search for a target is in progress. (5) In antennas, to sweep, i.e., rotate a beam about an axis not coincident with the axis of propagation. (6) One process by which a data acquisition system interrogates satellite locations for data. (7) To tune a receiver through a predetermined range of frequencies in prescribed increments or at prescribed times.

SCAN An acronym for <u>S</u>witched <u>C</u>ircuit <u>A</u>utomatic <u>N</u>etwork.

scan rate The number of remote stations, functions, or devices that can be polled by a master device in a given time period.

scan time The time required to examine the state of all remote device inputs.

scanning (1) In telephony, the periodic testing of the circuits under common control. For example, the *scanning* of subscriber line circuits for off-hook/on-hook transitions. (2) In facsimile, the process of sequentially analyzing each element of an input image for its level of darkness. The *scanning* normally proceeds on a line-by-line basis left to right and top to bottom. In reverse *scanning,* the procedure is right to left and top to bottom.

scattering The dispersion of wave energy due to a nonhomogeneous or nonlinear transmission medium or surface irregularities of a transmission boundary. It is a phenomenon in which the direction, frequency, and/or polarization of an electromagnetic wave is changed when the wave encounters discontinuities or interacts with the material at the atomic or molecular level in the medium which it is propagating. Scattering results in a disordered or random change in the incident energy distribution.

For example, in communications over a fiber optic cable, *scattering* is the signal loss that occurs when the light in the fiber core strike molecules or slight imperfections in the cladding-core boundary.

scattering angle The angle between the initial and final paths traveled by a scattered photon or particle.

scattering center A site, in the microstructure of a transmission medium, at which electromagnetic waves are scattered. Many scattering centers are locked into the medium's matrix at the time of solidification during manufacturing.

Examples of scattering centers are

- Vacancy defects.
- Interstitial defects.
- Inclusions, such as gas molecules, hydroxide ions, iron ions, and trapped water molecules.
  For example, in glass optical fibers, there is a high attenuation band at 0.95 μm, caused primarily by scattering and absorption by OH^- (hydroxyl) ions.
- Microcracks or fractures in dielectric waveguides.

See also *Rayleigh scattering.*

scattering coefficient The factor that expresses the attenuation caused by scattering of electromagnetic or acoustic energy, during its passage through a unit thickness of medium. *Scattering coefficient* is usually expressed in units of reciprocal distance.

scattering loss That portion of a signal lost in transmission due to scattering effects.

SCC (**1**) An abbreviation of $\underline{S}$atellite $\underline{C}$ommunications $\underline{C}$ontrol. (**2**) An abbreviation of $\underline{S}$atellite $\underline{C}$ontrol $\underline{C}$enter. (**3**) An abbreviation of $\underline{S}$pecialized $\underline{C}$ommon $\underline{C}$arrier. (**4**) An abbreviation of $\underline{S}$tandards Council of $\underline{C}$anada. (**5**) An abbreviation of $\underline{S}$witching $\underline{C}$ontrol $\underline{C}$enter. (**6**) An abbreviation of $\underline{S}$hort $\underline{C}$ircuit $\underline{C}$urrent. (**7**) An abbreviation of $\underline{S}$ingle $\underline{C}$onductor $\underline{C}$able. (**8**) An abbreviation of $\underline{S}$ingle $\underline{C}$otton $\underline{C}$overed wire insulation.

SCCP An abbreviation of $\underline{S}$ignaling $\underline{C}$onnection $\underline{C}$ontrol $\underline{P}$art.

SCFM An abbreviation of $\underline{S}$ub-$\underline{C}$arrier $\underline{F}$requency $\underline{M}$odulation.

schematic diagram A diagram that shows, by means of symbols, the electrical interconnection of circuit elements (components) in a specific circuit arrangement. A schematic diagram enables tracing out the circuit, its function, and analysis without regard to any physical characteristics (size, shape, component location, etc.).

The following schematic is of a third-order low-pass filter with a differential input and single-ended output and a gain of two in the passband. The cutoff frequency of the filter is unknown because the values of the three frequency-determining capacitors are not given.

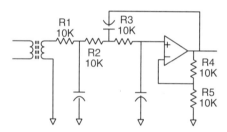

schlieren image A technique by which an image is created from the deviations of refracted light due to density variations in the transmission medium. The density variations may be generated by thermal, acoustic, or other phenomena. *Schlieren* is German for streaks or striae.

The principle is illustrated by the following figure. Light from a *slit source* is *collimated* by a mirror or lens system. After traversing the *test region*, it is focused at a *knife-edge* where approximately half of the light rays passing to the observer or observing screen are blocked. Changes in the refractive index near an object in the test area deviate the rays, causing changes in illumination at the corresponding points on the observing screen. Rays deviated downward (solid ray) pass over the knife-edge, while rays deviated upward (dotted ray) are blocked by the knife-edge.

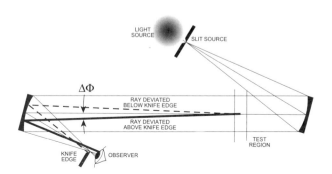

Schmitt trigger An electronic device whose output remains in the "off" state unless the input signal exceeds a specified input turn-on threshold. Similarly, the output remains at the "on" level until the input falls below a second specified turn-off threshold. The turn-off threshold is designed to be less than the turn-on threshold. This *hysteresis* provides noise immunity when the input signal is near the thresholds.

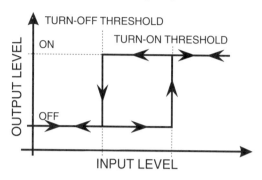

schrödinbug Jargon. A programming design or implementation bug in a program that isn't visible until someone reading source code or using the program in an unusual way notices that the program never should have worked (at which point the program promptly stops working for everybody until the bug is fixed). Although it seems impossible, some programs have had latent *schrödinbugs* for years. The name is derived from Schrödinger's Cat thought experiment in quantum physics. See also *Bohr bug, heisenbug, mandelbug,* and *software rot.*

SCI (**1**) An abbreviation of $\underline{S}$erial $\underline{C}$ommunications $\underline{I}$nterface. (**2**) An abbreviation of $\underline{S}$hort $\underline{CI}$rcuit. (**3**) An abbreviation of $\underline{S}$ingle $\underline{C}$hannel $\underline{I}$nterface. (**4**) An abbreviation of $\underline{S}$ystem $\underline{C}$ontrol $\underline{I}$nterface.

sci hierarchy One of the seven *world newsgroups* in USENet. The *sci* newsgroups report material relating to the sciences including *aeronautics, astronomy, biology, chemistry, engineering, mathematics, medicine, physics, psychology,* and many more. See also *newsgroup hierarchy.*

scintillation In electromagnetic wave propagation, a small random fluctuation of the received field strength about its mean value. This effect becomes more significant as the frequency of the propagating wave increases.

SCLOG An abbreviation of $\underline{SeC}$urity $\underline{LOG}$.

SCM (**1**) An abbreviation of $\underline{S}$tation $\underline{C}$lass $\underline{M}$ark. (**2**) An abbreviation of $\underline{S}$ingle $\underline{C}$hannel $\underline{M}$odem. (**3**) An abbreviation of $\underline{S}$ystem $\underline{C}$ontrol and $\underline{M}$onitor.

SCN (**1**) An abbreviation of $\underline{S}$atellite $\underline{C}$ommunications $\underline{N}$etwork. (**2**) An abbreviation of $\underline{S}$hortest $\underline{C}$onnected $\underline{N}$etwork.

SCP (**1**) An abbreviation of $\underline{S}$ervice $\underline{C}$ontrol $\underline{P}$oint or $\underline{S}$ignal $\underline{C}$ontrol $\underline{P}$oint. Part of Signaling System 7 (SS7) that supports carrier services, such as 800 and ANI. (**2**) An abbreviation of $\underline{S}$ystem $\underline{C}$ontrol $\underline{P}$oint.

SCPC An abbreviation of $\underline{S}$ingle $\underline{C}$hannel $\underline{P}$er $\underline{C}$arrier.

SCR (**1**) An abbreviation of $\underline{S}$ustainable $\underline{C}$ell $\underline{R}$ate. An asynchronous transfer mode (ATM) traffic parameter. (**2**) An abbreviation of $\underline{S}$ilicon $\underline{C}$ontrolled $\underline{R}$ectifier. (**3**) An abbreviation of $\underline{S}$ignal to $\underline{C}$rosstalk $\underline{R}$atio. (**4**) An abbreviation of $\underline{S}$ignal $\underline{C}$onversion $\underline{R}$elay. (**5**) An abbreviation of $\underline{S}$ingle $\underline{C}$haracter $\underline{R}$ecognition.

scrambler A device (or program) for reordering or encoding a signal at a transmitter in order to either:

- Render it indecipherable at a receiver without an appropriate descrambler, or
- To spread the transmitted energy over the available bandwidth.

See also *encryption* and *spread spectrum.*

screen (1) In a telecommunications, computing, or data processing system, to examine entities that are being processed to determine their suitability for further processing. For example, the ability to reject or accept call attempts by using:

1. Trunk or line class, or
2. Trunk or line number information.

(2) A nonferrous metallic mesh used to provide electromagnetic shielding. Also called a *shield.* (3) To reduce undesired electromagnetic signals and noise by enclosing devices in electrostatic or electromagnetic shields. (4) A viewing surface, such as a cathode ray tube (CRT) or liquid crystal display (LCD).

screening router A router configured to admit or deny traffic based on a set of rules assigned by the system administrator. See also *router.*

script An automated command sequence used to set up or control the operation of a program or device, such as a modem telecommunication program or a network logon procedure.

SCS (1) An abbreviation of Satellite Communications Systems. (2) An abbreviation of Silicon Controlled Switch. (3) An abbreviation of SNA Character String. A printing mode in IBM's Systems Network Architecture (SNA) that provides various printing and formatting capabilities. (4) An abbreviation of SWITCH Central System. (5) An abbreviation of System Communication Services. (6) An abbreviation of System Conformance Statement.

SCSA An abbreviation of Signal Computing System Architecture.

SCSI An acronym for Small Computer System Interface (pronounced "scuzzy"). A high-speed parallel interface defined by the American National Standards Insitute (ANSI) standard X3T9.2. A *SCSI* interface will support up to eight peripherals on a sequential connecting cable (called a *daisy chain* connection), although only one of the devices may transmit at a time. The last device on the daisy chain must be properly terminated. The device with the highest address on the chain has the highest priority. Major versions in use are *SCSI-1, SCSI-2,* or *SCSI-3* or *wide SCSI.*

- *SCSI-1*—the slower, less capable of the versions. It supports disk drives of up to 2 GB and transfer rates as high as 5 Mbps on an 8-bit bus.
- *SCSI-2*—supports drive capacities greater than 3 GB. Typical data transfer rates are up to 10 Mbps on a 16-bit bus.
- *SCSI-3*—a 32-bit interface on a 68-pin connector supporting data transfer rates to 40 Mbps. Defined in ANSI X3.302:1998.

A *SCSI* cable generally has either a D-type or Centronics-like connectors at one or both ends. The D-type is a 50-pin connector that looks like the DB-nn connectors used for serial ports on PCs, except it is somewhat smaller. The pin assignments for a *SCSI* connector are:

8 BIT SCSI PIN ASSIGNMENTS

Pin	Definition	Pin	Definition	Pin	Definition	Pin	Definition
1	Gnd	14	Data 6	27	Gnd	40	Reset
2	Data 0	15	Gnd	28	Gnd	41	Gnd
3	Gnd	16	Data 7	29	Gnd	42	Message
4	Data 1	17	Gnd	30	Gnd	43	Gnd
5	Gnd	18	Parity	31	Gnd	44	Select
6	Data 2	19	Gnd	32	Attention	45	Gnd
7	Gnd	20	Gnd	33	Gnd	46	C/D
8	Data 3	21	Gnd	34	Gnd	47	Gnd
9	Gnd	22	Gnd	35	Gnd	48	Request
10	Data 4	23	Gnd	36	Busy	49	Gnd
11	Gnd	24	Gnd	37	Gnd	50	I/O
12	Data 5	25	Key	38	Ack		
13	Gnd	26	Terminator power	39	Gnd		

SCSIF An abbreviation of Small Computer System Interface Fast.

SCSIW An abbreviation of Small Computer System Interface Wide.

SCSR An abbreviation of Single Channel Signaling Rate.

SCTO An abbreviation of Soft Carrier Turn-Off.

ScTP An abbreviation of Screened Twisted Pair. Generally refers to a type of 100-ohm cable, composed of four pairs of copper wire wrapped in a thin aluminum-foil shield.

SCTR An abbreviation of System Conformance Test Report.

SCUSA An abbreviation of Standards Council of the USA.

SD (1) An abbreviation of Starting Delimiter. (2) An abbreviation of Send Data. (3) An abbreviation of Send Digits. (4) An abbreviation of Sampled Data. (5) An abbreviation of Schematic Diagram. (6) An abbreviation of Seralizer/Deseralizer. (7) An abbreviation of Single Density.

SDA An abbreviation of Security Domain Authority.

SDAP An acronym for Standard Document Application Profile.

SDDN An abbreviation of Software Defined Data Network. A private virtual network built on a public data network.

SDE An abbreviation of Secure Data Exchange.

SDF An abbreviation of Sub-Distribution Frame.

SDH An abbreviation of Synchronous Digital Hierarchy. ITU-T's specification equivalent to Synchronous Optical Network (SONET).

SDI An abbreviation of Serial Data Interface.

SDIF An abbreviation of SGML Document Interchange Format.

SDK (1) An abbreviation of Software Developer's (or Development) Kit. (2) An abbreviation of System Design Kit.

SDL (1) An abbreviation of Specification and Description Language. (2) An abbreviation of Signaling Data Link.

SDL/GR An abbreviation of SDL/Graphical Representation.

SDL/PR An abbreviation of SDL/Phrase Representation.

SDLC (1) An abbreviation of Synchronous Data Link Control. (2) An abbreviation of System Data Link Control.

SDM (1) An abbreviation of Space Division Multiplex (deprecated). (2) An abbreviation of Subrate Digital Multiplex (alternatively Subrate Data Multiplex). A digital data system (DDS) service that multiplexes multiple low-speed data channels onto a DS-0 channel. (3) An abbreviation of Sequence Division Multiplexing. (4) An abbreviation of Subrate De-Multiplex.

SDMA (1) An abbreviation of Space Division Multiple Access. (2) An abbreviation of Shared Direct Memory Access.

SDN (1) An abbreviation of Software Defined Network. (2) An abbreviation of Subscriber's Directory Number. (3) An abbreviation of Synchronized Digital Network. (4) An abbreviation of the Korean System Development Network.

SDNS An abbreviation of Secure Data Network System.

SDO An abbreviation of Standards Development Organization.

SDP (1) An abbreviation of Signal Data Processor. (2) An abbreviation of Station Data Processing.

SDRAM An acronym for <u>S</u>ynchronous <u>D</u>ynamic <u>R</u>andom <u>A</u>ccess Memory. Features of *SDRAM* include:

- *SDRAM* clock synchronization with the microprocessor (μP). This allows the data to be delivered to the μP with less delay than conventional *DRAM*.
- A dual bank architecture, i.e., two interleaved memory banks. This structure allows one bank to be accessed while the other bank is preparing to read or write. By alternating bank usage, the effective throughput can be doubled.
- The ability to do *burst mode transfers.* Using burst mode, a *SDRAM* generates a block of consecutive data address fetches each time the μP requests one address. Because a μP tends to request sequential data addresses, the *SDRAM*'s anticipation of the next fetch speeds throughput.

SDS (**1**) An abbreviation of <u>S</u>witched <u>D</u>igital <u>S</u>ervice. A generic term for carrier systems such as Switched56. (**2**) An abbreviation of <u>S</u>pace <u>D</u>ivision <u>S</u>witching.

SDSL (**1**) An abbreviation of <u>S</u>ingle line <u>D</u>igital <u>S</u>ubscriber <u>L</u>ine. (**2**) An abbreviation of <u>S</u>ymmetric <u>D</u>igital <u>S</u>ubscriber <u>L</u>ine.

SDU An abbreviation of <u>S</u>ervice <u>D</u>ata <u>U</u>nit. An information packet or segment passed down the ISO/OSI protocol stack to become the payload of the next lower protocol layer.

SE (**1**) An abbreviation of <u>Sw</u><u>E</u>den. (**2**) An abbreviation of <u>S</u>ystem <u>E</u>xpansion. (**3**) An abbreviation of <u>S</u>ystems <u>E</u>ngineering.

SEA An abbreviation of <u>S</u>elf <u>E</u>xtracting <u>A</u>rchive.

SEAL An abbreviation of <u>S</u>imple and <u>E</u>fficient <u>A</u>daptation <u>L</u>ayer.

seal of approval (SOAP) In the Internet, a message attached to an article in the proposed Internet encyclopedia (*Interpedia*) which indicates that, in the opinion of the reviewer, the article is valid, accurate, and worthwhile.

Since anyone (in principle) can contribute an article to the encyclopedia, the articles will vary widely in depth, quality, and accuracy. The *SOAP* approach allows this freedom (of public contribution) while providing a means for a reader to determine its reliability. A *SOAP* provided by a competent practitioner of the subject matter would tend to attest to its accuracy. Seals of *disapproval* have been proposed. In the world of printed books and articles, a negative review by a peer can keep the manuscript out of print. Later it may be found that the reviewer's motives were based on professional jealousy, narrow-mindedness, or fear of innovation, but by then the damage has been done. In the Interpedia, the reader can review both the article and its SOAPs, and then reach his or her own conclusion.

search directory A World Wide Web (WWW) site designed to help users locate resources in cyberspace. In most *search directories,* users enter words or phrases into a search form that is submitted to the search engine. The search engine then looks for matching occurrences in its database and reports the Universal Resource Locator (URL) back to the user. Some provide not only the URL of the desired information but also reviews of the site.

search time The time interval required to locate a particular data element, record, or file in a storage device.

SEAS An abbreviation of <u>S</u>ignaling <u>E</u>ngineering and <u>A</u>dministration <u>S</u>ystem.

SEC (**1**) An abbreviation of <u>S</u>witching <u>E</u>quipment <u>C</u>ongestion. (**2**) An abbreviation of <u>SEC</u>ond. (**3**) An abbreviation of <u>S</u>ingle <u>E</u>rror <u>C</u>orrecting code. (**4**) An abbreviation of <u>S</u>witch <u>E</u>lement <u>C</u>ontroller.

SECAM An acronym for <u>SE</u>quential (or <u>Sy</u>stème <u>E</u>lectronique) Couleur <u>A</u>vec <u>M</u>emoire. A television signal standard used in France and some Eastern Europe, former Soviet Union countries, Middle East countries, and African countries. It uses a 625-line 50-Hz picture scan. SECAM 625 line, Phase Alternate Line (PAL) 625-line, and National Television Standards Committee (NTSC) 525-line systems are mutually incompatible without a converter.

SECDED An abbreviation of <u>S</u>ingle bit <u>E</u>rror <u>C</u>orrection and <u>D</u>ouble bit <u>E</u>rror <u>D</u>etection.

SECO (**1**) An acronym for <u>SE</u>quential <u>CO</u>ding. (**2**) An acronym for <u>SE</u>quential <u>CO</u>ntrol. (**3**) An acronym for <u>S</u>tation <u>E</u>ngineering <u>C</u>ontrol <u>O</u>ffice.

SECOBI An acronym for <u>SE</u>rvicio de <u>CO</u>nsulta a <u>B</u>ancos de <u>I</u>nformacion. Mexico's Database Consultation Service.

second The time duration of 9 192 631 770 periods of the radiation associated with the atomic resonance between the two hyperfine ground-state levels of the cesium-133 atom in a magnetically neutral environment. A diagram of a cesium "clock" is:

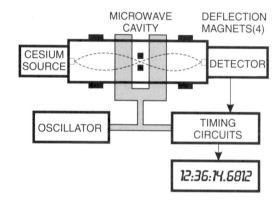

Realizable cesium standards use a strong electromagnet to deliberately introduce a magnetic field which overwhelms the field of the Earth. The presence of this strong magnetic field introduces a slight, but known, increase in the atomic resonance frequency (and a corresponding decrease in the second). Very small variations in the field strength due to mechanical tolerances and calibration of the electric current in the electromagnets introduce minuscule, but measurable, variations among different cesium oscillators.

second dial tone A call progress tone indicating that the caller has been given access to another dial-up switching system.

For example, a caller in an organization with a private automatic branch exchange (PABX) might dial an 8 or 9 to reach the public switched telephone network (PSTN). Upon seizing the PSTN, it will give the caller a dialtone indicating that the rest of the dial information may be entered. A third or fourth dialtone may be received if the caller reaches automatic switching equipment at the destination of a dial sequence. Also called an *outside dialtone.*

second window Of silica-based optical fibers, the transmission window at approximately 1300 nm. This is the *minimum-dispersion window* in silica-based glasses. See also *window.*

secondary channel In a system in which two channels share a common interface, a channel that has a lower data signaling rate (DSR) capacity than the primary channel.

secondary emission Particles or radiation, such as photons, Compton recoil electrons, and secondary electrons, that are produced by the action of primary radiation on matter.

secondary frequency standard A frequency standard that does not have inherent accuracy and therefore must be calibrated against a primary frequency standard. Examples of *secondary standards* include:

- *Crystal oscillators*—whose frequency depends on physical dimensions, which in turn vary with fabrication and environmental conditions.
- *Rubidium standards*—whose frequency is determined by atomic transitions in a gas cell through which an optical signal is passed. The gas cell has inherent inaccuracies because of
  - Gas pressure variations, including those induced by temperature variations.
  - Variations in the concentrations of the required buffer gases, which variations cause frequency deviations.

secondary link station (SLS) In environments that use IBM's Synchronous Data Link Control (SDLC) protocol, any node that responds to communications initiated by a primary link station (PLS). In SDLC, *SLS*s cannot initiate communications.

secondary protection A transient protective device that is associated with a particular device or piece of equipment. This is in contrast to primary protection which is generally at the entrance point to a facility.

Primary protection devices are intended to handle the bulk of the energy in a foreign voltage; however, some energy will pass. The job of the secondary protection is to limit this signal to a safe level for subsequent devices. In telephony, the local telephone company provides the primary protection at the demarcation point at the subscriber's premises. Secondary protection is provided by the subscriber.

secondary radiation See *secondary emission.*

secondary service area In radio broadcasting, a region in which satisfactory reception can be obtained only under favorable conditions. The area is served by the skywave, is not subject to objectionable interference, and is subject to intermittent variations in strength.

secondary time standard A time standard that requires periodic calibration against a primary time standard.

SECORD An acronym for SECure voice COrd boaRD. A desk-mounted patch panel capable of connecting and controlling communications facilities to the Defense Switched Network (DSN) or other narrowband facilities. The communications facilities can be:

- Sixteen 50-kbps wideband or sixteen 2400-bps narrowband user lines and
- Five narrowband trunks

SECTEL An acronym for SECure TELephone.

sectional center (SC) In telephony, a class 2 office (toll office) to which a number of primary centers (local offices), toll centers, or toll points may be connected. See also *class n office.*

sector In magnetic or optical disk or drum storage media, the division of tracks into a number of segments for the purpose of optimizing data storage capacity and data transfer rate. Each sector can be addressed and accessed by the read device in the course of a predetermined rotational displacement of the medium. See also *cylinder* and *track.*

sector sparing A method of improving the reliability of a hard disk drive. Before writing to the hard disk drive, it is checked for integrity. If bad sectors are found, they are marked unusable and the data are written to another location. Also called *bad sector remapping* and *sector fixing.*

secure electronic transfer (SET) A protocol that encrypts credit card transactions as they traverse the Internet and authenticates the identity of the sender and receiver.

secure link A communications path in which significant information is encrypted so as to make interception by a third-party impractical. For example, Netscape's browser (Navigator) encrypts credit card data between a buyer and seller.

secure transmission (1) In transmission security, telecommunications deriving security through use of type 1 products (products containing classified National Security Agency algorithms) and/or protected distribution systems. Also called *secure communications.* **(2)** In spread spectrum systems, the transmission of coded sequences that represent information that can be recovered only by systems or equipment that have the identical spread spectrum code sequence key as the transmitter.

security The protection of a computer system or communications network and its data from accidental or malicious loss, corruption, or unauthorized access. That is, ensuring that the data, circuits, and equipment on a network are used only by authorized users and in authorized ways.

Various methods are used to provide physical and logical security, for example:

- *Physically securing* hardware and software from theft, fire, flood, etc.
- Use of *power protection* devices, such as line conditioners to clean the electrical signals coming into the network components, and uninterruptible or standby power supplies (UPSs or SPSs) to keep the network running long enough to shut down properly in case of a power outage.
- *Logically securing* hardware, such as by using encryption chips on network interface cards.
  Encryption information must be stored in a separate location in memory that is not directly accessible to the computer. Hardware security measures are necessary for networks that comply with moderate *security levels* (such as C2) as specified by the National Security Agency.
- *Redundancy*
  - Use of system fault-tolerant servers.
  - Use of redundant cabling that provides multiple paths for information flow.
  - Use of redundant storage in which multiple copies of information are stored. See also *RAID.*
- Use of *diskless workstations,* to prevent users from copying files to removable media or logging transmissions to disk.
- Use of *encryption* of transmissions to prevent (or deter) unauthorized theft of information transferred across a network.
- Use of *callback modems* to prevent logins from unauthorized remote locations.
- Use of *traffic padding* to make the level of network traffic more constant, thus making it more difficult for an eavesdropper to infer network contents.
- Use of *packet filtering,* the transmission of packets only to the specified destination node.
- Use of *Virus Protection*
  - Use of detection and protection software.
  - Control of uploading privileges to minimize the likelihood that someone can deliberately or inadvertently load a virus or other damaging program onto the network.
- Use of *Backups.* Doing regular and frequent backups onto tape, disk, or optical media.
- Use of *transaction tracking.* All materials related to a transaction are kept in memory (or in temporary buffers on disk) and are written only once the transaction is completed. (This protects against data loss if the network goes down in the middle of a transaction.)

- Use of *audit trails,* in which all user actions are recorded and stored.
- Use of *verification,* such as message authentication codes (MACs), to determine whether a message has been received as sent. These codes are similar to ordinary cyclic redundancy check codes (CRCs) but differ because the checksum is encrypted. Other verification activities include the use of digital signatures, notarization, and origin and destination nonrepudiation.
- Use of *access control.*
  - Controlling *file access* to critical files or directories (e.g., the user account and password data).
  - Assigning users' *privileges* based on their "need-to-use."
  - Use of *passwords* and other user IDs to control access to the network. With *dynamic passwords,* users get new passwords (generated by a special device) every time a login occurs. Because the *password* is generally the front-line security defense, it is a very good idea to change the password frequently. The password should be a "good password." See also *password.*
  - Use of *host* and *key authentication* in addition to passwords to ensure that all parties involved in a network connection are allowed to be there.

security filter (1) In communications security, the hardware, firmware, and/or software used to prevent access to specified data by unauthorized persons or systems, such as by preventing transmission, preventing forwarding messages over unprotected lines or circuits, or requiring special codes for access to certain files. **(2)** An automated information system (AIS) trusted subsystem that enforces security policy on the data that passes through it.

security kernel (1) In computer and communications security, the central part of a computer or communications system's hardware, firmware, and software that implements the basic security procedures for controlling access to system resources. **(2)** A self-contained, usually small, collection of key security-related statements that works as part of an operating system to prevent unauthorized access to, or use of, the system and contain criteria that must be met before specified programs can be accessed. **(3)** Hardware, firmware, and software elements of a trusted computing base that implement the reference monitor concept.

security levels Four general security classes are defined in a government publication called the *Trusted Computer System Evaluation Criteria* but more commonly known as the *Orange Book.* The four classes are (in order of increasing security):

- *Class D* (minimal security)—any system that cannot meet any of the higher security criteria. Systems in this class cannot be considered secure. Examples include PC operating systems such as MS DOS or Macintosh's System 7.
- *Class C* (discretionary protection)—subdivided into two classes, labeled C1 and C2. Operating systems such as UNIX, Windows NT, or network operating systems that provide password protection and access rights can fall into either of these classes.
  - C1 security features include the use of passwords or other authentication measures, the ability to restrict access to files and resources, and the ability to prevent accidental destruction of system programs.
  - C2 systems include the capabilities of C1 in addition to the ability to audit or track all user activity, restrict operations for individual users, and make sure that data left in memory cannot be used by other programs or users.
- *Class B* (mandatory protection)—must in general, be able to provide mathematical documentation of security, actively seek out threats to security, and be able to maintain security even during system failure. Class B systems are divided into three subsystems:

- B1 systems must have all the security capabilities of a C2 system and must take all available security measures and separate the security-related system components from the ones that are not related to security. B1 documentation must include discussions of the security measures.
- B2 systems extend B1 requirements to be able to provide a mathematical description of the security system, manage all configuration changes (software updates, and so on) in a secure manner, and check explicitly to make sure new software does not have any backdoors or other ways through which an outsider might try to access the secure system.
- B3 systems extend B2 by requiring a system administrator in charge of security and must remain secure even if the system goes down.
- *Class A* (verified protection)—class A systems must be able to verify mathematically that their security system and policies match the security design specifications.

security management One of the five Open Systems Interconnection (OSI) network management domains specified by the ISO and ITU-T. The purpose is to provide a secure network as defined in the entry on security and to notify the system administrator of any efforts to compromise or breach this secure network. Generally, a secure network is one that is always accessible when needed and whose contents can be accessed (read and modified) only by authorized users.

Security Through Obscurity (STO) The philosophy that a system or message will remain secure as long as the would-be assailants do not know of its existence. This philosophy applies to both computer systems and cryptology. When applied to bugs, it implies that the vendor is simply ignoring the problem, hoping it will go away. It never does.

seed (1) The starting value for a pseudorandom number generator. **(2)** A small inclusion in an optical element.

seed router A router in an AppleTalk internetwork that defines the network number ranges for all other routers in the network. Each AppleTalk internetwork needs at least one seed router.

SEF An abbreviation of Source Explicit Forwarding. A *security* feature provided on networks that allows transmissions only from specified stations to be forwarded by bridges.

SEF/AIS An abbreviation of Severely Erred Framing/Alarm Indication Signal.

segment (1) In general, any of the parts into which an entity may be divided. **(2)** A contiguous portion of a communications network. **(3)** A local area network (LAN) term meaning an electrically contiguous piece of the bus. *Segments* can be joined together using repeaters or bridges. **(4)** In a distributed queue dual bus (DQDB) network, a protocol data unit (PDU) that

- Consists of 52 octets transferred between DQDB-layer peer entities as the information payload of a slot,
- Contains a header of 4 octets and a payload of 48 octets, and
- Is either a pre-arbitrated segment or a queued arbitrated segment.

(5) A 64-character block used as the basis billing on data networks in some countries. **(6)** A self-contained portion of a computer program that may be executed without the remainder of the program resident in memory.

segmentation (1) In networks that conform to the ISO/OSI Reference Model, the process by which a packet is broken into parts and encapsulated into several packets at a lower layer. (The process may be necessary because of packet length restrictions at certain layers or network nodes.)

When a packet is *segmented,* the data portion (payload) is broken in parts, and a header and segment sequence information is appended to each part. The packet is passed down to the layer below for further processing (e.g., for encapsulation into the lower-layer packets). The reverse process, removing redundant headers and recombining several segments into the original packet, is known as reassembly. Also called *fragmentation* in the Internet. **(2)** *Segmentation* is also used to describe the division of a large local area network (LAN) into multiple smaller, more manageable LANs.

segmentation and reassembly (SAR) The protocol, at an originating station, that cuts a protocol data unit (PDU) into appropriate lengths and formats them to fit into an asynchronous transfer mode (ATM) payload. At the receiving end station, the protocol that extracts the payload and reassembles the original PDU to be used by the intended application.

segmentation fault An error in which a running program attempts to access memory not allocated to it.

segmented encoding law An encoding law in which an approximation to a curve defined by a smooth encoding law is obtained by a number of linear segments. Also called *piecewise linear encoding law.* See also *A-law* and *mu-law.*

seize To access a circuit in preparation for use or to make it busy so that others cannot use it.

seizure signal In telephony, a signal transmitted from the sending end of a trunk to the far end to indicate a request for service. A seizure signal also locks out the trunk or line to other demands for service.

SEK An abbreviation of Swedish Electrical Commission.

selecting **(1)** Choosing a specific group of one or more servers in the establishment of a call. **(2)** In the Binary Synchronous Communications (BSC) polled protocol, the process of informing a particular station that the next transmission is directed to that station.

selective availability (SA) The name of the U.S. Department of Defense policy and the implementation method by which unauthorized users of the Global Positioning System (GPS) will have the position accuracy of NAVSTAR's Standard Positioning Service (SPS) *intentionally* degraded. Degradation is achieved by dithering the satellite clock (delta-process) and/or by corrupting the message ephemeris (epsilon-process). The horizontal accuracy under *SA* is 40 meters circular error probability (CEP), i.e., with a 50% confidence level or 100 meters at the 95% confidence level (2 dRMS). The altitude uncertainty is similarly degraded to 156 meters at the 95% confidence level.

The effects of *SA* can be removed by either encryption keys or a procedure called *differential GPS (DGPS).* With DGPS a CEP accuracy of 2 to 5 meters can be obtained. See also *differential GPS (DGPS)* and *Global Positioning System (GPS).*

selective calling The ability of a station on a multistation communications network to designate which station is to receive a transmission.

selective fading Fading in which the individual components of the received radio signal fluctuate independently.

selective reflection The reflection of different wavelengths by different amounts due to absorption, scattering, or interferences at different rates by the various wavelengths. When the wavelengths are in the visible light region, the effects produce various colors.

selective ringing In telephony, subscriber line ringing in which only the ringer at a designated station on a party line responds. This may be accomplished by several means including frequency selective, voltage biased, or selective wire techniques. Without selective ringing, all the instruments on the party line will ring at the same time, selection being made by the number of rings. See also *coded ringing, distinctive ringing,* and *semiselective ringing.*

selective signaling A method of in-band signaling used on private networks to direct switching entities to properly route the call.

selective transmission The transference of different wavelengths by different amounts due to absorption, scattering, or interferences at different rates by the various wavelengths. When the wavelengths are in the visible light region, the effects produce various colors.

selectivity **(1)** The characteristic of a filter that determines the extent to which it modifies the frequency spectrum of signals passing through it. It is a measure of the "steepness" of the transition band, that is, a highly selective filter has an abrupt transition between the passband and stopband. **(2)** The ability of a receiver or tuner to pass a desired channel's signal while rejecting those of adjacent channels. It is the ratio of the level of the desired signal to the level of adjacent channel signal at the receiver's output expressed in decibels (dB). Hence, the higher the value, the better the *selectivity.*

selector In the ISO/OSI Reference Model, a value used at a specific layer to distinguish each of the multiple service access points (SAPs) through which the entity at that level provides services to the layer above it.

self-adapting The process in which a device or system changes parameters or operating behavior in response to differing environmental conditions.

self-capacitance In multiconductor cables, the sum of all direct capacitances between a specified conductor and each of the other conductors in the cable. *Self-capacitance* can be directly measured by connecting all unspecified conductors together at one end and measuring the capacitance between this junction and the specified conductor.

self-checking code Any code in which one (or more) bit errors produce forbidden bit combinations. Also called *error-detecting codes.* See also *codes.*

self-clocking A data stream that is encoded in such a way that it provides the clock reference for itself.

In the diagram, the top waveform represents one possible self-clocking code where the rising edge can be used to synchronize a local oscillator. The output of the local oscillator is displayed in the second trace as the code-derived clock. The final trace illustrates the lack of clock information in an unencoded binary waveform, that is, the local oscillator receives no synchronizing pulses when long strings of 0s or 1s are transferred. See also *waveform codes.*

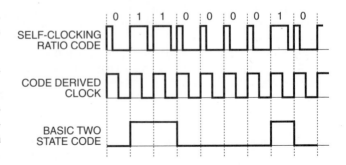

self-complement A binary coded value that becomes its own complement when each bit is complemented (inverted).

For example, the nine's complement of 3 is 6, of 4 is 5, and of 1 is 8. If the numbers in this example are encoded with a 2421 weighted binary code, then the complements are obtained simply by inverting the bits, that is,

EXAMPLE OF A SELF-COMPLEMENTING
BCD CODE

Original Value		Complement Value	
BCD	2421 Code	2421 Code (bits inverted)	BCD
0	0000	1111	9
1	0001	1110	8
2	0010 (1000)	1101 (0111)	7
3	0011 (1001)	1100 (0110)	6
4	0100 (1010)	1011 (0101)	5
5	0101 (1011)	1010 (0100)	4
6	0110 (1100)	1001 (0011)	3
7	0111 (1101)	1000 (0010)	2
8	1110	0001	1
9	1111	0000	0

Other examples of self-complementing codes include the *excess 3* code, 3321, 4311, 5211, 543-3, 84-2-1, and many more. See also *BCD, code, complement,* and *excess 3.*

self-delineating block A data block in which a specific bit pattern (flag) identifies the beginning or end of a block.

self-extinguishing The characteristic of a material that does not support combustion without an external source of heat, that is, its flame dies when the igniting flame is removed.

self-extracting archive (SEA) A compressed file (or set of files) concatenated with an executable decompression program, creating a single archive file. The archive file can be a stand-alone program, that is, it does not require any other support files or programs. When the archive file is run, it restores the compressed files to their original form.

For example, the archive format is used on the Apple Macintosh, and double-clicking a file of this type should extract its contents.

self-synchronizing code A code in which the symbol stream formed by a portion of one code word, or by the overlapped portion of any two adjacent code words, is not a valid code word.

A *self-synchronizing code* enables proper framing of transmitted code words if no uncorrected errors occur in the symbol stream. High-level data link control (HDLC) and Advanced Data Communication Control Procedures (ADCCP) frames are examples of *self-synchronizing code* words.

self-test A test or series of tests performed by a device upon itself. These tests, if passed, indicate the device is functional within the design limits. The tests may be automatically performed at power on (power on self-test—POST) or may be manually initiated.

semantics The relationship between symbols and their meaning.

semaphore A method by which multiple applications or devices can coordinate operations using messages or shared variables (in the case of software).

In operating systems, it is frequently a system variable used to synchronize concurrent processes by indicating whether an action has completed or a specific event has occurred.

semi-rigid cable A coaxial cable in which the inner conductor is surrounded by a relatively inflexible sheath or tube.

semiautomatic Combining manual and automatic features so that a manual operation is required to supply the automated feature with direction or initiation.

semiautomatic telephone system A telephone system in which the operator receives direction orally from calling parties and then establishes the connection by means of automated equipment.

semiconductor (1) Any partially conducting material without regard to any other property. For example, a semiconductive paint, tape, or wire is a material with significant resistance but conducts current in sufficient amounts to affect the circuits in which it is a part. (2) An electrically conducting solid with a resistivity between that of a metal (conductor) and an insulator. The resistance decreases with increasing temperature over some temperature range (called a *negative temperature coefficient*). The current carriers in semiconductors are negative *electrons* and positive *holes* depending on the doping of the pure (*intrinsic*) base material. The flow of current in such materials may be affected by external influences such as light, electric or magnetic fields, or other energy sources.

Intrinsic and doped semiconductor materials are used in the manufacture of transistors, integrated circuits, and semiconductor optical devices. Materials include:

Intrinsic semiconductor—a semiconductor material in which the charge-carrier concentration is essentially the same as that of an ideal crystal and are created solely by thermal excitation. The electron concentration must therefore equal the hole concentration. Examples of such semiconductor materials include silicon (Si) and germanium (Ge).

n-type semiconductor—a semiconductor material with impurities introduced such that the conduction electron concentration exceeds the mobile hole concentration. The impurity is a donor type material.

n^+*-type semiconductor*—an n-type material in which the excess conduction electron concentration exceeds that of n-type material, i.e., is very large.

p-type semiconductor—a semiconductor material with impurities introduced such that the mobile hole concentration exceeds the conduction electron concentration. The impurity is an acceptor-type material.

p^+*-type semiconductor*—a p-type material in which the excess mobile hole concentration exceeds that of p-type material, i.e., is very large.

semiconductor device A device in which the operating characteristic is determined within a *semiconductor* (2) material or at its boundary.

Examples of basic *semiconductor devices* include:

Insulated gate field effect transistor (IGFET)

Junction diode (rectifier)

Junction field effect transistor (JFET)

Junction transistor

Laser diode (injection laser diode—ILD)

Light emitting diode (LED)

Photo diode

Photo transistor

Schottky diode

Silicon bilateral switch (SBS)

Silicon controlled rectifier (SCR) (Thyristor)

Varistor (varicap)

semiselective ringing In telephony, subscriber line ringing in which all ringers on a party line respond simultaneously to an applied ringing signal. Differentiation at the called stations is accomplished by utilizing different ringing cadences or number of rings. See also *coded ringing, distinctive ringing,* and *selective ringing.*

semitone The interval between two frequencies which have a ratio of approximately the twelfth root of two. In equally tempered semitones, the ratio between any two adjacent frequencies is twelve times the log base two of the frequency ratio. Also called *semit* or *half step.*

$$r = 12 \cdot \log_2 \left(\frac{f_1}{f_2} \right)$$

SEND The clear to SEND signal bit in the STS-3 coded mark inversion (CMI) line signal.

send only A device or communications channel capable of only transmitting signals. It is not capable of receiving signals.

sender A device that generates and transmits signals in response to information received from another device or part of the system.

sendmail A UNIX program that serves as a *message transfer agent (MTA)* in an *e-mail system.*

In operation, a user generates a mail message using an e-mail client program such as Pine or Elm. The *sendmail* program, running in the background, receives the message from the user's mail program and routes to its destination. A system administrator can also use the program to *create mailing lists, aliases,* and *nicknames.*

sensitivity Generally, the ratio of response to stimulus, although in some fields the definition is reversed!

sensitivity analysis The determination of the variation of a specified system output variable in response to changes in other system variables about a selected operating or reference point. For example, the sensitivity of a particular system's frequency to changes in an element is defined mathematically as:

$$S_{E_1}^f = \frac{E_1}{f} \cdot \frac{\Delta f}{\Delta E_1}$$

where:

$S_{E_1}^f$ is the sensitivity of f with respect to E_1,
E_1 is the input variable reference point,
f is the output variable reference point,
Δf is the resultant change in output due to E_1, and
ΔE_1 is a small change of input value.

sensor A device that responds to a physical stimulus (such as light, temperature, field intensity, pressure) and may generate and transmit a signal corresponding to the stimulus or its change. *Sensors* may be active or passive. *Active sensors* require a source of power other than the signal being measured. *Passive sensors* require no power other than the input signal. See also *transducer.*

SEP An acronym for Spherical Error Probability.

separation The extent to which two communications channels on the same link are isolated, e.g., the left and right channel of a stereo broadcast or recording. It is the ratio of the desired signal level to the interfering signal level in a specified channel and is generally expressed in decibels (dB).

SEPT An acronym for Signaling End Point Translator. Part of Signaling System number 7 (SS7).

septet A byte composed of seven binary elements. Also called a *seven-bit byte,* naturally!

sequence An ordered arrangement of a set.

sequence number A number identifying the relative location of blocks or groups of blocks on a communications medium. The medium may be physical, such as a magnetic tape or transient such as a data packet.

sequencing receiver A single receiver that is sequentially switched between several transmitted carriers. These receivers are less expensive to build and consume less power than a multiple receiver system. However, because they can receive only one carrier at a time, some information may be lost.

Some Global Positioning System (GPS) receivers use this technique to monitor the position of constellation of visible satellites. The technique reduces cost but may limit the obtainable accuracy because simultaneous pseudoranges are not obtained.

sequential access A data access method in which every record from the beginning of a file must be read before arriving at the desired records location. For example, a magnetic tape drive required the tape be physically moved from the beginning through all locations up to the desired record before that desired record could be read. Also called *serial access.*

sequential logic A logic system with one or more input channels and one or more output channels and in which at least one output channel is influenced by not only the current state of the inputs but by previous input states. As a result, for a given set of input states, there may be more than one possible output state. See also *state diagram.*

sequential transmission A synonym for *serial transmission.*

SERDES An acronym for SERializer-DESerializer.

serial (1) Pertaining to the time sequencing of two or more processes or events. (2) Pertaining to a process in which all events occur sequentially, i.e., one after the other. An example is the serial transmission of the bits of a character according to the ITU-T V.25 protocol. *Serial* and *parallel* usually refer to devices, while *sequential* and *consecutive* usually refer to processes. Contrast with *parallel.* (3) Pertaining to the sequential or consecutive occurrence of multiple related activities in a single device or channel. (4) Pertaining to the sequential processing of the individual parts of a whole, such as the bits of a character or the characters of a word, using the same facilities for successive parts.

serial access (1) Pertaining to the sequential or consecutive transmission of data into or out of a device, such as a computer, transmission line, or storage device. (2) A process by which data are obtained from a storage device or entered into a storage device in such a way that the process depends on the location of those data and on a reference to data previously accessed. Also called *sequential access.*

Serial Bus See *IEEE P1394.*

serial communications The sending of information between computers or between a computer and a peripheral device over a single wire *one element (bit) at a time.* Serial communications can use either *synchronous* or *asynchronous* protocols. Compare with *parallel communications.*

serial computer A computer with some specified characteristic that is serial.

For example,

- A computer that has a single arithmetic and logic unit.
- A computer that manipulates all bits of a word serially.

serial interface An input/output port on a device that is designed for *serial communications*. Generally, the term *serial interface* refers to one of the Electronic Industries Association (EIA) or International Telecommunications Union-Telecommunitator (ITU-T) defined standards or recommendations, i.e., EIA-232, EIA-422, EIA-423, EIA-449, X.21, V.24, V.28, ISO 2110, ISO 4902, etc.

Serial Line Interface Protocol (SLIP) See *SLIP*.

serial parity See *longitudinal redundancy check*.

serial port A term that refers to the specific hardware entry channel to which a serial device (such as a modem or mouse) is attached to a data terminal equipment (DTE). It is an interface in which only one wire is available for data transmission in a given direction; hence, information must be transferred sequentially, i.e., one bit at a time. Contrast with *parallel port*.

serial to parallel conversion Conversion of a sequential stream of data elements, i.e., one element at a time, into a data stream consisting of multiple data elements transmitted simultaneously.

The inverse process is *parallel to serial conversion*.

serial transmission The transmission of a single signaling element one at a time from one device to another device. In the case of digital characters, each character's bits are transmitted one at a time. Also called *sequential transmission*.

serialize To change from a parallel format to a serial format. For example, to change 8-bit words to individual bits for modem transmission.

serializer A synonym for *parallel-to-serial conversion*.

series circuit A circuit in which two or more components or devices are connected "end to end." In a *series circuit,* the total current flows through each of the components or devices; however, the voltage is divided across the series string of components.

The circuit below contains five *series* elements: one voltage source and four impedances. Note that the same current passes through every device. See also *parallel circuit*.

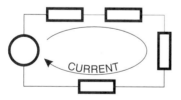

series-parallel circuit A circuit containing only two terminal elements that can be constructed by successively connecting series-connected groups in parallel.

For example, a series-parallel connected battery provides more voltage and current capability than is available from a single cell.

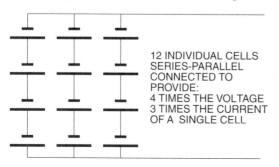

12 INDIVIDUAL CELLS
SERIES-PARALLEL
CONNECTED TO
PROVIDE:
4 TIMES THE VOLTAGE
3 TIMES THE CURRENT
OF A SINGLE CELL

series T junction A three-port waveguide junction in which the impedance of the branch waveguide is predominantly in series with the impedance of the main waveguide at the junction.

server (**1**) A node on a local area network (LAN) that runs the administrative software and/or provides shared services and resources for other computing stations. The administrative software controls access to the network and allocates the resources to requesting users. The shared resources can be files on a mass storage device, printer(s), modem(s), gateways to other networks, and so on.

Servers performing basically one function are called dedicated servers, e.g., file servers, print servers, backup server, post-office server. A *server* can also be non-dedicated, where the node can be used in other ways, such as a workstation. (**2**) A program running on a host computer which exists to assist clients in a specific way. On the Internet, programs such as Archie, Gopher, Veronica, WAIS, and whois are servers, as are name servers, domain servers, news servers, and e-mail servers.

server based network A network in which one or more nodes have special status as dedicated servers. Other nodes (workstations) must go through a server for resources on other machines.

This is in contrast to a peer-to-peer network, in which each node may be either a server or workstation as the need arises.

Server Message Block (SMB) A Microsoft-originated distributed system that enables access to another computer's files and peripherals over the network as if they were local.

server mirroring A network security method in which two or more redundant computers operate as servers and do exactly what the other is doing simultaneously. Should one server become unavailable, the other can supply the network workstation needs.

service In the International Standards Organization (ISO) Open Systems Interconnection Reference Model (ISO/OSI-RM), a capability of a given layer, and the layers below it, that is provided to the entities of the next higher layer.

service access code (SAC) North American Number Plan Area (NPA) codes that are set aside for special network use rather than for the definition of geographic areas. For example, 800 and 900 provide networkwide user paid access and caller paid access, respectively. Other codes assigned include 600, 700, and 888.

service access point (SAP) (**1**) A physical point at which a circuit may be accessed. (**2**) The actual interfaces between ISO/OSI Reference Model layers. They are unique addresses that the layers involved can use to exchange requests, replies, and data, i.e., the point at which a designated service may be obtained. Because multiple programs may be running at a given layer, each needs its own *SAP* for communicating with the layers above and below it.

SAPs represent the generic communications links between layers. To identify the layer under discussion, it is common practice to include a letter identifying the lower layer in the pair. For example, a *SAP* linking a presentation layer process to the session layer below it would be known as an SSAP. *SAP* addresses are assigned by the IEEE standards office. The table on page 464 illustrates a few of the common protocols with their addresses in hexadecimal form.

SAP ADDRESSES FOR COMMON PROTOCOLS

Protocol	Address
TCP/IP SNAP	AAh
ISO Network Layer	F5h
ARPAnet—IP	06h
Novell NetWare—IPX	E0h
IBM—NetBIOS	F0h
IBM—SNA Group Path Control	05h
IBM—SNA Individual Path Control	04h
3Com—XNS	80h

service access point identifier (SAPI) A field in the LAP-D (link access protocol-D channel) frame that indicates the *logical* address of the called user.

service advertising protocol (SAP) A transport layer protocol, in Novell's NetWare that servers can use to make their services known on a network. Servers advertise their services using *SAP* packets that are received and stored by other routers on the network. Each router maintains a database of all "nearby" servers. Each router periodically broadcasts this information to other routers, typically, every 60 seconds or when something changes.

Stations that need a service can broadcast *SAP* request packets. These packets will be answered by the "nearest" router with information about the requested service.

service bits Overhead bits within a communications packet that are neither information nor error control bits. They are used for providing a network service, such as a request for a repetition or for a numbering sequence.

service channel A synonym for an *orderwire circuit*.

service class See *class of service*.

service code In telephony, any of the three-digit destination codes for use by customers to obtain telephone company services, e.g., repair service (611), directory assistance (411), the business office (811), emergency service (911), and so on.

service data unit (SDU) In layered systems, data sent by a user of the services of a given layer and transmitted to a peer service user semantically unchanged. Information is delivered as a unit between adjacent entities. A *service data unit* may also contain a protocol data unit (PDU).

service feature In telephony, any of the many special functions that may be either specified initially or added to the user's basic service provided by the switching system. Such features include call waiting, call forwarding, distinctive ring, and automatic caller identification.

service life The time during which satisfactory performance of a device can be expected according to a specified set of conditions. For example, the *service life* of a primary battery at a designated current is the time before its working voltage falls below a specified cutoff voltage.

service line An exchange line associated with multiple data station installations. It is used to provide monitoring and testing of both customer and telephone company data equipment.

service point (1) In IBM's Network Management Architecture (NMA), software through which a non-SNA device or a network can communicate with the NMA network manager. **(2)** In telephony, the last point of service that is rendered by a carrier under applicable tariffs and usually corresponds to the demarcation point. The point, usually on the customer premises, where service channels or facilities are terminated in switching equipment used for communications with phones or customer-provided equipment located on the premises. The customer is responsible for equipment and operation from the service termination point to user end instruments. Also called the *service termination point*.

service profile identifier (SPID) A parameter used by an Integrated Services Digital Network (ISDN) central office switch to select which specific set is to be used when more than one set is attached to the same central office line.

service provider An organization that provides (generally for a price) connections to a communications network. Networks include the local telephone company and a multitude of companies that provide access to the Internet.

Examples of service providers include Prodigy, America Online, GEnie, CompuServe, and many others.

service signals Signals that enable data systems equipment to function correctly and possibly to provide ancillary facilities. Also called *housekeeping signals*.

service termination point See *service point* (2).

service voltage The root-mean-square (RMS) voltage at the point where the provider and the user circuits are connected. Sometimes called the *mains voltage* or the *ac mains*.

SES (1) An abbreviation of Severely Erred Second. **(2)** An abbreviation of Satellite Earth Station.

SESFE An abbreviation of Severely Erred Second, Far End.

SESP An abbreviation of Severely Erred Second, Path.

session (1) In general, the time during which a process is running or the time during which a program accepts inputs, responds to commands, or processes information. **(2)** In communications, the time during which a logical connection is maintained between two nodes (between servers and/or workstations) and usually are engaged in information transfers. This connection remains in effect until the task that caused the session is completed or some other constraint forces an end to the connection, e.g., an inactivity timeout.

session ID A unique number, in an AppleTalk network, associated with each session. The *ID* is used to identify the session and to distinguish it from other sessions.

session layer The fifth layer in the seven-layer ISO/OSI communications reference model. It provides the coordination of communications and the means for establishing a real-time connection to another network node. It establishes and terminates connections between processes, provides for the recovery at the transport level, provides virtual circuit service if the transport layer does not do so, and prevents two machines from trying to access the local area network (LAN) at the same time. See also *OSI*.

set (1) A group of objects, entities, or concepts, having one or more characteristics (or properties) in common. A collection of related data. **(2)** The placing of a device in one of two (or more) operating states. See also *reset*. **(3)** A contraction of telephone sub*set*.

SET An abbreviation of Secure Electronic Transfer.

setup The procedure involved in preparing software or hardware for operation. The procedure involves many of the following: connecting communication cables, connecting power, setting option switches, installing device drivers, and setting software parameters.

seven-bit byte A synonym for *septet*.

severely erred second (SES) An interval when the bit error rate (BER) exceeds a threshold (generally 10^{-3}), there are excessive cyclic redundancy check (CRC) errors in an extended super frame, there is a frame slip, or an alarm condition exists.

SEW An acronym for Surface Electromagnetic Wave.

sex changer See *gender changer*.

sexadecimal See *hexadecimal*.

sextet A byte composed of six binary elements. Also called a *six-bit byte* (*SBT*).

SF (1) An abbreviation of Single Frequency. (2) An abbreviation of Signal Frequency. (3) An abbreviation of SuperFrame. 12 T-1 frames combined. (4) An abbreviation of Scale Factor.

SF signaling From single frequency signaling. A method of in-band signaling that uses the presence or absence of a tone to indicate line state. See *single frequency (SF) signaling*.

SFC An abbreviation of Switch Fabric Controller.

SFD (1) An abbreviation of Simple Formattable Document. (2) An abbreviation of Start of Frame Delimiter.

SFET An abbreviation of Synchronous Frequency Encoding Technique.

SFJ An abbreviation of Swept Frequency Jamming.

SFS An abbreviation of Suomen Standardisoimisliitto Informaatiopalvelu (Standards Association of Finland).

SFT An abbreviation of System Fault Tolerance.

SFTP An abbreviation of Shielded Foil Twisted Pair wire.

SFTS An abbreviation of Standard Frequency and Time Signal.

SFU An abbreviation of Store and Forward Unit.

SG (1) An abbreviation of Signal Ground. (2) An abbreviation of Signal Generator. (3) An abbreviation of Study Group. (4) An abbreviation of Super Group. (5) An abbreviation of System Gain.

SGC An abbreviation of SuperGroup Connector.

SGD An abbreviation of Signal GrounD.

SGDF An abbreviation of SuperGroup Distribution Frame.

SGE An abbreviation of Subscriber Group Equipment.

SGML An abbreviation of Standard Generalized Markup Language.

SGML-B An abbreviation of Standard Generalized Markup Language—Binary.

SGMP An abbreviation of Simple Gateway Management (or Monitoring) Protocol. A predecessor of Simple Network Management Protocol (SNMP).

SGND An abbreviation of Signal GrouND.

SGNET An acronym for Sea Grant NETwork.

sgn(x) A mathematical shorthand way of expressing the sign of a number *x*. For example, sgn(80) is "+" and sgn(−32) is "−."

SGT An abbreviation of Satellite Ground Terminal.

SGX An abbreviation of Selector Group MatriX.

Sh The symbol for a unit of information, that is, a shannon.

SH (1) An abbreviation of Switch Hook. (2) An abbreviation of Send Hub. (3) An abbreviation of Session Handler. (4) An abbreviation of SHunt. (5) An abbreviation of Switch Handler. (6) An abbreviation of Sample and Hold, although it is more frequently written S/H.

shadow loss (1) In general, the attenuation of a signal caused by obstructions in the propagation path, i.e., between the transmitter and the receiver. (2) In a reflector antenna, the relative reduction in the effective aperture of the antenna caused by the masking effect of other antenna parts that obstruct a portion of the radiation path. Examples of such parts include the feed horn and the secondary reflector.

shadow ram A random access memory (RAM) area a computer uses to store frequently accessed read only memory (ROM) code. RAM, being faster than ROM, allows information to be accessed faster; hence, it speeds up computer operations.

shadow server Software or hardware that copies files so that there are always two current copies of each file. Also called a *backup server*.

shall Indicates that which is obligatory in any conforming specification; that which is necessary or essential to meet requirements.

shannon (Sh) The unit of information derived from the occurrence of one of two equiprobable, mutually exclusive, and exhaustive events. A bit may, with perfect formatting and source coding, contain 1 Sh of information. However, the information content of a bit is usually less than 1 Sh.

Shannon limit In 1948 Claude Elwood Shannon (1916–) demonstrated that the theoretical maximum capacity of an ideal bandwidth-limited communications channel with only white noise interference is related to bandwidth and the S/N ratio by the following equation:

$$C = Bw \cdot \log_2\left(1 + \frac{S}{N}\right)$$

where:
C is the maximum channel capacity (bps),
Bw is the channel bandwidth (Hz),
S is the signal power (watts), and
N is the noise power (watts).

Error-correction codes can improve the communications performance relative to uncoded transmission, but no practical error-correction coding scheme exists that can closely approach the theoretical performance limit set by Shannon's law. The diagram illustrates the effects of noise on the maximum channel capacity for several communications channels with various bandwidths (including 3200 Hz—the bandwidth of an ideal telephone channel). Also indicated on the dia-

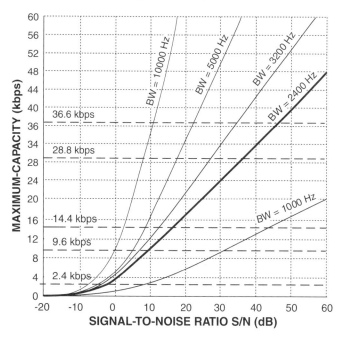

gram are several common modem data rates. From the diagram, one can observe that to achieve a 28.8-kbps link on a 2400-Hz channel, one must have an S/N ratio better than about 35 dB, while a 9600-bps connection requires only better than 12 dB. Also, extending the data rate beyond 28.8 kbps to, say, 36.6 kbps requires a channel with a 10-dB better S/N ratio. It should also be pointed out that this does not include any anomalous effects in the transmission path except white noise (group delay, frequency errors, etc.). Furthermore, the nonideal characteristics of the modem are also ignored. Also called *Shannon's law.*

shaping network An electronic circuit (network) inserted into a circuit or signal path to modify a waveform in a desirable manner.

shareable A file, device, or process that is available to multiple users and can be used simultaneously if requested.

shared lock An attribute on a database that allows multiple users to concurrently read from the database. No user may acquire an exclusive lock until all shared locks are released.

shared processing A network configuration in which a server processes tasks for multiple attached stations. The nodes must share the computing power of the central processor; hence, the busier the network becomes, the slower tasks will be processed.

Shared processing is the opposite of *distributed processing,* in which tasks are performed by multiple (possibly specialized) nodes somewhere on the network.

shareware Copyrighted software that is made available on a "try-before-buy" basis.

It is commercial software that a user is allowed to run and evaluate before paying for it—assuming it proves to be useful. Payment/registration for the software is on the honor system and generally directly to the author. Upon registration, the user generally is entitled to technical support, upgrade notices, and possibly printed copies of the users manual. *Shareware* is distributed by the author or owner via bulletin boards, online services, and disk vendors. Copying and passing the software among friends are generally encouraged. Again, registration is required for continued use of the product. See also *Association of Shareware Professionals* (*ASP*), *freeware, proprietary software,* and *public domain software.*

SHARP An acronym for Self Healing Alternate Route Protection. A system of redundant traffic carrying cables between two separate entities, routed along different paths.

sheath The outermost protective *jacket* or covering surrounding a cable (copper, fiber, or structural). The outer covering or coverings is a tough material, often plastic, that is resistant to environmental hazards such as abrasion, rodent damage, liquid intrusion, and solar radiation, and is used to protect interal cable components such as optical fibers or metallic conductors that transport the signal or power.

There may be more than one *sheath* surrounding a given cable. For example, some cables have an inner *sheath* surrounded by metallic armor, over which is an outer sheath. Also called a *jacket.*

sheath miles The actual length of cable installed between two separate points, not, the distance between the points.

shelfware Software that is purchased, frequently on a whim, and never used.

shell That portion of a computer *operating system* (*OS*) that interfaces with the user.

It is an OS *command interpreter,* i.e., software that reads an input specifying an operation and that may perform, direct, or control the specified operation. The *shell* may take its input from either a user terminal or from a file. For example, a shell may permit a user to switch among application programs without terminating any of them. See also *kernel.*

shell account A method of Internet access in which the user is given access to a *shell* of the host computer. *Shell accounts* generally require that the connection be established by way of a modem on a dial up connection.

From a *shell account* a user can run application programs such as FTP (file transfer protocol), Pine, Gopher, and others. There are three levels at which a user may be connected to the host on a shell account:

- *Basic Prompt*—the simplest and least informative *shell account.* There is simply a prompt, a cursor, and nothing else—nothing to provide a user with any suggestion as to what to do next. It is generally useful only to relatively knowledgeable users.
- *Menu selection*—the user is provided with a menu of options from which to choose. This is an easier interface to use but may not provide all of the options available at the host *basic prompt.*
- *Graphical user interface* (*GUI*)—a "front end" program that runs on the user's PC. It provides the easiest method of interfacing the Internet with a *shell account.* Generally, GUIs provide full mouse selection, pull down menus, and pictorial representations (rather than strictly textual as do the previous two methods).

See also *PPP* and *SLIP.*

SHF An abbreviation of Super High Frequency.

shield (**1**) In electrical or optical uses, a material impervious to specific frequencies in the electromagnetic spectrum that is used to protect an enclosed region from exposure to radiation or to prevent radiation from escaping the enclosure. Specific shielding includes:

- A conductive material placed around an insulated wire, circuit, system, or device to reduce the effects of electromagnetic interference (EMI) or radio frequency interference (RFI). The conductive material generally is not connected directly to the protected elements but is connected to some common ground point. When the protected element is a wire, the shield may be placed around a single conductor or a group of conductors. See also *coaxial cable, shielded cable shielded pair,* and *shielded twisted pair.*
- A magnetically susceptible material (mu-metal) placed around a region which prevents external magnetic fields from reaching the protected region.
- An optically opaque material used to prevent light from reaching a protected region.
- A reflective material used to prevent electromagnetic radiation from entering or exiting a specified volume.

(**2**) In mechanical uses, a protective cover that prevents the accidental contact of objects or persons with parts or components operating at hazardous voltage levels.

- A mechanical element used to protect the enclosed region from physical intrusion or to prevent entities in the physical region from escaping.

shield effectiveness The relative ability of a shield to screen out undesirable radiation.

shield factor The ratio of induced noise with a shield present to the noise without a shield.

shield percentage The ratio of actual coverage of a region to the maximum possible coverage.

shielded cable Cable in which the individual insulated wire(s) is (are) enclosed in a conductive envelope. Multiple shields, each insulated from the other, may be used in some instances.

The shield(s) provides protection against external electrical noise as well as preventing the signal from interfering with external devices. Two wires within a shield are called a *shielded pair.* Frequently, the pair is wrapped one around the other so as to further reduce the effects of external fields (a *twisted shielded pair*). A single shielded conductor with accurately controlled dimensions and a homogeneous dielectric is called a *coaxial cable.*

shielded pair   A two-conductor transmission line surrounded by an electrically conductive sheath that protects it from the effects of external electromagnetic fields and confines fields produced within the line.

shielded twisted pair (STP)   A transmission line composed of a twisted two-wire transmission line surrounded by a sheath of electrically conductive material. The sheath protects the pair from the effects of external electromagnetic fields and confines fields produced within the line. The twisting reduces the effects of any fields that penetrate the shield. See also *twisted pair.*

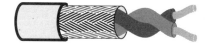

shift in <SI>   The character used on some keyboards to select an alternate character set. In ASCII it is decimal 15 (<Ctl-O>).

shift out <SO>   The character used on some keyboards to deselect the alternate character set. In ASCII it is decimal 14 (<Ctl-N>).

shift register   An electronic memory circuit in which all bits are shifted one physical position (up or down) on each clock cycle.

shock excitation   A synonym for *impulse excitation.*

short circuit   An abnormal condition that occurs when there is an unwanted electrical connection of relatively low impedance between two conductors. A *short circuit* may be either accidental or intentional and does not necessarily result in the flow of excess current, i.e., between points of equal potential. It is also called a *short,* and in telephony it may be called a *cross.*

short-haul modem   A modem or other communication device that connects a data terminal equipment (DTE) to a communication line but is capable of reliable transmission only over distances of a few miles or less. *Short-haul modems* frequently are not really modems in the strictest sense (i.e., they do not transport data using carrier modulation techniques) but are devices that transmit digital data directly as baseband signals such as T-1 or Integrated Service Digital Network (ISDN).

short-haul toll traffic   A general term describing message toll traffic between nearby points. In common usage, this term is applied to message toll traffic between points less than 20 to 50 miles apart.

short message service (SMS)   An add-on facility of the Global System for Mobile communications (GSM) digital cellular service that allows the transmission of limited text messages similar to those of pagers.

Using *SMS,* one can send a short alphanumeric message (up to 160 alphanumeric characters) to the cellular phone for display, much like an alphanumeric pager system. The message is buffered by the GSM network until the phone becomes active.

short wavelength   In optical communication, optical radiation having a wavelength less than approximately 1 μm.

shortest path first (SPF)   A routing strategy for passing packets between routers. A strategy used in Token Ring networks that may include connections to IBM mainframes.

shortwave   In radio communications, an unofficial designation of the band of radio frequencies between approximately 3 MHz and 30 MHz. The term is not officially recognized by the international community.

shot noise   **(1)** Noise generated by random fluctuations in the number and velocity of electrons in a conductor or emitter. **(2)** In optical systems, the term is often applied (loosely) when speaking of the mean square shot noise current (amperes^2) rather than noise power (watts).

shout   The act of using capital letters in an e-mail message (or on an Internet USENet posting) is considered shouting. Use normal upper and lower case letters for normal communications. To highlight a word, phrase, or comment use quotation marks (" or ') or asterisks (*).

shovelware   Extra software dumped onto a distribution medium, such as a CD-ROM, to fill up space or provide fodder for the marketing hype. *Shovelware* may have little to do with the product being sold.

showstopper   **(1)** A hardware or software bug that effectively makes a product unusable until it is fixed. **(2)** Hardware or software that is so stunningly good that all observers around the implementation are in awe.

SHR   An abbreviation of <u>S</u>elf <u>H</u>ealing <u>R</u>ing. A ring topology that can function with one failure in either the transmission medium or the node.

SHTTP or S-HTTP   An abbreviation of <u>S</u>ecure <u>H</u>yper<u>T</u>ext <u>T</u>ransfer <u>P</u>rotocol. An implementation of HTTP with the RSA (Rivest, Shamir, Allaman) Data Security encryption algorithm incorporated. *SHTTP* allows the transmission of secure data, signatures, and even payments over Internet.

shunt   See *parallel.*

SI   **(1)** The abbreviation of Le <u>S</u>ystème <u>I</u>nternational d'Units. Also called the *metric system,* the *SI* system (first proposed in 1670) provides three major features:

- *Decimalization* of values. All values regardless of the physical measure are represented in the same units and format. For example, 3 and ⁷⁄₁₆ of an inch would be expressed as 3.4375 inch.
- Standard *prefixes.* For convenience and ease of reading, a series of prefixes to the units is defined. These represent a power of 10 multiplier.

SI PREFIXES

Name	Term (U.S.)	Symbol	Multiplier Base 10	Multiplier Base 2
yotta			10^{24}	2^{80}
zetta			10^{21}	2^{70}
exa	quintillion	E	10^{18}	2^{60}
peta	quadrillion	P	10^{15}	2^{50}
tera (megamega)	trillion	T (MM)	10^{12}	2^{40}
giga (kilomega)	billion	G (kM)	10^{9}	2^{30}
mega	million	M	10^{6}	2^{20}
(myria)			10^{4}	
kilo	thousand	k	10^{3}	2^{10}
(hecto)		(h)	10^{2}	
(deka)		(da)	10	
(deci)		(d)	10^{-1}	
(centi)		(c)	10^{-2}	
milli	thousandth	m	10^{-3}	2^{-10}
(decimilli)		(dm)	10^{-4}	
micro	millionth	μ	10^{-6}	2^{-20}
nano (millimicro)	billionth	n (mμ)	10^{-9}	2^{-30}
pico (micromicro)	trillionth	p (μμ)	10^{-12}	2^{-40}
femto	quadrillionth	f	10^{-15}	2^{-50}
atto	quintillionth	a	10^{-18}	2^{-60}
zepto			10^{-21}	2^{-70}
yacto			10^{-24}	2^{-80}

Names and symbols in parentheses are to be avoided, however, they may be found in older literature.

- Standardized *base units.* All units of a given measurement are expressed in the same units. It is not necessary to express weight, for example, as 1 ton 202 pounds 8 ounces; it could be expressed as 1.1015 tons or 2202.5 pounds.

Also, all units are reduced to nine base units founded on invariable natural phenomena (not arbitrary measures such as the length of a Pharaoh's arm).

FUNDAMENTAL SI BASE UNITS

Unit Name	Symbol	Description
ampere(s)	A	Electric current
candela(s)	cd	Luminous intensity
kelvin(s)	K	Thermodynamic temperature
kilogram(s)	kg	Mass
meter	m	Length
mole(s)	mol	Amount of substance
radian	rad	Plane angle
second(s)	s	Time
steradian	sr	Solid angle

Derived units commonly used are given special names and symbols; however, they are still related to the nine Base SI Units as shown in the following table.

SI DERIVED UNITS

Unit Name	Symbol	Description	Relation to Base Units	
becquerel(s)	Bq	Radioactivity	s^{-1}	1/s
coulomb(s)	C	Electric charge	$A \cdot s$	$A \cdot s$
farad(s)	F	Capacitance	$m^{-2} \cdot kg^{-1} \cdot s^4 \cdot A^2$	C/V
gray(s)	Gy	Absorbed dose	$m^2 \cdot s^{-2}$	J/kg
henry(ies)	H	Inductance	$m^2 \cdot kg \cdot s^{-2} \cdot A^{-2}$	Wb/A
hertz	Hz	Frequency	s^{-1}	1/s
joule(s)	J	Energy	$m^2 \cdot kg \cdot s^{-2}$	$N \cdot m$
lumen(s)	lm	Luminous flux	$cd \cdot sr$	cd/sr
lux	lx	Illuminance	$m^{-2} \cdot cd \cdot sr$	lm/m^2
newton(s)	N	Force	$m \cdot kg \cdot s^{-2}$	$kg \cdot m/s^2$
ohm(s)	Ω	Resistance	$m^2 \cdot kg \cdot s^{-3} \cdot A^{-2}$	V/A
pascal(s)	Pa	Pressure; stress	$m^{-1} \cdot kg \cdot s^{-2}$	N/m^2
siemens	S	Conductance	$m^{-2} \cdot kg^{-1} \cdot s^3 \cdot A^2$	A/V
tesla(s)	T	Magnetic flux density	$kg \cdot s^{-2} \cdot A^{-1}$	
volt(s)	V	Electric potential	$m^2 \cdot kg \cdot s^{-3} \cdot A^{-1}$	W/A
watt(s)	W	Power	$m^2 \cdot kg \cdot s^{-3}$	J/s
weber(s)	Wb	Magnetic flux	$m^2 \cdot kg \cdot s^{-2} \cdot A^{-1}$	$V \cdot s$
		Area	m^2	
		Volume	m^3	
		Linear velocity	m/s	
		Radial velocity	rad/s	
		Acceleration	m/s^2	
		Torque	$N \cdot m$	
		Density	kg/m^3	

Some non-SI units in specialized fields and those of practical importance will remain in usage internationally, for example,

SOME NON-SI UNITS ACCEPTED

Name	Symbol	Relation to SI Units
minute(s)	min	1 min = 60 s
hour(s)	h	1 h = 3600 s
day(s)	d	1 d = 86400 s
degree(s)	°	$1° = (\pi/180)$ rad
minute(s)	'	$1' = (\pi/10800)$ rad
second(s)	"	$1'' = (\pi/648000)$ rad
bar(s)	bar	1 bar = $10^{5\ Pa}$
liter(s)	l	$1 l = 10^{-3}\ m^{-3}$

(2) An abbreviation of Sequenced Information. A LAP-D frame type. **(3)** An abbreviation of Secondary Input. The fiber optic port in Fiber Distributed Date Interface (FDDI) that receives signals from the secondary fiber ring. **(4)** An abbreviation of Step Index. **(5)** An abbreviation of Sample Interval. **(6)** An abbreviation of Serial Input. **(7)** An abbreviation of Shift In character (ASCII 15 0Fh). **(8)** An abbreviation of Speech Interpolation.

SIA (1) An abbreviation of Serial Input Adapter. **(2)** An abbreviation of Serial Interface Adapter.

SIB (1) An abbreviation of Serial Input Board. **(2)** An abbreviation of System Interconnect Bus. **(3)** An abbreviation of Systems Information Bulletin.

SIC (1) An abbreviation of Solar Induced Current. **(2)** An abbreviation of Serial Interface Circuit. **(3)** An abbreviation of Silicon Integrated Circuit.

SID (1) An abbreviation of Source IDentifier. **(2)** An acronym for Sudden Ionospheric Disturbance. **(3)** An acronym for Switch Interface Device. **(4)** An abbreviation of Service IDentification number. A five-digit number that identifies individual cellular carrier providers. **(5)** An abbreviation of Serial Input Data. **(6)** An abbreviation of Sound Interface Device.

side circuit Either of the two physical circuits used for deriving a *phantom circuit*. See also *phantom circuit*.

side frequency A sideband frequency.

side hour All hours other than the *busy hour*.

side lobe A radiation lobe in any direction other than that of the major lobe.

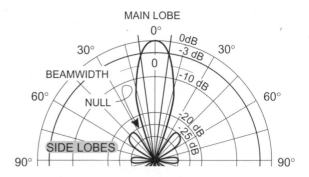

sidebands The band of frequencies on both sides of the carrier frequency that are generated by the process of modulation. Those frequencies higher than the carrier frequency are called the *upper sideband,* and those below the carrier are called the *lower sideband.*

The spectrum drawing depicts a pure carrier modulated by a single frequency tone. Each additional tone in the modulation will add another frequency to the upper and lower sidebands. With conventional amplitude modulation (AM), both sidebands are present and are located at the carrier plus the modulation frequency and at the carrier minus the modulation frequency. Transmission in which one sideband is removed is called single-sideband transmission.

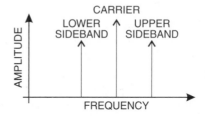

sideband suppression A process that attenuates one of the sidebands generated during modulation.

sidecar Any device added to an existing piece of equipment in an awkward manner, e.g., the IBM PC compatibility box that could be bolted onto the side of an Amiga. Designed and produced by Commodore, it is reported to have broken all of the company's own design rules. Another example is any of various devices designed to be connected to the expansion slot on the left side of the Amiga 500 (and later, 600 and 1200), which included a hard drive controller, a hard drive, and additional memory.

sidereal time A time standard based on the diurnal rotation of a star relative to the fixed stellar system. See also *time*.

sidetone In telephony, the sound-level output of a telephone subset due to the sound-level input on the same telephone.

Subjectively, the *sidetone* level affects how loud a speaker talks and how room ambient noise influences the far end listener's understanding. High *sidetone* levels tend to cause the speaker to talk quieter, while lower levels tend to make the speaker shout, and the absence of *sidetone* gives the speaker a feeling that the equipment is nonfunctional. Some telephone instruments have been designed to take advantage of the *sidetone* influence on speaker talking level. Specifically, the *sidetone* level is reduced on long loops, "causing" the speaker to talk louder than normal. It is increased on short loops to reduce the speaker's talk power. Also called a *talkback circuit*.

sidetone objective loudness rating (SOLR) A numerical rating of the sidetone loudness. It is defined as:

$$SOLR = -20 \cdot \log_{10} \frac{SPL_E}{SPL_M}$$

where SPL_E is the sound pressure level at the earpiece reference point and SPL_M is the sound pressure level at the microphone reference point.

SIDF An abbreviation of $\underline{S}$ystem $\underline{I}$ndependent $\underline{D}$ata $\underline{F}$ormat. A local area network (LAN) backup standard developed by Novell Incorporated.

siemens (S) The SI unit of *admittance* (conductance). One *siemen* is the conductance of a conductor in which a current of one ampere is produced by an electrical potential difference of one volt. The unit of conductance was formerly called the *mho* (℧).

SIF An abbreviation of $\underline{S}$ignaling $\underline{I}$nformation $\underline{F}$ield. The payload of a signaling packet or a Signaling System 7 (SS7) message signaling unit (MSU).

SIG (1) An acronym for $\underline{S}$pecial $\underline{I}$nterest $\underline{G}$roup, usually used as a prefix. For example:

- SIGCAPH—Special Interest Group on Computers and the Physically Handicapped.
- SIGCOMM—Special Interest Group on Data COMMunications.
- SIGCS—Special Interest Group on Computers and Society.

(2) An abbreviation of $\underline{SIG}$nal. (3) An abbreviation of $\underline{SIG}$nature.

sign The position of a number field where the direction of a number is located. If the direction is greater than or equal to zero, the *sign* is positive; if the direction is less than zero, the *sign* is negative.

Generally, in a binary number field the most significant bit position (leftmost) is the sign bit. It is zero if the number is positive and one if the number is negative.

sign off See *logoff*.

sign on See *logon*.

signal (1) A general term referring to any variation in a physical quantity used to convey information. Examples of quantities used as a carrier (and possible use) include:

Audible, telephone ringers, klaxons;

Current, the hook switch signaling and voice transport in a telephone;

Frequency, the carrier in a frequency shift key (FSK) modem;

Light, visual indicators, fiber optics;

Phase, the carrier in a phase shift key (PSK) modem;

Seismic, waves generated by an earthquake;

Smoke, impending disaster;

Tones, the error beep from a computer speaker;

Voltage, interface between data communication equipment (DCE) and data terminal equipment (DTE).

(2) As applied to electronics, any transmitted electrical impulse or wave. (3) Operationally, a type of message, the text of which consists of one or more letters, words, characters, signal flags, visual displays, or special sounds, with prearranged meaning. It is conveyed or transmitted by visual, acoustical, or electrical means.

signal compression (1) In analog systems (usually audio systems), reduction of the *dynamic range* of a signal by controlling its output level as an inverse function of the instantaneous input level relative to a specified reference level. It is usually expressed in decibels (dB).

In operation,

- The instantaneous values of the input signal that are low, relative to the reference level, are increased, and those that are high are decreased.
- The compression may be linear or nonlinear (but is always monotonic) with respect to signal level.
- Compression may be essentially instantaneous or have fixed or variable delay times.
- *Signal compression* always introduces distortion, which is usually not objectionable, if the compression is limited to a few dB.
- The original dynamic range of a compressed signal may be restored by the inverse process called expansion.

Signal compression is used for such purposes as:

- Improving signal-to-noise ratios prior to digitizing an analog signal for transmission over a digital carrier system.
- Preventing overload of succeeding elements of a system.
- Matching the dynamic ranges of devices.

See also *compressor, compandor,* and *data compression*. (2) In facsimile systems, a process in which the number of pels scanned on the original is larger than the number of encoded bits of picture information transmitted.

signal computing system architecture (SCSA) A set of common and open standards that both telecommunications equipment manufacturers and computer equipment manufacturers can use to create telephone systems. The family of standards extends from low-level hardware interfaces, such as buses, to high-level application programming and software interfaces. It defines how data modems, facsimile, voice, and video devices may operate together.

signal conditioner A device that modifies an electrical signal in order to make it more useful. For example, it filters out noise or amplifies the signal.

signal contrast The ratio, expressed in decibels (dB), of the white signal level to the black level of a facsimile signal.

signal conversion equipment A synonym for *modem*.

signal delay The transmission time of a signal through a network. The delay may be inherent in the transmission technology or may be purposefully inserted.

For example, the *signal delay* from an Earth station to a synchronous satellite back to another Earth station is about 0.24 seconds. (The satellite altitude is about 22 300 miles, so the signal path length is about 44 600 miles and the speed of light is about 186 000 mi/sec.)

signal distance A measure of the difference between a given signal and a reference signal.

- For analog signals, the *signal distance* is the root-mean-square difference between the given signal and a reference signal over a symbol period.
- For digital signals, it is the number of positions of difference between two binary words of equal length.

See also *codes* and *Hamming distance*.

signal droop In an otherwise essentially flat-topped rectangular pulse, *droop* is a distortion characterized by a decline of the pulse top. See also *pulse*.

signal element That part of a signaling code that occupies the shortest time interval.

Examples of *signal elements* include signal transitions, significant conditions, significant instants, and binary digits (bits). Also called the *unit duration* of the code and the *symbol time*.

signal expansion Restoration of the dynamic range of a compressed signal. See also *signal compression*.

signal ground (SG or SGD) A common reference point used as the return path for signal currents. The *signal ground* may or may not be identical to power ground. The two grounds are, however, generally tied together at one point by either a wire or a capacitor.

signal level The relative magnitude of a signal when compared to an arbitrary but specified reference.

Signal level may be expressed in absolute units (such as voltage or power) or in relative units, such as decibels (dB) or volume units (VU).

signal lines Passive transmission lines through which a signal passes from one system element to another.

signal parameter Those components of a transmission (electrical, optical, or other) whose values or sequence of values convey information.

signal plus noise plus distortion to noise plus distortion ratio (SINAD) **(1)** The ratio of total received power (i.e., the received signal plus noise plus distortion power) to the received noise plus distortion power. **(2)** The ratio of total recovered audio power (i.e., the original modulating audio signal plus noise plus distortion powers) from a modulated radio frequency carrier to the residual power (i.e., noise plus distortion powers) remaining after the original modulating audio signal is removed. *SINAD* is usually expressed in decibels (dB).

signal plus noise to noise ratio ((S+N)/N) At a given point in a communications system, the ratio of the power of the desired signal plus the noise $(S + N)$ to the power of the noise (N). *Note:* The ratio is usually expressed in decibels (dB).

signal processing The manipulation of signals that results in their transformation into other forms, such as other waveshapes, power levels, frequency distribution, and coding arrangements. Such processing includes detection, filtering, shaping, converting, coding, and time positioning.

signal processing gain **(1)** The ratio of the signal-to-noise ratio of a processed signal to the signal-to-noise ratio of the unprocessed signal. The gain is usually expressed in decibels (dB). **(2)** In a spread spectrum communications system, the signal gain, signal-to-noise ratio, signal shape, or other signal improvement obtained by coherent band spreading at the transmitter, remapping, and reconstitution of the desired signal at the signal receiver. See also *spread spectrum*.

signal regeneration Signal processing that restores a signal so that it conforms to its original characteristics. See also *regenerator* and *repeater*.

signal return circuit In a closed-loop delivering power to a load, the current carrying return path from the load back to the signal source. Also called the *low side of the loop*.

signal sample The value of a signal at a particular instant in time. The signal may be voltage, current, phase, etc., and the value of the signal may be an analog or digital representation. See also *sampling* and *Shannon limit*.

signal security A generic term that includes both communications security and electronics security.

signal strength indicator An indicator on a receiver that informs the user about the current relative signal level.

signal threshold The voltage, current, or time value at which a transition from one state to another state occurs. The threshold may be expressed as an absolute number (such as 24 volts or 5.0 mA) or as a percentage (e.g., 70% make break ratio).

If the detector has hysteresis, different thresholds will be specified for the transition in one direction than in the other. For example, the on-hook/off-hook threshold currents may be specified as:

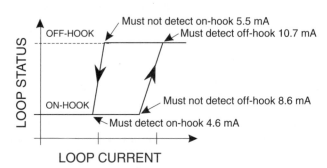

signal-to-crosstalk ratio (SCR) At a specified point in a multichannel circuit, the ratio of the power of the wanted signal to the power of the unwanted signal coupled from another channel. The ratio is generally expressed in decibels (dB), that is,

$$SCR = 10 \cdot \log_{10} \frac{P_{signal}}{P_{crosstalk}} \text{ dB}$$

When testing, both the wanted signal and the unwanted signal sources are adjusted in each channel so that they are of equal power at their respective zero transmission level point (0TLP). See also *far end crosstalk (FEXT)* and *near end crosstalk (NEXT)*.

signal-to-noise ratio (SNR) **(1)** The ratio of the signal magnitude (amplitude, power or energy) to the noise amplitude in a communication channel. The ratio may be expressed directly or in decibels, that is,

$$SNR = 20 \cdot \log_{10} \frac{V_{Signal}}{V_{Noise}} = 10 \cdot \log_{10} \frac{P_{Signal}}{P_{Noise}} \text{ dB}$$

V_{Signal} and V_{Noise} are generally peak values for impulse noise and the root mean square (RMS) for random noise. The bandwidth of the

SNR measurement may be limited with filters such as C-message and psophometric weighted filters to better approximate the performance of a telephone subset. *Signal-to-noise ratio* is often abbreviated as *S/N* or *S:N* ratio as well as *SNR*. (2) On the USENet newsgroups, *signal-to-noise ratio* refers to the ratio of the amount of useful information to the amount of irrelevant chatter bound to occur on an unmoderated forum. One of several things may happen when a newsgroup S/N becomes intolerable;

- The newsgroup may simply disappear, or
- The newsgroup may become moderated.

signal-to-noise ratio per bit (E_b/N_0) The ratio of the signal energy per bit (E_b) to the noise energy per hertz of noise bandwidth (N_0).

signal transfer point (STP) In telephony, a switching system, such as SS7, where common channel signaling facilities are interconnected. It provides for the transfer from one signaling link to another.

In nonassociated common channel signaling, the *STP* need not be the point through which the call (which is associated with the signaling being switched) passes.

signal transition In the modulation of a carrier (e.g., voltage, current, phase, RF, or optical), a change from one significant condition to another. *Signal transitions* are used to create signals that represent information, such as "1" and "0" or "mark" and "space."

Examples include:

- The change from one electrical voltage, current, or power level to another.
- A phase shift.
- A frequency shift.
- The change from one optical power level to another.
- The change from optical wavelength to another.

See also *significant condition*.

signaling The exchange of information specifically concerned with establishing, managing, and controlling communication connections in a telecommunications network. It includes the transmission of address and other switching or routing information between switching entities or between switching and circuit endpoint entities.

signaling path In a transmission system, a separate path from the data path used for system control, synchronization, checking, signaling, and service signals used in system management and operations rather than for the data, messages, or calls of the users.

signaling point interface (SPOI) The demarcation point on a signaling system number 7 (SS7) signaling link between a local exchange carrier (LEC) and a wireless service provider network. The point establishes the technical interface that can be used to establish the operational division of responsibility as well as a point to perform testing.

signaling system number 7 (SS7) An ITU-T standard common signaling channel protocol used for signaling among national and international central office switching systems. It is defined in the 1988 Blue Book, in Recommendations Q.771 through Q.774. *SS7* has a layered structure. Its major capabilities include:

- Fast call setup.
- Full calling and called party terminal information transfer. (This enables the caller ID feature for example.)
- A prerequisite for implementation of an Integrated Services Digital Network (ISDN).

signaling time slot In time division multiplex (TDM) systems, a specific time slot in each frame allocated for the transmission of signaling (supervisory and control) data.

signature (1) The complete set of electromagnetic and/or acoustic signals received from a source which may be analyzed to indicate the nature of the source and/or assist in its identification. (2) The complete set of electromagnetic and/or acoustic signals received from a source, which may be analyzed to indicate the nature of the medium through which it traversed or from which it was reflected. (3) A file appended to the end of a message to indicate the senders identity. Frequently abbreviated *sig*. See also *signature file*. (4) An encoded message used to authenticate the sender's identity.

signature file A short text file containing a user's identification and contact information (e-mail address, fax number, etc.). It is appended to e-mail and news articles posted on a network. Many newsreaders will automatically append the *signature file* if it exists. On most UNIX systems, the signature file is *.signature* and is placed in the user's home directory.

The composition of a signature can be quite an art form, including an ASCII logo or a witty saying, quip, or proverb. The *signature file* should be kept short, less than five lines or so.

significant condition In the modulation of a carrier, a possible value of a selected carrier parameter chosen to represent information.

- The carrier and modulation may be an electrical current, voltage, or power level; an optical power level; a phase value; or a frequency or wavelength.
- The possible values represent one of the symbols in the signal alphabet (including error control information). For example, in binary the values represent a "0" or "1," or a "mark" or "space," and in a quadbit the values represent any of the 16 possible conditions.
- The duration of a *significant condition*, called the *significant interval*, is the time interval between successive *significant instants*.
- The change from one *significant condition* to another is called a *signal transition*.

See also *significant instant*.

significant digits Any digit that contributes to the *accuracy* or *precision* of a number. The number of *significant digits* is counted beginning with the first nonzero digit from the left and ending with either the last nonzero digit to the left of the decimal point or the last digit to the right of the decimal point, whichever provides the highest count. (Zeros in-between the defined endpoints are significant digits.) The table gives some examples:

EXAMPLES OF SIGNIFICANT DIGITS

Number	Number of Significant Digits	Significant Digits
1234.	4	1234
12040.	4	1204
01204.	4	1204
1234.00	6	123400
0.00123	3	123
0.00120	3	120
01030.060	7	1030060

significant hour Any hour whose call rate influences the sizing of equipment.

significant instant In a signal, any instant at which a *significant condition* of a signal begins or ends. For a given signal, the significant condition may be arbitrarily chosen. However, it is commonly chosen to be the instant at which a signal crosses the baseline, reaches 10% or 90% of its maximum value, or reaches its peak value. See also *pulse* and *significant condition*.

significant interval The time interval between two consecutive *significant instants.*

SIL (1) An acronym for $\underline{S}$emiconductor $\underline{I}$njection $\underline{L}$aser. **(2)** An acronym for $\underline{S}$peech $\underline{I}$nterference $\underline{L}$evel.

silence suppression A term used in transmission channel reuse multiplexing methods wherein the silent periods of normal speech are filled with information from another channel.

Normal conversations on a full-duplex circuit use less than 40% of the transmission capacity. (AT&T claims 38% on the average.) (Fifty percent of the time one party is listening, and another 10% of the time the talker is also quiet; pauses between syllables, words, sentences, and to breathe, think, etc.) If the quiet periods are filled with another message, no user will suspect that the link is being shared, as long as it is switched properly at the terminating ends. With quiet periods as short as 300 ms, several hundred bits of digital information may be transmitted. Examples include TASI (Time Assigned Speech Interpolation) and Cellular Digital Packet Data (CDP).

silent answer A feature in some fax/modems that allows the fax/modem to differentiate an incoming facsimile call from a voice call. This is useful when both an answering machine and a fax/modem are connected to the same line.

When a ringing signal is received, the fax/modem allows an alternate device (e.g., answering machine or person) to answer the incoming call. The modem monitors the line for the calling station fax calling (CNG) tones. If the tones are not recognized, the fax/modem returns to the idle state, allowing the alternate device to retain control of the call. If the tones are recognized, the fax/modem goes off-hook (causing the alternate device to drop control) and attempts to negotiate a fax connection. See also *adaptive answer.*

silent zone A synonym for *skip zone.*

silica Silicon dioxide (SiO_2). The basic material from which glass and the most common communication grade optical fibers are fabricated.

Silica occurs naturally in impure forms such as quartz and sand and may take both crystalline and amorphous forms.

SILM An abbreviation of $\underline{S}$ingle $\underline{In}$-$\underline{L}$ine $\underline{M}$odule.

SILO A synonym for a first-in-first-out (FIFO) buffer.

SILS A acronym for $\underline{S}$tandard for $\underline{I}$nteroperable $\underline{L}$AN $\underline{S}$ecurity.

Silsbee effect The ability of an electric current flowing in a superconductor to destroy the superconductivity effect by means of the magnetic field generated by the current itself. The temperature of the material is not raised (until after resistance reappears). The effect is identical to application of an external magnetic field.

SIM (1) An acronym for $\underline{S}$ynchronous $\underline{I}$nterface $\underline{M}$odule. **(2)** An acronym for $\underline{S}$et $\underline{I}$nitialization $\underline{M}$ode. **(3)** An acronym for $\underline{S}$ervice $\underline{I}$nstructions $\underline{M}$essage. **(4)** An acronym for $\underline{S}$ubscriber $\underline{I}$dentity $\underline{M}$odule.

SIMD An abbreviation of $\underline{S}$ingle $\underline{I}$nstruction $\underline{M}$ultiple $\underline{D}$ata computer.

SIMM An acronym for $\underline{S}$ingle $\underline{In}$-line $\underline{M}$emory $\underline{M}$odule. An electronic module containing several random access memory (RAM) integrated circuits arranged to appear as a single memory block to the addressing device.

SIMMs are rated in several ways: memory size (both the number of words and number of bits per access), memory speed, and physical size.

- Memory size is always a power of two and generally a power of four, i.e., 256 k, 1 M, 4 M, or 16 M byte. The number of bits in the output word is 8, 9, 32, or 36 (8 or 32 if no parity bit is present and 9 or 36 if parity is supplied).

- Memory speed is usually specified at 50 ns, 60 ns, 70 ns, and 80 ns.
- Physical size is a combination of the number of electrical contacts on the module (either 30 or 72 pins) and the number of integrated circuits used to make up the output byte or word.

For example, eight 1-M RAM chips organized to deliver 1 bit when addressed will yield a *SIMM* that is 1 M by 8 bits; similarly, two 4-M RAM chips organized to deliver 4 bits when addressed will also yield a 1-M-by-8-bit *SIMM.* Some examples of available *SIMMs* include:

Descriptor (Part Number)	Word Size (bits)	Number of Words	Number of Pins	Speed (ns)
256 k × 8-80	8	256 k	30	80
1 M × 8-70	8	1 M	30	70
1 M × 9 60-9 chip	9	1 M	30	60
1 M × 9 60-3 chip	9	1 M	30	60
1 M × 32-70	8/32	1 M	72	70
1 M × 36-70	9/36	1 M	72	70
1 × 36-70	9/36	4 M	72	70
4 × 3-70	9	4 M	30	70
4 M × 9-70	9	4 M	30	70
4 M × 36-60	9/36	4 M	72	60
2 × 32-60	8/32	8 M	72	60
4 × 32-70	8/32	16 M	72	70
16 M × 36-70	9/36	16 M	72	70
32 M × 32-60	8/32	32 M	72	60
8 × 32-70	8/32	32 M	72	70

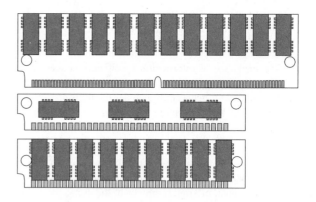

The sketch shows three *SIMMs;* the top is a 72-pin device, while the bottom two have 30 pins (3 and 9 chip version).

SIMP An acronym for $\underline{S}$atellite $\underline{I}$nformation $\underline{M}$essage $\underline{P}$rotocol.

Simple Mail Transport Protocol (SMTP) A Transmission Control Protocol/Internet Protocol (TCP/IP) standard protocol that specifies how two systems are to interact and the format of messages used to control the transfer of electronic mail on the Internet. It has become a de facto standard e-mail transfer protocol. *SMTP* is defined by the Internet standard STD10 and Request for Comment RFC821. Eudora, for example, uses *SMTP* to send e-mail.

SMTP describes messages as a rather limiting plain ASCII text message. Because of this limitation, other protocols are being explored, such as Multi-purpose Internet Mail Extension (MIME) and Privacy Enhanced Mail (PEM).

Simple Multicast Routing Protocol (SMRP) An Apple Computer network protocol for routing multimedia data streams on AppleTalk networks. See also *IP multicast.*

Simple Network Management Protocol (SNMP) An Internet Transmission Control Protocol/Internet Protocol (TCP/IP) standard protocol that provides a means to monitor network-related statistics and error conditions and to set network configuration and runtime parameters across the network, that is,

- It is used to manage and control IP gateways and the networks to which they are attached.
- It uses IP directly, bypassing the masking effects of TCP error correction.
- It has direct access to IP datagrams on a network that may be operating abnormally, thus requiring management.
- It defines a set of variables that the gateway must store.
- It specifies that all control operations on the gateway are a side effect of fetching or storing those data variables, i.e., operations that are analogous to writing commands and reading status.

The standard was created by the Internet Engineering Task Force (IETF) and was introduced in 1988 as an interim protocol until Common Network Management Information Protocol (CMIP) was finalized. It is defined in STD15 and RFC1157. The purpose of automated network management is to relieve the burden of mundane tasks on the network manager. See also *Common Network Management Information Protocol (CMIP)*.

Simple Network Management Protocol version 2 (SNMP v2) A revision of Simple Network Management Protocol (SNMP), which includes improvements in the areas of performance, security, confidentiality, and manager-to-manager communications.

The major components of SNMP v2 are defined in RFC1441 through 1452.

simple tone A sinusoidal sound pressure wave that is a single frequency. Also called a *pure tone*.

simplex circuit **(1)** A circuit that provides transmission in one direction only. **(2)** A contradictory and deprecated definition: A circuit using ground return and providing communication in either direction but in only one direction at a time. *Note:* Both definitions are in common use; hence, problems will arise if the user assumes the wrong one. **(3)** A circuit derived from one pair of wires and an Earth ground return or from two pairs. Also called a *phantom circuit*.

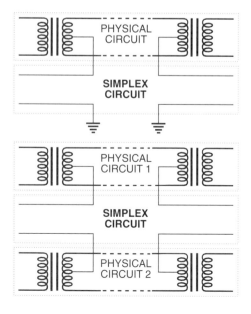

simplex operation **(1)** In telephony, the operation of a communications channel, protocol, or transmission facility in which information transmissions between two stations take place in only one preassigned direction. Examples of *simplex operation* include most CATV cable distribution systems, a tickertape machine, and radio or television transmissions. Also called *one-way operation*. **(2)** In radio, an operating method in which transmission is possible alternately in each direction of a telecommunications channel, e.g., by means of manual switching. *Note:* Both definitions, though contradictory, are in common use; hence, problems will arise if the user assumes the wrong one.

simplex (SX) signaling Signaling in which two conductors are used for a single channel and a center-tapped coil, or its equivalent, is used to split the signaling current equally between the two conductors.

SX signaling may be one-way, for intra-central-office use, or the simplex legs may be connected to form full-duplex signaling circuits that function like composite (CX) signaling circuits with E & M lead control. See also *simplex circuit (3)*.

simplex supervision A form of supervision that utilizes a simplex circuit.

simplified access control (SAC) One of two access methods in the ITU-T X.500 Directory Services model. It is the more restricted of the two sets. The other set is Basic Access Control (BAC).

simulate The attempt to represent certain features of the behavior of one physical or abstract system by the behavior of another system. This may be done by mathematical models in a computer or by substituting a physical device to represent another device.

For example, delay lines may be used to simulate propagation delay and phase shift caused by an actual transmission line. A simulator may imitate only a few of the operations and functions of the unit it simulates. Contrast with *emulate*.

simulator A device used to represent the behavior, inputs, or outputs of another device or system by its analogous characteristics.

SINAD An abbreviation of <u>SI</u>gnal plus <u>N</u>oise <u>A</u>nd <u>D</u>istortion to noise plus distortion ratio.

sine wave A periodic wave generated by an object vibrating or oscillating at a single frequency. A sinusoidal waveform is depicted in the drawing. It shows the relative position (amplitude) of a point as a function of time.

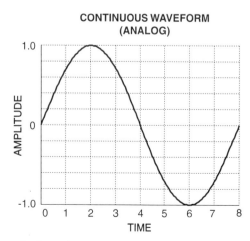

CONTINUOUS WAVEFORM (ANALOG)

Mathematically, a sine wave is defined by:

$$Ampl(t) = M \times \sin(\omega \times t + \Phi) + \text{Offset}$$

where:

Ampl(t)	is the amplitude of the *sine wave* at any specified time *t*,
M	is the maximum amplitude of the *sine wave* without regard to time,
ω	is the frequency determining term,
t	is time,
Φ	is the initial phase angle at time 0, and
Offset	is a constant term defining the dc offset if no *sine wave* were present.

singing An undesired self-sustaining oscillation in a transmission system. Oscillations at very low frequencies are frequently called *motor boating*. *Singing* is caused by positive feedback and excessive signal gain in one or more parts of the loop.

singing margin The excess of loss over gain around a potential singing path at any frequency or the minimum excess over a range of frequencies. (The difference in power levels between the singing point and the operating gain of a system.) *Singing margin* is generally expressed in dB. Also called *gain margin*.

singing point The threshold point at which additional gain in a system will cause self-oscillation.

singing return loss (SRL) A method of determining the singing margin of a circuit through the use of return loss measurements.

The return loss at a termination is measured in two frequency bands 260–500 Hz (which yields a measure of *SRL low*) and 2200–3400 Hz (which yields a measure of *SRL high*), the lower of the two *SRL* measurements is taken as a measure of the singing margin. The exciting signal used in the measurement is band limited noise with −3 dB points at the designated band edges.

single address message A message that is directed to only one specific destination. Other types of messages include group messages and broadcast messages.

single attachment concentrator (SAC) In the Fiber Distributed Data Interface (FDDI), a concentrator that serves as a termination point for single attachment stations (SASs) and that attaches to the FDDI ring through a dual attachment connector (DAC). (It is equivalent to a hub in other network architectures.)

single ended See *unbalanced*.

single ended amplifier An amplifier with a single wire output which is referenced to circuit common. Signal current flowing out of the amplifier returns via circuit common. The amplifier itself may be inverting or noninverting. See also *balanced amplifier*.

single ended synchronization Synchronization between two locations, in which the phase (and/or frequency) of the working clock is derived by comparing the phase of the incoming signals to the phase of the internal clock at that location. Also called *single ended control*. See also *phase lock loop (PLL)*.

single frequency interference Interference caused by a single frequency source.

One common example is the interference in a transmission channel induced by the 50/60 Hz mains power.

single frequency (SF) signaling In telephony, trunk signaling in which dial pulses and/or supervisory signals are conveyed by a single voice frequency tone in each direction. That is, dial pulse and supervisory information are sent from one switching office to another over a trunk by means of single frequency tone.

An *SF signaling unit* converts E & M wire signaling to a single voice frequency tone (*SF tone*) format that is suitable for transmission over an ac path (e.g., a carrier system) and converts received tones into dc signaling. Characteristics of the *SF signaling* system include:

- The *SF* tone is present in the idle state and absent during the seized state.
- Dial pulses are conveyed by bursts of *SF* tone, corresponding to the interruptions in dc continuity created by the dial pulse sending unit (rotary dial, etc.).
- *In-band SF* signaling tone is centered at either 1600 Hz or 2600 Hz
- *Out-of-band SF* signaling tone is centered at 3600 Hz.
- In either in-band or out-of-band signaling, the bandwidth is just wide enough to allow for the modulation sidebands and component tolerances.
- Usually, a notch filter is inserted in the transmission path between the user and the network to prevent the user from inadvertently disconnecting a call if user signals have sufficient energy at the *SF* frequency.

Also called *single frequency pulsing*.

single harmonic distortion Of a single frequency tone. The ratio of the power of a specified harmonic to the power of the fundamental frequency. *Single harmonic distortion* is measured at the output of a device under specified operating conditions and is expressed in decibels (dB).

single image random dot stereogram (SIRDS) A stereogram composed of either monochrome or colored entities (e.g., dots or patterns), which when viewed correctly appear to have depth into the page. That is, it appears as a three-dimensional image. The letters 3D are embedded in the following image—to view, stare at the image until the two dots become three.

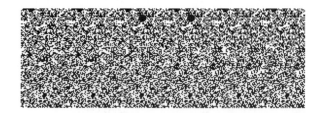

single instruction multiple data (SIMD) A parallel computing architecture wherein multiple arithmetic logic units (ALUs) operate on multiple data streams under the control of a single instruction and sequence controller. See also *multiple instruction multiple data (MIMD)*.

single mode optical fiber An optical fiber in which only the lowest order *bound mode* of the exciting wavelength can propagate the length of the cable. The core of a *single mode fiber* is extremely thin (less than 10 microns), which limits the number of light ray paths to exactly one. *Single mode fibers* have the potential bandwidth of 50–100 GHz per kilometer.

The lowest order bound mode is determined at a specified wavelength by solving *Maxwell's equations* for the boundary conditions imposed by the fiber, e.g., core (spot) size, and the refractive indices of the core and cladding. In step index guides, single mode operation occurs when the normalized frequency (V) is less than 2.405. For power law profiles, single mode operation occurs for a normalized frequency (V), less than approximately

$$2.405 \sqrt{\frac{g + 2}{g}}$$

where g is the *profile parameter*. Also called a *monomode optical fiber, single mode fiber, single mode optical waveguide*, and a *unimode fiber*. See also *optical fiber*.

single session CD-ROM drive CD-ROM drives that are capable of reading only the first session that was recorded on an ISO 9660 CD-ROM. See also *multi-session CD-ROM*.

single sideband (SSB) An amplitude modulation (AM) transmission method wherein one sideband (either the upper or lower sideband) is transmitted and the other sideband is suppressed. The carrier may or may not be transmitted or suppressed as well. See also *single sideband suppressed carrier (SSB-SC) transmission*.

single sideband suppressed carrier (SSB-SC) transmission Single sideband (SSB) transmission in which the carrier is reduced. In *SSB-SC* the carrier power level is reduced (suppressed) to the point where it is insufficient to demodulate the signal.

single step multimode fiber An optical fiber with a core large enough to allow multiple light paths (modes) through simultaneously. Unlike graded index multimode fiber, single step fiber has only a single layer of cladding, so that there is an abrupt difference in refractive index between fiber core and cladding. See also *optical fiber*.

single tone interference An undesired discrete frequency appearing in a transmission channel. The tone's frequency is the frequency that appears in the channel regardless of the nature of the source.

Contrast with *single frequency interference*.

sink A device that receives something (energy, information, . . .) from another device. Examples include:
- A *data sink*—that part of a terminal which receives data.
- A *heat sink*—that part that receives heat from other heat-producing components.
- A *sink node*—a point in a circuit or system that has only incoming branches.

SIO (1) An abbreviation of Serial Input Output. (2) An abbreviation of Service Information Octet. A field in the Signaling System 7 (SS7) Message Signaling Unit (MSU) used to identify individual users.

SIP (1) An acronym for Single Inline Package. (2) An acronym for Switched MultiMegabit Data Service (SMDS) Interface Protocol. (3) An acronym for Short Irregular Pulses.

SIPO (1) An acronym for Serial In-Parallel Out. (2) An acronym for Swiss Intellectual Property Office.

SIR (1) An abbreviation of Statistical Information Retrieval. A data search engine that locates information based on probability. (2) An abbreviation of Sustained Information Rate. (3) An abbreviation of Speaker Independent voice Recognition. (4) An abbreviation of Segment Identification Register.

SIRDS An acronym for Single Image Random Dot Stereogram.

SIS (1) An abbreviation of Structured Information Store. (2) An abbreviation of the Swedish Standards Committee Standardiseringskommissionen I Sverige. (3) An abbreviation of Signaling Interworking Subsystem.

.sit The default filename extension of the StuffIt file compression program.

SIT An acronym for Special Information Tone. An audible tone preceding a network broadcast announcement; frequently, a sequence of three notes with rising pitch.

SITD An abbreviation of Still In The Dark.

SIVR An abbreviation of Speaker Independent Voice Recognition.

six-bit byte A synonym for *sextet*.

six-bit transcode (SBT) A 6-bit communication code sometimes used with peripheral devices such as printers. The binary representation is frequently divided into two sets of three bits, then expressed in octal, for example, U (010 101) would be 25.

THE SBT CODE

Char	Code	Char	Code	Char	Code	Char	Code
@	000 000	P	010 000	□	100 000	0	110 000
A	000 001	Q	010 001	!	100 001	1	110 001
B	000 010	R	010 010	"	100 010	2	110 010
C	000 011	S	010 011	#	100 011	3	110 011
D	000 100	T	010 100	$	100 100	4	110 100
E	000 101	U	010 101	%	100 101	5	110 101
F	000 110	V	010 110	&	100 110	6	110 110
G	000 111	W	010 111	'	100 111	7	110 111
H	001 000	X	011 000	(	101 000	8	111 000
I	001 001	Y	011 001	)	101 001	9	111 001
J	001 010	Z	011 010	*	101 010	:	111 010
K	001 011	[	011 011	+	101 011	;	111 011
L	001 100	\	011 100	,	101 100	<	111 100
M	001 101	]	011 101	-	101 101	=	111 101
N	001 110	↑	011 110	.	101 110	>	111 110
O	001 111	←	011 111	/	101 111	?	111 111

skew (1) The deviation from synchronization (phase shift) of two or more signals. (2) The difference in arrival time of two signals transmitted simultaneously (in parallel). (3) The difference in position of multiple bits on a multichannel device (e.g., tape). (4) The difference in amplitude of the two tones in a dual tone multifrequency (DTMF) signal. (5) The difference in starting time of a two-tone transmission. The skew is positive when the low-frequency tone starts first and is negative when the high-frequency tone starts first. (6) In scanning devices (e.g., facsimile, printers, image scanners), the angle between a scan line or recording line and the perpendicular to the paper path. (7) The angular deviation from rectilinearity caused by asynchronism between a scanner and recorder. *Skew* is expressed numerically as the tangent of the deviation angle.

skew ray (1) A ray that does not lie in a plane containing the axis of a system having rotational symmetry. (2) In a multimode optical fiber, a bound ray that travels in a helical path along the fiber; hence, it:
- Is not parallel to the fiber axis,
- Does not lie in a meridional plane, and
- Does not intersect the fiber axis.

skin effect An effect where high frequencies, high current, or high magnetic fields tend to force the current flowing in a conductor to the outside surface of the conductor. This decreases the effective cross-sectional area of the conductor being used for moving electrons (current), and thereby increases the effective resistance. This, in turn, increases real signal attenuation, or loss.

SKIP An acronym for <u>S</u>imple <u>K</u>ey management for <u>I</u>nternet <u>P</u>rotocol. An IP layer security extension proposed by the IP Security Working Group of the Internet Engineering Task Force (IETF).

skip distance At a given azimuth, the minimum distance between a transmitting station and the closest point of return to the Earth of a transmitted wave reflected from the ionosphere. See also *ionosphere* and *sky wave.*

skip zone That region surrounding a transmitter which is beyond the furthest range of the *ground wave* and is inside the shortest range of the reflected *skywave;* hence, the signal cannot be received. Also called the *silent zone* and the *zone of silence.*

Skipjack The name of a dual key encryption/decryption method developed by the National Security Agency (NSA). The algorithm not only has the same type of key as does the Data Encryption Standard (DES) algorithm, but it has a second key related to the serial number of the device that the government can use to decrypt any communication. The circuit that performs the algorithm is contained in an integrated circuit called the *Clipper Chip.*

sky noise In communications satellite systems, background noise, seen at the ground station, that originates in astronomical sources (e.g., galaxies).

sky wave In radio wave transmission, a wave that is transmitted in a direction away from the Earth and may be reflected by one or more layers of the ionosphere.

Sky waves can be transmitted over a great distance before being reflected back to Earth. However, the ionosphere reflects high-frequency radio waves in a frequency-dependent manner, and the height of the reflecting layer varies with time. Therefore, the great transmission distances that can be achieved are often unreliable paths. It was formerly called an *ionospheric wave;* its use is now restricted to the description of the wavelike motions of the ionosphere itself.

SL (1) An abbreviation of <u>S</u>ession <u>L</u>ayer. (2) An abbreviation of <u>S</u>ink <u>L</u>oss. (3) An abbreviation of <u>S</u>ubscriber's <u>L</u>oop.

SLA (1) An abbreviation of <u>S</u>ynchronous <u>L</u>ine <u>A</u>dapter. (2) An abbreviation of <u>S</u>hared <u>L</u>ine <u>A</u>dapter.

slab dielectric waveguide Essentially the same as a circular optical waveguide (optical fiber), with shape being the major differentiator. Its characteristics include:

- The "active material" is constructed solely of dielectric materials.
- The dielectric propagation medium has a rectangular cross section.
- The operating wavelength and modes the guide will support, beyond the equilibrium length, are determined by the width, thickness, and refractive indices of the propagating medium.
- The "active materials" may be clad or otherwise protected, distributed, and electronically controllable.

Used in applications such as integrated optical circuits (IOCs) in which the rectangular shape is geometrically more convenient than the optical fibers that are circular in cross section. Also called a *planar waveguide.*

slamming The illegal act of transferring a subscriber from one long-distance carrier to another without the permission of the subscriber. It is performed at the request of the new long distance carrier.

slave clock A clock that is coordinated with a master clock. Coordination is usually achieved by phase locking the slave clock signal to a signal received from the master clock.

To adjust for the transit time of the signal from the master clock to the slave clock, the phase of the slave clock is adjusted with respect to the signal from the master clock such that both clocks are in phase. Therefore, time markers generated by both clocks occur essentially simultaneously at their respective outputs.

slave station (1) In a data network, a station that is selected and controlled by a master station. Usually a *slave station* can only communicate with a master station. (2) In navigation systems using precise time dissemination, a station having a clock is synchronized by a remote master station. Also called a *subordinate station.*

SLC An abbreviation of <u>S</u>ubscriber <u>L</u>oop <u>C</u>arrier.

SLC96 A <u>S</u>ubscriber <u>L</u>oop <u>C</u>arrier system with <u>96</u> transmission channels between the remote terminal and the terminal (pronounced "slick 96"). The system uses digital μ255 PCM coding and time division multiplexing to consolidate subscriber loops onto four 24 channel T-1 carrier links. See also *subscriber carrier.*

SLCC An abbreviation of <u>S</u>ubscriber <u>L</u>ine <u>C</u>arrier <u>C</u>ircuit (pronounced "slick"). A multiplexer that allows more subscribers to be connected to the central office switch than the number of available copper loops.

SLD An abbreviation of <u>S</u>ynchronous <u>L</u>ine <u>D</u>river.

SLDRAM An acronym for <u>S</u>ync<u>L</u>ink <u>DRAM</u> (Dynamic Random Access Memory). A memory standard proposed to the IEEE by nine memory manufacturers as an open standard for a high-speed DRAM architecture. The architecture uses two busses, a 10-bit upper bus which carries commands and addresses, and an 18-bit lower bus dedicated to transmission of packetized data signals.

SLE An abbreviation of <u>S</u>mall <u>L</u>ocal <u>E</u>xchange.

SLED (1) An acronym for <u>S</u>ingle <u>L</u>arge <u>E</u>xpensive <u>D</u>isk. (2) An acronym for <u>S</u>urface <u>L</u>ight <u>E</u>mitting <u>D</u>iode.

sleep To suspend operation of a process or program without terminating or exiting the software. Generally, operations are suspended due to inactivity. When this is the case, shared resources can be reassigned or nonessential equipment can be put into a low operating power mode.

For example, laptop computers go to *sleep* if they are not used for a specified amount of time, that is, the display and most electronics are turned off. Random access memory (RAM) and I/O remain active however so that data will not be lost and activity such as a keyboard operation may be detected. Because the computer is not using as much battery power, the battery life is extended.

sleeve (1) In telephony, a conductor that provides miscellaneous control or supervisory functions (such as E and M leads). The sleeve wire is usually accompanied by the tip and ring wires. (2) One of the terminations on the plug used in manual switchboards to connect the subscriber lines and interoffice trunks, for example,

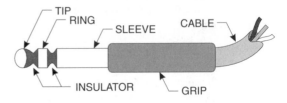

See also *ring* and *tip.*

slew rate The rate of change of a signal or physical entity at any instant.

Slew rate is dependent on frequency, signal amplitude, and wave shape. Higher frequencies, larger amplitudes, and wave shapes with edges steeper than a triangle wave lead to higher values of *slew rate*. Mathematically, *slew rate* is the *slope* of the waveform at the instant the measurement is taken. *Slew rate* is expressed as magnitude per time, e.g., volts/microsecond, megavolts/second, amps/millisecond. Because the *instantaneous slew rate* changes across the waveform, *slew rate* is generally expressed as the magnitude of the maximum slew rate. The figure illustrates these effects with a simple 1000-Hz sine wave. The sine wave varies from −6.28 kV/s through 0 kV/s to +6.28 kV/s and back during each period. The ±6.28 kV/s and 0 kV/s bounds are shown with dashed lines.

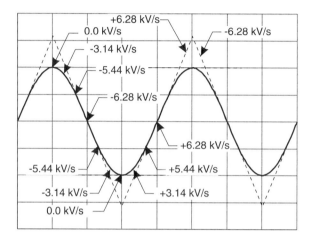

slewing (1) Rapidly rotating a directional device, such as an antenna or telescope, about one or more axes. The movement is generally thought of as redirecting rather than tracking a moving target. (2) Redirecting the beam of a fixed antenna array by changing the relative phases of the signals feeding the antenna elements. (3) Changing the frequency or pulse repetition rate of a signal source. (4) Changing the tuning of a receiver, usually by sweeping through many or all frequencies.

SLI An abbreviation of Synchronous Line Interface.

SLIC An acronym for Subscriber Line Interface Circuit (or Card). That part of a telephone switching system central office (CO) that connects the subscriber loop to the switching equipment or a line concentrator (such as a SLC 96).

slip (1) A phenomenon in re-timing the connection between two separately synchronous networks when there is a repetition or deletion of a frame.

Because the network clocks have slightly different frequencies, one network will, on the average, send data to the other slightly faster than it can be received. A frame buffer is interposed between the two networks to absorb the accumulating information. When the buffer is full, it overflows (is dumped), resulting in the loss of a frame of information. (The information is not truly lost, for the receiving station will signal the sending station to retransmit the frame.) In the other direction, data are not transmitted fast enough so that a duplicate frame is periodically inserted (by the network interfacing equipment) in order to maintain a continuous flow. Again, a buffer is used to smooth the flow, that is, the duplicate frame is transmitted from the buffer, not directly from the signal source. Slip-free operation can only occur on systems with sufficient phase locking of the clocks and

buffering depth in the transmission path. The average clock frequencies are maintained by the phase lock loop, while the short-term clock variations are absorbed by the buffers. (2) In rotating machinery, the difference between the synchronous speed and the actual speed. This difference is expressed in a number of ways, that is,

- The difference as an absolute number of revolution per unit time.
- The ratio of the difference to the synchronous speed as a decimal fraction.
- The ratio of the difference to the synchronous speed as a percentage.

SLIP An acronym for Serial Line Interface Protocol. An asynchronous protocol, defined in RFC1055, that allows dial-up modem users to access Internet using Transmission Control Protocol/Internet Protocol (TCP/IP) just as if the personal computer were a hard-wired mainframe. The modem node has both a name and an IP address. Because of the direct, albeit temporary, connection, users can use client programs such as Eudora, Mosaic, and Netscape's Navigator.

The *SLIP* protocol is older and less flexible than the newer Point-to-Point Protocol (PPP), for example,

- It modifies a standard Internet datagram by appending a special *SLIP END* character to it, which allows datagrams to be distinguished.
- It requires a port configuration of 8 data bits, no parity.
- It requires EIA or hardware flow control.
- It does not provide error detection (it relies on other high-layer protocols for error control); therefore, on a noisy link *SLIP* on its own would not be satisfactory.
- Unlike PPP, it does not provide for data compression.
- A *SLIP* connection needs to have its IP address configuration set each time before it is established whereas PPP can determine it automatically once it has started.

See also *PPP* and *shell account.*

slope (1) In a circuit, device, or system (such as a transmission line) in which there exists a frequency-dependent amplitude response, the rate of change of attenuation with respect to frequency over a specified frequency spectrum. *Slope* is usually expressed in dB/Hz, dB/octave, or dB/decade.

In metallic transmission lines, the *slope* is usually greater at high frequencies than at low frequencies. (2) A synonym for *droop* in a pulse.

slope equalizer A device or circuit used to achieve a specified slope in a metallic transmission line. See also *line conditioning.*

slot (1) A media access method used with time division multiplex and some ring protocols.

Ring protocols using this access method divide the ring into a number of fixed size slots, each of which circulates around the ring. The status of each slot (empty or full) is determined by the value of a control bit. When an empty slot passes a node in the ring, that node can access the network by setting the slot's control bit and inserting a packet (which contains address, data, and error-checking information) into the slot for transmission. As with every network access solution there are advantages and disadvantages. One advantage is that multiple packets can be on the network at the same time, thereby increasing the capacity of the network. A disadvantage is the potential for one node to hog all available slots, thus preventing other nodes from gaining access to the network. See also *CSMA/CA, CSMA/CD, polling, time division multiplex* (TDM), and *token passing.* (2) In a distributed queue dual-bus (DQDB) network, a protocol data unit (PDU) that

- Consists of 53 octets used to transfer segments of user information,
- Has the capacity to contain a segment of 52 octets and a 1-octet access control field, and
- May be either a pre-arbitrated (PA) slot or a queued arbitrated (QA) slot.

slot antenna A radiating element formed by a slot in a conducting surface or in the wall of a waveguide.

slot time (1) The amount of time available for a transmission in a time division multiplex (TDM) system. (2) In an Ethernet network, the maximum time that can elapse between the first and last node's receipt of a packet. To ensure that a node can detect whether the packet it transmitted has suffered a collision with another packet, the packet must be longer than the slot time. This is about half a microsecond, or more than about 512 bits. *Slot time,* however, varies from station to station.

slotted Aloha A media access method that extends pure Aloha to the case where access is allowed only in specified time intervals. See also *Aloha Network.*

slotted ring A local area network (LAN) protocol that continually carries a constant number of fixed-length packets or slots round the ring. The nodes then replace empty slots as they pass by to transmit data. All the nodes can recognize slots as empty or as addressed to them. Also called a *Cambridge Ring.*

slow operate relay A relay designed to have a delay between the application of the energizing current and the operation of the contacts. *Slow operate relays* may or may not also be slow release.

slow release relay A relay designed to have a delay between the removal of the energizing current and contact release. *Slow release relays* may or may not also be *slow operate.*

SLR An abbreviation of Send Loudness Rating.

SLS (1) An abbreviation of Signaling Link Selection. A field in the routing label of a Signaling System 7 (SS7) Signaling Unit (SU) that keeps related packets on the same path so as to preserve the delivery order. (2) An abbreviation of Sequential Logic System. (3) An abbreviation of Secondary Link Station.

SLSA An abbreviation of Single Line Switching Apparatus.

SLT An abbreviation of Single Line Telephone.

SM (1) An abbreviation of Standby Monitor. (2) An abbreviation of Signaling Module.

SMA 905/906 An abbreviation of *subminiature type A* connector. A standard fiber optic threaded connector with two variants. The *SMA 905* has a straight ferrule, while the *SMA 906* ferrule is stepped and uses a plastic sleeve for alignment.

SMAE An abbreviation of Systems Management Application Entity.

SMAP An acronym for Systems Management Application Process.

smart With respect to either software or hardware, the capability to process information. *Smartness* does not imply reason, understanding, or rationality; but *only* the ability to make decisions based on the comparison of two or more quantities. A synonym for *intelligent.*

smart card (1) In electronics, a printed circuit board with built-in logic, processor, or firmware that gives the assembly some independent decision-making ability. (2) In banking, a credit-card-size device containing not only the logic in (*1*) but also nonvolatile memory which contains some personal or monetary data.

smart jack Slang for the *RJ-48X connector.*

Smart Ring See *distinctive ringing.*

smart terminal A terminal that contains a microprocessor and memory and can do some rudimentary processing without assistance from the host computer. The *smart terminal* can limit keyboard input to ʾfic areas on the display as well as verify that the input is within ʾfined range and of a specific type (numbers only for example).

smart wiring hub A network concentrator enabling multiple media to be supported and managed from a central location. When supporting structured wiring systems, smart hubs provide port management.

SMAS An abbreviation of Switched Maintenance Access System.

SMASE An abbreviation of Systems Management Application Service Element.

smash case Jargon meaning to ignore the uppercase/lowercase distinction in text input. For example, DOS will *smash case* all file names.

SMB An abbreviation of Server Message Block.

SMC An abbreviation in telecommunications for Switch Maintenance Center.

SMD (1) An abbreviation of Surface Mount (or Mounted) Device. (2) An abbreviation of Synchronous Modulator Demodulator.

SMDA An abbreviation of Station Message Detail Accounting. Telephone call accounting.

SMDL An abbreviation of Standard Music Description Language.

SMDR An abbreviation of Station Message Detail Recording. The keeping of a detailed calling record of all phones served by a private automatic branch exchange (PABX). Formerly called *MDR* for Message Detail Recording.

SMDS An abbreviation of Switched MultiMegabit Data Service.

SMF (1) An abbreviation of Single Mode Fiber. (2) An abbreviation of Systems Management Function. (3) An abbreviation of S-band MultiFrequency.

SMFA (1) An abbreviation of Systems Management Functional Area. (2) An abbreviation of Simplified Modular Frame Assignment system.

SMI An abbreviation of Structure of Management Information.

smileys See *emoticons.*

SMLE An abbreviation of Small to Medium Local Exchange.

SMM (1) An abbreviation of Semiconductor Memory Module. (2) An abbreviation of Shared Multiport Memory.

smooth Earth An idealized Earth surface having radio horizons that are not formed by prominent ridges or mountains but are determined solely as a function of antenna height above ground and the effective Earth radius.

Examples of an approximately smooth Earth include water surfaces or very level terrain.

SMP (1) An abbreviation of Simple Management Protocol. (2) An abbreviation of Session Management Protocol. (3) Abbreviation of Standby Monitor Present. (4) Abbreviation of Symmetric MultiProcessing.

SMPDU (1) An abbreviation of Service Message Protocol Data Unit. (2) An abbreviation of System Management Protocol Data Unit.

SMR An abbreviation of Specialized Mobile Radio. An analog wireless communications technology that provides paging, fax, modem, and voice. Another feature claimed by the system is one phone number nationwide, regardless of geographic location. See also *ISM* and *DECT.*

FCC SMR SPECTRUM ALLOCATIONS

Channel "Block"	Frequency (GHz)
A	1.850-1.865 / 1.930-1.945
B	1.865-1.880 / 1.945-1.960
C	1.880-1.890 / 1.960-1.970
D	2.130-2.135 / 2.180-2.185
E	2.135-2.140 / 2.185-2.190
F	2.140-2.145 / 2.190-2.195
G	2.145-2.150 / 2.195-2.220

SMRP An abbreviation of Simple Multicast Routing Protocol.

SMRT An acronym for Single Message-unit Rate Timing.

SMS (**1**) An abbreviation of Storage Management Services. (**2**) An abbreviation of Service Management System. (**3**) An abbreviation of Short Message Service. A service defined by GSM.

SMSA An abbreviation of Standard Metropolitan Statistical Area.

SMSP An abbreviation of Storage Management Services Protocol.

SMT (**1**) An abbreviation of Surface Mount Technology. (**2**) An abbreviation of FDDI's Station ManagemenT document.

SMTP An abbreviation of Simple Mail Transport Protocol.

SMU (**1**) An abbreviation of Secondary Multiplex Unit. (**2**) An abbreviation of System Monitoring Unit.

SMX An abbreviation of Sub MultipleX unit.

SN (**1**) An abbreviation of Sequence Number. (**2**) An abbreviation of Signal Node. (**3**) An abbreviation of Serial Number.

SNA (**1**) An acronym for IBM's proprietary Systems Network Architecture (pronounced *snaw*). *SNA* as first introduced in 1974 was intended to connect dumb terminals to IBM mainframes. It is a design methodology that enables different models of computers to exchange and process data. It separates the network communications into five layers analogous to the seven-layer ISO/OSI model. The five layers are comparable to layers two through six, though are not compatible. *SNA* does not provide an equivalent to either layer 1 (physical or hardware connections layer) or layer 7 (application layer).

SNA has traditionally been a hierarchical network architecture for homogeneous networking between IBM systems, but in the new *SNA*,

COMPARISON OF SNA AND ISO/OSI LEVELS

SNA Layer	SNA Function	ISO/OSI Layer
		7. Application
5. Function Management	Responsible for "visible" tasks; such as, presentation of data and management of the interface between the network and the user.	6. Presentation
4. Data Flow Control	Addresses data flow during a communications session.	
3. Transmission Control	Handles the status and pacing of communication sessions.	5. Session
2. Path Control	Involves the data routing.	4. Transport
		3. Network
1. Data-Link Control	Addresses the actual transmission of data	2. Data-Link
		1. Physical

IBM has added increasing support for peer-to-peer networking through the development of Advanced Peer to Peer Communications. *SNA* is managed through the NetView network management system. (**2**) An acronym for Synchronous Digital Hierarchy (SDH) Network Aspects.

SNA/SDLC An abbreviation of Systems Network Architecture/Synchronous Data Link Control. A communications protocol used to transfer data between a host and a controller in an SNA environment.

SNAcP An abbreviation of SubNetwork Access Protocol.

SNADS An acronym for SNA Distribution Services. A standardized asynchronous distribution service architecture for the transmission of e-mail, files, or jobs around an IBM Systems Network Architecture (SNA) network. It is implemented as a transaction service of the SNA network.

SNAFU A military acronym for Situation Normal All Fouled Up—or something like that.

snail mail What the U.S. Postal Service is called on many networks. The term is used to emphasize the dramatic speed difference between e-mail and mail delivery (seconds on the networks vs. days by the Postal Service).

In general, any paper mail (*p-mail*) delivery service is referred to as *snail mail*. Also written *SnailMail* and *smail*.

SNAP (**1**) An acronym for Single Numbering Assignment Plan. (**2**) An acronym for Special Night Answering Position. (**3**) An acronym for Standard Network Access Protocol. (**4**) An acronym for SubNetwork Access Protocol.

SNARE An acronym for Subnetwork Address Resolution Entity.

snarf (**1**) To appropriate, with little regard for legality or etiquette (but not quite by stealing). For example, "They were giving away samples, so he *snarfed* up a bunch." (**2**) To fetch a file or set of files across a network.

SNCP An abbreviation of Single Node Control Point.

SND An abbreviation of SeND.

SNDCF An abbreviation of Subnetwork Dependent Convergence Facility.

SNDCP An abbreviation of Subnetwork Dependent Convergence Protocol.

sneak current Low-level, unwanted, anomalous, but persistent and steady current that flows into a device or circuit from an external source. It may cause improper operation or even damage.

sneakernet Probably the largest, most widespread, and slowest network available. A file is copied to a removable media such as a floppy disk or tape on one machine and then physically transported to another machine where it is copied to that second machine.

The implication may be either:
- The person carrying the physical medium is wearing sneakers, or
- The person is sneaking information from one location to another.

Sometimes also called *armpit-net, floppy-net, shoenet, tennis-net,* and *walknet*.

Snell's law A law of geometric optics that relates incident and refracted rays at the boundary of two materials. It states:
- The incident ray, the normal to the refracting surface at the point where the incident ray intercepts the surface, and the refracted ray below the surface all lie in a plane.

- The ratio of the sine of the angle between the incident ray and the normal to the sine of the angle between the refracted ray and the normal is constant.
- If a ray travels from a medium of lower refractive index into a medium of higher refractive index, it is bent toward the normal; if it travels from a medium of higher refractive index to a medium of lower index, it is bent away from the normal.
- If η_i is the refractive index of the material containing the incident ray and η_r is the index of the material with the refracted ray, then

$$\eta_i \sin \theta_i = \eta_r \sin \theta_r$$

- If the incident ray travels in a medium of higher refractive index toward a medium of lower refractive index at an angle that requires the sine of the refracted ray to be greater than unity (a mathematical impossibility), that is,

$$\sin \theta_r = \frac{\eta_i}{\eta_r} \sin \theta_i > 1$$

then the *refracted* ray becomes a *reflected* ray, i.e., is totally reflected back into the medium of higher refractive index (at an angle equal to the incident angle) and thus still obeys *Snell's Law*. This reflection occurs even in the absence of a reflective coating (e.g., aluminum or silver). This effect is called *total internal reflection*. The smallest angle of incidence which yields total internal reflection is called the *critical angle*. See also *critical angle*.

SNI (1) An abbreviation of Subscriber Network Interface. (2) An abbreviation of Signal to Noise Improvement. (3) An abbreviation of SNA Network Interconnect. The way in which autonomous SNA networks can be connected while still allowing them to be independently managed.

SNICF An abbreviation of Subnetwork Independent Convergence Facility.

SNICP An abbreviation of Subnetwork Independent Convergence Protocol.

SNMP An abbreviation of Simple Network Management Protocol. A transmission protocol defined by the Internet Engineering Task Force (IETF) and introduced in 1988 by the Internet Architecture Board (IAB) in RFC1157 for Transmission Control Protocol/Internet Protocol (TCP/IP) based network management as an interim protocol until Common Network Management Information Protocol (CMIP) was finalized. *SNMP* has become the accepted de facto standard for local area network (LAN) network management.

SNMP is used to monitor IP gateway and network-related statistics and error conditions and to set network configuration and runtime parameters across the network. It defines a set of variables that the gateway must keep, and it specifies that all operations on the gateway are

a side effect of fetching or storing to the data variables. *SNMP* consists of three parts:

- The structure of management information (SMI),
- The Management Information Base (MIB), and
- The protocol itself.

The SMI and MIB define and store the set of managed entities; *SNMP* itself conveys information to and from these entities. The public domain standard is based on the operational experience of TCP/IP internetworks within Darpa/NSFnet.

SNMP v2 Version 2 of the SNMP protocol with enhanced security (Secure SNMP) and improved management capabilities (Simple Management Protocol).

snow Random noise on the intensity-modulated signal of a display device (such as a cathode-ray tube—CRT) which manifests itself as a uniform distribution of fixed or moving spots, mottling, or speckling on the display.

SNP An abbreviation of Sequence Number Protection. The calculation of cyclic redundancy check code (CRC) or parity over both data and the sequence number (SN) field in the header.

SNPA An abbreviation of Subnetwork Point of Attachment.

SNR (1) An abbreviation of Signal to Noise Ratio. (2) An abbreviation of Saved Number Redial.

SNR psoph Signal-to-noise ratio measured with a psophometric weighted detector. See also *SNR*.

SNRM An abbreviation of Set Normal Response Mode.

SNRME An abbreviation of Set Normal Response Mode Extended.

SNS (1) An abbreviation of Satellite Navigation System. (2) An abbreviation of Secondary Network Server.

snubber A circuit or device used to reduce the effects of transients on other devices.

SO (1) An abbreviation of Serving Office. A central office in which the interexchange carrier (IXC) has a point of presence (POP). (2) An abbreviation of Secondary Output. The fiber optic port in Fiber Distributed Data Interface (FDDI) that transmits signals to the secondary fiber ring. (3) An abbreviation frequently found on the Internet meaning Significant Other. Always pronounced "Ess, Oh." Used to refer to one's primary relationship, especially a live-in to whom one is not married. (4) An abbreviation of Shift Out character (ASCII 14 0Eh).

SO-DIMM An acronym for Small Outline Dual Inline Memory Module. A type of *DIMM* with 72 pins instead of the 168 pins of a regular *DIMM*. It is designed to connect directly to 32-bit data buses and is intended for use in memory expansion applications in notebook computers.

SOA An abbreviation of Safe Operating Area.

SOAP An acronym for Seal Of APproval.

SOB An abbreviation of Start Of Block.

soc In USENet, one of the defined seven world newsgroup categories. They cover topics relating to social issues and social interaction. Subject matter is widely varied and includes material on culture, ethnic groups, human rights, bisexuality, homosexuality, feminism, and religion.

socket (1) That part of a connector designed to mechanically and electrically interface with a pin on a mating connector. See also *pin*.

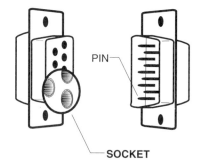

PIN

SOCKET

(2) A device into which a lamp is inserted. It provides mechanical support and electrical connections to the lamp. **(3)** The entity in a PC which accepts a PCMCIA (PC card) device and maps the host's internal bus to the device interface. **(4)** A synonym for a port. **(5)** In computer networking, a general-purpose interprocess communication (IPC) mechanism. The mechanism was originally implemented on the Berkeley Software Distribution (BSD) version of UNIX. *Sockets* are used as end points for sending and receiving data between applications. Each socket is associated with an address and usually with some other type of identification. In Internet addressing, it is the combination of the IP address (which identifies a host) and a port address (which identifies an application running on the host). Together, the *socket* is used to identify particular services throughout the Internet. Types of sockets frequently mentioned include:

- *Datagram socket*—for sending datagrams (packets used in connectionless delivery systems that do not guarantee delivery).
- *Stream socket*—a mechanism that provides a reliable connection (i.e., it guarantees delivery).
- *Raw socket*—used for access by low-level protocols and is available only to privileged programs.
- *Dynamically assigned socket* (*DAS*)—used for datagram delivery between nodes in an AppleTalk internetwork.
- *Statically assigned socket* (*SAS*)—used for datagram delivery between low-level processes in an AppleTalk internetwork.

socket client Any process or function associated with a socket in a particular network node. The client is said to "own" the socket. That is, it can make use of the socket to request and receive information and network services.

socket number In any of various networking environments, such as AppleTalk and Novell's NetWare, a unique value assigned to a socket. The maximum size of such a value depends on the number of bits allocated for the number.

For example, AppleTalk socket numbers are 8-bit values; hence, the range of values is 0–255. Values between 0 and 127 are reserved by Apple for system devices. (The 0–63 range is reserved for Apple's exclusive use.)

socket services The BIOS (basic input output system) level software layer directly above the hardware that provides a standardized interface.

SOCKS A security mechanism/package by which a secure data channel can be established between two computers in a client–server environment. For example, it allows a host behind a firewall to use finger, file transfer protocol (FTP), telnet, Gopher, and Mosaic to access resources outside the firewall while maintaining the security requirements.

SOD **(1)** An abbreviation of Start Of Data. **(2)** An abbreviation of Serial Output Data.

SODB An abbreviation of Start Of Data Block.

SOE An abbreviation of Start Of Entry.

SOF An abbreviation of Start Of Frame.

soft error **(1)** An error that is transient (intermittent) as opposed to a hard error that is permanent or repeatable. **(2)** An error from which a program, operating system, or device is able to recover, as opposed to a hard error that requires operator intervention (e.g., rebooting the computer). An error that is not considered serious or a threat to the performance or continued operation of the network, device, or system. See also *hard error.*

soft limiting A form of peak limiting or clipping in which the level of attenuation is progressively increased as the signal level is increased above a specified threshold. This is in contrast to hard limiting wherein the signal level is not allowed to exceed the threshold at all. See also *clipping* and *limiting.*

soft modem See *software upgradable modem.*

soft sectoring On magnetic disks, magnetic drums, and optical disks, the identification of sector boundaries by using recorded information. This is in opposition to hard sectoring in which a physical or mechanical method of identification is used.

software In computer programs, a set of machine readable instructions that cause hardware to do work. *Software* is divided into several categories and subcategories based on the function being performed. Examples of one top-level category based on task-based functions include:

- *Operating systems* (*OS*), also called system software—handles essential functions such as file management, display management, error detection and hardware control. The OS software provides a foundation from which application software is run.
- *Application program*—the set of specific job-related programs for which people use computers, e.g., word processors, spreadsheet, accounting, engineering analysis, drafting, and artwork programs.
- *Communications software*—enables multiple computers to share files and resources.
- *Development tools*—provide the tools needed to write, debug, and integrate new application software. Development tools include languages (BASIC, C, C++, ADA, PASCAL, FORTRAN, COBOL), debugging programs, programs to link several small programs into a single user program, editors, source code libraries, and code analyzers.

software maintenance The modification of delivered software to correct faults (bugs), improve performance, or adapt it to changing requirements.

software metering A program that observes the use (execution) of application programs. The usual purpose is to control the number of concurrent users so as to stay within the number allowed by the license.

For example, an organization may have 20 people who use a particular application; but the organization purchased a license for only five concurrent users. The *software metering* program would allow any five of the 20 to use the application simultaneously but only five at a time.

Software Publishing Association (SPA) An international trade association for the personal computer software industry formed in 1986. It has a worldwide membership of about 1200 companies, including large and small publishers and developers in the business, consumer, and education markets. Its association membership includes distributors, retailers, consultants, and other firms that provide services to publishers and are committed to the industry's growth.

Members benefit include:

- *Publications:* Member publications are on topics such as information on contracts, copyright protection, model licensing agreements, channel marketing, direct mail, public relations, packaging, and marketing.

  Cross-referenced membership directory, which provides a listing of member companies, key contacts, primary business focus, and products and services provided.

- *Anti-piracy:* Free anti-piracy material and participation in an industrywide litigation and education program protect members' intellectual property against copyright infringement.

- *Government contacts:* The *SPA* lobbies for legal protection on issues concerning such things as game ratings, competitiveness, cryptography, copyright legislation, and intellectual property protection (domestic and international).

- *Market-specific support:* Each member company can enroll in the business, consumer, and education sections. Each section is governed by a board that helps shape the focus of the activities and research projects so as to benefit these sectors of the software industry.

- *Peer recognition:* The Codie Awards held once each year recognize those products that have been determined by the industry to be the best in the business.

- *Special Interest Groups* (*SIGs*) such as:
  - Interactive Multimedia,
  - International,
  - Marketing,
  - Public Relations, and
  - Packaging.

  Each SIG is accompanied by resource guides and networking directories to support individual efforts in these areas.

- *Meetings and seminars:* Networking and educational opportunities at the fall (in the Midwest or on the East Coast) and spring conferences (West Coast) provide over 100 seminars, boot camps, and interactive sessions on software industry topics. In addition to these two annual conferences, a European conference is held each year in June.

software rot A term used to describe the effect of previously functional software, which has not been used in a while, to lose capability or fail. Frequently, *software rot* occurs when a program's assumptions become invalid or out of date. If the original design was insufficiently robust, the software may fail in mysterious ways. Another source of failure occurs when hardware is changed without a corresponding change in poorly written software. Such a failure may be semi-humorously ascribed to *bit rot.*

For example, many programs used a two-digit representation of a year. Most of these programs were expected to break down with *software rot* when their two-digit year counters wrapped around at the beginning of the year 2000. Without attention, they would have indicated the year 1900. Another example occurs when software timing loops are used to determine how fast certain parts of a program run (such as was done in early PC-based game programs). When the hardware speed was increased, the software no longer ran correctly. The BIOS (basic input output systems) in early PCs suffered the same fate when a faster crystal was used in the PC. See also *bit rot, Bohr bug, heisenbug, mandelbug,* and *schrödinbug.*

software upgradable modem A modem whose modulating and demodulating algorithms are stored in alterable memory rather than read only memory (ROM). This allows the user to load new programs into the existing modem hardware when new standards are developed rather than replacing the hardware.

Modulation and demodulation are accomplished through the use of digital signal processing (DSP) software rather than through classical hardware methods.

SOH (1) An abbreviation of Start Of Header. A control byte in Binary Synchronous Communications (BSC). (2) An abbreviation of Section OverHead.

SOHF An abbreviation of Sense Of Humor Failure.

SOHO An acronym for Small Office, Home Office.

SOL An acronym for Sure Out of Luck (or something like that).

solar array A photovoltaic power system. A collection of photovoltaic cells interconnected in a series parallel arrangement so as to produce a desired voltage and current when the collection (array) is exposed to sunlight.

solar constant The irradiance of the Sun at the Earth's mean orbital distance (1.5×10^{11} m or 92.9×10^6 mi), disregarding the effects of the atmosphere. The value is generally given as 1353 W/m^2 (or 125.7 W/ft^2).

solar induced current (SIC) A current in grounded power and telecommunication cables due to Earth's potential differences. The potential differences are in sporadic quasi-direct currents induced by geomagnetic storms resulting from the particles emitted from solar flares erupting on the surface of the Sun.

solar noise Electromagnetic radiated noise generated by the Sun. It exceeds other background sky noise sources by several orders of magnitude.

solar wind The constant outward flow of plasma from the Sun that is deflected by the magnetic field of the Earth. In flowing around the Earth, it creates the magnetosphere.

solder (1) A metal alloy, e.g., of tin and lead. (2) The act of joining together with solder (the alloy).

solenoid An electromagnetic device that converts electrical energy into linear mechanical motion. The basic physical construction of a solenoid is simply a coil of wire wrapped around a hollow tube into which is inserted a movable iron rod. When a current is passed through the coil, it creates a magnetic field, which in turn attracts the iron rod (pulling it into the tube). When the current ceases flowing, a spring or other means pushes the rod out to its resting position.

solid angle The ratio of the area on a sphere to the square of the radius of the sphere. Expressed in *steradians.*

solid-state component A component whose operation depends on the control of electric or magnetic phenomena in solids. Examples include all semiconductors (transistors, diodes, FETs), ferrite cores, metal oxide vaistors (MOVs), and so on.

solid-state memory Any of the semiconductor-based memories including EEPROM, EPROM, PROM, RAM, ROM, and UVPROM.

solid-state relay A relay that utilizes semiconductor components rather than mechanical components to open and close the controlled circuit.

soliton An optical pulse having a shape, spectral content, and power level designed to take advantage of nonlinear effects in an optical fiber, for the purpose of essentially negating dispersion over long distances.

SOLR An acronym for Sidetone Objective Loudness Rating.

SOM (1) An abbreviation of Start Of Message. (2) An abbreviation of System Object Model.

SOM-H An abbreviation of Start Of Message—High precedence.

SOM-L An abbreviation of Start Of Message—Low precedence.

SOM-P An abbreviation of <u>S</u>tart <u>O</u>f <u>M</u>essage—<u>P</u>riority.

SONet MCI's national fiber optic technology.

SONAD An acronym for <u>S</u>peech <u>O</u>perated <u>N</u>oise <u>A</u>djusting <u>D</u>evice.

SONET An acronym for <u>S</u>ynchronous <u>O</u>ptical <u>NET</u>work (pronounced "SO net" or "SAW net"). An American National Standards Institute (ANSI) & BellCore standard for connecting one high-bandwidth optical fiber communications system to another. The standards pertain to the lower layer of the ISO/OSI seven-layer Reference Model. *SONET* is designed to operate with a number of data link layer technologies, including Switched Multimegabit Data Services (SMDS) and Fiber Distributed Data Interface (FDDI). The data transfer rates can range from 51.84 Mbps (the OC1 rate) to 2.48832 Gbps (the OC48 rate) in multiples of the OC1 rate. An equivalent ITU-T international standard is specified by the Synchronous Digital Hierarchy (SDH) recommendation.

SONET supports drop and insert capabilities, which facilitates the identification and removal of channels to different destinations. This makes submultiplexing feasible, that is, channel capacities as low as 64 kbps may be multiplexed into *SONET* channels. The *SONET* layer can be broken down into a four-layer protocol stack as follows,

Function in Layer	Local DCE		Remote DCE	Overhead (embedded ops chan)
Payload to SPE mapping	Path layer		Path layer	
Maintenance & switching	Line layer		Line layer	576 kbps
Scrambling & framing	Section layer		Section layer	102 Kbps
Optical transmission	Photonic layer	fiber comm	Photonic layer	0 Kbps

The basic transmission unit is an 810-byte (6480-bit) frame, and 80,000 of these are transmitted per second. Both timing and framing can be adjusted during operation. See also *DS-n, OC-n, SDH, STM-n,* and *STS-n.*

SONET frame An 810-byte frame arranged as 9 rows of 90 bits. Four bytes in each row are overhead, leaving 86 for data.

Three overhead bytes in the first three rows are allocated for monitoring the section; three overhead bits in the remaining six rows are for the line; and one byte in each row is used for path overhead. The bytes containing the payloads for the nine rows and the path overhead byte are known as the *synchronous payload envelope* (*SPE*). The SPE can handle payloads in one of three ways:

- A continuous 50.11-Mbps envelope for carrying asynchronous data to 50.11-Mbps rates.
- A structured virtual tributary (VT) envelope to carry DS-1, DS-1C, DS-2, CEPT1, etc.
- A concatenation of multiple SPEs to yield higher capacity channels, for example, three SPEs may be combined to give a channel capacity of 155 Mbps.

SOP An abbreviation of <u>S</u>tandard <u>O</u>perating <u>P</u>rocedure.

SOR An abbreviation for <u>S</u>tart <u>O</u>f <u>R</u>ecord.

sound field A region containing sound waves.

sound powered telephone A telephone in which the only operating power is that which is derived from the speech input.

sound pressure Also called *instantaneous sound pressure,* it is the instantaneous difference between the static pressure and the pressure due to a sound wave at a point. *Sound pressure* is measured in pascals (newtons per square meter—N/m^2) or equivalent units. See also *effective sound pressure.*

sound pressure level (SPL) In telephony, the ratio of measured *sound pressure* to a reference pressure of 1 pascal expressed in decibels, that is,

$$SPL = 20 \cdot \log_{10} \left(\frac{Pressure_{measured} \, \text{Pa}}{1 \, \text{Pa}} \right) \text{dB}$$

source (1) The location or device from which data are considered to originate. Diskettes, host computers, files, a memory address, and documents all represent examples of sources. **(2)** The physical origin of electromagnetic radiation.

source address In packet communications systems, the *source address* is that part of a packet header that specifies the originator of the message. Depending on the type of address, this field may be 4 or 6 bytes or even longer. See also *Destination Address (DA).*

source code Human readable program statements written in a symbolic language (such as assembly) or a high-level language (such as BASIC, C), as opposed to object code which is designed to be readable by the computer. Object code is derived from *source code* through the use of a compiler or assembler program.

source data The original data on which an application program operates. For example, employee records might contain hourly pay rates; these records are one of the source data files for the payroll check printing program. (Another source data file contains the number of hours reported for the pay period.)

source efficiency The ratio of the power emitted by a source to the power supplied to that source.

source explicit forwarding A bridge feature that allows Medium Access Control (MAC)-layer bridges on local area networks (LANs) to forward packets from only source nodes specified by the network administrator.

source impedance The impedance presented by an energy source to the input terminals of a device, circuit, or system.

source node A node in a circuit or system with only outgoing branches.

source quench A flow-control technique in which a computer experiencing data traffic congestion sends a message back to the source of the messages or packets causing the congestion, requesting that the source stop transmitting.

source routing (SR) A packet routing strategy originally used in IBM Token Ring networks. Now specified by the IEEE standard for 802.5 token ring environments. In it, the route a packet will take between its source and destination is determined in advance by (or for) the source node.

Packet routes are determined by a discovery process in which the source node sends an *explorer packet* (or *discovery packet*) onto the network and then waits for return packets. The explorer packet is launched along all active paths in the network, eventually reaching the destination host; when received, the packet is sent back to the originator. Along the way, each bridge traversed adds its designator to the explorer packet. When the return packet arrives at the originating node, it contains complete routing information, which can be used in subsequent data transmission packets. This routing information is included when a packet is sent around a Token Ring network. *SR*-compliant bridges generally need less processing power since most of the work is done at the originating node. A liability of *SR*, however, is that the number of explorer packets traversing the network increases dramatically as the number of Token Ring local area network (LAN) segments and internetworking devices between the stations grows. See also *open shortest path first (OSPF), routing information protocol (RIP)*, and *spanning tree*.

Source Routing Transparent (SRT) An algorithm under consideration by IEEE combining Transparent Bridging (TB) for Ethernet networks and Source Routing (SR) of Token Ring networks for interconnectivity of the two local area networks (LAN) types. Upward migration to Fiber Distributed Data Interface (FDDI) is guaranteed. The bridge applies either TB or SR logic to each frame according to frame type.

source user The user providing the information to be transferred to a destination user during a particular information transfer transaction. Also called the *information source*.

Southeastern Universities Research Association Network (SURAnet) A regional Internet network and service provider that serves Alabama, Delaware, Florida, Georgia, Kentucky, Louisiana, Maryland, Mississippi, North Carolina, Puerto Rico, South America, South Carolina, Tennessee, Virginia, Washington, D.C., and West Virginia.

SP (1) An abbreviation of <u>SP</u>ace character. In ASCII <SP> is 020h. (2) An abbreviation of <u>S</u>tructure <u>P</u>ointer. A field in the asynchronous transfer mode (ATM) Adaptation Layer 1 cell. (3) An abbreviation of <u>S</u>ecurity <u>P</u>rotocol. (4) An abbreviation of <u>S</u>witch <u>P</u>ort. (5) An abbreviation of <u>S</u>emi<u>P</u>ublic. (6) An abbreviation of <u>S</u>inging <u>P</u>oint. (7) An abbreviation of <u>S</u>upervisory <u>P</u>rocess.

SPA (1) An abbreviation of <u>S</u>oftware <u>P</u>ublishers <u>A</u>ssociation. (2) An abbreviation of <u>S</u>catter <u>P</u>ropagation <u>A</u>ntenna. (3) An abbreviation of <u>S</u>emi<u>P</u>ermanently <u>A</u>ssociated.

space (1) The ASCII character 32 decimal (032h). (2) In binary communications, one of two *significant conditions* of a signal when a bit of information is being transmitted. In current loop signaling, the space is the absence of current flow (e.g., dial pulses). In binary systems, a *space* is the state of the zero (0) bit and the signaling start bit in asynchronous transmission. Also called *spacing pulse* and *spacing signal*. The complementary significant condition is called a mark or a one (1).

space diversity transmission A transmission system in which a multiplicity of physically separate communications paths are made available for message propagation. In the case of radio transmission systems, the transmitter and/or receiving antennas should be separated by one or more wavelengths.

For example, in some Earth satellite systems, multiple antennas are used to prevent *Sun transit outages*. By separating the receiving sites geographically, the Sun, satellite, and receiving site do not line up at the same time. Therefore, at least one antenna can always receive messages. On the Earth's surface, weather conditions (such as heavy rain) can cause outages. Here, multiple microwave paths may be used to provide continuous communications capability. On the Internet (as well as many other wide area networks), several paths through the system are available to nodes (hosts) to circumvent communications problems due to congestion, intermediate node failures, or an excessively noisy link.

space division multiplex (SDM) The routing of messages through physically distinct paths. The telephone central office switch is an example of a *space division multiplexer,* that is, a separate communication path is provided for each simultaneous call.

The illustration demonstrates how a set of input signals may be routed through a *space division switching system* to arbitrary output ports. See also *code division multiplex, frequency division multiplex, switch,* and *time division multiplex.*

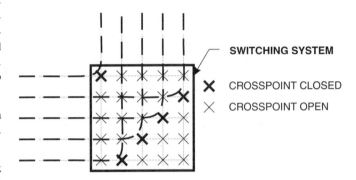

space division switch See *space division multiplex.*

space hold The normal "no traffic" condition of a serial communications line wherein a steady space (a binary 0) condition is transmitted. See also *mark hold.*

space parity See *parity.*

spacing bias The uniform lengthening of all spacing signal elements at the expense of the pulse width of all marking signal elements.

SPAG An abbreviation of <u>S</u>tandards <u>P</u>romotion and <u>A</u>pplication Group.

spam A slang term that originated on the Internet to describe unsolicited e-mail. Specifically, an uninvited message sent to a large number of unsuspecting users (or newsgroups). The term comes from the comparison to how offensive "SPAM®" thrown at the walls of a restaurant would be for diners viewing it.

On the USENet, *spam* is a synonym for *carpet bomb.*

span (1) Refers to that portion of a T-carrier link that interconnects two central offices. Repeated sections are called a *span line*. (2) Refers to a range of values. The algebraic difference between the upper and lower value.

SPAN An acronym for <u>S</u>ystem <u>P</u>erformance <u>AN</u>alysis.

spanning tree A technique used by bridges to create a logical topology that connects all network segments and ensures that only one path exists between two nodes at a time. If a path should become inoperative for any reason, the algorithm can find an alternative path (if one exists). The algorithm is detailed in the IEEE 802.1d standard.

A *minimal spanning tree* is one that covers all possible paths, does so with as few segments as possible, and makes sure there are no loops (closed paths) in the network. See also *open shortest path first (OSPF), routing information protocol (RIP)*, and *source routing.*

spanning tree algorithm (STA) An algorithm, documented in the IEEE 802.1d standard, for detecting and preventing loops from occurring in a multibridged environment. When three or more local area network (LAN) segments are connected by bridges, a loop can occur. As a bridge forwards all packets that are not recognized as being local, some packets can circulate for long periods of time (called a *broadcast storm*), eventually degrading system performance. This algorithm ensures that only one path connects any pair of stations, selecting one bridge as the root bridge from which all paths should radiate.

The algorithm is based on the original *spanning tree algorithm* developed by Digital Equipment Corporation (DEC). The two algorithms are not identical, however, nor are they even compatible.

SPARC (1) An acronym for <u>S</u>caleable <u>P</u>erformance <u>ARC</u>hitecture. (2) An acronym for <u>S</u>tandards <u>P</u>lanning <u>A</u>nd <u>R</u>eview <u>C</u>ommittee.

spark gap A gas dielectric separating two electrodes designed so as to carry an electrical current when the voltage exceeds a predetermined level (overvoltage protection). A *spark gap* may be exposed with air as the dielectric, or it may be enclosed in a container with selected gasses at pressures other than one atmosphere. *Spark gaps* are used to protect communications circuits from damage due to excessive voltages. See also *gas tube*.

spark killer A circuit or device connected across a relay or switch contact or in parallel with a switched inductive element for the purpose of reducing or absorbing the spark energy that occurs when an inductive circuit carrying a current is suddenly opened.

Frequently, the circuit is simply a series resistor and capacitor. The capacitor blocks the direct current being switched while passing the spark's transient energy and the resistor dissipates the energy of that spark. Also called a *snubber* or *surge arrestor*.

SPATA An acronym for <u>SP</u>eech and d<u>ATA</u>.

SPC (1) An abbreviation of <u>S</u>tored <u>P</u>rogram <u>C</u>ontrol. (2) An abbreviation of <u>S</u>ignal <u>P</u>rocessing <u>C</u>omponent.

SPCC An abbreviation of <u>S</u>trength, <u>P</u>ower, and <u>C</u>ommunications <u>Ca</u>ble.

SPCL An abbreviation of <u>SP</u>ectrum <u>CeL</u>lular error correction protocol.

SPD An abbreviation of <u>S</u>ynchronous <u>P</u>hase <u>D</u>emodulator.

SPDL An abbreviation of <u>S</u>tandard <u>P</u>age <u>D</u>escription <u>L</u>anguage.

SPDT An abbreviation of <u>S</u>ingle <u>P</u>ole <u>D</u>ouble <u>T</u>hrow switch.

SPDT-NC An abbreviation of <u>S</u>ingle <u>P</u>ole <u>D</u>ouble <u>T</u>hrow switch—<u>N</u>ormally <u>C</u>losed contact.

SPDT-NO An abbreviation of <u>S</u>ingle <u>P</u>ole <u>D</u>ouble <u>T</u>hrow switch—<u>N</u>ormally <u>O</u>pen contact.

SPDU An abbreviation of <u>S</u>ession layer <u>P</u>rotocol <u>D</u>ata <u>U</u>nit.

SPE (1) An abbreviation of <u>S</u>ynchronous <u>P</u>ayload <u>E</u>nvelope. That part of the Synchronous Optical Network (SONET), Switched Digital Service (SDS), or Synchronous Digital Hierarchy (SDH) frame that contains user data. (2) An abbreviation of <u>S</u>witch <u>P</u>rocessing <u>E</u>lement. (3) An abbreviation of <u>S</u>ignal <u>P</u>rocessing <u>E</u>lement.

speaker-dependent voice recognition A speech recognition technology/algorithm which requires that each user train the system to the user's speech characteristics. These systems will therefore recognize speech from only a given user or someone who sounds like that user. The system cannot, however, be used as a speaker verification device.

speaker-independent voice recognition (SIR or SIVR) A speech recognition system technology/algorithm that automatically recognizes utterances of an arbitrary user without prior training.

A typical application accepts input from callers where the callers are using rotary dial instead of dual tone multifrequency (DTMF) telephones and then use voice response to verbalized questions to route calls or provide information. These *SIR* products have deliberately limited vocabularies but are increasing due to the vast installed base of non-DTMF phones. Another developing application provides for the automated conversion of speech to accurate and meaningful textual information.

speaker phone A telephone instrument with a speaker and microphone arranged for hands-free, two-way conversation.

spec. An abbreviation of <u>SPEC</u>ification.

SPEC An acronym for <u>S</u>peech <u>P</u>redictive <u>E</u>ncoded <u>C</u>ommunications.

special access In telephony, a dedicated line, installed by the local exchange carrier (LEC) between a subscriber and a long-distance carrier.

special billing number A phone number assigned to a customer for billing purposes only. It cannot be called; however, it can be used as a bill to a third-party number or given to the operator as an outgoing paid call number.

special routing code In telephony, a three-digit code (of the form 0XX or 1XX) which is used within the network to modify call routing or handling. End users are blocked from its use because office codes starting with 0 or 1 are rejected by the local exchange carrier (LEC) switching system.

specialized common carrier (SCC) A common carrier offering a limited type of service or serving a limited market.

specification An essential technical requirements document for hardware and/or software items, materials, and/or services, including:

- A detailed, precise description of the operation, parameters, capabilities, input requirements, features, required environment, and limitations.
- The procedures to be used to determine whether the requirement has been met.
- The requirements for packaging, packing, and marking.

specific detectivity (D*) A figure of merit used to characterize the performance of a photodetector. It is equal to the reciprocal of *noise equivalent power* (*NEP*), normalized to unit area and unit bandwidth. Mathematically, it is expressed:

$$D* = \frac{\sqrt{A\Delta f}}{NEP}$$

where:

A is the area of the photosensitive region of the detector, and
Δf is the effective noise bandwidth.

Also called *D-Star* (*D**).

speckle noise A synonym for *modal noise*.

speckle pattern In optical systems, a field intensity pattern produced by the mutual interference of partially coherent beams that are subject to minute temporal and spatial fluctuations.

In a multimode fiber, a *speckle pattern* results from the superposition of mode field patterns. If the relative modal group velocities change with time, the *speckle pattern* will also change. If differential mode attenuation occurs, *modal noise* results.

spectral density For a specified bandwidth of radiation consisting of a continuous frequency spectrum, the total power in the specified bandwidth divided by the specified bandwidth. *Spectral density* is usually expressed in watts per hertz (W/Hz).

spectral irradiance Irradiance per unit wavelength interval at a given wavelength, usually expressed in watts per unit area per unit wavelength interval ($W/m^2/\Delta\lambda$).

spectral line A narrow range of emitted or absorbed wavelengths. See also *spectrum* (2).

spectral loss curve In an optical fiber, a plot of attenuation as a function of wavelength. *Spectral loss curves* must be normalized with respect to distance before meaningful comparison among fibers can be made.

spectral power distribution The relative power emitted by a source as a function of wavelength.

spectral purity The degree to which a signal is monochromatic.

spectral radiance Radiance per unit wavelength interval at a given wavelength, expressed in watts per steradian per unit area per wavelength interval.

spectral responsivity The ratio of an optical detector's electrical output to its optical input, as a function of optical wavelength.

spectral width In laser optics, the range of wavelengths (light frequencies) emitted by a single frequency, unmodulated laser with an optical power that exceeds a specified fraction of the maximum value. Generally, a narrower width is more desirable.

In optical communications applications, a common method of specifying *spectral width* ($\Delta\lambda$) is the full width at half maximum. (This method may be difficult to apply when the spectrum has a complex shape.) Also known as *laser line width* or *spectral purity*.

spectral window A wavelength region of relatively low transmission loss (high transmittance) surrounded by regions of higher loss (low transmittance). See also *window*.

spectrum (1) A graphical representation of the distribution of amplitudes of the constituent frequencies of a signal or waveform as a function of frequency. (2) The continuous range of frequencies available for a particular service, as in the 88 to 108 MHz *spectrum* allocated for the FM broadcast band.

The frequency spectrum diagram in the figure shows both:

• A line spectrum such as would be found in an unmodulated carrier, and
• A continuous band of frequencies such as would be found when the carrier is modulated by a range of frequencies.

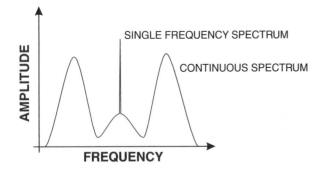

(3) The description of a portion of or the entire electromagnetic frequency range. See also *frequency band* and *frequency spectrum*.

specular In, or pertaining to, the direction in which a mirror reflects incident energy.

specular reflection Reflection from a smooth surface, such as a mirror, which maintains the integrity of the incident wavefront.

speech concatenation A voice processing term for economical digitized speech playback. It uses independently recorded phrases linked under application program control to produce a customized response in natural sounding language. Examples include bank balances or bus schedules. It is done for speed and economy, lending itself to limited, structured vocabularies that are best stored in RAM or are readily accessible from disk.

speech digit signaling A synonym for *bit robbing*.

speech interpolation A method of obtaining, on the average, more than one speech channel per communications channel. This is accomplished by giving each subscriber access to a half-channel (a one-way or simplex channel) only when there is activity (speech) at the terminal to be transmitted. See also *TASI*.

speech network An electric circuit containing a two- to four-wire hybrid that interfaces a telephone handset (microphone and earphone) to the telephone line. The network not only provides the desired transmit and receive levels to the two-wire line based on loop length, but it reduces the sidetone to an acceptable level. One *speech network* implementation is shown in the figure; it contains a transformer hybrid and sidetone balance circuit. Also called a *telephone network*. See also *hybrid*.

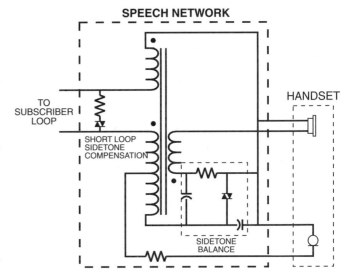

speech plus Pertaining to a circuit that was designed and used for speech transmission, but to which other uses, such as digital data transmission, facsimile transmission, telegraph, or signaling, have been added by means of multiplexing.

speech power See *volume unit (vu)*.

speech recognition The ability of a machine to convert speech (generally a word or phrase) into digital codes that can be processed by a computer as if the information had been entered via keyboard or mouse.

Although the details of the recognition process vary, the principle includes matching a digitized speech sample against a dictionary of known coded utterances. When a match occurs, a signal is sent to the computer program as if it were generated on the keyboard.

speech synthesis The ability of a machine to produce sounds that are understandable by a person as words. Generally, the term *synthesis* implies the generation of sounds by an algorithm, not by the playback of prerecorded speech fragments.

speed dialing (**1**) A feature of a telephone device or system that allows a user to quickly dial frequently used numbers. The repertory of stored numbers may be as few as 10 or as many as several hundred. Typically, it is about 30 to 50. The number of digits in each entry may be as many as 30 or more. This allows the complete sequence for international calls and any special calling codes to be saved in a single register. Also called *abbreviated dialing, memory dialing repertory dialing,* or *speed calling.* (**2**) Dialing at a speed greater than the normal 10 pulses per second.

speed of light (**c**) The speed of light is a constant in a uniform medium; however, it is not the same in all media. Therefore, the reference medium is generally taken to be the velocity in free space. In free space, the speed (*c*) is precisely:

$$c = 2.99792458 \times 10^8 \text{ m/s}$$

or

$$c \cong 186\ 282 \text{ mi/s}$$

The figure (in m/s) is precise because by international agreement the meter is defined in terms of the speed of light. In any physical medium, the speed of light is lower than in free space. Since the frequency is not changed, the wavelength is also decreased.

speed of service The time between a significant instant of a transmitted message at a sending station and the corresponding instant at the terminating station:

- As perceived by the end user, or
- As measured by the system.

Also called *originator to recipient speed of service.*

speed up tone A synonym for *camp-on busy signal.*

SPF (**1**) An abbreviation of <u>S</u>hortest <u>P</u>ath <u>F</u>irst. (**2**) An abbreviation of <u>S</u>ubscriber <u>P</u>lant <u>F</u>actor. (**3**) An abbreviation of <u>S</u>ystem <u>P</u>erformance <u>F</u>actor.

spherical error probability (**SEP**) A measure of navigational accuracy. It is the radius of a sphere, inside which there is a 50% probability that the true location will be found. See also *circular error probability* (*CEP*).

SPI (**1**) An abbreviation of <u>S</u>ubsequent <u>P</u>rotocol <u>I</u>dentifier. (**2**) An abbreviation of <u>S</u>hared <u>P</u>eripheral <u>I</u>nterface.

SPID An abbreviation of Integrated Services Digital Network's (ISDN's) <u>S</u>ervice <u>P</u>rofile <u>ID</u>entifier.

spike A transient electrical signal superimposed on a desired waveform. Usually very short in duration and high in amplitude. The drawing shows both a positive going and a negative going *spike* on an arbitrary waveform.

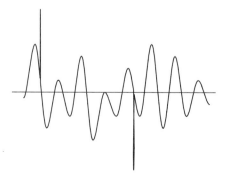

A transient is generally called a spike when the voltage is more than twice the normal signal's peak amplitude. See also *impulse, mov, surge,* and *transient.*

spike file See *last-in first-out* (*LIFO*).

spill forward In automatic switching. The transfer of full control on a call to the succeeding office by sending forward the complete called-party telephone address.

spillover In an antenna, that part of the radiated energy from the feed that does not intercept the reflectors.

spiral four See *star quad.*

SPL (**1**) An abbreviation of <u>S</u>ound <u>P</u>ressure <u>L</u>evel. (**2**) An abbreviation of <u>SPL</u>ice.

splice (**1**) To physically join together two (or more) like information or power-carrying media such that continuity is established. Examples include wire, coaxial cables, and optical fibers.

In the case of wire connection, the splice may be by means of twisting or soldering like conductors together and applying appropriate insulation to the junction. With optical fibers, the splice can be made by fusing the cores from the two cables together or by attaching the cores to each other with some mechanical means. In general, a fusion approach works better than a mechanical one. (**2**) A device used to join or splice. (**3**) The completed joint.

splice loss Losses that occur at a splice. See also *coupling loss* and *insertion loss.*

split cable system A broadband transmission system in which a portion of the bandwidth is allocated to transmission in one direction and the remaining bandwidth is used for transmission in the opposite direction. The bandwidth need not be distributed equally between the two directions. For example, cable television allocates a much larger part of the bandwidth to outgoing signals, since the subscribers need not communicate with the headend. For local area networks (LANs), the distribution is more even.

split homing The connection of a terminal facility to more than one switching center by separate access lines, each of which has a separate directory number.

split horizon routing A strategy for maintaining routing tables in an AppleTalk Phase 2 network. The strategy involves passing routing table updates only to nodes or routers that can and will actually use the information. See also *open shortest path first* (*OSPF*), *routing information protocol* (*RIP*), *source routing,* and *spanning tree.*

split pair In twisted pair wiring, refers to a type of wiring error in which wires from two different pairs are used for transmission instead of wires from the same pair.

This results in a significant reduction in noise immunity because the electromagnetic fields encountered by the *split pair* are no longer equal and opposite.

splitter A device that breaks a received signal into multiple derived output signals. One type of splitter is a wavelength (or frequency) selective coupler, which splits an incoming signal into outgoing signals based on wavelength (or frequency). See also *directional coupler* and *frequency division multiplex* (*FDM*).

SPM An abbreviation of <u>S</u>ession <u>P</u>rotocol <u>M</u>achine.

SPMF An abbreviation of Servo Play Mode Function.

SPN (1) An abbreviation of Signal Processor Network. (2) An abbreviation of Subscriber Premises Network.

SPOI An abbreviation of Signaling POint Interface.

spontaneous emission Radiation emitted when the internal energy of a quantum mechanical system drops from an excited level to a lower level without regard to the simultaneous presence of similar radiation. Examples of spontaneous emission include radiation from a light emitting diode (LED) and radiation from an injection laser diode (ILD) at currents below the lasing threshold.

spoofing (1) In general, a method of bypassing some aspect of a system by delivering deceptive messages, i.e., valid messages but from a source other than the one the destination is expecting. (2) A technique used by some "smart" modems to improve the data throughput when nonstreaming protocols are used by the local data terminal equipment (DTE).

The local modem returns a simulated acknowledge as soon as a packet is received from the local DTE. This allows the local DTE to send the next packet to the modem immediately rather than waiting for the acknowledge from the remote DTE. When the local modem receives the real acknowledge from the remote DTE, it is discarded (not sent on to the local DTE). Without *spoofing,* the local DTE would have to wait for the acknowledge from the remote DTE before releasing the next packet for transmission. The local DTE and modem are idle during this period; consequently, the throughput suffers. (3) A method of fooling network end stations into believing that keep-alive signals have come from and return to the host. Polls are normally received and returned locally at either end of the network and are transmitted over the network only if there is a status change. The result is a non-time critical network with a minimum of keep-alive traffic between deterministic end stations, while retaining the opportunity to send flags should an end station alter its state. (4) The interception, alteration, and retransmission of a cipher signal or data in such a way as to mislead the recipient. (5) An attempt to gain access to an automated information system (AIS) by posing as an authorized user.

spooler Application software (or a device) that allows a print request to be processed concurrently with another unrelated application.

The *spooler* manages multiple print requests (possibly according to priority) so that a job started may be completed while additional print jobs are placed into a waiting queue. At the completion of the print job in process, the next job on the queue is started. Although some say that *SPOOL* is an acronym from Simultaneous Peripheral Operation On Line, it is more likely that the term originally referred to the process of copying a print job to a spool of magnetic tape for off-line printing, e.g., spooling the job to tape.

sporadic E Irregular scattered patches of relatively dense ionization that develop seasonally within the ionospheric E region and that reflect and scatter frequencies up to 150 MHz. Reflections from these regions can reach 2400 km.

It occurs regularly over equatorial regions and is common in temperate latitudes during late spring, early summer and, to a lesser degree, in early winter. At polar latitudes, it can accompany auroras and associated disturbed magnetic conditions. Also called *sporadic E propagation.* See also *ionosphere.*

spot beam In satellite communications systems, a narrow beam from a satellite station antenna that illuminates a selected region of the Earth (called a footprint), with high irradiance. See also *footprint.*

SPP (1) An abbreviation of Sequential Packet Protocol. Xerox Network Systems (XNS) transport protocol governing sequential data. (2) An abbreviation of Signal Processing Peripheral.

SPR An abbreviation of SPaRe.

spread spectrum A modulation technique developed in the late 1930s that expands or spreads the information signal over a bandwidth much wider than required for transmission. There are several potential advantages of this modulation technique:

- Noise immunity improvement.
- Security improvement:
  - Low probability of detection, interception, or determination of the transmitter's location. (To an observer who does not possess information about the carrier, the transmission is indistinguishable from other sources of noise.)
  - High immunity against interference and jamming (intentional interference). The presence of (narrowband) interference signals only decreases the channel's signal-to-noise ratio (SNR); hence, it increases the error rate, which can be handled with appropriate error control.
- Reduction of adverse effects of multipath interference: Signals of certain frequencies cancel due to differences of path lengths; at other frequencies the same path does not cause cancellation. Hence, multipath does not cause the loss of all signal information.
- Multiple access capability (*code division multiple access—CDMA*): Increasing the number of transmit receive pairs only gradually increases each channel's error rate. In contrast, narrowband systems can only accommodate a fixed number of channels determined by available bandwidth and channel width (data rate).

There are several methods of arriving at a spread spectrum system, each with its own benefits and drawbacks. Five methods are:

- *Direct sequence* (*DS*) *modulation systems:* A high-speed randomizing code is used to directly modulate the data. Generates a sin *x/x spread spectrum.*
- *Frequency hopping* (*FH*) systems: The frequency of the data-modulated carrier is changed according to a randomizing code. Produces a flat *spread spectrum.*
- *Time hopping* systems: The data modulated carrier is keyed on and off according to the randomizing code.
- *Chirp* technique systems.
- *Hybrid* systems (using combinations of the above methods).

The fundamentals of a *direct sequence spread spectrum* modulation demodulation system are:

- Modulate the transmit data using a pseudorandom number (PRN) generator. This yields the very wide bandwidth *transmit signal spectrum* shown in the figure.
- Within the communications channel, nature adds unwanted noise. (In the figure on page 489 the noise is shown as an interfering modulated carrier.) The sum of the unwanted noise and the transmitted spectra is delivered to the receiver as is indicated in the receive signal spectrum.
- The demodulator compresses the modulated data spectrum by correlating it with the same PN sequence that the modulator used. However, because the noise is "uncorrelated," it is spread over the same bandwidth as was the original data.
- Following the demodulator is generally a *band-pass filter* (not shown) whose function it is to attenuate the noise outside the bandwidth of the wanted signal. This improves the received signal-to-noise ratio (SNR).

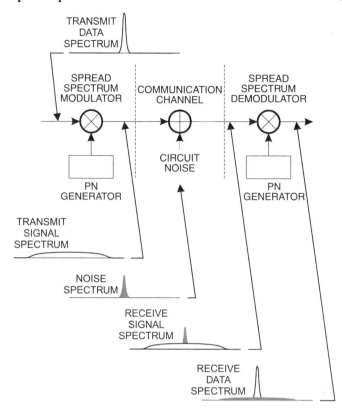

Sprintnet A public packet-switched network using the ITU-T X.25 protocols, which provides dial-up access to services like Delphi, Portal, GEnie, and Compuserve.

SPS (**1**) An abbreviation of Standard Positioning Service. (**2**) An abbreviation of Standby Power Supply. (**3**) An abbreviation of Samples Per Second. (**4**) An abbreviation of Symbols Per Second. (**5**) An abbreviation of System Performance Score.

SPST An abbreviation of Single Pole Single Throw switch or relay contact.

SPST-NC An abbreviation of Single Pole Single Throw switch—Normally Closed switch or relay contact.

SPST-NO An abbreviation of Single Pole Single Throw switch—Normally Open switch or relay contact.

spur (**1**) A secondary route having a junction to the primary route in a network. (**2**) An abbreviation of *spurious emission.*

spurious An unexpected, unwanted, and erroneous event.

spurious emission Any part of an electromagnetic output that is not a component of the theoretical output, as determined by the modulation technique and specified bandwidth constraints, i.e., frequencies that are outside the necessary signal bandwidth and which level may be reduced without affecting the corresponding transmission of information.

Spurious emissions include harmonic emissions, parasitic emissions, intermodulation products, and frequency conversion products but exclude out-of-band emissions. Frequently referred to as *spurs.*

spurious radiation Any unintentional emission.

SPX An abbreviation of Sequential Packet eXchange. Novell's NetWare implementation of *sequential packet protocol* (*SPP*), a connection-oriented protocol governing sequenced data.

SQD An abbreviation of Signal Quality Detector.

SQE An abbreviation of Signal Quality Error test function. In the older Ethernet systems, it is the defined signal quality test function.

IEEE 802.3 defines this for signals from the medium attachment unit (MAU) to the network interface card (NIC). The *SQE* signal indicates that the collision detection circuitry is functioning properly. Frequently called *heartbeat.* Also abbreviated *SQET.*

SQL An abbreviation of Structured Query Language.

square law detection An amplitude modulation (AM) detection method in which the output voltage is linearly proportional to the square of the input wave envelope.

square root ($\sqrt{x}$) A mathematical process that determines what number when multiplied by itself yields the specified value.

square wave A periodic wave that alternates between two fixed values (significant conditions) and remains at each value for the same period of time.

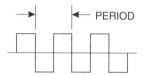

squelch (**1**) A circuit in a receiver that forces the output to a known state in the absence of a proper information carrier. The *squelch* function is activated in the absence of a sufficiently strong desired input signal, in order to exclude undesired lower-power input signals that may be present at or near the frequency of the desired signal. (**2**) An action taken by an Internet host system administrator to suspend or cancel a user's account privileges for noncompliance with the host's usage policies. Acts that may lead to *squelching* include harassing or threatening other users, engaging in software piracy, mail bombing, posting defamatory messages that attack the reputation of other users, using language not acceptable within the norms of the host (habitual or excessive profanity for example), and posting copyrighted material.

SQUID An acronym for Superconducting QUantum Interference Device.

sr An abbreviation of *steradian.*

SR (**1**) An abbreviation of Source Routing. (**2**) An abbreviation of Send Receive. (**3**) An abbreviation of Switch Register.

SRAM An acronym for Static Random Access Memory (pronounced "ESS-ram"). *SRAM* is a type of random access memory (RAM) with a permanent hold time. Therefore, unlike DRAM there is no need to refresh the information stored in it. *SRAM* is also generally faster than DRAM because it does not need to be refreshed. The contents of *SRAM* will be lost when operating power is removed.

SRDC An abbreviation of SubRate Digital Cross-connect.

SRDRAM An acronym for Self-Refreshing Dynamic Random Access Memory.

SREJ An abbreviation of Selective Reject.

SRF An abbreviation of Specifically Routed Frame.

SRL (**1**) An abbreviation of Singing Return Loss. (**2**) An abbreviation of Stability Return Loss.

SRP An abbreviation of Source Routing Protocol, an IBM routing specification that allows each local area network (LAN) station to append unique routing information to every packet. Each bridge the packet encounters complies with the path indicated for the packet.

SRT (1) An abbreviation of §ource §outing §ransparent. (2) An abbreviation of §tation §ing §ransfer. (3) An abbreviation of §tandard §emote §erminal.

SRTG An abbreviation of §ource §oute §ransparent §ateway.

SS (1) An abbreviation of §erver-to-§erver. (2) An abbreviation of §ession §ervice. (3) An abbreviation of §ignaling §ystem. (4) An abbreviation of §tart-§top.

SS7 An abbreviation of §ignaling §ystem number §.

SSAP (1) An abbreviation of §ource §ervice §ccess §oint. A field in a Logical Link Control (LLC) frame header to identify the sending session within a physical station. (2) An abbreviation of §ession §ervice §ccess §oint.

SSB An abbreviation of §ingle §ide§and.

SSB-SC An abbreviation of §ingle §ide§and §uppressed §arrier.

SSCF An abbreviation of §ervice §pecific §oordination §unction.

SSCOP An abbreviation of §ervice §pecific §onnection §riented §rotocol.

SSCP An abbreviation of §ystems §ervices §ontrol §oint.

SSCS An abbreviation of §ervice §pecific §onvergence §ublayer.

SSDU An abbreviation of §ession §ervice §ata §nit.

SSEL An abbreviation of §ession §ELector.

SSFDC An abbreviation of §olid-§tate §loppy §isk §ard. A memory device proposed by Toshiba in the spring of 1996; it is targeted at the digital camera market as the first major application. *Compact Flash* (*CF*) is a competing but less used memory device.

SSI (1) An abbreviation of §mall §cale §ntegration. (2) An abbreviation of §ubsystem §upport §nterface. (3) An abbreviation of §tart §ignal §ndicator.

SSL An abbreviation of §ecure §ockets §ayer. An Internet security program by Netscape Communications Corporation.

SSM An abbreviation of §ingle §egment §essage. A frame short enough to be carried in one cell.

SSN (1) An abbreviation of §witched-§ervices §etwork. See also *public switched telephone network* (PSTN). (2) An abbreviation of §ub§ystem §umber.

SSP An abbreviation of §ervice §witching §oint.

SSRAM An acronym for §ynchronous §tatic §andom §ccess §emory.

SSS An abbreviation of §erver §ession §ocket.

SST An abbreviation of §ingle §ideband §ransmitter.

SSTDMA An abbreviation of §pacecraft §witched §ime §ivision §ultiple §ccess.

SSU (1) An abbreviation of §ession §upport §tility. A Digital Equipment Corporation (DEC) proprietary protocol that allows multiple sessions to run simultaneously over a serial cable. *SSU* is used to allow terminals to provide two concurrent session windows, each displaying session output simultaneously. (2) An abbreviation of §ubscriber §witching §nit. (3) An abbreviation of §ingle §ignaling §nit. (4) An abbreviation of §ubsequent §ignal §nit.

ST (1) An abbreviation of §egment §ype. (2) An abbreviation of §traight §ip. (3) An abbreviation of §elf §est. (4) An abbreviation of §ystem §est. (5) An abbreviation of §ide§one.

STA (1) An abbreviation of §panning §ree §lgorithm. A technique for determining the most desirable path between segments of a multiloop, bridged network. (2) An abbreviation of §TAtion. (3) An abbreviation of §wedish §elecommunications §dministration.

stability (1) A condition of circuit, device, or system of behavior wherein a bounded input signal results in a bounded output signal. In the case of a steady-state input, the output will attain a steady state after a suitable period of time. (2) The sameness of a specified property of a substance, device, or apparatus with time, or under the influence of external parametric changes (e.g., temperature, voltage).

stack In software, a list that is accessed in a last-in first-out (LIFO) manner. Compare with *first-in first-out* (*FIFO*).

STAD An acronym for §Tart §Ddress.

stage One step of a multistep process, or the entity to accomplish the step, for example,

• An amplifying system with several gain *stages* required to achieve a specified gain-bandwidth product.
• The various boot programs a PC executes in order to load the complete operating system.

stagger tuning The adjustment of the center frequency of each stage of a multistage amplifier or filter to slightly different frequencies. The effect is to flatten the passband response.

stairstep wave A waveform with periodic and finite steps of equal magnitude. Also called a *staircase* waveform.

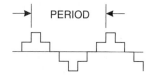

stand alone Any device that can perform its design function independently.

stand-alone hub An external hub requiring its own power supply, with connectors for the nodes to which it is attached, and possibly with connectors for linking to other hubs.

stand-alone server In LAN (local area network) Manager terminology, a server that maintains its own user accounts database and does not participate in logon security.

standard (1) A set of technical guidelines used to establish the uniformity of a specific area hardware or software product. Typically, it includes product requirements, limits, exceptions, reference documents, and definition of terms used. Standards are established in either of two ways:

• *De facto standard:* A product or standard is introduced into the marketplace and is so successful that others copy, emulate, or otherwise duplicate the essential characteristics of the product. Examples include the IBM personal computer, the Hayes modem command set, and Internet Transmission Control Protocol/Internet Protocol (TCP/IP) set.
• *De jure standard:* This standard is more formal in that a group or committee drafts a proposed standard after studying "real-world" limitations, existing methods, approaches, technology trends, and developments. These proposals are later ratified as standards by a recognized organization. Finally, over time, products based on these standards become commonplace. Examples include the ASCII character set, the RS-232 interface standard, the SCSI interface, and all ITU-T recommendations.

(2) A device having precisely defined and stable characteristics that may be used as a reference in testing or calibration procedures. **(3)** An exact value, a physical entity, or an abstract concept established and defined by authority, custom, or common consent to serve as a reference, model, or rule in measuring quantities or qualities, establishing practices or procedures, or evaluating results, i.e., a fixed quantity or quality.

standard cable In telephony, an approximation of a length of telephone cable. One such *standard cable* is fabricated with 88 ohms of series resistance and 0.054 μF shunt capacitance and represents one loop mile.

standard deviation (S or σ) In statistics, a measure of the dispersion of a group of data relative to the mean (average) of the group. The standard deviation (σ) of a total population can be estimated by:

$$s = \sqrt{\frac{\sum_{i=1}^{n}\left(x_i - \bar{x}\right)^2}{n - 1}} = \sqrt{\frac{n \cdot \sum_{i=1}^{n} x_i^2 - \left(\sum_{i=1}^{n} x_i\right)^2}{n \cdot (n - 1)}}$$

where:
- s is the estimate of the standard deviation,
- i is the index to a specific observation,
- n is the number of observations—generally less than the total number of entities in the population,
- x_i is a specific observation, and
- $\bar{x}$ is the *average* of all observations.

Also called *root-mean-square* (*RMS*) *deviation*. See also *mean*.

standard Ethernet cable The half-inch diameter coaxial cable used as the backbone of 10Base5 Ethernet networks. Also called *thickwire*.

Standard Generalized Markup Language (SGML) A markup language for plain ASCII text-based documents that allows them to be transferred among different manufacturers systems and yet display correct formatting. *SGML* is an international standard adopted in 1986 and defined by ISO 8879. It is widely used in electronic texts.

Although a *SGML* document is composed of ASCII characters, it must be viewed with a program called a *parser*. This is due to the interspersion of the *SGML* commands, called *tags*, throughout the document. The parser interprets the tags (*SGML* commands) and prepares the document for proper viewing or printing.

standard industry practice A term that indicates a normal rule developed over time and with frequent use. However, the rule lacks the formal support of a written standard supported by industry groups.

standard interface A methodology or a device that conforms to *accepted* guidelines or *standards*.

standard jack In telephony, generally refers to the RJ-nn series connectors used for connecting customer premises equipment to the subscriber loop. See also *RJ-nn*.

standard metropolitan statistical area (SMSA) An area consisting of at least one city, as defined by the U.S. Office of Management and Budget (OMB). It is used by the Federal Communication Commission (FCC) to allocate the cellular radio market.

standard noise temperature The temperature used in evaluating noise characteristics in signal transmission systems, specifically, 290 kelvins (27°C).

standard positioning service (SPS) In the Global Positioning System (GPS), the single receiver positioning service available to any user on a continuous, worldwide basis. It provides access to the C/A-code on the L1 carrier only and provides a horizontal positioning accuracy of 40 m at 50% confidence, 100 m at 95% confidence (2dRMS), and 300 m at 99.99% confidence. A vertical accuracy of 156 m and a time accuracy of 334 ns (at the 95% probability level) are attainable.

standard refraction The refraction that would take place in an idealized atmosphere where the refractive index is uniformly varied with respect to altitude at the rate of 39×10^{-6} per km.

standard source A reference source to which other devices are compared for calibration purposes.

In the United States, recognized *standard sources* must be traceable to the National Institute of Standards and Technology (NIST), formerly the National Bureau of Standards (NBS).

standard telegraph level (STL) The power per individual telegraph channel required to yield the standard composite data level.

For example, for a composite data level of -13 dBm at the 0 dBm transmission level point (0TLP), the *STL* would be approximately -25 dBm for a 16-channel voice frequency carrier telegraph (VFCT) terminal computed from

$$STL = -(13 + 10\log_{10} n)$$

where n is the number of telegraph channels and the *STL* is in dBm.

standard test tone A single frequency tone generator with 0 dBm output and at approximately 1 kHz. The actual frequency is slightly different throughout the world, as is the output impedance of the generator.

standard time A time standard base on local mean time (LMT) but positionally quantized into time zones of generally one hour (15° longitude). See also *time* (*3*).

standard time and frequency signal (STFS) service In the United States, standard time and frequency signals, broadcast on very precise carrier frequencies by the U.S. Naval Observatory and the National Institute of Standards and Technology (NIST), formerly the National Bureau of Standards (NBS).

The Radio Regulations (RR) define an identical international service as standard frequency and time signal service.

standby (STBY) **(1)** Pertaining to a dormant operating condition or state of a system or equipment that permits complete resumption of operation in a stable state within a short time, e.g., a power-saving condition. **(2)** Pertaining to spare equipment that is placed in operation only when other, in-use equipment becomes inoperative. It is usually classified in one of two ways:

- *Hot standby* equipment, which is warmed up, i.e., powered and ready for immediate service, and which may be switched into service automatically upon detection of a failure in the regular equipment, or
- *Cold standby* equipment, which is turned off or not connected, and which must be placed into service manually.

standby monitor (SM) In a Token Ring network, any node that is ready to take over as active monitor (AM) in the event the AM fails. (An active monitor is the dispenser of the token and de facto network manager.) A Token Ring network may have several *SMs*.

standby power supply (SPS) Similar to an uninterruptible power supply (UPS), except that the power does not go through the *SPS* battery during normal operation. When there is a power loss, the *SPS* will switch to the emergency battery, a process that generally occurs within a few milliseconds. Also called *off-line UPS*. See also *uninterruptible power supply* (*UPS*).

standing wave A wave in which the instantaneous value of any component of its field (at a physical location) does not vary with time. Also called a *stationary wave*.

standing wave ratio (SWR) The ratio of the maximum voltage to the minimum voltage along a transmission line in the direction of transmission. Current ratio as well as voltage ratios may be used for determining *SWR*.

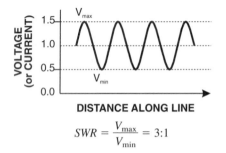

$$SWR = \frac{V_{\max}}{V_{\min}} = 3{:}1$$

SWR is a measure of the mismatch between the transmission line and the load impedance. *SWR* is, by custom, a positive number greater than one. When the line is perfectly matched to the load, the *SWR* is exactly one. *SWR* can be calculated from these impedances by

$$SWR = \frac{Z_O}{Z_R} \text{ or } = \frac{Z_R}{Z_O}$$

where:

Z_O is the *characteristic impedance* of the line, and

Z_R is the impedance of the load.

SWR can also be calculated or measured from the outward (or forward) voltage (V_O) and the reverse (or reflected) voltage (V_R) or from the *reflection coefficient* (ρ); that is,

$$SWR = \left| \frac{V_O + V_R}{V_O - V_R} \right| \text{ or } SWR = \frac{1 + \rho}{1 - \rho}$$

SWR indicates *return loss power* due to impedance mismatch, for example,

SWR	Power Loss	SWR	Power Loss
1:1	0 (perfect match)	3:1	25%
1.3:1	1.7%	4:1	36%
1.5:1	4.0%	5:1	44%
1.7:1	6.7%	6:1	51%
2:1	11%	10:1	67%

$$\text{Power Loss} = \left(\frac{SWR - 1}{SWR + 1} \right)^2$$

SWR is sometimes also called *voltage standing wave ratio* (VSWR). See also *return loss*.

STAR (1) An acronym for <u>S</u>elf <u>T</u>est <u>A</u>nd <u>R</u>epair. (2) An acronym for <u>S</u>tandard <u>T</u>elecommunications <u>A</u>utomatic <u>R</u>ecognizer.

star coupler A passive coupler that distributes signals from one or more inputs among a multiplicity of outputs without boosting the signal power.

star network (1) A local area network (LAN) topology in which each node is *individually* connected to a *master node* (host or hub) via its own signal transmission means. The master node (hub) provides a path (either physical or logical) between participating nodes without involving nonparticipating nodes. The topology (configuration) of the transmission structure resembles a star—with each node radiating from the hub (central computer).

The strength of a *star network* topology is also its weakness. Because no failure in a node can affect another node, the network can provide very reliable service. However, the failure of the hub will shut down the entire network. Furthermore, the cost of the individual wiring of each node can be higher than that of other methods. Also called *star*

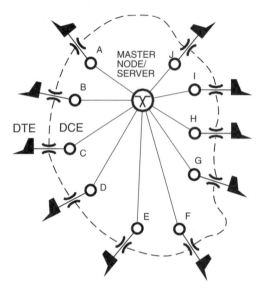

topology. See also *bus network, gateway, hierarchical network, hybrid network, mesh network,* and *ring network.* (2) A circuit with three or more branches and with one terminal of each branch connected to a common node.

star quad A cable consisting of four conductors that are twisted about a common axis. The two sets of opposing pairs are used as transmission pairs. Also called *spiral four.*

StarGroup A network operating system (NOS) from AT&T adapted from Microsoft's LAN (local area network) Manager to run on UNIX systems. It provides support for the many common protocols and devices, e.g., Transmission Control Protocol/Internet Protocol (TCP/IP), Systems Network Architecture (SNA), asynchronous gateways, routers, and X.25 networks.

StarLAN A 1-Mbps baseband network based on IEEE 802.3 and implemented by AT&T which communicates over two twisted pairs and uses the CSMA/CD (collision sense multiple access with carrier detection) protocol.

StarLAN10 A 10-Mbps baseband network developed by AT&T, conforming to the 10BaseT specification in IEEE 802.3. It operates over standard telephone twisted pairs emanating from a central hub (a star network topology) as is found in existing building wiring. Proper performance is specified with wire pairs up to 328 feet from the hub. Hubs may be cascaded and with optical fiber links may be separated by a kilometer or more.

Starlink A research network for astronomers that provides researchers with interactive computing facilities, including hardware and software for image analysis and enhancement and spectral analysis. With over 50 hosts, it facilitates the processing of data from a number of diverse sources including satellites, radio astronomy antennas, ground-based optical telescopes, and automatic photograph scanning equipment.

start bit The first bit in a block of serially transmitted data in an asynchronous system. It specifies the starting point for a single block of data bits and shows that the following "n" bits are data. In most systems, the block size "n" is 7 or 8 bits. The *start bit* is a *spacing signal.* Also called the *start element* or *start signal.*

start dialing signal In telephony, a signal transmitted from the incoming end of a circuit (following the reception of a seizing signal), which indicates that the necessary equipment and conditions are present in order that the call routing information may be received. Dial tone is an example of a *start dialing signal.*

start of header (SOH) A transmission control sequence used to delineate the beginning of a message header, e.g., the ASCII <SOH> 01h character.

start of message (SOM) A transmission control character used to delineate the beginning of a message.

start of text (STX) A transmission control sequence used to delineate the beginning of a message body, e.g., the ASCII <STX> 02h character.

start pulsing signal In telephony, a signal transmitted from the receiving end to the sending end of a trunk which indicates the receiving end is ready to accept pulse information.

start signal (1) In start/stop transmission, a signal that prepares the receiver for the reception and registration of a character or for the control of a function. See also *start bit.* (2) In telephony, a signal used to indicate that all routing digits have been transmitted. (3) A signal that prepares a device to receive data or to perform a function. Contrast with *A-condition.*

start/stop character A character that includes one start signal at the beginning and one or two stop signals at the end.

start/stop distortion In start/stop modulation, the ratio of the maximum absolute difference between the actual and the theoretical intervals that separate any significant instant of modulation or demodulation from the significant instant of the start signal element immediately preceding it to the unit interval.

start/stop margin In start/stop modulation, the maximum amount of overall *start/stop distortion* that is compatible with correct translation by the start/stop equipment of all the character signals that appear singly, at the maximum allowable speed, or at the standard modulation rate.

start/stop modulation A method of modulation in which the time of occurrence of the information bits within each character, or block of characters, relates to a fixed time frame established by the *start bit.*

start/stop transmission (1) *Asynchronous transmission* in which a start pulse and a stop pulse are used with each symbol. See also *asynchronous transmission.* (2) Signaling in which each group of code elements corresponding to an alphanumeric character is

• Preceded by a start signal that serves to prepare the receiving mechanism for the reception and registration of a character and
• Followed by a stop signal that serves to bring the receiving mechanism to rest in preparation for the reception of the next character.

starting frame delimiter A specified bit pattern that indicates the start of a transmission frame.

stat mux A contraction of STATistical MUltipleXor.

state diagram A type of diagram consisting of a number of nodes (describing a particular state or condition of a system, device, or program) and interconnecting lines (corresponding to action-causing events or inputs). The drawing illustrates a basic state diagram of the placing of a call. Also called *directed graph, Petri net,* and *transition diagram.*

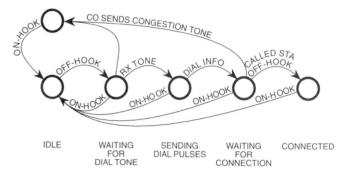

| IDLE | WAITING FOR DIAL TONE | SENDING DIAL PULSES | WAITING FOR CONNECTION | CONNECTED |

State Public Service Commission (PSC) The state legislative body that is responsible for regulating the operation of telephone companies and those involved in supplying telephone service. Regulated services include:

• The introduction of new services.
• The discontinuance of existing services.
• The pricing for services.

static (1) In general, having no motion. (2) Interference in reception caused by natural electric disturbances (random noise) or the electromagnetic phenomenon able to cause such interference. Also called *static noise.* See also *static electricity.*

static electricity An accumulation of charge on an insulated object, called *static* because there is no flow of electrons (current). A charge on an object gives rise to a potential difference (voltage) between the object with the charge and any other object. These voltages can be in the tens of thousands of volts but are generally harmless to people because the available energy is low. However, semiconductors are sensitive to voltage as well as energy and can easily be damaged by a static discharge.

static object An object that has been pasted into a document. It is not linked to any other entity, so the only way to edit a static object in a document is to delete it, create a new object (with the original program), and paste a new copy into the document.

static positioning In the Global Positioning System (GPS), location determination when the receiver antenna is at a fixed place (that is, not moving). Under these conditions, it is possible to improve the accuracy of the latitude and longitude measurement by a factor of several hundred to a thousand by averaging or other statistical techniques.

static RAM (SRAM) Pronounced "ESS-ram." A semiconductor memory element that retains information as long as power is applied to the device; *no memory refresh clocking is required.* Because no time is required to refresh the contents of an *SRAM,* they are generally much faster than dynamic RAMs.

static routing A routing system in which the routing table information is entered manually. The router will always attempt to use the same route, even if it is not the shortest. If there is no route address, the packet cannot be delivered.

statically assigned socket (SAS) Sockets that are allocated for use by various low-level protocols, such as Name Binding Protocol (NBP) and Routing Table Maintenance Protocol (RTMP) in the AppleTalk protocol suite. Values between 1 and 63 are used exclusively by Apple, and values between 64 and 127 can be used by whatever processes request the values.

staticizer See *serial to parallel conversion.*

station One of the input or output points on a communications system. Examples include telephone subsets, computer terminals, and facsimile machines.

station apparatus The equipment installed at a subscriber's premises, including telephones, private automatic branch exchanges (PABXs), and so on.

station battery A separate power source that provides the necessary power to operate critical station equipment. The batteries are typically −48 volt (−56 volt in Germany) wet lead-acid in a telephone central office and −24 volt nickel-cadmium or gelled lead-acid in a private automatic branch exchange (PABX). In all cases, the batteries are charged from the ac mains.

station class mark (SCM) A two-digit number that identifies particular capabilities of a cellular telephone. It informs the cellular system how to handle the call, that is,

- It informs the system whether the phone can transmit at high or low power levels.
- It informs the system if the phone can access the full 832 available channels or only the original 666 channels.
- It informs the system if the phone has VOX (voice-operated switching) capabilities.

station clock In a station, the principal clock, or alternate clock, that provides the timing reference at the station.

station code In telephony, the last four digits of the standard North American Numbering Plan (NANP). Also called the *line code.*

station equipment In telephony, a general term describing the telecommunications apparatus at a subscriber's premises. This equipment may belong to either the phone company or the subscriber. Also called *end office equipment.*

station hunting A switching system feature that allows an incoming call placed to a busy number to be automatically rolled over to the next idle station within the hunt group.

station keeping The process of maintaining an Earth-orbiting satellite in its prescribed orbit. The process is necessary for both geosynchronous and nongeosynchronous satellites because the gravitational field of the Earth is not uniform and the satellites tend to slide to the lowest energy points.

station line In telephony, the wires that carry the dc loop current from the switching equipment to a main station, subscriber's premises, private branch exchange (PBX), or other end office equipment.

station load The total power requirements of the integrated station facilities.

station loop resistance In telephony, the series resistance of the loop conductors and an off-hook subset.

station management (SMT) One of the documents defining Fiber Distributed Data Interface (FDDI). It defines monitoring, bandwidth allocation, and management of node's activity. The elements of *SMT* include the following:

- Connection management (CMT) controls access to the network.
- Frame services generate frames for diagnostics.
- Ring management (RMT) troubleshoots the network.

If there is a fault in the primary ring, the *SMT* facility redirects transmissions to use the secondary ring around the faulty section. Under some conditions, it can also use the secondary ring to concurrently transmit data, yielding a potential transmission rate of 200 Mbps. This component has no counterpart in the ISO/OSI Reference Model. See also *media access control (MAC)*, *physical (PHY)*, and *physical-layer medium dependent (PMD).*

station message detail recording (SMDR) A computer-generated record of all calls originated or received by a switching system.

station protector An overvoltage protecting device generally located at the subscriber demarcation point that routes potentially harmful transients to ground. The protecting device may be a gas discharge tube, carbon block, or other transient protecting device.

stationary orbit An orbit whose period is the same as the average rotational period of the primary body, is circular, and is above the equator or the primary body. For example, the Earth is in a *stationary orbit* around the Moon.

When the primary body is the Earth, it is called a *geostationary orbit.*

stationary wave See *standing wave.*

statistical multiplex A time division multiplexing (TDM) device that allocates communication bandwidth dynamically on an as-needed basis. If one "connected path" is not transmitting, the bandwidth can be *temporarily* reassigned to a channel that is transmitting. Also called a *statistical time division multiplex (STDM).* See also *TASI.*

statistics A branch of mathematics that deals with the relationships of groups of measurements and the relevance of similarities and differences in those relationships.

status The condition of a system, device, or program at a particular time. The *status* is reported in a number of ways, for example, the data set ready (DSR) signal from data communication equipment (DCE) is a voltage that data terminal equipment (DTE) can interpret as "I am ready to proceed." Another example is the "OH" indicator light on the front panel of many modems which indicates that the modem is off-hook.

statmux A contraction of STATistical MUltipleXer.

statute mile A unit of distance equal to 1.609 km (0.869 nmi, 5280 ft.).

STB An abbreviation of Start of Text Block.

STC (1) An abbreviation of Switching and Testing Center. (2) An abbreviation of Standard Transmission Code.

STD (1) An abbreviation of Subscriber Trunk Dialing. See also *direct distance dialing.* (2) An abbreviation of STandarD. (3) An abbreviation of Synchronous Time Division. (4) An abbreviation of Subscriber Toll Dialing. (5) An abbreviation of Subscriber Trunk Dialing.

STDA An abbreviation of StreetTalk Directory Assistance.

STDM (1) An abbreviation of Statistical Time Division Multiplexer. (2) An abbreviation of Synchronous Time Division Multiplexer.

STE (1) An abbreviation of Section (or Span) Terminating Equipment. (2) An abbreviation of Signal Terminal Equipment. (3) An abbreviation of Spanning Tree Explorer.

steady state (1) A characteristic of a device, system, or program in which a specified parameter such as amplitude, current, frequency, period, phase, rate, value, and voltage changes negligibly over an arbitrarily long time period. (2) In an electrical circuit, the condition that exists after all initial transients or fluctuating conditions have decayed, and all currents, voltages, or fields remain essentially constant or oscillate uniformly. (3) In fiber optics, the *steady-state* condition is a synonym for *equilibrium mode distribution.*

steganography A form of encryption that hides an encrypted message in the least significant bit position of a graphic, video, or sound file. The existence of the encrypted message is virtually impossible to detect, and its degradation of the graphic image or sound file is imperceptible (showing only a very slight increase in noise level).

step (1) An abrupt change (transition) in a waveform in which the change has negligible duration relative to the duration of the states immediately preceding and following the change. (2) One increment of an electromechanical switch; generally corresponding to a single dial pulse.

step-by-step (SXS) office An automated switching center that utilizes step-by-step switches as the means for routing calls. The switches may be operated directly by the signaling pulses or by stored control mechanisms that interpret the signaling in some manner.

step-by-step switch A multiposition switch in which the wipers are operated by an electromagnetic ratchet mechanism. The switch moves in synchronism with dial pulses. Each digit dialed causes the movement of successive switches to carry the connection forward until the addressed line is reached. Also called a *stepper, stepper switch,* or *stepping relay.*

step down (1) A term that refers to a voltage-reducing transformer or circuit. (2) A term that refers to the ability of a digital communications device (such as a fax or modem) to reduce the data transmission rate when it detects excessive transmission line noise.

step down transformer A transformer designed to reduce the output voltage as compared to the input voltage by the amount as the primary to secondary turns ratio.

step index fiber An optical fiber in which the core has a uniform index of refraction and a sharp decrease in refractive index at the core-cladding boundary. Also called *stepped index fiber.* See also *optical fiber.*

step response The observed time-dependent reaction to a step change of electromagnetic radiation upon or through a medium, network, or device.

step up transformer A transformer designed to increase the output voltage as compared to the input voltage by the amount as the primary to secondary turns ratio.

stepper (1) In telephony, slang for a step-by-step relay. See also *Strowger switch.* (2) In photolithography, slang for a step and repeat system. Used to copy a single image instance multiple times onto the target material. Used in the manufacturing of integrated circuits. (3) A type of motor that rotates a fixed number of degrees for each drive pulse. These motors can be found with step rotation angles of $360°/n$ (where $n \geq 3$).

steradian (sr) In the metric system, the solid angle subtended at the center of a sphere by an area on the surface of the sphere equal to the square of the radius of that sphere. When the shape of the *steradian* is a pure cone, the solid angle is approximately 66°. See also *SI.*

stereophonic crosstalk An undesired signal occurring in either the left or right output channel of a stereophonic transmission from the opposite channel, i.e., left channel information into the right channel or right channel information into the left channel.

In FM stereo transmission, the main channel carries L + R channel information and the subcarrier carries L − R information. The receiver signal processing, when functioning properly, generates:

$$L = (main + subcarrier)/2; i.e., ((L + R) + (L - R))/2$$

$$R = (main - subcarrier)/2; i.e., ((L + R) - (L - R))/2.$$

If the gain in the signal processor is not exactly the same in both paths, some of the opposite channel will leak into the output. For example, if the subcarrier channel's gain is off by 1% (i.e., 1.01 instead of 1), then the equations become

$$(L + R) + 1.01(L - R)/2 = 1.005 L - 0.005 R$$

which corresponds to a -46 dB crosstalk level, i.e., $-20 \log_{10}(.005/1.005)$.

stereophonic sound subcarrier A subcarrier within the FM broadcast baseband used for transmitting signals for stereophonic sound reception of the main broadcast program service. The subcarrier frequency is 68 kHz and carries the difference of the left and right channel information. (The main carrier is modulated with the sum of the left and right channel information.)

STF An abbreviation of Standard Transaction Format.

STFS An abbreviation of Standard Time and Frequency Signal (or Service).

STI An abbreviation of Single Tuned Interstage.

stimulated emission In a quantum mechanical system, the radiation emitted when the internal energy of the system drops from an excited level (induced by the presence of radiant energy at the same frequency) to a lower level. Both lasers and masers are examples of devices operating in a *simulated emission* manner.

STL (1) An abbreviation of Standard Telegraph Level. (2) An abbreviation of Studio to Transmitter Link.

STM An acronym for Synchronous Transfer Mode. One of several protocols used for Synchronous Optical Network (SONET) and Broadband Integrated Services Digital Network (BISDN).

STM-n An abbreviation of Synchronous Transport Mode. A high-speed communications standard used in the Synchronous Optical Network (SONET). The levels represent multiplexed 44.736-Mbps DS3 channels, plus overhead for signaling and framing. For example, the lowest *STM* capacity, *STM-1*, multiplexes three 51.84-Mbps channels into its 155.52-Mbps bandwidth. See also DS-n and OC-n.

Standard	Communications Rate
STM-1	155.52 Mbps
STM-4	622.08 Mbps
STM-16	2.4883 Gbps
STM-n	155.52×n Mbps

STO An abbreviation of Security Through Obscurity.

stochastic Based on random occurrences.

Stokes' law The wavelength of stimulated luminescence is always greater than the wavelength of the stimulating radiation.

Stone bridge One of many circuits to provide dc feed to a telephone loop. The *Stone bridge* injects the dc feed current in parallel with the voice-coupling transformer as is shown in the figure. Series resistors limit the current to the subscriber when the loop is short and a capacitor blocks dc current in the voice transformer path.

CONVENTIONAL SHUNT FEED (STONE BRIDGE)

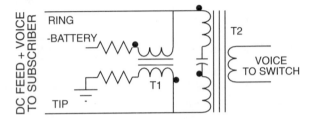

Some general characteristics of the circuit include:

- The transformer T2 can be a small "cheap" transformer as no dc flows through it.
- The two- to four-wire hybrid function (not shown) is nearly free as it only requires the addition of another winding on T2 and a line matching network.
- A large transformer core is required on T1 due to the high dc feed current flowing in the windings.
- With short loops, there is high power dissipation.
- Longitudinal balance is determined by both the feed resistors and transformer T1. The balance network must not only match the transmission line but also must take into account the inductance of T1.
- Ringing must be done from an external source which is switched onto the subscriber line during ringing.

See also *BORSCHT, Hayes bridge,* and *line circuit.*

stop bit The last element in an asynchronous serial transmission. It is the interval at the end of each character that allows the receiving computer to pause and resynchronize with the start of the next character. It can be thought of as extra "1" bits that follow the data and mark the end of the symbol.

The *stop bit* provides a degree of guard time between two successive bytes of data in an asynchronous transmission link. The guard time is necessary because the transmitting device and receiving device data clocks are not necessarily the exactly same rate. The minimum number of stop bits is generally set to 1, 1½, or 2. (Five-level Baudot machines used 1½ stop bits to resynchronize, older electronic devices used two, while newer devices require only one.) The *stop bit* is the marking or idle state of the transmission protocol. Most serial connections are described as *8N1*, which means 8 data bits, no parity, and 1 stop bit. *Stop bits* are not required for synchronous communications. Also called a *stop element* and *stop signal.*

stop element See *stop bit.*

stop pulse signaling In telephony, a method of trunk signaling whereby the sending end transmits signaling digits (pulses) until it receives a stop pulse signal. After a stop pulse signal is received, the remaining digits are held at the sending end until a start pulse signal is received. Also called *stop-go pulse signaling.*

stop signal (1) In *start/stop transmission,* a signal at the end of a character that prepares the receiving device for the reception of a subsequent character. It is usually limited to one signal element having any duration equal to or greater than a specified minimum value. See also *stop bit.* (2) A signal to a receiving mechanism to wait for the next signal.

stop/start transmission See *start/stop transmission.*

stopband A band of frequencies that a circuit, device, or system such as a filter, telephone circuit, or transmission system passes with substantial attenuation relative to another band of frequencies (the *passband*). The upper and/or lower band edge frequencies are those at which the transmitted power level is below a specified level, usually 3 dB below the maximum level in the passband.

Frequencies above the lower band edge and below the upper band edge are not transmitted, i.e., are not allowed to pass. The difference between the two band edges is the stopband bandwidth and is usually expressed in hertz (Hz). In the graph, the lower band edge is at 1.2 kHz and the upper band edge is at infinite frequency. Also called *rejection band.* See also *band-pass filter, band reject filter, filter, high-pass filter, low-pass filter, passband,* and *transition band.*

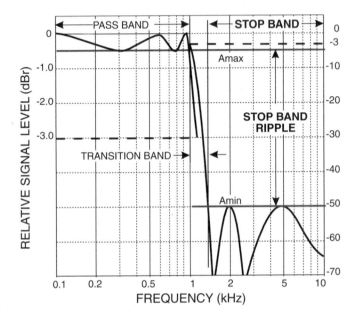

stop band ripple The difference between the maxima and minima of loss in a *stop band.* See also *stopband.*

STOQ An abbreviation of STOrage Queue.

storage (1) The retention of data in any form for the purpose of orderly retrieval or documentation. (2) A hardware device (electronic, magnetic, electrostatic, electrical, mechanical, optical, etc.) into which data may be entered and from which data may be retrieved, when desired.

storage cell (1) An addressable storage unit. (2) The smallest subdivision of storage into which a unit of data can be entered, stored, and retrieved. Also called a *storage element.*

Storage Management Services (SMS) A collection of services for managing data storage and retrieval in Novell's NetWare. Individual services are available in modular form and are independent of operating systems and hardware. *SMS* modules include:

- *Workstation Manager*—for identifying and keeping track of the stations waiting to be backed up.
- *SBACKUP*—the backup and restoration operations module.
- *SMDR* (from Storage Management Data Requester)—the module for passing commands and information between the backup program and target service agents (TSAs).
- *Storage device interfaces*—the modules for passing information between SBACKUP and the physical storage devices.

- *Device drivers*—modules to control the behavior of storage or other devices.
- *Target service agents* (*TSAs*)—server, database, and workstation agents for passing requests, commands, and data between SBACKUP and various other components on the network.

storage time **(1)** In opto-electronics, the time interval between the removal of an optical signal to a photoelectric device and the reduction of its output signal to a specified percentage of the no signal level. **(2)** In memory systems, the time interval between writing of information to the memory device and loss of a specified fraction of the charge in the device. The loss necessitates refreshing the memory contents. (In DRAM the time period is measured in ms, while in EPROMs the interval is measured in hundreds of years.)

store and forward (S-F) A communication network procedure in which message packets are received at an intermediate point and temporarily stored for later transmission to a further routing point or final destination. *Store-and-forward* message routing on a network may increase the delay time between sender and receiver, but it minimizes or eliminates *congestion* ("traffic jams"), and it allows messages to be sent to nodes temporarily off the network and in off-peak time periods.

In an Ethernet switch or bridge, for example, the whole packet is read before forwarding or filtering. The process is slower than *cut through,* but it ensures that all bad or misaligned packets are eliminated. In the FidoNet system, for example, entire messages are collected throughout the day by a bulletin board system (BBS). At a prescribed time, the BBS computer uses a modem and a dial-up connection to exchange messages (upload and download) with a central site. Although the system is not fast, it is effective. Over a period of only several days, a posting can propagate throughout the network. See also *cut through.*

store-and-forward switching center A message switching center in which a message from the originating user (the sender) is accepted when it is offered and held in physical storage until such time as it can be forwarded to the destination user (the receiver) in accordance with the message's priority (as assigned by the originating user) and the availability of an outgoing channel. The user may be a person, a computer, or another *store-and-forward switching center.*

STP **(1)** An abbreviation of <u>S</u>hielded <u>T</u>wisted <u>P</u>air. **(2)** An abbreviation of <u>S</u>panning <u>T</u>ree <u>P</u>rotocol. An IEEE 802.1 routing specification. **(3)** An abbreviation of <u>S</u>ignal <u>T</u>ransfer <u>P</u>oint. A packet switch for Signaling System 7 (SS7). **(4)** An abbreviation of <u>S</u>elf <u>T</u>est <u>P</u>rogram. **(5)** An abbreviation of <u>S</u>tandardized <u>T</u>est <u>P</u>rogram. **(6)** An abbreviation of the <u>ST</u>o<u>P</u> character. **(7)** An abbreviation of <u>S</u>tandard <u>T</u>emperature and <u>P</u>ressure.

STPL An abbreviation of <u>S</u>ide<u>T</u>one <u>P</u>ath <u>L</u>oss.

STPST An abbreviation of <u>ST</u>o<u>P</u>-<u>ST</u>art.

STR **(1)** An abbreviation of <u>S</u>ynchronous <u>T</u>ransmit <u>R</u>eceive. **(2)** An abbreviation of <u>S</u>ide<u>T</u>one <u>R</u>eduction. **(3)** An abbreviation of <u>ST</u>o<u>R</u>e.

STRAD An acronym for <u>S</u>ignal <u>T</u>ransmission <u>R</u>eception <u>A</u>nd <u>D</u>istribution.

straight through cable A cable with identical or mating connectors at each end and wired so that terminal 1 at one end is connected to terminal 1 at the other end. Similarly, wiring is performed with terminals 2, 3, 4, . . .

This is in contrast to a *crossover cable* in which the first terminal on one end is wired to the last terminal at the other cable end, the second is paired with the next to last, and so on.

strain Internal mechanical forces within a device, i.e., tension (pulling apart), compression (pushing together), shear (tearing), or torsion (twisting). The internal forces may be the result of externally applied forces or may be generated internally due to temperature or other effects.

STRAM An acronym for <u>S</u>ynchronous <u>T</u>ransmit <u>R</u>eceive <u>A</u>ccess <u>M</u>ethod.

strand Generally, a single uninsulated wire or optical fiber. One exception is *litz* wire, which is used for high-frequency circuits. Here each conductor is insulated from the other conductors except at the terminations.

strand lay The distance between like points in one turn of a spirally twisted stranded cable. (The cable may be used for mechanical support, electrical conductivity, or optical transmission.) Also called *lay, lay length,* and *pitch.* See also *lay length.*

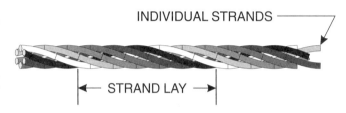

stranded conductor A group of multiple conductors twisted together to form a single conductive path.

stranded fiber cable An optical fiber cable in which the various fibers are twisted around one or more tensile strength elements.

strap See *cross connection.*

stratum The levels within the synchronization hierarchy which define the minimum accuracy for digital clocks of interconnected digital networks.

STRAW An acronym for <u>S</u>imultaneous <u>T</u>ape <u>R</u>ead <u>A</u>nd <u>W</u>rite.

stray Any element or circumstance occurring in a physical system but not in the theoretical design. For example, the capacitance developed between any two conductors in the same proximity.

stray current An electrical current through a path other than the intended path.

STRD An abbreviation of <u>ST</u>o<u>R</u>e<u>D</u>.

streaming protocol A file transfer protocol wherein the sending device makes the assumption that a packet transmitted is received correctly. It therefore can send the next packet immediately rather than wait for the receiving modem to send an acknowledge. If the receiving modem finds an error in a packet, it will send a message to the sending modem to retransmit the bad packet.

Examples of systems that use streaming protocols include tape backup hardware, audio or video transfers across the Internet, and file transfers using programs such as ZMODEM.

StreetTalk The global naming service for Banyan's VINES network operating system (NOS). It includes a database that contains all the necessary information about the network and each node or device on it and is updated every 90 seconds by every server on the network.

The naming service keeps track of which nodes and devices are attached to the network and assigns a global name to each node. The name is independent of the particular network in which the node is located and makes it possible for a user connected to one server to use resources attached to a different server, without knowing which specific server has the resources. A StreetTalk name may include three levels of identity: item, group, and organization where item is the most specific. A node or device may get a name at each of these levels, and these names will be separated by a @, for example,

node@group@item.

It is a protocol for determining and maintaining network resource information when the resources are distributed among several servers.

StreetTalk Directory Assistance (STDA) A global network naming system for Banyan's VINES. It provides a pop-up window in which a user can see the name of every node or device attached to the network, addressing facilities for e-mail, and other information about a specific node or device.

strength member Any component of a communications cable (electrical or optical) whose function is to protect the transport medium (wire or fiber) from excessive tensile and bending stresses either during installation and while in service.

stress relief A predetermined amount of slack to relieve tension in a component or other lead wire. The relief may be for tensile or bending stresses.

string Any group of successive characters arranged together for a common purpose and considered as a whole.

The command to instruct the modem to dial a number ATDT555-1212 is a *string* as is the transmitted data "Happy Birthday."

strip line A transmission line in which the main conductor is isolated from one or more conducting planes by a dielectric. Generally, the conducting planes are connected to circuit ground.

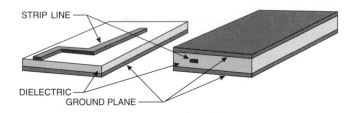

strobe An electronic timing signal that initiates an event or coordinates several events. A *strobe* from one device may inform a second device that data is present and stable on the input/output (I/O) port.

Strowger switch First patented in 1891 by Almond B. Strowger. A type of mechanical switch used to route calls through a telephone switching office without the aid of an operator.

By the late 1880s, Strowger had developed a model of an automatic exchange switch that consisted of:

- A cylinder—made from a collar box.
- A spindle—made from a pencil.
- A perpendicular arm—made from dressmaking pins.

Strowger's idea was to install switches like these at the telephone exchanges. It would be lined with rows of electrical contacts—one contact for each telephone line. To reach a particular contact, the spindle (controlled by a device associated with each telephone) would slide in and out of the cylinder. It would then stop and rotate, allowing the arm to sweep along one of the rows of contacts. Finally, it would stop at the one the caller required. As a result, callers could connect their line to any other line on the same exchange without the help of a manual operator. The first system was an unwieldy affair. Each telephone was connected to the exchange by five wires, of which only one (plus the

Earth return) was for the conversation itself. The other four were connected to four buttons on the telephone which transmitted the number to the exchange. (To dial a number, say 345, the caller had to press the buttons in sequence; i.e., button 1—three presses; button 2—four presses; button 3—five presses. To disconnect the call, the caller pressed button 4.) At the exchange, 1000 electrical contacts (one for each line) were arranged inside a cylinder with 10 rows, each row containing 10 groups of 10 contacts. The call was connected by operating selectors. With each button press, the selector takes a step, that is,

- Each press of button 1 moves the selector shaft UP one row.
- Each press of button 2 rotated the selector arm ALONG the selected row from one block of 10 contacts to the next.
- Each press of button 3 stepped the selector arm to the NEXT CONTACT within the selected group. At this point, the selector arm rested against the contact of the called line.

The first Strowger exchange opened in La Porte, Indiana, in 1892. Strowger improved his system, and by 1896, he had reduced the number of required wires to three and replaced the pushbutton arrangement by a dial. Eventually, the system needed only one pair of wires to carry both conversation and dialing information, and the complexity of the switch was reduced so that only two motions were required. Banks of switches were interconnected to provide the switching of tens of thousands of lines. This arrangement remained in use until replaced by crossbar and electronic switching systems.

Structure of Management Information (SMI) A component of the Internet Protocol (IP) network management model. The *SMI* specifies how information about managed objects is to be represented. The representation uses a restricted version of the International Standard Organization's Abstract Syntax Notation One (ASN.1) system. Documented in RFC 1155.

Structured Query Language (SQL) A standardized query language that can be used for interrogating and updating databases across a network in client–server applications.

STS (**1**) An abbreviation of <u>S</u>ynchronous <u>T</u>ransport <u>S</u>ignal; see STS-n. (**2**) An abbreviation of <u>S</u>atellite <u>T</u>ransmission <u>S</u>ystem. (**3**) An abbreviation of <u>S</u>pace <u>T</u>ime <u>S</u>pace switching. A telephone central office digital switch architecture. (**4**) An abbreviation of <u>S</u>hared <u>T</u>enant <u>S</u>ervices. (**5**) An abbreviation of <u>S</u>ystem <u>T</u>est <u>S</u>oftware.

STS-n An abbreviation of <u>S</u>ynchronous <u>T</u>ransport <u>S</u>ignal level "<u>n</u>." STS-1 is the electrical counterpart of the optical OC-1 protocol. It is the basic digital signal building block for Synchronous Optical Networks (SONET) with a rate of 51.84 Mbps. Higher order STS-ns are obtained by byte interleaving "n" STS-1 signals together. This yields a data stream with a rate of "n" times 51.84 Mbps. See OC-n for a comparison and listing of the various signaling rates. See also *DS-n, OC-n, SDH, SONET,* and *STM-n.*

STU (**1**) An abbreviation of <u>S</u>ignal <u>T</u>ransfer <u>U</u>nit. (**2**) An abbreviation of <u>S</u>ubscriber <u>T</u>runk <u>U</u>nit. (**3**) An abbreviation of <u>S</u>ecure <u>T</u>elephone <u>U</u>nit. A government-approved telecommunications terminal that protects the transmission of sensitive or classified information in voice, data, and facsimile systems.

stub network An autonomous network with only one connection path to a backbone network. The router performing the connection need not perform any intelligent routing because only one external path is available to it.

studio to transmitter link (STL) A communications link between a production studio and the radio transmitter used to convey program material. It may be a microwave, radio, or conditioned telephone landline link.

Stuffit A file archive and data compression program (or family of programs) originally developed as shareware by Raymond Lau for the Macintosh computers. The default file extension of the compressed output file is ".sit". *Stuffit* archives can be extracted with Stuffit Expander.

stunt box A device that controls the nonprinting functions of a printer (historically a teleprinter) at a terminal.

STX An abbreviation of STart of teXt.

SU (1) An abbreviation of Signaling Unit. A layer packet of Signaling System 7 (SS7). (2) An abbreviation of Segmentation Unit. In Switched Multimegabit Data Service (SMDS), an information field of a layer 2 Protocol Data Unit (PDU) or a layer 3 PDU with 44 or fewer octets. (3) An abbreviation of Service Unit. (4) An abbreviation of Switching Unit. (5) An abbreviation of Subscriber Unit.

SUA An abbreviation of Stored Upstream Address.

SUABORT An acronym for Session User ABORT (pronounced S-U-abort).

SUB A truncation of SUBstitute character. A control character used in place of a character found to be in error. In ASCII, it is 26 decimals (01Ah).

sub-address That part of a hierarchical addressing scheme that allows the location of or access to specific subcategories. Examples include:

- An apartment number within a building address.
- An extension in Integrated Services Digital Network (ISDN) that enables specific services (facsimile, telephone, PCs, etc.).

Sub-Area Network The original hierarchical approach used in the construction of IBM's Systems Network Architecture (SNA) backbone networks. The structure of a *Sub-Area Network* is predefined, that is, the relationship between the nodes of the network is designed into the software of the host systems. However, the network can be modified without having to rebuild the software of the entire network, that is, a new node can be added by defining it in each of the directly connected adjacent nodes. These networks are built on a backbone of communications controllers to which host systems are attached.

sub-directory See *directory*.

sub-distribution frame (SDF) An intermediate wiring location. For example, it may be used for all the equipment on a particular floor. *SDFs* are connected by backbone cable to a main distribution frame (MDF).

subcarrier Any carrier (potentially modulated with information) which is used to modulate the main carrier (or another subcarrier). Multiple subcarriers may be impressed onto the main carrier.

subdomain A set of two or more subcategories of hosts within the domain name which was assigned to the organization by the Network Information Center (NIC). The organization is free to assign subdomain names without consultation with the NIC.

For example, the NIC could assign the domain name *hc.com* to an organization. The organization is then free to assign subdomains such as:

new-prod.hc.com   for their new product announcements
tek-sup.hc.com   for their technical support
beta.hc.com   for their beta test programs

subharmonic A sinusoidal waveform with a frequency that is an exact submultiple of the fundamental frequency to which it is related. That is, the frequency of a *subharmonic* is $1/n$ times the fundamental frequency where n is an integer.

subject selector A program mode in a USENet newsreader that displays articles in the chronological order of receipt. In this mode, following a thread is difficult.

sublayer (1) In general, in a layered open communications system, a specified subset of the services, functions, and protocols included in a given layer. (2) In the Open Systems Interconnection—Reference Model, a subdivision of a given layer which is a conceptually complete group of the services, functions, and protocols included in the given layer.

subnet (1) As a verb, to divide a network into parts, each continuing to be a part of the same domain. The resulting *subnet* continues to be treated as a single network by outside networks. *Subnetting* allows a site to use a single network ID to refer to multiple physical networks. Also known as *Classless InterDomain Routing* (*CIDR*). (2) As a noun, used to refer to one of the smaller networks linked by a bridge or router.

subnet address In an Internet Protocol (IP) address, an extension that allows users in a network to use a single IP network address for multiple physical subnetworks.

The *IP address* is divided into two parts: the *network part* and the *local part*. The local part is further divided into a *subnet address* part and *host address* part. Inside the subnetwork, gateways and hosts divide the local portion of the IP address into a subnet address and a host address and forward packets onto the appropriate physical network. Outside of the subnetwork, routing continues as usual by dividing the destination address into a network portion and a local portion.

subnet layers The bottom three layers in the ISO/OSI Reference Model: physical, data link, and network. Intermediate systems (which are the devices that relay transmissions between other devices) use only these three layers to pass on transmissions.

subnet mask In the Internet Protocol (IP) addressing scheme, a group of bits whose value operates on the IP address to separate the network ID portion from the host ID to identify a subnetwork. All the members of the subnetwork share the mask value. Once identified using the mask, members of this subnet can be referenced more easily. Also called an *address mask*.

subnetwork A term for an autonomous network that is part of another network and is connected by way of a bridge, gateway, or router.

A *subnetwork* may include both terminating devices (nodes) and intermediate systems (routers). Nodes within each *subnetwork* use a single protocol to communicate among themselves. A *subnetwork* is connected to the higher level network through an intermediate system. The intermediate system may use a routing protocol to communicate with nodes outside the *subnetwork*. For example, departmental networks connected by bridges or routers to a company's local area network (LAN) are subnetworks. Similarly, a localized X.25 network may be a subnetwork in a larger wide area network (WAN). If the overall internetwork becomes too large or changes frequently, individual routers may not be able to maintain an accurate routing table. To alleviate this problem, the internetwork is divided into multiple levels, and routing tables are maintained for nodes only within its own domain. The highest level network is frequently called the *backbone*. The Internet, for example, is partitioned in this manner. An added advantage of the hierarchical structure is the ability to isolate individual *subnetworks* when problems occur.

Subnetwork Access Protocol (SNAP or SNAcP) In the ISO/OSI 8648 specifications for the Internal Organization of the Network Layer (IONL), the type of protocol used at the bottom of the network layer, the subnetwork access sublayer provides an interface over which data are sent across a network or subnetwork.

For example, in the Internet, it is a protocol that operates between a network entity in the subnetwork and a network entity in the end system and specifies a standard way of encapsulating Internet Protocol (IP) datagrams and address resolution protocol (ARP) messages. X.25 packet level protocol is another example of a subnetwork access protocol.

Subnetwork Dependent Convergence Protocol (SNDCP) In the ISO/OSI 8648 specifications for the Internal Organization of the Network Layer (IONL), the type of protocol used at the middle of the three sublayers into which the layer has been subdivided. The protocol must handle any details or problems relating to the subnetwork to which the data are being transferred.

Protocols operating at this sublayer assume a particular type of subnetwork, e.g., an Ethernet local area network.

Subnetwork Independent Convergence Protocol (SNICP) In the ISO/OSI 8648 specifications for the Internal Organization of the Network Layer (IONL), the type of protocol used at the highest of the three sublayers into which the layer has been subdivided. A *SNICP* must provide the routing and relaying capabilities needed to get data to its destination.

The Connectionless mode Network Protocol (CLNP) is an example.

subreflector Any reflector other than the main reflector of a multireflector antenna.

subordinate station A synonym for *slave station*.

subscribe The identification and selection of a USENet newsgroup as one that the user's newsreader will display.

The opposite of *subscribe* is *unsubscribe*.

subscriber In the telephone industry, a synonym for customer (the ultimate user; individuals, organizations, activities, etc.). Subscribers are usually subject to a tariff and use end instruments (e.g., telephones, modems, and facsimile machines) that are connected to a central office.

The term probably arose in the earliest days of telephony when a person could string his own wire among geographic locations for communications or *subscribe* to the telephone service.

subscriber carrier In telephony, a system that multiplexes customer signals in order to achieve higher utilization of the "loop plant" (the wiring between the central office and the subscriber). Carrier systems consist of two terminals: the central office terminal (COT) and the remote terminal (RT) separated by the transmission medium.

The remote terminal interfaces the subscriber loops, multiplexes the subscriber signals together, and conditions the signal for transmitting over the transmission link. The central office demultiplexes the signals and routes them to the appropriate line appearances on the switching equipment. *Subscriber carrier* systems may be either nonconcentrating or concentrating systems. In nonconcentrating systems, there are at least as many communications channels between the remote terminal and the central office as there are individual subscriber loops. In a concentrating system, there are fewer channels than subscriber loops; and the system reassigns channels to loops as the traffic pattern changes.

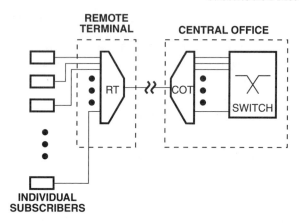

subscriber drop The cable that runs from a cable distribution point to a subscriber's premises.

subscriber loop A telephone line between the telephone company central office and a piece of equipment at the subscriber premises (equipment such as a telephone, answering machine, modem, fax, private branch exchange, or other end equipment). Also called a *subscriber line*. See also *local loop*.

subscriber number See *local number*.

subscriber trunk dialing (STD) See *direct distance dialing*.

subset (1) In telephony, a contraction of SUBscriber SET. A device used for originating and receiving telephonic calls on the premises of a telco subscriber. The *subset* contains at least a means for dialing a remote station number, a microphone, an earpiece, and an interface to the subscriber loop. The microphone and earpiece are combined into a single assembly called the *handset*. **(2)** A group of entities with similar characteristics within a larger group.

subset equalizer Circuits in a telephone subset that cause both the transmitter and receiver levels to be inversely proportional to loop current, thus, partially compensating for the transmission losses associated with loop length. The effect is to increase both the transmit level and the receiver sensitivity with increasing loop length.

Subsidiary Communications Authorization (SCA) A subcarrier modulation permit granted to broadcasters by the Federal Communications Commission (FCC) that allows the simultaneous transmission of private information (and/or control signals) along with the public broadcasting.

For example, an FM radio station transmitting a monophonic signal with *SCA* service may use multiple subcarriers within 20 to 75 kHz, providing the crosstalk into the main channel is attenuated more than 60 dB. If the station is transmitting stereophonic programming, the subcarriers are restricted to 53 to 75 kHz, again with the same crosstalk requirement. The typical *SCA* subcarrier frequency is 67 kHz. Sometimes called the *Supplemental Communications Authority*.

substitution method In fiber optics, a procedure for measuring the transmission loss of a fiber, that is,

1. Select a reference optical fiber 1–2 m in length with characteristics (physical and optical) matching those of the fiber to be tested.
2. Using a stable optical source at the wavelength of interest and passed through a mode scrambler, overfill the reference fiber.
3. Measure the power level at the output of the reference fiber.
4. Repeat steps 2 and 3 on the fiber under test (FUT).

5. Obtain the transmission loss of the fiber under test by subtracting the power level obtained at the output of the fiber under test from the power level obtained at the output of the reference fiber.

The procedure has some drawbacks with respect to accuracy, but its simplicity makes it a common method used in the field. The major "failure" is that it provides a conservative loss figure, that is, if it were used to measure the individual losses of several long fibers, and these long fibers were joined serially, the total loss obtained (excluding splice losses) would be expected to be less than the sum of the individual fiber measured losses.

subvoice grade channel A channel with a bandwidth narrower than that of a voice-grade channel, i.e., less than the nominal 300–3400 Hz band normally provided. A *subvoice grade* channel is usually a subchannel of a voice grade line.

sudden ionospheric disturbance (SID) An abnormally high ionization density in the ionospheric D region caused by an occasional sudden solar flare (outburst of ultraviolet light from the Sun). The ionospheric change result in a sudden increase in radio wave absorption, primarily at the upper medium-frequency (MF) and lower high-frequency (HF) bands.

suite A collection of entities assembled to serve a common goal. Examples include protocol suites, which are collections of rules brought together to provide reliable communication; or office suites which provide a collection of programs needed in a typical office (word processor, spreadsheet, database, and communications including fax and e-mail).

sum (1) The action of mathematically adding two or more numbers. (2) The resulting value from the addition of a set of numbers. (3) An acronym for S̲et-U̲p control M̲odule.

summation (Σ) A value found by adding together all numbers of a specified set or subset. The summation is indicated by the upper case Greek letter *sigma* (Σ). For example,

$$T = \sum_{i=1}^{10} N_i$$

indicates the total (T) is the sum of the 10 values associated with N_1, N_2, ... N_{10} [as specified by the index (i) above and below the summation symbol].

summation check (1) A validity-testing method in which the current sum of the numerical values of a set of data is compared to a previously computed sum (called the *checksum*). (2) A comparison of checksums on the same data on different occasions or on different representations of the data in order to verify data integrity. Also called a *sumcheck*.

summing amplifier A high-gain amplifier (called an operational amplifier) with accompanying circuits that allow it to perform as a *summing element*.

A block diagram symbol for a five-input summing amplifier with two of the five signals inverted before summing is shown here. Sometimes, a number accompanying the + and − symbols is used to indicate a multiplier coefficient associated with each input.

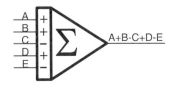

summing element A device that algebraically adds several signal inputs and delivers the resultant sum to the output. There may be any number of inputs, and each input may have a "+" or "−" sign associated with it.

A *summing element* is also called a *summer*. The block diagram symbol below is of a four-input *summing device*. Three of the inputs are added together, while the third is subtracted from the sum, that is,

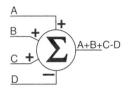

summing point (1) Any point at which multiple signals are algebraically added together. (2) The point in a feedback control system at which the algebraic sum of two or more signals is obtained.

Sun synchronous orbit Characterizes an Earth-orbiting satellite whose orbital plane is polar and whose altitude is such that it passes over every point on the Earth at an identical local Sun time twice per day.

Sun transit outage The condition that occurs when an Earth-orbiting satellite passes between the Sun and the Earth. For synchronous communications satellites, this occurs as much as on several days, twice a year.

The outage occurs when the pointing angle from the Earth station to the satellite and the pointing angle to the Sun are so nearly coincident that the receiving system is overloaded by the noise generated by the Sun. The actual length of the outage time is related to the antenna beam width (which in turn is a function of frequency and antenna size).

sundial time A timekeeping scheme based on the Sun's crossing of the local meridian at 12:00 noon. Also called *local apparent time* and *local apparent time (LAT)*. See also *time (3)*.

sunspots (1) A real phenomenon occurring on the surface of the Sun (appearing as a dark blotch in the photosphere) in which large numbers of charged particles and electromagnetic radiation are thrown into space at high velocity. Some of these particles reach the Earth and are channeled toward the poles by the Earth's magnetic field. Here they cause the aurora borealis and aurora australis (the northern and southern lights). Severe bombardment can induce extreme effects on the ionosphere, disrupt communications, and even cause failures in the power distribution grid.

Sunspot activity (the number of sunspots occurring at a given time or on a given day) seems to be periodic, and the period of the cycle from a maximum number to a minimum and back to a maximum is approximately 11 years. Some sunspots last several weeks. However, because the Sun rotates on its axis, the sunspot comes into view over one limb, only to later disappear over the opposite limb. (2) The humorous hypothesis as to the cause of an odd error. For example, a response to "Why did my machine crash?" might be, "Sunspots, I guess." (3) One of several "tongue-in-cheek" hypotheses as to the cause of bit rot, others being cosmic rays and the phase of the Moon.

super encryption The process of encrypting encrypted information.

super high frequency (SHF) See *frequency spectrum*.

superchromatic correction In optics, color correction of an optical system at four separate wavelengths. The correction requires the use of three separate glass types.

supercomputer Any computer that is among the largest and fastest few computers in the world. *Supercomputers* are used for complex and sophisticated calculations such as those found in scientific modeling and weather forecasting.

superconducting quantum interference device (SQUID) An extremely sensitive magnetic field detecting device. Its operation relies on three quantum effects:

- *Superconductivity*—the loss of all electrical resistance;
- *Electron tunneling*—effectively, current flow through a thin insulating layer, called the *Josephson effect;* and
- *Flux quantization*—only integer values of the product of magnetic field and loop area are permissible.

A direct current *SQUID* consists of two Josephson junctions in a superconducting ring with a bias current injected into the ring such that it divides between the two junctions. When the current is above a critical value, a voltage is developed across the structure. The critical current is affected by any external magnetic field; hence, the junction voltage is affected. This voltage varies in a manner such that fractions of a flux quantum can be measured.

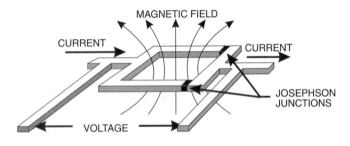

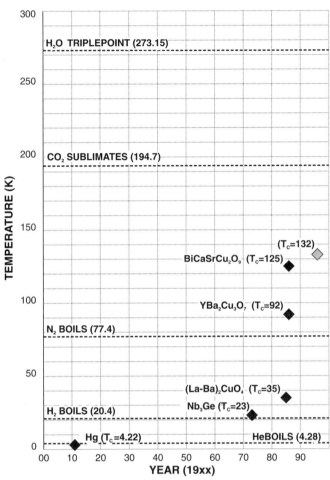

superconductor A material that has absolutely no resistance to the flow of an electrical current and, ideally, zero permeability. The term *superconductor* was coined in 1911 by the Dutch physicist Heike Kamerlingh-Onnes (1853–1926).

Normal conductors have small and finite measurable resistance; superconductors have none. All *superconductors* exist only at very cold temperatures; to date, the warmest temperature would be cold enough to turn air into a liquid. The temperature at which the material's resistance becomes zero is called the critical temperature (T_C). Because of the expense of creating an environment as cold as is needed, *superconductors* are used only in very special equipment.

superframe (SF) A data transmission format composed of 12 DS1 frames of 192 bits each and a 193rd bit used for link control and error checking. As an industry standard, *SF* has been superseded by the *Extended Super Frame* (*ESF*) format. However, because ESF is not backward compatible and there is a large installed base of channel banks and DS-1 Multiplexers that are based on *SF,* it is still the default private line formatting technique. Also known as *D4.*

supergroup The assembly of five 12-channel groups (60 channels) into a composite transmission group. See also *group.*

supergroup distribution frame (SGDF) In frequency division multiplex (FDM) systems, the distribution frame provided for terminating and interconnecting group modulator/demodulator circuits and supergroup modulator/demodulator circuits. See also *group.*

superheterodyne receiver A type of radio frequency receiver in which the heterodyne process is used to convert the received signal frequency to an intermediate frequency (IF) suitable for further processing.

superimposed ringing A form of selective ringing that uses the polarity of the direct current feed to obtain selectivity. See also *selective ringing.*

superluminescent light emitting diode (SLD) A diode light source that is based on stimulated emission with amplification, but with insufficient feedback for oscillation to build up. Its output has a narrower spectral width and a higher radiance than a normal light emitting diode (LED). See also *superradiance.*

superposed circuit A contraction of SUPERimPOSED circuit. A derived channel obtained from one or more circuits in such a manner that all channels can be used simultaneously and without interference.

superposition theorem A network theorem that states: "The current that flows through any branch of a linear network (or the voltage across that branch) due to the simultaneous application of any number of voltage sources is the same as the algebraic sum of the individual currents due to each source applied separately."

superradiance Amplification of spontaneously emitted radiation in a gain medium, characterized by moderate spectral line narrowing and moderate directionality. The process differs from lasing by the absence of positive feedback and the absence of well-defined modes of oscillation.

superregeneration A form of regenerative amplification in which oscillations are allowed to build up and are then quenched at an ultrasonic rate. Superregenerative detectors are sometimes used in radio receivers.

supervised circuit A connection between two points in which a device at one or both ends monitors the status of the connection for proper continuity. That is, it monitors for open circuit conditions and/or short circuits between the connection pairs or to ground.

supervision In telephony, the function of controlling and indicating the status of a connection.

supervisory control The use of signals (e.g., control characters) for the automatic actuation of equipment or indicators.

supervisory signal A signal used to indicate and/or control the various operating states of a system, device, or circuit.

supervisory tone In telephony, a tone or set of tones that indicate to equipment, an operator, or a customer that a specific call state has been reached or that an action is required. Also called *call progress tones*.

supply The immediate source of electrical energy to a system or device.

The supply may be the ac mains (117 Vac or 220 Vac), the telephone switching center battery (−48 Vdc typically), a 12-V lead acid battery used in an automobile, or the output of an ac to dc converter such as those used to convert the mains input to the 5 Vdc and 12 Vdc required by a computer. Also called a *power supply*.

suppressed carrier transmission Any AM modulation where any unmodulated carrier energy appearing at the modulator output is attenuated before being transmitted to a level unsuitable for direct use in demodulation. Characteristics include:

- One or both sidebands may be transmitted.
- Carrier suppression permits higher power levels in the sidebands, for a given total transmission power, than would be possible with conventional AM transmission.
- Carrier power must be restored by the receiving station to permit demodulation.
- Suppressed carrier transmission is a special case of reduced carrier transmission.

SURAnet An acronym for <u>S</u>outheastern <u>U</u>niversities <u>R</u>esearch <u>A</u>ssociation <u>net</u>work.

surface acoustic wave (SAW) A mechanical wave, opposed to an electromagnetic wave, that propagates along the surface along an interface between two different media (a surface) of a solid. It has both longitudinal and transverse (shear) components. Also called a *Rayleigh wave*.

surface electromagnetic wave An electromagnetic wave that propagates along the interface between two different media without radiation but with exponentially decaying short-lived fields on both sides of the interface.

surface leakage Current flow over the surface of a material rather than through the volume of the material itself.

surface mounted device (SMD) A device designed to be attached to one surface of a printed circuit board and has the entire body of the device projects from the mounting surface. This is in contrast to through-hole devices that have pins extending into or through the printed circuit board. See also *surface mounted technology* (*SMT*).

surface mount technology (SMT) A method of manufacturing printed circuit board assemblies (PBAs) such that components are soldered directly to the surface of the printed circuit board (PCB) instead of being soldered in holes drilled through the PCB. A component designed to be installed in this technology is called a surface mounted device (SMD). SMDs are much smaller, higher density, and more rugged than their through hole counterpart. See also *surface mounted device* (*SMD*).

surface reflection That portion of the incident wave radiation that is reflected from a surface of a refractive index material. The fraction of the incident radiation that is reflected (*r*) is related to the refractive index (*N*) by:

$$r = \left(\frac{N - 1}{N + 1}\right)^2$$

For example, for crown glass with a refractive index (*N*) of 1.5, the surface reflection (*r*) is 4%. Also called *Fresnel reflection*.

surface refractivity The refractive index of the Earth's atmosphere, calculated from observations of pressure, temperature, and humidity at the surface of the Earth.

surface refractivity gradient The difference in refractive index between the Earth's surface and a specified altitude.

surface wave A wave that is guided by and along the interface between two different media or by the refractive index gradient of the medium. Characteristics include:

- Energy is not converted from the surface wave field to another form of energy.
- The wave does not have a component directed normal to the interface surface, that is, the field components diminish with distance from the interface and become negligible within a finite and short distance of the interface.

Examples of *surface waves* include:

- In radio transmission, ground waves that propagate close to the surface of the Earth (the interface surface being the Earth with one refractive index and the atmosphere with another).
- In optical fiber transmission, evanescent waves. All guided modes, but not radiation modes, in an optical waveguide belong to this class.
- Surface acoustic waves.

surface wave antenna An antenna in which power radiates from discontinuities in the structure that interrupt a bound wave on the antenna surface.

surfave wave transmission line A transmission line in which propagation (other than a transverse electric and magnetic—TEM mode) is constrained to follow an interface boundary of a guiding structure. Also called a *wave guide*.

surge (1) In electrical circuits, a sudden short-term increase in line voltage. A power surge can damage a computer if the *surge* is sufficiently large and prolonged. Devices called *surge protectors* or *surge suppressors* can be inserted between the mains supply (wall outlet) and the computer to minimize the effects of *surges*.

A surge is frequently defined as a voltage greater than 10% of the nominal voltage and for a duration of more than 50% of a cycle. Also called an *impulse*. See also *transient*. (2) In packet switched networks, a temporary increase in required bandwidth. The increase is measured in relation to a guaranteed bandwidth, known as the *committed information rate* (*CIR*).

surge let through That part of an electrical surge that passes through a surge protector with little or no alteration. For example, segments that pass through the protective device before the device has time to turn on. See also *surge remnant*.

surge protector A voltage-protecting assembly that attenuates potentially damaging power line surges and transients from reaching the computer or other electronic devices. Surge protector work is done by dissipating the excess energy as heat. If the excess energy is too great, a fuse or circuit breaker in the protector will open.

Surge protectors are rated in several ways, that is,

- The way in which they deal with the excess voltage.
- The amount of voltage allowed to pass through during operation.
- The speed with which they can deal with the voltage.
- The amount of energy they can absorb.
- The number of surges they can withstand.
- The combinations of wires protected (tip, ring, hot, neutral, and ground, etc.).

The protection level for nonrepetitive lightning-induced surges is defined differently by various agencies and for various points in a network; however, they all tend to follow the format outlined in the accompanying figure. Typically in telephony, V_{peak} is ± 1000V to ±2000V, T_{peak} is 2 to 10 μs, and $T_{50\%}$ is 750 to 1000 μs. With mains power distribution circuits, the voltages are usually higher, for example, UL-1449 uses a 6000-V surge. The UL 1449 standard sends repeated high-voltage (6000 volts), high-current signals through the *surge protector* and monitors the voltage that the device lets through. To pass, a device must withstand these repeated pulses such that the first and last pulse generated pass through voltages less than the rating and differ by less than 10%. For example, a 6000/330 device allows 330 volts to pass through with the application of a 6000-volt surge. Other ratings include 6000/400, 6000/500, and so on. Several

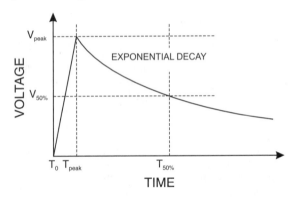

materials are used in manufacturing surge-protecting devices. A comparison of the capabilities and limitations of various devices used to protect against voltage surges is illustrated in the table. The volt-amp

COMPARISON OF SURGE PROTECTOR TECHNOLOGIES

	Energy capacity	P_{AVG} capacity	I_{PEAK}	Clamp ratio	Speed
Ideal	H	H	H	H	H
Carbon blocks	H	H	M-H	H	L
SiC Varistor	M	L-M	M	L	H
Zener diodes	L	M	L-M	H	M-H
Gas tubes	H	H	M-H	H	L
Spark gaps	H	L	H	L-M	M
MOVs	M-H	M-H	M-H	M-H	H
SCR crowbars	M			M-H	H
RC filter	H	H	H	L	L

H = high; M = medium; L = low.

characteristic of clamping devices in the active region can be compared using the approximation $I = kV^{\alpha}$ where I is the current through the device, V is the voltage across the device, k is the scaling factor, and α is the degree of nonlinearity of the device. α is 1 for a resistor and increases as the clamping factor increases. Surge protectors are also called *arresters, limiters, snubbers, surge arresters, surge diverters, surge suppressors, transient suppressors,* and sometimes *noise filters.* See also *gas tube, MOV,* and *zener diode.*

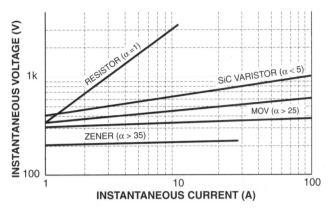

surge remnant Those parts of a surge that pass through a surge protector, i.e., are detectable downstream of the surge-protecting device(s).

surge suppressor See *surge protector.*

susceptance The imaginary part of admittance.

susceptibility (**1**) The lack of immunity. (**2**) The inability of a device or system to resist the influences of an electromagnetic disturbance. Also called *radio frequency interference (RFI).*

susceptibility threshold The amount of undesired signal power required at the input terminals of a receiver to cause barely perceptible interference at the receiver output terminals.

susceptiveness In telephone systems, the extent to which wire circuits pick up noise and low-frequency energy by induction from power distribution systems.

Susceptiveness depends on parameters such as:

- Circuit *balance;*
- Wire and connection transpositions, e.g., twisted pair or frogging;
- Wire spacing; and
- Isolation from ground.

suspend To temporarily halt a process or operation. See also *sleep.*

SUT An abbreviation of **S**ystem **U**nder **T**est.

SV An abbreviation of **S**ingle **V**alue.

SVC (**1**) An abbreviation of **S**witched **V**irtual **C**ircuit. (**2**) An abbreviation of **S**uper**V**isor **C**all.

SVD An abbreviation of **S**imultaneous **V**oice/**D**ata.

SVI An abbreviation of **S**er**V**ice **I**nterruption.

SVS (**1**) An abbreviation of **S**ecure **V**oice **S**witch. (**2**) An abbreviation of **S**ecure **V**oice **S**ystem.

SW (**1**) An abbreviation of **S**oft**W**are. (**2**) An abbreviation of **SW**itch. (**3**) An abbreviation of **S**outh **W**est.

SW56 An abbreviation of **SW**itched **56**kbps dial up data service.

SWA An abbreviation of **S**traight **W**ire **A**ntenna.

SWAK An acronym for **S**ealed **W**ith **A** **K**iss.

SWALK An acronym for **S**ealed **W**ith **A** **L**oving **K**iss.

SWAN An acronym for **S**atellite based **W**ide **A**rea **N**etwork.

SWAT An acronym for **S**imultaneous **W**ide **A**rea **T**elecommunications service.

SWBD An abbreviation of **SW**itch**B**oar**D**.

SWC An abbreviation of <u>S</u>erving <u>W</u>ire <u>C</u>enter.

sweep acquisition A technique of assuring that a phase lock loop (PLL) oscillator can lock onto the reference signal.

The technique slowly sweeps the frequency of the PLL local oscillator across the expected range of frequencies so that at some point it is within the pull-in (capture) range of the reference.

swell A persistent overvoltage that is not large enough to be considered a surge, but can damage sensitive equipment.

SWG An abbreviation of <u>S</u>tandard <u>W</u>ire <u>G</u>auge.

SWIFT An acronym for <u>SWI</u>tched <u>F</u>ractional <u>T</u>-1. A local exchange carrier (LEC) service, defined by Bellcore, that provides up to full T-1 capacity.

swim A measure of how much an on-screen image waivers over time. It is a slow oscillatory movement of objects on a monitor screen that are supposed to remain stationary. The movement may also manifest itself as a wavy pattern on the screen. *Swim* is slower than jitter and faster than drift.

Swim may be caused by the interaction of a magnetic field (from an electric motor or a transformer near the monitor) and the electronics of the monitor itself. It may also be due to poor translation methods from one video format to another. See also *drift* and *jitter.*

switch **(1)** In computer software, an argument used to control one of the optional parameters of a program; typically, the switch is designated with a hyphen (-) or slash (/). **(2)** An electronic device or circuit element that can be used to open, close, or transfer a communications or current path. When the path is open, the *switch* is said to be off; when the path is closed, the *switch* is on. **(3)** In communications, a *switch* is a device that provides a temporary circuit between nodes of a network or between two networks. The circuit connection may be established either throughout the duration of the requirement by the nodes (e.g., the telephone network) or only as required by activity (e.g., a packet network).

In digital networks two types of switches are common:

- A cross-point switch that routes or directs packets without checking for errors.
- A store-and-forward switch that checks each packet for errors before forwarding it to the selected network port. Because this methodology is time consuming, heavy traffic may overload the switch; in burst mode, a store-and-forward switch will almost certainly become saturated. In general, store-and-forward techniques are slower than cross-point switches.

The diagram shows several representations of a *communications switch.*

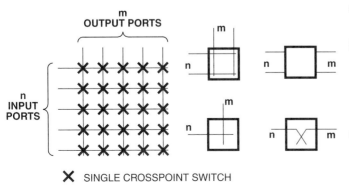

✗ SINGLE CROSSPOINT SWITCH

(4) In telephony, a *switch* may also be the switching system in the central office that controls, connects, and routes "telephone calls." In this instance, the switch element is frequently called a *crosspoint,* and the collection of crosspoints is called a *switch* or *switching system.* **(5)** A deprecated synonym for the central office itself.

switch busy hour In telephony, the busy hour for a single switch.

switch contact An electromechanical device that opens or closes an electrical circuit. The device may be manually operated (as with a switch) or electrically operated (as with a relay or semiconductor.) Simple *switch contacts,* single pole switches, are classified as normally open (NO) or "form A" and normally closed (NC) or "form B." Symbols for these contacts include those shown below.

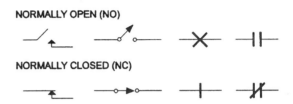

These two *switch contact* types may be used to create more complex arrangements. For example, one NO contact and one NC contact may be arranged with a common element and operated together to form a single pole double throw switch (SPDT). SPDT *switch contacts* are represented with one of the following symbols:

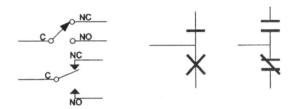

The SPDT *switch contact* arrangement may be designed as a break-before-make switch which is called a "form C" contact. Or it may be designed as a make-before-break switch which is called a "form D" contact. The complexity of *switch contact* designs is nearly limitless with multiple contacts and/or multiple poles on the switch, e.g., double pole double throw, single pole 8-position, or 10-pole 10-position. See also *relay.*

switch fabric In telephony, the means in a switching system by which transmitting circuits are connected to receiving circuits.

switch hook See *hook switch.*

switch over The transfer of control or communications to an alternate (redundant) entity. This can occur when the primary entity fails or during routine maintenance.

switchboard The attendant position at a private branch exchange (PBX) or private automatic branch exchange (PABX).

Switched 56 A 56-kbps, synchronous, circuit switched telecommunications, full-duplex service that can be leased from long-distance service providers. Switched 56 may be delivered on one pair or two pairs.

The service is being replaced by Integrated Services Digital Network (ISDN).

switched access A method of obtaining entry to telecommunications circuits by using (generally) electromechanical relays, e.g., the use of relays to allow test equipment to monitor or measure a circuit's performance.

switched attachment The IBM term for linking devices to host computers through a private branch exchange (PBX) capable of handling data switching.

switched circuit In a communications network, a circuit that is temporarily established at the request of one or more of the connected stations.

switched digital access In wide area networks (WANs), an arbitrated connection to long-distance lines by a local carrier. The user is connected directly to the local carrier and from there to the long-distance carrier.

switched Ethernet See *Ethernet*.

switched line The type of communication path found in a standard dial-up telephone connection, i.e., a line that passes through the central office switching system. See also *leased line*.

switched loop In telephony, a circuit that automatically releases the connection to the attendant once the connection to a third party is established. Also called a *released loop*.

Switched Multimegabit Data Services (SMDS) A BellCore developed public packet switched data service standard for the regional Bell operating companies (RBOCs). It is a high-speed, connectionless, broadband, packet switched, wide area network (WAN) data service network.

Currently, it supports data rates between 1.544 and 44.736 Mbps over public lines and can run over the physical wiring for a metropolitan area network (MAN) Distributed Queue Dual Bus (DQDB) or an asynchronous transfer mode (ATM). It is capable of supporting up to 155 Mbps data transfer rates. It offers variable data packet size, Virtual Private Network, and Closed User Group features. Access to the network is usually via T-1 or T-3 lines.

switched network **(1)** Any network providing switched communications service. **(2)** A communications network that establishes a connection from a calling station to a called station through the use of a circuit switching system. The switches may be left in place for the duration of a session or may be reassigned during breaks in information flow as in a packet network.

switched private line network A network derived by combining point-to-point private circuits with switches.

switched services network (SSN) See *public switched telephone network* (*PSTN*).

switched T-1 A circuit switched service that provides a 1.544-Mbps bandwidth channel (i.e., T1 line capability). Transmissions over this line may additionally go through a multiplexer, or channel bank, where they are broken down and transmitted across slower channels (generally, multiples of 64 kbps). See also *fractional T-1*.

switched virtual circuit (SVC) An implementation agreement (IA) from the Frame Relay Forum's Technical Committee that expands frame relay access options to include switched (dial) access. It is a temporary logical connection in a packet network that is established for the duration of a communications session and is terminated after the session is over.

The opposite of an *SVC* is a permanent virtual circuit (PVC), in which the connection is always established.

switching A communications method that uses temporary paths rather than permanent connections to establish links or to route messages between stations. In the normal telephone dial-up network, the switch path in the central office is established during the initial call setup process and is maintained throughout the session. In a message switching and packet switching networks, the connection is established only for the duration of the packet or message. (At other times, the circuits are made available to other users.)

switching array The aggregate of multiple crosspoints—generally arranged in a matrix.

switching center A site-containing equipment that terminates multiple circuits and is capable of interconnecting circuits or transferring traffic between circuits by manual or automated means. The interconnections may be on a circuit, message, or packet switching basis.

Synonyms in telephony include *central office (CO)*, *switching exchange, switching facility,* and the deprecated synonym *switch.* See also *class n office.*

switching circuit In telephony, the method of handling traffic through a switching center where connection is established between calling and called stations. The information for traffic handling may be supplied by the user or another switching system.

switching control center In telephony, the location where maintenance analysis and control activities are centralized for remotely located switching exchanges

switching element That part of a switching system that connects the input ports with output ports. See also *switch* and *switching system.*

switching entity In telephony, the switching network and its control.

switching exchange A synonym for *switching center.*

switching facility A synonym for *switching center.*

switching hierarchy In telephony, the levels of switch used for establishing connections for long-distance calls. See *class n office.*

switching hub A multiport hub that provides not only connectivity, but the full available bandwidth between any pair of ports. The hub may also provide bridging, routing, and protocol conversion services, in which case it is frequently called an *intelligent switching hub.*

switching network In telephony, the switching stages and their interconnections.

Within a switching system, there may be more than one *switching network.*

switching plan In telephony, a plan for the interconnection of *switching networks.*

switching point A synonym for end office.

switching stage In telephony, a set of *switching elements* arranged as a functional subgroup of the overall switching entity. By dividing the switching system into stages, substantially fewer switching points need to be included. However, the overall switch may become *blocking* if insufficient downstream stages are not provided.

In the diagram, two switching stages are shown in which each stage is composed of identical switch modules (each with "n" input and "n" output ports). Each signal path represents a bidirectional link. The inputs are connected to the individual line controllers, while the outputs may be connected to either output controllers, subsequent switching stages, or another stage 2 port (to create a folded switch matrix). For each stage 1 module used, "n" subscribers may be connected to the switching system. For each stage 2 module used, "n" simultaneous connections may be established.

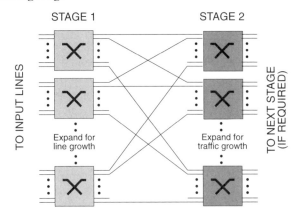

STAGE 1   STAGE 2

TO INPUT LINES

Expand for line growth

Expand for traffic growth

TO NEXT STAGE (IF REQUIRED)

switching system All necessary equipment and software required to connect data path from a source to a desired destination. The elements of a switching system include:

- *An input controller (IC)*—circuitry to recognize that the channel is being seized, destination address signaling, establishing clock synchronization where required, and so on. Its job is essentially to synchronize each input (which may be a message, packet, or cell, depending on the architecture being used) with the internal clock and control mechanism.
- *An interconnection means*—the fabric that allows the connection of input ports (lines) to output ports (lines). The switch may be a hardware matrix such as is used in telephony, or a computer node such as is found in a local area network (LAN). See also *switch.*
- *An output controller (OC)*—circuitry to provide outgoing signaling (e.g., ringing), packet queuing, buffering, transmission line conditioning and impedance matching, and so on.

Collectively, these elements provide services such as:

- Accepting requests for service (e.g., on-hook/off-hook condition).
- Accepting address information.
- Mapping input channels to appropriate output channels.
- Scheduling the sequence of packet transmission in appropriate time slots (based on circuit availability, priority, etc.).
- Data forwarding—the delivery of a packet to the appropriate output channel as prescribed by the scheduler.
- Transmitting address information toward the destination node.
- Notification of the destination node of an incoming message (e.g., ringing a telephone).

switching time The elapsed time between the application of the command to switch and the completion of the switching action. Some systems also require that the switching time include the time required for the switched signal to settle to within a specified tolerance.

SWO An abbreviation of <u>S</u>tandards <u>W</u>riting <u>O</u>rganization.

SWR An abbreviation of <u>S</u>tanding <u>W</u>ave <u>R</u>atio.

SX An abbreviation of <u>S</u>imple<u>X</u>.

SXS An abbreviation of Step-by-Step switch.

syllabic companding A compression-expansion (compander) scheme in which the gain variations occur at a rate similar to the syllabic rate of speech.

symbol (**1**) In general, a graphical representation of a physical entity or a relationship between quantities. (**2**) In communications, the smallest unit of information from which all messages are composed. For example, a binary system has two *symbols* designated 1 and 0, and a quaternary system has four *symbols,* each representing one of the four possible bit pair combinations 00, 01, 10, and 11. The set of all *symbols* in a given system is called the *alphabet.* (**3**) A character, letter, or combination of letters that may be used in place of a unit. Generally, the letters are taken from the name of the unit. See also *SI* for a list of units, their names, and their symbols.

symbolic link A file that points to either the relative or absolute path name of a file or application that is located on the same or another disk or even another network. Also called a *soft link.*

symbolic logic The treatment of logic by means of a formalized artificial language (a symbolic calculus) whose goal is to avoid the shortcomings of natural languages. The shortcomings include both ambiguous statements and logical inadequacies. See also *Boolean algebra.*

symmetric multiprocessing A type of multiprocessing in which more than one processor can execute kernel-level code concurrently.

symmetrical In transmission circuits, a term that refers to terminal (end-to-end) equivalence; that is, there is symmetry about a "vertical" centerline. The circuit in the illustration is *symmetrical;* it will also be *balanced* if the values $x = y$. See also *balanced.*

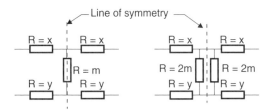

Line of symmetry

R = x   R = x   R = x   R = x

R = m   R = 2m   R = 2m

R = y   R = y   R = y   R = y

symmetrical alternating signal A periodic signal (voltage or current) in which all points one-half period apart are equal in magnitude and opposite in sign.

symmetrical channel A full-duplex communications channel in which the send and receive capacities, bandwidths, or data rates are the same.

symmetrical compression A data compression method in which the amount of computer power or time required to compress a data file is approximately equal to the power or time required to expand the compressed file. Compression methods such as these are used when the number of compression operations is similar to the number of expansions, for example, facsimile transmissions generally have an expansion for every compression.

symmetrical pair A balanced two-wire transmission line in which each conductor has the same resistance per unit length, the same impedance, and the same coupling to the environment in which it is immersed (e.g., the same coupling to all other conductors in the same cable bundle).

symmetrical periodic function A mathematical function $f(x)$ with period $2p$ which satisfies one or more of the following conditions:

$$f(x) = -f(\pm x), f(x) = f(p \pm x), f(x) = -f(p \pm x)$$

SYN An abbreviation of the <u>SYN</u>chronizing character used in synchronous data communications. The *SYN* character enables the receiver and transmitter to maintain the same timing relationship (remain in phase). It is used as a fill character when there is no user information or control information transfer. The ASCII <SYN> character is decimal 22 (16h).

sync An abbreviation of the word <u>SYNC</u>hronize, <u>SYNC</u>hronizing, or <u>SYNC</u>hronous. Although *sync* is an abbreviation, it has been used so extensively that it has become a word in its own right.

sync pulse A synonym for *synchronization pulse.*

synchronization (1) In general, the process or condition of bringing two entities into the same operating phase, i.e., the attaining of synchronism. (2) In communications, matching a receiver's clock and data timing to the same clock and data timing of the transmitter. In the more general case, the obtaining of a desired fixed relationship among corresponding significant instants of two or more signals. (3) In power transmission, bringing the ac voltage phase relationship of two mains to the same phase before they both can be used to deliver power to the load. (4) In audiovisual systems (multimedia), matching the sound track to the visual information. (5) The condition of bringing two data files to the same value (as would be the case in mirror servers). (6) A state of simultaneous occurrences of significant instants among two or more events.

synchronization bit A digital bit, character, or word used to achieve or maintain synchronization between a digital data transmitter and its corresponding receiver.

synchronization code A specified sequence of symbols used to achieve or maintain synchronization between a data transmitter and its corresponding receiver.

synchronizing pilot In frequency division multiplex (FDM) systems, a reference frequency used for achieving and maintaining synchronization of the oscillators of the system or for comparing the frequencies or phases of the signals controlled by those oscillators.

synchronizing pulse A pulse used to achieve or maintain synchronization between an analog transmitter and its corresponding receiver. Also called a *sync pulse.*

synchronous (1) In general, a term that describes signals in which corresponding significant events or entities occur at precisely the same average rate. See also *anisochronous, asynchronous, heterochronous, homochronous, isochronous,* and *plesiochronous.* Note that the terms *isochronous* and *anisochronous* pertain to characteristics, while *synchronous* and *asynchronous* pertain to relationships. (2) A communications strategy that uses precise timing to control transmission. A transmission consists of an initial synchronization sequence, followed by a predefined number of symbols transmitted at a constant rate. Except for the initial synchronization symbol or sequence (and fill symbols [SYN] if there is insufficient data for the defined period) synchronous transmissions do not require any additional periodic timing.

synchronous auto dial language (SADL) A public domain autodialing protocol developed by Racal Vadic for use in IBM's binary synchronous communications (BSC) protocol, IBM's Systems Network Architecture (SNA), synchronous data link control (SDLC), and the ISO/OSI high level data link control (HDLC). The protocol does for synchronous systems what the Hayes AT command set does for asynchronous modem systems.

synchronous data transfer A term that describes data transfers in which all bits are of equal duration, the interval between characters is fixed, and the data transmission signal is associated with a clocking signal. The clocking signal keeps the receiver in step with the transmitter. Because the characters occur at regular intervals, no start or stop bits are required.

synchronous data link control (SDLC) An IBM-developed bit-oriented, full- or half-duplex, synchronous data communication protocol for the Systems Network Architecture (SNA). It initiates, controls, checks, and terminates information exchanges on an individual link. It is a subset of the ITU-T high level data link control (HDLC) protocol.

synchronous digital hierarchy (SDH) The ITC-T international recommendation for high-bandwidth optical fiber data networks corresponding to the American National Standards Institute (ANSI) Synchronous Optical Network (SONET) standard. See also *OC-n, SONET,* and *STS-n.*

synchronous DRAM (SDRAM) A form of DRAM which adds a separate clock signal to the control signals. Synchronous Dynamic Random Access Memory (*SDRAM*) chips contain a more complex state machine, allowing them to support burst transfer modes, i.e., that clock out a series of successive bits.

synchronous graphics RAM (SGRAM) A type of *synchronous* Dynamic Random Access Memory (*DRAM*) optimized for use in graphics hardware. Features include burst operation and block write. *SGRAMs* are designed to provide the very high throughput needed for graphics-intensive operations such as 3-D rendering and full-motion video.

synchronous idle character <SYN> A control character used on synchronous transmission systems to maintain transmitter-receiver synchronism when no user or other system control information is to be transmitted.

synchronous network A network in which all communications links are synchronized to a common clock.

synchronous operation Any operation that proceeds under the control of a timing signal (clock). The timing signal may be embedded in a data stream or derived from a common external source.

synchronous optical network See *SONET.*

synchronous orbit A satellite orbit in which the period exactly matches the rotational period of the body being orbited. Nonequatorial orbits appear to move north and south, while noncircular orbits move east and west. Equatorial and circular orbits are also called *stationary orbits.*

For the Earth, satellites such as weather or communications satellites, which are at an altitude of about 35 800 km (above the Earth's surface) and over the equator, are in synchronous orbits (*geostationary orbit*).

synchronous protocol A set of standards devised to allow synchronous communications between terminals. A number of synchronous protocols exist, e.g., a *binary synchronous* (*BISYNC*) protocol, a character-based protocol, and a high-level data link control (HDLC) or *synchronous data link control* (*SDLC*) which are bit-oriented protocols.

synchronous satellite A satellite in a synchronous orbit.

synchronous TDM A time division multiplex (TDM) scheme in which timing is obtained from a clock that controls both the multiplexer and channel source.

synchronous transmission A transmission system in which the sending and receiving devices are operating continuously and at essentially the same frequency. The fixed relationship of sending and receiving equipment is maintained by means of a correction (synchronizing) signal (also called a clocking signal).

synchronous transport level "n" (STS-n) See *STS-n.*

syntax (1) In a language, the relationships among characters or groups of characters independent of their meanings or the manner of their interpretation and use. (2) The rules governing the structure of expressions in a language. (3) The rules specifying how commands (instructions) are constructed in a command line interface or a computer programming language. The rules specify everything typed, including "spelling," "punctuation," "spacing," and "order."

SYS An abbreviation of <u>SYS</u>tem.

SYSGEN A contraction of <u>SYS</u>tem <u>GEN</u>eration. The process of using a master control program (linker) to assemble and link all of the subparts that constitute another operating program.

sysop An acronym for <u>SYS</u>tem <u>OP</u>erator (pronounced "siss-op"). The person responsible for the maintenance, operation, and services of a system, network, bulletin board, or other online service. The *sysop* has complete control over the system. Control ranges from who can log on, the time available to each person, what files are available, what types of messages can be posted, and in general the character of the bulletin board system (BBS).

On mainframe systems the person is more frequently called the *system administrator* or *network administrator.*

system (1) Any collection of interconnected devices or elements assembled to achieve a specified objective or function. **(2)** A generic term for any computer and its peripherals. Some examples include a bulletin board system (BBS) or other online service, the mainframe computer at NASA, or the desktop computer and printer in your office.

system administrator A person who manages either a single computer or a collection of autonomous or networked computers. The administrator is responsible for installing hardware and software, upgrading hardware and software, ensuring hardware is repaired and functional, assigning new accounts, designing and implementing new applications, solving users' machine-related problems, performing miscellaneous maintenance functions such as screening systems for viruses, and doing routine backups. See also *network administrator.*

system attribute In a file system, such as the one used by DOS, a flag that marks a file or directory as usable only by the operating system. See also *attribute.*

system blocking A synonym for *access denial.*

system blocking signal A control message generated within a telecommunications system which indicates to a resource requestor that the resource is temporarily unavailable.

system clock A timing source designated as the primary reference for all timed operations in an electronic device, system, or network. The clock is the pulse train output of a stable oscillator, not a clock that reports time of day. The time of day clock is derived from the *system clock.*

system connect The physical connection to a network or a host computer.

Examples include the BNC (bayonet Neill-Concelman) "T" connector attached to the network interface card (NIC) in an Ethernet network, and the RJ-11 connector used to connect a telephone to the switched telephone network.

system error An error that causes the software operating system (program) to fail in a fatal manner. This failure can be caused by either hardware or software. Recovery usually requires rebooting the computer.

system fault tolerance (SFT) The ability of a system to be fully functional in the presence of failed elements.

Novell's NetWare strategy, for example, supports three levels of *SFT:*

Level 1 includes Hot Fix, read-after-write verify, and duplicate directory entry tables (DETs) to ensure data and directory integrity.

Level 2 includes disk mirroring or duplexing. See *RAID level 1.*

Level 3 uses server mirroring.

system gain The total of all gains and losses between a source and destination. It includes not only the electronic gain in amplifiers, but also the effective gain of an antenna and the free space path loss or the loss in a wire transmission system.

system loss The sum of all transmission losses and gains, including transmitting and receiving antenna gain and path loss.

system noise Noise that is generated within or by a system including recording media.

system operator See *sysop.*

system redundancy The duplication of some or all system components to achieve a desired level of system fault tolerance (SFT). In the event of a device failure, a redundant device is brought into operation, thus allowing the system to continue to perform its functions.

Redundant system elements frequently include alternate transmission paths, alternate power sources, mirror computers, Redundant Array of Inexpensive Disks (RAID) disk systems, and so on.

system reliability An expression of the probability that a system (including all hardware, firmware, and software) will satisfactorily perform the task for which it was designed or intended, for a specified time, and in a specified environment.

system routing code In telephony, a three-digit code consisting of the country code plus two additional digits that uniquely identify an international switching center in World Zone 1. See also *country code* and *world zone number.*

System Services Control Point (SSCP) A type of node in IBM's Systems Network Architecture (SNA) networks that provide the services needed to manage part or all of the entire network.

system side The cabling from a computer or network to the cross connect field of a distribution frame.

system utility Any computer program that aids either the user or administrator with the mechanics of using the computer, e.g., data compression, data encryption, virus checking, system backup, and file defragmentation.

systematic error The inherent bias or offset of a process or one of its components.

Systems Application Architecture (SAA) An IBM approach to standardizing the conventions, interfaces, and protocols used by applications in all IBM operating environments. The goal of *SAA* is to provide a unified, logical architecture for applications running on the family of machines ranging from a PS/2 up to a System/370 and operating systems such as MVS/ESA, VM/ESA, OS/400, and OS/2 EE.

The four main components of *SAA* are:

- *Common User Access* (*CUA*)—defines standard interfaces for applications that are either window based or character based. CUA includes specifications for screen and keyboard layout, and for selection methods using either a keyboard or mouse.
- *Common Program Interface* (*CPI*)—defines Application Program Interfaces (APIs) that are consistent across all systems. The CPI standards relating to languages and databases follow American National Standards Institute (ANSI) specifications.
- *Common Communications Support* (*CCS*)—defines a collection of communications protocols that systems can use to communicate with each other. The most commonly used protocols are LU 6.2 and HLLAPI.
- *Common Applications*—a product-oriented component concerned with developing common frameworks for the same kinds of applications running in different environments. It deals primarily with marketing and appearance; hence, it is sometimes not included as part of *SAA.*

Systems Management Application Entity (SMAE) In the Open System Interconnection (OSI) network management model, the component that implements the network management services and activities at the application level in a node.

Systems Management Application Process (SMAP) In the Open System Interconnection (OSI) network management model, the software that implements the network management capabilities in a single node of any type, e.g., a workstation, a router, a bridge, or a front-end processor (FEP).

Systems Management Application Service Element (SMASE) In the Open System Interconnection (OSI) network management model, the element that does the work for a systems management application entity (SMAE).

Systems Management Function (SMF) In the Open System Interconnection (OSI) network management model, any of several services available for managing particular network domains.

Systems Management Functional Area (SMFA) A term for any one of the five major domains that make up the Open System Interconnection (OSI) network management model:

- Accounting management
- Configuration management
- Fault management
- Performance management
- Security management

Systems Network Architecture See *SNA*.

Systems Services Control Point (SSCP) That program on a host computer in an IBM Systems Network Architecture (SNA) environment responsible for central control, directory services, and operational functions of the network. It may be used either alone or in combination with *SSCPs* of other hosts.

SYU An abbreviation of SYnchronization Signal Unit.

SZ An abbreviation of SeiZure.

T

T **(1)** The SI prefix for *tera* or 10^{12} in the decimal system. Generally, in the binary system *tera* indicates 2^{40}. **(2)** The SI symbol for units of *magnetic flux density* (*B*) measured in *teslas*. **(3)** A symbol for the Tip wire. One of the two wires in a standard twisted pair local loop. (The other wire is the Ring wire.) **(4)** The ending delimiter character in 4B/5B coding. **(5)** An abbreviation of Transparent. Meaning there is no robbed-bit signaling in a D4 extended super frame format. **(6)** A symbol for Temperature. **(7)** A symbol for Time or time duration

T carrier See *DS-n, T-carrier,* and *T-n.*

T interface point The reference point in an Integrated Services Digital Network (ISDN) customer premises terminal equipment set between the *NT2* (network terminator 2) and the *NT1* (network terminator 1). The *T interface* is functionally the same as the S interface; however, the protocols are different.

Its characteristics include:

- A four-wire, 144-kb/s (2B + D) user rate,
- Accommodations for the link access and transport layer function in the ISDN architecture,
- Being located at the customer premises,
- Distance sensitivity to the servicing network terminating equipment, and
- Functions analogous to that of the Channel Service Units (CSUs) and the Data Service Units (DSUs).

Also called the *T reference point.* See also *ISDN.*

T junction See *series T junction.*

T network A circuit with three branches and one common node to which one end of each branch is attached. One branch is designated the common branch, another the input branch and the third the output branch. An input signal is impressed across the input and common branch end points, while the output is taken from the output and common branch end points.

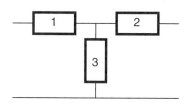

T pulse A $\sin^2$ pulse with a half-amplitude duration (HAD) equal to half of the period of the path cutoff frequency (f_c), that is,

$$T = \frac{1}{2 \cdot f_c}$$

For TV (with an f_c of 4 MHz) $T = 125$ ns and has a frequency spectrum envelope with the following characteristics:

Frequency	Relative Spectrum Envelope Amplitude
0	100 %
4 MHz	50 %
8 MHz and beyond	0

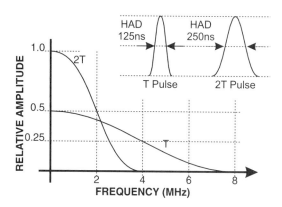

T reference point See *T interface point.*

T step A step waveform with a $\sin^2$ rise or fall and with a transition time (10% to 90%) equal to half of the period of the path cutoff frequency (f_c). For TV, $T = 125$ ns and the amplitude of the spectrum above 8 MHz is essentially zero.

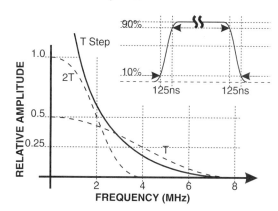

T tap A passive interface used to extract information from a circuit (electrical or optical).

T-carrier The generic designator for any of several digital time division multiplexed telecommunications carrier systems. The line types are of the form T-n, and the corresponding line signal standards are of the form DS-n with data rates at multiples of 1.544 MHz. See *DS-n* and *T-n.*

T-connector A "T"-shaped device used to join three circuit elements: generally two coaxial cables and a hardware element (such as a network interface card). The *T-connector* has two female connection points for the coaxial cable and a male connection to allow attachment to the hardware.

The diagram is of a BNC (bayonet-Neill-Concelman) type T-connector. Also called a "*tee.*"

T-coupler A bidirectional, passive optical coupler having three ports (three fibers). The device allows either:

- The two isolated inputs to be combined into one output; or
- One input to be split into two isolated outputs.

The amount of coupling loss, usually expressed in decibels (dB), between ports is determined by the design and construction of the coupler. Also called a *splitter.*

T-n Level of digital Transmission protocol at the ISO/OSI Reference Model physical layer (layer 1) used to transport voice and digital data between two fixed sites.

Originally, these four-wire digital communications methods (also called *T-carrier* systems) were used by the telephone operating companies as long-distance interconnects between switching offices and were intended to carry digitized voice. Current telco applications also include digital data transmission. End users are also using the circuits as the basis of private networks. These *T-carrier* systems are used primarily in the United States, Canada, and Japan. If an "F" precedes the "T," a fiber optic cable system is indicated at the same rates. See also *DS-n, E-n, OC-n, SDH, SONET, STM-n* and *STS-n.*

T-CARRIER DESIGNATIONS

Standard	Line Rate (bps)	Description
T-1	1 544 000	24 64 kbps channels plus one framing bit multiplexed together to form a single 193 bit frame.
T-1C	3 152 000	48 64 kbps channels plus one framing bit multiplexed together.
T-2	6 312 000	4 T-1 (96 channels) data streams plus synchronization and overhead bits multiplexed into a single serial stream.
T-3	44 736 000	7 T-2 (672 channels) data streams plus synchronization and overhead bits multiplexed into a single serial stream. Frequently called 45 Mbps.
T-4	274 176 000	6 T-3 (4032 channels) plus overhead.
T-5	400 352 000	5760 channels plus overhead

T.n recommendations An International Telecommunication Union-Telecommunication group (ITU-T) series of specifications concerning "Terminal characteristics and higher layer protocols for telematic services, document architecture."

For example, the T.50 recommendation is the joint ITU-T/ISO International Alphabet No. 5 specification. The ITU-T T.50/ISO 646 standard defines 128 7-bit symbols. These symbols include control characters, graphic characters (alphabetic letters and symbols), and numbers. (See ASCII for character definitions.) See Appendix E for a detailed list.

T1 An American National Standards Institute (ANSI) standards committee responsible for transmission issues in the United States, e.g., digital telephony such as Integrated Services Digital Network (ISDN). It corresponds to the European Telecommunications Standards Institute (ETSI) and the Japanese Telecommunications Technology Committee. The T1 committee is in no way responsible for or involved with the T-1 (1.5 Mbit/s) circuit standards. Subcommittees of T1 include:

- *T1D1*—the technical subcommittee for BRI U interfaces.
- *T1E1*—the technical subcommittee for subscriber network interfaces, for example, T1E1.4 is responsible for High speed Digital Subscriber Line (HDSL).
- *T1M1*—the technical subcommittee for network management and operations support.

- *T1X1*—the technical subcommittee for SS7 and Synchronous Optical Network (SONET).

TA (1) An abbreviation of Terminal Adapter. **(2)** An abbreviation of Technical Advisory. A Bellcore draft standard, prior to becoming a Technical Reference (TR). **(3)** An abbreviation of Test Access. **(4)** An abbreviation of Telecommunications Administration. An entity, usually a company or government agency, that provides public telecommunications services.

TAAS An abbreviation of Trunk Answer from Any Station.

TAB (1) An acronym for Telecommunications Advisory Board. **(2)** An acronym for Tape Automated Bonding. A printed circuit manufacturing technique.

table driven A computer process in which an acquired variable is matched with an array of predefined values. The information in the array associated with that value determines the computer's operation. The technique is frequently used in access security and network routing. Also called table *look-up.*

table look-up access An information retrieval system in which the input information and the associated output information are stored as a pair. In operation, a particular input causes the associated output information to be presented to the output register or port.

TABS An acronym for Telemetry Asynchronous Block Serial. A master/slave packet protocol used to control network elements and to obtain extended superframe network statistics.

TAC (1) An acronym for Technical Assistance Center. **(2)** An acronym for ARPAnet/MILnet Terminal Access Controller. Supersedes the term *terminal interface processor* (TIP). **(3)** An acronym for TELENET Access Controller.

TACACS An abbreviation of ARPAnet/MILnet Terminal Access Controller Access Control System.

TAD An acronym for Telephone Answering Device otherwise known as an answering machine.

TADIL An acronym for TActical Data Information Link. It is a standardized communications link, approved by the Joint Staff, that is suitable for digital information transmission, and has standardized message formats and transmission characteristics. Two systems are:

- *TADIL-A*—a system in which a control station polls units that respond by broadcasting the message on the net.
- *TADIL-B*—a point-to-point duplex system.

TADSS An abbreviation of Tactical Automatic Digital Switching System.

tag The embedded commands of an SGML (Standard Generalized Markup Language) document. The *tags* are embedded within a plain ASCII text string in order to delineate certain sections within the document, such as titles, headers, abstracts or the body. In order to properly read an SGML document, a program called a *parser* is required. The *parser* reads the *tags* and displays or prints the text with appropriate formatting. See also *flag* and *label.*

TAG An acronym for Technical Advisory Group. An IEEE committee whose task is to provide general recommendations and technical guidance for other committees. Examples include the 802.7 (which is concerned with issues relating to broadband networks) and 802.8 (which is concerned with the use of fiber optic cabling in networks) committees.

tag image file format (TIFF) A file format used to store an image using the particular data structure of the file.

TAI An abbreviation of International Atomic Time.

tail circuit A communications line, a feeder circuit, or an access line from the end of a major transmission link (such as a microwave link, satellite link, or local area network) to an end-user location. Usually used in a synchronous environment when the data communications equipment (DCE) to attachment point is long enough to require line drivers.

tail end hop off (TEHO) A term used to describe the call routing when a call is originated on a private network, travels to a distant node, leaves the private network, and uses the public network to complete the path to the destination.

tailing The excessive prolongation of the decay of a signal. Also called *hangover.*

takeoff angle A synonym for *departure angle.*

talk battery The dc voltage applied by a switching system to the user's telephone equipment. The voltage may be as low as 5 volts in some private automatic brand exchange (PABX) systems or as high as −60 volts in a German central office. Typical voltages, however, are −24 V in a PABX and −48 V in a central office.

talk hierarchy One of the seven *world newsgroups* in USENet. The categories of these newsgroups include anything about just about anything. This newsgroup invites controversy and debate. See also *newsgroup hierarchy.*

talk off An effect of in-band tone signaling where there is sufficient energy in a speaker's voice at the signaling frequency to cause the circuit to disconnect.

talk path The collection of circuits and devices that carry information from the sender to the receiver.

talker echo Echo that reaches the talker's ear. See also *NEXT* and *FEXT.*

tandem The series connection or sequencing of networks, circuits, or links, that is, the output of one entity is directly connected to the input of another like entity.

Examples include:

- The concatenation of microwave links constitutes a tandem connection.
- In the United States, those telephone-switching offices that connect class 5 offices.

See also *tandem exchange.*

tandem call A call routed through two or more switching systems.

tandem center See *tandem exchange.*

tandem exchange In telephony, an intermediate switching facility used primarily to switch traffic between end offices, other tandem exchanges, and with toll trunks within a given exchange. The *tandem center* completes all calls between end offices (connects trunks) but is not connected directly to subscriber loops. Also called a *tandem center.* See also *class n office* and *tandem office.*

tandem office In telephony, a central office (CO) that not only serves local subscriber loops, but also is used as an intermediate switching point for traffic between COs. See also *class n office* and *tandem exchange.*

tandem operation An operation that permits two or more trunks (direct or dial tie trunks) to be connected at a tandem center to form a through connection.

tandem switch In telephony, a switching system used to connect one trunk to another.

tandem tie trunk network (TTTN) An arrangement that permits sequential connection of tie trunks between private automatic branch exchange (PABX) and Centrex® locations by using tandem operation.

tandem trunks In telephony, circuits between end offices or those connected to a tandem exchange.

TANSTAAFL An abbreviation of There Ain't No Such Thing As A Free Lunch.

tap (1) A *device* for extracting a portion of the signal traveling down a transmission path. The *tap* may be electrical or optical in the case of fiber optic transmission. (2) A *point* on a device for extracting a portion of the total available energy, for example, a transformer may have the secondary *tapped* at several points in order to efficiently produce several voltages. (3) To *extract* energy from a circuit. (4) To *monitor,* with or without authorization, the information that is being transmitted via a communications circuit.

TAP An acronym from Telocator Alphanumeric paging Protocol.

tap key A pushbutton switch on a telephone instrument that generates a timed hook flash signal to the private automatic branch exchange (PABX). Also called *tap button* or *flash button.*

taper pad Any *pad* designed to operate between two circuits of *unequal* impedance. See also *pad.*

tapered fiber An optical fiber in which the cross-sectional diameter or area increases (or decreases) monotonically with distance.

TAPI An acronym for Telephony Application Programming Interface. *TAPI* is a series of application programming interfaces (APIs) that allow Windows-based programs (and supporting .DLLs) to interface a PC to common telephone equipment. The *TAPI* specification defines two function levels: simple *TAPI* and full *TAPI.*

Simple TAPI (also called Assisted Telephony) basically adds dialing capability to existing applications. *Full TAPI,* a broader specification, divides support into three levels: basic services, supplemental services, and extended services. Basic services, like *simple TAPI,* provide enough services to initiate a call. Supplemental services add functions such as the ability to:

- Select the proper communications protocol after a communication link is established.
- Monitor calls.
- Detect the type of data being sent.
- Generate and detect dual tone multifrequency (DTMF) signaling.
- Initiate call hold, conference, and transfer.

Extended services are features such as special modem commands and configuration change switches.

TAR An acronym for Tape ARchive. A program that groups several files together as a single archive file. It reads and writes archive files that conform to the Archive/Interchange File Format specified in IEEE Std. 1003.1-1988. Files inside a *tar* archive are called members. By convention (and default), the extension of a *tar* archive is ".tar".

The UNIX software was developed by Mark H. Colburn and is sponsored by the USENIX Associaion. *GNU tar* was originally written by John Gilmore and has subsequently been modified by many people. Most of the *tar* operations and options can be written in any of three forms: long (mnemonic) form, short form, and old style.

- *Long (mnemonic) form:* Each option has at least one long name starting with two hyphens, e.g., "--list". The long names are mnemonics of the desired operation and therefore are more easily remembered than their corresponding short or old names. The form of the long form is:

tar--command. . . [name]. . .

Examples include:

tar--create--file=/foo/bar.txt

tar--list

- *Short form:* Short options start with a single hyphen and are followed by a single character, e.g., "-t" (equivalent to "--list"). Multiple commands may be listed with or without spaces between the letter options. When no spaces are used, a single hyphen is used and only the last option may have an argument. The format of the short form is:

tar -letter(s). . . [name]. . .

Examples include:

tar -c -f /foo/bar.txt

tar -cf /foo/bar.txt

- *Old form:* The old form is included for compatibility with the older UNIX letter option style. Like the short form, the old form uses single letters. In the old form, however, the letters must be written together as a single set (that is, without spaces separating them), and no hyphens can precede them.

tar letter(s) [arg]. . . [option]. . . [name]. . .

Examples include:

tar cf/foo/bar.txt

A list of the operations and options is given in the following table. The short form and long form are both shown in the table.

-A | --catenate | --concatenate
> Appends other *tar archives* to the end of the archive.

-b n | --blocking-factor = n
> Sets the block length for tape-records to $512 \times n$. The default value of n is 1; and the maximum value is 20.

-B | --read-full-records
> Instructs tar to reblock its input.

-c | --create
> Creates a new archive; writing begins at the beginning of the archive, instead of after the last member.

-C dir |, --directory = dir
> Instructs tar to change its current directory to dir before performing any operations.

-d | --compare | --diff
> Causes tar to compare archive members with their counterparts in the file system and to report differences in file size, mode, owner, modification date and content.

-f name | --file = name
> Causes tar to use the file name "*name*" as the name of the archive instead of the default (usually a tape drive).
>
> If no name is specified *tar* writes to (or reads from) the standard output (or input), whichever is appropriate for the options given.

-F script | --info-script = script | --new-volume-script = script
> Causes tar to run the script file "*script*" and the end of each tape when it is performing multi-tape backups.

-G | --incremental
> Informs tar that it is working with an old GNU-format incremental backup archive. (It is intended primarily for backwards compatibility only.)

-h | --dereference
> When creating a tar archive, tar will archive the file that a symbolic link points to, rather than archiving the link.

-i | --ignore-zeros
> Causes tar to ignore zeroed blocks in the archive, which normally signals an EOF (end of file).

-k | --keep-old-files
> Instructs tar to not overwrite existing files when extracting members from an archive.

-K name | --starting-file = name
> Causes tar to skip members of an archive that do not match "*name*" when extracting.

-l | --one-file-system
> Prevents tar from recursing into directories that are on different file systems from the current directory when creating an archive. *tar* reports if it cannot resolve all of the links to the files being archived. (If -l is not specified, no error messages are written to the standard output device.)
>
> This modifier is only valid with the -c, -r and -u options.

-L n | --tape-length = n
> Specifies to tar that the tape length is to be n × 1024 bytes.

-m | --touch
> Causes tar to set the modification time to the time of extraction rather than the modification time stored in the archive.
>
> This option is invalid with the -t option.

-M | --multi-volume
> Informs tar that it should create (or operate) on a multi-volume archive.

-N date | --newer | --after-date = date
> Instructs tar to add files to an archive only if the file has changed any status since the specified date "*date.*"

-o | --portability | --old-archive
> Causes tar to create archives compatible with the UNIX V7 tar. Extracted files to take on the user and group identifier of the user running the program rather than those on the archive. (This option is only valid with the -x option.)

-O | --to-stdout
> Causes tar to extract files to *stdout* rather than to the file system.

-p | --preserve-permissions | --same-permissions
> Causes tar to use the permissions directly from the archive. When tar is extracting an archive, it normally subtracts the user's umask from the permissions specified in the archive and uses that number as the permissions to create the destination file.

-P | --absolute-names
> Causes tar to leave the initial "/" from file names, if present, when creating an archive. Normally it is stripped before becoming an archive member.

-r | --append
> Appends *files* to the end of the archive.

-R | --block-number
> Causes tar to print error messages with the block number for read errors in the archive file.

-s | --same-order |, --preserve-order
> Informs tar that the list of file arguments has already been sorted to match the order of files in the archive. (An optimization for tar when running on machines with small amounts of memory.)

-S | --sparse
> Invokes a GNU extension when adding files to an archive that handles sparse files efficiently.

-t | --list
> Causes tar to lists the names of all members in the archive.

-T list | --files-from = list
> Causes tar to use the contents of the file "list" as a list of archive members or files to operate on, in addition to those specified on the command line.

-u | --update
> Causes named files to be added to the archive if they are not already there, or if they have been modified since last written into the archive.

-U | --unlink-first
> Instructs tar to remove the corresponding file from the file system before extracting it from the archive.

-v | --verbose

Causes *tar* to print the name of each file it processes preceded by the option letter. Normally *tar* delivers no information. When combined with the -t option, -v gives more information about the archive entries than just the name.

-V name | --label = name

Instructs tar to write "*name*" as a name record in the archive when it is creating an archive.

When it is extracting or listing archives, tar will only operate on archives that have a label matching the pattern specified in "*name*."

-w | --interactive | --confirmation

Causes *tar* to print the requested action when it encounters a potentially destructive action (like overwriting an existing file), followed by the name of the file, and then to wait for the user's confirmation. If the user response is a word beginning with y is given, the action is performed. Any other response means "no."

This modifier is invalid with the -t option.

-W | --verify

Causes tar to verify that the archive was correctly written when creating an archive.

-x | --extract | --get

Extracts named members from the archive. If a named member matches a directory whose contents had been written onto the archive, that directory is recursively extracted. If a named member does not exist on the system, the file is created with the same mode as the one in the archive, except that the set-user-id and get-group-id modes are not set unless the user has appropriate privileges.

If the files exist, their modes are not changed except as described above. The owner, group, and modification time is restored if possible. If no member argument is given, the entire content of the archive is extracted. If several members with the same name are in the archive, the last physical one will overwrite all earlier files.

-X list | --exclude-from = file

Similar to "--exclude", except tar will use the list of patterns in the file *list*.

-z | --gzip | --gnuzip | --ungzip | --gunzip

Causes tar to use the compress program "gzip" when reading or writing the archive.

-Z | --compress | --uncompress

Causes *tar* to use the compress program when reading or writing the archive. (This allows direct action on archives while saving space.)

--backup = backup-type

Rather than deleting files from the file system, tar will back them up using simple or numbered backups, depending upon backup-type.

--checkpoint

Directs tar to print periodic checkpoint messages as it reads through the archive. Its intended to give a visual indication that tar is still running, but with less output than "--verbose" yields.

--delete

Deletes members from the archive.

Does not work on a magnetic tape!

--exclude = pat

Instructs tar to skip files that match *pat*.

--force-local

Causes tar to interpret the filename given to "--file" as a local file, even if it looks like a remote tape drive name.

--group = set

Causes files added to the tar archive to have a group id of "*set*", rather than the group from the source file. "*set*" is first decoded as a group symbolic name, but if this interpretation fails, it must be a decimal numeric group ID.

--help

Causes tar to print a short message summarizing the operations and options available, then to exit.

--ignore-failed-read

Instructs tar to exit successfully if it encounters an unreadable file.

--mode = permissions

Instructs tar to use permissions for the archive members, rather than the permissions from the files when appending. Permissions may be specified as an octal number or by generic symbols.

--newer-mtime

Modifies the "--newer" option so that tar will only add files whose contents have changed (as opposed to just "--newer", which will also back up files for which any status information has changed).

--no-recursion

Instructs tar not to recurse into directories unless a directory is explicitly named as an argument.

--null

Instructs tar to expect filenames terminated with <NUL> (so tar can correctly work with file names that contain newlines).

Used when tar is using the "--files-from" option.

--numeric-owner

Instructs tar to use numeric user and group IDs when creating a tar file, rather than names.

--owner = user

Instructs tar to use "*user*" as the owner of members when creating archives instead of the user associated with the source file.

"*user*" is first decoded as a user symbolic name, but if this interpretation fails, it has to be a decimal numeric user ID.

--posix

Instructs tar to create a POSIX compliant tar archive.

--preserve

Equivalent to specifying both the "--preserve-permissions" and "--same-order" options.

--record-size = n

Instructs tar to use "*n*" bytes per record when accessing the archive.

--recursive-unlink

Similar to the "--unlink-first" option, removes existing directory hierarchies before extracting directories of the same name from the archive.

--remove-files

Instructs tar to remove the source file from the file system after appending it to an archive.

--rsh-command = cmd

Instructs tar to use "cmd" to communicate with the remote device.

--same-owner

Instructs tar to preserve the owner specified in the tar archive.

--show-omitted-dirs

Instructs tar to indicate directories skipped when operating on an archive.

--suffix = suffix

Causes tar to use the suffix "*suffix*" rather than the default "~".

--totals

Displays the total number of bytes written after creating an archive.

--use-compress-program = prog

Instructs tar to access the archive through "*prog*", where "*prog*" is a compression program.

--version

Causes tar to print an informational message about its version, a copyright message, and some credits; and then to exit.

Details of the tar commands for a specific system and version can be obtained issuing a *tar--help* or *man tar* comand. See also *compression program* (2).

TARGA An acronym for AT&T's Truvision Advanced Raster Graphics Adapter.

target The device to which an action is being directed. For example, a *target computer* may be the computer that receives data from a communication device, a hardware add in, or a software package; and a *target diskette* is the diskette that is the destination of a copy command.

tariff In telecommunications, the published requirements for the types of services, the terms and conditions a common carrier may offer, and the vehicle by which the governmental regulatory agency details and approves rate schedules for them.

TARM An abbreviation of Telephone Answering and Recording Machines.

TARR An abbreviation of Test Action Request Receiver.

TAS An acronym for Telephone Answering Service (or System).

TASI An acronym for Time Assigned Speech Interpolation. A form of time division multiplex (TDM) where a pool of trunk circuits is assigned based on a *required by signal* basis rather than a *required by call* basis. That is, if an established connection between two stations has no traffic, a trunk circuit in the path can be temporarily reassigned to a different link with traffic. When the original path becomes active again, an available trunk from the pool is assigned for the duration of the traffic.

On average, the amount of free time on one pair of a connected four-wire circuit is greater than 60%. This permits a utilization increase of about 2.4 on a 96-channel trunk group.

TAT An acronym for Trans-Atlantic Telecommunications cable. Formerly an acronym for TransAtlantic Telephone cable.

TATC An abbreviation of TransAtlantic Telephone Cable. Also abbreviated TAT.

TAU (1) An abbreviation of Test Access Unit. (2) An abbreviation of Trunk Access Unit.

TAXI An acronym for Transparent Asynchronous Transmitter-Receiver (X) Interface.

Tb An abbreviation of Terabit, i.e., 10^{12} bits.

TB (1) An abbreviation of TeraByte, i.e., 10^{12} bytes. (2) An abbreviation of Transparent Bridge.

TBA An abbreviation of To Be Announced.

TBB An abbreviation of Transnational Broadband Backbone.

TBC (1) An abbreviation of Token Bus Controller. (2) An abbreviation of Trunk Block Connector.

TBD (1) An abbreviation of To Be Determined. (2) An abbreviation of To Be Done.

TBR An abbreviation of Timed BReak.

TC (1) An abbreviation of Transmission Convergence sublayer. (2) An abbreviation of Transport Connection. (3) An abbreviation of Transmission Control. (4) An abbreviation of Trunk Conditioning. (5) An abbreviation of Terminating Channel. (6) An abbreviation of Technical Committee. (7) An abbreviation of Terminal Controller. (8) An abbreviation of Test Conductor. (9) An abbreviation of Temperature Coefficient. (10) An abbreviation of Toll Center. (11) An abbreviation of Toll Completing. (12) An abbreviation of Trunk Control.

TCA An abbreviation of Tele-Communications Association.

TCAM An acronym for Telecommunications Access Method. The telecommunications access method software for IBM 3270 control.

TCAP An acronym for Transaction Capabilities Application Part. (1) The lower part of Signaling System 7 (SS7) layer 7. (2) An Integrated Service Digital Network (ISDN) application protocol.

TCAS An acronym for T-Carrier Administration System.

TCB (1) An abbreviation of Transmission Control Block. (2) An abbreviation of Task Control Block. (3) An abbreviation of Transfer Control Block. (4) An abbreviation of Trusted Computing Base.

TCC (1) An abbreviation of TeleCommunications Center. (2) An abbreviation of Technical Computing Center. (3) An abbreviation of Technical Control Center. (4) An abbreviation of Temperature Control Circuit. (5) An abbreviation of Through Connected Circuit. (7) An abbreviation of Transmission Control Character. (8) An abbreviation of Transfer Channel Control.

TCCC An abbreviation of Technical Committee for Computer Communications.

TCD An abbreviation of Telemetry and Command Data.

TCE An abbreviation of Telephone Company Engineered. As opposed to customer engineered.

TCF (1) An abbreviation of Training Check Frame. The last step in the series of handshake signals of a facsimile transmission training phase. *TCF* allows the receiver to adjust its equalizers to compensate for telephone line conditions. (2) An abbreviation of Terminal (or Technical) Communications Facility. (3) An abbreviation of Terminal (or Technical) Control Facility.

TCG An abbreviation of Test Call Generator.

TCL (1) An abbreviation of Terminal Control Language. (2) An abbreviation of Toll Circuit Layout. (3) An abbreviation of Troposcatter Communications Link.

TCLK An acronym for Transmit CLocK (EIA-232 signal DA/DB, EIA-449 signal TT/ST, and ITU-T signal 113/114). The transmit timing signal may be generated by either the data terminal equipment (DTE) (EIA-232 signal DA, etc.) or the data communications equipment (DCE) (EIA-232 signal DB, etc.). When generated by the DTE, the clock is an independent free-running source. When generated by the DCE, it may be derived from the received data stream or from an independent source in the DCE.

TCM (1) An abbreviation of Trellis Coded Modulation. (2) An abbreviation of Time Compression Multiplexing. (3) An abbreviation of TeleCommunications Monitor. (4) An abbreviation of Telemetry Code Modulation. (5) An abbreviation of Telephone Channel Monitor. (6) An abbreviation of Traveling Class Mark.

TCN (1) An abbreviation of Throughput Class Negotiation. (2) An abbreviation of TeleCommunications Networks.

TCNS An abbreviation of Thomas-Conrad Network System.

TCO An abbreviation of Total Cost of Ownership. The cost of an entity including required personnel, maintenance, upgrades, and consumable materials.

TCOS An abbreviation of Trunk Class Of Service.

TCP (1) An abbreviation of Transmission Control Protocol (also listed as Transport Control Protocol). An ARPAnet developed transport layer protocol (that corresponds to the ISO/OSI Reference Model layers 4 and 5). It is a standard full-duplex, connection-oriented, end-to-end protocol that provides reliable, sequenced, and unduplicated packet delivery to a remote or local user over a packet switched network. See also *Internet Protocol* (*IP*) and *TCP/IP*. (2) An abbreviation of Task Control Program. (3) An abbreviation of Terminal Control Program.

TCP/IP An abbreviation of Transmission Control Protocol and Internet Protocol. A software protocol suite developed at the University of California by the Advanced Research Projects Agency (ARPA) (for the U.S. Department of Defense) for communications between heterogeneous computers. The protocol suite includes:

ARP (*Address Resolution Protocol*)—protocol for determining a Data Link Control (DLC) address from an Internet Protocol (IP) address.

datagram—refers to a packet containing a destination address and data.

DNS (Domain Name Service)—protocol for determining and maintaining network resource information distributed among several servers.

FTP (File Transfer Protocol)—protocol governing file transfers.

ICMP—control message protocol.

IP (Internet Protocol)—protocol governing packet transmission.

RARP (Reverse Address Resolution Protocol)—protocol for determining an IP address from a DLC address.

SMTP (Simple Mail Transfer Protocol)—protocol governing electronic mail transfers.

SNMP (Simple Network Management Protocol)—protocol governing network monitoring.

TCP (Transmission Control Protocol)—defines the transport level protocol. It provides connection and stream-oriented, transport layer services. TCP uses the IP to deliver its packets.

Telnet—protocol governing terminal connections with character-oriented data. Provides terminal-emulation capabilities and allows users to log in to a remote network from their computers and operate as if they were hard wired to the host.

UDP— *(User Datagram Protocol)*—provides connectionless transport layer service. UDP also uses the IP to deliver its packets.

COMPARISON OF TCP/IP & OSI/ISO PROTOCOL LAYERS

ISO Reference Model		Equivalent TCP/IP Model	
Layer	Definition	Definition	Layer
7	Application	Applications (SMTP, FTP, TELNET)	5
6	Presentation		
5	Session		
4	Transport	TCP/UDP	4
3	Network	IP	3
2	Data Link MAC	Network Interface MAC	2
1	Physical	Physical	1

See also *Internet Protocol (IP)*.

TCR An abbreviation of <u>T</u>ransaction <u>C</u>onfirmation <u>R</u>eport. A report from a facsimile system listing the "faxes" sent and received including the date, time, number of pages, transmission speed, remote station identifier, total pages, and transmission result.

TCS (1) An abbreviation of <u>T</u>roposcatter <u>C</u>ommunications <u>S</u>ystem. **(2)** An abbreviation of <u>T</u>rusted <u>C</u>omputer <u>S</u>ystem.

TCT An abbreviation of <u>T</u>oll <u>C</u>ollecting <u>T</u>runk.

TCU (1) An abbreviation of <u>T</u>ransmission <u>C</u>ontrol <u>U</u>nit. **(2)** An abbreviation of <u>T</u>runk <u>C</u>oupling <u>U</u>nit. **(3)** An abbreviation of <u>T</u>iming <u>C</u>ontrol <u>U</u>nit. **(4)** An abbreviation of <u>T</u>erminal <u>C</u>luster <u>U</u>nit.

TD (1) An abbreviation of <u>T</u>ransmit <u>D</u>ata. **(2)** An abbreviation of <u>T</u>erminal <u>D</u>igit. **(3)** An abbreviation of <u>T</u>est <u>D</u>istributer.

TDA An abbreviation of <u>T</u>racking and <u>D</u>ata <u>A</u>cquisition. **(2)** An abbreviation of <u>T</u>oll <u>D</u>ial <u>A</u>ssistance. **(3)** An abbreviation of <u>T</u>unnel <u>D</u>iode <u>A</u>mplifier.

TDCS An abbreviation of <u>T</u>ime <u>D</u>ivision <u>C</u>ircuit <u>S</u>witching.

TDD (1) An abbreviation of <u>T</u>elecommunications <u>D</u>evice for the <u>D</u>eaf. A modem (and software) used to communicate over the switched network with other like devices. The data rate of these devices is 45.5 bps in the United States and 50 bps internationally. **(2)** An abbreviation of <u>T</u>ime <u>D</u>ivision <u>D</u>uplex.

TDDI An abbreviation of <u>T</u>wisted-pair <u>D</u>istributed <u>D</u>ata <u>I</u>nterface.

TDED An abbreviation of <u>T</u>rade <u>D</u>ata <u>E</u>lements <u>D</u>irectory.

TDF An abbreviation of <u>T</u>runk <u>D</u>istribution <u>F</u>rame.

TDG An abbreviation of <u>T</u>est <u>D</u>ata <u>G</u>enerator.

TDI (1) An abbreviation of <u>T</u>ransit <u>D</u>elay <u>I</u>ndication. **(2)** An abbreviation of <u>T</u>ransport <u>D</u>evice (or <u>D</u>river) <u>I</u>nterface. **(3)** An abbreviation of <u>T</u>elecommunications <u>D</u>ata <u>I</u>nterface.

TDID An abbreviation of <u>T</u>rade <u>D</u>ata <u>I</u>nterchange <u>D</u>irectory.

TDM (1) An abbreviation of <u>T</u>ime <u>D</u>ivision <u>M</u>ultiplex. **(2)** An abbreviation of <u>T</u>unnel <u>D</u>iode <u>M</u>ixer.

TDMA An abbreviation of <u>T</u>ime <u>D</u>ivision <u>M</u>ultiple <u>A</u>ccess.

TDMS (1) An abbreviation of <u>T</u>ime <u>D</u>ivision <u>M</u>ultiplex <u>S</u>ystem. **(2)** An abbreviation of <u>T</u>ransmission <u>D</u>istortion <u>M</u>easuring <u>S</u>et.

TDP An abbreviation of <u>T</u>elocator <u>D</u>ata <u>P</u>rotocol. An 8-bit protocol for sending binary files and messages to remote pagers. See also *Telocator Alphanumeric Protocol (TAP)*.

TDPSK An abbreviation of <u>T</u>ime <u>D</u>ifferential <u>P</u>hase <u>S</u>hift <u>K</u>eying.

TDR (1) An abbreviation of <u>T</u>ime <u>D</u>omain <u>R</u>eflectometry. **(2)** An abbreviation of <u>T</u>emporarily <u>D</u>isconnected at subscriber's <u>R</u>equest.

TDS An abbreviation of <u>T</u>ransit <u>D</u>elay <u>S</u>election.

TDSAI An abbreviation of <u>T</u>ransit <u>D</u>elay <u>S</u>ection <u>A</u>nd <u>I</u>ndication.

TE (1) An abbreviation of <u>T</u>erminal <u>E</u>quipment. **(2)** An abbreviation of <u>T</u>ransverse <u>E</u>lectric mode. See *transverse mode*. **(3)** An abbreviation of <u>T</u>ext <u>E</u>ditor.

TE1 An abbreviation of <u>T</u>erminal <u>E</u>quipment type <u>1</u>. A data terminal equipment (DTE) whose interface is Integrated Services Digital Network (ISDN) compliant. That is, the DTE that can be connected directly to the digital network.

TE2 An abbreviation of <u>T</u>erminal <u>E</u>quipment type <u>2</u>. A data terminal equipment (DTE) whose interface is *not* ISDN compliant. That is, the DTE requires an adapter to interface to the digital network. The adapter is called a TA (terminal adapter).

TEC An acronym for <u>T</u>elephone (or <u>T</u>echnical) <u>E</u>ngineering <u>C</u>enter.

tech (1) An abbreviation of <u>TECH</u>nical. **(2)** An abbreviation of <u>TECH</u>nology.

technical and office protocol (TOP) A version of the ISO/OSI Reference Model, developed by Boeing, for use in the office. It provides standards for the representation and exchange of messages, documents, and other files in office settings. It has been merged with Manufacturing Automation Protocol (MAP), and the two functional profiles share a common integration strategy and a single (Map/Top) user group.

technical control center (TCC) A testing facility for telecommunications circuits. It allows telecommunications systems control personnel to exercise operational control of communications paths and facilities, make quality analyses of communications and communications channels, monitor operations and maintenance functions, recognize and correct deteriorating conditions, restore disrupted communications, provide requested on-call circuits, and take or direct such actions as may be required and practical to provide effective telecommunications services.

Features and capabilities include:

- Equipment necessary for ensuring fast, reliable, and secure exchange of information.
- Typically includes distribution frames and associated panels, jacks, and switches and monitoring, test, conditioning, and order-wire equipment, that is,
  - Provision for network access.
  - Provision for automated testing to aid in the diagnostics of the switched network facilities and the interconnecting transmission lines.

Also called a *technical control facility* (*TCF*).

technical control hubbing repeater A synonym for *data conferencing repeater*.

technical load The operational load required for full capability of a communications facility. Such loads include the communications equipment and necessary lighting, air conditioning, and/or ventilation.

TED An acronym for Trunk Encryption Device.

TEDIS An acronym for Trade Electronic Data Interchange System.

tee See *T-connector*.

tee coupler A passive coupler with three ports, generally one input and two output ports. See also *T-connector*.

TEHO An acronym for Tail End Hop Off.

TEI (1) An abbreviation of Terminal Equipment Identifier. (2) An abbreviation of Terminal Endpoint Identifier. A subfield in the second octet of the LAP-D address field in the Integrated Services Digital Network (ISDN).

TEK An acronym for Traffic Encryption Key.

TEL An abbreviation of telecommunications, telephone, telegraph, telegram, telemetry, or teletypewriter.

telco (1) An acronym for TELephone COmpany. (2) An acronym for TELephone Central Office.

telecommunication (1) The literal meaning is "communication at a distance"; however, it is generally used as a term to describe the communication of information by electronic means, i.e., *electronic communications*. Examples include the transfer of voice, digital data (via modem), or pictures (via facsimile) over the telephone network, radio, television, and so on. (2) Any transmission, emission, or reception of symbols, signals, writing, images, sounds, or other intelligence of any nature by wire, radio, optical, or other electromagnetic systems.

telecommunication loop The communication channel between a communication station and a switching entity. See also *subscriber loop*.

telecommunications The art and science of communicating information of a user's choosing (including voice, data, image, graphics, and video) at a distance by means of electrical (through wire), optical (through optical fiber), or electromagnetic radiation (wireless) methods without change in the form or content of the information, between or among points specified by the user. Examples include telephone, radio, television, facsimile, and telegraph. Also called *telephony*.

telecommunications carrier Any provider of telecommunications services.

telecommunications customer The person for whom the telecommunications service is provided. Also called a *subscriber*.

Telecommunications Device for the Deaf (TDD) A machine that uses typed input and output, usually with a visual text display, to enable individuals with hearing or speech impairments to communicate over a telecommunications network.

telecommunications equipment Any equipment (other than customer premises equipment) used by a telecommunications carrier to provide telecommunications services. *Telecommunications equipment* includes both hardware and the software (including upgrades) integral to the hardware. Examples include transmitters, receivers, telephones, cables, antennas, teletypewriters, facsimile equipment, data equipment, and network switches.

telecommunications exchange A means of providing *telecommunications service* to subscribers within some geographic area.

telecommunications facilities The total set of *telecommunications equipment* used for providing telecommunications services. See also *telecommunications equipment*.

telecommunications management network (TMN) An auxiliary network that interfaces with a telecommunications network at critical points in order to obtain information from, and to control the operation of, the telecommunications network. The auxiliary *TMN* need not be totally separate from the primary network. It may, in fact, use parts of the managed telecommunications network to provide for the *TMN* communications.

telecommunications service Any offering of telecommunications for a fee either directly to the public or to such classes of users as to be effectively available directly to the public, regardless of the facilities used. The responsibilities are divided as:

- The service provider has the responsibility for the acceptance, transmission, and delivery of the message.
- The service user is responsible for the information content of the message.

telecommunications switching A means of selectively establishing and disconnecting communication paths between calling and called stations.

telecommunications system An assemblage of telecommunications stations, station service lines, a routing and switching means (switching offices and trunks for interconnecting switching offices), and all the accessories required to provide a telecommunication service.

telecommuting The practice of working in one location (generally, at home) and communicating with an office or computing facility in a different location through a local computer and a communication device such as a modem.

In a general sense, telecommuting is a substitute for transportation.

telecon A contraction of TELEphone CONference or TELEphone CONversation.

teleconference A contraction of TELEphone CONFERENCE. A type of communication utilizing *telecommunications equipment* (such as the public switched telephone network or a satellite link) and subscriber equipment (such as, computers, telephones, video cameras, etc.) which allows live information exchange among multiple participants from separate geographic locations.

A *teleconference* allows a collaboration in real time. It enables (with proper software) audio, video, and scratchpad communication. (Scratchpad communications is the ability to type messages or draw sketches at any terminal in the teleconference and have all other participants see, edit, or add to the display.) Often further abbreviated *telecon*.

telecopier A synonym for a facsimile machine.

tele-experimentation The use of a long-distance communications link to provide an experimenter access to a geographically distant research tool or facility.

Many research tools today are computer-controlled devices that provide result data only in machine readable form. The machine readable data are manipulated by a computer to generate a graphical or pictorial display of the experimental results. Because both the input and output of these experiments are via a computer, the experimenter need not collocate with the experiment. Hence, the experimenter may use communications facilities such as a dial-up connection or the Internet to gain access when time is available or the equipment. Examples of such facilities include earthquake monitoring stations, astronomical telescopes, high-energy particle accelerators, and super computing facilities.

telefax (1) An Integrated Services Digital Network (ISDN) service that provides high-speed fax transmission and store-and-forward capabilities. It uses one "B-channel" (64 kbps) of the ISDN basic rate interface (BRI) to provide about a 10-page per minute transfer rate on Group 4 fax machines. (2) In Europe, a synonym for a fax machine.

telegram Written information intended to be transmitted by telegraphy for delivery to the addressee. This term generally includes radiotelegrams unless otherwise specified.

telegraph (1) A communications system concerned with the transmission of written matter by using a signal code. (2) A communications system employing either the interruption of current flow or the change of polarity of current flow in the transmission medium to convey coded information. See also *teleprinter codes*.

telegraph concentrator A switching device that allows a number of subscriber lines to be connected to a smaller number of trunk lines or operating positions. The switching means may be automatic or manual. See also *telegraph distributor* and *telegraph selector*.

telegraph distortion The shifting of the transition points of asynchronous digital signaling elements from their proper position relative to the start pulse. The distortion is expressed as a percentage of the perfect pulse locations. The reference point for measuring telegraph distortion is the space to mark the transition of each character. The sampling point on the waveform is the 50% point on the rising (or falling) edge of the pulse. Mathematically, *telegraph distortion* is given by:

$$distortion = \frac{|t_e - t_a|}{t_e} \cdot 100\%$$

where t_e is the expected or theoretical time duration of a signaling element and t_a is the actual duration.

telegraph distributor A switching device that associates one telegraph channel with a multiplicity of sending or receiving devices. See also *telegraph concentrator* and *telegraph selector*.

telegraph key A hand-operated telegraph sender (transmitter). Physically, it is a switch for making and breaking a circuit so as to generate the dots and dashes of individual Morse character codes. See also *teleprinter codes*.

telegraph regenerator A device for receiving telegraph signals from a line or line segment, retiming, reconstituting the waveshape, and retransmitting the digital signal on another line or line segment. See also *telegraph repeater*.

telegraph repeater A device for receiving telegraph signals from one line or line segment and retransmitting the signal on another line or line segment. See also *telegraph regenerator*.

telegraph sender A transmitting device for forming telegraph signals. Also called a *telegraph transmitter*.

telegraph signal The set of elements defined by a particular code to enable the transmission of written characters, certain punctuation marks, and control codes. See also *teleprinter codes*.

telegraph word A unit of five letters and a space. Used to calculate words per minute.

telegraphy (1) A telecommunications system for the transmission of graphic symbols (usually letters, numbers, and some punctuation) by means of a signal code.

Historically, the data transmission technique is one in which the maximum data rate is 75 bps and the line signaling method is either current polarity reversal or the interruption of current flow. Currently, however, it includes most digital transmission means. (2) The term is sometimes extended to include any system for the transmission of graphic symbols or images (usually without grayscale content) in record form.

telemetry The use of telecommunications methods and means to transmit measurement information from a remote transducer (measuring device) to a local indicating and/or recording device. The transmission means may be analog or digital, and the information may be status, location, or some other aspect of a remote entity.

telephone Primarily an end-user voice communications device that:

- Is used to transmit and receive voice-frequency signals, and
- Provides access to the worldwide phone system.

The telephone provides the signaling to the phone system that allows communications with a specific station anywhere within the system. It notifies a called user that a call is coming in, and it converts the acoustic signal to an electrical signal suitable for transmission to the destination.

telephone amplifier A device used to amplify the received signal level to enable people with poor hearing to use the telephone. Typical gains are between 10 and 20 dB.

telephone channel A communications link with characteristics suitable for the transmission of telephonic signals. The bandwidth of the channel is typically 300 to 3400 Hz. See also *attenuation distortion* and *bandwidth*.

telephone connection A two-way telephone channel established between two stations by means of a switching network and signal transmission facilities. Each station associated with the channel is provided:

- Communication bandwidth.
- Dc current to operate the terminal device.
- Signaling means to control the switching equipment, such as pulse or tone dialing.
- Signaling means to inform a called station that an attempt to establish a path is in progress, such as ringing.

telephone equalization See *subset equalization*.

telephone exchange A switching center to which local users are connected. Called a *central office* in North America.

telephone feed circuit That portion of a local central office that provides *BORSCHT* functions to a subscriber line. That is, it provides:

- (B) Battery feed, the application of dc loop current to operate the subscriber's subset;
- (O) Overvoltage protection, a circuit to protect the central office from power cross and lightning induced transients;

- (R) Ringing, the incoming call-alerting signal sent to the subscriber equipment;
- (S) Signaling, dial pulse, or tone station signaling;
- (C) Codec, the conversion of subscriber information between the analog line format and digital transmission format;
- (H) Hybrid, the conversion of the two-wire subscriber loop to a four-wire communications link; and
- (T) Transmission, the conveyance of subscriber equipment generated signals from the subscriber loop to the next stage of the network transmission equipment.

See also *feed, feed current, Hayes bridge, hybrid,* and *Stone bridge.*

telephone frequency The band of audio or voice frequency from about 300 Hz to 3400 Hz.

telephone handset A telephone transmitter and receiver within a single handheld package.

telephone line A general term used to depict either the path between a station and a central switching office or between switching offices. The path includes not only the copper conductors but the circuits and apparatus associated with the channel. See also *subscriber line* and *trunk.*

telephone number The unique network address that is assigned to a telephone user (subscriber) for routing telephone calls.

telephone receiver A device that converts the received voiceband signals to audible signals. Also called an *earphone* or *earpiece.*

telephone ringer A device that converts the *ringing signal* to an audible signal.

telephone sidetone See *sidetone.*

telephone transmitter A device that converts sounds to electrical signals for transmission over the telephone network. Also called a *microphone.*

telephony (1) Originally, the term was used to indicate transmission of the voice, as distinguished from telegraphy (using Morse code) and radio teletypewriter (RTTY) transmissions. Later, the term's scope expanded such that it referred to the business of the telephone companies, i.e., the conversion of sound into electrical signals (specifically speech signals), its transmission to another location (by any means with or without wire), its restoration to a sound, and the management of the equipment to provide these services.

It has now been expanded to include the combination and integration of all telephone and networking services, such as the transmission of graphics information (facsimile) and data (via modems or other means, e.g., Integrated Services Digital Network). Analog representations of sounds may be digitized, transmitted, and, on reception, converted back to analog form. Transmission may be by any means such as wire (traditional), optical fiber, radio, or satellite. (2) The branch of science devoted to the transmission, reception, and reproduction of sounds, such as speech and tones that represent digits for signaling. (3) A form of telecommunication set up for the transmission of speech or, in some cases, other sounds. Also called *telecommunications.*

Telephony Application Programming Interface (TAPI) A Windows 95 (or later) Application Program Interface enabling hardware independent access to telephone-based communication. *TAPI* includes a wide range of services from initializing equipment (such as a modem) and placing a call to voice mail or control of a remote computer. Sometimes also called *Telephone Application Program Interface.* See also *TAPI.*

telephoto (1) Pertaining to pictures transmitted via a telecommunications system. (2) A compound lens constructed such that its overall length is less than or equal to its effective focal length.

teleprinter A printing telegraph instrument having a signal-actuated mechanism for automatically printing received messages.

A *teleprinter* does not have a keyboard for entering messages to be transmitted. Also called a *printer* and *receive only teletypewriter.* See also *teletypewriter.*

teleprinter codes Several teleprinter codes are still in use, the most recent being the *ASCII* code. Other widely used codes are based on the first five-level (5-bit) teleprinter code, the *Baudot code,* invented in the 1880s by Jean-Maurice-Emile Baudot (1845–1903).

The *Baudot code* (also called the International Telegraph Alphabet No. 1 or ITA1) contains only 32 symbols (as do all five-level codes), not enough to represent the alphabet (26 symbols), numbers (10 symbols), and required punctuation needed for telegraph messages. The solution was to reuse a part of the code much like the "caps lock" key on a typewriter redefines individual keys. Two of the 32 symbols were defined as *letter shift* (*LS*) and *figures shift* (*FS*) keys. In this way 60 symbols are available (62 counting FS and LS). In use, all characters following the LS symbol are interpreted as "letters," and all symbols following the FS symbol are assumed to be "figures." The Murray code, a modification of the original Baudot code, was adopted by ITU-T as the standard International Telegraph Alphabet number 2 (abbreviated ITA2). Other adaptations of this code persist in specialized fields or communication disciplines. A few of these codes are shown in the chart on next page.

teletax A European system of informing a calling station how much the current call in progress is costing. The telephone exchange transmits an out-of-band tone pulse stream to a receiver at the calling station's location; the receiver increments a counter each time a tone pulse is received. The rate at which tone pulses are transmitted is dependent on both the state of call progress and the distance between the calling and called stations.

Typical tone frequencies are 12 kHz and 16 kHz. Because both frequencies are audible, they must be filtered out before being presented the telephone instrument. Also called *metering.*

teletex An international store-and-forward, essentially error-free, 2400 bps data rate communications service defined by ITU-T (CCITT) for use with the Integrated Services Digital Network (ISDN) and the switched telephone networks. It provides:

- Text communication capabilities using standardized character sets (ITU T.50 or ASCII), formats, and communication protocols, and
- A communications protocol that supports the Group 4 facsimile service.

teletext A one-way text-only information retrieval service. A fixed number of all text information pages are repetitively broadcast on an unused subcarrier of a television channel. A proprietary decoder (video adapter) at the subscriber's television receiver is used to select and display pages. Originally intended as a service to provide general information (weather, sports updates, and so on), it has not yet become popular with the general public. See also *viewdata.*

Teletype® A trademark of Teletype Corporation, usually referring to equipment, such as, teleprinters, tape readers, tape punches, and page printers, used in communications systems.

Teletype mode A mode of operation in which input/output operations are limited to those characteristics of a teletypewriter (monospaced, alphanumeric characters only—no graphics, no character formatting, no cursor control, no color, and limited page formatting—carriage return, and line feed).

FIVE LEVEL TELEPRINTER CODE ALPHABETS

Original Baudot Code — ITU-T Standard International Telegraph Alphabet 1			Currently Used Codes		CCITT Standard International Telegraph Alphabet 2	United States Teletype Commercial Keyboard	AT&T Fractions Keyboard	TWX	Telex	Weather
Code	Letters	Figures	Code	Letters	Figures					
10000	A	1	11000	A	–	–	–	–	–	↑
00110	B	8	10011	B	?	?	5/8	5/8	?	⊕
10110	C	9	01110	C	:	:	1/8	WRU	:	○
11110	D	0	10010	D	WRU	$	$	$	WRU	/
01000	E	2	10000	E	3	3	3	3	3	3
01110	F	N/A	10110	F	②	!	1/4	1/4	② (US $)	→
01010	G	7	01011	G	②	&	&	&	② (US &)	\
11010	H	+	00101	H	②	#			② (US #)	↓
01100	I	N/A	01100	I	8	8	8	8	8	8
10010	J	6	11010	J	Bell	Bell	,	,	Bell	/
10011	K	(	11110	K	(	(	1/2	1/2	(	←
11011	L	=	01001	L	)	)	3/4	3/4	)	\
01011	M	)	00111	M	.	.	.	.	.	
01111	N	N/A	00110	N	,	,	7/8		,	⊕
11100	O	5	00011	O	9	9	9	9	9	9
11111	P	%	01101	P	0	0	0	0	0	∅
10111	Q	/	11100	Q	1	1	1	1	1	1
00111	R	–	01010	R	4	4	4	4	4	4
00101	S	.	10100	S	'	'	Bell	Bell	'	Bell
10101	T	N/A	00001	T	5	5	5	5	5	5
10100	U	4	11100	U	7	7	7	7	7	7
11101	V	'	01111	V	=	;	3/8	3/8	;	⊙
01101	W	?	11001	W	2	2	2	2	2	2
01001	X	,	10111	X	/	/	/	/	/	/
00100	Y	3	10101	Y	6	6	6	6	6	6
11001	Z	:	10001	Z	+	"	"	"	"	+
00000	N/A	N/A	00000				Blank ③			–
00001	LS	LS	11111				Letters Shift (↑)			
00010	FS	FS	11011				Figures Shift (↓)			
00011	ER	ER	00100				Space (■)			
11000	CR	CR	00010				Carriage Return (<)			
10001	LF	LF	01000				Line Feed (=)			

① CR = carriage return, ER = error, FS = figures shift, LF = line feed, LS = letters shift, N/A = not assigned, and WRU = Who Are You?. A space is printed for both LS and FS symbols.

② Unassigned (domestic variation, not used internationally).

③ Blank in the United States, "No Action" in the International Alphabet 2.

teletypewriter (TTY) A printing telegraph instrument that has a signal-actuated mechanism for automatically printing received messages. It may also have a keyboard similar to that of a typewriter for entering and sending messages. If no keyboard is present, a *teletypewriter* is also called a *teleprinter* or *receive only teletypewriter*.

Radio circuits carrying *TTY* traffic are called *RTTY circuits* or *RATT circuits*.

teletypewriter exchange service (TWX) An AT&T public switched service that allows teletypewriter stations to be connected to a central office for communication with other like stations. Both Baudot (5 level codes) and ASCII (7 level codes) machines are used.

teletypewriter signal distortion The shifting of signal pulse transitions from their proper positions relative to the beginning of the start pulse. The magnitude of the distortion is expressed in a percentage of a perfect unit pulse length. Also called *start stop TTY distortion*.

television (TV) A form of telecommunications designed for the transmission of transient images of fixed or moving objects and the associated sound signal. The format of the signal varies with country; the most prevalent standards are National Television Standards Committee (NTSC) in the United States, Programmable Array Logic (PAL) in Europe and many other places in the world, and Système Electronique Couleur Avec Memoire (SECAM) in France.

Telex® (TEX) A worldwide dial-up communications service, offered by Western Union and the International Record Carriers, which enables its subscribers to communicate among themselves over the public telegraph network using the Baudot code.

telluric current Currents circulating in the Earth or in conductors connecting multiple points on the Earth (grounded points) due to voltages induced in the Earth.

Telnet (1) An acronym for TELephone NETwork. (2) A standardized application that provides a terminal interface with hosts using the Transmission Control Protocol/Internet Protocol (TCP/IP) network protocol. *Telnetting* to any host provides an interactive terminal session regardless of the remote host type or its operating system. It is defined in Internet STD8, RFC854 and extended with options by many other Request for Comments (RFCs).

Telnet provides a standard protocol (the *Network Virtual Terminal—NVT—*protocol) for character-based remote terminal connections. It allows users to log onto remote hosts as if they were a terminal connected directly to the host. However, the NVT protocol includes only half-duplex communications and that, only in line mode. See also *TCP/IP*.

Telocator Alphanumeric Protocol (TAP) A protocol for submitting messaging requests to a pager service by means of a modem. It uses a 7-bit ASCII, even parity, <XON>/<XOFF> protocol. Doug Morrison of Radiofone Corporation wrote the original version.

RFC1645 and RFC1861 also address methods of delivering both alphanumeric and numeric messages to radio paging terminals. Also called *Personal Entry Terminal* (*PET*) by Motorola, *IXO* protocol by the IXO Company, and *IXO/TAP* by most users. See also *TDP*.

TELSET An acronym for TELephone SET.

TEM An abbreviation from Transverse Electric and Magnetic (Transverse ElectroMagnetic). See also *transverse mode*.

temperature coefficient The ratio of the change of a particular parameter value of a device to the change in temperature of the device. Generally, temperature coefficients are expressed as a percentage.

temperature rating The range of temperatures to which a device may be subjected, under a specified set of operating conditions, such that no operating property is degraded more than allowed.

tempest (1) In encryption, the name referring to the investigation and control of security jeopardizing emanations from electrical, electronic, and other sources. *Tempest* secure devices provide complete isolation between the unencrypted and encrypted environments and signals. (2) To shield against compromising emanations.

temporal Pertaining to the relation of a process, action, or other parameter with time.

temporal database A database that can store and retrieve temporal data, that is, data that depends on time in some way.

TEN An acronym for Telephone Equipment Network.

ten finger interface A term that refers to the interface between two networks that cannot be directly connected for security reasons. The derivation refers to the practice of placing two terminals side by side and having an operator read from one and type into the other. See also *tempest*.

ten's complement See *complement*.

ter Meaning third in Latin. The term is used as a suffix to denote a third version of a standard released by ITU-T.

tera- (T) (1) The SI prefix for 10^{12} (one trillion). (2) With computers and other binary systems, a multiplier of 2^{40}.

terabyte 2^{40} bytes (1 099 511 627 776 bytes).

terminal (1) A device used for data input and/or output. *Terminals* can be input only (called *send only terminals*), output only (called *receive only terminals* or *display terminals*), or both. A terminal may have some degree of computing power; hence, there are several categories possible. That is,

- *Dumb terminal*—lacks any memory or other components needed for doing computations. All processing for the terminal is done by the host or by the host's front-end processor (FEP). They have limited flexibility for use because they are not addressable. This means they cannot be shared on a bus.
- *Smart terminal*—has some processing capabilities and can be addressed.
- *Intelligent terminal*—has its own processor, can do its own processing, and can run programs. PCs often are used as intelligent terminals.

(2) In electronics, an electrical termination that can be connected to something else, e.g., a wire or connector.

terminal access controller (TAC) A host computer that accepts terminal connections, usually from dial-up lines, and that allows the user to invoke Internet remote logon procedures, such as Telnet.

terminal adapter (TA) A device that is used to permit non-Integrated Services Digital Network equipment (such as fax machines, computers, and analog telephones) to be connected to the ISDN digital telephone line (at the "R" interface point). Like modems, *TAs* may be designed for internal installation to a computer or may be external with their own power supply. A *TA*'s output conforms to one of the four ITU-T standards V.110, V.120, X.30, or X.31. Typically, it will support standard RJ-11 telephone connection plugs for voice and RS-232, V.35 and RS-449 interfaces for data.

terminal block An insulating base with a means for connecting wiring paths together. The connecting means may be screw terminals, insulation displacement terminals, quick release clips, and so on.

One example of a terminal block is Western Electric's 66-type telephone wire terminal block.

terminal cluster controller A device that connects one or more PCs to a front-end processor of a mainframe computer, e.g., an IBM mainframe network. An alternative uses a gateway to the mainframe.

terminal emulation The imitation of a terminal through the use of software in a computer. Terminal emulation software is used to make a local microcomputer behave as if it were a specific terminal type while it is communicating with a host.

Common terminals emulated include IBM's 3270 and Digital Equipment Corporation's VT-52, VT-100, and VT-200. (The VT-100 is by far the most common.) Host computers can only communicate with a finite number of terminal types; therefore, if the user's terminal is not one of the expected types, only two choices are possible:

- First, don't communicate, or
- Second, use software to fool the host into thinking that the terminal is one of the acceptable types.

terminal equipment (1) Communications equipment at either end of a communications link. It permits the involved terminal stations to interface with the communications link and to accomplish the task for which the link was established. (2) In radio relay systems, equipment used at points where data are inserted or delivered (as distinct from equipment used only to relay a reconstituted signal). (3) Telephone and telegraph switchboards and other centrally located equipment at which communications circuits are terminated.

terminal equipment identifier (TEI) A field of the LAPD (Link Access Protocol D-channel) frame that indicates the *physical* address of the called user.

terminal impedance (1) The impedance as measured at the unloaded output terminals of transmission equipment or a line that is otherwise in normal operating condition. (2) The ratio of voltage to current at the output terminals of a device, including the connected load.

terminal mode See *command state.*

terminal server A concentrator that aids communication between hosts and terminals.

terminal trunk A trunk circuit connecting two switching centers that is used only in conjunction with those switching centers.

terminate (1) To connect a wire or cable to an entity. (2) To end a communications session; to break the communications link.

terminate and stay resident (TSR) program A program that remains in random access memory (RAM) after it has relinquished control back to the operating system. It generally remains dormant until a specified sequence of events occurs. It then awakens, performs a predetermined task, and returns to the dormant state. Well-written *TSR* programs can be removed by an operator; others can be removed only by rebooting the computer.

terminated (1) The condition in which a wire or cable is connected to an entity. (2) The condition in which a transmission line is connected to an entity that has the same impedance as the characteristic impedance of the transmission line.

terminated line A transmission line in which the resistance at the cable ends is the same as the characteristic impedance of the cable. Under these conditions, no reflections or standing waves are present when a signal is injected at the transmitting end.

terminating office The telephone exchange to which the destination station is attached.

terminating resistor A resistor placed at the end of a transmission line to minimize reflections. The value of the resistor must be the same as the characteristic impedance of the transmission line. See *terminator.*

terminating traffic In telephony, traffic delivered directly to subscriber lines.

termination (1) In transmission systems, a load connected to a transmission line, circuit, or device that acts like a pure resistance *in the operating frequency range* of the driving source.

Terminating a uniform transmission line with its characteristic impedance causes the *terminator* to absorb all forward energy, thereby causing no reflection of energy back toward the source. (2) In hollow metallic waveguides, the point at which energy propagating in the waveguide continues in a nonwaveguide propagation mode into a load. (3) An impedance, often resistive, that is connected to a transmission line or piece of equipment as a *dummy load,* for test purposes.

terminator A device used in networks and transmission systems to reduce noise. It is generally placed at the end of the communications cable. The *terminator* prevents a signal from being reflected from the end of the cable back to the signal source.

The resistance of a *terminator* must match the *characteristic impedance* of the communications cable—e.g., when using RG-58 cable (50-ohm cable), the *terminator* must be 50 ohms. See also *termination.*

terminus A device used to terminate, position, and hold an optical fiber within a connector. It provides the equivalent function of a resistive terminator on an electrical transmission line.

ternary (1) Pertaining to the base three number system. (2) Pertaining to a characteristic, property or condition in which there exists three possible selections, choices, or significant conditions. A *ternary signal,* for example, may be a:

- Pulse with positive, zero, and negative values.
- Waveform that can assume 0°, 120°, or 240° relative to a reference waveform.
- Carrier wave that can assume any one of three different frequencies.

ternary code (1) A code whose alphabet contains exactly three symbols. (2) A multilevel waveform code with three states, generally 0 and a symmetric signal pair designated "+" and "−". Ternary codes are useful as:

- A mechanism for eliminating the dc content in a data stream, e.g., an alternate mark inversions (AMI) code. If the "+" and "−" states are equal but of opposite polarity, the average of a statistically random signal will be zero.
- A method of reducing the number of bauds to be transported for a given bit rate. For example, the 4B3T code encodes four binary bits into three ternary states for transmission.

See also *AMI, 4B3T,* and *waveform codes.*

tesla The SI unit of magnetic induction, equal to 1 weber per square meter.

test (1) In general, a check and/or diagnostic procedure intended to evaluate an entity for conformance with an applicable specification. (2) An action or group of actions, performed on a device under test (DUT), designed to evaluate a particular parameter or characteristic under specified, controlled operating conditions. (3) A check on an entity to determine its current operating state. (4) A command in a Logical Link Control (LLC) unnumbered information frame to create a loopback.

test antenna An antenna with known performance characteristics used in determining the transmission characteristics of equipment and/or associated propagation paths.

test call A call made to determine whether circuits or equipment are functioning properly.

test desk In telephony, a position equipped with testing apparatus so arranged that connection can be made to telephone lines or central office equipment for testing purposes.

test frequency (1) The frequency in hertz (Hz) of a test signal. (2) The number of tests (of the same type) performed per unit time interval, i.e., the reciprocal of the test interval.

test interval The time between the initiation of identical tests on the same entity.

test loop A circuit inserted between transmission equipment terminals to simulate a real transmission line, e.g., between a telephone feed circuit and a telephone set.

test plug See *loopback plug.*

test point A point within a piece of equipment, device, or system that provides access to signals for the purpose of calibration and/or fault isolation.

test posting A posting (message) sent to USEnet for the sole purpose of verifying that the process of placing the message on a particular newsgroup actually works. It is generally considered bad netiquette to *test post* to any USEnet newsgroup other than *alt.test* or *misc.test*. *Test posting* to other groups may lead members of the group to conclude that the poster is new to the Internet (a newbie).

test set A telephone test handset designed to test telephone circuits. It has the microphone and earphone of a normal handset, plus both a rotary dial and dual tone multifrequency (DTMF) sender of a telephone subset. It also has several switches that allow it to monitor line quality without disturbing a connected call. Also called a *butt set* or a *test handset*.

test shoe A device that is attached to a distributing frame circuit appearance so as to allow testing of the circuit.

test tone A signal sent through a transmission system or device for testing or alignment purposes, e.g., for aligning gains and losses in the system. The tone is transmitted at a prescribed level and frequency.

testing The process of examining a design (hardware or software) or an idea with the intent of

- Locating deviations from a predefined specification or expectation,
- Determining the margins of actual performance with respect to the specification,
- Confirming the lack of deviations from that specification, or in some cases,
- Determining what the actual performance is under various operating conditions.

TEX (1) An acronym for TELEX. (2) An acronym from TEletype eXchange.

text Data that consists of only words and symbols *representing human speech;* usually encoded to meet the ASCII or ITU-T T.50 standard. Numeric symbols are not taken as values on which any math should be performed.

text editor A program that allows a user to create or edit a text file. It is similar to a word processor with the exception of the extensive formatting capabilities normally found in word processors. Text editors are designed to assist users in the preparation of documents and control files to be used within the computer environment and are rarely printed for general distribution.

text file A file composed of only ASCII text characters. See *ASCII file* and *binary file*.

TF (1) An abbreviation of Test Frame. (2) An abbreviation of Trunk Frame. (3) An abbreviation of Transmitter Frequency. (4) An abbreviation of Triple Frequency.

TFA An acronym for Transparent File Access.

TFB An abbreviation of Too F—ing Bad. Seen in flame mail and postings, indicating sneering, sarcastic "pity."

TFC An abbreviation of Transmission Fault Control.

TFOM An acronym for Time Figure Of Merit.

TFP An abbreviation of TOPS Filing Protocol.

TFT An abbreviation of Thin Film Transistor. A liquid crystal display (LCD) display technology in which each pixel of the display is controlled by its own transistor. The display is brighter and provides more contrast than passive displays; such as the *DSTN* (dual super twist nematic). Also called an *active matrix* display. See also *DSTN* and *LCD*.

TFR (1) An abbreviation of Theoretical Final Route. See also *class n office*. (2) An abbreviation of Tunable Frequency Range.

TFS An abbreviation of Traffic Flow Security.

TFTP An abbreviation of Trivial File Transfer Protocol.

TG (1) An abbreviation of Transmission Group. In IBM's Systems Network Architecture (SNA), one or more links between adjacent nodes. (2) An abbreviation of Task Group. (3) An abbreviation of TeleGraph. (4) An abbreviation from Terminator Group.

TGB An abbreviation of Trunk Group Busy.

TGC An abbreviation of Transmission Group Control in IBM's Systems Network Architecture (SNA).

TGF An abbreviation of Through Group Filter.

TGID An abbreviation of Trunk Group IDentification.

TGIF An abbreviation of Thank God Its Friday.

TGM An abbreviation of Trunk Group Multiplex.

TGN An abbreviation of Trunk Group Number.

TGS An abbreviation of Telemetry Ground Station.

TGW An abbreviation of Trunk Group Warning.

TH An abbreviation of Transmission Header. In IBM's Systems Network Architecture (SNA), two bytes in the framing format added to a basic information unit (BIU) at the path control layer (layer 4). The BIU, together with the *TH*, form a path information unit (PIU).

THD An abbreviation of Total Harmonic Distortion.

THE An acronym for the Technical Help for Exporters group within the British Standards Institute (BSI). *THE* publishes many English translations of international standards, surveys of international requirements, and a periodic newsletter.

THEnet An acronym for Texas Higher Education network. A service provider that furnishes interconnectivity statewide for Texas and in Mexico as well as access to the Internet. Members include both universities and businesses, such as the campuses of the University of Texas, Texas A & M, Rice University, Texas Instruments, Lockheed, Sematech, and Schlumberger.

thermal conductivity The ratio of heat conducted per unit area per unit time to the component of the temperature gradient normal to that area along the path of conduction.

thermal equilibrium The condition or state that a device achieves when the power into the device is equal to the power removed in the form of heat. That is, the device temperature remains constant.

thermal noise Noise occurring in any noninsulating material resulting from the random motion of free charge carriers (e.g., electrons) within the material.

The *thermal noise* power (P_N) in a resistor is given by:

$$P_N = k\,T\,B_w$$

where:

k is Boltzmann's constant (1.38×10^{-23} J/K),

T is the conductor temperature in $^\circ$K, and

B_w is the bandwidth of interest in Hz.

Thermal noise has a flat power spectrum to extremely high frequencies, that is, *thermal noise power,* per hertz, is equal throughout the frequency spectrum, depending only on k and T. At room temperature

(293°K) and in a bandwidth of 1 Hz, the *thermal noise* power is approximately −174 dBm/Hz. When P_N is expressed in a 1-Hz bandwidth, it is often called *noise power density* (*NPD*). Also called *Johnson noise*.

thermal radiation Electromagnetic radiation (essentially ultraviolet, visible, and infrared radiations) emitted from a source (heat or light) as a consequence of its temperature.

thermal resistance The impeding of a material to the flow of thermal (heat) energy. Generally expressed as the temperature rise at a designated point on one side of the material (the heat source) as compared to a second designated point on the other side of the material per watt applied at the source. Examples of *thermal resistance* include:

- *Of a cable*—the temperature rise of the conductor over the outer surface temperature of the aggregate coverings per foot per watt dissipated in the conductor. Expressed in degrees Celsius per watt per foot (°C/W/ft).
- *Case to ambient* ($\Re_{\theta CA}$)—the difference between the case and ambient temperatures divided by the power dissipated by the device.
- *Semiconductor device* ($\Re_{\theta JC}$)—the temperature rise of the junction compared to a designated point in the case divided by the power dissipation in the junction.

thermistor A bulk effect metal oxide semiconductor that dramatically changes resistance with temperature. Over a limited working temperature range, it exhibits a negative temperature coefficient, that is, the resistance decreases with increasing temperature, as is shown in the drawing.

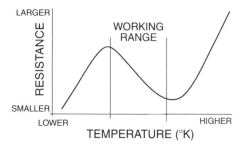

Over this working range, the temperature-resistance relationship follows an approximately logarithmic relationship, e.g., the J. S. Steinhart and S. R. Hart approximation equation:

$$\frac{1}{T} = A + B \cdot \ln R + C \cdot (\ln R)^3$$

where:

T is the temperature in kelvins,

R is the resistance at temperature T,

A, B, and C are curve-fitting constants determined experimentally for each manufacturing recipe, that is, from three resistance measurements (R_1, R_2, R_3) at three distinct temperatures (T_1, T_2, T_3) one can derive the equations:

$$C = \frac{\dfrac{T_3 - T_2}{T_2(L_2 - L_3)} - \dfrac{T_3 - T_1}{T_1(L_1 - L_3)}}{T_3(L_2 - L_1)(L_1 + L_2 + L_3)}$$

$$B = \frac{T_3 - T_2}{T_2 T_3(L_2 - L_3)} - C(L_2^2 + L_2 L_3 + L_3^2)$$

$$A = \frac{1}{T_1} + BL_1 - CL_1^3$$

where:

L_n is $\ln R_n$.

thermocouple A pair of dissimilar conductors joined at two points in such a manner that a potential difference is created when the two junctions are at different temperatures. The generated voltage is proportional to the temperature *difference* and to the material from which the dissimilar conductors are made.

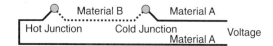

thermodynamic temperature A measure proportional to the energy of molecular activity (thermal energy) of a given body at equilibrium and indicated in kelvins (K). Minimum thermal energy occurs with minimum molecular activity and is defined as a temperature of *absolute zero* (0 K).

The International Temperature Scale of 1990 (ITS-90) is the basis for high-accuracy temperature measurements in science and technology. Formerly called *absolute temperature*.

thermoluminescence A synonym for *incandescence*.

thermopile A group of thermocouples connected in series and/or parallel so as to generate a higher voltage and/or higher current capability.

Thevenin's theorem A circuit principle that states any complex circuit containing only resistive elements, voltage sources, and current sources can be reduced to a single equivalent voltage source and an equivalent series resistance. See also *Norton's theorem*.

THF An abbreviation of Tremendously High Frequency; 300–3000 GHz. See also *frequency spectrum*.

Thick Ethernet See *Ethernet*.

thick film In electronics, a technology in which a film or ink about 1 mil thick is printed (usually with a silk screen process) onto an insulating substrate and then fused to the substrate by firing. The process is used to create resistors, conductors, and even capacitors.

Thicknet See *Ethernet*.

thickwire See *standard Ethernet cable*.

thin client See *network computer* (*NC*).

thin film (1) In electronics, a technique in which a layer of material from 10s to 100s of nanometers thick are deposited in a vacuum on an insulating substrate. Resistors, capacitors, conductors, and magnetic memory (e.g., computer hard disks) are commonly manufactured by this technique. (2) In optics, the deposition of one or more layers of transparent dielectric film, each with a specifically chosen thickness and refractive index. Thin films are used in a number of passive and active applications, including:

- Anti-reflection coatings,
- Achromatic beam splitters,
- Semitransparent mirrors,
- High reflectivity mirrors,
- Wavelength transmission and reflection filters (e.g., filters for heat control, color exclusion, contrast enhancement, color selection),
- Optical waveguides,
- Lasers,
- Optical modulators, and
- Optical switches and multiplexers.

thin film laser A laser constructed by thin film deposition techniques on a substrate. Generally used as a light source to drive thin film optical waveguides and may be used in integrated optical circuits (IOCs).

thin film optical modulator A modulator constructed of multiple thin film layers, each of a different optical characteristic. It is capable of modulating transmitted light by electro-optic, electro-acoustic, or magneto-optic effects, thus impressing a signal on the optical carrier. It is frequently used as a component in integrated optical circuits (IOCs).

thin film optical multiplexer A multiplexer constructed of multiple thin film layers, each of a different optical characteristic. It is capable of multiplexing transmitted light by electro-optic, electro-acoustic, or magneto-optic effects to obtain signal multiplexing. It is frequently used as a component in integrated optical circuits (IOCs).

thin film optical switch A switch constructed of multiple thin film layers, each of a different optical characteristic. It is capable of switching transmitted light by electro-optic, electro-acoustic, or magneto-optic effects to obtain signal switching. It is frequently used as a component in integrated optical circuits (IOCs).

thin film optical waveguide One or more thin transparent dielectric films bounded by material of a lower refractive index which is capable of guiding light (an optical signal). Such a construction is frequently used as a component in integrated optical circuits (IOCs).

Thinnet See *Ethernet.*

thinwire RG-58 coaxial cable used to interconnect 10Base2 Ethernet nodes.

third octave The interval between any two frequencies having a ratio of $\sqrt[3]{2}{:}1$ (about 1.26:1), e.g., 126 Hz and 100 Hz, 5000 Hz and 3969 Hz. The *center frequency* is defined as the *geometric mean* of the *bandedge* frequencies. For example, bandedges of 126 Hz and 100 Hz yield a center frequency of 112.25 Hz. In equation form, these relationships are:

$$f_c = \sqrt{f_l \cdot f_h}, \; f_h = f_c \cdot \sqrt[6]{2}, \; f_l = f_l = \frac{f_c}{\sqrt[6]{2}}$$

where:
- f_c is the center frequency,
- f_l is the lower band edge frequency, and
- f_h is the upper band edge frequency.

See also *octave.*

third-party call control A call control methodology that allows management of a communications path (including setup, forwarding, transfer, and disconnecting) *without being a party to the call.* See also *first-party call control.*

third window The minimum loss window in silica-based fibers with transmission at approximately 1530–1560 nm. See also *window.*

THL An abbreviation of <u>T</u>rans<u>H</u>ybrid <u>L</u>oss.

Thomas-Conrad Network System (TCNS) A 100-Mbps proprietary implementation of the ARCnet local area network (LAN) architecture, developed by Thomas-Conrad. *TCNS* can use existing ARCnet drivers, but it also includes drivers for use with other operating environments, such as Novell's NetWare, Microsoft's LAN Manager, or Banyan's VINES.

TCNS requires special network interface cards (NICs) because of the higher transmission speed and because it uses a different encoding scheme than standard ARCnet. Specifically, it uses a 4B/5B translation scheme (which converts 4 signal bits into a 5-bit symbol) and then uses a nonreturn to zero, inverted (NRZI) waveform encoding scheme. The *TCNS* can use coaxial, shielded twisted pair (STP), or fiber optic cable, but it does not support unshielded twisted pair (UTP).

thrashing The collapse of throughput of a system as the load increases. (Here load is defined as a backlog of unfinished tasks.)

On a system capable of processing multiple requests for service, such as a local area network (LAN), one might think that the throughput of the system would increase, up to some maximum level, as more requests were made. As it turns out, this is only partially correct: The throughput increases to a point and then falls off to zero as additional requests for service are made. The solution to the *thrashing* problem is solved by some form of a load controller. In a communications packet network, two frequent solutions are token passing and backoff algorithms.

thread **(1)** In bulletin board systems, a set of messages in a discussion group that all share a common subject or topic. Generally, the thread contains:

- An initial comment or question followed by
- Both direct responses and comments and
- Responses and comments to other responses and comments.

In this respect, a *thread* can be loosely described as a conversation. On the Internet a *thread* is more easily read with a *threaded newsreader.* **(2)** In *software,* individual processes within a single application. That is, an executable object (with its own stacks, registers, and instruction counter), which belongs to a single process or program.

thread selector In a threaded newsreader, the program mode that displays posted articles organized by topic rather than in chronologically received order.

threaded newsreader A program for reading USENet newsgroup postings organized by topic (in opposition to chronologically received order).

Examples of UNIX-based threaded newsreaders include nn, tin, and trn. Others include WinVin, Navigator, and Explorer.

three bit byte A synonym for a *triplet.*

three finger salute Rebooting a PC by hitting the CTL-ALT-DEL keys simultaneously.

three letter acronym (TLA) Used as a generic description of any abbreviation or acronym. *TLA* is a self-describing abbreviation of itself.

Because many people believe there are too many abbreviations, they use the phrase, with disgust, to describe any abbreviation or acronym and claim that "not all *TLAs* have three letters, just as not all four letter words have four letters." There are only 17 576 (26^3) true *TLAs*. This can be expanded by including numbers or by going to four or more letters—hence the expression *extended three letter acronyms* (*ETLA*) or *stupid four letter acronyms* (*SFLA*).

three phase circuit An ac power distribution system in which three circuits deliver ac power and each circuit is phase shifted one-third of a cycle (120°) from the other two. The power may be delivered in either of two systems:

- A wye (Y) connection using four wires, one for each phase and one common return.
- A delta (Δ) connection using three wires in which power may be taken between any two conductors.

three-way calling A switching system service feature that permits users to add a third party at a different number during a call, without the assistance of an attendant. See also *conference call.*

three-way handshake A reference to the three steps required to establish a connection and synchronize activities. Specifically:

- The calling station requests a connection.
- The called station responds with a ready to receive.
- The calling station acknowledges the called station response.

threshold (1) That point of a transfer function where a change in the input causes an abrupt change in the output. See also *hysteresis* and *Schmitt trigger.* (2) In network management. An attribute value that is used as a cutoff point between significant (or critical) and non-significant events. (3) The smallest value of a parameter that can be detected or causes a specified action by a particular device. For example, the smallest value an analog-to-digital converter can discern. (4) The minimum value a stimulus may have to create a desired effect, e.g., the minimum current necessary for a laser diode to produce coherent emission. (5) A value used to denote predetermined levels. For example:

- The levels in a UART (Universal asynchronous receiver transmitter) transmit or receive buffer.
- Those pertaining to volume of message storage, i.e., in-transit storage or queue storage, used in a message-switching center.

threshold current (1) In a laser, the driving current corresponding to lasing threshold. Currents below this threshold may cause emission; however, it will not be coherent. (2) Of a fuse. The current at which the device becomes current limiting, that is, it becomes an open circuit in a specified period of time.

threshold extension See *FM threshold extension.*

threshold frequency In opto-electronics, the frequency of incident radiant energy below which there is no photoemissive effect.

threshold signal to interference (TSI) ratio The minimum signal to interference power required by a circuit, device, or system to provide a specified performance level.

through group A group of 12 voice-frequency channels transmitted as a unit through a carrier system.

through supergroup An aggregate of 60 voice-frequency channels (five groups) transmitted as a unit through a carrier system.

throughput (1) The *average* transfer rate of actual, error-free, useful information (not including overhead, error bits, or retransmitted bits) through a device or series of devices. See also *transfer rate of information bits (TRIB).* (2) The maximum capacity of a communications channel or system. (3) A measure of the number of tasks performed by a system over a period of time.

THT An abbreviation of Token Holding Time.

THz The SI symbol for terahertz (10^{12} Hz).

TI (1) An abbreviation of Time In. (2) An abbreviation of Texas Instruments Inc. (3) An abbreviation of Terminal Interface.

TIA (1) An acronym for Telecommunications Industry Association. A trade group representing about 600 manufacturers. A key player in setting telecommunications standards both in the United States and abroad. (2) An acronym for Thanks In Advance. (3) An acronym for The Internet Adapter. (4) An acronym for Telematic Internetworking Application.

TIA/EIA IS-n See *IS-n.*

TIC (1) An acronym for Tongue In Cheek. (2) An acronym for Token ring Interface Coupler. (3) An acronym for Terminal Identification Code.

TICS An acronym for Telecommunication Information Control System.

tick tone A signal generated by some auto dialers and private automatic branch exchanges (PABXs) that indicates digits are being outpulsed to the local exchange.

ticker A receive only telex machine typically used to report stock and commodity prices. Information is printed on a one-inch-wide paper tape.

TID (1) An abbreviation of Transaction ID. (2) An abbreviation of Traveling Ionospheric Disturbance.

TIDF An abbreviation of Trunk Intermediate Distribution Frame.

TIDS An acronym for Technical Information Distribution Service.

TIE (1) An acronym for Time Interval Error. (2) An acronym for Trusted Information Environment—an encoding method. (3) An acronym for Terminal Interface Equipment.

tie line (1) TIE is an acronym for Terminal Interface Equipment. A *TIE line* is a long-distance line leased from a communications carrier for private, nonswitched use. The *TIE line* may be a single voice grade line or may have the capacity of a T-1 or greater link. *TIE lines* are frequently used to connect two or more buildings, facilities, or private branch exchanges (PBXs) of an organization. Also called a *tie trunk.* (2) A transmission line connecting two or more power systems.

tie trunk A dedicated telephone line or channel directly connecting two private branch exchanges (PBXs). See also *tie line.*

TIES An acronym for Time Independent Escape Sequence. *TIES* is an *escape sequence,* which is not the same as the escape sequence with guard time (patented by Hayes Computer products) that is now used as the de facto standard for reliable modem operation by modem manufacturers worldwide. A modem supporting only *TIES* will respond to the escape sequence with guard time but will also escape into *command mode* if the chosen escape sequence is embedded in the data being transmitted.

.TIF A graphics file adhering to the Tagged Image File Format. See also *TIFF.*

TIF (1) An acronym for Telecommunications Interference Filter. (2) An acronym for Telephone Influence (or Interference) Factor.

TIFF An acronym for Tagged Image File Format. The standard was developed by Microsoft, Aldus, and major scanning hardware vendors as a means to capture and transfer images into publishing applications. There are several image types supported in the standard, that is,

- Black and white,
- Halftone or dithered, and
- Grayscale.

tight buffer In fiber optic cabling, a protective layer that is extruded over the cladding to keep the fiber from moving around or bending too sharply. Tight buffers are commonly used in patch cords and other areas where the cable is likely to be "carelessly" moved.

time (1) The interval between two events, or the duration of an event. (2) An apparently irreversible continuum of ordered events. (3) The designation of an instant on a selected time scale, for example.

- *Local Apparent Time (LAT)*—an early timekeeping scheme based on the Sun's crossing the meridian, that is, when the Sun is due south (for positions in northern temperate latitudes), the time is 12:00 noon. (For positions in the southern temperate latitudes, the Sun will be due north, and for regions between these latitudes, the Sun may be due north or south depending on the season.) Because of the Earth's elliptical orbit and its tilted axis, the time scale is not linear. In fact, it can be off as much as 16 minutes in some seasons. Another problem is caused by the fact that the Earth is round. When it is noon at a specified location, it is past noon to the east and not yet noon to the west. At the equator, this effect produces a 1-minute time difference for each 27 km east or west; at 40° latitude the 1-minute difference occurs at 21 km. Also called *apparent solar time* and *sundial time.*

- *Local Mean Time* (*LMT*)—essentially the same as Local Apparent Time but is based on a hypothetical well-behaved situation in which the Earth's orbit is circular and the Sun travels along the equator at a uniform rate. The difference in time of this *mean Sun* to the real Sun position is called the *equation of time.* (The value for any date can be found in an almanac.) This *mean Sun* has the average (mean) right ascension of the real Sun. Hence, noon became the moment when the *mean Sun* crossed the meridian. Also called *Mean Solar Time.*

- *Standard Time*—On November 18, 1883, the United States was divided into Standard Time zones, each 15° wide. (This is 1/24 of the way around the Earth; therefore, each time zone differs from its neighbor by one hour.) In each zone, all clocks are set to the Local Mean Time of a standard longitude, i.e., 75° west for the Eastern Standard Time (EST), 90° for Central (CST), 105° for Mountain (MST), and 120° for Pacific (PST). The rest of the world essentially followed suit; however, some political regions placed the time zones on half-hour boundaries, and other political regions are either wider or narrower than the 15° time zone. This was done in order to keep the entire region in the same time zone. Correcting from Standard Time to Local Mean Time is accomplished by subtracting 4 minutes for each degree of longitude the local position is west of the time zone reference; or in equation form:

$$LMT = ST - 4n°$$

- *Daylight Savings Time* (*DST*)—a variant of Standard Time used in some parts of the world to shift sunrise time during some seasons. In the United States most states (except most of Arizona, Hawaii, and a few individual counties) add one hour to Standard Time at 2:00 A.M. on the first Sunday in April and return to Standard Time at 2:00 A.M. on the last Sunday in October.

$$DST = ST + 1$$

Not all countries used the same dates for DST. One country even voted on whether to have DST each year and what dates to have it on!

- *Universal Time* (*UT*)— a time reference used when a time applies worldwide, such as with astronomical events. It is a measure of time that conforms, within a close approximation, to the mean diurnal rotation of the Earth and serves as the basis of civil timekeeping. UT is Standard Time at the 0° longitude. The 0° longitude is defined by a line engraved in a brass plate in the floor of the Old Royal Observatory at Greenwich, England. Conversion of Universal Time into Standard Time is accomplished by subtracting one hour for each time zone west of 0°. In the United States, for example, to get Eastern Standard Time, subtract 5 hours from UT (EST = UT · 5), for Central Standard Time CST = UT · 6, for Mountain Standard Time MST = UT · 7, and for Pacific Standard Time PST = UT · 8. (To get Daylight Saving Time, subtract one hour less than these values.) If subtraction yields a negative number, add 24 hours and subtract one from the date, for example, 3:00 UT on January 1, 1996 is 10:00 P.M. EST on December 31, 1995. The determination of UT by direct observations of celestial objects without regard for the physical location of the observation is called *Universal Time Observed* (*UTO*). When UTO

is corrected for the small but measurable longitude shifts in the Earth's crust with respect to its axis (polar motion) and for astronomical-atomic second differences, it is labeled UT1. Formerly called *Greenwich Mean Time* (*GMT*) and *Greenwich Civil Time* (*GCT*).

- *Terrestrial Dynamical Time* (*TDT* or *TT*)—originally called Ephemeris Time (ET). It is a refinement of Universal Time (UT) to compensate for the fact that the Earth's rotation is not uniform. The Earth's rotation slows down and speeds up by small, essentially unpredictable amounts, and it undergoes a very long-term slowing trend. The gradual slowing is caused by the friction of tides raised by the Moon and Sun. Irregular changes have two classes:
  - Long-term changes probably involve the motion of the Earth's fluid interior.
  - Short-term changes are caused by changes in winds, air masses, snow packs, and other factors.

  In 1960 a system (Ephemeris Time or ET) was instituted which runs absolutely steady regardless of the Earth's irregular rotation. ET was the same as UT around 1902; however, UT gradually drifts behind ET such that in 1996 it lagged by about 62 seconds. Conversion of ET to UT is simply

$$UT = ET - Delta\ T$$

  where *Delta T* is the time difference and can be found in an Almanac. However, because of the unpredictability of the Earth's rotation rate, accurate predictions of Delta T are impossible. In 1984 ET was renamed Terrestrial Dynamical Time (TDT or TT).

- *Barycentric Dynamical Time* (*TDB*)—defined in 1984, TDB is similar to Terrestrial Dynamic Time (TDT). However, it refers to the solar system's center of mass. The difference between TDT and TDB is only milliseconds.

- *Coordinated Universal Time* (*UTC*)—a time based on an atomic second rather than an astronomical second. An astronomical second is 1/86 400 of a mean solar day, while an atomic second is the time it takes a cesium-133 atom to emit 9 192 631 770 cycles of radiation under certain conditions.

  Because the atomic second does not vary, as does an astronomical second, a day is no longer exactly 24 hours. That is, in 1983 an average day was 24.000 000 63 hours, and in 1986 the average day was 24.000 000 34 hours. To keep astronomical-based time and atomic-based time synchronized, UTC provides for a one-second correction (called a leap second) to be applied, as necessary, in the last minute of June 30 and/or December 31. This corrected version of UTC is always within 0.9 second of UT1. UT1 includes a further correction, that is, it allows for continental drift: it includes the small effect of the Earth's crust moving with respect to its axis (polar motion). *UTC* is maintained by the Bureau International des Poids et Mesures (BIPM). Maintenance by BIPM requires cooperation among various national laboratories around the world, each of which maintains its version of UTC called UTC(*x*) where *x* is the designation of the laboratory. For example, UTC (USNO) represents the U.S. Naval Observatory, the official laboratory in the United States. The full definition of UTC is contained in CCIR Recommendation 460-4. Also called *atomic time* (*AT*), *World Time*, *Z time*, and *Zulu.*

- *Sidereal Time*—time based on the position of the stars rather than the position of the Sun. Because of the Earth's rotation, there is exactly one less sidereal day in a year than a year based on (or synchronized to) the Sun's position. Sidereal time, therefore, runs approximately four minutes a day faster than any of the systems described above based on the Sun.

- *GPS Time*—a time based on the atomic second and synchronized to within 1 μs of UTC but without the leap second corrections. Global Positioning System (GPS) time zero is 0000 UT (midnight) on January 6, 1980.

  Because GPS time is not leap second adjusted, there is an offset from UTC by an integer number of seconds. The offset remains constant between leap second occurrences. This offset is given in the GPS navigation (NAV) message, and GPS receivers generally apply the correction automatically. On January 1, 1999, GPS time was ahead of UTC by 13 seconds.

  See also *Julian date* and *second*.

time ambiguity A situation in which more than one time or time measurement can be obtained under the stated conditions.

time-based billing A method of recovering the cost of operating a system by billing each user for the amount of time (and other resources) used. Another method of cost recovery is the monthly flat rate billing, that is, each user is billed a flat rate regardless of the amount of resources used. A third method is a combination of the two previous methods.

time code A code used for the transmission and identification of time signals. Because several different time code formats are in use, the format of the time code must be specified.

time code ambiguity The shortest interval between successive repetitions of the same time code value. For example:

- For a digital clock in which hours and minutes up to a maximum of 11:59 are displayed, the time code ambiguity is 12 hr.
- For a code in which year-of-century is the most slowly changing field, the time code ambiguity would be 100 years; hence, the year 2000 (Y2K) problem arose.

time code resolution The interval between two successive time code states—i.e., the elapsed time between the changes of the most rapidly changing symbol position of the specific code. For example, a digital clock that displays minutes and seconds (MM:SS) has a *time code resolution* of 1 second.

time compression multiplex (TCM) A digital transmission technique that provides full-duplex services by sending compressed bursts of data in a "ping-pong"-like manner.

time constant The natural time response of a system. The time required for the system output to change to 63.2 % of its final value in response to a step change input. The following graph shows the curve of a natural response and the values at 1, 2, 3, 4, and 5 time constants.

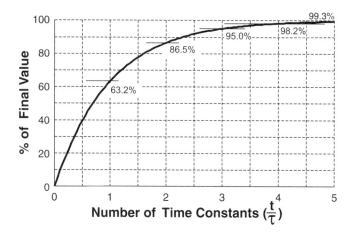

Number of Time Constants ($\frac{t}{\tau}$)

The output response (*P*) at other times (*t*) may be calculated by

$$P = F - (F - I) \cdot e^{-t/\tau}$$

where:

τ is the actual time of one *time constant,*
I is the *initial* value, and
F is the *final* value.

time derived channel Any of the channels obtained from the time division multiplexing (TDM) of a communications channel.

time diversity An error control technique in which signals representing the same information are sent over the same channel at different times. The method is used over systems subject to burst error conditions, in which case the transmit intervals are adjusted to be longer than expected error burst.

time division multiple access (TDMA) A form of time division multiplexing (TDM) in which individual terminals are assigned a time to transmit a burst of information, that is, the information from each signal source is transmitted sequentially across the common link. The assignment may be either permanent or based on demand. If the assignment is based on demand, more terminals may exist on the network than time slots (a blocking network). Several *TDMA* schemes include:

- *Slotted:* All time slots are of equal duration. When times slots are uniquely assigned, a node must wait for its assigned time slot before any transmission can take place. When time slots are allocated on demand, some form of CSMA/CD is implemented.
- *CSMA/CD:* Carrier sense multiple access with collision detection multiplex methods allow a node to transmit essentially whenever the local area network (LAN) is idle. If two stations present a carrier, a collision occurs. Both the nodes detect the collision and retry after a delay.
- *Token passing:* Unique time slot positions are assigned to each node. Each node is allowed to transmit only when it receives an "OK to transmit" word (token). If a node has no traffic, it passes the token on to the next node in on the LAN. If the node has traffic, it transmits a packet of information, followed by the token allowing the next node access.

The diagram compares various methods of multiple access using a LAN with six nodes.

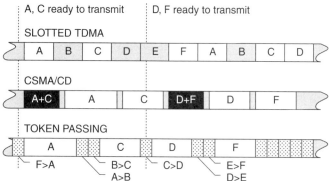

time division multiplex (TDM) A form of multiplexing wherein separate low-speed signals are combined onto a single high-speed transmission facility. The high-speed transmission facility allocates a fraction of its total time to transmitting each of the low-speed signals. The allocated time, called a *time slot,* is made available periodically. (For telephony applications, the period is generally 125 μs.) Each time slot carries one element (or sample) of a signal. Hence, each period a new sample of the low-speed signal is transmitted.

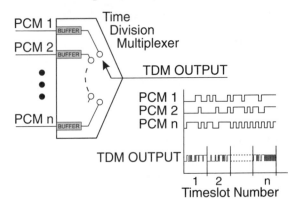

time division switching In telephony, a method of switching that provides a common path with separate time intervals assigned to each of the simultaneous calls. It switches time division multiplexed channels by moving them between time slots in a time division multiplexing frame. See also *time division multiplex.*

time domain reflectometry (TDR) A diagnostic method in which a signal pulse of known amplitude and duration is transmitted into a cable segment and a receiver listens for echoes. Depending on the amount of time the echo takes to return, the magnitude of the echo, and the cable's nominal velocity of propagation, a determination of the distance to any anomaly (impedance mismatch, short, open, "kink," or nonhomogeneous inclusion) in the cable may be ascertained.

A time domain reflectometer can be used to test the integrity of a section of cable before the cable is removed from the spool. *TDR* may be used in acoustic, electrical, or optical transmission systems.

time gated direct sequence spread spectrum Direct sequence spread spectrum in which the transmitter is on only for a fraction of a time interval. The on time may be periodic or random within a time interval.

time guard band A time interval left vacant between sequential events (such as time division multiplexing, encoding, decoding, switching, transmission, etc.), used to prevent mutual interference of information in adjacent time slots.

time instability The fluctuation of the time interval error caused by the instability of a realizable (real) clock.

time interval error (TIE) The time difference between a realizable (real) clock and an ideal uniform time scale, after a time interval following perfect synchronization between the clock and the scale.

time jitter Short-term variations of a specified signal instance when compared to a reference.

time marker Generally, a periodic reference signal that enables the synchronization of events or correlation of events with time.

time of day routing In telephony, the choice of transmission path based on the most economical route at that particular time of day.

time of occurrence The date of an event, i.e., the instant an event occurs, with reference to a specified time scale and resolution (year, month, day, hour, second, millisecond).

time of week (TOW) In the Global Positioning System (GPS),
- A 19-bit field in the z-count with a resolution of 1.5 seconds.
- A field in the navigation message handover word with a resolution of 6 seconds.

time scale **(1)** A time-measuring system that relates the passage of temporal events to a selected reference time. *Time scales* are graduated in intervals such as eons, centuries, years, months, weeks days, hours, minutes, seconds, and fractions thereof. **(2)** Time coordinates placed on the abscissa (*x*-axis) of Cartesian-coordinate graphs used for depicting waveforms or events.

time sequence diagram A method of graphically representing events over time. Time is generally represented on a vertical axis, with the oldest event at the top and the most recent event at the bottom. The information presented horizontally depends on the events to be depicted.

For example, the *time sequence diagram* illustrates the basic operations in establishing a telephone call.

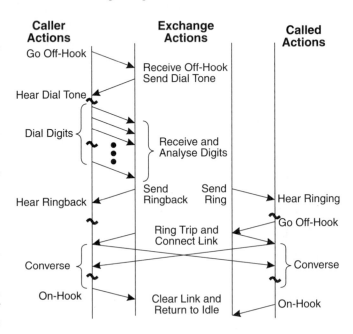

time sharing The use of a computer by more than one application, device, or user at the same time. *Time sharing* is accomplished by interleaving portions of processing time to each program. *Time sharing* is essentially the same as multitasking; however, it is generally associated with the situation of multiple users accessing a large mainframe computer, while multitasking generally implies performing multiple tasks for a single user. Also spelled *timesharing.*

time slice The period of time allocated to a particular program, process, or user in a timesharing or multitasking system.

time slot **(1)** In time division multiplexing (TDM) or time division switching, the period of time "belonging" to a particular communications source. The *time slot* will always be present, even if the source has no information to be transmitted. **(2)** A time interval that can be recognized and uniquely defined.

time space time (TST) One type of switching matrix in which three stages are used; the first stage switches an incoming signal on a selected time slot of a time division multiplexing (TDM) stream, the second stage routes the TDM stream to one of several possible outputs, and the third stage accepts the TDM stream and reassigns individual slot information to the destination's time slot.

Other switching schemes include space time space (STS), time only, and space only.

time standard A stable reference device that emits signals at precise and accurate equal intervals such that their count may be used as a clock.

time switch A switch that changes routing based on clock time rather directly upon demand.

time tick A time mark output of a clock system.

time to live (TTL) A field in the Internet Protocol header that indicates how many more hops the packet should be allowed to make before being discarded or returned.

timeout **(1)** A network parameter related to a forced event designed to occur at the conclusion of a predetermined elapsed time.

For example, on digital communications circuits, periods of inactivity (or idle time) are frequently measured, and after excessive periods of idle time (the *timeout* period) the circuit is disconnected. **(2)** A specified period of time that will be allowed to elapse in a device, circuit, or system before a specified event is to take place. The *timeout* timer is reset or canceled if:

- The *timeout* occurs.
- Another specified event occurs before the *timeout* period expires.
- An appropriate cancellation signal is received by the timer.

See also *watchdog timer.*

timing jitter Short-term variations or timing discrepancies between two digital signal elements.

timing recovery The derivation of a reference clock or timing signal from a received signal. Also called *timing extraction.*

timing signal **(1)** A signal used to synchronize two events. **(2)** The output of a clock.

TIMS An acronym for Transmission Impairment Measuring Set.

tin A UNIX threaded newsreader written by Iain Lea. See also *newsreader.*

tinsel wire An extremely flexible wire in which the conductor is a thin metal ribbon spiraly wound on a thread core. *Tinsel wire* is used in cords and cables that must withstand tremendous amounts of flexing.

tip **(1)** In telephony, one of the two wires connecting the central office to the subscriber's equipment. The tip has the more positive voltage of the pair (although it is generally still a negative voltage). The second wire is the "ring." See also *color code, ring,* and *sleeve.* **(2)** One of the terminations on the plug used in manual switchboards to connect the subscriber lines and interoffice trunks, that is,

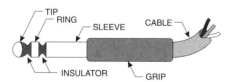

TIP An acronym for Terminal Interface Package. **(2)** An acronym for Terminal Interface Processor.

tired iron Hardware (generally, an IBM mainframe computer) that is perfectly functional but far enough behind the state of the art to have been superseded by new products with significant performance improvements. Also called a *dinosaur.*

TIRKS An acronym for Trunk Inventory Record Keeping System. The telephone company's line database system.

TIS **(1)** An acronym for Technical Information Sheet. **(2)** An acronym for Telephone Information Services. **(3)** An acronym for Terminal Interface Subsystem.

TIU **(1)** An abbreviation of Terminal Interface Unit. A channel service unit/data service unit (CSU/DSU) or network terminal device (NT1) for switched 56 service that handles dialing. **(2)** An abbreviation of Telematic Internetworking Unit. **(3)** An abbreviation of Telecommunications International Unit.

TJ **(1)** An abbreviation of Terminal Jack. **(2)** An abbreviation of Telephone Jack.

TJF An abbreviation of Test Jack Frame.

TJR An abbreviation of Trunk and Junction Routing.

TJS An abbreviation of Terminal Junction System.

TJT An abbreviation of Trunk and Junction Routing.

TKO An abbreviation of TrunK Offering.

TL **(1)** An abbreviation of Transport Layer. **(2)** An abbreviation of Tie Line. **(3)** An abbreviation of Transmission Level.

TLA **(1)** An abbreviation of Transmission Line Adapter (or Assembly). **(2)** A canonic abbreviation of Three Letter Acronym.

TLAP An acronym for TokenTalk Link Access Protocol.

TLB An abbreviation of Test LoopBack.

TLF An abbreviation of Trunk Link Frame.

TLI **(1)** An abbreviation of UNIX's Transport Level Interface. **(2)** An abbreviation of Telephone Line Interface.

TLK An abbreviation of TaLK or TaLKing.

TLM An abbreviation of TeLeMetry.

TLN An abbreviation of Trunk Line Network.

TLP **(1)** An abbreviation of Transmission Level Point. The system reference location to which relative signal gain or loss levels are measured. **(2)** An abbreviation of Telephone Line Patch. **(3)** An abbreviation of Telephone (or Telegraph) Line Pair.

TLSA An abbreviation of Transparent Line Sharing Adapter.

TLSPP An abbreviation of Transport Layer Sequenced Packet Protocol.

TLTP An abbreviation of Trunk Line Test Panel.

TLV An abbreviation of Type-Length-Value.

TLX An abbreviation of TeLeX.

TLXAU An abbreviation of TeLeX Access Unit.

TM **(1)** An abbreviation of Test Mode (RS-232 signal TM, RS-449 signal TM, and ITU-T signal 142). A signal generated by the local data communications equipment (DCE) to indicate it is in a test condition. **(2)** An abbreviation of Transverse Magnetic mode. See also *transverse mode.* **(3)** An abbreviation of TeleMetry. **(4)** An abbreviation of Time Modulation.

T$_{MAX}$ A symbol for maximum time.

TMF An abbreviation of Test Management Function.

T$_{MIN}$ A symbol for minimum time.

TMIS An acronym for Telecommunications Management Information System.

TML **(1)** An abbreviation of Terrestrial Microwave Link. **(2)** An abbreviation of Tandem Matching Loss.

TMN An abbreviation of <u>T</u>elecommunications <u>M</u>anagement <u>N</u>etwork. A Synchronous Optical Network (SONET) support network.

TMP (1) An abbreviation of <u>T</u>est-<u>M</u>anagement <u>P</u>rotocol. **(2)** A frequent abbreviation for <u>TeMP</u>orary. **(3)** An abbreviation of <u>TeMP</u>erature.

TMPDU An abbreviation of <u>T</u>est-<u>M</u>anagement <u>P</u>rotocol <u>D</u>ata <u>U</u>nit.

TMS An abbreviation of <u>T</u>elecommunications <u>M</u>essage <u>S</u>witcher. **(2)** An abbreviation of <u>T</u>elephone <u>M</u>anagement <u>S</u>ystem. **(3)** An abbreviation of <u>T</u>ime-<u>M</u>ultiplexed <u>S</u>witching. **(4)** An abbreviation of <u>T</u>ransmission <u>M</u>easuring <u>S</u>et. **(5)** An abbreviation of <u>T</u>elemetry <u>M</u>ultiplex <u>S</u>ystem.

TMU An abbreviation of <u>T</u>ransmission <u>M</u>essage <u>U</u>nit.

TN (1) An abbreviation of <u>T</u>ransit <u>N</u>etwork. **(2)** An abbreviation of <u>T</u>erminal (or <u>T</u>elephone) <u>n</u>umber. **(3)** An abbreviation of <u>T</u>erminal <u>N</u>ode.

TN3270 A protocol that emulates the IBM SNA 3270 terminal protocol data stream over a network such as Ethernet.

On the Internet, the TN3270 is a variant of Telnet which uses page mode instead of the normal line mode.

TNC (1) A standard connector used with coaxial cables similar to the BNC (bayonet Neill Concelman); however, *TNC* connectors are threaded for a more secure fastening in high-vibration environments. The name may be derived from <u>T</u>hreaded <u>N</u>eill-<u>C</u>oncelman, <u>T</u>hreaded <u>N</u>ut <u>C</u>onnector, <u>T</u>hreaded <u>N</u>avy <u>C</u>onnector or, as some claim, <u>T</u>iny "<u>N</u>" Connector.

The drawing shows the attachment of a coaxial cable to a *TNC* connector. The coax has each layer progressively exposed per manufacturer's instructions. The coax center conductor is soldered to the connector pin; then the cable is inserted into the connector body. The ferrule is crimped over the sheath holding the assembly together. **(2)** An abbreviation of <u>T</u>erminal <u>N</u>ode <u>C</u>ontroller. **(3)** An abbreviation of <u>T</u>he <u>N</u>etworking <u>C</u>enter. **(4)** An abbreviation of <u>T</u>ransport <u>N</u>etwork <u>C</u>ontroller.

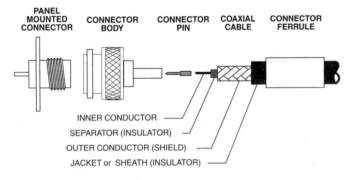

PANEL MOUNTED CONNECTOR   CONNECTOR BODY   CONNECTOR PIN   COAXIAL CABLE   CONNECTOR FERRULE

INNER CONDUCTOR
SEPARATOR (INSULATOR)
OUTER CONDUCTOR (SHIELD)
JACKET or SHEATH (INSULATOR)

TNIC An acronym for <u>T</u>ransit <u>N</u>etwork <u>I</u>dentification <u>C</u>ode.

TNL (1) An abbreviation of <u>T</u>erminal <u>N</u>et <u>L</u>oss. **(2)** An abbreviation of <u>T</u>echnical <u>N</u>ews<u>L</u>etter.

TNN An abbreviation of <u>T</u>runk <u>N</u>etwork <u>N</u>umber.

TNPC An abbreviation of <u>T</u>raffic <u>N</u>etwork <u>P</u>lanning <u>C</u>enter.

TNS (1) An abbreviation of <u>T</u>ransaction <u>N</u>etwork <u>S</u>ervice. **(2)** An abbreviation of <u>T</u>elecommunications <u>N</u>etwork <u>S</u>ervices.

TNX An abbreviation for thanks.

TNXE6 Jargon for *thanks a million*. TNX is the abbreviation for "thanks," and the "E," used in much of the computer world, means "exponent follows" and the 6 is the power of ten; hence, E6 represents 1 000 000. Also written *TNX 1.0E6*.

TO (1) An abbreviation of <u>T</u>elephone (or <u>T</u>elegraph) <u>O</u>ffice. **(2)** An abbreviation of <u>T</u>ime<u>O</u>ut. **(3)** An abbreviation of <u>T</u>ransmit <u>O</u>nly.

TOA (1) An abbreviation of <u>T</u>ype <u>O</u>f <u>A</u>ddress. A 1-bit field that indicates X.121 type or not. **(2)** An abbreviation of <u>T</u>ime <u>O</u>f <u>A</u>rrival.

toast (1) Jargon meaning any completely inoperable system or component, especially one that has just failed irreparably. For example, one might say, "Uh, oh . . . I think that lightning bolt just toasted my CPU." **(2)** Jargon meaning to cause a system to crash accidentally, especially in a manner that requires manual rebooting.

TOC (1) An abbreviation of <u>T</u>echnical <u>O</u>perations <u>C</u>enter. **(2)** An acronym for <u>T</u>able <u>O</u>f <u>C</u>oincidences.

TOCC An abbreviation of <u>T</u>echnical <u>O</u>perations and <u>C</u>ontrol <u>C</u>enter. Communications Satellite operations center.

TOD An abbreviation of <u>T</u>ime <u>O</u>f <u>D</u>ay. See also *time of occurrence*.

toeprint Jargon meaning an especially small-sized *footprint*.

TOF (1) An abbreviation of <u>T</u>one <u>OF</u>f. **(2)** An abbreviation of <u>T</u>op <u>O</u>f <u>F</u>ile. **(3)** An abbreviation of <u>T</u>op <u>O</u>f <u>F</u>orm.

TOG An abbreviation of <u>TOG</u>gle.

toggle (1) To change the state of a two-state device from the current state to the opposite state. A toggle switch, for example, has only two positions: on and off. A binary bit has two states: "0" and "1." There are four operations possible on a two state device:

- Do nothing ($0 \rightarrow 0$ and $1 \rightarrow 1$).
- Force the "0" state ($0 \rightarrow 0$ and $1 \rightarrow 0$).
- Force the "1" state ($0 \rightarrow 1$ and $1 \rightarrow 1$).
- Toggle it ($0 \rightarrow 1$ and $1 \rightarrow 0$).

(2) A type of electronic circuit (flip-flop) that changes state each time it receives a clock pulse. Generally designated a "T flip-flop."

token A unique data message (or "object") that circulates continuously among the nodes of a token network and describes the current state of the network. It is passed in sequence from node to node and indicates that the receiving node has the authority to transmit for a prescribed amount of time. Before any node can transmit, it must receive the *token* and the *token* must indicate the network is available (idle). See also *token ring packet*.

token bus network A local area network (LAN) whose node connection topology is a bus and uses a token passing method for regulating message flow. That is, it is a method of avoiding *contention*. In operation, a *token* governing transmission rights is passed from one station to the next logical station. Each station retains the token for a brief time, during which it alone has the privilege to transmit messages onto the network. The token is passed from the highest priority node to the next highest, and so on, until it reaches the lowest priority node. The token is then passed to the highest priority node again where the process starts again. The token "circles" through the network in a logical ring. (Node priority levels do not necessarily correspond to physical position on the bus.)

Token bus networks are defined in the IEEE 802.4 specifications, which include physical layer and media access control (MAC) sublayer details for networks that use a bus topology and use token passing as the media-access method. Some token bus architecture features include:

- Network speeds of 1, 5, 10, and 20 Mbps.
- Operation over 75-ohm coaxial cable or optical fiber.
- Two single-channel carrier band configurations:
  - The easiest to implement and the least expensive of the four configurations-supported architectures is the phase continuous frequency shift key (FSK) system. It can be used even with older cable that may already be installed in a building. However, a disadvantage is that the top speed is only 1 Mbps.
  - The faster but more difficult and expensive to implement is the phase-coherent FSK system known as minimum shift key (MSK). Two network speeds are possible, that is,

Data rate	Frequency 1	Frequency 2
5 Mbps	5 MHz	10 MHz
10 Mbps	10 MHz	20 MHz

  Both single-channel, carrier-band configurations use a bus in which all signals are broadcast in all directions. Cable segments are joined using BNC (bayonet-Neill-Concelman) connectors.
- A broadband configuration:

  The broadband directed bus (tree topology) assumes that transmissions originate in a special node, known as the *head-end.* From the head-end, signals are sent to other nodes along the network bus or tree. The data stream may be scrambled before modulation to avoid loss of clock synchronization during periods in which the signal does not change. The modulation method includes both amplitude and phase. Network transmission speeds of 1, 5, or 10 Mbps are supported. Cable segments are joined using Type F connectors.
- An optical configuration:

  The physical implementation is a star (active or passive), and the modulation method is amplitude shift keying (ASK). In ASK, a logical one is coded as a pulse of light, while a logical zero is no light. As with the broadband configuration, the data signal is encoded (using Manchester encoding) to prevent clock synchronization loss when long periods of data inactivity occur. Data rates can be 5, 10, or 20 Mbps.
- Four priority levels to regulate medium access. The priority levels supported by the token bus architecture are named 6, 4, 2, and 0, where 6 is the highest and 0 is the lowest priority. To ensure that no node hogs the token, restrictions are placed on the amount of time that any node may hold a token. This restriction is called the token holding time (THT).

The architecture uses data and control frames for transmitting information and managing the network. The data structure for a token bus frame is:

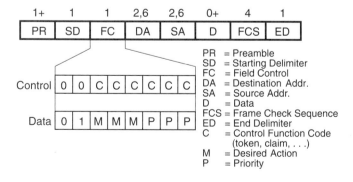

The various fields within the frame (or packet) include:
- *The preamble* (*PR*)—used to synchronize the receiver to the transmitter. At 1 Mbps only 1 byte is needed, while at 10 Mbps 3 bytes are needed.

- The *start delimiter* (*SD*)—one unique byte used to indicate the beginning of the frame. The byte is designed so that it will not occur in a data field.
- The *destination address* (*DA*)—depending on the type of addresses being used, 2 or 6 bytes which specify the node to which the token is being sent.
- The *source address* (*SA*)—depending on the type of addresses being used, 2 or 6 bytes which specify the node from which the token is being sent.
- *Data* (*D*)—0 bytes for tokens to more than 8000 bytes for data. Some control frames use this field for command or setting information.
- The *frame check sequence* (*FCS*)—a 4-byte sequence used to verify error-free data reception.
- The *end delimiter* (*ED*)—one unique byte used to indicate the end of the frame. The byte is designed so that it will not occur in a data field.

See also *frequency shift key* (*FSK*), *IEEE802.x, minimum shift key* (*MSK*), *star coupler, star network,* and *token ring network.*

token holding time (THT) In token networks, a parameter whose value can be used to adjust access to the network. A high value allows a node to keep the token for a long time, which is useful if network activity consists mainly of large file transfers and if rapid access to the network is not critical. A small value gives all nodes more equal access to the network.

THT works in conjunction with the token rotation time (TRTn) parameter. For each *THT* level, a maximum token rotation time (TRTn) is specified. For example, a value of TRT2 represents how long the token can take to make its way around the ring while still being able to ensure that packets at priority level 2 will be transmitted.

token latency The time it takes for a token to be passed completely around a token ring network, i.e., to all registered nodes.

token passing A deterministic method of controlling which node in a local area network (LAN) is allowed to transmit at any given time through the use of a control message (token) passed from terminal to terminal. The token, received by only one node at a time, enables message transmission if that node has information to be transmitted. If no information is to be transmitted, the node will transmit the token on to the next node. The technique was described by W. D. Farmer and E. E. Newhall in their 1969 paper "An experimental distributed switching system to handle bursty computer traffic."

A deterministic access method guarantees that every registered node will get access to the network within a predictable length of time. This is in contrast to a probabilistic access method (such as CSMA/CD) in which access time is only statistically known. Because each node gets its turn within a fixed period, deterministic access methods are more efficient on networks that have heavy traffic. Nodes on networks using probabilistic access methods waste much of their time trying to gain access and relatively little time actually transmitting data over a busy network. When token passing is used as a medium access method, monitoring to keep track of the token health is required. If a token should be lost or corrupted, the network active monitor (or a standby monitor) will attempt a token recovery and, failing that, will generate a new token. To enable new nodes to connect to the ring, "registration" opportunities are provided periodically. Each node will occasionally ask whether any nodes with lower addresses are interested in joining the ring. Network architectures that support token passing as an access method include ARCnet, Fiber Distributed Data Interface (FDDI), and IBM's Token Ring. See also *ALOHA, CSMA/CA,* and *CSMA/CD.*

Token Ring IBM's mid-1980 proprietary token passing baseband specification on which the IEEE 802.5 specification is based. It operates at 4 or 16 Mbps. The waveform protocol is differential Manchester encoding.

Although they use a logical ring structure, Token Ring networks are actually arranged in a physical star topology, with each node connected to a central hub (the multistation access unit—MAU), as illustrated in the following figure. The MAU arranges the lobes (nodes or workstations) attached to it in a ring.

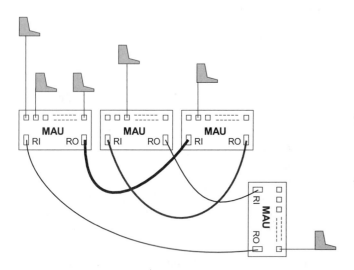

MAUs connected to each other use special ring in (RI) and ring out (RO) ports on the MAUs. These connections maintain the ring structure across the MAUs. The RO port from one MAU is connected to the RI port of another. Several MAUs may be linked this way. If there are multiple MAUs, the RO port of the last MAU in the series is connected to the RI port of the first MAU to complete the ring. The MAU-to-MAU connection actually creates both a primary, or main, ring and a secondary or backup ring. In a properly connected network, the logical layout results in each node (*X*) having exactly two adjacent nodes:

- The *nearest active upstream neighbor (NAUN)*, the node that passes frames and the token to node *X* in the ring.
- The *nearest active downstream neighbor (NADN)*, the node to which *X* passes frames and the token.

Cable lengths and between the lobe (node) and MAU, as well as the length between MAUs, are somewhat complex calculations and do not readily lend themselves to tabular layout; however, a conservative estimation can be derived from the following table. See also *token ring network*.

Description	Node-MAU	MAU-MAU
Type 1 & 2 Cable	100 m (330 ft)	200 m (660 ft)
Type 3 Cable	45 m (150 ft)	120 m (400 ft)
Type 6 Cable	66 m (220 ft)	
Type 9 Cable	66 m (220 ft)	
Type 5 Fiber Optic		1 km (0.6 mi)
Minimum distance		2.5 m (8 ft)
Inter-MAU segments		3
Maximum 802.5 node	250	
Maximum IBM lobes	260 w/STP	
	72 w/UTP	
Maximum MAUs		33

token ring adapter A network interface card (NIC) used to attach a node to a token ring network and operate as a token passing interface.

token ring cable Although the IEEE 802.5 specifications do not specify a particular type of cabling, IBM specifications do and the categories defined by the IBM Cabling System are generally used.

In a token ring network, cable is used for two purposes: for the main ring path (which interconnects multistation access units—MAUs) and for short runs (lobe to MAU or MAU to patch panel). IBM Type 1, 2, or possibly 9 shielded twisted pair (STP) cable is generally used for the main ring path. However, the token ring specifications also support Type 3 unshielded twisted pair (UTP) and fiber optic, Type 5, cables. For patch or jumper cables, Type 6 cable is commonly used. Token ring network interface cards (NICs) generally have a DB-9 connector for STP cable and may have a modular RJ-45 plug for UTP cable. MAUs have IBM Data Connectors (a special type of connector that self shorts when the mating plug is removed, so that the ring inside the MAU is not broken when a lobe is disconnected). See also *type n cable*.

Token Ring interface coupler (TIC) A device that enables direct connections from a Token Ring network to various types of mainframe equipment, including front-end processors, AS/400s, and 3174 terminal cluster controllers.

token ring network A local area network (LAN) whose node connection topology is a ring and uses a token passing method for regulating message flow. (IEEE 802.5 standards define the *token ring protocols*.) On a *token ring network*, a token controlling the right to transmit is passed from one node to the next physical node on the circular network. When a node has information to transmit, it seizes the passing node, marks it busy, and attaches the message. The busy token and message are then passed around the network until it reaches the destination node. The destination node receives the message and passes it on. Eventually, the sending node receives the original transmitted package, strips the message off, marks the token idle, and sends it on.

When Token Ring is capitalized, it refers specifically to IBM's Token Ring network. When it is spelled with all lower case (token ring), it refers to any ring network using a token passing protocol. Token ring networks have the following characteristics:

- They use a ring as the logical topology but a star as the physical topology (wiring).
- IEEE 802.5 networks operate at either 1 or 4 Mbps; IBM's Token Ring operates at either 4 or 16 Mbps.
- They use baseband signaling, meaning that only one signal is present at a time.
- They use the differential Manchester waveform-coding method.
- They use shielded twisted pair (STP), unshielded twisted pair (UTP) or fiber optic cable, but not coaxial cable. The characteristic impedance of STP is 150 ohm, while UTP is 100 ohm.
- They use a four-wire cable, with two of the wires used for the main ring and two for the redundant secondary ring.
- Each node (called a lobe by IBM) is connected to a wiring center, called a multistation access unit (MAU). The wiring inside an MAU creates a ring of the attached nodes.
- MAUs can be connected to each other to create larger rings. Each MAU includes two dedicated connectors for making a MAU–MAU connection.
- They allow the use of patch panels, which sit between nodes and MAUs and make it easier to reconfigure the network.
- The network interface cards (NICs) implement an agent in the chip set for network monitoring, i.e., to determine whether a token has been corrupted or lost.
- They are controlled by the node that generates the token (the active monitor).

See also *IEEE 802.x*.

token ring packet A variable-length frame initially generated by the active monitor (AM) and is composed of up to 10 variable-length fields. The basic form of all frames as well as the specific token and abort frames are illustrated here.

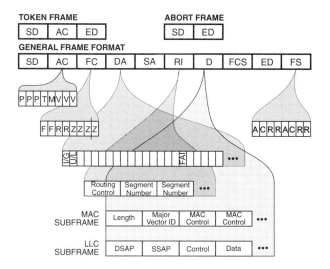

- *Starting delimiter* (*SD*)—1 byte. A unique byte used to indicate the beginning of the frame. The byte is designed so that it will not occur in a data field.
- *Access control* (*AC*)—1 byte. Indicates the type of frame, its priority level and its status, that is:
  - Three *Priority* bits (P) specify the frame's priority value, 0 being the lowest and 7 the highest.
  - A *Token* bit (T) is set to 0 if the frame is a token and 1 otherwise.
  - A *Monitor* bit (M) is set to 1 by the active monitor (AM) and to 0 whenever a lobe grabs the token.
  - Three *Reservation* (V) bits can be used by a lobe to request the priority level that is required to get access to the network.
- *Frame control* (*FC*)—1 byte. Distinguishes logical link control (LLC) and medium access control (MAC) data frames and defines how MAC frames should be processed.
  - Two *Frame* (*F*) bits define the data type.

F Subfield	Definition
00	MAC
01	LLC
10	Reserved
11	Reserved

  - Two *reserved* bits (R).
  - Four *control* bits (Z) are reserved for future use in LLC frames. For MAC frames, they indicate whether the frame should be copied to the lobe's regular input buffer (Z = 0000) for normal handling or to an "express" buffer (Z ≠ 0000) indicating the frame is to be processed immediately by the MAC sublayer.
- *Destination address* (*DA*)—6 bytes. Indicates the address of the lobe to which the frame is being sent and the type of address.
  - Byte 0 bit 0 (I/G) indicates whether the address is an *individual* (0) or a *group* (1) address. Generally, each lobe has its own unique (individual) address. However, in group addressing, multiple lobes share the same address for the purpose of communication (so that a frame sent to that location will be received by each lobe that belongs to the group).
  - Byte 0 bit 1 (U/L) indicates whether the address is administered *universally* (0) or *locally* (1). In universal administration, NIC

hardware addresses (those assigned by the IEEE and the board's manufacturer) are used. In local administration, software or switch configurable addresses are used.
  - Byte 2 bit 0 (FAI). The *functional address indicator* is 0 if the address is a functional one and 1 otherwise. A functional address specifies a lobe with a particular function (Token Ring management or user-defined). The table lists predefined functional addresses.

Address	Server with Address
C00000000001	Active Monitor (AM)
C00000000002	Ring Parameter Server (RPS)
C00000000008	Ring Error Monitor (REM)
C00000000010	Configuration Report Server (CPS)
C00000000100	Bridge
C00000002000	LAN Manager
C00000800000 to	User-Defined Servers
C00040000000	

- *Source address* (*SA*)—6 bytes. Indicates the frame's originating location. The I/G and U/L bits are used in the same way as the destination address (DA) uses them.
- *Routing information* (*RI*)—0+ bytes. Information regarding the bridges or routers through which the frame must pass if the destination is on another network. If present, the first 2 bytes are routing control; the remaining bytes are grouped into pairs, each of which identifies a bridge or router in the path.
- *Data* (*D*)—0+ bytes. The data field contains either LLC information or MAC information.
  - An *LLC frame* is received from the LLC sublayer (defined in the IEEE 802.2 standard). It contains a packet from the higher layer protocol, which is being sent as data to another node. For such a frame, the data field (D) is known as the protocol data unit (PDU). The PDU is broken down into the destination service access point (DSAP) address, source service access point (SSAP) address, and control subfields. The 1-byte DSAP provides information about the process running at the layer that will be receiving the packet. The one byte SSAP provides information about the process running at the layer that is sending the packet. And the 1- or 2-byte control field indicates the type of subsequent data bytes.
  - A *MAC frame* issues commands and provides status information. There are 25 MAC frame types defined; 15 can be used by workstations, and the remaining 10 are used by the active monitor (AM) or by special management servers. The MAC field has three subfields: a 2-byte length indicator that specifies the length of the control subfield, a 2-byte Major Vector Identifier (MVID) that identifies the frame function, and the control information itself (zero or more bytes).
- *Frame check status* (*FCS*)—4 bytes. A 32-bit cyclic redundancy check (CRC) value computed by the sender. Upon reception, the receiving lobe also computes the value and compares it with the field's value. If the values match, the frame is assumed to be properly received and intact.
- *End delimiter* (*ED*)—1 byte. A unique byte used to indicate the end of the frame. The byte is the same as the starting delimiter (SD).
- *Frame status* (*FS*)—1 byte. Contains information about what happened to the frame as it traversed the ring. It is also used in error control.
  - Bits 0 and 4 are *Address Recognized* (A) bits. They are set to 0 by the sender and are changed to 1 when the destination lobe recognizes the source address. If the frame returns to the sender with these bits still equal to 0, the sender assumes the destination node is not on the ring.

- Bits 1 and 5 are Copied (C) bits. They are set to 0 by the sender and are changed to 1 when the destination lobe copies the frame's contents to its input buffer. If the frame is not received correctly, the destination node sets the Address Recognized bits to 1 but leaves the Copied bits set to 0. The sender then knows that the destination is on the ring and that the frame was not received correctly.
- Bits 3, 4, 6, and 7 are reserved for future use.

token rotation time (TRTn) A token passing network parameter that specifies the maximum time before every node on a network gets access to the token.

TokenTalk Apple Computer's token ring protocol for transmissions in its own AppleTalk environments.

TokenTalk characteristics include:

- Definition at the lowest two ISO/OSI Reference Model layers, i.e., the physical and data link.
- Use of the TokenTalk Link Access Protocol (TLAP) to gain access to the network.
- Support for both 4- and 16-Mbps IEEE 802.5 Token Ring networks.

tolerance The permissible range a quantity is allowed to vary from its nominal value before it is considered out of specification. The *tolerance* is the difference between the maximum and minimum limits. The limits need not be symmetrical, that is, an upper limit may be specified as one number (+5 for example), while the lower limit is specified at another (−1 for example).

tolerance field (**1**) The region between two curves, such as circles or rectangles, used to specify the tolerance on component size and geometry.

For example, the tolerance field in the diagram ($D \pm \Delta D$ and $d \pm \Delta d$) is used to specify the cross section of an optical fiber, that is,

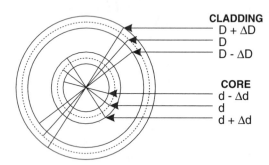

Together, the tolerance field
- Specifies respective diameters.
- Specifies respective ovality.
- Specifies core-cladding concentricity.

When both the core and cladding boundaries of the cross section of the fiber in question simultaneously fall entirely within their respective defined areas, the fiber meets the specification.

toll call A call for a destination outside the service area of the local central office to which the calling subscriber is connected.

toll center (TC) A switching office where trunks from end offices (class 5 offices) are connected to intertoll trunks and operators provide assistance with the completion of incoming calls.

Toll centers are class 4C offices. See also *class n office* and *toll point.*

toll circuit A circuit connecting two exchanges in different localities (generally greater than 15 miles apart). Exchanges less than 15 miles apart are generally called *junction circuits.* Also called *trunk circuit* (in Great Britain).

toll connecting trunk A trunk connecting a local office (class 5 office) and a toll office (class 4 office) or switchboard.

toll denial A feature on a private automatic branch exchange (PABX) that allows a user to dial local numbers but denies access to toll numbers or the toll operator.

toll diversion A switching system service feature by which users are denied the ability to place toll calls without the assistance of an attendant.

toll line A *toll line* is a telephone line (or channel) connecting two different telephone exchanges.

toll office Any intermediate office serving toll calls, i.e., an office controlling the switching of toll trunks. Also called *trunk exchange* in Great Britain.

toll point (TP) A switching office where end offices (class 5 offices) are connected to the distance dialing network, and, if present, operators assist with outgoing calls.

Toll points are class 4P offices. See also *class n office* and *toll center.*

toll quality The voice quality resulting from use of a nominal 4-kHz telephone channel. It may be specified in analog systems in terms of S/N ratio and bandwidth or in digital systems in terms of a specified bit error rate (BER).

toll restriction A method that prevents a private automatic branch exchange (PABX) from completing selected toll calls (or even reaching the toll operator) without the assistance of an attendant. Also called a *classmark.*

toll switching trunk A circuit (trunk) connecting one or more end offices (local offices) to a toll center as the first stage of concentration for intertoll traffic.

In the United States, a *toll center* in which assistance in completing incoming calls is provided in addition to other traffic is called a Class 4C office. A center in which assistance in completing outgoing calls is provided or where switching is performed without operator assistance is called a Class 4P office. Also called a *trunk junction* in Great Britain.

TOLR (**1**) Abbreviation of <u>TOL</u>l <u>R</u>estricted. (**2**) An abbreviation of <u>T</u>ransmitting <u>O</u>bjective <u>L</u>oudness <u>R</u>ating.

TON (**1**) An acronym for <u>T</u>ype <u>O</u>f <u>N</u>umber. Part of an Integrated Services Digital Network (ISDN) address that indicates nation or international destination. (**2**) An acronym for <u>T</u>one <u>ON</u>.

tone (**1**) A sound wave that can be heard and has pitch. (**2**) Any audible collection of frequencies transmitted over a telecommunications network. Generally, the number of frequencies is limited to one or two, however.

tone burst See *tone pulse.*

tone decay time (**1**) At a tone detector, the time interval between the end of the tone present condition and the beginning of the tone off condition at the end of a tone pulse. See also *tone pulse.* (**2**) At a tone transmitter, the time interval between command to remove a transmitted tone and the instant the tone level falls below a prescribed threshold value.

tone dialing A dialing method that places a unique pair of tones on the telephone line for each digit dialed by a telephone or data communications equipment (DCE).

Tone dialing is also called DTMF dialing (from Dual Tone MultiFrequency). See also *DTMF dialing* and *rotary dialing*.

tone disabling A method of controlling the operation of telecommunications equipment by transmitting a tone signal down the transmission line.

tone duration The time interval during which a tone is continuously present above a specified threshold value.

tone fall time The time interval between the end of the tone-on condition and the beginning of the tone-off condition at the end of the tone pulse under consideration. See also *tone pulse*.

tone leak In telephony, the occurrence of any unintended address or other control tone signal either in the presence of another address or control tone, or during signal off intervals.

tone off The condition in which the level of a tone under consideration is less than a specified "off" threshold value. See also *tone pulse*.

tone pulse A burst of tone energy rising above a specified threshold level for a specified period of time. Also called a *tone burst*.

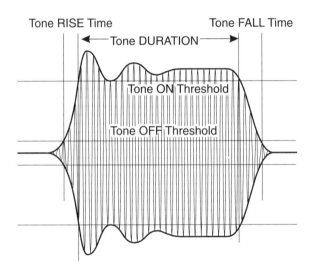

tone present The condition in which the level of a tone under consideration is greater than or equal to a specified "on" threshold value. Also called *tone-on*. See also *tone pulse*.

tone rise time The time interval between the end of the tone-off condition and the beginning of the tone-on condition at the beginning of the tone pulse under consideration. See also *tone pulse*.

tone set In telephony, the collection of in-band tone signals used to set up and tear down a link. The tone set includes: DTMF, R1 MF, and R2 MF.

tone signaling In telephony, the use of in-band tone signals for supervisory, address, and alerting signals. See also *DTMF, DTMF dialing, multifrequency (MF) signaling, single frequency (SF) signaling*, and *trunk multifrequency (MF) signaling*.

tonlar An acronym for Tone Operated Net Loss AdjusteR. In telephony, a system for stabilizing the net loss of a circuit by means of a tone transmitted during idle time.

TOO An acronym for Time Of Origin.

tool A synonym for *utility computer program*.

TOOL An acronym for Test Oriented Operated Language.

TOP An acronym for Technical and Office Protocol. Boeing's version of the manufacturing automation protocol (MAP) protocol suite for office and engineering applications.

TOPES An acronym for Telephone Office Planning and Engineering System.

topography The specification and arrangement of physical locations of actual communication and information system components which implement the topology.

topology The physical and/or logical configuration formed by the host computer and nodes of a local area network (LAN). Basic LAN *topologies* are:

- *bus:* Physically, all nodes and the host share the same communication lines. Logically, packets are broadcast so that every node receives the message at approximately the same time. Also called a *linear topology.*

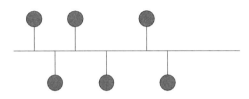

- *daisy chain:* Physically, each node receives all information on one port and retransmits information not directed to itself on another port.

- *ring:* Physically, a separate communication line is provided between each node and the next node as with a daisy chain; and there is a line from the last node back to the first node. That is, every node has exactly two branches connected to it. Logically, each node hears from exactly one node and talks to exactly one other node. Information is passed sequentially, in a previously determined order, from node to node.

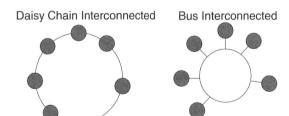

Daisy Chain Interconnected   Bus Interconnected

- *dual ring:* Similar to a ring network except two separate communications paths are involved; each path has its own send and receive path. One ring carries information clockwise while the other ring carries it counterclockwise.

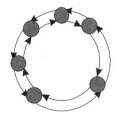

- *star:* Physically, a separate communication line is routed from a central node (or host) to each peripheral node. The central node (or host) rebroadcasts all transmissions received from any path to all connected paths. All peripheral nodes may communicate with all other nodes by transmitting to, and receiving from, the central node only. A failure of a transmission line, i.e., the channel, linking any peripheral node to the central node will result in the isolation of only that peripheral node from all others. The remainder of the network continues to perform normally.

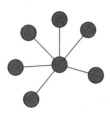

- *tree:* Physically, an extension of the star topology wherein the host is connected to intermediate nodes by star connections; the intermediate nodes are extended to additional intermediate nodes or terminal nodes also in a star fashion. Also called a *distributed bus.*

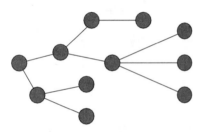

- *mesh:* Physically, each node has a duplex connection with two or more other nodes. In a *full mesh* network, every node is connected to every other node. This is not practical in a network of any size, that is, for a network of *n* nodes there are $n(n-1)/2$ direct paths (branches).

A full mesh topology is also called a *fully connected mesh topology.*

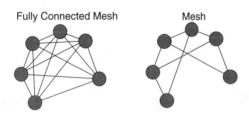

Fully Connected Mesh   Mesh

Hybrid topologies are formed when two or more of the basic topologies are combined in a single network. For example,

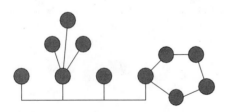

Networks may be physically interconnected with one topology and logically operated with another topology. For example, the Token Ring is a star wired but logical ring topology.

TOPS (**1**) An acronym for $\underline{T}$ask $\underline{O}$riented $\underline{P}$rocedure$\underline{S}$. (**2**) In telephony, an acronym for $\underline{T}$raffic $\underline{O}$perator $\underline{P}$osition $\underline{S}$ystem. (**3**) An acronym for $\underline{T}$imesharing $\underline{OP}$erating $\underline{S}$ystem.

TOPSMP An acronym for $\underline{T}$raffic $\underline{O}$perator $\underline{P}$osition $\underline{S}$ystem $\underline{M}$ulti-$\underline{p}$urpose.

TORC An acronym for $\underline{T}$raffic $\underline{O}$verload $\underline{R}$eroute $\underline{C}$ontrol.

TORES An acronym for $\underline{T}$ext $\underline{OR}$iented $\underline{E}$diting $\underline{S}$ystem.

toroid A geometric shape in the form of a tube with the two ends joined together.

TOS (**1**) An acronym for $\underline{T}$ype $\underline{O}$f $\underline{S}$ervice An attribute in the shortest path first (SPF) algorithm used to select a local area network (LAN) route. (**2**) An acronym for $\underline{T}$echnical and $\underline{O}$ffice $\underline{S}$ystem. (**3**) An acronym for $\underline{T}$aken $\underline{O}$ut of $\underline{S}$ervice. (**4**) An acronym for $\underline{T}$emporarily $\underline{O}$ut of $\underline{S}$ervice.

TOSD An abbreviation of the Rural Electrification Administration's (REA) $\underline{T}$elephone $\underline{O}$perations and $\underline{S}$tandards $\underline{D}$ivision.

total channel noise The sum of random noise, intermodulation noise, and crosstalk. It excludes impulse noise because different techniques are required for its measurement.

total harmonic distortion (THD) The ratio of the sum of all output powers due to the input signal excluding the signal fundamental to the power of the signal fundamental itself. Typically, the ratio is expressed in decibels (dB), that is,

$$THD = 10 \cdot \log_{10}\left(\frac{P_{out} - P_{fundamental}}{P_{fundamental}}\right).$$

total internal reflection Reflection that occurs when wave energy in a material strikes an interface with a medium of lesser refractive indexes and at an angle of incidence equal to or greater than the *critical angle.* See also *critical angle* and *Snell's law.*

total power loss The sum of all forward and reverse power losses.

touch-tone Formerly an AT&T trademarked term for a pushbutton telephone that uses dual tone multifrequency (DTMF) tone signaling. However, at divestiture in 1984, AT&T gave the term to the public. It is now a generic term.

TP (**1**) An abbreviation of $\underline{T}$est $\underline{P}$oint (or Position). (**2**) An abbreviation of $\underline{T}$ransaction $\underline{P}$rocessing. (**3**) An abbreviation from $\underline{T}$wisted $\underline{P}$air. (**4**) An abbreviation of $\underline{T}$ransport $\underline{P}$rotocol. (**5**) An abbreviation of $\underline{T}$oll $\underline{P}$oint. (**6**) An abbreviation of $\underline{T}$oll $\underline{P}$refix.

TP-4/IP A term given to an International Standards Organization (ISO) protocol suite that closely resembles Transmission Control Protocol Internet/Protocol (TCP/IP).

TP-n An acronym for the ISO/OSI Reference Model layer 4 $\underline{T}$ransport $\underline{P}$rotocol class $\underline{n}$ (n =0 to 4).

TRANSPORT PROTOCOL CLASSES

Class	Description
0	Connectionless (ISO 8602)
1	
2	
3	
4	Connection Oriented (ISO 8073)

TP-PMD An abbreviation of $\underline{T}$wisted $\underline{P}$air, Physical $\underline{M}$edium $\underline{D}$ependent.

TPAD An acronym for Terminal Packet Assembler/Disassembler.

TPC (1) An abbreviation of Twisted Pair Cable. (2) An abbreviation of Telephone Pickup Coil.

TPDDI An abbreviation of Twisted Pair Distributed Data Interface.

TPDT An abbreviation of Triple Pole Double Throw switch.

TPDU An abbreviation of Transport Protocol Data Unit.

TPE An abbreviation of Twisted Pair Ethernet.

TPEX An abbreviation of Twisted Pair Ethernet transceiver (Xceiver).

TPF An abbreviation of Transaction Processing Facility. IBM host software for online transaction processing (OLTP).

TPFSE An abbreviation of Transport Processing Service Element.

TPON (1) An acronym for Telephone Passive Optical Network. (2) An acronym for Open Systems Interconnect (OSI's) Transport Protocol Class O. The simplest of the OSI transport protocols. It is generally used only with networks that provide other means of error control.

TPST An abbreviation of Triple Pole Single Throw switch.

TPTB An abbreviation of The Powers That Be.

TQFP An abbreviation of Thin Quad Flat Pack.

TR (1) An abbreviation of Token Ring. (2) An abbreviation of Technical Reference (or Technical Requirement). A final Bellcore standard. (3) An abbreviation of Trouble Report. (4) An abbreviation of Transmission Report. (5) An abbreviation of TransmitteR. (6) An abbreviation of Traffic Route. (7) An abbreviation of Test Run.

TR29.2 See *FAX class 2.*

trace packet A special kind of packet in a packet switching network that causes each node encountered on its journey from source to destination to generate a report that is sent to the network control center.

track (1) That portion of a moving data medium, such as a magnetic disk or drum, that is accessible to a given head. See also *cylinder* and *sector.* (2) The conductors on a printed circuit board (PCB). Also called a *trace.*

track density The number of tracks per unit distance, measured in a direction perpendicular to the direction in which the tracks are read.

tracking In documents or text, the spacing between characters in a line of text. This is defined when a font is designed but can often be altered in order to change the appearance of the text or for special effects. Tracking should not be confused with *kerning,* which deals with the spacing between certain pairs of characters. See also *kerning* and *leading.*

tracking A/D converter See *delta sigma modulator.*

tracking error The deviation of a dependent variable with respect to a reference function.

For example, in a phase lock loop (PLL), the phase error between the local oscillator and the incoming reference signal when the frequencies are the same (i.e., the PLL is locked).

tracking mode An operational mode during which a system is operating within specified movement limits relative to a reference. Also called *tracking phase.*

trade-off link In Internet, a tie from one home page to a related home page which allows easy exploration (and possibly a return through a reciprocal link).

traffic (1) In networking, the information moved over a communication channel. (2) In telephony, a quantitative measurement of the total messages and their length, expressed in hundred call seconds (CCS) or other units, during a specified period of time. The combined connect time or occupancy time of one or more communications paths. Also called *traffic quantity.*

traffic analysis (1) In a communications system, the analysis of traffic rates, volumes, densities, capacities, and patterns specifically for system performance improvement. See also *erlang (E), erlang B formula,* and *erlang C formula.* (2) The analysis of electronic and optical communicaitons system environments for use in design, development, and operation of new communications systems. (3) The study of communications characteristics external to the text.

traffic capacity The maximum traffic per unit of time that a given telecommunications link, device, subsystem, or system can carry under specified conditions. Usually expressed in hundred call seconds (CCS) per hour.

traffic carried The number of call attempts that are successfully connected to the destination. Those not carried are lost or delayed, generally due to congestion. See also *traffic offered.*

traffic concentration The ratio of the average busy hour traffic to the total traffic during the day.

traffic descriptor An element, in the asynchronous transfer mode (ATM) architecture, that specifies parameters for a virtual channel connection (VCC) or virtual path connection (VPC). These parameter values can be negotiated by the entities involved in the connection. Also called a *user network contract.*

traffic encryption key (TEK) A key used to encrypt plaintext or to superencrypt previously encrypted text and/or to decrypt cipher text.

traffic engineering The design of a network or network expansion to adequately meet the anticipated traffic loads throughout a communications system. It includes the determination of:

- The numbers and kinds of circuits required, e.g., lines and trunks.
- The quantities of related terminating and switching equipment required.

traffic intensity The ratio of the time during which a facility is occupied (continuously or cumulatively) to the time this facility is available for occupancy over a specified period of time. It is therefore a measure of the average occupancy of a facility during a specified period of time (normally a busy hour) measured in traffic units of erlangs (E).

A *traffic intensity* of one traffic unit (one erlang) means continuous occupancy of a facility during the time period under consideration, regardless of whether or not information is transmitted. Also called *call intensity.*

traffic load The total traffic carried by a link (trunk or trunk group) during a certain time interval.

traffic measure Two basic ways of quantifying traffic are in erlangs and hundred call seconds (CCS).

traffic offered The number of attempts received by a switching system to place calls. See also *traffic carried.*

traffic overflow (1) The condition that occurs when the *traffic offered* to a link, device, or system exceeds its *traffic capacity.* Frequently, links are designed so that excess is re-routed on alternate paths; however, it may just be blocked or lost. (2) The excess traffic itself.

traffic path A channel, circuit, frequency band, line, switch, time slot, or trunk over which individual communications pass.

traffic shaping Allows the station sending information into an asynchronous transfer mode (ATM) network to:

- Specify the priority and throughput of information going into the network, and
- Monitor performance to meet required service levels.

traffic unit A synonym for *erlang* (*E*).

traffic usage recorder A device (or system) used to sample and record the occupancy of equipment, i.e., the amount of telephone traffic carried by a group, or several groups, of switches or trunks.

.trailer Information occupying the last several bytes of a block or packet. *Trailers* often contain checksums or other error control information.

train (**1**) A sequence of events or items, such as a *pulse train.* (**2**) To modify the behavior of a device based on external conditions, as in the *training* of an echo canceler for each new connected path.

transaction (**1**) An interaction between a client and a server. A sequence of messages between a master and client station required to perform a specific function. (**2**) The smallest complete action when using the Structured Query Language (SQL) to search or modify a database. (If any step in the transaction cannot be completed, the entire transaction fails, and all the intermediate steps in the transaction are undone.) (**3**) An entry in a database.

transaction code An identifier or symbol associated with a specific transaction and representing the action to be carried out.

For example, the letter *A* may be used as a transaction code for the operation "add," *D* may be "delete," and so on.

transceiver (**1**) A contraction of TRANSmitter and reCEIVER. A device that can both transmit and receive signals, such as cellular telephones, modems, and network interface controllers (NICs). Often NICs provide some form of collision detection as well. Also known as a *medium attachment unit* (*MAU*) in IEEE specifications. (**2**) In military communications, the combination of transmitting and receiving equipment which:

- Is housed in a common chassis or enclosure,
- Is usually designed for portable or mobile use,
- Uses common circuit components for both transmitting and receiving, and
- Provides half-duplex operation.

transceiver cable In Ethernet, a cable that attaches a terminal device to an Ethernet backbone cable (either 10Base2 or 10Base5).

transcoder A device that directly converts one digital code into another digital code, i.e., without returning the original code to an analog form before generating the new code.

For example, the conversion of μ255-law encoded pulse code modulation (PCM) to A-law PCM for transmission from the United States to countries in Europe.

transducer A device for converting energy from one form (heat, light, sound, temperature, electrical, etc.) to another for either measurement of a physical quantity or information transfer.

Examples of transducers include devices that:

- Convert sound pressure levels into electrical signals (microphones).
- Convert electrical signals into sound pressure waves (speakers).

transfer (**1**) The *movement of data* from one location to another. (**2**) The *passing of control* from one device to another. (**3**) In telephony, a switching system feature that allows a user to reassign a call to different end station.

transfer characteristics Those intrinsic parameters of an entity (system, subsystem, or device) which, when applied to the input of the entity, will fully describe its output. See also *transfer function.*

transfer delay A performance characteristic that expresses the amount of elapsed time required to send a message through a system. It includes not only the link's propagation time but any signal processing time required at either end of the link.

transfer function (**1**) A mathematical statement that describes the *transfer characteristics* of a system, subsystem, or device. (**2**) A rule (the *transfer characteristic*) describing how the output signal of circuit, device or system responds to an input signal. The rule may be stated in mathematical, graphical, or tabular terms. A *transfer function* is essentially the complex ratio of the output signal of the entity to the input signal.

When the *transfer function* (*T*) operates on the input (e_i), the output (e_o) is obtained. Given any two of these three entities (*T*, e_i, and e_o), the third can be obtained, that is,

$$T = e_o / e_i, \ e_o = T \cdot e_i, \ \text{or} \ e_i = e_o / T$$

Simple transfer functions express the ratio of output to input signals when the imaginary part of the signals can be ignored, examples are voltage and current gains, reflection coefficients, transmission coefficients, and efficiency ratios. *Complex transfer functions* include the imaginary part and are frequency dependent, examples include filter response and envelope delay distortion. *Transfer functions* are frequently expressed in terms of amplitude vs. frequency and phase vs. frequency. (**3**) In an optical fiber, the complex mathematical function that expresses the ratio of the variation of the instantaneous power of the optical signal at the output of the fiber (P_o) to the instantaneous power of the optical signal that is launched into the fiber (P_i), as a function of modulation frequency.

transfer mode In telecommunications, the manner in which data are transmitted and/or switched in a network, i.e., synchronous vs. asynchronous, circuit switched vs. packet switched, and so on.

transfer rate The rate at which information is conveyed across a communications channel or circuit. *Transfer rate* is expressed in units per second (e.g., bits per second, characters per second, bytes per second, and so on). It may represent either the maximum number of units per second possible or the average number of units per second (including headers, trailers, and gaps between blocks). See also *baud.*

transfer rate of information bits (TRIB) The *average* transfer rate of actual, error-free, useful information (not including overhead, error bits, or retransmitted bits) through a device or series of devices. Mathematically, *TRIB* can be expressed as:

$$\text{TRIB} = \frac{\text{Number of information bits properly received}}{\text{Total time required to get the bits}}$$

Data compression increases the transfer rate by reducing the number of bits to be transmitted so that more information can be transferred in the same time. *TRIB* is also called the *throughput* or *data transfer rate.*

transfer ratio A dimensionless transfer function.

transfer time (**1**) The time it takes to switch from one process or device to an alternate. (**2**) In gas tube surge protectors, the amount of time required for the voltage across the gap to drop into the arc region after the initial gap conduction begins. (**3**) The amount of time required to transmit and receive a complete message.

transformer An electric device that transfers electrical energy from one circuit to another without a direct connection. This transfer is accomplished by coupling the alternating magnetic field generated in one coil of wire (the primary coil) to a second coil of wire (the secondary). Shown schematically, the basic transformer is a four-terminal device with an input (primary) and an output (secondary). The arrows indicate the direction of current flow and the dots indicate like voltage polarities, positive, for example.

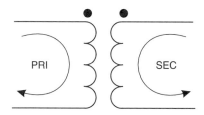

The voltage, current, and impedance ratio between the primary and secondary coils can be determined from the ratio of the number of turns on the primary (n_p) and the number of turns on the secondary (n_s). Specifically:

$$V_s = \frac{n_s}{n_p} V_p$$

$$I_s = \frac{n_p}{n_s} I_p$$

$$Z_s = \frac{n_s^2}{n_p^2} Z_p$$

where the symbols are defined by:

	Primary	Secondary
Voltage	V_p	V_s
Current	I_p	I_s
Impedance	Z_p	Z_s
Turns	n_p	n_s

transhybrid loss The balance return loss (BRL) of a *hybrid* plus 6 dB plus the losses intrinsic to the parts used in constructing the hybrid. The 6-dB loss is due to the division of the transmit signal among the branches of an ideal hybrid bridge.

Intrinsic loss, for example, may be illustrated by a hybrid implemented with transformers. Here, intrinsic losses occur due to the resistance of the wires in the transformer as well as magnetic coupling losses. Typical analog telephone systems will have *transhybrid losses* defined by a mask similar to the one in the chart. See also *balanced return loss (BRL), hybrid,* and *sidetone.*

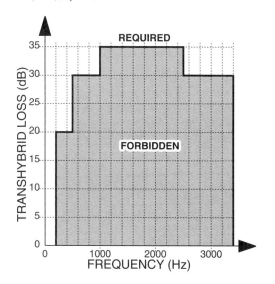

transient (1) In general, fleeting, brief, temporary, or unpredictable. **(2)** That part of a signal that disappears after the completion of a transition from one steady-state operating condition to another. **(3)** In electronics, a short-duration, abnormal, and unpredictable increase in the signal level on a data line or power supply distribution line. *Transients,* if sufficiently large, can damage electronic circuits to which the conducting lines are attached. Also called a *spike.* See also *surge* and *surge protector.*

transient suppression network A circuit composed of capacitors, resistors, inductors, and/or diodes so placed as to control the discharge of stored energy in switched devices.

transient suppressor A device or circuit designed to attenuate or dissipate transients to safe levels, that is, to levels that do not damage the electronics to which the line is attached. See also *gas tube, MOV, surge protector,* and *zener diode.*

transistor A three (or more) terminal semiconductor device that can provide *power gain* to a signal. It was invented at Bell Laboratories by John Bardeen, Walter H. Battain, and William Shockley in 1947. A selection of various transistor types and their symbols is shown in the figure.

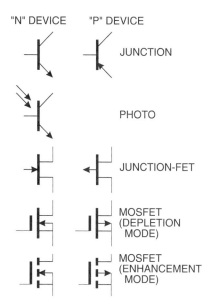

Multiple transistors (and other components) are interconnected on a common substrate to form complex circuits called *integrated circuits.* Examples of integrated circuits include operational amplifiers, logic gates, microprocessors, memory, and other "chips."

transit center International exchange CT1, CT2, CT3 (CT from the French *Centres de Transit*). At the highest level, there are only 7 CT1s in the world which, when fully implemented, will form a full mesh network.

transit delay The elapsed time between the first occurrence of a packet as it passes a given point and the moment that same occurrence passes another point, plus the transmission time (length) of the packet itself. The time required for complete transmission and reception of the packet.

transit network A network that routes traffic among other networks possibly in addition to carrying traffic for its own hosts. A transit network, therefore, must have paths to at least two other networks.

One example of a *transit network* is an Internet backbone.

transit network identification A network service feature that specifies the sequence of networks used to establish or partially establish a virtual circuit.

transit time A synonym for *phase delay.*

transition (1) A signal change from one condition or state to another. The change can occur in a voltage, current, phase, frequency, or other waveform property. (2) In pulse terms, a transition is that portion of a wave between one normal state and the next normal state. (3) In filter terms, the transition is the band of frequencies between the passband and the stopband.

transition band Those frequencies in a circuit, network, or system between the passband and stopband frequencies. See also *band-pass filter, band reject filter, high-pass filter, low-pass filter, passband,* and *stopband.*

transition frequency The frequency associated with the difference between two discrete energy levels in a quantum mechanical system, given by

$$f_{2,1} = \frac{E_2 - E_1}{h}$$

where:

E_1 is the lower energy level,
E_2 is the higher energy level,
h is Planck's constant, and
$f_{2,1}$ is the frequency associated with the difference between two energy levels, E_2 and E_1.

If the energy level of the quantum system "falls" from E_2 to E_1, a photon, phonon, or particle with frequency $f_{2,1}$ will be emitted. If the atomic system is at energy level E_1 and a photon of frequency $f_{2,1}$ is absorbed, the energy level will be raised to E_2. Transition concerned with only photons are called *direct radiative transitions;* otherwise it is called indirect.

translating bridge A bridge designed to interconnect two different networks at the physical and datalink layers, that is, with dissimilar medium access control (MAC) sublayer protocols. When a packet from one network is forwarded to the other network, the bridge translates the protocols to the appropriate packet format.

An example is a bridge used to interconnect Ethernet and Token Ring networks. Also called a *transparent source routing bridge.* See also *bridge, gateway,* and *router.*

translator (1) In general, a device or equipment capable of interpreting or converting information or a signal from one form (representation) to another form. (2) In telephony, a device that translates destination addresses (dialed digits) into call routing path information. (3) A computer program that translates from one language into another language and in particular from one programming language into another programming language. Also called a *translating program.* (4) A repeater station that receives a primary station's signal, amplifies it, shifts it in frequency, and retransmits it. See also *transponder.* (5) A device that converts one frequency to another.

transliterate To convert the characters of one alphabet to the corresponding characters of another alphabet without regard for meaning.

transmission (1) The transferral, for reception elsewhere, of a signal, message, or other form of information. (2) The propagation of a signal, message, or other form of information by any means (such as whistles, telegraph, telephone, radio, television, smoke, mirrors, or facsimile) via any medium (such as wire, coaxial cable, microwave, optical fiber, radio frequency, air, water, or smoke). Data delivery can take place with several different *transmission protocols.* (3) In communications systems, a series of data units, such as blocks, messages, or frames. (4) The transfer of electrical power from one location to another via conductors. (5) The passage of electromagnetic radiation through a medium, e.g., light through glass.

transmission band See *passband.*

transmission block (1) A group of bits, bytes, or characters transmitted as a unit and usually containing additional error control information (parity, CRC, etc.). (2) In data transmission, a group of records sent, processed, or recorded as a unit. A transmission block is usually terminated by an end-of-block character (EOB), end-of-transmission-block character (ETB), or end-of-text character (EOT or ETX).

transmission code A set of rules for representing data, usually, characters. Commonly used transmission codes include EBCDIC (an 8-bit code used on IBM mainframes), ASCII (a 7-bit code commonly used on PCs), and Baudot. See also *ASCII* and *codes.*

transmission coefficient (1) In optics, the ratio of the transmitted field strength to the incident field strength of an electromagnetic wave when it is incident upon an interface surface between media with two different refractive indices. (2) In a transmission line, the ratio of the amplitude of the complex transmitted wave to that of the incident wave at a discontinuity in the transmission line. (3) The probability that a portion of a communications system, such as a line, circuit, channel, or trunk, will meet specified performance criteria. The value of the *transmission coefficient* is inversely related to the quality of the line, circuit, channel, or trunk.

transmission control character See *control character.*

transmission control protocol (TCP) A transport layer protocol for packetizing data, managing the transmission of the packets, and performing error control. *TCP* is built on top of the Internet Protocol (IP) and is generally seen in the combination TCP/IP (TCP over IP). It adds reliable communication, flow-control, multiplexing, connection-oriented communication, and it provides full-duplex, process-to-process connections. It is that part of the TCP/IP protocol that governs the transfer of sequential data.

TCP is defined in the Internet documents STD7 and RFC793. See also *TCP/IP.*

transmission delay The length of time it takes an arbitrary data element (or packet) to traverse a specified link. The delay is the sum of all sources, e.g., propagation delay and processing delays.

transmission facility The transmission medium and all associated equipment required to send and receive signals on the medium.

transmission frame A data structure consisting of multiple fields as specified by a particular protocol. The frame contains a minimum of two fields: a starting delimiter and information, although, generally at least a destination address, a frame type, and some form of error control are also present. See also *packet format* and *token ring packet.*

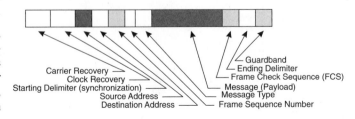

Carrier Recovery
Clock Recovery
Starting Delimiter (synchronization)
Source Address
Destination Address
Guardband
Ending Delimiter
Frame Check Sequence (FCS)
Message (Payload)
Message Type
Frame Sequence Number

transmission gain A general term describing the increase in signal power during conveyance from one point to another. *Transmission gain* is the algebraic sum of all gains and losses encountered by the signal. *Transmission gain* is generally expressed in decibels (dB).

transmission header (TH) In IBM's Systems Network Architecture (SNA) networks, an element added to a basic information unit (BIU) at the path control layer. The BIU, together with the *TH*, forms a path information unit (PIU). See also *header*.

transmission level The ratio of electrical or optical signal power at a point in a transmission system to the power at a corresponding reference point (transmission level point, TLP). *Transmission level* is generally expressed in decibels (dB).

transmission level point (TLP) In a telecommunications system, a specified location in a circuit or system where the nominal level of a test signal is specified and to which measurements in other parts of the system are referenced. It may be a physical test point (a point where a signal may be inserted or measured).

Generally, the abbreviation *TLP* is used and modified to indicate the nominal level at the point in question and is usually expressed in decibels (dB). For example,

- 0 dBm TLP (or 0TLP) is used to indicate the nominal level is 0 dBm.
- −16 dBm TLP (or −16TLP) is used to indicate the nominal level is −16 dBm.

The nominal level at a specified *TLP* is a function of the particular system design. In telephony, the frequency is approximately 1000 Hz (1004 Hz in the United States and 800 Hz per CCITT / ITU-T).

transmission limit The wavelengths above and below which an optical fiber ceases to be a practical waveguide and therefore is no longer useful as a transmission medium.

transmission line The material medium or structure that forms a path from one place to another for directing the transmission of energy (such as electric currents, magnetic fields, acoustic waves, or electromagnetic waves).

Examples of transmission lines include twisted pairs, optical fibers, coaxial cables, rectangular closed waveguides, dielectric slabs, and the parallel power distribution wires.

transmission loss A general term describing the decrease in signal power during conveyance from one point to another. *Transmission loss* is the algebraic sum of all gains and losses encountered by the signal and is generally expressed in decibels (dB).

transmission medium Any physical substance that can be used for the propagation of signals (such as modulated electromagnetic or acoustic waves). Common *transmission media* include: fiber optic cable, twisted-wire pair, coaxial cable, dielectric-slab waveguide, water, and air. By extension, free space is considered a *transmission medium* for electromagnetic waves, although it is not a physical medium.

transmission mode (1) A characterization of the manner in which a communication between a sender and a receiver can take place. Classification is based on data flow, physical connection, and timing.

The following data flow modes are defined:

- *Simplex:* Communication goes in one direction only, and the sender can use the entire communication channel bandwidth. Broadcast radio is an example.
- *Half duplex:* Bidirectional communication is possible but only in one direction at a time. The sender can use nearly the entire chan-

nel bandwidth. In order to change direction, a special signal must be transmitted by the receiver and acknowledged by the sender. The time required to turn control over to the other side is called the transmission turnaround time (or just turnaround) and can sometimes become significant.

- *Full duplex:* Communication travels in both directions simultaneously, but each direction may only have a fraction of the channel's bandwidth. Modem connections are an example.

Physical connection modes include:

- *Parallel transfer,* and
- *Serial transfer.*

The following timing modes are defined:

- *Asynchronous,* and
- *Synchronous.*

(2) A form of propagation along a transmission line characterized by the presence of one of the elemental waves, that is,

- Transverse Electric (TE),
- Transverse Magnetic (TM), or
- Transverse ElectroMagnetic (TEM).

transmission objectives The stated set of performance characteristics for a transmission system. They include parameters such as noise, S/N, echo, crosstalk, attenuation distortion, data rate, .frequency shifts, bit error rate, and so on.

transmission protocol The methods and rules adhered to during the transfer of information from one device to another. The protocol involves several parameters such as:

- *Asynchronous* (variable) or *synchronous* (exact) clocking.
- *Serial* (bit by bit) or *parallel* (byte by byte) data transfers.
- *Duplex* or *full-duplex* (simultaneous two way transfers), *half duplex* (alternate two way transfers), or *simplex* (one-way transfers only).
- *Streaming* (continuous data flow) or *burst* (packetized or block data) line transfers.

transmission rate The instantaneous rate at which a transmission facility is processing information. *Transmission rate* is usually expressed in characters per second. Also called *transmission speed*. See also *baud* and *connection rate*.

transmission turnaround time The delay between the transmissions from data communications equipment (DCE) 1 and DCE 2 in half-duplex circuits. It is the time required to:

1. Turn off the transmitter at one end of a communication channel (DCE 1),
2. Allow residual signals to decay in the transmission channel, and finally
3. Turn the transmitter on at the opposite end of the channel (DCE 2).

transmission window A synonym for *spectral window*. See *window*.

transmissivity Obsolete. The internal transmittance per unit thickness of a nondiffusing material. See *transmittance*.

transmittance The ratio of the total transmitted power (or flux) exiting a device to the incident power (or flux) entering the device. In an optical waveguide, therefore, *transmittance* includes losses due to reflections as well as those from absorption.

In an optical system, *transmittance* is generally expressed as a percentage or as optical density. In communication applications it is generally expressed in decibels (dB). Formerly called *transmission*.

transmitter Any device that is able to send encoded data to another location.

transmitting objective loudness rating (TOLR) A loudness rating of telephone connections. It is defined mathematically as:

$$TOLR = -20 \cdot \log_{10}\left(\frac{V_T}{S_M}\right)$$

where V_T is the output voltage of the transmitting element in mV and S_M is the sound pressure level at the mouth reference point, measured in pascals. See also *ROLR (receiving objective loudness rating)*.

transmitting station identifier (TSID) The information that identifies the sending station to the receiving station. In a fax transmission, the *TSID* information is usually appended to the message as a header on the fax pages.

transmultiplexer Equipment that transforms signals derived from *frequency division multiplex (FDM)* equipment (such as group or supergroups) to *time division multiplexed (TDM)* signals having the same structure as those derived from pulse code modulation PCM multiplex equipment (such as PCM multiplex signals) and vice versa.

transparency **(1)** The property of an entity that allows another entity to pass thorough it without altering either of the entities. **(2)** The property in which transmission of electromagnetic radiation occurs without appreciable scattering or diffusion. **(3)** In telecommunications, the property that allows a transmission system or channel to accept, at its input, user information, and deliver corresponding user information at its output, unchanged in form or information content. (The user information may be changed internally within the transmission system, but it is restored to its original form prior to delivery without the involvement of the user.) **(4)** The quality of a data communications system or device that uses a bit-oriented link protocol that does not depend on the bit sequence structure used by the data source. **(5)** The effect in which, to an observer or a user, an entity does not appear to be present but actually is. Contrast with *virtual* (apparently there, but actually not).

transparent adaptive routing The ability of a network to respond to dynamic conditions, such as the temporary loss or destruction of part of the network in such a way that alternative routes are automatically found and selected. This is one of the cornerstones of the Internet.

Transparent Asynchronous Transmitter Receiver Interface (TAXI) A 100-Mbps asynchronous transfer mode (ATM) physical interface specification based on Fiber Distributed Digital Interface (FDDI) PHY.

transparent bridge A network bridge with sufficient "intelligence" to make routing decisions itself; it is therefore "transparent" to both end nodes. In the method, each bridge along the path passes frames forward one hop at a time (based on tables associating end nodes with bridge ports). A *transparent bridge* involves at least learning node addresses, preventing topological loops, and packet forwarding. Also called a *learning bridge.*

transparent mode A communications mode in which a device passes received information on to the next element without interpreting or acting upon it.

Examples include:

- A modem, when in transparent mode does not respond to the flow control characters XON and XOFF.
- A terminal display mode in which control characters are displayed literally, rather than being interpreted as commands (e.g., the <BEL> character is displayed as Ctl-G rather than causing the bell to ring or beep).

- In binary synchronous communications (BSC), the transmission of binary data with the recognition of most control characters supressed. Entry to and exit from the transparent mode are indicated by the sequences DLE-STX and DLE-ETX (or DLE-ETC, DLE-ITB, ENQ).

transponder **(1)** A transceiver that receives a signal on one frequency, amplifies it, and retransmits it on another frequency, e.g., a communications satellite relay. See also *repeater.* **(2)** A transceiver that receives a signal and transmits a response signal on the same or different frequency. If the response is on a different frequency, it is called a *crossband transponder.* For example, an aircraft transponder responds with an ID when interrogated by air-traffic-control secondary radar (beacon radar).

transport driver A network device driver that implements a protocol for communicating between LAN (local area network) Manager and one or more media access control drivers. The transport driver transfers LAN Manager events between computers on the local area network.

transport efficiency The ability of a communications system to carry information using no more resources than necessary. Circuit switched communications links are among the least efficient, while packet switched or statistically switched systems tend to be the most efficient because they remove silent intervals from information transfers.

transport layer The fourth layer in the ISO/OSI communications Reference Model which manages the transmission of data between end points of a channel. (A channel may be composed of more than one link; link management is handled in layer 3.) All layers above the transport layer are network independent.

Two methods of establishing a connection are possible: *connection oriented* and *connectionless protocols.*

- Connection-oriented protocols (such as Internet's transmission control protocol/Internet control protocol) keep transmitting messages until receipt is acknowledged by the destination.
- Connectionless protocols (such as the Internet User Datagram Protocol—UDP) transmit the information once and assume it made it without error.

See *connection oriented, connectionless mode transfer,* and *OSI.*

transport medium A synonym for *transmission medium.*

transposition **(1)** In general, the interchange of two elements. **(2)** Interchanging the relative position of two conductors at regular intervals to reduce inductive crosstalk. A form of a twisted pair. **(3)** In data transmission, the interchanging of the sense of individual signaling elements, that is, a logical 1 becomes a logical 0 or the reverse.

transversal filter See *finite impulse response (FIR) filter.*

transverse interferometry A method of measuring the refractive index profile of an optical fiber. The technique involves placing the fiber in an interferometer, illuminating it transversally to the fiber axis, and analyzing the resultant interference pattern with a computer.

transverse mode Modes that have a field vector normal to the direction of propagation. Three modes are possible:

- The *transverse electric (TE)* mode in which the electric field vector is normal to the direction of propagation. In an optical fiber, the TE modes correspond to meridional rays.
- The *transverse magnetic (TM)* mode in which the magnetic field vector is normal to the direction of propagation. In an optical fiber, the TM modes correspond to meridional rays.

- The *transverse electromagnetic* (*TEM*) mode in which both electric and magnetic field vectors are normal to the direction of propagation. The TEM mode is the most useful mode in a coaxial cable.

See also *transmission mode*.

transverse offset loss In fiber optics, a synonym for *lateral offset loss*.

transverse parity check A parity check performed on a group of binary digits in a synchronized parallel data stream and in a transverse direction to the data stream, such as the parallel tracks of magnetic tape. See also *checksum, cyclic redundancy check* (*CRC*), *longitudinal redundancy check* (*LRC*), *parity,* and *vertical redundancy check* (*VRC*).

transverse redundancy check (TRC) In synchronized parallel bit streams, a type of redundancy check or error check. It is based on rules, such as parity or cyclic redundancy checking (CRC), applied across the parallel bit streams and blocks.

When the *TRC* is based on a parity bit applied to each character and block, it is a weak error detector and cannot correct any errors. This is because an even number of errors in the same character (or block) does not affect parity of character or block. If however, *TRC* is combined with longitudinal redundancy checking (LRC) individual erroneous bits can be corrected. Also called *vertical redundancy check* (*VRC*). See also *checksum, cyclic redundancy check* (*CRC*), *longitudinal redundancy check* (*LRC*), *parity,* and *vertical redundancy check* (*VRC*).

transverse scattering A method of measuring the refractive index profile of an optical fiber. The technique involves illuminating the fiber with a coherent source transversally to its axis and analyzing the far-field irradiance pattern. (A computer is required to interpret this pattern of scattered light.)

trap door function A mathematical function that is easy to compute but whose inverse is very difficult to compute. Such functions are important to applications in cryptography, specifically in public key cryptography.

trapped electromagnetic wave An electromagnetic wave bounded in a region by virtue of

- The region in which the wave entered is surrounded by material of a lower refractive index.
- The direction of propagation of the wave is at an incident angle greater than the critical angle at all surfaces of that region. Hence, total internal reflection will occur, thus bounding the wave.

Dielectric slabs, optical fibers, waveguides, and layers of air can serve as an electromagnetic wave trap, thus confining the wave to a specific direction of propagation.

trapped mode A synonym for *bound mode.*

trapped ray A synonym for *guided ray.*

traveling class mark (TCM) A code that accompanies a long-distance call and is used by the distant system to determine the best available long-distance line consistent with the calling privileges. The *TCM* indicates the level of restriction to be used based on the phone, trunk, attendant, or access code used when initiating the call.

traveling ionospheric disturbance A wavelike disturbance in the electron density of the ionosphere.

traveling wave A wave of energy that:

- Propagates in a *transmission medium* (including free space for electromagnetic waves),
- Has a propagation velocity determined by the launching conditions and the physical properties of the medium, and
- May be a longitudinal or transverse wave or have components of both.

Examples of traveling waves include: radio waves propagating in free space, lightwaves propagating in optical fibers, water waves on the surface of the ocean, sound waves in the air, and seismic waves in the Earth. A traveling wave is not a wave that is reduced to a standing wave by reflections from a distant boundary.

trawl (1) To sift through large volumes of data on the Internet (e.g., USEnet postings, FTP (file transfer protocol) archives, or WWW pages). (2) See *trolling.*

TRC (1) An abbreviation of Transverse Redundancy Check. (2) An abbreviation of Transmit-Receive Control unit. Related categories include:

- *TRC-AS* from Transmit-Receive Control unit—Asynchronous Start/stop.
- *TRC-SC* from Transmit-Receive Control unit—Synchronous Character.
- *TRC-SF* from Transmit-Receive Control unit—Synchronous Frame.

TREAT An acronym for Trouble Report Evaluation Analysis Tool.

tree A hierarchical structure consisting of nodes connected by branches in which:

- Each node is connected to one and only one hierarchical superior node by a branch (i.e., there is one and only one path between any two nodes).
- A given node is considered to be an *ancestor* of all the lower level nodes to which the given node is connected.
- There is a unique node, called the root node, to which only branches to subsidiary nodes exist. (It is the ancestor of all other nodes.)

tree network A network topology where there is only one route between any two nodes. See also *network topology* and *topology.*

tree search A search in which it is possible to decide, at each step, which part of the tree may be rejected without a further search.

tree structure See *tree.*

treeware Printouts, books, and other information transport media made from paper (pulped trees).

trellis code modulation (TCM) A forward error-correcting technique utilizing an enhanced form of quadrature amplitude modulation (QAM). *TCM* is a specific data-encoding algorithm wherein one or more redundancy bits are added to each data block at the sending data communications equipment (DCE) creating a constellation with more points than are necessary to represent the message. The added bits permit only certain *sequences* of signal points to be valid in the constellation. The receiving DCE decodes the signal. If the decoded signal is not in an allowable position (by the rules of *TCM*), the receiver takes the closest allowable state on the trellis (thereby correcting the received error). The result of this process is an improvement in the *bit error rate* (*BER*).

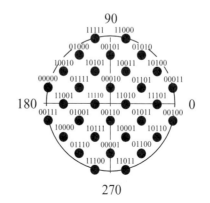

The ITU-T V.32 signal structure with trellis coding for 9600 bps transmission. The code is a 4-data bit plus 1 redundancy bit code. The 5 bits yield the 32 possible points on the phase amplitude map shown on page 545.

TRF Abbreviation of <u>T</u>uned <u>R</u>adio <u>F</u>requency.

triaxial cable A three-conductor cable construction in which all conductors have the coincident axes.

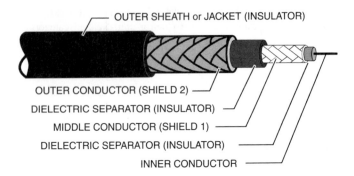

OUTER SHEATH or JACKET (INSULATOR)
OUTER CONDUCTOR (SHIELD 2)
DIELECTRIC SEPARATOR (INSULATOR)
MIDDLE CONDUCTOR (SHIELD 1)
DIELECTRIC SEPARATOR (INSULATOR)
INNER CONDUCTOR

TRIB An abbreviation of <u>T</u>ransfer <u>R</u>ate of <u>I</u>nformation <u>B</u>its.

tribit Literally means 3 bits. It generally refers to any modulation technique that encodes 3 bits at a time to generate 1 of 8 symbols for transmission (1 baud). For example, in a phase modulation scheme the 000 *tribit* could be transmitted as a single frequency carrier, the *tribit* 001 could be the same frequency but phase shifted by 45° from *tribit* 000, *tribit* 010 could be transmitted at a phase shift of 90°, *tribit* 011 at a phase shift of 135°, and so on. Tribit encoding is a specific form of *multilevel encoding*. See also *dibit* and *quadbit*.

tributary circuit A circuit connecting an individual phone to a switching center.

tributary office In telephony, a central office that has a different rate center from the toll office with which it communicates.

tributary station In a data network, any station other than the control station.

TRICKLE A utility program that allows a user to subscribe to an archive site in order to obtain periodic summaries of the new files available at the site.

trickle charge To charge a battery system continuously at a low rate. The purpose is to provide sufficient charging energy to replace the naturally occurring internal losses an unused battery experiences over time.

trigger The combination of events that satisfy the criteria required to initiate some action. A trigger may initiate hardware or software activities. For example, the trigger for a virus, may be a specific date, number of files infected, or any other activity.

trillion In the United States and Canada, a number represented by 10^{12}, in Germany, Great Britain, and France by 10^{18}. Because of the possible confusion in meaning, one should use the SI prefix *tera-* for 10^{12} and *exa-* for 10^{18}.

trimmer A device used to make small adjustments of a variable in order to optimize the performance of a device.

triplet A byte composed of 3 bits. Also called a *three-bit byte*.

trit A contraction of <u>TR</u>inary digi<u>T</u>. It is a single symbol or a digit with three unique values. (A bit has 2 values and a decimal digit has 10.) Sometimes called a *3-state bit* or a *bit and a half*.

trivial file transfer protocol (TFTP) A file transfer protocol with minimal capability and minimal overhead used on the Internet. It is a subset of FTP but does not provide password protection or user directory capability. It uses the *User Datagram Protocol* (*UDP*) to quickly send files across the network.

The message is broken into numbered blocks that are transmitted to the destination one at a time. Upon receiving a block, the destination sends an acknowledge to the sending station, which then sends the next block to the destination.

TRMTR An abbreviation of <u>TR</u>ans<u>M</u>i<u>T</u>te<u>R</u>.

trn A popular UNIX threaded newsreader. In addition to the grouping capabilities of a threaded newsreader, *trn* provides the facilities for automatic file extraction when the binary posting was encoded with *uuencode*. See also *newsreader*.

trojan horse See *virus*.

trolling The act of posting a frivolous, idiotic, or facetious message on a bulletin board or newsgroup (baiting) in the hope an unsuspecting newbie, gullible, arrogant, or egotistical person would respond (the catch). The response more often than not demonstrates that person's ignorance—often in a hilarious manner to all—except the target.

Don't respond to trolling: it is exactly their desire, i.e., to obtain any response.

troposcatter A contraction of <u>TROPO</u>spheric <u>SCATTER</u>.

troposphere The lower region of the atmosphere (between the surface and the stratosphere), in which the temperature decreases with height.

- It contains approximately 80% of the total air mass.
- Its thickness varies with both season and latitude. It is usually 16 km to 18 km thick over tropical regions and 6 km to 10 km thick over the poles.
- It is the region where clouds form, convection is active, and mixing is continuous and more or less complete.

tropospheric duct See *atmospheric duct*.

tropospheric scatter propagation The propagation of radio waves in the troposphere over distances greater than expected line of sight distances. The effect is primarily evident at frequencies between 350 MHz and 8 GHz.

The full propagation mechanism is not fully understood but includes mechanisms such as:

- Propagation by means of random reflections and scattering from irregularities in the dielectric gradient density (refractive index) of the troposphere,
- Smooth-Earth diffraction, and
- Diffraction over isolated obstacles such as mountains (knife-edge diffraction).

Also called *troposcatter*.

tropospheric wave An electromagnetic wave that is propagated by reflection from a place of abrupt change (or a gradient) in the dielectric constant.

In some cases, the ground wave may be so altered that new components appear to arise from reflections in regions of rapidly changing dielectric constant. When these components are distinguishable from other components, they are called *tropospheric waves*.

trouble ticket (1) A piece of paper or a record in a computer system used to report and manage the resolution of network or circuit outages. (2) An error log in network fault management. When a fault or problem occurs in a network, an administrator addresses it. This administrator fills out a *trouble ticket* to indicate that the fault has been detected and is being worked on. When the fault is resolved, the administrator adds the resolution and date to the trouble ticket. *Trouble tickets* can be stored in a problem library where they would serve as both a reference and performance data.

TRR An abbreviation of Tip-Ring Reverse.

TRS (1) An abbreviation of Terrestrial Radio System. (2) An abbreviation of Toll Room Switch.

TRT An abbreviation of Traffic Route Testing.

TRTn An abbreviation of Token Rotation Timer.

TRU An abbreviation of Transmit-Receive Unit.

true complement A number representation that can be derived from another number by subtracting each digit position from one less than the number base and adding one to the resulting composite value. See also *complement*.

true power A synonym for *effective power*.

truncate (1) To terminate a computational process in accordance with a specified criteria, e.g., to end the evaluation of an infinite series at a specified term. (2) The deletion or omission of a leading or trailing portion of a string. The deletion is generally by a specified rule; however, in communications systems, packet *truncations* can occur randomly. (3) To discard all digits with significance less than a specified value, for example, truncating the decimal value of ⅔ to 2 significant places is 0.66. See also *round off* and *truncation error*.

truncated binary exponential backoff See *binary exponential backoff* and *CSMA/CD*.

truncation error The error introduced when one or more digits are truncated from a number. See also *truncate*.

trunk (1) In telephony, a communications link between two central offices or switching devices which, when required, is utilized as part of the communications link between subscribers. (2) A link between private automatic branch exchanges (PABXs) or other switching equipment, as distinguished from circuits that extend between central office switching equipment and information origination/termination equipment. (3) In a token ring, the cable running between multistation access unit (MSAUs). It can be either fiber or shielded twisted pair (STP) cable. STP uses two positive transmit wires in normal mode, with no crossover, while fiber has one transmit fiber and one receiver fiber. In normal mode, the second pair of wires is not used. It acts as backup and implements the Wrap feature. *Trunks* (1) and (2) may be used to interconnect switches to form networks.

trunk circuit (1) In telephony, the interface circuit between a trunk and the switching system. It is similar to a *line circuit* on the subscriber side of the switch. (2) In Great Britain, a synonym for the American *toll circuit*.

trunk encryption device (TED) A bulk encryption device used to provide secure communications over a wideband digital transmission link. It is usually located between the output of a trunk group multiplexer and a wideband radio or cable facility.

trunk exchange The British equivalent of the American term *toll office*.

trunk finder See *finder switch*.

trunk group A collection of trunks between two switching systems that can be used interchangeably.

trunk group multiplexer (TGM) A time division multiplexer/demultiplexer sustem that combines individual digital trunk groups into a higher rate bit stream for transmission over wideband digital communications links.

trunk junction The British equivalent of the American term *toll switching trunk*.

trunk member A single circuit in a trunk group.

trunk multifrequency (MF) signaling A means of sending register and address signaling over a trunk by using combinations of two of six frequencies to represent digits. *MF signaling* is similar to dual tone multifrequency (DTMF) signaling; however, it uses different frequencies, that is,

Signal	Frequency (Hz)					
	700	900	1100	1300	1500	1700
Code 11	X					
Code 12		X				
KP-start pulsing			X			
KP2 - transit traffic				X		X
End pulsing					X	X
Digit 1	X	X				
Digit 2	X		X			
Digit 3		X	X			
Digit 4	X			X		
Digit 5		X		X		
Digit 6			X	X		
Digit 7	X				X	
Digit 8		X			X	
Digit 9			X		X	
Digit 0				X	X	

trunk occupancy A measure of the time that a trunk or group of trunks is busy. It is generally over a one-hour period and is expressed as a percentage or in hundred call seconds (ccs) per trunk.

trunk queuing A feature of a private automatic branch exchange (PABX) in which requests for outgoing trunk circuits are automatically stacked and processed as circuits become available. The queuing is generally on a first-in-first-out (FIFO) basis.

trunk reservation A feature of a private automatic branch exchange (PABX) in which the attendant can hold a trunk and connect it to a specific station.

trunk restriction A feature of a private automatic branch exchange (PABX) in which some stations are not allowed access to certain trunks or possibly only at certain times.

trunk segment The main distribution cable in an Ethernet network.

trunk side In telephony, connections that are attached to other switching systems.

trunk-to-trunk by station A private automatic branch exchange (PABX) feature that enables the user who established a three-way conference involving the user's station and two trunks to drop from the call without disconnecting the trunk-to-trunk connection.

trunk-to-trunk connections A private automatic branch exchange (PABX) feature that allows the connection between two outside parties on separate trunks. For example, an attendant could route an incoming call to another location by dialing the number of that location and then connecting the two trunks. Some PABXs have the ability to allow incoming calls to dial outside numbers without attendant intervention.

trunk-to-trunk consultations A private automatic branch exchange (PABX) feature that allows a station connected to a trunk to gain access to a second trunk for consultation. (Conferencing is not possible and neither trunk party can hear the other.)

trunks in service The number of trunks in a group that are either carrying traffic or are able to carry traffic. It does not include faulty trunks or those *made busy* for any reason.

truth table (1) A table of all possible input conditions of a Boolean expression and the corresponding output value. The table has two basic sets of entries, all of the independent variables (inputs), and the dependent variable(s) (output).

For example, the expression $Y = (A$ AND $B)$ OR C yields the following table [there are 8 possible input conditions ($2^3 = 8$) and one output or result]:

TRUTH TABLE FOR $(A \cdot B) + C$

Inputs			Output
A	B	C	Y
0	0	0	0
0	0	1	1
0	1	0	0
0	1	1	1
1	0	0	0
1	0	1	1
1	1	0	1
1	1	1	1

See also *Venn diagram.* (2) An operation table for a logic operation.

TS (1) An abbreviation of <u>T</u>ransaction <u>S</u>ervices. Level 7 of IBM's Systems Network Architecture (SNA) protocol stack. (2) An abbreviation of <u>T</u>ime <u>S</u>lot. A channel in many implementations of time division multiplex (TDM). (3) An abbreviation of <u>T</u>ime <u>S</u>witch. (4) An abbreviation of <u>T</u>elecommunications <u>S</u>ystem. (5) An abbreviation of <u>T</u>oll <u>S</u>witch (or Switching). (6) An abbreviation of <u>T</u>ransmission <u>S</u>et.

TSA An abbreviation of Time Slot Access.

TSAP An acronym for <u>T</u>ransport <u>S</u>ervice <u>A</u>ccess <u>P</u>oint.

TSAPI An acronym for <u>T</u>elephone <u>S</u>ervices <u>A</u>pplication <u>P</u>rogramming <u>I</u>nterface. An AT&T and Novell protocol specification.

TSB An abbreviation of <u>T</u>elecommunications <u>S</u>tandardization <u>B</u>ureau. Formed by the International Telecommunications Union (ITU) in 1993 by consolidating the Comité Consultatif Internationale des Radiocommunications (CCIR) and the Comité Consultatif Internationale Téléphonique et Télégraphique (CCITT).

TSC (1) An abbreviation of <u>T</u>ransit <u>S</u>witching <u>C</u>enter. (2) An acronym for <u>T</u>ransmitter <u>S</u>tart <u>C</u>ode.

TSCPF (1) An abbreviation of <u>T</u>ime <u>S</u>witch and <u>C</u>all <u>P</u>rocessor <u>F</u>rame. (2) An abbreviation of <u>T</u>ime <u>S</u>witch and <u>C</u>entral <u>P</u>rocessor <u>F</u>rame.

TSDU An abbreviation of <u>T</u>ransport <u>S</u>ervice <u>D</u>ata <u>U</u>nit.

TSI (1) An abbreviation of <u>T</u>ime<u>S</u>lot <u>I</u>nterchange (or <u>I</u>nterchanger). (2) An abbreviation of <u>T</u>ransmitting <u>S</u>ubscriber <u>I</u>nformation.

TSID An abbreviation of <u>T</u>ransmitting <u>S</u>tation <u>ID</u>entifier.

TSIU (1) An abbreviation of <u>T</u>elephone <u>S</u>ystem <u>I</u>nterface <u>U</u>nit. (2) An abbreviation of <u>T</u>ime <u>S</u>lot <u>I</u>nterchange <u>U</u>nit.

TSK (1) An abbreviation of <u>T</u>ime <u>S</u>hift <u>K</u>ey. (2) An abbreviation of <u>T</u>ransmission <u>S</u>ecurity <u>K</u>ey.

TSO An acronym for <u>T</u>erminating <u>S</u>ervice <u>O</u>ffice.

TSP (1) An abbreviation of <u>T</u>ime <u>S</u>ynchronization <u>P</u>rotocol (2) An abbreviation of <u>T</u>eleprocessing <u>S</u>ervices <u>P</u>rogram. (3) An abbreviation of or <u>T</u>elecommunications <u>S</u>ervice <u>P</u>riority. (4) An abbreviation of <u>T</u>est <u>S</u>upervisor <u>P</u>osition. (5) An abbreviation of <u>T</u>est <u>SuP</u>ervisor.

TSPEC An acronym for <u>T</u>raffic <u>SPEC</u>ification. See also *resource reservation protocol (RSVP).*

TSPS An abbreviation of <u>T</u>raffic <u>S</u>ervice <u>P</u>osition <u>S</u>ystem.

TSR An abbreviation of <u>T</u>erminate and <u>S</u>tay <u>R</u>esident. Any program that remains resident in random access memory (RAM) when the user exits it. The program can be reactivated by a keyboard key stroke or other external event such as a modem detecting a ring.

TSS (1) An abbreviation of the <u>T</u>elecommunications <u>S</u>tandardization <u>S</u>ector, part of International Telecommunications Union (ITU) Telecommunications Standardization Bureau (TSB). Frequently called the *ITU-T*, the *TSS* was formerly called the CCITT. (2) An abbreviation of <u>T</u>runk <u>S</u>ervicing <u>S</u>ystem.

TSST An abbreviation of <u>T</u>ime <u>S</u>pace <u>S</u>pace <u>T</u>ime network or switch.

TST (1) An abbreviation of <u>Te</u>ST. (2) An abbreviation of <u>T</u>ime <u>S</u>pace <u>T</u>ime network or switch.

TSTA An abbreviation of <u>T</u>ransmission, <u>S</u>ignaling and <u>T</u>est <u>A</u>ccess.

TSTN An abbreviation of <u>T</u>riple <u>S</u>uper <u>T</u>wisted <u>N</u>ematic. A liquid crystal display (LCD) technology that gives better contrast and gray scale displays.

TSTPAC An abbreviation of <u>T</u>ransmission and <u>S</u>ignaling <u>T</u>est <u>P</u>lan and <u>A</u>nalysis <u>C</u>oncept.

TSTS (1) An abbreviation of <u>T</u>ransaction <u>S</u>witching and <u>T</u>ransport <u>S</u>ervices. The uniform transaction processing capabilities agreed to by the Regional bell operating companies (RBOCs) in 1992. (2) An abbreviation of <u>T</u>ime <u>S</u>pace <u>T</u>ime <u>S</u>pace Network or switch.

TSU An abbreviation of <u>T</u>erminal <u>S</u>ervice <u>U</u>nit. A high-speed modem typically used with T-1 (DS0) or fractional T-1 circuits.

TT An abbreviation of <u>T</u>runk <u>T</u>ype.

TT/N An abbreviation of <u>T</u>est <u>T</u>one to <u>N</u>oise Ratio.

TTC (1) An abbreviation of <u>T</u>elecommunications <u>T</u>echnology <u>C</u>ommittee, Japan's standards body. (2) An abbreviation of <u>T</u>erminating <u>T</u>oll <u>C</u>enter.

TTD An abbreviation of <u>T</u>emporary <u>T</u>ext <u>D</u>elay. A control sequence (<STX> <ENQ>) sent by a station that is in the message transfer state when it is not ready to transmit but wants to retain the line.

TTFN An abbreviation of <u>T</u>a <u>T</u>a <u>F</u>or <u>N</u>ow.

TTL (1) An abbreviation of <u>T</u>ransistor <u>T</u>ransistor <u>L</u>ogic. A description of a particular integrated circuit technology. (2) An abbreviation of <u>T</u>ime <u>T</u>o <u>L</u>ive. Refers to the time a packet can exist in a network before it is discarded.

TTO (1) An abbreviation of Traffic Trunk Order. (2) An abbreviation of Transmitter Turn-Off.

TTP An abbreviation of Trunk Test Panel.

TTR (1) An abbreviation of Timed Token Rotation. A particular type of token passing protocol found in Fiber Distributed Data Interface (FDDI), for example. See *token rotation time* (*TRTn*). (2) An abbreviation of Touch Tone Receiver.

TTRT An abbreviation of Target Token Rotation Time. The expected or allowed time period for a token to circulate once around a Fiber Distributed Data Interface (FDDI) ring. See also *token rotation time* (*TRTn*).

TTS (1) An abbreviation of Transaction Tracking System. (2) An abbreviation of Text-To-Speech. (3) An abbreviation of Trunk Time Switch. (4) An abbreviation of Transmission Test Set.

TTTN An abbreviation of Tandem Tie-Trunk Network.

TTX An abbreviation of TeleTeX.

TTXAU An abbreviation of TeleTeX Access Unit.

TTY An abbreviation of TeleTYpewriter equipment. A low-speed electromechanical communications device generally consisting of a keyboard and printer (called a KSR for Keyboard Send and Receive) or sometimes a printer only (called a RO for Receive Only). The *TTY* emulation mode on many communications programs causes the computer to emulate the KSR teletypewriters.

TTYL An abbreviation of Talk To You Later.

TU (1) An abbreviation of Tributary Unit. In the synchronous digital hierarchy (SDH), a virtual container plus the path overhead. (2) An abbreviation of Thank yoU. (3) An abbreviation of Timing Unit. (4) An abbreviation of Traffic Unit. (5) An abbreviation of Transmission Unit.

TUA An abbreviation of Telecommunications Users Association of the United Kingdom.

TUBA An acronym for TCP and UDP with Bigger Addresses. Defined in RFC1347, 1526, and 1561. It is based on the OSI Connectionless Network Protocol (CNLP).

TUC An abbreviation of Total User Cells.

TUG (1) An acronym for Tributary Unit Group. (2) An acronym for Telecommunications Users Group.

tune (1) To adjust the operating frequency of a transmitter, receiver, or subsystem of such a device. The adjustment of a circuit's resonant frequency. (2) To optimize the performance of a device, system or software. (3) To adjust a circuit, device, or system parameter so as to maximize or minimize a second selected parameter (voltage, current, power, etc.).

tunneling A "translation" technique for transferring packets of information through an incompatible system. The translator simply takes the original message packet and wraps it in the frame protocol of the incompatible system. After routing, another translator removes the wrapper, yielding the original, unmodified packet.

This technique is frequently used to ship Internet datagrams across X.25 networks. Similarly, the Apple Internet Router (AIR) can wrap an AppleTalk packet inside X.25 or TCP/IP packets. Also called *protocol encapsulation* and *synchronous pass-through*.

tunneling mode A synonym for *leaky mode* in fiber optics.

tunneling ray A synonym for *leaky ray*.

tunneling router (1) A router that is capable of routing foreign protocols not by protocol conversion but by protocol encapsulation. That is, the router wraps the entire received packet in a protocol format that the network can handle (encapsulation) and then launches it toward the destination. At the far end of the network, another router deencapsulates the packet and passes it on. (2) A router that encrypts an incoming packet and then encapsulates it in a protocol format suitable for transmission over an untrusted network. At the edge of the untrusted network, another router deencapsulates and decrypts the packet for further routing if necessary. See also *router.*

TUR (1) An acronym for Traffic Usage Recording. (2) An acronym for Trunk Utilization Report.

turn ratio In transformers, the ratio between the number of turns of one winding to the number of turns on another winding. A transformer may have more than one winding, hence, more than one ratio.

Note: When using *turn ratio,* one must specify which winding order is used in the ratio. For example, one transformer's primary to secondary *turn ratio* may be specified, while another transformer's secondary *turn ratio* is specified. See also *transformer.*

turnaround time The total time required to reverse the direction of information flow on a half-duplex communication path from transmit mode to receive mode. On a digital circuit, the time includes transmission propagation delay, modem delay, software execution time delay, and so on.

turnpike effect An effect caused by heavy traffic over a communications system or network wherein a node cannot gain access to the communications facility in order to launch messages. The effect is similar to that of automobile highway gridlock.

turns ratio In transformers, the ratio of the number of turns of a given secondary to the number in the primary. Thus, a turns ratio less than one is a voltage step-down transformer, while a ratio greater than one is a step up. See also *transformer.*

turnup Completing the installation of a circuit and making it available to the customer that requested it.

TUV An abbreviation of Technischer Uberwachungs-Verein. A German electrical testing and certification organization similar to the UL laboratories in the United States.

TV (1) An abbreviation of TeleVision. (2) An abbreviation of Transfer Vector.

TVI An abbreviation of TeleVision Interference.

TVS An abbreviation of Transient Voltage Suppressor.

TVS diode An abbreviation of Transient Voltage Suppression diode. An avalanche diode (zener diode) that has been constructed with a large junction area. The large junction area allows the device to handle more joules of energy than a standard zener diode.

TVX An abbreviation for Timer, Valid Transmission (X).

TW An abbreviation of Traveling Wave.

TWA An abbreviation of Two-Way Alternate.

twain Said to be an acronym for Technology Without An Important (Interesting) Name. It is a machine-independent open application interface standard for image capture. It was developed by Hewlett-Packard, Logitech, Eastman Kodak, Aldus, and other hardware and software vendors as a means to bring images into user applications from scanners, frame grabbers, cameras, and so on.

tweak To change a parameter slightly or to optimize, usually in reference to a value. Also called *twiddle*.

TWIMC An acronym for <u>T</u>o <u>W</u>hom <u>I</u>t <u>M</u>ay <u>C</u>oncern.

twin cable A cable composed of two insulated conductors laid parallel and bound together as a single unit in a ribbonlike structure. Also called *twin-lead*.

twin T network A three-terminal network composed of two separate T networks with their terminals connected in parallel. See also *pad*.

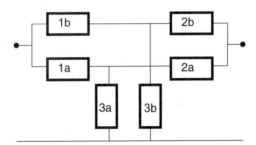

twinaxial cable A cable containing two coaxial cables bonded together in a single outer sheath. Sometimes shortened further to *twinax*.

twinning (**1**) In optics, a defect in a crystal such that both left-handed and right-handed versions are present in the same crystal. (**2**) Keeping a mirror of a magnetic tape.

twinplex A frequency shift keyed (FSK) carrier telegraphy system in which two *tone pairs* are transmitted on a single transmission channel (such as a twisted pair). One tone of each pair represents a *mark* and the other a *space*.

twist (**1**) The difference of the levels of a two-tone signal; specifically, the level of the higher frequency minus the level of the lower frequency. An example is the difference of the transmission levels of the two tones in dual tone multifrequency (DTMF) signaling used in telephone sets. (**2**) In telephony, the change of a transmission characteristic (such as the shape of the frequency-attenuation curve) as a function of temperature of a transmission line.

twisted pair (TWP) A cable containing two separate insulated conductors mechanically twisted as a pair. Twist rates may be as low as one or two per meter on some cable types to as high as several per centimeter on other types. The conductor pairs are connected to the same signal source, one wire for current going to the sink and the other wire for current returning. *Twisted pair* wiring is used to reduce both noise susceptibility and radiation. The *twisted pair* may be either shielded or nonshielded.

As is illustrated in the figure, the electromagnetic fields generated by each of the wires in a twisted pair tend to cancel, thereby reducing the effects of interference on nearby circuits. Similarly, external electromagnetic fields (noise sources) will induce equal signals in each of the wires of the twisted pair. These currents will be rejected by the signal receiving equipment as common mode (or longitudinal) signals. When multiple twisted pair are bound within a single cable, the amount of twist or lay among pairs is sometimes varied to further reduce the amount of coupling between individual pairs. In telephony, where cables comprise more than 25 pairs, they are usually bundled and wrapped in a cable sheath. Twisted pair is the most common medium for connecting phones, computers and terminals to private automatic branch exchanges (PABXs) as well as interconnecting 10 Mbps 10BaseT Ethernet networks. Newer data grade cables support bit rates to 100 Mbps or higher. See also *cable, lay, strand lay,* and *wire*.

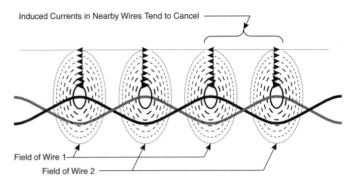

twisted pair distributed data interface (TPDDI) A network architecture that implements 100-Mbps Fiber Distributed Data Interface (FDDI) capabilities and protocols on unshielded twisted pair, copper-based cable. The maximum cable length is limited to about 50 m (164 ft). Also called ANSI X3T9.5. Sometimes the abbreviation *TDDI* is used. See also *copper distributed data interface (CDDI)*.

twisted pair, physical media dependent (TP-PMD) The 100 megabit per second, Fiber Distributed Data Interface (FDDI) standard as implemented on unshielded twisted pair (UTP) cable.

two-dimensional coding A data compression scheme designed to deal with data such as from an image where there is a high degree of correlation both horizontally and vertically. For example, in a facsimile system the previous scan line is used as a reference when scanning and encoding a subsequent line. See also *codes* and *waveform codes*.

two- of five code An error-detecting code for data transmission where each of the 10 decimal digits (0 through 9) is encoded as a set of five bits with either two 1s and three 0s or two 0s and three 1s. See also *code*.

POSSIBLE 2 OF 5 CODES

Value	7 4 2 1 0	6 3 2 1 0	Gray Like
0	11000	00110	10001
1	00011	00011	10010
2	00101	00101	10100
3	00110	01001	11000
4	01001	01010	01100
5	01010	01100	01010
6	01100	10001	00110
7	10001	10010	00011
8	10010	10100	00101
9	10100	11000	01001

two-pilot regulation In frequency division multiplex (FDM) systems, the use of two in-band pilot signals so that transmission path *twist* (frequency response changes with respect to temperature) may be detected and compensated by a regulator.

two- to four-wire hybrid (2–4 wire hybrid) See *hybrid*.

two-tone keying A synonym for frequency shift keying (FSK).

two-way alternate operation A synonym for *half-duplex operation*.

two-way simultaneous operation A synonym for *duplex operation*.

two-way splitting A private automatic branch exchange (PABX) feature that enables a user to split up a three-way conference into two separate communications links in which the user may select which of the two destination stations is to be connected at any time. (The other station is effectively put on hold.)

two-way trunk In telephony, a trunk between two switching offices in which calls may originate at either end.

two-wire circuit (1) An electric circuit in which all current delivered on one wire returns on the other wire of the pair. **(2)** A transmission facility in which both the transmitted signal to a remote terminal and received signals from the remote terminal are routed over one twisted pair of wires, e.g., the subscriber loop connecting a central office to a telephone subset. See also *four-wire circuit*.

two-wire repeater In telephony, a device used to amplify signals traveling in both directions on a two-wire telephone circuit.

The block diagram shows one possible way to achieve such a device.

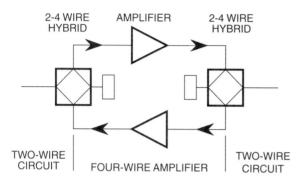

two-wire switching In telephony, switching wherein the same path, frequency, or time interval is used for signal transmission in both directions.

twos complement See *complement*.

TWP An acronym for TWisted Pair.

TWPS An acronym for Traveling Wave Phase Shifter.

TWS An abbreviation of Two-Way Simultaneous.

TWT An abbreviation of Traveling Wave Tube.

TWX® An acronym for TeletypeWriter eXchange service.

TX (1) An abbreviation of transmit or transmitter. **(2)** An abbreviation of Telephone eXchange.

TXD (1) An abbreviation of Transmit Data (RS-232 signal BA, RS-449 signal SD, and ITU-T signal 103). *TXD* is the data signal stream generated by the local data terminal equipment (DTE) and sent to the local data communication equipment (DCE) for transmission to the remote equipment. **(2)** An abbreviation of Telephone eXchange, Digital.

TXE An abbreviation of Telephone eXchange, Electronic.

TXK An abbreviation of Telephone Exchange Crossbar.

TXS An abbreviation of Telephone Exchange Strowger.

.TXT The filename extension of an unencoded ASCII file, i.e., a pure text file that can be directly read by most viewers. See also *filename extension*.

Tymnet A common carrier offering an X.25 public data network (PDN).

Tyndall cone The shape taken by scattered light from a beam as a result of the Tyndall effect.

Tyndall effect The effect by which sufficiently small particles dispersed in a transmission medium will scatter radiated energy (e.g., light) at right angles to the incident beam. The scattered energy is completely linearly polarized. The effect is named after the Irish physicist John Tyndall (1820–1893).

typ An abbreviation of TYPical.

type 3 encryption algorithm A cryptographic algorithm for use in protecting unclassified sensitive information or commercial information which has been registered by the National Institute of Standards and Technology (NIST) and has been published as a Federal Information Processing Standard (FIPS).

type 4 encryption algorithm An unclassified cryptographic algorithm which has been registered by the National Institute of Standards and Technology (NIST) but is not a Federal Information Processing Standard (FIPS).

type A IN service An Intelligent Network (IN) term describing services evoked by, and affecting, a single user. Most of them can only be invoked during call setup or teardown.

type approval A concept in which a design prototype is approved by an agency and all devices are subsequently manufactured to the same set of drawings (i.e., the same design) are automatically approved.

type B IN service An Intelligent Network (IN) term describing services invoked at any point by, and affecting directly, several users.

type n cable A cable system designed by IBM for use in its Token Ring networks and for general-purpose premises wiring. There are seven types specified in the system; they are identified by the numbers 1 through 9 (types 4 and 7 are reserved for future use).

Although not required by the specifications, a plenum version of most cable types is also available. The system, called IBM Cabling System (IBMCS or ICS), is defined as:

Type 1 Cable: Shielded twisted pair (STP), with two pairs of 22-gauge solid wire. *Type 1 cable* used for data quality transmission in IBM's Token Ring network. It can be used for either the main ring or to connect lobes (nodes) to multistation attachment units (MAUs), which are wiring centers.

Type 2 Cable: A hybrid consisting of four pairs of unshielded 22-gauge solid wire (for voice transmission) and two pairs of shielded 22-gauge solid wire (for data).

Type 3 Cable: Unshielded twisted pair (UTP), with two, three, or four pairs of 22- or 24-gauge solid wire, each pair having at least two twists per foot. Because this cable type is intended for only voice grade capabilities, and therefore may be used as telephone wire for voice transmissions, it is not recommended for 16 Mbps Token Ring networks.

Type 3 cable is sometimes used as an adapter cable to connect a node to a medium attachment unit (MAU). A media filter must be used when using Type 3 cable to connect a node to a MAU or when there is a switch between UTP and STP in a Token Ring network. Type 1 and type 3 cables should not be mixed in the same ring.

Some manufacturers offer higher quality Type 3 cable for greater reliability. These cables have more twists per foot, thereby reducing the incidence of crosstalk interference.

Type 4 Cable: Unspecified.

Type 5 Cable: Fiber optic cable, with two glass fibers, each with a 100-micron diameter core and a 140-micron cladding diameter. (IBM also allows the more widely used 62.5/125-micron fiber.) *Type 5 cable* is used for the Token Ring network's main ring in order to connect MAUs over greater distances or to connect network segments between buildings.

Type 6 Cable: An STP cable, with two pairs of 26-gauge stranded wire. Commonly, uses include:

Adapter cables, used to connect a node to an MAU. Here, the PC end of the cable is terminated with a DB-9P or DB-25P connector, and the MAU end has a specially designed IBM data connector.
Patch cables, used to connect MAUs. In this application, the cable has IBM data connectors at each end.

Type 7 Cable: Unspecified.

Type 8 Cable: An STP cable specially designed to be run under carpets, it contains two pairs of flat, 26-gauge solid wire. Because of this, the cable typically has a higher signal loss than Type 1 or Type 2 cable. However, the performance of *Type 8 cable* is satisfactory for the short distances usually involved in "under the carpet" cabling.

Type 9 Cable: An STP cable with two pairs of 26-gauge solid or stranded wire and covered with a plenum jacket. It is designed to be run between building floors.

See also *cable, category n cable, level n cable,* and *wire.*

type n PC Card See *PCMCIA.*

type n product A classification of cyptographic equipment, components, or assemblies by the National Security Agency (NSA). The term is applied only to products (i.e., not to information, keys, key management, control, or services). Because the products contain classified cryptographic algorithms, they are subject to the export restrictions described in the International Traffic in Arms Regulation (ITAR).

- *Type 1 products*—classified and controlled cryptographic entities authorized for use in secure communications (handling classified or sensitive material) by the NSA.
  Type 1 products are only available to qualified U.S. government users and contractors. Non-U.S. but federally sponsored activities may also qualify, subject to NSA and ITAR restrictions.
- *Type 2 products*—unclassified cyptographic products not to be used for classified information. Type 2 products contain classified NSA algorithms that differentiate them from products containing unclassified data algorithms.

type of service (ToS) A field in an Internet Protocol (IP) packet, or datagram, header. The contents of this byte specify the kind of transmission desired, with respect to delay, throughput, and reliability. Part of this byte specifies a priority for the datagram's handling.

TZ An abbreviation of Transmitter Zone.

U

u A representation of the greek letter μ (mu) (generally indicating the prefix micro) in character sets not containing Greek symbols.

U interface A user-to-network interface reference point in an Integrated Services Digital Network (ISDN) environment between the network terminator 1 (NT1) and the central office. It:

- Is a point-to-point, two-wire twisted pair subscriber loop that provides basic rate access to the ISDN from the NT1.
- Conveys information between the four-wire user-to-network interface (S/T reference point) and the local exchange.
- Is located on the carrier side of the demarcation point.
- Is not as sensitive to distance as a service using the T interface.

See also *ISDN*.

u law See *mu-law*.

U-PCS An abbreviation of Underlicensed Personal Communications Service. Here, *unlicensed* refers to the frequency bands used by wireless local area networks (LANs) in the United States. The bands include 1.88–1.89 GHz, 1.91–1.92 GHz, and 1.92–2.15 GHz. Various modulation and data-encoding schemes are used depending on the manufacturer, so compatibility problems may occur.

U.n Recommendations A series of International Telecommunications Union (ITU) standards covering "Telegraph switching." See Appendix E for a complete list.

U.S.P.S. or USPS An abbreviation of the United States Postal Service.

UA (**1**) An abbreviation of User Agent. (**2**) An abbreviation of Universal Access. (**3**) An abbreviation of Unnumbered Acknowledgment. (**4**) An abbreviation of User Account.

UAE An abbreviation of User Agent Entity.

UAL (**1**) An abbreviation of User Agent Layer. (**2**) An abbreviation of User Access Line.

UAM An abbreviation of User Authentication Method.

UAOS An abbreviation of User Alliance for Open Systems.

UAPDU An abbreviation of User Agent Protocol Data Unit.

UART An acronym for Universal Asynchronous Receiver/Transmitter (pronounced "YOU-art"). A hardware device that handles the serial data interface between a data terminal equipment (DTE) and data communications equipment (DCE). The *UART* adds both the framing bits (start and stop bits) and the *parity* bit to the serial data stream in the transmit direction. In the receive direction, it extracts data from the serial data stream and checks it for proper parity if required. Several popular *UART* chips are the 8250, A, B, the 16450, and the 16550; the prefix to these chip numbers is manufacturer dependent.

The *UART* performs the following tasks:

- Converts parallel input from a computer bus to a serial form for transmission at the transmitting node and converts the serial input to parallel at the receiving node.
- Adds any required start, stop, and parity bits to the byte at the transmitting node and strips these framing bits off at the receiving end.

- Controls the timing for communication.
- Maintains and administrates the transmit and receive buffers.
- Monitors and reports the serial port's status.

See also *8250, 16450,* and *16550.*

UAS An abbreviation of Unavailable Second.

UASFE An abbreviation of UnAvailable Second, Far End.

UASL An abbreviation of User Agent SubLayer.

UBC An abbreviation of Universal Block Channel.

UBR An abbreviation of Unspecified Bit Rate. See also *CBR* and *VBR*.

UC An abbreviation of Up Converter.

UCC An abbreviation of University Computing Center.

UCD An abbreviation of Uniform Call Distribution.

UCS (**1**) An abbreviation of Uniform Communications Standard. (**2**) An abbreviation of Universal Component System. (**3**) An abbreviation of Universal Character Set. (**4**) An abbreviation of Underwater Communication System. (**5**) An abbreviation of Underwater Cable System.

UD An abbreviation of Unit Data.

UDLC An abbreviation of Universal Data Link Control.

UDP An abbreviation of User Datagram Protocol. An Internet protocol for datagram service.

UE An abbreviation of User Element.

UERE An abbreviation of User Equivalent Range Error.

uF The symbol for micro farad, a unit of capacitance. Used when the Greek letter mu (μ) is not available, e.g., a typewriter.

UFI An abbreviation of User Friendly Interface.

UFS An abbreviation of UNIX File System.

UHF An abbreviation of Ultra High Frequency. See also *frequency bands.*

UI (**1**) An abbreviation of Unit Interval. (**2**) An abbreviation of UNIX International. (**3**) An abbreviation of Unnumbered Information. (**4**) An abbreviation of Unnumbered Interrupt. (**5**) An abbreviation of User Interface.

UIC An abbreviation of User Identification Code.

UID (**1**) An abbreviation of User ID. (**2**) An abbreviation of User Interactive Data.

UIS An abbreviation of Universal Information Services.

UJT An abbreviation of UniJunction Transistor.

UKRA An abbreviation of United Kingdom Registration Authority.

UL (**1**) An abbreviation of the Underwriters Laboratory®. UL® is a private testing organization in the United States concerned primarily with product safety. (**2**) An abbreviation of User Location. (**3**) An abbreviation of UpLink.

UL-13 An Underwriters Laboratories specification for the "Standard for Safety for Power-Limited Circuit Cable."

UL-444 An Underwriters Laboratories specification for the "Standard for Communications Cable."

UL-497 An Underwriters Laboratories specification for primary transient protectors to be used on telephone circuits.

UL-497A An Underwriters Laboratories specification for secondary transient protectors to be used on telephone circuits.

UL-1449 An Underwriters Laboratories specification, method of rating, and approval for surge suppressors. It specifies the method of testing a surge suppressor and the minimum results for passing the test.

Surge suppressors are also rated by the amount of pass-through voltage when the device is repetitively subjected to a 6000-volt transient. The ratings range from about 300 volts to 600 volts.

UL-1459 An Underwriters Laboratories specification dictating the protection of telephone equipment to transients on lines that enter the system from the public network. The requirement includes single and multiline telephones, private automatic branch exchanges (PABXs), key systems, and central offices.

UL certified cable levels A program to rate communications (and power-limited) cable according to performance capabilities. The cables are evaluated for compliance with UL-13 and/or UL-444. Five levels of performance are identified and are marked Level I through Level V; that is

* *LEVEL I*—a basic communications and power-limited circuit cable. There are no performance criteria for these cables.
* *LEVEL II*—covers cable bundles with 2 to 25 twisted pairs. The performance specifications are similar to those defined in IBM's Level 3 cable specification GA27-3773-1.
* *LEVEL III*—a standard for horizontal twisted pair (both shielded twisted pair—STP; and unshielded twisted pair—UTP) wiring similar to the EIA/TIA Category 3 cable listed in the Technical Systems Bulletin PN-2841.
* *LEVEL IV*—a low-loss premises wire standard that is compliant with the National Manufacturers Association (NEMA) standard "Low-Loss Extended Frequency Premises Telecommunications Cable." It is also similar to the EIA/TIA Category 4 cable listed in the Technical Systems Bulletin PN-2841.
* *LEVEL V*—a low-loss premises wire standard that is compliant with the National Manufacturers Association (NEMA) standard "Low-Loss Extended Frequency Premises Telecommunications Cable." It is also similar to the EIA/TIA Category 5 cable listed in the Technical Systems Bulletin PN-2841.

ULA (1) An abbreviation of Upper Layer Architecture. (2) An abbreviation of Uncommitted Logic Array.

ULB An abbreviation of Universal Logic Block.

ULCT An abbreviation of Upper Layer Conformance Testing.

ULF (1) An abbreviation of Ultra Low Frequency. See also *frequency bands*. (2) An abbreviation of Upper Limiting Frequency.

ULP (1) An abbreviation of Upper Layer Protocol. (2) An abbreviation of Upper Layer Process.

ultra high frequency (UHF) Frequencies between 300 MHz and 3 GHz (wavelengths from 100 to 10 cm, respectively). See also *frequency bands* and *frequency spectrum.*

ultra low frequency (ULF) Frequencies between 300 Hz and 3 kHz. See also *frequency bands* and *frequency spectrum.*

ultraphotic rays Electromagnetic rays lying beyond the visible portion of the spectrum, e.g., ultraviolet rays.

ultrasonic frequency Any sound frequency above the audible frequency range. It is generally applied to elastic wave propagation (in gasses, liquids, or solids).

ultraviolet (UV) Any electromagnetic radiation with a wavelength in the range of 4 to 400 nm. The ultraviolet wavelengths are arbitrarily divided into bands based on both effect and practical application, that is,

Band and Effect	Wavelength (nm)
Vacuum ultraviolet (VUV)	<10*
Extreme ultraviolet (EUV)	10*–100
Far ultraviolet	100–200
Middle ultraviolet	(200–300)
Ozone producing	180–220
Bactericidal	220–300
Near ultraviolet	(300–380)
Erythermal	280–320
"Black light"	320–400

*The lower edge of the UV band is the upper edge of the x-ray band and is sometimes specified at other points in the 1–40 nm range.

See also *frequency spectrum.*

um A symbol for micron (micrometer, 10^{-6}, and μm).

UMA An abbreviation of Upper Memory Area.

UMB An abbreviation of Upper Memory Block.

UMPDU An abbreviation of User Message Protocol Data Unit.

UN (1) An abbreviation of United Nations. (2) An abbreviation of UNassigned. (3) An abbreviation of UNknown.

UNA (1) An abbreviation of Upstream Neighbor's Address. (2) An abbreviation of Universal Network Architecture. (3) An abbreviation of Universal Night Answer.

unallowable character A synonym for *illegal character.*

unattended call A call placed by a computerized dialing system so as to deliver a message (usually an advertisement) to the called party.

unauthorized access The gaining of access into a secured system without the permission of the system owner. Access is usually by means of stolen or guessed passwords or through a loophole in the systems operating system.

unbalanced A circuit in which the current path in one direction is inherently different from the current path in the return direction with respect to a common reference (generally ground). Many times an unbalanced circuit signifies the return path is ground. See also *balanced* and *symmetrical.*

unbalanced line A transmission line in which the magnitudes of the signals on the two conductors are not equal with respect to ground, e.g., a coaxial cable.

unbalanced modulator A modulator in which the modulation factor is different for the alternate half-cycles of the carrier. Also called an *asymmetrical modulator.*

unbalanced ringing Signaling applied only on the Ring side of the subscriber loop. See also *balanced ringing, ring,* and *ringer.*

unbalanced wire circuit A circuit in which the two sides are inherently electrically dissimilar.

unbiased ringer A mechanical telephone ringer whose clapper is not normally held in a preferred position so that the ringer will operate on ac. *Unbiased ringers* do not operate reliably on pulsed dc.

A ringer that is only weakly biased so that it does not respond to dial pulses is sometimes also called an *unbiased ringer.*

unbound mode In fiber optics, any mode that is not a bound mode, i.e., a leaky or radiation mode of the waveguide.

UNC An abbreviation of <u>U</u>niform <u>N</u>aming <u>C</u>onvention.

uncertainty The limits of the confidence interval of a measured or calculated value. The probability of the confidence limits is generally specified at one standard deviation.

underdamped A damping factor insufficient to prevent ringing or oscillations in the output of a device following a transient or step change at the input. See also *damped response.*

underfill In fiber optics, a condition for launching light into an optical fiber in which not all of the transmission modes supported by the fiber are excited.

underflow In computing, a condition that occurs when a mathematical operation on a number delivers a result smaller than the smallest value (other than zero itself) that can be represented in the computer.

For example, suppose the smallest number possible in a particular computer is exactly 1; then, dividing 5 by 10 will give a result of 0. (The true result, 0.5, is less than the smallest number the machine can represent.)

underground cable A cable designed to be placed under the surface of the Earth in a duct system that isolates it from direct contact with the soil. Contrast with *direct-buried cable.*

underlap In facsimile, the amount by which the scanning spot size is less than the pitch of the lines being scanned.

Underlap can occur either in the X-dimension—i.e., adjacent spots on the same line (called X underlap) or in the Y-dimension; adjacent spots in adjacent lines (called Y underlap).

Undernet An Internet Relay Chat network that broke away from the main IRC network (EFNet) circa 1990.

underscore (_) An ASCII character (95, 5Fh). A frequently used substitute for the space character in file names in file naming systems that do not accept spaces, e.g., MS-DOS. For example, CHAP_1.TXT is a legal file name while CHAP 1.TXT is not. Also called *underline, underbar,* or simply *under.* Less commonly called but sometimes heard are *backarrow, flatworm, score,* or *skid.*

undershoot See *pulse.*

undervoltage A mains or supply voltage less than the minimum value specified. See also *Uninterruptible power supply (UPS).*

undesired signal Any signal that interferes with or produces a degradation in the operation of a system.

undetected error rate A synonym for undetected error ratio.

undetected error ratio The ratio of the number of bits (or characters or blocks) received in error (undetected or uncorrected) to the total number of bits (or characters or blocks) transmitted. Also called *residual error rate* and formerly called *undetected error rate.*

undirected information Data broadcast without regard as to who will read it.

Examples include USENet on the Internet and public bulletin board postings.

undisturbed day A day in which neither sunspot activity nor ionospheric disturbance interferes with radio communications.

unguarded release In telephony, a condition that may occur during the restoration of a circuit to its idle state when it can be prematurely seized.

UNI An acronym for <u>U</u>ser to <u>N</u>etwork <u>I</u>nterface. See also *ATM.*

unicast The transmission of a message from one node on a network to a specific second node on the network. All other network nodes ignore the message. See also *broadcast* and *multicast.*

Unicode A 16-bit character code specification developed by the Unicode Consortium, which supports up to 65,536 different symbols. The code is most useful for languages with large alphabets or other basic units (e.g., the Han characters used in Asian languages). Most of the commonly used character codes (such as ASCII or EBCDIC) are encoded within the *Unicode*'s structure. ASCII symbols, for example, are the first 128 symbols of the *Unicode.*

unidirectional channel A one-way-only communications channel.

unidirectional operation Operation in which data are transmitted across a communications link in only one direction.

Unified Network Management Architecture (UNMA) An Open Systems Interconnection (OSI) protocol-based architecture developed by AT&T to provide a unified framework for AT&T's conception of network management tasks. It is medium and vendor independent, and it relies on distributed (rather than centralized or mainframe based) processing.

It serves as an operating environment for AT&T's Accumaster Integrator network management package and provides a framework for dealing with the nine major management functions in AT&T's model:

- Accounting management
- Configuration management
- Fault management
- Performance management
- Security management
- Integrated control
- Operations support
- Planning capability
- Programmability

The first five of these function areas are identical to those specified in the OSI network management model. *UNMA* consists of five major elements:

- The *unified user interface (UUI)*—which provides a graphical-based summary of the network's operation and where the user interacts.
- An *integrated network management system*—which does the actual network management. Actually accomplished by the Accumaster Integrator.
- *Element management systems (EMSs)*—which serve as local network managers for a part of the entire network, such as a local area network (LAN), a mainframe, or a telecommunications link. Each EMS's operation is supervised by the integrated management system.

- *Network elements*—the components operating at the user level. Network elements include nodes (PCs, mainframes, modems), LANs, PABXs (private automatic branch exchanges), IXCs (interexchange carriers), or entire PTTs (Post, Telephone, and Telegraph).
- *Network management protocol* (*NMP*)—the OSI-based protocol used to communicate between an EMS and network element.

uniform call distributor (UCD) A device for routing incoming calls evenly among a group of agents. It is a very simple *automatic call distributor* (*ACD*) in that it follows a predetermined pattern for apportioning incoming calls, and it will not adapt to real-time load changes. Nor will it take into account which agent has been idle the longest or the busiest.

uniform encoding An analog-to-digital (A/D) conversion process in which all of the quantization subrange values are equal (except perhaps for the highest and lowest quantization steps). Also called *uniform quantizing*.

uniform linear array An antenna composed of a number of identical elements, arranged in a single line or in a plane with uniform spacing and usually with a uniform feed system.

Uniform Naming Convention (UNC) Used in IBM PC networking to completely specify a directory on a file server. The basic format is:

$$\backslash\backslash servername\backslash sharename$$

where
servername	is the hostname of a network file server, and
sharename	is the name of a networked or shared directory.

This is not the same as the conventional MS-DOS "C:\windows" directory name. For example, the *UNC* location

$$\backslash\backslash server1 \backslash dave$$

might point to the "real" location

$$C:\backslash users\backslash homedirs\backslash dave$$

on a server called "server1." It is possible to execute a program using this convention without having to specifically link a drive, by running:

$$\backslash\backslash server\backslash share\backslash directory\backslash program.exe$$

The *undocumented* DOS command, TRUENAME, can be used to find out the *UNC* name of a file or directory on a network drive.

uniform numbering plan A network station identification scheme in which all locations have the same number of digits.

For example, the United States' long-distance network has a *uniform numbering plan* (10 digits). Some countries without a *uniform numbering plan* need to dial four digits to reach one city, six digits to reach another and eight to reach a third.

uniform quantizing A synonym for *uniform encoding*.

Uniform Resource Locator (URL) In the Internet World Wide Web (WWW) *URL* is pronounced "you-are-ell," or "earl" which rhymes with "hurl." Both pronunciations are widely used. It is a standardized way of representing a location and type of documents, media, and network services. It provides a standard for jumping to remote hosts and their resources.

The *URL* is divided into parts.
- The first part identifies the protocol access method. Examples include:

- http:// . . . indicates that the access method is HyperText Transport Protocol and documents will use HyperText Markup Language (HTML), CGI, a Java applet or other files supported by HTTP.
- ftp:// . . . indicates File Transport Protocol.
- gopher:// . . . indicates Gopher protocol.
- wais:// . . . indicates Wide Area Information Servers.
- file:// . . . a host-specific file.
- The second part cites a machine name, organization, and domain name and possibly a port number. For example,

$$http://www.nasa.gov:80/$$

indicates the protocol is HTTP, the organization and machine are identified by www.nasa, the domain is .gov, and the port number is 80.
- The third part is a hierarchical description to the location of the desired file or program on the cited domain and computer. For example,

$$http://www.w3.org/addressing/rfc1808.txt$$

where:

$$/addressing/is \text{ the subdirectory and}$$

rfc 1808.txt is the file being viewed.
- The next part is a set of parameters transferred to the desired program and strictly speaking is not part of the URL. For example,

$$http://www.google.com/search?q5URL$$

URLs are described in RFC1738, 1808, and 2396. Sometimes referred to by its former name *Universal Resource Locator* (*URL*).

uniform spectrum random noise See *white noise*.

uniform time scale A time scale with equal intervals.

uniform transmission line A transmission line in which electrical properties are distributed equally along the line length. (That is, resistance, inductance, and capacitance per unit length are constant along the line.) Properties of a *uniform transmission line* include:

- The voltage-to-current ratio does not vary with distance along the line, if the line is terminated in its characteristic impedance.
- Signal attenuation is a function of the length of the line and frequency.

Examples of *uniform transmission lines* include coaxial cables, twisted pairs, and single wires at constant height above ground, all of which have no changes in geometry, materials, or construction along their length.

unilateral synchronization system A system synchronization scheme in which signals from a single node are used to synchronize clocks at all other nodes. Also called *unilateral control system*.

unimode fiber A synonym for *single mode optical fiber*.

unintelligible crosstalk Crosstalk that does not yield understandable signals in the adjacent channel, that is, information cannot be derived from the crosstalk signal power.

uninterruptible power supply (UPS) A battery backup system that provides continuous power to electronic equipment even in the event of momentary mains failure. The device is designed to provide short-term electrical power in the event the mains power (115V, 220V, etc.) is lost.

The fundamental structure of a *UPS* is shown in the block diagram. Here, the local power company supplies the raw *mains input,* it is converted to dc in order to charge a battery and supply power to the dc *to ac converter* (called an *inverter*). The output of the inverter delivers power to the user's devices (computers, modems, etc.) In the event of a mains power loss, the battery supplies power to the dc to ac converter for distribution to the user devices, generally, without the loss of even a fraction of a cycle in the ac output. The battery also provides protection against transients. Also called an *on-line UPS*. See also *line conditioner, standby power supply (SPS)*, and *surge protector*.

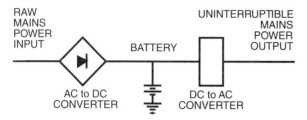

unipolar A term meaning having one active state. In electronics, a *unipolar* signal is one having either no voltage, current, or tone representing one signaling state and a single nonzero voltage, current, or tone representing the other state (active state). The active state voltage or current may be of either polarity.

unipolar signal A two-state signal in which one state is represented by voltage or current (of either polarity), and the other state is represented by no voltage or current.

unipolar signaling See *neutral transmission*.

unit cable construction That method of cable construction in which a number of wire pairs are bundled together (called units) and groups of these bundles are wrapped together to form the cable.

unit distance code An unweighted code that changes at only one digit position when going from any value to the next or previous consecutive value in a sequence of numbers. This type of code is frequently used when converting mechanical analog values to digital values (such as with shaft position encoders). One example of a *unit distance code* is the *Gray code*. See also *code (6)*.

unit element In the representation of a character, a signal element that has a duration equal to the unit interval.

unit impulse (1) A mathematical entity consisting of an impulse with infinite amplitude and zero width, and having an area under the curve equal to unity.

The *unit impulse* is a useful tool for analyzing circuits, networks, and systems describing their transfer function. For example, by injecting an approximation of a unit impulse into a system and recording the device's response, a complete characterization of the systems transfer function may be determined. Also called the *Dirac delta function*. (2) A transient (pulse) signal whose enclosed area is one. Generally, the width of the impulse is significantly smaller than other time-related features of the system.

unit interval In isochronous systems, the shortest time interval between two consecutive significant instants of an undistorted signal.

unit under test (UUT) A component, circuit, device, or system being tested. Also called a *device under test*.

United States digital cellular (USDC) One of the cellular telephone standards used in the United States. See also *GSM*.

units See *SI*.

unity gain The condition in which a signal is neither attenuated nor amplified.

unit interval In isochronous transmission, the shortest time interval in which the theoretical duration of a significant signal element is wholly contained.

unity gain bandwidth A measure of the amplification frequency product of an amplifier. It is the frequency at which the amplification factor reduces to one (0 dB).

The figure shows a straight-line approximation of the gain-frequency response (Bode plot) of an amplifier whose *unity gain bandwidth* point is 1 MHz. The gain-frequency response line must cross the unity gain-frequency point with a slope not greater than -20 dB per decade (-6 dB per octave) if feedback systems are to be stable.

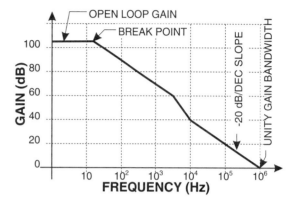

Universal Asynchronous Receiver/Transmitter (UART) A serial communications integrated circuit, containing a transmitter (parallel-to-serial converter) and a receiver (serial-to-parallel converter), each clocked separately. See also *UART*.

Universal Character Set (USC) A 1993 ISO and IEC standard character set (ISO/IEC 10646). Part 1 is Architecture and *Basic Multilingual Plane (BMP)*, where a plane is 256×256 locations (65 536). The standard results from the combining of Unicode and an ISO draft for 16-bit character codes. BMP is the first and only plane defined to date. UCS/BMP features include:

- Combined characters, which makes the ultimate symbol set virtually unlimited.
- ISO 646 (ASCII) and ISO 8859-1 (Latin 1) as subsets, with binary values unchanged.

Also known as *Universal Multiple-Octet Coded Character Set*.

universal gas constant A constant 8.314 joules per kelvin (J/K). The product of the volume and pressure of 1-gram molecule of an ideal gas divided by the absolute temperature.

Universal International Freephone Number (UIFN) A standardized plan for international toll free service (ITFS) described in the ITU-T recommendation E.169. The *UIFN* enables a customer to register a single toll free number with multiple countries. The format of the number is:

$$+ 800 \text{ XXXX XXXX}$$

Where "+" indicates the international direct dial (IDD) prefix of the country from which the call is originated. The "800" is a special country code assigned by the International Telecommunications Union (ITU) as nongeographically specific (i.e., the entire world). "XXXX XXXX" is the subscriber's unique access number. Also called *Universal 800* or *Global 800 numbers*. See also *area code, city code, country code, domain, international access code (IAC), international toll free service (ITFS), local access code,* and *local number*.

universal naming convention (UNC) A method of defining a unique name for a specific file on any network. The convention includes a hierarchical structure of entities from the server to the local file name itself, that is,

$$\backslash\backslash servername\backslash sharename\backslash path\backslash filename$$

universal night answer (UNA) A feature of private automatic branch systems (PABX) systems that permits any station to pick up an incoming trunk call when the attendant's console is unattended and the feature is enabled.

universal numbering plan A network station addressing method that provides nonconflicting station codes arranged so that all main stations can be reached from any point within the network.

Universal Personal Telecommunications (UPT) service A telecommunications service that provides personal mobility and service profile management. The principles are given in ITU-T recommendation F.850.

universal power supply A power supply that will deliver the required system voltages when connected to the mains anywhere in the world. That is, it will operate from 117 V 60 Hz to 240 V 50 Hz mains.

Universal Resource Identifier (URI) The generic set of all names and addresses that are short strings that refer to objects. *URIs* are defined in the Internet RFC1630. Originally called *UDI* in some World Wide Web documents.

Universal Resource Locator (URL) See *Uniform Resource Locator* (*URL*).

Universal Serial Bus (USB) A specification authored by Compaq, DEC, IBM, Intel, Microsoft, NEC, and Northern Telecom defining a serial interface. Version 1.0 was released in the Fall of 1995. Its features and characteristics include:

- Reduced need for card slots.
- Transfer rates up to 12 Mbps.
- Support for up to 127 physical devices.
- Self-identifying peripherals.
- Only one connector type.
- Tiered-star multidrop physical interconnection topology.
- Each cable segment up to 5 meters.
- Capability to dynamically attach or reconfigure peripherals without rebooting.
- Full support for real-time data for voice, audio, and compressed video applications.
- Protocol flexibility for mixed mode isochronous data transfers and asynchronous messaging.
- Various PC configurations.
- Built-in power distribution for low power devices.
- A standard interface.

universal service In telephony, the concept of making basic local telephone service (and certain other telecommunications and information services) available at an affordable price for everyone within a country or specified jurisdictional area.

Universal Time (UT) **(1)** A measure of time based on Standard Time at the 0° longitude mark in the floor of the Old Royal Observatory at Greenwich, England, which provides a close approximation to the mean diurnal rotation of the Earth and serves as the basis of civil timekeeping. When *Universal Time* is corrected for observing location and set to the atomic second standard, it is designated UT1. Formerly called *Greenwich Mean Time* (*GMT*). See also *time* (*3*) and *universal time, coordinated* (*UTC*). **(2)** The official civil time of the United Kingdom.

universal time, coordinated (UTC) An atomic time standard/system maintained by the U.S. Naval Observatory which defines the mean solar time at the Greenwich Meridian. The time is occasionally corrected by the insertion of "leap seconds"; this keeps it in approximate synchronization with time based on the Earth's rotation (UT1 time). The leap second adjustments keep UTC within 0.9 second of UT1. Also called *coordinated universal time* and *universal coordinated time*. See also *time* (*3*) and *UTC(x)*.

UNIX™ A multi-user *operating system* originally developed by AT&T's Bell Laboratories in 1969 for large computing machines. Generally, Ken Thompson and Dennis Ritchie are considered the inventors. The name *UNIX* was a pun on the earlier operating system, Multics. The trademarked *UNIX* is now owned by the Open Group, an industry organization that certifies *UNIX* implementations.

It is the predominant operating system used on the Internet host computers. *UNIX* is a very complete and powerful system, but as is often stated, "It has never been known for its user-friendliness." There are two major variants of UNIX: the AT&T System releases and the University of California Berkeley System Distribution (BSD). A version that combines features of various versions is also available from Novell's UNIX Systems Group. It has been adapted to all manner of hardware from PCs to Crays. A few hardware specific versions include A/UX (Macintosh), AIX (IBM), LINUX (Intel), NeXTSTEP (NeXT and Intel), UnixWare (Intel), Xenix (Intel), and so on. Some of the many UNIX commands and pseudo-commands (programs) include:

cd	change directory
finger *name*	gives information about a "name"
irc	International Relay Chat allows conversation with people over the world
logout	logs off the machine (logoff, exit, & ^d)
ls	list directories and files
man *command*	display manual for a command
mail *address*	with an address to send mail, without an address to receive mail
more *filename*	displays one screen of a file at a time
mv *file1 file2*	rename command
nn	a net news interface program
rm *filename*	remove a file
rn	a net news interface program
rz *filename*	receive with ZMODEM protocol
sz *filename*	send with ZMODEM protocol
talk *username*	allows online conversation with someone currently logged onto the system
tar *option*	archive files

There are a number of switches associated with many commands. To discover what switches are associated with a command, type *man command*. To discover what commands might be available to do a particular function, type *man-k keyword* where "keyword" is the term to be explored.

UNIX-to-UNIX copy (uucp) An older store-and-forward network protocol that allows UNIX users to exchange files, mail, and USENet news via dial-up connections.

unloaded **(1)** A telephone line without any load coils. Also called *nonloaded line*. See also *load coil* and *loaded line*. **(2)** Pertaining to an electronic circuit or device that does not have any external dissipative element. For example, the unloaded Q of a parallel inductor and capacitor circuit might be a number like 20. Adding a load (resistor) can bring the Q as close to 1 as desired.

unlock code With cellular telephones, a three-digit code used to enable a cellular telephone. Without the code unauthorized use is much more difficult. A frequent factory default code is 123.

UNMA An abbreviation of Unified Network Management Architecture. AT&T's network management scheme.

unnumbered command A command in data transmission that does not have a sequence number in the control field.

unnumbered response A response in data transmission that does not have a sequence number in the control field.

unpack The operation necessary to restore a packed file to its original format. Also called *decompressing* or *uncompressing*.

UNR An abbreviation of Uncontrolled Not Ready.

unreliable protocol A data transfer protocol that does not provide any error-detection or correction capability. In use, an ureliable protocol transmits its message packets and assumes they arrived intact.

This does not mean the protocol is bad or shouldn't be used because, error detection or correction is generally accomplished in a higher layer of the communications protocol stack. Multiple error control schemes may actually degrade overall system performance.

unshielded An entity not covered or protected from something else.

In wire, for example, there is no outer conductive sheathing to protect the wire from electromagnetic radiation.

unsubscribe The opposite of *subscribe*. The removal of a newsgroup from the list of newsgroups a USENet user is following. After unsubscribing, the newsreader will not present the newsgroup again. The user may subscribe and unsubscribe as many times as desired.

unsuccessful call attempt A call attempt that does not result in the establishment of a connection. Also called an *unsuccessful call*.

unsupervised transfer A call transferred to a new station without informing that person who is calling. Also called *blind transfer*.

unthreaded newsreader A USENet newsreader that organizes the presentation in the chronological order in which it was received. See also *threaded newsreader*.

unused character A synonym for an *illegal character*.

unzip (1) To extract files from an archive created with PKWare's PKZIP archiver. (2) A program to list, test, or extract files from a PKWare ZIP archive, commonly found on PC-based systems.

UOA An abbreviation of U-interface Only Active. An ISDN basic rate interface control bit.

UP An abbreviation of Unnumbered Poll.

up converter A device for translating a signal from one frequency range to a higher one. Accomplished through the use of heterodyning (mixing with another carrier).

The opposite is a *down converter*.

UPC (1) An abbreviation of Universal Product Code. Also called a *barcode*. (2) An abbreviation of Usage Parameter Control. An asynchronous transfer mode (ATM) flow-control signal. (3) An abbreviation of Universal Peripheral Controller.

update (1) The regeneration (by manual or automatic means) of a display to show the current status of displayed data. Also called *refresh*. (2) To install a newer version of an entity (hardware or software). For example, one can update a communications system by replacing a V.34 modem with a V.90 modem. See also *upgrade*. (3) The process of modifying or reestablishing data with more current information, i.e., entering corrected or new information into a database.

UPF An abbreviation of Uni Picture Format. A universal format used to facilitate viewing picture information on dissimilar devices, e.g., video printers, multimedia kiosk terminals, public pay phones with Integrated Services Digital Network (ISDN) and IrDA interfaces, digital still cameras (DSC), and image capture and display software on PCs.

upgrade A newer, later, faster, or more featured version of hardware or software previously manufactured by a company. (An upgrade that requires complete replacement is sometimes called a *forklift upgrade*.)

uplink (U/L) (1) That portion of a communications link used for the transmission of signals from a ground station (Earth station) to a satellite or an airborne platform. The inverse of an *uplink* is a *downlink (D/L)*. (2) Pertaining to data transmission from a data station toward the headend. Also spelled *up link*.

upload The process of transferring information (usually in the form of a data file) from a local data terminal equipment (DTE) to a remote DTE, as in *uploading* a file to a bulletin board system (BBS).

To upload a file, a *communications program* must be used. The communications program provides the necessary flow-control and error control protocols to allow a successful transfer. Many communications programs are available; however, they all offer one or more protocols such as Kermit, XMODEM, YMODEM, or ZMODEM. See also *download* and *FTP*.

upper frequency limit The highest usable frequency determined by the cutoff frequency or higher order waveguide modes of propagation and the effect they have on impedance and transmission characteristics.

upper limit The maximum acceptable limit of a variable or parameter.

upper memory block (UMB) In DOS environments, a reference to the memory in the area between 640 KB and 1 MB.

With memory managers, *UMBs* are allocated for storing drivers, buffers, and other items, which frees conventional memory (memory between 0 and 640 KB) and gives programs more room to execute. See also *high memory area* (*HMA*).

upper sideband The higher frequency band of the two frequency bands produced by a modulation process. See also *sideband*.

UPS (1) An abbreviation of United Parcel Service. (2) An abbreviation of Uninterruptible Power Supply.

upstream (1) The direction opposite of the data flow. (2) All network nodes that pass information serially from a source node to a specified node are said to be *upstream* from the specified node. See also *downstream*.

upstream bandwidth A representation of the amount of information a system can support in the direction from a user toward the host node. This capacity must be sufficient to allow a user to become an originator of information, not just a sink.

uptime The time interval during which a functional unit is fully operational and available for its intended purpose.

UPT An abbreviation of Universal Personal Telecommunications.

URA An abbreviation of User Range Accuracy.

URE An abbreviation of User Range Error. See *user equivalent range error* (*UERE*).

URL An abbreviation of <u>U</u>niform (or <u>U</u>niversal) <u>R</u>esource <u>L</u>ocator.

us The SI abbreviation for microsecond (10^{-6} s).

US classification levels A set of security (classification) categories specified by the United States government and used in datagrams transmitted across the Internet. A datagram's classification level is specified in an 8-bit value; classification levels include:

US CLASSIFICATION LEVELS

Code	Definition
0000 0001 (01h)	Reserved
0011 1101 (3Dh)	Top secret
0101 1010 (5Ah)	Secret
1001 0110 (96h)	Confidential
0110 0110 (66h)	Reserved
1100 1100 (CCh)	Reserved
1010 1011 (ABh)	Unclassified
1111 0001 (F1h)	Reserved

With a minimum distance of four between any code values, an error-correction scheme can identify the proper level even with multiple errors.

USART An acronym for <u>U</u>niversal <u>S</u>ynchronous/<u>A</u>synchronous <u>Re</u>ceiver/<u>T</u>ransmitter (pronounced "U-sart"). A UART with the additional capability of *synchronous* communication.

USAT An acronym for <u>U</u>ltra<u>S</u>mall <u>A</u>perture <u>S</u>atellite antenna.

USB (1) An abbreviation of <u>U</u>pper <u>S</u>ide<u>B</u>and. (2) An abbreviation of <u>U</u>niversal <u>S</u>erial <u>B</u>us.

USDC An abbreviation of <u>U</u>nited <u>S</u>tates <u>D</u>igital <u>C</u>ellular.

useful life (1) The period of time that a device has an acceptable failure rate or an unrepairable failure occurs. Generally, the *useful life* is stated only for a given set of operating conditions. Also called *useful service life*. (2) The period of time that a computer can provide useful service. Although the computer may be fully functional, it may not be capable of supporting required software upgrades.

USENet Also written USEnet. A contraction of <u>USE</u>r's <u>Net</u>work. Started in 1979 by graduate students at Duke University and the University of North Carolina (Steve Bellovin, Jim Ellis, Tom Truscott, and Steve Daniel) when they decided to link several UNIX computers together in an attempt to better communicate with other UNIX system users. The system created included software to read, write, and post news. From this beginning, *USENet* has grown to include innumerable topics (about 20 000—not just UNIX), is international in scope and size (over a million users), and runs on most every computing platform available. *USENet* has provided much of the attitude of free information sharing that exists on the Internet today.

Groups are established by the users for the purpose of discussing a particular subject. Each group has its own Internet address, so membership (discussions) is limited to those who seek out the group. See also *newsgroup hierarchies*.

USENet newsgroups There are supposed to be seven fundamental (or core) groups and a number of interest groups within each core group. Additional subgroups are below these. See also *newsgroups*.

user A generic term referring to any person, organization, hardware, or program that operates, manipulates, or employs a computer or communications system to accomplish something. Also called a *subscriber*, especially in telephony.

user access line (UAL) In an X.25 network, the line that provides a connection between a data terminal equipment (DTE) (computer) and a network, with the user's data communications equipment (DCE) (digital service unit, modem, or multiplexer) as the interface to the network.

user account A means for establishing who can gain access to a multi-user computing system and what resources are available to that user. The account contains a *user profile* which includes information such as the user's name, password, time allowed per session, rights, and permissions. A *user account* is established when one registers onto a host.

user agent (UA) The client program in an e-mail application, as defined in the ITU X.400 Message Handling System (MHS) recommendation, that helps the user compose, send, and receive mail. The *user agent* communicates with a local *message transfer agent* (*MTA*) which relays the message onto remote hosts. Also called an *e-mail client*.

user agent layer (UAL) The upper sublayer of the ISO/OSI application layer in the 1984 version of the X.400 *Message Handling System* (*MHS*) recommendations. Users interact with it, and the *UAL,* in turn communicates with the *message transfer layer* (*MTL*) below it.

user authentication method (UAM) Identifies users for a file server before giving the users access to services in an AppleTalk network. Depending on the authentication method being used, this can be done on the basis of:

- An unencrypted password over the network,
- A random number from which the user's password can be derived by decrypting at the server's end.

User Datagram Protocol (UDP) A host-to-host protocol in Internet's Internet Protocol (IP) suite at the transport layer which provides a *connectionless mode transfer, unreliable* link. (*Note:* Unreliable means "unverified," not "prone to error.") It is defined in STD6 (and RFC768).

A *UDP* datagram makes far fewer demands on the network bandwidth than the connection-oriented reliable Transmission Control Protocol/Internet Protocol (TCP/IP). For this reason, the *UDP* is used whenever there is a higher layer of the protocol stack "worrying" about error control. See also *connectionless mode transfer* and *unreliable*.

user equivalent range error (UERE) In the Global Positioning System (GPS), any error contributing to the total error of the stand-alone GPS receiver positioning, expressed as an equivalent range error between the receiver and a satellite. Individual errors are from different sources and are therefore independent. The total error is expressed as the root mean square (RMS) of individual contributors (the square root of the sum of the squares of individual contributors). Within each satellite navigation message is a prediction of the maximum anticipated *UERE* (exclusive of path errors, such as ionospheric or multipath errors). The prediction is the *user range accuracy* (*URA*). Also called the *user range error* (*URE*).

user group A group of people drawn together through common interest, e.g., in the same computer system, software, or skill. User groups provide support for new entrants and a forum where members can exchange ideas or information and discuss problems.

user ID The name a user uses to log onto a computer system. On the network, the name must be unique. (It cannot be the same as another user, group, or domain name.) Also called *user name* and *login name*.

user information Information transferred across the functional interface between a source user and a telecommunications system for delivery to a destination user. User information includes user overhead information but not telecommunication control overhead.

user information bit A bit transferred from a source user to a telecommunications system for delivery to a destination user. They are encoded to form channel bits.

User information bits do not include the overhead bits originated by, or having their primary functional effect within, the telecommunications system.

user information block A block that contains at least one user information bit.

user interface That portion of a device or program with which a person interacts.

A program that requires data entry via a keyboard is said to have a *command line interface*. A program that displays a list of options and that allows the user to select a specific option is said to have a *menu interface*. A program that displays a set of icons and requires a pointing device is said to have a *graphical user interface (GUI)*.

user level security A network in which a central server maintains a database of all user accounts and provides authentication of passwords. After logon, users are allowed access to all network resources for which the password or group has been granted permission. Usually, only one password is required to gain access to the resources. See also *share level security*.

user line A synonym for *user loop*.

user loop A two- or four-wire circuit connecting a user to a private automatic branch exchange (PABX) or to other telephone equipment. See also *loop*.

user name See *user ID*.

user network contract An element that specifies parameters for a virtual channel connection (VCC) or virtual path connection (VPC) in the asynchronous transfer mode (ATM) architecture. These parameter values are negotiated by the entities involved in the connection. Also called a *traffic descriptor*.

user profile The computer-based record of all of the attributes about an authorized user of a multiuser computer system. The *user profile* contains information such as user name, user password, security level, resources available, time allowed per session, time used, file count uploaded, file count downloaded, mailbox location, terminal type, communication attributes, and so on.

user range accuracy (URA) A statistical indicator of the effects of apparent clock and ephemeris prediction accuracy to the ranging accuracy possible with each specific Global Positioning System (GPS) satellite based on historical data. See also *user equivalent range error (UERE)*.

User Range Error (URE) In the Global Positioning System (GPS), an estimated error in the calculated distance from the receiving antenna to the satellite due to factors such as unmodeled atmospheric effects, orbital calculation errors, satellite clock bias, multipath, and selective availability (S/A). This value is transmitted by the NAV-STAR satellites and may be displayed by some GPS receivers. *URE* is expressed in meters. (32 is common with S/A on and 5 meters is possible with S/A off.) See also *user equivalent range error (UERE)*.

user to network interface (UNI) *UNI* (pronounced YOU-knee) is one of three levels of interface in asynchronous transfer mode (ATM) networks. The other two are network-to-network (NNI) and user-to-user (UUI). It is the protocol to define connections between ATM end stations and the ATM switch (including signaling, cell structure, addressing, traffic management, and adaptation layers). It interfaces:

- User ATM equipment (e.g., hosts) and private or public ATM network equipment (e.g., ATM switches)
- Customer Premises Equipment (e.g., private ATM switch) and public network equipment (e.g., provider ATM switch)

The ATM Forum has defined two contexts in which the *UNI* appears:

- Public *UNI*, connecting ATM equipment within a private network (either hosts or switches) with public network provider (carrier) networks.
- Private *UNI*, used exclusively to connect hosts to switches where both are managed by the same administrative entity.

user to user interface (UUI) One of three levels of interface in asynchronous transfer mode (ATM) networks. The other two are network-to-network (NNI) and user-to-network (UNI).

USITA An acronym for United States Independent Telephone Association.

USL An abbreviation of UNIX System Laboratories.

USNC An abbreviation of United States National Committee.

USNO An abbreviation of United States Naval Observatory.

USO An abbreviation of UNIX Software Operation.

USOC An acronym for Uniform Service Order Code. A set of telephone company standards for equipment, connectors, and interfaces.

USPS An abbreviation of United States Postal Service.

USR An abbreviation of User Service Routines.

USRT An acronym for Universal Synchronous Receiver Transmitter.

USTA An abbreviation for United States Telephone Association.

USV An abbreviation of User SerVices.

UT An abbreviation of Universal Time. See also *time, UT1, UTC,* and *UTC(x)*.

UT1 The time scale based on the Earth's rotation. It is a measure of the true angular position of the Earth in space; however, because the Earth's rotation is not constant, *UT1* is not a uniform time scale.

UTC **(1)** An abbreviation of Universal Time, Coordinated. **(2)** An abbreviation of Universal Time Code.

UTC(x) Universal Time, Coordinated (UTC), as kept by the "x" laboratory, where "x" is the designation of any laboratory cooperating in the determination of UTC. In the United States, the official UTC is kept by the U.S. Naval Observatory; hence, *UTC(x)* is UTC (USNO).

UTE An abbreviation of Union Technique de l'Électricité. The French standards-setting and compliance certification agency.

utility A program used to assist a computer operator with maintenance and housekeeping, or to provide features not included in the machine's operating system. Examples include file compression programs, screen savers, hard disk drive defragmentation programs, and spell checkers.

utility load Any load not required by a facility to accomplish its primary mission, e.g., administrative, support, and housing power requirements. Also called a *nonoperational load*.

utility program A computer program that is in general support of the operations and processes of a computer.

Examples include diagnostic programs, trace programs, input routines, disk defragmentation programs, and programs used to perform routine tasks, i.e., performing everyday tasks, such as copying data from one storage location to another. Also called *service program, service routine, tools,* and *utility routine.*

utilization factor The ratio of the maximum demand on a system to the rated capacity of the system.

UTP-n An abbreviation of Unshielded Twisted Pair cable category n. "n" currently ranges from 1 to 5. See *category n cable.*

utterance In speech recognition, any vocalized sound. Typically a word.

UTTP An abbreviation of Unshielded Telephone Twisted Pair.

.UU One default filename extension of *uuencoded* files. Another is .UUE.

uucp An abbreviation of UNIX to UNIX CoPy. Sometimes listed as an abbreviation of UNIX to UNIX Copy Program.

uudecode In Internet, an acronym from UNIX Uniform decode, a program that extracts the original character set (generally a binary file) encoded by *uuencode.*

The format of the *uudecode* command is:

uudecode [sw] source.file

where the optional switches [sw] are:

- -L   Disables encoder line checksums.
- -u   Uses UU decoding.
- -x   Uses XX decoding.
- -Z"x"   Defines "x" as the separator (cut line) between the header and data portions of an encoded file. Used when *uudecode* does not automatically recognize the separator: such as the cut line "---" generated by some DOS bulletin board systems (BBSs). For example,

         uudecode -Z "---" name

     tells the decoder that the cut line is "---".

UUE An abbreviation of UNIX Uniform encode.

uuencode In Internet, an acronym for UNIX Uniform encode, one of several programs that encode a binary file (which may contain any character) into another file containing only characters that can be transmitted over diverse networks.

The basic scheme groups three 8-bit characters together and then divides the 24 bits into four 6-bit characters. To each of the 6-bit characters 32 is added. This creates a new symbol set with values ranging from 32 to 95—or in ASCII

$$‡ ! '' \# \$ \% \& ' ( ) * + , - . / 0\ 1\ 2\ 3\ 4\ 5\ 6\ 7$$
$$8\ 9 : ; < = > ? @ A\ B\ C\ D\ E\ F\ G\ H\ I\ J\ K\ L$$
$$M\ N\ O\ P\ Q\ R\ S\ T\ U\ V\ W\ X\ Y\ Z\ [\backslash] \char`^ \_$$

where ‡ represents a space. Several characters generated by this encoding scheme do not get through an EBCDIC to ASCII translation correctly. This potential problem can be resolved by using *xxencode.* The format of the *uuencode* command is:

     uuencode [sw] source.file [destination.file]

where the optional switches [sw] are:

- -h n   Leaves space for "n" lines in the beginning of the first split section. This space is frequently used for comments about the file.
- -L   Enables encoder line checksums.
- -s   Disables destination file splitting. Normally, long files are split into sections of about 59 kbytes (950 lines). This is required because some network nodes require files to be less than 64 kbytes.
- -s n   Splits the destination file into sections of "n" lines each.
- -t   Inserts the character mapping table at the front of the converted file.
- -X   Uses XX encoding.

File names generated by the *UU-encoder* have the default extension .UU, .UUD, or .UUE. This may be overridden simply by specifying an extension in the destination file name. Also abbreviated *UUE.* See also *BinHex, uudecode,* and *XXENCODE.*

UUG An abbreviation of UNIX User Group.

UUI (1) An abbreviation of User to User Interface. **(2)** An abbreviation of Unified User Interface.

UUSCC An abbreviation of User to User Signaling with Call Control. An Integrated Services Digital Network (ISDN) feature that passes user data with some signaling messages.

UUT An abbreviation of Unit Under Test.

UV (1) An abbreviation of UltraViolet. **(2)** A symbol for microvolt $(10^{-6}$ V, μV).

UVPROM An acronym for Ultra Violet erasable Programmable Read Only Memory. See *EPROM.*

V

V The SI symbol for *electric potential* measured in *volts.*

V interface The two-wire Integrated Services Digital Network (ISDN) physical interface used for a single termination from a remote terminal. Also called the *V reference point.* See also *ISDN.*

V number A synonym for *normalized frequency.*

V.x Recommendations A series of ITU-T recommended standards covering interchange circuit characteristics, signaling, signal coding, and modem design and operation, e.g., data transmission over telephone lines. Some of the V-series recommendations are:

- *V.10*—electrical characteristics for unbalanced double-current interchange circuits at data signaling rates nominally up to 100 kbit/s.
- *V.11*—electrical characteristics for balanced double-current interchange circuits operating at data signaling rates up to 10 Mbit/s.
- *V.14*—transmission of asynchronous (start-stop) characters over synchronous bearer channels.
- *V.15*—use of acoustic coupling for data transmission.
- *V.17*—ITU-T 14,400 bps FAX transmission standard, including many fallback speeds and a short retraining sequence.
- *V.21*—A 300-bps full-duplex data transmission standard for the public switched telephone network (PSTN) specifying frequency shift key (FSK) modulation. (Similar to but different from the North American Bell 103 standard.)

ITU-T V.21 FREQUENCIES (HZ)

Mode	Transmit		Receive		Answer Tone
	Space	Mark	Space	Mark	
Orig.	1180	980	1850	1650	—
Ans.	1850	1650	1180	980	2100

- *V.22*—a 1200-bps data transmission standard for the PSTN specifying DPSK modulation and data scrambling. (Similar to but different from the North American Bell 212A standard.)
- *V.22 bis*—a 2400-bps duplex data transmission standard for the PSTN specifying quadrature amplitude modulation (QAM), data scrambling (trellis coding), and telco line equalization.
- *V.23*—a 1200/600/300-bps asynchronous/synchronous half-duplex transmission standard for PSTN specifying FSK modulation. *V.23* has a 75-bps FSK reverse channel for signaling and error control.

ITU-T 600/1200 BAUD FREQUENCIES (HZ)

Channel	Transmit		Receive		Answer Tone
	Space	Mark	Space	Mark	
Mode 1–600	1700	1300	1700	1300	2100
Mode 2–1200	2100	1300	2100	1300	2100
75 bps Reverse	450	390	450	390	—

- *V.24*—a standard describing the set of data communication equipment (DCE)/data terminal equipment (DTE) interface signals. The document covers only the functional characteristics of the signals. Physical characteristics are covered in ISO 2110 and ISO 4902. Electrical characteristics are described in ITU-T V.28. (Similar to the EIA/RS-232 standard.)
- *V.25*—an automatic calling (dialing) and answering equipment standard for the PSTN. It includes procedures for disabling echo suppressors.
- *V.25 bis*—an automatic calling and answering equipment standard for the PSTN using 100 series interchange circuits.

V.25 BIS DIALER COMMANDS

Command	Function
CIC	Connect incoming call (auto answer).
CRNxxxxxxx	Call requested to xxxxxxx.
CRSyy	Call requested from memory location yy.
DIC	Disregard incoming calls (no auto-answer).
PRNyy,xxxxxxx	Program memory location yy with xxxxxxx.
RLD	Request list of delayed numbers.
RLF	Request list of forbidden (blacklisted) numbers.
RLN	Request list of stored numbers.

V.25 BIS DIALER RESPONSES

Terse	Verbose	Description
CFIAB	ABORT CALL	The current call is aborted
CFICB	DCE BUSY	Modem unable to dial
CFIET	ENGAGED TONE	Busy tone detected
CFIFC	FORBIDDEN	Telephone number is forbidden
CFIND	NO DIAL TONE	No dial tone detected
CFINS	NOT STORED	Requested number not in memory
CFINT	NO ANSWER TONE	No answer tone detected
DLCx	DELAYED X	Number cannot be dialed for X minutes
INC	RINGING	Incoming ring signal
INV	INVALID	Invalid command
VAL	VALID	Valid command

- *V.26 ter*—a 2400-bps full-duplex data transmission standard for the PSTN specifying QAM, data scrambling, telco line equalization, and echo cancellation techniques. Used primarily in France.
- *V.27 ter*—a 4800/2400-bps half-duplex Group III FAX transmission standard for the PSTN specifying 8 PSK with fallback to 4 PSK, automatic adaptive equalizer, and a 75-bps reverse communication channel. (Similar to Bell 208 B.)
- *V.28*—a standard describing the electrical characteristics of the interface signals between the DTE and DCE. The mechanical standards are described in ISO 2110 and ISO 4902. The list of signals and their definitions between DTE and DCE is given in ITU-T V.24. (Similar to the RS-232 standard.)
- *V.29*—a 4800, 7200, 9600-bps half-duplex standard used in Group III FAX transmission. (Similar to Bell 209.)
- *V.32*—a 9600, 4800, and 2400-bps full-duplex data transmission standard for the PSTN specifying QAM, data scrambling or trellis coding, and echo cancellation techniques.

- *V.32 bis*—a data transmission standard which allows 14,400, 12,000, 9600, 7200, and 4800-bps duplex data communication. *V.32 bis* also defines procedures enabling the modems to fall back to slower data rates when the transmission errors become excessive and to step up to higher rates when conditions improve. The protocol employs trellis modulation.
- *V.32 terbo*—see *V.terbo*.
- *V.34*—an ITU-T 33.6-kbps modem standard with fallback to many possible lesser speeds with degraded telephonic connections. The 28.8-kbps specification was released in 1994.
- *V.42*—an error-detection and correction specification link access protocol for modems (LAPM) that use *V.22, V.22 bis, V.26 ter, V.32,* or *V.32 bis* modulation methods. The specification provides an alternative protocol for compatibility with Microcom Networking Protocol (MNP) Class 2 through MNP Class 4 modems. Fallback to V.14 is included for compatibility with nonerror-correcting modems.
- *V.42 bis*—a data compression standard that yields up to 4:1 compression on data *not previously compressed. V.42 bis* uses a Lempel-Ziv data compression technique.
- *V.54*—defines loop test devices for modems.
- *V.90*—a 56-kbps modem specification from the ITU Telecommunication Standardization Sector (ITU-T) Study Group 16. Work began on the development of V.90 in the ITU-T in March 1997 and was ratified on February 6, 1998 in Geneva, Switzerland.

The technique takes advantage of the normal way a dial-up connection to networks such as the Internet, AOL, Prodigy, GEnie, Compuserve, and others are attached to the PSTN, i.e., via a digital link (T-carrier or ISDN). Under these conditions there is only one analog-to-digital (A/D) converter in the signal path, hence, only one point in which significant amounts of quantizing noise is injected. The figure illustrates this connection. In the down-

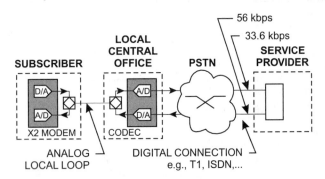

stream direction, only a digital-to-analog converter (DAC) is present. The service provider injects pulse code modulation (PCM) words directly. Because each 8 bit PCM codec sample generates one of 256 analog levels (signaling symbols) at 8 kilosamples per second, one would think that a throughput of 64 kbps was possible, that is,

Signaling Bits (n)	Signaling Symbols (2^n)	Signaling Rate (bps)
1	2	8 000
2	4	16 000
3	8	24 000
4	16	32 000
5	32	40 000
6	64	48 000
7	128	56 000
8	256	64 000

Unfortunately, there are some limitations with digital signals that are not significant with voice signals; notably,

- Low-level noise,
- An unknown dc bias conferred by the network,
- Nonlinear distortions in the circuit, and
- Robbed bit signaling used in T-carrier circuits. (Robbed bit signaling "steals" the low-order sample bit in two of the samples of each T-1 frame to indicate call status.)

The first three make it difficult for the modem to determine the exact quantizing point transmitted by the service provider. The use of bit 8 as network overhead (robbed bit) means that not all bits are always available for data. Taken together, these limitations reduce the number of usable bits to 7, hence, the achievable data rate to about 56 kbps. The Federal Communications Commission (FCC) and other regulatory agencies, in order to minimize noise and crosstalk between adjacent telephone lines, have specified maximum transmit signal power levels. To meet these requirements, further speed reductions are imposed. Collectively, the maximum downstream speed is 53–54 kbps. A trellis8 code, which recognizes the nonuniform spacing of the codec symbols, provides forward error correction and enables discrimination between the service provider generated quantization levels, especially those near the origin. The upstream rate from the subscriber to the service provider is still limited to rates below about 35 kbps because of the A/D conversion. The *V.90* modem standard is likely to persevere as the ultimate in analog modem speed for a much longer time than its predecessor, the V.34 (33.6 kbps) standard. The next step in transmission speed is likely to be some form of digital subscriber line (xDSL). V.90 replaces the *K56FLEX* specification by Rockwell Semiconductor Systems, Lucent Technology, 3Com and Motorola, and the 3Com/U.S. Robotics X2 specification. *V.90* was called *V.pcm* during development at the ITU. See also *mu-law, Shannon limit, trellis code,* and *xDSL.*

- *V.110*—Support of DTEs with V-series type interfaces by an ISDN.

For a complete list, see Appendix E.

V.fast class (V.fc) A Rockwell standard defining modem line rates up to 28.8 kbps. *V.fc* is a forerunner of V.34. Although *V.fc* is incompatible with V.34, modems with *V.fc* firmware are generally upgradable by the modem manufacturer to V.34.

V.fc A contraction of *V.fast class.*

V.pcm The working name for the ITU-T V.90 modem standard.

V.terbo A 19.2-kbps modem standard with fallback to V.32-bis protocols. It was developed jointly by AT&T, Penril, and Data Race. The *V.terbo* standard predates and is incompatible with the V.34 standard.

V/D An abbreviation of <u>V</u>oice over <u>D</u>ata.

V/m The symbol for <u>V</u>olts per <u>M</u>eter.

VA The symbol for <u>v</u>olt-<u>a</u>mp.

Vac An abbreviation of <u>V</u>olts <u>AC</u> (alternating current).

vacant code An unassigned digit or combination of digits, e.g., unassigned area codes or central office codes.

vacant number An unassigned station code or the code of an unequipped circuit.

validation **(1)** Tests to determine whether a specific system implementation meets its specifications **(2)** The checking of data for correctness or for compliance with applicable standards, rules, and conventions.

(3) In Universal Personal Telecommunications (UPT), the process of verifying that a user or terminal is authorized to access UPT services. **(4)** The process of applying specialized security test and evaluation procedures, tools, and equipment needed to establish acceptance for joint usage of an alarm indication signal (AIS) by one or more departments or agencies and their contractors. This action will include, as necessary, final development, evaluation, and testing, preparatory to acceptance by senior security test and evaluation staff specialists.

vampire tap In a thick Ethernet network, a horizontal cable attachment method that pierces the main local area network (LAN) cable's outer shield in order to connect to the inner conductor. The node attachment method. Also called a *piercing tap.*

VAN (Value Added Network)/VANS (Value Added Network Service) A data transmission network that guarantees data security and integrity through added computer control and communications, from the sender to the recipient—often in the manner of a door-to-door courier or freight forwarder.

vanilla Something plain, that is, without options or frills. It may apply to either hardware or software.

vaporware Software that appears much later in the marketplace than promised (many delivery dates missed) or never is made available at all.

var A symbol for <u>v</u>olt-<u>a</u>mp-<u>r</u>eactive. The SI unit of reactive power.

VAR **(1)** An abbreviation of <u>VAR</u>iable. **(2)** An abbreviation of <u>V</u>alue <u>A</u>dded <u>R</u>eseller.

varactor A semiconductor device (diode) whose capacitance is a function of applied voltage.

variable bit rate (VBR) In asynchronous transfer mode (ATM) networks, ATM packetized bandwidth on demand, not dedicated. A connection in which transmissions occur at varying rates, such as in bursts. They employ class B, C, or D services, and are used for data transmissions (as opposed to voice), whose contents are not constrained by timing restrictions. See also *constant bit rate (CBR).*

variable bit rate/non-real time (VBR/nrt) One of five asynchronous transfer mode (ATM) Forum defined service types. It supports variable bit rate traffic that average and peak traffic parameters which can tolerate variable but predictable delays.

variable bit rate/real time (VBR/rt) One of five asynchronous transfer mode (ATM) Forum defined service types. It supports variable bit rate traffic that requires strict timing control, e.g., packetized voice or video.

variable length buffer A buffer into which data may be written at a different instantaneous rate than the data extraction rate, e.g., a first-in first-out (FIFO) device.

Some form of overflow (buffer full) and underflow (buffer empty) detection is generally implemented as part of the flow control to assist in lossless data transfer.

variable length field In a record, a field whose length can vary depending on the contents.

variable length record A record whose length can vary because either it contains fields of variable length or it contains optional fields that may or may not be present, or for both reasons.

variance In statistics, in a population of samples, the mean of the squares of the differences between the respective samples and their mean, expressed mathematically as:

$$\sigma^2 = \frac{1}{n} \sum_{i=1}^{n} \left(x_i - \bar{x} \right)^2$$

where:

i	is the index to a specific value,
n	is the number of samples,
x_i	is the value of sample i,
$\bar{x}$	is the mean of the samples, and
σ^2	is the variance.

The square root of the variance (σ^2) is the *standard deviation* (σ).

variant **(1)** Any of several versions of an entity that are equivalent in most aspects. For example, there are several variants of UNIX, the original from AT&T and others (UCBs, Linex, etc.). **(2)** One of two or more code symbols that have the same plaintext equivalent. **(3)** One of several plaintext meanings that are represented by a single code group.

varindor An inductor whose inductance changes significantly with current. Also called a *swinging choke.*

variolosser A device whose loss can be controlled by external voltages or currents.

varistor An abbreviation of <u>vari</u>able resi<u>stor</u>. A two-terminal semiconductor device having a voltage-dependent nonlinear resistance. See also *MOV.*

VARP An acronym for <u>V</u>INES <u>A</u>ddress <u>R</u>esolution <u>P</u>rotocol.

vars An abbreviation of *volt-amperes reactive.*

VAX **(1)** An abbreviation of <u>V</u>irtual <u>A</u>ccess e<u>X</u>tended. **(2)** An abbreviation of <u>V</u>irtual <u>A</u>ddress e<u>X</u>tension. **(3)** The trademarked name of a line of minicomputers manufactured by Digital Equipment Corporation (DEC).

VB **(1)** An abbreviation of <u>V</u>oice <u>B</u>and. **(2)** An abbreviation of <u>V</u>oice <u>B</u>ank.

VBD An abbreviation of <u>V</u>oice <u>B</u>and <u>D</u>ata.

VBR An abbreviation of <u>V</u>ariable <u>B</u>it <u>R</u>ate.

vBNS An abbreviation of <u>v</u>ery high speed <u>B</u>ackbone <u>N</u>etwork <u>S</u>ervices. The National Science Foundation (NSF) proposed successor to Internet, a 155-Mbps to 622-Mbps network interconnect for high-speed computing facilities. The MCI-sponsored test bed began carrying small amounts of traffic in April 1995. The eventual data rate is expected to be 2.5 Gbps. (Current transmission speeds, on NSFNet, are limited to about 45 Mbps.)

Commercial participants include AT&T, Cisco, IBM, MCI, Microsoft, and Sun. Governmental agency participants include the Defense Advanced Research Projects Agency (DARPA), Department of Energy (DOE), and NSF. In 1996, 12 universities were participating with expectations of growth to 100. Also called *Internet2.*

VC **(1)** An abbreviation of <u>V</u>irtual <u>C</u>all. **(2)** An abbreviation of <u>V</u>irtual <u>C</u>hannel or <u>V</u>irtual <u>C</u>ircuit. **(3)** An abbreviation of <u>V</u>irtual <u>C</u>ontainer.

VC-n An abbreviation of <u>V</u>irtual <u>C</u>ontainer <u>n</u>. In the synchronous digital hierarchy (SDH), a cell of slower bytes carrying a slower channel to define a path. "n" corresponds to the "n" in the DS-n standard.

VCC **(1)** An abbreviation of <u>V</u>irtual <u>C</u>hannel <u>C</u>onnection. **(2)** A symbol of the positive supply voltage to an element, device, or system. **(3)** An abbreviation of <u>V</u>irtual <u>C</u>ontrol <u>C</u>hannel.

VCD An abbreviation of <u>V</u>ariable <u>C</u>apacitance <u>D</u>iode. See also *varactor.*

VCI (1) An abbreviation of <u>V</u>irtual <u>C</u>hannel <u>I</u>dentifier. Part of the packet/frame/cell header address in IEEE 802.6 asynchronous transfer mode (ATM). (2) An abbreviation of <u>V</u>irtual <u>C</u>ircuit <u>I</u>dentifier.

VCL An abbreviation of <u>V</u>irtual <u>C</u>hannel <u>L</u>ink.

VCO An abbreviation of <u>V</u>oltage <u>C</u>ontrolled <u>O</u>scillator. An oscillator whose output frequency can be set by a control signal (voltage).

VCPI An abbreviation of <u>V</u>irtual <u>C</u>ontrol <u>P</u>rogram <u>I</u>nterface.

VCR An abbreviation of <u>V</u>ideo <u>C</u>assette <u>R</u>ecorder.

VCS An abbreviation of <u>V</u>irtual <u>C</u>ircuit <u>S</u>witch.

VCSEL An abbreviation of <u>V</u>ertical <u>C</u>avity <u>S</u>urface <u>E</u>mitting <u>L</u>aser.

VCX An abbreviation of <u>V</u>irtual <u>C</u>hannel <u>C</u>ross connect. A device that switches packets/frames/cells on a logical connection basis.

VCXO An abbreviation of <u>V</u>oltage-<u>C</u>ontrolled crystal (<u>X</u>) <u>O</u>scillator.

Vdc An abbreviation of <u>V</u>oltage <u>DC</u>. Also written *VDC*.

VDD A symbol of the negative supply voltage to an element, device, or system.

VDE An abbreviation of <u>V</u>erband <u>D</u>eutscher <u>E</u>lectrotechniker e. V., the German national testing and component safety agency. *VDE* issues standards in several forms:

Regulations—VDE Bestimmungen

Leaflets—VDE Druckschriften

Instructions—VDE Merkblaetter

Guiding principles—VDE Richtlinien

VDI (1) An abbreviation of <u>V</u>irtual <u>D</u>evice <u>I</u>nterface. (2) An abbreviation of the German <u>V</u>ereins <u>D</u>eutscher <u>I</u>ngenieure.

VDM An abbreviation of <u>V</u>irtual <u>DOS</u> <u>M</u>achine.

VDOP An acronym for <u>V</u>ertical <u>D</u>ilution <u>O</u>f <u>P</u>recision. A measure of how much the geometry of satellite positions affects the vertical position estimate of a Global Positioning System (GPS) receiver. The estimate is computed from satellite range measurements. See also *DOP (dilution of precision)* and *Global Positioning System (GPS)*.

VDSL An abbreviation of <u>V</u>ery high speed <u>D</u>igital <u>S</u>ubscriber <u>L</u>ine.

VDT An abbreviation of <u>V</u>ideo <u>D</u>isplay <u>T</u>erminal.

VDU An abbreviation of <u>V</u>ideo (or <u>V</u>isual) <u>D</u>isplay <u>U</u>nit.

vector Any physical quantity that requires two or more variables to describe its value. For example, the magnitude and phase of a signal or the latitude, longitude, and altitude of a position.

vector image A method of generating images based on mathematical descriptions to determine position, length, and direction of a line or line segment. Graphics are a collection of line segments rather than a collection of individual dots as with a *raster image*. See also *raster image*.

vector processor A synonym for *array processor*.

velocity of light See *speed of light*.

velocity of propagation (VOP) In a transmission medium, a value that indicates the speed that a signal travels in that medium, as a proportion of the maximum speed that is theoretically possible. This value varies with both the medium and with the physical architecture. Values for electrically based local area networks range from about 60 to 85% of maximum. See also *propagation velocity*.

velocity of sound The speed of transmission of mechanical waves in any given medium. It is expressed as the square root of the ratio of the material's elasticity to its density.

Vendor Independent Messaging (VIM) A standard for the application interface to e-mail from Lotus, WordPerfect, and others. It includes Messaging Application Programming Interface (MAPI) compliance.

Venn diagram A diagram of multiple entities in which each entity is represented by a closed region and the union of entities by the overlap of the regions. The associated diagram illustrates a three-variable (A, B, and C) system with A · B, B · C, A · C, and A · B · C regions highlighted.

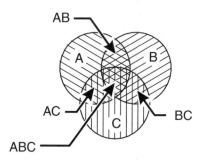

verbose A mode in which narrative type messages are provided to the user rather than cryptic numbers.

Setting ATV1 will turn on the verbose mode of most Hayes compatible modems. ATV0 sets the modem in the numeric reporting mode.

verified off-hook In telephony, a service provided by a unit inserted on each end of a transmission circuit for verifying supervisory signals on the circuit.

Veronica In Internet. Said to be an acronym from <u>V</u>ery <u>E</u>asy <u>R</u>odent-<u>O</u>riented <u>N</u>et-wide <u>I</u>ndex to <u>C</u>omputerized <u>A</u>rchive. A server program or utility (within Gopher) that will find occurrences of something based on some defined search criteria. The search criteria may be a complex string of Boolean expressions. *Veronica* actually scans *lists* of resources (such as text files, programs, and graphics) rather than the directories themselves. A list is a compiled index of the titles of all of the items that can be retrieved in *gopherspace*. See also *Jughead*.

version A variant of an original entity.

vertex angle In an optical fiber, the angle formed by:

- Twice the *acceptance angle* (the angle formed by the largest cone of light accepted or emitted by the fiber).
- The extreme bound meridional rays accepted by the fiber or emerging from it.

See also *acceptance angle* and *optical fiber*.

vertical redundancy check (VRC) An error-checking method wherein an extra bit is appended to each byte transmitted and that bit is set to either a "0" or a "1" in a manner that maintains either even parity or odd parity across the character. A character with odd parity contains an odd number of "1s" and a character with even parity contains an even number of "1s." Also called *vertical parity check* or *transverse redundancy check (TRC)*. See also *longitudinal redundancy check (LRC), parity,* and *transverse redundance check (TRC)*.

very high frequency (VHF) The range of frequencies from 30 MHz to 300 MHz. See also *frequency bands*.

very low frequency (VLF) The range of frequencies from 3 kHz to 30 kHz. See also *frequency bands*.

very small aperture terminal (VSAT) A relatively small satellite dish (less than about 2 meters) used for digital satellite communications.

VESA An acronym for Video Electronics Standards Association.

vestigial sideband (VSB) transmission Modified AM transmission in which one sideband, the carrier, and only a portion of the other sideband power are transmitted.

VEX An abbreviation of Video Extensions for X-Windows.

VF (1) An abbreviation of Voice Frequency. In telephony, the frequency band is 300–3400 Hz. See also *frequency bands*. (2) An abbreviation of Vacuum Florescent.

VFC (1) An abbreviation of Video Frequency Channel (or Carrier). (2) An abbreviation of Voice Frequency Channel (or Carrier). (3) An abbreviation of Voltage to Frequency Converter.

VFCT An abbreviation of Voice Frequency Carrier Telegraph. See *voice-frequency telegraph*.

VFCTG An abbreviation of Voice Frequency Carrier TeleGraph. See *voice-frequency telegraph*.

VFDF An abbreviation of Voice Frequency Distribution Frame.

VFL An abbreviation of Voice Frequency Line.

VFO An abbreviation of Variable Frequency Oscillator.

VFRP An abbreviation of VINES FRagmentation Protocol.

VFS (1) An abbreviation of VINES File System. (2) An abbreviation of Virtual File Store (3) An abbreviation of Virtual File System.

VFSS An abbreviation of Voice Frequency Signaling System.

VFT An abbreviation of Voice Frequency Telegraph.

VFTG An abbreviation of Voice-Frequency TeleGraph.

VfW An abbreviation of Video for Windows.

VG An abbreviation of Voice Grade.

VGA An abbreviation of Video Graphics Array.

VGPL An abbreviation of Voice Grade Private Line.

VHDSL An abbreviation of Very High speed Digital Subscriber Loop.

VHF An abbreviation of Very High Frequency. See also *frequency bands*.

vi A full-screen UNIX system text editor. It is simple but cumbersome. It is difficult to learn by those used to the high-quality word processors found on most PCs.

via net loss (VNL) (1) In telephony, the overall (or net) losses of trunks in the long-distance switched network. (A trunk is in the "via" condition when it is an intermediate link in a switched connection.) (2) Pertaining to circuit performance prediction and description that allows circuit parameters to be predetermined and the circuit to be designed to meet established criteria by analyzing actual, theoretical, and calculated losses.

VICP An abbreviation of VINES Internet Control Protocol.

video (1) The composite electrical signal containing timing (synchronization), luminance (intensity), and optionally chrominance (color) information which, when displayed on by an appropriate device, gives a visual image or representation of the original image sequence.

(2) Pertaining to the sections of a television system that carry either modulated or unmodulated television signals. (3) Pertaining to the bandwidth or data rate necessary for the transmission of real-time television pictures. (The bandwidth required for the transmission of television pictures, excluding audio carriers, is approximately 6 MHz.) (4) Pertaining to the demodulated radar signal that is applied to a radar display device.

Video CD A format that allows the viewing of Motion Picture Experts Group (MPEG) 1 video on CD-ROM. Playback of these CDs requires a computer with MPEG hardware and software, a *Video CD player*, or a CD-i player.

Video CD is based on the *White Book* standard developed by Philips and others.

video compression A data compression scheme optimized for video. Many algorithms use the fact that there are usually only small changes from one *frame* to the next, so they only need to encode the starting frame and a sequence of differences between frames. The process is called interframe coding or 3-D coding.

Two *video compression* algorithms are defined by Motion Picture Experts Group (MPEG) and International Telecommunications Union Telecommunications section (ITU-T's) H.261 recommendation.

video conferencing A term that refers to multiparty communications involving both audio and video. It includes equipment that permits two or more persons in different locations to engage in the equivalent of face-to-face audio and video communications, that is, it permits participants to conduct business as if they were all in the same room.

Because of the huge bandwidth required, special-purpose hardware is generally needed. With efficient voice and video compression techniques, this problem may become less of an obstacle. Standards and specifications for codecs, the information they must process, and *video conferencing* are delineated in the ITU-T H.200 and H.300 series recommendations. Some particularly important recommendations include:

- H.221—communications framing.
- H.230—control and indication signals.
- H.242d—call setup and disconnect.
- H.261—announced in 1990 and originally defined by Compression Labs Inc., relating to the decoding process used when decompressing *video conferencing* pictures.

Also called *video teleconferencing*.

video dial-tone A service planned by various telephone companies that will deliver video to homes as an alternative to the existing cable networks. The deliver system of choice is fiber optics, and many local operating companies are now building the network infrastructure to support the service.

video display terminal (VDT) See *monitor*.

video display unit (VDU) See *monitor*.

Video Electronics Standards Association (VESA) An association of video adapter and display manufacturers. They have developed standards for

- Display formats (such as the Super VGA graphics standard) and
- A system bus, called the VL bus, or VESA local bus which is capable of 64-bit operation and can support much faster clock speeds than earlier bus designs.

See also *EISA, ISA, MicroChannel,* and *Peripheral Component Interconnect (PCI)*.

Video for Windows (VfW) A standard established by Microsoft for the integration of digital video, animation, and sound. The Audio-Video Interleaved (AVI) file format is used by *VfW*. Software drivers once were required to view AVI files in Windows. These drivers are incorporated into the latest versions of Windows.

video on demand A television program delivery service that allows a consumer to choose and view programming material from a library of material at any time. This is in contrast with "pay-per-view" in which the consumer chooses from material broadcast only at a predetermined time.

video teleconference See *video conferencing*.

videophone A telephone coupled to an imaging device that enables the call receiver and/or the call originator to view one another or other picture information, if they so desire.

videotex A term invented by the International Telecommunications Union Telecommunication section (ITU-T) to describe TV equipment used to display computer-based data, whether sent via a telephone (often called *viewdata*) or broadcasting channel (*teletext*).

It is an interactive information retrieval service accessed by a user over a standard telephone line. The accessed information is displayed on a standard television or a *videotex* terminal. The system provides access to its various services and databases via keyboard selection of displayed menu items. See also *teletext*.

viewdata A type of information retrieval service in which a subscriber can:

- Access a remote database via a common carrier channel,
- Request data, and
- Receive requested data on a video display over a separate channel.

The access, request, and reception are usually via common carrier broadcast channels. See also *teletext*.

VIM An acronym for <u>V</u>endor <u>I</u>ndependent <u>M</u>essaging.

V_{in} The symbol for voltage input.

VINES™ An acronym for <u>VI</u>rtual <u>NE</u>twork <u>S</u>oftware. Banyan®'s network operating system (NOS) based on UNIX system V and its protocols. Features of VINES include:

- Support for up to four network cards (NICs) per server.
- Automatic service of any necessary protocol binding or translation when routing packets between different LAN topologies. Protocol binding is accomplished using Microsoft's Network Driver Interface Specification (NDIS).
- Support for multiple servers and enterprise networks.
- Server-to-server connections for local area networks (LANs) or wide area networks (WANs) via direct, X.25, ISDN, T-1, SNA, dial-up, or leased lines connections.
- Support for various protocol stacks, including: VINES and optional support for Open System Interconnection (OSI), Transmission Control Protocol/Internet Protocol (TCP/IP), and AppleTalk stacks. In addition, it supports NetBIOS emulation, which provides generic support for other layered networking environments.
- Asynchronous communication capabilities for remote networking.
- Gateway services for communicating with other networking environments.
- Networkwide security services that provide user authentication services and that use access rights lists associated with files and resources to determine who is allowed to use which resources.

- Symmetric multiprocessing capabilities, which support multiple processors working independently of each other, but all communicating with the NOS. This allows the NOS to allocate different tasks to different processors.
- Both local and network wide management and monitoring capabilities, including the optional ability to monitor (in real time) the network from the server console or from any network PC.
- And, of course, the basic network facilities (file and directory access, archival capabilities, mail and message handling services, etc.).

The architecture as compared to the ISO/OSI Reference Model is illustrated in the table.

Layer 7: Application Layer	VINES File Service		VINES Applications Services	
Layer 6: Presentation Layer	VINES Remote Procedure Calls (RPC)		Server Message Block (SMB)	
Layer 5: Session Layer	Socket Interface			
Layer 4: Transport Layer	VINES Interprocess Communications (VIPC)	VINES Sequenced Packet Protocol (VSPP)	Transmission Control Protocol (TCP)	User Datagram Protocol (UDP)
Layer 3: Network Layer	VINES Internet Protocol (VIP)	VINES Internet Control Protocol (VICP)	Internet Protocol (IP)	X.25
Layer 2: Link Layer	Network Driver Interface Specification (NDIS)		HDLC	
Layer 1: Physical Layer	Network Interface Cards (NIC)			
Layer 0: Media Layer	Cabling			

violation See *AMI violation*.

VIP (1) An acronym for <u>VI</u>NES <u>I</u>nternet <u>P</u>rotocol. **(2)** An acronym for <u>V</u>isual <u>I</u>nformation <u>P</u>rocessing. **(3)** An acronym for <u>V</u>ideo <u>I</u>nput <u>P</u>ort.

VIPC An abbreviation of <u>VI</u>NES <u>I</u>nter<u>P</u>rocess <u>C</u>ommunications.

virtual A description of an emulated device or process that is perceived to be something that it is not. A *virtual* entity can be treated in every aspect as if it were the entity being emulated—an entity that apparently exists but in reality does not. For example, virtual memory in a PC is actually hard disk space holding working memory contents not needed in the current process.

Contrast with *transparent* (actually there but apparently not).

virtual call A call, established over a network, that uses the capabilities of either a real or virtual circuit by sharing all or any part of the resources of the circuit for the duration of the call.

virtual call capability A network service feature in which:

- A call setup procedure and a call disengagement procedure determine the period of communication between two data terminal equipments (DTEs) in which user data are transferred by the network in the packet mode of operation.
- End-to-end transfer control of packets within the network is required.
- Data may be delivered to the network by the call originator before the call access phase is completed. However, data are not delivered to the call receiver if the call attempt is unsuccessful.
- The network delivers all the user data to the call receiver in the same sequence in which the data are received by the network.

Multi-access DTEs may have several virtual calls in progress at the same time. Also called *virtual call facility*.

virtual carrier frequency In radio or carrier systems in which no carrier is transmitted (e.g., single sideband or double sideband with suppressed carrier), the location in the frequency spectrum that the carrier would occupy if it were present.

virtual channel connection (VCC) The basic switching level for asynchronous transfer mode (ATM) that forms a logical connection between two entities (which may be users or networks). Some of the characteristics include:

- *Selectable circuit type*—switched circuits (established as required or demanded) or dedicated circuits (semi-permanent links) as required.
- *Packet order preservation*—cells transmitted at the source and delivered to the destination in the same order.
- *Negotiated performance parameters*—entities involved in the connection to negotiate performance parameters. The parameters are specified in a *traffic descriptor* (or user network contract).
- *Quality of service (QoS)* as required—a quality of service provided as specified by parameters governing cell delays, losses, peak transmission rate, maximum burst length, and so on.

virtual channel identifier (VCI) In asynchronous transfer mode (ATM) networks, a value, associated with a single virtual channel connection (VCC) for a particular user, used to route a cell to and from that user. A given VCC may have different *VCIs* at the sending and receiving ends. See also *ATM* and *virtual path identifier (VPI)*.

virtual circuit (VC) In packet switching networks, a temporary communications link that, to a user, appears to be a direct connection between the sender and receiver but may be a collection of meandering paths. This type of circuit may appear as a dedicated line to the user, but it will actually be connected only for the duration of any packets routed between the nodes. Each virtual connection exists only for the duration of one packet transmission.

To a user, a *virtual circuit* appears to be a communications link between points A and B, but it in fact may take a path from A to C to D to E and finally to B. It may also be rerouted during the session based on traffic flow requirements. However, the user is unaware of any of this, knowing only that data are flowing from A to B. A virtual circuit is a logical connection. The connection can be a *switched virtual circuit (SVC)*, which can connect to a different data terminal equipment (DTE) each session; or a *permanent virtual circuit (PVC)* in which all sessions connect to the same DTE at the other end. Also called a *logical circuit* and *logical route*.

virtual circuit capability A network-provided service feature in which a user is provided with a virtual circuit.

Virtual circuit capability is not necessarily limited to packet mode transmission. For example, an analog signal may be converted to a digital signal and then be routed over the network via any available route—e.g., voice over Internet Protocol (IP) (VoIP).

virtual company A company built on the basis of teleworking with limited central office administration. Made possible by improved communications and groupware software.

virtual connection A logical connection that is made to a virtual circuit.

virtual container (VC) The synchronous digital hierarchy (SDH) defines a number of *containers,* each corresponding to an existing plesiochronous rate. Information from a plesiochronous signal is mapped into the relevant container along with control information known as the "path overhead." The container plus path overhead form a *VC*.

Virtual Control Program Interface (VCPI) A DOS-protected mode interface developed by Quarterdeck Systems, Phar Lap Software, and other vendors which became a de facto standard.

It enables DOS programs to run in protected mode on 80386 and higher machines and to execute cooperatively with other operating environments (for example, DESQview). *VCPI,* however, is incompatible with Microsoft's DOS extender, *DOS Protected Mode Interface (DPMI)*.

virtual device A device that can be referenced but does not physically exist. A terminal emulator program makes a computer appear to be a specific terminal (a *virtual terminal*) to the host computer.

virtual height A term referring to the ionosphere. The apparent height of the ionosphere as determined by the time interval between a transmitted radio pulse and its echo at vertical incidence and assuming the velocity of propagation is constant over the entire path.

virtual LAN (1) A logical subgroup of users and devices (independent of their physical locations) from the total population of a physical local area network (LAN) or wide area network (WAN) which are treated as a distinct and independent LAN. Multiple *virtual LANs* may reside on a single physical network.

Benefits of the *virtual LAN* include:

- The ability to move the physical location of a user without rewiring.
- The ability to broadcast messages to members of the *virtual LAN* without others on the LAN seeing them.
- When multiple *virtual LANs* are present on a single physical LAN, reduced communications cost to the LAN owner (and users).

(2) The connection of a number of physical segments into a single virtual segment that functions as a self-contained traffic domain. They are switched at the ISO/OSI data link layer (layer 2). Also called *Virtual Segment Virtual LAN* or a *Virtual Port Group Virtual LAN*.

virtual memory In computer systems, the memory as it appears (or is available) to the programs running on the central processing unit (CPU). The virtual memory may be smaller, equal to, or larger than the real memory present in the system.

Although virtual memory may be smaller or the same size as physical memory, it is usually larger than physical memory. Paging allows the excess to be stored in a reservoir (disk or external random access memory—RAM) and copied to physical RAM address space as required. This makes it possible to run programs whose code plus data size is greater than the amount of physical RAM available. *Demand paged virtual memory access* is the most common technique of accessing information in the reservoir memory. In the method, a page is copied (paged) from the reservoir to main memory when an access attempt is made and the information is not already in physical RAM. This paging is performed automatically by collaboration between the hardware (CPU), memory management unit (MMU), and the operating system. The program is unaware of it, except perhaps in terms of access time. A program's actual performance depends dramatically on how its access to virtual memory pages interacts with the paging scheme. If accesses tend to be close together, they tend to be in the same page; hence, no page swapping is required (producing a faster program). If, on the other hand, accesses are randomly distributed over the program's address space, more paging is required, thus, slowing the program.

virtual network (1) A subnetwork of a larger network defined in scope and attributes by software rather than physical hardware and wiring. **(2)** A network that provides virtual circuits and that is established by using the facilities of a real network.

virtual path See *virtual circuit* (*VC*).

virtual path identifier (VPI) A value associated with a particular virtual path connection (VPC) and used to route a cell across a network. See also *ATM* and *virtual channel identifier* (*VCI*).

virtual private network (VPN) A network tunnel created between two or more authenticated terminations (users). It is an advanced provision for private voice and data networking via a public or other multiply accessed switched networks. The network connection appears to the user as an end-to-end, nailed-up circuit without actually involving a permanent physical connection, as in the case of a leased line. *VPNs* retain the advantages of private networks but add benefits like capacity on demand.

Capabilities of secure *VPNs* may include:

- *Tunneling*—the ability to pass a foreign protocol through a network in such a manner that the network is transparent to the end nodes.
  Standards used include L2TP, PPTP, and the Internet Protocol Security (IPSec) suite.
- *Remote access*—enables a user to initiate a connection from a dial-up terminal.
- *Encryption*—used to provide user authentication and data confidentiality. Multiple encryption algorithms may be available so that an appropriate level of security and speed may be selected. Both public and secret key systems are used (e.g., DES and RSA respectively).
- *Authentication*—a means to verify the identity of a user or terminal. Examples of authentication means include passwords and digital certificates (such as ITU X.509 or Simple Key management for Internet Protocol—SKIP).
- *Automated key management*—defines the time periods for which session keys and digital certificates are valid.

virtual reality (VR) An interactive, computer-generated simulated environment with which users can interact. The technology allows the user to experience 3-D interaction with the computer. Most *VR* systems incorporate special peripherals such as visors or helmets (with computer-graphic displays), gloves, and special 3-D graphics technology to simulate the real-world environment.

virtual reality modeling language (VRML) A programming language for creating 3-D World Wide Web (WWW) sites. VRML 1.0 was released in May 1995.

virtual storage Storage space that may be viewed as addressable main storage but is actually auxiliary storage (usually peripheral mass storage such as a hard disk drive) mapped into read addresses. The amount of *virtual storage* is limited by the addressing scheme of the computer, the amount of auxiliary storage allocated, and the amount of auxiliary actually available. It is not a function of the size of the main storage. See also *virtual memory*.

virtual teams Ad hoc groups of users formed to solve particular problems without taking them away from their desks. A useful option made feasible with groupware.

virtual terminal (VT) An application layer service, in the ISO/OSI Reference Model, that makes it possible to emulate the behavior of a particular terminal. This emulation enables an application to communicate with a remote system, such as a mainframe, without needing to worry about the type of hardware sending or receiving the communications. It acts as a translator for the two different protocols (host and remote).

The host will use the host's native protocol to communicate with the *virtual terminal*. The *virtual terminal* will convert the communications from the host into a standardized intermediate protocol and finally into the compatible protocol of the PC.

virus An unwanted computer program with two characteristics:

- It is a program, that is, it can run or execute commands.
- It can duplicate itself.

Generally, however, it is also a hidden computer program that lies in a system until activated. When activated, it attaches copies of itself to other computer programs (*infects* them) and/or executes a payload. The "infected" computer program, when run, will infect another programs and so on. *Viruses* often have secondary effects (from the payload portion), sometimes have just a simple message, and sometimes are massively damaging. The only way a *virus* can infect a program is by *running* an infected program. This means only *executable programs, batch programs, macros,* etc., can be carriers. Simply *viewing* a file (even if it contains a *virus*) cannot infect a system. Note that many word processors and spreadsheet programs have macros imbedded in their data files. These macros are candidates for attack by virus programmers; hence, it is wise to be cautious even with "data only" files. *Virus* programs are often classified in a number of ways based on such things as where they locate themselves, how they operate, and how they defend themselves. Some examples of classification include:

- *Armored*—a *virus* that attempts to protect itself by making analysis difficult. For example, it may encrypt selected code bytes in a way that scrambles pieces of code when it is simply disassembled for analysis.
- *Benign*—a *virus* that **tends** to do nothing more than display a message or degrade system performance. Its intent is not data corruption, although corruption can occur unintentionally.
- *Bomb*—see *Trojan horse*.
- *Boot & file*—see *Multipartite virus*.
- *Boot sector*—a *virus* that resides in the boot sector of a disk. The infection is spread when a computer is booted from a disk containing the *virus*.
- *Encrypted*—a *virus* in which the main code is encrypted with a key that is varied. Because the key and main code are variable, they present no fixed signature that is easily detectable. The decoding program cannot itself be encoded so it does have a fixed signature. However, the routine (and therefore its signature) may be the same as a legitimate program. An alternate approach uses an extraneous nonexecuted code segment that is modified at each replication.
- *File infectors (parasitic)*—a *virus* that resides in an executable program. When the infected program is run, the infection is passed on to other programs in the system.
- *Malignant*—a *virus* that has the intent of causing harm to data in as short a time as possible.
- *Multipartite*—a *virus* that uses both the boot sectors and an executable files for its residence.
- *Polymorphic*—a *virus* that attempts to avoid detection by modifying itself each time it replicates. This makes detection by direct signature analysis difficult, if not impossible.
- *Stealth*—a *virus* that attempts to avoid detection by camouflage, for example, a boot sector *virus* may redirect read attempts to other sectors containing an uninfected copy of the boot sectors.
- *Trojan horse*—a *virus* program disguised as a "useful" program (such as a game, utility, other application) that when run intentionally does harm to some aspect of the computer, network, or system. Also called a *bomb* or *content virus*.

- *Worm*—a *virus* that after entering a computer system duplicates itself multiple times, filling up all available memory (and possible disk space), thus bringing the system to a halt (crashing the system).

Many virus detection programs are available. Some of them run in the background at all times, and others must be manually initiated. If you download many files from public bulletin boards or the Internet, you will eventually download a file that contains a virus. The only defenses are:

- Test every file downloaded with a *current version* of a good virus detection program.
- Make sure that the machine on which the file is to be tested has a *current backup.*
- Test the file only on a machine with data that can be lost.
- Download only from known sources, and even these can be dangerous.
- Keep a write protected boot diskette with the current system configuration available at all times.
- Know *ahead of time* how to use your virus removal programs.

VISCA An abbreviation of VIdeo System Control Architecture. A platform-independent protocol designed by Sony to provide computer control of multiple video devices.

visible spectrum The region of the electromagnetic spectrum that can be perceived by human vision, approximately the wavelength range of 400 nm to 700 nm. See also *frequency spectrum.*

visual display unit (VDU) See *monitor.*

VL (1) An abbreviation of VESA Local bus. **(2)** An abbreviation of Virtual Link.

VLAN An acronym for Virtual LAN.

VLB An abbreviation of VESA Local Bus.

VLF An abbreviation of Very Low Frequency. The range of frequencies is between 3 kHz and 30 kHz. See also *frequency bands.*

VLIW An abbreviation of Very Long Instruction Word.

VLM An abbreviation of Virtual Loadable Module.

VLSI An abbreviation of Very Large Scale Integration.

VLSM An abbreviation of Variable Length Subnet Mask.

VM (1) An abbreviation of Virtual Machine. **(2)** An abbreviation of Virtual Memory. **(3)** An abbreviation of Voice Messaging.

VM/CMS An abbreviation of Virtual Machine/Conversation Monitor System.

VMD An abbreviation of Virtual Manufacturing Device.

VMS (1) The main operating system on Digital Equipment Corporation's (DEC) VAX computers. **(2)** An abbreviation of Virtual Memory operating System. **(3)** An abbreviation of Voice Message System.

VMSnet One of several USENet alternative newsgroups. The newsgroups discuss information relevant to those with interests in Digital Equipment Corporation (DEC) VAX computers. See also *ALT* and *alternative newsgroup hierarchies.*

VMTP An abbreviation of Versatile Message Transaction Protocol. Designed at the University of California at Stanford to replace the TCP and TP4 protocols in high-speed networks.

VNA An abbreviation of Virtual Network Architecture.

VNET An abbreviation of Virtual NETwork.

VNL An algorithm for Via Net Loss.

VNLF An abbreviation of Via Net Loss Factor.

VNN An abbreviation of Vacant National Number.

vocoder A contraction of VOice-CODER. A device that usually consists of a speech analyzer, which converts analog speech waveforms into narrowband digital signals, and a speech synthesizer, which converts the digital signals into artificial speech sounds.

A *vocoder* may be used in conjunction with encryption so as to allow the transmission of encrypted speech signals over narrowband voice communications channels.

VODAS An acronym for Voice Operated Device Anti-Sing. A voice-operated system that prevents a two-way voice circuit from singing (oscillating) by disabling one of the transmission directions at all times. The active direction is unimpeded while the quiet path is disabled.

VOGAD An acronym for Voice Operated Gain Adjusting Device. A device that gives, to a large extent, the same volume of signal for a wide range of input levels. It is similar in function to a compressor; however, no expander is required at the receiving station.

voice/data PABX A switching system that combines the functions of both a *voice PABX* (private automatic branch exchange) and a *data PABX,* often with emphasis on the voice facilities.

voice band Depending on the specification, a range of frequencies from less than 100 Hz to about 4000 Hz, including guard bands. Some example specifications are:

Service Specification	Frequency	
	Low	High
ANSI/IEEE Std 455-1985	50	4000
FCC Mobil Radio	250	3000
Telephone (typical)	200 (300)	3400

In telephony, the bandwidth allocated for a single voice-frequency transmission channel is usually 4 kHz, including any guard bands. Also called *voice frequency* (*VF*).

voice data signal See *quasi-analog signal.*

voice frequency (VF) See *voice band.*

voice-frequency (VF) channel A communications channel capable of carrying voice band analog and quasi-analog signals. The term *voice-frequency channel* does not imply any specific signaling or supervisory scheme, only that the channel has sufficient bandwidth for voice grade transmissions.

In telephony, the channel bandwidth is approximately 300 Hz to 3400 Hz usable in an allocated 4-kHz band (including guardband).

voice-frequency telegraph (VFT or **VFTG)** A method of multiplexing one or more dc telegraph channels onto a nominal 4-kHz voice frequency channel. Also called *voice-frequency carrier telegraph* (*VFCT* or *VFCTG*).

voice grade A telecommunications service grade defined by the U.S. Code of Federal Regulations (CFR), Title 47, part 68. *Voice grade* service does not imply any specific signaling, supervisory scheme, or even the information type.

voice grade channel A communications channel suitable for carrying speech but not necessarily good enough for high-speed data communications. *Voice grade* telephone lines have a usable bandwidth of only 3100 Hz, sufficient for only 2400- to 3000-baud transmissions. Because of the quantizing noise introduced by the analog-to-digital conversion in the coder-decoder (CODEC), multibit encoding techniques can increase the throughput to 33 600 bps in ITU-T V.34 modems, with only somewhat higher capabilities in the future. ITU-T V.90 modems can operate to 56 kbps in the downstream direction because they avoid the analog-to-digital conversion process in one direction. Also called a *voice grade line*. See also *Shannon limit* and *V.90*.

voice mail A system for recording, storing, retrieving, and delivering voice messages. It may be either a stand-alone device or integrated, to some extent, with a user's phone system. If the phone rings for a specified number of rings, it can default to a mailbox which delivers its prerecorded invitation to leave a message and records the results. Messages can be delivered at a prearranged time, tagged and edited.

Stand-alone *voice mail* is similar to a collection of answering machines but has added features such as call forwarding. Integrated systems indicate messages waiting via a light on a user's phone and/or an alphanumeric display.

voice-operated switch (VOX) A switching device that monitors the signal level on a transmitter's input. When the level exceeds a specified threshold, the transmitter is turned on and the receiver is turned off. When the level falls below that threshold, the transmitter is turned off. Also called *voice-operated transmit*.

voice over data (VOD) A method of sending voice and data simultaneously over a single telephone line.

There are several methods of accomplishing this task, e.g., frequency division multiplex (FDM) where the lower frequencies of a band are used for data transmission and the upper part of the band is used for voice. A second method is time division multiplex (TDM). The firmware to accomplish this is included in some modems. *Note:* The total message capacity of the telephone line is bounded by Shannon's limit. Therefore, when the voice option is in effect, the digital transport rate is reduced. See also *Shannon limit*.

voice PABX A private automatic branch exchange (PABX) for voice only circuits, e.g., a telephone exchange.

voice plus circuit A circuit carrying both voice and other services. Also called a *composited circuit*.

voice recognition See *speech recognition*.

VoIP An abbreviation of Voice Over Internet Protocol (IP).

VOIS An abbreviation of Voice Operated Information System.

VOL An abbreviation of VOLume.

volatile memory Memory that loses data when power is removed. Both dynamic random access memory—RAM (DRAM) and static RAM (SRAM) memories will lose data when power is removed. See also *nonvolatile memory*.

volatile storage A term that refers to any storage device (memory) in which the contents are lost when electrical power is removed.

volcas An acronym for Voice Operated Loss Control And Suppressor. A voice-operated system that prevents a two-way voice circuit from singing (oscillating) by attenuating one of the transmission directions at all times. The active direction is unimpeded, while the quiet path is attenuated.

volt (V) The unit of voltage, electromagnetic force, or potential difference.

One *volt* is defined as the potential difference across which 1 coulomb of charge will do 1 joule of work. Ohm's law relates voltage, current, and resistance with the statement, "One *volt* of potential difference is generated when one ampere of current flows through a resistance of one Ohm," or mathematically,

$$V = I \cdot R$$

where

V is the voltage,
I is the current, and
R is the resistance.

The SI symbol for the *volt* is V.

volt-ampere (VA) The unit of apparent power in the SI system. It is the product of the root-mean-square (RMS) voltage, the RMS current, and the cosine of the angle between them. Frequently abbreviated *voltamp*.

volt amperes reactive (vars) In alternating-current (ac) power transmission and distribution, the product of the root-mean-square (RMS) voltage and amperage (i.e., the *apparent power*), multiplied by the sine of the phase angle between the voltage and the current.

Var is properly expressed only in volt-amperes (VA)—not watts (W). (Only *effective power*, i.e., the actual power delivered to or consumed by the load, is expressed in watts.) *Vars* represents the power *not* consumed by a reactive load. To maximize transmission efficiency, therefore, *vars* must be minimized. This is accomplished by balancing capacitive and inductive loads, or by adding an appropriate capacitive or inductive reactance elements to compensate for reactive loads.

voltage The amount of energy available to move a certain number of electrons from one point to another in an electrical circuit. Also called *electromotive force* (*EMF*).

voltage breakdown impulse ratio The ratio of the impulse voltage breakdown (V_{IBD}) of an entity to the dc breakdown voltage (V_{DCBD}), that is,

$$\delta_{RATIO} = \frac{V_{IBD}}{V_{DCBD}}$$

Note that this ratio is never less than unity.

voltage-controlled oscillator (VCO) Any oscillator in which the output frequency can be set by a control signal (voltage). Neither amplitude nor waveshape of the oscillator output is intentionally changed by the control signal. See also *oscillator*.

voltage delay In electrochemical cells, a time delay between the application of a load to a battery source and the full operating voltage. The delay is dependent on the percentage of the battery capacity the load requires, the ambient temperature, and the cell chemistry.

voltage depression In electrochemical cells, an abnormal drop in terminal voltage, i.e., below the expected values during the discharge of a battery.

voltage keyed A term that refers to a system which incorporates a mechanical identifier on battery packs and devices to ensure only batteries of the correct voltage and polarity are connected to the device.

voltage reversal In electrochemical cells, a changing of the normal polarity of one or more cells due to overdischarge of the battery.

voltage standing wave ratio (VSWR) See *standing wave ratio.*

volume (1) A term that refers to the magnitude of an audible signal, i.e., signal level. (2) In computer hierarchical file structures, the highest level in a file computer's directory and file structure. (3) A portion of data, with its physical storage medium, that can be handled conveniently as a unit. Examples include a "floppy" diskette or a magnetic tape.

volume limiter A device that automatically limits the signal level in a circuit or portion of a circuit. The device may be hard limiting (clipping) or soft limiting (compression). See also *clipping* and *commander.*

volume unit (vu) A quantitative measure of audio signal level (volume) in an electric circuit. The *volume unit* is numerically equal to the ratio of the signal to a reference volume expressed in decibels (dB). For sine waves, 0 *vu* is equal to 0 dBm; however, the term *vu* should not be used to express the results of measurements of a complex waveform made with a device whose characteristics differ from those of a standard volume indicator.

A *vu meter* is built and used in accordance with American National Standard C16.5-1942.

VOP An abbreviation of $\underline{V}$elocity $\underline{O}$f $\underline{P}$ropagation.

VOTS An abbreviation of $\underline{V}$AX $\underline{O}$SI $\underline{T}$ransport $\underline{S}$ervice.

VOM (1) An abbreviation of $\underline{V}$olt-$\underline{O}$hm-$\underline{M}$illiammpmeter. A device for measuring circuit voltages, resistances, or current. (2) An abbreviation of $\underline{V}$olt-$\underline{O}$hm-$\underline{M}$eter.

V$_{out}$ A symbol for output voltage.

VOX An acronym for $\underline{V}$oice $\underline{O}$perated Switch ($\underline{X}$) or $\underline{V}$oice $\underline{O}$perated transmit ($\underline{X}$mit).

VP An abbreviation of $\underline{V}$irtual $\underline{P}$ath.

VPC An abbreviation of $\underline{V}$irtual $\underline{P}$ath $\underline{C}$onnection.

VPI An abbreviation of $\underline{V}$irtual $\underline{P}$ath $\underline{I}$dentifier. See *ATM.*

VPL Generally, an abbreviation of $\underline{V}$irtual $\underline{P}$ath $\underline{L}$ink.

VPN (1) An abbreviation of $\underline{V}$irtual $\underline{P}$rivate $\underline{N}$etwork. (2) Sometimes an abbreviation from $\underline{V}$irtual $\underline{P}$ublic $\underline{N}$etwork.

VPX An abbreviation of Virtual Path Cross connect.

VQC An abbreviation of $\underline{V}$ector $\underline{Q}$uantizing $\underline{C}$ode. A voice compression technique that reduces speech transmission rates to 16 or 32 kbps.

VQL An abbreviation of $\underline{V}$ariable $\underline{Q}$uantizing $\underline{L}$evel. A voice-encoding method.

VR (1) An abbreviation of $\underline{V}$irtual $\underline{R}$oute. (2) An abbreviation of $\underline{V}$oltage $\underline{R}$egulator.

VRAM An acronym for $\underline{V}$ideo $\underline{RAM}$. Basically, *VRAM* is normal RAM but is optimized for video applications. *VRAM* is generally dual ported, which enables the central processing unit (CPU) to load information into memory via the parallel port while the video controller is reading information via the serial port. Increasing *VRAM* size in a graphics interface card increases the number of colors possible and/or the number of pixels that can be displayed.

VRC An abbreviation of $\underline{V}$ertical $\underline{R}$edundancy $\underline{C}$heck.

VRML An abbreviation of $\underline{V}$irtual $\underline{R}$eality $\underline{M}$odeling $\underline{L}$anguage.

VRTP An abbreviation of $\underline{V}$INES $\underline{R}$ou$\underline{T}$ing update $\underline{P}$rotocol.

VRU An abbreviation of $\underline{V}$oice $\underline{R}$esponse $\underline{U}$nit.

VRUP An abbreviation of $\underline{V}$INES $\underline{R}$outing $\underline{U}$pdate $\underline{P}$rotocol.

VS An abbreviation of $\underline{V}$irtual $\underline{S}$torage.

VSAM An acronym for $\underline{V}$irtual index $\underline{S}$equential $\underline{A}$ccess $\underline{M}$ethod.

VSAT An acronym for $\underline{V}$ery $\underline{S}$mall $\underline{A}$perature $\underline{T}$erminal. A very small-diameter satellite receiving antenna made possible by increasing the *effective isotropic radiated power* (*EIRP*) of the satellite transmitter.

VSB An abbreviation of $\underline{V}$estigial $\underline{S}$ide$\underline{B}$and.

VSE An abbreviation of $\underline{V}$irtual $\underline{S}$torage $\underline{E}$xtended.

VSF An abbreviation of $\underline{V}$oice $\underline{S}$tore and $\underline{F}$orward.

VSIA An abbreviation of $\underline{V}$irtual $\underline{S}$ocket $\underline{I}$nterface $\underline{A}$lliance.

VSM An abbreviation for $\underline{V}$estigial $\underline{S}$ideband $\underline{M}$odulation.

VSPC An abbreviation of $\underline{V}$isual $\underline{S}$torage $\underline{P}$ersonal $\underline{C}$omputing.

VSPP An abbreviation of $\underline{V}$INES $\underline{S}$equenced $\underline{P}$acket $\underline{P}$rotocol.

VSWR An abbreviation of $\underline{V}$oltage $\underline{S}$tanding $\underline{W}$ave $\underline{R}$atio. See also *standing wave ratio.*

VSX An abbreviation of X/open Verification Suite.

VT (1) An abbreviation of $\underline{V}$ertical $\underline{T}$ab. (2) An abbreviation of $\underline{V}$irtual $\underline{T}$erminal. (3) An abbreviation of $\underline{V}$irtual $\underline{T}$ributary. A logical channel composed of a sequence of cells.

VT-100 A terminal designed by Digital Equipment Corporation (DEC) for its mainframe computers. Although the terminal itself has become obsolete, the protocol is now one of the major terminal standards that most modem communications programs emulate. Other terminals in the series include VT-54 and VT-102.

VTAM An acronym for $\underline{V}$irtual $\underline{T}$elecommunication $\underline{A}$ccess $\underline{M}$ethod. IBM's Systems Network Architecture (SNA) protocol and host communications program. The virtual access method for 3270 systems.

VTE An abbreviation of $\underline{V}$irtual $\underline{T}$ributary $\underline{E}$nvelope. The real payload plus any path overhead within a virtual tributary (VT) channel.

VTNS An abbreviation of $\underline{V}$irtual $\underline{T}$elecommunications $\underline{N}$etwork Services.

VTOC An acronym for $\underline{V}$olume $\underline{T}$able $\underline{O}$f $\underline{C}$ontents.

VTP An abbreviation of $\underline{V}$irtual $\underline{T}$erminal $\underline{P}$rotocol.

VTS An abbreviation of $\underline{V}$irtual $\underline{T}$erminal $\underline{S}$ervice.

VTU An abbreviation of $\underline{V}$ideo $\underline{T}$eleconferencing $\underline{U}$nit.

vu An abbreviation of $\underline{V}$olume $\underline{U}$nit.

VxD An abbreviation of $\underline{V}$irtual $\underline{D}$evice $\underline{D}$river.

W

W The SI symbol for the units of *power* (*P*) measured in *watts* (*W*).

W-code A secret code used to encrypt the real Global Positioning System (GPS) P-code so as to prevent a receiver from reporting false locations in the event it receives bogus, enemy P-code. The encrypted P-code is called *Y-code.*

W-DCS An abbreviation of Wideband Digital Cross connect System. For OC-1, STS-1, DS-3 and below.

W-profile fiber A synonym for *doubly clad fiber.*

W3 A symbol for *World Wide Web* (*WWW*).

WAAS An abbreviation of Wide Area Augmentation System.

WACK An acronym for Wait for positive ACKnowledgment. A two-character control sequence.

WADGPS An abbreviation of Wide Area Differential GPS.

wafer A thin slice of semiconducting material such as a crystal of silicon, gallium, or germanium. A wafer may be 25 cm in diameter or more. Onto this *wafer* multiple copies of a desired microcircuit are constructed by essentially photographic, chemical diffusion, and deposition techniques. Individual microcircuits are separated by scoring and breaking the wafer into individual pieces called *chips* or *dice.*

Waffle A bulletin board system (BBS) shareware program that allows the BBS to become a USENet site. The BBS dials up a (UUCP) site periodically and transfers the latest USENet postings to its local memory for the BBS subscribers to access.

WAIS An acronym for Wide Area Information Server (pronounced "weighs").

wait on busy The English term for camp-on or call waiting.

wait state A delay of one or more clock cycles added to a processor's instruction execution time to allow communication with slow external devices (for example, random access memory). The number and duration of *wait states* may be preconfigured, or they may be controlled dynamically via certain control lines.

WAITS An acronym for Wide Area Information Transfer System.

wall A contraction of Write ALL. A command in UNIX to send a message to all users currently logged on.

wall time (1) The time as displayed by the "clock on the wall," as opposed to the system clock's time. Also called *people time, real time,* and *wall clock time.* (2) The real elapsed time a program takes to run, as opposed to the number of ticks required to execute it. This is important in multitasking and time-shared systems as no one program is allowed to hog all of the system resources (including available time); hence, the *wall time* for execution will be longer than the number of ticks imply. Conversely, on a multiprocessor system a task may get more processor time than *real time.*

WAN An acronym for Wide Area Network. A network capable of providing data communications over a large geographic area, typically linking cities. Generally, it uses transmission facilities provided by a common carrier. Also called an *enterprise network.*

wander A long-term random variation of a system or device parameter around the ideal value.

Although not a specification, the term *wander* is generally applied to variations over time periods of about a second or more. See also *drift, jitter,* and *swim.*

WARC An acronym for World Administrative Radio Conference.

warez A term used by software pirates to describe a cracked game or application that is made available on the Internet, usually via file transfer protocol (FTP), telnet, or USEnet. Often the pirate will make use of a site with lax security.

The practice is illegal and a very good way to acquire a virus. Software piracy is illegal and should be reported to the Business Software Alliance (BSA) or a similar agency.

warm start (1) Restarting or resetting a previously running sequential control device system without power cycling or clearing certain register values. Also called *warm restart.* (2) In computer systems, the restarting of equipment, after a sudden shutdown, that allows reuse of previously retained initialized input data, retained programs, and retained output queues. A *warm start* cannot occur if initial data, programs, and files are not retained after closedown. Also called *hot boot, soft boot,* and *warm boot.*

WARN An acronym for Weather Amateur Radio Network.

WAS An acronym for Wideband Antenna System.

watchdog timer An independent timer on a device or system that indicates when an activity has taken too long; then it sets an alarm or forces some corrective action. See also *timeout.*

WATS An acronym for Wide Area Telecommunications (formerly Telephone) Service. A service provided by the telephone companies in the United States which allows a customer (by use of an access line) to place calls to stations into one of several specific zones for a flat monthly rate (independent of the number of calls placed or the duration of the connection).

watt (W) The unit of power equal to the expenditure of energy at the rate of one joule per second. In electronics, *wattage* (*W*) can be calculated from the voltage (*V*), current (*I*), and resistance (*R*) using the following formulas:

$$= V \cdot I$$
$$= I^2 \cdot R$$
$$= \frac{V^2}{R}$$

watt-second The units of energy (also *joule*).

.WAV A name extension for a multimedia file format for a particular sound format developed by Microsoft and used extensively in its Windows products.

wave Any periodic variation of an electrical signal with respect to time.

When the variations of the electrical signal repeat exactly from any time segment to the next time segment, the wave is said to be a *con-*

tinuous wave. The *amplitude* of a wave is a measure of the size of the variation of the signal. The amplitude may be of a discrete point or a measure of the variation over time. The maximum amplitude (without regard to sign) is called the *peak* or *crest* of the wave. If the amplitude is measured over time, it may be expressed as an *average* or as *RMS* short for *root mean square*. The *period* of a wave is the time from any point on a continuous wave to the next identical point in the wave, e.g., from one positive peak to the next positive peak. Each repeating period of a wave is called a *cycle.* (Generally, a cycle is from a peak or zero crossing to the corresponding peak or zero crossing.) *Wavelength* is related to period by the speed at which the wave is moving through the medium, that is,

$$\text{Wavelength} = \text{Velocity (m/s)} \cdot \text{Period (s)}.$$

The *frequency* of a continuous wave is the reciprocal of the period, that is,

$$\textit{frequency} = \textit{1/period.}$$

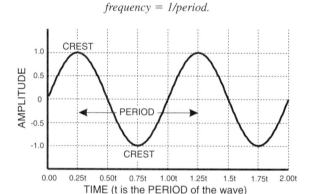

TIME (t is the PERIOD of the wave)

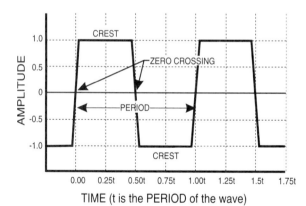

TIME (t is the PERIOD of the wave)

The upper figure is that of a *sine wave* with a period of *t* seconds and a peak amplitude of 1. The lower figure shows a *square wave* with the same period, frequency, and peak amplitude. See also *pulse.*

wave equation See *Maxwell's equations.*

wave impedance At a point in an electromagnetic wave, the ratio of the electric field strength to the magnetic field strength.

If the electric field strength is expressed in volts per meter (V/m) and the magnetic field strength is expressed in ampere-turns per meter (A-T/m), the *wave impedance* will have the units of ohms. The *wave impedance* (Z) of an electromagnetic wave is given by:

$$Z = \sqrt{\mu/\epsilon}$$

where:

μ   is the magnetic permeability and
ε   is the electric permittivity.

For free space, $\mu = 4\pi \times 10^{-7}$ henries per meter (H/m) and $\epsilon = 1/36\pi$ farads per meter (F/m), which yields $Z = 120\pi$ (~377) ohms. In dielectric materials, the *wave impedance* becomes $377/n$, where *n* is the material's refractive index. Although the ratio is called the *wave impedance,* it is also the impedance of the free space or the material medium.

wave length See *wavelength.*

wave number The number proportional to the inverse of the wavelength in a harmonic wave, i.e., the number of waves per unit length.

wave optics That branch of optics concerned with radiant energy and related phenomena, as defined by wave characteristics. Also called *physical optics.* See also *geometric optics.*

wave trap A device used to exclude unwanted frequency components of a complex wave. Traps may be fixed or tunable to permit the selection of unwanted or interfering signals. Also called a *trap.*

waveform A general term used to describe the manner in which a wave's amplitude (or other specified parameter) varies over time.

waveform codes The way a digital signal is represented and presented to a transmission line. Different *waveform codes* are used to optimize the energy distribution for some purpose or to increase the data transfer rate. A comparison of several waveform codes is shown on page 576. Also called *line coding.*

wavefront Of a wave. The surface defined by all points that have the same phase, i.e., have the same path length from the source in a uniform medium. Characteristics of an electromagnetic *wavefront* include:

- It is perpendicular to the ray that represents the direction of propagation of the wave.
- The plane in which the electric and magnetic field vectors lie is tangential to the wavefront at every point.
- For parallel (collimated) rays, the wavefront is plane.
- For rays diverging from a point source, or converging toward a point sink, the wavefront is spherical.

waveguide A system of materials designed to confine and direct electromagnetic waves in a direction determined by its physical boundaries. Different materials are used for different portions of the electromagnetic spectrum, for example:

- With microwaves, the waveguide is normally a hollow metallic conductor, usually of rectangular, elliptical, or circular cross section. The waveguide may contain a dielectric material in some cases.
- In optics, the waveguide consists of a series of dielectric films or layers. In optical transmission lines, the layers are generally concentric cylinders of glass or plastic. In integrated optical circuits an optical waveguide may consist of a thin dielectric film.

The ionized layers of the stratosphere and refractive surfaces of the troposphere may also act as a waveguide to RF energy. See also *ionosphere.*

waveguide dispersion See *dispersion.*

waveguide scattering Scattering in a waveguide from sources other than material scattering, i.e., from variations in geometry and the refractive index in an optical waveguide.

wavelength (λ) The distance between points of corresponding phase of two consecutive cycles of a wave, e.g., the distance from the crest of a wave to the crest of the next wave, the length of one cycle. (Any point on the wave may be used as the starting point as long as the corresponding point on the next wave is used as the ending point.)

CODING RULE OR ALGORITHM

CODED WAVEFORM

CODE NAME	Binary data	0	1	0	1	1	0	0	1	1	0	0	CODING RULE OR ALGORITHM

Binary data — Source information to be encoded.

NRZ, NRZ-L — Nonreturn to Zero-Level (Basic binary waveform) Waveform is at *one level* while the source data = "1" and is at the *other level* while the source = "0."

NRZ-M, NRZ-I — Nonreturn to Zero-Mark Waveform *changes state* while the source is "1," no change for a "0."

NRZ-S — Nonreturn to Zero-Space Waveform *changes state* while the source is "0," no change for a "1."

Differential Manchester — Waveform *changes state* at leading edge only when the source = "0," and always *changes state* mid-bit.

BIΦ-L Manchester II — Biphase Level (Manchester II) Mid-bit change from "1 to 0" when the source = "1" and from "0 to 1" when source = "0," transition at leading edge if successive source bits same.

BIΦ-M — Biphase Mark Transition occurs at the beginning of every bit cell. Source = "1" causes a second transition half a bit period later, source = "0" no second transition.

BIΦ-S — Biphase Space Transition occurs at the beginning of every bit cell. Source = "0" causes a second transition half a bit period later, source = "1" no second transition.

DM Miller — Delay Modulation (Miller Coding) Source = "1" causes a waveform transition at mid bit, source = "0" causes a waveform transition at the start of the bit cell except after source = "1."

RZ — Return to Zero Source = "1" causes a pulse less than 1 bit long, source = "0" is no pulse.

Bipolar RZ — Bipolar Return to Zero Source = "1" causes a pulse less than 1 bit long, source = "0" causes a pulse in the opposite polarity and of the same duration.

Ratio — Ratio Source = "1" causes a pulse approx ⅔ bit long, source = 0 generates a pulse 1/3 bit long.

AMI (T-1) — Alternate Mark Inversion Source = "1" causes a pulse, each succeeding pulse of opposite polarity, source = "0" causes no pulse.

MLT-3 — Multilevel Source = "1" causes change to next available level source = "0" causes no change.

2B1Q — 2B1Q Source = "00" sets level 1. Source = "01" sets level 2. Source = "10" sets level 4. Source = "11" sets level 3.

Differential NRZ — Differential NRZ Source = "1" causes transition at beginning of bit cell source = "0" causes no change.

Dipulse — Multilevel Source = "1" causes two pulses of opposite polarity source = "0" causes no change.

Mathematically, wavelength (λ) is the propagation velocity (v, in m/s, ft/s, etc.) divided by the frequency (f, in Hz) of the wave within the medium in which it is propagating, that is,

$$\lambda = \frac{v}{f}$$

Wavelength is generally expressed in meters with appropriate SI prefixes, for example, see table. See also *frequency spectrum, period,* and *wave.*

Name	Symbol	Value
millimeter	mm	10^{-3}m
micrometer	μm	10^{-6}m
nanometer	nm	10^{-9}m
Angstrom (non-SI unit—found in older literature)	Å	10^{-10}m

wavelength division multiplex (WDM) An optical multiplexing technique using light of different wavelengths for each channel. Essentially the same as frequency division multiplex (FDM). The term *WDM* is used in optical systems rather than *FDM.*

wavelength selective coupler (WSC) In fiber optics, a type of splitter/coupler that breaks an incoming light signal into multiple-derived output signals based on wavelength and couples energy at a specific wavelength from one fiber to a second fiber.

wavelength stability In an optical source, the maximum deviation of the peak wavelength from its mean value during a specified period.

wavelet A waveform that is bounded in both frequency and duration.

wavelet transform An alternative to the more traditional Fourier transforms used to analyze waveforms. *Wavelet transforms* convert a signal into a series of wavelets. In theory, signals processed by the *wavelet transform* can be stored more efficiently than signals processed by Fourier transform. Wavelets can also be constructed with rough edges, which better approximate real-world signals.

By contrast, the Fourier transform converts an arbitrary signal into a set of continuous sine waves, each of which has a constant frequency and amplitude and an infinite duration. Most real-world signals (such as music or images) have finite durations and abrupt changes in frequency and amplitude.

waZOO An acronym for Warp-Zillion Opus-to-Opus. Fidonet's network session layer protocol. A more efficient protocol than others, such as file transfer protocol (FTP), it is also sometimes used in on the Internet in automated or batch communications. (Opus is the name of a specific, vintage 1980s bulletin board system (BBS).)

WBC An abbreviation of WideBand Channel. One of 16 Fiber Distributed Digital Interface (FDDI) subframes assignable to packet or circuit connections.

WC An abbreviation of Wire Center.

WCC An abbreviation of World Congress on Computing.

WD (1) An abbreviation of Working Document. (2) An abbreviation of Working Draft.

WDM An abbreviation of Wavelength Division Multiplexing.

WDT An abbreviation of Watch Dog Timer.

weakly guiding fiber An optical fiber in which the refractive index contrast is small (substantially less than 1%).

Web An abbreviation of World Wide Web (WWW).

web browser In Internet, a graphical user interface program that allows users to access documents on that part of the Internet called the World Wide Web (WWW) via hypertext links and to review these documents online. Most *browsers* incorporate the graphical program *Mosaic* into their structure.

Web zine See *zine.*

weber (Wb) The SI unit of magnetic flux (ϕ).

Weber's law The premise that the "just perceptible" difference between two stimuli is a constant fraction of the stimulus.

webmaster In Internet, the administrator responsible for management of a World Wide Web site.

WECO An acronym for Western Electric COmpany. The equipment manufacturing arm of AT&T now called AT&T Technologies.

weighting network An electronic network with loss or gain that varies with frequency in a prescribed manner. This network is used for compensation of transmission characteristics or for characterization of noise measurements. See also *C-message, conditioning,* and *psophometric.*

well-known port On the Internet, a port address with a fixed, standardized number because it provides access to *well-known services.* When accessing these services from a remote system, the user need not specify the address.

well-known service A standard Internet program such as file transfer protocol (FTP) or Telnet that is assigned a fixed *protocol number* and *port address.* Because these addresses are fixed, local client software will be able to locate the desired server program on any remote host. See also *port address* and *protocol number.*

WEP An abbreviation of Well-known Entry Point.

wet contact In telephony, a mechanical switch contact through which a direct current flows.

wet loop An electrical circuit which, on the average, carries enough current to keep oxides from forming on the splices and junction of the conducting circuits. In telephony, 10 to 30 mA is generally sufficient. See also *dry loop.*

wetware Slang for "people." Also called *jellyware* and *liveware.*

WG An abbreviation of Working Group.

whatis A command in an Archie server that provides descriptive information about files.

whip antenna A flexible rod antenna, usually between 1/10 and 5/8 wavelength long, supported on a base insulator. The physical length is determined by a series load coil.

white area The area or population that does not receive interference-free primary service from an authorized AM station or does not receive a signal strength of at least 1 mV/m from an authorized FM station.

White Book (1) The 1992 compilation of the Comité Consultatif Internationale Téléphonique et Télégrahique (CCITT) now called International Telecommunications Union-Telecom section (ITU-T's) standards for telecommunications. (2) Specifications of the Video CD™ standard.

A major extension to the audio CD (Red Book) standard specific to the Sony/Philips Video CD. Unlike standards for CD-ROM, CD-I, and CD-R, the White Book designates a very medium-specific format (similar to the original audio only CD format). The White Book standard allows for 74 minutes of audio and video in MPEG (Motion Picture Experts Group) format on a compact disc. See also *Video CD.* **(3)** The fourth book in Adobe's PostScript series, describing the previously secret format of Type 1 fonts. The other three official guides are known as the Blue, Green, and Red Books.

white noise Noise (acoustic, optic, or electric) having a frequency spectrum that is continuous and uniform over a specified range of interest. It has equal power per hertz over the specified frequency band.

From an analogy with white light. Also called *additive white gaussian noise.*

white pages **(1)** In telephony, a hard copy directory of subscriber names listed in alphabetical order along with a telephone number and address. **(2)** An electronic database containing name and address information for users on a server or network. White pages directories may be found through the Gopher and the Whois servers. The user-based *white pages* are in contrast to the business-oriented yellow pages (YP).

Electronic *white pages* usually contain additional information, such as office location, phone number, e-mail address, and mailstop. Such databases are frequently found on the Internet, at educational institutions, and in large corporations.

who are you (WRU) A control character that invokes an answer-back response (ID and equipment type) in a distant terminal (generally a telex terminal). The *WRU* signal corresponds to the 7-bit code assigned to the *WRU* symbol. Also called a *WRU signal.*

whois A directory on the Internet of people who are responsible for the actual workings of the network. It is maintained by the DDN Network Information Center. It is described in RFC954.

wide area augmentation system (WAAS) In the Global Positioning System (GPS), a system that enhances the "performance" of the Standard Positioning Service (SPS) of the GPS facilities over a wide geographical area. The *WAAS* is being developed by the Federal Aviation Administration (FAA) along with several other agencies. The system provides wide area differential GPS (WADGPS) corrections via geosynchronous satellites.

wide area differential GPS (WADGPS) In the Global Positioning System (GPS), a specific form of differential GPS wherein the receiver obtains correction information from a network of reference stations over a wide geographical area. Corrections for each error source are provided and can be applied at either the receiver in real time or in a data processing center for post processing. Typical correction information is available for the satellite clock, ephemeris, ionospheric propagation delay, and so on. See also *local area DGPS (LADGPS)* and *wide area augmentation system (WAAS).*

Wide Area Information Server (WAIS) Pronounced "weighs." A distributed text searching system that uses the client–server model and the ANS Z39.50 protocol standard to search index databases on remote computers. It is a service for looking up and retrieving specific information on a network (frequently the Internet) using natural language queries and without regard to the information's physical location. It can search specified locations (sources) for files that contain specified terms (keywords) and return a list of files that satisfy the search criteria. WAIS allows the use of one or more keywords, which can be combined using simple relationships (AND, OR, or NOT).

wide area network (WAN) A network that provides communication services over areas larger than those areas serviced by a local area network (LAN). A WAN typically interconnects several LAN subnetworks via a serial interface or gateway. See also *LAN* and *WAN.*

Wide Area Telephone Service (WATS) In telephony, a toll service offering for a dial-type single line in which a customer can purchase fixed rate communications services with other stations based on:

- Inter- or intra-LATA geographic rate area called zones.
- Calling direction, outward (OUT-WATS) and/or inward (IN-WATS) service.
- Calling time.
- Call duration.

wide character A 16-bit *Unicode* character.

wideband "Wideband" has many meanings depending on application. The term is often used to distinguish it from *narrowband,* where both terms are subjectively defined relative to an implied context. The wider bandwidth enables higher data transmission rates.

Some specific definitions include:

1. A signal that pertains to or occupies a broad frequency spectrum.
2. The property of any communications facility, device, channel, or system in which the range of frequencies used for transmission is greater than 0.1% of the midband frequency.
3. A communications channel having a bandwidth greater than a reference channel (e.g., frequently a voice grade circuit is used as the reference).
4. The property of a circuit that has a bandwidth wider than normal for the type of circuit, frequency of operation, or type of modulation.
5. In telephony, the property of a circuit that has a bandwidth greater than 4 kHz.
6. In communications security systems, a bandwidth exceeding that of a nominal 4-kHz telephone channel.

Also called *broadband.*

wideband channel (WBC) **(1)** In a Fiber Distributed Digital Interface (FDDI) network, a channel with a bandwidth of 6.144 Mbps. The total FDDI bandwidth can support 16 *WBCs.* Each channel has a minimum bandwidth of 768 kbps and a maximum of about 99 Mbps.

In FDDI-II, a *WBC* can be allocated either for packet switched or circuit switched service. If it is used for packet switched service, the channel is merged with the other *WBCs* allocated similarly and the aggregate is called the packet data channel. If a *WBC* is used for circuit switched service, it may be allocated entirely to a single connection, or the *WBC* may be broken into slower channels, each of which can then be used to connect a different pair of nodes. **(2)** A communication channel of a bandwidth equivalent to 12 or more voice grade channels.

wideband modem **(1)** Any modem whose modulated output signal has an essential frequency spectrum that is broader than that which can be wholly contained within, and faithfully transmitted through, a voice channel with a nominal 4-kHz bandwidth. **(2)** A modem whose bandwidth capability is greater than that of a narrowband modem.

Wiener filtering A method that accepts the classical approach to signal restoration and attempts to minimize the mean square difference between the original degraded signal and the restoration.

wildcard account An account that gives all uses on a local area network (LAN) full access to all resources. To establish network security one must use individual and access control list (ACL) group accounts.

wildcard character A character that may be substituted for any of a defined subset of all possible characters. Whether the wildcard character represents a single character or a string of characters must be specified. Examples include:

- In high-frequency (HF) radio automatic link establishment, the wildcard character "?" may be substituted for any one of the 36 characters, A–Z and 0–9.
- In computer (software) technology, a character that can be used to substitute for any other character or characters in a string. Common Wildcard characters include:
    - The question mark (?), which usually substitutes as a wildcard character for any single ASCII character position. For example, given the set of directories *file1.txt, file1a.txt, file2.txt, file3.doc, fileX.txt,* and *filename.txt.* The search command DIR file?.txt would find only *file1.txt, file2.txt,* and *fileX.txt.*
    - The asterisk (*) which usually substitutes as a wildcard character for a string of any one or more ASCII characters. The string is terminated by a defined delimiter. For example, given the set of file from above, the command DIR file*.txt would find *file1.txt, file1a.txt, file2.txt, fileX.txt,* and *filename.txt.* Also called *don't care* characters.

WIMP An acronym for <u>W</u>indows, <u>I</u>cons, <u>M</u>enus, and <u>P</u>ictures. See *graphical user interface (GUI)*.

WIN An acronym for <u>Wi</u>reless <u>I</u>n-building <u>N</u>etwork.

winding Coils of wire wrapped around a volume. The volume may be filled with air or a magnetic material, such as iron.

window (**1**) That band of electromagnetic frequencies (wavelengths) which will pass through a device or system. (**2**) In fiber optics, a band of wavelengths at which an optical fiber is sufficiently transparent for practical use in communications. Also called *spectral window* and *transmission window.* See also *first window, second window, third window,* and *fourth window.* (**3**) A device designed to allow a particular band of electromagnetic frequencies to pass. (**4**) A flow-control mechanism in communication systems that determines the number of frames or packets that can be transmitted before receiving an acknowledgment. Also called a *pacing group* in IBM's Systems Network Architecture (SNA). (**5**) A generic method of displaying data on a screen, mimicking the effect of looking at several pieces of paper at once. The technique allows viewing activities and information of several programs running concurrently. Each program is given a box (a *window*) into which it places its information. Each *window* may be placed anywhere on the video display even if it means one window partially or totally overlaps another. Because a user can enter data from the keyboard into only one program at a time, only that *window* need be actively displayed. To accomplish this, the active program's window is placed on top of all others. (**6**) The gap between an upper threshold and a lower threshold of a device. (**7**) A period during which an event can occur, can be expected to occur, or is allowed to occur.

window control A scheme that limits the number of packets (frames) or calls into a system.

window size The maximum number of packets or frames that are allowed to be transmitted before the receiver issues an acknowledgment signal.

Windows® The now ubiquitous Microsoft stand-alone operating system (or shell) with integral graphical user interface (GUI), running on top of DOS. Versions include:

- *Windows® 1.0* and *2.0*—a graphical user interface shell used primarily to launch DOS applications; although some applications were written to take advantage of Windows' built-in features.
- Windows® 3.0—a complete reworking of Windows with many new facilities such as the ability to address memory beyond 640k. It was released in 1990, and the development of a plethora of applications by third parties helped Microsoft establish it as a standard.
- *Windows® 3.1*—a version of Windows 3.0 with improvements such as True Type Fonts (TTF) and Object Linking and Embedding (OLE). It also ceased to support Real Mode (which means it would no longer run on Intel 8086 processors). Also called *stand-alone Windows.*
- *Windows® 3.11*—a free upgrade to Windows 3.1 containing mainly minor bug fixes.
- *Windows® for Workgroups (WFWG)*—an extension of Windows 3.1 that allows between 2 and 20 users to form a peer-to-peer network. It allows sharing information such as files and e-mail by clicking on icons. WFWG is compatible with NetWare and LAN Manager. Windows for Workgroups 3.11 is a significant upgrade to Windows for Workgroups 3.1 adding 32-bit file access, fax capability, and higher performance.
- *Windows® 95 (Win95)*—a true operating system and successor to the Windows 3.11 shell. It provides both 16-bit and 32-bit application support, preemptive multitasking (for 32-bit applications), threading and built-in networking with TCP/IP, IPX, SLIP/PPP, and Windows Sockets. It includes MS-DOS 7.0 but takes over completely after booting. It was known as "Chicago" during development. Its release was originally scheduled for late 1994, but it was not released until August 24, 1995.
- *Windows NT®*—NT is an abbreviation of <u>N</u>ew <u>T</u>echnology. Windows NT is Microsoft's scalable 32-bit version of Windows aimed at high-end workstation users. It is a stand-alone operating system (OS)/network operating system (NOS) capable of being a small application server for a workgroup of Windows® based PCs.
- *Windows NT® Advanced Server (NTAS)*—an extension of Microsoft's Windows NT incorporating all of its features. NTAS is a server operating system offering centralized management and security, fault tolerance, and multiple connectivity options. Designed for client–server computing on practically any network, including NetWare, VINES, and LAN Manager.
- *Windows 2000®*—essentially an upgrade of Windows NT. It includes new features such as the ability to reconfigure without shutting down. Plug-and-Play, better security, and the Advanced Server version supports clustered multiple servers. Currently there is no easy way to remove Windows 2000 after the upgrade is installed.
- *Windows Millennium Edition® (Windows (ME)*—an upgrade of the Windows 95/98 operating system. Improvements include new system file protection, faster start-up, better methods of old configuration restoration.

Windows Open Service Architecture (WOSA) A framework of open-ended interfaces allowing Microsoft Windows and applications running under it to integrate with enterprise computing environments. It includes Application Program Interfaces (APIs) for messaging (MAPI), standard access to databases via Open Database Connectivity (ODBC) and extensions to financial services.

wink (**1**) In telephony, the rate at which an indicator lamp flashes to signal that the associated line is on hold. The rate is about 450–475 ms on and 25–50 ms off. (**2**) A momentary signal sent between two telecommunications devices as part of the handshaking protocol. It varies in form, depending on the signaling method, for example,

- In single-frequency (SF) signaling, it is a short interruption in the transmitted tone indicating that the distant central office is ready to receive dialed digits.
- On T-1 circuits, it is a brief change in the state of the A and B signaling bits.
- On an analog line, it is the brief polarity reversal of the loop current.

wink release A tone pulse transmitted to the off-hook station when a station goes on-hook.

wink start pulsing In telephony, a method of trunk pulsing control and integrity checking wherein the sender waits for a momentary off-hook signal from the far end before sending the address pulses.

WINS An acronym for <u>W</u>indows <u>I</u>nternet <u>N</u>aming <u>S</u>ervice. A name resolution service that resolves Windows networking computer names to Internet Protocol (IP) addresses in a routed environment. It handles name registration, queries, and releases.

WinSock A contraction of <u>Win</u>dows <u>Sock</u>et. A Transmission Control Protocol/Internet Protocol (TCP/IP) extension of the Windows Applications Program Interface (API) which allows Windows applications to run independently of the hardware underneath. It is a dynamic link library (DLL) file that contains the information and procedures necessary for Windows to interface with TCP/IP. *WinSock* is required on Windows systems if client programs such as Eudora or Mosaic are to communicate with the Internet via Ethernet or a dial-up connection.

WinTel An acronym for <u>WIN</u>dows-In<u>TEL</u>. A PC with the Microsoft Windows operating system and an Intel microprocessor.

WinVn A threaded newsreader program used to download and display USENet news postings.

wire One or more slender filaments of semi-rigid or flexible metal woven or twisted into a single structure. When only a single filament, the structure is said to be a *solid wire*. With multiple filaments, the structure is called *stranded wire*.

In the strictest sense, the term *wire* refers to the metal, and a *wire* covered with insulation is properly called an *insulated wire*. However, when no confusion can arise in the context of usage, the term *wire* may include the insulation. See also *color code* and *wire gage*.

wire center (WC) In telephony, a central point from which subscriber loops emanate in a treelike manner into the service area associated with one or more class 5 switching offices.

wire gage The specification of the diameter (or equivalent diameter) of wire. Several standards exist for specifying the size of wire, with the American Wire Gage (AWG) and millimeter gages being the most common for most materials. Steel wire is specified by the Steel Wire Gage (Stl WG).

The Brown & Sharpe (B&S) and the Birmingham Wire Gage are considered obsolete. The B&S is now known as the AWG. Several useful approximations relating AWG and the wire's diameter (*D*) in mils are:

$$AWG = Integer \left[20 \cdot \log_{10} \left(\frac{D}{325} \right) \right]$$

and

$$D = \frac{325}{10^{\left(\frac{AWG}{20} \right)}}$$

See also *cable* and *color code*.

wire gauge A device for determining the gage (diameter) of noninsulated wire.

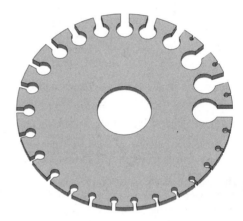

wire pair Two individual conductors, generally twisted together, that serve as a communications path between two points. See also *balanced line*.

wire stripper A tool that removes the insulation from a wire without nicking, scratching, or otherwise damaging the internal conductor.

wire wrap A method of terminating (connecting) individual wires to a device. The connection consists of a number of helical turns of solid uninsulated wire tightly wrapped around a terminal post so as to form an electrically and mechanically stable connection.

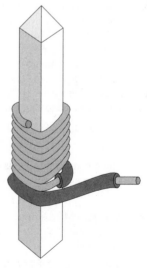

wired logic A logic function performed in hardware rather than software. Generally, it also implies that the function is nonchangeable.

BARE ANNEALED COPPER WIRE TABLE

AWG	Diameter in Mils	Cross Section in Circular Mils	Ohms per 1000 ft @ 20 C	Current Rating ①	Fusing Current ①
10	101.9	10380	0.9989	10-40	330
12	80.81	6530	1.588	8-25	240
14	64.08	4107	2.525	5-20	170
16	50.82	2583	4.016	4-12	120
18	40.30	1624	6.385	2.5-10	83
20	31.96	1022	10.15	2-6	58
22	25.35	642.4	16.14	1.3-5	41
24	20.10	404.0	25.67	1-3	29
26	15.94	254.1	40.81	0.7-2.5	20
28	12.64	159.8	64.90	0.5-1.6	14
30	10.03	100.5	103.2	0.3-1.3	10
32	7.950	63.21	164.1	0.2-0.9	7.2
34	6.305	39.75	260.9	0.15-0.7	5.1

Values are very approximate since they are a function of both the environment and application. For example, the wire in a transformer will get hotter because it cannot radiate its heat as easily as a wire in the open, hence the resistance is higher resulting in higher losses.

wirehead Jargon (1) A hardware expert, especially one who concentrates on communications hardware. (2) An expert in local area networks (LANs).

A *wirehead* can be a network software wizard too but will always have the ability to deal with network hardware, down to the smallest component.

wireless access mode In a personal communications service (PCS), interfacing with a network access point by means of a standardized air interface protocol without the use of a physical connection to the network.

wireless cable A term referring to the use of radio networks to transmit cable-like TV programming to homes and businesses.

wireless in-building network (WIN) A wireless network that is confined to a single building.

wireless LAN (WLAN) A local area network (LAN) architecture radiated transmissions (such as radio or light) rather than conducted transmissions (including twisted pair, coax, or fiber optic cables).

wireless local loop (WLL) The use of a radio link to provide service to a remote location or to locations where it is too expensive to install copper loops.

wireless modem A modem that transmits over a wireless network rather than over telephone lines.

wireless packet network A switching network designed specifically for handling packet data communications. The network, like most telephonic systems, is a hierarchical topology. That is, at the lowest level, a base station exchanges wireless packets with nearby devices. The base stations then route these packets via traditional local area network (LAN) techniques to the destination user.

wireless switching center (WSC) The second of the two cellular systems in every market, it always belongs to the local telephone company. The frequencies used are in the 869–894 MHz band. Also called the *Block B carrier.*

wireline A term for a communications system that uses wire cable rather than radio for communications links to its customers.

wireline common carrier Common carriers that are in the business of providing landline local exchange telephone service.

wiring center (1) In general, a term for any of several components that serve as common termination points for one or more nodes and/or other wiring centers. The *wiring center* frequently connects to either an intermediate distribution frame (IDF) or a main distribution frame (MDF). (2) The IEEE 802.5 name for a token ring node concentrator. Called a *multistation access unit (MAU)* by IBM. See also *multistation access unit (MAU).*

wiring closet The termination point for customer premises cabling. These cables connect the various areas in an office or building to the central wiring, and to the telephone or power company wiring. Usually a physical box, room, or even a closet in which the cabling on a particular floor is terminated, typically on a *wiring frame.*

wiring frame A frame used to organize and manage the termination and connection and cross connection of multiple cables.

WLAN An acronym for <u>W</u>ireless <u>LAN</u>.

WLL An abbreviation of <u>W</u>ireless <u>L</u>ocal <u>L</u>oop.

WN An abbreviation of <u>W</u>rong <u>N</u>umber.

WOM An acronym for <u>W</u>rite <u>O</u>nly <u>M</u>emory. Useful in FINO (First In Never Out) buffers.

WOMBAT An acronym for <u>W</u>aste <u>O</u>f <u>M</u>oney, <u>B</u>rains, <u>A</u>nd <u>T</u>ime. Describes problems that are both profoundly uninteresting in themselves and unlikely to benefit anyone, even if solved.

word A set of bits or characters treated as a single unit within a system.
- In computers, although a binary word length can be any fixed or variable length, in computer-based systems, it is generally fixed at 8, 16, or 32 bits depending on the particular computer.
- In telegraphy, six character intervals are defined as a *word* when computing traffic capacity in words per minute, which is computed by multiplying the data signaling rate in baud by 10 and dividing the resulting product by the number of unit intervals per character.

word length The total number of bits in a symbol (or character) without start, stop, or parity bits.

work space In computers and data processing systems, that portion of main storage used by a computer program for temporarily storing data.

workflow automation The flow of documents around an organization in a prescribed order (workflow) which can be automated, delivering an hierarchical and controlled form of *workgroup computing.*

workgroup A collection of network nodes grouped for administrative purposes but not sharing security information. Each workgroup has a unique name. See also *domain* and *user.*

workgroup computing A method of organizing a business around productive teams using computer support to enable cooperative working and to eliminate time/space restrictions. An extension of conventional local area network (LAN) working.

workstation (1) A networked computer typically dedicated for end-user applications (i.e., it does not provide any resources that can be shared on a network). See also *mainframe, microcomputer, microprocessor, minicomputer,* and *personal computer.* (2) A term used freely to mean a PC, node, terminal, or high-end desktop processor (for computer aided design or computer aided manufacturing (CAD/CAM) and similar computer intensive applications). (3) In automated systems, such as computer, communications, and control systems, the input/output, display, and processing equipment that provides the operator to system interface.

World Geodetic System 1984 (WGS 84) A geodetic reference system that includes a geocentric reference ellipsoid, a coordinate system, and a gravity field model. The ellipsoid attempts to fit the shape of the entire Earth as well as is possible with a single ellipsoid. Although *WGS 84* is often used as the worldwide reference, other models are used locally to provide a better fit to the Earth in a specific region.

WGS 84 defines the Earth-centered ellipsoid coordinates such that the Z-axis is collinear with the Earth's spin axis, the X-axis is through the intersection of the prime meridian and the equator, and the Y-axis is rotated 90° east of the X-axis about the Z-axis. *WGS84* is the geodetic coordinate system and ellipsoid commonly used by the Global Positioning System (GPS). See also *ECEF (Earth Centered Earth Fixed)* and *Global Positioning System (GPS).*

world newsgroups The seven major newsgroup categories that are automatically distributed to every USENet site on the Internet. The seven categories are *comp, news, rec, sci, soc, talk,* and *misc.*

Two other classifications of newsgroups are also available in USENet: *alternative newsgroups* and *local newsgroups.* The alternative newsgroups (which are not carried by all USENet sites) are created by the users themselves; hence, they cover almost every conceivable topic, but at the expense of a low signal-to-noise ratio. Local newsgroups provide information of interest regionally. See also *ALT* and *newsgroup hierarchies*.

World Time A synonym for Coordinated Universal Time (UTC).

world zone number The first digit of a country code. It specifies a large geographic area within the world number plan. The zone numbers and their locations are:

Zone	Geographic Area
0	Spare
1	United States, Canada, and some Caribbean countries
2	Africa
3	Europe
4	Europe
5	South America, Cuba, Central America, and parts of Mexico
6	South Pacific (Australia-Asia)
7	Countries of the former USSR
8	North Pacific (Eastern Asia)
9	Far East and Middle East

See also *country code*.

World Wide Web (WWW) In Internet, a hypertext-based service used for browsing Internet resources. The graphical user interface enables users to access different databases and resources by clicking on displayed keywords. The WWW got its start at CERN in 1991 when the original software, written by Tim Berners-Lee, was released. Also called *WWW, W3,* and *W³*.

Worm See *virus*.

WORM An acronym for Write Once Read Many (Mostly or Mainly). A writeable but nonerasable CD-ROM. That is, an optical disc format that can be written to only one time, but it can be read many times. *WORM* systems are often used for backup and archiving of data.

Based on the *Orange Book Standard* published by Philips.

WORMFACE An acronym for Write, Read, Modify, File scan, Access control, Create, Erase, Supervisor (the "O" is needed to make the acronym but is not used). The Novell file system-access rights which control what a user is allowed to do to a file.

WORN An acronym for Write Once Read Never. See also WOM.

worst hour of the year That hour of the year during which the median noise over any radio path is at a maximum. (This hour is considered to coincide with the hour during which the greatest transmission loss occurs.)

WOS An acronym for Workstation Operating System.

WOSA An acronym for Windows Open Service Architecture.

.WPG The filename extension for a WordPerfect Graphics file.

WPM (1) An abbreviation of Words Per Minute. **(2)** An abbreviation of Write Protected Memory.

WPS An abbreviation of Words Per Second.

Wrap A redundancy measure in IEEE 803 / IBM Token Ring local area networks. The trunk cabling used in Token Ring TCUs contains two data paths: a main (normally used) and back-up (normally unused). If the trunk cable fails, the physical disconnection of the connector at a transmission control unit (TCU) causes the signal from the main path to connect (wrap) onto the backup and maintain the loop.

wrapping (1) In a network using a dual counter-rotating ring architecture, reconfiguration to circumvent a failed link or node. **(2)** In open systems architecture, the use of a network to connect two other networks, thus providing increased interaction capability between the two connected networks. Recurring application of *wrapping* usually results in a hierarchical structure.

write To make a temporary or permanent recording of information in a data storage device or on a data storage medium.

write precompensation The purposeful variation of the signal amplitude to the write heads of magnetic disk drive from the inner tracks to the outer tracks. This variation is designed to compensate for the higher linear speed of the outer tracks as compared to the inner tracks. (If the outer track is 5 inches and the inner track is 2 inches in diameter, then the linear speed of the outer track is 2.5 times faster than the inner track.)

WRT An abbreviation of With Respect To.

WRTC An abbreviation of Working Reference Telephone Circuit.

WRU (1) An abbreviation of Who aRe yoU. **(2)** A specific character in some digital coding schemes.

WSC An abbreviation of Wireless Switching Center.

wv An abbreviation of working voltage.

WWW An abbreviation of World Wide Web.

WWWW An abbreviation of World Wide Web Worm.

WYSIAYG An abbreviation of What You See Is All You Get. Describes a user interface that has easy initial learning curves, but lacks the depth to do complex or very detailed functions.

WYSIWYG An abbreviation of What You See Is What You Get.

X

X (1) The symbol for reactance, the imaginary part of impedance. Inductive reactance is written X_L while capacitive reactance is X_C. **(2)** In mathematics, an unknown quantity that may take on any value in an equation. **(3)** Used in writing to approximate the algebraic sense of *unknown* within a set defined by the context. For example, the term 80×86 indicates the total family of Intel microprocessors based in the 8086 (i.e., 8086, 8088, 80186, 80286, 80386, 80486, Pentium, Pentium Pro, etc.).

X band The International Telecommunications Union (ITU) designation of electromagnetic radiation in the 8- to 12-GHz frequency range.

X terminal A networked desktop computer that displays application information run on a central server. They are cheaper and easier to network than PCs or workstations.

X-10 protocol A power line (carrier current) signaling protocol used in home automation remote control systems such as those used to turn lights on and off or control a security system. The transmitted signal is a series of pulse position modulated 120-kHz tone bursts synchronized to the zero crossings of the mains power. The transmitted packet is capable of addressing 256 nodes and carries a payload of one command (which may be any of the possible commands).

x-y mount A synonym for *altazimuth mount.*

X.n Recommendations A series of ITU-T recommended standards for terminal equipment (DTE), data circuit terminating equipment (DCE), communication lines, and protocols for use in public and private digital networks (PDN). The titles of some of the X-series recommendations include:

- *X.3*—packet assembly/disassembly facility (PAD) in a public data network.
- *X.4*—general structure of signals of the International Alphabet No. 5 (IA5) for transmission over public data networks.
- *X.20*—asynchronous communications interface definitions for use over the public switch telephone network (PSTN).
- *X.20 bis*—V-series-compatible modem, asynchronous communications interface definitions for use over the PSTN.
- *X.21*—synchronous communications interface definitions for use over the PSTN.
- *X.21 bis*—V-series-compatible modem, synchronous communications interface definitions for use over the PSTN.
- *X.24*—list of definitions for interchange circuits between data terminal equipment (DTE) and data circuit terminating equipment (DCE) on public data networks.
- *X.25*—a 64-kbps, connection-oriented, reliable protocol for communications between two points. It provides an interface between DTE and DCE for terminals operating in the packet mode and connected to public data networks by a dedicated circuit. *X.25* is a standard packet switch data communication protocol defined by the ITU-T for implementing the lowest layer functions specified in the ISO/OSI network model. It describes how data are handled in a packet switched network and how to access the network. It is often used in wider area network (WAN) communication because it provides extensive error correction at every node. X.25 defines

a synchronous, full-duplex, terminal to network packet connection, that is,

- The *electrical connection* between the terminal and network,
- The *link access or transmission protocol,* and
- The implementation of *virtual circuits* between network users. As *X.25* is the protocol standard governing most packet switched networks, it allows a node not directly connected to a public data network to access the facilities of that network through an intermediate node.
- *X.110*—international routing principles and routing plan for PDNs.
- *X.121*—international numbering plan for X.25 networks and PDNs.
- *X.130*—call setup and clear-down times for international connection to synchronous PDNs.
- *X.200*—defines the reference model of Open Systems Interconnection (OSI) for ITU-T applications.
- *X.400*—governs message handling systems, e.g., e-mail graphics and fax. The X.400 standard is an overview that is broken down under these and other numbers:

X.402	Overall architecture.
X.403	Conformance testing.
X.407	Abstract service definition conventions.
X.408	Encoded information type conversion rules.
X.411	Message transfer system.
X.413	Message store.
X.419	Protocol specifications.
X.420	Interpersonal messaging system.
X.440	Voice messaging systems.

- *X.500*—The ITU-T X.500 series of protocol specifications govern the maintenance of files and directories on multiple systems. The protocol is a hierarchical model that tracks layered structures of names, job titles, company names, company addresses, and so on. It lets applications like e-mail access information that can either be central or distributed. The standard is broken down under these and other numbers:

X.501	Models.
X.509	Authentication framework.
X.511	Abstract service definition.
X.518	Procedures for distributed operation.
X.519	Protocol specifications.
X.520	Selected attribute types.
X.521	Selected object types.

See Appendix E for a complete list.

X.Windows A networked graphical user interface (GUI) based on a client–server architecture, it displays information from multiple networked hosts on a single workstation. Available on PCs as X.terminal emulation and emulation on local area network (LAN) servers.

X2 A modem data encoding/transmission scheme developed by U.S. Robotics that can achieve raw transmission speeds of 56 kbps through the public switched telephone network (PSTN). Further, this rate is before any compression is applied to the data. See also *V.90.*

X/Open A body of computer vendors responsible for researching, defining, and publicizing open systems.

583

XALS An abbreviation of E̲xtended A̲pplication L̲ayer Structure.

XAPIA An acronym for the X̲.400 A̲pplication P̲rogramming I̲nterface A̲ssociation. A group attempting to resolve the X.400 cross-system directory synchronization interface problem, e.g., standardizing the interface to X.400 e-mail services.

XAS An abbreviation of X̲-band A̲ntenna S̲ystem.

xB/yB encoding The general label for any of several schemes that encode groups of x bits into a y bit symbol prior to waveform encoding. See also *4B/5B, 5B/6B,* and *8B/10B.*

XBT An abbreviation of C̲rossB̲ar T̲andem.

X$_c$ The symbol for capacitive reactance.

XC An abbreviation of C̲ross C̲onnect.

xDSL An abbreviation of x̲ (the unknown variable) D̲igital S̲ubscriber L̲ine. Several types of *digital subscriber lines* (*DSLs*) are referred to simultaneously, including:

DSL TYPES

Service	Std.	Rate (bps)	Dist. *, **
ADSL (generic)			1 pair
Asymmetric Digital		8M down	5.5 km
Subscriber Line		640k up	18 kft
CAP ADSL			
Carrierless Amplitude	Proposed ANSI	6.144M	3670 m
and Phase ADSL			12000 ft
CDSL			
Consumer Digital	Proposed by		
Subscriber Line	Rockwell		
DMT ADSL			
Discrete MultiTone	ANSI/ETSI	6.144M	3670 m
ADSL			12000 ft
EtherLoop	Nortel	800k to 10M	21000 to 2000 ft
HDSL			
High speed Digital	ANSI/ETSI	1.544M or	2 pair 3670 m
Subscriber Line (2B1Q)		2.048M	12000 ft
HDSL2			
High speed Digital	Proposed ANSI	1.544M or	2 pair 3670 m
Subscriber Line ver. 2		2.048M	12000 ft
IDSL			
ISDN Digital		128k	
Subscriber Line		FDX	
ISDN			
Integrated Services	ANSI/ITU-T	160k	1 pair 5500 m
Digital Network			18000 ft
MDSL		272k - 1168k	
MVL			
Multiple Virtual Line	Paradyne	384k FDX	
RADSL			
Rate Adaptive Digital		to 9M	
Subscriber Line			
SDSL			
Single line Digital	ETSI	1.544M or	1 pair 2-2.8 km
Subscriber Line		2.048M	7-9 kft

DSL TYPES (*CONTINUED*)

Service	Std.	Rate (bps)	Dist. *, **
SDSL			1 pair
Symmetric Digital	Ascend	1.544M	6.3 km
Subscriber Line			21 kft
UDSL			
(*G.Lite, DSL.Lite*)	UAWG		
Universal Digital	ITU-T		
Sunscriber Line			
VDSL			
Very high speed Digital	Proposed ANSI	13-51M	90-305 m
Subscriber Line			0.3-1 kft

*Maximum distance between repeaters with 24 AWG copper. Some manufacturers claim longer distances.

**Variable, depends on data rate.

Most provide a high-bandwidth downstream data path (several Mbps) and a medium-bandwidth upstream path (less than 1 Mbps). The spectrum on the subscriber loop is similar to the following diagram. See also *ADSL, DSL, HDSL, RADSL, UDSL,* and *VDSL.*

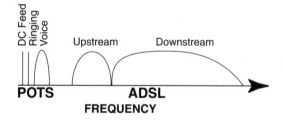

XDR An abbreviation of eX̲ternal D̲ata R̲epresentation.

Xenix™ A multiuser operating system developed by Microsoft, Inc.; a subset of UNIX.

XFER An abbreviation of *transfer.*

XFR An abbreviation of *transfer.*

XID An abbreviation of E̲xchange I̲D̲.

XIP An abbreviation of eX̲ecute I̲n P̲lace.

X$_L$ The symbol for inductive reactance.

XMA An abbreviation of eX̲tended M̲emory A̲rchitecture (or specificA̲tion).

XMH An abbreviation of X̲ M̲ail H̲andler.

XMIT An abbreviation of transmit.

XML An abbreviation of eX̲tended M̲arkup L̲anguage. *XML* is a simplified subset of the Standard Generalized Markup Language (SGML) developed under the auspices of the World Wide Web Consortium (W3C). It allows developers to define their own tags and document types.

XMODEM One of the oldest block file transfer protocols (developed by Ward Christensen in 1978). *XMODEM* uses a 1-byte control sequence, a simple checksum for error checking, and a 128-byte block size. Because of its ease and simplicity, it is one of the most common transfer protocols. However, it is neither particularly fast nor good at error checking (error detection is about 95%).

The layout of an *XMODEM* packet is:

SOH (*01h*) (1 byte)

Packet number (1 byte)

1's complement of packet number (1 byte)

Payload (128 bytes)

Arithmetic Checksum (1 byte).

The *XMODEM* messages are:

ACK (*06h*) packet received OK

NAK (*15h*) packet received NOT OK and start character

CAN (*18h*) cancel transfer

EOT (*04h*) end of transmission

The table demonstrates the steps taken to transfer a file from a sending terminal to a receiving terminal after the basic connection has been established. The start character is the NAK (15h). Sometimes called the *Christensen Protocol* after Ward Christensen, its inventor.

XMODEM FILE TRANSFER SEQUENCE

Sending DCE	Receiving DCE
Idle	Send start character
Send packet 1	Idle
Idle	Send ACK
Send packet 2	Idle
Idle	Send NAK
Resend packet 2	Idle
Idle	Send ACK
Send packet 3	Idle
·	Send ACK
·	·
·	·
Send packet n	·
Idle	Send ACK
Send EOT	Idle
Idle	Send ACK

XMODEM-1K A variation of the standard XMODEM protocol with payload block-length capability increased to 1024 bytes (from 128 bytes). Once the payload size has been selected for a packet, it cannot be changed in retransmissions. The layout of an XMODEM-1K packet is:

STX (*02h*) or *SOH* (*01h*) (1 byte)

Packet number (1 byte)

1's complement of packet number (1 byte)

Payload (128 or 1024 bytes)

Arithmetic Checksum (1 byte).

The XMODEM-1K messages are:

SOH (*01h*) (send 128 byte payload)

STX (*02h*) (send 1024 byte payload)

ACK (*06h*) packet received okay

NAK (*15h*) packet received NOT okay and start character

CAN (*18h*) cancel transfer

EOT (*04h*) end of transmission

XMODEM/CRC A variation of the XMODEM protocol wherein the payload capability is increased to 1K (as in XMODEM-1K) and the simple 8-bit arithmetic checksum is replaced by a 16-bit cyclic redundancy check (CRC). This improves the error detection from about 95% to 99.997%.

The layout of an XMODEM/CRC packet is:

STX (*02h*) or *SOH* (*01h*) (1 byte)

Packet number (1 byte)

1's complement of packet number (1 byte)

Payload (1024 bytes)

Checksum (depends on start character)

  CRC (2 bytes) if start is a C

  Arithmetic (1 byte) if strt is a NAK.

The XMODEM/CRC messages are:

C (*43h*) start character (XMODEM/CRC)

SOH (*01h*) (send 128-byte payload)

STX (*02h*) (send 1024-byte payload)

ACK (*06h*) packet received okay

NAK (*15h*) packet received not okay and start character (XMODEM-1K)

CAN (*18h*) cancel transfer

EOT (*04h*) end of transmission

The CRC generating polynomial is:

$$X^{16} + X^{12} + X^5 + 1$$

XMP An abbreviation of <u>X</u>/Open <u>M</u>anagement <u>P</u>rotocol. The Open Software Foundation's software interface specification in its Distributed Management Environment.

XMS An abbreviation of <u>E</u>xtended <u>M</u>emory <u>S</u>pecification. The 64-KB immediately above the 1-M boundary.

XMSN An abbreviation of transmission.

XMTD An abbreviation of transmitted.

XMTR An abbreviation of transmitter.

XNA An abbreviation of <u>X</u>erox <u>N</u>etwork <u>A</u>rchitecture.

XNS **(1)** An abbreviation of <u>X</u>erox <u>N</u>etwork <u>S</u>ervices. A local area network (LAN) protocol stack. **(2)** An abbreviation of <u>X</u>erox <u>N</u>etwork <u>S</u>ystem, a peer-to-peer protocol developed by Xerox® that has been used by several LAN schemes such as 3Com®, 3+®, and 3+Open® network operating system.

COMPARISON OF XNS & OSI/ISO PROTOCOL LAYERS

ISO Reference Model		Equivalent TCP/IP Model	
Layer	Definition	Definition	Layer
7	Application	Application	4
6	Presentation	Printing, Filing, Clearinghouse	3
5	Session		
4	Transport	RIP, Echo, Error, PXP, SPP	2
3	Network	IDP	1
2	Data Link MAC	Ethernet, X.25, HDLC, Leased Lines, etc.	0
1	Physical		

- *XNS* Level 0 protocols correspond to ISO Layers 1, 2, and part of 3. It is essentially left open, implicitly allowing any adequate protocol to transport *XNS* packets over the physical medium.
- *XNS* Level 1 protocol, the Internet Datagram Protocol (IDP), performs the standard ISO internetworking functions contained in ISO Layer 3.
- *XNS* Level 2 protocol corresponds to the ISO Layer 4 protocol. It includes:
  - *The Sequences Packet Protocol* (*SPP*)—provides reliable, connection-based flow control.
  - *The Routing Information Protocol* (*RIP*)—provides dynamic routing.
  - *The Packet Exchange Protocol* (*PXP or PEP*)—a protocol with greater reliability than simple datagram service (such as IDP) but is less reliable than SPP.
  - *An Error Protocol* (*EP*)—used to notify another client process that a network error has occurred.
  - *An Echo Protocol*—provides functions such as the *ping* command and is used to test the accessibility of *XNS* network nodes.
- *XNS* Level 3 protocol corresponds to the ISO Layer 6 protocol. It provides print services via the Printing Protocol, file-access services via the Filing Protocol, and name services via the Clearinghouse protocol.
- *XNS* Level 4 protocol corresponds to ISO Layer 7 protocol and has no specified protocols as it has little to do with communications functions.

XO An abbreviation of Crystal (X) Oscillator.

XOFF Standing for *transmitter off, XOFF* is one of the control characters in the ASCII character set (DC1 <CTL-s>, 017) used to control data transmissions. When a data communication equipment (DCE) is using XON/XOFF flow control, the character will stop data transmission until the XON character is received.

XON Standing for *transmitter on, XON* is one of the control characters in the ASCII character set (DC3 <CTL-q>, 019) used to control data transmissions. When a DCE is using XON/XOFF flow control, the character will transmit data until the XOFF character is received.

XON/XOFF An asynchronous communication protocol utilizing software data flow control. The ASCII characters DC1/DC3 (017 and 019) are used to start or stop data transmission. Generally, these control characters are embedded in the data stream by the data terminal equipment (DTE). The receiving DTE communicates with the sending DTE by transmitting an XOFF when its buffer is full and an XON signal when it is ready to receive mode data. The characters DC1 and DC3 may be entered manually by typing <CTL-s> and <CTL-q>.

XOR An acronym for eXclusive OR (pronounced "EX-or").

XPAD An abbreviation of eXternal Packet Assembler Disassembler.

XRB An abbreviation of transmit (Xmit) Reference Burst.

xref An acronym for cross (X) REFerence.

XT An abbreviation of crosstalk.

XTAL An abbreviation of crystal.

XTP An abbreviation of eXpress Transfer Protocol. A simplified low-processing protocol for broadband networks.

XXENCODE In Internet, one of several encoding methods that convert a binary file (which may contain any character) into another file containing only characters that can be transmitted over diverse networks.

The basic scheme groups three 8-bit characters together and then divides the 24 bits into four 6-bit characters. These 6-bit characters are mapped into the ASCII characters:

```
+ - 0 1 2 3 4 5 6 7 8 9 A B C D E F G H I
J K L M N O P Q R S T U V W X Y Z a b
c d e f g h i j k l m n o p q r s t u v w x y z.
```

Unlike *uuencode,* all of these characters can be mapped in an EBCDIC to ASCII translation. See also *uuencode.*

Y

Y (1) The SI symbol for *admittance*. **(2)** An abbreviation of the Yellow alarm control bit in the sync byte of time slot 24 of a T-1 multiplexer. Y = 0 in bit two of all time slots indicates an alarm.

Y-code Encrypted P-code. See *W-code.*

Y2k problem An abbreviation of Year 2000 (2k). The situation that was projected to occur for the year after the year 1999 when computer dates represented with only two digits for the year could not determine the proper century. That is, did 00 represent the year 1900 or the year 2000? The real problem was the expense industries had to incur to fix/rewrite system programs.

ISO 8601:1988, the international date standard, specifies the numeric format to be yyyy-mm-dd.

YA- A prefix meaning Yet Another, for example,

YAA   Yet Another Assembler

YABA   Yet Another Bloody Acronym

YAC   Yet Another Compiler

Yagi-Uda antenna A linear end-fire antenna consisting of three or more elements (one driven, one reflector, and one or more directors). The design provides very high directivity and gain.

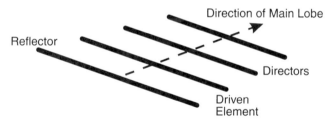

The length of the driven element is approximately a half-wave, the directors are generally a little longer, and the reflectors are somewhat shorter than the driven element. Each director is progressively shorter than its predecessor. Also called a *Yagi antenna.*

Yahoo One of the biggest hierarchical indexes of the World Wide Web. Originally at Stanford University, Yahoo moved to its own site in April 1995. It allows a user to move up and down the hierarchy to search it and to suggest additions. It also features, as do most other search sites, a *What's New, What's Popular, What's Cool,* and a *random link.*

yellow alarm (Y) An indication provided to remote T-1 terminals that the local terminal is out of synchronization. (Every second bit position is zero in every channel in a T1 frame.) The yellow alarm is sent to the local site after a red alarm at the remote site.

Yellow Book The industry standard that defines the format of CD-ROMs. It was the first extension of the audio CD (Red Book) stan-

dard, and it enables CDs to contain 650MB of computer data instead of only digital sound. (Because neither of the standard's defining companies, Philips and Sony, was a major player in the personal computer market, the standard had some deficiencies. Improvements to the standard include:

- *High Sierra*—an early attempt to make the *Yellow Book* standard readable on any computer platform. But it worked only if the data was in a universal format such as ASCII.
- *ISO 9660*—an international standard that refined High Sierra version to be readable on any platform, regardless of the data format (providing the right device driver is installed).
- *Mac HFS*—a format that can be read only by Macintosh computers.
- *Hybrid CD-ROM*—defines a CD-ROM with a format, or file structure, and programs that can be read by either Macs or PCs.

It forms the basis of the ISO 10149 international standard.

yellow cable An alias for Ethernet 10Base5.

yellow pages (YP) A directory of businesses arranged by category. It may be printed on paper (historically yellow paper), recorded in a voice processing and response systems, distributed on the Internet, or recorded on CD-ROMs.

YMMV An abbreviation of Your Mileage May Vary. Translation: "It works this way for me, but it may be different on your system!"

YMODEM A variation of the standard *XMODEM* protocol with:

- cyclic redundancy code (CRC) error checking,
- The ability to send multiple files (batch file transmission),
- The ability to abort a transmission by transmitting two <CAN> characters in a row,
- 1024 bytes per block, and
- An extra field at the beginning of the transfer containing the file-name, file size, and date.

YMODEM-G extends the *YMODEM* protocol to include support modems that have a built-in error control protocol. It also allows streaming, that is, the software transfers blocks continuously to the modem (not waiting for a positive acknowledgment after each block). If any block fails to transfer to the destination correctly, the entire transfer is canceled.

YMU An abbreviation of Y-Net Management Unit.

yotta (1) The SI prefix for 10^{24}. **(2)** In computer terms, the prefix 2^{80} or 1 208 925 819 614 629 174 706 176.

YP An abbreviation of Yellow Pages.

YTD An abbreviation of Year To Date.

YWIA An abbreviation of Your Welcome In Advance.

Z

Z (1) The symbol for impedance. **(2)** An abbreviation of Zulu time. See also *time (3)*. **(3)** The symbol for Zenith. **(4)** The symbol for Zero.

Z Time An abbreviation of Zulu Time. See *time (3)*.

z-count The fundamental unit of time in the Global Positioning System (GPS). It is a 29-bit binary number divided so that the first 10 most significant bits (MSBs) represent the GPS week and the remaining 19 least significant bits (LSBs) are the time-of-week (TOW) in units of 1.5 seconds.

A truncated version of the TOW is included in the navigation message handover word. The resolution of the truncated time is 6 seconds.

Z.n Recommendations A series of ITU-T standards covering "Programming Languages." See Appendix E for a complete list.

zap (1) To modify, usually to correct; especially used when the action is performed with a debugger or binary patching tool.

In the IBM mainframe world, binary patches are applied to programs or to the operating system with a program called superzap, whose file name is IMASPZAP (possibly contrived from I M A SuPerZAP). **(2)** To destroy an electronic part with static electricity, lightning, or another severe transient overvoltage condition. For example, "that lightning strike just *zapped* the modem."

ZBTSI An abbreviation of Zero Byte Time Slot Interchange.

ZCR An abbreviation of Zero Crossing Rate.

zener diode A type of diode that tends to maintain a constant voltage drop regardless of the current through it.

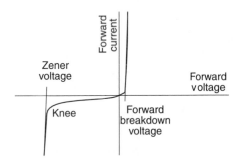

zero bit insertion A technique used with bit-oriented protocols to ensure that six consecutive 1-bits never appear between the start-of-frame and end-of-frame flags delineating a transmission frame. In operation, when five consecutive 1-bits occur in any part of the frame other than the beginning and ending flag, the sending station inserts an extra 0-bit. When the receiving station detects five 1-bits followed by a 0-bit, it removes the extra 0-bit, thereby restoring the original bit stream value. See also *BnZS*.

zero code suppression The insertion of a "one" bit into a long string of "zero" bits so as to force a transition in the transmitted waveform periodically. This allows the clock recovery circuit in the receiver to maintain its synchronization with the transmitter. See also *BnZS*.

zero dBm transmission level point (0 dBm TLP) In a communication system, any point at which the reference level is 1 mW (0 dBm). This does not imply that the actual power level of the carried traffic is 0 dBm. The 0 dBm level is a reference for system design and test purposes. Actual traffic is usually well below these test levels. Also called the *zero transmission level point (0TLP)*.

zero dispersion slope In fiber optics, the rate of change of dispersion, with respect to wavelength, at the fiber's zero dispersion wavelength.

In a pure silica-based optical fiber, the *zero-dispersion wavelength* occurs naturally at about 1.3 μm. However, this wavelength may be shifted somewhat toward the minimum-loss window through the addition of dopants to the fiber material. The drawing shows the *zero dispersion slopes* for singly, doubly, and quadruply clad single mode fibers.

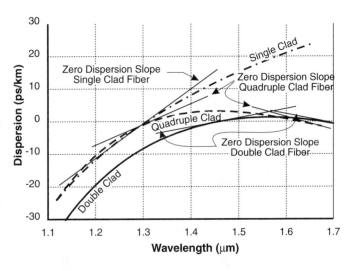

zero dispersion wavelength (1) In a single mode optical fiber, the wavelength (or wavelengths) at which material dispersion and waveguide dispersion cancel one another.

In all silica-based optical fibers, minimum material dispersion occurs naturally at a wavelength of approximately 1.3 μm. With the addition of dopants, this wavelength can be shifted toward the minimum loss window at approximately 1.55 μm. The penalty for this is a slight increase in the minimum attenuation coefficient. **(2)** Loosely, in a multimode optical fiber, the wavelength at which material dispersion is minimum, i.e., essentially zero. Also called *minimum dispersion wavelength*.

zero dispersion window A synonym for *minimum dispersion window*.

zero fill (1) The writing of the character zero to unused storage locations. **(2)** The insertion of one or more "zero symbols" into a transmission stream as a means of maintaining synchronization. For example, when a facsimile machine with a slow mechanical printer is receiving from a fast transmitter, the transmitter must add "fill bits" (called *zero fill*) to the end of each scan line to allow the receiver time to reset the mechanical mechanism.

zero level decoder A decoder that yields an analog level of 0 dBm at its output when the input is the digital milliwatt signal.

zero minus call (0-) In telephony, a call in which only the digit "0" is dialed, indicating the caller desires operator assistance.

zero modulation (1) The state where a transmission, recording, or playback system is connected and processing a signal stream with zero amplitude (no signal applied to the input). (2) A waveform encoding method in which one binary sequence is mapped into a second binary sequence with a reduced zero frequency (dc) content. This encoding technique is useful on systems that cannot tolerate any dc content in the signal, for example, magnetic recording systems or signal transmission systems that include capacitors or transformers.

zero plus call (0+) In telephony, a call in which the digit "0" is dialed as a prefix to another number indicating that operator intervention is necessary.

zero power modem A modem that draws operating power from the telephone line to which it is attached and/or from the communications signals of data terminal equipment (DTE) to which it is attached. Because of the low power available, the range and speed are generally somewhat limited.

zero slot LAN A local area network (LAN) that uses a serial or parallel port on the computer rather than a special network interface card (*NIC*) plugged into the computer's expansion bus. Because *zero slot LANs* can transmit only as fast as the computer's output port, they are considerably slower than networks that use specialized hardware and software. Furthermore, the maximum length of each cable segment is also severely limited, so *zero slot LANs* can connect only two or three computers. The major (and perhaps the only) advantage of a *zero slot LAN* is its low cost compared with dedicated network systems. However, the prices of peer-to-peer networks are eroding and tend to negate even this advantage.

zero stuffing In transmission systems, the addition of a "zero" bit into a long string of "one" bits for one of two reasons:

- To force transitions in the waveform (so that the receiver can maintain clock synchronization), or
- To prevent the occurrence of the start (or end) of frame (packet) synchronization word in the middle of a data sequence. The frame delineation marker is 01111110; hence, any data sequence with more than five consecutive ones is not allowed. The transmitter always sends a zero following a string of five ones, and the receiver always strips the zero following five ones. Therefore, the receiver knows that when it receives a string of six ones it is a frame marker.

zero suppression The elimination of nonsignificant zeros (leading zeros) in a displayed number. Sometimes, the zeros are replaced by spaces rather than simply eliminating the physical position. For example, the number 00123 could be displayed as 123 or   123.

zero transmission level point (0 TLP) That point in a circuit or system which is used as a reference point for measuring for *relative* transmission measurements. The signal level at that point is always 0 dBr, by definition. Also called *zero dBm transmission level point.*

zeroth order approximation Jargon. A *really* sloppy (and probably unusable) approximation, i.e., a wild guess.

zif socket An acronym for Zero Insertion Force socket. A kind of socket for integrated circuits (IC). A *Zif* socket can be opened and closed by means of a lever or screwdriver. When open, the IC may be placed in the socket without applying any significant pressure. When the socket is closed, its contacts grip the pins of the IC, thereby holding the device in place and making electrical contact. Such sockets are used only when the cost of the IC is high or in test equipment (when many insertions are expected).

zine A magazine designed for use on electronic networks. It is not just an electronic or online version of a magazine posted on the network; it is designed specifically for the network on which it is used. In general, the articles tend to be shorter, contain hot links to other related material and sites, and sometimes contain multimedia capabilities. Those found on the Internet are called *Web zines.*

.ZIP The file extension of one of the many file compression types. The compression program implemented by PKWARE Inc. is the shareware program PKZIP, and the expansion program is PKUNZIP, both available as downloads on most bulletin board systems (BBSs). See also *compression* (2).

ZIP An acronym for Zone Information Protocol.

zip code A five-digit code assigned to each U.S. post office to facilitate quicker delivery. A four-digit extender to the five digits allows sorting to the individual delivery route.

zip-cord (1) A two-conductor electrical cable in which the insulation or jacket of the cable is manufactured in such a way that the conductors are joined only by a thin strip. The two conductors are easily separated to facilitate termination. Also called *lamp cord.* (2) In fiber optics, a cable consisting of two single fiber cables whose jackets are joined by a thin strip of jacket material. As with electrical *zip-cord,* optical *zip-cord* may be separated to permit the installation of connectors. Both loose buffer and tight buffer *zip-cord* cables are manufactured.

ZIS An abbreviation of Zone Information Socket.

ZIT An abbreviation of Zone Information Table.

ZMA An abbreviation of Zone Multicast Access.

ZMODEM *ZMODEM* is a file transfer protocol, developed by Chuck Forsberg, similar to XMODEM and YMODEM but with better performance. ZMODEM provides faster transfers, better error checking (uses CRC-32), and the ability to restart interrupted transfers (a feature called *checkpoint restart*). Unlike XMODEM or YMODEM, *ZMODEM* is a streaming protocol.

Zobel filter A filter designed according to image parameter rules. See also *attenuator, band reject filter, band-pass filter, filter, high-pass filter, low-pass filter,* and *notch filter.*

zone (1) In an internetwork, a subset of nodes which, together, form a logical subdivision. A node can be part of one or more *zones.* A *zone* can encompass multiple networks and can cross network boundaries. (That is, it can apply to parts of several networks.) A *zone* may have a name associated with it that is used to simplify routing and service advertising. (2) In AppleTalk. A logical subset of nodes which together form a subdivision. It can have an associated name, and a node can be part of one or more zones. The zone name is used to simplify routing and service advertising. A zone can encompass multiple networks and can cross network boundaries (that is, apply to parts of several networks).

zone information socket (ZIS) A socket (access point) associated with the zone information protocol (ZIP) services in an AppleTalk network.

zone information table (ZIT) Maps the zone name(s) associated with each subnetwork in a network or internetwork in an AppleTalk network.

zone of silence That region on the Earth that is out of range of a transmitter's ground wave and closer to the transmitter than the first reflection from the ionosphere (the skywave). Also called a *skip zone*.

zone paging The ability of a device, such as a private automatic branch exchange (PABX), to page specific areas of a facility. *Zone paging* is intermediate to facility paging (where the entire facility is paged, i.e., only one zone) and individual paging (where each person is paged individually, e.g., by a beeper or the telephone at their desk.)

Zulu Time (Z) A synonym for *time (3)*. Formerly called *Greenwich Mean Time.*

zyx Pronounced *zikes,* a term addressing the collection of all of the abbreviations, acronyms, phrases, and terms of a specified discipline or field. Obviously, for an active field the collection will never be complete. For a field as broad and rapidly changing as communications/computing there is no hope of just listing all of the elements of the collection.

Appendix A
"AT" Fax/Modem Commands

MODEM COMMANDS

Modem commands can be thought of as a number of user selectable switches. Each switch has the ability to control some aspect of the operation of the modem, and each switch generally has several settings. For example, one command sets the speaker level at one of three values. The original command set was developed by Hayes Computer Products, Inc. for its "Smartmodem 300" many years ago. It was so successful that most modem manufacturers adopted this "Hayes compatible" command set to operate their own modems. Hence, the command set used by the SmartModem 300 is used by most modems today (with additional advanced commands). It is also known as the AT command set. (AT stands for attention and precedes all [with few exceptions] commands directing the modem.)

The following list is a compilation of the "AT" commands from many modem models and manufacturers. All of the commands are an extension of the original set of "AT" commands developed by Hayes for their SmartModem some years ago. Because each manufacturer developed his product separately, these extended commands were developed individually rather than from a common standard. Hence, not all modems use all of the listed commands, nor do the extended commands mean the same thing to every modem manufactured. Where the "AT" command differs between manufacturers, a separate description is provided for each. The Telecommunications Industry Association has adopted a number of commands in its TIA 602 specification.

The American Standard Code for Information Interchange (ASCII) character set (ISO 646 or ITU-T T.50 IA5) is used for all commands and responses. Only the low-order 7 bits of each character are used for commands or parameters. The high-order bit is generally ignored (although, for facsimile Phase C data transmission or reception, all 8 bits are used).

A command line is a string of characters sent from data terminal equipment (DTE) to data communications equipment (DCE), while the DCE is in COMMAND mode. Command lines have a prefix, a body, and a terminator. The prefix consists of the ASCII characters "AT." The body is a string of commands restricted to printable ASCII characters, (032–126). Space characters (ASCII 032) and control characters other than <CR> (ASCII 013) and <BS> (ASCII 008) in the command string are ignored. The default terminator is the ASCII <CR> character. All characters preceding the "AT" prefix are ignored. On most modems either upper or lower case may be used for the prefix and/or body of a command; therefore "*ATDT*" will perform the same as "*atdt*", and "*ATdt*." Many modems restrict the "AT" to be of the same case, that is, neither "At" nor "aT" is allowed. A command line may be aborted at any time by entering <CTL> X.

The basic command consists of a single letter character (ASCII 065–090) possibly preceded by a prefix character (e.g., '&', '%', '#', '*', '-', or '\') and possibly followed by a decimal parameter "n". If a numeric parameter is included in the command definition but is omitted on the "AT" command line, the missing parameter is evaluated as n = 0.

Multiple commands and their associated parameters may be entered on a single command line in most cases. The maximum command line length, however, is generally limited to about 55 to 60 characters. The format for transmitting single or multiple commands from the DTE to the DCE is:

```
AT [command[parameter(s)]] [command[parameter(s)]] . . . <return>
```

+++ . . . . *Default Escape Code*

The "+++" escape code must be preceded and followed by a 1-second guard time (a period of time with no data transmission) to be a "Hayes compliant" escape sequence. The entire escape sequence, "<guard time> +++ <guard time>," will put the modem into the COMMAND mode. Only while in the COMMAND mode will the modem accept any AT command strings. The command ATO is used to put the modem back online.

Some manufacturers, in order to avoid patent infringement, have eliminated the guard time before and after the escape code. The modems using this *time independent escape sequence* (*TIES*) have a "Hayes compatible" escape sequence, though not a "Hayes compliant" one. These modems have the potential to accidentally enter the COMMAND mode due to the presence of an escape sequence embedded in a transmitted data stream. The probability of this causing a problem is a function of the escape code chosen, the type of data being transmitted, and the amount (and type) of data transmitted. Some manufacturers have added more constraints to TIES. For example, the characters immediately following the "+++" escape code must be a VALID "AT" command in order to put the modem into COMMAND mode. Others have chosen methods such as including a BREAK in the escape sequence.

The "+++" default sequence can be changed by modifying the value in the S2 register. The guard time duration is set by the value in the S12 register. The modem will respond to the escape sequence with "OK," indicating it is ready to accept commands.

Result Codes:
OK . . . . For an accepted command.

? . . . . *Report value of S-Register*

"AT?" reports the value of the *last S-Register* used. A specific S-Register can be selected by using the ATSn? command. Some registers may not be written due to country specific PTT limitations.

See also Sr?.

Result Codes:
OK

591

=v . . . . *Writes n to S-Register*

"AT = v" writes n into the *last S-Register* used. A specific S-Register can be selected by using the ATSn=v command.

See also Sr = v.

Result Codes:
OK

$ . . . . *HELP, Command Summary*

The "AT$" command instructs a U.S. Robotics modem to display a quick reference of the commands available to the modem.

Example:
```
AT$<return>

HELP, Command Quick Reference (CTRL-S to Stop, CTRL-C to Cancel)

A      Answer Call                    Kn     n=0  Call Duration Mode
A/     Repeat Last Command                   n=1  Real Time Clock Mode
Cn     n=0  Transmitter Off           Mn     n=0  Speaker Off
       n=1  Transmitter On                   n=1  On Until CD
Ds     Dial a Telephone Number               n=2  Always On
       s=0..9 # * T P R , ; " W @ !           n=3  Off during Dial
D$     HELP, Dial Commands            O      Return On Line
En     n=0  Do Not Echo Chars         Qn     n=0  Result Codes Sent
       n=1  Echo Command Chars               n=1  Quiet (No Result Codes)
Fn     n=0  Half Duplex               Sr=n   Sets Register "R" to "n"
       n=1  Full Duplex               Sr?    Query Register"r"
Hn     n=0  On Hook (Hang Up)         S$     Help, S Registers
       n=1  Off Hook                  Vn     n=0  Non-Verbal Responses
In     n=0  Product Code                     n=1  Verbal responses
       n=1  Checksum                  Xn     n=0  Basic Result Codes
       n=2  Ram Test                         n=1  Extended Result Codes
       n=3  Call Duration/Clock              n=2-6 Advanced Result Codes
       n=4  Register Settings         Z      Software Reset
I3=s   Sets the Clock                 >      Repeat Command String
       s=Hours:Mins:Secs              $      HELP, Command Summary
OK
```

A . . . . *Answer Phone*

The "*ATA*" command instructs the local modem to proceed with the connection sequence (answer a call); that is, the local modem will go off-hook and attempt to complete the handshake procedure. Specifics of this command are affected by the state of +FCLASS, #CLS, #VTD, Line Current Sense–if enabled (most countries do not require Line Current Sense), and other country-specific requirements.

If the modem is in the data mode (i.e., +FCLASS=0), the modem will enter the connect state after detecting carrier and establishing a common data protocol with the remote modem. If no carrier is detected within a period specified in register S7, the modem will disconnect and return the result code NO CARRIER. Any character entered during the connect sequence will abort the connection attempt.

If the modem is in the facsimile mode (i.e., +FCLASS≠0), the modem will go off-hook in V.21 answer mode, generate the V.21 2100 Hz answer tone for 3 ± 0.5 seconds and, following a delay of 70 ms, will proceed as if the +FTH=3 command were issued. At any stage up to (but excluding) the +FTH=3 command state, any character will abort the communication. (See the description of the +FTH command for details.)

Result Codes:
CONNECT XXXX . . . . If a connection is established (XXXX = telco rate)
NO CARRIER . . . . If a connection cannot be established, if the S7 timer expires before a carrier is detected, or if the command is aborted.
VCON . . . . Issued in Voice Mode (#CLS=8) immediately after going off-hook.

A/ . . . . *Reexecute Last Command*

The "*A/*" command instructs the modem to repeat the last AT command line string entered into the command line buffer. The principal use of this command is to initiate another call (using the Dial command) that failed to connect due to a busy line or no answer condition.

NOTE: This command is *not* preceded by AT and must not be followed by a carriage return; therefore, it must be the only command on the command line. Typing any character after the *A/* will terminate the command.

A> . . . . *Reexecute Last Command Continuously*

"*A>*" causes the modem to execute the last AT command line string entered *continuously* or until canceled by the receipt of any character. Dial strings will repeat up to 10 times.

NOTE: This command *is not* preceded by AT and must not be followed by a carriage return; furthermore, it must be the only command on the command line. Typing any character after the A> will terminate the command.

AT . . . . *<u>AT</u>tention Command Line Prefix*

"*AT*" is the command line prefix and is the first entry of all command lines sent to a modem (except for the $, A/, and A> commands). All command lines must end with a carriage return <CR> (again, except for the A/ and A> command). Only while in COMMAND mode will the modem accept any AT command sequences. If the modem is in the DATA mode (also called online mode), an escape sequence must be sent first. See +++.

Commands are not executed until a carriage return <CR> is entered; therefore, a backspace <BS> may be used to correct command line entry errors as long as the carriage return has not been entered. Note that the backspace may not delete the AT at the beginning of the command line.

While the modem is in the DATA mode, entering the command line "*AT<CR>*" will only cause the modem to return an "OK" response (or a "0" if the modem is set for numeric responses).

Result Codes:
OK . . . . For accepted command (or no command).
ERROR . . . . For commands not recognized or inappropriate for the current operating state.
Others . . . . See Appendix C for a complete list.

Bn . . . . *Modulation Mode Select (BELL/ITU-T)*

The "*ATBn*" command selects between ITU-T and Bell carrier frequencies for 300 and 1200 bps operation. The command is not valid when N1 (automode detection) is selected. For most modems the range of n is limited to 0 or 1.

Command Option Details:
B0 . . . . **ITU-T standard frequencies** (International)
                                  V.21 (300 bps), V.22 (1200 bps), V.22 bis (2400 bps), V.25 answer sequence, V.32
B1 . . . . Bell standard frequencies (default in North America)
                                   103 (300 bps), 212A (1200 bps).
B2, B3 . . . . Selects ITU-T V.23 R1200/T75 ASB when the modem is originating,
                                   and ITU-T V.23 T1200/R75 when the modem is answering.
B15 . . . . Selects ITU-T V.21 when the modem is at 300 bps.
B16 . . . . Selects Bell 103 when the modem is at 300 bps.
B30 . . . . Selects ITU-T V.22 bis when the modem is at 2400 bps.
B41 . . . . Selects ITU-T V.32 when the modem is at 4800 bps.
B52 . . . . Selects ITU-T V.32 bis when the modem is at 7200 bps
B60 . . . . Selects ITU-T V.32 when the modem is at 9600 bps
B70 . . . . Selects ITU-T V.32 bis when the modem is at 12000 bps
B75 . . . . Selects ITU-T V.32 bis when the modem is at 14400 bps

Result Codes:
OK . . . . For n valid number.
ERROR . . . . Otherwise.

Cn . . . . *Carrier On/Off Control*

"*ATCn*" controls the transmit carrier. The modem will turn the carrier on and off as required in normal operation when C1 is selected. When *C0* is asserted, the carrier is turned off providing a "*receive only*" modem. This feature is not included in most modems.

Command Option Details:
C0 . . . . Disable carrier and returns ERROR for most modems. Receive only mode on others.
C1 . . . . **Enable normal transmit carrier operation.**

Result Codes:
OK . . . . For n = 1 (for most modems).
ERROR . . . . Otherwise.

Dp . . . .*Dialing Command*

The "*ATDp*" command must be the last command on the command line. It causes the modem to go off-hook, wait for a dial tone (for a maximum time specified in the S6-Register), dial according to the parameter string "p" following the D, and attempt to establish a connection (in the data mode if +FCLASS=0 or in the facsimile mode if +FCLASS≠0). If there are no parameters following the D, the modem goes off-hook in originate mode and waits for either a carrier or the expiration of the abort timer (register S7). Up to 40 symbols may follow the D command (the modem truncates all symbols). The symbols may be any mix of 0123456789ABCD#* and the dial command modifiers listed below. The symbols ABCD#* are valid only in *tone dialing*.

Command Option Details (Dial Command Modifiers):

! . . . . Hook flash. The exclamation point "!" forces a hook flash (the modem goes on-hook for the time specified in register S29 (typically 700 ms) then return to off-hook).

" . . . . Dial Alpha. The quote " ″ " dials the numeric equivalent of the letters that follow, e.g., ATDT1(800)″CALLIBM″ is the same as ATDT1(800)225-5422 (Courier specific command).

$ . . . . Help. The dollar symbol ($) causes U.S. Robotics modems to display a dial command summary, e.g.,

Example:
```
ATD$ <return>

HELP, Dial Commands (CTRL-S to Stop, CTRL-C to Cancel)

0-9    Digits to dial
* #    Auxiliary Digits
T      Tone Dialing
P      Pulse Dialing
R      Call an Originate Only Modem
,      Pause
;      Remain in Command Mode
"      Used to Dial Alpha Phone #'s
W      Wait for 2nd Dial Tone (X3-X6)
@      Wait for Answer (X3-X6)
!      Flash Switch Hook
OK
```

& . . . . Wait for "Bong" tone for credit card dialing before continuing with the remaining dial string.

, ... Pause. The comma "," instructs the modem to pause before continuing the dialing string. Multiple commas may be used to extend the pause time80. The pause time for each comma is set in register *S8*.

/ . . . . Delay 125 ms. The slash "/" inserts a 125 ms delay before proceeding with the dial string.

; . . . . Return to COMMAND mode. The semicolon ";" instructs the modem to return to COMMAND mode after dialing (must be the last character in the command line). Useful if more than 40 symbols are to be processed. The *ATH0* command will abort the dial command.

> . . . . Generate a grounding pulse on the EARTH relay output. (Valid only if the country code settings support it.)

@ . . . . Wait for "*quiet answer.*" The "@" symbol causes the modem to wait for 5 seconds of silence before proceeding to the next step in the dial string. If a 5-second silence is not detected in the time period defined by *S7*, the modem disconnects and returns either a BUSY or a NO ANSWER result code. If a 5-second silence is detected, the modem continues with the dial string.

^ . . . . Turn on the calling tone. The caret "^" causes the modem to turn on a periodic 1300 Hz call originating tone.

L . . . . Re-dial the last string dialed. Repeats the entire dial string after the D in the ATD. . . command.

J . . . . Causes MNP10 link negotiation at the highest rate for the current call only. See also the *H command.

K . . . . Enable power level adjustment during the MNP 10 link negotiation for the current call only. See also the Mn command.

P . . . . Pulse dial. Dial using pulse dialing (must be placed before the string to be pulse dialed). Only the digits following the P in the current string are pulse dialed; subsequent dial commands revert to the previous dial mode. Some countries prevent changing dialing modes after the first digit is dialed.

R . . . . Originate call in answer mode. Reverse the modem's mode from originate to answer mode (allows calling an originate only modem). This command MUST be the last command in the dial string.

Sx . . . . Dial the number previously stored in memory location x (x = 0 − 3). Some modems only support two or three stored numbers.

T . . . . Tone dial. Dial using tone dialing (must be placed before the string to be dialed). Only the digits following the T in the current string are tone dialed; subsequent dial commands revert to the previous dial mode. Some countries prevent changing dialing modes after the first digit is dialed.

W . . . . Wait for dial tone. The "W" dial modifier causes the modem to wait for the dial tone before processing the next character in the dial string (used with PBX "dial 9 to get an outside line"). The modem will wait for the time specified in *S7;* if no dial tone is detected the modem will disconnect.

- ( ) . . . . Spaces, hyphens, and parentheses are ignored by most modems so they can usually be added to the dial string to aid the readability of the command line.

Result Codes:

OK. . . . . If either a ";" dial modifier is used or an abort (*ATH0*) during dialing is entered.

NO DIALTONE. . . . . If X2 or X4 is asserted and 1 second of dial tone is not detected within 5 seconds; or if the "W" dial modifier is used and 3 seconds of dial tone is not detected within the time specified in *S7.*

BUSY . . . . If X3, X4, or the dial modifier "W" is asserted and a busy tone is detected.

NO ANSWER . . . . 1. If the "@" dial modifier is used and 5 seconds of silence are not detected in the time specified by *S7.*
2. Issued in Voice/Audio Mode (#CLS=8) when the modem determines that the remote has not picked up the line before the S7 timer expires. Although the modem reports "NO ANSWER," the DTE must issue an ATH (or ATZ) to disconnect the telephone line while #CLS=8.

CONNECT XXXX . . . . If a connection is established.

VCON . . . . Issued in Voice/Audio Mode (#CLS=8) when the modem determines that the remote modem or handset has gone off-hook, or when returning to the Online Voice Command Mode. (See also #VRA and #VRN commands.)

NO CARRIER. . . . . If a connection cannot be established, or the abort timer *S7* expires.

ERROR . . . . If the modem is in the DATA mode.

En . . . . *Command Mode Echo Control*

        "*ATEn*" controls the local echo of characters sent to the modem by the DTE while the modem is in the COMMAND mode. When enabled, the modem echoes the character it receives from the DTE back to the DTE. The value is written to register S14 bit 1 if the result code is OK.

Command Option Details:

E0 . . . . No echo.

E1 . . . . **Echo command** (default).

Result Codes:

OK. . . . . If for n = 0 or 1.

ERROR . . . . Otherwise.

Fn . . . . *(1) On-Line State Echo Control*

"*ATFn*" determines if the modem will echo data received from the DTE back to the DTE while in the DATA mode (local echo). Used in half-duplex communication channels to enable the DTE to "see" the characters that were transmitted.

Command Option Details:

F0 . . . . Enable echo at modem (returns ERROR on many modems that do not support the feature).

F1 . . . . **Disables data echo at modem.**

Result Codes:

OK. . . . . For n = 0 or 1 if supported.

ERROR . . . . Otherwise.

Fn . . . . *(2) Select Line Modulation*

The "*ATFn*" command selects the line modulation according to the parameter supplied. The line modulation is fixed unless Automode is selected (n=0). The parameter value, if valid, is written to S37 bits 0-4. To select line modulation, it is generally recommended that either the F command, or a combination of the N command and setting the S37 register be used, but not both, as the Fn command interacts with the S37.

Command Option Details:

F0 . . . . Selects auto-detect mode. Sets N1 and sets register S31 bit 1. In this mode, the modem configures for automode operation. All connect speeds supported by the modem are possible according to the remote modem's preference. The contents of register S37 are ignored, as is the sensed DTE speed.

F1 . . . . Selects V.21 or Bell 103 according to the Bn setting as the only acceptable line speed resulting in a subsequent connection. Sets N0, sets register S37 to 1, and clears register S31 bit 1. *ATF1* is equivalent to the command string ATN0S37=1.

F2 . . . . Some Rockwell chipsets set speed to 600 bps.

F3 . . . . Selects ITU-T V.23 as the only acceptable line modulation for subsequent connections, that is, the originator is at 75 bps and answer is at 1200 bps. Sets N0, sets register S37 to 7, and clears register S31 bit 1. *ATF3* is equivalent to the command string ATN0S37=7.

F4 . . . . Selects ITU-T V.22 1200 or Bell 212A according to the Bn command setting as the only acceptable line speed for a subsequent connection. Sets N0, sets register S37 to 5, and clears register S31 bit 1. *ATF4* is equivalent to the command string ATN0S37=5.

F5 . . . . Selects ITU-T V.22 bis as the only acceptable line modulation for subsequent connections. It sets N0, sets register S37 to 6, and clears register S31 bit 1. *ATF5* is equivalent to the command string ATN0S37=6.

F6 . . . . Selects ITU-T V.32 bis 4800 or V.32 4800 as the only acceptable line modulation for subsequent connections. It sets N0, sets register S37 to 8, and clears register S31 bit 1. *ATF6* is equivalent to the command string ATN0S37=8.

F7 . . . . Selects ITU-T V.32 bis 7200 as the only acceptable line modulation for subsequent connections. It sets N0, sets register S37 to 12, and clears register S31 bit 1. *ATF7* is equivalent to the command string ATN0S37=12.

F8 . . . . Selects ITU-T V.32 bis 9600 or V.32 9600 as the only acceptable line modulations for subsequent connections. It sets N0, sets register S37 to 9, and clears register S31 bit 1. *ATF8* is equivalent to the command string ATN0S37=9.

F9 . . . . Selects ITU-T V.32 bis 12000 as the only acceptable line modulation for subsequent connections. It sets N0, sets register S37 to 10, and clears register S31 bit 1. *ATF9* is equivalent to the command string ATN0S37=10.

F10 . . . . Selects V.32 bis 14400 as the only acceptable line modulation for subsequent connections. It sets N0, sets register S37 to 11, and clears register S31 bit 1. *ATF10* is equivalent to the command string ATN0S37=11.

F13 . . . . Selects ITU-T V.34 14400 as the only acceptable line modulation for subsequent connections. It sets N0, sets register S37 to 15, and clears register S31 bit 1. *ATF13* is equivalent to the command string ATN0S37=10.

F14 . . . . Selects ITU-T V.34 16800 as the only acceptable line modulation for subsequent connections. It sets N0, sets register S37 to 16, and clears register S31 bit 1. *ATF14* is equivalent to the command string ATN0S37=10.

F15 . . . . Selects ITU-T V.34 19200 as the only acceptable line modulation for subsequent connections. It sets N0, sets register S37 to 17, and clears register S31 bit 1. *ATF15* is equivalent to the command string ATN0S37=10.

F16 . . . . Selects ITU-T V.34 21600 as the only acceptable line modulation for subsequent connections. It sets N0, sets register S37 to 18, and clears register S31 bit 1. *ATF16* is equivalent to the command string ATN0S37=10.

F17 . . . . Selects ITU-T V.34 24000 as the only acceptable line modulation for subsequent connections. It sets N0, sets register S37 to 19, and clears register S31 bit 1. *ATF17* is equivalent to the command string ATN0S37=10.

F18 . . . . Selects ITU-T V.34 26400 as the only acceptable line modulation for subsequent connections. It sets N0, sets register S37 to 20, and clears register S31 bit 1. *ATF18* is equivalent to the command string ATN0S37=10.

F19 . . . . Selects ITU-T V.34 28800 as the only acceptable line modulation for subsequent connections. It sets N0, sets register S37 to 21, and clears register S31 bit 1. *ATF19* is equivalent to the command string ATN0S37=10.

F20 . . . . Selects ITU-T V.34 31200 as the only acceptable line modulation for subsequent connections. It sets N0, sets register S37 to 22, and clears register S31 bit 1. *ATF20* is equivalent to the command string ATN0S37=10.

F21 . . . . Selects ITU-T V.34 33600 as the only acceptable line modulation for subsequent connections. It sets N0, sets register S37 to 23, and clears register S31 bit 1. *ATF21* is equivalent to the command string ATN0S37=10.

Result Codes:

OK. . . . . For n = 0 through 10.
ERROR . . . . Otherwise.

Hn . . . .*Switch-Hook Control*

The "*ATHn*" command controls the state of the hook switch in countries where its use is not restricted by the PTT.

Command Option Details:

H0 . . . . Goes on-hook (hang up the phone—disconnect from phone line). It also terminates any test that is in progress (AT&Tn commands), sets #CLS=0 (data mode), and sets #BDR=0 (autobaud).

H1 . . . . Goes off-hook (pick up phone—connect to phone line). For U.S. models, the modem will generally remain off-hook. With other models, the modem will return on-hook after a period of time determined by the value in register S7.

Result Codes:
OK. . . . . For n = 0 or 1.
ERROR . . . . Otherwise (see also S38).

In . . . . *(1) Identification or Inquiry*

"ATIn" returns information about the modem firmware and capabilities. The details of the report are ***MANUFACTURER DEPENDENT.***

Command Option Details:
I0 . . . . Displays the product ID code.
I1 . . . . Displays the precomputed ROM checksum.
I2 . . . . Computes checksum and compare to ROM's precomputed checksum (return OK if the same, ERROR if not).
I3 . . . . Displays firmware (ROM) revision number, model, and interface type (e.g., S for serial or P for parallel). In addition, some modems report an application code, such as D for desktop or L for low power (PCMCIA).
I4 . . . . Encrypted report of supported protocols (or OK).
I5 . . . . Firmware copyright information (alternatively, a country code parameter).
I6 . . . . Returns the modem data pump model and firmware revision.
I7 . . . . Reports the DAA code resulting from MCU interrogation of the DAA for auto DAA recognition. Examples: 000 for US or Canada, 016 for Japan, 033 for Belgium, 034 for Finland, 035 for France, 037 for Italy, 038 for Netherlands, 039 for Sweden, 040 for Switzerland, and 041 for UK.
I10 . . . . Supported features (encoded). For example, the SupraFax codes are

Code #	Description	Code #	Description
101	FAX CLASS 1	140	ITU-T V.32 bis (14400 bps data)
104	FAX CLASS 2	141	Silent Answer
105	FAX CLASS 2.0	142	Caller ID
115	ITU-T V.17 (14400 bps FAX)	143	MNP 10
120	WORLDWIDE SUPPORT	144	MNP cellular
121	ITU-T V.23	145	Autosync
122	ITU-T V.25	146	Enhanced Configuration Display
123	Blacklisting	160	Voice Processing
124	Access Security		

I99 . . . . An electronic serial number.

Result Codes:
OK. . . . . For successful completion of supported "n."
ERROR . . . . Otherwise.

In . . . . *(2) Identification or Inquiry (U.S. Robotics)*

"ATIn" returns information about the modem firmware and capabilities. The details of the report are manufacturer dependent; this list is for the Courier modem.

I0 . . . . Displays product ID code.
I1 . . . . Displays precomputed ROM checksum.
I2 . . . . Displays results of RAM test.
I3 . . . . Displays call duration or real time, selected by Kn (Courier modem).
. . . . Displays product type (Sportster modem).
I4 . . . . Displays current modem settings.
I5 . . . . Displays NVRAM.
I6 . . . . Displays Link Diagnostics.
I7 . . . . Displays product configuration.
I11 . . . . Displays V.FC link screen.

Result Codes:
OK. . . . . For successful completion of supported "n."
ERROR . . . . Otherwise.

ATI4 Example:

```
ATI4<return>
USRobotics Sportster 28800 V.34 Fax Settings...

   B0 E1 F1 M1 Q0 V1 X4 Y0
   BAUD=57600 PARITY=N WORDLEN=8
   DIAL=PULSE ON HOOK

   &A3 &B1 &C1 &D2 &G0 &H1 &I0 &K1
   &M4 &N0 &P0 &R2 &S0 &T5 &U0 &Y1

   S00=000  S01=000  S02=043  S03=013  S04=010  S05=008  S06=002
   S07=060  S08=002  S09=006  S10=007  S11=070  S12=050  S13=000
   S14=000  S15=000  S16=000  S17=000  S18=000  S19=000  S20=000
   S21=010  S22=017  S23=019  S24=000  S25=005  S26=000  S27=000
   S28=008  S29=020  S30=000  S31=000  S32=000  S33=000  S34=000
   S35=000  S36=014  S37=000  S38=000  S39=000  S40=000  S41=000
   S42=000  S43=200  S44=015  S45=000  S46=000  S47=000  S48=000
   S49=000  S50=000  S51=000  S52=000  S53=000  S54=064  S55=000
   S56=016  S57=000

   LAST DIALED #:

OK
```

Kn . . . .*Modem Clock Operation (Courier modems)*

"ATKn" used by the Courier modem to control the behavior of the real-time clock on the modem itself. The clock is capable of displaying either real-time or call-elapsed time. Kn is used by some modem manufacturers for a proprietary diagnostic.

Command Option Details:
K0 . . . . *ATI3* returns call duration (default for Courier modem).
K1 . . . . *ATI3* returns actual time in HH:MM:SS format (Courier modem).

Result Codes:
OK. . . . . For n = 0 or 1.
ERROR . . . . Otherwise.

L . . . .*Re-dial Last Number*

A dial command modifier—see *Dialing Command.*

Ln . . . .*Speaker Volume*

📁   *"ATLn"* sets the volume of the speaker built into the modem. The value, if valid, is written into register S22 bits 0 and 1.

Command Option Details:
L0 & L1 . . . . Low. On some modems, L0 is the lowest volume setting of four levels.
L2 . . . . **Medium (default).**
L3 . . . . High.

Result Codes:
OK. . . . . For n = 0, 1, 2, or 3.
ERROR . . . . Otherwise.

Mn . . . .*Speaker Control*

📁   *"ATMn"* controls when the speaker is on during a call. The value, if valid, is written into register S22 bits 2 and 3.

Command Option Details:
M0 . . . . Always OFF.
M1 . . . . **On until carrier is received (default).**
M2 . . . . Always On.
M3 . . . . On when answering.
                                     On after dialing and until a carrier is detected.

Result Codes:
OK. . . . . For n = 0, 1, 2, or 3.
ERROR . . . . Otherwise.

Nn . . . .*Connection Rate When Originating Call*

The "*ATNn*" command determines if a connection is to be made a specific rate or if the modem can connect at the fastest rate the remote modem, local modem, and telephone line can support.

Command Option Details:

N0 . . . . Connect ONLY at rate specified in register *S37* or if *S37=0*, connect at the rate at which the last AT command was issued. Only attempts connection protocols specified in the Bn command.

N1 . . . . **Connect Normally (at the highest rate supported by *both* modems).** The Bn command setting is generally ignored when *ATN1* is asserted.

Result Codes:

OK. . . . . For n = 0, or 1.

ERROR . . . . Otherwise.

On . . . .*Return to the ONLINE State from the COMMAND Mode*

The "*ATOn*" command applies to asynchronous operation only. The modem will be put back into the online mode (with or without retraining) from the COMMAND mode.

Command Option Details:

O0 . . . . Returns to ONLINE state.

O1 . . . . Requests a retrain and return to ONLINE state.

O2 . . . . Returns ONLINE and speed shift.

O3 . . . . Requests rate renegotiation sequence and return to ONLINE state.

Result Codes:

CONNECT XXXX . . . . If a connection is established.

NO CARRIER. . . . . If retrain is not successful in the time specified in S7.

ERROR . . . . Otherwise.

P . . . .*Set Pulse Dialing as Default for All Subsequent Dial Commands*

🗁   The P command may be issued either as a stand-alone command or as a dial command modifier. As a stand-alone command ("ATP"), the command causes the modem to assume that all subsequent dial commands are pulse dial. If the command is issued within a dial string ("*ATDP. . .*"), only the digits following the P in the current string are pulse dialed; subsequent dial commands revert to the previous dial mode with some modems. The command sets register S14 bit 5 on.

The *ATDP* commands may not be permitted in some countries.

Result Codes:

OK. . . . . If issued as a stand-alone command, i.e., not as part of a dial string.

Qn . . . .*Result Code Control*

🗁   "*ATQn*" determines how the modem will report result codes. ATQ0 clears register 14 bit 2 to 0, while ATQ1 sets the bit to 1.

Command Option Details:

Q0 . . . . **Enables result code reporting to DTE .**

Q1 . . . . Disables all result code reporting.

Q2 . . . . Enables reporting when originating; disables reporting when answering with the exception of OK and ERROR.

Result Codes:

OK. . . . . For n = 0 or 2.

none. . . . . For n = 1.

ERROR . . . . Otherwise.

R . . . .*Reverse to Answer Mode.*

A dial command modifier—see Dialing Command *ATDp*.

S$ . . . .*Help, S Register Functions*

The "*ATS$*" command is a U.S. Robotics specific command, it instructs the modem to display the definitions of its S-Registers.

Example:

```
ATS$<return>

HELP, S Register Functions (CTRL-S to Stop, CTRL-C to Cancel)

S0    Ring to Answer              S9    Carrier Detect Time (1/10 sec)
S1    Count # of Rings            S10   Carrier Loss Time (1/10 sec)
S2    Escape Code Char            S11   Touch Tone Spacing (msec)
S3    Carriage Return Char        S12   Escape Code Time (1/50 sec)
S4    Line Feed Char              S16   0=Data Mode
S5    Backspace Char                    1=Analog Loopback
S6    Wait Time/Dial Tone (sec)         2=Dial Test
S7    Wait Time/Carrier (sec)           4=Test Pattern
S8    Comma and Repeat Pause (sec)      5=Analog Loopback & Test Pattern

OK
```

SCr? . . . . Report Value in SC Register r

"*ATSCr?*" reads and displays the contents of register SCr where r is one of the defined registers in the modem.

Result Codes:
OK. . . . . For n = defined register numbers.
ERROR . . . . Otherwise.

SCr = n . . . . Write Value n into SC Register r

"*ATSCr=n*" writes the value n into register Sr where r is one of the defined register numbers in the modem.

Result Codes:
OK. . . . . For n = defined register numbers or no argument.
ERROR . . . . Otherwise.

Sr . . . . Set Register "r" as Default

"ATSr" defines the register used by the "*AT?*" and "*AT=v*" commands.

Sr.b = v . . . . Write Value v into S Register r Bit b

"*ATS.b=v*" writes the value "v" (limited to a 0 or 1) into bit position "b" register Sr where r is one of the bit mapped registers of the modem. The range of "v" is limited to the binary 0 (off) or 1 (on).

Result Codes:
OK. . . . . For b = 0 through 7 and r = defined register and v = 0 or 1.
ERROR . . . . Otherwise.

Sr? . . . . Report Value in S Register r

"*ATSr*" reads and displays the contents of register Sr where r is one of the defined registers in the modem. The range of r varies from modem to modem. If "r" is not specified (i.e., *ATS?*), the modem will report the last S-register addressed.

Result Codes:
OK. . . . . For n = defined register numbers.
ERROR . . . . Otherwise.

Sr = v . . . . Write Value v into S Register r

"*ATSr=v*" writes the value *v* into register Sr where r is one of the defined register numbers in the modem. Although some registers are read only registers (e.g., register S1), on some modems the result code of OK is reported, even though nothing was written into the register. The range of n is 0 to 255. If "r" is not specified (i.e., *ATS=v*), the modem will write to the last S-register addressed.

If the number "r" is beyond the allowable range of the S-Registers available, the modem will return the ERROR message. The value "v" may be stored MOD-256. If the result is outside the range permitted for a given S-Register, the values may still be stored, but functionally the lower and higher limits will be observed. Input and output values "v" are always in decimal form.

Due to country restrictions, some commands will be accepted, but the value may be limited and replaced by a maximum or minimum value.

Result Codes:
OK. . . . . For v = defined register numbers or no argument.
ERROR . . . . Otherwise.

Sx . . . . *Dial the Number Stored in Memory Location x*

> A dial command modifier—see Dialing Command *ATDp.*

T . . . . *Use Tone Dialing for all Subsequent Dial Commands*

📁 The "*ATT*" command may be issued either as a stand-alone command or as a dial command modifier. As a stand-alone command ("*ATT*"), this command causes the modem to assume that all subsequent dial commands are to DTMF tone dial. If the command is issued within a dial string ("*ATDT . . .*"), only the digits following the T in the current string are dialed as tones, subsequent dial commands may revert to the previous dial mode. The ATT command clears register S14 bit 5.

The ATT and ATDT commands may not be allowed by some country's PTT.

Result Codes:
OK. . . . . If issued as a stand-alone command, i.e., not as part of a dial string.

Un . . . . *User Help Screen*

> On the Silicon Systems 73D2404, the "*ATUn*" provides S-register, dial string, and data formats.

Vn . . . . *Result Code Format Select*

The "*ATVn*" command determines whether the modem sends result codes in the short form (numeric) or the long form (words) to the DTE. Short-form responses are ASCII values and are followed by carriage return. The long-form or verbose responses are preceded and followed by both a carriage return and a line feed. See Appendix C for a description of the individual result codes.

Command Option Details:
V0 . . . . Short form (Terse—numeric format) clears register S13 bit 3.
V1 . . . . **Long form** (Verbose format—word format) sets register S13 bit 3.

Result Codes:
0 . . . . If n = 0.
OK. . . . . If n = 1.
ERROR . . . . Otherwise.

W . . . . *Wait for Dial Tone*

> A dial command modifier—see Dialing Command *ATDp.*

Wn . . . . *Error-Correction Result Code Reporting*

📁 "*ATWn*" controls the reporting of error-correction call progress. These codes report the carrier rate and the error-correction protocol. The codes will be either numeric or verbose as selected by the "*ATVn*" command. The value of "n" is written into register S31 bits 2 and 3. Register *S95 = 0* will disable all extended result codes.

Command Option Details:

W0	**Report only**	**CONNECT XXXX (DTE rate).**
W1.	Reports:	
	Computer connection rate	CONNECT XXXX (DTE rate)
	Error-Correction Mode	PROTOCOL: XXXX
	Connection rate as	CARRIER XXXX (DCE rate)
W2.	Reports connection rate as	CONNECT XXXX (DCE rate)

Note: The DTE rate is the computer/modem speed, and the DCE rate is the modem/modem speed.

Result Codes:
OK. . . . . If n = 0, 1, or 2.
ERROR . . . . Otherwise.

Xn . . . . *(1) Dialing Type and Connect Result Codes*

📁 The "*ATXn*" command determines how the modem responds to telco call progress tones and how it displays codes for CONNECT messages. When the modem ignores dial tone, it waits the period of time specified in the *S6* register; then it dials whether dial tone is present or not. This is called blind dialing. When the modem monitors dial tone, it must detect dial tone within 5 seconds of going off-hook, or it will report NO DIAL TONE. The modem will send the message BUSY when it detects a busy tone after dialing.

If the modem is in facsimile mode (+FCLASS ≠ 0), the only message sent to indicate a connection is CONNECT, without a speed indication.

Command Option Details:

		Xn Settings							
	Result Codes	X0	X1	X2	X3	X4	X5	X6	X7
0	OK	X	X	X	X	X			
1	CONNECT	X	X	X	X	X			
2	RING	X	X	X	X	X			
3	NO CARRIER	X	X	X	X	X			
4	ERROR	X	X	X	X	X			
5	CONNECT XXXX		X		X	X			
6	NO DIAL TONE			X		X			
7	BUSY				X	X			
8	NO ANSWER (replaces NO CARRIER if the @ is included in the dial string)	X	(X)	(X)	X	X			
–	CONNECT XXXX full range enabled		X	X	X	X			

Call Progress Tone Monitoring	X0	X1	X2	X3	X4	X5	X6	X7
Blind Dial	X	X		X				
Dial Tone detect			X		X			
Busy Tone detect				X	X			
Register S22 value (bits 6, 5, and 4)	.000....	.100....	.101....	.110....	.111....			

Result Codes:

OK. If n = value defined for the modem. The value is written into register S22 bits 6, 5, and 4.
ERROR . . . . Otherwise.

Xn . . . . ***(2) Dialing Type and Connect Result Codes (U.S. Robotics)***

The "*ATXn*" command determines how the modem responds to telco call progress tones and how it displays codes for CONNECT messages. When the modem ignores dial tone, it waits the period of time specified in the *S6* register; then it dials whether dial tone is present or not. This is called blind dialing. When the modem monitors dial tone, it must detect dial tone within 5 seconds of going off-hook, or it will report NO DIAL TONE. The modem will send the message BUSY when it detects a busy tone after dialing.

Command Option Details:

		Xn Settings							
	Result Codes	X0	X1	X2	X3	X4	X5	X6	X7
0	OK	X	X	X	X	X	X	X	X
1	CONNECT	X	X	X	X	X	X	X	X
2	RING	X	X	X	X	X	X	X	X
3	NO CARRIER	X	X	X	X	X	X	X	X
4	ERROR	X	X	X	X	X	X	X	X
5	CONNECT 1200		X	X	X	X	X	X	X
6	NO DIAL TONE			X		X		X	X
7	BUSY				X	X	X	X	X
8	NO ANSWER (if @ in dial string)				X	X	X	X	X
9	RESERVED								
10	CONNECT 2400		X	X	X	X	X	X	X
11	RINGING						X	X	X
12	VOICE						X	X	
13	CONNECT 9600		X	X	X	X	X	X	X
18	CONNECT 4800		X	X	X	X	X	X	X
20	CONNECT 7200		X	X	X	X			
21	CONNECT 12000		X	X	X	X			
25	CONNECT 14400		X	X	X	X			
43	CONNECT 16800 (on 28.8 modem)		X	X	X	X			
85	CONNECT 19200 (on 28.8 modem)		X	X	X	X			
91	CONNECT 21600 (on 28.8 modem)		X	X	X	X			
99	CONNECT 26400 (on 28.8 modem)		X	X	X	X			
103	CONNECT 16800 (on 28.8 modem)		X	X	X	X			
107	CONNECT 28800 (on 28.8 modem)		X	X	X	X			
151	CONNECT 31200 (on 33.6 modem)		X	X	X	X			
155	CONNECT 33600 (on 33.6 modem)	X	X	X	X	X			
180	CONNECT 33333 (on 56K modem)	X	X	X	X	X			

	Result Codes	X0	X1	X2	X3	X4	X5	X6	X7
184	CONNECT 37333 (on 56K modem)	X	X	X	X	X			
188	CONNECT 41333 (on 56K modem)	X	X	X	X	X			
192	CONNECT 42666 (on 56K modem)	X	X	X	X	X			
196	CONNECT 44000 (on 56K modem)	X	X	X	X	X			
200	CONNECT 45333 (on 56K modem)	X	X	X	X	X			
204	CONNECT 46666 (on 56K modem)	X	X	X	X	X			
208	CONNECT 48000 (on 56K modem)	X	X	X	X	X			
212	CONNECT 49333 (on 56K modem)	X	X	X	X	X			
216	CONNECT 50666 (on 56K modem)	X	X	X	X	X			
220	CONNECT 52000 (on 56K modem)	X	X	X	X	X			
224	CONNECT 53333 (on 56K modem)	X	X	X	X	X			
228	CONNECT 54666 (on 56K modem)	X	X	X	X	X			
232	CONNECT 56000 (on 56K modem)	X	X	X	X	X			
236	CONNECT 57333 (on 56K modem)	X	X	X	X	X			

Call Progress Tone Monitoring	X0	X1	X2	X3	X4	X5	X6	X7
Blind Dialing	X	X		X		X		
Dial Tone			X		X		X	X
Busy Tone				X	X	X	X	X
Adaptive Dialing			X	X	X	X	X	X
Wait for second dial tone (W)			X		X	X	X	X
Wait for answer (@)				X	X	X	X	X
Fast dial			X		X		X	X

Result Codes:

OK. . . . . If n = value defined for the modem.
ERROR . . . . Otherwise.

Yn . . . . *(1) Long Space Disconnect*

"ATY*n*" determines if the local modem will disconnect when it receives a 1.6-second (or greater) BREAK (or "space") signal from the remote mode. It also determines whether the modem will send a BREAK to the remote modem when either an *H0* command is received or an ON-to-OFF transition of DTR occurs while *&D2* is asserted.

Command Option Details:

Y0 . . . . **Disable. The modem does not respond to break signal** (default).
Y1 . . . . Enable. The modem will disconnect if it receives a BREAK for 1.6-seconds or greater. It will send a 4-second BREAK before disconnecting due to either an ON-to-OFF transition of DTR (with &D2 asserted) or upon receiving an H0 command.

Result Codes:

OK. . . . . If n = 0 or 1.
ERROR . . . . Otherwise.

Yn . . . . *(2) Power-on Reset Profile (U.S. Robotics)*

Command Option Details:

Y0 . . . . **Default profile "0" setting in NVRAM is** (default).
Y1 . . . . Default profile "1" settings in NVRAM.

Zn . . . . *Soft Reset and Load Stored Profile n*

"AT*Zn*" causes the modem to disconnect if it is connected and perform a warm start. The command must be the last command on the AT command line. The reset actions are:

1. Clears the serial port buffers.
2. Sets the data rate and parity to match the local DTE.
3. Restores the active configuration with the user profile stored in NVRAM n or with the factory defaults if no NVRAM exists.
4. Resets all voice parameters to their default values.
5. Sets #BDR=0 (autobaud).
6. Goes on-hook if #CLS=8.

Command Option Details:

Z0 . . . . Resets modem and load user profile 0.

. . . . Resets modem and load user profile specified by *Yn* (U.S. Robotics).

Z1 . . . . Resets modem and load user profile 1.

. . . . Resets modem and load user profile 0 (U.S. Robotics).

Z2 . . . . Resets modem and load user profile 1 (U.S. Robotics).

Z3 . . . . Resets modem and load factory profile 0 (same as *&F0*).

Z4 . . . . Resets modem and load factory profile 1 (same as *&F1*).

Z5 . . . . Resets modem and load factory profile 2 (same as *&F2*).

Result Codes:

OK. . . . . If n = 0 or 1.

ERROR . . . . Otherwise.

"Hn . . . .*V.42 bis Compression Control*

The *"AT"Hn"* command determines if and where data compression is used.

Command Option Details:

"H0 . . . . Disables V.42 bis.

"H1 . . . . Enables V.42 bis only when transmitting.

"H2 . . . . Enables V.42 bis only when receiving.

"H3 . . . . Enables V.42 bis for both transmitting and receiving.

"On . . . .*V.42 bis String Length*

"AT"On" defines the number of characters in the string. Default n = 16.

#BDR <? | =? | =n> . . . .*Select Baud Rate (default = 0)*

There are three forms of the #BDR command: one form (AT#BDR=?) allows the DTE to determine the range of acceptable communication rates; another form (AT#BDR?) reports the current modem setting; and the last form (AT#BDR=n) allows the DTE to either select a particular baud rate and turn off autobaud or to instruct the modem to determine the rate itself (autobaud).

"AT#BDR=0" enables autobaud detection on the DTE interface.

"AT#BDR=n" with n > 0 causes the modem to select a specific DTE/DCE baud rate without further speed sensing on the interface. When a valid #BDR=n command is entered (i.e., $1 \leq n \leq 48$), the OK result code is sent at the current assumed speed. After the OK has been sent, the modem switches to the speed indicated by the #BDR=n command it has just received. The selected baud rate is n × 2400 bps.

When in Online Voice Command Mode and the #BDR setting is nonzero, the modem supports a full-duplex DTE interface. This means that the DTE can send commands to the modem at any time, even if the modem is in the process of sending a shielded code indicating DTMF detection to the DTE.

Command Options:

#BDR?. . . . . Returns the current setting of the #BDR command as an ASCII decimal value in result code format.

#BDR=? . . . . Returns a message indicating the range of speeds that are supported.

#BDR=0 . . . . **Enables autobaud detection on the DTE interface** (Default).

#BDR=n . . . . Where 1<n<48. Sends OK message at current speed, then switches to the new speed defined by n*2400 bps unless and until another #BDR=n command is received. Autobaud is disabled, and the character format is maintained at the format most recently detected. When #BDR has been set nonzero, the modem employs the S30 Disconnect Inactivity timer, and this timer starts at the point where #BDR is set nonzero. If this period expires (nominally 60 seconds) with no activity on the DTE interface, the modem reverts to #BDR=0, #CLS=0, and #VLS=0.

Result Codes:

n

OK. . . . . For AT#BDR?

OK. . . . . For AT#BDR=n with $0 \leq n \leq 48$.

ERROR . . . . Otherwise.

#CID <? | =? | =n> . . . .*Caller ID Mode (default = 0)*

There are three forms of the #CID command: one form (AT#CID=?) allows the DTE to determine the range of acceptable values; another form (AT#CID?) reports the current modem setting; and the last form (AT#CID=n) allows the DTE to set the modem in a particular Caller ID mode.

The "*AT#CID=n*" command determines if and how caller ID information is reported to the local DTE. Caller ID works in conjunction with telephone company equipment to provide the date, time, phone number, and possibly the name of the calling party, OR if the calling party has blocked caller ID, a message stating the information is restricted may be reported. This information is provided by the telephone company between the first and second ring signal while the modem is still on-hook.

Command Option:

#CID=0 . . . . **Disables Caller ID** (Default).

#CID=1. . . . . Enables Formatted Caller ID Mode. The *caller ID* information is received in either a single data message (SDM) (sometimes called single-page mode) or a multiple data message (MDM) (sometimes called multiple-page mode) format. The format for each of mode is:

SDM - Single Data Message (Single Page Mode):

> RING
> DATE=1225
> TIME=1801
> NMBR=2135551212
> NAME=INFORMATION (optional, not always supported)
> RING

Where:

DATE=MMDD and MM is the month (01-12) & DD day of the month (01-31).

TIME=HHMM and HH is the hour (00-23) & MM is the minute (00-59).

NMBR=number code OR a statement. The number code is either the area code, the office code and the subscriber code, or a unique number to the individual subscriber. If the subscriber has requested that the number not be given out, a "P" indicating Private is displayed. If the calling number is in an area not providing the feature (e.g., out of service), an "O" will be displayed.

NAME=The listed subscriber name. (Optional parameter and not always supported.)

MDM—Multiple Data Mode (Multiple Page Mode):

> *DATE=MMDD* and MM is the month (01-12) & DD day of the month (01-31).
> *TIME=HHMM* and HH is the hour (00-23) & MM is the minute (00-59).
> *MESG=Message string with header.* The message string's format is

> *CCLLd0d1d2d3d4d5d6d7d8d9CS*

where:

CC is the message type identifier (03),
LL is the length of the data in hexadecimal,
dn is the data (dialed digits) in hexadecimal, and
CS is the checksum.

#CID=2. . . . . Enables Raw Mode (ASCII printable HEX number).

The modem presents all information and packet control information (including SDM, MDM, or call waiting packets) found in the message except channel seizure information (leading U's). The packet is presented in ASCII printable hex numbers, without inserted spaces, line feeds, or formatting between bytes or words of the packet. The modem does not detect the checksum of the packet.

Example of the un-formatted form

> RING
>
> 04123033323232343035393134353531323334 35
>
> RING

#CID? . . . . Returns the current setting (0, 1, or 2) of the #CID parameter as an ASCII decimal value in result code format.

#CID=?. . . . . Returns the range of allowable values.

Result Codes:

n
OK. . . . . For AT#CID? (n = 0, 1, or 2).
0, 1, 2
OK. . . . . For AT#CIS=?
OK. . . . . For AT#CID=n with n = 0, 1, or 2.
ERROR . . . . Otherwise.

#CLS <? | =? | =n> . . . .*Select Data, Fax, or Voice/Audio (default = 0)*

There are three forms of the *#CLS* command: one form (AT#CLS=?) allows the DTE to determine the range of acceptable values; another form (AT#CLS?) reports the current modem setting; and the last form (AT#CLS=n) allows the DTE to set the modem in a specified Data, Fax, or Voice/Audio Mode from operation in any mode.

Command Options:

#CLS? . . . . Returns the current setting (0, 1, 2, or 8) of the #CLS parameter as an ASCII decimal value in result code format.

#CLS=? . . . . Returns the range of values allowable, i.e., "0, 1, 2, or 8."

#CLS=0 . . . . **Sets Data mode.** This is similar to setting +FCLASS=0. It instructs the modem to act like a data modem on subsequent answer or originate operations. (When a disconnect or inactivity time out in the non-autobaud mode is detected, the modem automatically sets the #CLS setting to 0 and disconnects from the telephone line. This ensures that the modem is always in a known state despite disorderly DTE behavior.)

#CLS=1 . . . . Sets Class 1 fax. This is similar to setting +FCLASS=1. It instructs the modem to be a Class 1 fax modem. Once this is set, either the +FAA or the +FAE command can be used to force subsequent answers to be Class 1 adaptive answers.

#CLS=2 . . . . Sets Class 2 fax. This is similar to setting +FCLASS=2. It instructs the modem to be a Class 2 fax modem. Once this is set, the +FAA command can be used to force subsequent answers to be Class 2 adaptive answers.

#CLS=8 . . . . Sets Voice/Audio Mode. This is the setting which the DTE uses to effect directed or adaptive answer or originate sequences involving voice modes. All calls initialized by *AT#CLS=8* result in the modem in Online Voice Command Mode after answer or successful call progress.

Result Codes:

n

OK. . . . . For AT#CLS?

0, 1, 2, 8

OK. . . . . For AT#CLS=?

OK. . . . . For AT#CLS=n, if n = 0, 1, 2, or 8.

ERROR . . . . Otherwise.

#MDL? . . . . *Identify Model*

The AT#MDL? command causes the modem to report the model number of the chipset. It is similar to the ATI0 command.

Command Option:

#MDL? . . . . Requests model number.

Result Codes:

<model> . . . . RC144ACi for example.

OK

#MFR? . . . . *Identify Manufacturer*

The AT#MFR? command causes the modem to report the manufacturer of the chipset.

Command Option:

#MFR?. . . . . Request manufacturer.

Result Codes:

<manufacturer> . . . . Rockwell, for example.

OK

#REV? . . . . *Identify Revision Level*

The AT#REV? command causes the modem to report the revision level of the chipset.

Command Option:

#REV?. . . . . Request revision level.

Result Codes:

<xxxx-xxxx-xxx yyy>

OK. . . . . x's represent the part number and y's the revision at the time of manufacture.

#SPK=<mute>,<spkr>,<mic> . . . . *Speakerphone (default 1, 5, 1)*

The #SPK command can be used to control the microphone state (mute or on), adjust the speaker volume, and set microphone gain. The #SPK parameters are valid only after the modem has entered the Voice Online (VCON) mode while in the #VLS=6 setting (after a VCON during originate or answer). The modem will respond "ERROR" otherwise.

<MUTE>, <SPKR>, AND <MIC> PARAMETER VALUES

	<mute> Mode	<spkr> Output Attenuation	<mic> Input Gain
0	Mute microphone	0.0 dB	0 dB
1	**Microphone on**	2.0 dB	**6 dB**
2	Room monitor mode (mic on, mic gain to 50 dB, spkr off)	4.0 dB	9.5 dB
3		6.0 dB	12 dB
4		8.0 dB	
5		**10.0 dB**	
6		12.0 dB	
7		14.0 dB	
8		16.0 dB	
9		18.0 dB	
10		20 dB	
11		22 dB	
12		24 dB	
13		26 dB	
14		28 dB	
15		30 dB	
16		Muted	

Command Options:

#SPK=[<mute>],[<spkr>],[<mic>]

. . . . Sets one or more speakerphone parameters; if a parameter is omitted, a place holder comma must be used. For example,

AT#SPK=0,, ... Mutes the microphone without changing speaker volume or microphone gain.

AT#SPK=2,, ... Sets the microphone gain to maximum (50 dB) and the speaker off.

AT#SPK=,6,,. . . . . Changes the speaker volume to 6 (12 dB attenuation) without affecting microphone settings.

AT#SPK=,6,2 . . . . Changes the speaker volume to 6 (12 dB attenuation) and the microphone gain to 2 (9.5 dB).

Result Codes:

OK. . . . . For defined parameter values.

ERROR . . . . Otherwise.

#TL = n . . . .*Audio Output Transmit Level*

When the modem is configured for Voice/Audio Mode with the #CLS=8 command, the *AT#TL=n* command allows adjustment of the transmit level of the data pump audio output. It affects the playback level of the handset, speaker, and telephone line. The command may be accepted by some chipsets for compatibility; however, no action may occur.

Command Option:

#TL=n. . . . . n=0000 to 7FFF if the handset or speaker is selected.

　　　　　　　　　　　　Default: 7FFF, if local handset or speaker is selected or

　　　　　　　　　　　　　　　3F44, if the phone line is selected

Result Codes:

OK. . . . . For n=0000 to 7FFF.

#VBQ? . . . .*Query Buffer Size*

When the modem is configured for Voice/Audio Mode with the #CLS=8 command, the AT#VBQ command returns the size of the modem's voice transmit and voice receive buffers.

Command Option:

#VBQ?. . . . . Requests the modem to return the voice transmit and voice receive buffer size.

Result codes:

n

OK. . . . . The size of the modem voice transmit and voice receive buffers in ASCII decimal.

#VBS <? | =? | =n> . . . .*Bits per Sample (ADPCM or PCM)(default = 4)*

There are three forms of the *#VBS* command: one form (*AT#VBS=?*) allows the DTE to determine the range of acceptable values; another form (*AT#VBS?*) reports the current modem setting; and the last form (*AT#VBS=n*) allows the DTE to select the degree of ADPCM voice compression or format of linear PCM to be used. These commands are active when the modem is configured for Voice/Audio Mode with the #CLS=8 command.

Command Options:

#VBS?. . . . . Returns the current setting of the #VBS command as an ASCII decimal value in result code format.

#VBS=? . . . . Returns "2" or "4," which are the ADPCM compression bits/sample rates available, or "8" which is the PCM bits/sample rate. The 2 and 4 bits/sample rates are correlated with the #VCI query command response, which provides the single compression method available. The "8" represents the 8-bit unsigned linear PCM format.

#VBS=2 . . . . Selects 2 bits per sample ADPCM compression (valid for #VSR=7200).

#VBS=4 . . . . Selects 4 bits per sample ADPCM compression (valid for #VSR=7200). Default.

#VBS=8 . . . . Selects 8 bits per sample PCM (valid for #VSR=7200 or #VSR=11025).

#VBS=16 . . . . Selects 16 bits per sample PCM (valid for #VSR=7200 or #VSR=11025).

Result Codes:

n

OK. . . . . For AT#VBS?. n = 2, 4, or 8.

2, 4, 8

OK. . . . . For AT#VBS=?

OK. . . . . For AT#VBS=n, if n=2, 4, or 8.

ERROR . . . . Otherwise.

#VBT <? | =? | =n> . . . . *Beep Tone Timer (default = 10)*

There are three forms of the *#VBT* command: one form (*AT#VBT=?*) allows the DTE to determine the range of acceptable values; another form (*AT#VBT?*) reports the current modem setting; and the last form (*AT#VBT=n*) allows the DTE to define the time period, in tenths of a second up to 4 seconds (40), which is used by the modem as the DTMF or fixed tone duration for generating tones while in Online Voice Command Mode. These commands are active when the modem is configured for Voice/Audio Mode with the #CLS=8 command.

Command Options:

#VBT?. . . . . Returns the current #VBT parameter value as an ASCII decimal value.

#VBT=? . . . . Returns the message, "0-40."

#VBT=0 . . . . Disables the tone generation capability.

#VBT=n . . . . Sets tone duration time from 0.1 to 4.0 seconds (n=1 to 40).

Result Codes:

n

OK. . . . . For AT#VBT?

0-40

OK. . . . . For AT#VBT=?

OK. . . . . For AT#VBT=n, if n = 0 – 40.

ERROR . . . . Otherwise.

#VCI? . . . . *Identify Compression Method (ADPCM)*

When the modem is configured for Voice/Audio Mode with the #CLS=8 command, the *AT#VCI* command provides a unique character string which may be used for identifying the proprietary compression method of the modem (Rockwell ADPCM).

Command Option:

#VCI? . . . . Requests the Compression Identifier string, which identifies the proprietary voice data encoding of the modem.

Result Codes:

<string>

OK. . . . . Returned after the Compression Identifier string.

ERROR . . . . Returned if attempting to assign a parameter value.

#VGT <? | = ? | = n> . . . . *Playback Volume in Command State (default = 153)*

There are three forms of the *#VGT* command: one form (*AT#VGT=?*) allows the DTE to determine the range of acceptable values; another form (AT#VGT?) reports the current modem setting; and the last form (*AT#VGT=n*) allows the DTE to set the playback volume of the headphone and speaker outputs via the on-board digital potentiometer while in the command state. These commands are active when the modem is configured for Voice/Audio Mode with the #CLS=8 command.

Command Options:

#VGT=? . . . . Returns "128-228" if the external volume control is present; otherwise "128." The lowest value in the range means the lowest transmit level, while the highest in the range means the highest transmit level.

#VGT?. . . . . Returns the current #VBT parameter value as an ASCII decimal value. 153 (default).

#VGT=n . . . . Sets the playback level if 128 ≤ n ≤ 288.

Result Codes:

n
OK. . . . . For AT#VGT?
128-288
OK. . . . . For AT#VGT=?
OK. . . . . For AT#VGT=n, if n=128 to 228.
ERROR . . . . Otherwise.

#VLS <? | =? | =n> . . . .*Voice Line Select (ADPCM or PCM)(default = 0)*

There are three forms of the ***#VLS*** command: one form (***AT#VLS=?***) allows the DTE to determine the range of acceptable values; another form (***AT#VLS?***) reports the current modem setting; and the last form (***AT#VLS=n***) can be used by the DTE to select which devices are routed through the modem. These commands are active when the modem is configured for Voice/Audio Mode with the #CLS=8 command.

Command Options:

#VLS? . . . . Returns the current #VLS parameter value as an ASCII decimal value.

#VLS=? . . . . Requests a report of the device types available for selection. The response is a number from 0 to 9 corresponding to each device position number.

#VLS=0 . . . . **Instructs the modem that when entering any of the three voice operating submodes (Online Command, Transmit, or Receive), the telephone line interface should be routed through the modem.** The OK response is sent to the DTE, and any previous connection is lost (i.e., the modem ends up on-hook as a result of issuing this command to connect to the telephone line). **Default mode.**

#VLS=1 . . . . Instructs the modem to route only the handset through the modem. This setting can be chosen before recording a greeting message.

#VLS=2 . . . . Instructs the modem to route only the speaker through the modem. This setting can be chosen before playing back any message. The modem immediately switches to Online Voice Command Mode, and the VCON response is generated for completeness. However, since this is an output only device, nothing of consequence can happen until the DTE sends the #VTX command.

#VLS=3 . . . . Instructs the modem that only the auxiliary input device (microphone) should be routed through the modem. This setting can be chosen before recording a greeting message.

#VLS=4 . . . . The same as #VLS=0, except that the modem enables the internal speaker as well as the telephone line/handset circuit.

#VLS=5 . . . . Selects telephone emulation mode while in #CLS=8 mode. Speakerphone operation is entered during Voice Online (VCON) mode after completing dialing (ATD) or answering (ATA). #VLS=5 will respond ERROR if speakerphone is supported.

#VLS=6 . . . . Selects speakerphone mode while in #CLS=8 mode. Telephone emulation operation is entered during Voice Online (VCON) mode after completing dialing (ATD) or answering (ATA). #VLS=6 will respond ERROR if speakerphone is not supported.

#VLS=7 . . . . Mutes the local handset by switching the handset or speakerphone out of the telephone line path if in #VLS=0 or #VLS=6 mode. To unmute, the host software should issue AT#VLS=0 or AT#VLS=6 to return to phone line or speakerphone mode, respectively. On hardware designs which incorporate the audio codec, #VLS=7 can be used to route the codec's output through the phone line to provide music-on-hold feature. (Valid after Voice Online mode.)

#VLS=8 . . . . Enables the Caller ID relay to allow recording of conversation when using a handset by routing the signal to the audio codec (if populated). To deactivate the Caller ID relay, the host should issue an AT#VLS=0 or AT#VLS=6 to return to phone line or speakerphone mode, respectively. (Valid after Voice Online mode.)

#VLS=9 . . . . Routes the handset to the sound codec to allow recording/playback of audio (e.g., greeting messages) through the local handset.

Result Codes:

n
OK. . . . . For AT#VLS?
OK. . . . . For AT#VLS=n and if n = 0 to 9 as supported by the modem model.
VCON . . . . If device selected does not connect to the telephone line. (A speaker is such a device, but a telephone line with speaker ON is not such a device, and generates OK.)
ERROR . . . . If n does not equal 0 to 9 or if the telephone line or local handset is already off-hook.

#VRA <? | =? | =n> . . . .*Ringback Goes Away Timer (default = 70)*

There are three forms of the ***#VRA*** command: one form (***AT#VRA=?***) allows the DTE to determine the range of acceptable values; another form (***AT#VRA?***) reports the current modem setting; and the last form (***AT#VRA=n***) is used when originating a call to set the "Ringback Goes Away" timer value to a value between 0 and 25.5 seconds (in 100 ms units and measured from when the ringback cadence stops once detected). These commands are active when the modem is configured for Voice/Audio Mode with the #CLS=8 command.

Every time a ringback cycle is detected, this timer is reset. If ringback is not detected within the period set, the modem assumes that the remote station has picked up the line and switches to Online Voice Command Mode.

Command Options:

#VRA?. . . . . Returns the current #VRA parameter value as an ASCII decimal value.

#VRA=?. . . . . Returns the message, "0-255".

#VRA=0. . . . . Turns off the "ringback goes away timer."
                                After one ringback ring, the dialing modem sends VCON and enters Online Voice Command Mode immediately.

#VRA=n. . . . . Where n defines the period without ringback (after at least one ringback has been detected) in 100 ms units.

Result Codes:

n

OK. . . . . For AT#VRA?

0-255

OK. . . . . For AT#VRA=?

VCON . . . . FOR AT#VRA=0 and after one ringback ring.

OK. . . . . For AT#VRA=n, if n = 0 − 255 (corresponding to 0 to 25.5 seconds).

ERROR . . . . Otherwise.

#VRN . . . .*Ringback Never Came Timer (default = 100)*

There are three forms of the *#VRN* command: one form (*AT#VRN=?*) allows the DTE to determine the range of acceptable values; another form (*AT#VRN?*) reports the current modem setting; and the last form (*AT#VRN=n*) is used when originating a call to set the "Ringback Never Came" timer value to a value between 0 and 25.5 seconds (in 100 ms units and measured from the completion of dialing). These commands are active when the modem is configured for Voice/Audio Mode with the #CLS=8 command.

If ringback is not detected within the defined period, the modem assumes the remote has picked up the line and switches to Online Voice Command Mode.

Command Options:

#VRN?. . . . . Returns the current #VRN parameter value as an ASCII decimal value.

#VRN=?. . . . . Returns the message, "0-255".

#VRN=0. . . . . Turns off the "ringback never came timer." After dialing, the modem sends VCON and immediately enters Online Voice Command Mode.

#VRN=n. . . . . Where n defines the period without ringback after dialing in 100 ms units.

Result Codes:

n

OK. . . . . For AT#VRN?

0-255

OK. . . . . For AT#VRN=?

VCON . . . . For AT#VRN=0 and after dialing.

OK. . . . . For AT#VRN=n, if n = 0 − 255 (corresponding to 0 to 25.5 seconds.)

ERROR . . . . Otherwise.

#VRX . . . .*Voice Receive Mode*

The *AT#VRX* action command is only valid if the modem is in the Online Voice Command Mode (indicated by a previously issued VCON message) and is the switch to the Voice Receive Mode. This command is used when a voice file is to be received from the line, microphone, or handset. The #VLS command should have been previously issued to select the input source. The command is valid for both ADPCM and PCM.

Command Option:

#VRX . . . . Causes the modem to switch to voice/audio mode.

Result Codes:

CONNECT . . . . When voice/audio transfer from modem to DTE can begin.

ERROR . . . . If #VLS=0 and not connected to any input device.

#VSD <? | =? | =n> . . . .*Enable Silence Deletion (default = 0)*

There are three forms of the *#VSD* command: one form (*AT#VSD=?*) allows the DTE to determine the range of acceptable values, another form (*AT#VSD?*) reports the current modem setting, and the last form (*AT#VSD=n*) is used to enable or disable the removal of quiet (no speech) periods in ADPCM compression. These commands are active when the modem is configured for Voice/Audio Mode with the #CLS=8 command.

On many modems, this command provides no function other than command response compatibility.

Command Options:

#VSD? . . . . Returns the current #VSD parameter value as an ASCII decimal value.

#VSD=? . . . . Returns the message, "0,1."

#VSD=0 . . . . Disables silence deletion.

#VSD=1 . . . . Enables silence deletion if the silence sensitivity setting #VSS is nonzero. The aggressiveness of the silence deletion is controlled by #VSS.

Result Codes:

n
OK . . . . For AT#VSD?
0, 1
OK . . . . For AT#VSD=?
OK . . . . For AT#VSD=n, if n = 0 or 1.
ERROR . . . . Otherwise.

#VSK <? | =? | =n> . . . .*Buffer Skid Setting (default = 255)*

There are two forms of the #VSK command: one form (AT#VSK=?) which reports 255 and the form (AT#VSK?) reports the current modem setting. These commands are active when the modem is configured for Voice/Audio Mode with the #CLS=8 command.

The command queries and sets the number of bytes of spare space, after the XOFF threshold is reached, in the modem's buffer during Voice Transmit Mode. This equates to the "skid" spare buffer space or the amount of data the DTE can continue to send after being told to stop sending data by the modem, before the modem voice transmit buffer overflows.

Command Options:

#VSK? . . . . Returns the current #VSK parameter value as an ASCII decimal value.

#VSK=? . . . . Returns the message, "255."

Result Codes:

OK . . . . If n = 255.
ERROR . . . . Otherwise.

#VSP <? | =? | =n> . . . .*Silence Detection Period (default = 55)*

There are three forms of the *#VSP* command: one form (*AT#VSP=?*) allows the DTE to determine the range of acceptable values; another form (*AT#VSP?*) reports the current modem setting; and the last form (*AT#VSP=n*) sets the Voice Receive Mode silence detection period (inactivity timer) to value a value between 0 and 25.5 seconds. These commands are active when the modem is configured for Voice/Audio Mode with the #CLS=8 command.

The parameter, in 100 ms units, can be used when receiving voice data. This is an amount of time, which if elapsed without receiving any ADPCM data, causes the modem to send the <DLE>s or <DLE>q codes after ensuring that the buffer is empty.

The modem determines what constitutes silence. This involves monitoring and debouncing the modem value for average energy. If this debounced value is less than an arbitrary threshold constituting the modem's definition of silence for a period greater than that defined by the #VSP setting, the modem sends the <DLE>q or <DLE>s shielded code to the DTE.

Command Options:

#VSP? . . . . Returns the current #VSP parameter value as an ASCII decimal value.

#VSP=? . . . . Returns the message, "0-255."

#VSP=0 . . . . Turns off the silence period detection timer.

#VSP=n . . . . Where 1 ≤ n ≤ 255 defines the period (0.1 to 25.5 seconds) without received voice data in 100 ms units.

Result Codes:

n
OK . . . . For AT#VSP?
0-255
OK . . . . For AT#VSP=?
OK . . . . For AT#VSP, if n = 0–255.
ERROR . . . . Otherwise.

#VSR <? | =? | =n> . . . .*Sampling Rate Selection (default = 7200)*

There are three forms of the *#VSR* command: one form (*AT#VSR=?*) allows the DTE to determine the range of acceptable values; another form (*AT#VSR?*) reports the current modem setting; and the last form (*AT#VSR=n*) is used along with the bits per sample (#VBS) command, determines the necessary DTE interface speed to transmit and receive in Voice/Audio Mode. These commands are active when the modem is configured for Voice/Audio Mode with the #CLS=8 command.

Command Options:

#VSR?................ Returns the current #VSR parameter value as an ASCII decimal value.
#VSR=? .... Returns the message "7200, 11025."
#VSR=7200 .... **Selects 7200 Hz sampling rate** (valid for ADPCM or PCM).
#VSR=11025 .... Selects 11.025 kHz sampling rate (valid for PCM only).

Result Codes:

n
OK................... For AT#VSR?
7200, 11025
OK................... For AT#VSR=?
OK................... For AT#VSR=n, if n=7200, or if n=11025 and #VBS=8
ERROR .... Otherwise.

#VSS <? | =? | =n>..*Silence Detection Tuner (default = 0)*

There are three forms of the *#VSS* command: one form (*AT#VSS=?*) allows the DTE to determine the range of acceptable values; another form (*AT#VSS?*) reports the current modem setting; and the last form (*AT#VSS=n*) enables or disables the Voice Receive mode silence detection, and controls the sensitivity employed by the modem in ADPCM compressing periods of silence. These commands are active when the modem is configured for Voice/Audio Mode with the #CLS=8 command.

Command Options:

#VSS?................ Returns the current #VSS parameter value as an ASCII decimal value.
#VSS=? .... Returns the message "0,1,2,3."
#VSS =0.............. **Disables silence detection when in Voice Receive Mode** (default).
#VSS =1.............. Least sensitive setting. When this command is received by the modem, the system is configured to a state which is least likely to detect and compress periods of silence, but still able to do so if the line is really quiet.
#VSS =2.............. Midrange setting. When this command is received by the modem, the system is configured to a state which is likely to be the best overall compromise on normal telephone lines.
#VSS =3.............. Most sensitive setting. When this command is received, the system is configured so that it is most likely to detect and compress silent periods.

Result Codes:

n
OK................... For AT#VSS?
0, 1, 2, 3
OK................... For AT#VSS=?
OK................... For AT#VSS=n, if n = 0–3.
ERROR .... Otherwise.

#VTD <? | =? | =i,j,k> ....*DTMF Reporting Capability (default 3Fh, 3Fh, 3Fh)*

There are three forms of the *#VTD* command: one form (*AT#VTD=?*) allows the DTE to determine the range of acceptable values; another form (*AT#VTD?*) reports the current modem setting; and the last form (*AT#VTD=n*) queries and controls which types of tones can be detected and reported to the DTE via shielded codes in Voice Transmit, Voice Receive, and Online Voice Command Modes. These commands are active when the modem is configured for Voice/Audio Mode with the #CLS=8 command.

Command Options:

#VTD?................ Returns the current #VTD parameter values as an ASCII string.
#VTD=?.............. Returns the tone reporting capabilities of the modem.
#VTD=i,j,k .... Where i,j,k corresponds (in ASCII hexadecimal) to the desired capabilities for Voice Transmit, Receive, and Online Command Modes, respectively.

I, J, K TONE DETECTION/REPORTING BIT SETTINGS

Bit #	Definition	Setting = 0	Setting = 1
0	DTMF tone capability	Disable	**Enable**
1	V.25 1300 Hz Calling tone capability	Disable	**Enable**
2	T.30 1100 Hz Facsimile Calling tone capability	Disable	**Enable**
3	V.25/T.30 2100 Hz Answer tone capability	Disable	**Enable**
4	Bell 2225 Hz Answer tone capability	Disable	**Enable**
5	Call progress tone and cadence (busy and dial tone) capability	Disable	**Enable**
6-7	Reserved		

Notes: 1. The modem detects these tone/cadence and reports it via a shielded code to the DTE.
2. Tone detection is only performed when the sampling rate is 7.2 kHz.

Result Codes:
OK . . . . If bits supported with #VTD=? are selected.
ERROR . . . . Otherwise.

#VTM <? | =? | =n> . . . .*Enable Timing Mark Placement (default = 0)*

There are three forms of the ***#VTM*** command: one form (***AT#VTM=?***) allows the DTE to determine the range of acceptable values; another form (***AT#VTM?***) reports the current modem setting; and the last form (***AT#VTM=n***) controls the placement of <DLE><T> timing marks by the DCE in the data stream during ADPCM recording. These commands are active when the modem is configured for Voice/Audio Mode with the #CLS=8 command.

Command Options:
#VTM? . . . . Returns the current #VTM parameter value as an ASCII decimal value.
#VTM=? . . . . Returns "0, 10."
#VTM=0 . . . . **Disables time mark placement** (default)
#VTM=10 . . . . Enables time mark placement at 1-second intervals.

Result Codes:
OK . . . . For AT#VTM=n, if n=0 or 10.
ERROR . . . . Otherwise.

#VTS=<x | {x,z} | [x,y,z]> . . . .*Generate Tone Signals*

The ***AT#VTS=*** action command can be issued to play one or more tones including DTMF tones if and only if the modem is in the Online Voice Command Mode and the sampling rate (see the #VSR command) is set to 7.2 kHz. The modem parses and plays the tones defined in the parameter in the order listed, and no key abort is accepted. The parameter can have three types of elements separated by commas, and following "#VTS=":

x . . . . DTMF tones with duration defined by #VBT: x represents the DTMF digit (0-9,A-D,*,#), with the duration defined by #VBT.
{x, z} . . . . DTMF Digits with variable duration: x represents the DTMF digit (0-9,A-D,*,#) and z represents the tone duration (represented in ASCII decimal in units of 100 ms).
[x, y, z] . . . . Dual or Single Tones with Variable Duration:
x represents the first frequency (0 or 200–3000 Hz).
y represents the second frequency (0 or 200–3000 Hz).
z represents the tone duration (represented in ASCII decimal in units of 100 ms).

A tone generation sequence consists of one or more elements in a list with each element separated by commas. Multiple element types may be used in a single list.

Command Options:
#VTS=x . . . . Plays DTMF tone, the duration specified by the #VBT parameter.
#VTS=<x,z> . . . . Plays DTMF tone "x" for "z/10" seconds.
#VTS=[x,y,z] . . . . Plays the tone pair x,y for "z/10" seconds.
#VTS=[x1,y1,z1], [x2,y2,z2], x3, {x4,z4}, . . .
Plays the tone sequence consisting of tone pair x1,y1 for z1/10 seconds, the tone pair x2,y2 for z2/10 seconds, the DTMF tone x3 for #VBT seconds, the DTMF tone x4 for z4/10 seconds, etc.

Result Codes:
OK . . . . Command to play tones on currently selected device is accepted.
ERROR . . . . Command was not issued during Online Voice Command Mode or string is grammatically incorrect.

#VTX . . . .*Voice Transmit Mode*

The ***AT#VTX*** action command can be issued if and only if the modem is in the Online Voice Command Mode (indicated previously with the VCON message) and is the switch to Voice Transmit Mode. AT#VTX is used when a voice file is to be transmitted to the line, speaker, or handset. The #VLS command should have been previously issued to select the output source.

Command Option:
#VTX . . . . Request

Result Codes:
CONNECT . . . . When voice transmission by DTE can begin.
ERROR . . . . If #VLS =0 and output device not connected.

%An . . . .*Set Auto-Reliable Fallback Character*

The "***AT%An***" command determines the Auto Reliable fallback character. The ASCII character n can be any value 0 through 127. Default = 13.

%Cn . . . . *MNP5 and BTLZ Data Compression Enable/Disable*

📁 "*AT%Cn*" enables data compression in both ITU-T V.42 bis and MNP class 5 operation. On some modems it controls bit 1 of register S46 .

Command Option Details:

%C0. . . . . No data compression. Clear register S46 bit 1 to 0.

%C1 . . . . **Enables data compression using V.42 bis in LAP-M and MNP5** in a Reliable Link connection. (On some modems, %C1 enables MNP 5 only.) Sets register S46 bit 1 to 1.

%C2. . . . . Enables V.42 bis. Sets register S46 bit 1 to 1.

%C3. . . . . Enables both V.42 and MNP 5. Sets register S46 bit 1 to 1.

Result Codes:

OK. . . . . n = 0, 1, 2, or 3.

ERROR . . . . Otherwise.

%Dn . . . .*BTLZ Dictionary Size*

"*AT%Dn*" sets the dictionary size used in data compression.

Command Option Details:

%D0 . . . . 512 bytes

%D1 . . . . 1024 bytes

%D2 . . . . **2048 bytes** (default).

%D3 . . . . 4096 bytes if one-way compression (%M1 or %M2) is enabled.

2048 bytes if two-way compression (%M3).

%En . . . .*Auto Retrain*

📁 "*AT%En*" controls whether or not the modem will automatically monitor the line quality and request a retrain (%E1) or fall back when line quality is low and fall forward when line quality is high (%E2). The parameter value, if valid, is written to register S41.

If enabled, the modem attempts to retrain for a maximum of 30 seconds.

Command Option Details:

%E0. . . . . Disables line quality monitor and auto retrain.

%E1. . . . . Enables line quality monitor and auto retrain.

%E2. . . . . Enables line quality monitor fallback and fall forward. When AT%E2 is asserted, the modem monitors the line quality (EQM). When line quality is inadequate, the modem will initiate a rate renegotiation to a lower speed within the V.34/V.FC/V.32 bis/V.32 (28800 modems) or V.32 bis/V.32 (14400 modems) modulation speeds. The modem will keep falling back within the current modulation protocol if necessary until the speed reaches 2400 bps (V.34), 14400 bps (V.FC), or 4800 bps (V.32). Below this rate, the modem will only do retrains if EQM thresholds are exceeded. If the EQM is better than the required level for at least one minute, the modem will initiate a rate renegotiation to a higher speed within the current available modulation protocol speeds. The rate renegotiation will be done without a retrain if a V.32 bis connection is established.

Speeds attempted during fallback and fall forward are those shown to be available during the initial handshake sequence. Fallback and fall forward is available in error correction and normal modes but not in direct mode or synchronous mode with external clocks.

Result Codes:

OK. . . . . n = 0, 1, or 2.

ERROR . . . . Otherwise.

%Fn . . . .*Split Speed Direction Select*

Command Option Details:

%F1. . . . . 75 bps transmit 1200 bps receive.

%F2. . . . . 1200 bps transmit 75 bps receive.

%F3. . . . . V.23 Half-duplex operation (independent of \Wn).

%Gn . . . .*Rate Renegotiation Control*

📁 "*AT%Gn*" determines whether the modem automatically monitors the line quality and renegotiates the rate on V.32 and V.32 bis connections.

Command Option Details:

%G0 . . . . Disables rate renegotiation on V.32 bis.

%G1 . . . . Enables rate renegotiation on V.32 bis and enables rate renegotiation with retrain on V.32.

%L . . . .*Report Received Signal Level nnn in -dBm*

The "*AT%L*" command reports the measured receive carrier level in decibels (dBm) as an unsigned (positive) number (the real value is always negative). FCC specifications state that a modem will transmit at a maximum level no greater than -10 dBm. Readings from %L that are less than 10 (i.e. 0 to 9) show that the signal is being amplified to a level beyond the original transmit signal level. Such amplification will cause distortion and may have a very high level of noise mixed with the original signal, hence, a very high bit error rate (BER).

Command Option Details:

AT%L . . . . Reports levels from 0 to -9 dBm as 009,

           levels from -10 to -42 dBm as 010 to 042,

           levels -43 dBm or less as 043.

How to use %L to measure signal level:

1 . . . . Dial and establish a connection to another modem.
2 . . . . Escape to the COMMAND mode by issuing <guard> <+++> <guard> and then wait for OK.
3 . . . . Type *AT%L* <CR>. Repeat the *AT%L* <CR> a few times and average the values together to get a "middle of the road" value.
4 . . . . Issue *ATO* to return to "ONLINE" mode, or *ATH* to hang up.

%Mn . . . .*BTLZ Compression Mode*

"*AT%Mn*" determines that the if and where data compression is used.

Command Option Details:

%M0 . . . . Disabled.
%M1 . . . . Transmit compression only.
%M2 . . . . Receive compression only.
%M3. . . . . **Two-way compression** (default).

%Mn . . . .*AUXCTL Output Line Control*

Command Option Details:

%M0. . . . . **Line high only while modem is in synchronous data mode** (default).
%M1 . . . . Line low only while modem is in the synchronous mode.

%P . . . .*Clear Dictionaries*

"*AT%P*" clear local and remote modems' BTLZ dictionaries. The command returns an error if the BTLZ dictionary is not in use.

%Q . . . .*Report Received Signal Quality*

The "*AT%Q*" command returns an *instantaneous* reading of the received carrier signal quality as a value between 0 and 127. Values of greater than 20 are generally considered to be poor. When poor *%Q* values occur, the carrier rate should be reduced by one setting, then the %Q value should be rechecked. This process should be repeated until a good %Q reading is obtained. Values less than 8 show that the line quality is very good at the present carrier rate and that a higher rate may also be supported under the present conditions.

%Q readings in many modems are based on the root-mean-square of the ERROR VECTOR value *currently* generated in the modem. (The larger the vector, the larger the %Q value reported.) The ERROR VECTOR is the average of the vector error for all the points in the carrier encoding pattern currently in use (either TCM or QAM). The number of points in the carrier increases with the carrier rate. The ERROR VECTOR is a measure of the amount of separation between the current pattern of the points on the carrier and the ideal location of these points. The more the signal is corrupted, the greater the uncertainty these points will become, hence the greater the error rate. The measurement is not a direct measurement of the telephone connection; however, the telephone connection between one modem and the other is the major source of signal distortions.

nnn = 001 to 008. . . . . current signal quality excellent (higher rate may be used).
nnn = 009 to 020. . . . . current signal good (best carrier rate for line condition).
nnn = 020 to 030. . . . . current signal poor (lower carrier rate should be used).
nnn = 031 to 127. . . . . signal very bad (or gone).
NOTE: Many modems use 8 and 20 as rate renegotiation thresholds.

Steps to use %Q to determine the received signal quality:

1 . . . . Dial and connect to a remote modem.
2 . . . . Escape to the COMMAND mode by issuing "<guard time> <+++> <guard time>" and then wait for OK.
3 . . . . Type "*AT%Q*" and press return. Repeat the "*AT%Q*" several times and average the values together to get an average value.
4 . . . . Issue "*ATO*" to return to "ONLINE" mode, or "*ATH*" to hang up.

Result Codes:
ERROR . . . . If not connected, if connected at 300 bps, in V.23 mode, or in a fax mode.
OK. . . . . Otherwise.

%Sn . . . . *Set Maximum BTLZ String Length*

"*AT%Sn*" sets the maximum number of characters that can be compressed into one word. The maximum BTLZ string length "n" can be set from 6 to 250; the default is 32.

&$ . . . . *Display List of Ampersand (&) Commands (U.S. Robotics)*

For example, a U.S. Robotics Sportster 28,800 fax/modem reports the following:

```
AT&$
HELP, Ampersand Commands (CTRL-S to Stop, CTRL-C to Cancel)
&An   n=0  Disable /ARQ Result Codes    &Pn   n=0  N.American Pulse Dial
      n=1  Enable /ARQ Result Codes           n=1  UK Pulse Dial
      n=2  Enable /Modulation Codes     &Rn   n=1  Ignore RTS
      n=3  Enable /Extra Result Codes         n=2  RX to DTE/RTS high
&Bn   n=0  Floating DTE Speed           &Sn   n=0  DSR Always On
      n=1  Fixed DTE Speed                    n=1  Modem Controls DSR
      n=2  DTE Speed Fixed When ARQ     &Tn   n=0  End Test
&Cn   n=0  CD Always On                       n=1  Analog Loopback (ALB)
      n=1  Modem Controls CD                  n=3  Digital Loopback (DLB)
&Dn   n=0  Ignore DTR                         n=4  Grant Remote DLB
      n=1  Online Command Mode                n=5  Deny Remote DLB
      n=2  DTE Controls DTR                   n=6  Remote Digital Loopback
&Fn   n=0  Load Factory Configuration         n=7  Remote DLB With Self Test
      n=1  Hardware Flow Control Cnfg.        n=8  ALB With Self Test
      n=2  Software Flow Control Cnfg.  &Un   Lowest Link Speed Limit
&Gn   n=0  No Guard Tone                      n=0  Disabled
      n=1  550 Hz Guard Tone                  n=1  300 bps
      n=2  1800 Hz Guard Tone                 n=2  1200 bps
&Hn   n=0  Disable TX Flow Control            n=3  2400 bps
      n=1  CTS                                n=4  4800 bps
      n=2  Xon/Xoff                           n=5  7200 bps
      n=3  CTS and Xon/Xoff                   n=6  9600 bps
&In   n=0  Disable RX Flow Control            n=7  12000 bps
      n=1  Xon/Xoff                           n=8  14400 bps
      n=2  Xon/Xoff Chars Filtered            n=9  16800 bps
      n=3  HP Enq/Ack Host Mode               n=10 19200 bps
      n=4  HP Enq/Ack Terminal Mode           n=11 21600 bps
      n=5  Xon/Xoff for non-ARQ Mode          n=12 24000 bps
      n=5  Xon/Xoff for non-ARQ Mode          n=12 24000 bps
      n=1  Auto Data Compression              n=14 28800 bps
      n=2  Enable Data Compression      &W    Store Configuration
      n=3  Selective Data Compression   &Yn   n=0  Destructive
&Mn   n=0  Normal Mode                        n=1  Destructive/Expedited
      n=4  ARQ/Normal Mode                    n=2  Nondest./Expedited
      n=5  ARQ Mode                           n=3  Nondest./Unexpedited
&Nn   n=0  Highest Link Speed           &Zn=s Store Phone Number
      n=1  300 bps                      &Zn?  Query Phone Number
      n=2  1200 bps
      n=3  2400 bps
      n=4  4800 bps
      n=5  7200 bps
      n=6  9600 bps
      n=7  12000 bps
      n=8  14400 bps
      n=9  16800 bps
      n=10 19200 bps
      n=11 21600 bps
      n=12 24000 bps
      n=13 26400 bps
      n=14 28800 bps
OK
```

&An . . . . *Enable/Disable Additional Result Code*

The "*AT&An*" command enables and disables the additional result code subsets.

Command Option Details:
&A0. . . . . /ARQ result codes disabled.
&A1 . . . . **/ARQ result codes enabled** (default).
&A2. . . . . Additional HST or V32 indicator for calls at 4800 & 9600 bps enabled..
&A3. . . . . Protocol indicator added—LAP-M, MNP, NONE (error control) and V.42 bis or MNP (data compression).

&Bn . . . . *(1) V.32 Auto Retrain (ATT modem chipset based)*

Command Option Details:
&B0. . . . . Return to ON-HOOK if bad MSE.
&B1. . . . . **Retrain.**
&B2. . . . . Remain OFF-HOOK; do not retrain.

&Bn . . . . *(2) DTE to DCE Date Rate Control (U.S. Robotics)*

In US Robotics modems the "*AT&Bn*" command, determines the DTE - DCE data rate.

Command Option Details:
&B0. . . . . DTE-DCE rate switches to follow the connection rate.
&B1. . . . . **DTE-DCE rate remains fixed at DTE rate (default).**
&B2. . . . . Fixed in ARQ mode, variable in non-ARQ mode.

&Cn . . . . *DCD (RLSD) Signal Control*

The "*AT&Cn*" command controls how the carrier detect signal is reported to the serial port (DCD). When synchronous mode is selected, this command is ignored and the modem functions as if the &C1 command is selected. The value is written into register S21 bit 5.

Command Option Details:
&C0. . . . . DCD always on.
&C1 . . . . **DCD follows carrier state** (ON when carrier present).

Result Codes:
OK. . . . . If n = 0 or 1.
ERROR . . . . Otherwise.

&Dn . . . .*DTR Options*

"*AT&Dn*" (in conjunction with the &Qn state) determines the modem's response to a DTR ON-to-OFF transition (with an off time greater than time specified in the S25 register). The default is dependent on the &F setting.

Command Option Details:

	&D0	&D1	&D2	&D3	&D4
&Q0, &Q5, &Q6, &Q8, &Q9, or + FCLASS ≠ 0	N	2	3	4	5
&Q1	1	2	3	4	
&Q2	3	3	3	3	
&Q3	3	3	3	3	
&Q4	1	2	3	4	

N . . . . Ignore DTR. Allows operation with DTEs that do not provide DTR.
1 . . . . Disconnect and send OK result code. Auto answer is unaffected.
2 . . . . Switch to asynchronous COMMAND mode, maintain link connection, and send OK result code.
3 . . . . Disconnect (go on-hook), return to COMMAND mode, send OK result code, and disable auto answer while DTR is off.
4 . . . . Perform warm start (*ATZ0*). *&Yn* setting determines loaded profile.
5 . . . . Reset and enter low power mode (Hayes V-Series Smartmodem 2400).

US Robotics
&D0. . . . . DTR always on.
&D1. . . . . DTR toggle causes online COMMAND mode.
&D2 . . . . **DTR = 1 enables DATA and COMMAND modes;**
                         DTR = 0 disables and terminates a connection.
&D3. . . . . DTR = 1 resets modem.

Result Codes:
OK . . . . If n = 0, 1, 2, 3, or 4.
ERROR . . . . Otherwise.

&Fn . . . .*Load Factory Default Configuration*

The "*AT&Fn*" command loads the active user profile with one of the available factory preset default profiles. *AT&F0* generally loads a generic template while &F1 and &F2 load specialized templates (e.g., Macintosh vs. PC or H/W vs. S/W flow control).

Command Option Details:
&F0 . . . . Set Factory configuration 0.
&F1 . . . . Set Factory configuration 1.

Result Codes:
OK . . . . If n = 0, 1, or 2 (n > 0 varies by modem manufacturer).
ERROR . . . . Otherwise or if modem is online.

&Gn . . . .*Set Guard Tone*

The "*AT&Gn*" command controls the generation of guard tones. This command is not implemented on many modems. When it is implemented, it may require a *B0* setting. The value is stored in register S23 bits 6 and 7.

Command Option Details:
&G0 . . . . **No Guard tone** (US and Canada).
&G1 . . . . Generates 550 Hz guard tone (some European countries).
&G2 . . . . Generates 1800 Hz guard tone (UK and some Commonwealth countries).

Result Codes:
OK . . . . If n = 0, 1, or 2.
ERROR . . . . Otherwise.

&Hn . . . .*Transmit Data Flow Control*

The "*AT&Hn*" command selects the method of transmit data flow control. Not all methods are implemented in all modems that accept this command.

Command Option Details:
&H0 . . . . **Disables flow control** (default).
&H1 . . . . Hardware flow control RTS/CTS flow control enabled.
&H2 . . . . Software flow control XON/XOFF flow control enabled.
&H3 . . . . Hardware and software flow control enabled.

Result Codes:
OK . . . . If n = 0, 1, 2, or 3.
ERROR . . . . Otherwise.

&In . . . .*Receive Data Flow Control*

The "*AT&In*" command selects the method of receive data flow control. Not all methods are implemented in all modems that accept this command.

Command Option Details:
&I0 . . . . **Disables flow control** (default).
&I1 . . . . Transparent flow control, XON/XOFF to local DCE and remote DTE.
&I2 . . . . Software flow control, XON/XOFF flow control enabled local to modem only.
&I3 . . . . Host mode, Hewlett-Packard protocol.
&I4 . . . . Terminal mode, Hewlett-Packard protocol.
&I5 . . . . Software flow control, XON/XOFF enabled local to modem only in ARQ mode. In non-ARQ mode XON/XOFF to remote DCE for link flow control.

Result Codes:
OK . . . . If n = 0, through 5 (or maximum value implemented).
ERROR . . . . Otherwise.

&Jn . . . .*Phone Jack Selection*

The "*AT&Jn*" command determines how the auxiliary telco relay is controlled (and the subsequent connection of the A and A1 leads). For &J0, the relay never operates. With &J1 and the modem off-hook, the relay is operated and lead A is shorted to lead A1. The setting is stored in register S21 bit 0.

Command Option Details:

&J0 . . . . **RJ-11, RJ-41S, or RJ-45S type telco phone jacks.** The auxiliary relay is never operated (default).

&J1 . . . . RJ-12 and RJ-13 type phone jacks. The auxiliary relay connects the A lead to the A1 lead while the modem is off-hook. For multiple phones or modems on the line.

Result Codes:

OK . . . . If n = 0 or 1.

ERROR . . . . Otherwise.

&Kn . . . .*Serial Port Flow Control*

The "*AT&Kn*" command determines how the flow of data between the DTE and modem is controlled. When the local modem transmit buffer is almost full, the local modem will send an XOFF character or drop CTS to the DTE in order to stop the flow of data. When the buffer is nearly empty, the local modem will send an XON character or raise CTS to restart data flow. When the DTE sends XOFF/XON characters or lowers/raises RTS to the local modem, the local modem responds by suspending or resuming data transmission to the DTE. The local modem will pass the XON/XOFF information to the remote modem if transparent flow control is selected. When the modem is in Direct mode (selected by &Q0), flow control is not used, that is, the &Kn command is ignored.

Command Option Details:

&K0 . . . . Disables flow control.

&K1 . . . . Auto enables/disables. Disables flow control if modem is B0 selected.

&K2 . . . . Enables flow control, ignores Bn setting.

&K3 . . . . **Enables bidirectional hardware (RTS/CTS) flow control** (default with data mode).

&K4 . . . . Enables software (XON/XOFF) flow control.

&K5 . . . . Enables transparent software flow control (XON/XOFF).

&K6 . . . . Enables software (XON/XOFF) and hardware (RTS/CTS) flow control (default in fax and voice modes).

Result Codes:

OK . . . . If n = defined value.

ERROR . . . . Otherwise.

&Kn . . . .*Enable/Disable Data Compression (U.S. Robotics)*

Command Option Details:

&K0 . . . . Data compression disabled.

&K1 . . . . Auto enable/disable.

&K2 . . . . Data compression enabled.

&K3 . . . . MNP-5 compression disabled.

Result Codes:

OK . . . . If n = defined value.

ERROR . . . . Otherwise.

&Ln . . . .*Line Type Selection*

"*AT&Ln*" controls the selection of dial-up or leased line characteristics. The power level is the primary characteristic changed when setting leased line operation. The value is written to register S27 bit 2.

Command Option Details:

&L0 . . . . **Dial-up line** (default).

&L1 . . . . Leased line (not supported by most modems. For compatibility the modem may respond with OK; however, it will take no action).

Result Codes:

OK . . . . If n = 0 or 1.

ERROR . . . . Otherwise.

&Mn . . . .*Communication Mode*

The "*AT&Mn*" command is essentially the same as the "*AT&Qn*" command for most modems. Some differences are listed below.

Command Option Details:

&M0 . . . . *Asynchronous Direct mode,* error control disabled. The line rate must match the DTE rate. The value xxxx0x00b is written to register S27.

&M1 . . . . *Synchronous Mode 1,* error control disabled. (Terminal must support both synchronous and asynchronous modes.) After call placed in asynchronous mode, the modem switches to synchronous mode. When DTR is dropped or the loss of carrier from the remote modem is detected for greater than the time specified in register *S10,* the modem returns to asynchronous mode. The value xxxx0x01b is written to register S27.

&M2 . . . . *Synchronous Mode 2;* the modem dials a previously stored number, then establishes a synchronous connection as in &M1. The value xxxx0x10b is written to register S27.

&M3 . . . . *Synchronous Mode 3;* a telephone instrument dials the number of the remote location. The call is manually initiated while DTR is inactive. When DTR becomes active, the handshake proceeds in originate or answer mode according to S14 bit 7. The value xxxx0x11b is written to register S27.

&M4 . . . . **Normal/ARQ asynchronous mode with fallback to NORMAL asynchronous mode** when an ARQ connection cannot be established. The value xxxx1x00b is written to register 27.

&M5 . . . . *ARQ asynchronous reliable mode.* The modem terminates the connection if an ARQ connection cannot be established. The value xxxx1x01b is written to register S27.

&M6 . . . . *Asynchronous Normal mode.* The value xxxx1x10b is written to register S27.

&M7 . . . . Reserved. The value xxxx1x11b is written to register S27.

Result Codes:

OK. . . . . If n = defined.

ERROR . . . . Otherwise.

&Nn . . . . *Link Rate*

The "*AT&Nn*" parameter sets the DCE to DCE transmission rate (line rate) at either a particular fixed value (N>0) or allows the DCEs to negotiate the rate (N=0). If the rate is negotiated, it is set to the highest possible value both modems can support, with the prevailing line conditions established by the connection.

If "*AT&Un*" is greater than 0, then "*AT&Nn*" sets a ceiling rate.

Command Option Details:

&N0	**Variable link rate, The DCEs negotiate the highest attainable rate.**	&N16	Set rate to 33600 bps (33.6 Kbps).
&N1	Set rate to 300 bps.	&N17	Set rate to 33333 bps (56 Kbps).
&N2	Set rate to 1200 bps.	&N18	Set rate to 37333 bps (56 Kbps).
&N3	Set rate to 2400 bps.	&N19	Set rate to 41333 bps (56 Kbps).
&N4	Set rate to 4800 bps.	&N20	Set rate to 42666 bps (56 Kbps).
&N5	Set rate to 7200 bps.	&N21	Set rate to 44000 bps (56 Kbps).
&N6	Set rate to 9600 bps.	&N22	Set rate to 45333 bps (56 Kbps).
&N7	Set rate to 12000 bps (HST or 14.4 Kbps).	&N23	Set rate to 46666 bps (56 Kbps).
&N8	Set rate to 14.4 Kbps (HST or 14.4 Kbps).	&N24	Set rate to 48000 bps (56 Kbps).
&N9	Set rate to 16.8 Kbps (28.8 Kbps).	&N25	Set rate to 49333 bps (56 Kbps).
&N10	Set rate to 19.2 Kbps (28.8 Kbps).	&N26	Set rate to 50666 bps (56 Kbps).
&N11	Set rate to 21.6 Kbps (28.8 Kbps).	&N27	Set rate to 52000 bps (56 Kbps).
&N12	Set rate to 24.0 Kbps (28.8 Kbps).	&N28	Set rate to 53333 bps (56 Kbps).
&N13	Set rate to 26.4 Kbps (28.8 Kbps).	&N29	Set rate to 54666 bps (56 Kbps).
&N14	Set rate to 28.8 Kbps (28.8 Kbps).	&N30	Set rate to 56000 bps (56 Kbps).
&N15	Set rate to 31.2 Kbps (33.6 Kbps).	&N31	Set rate to 57333 bps (56 Kbps).

Result Codes:

OK. . . . . If n = defined.

ERROR . . . . Otherwise.

&Pn . . . . *Pulse Dialing MAKE/BREAK Ratio*

The "*AT&Pn*" command selects the ratio of off-hook (make) to on-hook (break) timing and the number of dial pulses per second (pps) when the modem is pulse dialing.

Command Option Details:

&P0. . . . . **39%/61% make/break ratio at 10 pps** (US and CANADA) (Default).

&P1. . . . . 33%/67% make/break ratio at 10 pps (UK and Hong Kong).

&P2. . . . . 33%/67% make/break ratio at 20 pps (Japan), or on some modems. 39%/61% make/break ratio at 20 pps.

&P3. . . . . 39%/61% make/break ratio at 20 pps, or on some modems. 33%/67% make/break ratio at 20 pps.

Result Codes:

OK. . . . . If n = defined.

ERROR . . . . Otherwise.

&Qn . . . . *Asynchronous/Synchronous Mode Selection*

"*AT&Qn*" selects one of three basic communication modes: asynchronous, synchronous, and error correction. It is used in conjunction with registers S36 and S48 .

Note: Use caution when saving &Q1, &Q2, and &Q3 settings to memory. Once issued, asynchronous mode may be again selected by either (1) issuing an AT&Q0&W (or &Q5&W through &Q9&W) command from a *synchronous* terminal, or (2) connecting the modem to an *asynchronous* terminal with DTR disabled and issuing the an AT&Q0&W (or &Q5&W through &Q9&W) command.

Command Option Details:

&Q0. . . . .*Asynchronous Direct mode,* error control disabled. The line rate must match the DTE rate. The value xxxx0x00b is written to register S27.

&Q1. . . . .*Synchronous Mode 1,* error control disabled. (Terminal must support both synchronous and asynchronous modes.) After call placed in asynchronous mode, the modem switches to synchronous mode. When DTR is dropped or the loss of carrier from the remote modem is detected for greater than the time specified in register *S10,* the modem returns to asynchronous mode. The value xxxx0x01b is written to register S27.

&Q2. . . . .*Synchronous Mode 2.* Modem dials number in stored location 0 when DTR goes from LOW to HIGH. The modem returns to asynchronous mode when DTR transitions to LOW (OFF). The value xxxx0x10b is written to register S27.

&Q3. . . . .*Synchronous Mode 3.* With DTR LOW (OFF), dial the number manually on a telephone. After the connection is established (and before the telephone is returned to the on-hook state), bring DTR HIGH (ON) to establish the modem connection. The value xxxx0x11b is written to register S27.

&Q4. . . . .*Selects AutoSync mode.* AutoSync operation, when used in conjunction with the Hayes Synchronous Interface (HSI) capability in the DTE, provides synchronous communication capability from an asynchronous terminal. The value xxxx1x00b is written to register S27.

To start AutoSync operation, set registers S19, S20, and S25 to the desired values before issuing the AT&Q4 command. After the CONNECT message is issued, the modem waits the period of time specified by S25 before examining DTR. If DTR is on, the modem enters the synchronous operating state; if DTR is off, the modem terminates the line connection and returns to the asynchronous command state.

AutoSync operation is terminated with either the loss of carrier or the on-to-off transition of DTR. Loss of carrier will cause the modem to return to the asynchronous command state. An on-to-off transition of DTR will cause the modem to return to the asynchronous command state and either not terminate the line connection (&D1 active) or terminate the line connection (any other &Dn value).

&Q5 . . . .**Asynchronous Reliable Mode.** The modem negotiates an error-correction link based on settings in registers *S36* and *S48.* (most common default). The value xxxx1x01b is written to register S27.

&Q6. . . . .*Asynchronous Normal Mode.* The value xxxx1x10b is written to register S27.

&Q7. . . . .*Reserved.* The value xxxx1x11b is written to register S27.

&Q8. . . . .*Asynchronous Reliable Mode*—causes the modem to act as if *S48=128*
If *S36 = odd* number (bit 0 = 1), then modem acts as if *S36=7,* that is, the modem will attempt an MNP connection, then normal mode.
If *S36 = even* (bit 0=0), then modem acts as if *S36=5,* that is, the modem will attempt an MNP connection, then direct mode.

&Q9. . . . .*Asynchronous Reliable Mode*—attempts V.42 bis, then V.42, then NORMAL mode (without attempting MNP 2-5).
Modem acts like &Q5 with *S48=7* and *S46=138.*

Result Codes:

OK. . . . . .If n = 0 through 9.
ERROR . . . . .Otherwise.

&Rn . . . . *RTS/CTS Control*

"*AT&Rn*" controls how the clear to send (CTS) signal responds to the request to send (RTS) signal in synchronous mode. (In *asynchronous mode,* CTS is always on unless &K3 is set.) Default value is modem dependent. The value of "n=0 or 1" is written to register 21 bit 2.

Command Option Details:

&R0. . . . . CTS follows RTS after a delay specified by register *S26.* (V.25 bis sync.)
&R1. . . . . CTS always "on" while the modem is online unless &K3 is set.
&R2. . . . . Receive data is sent to DTE only when RTS is high.

Result Codes:

OK. . . . . .If n = defined value.
ERROR . . . . .Otherwise.

&Sn . . . . *Data Set Ready (DSR) Action Select*

"*AT&Sn*" determines whether data set ready (DSR) is always on or operates in accordance with EIA RS-232 specifications. The value is written into register S21 bit 6.

Command Option Details:

&S0 . . . .   **DSR always on** (default).

&S1 . . . .   DSR follows EIA RS-232 (or ITU-T V.25) specification (active following carrier tone detection and until the carrier is lost).

&S2 . . . .   Upon detecting loss of carrier, the DCE sends the DTE a pulsed DSR. CTS follows the carrier detect signal CD.

&S3 . . . .   Upon detecting loss of carrier, the DCE sends the DTE a pulsed DSR.

Result Codes:

OK . . . .   If n = defined value.

ERROR . . . .   Otherwise.

&Tn . . . . *Test and Diagnostics*

The *"AT&Tn"* test and diagnostic command selects various test modes of operation. Test commands must be initiated in the COMMAND mode with asynchronous direct mode selected, no flow control, no error correction, no data compression, and with DTE data at a specified rate (generally 2400 bps or 9600 bps). A telco line connection must be established prior to initiating a remote mode loopback test. Tests are performed for the period of time defined in register *S18* (if *S18=0*; then the tests will run continuously and must be terminated with the *&T0* command).

Command Option Details:

&T0 . . . .   Terminates test in progress (command must be the last command on the command line). Clears register S16.

&T1 . . . .   Local analog loopback, ITU-T V.54 Loop 3. If a connection exists when *AT&T1* is issued, the modem goes on-hook. The CONNECT XXXX message is displayed upon the start of the test. The test verifies the operation between DTE and the local modem. Sets register S16 bit 0 = 1. (Issue +++ *AT&T0* to terminate test.)

&T2 . . . .   Generally returns ERROR. Reserved function.

&T3 . . . .   Local digital loopback, ITU-T V.54 Loop 2. Characters from the local DTE are looped back in the local modem to the local DTE. Sets register S16 bit 2 = 1.

&T4 . . . .   Allows local modem to grant a request for remote digital loopback from a remote modem, that is, an RDL request from a remote modem is allowed. Sets register S23 bit 0 = 1 (default on some modems).

&T5 . . . .   **Prohibits local modem from granting a request for remote digital loopback,** that is, an RDL request from a remote modem is denied. Clears register S23 bit 0 = 0 (default).

&T6 . . . .   Initiates a remote digital loopback (RDL), ITU-T V.54 Loop 2, without self-test. Characters from the local DTE are sent to the remote modem, looped back from the remote modem back to the local DTE. This test verifies the functionality of the equipment and path from the local equipment. A CONNECT XXXX message is displayed upon the start of the test. If no connection exists, ERROR is returned. Sets register S16 bit 4 = 1.

&T7 . . . .   Remote digital loopback (RDL), ITU-T V.54 Loop 2, with self-test. This test is similar to &T6 except the local modem sends a continuous test pattern to the remote modem and counts the errors in the returned pattern. If no connection exists, ERROR is returned. At the end of the test (register S18 time-out, +++AT&T0, or +++ATH0), this error count is sent to the DTE. Sets register S16 bit 5.

&T8 . . . .   Local analog loopback, V.54 Loop 3, with self-test. This test is similar to &T7 except the local modem sends a continuous test pattern through the local analog loopback and counts the errors in the returned pattern. If a connection exists, the modem returns to the on-hook state before the test is initiated. At the end of the test (register S18 time-out, +++AT&T0, or +++ATH0), this error count is sent to the DTE.

This test DOES NOT evaluate performance with the actual telco line connection, that is, the test verifies only the functionality of the local modem.

The command may not be available in some countries due to PTT restrictions.

Result Codes:

OK . . . .   If n = 0, 1, or 3 through 8.

ERROR . . . .   Otherwise.

&Un . . . . *Trellis Coding*

The *"AT&Un"* command enables and disables trellis coding.

Command Option Details:

&U0 . . . .   **Enables trellis coding.**

&U1 . . . .   Disables trellis coding.

Result Codes:

OK . . . .   If n = 0, or 1.

ERROR . . . .   Otherwise.

&Un . . . . *Sets Floor Connect Speed (U.S. Robotics)*

"*AT&Un*" sets the floor connection speed if "*AT&Nn*" is greater than 0 (i.e., a ceiling rate is set).

Command Option Details:

&U0	**Disabled**	&U16	33600 bps
&U1	300 bps	&U17	33333 bps
&U2	1200 bps	&U18	37333 bps
&U3	2400 bps	&U19	41333 bps
&U4	4800 bps	&U20	42666 bps
&U5	7200 bps	&U21	44000 bps
&U6	9600 bps	&U22	45333 bps
&U7	12000 bps	&U23	46666 bps
&U8	14400 bps	&U24	48000 bps
&U9	16800 bps	&U25	49333 bps
&U10	19200 bps	&U26	50666 bps
&U11	21600 bps	&U27	52000 bps
&U12	24000 bps	&U28	53333 bps
&U13	26400 bps	&U29	54666 bps
&U14	28800 bps	&U30	56000 bps
&U15	31200 bps	&U31	57333 bps

Result Codes:
OK. . . . . If n = defined in the particular modem.
ERROR . . . . Otherwise.

&Vn . . . . *Display Configuration Profiles*

The "AT&Vn" command displays the active (current) configuration, stored (user) profiles, and stored phone numbers. If no NVRAM is present, the stored profiles and phone numbers will not be displayed and an error result code may be given. On many modems only the n=0 option is valid. On these modems, all stored information is displayed upon assertion of the "AT&V" command.

Command Option Details:
&V0. . . . . Views user stored user profile 0. On many modems, the &V (&V0) command will display all user stored
           information as well as the current configuration.
&V1. . . . . Views user stored user profile 1.

Result Codes:
OK. . . . . If n = 0, or 1.
ERROR . . . . Otherwise (*NVRAM FAILED OR NOT INSTALLED*).

For example:

```
at&V<return>

ACTIVE PROFILE:
B1 E1 L2 M1 N1 Q0 T V1 W0 X4 Y0 %C1 %E0 %G1 \N3 &C1 &D2 &K3 &Q5 &R0 &S0 &X0 &Y0
S00:000 S01:000 S02:043 S03:013 S04:010 S05:008 S06:002 S07:050 S08:002 S09:006
S10:014 S11:095 S12:050 S18:000 S23:053 S25:005 S26:001 S36:007 S37:000 S38:020
S40:087 S46:138 S48:007 S95:000 S109:062 S110:001

STORED PROFILE 0:
B1 E1 L2 M1 N1 Q0 T V1 W0 X4 Y0 %C1 %E0 %G1 \N3 &C1 &D2 &K3 &Q5 &R0 &S0 &X0
S00:000 S02:043 S06:002 S07:050 S08:002 S09:006 S10:014 S11:095 S12:050 S18:000
S23:059 S25:005 S26:001 S36:007 S37:000 S38:020 S40:087 S46:138 S48:007 S95:000

STORED PROFILE 1:
B1 E1 L2 M0 N1 Q0 T V1 W1 X4 Y0 %C1 %E0 %G1 \N3 &C1 &D2 &K3 &Q5 &R0 &S0 &X0
S00:000 S02:043 S06:002 S07:050 S08:002 S09:006 S10:014 S11:095 S12:050 S18:000
S23:061 S25:005 S26:001 S36:007 S37:000 S38:020 S40:087 S46:138 S48:007 S95:000

TELEPHONE NUMBERS:
0=                    1=
2=                    3=
```

&Wn . . . . *Write ACTIVE Profile to Stored Profile n*

"*AT&Wn*" saves the current modem configuration in the selected NVRAM location. The settings are restored with the "*ATZn*" command or at power up if selected by the "*AT&Yn*" command.

Command Option Details:

&W0 . . . . Writes current profile to NVRAM location 0 (restored with Y0).

&W1 . . . . Writes current profile to NVRAM location 1 (restored with Y1).

Result Codes:

OK. . . . . If n = 0, or 1.

ERROR . . . . Otherwise or if the NVRAM is not installed or is not functional.

&Xn . . . . *Synchronous Transmit Clock Source*

📁 "*AT&Xn*" selects the source of the synchronous transmit clock (TCLK) for the modem during synchronous operation. In asynchronous mode, the transmit and receive clocks are turned off.

Command Option Details:

&X0 . . . . **A modem generated clock is applied to TCLK on the serial interface** (default).

&X1. . . . . DTE supplies external clock. The DTE supplies a clock to the TCLK input of the serial interface. The modem applies this clock to the TCLK output of the serial interface.

&X2. . . . . The transmit clock is derived from the data carrier received from the remote modem and applies it to the TCLK on the serial interface.

Result Codes:

OK. . . . . If n = 0, 1, or 2.

ERROR . . . . Otherwise.

&Yn . . . . *(1) Select Configuration Loaded at Power Up*

"*AT&Yn*" designates which of the saved user profiles is to be loaded into the active configuration at power up or hard reset.

Command Option Details:

&Y0 . . . . **Loads profile 0 at power up** (default).

&Y1. . . . . Loads profile 1 at power up.

Result Codes:

OK. . . . . If n = 0, or 1.

ERROR . . . . Otherwise or if NVRAM fails during reset testing.

&Yn . . . . *(2) Break Handling (U.S. Robotics)*

"*AT&Yn*" determines how the DCE will respond when a BREAK is received from the remote DCE or the local DTE.

Command Option Details:

&Y0. . . . . Destructive, break not forwarded to remote DCE.

&Y1 . . . . **Destructive, break forwarded immediately to remote modem** (default).

&Y2. . . . . Nondestructive, break forwarded immediately to remote DCE.

&Y3. . . . . Nondestructive, break sent in sequence with buffered data to the remote DCE.

. . . . Destructive breaks clear DCE buffers.

. . . . In data compression modes, the compression tables are cleared.

Result Codes:

OK. . . . . If n = 0, through 4.

ERROR . . . . Otherwise.

&ZL? . . . . *Display Last Executed Dial String*

&Zn? . . . . *Display Stored Phone Numbers*

The "*ATn?*" command displays the phone number stored at location n (n = 0 - 3).

&Zx = L . . . . *Store Last Executed Dial String*

📁 The "*AT&Zx=L*" command writes the last executed dial string into the NVRAM memory location "x" (x = 0 to 3). Any of the dial modifiers may be included in the string. The maximum string length is modem dependent and usually in the range of 35 to 60 characters.

Result Codes:

OK. . . . . If x = 0, 1, 2, or 3 and the length of L < maximum allowable.

ERROR . . . . Otherwise.

&Zx = s . . . .*Store Phone Numbers*

The "*AT&Zx=s*" command writes a dial string "s" in the NVRAM memory location "x" (x = 0 to 3). The maximum length of the dial string is modem dependent but is generally greater than 36. Any of the dial modifiers may be included in the string.

Result Codes:
OK. If n = 0, 1, 2, or 3.
ERROR . . . . Otherwise.

&Zx? . . . .*Display Phone Number Memory*

The "*AT&Zx?*" command displays the phone number stored in NVRAM location "x."

)Mn . . . .*Transmit Level Adjustment for Cellular Connection*

AT)Mn sets the cellular transmit power level.

Command Option Details:
)M0 . . . . Automatically adjusts the transmit level only if the remote modem is set to)M1. Otherwise, disables the transmit power level adjustment during MNP10 link negotiation (sets the transmit level to −10 dBm). [On many modems)M0 disables the transmit level adjustment even if the remote modem)M1 is set]. (Default)
)M1 . . . . Enables transmit power level adjustment during MNP10 link negotiation. Uses the @Mn value to establish initial cellular connection. After the connection is established, the power level is determined by the modem. Both modems will attempt to reduce the transmit power to less than −10 dBm.
)M2 . . . . Enables transmit power level adjustment during MNP10 link negotiation. Uses the @Mn value to establish initial cellular connection. After connection power level is fixed.

Result Codes:
OK. n = 0, 1, or 2.
ERROR . . . . Otherwise.

*B . . . .*Display Blacklisted Numbers*

The "*AT*B*" command causes the modem to return a list of blacklisted numbers to the DTE. Permanently forbidden numbers as defined by country requirements do not appear on this list. If no numbers are blacklisted, only the OK result code is issued.

For example:
```
AT*B <CR>
NO. - PHONE NUMBER -
-------------------
1; 2125551212
2; 9001234567
3; 9007654321
4; 011496832453
5; 2837004
6; 213
OK
```

*D . . . .*Display Delayed Numbers*

The "*AT*D*" command causes the modem to send a list of the blacklist delayed numbers and the delay associated with each (used in conjunction with the *AT*B* command). The format of the response is shown by the example below (delay times are shown as hours:minutes:seconds). If no numbers are delayed, only the OK result code is issued.

For example:
```
AT*D <CR>
--------------------
1; 2125551212 2:10:00
2; 9001234567 2:30:00
3; 011496832453 0:02:00
4; 2837004 0:01:45
OK
```

*Hn . . . .*MNP-10 Link Negotiation Rate*

"*AT*Hn*" sets the connection rate for link negotiation before rate increase occurs between modems.

Command Option Details:
*H0 . . . . **Negotiates link at the highest supported speed** (default).
*H1 . . . . Negotiates link at 1200 bps. If the modems make a LAP-M connection, then the connection rate will be fixed at 1200 bps. (This command is useful with noisy or poor quality connections.)
*H2 . . . . Negotiates link at 4800 bps.

Result Codes:

OK . . . . n = 0, 1 or 2.
ERROR . . . . Otherwise.

+FCLASS = n . . . . *Select Service Class*

The +FCLASS=n command sets the active service class of the modem. The service class may be set by the DTE from the choices available within a particular modem using the AT+FCLASS=n command.

Command Options:

+FCLASS = 0 . . . . **Selects Data Mode** (default).
+FCLASS = 1 . . . . Selects Facsimile class 1 operation.
+FCLASS = 2 . . . . Selects Facsimile class 2 operation.
+FCLASS = 2.0 . . . . Selects Facsimile class 2.0 operation.

Result Codes:

OK . . . . For n = 0, 1, 2, or 2.0 (some FAX/modems may allow only Classes 0 and 1 service, while others may allow only Classes 0 and 2 or 2.0).
ERROR . . . . Otherwise.

+Hn . . . . *Enable/Disable RPI and DTE Speed*

This command enables or disables Rockwell Protocol Interface (RPI) processing and sets the DTE speed. (Applicable only to modems supporting RPI and RPI+.)

Command Option Details:

+H0 . . . . Disables RPI/RPI+.
+H1 . . . . Enables RPI mode and set DTE speed to 19200 bps.
+H2 . . . . Enables RPI mode and set DTE speed to 38400 bps.
+H3 . . . . Enables RPI mode and set DTE speed to 57600 bps.
+H11 . . . . Enables RPI+ mode (applicable only to modems supporting RPI and PC hosts with appropriate software drivers installed). When in RPI+ mode, a link is established between the modem and the host PC software driver (WinRPI or WinRPI95, for example) to allow the modem to support protocol (V.42 bis/LAP-M/MNP2-5) connections with a remote modem.

Result Codes:

OK . . . . n = 0 to 3, 11.
ERROR . . . . Otherwise.

+MS = x . . . . *Modulation Select*

The "*AT+MS = x*" command selects the modulation protocol <MOD>, optionally enables or disables automode <AUTOMODE>, and optionally specifies the lowest <MIN_RATE> and highest <MAX_RATE> connection rates using one to four subparameters. Subparameters not entered remain at their current values (enter a comma as a place holder to skip a value). For speeds at or below 14400 bps, the *ATNnS37n=x* command can alternatively be used, in which case the +MS subparameters will be modified to reflect the Nn and S37=x settings. Use of the ATNn and S37=x commands is generally discouraged but is provided for compatibility with existing communication software. Register S37 is not updated by the +MS command.

Command Option Details:

+MS = <MOD> [,[<AUTOMODE>][,[<MIN_RATE>][,[<MAX_RATE>]]]]<CR>

where:

<MOD> . . . . Is a decimal number which specifies the preferred modulation (automode enabled) or the modulation (automode disabled) to use in originating or answering a connection. The options are:

	<mod>	Modulation	Possible Rates (bps)	Notes
	0	V.21	300	
	1	V.22	1200	
	2	V.22 bis	2400 or 1200	
	3	V.23	1200	See note
	9	V.32	9600 or 4800	
	10	V.32 bis	14400, 12000, 9600, 7200, or 4800	Default 14400
	11	V.34	28800, 26400, 24000, 21600, 19200, 16800, 14400, 12000, 9600, 7200, 4800, or 2400	
	64	Bell 103	300	

Continued

	<mod>	Modulation	Possible Rates (bps)	Notes
69	Bell 212	1200		
74	V.FC	28800, 26400, 24000, 21600,		
		19200, 16800, or 14400		

Note: For V.23, originating modes transmit at 75 bps and receive at 1200 bps; answering modes transmit at 1200 bps and receive at 75 bps. In both cases, the rate is specified as 1200 bps.

<AUTOMODE> . . . . <AUTOMODE>=0 disables automode.

    1. If <MAX_RATE> is in the allowable range specified by <MOD>, the AT+MS<MOD>, 0,<MIN_RATE>,<MAX_RATE> command disables automode and sets the modem to <MAX_RATE>.

    2. If <MAX_RATE> is not in the allowable range specified by <MOD>, the AT+MS<MOD>, 0,<MIN_RATE>,<MAX_RATE> command disables automode and sets the modem to the maximum rate allowed by <MOD>.

    <AUTOMODE>=1 enables automode and the modem connects at the highest possible rate in accordance with V.8, or V.32 bis Annex A if V.8 is not supported by the remote modem.

    1. If <MAX_RATE> is greater than the highest rate supported by the modulation specified by <MOD>, the modem automodes down from the highest allowable rate of the selected modulation protocol. For example, +MS=10,1,1200,24000 selects automoding down from V.32 bis 14400 bps.

    Notes:

    1. There are modulation protocols for which there is no automatic negotiation, e.g., Bell 103 or Bell 212 (<MOD> = 64 and 69, respectively).

    2. The DTE may disable automode operation.

    3. The DTE may constrain the range of modulations available by specifying the lowest and highest rates (see <MIN_RATE> and <MAX_RATE>).

    4. By setting the <MIN_RATE> = <MAX_RATE> the command can emulate an ATNnS37n command (even for values not specified in register S37). For example AT+MS11,0,16800,16800 has no equivalent S37 register value; AT+MS10,0,12000,12000 is equivalent to ATN0S37=10.

<MIN_RATE> . . . . Is an optional decimal value which specifies the lowest rate at which the modem is allowed to establish a connection. The value is in units of bps, e.g., 2400 specifies the lowest rate to be 2400 bps. The default is 300 for 300 bps.

<MAX_RATE> . . . . Is an optional decimal value which specifies the highest rate at which the modem is allowed to establish a connection. The value is in units of bps, e.g., 9600 specifies the highest rate to be 9600 bps. The default is 14400 (14400 bps) for 14400 modems and 28800 (28800 bps) for 28800 modems.

+MS = ? . . . . *Report Modulation Capabilities*

The *AT+MS = ?* command causes the modem to report to the DTE all supported value ranges.

Command Option Details:

+MS = ? . . . . Causes the modem to respond with the string:

    +MS: (<MOD> value list), (<AUTOMODE> value list), (<MIN_RATE> value list), (<MAX_RATE> value list).

For Example:

+MS? . . . . Causes a Rockwell RC288 modem chipset to respond with the string:

    +MS: (0, 1, 2, 3, 9, 10, 11, 64, 69, 74), (0,1) (300-28800), (300-28800)

+MS? . . . . *Report Modulation Setting*

The *AT+MS?* command causes the modem to send the current modulation profile to the DTE.

Command Option Details:

+MS? . . . . Causes the modem to respond with the string:

    +MS: <MOD>,<AUTOMODE>,<MIN_RATE>,<MAX_RATE>

For example:

AT+MS? . . . . Gives: "+MS: 11,1,300,28800" the default for Rockwell's RC288 chipset.

-B . . . . *Forced Fallback to 1200 bps*

Fallback to 1200 bps at −10 dBm.

-Jn . . . .*Set V.42 Detect Phase*

Command Option Details:
-J0 . . . . Disables the V.42 detect phase.
-J1 . . . . Enables the V.42 detect phase.

-Kn . . . .*LAP-M to MNP-10 Conversion*

"*AT-Kn*" directs a connection to be converted from a LAP-M connection to a MNP10 connection if both modems support the feature.

Command Option Details:
-K0 . . . . **Does not convert a V.42 LAP-M connection to an MNP-10 connection** (default).
-K1 . . . . Converts a V.42 LAP-M connection to an MNP-10 connection if -K1 is used on both modems.
-K2 . . . . Enables V.42 LAPM to MNP 10 conversion; inhibits MNP Extended Services initiation during V.42 LAPM answer mode detection phase.

Result Codes:
OK. . . . . n = 0, 1, or 2.
ERROR . . . . Otherwise.

-Qn . . . .*MNP-10 Fallback to V.22/V.22 bis*

"AT-Qn" determines whether V.22/V.22 bis is allowed in MNP-10 mode.

Command Option Details:
-Q0 . . . . Disables fallback (fallback allowed only to 4800 bps).
-Q1 . . . . **Enables fallback–default (fallback allowed to V.22/V.22 bis).**

Result Codes:
OK. . . . . n = 0, 1, or 2.
ERROR . . . . Otherwise.

-SDR = n . . . .*Distinctive Ring Control*

The "*AT-SDR = n*" command enables or disables the detection and reporting of distinctive ring. One, two, or three distinctive ring types can be simultaneously enabled depending on the value of n (bit mapped). The detected ring type is reported in the long form (verbose) of the result code by appending the ring type number to the end of the RING message.

The modem's ring output (RI) does not change state on the first ring if "n" in AT-SDRn is greater than zero. The output does not follow the different ring cadences; it remains on during the entire ring period.

Ring Type	Ring Cadence				
1	2.0 sec ON	4.0 sec OFF			
2	0.8 sec ON	0.8 sec OFF	0.8 sec ON	4.0 sec OFF	
3	0.4 sec ON	0.2 sec OFF	0.4 sec ON	0.2 sec OFF	0.8 sec ON	4.0 sec OFF

Command Option Details:
-SDR=1 . . . . Enables Distinctive Ring Type 1 detection.
-SDR=2 . . . . Enables Distinctive Ring Type 2 detection.
-SDR=3 . . . . Enables Distinctive Ring Type 1 and 2 detection.
-SDR=4 . . . . Enables Distinctive Ring Type 3 detection.
-SDR=5 . . . . Enables Distinctive Ring Type 1 and 3 detection.
-SDR=6 . . . . Enables Distinctive Ring Type 2 and 3 detection.
-SDR=7 . . . . Enables Distinctive Ring Type 1, 2, and 3 detection.

Result Codes:
OK. . . . . n = 0 to 7.
ERROR . . . . Otherwise.

-SEC? . . . .*Reports the Current -SEC Settings*

Command Option Details:
-SEC? . . . . Reports if MNP10-EC is enabled or not and what the transmit level is set to, e.g., 1,12 (indicating MNP10-EC enabled and transmitting at -15 dBm).

-SEC = n,[L] . . . .*Enable/Disable MNP10-EC*

The "*AT-SEC = n,<L>*" command enables or disables MNP10-EC operation, and the optional parameter <tx level>sets the transmit level used in cellular connections.

Command Option Details:

-SEC = 0............... Disables MNP-EC operation. The transmit level is set by the value in register S91.

If the modem is set with the "*AT-SEC = 0*" command, may automatically be set with "*AT-SEC = 1*" if the remote modem indicates Cellular in the V.8 phase or if a Cellular Driver is loaded and the Cell Phone is attached.

-SEC=1,[<L>]........... Enables MNP10-EC; the transmit level will be defined by the <L> parameter. The range of range of <L> is 0 to 30 (corresponding to a range of 0 dBm to -30 dBm). If <L> is omitted, the default transmit level defined in register S91 is used.

For example, the command "*AT-SEC = 1,15*" enables MNP10-EC and sets the transmit level to -15 dBm.

Result Codes:

OK..................... n=0, 1, or 1 and <tx level>=0 to 30.
ERROR .... Otherwise.

-SSE = n .... Simultaneous Voice/Data Control

The "*AT-SSE = n*" command enables or disables digital simultaneous voice and data (DSVD) in modem models supporting DSVD.

Command Option Details:

-SSE = 0............... **Disables simultaneous voice and data** (default).
-SSE = 1............... Enables simultaneous voice and data.

Result Codes:

OK..................... n = 0 or 1.
ERROR .... Otherwise.

-U .... Forced Transmit Level Change During Fall Forward

"*AT-U*" forces the transmit level to change to a predetermined level during a fall forward. This command has no effect if the line speed is already 2400 bps.

Command Option Details:

-U0 .... Allows automatic attenuation to -10 dBm.
-U1 .... Allows automatic attenuation to as low as -18 dBm.
-U2 .... Allows automatic attenuation to as low as -22 dBm.
-U3 .... Allows automatic attenuation to as low as -25 dBm.
-U4 .... Forces 2400 bps transmit level to -10 dBm.
-U5 .... Forces 2400 bps transmit level to -18 dBm.
-U6 .... Forces 2400 bps transmit level to -22 dBm.
-U7 .... Forces 2400 bps transmit level to -25 dBm.

:En .... Compromise Equalizer Setting

@Mn .... MNP-10 Power Level

AT@Mn sets the initial cellular transmit power level. The range of "n" is 0 to 30 corresponding to a transmit level of 0 to -30 dBm.

Result Codes:

OK..................... n = 0 to 30.
ERROR .... Otherwise.

\An .... Maximum MNP Block Size for Stream Links

"*AT\An*" is an ECC command that sets the maximum block size for MNP connections. The parameter value, if valid, is written into register S40 bits 5 and 6 in some modems and 6 and 7 in others.

Command Option Details:

\A0................... Maximum block size = 64 characters.
\A1................... **Maximum block size = 128 characters** (default on most modems).
\A2................... Maximum block size = 192 characters.
\A3................... Maximum block size = 256 characters.

Result Codes:

OK..................... If n = 0, 1, 2, or 3 (on some modems n>3 will report OK and n will be set to 3).
ERROR .... Otherwise.

\Bn . . . .*Transmit Break Control*

"*AT\Bn*" is an ECC command that sets the line break time duration. In a nonerror correction mode, the modem transmits a break of length 100 ms to 900 ms corresponding to n = 1 to 9. If n=0, the modem will use the default 3; if n>9, the modem will set n=9. In an error-correction mode, the modem sends a Link Attention PDU to the remote modem.

Command Option Details:
\B1 thru \B9. . . . . Sets the transmitted break time to 100ms, 200ms, . . . , 900ms. (Default = 300ms).

Result Codes:
OK. . . . . If connected.
NO CARRIER. . . . . If not connected or connected in fax modem mode.
ERROR . . . . Otherwise.

See also the \Kn command.

\En . . . .*Optimize Echo Cancellation*

The "*AT\En*" command causes the modem to optimize its echo cancellation networks. The command is issued before a call is originated. When the next call is originated, the modem will try to optimize internal networks so that the echo is minimized for that connection. The value obtained in this optimization is stored in NVRAM and used for all subsequent connections or until another \En command and call origination attempt is issued.

Result Codes:
OK. . . . . At successful completion of operation.
ERROR . . . . .Otherwise.

\Gn . . . .*Modem-to-Modem XON/XOFF Flow Control*

The "*AT\Gn*" command sets the modem to modem software flow control for a NORMAL connection. This command is ignored when error correction is enabled, e.g., &Q5, &Q8 or &Q9.

Command Option Details:
\G0 . . . . **Disables XON/XOFF flow control** (default).
\G1. . . . . Enables XON/XOFF flow control.

Result Codes:
OK. . . . . If n = 0, or 1 (on some modems n>1 will report OK and n=1 is used).
ERROR . . . . Otherwise.

\Jn . . . .*Port Rate Adjustment Control*

The "*AT\Jn*" command selects if the communication rate between the DCE and DTE is adjusted.

Command Option Details:
\J0 . . . . **Disables port rate adjustment.** Serial port is independent of the connection.
\J1 . . . . Enables port rate adjustment. After a connection is made, the serial port adjusts to the connection rate.

\Kn . . . .*Break Processing Control*

"*AT\Kn*" determines how the local modem will respond when a BREAK is received from the remote modem or the DCE. While in MNP mode, the remote modem's BREAK setting determines how the local modem will handle the BREAK. The value is written in register S40.

Command Option Details:
When a BREAK is received from DTE in the data transfer mode, the modem response is:
\K0, \K2, or \K4. . . . . Enters COMMAND mode without sending BREAK to remote.
\K1. . . . . Clears modem and terminal buffers; sends a BREAK to remote modem.
\K3. . . . . Does not clear buffers but sends a BREAK to the remote modem.
\K5 . . . . **Sends BREAK to the remote modem in sequence with any transmitted data** (default).

When a BREAK is received from a remote modem during Normal mode or nonerror-corrected mode, the modem response is:
\K0 or \K1 . . . . Clears the modem and terminal buffers; sends a BREAK to the DTE.
\K2 or \K3 . . . . Does not clear the buffers but sends a BREAK to the DTE.
\K4 or \K5. . . . . **Sends BREAK in sequence with any data in buffer** (default).

When a BREAK is received from the DTE during Direct mode, the modem response is:
\K0, or \K2, or \K4. . . . . . . Enters COMMAND mode without sending BREAK to remote.
\K1, or \K3, or **\K5** . . . . . . **Sends a BREAK to the remote modem** (default).

In LAP-M, BREAKs are timed, and the modem attempts to preserve the duration of the BREAK when transmitting to the remote modem.

In MNP-4, BREAKs are not timed; hence, long BREAKs are the same as short BREAKs.

Result Codes:
OK . . . . If n = 0 through 5 (on some modems n>5 will report OK and n = 5 is used).
ERROR . . . . Otherwise.

\Ln . . . .*MNP Block Transfer Control*

"AT\Ln" determines whether the modem will use Block or Stream mode for MNP link.

Command Option Details:
\L0 . . . . **Uses stream mode for MNP links** (default).
\L1 . . . . Uses block mode for MNP links.

Result Codes:
OK . . . . If n = 0, or 1 (on some modems n>1 will report OK and n=1 is used).
ERROR . . . . Otherwise.

\Wn . . . .*Operating State While Connecting*

"AT\Wn" selects the preferred error-correcting mode to be negotiated in subsequent data connections. This command is manufacturer and model dependent.

Command Option Details:
\N0 . . . . Normal mode. Disables error correction. (*AT\W0* forces &Q6.)
\N1 . . . . Directs mode with serial interface (*AT\W1* forces &Q0). Same as \N0 with parallel interface selection.
\N2 . . . . MNP Reliable (error correction) mode. If the modem attempts to establish a connection and fails, the modem disconnects. (*AT\W2* forces %Q5, S36 = 4, and S48 = 7.)

                        To establish MNP after a connection has been established, use the commands AT\O, AT\U, and AT\Y.

\N3 . . . . **AutoReliable mode.** The modem will attempt to establish an error-correcting connection but will fall back to Normal mode if it fails to establish a reliable link. (AT\N3 forces &Q5, S36 = 4, and S48 = 7.)

                        To establish an MNP link after a connection has been established, use the commands AT\O, AT\U, and AT\Y.

\N4 . . . . Forces LAP-M / MNP Reliable mode. Failure to establish an error-correction connection results in the modem disconnecting. (*AT\W4* forces &Q5 and S48=0.)

                        Note: The -K1 command can override the \N4 command.

\N5 . . . . Forces MNP reliable mode. Failure to establish an MNP error-correction link results in the modem disconnecting. (*AT\W5* forces &Q5, S36=4, and S48=128.)

Result Codes:
OK . . . . If n = defined n (on some modems n > maximum defined value will report OK and the maximum value will be used).
ERROR . . . . Otherwise.

\O . . . .*Originate MNP Reliable Mode Request*

"AT\O" forces the modem to initiate a Reliable connection regardless of whether the local modem originated or answered the call.

If a Reliable connection already exists, the modem returns to the online state.

If the modem fails to establish a Reliable link, the modem returns to the previous state.

Result codes:
The CONNECT result code indicates the type of connection established.

\Qn . . . .*Serial Port Flow Control*

"AT\Qn" controls if local DTE to DCE data flow control is present and whether it is hardware flow control or software flow control.

Command Option Details:
\Q0 . . . . Disables flow control.
\Q1 . . . . Enables bidirectional XON/XOFF flow control.
\Q2 . . . . Enables unidirectional hardware flow control.
\Q3 . . . . **Enables bidirectional hardware (RTS/CTS) flow control.**

Result Codes:
OK . . . . If n = 0, 1, 2, or 3.
ERROR . . . . Otherwise.

\Tn . . . . *Inactivity Timer*

"*AT\Tn*" establishes how long the modem will remain connected with no activity. The modem disconnects if no activity in time n minutes. In MNP mode, the activity timer is reset by both transmit and receive data. In non-MNP mode, the timer is reset only by transmitted data. The timer is ignored in Direct mode.

n = 0 . . . . **Timer disabled** (default).
n=1 to x . . . . Modem disconnects after n minutes without activity. The maximum "x" varies between 30 and 90 for different modems.

\U . . . . *Accept MNP Reliable Request*

"*AT\U*" directs the local modem to accept an MNP Reliable connection regardless of whether the modem originated or answered the call. If a Reliable connection already exists, the modem returns online.

If it fails, or if no request is received in 12 seconds, the modem returns to its previous state.

The CONNECT code indicates the type of connection established.

\Vn . . . . *(1) Modify Result Codes*

"*AT\Vn*" selects which set of result codes is reported: normal or a modified set for MNP.

Command Option Details:
\V0 . . . . **Enables codes defined by ATV command.**
\V1 . . . . Enables modified MNP codes.

Result Codes:
OK . . . . If n = 0, or 1.
ERROR . . . . Otherwise.

\Vn . . . . *(2) Single Line Connect Message*

AT\Vn switches the normal multiline connect message to a single-line format, that is, *CONNECT <DTE speed></Modulation></Protocol></Compression></Line Speed>*

Where:
DTE Speed . . . . Is DTE to DCE speed.
Modulation . . . . Is V32 for ITU-T V.32 or V.32 bis, VFC for proprietary V.FC™, or V34 for ITU-T V.34.
Protocol . . . . Is LAPM for LAP-M, ALT for Microcom's MNP, or NONE.
Compression . . . . Is V42 bis for ITU-T V.42 bis compression, or CLASS5 for Microcom's MNP-5 compression. (Omitted parameter if no compression is used.)
Line Speed . . . . Is a single numeric value for symmetric rates (e.g., 28800) or
A pair of values for asymmetric links (e.g., 1200TX/75RX).

Command Option Details:
\V0 . . . . Codes are controlled by the ATX and ATW commands and the value in register S95.
\V1 . . . . Connect messages are displayed in the single-line format. The display is subject to the command settings V (Verbose) and Q (Quiet). In Non-Verbose mode (V0), single-line connect messages are disabled, and a single numeric result code is generated for CONNECT DTE.

Result Codes:
OK . . . . If n = 0, or 1.
ERROR . . . . Otherwise.

\Wn . . . . *Split Speed Operation (V.23)*

Command Option Details:
\W0 . . . . **Disables split speed mode** (default).
\W1 . . . . Enables split speed mode—V.23 (forces ATF3).

Result Codes:
OK . . . . If n = 0, or 1.
ERROR . . . . Otherwise.

\Xn . . . .*Set XON/XOFF Pass-through*

"*AT\Xn*" determines whether or not the software data flow characters are passed on to the remote modem.

Command Option Details:
\X0. . . . . **Processes flow-control characters.**
\X1. . . . . Processes flow-control characters and passes them through to the local and remote equipment so they
                                       can process them.

Result Codes:
OK. . . . . If n = 0, or 1.
ERROR . . . . Otherwise.

\Y . . . .*Switch to MNP Reliable Mode*

"*AT\Y*" directs the modem to try to establish an MNP Reliable connection.

If it fails, the modem returns to its previous state.

If a Reliable connection already exists, the modem returns online.

\Z . . . .*Switch from MNP to Normal Mode*

"*AT\Z*" causes both the local and remote modem to switch back to the NORMAL mode from an MNP Reliable connection. Data in
the buffers is lost. Either the local or the remote modem may issue the \Z command to fall back to Normal mode.

^C2 . . . .*Download Cellular Phone Driver*

The data interface to cellular phones differs among manufacturers and models; hence, a unique cellular phone driver for each phone
or group of phones is required. Therefore, the particular phone driver needed must be downloaded from the PC into the modem's RAM
before the modem can be used directly with the cellular phone. If a driver is not loaded, the modem will operate as a normal landline
modem.

The "*AT^C2*" command initiates the cellular phone driver download function. Upon receipt of the command, the modem responds
with the "OK" message. The user then performs an ASCII transfer of the driver from the host to the modem, typically using a communications software package (with transmit flow control turned off).

Once the driver is downloaded to a modem, the modem is connected to a cellular phone, and the phone is powered on, dial & answer
functions will be routed through the phone instead of the landline digital access arrangement (DAA). This means no special commands
are needed to place or answer calls; the same AT commands and software packages that are used for landline communication sessions
can be used. If the cellular phone is not connected or is powered off, dial & answer functions will be routed through the landline DAA.
Further, if a V.42 bis connection is established, the cellular phone driver will be purged so that the V.42 bis dictionaries can be increased.

While the modem is being used with a cellular phone, it will respond with normal result messages with the meaning modified on NO
DIALTONE and RING, i.e.:
NO DIALTONE. . . . . Indicates that cellular service is not currently available.
RING. . . . . Indicates that the cellular phone is receiving an incoming call.

Command Option Details:
^C2 . . . . Downloads Cellular Phone command. Following the initial response code OK, the download process is
                                       initiated.

Result Codes:
OK. . . . . Response to AT^C2 and successful download.
ERROR . . . . Cellular phone driver download not completed successfully, e.g.,
                                       the modem is connected,
                                       the checksum of record is not correct,
                                       the driver size is larger than 2k bytes, or
                                       an invalid driver is downloaded.

^I . . . .*Identify Cellular Phone Driver*

The "AT^I" command causes the modem to report the identification of the loaded cellular phone driver.

Command Option Details:
^I . . . . Requests driver identity.

Result Codes:

CELLULAR DRIVER: <Manufacturer and identity>
<(c) Copyright information>
<Version> <date>
OK. . . . . The exact response is dependent on the loaded driver; however, it is generally similar to the presented form.
ERROR . . . . Cellular phone driver is not loaded.

^T6 . . . .*Report Status of Cellular Phone*

The "AT^T6" command causes the status of the cellular phone connected to the modem to be reported to the DTE. The status is reported in a single-bit-mapped byte expressed as a decimal value.

Command Option Details:

^T6 . . . . Reports cellular telephone status.

Result Codes:

<value> . . . . Value represents the status of the cellular phone expressed as a bit-mapped byte; each bit defined as:

bit 0	1 = Cellular phone is receiving an incoming call.
bit 1	1 = Cellular phone is in use.
bit 2	1 = Cellular phone is locked (cannot be used).
bit 3	1 = No service for cellular phone (not a signal strength indicator).
bit 4	1 = Cellular phone is powered on.
bit 5	1 = Cellular driver is initialized.
bit 6	0 = Reserved (0).
bit 7	1 = Cellular cable detected.

OK

FAX COMMANDS

Command Structure and Syntax

The facsimile commands follow the same principle as the basic AT commands, that is, each command line has a prefix, a body, and a terminator. However, in addition to the prefix "AT" at the beginning of each command line, a "+F" must precede each command, e.g., "*AT+FCLASS?*" Individual commands may be terminated by the semicolon character ";" or by the <CR>. The <CR> must be the last terminator on a command line. The semicolon may be used to separate commands when more than one command is in the command line. The other difference is in the command itself; it may be several characters in length.

Some general FAX command rules include:

1. The AT+F<command> commands must be entered completely; otherwise an ERROR response is sent by the modem.

2. All response messages are preceded <u>and</u> followed by <CR><LF>. Multiple response commands on a single command line will display a blank line between them. For example: the command *AT+FCLASS? +FCLASS=?* on a modem in the data mode will appear as:

```
at+fclass?+class=?
0
0, 1, 2.0
OK
```

3. Fax Class 2 commands may be separated by the ";" character. The ";" may be omitted if desired.

4. All Class 2 commands are assumed to be the final command on a command line. Additional characters will be ignored (although some modems will produce an ERROR message result).

5. An ERROR message will be generated for any of the following conditions:
   a. A Class 1 command is received while the modem is in a Class 2 state.
   b. A Class 2 command is received while the modem is in a Class 1 state.
   c. A Class 1 or Class 2 action command is received while in the modem is in a data mode.
   d. A Class 2 read-only parameter is given the "=" form of a +F command, e.g., AT+FAXERR=5.
   e. A Class 2 action command is given the inappropriate "=" or "=?", e.g., AT+FDR=?

Many commands can be expressed in three forms: a form to report the range of values possible, a form to set a value, and a form to report the current value. The structure of these forms is:

+F<command>=? . . . .*Report Operating Capabilities*

The *+F<command>=?* command instructs the modem to report its range of permissible values for the parameter <command> back to the DTE.

Some examples are:

+FCLASS=?	0,1	(if data and CLASS 1 modes possible)
	0,2	(if data and CLASS 2 modes possible)
	0,1,2	(if data, CLASS 1 and CLASS 2 modes possible)
	0,1,2.0	(if data, CLASS 1 and CLASS 2.0 modes possible)

+FTM=? 3,24,48,72,96   (for a Rockwell RC96AC)
+FTM=? 3,24,48,72,73,74,96,97,98,121,122,145,146   (for a Rockwell RC144AC)
+FRM=? 3,24,48,72,96   (for a Rockwell RC96AC)

+F<command>=n . . . . .Set a Parameter Value

The *+F<command>=n* command instructs the modem to set the specified parameter <command> to the value "n."

For example:
+FCLASS=0 . . . . Instructs the modem to go to data mode.
+FCLASS=1 . . . . Instructs the modem to go to fax Class 1 mode.
+FCLASS=2 . . . . Instructs the modem to go to fax Class 2 mode.

+F<command>? . . . . .Report Active Configuration

The *+F<command>?* command instructs the modem to report its current active value for the parameter <command> back to the DTE.

For example:
+FCLASS? . . . . 0 if the modem is in the data mode.
                               1 if the modem is in the fax Class 1 state.
                               2 if the modem is in the fax Class 2 state.
                               2.0 if the modem is in the fax Class 2 state.

COMMON FAX COMMANDS

A . . . . .Answer an Incoming Call

See description on page 592.

D . . . . .Dial an Outgoing Call

See description on page 594.

+FCLASS=n . . . . .Select Service Class

The +FCLASS=n command sets the active service class. The service class may be set by the DTE from the choices available within a particular modem using the *AT+FCLASS=n* command.

Command Options:
+FCLASS = 0 . . . . **Selects Data Mode** (default).
+FCLASS = 1 . . . . Selects Facsimile class 1 operation.
+FCLASS = 2 . . . . Selects Facsimile class 2 operation.
+FCLASS = 2.0 . . . . Selects Facsimile class 2.0 operation.

Result Codes:
OK. . . . . For n = 0, 1, or 2 (some FAX/modems may allow only Classes 0 and 1 service, while others may allow
                                only Classes 0 and 2).
ERROR . . . . Otherwise.

CLASS 1 FAX COMMANDS

Class 1 fax commands are defined in the EIA 578 specification.

+FAEn . . . . .Data/Fax Auto Answer

The +FAEn command instructs the modem either to restrict answering to class 1 or to automatically detect whether the calling station is a fax class 1 modem or data modem, and answer accordingly.

Command Options:
n = 0. . . . . **Disables data/fax auto answer mode.** The modem answers as a fax modem only (default).
n = 1 . . . . Enables data/fax auto answer mode. The modem answers as a fax or data modem.

+FRH=n . . . . .Receive Data with HDLC Framing

The +FRH=n command causes the modem to receive frames using the HDLC protocol and to use the modulation selected by n as defined in command options.

Command Options:

n = 3	V.21 ch 2	300 bps	n = 96	V.29	9600 bps
n = 24	V.27 ter	2400 bps	n = 97	V.17	9600 bps long
n = 48	V.27 ter	4800 bps	n = 98	V.17	9600 bps short
n = 72	V.29	7200 bps	n = 121	V.17	12000 bps long
n = 96	V.29	9600 bps	n = 122	V.17	12000 bps short
n = 73	V.17	7200 bps long	n = 145	V.17	14400 bps long
n = 74	V.17	7200 bps short	n = 146	V.17	14400 bps short

Response Codes:

ERROR . . . . If the command is issued while the modem is on-hook.

+FRM=n . . . .*Receive Data*

The +FRM=n command causes the modem to enter the receive mode and to use the modulation selected by n as defined in command options. If the modem detects the carrier specified by "n," the modem enters the data receive mode; otherwise an error condition is reported, and the modem remains in the command mode.

Command Options:

n = 3	V.21 ch 2	300 bps	n = 96	V.29	9600 bps
n = 24	V.27 ter	2400 bps	n = 97	V.17	9600 bps long
n = 48	V.27 ter	4800 bps	n = 98	V.17	9600 bps short
n = 72	V.29	7200 bps	n = 121	V.17	12000 bps long
n = 96	V.29	9600 bps	n = 122	V.17	12000 bps short
n = 73	V.17	7200 bps long	n = 145	V.17	14400 bps long
n = 74	V.17	7200 bps short	n = 146	V.17	14400 bps short

Response Codes:

CONNECT . . . . If the specified carrier is detected.
ERROR . . . . If the command is issued while the modem is on-hook.
NO CARRIER . . . . If no carrier is detected before the time specified in register S7.
+FCERROR . . . . If a carrier is detected that is not the one specified by "n."

+FRS=n . . . .*Receive Silence*

The +FRS=n command causes the modem to report back to the DTE with an OK result code after n 10-millisecond intervals of silence have been detected on the line.

Response Codes:

OK. . . . . 1. After "n" ms intervals of silence have elapsed.
                          2. If a character is received from the DTE (modem discards the DTE character and aborts the command).
ERROR . . . . If the command is issued while the modem is on-hook.

+FTH=n . . . .*Transmit Data with HDLC Framing*

The +FTH=n command causes the modem to transmit data using HDLC protocol and to use the modulation selected by n as defined in command options.

Command Options:

n = 3	V.21 ch 2	300 bps	n = 96	V.29	9600 bps
n = 24	V.27 ter	2400 bps	n = 97	V.17	9600 bps long
n = 48	V.27 ter	4800 bps	n = 98	V.17	9600 bps short
n = 72	V.29	7200 bps	n = 121	V.17	12000 bps long
n = 96	V.29	9600 bps	n = 122	V.17	12000 bps short
n = 73	V.17	7200 bps long	n = 145	V.17	14400 bps long
n = 74	V.17	7200 bps short	n = 146	V.17	14400 bps short

Response Codes:

ERROR . . . . If the command is issued while the modem is on-hook.

+FTM=n . . . .*Transmit Data*

The +FTM=n command causes the modem to transmit data and to use the modulation selected by n as defined in command options.

Command Options:

n = 3	V.21 ch	2300 bps	n = 96	V.29	9600 bps
n = 24	V.27 ter	2400 bps	n = 97	V.17	9600 bps long
n = 48	V.27 ter	4800 bps	n = 98	V.17	9600 bps short
n = 72	V.29	7200 bps	n = 121	V.17	12000 bps long
n = 96	V.29	9600 bps	n = 122	V.17	12000 bps short
n = 73	V.17	7200 bps long	n = 145	V.17	14400 bps long
n = 74	V.17	7200 bps short	n = 146	V.17	14400 bps short

Response Codes:

ERROR . . . . 1. If the command is issued while the modem is on-hook.

                      2. If the last character transmitted while the transmit buffer is empty is a <NUL> and 5 seconds elapse with no data transfers from the DTE.

OK. . . . . After the last character is transmitted, the transmit buffer is empty and the last character is not a <NUL>.

CONNECT . . . . Otherwise.

+FTS =n . . . . *Stop Transmission and Wait*

The +FTS=n command causes the modem to terminate transmission and to wait for n 10-millisecond intervals before responding with the OK result code.

Response Codes:

OK. . . . . After "n" ms intervals of silence have elapsed.

ERROR . . . . If the command is issued while the modem is on-hook.

CLASS 2 FAX COMMANDS

In Class 2, the data communications equipment (DCE) makes and terminates calls, manages the communication session, negotiates (T.30 protocol), and transports the image date to the data terminal equipment (DTE). The T.4 protocol management of image data, etc. is done by DTE.

FAX CLASS 2 COMMANDS

Command	Function	Page
DTE ACTION COMMANDS		
A	Answer an incoming call	592
D	Originate an outgoing call	594
+FCLASS = n	Set service class	635
+FDR	Begin or continue phase C data reception	641
+FDT =	Transmit data	643
+FET = n	Transmit page punctuation	643
+FK	Terminate session	644
DCE RESPONSES		
+FCFR	Indicate confirmation to receive	640
+FCON	Facsimile connection response	640
+FCSI :	Report called station identification	640
+FDCS :	Report current session	641
+FDIS :	Report remote identification	642
+FET :	Post page message response	643
+FHNG	Call termination with status	643
+FPTS :	Page transfer status	646
+FTSI :	Report transmit station identification	647
SESSION PARAMETERS		
+FAA	Adaptive answer enable	638
+FAXERR	Fax error value	638
+FBOR	Phase C data bit transmission order	639
+FBUF ?	DCE buffer size (read only)	639
+FCR	Capability to receive	640
+FDCC =	DCE capabilities parameters	640
+FDCS =	Current session results	641
+FDIS =	Current session parameters	642
+FLID =	Local identification string	645
+FMDL ?	Identify DCE model	645

Continued

FAX CLASS 2 COMMANDS (*CONTINUED*)

Command	Function	Page
+FMFR ?	Identify DCE manufacturer	645
+FPHCTO	Phase C timeout	645
+FPTS =	Page transfer status	646
+FREV ?	Identify revision	646

+FAA = <0 | 1> ....*Auto Answer*

The "*AT+FAA = n*" session parameter command enables the DCE to adaptively answer a call as a data or fax terminal.

When enabled (+FAA=1), an answered call (via the ATA command) is first assumed to be a data call and the DCE attempts a data mode handshake—a process that may take several seconds. If the connection is successful, FCLASS is set to 0 and the DTE is notified of the connection type via a "D" result code. If the connection is unsuccessful, a facsimile Class 2 connection is then assumed. A successful handshake results in a "FAX" result code from the DCE to the DTE.

Command Options:

+FAA = 0 .... Auto answer is disabled, i.e., constrains the DCE to answer as set by the +FCLASS command.

+FAA = 1 .... Auto answer as a facsimile modem or a data modem depending on the call signaling tones.

When "*AT+FAA=1*" is issued, the DCE will adaptively answer as a facsimile DCE or as a data DCE and set the +FCLASS parameter. For example, if the DCE answers as a data DCE, it resets the +FCLASS parameter to 0 and issues the appropriate final result code (e.g., CONNECT or NO CARRIER) to the DTE.

Result Codes:

CONNECT XXXX .... DCE status response if the call is detected to be a data call.

FCON .... DCE status response if the call is detected to be a fax call.

NO CARRIER .... DCE status response if no carrier is detected.

+FAXERR = ? ....*Fax Error Value*

A read only session parameter indicating the cause of the call termination (both normal and abnormal terminations). The DCE clears +FAXERR to 0 at the beginning of Phase A at off-hook time and sets it at the conclusion of a fax session, e.g., by +FHNG.

FAXERR STATUS CODES

Code	Description	Code	Description
0-9	CALL PLACEMENT AND TERMINATION	52	No response to MPS repeated 3 times
		53	Invalid response to MPS
0	Normal and proper end of connection	54	No response to EOP repeated 3 times
1	Ring Detect without successful handshake	55	Invalid response to EOM
2	Call aborted, from +FK or AN	56	No response to EOM repeated 3 times
3	No Loop Current	57	Invalid response to EOM
		58	Unable to continue after PIN or PIP
10-19	TRANSMIT PHASE A & MISCELLANEOUS ERRORS	70-89	RECEIVE PHASE B HANG-UP CODES
10	Unspecified Phase A error		
11	No Answer (T.30 T1 timeout)	70	Unspecified Receive Phase B error
		71	RSPREC error
20-39	TRANSMIT PHASE B HANG-UP CODES	72	COMREC error
		73	T.30 T2 timeout, expected page not received
20	Unspecified Transmit Phase B error	74	T.30 T1 timeout after EOM received
21	Remote cannot receive or send		
22	COMREC error in transmit Phase B	90-99	RECEIVE PHASE C HANG-UP CODES
23	COMREC invalid command received		
24	RSPEC error	90	Unspecified Receive Phase C error
25	DCS sent three times without response	91	Missing EOL after 5 seconds
26	DIS/DTC received 3 times; DCS not recognized	92	Unused code
27	Failure to train at 2400 bps or +FMINSP value	93	DCE to DTE buffer overflow
28	RSPREC invalid response received	94	Bad CRC or frame (ECM or BFT modes)
40-49	TRANSMIT PHASE C HANG-UP CODES	100-119	RECEIVE PHASE D HANG-UP CODES
40	Unspecified Transmit Phase C error	100	Unspecified Receive Phase D errors
43	DTE to DCE data underflow	101	RSPREC invalid response received
		102	COMREC invalid response received
50-59	TRANSMIT PHASE D HANG-UP CODES	103	Unable to continue after PIN or PIP
50	Unspecified Transmit Phase D error		
51	RSPREC error	120-255	RESERVED CODES

+FBOR = <0 | 1> . . . . *Phase C Data Bit Order*

The "*AT+FBOR=n*" parameter controls the mapping of facsimile data between the public switched telephone network (PSTN) and the DTE-DCE links by one of two methods:

DIRECT: . . . . The first bit transferred to each byte on the DTE-DCE link is the first bit transferred on the PSTN data carrier.

REVERSED: . . . . The last bit transferred of each byte on the DTE-DCE link is the first bit transferred on the PSTN data carrier.

The mapping is effective for Phase C facsimile data (ITU-T T.4 encoded data) transferred during execution of +FDT (data transmit) or +FDR (begin/continue phase C data receive) commands.

+FBOR=<0 | 1> does not affect the bit order of control characters generated by the DCE.

Command Options:
+FBOR=0 . . . . **Selects direct bit order for Phase C data.** Default value.
+FBOR=1 . . . . Selects reversed bit order for Phase C data.

+FBUF? . . . .*Buffer Size*

The *AT+FBUF* read only parameter allows the DTE to determine the characteristics of the DCE's data buffer. This is useful in optimizing the DTE-DCE flow control.

Command Options:
+FBUF? . . . . Requests DCE buffer characteristics.

Result Codes:
<bs>,<xoft>,<xont>,<bc> . . where:

                         <bs> = total buffer size
                         <xoft> = XOFF threshold
                         <xcont> = XON threshold
                         <bc> = current buffer byte count

+FCFR . . . . *Indicate Confirmation to Receive*

The DCE sends a "*+FCFR*" response to the DTE upon reception of an acceptable training burst and a valid FDCS (current session report) signal from the remote machine. This indicates that the DCE will receive Phase C data after the remote station receives the local DCE's +FCFR message. The +FCFR message is generated by the execution of an AT+FDR (begin/continue phase C data receive) command.

+FCIG = <string> . . . .*Set Polling ID*

+FCIG = allows setting and reporting a 20-character Local PollingID String. The string may be either numeric (<space>, "+", and the digits 0-9) or alphanumeric as specified in ITU-T T.30. If *+FCIG* is not a null string, it generates a CIG frame (the default value is empty or null string).

If the DCE supports use of numeric values only, the response to an "*AT+FCIG=?*" command is "(20)(32, 43, 48-57)." If the DCE supports all printable ASCII, the response is "(20)(32-127)<CR><LF>". The first "(20)" represents string length, and the following field reports the range of supported string values (decimal ASCII character values).

Command Options:
AT+FCIG=?. . . . . (20)(32, 43, 48-57) if the DCE only supports numeric strings or
. . . . (20)(32-127)<CR><LF> if the DCE supports alphanumeric strings.
AT+FCIG=<string> . . . . . . Default value is empty or null string.

+FCON . . . .*Facsimile Connection Response*

The DCE response "*+FCON*" indicates connection with a fax machine. It is generated in response to an originate (ATD) or answer (ATA) command and is released by the detection of HDLC flags in the first received frame.

+FCQ = n . . . .*Copy Quality Checking*

The *+FCQ* parameter controls Copy Quality checking by a receiving facsimile DCE. The DCE will generate Copy Quality OK (MCF) responses to complete pages and sets +FPTS = 1.

Result Codes:
+FCQ = 0 . . . . DCE generates Copy Quality OK (MCF) responses to complete pages and sets +FPTS = 1.

+FCR = <0 | 1> . . . .*Capability to Receive*

The "*AT+FCR = 0*" command indicates that the DCE will not receive message data. This flow control command can be issued when the DTE has insufficient storage. AT+FCR is sampled in both ITU-T T.30 Phase A and Phase D.

Command Options:
AT+FCR = 0 . . . . **Indicates message data will not be received.** Default value.
AT+FCR = 1 . . . . Indicates message data will be received.

+FCSI:<string> . . . .*Report the Called Station ID*

The DCE response which reports the received called station ID string, if any. This message is generated by the execution of the originate (ATD), answer (ATA), AT+FDT (data transmit), or AT+FDR (begin/continue phase C data receive) commands.

Result codes:

+FCSI:<CSI ID string>. . . . Called Station ID.

+FDCC:<string> . . . .*DCE Capabilities Parameters*

+FDCC<string> allows the DTE to sense and/or constrain the capabilities of the DCE to the choices defined by ITU-T T.30. The values of the <string> subparameters VR, BR, WD, LN, DF, EC, BF, and ST are defined in the following table. When +FDCC is modified by the DTE, the DCE copies *+FDCC* into +FDIS (session parameters).

SESSION SUBPARAMETER CODES

Label	Function	Value	Description	
VR	Vertical Resolution	0 1	Normal, 98 lpi Fine, 196 lpi	
BR	Bit Rate[1]	0 1 2 3 4 5	2400 bit/s V.27 ter 4800 bit/s V.27 ter 7200 bit/s V.29 or v.17 9600 bit/s V.29 or v.17 12000 bit/s V.33 or v.17 14400 bit/s V.33 or v.17	
WD	Page Width[2]	0 1 2 3 4	1728 pixels in 215 mm 2048 pixels in 255 mm 2432 pixels in 303 mm 1216 pixels in 151 mm 864 pixels in 107 mm	
LN	Page Length[3]	0 1 2	A4, 297 mm B4, 364 mm unlimited length	
DF	Data Compression[4]	0 1 2 3	1-D modified Huffman 2-D modified Read 2-D uncompressed mode 2-D modified Read	
EC	Error Correction (Annex A/T.30)[5]	0 1 2	Disable ECM Enable ECM, 64 bytes/frame Enable ECM, 256 bytes/frame	
BF	Binary File Transfer (BFT)[6]	0 1	Disable BFT Enable BFT	
ST	Scan Time/Line		VR = normal	VR = fine
		0	0 ms	0 ms
		1	5 ms	5 ms
		2	10 ms	5 ms
		3	10 ms	10 ms
		4	20 ms	10 ms
		5	20 ms	20 ms
		6	40 ms	20 ms
		7	40 ms	40 ms

Notes: [1]ITU-T T.30 does not provide for the answering station to specify all speeds exactly using the DIS frame. Implementation of some BR codes (e.g., codes 2, 4, and 5) by an answering DCE is manufacturer specific.

[2]Page width modes 3 and 4 are not implemented in all fax/modems.

[3]Page length modes 2 and 3 are not implemented in all fax/modems.

[4]Not all data compression modes are implemented (i.e., modes 2 and 3 are frequently omitted).

[5]ECM is not implemented in all fax modems; further, although ECM transmit and receive may work correctly, the modem may not tell the transmitting machine that the fax was received correctly.

[6]BFT is not implemented in all fax modems; further, although BFT transmit and receive may work correctly, the modem may not tell the transmitting machine that the fax was received correctly.

+FDR . . . .*Begin/Continue Phase C Receive Data*

Class 2 action command.

+FDCS:<string> . . . .*Report Current Session Capabilities*

The DCE response "+FDCS:<string>" reports the negotiated parameters. Phase C data is formatted as described by the <string> subparameters (VR, BR, WD, LN, DF, EC, BF, and ST). This message may be generated by the execution of AT+FDT (data transmit) or AT+FDR (begin/continue phase C data receive) commands before the CONNECT result code if new DCS frames are generated or received.

Result Codes:

+FDCS:VR,BR,WD,LN,DF,EC,BF,ST

SESSION SUBPARAMETER CODES

Label	Function	Value	Description	
VR	Vertical Resolution	0	Normal, 98 lpi	
		1	Fine, 196 lpi	
BR	Bit Rate[1]	0	2400 bit/s V.27 ter	
		1	4800 bit/s V.27 ter	
		2	7200 bit/s V.29 or v.17	
		3	9600 bit/s V.29 or v.17	
		4	12000 bit/s V.33 or v.17	
		5	14400 bit/s V.33 or v.17	
WD	Page Width[2]	0	1728 pixels in 215 mm	
		1	2048 pixels in 255 mm	
		2	2432 pixels in 303 mm	
		3	1216 pixels in 151 mm	
		4	864 pixels in 107 mm	
LN	Page Length[3]	0	A4, 297 mm	
		1	B4, 364 mm	
		2	unlimited length	
DF	Data Compression[4]	0	1-D modified Huffman	
		1	2-D modified Read	
		2	2-D uncompressed mode	
		3	2-D modified Read	
EC	Error Correction (Annex A/T.30)[5]	0	Disable ECM	
		1	Enable ECM, 64 bytes/frame	
		2	Enable ECM, 256 bytes/frame	
BF	Binary File Transfer (BFT)[6]	0	Disable BFT	
		1	Enable BFT	
ST	Scan Time/Line		VR = normal	VR = fine
		0	0 ms	0 ms
		1	5 ms	5 ms
		2	10 ms	5 ms
		3	10 ms	10 ms
		4	20 ms	10 ms
		5	20 ms	20 ms
		6	40 ms	20 ms
		7	40 ms	40 ms

Notes: [1]ITU-T T.30 does not provide for the answering station to specify all speeds exactly using the DIS frame. Implementation of some BR codes (e.g., codes 2, 4, and 5) by an answering DCE is manufacturer specific.

[2]Page width modes 3 and 4 are not implemented in all fax/modems.

[3]Page length modes 2 and 3 are not implemented in all fax/modems.

[4]Not all data compression modes are implemented (i.e., modes 2 and 3 are frequently omitted).

[5]ECM is not implemented in all fax modems; further, although ECM transmit and receive may work correctly, the modem may not tell the transmitting machine that the fax was received correctly.

[6]BFT is not implemented in all fax modems; further, although BFT transmit and receive may work correctly, the modem may not tell the transmitting machine that the fax was received correctly.

+FDIL = \<string\> ....*Local String ID*

    \<string\> is a valid 20-character ASCII string (typically, the characters 0-9, "+", and \<space\>). The default value is a null string, i.e., empty. If \<string\> is not null, it generates a Transmit Station ID (TSI) or Called Station ID (CSI) frame.

    If the DCE supports the full printable ASCII character set, the response to **AT+FDIL = ?** is "n \<string\> \<CR\> \<LF\>" where "n" is string length and \<string\> is composed from the ASCII characters 32-127. Otherwise, the response is "n \<string\>" where "n" is the string length and \<string\> is composed from the ASCII characters 32, 43, and 48-57.

+FDIS:\<string\> ....*Report Remote Station Capabilities*

    The DCE response "+FDIS:\<string\>" reports remote station capabilities and intentions. The string consists of the subparameters code VR, BR, WD, LN, DF, EC, BF, and ST. The message is generated by the execution of originate (ATD), answer (ATA), AT+FDT (data transmit), or AT+FDR (begin/continue phase C data receive) commands.

    Details of the session subparameter codes (VR, BR, WD, LN, DF, EC, BF, and ST) are the same as those listed in the FDCS:\<string\> command.

+FDIS\<string\> ....*Current Session Parameters*

    The +FDIS\<string\> parameter allows the DTE to sense and constrain the capabilities used for the current session. The string consists of the subparameters VR, BR, WD, LN, DF, EC, BF, and ST. The DCE uses **+FDIS** to generate DIS or DTC messages directly and uses **+FDIS** and received DIS messages to generate DCS messages. Default values are product dependent. The DCE initializes the **+FDIS** parameter from the +FDCC parameter on initialization, when +FDCC is written, and at the end of a session.

    Details of the session subparameter codes (VR, BR, WD, LN, DF, EC, BF, and ST) are the same as those listed in the FDCS:\<string\> command.

+FDT ....*Data Transmission*

    In Phase B, the "AT+FDT" command releases the DCE to proceed with negotiation and releases the DCS message (session capabilities message) to the remote station. In Phase C, the "AT+FDT" command resumes transmission after the end of a prior transmit data stream.

    The "AT+FDT" command prefixes Phase C data transmission. When the DCE is ready to accept Phase C data, it issues the negotiation responses and the CONNECT result code to the DTE.

+FET = \<ppm\>[,\<pc\>,\<bc\>,\<fc\>] ....*Transmit Page Punctuation*

    The "AT+FET=" command is used to punctuate page and document transmission after one or more +FDT commands. This command generates ITU-T T.30 Post Page Messages selected by the \<ppm\> codes (see table).

    The **+FET=\<ppm\>** command indicates that the current page is complete; no more data will be appended to it. The value indicates if there are any additional pages are to be sent and, if so, whether there is a change in any of the document parameters.

    This command must be sent within the timeout specified by +FPHCTO (DTE phase C timeout) after sending Phase C data, or else the DCE will end the page and document transmission. If the Phase C timeout is reached, the DCE sends an EOP post-page message and terminates the session.

Command Options:
+FET=\<ppm\>[,\<pc\>,\<bc\>,\<fc\>]

POST-PAGE MESSAGE CODES

Code	Mnemonic	Description
1	[PPS]-MPS	Another page next, same document
2	[PPS]-EOM	Another page next
3	[PPS]-EOP	No more pages or documents
4	[PPS-]PRI-MPS	Another page, procedure interrupt
5	[PPS-]PRI-EOM	Another document, procedure interrupt
6	[PPS-]PRI-EOP	All done, procedure interrupt
7	CTC	Continue to correct
8–15	EOR-	End of retransmission (8)+
8+ppm		Post-page message (ppm code)

+FET= . . . . Causes the DCE to append an ETC (6 EOL) pattern as needed and enter Phase D by sending the se-lected T.30 Post-Page message.

+FET=1 . . . . The EOM command signals the remote station that the next document will have a new DCS negotiated; this causes the session to reenter Phase B.

+FET=4, 5, or 6 . . . . Causes the DCE to generate PRI-messages.

Result Codes:

+FPTS:<ppr> . . . . DCE response when received from remote OK.

The remote station should respond to the post-page message with a post-page response. The local DCE will report this using a +FPTS:<ppr> response.

+FET:<ppm> . . . . *Post-Page Message Response*

The **+FET:<post-page message>** response is generated by a receiving DCE after the end of Phase C reception upon receipt of the post-page message from the transmitting station. The **+FET:<ppm>** response is generated by the execution of an AT+FDR () com-mand. The <ppm> codes respond to the T.30 messages described in the *AT+FET = <ppm>[,<pc>,<bc>,<fc>]* command above.

+FHNG:<hsc> . . . . *Call Termination with Status*

The **+FHNG:<hang-up status code (hsc)>** DCE response indicates that the call has been terminated and the cause of the termina-tion. The disconnect cause is reported and stored in the +FAXERR parameter for later inspection. The values are in decimal notation with leading zero characters optional. *+FHNG:<hsc>* is a possible intermediate result code to any DTE action command. The hang-up values are organized according to the phases of the facsimile transaction as defined by ITU-T T.30. A COMREC error or RSPREC error indicates that one of two events occurred, that is,

• A DCN (disconnect) signal was received, or
• An FCS error was detected and the incoming signal was still present after 3 seconds.

+FHNG:<HSC> is always followed by the OK final result code.

FAXERR STATUS CODES (hsc)

Code	Description	Code	Description
0-9	CALL PLACEMENT AND TERMINATION	52	No response to MPS repeated 3 times
		53	Invalid response to MPS
0	Normal and proper end of connection	54	No response to EOP repeated 3 times
1	Ring Detect without successful handshake	55	Invalid response to EOM
2	Call aborted, from +FK or AN	56	No response to EOM repeated 3 times
3	No Loop Current	57	Invalid response to EOM
		58	Unable to continue after PIN or PIP
10-19	TRANSMIT PHASE A & MISCELLANEOUS ERRORS		
		70-89	RECEIVE PHASE B HANG-UP CODES
10	Unspecified Phase A error		
11	No Answer (T.30 T1 timeout)	70	Unspecified Receive Phase B error
		71	RSPREC error
20-39	TRANSMIT PHASE B HANG-UP CODES	72	COMREC error
		73	T.30 T2 timeout, expected page not received
20	Unspecified Transmit Phase B error	74	T.30 T1 timeout after EOM received
21	Remote cannot receive or send		
22	COMREC error in transmit Phase B	90-99	RECEIVE PHASE C HANG-UP CODES
23	COMREC invalid command received		
24	RSPEC error	90	Unspecified Receive Phase C error
25	DCS sent three times without response	91	Missing EOL after 5 seconds
26	DIS/DTC received 3 times; DCS not recognized	92	Unused code
27	Failure to train at 2400 bps or +FMINSP value	93	DCE to DTE buffer overflow
28	RSPREC invalid response received	94	Bad CRC or frame (ECM or BFT modes)
40-49	TRANSMIT PHASE C HANG-UP CODES	100-119	RECEIVE PHASE D HANG-UP CODES
40	Unspecified Transmit Phase C error	100	Unspecified Receive Phase D errors
43	DTE to DCE data underflow	101	RSPREC invalid response received
		102	COMREC invalid response received
50-59	TRANSMIT PHASE D HANG-UP CODES	103	Unable to continue after PIN or PIP
50	Unspecified Transmit Phase D error		
51	RSPREC error	120-255	RESERVED CODES

+FK . . . .*Session Termination*

The "*AT+FK*" command causes the DCE to terminate the session in an orderly manner. The DCE will send a DCN message at the next opportunity and hang up. (The DCE will wait until the current page completes, unless the reception is of unlimited length, in which case, the DCE may halt reception and terminate the session immediately.) At the end of the termination process, the DCE will report the +FHNG:<hsc> (hang-up status code) and disconnect.

This operation may also be invoked by using the cancel <CAN> character during Phase C data reception. Upon receipt of the <CAN> character, the DCE will cease reporting receive data by sending <DLE> <ETX> characters to the DTE. It will then execute an implied *AT+FK* command to terminate the session.

+FLID = ? | <string> . . . .*Local ID String*

+FLID = allows setting and reporting a 20-character Local ID String. The string may be either numeric (<space>, "+", and digits 0-9) or alphanumeric as specified in ITU-T T.30. If *FLID* is not a null string, it generates a TSI (transmit station ID) or CSI (called station ID) frame (the default value is empty or null string).

Command Options:
AT+FLID=? . . . . (20)(32, 43, 48-57) if the DCE supports only numeric strings or
. . . . (20)(32-127) <CR> <LF> if the DCE supports alphanumeric strings. The "(20)" represents string length
                 and the following field reports the supported string values (ASCII character decimal values).
AT+FLID? . . . . n <string> where "n" is the string length.
AT+FLID=<string> . . . . . . default value is empty or null string.

+FLPL = ? | 0 | 1 . . . .*Indicate a Document for Polling*

The "*AT+FLPL=*" command notifies a DCE that there is a pending fax to be transmitted when the DCE is polled. This allows a calling modem to receive this fax from the answering station.

Command Options:
AT+FLPL=? . . . . Interrogates the DCE for the range of values allowable.
AT+FLPL=n . . . . Sets the value in the DCE.
AT+FLPL? . . . . Interrogates the DCE for the current setting.

Result Codes:
0,1
OK. . . . . Response to AT+FLPL=?
OK. . . . . Response to AT+FLPL=n.
n
OK. . . . . Response to AT+FLPL?

+FMDL? . . . .*Request Product Model*

"*AT+FMDL?*" is a class 2 session command that causes the DCE to send a 20-byte, ASCII encoded message identifying the DCE product model (generally the basic chip set's number).

Command Options:
AT+FMDL? . . . . Identify equipment model.

Result Codes:
<model number> . . . . For example, V.32AC.
OK. . . . . Completed command.

+FMFR? . . . .*Request Manufacturer ID*

"*AT+FMFR?*" is a class 2 session command that causes the DCE to send a 20 byte, ASCII encoded message identifying the DCE's manufacturer.

Command Options:
AT+FMFR?. . . . . Identifies equipment manufacturer.

Result Codes:
<manufacturer> . . . . Examples include ATT, ROCKWELL, and US ROBOTICS.
OK. . . . . Completed command.

+FPHCTO = n . . . .*DTE Phase C Response Timeout*

The "*AT+FPHCTO=n*" command determines how long the DCE will wait for a command after reaching the end of data when transmitting in Phase C. When the *+FPHCTO* timeout expires, the DCE assumes there are no additional items to transmit (neither pages nor documents). It then sends the ITU-T T.30 EOP response to the remote device.

The range of values for "n" is 0 to 255, which corresponds to a time range of 0 to 25.5 seconds with 1/10 second resolution. The default value is n=30 or 3 seconds.

Command Options:

AT+FPHCTO = n. . . . . . . Wait n/10's of a second before terminating the call.

+FPOLL: . . . .*Indicate Polling Request*

The "*+FPOLL:*" message is sent to the DTE if the received DIS (remote station capabilities) message indicates that the remote station has a document to poll (by bit 9 in the DIS) and polling has been enabled with the +FSPL (enable polling) command. The DTE may then elect to receive or transmit.

+FPTS:<ppr>,<lc>[,<blc>,<cblc>] . . . .*Receive Page Transfer Status*

The *+FPTS:<post-page response (ppr)>,<lc>[,<blc>,<cblc>]* response is generated by the DCE at the end of Phase C data reception by the execution of an AT+FDR (begin/continue phase C data receive) command. The value of <ppr> is dependent on the DCE's ITU-T T.4 error-checking capabilities.

POST PAGE RESPONSE MESSAGE CODES

Code	Mnemonic	Description
0	PPR	Partial page errors
1	MCF	Page good
2	RTN	Page bad, retrain requested
3	RTP	Page good, retrain requested
4	PIN	Page bad, interrupt requested
5	PIP	Page good, interrupt requested

The receiving DCE will count the lines and may optionally generate bad line counts, that is,

<lc> . . . . line count
<blc> . . . . bad line count
<cblc> . . . . consecutive bad line count

A receiving DTE may inspect <ppr> and write a modified value into the +FPTS parameter. The DCE will hold the corresponding post-page response message until released by a +FDR command from the DTE.

+FPTS:<ppr> . . . .*Transmit Page Transfer Status*

The "*AT+FPTS:<ppr>*" DCE response reports a <ppr> number that represents the copy quality and related post-page message responses received from the remote DCE. The response is generated by the execution of an AT+FET=<ppm> command. The set of valid <ppr> values are defined in the *+FPTS:<(ppr>,<lc>[,<blc>,<cblc>]* response above.

+FPTS = <ppr> . . . .*Page Transfer Status*

The "*AT+FPTS = n*" command sets the page transfer status to one of the values defined in the table defined in the *+FPTS:<(ppr>,<lc>[,<blc>,<cblc>]* response above.

Command Options:

+FPTS = <ppr> . . . . Sets the page transfer status code to a value of 0, 1, 2, 3, 4, or 5. The default value is 0.

+FREV? . . . .*Request Product Revision*

"*AT+FREV?*" is a class 2 session command that causes the DCE to send a 20 byte, ASCII encoded message identifying the DCE product model revision number. The message format is generally the same as the ATI3 message (see page 597).

Command Options:

AT+FREV? . . . . Identifies equipment firmware revision level.

Result Codes:

<revision number> . . . .
OK . . . . Completed command.

+FSPL = ? | 0 | 1 . . . .*Enable Polling*

This "*AT+FSPL=*" command allows setting up an originating modem to be able to request to receive a document from a polled station.

Command Options:

AT+FSPL=?. . . . . Interrogates the DCE for the range of values allowable.

AT+FSPL=n. . . . . Sets the value in the DCE.

AT+FSPL? . . . . Interrogates the DCE for the current setting.

Result Codes:

0,1

OK. . . . . Response to AT+FSPL=?

OK. . . . . Response to AT+FSPL=n.

n

OK. . . . . Response to AT+FSPL?

+FTSI:<string> . . . .*Report the Transmit Station ID*

The DCE response that reports the received transmit station ID string, if any. This message is generated by the execution of originate (ATD), answer (ATA), AT+FDT (data transmit), or AT+FDR (begin/continue phase C data receive) commands.

Result Codes:

+FTSI:<TSI ID string> . . . . Transmit Station ID

Appendix B
"AT" Fax/Modem S-Registers

GENERAL INFORMATION

There are a number of memory locations in a modem called S-registers. These registers are used either to set an operating mode or parameter, or to report conditions and status. In most cases, the register values may be both read and changed via an AT command string. A few registers, however, are read only registers; that is, they may be read with an AT command, but the value is changed only by some change in hardware condition.

If a register can be changed, it is listed as *Sr = n* where "*r*" is the register number and "*n*" is the value to be written into the register. If the register is a read only register, it is listed only as *Sr;* again the "*r*" is the register number—the meaning of the value "*n*" in the register is given in the text of the description of the register. The default values given are the most common values. Values marked with an "↕" are the most likely to be different from manufacturer to manufacturer or even model to model.

Some registers must be changed with an *ATSr = n* command; others may be changed with the more easily used AT letter command. For example, *ATL3* (set speaker to maximum volume) is the same as setting bits 0 and 1 of register *S22* to "1" or "on," and some registers (such as the S1 ring count register) are changed only by the hardware. If the *ATSr = n* command is used, all 8 bits of that register are set. In the preceding example, *S22* contains not only the Speaker Volume information, but also Speaker Control, Dial Signaling Response & Result Code Control, and pulse dialing Make/Break Ratio control. Simply issuing the command *ATS22 = 3* (to set bits 0 and 1 equal 1) will turn the Speaker Volume to maximum; however, bits 2 and 3 are set to "0," which tells the speaker to always be off! The command *ATL3* will change bits 0 and 1 to "1" but will not change bits 2 through 7.

Registers marked with a "☐" may have their contents saved in the modem's NVRAM using the *AT&Wn* command, assuming NVRAM memory is installed and functioning. This value may then be restored at a later date with the *AT&Zn* command or on power-up after an *AT&Yn* has been issued.

NOTES:

1. Not all of the registers listed are in all modems.

2. Some modems that do not have a particular register may still give a value if read (albeit the value may always be 0) and may not give an error if an attempt to change the value is made.

3. In a few cases, the function of a register in a particular model modem is different from the "normal" use of that register. Further, the use of the register may be different for different models from the same company; for example, the use register S13 of a U.S. Robotics 14.4 Kbps Sportster modem is different from that of the 28.8 Kbps Sportster modem!

4. The settings of manual controls such as DIP switches and jumper plugs may alter the way an S-register operates.

5. Default values are typical of most modems; however, some specific modems use different values.

S0 = n . . . .*Number of Rings to Auto Answer (default = 0)*

☐   The value stored in register *S0* determines the number of rings that must be detected before the modem automatically answers an incoming call. Setting n=0 disables the *auto-answer mode.*

        n = 0 . . . . **Disables auto answer,** (default).
        n = 1 to 255 . . . . Enables auto answer and goes off-hook after n rings.

S1 . . . .*Ring Counter*

        The number of telephone rings detected. The counter is automatically reset to 0 if no rings are detected for an 8-second period.

        n = 0 to 255 . . . . "n" rings have been detected since the last 8-second silent period.

S2 = n . . . .*Escape Code Character (default = 43, "+")*

☐   The value in *S2* sets the ASCII value of the character used as the escape code in the escape sequence. Although the register will accept values from 0 to 255, values greater than 127 will disable the escape feature, preventing the return to the COMMAND mode. If n > 127 and a call is connected, it may be disconnected only if the carrier is lost or DTR from the DTE is turned off.

        n = 0 to 127 . . . . The ASCII value of the escape code character used in the escape sequence (default = 43, an ASCII "+").
        n = 128 to 255. . . . . Disables the escape sequence.

S3 = n . . . .*Carriage Return Character (default = 13, <CR>)*

        The value in *S3* sets the ASCII value to be used as the command line and result code terminator code.

        n = 0 to 127 . . . . The ASCII value used as the return character (default = 13).

S4 = n . . . .*Line Feed Character (default = 10, <LF>)*

        The value in *S4* sets the ASCII value of the line feed character. This line feed character is sent after a carriage return is sent.

        n = 0 to 127 . . . . The ASCII value used as the line feed character (default = 10).

647

S5 = n . . . .*Backspace Character (default = 8, <BS>)*

> The value in *S5* sets the ASCII value of the backspace character. This character is created by pressing the BACKSPACE key on the keyboard and is the character echoed when the cursor is moved to the left.
>
> When the modem is set to echo commands (ATE1), the modem will send the three characters <BS>, <SPACE>, <BS> to the DTE when it receives a <BS> from the DTE.
>
> n = 0 to 32, 127 . . . . The ASCII value used as the backspace character. Some modems allow the full range of 0 to 127, others restrict the range, e.g., 0 to 32.
>
> n = 128 to 255 . . . . On some modems, disables the backspace key's delete function.

S6 = n . . . .*Time to Wait before Dial (default = 2)*

> The value in S6 defines the delay time (in seconds) after going off-hook and before dialing a number when *Blind Dial* mode is selected (X0, X1, and X3 dialing modes). The minimum delay is 2 seconds even if n is less than 2. The value in register S6 is superseded by the W dial command modifier.
>
> If X2 or X4 is asserted, register *S6* is ignored by many modems, but it is treated as the maximum delay before dial tone by other modems.
>
> n = 0, 1 . . . . Wait 2 seconds after going off-hook to dial the first digit.
>
> n = 2 to 255 . . . . Wait n seconds after going off-hook to dial the first digit.

S7 = n . . . .*Time to Wait for Carrier (default = 50 ↕)*

> 1. The maximum time (in seconds) the originating modem waits for a carrier from the remote modem before going on-hook and issuing a NO CARRIER message.
> 2. The maximum time the modem waits for dial tone when the W (wait for dial tone) dial command modifier is asserted.
> 3. The maximum time the modem waits for silence when the @ (wait for silence) dial command modifier is asserted.
> 4. The maximum time the modem waits for ringback when X3 or X4 dial modes are asserted.
>
> n = 0 . . . . Accepted but meaningless; don't wait, just proceed.
>
> n = 1 to 255 . . . . Wait n seconds.

S8 = n . . . .*Pause Time for Comma (default = 2)*

> The time (in seconds) which the "comma (,) dial command modifier" waits before going on to the next item in the dial command list.
>
> Frequently, the value in register *S8* is left at 2 seconds, and more than one comma is used to make longer delays, for example, the command "ATD9,123-4567,,,234" will
>> dial a 9 (to get an outside line from a PABX for example),
>> wait 2 seconds,
>> then it will dial 123-4567 to connect to the remote switching equipment,
>> wait 6 seconds (3 2 second waits), and finally
>> dial 234 to access the remote modem.
>
> n = 0 to 255 . . . . Wait n seconds; then proceed to the next item in the dial command list.

S9 = n . . . . *(1) Carrier Detect Response Time (default = 6)*

> The persistence time (in tenths of a second) required between the first detection of a carrier and the turn on of DCD. This delay allows the modem to ignore spurious noise signals that would result in false carrier detection.
>
> Larger values of n will reduce the probability of falsely detecting a carrier but will slow the time to connect.
>
> n = 0 . . . .
>
> n = 1 to 255 . . . . Wait n tenths of a second (0.1 to 25.5 seconds) to assert DCD after detecting a carrier.

S9 = n . . . . *(2) Bit-Mapped Option Codes (default = 317)*
(Cisco MICA Six-Port Modules)

> A bit-mapped storage register containing the values set with the E, Q, V, X, and &D commands. Generally, it is recommended that the letter commands be used rather than directly setting this register.

Bit #	Definition	Reference Command	Setting = 0	Setting = 1
0	Command echo	E	No Echo	**Echo (default)**
1, 2	Result codes	Q	Q0 = Result codes	
			Q1 = No result codes	
			Q2 = No result codes in answer mode (default)	
3	Result type	V	Short form	**Long form (default)**
4,5	Action on loss of data terminal ready (DTR)	&D	&D0 = Ignore DTR	
			&D1 = Modem goes to command mode	
			&D2 = Modem goes on hook	
			&D3 = Reset modem (default)	
6,7,8	Connect result code and call progress format	X	X0 = Sends connect result code, ignores dial and busy signals	
			X1 = Sends verbose connect result code, ignores dial tones and busy signals	
			X2 = Sends verbose connect and no dial tone result codes, ignores busy signals	
			X3 = Sends verbose connect and busy result codes, ignores dial tone	
			X4 = Sends verbose connect, no dial tone, and busy result codes (default)	

S10 = n . . . .*Delay from Lost Carrier to Hang Up (default = 14 ↕)*

📂   The time (in tenths of a second) the modem waits after the loss of the carrier is detected before disconnecting. This time allows the remote signal to momentarily disappear and return without causing the local modem to disconnect. The maximum lost signal time is actually the difference between the *lost carrier delay* (*S10*) and the *carrier detect time* (*S9*). Using the listed default values yields a maximum momentary signal loss of 1.4–0.6 = 0.8 seconds can be tolerated before the modem will return to the on-hook state. If *S10* = 255 the modem behaves as if the carrier were present at all times (unless the modem is in an error-correcting mode, then the modem will disconnect). If *S10* < *S9*, then any loss of carrier will result in the modem hanging up (i.e., it disconnects before carrier is recognized).

Setting *S10* to a large number can be used to allow the modem to tolerate the tone burst introduced by *call waiting* for both incoming and outgoing calls. A better method, however, is to issue the temporary *cancel call waiting* command to the telephone company (usually *70 with DTMF signaling or 1170 with pulse signaling) for outgoing calls.

n = 1 to 254 . . . . Wait n tenths of seconds (0.1–25.4 seconds) after the loss of carrier before returning on-hook.

n = 0, 255 . . . . The modem acts as though a carrier is present at all times (unless in an error-correcting mode, then carrier loss will cause a disconnect). Dropping DTR will also cause the modem to go on hook.

S11 = n . . . . *(1) DTMF Tone Duration/Spacing (default = 95 ↕)*

📂   Determines the DTMF signaling rate, i.e., the tone duration and the spacing between tones during tone dialing. The duration of the DTMF tone is equal to the spacing between the tones.

n < 50 . . . . The values may be accepted by the modem; however, the telco may not correctly recognize the digits being sent. The minimum time a tone must be present is 50 ms according to U.S. and international standards.

n = 50 to 255 . . . . The DTMF tone is on for "n" ms and off for "n" ms.

S11 = n . . . . *(2) Link Initiate Action (default = 0)*
                                                                (Cisco MICA Six-Port Modules)

If the modem is idle (i.e., in escape or command mode) and receives the ATO (on line) command, it originates or answers a call, based on the contents of this register.

n = 0 . . . . **Answer (default).**

n = 1 . . . . Originate

S12 = n . . . . *(1) Escape Sequence Guard Time (default = 50)*

📂   The value in *S12* defines the minimum guard time (or idle time) before (and after) the escape code (defined by the value in register *S2*) is entered for the entire sequence to be recognized as an escape sequence. "n" is in 1/50 seconds units; therefore, the range of guard time is 1/50 of a second to 5.1 seconds (20 ms to 5.1 seconds). If n = 0, the guard time is to 0 seconds and therefore has been disabled.

If small values of n are used, it may not be possible to manually enter the three-character escape code before the end guard timer times out.

n = 0 . . . . Disables the timer.

n = 1 to 255 . . . . Sets the guard time to n × 20 ms. Using the default value of 50, the guard time is 50 × 20 ms = 1000 ms (or 1 second)!

S12 = n .... *(2) Number of Asynchronous Data Bits (default = 8)*
(Cisco MICA Six-Port Modules)

Specifies the number of data bits in each transmitted character in an asynchronous (start-stop mode) call. The modem always uses 8-bit data, regardless of how this register is set unless error correction is successfully negotiated (see registers S15, S22, S23, and S24), and bit 6 in register S54 is set, then this register must be set to 7.

n = 7, 8, or 9 .... Character length

S13 = n .... *(1) Bit-Mapped Option Codes (default = 0↕)*

The individual bits in register *S13* are treated as switches by the modem. If the bit is a "1" the specified function is on or enabled; if the bit is a "0," the function is off or disabled. The register may set using the *ATS13=n* command format.

U.S. ROBOTICS (14.4 KBPS SPORTSTER MODEMS)

Bit #	Definition	Reference Command	Setting = 0	Setting = 1
0	Reset on DTR negation			Reset
1	Set non-MNP buffer size		1500 byte transmit buffer	128-byte transmit buffer
2	Define backspace key		Backspace	Delete
3	Auto dial on DTR assertion		No action	Dial number stored in NVRAM position 0
4	Auto dial at power on		No action	Dial number stored in NVRAM position 0
5	Reserved			
6	Reserved			
7	Disconnect on escape code		No action	Disconnect on escape code

The 1500-byte buffer size (*S13 bit 2 = 0*) generally allows data transfers with XMODEM or YMODEM protocols without using flow control. The 128-byte buffer size option provides quicker response to the XON/XOFF flow control of the remote station. For example, when the remote user sends an XOFF character (<CTL-S>), the local DTE stops transmitting; however, the modem continues to transmit until the buffer is empty. With a 128-byte buffer only a few more lines would be written to the screen of the remote terminal.

U.S. ROBOTICS (SPORTSTER MODEMS)

Bit #	Definition	Reference Command	Setting = 0	Setting = 1
0	Reset on DTR negation			Reset
1	Reserved or			
	Set non-MNP buffer size		1500 byte transmit buffer	128 byte transmit buffer
2	Reserved or			
	Define backspace key		Backspace	Delete
3	Auto dial on DTR assertion		No action	Dial number stored in NVRAM position 0
4	Auto dial at power on		No action	Dial number stored in NVRAM position 0
5	Reserved			
6	Disable MNP level 3		Enabled	Disable MNP level 3 (used while testing level 2)
7	Reserved (Disconnect on escape code on 33.6 Vi modem)			Enable disconnect

S13 = n .... *(2) Parity Configuration (default = 0)*
(Cisco MICA Six-Port Modules)

Specifies the character parity for the call in start-stop (asynchronous) mode. The modem sets this value on outgoing (to line) characters and checks it on incoming characters. The modem always uses no parity (n = 0) regardless of how this register is set unless error correction is successfully negotiated (see registers S15, S22, S23, and S24), and bit 6 in register S54 is set.

Bit #	Definition	Reference Command	Setting = 0	Setting = 1
0,1,2	Parity Configuration		S13 = 0 No parity (default)	
			S13 = 1 Even parity	
			S13 = 2 Odd parity	
			S13 = 3 Mark parity	
			S13 = 4 Space parity	

S14 = n . . . . *(1) Bit Mapped Option Codes (default = 174, AEh ↕)*

The individual bits in register *S14* are treated as switches by the modem. If the bit is a "1," the specified function is on or enabled; if the bit is a "0," the function is off or disabled. Individual bits may be altered by using the referenced AT letter command, or the entire register may set using the ***ATS14=n*** command format.

Bit #	Definition	Reference Command	Setting = 0	Setting = 1
0	On-line state echo	Fn	Disable (default)	Enable
1	Command echo	En	Disable	Enable (default)
2	Result codes	Qn	Disable	Enable (default)
3	Result codes	Vn	Numeric	Verbose (default)
4	Dumb mode		Off (default)	On
5	Dial method	P, T	Tone	Pulse (default)
6	Result codes	Qn	Q0 or Q1 (default)	Q2
7	Originate/Answer	On	Answer	Originate (default)

ROCKWELL CHIPSET BASED

Bit #	Definition	Reference Command	Setting = 0	Setting = 1
0	Ignored		(default)	
1	Command echo	En	Disable	Enable (default)
2	Quiet mode	Qn	Send result codes (default)	Do not send result codes
3	Result codes	Vn	Numeric	Verbose (default)
4	Reserved		(default)	
5	Dial method	P, T	Tone (default)	Pulse
6	Reserved		(default)	
7	Originate/Answer	On	Answer	Originate (default)

U.S. ROBOTICS (28.8 KBPS SPORTSTER MODEMS)

Bit #	Definition	Reference Command	Setting = 0	Setting = 1
0	On-line reaction to "+++"		No action	Modem goes on-hook, returns to COMMAND mode, and issues a NO CARRIER result code.
1–7	Reserved			

S14 = n . . . . *(2) Stop Bit Configuration (default = 1)*
(Cisco MICA Six-Port Modules)

Specifies the number of transmitted stop bits in start-stop (asynchronous) mode. If error correction is successfully negotiated (see registers S15, S22, S23, and S24), stop bits are meaningless, and the setting of this register has no effect.

n = 1 or 2 . . . . Number of stop bits in asynchronous mode

S15 = n . . . . *(1) Bit-Mapped Option Codes (default = 0)*

The individual bits in register *S15* are treated as switches by the modem. If the bit is a "1," the specified function is on or enabled; if the bit is a "0," the function is off or disabled. The register may set using the ***ATS15=n*** command format.

U.S. ROBOTICS (14.4 KBPS AND 56K SPORTSTER MODEMS)

Bit #	Definition	Reference Command	Setting = 0	Setting = 1
0	Disable ARQ/MNP for V.22		Enabled	Disable ARQ/MNP for V.22
1	Disable ARQ/MNP for V.22 bis		Enabled	Disable ARQ/MNP for V.22 bis
2	Disable ARQ/MNP for V.22, V.32 bis, and V.32 ter.		Enabled	Disable ARQ/MNP for V.22, V.32 bis, and V.32 ter.
3	Disable MNP handshake		Enabled	Disable MNP handshake
4	Disable MNP level 4		Enabled	Disable MNP level 4
5	Disable MNP level 3		Enabled	Disable MNP level 3
6	MNP incompatibility			MNP incompatibility
7	Disable V.42		Enabled	Disable V.42

On the U.S. Robotics 56K Sportster fax/modem, a value of 136 disables the V.42 detect phase.

U.S. ROBOTICS (28.8 KBPS SPORTSTER MODEMS)

Bit #	Definition	Reference Command	Setting = 0	Setting = 1
0–2	Reserved			
3	Set non-ARQ buffer size		1500 byte transmit buffer	128 byte transmit buffer
4	Disable MNP level 4			Disable MNP level 4
5	Define backspace key		Backspace	Delete
6–7	Reserved			

The 1500-byte buffer size (*S15 bit 3 = 0*) generally allows data transfers with XMODEM or YMODEM protocols without using flow control. The 128-byte buffer size option provides quicker response to the XON/XOFF flow control of the remote station. For example, when the remote user sends an XOFF character (<CTL-S>), the local DTE stops transmitting; however, the modem continues to transmit until the buffer is empty. With a 128-byte buffer only a few more lines would be written to the screen of the remote terminal.

On noisy communications links, it may be beneficial to limit the size of retransmitted blocks by disabling MNP level 4 (*S15 bit 4 = 1*).

S15 = n . . . . *(2) V.42 Detect Phase Operation (default = 1)*
(Cisco MICA Six-Port Modules)

Enables or disables V.42 detect phase, during which two modems prepare to negotiate a common error correction protocol. This register controls the originating modem's behavior only; in answering mode, detect phase is automatic. The length of the negotiation period is controlled by S16.

The settings of this register are ignored and V.42 detect phase is disabled when

• LAP-M error correction is disabled by setting S23 to 0, or

• The modulation standard in effect uses FSK (frequency shift keying). For example, with V.21, V.23, and Bell103.

n = 0 . . . . Disable V.42 detect phase
n = 1 . . . . Enable V.42 detect phase (default)

S16 = n . . . . *(1) Diagnostic Test Mode Setting—Bit Mapped (default = 0)*

The individual bits in register *S16* are treated as switches by the modem. If the bit is a "1," the specified function is on or enabled; if the bit is a "0," the function is off or disabled. Individual bits may be altered by using the referenced ATT*n* letter command. The register may also be set using the ***ATS16=n*** command format; however, although *S16* is bit mapped, only one test can be performed at a time.

Bit #	Definition	Reference Command	Setting = 0	Setting = 1
0	Local analog loopback (LAL)	&T1	Disable	Enable
1	Touch tone dialing test (US Robotics Sportster)		Off	
2	Local digital loopback (LDL)	&T3	Disable	Enable
3	Remote digital loopback (RDL) status	&T4, &T5	Off	RDL in process
4	Remote digital loopback request (RDL)	&T6	Not requested	Requested
5	Remote digital loopback with self test	&T7	Disable	Enable
6	Local analog loopback with self test	&T8	Disable	Enable
7	Not used			

If there is a test mode indicator on the modem, it will be active any time the register is non-zero.

The &T0 command terminates any test in progress and sets register *S16=0*.

S16 = n . . . . *(2) Bit-Mapped Option Codes (default = 0)(U.S. Robotics Sportster)*

U.S. ROBOTICS (56K SPORTSTER MODEMS)

Bit #	Definition	Reference Command	Setting = 0	Setting = 1
0	Reserved			
1	Touch tone dialing test			
2–7	Reserved			

S16 = n . . . . *(3) Error Correction Autodetect Timeout (default=50)*
(Cisco MICA Six-Port Modules)

Specifies the period that the modem can remain in V.42 detect phase. This timeout period is used only if V.42 detect phase is enabled (S15 = 1). S25 (Link Protocol Fallback) determines the fallback action invoked by the modem if the detect phase times out. The value of n is in 100-milliseconds.

n = 0 . . . . Remain in autodetection until manual intervention occurs or a pattern match is found.

n = 1–8589 . . . . Remain in autodetection for 0.1 to 858.9 seconds or until manual intervention occurs or a pattern match is found. (Values larger than 8589 will be accepted but will yield unpredictable results.) The value in register S16 must be less than the value set in register S17. The default value is 50 (5.0 seconds).

S17 . . . . *Error Correction Negotiation Timeout (default=100)*
(Cisco MICA Six-Port Modules)

Specifies the period that the modem tries to detect and negotiate an error correction protocol. The value of n is in 100-milliseconds. The clock for this timeout starts at the same time as the clock for the error correction autodetect timeout (S16). The time allowed for error correction negotiation is the difference between the values of S16 and S17. For example, if S16 is set to 50 (the default, 5 seconds) and S17 is set to 100 (the default, 10 seconds), then the modem may have as much as 10 seconds for error correction negotiation (if the autodetect phase completes instantaneously) or as little as 5 seconds (if the autodetect phase times out). If negotiation is unsucessful, the fallback action specified in register S25 is implemented.

n = 0 . . . . Wait indefinitely for error correction protocol negotiation or until manual intervention.

n = 1–8589 . . . . Remain in error correction negotiation for 0.1 to 858.9 seconds or until manual intervention occurs. (Values larger than 8589 will be accepted but will yield unpredictable results.) The value in register S17 must be greater than the value set in register S16. The default value is 100 (10.0 seconds).

S18 = n . . . . *(1) Test Mode Timer (default=0)*

The value in register *S18* specifies the duration (in minutes) the modem diagnostic tests are to run. A diagnostic test is active for the period of time n and then automatically terminates. When n = 0, a test will run continuously or until it is terminated by the AT&T0 or ATH0 command, that is, the timer is disabled.

n = 0 . . . . **Runs tests continuously (default).**

n=1–255 . . . . Runs test n seconds, then terminates the test.

S18 = n . . . . *(2) Error Correction Fallback Character (default=13)*
(Cisco MICA Six-Port Modules)

Specifies the ASCII value of the error correction fallback character. The far-end modem can send this character during the V.42 detect phase (during call establishment) to force this modem to stop negotiating and use the fallback option specified in register S25.

U.S. ROBOTICS (56K SPORTSTER MODEMS)

Bit #	Definition	Reference Command	Setting = 0	Setting = 1
0–7	ASCII decimal code for fallback character		0 : **13** : 255	No fallback character **Carriage return as fallback character**
8	Number of times fallback character must be received		**3 (default)**	1

S19 = n . . . . *(1) AutoSync Options (default=0)*

Register S19 defines the options for AutoSync operation (see &Q4 command). The register must be set to the desired value before &Q4 is issued.

Bit #	Definition	Reference Command	Setting = 0	Setting = 1
0	Reserved			
1	Protocol Select		BSC	HDLC
2	Address detection		Disable	Enable
3	Coding selection		NRZI	NZI
4	Idle indicator select		Mark idle	Flag or sync idle
5–7	Reserved			

S19 = n . . . . *(2) Inactivity Timer (default=0)(U.S. Robotics Sportster)*

The value in register *S19* determines the time duration (in minutes) for which the modem will tolerate no activity before it returns to the on-hook state. A value of *S19 = 0* disables the timer.

S19 = n . . . . *(3) Error Correction Retransmission Limit (default=12)*
(Cisco MICA Six-Port Modules)

Specifies the number of successive data frame retransmissions that can take place in a call before the modem performs a recovery action and repeats the specified number of retransmissions. If the retransmissions fail, the modem disconnects.

n = 0 . . . . Do not disconnect on excessive retransmissions and do not perform recovery action before retries.

n = 1–65535 . . . . The number of successive frame retransmissions before recovery attempt and then disconnect. (Default 12 retransmissions.)

S20 = n . . . . *(1) AutoSync HDLC Address or BSC Sync Character (default=0)*

Defines the *HDLC address* (S19 bit 1 = 1) or *BSC Sync Character* (S19 bit 1 = 0) for AutoSync operation (see &Q4 command). Register S20 must be set to the desired value before &Q4 is issued.

n = 0–255 . . . . Defined HDLC address or BSC character.

S20 = n . . . . *(2) Error Correction Max Frame Length (default=256)*
(Cisco MICA Six-Port Modules)

Specifies the maximum length for error correction frames (MNP, LAP-M, and ARA [AppleTalk Remote Access] frames). (Use higher values on links with good transmission quality as they are more efficient – lower values are better on a line with poor transmission quality.)

n = 64–1024 . . . . Octets of data (default is 256 octets).

S21 = n . . . . *(1) Bit-Mapped Option Codes (default=0↕)*

The individual bits in register *S21* are treated as switches by the modem. If the bit is a "1," the specified function is on or enabled; if the bit is a "0," the function is off or disabled. Individual bits may be altered by using the referenced AT letter command, or the entire register may set using the *ATS21=n* command format.

Bit #	Definition	Reference Command	Setting = 0	Setting = 1
0	Telephone jack	&Jn	RJ-11/41/45S	RJ-12/13
1	Stored configuration	&Yn	Location 0	Location 1
2	Synchronous CTS	&Rn	CTS follows RTS	CTS always on
3,4	DTR	&D0	0,0	See &D command for
		&D1	0,1	definition. Operates
		&D2	1,0	in conjunction with
		&D3	1,1	&Q command.
5	DCD	&Cn	Always on	Tracks carrier
6	DSR	&Sn	Always on	Per EIA spec.
7	Long space disconnect	Yn	Disabled	Enabled

S21 = n . . . . *(2) Break Duration (default = 10)(U.S. Robotics Sportster)*

S21 = n on the U.S. Robotics Sportster series of modems sets the length of breaks (in tens of milliseconds—0.01 s) from the modem to the computer in ARQ mode. The range is from 10 ms to 2.55 s.

S21 = n . . . . *(3) Data Compression (default = 3)*
(Cisco MICA Six-Port Modules)

Specifies the methods of data compression that can be negotiated between two modems. (Both modems must permit a particular compression scheme before it can be used.)

n = 0 . . . . Data compression disabled.
n = 1 . . . . V.42bis.
n = 2 . . . . MNP 5.
n = 3 . . . . V.42bis then MNP 5 if V.42bis fails (default).

S22 = n . . . . *(1) Bit-Mapped Option Codes (default = 118 ↕)*

The individual bits in register *S22* are treated as switches by the modem. If the bit is a "1," the specified function is on or enabled; if the bit is a "0," the function is off or disabled. Individual bits may be altered by using the referenced AT letter command, or the entire register may set using the *ATS22=n* command format.

Bit #	Definition	Reference Command	Setting = 0	Setting = 1
1,0	Speaker volume	L0	0,0	Low (with some modems T0 is off)
		L1	0,1	Low
		L2	**1,0**	**Medium (default)**
		L3	1,1	High
2,3	Speaker control	M0	0,0	Off
		M1	**0,1**	**On until carrier (default)**
		M2	1,0	Always on
		M3	1,1	On after dialing and before carrier
4,5,6	Result codes	X0	0,0,0	
		X1	1,0,0	
		X2	1,0,1	See Xn command
		X3	1,1,0	
		X4	**1,1,1**	
7	Make/Break ratio	&Pn	**39/61 @ 10 pps (default)**	33/67 @ 10 pps

Note that the ATXn commands do not map into the respective values of n.

S22 = n . . . . *(2) XON Character (default = 17)(U.S. Robotics Sportster)*

S22 = n on the U.S. Robotics Sportster series of modems sets the ASCII code for the XON symbol. The valid range is 0 - 127. The default is 17.

S22 = n . . . . *(3) ARA Error Correction Options (default=0)*
                                                                                                     (Cisco MICA Six-Port Modules)

Specifies the AppleTalk Remote Access (ARA) error correction method. If more than one method of error correction is enabled at both ends of the connection, the choice of method is based on the order of precedence ARA originate only (highest precedence), LAP-M, MNP/ARA answer only, then fall back specified in register *S25* (lowest precedence).

n = 0 . . . . **ARA 1.0 and ARA 2.0 disabled (default).**
n = 1 . . . . ARA 1.0 and ARA 2.0 enabled for answer only.
n = 2 . . . . ARA 1.0 and ARA 2.0 enabled for answer, ARA 1.0 enabled for call origination.
n = 3 . . . . ARA 1.0 and ARA 2.0 enabled for answer, ARA 2.0 enabled for call origination.

S23 = n . . . . *(1) Bit-Mapped Option Codes (default = 62 ↕)*

☞ The individual bits in register *S23* are treated as switches by the modem. If the bit is a "1," the specified function is on or enabled; if the bit is a "0," the function is off or disabled. Individual bits may be altered by using the referenced AT letter command or the entire register may set using the *ATS23=n* command format. In order to set the DTE rate or the parity mode, the *ATS23=n* format must be used.

Bit #	Definition	Reference Command	Setting = 0	Setting = 1
0	Remote digital loopback (RDL)	&T4, &T5	&T5 = dis-allow RDL (default)	&T4 allow RDL
1,2,3	DTE rate		0,0,0 0 to 300 bps	
			0,0,1 600 bps	
			0,1,0 1200 bps	
			0,1,1 2400 bps	
			1,0,0 4800 bps	
			1,0,1 7200, 9600, or 14400 bps	
			1,1,0 19200 bps	
			1,1,1 38400 or 57600 bps	
4,5	Parity		0,0 Even	
			0,1 Not used	
			1,0 Odd	
			1,1 None	
6,7	Guard tones	&G0	0,0 Guard tone disabled	
		&G1	0,1 550 Hz guard tone	
		&G2	1,0 1800 Hz guard tone	

S23 = n . . . . *(2) XOFF Character (default = 19)(U.S. Robotics Sportster)*

S23 = n on the U.S. Robotics Sportster series of modems sets the ASCII code for the XOFF symbol. The valid range is 0–127. The default is 19.

S23 = n ....*(3) V.42 LAP-M Error Correction (default=1)*
(Cisco MICA Six-Port Modules)

Enables or disables V.42 (LAP-M) error correction. If more than one method of error correction is enabled at both ends of the connection, the choice of method is based on the order of precedence ARA originate only (highest precedence), LAP-M, MNP/ARA answer only, then fall back specified in register *S25* (lowest precedence).

n = 0 .... V.42 (LAP-M) disabled.
n = 1 .... **V.42 (LAP-M) originate and answer enabled (default).**

S24 = n .... *(1) Sleep Inactivity Timer (default 0)*

Register S24 sets the length of time (in seconds) that the modem will operate in normal mode with no detected telephone line or DTE line activity before entering low-power sleep mode. The timer is reset with any DTE line or telephone line activity. Setting register S24 to zero disables the timeout feature.

n = 0 .... Disables timeout (neither DTE line nor telephone inactivity will cause the modem to enter the sleep mode).
n = 1–255 .... Puts the modem in sleep mode after "n" seconds of no activity. Some modems use 5 seconds for n = 1–4.

S24 = n .... *(2) MNP Error Correction (default 1)*
(Cisco MICA Six-Port Modules)

Enables or disables MNP error correction. If more than one method of error correction is enabled at both ends of the connection, the choice of method is based on the order of precedence LAP-M, (highest precedence), ARA, MNP, then fall back specified in register *S25* (lowest precedence).

n = 0 .... MNP disabled.
n = 1 .... **MNP originate and answer enabled (default).**

S25 = n .... *(1) DTR Delay (default = 5 ↕)*

* The value in register *S25* defines the time the modem ignores DTR ON to OFF transitions. The units of n are either seconds or tens of milliseconds (0.01 second), depending on the modem operating mode, that is,

1. When the modem is operating in synchronous mode 1 (**AT&M1** or **AT&Q1**), the value in *S25* specifies the length of time (in seconds) that the modem waits after a connection has been established before testing the state of DTR. This allows time to transfer the communication connection at the modem from an asynchronous DTE to a synchronous DTE without the modem reverting to asynchronous mode.

2. In all other modes (and after the connection is established in synchronous mode 1), the time delay is 100 times faster, that is, n is in 1/100ths of a second.

n = 0 to 255 .... Synchronous mode, n is in *seconds*. With a default = 5, the shortest transition of DTR that the modem will recognize is 5 seconds.
n = 0 to 255 .... All other modes n is in *1/100ths of a second* (0 - 2.55 seconds). With a default = 5, the shortest transition of DTR that the modem will recognize is 50 ms.

S25 = n .... *(2) Link Protocol Fallback (default=0)*
(Cisco MICA Six-Port Modules)

Specifies the fall back action if either the V.42 detect phase (enabled by *S15* = 1) or error correction negotiation phase (timed by *S16* and *S17*) ends in a failure to agree on an error correction protocol. The far-end modem can also force the local modem to fall back using the character specified in register *S18*. In asynchronous framing mode (*S25* = 0), the modem uses the settings of registers *S12, S13,* and *S14*.

n = 0 .... **Use asynchronous framing mode (start/stop/parity bits) (default).**
n = 1 .... Use synchronous framing mode (raw 8 bits to line).
n = 2 .... Disconnect.

S26 = n .... *(1) RTS to CTS Delay (default = 1 ↕)*

When the modem is in synchronous mode and CTS tracks RTS (**AT&R0** asserted), the value in register *S26* determines how long CTS is held OFF after RTS is turned ON. "n" is in 1/100ths of a second; therefore, the range of holdoff delay is 0 to 2.55 seconds.

n=0–255 .... Delay CTS OFF to On transition by n/100 seconds (0 - 2.55 seconds) from RTS OFF to ON transition. The default = 1 or 10 ms.

S26 .... *(2) DSP Processor MVIP TDM Slice*
(Cisco MICA Six-Port Modules)

A read only register indicating the time-division multiplexing (TDM) pair assigned to the digital signal processors (DSP) on the modem card. Each of the three DSPs per modem card is assigned one TDM pair. The value of this register is set automatically to match a value determined by the router or access server. The value of S26 is used for diagnostic purposes. An attempt to manually change the value causes the message ERROR to be returned.

n = 0–15 ....

S27 = n . . . . *(1) Bit-Mapped Option Codes (default = 73 ↕)*

The individual bits in register *S27* are treated as switches by the modem. If the bit is a "1," the specified function is on or enabled; if the bit is a "0," the function is off or disabled. Individual bits may be altered by using the referenced AT letter command, or the entire register may set using the *ATS27=n* command format.

Bit #	Definition	Reference Command	Setting = 0	Setting = 1
0,1,3	Async/Sync	&M0, &Q0	0,x,0,0	Async direct mode
		&M1, &Q1	0,x,0,1	Sync mode 1
		&M2, &Q2	0,x,1,0	Sync mode 2
		&M3, &Q3	0,x,1,1	Sync mode 3
		&M4, &Q4	1,x,0,0	Reserved
		&M5, &Q5	**1,x,0,1**	**Async Reliable mode**
		&M6, &Q6	1,x,1,0	Async Normal mode
		&M7, &Q7	1,x,1,1	Reserved
2	Dial-up / Leased line	&Ln	Dialup (default)	Leased
4,5	Clock source	**&X0**	**0,0**	**Modem generates**
		&X1	0,1	DTE supplies
		&X2	1,0	Derived from carrier
6	300/1200 protocol	Bn	CCITT V.21/.22	Bell 103/212
7	Reserved			

S27 = n . . . . *(2) Bit-Mapped Option Codes (default = 0)(U.S. Robotics Sportster)*

The individual bits in register *S27* are treated as switches by the modem. If the bit is a "1," the specified function is on or enabled; if the bit is a "0," the function is off or disabled. Individual bits may be altered by using the referenced AT letter command, or the entire register may set using the *ATS27=n* command format.

U.S. ROBOTICS SPORTSTER 14.4 KBPS AND 56K MODEMS

Bit #	Definition	Reference Command	Setting = 0	Setting = 1
0	Enables V.21 300 bps		Disabled	Enabled
1	Enables non-trellis coded V.32		Disabled	Enabled
2	Disables V.32 modulation		Enabled	Disabled
3	Disables 2100 Hz answer tone		Enabled	Disabled
4	Enables V.23 fallback mode		Disable	Enable
5	Disables V.32 bis mode		Enabled	Disable
6	Reserved on 14.4 Kbps			
	Disables V.42 selective reject		Enabled	Disable
7	S/W compatibility mode		Normal	Disables actual code & displays 9600 code.

U.S. ROBOTICS SPORTSTER 28.8 KBPS MODEMS

Bit #	Definition	Reference Command	Setting = 0	Setting = 1
0	Enables V.21 300 bps		Disabled	Enabled
1	Enables nontrellis coded V.32		Disabled	Enabled
2	Disables V.32 modulation		Enabled	Disabled
3	Disables 2100 Hz answer tone		Enabled	Disabled
4,5	Error-control (ec) handshaking option		0 0 Complete handshake sequence V.42 detect, LAP-M ec, MNP. 0 1 Disable MNP 1 0 Disable V.42 detect & LAP-M 1 1 Disable V.42 detect. Used if remote mode uses LAP-M but does not support detection phase.	
6	Reserved			
7	S/W compatibility mode		Normal	Disables actual codes and displays 9600 code instead.

S27 bit 7 is used where 7200, 12000, 14400, or 28800 result codes are not accepted or recognized by the software. The actual connection rate can be retrieved by using the *ATI6* command.

S27 = n . . . . ***(3) Calling Tones (default=0)***
(Cisco MICA Six-Port Modules)

Enables or disables the V.25 optional calling tone, a 1300-Hz signal that allows the called party to determine whether the calling device is a modem. (Some PTTs outside the United States and Canada require a modem to send a calling tone when it originates a connection.)

n = 0. . . . . **Calling tone disabled (default).**
n = 1. . . . . Send 1300 Hz calling tone.

S28 = n . . . . ***(1) Bit-Mapped Option Codes (default = 0)***

Register S28 sets the dial pulse ration and rate.

Bit #	Definition	Reference Command	Setting = 0			Setting = 1
0,1	Reserved					
2	Reserved (always 0)					
3,4	Pulse dialing ratio and speed		bits	make/break	pps	
		&P0	**00**	**39%-61%**	**10**	
		&P1	01	33%-67%	10	
		&P2	02	39%-61%	20	
		&P3	02	33%-67%	20	
5,6,7	Reserved					

S28 = n . . . . ***(2) V.32 Handshake Time Allowed (default = 8)(U.S. Robotics Sportster)***

S28 = n sets the allowable time (in tenths of a second) for V.32 bis handshaking to be completed before falling back to V.23 or other modes.

n = 0. . . . . Eliminates the V.32 answer tones (for faster dialing).
n = 1–254. . . . . Sets timer value in 100 ms increments (range is 0–25.4 seconds).
n = 255. . . . . Disables all connection protocols except V.32 at 9600 bps.

S28 = n . . . . ***(3) V.34 Modulation Enable/Disable (default=1)***
(Zoltrix)

Enables or disables V.34 modulation.

n = 0. . . . . Disable V.34
n = 1–255. . . . . **Enable V.34 (default = 1).**

S28 = n. . . . . ***(4) Guard Tone (default=0)***
(Cisco MICA Six-Port Modules)

Enables or disables the 1800-Hz guard tone in V.22 and V.22bis modes. Some PTTs outside the United States and Canada require a modem to send a guard tone.

n = 0. . . . . **Guard tone disabled (default).**
n = 1. . . . . Use 1800 Hz guard tone in V.22 and V.22bis.

S29 = n. . . . . ***(1) Flash Dial Modifier Time (default = 70)***

Sets the length of time (in units of 10 ms) that the modem will go on-hook when it encounters the flash (!) dial modifier in the dial string. The time range may be limited in some countries.

n = 0–255. . . . . Flash time in increments of 10 ms intervals, i.e., total time from 0 to 2.55 seconds.

S29 = n. . . . . ***(2) V.21 Fallback Timer (default = 20)(U.S. Robotics Sportster)***

S29 = n sets the time duration (in tenths of a second) of the V.21 answer mode fallback timer for the 14.4 Kbps and 56K modem.

S29 = n . . . . ***(3) Modulation Standard (default=6)***
(Cisco MICA Six-Port Modules)

Register *S29* specifies the modulation standards available for use. When negotiating a connection, the modems attempt to establish a connection based on a standard acceptable to both. This is based on the sequence and timing of generated and detected tones. (For example, when *S29* = 0, the modem opens negotiations by generating a specific tone that is a valid starting point for many standards, including K56FLEX, V.90, V.34+, V.34, V.32bis, V.32, V.22bis, V.22, and V.21. If there is no response, the modem generates a sequence specific to V.32bis. Then the modem tries V.22/V.22bis, then V.32 again, then V.21, and so on.)

If both modems are MICA (modem ISDN channel aggregation), they will at a minimum be able to agree on one of the standards set with n = 4. If one modem is not MICA, the modems may fail to negotiate a standard and disconnect.

n = 0, 1 . . . . V.34+, V.34, V.32bis, V.32, V.23, V.22bis, V.22, V.21, Bell 212, Bell 103
n = 2 . . . . V.32ter, V.32bis, V.32, V.23, V.22bis, V.22, V.21, Bell 212, Bell 103
n = 3 . . . . V.32bis, V.32, V.23, V.22bis, V.22, V.21, Bell 212, Bell 103
n = 4 . . . . V.23, V.22bis, V.22, V.21, Bell 212, Bell 103
n = 5 . . . . K56Flex 1.1, V.34+, V.34, V.32bis, V.32, V.23, V.22bis, V.22, V.21, Bell 212, Bell 103
n = 6 . . . . V.90, K56Flex 1.1, V.34+, V.34, V.32bis, V.32, V.23, V.22bis, V.22, V.21, Bell 212, Bell 103
n = 7 . . . . SS7/ COT
n = 8 . . . . V.110

S30 = n . . . . *(1) Inactivity Timer (default = 0 ↕)*

The value in the *S30* register determines the time that the modem waits before disconnecting when no data is transferred. In *reliable mode* any data transfer resets the timer. In *normal mode* only transmit data resets the timer. The timer is not used in synchronous mode.

n = 0 . . . . Disables the inactivity timer.
n = 1–255 . . . . Forces a disconnect after "n" 10-second intervals with no data transferred (range is 0–2550 seconds or 42.5 minutes).

Note: In a few modems, n is in seconds rather than tens of seconds.

S30 = n . . . . *(2) Maximum Connection Rate (default=33600)*
(Cisco MICA Six-Port Modules)

Sets the maximum rate at which the modem will receive data, in bits per second.

n = 75–33600 . . . . Maximum connection rate (default = 33600). The value in this register must be equal to or larger than the value in register *S31*

	S-Registers Controlling	
Modulation Standard	Transmit Rate	Receive Rate
V.34 and lower	S30 and S31	S30 and S31
K56Flex , V.90	S50 and S51	S30 and S31

S31 = n . . . . *(1) Bit-Mapped Option Codes (default = 194, C2h)*

Bit #	Definition	Reference Command	Setting = 0	Setting = 1
0	Single-line connect message	\Vn	Disabled (use Wn and S95)	Enabled
1	Auto line speed detection	Nn	Disabled	Enabled
2,3	Error-correction progress messages	**W0**	**00 DTE speed only (default)**	
		W1	01 Full reporting	
		W2	10 DCE speed only	
4,5	Caller ID control	#CID=0	**00 Disabled (default)**	
		#CID=1	01 Short (formatted) Caller ID enabled	
		#CID=2	10 Long (unformatted) Caller ID enabled	
6,7	Reserved (set to 11b)		Default	

S31 = n . . . . *(2) Minimum Connection Rate (default=300)*
(Cisco MICA Six-Port Modules)

Sets the minimum rate at which the modem will receive data, in bits per second.

n = 75 – 33600 . . . . minimum connection rate (default = 300). The value in this register must be equal to or less than the value in register *S30*

	S-Registers Controlling	
Modulation Standard	Transmit Rate	Receive Rate
V.34 and lower	S30 and S31	S30 and S31
K56Flex , V.90	S50 and S51	S30 and S31

S32 . . . . *(1) XON Character (default = 17, 11h)*

Sets the value of the XON character.

n = 0–255 . . . . ASCII XON character in decimal format.

S32 = n . . . . *(2) Bit-Mapped Option Codes (default = 2)(U.S. Robotics Sportster)*

U.S. ROBOTICS SPORTSTER 56K MODEMS

Bit #	Definition	Reference Command	Setting = 0	Setting = 1
0	V.8 Call indicator enable		Disable	Enable
1	Enable V.8 mode		Disable	Enable
2	Reserved			
3	Disable V.34 modulation		Disable	Enable
4	Disable V.34+ modulation		Disable	Enable
5	Disable X2 modulation		Disable	Enable
6–7	Reserved			

In the 33.6 Vi modem, only bits 0, 1, & 3 are defined; bits 0–5 are defined in the 56K.

S32 = n . . . . *(3) Synthetic Ring Volume (default=10)*
(Zoltrix)

Sets the synthetic ring volume in dB with an implied minus sign

n = 0–255 . . . . Range of volume from 0 to -255dB (default$= -10$ dB).

S32 = n . . . . *(4) 56K Protocol Selection*
(US Robotics/3Com)

Bit #	Definition	Reference Command	Setting = 0	Setting = 1
0	V.8 Call Indicate enabled		Disable	Enable
1	Enables V.8 mode		Disable	Enable
2	Reserved			
3	Disable V.34 modulation		Enable	Disable
4	Disable V.34+ modulation		Enable	Disable
5	Disable X2 modulation only		Enable	Disable
6	Disable V.90 modulation		Enable	Disable
7	Reserved			

Some useful examples of combined values include:

n = 2 . . . . Enable X2 and V.90 (Sportster)
n = 32 . . . . Disable X2 mode (X2 only modem)
n = 34 . . . . Disable X2, enable V.8 mode (X2 only modem)
n = 34 . . . . Disable X2, enable V.90 (Sportster)
n = 66 . . . . Disable V.90, enable X2 (Sportster)
n = 98 . . . . Disable X2 and V.90 (Sportster)

S32 = n . . . .*(5). Signal Quality Threshold (default=2)*
(Cisco MICA Six-Port Modules)

Register *S32,* in conjunction with register *S54* bit-7, specifies the minimum bit error rate at which the modem initiates recovery, and below which the modem considers the line to be good. Recovery consists of a retrain and/or falling back to a lower data rate, depending on the modulation standard currently in use. With *S54* bit-7 = 0, the value of n is the *signal quality threshold (SQT)*. With bit-7 = 1 the SQT is n + 0.5. Bit error rate threshold (*BER_threshold*) and SQT are related by

$$BER\_threshold = 1:10^{SQT+1}$$

S-Register Settings		Effective SQT	BER Threshold
S32	S54		
0	0	None	N/A
0	1	None	N/A
1	0	1	1 : 100
1	1	1.5	1 : 316
2 (Default)	0	2	1 : 1000
2 (Default)	**1 (default)**	**2.5**	**1 : 3160**
3	0	3	1 : 10 000
3	1	3.5	1 : 31 600
4	0	4	1 : 100 000
4	1	4.5	1 : 316 000
5	0	5	1 : 1 000 000
5	1	5.5	1 : 3 160 000

S33 .... *(1) XOFF Character (default = 19, 13h)*

Sets the value of the XOFF character.

n = 0–255 . . . . ASCII XOFF character in decimal format.

S33 = n .... *(2) Bit-Mapped Option Codes (default = 0)(U.S. Robotics 56K Sportster)*

U.S. ROBOTICS SPORTSTER 56K MODEMS

Bit #	Definition	Reference Command	Setting = 0	Setting = 1
0	Disable 2400 symbol rate		Enable	Disable
1	Disable 2743 symbol rate		Enable	Disable
2	Disable 2800 symbol rate		Enable	Disable
3	Disable 3000 symbol rate		Enable	Disable
4	Disable 3200 symbol rate		Enable	Disable
5	Disable 3429 symbol rate		Enable	Disable
6	Reserved			
7	Disable shaping		Enable	Disable

S33 = n .... *(3) Synthetic Ring Frequency (default=0)*
(Zoltrix)

Register *S33* sets a synthetic ring frequency.

n = 0 . . . . Disabled (default).
n = 1–5 . . . . Selects 1 of 5 ring frequencies.

S33 .... *(4) Speed Change Squelch Timer (default=500)*
(Cisco MICA Six-Port Modules)

Sets the time to delay (in milliseconds) after a speed shift before another speed shift is allowed.

n = 0 . . . . Timer disabled
n = 1 – 65535 . . . . Delay time (0.001–65.535 seconds); default =500 (0.5 seconds).

S34 = n .... *(1) Bit-Mapped Option Codes (default = 6)(U.S. Robotics Sportster)*

U.S. ROBOTICS SPORTSTER 28.8 KBPS MODEMS

Bit #	Definition	Reference Command	Setting = 0	Setting = 1
0	Disable V.32 bis		Enable V.32 bis	Disable V.32 bis
1-2	Reserved			
3	Enable V.23 modulation		Disable V.23	Enable V.23
4-7	Reserved			

To enter a value into register *S34*, one must add six to the bit value desired; for example, to set bit 3 =1 (bit value 8), one must issue the command "*ATS34 = 14*" (6 + 8 = 14). This implies that allowable values for register *S34* include only 6, 7, 14, and 15. *S34* is used primarily for troubleshooting by the manufacturer.

S34 = n .... *(2) Bit-Mapped Option Codes (default = 0)(U.S. Robotics 56K Sportster)*

U.S. ROBOTICS SPORTSTER 56K MODEMS

Bit #	Definition	Reference Command	Setting = 0	Setting = 1
0	Disable 8S-2D trellis encoding		Enable	Disable
1	Disable 16S-4D trellis encoding		Enable	Disable
2	Disable 32S-2D trellis encoding		Enable	Disable
3	Disable 64S-4D trellis encoding		Enable	Disable
4	Disable non-linear coding		Enable	Disable
5	Disable transmit level deviation		Enable	Disable
6	Disable pre-emphasis		Enable	Disable
7	Disable precoding		Enable	Disable

S34 = n .... *(3) Fall-forward Timer (default=2000)*
(Cisco MICA Six-Port Modules)

Sets the elapsed time (in 10-milliseconds) during which the signal quality must be good before the modem attempts to increase transmission speed.

n = 0 . . . . Fall-forward disabled
n = 0 – 65535 . . . . Delay time (0.01 – 655.35 seconds), maximum value is almost 11 minutes; default = 2000 (20 seconds).

S35 = n . . . . *(1) Calling Tones (default=0)*
(Zoltrix)

Enables or disables the V.25 optional calling tone, a 1300-Hz signal that allows the called party to determine whether the calling device is a modem. (Some PTTs outside the United States and Canada require a modem to send a calling tone when it originates a connection.)

n = 0. . . . . **Calling tone disabled (default).**
n = 1 . . . . Send 1300 Hz calling tone.

S35 = n . . . . *(2) Fallback Timer (default=50)*
(Cisco MICA Six-Port Modules)

Sets the elapsed time (in 10-milliseconds) during which the signal quality must be poor for the modem to decrease transmission speed.

n = 0 . . . . Fallback disabled
n = 1 – 65535 . . . . Delay time (0.01 – 655.35 seconds); default =50 (0.5 seconds).

S36 = n . . . . *(1) Negotiation Failure Handling (default = 7)*

Register *S36* defines the fallback procedure when an attempted error-correction link (LAP-M) fails or immediately upon connection if register *S48* = 128. Although the range of n is 0 through 255, the only valid numbers are 0, 1, 3–5, and 7. If a number greater than 7 is entered, it may be accepted but the modem will behave as if the default had been entered.

n	Definition	Reference Command
0	Modem disconnect.	
1	Modem stays online and establishes a Direct connection.	
2	Undefined.	
3	Modem stays online and establish a Normal connection.	
4	Establish an MNP connection if possible, otherwise modem disconnects.	
5	Establish an MNP connection if possible, otherwise establish a direct connection.	
6	Undefined	
7	**Establish an MNP connection if possible, otherwise a Normal connection (default)**	

S36 = n . . . . *(2) Terminate Timeout (default=20)*
(Cisco MICA Six-Port Modules)

Sets the maximum time (in seconds) to delay after a host disconnect request before forcing the link to disconnect. During this period, the modem sends buffered data and then clears the link. Low values cause the modem to disconnect faster, but may result in some final data being lost and in the remote modem hanging on to a dead line for a while. A disconnect request can be an ATH (hang up) command or a request from the router or access server.

n = 0–858 . . . . Delay time (0–858 seconds), maximum value is more than 14 minutes; default = 20 (20 seconds).

S37 = n . . . . *(1) Desired Telco Carrier Rate (default = 0)*

The value in register *S37* defines the highest rate the local modem attempts to connect with the remote modem. If a value other than one of the valid numbers is entered, it may be accepted by the modem, but the modem will behave as if the default had been entered.

n	Definition	Reference Command
0	**Use the same carrier rate as the last AT rate command issued, or the maximum carrier rate the modem supports if the last command rate was greater than the modem's maximum carrier rate.**	Fn, Nn, S37, +MS
1	Attempt to connect at 300 bps	F1, Nn, S37, +MS
2	Attempt to connect at 300 bps	F1, Nn, S37, +MS
3	Attempt to connect at 300 bps	F1, Nn, S37, +MS
4	Undefined	
5	Attempt to connect at ITU-T V.21 1200 bps	F4, Nn, S37, +MS
6	Attempt to connect at ITU-T V.22 bis 2400 bps	F5, Nn, S37, +MS
7	Attempt to connect at ITU-T V.23 4800 bps	F3, Nn, S37, +MS
8	Attempt to connect at ITU-T V.32 / 32 bis 7200 bps	F6, Nn, S37, +MS
9	Attempt to connect at ITU-T V.32 / 32 bis 9600 bps	F8, Nn, S37, +MS
10	Attempt to connect at ITU-T V.32 bis 12,000 bps	F9, Nn, S37, +MS
11	Attempt to connect at ITU-T V.32 bis 14,400 bps	F10, Nn, S37, +MS
12	Attempt to connect at ITU-T V.34	F7, Nn, S37, +MS

1. When the Nn command is issued or the *S37* register value is modified, the +MS command subparameters are updated to reflect the speed and modulation specified by the *S37* value (see +MS command).

   For example:
   ATN0S37=10. . . . . Updates the +MS command subparameters to +MS=10,1,300,12000.
   ATN1S37=10. . . . . Updates the +MS command subparameters to +MS=10,0,1200,12000.
   ATN1S37=0. . . . . Updates the +MS command subparameters to +MS=11,1,300,28800.

2. *S37;* however, is not updated by the +MS command.

3. Use of the +MS command is generally recommended instead of the Nn and *S37*=x commands. Nn and S37=x commands are supported for compatibility with existing communication software.

S37 = n . . . . *(2) 56K Upstream Rate*

The S37 register is used to control the upstream V.34 rate.

n = 0 . . . . Auto-rate select	n = 10 . . . . 12000 bps
n = 1 . . . . Reserved	n = 11 . . . . 14400 bps
n = 2 . . . . 1200/75 bps	n = 12 . . . . 16800 bps
n = 3 . . . . 300 bps	n = 13 . . . . 19200 bps
n = 4 . . . . Reserved	n = 14 . . . . 21600 bps
n = 5 . . . . 1200 bps	n = 15 . . . . 24000 bps
n = 6 . . . . 2400 bps	n = 16 . . . . 26400 bps
n = 7 . . . . 4800 bps	n = 17 . . . . 28800 bps
n = 8 . . . . 7200 bps	n = 18 . . . . 31200 bps
n = 9 . . . . 9600 bps	n = 19 . . . . 33600 bps

S37 = n . . . . *(3) Wait for Carrier After Dial (default = 60)*
(Cisco MICA Six-Port Modules)

The value in register *S37* is linked to register *S7* so that the two registers share a single value. Changing the value in one changes the value in the other. See register *S7*.

S38 = n . . . . *(1) Delay Before Forced Disconnect (default = 20)*

☞* The value in register *S38* generally sets the maximum delay to disconnect from either the receipt of an ***ATH0*** command or from an ON to OFF transition of DTR—although some modems disconnect immediately (ignoring *S38*) upon receipt of ***ATH0***. If n is less than 255, the modem will wait n seconds for the remote modem to acknowledge the receipt of the data in the local modem's transmit buffer before disconnecting. If n = 255, the local modem will wait indefinitely for the remote modem to acknowledge the receipt of the data in the local modem buffer or until the connection is lost.

For an error-correction connection, register *S38* can be used to ensure that all data in the modem's transmit buffer are sent to the remote modem before disconnecting

n = 0–254 . . . . Wait n seconds for the remote modem to acknowledge all data received or until the connection is lost.
n = 255. . . . . Wait indefinitely for the remote modem to acknowledge all data received or until the connection is lost.

n	Disconnect delay conditions
0–254	Delay in seconds from H command, or DTR ON to OFF toggle (if modem is set to follow DTR), before modem disconnects.
255	Delay until data in buffer is delivered or until connection is lost.

ATH0 Result Codes:
OK. . . . . If all data were transmitted.
NO CARRIER . . . . If n < 255 and the time expired before the transmit data were sent.
ERROR . . . . Otherwise.

S38 = n . . . . *(2) 56K Downstream Rate*

The *S38* register sets the maximum 56K downstream speed that the modem attempts to connect. To disable 56K, set *S38* to 0. (*S37* register is used to control the upstream V.34 rate.)

n	K56Flex Description	V.90 Description		n	K56Flex Description	V.90 Description
0	K56Flex disabled	V.90 Disabled		12	52000 bps	41333 bps
1	K56Flex enabled—	V.90 Enabled—		13	54000 bps	42666 bps
	Auto-rate select (default)	Auto-rate select		14	56000 bps	44000 bps
2	32000 bps	28000 bps		15	58000 bps	45333 bps
3	34000 bps	29333 bps		16	60000 bps	46666 bps
4	36000 bps	30666 bps		17		48000 bps
5	38000 bps	32000 bps		18		49333 bps
6	40000 bps	33333 bps		19		50666 bps
7	42000 bps	34666 bps		20		52000 bps
8	44000 bps	36000 bps		21		53333 bps
9	46000 bps	37333 bps		22		54666 bps
10	48000 bps	38666 bps		23		56000 bps
11	50000 bps	40000 bps				

S38 = n . . . . *(3) Wait for Carrier After Dial (default=2)*
(Cisco MICA Six-Port Modules)

The value in register *S38* is linked to register *S10* so that the two registers share a single value. Changing the value in one changes the value in the other. See register *S10*.

S39 . . . . *(1) Current Flow Control Setting (SSI 73D2247)*
🗁

S39 = n . . . . *(2) Flow Control Bit Mapped Options Status (default=3)*

Bit #	Definition	Reference Command	Setting = 0	Setting = 1
0–2	Flow control method		0, 0, 0	No flow control
		&K3	0, 1, 1	RTS/CTS (default)
		&K4	1, 0, 0	XON/XOFF
		&K5	1, 0, 1	Transparent XON
		&K6	1, 1, 0	Both methods
3–7	Reserved			

S39 = n . . . . *(3) Transmit Level Setting (default=7)*
(Cisco MICA Six-Port Modules)

Sets the transmission level of the modem in dB.

n	Transmit level		n	Transmit level
0	–6 dBm		8	–14 dBm
1	–7 dBm		9	–15 dBm
2	–8 dBm		10	–16 dBm
3	–9 dBm		11	–17 dBm
4	–10 dBm		12	–18 dBm
5	–11 dBm		13	–19 dBm
6	–12 dBm		14	–20 dBm
7 (default)	**–13 dBm (default)**		15	–21 dBm

The actual range of transmit levels you can use is specified by the PTT of the country in which the modem is operating. The modem will limit the level to a value related to the country code specified at setup. If the modem is operating to V.90 specifications, the levels are controlled by register *S59*.

S40 = n . . . . *(1) Bit-Mapped MNP Option Codes (default = 55 ↕)*

📁 The individual bits in register *S40* are treated as switches by the modem. If the bit is a "1," the specified function is on or enabled; if the bit is a "0," the function is off or disabled. Individual bits may be altered by using the referenced AT letter command, or the entire register may be set using the *ATS40=n* command format.

Bit #	Definition	Reference Command	Setting = 0	Setting = 1
1,0	Mode	\N0	0,0	Normal mode
		\N1	0,1	Direct mode
		\N2	1,0	MNP Reliable mode
		\N3	1,1	AutoReliable mode
2,3,4	Break processing	\K0	0,0,0	Enter command state w/o sending BREAK
		\K1	0,0,1	Clear all buffers, send BREAK
		\K2	0,1,0	Enter command state w/o sending BREAK
		\K3	0,1,1	Hold buffers, send BREAK
		\K4	1,0,0	Enter command state w/o sending BREAK
		\K5	1,0,1	Send BREAK in sequence with data
5,6	MNP maximum block size	\A0	0,0	64 characters
		\A1	0,1	128 characters (default)
		\A2	1,0	192 characters
		\A3	1,1	256 characters
7	Modem-to-modem XON/XOFF flow control	\Gn	Disable (default)	Enable

S40 = n . . . . *(2) ETC Startup Autorating (default = 0)*
(Zoltrix)

n = 0. . . . . **Start with normal auto-rating (default).**
n = 1. . . . . Start up at initial rate of 4800 or below.
n = 2. . . . . Start up at initial rate of 9600 or below.

S40 = n . . . . *(3) Consecutive Retrain Disconnect Threshold (default = 4)*
(Cisco MICA Six-Port Modules)

The number of consecutive retrain failures that cause a modem disconnect.

n = 0. . . . . Never disconnect because of failed retrains
n = 1–255 . . . . Disconnect after "n" failed retrains (default = 4)

S41 = n . . . . *(1) Bit-Mapped MNP Option State Codes (default = 1)*

📁 The individual bits in register *S41* are treated as switches by the modem. If the bit is a "1," the specified function is on or enabled; if the bit is a "0," the function is off or disabled. Individual bits may be altered by using the referenced AT letter command, or the entire register may set using the *ATS41=n* command format.

Bit #	Definition	Reference Command	Setting = 0	Setting = 1
0	Data compression	%Cn	Disable	Enable
1	Auto-retrain	%En	Disable	Enable
2	MNP stream/block link	\Ln	Stream	Block
3–7	Not used			

S41 = n . . . . *(2) Bit-Mapped Option (default = 0)*

U.S. ROBOTICS SPORTSTER 56K MODEMS

Bit #	Definition	Reference Command	Setting = 0	Setting = 1
0	Enable distinctive ring		Disable	Enable
1–7	Reserved			

S41 = n . . . . *(3) General Bit-Mapped Options (default = 195↕)*

Bit #	Definition	Reference Command	Setting = 0		Setting = 1
1–0	Compression Selection	%C0	0, 0	Disabled	
		%C1	0, 1	MNP 5	
		%C2	1, 0	V.42 bis	
		%C3	**1, 1**	**MNP 5 and V.42 bis (default)**	
6,2	Auto-retrain and fall back/forward	%E0	0, x, x, x, 0	Retrain and fallback/fall forward disabled	
		%E1	0, x, x, x, 1	Retrain enabled	
		%E2	**1, x, x, x, 0**	**Fallback/fall forward enabled (default)**	
7, 5-3	Reserved				

S41 = n . . . . *(4) V.34 Maximum Symbol Rate (default = 5)*
(Cisco MICA Six-Port Modules)

Sets the maximum symbol rate (bauds) that pertains when the modem is using V.34 modulation. This register is read-only unless the DEBUGTHISMODEM command has been asserted.

n = 0 . . . . 2400 baud
n = 1 . . . . 2743 baud
n = 2 . . . . 2800 baud
n = 3 . . . . 3000 baud
n = 4 . . . . 3200 baud
n = 5 . . . . 3429 baud (default)

S42 . . . . *(1) Auto Rate (default = 1)*

Register *S42* is read only and is used for testing and debugging only. V.32bis and V.22bis auto rate is disabled. Retrain operation is disabled or enabled in date mode, and fallback is disabled in data mode.

n = 0 . . . . Auto-rate disable
n = 1 . . . . Auto-rate enable (default)

S42 = n . . . . *(2) V.34 minimum Symbol Rate (default = 0)*
(Cisco MICA Six-Port Modules)

Sets the minimum symbol rate (bauds) that pertains when the modem is using V.34 modulation. This register is read-only unless the DEBUGTHISMODEM command has been asserted.

n = 0 . . . . 2400 baud (default)
n = 1 . . . . 2743 baud
n = 2 . . . . 2800 baud
n = 3 . . . . 3000 baud
n = 4 . . . . 3200 baud
n = 5 . . . . 3429 baud

S43 . . . . *(1) Current DCE rate (SSI 73D2247)*

S43 . . . . *(2) Auto Mode (default = 1)*

Register *S43* is read only and is used for testing and debugging only. V.32bis startup auto mode is disabled.

n = 0 . . . . Auto-mode disable
n = 1 . . . . Auto-mode enable (default)

S43 . . . . *(3) V.34 Carrier Frequency (default = 2)*
(Cisco MICA Six-Port Modules)

Sets the carrier frequency that pertains when the modem is using V.34 modulation. This register is read-only unless the DEBUG-THISMODEM command has been asserted.

n = 0 . . . . Low carrier frequency
n = 1 . . . . High carrier frequency
n = 2 . . . . Automatic carrier frequency selection (default)

S44 . . . .*V.34 Preemphasis Filter Selection (default = 11)*
(Cisco MICA Six-Port Modules)

Sets the preemphasis filter to use on the transmit signal when the modem is using V.34 modulation. For more information on preemphasis filter selection, see CCITT Recommendation V.34. This register is read-only unless the DEBUGTHISMODEM command has been asserted.

n = 0–10 . . . . Select a specific preemphasis filter.
n = 11 . . . . **Automatic preemphasis filter selection (default).**

S45 . . . .*Signaling type (default = 0)*
(Cisco MICA Six-Port Modules)

Sets the type of transmit and receive signaling the modem uses on T1/R2 trunks. If *S29* = 8 (V.110), the modem ignores values in this register and uses null signaling (0)

n = 0 . . . . **Null signaling**	n = 4 . . . . .High-band R2 signaling
n = 1 . . . . MF signaling	n = 5 . . . . .Modified R1 signaling
n = 2 . . . . DTMF signaling	n = 6 . . . . .SS7/COT signaling
n = 3 . . . . Low-band R2 signaling	

S46 = n . . . . *(1) Compression Protocol Selection (default = 138)*

🗁   The value in register *S46* is treated as a switch by the modem. For example, with some modems, if n=136, the error-correction protocol *without* data compression is selected. If n=138 (the default value), the error-correction protocol *with* data compression is selected. With other modems, bit one is controlled by the *AT%Cn* command. In both cases, the default is generally to enable compression.

V.42bis data compression is used only with LAP-M protocol and MNP5 data compression is used only with MNP4 protocol.

n = 0 . . . . Disable compression (See also %Cn command)
n = 1 . . . . Enable compression (See also %Cn command)
n = 136 . . . . Disable V.42bis and MNP5 data compression
n = 138 . . . . Enable V.42bis and MNP5 data compression

S46 = n . . . . *(2) Call Progress Tone Detection (default = 0)*
(Cisco MICA Six-Port Modules)

Register *S46* is a bit-mapped register that determines whether the modem detects dial, ring-back, and busy call progress tone signals generated by other devices during link establishment.

n = 0–7 . . . . Range of bit map.

Bit #	Definition	Reference Command	Setting = 0	Setting = 1
0	Dial tone		Ignore (default)	Detect
1	Ring back tone		Ignore (default)	Detect
2	Busy tone		Ignore (default)	Detect

S47 . . . .*Hayes +++ Escape Detection (default = 2)*
(Cisco MICA Six-Port Modules)

Along with register *S48*, *S47* enables or disables the detection of the Hayes escape sequence (default = +++). The escape sequence allows a user to put the modem into command mode while it is connected to a far-end modem. Registers *S2* and *S47* determine the exact nature of the escape sequence.

(If escape code detection is enabled on the modem for a host that echoes the user's input, the host modem may inadvertently be placed in online command mode when a user enters the escape code on the local modem. Therefore escape detection should be disabled on host/server modems, i.e., set *S47* = 0.)

n = 0 . . . . Disable
n = 1 . . . . Enabled
n = 2 . . . . **Enabled in originate mode only (default)**

S48 = n . . . . *(1) V.42 Negotiation (default = 7)*

🗁   The value in *S48* determines which of three negotiation strategies will be applied to a connection. The full V.42 negotiation determines the capabilities of the remote modem. If the capabilities of the remote modem are known a priori, then negotiation can be disabled and the protocol procedure can be initiated directly.

n = 0 . . . . Disable negotiation. Bypass detection and negotiation phases; proceed with LAP-M.
n = 7 . . . . **Enable negotiation (default)**
n = 128 . . . . Disable Negotiation. Bypass detection and negotiation phases; proceed with fallback per S36 setting.

S48 = n . . . . *(2) AT Command Processor (default = 1)*
(Cisco MICA Six-Port Modules)

Enables or disables the AT command processor.

n = 0 . . . . Disable
n = 1 . . . . **Enabled (default)**

S49 = n . . . . *(1) ASB Buffer Low Limit (default = 8)*

🖚   The *S49* and *S50* registers are used to determine the lower and upper limit of characters that can be stored in the modem buffer. When the number of characters in the buffer is less than or equal to the number in *S49*, data flow is enabled (XON or DTR-ON). If the value in *S49* is greater than the value in *S50* or the value in *S49* is greater than 249, the modem will use values to match the modem buffer.

n = 0 or n > S50 . . . . Modem uses values to match those of the modem buffer.
n = 1 to 249 . . . . Modem enables flow when character count is less than n.

S49 = n . . . . *(2) Call Setup Delay (default = 0)*
(Cisco MICA Six-Port Modules)

Sets the amount of time in 100-milliseconds that the modem waits before initiating a new link.

n = 0 . . . . **No delay before link initiation (default)**
n = 1–255 . . . . Wait 0.1 to 25.5 seconds

S50 = n . . . . *(1) ASB Buffer High Limit (default = 255)*

🖚   The *S49* and *S50* registers are used to determine the lower and upper limit of characters that can be stored in the modem buffer. When the number of characters in the buffer is greater than or equal to the number in *S50*, data flow is disabled (XOFF or DTR-OFF). If the value in *S50* is not between 2 and 250 and is not greater than the value in *S49*, the modem will use values to match the modem buffer. Note the default is n = 255 (greater than the maximum 250); this forces the modem to use the maximum available buffer.

n = 0, 1 or n < S49 . . . . . . Modem uses values to match those of the modem buffer.
n = 2 to 250 . . . . Modem disables flow when character count is greater than n.

S50 = n . . . . *(2) Maximum PCM Connect Rate (default = 60000)*
(Cisco MICA Six-Port Modules)

Sets the maximum rate, in bits per second, at which the modem will transmit data when it is using the K56Flex or V.90 modulation standards (set by register *S29*).

Pulse code modulation (PCM) is digital; the other types of modulation available on MICA modems (V.34, V.22, etc.) are analog.

n = 28000–60000 . . . . Maximum connection rate (default = 60000). The value in this register must be equal to or greater than the value in register *S51*

Modulation Standard	S-Registers Controlling	
	Transmit Rate	Receive Rate
V.34 and lower	S30 and S31	S30 and S31
K56Flex , V.90	S50 and S51	S30 and S31

S51 . . . . *(1) Bit-Mapped Option Codes (default = 0)(U.S. Robotics Sportster)*

U.S. ROBOTICS SPORTSTER 28.8 KBPS MODEMS

Bit #	Definition	Reference Command	Setting = 0	Setting = 1
0	MNP/V.42 disable in V.22		**Enable**	Disable
1	MNP/V.42 disable in V.22 bis		**Enable**	Disable
2	MNP/V.42 disable in V.32 bis		**Enable**	Disable
3-7	Reserved			

S51 = n . . . . *(2) Minimum PCM Connect Rate (default = 28000)*
(Cisco MICA Six-Port Modules)

Sets the minimum rate, in bits per second, at which the modem will transmit data when it is using the K56Flex or V.90 modulation standards (set by register *S29*).

Pulse code modulation (PCM) is digital; the other types of modulation available on MICA modems (V.34, V.22, etc.) are analog.

n = 28000–60000 . . . . Minimum connection rate (default = 28000). The value in this register must be equal to or less than the value in register *S50*

Modulation Standard	S-Registers Controlling	
	Transmit Rate	Receive Rate
V.34 and lower	S30 and S31	S30 and S31
K56Flex , V.90	S50 and S51	S30 and S31

S52 = n . . . .*Digital Pad Compensation (default = 1)*
(Cisco MICA Six-Port Modules)

This register controls whether the MICA modem allows the partner analog modem to compensate on circuits where a digital pad is detected. (For V.90 transmissions, MICA can boost the signal it transmits when the partner analog modem detects a digital pad (attenuator) in the circuit. Boosting the signal to compensate for digital pad attenuation can improve throughput.) This register has no effect on transmissions in modes other than V.90.

n = 0 . . . . No compensation
n = 1 . . . . Modem supports digital pad compensation for V.90 transmissions

S53 = n . . . .*V.8bis Capability (default = 3)*
(Cisco MICA Six-Port Modules)

Register *S53* is a bit-mapped register that enables/disables V.8bis negotiation. V.8bis is a protocol used for exchanging K56Flex information. It is used during modem training prior to all other exchanges/tones. If K56Flex clients do not exist, set *S53* = 0 to shorten modem train-up times for other protocols by about 3 seconds. Enabling V.8bis has no effect unless you are using a modulation standard that includes K56Flex. Setting *S53* to 0 effectively disables K56Flex, because K56Flex cannot operate without V.8bis.

n = 0–7 . . . . Range of bit map.

Bit #	Definition	Reference Command	Setting = 0	Setting = 1
0	V.8bis negotiation		Disable	**Enable (default)**
1	V.90 negotiation in V.8bis		Disable	**Enable (default)**

S54 . . . . *(1) Bit-Mapped Option Codes (default = 0)(U.S. Robotics Sportster)*

U.S. ROBOTICS SPORTSTER 28.8 KBPS MODEMS

Bit #	Definition	Reference Command	Setting = 0	Setting = 1
0	Disable 2400 symbol rate		**Enable**	Disable
1	Disable 2743 symbol rate		**Enable**	Disable
2	Disable 2800 symbol rate		**Enable**	Disable
3	Disable 3000 symbol rate		**Enable**	Disable
4	Disable 3200 symbol rate		**Enable**	Disable
5	Disable 3429 symbol rate		**Enable**	Disable
6	Disable call indicator for V.34		**Enable**	Disable
7	Disable V.8 for V.34		**Enable**	Disable

S54 = n . . . . *(2) Bit-Mapped Option Codes (default = 136)*
(Cisco MICA Six-Port Modules)

n = 0–511 . . . . Range of values

Bit #	Definition	Reference Command	Setting = 0	Setting = 1
0	Sends automatic messages to host router when a modem session has debug information to log		**Disable (default)**	Enable
1	Aggressive client capping		**Disable (default)**	Enable

Continued

(CONTINUED)

Bit #	Definition	Reference Command	Setting = 0	Setting= 1
2	Force 4-point training		**Disable (default)**	Enable
3	Power control		Disable	**Enable (default)**
4	Error correction (EC) quality checking		**Disable (default)**	Enable
5	Cap the receive rate at 26400 when detecting a PC-Tel modem		**Disable (default)**	Enable
6	Error correction (EC) parity. Enables 7E, 7O, and 7N data to be passed over a modem connection that has EC. Note that S12 and S13 apply to modem connections that have EC when EC parity is enabled. Set S12 = 7 in order for 7E, 7O, and 7N data to be passed on an EC modem connection when EC parity is enabled in S54.		**Disable (default)**	Enable
7	Enable 0.5 shift in signal quality (SQ) threshold, to lower the receive rate. This parameter works in conjunction with S32		Disable	**Enable (default)**
8	DSP reset		**Disable (default)**	Enable

S55 . . . . *(1) Bit-Mapped Option Codes (default = 0)(U.S. Robotics Sportster)*

U.S. ROBOTICS SPORTSTER 28.8 KBPS MODEMS

Bit #	Definition	Reference Command	Setting = 0	Setting = 1
0	Disable 8S-2D map		**Enable**	Disable
1	Disable 16S-4D map		**Enable**	Disable
2	Disable 32S-2D map		**Enable**	Disable
3	Disable 64S-4D map		**Enable**	Disable
4–7	Reserved			

S55 . . . . *(2) SS7/COT Control (default = 0)*
(Cisco MICA Six-Port Modules)

A read only register provided for debugging. It is reserved for SS7/COT control, and is set by the host router or access server. (SS7/COT is in effect only when *S45 = 6*)

n = 0–2009 . . . . Range of values

Bit #	Definition	Reference Command	Setting = 0	Setting = 1
0–10	Timeout period, in 10 ms increments		**0**	**No timeout (default)**
			1	10 ms
			2	20 ms
			:	
			2000	20 000 ms (20 seconds)
11–13	Operation		**0**	**Transmit then receive (default)**
			1	Receive then transmit
			2	Receive monitor only (test mode)
			3	Transmit generate only (test mode)
			4	Receive monitor & transmit generate (test mode)
			5–7	Reserved
14	Transmit frequency		**1780 Hz (default)**	2010 Hz
15	Receive frequency		**1780 Hz (default)**	2010 Hz

S56 . . . . *(1) Bit-Mapped Option Codes (default = 0)(U.S. Robotics Sportster)*

U.S. ROBOTICS SPORTSTER 28.8 KBPS MODEMS

Bit #	Definition	Reference Command	Setting = 0	Setting = 1
0	Disable nonlinear coding		**Enable**	Disable
1	Disable transmit level deviation		**Enable**	Disable
2	Disable pre-emphasis		**Enable**	Disable
3	Disable precoding		**Enable**	Disable
4	Disable shaping		**Enable**	Disable
5	Reserved			
6	Disable V.34		**Enable**	Disable
7	Disable V.FC		**Enable**	Disable

S56 . . . . *(2) Set Maximum V.34+ Transmit Rate (default = 33600)*
(Cisco MICA Six-Port Modules)

The maximum allowable V.34+ transmit rate in bits per second (bps). This register is read-only unless the DEBUGTHISMODEM command has been executed.

n = 4800, 7200, 9600, 12000, 14400, 16800, 19200, 21600, 24000, 26400, 28800, 31200, or 33600

S57 . . . . *V.110 User Rate (default = 9600)*
(Cisco MICA Six-Port Modules)

Register *S57* specifies the user rate (send and receive speed) for originating V.110 connections. The user rate is communicated to the remote modem. The register is set by the host router or access server.

n = 600, 1200, 2400, 4800, 7200, **9600,** 14400, 19200, 38400

S58 = n . . . . *(1) 56K Protocol Selection*
(Lucent & US Robotics/3Com)

n = 0 . . . . Enable X2 and V.90 (Courier). Disable K56Flex (Lucent)
n = 1 . . . . Enable V.90, disable X2 (Courier). Enable K56Flex (Lucent)
n = 32 . . . . Enable X2, disable V.90 (Courier)
n = 33 . . . . Disable X2, and V.90 (Courier)

S58 . . . . *(2) V.110 Flow Control and Clock Bits (default = 0)*
(Cisco MICA Six-Port Modules)

Register *S58* is set by the host. If the modem is in originate mode, *S58* is set to 0. If the modem is in answer mode, the host sets *S58* according to Q.931 call setup standards. The information is useful to the remote modem for debugging.

S59 = n . . . . *V.90 Transmit Level Setting (default = 7)*
(Cisco MICA Six-Port Modules)

Sets the V.90 transmission level of the modem in dB.

n	Transmit level	n	Transmit level
0	–6 dBm	8	–14 dBm
1	–7 dBm	9 (default in Japan)	–15 dBm (default in Japan)
2	–8 dBm	10	–16 dBm
3	–9 dBm	11	–17 dBm
4	–10 dBm	12	–18 dBm
5	–11 dBm	13	–19 dBm
6 (default)	**–12 dBm (default)**	14	–20 dBm
7	–13 dBm	15	–21 dBm

The actual range of transmit levels you can use is specified by the PTT of the country in which the modem is operating. The modem will limit the level to a value related to the country code specified at setup. For example, the maximum transmit level for the United States is –12 dBm, for Japan –15 dBm, and for all other countries it is –10 dBm.

If the modem is not operating with V.90 specifications, the levels are controlled by register *S39*.

S60 . . . . *Sticky Flags (default = 0)*
(Cisco MICA Six-Port Modules)

A read only bit-mapped register provided for debugging. It is reserved for use by the host router or access server.

n = 0–15 . . . . Range of values

Bit #	Definition	Reference Command	Setting = 0	Setting = 1
0	Specifies whether the host accepts expedited Terminate_Event messages, which convey accounting information about call termination from the modem to the host.		**Disable Terminate_Event messages (default)**	Enable Terminate_Event messages
1	Specifies whether the modem will send a capabilities mask, which identifies the modem's call type capabilities, to the host. The capabilities mask is sent once in the lifespan of a Portware release. Based on the information sent, the host sets various S registers, including S12, S13, S14, S29 (V.110 only) S55, and S57.		**Disable capabilities mask (default)**	Enable capabilities mask
2	Activates in-band PPP mode switches.		**Disable switches (default)**	Enable switches
3	Specifies whether the ConnectInfo and TerminateInfo messages contain data. If the messages contain no data, the host reads the information directly from the shared memory area.		**Messages contain data (default)**	Messages contain no data

S61 through S81 ....*Reserved*

S82 = n ....*LAP-M Break Handling Options (default = 128)*

📁*   Break signals provide a way for an application (or user) to get the attention of the remote modem. The value in register *S82* selects the method the local modem uses when sending a break to the remote mode. LAP-M specifies the three break signaling procedures: *expedited, destructive,* and *in sequence.* The break type selection depends on the requirements of the specific application.

The values of n corresponding to the break procedure are given in the table below. Although the modem will accept all values in the range of 0 to 255, any value other than 3, 7, or 128 will cause the modem to behave as if the default value had been entered.

n = 3................. Expedited: Modem sends break immediately; data integrity is maintained before and after break.

n = 7................. Destructive: Modem sends break immediately; data being processed by each modem at that time is destroyed.

n = 128............... **In Sequence: Modem sends break in sequence with transmitted data; data integrity is maintained before and after the break.**

S83 through S85 ....*Reserved*

S86..*Connection Failure Cause Code (default = 0)*

Whenever the modem issues a NO CARRIER result code, a code is written into register *S86* to assist in determining the cause for the failed connection. *S86* records only the first event that contributes to the NO CARRIER message.

n = 0................. **Normal Disconnect (no error)**

n = 1–3............... Undefined

n = 4................. Loss of Carrier

n = 5................. V.42 negotiation failure (no error-correction remote modem)t

n = 6-8 .... Undefined

n = 9................. Modems could not find a common protocol

n = 10–11 .... Undefined

n = 12................. Normal Disconnect initiated by the remote modem

n = 13................. The remote modem did not respond after 10 retransmissions of the same message

n = 14................. Protocol violation

S87 through 88

S89 . . . . *(1) Power*

> *S89* = 0 in Silicon Systems 73D240/241 chips disables Power Down.

S89 = n . . . . *(2) Sleep Mode Control Timer (default = 60)(Zoltrix)*

> The *S89* register sets/displays the number of seconds of inactivity (no characters sent from DTE and no RING) in the off-line command state before the modem places itself into standby mode.
>
> n = 0 . . . . Disable standby mode
> n = 1–255 . . . . Number of seconds of inactivity before standby mode. Entered values of 1 – 4 are accepted for compatability, but are interpreted as 5 seconds.

S90 . . . . *Local Line Status*

> A read only register that indicates the hook-switch status of the local phone line..
>
> n = 0 . . . . Local phone on-hook
> n = 1 . . . . Local phone off-hook

S91 = n . . . . *Programmable Transmit Level (default = 10)*

> The value "n" in register *S91* represents the amount the transmit signal is reduced from the normal transmit level. The values are expressed in dB and have a possible range of 0 to 15. The selected transmit level is used for both synchronous and asynchronous modes. Once the connection to the remote modem is established, the transmit level may not be changed.
>
> **n = 0 . . . . 0 dB loss (US default value)**
> n = 1 . . . . 1 dB loss (sometimes listed as -1 dBr)
> n = 2 . . . . 2 dB loss (sometimes listed as -2 dBr)
> n = 3 . . . . 3 dB loss (sometimes listed as -3 dBr)
> n = 4 . . . . 4 dB loss (sometimes listed as -4 dBr)
> n = 5 . . . . 5 dB loss (sometimes listed as -5 dBr)
> n = 6 . . . . 6 dB loss (sometimes listed as -6 dBr)
> n = 7 . . . . 7 dB loss (sometimes listed as -7 dBr)
> n = 8 . . . . 8 dB loss (sometimes listed as -8 dBr)
> n = 9 . . . . 9 dB loss (sometimes listed as -9 dBr)
> n = 10 . . . . 10 dB loss (sometimes listed as -10 dBr) (default)
> n = 11 . . . . 11 dB loss (sometimes listed as -11 dBr)
> n = 12 . . . . 12 dB loss (sometimes listed as -12 dBr)
> n = 13 . . . . 13 dB loss (sometimes listed as -13 dBr)
> n = 14 . . . . 14 dB loss (sometimes listed as -14 dBr)
> n = 15 . . . . 15 dB loss (sometimes listed as -15 dBr) (Default value in Japan)

S92 . . . . *FAX Transmit Level (default = 10)*

S93 through S94

S95 = n . . . . *Extended Result Code Control (default = 0).*

> The individual bits in register *S95* are treated as switches by the modem. If the bit is a "1," the specified function is on or enabled, overriding the conditions specified by the *ATWn* command. If the bit is a "0," the function is treated as specified in the last *ATWn* command.
>
> Each bit set high in this register enables the corresponding result code regardless of the *ATWn* command setting. Therefore, *ATS95=44* will enable CARRIER, PROTOCOL, COMPRESSION, and CONNECT messages regardless of the setting specified by *ATWn.*

Bit #	Decimal Value	Description
0	1	CONNECT XXXX result code indicates DCE to DCE rate instead of local DTE to DCE rate
1	2	Append /ARQ to verbose CONNECT result code if protocol is not NONE
2	4	Enable CARRIER XXXX result code
3	8	Enable PROTOCOL XXXX result code
4	16	Reserved
5	32	Enable COMPRESSION XXXX result code
6	64	Enable Protocol result codes 81-83 for MNP connections (replaces result code 80)
7	128	Reserved

S96 through S107

S108 = n . . . .*Signal Quality Selector (default = 1)*

      n = 0. . . . . No limit
      n = 1. . . . . **Low quality**
      n = 2. . . . . Medium quality
      n = 3. . . . . High quality

S109 = n . . . . *(1) V.32/V.32 bis Negotiation Rate Selection (default = 62)*

The individual bits in register *S109* are treated as switches by the modem. If the bit is a "1," the specified function is on or enabled, if the bit is a "0" the function is off or disabled. Individual bits may be set using the **ATS109=n** command.

Each bit set high in this register enables the corresponding rate as a valid rate to be used during rate negotiation. For example, if bits 1, 3, and 5 are set (**ATS109=42**), then the only rates used during up-shifting will be 4800 bps, 9600 bps, and 14400 bps. This may be useful if the modems are not capable of up-shifting many steps. But it may force a rate lower than the communications channel could support; for example, the channel would fail at 14400 bps and be acceptable at 12000 bps. However, with bit 4 = 0, the modem would be forced to operate at 9600 bps.

Bit #	Decimal Value	Permissible Transmission Data Rates
0	1	Reserved
1	2	4800 bps
2	4	7200 bps
3	8	9600 bps
4	16	12000 bps
5	32	14400 bps
6	64	Reserved
7	128	Reserved

S109 = n . . . . *(2) Automode Selection*
(Zoltrix)

      n = 0. . . . . V.PCM disabled
      n = 1. . . . . K56Flex or V.90 enabled
      n = 2. . . . . Enable V.90, disable K56Flex

S110 = n . . . .*V.32 / V.32 bis Mode and Rate Negotiation Control (default-2)*

The V.32 bis standard specifies that negotiation will take place at 4800 bps and then jump to the agreed carrier rate to negotiate error-correction and data compression protocols. This technique has difficulties when the telephone lines are noisy. An alternative method is chosen when n = 3; that is, the modems connect at the slowest allowable V.32 or V.32 bis rate, and then up-shift to the next allowable common rate if the telephone line is providing a high enough quality connection. The rates that are allowable are specified in register *S109*.

      n = 0. . . . . Normal V.32 enabled (no V.32 bis support)
      n = 1. . . . . Normal V.32 bis enabled
      n = 2. . . . . **V.32 bis mode with automatic rate renegotiation enabled (default)**
      n = 3. . . . . V.32 bis mode with automatic rate renegotiation starting with the lowest rate set in S109 and working up one defined rate at a time toward the highest rate set in S109 (based on %Q level at each rate prior to stepping up.).
      . . . . If the modem steps back down, it will also follow the values specified by the S109 settings.

S111 . . . .*Reserved*

S112 = n . . . .*DTE Rate Select During Data Transfer*

      n = 0. . . . . Use the same rate of the last AT rate command issued, or the maximum rate the modem supports if the last command rate was greater than the modem's maximum rate.

n = 1. . . . . Reserved		n = 8. . . . . .12000 bps
n = 2. . . . . 300 bps		n = 9. . . . . .14400 bps
n = 3. . . . . 1200 bps		n = 10. . . . . .16800 bps
n = 4. . . . . 2400 bps		n = 11. . . . . .19200 bps
n = 5. . . . . 4800 bps		n = 12. . . . . .38400 bps
n = 6. . . . . 7200 bps		n = 13. . . . . .57600 bps
n = 7. . . . . 9600 bps		n = 14. . . . . .600 bps

SCr = n ....*BUSY Detect Watermark Controls*

Each register pair holds the value for each setting. To set a value, divide the desired value by 256. The integer portion (quotient) goes in the second register (the odd-numbered register), while the remainder goes in the first (the even-numbered register). Settings are in 1/100ths of a second (10 ms increments).

As an example: A setting of 516 ms would convert to 2 remainder 4; that is,

the integer part of $\dfrac{516}{256} = 2$, and $516 - 2 \cdot 256 = 4$.

This setting would be entered by sending *ATSC0=4SC1=2* to the modem.

Default Values		Definition
SC0=30	SC1=0	Minimum BUSY ON Time (30 ms)
SC2=75	SC3=0	Maximum BUSY ON Time (75 ms)
SC4=30	SC5=0	Minimum BUSY OFF Time (30 ms)
SC6=75	SC7=0	Maximum BUSY OFF Time (75 ms)
SC8=n		Set number of valid BUSY Pulses before reporting BUSY (4 is default)

Appendix C
"AT" Fax/Modem Result Messages

The messages listed are a composite from many different modem manufacturers and models; therefore, not all messages will exist on all modems. In particular, messages above 10 may have different meanings among the available hardware (for example, code 25 has 4 meanings listed). Some hardware may only deliver terse codes or only verbose messages for some of the result messages. For information specific to a particular modem, see the manufacturer's documentation. Generally, messages are reported to the DTE in the order CARRIER, PROTOCOL, COMPRESSION, and CONNECTION.

AT COMMAND RESULT MESSAGES AND DEFINITIONS

Terse	Verbose	Message Definition	Ref.	Xn[1]
0	OK	Command executed without errors		xxxxx
1	CONNECT	1. The line speed is 300 bps and the modem has been instructed to report the line speed to the DTE upon connecting, 2. The DTE speed is 300 bps and the modem has been instructed to report the DTE speed to the DTE upon connecting, or 3. The range of result code responses is restricted by the X command such that no speed reporting is allowed.	Wn	xxxxx
2	RING	1. Indicates incoming ringing is detected on the telephone line. (What qualifies as a ring signal is country-dependent.) 2. When the cellular interface is selected, RING indicates that the cellular phone is receiving an incoming call.		xxxxx
3	NO CARRIER	1. On call origination, ringback is detected and later ceases but no carrier is detected within the period of time determined by register S7, 2. On call origination, no ringback is detected within the period of time determined by register S7, or 3. The modem auto-disconnects due to loss of carrier.	S7	xxxxx
4	ERROR	1. Error in command line syntax or command too long, 2. The command line was not able to be executed, or 3. A command parameter is outside the permitted range.		xxxxx
5	CONNECT 1200	1. The line speed is 1200 bps and the modem has been instructed to report the line speed to the DTE upon connecting, or 2. The DTE speed is 1200 bps and the modem has been instructed to report the DTE speed to the DTE upon connecting.	Wn	1xxxx
6	NO DIALTONE	1. A landline modem sends this result code if it has been instructed to wait for dial tone during dialing but none is received within the period specified by register S7. 2. When cellular phone interface is selected, NO DIALTONE indicates that cellular service is not currently available. 3. Number blacklisted.	S6, S7	33xxx
7	BUSY	Busy tone detected		333xx
8	NO ANSWER	1. No response from remote modem before time-out of register S7. 2. The @ Wait For Quiet Answer dial modifier was executed, and 5 s of silence was not detected.	S7	xxxxx
9	CONNECT 0600	DTE to modem (if W0 or W1) or line (if W2) rate at 600 bps	Wn	1xxxx
10	CONNECT 2400	DTE to modem (if W0 or W1) or line (if W2) rate at 2400 bps	Wn	1xxxx
11	CONNECT 4800 RINGING	DTE to modem (if W0 or W1) or line (if W2) rate at 4800 bps U.S. Robotics/3Com	Wn	1xxxx
12	CONNECT 9600 VOICE	DTE to modem (if W0 or W1) or line (if W2) rate at 9600 bps U.S. Robotics	Wn	1xxxx
13	CONNECT 7200 CONNECT 9600 CONNECT 14400	DTE to modem (if W0 or W1) or line (if W2) rate at 7200 bps U.S. Robotics/3Com Hayes	Wn	1xxxx 1xxxx
14	CONNECT 12000 CONNECT 19200	DTE to modem (if W0 or W1) or line (if W2) rate at 12000 bps Hayes	Wn	1xxxx
15	CONNECT 14400 CONNECT 1200 CONNECT 28800	DTE to modem (if W0 or W1) or line (if W2) rate at 14400 bps 3Com Hayes	Wn	1xxxx ~xxxx
16	CONNECT 19200 TIMEOUT	DTE to modem (if W0 or W1) or line (if W2) rate at 19200 bps Hayes	Wn	1xxxx
17	CONNECT 38400 CONNECT 56000	DTE to modem (if W0 or W1) or line (if W2) rate at 38400 bps Hayes	Wn	1xxxx
18	CONNECT 57600 CONNECT 4800	DTE to modem (if W0 or W1) or line (if W2) rate at 57600 bps U.S. Robotics/3Com	Wn	1xxxx 1xxxx

AT COMMAND RESULT MESSAGES AND DEFINITIONS (*CONTINUED*)

Terse	Verbose	Message Definition	Ref.	Xn[1]
19	CONNECT 115200	DTE to modem (if W0 or W1) or line (if W2) rate at 115200 bps	Wn	1xxxx
	CONNECT 64000	Hayes		
20	CONNECT 7200	U.S. Robotics/3Com		1xxxx
	CONNECT 0600/75	Hayes: DTE / modem rate at 600 (send)/75 (receive) bps		
	CONNECT 28800	Zoltrix		
	CONNECT 230400	Rockwell		xxxxx
21	CONNECT 12000	U.S. Robotics/3Com		1xxxx
	CONNECT 75/0600	Hayes: DTE / modem rate at 75 (send)/600 (receive) bps		
	CONNECT 300	Zoltrix		
22	CONNECT 75TX/1200RX	DTE / modem rate at 75 (send)/1200 (receive) bps		1xxxx
23	CONNECT 1200TX/75RX	DTE / modem rate at 1200 (send)/75 (receive) bps		1xxxx
24	DELAYED	The modem returns this result code when a call fails to connect and the number dialed is considered "delayed" due to country blacklisting requirements.		4444x
	CONNECT 7200	Hayes: DTE to modem rate at 7200 bps.		
	CONNECT 110	Zoltrix		
25	CONNECT 14400	U.S. Robotics/3Com		1xxxx
	CONNECT 12000	Hayes: DTE to modem rate at 12000 bps.		
	CONNECT 16800	Hayes		
	RING BACK	Zoltrix: Ringback signal detected		
26	CONNECT 38400	Hayes: DTE to modem rate at 38400 bps.		
27	CALL WAITING	Hayes		
28	CONNECT 38400	Hayes: DTE to modem rate at 38400 bps.		
29	CONNECT 24000	Hayes: DTE to modem rate at 24000 bps.		
30	CONNECT 33600	Hayes: DTE to modem rate at 33600 bps.		
31	CONNECT 115200	Hayes: DTE to modem rate at 115200 bps.		
32	BLACKLISTED	The modem returns this result code when a call fails to connect and the number dialed is considered 'blacklisted'.		4444x
	CONNECT 48000	Hayes: DTE to modem rate at 48000 bps.		
33	FAX	A fax modem connection is established in a facsimile mode.		xxxxx
34	FCERROR (+FCERROR)			
	CONNECT 26400	Hayes: DTE to modem rate at 26400 bps.		
35	DATA	A data modem connection is established.		xxxxx
36	CONNECT 26400	Hayes		
	CARRIER 26400	Hayes: Telephone line rate at 26400 bps.		
37	CONNECT 24000	Hayes		
	CARRIER 24000	Hayes: Telephone line rate at 24000 bps.		
38	CONNECT 21600	Hayes		
	CARRIER 21600	Hayes: Telephone line rate at 21600 bps.		
39	CONNECT 48000	Hayes		
	CARRIER 48000	Hayes: Telephone line rate at 48000 bps.		
40	CARRIER 300	Telephone line rate at 300 baud	S95	xxxxx
41	CARRIER 600	Hayes: Telephone line rate at 600 bps.		
42	CARRIER 600/75	Hayes: Telephone line rate at 600 (Tx) and 75 (Rx) bps.		
43	CONNECT 16800	U.S. Robotics/3Com		~xxxx
	CARRIER 75/600	Hayes: Telephone line rate at 75 (Tx) and 600 (Rx) bps.		
44	CARRIER 1200/75	Telephone line rate at 1200 (send)/75 (receive) bps and ITU-T V.23 backward channel carrier detected.	S95	xxxxx
45	CARRIER 75/1200	Telephone line rate at 75 (send)/1200 (receive) bps and ITU-T V.23 forward channel carrier detected.	S95	xxxxx
46	CARRIER 1200	Telephone line rate at 1200 bps	S95	xxxxx
47	CARRIER 2400	Telephone line rate at 2400 bps	S95	xxxxx
48	CARRIER 4800	Telephone line rate at 4800 bps	S95	xxxxx
49	CARRIER 7200	Telephone line rate at 7200 bps	S95	xxxxx
50	CARRIER 9600	Telephone line rate at 9600 bps	S95	xxxxx
51	CARRIER 12000	Telephone line rate at 12000 bps	S95	xxxxx
52	CARRIER 14400	Telephone line rate at 14400 bps	S95	xxxxx
53	CARRIER 16800	Telephone line rate at 16800 bps	S95	xxxxx
54	CARRIER 19200	Telephone line rate at 19200 bps	S95	xxxxx
55	CARRIER 21600	Telephone line rate at 21600 bps	S95	xxxxx
	CARRIER 28800	Hayes: Telephone line rate at 28800 bps.		
56	CARRIER 24000	Telephone line rate at 24000 bps	S95	xxxxx
	CARRIER 38400	Hayes: Telephone line rate at 38400 bps.		
57	CARRIER 26400	Telephone line rate at 26400 bps	S95	xxxxx
	CARRIER 56000	Hayes: Telephone line rate at 56000 bps.		

Continued

AT COMMAND RESULT MESSAGES AND DEFINITIONS (*CONTINUED*)

Terse	Verbose	Message Definition	Ref.	Xn[1]
58	CARRIER 28800	Telephone line rate at 28800 bps	S95	xxxxx
	CARRIER 57600	Hayes: Telephone line rate at 57600 bps.		
59	CARRIER 16800	Telephone line rate at 16800 bps		1xxxx
	CONNECT 16800	Rockwell: DTE–modem rate at 16800bps		1xxxx
	CARRIER 64000	Hayes: Telephone line rate at 64000 bps.		
60	CONNECT 21600	Hayes: DTE to modem rate at 21600 bps.		
	CARRIER 31200	AltoCom:		xxxxx
61	CARRIER 21600	Telephone line rate at 21600 bps		1xxxx
	CONNECT 21600	Rockwell: DTE–modem rate at 21600		1xxxx
	CONNECT 20800	Hayes: DTE to modem rate at 20800 bps.		
62	CARRIER 24000	Telephone line rate at 24000 bps		1xxxx
	CONNECT 24000	Rockwell: DTE–modem rate at 24000		1xxxx
	CONNECT 41600	Hayes: DTE to modem rate at 41600 bps.		
63	CARRIER 26400	Telephone line rate at 26400 bps		1xxxx
	CONNECT 26400	Rockwell: DTE–modem rate at 26400		1xxxx
	CONNECT 51200	Hayes: DTE to modem rate at 51200 bps.		
64	CARRIER 28800	Telephone line rate at 28800 bps		1xxxx
	CONNECT 28800	Rockwell: DTE–modem rate at 2880 bps		1xxxx
	CONNECT 62400	Hayes: DTE to modem rate at 62400 bps.		
65	CONNECT 230400	Hayes: DTE to modem rate at 230400 bps.		
	CARRIER 33600	AltoCom:		xxxxx
66	COMPRESSION: CLASS 5	Data compression is MNP- 5 (2:1)	S95	xxxxx
67	COMPRESSION: V.42BIS	Data compression is V.42bis - BTLZ (4:1)	S95	xxxxx
68	COMPRESSION: ADC	Hayes: Using Hayes Adaptive Data Compression (ADC).		
69	COMPRESSION: NONE	No data compression is in effect	S95	xxxxx
70	PROTOCOL:NONE	No data protocol. A standard asynchronous connection is used	S95	xxxxx
	CONNECT 33600			1xxxx
71	PROTOCOL: ERROR-CONTROL/LAP-B	Hayes: Using MNP Class 5 data compression.		
72	PROTOCOL: ERROR-CONTROL/LAP-B/HDX	Hayes: Using ITU-T V.42bis data compression.		
73	PROTOCOL: ERROR-CONTROL/AFT	Hayes: Using the Hayes Asynchronous Framing Technique (AFT). (AFT connects modems such as the Hayes Smartmodem 1200 that do not communicate synchronously across the telephone line. AFT enables an error-control protocol.)		
74	PROTOCOL: X.25/LAP-B	Hayes: Using the X.25 protocol with a carrier speed of 1200, 2400, 4800, or 9600 bps.		
75	PROTOCOL: X.25/LAP-B/HDX	Hayes: Using a half-duplex X.25 error-control connection with a carrier speed of 4800 or 9600 bps		
76	PROTOCOL: NONE			xxxxx
	PROTOCOL: X.25/LAP-B/AFT	Hayes: Using an asynchronous X.25 error-control connection with a carrier speed of 1200 bps and Hayes Asynchronous Framing Technique (AFT).		
77	PROTOCOL: LAPM	Data protocol is V.42 LAP-M with a carrier speed of 1200, 2400, 4800, or 9600 bps	S95	xxxxx
78	PROTOCOL: LAP-M/HDX	Hayes: Using a half-duplex V.42 LAP-M error-control connection with a carrier speed of 4800 or 9600 bps.		
	CARRIER 31200	Rockwell: Telephone line rate at 31200 bps.		xxxxx
79	PROTOCOL: LAP-M/AFT	Hayes: Using an asynchronous V.42 LAP-M error-control connection with a carrier speed of 1200 bps and Hayes Asynchronous Framing Technique (AFT).		
	CARRIER 33600	Rockwell: Telephone line rate at 33600 bps.		xxxxx
80	PROTOCOL: ALT	Using a LAP-M error-control connection with a carrier speed of 1200, 2400, 4800, or 9600 bps. Data protocol is MNP class 2, 3, and 4 compatible.	S95	xxxxx
81	PROTOCOL: V.42BIS	Data protocol is V.42 LAP-M with V.42bis		
	PROTOCOL: ALT-CELLULAR	Sent to the DTE when the modem has connected in the MNP 10 mode and cellular power level adjustment is enabled; i.e., AT)M1 or AT)M2.	)Mn, S95	xxxxx
	PROTOCOL: ISDN	Hayes		
82	PROTOCOL: X.25	Hayes		
83	PROTOCOL: V.120	Hayes		
84	PROTOCOL: T-LINK	Hayes		1xxxx
	CONNECT 33600	Rockwell: DTE–modem rate at 33600 bps.		1xxxx
85	CONNECT 19200	U.S. Robotics/3Com		~xxxx
	PROTOCOL: DTMF	Hayes		
	RRING	AltoCom		~~~~x
86	PROTOCOL: FAX	Hayes		
	NO BONGTONE	AltoCom: No second "credit card" dial tone		~~~~x
87	REORDER	AltoCom:		3333x
88	WARBLE	AltoCom:		3333x
89				

AT COMMAND RESULT MESSAGES AND DEFINITIONS (*CONTINUED*)

Terse	Verbose	Message Definition	Ref.	Xn[1]
90				
91	CONNECT 21600	U.S. Robotics/3Com		1xxxx
	CONNECT 31200	Rockwell: DTE–modem rate at 31200 bps.		1xxxx
92				
93				
94				
95	CARRIER 14400/VFC	SupraFAX: Telephone line rate at 14400 bps V.FC modulation		
96	CARRIER 16800/VFC	SupraFAX: Telephone line rate at 16800 bps V.FC modulation		
97	CARRIER 19200/VFC	SupraFAX: Telephone line rate at 19200 bps V.FC modulation		
98	CARRIER 21600/VFC	SupraFAX: Telephone line rate at 21600 bps V.FC modulation		
	CONNECT 25333	AltoCom:		1xxxx
99	CONNECT 26400	U.S. Robotics/3Com		~xxxx
	CONNECT 26666	AltoCom:		1xxxx
	CARRIER 24000/VFC	SupraFAX: Telephone line rate at 24000 bps V.FC modulation		
100	CARRIER 26400/VFC	SupraFAX: Telephone line rate at 26400 bps V.FC modulation		
	CONNECT 28000	AltoCom:		1xxxx
101	CONNECT 76800	Hayes: DTE to modem rate at 76800 bps.		
	CONNECT 29333	AltoCom:		1xxxx
	CARRIER 28800/VFC	SupraFAX: Telephone line rate at 28800 bps V.FC modulation		
102	CONNECT 124800	Hayes: DTE to modem rate at 124800 bps.		
	CONNECT 30666	AltoCom:		1xxxx
103	CONNECT 16800	U.S. Robotics/3Com		~xxxx
	CONNECT 32000	AltoCom:		1xxxx
	CONNECT 153600	Hayes: DTE to modem rate at 153600 bps.		
104	CONNECT 31200	Hayes: DTE to modem rate at 31200 bps.		
	CONNECT 33333	AltoCom:		1xxxx
105	CARRIER 115200	Hayes: Telephone line rate at 115200 bps.		
	CONNECT 34666	AltoCom:		1xxxx
106	CARRIER 31200	Hayes: Telephone line rate at 31200 bps.		
	CONNECT 36000	AltoCom:		1xxxx
107	CONNECT 28800	U.S. Robotics/3Com		~xxxx
	CONNECT 37333	AltoCom:		1xxxx
	CARRIER 33600	Hayes: Telephone line rate at 33600 bps.		
108	CONNECT 38666	AltoCom:		1xxxx
109	CONNECT 40000	AltoCom:		1xxxx
110	CONNECT 41333	AltoCom:		1xxxx
111	DELAYED	Hayes: Must wait to dial because of blacklist.		
	CONNECT 42666	AltoCom:		1xxxx
112	BLACKLISTED	Hayes: Number is blacklisted, reset to clear.		
	CONNECT 42666	AltoCom:		1xxxx
113	CONNECT 45333	AltoCom:		1xxxx
114	CONNECT 46666	AltoCom:		1xxxx
115	CONNECT 48000	AltoCom:		1xxxx
116	CONNECT 49333	AltoCom:		1xxxx
117	CONNECT 50666	AltoCom:		1xxxx
118	CONNECT 52000	AltoCom:		1xxxx
119	MODULATION: Express	Hayes: Using Express modulation.		
	CONNECT 53333	AltoCom:		1xxxx
120	MODULATION: Bell 103	Hayes: Using Bell 103 modulation.		
	CONNECT 54666	AltoCom:		1xxxx
121	MODULATION: Bell 212	Hayes: Using Bell 212 modulation.		
	CONNECT 56000	AltoCom:		1xxxx
122	MODULATION: V.21	Hayes: Using V.21 modulation.		
	CONNECT 57333	AltoCom:		1xxxx
123	MODULATION: V.22	Hayes: Using V.22 modulation.		
124	MODULATION: V.22bis	Hayes: Using V.22bis modulation.		
125	MODULATION: V.23	Hayes: Using V.23 modulation.		
126	MODULATION: V.32	Hayes: Using V.32 modulation.		
127	MODULATION: V.32bis	Hayes: Using V.32bis modulation.		
128	MODULATION: V.FC	Hayes: Using V.FC modulation.		
	CARRIER 25333	AltoCom:		1xxxx
129	MODULATION: V.34	Hayes: Using V.34 modulation.		
	CARRIER 26666	AltoCom:		1xxxx
130	CARRIER 28000	AltoCom:		1xxxx

AT COMMAND RESULT MESSAGES AND DEFINITIONS (*CONTINUED*)

Terse	Verbose	Message Definition	Ref.	Xn[1]
131	CARRIER 29333	AltoCom:		1xxxx
132	CARRIER 30666	AltoCom:		1xxxx
133	CARRIER 32000	AltoCom:		1xxxx
134	CARRIER 33333	AltoCom:		1xxxx
135	CARRIER 34666	AltoCom:		1xxxx
136	CARRIER 36000	AltoCom:		1xxxx
137	CARRIER 37333	AltoCom:		1xxxx
138	CARRIER 38666	AltoCom:		1xxxx
139	CARRIER 40000	AltoCom:		1xxxx
140	CARRIER 41333	AltoCom:		1xxxx
141	CARRIER 42666	AltoCom:		1xxxx
142	CARRIER 44000	AltoCom:		1xxxx
143	CARRIER 46666	AltoCom:		1xxxx
144	CARRIER 48000	AltoCom:		1xxxx
145	CARRIER 49333	AltoCom:		1xxxx
146	CARRIER 50666	AltoCom:		1xxxx
147	CARRIER 52000	AltoCom:		1xxxx
148	CARRIER 53333	AltoCom:		1xxxx
149	CARRIER 54666	AltoCom:		1xxxx
150	CARRIER 32000	Telephone line rate is 32000 bps	S95	xxxxx
	CARRIER 54666	AltoCom:		1xxxx
151	CARRIER 34000	Telephone line rate is 34000 bps	S95	xxxxx
	CARRIER 56000	AltoCom:		1xxxx
	CONNECT 31200	3Com		~xxxx
152	CARRIER 36000	Telephone line rate is 36000 bps	S95	xxxxx
	CARRIER 57333	AltoCom:		1xxxx
153	CARRIER 38000	Telephone line rate is 38000 bps	S95	xxxxx
154	CARRIER 40000	Telephone line rate is 40000 bps	S95	xxxxx
155	CARRIER 42000	Telephone line rate is 42000 bps	S95	xxxxx
	CONNECT 33600	3Com		~xxxx
156	CARRIER 44000	Telephone line rate is 44000 bps	S95	xxxxx
157	CARRIER 46000	Telephone line rate is 46000 bps	S95	xxxxx
158	CARRIER 48000	Telephone line rate is 48000 bps	S95	xxxxx
159	CARRIER 50000	Telephone line rate is 50000 bps	S95	xxxxx
160	CARRIER 52000	Telephone line rate is 52000 bps	S95	xxxxx
161	CARRIER 54000	Telephone line rate is 54000 bps	S95	xxxxx
162	CARRIER 56000	Telephone line rate is 56000 bps	S95	xxxxx
163				
164				
165	CONNECT 32000	DCE speed is 32000 bps		xxxxx
166	CONNECT 34000	DCE speed is 34000 bps		xxxxx
167	CONNECT 36000	DCE speed is 36000 bps		xxxxx
168	CONNECT 38000	DCE speed is 38000 bps		xxxxx
169	CONNECT 40000	DCE speed is 40000 bps		xxxxx
170	CONNECT 42000	DCE speed is 42000 bps		xxxxx
171	CONNECT 44000	DCE speed is 44000 bps		xxxxx
172	CONNECT 46000	DCE speed is 46000 bps		xxxxx
173	CONNECT 48000	DCE speed is 48000 bps		xxxxx
174	CONNECT 50000	DCE speed is 50000 bps		xxxxx
175	CONNECT 52000	DCE speed is 52000 bps		xxxxx
176	CONNECT 54000	DCE speed is 54000 bps		xxxxx
177	CONNECT 56000	DCE speed is 56000 bps		xxxxx
178	CONNECT 230400	Zoltrix		xxxxx
179				
180	CONNECT 28000	Zoltrix		~xxxx
181	CONNECT 29333	Zoltrix		~xxxx
182	CONNECT 33333	3Com		~xxxx
	CONNECT 30667	Zoltrix		~xxxx
183	CONNECT 33333	Zoltrix		~xxxx
184	CONNECT 34667	Zoltrix		~xxxx
185	CONNECT 37333	Zoltrix		~xxxx
186	CONNECT 37333	3Com		~xxxx
	CONNECT 38667	Zoltrix		~xxxx
187	CONNECT 41333	Zoltrix		~xxxx
188	CONNECT 42667	Zoltrix		~xxxx

AT COMMAND RESULT MESSAGES AND DEFINITIONS (*CONTINUED*)

Terse	Verbose	Message Definition	Ref.	Xn[1]
189	CONNECT 45333	Zoltrix		~xxxx
190	CONNECT 41333	3Com		~xxxx
	CONNECT 46667	Zoltrix		~xxxx
191	CONNECT 49333	Zoltrix		~xxxx
192	CONNECT 50667	Zoltrix		~xxxx
193	CONNECT 53333	Zoltrix		~xxxx
194	CONNECT 42666	3Com		~xxxx
	CONNECT 54667	Zoltrix		~xxxx
198	CONNECT 44000	3Com		~xxxx
202	CONNECT 45333	3Com		~xxxx
206	CONNECT 46666	3Com		~xxxx
210	CONNECT 48000	3Com		~xxxx
214	CONNECT 49333	3Com		~xxxx
218	CONNECT 50666	3Com		~xxxx
222	CONNECT 52000	3Com		~xxxx
226	CONNECT 53333	3Com		~xxxx
230	CONNECT 54666	3Com		~xxxx
234	CONNECT 56000	3Com		~xxxx
238	CONNECT 57333	3Com		~xxxx
+F4	+FCERROR	Sent to the DTE when high speed fax data (V.27, V.29, V.33, or V.17) is expected and a V.21 signal is received.		xxxxx

[1]The column indicates what message will be displayed for the various values of "n" of the **ATFn** command. The "x" indicates the message on that line will be displayed while a number indicates the lesser message that will be displayed with a particular Fn value. For example, the terse message 7 indicating that a BUSY tone was detected will display 3 (terse) NO CARRIER (verbose) for ATF0, ATF1, and ATF2; it will indicate 7 (terse) BUSY (verbose) for ATF3 and ATF4.

Appendix D
Greek Alphabet

THE GREEK ALPHABET AND SOME COMMON USAGE

Symbol Name	Small Symbol	Common Usage	Capital Symbol	Common Usage
Alpha	α	Angle, attenuation constant, coefficients, absorption factor.	A	
Beta	β	Angle, phase constant, luminous intensity, coefficients.	B	
Gamma	γ	Specific quantity, angles, electrical conductivity, propagation constant.	Γ	Complex propagation constant.
Delta	δ	Increment/decrement (differences), density, angle.	Δ	Increment/decrement (differences), determinant.
Epsilon	ϵ	Dielectric constant, base of natural (Naperian) log, the mathematical constant 2.718281828459045 . . .	E	
Zeta	ζ	Coordinates, coefficients, damping factor.	Z	
Eta	η	Efficiency, intrinsic impedance, hysteresis, coordinates.	H	
Theta	θ, ϑ	Angles, angular phase displacement, time constants.	Θ	
Iota	ι		I	
Kappa	κ	Coupling coefficient, thermal conductivity.	K	
Lambda	λ	Wavelength	Λ	Permeance.
Mu	μ	SI prefix for micro, amplification factor.	M	
Nu	ν	Frequency.	N	
Xi	ξ	Coordinates.	Ξ	
Omicron	o		O	
Pi	π	A mathematical constant 3.141 592 653 589 . . .	Π	Series product.
Rho	ρ	Resistivity	P	
Sigma	σ	Complex propagation constant, electrical conductivity, leakage coefficient.	Σ	Summation.
Tau	τ	Time constant, volume resistivity, transmission factor, density.	T	
Upsilon	υ		Υ	
Phi	ϕ, φ	Angles, magnetic flux.	Φ	Scalar potential.
Chi	χ	Angles.	X	
Psi	ψ	Angles, coordinates, phase difference, dielectric flux.	Ψ	
Omega	ω	Angular velocity	Ω	Resistance, solid angle.

Appendix E
ITU-T Recommendation Titles

The following is a list of International Telecommunications Union—Telecommunications sub-section (ITU-T) recommendations. Markings in the recommendation number column (Rec. No.) indicate the status of the document (see below for details). Those with a check "✗" in the "S" column have a separate summary available. Because from time to time the recommendations are updated or have new sections added, one should obtain the most current issue before committing a design to them. ITU's URL address is http://www.itu.ch/. Many of the publications can be purchased through the ITU online subscription or from their bookstore. In the United States, another source for these and many other organization's standards is Global Engineering Documents™ whose URL is http://www.ihs.com/.

*	Recommendation in force as of June 2000
x93	Recommendation cancelled by the WTSC-93
x93-6	Recommendation cancelled during the study period 1993–1996
x96	Recommendation cancelled by the WTSC-96
x00	Recommendation cancelled by the WTSC-2000

Series A Recommendations
Organization of the work of the ITU-T

Rec. No.	Description/Title	S
A.1	Work methods for study groups of the ITU Telecommunication Standardization Sector (ITU-T)	
A.2	Presentation of contributions relative to the study of questions assigned to the ITU-T	
A.3	Elaboration and presentation of texts and development of terminology and other means of expression for Recommendations of the ITU Telecommunication Standardization Sector	
A 3Amend 1	Modifications of document type definitions	
A.4	Communication process between ITU-T and forums and consortia	
A 5	Generic procedures for including references to documents of other organizations in ITU-T Recommendations	
A 5 Imp	Implementers Guide	
A 6	Cooperation and exchange of information between ITU-T and national and regional standards development organizations	
A.10	Terms and definitions	
A.12	Collaboration with the international electrotechnical commission on the subject of definitions for telecommunications	
A.13	Collaboration with the international electrotechnical commission on graphical symbols and diagrams used in telecommunications	
A.14	Production maintenance and publication of ITU-T terminology	
A.15	Elaboration and presentation of texts for Recommendations of the ITU Telecommunication Standardization Sector	
A.20	Collaboration with other international organizations over data transmission	
A.21	Collaboration with other international organizations on ITU-T defined telematic services	
A.22	Collaboration with other international organizations on information technology	
A.23	Collaboration with other international organizations on information technology, telematic services and data transmission	
A.23 Annex A	Guide for ITU-T and ISO/IEC JTC 1 cooperation	
A.30	Major degradation or disruption of service	

Series B Recommendations

Means of expression (definitions, symbols, classification)

Rec. No.	Description/Title	S
B.1	Letter symbols for telecommunications	
B.3	Use of the international system of units (SI)	
B.10	Graphical symbols and rules for the preparation of documentation in telecommunications	
B.11	Use of the term UTC	
B.12	Use of the decibel and the neper in telecommunications	
B.13	Terms and definitions	
B.14	Terms and abbreviations for information quantities in telecommunications	
B.15	Nomenclature of the frequency and wavelength bands used in telecommunications	
B.16	Use of certain terms linked with physical quantities	
B.17	Adoption of the ITU-T Specification and Description Language (SDL)	
B.18	Traffic intensity unit	
B.19	Abbreviations and initials used in telecommunications	

Series C Recommendations

General telecommunication statistics

Rec. No.	Description/Title	S
C.1	ITU Statistical Yearbook	
C.2	Collection and publication of official service information	
C.3	Instructions for international telecommunication services	

Series D Recommendations

General tariff principles

Rec. No.	Description/Title	S
D.000	Terms and definitions for the Series D Recommendations	
D.1	General principles for the lease of international (continental and intercontinental) private telecommunication circuits and networks	✗
D.3	Principles for the lease of analogue international circuits for private service	
D.4	Special conditions for the lease of international (continental and intercontinental) sound- and television-programme circuits for private service	✗
D.5	Cost and values of services rendered as factors in the fixing of rates	
D.6	General principles for the provision of international telecommunications facilities to organizations formed to meet the specialized international communication needs of their members	x93
D.7	Concept and implementation of "one-stop shopping" for international private leased telecommunication circuits	✗
D.8	Special conditions for the lease of international end-to-end digital circuits for private service	
D.9	Private leasing of transmitters or receivers	
D.10	General tariff principles for international public data communication services	✗
D.11	Special tariff principles for international packet-switched public data communication services by means of the virtual call facility	✗
D.12	Measurement unit for charging by volume in the international packet-switched data communication service	
D.13	Guiding principles to govern the apportionment of accounting rates in international packet-switched public data communication relations	
D.15	General charging and accounting principles for non-voice services provided by interworking between public data networks	
D.20	Special tariff principles for the international circuit-switched public data communication services	
D.21	Special tariff principles for short transaction transmissions on the international packet switched public data networks using the fast select facility with restriction	
D.30	Implementation of reverse charging on international public data communication services	
D.35	General charging principles in the international public message handling services and associated applications	✗
D.36	General accounting principles applicable to message handling services and associated applications	

Series D Recommendations

General tariff principles

Rec. No.	Description/Title	S
D 37	Accounting and settlement principles applicable to the provision of public directory services between interconnected Directory Management Domains	
D39	Charging and accounting in the international land mobile telephone service (provided via cellular radio system)	
D.40 rev.1	General tariff principles applicable to telegrams exchanged in the international public telegram service	✗
D.41	Introduction of accounting rates by zones in the international public telegram service	
D.42	Accounting in the international public telegram service	
D.43	Partial and total refund of charges in the international telex service	
D.45 rev. 1	Charging and accounting principles for the international telemessage service	✗
D.50	Tariff and international accounting principles for the international Teletex service	x96
D.60	Guiding principles to govern the apportionment of accounting rates in intercontinental telex relations	✗
D.61	Charging and accounting provisions relating to the measurement of the chargeable duration of a telex call	
D.65	General charging and accounting principles in the international telex service for multi-address messages via store-and-forward units	
D.67	Charging and accounting in the international telex service	✗
D.70 rev. 1	General tariff principles for the international public facsimile service between public bureaux (bureaufax service)	✗
D.71 rev. 1	General tariff principles for the public facsimile service between subscriber stations (telefax service)	✗
D.73 rev. 1	General tariff and international accounting principles for interworking between the international bureaufax and telefax services	✗
D.79	Charging and accounting principles for the international videotex service	✗
D.80	Accounting and refunds for phototelegrams	
D.81	Accounting and refunds for private phototelegraph calls	
D.83	Rates for phototelegrams and private phototelegraph calls	
D.85	Charging for international phototelegraph calls to multiple destinations	
D.90	Charging, billing, international accounting and settlement in the maritime mobile service	
D.91	Transmission in encoded form of maritime telecommunications accounting information	✗
D.93	Charging and accounting in the international land mobile telephone service (provided via cellular radio systems)	
D.94	Charging, billing and accounting principles for international aeronautical mobile service, and international aeronautical mobile-satellite service	✗
D.95	Charging, billing, accounting and refunds in the data messaging land/maritime mobile-satellite service	
D.98	Charging and accounting provisions relating to the transferred account telegraph and telematic services	x96
D.100	Charging for international calls in manual or semi-automatic operating	
D 101	Charging in automatic international telephone service	x93
D.103 rev. 1	Charging in automatic service for calls terminating on a recorded announcement stating the reason for the call not being completed	✗
D.104	Charging for calls to subscriber's station connected either to the absent subscriber's service or to a device substituting a subscriber in his absence	
D.105	Charging for calls from or to a public call office	
D.106	Introduction of reduced rates during periods of light traffic in the international telephone service	
D.110 rev. 1	Charging and accounting for conference calls	✗
D.115	Tariff principles and accounting for the international freephone service (IFS)	
D.116	Charging and accounting principles relating to the home country direct telephone service	✗
D.120	Charging and accounting principles for automated telephone credit card service	
D.140	Accounting rate principles for international telephone services	
D.150	New system for accounting in international telephony	
D.151	Old system for accounting in international telephony	
D.155	Guiding principles governing the apportionment of accounting rates in intercontinental telephone relations	
D.160	Mode of application of the flat-rate price procedure set forth in Recommendation D.67 and Recommendation D.150 for remuneration of facilities made available to the Administrations of other countries	

Series D Recommendations
General tariff principles

Rec. No.	Description/Title	S
D.170	Monthly telephone and telex accounts	
D.171	Adjustments and refunds in the international telephone service	
D.172	Accounting for calls circulated over international routes for which accounting rates have not been established	
D.173	Defaulting subscribers	
D.174	Conventional transmission of information necessary for billing and accounting regarding collect and credit card calls	
D.176	Transmission in encoded form of telephone reversed charge billing and accounting information	✗
D.176	NOTICE relating to D.176	
D.177	Adjustment of charges and refunds in the international telex service	
D.178	Monthly accounts for semi-automatic telephone calls (ordinary and urgent calls, with or without special facilities)	
D.180	Occasional provision of circuits for international sound- and television-programme transmissions	
D.185	General tariff and accounting principles for international one-way point-to-multipoint satellite services	
D.186	General tariff and accounting principles for international two-way multipoint telecommunication service via satellite	
D.188	General charging and accounting principles applicable to an international videoconferencing service	
D.190	Transmission in encoded form of monthly international accounting information	
D.192	Principles for charging and accounting of service telecommunications	✗
D.193	Special tariff principles for privilege telecommunications	
D.195	Settlement of international telecommunication balances of accounts	x93
D.196	Clearing of international telecommunication balances of accounts	✗
D.197	Notification of change of address(es) for accounting and settlement purposes	✗
D.210	General charging and accounting principles for international telecommunication services provided over the integrated services digital network (ISDN)	
D.211	International accounting for the use of the signal transfer point (STP) in CCITT Signalling System No. 7	
D.212	Charging and accounting principles for the use of Signalling System No.7	
D.220	Charging and accounting principles to be applied to international circuit-mode demand bearer services provided over the integrated services digital network (ISDN)	✗
D 224	Charging and accounting principles for ATM/B-ISDN	
D.225	Charging and accounting principles to be applied to frame relay data transmission service	
D.230	General charging and accounting principles for supplementary services associated with international telecommunication services provided over the integrated services digital network (ISDN)	
D.231	Charging and accounting principles relating to the user-to-user information (UUI) supplementary service	
D.232	Specific tariff and accounting principles applicable to ISDN supplementary services	
D.233	Charging and accounting principles to be applied to the reversed charge supplementary service	
D.240	Charging and accounting principles for teleservices supported by the ISDN	✗
D.250	General charging and accounting principles for non-voice services provided by interworking between the ISDN and existing public data networks	✗
D.251	General charging and accounting principles for the basic telephone service provided over the ISDN or by interconnection between the ISDN and the public switched telephone network	
D.260	Charging and accounting capabilities to be applied on the ISDN	✗
D.280	Principles for charging and billing, accounting and reimbursements for universal personal telecommunication	
D.285	Guiding principles for charging and accounting for intelligent network supported services	
D.286	Charging and accounting principles for the global virtual network service	
D.300 R	Determination of accounting rate shares in telephone relations between countries in Europe and the Mediterranean Basin	
D.301 R	Determination of accounting rate shares and collection charges in telex relations between countries in Europe and the Mediterranean Basin	
D.302 R	Determination of the accounting rate shares and collection charges for the international public telegram service applicable to telegrams exchanged between countries in Europe and the Mediterranean Basin	
D.303 R	Determination of accounting rate shares and collection charges applicable by countries in Europe and the Mediterranean Basin to the occasional provision of circuits for sound- and television-programme transmissions	

Series D Recommendations

General tariff principles

Rec. No.	Description/Title	S
D.306 R	Remuneration of public packet-switched data transmission networks between the countries of Europe and the Mediterranean Basin	✗
D.305 R	Remuneration for facilities used for the switched-transit handling of intercontinental telephone traffic in a country in Europe or the Mediterranean Basin	x93-6
D.307 R	Remuneration of digital systems and channels used in telecommunication relations between the countries of Europe and the Mediterranean Basin	
D.310 R	Determination of rentals for the lease of international programme (sound- and television-) circuits and associated control circuits for private service in relations between countries in Europe and the Mediterranean	
D.390 R	Accounting system in the international automatic telephone service	x93-6
D.400 R	Accounting rates applicable in telephone relations between countries in Latin America	x96
D.401 R	Accounting rates applicable to telex relations between countries in Latin America	x96
D.500 R	Accounting rates applicable to telephone relations between countries in Asia and Oceania	
D.501 R	Accounting rates applicable to telex relations between countries in Asia and Oceania	
D.600 R	Determination of accounting rate shares and collection charges in telephone relations between countries in Africa	
D.601 R	Determination of accounting rate shares and collection charges in telex relations between countries in Africa	
D.606 R	Preferential rates in telecommunication relations between countries in Africa	
Supp-3	Supplement 3 to ITU-T Series D Recommendations (03/93) - Handbook on the methodology for determining costs and establishing national tariffs	

Series E Recommendations

Overall network operation (numbering, routing, network management, operational performance and traffic engineering); telephone service, service operation and human factors

Rec. No.	Description/Title	S
E.100	Definitions of terms used in international telephone operation	
E.104	International telephone directory assistance service and public access	✗
E.105	International telephone service	
E.109	International billed number screening procedures for collect and third-party calling	✗
E.110	Organization of the international telephone network	
E.111	Extension of international telephone services	
E.112	Arrangements to be made for controlling the telephone services between two countries	
E.113	Validation procedures for the international telecommunications charge card service	
E.114	Supply of lists of subscribers (directories and other means)	
E.115	Computerized directory assistance	✗
E.116	International telecommunication charge card service	✗
E.117	Terminal devices used in connection with the public telephone service (other than telephones)	✗
E.118	The international telecommunication charge card	
E.119	Instruction of staff operating international positions	x93
E.120	Instructions for users of the international telephone service	
E.121	Pictograms, symbols and icons to assist users of the telephone service	✗
E.122	Measures to reduce customer difficulties in the international telephone service	
E.123	Notation for national and international telephone numbers	
E.124	Discouragement of frivolous international calling to unassigned or vacant numbers answered by recorded announcements without charge	
E.125	Inquiries among users of the international telephone service	
E.126	Harmonization of the general information pages of the telephone directories published by Administrations	
E.127	Pages in the telephone directory intended for foreign visitors	
E.128	Leaflet to be distributed to foreign visitors	
E.130	Choice of the most useful and desirable supplementary telephone services	
E.131	Subscriber control procedures for supplementary telephone services	
E.132	Standardization of elements of control procedures for supplementary telephone services	
E.133	Operating procedures for cardphones	

Series E Recommendations
Overall network operation (numbering, routing, network management, operational performance and traffic engineering); telephone service, service operation and human factors

Rec. No.	Description/Title	S
E.134	Human factors aspects of public terminals: Generic operating procedures	
E.135	Human factors aspects of public telecommunication terminals for people with disabilities	X
E.136	Specification of a tactile identifier for use with telecommunication cards	
E.137	User instructions for payphones	
E.140	Operator-assisted telephone service	
E 141	Instructions for the international telephone service	
E.142	Time-to-answer by operators	x93
E.143	Demand operating of international circuits	x93
E.144	Advantages of semiautomatic international service	x93
E.145	Advantages of international automatic service	x93
E.146	Division of circuits into outgoing and incoming	x93
E.147	Manually operated international transit traffic	x93
E.148	Routing of traffic by automatic transit exchanges	
E.149	Presentation of routing data	
E.150	Publication of a "List of international telephone routes"	x93-6
E.151	Telephone conference calls	
E.152	International Freephone Service	X
E.153	Home country direct	X
E 155	International premium rate service	
E.160	Definitions relating to national and international numbering plans	
E.161	Arrangement of digits, letters and symbols on telephones and other devices that can be used for gaining access to a telephone network	
E.162	Capability for seven digit analysis of international E.164 numbers at Time T	X
E.164/I.331	The international public telecommunication numbering plan for the ISDN era	X
E.165	Timetable for coordinated implementation of the full capability of the numbering plan for the ISDN era (Recommendation E.164)	
E.165.1	Use of escape code "0" within the E.164 numbering plan during the transition period to implementation of NPI mechanism	X
E.166/X.122	Numbering plan interworking for the E.164 and X.121 numbering plans	X
E.167	ISDN Network Identification Codes	
E.168	Application of E.164 numbering plan for UPT	
E 169	Application of Recommendation E.164 numbering plan for universal international freephone numbers for international freephone service	
E.169.2	Application of Recommendation E.164 Numbering Plan for universal international premium rate numbers for the international premium rate service	
E.169.3	Application of Recommendation E.164 Numbering Plan for universal international shared cost numbers for international shared cost service	
E.170	Traffic routing	
E.171	International telephone routing plan	
E.172	ISDN routing plan	
E.173	Routing plan for interconnection between public land mobile networks and fixed terminal networks	X
E.174	Routing principles and guidance for universal personal telecommunications (UPT)	X
E.175	Models for international network planning	
E.177	B-ISDN routing	X
E.180	Technical characteristics of tones for the telephone service	
E.180/Q.35 Supplement 2	Various tones used in national networks	
E.181	Customer recognition of foreign tones	
E.182	Application of tones and recorded announcements in telephone services	
E.183	Guiding principles for telephone announcements	
E.184	Indications to users of ISDN terminals	
E.190	Principles and responsibilities for the management, assignment, and reclamation of E-Series international numbering resources	X
E.191	B-ISDN numbering and addressing	X

Series E Recommendations

Overall network operation (numbering, routing, network management, operational performance and traffic engineering); telephone service, service operation and human factors

Rec. No.	Description/Title	S
E.200	Operational provisions for the maritime mobile service	x93-6
E.201	Reference recommendation for mobile services	✗
E.202	Network operational principles for future public mobile systems	
E.210	Ship station identification for VHF/UHF and maritime mobile-satellite services	
E.211	Selection procedures for VHF/UHF maritime mobile services	
E.212	Identification plan for land mobile stations	
E.213	Telephone and ISDN numbering plan for the land mobile stations in public land mobile networks (PLMN)	
E.214	Structure of the land mobile global title for the signalling connection control part (SCCP)	
E.215	Telephone/ISDN numbering plan for the mobile-satellite services of INMARSAT	
E.216	Selection procedures for Inmarsat mobile-satellite telephone and ISDN services	x96
E.220	Interconnection of public land mobile networks (PLMN)	
E.230	Chargeable duration of calls	
E.231	Charging in automatic service for calls terminating on special services for suspended, cancelled or transferred subscribers	
E.232	Charging for calls to subscriber's station connected either to the absent subscriber's service or to a device substituting a subscriber in his absence	
E.250	See recommendation D.150	
E.251	See recommendation D.151	
E.252	See recommendation D.152	
E.260	Basic technical problems concerning the measurement and recording of call durations	
E.261	Devices for measuring and recording call durations	
E.270	See recommendation D.170	
E.275	See recommendation D.190	
E.276	See recommendation D.176	
E.277	See recommendation D.174	
E.300	Special uses of circuits normally employed for automatic telephone traffic	
E.301	Impact of non-voice applications on the telephone network	
E.320	Speeding up the establishment and clearing of phototelegraph calls	
E.323	See recommendation F.107	
E.330	User control of ISDN-supported services	
E.331	Minimum user-terminal interface for a human user entering address information into an ISDN terminal	✗
E.333	See recommendation Z.323	
E 401	Statistics for the international telephone service (number of circuits in operation and volume of traffic)	
E.410	International network management - General information	✗
E.411	International network management - Operational guidance	
E.412	Network management controls	✗
E.413	International network management - Planning	
E.414	International network management - Organization	
E.415	International network management guidance for common channel signalling system no.7	✗
E.420	Checking the quality of the international telephone service - General considerations	
E.421	Service quality observations on a statistical basis	
E.422	Observations on international outgoing telephone calls for quality of service	✗
E.423	Observations on traffic set up by operators	
E.424	Test calls	
E.425	International automatic observations	
E.426	General guide to the percentage of effective attempts which should be observed for international telephone calls	
E.427	Collection and statistical analysis of special quality of service observation data for measurements of customer difficulties in the international automatic service	
E.428	Connection retention	
E.430	Quality of service framework	✗
E.431	Service quality assessment for connection set-up and release delays	✗

Series E Recommendations

Overall network operation (numbering, routing, network management, operational performance and traffic engineering); telephone service, service operation and human factors

Rec. No.	Description/Title	S
E.432	Connection quality	X
E.433	Billing integrity	X
E.434	Subscriber-to-subscriber measurement of public switched telephone network	X
E 436	Customer Affecting Incidents and blocking Defects Per Million	
E 437	Comparative metrics for network performance management	
E.440	Customer satisfaction point	X
E.450	Facsimile quality of service on PSTN - General aspects	
E.451	Facsimile call cut-off performance	
E.452	Facsimile modem speed reductions and transaction time	
E.453	Facsimile image quality as corrupted by transmission-induced scan line errors	X
E.454	Transmission performance metrics based on Error Correction Mode (ECM) facsimile	X
E.456	Test transaction for facsimile transmission performance	X
E.457	Facsimile measurement methodologies	X
E.458	Figure of merit for facsimile transmission performance	X
E 459	Measurements and metrics for characterizing facsimile transmission performance using non-intrusive techniques	
E.490	Traffic measurement and evaluation - General survey	X
E.491	Traffic measurement by destination	X
E.492	Traffic reference period	X
E.493	Grade of service (GOS) monitoring	X
E.500 rev.1	Traffic intensity measurement principles	X
E.501 rev.1	Estimation of traffic offered in the network	X
E.502 rev.1	Traffic measurement requirements for digital telecommunication exchanges	X
E.503 rev.1	Traffic measurement data analysis	X
E.504	Traffic measurement administration	
E.505	Measurements of the performance of common channel signalling network	X
E.506 rev. 1	Forecasting international traffic	X
E.507	Models for forecasting international traffic	
E.508	Forecasting new telecommunication services	
E.510	Determination of the number of circuits in manual operation	
E.520	Number of circuits to be provided in automatic and/or semiautomatic operation, without overflow facilities	
E.521	Calculation of the number of circuits in a group carrying overflow traffic	
E.522	Number of circuits in a high-usage group	
E.523	Standard traffic profiles for international traffic streams	
E.524	Overflow approximations for non-random inputs	X
E.525	Designing networks to control grade of service	X
E.526	Dimensioning a circuit group with multi-slot bearer services and no overflow inputs	
E.527	Dimensioning at a circuit group with a multi-slot bearer services and overflow traffic	
E.528	Dimensioning of digital circuit multiplication equipment (DCME) systems	X
E.529	Network dimensioning using end-to-end GOS objectives	X
E.540	Overall grade of service of the international part of an international connection	
E.541	Overall grade of service for international connections (subscriber-to-subscriber)	
E.543	Grades of service in digital international telephone exchanges	
E.550	Grade of service and new performance criteria under failure conditions in international telephone exchanges	
E.600	Terms and definitions of traffic engineering	
E.700	Framework of the E.700-series Recommendations	
E.701	Reference connections for traffic engineering	
E.710	ISDN traffic modelling overview	x96
E.711	User demand modelling	X
E.712	User plane traffic modelling	
E.713	Control plane traffic modelling	
E.716	User demand modelling in Broadband-ISDN	X

Series E Recommendations

Overall network operation (numbering, routing, network management, operational performance and traffic engineering); telephone service, service operation and human factors

Rec. No.	Description/Title	S
E.720	ISDN grade of service concept	
E.721	Network grade of service parameters and target values for circuit-switched services in the evolving ISDN	✗
E.723	Grade-of-service parameters for Signalling System No. 7 networks	✗
E.724	Recommendation E.724 (02/96) – GOS parameters and target GOS objectives for in service	✗
E 728	Grade-of-service parameters for B-ISDN signalling	
E.730	ISDN dimensioning methods overview	x96
E.731	Methods for dimensioning resources operating in circuit switched mode	
E.733	Methods for dimensioning resources in Signalling System No. 7 networks	✗
E.734	Methods for allocating and dimensioning Intelligent Network (IN) resources	✗
E.735	Framework for traffic control and dimensioning in B-ISDN	✗
E 736	Methods for cell level traffic control in B-ISDN	
E 737	Dimensioning methods for B-ISDN	
E.743	Traffic measurements for SS No.7 dimensioning and planning	
E.744	Traffic and congestion control requirements for SS No. 7 and IN-structured networks	✗
E.750	Introduction to the E.750-series of Recommendations on traffic engineering aspects of mobile networks	✗
E.751	Reference connections for traffic engineering of land mobile networks	✗
E.752	Reference connections for traffic engineering of maritime and aeronautical systems	✗
E.755	Reference connections for UPT traffic performance and GOS	✗
E.770	Land mobile and fixed network interconnection traffic grade of service concept	
E.771	Network grade of service parameters and target values for circuit-switched land mobile services	✗
E.773	Maritime and aeronautical mobile grade of service concept	✗
E.774	Network grade of service parameters and target values for maritime and aeronautical mobile services	✗
E.775	UPT Grade of service concept	✗
E.776	Network grade of service parameters for UPT	✗
E.800	Quality of service and dependability vocabulary	
E.801	Framework for service quality agreement	✗
E.810	Framework of the Recommendations on the serveability performance and service integrity for telecommunication services	
E.820	Call models for serveability and service integrity performance	
E.830	Models for the specification, evaluation and allocation of serveability and service integrity	
E.845	Connection accessibility objective for the international telephone service	
E.846	Accessibility for 64 kbits circuit switched international end to end ISDN connection types	
E.850	Connection retainability objective for the international telephone service	
E.855	Connection integrity objective for international telephone service	
E.862 rev.1	Dependability planning of telecommunication networks	✗
E.880	Field data collection and evaluation on the performance of equipment, networks and services	

Series F Recommendations

Telecommunication services other than telephone (operations, quality of service, service definitions and human factors)

Rec. No.	Description/Title	S
F.1	Operational provisions for the international public telegram service	
F.2	Operational provisions for the collection of telegram charges	
F.4	Plain and secret language	
F.10	Character error rate objective for telegraph communication using 5-unit start-stop equipment	
F.11	Continued availability of traditional services	✗
F.12	See Recommendation A.30	
F.13	Operational provisions for participation in the transferred account telegraph and telematic service	x93-6
F.14	General provisions for one-stop-shopping arrangements	
F.15	Evaluating the success of new services	
F.16	Global virtual network service	✗
F.17	Operational aspects of service telecommunications	
F.18	Guidelines on harmonization of international public bureau services	✗

Series F Recommendations

Telecommunication services other than telephone (operations, quality of service, service definitions and human factors)

Rec. No.	Description/Title	S
F.20	The international gentex service	
F.21	Composition of answer-back codes for the international gentex service	
F.23	Grade of service for long-distance international gentex circuits	
F.24	Average grade of service from country to country in the gentex service	
F.30	Use of various sequences of combinations for special purposes	
F.31	Telegram retransmission system	
F.32	Telegram destination indicators	✗
F.35	Provisions applying to the operation of an international public automatic message switching service for equipments utilizing the International Telegraph Alphabet No. 2	
F.40	International public telemessage service	✗
F.41	Interworking between the telemessage service and the international public telegram service	✗
F.59	General characteristics of the international telex service	✗
F.60	Operational provisions for the international telex service	
F.61	Operational provisions relating to the chargeable duration of a telex call	
F.62	Duplex operation in the telex service	x93
F.63	Additional facilities in the international telex service	
F.64	Determination of the number of international telex circuits required to carry a given volume of traffic	
F.65	Time-to-answer by operators at international telex positions	
F.68	Establishment of the automatic intercontinental telex network	
F.69	The international telex service – Service and operational provisions of telex destination codes and telex network identification codes	✗
F.70	Evaluating the quality of the international telex service	
F.71	Interconnection of private teleprinter networks with the telex network	
F.72	The international telex service – General principles and operational aspects of a store and forward facility	
F.73	Operational principles for communication between terminals of the international telex service and data terminal equipment on packet switched public data networks	✗
F.74	Intermediate storage devices accessed from the international telex service using single stage selection – answerback format	
F.80	Basic requirements for interworking relations between the international telex service and other services	✗
F.82	Operational provisions to permit interworking between the international telex service and the intex service	
F 83	Operational principles for communication between terminals of the international telex service and data terminal equipment on packet switched public data networks	
F.85	See Recommendation F.421	
F.86	Interworking between the international telex service and the videotex service	✗
F.87	Operational principles for the transfer of messages from terminals on the telex network to Group 3 facsimile terminals connected to the public switched telephone network	✗
F.89	Status inquiry function in the international telex service	
F.91	General statistics for the telegraph services	x93-6
F.92	Services codes	x93-6
F.93	Routing tables for offices connected to the gentex service	x93-6
F.95	Table of international telex relations and traffic	x93-6
F.96	List of destination indicators	x93-6
F.100	Scheduled radiocommunication services	
F.104	International leased circuit services – Customer circuit designations	✗
F.105	Operational provisions for phototelegrams	
F.106	Operational provisions for private phototelegraph calls	
F.107	Rules for phototelegraph calls established over circuits normally used for telephone traffic	
F.108	Operating rules for international phototelegraph calls to multiple destinations	
F.110	Operational provision for the maritime mobile service	
F.111	Principles of service for mobile systems	✗
F.112	Quality objectives for 50-baud start-stop telegraph transmission in the maritime mobile-satellite service	
F.113	Service provisions for aeronautical passenger communications supported by mobile-satellite systems	
F.115	Service objectives and principles for future public land mobile telecommunication systems	✗

Series F Recommendations

Telecommunication services other than telephone (operations, quality of service, service definitions and human factors)

Rec. No.	Description/Title	S
F.120	Ship station identification for VHF/UHF and maritime mobile-satellite services	
F.122	Operational procedures for the maritime satellite data transmission service	
F.125	Numbering plan for access to the mobile-satellite services of INMARSAT from the international telex service	
F.126	Selection procedures for the INMARSAT mobile-satellite telex service	x93-6
F.127	Operational procedures for interworking between the international telex service and the service offered by INMARSAT-C system	
F.130	Maritime answer-back codes	
F.131	Radiotelex service codes	
F.140	Point-to-multipoint telecommunication service via satellite	
F.141	International two-way multipoint telecommunication service via satellite	✗
F.150	Service and operational provisions for the intex service	✗
F.160	General operational provisions for the international public facsimile services	
F.162	Service and operational requirements of store-and-forward facsimile service	
F.163	Operational requirements of the interconnection of facsimile store-and-forward units	
F.166	Service and operational requirements for a fax database service (FaxDB)	
F.170	Operational provisions for the international public facsimile service between public bureaux (bureaufax)	✗
F.171	Operational provisions relating to the use of store-and-forward switching nodes within the bureaufax service	
F.180	General operational provisions for the international public facsimile service between subscriber stations (telefax)	✗
F.182	Operational provisions for the international public facsimile service between subscribers' stations with Group 3 facsimile machines (Telefax 3)	✗
F182 bis	Guidelines for the support of the communication of documents using Group 3 facsimile between user terminals via public networks	✗
F.184	Operational provisions for the international public facsimile service between subscriber stations with Group 4 facsimile machines (Telefax 4)	✗
F 185	Internet facsimile: Guidelines for the support of the communication of facsimile documents	
F.190	Operational provisions for the international facsimile service between public bureaux and subscriber stations and vice versa (bureaufax-telefax and vice versa)	
F.200	Teletex service	x93-6
F.201	Interworking between teletex service and telex service – General principles	x93-6
F.202	Interworking between the telex service and the teletex service – General procedures and operational requirements for the international interconnection of telex/teletex conversion facilities	x93-6
F.203	Network based storage for the teletex service	x93-6
F.220	Service requirements unique to the processable mode number eleven (PM11) used within teletex service	x93-6
F.230	Service requirements unique to the mixed mode (MM) used within the teletex service	x93-6
F.300	Videotex service	
F.301	Fast speed PSTN videotex	✗
F.350	Application of Series T Recommendations	
F.351	General principles on the presentation of terminal identification to users of the telematic services	
F.353	Provision of telematic and data transmission services on integrated services digital network (ISDN)	
F.400/X.400	Message handling services: Message Handling System and service overview	
F.401	Message handling services: naming and addressing for public message handling services	
F.410	Message handling Services: the public message transfer service	
F.415	Message handling services: Intercommunication with public physical delivery services	
F.420	Message handling services: the public interpersonal messaging service	
F.421	Message handling services: Intercommunication between the IPM service and the telex service	
F.421 err	Covering F.421	
F.422	Message handling services: Intercommunication between the IPM service and the teletex service	x93-6
F.423	Message handling services: Intercommunication between the interpersonal messaging service and the telefax service	
F.435	Message handling: electronic data interchange messaging service	✗
F.440	Message handling services: the voice messaging service	
F.471	Operational requirements for the interconnection of voice-mail store-and-forward units	✗
F.471 corr 1	Corrigendum to F.471	

Series F Recommendations

Telecommunication services other than telephone (operations, quality of service, service definitions and human factors)

Rec. No.	Description/Title	S
F.472	Service and operational requirements of the voice-mail store-and-forward service	
F.500	International public directory services	
F.510	Automated directory assistance – White pages service definition	
F.551	Service Recommendation for the telematic file transfer within Telefax 3, Telefax 4, Teletex services and message handling services	
F.581	Guidelines for programming communication interfaces (PCIs) definition: Service Recommendation	
F.600	Service and operational principles for public data transmission services	✗
F.601	Service and operational principles for packet-switched public data networks	x93
F.700	Framework Recommendation for audiovisual/multimedia services	
F.701	Teleconference service	
F.702	Multimedia conference services	
F.710	General principles for audiographic conference service	✗
F.711	Audiographic conference teleservice for ISDN	
F.720	Videotelephony services - general	
F.721	Videotelephony teleservice for ISDN	
F.723	Videophone service in the Public Switched Telephone Network (PSTN)	
F.730	Videoconference service - general	
F.731	Multimedia conference services in the ISDN	
F.732	Multimedia conference services in the B-ISDN	✗
F.740	Audiovisual interactive services	
F.761	Service-oriented requirements for telewriting applications	
F.811	Broadband connection-oriented bearer service	✗
F.812	Broadband connectionless data bearer service	
F.813	Virtual path service for reserved and permanent communications	✗
F.850	Principles of Universal Personal Telecommunication (UPT)	
F.851	Universal personal telecommunication (UPT) - Service description (service set 1)	✗
F.853	Supplementary services in the Universal Personal Telecommunication (UPT) environment	
F.862	Universal Personal Telecommunication (UPT) – Service description (service set 2)	
F.901	Usability evaluation of telecommunication services	
F.902	Interactive services design guidelines	✗
F.910	Procedures for designing, evaluating and selecting symbols, pictograms and icons	✗

Series G Recommendations

Transmission systems and media, digital systems and networks

Rec. No.	Description/Title	S
Supplement 29	Planning of mixed analogue-digital circuits (chains, connections)	
Supplement 31	Principles of determining an impedance strategy for the local network	
Supplement 32	Transmission aspects of digital mobile radio systems	
Supplement 37	Supplement 37 to ITU-T Series G Recommendations - ITU-T Recommendation G.763 digital circuit multiplication equipment (DCME) tutorial and dimensioning	
Supplement 38	Supplement 38 to ITU-T Series G Recommendations – Variable bit rate calculations for ITU-T Recommendation G.767 Digital Circuit Multiplication Equipment (DCME)	
G.100	Definitions used in Recommendations on general characteristics of international telephone connections and circuits	
G.101	The transmission plan	✗
G.102	Transmission performance objectives and Recommendations	
G.103	Hypothetical reference connections	
G.105	Hypothetical reference connection for crosstalk studies	
G.107	The E-model, a computational model for use in transmission planning	
G.108.01	Conversational impacts on end-to-end speech transmission quality – Evaluation of effects not covered by the E-model	
G.109	Definition of categories of speech transmission quality	✗

Series G Recommendations
Transmission systems and media, digital systems and networks

Rec. No.	Description/Title	S
G.111	Loudness ratings (LRs) in an international connection	
G.113	Transmission impairments	✗
G.114	One-way transmission time	✗
G.115	Mean active speech level for announcements and speech synthesis systems	✗
G.116	Transmission performance objectives applicable to end-to-end international connections	✗
G.117	Transmission aspects of unbalance about earth (definitions and methods)	✗
G.120	Transmission characteristics of national networks	✗
G.121	Loudness ratings (LRs) of national systems	
G.122	Influence of national systems on stability and talker echo in international connections	
G.123	Circuit noise in national networks	x96
G.125	Characteristics of national circuits on carrier systems	x96
G.126	Listener echo in telephone networks	
G.131	Control of talker echo	✗
G.131 appdx	Appendix II to G.131 Relation between echo disturbances under single talk and double talk conditions (evaluated for one-way transmission time of 100 ms)	
G.132	Attenuation distortion	x96
G.133	Group-delay distortion	x96
G.134	Linear crosstalk	x96
G.135	Error on the reconstituted frequency	x96
G.136	Application rules for automatic level control devices	
G.136 err	Covering Note to Recommendation G.136	
G.141	Attenuation distortion	x96
G.142	Transmission characteristics of exchanges	✗
G.143	Circuit noise and the use of compandors	x00
G.151	General performance objectives applicable to all modern international circuits and national extension circuits	x96
G.152	Characteristics appropriate to long-distance circuits of a length not exceeding 2500 km	x96
G.153	Characteristics appropriate to international circuits more than 2500 km in length	x96
G.161	Echo-suppressors suitable for circuits having either short or long propagation times	x93
G.162	Characteristics of compandors for telephony	x96
G.163	Call concentrating systems	x93
G.164	Echo suppressors	
G.165	Echo cancellers	
G.166	Characteristics of syllabic compandors for telephony on high capacity long distance systems	x96
G.167	Acoustic echo controllers	
G.168	Digital network echo cancellers	✗
G.169	Automatic level control devices	✗
G.171	Transmission plan aspects of privately operated networks	
G.172	Transmission plan aspects of international conference calls	
G.173	Transmission planning aspects of the speech service in digital public land mobile networks	
G.174	Transmission performance objectives for terrestrial digital wireless systems using portable terminals to access the PSTN	
G.175	Transmission planning for private/public network interconnection of voice traffic	✗
G.176	Planning guidelines for the integration of ATM technology into networks supporting voiceband services	✗
G.177	Transmission planning for voiceband services over hybrid Internet/PSTN connections	
G.180	Characteristics of N + M type direct transmission restoration systems for use on digital and analogue sections, links or equipment	
G.181	Characteristics of 1 + 1 type restoration systems for use on digital transmission links	
G.191	Software tools for speech and audio coding standardization	✗
G.192	A common digital parallel interface for speech standardisation activities	
G.211	Make-up of a carrier link	
G.212	Hypothetical reference circuits for analogue systems	
G.213	Interconnection of systems in a main repeater station	
G.214	Line stability of cable systems	

Series G Recommendations

Transmission systems and media, digital systems and networks

Rec. No.	Description/Title	S
G.215	Hypothetical reference circuit of 5000 km for analogue systems	
G.221	Overall recommendations relating to carrier-transmission systems	
G.222	Noise objectives for design of carrier-transmission systems of 2500 km	
G.223	Assumptions for the calculation of noise on hypothetical reference circuits for telephony	
G.224	Maximum permissible value for the absolute power level (power referred to one milliwatt) of a signalling pulse	
G.225	Recommendations relating to the accuracy of carrier frequencies	
G.226	Noise on a real link	
G.227	Conventional telephone signal	
G.228	Measurement of circuit noise in cable systems using a uniform-spectrum random noise loading	
G.229	Unwanted modulation and phase jitter	
G.230	Measuring methods for noise produced by modulating equipment and through-connection filters	
G.231	Arrangement of carrier equipment	
G.232	12-channel terminal equipments	
G.233	Recommendations concerning translating equipments	
G.234	8-channel terminal equipments	x93
G.235	16-channel terminal equipments	x93
G.241	Pilots on groups, supergroups, etc.	
G.242	Through-connection of groups, supergroups, etc.	
G.243	Protection of pilots and additional measuring frequencies at points where there is a through-connection	
G.311	General characteristics of systems providing 12 carrier telephone circuits on an open-wire pair	x93
G.312	Intermediate repeaters for open-wire carrier systems conforming to Recommendation G.311	x93
G.313	Open-wire lines for use with 12-channel carrier systems	x93
G.314	General characteristics of systems providing eight carrier telephone circuits on an open-wire pair	x93
G.322	General characteristics recommended for systems on symmetric pair cables	
G.323	A typical transistorized system on symmetric cable pairs	x93
G.324	General characteristics for valve-type systems on *symmetric cable pairs	x93
G.325	General characteristics recommended for systems providing 12 telephone carrier circuits on a symmetric cable pair [(12+12) systems]	
G.326	Typical systems on symmetric cable pairs [(12 + 12) systems]	x93
G.327	Valve-type systems offering 12 carrier telephone circuits on a symmetric cable pair [(12 + 12) systems]	x93
G.332	12 MHz systems on standardized 2.6/9.5 mm coaxial cable pairs	
G.333	60 MHz systems on standardized 2.6/9.5 mm coaxial cable pairs	
G.334	18 MHz systems on standardized 2.6/9.5 mm coaxial cable pairs	
G.337	General characteristics of systems on 2.6/9.5 mm coaxial cable pairs	x93
G.338	4 MHz valve-type systems on standardized 2.6/9.5 mm coaxial cable pairs	x93
G.339	12 MHz valve-type systems on standardized 2.6/9.5 mm coaxial cable pairs	x93
G.341	1.3 MHz systems on standardized 1.2/4.4 mm coaxial cable pairs	
G.343	4 MHz systems on standardized 1.2/4.4 mm coaxial cable pairs	
G.344	6 MHz systems on standardized 1.2/4.4 mm coaxial cable pairs	
G.345	12 MHz systems on standardized 1.2/4.4 mm coaxial cable pairs	
G.346	18 MHz systems on standardized 1.2/4.4 mm coaxial cable pairs	
G.352	Interconnection of coaxial carrier systems of different designs	
G.356	(120 + 120) channel systems on a single coaxial pair	x93
G.361	Systems providing three carrier telephone circuits on a pair of open-wire lines	x93
G.371	FDM carrier systems for submarine cable	x93
G.411	Use of radio-relay systems for international telephone circuits	
G.412	Terminal equipments of radio-relay systems forming part of a general telecommunication network	x93
G.421	Methods of interconnection	
G.422	Interconnection at audio-frequencies	
G.423	Interconnection at the baseband frequencies of frequency-division multiplex radio-relay systems	
G.431	Hypothetical reference circuits for frequency-division multiplex radio-relay systems	
G.433	Hypothetical reference circuit for trans-horizon radio-relay systems for telephony using frequency-division multiplex	x93

Series G Recommendations

Transmission systems and media, digital systems and networks

Rec. No.	Description/Title	S
G.434	Hypothetical reference circuit for systems using analogue transmission in the fixed-satellite service	x93
G.441	Permissible circuit noise on frequency-division multiplex radio-relay systems	
G.442	Radio-relay system design objectives for noise at the far end of a hypothetical reference circuit with reference to telegraphy transmission	
G.444	Allowable noise power in the hypothetical reference circuit of trans-horizon radio-relay systems for telephony using frequency-division multiplex	x93
G.445	Allowable noise power in the hypothetical reference circuit for frequency-division multiplex telephony in the fixed-satellite service	x93
G.451	Use of radio links in international telephone circuits	
G.453	Improved transmission system for HF radio-telephone circuits	x93
G.464	Principles of the devices used to achieve privacy in radiotelephone conversations	x93
G.471	Conditions necessary for interconnection of mobile radiotelephone stations and international telephone lines	x93
G.473	Interconnection of a maritime mobile satellite system with the international automatic switched telephone service; transmission aspects	x00
G.511	Test methodology for Group 3 facsimile processing equipment in the Public Switched Telephone Network	✗
G.541	Specification of factory lengths of loaded telecommunication cable	x93
G.542	Specification of loading coils for loaded telecommunication cables	x93
G.543	Specification for repeater sections of loaded telecommunication cable	x93
G.544	Specifications for terminal equipment and intermediate repeater stations	x93
G.601	Terminology for cables	
G.602	Reliability and availability of analogue cable transmission systems and associated equipments	
G.611	Characteristics of symmetric cable pairs for analogue transmission	
G.612	Characteristics of symmetric cable pairs designed for the transmission of systems with bit rates of the order of 6 to 34 Mbit/s	
G.613	Characteristics of symmetric cable pairs usable wholly for the transmission of digital systems with a bit rate of up to 2 Mbits	
G.614	Characteristics of symmetric pair star-quad cables designed earlier for analogue transmission systems and being used now for digital system transmission at bit rates of 6 to 34 Mbit/s	
G.621	Characteristics of 0.7/2.9 mm coaxial cable pairs	
G.622	Characteristics of 1.2/4.4 mm coaxial cable pairs	
G.623	Characteristics of 2.6/9.5 mm coaxial cable pairs	
G.631	Types of submarine cable to be used for systems with line frequencies of less than about 45 MHz	
G.641	Waveguide diameters	x93
G.650	Definition and test methods for the relevant parameters of single-mode fibres	
G.651	Characteristics of a 50/125 um multimode grades index optical fibre cable	
G.652	Characteristics of a single-mode optical fibre cable	✗
G.653	Characteristics of a dispersion-shifted single-mode optical fibre cable	
G.654	Characteristics of a 1550 nm wavelength loss-minimized single-mode optical fibre cable	
G.655	Characteristics of a non-zero dispersion shifted single-mode optical fibre cable	
G.661	Definition and test methods for relevant generic parameters of optical fibre amplifiers	
G.662	Generic characteristics of optical fibre amplifier devices and sub-systems	
G.663	Application related aspects of optical fibre amplifier devices and sub-system	✗
G.664	General automatic power shutdown procedures for optical transport systems	
G.665	Optical network components and subsystems	
G.671	Transmission characteristics of passive optical components	✗
G.681	Functional characteristics of interoffice and long-haul line systems using optical amplifiers, including optical multiplexing	✗
G.691	Optical interfaces for single-channel SDH systems with optical amplifiers	
G.692	Optical interfaces for multi-channel SDH systems with optical amplifiers	✗
G.700	Framework of the series G.700, G.800 and G.900 Recommendations	x93
G.701	Vocabulary of digital transmission and multiplexing, and pulse code modulation (PCM) terms	
G.702	Digital hierarchy bit rates	
G.703	Physical/electrical characteristics of hierarchical digital interfaces	✗

Series G Recommendations

Transmission systems and media, digital systems and networks

Rec. No.	Description/Title	S
G.704	Synchronous frame structures used at primary and secondary hierarchical levels	✗
G.705	Characteristics required to terminate digital links on a digital exchange	x96
G.706	Frame alignment and cyclic redundancy check (CRC) procedures relating to basic frame structures defined in Recommendation G.704	✗
G.707	Synchronous digital hierarchy bit rates	✗
G.708	Sub STM-0 network node interface for the synchronous digital hierarchy (SDH)	✗
G.709	Synchronous multiplexing structure	✗
G.711	Pulse code modulation (PCM) of voice frequencies	
G.712	Transmission performance characteristics of pulse code modulation	✗
G.720	Characterization of low-rate digital voice coder performance with non-voice signals	
G.722	7 kHz audio-coding within 64 kbit/s	
G.722 Annex A	7 kHz audio-coding within 64 kbit/s; Annex A: Testing signal-to-total distortion ratio for audio-codecs at 64 kbit/s Recommendation G.722 connected back-to-back	
G.722 Encl 1	Digital test sequences for the verifications of the algorithms in Rec. G.722	
&.723.1	Dual rate speech coder for multimedia communications transmitting at 5.3 and 6.3 kbit/s	✗
G.723.1 Annex A Encl 1	Annex A – Silence compression scheme Reference codes and test vectors	
G.723.1 Annex B Encl 1	Annex B – Alternative specification based on floating point arithmetic Reference codes and test vectors	
G.723.1 Annex C Encl 1	Annex C – Scalable channel coding scheme for wireless applications Reference codes and test vectors	
G.724	Characteristics of a 48-channel low bit rate encoding primary multiplex operating at 1544 kbit/s	
G.725	System aspects for the use of the 7 kHz audio codec within 64 kbit/s	
G.726	40, 32, 24, 16 kbit/s Adaptive Differential Pulse Code Modulation (ADPCM)	✗
G.726 Annex A	Extensions of Recommendation G.726 for use with uniform-quantized input and output	
G.726 App II	Appendix II Test Vectors to Recommendation G.726 (03/91) – Description of the digital test sequences for the verification of the G.726 40, 32, 24 and 16 kbit/s ADPCM algorithm	
G.726/G.727	Appendix III (Rec. G.726)/Appendix II (Rec. G.727) – Comparison of ADPCM algorithms	
G.726 Encl 1	Digital test sequences for the verification of the algorithms in Rec. G.726	
G.727	5-, 4-, 3- and 2-bits sample embedded adaptative differential pulse code modulation (ADPCM)	✗
G.727 Annex A	Extensions of Recommendation G.727 for use with uniform-quantized input and output	
G.727 App I	Appendix I to Recommendation G.727 (03/91) – Description of the digital test sequences for the verification of the G.727 5-, 4-, 3- and 2-bit/sample embedded ADPCM algorithm	
G.727 Encl. 1	Digital test sequences for the verification of the algorithms in Rec. G.727	
G.728	Coding of speech at 16 kbit/s using low-delay code excited linear prediction	
G.728 App II	Appendix II (11/95) to Recommendation G.728 – Speech performance	
G.728 Annex G	Coding of speech at 16 kbit/s using low-delay code excited linear prediction: 16 kbit/s fixed point specification	
G.728 Annex H	Annex G (11/94) to Recommendation G.728 – Coding of speech at 16 kbit/s using low-delay code excited linear prediction – Annex G: 16 kbit/s fixed point specification	
G.728 Encl. 1	Recommendation G.728 test vectors (4)	
G.729	Coding of speech at 8 kbit/s using conjugate-structure algebraic-code-excited linear-prediction (CS-ACELP)	
G.729 Encl. 1	Source code file and test vectors	
G.729 Annex A Encl. 1	Reduced complexity 8 kbit/s CS-ACELP speech codec Source code and test vectors	
G.729 Annex B Encl. 1	A silence compression scheme for G.729 optimized for terminals conforming to Recommendation V.70 Source code and test vectors	
G.729 Annex C	Reference floating-point implementation for G.729 CS-ACELP 8 kbit/s speech coding	
G.729 Annex C+	Reference floating-point implementation for integrating G.729 CS-ACELP speech coding main body with Annexes B, D and E	
G.729 Annex D	Coding of speech at 8 kbit/s using Conjugate Structure Algebraic-Code-Excited Linear-Prediction (CS-ACELP) Annex D: 6.4 kbit/s CS-ACELP speech coding algorithm	
G.729 Annex E	Coding of speech at 8 kbit/s using Conjugate-Structure Algebraic-Code-Excited Linear-Prediction (CS-ACELP) Annex E: 11.8 kbit/s CS-ACELP speech coding algorithm	

Series G Recommendations
Transmission systems and media, digital systems and networks

Rec. No.	Description/Title	S
G.729 Annex F	Reference implementation of G.729 Annex B DTX functionality for Annex D	
G.729 Annex G		
G.729 Annex H	Reference implementation of switching procedure between G.729 Annexes D and E	
G.729 Annex I	Reference fixed-point implementation for integrating G.729 CS-ACELP speech coding main body with Annexes B, D and E	
G.731	Primary PCM multiplex equipment for voice frequencies	
G.732	Characteristics of primary PCM multiplex equipment operating at 2048 kbit/s	
G.733	Characteristics of primary PCM multiplex equipment operating at 1544 kbit/s	
G.734	Characteristics of synchronous digital multiplex equipment operating at 1544 kbit/s	
G.735	Characteristics of primary PCM multiplex equipment operating at 2048 kbit/s and offering synchronous digital access at 384 kbit/s and/or 64 kbit/s	
G.736	Characteristics of a synchronous digital multiplex equipment operating at 2048 kbit/s	
G.737	Characteristics of an external access equipment operating at 2048 kbit/s offering synchronous digital access at 384 kbit/s and/or 64 kbit/s	
G.738	Characteristics of primary PCM multiplex equipment operating at 2048 kbit/s and offering synchronous digital access at 320 kbit/s and/or 64 kbit/s	
G.739	Characteristics of an external access equipment operating at 2048 kbit/s offering synchronous digital access at 320 kbit/s and/or 64 kbit/s	
G.741	General considerations on second order multiplex equipments	
G.742	Second order digital multiplex equipment operating at 8448 kbit/s and using positive justification	
G.743	Second order digital multiplex equipment operating at 6312 kbit/s and using positive justification	
G.744	Second order PCM multiplex equipment operating at 8448 kbit/s	
G.745	Second order digital multiplex equipment operating at 8448 kbit/s and using positive/zero/negative justification	
G.746	Characteristics of second order PCM multiplex equipment operating at 6312 kbit/s	
G.747	Second order digital multiplex equipment operating at 6312 kbit/s and multiplexing three tributaries at 2048 kbit/s	
G.751	Digital multiplex equipments operating at the third order bit rate of 34 368 kbit/s and the fourth order bit rate of 139 264 kbit/s and using positive justification	
G.752	Characteristics of digital multiplex equipments based on a second order bit rate of 6312 kbit/s and using positive justification	
G.753	Third order digital multiplex equipment operating at 34 368 kbit/s and using positive/zero/negative justification	
G.754	Fourth order digital multiplex equipment operating at 139 264 kbit/s and using positive/zero/negative justification	
G.755	Digital multiplex equipment operating at 139 264 kbit/s and multiplexing three tributaries at 44 736 kbit/s	
G.761	General characteristics of a 60-channel transcoder equipment	
G.762	General characteristics of a 48-channel transcoder equipment	
G.763	Digital circuit multiplication equipment using ADPCM (Recommendation G.726) and digital speech interpolation	
G.763 err	Covering Note to Recommendation G.763	
G.764	Voice packetization – packetized voice protocols	✗
G.764 App I	Packetization guide	
G.765	Packet circuit multiplication equipment	
G.765 App I	Packet circuit multiplication equipment	
G.766	Facsimile demodulation/remodulation for DCME	
G.771	Q-interfaces and associated protocols for transmission equipment in the TMN	x93
G.772	Protected monitoring points provided on digital transmission systems	
G.773	Protocol suites for Q-interfaces for management of transmission systems	✗
G.774	Synchronous Digital Hierarchy (SDH) management information model for the network element view	
G.774 Corr.1	Corrigendum 1 (11/96) to Recommendation G.774	
G.774.01	Synchronous digital hierarchy (SDH) performance monitoring for the network element view	
G.774.01 Corr.1	Corrigendum 1 (11/96) to Recommendation G.774.01	
G.774.02	Synchronous digital hierarchy (SDH) configuration of the payload structure for the network element view	

Series G Recommendations

Transmission systems and media, digital systems and networks

Rec. No.	Description/Title	S
G.774.02 Cor.1	Corrigendum 1 (11/96) to Recommendation G.774.02	
G.774.03	Synchronous digital hierarchy (SDH) management of multiplex-section protection for the network element view	
G.774.03 Cor.1	Corrigendum 1 (11/96) to Recommendation G.774.03	
G.774.04	Synchronous digital hierarchy (SDH) management of the subnetwork connection protection for the network element view	
G.774.04 Cor.1	Corrigendum 1 (11/96) to Recommendation G.774.04	
G.774.05	Synchronous digital hierarchy (SDH) management of connection supervision functionality (HCS/LCS) for the network element view	
G.774.05 Cor.1	Corrigendum 1 (11/96) to Recommendation G.774.05	
G.774.6	Synchronous Digital Hierarchy (SDH) – Unidirectional performance monitoring for the network element view	X
G.774.7	Synchronous Digital Hierarchy (SDH) management of lower order path trace and interface labelling for the network element view	X
G.774.8	Synchronous digital hierarchy (SDH) management of radio-relay systems for the network element of view	X
G.774.9	Synchronous digital hierarchy (SDH) configuration of linear multiplex section protection for the network element view	X
G.775	Loss of signal (LOS) and alarm indication signal (AIS) defect detection and clearance criteria	
G.776.1	Managed objects for signal processing network elements	X
G.780	Vocabulary of terms for synchronous digital hierarchy (SDH) networks and equipment	X
G.781	Structure of Recommendations on equipment for the synchronous digital hierarchy (SDH)	X
G.782	Types and general characteristics of synchronous digital hierarchy (SDH) equipment	
G.783	Characteristics of synchronous digital hierarchy (SDH) equipment functional blocks	X
G.784	Synchronous digital hierarchy (SDH) management	X
G.785	Characteristics of a flexible multiplexer in a synchronous digital hierarchy environment	X
G.791	General considerations on transmultiplexing equipments	
G.792	Characteristics common to all transmultiplexing equipments	
G.793	Characteristics of 60-channel transmultiplexing equipments	
G.794	Characteristics of 24-channel transmultiplexing equipments	
G.795	Characteristics of codecs for FDM assemblies	
G.796	Characteristics of a 64 kbit/s cross-connect equipment with 2048 kbit/s access ports	
G.796.1	Corrigendum 1 to G.796	
G.797	Characteristics of a flexible multiplexer in a plesiochronous digital hierarchy environment	X
G.798	Characteristics of optical transport network hierarchy equipment functional blocks	
G.801	Digital transmission models	
G.802	Interworking between networks based on different digital hierarchies and speech encoding laws	
G.803	Architectures of transport networks based on the synchronous digital hierarchy (SDH)	X
G.804	ATM cell mapping into plesiochronous digital hierarchy (PDH)	X
G.805	Generic functional architecture of transport networks	
G.810	Definitions and terminology for synchronization networks	X
G.811	Timing requirements at the outputs of primary reference clocks suitable for plesiochronous operation of international digital links	
G.812	Timing requirements at the outputs of slave clocks suitable for plesiochronous operation of international digital links	X
G.813	Timing characteristics of SDH equipment slave clocks (SEC)	X
G.821	Error performance of an international digital connection forming part of an integrated services digital network	X
G.822	Controlled slip rate objectives on an international digital connection	
G.823	The control of jitter and wander within digital networks which are based on the 2048 kbit/s hierarchy	
G.824	The control of jitter and wander within digital networks which are based on the 1544 kbit/s hierarchy	
G.825	The control of jitter and wander within digital networks which are based on the synchronous digital hierarchy (SDH)	
G.826	Error performance parameters and objectives for international, constant bit rate digital paths at or above the primary rate	X
G.827	Availability parameters and objectives for path elements of international constant bit-rate digital paths at or above the primary rate	X

Series G Recommendations
Transmission systems and media, digital systems and networks

Rec. No.	Description/Title	S
G.831	Management capabilities of transport networks based on the Synchronous Digital Hierarchy (SDH)	✗
G.832	Transport of SDH elements on PDH networks: Frame and multiplexing structures	✗
G.841	Types and characteristics of SDH network protection architectures	
G.842	Interworking of SDH network protection architectures	
G.851.1	Management of the transport network – Application of the RM-ODP framework	
G.852.1	Management of the transport network – Enterprise viewpoint for simple subnetwork connection management	✗
G.852.2	Enterprise viewpoint description of transport network resource model	
G.852.3	Enterprise viewpoint for topology management	✗
G.852.6	Enterprise viewpoint for trail management	✗
G.852.8	Enterprise viewpoint for pre-provisioned adaptation management	✗
G.852.10	Enterprise viewpoint for pre-provisioned link connection management	✗
G.852.12	Enterprise viewpoint for pre-provisioned link management	✗
G.853.1	Common elements of the information viewpoint for the management of a transport network	✗
G.853.2	Subnetwork connection management information viewpoint	✗
G.853.3	Information viewpoint for topology management	✗
G.853.6	Information viewpoint for trail management	✗
G.853.8	Information viewpoint for pre-provisioned adaptation management	✗
G.853.10	Information viewpoint for pre-provisioned link connection management	✗
G.853.12	Information viewpoint for pre-provisioned link management	✗
G.854.1	Management of the transport network – Computational interfaces for basic transport network model	✗
G.854.3	Computational viewpoint for topology management	✗
G.854.6	Computational viewpoint for trail management	✗
G.854.7	Computational viewpoint for pre-provisioned adaptation management	✗
G.854.10	Computational viewpoint for pre-provisioned link connection management	✗
G.854.12	Computational viewpoint for pre-provisioned link management	✗
G.855.1	GDMO engineering viewpoint for the generic network level model	✗
G.861	Principles and guidelines for the integration of satellite and radio systems in SDH transport networks	✗
G.871	Framework for optical transport network recommendations	
G.872	Architecture of optical transport networks	✗
G.873	Optical transport network requirements	
G.874	Management aspects of the optical transport network element	
G.875	Optical transport Network (OTN) management information model for the network element view	
G.901	General considerations on digital sections and digital line systems	✗
G.902	Framework recommendation on functional access networks (AN) Architecture and functions, access types, management and service node aspects	✗
G.911	Parameters and calculation methodologies for reliability and availability of fibre optic systems	✗
G.921	Digital sections based on the 2048 kbit/s hierarchy	
G.931	Digital line sections at 3152 kbit/s	
G.941	Digital line systems provided by FDM transmission bearers	
G.950	General considerations on digital line systems	
G.951	Digital line systems based on the 1544 kbit/s hierarchy on symmetric pair cables	
G.952	Digital line systems based on the 2048bit/s hierarchy on symmetric pair cables	
G.953	Digital line systems based on the 1544 kbit/s hierarchy on coaxial pair cables	
G.954	Digital line systems based on the 2048 kbit/s hierarchy on coaxial pair cables	
G.955	Digital line systems based on the 1544 kbit/s and the 2048 kbit/s hierarchy on optical fibre cables	✗
G.956	Digital line systems based on the 2048 kbit/s hierarchy on optical fibre cables	x93
G.957	Optical interfaces for equipments and systems relating to the synchronous digital hierarchy	✗
G.958	Digital line systems based on the synchronous digital hierarchy for use on optical fibre cables	
G.959.1	Optical networking physical layer interfaces	
G.960	Access digital section for ISDN basic rate access	
G.961	Digital transmission system on metallic local lines for ISDN basic rate access	
G.962	Access digital section for ISDN primary rate at 2048 kbit/s	

Series G Recommendations
Transmission systems and media, digital systems and networks

Rec. No.	Description/Title	S
G.962 Amend 1	Amendment 1 (06/97) to Recommendation G.962 - Access digital section for ISDN primary rate at 2048 kbit/s	
G.963	Access digital section for ISDN primary rate at 1544 kbit/s	
G.964	V-Interfaces at the digital local exchange (LE) - V5.1-Interface (based on 2048 kbit/s) for the support of access network (AN)	
G.965	V-Interfaces at the digital local exchange (LE) - V5.2 interface (based on 2048 kbit/s) for the support of Access Network (AN)	
G.966	Access digital section for B-ISDN	✗
G.971	General features of optical fibre submarine cable systems	✗
G.972	Definition of terms relevant to optical fibre submarine cable systems	
G.973	Characteristics of repeaterless optical fibre submarine cable systems	✗
G.974	Characteristics of regenerative optical fibre submarine cable systems	
G.975	Forward error correction for submarine systems	✗
G.976	Test methods applicable to optical fibre submarine cable systems	✗
G.981	PDH optical line systems for the local network	
G.982	Optical access networks to support services up to the ISDN primary rate or equivalent bit rates	✗
G.983.1	Broadband optical access systems based on Passive Optical Networks (PON)	✗
G.983.1 corr	Corrigendum 1 to G.983.1	
G.991.1	High bit rate Digital Subscriber Line (HDSL) transceivers	✗
G.992.1	Asymmetric digital subscriber line (ADSL) transceivers	✗
G.992.2	Splitterless asymmetric digital subscriber line (ADSL) transceivers	✗
G.994.1	Handshake procedures for digital subscriber line (DSL) transceivers	✗
G.995.1	Overview of digital subscriber line (DSL) Recommendations	✗
G.996.1	Test procedures for digital subscriber line (DSL) transceivers	
G.997.1	Physical layer management for digital subscriber line (DSL) transceivers	✗

Series H Recommendations
Line transmission of non-telephone signals

Rec. No.	Description/Title	S
Supplement 1	Supplement 1 (05/99) to Series H - Application profile – Sign language and lip-reading real-time conversation using low bit-rate video communication	
H.11	Characteristics of circuits in the switched telephone network	x00
H.12	Characteristics of telephone-type leased circuits	x00
H.13	See Recommendation O.71	x00
H.14	Characteristics of group links for the transmission of wide-spectrum signals	x00
H.15	Characteristics of supergroup links for the transmission of wide-spectrum signals	x00
H.16	Characteristics of an impulsive-noise measuring instrument for wideband data transmission	x00
H.21	Composition and terminology of international voice-frequency telegraph systems	x00
H.22	Transmission requirements of international voice-frequency telegraph links (at 50, 100 and 200 bauds)	x00
H.23	Basic characteristics of telegraph equipments used in international voice-frequency telegraph systems	x00
H.32	See Recommendation R.43	x00
H.34	Subdivision of the frequency band of a telephone-type circuit between telegraphy and other services	x00
H.43	See Recommendation T.10	x00
H.51	See Recommendation V.2	x00
H.52	Transmission of wide-spectrum signals (data, facsimile, etc.) on wideband group links	x00
H.53	Transmission of wide-spectrum signals (data,etc.) over wideband supergroup links	x00
H.100	Visual telephone systems	
H.110	Hypothetical reference connections for videoconferencing using primary digital group transmission	
H.120	Codecs for videoconferencing using primary digital group transmission	
H.130	Frame structures for use in the international interconnection of digital codecs for videoconferencing or visual telephony	
H.140	A multipoint international videoconference system	
H.200	Framework for Recommendations for audiovisual services	

Series H Recommendations
Line transmission of non-telephone signals

Rec. No.	Description/Title	S
H.221	Frame structure for a 64 to 1920 kbit/s channel in audiovisual teleservices	✗
H.221 err	Covering Note to Recommendation H.221 (05/99)	
H222.0	Information technology - Generic coding of moving pictures and associated audio information: Systems	✗
H222.0 Amend	Amendments 1 through 2 (11/96-2/00) to Recommendation H.222.0	
H222.1	Multimedia multiplex and synchronization for audiovisual communication in ATM environments	✗
H223	Multiplexing protocol for low bit rate multimedia communication	✗
H.223 Annex A	Annex A (02/98) – Multiplexing protocol for low bit-rate multimedia mobile communication over low error-prone channels	
H.223 Annex B	Annex B (02/98) – Multiplexing protocol for low bit rate multimedia mobile communication over moderate error-prone channels	
H.223 Annex C	Annex C (02/98) – Multiplexing protocol for low bit rate multimedia mobile communication over highly error-prone channels	
H.223 Annex D	Annex D (05/99) – Optional multiplexing protocol for low bit rate multimedia mobile communication over highly error-prone channels	
H.224	A real time control protocol for simplex application using the H.221 LSD/HSD/MLP channels	✗
H225.0	Media stream packetization and synchronization on non-guaranteed quality of service LANs	✗
H225.0 Annex	Annex G (05/99) to Recommendation H.225.0 - Communication between administrative domains	
H225.0 Annex	Annex I (09/98) to Recommendation H.225.0 - Call signalling protocols and media stream packetization for packet-based multimedia communication systems Annex I: H.263 + video packetization	
H226	Channel aggregation protocol for multilink operation on circuit-switched networks	
H.230	Frame-synchronous control and indication signals for audiovisual systems	✗
H.231	Multipoint control units for audiovisual systems using digital channels up to 2 Mbit/s	✗
H.233	Confidentiality system for audiovisual services	
H.234	Encryption key management and authentication system for audiovisual services	✗
H.235	Security and encryption for H-Series (H.323 and other H.245-based) multimedia terminals	
H.242	System for establishing communication between audiovisual terminals using digital channels up to 2 Mbit/s	✗
H.243	Procedures for establishing communication between three or more audiovisual terminals using digital channels up to 2 Mbit/s	✗
H.244	Synchronized aggregation of multiple 64 or 56 kbit/s channels	
H.245	Control protocol for multimedia communication	✗
H.246	Interworking of H-Series multimedia terminals with H-Series multimedia terminals and voice/voiceband terminals on GSTN and ISDN	
H.246 Annex	Annex C (02/00) to Recommendation H.246 - To be published	
H.247	Multipoint extension for broadband audiovisual communication systems and terminals	
H.248	Gateway control protocol (published in COM 16-107)	
H.261	Video codec for audiovisual services at p x 64 kbit/s	
H.262	Information technology - Generic coding of moving pictures and associated audio information: Video	✗
H.262 Amend.1	Amendment 1 (11/96) to Recommendation H.262	
H.262 Corr.1	Corrigendum 1 (11/96) to Recommendation H.262	
H.262 Amend.2	Amendment 2 (11/96) to Recommendation H.262	
H.262 Corr.2	Corrigendum 2 (11/96) to Recommendation H.262	
H.263	Video coding for low bit rate communication	✗
H.263 Append.I	Appendix I (03/97) to Recommendation H.263	
H.281	A far end camera control protocol for videoconferences using H.224	✗
H.282	Remote device control protocol for multimedia applications	
H.283	Remote device control logical channel transport	✗
H.310	Broadband audiovisual communication systems and terminals	✗
H.310 corr	Corrigendum 1 (02/98) to Recommendation H.310 - Broadband audiovisual communication systems and terminals	
H.320	Narrow-band visual telephone systems and terminal equipment. A video conferencing standards suite that allows dissimilar videoconferencing systems to communicate with each other. See also H.711 and T.120	✗
H.321	Adaptation of H.320 visual telephone terminals to B-ISDN environments	✗

Series H Recommendations
Line transmission of non-telephone signals

Rec. No.	Description/Title	S
H.322	Visual telephone systems and terminal equipment for local area networks which provide a guaranteed quality of service	✗
H.323	Visual telephone systems and equipment for local area networks which provide a non-guaranteed quality of service	✗
H.323 Annex D	Annex D (09/98) to Recommendation H.323 - Packet-based multimedia communications systems Annex D: Real-time facsimile over H.323 systems	
H.323 Annex E	Annex E (05/99) to Recommendation H.323 – Framework and wire-protocol for multiplexed call signalling transport	
H.323 Annex F	Annex F (05/99) to Recommendation H.323 – Simple endpoint types	
H.323 Annex G	Annex G (02/00) to Recommendation H.323	
H.324	Terminal for low bit rate Multimedia Communication	✗
H.324 Annex F	Annex F (09/98) to Recommendation H.324 - Terminal for low bit-rate multimedia communication Annex F: Multilink operation	✗
H.324 Annex G	Annex G (02/00) to Recommendation H.324	
H.331	Broadcasting type audiovisual multipoint systems and terminal equipment	
H.332	Extended for loosely coupled conferences	✗
H.341	Multimedia management information base	
H450.1	Generic functional protocol for the support of supplementary services in H.323	✗
H450.2	Call transfer supplementary service for H.323	✗
H450.3	Call diversion supplementary service for H.323	✗
H450.4	Call hold supplementary service for H.323	✗
H450.5	Call park and call pickup supplementary services for H.323	✗
H450.5 err	Covering Note to Recommendation H.450.5 (05/99)	
H450.6	Call waiting supplementary service for H.323	✗
H450.7	Message waiting indication supplementary service for H.323	✗
H.450.8	Name identification supplementary service for H.323 (published in COM 16-106)	

Series I Recommendations
Integrated Services Digital Networks (ISDN)

Rec. No.	Description/Title	S
I.110	Preamble and general structure of the I-Series Recommendations for the integrated services digital network (ISDN)	x93
I.111	Relationship with other Recommendations relevant to ISDNs	x93
I.112	Vocabulary of terms for ISDNs	
I.113	Vocabulary of terms for broadband aspects of ISDN	✗
I.114	Vocabulary of terms for universal personal telecommunication	
I.120	Integrated services digital networks (ISDNs)	
I.121	Broadband aspects of ISDN	✗
I.122	Framework for frame mode bearer services	
I.130	Method for the characterization of telecommunication services supported by an ISDN and network capabilities of an ISDN	
I.140	Attribute technique for the characterization of telecommunication services supported by an ISDN and network capabilities of an ISDN	
I.141	ISDN network charging capabilities attributes	
I.150	B-ISDN asynchronous transfer mode functional characteristics	✗
I.200	Guidance to the I.200-series of Recommendations	
I.210	Principles of telecommunication services supported by an ISDN and the means to describe them	
I.211	B-ISDN service aspects	
I.220	Common dynamic description of basic telecommunication services	
I.221	Common specific characteristics of services	
I.230	Definition of bearer service categories	
I.231	Circuit-mode bearer service categories	
I.231.1	Circuit-mode bearer service categories - Circuit-mode 64 kbit/s unrestricted, 8 kHz structured bearer service	

Series I Recommendations
Integrated Services Digital Networks (ISDN)

Rec. No.	Description/Title	S
I.231.2	Circuit-mode bearer service categories - Circuit-mode 64 kbit/s, 8 kHz structured bearer service usable for speech information transfer	
I.231.3	Circuit-mode bearer service categories - Circuit-mode 64 kbit/s, 8 kHz structured bearer service usable for 3.1 kHz audio information transfer	
I.231.4	Circuit-mode bearer service categories - Circuit-mode, alternate speech / 64 kbit/s unrestricted, 8 kHz structured bearer service	
I.231.5	Circuit-mode bearer service categories - Circuit-mode 2 x 64 kbit/s unrestricted, 8 kHz structured bearer service	
I.231.6	Circuit-mode bearer service categories: Circuit-mode 384 kbit/s unrestricted, 8 kHz structured bearer service	
I.231.7	Circuit-mode bearer service categories: Circuit-mode 1536 kbit/s unrestricted, 8 kHz structured bearer service	
I.231.8	Circuit-mode bearer service categories: Circuit-mode 1920 kbit/s unrestricted, 8 kHz structured bearer service	
I.231.9	Circuit mode 64 kbit/s 8 kHz structured multi-use bearer service category	
I.231.10	Circuit-mode multiple-rate unrestricted 8 kHz structured bearer service category	
I.232	Packet-mode bearer services categories	
I.232.1	Packet-mode bearer service categories - Virtual call and permanent virtual circuit bearer service category	
I.232.2	Packet-mode bearer service categories - Connectionless bearer service category	
I.232.3	User signalling bearer service category (USBS)	
I.233	Frame mode bearer services	✗
I.233.1	ISDN frame relaying bearer service	✗
I.233.1 Annex F	Frame relay multicast	
I.233.2	ISDN frame switching bearer service	✗
I.240	Definition of teleservices	
I.241.1	Teleservices supported by an ISDN: Telephony	
I.241.2	Teleservices supported by an ISDN: Teletex	
I.241.3	Teleservices supported by an ISDN: Telefax	
I.241.4	Teleservices supported by an ISDN: Mixed mode	
I.241.5	Teleservices supported by an ISDN: Videotex	
I.241.6	Teleservices supported by an ISDN: Telex	
I.241.7	Telephony 7 kHz teleservice	
I.241.8	Teleaction stage one service description	✗
I.250	Definition of supplementary services	
I.251.1 (rev.1)	Direct-dialling-in	
I.251.2 (rev.1)	Multiple subscriber number	
I.251.3 (rev.1)	Calling line identification presentation	
I.251.4	Calling line identification restriction	
I.251.5	Connected Line Identification Presentation (COLP)	
I.251.6	Connected Line Identification Restriction (COLR)	
I.251.7	Malicious call identification	
I.251.8	Sub-addressing supplementary service	
I.251.9	Number identification supplementary services: Calling name identification presentation	
I.251.10	Number identification supplementary services: Calling name identification restriction	
I.252.1	Call offering supplementary services: Call Transfer	
I.252.2	Call forwarding busy	
I.252.3	Call forwarding no reply	
I.252.4	Call forwarding unconditional	
I.252.5	Call deflection	
I.252.6	Call offering supplementary services - Line hunting (LH)	
I.252.7	Call offering supplementary services: Explicit call transfer	✗
I.253.1	Call waiting (CW) supplementary service	✗
I.253.2	Call hold	
I.253.3	Call completion supplementary services: Completion of calls to busy subscribers	
I.253.4	Call completion supplementary services: Completion of calls on no reply	✗
I.254.1	Multiparty supplementary services: conference calling (CONF)	

Series I Recommendations
Integrated Services Digital Networks (ISDN)

Rec. No.	Description/Title	S
I.254.2	Three-party supplementary service	
I.254.5	Multiparty supplementary services: Meet-me conference	✗
I.255.1	Closed user group	
I.255.2	Community of interest supplementary services: Support of Private Numbering Plans	
I.255.3	Multi-level precedence and preemption service (MLPP)	✗
I.255.4	Priority service	✗
I.255.5	Outgoing call barring	
I.256.2a	Advice of charge: Charging information at call set-up time (AOC-S)	
I.256.2b	Advice of charge: Charging information during the call (AOC-D)	
I.256.2c	Advice of charge: Charging information at the end of the call (AOC-E)	
I.256.3	Reverse charging	
I.257.1	User-to-user signalling	
I.258.1	Terminal portability (TP) supplementary service	
I.258.2	In-call modification (IM)	✗
I.259.1	Screening supplementary services: Address screening (ADS)	✗
I.310	ISDN - Network functional principles	
I.311	B-ISDN general network aspects	✗
I.312/Q.1201	Principles of intelligent network architecture	
I.313	B-ISDN network requirements	✗
I.320	ISDN protocol reference model	
I.321	B-ISDN protocol reference model and its application	✗
I.324	ISDN network architecture	✗
I.325	Reference configurations for ISDN connection types	✗
I.326	Reference configuration for relative network resource requirements	x93
I.326	Functional architecture of transport networks based on ATM	✗
I.327	B-ISDN functional architecture	
I.328/Q.1202	Intelligent Network - Service plane architecture	
I.329/Q.1203	Intelligent Network - Global functional plane architecture	
I.330	ISDN numbering and addressing principles	
I.331/E.164	Numbering plan for the ISDN era	
I.333	Terminal selection in ISDN	
I.334	Principles relating ISDN numbers/subaddresses to the OSI reference model network layer addresses	
I.340	ISDN connection types	
I.350	General aspects of quality of service and network performance in digital networks, including ISDNs	
I.351	Relationships among ISDN performance Recommendations	✗
I.352	Network performance objectives for connection processing delays in an ISDN	
I.353	Reference events for defining ISDN performance parameters	✗
I.354	Network performance objectives for packet mode communication in an ISDN	
I.355	ISDN 64 kbit/s connection type availability performance	✗
I.356	B-ISDN ATM layer cell transfer performance	✗
I.357	B-ISDN semi-permanent connection availability	✗
I.358	All processing performance for switched Virtual Channel Connections (VCCs) in a B-ISDN	✗
I.359	Accuracy and dependability performance of ISDN 64 kbit/s circuit mode connection types	✗
I.361	B-ISDN ATM layer specification	
I.362	B-ISDN ATM Adaptation Layer (AAL) functional description	
I.363	B-ISDN ATM Adaptation Layer (AAL) specification	
I.363.1	B-ISDN ATM Adaptation: Type 1 AAL	
I.363.3	B-ISDN ATM Adaptation Layer specification: Type 3/4 AAL	✗
I.363.5	B-ISDN ATM Adaptation Layer specification: Type 5 AAL	✗
I.364	Support of broadband connectionless data service on B-ISDN	✗
I.365.1	Frame relaying service specific convergence sublayer (FR-SSCS)	
I.365.2	B-ISDN ATM adaptation layer sublayers: service specific coordination function to provide the connection oriented network service	✗

Series I Recommendations
Integrated Services Digital Networks (ISDN)

Rec. No.	Description/Title	S
I.365.3	B-ISDN ATM adaptation layer sublayers: service specific coordination function to provide the connection-oriented transport service	✗
I.365.4	B-ISDN ATM adaptation layer sublayers: Service specific convergence sublayer for HDLC applications	✗
I.366.1	Segmentation and Reassembly Service Specific Convergence Sublayer for the AAL type 2	✗
I.366.2	AAL type 2 service specific convergence sublayer for trunking	✗
I.370	Congestion management for the ISDN frame relaying bearer service	✗
I.371	Traffic control and congestion control in B-ISDN	
I.371.1	Traffic control and congestion control in B-ISDN: Conformance definitions for ABT and ABR	✗
I.371.2	Frame relaying bearer service network-to-network interface requirements	
I.372	Frame relaying bearer service network-to-network interface requirements	
I.373	Network capabilities to support Universal Personal Telecommunication (UPT)	
I.374	Framework Recommendation on "Network capabilities to support multimedia services"	
I.375.1	Network capabilities to support multimedia services: General aspects	✗
I.375.2	Network capabilities to support multimedia services: Example of multimedia retrieval service class - Video-on-demand service using an ATM based network	✗
I.375.3	Network capabilities to support multimedia services – Examples of multimedia distribution service class, switched digital broadcasting	
I.376	ISDN network capabilities for the support of the teleaction service	✗
I.380	Internet protocol data communication service - IP packet transfer and availability performance parameters	✗
I.410	General aspects and principles relating to Recommendations on ISDN user-network interfaces	
I.411	ISDN user-network interfaces - References configurations	
I.412	ISDN user-network interfaces - Interface structures and access capabilities	
I.413	B-ISDN user-network interface	✗
I.414	Overview of Recommendations on layer 1 for ISDN and B-ISDN customer accesses	✗
I.420	Basic user-network interface	
I.421	Primary rate user-network interface	
I.430	Basic user-network interface - Layer 1 specification	✗
I.431	Primary rate user-network interface - Layer 1 specification	
I.431 Amend.1	Amendment 1 (06/97) to Recommendation I.431	
I.432	B-ISDN user-network interface - Physical layer specification	
I.432.1	B-ISDN user-network interface - Physical layer specification: General characteristics	✗
I.432.2	B-ISDN user-network interface - Physical layer specification: 155 520 kbit/s and 622 080 kbit/s operation	✗
I.432.3	B-ISDN user-network interface - Physical layer specification: 1544 kbit/s and 2048 kbit/s operation	✗
I.432.4	B-ISDN user-network interface - Physical layer specification: 51 840 kbit/s operation	✗
I.432.5	B-ISDN user-network interface - Physical layer specification: 25 600 kbit/s operation	✗
I.440	See Recommendation Q.920	
I.441	See Recommendation Q.921	
I.450	See Recommendation Q.930	
I.451	See Recommendation Q.931	
I.452	See Recommendation Q.932	
I.460	Multiplexing, rate adaption and support of existing interfaces	
I.461	See Recommendation X.30	
I.462	See Recommendation X.31	
I.463	See Recommendation V.110	
I.464	Multiplexing, rate adaption and support of Existing interfaces for restricted 64 kbit/s transfer capability	✗
I.465	See Recommendation V.120	
I.470	Relationship of terminal functions to ISDN	
I.500	General structure of the ISDN interworking Recommendations	
I.501	Service interworking	
I.510	Definitions and general principles for ISDN interworking	
I.511	ISDN-to-ISDN layer 1 internetwork interface	
I.515	Parameter exchange for ISDN interworking	

Series I Recommendations

Integrated Services Digital Networks (ISDN)

Rec. No.	Description/Title	S
I.520	General arrangements for network interworking between ISDNs	
I.525	Interworking between ISDN and networks which operate at bit rates of less than 64 kbit/s	
I.530	Network interworking between an ISDN and a public switched telephone network (PSTN)	
I.540	See Recommendation X.321	
I.550	See Recommendation X.325	
I.555	Frame relaying bearer service interworking	✗
I.560	See Recommendation U.202	
I.570	Public/private ISDN interworking	
I.571	Connection of VSAT based private networks to the public ISDN	
I.580	General arrangements for interworking between B-ISDN and 64 kbit/s based ISDN	✗
I.581	General arrangements for B-ISDN interworking	✗
I.601	General maintenance principles of ISDN subscriber access and subscriber installation	
I.610	B-ISDN operation and maintenance principles and functions	
I.620	Frame relay operation and maintenance principles and functions	✗
I.620 adden	Addendum 1 to ITU-T Recommendation I.620	
I.630	ATM protection switching	✗
I.731	Types and general characteristics of ATM equipment	✗
I.732	Functional characteristics of ATM equipment	✗
I.741	Interworking and interconnection between ATM and switched telephone networks for the transmission of speech, voiceband data and audio signals	✗
I.751	Asynchronous transfer mode management of the network element view	✗

Series J Recommendations

Transmission of sound programme and television signals

Rec. No.	Description/Title	S
J.Supp 1	Supplement 1 to ITU-T J-series Recommendations (11/98) - Example of linking options between annexes of ITU-T Recommendation J.112 and annexes of ITU-T Recommendation J.83	✗
J.Supp 2	Supplement 2 to ITU-T J-series Recommendations (11/98) - Guidelines for the implementation of Annex A of Recommendation J.112, "Transmission systems for interactive cable television services" - Example of Digital Video Broadcasting (DVB) interaction	✗
J.Supp 3	Supplement 3 to ITU-T J-series Recommendations (11/98) - Guidelines for the implementation of Recommendation J.111 "Network Independent Protocols" - Example of digital video broadcasting (DVB) systems for interactive services	✗
J.1	Terminology for new services in television and sound-programme transmission	
J.1 Amend.1	Amendment 1 (04/97) to Recommendation J.1 -	✗
J.11	Hypothetical reference circuits for sound-programme transmissions	
J.12	Types of sound-programme circuits established over the international telephone network	
J.13	Definitions for international sound-programme circuits	
J.14	Relative levels and impedances on an international sound-programme connection	
J.15	Lining-up and monitoring an international sound-programme connection	
J.16	Measurement of weighted noise in sound-programme circuits	
J.17	Pre-emphasis used on sound-programme circuits	
J.18	Crosstalk in sound-programme circuits set up on carrier systems	
J.19	A conventional test signal simulating sound-programme signals for measuring interference in other channels	
J.21	Performance characteristics of 15 kHz-type sound-programme circuits - circuits for high quality monophonic and stereophonic transmissions	✗
J.22	Performance characteristics of 10 kHz type sound-programme circuits	x93
J.23	Performance characteristics of 7 kHz type (narrow bandwidth) sound-programme circuits	
J.24	Modulation of signals carried by sound-programme circuits by interfering signals from power supply sources	
J.25	Estimation of transmission performance of sound-programme circuits shorter or longer than the hypothetical reference circuit	
J.26	Test signals to be used on international sound-programme connections	

Series J Recommendations

Transmission of sound programme and television signals

Rec. No.	Description/Title	S
J.27	Signals for the alignment of international sound-programme connections	
J.31	Characteristics of equipment and lines used for setting up 15 kHz type sound-programme circuits	x96
J.32	Characteristics of equipment and lines used for setting up 10 kHz type sound-programme circuits	x93
J.33	Characteristics of equipment and lines used for setting up 6.4 kHz type sound-programme circuits	x96
J.34	Characteristics of equipment used for setting up 7 kHz type sound-programme circuits	x96
J.41	Charateristics of equipment for the coding of analogue high quality sound programme signals for transmission on 384 kbit/s channels	
J.42	Characteristics of equipment for the coding of analogue medium quality sound-programme signals for transmission on 384-kbit/s channels	
J.43	Characteristics of equipment for the coding of analogue high quality sound programme signals for transmission on 320 kbit/s channels	
J.44	Characteristics of equipment for the coding of analogue medium quality sound-programme signals for transmission on 320 kbit/s channels	x96
J.51	General principles and user requirements for the digital transmission of high quality sound programmes	✗
J.52	Digital transmission of high-quality sound-programme signals using one, two, or three 64 kbit/s channels per mono signal (and up to six per stereo signal)	✗
J.53	Sampling frequency to be used for the digital transmission of high-quality sound-programme signals	
J.54	Transmission of analogue high-quality sound-programme signals on mixed analogue-and-digital circuits using 384 kbit/s channels	
J.55	Digital transmission of high-quality sound-programme signals on distribution circuits using 480 kbit/s (496 kbit/s) per audio channel	
J.56	Transmission of high-quality sound-programme analogue signals over mixed analogue/digital circuits at 320 kbit/s	x96
J.57	Transmission of digital studio quality sound signals over H1 channels	
J.57 info	Information Note relating to Recommendation J.57	
J.61	Transmission performance of television circuits designed for use in international connections	
J.62	Single value of the signal-to-noise ratio for all television systems	
J.63	Insertion of test signals in the field-blanking interval of monochrome and colour television signals	
J.64	Definitions of parameters for simplified automatic measurement of television insertion test signals	
J.65	Standard test signal for conventional loading of a television channel	
J.66	Transmission of one sound programme associated with analogue television signal by means of time division multiplex in the line synchronizing pulse	
J.67	Test signals and measurement techniques for transmission circuits carrying MAC/packet signals for HD-MAC signals	✗
J.68	Hypothetical reference chain for television transmissions over very long distances	
J.73	Use of a 12-MHz system for the simultaneous transmission of telephony and television	x96
J.74	Methods for measuring the transmission characteristics of translating equipments	x96
J.75	Interconnection of systems for television transmission on coaxial pairs and on radio-relay links	x96
J.77	Characteristics of the television signals transmitted over 18 MHz and 60-MHz systems	x96
J.80	Transmission of component-coded digital television signals for contribution-quality applications at bit rates near 140 Mbit/s	
J.81	Transmission of component-coded television signals for contribution-quality applications at the third hierarchical level of ITU-T Recommendation G.702	
J.81 Amend.1	Amendment 1 to Recommendation J.81 (10/95) - Appendix II to Annex A: Guidelines for implementation of a complete television codec	
J.81 Amend.2	Amendment 2 to Recommendation J.81 (03/98) – Appendix IV to Annex A - Results of 34 Mbit/s codec interworking tests (February 1996)	
J.81 Corr.1	Corrigendum 1 (10/96) to Recommendation J.81 -T ransmission of component-coded digital television signals for contribution-quality applications at the third hierarchical level of ITU-T Recommendation G.702	
J.82	Transport of MPEG-2 constant bit rate television signals in B-ISDN	✗
J.83	Digital multi-programme systems for television sound and data services for cable distribution	✗
J.83 Info	Information Note relating to Recommendation J.83 (04/97) - Digital multi-programme systems for television and data services for cable distribution	
J.84	Distribution of digital multi-programme signals for television, sound and data services through SMATV networks	✗

Series J Recommendations

Transmission of sound programme and television signals

Rec. No.	Description/Title	S
J.85	Digital television transmission over long distances - general principles	
J.86	Mixed analogue-and-digital transmission of analogue composite television signals over long distances	
J.87	Use of hybrid cable television links for the secondary distribution of television into the user's premises	X
J.90	Electronic programme guides for delivery by digital cable television and similar methods	X
J.91	Technical methods for ensuring privacy in long-distance international television transmission	X
J.92	Recommended operating guidelines for point-to-point transmission of television programs	X
J.93	Requirements for conditional access in the secondary distribution of digital television on cable television systems	X
J.94	Service information for digital broadcasting in cable television systems	X
J.100	Tolerances for transmission time differences between the vision and sound components	
J.101	Measurement methods and test procedures for teletext signals	
J.110	Basic principles for a worldwide common family of systems for the provision of interactive television services	
J.111	Network independent protocols for interactive systems	X
J.112	Transmission systems for interactive cable television services	X
J.113	Digital video broadcasting interaction channel through the PSTN/ISDN	X
J.115	Interaction channel using the global system for mobile communications	X
J.116	(J.isl) Interaction channel for local multipoint distribution systems	
J.118	(J.smatv/matv) Access systems for interactive services on SMATV/MATV networks	
J.120	(J.web) Distribution of sound and television programs over the IP network	
J.131	Transport of MPEG-2 signals in PDH networks	X
J.132	Transport of MPEG-2 signals in SDH networks	X
J.140	Subjective picture quality assessment for digital cable television systems	X
J.141	Performance indicators for data services delivered over digital cable television systems	X
J.142	(J.mdt) Methods for the measurement of parameters in the transmission of digital cable television signals	
J.143	(J.ovq-req) User requirements for objective perceptual video quality measurements in digital cable television	
J.150	Operational functionalities for the delivery of digital multiprogramme television, sound and data services through multichannel, multipoint distribution systems (MMDS)	X
J.151	Amendment 1 (09/99) to Recommendation J.150 - Additions to Recommendation J.150 to also encompass local multipoint distribution systems (LMDS)	
J.180	(J.mux) User requirements for statistical multiplexing of several programmes on a transmission channel	

Series K Recommendations

Protection against interference

Rec. No.	Description/Title	S
K.1	Connection to earth of an audio-frequency telephone line in cable	x96
K.2	Protection of repeater power-feeding systems against interference from neighbouring electricity lines	x96
K.3	Interference caused by audio-frequency signals injected into a power distribution network	x96
K.4	Disturbance to signalling	x96
K.5	Joint use of poles for electricity distribution and for telecommunications	
K.6	Precautions at crossings	
K.7	Protection against acoustic shock	
K.8	Separation in the soil between telecommunication cables and earthing system of power facilities	
K.9	Protection of telecommunication staff and plant against a large earth potential due to a neighbouring electric traction line	
K.10	Unbalance about earth of telecommunication installations	X
K.11	Principles of protection against overvoltages and overcurrents	
K.12	Characteristics of gas discharge tubes for the protection of telecommunications installations	
K.13	Induced voltages in cables with plastic-insulated conductors	
K.14	Provision of a metallic screen in plastic-sheathed cables	

Series K Recommendations
Protection against interference

Rec. No.	Description/Title	S
K.15	Protection of remote-feeding systems and line repeaters against lightning and interference from neighbouring electricity lines	
K.16	Simplified calculation method for estimating the effect of magnetic induction from power lines on remote-fed repeaters in coaxial pair telecommunication systems	
K.17	Tests on power-fed repeaters using solid-state devices in order to check arrangements for protection from external inteference	
K.18	Calculation of voltage induced into telecommunication lines from radio station broadcasts and methods of reducing interference	
K.19	Joint use of trenches and tunnels for telecommunication and power cables	
K.20	Resistibility of telecommunication switching equipment to overvoltages and overcurrents	✗
K.21	Resistibility of subscribers' terminals to overvoltages and overcurrents	✗
K.22	Overvoltage resistibility of equipment connected to an ISDN T/S bus	
K.23	Types of induced noise and descriptions of noise voltage parameters for ISDN basic user networks	
K.24	Method for measuring radio-frequency induced noise on telecommunications pairs	
K.25	Lightning protection of optical fibre cables	
K.26	Protection of telecommunication lines against harmful effects from electric power and electrified railway lines	
K.27	Bonding configurations and earthing inside a telecommunication building	✗
K.28	Characteristics of semi-conductor arrester assemblies for the protection of telecommunications installations	✗
K.29	Coordinated protection schemes for telecommunication below ground	✗
K.30	Positive temperature coefficient (PTC) thermistors	
K.31	Bonding configurations and earthing of telecommunication installations inside a subscriber's building	
K.32	Immunity requirements and test methods for electrostatic discharge to telecommunication equipment - Generic EMC recommendation	
K.33	Limits for people safety related to coupling into telecommunications system from a.c. electric power and a.c. electrified railway installations in fault conditions	✗
K.34	Classification of electromagnetic environmental conditions for telecommunication equipment - fast transient and radio-frequency phenomenon	
K.35	Bonding configurations and earthing at remote electronic sites	
K.36	Selection of protective devices	
K.37	High frequency EMC mitigation techniques for telecommunication installations (Basic EMC Recommendation)	✗
K.38	Radiated emission test procedure for physically large systems	✗
K.39	Risk assessment of damages to telecommunication sites due to lightning discharges	✗
K.40	Protection against LEMP in telecommunications centres	✗
K.41	Resistibility of internal interfaces of telecommunication centres to surge overvoltages	✗
K.42	Preparation of emission and immunity requirements for telecommunication equipment - General principles	✗
K.43	Immunity requirements for telecommunication equipment	✗
K.46	(K.pair) Protection of telecommunication lines using metallic symmetric conductors against lightning induced surges	
K.47	(K.light) Protection of telecommunication lines using metallic conductors against direct lightning discharges	

Series L Recommendations
Construction, installation and protection of cable and other elements of outside plant

Rec. No.	Description/Title	S
L.1	Construction, installation and protection of telecommunication cables in public networks	
L.2	Impregnation of wooden poles	
L.3	Armoring of cables	
L.4	Aluminium cable sheaths	
L.5	Cable sheaths made of metals other than lead or aluminium	

Series L Recommendations
Construction, installation and protection of cable and other elements of outside plant

Rec. No.	Description/Title	S
L.6	Methods of keeping cables under gas pressure	
L.7	Application of joint cathodic protection	
L.8	Corrosion caused by alternating current	
L.9	Methods of terminating metallic cable conductors	
L.10	Optical fibre cables for duct, tunnel, aerial and buried application	
L.11	Joint use of tunnels by pipelines and telecommunication cables, and the standardization of underground duct plans	
L.12	Optical fibre joints	X
L.13	Sheath joints and organizers of optical fibre cables in the outside plant	
L.14	Measurement method to determine the tensile performance of optical fibre cables under load	
L.15	Optical local distribution networks - Factors to be considered for their construction	
L.16	Conductive plastic material (CPM) as protective covering for metal cable sheaths	
L.17	Implementation of connecting customers into the public switched telephone network (PSTN) via optical fibres	
L.17 App 1	Appendix 1 to Recommendation L.17 (02/97) – Implementation of connecting customers into the public switched telephone networks (PSTN) via optical fibres, Appendix 1: Examples of possible applications	X
L.18	Sheath closures for terrestrial copper telecommunication cables	
L.19	Outside plant copper networks for ISDN services	
L.20	Creation of a fire security code for telecommunication facilities	
L.21	Fire detection and alarm systems, detector and sounder devices	
L.22	Fire protection	
L.23	Fire extinction - Classification and location of fire extinguishing installations and equipment on premises	
L.24	Classification of outside plant waste	
L.25	Optical fibre cable network maintenance	
L.26	Optical fibre cables for aerial application	
L.27	Method for estimating the concentration of hydrogen in optical fibre cables	
L.28	External additional protection for marinized terrestrial cables	
L.29	As-laid report and maintenance/repair log for marinized terrestrial cable installation	
L.30	Markers on marinized terrestrial cables	X
L.31	Optical fibre attenuators	X
L.32	Protection devices for through-cable penetrations of fire-sector partitions	X
L.33	Periodic control of fire extinction devices in telecommunication buildings	X
L.34	Installation of Optical Fibre Ground Wire (OPGW) cable	X
L.35	Installation of optical fibre cables in the access network	X
L.36	Single mode fibre optic connector	X
L.37	Fibre optic (non-wavelength selective) branching devices	X
L.38	Use of trenchless techniques for the construction of underground infrastructures for telecommunication cable installation	X

Series M Recommendations
Maintenance: international transmission systems, telephone circuits, telegraphy, facsimile and leased cicuits

Rec. No.	Description/Title	S
M.10	Scope and application of Recommendations for maintenance of telecommunication networks and services	
M.15	Maintenance considerations for new systems	
M.20	Maintenance philosophy for telecommunication networks	
M.21	Maintenance philosophy for telecommunication services	
M.32	Principles for using alarm information for maintenance of international transmission systems and equipment	
M.34	Performance monitoring on international transmission systems and equipment	
M.35	Principles concerning line-up and maintenance limits	
M.50	Use of telecommunication terms for maintenance	

Series M Recommendations

Maintenance: international transmission systems, telephone circuits, telegraphy, facsimile and leased cicuits

Rec. No.	Description/Title	S
M.60	Maintenance terminology and definitions	
M.70	Guiding principles on the general maintenance organization for telephone-type international circuits	
M.75	Technical service	
M.80	Control stations	
M.85	Fault report points	
M.90	Sub-control stations	
M.100	Service circuits	
M.110	Circuit testing	
M.120	Access points for maintenance	
M.125	Digital loopback mechanisms	
M.160	Stability of transmission	
M.320	Numbering of the channels in a group	
M.330	Numbering of groups within a supergroup	
M.340	Numbering of supergroups within a mastergroup	
M.350	Numbering of mastergroups within a supermastergroup	
M.380	Numbering in coaxial systems	
M.390	Numbering in systems on symmetric pair cable	
M.400	Numbering in radio-relay links or open-wire line systems	
M.410	Numbering of digital blocks in transmission systems	
M.450	Bringing a new international transmission system into service	
M.460	Bringing international group, subgroup, etc. links into service	
M.470	Setting up and lining up analogue channels for international telecommunication services	
M.475	Setting up and lining up mixed analogue/digital channels for international telecommunication services	
M.495	Transmission restoration and transmission route diversity: terminology and general principles	
M.496	Functional organization for automatic transmission restoration	
M.500	Routine maintenance measurements to be made on regulated line sections	
M.510	Readjustment to the nominal value of a regulated line section (on a symmetric pair line, a coaxial line or a radio-relay link)	
M.520	Routine maintenance on international group, supergroup, etc., links	
M.525	Automatic maintenance procedures for international group, supergroup, etc., link	
M.530	Readjustment to the nominal value of an international group, supergroup, etc., link	
M.535	Special maintenance procedures for multiple destination unidirectional (MU) group and supergroup links	
M.540	Routine maintenance of carrier and pilot generating equipment	
M.556	Setting up and initial testing of digital channels on an international digital path or block	
M.560	International telephone circuits - principles, definitions and relative transmission levels	
M.562	Types of circuit and circuit section	
M.565	Access points for international telephone circuits	
M.570	Constitution of the circuit; preliminary exchange of information	
M.580	Setting up and lining up an international circuit for public telephony	
M.585	Bringing an international digital circuit into service	
M.590	Setting up and lining up a circuit fitted with a compandor	
M.600	Organization of routine maintenance measurements on circuits	
M.605	Routine maintenance schedule for international public telephony circuits	
M.610	Periodicity of maintenance measurements on circuits	
M.620	Methods for carrying out routine measurements on circuits	
M.630	Maintenance of circuits using control chart methods	
M.650	Routine line measurements to be made on the line repeaters of audio-frequency sections or circuits	
M.660	Periodical in-station tests of echo suppressors complying with Recommendations G.161 and G.164	
M.665	Testing of echo cancellers	
M.670	Maintenance of a circuit fitted with a compandor	
M.675	Lining up and maintaining international demand assignment circuits (SPADE)	
M.710	Performance monitoring on international transmission systems and equipment	
M.715	Fault report point (circuit)	
M.716	Fault report point (network)	

Series M Recommendations
Maintenance: international transmission systems, telephone circuits, telegraphy, facsimile and leased cicuits

Rec. No.	Description/Title	S
M.717	Testing point (transmission)	
M.718	Testing point (line signalling)	
M.719	Testing point (switching and interregister signalling)	
M.720	Network analysis point	
M.721	System availability information point	
M.722	Network Management Point	
M.723	Circuit control station	
M.724	Circuit sub-control station	
M.725	Restoration control point	
M.726	Maintenance organization for the wholly digital international automatic and semi-automatic telephone service	
M.729	Organization of the maintenance of international public switched telephone circuits used for data transmission	
M.730	Maintenance methods	
M.731	Subjective testing	
M.732	Signalling and switching routine maintenance tests and measurements	
M.733	Transmission routine maintenance measurements on automatic and semi-automatic telephone circuits	
M.734	Exchange of information on incoming test facilities at international switching centres	
M.760	Transfer link for common channel Signalling System No. 6	
M.762	Maintenance of common channel Signalling System No. 6	
M.782	Maintenance of common channel Signalling System No. 7	x93
M.800	Use of circuits for voice-frequency telegraphy	
M.810	Setting up and lining up an international voice-frequency telegraph link for public telegraph circuits (for 50, 100 and 200 baud modulation rates)	
M.820	Periodicity of routine tests on international voice-frequency telegraph links	
M.830	Routine measurements to be made on international voice-frequency telegraph links	
M.850	International time division multiplex (TDM) telegraph systems	
M.880	International phototelegraph transmission	
M.900	Use of leased group and supergroup links for wide-spectrum signal transmission (data, facsimile, etc.)	
M.910	Setting up and lining up an international leased group link for wide-spectrum signal transmission	
M.1010	Constitution and nomenclature of international leased circuits	
M.1012	Circuit control station for leased and special circuits	
M.1013	Sub-control station for leased and special circuits	
M.1014	Transmission maintenance point (international line) (TMP-IL)	
M.1015	Types of transmission on leased circuits	
M.1016	Assessment of the service availability performance of international leased circuits	
M.1020	Characteristics of special quality international leased circuits with special bandwidth conditioning	X
M.1025	Characteristics of special quality international leased circuits with basic bandwidth conditioning	X
M.1030	Characteristics of ordinary quality international leased circuits forming part of private switched telephone networks	
M.1040	Characteristics of ordinary quality international leased circuits	
M.1045	Preliminary exchange of information for the provision of international leased circuits	
M.1050	Lining up an international point-to-point leased circuit	
M.1055	Lining up an international multiterminal leased circuit	
M.1060	Maintenance of international leased circuits	
M.1130	General definitions and general principles of operation/maintenance procedures to be used in satellite mobile systems	
M.1140	Maritime mobile telecommunication services via satellite	
M.1150	Maritime mobile telecommunication store-and-forward services (packet mode) via satellite	X
M.1160	Aeronautical mobile telecommunication service via satellite	X
M.1170	Maintenance aspects of mobile digital telecommunication service via satellite	X
M.1230	Method to improve the management of operations and maintenance processes in the International Telephone Network	X
M.1235	Use of automatically generated test calls for assessment of network performance	
M.1300	International data transmission systems operating in the range 2.4 kbit/s to 2048 kbit/s	

Series M Recommendations

Maintenance: international transmission systems, telephone circuits, telegraphy, facsimile and leased cicuits

Rec. No.	Description/Title	S
M.1320	Numbering of channels in data transmission systems	
M.1340	Performance allocations and limits for international data transmission links and systems	✗
M.1350	Setting up, lining up and characteristics of international data transmission systems operating in the range 2.4 kbit/s to 14.4 kbit/s	
M.1355	Maintenance of international data transmission systems operating in the range 2.4 to 14.4 kbit/s	
M.1370	Bringing-into-service of international data transmission systems	✗
M.1375	Maintenance of international data transmission systems	
M.1380	Bringing-into-service of international leased circuits that are supported by international data transmission systems	✗
M.1385	Maintenance of international leased circuits that are supported by international data transmission systems	✗
M.1400	Designations for international networks	
M.1400 Amend	Amendment 1 (06/98) to Recommendation M.1400 - Designations for international networks	✗
M.1510	Exchange of contact point information for the maintenance of international services and the international network	
M.1520	Standardized information exchange between administrations	
M.1530	Network maintenance information	
M.1535	Network maintenance information	✗
M.1537	Definition of maintenance information to be exchanged at customer contact point (MICC)	✗
M.1539	Management of the grade of network maintenance services at the maintenance service customer contact point (MSCC)	✗
M.1540	Exchange of information for planned outages of transmission systems	
M.1550	Escalation procedure	
M.1560	Escalation procedure for international leased circuits	
M.2100	Performance limits for bringing-into-service and maintenance of international digital paths, sections and transmission systems	
M.2101	Performance limits for bringing-into-service and maintenance of international SDH paths and multiplex sections	
M.2101.1	Performance limits for bringing-into-service and maintenance of international SDH paths and multiplex sections	✗
M.2110	Bringing into service international digital paths, sections and transmission systems	✗
M.2120	Digital path, section and transmission system fault detection and localization procedures	✗
M.2130	Operational procedures in locating and clearing transmission faults	
M.3000	Overview of TMN Recommendations	✗
M.3010	Principles for a telecommunications management network	✗
M.3010 Adden	Addendum 1 (06/98) Recommendation M.3010 – Principles for a Telecommunications management network Addendum 1: TMN conformance and TMN compliance	
M.3016	TMN security overview	✗
M.3020	TMN interface specification methodology	
M.3100	Generic network information model	✗
M.3100 Corr	Corrigendum 1 to Recommendation M.3100 (07/98) - Generic network information model	
M.3100 Amend	Amendment 1 (03/99) to Recommendation M.3100 - Generic network information model	
M.3101	Managed object conformance statements for the generic network information model	
M.3108.1	Information model for management of leased circuit and reconfigurable service	✗
M.3180	Catalogue of TMN management information	
M.3200	TMN management services: Overview	✗
M.3207.1	TMN management service: maintenance aspects of B-ISDN management	✗
M3208.1	TMN management services for dedicated and reconfigurable circuits network: Leased circuit services	✗
M3208.2	TMN management services for dedicated and reconfigurable circuits network: Connection management of pre-provisioned service link connections to form a leased circuit service	
M.3211.1	TMN management service: Fault and performance management of the ISDN access	✗
M.3300	TMN management capabilities presented at the F interface	
M.3320	Management requirements framework for the TMN X-interface	✗
M.3400	TMN management functions	✗
M.3600	Principles for the management of ISDNs	
M.3602	Application of maintenance principles to ISDN subscriber installations	

Series M Recommendations
Maintenance: international transmission systems, telephone circuits, telegraphy, facsimile and leased cicuits

Rec. No.	Description/Title	S
M.3603	Application of maintenance principles to ISDN basic rate access	
M.3604	Application of maintenance principles to ISDN primary rate access	
M.3605	Application of maintenance principles to static multiplexed ISDN basic rate access	
M.3610	Principles for applying the TMN concept to the management of B-ISDN	✗
M.3620	Principles for the use of ISDN test calls, systems and responders	
M.3621	Integrated management of the ISDN customer access	
M.3640	Management of the D-channel - Data link layer and network layer	
M.3641	Management information model for the management of the data link and network layer of the ISDN D channel	✗
M.3650	Network performance measurements of ISDN calls	
M.3660	ISDN interface management services	
M.4010	Inter-administration agreements on common channel Signalling System No. 6	
M.4030	Transmission characteristics for setting up and lining up a transfer link for common channel Signalling System No. 6 (analogue version)	
M.4100	Maintenance of common channel Signalling System No. 7	✗
M.4110	Inter-administration agreements on common channel signalling system no. 7	✗

Series N Recommendations
Maintenance: international sound programme and television transmission circuits

Rec. No.	Description/Title	S	
N.1	Definitions for application to international sound-programme and television-sound transmission	✗	
N.2	Different types of sound-programme circuit		
N.3	Control circuits		
N.4	Definition and duration of the line-up period and the preparatory period		
N.5	Sound-programme control, sub-control and send reference stations		
N.10	Limits for the lining-up of international sound-programme links and connections	✗	
N.11	Essential transmission performance objectives for international sound-programme centres		
N.12	Measurements to be made during the line-up period that precedes a sound-programme transmission		
N.13	Measurements to be made by the broadcasting organizations during the preparatory period		
N.15	Maximum permissible power during an international sound-programme transmission		
N.16	Identification signal		
N.17	Monitoring the transmission		
N.18	Monitoring for charging purposes, releasing		
N.21	Limits and procedures for the lining-up of a sound-programmae circuit		
N.23	Maintenance measurements to be made on international sound-programme circuits		
N.51	Definitions for application to international television transmissions		
N.52	Multiple destination television transmissions and coordination centres		
N.54	Definition and duration of the line-up period and the preparatory period		
N.55	Organisation, responsibilities and functions of control and sub-control international television centres and control and sub-control stations for international television connections, links, circuits and circuit section	✗	
N.60	Nominal amplitude of video signals at video interconnection points	✗	
N.61	Measurements to be made before the line-up period that precedes a television transmission		
N.62	Tests to be made during the line-up period that precedes a television transmission	✗	
N.63	Test signals to be used by the broadcasting organizations during the preparatory period		
N.64	Quality and impairment assessment		
N.67	Monitoring television tranmissions - use of the field blanking interval	✗	
N.73	Maintenance of permanent international television circuits, links and connections		
N.81	Note		x96
N.86	Line-up and service commissioning of international videoconference systems operating at transmission bit rates of 1544 and 2048 kbit/s	✗	x96
N.90	Maintenance of international videoconference systems operating at transmission bit rates of 1544 and 2048 kbit/s	✗	x96

Series O Recommendations
Specifications of measuring equipment

Rec. No.	Description/Title	S
O.1	Scope and Application of Measurement Equipment Specifications covered in the O-series Recommendations	✗
O.3	Climatic conditions and relevant tests for measuring equipment	
O.6	1020 Hz reference test frequency	
O.9	Measuring arrangements to assess the degree of unbalance about earth	
O.11	Maintenance access lines	
O.22	CCITT Automatic transmission measuring and signalling testing ATME No. 2	
O.25	Semiautomatic in-circuit echo suppressor testing system (ESTS)	x93
O.27	In-station echo canceller test equipment	
O.31	Automatic measuring equipment for sound-programme circuits	x93
O.32	Automatic measuring equipment for stereophonic pairs of sound-programme circuits	x93
O.33	Automatic equipment for rapidly measuring stereophonic pairs and monophonic sound-programme circuits, links and connections	
O.41	Psophometer for use on telephone-type circuits	✗
O.42	Equipment to measure nonlinear distortion using the 4-tone intermodulation method	
O.51	Volume meters	x93
O.61	Simple equipment to measure interruptions on telephone-type circuits	
O.62	Sophisticated equipment to measure interruptions on telephone-type circuits	
O.71	Impulsive noise measuring equipment for telephone-type circuits	
O.72	See Recommendation H.16	
O.81	Group-delay measuring equipment for telephone-type circuits	
O.81 appndx I	Appendix I (06/98) to Recommendation O.81 - A measuring signal (multitone test signal) for fast measurement of amplitude and phase for telephone type circuits	
O.81 ap I err	Covering Note to Appendix I to Recommendation O.81 (06/98)	
O.82	Group-delay measuring equipment for the range 5 to 600 kHz	
O.91	Phase jitter measuring equipment for telephone-type circuits	
O.95	Phase and amplitude hit counters for telephone-type circuits	
O.111	Frequency shift measuring equipment for use on carrier channels	
O.131	Quantizing distortion measuring equipment using a pseudo-random noise test signal	
O.132	Quantizing distortion measuring equipment using a sinusoidal test signal	
O.133	Equipment for measuring the performance of PCM encoders and decoders	✗
O.150	Digital test patterns for performance measurements on digital transmission equipment	✗
O.151	Error performance measuring equipment operating at the primary rate and above	
O.152	Error performance measuring equipment for bit rates of 64 kbit/s and N x 64 kbit/s	
O.153	Basic parameters for the measurement of error performance at bit rates below the primary rate	
O.161	In-service code violation monitors for digital systems	
O.162	Equipment to perform in-service monitoring on 2048, 8448, 34 368 and 139 264 kbit/s signals	
O.163	Equipment to perform in-service monitoring on 1544 kbit/s signals	
O.171	Timing jitter measuring equipment for digital systems	✗
O.181	Equipment to assess error performance on STM-N interfaces	✗
O.191	Equipment to assess ATM layer cell transfer performance	✗

Series P Recommendations
Telephone transmission quality, telephone installations, local line networks

Rec. No.	Description/Title	S
Supplement 3	Models for predicting transmission quality from objective measurements	
Supplement 10	Supplement 10 to ITU-T Series P Recommendations (11/88) - Consideration relating to transmission characteristics for analogue handset telephones	
Supplement 11	Some effect of sidetone	
Supplement 16	Supplement 16 to ITU-T Series P Recommendations (11/88) - Guidelines for placement of microphones and loudspeakers in telephone conference rooms [1] and for Group Audio Terminals (GATS)	

Series P Recommendations

Telephone transmission quality, telephone installations, local line networks

Rec. No.	Description/Title	S
Supplement 20	Examples of measurements of handset receive-frequency responses: Dependence one earcap leakage losses	
Supplement 21	The principles of a composite source signal as an example of a measurement signal to determine the transfer characteristics of terminal equipment	
Supplement 22	Transmission characteristics of wideband audio telephones	
Supplement 23	Supplement 23 to ITU-T Series P Recommendations (02/98) - ITU-T coded-speech database	
P.10	Vocabulary of terms on telephone transmission quality and telephone sets	✗
P.11	Effect of transmission impairments	
P.16	Subjective effects of direct crosstalk; thresholds of audibility and intelligibility	
P.30	Transmission performance of group audio terminals (GATs)	
P.31	Transmission characteristics for digital telephones	x93-6
P.32	Evaluation of the efficiency of telephone booths and acousting hoods	
P.33	Subscriber telephone sets containing either loudspeaking receivers or microphones associated wtih amplifiers	x00
P.34	Transmission characteristics of hands-free telephones	x93-6
P.35	Handset telephones	
P.36	Efficiency of devices for preventing the occurrence of excessive acoustic pressure by telephone receivers	
P.37	Coupling hearing aids to telephone set	x93-6
P.38	Transmission characteristics of operator telephone systems (OTS)	
P.48	Specification for an intermediate reference system	
P.50	Artificial voices	
P.50 appndx	Appendix 1 to Recommendation P.50 (02/98) - Test signals	
P.50 app err	Covering Note to Recommendation P.50 (09/99)	
P.51	Artificial mouth	✗
P.52	Volume meters	
P.53	See Recommendation O.41	
P.54	Sound level meters (Apparatus for the objective measurement of room noise)	
P.55	Apparatus for the measurement of impulsive noise	
P.56	Objective measurement of active speech level	
P.57	Artificial ears	✗
P.58	Head and torso simulator for telephonometry	✗
P.59	Artificial conversational speech	
P.61	Methods for the calibration of condenser microphones	
P.62	Measurements on subscribers' telephone equipment	
P.63	Methods for the evaluation of transmission quality on the basis of objective measurements	x00
P.64	Determination of sensitivity/frequency characteristics of local telephone systems	✗
P.64 err	Covering Note to Recommendation P.64 (09/99)	
P.65	Objective instrumentation for the determination of loudness ratings	
P.66	Methods for evaluating the transmission performance of digital telephone sets	x93-6
P.75	Standard conditioning method for handsets with carbon microphones	
P.76	Determination of loudness ratings; fundamental principles	
P.78	Subjective testing method for determination of loudness ratings in accordance with Recommendation P.76	✗
P.79	Calculation of loudness ratings for telephone sets	✗
P.80	Methods for subjective determination of transmission quality	x93-6
P.81	Modulated noise reference unit (MNRU)	x93-6
P.82	Method for evaluation of service from the standpoint of speech transmission quality	
P.83	Subjective performance assessment of telephone-band and wideband digital codecs	x93-6
P.84	Subjective listening test method for evaluating digital circuit multiplication and packetized voice systems	
P.85	A method for subjective performance assessment of the quality of speech voice output devices	✗
P.310	Transmission characteristics for telephone band (300-3400 Hz) digital telephones	✗
P.311	Transmission characteristics of wideband handset telephones	✗
P.313	Transmission characteristics for cordless and mobile digital terminals	✗
P.340	Transmission characteristics of hands-free telephones	✗
P.341	Transmission characteristics of wideband handsfree telephone	✗

Series P Recommendations

Telephone transmission quality, telephone installations, local line networks

Rec. No.	Description/Title	S
P.341 corr	Corrigendum 1 (09/99) to Recommendation P.341	
P.342	Transmission characteristics for telephone band (300 - 3400 Hz) digital loudspeaking and hands-free telephony terminals	✗
P.360	Efficiency of devices for preventing the occurrence of excessive acoustic pressure by telephone receivers	✗
P.370	Coupling Hearing Aids to Telephone sets	✗
P.501	Test signals for use in telephonometry	✗
P.502	Objective test methods for speech communication systems using complex test signals	
P.561	In-service, non-intrusive measurement device - Voice service measurements	✗
P.561 appndx	Appendix 3 to Recommendation P.561 (02/98) – Digital speech recordings	
P.562	Analysis and interpretation of INMD voice-services measurements	
P.581	(HATSHFT) – Use of head and torso simulator (HATS) for hands-free terminal testing	
P.800	Methods for subjective determination of transmission quality	✗
P.810	Modulated noise reference unit (MNRU)	✗
P.830	Subjective performance assessment of telephone-band and wideband digital codecs	✗
P.831	Subjective performance evaluation of network echo cancellers	✗
P832	Subjective performance evaluation of hands-free terminals	
P.861	C reference code of Perceptual Speech Quality Measure (PSQM)	✗
P.910	Subjective video quality assessment methods for multimedia applications	✗
P.911	Subjective audiovisual quality assessment methods for multimedia applications	✗
P.911 corr	Corrigendum 1 (09/99) to Recommendation P.911	
P.920	Interactive test methods for audiovisual communications	✗
P.930	Principles of a reference impairment system for video	✗
P.931	Multimedia communications delay, synchronization and frame rate measurement	✗

Series Q Recommendations

Switching and Signalling

Rec. No.	Description/Title	S
Supplement 1	Definition of relative levels, transmission loss and attenuation/frequency distortion for digital exchanges with complex impedances at Z interfaces	
Supplement 2	Impedance strategy for telephone instruments and digital local exchanges in the British Telecom Network	
Supplement 3	Supplement 3 (05/98) to Series Q - Number portability – Scope and capability set 1 architecture	✗
Supplement 4	Supplement 4 (05/98) to Series Q - Number portability - Capability set 1 requirements for service provider portability (All call query and Onward routing)	✗
Supplement 5	Supplement 5 (03/99) to Series Q - Number portability - Capability set 2 requirements for service provider portability (Query on release and Dropback) (also E.100 supplement 3)	✗
Supplement 6	Supplement 6 (03/99) to Series Q - Technical report TRQ.2000: Roadmap for the TRQ.2xxx-series technical reports. Treatment of calls considered as terminating abnormally (also E.100 supplement 4)	✗
Supplement 7	Supplement 7 (03/99) to Series Q - Technical report TRQ.2001: General aspects for the development of unified signalling requirements	✗
Supplement 8	Supplement 8 (03/99) to Series Q - Technical report TRQ.2400: Transport control signalling requirements - Signalling requirements for AAL type 2 link control capability set 1	✗
Q.1	Signal receivers for manual working	
Q.2	Signal receivers for automatic and semi-automatic working, used for manual working	
Q.4	Automatic switching functions for use in national networks	
Q.5	Advantages of semi-automatic service in the international telephone service	
Q.6	Advantages of international automatic working	
Q.7	Signalling systems to be used for international automatic and semi-automatic telephone working	
Q.8	Signalling systems to be used for international manual and automatic working on analogue lead circuits	
Q.9	Vocabulary of switching and signalling terms	
Q.10	Definitions relating to national and international numbering plans	x93
Q.11 bis	Numbering plan for the ISDN era	x93
Q.11 ter	Timetable for coordinated implementation of the full capability of the numbering plan for the ISDN era (See recommendation E.164)	x93

Series Q Recommendations
Switching and Signalling

Rec. No.	Description/Title	S
Q.12	Overflow - Alternative routing - Rerouting - Automatic repeat attempt	
Q.13	International telephone routing plan	x93
Q.14	Means to control the number of satellite links in an international telephone connection	
Q.15	Nominal mean power during the busy hour	x93
Q.16	Maximum permissible value for the absolute power level of a signalling pulse	x93
Q.20	Comparative advantages of "in-band" and "out-band" systems	
Q.21	Systems recommended for out-band signalling	
Q.22	Frequencies to be used for in-band signalling	
Q.23	Technical features of push-button telephone sets	
Q.24	Multifrequency push-button signal reception	
Q.25	Splitting arrangements and signal recognition times in "in-band" signalling systems	
Q.26	Direct access to the international network from the national network	
Q.27	Transmission of the answer signal	
Q.28	Determination of the moment of the called subscriber's answer in the automatic service	
Q.29	Causes of noise and ways of reducing noise in telephone exchanges	
Q.30	Improving the reliability of contacts in speech circuits	
Q.31	Noise in a national 4-wire automatic exchange	
Q.32	Reduction of the risk of instability by switching means	
Q.33	Protection against effects of faulty transmission on groups of circuits	
Q.35/E.180	Technical characteristics of tones for the telephone service	✗
Q.36	Customer recognition of foreign tones	x93
Q.40	The transmission plan	x93
Q.41	Mean one-way propagation time	x93
Q.42	Stability and echo (echo suppressors)	x93
Q.43	Transmission losses, relative levels	x93
Q.44	Attenuation distortion	
Q.45	Transmission characteristics of an analogue international exchange	
Q45 bis	Transmission characteristics of an analogue international exchange	
Q.48	Demand assignment signalling systems	
Q.49	Specification for the CCITT automatic transmission measuring and signalling testing equipment ATME No. 2	x93
Q.50	Signalling between circuit multiplication equipments (CME) and international switching centres (ISC)	✗
Q.65	Stage 2 of the method for the characterization of services supported by an ISDN	
Q.68	Overview of methodology for developing management services	✗
Q.71	ISDN Circuit mode switched bearer services	✗
Q.72	Stage 2 description for packet mode	✗
Q.76	Service procedures for universal personal telecommunication - functional modelling and information flows	✗
Q.80	Introduction to Stage 2 service descriptions for supplementary services	
Q.81	Stage 2 description for number identification supplementary services	
Q.81.1	Stage 2 description for number identification supplementary services - Direct dialing-in	
Q.81.2	Stage 2 description for number identification supplementary services (Addenda to Rec. Q.81) - Section 2 - Multiple subscriber number	✗
Q.81.3	Stage 2 description for number identification supplementary services (Modifications to Rec. Q.81) - Calling line identification presentation (CLIP) and calling line identification restriction (CLIR)	✗
Q.81.5	Stage 2 description for number identification supplementary services - Connected line identification, presentation and restriction (COLP) and (COLR)	✗
Q.81.7	Stage 2 description for number identification supplementary services - Malicious call identification (MCID)	
Q.81.8	Stage 2 description for number identification supplementary services – Sub addressing (SUB)	✗
Q.82	Stage 2 description for call offering supplementary services	
Q.82.1	Stage 2 description for call offering supplementary services - Call transfer	
Q.82.2	Stage 2 description for call offering supplementary services; - Call forwarding	
Q.82.3	Stage 2 description for call offering supplementary services; - Call deflection	
Q.82.4	Stage 2 description for call offering supplementary services - Line hunting	
Q.82.7	Stage 2 description for call offering supplementary services - Explicit call transfer	✗

Series Q Recommendations
Switching and Signalling

Rec. No.	Description/Title	S
Q.83	Stage 2 description for call completion supplementary services	
Q.83.1	Sections 1 and 4 - Stage 2 description for call completion supplementary services (Modifications and addenda to: Recommendation Q.83) – Call waiting	✗
Q.83.2	Section 2 - Stage 2 description for call completion supplementary services (Modifications to: Recommendation Q.83) - Section 2 - Call hold	✗
Q.83.3	Stage 2 description for call completion supplementary services - Completion of call to busy subscriber	
Q.83.4	Stage 2 description for call completion supplementary services - Terminal portability	✗
Q.84	Stage 2 description for multiparty supplementary services	
Q.84.1	Clause 1 - Stage 2 description for multiparty supplementary services – conference calling	✗
Q.84.2	Section 2 - Stage 2 description for multiparty supplementary services - Section 2 - Three-party service	✗
Q.85	Stage 2 description for community of interest supplementary services	
Q.85.1	Sections 1 and 3 - Stage 2 description for community of interest supplementary services (Modifications and addenda to Rec. Q.85) - Section 1 - Closed User Group; Section 3 - Multi-level Precedence and Preemption	✗
Q.85.3	Stage 2 description for community of interest supplementary services - Multi-level precedence and preemption (MLPP)	✗
Q.85.6	Stage 2 description for community of interest supplementary services - Global virtual network service	✗
Q.85.6 Annex A	Stage 2 description for community of interest supplementary services: Global virtual network service (GVNS) - Annex A: Service procedures and information flows based on Intelligent Network CS-1 capabilities	✗
Q.86	Stage 2 description for charging supplementary services	
Q.86.1	Stage 2 description for charging supplementary services - Credit card call (Note)	
Q.86.2	Stage 2 description for charging supplementary services - advice of charge	✗
Q.86.3	Stage 2 description for charging supplementary services - Reverse charging (Rev)	✗
Q.86.4	Stage 2 description for charging supplementary services - International Freephone Service (IFS)	✗
Q.86.7	Stage 2 description for charging supplementary services - Clause 7 - International Telecommunication Charge Card (ITCC)	
Q.87	Stage 2 description for additional information transfer supplementary services	
Q.87.1	Stage 2 description for additional information transfer supplementary services; Clause 1 - User-to-user signalling (UUS)	✗
Q.87.2	Stage 2 description for additional information transfer supplementary services - User signalling bearer services (Note)	
Q.101	Facilities provided in international semi-automatic working	
Q.102	Facilities provided in international automatic working	
Q.103	Numbering used	
Q.104	Language digit or discriminating digit	
Q.105	National (significant) number	
Q.106	The sending-finished signal	
Q.107	Standard sending sequence of forward address information	
Q.107bis	Analysis of forward address information for routing	
Q.108	One-way or both-way operation of international circuits	
Q.109	Transmission of the answer signal in international exchanges	
Q.110	General aspects of the utilization of standardized CCITT Signalling Systems on PCM links	
Q.112	Signal levels and signal receiver sensitivity	
Q.113	Connection of signal receivers in the circuit	
Q.114	Typical transmission requirements for signal senders and receivers	
Q.115	Control of echo suppressors - control of echo suppressors and echo cancellers	
Q.116	Indication given to the outgoing operator or calling subscriber in case of an abnormal condition	
Q.117	Alarms for technical staff and arrangements in case of faults	
Q.118	Abnormal conditions special release arrangements	✗
Q.118bis	Indication of congestion conditions at transit exchanges	
Q.120	(Signalling System No. 4) Definition and function of signals	
Q.121	(Signalling System No. 4) Signal code	
Q.122	(Signalling System No. 4) Signal sender	
Q.123	(Signalling System No. 4) Signal receiver	
Q.124	(Signalling System No. 4) Splitting arrangements	

Series Q Recommendations
Switching and Signalling

Rec. No.	Description/Title	S
Q.125	(Signalling System No. 4) Speed of switching in international exchanges	
Q.126	(Signalling System No. 4) Analysis and transfer of digital information	
Q.127	(Signalling System No. 4) Release of registers	
Q.128	(Signalling System No. 4) Switching to the speech position	
Q.129	(Signalling System No. 4) Maximum duration of a blocking signal	
Q.130	(Signalling System No. 4) Special arrangements in case of failures in the sequence of signals	
Q.131	(Signalling System No. 4) Abnormal release conditions of the outgoing register causing release of the international circuit	
Q.133	(Signalling System No. 4) Numbering for access to automatic measuring and testing devices	
Q.134	(Signalling System No. 4) Routine testing of equipment (local maintenance)	
Q.135	(Signalling System No. 4) Principles of rapid transmission testing equipment	
Q.136	(Signalling System No. 4) Loop transmission measurements	
Q.137	(Signalling System No. 4) Automatic testing equipment	
Q.138	(Signalling System No. 4) Instruments for checking equipment and measuring signals	
Q.139	(Signalling System No. 4) Manual testing	
Q.140	(Signalling System No. 5) Definition and function of signals	
Q.141	(Signalling System No. 5) Clause 2 - Line signalling; 2.1 Signal code for line signalling	
Q.142	(Signalling System No. 5) Double seizing with both-way operation	
Q.143	(Signalling System No. 5) Line signal sender	
Q.144	(Signalling System No. 5) Clause 2 - Line signalling; 2.4 - Line signal receiver	
Q.145	(Signalling System No. 5) Splitting arrangements	
Q.146	(Signalling System No. 5) Speed of switching in international exchanges	
Q.151	(Signalling System No. 5) Signal code for register signalling	
Q.152	(Signalling System No. 5) End-of-pulsing conditions - Register arrangements concerning ST (end-of-pulsing) signal	
Q.153	(Signalling System No. 5) Multifrequency signal sender	
Q.154	(Signalling System No. 5) Multifrequency signal receiver	
Q.155	(Signalling System No. 5) Analysis of digital information for routing	
Q.156	(Signalling System No. 5) Release of international registers	
Q.157	(Signalling System No. 5) Switching to the speech position	
Q.161	(Signalling System No. 5) General arrangements for manual testing	
Q.162	(Signalling System No. 5) Routing testing of equipment (local maintenance)	
Q.163	(Signalling System No. 5) Manual testing	
Q.164	(Signalling System No. 5) Test equipment for checking equipment and signals	
Q.180	Interworking of systems No. 4 and No. 5	
Q.251	(Signalling System No. 6 - Functional description of the signalling system) General	
Q.252	(Signalling System No. 6 - Functional description of the signalling system) Signal transfer time definitions	
Q.253	(Signalling System No. 6 - Functional description of the signalling system) Association between signalling and speech networks	
Q.254	(Signalling System No. 6 - Definition and function of signals) Telephone signals	
Q.255	(Signalling System No. 6 - Definition and function of signals) Signalling-system-control signals	
Q.256	(Signalling System No. 6 - Definition and function of signals) Management signals	
Q.257	(Signalling System No. 6 - Signal unit formats and codes) General	
Q.258	(Signalling System No. 6 - Signal unit formats and codes) Telephone signals	
Q.259	(Signalling System No. 6 - Signal unit formats and codes) Signalling-system-control signals	
Q.260	(Signalling System No. 6 - Signal unit formats and codes) Management signals	
Q.261	(Signalling System No. 6 - Signalling procedures) Normal call set-up	
Q.262	(Signalling System No. 6 - Signalling procedures) Analysis of digital information for routing	
Q.263	(Signalling System No. 6 - Signalling procedures) Double seizing with both-way operation	
Q.264	(Signalling System No. 6 - Signalling procedures) Potential for automatic repeat attempt and re-routing	
Q.265	(Signalling System No. 6 - Signalling procedures) Speed of switching and signal transfer in international exchanges	
Q.266	(Signalling System No. 6 - Signalling procedures) Blocking and unblocking sequences and control of quasi-associated signalling	

Series Q Recommendations
Switching and Signalling

Rec. No.	Description/Title	S
Q.267	(Signalling System No. 6 - Signalling procedures) Unreasonable and superfluous messages	
Q.268	(Signalling System No. 6 - Signalling procedures) Release of international connections and associated equipment	
Q.271	(Signalling System No. 6 - Continuity check of the speech path) General	
Q.272	(Signalling System No. 6 - Signalling link) Requirements for the signalling data link	
Q.273	(Signalling System No. 6 - Signalling link) Data transmission rate	
Q.274	(Signalling System No. 6 - Signalling link) Transmission methods; Modem and interface requirements	
Q.275	(Signalling System No. 6 - Signalling link) Data channel failure detection	
Q.276	(Signalling System No. 6 - Signalling link) Service dependability	
Q.277	(Signalling System No. 6 - Signalling link) Error control	
Q.278	(Signalling System No. 6 - Signalling link) Synchronization	
Q.279	(Signalling System No. 6 - Signalling link) Drift compensation	
Q.285	(Signalling System No. 6 - Signal traffic characteristics) Signal priority categories	
Q.286	(Signalling System No. 6 - Signal traffic characteristics) Signalling channel loading and queuing delays	
Q.287	(Signalling System No. 6 - Signal traffic characteristics) Signal transfer time requirements	
Q.291	(Signalling System No. 6 - Security arrangements) General	
Q.292	(Signalling System No. 6 - Security arrangements) Reserve facilities provided	
Q.293	(Signalling System No. 6 - Security arrangements) Intervals at which security measures are to be invoked	
Q.295	(Signalling System No. 6 - Testing and maintenance) Overall tests of Signalling System No. 6	
Q.296	(Signalling System No. 6 - Testing and maintenance) Monitoring and maintenance of the common signalling channel	
Q.297	(Signalling System No. 6) Network management	
Q.300	(Signalling System No. 6) Interworking between ITU-T Signalling System No. 6 and national common channel signalling systems	
Q.310	(Signalling System R1) Definition and function of signals	
Q.311	(Signalling System R1) 2600 Hz line signalling	
Q.312	(Signalling System R1) 2600 Hz line signal sender (Transmitter)	
Q.313	(Signalling System R1) 2600 Hz line signal receiving equipment	
Q.314	(Signalling System R1) PCM line signalling	
Q.315	(Signalling System R1) PCM line signal sender (Transmitter)	
Q.316	(Signalling System R1) PCM line signal receiver	
Q.317	(Signalling System R1) Further specification clauses relative to line signalling	
Q.318	(Signalling System R1) Double seizing with both-way operation	
Q.319	(Signalling System R1) Speed of switching in international exchanges	
Q.320	(Signalling System R1) Signal code for register signalling	
Q.321	(Signalling System R1) End-of-pulsing conditions - Register arrangements concerning ST signal	
Q.322	(Signalling System R1) Multifrequency signal sender	
Q.323	(Signalling System R1) Multifrequency signal receiving equipment	
Q.324	(Signalling System R1) Analysis of address information for routing	
Q.325	(Signalling System R1) Release of registers	
Q.326	(Signalling System R1) Switching to the speech position	
Q.327	(Signalling System R1) General arrangements	
Q.328	(Signalling System R1) Routing testing of equipment (Local maintenance)	
Q.329	(Signalling System R1) Manual testing	
Q.330	(Signalling System R1) Automatic transmission and signalling testing	
Q.331	(Signalling System R1) Test equipment for checking equipment and signals	
Q.332	(Signalling System R1) Interworking	
Q.400	(Signalling System R2) Forward line signals	
Q.411	(Signalling System R2) Line signalling code	
Q.412	(Signalling System R2) Clauses for exchange line signalling equipment	
Q.414	(Signalling System R2) Signal sender	
Q.415	(Signalling System R2) Signal receiver	
Q.416	(Signalling System R2) Interruption control	
Q.421	(Signalling System R2) Digital line signalling code	

Series Q Recommendations
Switching and Signalling

Rec. No.	Description/Title	S
Q.422	(Signalling System R2) Clauses for exchange line signalling equipment	
Q.424	(Signalling System R2) Protection against the effects of faulty transmission	
Q.430	(Signalling System R2) Conversion between analogue and digital versions of system R2 line signalling	
Q.440	(Signalling System R2 - Interregister signalling) General	
Q.441	(Signalling System R2 - Interregister signalling) Signalling code	
Q.442	(Signalling System R2 - Interregister signalling) Pulse transmission of backward signals A-3, A-4, A-6 or A-15	
Q.450	(Signalling System R2 - Multifrequency signalling equipment) - General	
Q.451	(Signalling System R2 - Multifrequency signalling equipment) - Definitions	
Q.452	(Signalling System R2 - Multifrequency signalling equipment) - Requirements relating to transmission conditions	
Q.454	(Signalling System R2 - Multifrequency signalling equipment) - The sending part of the multifrequency signalling equipment	
Q.455	(Signalling System R2 - Multifrequency signalling equipment) - The receiving part of the multifrequency equipment	
Q.457	(Signalling System R2) Range of interregister signalling	
Q.458	(Signalling System R2) Reliability of interregister signalling	
Q.440-Q.458 Annex A, B & C	Annes A, B, and C to Recommendations Q.440 through Q.458	
Q.460	(Signalling System R2) Normal call set-up procedures for international working	
Q.462	(Signalling System R2) Signalling between the outgoing international R2 register and an incoming R2 register in an international exchange	
Q.463	(Signalling System R2) Signalling between the outgoing international R2 register and an incoming R2 register in a national exchange in the destination country	
Q.464	(Signalling System R2) Signalling between the outgoing international R2 register and the last incoming R2 register	
Q.465	(Signalling System R2) Particular cases	
Q.466	(Signalling System R2) Supervision and release of the call	
Q.468	(Signalling System R2) See Recommendations Q.107 and Q.107 bis	
Q.470	(Signalling System R2) At an incoming R2 register situated in a transfer exchange	
Q.471	(Signalling System R2) At the last incoming R2 register situated in the exchange to which the called subscriber is connected	
Q.472	(Signalling System R2) At the last incoming R2 register situated in a transit exchange	
Q.473	(Signalling System R2) Use of end-of-pulsing signal I-15 in international working	
Q.474	(Signalling System R2) Use of Group B signals	
Q.475	(Signalling System R2) Normal release of outgoing and incoming R2 registers	
Q.476	(Signalling System R2) Abnormal release of outgoing and incoming R2 registers	
Q.478	(Signalling System R2) Relay and regeneration of R2 interregister signals by an outgoing R2 register in a transit exchange	
Q.479	(Signalling System R2) Echo-suppressor control - Signalling requirements	
Q.480	(Signalling System R2) Miscellaneous procedures	
Q.490	(Signalling System R2) Testing and maintenance	
Q.400-Q.490 Annex A-R2	(Signalling System R2) Specifications of Signalling System R2 - Provision of a forward transfer signalling facility	
Q.500	Digital local, combined, transit and international exchanges - Introduction and field of application	
Q.511	Exchange interfaces towards other exchanges	
Q.512	Digital exchange interfaces for subscriber access	✗
Q.513	Digital exchange interfaces for operations, administrations and maintenance	
Q.521	Digital exchange functions	
Q.522	Digital exchange connections, signalling and ancillary functions	
Q.541	Digital exchange design objectives - General	
Q.542	Digital exchange design objectives - Operations and maintenance	
Q.543	Digital exchange performance design objectives	
Q.544	Digital exchange measurements	
Q.551	Transmission characteristics of digital exchanges	✗

Series Q Recommendations
Switching and Signalling

Rec. No.	Description/Title	S
Q.552	Transmission characteristics at 2-wire analogue interfaces of digital exchanges	✗
Q.553	Transmission characteristics at 4-wire analogue interfaces of digital exchanges	✗
Q.554	Transmission characteristics at digital interfaces of digital exchanges	✗
Q.601	Interworking of signalling systems - General	
Q.601 to Q.695 Annex A	Interworking of signalling systems - Lists and meanings of Fites, Bites and Spites - Representation of information contents of signals of the signalling systems	
Q.601 to Q.695 Annex B	Interworking of signalling systems - Narrative presentation of interworking	
Q.602	Interworking of signalling systems - Introduction	
Q.603	Interworking of signalling systems - Events	
Q.604	Interworking of signalling systems - Information analysis tables	
Q.605	Interworking of signalling systems - Drawing conventions	
Q.606	Interworking of signalling systems - Logic procedures	
Q.607	Interworking of signalling systems - Interworking requirements for new signalling systems	
Q.608	Interworking of signalling systems - Miscellaneous interworking aspects	
Q.611	Logic procedures for incoming Signalling System No. 4	
Q.612	Logic procedures for incoming Signalling System No. 5	
Q.613	Logic procedures for incoming Signalling System No. 6	
Q.614	Interworking of signalling systems - Logic procedures for incoming signalling system no 7 (TUP)	
Q.615	Logic procedures for incoming Signalling System R1	
Q.616	Logic procedures for incoming Signalling System R2	
Q.617	Interworking of signalling systems - Logic procedures for incoming signalling system no 7 (ISUP)	
Q.621	Logic procedures for outgoing Signalling System No. 4	
Q.622	Logic procedures for outgoing Signalling System No. 5	
Q.623	Logic procedures for outgoing Signalling System No. 6	
Q.624	Interworking of signalling systems - logic procedures for outgoing signalling system no 7 (TUP)	
Q.625	Logic procedures for outgoing Signalling System R1	
Q.626	Logic procedures for outgoing Signalling Systems R2	
Q.627	Interworking of signalling systems - logic procedures for outgoing signalling system no 7 (ISUP)	
Q.634	Logic procedures for interworking of Signalling System No. 4 to R2	
Q.642	Logic procedures for interworking of Signalling System No. 5 to No. 6	
Q.643	Logic procedures for interworking of Signalling System No. 5 to No. 7 (TUP)	
Q.644	Logic procedures for interworking of Signalling System No. 5 to R1	
Q.645	Logic procedures for interworking of Signalling System No. 5 to R2	
Q.646	Interworking of signalling systems - logic procedures for interworking of signalling system no. 5 to signalling system no. 7 (ISUP)	
Q.652	Logic procedures for interworking of Signalling System No. 6 to No. 5	
Q.653	Logic procedures for interworking of Signalling System No. 6 to No. 7 (TUP)	
Q.654	Logic procedures for interworking of Signalling System No. 6 to R1	
Q.655	Logic procedures for interworking of Signalling System No. 6 to R2	
Q.656	Interworking of signalling systems - Logic procedures for interworking of signalling system no. 6 to signalling system no. 7 (ISUP)	
Q.662	Logic procedures for interworking of Signalling System No. 7 (TUP) to No. 5	
Q.663	Logic procedures for interworking of Signalling System No. 7 (TUP) to No. 6	
Q.664	Logic procedures for interworking of Signalling System No. 7 (TUP) to No. 7 (TUP)	
Q.665	Logic procedures for interworking of Signalling System No. 7 (TUP) to R1	
Q.666	Logic procedures for interworking of Signalling System No. 7 (TUP) to R2	
Q.667	Interworking of signalling systems - Logic procedures for interworking of Signalling System No. 7 (TUP) to Signalling System No. 7 (ISUP)	
Q.671	Logic procedures for interworking of Signalling System R1 to No. 5	
Q.672	Logic procedures for interworking of Signalling System R1 to No. 6	
Q.673	Logic procedures for interworking of Signalling System R1 to No. 7 (TUP)	
Q.674	Logic procedures for interworking of Signalling System R1 to R2	
Q.675	Interworking of signalling systems - Logic procedures for interworking of signalling system R1 to signalling system no 7 (ISUP)	

Series Q Recommendations
Switching and Signalling

Rec. No.	Description/Title	S
Q.681	Logic procedures for interworking of Signalling System R2 to No. 4	
Q.682	Logic procedures for interworking of Signalling System R2 to No. 5 (TUP)	
Q.683	Logic procedures for interworking of Signalling System R2 to No. 6	
Q.684	Logic procedures for interworking of Signalling System R2 to No. 7 (TUP)	
Q.685	Logic procedures for interworking of Signalling System R2 to R1	
Q.686	Interworking of signalling systems - logic procedures for interworking of signalling system R2 to signalling system no. 7 (ISUP)	
Q.690	Interworking of signalling systems - logic procedures for interworking of signalling system no. 7 (ISUP) to No. 5	
Q.691	Interworking of signalling systems - logic procedures for interworking of Signalling System No. 7 (ISUP) to No. 6	
Q.692	Interworking of signalling systems - logic procedures for interworking of signalling system no. 7 (ISUP) to no. 7 (TUP)	
Q.694	Interworking of signalling systems - logic procedures for interworking of Signalling System No. 7 (ISUP) to R1	
Q.695	Interworking of signalling systems - logic procedures for interworking of signalling system no. 7 (ISUP) to R2	
Q.696	Interworking between the Signalling System No. 7 ISDN User Part (ISUP) and Signalling Systems No. 5, R2 and Signalling System No. 7 TUP	✗
Q.698	Interworking of signalling system No 7 ISUP, TUP and signalling system No 6 using arrow diagrams	
Q.699	Interworking DSS1 - SS NO.7 for Completion of call on No reply	
Q.699 Adden	Addendum 1 to Interworking DSS1 - SS NO.7 for Completion of call on No reply	
Q.699.1	Interworking between ISDN access and non-ISDN access over ISDN user part of signalling system No. 7: Support of VPN applications with PSS1 information flows	✗
Q.700	Introduction to CCITT Signalling System No. 7	
Q.701	Functional description of the message transfer part (MTP) of Signalling System No. 7	
Q.702	Signalling System No. 7 - Signalling data link	
Q.703	Signalling System No. 7 - Signalling link	✗
Q.704	Signalling System No. 7 - Signalling network functions and messages	✗
Q.705	Signalling System No. 7 - Signalling network structure	
Q.706	Signalling System No. 7 - Message transfer part signalling performance	
Q.707	Signalling System No. 7 - Testing and maintenance	
Q.708	Numbering of international signalling point codes	
Q.709	Signalling System No. 7 - Hypothetical signalling reference connection	
Q.710	Signalling System No. 7 - Simplified MPT version for small systems	
Q.711	Signalling System No. 7 - Functional description of the signalling connection control part	✗
Q.712	Signalling System No. 7 - Definition and function of SCCP messages	✗
Q.713	Signalling System No. 7 - SCCP formats and codes	✗
Q.714	Signalling System No 7 - Signalling connection control part procedures	✗
Q.715	Signalling connection control part user guide	✗
Q.716	Signalling System No. 7 - Signalling connection control part (SCCP) performance	
Q.721	Signalling System No. 7 - Functional description of the Signalling System No. 7 Telephone User Part (TUP)	
Q.722	General function of telephone messages and signals	
Q.723	Formats and codes	
Q.724	Specifications of Signalling System No. 7 - Telephone user part (Signalling procedures)	
Q.725	Signalling System No. 7 - Signalling performance in the telephone application	
Q.730	Signalling System No. 7 - ISDN supplementary services	
Q.731	Stage 3 description for number identification supplementary services using Signalling System No. 7	
Q.731.1	Stage 3 description for number identification supplementary services using Signalling System No. 7 - Direct dialling in (DDI)	✗
Q.731.3	Stage 3 description for number identification supplementary services using Signalling System No. 7 - Calling line identification presentation (CLIP)	
Q.731.4	Stage 3 description for number identification supplementary services using Signalling System No. 7 - Calling line identification restriction (CLIR)	

Series Q Recommendations
Switching and Signalling

Rec. No.	Description/Title	S
Q.731.5	Stage 3 description for number identification supplementary services using Signalling System No. 7 - Connected line identification presentation (COLP)	
Q.731.6	Stage 3 description for number identification supplementary services using Signalling System No. 7 - Connected line identification restriction (COLR)	
Q.731.7	Stage 3 description for number identification supplementary services using Signalling System No.7: Malicious call identification (MCID)	✗
Q.731.8	Stage 3 description for number identification supplementary services using Signalling System No. 7 - Sub-addressing (SUB)	✗
Q.732	Stage 3 description for call offering supplementary services using Signalling System No. 7	
Q.732.2	Stage 3 description for call offering supplementary services using signalling system no. 7 (Call diversion services)	✗
Q.732.5	Stage 3 description for call offering supplementary services using signalling system No. 7 - Call diversion services	✗
Q.732.7	Stage 3 description for call offering supplementary services using Signalling System No. 7 - Explicit Call Transfer	✗
Q.733	Stage 3 description for call completion supplementary services using Signalling System No. 7	
Q.733.1	Section 1 - Stage 3 description for call completion supplementary services using No.7 Signalling System - Call Waiting	✗
Q.733.2	Clauses 2 and 4 - Stage 3 description for call completion supplementary services using SS No. 7	
Q.733.3	Stage 3 description for call completion supplementary services using Signalling System No. 7 - Completion of calls to busy subscriber (CCBS)	✗
Q.733.4	Stage 3 description for call completion supplementary services using Signalling System No. 7 - Terminal portability (TP)	
Q.734	Stage 3 description for multiparty supplementary services using Signalling System No. 7	
Q.734.1	Clauses 1 and 2 - Stage 3 description for multiparty supplementary services using Signalling System No. 7	
Q.734.2	Stage 3 description for multiparty supplementary services using Signalling System No. 7 - Three-party service	
Q.735	Stage 3 description for community of interest supplementary services using Signalling System No. 7	
Q.735.1	Clauses 1 and 3 - Stage 3 description for community of interest supplementary services using SS No. 7	
Q.735.3	Completion of calls to busy subscriber (CCBS)	
Q.735.6	Stage 3 description for community of interest supplementary services using Signalling System No.7 - Global Virtual Network Service (GVNS)	✗
Q.736	Stage 3 description for charging supplementary services using Signalling System No. 7	
Q.736.1	Stage 3 description for charging supplementary services using signalling system no. 7 - Clause 1 - International Telecommunication Charge Card (ITCC)	✗
Q.736.3	Stage 3 description for charging supplementary services using signalling system no. 7	
Q.737	Stage 3 description for additional information transfer supplementary services using SS No 7 -	
Q.737.1	User-to-user signalling (UUS)	
Q.741	See Recommendation X.61	
Q.750	Overview of Signalling System No. 7 management	
Q.751.1	Network element management information model for the message transfer part (MTP)	✗
Q.751.2	Network element management information model for the Signalling Connection Control Part	✗
Q.751.3	Network element information model for MTP accounting	✗
Q.751.4	Network element information model for SCCP accounting and accounting verification	✗
Q.752	Monitoring and measurements for Signalling System No. 7 networks	✗
Q.753	Signalling System No. 7 management functions MRVT, SRVT and CVT and definition of the OMASE-user	✗
Q.754	Signalling System No. 7 management application service element (ASE) definitions	✗
Q.755	Signalling System No. 7 protocol tests	
Q.755.1	MTP protocol tester	✗
Q.755.2	Transaction capabilities test responder	✗
Q.756	Guidebook to Operations, Maintenance and Administration Part (OMAP)	✗
Q.761	Functional description of the ISDN user part of Signalling System No. 7	
Q.762	General function of messages and signals of the ISDN user part of signalling system no. 7	
Q.762 Adden	ADDENDUM 1 Signalling System No.7 – ISDN user part general functions of messages and signals	
Q.763	Formats and codes of the ISDN user part of Signalling System No. 7	

Series Q Recommendations
Switching and Signalling

Rec. No.	Description/Title	S
Q.763 Adden	ADDENDUM 1 Signalling System No. 7 – ISDN user part formats and codes	
Q.764	Signalling System No. 7 - ISDN user part signalling procedures	
Q.764 Adden	Addendum to Q.764	
Q.764 Annex H	Annex H to Recommendation Q.764	
Q.765	Signalling System No. 7 – application transport mechanism	
Q.765.1	Signalling System No. 7 - Application transport mechanism: Support of VPN applications with PSS1 information flows	✗
Q.765.4	Signalling System No. 7 – APPLICATION TRANSPORT mechanism – support of THE GENERIC ADDRESSING AND TRANSPORT PROTOCOL	
Q.765.5	Q.765.5 application transport MECHANISM – Bearer Independent Call Control (BICC)	
Q.766	Performance objectives in the integrated services digital network application	
Q.767	Application of the ISDN user part of ITU-T Signalling System No. 7 for international ISDN interconnections	✗
Q.768	Signalling interface between an international switching centre (ISC) and an ISDN satellite subnetwork	✗
Q.769.1	Signalling System No. 7 - ISDN user part enhancements for the support of number portability	
Q.771	Signalling System No. 7 – Functional description of transaction capabilities	✗
Q.772	Signalling System No. 7 – Transaction capabilities information element definitions	✗
Q.773	Signalling System No. 7 – Transaction capabilities formats and encoding	✗
Q.774	Signalling System No. 7 – Transaction capabilities procedures	✗
Q.775	Signalling System No. 7 – Guidelines for using transaction capabilities	✗
Q.780	Signalling System No. 7 test specification – General description	✗
Q.780 supp	Supplement 1 to Series Q.780 Recommendations (10/95) - Signalling System No. 7 testing and planning tools	
Q.781	Signalling System No. 7 – MTP level 2 test specification	✗
Q.781 Supp	Supplement 1 to Series Q.780 Recommendations (10/95) – Signalling System No. 7 testing and planning tools	
Q.782	Signalling System No. 7 – MTP level 3 test specification	✗
Q.783	TUP test specification	
Q.784	ISUP basic call test specification	✗
Q.784.1	ISUP basic call test specification: Validation and compatibility for ISUP'92 and Q.767 protocols	✗
Q.784.1 corr 1	Corr.1 ISUP Basic call test specification validation and compatibility for ISUP '92 and Q.767 Protocols	
Q.784.2	ISUP basic call test specification: Abstract test suite for ISUP'92 basic call control procedures	✗
Q.784 Annex A	TTCN version of Recommendation Q.784	
Q.785	ISUP protocol test specification for supplementary services	✗
Q.785.2	ISUP'97 supplementary services - Test suite structure and test purposes (TSS & TP)	
Q.786	Signalling System No. 7 – SCCP test specification	
Q.787	Transaction capabilities (TC) test specification	✗
Q.788	User-network-interface to user-network-interface compatibility test specifications for ISDN, non-ISDN and undetermined accesses interworking over international ISUP	✗
Q.788 supp	Suppl. 1 (10/95) Signalling System No. 7 testing and planning tools	
Q.791	Specifications of Signalling System No. 7 – Monitoring and measurements for Signalling System No. 7 networks	x93
Q.795	Specifications of Signalling System No. 7 – Operations, Maintenance and Administration Part (OMAP)	x93
Q.811	Lower layer protocol profiles for the Q3 interface	✗
Q.812	Upper layer protocol profiles for the Q3 interface	✗
Q.812 Amend	Amendment 1 (03/99) to Recommendation Q.812 - Upper layer protocol profiles for the Q3 and X interfaces - Amendment 1: Additional X interface protocols for the service management layer (SML)	
Q.812 Appndx	Appendix I to Recommendation Q.812 (03/99) – Guidance on using allomorphic management	
Q.813	Security Transformations Application Service Element for Remote Operations Service Element (STASE-ROSE)	✗
Q.821	Stage 2 and Stage 3 description for the Q3 interface – Alarm surveillance	✗
Q.821 corr	Corr 1 to Q.821	
Q.822	Stage 1, Stage 2 and Stage 3 description for the Q3 interface – Performance management	✗
Q.823	Stage 2 and Stage 3 functional specifications for traffic management	✗
Q.823.1	Management Conformance Statement Proformas	✗
Q.824	Stage 2 and Stage 3 description for the Q3 interface - Customer administration	

Series Q Recommendations
Switching and Signalling

Rec. No.	Description/Title	S
Q.824.0	Stages 2 and 3 description for the Q3 interface – Customer administration – Common information	✗
Q.824.1	Stages 2 and 3 description for the Q3 interface – Customer administration – Integrated services digital network (ISDN) basic and primary rate access	✗
Q.824.2	Stages 2 and 3 description for the Q3 interface – Customer administration – Integrated services digital network (ISDN) supplementary services	✗
Q.824.3	Stages 2 and 3 description for the Q3 interface – Customer administration – Integrated Services Digital Network (ISDN), optional user facilities	✗
Q.824.4	Stages 2 and 3 description for the Q3 interface – Customer administration – Integrated services digital network (ISDN) – Teleservices	✗
Q.824.5	Configuration management of V5 interface environments and associated customer profiles	✗
Q.824.6	Broadband Switch Management	✗
Q.825	Specification of TMN applications at the Q3 interface: Call detail recording	✗
Q.831	Fault and performance management of V5 interface environments and associated customer profiles	✗
Q.832.1	VB5.1 Management	✗
Q.832.2	VB5.2 Management	✗
Q.835	Line and line circuit test management of ISDN and analogue customer accesses	✗
Q.850	Usage of cause and location in the digital subscriber signalling system no 1 and the signalling system no 7 ISDN user part	
Q.850 Adden	Addendum 1 to Usage of cause and location in the digital subscriber signalling system No.1 and the signalling system No. 7 ISDN user part.	
Q.860	Integrated Services Digital Network (ISDN) and Broadband Integrated Services DIGITAL NETWORK (B-ISDN) – Generic ADDRESSING AND TRANSPORT (GAT) Protocol	
Q.920	Digital Subscriber Signalling System No. 1 (DSS1) – ISDN user-network interface data link layer – General aspects	
Q.921	ISDN user-network interface-data link layer specification	
Q.921 Amend	Amendment 1 ISDN User-network interface – data link layer specification	
Q.921 bis	Abstract test suite for LAPD conformance testing	
Q.921 bis Encl.1.1	Encl. 1.1 (03/93) – Abstract test suite – Part 1	
Q.922	ISDN data link layer specification for frame mode bearer services	✗
Q.923	Specification of a synchronization and coordination function for the provision of the OSI connection-mode network service in an ISDN environment	✗
Q.930	Digital Subscriber Signalling System No. 1 (DSS 1) – ISDN user-network interface layer 3 – General aspects	
Q.931	Digital Subscriber Signalling System No. 1 (DSS 1) – ISDN user-network interface layer 3 specification for basic call control	
Q.931 bis	PICS and abstract test suite for ISDN DSS 1 layer 3 – Circuit and circuit mode, basic call control conformance testing	
Q.932	Digital Subscriber Signalling System No. 1 (DSS 1) – Generic procedures for the control of ISDN supplementary services	
Q.932 Amend	AMENDMENT 1 digital subscriber signalling system No. 1 – generic procedures for the control of ISDN supplementary services	
Q.933	Digital Subscriber Signalling System No. 1 (DSS 1) – Signalling specification for frame mode, Basic call control	✗
Q.933 bis	PICS and abstract test suite for frame mode basic call control conformance testing of PVCs – Section 1: user and network sides of user-network interface	✗
Q.939	Digital Subscriber Signalling System No. 1 (DSS 1) - Typical DSS 1 service indicator codings for ISDN telecommunications services	
Q.940	ISDN user-network interface protocol for management - General aspects	
Q.941	Digital Subscriber Signalling System No. 1 (DSS 1) - ISDN user-network interface protocol profile for management	
Q.950	Digital Subscriber Signalling System No. 1 (DSS 1) - Supplementary services protocols, structure and general principles	✗
Q.951.1	Stage 3 description for number identification supplementary services using DSS 1 - Direct-dialling-in (DDI)	✗
Q.951.2	Stage 3 description for number identification supplementary services using DSS 1 - Multiple subscriber number (MSN)	✗

Series Q Recommendations
Switching and Signalling

Rec. No.	Description/Title	S
Q.951.3	Stage 3 description for number identification supplementary services using DSS 1 - Calling line identification presentation (CLIP)	
Q.951.4	Stage 3 description for number identification supplementary services using DSS 1 - Calling line identification restriction	
Q.951.5	Stage 3 description for number identification supplementary services using DSS 1 - Connected line identification presentation	
Q.951.6	Stage 3 description for number identification supplementary services using DSS 1 - Connected line identification restriction	
Q.951.7	Stage 3 description for number identification supplementary services using DSS 1: Malicious Call Identification (MCID)	
Q.951.8	Stage 3 description for number identification supplementary services using DSS 1	X
Q.952	Stage 3 service description for call offering supplementary services using DSS 1 - Diversion supplementary services	
Q.952.7	Stage 3 description for call offering supplementary services using DSS 1: Explicit Call Transfer (ECT)	X
Q.953.1	Stage 3 description for call completion supplementary services using DSS 1 - Call waiting	X
Q.953.2	Stage 3 description for call completion supplementary services using DSS 1; Clause 2 - Call hold	
Q.953.3	Stage 3 description for call completion supplementary services using DSS 1: Completion of Calls to Busy Subscribers (CCBS)	X
Q.953.4	Integrated services digital network (ISDN) – Stage 3 description for call completion supplementary services using DSS 1: Clause 4 – Terminal Portability (TP)	X
Q.953.5	Stage 3 description for call completion supplementary services using DSS1: Call Completion on No Reply (CCNR)	X
Q.954.1	Stage 3 description for multiparty supplementary services using DSS 1; Conference calling; Three-party service	
Q.954.1 note	Notice to Recommendation Q.954.1 (03/93) - Stage 3 description for multiparty supplementary services using DSS 1 - Clause 1: Conference calling, Clause 2: Three-party service	
Q.954.2	Intergrated services digital network (ISDN) - Stage 3 description for multiparty supplementary services using DSS 1 - Three party (3PTY)	X
Q.955.1	Stage 3 description for community of interest supplementary services using DSS 1 - Closed user group	X
Q.955.3	Stage 3 description for community of interest supplementary services using DSS 1; Multi-level precedence and preemption (MLPP)	
Q.956.2	Integrated services digital network (ISDN) - Stage 3 service description for charging supplementary services using DSS 1: Clause 2 - Advice of charge (AOC)	X
Q.956.3	Integrated services digital network (ISDN) - Stage 3 description for charging supplementary services using DSS 1: Clause 3 - Reverse charging	X
Q.957.1	Stage 3 description for additional information transfer supplementary services using DSS 1; Clause 1 - User-to-user signalling (UUS)	X
Q.1000	Structure of the Q.1000-series Recommendations for public land mobile networks	
Q.1001	General aspects of public land mobile networks	
Q.1002	Network functions	
Q.1003	Location registration procedures	
Q.1004	Location register restoration procedures	
Q.1005	Handover procedures	
Q.1031	General signalling requirements on interworking between the ISDN or PSTN and the PLMN	
Q.1032	Signalling requirements relating to routing of calls to mobile subscribers	
Q.1061	General aspects and principles relating to digital PLMN access signalling reference points	
Q.1062	Digital PLMN access signalling reference configuration	
Q.1063	Digital PLMN channel structures and access capabilities at the radio interface (Um reference point)	
Q.1100	Interworking with Standard A INMARSAT system - Structure of the Recommendations on the INMARSAT mobile satellite systems	
Q.1101	General requirements for the interworking of the terrestrial telephone network and INMARSAT Standard A system	
Q.1102	Interworking between Signalling System R2 and INMARSAT Standard A system	
Q.1103	Interworking between Signalling System No. 5 and INMARSAT Standard A system	
Q.1111	Interfaces between the INMARSAT standard B system and the international public switched telephone network/ISDN	

Series Q Recommendations
Switching and Signalling

Rec. No.	Description/Title	S
Q.1112	Procedures for interworking between INMARSAT standard-B system and the international public switched telephone network/ISDN	
Q.1151	Interfaces for interworking between the INMARSAT aeronautical mobile-satellite system and the international public switched telephone network/ISDN	
Q.1152	Procedures for interworking between INMARSAT aeronautical mobile satellite system and the international public switched telephone network/ISDN	
Q.1200	Q-Series intelligent network Recommendation structure	
Q.1201/I.312	Principles of intelligent network architecture	
Q.1202/I.328	Intelligent Network - Service plane architecture	
Q.1203/I.329	Intelligent Network - Global functional plane architecture	
Q.1204	Intelligent network distributed functional plane architecture	✗
Q.1205	Intelligent network physical plane architecture	✗
Q.1208	General aspects of the intelligent network application protocol	✗
Q.1210	Series intelligent network recommendation structure	✗
Q.1211	Introduction to intelligent network capability set 1	✗
Q.1213	Global functional plane for intelligent network CS-1	✗
Q.1214	Distributed functional plane for intelligent network CS-1	
Q.1215	Physical plane for intelligent network CS-1	✗
Q.1218	Interface Recommendation for intelligent network CS-1	✗
Q.1218 adden	Addendum 1 to Recommendation Q.1218 (09/97) - Interface Recommendation for Intelligent Network CS-1; Addendum 1: Definition for two new contexts in the SDF data model	✗
Q.1219	Intelligent network user's guide for capability set 1	✗
Q.1219 Supp	Supplement 1 to Recommendation Q.1219 (09/97) - Intelligent Network user's guide: Supplement for IN CS-1	✗
Q.1220	Series Intelligent Network Capability Set 2 Recommendation structure	✗
Q.1221	Introduction to Intelligent Network Capability Set 2	✗
Q.1222	Service plane for Intelligent Network Capability Set 2	✗
Q.1223	Global functional plane for Intelligent Network Capability Set 2	✗
Q.1224	Distributed functional plane for intelligent network Capability Set 2: Part 1, 2, 3	
Q.1225	Physical plane for Intelligent Network Capability Set 2	✗
Q.1228	Interface Recommendation for intelligent network Capability Set 2	✗
Q.1229	Intelligent Network user's guide for capability set 2	
Q.1237	Extensions to Intelligent Network Capability Set 3 in Support of B-ISDN	
Q.1238.1	Interface Recommendation for intelligent network capability set 3: Common aspects	
Q.1238.2	Interface Recommendation for intelligent network capability set 3: SCF-SSF Interface	
Q.1238.3	Interface Recommendation for Intelligent network Capability Set 3: SCF-SRF Interface	
Q.1238.4	Interface Recommendation for intelligent network Capability Set 3: SCF-SDF Interface	
Q.1238.5	Interface Recommendation for intelligent network Capability Set 3: SDF-SDF Interface	
Q.1238.6	Interface Recommendation for intelligent network Capability Set 3: SCF-SCF Interface	
Q.1238.7	Interface Recommendation for intelligent network Capability Set 3: SCF-CUSF Interface	
Q.1290	Glossary of terms used in the definition of intelligent networks	✗
Q.1300	Telecommunication applications for switches and computers (TASC) - General overview	✗
Q.1301	Telecommunication applications for switches and computers (TASC) - TASC Architecture	✗
Q.1302	Telecommunication applications for switches and computers (TASC) - TASC functional services	✗
Q.1303	Telecommunication applications for switches and computers (TASC) - TASC management: Architecture, methodology and requirements	✗
Q.1400	Architecture framework for the development of signalling and OA&M protocols using OSI concepts	✗
Q.1400 Add. 1	Architecture framework for the development of signalling and OAM protocols using OSI concepts	
Q.1521	Requirements on underlying networks and signalling protocols to support UPT	
Q.1531	UPT security requirements for Service Set 1	
Q.1541	UPT stage 2 for Service Set 1 on IN CS-1 - Procedures for universal personal telecommunication: Functional modelling and information flows	
Q.1542	UPT stage 2 for service set 1 ON CS2 - Procedures for universal personal telecommunication functional modelling and information flows	
Q.1551	Application of Intelligent Network Application Protocols (INAP) CS-1 for UPT service set 1	✗
Q.1600	Signalling System No. 7 - Interaction between ISUP and INAP	✗

Series Q Recommendations
Switching and Signalling

Rec. No.	Description/Title	S
Q.1701	Framework for IMT-2000 networks	✗
Q.1711	Network functional model for IMT-2000	✗
Q.1721	Information flows for IMT-2000 capability set 1	
Q.1731	Radio-technology Independent Requirements for IMT-2000 Layer 2 Radio Interface	
Q.1751	Internetwork signalling requirements for IMT-2000 capability set 1	
Q.1831.1	SSF Management Information Model.	
Q.1901	Bearer independent call control protocol	
Q.2010	Broadband integrated services digital network overview - signalling capability set, release 1	✗
Q.2100	B-ISDN signalling ATM adaptation layer (SAAL) overview description	✗
Q.2110	B-ISDN ATM adaptation layer - service specified connection oriented protocol (SSCOP)	✗
Q.2119	B-ISDN ATM adaptation layer - Convergence function for SSCOP above the frame relay core service	✗
Q.2120	B-ISDN meta-signalling protocol	
Q.2130	B-ISDN signalling ATM adaptation layer - service specific coordination function for support of signalling at the user network interface (SSFC At UNI)	✗
Q.2140	B-ISDN ATM adaptation layer - service specific coordination function for signalling at the network node interface (SSCF at NNI)	✗
Q.2144	B-ISDN Signalling ATM adaptation layer (SAAL) - Layer management for the SAAL at the network node interface (NNI)	✗
Q.2210	Message transfer part level 3 functions and messages using the services of ITU-T Recommendation Q.2140	✗
Q.2610	Broadband integrated services digital network (B-ISDN) - usage of cause and location in B-ISDN user part and DSS 2	✗
Q.2650	Broadband-ISDN, interworking between Signalling System No. 7 broadband ISDN user part (B-ISUP) and Digital Subscriber Signalling System No. 2 (DSS 2)	✗
Q.2660	Broadband integrated services digital network (B-ISDN) - interworking between Signalling System No. 7 - Broadband ISDN user part (B-ISUP) and narrow-band ISDN user part (N-ISUP)	✗
Q.2721.1	B-ISDN user part - Overview of the B-ISDN network node interface signalling capability set 2, step 1	✗
Q.2722.1	AMENDMENT 1 B-ISDN User Part – Network Node Interface Specification for Point-to-Multipoint call/Connection Control	
Q.2723.1	B-ISDN User Part - Support of additional traffic parameters for Sustainable Cell Rate and Quality of Service	✗ x96
Q.2723.2	Extensions to the B-ISDN User Part – Support of ATM transfer capability in the broadband bearer capability parameter	x96
Q.2723.3	Extensions to the B-ISDN User Part – Signalling capabilities to support traffic parameter for the Available Bit Rate (ABR) ATM Transfer Capability	x96
Q.2723.3 corr	Corr.1 Extensions to the B-ISDN user part – signalling capabilities to support traffic parameters for the available bit rate (abr) ATM transfer capability	
Q.2723.4	Extensions to the B-ISDN User Part – Signalling capabilities to support traffic parameters for the ATM Block Transfer (ABT) ATM transfer capability	x96
Q.2723.5	B-ISDN User Part –Support of cell delay variation tolerance indication	x96
Q.2723.6	Extensions to the Signalling System No. 7 B-ISDN User Part – Signalling capabilities to support the indication of the Statistical Bit-Rate configuration 2 (SBR 2) and 3 (SBR 3) ATM transfer capabilities	x96
Q.2724.1	B-ISDN User Part - Look-ahead without state change for the Network Node Interface (NNI)	✗ x96
Q.2725.1	B-ISDN User Part - Support of negotiation during connection setup	✗ x96
Q.2725.2	B-ISDN User Part - Modification procedures	✗ x96
Q.2725.3	Extensions to the B-ISDN User Part – Modification procedures for sustainable cell rate parameters	x96
Q.2725.4	Extensions to the Signalling System No. 7 B-ISDN User Part – Modification procedures with negotiation	x96
Q.2726.1	B-ISDN user part - ATM end system address	✗ x96
Q.2726.2	B-ISDN user part - Call priority	✗
Q.2726.3	B-ISDN user part - Network generated session identifier	✗
Q.2726.4	EXTENSIONS TO THE B-ISDN USER PART – APPLICATION GENERATED IDENTIFIERS	
Q.2727	B-ISDN User Part - support of frame relay	✗ x96
Q.2727.1	B-ISDN User Part –Overview of the B-ISDN Network Node Interface Signalling Capability Set 2, Step 1	x96
Q.2730	Broadband integrated services digital network (B-ISDN) signalling system no. 7 B-ISDN user part (B-ISUP) - supplementary services	✗

Series Q Recommendations
Switching and Signalling

Rec. No.	Description/Title	S
Q.2731	Broadband Integrated Services Digital Network (B-ISDN) - Digital subscriber signalling system no. 2 (DSS 2) - User-network interface (UNI) - Layer 3 specification for basic call/connection control	✗
Q.2731 Amend.1	Digital Subscriber Signalling System No. 2 - User-network interface (UNI) layer 3 specification for basic call/connection control Amendment 1	✗
Q.2731 Amend.2	AMENDMENT 2 CORRIGENDUM 1 digital subscriber signalling system no. 2 - user-network interface (UNI) layer 3 specification for basic call/connection control	
Q.2932.1	Digital Subscriber Signalling System No. 2 - Generic functional protocol: Core functions	✗
Q.2933	Digital Subscriber Signalling System No. 2 (DSS 2) - Signalling specification for Frame Relay service	✗
Q.2735	Stage 3 description for community of interest supplementary services for B-ISDN using SS No. 7	
Q.2735.1	Stage 3 description for community of interest supplementary services for B-ISDN using SS No. 7: Closed User Group (CUG)	✗
Q.2951	(Clauses 1, 2, 3, 4, 5, 6 and 8) - Stage 3 description for number identification supplementary services using B-ISDN Digital Subscriber Signalling System No. 2 (DSS 2) - Basic call	✗
Q.2751.1	Extension of Q.751.1 for SAAL signalling links	✗
Q.2761	Broadband integrated services digital network (B-ISDN) - Functional description of the B-ISDN user part (B-ISUP) of signalling system No. 7	✗
Q.2762	Broadband integrated services digital network (B-ISDN) - General functions of messages and signals of the B-ISDN user part (B-ISUP) of Signalling System No. 7	✗
Q.2763	Broadband Integrated Services Digital Network (B-ISDN) - Signaling System No. 7 B-ISDN user part (B-ISUP) - formats and codes	✗
Q.2764	Broadband Integrated Services Digital Network (B-ISDN) - Signalling system no. 7 B-ISDN user part (B-ISUP) - Basic call procedures	✗
Q.2766.1	Switched virtual path capability	✗
Q.2767.1	Soft PVC capability	
Q.2769.1	SUPPORT OF NUMBER PORTABILITY INFORMATION ACROSS B-ISUP	
Q.2931	Broadband Integrated Services Digital Network (B-ISDN) - Digital Subscriber Signalling System No. 2 (DSS 2) - User-Network Interface (UNI) - Layer 3 specification for basic call/connection control	✗
Q.2931 ammend 1	Amendment 1 to Recommendation Q.2931 (06/97) - Digital Subscriber Signalling System No. 2 - User-network interface (UNI) layer 3 specification for basic call/connection control Amendment 1	✗
Q.2931 ammend 2	Amendment 2 (03/99) to Recommendation Q.2931 - Digital subscriber signalling system No. 2 - User-network interface (UNI) layer 3 specification for basic call/connection control	✗
Q.2931 ammend 3	Amendment 3 to Recommendation Q.2931 (03/99) - User-network interface (UNI) layer 3 specification for basic call/connection control	✗
Q.2932.1	Digital Subscriber Signalling System No. 2 - Generic functional protocol: Core functions	✗
Q.2933	Digital Subscriber Signalling System No. 2 - (DSS 2) - Signalling specification for Frame Relay service	✗
Q.2934	Digital subscriber Signalling System No. 2 - Switched virtual path capability	✗
Q.2939.1	Digital Subscriber Signalling System No. 2 - Application of DSS 2 service-related information elements by equipment supporting B-ISDN services	✗
Q.2941.1	Digital Subscriber Signalling System No. 2 - Generic identifier transport	✗
Q.2941.3	Broadband integrated services digital network (B-ISDN) digital subscriber signalling system no. 2 (dss 2): GENERIC IDENTIFIER TRANSPORT EXTENSION for support of Bearer Independent Call Control	
Q.2951	Stage 3 description for number identification supplementary services using B-ISDN Digital Subscriber Signalling System No. 2 (DSS 2) - Basic call	✗
Q.2951.1	Direct-Dialling-In (DDI)	✗
Q.2951.1 corr	Corrigendum 1 (05/98) Recommendation Q.2951 - Stage 3 description for number identification supplementary services using B-ISDN Digital Subscriber Signalling System No.2 (DSS 2) - Basic call Corrigendum 1	
Q.2951.2	Multiple Subscriber Number (MSN)	
Q.2951.3	Calling Line Identification Presentation (CLIP)	
Q.2951.4	Calling Line Identification Restriction (CLIR)	
Q.2951.5	Connected Line Identification Presentation (COLP)	
Q.2951.6	Connected Line Identification Restriction (COLR)	
Q.2951.8	Sub-addressing (SUB)	
Q.2955.1	Stage 3 description for community of interest supplementary services using B-ISDN Digital Subscriber Signalling System No. 2 (DSS 2): Closed User Group (CUG)	✗

Series Q Recommendations
Switching and Signalling

Rec. No.	Description/Title	S
Q.2957.1	Stage 3 description for additional information transfer supplementary services using B-ISDN digital subscriber signalling system no. 2 (DSS 2) - Basic call; Clause 1 - User-to-user signalling (UUS)	✗
Q.2959	Digital subscriber signalling system No.2 - Call priority	✗
Q.2961	Broadband integrated services digital network (B-ISDN) - Digital subscriber signalling system No. 2 (DSS 2) - additional traffic parameters	✗
Q.2961.2	Digital subscriber signalling system No. 2 - Additional traffic parameters: Support of ATM transfer capability in the broadband bearer capability information element	✗
Q.2961.2 corr	Corrigendum 1 (03/99) to Recommendation Q.2961.2 - Digital subscriber signalling system No. 2 - Additional traffic parameters: Support of ATM transfer capability in the broadband bearer capability information element	✗
Q.2961.3	Digital Subscriber Signalling System No. 2 - Additional traffic parameters: Signalling capabilities to support traffic parameters for the available bit rate (ABR) ATM transfer capability	✗
Q.2961.4	Digital Subscriber Signalling System No. 2 - Additional traffic parameters: Signalling capabilities to support traffic parameters for the ATM Block Transfer (ABT) ATM transfer capability	✗
Q.2961.5	Digital subscriber signalling system No. 2 - Additional traffic parameters: Additional traffic parameters for cell delay variation tolerance indication	✗
Q.2961.6	Digital Subscriber Signalling System No. 2 - Additional traffic parameters: Additional signalling procedures for the support of the SBR2 and SBR3 ATM transfer capabilities	✗
Q.2962	Digital subscriber signalling system No. 2 - Connection characteristics negotiation during call/connection establishment phase	✗
Q.2963.1	Peak cell rate modification by the connection owner	✗
Q.2963.2	Modification procedures for sustainable cell rate parameters	✗
Q.2963.3	ATM traffic descriptor modification with negotiation by the connection owner	✗
Q.2964.1	Digital Subscriber Signalling No. 2: Basic Look-Ahead	✗
Q.2965.1	AMENDMENT 1 Digital Subscriber Signalling System No. 2 - Support of Quality of Service classes	✗
Q.2971	B-ISDN - DSS 2 - User-Network interface layer 3 specification	
Q.2971 corr	Corrigendum 1 (12/99) to Recommendation Q.2971 - Digital subscriber signalling system No. 2 (DSS2) - User-Network Interface Layer 3 specification for point-to-multipoint Call/Connection Control	
Q.2982	Broadband integrated services digital network (B-ISDN) Digital subscriber signalling system No. 2 (DSS 2) Q.2931-based separated call control protocol	
Q.2983	Broadband integrated services digital network (B-ISDN) Digital subscriber signalling system No. 2 (DSS 2): Bearer control protocol	

Series R Recommendations
Telegraph transmission

Rec. No.	Description/Title	S
R.2	Element error rate	
R.4	Methods for the separate measurements of the degrees of various types of telegraph distortion	
R.5	Observation conditions recommended for routine distortion measurements on international telegraph circuits	
R.9	How the laws governing distribution of distortion	
R.11	Calculation of the degree of distortion of a telegraphic circuit in terms of the degrees of distortion of the component links	
R.20	Telegraph modem for subscriber lines	
R.21	9600 bit/s modem standardized for use in the telegraph TDM system	
R.22	Data over voice 19 200 bit/s modem standardized for use on telephone network subscriber lines	
R.30	Transmission characteristic for international VFT links	
R.31	Standardization of AMVFT systems for a modulation rate of 50 bauds	
R.35	Standardization of FMVFT systems for a modulation rate of 50 bauds	
R.35 bis	50-baud wideband VFT systems	
R.36	Coexistence of 50-baud/120-Hz channels, 100-baud/240-Hz channels, 200-baud/360-Hz or 480-Hz channels on the same voice-frequency telegraph system	
R.37	Standardization of FMVFT systems for a modulation rate of 100 bauds	
R.38 A	Standardization of FMVFT system for a modulation rate of 200 bauds with channels spaced at 480 Hz	

Series R Recommendations

Telegraph transmission

Rec. No.	Description/Title	S
R.38 B	Standardization of FMVFT systems for a modulation rate of 200 bauds with channels spaced at 360 Hz usable on long intercontinental bearer circuits generally used with a 3 Hz spacing	
R.39	Voice-frequency telegraphy on radio circuits	
R.40	Coexistence in the same cable of telephony and super-telephone telegraphy	
R.43	Simultaneous communication by telephone and telegraph on a telephone-type circuit	
R.44	6-unit synchronous time-division 2-3-channel multiplex telegraph system for use over FMVFT channels spaced at 120 Hz for connection to standardized teleprinter networks	
R.49	Interband telegraphy over open-wire 3-channel carrier systems	
R.50	Tolerable limits for the degree of isochronous distortion of code-independent 50-baud telegraph circuits	
R.51	Standardized text for distortion testing of the code-independent elements of a complete circuit	
R.51 bis	Standardized text for testing the elements of a complete circuit	
R.52	Standardization of international texts for the measurement of the margin of start-stop equipment	
R.53	Permissible limits for the degree of distortion on an international 50-baud/120-Hz VFT channel (Frequency and amplitude modulation)	
R.54	Conventional degree of distortion tolerable for standardized start-stop 50-baud systems	
R.55	Conventional degree of distortion	
R.56	Telegraph distortion limits to be quoted in Recommendations or equipment and transmission plans	
R.57	Standard limits of transmission quality for planning code-independent international point-to-point telegraph communications and switched networks using 50-baud start-stop equipment	
R.58	Standard limits of transmission quality for the gentex and telex networks	
R.58 bis	Limits on signal transfer delay for telegraph, telex and gentex networks	
R.59	Interface requirements for 50-baud start-stop telegraph transmission in the maritime mobile satellite service	
R.60	Conditions to be fulfilled by regenerative repeaters for start-stop signals of International Telegraph Alphabet No. 2	
R.62	Siting of regenerative repeaters in international telex circuits	
R.70	Designation of international telegraph circuits	
R.70 bis	Numbering of international VFT channels	
R.71	Organization of the maintenance of international telegraph circuits	
R.72	Periodicity of maintenance measurements to be carried out on the channels of international VFT systems	
R.73	Maintenance measurements to be carried out on VFT systems	
R.74	Choice of type of telegraph distortion-measuring equipment	
R.75	Maintenance measurements on code-independent international sections of international telegraph circuits	
R.75 bis	Maintenance measurements of character error rate on international sections of international telegraph circuits	
R.76	Reserve channels for maintenance measurements on channels of international VFT systems	
R.77	Use of bearer circuits for voice-frequency telegraphy	
R.78	Pilot channel for AMVFT systems	
R.79	Automatic tests of transmission quality on telegraph circuits between switching centres	
R.80	Causes of disturbances to signals in VFT channels and their effect on telegraph distortion	
R.81	Maximum acceptable limit for the duration of interruption of telegraph channels arising from failure of the normal power supplies	
R.82	Appearance of false calling and clearing signals in circuits operated by switched teleprinter services	
R.83	Changes of level and interruptions in VFT channels	
R.90	Organization for locating and clearing faults in international telegraph switched networks	
R.91	General maintenance aspects for the maritime satellite telex service	
R.100	Transmission characteristics of international TDM links	
R.101	Code and speed dependent TDM system for anisochronous telegraph and data transmission using bit interleaving	
R.102	4800 bit/s code and speed dependent and hybrid TDM systems for anisochronous telegraph and data transmission using bit interleaving	
R.103	Code and speed-dependent TDM 600 bit/s system for use in point-to-point or branch-line muldex configurations	
R.105	Duplex muldex concentrator, connecting a group of gentex and telex subscribers to a telegraph exchange by assigning virtual channels to time slots of a bit-interleaved TDM system	
R.106	Muldex unit for telegraph and low speed data transmission using TDM bit interleaving with an aggregate bit rate higher than 4800 bit/s	

Series R Recommendations

Telegraph transmission

Rec. No.	Description/Title	S
R.111	Code and speed independent TDM system for anisochronous telegraph and data transmission	
R.112	TDM hybrid system for anisochronous telegraph and data transmission using bit interleaving	
R.113	Combined muldex for telegraphy and synchronous data transmission	
R.114	Numbering of international TDM channels	
R.115	Maintenance loops for TDM-systems	
R.116	Maintenance tests to be carried out on international TDM systems	
R.117	End-to-end performance for telegraph, telex and gentex connections involving regenerative equipment	
R.118	Performance and availability monitoring in regenerative TDM	
R.120	Tolerable limits for the degree of isochronous distortion of code-independent telegraph circuits operating at modulation rates of 75, 100 and 200 bauds	
R.121	Standard limits of transmission quality for start-stop user classes of service 1 and 2 on anisochronous data networks	
R.122	Summary of transmission plans for rates up to 300 bauds	
R.140	Definitions of essential technical terms in the field of telegraph transmission	
R.150	Automatic protection switching of dual diversity bearers	

Series S Recommendations

Telegraph services terminal equipment

Rec. No.	Description/Title	S
S.1	International telegraph alphabet no 2	
S.2	Coding scheme using International Telegraph Alphabet No. 2 (ITA2) to allow the transmission of capital and small letters	
S.3	Transmission characteristics of the local end with its termination (ITA2)	
S.4	Special use of certain characters of the international telegraph alphabet No. 2	
S.5	Standardization of page-printing start-stop equipment and cooperation between page-printing and tape-printing start-stop equipment (ITA2)	
S.6	Characteristics of answer back units (ITA2)	
S.7	Control of teleprinter motors	
S.8	Intercontinental standardization of the modulation rate of start-stop apparatus and of the use of combination No. 4 in figure-shift	
S.9	Switching equipment of start-stop apparatus	
S.10	Transmission at reduced character transfer rate over a standardized 50-baud telegraph channel	
S.11	Use of start-stop reperforating equipment for perforated tape retransmission	
S.12	Conditions that must be satisfied by synchronous systems operating in connection with standard 50-baud teleprinter circuits	
S.13	Use of radio circuits of 7-unit synchronous systems giving error correction by automatic repetition	
S.14	Suppression of unwanted reception in radiotelegraph multi-destination teleprinter systems	
S.15	Use of the telex network for data transmission at 50 bauds	
S.16	Connection to the telex network of an automatic terminal using a V.24 DCE/DTE interface	
S.17	Answer-back unit simulators	
S.18	Conversion between International Telegraph Alphabet No. 2 and International Alphabet No. 5	
S.19	Calling and answering in the telex network with automatic terminal equipment	
S.20	Automatic clearing procedure for a telex terminal	
S.21	Use of display screens in telex machines	
S.22	"Conversation impossible" and or pre-recorded message in response to J/BELL signals from a telex terminal	
S.23	Automatic request of the answer-back of the terminal of the calling party, by the telex terminal of the called party or by the international network	
S.30	Standardization of basic model page-printing machine using International Alphabet No. 5	
S.31	Transmission characteristics for start-stop data terminal equipment using International Alphabet No. 5	
S.32	Answer-back units for 200- and 300-baud start-stop machines in accordance with Recommendation S.30	
S.33	Alphabets and presentation characteristics for the intex service	
S.34	Intex terminals – requirements to effect interworking with the international telex service	

Series S Recommendations
Telegraph services terminal equipment

Rec. No.	Description/Title	S
S.35	Answer-back coding for the intex service	
S.36	INTEX and similar services – Terminal requirements to effect interworking between terminals operating at different speeds	✗
S.40	Definitions of essential technical terms relating to apparatus for alphabetic telegraphy	

Series T Recommendations
Terminal characteristics and higher layer protocols for telematic services, document architecture

Rec. No.	Description/Title	S
T.0	Classification of facsimile apparatus for document transmission over the public networks	
T.1	Standardization of phototelegraph apparatus	
T.2	Standardization of Group 1 facsimile apparatus for document transmission	x93-6
T.3	Standardization of Group 2 facsimile apparatus for document transmission	x93-6
T.4	Standardization of Group 3 facsimile apparatus for document transmission	✗
T.4 Amendment 1	Standardization to Group 3 facsimile apparatus for document transmission	
T.6	Facsimile coding schemes and coding control functions for Group 4 facsimile apparatus	
T.10	Document facsimile transmissions over leased telephone-type circuits	
T.10bis	Document facsimile transmissions in the general switched telephone network	
T.11	Phototelegraph transmissions on telephone-type circuit	x96
T.12	Range of phototelegraph transmissions on a telephone-type circuit	x96
T.15	Phototelegraph transmission over combined radio and metallic circuits	x96
T.20	Standardized test chart for facsimile transmission	x93-6
T.20	Standardized test charts for document facsimile transmissions	x93-6
T.22	Standardized test charts for document facsimile transmissions	
T.23	Standardized colour test chart for document facsimile transmissions	✗
T.24	Standardized digitized image set	
T.30	Procedures for document facsimile transmission in the general switched telephone network	✗
T.30 Amend.1	Procedures for document facsimile transmission in the general switched telephone network	
T.31	Asynchronous facsimile DCE control – Service class 1	✗
T.31 Amend.1	Asynchronous facsimile DCE control – Service class 1 – Annex B – Procedure for service class 1 support of V.34 Modems	✗
T.32	Asynchronous facsimile DCE control – Service class 2	✗
T.32 Amend.1	Asynchronous facsimile DCE control – Service class 2	✗
T.32 Corr.1	Corrigendum 1 (10/97) to Recommendation T.32	
T.33	Facsimile routing utilizing the Subaddress	✗
T.35	Procedure for the allocation of CCITT defined codes for non-standard facilities	✗
T.36	Security capabilities for use with Group 3 facsimile terminals	✗
T.36 Amend	Amendment 1 (04/99) to Recommendation T.36 – Security capabilities for use with Group 3 facsimile terminals	✗
T.37	Procedures for the transfer of facsimile data via store-and-forward on the Internet	✗
T.38	Procedures for real-time Group 3 facsimile communication over IP networks	✗
T.38 Imp Guide	Implementor's Guide (06/00) for Recommendation T.38	
T.38 amend 1	Amendment 1 (04/99) to Recommendation T.38 - Procedures for real-time Group 3 facsimile communication over IP networks - Amendment 1	✗
T.38 amend 2	Amendment 2 (02/00) to Recommendation T.38	
T.39	Application profiles for simultaneous voice and facsimile terminals	✗
T.42	Continuous colour representation method for facsimile	✗
T.43	Colour and gray-scale image representations using lossless coding scheme for facsimile	✗
T.43 amend 1	Amendment 1 (02/00) to Recommendation T.43	
T.44	Mixed raster content (MRC)	✗
T.44 amend 1	Amendment 1 (02/00) to Recommendation T.44	
T.45	Recommendation T.45 (02/00)	
T.50	Information technology - 7-bit coded character set for information interchange	✗

Series T Recommendations
Terminal characteristics and higher layer protocols for telematic services, document architecture

Rec. No.	Description/Title	S
T.51	Latin based coded character sets for telematic services	
T.51 Amend.1	Amendment 1 to Recommendation T.51 (08/95)	
T.52	Non-latin coded character sets for telematic services	
T.52 Amend.1	Amendment 1 (10/96) to Recommendation T.52	✗
T.53	Character coded control functions for telematic services	✗
T.60	Terminal equipment for use in the teletex service	
T.61	Standardized test charts for document facsimile transmissions	x93-6
T.62	Control procedures for teletex and Group 4 facsimile services (3)	
T.62 bis	Control procedures for teletex and G4 facsimile services based on Recommendations X.215 and X.225	
T.63	Provisions for verification of teletex terminal compliance	
T.64	Conformance testing procedures for the teletex recommendations	
T.65	Applicability of telematic protocols and terminal characteristics to computerized communication terminals (CCTs)	
T.66	Facsimile code points for use with Recommendations V.8 and V.8 bis	✗
T.70	Network-independent basic transport service for the telematic services	
T.71	Link Access Protocol Balanced (LAPB) extended for half-duplex physical level facility	
T.80	Common components for image compression and communication – Basic principles	
T.81	Information technology – digital compression and coding of continuous-tone still images – requirements and guidelines	
T.82	Information technology – Coded representation of picture and audio information – progressive bi-level image compression	
T.82 Corr.1	Technical Corrigendum 1 – Information technology – Coded representation of picture and audio information – progressive bi-level image compression	
T.83	Information technology – digital compression and coding of continuous-tone still images: compliance testing	✗
T.83 Encl.1	Compliance test data	
T.84	Information Technology – Digital compression and coding of continuous-tone still images: Extensions	✗
T.84 amend 1	Recommendation T.84 (07/96) - Information Technology – Digital compression and coding of continuous-tone still images: Extensions	✗
T.85	Application profile for Recommendation T.82 - progressive bi-level image compression (JBIG coding scheme) for facsimile apparatus	✗
T,85 Amend.1	Amendment 1 (10/96) to Recommendation T.85	
T.85 Corr.1	Corrigendum 1 (02/97) to Recommendation T.85	✗
T.85 Amend 2	Amendment 2 (10/97) to Recommendation T.85 - Application profile for Recommendation T.82 - Progressive bi-level image compression (JBIG coding scheme) for facsimile apparatus Amendment 2	✗
T.86	Information Technology – Digital compression and coding of continuous-tone still images: Registration of JPEG Profiles, SPIFF Profiles, SPIFF Tags, SPIFF colour Spaces, APPn Markers, SPIFF Compression types and Registration..	✗
T.87	Information technology – Lossless and near-lossless compression of continuous-tone still images - Baseline	✗
T.88	Recommendation T.88 (02/00)	
T.90	Characteristics and protocols for terminals for telematic services in ISDN	✗
T.90 Amend.1	Amendment 1 to Recommendation T.90 (11/94)	
T.90 Amend.2	Amendment 2 to Recommendation T.90 (07/96)	✗
T.90 Amend.3	Amendment 3 to Recommendation T.90 (06/98) - Characteristics and protocols for terminals for telematic services in ISDN Amendment 3: Cause value for a G4 Fax fall-back	✗
T.100	International information exchange for interactive Videotex	
T.101	International interworking for videotex services	
T.102	Syntax-based videotex end-to-end protocols for the circuit mode ISDN	
T.103	Syntax-based videotex end-to-end protocols for the packet mode ISDN	
T.104	Packet mode access for syntax-based videotex via PSTN	
T.105	Syntax-based videotex application layer protocol	✗
T.106	Framework of videotex terminal protocols	
T.107	Enhanced man machine interface for videotex and other retrieval services (VEMMI)	✗
T.120	Document Conferencing	✗

Series T Recommendations

Terminal characteristics and higher layer protocols for telematic services, document architecture

Rec. No.	Description/Title	S
T.120 annex C	Annex C to Recommendation T.120 (02/98) – Lightweight profiles for the T.120 architecture	x
T.121	Generic application template	x
T.122	Multipoint communication service for audiographics and audiovisual conferencing service definition	x
T.123	Protocol stacks for audiographic and audiovisual teleconference applications	x
T.124	Generic conference control	
T.125	Multipoint communication service protocol specification	x
T.126	Multipoint still image and annotation protocol	x
T.127	Multipoint binary file transfer protocol	x
T.128	Multipoint application sharing	x
T.134	Text chat application entity	x
T.135	User-to-reservation system transactions within T.120 conferences	x
T.136	Remote device control application protocol	x
T.137	Recommendation T.137 (02/00)	
T.140	Protocol for multimedia application text conversation	x
T.140 Adden 1	Addendum 1 (02/00) to Recommendation T.140	
T.150	Telewriting terminal equipment	
T.170	Framework of the T.170-Series of Recommendations	x
T.171	Protocols for interactive audiovisual services: Coded representation of multimedia and hypermedia objects	
T.172	MHEG-5 - Support for base-level interactive applications	x
T.173	MHEG-3 script interchange representation	
T.174	Application Programming Interface (API) for MHEG-1	x
T.175	Application Programming Interface (API) for MHEG-5	x
T.176	Application Programming Interface (API) for Digital Storage Media Command and Control (DSM-CC)	x
T.180	Homogeneous access mechanism to communication services	x
T.190	Cooperative document handling (CDH) – Framework and basic services	x
T.191	Cooperative document handling (CDH) – Joint synchronous editing (point-to-point)	
T.192	Cooperative document handling (CDH) – Complex services: Joint synchronous editing and joint document presentation/viewing	x
T.200	Programmable communication interface for terminal equipment connected to ISDN	
T.300	General Principles of telematic interworking	
T.330	Telematic access to interpersonal message system	
T.351	Imaging process of character information on facsimile apparatus	
T.390	Teletex requirements for interworking wtih the telex service	
T.400	Introduction to document architecture, transfer and manipulation	x96
T.400 S1	First extension to the T.410-series (1988) of Recommendations contained in the CCITT Blue Book, Fascicle VII.6: I - Tiled raster graphics; II - Annex E to T.411; III - Alternative representation; IV - Styles extension; V - Security	x96
T.400 S2	Revision of the T.410-series (1988) of Recommendations contained in the CCITT Blue Book, Fascicle VII.6, on the subject of "colour"	x96
T.400 S3	Amendments to the T.410-series (1988): I - Streams; II - Support for additional bit order mapping	x96
T.411	Information technology - Open document architecture (ODA) and interchange format: Introduction and general principles	
T.411 corr 1	Corrigendum 1 (10/97) Recommendation T.411 - Information technology - Open document Architecture (ODA) and interchange format: Introduction and general principles Technical Corrigendum 1	
T.412	Information Technology - Open Document Architecture (ODA) and interchange format - Document structures	
T.412 corr 1	Corrigendum 1 (10/97) Recommendation T.412 - Information technology - Open Document Architecture (ODA) and interchange format: Document structures Technical Corrigendum 1	
T.412 corr 2	Corrigendum 2 (10/97) Recommendation T.412 - Information technology - Open Document Architecture (ODA) and interchange format: Document structures Technical Corrigendum 2	
T.413	Information technology - Open document architecture (ODA) and interchange format: abstract interface for the manipulation of ODA documents	x
T.414	Information technology - Open Document Architecture (ODA) and interchange format: Document profile	

Series T Recommendations

Terminal characteristics and higher layer protocols for telematic services, document architecture

Rec. No.	Description/Title	S
T414 corr 1	Corrigendum 1 to Recommendation T.414 (10/97) - Information technology - Open Document Architecture (ODA) and interchange format: Document profile; Technical Corrigendum 1	
T414 corr 2	Corrigendum 2 (10/97) Recommendation T.414 - Information technology - Open Document Architecture (ODA) and interchange format: Document profile Technical Corrigendum 2	
T.415	Information technology - Open Document Architecture (ODA) and interchange format: Open document interchange format	
T.415 corr 1	Corrigendum 1 to Recommendation T.415 (10/97) - Information technology - Open Document Architecture (ODA) and interchange format: Open document interchange format; Technical Corrigendum 1	✗
T.415 corr 2	Corrigendum 2 (10/97) to Recommendation T.415 - Information technology - Open Document Architecture (ODA) and interchange format: Open Document interchange format	✗
T.416	Information Technology - Open Document Architecture (ODA) and interchange format: Character content architectures	
T.416 corr 1	Corrigendum 1 (10/97) to Recommendation T.416 - Information technology - Open Document Architecture (ODA) and interchange format: Character content architectures	
T.417	Information technology - Open document architecture (ODA) and interchange format: Raster graphics content architectures	✗
T.417 corr 1	Corrigendum 1 (10/97) Recommendation T.417 - Information technology - Open Document Architecture (ODA) and interchange format: Raster graphics content architectures Technical Corrigendum 1	
T.417 corr 2	Amendment 2 (02/00) to Recommendation T.417	
T.418	Information technology – Open document architecture (ODA) and interchange format: Geometric graphics content architecture	
T.419	Information technology – Open document architecture (ODA) and interchange format: Audio content architectures	✗
T.421	Information technology – Open document architecture (ODA) and interchange format: tabular structures and tabular layout	✗
T.422	Information technology – Open document architecture (ODA) and interchange format – Identification of document fragments	✗
T.424	Information technology – Open Document Architecture (ODA) and interchange format – Temporal relationships and non-linear structures	✗
T.431	Document transfer and manipulation (DTAM) – Services and protocols – Introduction and general principles	
T.432	Document transfer and manipulation (DTAM) services and protocols – Service definition	
T.432 Amend.1	Amendment 1 to Recommendation T.432 (08/95)	
T.433	Document Transfer, Access and Manipulation (DTAM) – Services and protocols – Protocol specification	✗
T.433 Amend.1	Amendment 1 to Recommendation T.433 (08/95)	
T.434	Binary file transfer format (BFT) for the telematic services	
T.434 Amend 1	Amendment 1 to Recommendation T.433 (08/95) - Revisions of T.433 to support G4 colour and file transfer	✗
T.435	Document transfer and manipulation (DTAM) - Services and protocols - Abstract service definition and procedures for confirmed document manipulation	✗
T.436	Document transfer and manipulation (DTAM) - Services and protocols - Protocol specifications for confirmed document manipulation	✗
T.441	Document transfer and manipulation (DTAM) - Operational structure	
T.501	Document application profile MM for the interchange of formatted mixed mode documents	
T.502	Document application profile PM-11 for the interchange of simple structure, character content documents in processable and formatted forms	✗
T.503	A document application profile for the interchange of Group 4 facsimile documents	✗
T.503 annex B	Extension for continuous-tone colour and gray-scale image documents	✗
T.503 Amend 2	Amendment 2 to Recommendation T.503 (08/95)	✗
T.503 Amend 3	Amendment 3 to Recommendation T.503 (07/97)	✗
T.503 Amend 4	Amendment 4 (04/99) to Recommendation T.503 – A document application profile for the interchange of Group 4 facsimile documents - Amendment 4	
T.503 Amend 5	Amendment 5 (02/00) to Recommendation T.503	
T.504	Document application profile for videotex interworking	

Series T Recommendations

Terminal characteristics and higher layer protocols for telematic services, document architecture

Rec. No.	Description/Title	S
T.505	Document application profile PM-26 for the interchange of enhanced structure, mixed content documents in processable and formatted forms	
T.506	Document application profile PM-36 for the interchange of extended document structures and mixed content documents in processable and formatted forms	
T.510	General overview of the T.510-series Recommendations	
T.521	Communication application profile BT0 for document bulk transfer based on the session service	✗
T.521 Amend.1	Amendment 1 to Recommendation T.521 (08/95)	
T.522	Communication application profile BT1 for document bulk transfer	
T.523	Communication application profile DM-1 for videotex interworking	
T.541	Operational application profile for videotex interworking	
T.561	Terminal characteristics for mixed mode of operation MM	
T.562	Terminal characteristics for teletex processable mode PM.1	
T.563	Terminal characteristics for Group 4 facsimile apparatus	✗
T.563 Amend.1	Amendment 1 to Recommendation T.563 (07/97)	
T.563 Amend.2	Amendment 2 to Recommendation T.563 (10/97) - Terminal characteristics for Group 4 facsimile apparatus Amendment 2: Annex C - T.30 frames for G4 facsimile	✗
T.563 Amend.3	Amendment 3 (04/99) to Recommendation T.563 - Terminal characteristics for Group 4 facsimile apparatus	✗
T.563 corr 1	Corrigendum 1 (06/98) Recommendation T.563 – Terminal characteristics for Group 4 facsimile apparatus Corrigendum 1	
T.564	Gateway characteristics for videotex interworking	
T.571	Terminal characters for the telematic file transfer within teletex service	
T.611	Programming communication interface (PCI) APPLI/COM for facsimile Group 3, facsimile Group 4, teletex, telex, e-mail and file transfer services	✗

Series U Recommendations

Telegraph switching

Rec. No.	Description	S
U.1	Signalling conditions to be applied in the international telex service	
U.2	Standardization of dials and dial pulse generators for the international telex service	
U.3	Arrangements in switching equipment to minimize the effects of false calling signals	
U.4	Exchange of information regarding signals destined to be used over international circuits concerned with switched teleprinter networks	
U.5	Requirements to be met by regenerative repeaters in international connections	
U.6	Prevention of fraudulent transit traffic in the fully automatic international telex service	
U.7	Numbering schemes for automatic switching networks	
U.8	Hypothetical reference connections for telex and gentex networks	
U.10	Equipment of an international telex position	
U.11	Telex and gentex signalling on intercontinental circuits used for intercontinental automatic transit traffic (type C signalling)	
U.12	Terminal and transit control signalling system for telex and similar services on international circuits (type D signalling)	
U.15	Interworking rules for international signalling systems according to Recommendations U.1, U.11 and U.12	
U.20	Telex and gentex signalling on radio channels (synchronous 7-unit systems affording error correction by automatic repetition)	
U.21	Operator recall on a telex call set up on a radiotelegraph circuit	
U.22	Signals indicating delay in transmission on calls set up by means of synchronous systems wtih automatic error correction by repetition	
U.23	Use of radiotelegraph circuits with ARQ equipment for fully automatic telex calls charged on the basis of elapsed time	
U.24	Requirements for telex and gentex operation to be met by synchronous multiplex equipment described in Recommendation R.44	

Series U Recommendations

Telegraph switching

Rec. No.	Description	S
U.25	Requirements for telex and gentex operation to be met by code- and speed-dependent TDM systems conforming to Recommendation R.101	
U.30	Signalling conditions for use in the international gentex network	
U.31	Prevention of connection to faulty stations and/or station lines in the gentex service	
U.40	Reactions by automatic terminals connected to the telex network in the event of ineffective call attempts or signalling incidents	
U.41	Changed address interception and call redirection in the telex service	
U.43	Follow-on calls	
U.44	Multi-address calls in real time for broadcast purposes in the international telex service	
U.45	Response to the not-ready condition of the telex terminal	
U.46	Interruption of automatic transmission and flow control in the international telex service	
U.60	General requirements to be met in interfacing the international telex network with the maritime satellite systems	
U.61	Detailed requirements to be met in interfacing the international telex network with maritime satellite systems	
U.62	General requirements to be met in interfacing the international telex network with the fully automated maritime VHF/UHF radio system	
U.63	General requirements to be met in interfacing the international telex network with the maritime "direct printing" system	
U.70	Telex service signals for telex to Teletex interworking	
U.74	Extraction of telex selection information from a calling telex answer-back	
U.75	Automatic called telex answerback check	
U.80	International telex store and forward access from a telex subscriber	
U.81	International telex store-and-forward - Delivery to a telex subscriber	
U.82	International telex store and forward - Interconnection of telex store and forward units	x93
U.101	Signalling systems for the intex service (Types E and F signalling)	
U.102	Intex and similar services - Network requirements to effect interworking between terminals operating at different speeds	✗
U.140	Definitions of essential technical terms relating to telegraph switching and signalling	
U.200	The international telex service-General technical requirements for interworking	
U.201	Interworking between the teletex service and the international telex service	
U.202	Technical requirements to be met in providing the international telex service within an integrated services digital network	
U.203	Technical requirements to be met when providing real-time both way communications between terminals of the international telex service and data terminal equipments on a PSPDN or via the PSTN	
U.204	Interworking between the international telex service and the public interpersonal messaging service	
U.205	Store-and-retrieve facility for the delivery of messages from a terminal of the international telex service to a data terminal equipment which connects to a packet-switched public data network over the public switched	
U.206	Technical requirements for interworking between the international telex service and the videotex service	
U.207	Technical requirements to be met for the transfer of messages between terminals of the international telex service and Group 3 facsimile terminals connected to PSTN	
U.208	The international telex service interworking with the INMARSAT C system using one-stage selection	
U.210	Intex service network requirements to effect interworking with the international telex service	
U.220	The international telex service - Technical requirements for a status enquiry function in an interworking scenario	

Series V Recommendations

Data communication over the telephone network

Rec. No.	Description/Title	S
V.1	Equivalence between binary notation symbols and the significant conditions of a two-condition code	
V.2	Power levels for data transmission over telephone lines	
V.4	General structure of signals of International Alphabet No. 5 code for character oriented data transmission over public telephone networks	

Series V Recommendations

Data communication over the telephone network

Rec. No.	Description/Title	S
V.5	Standardization of data signalling rates for synchronous data transmission in the general switched telephone network	x93
V.6	Standardization of data signalling rates for synchronous data transmission on leased telephone-type circuits	x93
V.7	Definitions of terms concerning data communication over the telephone network	
V.8	Procedures for starting sessions of data transmission over the general switched telephone network	
V.8 bis	Procedures for the identification and selection of common modes of operation between data circuit-terminating equipments (DCEs) and between data terminal equipments (DTEs) over the general switched telephone	
V.10	Electrical characteristics for unbalanced double-current interchange circuits operating at data signalling rates nominally up to 100 kbit/s	
V.11	Electrical characteristics for balanced double-current interchange circuits operating at data signalling rates up to 10 Mbit/s	
V.12	Electrical characteristics for balanced double-current interchange circuits for interfaces with data signalling rates up to 52 Mbit/s	
V.13	Simulated carrier control	
V.14	Transmission of start-stop characters over synchronous bearer channels	
V.14 corr 1	Corrigendum 1 (06/98) Recommendation T.563 – Terminal characteristics for Group 4 facsimile apparatus Corrigendum 1	
V.15	Use of acoustic coupling for data transmission	
V.16	Medical analogue data transmission modems	
V.17	A 2-wire modem for facsimile applications with rates up to 14 400 bit/s	✗
V.17 corr 1	Corrigendum 1 (09/98) to Recommendation V.17 - A 2-wire modem for facsimile applications with rates up to 14 400 bit/s – Corrigendum 1	
V.18	Operational and interworking requirements for modems operating in the text telephone mode	
V.18 app III	Appendix III (09/98) to Recommendation V.18 – Connection procedures for terminals including V.18 functionality	✗
V.18 app IV	Appendix IV (09/98) to Recommendation V.18 – Specification of V.18 implementation tests	✗
V.19	Modems for parallel data transmission using telephone signalling frequencies	
V.20	Parallel data transmission modems standardized for universal use in the general switched telephone network	x93
V.21	300 bits per second duplex modem standardized for use in the general switched telephone network	
V.22	1200 bits per second duplex modem standardized for use in the general switched telephone network and on point-to-point 2-wire leased telephone-type circuits	
V.22 bis	2400 bits per second duplex modem using the frequency division technique standardized for use on the general switched telephone network and on point-to-point 2-wire leased telephone-type circuits	
V.23	600/1200-baud modem standardized for use in the general switched telephone network	
V.24	List of definitions for interchange circuits between terminal equipment (DTE) and data circuit-terminating equipment (DCE)	
V.25	Automatic answering equipment and/or parallel automatic calling equipment on the general switched telephone network including procedures for disabling of echo control devices for both manually and automatically established calls.	
V.25 bis	Automatic calling and/or answering equipment on the general switched telephone network (GSTN) using the 100-series interchange circuits	
V.25 ter	Serial asynchronous automatic dialing and control	✗
V.25 ter supp	Supplement to Recommendation V.25ter	
V.25 terr annex	Recommendation V.25 ter Annex A (08/96) - Annex A: Procedure for DTE-controlled call negotiation	
V.26	2400 bits per second modem standardized for use on 4-wire leased telephone-type circuits	
V.26 bis	2400/1200 bits per second modem standardized for use in the general switched telephone network	
V.26 ter	2400 bits per second duplex modem using the echo cancellation technique standardized for use on the general switched telephone network and on point-to-point 2-wire leased telephone-type circuits	
V.27	4800 bits per second modem with manual equalizer standardized for use on leased telephone-type circuits	
V.27 bis	4800/2400 bits per second modem with automatic equalizer standardized for use on leased telephone-type circuits	
V.27 ter	4800/2400 bits per second modem standardized for use in the general switched telephone network	

Series V Recommendations
Data communication over the telephone network

Rec. No.	Description/Title	S
V.28	Electrical characteristics for unbalanced doubled-current interchange circuits	
V.29	9600 bits per second modem standardized for use on point-to-point 4-wire leased telephone-type circuits	
V.31	Electrical characteristics for single-current interchange circuits controlled by contact closure	
V.31 bis	Electrical characteristics for single-current interchange circuits using optocouplers	
V.32	A family of 2-wire, duplex modems operating at data signalling rates of up to 9600 bit/s for use on the general switched telephone network and on leased telephone-type circuits	
V.32 bis	A duplex modem operating at data signalling rates of up to 14 400 bit/s for use on the general switched telephone network and on leased point-to-point 2-wire telephone-type circuit	✗
V.33	14 400 bits per second modem standardized for use on point-to-point 4-wire leased telephone-type circuits	
V.34	A modem operating at data signalling rates of up to 33 600 bit/s for use on the general switched telephone network and on leased point-to-point 2-wire telephone-type circuits	
V.34 Corr	Corrigendum to Recommendation V.34 (10/96)	
V.35	Data transmission at 48 kbit/s using 60-108 kHz group band circuits	x93
V.36	Modems for synchronous data transmission using 60-108 kHz group band circuits	
V.37	Synchronous data transmission at a data signalling rate higher than 72 kbit/s using 60-108 kHz group band circuits	
V.38	A 48/56/64 kbit/s data circuit terminating equipment standardized for use on digital point-to-point leased circuits	
V.40	Error indication with electromechanical equipment	x93
V.41	Code-independent error-control system	
V.42	Error-correcting procedures for DCEs using asynchronous-to-synchronous conversion	
V.42 bis	Data compression procedures for data circuit terminating equipment (DCE) using error correction procedures	✗
V.43	Data flow control	✗
V.44	Data compression procedures	
V.50	Standard limits for transmission quality of data transmission	
V.51	Organization of the maintenance of international telephone-type circuits used for data transmission	
V.52	Characteristics of distortion and error-rate measuring apparatus for data transmission	
V.53	Limits for the maintenance of telephone-type circuits used for data transmission	
V.54	Loop test devices for modems	
V.55	See Recommendation O.71	
V.56	Comparative tests of modems for use over telephone-type circuits	
V.56 bis	Network transmission model for evaluating modem performance over 2-wire voice grade connections	✗
V.56 ter	Test procedure for evaluation of 2-wire 4 KHz voiceband duplex modems (2)	
V.57	Comprehensive data test set for high data signalling rates	
V.58	Management information model for V-series DCE's	
V.61	A simultaneous voice plus data modem, operating at a voice plus data signalling rate of 4800 bit/s, with optional automatic switching to data-only signalling rates of up to 14 400 bit/s, for use on the general switched	
V.70	Procedures for the simultaneous transmission of data and digitally encoded voice signals over the GSTN, or over 2-wire leased point-to-point telephone type circuits	
V.75	DSVD terminal control procedures	
V.75 appndx	Recommendation V.75 (Appendix II) (02/98) - DSVD terminal control procedures: Session establishment using V.75/H.245 procedures	
V.76	Generic multiplexer using V.42 LAPM-based procedures	
V.80	In-band DCE control and synchronous data modes for asynchronous DTE	
V.90	A digital modem and analogue modem pair for use on the Public Switched Telephone Network (PSTN) at data signalling rates of up to 56 000 bit/s downstream and up to 33 600 bit/s upstream	✗
V.91	A digital modem operating at data signalling rates of up to 64 000 bit/s for use on a 4-wire circuit switched connection and on leased point-to-point 4-wire digital circuits	
V.92		
V.100	Interconnection between public data networks (PDNs) and the public switched telephone networks (PSTN)	
V.110	Support of data terminal equipments with V-Series type interfaces by an integrated services digital network	

Series V Recommendations

Data communication over the telephone network

Rec. No.	Description/Title	S
V.120	Support by an ISDN of data terminal equipment with V-series type interfaces with provision for statistical multiplexing	
V.120 corr 1	Corrigendum 1 to Recommendation V.120 (05/99) – Support by an ISDN of data terminal equipment with V-series type interfaces with provision for statistical multiplexing	
V.130	ISDN terminal adaptor framework	✗
V.140	Procedures for establishing communication between two multiprotocol audiovisual terminals using digital channels at a multiple of 64 or 56 kbit/s	✗
V.230	General data communications interface layer 1 specification	
V.250	Serial asynchronous automatic dialling and control	
V.250 supp 1	Supplement 1 (09/98) to Recommendation V.250 – Serial asynchronous automatic dialling and control - Supplement 1: Various extensions to V.250 basic command set	
V.251	Procedure for DTE-controlled call negotiation	✗
V.252	Procedure for control of V.70 and H.324 terminals by a DTE	
V.253	Control of voice-related functions in a DCE by an asynchronous DTE	✗
V.300	A 128 (144) kbit/s data circuit-terminating equipment standardized for use on digital point-to-point leased circuit	✗

Series X Recommendations

Data networks and open system communication

Rec. No.	Description/Title	S
X.1	International user classes of service in, and categories of access to, public data networks and integrated services digital networks (ISDNs)	✗
X.2	International data transmission services and optional user facilities in public data networks and ISDNs	✗
X.3	Packet assembly/disassembly facility (PAD) in a public data network	
X.4	General structure of signals of international alphabet No.5 code for character oriented data transmission over public data networks	
X.5	Facsimile packet assembly/disassembly facility (FPAD) in a public data network	✗
X.6	Multicast service definition	
X.7	Technical characteristics of data transmission services	✗
X.8	Multi-aspect pad (MAP) framework and service definition	
X.10	Categories of access for data terminal equipment (DTE) to public data transmission services	x93-6
X.20	Interface between data terminal equipment (DTE) and data circuit-terminating equipment (DCE) for start-stop transmission services on public data networks	
X.20 bis	Use on public data networks of data terminal equipment (DTE) which is designed for interfacing to asynchronous duplex V-series modems	
X.21	Interface between data terminal equipment and data circuit-terminating equipment for synchronous operation on public data networks	
X.21 bis	Use on public data networks of data terminal equipment (DTE) which is designed for interfacing to synchronous V-series modems	
X.22	Multiplex DTE/DCE interface for user classes 3-6	
X.24	List of definitions for interchange circuits between data terminal equipment (DTE) and data circuit-terminating equipment (DCE) on public data networks	
X.25	Interface between data terminal equipment (DTE) and data circuit-terminating equipment (DCE) for terminals operating in the packet mode and connected to public data networks by dedicated circuit	✗
X.25 corr 1	Recommendation X.25 Corrigendum 1 (09/98) – Interface between Data Terminal Equipment (DTE) and Data Circuit-terminating Equipment (DCE) for terminals operating in the packet mode and connected to public data networks by dedicated circuit - Corrigendum 1	
X.26	See Recommendation V.10	
X.27	See Recommendation V.11	
X.28	DTE/DCE interface for a start-stop mode data terminal equipment accessing the packet assembly/disassembly facility (PAD) in a public data network situated in the same country	✗

Series X Recommendations
Data networks and open system communication

Rec. No.	Description/Title	S
X.28 Addm 1	to enable map support in accordance with Recommendation X.8	
X.29	Procedures for the exchange of control information and user data between a packet assembly/disassembly (PAD) facility and a packet mode DTE or another PAD	✗
X.30	Support of X.21, X.21 bis and X.20 bis based data terminal equipments (DTEs) by an integrated services digital network (ISDN)	
X.31	Support of packet mode terminal equipment by an ISDN	✗
X.32	Interface between data terminal equipment (DTE) and data circuit-terminating equipment (DCE) for terminals operating in the packet mode and accessing a packet switched public data network through a public switched tele	
X.33	Access to packet switched data transmission services via frame relaying data transmission services	✗
X.34	Access to packet switched data transmission services via B-ISDN	✗
X.35	Interface between a PSPDN and a private PSDN which is based on X.25 procedures and enhancements to define a gateway function that is provided in the PSPDN	
X.36	Interface between data terminal equipment (DTE) and data circuit-terminating equipment (DCE) for public data networks providing frame relay data transmission service by dedicated circuit	✗
X.36 Amend.1	Amendment 1 to Recommendation X.36 (10/96)	✗
X.36 Amend.2	Frame transfer priority	
X.36 Amend.3	Frame discard priority, service classes, NSAP signalling and protocol encapsulation	
X.37	Encapsulation in X.25 packets of various protocols including frame relay	✗
X.38	G3 facsimile equipment/DCE interface for G3 facsimile equipment accessing the facsimile packet assembly/disassembly facility (FPAD) in a public data network situated country	✗
X.39	Procedures for the exchange of control information and user data between a facsimile packet assembly/disassembly (FPAD) facility and a packet mode data terminal equipment (DTE) or another FPAD	✗
X.40	Standardization of frequency-shift modulated transmission systems for the provision of telegraph and data channels by frequency division of a group	x93
X.42	Procedures and methods for accessing a Public Data Network from a DTE operating under control of a generalized polling protocol	✗
X.45	Interface between Data Terminal Equipment (DTE) and Data Circuit-terminating Equipment (DCE) for terminals operating in the packet mode and connected to public data networks, designed for efficiency at higher speeds	✗
X.46	Access to FRDTS via B-ISDN	✗
X.48	Procedures for the provision of a basic multicast service for Data Terminal Equipments (DTEs) using Recommendation X.25	✗
X.49	Procedures for the provision of an extended multicast service for Data Terminal Equipments (DTEs) using Recommendation X.25	✗
X.50	Fundamental parameters of a multiplexing scheme for the international interface between synchronous data networks	
X.50 bis	Fundamental parameters of a 48-kbit/s user data signalling rate transmission scheme for the international interface between synchronous data networks	
X.51	Fundamental parameters of a multiplexing scheme for the international interface between synchronous data networks using 10-bit envelope structure	
X.51 bis	Fundamental parameters of a 48-kbit/s user data signalling rate transmission scheme for the international interface between synchronous data networks using 10-bit envelope structure	
X.52	Method of encoding anisonchronous signals into a synchronous user bearer	
X.53	Numbering of channels on international multiplex links at 64 kbit/s	
X.54	Allocation of channels on international multiplex links at 64 kbit/s	
X.55	Interface between synchronous data networks using a 6 + 2 envelope structure and single channel per carrier (SCPC) satellite channels	
X.56	Interface between synchronous data networks using an 8 + 2 envelope structure and single channel per carrier (SCPC) satellite channels	
X.57	Method of transmitting a single lower speed data channel on a 64kbit/s data stream	
X.58	Fundamental parameters of a multiplexing scheme for the international interface between synchronous non-switched data networks using no envelope structure	
X.60	Common channel signalling for circuit switched data applications	

Series X Recommendations
Data networks and open system communication

Rec. No.	Description/Title	S
X.61	Signalling System No. 7 - Data user part	
X.70	Terminal and transit control signalling system for start-stop services on international circuits between anisochronous data networks	
X.71	Decentralized terminal and transit control signalling system on international circuits between synchronous data networks	
X.71 corr 1		
X.75	Packet-switched signalling system between public networks providing data transmission services	✗
X.75 corr 1	Corrigendum 1 (09/98) to Recommendation X.75 - Packet-switched signalling system between public networks providing data transmission services Corrigendum 1	
X.76	Network-to-network interface between public data networks providing the frame relay data transmission service	
X.76 amend1	Switched virtual circuits	
X.76 amend2	Frame relay service classes and priorities	
X.77	Interworking between PSPDNs via B-ISDN	✗
X.78	Interworking procedures between networks providing frame relay data transmission services via B-ISDN	✗
X.80	Interworking of interexchange signalling systems for circuit switched data services	
X.81	Interworking between an ISDN circuit-switched and a circuit-switched public data network (CSPDN)	
X.82	Detailed arrangements for interworking between CSPDNs and PSPDNs based on Recommendation T.70	
X.92	Hypothetical reference connections for public synchronous data networks	
X.96	Call progress signals in public data networks	
X.110	International routing principles and routing plan for public data networks	✗
X.115	Definition of address translation capability in public data networks	✗
X.115 Amend.1	Amendment 1 to Recommendation X.115 (10/96)	✗
X.116	Address translation registration and resolution protocol	✗
X.121	International numbering plan for public data networks	✗
X.122	Numbering plan interworking for the E.164 and X.121 numbering plans	
X.123	Mapping between escape codes and TOA/NPI for E.164/X.121 numbering plan interworking during the transition period	✗
X.124	Arrangements for the interworking of the E.164 and X.121 numbering plans for frame relay and ATM networks	
X.125	Procedure for the notification of the assignment of international network identification codes for public frame relay data networks and ATM networks numbered under the E.164 numbering plan	✗
X.130	Call processing delays in public data networks when providing international synchronous circuit-switched data services	
X.131	Call blocking in public data networks when providing international synchronous circuit-switched data services	
X.134	Portion boundaries and packet layer reference events: basis for defining packet-switched performance parameters	✗
X.135	Speed of service (delay and throughput) performance values for public data networks when providing international packet-switched services	✗
X.135 supp 1	Some test results from specific national and international portions	
X.136	Accuracy and dependability performance values for public data networks when providing international packet-switched services	✗
X.137	Availability performance values for public data networks when providing international packet-switched services	✗
X.138	Measurement of performance values for public data networks when providing international packet-switched services	✗
X.139	Echo, drop, generator and test DTEs for measurement of performance values in public data networks when providing international packet switched services	✗
X.140	General quality of service parameters for communication via public data networks	
X.141	General principles for the detection and correction of errors in public data networks	
X.144	User information transfer performance parameters for data networks providing international frame relay PVC service	✗
X.144 amend 1	Annex C - Some relations between frame-level and ATM-level performance parameters	✗
X.145	Performance for data networks providing international frame relay SVC service	✗

Series X Recommendations

Data networks and open system communication

Rec. No.	Description/Title	S
X.146	Performance objectives and quality of service classes applicable to frame relay	✗
X.146 note	Notice to Recommendation X.146 (09/98) – Performance objectives and quality of service classes applicable to frame relay	
X.150	Principles of maintenance testing for public data networks using data terminal equipment (DTE) and data circuit-terminating equipment (DCE) test loops	
X.160	Architecture for customer network management service for public data networks	✗
X.161	Definition of customer network management services for public data networks	✗
X.162	Definition of management information for customer network management service for public data networks to be used with the CNMc interface	✗
X.162 amend 1	Implementation conformance statement proformas	
X.163	Definition of management information for customer network management service for public data networks to be used with the CNMe interface	✗
X.180	Administrative arrangements for international closed user groups (CUGs)	
X.181	Administrative arrangements for the provision of international permanent virtual circuits (PVCs)	
X.200	Information technology - Open Systems Interconnection - Basic reference model: The basic model	
X.207	Information technology - Open Systems Interconnection - Application layer structure	
X.208	Specification of Abstract Syntax Notation One (ASN.1)	
X.209	Specification of basic encoding rules for Abstract Syntax Notation One (ASN.1)	
X.210	Information technology - Open Systems Interconnection - Basic reference model: Conventions for the definition of OSI services	
X.211	Physical service definition of Open Systems Interconnection for CCITT applications	✗
X.212	Data link service definition for Open Systems Interconnection for CCITT applications	✗
X.213	Information technology - Network service definition for open systems interconnection	✗
X.213 Corr.1	Information technology - Open Systems Interconnection - network service definition	
X.213 amend 1	Addition of the Internet protocol address format identifier	✗
X.214	Information technology - Open Systems Interconnection - Transport service definition	✗
X.215	Information technology - Open Systems Interconnection - Session service definition	✗
X.215 Adden 1	Addendum No. 1 to Recommendation X.215 (11/95) - Efficiency enhancements	✗
X.215 Adden 2	Nested connections functional unit	✗
X.216	Information technology - Open Systems Interconnection - Presentation service definition	✗
X.216 Adden.1	Addendum No. 1 to Recommendation X.216 (11/95)	✗
X.216 Adden.2	Amendment 2 to Recommendation X.216 (12/97) - Information technology - Open Systems Interconnection - Presentation service definition Amendment 2: Nested connections function	✗
X.217	Information technology - Open Systems Interconnection - Service definition for the association control service element	✗
X.217 Amend.1	Amendment 1 to Recommendation X.217 (10/96) -Support of authentication mechanisms for the connectionless mode	✗
X.217 Amend.2	Fast associate mechanism	✗
X.217 bis	Information technology - Open Systems Interconnection - Service definition for the Application Service Object Association Control Service Element	✗
X.218	Reliable transfer: Model and service definition	
X.219	Remote operations: Model, notation and service definition	
X.220	Use of X.200-series protocols in CCITT applications	
X.222	Use of X.25 LAPB compatible data link procedures to provide the OSI connection-mode data link service	
X.222 Amend.1	Amendment 1 to Recommendation X.222 (10/96)	✗
X.223	Use of X.25 to provide the OSI connection-mode network service for the ITU-T applications	
X.223 Amend.1	Amendment 1 to Recommendation X.223 (10/96)	✗
X.224	Protocol for providing the OSI connection-mode transport service	✗
X.224 Amend 1	Relaxation of class conformance requirements and expedited data service feature negotiation	✗
X.225	Information technology - Open Systems Interconnection - Connection-oriented session protocol: Protocol specification	✗
X.225 Adden.1	Addendum 1 to Recommendation X.225 (11/95) - Efficiency enhancements	✗
X.225 Adden.2	Nested connections functional unit	✗
X.226	Information technology - Open Systems Interconnection - Connection-oriented presentation protocol: Protocol specification	

Series X Recommendations

Data networks and open system communication

Rec. No.	Description/Title	S
X.226 Adden.1	Addendum 1 to Recommendation X.226 (11/95) - Efficiency enhancements	✗
X.226 Adden.2	Nested connections functional unit	✗
X.227	Information technology - Open systems interconnection - Connection-oriented protocol for the association control service	✗
X.227 Adden.1	Amendment 1 to Recommendation X.227 (10/96) - Incorporation of extensibility markers	✗
X.227 Adden.2	Fast-associate mechanism	✗
X.227 bis	Information technology - Open Systems Interconnection - Connection-mode protocol for the Application Service Object Association Control Service Element	✗
X.228	Reliable transfer: Protocol specification	
X.229	Remote operations: Protocol specification	
X.233	Information technology - Protocol for providing the connectionless-mode network service: protocol specification	
X.233 Amend.1	Amendment 1 to Recommendation X.233 (04/95)	
X.233 Amend.2	Amendment 2 to Recommendation X.233 (11/95)	✗
X.233 Amend.3	Amendment 3 to Recommendation X.233 (11/95)	✗
X.234	Information technology - Protocol for providing the OSI connectionless-mode transport service	✗
X.234 Amend.1	Amendment 1 to Recommendation X.234 (11/95)	✗
X.235	Information technology - Open Systems Interconnection - Connectionless session protocol: Protocol specification	✗
X.235 Amend 1	Amendment 1 to Recommendation X.235 (06/99) – Efficiency enhancements	✗
X.236	Information technology – Open Systems Interconnection – Connectionles presentaition protocol: Protocol specification	
X.236 Amend 1	Amendment 1 (06/99) to Recommendation X.236 - Efficiency enhancements	
X.237	Information technology - Open Systems Interconnection - Connectionless protocol for the association control service element: Protocol specification	
X.237 Amend.1	Information technology - Open Systems Interconnection - Connectionless protocol for the association control service element: Protocol specification	✗
X.237 bis	Information technology - Open Systems Interconnection - Connectionless protocol for the Application Service Object Association Control Service Element	✗
X.244	Procedure for the exchange of protocol identification during virtual call establishment on packet switched public data networks	x93-6
X.245	Information technology - Open Systems Interconnection - Connection-oriented session protocol: Protocol implementation conformance statement (PICS) proforma	✗
X.246	Information technology - Open Systems Interconnection - Connection-oriented presentation protocol: Protocol Implementation Conformance Statement (PICS) proforma	✗
X.247	Information technology - Open Systems Interconnection - Protocol specification for the association control service element: protocol implementation conformance statement (PICS) proforma	✗
X.248	Reliable transfer service element - Protocol implementation conformance statement (PICS) proforma	✗
X.249	Remote operations service element - Protocol implementation conformance statement (PICS) proforma	✗
X.255	Information technology - Open Systems Interconnection - Connectionless session protocol: Protocol Implementation Conformance Statement (PICS) proforma	✗
X.256	Information technology - Open Systems Interconnection - Connectionless presentation protocol: Protocol Implementation Conformance Statement (PICS) proforma	✗
X.257	Information technology - Open Systems Interconnection - Connectionless protocol for the association control service element: Protocol Implementation Conformance Statement (PICS) proforma	✗
X.257 Amend.1	Amendment 1 to Recommendation X.257 (10/96)	✗
X.260	Information technology - Framework for protocol identification and encapsulation	✗
X.263	Information technology - Protocol identification in the network layer	✗
X.264	Transport protocol identification mechanism	✗
X.273	Information technology - Open Systems Interconnection - Network layer security protocol	✗
X.274	Information technology - Telecommunication and information exchange between systems - Transport layer security protocol	✗
X.281	Information technology - Elements of management information related to the OSI physical layer	✗
X.282	Elements of management information related to the OSI data link layer	✗
X.282 Amend.1	Amendment 1 to Recommendation X.282 (10/96) - Implementation conformance statements (ICSs) proformas	✗
X.282 Amend.2	Addition of new counter attributes	
X.283	Elements of management information related to the OSI network layer	✗

Series X Recommendations
Data networks and open system communication

Rec. No.	Description/Title	S
X.283 Corr.1	Corrigendum 1 to Recommendation X.283 (04/96)	
X.283 Amend.1	Amendment 1 to Recommendation X.283 (10/96)	✗
X.284	Elements of management information related to the OSI transport layer	✗
X.284 Amend.1	Amendment 1 to Recommendation X.284 (11/95)	✗
X.284 Amend.2	Amendment 1 to Recommendation X.284 (10/96)	✗
X.290	OSI conformance testing methodology and framework for protocol recommendations for ITU-T applications - General concepts	✗
X.291	OSI conformance testing methodology and framework for protocol Recommendations for ITU-T applications - Abstract test suite specification	✗
X.292	OSI conformance testing methodology and framework for protocol Recommendations for CCITT applications - The Tree and Tabular Combined Notation (TTCN)	
X.293	OSI conformance testing methodology and framework for protocol Recommendations for ITU-T applications - Test realization	✗
X.294	OSI conformance testing methodology and framework for protocol recommendations for CCITT applications - Requirements on test laboratories and clients for the conformance assessment process	✗
X.295	OSI conformance testing methodology and framework for protocol recommendations for ITU-T applications - Protocol profile test specification	✗
X.296	OSI conformance testing methodology and framework for protocol recommendations for ITU-T applications - Implementation conformance statements	✗
X.300	General principles for interworking between public networks, and between public networks and other networks for the provision of data transmission services	✗
X.301	Description of the general arrangements for call control within a subnetwork and between subnetworks for the provision of data transmission services	
X.302	Description of the general arrangements for internal network utilities within a subnetwork and intermediate utilities between subnetworks for the provision of data transmission services	
X.305	Functionalities of subnetworks relating to the support of the OSI connection-mode network service	
X.320	General arrangements for interworking between Integrated Services Digital Networks (ISDNs) for the provision of data transmission services	✗
X.321	General arrangements for interworking between Circuit Switched Public Data Networks (CSPDNs) and Integrated Services Digital Networks (ISDNs) for the provision of data transmission services	✗
X.322	General arrangements for interworking between Packet Switched Public Data Networks (PSPDNs) and Circuit Switched Public Data Networks (CSPDNs) for the provision of data transmission services	
X.323	General arrangements for interworking between Packet Switched Public Data Networks (PSPDNs)	
X.324	General arrangements for interworking between Packet Switched Public Data Networks (PSPDNs) and public mobile systems for the provision of data transmission services	
X.325	General arrangements for interworking between Packet Switched Public Data Networks (PSPDNs) and Integated Services Digital Networks (ISDNs) for the provision of data transmission services	✗
X.326	General arrangements for interworking between Packet Switched Public Data Networks (PSPDNs) and Common Channel Signalling Network (CCSN)	
X.327	General arrangements for interworking between packet switched public data networks (PSPDNs) and private data networks for the provision of data transmission services	✗
X.328	General arrangements for interworking between public data networks providing frame relay data transmission services and, Integrated Services Digital Networks (ISDNs) for the provision of data transmission services	✗
X.340	General arrangements for interworking between a packet switched public data network (PSPDN) and the international telex network	
X.350	General interworking requirements to be met for data transmission in international public mobile satellite systems	
X.351	Special requirements to be met for packet assembly/disassembly facilities (PADs) located at or in association with coast earth stations in the public mobile satellite service	
X.352	Interworking between packet switched public data networks and public maritime mobile satellite data transmission systems	
X.353	Routing principles for interconnecting public maritime mobile satellite data transmission systems with public data networks	
X.361	Connection of VSAT systems with Packet-Switched Public Data Networks based on X.25 procedures	✗

Series X Recommendations

Data networks and open system communication

Rec. No.	Description/Title	S
X.370	Arrangements for the transfer of internetwork management information	x93-6
X.400	Message handling services: Message Handling System and service overview	
X.400 amend 1	Amendment 1 (09/98) to Recommendation F.400/X.400 - Message handling system and service overview Amendment 1	✗
X.402	Message Handling Systems - Overall architecture	✗
X.402 corr 1		
X.402 amend 1		
X.403	Message Handling Systems - Conformance testing	x93-6
X.407	Message Handling Systems - Abstract service definition conventions	x93-6
X.408	Message Handling Systems: Encoded information type conversion rules	
X.411	Message Handling Systems - Message transfer system: Abstract service definition and procedures	✗
X.411 corr 1		
X.411 corr 2		
X.411 corr 3		
X.411 amend 1	Additional correlation attribute and security error code	
X.411 amend 2		
X.412	Information technology – Message Handling System (MHS) - MHS Routing	
X.413	Information technology - Message Handling Systems (MHS): Message store: Abstract service definition	✗
X.419	Message Handling Systems - Protocol specifications	✗
X.419 corr 1		
X.419 amend 1		
X.419 amend 2		
X.420	Message Handling Systems - Interpersonal messaging system	✗
X.420 corr 1		
X.420 corr 2		
X.420 corr 3		
X.420 amend 1		
X.420 amend 2		
X.420 amend 3		
X.421	Message Handling Systems - Comfax use of MHS	✗
X.421 amend 1		
X.435	Message Handling Systems - Electronic data interchange messaging system	✗
X.435 corr 1		
X.435 amend 1		
X.440	Message Handling Systems - Voice messaging system	
X.440 Amend.1	Amendment 1 to Recommendation X.440 (11/95)	✗
X.445	Asynchronous protocol specification - Provision of OSI connection mode network service over the telephone network	✗
X.446	Common messaging call API	✗
X.460	Information technology - Message Handling Systems (MHS) management: Model and architecture	✗
X.462	Information technology - Message Handling Systems (MHS) management: Logging information	✗
X.467	Information technology - Message Handling Systems (MHS) management: Message transfer agent management	✗
X.480	Message Handling Systems and directory services - Conformance testing	
X.481	P2 protocol - Protocol Implementation Conformance Statement (PICS) proforma	✗
X.482	P1 Protocol - Protocol Implementation Conformance Statement (PICS) proforma	✗
X.483	P3 protocol - Protocol Implementation Conformance Statement (PICS) proforma	✗
X.484	P7 protocol - Protocol Implementation Conformance Statement (PICS) proforma	✗
X.485	Message Handling Systems: Voice messaging system Protocol Implementation Conformance Statement (PICS) proforma	
X.486	Messaging Handling Systems - Pedi protocol PICS proforma	✗
X.487	Message Handling Systems - IPM-MS attributes PICS proforma	✗
X.488	Message Handling Systems – EDI-MS attributes PICS proforma	
X.500	Information technology – Open Systems Interconnection – The directory: overview of concepts, models, and services	✗

Series X Recommendations

Data networks and open system communication

Rec. No.	Description/Title	S
X.501	Information technology – Open Systems Interconnection – The Directory: Models	✗
X.509	Information technology – Open Systems Interconnection – The directory: Authentication framework	✗
X.511	Information technology – Open Systems Interconnection – The Directory: Abstract service definition	✗
X.518	Information technology – Open Systems Interconnection – The Directory: Procedures for distributed operation	✗
X.519	Information technology – Open Systems Interconnection – The directory: Protocol specifications	✗
X.520	Information technology – Open Systems Interconnection – The Directory: Selected attribute types	✗
X.521	Information technology – Open Systems Interconnection – The Directory: Selected object classes	✗
X.525	Information technology – Open Systems Interconnection – The Directory: Replication	✗
X.530	Information technology – Open Systems Interconnection - The Directory: Use of systems management for administration of the Directory	✗
X.581	Directory access protocol - Protocol implementation conformance statement (PICS)	✗
X.582	Directory system protocol - Protocol implementation conformance statement (PICS)	✗
X.583	Information technology - Open Systems Interconnection - The Directory: Protocol Implementation Conformance Statement (PICS) Proforma for the Directory access protocol	✗
X.584	Information technology - Open Systems Interconnection - The Directory: Protocol Implementation Conformance Statement (PICS) Proforma for the Directory system protocol	✗
X.585	Information technology - Open Systems Interconnection - The Directory: Protocol Implementation Conformance Statement (PICS) Proforma for the Directory operational binding management protocol	✗
X.586	Information technology - Open Systems Interconnection - The Directory: Protocol Implementation Conformance Statement (PICS) Proforma for the Directory information shadowing protocol	✗
X.605	Information technology – Enhanced communications transport service definition	✗
X.610	Provision and support of the OSI connection-mode network service	✗
X.612	Information technology – Provision of the OSI connection-mode network service by packet-mode terminal equipment connected to an integrated services digital network (ISDN) for CCITT applications	✗
X.613	Information technology – Use of X.25 packet layer protocol in conjunction with X.21/X.21 bis to provide the OSI connection-mode network service	
X.614	Information technology – Use of X.25 packet layer protocol to provide the OSI connection-mode network service over the telephone network	
X.622	Information technology – Protocol for providing the connectionless-mode network service: Provision of the underlying service by an X.25 subnetwork	✗
X.623	Information technology – Protocol for providing the connectionless-mode network service: Provision of the underlying service by a subnetwork that provides the OSI data link service	✗
X.625	Information technology – Protocol for providing the connectionless-mode network service: Provision of the underlying service by ISDN circuit-switched B-channels	✗
X.630	Efficient Open Systems Interconnection (OSI) operations	✗
X.633	Information technology - Open systems interconnection - Network Fast Byte Protocol	✗
X.633 adden 1	SDL specifications	✗
X.634	Information technology - Open Systems Interconnection - Transport Fast Byte Protocol	✗
X.634 adden 1	SDL specifications	✗
X.637	Basic connection-oriented common upper layer requirements	✗
X.638	Minimal OSI facilities to support basic communications applications	✗
X.639	Basic connection-oriented requirements for ROSE-based profiles	✗
X.641	Quality of service: framework	✗
X.642	Information technology - Quality of service - Guide to methods and mechanisms	✗
X.650	Open systems interconnection (OSI) - reference model for naming and addressing	✗
X.660	Information technology - Open Systems Interconnection - Procedures for the operation of OSI Registration Authorities - General procedures	
X.660 Amend.1	Amendment 1 to Recommendation X.660 (10/96) - Incorporation of object identifiers components	✗
X.660 Amend.2	Incorporation of the root arcs of the object identifier tree	✗
X.662	Information technology - Open Systems Interconnection - Procedures for the operation of OSI Registration Authorities: Registration of values of RH-name-tree components for joint ISO and ITU-T use	✗
X.665	Information technology - Open Systems Interconnection - procedures for the operation of OSI registration authorities: application processes and application entities	
X.666	Procedures for registration of international and multinational organization names	✗

Series X Recommendations
Data networks and open system communication

Rec. No.	Description/Title	S
X.669	Procedures for the operation of OSI registration authorities: Registration procedures for the ITU-T subordinate arcs	✗
X.669 corr 1	Corrigendum 1 (06/99) to Recommendation X.669 - Procedures for the operation of OSI registration authorities: Registration procedures for the ITU-T subordinate arcs	✗
X.670	Procedures for registration agents operating on behalf of organizations to register organization names subordinate to country names	✗
X.671	Procedures for a registration authority operating on behalf of countries to register organization names subordinate to country names	✗
X.680	Information technology - Abstract Syntax Notation One (ASN.1): Specification of basic notation	✗
X.680 Amend.1	Amendment 1 to Recommendation X.680 (04/95) - Rules of extensibility	✗
X.680 Corr.1	Corrigendum 1 (11/95) to Recommendation X.680	✗
X.681	Information technology - Abstract Syntax Notation One (ASN.1): Information object specification	✗
X.681 Amend.1	Amendment 1 to Recommendation X.681 (04/95)	✗
X.681 corr 1	Corrigendum 1 (06/99) to Recommendation X.681 - Information technology - Abstract Syntax Notation One (ASN.1): Information object specification	
X.682	Information technology - Abstract Syntax Notation One (ASN.1): constraint specification	✗
X.683	Information technology - Abstract Syntax Notation One (ASN.1): parameterization of ASN.1 specifications	✗
X.683 Amend 1	Amendment 1 (06/99) to Recommendation X.683 – ASN.1 semantic model	
X.690	Information technology – ASN.1 encoding rules: specification of basic encoding rules (BER), canonical encoding rules (CER) and distinguished encoding rules (DER)	✗
X.690 Amend 1	Amendment 1 (06/99) to Recommendation X.690 - Relative object identifiers	
X.690 Corr.1	Corrigendum 1 (06/99) to Recommendation X.690 - Information technology - ASN.1 encoding rules: Specification of Basic Encoding Rules (BER), Canonical Encoding Rules (CER) and Distinguished Encoding Rules (DER)	✗
X.691	Information technology - ASN.1 encoding rules - Specification of Packed Encoding Rules (PER)	✗
X.691 Amend 1	Amendment 1 (06/99) to Recommendation X.691 – Relative object identifiers	
X.691 corr 1	Corrigendum 1 (06/99) to Recommendation X.691 - Specification of Packed Encoding Rules (PER)	
X.700	Management framework for Open Systems Interconnection (OSI) for CCITT applications	
X.701	Information technology - Open Systems Interconnection - Systems management overview	
X.701 Amend 1	Management knowledge management	✗
X.701 Corr.1	Corrigendum 1 to Recommendation X.701 (02/94)	
X.701 Corr.2	Corrigendum 2 to Recommendation X.701 (04/95)	✗
X.701 Corr.3	Corrigendum 3 to Recommendation X.701 (10/96)	✗
X.702	Information technology - Open Systems Interconnection - Application context for systems management with transaction processing	✗
X.703	Information technology - Open Distributed Management Architecture	✗
X.703 amend 1	Support using Common Object Request Broker Architecture (CORBA)	
X.710	Information technology – Open Systems Interconnection - Common Management Information Service	✗
X.711	Information technology – Open Systems Interconnection - Common Management Information Protocol: Specification	✗
X.711 corr 1	Corrigendum 1 (03/99) to Recommendation X.711 - Information technology - Open Systems Interconnection – Common management information protocol: Specification - Technical Corrigendum 1	
X.712	Information technology - Open Systems Interconnection - Common management information protocol: Protocol implementation conformance statement proforma	
X.712 Corr 1&2	Corrigenda 1 and 2 (10/96) to Recommendation X.712 - Information technology - Open Systems Interconnection – Common management information protocol: Protocol Implementation Conformance Statement (PICS) proforma	✗
X.712 Corr 3	Corrigendum 3 (06/98) to Recommendation X.712 - Information technology - Open Systems Interconnection – Common management information protocol: Protocol Implementation Conformance Statement (PICS) proforma Technical Corrigendum 3	
X.720	Information technology - Open Systems Interconnection - Structure of management information: management information model	
X.720 Corr.1	Information technology - Open Systems Interconnection - Structure of management information: Management information model	
X.720 Amend.1	Amendment 1 to Recommendation X.720 (11/95) - Generalization of terms	✗

Series X Recommendations
Data networks and open system communication

Rec. No.	Description/Title	S
X.721	Information technology - Open Systems Interconnection - Structure of management information: definition of management information	
X.721 Corr.1	Corrigendum 1 to Recommendation X.721 (02/94)	
X.721 Corr.2	Corrigendum 2 to Recommendation X.721 (10/96)	✗
X.721 Corr 3	Corrigendum 3 (06/98) to Recommendation X.721 - Information technology - Open Systems Interconnection - Structure of management information: Definition of management information Technical corrigendum 3	
X.722	Information technology - Open Systems Interconnection - Structure of management information: Guidelines for the definition of managed objects	
X.722 Amend.1	Amendment 1 to Recommendation X.722 (11/95) - Set by create and components registration	✗
X.722 Amend.2	Addition of the NO-MODIFY syntax element and guidelines extension	✗
X.722 Amend.3	Information technology - Open Systems Interconnection - Structure of management information: Generic management information	✗
X.722 Corr.1	Corrigendum 1 to Recommendation X.722 (10/96)	✗
X.723	Information technology - Open Systems Interconnection - Structure of management information: Generic management information	✗
X.724	Information technology - Open Systems Interconnection - Structure of management information: Requirements and guidelines for implementation conformance statement proformas associated with OSI management	✗
X.725	Information technology - Open Systems Interconnection - Structure of management information: General relationship model	✗
X.726	Information technology - Open Systems Interconnection - Structure of management information: General Relationship Model	
X.727	Information technology - Open Systems Interconnection - Structure of management information: Systems management application layer managed objects	
X.730	Information technology - Open Systems Interconnection - Systems Management: Object Management Function	✗
X.730 Amend.1	Amendment 1 to Recommendation X.730 (04/95)	✗
X.730 Corr.1 Amend.1	Corrigendum 1 to Amendment 1 of Recommendation X.730 (10/96) Implementation conformance statement proformas	✗
X.731	Information technology - Open Systems Interconnection - Systems management: State management function	
X.731 Amend.1	Amendment 1 to Recommendation X.731 (04/95)	✗
X.731 Corr.1	Corrigendum 1 to Recommendation X.731 (04/95)	✗
X.731 Corr.1 Amend.1	Corrigendum 1 to Amendment 1 of Recommendation X.731 (10/96) - Implementation conformance statement proformas	
X.732	Information technology - Open Systems Interconnection - Systems management: Attributes for representing relationships	
X.732 Amend.1	Amendment 1 to Recommendation X.732 (04/95)	✗
X.732 Corr.1 Amend.1	Corrigendum 1 to Amendment 1 of Recommendation X.732 (10/96)	✗
X.733	Information technology - Open Systems Interconnection - Systems management: Alarm reporting function	
X.733 Corr.1	Corrigendum 1 to Recommendation X.733 (02/94)	
X.733 Amend.1	Amendment 1 to Recommendation X.733 (04/95)	
X.733 Corr.1 Amend.1	Corrigendum 1 to Amendment 1 of Recommendation X.733 (10/96) - Implementation conformance statement proformas	✗
X.733 corr 2	Corrigendum 2 (03/99) to Recommendation X.733 - Information technology - Open Systems Interconnection – Systems Management: Alarm reporting function - Technical Corrigendum 2	
X.734	Information technology - Open Systems Interconnection - Systems management: Event report management function	
X.734 Corr.1	Corrigendum 1 (02/94) to Recommendation X.734 - Information technology - Open Systems Interconnection – Systems management: Event report management functions	
X.734 Amend.1	Amendment 1 (04/95) to Recommendation X.734 - Information technology - Open Systems Interconnection – Systems management: Event report management function	

Series X Recommendations
Data networks and open system communication

Rec. No.	Description/Title	S
X.734 Corr.1 Amend.1	Corrigendum 1 (10/96) to Amendment 1 of Recommendation X.734 - Information technology - Open Systems Interconnection - Systems management: Event report management function - Amendment 1: Implementation conformance statement proformas	✗
X.734 Corr.2	Corrigendum 2 (03/99) to Recommendation X.734 - Information technology - Open Systems Interconnection – Systems Management: Event Report Management Function - Technical Corrigendum 2	
X.735	Information technology – Open Systems Interconnection – Systems management: Log control function	
X.735 Amend.1	Amendment 1 to Recommendation X.735 (04/95)	
X.735 Corr.1 Amend.1	Corrigendum 1 to Amendment 1 of Recommendation X.735 (10/96) – Implementation conformance statement proformas	✗
X.736	Information technology – Open Systems Interconnection – Systems management: Security alarm reporting function	✗
X.736 Amend.1	Amendment 1 to Recommendation X.736 (04/95)	
X.736 Corr.1 Amend.1	Corrigendum 1 to Amendment 1 of Recommendation X.736 (10/96) – Implementation conformance statement proformas	✗
X.737	Information Technology – Open Systems Interconnection – Systems Management: Confidence and diagnostic test categories	✗
X.737 Corr 1	Corrigendum 1 (06/98) to Recommendation X.737 - Information technology - Open Systems Interconnection – Systems management: Confidence and diagnostics test categories: Technical corrigendum 1	
X.738	Information Technology - Open Systems Interconnection - Systems Management: Summarization Function	✗
X.738 Amend.1	Amendment 1 to Recommendation X.738 (11/93)	✗
X.738 Corr 1	Corrigendum 1 (06/98) to Recommendation X.738 - Information technology - Open Systems Interconnection – Systems management: Summarization function: Technical corrigendum 1	
X.739	Information Technology - Open Systems Interconnection - Systems Management: Metric objects and attributes	✗
X.739 Corr 1	Corrigendum 1 (06/98) to Recommendation X.739 - Information technology - Open Systems Interconnection – Systems management: Metric objects and attributes: Technical corrigendum 1	
X.740	Information technology - Open Systems Interconnection - Systems management: Security audit trail function	
X.740 Corr.1	Corrigendum 1 (04/95) to Recommendation X.740 - Information technology - Open Systems Interconnection – Systems management: Security audit trail function	
X.740 Corr.2	Corrigendum 2 (10/96) to Recommendation X.740 - Information technology - Open Systems Interconnection – Systems management: Security audit trail function - Technical corrigendum 2	
X.740 Corr.3	Corrigendum 3 (06/98) to Recommendation X.740 - Information technology - Open Systems Interconnection – Systems management: Security audit trail fuction: Technical corrigendum 3	
X.741	Information technology - Open Systems Interconnection - Systems management: Objects and attributes for access control	
X.741 Corr.1	Corrigendum 1 (10/96) to Recommendation X.741 - Information technology - Open systems interconnection – Systems management: Objects and attributes for access control – Technical corrigendum 1	
X.741 Corr.2	Corrigendum 2 (06/98) to Recommendation X.741 - Information technology - Open Systems Interconnection – Systems management: Objects and attributes for access control: Technical corrigendum 2	
X.742	Information technology - Open Systems Interconnection - Systems management: Usage metering function for accounting purposes	
X.742 Corr.1	Amendment 1 to Recommendation X.742 (10/97) - Information technology - Open Systems Interconnection – Systems management: Usage metering function for accounting purposes; Amendment 1: Implementation conformance statement proformas	
X.743 Corr.2	Information technology - Open Systems Interconnection – Systems management: Usage metering function for accounting purposes: Technical corrigendum 1	
X.744	Information technology – Open Systems Interconnection – Systems management: Software management function	✗
X.744 Corr 1	Corrigendum 1 (06/98) to Recommendation X.744 - Information technology - Open Systems Interconnection – Systems management: Software management function; Technical corrigendum 1	

Series X Recommendations
Data networks and open system communication

Rec. No.	Description/Title	S
X.745	Information technology - Open Systems Interconnection - Systems management: Test management function	✗
X.745 Corr 1	Corrigendum 1 (08/97) to Recommendation X.745 - Information technology - Open Systems Interconnection – Systems management: Test management function Technical corrigendum 1	
X.745 Corr 2	Corrigendum 2 (06/98) to Recommendation X.745 - Information technology - Open Systems Interconnection – Systems management: Test management function: Technical corrigendum 2	
X.746	Information technology - Open Systems Interconnection - Systems management: Scheduling function	✗
X.748	Information technology – Open Systems Interconnection - Systems management: Response Time Monitoring Function	
X.749	Information technology – Open Systems Interconnection – Systems Management: Management domain and management policy management function	✗
X.750	Information technology - Open Systems Interconnection - Systems management: Management knowledge management function	✗
X.750 Amend 1	Amendment 1 to Recommendation X.750 (10/97) - Information technology - Open Systems Interconnection – Systems management: Management knowledge management function; Amendment 1: Extension for General Relationship Model	
X.751	Information technology - Open Systems Interconnection - Systems management: Change over function	✗
X.751 Corr 1	Corrigendum 1 (06/98) to Recommendation X.751 - Information technology - Open Systems Interconnection – Systems management: Changeover function: Technical corrigendum 1	
X.753	Information technology – Open Systems Interconnection - Systems management: Command sequencer for systems management	
X.790	Trouble management function for ITU-T applications	✗
X.790 Amend.1	Amendment 1 to Recommendation X.790 (10/96) - Implementation conformance statement proformas	✗
X.790 Corr 1	Corrigendum 1 (03/99) to Recommendation X.790 - Trouble management function for ITU-T applications - Corrigendum 1	
X.791	Profile for trouble management function for ITU-T applications	✗
X.792	Configuration audit support function for ITU-T applications	
X.800	Security architecture for Open Systems Interconnection for CCITT applications	✗
X.800 Amend.1	Amendment 1 to Recommendation X.800 (10/96)	✗
X.802	Information technology - Lower layers security model	✗
X.803	Information technology - Open Systems Interconnection - Upper layers security model	✗
X.810	Information technology - Open Systems Interconnection - Security frameworks for open systems: Overview	✗
X.811	Information technology - Open Systems Interconnection - Security frameworks for open systems: Authentication framework	✗
X.812	Information technology - Open Systems Interconnection - Security frameworks for open systems: Access control framework	✗
X.813	Information technology - Open Systems Interconnection - Security frameworks in open systems: Non-repudiation framework	✗
X.814	Information technology - Open Systems Interconnection - Security frameworks for open systems: Confidentiality framework	✗
X.815	Information technology - Open Systems Interconnection - Security frameworks for open systems: Integrity frameworks	✗
X.816	Information technology - Open Systems Interconnection - Security frameworks for open systems: Security audit and alarms framework	✗
X.830	Information technology - Open Systems Interconnection - Generic upper layers security: Overview, models and notation	✗
X.831	Information technology - Open Systems Interconnection - Generic upper layers security: Security Exchange Service Element (SESE) service definition	✗
X.832	Information technology - Open Systems Interconnection - Generic upper layers security: Security Exchange Service Element (SESE) protocol specification	✗
X.833	Information technology - Open Systems Interconnection - Generic upper layers security: Protecting transfer syntax specification	✗
X.834	Information technology - Open systems interconnection - Generic upper layers security: Security exchange service element (SESE) protocol implementation conformance statement (PICS) proforma	✗
X.835	Information technology - Open Systems Interconnection - Generic upper layers security: Protecting transfer syntax protocol implementation conformance statement (PICS) proforma	✗

Series X Recommendations

Data networks and open system communication

Rec. No.	Description/Title	S
X.851	Information technology - Open Systems Interconnection - Service definition for the commitment, concurrency and recovery service element	✗
X.852	Information technology - Open Systems Interconnection - Protocol for the commitment, concurrency and recovery service element: Protocol specification	✗
X.853	Information technology - Open Systems Interconnection - Protocol for the commitment, concurrency and recovery service element: Protocol implementation conformance statement (PICS) proforma	✗
X.860	Open Systems Interconnection - Distributed transaction processing	✗
X.861	Open Systems Interconnection - Distributed transaction processing: Service definition	✗
X.862	Open Systems Interconnection - Distributed transaction processing: Protocol specification	✗
X.863	Information technology - Open Systems Interconnection - Distributed transaction processing: Protocol implementation conformance statement (PICS) proforma	✗
X.880	Information technology - Remote operations: Concepts, model and notation	✗
X.880 Corr.1	Corrigendum 1 to Recommendation X.880 (11/95)	✗
X.880 Amend.1	Amendment 1 to Recommendation X.880 (11/95) - Built-in operations	✗
X.881	Information technology - Remote operations: OSI realizations - Remote Operations Service Element (ROSE) service definition	✗
X.881 Amend.1	Amendment 1 to Recommendation X.881 (11/95): Mapping to A-UNIT-DATA and built-in operations	✗
X.882	Information technology - Remote operations: OSI realizations - Remote Operations Service Element (ROSE) protocol specification	✗
X.882 Corr.1	Corrigendum 1 (11/95) to Recommendation X.882 - Information technology - Remote operations: OSI realizations – Remote Operations Service Element (ROSE) protocol specification: Technical Corrigendum 1	✗
X.882 Amend.1	Amendment 1 to Recommendation X.882 (11/95): Mapping to A-UNIT-DATA and built-in operations	✗
X.901	Information technology – Open distributed processing - Reference Model: Overview	
X.902	Information technology - Open distributed processing - Reference model: Foundations	✗
X.903	Information technology - Open distributed processing - Reference model: Architecture	✗
X.904	Information technology – Open distributed processing - Reference Model: Architectural semantics	✗
X.910	Information technology – Open distributed processing - Naming framework	✗
X.920	Information technology – Open distributed processing - Interface definition language	✗
X.930	Information technology – Open distributed processing - Interface references and binding	✗
X.931	Information technology – Open distributed processing - Protocol support for computational interactions	✗
X.950	Information technology – Open distributed processing - Trading function: Specification	✗
X.952	Information technology – Open distributed processing - Trading function: Provision of trading function using OSI Directory service	✗

Series Y Recommendations

Global information infrastructure and Internet protocol aspects

Rec. No.	Description/Title	S
Y.100	General overview of the Global Information Infrastructure standards development	✗
Y.110	Global Information Infrastructure principles and framework architecture	✗
Y.120	Global Information Infrastructure scenario methodology	✗
Y.120 Annex A	Annex A to Recommendation Y.120 (02/99) – Global information infrastructure scenario methodology - Annex A: Examples of use	✗
Y.1000	Recommendation Series on IP Related Issues - Full Featured IP Integrated Networks	
Y.1010	Recommendation Series on IP Related Issues - Vocabulary	
Y.1100	Recommendation Series on IP Related Issues - Services and Applications (including Multimedia)	
Y.1200	Recommendation Series on IP Related Issues - General Network Considerations	
Y.1210	Recommendation Series on IP Related Issues - Reference Models	
Y.1220	Recommendation Series on IP Related Issues - Functional Architecture	
Y.1230	Recommendation Series on IP Related Issues - Access Architectures	

Series Y Recommendations

Global information infrastructure and Internet protocol aspects

Rec. No.	Description/Title	S
Y.1231	Recommendation Series on IP Related Issues - Access Capabilities	
Y.1232	Recommendation Series on IP Related Issues - Access Interfaces	
Y.1240	Recommendation Series on IP Related Issues - Network Capabilities	
Y.1250	Recommendation Series on IP Related Issues - Resource Management	
Y.1260	Recommendation Series on IP Related Issues - Traffic Engineering	
Y.1270	Recommendation Series on IP Related Issues - IP Network Security	
Y.1300	Recommendation Series on IP Related Issues - General Transport Considerations	
Y.1310	Recommendation Series on IP Related Issues - IP over ATM	
Y.1320	Recommendation Series on IP Related Issues - IP over SDH	
Y.1330	Recommendation Series on IP Related Issues - IP over Optical (WDM)	
Y.1340	Recommendation Series on IP Related Issues - IP over Satellite	
Y.1350	Recommendation Series on IP Related Issues - IP over Cable	
Y.1360	Recommendation Series on IP Related Issues - IP over Wireless	
Y.1400	Recommendation Series on IP Related Issues - General Interworking Considerations	
Y.1410	Recommendation Series on IP Related Issues - Narrowband ISDN	
Y.1420	Recommendation Series on IP Related Issues - Broadband ISDN	
Y.1430	Recommendation Series on IP Related Issues - Wireless	
Y.1440	Recommendation Series on IP Related Issues - Satellite	
Y.1450	Recommendation Series on IP Related Issues - Cables	
Y.1500	Recommendation Series on IP Related Issues - General QoS and NP Considerations	
Y.1510	Recommendation Series on IP Related Issues - Customer-Perceived QoS Including Customer Equipment Effects	
Y.1520	Recommendation Series on IP Related Issues - Reliability, Availability, Survivability, and Emergency Services	
Y.1530	Recommendation Series on IP Related Issues - Signalling, Call, and Connection Processing Performance	
Y.1540	Recommendation Series on IP Related Issues - User Information Transfer Performance	
Y.1550	Recommendation Series on IP Related Issues - Timing and Synchronization Performance	
Y.1560	Recommendation Series on IP Related Issues - QoS and NP Across Heterogeneous Networks	
Y.1580	Recommendation Series on IP Related Issues - Performance of Network Components	
Y.1580	Recommendation Series on IP Related Issues - Performance Monitoring and Measurement	
Y.1600	Recommendation Series on IP Related Issues – Signalling	
Y.1700	Recommendation Series on IP Related Issues – OAM	
Y.1800	Recommendation Series on IP Related Issues - Charging	

Series Z Recommendations

Programming languages

Rec. No.	Description/Title	S
Z.100	CCITT specification and description language (SDL)	✗
Z.100 Adden 1	Addendum 1 (10/96) to Recommendation Z.100 – CCITT Specification and Description Language (SDL)	✗
Z.100 Supp 1	Supplement 1 to Recommendation Z.100 (05/97) - SDL + methodology: Use of MSC and SDL (with ASN.1)	✗
Z.105	SDL Combined with ASN.1 (SDL/ASN.1)	✗
Z.106	Common interchange format for SDL	✗
Z.107	SDL with embedded ASN.1	✗
Z.109	SDL combined with UML	✗
Z.110	Criteria for the use and applicability of formal description techniques	✗
Z.120	Message sequence chart (MSC)	
Z.130	ITU object definition language	✗
Z.200	CCITT High Level Language (CHILL)	
Z.301	Introduction to the CCITT man-machine language	
Z.302	The meta-language for describing MML syntax and dialogue procedures	

Series Z Recommendations
Programming languages

Rec. No.	Description/Title	S
Z.311	Introduction to syntax and dialogue procedures	
Z.312	Basic format layout	
Z.314	The character set and basic elements	
Z.315	Input (command) language syntax specification	
Z.316	Output language syntax specification	
Z.317	Man-machine dialogue procedures	
Z.321	Introduction to the extended MML for visual display terminals	
Z.322	Capabilities of visual display terminals	
Z.323	Man-machine interaction	
Z.331	Introduction to the specification of the man-machine interface	
Z.332	Methodology for the specification of the man-machine interface – General working procedure	
Z.333	Methodology for the specification of the man-machine interface – Tools and methods	
Z.334	Subscriber administration	
Z.335	Routing administration	
Z.336	Traffic measurement administration	
Z.337	Network management administration	
Z.341	Glossary of terms	
Z.351	Data oriented human-machine interface specification technique – Introduction	X
Z.352	Data oriented human-machine interface specification technique – Scope, approach and reference model	X
Z.360	Graphic GDMO: A graphic notation for the Guidelines for the Definition of Managed Objects	X
Z.361	Design guidelines for Human-Computer Interfaces (HCI) for the management of telecommunications networks	X
Z.400	Structure and format of quality manuals for telecommunications software	
Z.500	Framework on formal methods in conformance testing	X

Appendix F
ITU-R Recommendation Titles

The following is a list of the International Telecommunications Union—Radio sub-section (ITU-R) recommendations that are in force as of December 1997. Because from time to time the recommendations are updated or have new sections added, one should obtain the most current issue before committing a design to them. ITU's URL address is *http://www.itu.ch/*. Many of the publications can be purchased through the ITU on-line subscription or from their bookstore. In the United States, another source for these and many other organizations' standards is Global Engineering Documents™ whose URL is *http://www.ihs.com*.

BO Series Recommendations
Broadcasting satellite service (sound and television)

Rec. No.	Description/Title
BO.566-3	Terminology relating to the use of space communication techniques for broadcasting
BO.600-1	Standardized set of test conditions and measurement procedures for the subjective and objective determination of protection ratios for television in the terrestrial broadcasting and the broadcasting-satellite services
BO.650-2	Standards for conventional television systems for satellite broadcasting in the channels defined by Appendix 30 of the Radio Regulations
BO.651	Digital PCM coding for the emission of high-quality sound signals in satellite broadcasting (15 kHz nominal bandwidth)
BO.652-1	Reference patterns for earth-station and satellite antennas for the broadcasting-satellite service in the 12 GHz band and for the associated feeder links in the 14 GHz and 17 GHz bands
BO.712-1	High-quality sound/data standards for the broadcasting-satellite service in the 12 GHz band
BO.786	MUSEsystem for HDTV broadcasting-satellite services
BO.787	MAC/packet based system for HDTV broadcasting-satellite services
BO.788-1	Coding rate for virtually transparent studio quality HDTV emissions in the broadcasting-satellite services
BO.789-2	Service for digital sound broadcasting to vehicular, portable and fixed receivers for broadcasting-satellite service (sound) in the frequency range 1 400 to 2 700 MHz (Revised)
BO.790	Characteristics of receiving equipment and calculation of receiver figure-of-merit (G/T) for the broadcasting-satellite service
BO.791	Choice of polarization for the broadcasting-satellite service
BO.792	Interference protection ratios for the broadcasting-satellite service (television) in the 12 GHz band
BO.793	Partitioning of noise between feeder links for the broadcasting-satellite service (BSS) and BSS down links
BO.794	Techniques for minimizing the impact on the overall BSS system performance due to rain along the feeder-link path
BO.795	Techniques for alleviating mutual interference between feeder links to the BSS
BO.1130-1	System for digital sound broadcasting to vehicular, portable and fixed receivers for broadcasting service satellite (sound) bands in the frequency range 1 400 to 2 700 MHz (revised)
BO.1211	Digital multi-programme emission systems for television, sound and data services for satellites operating in the 11/12 GHz frequency range
BO.1212	Calculation of total interference between geostationary-satellite networks in the broadcasting-satellite service
BO.1213	Reference receiving earth station antenna patterns for replanning purposes to be used in the revision of the WARC-77 BSS plans for Regions 1 and 3

BR Series Recommendations
Sound and television recording

Rec. No.	Description/Title
BR.265-7	Standards for the international exchange of programmes on film for television use
BR.407-4	International exchange of sound programmes recorded in analogue form
BR.408-6	Standards of sound recording on magnetic tape for the international exchange of programmes

BR Series Recommendations
Sound and television recording

Rec. No.	Description/Title
BR.469-6	Analogue composite television tape recording. Standards for the international exchange of television programmes on magnetic tape
BR.602-3	Exchange of television recordings for programme evaluation
BR.648	Digital recording of audio signals
BR.649-1	Measuring methods for analogue audio tape recordings
BR.657-2	Digital television tape recording. Standards for the international exchange of television programmes on magnetic tape
BR.713	Recording of HDTV images on film
BR.714-1	International exchange of programmes electronically produced by means of high-definition television
BR.715	International exchange of ENG recordings
BR.777-2	International exchange of digital audio recordings (Revised)
BR.778-1	Analogue component television tape recording. Standards for the international exchange of television programmes on magnetic tapes
BR.779	Operating practices for 4:2:2 digital television recording
BR.780	Time and control code standards for the international exchange of television programmes on magnetic tapes
BR.785	The release of programmes in a multimedia environment
BR.1214	Studio recording of sound-broadcasting programmes on magnetic tape for release on multi-programme digital channels
BR.1215	Handling and storage of television and sound recordings on magnetic tape
BR.1216	Recording of television programmes on magnetic tape in the case when several programmes are broadcast in the same digital multiplex
BR.1218	Recording of teletext on future digital recorder for consumer use
BR.1219	Handling and storage of cinematographic film recording
BR.1220	Requirements for the generation, recording and presentation of HDTV programmes intended for release in the "electronic cinema"

BS Series Recommendations
Broadcasting service (sound)

Rec. No.	Description/Title
BS.48-2	Choice of frequency for sound broadcasting in the Tropical Zone
BS.80-3	Transmitting antennas in HF broadcasting
BS.139-3	Transmitting antennas for sound broadcasting in the Tropical Zone
BS.215-2	Maximum transmitter powers for broadcasting in the Tropical Zone
BS.216-2	Protection ratio for sound broadcasting in the Tropical Zone
BS.411-4	Fading allowances in HF broadcasting
BS.412-7	Planning standards for FM sound broadcasting at VHF (Revised)
BS.415-2	Minimum performance specifications for low-cost sound-broadcasting receivers
BS.450-2	Transmission standards for FM sound broadcasting at VHF (Revised)
BS.467	Technical characteristics to be checked for frequency-modulation stereophonic broadcasting. Pilot-tone system
BS.468-4	Measurement of audio-frequency noise voltage level in sound broadcasting
BS.498-2	Ionospheric cross-modulation in the LF and MF broadcasting bands
BS.559-2	Objective measurement of radio-frequency protection ratios in LF, MF and HF broadcasting
BS.560-3	Radio-frequency protection ratios in LF, MF, and HF broadcasting
BS.561-2	Definitions of radiation in LF, MF and HF broadcasting bands
BS.562-3	Subjective assessment of sound quality
BS.597-1	Channel spacing for sound broadcasting in band 7 (HF)
BS.598-1	Factors influencing the limits of amplitude-modulation sound-broadcasting coverage in band 6 (MF)
BS.599	Directivity of antennas for the reception of sound broadcasting in band 8 (VHF)
BS.638	Terms and definitions used in frequency planning for sound broadcasting
BS.639	Necessary bandwidth of emission in LF, MF and HF broadcasting
BS.640-2	Single sideband (SSD) system for HF broadcasting
BS.641	Determination of radio-frequency protection ratios for frequency-modulation sound broadcasting

BS Series Recommendations
Broadcasting service (sound)

Rec. No.	Description/Title
BS.642-1	Limiters for high-quality sound-programme signals
BS.643-2	System for automatic tuning and other applications in FM radio receivers for use with the pilot-tone system (Revised)
BS.644-1	Audio quality parameters for the performance of a high-quality sound-programme transmission chain
BS.645-2	Test signals and metering to be used on international sound-programme connections
BS.646-1	Source encoding for digital sound signals in broadcasting studios
BS.647-2	A digital audio interface for broadcasting studios
BS.702-1	Synchronization and multiple frequency use per programme in HF broadcasting
BS.703	Characteristics of AM sound broadcasting reference receivers for planning purposes
BS.704	Characteristics of FM sound broadcasting reference receivers for planning purposes
BS.705-1	HF transmitting and receiving antennas characteristics and diagrams (Revised)
BS.706-1	Data system in monophonic AM sound broadcasting (AMDS)
BS.707-2	Transmission of multi-sound in terrestrial television systems PAL B, G, H, and I, and SECAM L (Revised)
BS.708	Determination of the electro-acoustical properties of studio monitor headphones
BS.773	Radio-frequency protection ratios required by FM sound broadcasting in the band between 87.5 MHz and 108 MHz against interference from D/SECAM television transmissions
BS.774-2	Service requirements for digital sound broadcasting to vehicular, portable and fixed receivers using terrestrial transmitters in the VHF/UHF bands (Revised)
BS.775-1	Multichannel stereophonic sound systems with and without accompanying picture
BS.776	Format for user data channel of the digital audio interface
BS.1114-1	Systems for terrestrial digital sound broadcasting to vehicular, portable and fixed receivers in the frequency range 30-3 000 MHz (Revised)
BS.1115	Low bit-rate audio coding
BS.1116	Methods for the subjective assessment of small impairments in audio systems including multichannel sound systems
BS.1194	System for multiplexing FM sound broadcasts with a sub-carrier data channel having a relatively large transmission capacity for stationary and mobile reception
BS.1195	Transmitting antenna characteristics at VHF and UHF
BS.1196	Audio coding for digital terrestrial television broadcasting

BT Series Recommendations
Broadcasting service (television)

Rec. No.	Description/Title
BT.266-1	Phase pre-correction of television transmitters
BT.417-4	Minimum field strengths for which protection may be sought in planning a television service
BT.419-3	Directivity and polarization discrimination of antennas in the reception of television broadcasting
BT.470-4	Television Systems (Revised)
BT.471-1	Nomenclature and description of colour bar signals
BT.472-3	Video-frequency characteristics of a television system to be used for the international exchange of programmes between countries that have adopted 625-line colour or monochrome systems
BT.500-7	Methodology for the subjective assessment of the quality of television pictures (Revised)
BT.565	Protection ratios for 625-line television against radio navigation transmitters operating in the shared bands between 582 and 606 MHz
BT.601-5	Studio encoding parameters of digital television for standard 4:3 and wide-screen 16:9 aspect ratios (Revised)
BT.653-2	Teletext Systems
BT.654	Subjective quality of television pictures in relation to the main impairments of the analogue composite television signal
BT.655-4	Radio-frequency protection ratios for AM vestigial sideband terrestrial television systems (Revised)
BT.656-3	Interfaces for digital component video signals in 525-line and 625-line television systems operating at the 4:2:2 level of Recommendation ITU-R BT.601 (Part A) (Revised)
BT.709-2	Parameter values for the HDTV standards for production and international programme exchange (Revised)

BT Series Recommendations
Broadcasting service (television)

Rec. No.	Description/Title
BT.710-2	Subjective assessment methods for image quality in high-definition television
BT.711-1	Synchronizing reference signals for the component digital studio
BT.796	Parameters for enhanced compatible coding systems based on 625-line PAL and SECAM television systems
BT.797-1	Parameters for 4:3 enhanced television systems that are NTSC-compatible
BT.798-1	Digital television terrestrial broadcasting in the VHF/UHF bands
BT.799-2	Interfaces for digital component video signals in 525-line and 625-line television systems operating at the 4:4:4 level of Recommendation ITU-R BT.601 (Part A) (Revised)
BT.800-2	User requirements for the transmission through contribution and primary distribution networks of digital television signals defined according to the 4:2:2 standard of Recommendation ITU-R BT.601 (Part A) (Revised)
BT.801-1	Test signals for digitally encoded colour television signals conforming with Recommendations ITU-R BT.601 (Part A) and ITU-R BT.656
BT.802-1	Test pictures and sequences for subjective assessments of digital codecs conveying signals produced according to Recommendation ITU-R BT.601
BT.803	The avoidance of interference generated by digital television studio equipment
BT.804	Characteristics of TV receivers essential for frequency planning with PAL/SECAM/NTSC television systems
BT.805	Assessment of impairment caused to television reception by a wind turbine
BT.806	Common channel raster for the distribution of D-MAC, D2-MAC and HD-MAC signals in collective antenna and cable distribution systems
BT.807	Reference model for data broadcasting
BT.808	The broadcasting of time and date information in coded form
BT.809	Programme delivery control (PDC) system for video recording
BT.810	Conditional-access broadcasting systems
BT.811-1	The subjective assessment of enhanced PAL and SECAM systems
BT.812	Subjective assessment of the quality of alphanumeric and graphic pictures in teletext and similar services
BT.813	Methods for objective picture quality assessment in relation to impairments from digital coding of television signals
BT.814-1	Specifications and alignment procedures for setting of brightness and contrast of displays
BT.815-1	Specification of a signal for measurement of the contrast ratio of displays
BT.1117-1	Studio format parameters for enhanced 16:9 625-line television systems (D-and D2-MAC, PALplus, enhanced SECAM) (Revised)
BT.1118	Enhanced compatible widescreen television based on conventional television systems
BT.1119-1	Widescreen signalling for broadcasting. Signalling for widescreen and other enhanced television parameters
BT.1120	Digital Interfaces for 1125/60/2:1 and 1250/50/2:1 HDTV studio signals
BT.1121-1	User requirements for the transmission through contribution and primary distribution network of digital HDTV signals (Revised)
BT.1122-1	User requirements for emission and secondary distribution systems for SDTV, HDTV and hierarchical coding schemes (Revised)
BT.1123	Planning methods for 625-line terrestrial television in VHF/UHF bands
BT.1124-1	Reference signals for ghost cancelling in television (Revised)
BT.1125	Basic objectives for the planning and implementation of digital terrestrial television broadcasting systems
BT.1126	Data transmission protocols and transmission control scheme for data broadcasting systems using a data channel in satellite television broadcasting
BT.1127	Relative quality requirements of television broadcast systems
BT.1128-1	Subjective assessment of conventional television systems (Revised)
BT.1129-1	Subjective assessment of standard definition digital television (SDTV) systems (Revised)
BT.1197	Enhanced wide-screen PAL TV transmission system (the PALplus system)
BT.1198	Stereoscopic television based on R- and L-eye two channel signals
BT.1199	Use of bit-rate reduction in the HDTV studio environment
BT.1200	Target standard for digital video systems for the studio and for international programme exchange
BT.1201	Extremely high resolution imagery
BT.1202	Displays for future television systems
BT.1203	User requirements for generic bit-rate reduction coding of digital TV signals (SDTV, EDTV and HDTV) for an end-to-end television system

BT Series Recommendations
Broadcasting service (television)

Rec. No.	Description/Title
BT.1204	Measuring methods for digital video equipment with analogue input/output
BT.1205	User requirements for the quality of baseband SDTV and HDTV signals when transmitted by digital satellite news gathering (SNG)
BT.1206	Spectrum shaping limits for digital terrestrial television broadcasting
BT.1207	Data access methods for digital terrestrial television broadcasting
BT.1208	Video coding for digital terrestrial television broadcasting
BT.1209	Service multiplex methods for digital terrestrial television broadcasting
BT.1210	Test materials to be used in subjective assessment

F Series Recommendations
Fixed service

Rec. No.	Description/Title
F.106-1	Voice-frequency telegraphy on radio circuits
F.162-3	Use of directional transmitting antennas in the fixed service operating in bands below about 30 MHz
F.240-6	Signal-to-interference protection ratios for various classes of emission in the fixed service below about 30 MHz
F.246-3	Frequency-shift keying
F.268-1	Interconnection at audio frequencies of radio-relay systems for telephony
F.270-2	Interconnection at video signal frequencies of radio-relay systems for television
F.275-3	Pre-emphasis characteristic for frequency modulation radio-relay systems for telephony using frequency-division multiplex
F.276-2	Frequency deviation and the sense of modulation for analogue radio-relay systems for television
F.283-5	Radio-frequency channel arrangements for low and medium capacity analogue or digital radio-relay systems operating in the 2 GHz band
F.290-3	Maintenance measurements on radio-relay systems for telephony using frequency-division multiplex
F.302-3	Limitation of interference from trans-horizon radio-relay systems
F.305	Stand-by arrangements for radio-relay systems for television and telephony
F.306	Procedure for the international connection of radio-relay systems with different characteristics
F.335-2	Use of radio links in international telephone circuits
F.338-2	Bandwidth required at the output of a telegraph or telephone receiver
F.339-6	Bandwidths, signal-to-noise ratios and fading allowances in complete systems
F.342-2	Automatic error-correcting system for telegraph signals transmitted over radio circuits
F.345	Telegraph distortion
F.347	Classification of multi-channel radiotelegraph systems for long-range circuits operating at frequencies below about 30 MHz and the designation of the channels in these systems
F.348-4	Arrangements of channels in multi-channel single-sideband and independent-sideband transmitters for long-range circuits operating at frequencies below about 30 MHz
F.349-4	Frequency stability required for systems operating in the HF fixed service to make the use of automatic frequency control superfluous
F.380-4	Interconnection at baseband frequencies of radio-relay systems for telephony using frequency-division multiplex
F.381-2	Conditions relating to line regulating and other pilots and to limits for the residues of signals outside the baseband in the interconnection of radio-relay and line systems for telephony
F.382-6	Radio-frequency channel arrangements for radio-relay systems operating in the 2 and 4 GHz bands
F.383-5	Radio-frequency channel arrangements for high capacity radio-relay systems operating in the 6 GHz band
F.384-6	Radio-frequency channel arrangements for medium and high capacity analogue or digital radio-relay systems operating in the upper 6 GHz band (Revised)
F.385-6	Radio-frequency channel arrangements for radio-relay systems operating in the 7 GHz band
F.386-4	Radio-frequency channel arrangements for radio-relay systems operating in the 8 GHz band
F.387-7	Radio-frequency channel arrangements for radio-relay systems operating in the 11 GHz band (Revised)
F.388	Radio-frequency channel arrangements for trans-horizon radio-relay systems
F.389-2	Preferred characteristics of auxiliary radio-relay systems operating in the 2, 4, 6 or 11 GHz bands

F Series Recommendations
Fixed service

Rec. No.	Description/Title
F.390-4	Definitions of terms and references concerning hypothetical reference circuits and hypothetical reference digital paths for radio-relay systems
F.391	Hypothetical reference circuit for radio-relay systems for telephony using frequency-division multiplex with a capacity of 12 to 60 telephone channels
F.392	Hypothetical reference circuit for radio-relay systems for telephony using frequency-division multiplex with a capacity of more than 60 telephone channels
F.393-4	Allowable noise power in the hypothetical reference circuit for radio-relay systems for telephony using frequency-division multiplex
F.395-2	Noise in the radio portion of circuits to be established over real radio-relay links for FDM telephony
F.396-1	Hypothetical reference circuit for trans-horizon radio-relay systems for telephony using frequency-division multiplex
F.397-3	Allowable noise power in the hypothetical reference circuit of trans-horizon radio-relay systems for telephony using frequency-division multiplex
F.398-3	Measurements of noise in actual traffic over radio-relay systems for telephony using frequency-division multiplex
F.399-3	Measurement of noise using a continuous uniform spectrum signal on frequency-division multiplex telephony radio-relay systems
F.400-2	Service channels to be provided for the operation and maintenance of radio-relay systems
F.401-2	Frequencies and deviations of continuity pilots for frequency modulation radio-relay systems for television and telephony
F.402-2	The preferred characteristics of a single sound channel simultaneously transmitted with a television signal on an analogue radio-relay system
F.403-3	Intermediate-frequency characteristics for the interconnection of analogue radio-relay systems
F.404-2	Frequency deviation for analogue radio-relay systems for telephony using frequency-division multiplex
F.405-1	Pre-emphasis characteristics for frequency modulation radio-relay systems for television
F.436-4	Arrangement of voice-frequency, frequency-shift telegraph channels over HF radio circuits (Revised)
F.444-3	Preferred characteristics for multi-line switching arrangements of analogue radio-relay systems
F.454-1	Pilot carrier level for HF single-sideband and independent-sideband reduced-carrier systems
F.455-2	Improved transmission system for HF radiotelephone circuits
F.463-1	Limits for the residues of signals outside the baseband of radio-relay systems for television
F.480	Semi-automatic operation on HF radiotelephone circuits. Devices for remote connection to an automatic exchange by radiotelephone circuits
F.497-5	Radio-frequency channel arrangements for radio-relay systems operating in the 13 GHz frequency band (Revised)
F.518-1	Single-channel simplex ARQ telegraph system
F.519	Single-channel duplex ARQ telegraph system
F.520-2	Use of high frequency ionospheric channel simulators
F.555-1	Permissible noise in the hypothetical reference circuit of radio-relay systems for television
F.556-1	Hypothetical reference digital path for radio-relay systems which may form part of an integrated services digital network with a capacity above the second hierarchical level
F.557-3	Availability objective for radio-relay systems over a hypothetical reference circuit and a hypothetical reference digital path
F.592-2	Terminology used for radio-relay systems
F.593	Noise in real circuits of multi-channel trans-horizon FM radio-relay systems of less than 2500 km
F.594-3	Allowable bit error ratios at the output of the hypothetical reference digital path for radio-relay systems which may form part of an integrated services digital network
F.595-4	Radio-frequency channel arrangements for radio-relay systems operating in the 18 GHz frequency band (Revised)
F.596-1	Interconnection of digital radio-relay systems
F.612	Measurement of reciprocal mixing in HF communication receivers in the fixed service
F.613	The use of ionospheric channel sounding systems operating in the fixed service at frequencies below about 30 MHz
F.634-3	Error performance objectives for real digital radio-relay links forming part of a high-grade circuit within an integrated services digital network
F.635-3	Radio-frequency channel arrangements based on a homogeneous pattern for radio-relay systems operating in the 4 GHz band (Revised)
F.636-3	Radio-frequency channel arrangements for radio-relay systems operating in the 15 GHz band

F Series Recommendations
Fixed service

Rec. No.	Description/Title
F.637-2	Radio-frequency channel arrangements for radio-relay systems operating in the 23 GHz band
F.695	Availability objectives for real digital radio-relay links forming part of a high-grade circuit within an integrated services digital network
F.696-1	Error performance and availability objectives for hypothetical reference digital sections utilizing digital radio-relay systems forming part or all of the medium-grade portion of an ISDN connection
F.697-1	Error performance and availability objectives for the local-grade portion at each end of an ISDN connection utilizing digital radio-relay systems
F.698-2	Preferred frequency bands for trans-horizon radio-relay systems
F.699-4	Reference radiation patterns for line-of-sight radio-relay system antennas for use in coordination studies and interference assessment in the frequency range from 1 to about 40 GHz
F.700-2	Error performance and availability measurement algorithm for digital radio-relay links at the system bit-rate interface
F.701-1	Radio-frequency channel arrangements for analogue and digital point-to-multipoint radio systems operating in frequency bands in the range 1.427 to 2.690 GHz
F.745	CCIR Recommendations for analogue radio-relay systems
F.746-3	Radio-frequency channel arrangements for radio-relay systems
F.747	Radio-frequency channel arrangements for radio-relay systems operating in the 10 GHz band
F.748-2	Radio-frequency channel arrangements for radio-relay systems operating in the 25, 26 and 28 GHz bands (Revised)
F.749-1	Radio-frequency channel arrangements for radio-relay systems operating in the 38 GHz band
F.750-2	Architectures and functional aspects of radio-relay systems for SDH-based networks (Revised)
F.751-1	Transmission characteristics and performance requirements of radio-relay systems for SDH-based networks
F.752-1	Diversity techniques for radio-relay systems
F.753	Preferred methods and characteristics for the supervision and protection of digital radio-relay systems
F.754	Radio-relay systems in bands 8 and 9 for the provision of telephone trunk connections in rural areas
F.755-1	Point-to-multipoint systems used in the fixed service
F.756	TDMA point-to-multipoint systems used as radio concentrators
F.757	Basic system requirements and performance objectives for cellular type mobile systems used as fixed systems
F.758	Considerations in the development of criteria for sharing between the terrestrial fixed service and other services
F.759	Use of frequencies in the band 500 to 3 000 MHz for radio-relay systems
F.760-1	Protection of terrestrial line-of-sight radio-relay systems against interference from the broadcasting-satellite service in the bands near 20 GHz
F.761	Frequency sharing between the fixed service and passive sensors in the band 18.6-18.8 GHz
F.762-2	Main characteristics of remote control and monitoring systems for HF receiving and transmitting stations (Revised)
F.763-2	Data transmission over HF circuits using phase-shift keying (Revised)
F.764-1	Minimum requirements for HF radio systems using a packet transmission protocol
F.1092	Error performance objectives for constant bit rate digital path at or above the primary rate carried by digital radio-relay systems which may form part of the international portion of a 27 500 KM hypothetical reference path
F.1093	Effects of multipath propagation on the design and operation of line-of-sight digital radio-relay systems
F.1094-1	Maximum allowable error performance and availability degradations to digital radio-relay systems arising from interference from emissions and radiations from other sources (Revised)
F.1095	A procedure for determining coordination area between radio-relay stations of the fixed service
F.1096	Methods of calculating line-of-sight interference into radio-relay systems to account for terrain scattering
F.1097	Interference mitigation options to enhance compatibility between radar systems and digital radio-relay systems
F.1098-1	Radio-frequency channel arrangements for radio-relay systems in the 1 900-2 300 MHz band (revised)
F.1099-1	Radio-frequency channel arrangements for high-capacity digital radio-relay systems in the 5 GHz (4 400 to 5 000 MHz) band
F.1100	Radio-frequency channel arrangements for radio-relay systems operating in the 55 GHz band
F.1101	Characteristics of digital radio-relay systems below about 17 GHz
F.1102	Characteristics of radio-relay systems operating in frequency bands above about 17 GHz
F.1103	Radio-relay systems operating in bands 8 and 9 for the provision of subscriber telephone connections in rural areas

F Series Recommendations
Fixed service

Rec. No.	Description/Title
F.1104	Requirements for point-to-multipoint radio systems used in the local grade portion of an ISDN connection
F.1105	Transportable fixed radiocommunications equipment for relief operations
F.1106	Effects of propagation on the design and operation of trans-horizon radio-relay systems
F.1107	Probabilistic analysis for calculating interference into the fixed service from satellites occupying the geostationary orbit
F.1108-1	Determination of the criteria to protect fixed service receivers from the emissions of space stations operating in non-geostationary orbits in shared frequency bands (Revised)
F.1109	ITU-R Recommendations relating to systems in the fixed service service operating at frequencies below about 30 MHz which are not reprinted
F.1110-1	Adaptive radio systems for frequencies below about 30 MHz (Revised)
F.1111-1	Improved Lincompex system for HF radiotelephone circuits (Revised)
F.1112-1	Digitized speech transmissions for systems operating below about 30 MHz (Revised)
F.1113	Radio systems employing meteor-burst propagation
F.1189	Error-performance objectives for constant bit rate digital paths at or above the primary rate carried by digital radio-relay systems which may form part or all of the national portion of a 27 500 km hypothetical reference path
F.1190	Protection criteria for digital radio-relay systems to ensure compatibility with radar systems in the radiodetermination service
F.1191	Bandwidths and unwanted emissions of digital radio-relay systems
F.1192	Traffic capacity of automatically controlled radio systems and networks in the HF fixed service
F.1241	Performance degradation due to interf. from other services sharing the same freq. bands on a primary basis with digital radio-relay syst. oper. at or above the primary rate and which may form part of the intern. portion of a 27 500 km hypoth. ref. path
F.1242	Radio-frequency channel arrangements for digital radio systems operating in the range 1 350 to 1 530 MHz
F.1243	Radio-frequency channel arrangements for digital radio systems operating in the range 2 290-2 670 MHz
F.1244	Radio local area networks (RLANs)
F.1245	Mathematical model of average radiation patterns for line-of-sight point-to-point radio-relay system antennas for use in certain coordination studies and interference assessment in the frequency range from 1 to about 40 GHz
F.1246	Reference bandwidth of receiving stations in the fixed service to be used in coordination of frequency assignments with transmitting space stations in the mobile-satellite service in the 1-3 GHz range
F.1247	Technical and operational characteristics of systems in the fixed service to facilitate sharing with the space research, space operation and Earth exploration-satellite services operating in the bands 2 025 to 2 110 MHz and 2 200 to 2 290 MHz
F.1248	Limiting interference to satellites in the space science services from the emissions of trans-horizon radio-relay systems in the bands 2 025 to 2 110 MHz and 2 200 to 2 290 MHz
F.1249	Maximum equivalent isotropically radiated power of transmitting stations in the fixed service operating in the frequency band 25.25 to 27.5 GHz shared with the inter-satellite service

IS Series Recommendations
Inter-service sharing and compatibility

Rec. No.	Description/Title
IS.847-1	Determination of the coordination area of an earth station operating with a geostationary space station and using the same frequency band as a system in a terrestrial service
IS.848-1	Determination of the coordination area of a transmitting earth station using the same frequency band as receiving earth stations in bidirectionally allocated frequency bands
IS.849-1	Determination of the coordination area for earth stations operating with non-geostationary spacecraft in bands shared with terrestrial services
IS.850-1	Coordination areas using predetermined coordination distances (Revised)
IS.851-1	Sharing between the broadcasting service and the fixed and/or mobile services in the VHF and UHF bands
IS.1009-1	Compatibility between the sound-broadcasting service in the band of about 87-108 MHz and the aeronautical services in the band 108-137 MHz (Revised)

IS Series Recommendations	
Inter-service sharing and compatibility	
Rec. No.	Description/Title
IS.1140	Test procedures for measuring aeronautical receiver characteristics used for determining compatibility between the sound-broadcasting service in the band of about 87-108 MHz and the aeronautical services in the band 108-118 MHz
IS.1141	Sharing in the frequency bands in the 1-3 GHz frequency range between the non-geostationary space stations operating in the mobile-satellie service and the fixed service
IS.1142	Sharing in the frequency bands in the 1-3 GHz frequency range between geostationary space stations operating in the mobile-satellite service and the fixed service
IS.1143	System specific methodology for coordination of non-geostationary space stations (space-to-Earth) operating in the mobile-satellite service with the fixed service

M Series Recommendations	
Mobile, radiodetermination, amateur and related satellite services	
Rec. No.	Description/Title
M.218-2	Prevention of interference to radio reception on board ships
M.219-1	Alarm signal for use on the maritime radiotelephony distress frequency of 2182 kHz
M.257-3	Sequential single frequency selective-calling system for use in the maritime mobile service (Revised)
M.428-3	Direction-finding and/or homing in the 2 MHz band on board ships
M.441-1	Signal-to-interference ratios and minimum field strengths required in the aeronautical mobile (R) service above 30 MHz
M.476-5	Direct-printing telegraph equipment in the maritime mobile service (Revised)
M.478-5	Technical characteristics of equipment and principles governing the allocation of frequency channels between 25 and 3 000 MHz for the FM land mobile service
M.488-1	Equivalent powers of double-sideband and single-sideband radio telephone emissions in the maritime mobile service
M.489-2	Technical characteristics of VHF radiotelephone equipment operating in the maritime mobile service in channels spaced by 25 kHz (Revised)
M.490	The introduction of direct-printing telegraph equipment in the maritime mobile service. Equivalence of terms
M.491-1	Translation between an identity number and identities for direct-printing telegraphy in the maritime mobile service
M.492-6	Operational procedures for the use of direct-printing telegraph equipment in the maritime mobile service (Revised)
M.493-8	Digital selective-calling system for use in the maritime mobile service
M.496-3	Limits of power flux-density of radionavigation transmitters to protect space station receivers in the fixed-satellite service in the 14 GHz band
M.539-3	Technical and operational characteristics of international radio-paging systems
M.540-2	Operational and technical characteristics for an automated direct-printing telegraph system for promulgation of navigational and meteorological warnings and urgent information to ships
M.541-7	Operational procedures for the use of digital selective-calling (DSC) equipment in the maritime mobile service
M.542-1	On-board communications by means of portable radiotelephone equipment
M.546-2	Hypothetical telephone reference circuit in the aeronautical, land and maritime mobile-satellite services
M.547	Noise objectives in the hypothetical reference circuit for systems in the maritime mobile-satellite service
M.548	Overall transmission characteristics of telephone circuits in the maritime mobile-satellite service
M.549-1	Side tone reference equivalent of handset used on board a ship in the maritime mobile-satellite service and in automated VHF/UHF maritime mobile radiotelephone systems
M.550-1	Use of echo suppressors in the maritime mobile-satellite service
M.552	Quality objectives for 50-baud start-stop telegraph transmission in the maritime mobile-satellite service
M.553	Interface requirements for 50-baud start-stop telegraph transmission in the maritime mobile-satellite service
M.584-1	Standard codes and formats for international radio paging
M.585-2	Assignment and use of maritime mobile service identities
M.586-1	Automated VHF/UHF maritime mobile telephone system

M Series Recommendations
Mobile, radiodetermination, amateur and related satellite services

Rec. No.	Description/Title
M.587-1	Coast station identities and initiation of location registration in an automated VHF/UHF maritime mobile telephone system
M.588	Characteristics of maritime radio beacons (Region 1)
M.589-2	Interference to radionavigation services from other services in the frequency bands between 70 and 130 kHz
M.622	Technical and operational characteristics of analogue cellular systems for public land mobile telephone use
M.623	Data transmission bit rates and modulation techniques in the land mobile service
M.624	Public land mobile communication systems location registration
M.625-3	Direct-printing telegraph equipment employing automatic identification in the maritime mobile service (Revised)
M.626	Evaluation of the quality of digital channels in the maritime mobile service
M.627-1	Technical characteristics for HF maritime radio equipment using narrow-band phase-shift keying (NBPSK) telegraphy (Revised)
M.628-3	Technical characteristics for search and rescue radar transponders
M.629	Use of the radionavigation service of the frequency bands 2 900 to 3 100 MHz, 5 470 to 5 650 MHz, 9 200 to 9 300 MHz, 9 300 to 9 500 MHz and 9 500 to 9 800 MHz
M.630	Main characteristics of two frequency shipborne interrogator transponders (SIT)
M.631-1	Use of hyperbolic maritime radionavigation systems in the band 283.5 to 315 kHz
M.632-3	Transmission characteristics of a satellite emergency position-indicating radio beacon (satellite EPIRB) system operating through geostationary satellites in the 1.6 GHz band
M.633-1	Transmission characteristics of a satellite emergency position-indicating radiobeacon (satellite EPIRB) system operating through a low polar-orbiting satellite system in the 406 MHz band
M.687-2	Future Public Land Mobile Telecommunication Systems (FPLMTS)
M.688	Technical characteristics for a high frequency direct-printing telegraph system for promulgation of high seas and NAVTEX-type maritime safety information
M.689-2	International maritime VHF radiotelephone system with automatic facilities based on DSC signalling format
M.690-1	Technical characteristics of emergency position-indicating radio beacons (EPIRBs) operating on the carrier frequencies of 121.5 MHz and 243 MHz (Revised)
M.691-1	Technical characteristics and compatibility criteria of maritime radiolocation systems operating in the medium frequency band and using spread-spectrum techniques
M.692	Narrow-band direct-printing telegraph equipment using a single-frequency channel
M.693	Technical characteristics of VHF emergency position-indicating radio beacons using digital selective calling (DSC VHF EPIRB)
M.694	Reference radiation pattern for ship earth station antennas
M.816	Framework for services supported on future public land mobile telecommunication systems (FPLMTS)
M.817	Future public land mobile telecommunication systems (FPLMTS). Network architectures
M.818-1	Satellite operation within future public land mobile telecommunication systems (FPLMTS)
M.819-2	Future Public Land Mobile Telecommunication Systems (FPLMTS) for developing countries
M.820	Use of 9-digit identities for narrow-band direct-printing telegraphy in the maritime mobile service
M.821-1	Optional expansion of the digital selective-calling system for use in the maritime mobile service
M.822-1	Calling-channel loading for digital selective-calling (DSC) for the maritime mobile service
M.823-1	Technical characteristics of differential transmissions for Global Navigation Satellite Systems (GNSS) from maritime radio beacons in the frequency band 283.5 to 315 kHz in Region 1 and 285 to 325 kHz in Regions 2 & 3
M.824-2	Technical parameters of radar beacons (RACONS) (Revised)
M.825-1	Characteristics of a transponder system using digital selective calling techniques for use with vessel traffic services and ship-to-ship identification (Revised)
M.826	Transmission of information for updating electronic chart display and information systems (ECDIS)
M.827	Hypothetical reference digital path for systems in the mobile-satellite service using feeder links
M.828-1	Definition of availability for communication circuits in the mobile-satellite service (MSS)
M.830	Operational procedures for mobile-satellite networks or systems in the bands 1 530 to 1 544 MHz and 1 626.5 to 1 645.5 MHz which are used for distress and safety purposes as specified for GMDSS
M.831	Frequency sharing between services in the band 4 to 30 MHz
M.1032	Technical and operational characteristics of land mobile systems using multi-channel access techniques without a central controller
M.1033-1	Technical and operational characteristics of cordless telephones and cordless telecommunication systems

M Series Recommendations
Mobile, radiodetermination, amateur and related satellite services

Rec. No.	Description/Title
M.1034-1	Requirements for the radio interface(s) for Future Public Land Mobile Telecommunication Systems (FPLMTS)
M.1035	Framework for the radio interface(s) and radio sub-system functionality for future public land mobile telecommunication systems (FPLMTS)
M.1036	Spectrum considerations for implementation of future public land mobile telecommunication systems (FPLMTS) in the bands 1 885 to 2 025 MHz AND 2 110 to 2 200 MHz
M.1037	Bit error performance objectives for aeronautical mobile-satellite (R) service (AMS(R)S) radio link
M.1038	Efficient use of the geostationary-satellite orbit and spectrum in the 1-3 GHz frequency range by mobile-satellite systems
M.1039-1	Co-frequency sharing between stations in the mobile service below 1 GHz and FDMA non-geostationary-satellite orbit (non-GSO) mobile earth stations
M.1040	Public mobile telecommunication service with aircraft using the bands 1 670 to 1 675 MHz and 1 800 to 1 805 MHz
M.1041	Future amateur radio systems (FARS)
M.1042	Disaster communications in the amateur and amateur-satellite services
M.1043	Use of the amateur and amateur-satellite services in developing countries
M.1044	Frequency sharing criteria in the amateur and amateur-satellite services
M.1072	Interference due to intermodulation products in the land mobile service between 25 and 3 000 MHz
M.1073-1	Digital cellular land mobile telecommunication systems
M.1074	Integration of public mobile radiocommunication systems
M.1075	Leaky feeder systems in the land mobile services
M.1076	Wireless communication systems for persons with impaired hearing
M.1077	Multi-transmitter radio systems using quasi-synchronous (simulcast) transmission for analogue speech
M.1078	Security principles for future public land mobile telecommunication systems (FPLMTS)
M.1079	Speech and voiceband data performance requirements for future public land mobile telecommunication systems (FPLMTS)
M.1080	Digital selective calling system enhancement for multiple equipment installations
M.1081	Automatic HF facsimile and data system for maritime mobile users
M.1082	International maritime MF/HF radiotelephone system with automatic facilities based on DSC signalling format
M.1083	Interworking of maritime radiotelephone systems
M.1084-1	Improved efficiency in the use of the band 156-174 MHz by stations in the maritime mobile service (Revised)
M.1085-1	Technical and operational characteristics of wind profiler radars for bands in the vicinity of 400 MHz
M.1086	Determination of the need for coordination between geostationary mobile satellite networks sharing the same frequency bands
M.1087	Methods for evaluating sharing between systems in the land mobile service and spread-spectrum low-earth orbit (LEO)
M.1088	Considerations for sharing with systems of other services operating in the bands allocated to the radionavigation-satellite service
M.1089	Technical considerations for the coordination of mobile-satellite systems supporting the aeronautical mobile-satellite (R) service (AMS(R)S)
M.1090	Frequency plans for satellite transmission of single channel per carrier (SCPC) carriers using non-linear transponders in the mobile-satellite service
M.1091	Reference off-axis radiation patterns for mobile earth station antennas operating in the land mobile-satellite service in the frequency range 1 to 3 GHz
M.1167	Framework for the satellite component of future public land mobile telecommunication systems (FPLMTS)
M.1168	Framework of future public land mobile telecommunication systems management (FPLMTS)
M.1169	Hours of service of ship stations
M.1170	Morse telegraphy procedures in the maritime mobile service
M.1171	Radiotelephony procedures in the maritime mobile service
M.1172	Miscellaneous abbreviations and signals to be used for radiocommunications in the maritime mobile service
M.1173	Technical characteristics of single-sideband transmitters used in the maritime mobile service for radiotelephony in the bands between 1 606.5 kHz (1 605 kHz Region 2) and 4 000 kHz and between 4 000 and 27 500 kHz
M.1174	Characteristics of equipment used for on-board communications in the bands between 450 and 470 MHz

M Series Recommendations
Mobile, radiodetermination, amateur and related satellite services

Rec. No.	Description/Title
M.1175	Automatic receiving equipment for radiotelegraph and radiotelephone alarm signals
M.1176	Technical parameters of radar target enhancers
M.1177	Techniques for measurement of spurious emissions of maritime radar systems
M.1178	Use of the maritime radionavigation band 283.5-315 kHz in Region 1 and 285-325 kHz in Regions 2 and 3
M.1179	Procedures for determining the interference coupling mechanisms and mitigation options for systems operating in bands adjacent to and in harmonic relationships with radar stations in the radio determination service
M.1180	Availability of communication circuits in the aeronautical mobile-satellite (R) services (AMS(R)S)
M.1181	Minimum performance objectives for narrow-band digital channels using geostationary satellites to serve transportable and vehicular mobile earth stations in the 1-3 GHz range, not forming part of the ISDN
M.1182	Integration of terrestrial and satellite mobile communication systems
M.1183	Permissible levels of interference in a digital channel of a geostationary network in mobile-satellite service in 1-3 GHz caused by other networks of this service and fixed-satellite service
M.1184	Technical characteristics of mobile satellite systems in the 1-3 GHz range for use in developing criteria for sharing between the mobile-satellite service (MSS) and other services using common frequencies
M.1185	Method for determining coordination distance between ground based mobile earth stations and terrestrial stations operating in the 148.0 to 149.9 MHz band
M.1186	Technical considerations for the coordination between MSS networks utilizing code division multiple access (CDMA) and other spread spectrum techniques in the 1-3 GHz band
M.1187	A method for the calculation of the potentially affected region for a mobile-satellite service (MSS) network in the 1-3 GHz range using circular orbits
M.1188	Impact of propagation on the design of non-GSO mobile-satellite systems not employing satellite diversity which provide service to handheld equipment
M.1221	Technical and operational requirements for cellular multimode mobile radio stations
M.1222	Transmission of data messages on shared private land mobile radio channels
M.1223	Evaluation of security mechanims for IMT-2000
M.1224	Vocabulary of terms for Future Public Land Mobile Telecommunication Systems (FPLMTS)
M.1225	Guidelines for evaluation of radio transmission technologies for IMT-2000
M.1226	Technical and operational characteristics of wind profiler radars in bands in the vicinity of 50 MHz
M.1227	Technical and operational characteristics of wind profiler radars in bands in the vicinity of 1 000 MHz
M.1228	Methodology for determining performance objectives for narrow-band channels in mobile satellite systems using geostationary satellites not forming part of the ISDN
M.1229	Performance objectives for the digital aeronautical mobile-satellite service (AMSS) channels operating in the bands 1 525 to 1 559 MHz and 1 626.5 to 1660.5 MHz not forming part of the ISDN
M.1230	Performance objectives for space-to-Earth links operating in the mobile-satellite service with non-geostationary satellites in the 137 to 138 MHz band
M.1231	Interference criteria for space-to-Earth links operating in the mobile-satellite service with non-geostationary in the 137 to 138 MHz band
M.1232	Sharing criteria for space-to-Earth links operating in the mobile-satellite service with non-geostationary satellites in the 137 to 138 MHz band
M.1233	Technical considerations for sharing satellite network resources between the mobile-satellite service (MSS) (other than the aeronautical mobile-satellite (R) service (AMS(R)S)) and AMS(R)S
M.1234	Permissible level of interf. in dig. chan. of geostat. sat. netw. in the aero. mob.-sat. (R) serv. (AMS(R)S) in the band. 1 545-1 555 MHz to 1 646.5-1 656.5 MHz ands its ass. feeder links caus. by other netw. of this serv. and the fixed-sat.-serv.

P Series Recommendations
Radiowave Propagation

Rec. No.	Description/Title
P.310-9	Definitions of terms relating to propagation in non-ionized media
P.311-8	Acquisition, presentation and analysis of data in studies of tropospheric propagation
P.313-8	Exchange of information for short-term forecasts and transmission of ionospheric disturbance warnings (Revised)
P.341-4	The concept of transmission loss for radio links (Revised)
P.368-7a	Ground-wave propagation curves for frequencies between 10 kHz and 30 MHz

P Series Recommendations	
Radiowave Propagation	
Rec. No.	Description/Title
P.368-7b	Ground-wave propagation curves for frequencies between 10 kHz and 30 MHz
P.368-7c	Ground-wave propagation curves for frequencies between 10 kHz and 30 MHz
P.370-7	VHF and UHF propagation curves for the frequency range from 30 MHz to 1000 MHz. Broadcasting services (Revised)
P.371-7	Choice of indices for long-term ionospheric predictions (Revised)
P.372-6a	Radio noise
P.372-6b	Radio noise
P.372-6c	Radio noise
P.372-6d	Radio noise
P.372-6e	Radio noise
P.373-7	Definitions of maximum and minimum transmission frequencies (Revised)
P.452-8	Prediction procedure for the evaluation of microwave interference between stations on the surface of the Earth at frequencies above about 0.7 GHz
P.453-6	The radio refractice index: its formula and refractivity data
P.525-2	Calculation of free space attenuation
P.526-5	Propagation by diffraction
P.527-3	Electrical characteristics of the surface of the Earth
P.528-2	Propagation curves for aeronautical mobile and radionavigation services using the VHF, UHF and SHF bands
P.529-2	Prediction methods for the terrestrial land mobile service in the VHF and UHF bands (Revised)
P.530-7	Propagation data and prediction methods required for the design of terrestrial line-of-sight systems
P.531-4	Ionospheric propagation data and prediction methods required for the design of satellite services and systems
P.532-1	Ionospheric effects and operational considerations associated with artificial modification of the ionosphere and the radio-wave channel
P.533-5	HF propagation prediction method (Revised)
P.534-3	Method for calculating sporadic-E field strength
P.581-2	The concept of "worst month"
P.616	Propagation data for terrestrial maritime mobile services operating at frequencies above 30 MHz
P.617-1	Propagation prediction techniques and data required for the design of trans-horizon radio-relay systems
P.618-5	Datos de propagación y métodos de predicción necesarios para el diseño de sistemas de telecomunicación Tierra-espacio
P.619-1	Propagation data required for the evaluation of interference between stations in space and those on the surface of the Earth
P.620-3	Propagation data required for the evaluation of coordination distances in the frequency range 0.85 to 60 GHz
P.676-3	Attenuation by atmospheric gases
P.678-1	Characterization of the natural variability of propagation phenomena
P.679-1	Propagation data required for the design of broadcasting-satellite systems
P.680-2	Propagation data required for the design of Earth-space maritime mobile telecommunication systems
P.681-3	Propagation data required for the design of Earth-space land mobile telecommunication systems
P.682-1	Propagation data required for the design of Earth-space aeronautical mobile telecommunication systems
P.684-1	Prediction of field strength at frequencies below about 500 kHz
P.832-1a	World atlas of ground conductivities
P.832-1b	World atlas of ground conductivities
P.832-1c	World atlas of ground conductivities
P.832-1d	World atlas of ground conductivities
P.833-1	Attenuation in vegetation
P.834-2	Effects of tropospheric refraction on radiowave propagation
P.835-2	Reference standard atmospheres
P.836-1	Water vapour: surface density and total columnar content
P.837-1	Characteristics of precipitation for propagation modelling
P.838	Specific attenuation model for rain for use in prediction methods
P.839-1	Rain height model for prediction methods
P.840-2	Attenuation due to clouds and fog

P Series Recommendations
Radiowave Propagation

Rec. No.	Description/Title
P.841	Conversion of annual statistics to worst-months statistics
P.842-1	Computation of reliability and compatibility of HF radio systems
P.843-1	Communication by meteor-burst propagation
P.844-1	Ionospheric factors affecting frequency sharing in the VHF and UHF bands (30 MHz - 3 GHz)
P.845-3	HF field-strength measurement
P.846-1	Measurements of ionospheric and related characteristics (Revised)
P.1057	Probability distributions relevant to radiowave propagation modelling
P.1058-1	Digital topographic databases for propagation studies
P.1060	Propagation factors affecting frequency sharing in HF terrestrial systems
P.1144	Guide to the application of the propagation methods of Study Group 3
P.1145	Propagation data for the terrestrial land mobile service in the VHF and UHF bands
P.1146	The prediction of field strength for land mobile and terrestrial broadcasting services in the frequency range from 1 to 3 GHz
P.1147	Prediction of sky-wave field strength at frequencies between about 150 and 1 700 kHz
P.1148-1	Standardized procedure for comparing predicted and observed HF sky-wave signal intensities and the presentation of such comparisons
P.1238	Propagation data and prediction models for the planning of indoor radiocommunication systems and radio local area networks in the frequency range 900 MHz to 100 GHz
P.1239	ITU-R Reference ionospheric characteristics
P.1240	ITU-R Methods of basic MUF, operational MUF and ray-path prediction
P.1321	Propagation factors affecting systems using digital modulation techniques at LF and MF
P.1322	Radiometric estimation of atmospheric attenuation

RA Series Recommendations
Radioastronomy

Rec. No.	Description/Title
RA.314-8	Preferred frequency bands for radioastronomical measurements
RA.479-4	Protection of frequencies for radioastronomical measurements in the shielded zone of the Moon (Revised)
RA.517-2	Protection of the radioastronomy service from transmitters in adjacent bands
RA.611-2	Protection of the radioastronomy service from spurious emissions
RA.769-1	Protection criteria used for radioastronomical measurements (Revised)
RA.1031-1	Protection of the radioastronomy service in frequency bands shared with other services (Revised)
RA.1237	Protection of the radioastronomy service from unwanted emissions resulting from applications of wideband digital modulation

S Series Recommendations
Fixed satellite service

Rec. No.	Description/Title
S.352-4	Hypothetical reference circuit for systems using analogue transmission in the fixed-satellite service
S.353-8	Allowable noise power in the hypothetical reference circuit for frequency-division multiplex telephony in the fixed-satellite service
S.354-2	Video bandwidth and permissible noise level in the hypothetical reference circuit for the fixed-satellite service
S.446-4	Carrier energy dispersal for systems employing angle modulation by analogue signals or digital modulation in the fixed-satellite service
S.464-2	Pre-emphasis characteristics for frequency-modulation systems for frequency-division multiplex telephony in the fixed-satellite service
S.465-5	Reference earth-station radiation pattern for use in coordination and interference assessment in the frequency range from 2 to about 30 GHz
S.466-6	Maximum permissible level of interference in a telephone channel of a geostationary-satellite network in the fixed-satellite service employing frequency modulation with frequency-division multiplex, caused by other networks of this service

S Series Recommendations
Fixed satellite service

Rec. No.	Description/Title
S.481-2	Measurement of noise in actual traffic for systems in the fixed-satellite service for telephony using frequency-division multiplex
S.482-2	Measurement of performance by means of a signal of a uniform spectrum for systems using frequency-division multiplex telephony in the fixed-satellite service
S.483-3	Maximum permissible level of interference in a television channel of a geostationary-satellite network in the fixed- satellite service employing frequency modulation, caused by other networks of this service
S.484-3	Station-keeping in longitude of geostationary satellites in the fixed satellite service
S.521-3	Hypothetical reference digital path for systems using digital transmission in the fixed-satellite service
S.522-5	Allowable bit-error ratios at the output of the hypothetical reference digital path for systems in the fixed-satellite service using pulse-code modulation for telephony
S.523-4	Maximum permissible levels of interference in a geostationary-satellite network in the fixed-satellite service using 8 bit PCM encoded telephony, caused by other networks of this service
S.524-5	Maximum permissible levels of off-axis e.i.r.p. density from earth stations in the fixed-satellite service transmitting in the 6 and 14 GHz frequency bands
S.579-4	Availability objectives for a hypothetical reference circuit and a hypothetical reference digital path when used for telephony using pulse-code modulation, or as part of an integrated services digital network hypothetical reference connection
S.580-5	Radiation diagrams for use as design objectives for antennas of earth stations operating with geostationary satellites
S.614-3	Allowable error performance for a hypothetical reference digital path in the fixed-satellite service operating below 15 GHz when forming part of an international connection in an integrated services digital network
S.670-1	Flexibility in the positioning of satellites as a design objective
S.671-3	Necessary protection ratios for narrow-band single channel-per-carrier transmissions interfered with by analogue television carriers
S.672-4	Satellite antenna radiation pattern for use as a design objective in the fixed-satellite service employing geostationary satellites
S.673	Terms and definitions relating to space radiocommunications
S.725	Technical characteristics for very small aperture terminals (VSATs)
S.726-1	Maximum permissible level of spurious emissions from very small aperture terminals (VSATs)
S.727	Cross-polarization isolation from very small aperture terminals (VSATs)
S.728-1	Maximum permissible level of off-axis e.i.r.p. density from very small aperture terminals (VSATs) (Revised)
S.729	Control and monitoring function of very small aperture terminals (VSATs)
S.730	Compensation of the effects of switching discontinuities for voice band data and of doppler frequency-shifts in the fixed-satellite service
S.731	Reference earth-station cross-polarized radiation pattern for use in frequency coordination and interference assessment in the frequency range from 2 to about 30 GHz
S.732	Method for statistical processing of earth-station antenna side-lobe peaks
S.733-1	Determination of the G/T ratio for earth stations operating in the fixed-satellite service
S.734	The application of interference cancellers in the fixed-satellite service
S.735-1	Maximum permissible levels of interference in a geostationary-satellite network for an HRDP when forming part of the ISDN in the fixed-satellite service caused by other networks of this service below 15 GHz
S.736-3	Estimation of polarization discrimination in calculations of interference between geostationary-satellite networks in the fixed-satellite service
S.737	Relationship of technical coordination methods within the fixed satellite service
S.738	Procedure for determining if coordination is required between geostationary-satellite networks sharing the same frequency bands
S.739	Additional methods for determining if detailed coordination is necessary between geostationary satellite networks in the fixed-satellite service sharing the same frequency bands
S.740	Technical coordination methods for fixed-satellite networks
S.741-2	Carrier-to-interference calculations between networks in the fixed-satellite service
S.742-1	Spectrum utilization methodologies
S.743-1	The coordination between satellite networks using slightly inclined geostationary-satellite orbits (GSOs) and between such networks and satellite networks using non-inclined GSO satellites
S.744	Orbit spectrum improvement measures for satellite networks having more than one service in one or more frequency bands

S Series Recommendations
Fixed satellite service

Rec. No.	Description/Title
S.1001	Use of systems in the fixed-satellite service in the event of natural disasters and similar emergencies for warning and relief operations
S.1002	Orbit management techniques for the fixed-satellite service
S.1003	Environmental protection of the geostationary orbit
S.1061	Utilization of fade countermeasures strategies and techniques in the fixed-satellite service
S.1062-1	Allowable error performance for a hypothetical reference digital path operating at or above the primary rate (Revised)
S.1063	Criteria for sharing between BSS feeder links and other earth-to-space or space-to-earth links of the FSS
S.1064-1	Pointing accuracy as a design objective for earthward antennas on board geostationary satellites in the FSS (Revised)
S.1065	Power flux-density values to facilitate the application of RR article 14 for the FSS in Region 2 in relation to the BSS in the band 11.7-12.2 GHz
S.1066	Ways of reducing the interference from the broadcasting-satellite service of one region into the fixed-satellite service of another region around 12 GHz
S.1067	Ways of reducing the interference from the broadcasting-satellite service into the fixed-satellite service in adjacent frequency bands around 12 GHz
S.1068	Fixed-satellite and radiolocation/radionavigation services sharing in the band 13.75 to 14 GHz
S.1069	Compatibility between the fixed-satellite service and the space science services in the band 13.75 to 14 GHz
S.1149-1	Network architecture and equipment functional aspects of digital satellite systems in the fixed-satellite service forming part of synchronous digital hierarchy transport networks
S.1150	Technical criteria to be used in examinations relating to the probability of harmful interference between frequency assignments in the FSS as required in No.1506 of the Radio Regulations
S.1151	Sharing between the inter-satellite service involving geostationary satellites in the fixed-satellite service and the radionavigation service at 33 GHz
S.1250	Network management architecture for digital satellite systems forming part of SDH transport networks in the fixed-satellite service
S.1251	Network management - Performance management object class definitions for satellite systems network elements forming part of SDH transport networks in the fixed-satellite service
S.1252	Network management - Payload configuration object class definitions for satellite system network elements forming part of SDH transport networks in the fixed-satellite service forming part of SDH transport networks in the fixed-satellite service
S.1253	Technical options to facilitate coordination of fixed-satellite service networks in certain orbital arc segments and frequency bands
S.1254	Best practices to facilitate the coordination process of fixed-satellite service satellite networks
S.1255	Use of adapt. uplink power cont to mitig. codirect. interf. between geo. sat. orbit/fixed-sat. serv. (GSO/FSS) networks and feeder links of non-geostationary sat. orbit/mobile sat. serv. (NON-GSO/FSS) netw. and between GSO/FSS netw. and NON-GSO/FSS netw.
S.1256	Methodology for determining the maximum aggregate power flux-density at the geostationary-satellite orbit in the band 6 700-7 075 MHz from feeder links of non-geostationary satellite systems in the mobile-satellite service in the space-to-Earth direction
S.1257	Analytical method to calculate visibility statistics for non-geostationary satellite orbit satellites as seen from a point on the Earth's surface
S.1323	Maximum permissible levels of interference in a satellite network (GSO/FSS; non-GSO/FSS; non-GSO/MSS feeder links) for a hypothetical reference digital path in the fixed-satellite service caused by other codirectional networks below 30 GHz
S.1324	Analytical method for estimating interference between non-geostationary mobile-satellite feeder links and geostationary fixed-satellite networks operating co-frequency and codirectionally
S.1325	Simulation methodology for assessing short-term interference between co-frequency, codirectional non-GSO FSS networks and other non-GSO FSS or GSO FSS networks
S.1326	Feasibility of sharing between the inter-satellite service and the fixed-satellite service in the frequency band 50.4 to 51.4 GHz
S.1327	Requirements and suitable bands for operation of the inter-satellite service within the range 50.2 to 71 GHz
S.1328	Satellite system characteristics to be considered in frequency sharing analyses between GSO and non-GSO satellite systems in the fixed-satellite service including feeder links for the mobile-satellite service
S.1329	Frequency sharing of the bands 19.7 to 20.2 GHz and 29.5 to -30.0 GHz between systems in the mobile-satellite service and systems in the fixed-satellite service

SA Series Recommendations

Space applications and meteorology

Rec. No.	Description/Title
SA.363-5	Space operation systems frequencies, bandwidths and protection criteria
SA.364-5	Preferred frequencies and bandwidths for manned and unmanned near-Earth research satellites
SA.509-1	Generalized space research earth station antenna radiation pattern for use in interference calculations, including coordination procedures
SA.510-1	Feasibility of frequency sharing between the space research service and other services in band 10. Potential interference from data relay satellite systems
SA.514-2	Interference criteria for command and data transmission systems operating in the earth exploration-satellite and meteorological-satellite services
SA.515-3	Frequency bands and bandwidths used for satellite passive sensing
SA.516-1	Feasibility of sharing between active sensors used on earth exploration and meteorological satellites and the radiolocation service
SA.577-5	Preferred frequencies and necessary bandwidths for spaceborne active remote sensors
SA 609-1	Protection criteria for telecommunication links for manned and unmanned near earth research satellites
SA.1012	Preferred frequency bands for deep-space research in the 1-40 GHz range
SA.1013	Preferred frequency bands for deep-space research in the 40-120 GHz range
SA.1014	Telecommunication requirements for manned and unmanned deep-space research
SA.1015	Bandwidth requirements for deep-space research
SA.1016	Sharing considerations relating to deep-space research
SA.1017	Preferred method for calculating link performance in the space research service
SA.1018	Hypothetical reference system for systems comprising data relay satellites in the geostationary orbit and user spacecraft in low earth orbits
SA.1019	Preferred frequency bands and transmission directions for data relay satellite systems
SA.1020	Hypothetical reference system for the earth exploration-satellite and meteorological-satellite services
SA.1021	Methodology for determining performance objectives for systems in the earth exploration-satellite and meteorological-satellite services
SA.1022	Methodology for determining interference criteria for systems in the earth exploration-satellite and meteorological-satellite services
SA.1023	Methodology for determining sharing and coordination criteria for systems in the earth exploration-satellite and meteorological-satellite services
SA.1024-1	Necessary bandwidths and preferred frequency bands for data transmission from Earth exploration satellites (not including meteorological satellites)
SA.1025-2	Performance criteria for space-to-Earth data transmission systems operating in the Earth exploration-satellite and meteorological-satellite services using satellites
SA.1026-2	Interference criteria for space-to-Earth data transmission systems operating in the Earth exploration-satellite and meteorological-satellite services using satellites in low-Earth orbit
SA.1027-2	Sharing and coordination criteria for space-to-Earth data transmission systems in the Earth exploration-satellite and meteorological-satellite services using satellites in low-Earth orbit
SA.1028-1	Performance criteria for satellite passive remote sensing
SA.1029-1	Interference criteria for satellite passive remote sensing
SA.1030	Telecommunication requirements of satellite systems for geodesy and geodynamics
SA.1071	Use of the 13.75 to 14.0 GHz band by the space science services and the fixed-satellite service
SA.1154	Provisions to protect the Space Research (SR), Space Operations (SO) and Earth-Exploration Satellite Services (EES) and to facilitate sharing with the mobile service in the 2 025 to 2 110 and 2 200 to - 2 290 MHz bands
SA.1155	Protection criteria related to the operation of data relay satellite systems
SA.1156	Methods for calculating low-orbit satellite visibility statistics
SA.1157	Protection criteria for deep-space research
SA.1158-1	Sharing of the 1 675-1 710 MHz band between the meteorological-satellite service (space-to-Earth) and the mobile-satellite service (Earth-to-space)
SA.1159-1	Performance criteria for data dissemination and direct data readout systems in the Earth exploration-satellite service and meteorological-satellite services using satellites in geostationary orbit
SA.1160-1	Interference criteria for data dissemination and direct data readout systems in the Earth exploration-satellite and meteorological-satellite services using satellites in the geostationary orbit
SA.1161	Sharing and coordination criteria for data dissemination and direct data readout systems in the meteorological-satellite service using satellites in geostationary orbit

SA Series Recommendations

Space applications and meteorology

Rec. No.	Description/Title
SA.1162-1	Telecommunication requirements and performance criteria for service links in data collection and platform location systems in the Earth exploration- and meteorological-satellite services
SA.1163-1	Interference criteria for service links in data collection systems in the Earth exploration- and meteorological-satellite service
SA.1164-1	Sharing and coordination criteria for service links in data collection systems in the Earth exploration- and meteorological-satellite services
SA.1165-1	Technical characteristics and performance criteria for radiosonde systems in the meteorological aids service
SA.1166-1	Performance and interference criteria for active spaceborne sensors
SA.1236	Frequency sharing between space research service extra-vehicular activity (EVA) links and fixed and mobile service links in the 410 to 420 MHz band
SA 1258	Sharing of the frequency band 401 to 403 MHz between the meteorological satellite service, Earth exploration satellite service and meteorological aids service
SA 1259	Feasibility of sharing between spaceborne passive sensors and the fixed service from 50 to 60 GHz
SA 1260	Feasibility of sharing between active spaceborne sensors and other services in the vicinity of 410 to 470 MHz
SA 1261	Feasibility of sharing between spaceborne cloud radars and other services in the range of 92 to 95 GHz
SA 1262	Sharing and coordination criteria for meteorological aids in the 400.15 to 406 and 1 668.4 to 1 700 MHz bands
SA 1263	Interference criteria for meteorological aids operated in the 400.15 to 406 and 1 668.4 to 1 700 MHz bands
SA 1264	Frequency sharing between the meteorological aids service and the mobile-satellite service (Earth-to-space) in the 1 675 to 1 700 MHz band

SF Series Recommendations

Frequency sharing between the fixed satellite service and the fixed service

Rec. No.	Description/Title
SF.355-4	Frequency sharing between systems in the fixed-satellite service and radio-relay systems in the same frequency bands
SF.356-4	Maximum allowable values of interference from line-of-sight radio-relay systems in a telephone channel of a system in the fixed-satellite service employing frequency modulation, when the same frequency bands are shared by both systems
SF.357-4	Maximum allowable values of interference in a telephone channel of an analogue angle-modulated radio-relay system sharing the same frequency bands as systems in the fixed-satellite service
SF.358-5	Maximum permissible values of power flux-density at the surface of the Earth produced by satellites in the fixed-satellite service using the same frequency bands above 1 GHz as line-of-sight radio-relay systems (Revised)
SF.406-8	Maximum equivalent isotropically radiated power of radio-relay system transmitters operating in the frequency bands shared with the fixed-satellite service
SF.558-2	Maximum allowable values of interference from terrestrial radio links to systems in the fixed-satellite service employing 8-bit PCM encoded telephony and sharing the same frequency bands
SF.615-1	Maximum allowable values of interference from the fixed-satellite service into terrestrial radio-relay systems which may form part of an ISDN and share the same frequency band below 15 GHz
SF.674-1	Power flux-density values to facilitate the application of Article 14 of the Radio Regulations for FSS in relation to the fixed-satellite service in the 11.7-12.2 GHz band in region 2
SF.675-3	Calculation of the maximum power density (averaged over 4 kHz) of an angle-modulated carrier
SF.765	Intersection of radio-relay antenna beams with orbits used by space stations in the fixed-satellite service
SF.766	Methods for determining the effects of interference on the performance and the availability of terrestrial radio-relay systems and systems in the fixed-satellite service
SF.1004	Maximum equivalent isotropically radiated power transmitted towards the horizon by earth stations of the fixed satellite service sharing frequency bands with the fixed service
SF.1005	Sharing between the fixed service and the fixed-satellite service with bidirectional usage in bands above 10 GHz currently unidirectionally allocated
SF.1006	Determination of the interference potential between earth stations of the fixed-satellite service and stations in the fixed service

SF Series Recommendations
Frequency sharing between the fixed satellite service and the fixed service

Rec. No.	Description/Title
SF.1008-1	Possible use by space stations in the fixed-satellite service of orbits slightly inclined with respect to the geostationary-satellite orbit in bands shared with the fixed service (Revised)
SF.1193	Carrier-to-interference calculations between earth stations in the fixed-satellite service and radio-relay systems
SF.1320	Maximum allowable values of power flux-density at the surface of the Earth produced by non-geostationary satellites in the fixed-satellite service used in feeder links for the mobile-satellite service and sharing the same frequency bands with radio

SM Series Recommendations
Spectrum management

Rec. No.	Description/Title
SM.182-4	Automatic monitoring of occupancy of the radio-frequency spectrum
SM.239-2	Spurious emissions from sound and television broadcast receivers
SM.326-6	Determination and measurement of the power of radio transmitters
SM.328-8	Spectra and bandwidth of emissions
SM.329-7	Spurious emissions
SM.331-4	Noise and sensitivity of receivers
SM.332-4	Selectivity of receivers
SM.337-3	Frequency and distance separations
SM.377-3	Accuracy of frequency measurements at stations for international monitoring
SM.378-6	Field-strength measurements at monitoring stations (Revised)
SM.433-5	Methods for the measurement of radio interference and the determination of tolerable levels of interference
SM.443-2	Bandwidth measurement at monitoring stations (Revised)
SM.575	Protection of fixed monitoring stations against radio-frequency interference
SM.667	National spectrum management data
SM.668-1	Electronic exchange of information for spectrum management purposes
SM.669-1	Protection ratios for spectrum sharing investigations
SM.852	Sensitivity of radio receivers for class of emissions F3E
SM.853	Necessary bandwidth
SM.854	Direction finding at monitoring stations of signals below 30 MHz
SM.855	Multi-service telecommunication systems
SM.856-1	New spectrally efficient techniques and systems
SM.1045-1	Frequency tolerance of transmitters
SM.1046	Definition of spectrum use and efficiency of a radio system
SM.1047	National spectrum management
SM.1048	Design guidelines for a basic automated spectrum management system (BASMS)
SM.1049-1	A method of spectrum management to be used for aiding frequency assignment for terrestrial services in border areas (Revised)
SM.1050	Tasks of a monitoring service
SM.1051-2	Priority of identifying and eliminating harmful interference in the band 406 to 406.1 MHz
SM.1052	Automatic identification of radio stations
SM.1053	Methods of improving HF direction finding accuracy at fixed stations
SM.1054	Monitoring of radio emissions from spacecraft at monitoring stations
SM.1055	The use of spread spectrum techniques
SM.1056	Limitation of radiation from industrial, scientific and medical (ISM) equipment
SM.1131	Factors to consider in allocating spectrum on a worldwide basis
SM.1132	General principles and methods for sharing between radio services
SM.1133	Spectrum utilization of broadly defined services
SM.1134	Intermodulation interference calculations in the land-mobile service
SM.1135	SINPO and SINPFEMO codes
SM.1138	Determination of necessary bandwidths including examples for their calculation and associated examples for the designation of emissions

SM Series Recommendations
Spectrum management

Rec. No.	Description/Title
SM.1139	International monitoring system
SM.1235	Performance functions for digital modulation systems in an interference environment
SM.1265	Alternative allocation methods
SM.1266	Adaptive MF/HF systems
SM.1267	Collection and publication of monitoring data to assist frequency assignment for geostationary satellite systems
SM.1268	Method of measuring the maximum frequency deviation of FM broadcast emissions at monitoring stations
SM.1269	Classification of direction finding bearings
SM.1270	Additional information for monitoring purposes related to classification and designation of emission

SNG Series Recommendations
Satellite news gathering

Rec. No.	Description/Title
SNG.722-1	Uniform technical standards (analogue) for satellite news gathering (SNG)
SNG.770-1	Uniform operational procedures for satellite news gathering (SNG)
SNG.771-1	Auxiliary coordination satellite circuits for SNG terminals
SNG.1007-1	Uniform technical standards (digital) for satellite news gathering (SNG) (Revised)
SNG.1070	An automatic transmitter identification system (ATIS) for analogue-modulation transmissions for satellite news gathering and outside broadcasts
SNG.1152	Use of digital transmission techniques for satellite news gathering (sound)

TF Series Recommendations
Time signals and frequency standards emissions

Rec. No.	Description/Title
TF.374-3	Standard-frequency and time-signal emissions
TF.457-1	Use of the modified Julian date by standard-frequency and time-signal services
TF.458-2	International comparisons of atomic time scales
TF.460-4	Standard-frequency and time-signal emissions
TF.486-1	Reference of precisely controlled frequency generators and emissions to the international atomic time scale
TF.535-1	Use of the term UTC
TF.536	Time-scale notations
TF.538-3	Measures for random instabilities in frequency and time (phase)
TF.582-1	Time and frequency reference signal dissemination and coordination using satellite methods
TF.583-4	Time codes
TF.686	Glossary
TF.767	Use of the global positioning system (GPS) and the global navigation satellite system (GLONASS) for high-accuracy time transfer
TF.768-3	Standard frequencies and time signals
TF.1010	Relativistic effects in a coordinate time system in the vicinity of the earth
TF.1011	Systems, techniques and services for time and frequency transfer
TF.1153-1	The operational use of two-way satellite time and frequency transfer employing PN codes

Rec. No.	Description/Title
TF Series Recommendations	
Time signals and frequency standards emissions	
V.430-3	Use of international system of units (SI)
V.431-6	Nomenclature of the frequency and wavelength bands used in telecommunications
V.461-5	Graphical symbols and rules for the preparation of documentation in telecommunications
V.573-3	Radiocommunication vocabulary
V.574-3	Use of decibel and the neper in telecommunications
V.607-2	Terms and symbols for information quantities in telecommunications
V.608-2	Letter symbols for telecommunications
V.662-2	Terms and definitions
V.663-1	Use of certain terms linked with physical quantities
V.664	Adoption of the CCITT Specification and Description Language (SDL)
V.665-1	Traffic intensity unit
V.666-2	Abbreviations and initials used in telecommunications

Appendix G
Internet RFCs

RFC No.	Authors	Title	Date	Pgs	Updated By	Updates	Obsoleted By	Obsoletes	Notes / Status
0001	S. Crocker	Host software	04/07/1969	7					
0002	B. Duvall	Host software	04/09/1969	10					
0003	S. Crocker	Documentation conventions	04/09/1969	2			10		
0004	E. Shapiro	Network timetable	03/24/1969	5					
0005	J. Rulifson	Decode Encode Language	06/02/1969	18					
0006	S. Crocker	Conversation with Bob Kahn	04/10/1969	1					
0007	G. Deloche	Host-IMP interface	05/01/1969	4					
0008	G. Deloche	Functional specifications for the ARPA Network	05/05/1969	4					
0009	G. Deloche	Host software	05/01/1969	14					
0010	S. Crocker	Documentation conventions	07/29/1969	3			16	3	
0011	G. Deloche	Implementation of the Host-Host software procedures in GORDO	08/01/1969	52			33		
0012	M. Wingfield	IMP-Host interface flow diagrams	08/26/1969	5					
0013	V. Cerf	[Referring to NWG/RFC 11]	08/20/1969	1					
0014		Not Issued							
0015	C. Carr	Network subsystem for time sharing hosts	09/25/1969	8					
0016	S. Crocker	M.I.T	09/27/1969	1			24	10	
0017	J. Kreznar	Some questions re: Host-IMP Protocol	08/27/1969	1					
0018	V. Cerf	[Link assignments]	09/01/1969	1					
0019	J. Kreznar	Two protocol suggestions to reduce congestion at swap bound nodes	10/07/1969	1					
0020	V. Cerf	ASCII format for network interchange	10/16/1969	6					
0021	V. Cerf	Network meeting	10/17/1969	3					
0022	V. Cerf	Host-host control message formats	10/17/1969	2					
0023	G. Gregg	Transmission of multiple control messages	10/16/1969	1					
0024	S. Crocker	Documentation conventions	11/21/1969	3	27, 30	10, 16		16	
0025	S. Crocker	No high link numbers	10/30/1969	1					
0026		Not Issued							
0027	S. Crocker	Documentation conventions	12/09/1969	3		10, 16, 24	30	24	
0028	W. English	Time standards	01/13/1970	1					
0029	R. Kahn	Response to RFC 28	01/19/1970	1					
0030	S. Crocker	Documentation conventions	02/04/1970	1				27	
0031	D. Bobrow, W. Sutherland	Binary message forms in computer	02/01/1968	6					
0032	D. Vedder	Connecting M.I.T. computers to the ARPA Computer-to-computer communication network	01/31/1969	23					
0033	S. Crocker	New Host-Host Protocol	02/12/1970	31	47, 36			11	
0034	W. English	Some brief preliminary notes on the Augmentation Research Center clock	02/26/1970	1					
0035	S. Crocker	Network meeting	03/03/1970	1					
0036	S. Crocker	Protocol notes	03/16/1970	7	44, 39	33			
0037	S. Crocker	Network meeting epilogue, etc	03/20/1970	4					
0038	S. Wolfe	Comments on network protocol from NWG/RFC #36	03/20/1970	2					
0039	E. Harslem, J. Heafner	Comments on protocol re: NWG/RFC #36	03/25/1970	3		36			
0040	E. Harslem, J. Heafner	More comments on the forthcoming protocol	03/27/1970	3					
0041	J. Melvin	IMP-IMP teletype communication	03/30/1970	1					
0042	E. Ancona	Message data types	03/31/1970	3					
0043	A. Nemeth	Proposed meeting [LIL]	04/08/1970	1					
0044	A. Shoshani, R. Long, A. Landsberg	Comments on NWG/RFC 33 and 36	04/10/1970	5		36			
0045	J. Postel, S. Crocker	New protocol is coming	04/14/1970	1					
0046	E. Meyer	ARPA Network protocol notes	04/17/1970	27					
0047	W. Crowther	BBN's comments on NWG/RFC #33	04/20/1970	4		33			
0048	J. Postel, S. Crocker	Possible protocol plateau	04/21/1970	16					
0049	E. Meyer	Conversations with S. Crocker UCLA	04/23/1970	5					
0050	E. Harslem, J. Haverty	Comments on the Meyer proposal	04/30/1970	3					
0051	M. Elie	Proposal for a Network Interchange Language	05/04/1970	19					
0052	J. Postel, S. Crocker	Updated distribution list	07/01/1970	3	69				
0053	S. Crocker	Official protocol mechanism	06/09/1970	1					
0054	S. Crocker, M. Kraley, J. Postel, J. Newkirk	Official protocol proffering	06/18/1970	16	57				
0055	S. Crocker, M. Kraley, J. Postel, J. Newkirk	Prototypical implementation of the NCP	06/19/1970	32					

RFC No.	Authors	Title	Date	Pgs	Updated By	Updates	Obsoleted By	Obsoletes	Notes / Status
0056	E. Belove, D. Black, R. Flegal, L. Farquar	Third level protocol: Logger Protocol	06/01/1970	8					
0057	M. Kraley, J. Newkirk	Thoughts and reflections on NWG/RFC 54	06/19/1970	5		54			
0058	T. Skinner	Logical message synchronization	06/26/1970	3					
0059	E. Meyer	Flow control - fixed versus demand allocation	06/27/1970	7					
0060	R. Kalin	Simplified NCP Protocol	07/15/1970	8					
0061	D. Walden	Note on interprocess communication in a resource sharing computer network	07/17/1970	26			62		
0062	D. Walden	Systems for interprocess communication in a resource sharing computer network	08/03/1970	37				61	
0063	V. Cerf	Belated network meeting report	07/31/1970	2					
0064	M. Elie	Getting rid of marking	07/01/1970	7					
0065	D. Walden	Comments on Host/Host Protocol document #1	08/29/1970	2					
0066	S. Crocker	NIC - third level ideas and other noise	08/26/1970	3	80, 93		123		
0067	W. Crowther	Proposed change to Host/IMP spec to eliminate marking	01/01/1970	1					
0068	M. Elie	Comments on memory allocation control commands: CEASE, ALL, GVB, RET, and RFNM	08/31/1970	2					
0069	A. Bhushan	Distribution list change for MIT	09/22/1970	1		52			
0070	S. Crocker	Note on padding	10/15/1970	8	228				
0071	T. Schipper	Reallocation in case of input error	09/25/1970	1					
0072	R. Bressler	Proposed moratorium on changes to network protocol	09/28/1970	3					
0073	S. Crocker	Response to NWG/RFC 67	09/25/1970	1					
0074	J. White	Specifications for network use of the UCSB On-Line System	10/16/1970	11	225, 217				
0075	S. Crocker	Network meeting	10/14/1970	1					
0076	J. Bouknight, J. Madden, G. Grossman	Connection by name: User oriented protocol	10/28/1970	8					
0077	J. Postel	Network meeting report	11/20/1970	9					
0078	E. Harslem, J. Heafner, J. White	NCP status report: UCSB/Rand	10/01/1970	1					
0079	E. Meyer	Logger Protocol error	11/16/1970	1					
0080	E. Harslem, J. Heafner	Protocols and data formats	12/01/1970	9	93	66	123		
0081	J. Bouknight	Request for reference information	12/03/1970	1					
0082	E. Meyer	Network meeting notes	12/09/1970	16					
0083	R. Anderson, E. Harslem, J. Heafner	Language-machine for data reconfiguration	12/18/1970	12					
0084	J. North	List of NWG/RFC's 1-80	12/23/1970	8					
0085	S. Crocker	Network Working Group meeting	12/28/1970	1					
0086	S. Crocker	Proposal for a network standard format for a data stream to control graphics display	01/05/1971	5	125				
0087	A. Vezza	Topic for discussion at the next Network Working Group meeting	01/12/1971	3					
0088	R. Braden, S. Wolfe	NETRJS: A third level protocol for Remote Job Entry	01/13/1971	10			189		
0089	R. Metcalfe	Some historic moments in networking	01/19/1971	12					
0090	R. Braden	CCN as a network service center	01/15/1971	6					
0091	G. Mealy	Proposed User-User Protocol	12/27/1970						
0092		Not Issued							
0093	A. McKenzie	Initial Connection Protocol	01/27/1971	1		66, 80			
0094	E. Harslem, J. Heafner	Some thoughts on network graphics	02/03/1971	8					
0095	S. Crocker	Distribution of NWG/RFC's through the NIC	02/04/1971	4			155		
0096	R. Watson	Interactive network experiment to study modes of access to the Network Information Center	02/12/1971	4					I
0097	J. Melvin, R. Watson	First cut at a proposed Telnet Protocol	02/15/1971	10					
0098	E. Meyer, T. Skinner	Logger Protocol proposal	02/11/1971	12	123				
0099	P. Karp	Network meeting	02/22/1971	1	116				
0100	P. Karp	Categorization and guide to NWG/RFCs	02/26/1971	43					
0101	R. Watson	Notes on the Network Working Group meeting, Urbana, Illinois, February 17, 1971	02/23/1971	14	108, 123				
0102	S. Crocker	Output of the Host-Host Protocol glitch cleaning committee	02/22/1971	6	107				
0103	R. Kalin	Implementation of interrupt keys	02/24/1971	3					
0104	J. Postel, S. Crocker	Link 191	02/25/1971	1					
0105	J. White	Network specifications for Remote Job Entry and Remote Job Output Retrieval at UCSB	03/22/1971	8	217				
0106	T. O'Sullivan	User/Server Site Protocol network host questionnaire	03/03/1971	3					
0107	R. Bressler, S. Crocker, W. Crowther, G. Grossman, R. Tomlinson	Output of the Host-Host Protocol glitch cleaning committee	03/23/1971	11	179, 154, 132, 124, 111	102			

RFC No.	Authors	Title	Date	Pgs	Updated By	Updates	Obsoleted By	Obsoletes	Notes / Status
0108	R. Watson	Attendance list at the Urbana NWG meeting, February 17-19,1971	03/25/1971	3		101			
0109	J. Winett	Level III Server Protocol for the Lincoln Laboratory NIC 360/67 Host	03/24/1971	12					393
0110	J. Winett	Conventions for using an IBM 2741 terminal as a user console for access to network server hosts	03/25/1971	4	135				
0111	S. Crocker	Pressure from the chairman	03/31/1971	2	130	107			
0112	T. O'Sullivan	User/Server Site Protocol: Network host questionnaire responses	04/01/1971	3					
0113	E. Harslem, J. Heafner, J. White	Network activity report: UCSB Rand	04/05/1971	2	227				
0114	A. Bhushan	File Transfer Protocol	04/10/1971	24	172, 171, 141				
0115	R. Watson, J. North	Some Network Information Center policies on handling documents	04/16/1971	12					
0116	S. Crocker	Structure of the May NWG meeting	04/12/1971	1	156, 131	99			
0117	J. Wong	Some comments on the official protocol	04/07/1971	5					
0118	R. Watson	Recommendations for facility documentation	04/16/1971	3					
0119	M. Krilanovich	Network Fortran subprograms	04/21/1971	17					
0120	M. Krilanovich	Network PL1 subprograms	04/21/1971	16					
0121	M. Krilanovich	Network on-line operators	04/21/1971	14					
0122	J. White	Network specifications for UCSB's Simple-Minded File System	04/26/1971	21	431, 399, 269. 217				
0123	S. Crocker	Proffered official ICP	04/20/1971	4	148, 143	98	165	66	
0124	J. Melvin	Typographical error in RFC 107	04/19/1971	1		107			
0125	J. McConnell	Response to RFC 86: Proposal for network standard format for a graphics data stream	04/18/1971	4	177	86			
0126	J. McConnell	Graphics facilities at Ames Research Center	04/18/1971	2					
0127	J. Postel	Comments on RFC 123	04/20/1971	1	151	123	145		
0128	J. Postel	Bytes	04/21/1971	2					
0129	E. Harslem, J. Heafner, E. Meyer	Request for comments on socket name structure	04/22/1971	6	147				
0130	J. Heafner	Response to RFC 111: Pressure from the chairman	04/22/1971	2		111			
0131	E. Harslem, J. Heafner	Response to RFC 116: May NWG meeting	04/22/1971	4		116			
0132	J. White	Typographical error in RFC 107	04/28/1971	11		107	154		
0133	R. Sundberg	File transfer and recovery	04/27/1971	5					
0134	A. Vezza	Network Graphics meeting	04/29/1971	2					
0135	W. Hathaway	Response to NWG/RFC 110	04/29/1971	2		110			
0136	R. Kahn	Host accounting and administrative procedures	04/29/1971	6					
0137	T. O'Sullivan	Telnet Protocol - a proposed document	04/30/1971	6	139				
0138	R. Anderson, V. Cerf, E. Harslem, J. Heafner, J. Madden	Status report on proposed Data Reconfiguration Service	04/28/1971	30					
0139	T. O'Sullivan	Discussion of Telnet Protocol	05/07/1971	11	158	137			393
0140	S. Crocker	Agenda for the May NWG meeting	05/04/1971	3	149				
0141	E. Harslem, J. Heafner	Comments on RFC 114: A File Transfer Protocol	04/29/1971	2		114			
0142	C. Kline, J. Wong	Time-out mechanism in the Host-Host Protocol	05/03/1971	3					
0143	W. Naylor, C. Kline, J. Wong, J. Postel	Regarding proffered official ICP	05/03/1971	4		123, 145	165		
0144	A. Shoshani	Data sharing on computer networks	04/1971						
0145	J. Postel	Initial Connection Protocol control commands	05/04/1971	1	143		165	127	
0146	P. Karp, D. McKay, D. Wood	Views on issues relevant to data sharing on computer networks	05/12/1971	7					
0147	J. Winett	Definition of a socket	05/07/1971	2		129			
0148	A. Bhushan	Comments on RFC 123	05/07/1971	1		123			
0149	S. Crocker	Best laid plans	05/10/1971	1		140			
0150	R. Kalin	Use of IPC facilities: A working paper	05/05/1971	16					
0151	A. Shoshani	Comments on a proffered official ICP: RFCs 123, 127	05/10/1971	3		127			
0152	M. Wilber	SRI Artificial Intelligence status report	05/1971	2726 b					
0153	J. Melvin, R. Watson	SRI ARC-NIC status	05/15/1971	4					
0154	S. Crocker	Exposition style	05/12/1971	1		107		132	
0155	J. North	ARPA Network mailing lists	05/01/1971	5			168	95	
0156	J. Bouknight	Status of the Illinois site: Response to RFC 116	04/26/1971	1		116			
0157	V. Cerf	Invitation to the Second Symposium on Problems in the Optimization of Data Communications Systems	05/12/1971	1					
0158	T. O'Sullivan	Telnet Protocol: A proposed document	05/19/1971	11	318	139	495		393
0159		Not Issued							
0160	Stanford Research Inst., NIC	RFC brief list	05/18/1971	4		NIC6716	200, 999		
0161	A. Shoshani	Solution to the race condition in the ICP	05/19/1971	2					
0162	M. Kampe	NETBUGGER3	05/22/1971	1					
0163	V. Cerf	Data transfer protocols	05/19/1971	3					
0164	J. Heafner	Minutes of Network Working Group meeting, 5/16 through 5/19/71	05/25/1971	38					

RFC No.	Authors	Title	Date	Pgs	Updated By	Updates	Obsoleted By	Obsoletes	Notes / Status
0165	J. Postel	Proffered official Initial Connection Protocol	05/25/1971	6	NIC7101			145, 143, 123	
0166	R. Anderson, V. Cerf, E. Harslem, J. Heafner, J. Madden	Data Reconfiguration Service: An implementation specification	05/25/1971	24					
0167	A. Bhushan, R. Metcalfe, J. Winett	Socket conventions reconsidered	05/24/1971	7					147, 129
0168	J. North	ARPA Network mailing lists	05/26/1971	5			211	155	
0169	S. Crocker	Computer networks	05/27/1971	5					
0170	Stanford Research Inst., NIC	RFC list by number	06/01/1971	2			200		
0171	A. Bhushan, R. Braden, J. Heafner, W. Crowther, E. Harslem	Data Transfer Protocol	06/23/1971	13	238	114	264		
0172	A. Bhushan, R. Braden, J. Heafner, W. Crowther, E. Harslem	File Transfer Protocol	06/23/1971	15	238	114	265		
0173	D. Karp, D. McKay	Network data management committee meeting announcement	06/04/1971	3					
0174	J. Postel, V. Cerf	UCLA - computer science graphics overview	06/08/1971	3					
0175	E. Harslem, J. Heafner	Comments on "Socket conventions reconsidered"	06/11/1971	1					
0176	A. Bhushan, R. Kanodia, R. Metcalfe, J. Postel	Comments on "Byte size for connections"	06/14/1971	5					
0177	J. McConnell	Device independent graphical display description	06/15/1971	10	181	125			
0178	I. Cotton	Network graphic attention handling	06/27/1971	18					
0179	A. McKenzie	Link number assignments	06/22/1971	1		107			
0180	A. McKenzie	File system questionnaire	06/25/1971	8					
0181	J. McConnell	Modifications to RFC 177	07/21/1971	2		177			
0182	J. North	Compilation of list of relevant site reports	06/25/1971	1					
0183	J. Winett	EBCDIC codes and their mapping to ASCII	07/21/1971	15					
0184	K. Kelley	Proposed graphic display modes	07/06/1971	7					
0185	J. North	NIC distribution of manuals and handbooks	07/07/1971	1					
0186	J. Michener	Network graphics loader	07/12/1971	21					
0187	D. McKay, D. Karp	Network/440 protocol concept	07/01/1971	15					
0188	D. Karp, D. McKay	Data management meeting announcement	01/28/1971	2					
0189	R. Braden	Interim NETRJS specifications	07/15/1971	19	283		599	88	
0190	L. Deutsch	DEC PDP-10-IMLAC communications system	07/13/1971	15					
0191	C. Irby	Graphics implementation and conceptualization at Augmentation Research Center	07/13/1971	4					
0192	R. Watson	Some factors which a Network Graphics Protocol must consider	07/12/1971	21					
0193	E. Harslem, J. Heafner	Network checkout	07/14/1971				198		
0194	V. Cerf, E. Harslem, J. Heafner, R. Metcalfe, J. White	Data Reconfiguration Service - compiler/interpreter implementation notes	07/01/1971	22					
0195	G. Mealy	Data computers-data descriptions and access language	07/16/1971	4					
0196	R. Watson	Mail Box Protocol	07/20/1971	4			221		
0197	A. Shoshani, E. Harslem	Initial Connection Protocol - Reviewed	07/14/1971	4					
0198	J. Heafner	Site certification - Lincoln Labs 360/67	07/20/1971	1		193	214	193	
0199	T. Williams	Suggestions for a network data-tablet graphics protocol	07/15/1971	13					
0200	J. North	RFC list by number	08/01/1971	2			NIC7724	170, 160	
0201		Not Issued							
0202	S. Wolfe, J. Postel	Possible deadlock in ICP	07/26/1971	2					
0203	R. Kalin	Achieving reliable communication	08/10/1971	14					
0204	J. Postel	Sockets in use	08/05/1971	1		234			
0205	R. Braden	NETCRT - a character display protocol	08/06/1971	14					
0206	J. White	User Telnet - description of an initial implementation	08/09/1971	17					
0207	A. Vezza	September Network Working Group meeting	08/09/1971	2			212		
0208	A. McKenzie	Address tables	08/09/1971	4					
0209	B. Cosell	Host/IMP interface documentation	08/13/1971	2					
0210	W. Conrad	Improvement of flow control	08/16/1971	3					
0211	J. North	ARPA Network mailing lists	08/18/1971	5			300	168	
0212	University of Southern California, Information Sciences Inst	NWG meeting on network usage	08/23/1971	2	222			207	
0213	B. Cosell	IMP System change notification	08/20/1971						
0214	E. Harslem	Network checkpoint	08/21/1971	2				198	
0215	A. McKenzie	NCP, ICP, and Telnet: The Terminal IMP implementation	08/30/1971	7					

RFC No.	Authors	Title	Date	Pgs	Updated By	Updates	Obsoleted By	Obsoletes	Notes / Status
0216	J. White	Telnet access to UCSB's On-Line System	09/08/1971	27					
0217	J. White	Specifications changes for OLS, RJE/RJOR, and SMFS	09/08/1971	2		74, 105, 122			
0218	B. Cosell	Changing the IMP status reporting facility	09/08/1971	1					
0219	R. Winter	User's view of the datacomputer	09/03/1971	10					
0220		Not Issued							
0221	R. Watson	Mail Box Protocol: Version 2	08/27/1971				278	196	
0222	R. Metcalf	System programmer's workshop	09/13/71	2	234	212			
0223	J. Melvin, R. Watson	Network Information Center schedule for network users	09/14/1971	3					
0224	A. McKenzie	Comments on Mailbox Protocol	09/14/1971	2					
0225	E. Harslem, R. Stoughton	Rand/UCSB network graphics experiment	09/13/1971	6		74			
0226	P. Karp	Standardization of host mnemonics	09/20/1971	1			247		
0227	J. Heafner, E. Harslem	Data transfer rates Rand/UCLA	09/17/1971	2		113			
0228	D. Walden	Clarification	09/22/1971	1		70			
0229	J. Postel	Standard host names	09/22/1971	2			236		
0230	T. Pyke	Toward reliable operation of minicomputer-based terminals on a TIP	09/24/1971	3					
0231	J. Heafner, E. Harslem	Service center standards for remote usage: Auser's view	09/21/1971	5					
0232	A. Vezza	Postponement of network graphics meeting	09/23/1971	1					
0233	A. Bhushan, R. Metcalfe	Standardization of host call letters	09/28/1971	1					
0234	A. Vezza	Network Working Group meeting schedule	10/05/1971	1		222, 204			
0235	E. Westheimer	Site status	09/27/1971	4			240		
0236	J. Postel	Standard host names	09/27/1971	2				229	
0237	R. Watson	NIC view of standard host names	09/29/1971	1			273		
0238	R. Braden	Comments on DTP and FTP proposals	09/29/1971	1	269	171, 172			
0239	R. Braden	Host mnemonics proposed in RFC 226 NIC 7625	09/23/1971	1					226, 229, 236
0240	A. McKenzie	Site status	09/30/1971	5			252	235	
0241	A. McKenzie	Connecting computers to MLC ports	09/29/1971	2					
0242	L. Haibt, A. Mullery	Data descriptive language for shared data	07/19/1971	12					
0243	A. Mullery	Network and data sharing bibliography	10/05/1971	6			290		
0244		Not Issued							
0245	C. Falls	Reservations for Network Group meeting	10/05/1971	1					
0246	A. Vezza	Network Graphics meeting	10/05/1971	1					
0247	P. Karp	Proffered set of standard host names	10/12/1971	4				226	
0248		Not Issued							
0249	R. Borelli	Coordination of equipment and supplies purchase	10/08/1971	2					
0250	H. Brodie	Some thoughts on file transfer	10/07/1971						
0251	D. Stern	Weather data	10/13/1971	2					
0252	E. Westheimer	Network host status	10/08/1971	3			255	240	
0253	J. Moorer	Second Network Graphics meeting details	10/19/1971	1					
0254	A. Bhushan	Scenarios for using ARPANET computers	10/29/1971	32					
0255	E. Westheimer	Status of network hosts	10/26/1971	2			266	252	
0256	B. Cosell	IMPSYS change notification	11/03/1971	1					
0257		Not Issued							
0258		Not Issued							
0259		Not Issued							
0260		Not Issued							
0261		Not Issued							
0262		Not Issued							
0263	A. McKenzie	"Very Distant" Host interface	12/17/1971	2					
0264	A. Bhushan, R. Braden, W. Crowther, E. Harslem, J. Heafner	Heafner, J.F., McKenzie, A.M., Melvin, J.T., Sundberg, R.L., Watson, R.W., White, J.E. Data Transfer Protocol	12/15/1971	8			354	171	
0265	A. Bhushan, R. Braden, W. Crowther, E. Harslem, J. Heafner	Heafner, J.F., McKenzie, A.M., Melvin, J.T., Sundberg, R.L., Watson, R.W., White, J.E. File Transfer Protocol	11/17/1971	11	294		354	172	
0266	E. Westheimer	Network host status	11/08/1971	2			267	255	
0267	E. Westheimer	Network host status	11/22/1971	3			287	266	
0268	J. Postel	Graphics facilities information	11/24/1971	1					
0269	H. Brodie	Some experience with file transfer	12/06/1971	3		122, 238			
0270	A. McKenzie	Correction to BBN Report No. 1822 NIC 7958	01/01/1972	3		NIC7959			
0271	B. Cosell	IMP System change notifications	01/03/1972	2					
0272		Not Issued							
0273	R. Watson	More on standard host names	10/18/1971					237	
0274	E. Forman	Establishing a local guide for network usage	11/01/1971	5					
0275		Not Issued							
0276	R. Watson	NIC course	11/08/1971	2					
0277		Not Issued							

RFC No.	Authors	Title	Date	Pgs	Updated By	Updates	Obsoleted By	Obsoletes	Notes / Status
0278	A. Bhushan, R. Braden, E. Harslem, J. Heafner, A. McKenzie	McKenzie, A.M., Melvin, J.T., Sundberg, R.L., Watson, R.W., White, J.E. Revision of the Mail Box Protocol	11/17/1971	4				221	
0279		Not Issued							
0280	R. Watson	Draft of host names	11/17/1971	4					
0281	A. McKenzie	Suggested addition to File Transfer Protocol	12/08/1971	5					
0282	M. Padlipsky	Graphics meeting report	12/08/1971	8					
0283	R. Braden	NETRJT: Remote Job Service Protocol for TIPS	12/20/1971	9		189			
0284		Not Issued							
0285	D. Huff	Network graphics	12/15/1971	13					
0286	E. Forman	Network library information system	12/21/1971	1					
0287	E. Westheimer	Status of network hosts	12/22/1971	3			288	267	
0288	E. Westheimer	Network host status	01/06/1972	6	293		293	287	
0289	R. Watson	What we hope is an official list of host names	12/21/1971	3			384		
0290	A. Mullery	Computer networks and data sharing: A bibliography	01/11/1972	15				243	
0291	D. McKay	Data management meeting announcement	01/14/1972	2					
0292	J. Michener, I. Cotton, K. Kelley, D. Liddle, E. Meyer	E.W., Jr Graphics Protocol: Level 0 only	01/12/1972	9					
0293	E. Westheimer	Network host status	01/18/1972	3			298	288	
0294	A. Bhushan	On the use of "set data type" transaction in File Transfer Protocol	01/25/1972	2		265			
0295	J. Postel	Report of the Protocol Workshop, 12 October 1971	01/02/1972	4					
0296	D. Liddle	DS-1 display system	01/27/1972	23					
0297	D. Walden	TIP message buffers	01/31/1972	5					
0298	E. Westheimer	Network host status	02/11/1972	3			306	293	
0299	D. Hopkin	Information management system	02/11/1972	1					
0300	J. North	ARPA Network mailing lists	01/25/1972	6			303	211	
0301	R. Alter	BBN IMP #5 and NCC schedule March 4, 1971	02/11/1972	1					
0302	R. Bryan	Exercising the ARPANET	02/08/1972	3					
0303	Stanford Research Inst., NIC	ARPA Network mailing lists	02/23/1972	6			329	300	
0304	D. McKay	Data management system proposal for the ARPA network	02/17/1972	12					
0305	R. Alter	Unknown host numbers	02/23/1972	1					
0306	E. Westheimer	Network host status	02/15/1972	3			315	298	
0307	E. Harslem	Using network Remote Job Entry	02/24/1972	6					
0308	M. Seriff	ARPANET host availability data	03/13/1972	3					
0309	A. Bhushan	Data and File Transfer workshop announcement	03/17/1972	5					
0310	A. Bhushan	Another look at Data and File Transfer Protocols	04/03/1972	7					
0311	R. Bryan	New console attachments to the USCB host	02/29/1972	2					
0312	A. McKenzie	Proposed change in IMP-to-Host Protocol	03/22/1972	2					
0313	T. O'Sullivan	Computer based instruction	03/06/1972						
0314	I. Cotton	Network Graphics Working Group meeting	03/14/1972	1					
0315	E. Westheimer	Network host status	03/08/1972	3			319	306	
0316	D. McKay, A. Mullery	ARPA Network Data Management Working Group	02/23/1972	10					
0317	J. Postel	Official Host-Host Protocol modification: Assigned link numbers	03/20/1972	1			604		
0318	J. Postel	[Ad hoc Telnet Protocol]	04/03/1972	23	435	158			139, 158
0319	E. Westheimer	Network host status	03/21/1972	3			326	315	
0320	R. Reddy	Workshop on hard copy line printers	03/27/1972	4					
0321	P. Karp	CBI networking activity at MITRE	03/24/1972	13					
0322	V. Cerf, J. Postel	Well known socket numbers	03/26/1972	1					
0323	V. Cerf	Formation of Network Measurement Group NMG	03/23/1972	9	388				
0324	J. Postel	RJE Protocol meeting	04/03/1972	1					
0325	G. Hicks	Network Remote Job Entry program - NETRJS	04/06/1972	9					
0326	E. Westheimer	Network host status	04/03/1972	3	330		330	319	
0327	A. Bhushan	Data and File Transfer workshop notes	04/27/1972	7					
0328	J. Postel	Suggested Telnet Protocol changes	04/29/1972	1					
0329	Stanford Research Inst., NIC	ARPA Network mailing lists	05/17/1972	7			363	303	
0330	E. Westheimer	Network host status	04/13/1972	3	332	326	332	326	
0331	J. McQuillan	IMP System change notification	04/19/1972	1			343		
0332	E. Westheimer	Network host status	04/25/1972	3	342	330	342	330	
0333	R. Bressler, D. Murphy, D. Walden	Proposed experiment with a Message Switching Protocol	05/15/1972	52					
0334	A. McKenzie	Network use on May 8	05/01/1972	1					
0335	R. Bryan	New interface - IMP/360	05/01/1972	1					
0336	I. Cotton	Level 0 Graphic Input Protocol	05/05/1972	2					
0337		Not Issued							
0338	R. Braden	EBCDIC/ASCII mapping for network RJE	05/17/1972	6					
0339	R. Thomas	MLTNET: A "Multi Telnet" subsystem for Tenex	05/05/1972	8					

RFC No.	Authors	Title	Date	Pgs	Updated By	Updates	Obsoleted By	Obsoletes	Notes / Status
0340	T. O'Sullivan	Proposed Telnet changes	05/15/1972	1					328
0341		Not Issued							
0342	E. Westheimer	Network host status	05/15/1972	3	344	332	344	332	
0343	A. McKenzie	IMP System change notification	05/19/1972	2			359	331	
0344	E. Westheimer	Network host status	05/22/1972	3		342	353	342	
0345	K. Kelley	Interest in mixed integer programming MPSX on NIC 360/91 at CCN	05/26/1972	1					
0346	J. Postel	Satellite considerations	05/30/1972	1					
0347	J. Postel	Echo process	05/30/1972	1					
0348	J. Postel	Discard process	05/30/1972	1					
0349	J. Postel	Proposed standard socket numbers	05/30/1972	1			433		322, 204
0350	R. Stoughton	User accounts for UCSB On-Line System	05/18/1972	3					
0351	D. Crocker	Graphics information form for the ARPANET graphics resources notebook	06/05/1972	3					
0352	D. Crocker	TIP site information form	06/05/1972	3					
0353	E. Westheimer	Network host status	06/12/1972	3			362	344	
0354	A. Bhushan	File Transfer Protocol	07/08/1972	29	385		542	264, 265	
0355	J. Davidson	Response to NWG/RFC 346	06/09/1972	2					
0356	R. Alter	ARPA Network Control Center	06/21/1972	1					
0357	J. Davidson	Echoing strategy for satellite links	06/26/1972	15					
0358		Not Issued							
0359	D. Walden	Status of the release of the new IMP System	06/22/1972	1				343	
0360	C. Holland	Proposed Remote Job Entry Protocol	06/24/1972	16			407		
0361	R. Bressler	Deamon processes on host 106	07/05/1972	1					
0362	E. Westheimer	Network host status	06/28/1972	3			366	353	
0363	SRI., NIC	ARPA Network mailing lists	08/08/1972	6			402	329	
0364	M. Abrams	Serving remote users on the ARPANET	07/11/1972	7					
0365	D. Walden	Letter to all TIP users	07/11/1972	5					
0366	E. Westheimer	Network host status	07/11/1972	3			367	362	
0367	E. Westheimer	Network host status	07/19/1972	3	370		370	366	
0368	R. Braden	Comments on Proposed Remote Job Entry Protocol	07/21/1972	2					
0369	J. Pickens	Evaluation of ARPANET services January-March, 1972	07/25/1972	14					
0370	E. Westheimer	Network host status	07/31/1972	3			376	367	
0371	R. Kahn	Demonstration at International Computer Communications Conference	07/12/1972	2					
0372	R. Watson	Notes on a conversation with Bob Kahn on the ICCC	07/12/1972	3					
0373	J. McCarthy	Arbitrary character sets	07/14/1972	4					
0374	A. McKenzie	IMP system announcement	07/19/1972	2					
0375		Not Issued							
0376	E. Westheimer	Network host status	08/08/1972	3			370	370	
0377	R. Braden	Using TSO via ARPA Network Virtual Terminal	08/10/1972	8					
0378	A. McKenzie	Traffic statistics July 1972	08/10/1972	3			391		
0379	R. Braden	Using TSO at CCN	08/11/1972	5					
0380		Not Issued							
0381	J. McQuillan	Three aids to improved network operation	07/26/1972	4	394				
0382	L. McDaniel	Mathematical software on the ARPA Network	08/03/1972	1					
0383		Not Issued							
0384	J. North	Official site idents for organizations in the ARPA Network	08/28/1972	5				289	
0385	A. Bhushan	Comments on the File Transfer Protocol	08/18/1972	5	414	354			
0386	B. Cosell, D. Walden	Letter to TIP users-2	08/16/1972	7					
0387	K. Kelley, J. Meir	Some experiences in implementing Network Graphics Protocol Level 0	08/10/1972	6	401				
0388	V. Cerf	NCP statistics	08/23/1972	4		323			
0389	B. Noble	UCLA Campus Computing Network liaison staff for ARPA Network	08/30/1972	2			423		
0390	R. Braden	TSO scenario	09/12/1972	3					
0391	A. McKenzie	Traffic statistics August 1972	09/15/1972	3				378	
0392	G. Hicks, B. Wessler	Measurement of host costs for transmitting network data	09/20/1972	9					
0393	J. Winett	Comments on Telnet Protocol changes	10/03/1972	5					109, 139, 158, 318, 328
0394	J. McQuillan	Two proposed changes to the IMP-Host Protocol	09/27/1972	3		381			
0395	J. McQuillan	Switch settings on IMPs and TIPs	10/03/1972	1					
0396	S. Bunch	Network Graphics Working Group meeting - second iteration	11/13/1972	1	474				
0397		Not Issued							
0398	J. Pickens, E. Faeh	ICP sockets	09/22/1972	2					
0399	M. Krilanovich	SMFS login and logout	09/26/1972	2		122	431		
0400	A. McKenzie	Traffic statistics September 1972	10/18/1972	3					
0401	J. Hansen	Conversion of NGP-0 coordinates to device specific coordinates	10/23/1972	2		387			
0402	J. North	ARPA Network mailing lists	10/26/1972	8				363	
0403	G. Hicks	Desirability of a network 1108 service	01/10/1973	5					
0404	A. McKenzie	Host address changes involving Rand and ISI	10/05/1972	1			405		

RFC No.	Authors	Title	Date	Pgs	Updated By	Updates	Obsoleted By	Obsoletes	Notes / Status
0405	A. McKenzie	Correction to RFC 404	10/10/1972	1				404	
0406	J. McQuillan	Scheduled IMP software releases	10/10/1972	2					
0407	R. Bressler, R. Guida, A. McKenzie	Remote Job Entry Protocol	10/16/1972	24				360	H
0408	A. Owen, J. Postel	NETBANK	10/25/1972	1					
0409	J. White	Tenex interface to UCSB's Simple-Minded File System	12/08/1972	8					
0410	J. McQuillan	Removal of the 30-second delay when hosts come up	11/10/1972	2					
0411	M. Padlipsky	New MULTICS network software features	11/14/1972	1					
0412	G. Hicks	User FTP documentation	11/27/1972	10					
0413	A. McKenzie	Traffic statistics October 1972	11/13/1972	8					
0414	A. Bhushan	File Transfer Protocol FTP status and further comments	12/29/1972	5		385			
0415	H. Murray	Tenex bandwidth	11/29/1972	2					
0416	J. Norton	ARC system will be unavailable for use during Thanksgiving week	11/07/1972	1					
0417	J. Postel, C. Kline	Link usage violation	12/06/1972	1					
0418	W. Hathaway	Server file transfer under TSS/360 at NASA Ames	11/27/1972	10					
0419	A. Vezza	To: Network liaisons and station agents	12/12/1972	1					
0420	H. Murray	CCA ICCC weather demo	01/04/1973	11					
0421	A. McKenzie	Software consulting service for network users	11/27/1972	1					
0422	A. McKenzie	Traffic statistics November 1972	12/11/1972	4					
0423	B. Noble	UCLA Campus Computing Network liaison staff for ARPANET	12/12/1972	1				389	
0424		Not Issued							
0425	R. Bressler	"But my NCP costs $500 a day"	12/19/1972	1					
0426	R. Thomas	Reconnection Protocol	01/26/1973	16					
0427		Not Issued							
0428		Not Issued							
0429	J. Postel	Character generator process	12/12/1972	1					
0430	R. Braden	Comments on File Transfer Protocol	02/07/1973	8					
0431	M. Krilanovich	Update on SMFS login and logout	12/15/1972	3		122		399	
0432	N. Neigus	Network logical map	12/29/1972	2					
0433	J. Postel	Socket number list	12/22/1972	8			503	349	
0434	A. McKenzie	IMP/TIP memory retrofit schedule	01/04/1973	2			447		
0435	B. Cosell, D. Walden	Telnet issues	01/05/1973	14		318			
0436	M. Krilanovich	Announcement of RJS at UCSB	01/10/1973	2					
0437	E. Faeh	Data Reconfiguration Service at UCSB	06/30/1973	9					
0438	R. Thomas, R. Clements	FTP server-server interaction	01/15/1973	5					
0439	V. Cerf	PARRY encounters the DOCTOR	01/21/1973	5					
0440	D. Walden	Scheduled network software maintenance	01/01/1973	1					
0441	R. Bressler, R. Thomas	Inter-Entity Communication - an experiment	01/19/1973	10					
0442	V. Cerf	Current flow-control scheme for IMPSYS	01/24/1973	7	449				
0443	A. McKenzie	Traffic statistics December 1972	01/18/1973	3					
0444		Not Issued							
0445	A. McKenzie	IMP/TIP preventive maintenance schedule	01/22/1973	3					
0446	L. Deutsch	Proposal to consider a network program resource notebook	01/25/1973	1					
0447	A. McKenzie	IMP/TIP memory retrofit schedule	01/29/1973	2			476	434	
0448	R. Braden	Print files in FTP	02/27/1973	4					
0449	D. Walden	Current flow-control scheme for IMPSYS	01/06/1973	1		442			
0450	M. Padlipsky	MULTICS sampling timeout change	02/08/1973	1					
0451	M. Padlipsky	Tentative proposal for a Unified User Level Protocol	02/22/1973	3					
0452		Not Issued							
0453	M. Kudlick	Meeting announcement to discuss a network mail system	02/07/1973	3					
0454	A. McKenzie	File Transfer Protocol - meeting announcement and a new proposed document	02/16/1973	38					
0455	A. McKenzie	Traffic statistics January 1973	02/12/1973	4					
0456	M. Kudlick	Memorandum: Date change of mail meeting	02/13/1973	1					
0457	D. Walden	TIPUG	02/15/1973	1					
0458	R. Bressler, R. Thomas	Mail retrieval via FTP	02/20/1973	2					
0459	W. Kantrowitz	Network questionnaires	02/26/1973	1					
0460	C. Kline	NCP survey	02/13/1973	7					
0461	A. McKenzie	Telnet Protocol meeting announcement	02/14/1973	1					
0462	J. Iseli, D. Crocker	Responding to user needs	02/22/1973	2					
0463	A. Bhushan	FTP comments and response to RFC 430	02/21/1973	3					
0464	M. Kudlick	Resource notebook framework	02/27/1973	2					
0465		Not Issued							
0466	J. Winett	Telnet logger/server for host LL-67	02/27/1973	8					
0467	J. Burchfiel, R. Tomlinson	Proposed change to Host-Host Protocol:Resynchronization of connection status	02/20/1973	13		492			

RFC No.	Authors	Title	Date	Pgs	Updated By	Updates	Obsoleted By	Obsoletes	Notes / Status
0468	R. Braden	FTP data compression	03/08/1973	5					
0469	M. Kudlick	Network mail meeting summary	03/08/1973	9					
0470	R. Thomas	Change in socket for TIP news facility	03/13/1973	1					
0471	R. Thomas	Workshop on multi-site executive programs		2					
0472	S. Bunch	Illinois' reply to Maxwell's request for graphics information NIC 14925	03/01/1973	2					
0473	D. Walden	MIX and MIXAL?	02/28/1973	1					
0474	S. Bunch	Announcement of NGWG meeting: Call for papers	03/01/1973	1					
0475	A. Bhushan	FTP and network mail system	03/06/1973	8					
0476	A. McKenzie	IMP/TIP memory retrofit schedule rev. 2	03/07/1973	2				447	
0477	M. Krilanovich	Remote Job Service at UCSB	05/23/1973	18					
0478	R. Bressler, R. Thomas	FTP server-server interaction - II	03/26/1973	2					
0479	J. White	Use of FTP by the NIC Journal	03/08/1973	6					
0480	J. White	Host-dependent FTP parameters	03/08/1973	1					
0481		Not Issued							
0482	A. McKenzie	Traffic statistics February 1973	03/12/1973	4					
0483	M. Kudlick	Cancellation of the resource notebook framework meeting	03/14/1973	1					
0484		Not Issued							
0485	J. Pickens	MIX and MIXAL at UCSB	03/19/1973	1					
0486	R. Bressler	Data transfer revisited	03/20/1973	2					
0487	R. Bressler	Free file transfer	04/06/1973	2					
0488	M. Auerbach	NLS classes at network sites	03/23/1973	2					
0489	J. Postel	Comment on resynchronization of connection status proposal	03/26/1973	1					
0490	J. Pickens	Surrogate RJS for UCLA-CCN	03/06/1973	5					
0491	M. Padlipsky	What is "Free"?	04/12/1973	2					
0492	E. Meyer	Response to RFC 467	04/18/1973	9		467			
0493	J. Michener, I. Cotton, K. Kelley, D. Liddle, E. Meyer	E.W., Jr Graphics Protocol	04/26/1973	30					
0494	D. Walden	Availability of MIX and MIXAL in the Network	04/20/1973	1					
0495	A. McKenzie	Telnet Protocol specifications	05/01/1973					158	
0496	M. Auerbach	TNLS quick reference card is available	04/05/1973	2					
0498	R. Braden	On mail service to CCN	04/17/1973	2					
0499	B. Reussow	Harvard's network RJE	04/01/1973	7					
0500	A. Shoshani, I. Spiegler	Integration of data management systems on a computer network	04/16/1973	6					
0501	K. Pogran	Un-muddling "free file transfer"	05/11/1973	5					
0502		Not Issued							
0503	N. Neigus, J. Postel	Socket number list	04/12/1973	9			739	433	
0504	R. Thomas	Distributed resources workshop announcement	04/30/1973	4					
0505	M. Padlipsky	Two solutions to a file transfer access problem	06/25/1973	3					
0506	M. Padlipsky	FTP command naming problem	06/26/1973	1					
0507		Not Issued							
0508	L. Pfeifer, J. McAfee	Real-time data transmission on the ARPANET	05/07/1973	11					
0509	A. McKenzie	Traffic statistics April 1973	04/07/1973	3					
0510	J. White	Request for network mailbox addresses	05/30/1973	3					
0511	J. North	Enterprise phone service to NIC from ARPANET sites	05/23/1973	4					
0512	W. Hathaway	More on lost message detection	05/23/1973	1					
0513	W. Hathaway	Comments on the new Telnet specifications	05/30/1973	3					
0514	W. Kantrowitz	Network make-work	06/05/1973	3					
0515	R. Winter	Specifications for data language: Version 0/9	06/06/1973	35					
0516	J. Postel	Lost message detection	05/18/1973	2					
0517		Not Issued							
0518	N. Vaughan, E. Feinler	ARPANET accounts	06/19/1973						
0519	J. Pickens	Resource evaluation	06/01/1973	6					
0520	J. Day	Memo to FTP group: Proposal for File Access Protocol	06/25/1973	8					
0521	A. McKenzie	Restricted use of IMP DDT	05/30/1973	2					
0522	A. McKenzie	Traffic statistics May 1973	06/05/1973						
0523	A. Bhushan	SURVEY is in operation again	06/05/1973	1					
0524	J. White	Proposed Mail Protocol	06/13/1973	44					
0525	W. Parrish, J. Pickens	MIT-MATHLAB meets UCSB-OLS -an example of resource sharing	06/01/1973	10					
0526	W. Pratt	Technical meeting: Digital image processing software systems	06/25/1973	3					
0527	D. Covill	ARPAWOCKY	05/01/1973	1					
0528	J. McQuillan	Software checksumming in the IMP and network reliability	06/20/1973	11					
0529	A. McKenzie, R. Thomas, K. Pogran, R. Tomlinson	Note on protocol synch sequences	06/29/1973	6					
0530	A. Bhushan	Report on the Survey project	06/22/1973	9					

RFC No.	Authors	Title	Date	Pgs	Updated By	Updates	Obsoleted By	Obsoletes	Notes / Status
0531	M. Padlipsky	Feast or famine? A response to two recent RFC's about network information	06/26/1973	2					
0532	R. Merryman	UCSD-CC Server-FTP facility	07/12/1973	3					
0533	D. Walden	Message-ID numbers	07/17/1973	1					
0534	D. Walden	Lost message detection	07/17/1973	2					
0535	R. Thomas	Comments on File Access Protocol	07/25/1973						
0536		Not Issued							
0537	S. Bunch	Announcement of NGG meeting July 16-17	06/27/1973	2					
0538	A. McKenzie	Traffic statistics June 1973	07/05/1973	4					
0539	D. Crocker, J. Postel	Thoughts on the mail protocol proposed in RFC524	07/07/1973	3					
0540		Not Issued							
0541		Not Issued							
0542	N. Neigus	File Transfer Protocol	07/12/1973	52			765	354	354, 454, 495
0543	N. Meyer	Network journal submission and delivery	07/24/1973	8					
0544	N. Meyer, K. Kelley	Locating on-line documentation at SRI-ARC	07/13/1973	1					
0545	J. Pickens	Of what quality are the UCSB resources evaluators?	07/23/1973	2					
0546	R. Thomas	Tenex load averages for July 1973	08/10/1973	4					
0547	D. Walden	Change to the Very Distant Host specification	08/13/1973	4					
0548	D. Walden	Hosts using the IMP Going Down message	08/16/1973	1					
0549	J. Michener	Minutes of Network Graphics Group meeting, 15-17 July 1973	07/17/1973	13					
0550	L. Deutsch	NIC NCP experiment	08/24/1973	2					
0551	Y. Feinroth, R. Fink	[Letter from Feinroth re: NYU, ANL, and LBL entering the net, and FTP protocol]	08/27/1973	1					
0552	A. Owen	Single access to standard protocols	07/13/1973	1					
0553	C. Irby, K. Victor	Draft design for a text/graphics protocol	07/14/1973	17					
0554		Not Issued							
0555	J. White	Responses to critiques of the proposed mail protocol	07/30/1973	14					
0556	A. M. McKenzie	Traffic statistics (July 1973)	08/1973						
0557	B. Wessler	Revelations in network host measurements	08/30/1973	2					
0558		Not Issued							
0559	A. Bhushan	Comments on the new Telnet Protocol and its implementation	08/15/1973	5					
0560	D. Crocker, J. Postel	Remote Controlled Transmission and Echoing Telnet option	08/18/1973	11					
0561	A. Bhushan, K. Pogran, J. White, R. Tomlinson	Standardizing network mail headers	09/05/1973	2	680				
0562	A. McKenzie	Modifications to the Telnet specification	08/28/1973	1					
0563	J. Davidson	Comments on the RCTE Telnet option	08/28/1973	4					
0564		Not Issued							
0565	D. Cantor	Storing network survey data at the data computer	08/28/1973	6					
0566	A. McKenzie	Traffic statistics August 1973	09/04/1973	4					
0567	L. Deutsch	Cross country network bandwidth	09/06/1973	1	568				
0568	J. McQuillan	Response to RFC 567 - cross country network bandwidth	09/18/1973	2		567			
0569	M. Padlipsky	NETED: A common editor for the ARPA network	10/15/1973	7					H
0570	J. Pickens	Experimental input mapping between NVT ASCII and UCSB On Line System	10/30/1973	10					
0571	R. Braden	Tenex FTP problem	11/15/1973	1					
0572		Not Issued							
0573	A. Bhushan	Data and file transfer: Some measurement results	09/14/1973	12					
0574	M. Krilanovich	Announcement of a mail facility at UCSB	09/26/1973	1					
0575		Not Issued							
0576	K. Victor	Proposal for modifying linking	09/26/1973	2					
0577	D. Crocker	Mail priority	10/18/1973	2					
0578	A. Bhushan, N. Ryan	Using MIT-Mathlab MACSYMA from MIT-DMS Muddle	10/29/1973	13					
0579	A. McKenzie	Traffic statistics September 1973	11/26/1973	4					
0580	J. Postel	Note to protocol designers and implementers	10/25/1973	1	582				
0581	D. Crocker, J. Postel	Corrections to RFC 560: Remote Controlled Transmission and Echoing Telnet option	11/02/1973	4					
0582	R. Clements	Comments on RFC 580: Machine readable protocols	11/05/1973			580			
0583		Not Issued							
0584	J. Iseli, D. Crocker, N. Neigus	Charter for ARPANET Users Interest Working Group	11/06/1973	2					
0585	D. Crocker, N. Neigus, E. Feinler, J. Iseli	ARPANET users interest working group meeting	11/06/1973	9					
0586	A. McKenzie	Traffic statistics October 1973	11/08/1973	4					
0587	J. Postel	Announcing new Telnet options	11/13/1973	1					
0588	A. Stokes	London node is now up	10/29/1973	3					
0589	R. Braden,	CCN NETRJS server messages to remote user	11/26/1973	4					

RFC No.	Authors	Title	Date	Pgs	Updated By	Updates	Obsoleted By	Obsoletes	Notes / Status
0590	M. Padlipsky	MULTICS address change	11/19/1973	1					
0591	D. Walden	Addition to the Very Distant Host specifications	11/29/1973	1					
0592	R. Watson	Some thoughts on system design to facilitate resource sharing	11/20/1973	5					
0593	A. McKenzie, J. Postel	Telnet and FTP implementation schedule change	11/29/1973	1					
0594	J. Burchfiel	Speedup of Host-IMP interface	12/10/1973	3					
0595	W. Hathaway	Second thoughts in defense of the Telnet Go-Ahead	12/12/1973	5					
0596	E. Taft	Second thoughts on Telnet Go-Ahead	12/08/1973	6					
0597	N. Neigus, E. Feinler	Host status	12/12/1973	9	603				
0598	Stanford Research Inst., NIC	RFC index - December 5, 1973	12/05/1973	8					
0599	R. Braden	Update on NETRJS	12/13/1973	8			740	189	
0600	A. Berggreen	Interfacing an Illinois plasma terminal to the ARPANET	11/26/1973	4					
0601	A. McKenzie	Traffic statistics November 1973	12/14/1973	5					
0602	R. Metcalfe	"The stockings were hung by the chimney with care"	12/27/1973	2					
0603	J. Burchfiel	Response to RFC 597: Host status	12/31/1973	1	613	597			
0604	J. Postel	Assigned link numbers	12/26/1973	2			739	317	
0605		Not Issued							
0606	L. Deutsch	Host names on-line	12/29/1973	3					
0607	M. Krilanovich, G. Gregg	Comments on the File Transfer Protocol	01/07/1974	4	614		624		
0608	M. Kudlick	Host names on-line	01/10/1974	3			810		
0609	B. Ferguson	Statement of upcoming move of NIC/NLS service	01/10/1974	1					
0610	R. Winter, J. Hill, W. Greiff	Further data language design concepts	12/15/1973	79					
0611	D. Walden	Two changes to the IMP/Host Protocol to improve user/network communications	02/14/1974	4					
0612	A. McKenzie	Traffic statistics December 1973	01/16/1974	5					
0613	A. McKenzie	Network connectivity: A response to RFC 603	01/21/1974	1		603			
0614	K. Pogran, N. Neigus	Response to RFC 607: "Comments on the File Transfer Protocol"	01/28/1974	5		542, 607			
0615	D. Crocker	Proposed Network Standard Data Pathname Syntax	03/01/1974	5					
0616	D. Walden	Latest network maps	02/11/1973	3					
0617	E. Taft	Note on socket number assignment	02/19/1974	4					
0618	E. Taft	Few observations on NCP statistics	02/19/1974	4					
0619	W. Naylor, H. Opderbeck	Mean round-trip times in the ARPANET	03/07/1974	13					
0620	B. Ferguson	Request for monitor host table updates	03/01/1974	2					
0621	M. Kudlick	NIC user directories at SRI ARC	03/06/1974	1					
0622	A. McKenzie	Scheduling IMP/TIP down time	03/13/1974	3					
0623	M. Krilanovich	Comments on on-line host name service	02/22/1974	2					
0624	M. Krilanovich, G. Gregg, J. White, W. Hathaway	Comments on the File Transfer Protocol	02/28/1974	4				607	
0625	M. Kudlick, E. Feinler	On-line hostnames service	03/07/1974	1					
0626	L. Kleinrock, H. Opderbeck	On a possible lockup condition in IMP subnet due to message sequencing	03/14/1974	6					
0627	M. Kudlick, E. Feinler	ASCII text file of hostnames	03/25/1974	1					
0628	M. Keeney	Status of RFC numbers and a note on pre-assigned journal numbers	03/27/1974	1					
0629	J. North	Scenario for using the Network Journal	03/27/1974	2					
0630	J. Sussmann	FTP error code usage for more reliable mail service	04/10/1974	2					
0631	A. Danthine	International meeting on minicomputers and data communication: Call for papers	04/17/1974	1					
0632	H. Opderbeck	Throughput degradations for single packet messages	05/20/1974	6					
0633	A. McKenzie	IMP/TIP preventive maintenance schedule	03/18/1974	4			638		
0634	A. McKenzie	Change in network address for Haskins Lab	04/10/1974	1					
0635	V. Cerf	Assessment of ARPANET protocols	04/22/1974						
0636	J. Burchfiel, B. Cosell, R. Tomlinson, D. Walden	TIP/Tenex reliability improvements	06/10/1974	9					
0637	A. McKenzie	Change of network address for SU-DSL	04/23/1974	1					
0638	A. McKenzie	IMP/TIP preventive maintenance schedule	04/25/1974	4				633	
0639		Not Issued							
0640	J. Postel	Revised FTP reply codes	06/05/1974	16					
0641		Not Issued							
0642	J. Burchfiel	Ready line philosophy and implementation	07/05/1974	5					

RFC No.	Authors	Title	Date	Pgs	Updated By	Updates	Obsoleted By	Obsoletes	Notes / Status
0643	E. Mader	Network Debugging Protocol	07/01/1974	8					
0644	R. Thomas	On the problem of signature authentication for network mail	07/22/1974	6					
0645	D. Crocker	Network Standard Data Specification syntax	06/26/1974	9					
0646		Not Issued							
0647	M. Padlipsky	Proposed protocol for connecting host computers to ARPA-like networks via front end processors	11/12/1974	20					
0648		Not Issued							
0649		Not Issued							
0650		Not Issued							
0651	D. Crocker	Revised Telnet status option	10/25/1974	3			859		
0652	D. Crocker	Telnet output carriage-return disposition option	10/25/1974	3					H
0653	D. Crocker	Telnet output horizontal tabstops option	10/25/1974						H
0654	D. Crocker	Telnet output horizontal tab disposition option	10/25/1974	4					H
0655	D. Crocker	Telnet output formfeed disposition option	10/25/1974	4					H
0656	D. Crocker	Telnet output vertical tabstops option	10/25/1974	3					H
0657	D. Crocker	Telnet output vertical tab disposition option	10/25/1974	4					H
0658	D. Crocker	Telnet output linefeed disposition	10/25/1974	4					H
0659	J. Postel	Announcing additional Telnet options	10/18/1974	1					
0660	D. Walden	Some changes to the IMP and the IMP/Host interface	10/23/1974	2					
0661	J. Postel	Protocol information	11/23/1974	23					
0662	R. Kanodia	Performance improvement in ARPANET file transfers from Multics	11/26/1974	3					
0663	R. Kanodia	Lost message detection and recovery protocol	11/29/1974	17					
0664		Not Issued							
0665		Not Issued							
0666	M. Padlipsky	Specification of the Unified User-Level Protocol	11/26/1974	17					
0667	S. Chipman	BBN host ports	12/17/1974	1					
0668		Not Issued							
0669	D. Dodds	November, 1974, survey of New-Protocol Telnet servers	12/04/1974	4					
0670		Not Issued							
0671	R. Schantz	Note on Reconnection Protocol	12/06/1974	8					
0672	R. Schantz	Multi-site data collection facility	12/06/1974	10					
0673		Not Issued							
0674	J. Postel, J. White	Procedure call documents - version 2	12/12/1974	4					
0675	V. Cerf, Y. Dalal, C. Sunshine	Specification of Internet Transmission Control Program	12/01/1974	70					
0676		Not Issued							
0677	p. Johnson, R. Thomas	Maintenance of duplicate databases	01/27/1975	9					
0678	J. Postel	Standard file formats	12/19/1974	8					
0679	D. Dodds	February, 1975, survey of New-Protocol Telnet servers	02/21/1975	2					
0680	T. Myer, D. Henderson	Message Transmission Protocol	04/30/1975	6		561			
0681	S. Holmgren	Network UNIX	03/18/1975	6					
0682		Not Issued							
0683	R. Clements	FTPSRV - Tenex extension for paged files	04/03/1975	9					
0684	R. Schantz	Commentary on procedure calling as a network protocol	04/15/1975	7					
0685	M. Beeler	Response time in cross network debugging	04/16/1975	4					
0686	B. Harvey	Leaving well enough alone	05/10/1975	9					
0687	D. Walden	IMP/Host and Host/IMP Protocol changes	06/02/1975	3	690		704		
0688	D. Walden	Tentative schedule for the new Telnet implementation for the TIP	06/04/1975	1					
0689	R. Clements	Tenex NCP finite state machine for connections	05/23/1975	6					
0690	J. Postel	Comments on the proposed Host/IMP Protocol changes	06/06/1975	4	692	687	692	687	
0691	B. Harvey	One more try on the FTP	05/28/1975	13					
0692	S. Wolfe	Comments on IMP/Host Protocol changes RFCs 687 and 690	06/20/1975	2		690			
0693		Not Issued							
0694	J. Postel	Protocol information	06/18/1975	36					
0695	M. Krilanovich	Official change in Host-Host Protocol	07/05/1975	2					
0696	V. Cerf	Comments on the IMP/Host and Host/IMP Protocol changes	07/13/1975	2					
0697	J. Lieb	CWD command of FTP	07/14/1975	2					
0698	T. Mock	Telnet extended ASCII option	07/23/1975	4					
0699	J. Postel, J. Vernon	Request For Comments summary notes: 600-699	11/01/1982	9					
0700	E. Mader, W. Plummer, R. Tomlinson	Protocol experiment	08/01/1974	6					
0701	D. Dodds	August, 1974, survey of New-Protocol Telnet servers	08/01/1974	2					
0702	D. Dodds	September, 1974, survey of New-Protocol Telnet servers	09/25/1974	2					
0703	D. Dodds	July, 1975, survey of New Protocol Telnet Servers	07/11/1975	2					

RFC No.	Authors	Title	Date	Pgs	Updated By	Updates	Obsoleted By	Obsoletes	Notes / Status
0704	P. Santos	IMP/Host and Host/IMP Protocol change	09/15/1975					687	
0705	R. Bryan	Front-end Protocol B6700 version	11/05/1975	40					
0706	J. Postel	On the junk mail problem	11/08/1975	1					
0707	J. White	High-level framework for network-based resource sharing	12/23/1975	27					
0708	J. White	Elements of a distributed programming system	01/28/1976	29					
0709		Not Issued							
0710		Not Issued							
0711		Not Issued							
0712	J. Donnelley	Distributed Capability Computing System DCCS	02/05/1976	38					
0713	J. Haverty	MSDTP-Message Services Data Transmission Protocol	04/06/1976	29					
0714	A. McKenzie	Host-Host Protocol for an ARPANET-type network	04/21/1976	43					
0715		Not Issued							
0716	D. Walden, J. Levin	Interim revision to Appendix F of BBN 1822	05/24/1976	2					
0717	J. Postel	Assigned network numbers	07/01/1976	2					
0718	J. Postel	Comments on RCTE from the Tenex implementation experience	06/30/1976	2					
0719	J. Postel	Discussion on RCTE	07/22/1976	2					
0720	D. Crocker	Address specification syntax for network mail	08/05/1976	4					
0721	L. Garlick	Out-of-band control signals in a Host-to-Host Protocol	09/01/1976	7					
0722	J. Haverty	Thoughts on interactions in distributed services	09/16/1976	20					
0723		Not Issued							
0724	D. Crocker, K. Pogran, J. Vittal, D. Henderson	Proposed official standard for the format of ARPA Network messages	05/12/1977				733		
0725	J. Day, G. Grossman	RJE protocol for a resource sharing network	03/01/1977	26					
0726	J. Postel, D. Crocker	Remote Controlled Transmission and Echoing Telnet option	03/08/1977	16					
0727	M. Crispin	Telnet logout option	04/27/1977	3					
0728	J. Day	Minor pitfall in the Telnet Protocol	04/27/1977	1					
0729	D. Crocker	Telnet byte macro option	05/13/1977	4			735		
0730	J. Postel	Extensible field addressing	05/20/1977	5					
0731	J. Day	Telnet Data Entry Terminal option	06/27/1977	28			732		
0732	J. Day	Telnet Data Entry Terminal option	09/12/1977	30	1043			731	
0733	D. Crocker, J. Vittal, K. Pogran, D. Henderson	Standard for the format of ARPA network text messages	11/21/1977	38			822	724	
0734	M. Crispin	SUPDUP Protocol	10/07/1977	14					H
0735	D. Crocker, R. Gumpertz	Revised Telnet byte macro option	11/03/1977	5				729	PS
0736	M. Crispin	Telnet SUPDUP option	10/31/1977	2					PS
0737	K. Harrenstien	FTP extension: XSEN	10/31/1977	1					
0738	K. Harrenstien	Time server	10/31/1977	1					
0739	J. Postel	Assigned numbers	11/11/1977	11			750	604, 503	H
0740	R. Braden	NETRJS Protocol	11/22/1977	19				599	H
0741	D. Cohen	Specifications for the Network Voice Protocol NVP	11/22/1977	30					
0742	K. Harrenstien	NAME/FINGER Protocol	12/30/1977	7				1194, 1288	PS
0743	K. Harrenstien	FTP extension: XRSQ/XRCP	12/30/1977	8					
0744	J. Sattley	MARS - a Message Archiving and Retrieval Service	01/08/1978	6					
0745	M. Beeler	JANUS interface specifications	03/30/1978	22042					
0746	R. Stallman	SUPDUP graphics extension	03/17/1978	15					
0747	M. Crispin	Recent extensions to the SUPDUP Protocol	03/21/1978	3					
0748	M. Crispin	Telnet randomly-lose option	04/01/1978	2					
0749	B. Greenberg	Telnet SUPDUP-Output option	09/18/1978	4					
0750	J. Postel	Assigned numbers	09/26/1978	10			755	739	H
0751	P. Lebling	Survey of FTP mail and MLFL	12/10/1978	5					
0752	M. Crispin	Universal host table	01/02/1979	13					
0753	J. Postel	Internet Message Protocol	03/01/1979	62					
0754	J. Postel	Out-of-net host addresses for mail	04/06/1979	10					
0755	J. Postel	Assigned numbers	05/03/1979	12			758	750	H
0756	J. Pickens, E. Feinler, J. Mathis	NIC name server - a datagram-based information utility	07/01/1979	11					
0757	D. Deutsch	Suggested solution to the naming, addressing, and delivery problem for ARPANET message systems	09/10/1979	17					
0758	J. Postel	Assigned numbers	08/01/1979	12			762	755	H
0759	J. Postel	Internet Message Protocol	08/01/1980	71					H
0760	J. Postel	DoD standard Internet Protocol	01/01/1980	41			791, 777	IEN123	
0761	J. Postel	DoD Standard Transmission Control Protocol	01/01/1980	84	793				STD 7
0762	J. Postel	Assigned numbers	01/01/1980	13			770	758	
0763	M. Abrams	Role mailboxes	05/07/1980	1					
0764	J. Postel	Telnet Protocol specification	06/01/1980	15			854		
0765	J. Postel	File Transfer Protocol specification	06/01/1980	70			959	542	
0766	J. Postel	Internet Protocol Handbook: Table of contents	07/01/1980				774		

RFC No.	Authors	Title	Date	Pgs	Updated By	Updates	Obsoleted By	Obsoletes	Notes / Status
0767	J. Postel	Structured format for transmission of multi-media documents	08/01/1980	33					
0768	*J. Postel*	*User Datagram Protocol*	*08/28/1980*	*3*					*STD 6*
0769	J. Postel	Rapicom 450 facsimile file format	09/26/1980	2					
0770	J. Postel	Assigned numbers	09/01/1980	15			776	762	
0771	V. Cerf, J. Postel	Mail transition plan	09/01/1980	9					
0772	S. Sluizer, J. Postel	Mail Transfer Protocol	09/01/1980	31			780		
0773	V. Cerf	Comments on NCP/TCP mail service transition strategy	10/01/1980	11					H
0774	J. Postel	Internet Protocol Handbook: Table of contents	10/01/1980	3				766	
0775	D. Mankins, D. Franklin, A. Owen	Directory oriented FTP commands	12/01/1980	6					
0776	J. Postel	Assigned numbers	01/01/1981	13			790	770	
0777	J. Postel	Internet Control Message Protocol	04/01/1981	14			792	760	
0778	D. Mills	DCNET Internet Clock Service	04/18/1981	5					H
0779	E. Killian	Telnet send-location option	04/01/1981	2					PS
0780	S. Sluizer, J. Postel	Mail Transfer Protocol	05/01/1981	43			788	772	
0781	Z. Su	Specification of the Internet Protocol IP timestamp option	05/01/1981	2					
0782	J. Nabielsky, A. Skelton	Virtual Terminal management model	01/01/1981	20					
0783	K. Sollins	TFTP Protocol revision 2	06/01/1981	18				IEN133	DS
0784	S. Sluizer, J. Postel	Mail Transfer Protocol: ISI TOPS20 implementation	07/01/1981	3			1350		
0785	S. Sluizer, J. Postel	Mail Transfer Protocol: ISI TOPS20 file definitions	07/01/1981	3					
0786	S. Sluizer, J. Postel	Mail Transfer Protocol: ISI TOPS20 MTP-NIMAIL interface	07/01/1981	2					
0787	A. Chapin	Connectionless data transmission survey/tutorial	07/01/1981	41					
0788	J. Postel	Simple Mail Transfer Protocol	11/01/1981	62			821	780	
0789	E. Rosen	Vulnerabilities of network control protocols: An example	07/01/1981	15					
0790	J. Postel	Assigned numbers	09/01/1981	15			820	776	H
0791	*J. Postel*	*Internet Protocol*	*09/01/1981*	*45*				*760*	*STD 5*
0792	*J. Postel*	*Internet Control Message Protocol*	*09/01/1981*	*21*	*950*			*777*	*STD 5*
0793	*J. Postel*	*Transmission Control Protocol*	*09/01/1981*	*85*		*761*			*STD 7*
0794	V. Cerf	Pre-emption	09/01/1981	4					
0795	J. Postel	Service mappings	09/01/1981	7					
0796	J. Postel	Address mappings	09/01/1981	7					
0797	A. Katz	Format for Bitmap files	09/01/1981	2					
0798	A. Katz	Decoding facsimile data from the Rapicom 450	09/01/1981	17					
0799	D. Mills	Internet name domains	09/01/1981	6					
0800	J. Postel, JVernon	Request For Comments summary notes: 700-799	11/01/1982	10					
0801	J. Postel	NCP/TCP transition plan	11/01/1981	21					
0802	A. Malis	ARPANET 1822L Host Access Protocol	11/01/1981	43			851		
0803	A. Agarwal, D. Mills, M. O'Connor	Dacom 450/500 facsimile data transcoding	11/02/1981	14					
0804	ITU	CCITT draft recommendation T.4 [Standardization of Group 3 facsimile apparatus for document transmission]	01/01/1981	12					
0805	J. Postel	Computer mail meeting notes	02/08/1982	6					
0806	National Bureau of Standards	Proposed Federal Information Processing Standard: Specification for message format for computer based message systems	09/01/1981	99			841		
0807	J. Postel	Multimedia mail meeting notes	02/09/1982	6					
0808	J. Postel	Summary of computer mail services meeting held at BBN on 10 January 1979	03/01/1982	8					
0809	T. Chang	UCL facsimile system	02/01/1982	96					
0810	E. Feinler, K. Harrenstien, Z. Su, V. White	DoD Internet host table specification	03/01/1982	9			952	608	
0811	K. Harrenstien, V. White, E. Feinler	Hostnames Server	03/01/1982	5			953		
0812	K. Harrenstien, V. White	NICNAME/WHOIS	03/01/1982	3			954		
0813	D. Clark	Window and acknowledgement strategy in TCP	07/01/1982	22					
0814	D. Clark	Name, addresses, ports, and routes	07/01/1982	14					
0815	D. Clark	IP datagram reassembly algorithms	07/01/1982	9					
0816	D. Clark	Fault isolation and recovery	07/01/1982	12					
0817	D. Clark	Modularity and efficiency in protocol implementation	07/01/1982	26					
0818	J. Postel	Remote User Telnet service	11/01/1982	2					H
0819	Z. Su, J. Postel	Domain naming convention for Internet user applications	08/01/1982	18					
0820	J. Postel	Assigned numbers	08/14/1982	1			870	790	H
0821	*J. Postel*	*Simple Mail Transfer Protocol*	*08/01/1982*	*58*				*788*	*STD 10*

RFC No.	Authors	Title	Date	Pgs	Updated By	Updates	Obsoleted By	Obsoletes	Notes / Status
0822	D. Crocker	*Standard for the format of ARPA Internet text messages*	*08/13/1982*	*47*	*2156, 1327, 1148, 1138*			*733*	*STD 11*
0823	R. Hinden, A. Sheltzer	DARPA Internet gateway	09/01/1982	33					H
0824	W. MacGregor, Tappan D.C	CRONUS Virtual Local Network	08/25/1982	41					
0825	J. Postel	Request for comments on Requests For Comments	11/01/1982	2			1111, 1543		
0826	D. Plummer	*Ethernet Address Resolution Protocol: Or converting network protocol addresses to 48.bit Ethernet address for transmission on Ethernet hardware*	*11/01/1982*	*10*					*STD 37*
0827	E. Rosen	Exterior Gateway Protocol EGP	10/01/1982	44	904				
0828	K. Owen	Data communications: IFIP's international "network" of experts	08/01/1982	11					
0829	V. Cerf	Packet satellite technology reference sources	11/01/1982	5					
0830	Z. Su	Distributed system for Internet name service	10/01/1982	16					
0831	R. Braden	Backup access to the European side of SATNET	12/01/1982	5					
0832	D. Smallberg	Who talks TCP?	12/07/1982	13			833		
0833	D. Smallberg	Who talks TCP?	12/14/1982	13			834	832	
0834	D. Smallberg	Who talks TCP?	12/22/1982	13			835	833	
0835	D. Smallberg	Who talks TCP?	12/29/1982	13			836	834	
0836	D. Smallberg	Who talks TCP?	01/05/1983	13			837	835	
0837	D. Smallberg	Who talks TCP?	01/12/1983	14			838	836	
0838	D. Smallberg	Who talks TCP?	01/20/1983	14			839	837	
0839	D. Smallberg	Who talks TCP?	01/26/1983	14			842	838	
0840	J. Postel	Official protocols	04/13/1983	23			880		H
0841	National Bureau of Standards	Specification for message format for Computer Based Message Systems	01/27/1983	110				806	
0842	D. Smallberg	Who talks TCP? - survey of 1 February 83	02/03/1983	14			843	839	
0843	D. Smallberg	Who talks TCP? - survey of 8 February 83	02/09/1983	14	844		845	842	
0844	R. Clements	Who talks ICMP, too? - Survey of 18 February 1983	02/18/1983	5		843			
0845	D. Smallberg	Who talks TCP? - survey of 15 February 1983	02/17/1983	14			846	843	
0846	D. Smallberg	Who talks TCP? - survey of 22 February 1983	02/23/1983	14			847	845	
0847	D. Smallberg, A. Westine, J. Postel	Summary of Smallberg surveys	02/01/1983	2				846	
0848	D. Smallberg	Who provides the "little" TCP services?	03/14/1983	5					
0849	M. Crispin	Suggestions for improved host table distribution	05/01/1983	2					
0850	M. Horton	Standard for interchange of USENET messages	06/01/1983	18			1036		
0851	A. Malis	ARPANET 1822L Host Access Protocol	04/18/1983	44			878	802	
0852	A. Malis	ARPANET short blocking feature	04/01/1983	13					
0853		Not Issued							
0854	J. Postel, J. Reynolds	*Telnet Protocol specification*	*05/01/1983*	*15*				*764*	*STD 8*
0855	J. Postel, J. Reynolds	*Telnet option specifications*	*05/01/1983*	*4*				*NIC18640*	*STD 8*
0856	J. Postel, J. Reynolds	*Telnet binary transmission*	*05/01/1983*	*4*				*NIC15389*	*STD 27*
0857	J. Postel, J. Reynolds	*Telnet echo option*	*05/01/1983*	*5*				*NIC15390*	*STD 28*
0858	J. Postel, J. Reynolds	*Telnet Suppress Go Ahead option*	*05/01/1983*	*3*				*NIC15392*	*STD 29*
0859	J. Postel, J. Reynolds	*Telnet status option*	*05/01/1983*	*3*				*651*	*STD 30*
0860	J. Postel, J. Reynolds	*Telnet timing mark option*	*05/01/1983*	*4*				*NIC16238*	*STD 31*
0861	J. Postel, J. Reynolds	*Telnet extended options: List option*	*05/01/1983*	*1*				*NIC16239*	*STD 32*
0862	J. Postel	*Echo Protocol*	*05/01/1983*	*1*					*STD 20*
0863	J. Postel	*Discard Protocol*	*05/01/1983*	*1*					*STD 21*
0864	J. Postel	*Character Generator Protocol*	*05/01/1983*	*3*					*STD 22*
0865	J. Postel	*Quote of the Day Protocol*	*05/01/1983*	*1*					*STD 23*
0866	J. Postel	*Active users*	*05/01/1983*	*1*					*STD 24*
0867	J. Postel	*Daytime Protocol*	*05/01/1983*	*2*					*STD 25*
0868	K. Harrenstien, J. Postel	*Time Protocol*	*05/01/1983*	*2*					*STD 26*
0869	R. Hinden	Host Monitoring Protocol	12/01/1983	70					H
0870	J. Reynolds, J. Postel	Assigned numbers	10/01/1983	26			900	820	
0871	M. Padlipsky	Perspective on the ARPANET reference model	09/01/1982	25					
0872	M. Padlipsky	TCP-on-a-LAN	09/01/1982	8					
0873	M. Padlipsky	Illusion of vendor support	09/01/1982	8					
0874	M. Padlipsky	Critique of X.25	09/01/1982	13					
0875	M. Padlipsky	Gateways, architectures, and heffalumps	09/01/1982	8					
0876	D. Smallberg	Survey of SMTP implementations	09/01/1983	13					
0877	J. Korb	Standard for the transmission of IP datagrams over public data networks	09/01/1983	2			1356		
0878	A. Malis	ARPANET 1822L Host Access Protocol	12/01/1983	48				851	

RFC No.	Authors	Title	Date	Pgs	Updated By	Updates	Obsoleted By	Obsoletes	Notes / Status
0879	J. Postel	TCP maximum segment size and related topics	11/01/1983	11					
0880	J. Reynolds, J. Postel	Official protocols	10/01/1983	26			901	840	(STD 1)
0881	J. Postel	Domain names plan and schedule	11/01/1983	10			897		
0882	P. Mockapetris	Domain names: Concepts and facilities	11/01/1983	31	973		1034, 1035		
0883	P. Mockapetris	Domain names: Implementation specification	11/01/1983	73	973		1034, 1035		
0884	M. Solomon, E. Wimmers	Telnet terminal type option	12/01/1983	5			930		
0885	J. Postel	Telnet end of record option	12/01/1983	2					PS
0886	M. Rose	Proposed standard for message header munging	12/15/1983	16					
0887	M. Accetta	Resource Location Protocol	12/01/1983	16					E
0888	L. Seamonson, E. Rosen	"STUB" Exterior Gateway Protocol	01/01/1984	38	904				
0889	D. Mills	Internet delay experiments	12/01/1983	12					
0890	J. Postel	Exterior Gateway Protocol implementation schedule	02/01/1984	3					
0891	*D. Mills*	*DCN local-network protocols*	*12/01/1983*	*26*					*STD 44*
0892	ISO	ISO Transport Protocol specification [Draft]	12/01/1983	82			905		
0893	S. Leffler, M. Karels	Trailer encapsulations	04/01/1984	3					
0894	*C. Hornig*	*Standard for the transmission of IP datagrams over Ethernet networks*	*04/01/1984*	*3*					*STD 41*
0895	*J. Postel*	*Standard for the transmission of IP datagrams over experimental Ethernet networks*	*04/01/1984*	*3*					*STD 42*
0896	J. Nagle	Congestion control in IP/TCP internetworks	01/06/1984	9					
0897	J. Postel	Domain name system implementation schedule	02/01/1984	8	921	881			
0898	R. Hinden, J. Postel, M. Muuss, J. Reynolds	Gateway special interest group meeting notes	04/01/1984	24					
0899	J. Postel, A. Westine	Request For Comments summary notes: 800-899	05/01/1984	18					I
0900	J. Reynolds, J. Postel	Assigned Numbers	06/01/1984	43			923	870	H
0901	J. Reynolds, J. Postel	Official ARPA-Internet protocols	06/01/1984	28			0924	880	(STD 1)
0902	J. Reynolds, J. Postel	ARPA Internet Protocol policy	07/05/1984						*
0903	*R. Finlayson, T. Mann, J. Mogul, M. Theimer*	*Reverse Address Resolution Protocol*	*06/01/1984*	*4*					*STD 38*
0904	*ITT, D. Mills*	*Exterior Gateway Protocol formal specification*	*04/01/1984*	*30*		*827, 888*			*STD 18*
0905	A. McKenzie	ISO Transport specification ISO DP 8073	04/01/1984	154				892	
0906	R. Finlayson	Bootstrap loading using TFTP	06/01/1984	4					
0907	*Bolt Beranek and Newman*	*Host Access Protocol specification*	*07/01/1984*	*75*			*1221*		*STD 40, H*
0908	R. Hinden, J. Sax, D. Velten	Reliable Data Protocol	07/01/1984	56	1151				E
0909	W. Milliken, C. Welles	Loader Debugger Protocol	07/01/1984	127					E
0910	H. Forsdick	Multimedia mail meeting notes	08/01/1984	11					
0911	P. Kirton	EGP Gateway under Berkeley UNIX 4.2	08/22/1984	22					
0912	M. St. Johns	Authentication service	09/01/1984	3			931		
0913	M. Lottor	Simple File Transfer Protocol	09/01/1984	15					H
0914	T. Conte, G. Delp, D. Farber	Thinwire protocol for connecting personal computers to the Internet	09/01/1984	22					H
0915	M. Elvy, R. Nedved	Network mail path service	12/01/1984	11					
0916	G. Finn	Reliable Asynchronous Transfer Protocol RATP	10/01/1984	54					H
0917	J. Mogul	Internet subnets	10/01/1984	22					
0918	J. Reynolds	Post Office Protocol	10/01/1984	5			937		
0919	*J. Mogul*	*Broadcasting Internet datagrams*	*10/01/1984*	*8*					*STD 5*
0920	J. Postel, J. Reynolds	Domain requirements	10/01/1984	14					
0921	J. Postel	Domain name system implementation schedule - revised	10/01/1984	13		897			
0922	*J. Mogul*	*Broadcasting Internet datagrams in the presence of subnets*	*10/01/1984*	*12*					*STD 5*
0923	J. Reynolds, J. Postel	Assigned numbers	10/01/1984	47			943	900	H
0924	J. Reynolds, J. Postel	Official ARPA-Internet protocols for connecting personal computers to the Internet	10/01/1984	35			944	901	
0925	J. Postel	Multi-LAN address resolution	10/01/1984	15					
0926	ISO	Protocol for providing the connectionless mode network services	12/01/1984	101			994		
0927	B. Anderson	TACACS user identification Telnet option	12/01/1984	4					PS
0928	M. Padlipsky	Introduction to proposed DoD standard H-FP	12/01/1984	21					
0929	J. Lilienkamp, R. Mandell, M. Padlipsky	Proposed Host-Front End Protocol	12/01/1984	52					H

RFC No.	Authors	Title	Date	Pgs	Updated By	Updates	Obsoleted By	Obsoletes	Notes / Status
0930	M. Solomon, E. Wimmers	Telnet terminal type option	01/01/1985	4			1091	884	
0931	M. St. Johns	Authentication server	01/01/1985	4			1413	912	
0932	D. Clark	Subnetwork addressing scheme	01/01/1985	4					
0933	S. Silverman	Output marking Telnet option	01/01/1985	4					PS
0934	M. Rose, E. Stefferud	Proposed standard for message encapsulation	01/01/1985	10					
0935	J. Robinson	Reliable link layer protocols	01/01/1985	13					
0936	M. Karels	Another Internet subnet addressing scheme	02/04/1985						
0937	M. Butler, D. Chase, J. Goldberger, J. Postel, J. Reynolds	Post Office Protocol - version 2	02/01/1985	24				918	' H
0938	T. Miller	Internet Reliable Transaction Protocol functional and interface specification	02/01/1985	16					E
0939	National Research Council	Executive summary of the NRC report on transport protocols for Department of Defense data networks	02/01/1985	20					
0940	D. Estrin, T. Li, Y. Rekhter, K. Varadhan, D. Zappala	Gateway Algorithms and Data Structures Toward an Internet standard scheme for subnetting	04/01/1985	3					
0941	ISO	Addendum to the network service definition covering network layer addressing	04/01/1985	34					
0942	National Research Council,	Transport protocols for Department of Defense data networks	02/01/1985	68					
0943	J. Reynolds, J. Postel	Assigned numbers	04/01/1985	50			923	960	H
0944	J. Reynolds, J. Postel	Official ARPA-Internet protocols	04/01/1985	40			961	924	(STD 1)
0945	J. Postel	DoD statement on the NRC report	05/01/1985	2			1039		
0946	R. Nedved	Telnet terminal location number option	05/01/1985	4					
0947	K. Lebowitz, D. Mankins	Multi-network broadcasting within the Internet	06/01/1985	5					
0948	*I. Winston*	*Two methods for the transmission of IP datagrams over IEEE 802.3 networks*	*06/01/1985*	*5*			*1042*		*STD 43, H*
0949	M. Padlipsky	FTP unique-named store command	07/01/1985	2					
0950	**J. Mogul, J. Postel**	**Internet standard subnetting procedure**	**08/01/1985**	**18**		**792**			**STD 5**
0951	W. Croft, J. Gilmore	Bootstrap Protocol	09/01/1985	12			1542, 1532, 1497, 1395		DS
0952	E. Feinler, M. Stahl, K. Harrenstien	DoD Internet host table specification	10/01/1985	6				810	
0953	E. Feinler, M. Stahl, K. Harrenstien	Hostname Server	10/01/1985	5				811	H
0954	E. Feinler, M. Stahl, K. Harrenstien	NICNAME/WHOIS	10/01/1985	4				812	DS
0955	R. Braden	Towards a transport service for transaction processing applications	09/01/1985	10					
0956	D. Mills	Algorithms for synchronizing network clocks	09/01/1985	26					
0957	D. Mills	Experiments in network clock synchronization	09/01/1985	27					
0958	D. Mills	Network Time Protocol NTP	09/01/1985	14			1059		
0959	**J. Postel, J. Reynolds**	**File Transfer Protocol**	**10/01/1985**	**69**	**2228, 2640**			**765**	**STD 9**
0960	J. Reynolds, J. Postel	Assigned numbers	12/01/1985	60			990	943	
0961	J. Reynolds, J. Postel	Official ARPA-Internet protocols	12/01/1985	38			991	944	(STD 1)
0962	M. Padlipsky	TCP-4 prime	11/01/1985	2					
0963	D. Sidhu	Some problems with the specification of the Military Standard Internet Protocol	11/01/1985	19					
0964	D. Sidhu	Some problems with the specification of the Military Standard Transmission Control Protocol	11/01/1985	10					
0965	L. Aguilar	Format for a graphical communication protocol	12/01/1985	51					
0966	S. Deering, D. Cheriton	Host groups: A multicast extension to the Internet Protocol	12/01/1985	27			988		
0967	M. Padlipsky	All victims together	12/01/1985	2					
0968	V. Cerf	Twas the night before start-up	12/01/1985	2					
0969	D. Clark, M. Lambert, L. Zhang	NETBLT: A bulk data transfer protocol	12/01/1985	15			998		
0970	J. Nagle	On packet switches with infinite storage	12/01/1985	9					
0971	A. DeSchon	Survey of data representation standards	01/01/1986	9					
0972	F. Wancho	Password Generator Protocol	01/01/1986	2					
0973	P. Mockapetris	Domain system changes and observations	01/01/1986	10		882, 883	1035, 1034		
0974	**C. Partridge**	**Mail routing and the domain system**	**01/01/1986**	**7**					**STD 14**
0975	D. Mills	Autonomous confederations	02/01/1986	10					
0976	M. Horton	UUCP mail interchange format standard	02/01/1986	12	1137				
0977	B. Kantor, P. Lapsley	Network News Transfer Protocol: A Proposed Standard for the Stream-Based Transmission of News	02/01/1986	27					PS
0978	J. Reynolds, R. Gillman, J. Postel, W. Brackenridge, A. Witkowski	Postel, J.B. Voice File Interchange Protocol VFIP	02/01/1986	5					

RFC No.	Authors	Title	Date	Pgs	Updated By	Updates	Obsoleted By	Obsoletes	Notes / Status
0979	A. Malis	PSN End-to-End functional specification	03/01/1986	15					
0980	Jacobsen, J. Postel	Protocol document order information	03/01/1986	12					
0981	D. Mills	Experimental multiple-path routing algorithm	03/01/1986	22					
0982	H. Braun	Guidelines for the specification of the structure of the Domain Specific Part DSP of the ISO standard NSAP address	04/01/1986	11					
0983	D. Cass, M. Rose	ISO transport arrives on top of the TCP	04/01/1986	27			1006		
0984	D. Clark, M. Lambert	PCMAIL: A distributed mail system for personal computers	05/01/1986	31			993		
0985	NSF, Network Technical Advisory Group	Requirements for Internet gateways - draft	05/01/1986	23			1009		
0986	R. Callon, H. Braun	Guidelines for the use of Internet-IP addresses in the ISO Connectionless-Mode Network Protocol [Working draft]	06/01/1986	7			1069		
0987	S. Kille	Mapping between X.400 and RFC 822	06/01/1986	69	1026, 1138, 1148	882	2156		PS
0988	S. Deering	Host extensions for IP multicasting	07/01/1986	20			1112, 1054	966	
0989	J. Linn	Privacy enhancement for Internet electronic mail: Part I: Message encipherment and authentication procedures	02/01/1987	23			1113, 1040		
0990	J. Postel, J. Reynolds	Assigned numbers	11/01/1986	75	997		1010	960	
0991	J. Reynolds, J. Postel	Official ARPA-Internet protocols	11/01/1986	46			1011	961	(STD 1)
0992	K. Birman, T. Joseph	On communication support for fault tolerant process groups	11/01/1986	18					
0993	D. Clark, M. Lambert	PCMAIL: A distributed mail system for personal computers	12/01/1986	28			1056	984	
0994	ISO	Final text of DIS 8473, Protocol for Providing the Connectionless-mode Network Service	03/01/1986	52				926	
0995	ISO	End System to Intermediate System Routing Exchange Protocol for use in conjunction with ISO 8473	04/01/1986	41					
0996	D. Mills	Statistics server	02/01/1987	3					H
0997	J. Reynolds, J. Postel	Internet numbers	03/01/1987	42		990	1020, 1117		
0998	D. Clark, M. Lambert, L. Zhang	NETBLT: A bulk data transfer protocol	03/01/1987	21				969	E
0999	A. Westine, J. Postel	Requests For Comments summary notes: 900-999	04/01/1987	22			1000		
1000	J. Postel, J. Reynolds	Request For Comments reference guide	08/01/1987	149				999	
1001	*DARPA, End-to-End Services Task Force, IAB, NetBIOS Working Group*	*Protocol standard for a NetBIOS service on a TCP/UDP transport: Concepts and methods*	*03/01/1987*	*68*					*STD 19*
1002	*DARPA, End-to-End Services Task Force, IAB, NetBIOS Working Group*	*Protocol standard for a NetBIOS service on a TCP/UDP transport: Detailed specifications*	*03/01/1987*	*85*					*STD 19*
1003	A. Katz	Issues in defining an equations representation standard	03/01/1987	7					
1004	D. Mills	Distributed-protocol authentication scheme	04/01/1987	8					E
1005	A. Khanna A. Malis	ARPANET AHIP-E Host Access Protocol enhanced AHIP	05/01/1987	31					
1006	*D. Cass, M. Rose*	*ISO transport services on top of the TCP: Ver. 3*	*05/01/1987*	*17*	*2126*			*983*	*STD 35*
1007	W. McCoy	Military supplement to the ISO Transport Protocol	06/01/1987	23					
1008	W. McCoy	Implementation guide for the ISO Transport Protocol	06/01/1987	73					
1009	R. Braden, J. Postel	Requirements for Internet gateways	06/01/1987	55			1812	985	(STD 4), H
1010	J. Postel, J. Reynolds	Assigned numbers	05/01/1987	44			1060	990	(STD 2)
1011	J. Postel, J. Reynolds	Official Internet protocols	05/01/1987	52				991	(STD 1)
1012	J. Reynolds, J. Postel	Bibliography of Request For Comments 1 through 999	06/01/1987	64					I
1013	R. Scheifler	X Window System Protocol, version 11: Alpha update April 1987	06/01/1987	101					
1014	Sun Microsystems, Inc.	XDR: External Data Representation standard	06/01/1987	20					
1015	B. Leiner	Implementation plan for interagency research Internet	07/01/1987	24					
1016	W. Prue, J. Postel	Something a host could do with source quench: The Source Quench Introduced Delay SQuID	07/01/1987	18					
1017	B. Leiner	Network requirements for scientific research: Internet task force on scientific computing	08/01/1987	19					
1018	A. McKenzie	Some comments on SQuID	08/01/1987	3					

RFC No.	Authors	Title	Date	Pgs	Updated By	Updates	Obsoleted By	Obsoletes	Notes / Status
1019	D. Arnon	Report of the Workshop on Environments for Computational Mathematics	09/01/1987	8					
1020	S. Romano, M. Stahl	Internet numbers	11/01/1987	51			1062, 1117	997	
1021	C. Partridge, G. Trewitt	High-level Entity Management System HEMS	10/01/1987	5					H
1022	C. Partridge, G. Trewitt	High-level Entity Management Protocol HEMP	10/01/1987	12					
1023	G. Trewitt, C. Partridge	HEMS monitoring and control language	10/01/1987	17			1076		
1024	C. Partridge, G. Trewitt	HEMS variable definitions	10/01/1987	74					
1025	J. Postel	TCP and IP bake off	09/01/1987	6					
1026	S. Kille	Addendum to RFC 987: Mapping between X.400 and RFC-822	09/01/1987	4	1138, 1148	987	1327, 1495, 2156		PS
1027	S. Carl-Mitchell, J. Quarterman	Using ARP to implement transparent subnet gateways	10/01/1987	8					
1028	J. Case, J. Davin, M. Fedor, M. Schoffstall	Simple Gateway Monitoring Protocol	11/01/1987	38					H
1029	G. Parr	More fault tolerant approach to address resolution for a Multi-LAN system of Ethernets	05/01/1988	17					*
1030	M. Lambert	On testing the NETBLT Protocol over divers networks	11/01/1987	16					*
1031	W. Lazear	MILNET name domain transition	11/01/1987	10					*
1032	M. Stahl	Domain administrators guide	11/01/1987	14					*
1033	M. Lottor	Domain administrators operations guide	11/01/1987	22					*
1034	P. Mockapetris	Domain names - concepts and facilities	11/01/1987	55	1101, 1183, 1348, 1876, 1982, 2065, 2181, 2308		1065, 2308	973, 882, 883	(STD 13)
1035	*P. Mockapetris*	*Domain names - implementation and specification*	*11/01/1987*	*55*	*1101, 1183, 1348, 1876, 1982, 1995, 1996, 2065, 2181, 2136, 2137, 2308, 2535*			*973, 882, 883*	*STD 13*
1036	M. Horton, R. Adams	Standard for interchange of USENET messages	12/01/1987	19				850	
1037	B. Greenberg, S. Keene	NFILE - a file access protocol	12/01/1987	86					H
1038	M. St. Johns	Draft revised IP security option	01/01/1988	7			1108		
1039	D. Latham	DoD statement on Open Systems Interconnection protocols	01/01/1988	3				945	
1040	J. Linn	Privacy enhancement for Internet electronic mail: Part I: Message encipherment and authentication procedures	01/01/1988	29			1113	989	
1041	Y. Rekhter	Telnet 3270 regime option	01/01/1988	6					PS
1042	*J. Postel, J. Reynolds*	*Standard for the transmission of IP datagrams over IEEE 802 networks*	*02/01/1988*	*15*				*948*	*STD 43*
1043	A. Yasuda, T. Thompson	Telnet Data Entry Terminal option: DODIIS implementation	02/01/1988	26		732			
1044	*K. Hardwick, J. Lekashman*	*Internet Protocol on Network System's HYPERchannel: Protocol specification*	*02/01/1988*	*43*					*STD 45*
1045	D. Cheriton	VMTP: Versatile Message Transaction Protocol: Protocol specification	02/01/1988	123					E
1046	W. Prue, J. Postel	Queuing algorithm to provide type-of-service for IP links	02/01/1988	11					
1047	C. Partridge	Duplicate messages and SMTP	02/01/1988	3					
1048	P. Prindeville	BOOTP vendor information extensions	02/01/1988	7			1084		DS
1049	M. Sirbu	Content-type header field for Internet messages	03/01/1988	8	1154				H
1050	Sun Microsystems, Inc.	RPC: Remote Procedure Call Protocol specification	04/01/1988	24			1057		H
1051	P. Prindeville	Standard for the transmission of IP datagrams and ARP packets over ARCNET networks	03/01/1988	4			1201		(STD 46), H
1052	V. Cerf	IAB recommendations for the development of Internet network management standards	04/01/1988	14					
1053	S. Levy, T. Jacobson	Telnet X.3 PAD option	04/01/1988	21					PS
1054	S. Deering	Host extensions for IP multicasting	05/01/1988	19			1112	988	
1055	*J. Romkey*	*Nonstandard for transmission of IP datagrams over serial lines: SLIP*	*06/01/1988*	*6*					*STD 47*
1056	M. Lambert	PCMAIL: A distributed mail system for personal computers	06/01/1988	38				993	I
1057	Sun Microsystems, Inc.	RPC: Remote Procedure Call Protocol specification version 2	06/01/1988	25				1050	I
1058	C. Hedrick	Routing Information Protocol	06/01/1988	33	1723, 1388				(STD34), H
1059	D. Mills	Network Time Protocol version 1 specification and implementation	07/01/1988	58			1119, 1305	958	
1060	J. Postel, J. Reynolds	ASSIGNED NUMBERS	03/20/1990	86			1340	1010	(STD 2), H

RFC No.	Authors	Title	Date	Pgs	Updated By	Updates	Obsoleted By	Obsoletes	Notes / Status
1061		Not Issued							
1062	S. Romano, M. Stahl, M. Recker	Internet numbers	08/01/1988	65			1117, 1166	1020	
1063	C. Kent, J. Mogul, K. McCloghrie, C. Partridge	IP MTU Discovery options	07/01/1988	11			1191		
1064	M. Crispin	Interactive Mail Access Protocol: Version 2	07/01/1988	26			1176, 1203		H
1065	*K. McCloghrie, M. Rose*	*Structure and identification of management information for TCP/IP-based internets*	*08/01/1988*	*21*			*1155*	*1034*	*STD 16, H*
1066	K. McCloghrie, M. Rose	Management Information Base for network management of TCP/IP-based internets	08/01/1988	90			1156		H
1067	J. Case, M. Fedor, M. Schoffstall, J. Davin	Simple Network Management Protocol	08/01/1988	33			1098		
1068	A. DeSchon, R. Braden	Background File Transfer Program BFTP	08/01/1988	27					
1069	R. Callon, H. Braun	Guidelines for the use of Internet-IP addresses in the ISO Connectionless-Mode Network Protocol	02/01/1989	10				986	
1070	R. Hagens, N. Hall, M. Rose	Use of the Internet as a subnetwork for experimentation with the OSI network layer	02/01/1989	17					
1071	R. Braden, D. Borman, C. Partridge	Computing the Internet checksum	09/01/1988	24		1141			
1072	R. Braden, V. Jacobson	TCP extensions for long-delay paths	10/01/1988	16			1323, 2018		
1073	D. Waitzman	Telnet window size option	10/01/1988	4					PS
1074	J. Rekhter	NSFNET backbone SPF based Interior Gateway Protocol	10/01/1988	5					
1075	S. Deering, C. Partridge, D. Waitzman	Distance Vector Multicast Routing Protocol	11/01/1988	24					E
1076	G. Trewitt, C. Partridge	HEMS monitoring and control language	11/01/1988	42				1023	
1077	B. Leiner	Critical issues in high bandwidth networking	11/01/1988	46					
1078	M. Lottor	TCP port service Multiplexer TCPMUX	11/01/1988	2					
1079	C. Hedrick	Telnet terminal speed option	12/01/1988	3					
1080	C. Hedrick	Telnet remote flow control option	11/01/1988	4			1372		
1081	M. Rose	Post Office Protocol - version 3	11/01/1988	16			1225		PS
1082	M. Rose	Post Office Protocol - version 3: Extended service offerings	11/01/1988	11					H
1083	DARPA, IAB	IAB official protocol standards	12/01/1988	12			1100		(STD 1)
1084	J. Reynolds	BOOTP vendor information extensions	12/01/1988	8			1395	1048	
1085	M. Rose	ISO presentation services on top of TCP/IP based internets	12/01/1988	32					
1086	J. Onions, M. Rose	ISO-TP0 bridge between TCP and X.25	12/01/1988	9					
1087	DARPA, IAB	Ethics and the Internet	01/01/1989	2					
1088	*L. McLaughlin*	*Standard for the transmission of IP datagrams over NetBIOS networks*	*02/01/1989*	*3*					*STD 48*
1089	M. Schoffstall, C. Davin, M. Fedor, J. Case	SNMP over Ethernet	02/01/1989	3					
1090	R. Ullmann	SMTP on X.25	02/01/1989	4					
1091	J. VanBokkelen	Telnet terminal-type option	02/01/1989	7				930	
1092	J. Rekhter	EGP and policy based routing in the new NSFNET backbone	02/01/1989	5					
1093	H. Braun	NSFNET routing architecture	02/01/1989	9					
1094	Sun Microsystems Inc	NFS: Network File System Protocol specification	03/01/1989	27					1813, H
1095	U. Warrier, L. Besaw	Common Management Information Services and Protocol over TCP/IP CMOT	04/01/1989	67			1189		
1096	G. Marcy	Telnet X display location option	03/01/1989	3					PS
1097	B. Miller	Telnet subliminal-message option	04/01/1989	3					
1098	J. Case, C. Davin, M. Fedor	Simple Network Management Protocol SNMP	04/01/1989	34	1157			1067	
1099	J. Reynolds	Request for Comments Summary RFC Numbers 1000-1099	12/19/1991	22					I
1100	DARPA, IAB	IAB official protocol standards	04/01/1989	14			1130	1083	(STD 1)
1101	P. Mockapetris	DNS encoding of network names and other types	04/01/1989	14		1034, 1035			
1102	D. Clark	Policy routing in Internet protocols	05/01/1989	22					
1103	D. Katz	Proposed standard for the transmission of IP datagrams over FDDI Networks	06/01/1989	9			1188		
1104	H. Braun	Models of policy based routing	06/01/1989	10					
1105	K. Lougheed, Y. Rekhter	Border Gateway Protocol BGP	06/01/1989	17			1163		E
1106	R. Fox	TCP big window and NAK options	06/01/1989	13					
1107	K. Sollins	Plan for Internet directory services	07/01/1989	19					I
1108	S. Kent	U.S. Department of Defense Security Options for the Internet Protocol	11/27/1991	17				1038	H

RFC No.	Authors	Title	Date	Pgs	Updated By	Updates	Obsoleted By	Obsoletes	Notes / Status
1109	V. Cerf	Report of the second Ad Hoc Network Management Review Group	08/01/1989	8					
1110	A. McKenzie	Problem with the TCP big window option	08/01/1989	3					
1111	J. Postel	Request for comments on Request for Comments: Instructions to RFC authors	08/01/1989	6			1543, 2223	825	I
1112	S. Deering	Host extensions for IP multicasting	08/01/1989	17	2236			988, 1054	(STD 5)
1113	J. Linn	Privacy enhancement for Internet electronic mail: Part I - message encipherment and authentication procedures [Draft]	08/01/1989	34			1421	989, 1040	H
1114	S. Kent, J. Linn	Privacy enhancement for Internet electronic mail: Part II - certificate-based key management [Draft]	08/01/1989	25			1422		H
1115	J. Linn	Privacy enhancement for Internet electronic mail: Part III - algorithms, modes, and identifiers [Draft]	08/01/1989	8	1319		1423		H
1116	D. Borman	Telnet Linemode option	08/01/1989	21			1184		PS
1117	M. Stahl, S. Romano, M. Recker	Internet numbers	08/01/1989	109			1166	997, 1020, 1062	I
1118	E. Krol	Hitchhikers guide to the Internet	09/01/1989	24					I
1119	D. Mills	Network Time Protocol version 2 specification and implementation	09/01/1989	64			1305	958, 1059	(STD 12)
1120	V. Cerf	Internet Activities Board	09/01/1989	11			1160		I
1121	J. Postel, L. Kleimock, V. Cerf, B. Boehm	Act one - the poems	09/01/1989	6					I
1122	*R. Braden*	*Requirements for Internet hosts - communication layers*	*10/01/1989*	*116*					*STD 3*
1123	*R. Braden*	*Requirements for Internet hosts - application and support*	*10/01/1989*	*98*	*2181*	*822*			*STD 3*
1124	B. Leiner	Policy issues in interconnecting networks	09/01/1989	54					
1125	D. Estrin	Policy requirements for inter Administrative Domain routing	11/01/1989	18					
1126	M. Little	Goals and functional requirements for inter-autonomous system routing	10/01/1989	25					
1127	R. Braden	Perspective on the Host Requirements RFCs	10/01/1989	20					I
1128	D. Mills	Measured performance of the Network Time Protocol in the Internet system	10/01/1989	20					
1129	D. Mills	Internet time synchronization: The Network Time Protocol	10/01/1989	29					
1130	IAB, DARPA	IAB official protocol standards	10/01/1989	17			1140	1100	(STD 1)
1131	J. Moy	OSPF specification	10/01/1989	107			1247		PS
1132	*L. McLaughlin*	*Standard for the transmission of 802.2 packets over IPX networks*	*11/01/1989*	*4*					*STD 49*
1133	J. Yu, H. Braun	Routing between the NSFNET and the DDN	11/01/1989	10					I
1134	D. Perkins	Point-to-Point Protocol: A proposal for multi-protocol transmission of datagrams over Point-to-Point links	11/01/1989	38			1171		PS
1135	J. Reynolds	Helminthiasis of the Internet	12/01/1989	33					I
1136	S. Hares, D. Katz	Administrative Domains and Routing Domains: A model for routing in the Internet	12/01/1989	10					I
1137	S. Kille	Mapping between full RFC 822 and RFC 822 with restricted encoding	12/01/1989	3		976			H
1138	S. Kille	Mapping between X.400(1988) / ISO 10021 and RFC 822	12/01/1989	92	1148	822, 987, 1026	1327, 1495, 2156		E
1139	R. Hagens	Echo function for ISO 8473	01/30/1990	6			1575, 1574		PS
1140	J. Postel	IAB Official Protocol Standards	05/11/1990	27		1130	1200		(STD 1)
1141	T. Mallory, A. Kullberg	Incremental Updating of the Internet Checksum	01/01/1990	2	1624	1071			I
1142	D. Oran	OSI IS-IS Intra-domain Routing Protocol	12/30/1991	117					I
1143	D. Bernstein	The Q Method of Implementing TELNET Option Negotiation	02/01/1990	10					E
1144	V. Jacobson	Compressing TCP/IP headers for low-speed serial links	02/01/1990	43					PS
1145	J. Zweig, C. Partridge	TCP Alternate Checksum Options	02/01/1990	5			1146		E
1146	J. Zweig, C. Partridge	TCP Alternate Checksum Options	03/01/1991	5				1145	E
1147	R. Stine	FYI on a Network Management Tool Catalog: Tools for Monitoring and Debugging TCP/IP Internets and Interconnected Devices	04/04/1990	126			1470		FYI 2
1148	B. Kantor, S. Kille, P. Lapsley	Mapping between X.400 (1988) / ISO 10021 and RFC 822	03/01/1990	94		822, 987, 1026, 1138	1327, 1495, 2156		E
1149	D. Waitzman	A Standard for the Transmission of IP Datagrams on Avian Carriers	04/01/1990	2					E
1150	G. Malkin, J. Reynolds	F.Y.I. on F.Y.I.: Introduction to the F.Y.I. notes	03/01/1990	4					FYI 1
1151	R. Hinden, C. Partridge	Version 2 of the Reliable Data Protocol (RDP)	04/05/1990	4		908			E
1152	C. Partridge	Workshop Report: Internet Research Steering Group Workshop on Very-High-Speed Networks	04/06/1990	23					I
1153	F. Wancho	Digest Message Format	04/01/1990	4					E

RFC No.	Authors	Title	Date	Pgs	Updated By	Updates	Obsoleted By	Obsoletes	Notes / Status
1154	R. Ullmann, D. Robinson	Encoding Header Field for Internet Messages	04/16/1990	7		1049	1505		E
1155	K. McCloghrie, M. Rose	Structure and Identification of Management Information for TCP/IP-based Internets	05/10/1990	22				1065	STD 16
1156	K. McCloghrie, M. Rose	Management Information Base for Network Management of TCP/IP-based internets	05/10/1990	91		1066	1158	1066	H
1157	M. Schoffstall, M. Fedor, J. Davin, J. Case	A Simple Network Management Protocol (SNMP)	05/10/1990	36		1098		1098	STD 15
1158	M. Rose	Management Information Base for Network Management of TCP/IP-based internets: MIB-II	05/23/1990	133			1213	1156	PS
1159	R. Nelson	Message Send Protocol	06/25/1990	2			1312		E
1160	V. Cerf	The Internet Activities Board	05/25/1990	11		1120		1120	I
1161	M. Rose	SNMP over OSI	06/05/1990	8			1418		E
1162	G. Satz	Connectionless Network Protocol (ISO 8473) and End System to Intermediate System (ISO 9542) Management Information Base	06/05/1990	70			1238		E
1163	K. Lougheed, Y. Rekhter	A Border Gateway Protocol (BGP)	06/20/1990	29			1267	1105	H
1164	J. Honig, D. Katz, M. Mathis, Y. Rekhter, J. Yu	Application of the Border Gateway Protocol in the Internet	06/20/1990	23			1268		H
1165	J. Crowcroft, J. Onions	Network Time Protocol (NTP) over the OSI Remote Operations Service	06/25/1990	9					E
1166	S. Kirkpatrick, M. Recker	Internet Numbers	07/11/1990	182				1117, 1062, 1020	I
1167	V. Cerf	Thoughts on the National Research and Education Network	07/12/1990	8					I
1168	Ward, J. Postel, DeSchon, A. Westine	Intermail and Commercial Mail Relay Services	07/17/1990	23					I
1169	K. Mills, V. Cerf	Explaining the Role of GOSIP	08/09/1990	15					I
1170	R. Fougner	Public Key Standards and Licenses	01/11/1991	2					I
1171	D. Perkins	The Point-to-Point Protocol for the Transmission of Multi-Protocol Datagrams Over Point-to-Point Links	07/24/1990	48			1331	1134	DS
1172	R. Hobby, D. Perkins	The Point-to-Point Protocol (PPP) Initial Configuration Options	07/24/1990	40			1331		PS
1173	J. Van Bokkelen	Responsiblilities of Host and Network Managers. A Summary of the "Oral Tradition" of the Internet	08/07/1990	5					I
1174	V. Cerf	IAB Recommended Policy on Distributing Internet Identifier Assignment and IAB Recommended Policy Change to Internet "Connected" Status	08/09/1990	9					I
1175	M. A. Yuan, J. K. Roubicek, K. T. LaQuey	FYI on Where to Start - A Bibliography of Internetworking Information	08/16/1990	43					FYI 3
1176	M. Crispin	Interactive Mail Access Protocol - Version 2	08/20/1990	30				1064	E
1177	G. Malkin, A. Marine, J. Reynolds	FYI on Questions and Answers - Answers to Commonly Asked ``New Internet User'' Questions	09/04/1990	24			1206		FYI 4
1178	D. Libes	Choosing a Name for Your Computer	09/04/1990	8					FYI 5
1179	L. McLaughlin, III	Line Printer Daemon Protocol	09/04/1990	14					I
1180	T. Socolofsky, C. Kale	A TCP/IP Tutorial	01/15/1991	28					I
1181	R. Blokzijl	RIPE Terms of Reference	09/26/1990	2					I
1182		Not Issued							
1183	R. Ullman, P. Mockapetris, L. Mamakos, C. Everhart	New DNS RR Definitions	10/08/1990	11		1034, 1035			E
1184	D. Borman	Telnet Linemode Option	10/15/1990	23				1116	DS
1185	R. Braden, V. Jacobson, L. Zhang	TCP Extension for High-Speed Paths	10/15/1990	21			1323		E
1186	R. Rivest	The MD4 Message Digest Algorithm	10/18/1990	18			1320		I
1187	J. Davin, M. Rose, K. McCloghrie	Bulk Table Retrieval with the SNMP	10/18/1990	12					E
1188	D. Katz	A Proposed Standard for the Transmission of IP Datagrams over FDDI Networks	10/30/1990	10			1390	1103	DS
1189	L. Besaw, B. Handspicker, L. LaBarre, U. Warrier	The Common Management Information Services and Protocols for the Internet	10/26/1990	15				1095	H
1190	C. Topolcic	Experimental Internet Stream Protocol, Version 2 (ST-II)	10/30/1990	148			1819		E
1191	J. Mogul, S. Deering	Path MTU Discovery	11/16/1990	19				1063	DS
1192	B. Kahin	Commercialization of the Internet Summary Report	11/12/1990	13					I

RFC No.	Authors	Title	Date	Pgs	Updated By	Updates	Obsoleted By	Obsoletes	Notes / Status
1193	D. Ferrari	Client Requirements for Real-Time Communication Services	11/15/1990	24					I
1194	D. Zimmerman	The Finger User Information Protocol	11/21/1990	12			1196, 1288	742	DS
1195	R. Callon	Use of OSI IS-IS for Routing in TCP/IP and Dual Environments	12/19/1990	68					PS
1196	D. Zimmerman	The Finger User Information Protocol	12/26/1990	12			1288	742, 1194	DS
1197	M. Sherman	Using ODA for Translating Multimedia Information	12/31/1990	2					I
1198	B. Scheifler	FYI on the X Window System	01/01/1991	3					I
1199	J. Reynolds	Request for Comments Summary RFC Numbers 1100-1199	12/31/1991	22					I
1200	J. Postel	IAB Official Protocol Standards	04/01/1991	31			1250	1140	(STD 1)
1201	*D. Provan*	*Transmitting IP Traffic over ARCNET Networks*	*02/01/1991*	*7*				*1051*	*STD 46*
1202	M. Rose	Directory Assistance Service	02/07/1991	11					I
1203	J. Rice	Interactive Mail Access Protocol - Version 3	02/08/1991	49				1064	H
1204	D. Lee, S. Yeh	Message Posting Protocol (MPP)	02/15/1991	6					E
1205	P. Chmielewski	5250 Telnet Interface	02/21/1991	12					I
1206	G. Malkin, A. Marine	FYI on Questions and Answers - Answers to Commonly asked "New Internet User" Questions	02/26/1991	32			1325	1177	FYI 4
1207	G. Malkin, A. Marine, J. Reynolds	Answers to Commonly asked "Experienced Internet User" Questions	02/28/1991	15					FYI 7
1208	O. Jacobsen, D. Lynch	A Glossary of Networking Terms	03/04/1991	18					I
1209	*J. Lawrence, D. Piscitello*	*The Transmission of IP Datagrams over the SMDS Service*	*03/06/1991*	*11*					*STD 52*
1210	V. Cerf, P. Kirstein, B. Randell	Network and Infrastructure User Requirements for Transatlantic Research Collaboration - Brussels, July 16-18, and Washington July 24-25, 1990	03/21/1991	36					I
1211	A. Westine, J. Postel	Problems with the Maintenance of Large Mailing Lists	03/22/1991	54					I
1212	*K. McCloghrie, M. Rose*	*Concise MIB Definitions*	*03/26/1991*	*19*					*STD 16*
1213	*K. McCloghrie, M. Rose*	*Management Information Base for Network Management of TCP/IP-based internets: MIB-II*	*03/26/1991*	*70*	*2011, 2012, 2013*			*1158*	*STD 17*
1214	L. Labarre	OSI Internet Management: Management Information Base	04/05/1991	83					H
1215	M. Rose	A Convention for Defining Traps for use with the SNMP	03/27/1991	9					I
1216	P. Kunikos, P. Richard	Gigabit Network Economics and Paradigm Shifts	03/30/1991	4					I
1217	V. Cerf	Memo from the Consortium for Slow Commotion Research (CSCR)	04/01/1991	5					I
1218	N. Directory Forum	A Naming Scheme for c=US	04/03/1991	23			1255, 1417		I
1219	P. Tsuchiya	On the Assignment of Subnet Numbers	04/16/1991	13					I
1220	F. Baker	Point-to-Point Protocol Extensions for Bridging	04/17/1991	18			1638		PS
1221	*W. Edmond*	*Host Access Protocol (HAP) Specification - Version 2*	*04/16/1991*	*68*				*907*	*STD 40*
1222	H. Braun, Y. Rekhter	Advancing the NSFNET Routing Architecture	05/08/1991	6					I
1223	J. Halpern	OSI CLNS and LLC1 Protocols on Network Systems HYPERchannel	05/09/1991	12					I
1224	L. Steinberg	Techniques for Managing Asynchronously Generated Alerts	05/10/1991	22					E
1225	M. Rose	Post Office Protocol - Version 3	05/14/1991	16			1460	1081	DS
1226	B. Kantor	Internet Protocol Encapsulation of AX.25 Frames	05/13/1991	2					E
1227	M. Rose	SNMP MUX Protocol and MIB	05/23/1991	13					E
1228	G. Carpenter, B. Wijnen	SNMP-DPI - Simple Network Management Protocol Distributed Program Interface	05/23/1991	50			1592		E
1229	K. McCloghrie	Extensions to the Generic-Interface MIB	08/03/1992	16	1239		1573		PS
1230	R. Fox, K. McCloghrie	IEEE 802.4 Token Bus MIB	05/23/1991	23	1239				H
1231	E. Decker, R. Fox, K. McCloghrie	IEEE 802.5 Token Ring MIB	02/11/1993	23	1239		1743, 1748		DS
1232	F. Baker, C. Kolb	Definitions of Managed Objects for the DS1 Interface Type	05/23/1991	28	1239		1406		PS
1233	T. Cox, K. Tesink	Definitions of Managed Objects for the DS3 Interface Type	05/23/1991	23	1239		1407		PS
1234	D. Provan	Tunneling IPX Traffic through IP Networks	06/20/1991	6					PS
1235	J. Ioannidis, G. Maguire, Jr.	The Coherent File Distribution Protocol	06/20/1991	12					E
1236	L. Morales, P. Hasse	IP to X.121 Address Mapping for DDN	06/25/1991	7					I
1237	R. Colella, R. Callon, E. Gardner	Guidelines for OSI NSAP Allocation in the Internet	07/23/1991	49			1629		PS
1238	G. Satz	CLNS MIB - for use with Connectionless Network Protocol (ISO 8473) and End System to Intermediate System (ISO 9542)	06/25/1991	32				1162	E

RFC No.	Authors	Title	Date	Pgs	Updated By	Updates	Obsoleted By	Obsoletes	Notes / Status
1239	J. Reynolds	Reassignment of Experimental MIBs to Standard MIBs	06/25/1991	2		1229, 1230, 1231, 1232, 1233			PS
1240	K. Dobbins, W. Haggerty, C. Shue	OSI Connectionless Transport Services on top of UDP - Version: 1	06/26/1991	8					H
1241	D. Mills, R. Woodburn	A Scheme for an Internet Encapsulation Protocol: Version 1	07/02/1991	15					E
1242	S. Bradner	Benchmarking Terminology for Network Interconnection Devices	07/02/1991	12					
1243	S. Waldbusser	AppleTalk Management Information Base	07/08/1991	29			1742		PS
1244	P. Holbrook, J. Reynolds	Site Security Handbook	07/23/1991	101			2196		FYI 8
1245	J. Moy	OSPF Protocol Analysis	08/08/1991	12					1246, 1247, I
1246	J. Moy	Experience with the OSPF Protocol	08/08/1991	31					1245, 1247, I
1247	J. Moy	OSPF Version 2	08/08/1991	189			1583	1131	1245, 1246, DS
1248	F. Baker, R. Coltun	OSPF Version 2 Management Information Base	08/08/1991	42	1349		1252		PS
1249	T. Howes, M. Smith, B. Beecher	DIXIE Protocol Specification	08/09/1991	10					1202, I
1250	J. Postel	IAB Official Protocol Standards	08/26/1991	28			2200	1200	(STD 1)
1251	G. Malkin	Who's Who in the Internet: Biographies of IAB, IESG and IRSG Members	08/19/1991	26			1336		FYI 9
1252	F. Baker, R. Coltun	OSPF Version 2 Management Information Base	08/21/1991	42			1253	1248	PS
1253	F. Baker, R. Coltun	OSPF Version 2 Management Information Base	08/30/1991	42			1850	1252	PS
1254	A. Mankin, K. Ramakrishnan	Gateway Congestion Control Survey	08/30/1991	25					I
1255	T. Directory Forum	A Naming Scheme for c=US	09/05/1991	25			1417	1218	I
1256	S. Deering	ICMP Router Discovery Messages	09/05/1991	19					792, PS
1257	C. Partridge	Isochronous Applications Do Not Require Jitter-Controlled Networks	09/09/1991	4					I
1258	B. Kantor	BSD Rlogin	09/11/1991	5			1282		I
1259	M. Kapor	Building The Open Road: The NREN As Test-Bed For The National Public Network	09/17/1991	23					I
1260									
1261	S. Williamson, L. Nobile	Transition of NIC Services	09/19/1991	3					I
1262	IAB	Guidelines for Internet Measurement Activities	10/15/1991	3					
1263	L. Peterson, S. O'Malley	TCP Extensions Considered Harmful	10/22/1991	19					I
1264	B. Hinden	Internet Routing Protocol Standardization Criteria	10/25/1991	8					I
1265	Y. Rekhter	BGP Protocol Analysis	10/28/1991	8					I
1266	Y. Rekhter	Experience with the BGP Protocol	10/28/1991	9					I
1267	K. Lougheed, Y. Rekhter	A Border Gateway Protocol 3 (BGP-3)	10/25/1991	35				1163	H
1268	P. Gross, Y. Rekhter	Application of the Border Gateway Protocol in the Internet	10/25/1991	13			1655	1164	H
1269	J. Burruss, S. Willis	Definitions of Managed Objects for the Border Gateway Protocol (Version 3)	10/26/1991	13					PS
1270	F. Kastenholz	SNMP Communications Services	10/30/1991	11					I
1271	S. Waldbusser	Remote Network Monitoring Management Information Base	11/12/1991	81	1513		1757		PS
1272	D. Hirsh, C. Mills, G. Ruth	Internet Accounting: Background	11/11/1991	19					I
1273	M. Schwartz	A Measurement Study of Changes in Service-Level Reachability in the Global TCP/IP Internet: Goals, Experimental Design, Implementation, and Policy Considerations	11/14/1991	8					I
1274	P. Barker, S. Kille	The COSINE and Internet X.500 Schema	11/27/1991	60					PS
1275	S. Hardcastle-Kille	Replication Requirements to provide an Internet Directory using X.500	11/27/1991	17					I
1276	S. Hardcastle-Kille	Replication and Distributed Operations extensions to provide an Internet Directory using X.500	11/27/1991	17					PS
1277	S. Hardcastle-Kille	Encoding Network Addresses to Support Operation Over Non-OSI Lower Layers	11/27/1991	10					PS
1278	S. Hardcastle-Kille	A String Encoding of Presentation Address	11/27/1991	5					I
1279	S. Kille	X.500 and Domains	11/27/1991	13					E
1280	IAB, A. Chapin	IAB OFFICIAL PROTOCOL STANDARDS	03/14/1992	32			2200	1360	(STD 1)
1281	S. Crocker, B. Fraser, R. Pethia	Guidelines for the Secure Operation of the Internet	11/27/1991	10					I
1282	B. Kantor	BSD Rlogin	12/04/1991	5				1258	I
1283	M. Rose	SNMP over OSI	12/06/1991	8			1418	1161	E
1284	J. Cook	Definitions of Managed Objects for the Ethernet-like Interface Types	12/04/1991	21			1398		PS
1285	J. Case	FDDI Management Information Base	01/24/1992	46	1512				PS

RFC No.	Authors	Title	Date	Pgs	Updated By	Updates	Obsoleted By	Obsoletes	Notes / Status
1286	K. McCloghrie, E. Decker, P. Langille, A. Rijsinghani	Definitions of Managed Objects for Bridges	12/11/1991	40			1525, 1493		PS
1287	R. Braden, V. Cerf, L. Chapin, D. Clark, R. Hobby	Towards the Future Internet Architecture	12/12/1991	29					I
1288	D. Zimmerman	The Finger User Information Protocol	12/19/1991	12				742, 1194, 1196	DS
1289	J. Saperia	DECnet Phase IV MIB Extensions	12/20/1991	64			1559		PS
1290	J. Martin	There's Gold in them thar Networks! or Searching for Treasure in all the Wrong Places	12/31/1991	27			1402		FYI 10
1291	V. Aggarwal	Mid-Level Networks: Potential Technical Services	12/30/1991	10					I
1292	R. Lang, R. Wright	A Catalog of Available X.500 Implementations	01/03/1992	103			1632		FYI 11
1293	T. Bradley, C. Brown	Inverse Address Resolution Protocol	01/17/1992	6			2390		PS
1294	T. Bradley, C. Brown, A. Malis	Multiprotocol Interconnect over Frame Relay	01/17/1992	28			1490, 2427		STD55, H
1295	NADF	User Bill of Rights for entries and listings in the Public Directory	01/29/1992	2			1417		I
1296	M. Lottor	Internet Growth (1981-1991)	01/29/1992	9					I
1297	D. Johnson	NOC Internal Integrated Trouble Ticket System Functional Specification Wishlist ("NOC TT REQUIREMENTS")	01/31/1992	12					I
1298	R. Wormley, S. Bostock	SNMP over IPX	02/07/1992	5			1420		I
1299	M. Kennedy	Summary of 1200-1299	1/1997						I
1300	S. Greenfield	Remembrances of Things Past	02/07/1992	4					I
1301	S. Armstrong, A. Freier, K. Marzullo	Multicast Transport Protocol	02/19/1992	38					I
1302	D. Sitzler, P. Smith, A. Marine	Building a Network Information Services Infrastructure	02/25/1992	13					FYI 12
1303	K. McCloghrie, M. Rose	A Convention for Describing SNMP-based Agents	02/26/1992	12				1155, 1212, 1213, 1157, I	
1304	T. Cox, K. Tesink	Definitions of Managed Objects for the SIP Interface Type	02/28/1992	25			1694		PS
1305	*D. Mills*	*Network Time Protocol (v3)*	*04/09/1992*	*120*				*1119*	*STD 12*
1306	A. Nicholson, J. Young	Experiences Supporting By-Request Circuit-Switched T3 Networks	03/12/1992	10					I
1307	A. Nicholson, J. Young	Dynamically Switched Link Control Protocol	03/12/1992	13					E
1308	J. Reynolds, C. Weider	Executive Introduction to Directory Services Using the X.500 Protocol	03/12/1992	4					FYI 13
1309	S. Heker, J. Reynolds, C. Weider	Technical Overview of Directory Services Using the X.500 Protocol	03/12/1992	16					FYI 14
1310	IAB, A. Chapin	The Internet Standards Process	03/14/1992	23			1602		I
1311	IAB, A. Chapin	Introduction to the STD Notes	03/14/1992	5					I
1312	R. Nelson, G. Arnold	Message Send Protocol	04/01/1992	8				1159	E
1313	C. Partridge	Today's Programming for KRFC AM 1313 Internet Talk Radio	04/01/1992	3					I
1314	D. Cohen, A. Katz	A File Format for the Exchange of Images in the Internet	04/10/1992	23					PS
1315	C. Brown, F. Baker, C. Carvalho	Management Information Base for Frame Relay DTEs	04/09/1992	19			2115		PS
1316	B. Stewart	Definitions of Managed Objects for Character Stream Devices	04/16/1992	17			1658		PS
1317	B. Stewart	Definitions of Managed Objects for RS-232-like Hardware Devices	04/16/1992	17			1659		PS
1318	B. Stewart	Definitions of Managed Objects for Parallel-printer-like Hardware Devices	04/16/1992	11			1660		PS
1319	B. Kaliski	The MD2 Message-Digest Algorithm	04/16/1992	17		1115			I
1320	R. Rivest	The MD4 Message-Digest Algorithm	04/16/1992	20				1186	I
1321	R. Rivest	The MD5 Message-Digest Algorithm	04/16/1992	21					I
1322	D. Estrin, S. Hotz, Y. Rekhter	A Unified Approach to Inter-Domain Routing	05/11/1992	38					I
1323	D. Borman, R. Braden, V. Jacobson	TCP Extensions for High Performance	05/13/1992	37				1072, 1185	PS
1324	D. Reed	A Discussion on Computer Network Conferencing	05/13/1992	11					I
1325	G. Malkin, A. Marine	FYI on Questions and Answers to Commonly asked "New Internet User" Questions	05/15/1992	42			1594	1206	FYI 4
1326	P. Tsuchiya	Mutual Encapsulation Considered Dangerous	05/15/1992	5					I
1327	S. Hardcastle-Kille	Mapping between X.400(1988) / ISO 10021 and RFC 822	05/18/1992	113	1495	822	2156	1148	PS

RFC No.	Authors	Title	Date	Pgs	Updated By	Updates	Obsoleted By	Obsoletes	Notes / Status
1328	S. Hardcastle-Kille	X.400 1988 to 1984 downgrading	05/18/1992	5	1496				PS
1329	P. Kuehn	Thoughts on Address Resolution for Dual MAC FDDI Networks	05/19/1992	28					I
1330	ESCC X.500/X.400 Task Force	Recommendations for the Phase I Deployment of OSI Directory Services (X.500) and OSI Message Handling Services (X.400) within the ESnet Community	05/22/1992	87					I
1331	W. Simpson	The Point-to-Point Protocol (PPP) for the Transmission of Multi-protocol Datagrams over Point-to-Point Links	05/26/1992	69			1548	1171, 1172	PS
1332	G. McGregor	The PPP Internet Protocol Control Protocol (IPCP)	05/26/1992	14				1172	PS
1333	W. Simpson	PPP Link Quality Monitoring	05/26/1992	17			1989		PS
1334	B. Lloyd, W. Simpson	PPP Authentication Protocols	10/20/1992	16			1994		PS
1335	Z. Wang, J. Crowcroft	A Two-Tier Address Structure for the Internet: A Solution to the Problem of Address Space Exhaustion	05/26/1992	7					I
1336	G. Malkin	Who's Who in the Internet Biographies of IAB, IESG and IRSG Members	05/27/1992	33				1251	FYI 9
1337	R. Braden	TIME-WAIT Assassination Hazards in TCP	05/27/1992	11					I
1338	V. Fuller, T. Li, K. Varadhan, J. Yu	Supernetting: an Address Assignment and Aggregation Strategy	06/26/1992	20			1519		I
1339	S. Dorner, P. Resnick	Remote Mail Checking Protocol	06/29/1992	5					E
1340	J. Reynolds, J. Postel	ASSIGNED NUMBERS	07/10/1992	139			1700	1060	STD 2
1341	N. Borenstein, N. Freed	MIME (Multipurpose Internet Mail Extensions): Mechanisms for Specifying and Describing the Format of Internet Message Bodies	06/11/1992	80			1521		PS
1342	K. Moore	Representation of Non-ASCII Text in Internet Message Headers	06/11/1992	7			1522		PS
1343	N. Borenstein	A User Agent Configuration Mechanism For Multimedia Mail Format Information	06/11/1992	10					I
1344	N. Borenstein	Implications of MIME for Internet Mail Gateways	06/11/1992	9					I
1345	K. Simonsen	Character Mnemonics & Character Sets	06/11/1992	103					I
1346	P. Jones	Resource Allocation, Control, and Accounting for the Use of Network Resources	06/19/1992	6					I
1347	R. Callon	TCP and UDP with Bigger Addresses (TUBA), A Simple Proposal for Internet Addressing and Routing	06/19/1992	9					I
1348	B. Manning	DNS NSAP RRs	07/01/1992	4		1034, 1035	1637		E
1349	P. Almquist	Type of Service in the Internet Protocol Suite	07/06/1992	28		1248		2474	PS
1350	*K. Sollins*	*THE TFTP PROTOCOL (REVISION 2)*	*07/10/1992*	*11*	*1350, 1782, 1783, 1784, 1785, 2348, 2349, 2547*	*1350, 1783*		*784*	*STD 33*
1351	J. Davin, J. Galvin, K. McCloghrie	SNMP Administrative Model	07/06/1992	35					H
1352	J. Davin, J. Galvin, K. McCloghrie	SNMP Security Protocols	07/06/1992	41					H
1353	K. McCloghrie, J. Davin, J. Galvin	Definitions of Managed Objects for Administration of SNMP Parties	07/06/1992	26					H
1354	F. Baker	IP Forwarding Table MIB	07/06/1992	12			2096		PS
1355	J. Curran, A. Marine	Privacy and Accuracy Issues in Network Information Center Databases	08/04/1992	4					FYI 15
1356	A. Malis, D. Robinson, R. Ullmann	Multiprotocol Interconnect on X.25 and ISDN in the Packet Mode	08/06/1992	14				877	PS
1357	D. Cohen	A Format for E-mailing Bibliographic Records	07/10/1992	13			1807		I
1358	L. Chapin	Charter of the Internet Architecture Board (IAB)	08/07/1992	5			1601		I
1359	ACM SIGUCCS	Connecting to the Internet What Connecting Institutions Should Anticipate	08/14/1992	25					FYI 16
1360	IAB, A. Chapin	IAB Official Protocol Standards	09/09/1992	33			1280, 1410		(STD 1)
1361	D. Mills	Simple Network Time Protocol (SNTP)	08/10/1992	10			1769		I
1362	M. Allen	Novell IPX Over Various WAN Media (IPXWAN)	09/10/1992	13			1551, 1634		I
1363	C. Partridge	A Proposed Flow Specification	09/10/1992	20					E
1364	K. Varadhan	BGP OSPF Interaction	09/11/1992	14			1403		PS
1365	K. Siyan	An IP Address Extension Proposal	09/10/1992	6					I
1366	E. Gerich	Guidelines for Management of IP Address Space	10/22/1992	8			1466		I
1367	C. Topolcic	Schedule for IP Address Space Management Guidelines	10/22/1992	3			1467		I
1368	D. McMaster, K. McCloghrie	Definitions of Managed Objects for IEEE 802.3 Repeater Devices	10/26/1992	40			1516		PS
1369	F. Kastenholz	Implementation Notes and Experience for the Internet Ethernet MIB	10/23/1992	7					I
1370	IAB	Applicability Statement for OSPF	10/23/1992	2					PS
1371	P. Gross	Choosing a "Common IGP" for the IP Internet (The IESG's Recommendation to the IAB)	10/23/1992	9					I

RFC No.	Authors	Title	Date	Pgs	Updated By	Updates	Obsoleted By	Obsoletes	Notes / Status
1372	D. Borman, C. Hedrick	Telnet Remote Flow Control Option	10/23/1992	6				1080	PS
1373	T. Tignor	PORTABLE DUAs	10/27/1992	12					I
1374	J. Renwick, A. Nicholson	IP and ARP on HIPPI	11/02/1992	43			2067		PS
1375	P. Robinson	Suggestion for New Classes of IP Addresses	11/03/1992	7					I
1376	S. Senum	The PPP DECnet Phase IV Control Protocol (DNCP)	11/05/1992	6			1762		PS
1377	D. Katz	The PPP OSI Network Layer Control Protocol (OSINLCP)	11/05/1992	10					PS
1378	B. Parker	The PPP AppleTalk Control Protocol (ATCP)	11/05/1992	16					PS
1379	R. Braden	Extending TCP for Transactions -- Concepts	11/05/1992	38					I
1380	P. Gross, P. Almquist	IESG Deliberations on Routing and Addressing	11/09/1992	22					I
1381	D. Throop, F. Baker	SNMP MIB Extension for X.25 LAPB	11/10/1992	33					PS
1382	D. Throop	SNMP MIB Extension for the X.25 Packet Layer	11/10/1992	69					PS
1383	C. Huitema	An Experiment in DNS Based IP Routing	12/28/1992	14					I
1384	P. Barker, S. Hardcastle-Kille	Naming Guidelines for Directory Pilots	02/11/1993	12			1617		I
1385	Z. Wang	EIP: The Extended Internet Protocol A Framework for Maintaining Backward Compatibility	11/13/1992	17					I
1386	A. Cooper, J. Postel	The US Domain	12/28/1992	31			1480		I
1387	G. Malkin,	RIP Version 2 Protocol Analysis	01/06/1993	3			1721		I
1388	*G. Malkin*	*RIP Version 2 Carrying Additional Information*	*01/06/1993*	*7*	*2453*	*1058*	*1723*		*STD56, H*
1389	G. Malkin, F. Baker	RIP Version 2 MIB Extension	01/06/1993	13			1724		PS
1390	*D. Katz*	*Transmission of IP & ARP over FDDI Networks*	*01/05/1993*	*12*				*1188*	*STD 36*
1391	G. Malkin	The Tao of IETF: A Guide for New Attendees of the Internet Engineering Task Force	01/06/1993	19			1539		FYI 17
1392	G. Malkin, T. Parker	Internet Users' Glossary	01/12/1993	53			1983		FYI 18
1393	G. Malkin	Traceroute Using an IP Option	01/11/1993	7					E
1394	P. Robinson	Relationship of Telex Answerback Codes to Internet Domains	01/08/1993	15					I
1395	J. Reynolds	BOOTP Vendor Information Extensions	01/11/1993	8		951	1497, 1533	1048, 1084	DS
1396	S. Crocker	The Process for Organization of Internet Standards Working Group (POISED)	01/11/1993	10					I
1397	D. Haskin	Default Route Advertisement In BGP2 And BGP3 Versions Of The Border Gateway Protocol	01/13/1993	2					PS
1398	*F. Kastenholz*	*Definitions of Managed Objects for the Ethernet-like Interface Types*	*01/14/1993*	*24*			*1623, 1643*	*1284*	*STD 50, H*
1399	J. Elliott	Summary of 1300–1399	01/1997	43662 b					I
1400	A. Williamson	Transition and Modernization of the Internet Registration Service	03/25/1993	7					DS
1401	IAB, L. Chapin	Correspondence between the IAB and DISA on the use of DNS throughout the Internet	01/13/1993	8					I
1402	J. Martin	There's Gold in them thar Networks! Searching for Treasure in all the Wrong Places	01/14/1993	39			1290		FYI 10
1403	K. Varadhan	BGP OSPF Interaction	01/14/1993	17				1364	PS
1404	B. Stockman	A Model for Common Operational Statistics	01/20/1993	27			1857		I
1405	C. Allocchio	Mapping between X.400(1984/1988) and Mail-11 (DECnet mail)	01/20/1993	19			2162		E
1406	F. Baker, J. Watt	Definitions of Managed Objects for the DS1 and E1 Interface Types	01/26/1993	50			2495	1232	PS
1407	T. Cox, K. Tesink	Definitions of Managed Objects for the DS3/E3 Interface Type	01/26/1993	55			2496	1233	PS
1408	D. Borman	Telnet Environment Option	01/26/1993	7	1571				H
1409	D. Borman	Telnet Authentication Option	01/26/1993	7			1416		E
1410	IAB, L. Chapin	IAB OFFICIAL PROTOCOL STANDARDS	03/24/1993	35			1500	1360	(STD 1)
1411	D. Borman	Telnet Authentication: Kerberos Version 4	01/26/1993	4					E
1412	K. Alagappan	Telnet Authentication : SPX	01/27/1993	4					E
1413	M. St. Johns	Identification Protocol	02/04/1993	10				931	PS
1414	M. St. Johns, M. Rose	Ident MIB	02/04/1993	13					PS
1415	J. Mindel, R. Slaski	FTP-FTAM Gateway Specification	01/27/1993	58					PS
1416	D. Borman	Telnet Authentication Option	02/01/1993	7				1409	E
1417	T. Myer, NADF	NADF Standing Documents: A Brief Overview	02/04/1993	4			1758	1295, 1255, 1218	I
1418	M. Rose	SNMP over OSI	03/03/1993	4				1161, 1283	PS
1419	G. Minshall, M. Ritter	SNMP over AppleTalk	03/03/1993	7					PS
1420	S. Bostock	SNMP over IPX	03/03/1993	4				1298	PS
1421	J. Linn	Privacy Enhancement for Internet Electronic Mail: Part I: Message Encryption and Authentication Procedures	02/10/1993	42				1113	PS
1422	S. Kent	Privacy Enhancement for Internet Electronic Mail: Part II: Certificate-Based Key Management	02/10/1993	32				1114	PS

RFC No.	Authors	Title	Date	Pgs	Updated By	Updates	Obsoleted By	Obsoletes	Notes / Status
1423	D. Balenson	Privacy Enhancement for Internet Electronic Mail: Part III: Algorithms, Modes, and Identifiers	02/10/1993	14				1115	PS
1424	B. Kaliski	Privacy Enhancement for Internet Electronic Mail: Part IV: Key Certification and Related Services	02/10/1993	9					PS
1425	J. Klensin, D. Crocker, N. Freed, M. Rose, E. Stefferud	SMTP Service Extensions	02/10/1993	10			1651		PS
1426	J. Klensin, D. Crocker, N. Freed, M. Rose, E. Stefferud	SMTP Service Extension for 8bit-MIME transport	02/10/1993	6			1652		PS
1427	K. Moore, N. Freed, J. Klensin	SMTP Service Extension for Message Size Declaration	02/10/1993	8			1653		PS
1428	G. Vaudreuil	Transition of Internet Mail from Just-Send-8 to 8Bit-SMTP/MIME	02/10/1993	6					I
1429	E. Thomas	Listserv Distribute Protocol	02/24/1993	8					I
1430	S. Kille, E. Huizer, V. Cerf, R. Hobby, S. Kent	A Strategic Plan for Deploying an Internet X.500 Directory Service	02/26/1993	20					I
1431	P. Barker	DUA Metrics	02/26/1993	19					I
1432	J. Quarterman	Recent Internet Books	03/03/1993	15					I
1433	J. Garrett, J. Hagan, J. Wong	Directed ARP	03/05/1993	17					E
1434	R. Dixon, D. Kushi	Data Link Switching: Switch-to-Switch Protocol	03/17/1993	33			1795		I
1435	S. Knowles	IESG Advice from Experience with Path MTU Discovery	03/17/1993	2					I
1436	F. Anklesaria, M. McCahill, P. Lindner, D. Torrey, D. Johnson, D. John, B. Alberti	The Internet Gopher Protocol (a distributed document search and retrieval protocol)	03/18/1993	16					I
1437	N. Borenstein, M. Linimon	The Extension of MIME Content-Types to a New Medium	04/01/1993	6					I
1438	A. Chapin, C. Huitema	Internet Engineering Task Force Statements Of Boredom (SOBs)	03/31/1993	2					I
1439	C. Finseth	The Uniqueness of Unique Identifiers	03/25/1993	11					I
1440	R. Troth	SIFT/UFT: Sender-Initiated/Unsolicited File Transfer	07/23/1993	9					E
1441	J. Case, M. Rose, K. McCloghrie, S. Waldbusser	Introduction to version 2 of the Internet-standard Network Management Framework	05/03/1993	13					PS
1442	J. Case, M. Rose, K. McCloghrie, S. Waldbusser	Structure of Management Information for version 2 of the Simple Network Management Protocol (SNMPv2)	05/03/1993	55			1902		PS
1443	J. Case, M. Rose, K. McCloghrie, S. Waldbusser	Textual Conventions for version 2 of the Simple Network Management Protocol (SNMPv2)	05/03/1993	31			1903		PS
1444	J. Case, M. Rose, K. McCloghrie, S. Waldbusser	Conformance Statements for version 2 of the Simple Network Management Protocol (SNMPv2)	05/03/1993	33			1904		PS
1445	J. Davin, K. McCloghie	Administrative Model for version 2 of the Simple Network Management Protocol (SNMPv2)	05/03/1993	47					PS
1446	J. Galvin, K. McCloghrie	Security Protocols for version 2 of the Simple Network Management Protocol (SNMPv2)	05/03/1993	51					PS
1447	K. McCloghrie, J. Galvin	Party MIB for version 2 of the Simple Network Management Protocol (SNMPv2)	05/03/1993	50					PS
1448	J. Case, M. Rose, K. McCloghrie, S. Waldbusser	Protocol Operations for version 2 of the Simple Network Management Protocol (SNMPv2)	05/03/1993	36			1905		PS
1449	J. Case, M. Rose, K. McCloghrie, S. Waldbusser	Transport Mappings for version 2 of the Simple Network Management Protocol (SNMPv2)	05/03/1993	24			1906		PS
1450	J. Case, M. Rose, K. McCloghrie, S. Waldbusser	Management Information Base for version 2 of the Simple Network Management Protocol (SNMPv2)	05/03/1993	27			1907		PS
1451	J. Case, M. Rose, K. McCloghrie, S. Waldbusser	Manager to Manager Management Information Base	05/03/1993	36					PS
1452	J. Case, M. Rose, K. McCloghrie, S. Waldbusser	Coexistence between version 1 and version 2 of the Internet-standard Network Management Framework	05/03/1993	17			1908		PS
1453	W. Chimiak	A Comment on Packet Video Remote Conferencing and the Transport/Network Layers	04/15/1993	10					I
1454	T. Dixon	Comparison of Proposals for Next Version of IP	05/08/1993	15					I
1455	D. Eastlake, III	Physical Link Security Type of Service	05/26/1993	6			2474		E

RFC No.	Authors	Title	Date	Pgs	Updated By	Updates	Obsoleted By	Obsoletes	Notes / Status
1456	C. Nguyen, H. Ngo, C. Bui, T. van Nguyen	Conventions for Encoding the Vietnamese Language VISCII: VIetnamese Standard Code for Information Interchange VIQR: Vietnamese Quoted-Readable Specification	05/08/1993	7					I
1457	R. Housley	Security Label Framework for the Internet	05/26/1993	14					I
1458	R. Braudes, S. Zabele	Requirements for Multicast Protocols	05/26/1993	19					I
1459	J. Oikarinen, D. Reed	Internet Relay Chat Protocol	05/26/1993	65					E
1460	M. Rose	Post Office Protocol - Version 3	06/16/1993	19			1725	1225	DS
1461	D. Throop	SNMP MIB extension for MultiProtocol Interconnect over X.25	05/27/1993	30					PS
1462	E. Krol, E. Hoffman	FYI on "What is the Internet?"	05/27/1993	11					FYI 20
1463	E. Hoffman, L. Jackson	FYI on Introducing the Internet--A Short Bibliography of Introductory Internetworking Readings for the Network Novice	05/27/1993	4					FYI 19
1464	R. Rosenbaum	Using the Domain Name System To Store Arbitrary String Attributes	05/27/1993	4					E
1465	D. Eppenberger	Routing coordination for X.400 MHS services within a multi protocol / multi network environment Table Format V3 for static routing	05/26/1993	31					E
1466	E. Gerich	Guidelines for Management of IP Address Space	05/26/1993	10			2050	1366	I
1467	C. Topolcic	Status of CIDR Deployment in the Internet	08/06/1993	9				1367	I
1468	J. Murai, M. Crispin, E. van der Poel	Japanese Character Encoding for Internet Messages	06/04/1993	6					I
1469	T. Pusateri	IP Multicast over Token-Ring Local Area Networks	06/17/1993	4					PS
1470	R. Enger, J. Reynolds	FYI on a Network Management Tool Catalog: Tools for Monitoring and Debugging TCP/IP Internets and Interconnected Devices	06/25/1993	216				1147	FYI 2
1471	F. Kastenholz	The Definitions of Managed Objects for the Link Control Protocol of the Point-to-Point Protocol	06/08/1993	25					PS
1472	F. Kastenholz	The Definitions of Managed Objects for the Security Protocols of the Point-to-Point Protocol	06/08/1993	11					PS
1473	F. Kastenholz	The Definitions of Managed Objects for the IP Network Control Protocol of the Point-to-Point Protocol	06/08/1993	9					PS
1474	F. Kastenholz	The Definitions of Managed Objects for the Bridge Network Control Protocol of the Point-to-Point Protocol	06/08/1993	15					PS
1475	R. Ullmann	TP/IX: The Next Internet	06/17/1993	35					E
1476	R. Ullmann	RAP: Internet Route Access Protocol	06/17/1993	20					E
1477	M. Steenstrup	IDPR as a Proposed Standard	07/26/1993	13					I
1478	M. Lepp, M. Steenstrup	An Architecture for Inter-Domain Policy Routing	07/26/1993	35					PS
1479	M. Steenstrup	Inter-Domain Policy Routing Protocol Specification: Version 1	07/26/1993	108					PS
1480	A. Cooper, J. Postel	The US Domain	06/28/1993	47				1386	I
1481	C. Huitema, IAB	IAB Recommendation for an Intermediate Strategy to Address the Issue of Scaling	07/02/1993	2					I
1482	M. Knopper, S. Richardson	Aggregation Support in the NSFNET Policy Routing Database	07/20/1993	7					I
1483	J. Heinanen	Multiprotocol Encapsulation over ATM Adaptation Layer 5	07/20/1993	16			2684		PS
1484	S. Hardcastle-Kille	Using the OSI Directory to achieve User Friendly Naming (OSI-DS 24 (v1.2))	07/28/1993	25			1781		E
1485	S. Hardcastle-Kille	A String Representation of Distinguished Names (OSI-DS 23 (v5))	07/28/1993	7			1779		PS
1486	M. Rose, C. Malamud	An Experiment in Remote Printing	07/30/1993	14			1528, 1529		E
1487	W. Yeong, T. Howes, S. Hardcastle-Kille	X.500 Lightweight Directory Access Protocol	07/29/1993	21			1777		PS
1488	T. Howes, S. Hardcastle-Kille, W. Yeong, C. Robbins	The X.500 String Representation of Standard Attribute Syntaxes	07/29/1993	11			1778		PS
1489	A. Chernov	Registration of a Cyrillic Character Set	07/23/1993	5					I
1490	T. Bradley, C. Brown, A. Malis	Multiprotocol Interconnect over Frame Relay	07/26/1993	35			2427	1294	DS
1491	C. Weider, R. Wright	A Survey of Advanced Usages of X.500	07/26/1993	18					FYI 21
1492	C. Finseth	An Access Control Protocol, Sometimes Called TACACS	07/23/1993	21					I
1493	E. Decker, P. Langille, A. Rijsinghani, K. McCloghrie	Definitions of Managed Objects for Bridges	07/28/1993	34				1286	DS

RFC No.	Authors	Title	Date	Pgs	Updated By	Updates	Obsoleted By	Obsoletes	Notes / Status
1494	H. Alvestrand, S. Thompson	Equivalences between 1988 X.400 and RFC-822 Message Bodies	08/26/1993	26					PS
1495	H. Alvestrand, S. Kille, R. Miles, M. Rose, S. Thompson	Mapping between X.400 and RFC-822 Message Bodies	08/26/1993	15			2156	987, 1026, 1138, 1148, 1327	PS
1496	H. Alvestrand, J. Romaguera, K. Jordan	Rules for downgrading messages from X.400/88 to X.400/84 when MIME content-types are present in the messages	08/26/1993	7		1328			PS
1497	J. Reynolds	BOOTP Vendor Information Extensions	08/04/1993	8		951	1533	1395, 1084, 1048	DS
1498	J. Saltzer	On the Naming and Binding of Network Destinations	08/04/1993	10					I
1499	J. Elliott	Summary of 1400–1499	01/1997	40923 b					I
1500	J. Postel	Internet official protocol standards	08/30/1993	36			1540	1410	(STD 1)
1501	E. Brunsen	OS/2 User Group	08/06/1993	2					I
1502	H. Alvestrand	X.400 Use of Extended Character Sets	08/26/1993	16					PS
1503	K. McCloghrie, M. Rose	Algorithms for Automating Administration in SNMPv2 Managers	08/26/1993	19					I
1504	A. Oppenheimer	Appletalk Update-Based Routing Protocol: Enhanced Appletalk Routing	08/27/1993	82					I
1505	A. Costanzo, D. Robinson, R. Ullmann	Encoding Header Field for Internet Messages	08/27/1993	36				1154	E
1506	J. Houttuin	A tutorial on gatewaying between X.400 and Internet mail	09/23/1993	39					RTR 6
1507	C. Kaufman	DASS - Distributed Authentication Security Service	09/10/1993	119					E
1508	J. Linn	Generic Security Service Application Program Interface	09/10/1993	49			2078		PS
1509	J. Wray	Generic Security Service API: C-bindings	09/10/1993	48			2744		PS
1510	J. Kohl, B. Neuman	The Kerberos Network Authentication Service (V5)	09/10/1993	112					PS
1511	J. Linn	Common Authentication Technology Overview	09/10/1993	2					I
1512	J. Case, A. Rijsinghani	FDDI Management Information Base	09/10/1993	51		1285			PS
1513	S. Waldbusser	Token Ring Extensions to the Remote Network Monitoring MIB	09/23/1993	55		1271			PS
1514	P. Grillo, S. Waldbusser	Host Resources MIB	09/23/1993	33					PS
1515	D. McMaster, K. McCloghrie, S. Roberts	Definitions of Managed Objects for IEEE 802.3 Medium Attachment Units (MAUs)	09/10/1993	25					PS
1516	D. McMaster, K. McCloghrie	Definitions of Managed Objects for IEEE 802.3 Repeater Devices	09/10/1993	40			2108	1368	DS
1517	R. Hinden	Applicability Statement for the Implementation of Classless Inter-Domain Routing (CIDR)	09/24/1993	4					PS
1518	Y. Rekhter, T. Li	An Architecture for IP Address Allocation with CIDR	09/24/1993	27					PS
1519	V. Fuller, T. Li, J. Yu, K. Varadhan	Classless Inter-Domain Routing (CIDR): an Address Assignment and Aggregation Strategy	09/24/1993	24				1338	PS
1520	Y. Rekhter, C. Topolcic	Exchanging Routing Information Across Provider Boundaries in the CIDR Environment	09/24/1993	9					I
1521	N. Borenstein, N. Freed	MIME (Multipurpose Internet Mail Extensions) Part One: Mechanisms for Specifying and Describing the Format of Internet Message Bodies	09/23/1993	81	1590		2045, 1046, 2047, 2048, 2049	1341	DS
1522	K. Moore	MIME (Multipurpose Internet Mail Extensions) Part Two: Message Header Extensions for Non-ASCII Text	09/23/1993	10			2045, 2046, 2047, 2048, 2049	1342	DS
1523	N. Borenstein	The text/enriched MIME Content-type	09/23/1993	15			1563, 1896		I
1524	N. Borenstein	A User Agent Configuration Mechanism For Multimedia Mail Format Information	09/23/1993	12					I
1525	E. Decker, P. Langille, K. McCloghrie, A. Rijsinghani	Definitions of Managed Objects for Source Routing Bridges	09/30/1993	18				1286	PS
1526	D. Piscitello	Assignment of System Identifiers for TUBA/CLNP Hosts	09/30/1993	8					I
1527	G. Cook	What Should We Plan Given the Dilemma of the Network?	09/30/1993	17					I
1528	C. Malamud, M. Rose	Principles of Operation for the TPC.INT Subdomain: Remote Printing -- Technical Procedures	10/06/1993	12				1486	E
1529	C. Malamud, M. Rose	Principles of Operation for the TPC.INT Subdomain: Remote Printing -- Administrative Policies	10/06/1993	5				1486	I
1530	C. Malamud, M. Rose	Principles of Operation for the TPC.INT Subdomain: General Principles and Policy	10/06/1993	7					I
1531	R. Droms	Dynamic Host Configuration Protocol	10/07/1993	39			1541		PS

RFC No.	Authors	Title	Date	Pgs	Updated By	Updates	Obsoleted By	Obsoletes	Notes / Status
1532	W. Wimer	Clarifications and Extensions for the Bootstrap Protocol	10/08/1993	22		951	1542		PS
1533	S. Alexander, R. Droms	DHCP Options and BOOTP Vendor Extensions	10/08/1993	30			2132	1497, 1395, 1084, 1048	PS
1534	R. Droms	Interoperation Between DHCP and BOOTP	10/08/1993	4					PS
1535	E. Gavron	A Security Problem and Proposed Correction With Widely Deployed DNS Software	10/06/1993	5					I
1536	A. Kumar, J. Postel, C. Neuman, P. Danzig, S. Miller	Common DNS Implementation Errors and Suggested Fixes.	10/06/1993	12					I
1537	P. Beertema	Common DNS Data File Configuration Error	10/06/1993	9			1912		I
1538	W. Behl, B. Sterling, W. Teskey	Advanced SNA/IP : A Simple SNA Transport Protocol	10/06/1993	10					I
1539	G. Malkin	The Tao of IETF - A Guide for New Attendees of the Internet Engineering Task Force	10/07/1993	22			1718	1391	FYI 17
1540	J. Postel	Internet official protocol standards	10/22/1993	34			1600, 2790	1500	(STD 1)
1541	R. Droms	Dynamic Host Configuration Protocol	10/27/1993	39			2131	1531	PS
1542	W. Wimer	Clarifications and Extensions for the Bootstrap Protocol	10/27/1993	23		951		1532	PS
1543	J. Postel	Instructions to RFC Authors	10/28/1993	16			2223	825, 1111	I
1544	M. Rose	The Content-MD5 Header Field	11/16/1993	3			1864		PS
1545	D. Piscitello	FTP Operation Over Big Address Records (FOOBAR)	11/16/1993	5			1639		E
1546	C. Partridge, T. Mendez, W. Milliken	Host Anycasting Service	11/16/1993	9					I
1547	D. Perkins	Requirements for an Internet Standard Point-to-Point Protocol	12/09/1993	21					I
1548	W. Simpson	The Point-to-Point Protocol (PPP)	12/09/1993	62	1570		1661	1331	DS
1549	*W. Simpson*	*PPP in HDLC Framing*	*12/09/1993*	*20*			*1662*		*STD 51, H*
1550	S. Bradner, A. Mankin	IP: Next Generation (IPng) White Paper Solicitation	12/16/1993	6					I
1551	M. Allen	Novell IPX Over Various WAN Media (IPXWAN)	12/09/1993	22			1634	1362	I
1552	W. Simpson	The PPP Internetwork Packet Exchange Control Protocol (IPXCP)	12/09/1993	19					PS
1553	S. Mathur, M. Lewis	Compressing IPX Headers Over WAN Media (CIPX)	12/09/1993	27					PS
1554	M. Ohta, K. Handa	ISO-2022-JP-2: Multilingual Extension of ISO-2022-JP	12/23/1993	6					I
1555	H. Nussbacher, Y. Bourvine	Hebrew Character Encoding for Internet Messages	12/23/1993	5					I
1556	H. Nussbacher	Handling of Bi-directional Texts in MIME	12/23/1993	3					I
1557	K. Chon, H. Je Park, U. Choi	Korean Character Encoding for Internet Messages	12/27/1993	5					I
1558	T. Howes	A String Representation of LDAP Search Filters	12/23/1993	3			1960		I
1559	J. Saperia	DECnet Phase IV MIB Extensions	12/27/1993	69				1289	DS
1560	B. Leiner, Y. Rekhter	The MultiProtocol Internet	12/23/1993	5					I
1561	D. Piscitello	Use of ISO CLNP in TUBA Environments	12/23/1993	25					E
1562	G. Michaelson, M. Prior	Naming Guidelines for the AARNet X.500 Directory Service	12/29/1993	4					I
1563	N. Borenstein	The text/enriched MIME Content-type	01/10/1994	16			1896	1523	I
1564	P. Barker, R. Hedberg	DSA Metrics (OSI-DS 34 (v3))	01/14/1994	20					I
1565	N. Freed, S. Kille	Network Services Monitoring MIB	01/11/1994	18			2248		PS
1566	N. Freed, S. Kille	Mail Monitoring MIB	01/11/1994	21			2249		PS
1567	G. Mansfield, S. Kille	X.500 Directory Monitoring MIB	01/11/1994	19			2605		PS
1568	A. Gwinn	Simple Network Paging Protocol - Version 1(b)	01/07/1994	4			1645		I
1569	M. Rose	Principles of Operation for the TPC.INT Subdomain: Radio Paging -- Technical Procedures	01/07/1994	6			1703		I
1570	W. Simpson	PPP LCP Extensions	01/11/1994	22	2484	1548			PS
1571	D. Borman	Telnet Environment Option Interoperability Issues	01/14/1994	4		1408			I
1572	S. Alexander	Telnet Environment Option	01/14/1994	7					PS
1573	K. McCloghrie, F. Kastenholz	Evolution of the Interfaces Group of MIB-II	01/20/1994	55			2233	1229	PS
1574	S. Hares, C. Wittbrodt	Essential Tools for the OSI Internet	02/18/1994	14				1139	I
1575	S. Hares, C. Wittbrodt	An Echo Function for CLNP (ISO 8473)	02/18/1994	10				1139	DS
1576	J. Penner	TN3270 Current Practices	01/20/1994	10					I
1577	M. Laubach	Classical IP and ARP over ATM	01/20/1994	17			2225		PS
1578	J. Sellers	FYI on Questions and Answers: Answers to Commonly Asked "Primary and Secondary School Internet User" Questions	02/18/1994	53			1941		FYI 22

RFC No.	Authors	Title	Date	Pgs	Updated By	Updates	Obsoleted By	Obsoletes	Notes / Status
1579	S. Bellovin	Firewall-Friendly FTP	02/18/1994	4					I
1580	E. EARN Staff	Guide to Network Resource Tools	03/22/1994	107					FYI 23
1581	G. Meyer	Protocol Analysis for Extensions to RIP to Support Demand Circuits	02/18/1994	5					I
1582	G. Meyer	Extensions to RIP to Support Demand Circuits	02/18/1994	32					PS
1583	J. Moy	OSPF Version 2	03/23/1994	212			2178	1247	STD 1
1584	J. Moy	Multicast Extensions to OSPF	03/24/1994	102					PS
1585	J. Moy	MOSPF: Analysis and Experience	03/24/1994	13					I
1586	O. deSouza, M. Rodrigues	Guidelines for Running OSPF Over Frame Relay Networks	03/24/1994	6					I
1587	R. Coltun, V. Fuller	The OSPF NSSA Option	03/24/1994	17					PS
1588	J. Postel, C. Anderson	WHITE PAGES MEETING REPORT	02/25/1994	35					I
1589	D. Mills	A Kernel Model for Precision Timekeeping	03/03/1994	37					I
1590	J. Postel	Media Type Registration Procedure	03/02/1994	7		1521	2045, 2046, 2047, 2048, 2049		I
1591	J. Postel	Domain Name System Structure and Delegation	03/03/1994	7					I
1592	B. Wijnen, G. Carpenter, K. Curran, A. Sehgal, G. Waters	Simple Network Management Protocol Distributed Protocol Interface Version 2.0	03/03/1994	54				1228	E
1593	W. McKenzie, J. Cheng	SNA APPN Node MIB	03/10/1994	120					I
1594	A. Marine, J. Reynolds, G. Malkin	FYI on Questions and Answers to Commonly asked "New Internet User" Questions	03/11/1994	44			2664	1325	FYI 4
1595	T. Brown, K. Tesink	Definitions of Managed Objects for the SONET/SDH Interface Type	03/11/1994	59			2558		PS
1596	T. Brown	Definitions of Managed Objects for Frame Relay Service	03/17/1994	46			1604		PS
1597	Y. Rekhter, R. Moskowitz, D. Karrenberg, G. de Groot	Address Allocation for Private Internets	03/17/1994	8			1918		I
1598	W. Simpson	PPP in X.25	03/17/1994	8					PS
1599	M. Kennedy	Summary of 1500–1599	01/1997	43761 b					I
1600	J. Postel	Internet official protocol standards	03/14/1994	36			1610	1540	(STD 1)
1601	C. Huitema, IAB	Charter of the Internet Architecture Board (IAB)	03/22/1994	6				1358	I
1602	IAB, IESC, C. Huitema, P. Gross	The Internet Standards Process -- Revision 2	03/24/1994	37	1871		2026	1310	I
1603	E. Huizer, D. Crocker	IETF Working Group Guidelines and Procedures	03/24/1994	29	1871		2418		I
1604	T. Brown	Definitions of Managed Objects for Frame Relay Service	03/25/1994	46				1596	PS
1605	W. Shakespeare	SONET to Sonnet Translation	04/01/1994	3					I
1606	J. Onions	A Historical Perspective On The Usage Of IP Version 9	04/01/1994	4					I
1607	V. Cerf	A VIEW FROM THE 21ST CENTURY	04/01/1994	13					I
1608	T. Johannsen, G. Mansfield, M. Kosters, S. Sataluri	Representing IP Information in the X.500 Directory	03/25/1994	20					E
1609	G. Mansfield, T. Johannsen, M. Knopper	Charting Networks in the X.500 Directory	03/25/1994	15					E
1610	J. Postel, IAB	Internet official protocol standards	07/08/1994	37			1720	1600	(STD 1)
1611	R. Austein, J. Saperia	DNS Server MIB Extensions	05/17/1994	32					PS
1612	R. Austein, J. Saperia	DNS Resolver MIB Extensions	05/17/1994	36					PS
1613	J. Forster, G. Satz, G. Glick, R. Day	Cisco Systems X.25 over TCP (XOT)	05/13/1994	13					I
1614	C. Adie	Network Access to Multimedia Information	05/20/1994	79					RTR 8
1615	J. Houttuin, J. Craigie	Migrating from X.400(84) to X.400(88)	05/19/1994	17					RTR 9
1616	E. Huizer, J. Romaguera, RARE WG-MSG Task Force 88	X.400(1988) for the Academic and Research Community in Europe	05/19/1994	44					RTR 10
1617	P. Barker, S. Kille, T. Lenggenhager	Naming and Structuring Guidelines for X.500 Directory Pilots	05/20/1994	28				1384	RTR 11
1618	W. Simpson	PPP over ISDN	05/13/1994	7					PS
1619	W. Simpson	PPP over SONET/SDH	05/13/1994	5			2615		PS
1620	R. Braden, J. Postel, Y. Rekhter	Internet Architecture Extensions for Shared Media	05/20/1994	19					I
1621	P. Francis	Pip Near-term Architecture	05/20/1994	51					I
1622	P. Francis	Pip Header Processing	05/20/1994	16					I
1623	F. Kastenholz	Definitions of Managed Objects for the Ethernet-like Interface Types	05/24/1994	23			1643	1398	STD 50, H

RFC No.	Authors	Title	Date	Pgs	Updated By	Updates	Obsoleted By	Obsoletes	Notes / Status
1624	A. Rijsinghani	Computation of the Internet Checksum via Incremental Update	05/20/1994	6		1141			I
1625	M. J. Fullton, K. J. Goldman, B. J. Kunze, H. Morris, F. Schiettecatte	WAIS over Z39.50-1988	06/09/1994	7					I
1626	R. Atkinson	Default IP MTU for use over ATM AAL5	05/19/1994	5			2225		PS
1627	E. Lear, E. Fair, D. Crocker, T. Kessler	Network 10 Considered Harmful (Some Practices Shouldn't be Codified)	07/01/1994	8			1918		I
1628	J. Case	UPS Management Information Base	05/19/1994	45					PS
1629	R. Colella, R. Callon, E. Gardner, Y. Rekhter	Guidelines for OSI NSAP Allocation in the Internet	05/19/1994	52				1237	DS
1630	T. Berners-Lee	Universal Resource Identifiers in WWW: A Unifying Syntax for the Expression of Names and Addresses of Objects on the Network as used in the World-Wide Web	06/09/1994	28					I
1631	P. Francis, K. Egevang	The IP Network Address Translator (Nat)	05/20/1994	10					I
1632	A. Getchell, S. Sataluri	A Revised Catalog of Available X.500 Implementations	05/20/1994	94			2116	1292	FYI 11
1633	R. Braden, D. Clark, S. Shenker	Integrated Services in the Internet Architecture: an Overview.	06/09/1994	33					I
1634	M. Allen	Novell IPX Over Various WAN Media (IPXWAN)	05/24/1994	23				1362, 1551	I
1635	P. Deutsch, A. Emtage, A. Marine	How to Use Anonymous FTP	05/25/1994	13					FYI 24
1636	IAB, R. Braden, D. Clark, S. Crocker, C. Huitema	Report of IAB Workshop on Security in the Internet Architecture - February 8-10, 1994	06/09/1994	52					I
1637	B. Manning, R. Colella	DNS NSAP Resource Records	06/09/1994	11	1348		1706	1348	E
1638	F. Baker, R. Bowen	PPP Bridging Control Protocol (BCP)	06/09/1994	28				1220	PS
1639	D. Piscitello	FTP Operation Over Big Address Records (FOOBAR)	06/09/1994	5				1545	E
1640	S. Crocker	The Process for Organization of Internet Standards Working Group (POISED)	06/09/1994	10					I
1641	D. Goldsmith, M. Davis	Using Unicode with MIME	07/13/1994	6					E
1642	D. Goldsmith, M. Davis	UTF-7 - A Mail-Safe Transformation Format of Unicode	07/13/1994	14			2152		E
1643	*F. Kastenholz*	*Definitions of Managed Objects for the Ethernet-like Interface Types*	*07/13/1994*	*19*				*1623, 1398*	*STD 50*
1644	R. Braden	T/TCP -- TCP Extensions for Transactions Functional Specification	07/13/1994	38					E
1645	A. Gwinn	Simple Network Paging Protocol - Version 2	07/14/1994	15			1861, 1863	1568	I
1646	C. Graves, T. Butts, M. Angel	TN3270 Extensions for LU name and Printer Selection	07/14/1994	13					I
1647	B. Kelly	TN3270 Enhancements	07/15/1994	33			2355		PS
1648	C. Cargille	Postmaster Convention for X.400 Operations	07/18/1994	4					PS
1649	R. Hagens, A. Hansen	Operational Requirements for X.400 Management Domains in the GO-MHS Community	07/18/1994	14					I
1650	F. Kastenholz	Definitions of Managed Objects for the Ethernet-like Interface Types using SMIv2	08/23/1994	20			2358		PS
1651	J. Klensin, N. Freed, M. Rose, E. Stefferud, D. Crocker	SMTP Service Extensions	07/18/1994	11			1869	1425	DS
1652	J. Klensin, N. Freed, M. Rose, E. Stefferud, D. Crocker	SMTP Service Extension for 8bit-MIME transport	07/18/1994	6				1426	DS
1653	J. Klensin, N. Freed, K. Moore	SMTP Service Extension for Message Size Declaration	07/18/1994	8			1870	1427	DS
1654	Y. Rekhter, T. Li	A Border Gateway Protocol 4 (BGP-4)	07/21/1994	56			1771		PS
1655	Y. Rekhter, P. Gross	Application of the Border Gateway Protocol in the Internet	07/21/1994	19			1772	1268	PS
1656	P. Traina	BGP-4 Protocol Document Roadmap and Implementation Experience	07/21/1994	4			1773		I
1657	S. Willis, J. Burruss, J. Chu	Definitions of Managed Objects for the Fourth Version of the Border Gateway Protocol (BGP-4) using SMIv2	07/21/1994	21					PS
1658	B. Stewart	Definitions of Managed Objects for Character Stream Devices using SMIv2	07/20/1994	18				1316	DS
1659	B. Stewart	Definitions of Managed Objects for RS-232-like Hardware Devices using SMIv2	07/20/1994	21				1317	DS

RFC No.	Authors	Title	Date	Pgs	Updated By	Updates	Obsoleted By	Obsoletes	Notes / Status
1660	B. Stewart	Definitions of Managed Objects for Parallel-printer-like Hardware Devices using SMIv2	07/20/1994	10				1318	DS
1661	*W. Simpson*	*The Point-to-Point Protocol (PPP)*	*07/21/1994*	*54*	*2153*			*1548*	*STD 51*
1662	*W. Simpson*	*PPP in HDLC-like Framing*	*07/21/1994*	*27*				*1549*	*STD 51*
1663	D. Rand	PPP Reliable Transmission	07/21/1994	7					PS
1664	C. Allocchio, A. Bonito, B. Cole, S. Giordano, R. Hagens	Using the Internet DNS to Distribute RFC1327 Mail Address Mapping Tables	08/11/1994	23			2163		E
1665	Z. Kielczewski, D. Kostick, K. Shih	Definitions of Managed Objects for SNA NAUs using SMIv2	07/22/1994	67			1666		PS
1666	Z. Kielczewski, D. Kostick, K. Shih	Definitions of Managed Objects for SNA NAUs using SMIv2	08/11/1994	68				1665	PS
1667	S. Symington, D. Wood, J. Pullen	Modeling and Simulation Requirements for IPng	08/08/1994	7					I
1668	D. Estrin, T. Li, Y. Rekhter	Unified Routing Requirements for IPng	08/08/1994	3					I
1669	J. Curran	Market Viability as a IPng Criteria	08/08/1994	4					I
1670	D. Heagerty	Input to IPng Engineering Considerations	08/08/1994	3					I
1671	B. Carpenter	IPng White Paper on Transition and Other Considerations	08/08/1994	8					I
1672	J. Brownlee	Accounting Requirements for IPng	08/08/1994	2					I
1673	R. Skelton	Electric Power Research Institute Comments on IPng	08/08/1994	4					I
1674	M. Taylor	A Cellular Industry View of IPng	08/08/1994	3					I
1675	S. Bellovin	Security Concerns for IPng	08/08/1994	4					I
1676	A. Ghiselli, D. Salomoni, C. Vistoli	INFN Requirements for an IPng	08/11/1994	4					I
1677	B. Adamson	Tactical Radio Frequency Communication Requirements for IPng	08/08/1994	9					I
1678	E. Britton, J. Tavs	IPng Requirements of Large Corporate Networks	08/08/1994	8					I
1679	D. Green, P. Irey, D. Marlow, K. O'Donoghue	HPN Working Group Input to the IPng Requirements Solicitation	08/08/1994	10					I
1680	C. Brazdziunas	IPng Support for ATM Services	08/08/1994	7					I
1681	S. Bellovin	On Many Addresses per Host	08/08/1994	5					I
1682	J. Bound	IPng BSD Host Implementation Analysis	08/11/1994	10					I
1683	R. Clark, M. Ammar, K. Calvert	Multiprotocol Interoperability In IPng	08/11/1994	12					I
1684	P. Jurg	Introduction to White Pages services based on X.500	08/11/1994	10					I
1685	H. Alvestrand	Writing X.400 O/R Names	08/11/1994	11					RTR 12
1686	M. Vecchi	IPng Requirements: A Cable Television Industry Viewpoint	08/11/1994	14					I
1687	E. Fleischman	A Large Corporate User's View of IPng	08/11/1994	13					I
1688	W. Simpson	IPng Mobility Considerations	08/11/1994	9					I
1689	J. Foster	A Status Report on Networked Information Retrieval: Tools and Groups	08/17/1994	204					FYI 25 RTR 13
1690	G. Huston	Introducing the Internet Engineering and Planning Group (IEPG)	08/17/1994	2					I
1691	W. Turner	The Document Architecture for the Cornell Digital Library	08/17/1994	10					I
1692	P. Cameron, D. Crocker, D. Cohen, J. Postel	Transport Multiplexing Protocol (TMux)	08/17/1994	12					PS
1693	T. Connolly, P. Amer, P. Conrad	An Extension to TCP: Partial Order Service	11/01/1994	36					E
1694	T. Brown, K. Tesink	Definitions of Managed Objects for SMDS Interfaces using SMIv2	08/23/1994	35				1304	PS
1695	M. Ahmed, K. Tesink	Definitions of Managed Objects for ATM Management Version 8.0 using SMIv2	08/25/1994	73			2515		PS
1696	J. Barnes, L. Brown, R. Royston, S. Waldbusser	Modem Management Information Base (MIB) using SMIv2	08/25/1994	31					PS
1697	D. Brower, R. Purvy, A. Daniel, M. Sinykin, J. Smith	Relational Database Management System (RDBMS) Management Information Base (MIB) using SMIv2	08/23/1994	38					PS
1698	P. Furniss	Octet Sequences for Upper-Layer OSI to Support Basic Communications Applications	10/26/1994	29					I
1699	J. Elliott	Summary of 1600-1699	1/1997						I
1700	*J. Reynolds, J. Postel*	*ASSIGNED NUMBERS*	*10/20/1994*	*230*				*1340*	*STD 2*
1701	S. Hanks, T. Li, D. Farinacci, P. Traina	Generic Routing Encapsulation (GRE)	10/21/1994	8					I

RFC No.	Authors	Title	Date	Pgs	Updated By	Updates	Obsoleted By	Obsoletes	Notes / Status
1702	S. Hanks, T. Li, D. Farinacci, P. Traina	Generic Routing Encapsulation over IPv4 networks	10/21/1994	4					I
1703	M. Rose	Principles of Operation for the TPC.INT Subdomain: Radio Paging -- Technical Procedures	10/26/1994	9				1569	I
1704	N. Haller, R. Atkinson	On Internet Authentication	10/26/1994	17					I
1705	R. Carlson, D. Ficarella	Six Virtual Inches to the Left: The Problem with IPng	10/26/1994	23					I
1706	B. Manning, R. Colella	DNS NSAP Resource Records	10/26/1994	10				1637	I
1707	M. McGovern, R. Ullmann	CATNIP: Common Architecture for the Internet	11/02/1994	16					I
1708	D. Gowin	NTP PICS PROFORMA For the Network Time Protocol Version 3	10/26/1994	13					I
1709	J. Gargano, D. Wasley	K-12 Internetworking Guidelines	12/23/1994	26					FYI 26
1710	R. Hinden	Simple Internet Protocol Plus White Paper	10/26/1994	23					I
1711	J. Houttuin	Classifications in E-mail Routing	10/26/1994	19					I
1712	C. Farrell, S. Pleitner, M. Schulze, D. Baldoni	DNS Encoding of Geographical Location	11/01/1994	7					E
1713	A. Romao	Tools for DNS debugging	11/03/1994	13					FYI 27
1714	S. Williamson, M. Kosters	Referral Who is Protocol (RWhois)	12/15/1994	46			2167		I
1715	C. Huitema	The H Ratio for Address Assignment Efficiency	11/03/1994	4					I
1716	P. Almquist, F. Kastenholz	Towards Requirements for IP Routers	11/04/1994	186			1812	1009	I
1717	K. Sklower, B. Lloyd, D. Carr, G. McGregor	The PPP Multilink Protocol (MP)	11/21/1994	21			1990		PS
1718	T. IETF Secretariat, G. Malkin	The Tao of IETF - A Guide for New Attendees of the Internet Engineering Task Force	11/23/1994	23				1539, 1391	FYI 17
1719	P. Gross	A Direction for IPng	12/16/1994	5					I
1720	J. Postel, IAB	Internet official protocol standards	11/23/1994	41			1780	1610	(STD 1)
1721	G. Malkin	RIP Version 2 Protocol Analysis	11/15/1994	4				1387	I
1722	G. Malkin	RIP Version 2 Protocol Applicability Statement	11/15/1994	5					DS
1723	G. Malkin	RIP Version 2 Carrying Additional Information	11/15/1994	9		1058	2453	1388	(STD56), H
1724	G. Malkin, F. Baker	RIP Version 2 MIB Extension	11/15/1994	18				1389	DS
1725	J. Myers, M. Rose	Post Office Protocol - Version 3	11/23/1994	18			1939	1460	(STD 53), H
1726	F. Kastenholz, C. Partridge	Technical Criteria for Choosing IP:The Next Generation (IPng)	12/20/1994	31					I
1727	C. Weider, P. Deutsch	A Vision of an Integrated Internet Information Service	12/16/1994	11					I
1728	C. Weider	Resource Transponders	12/16/1994	6					I
1729	C. Lynch	Using the Z39.50 Information Retrieval Protocol in the Internet Environment	12/16/1994	8					I
1730	M. Crispin	INTERNET MESSAGE ACCESS PROTOCOL - VERSION 4	12/20/1994	77			2060, 2061		PS
1731	J. Myers	IMAP4 Authentication mechanisms	12/20/1994	6					PS
1732	M. Crispin	IMAP4 COMPATIBILITY WITH IMAP2 AND IMAP2BIS	12/20/1994	5					I
1733	M. Crispin	DISTRIBUTED ELECTRONIC MAIL MODELS IN IMAP4	12/20/1994	3					I
1734	J. Myers	POP3 AUTHentication command	12/20/1994	5					PS
1735	J. Heinanen, R. Govindan	NBMA Address Resolution Protocol (NARP)	12/15/1994	11					E
1736	J. Kunze	Functional Requirements for Internet Resource Locators	02/09/1995	10					I
1737	K. Sollins, L. Masinter	Functional Requirements for Uniform Resource Names	12/20/1994	7					I
1738	T. Berners-Lee, L. Masinter, M. McCahill	Uniform Resource Locators (URL)	12/20/1994	25	1808, 2368				PS
1739	G. Kessler, S. Shepard	A Primer On Internet and TCP/IP Tools	12/22/1994	32			2151		I
1740	P. Faltstrom, D. Crocker, E. Fair	MIME Encapsulation of Macintosh files - MacMIME	12/22/1994	16					PS
1741	P. Faltstrom, D. Crocker, E. Fair	MIME Content Type for BinHex Encoded Files	12/22/1994	6					DS
1742	S. Waldbusser, K. Frisa	AppleTalk Management Information Base II	01/05/1995	84				1243	I
1743	K. McCloghrie, E. Decker	IEEE 802.5 MIB using SMIv2	12/27/1994	25			1748	1231	PS
1744	G. Huston	Observations on the Management of the Internet Address Space	12/23/1994	12					I

RFC No.	Authors	Title	Date	Pgs	Updated By	Updates	Obsoleted By	Obsoletes	Notes / Status
1745	K. Varadhan, S. Hares, Y. Rekhter	BGP4/IDRP for IP---OSPF Interaction	12/27/1994	19					PS
1746	B. Manning, D. Perkins	Ways to Define User Expectations	12/30/1994	18					I
1747	J. Hilgeman, S. Nix, A. Bartky, W. Clark	Definitions of Managed Objects for SNA Data Link Control: SDLC	01/11/1995	67					DS
1748	K. McCloghrie, F. Baker, E. Decker	IEEE 802.5 MIB using SMIv2	12/29/1994	25	1749			1743, 1231	DS
1749	K. McCloghrie, F. Baker, E. Decker	IEEE 802.5 Station Source Routing MIB using SMIv2	12/29/1994	10		1748			PS
1750	D. Eastlake, S. Crocker, J. Schiller	Randomness Recommendations for Security	12/29/1994	25					I
1751	D. McDonald	A Convention for Human-Readable 128-bit Keys	12/29/1994	15					I
1752	S. Bradner, A. Mankin	The Recommendation for the IP Next Generation Protocol	01/18/1995	52					PS
1753	J. Chiappa	IPng Technical Requirements Of the Nimrod Routing and Addressing Architecture	01/05/1995	18					I
1754	M. Laubach	IP over ATM Working Group's Recommendations for the ATM Forum's Multiprotocol BOF Version 1	01/19/1995	7					I
1755	M. Perez, A. Malis, F. A. Mankin, E. Hoffman, G. Grossman	ATM Signaling Support for IP over ATM	02/17/1995	32					PS
1756	T. Rinne	Remote write protocol - version 1.0	01/19/1995	11					E
1757	S. Waldbusser	Remote Network Monitoring Management Information Base	02/10/1995	91				1271	DS
1758	T. American Directory Forum	NADF Standing Documents: A Brief Overview	02/09/1995	4				1417	I
1759	R. Smith, F. Wright, T. Hastings, S. Zilles, J. Gyllenskog	Printer MIB	03/28/1995	113					PS
1760	N. Haller	The S/KEY One-Time Password System	02/15/1995	12					I
1761	B. Callaghan, R. Gilligan	Snoop Version 2 Packet Capture File Format	02/09/1995	6					I
1762	S. Senum	The PPP DECnet Phase IV Control Protocol (DNCP)	03/01/1995	7				1376	DS
1763	S. Senum	The PPP Banyan Vines Control Protocol (BVCP)	03/01/1995	10					PS
1764	S. Senum	The PPP XNS IDP Control Protocol (XNSCP)	03/01/1995	5					PS
1765	J. Moy	OSPF Database Overflow	03/02/1995	9					E
1766	H. Alvestrand	Tags for the Identification of Languages	03/02/1995	9					PS
1767	D. Crocker	MIME Encapsulation of EDI Objects	03/02/1995	7					PS
1768	D. Marlow	Host Group Extensions for CLNP Multicasting	03/03/1995	42					E
1769	D. Mills	Simple Network Time Protocol (SNTP)	03/17/1995	14			2030	1361	I
1770	C. Graff	IPv4 Option for Sender Directed Multi-Destination Delivery	03/28/1995	6					I
1771	Y. Rekhter, T. Li	A Border Gateway Protocol 4 (BGP-4)	03/21/1995	57				1654	DS
1772	Y. Rekhter, P. Gross	Application of the Border Gateway Protocol in the Internet	03/21/1995	19				1655	DS
1773	P. Traina	Experience with the BGP-4 protocol	03/21/1995	9				1656	I
1774	P. Traina	BGP-4 Protocol Analysis	03/21/1995	10					I
1775	D. Crocker	To Be "On" the Internet	03/17/1995	4					I
1776	S. Crocker	The Address is the Message	04/01/1995	2					I
1777	W. Yeong, T. Howes, S. Kille	Lightweight Directory Access Protocol	03/28/1995	22				1487	DS
1778	T. Howes, S. Kille, W. Yeong, C. Robbins	The String Representation of Standard Attribute Syntaxes	03/28/1995	12.			2559	1488	DS
1779	S. Kille	A String Representation of Distinguished Names	03/28/1995	8			2253	1485	DS
1780	J. Postel	Internet official protocol standards	03/28/1995	39			1800	1720	(STD 1)
1781	S. Kille	Using the OSI Directory to Achieve User Friendly Naming	03/28/1995	26	1484			1484	PS
1782	G. Malkin, A. Harkin	TFTP Option Extension	03/28/1995	6		1350	2347		PS
1783	G. Malkin, A. Harkin	TFTP Blocksize Option	03/28/1995	5		1350	2348		PS
1784	G. Malkin, A. Harkin	TFTP Timeout Interval and Transfer Size Options	03/28/1995	5		1350	2349		PS
1785	G. Malkin, A. Harkin	TFTP Option Negotiation Analysis	03/28/1995	2		1350			I
1786	T. Bates, E. Gerich, L. Joncheray, J. Jouanigot, . . .	Representation of IP Routing Policies in a Routing Registry (ripe-81++)	03/28/1995	83					I
1787	Y. Rekhter	Routing in a Multi-provider Internet	04/14/1995	8					I
1788	W. Simpson	ICMP Domain Name Messages	04/14/1995	7					E
1789	C. Yang	INETPhone: Telephone Services and Servers on Internet	04/17/1995	6					I
1790	V. Cerf	An Agreement between the Internet Society and Sun Microsystems, Inc. in the Matter of ONC RPC and XDR Protocols	04/17/1995	6					I

RFC No.	Authors	Title	Date	Pgs	Updated By	Updates	Obsoleted By	Obsoletes	Notes / Status
1791	T. Sung	TCP And UDP Over IPX Networks With Fixed Path MTU	04/18/1995	12					E
1792	T. Sung	TCP/IPX Connection Mib Specification	04/18/1995	9					E
1793	J. Moy	Extending OSPF to Support Demand Circuits	04/19/1995	31					PS
1794	T. Brisco	DNS Support for Load Balancing	04/20/1995	7					I
1795	L. Wells, A. Bartky	Data Link Switching: Switch-to-Switch Protocol AIW DLSw RIG: DLSw Closed Pages, DLSw Standard Version 1.0	04/25/1995	91				1434	I
1796	C. Huitema, J. Postel, S. Crocker	Not All RFCs are Standards	04/25/1995	4					I
1797	I. Assigned Numbers Authority (IANA)	Class A Subnet Experiment	04/25/1995	4					E
1798	A. Young	Connection-less Lightweight Directory Access Protocol	06/07/1995	9					PS
1799	M. Kennedy	Request for Comments Summary RFC Numbers 1700-1799	01/1997	42038 b					I
1800	J. Postel	Internet official protocol standards	07/11/1995	36			1880	1780	(STD 1)
1801	S. Kille	MHS use of the X.500 Directory to support MHS Routing	06/09/1995	73					E
1802	H. Alvestrand, K. Jordan, S. Langlois, J. Romaguera	Introducing Project Long Bud: Internet Pilot Project for the Deployment of X.500 Directory Information in Support of X.400 Routing	06/12/1995	11					I
1803	R. Wright A. Malis, A. Getchell, S. Sataluri, P. Yee, W. Yeong	Recommendations for an X.500 Production Directory Service	06/07/1995	8					I
1804	G. Mansfield, P. Rajeev, T. Howes, S. Raghavan	Schema Publishing in X.500 Directory	06/09/1995	10					E
1805	A. Rubin	Location-Independent Data/Software Integrity Protocol	06/07/1995	6					I
1806	R. Troost, S. Dorner	Communicating Presentation Information in Internet Messages: The Content-Disposition Header	06/07/1995	8	2183				E
1807	R. Lasher, D. Cohen	A Format for Bibliographic Records	06/21/1995	16				1357	I
1808	R. Fielding	Relative Uniform Resource Locators	06/14/1995	16	2368	1738			PS
1809	C. Partridge	Using the Flow Label Field in IPv6	06/14/1995	6					I
1810	J. Touch	Report on MD5 Performance	06/21/1995	7					I
1811	F. Networking Council (FNC)	U.S. Government Internet Domain Names	06/21/1995	3			1816		I
1812	F. Baker	Requirements for IP Version 4 Routers	06/22/1995	175	2644			1009, 1716	PS
1813	B. Callaghan, B. Pawlowski, P. Staubach	NFS Version 3 Protocol Specification	06/21/1995	126					1094, I
1814	E. Gerich	Unique Addresses are Good	06/22/1995	3					I
1815	M. Ohta	Character Sets ISO-10646 and ISO-10646-J-1	08/01/1995	6					I
1816	F. Networking Council (FNC)	U.S. Government Internet Domain Names	08/03/1995	8			2146	1811	I
1817	Y. Rekhter	CIDR and Classful Routing	08/04/1995	2					I
1818	J. Postel, T. Li, Y. Rekhter	Best Current Practices	08/04/1995	3					BCP1
1819	L. Delgrossi, L. Berger	Internet Stream Protocol Version 2 (ST2) Protocol Specification - Version ST2+	08/11/1995	109				1190, IEN119	E
1820	E. Huizer	Multimedia E-mail (MIME) User Agent Checklist	08/22/1995	8			1844		I
1821	M. Borden, E. Crawley, B. Davie, S. Batsell	Integration of Real-time Services in an IP-ATM Network Architecture	08/11/1995	24					I
1822	J. Lowe	A Grant of Rights to Use a Specific IBM patent with Photuris	08/14/1995	2					I
1823	T. Howes, M. Smith	The LDAP Application Program Interface	08/09/1995	22					I
1824	H. Danisch	The Exponential Security System TESS: An Identity-Based Cryptographic Protocol for Authenticated Key-Exchange (E.I.S.S.-Report 1995/4)	08/11/1995	21					I
1825	R. Atkinson	Security Architecture for the Internet Protocol	08/09/1995	22			2401		PS
1826	R. Atkinson	IP Authentication Header	08/09/1995	13			2406		PS
1827	R. Atkinson	IP Encapsulating Security Payload (ESP)	08/09/1995	12					PS
1828	P. Metzger, W. Simpson	IP Authentication using Keyed MD5	08/09/1995	5					PS
1829	P. Metzger, P. Karn, W. Simpson	The ESP DES-CBC Transform	08/09/1995	10					PS
1830	G. Vaudreuil	SMTP Service Extensions for Transmission of Large and Binary MIME Messages	08/16/1995	8					E
1831	R. Srinivasan	RPC: Remote Procedure Call Protocol Specification Version 2	08/09/1995	18					PS
1832	R. Srinivasan	XDR: External Data Representation Standard	08/09/1995	24					PS
1833	R. Srinivasan	Binding Protocols for ONC RPC Version 2	08/09/1995	14					PS
1834	J. Gargano, K. Weiss	Whois and Network Information Lookup Service Whois++	08/16/1995	7					I

RFC No.	Authors	Title	Date	Pgs	Updated By	Updates	Obsoleted By	Obsoletes	Notes / Status
1835	P. Deutsch, R. Schoultz, P. Faltstrom, C. Weider	Architecture of the WHOIS++ service	08/16/1995	41					PS
1836	S. Kille	Representing the O/R Address hierarchy in the X.500 Directory Information Tree	08/22/1995	11			2294		E
1837	S. Kille	Representing Tables and Subtrees in the X.500 Directory	08/22/1995	7			2293		E
1838	S. Kille	Use of the X. 500 Directory to support mapping between X.400 and RFC 822 Addresses	08/22/1995	8			2164		E
1839									*
1840									*
1841	J. Chapman, D. Coli, A. Harvey, B. Jensen, K. Rowett	PPP Network Control Protocol for LAN Extension	09/29/1995	66					I
1842	Y. Wei, Y. Zhang, J. Li, J. Ding, Y. Jiang	ASCII Printable Characters-Based Chinese Character Encoding for Internet Messages	08/24/1995	5					I
1843	F. Lee	HZ-A Data Format for Exchanging Files of Arbitrarily Mixed Chinese and ASCII characters	08/24/1995	5					I
1844	E. Huizer	Multimedia E-mail (MIME) User Agent checklist	08/24/1995	8				1820	I
1845	D. Crocker, N. Freed, A. Cargille	SMTP Service Extension for Checkpoint/Restart	10/02/1995	8					E
1846	A. Durand, F. Dupont	SMTP 521 reply code	10/02/1995	4					E
1847	J. Galvin, S. Murphy, S. Crocker, N. Freed	Security Multiparts for MIME: Multipart/Signed and Multipart/Encrypted	10/03/1995	11					PS
1848	S. Crocker, N. Freed, J. Galvin, S. Murphy	MIME Object Security Services	10/03/1995	48					PS
1850	F. Baker, R. Coltun	OSPF Version 2 Management Information Base	11/03/1995	80				1253	DS
1851	P. Metzger, P. Karn, W. Simpson	The ESP Triple DES-CBC Transform	10/02/1995	9					E
1852	P. Metzger, W. Simpson	IP Authentication using Keyed SHA	10/02/1995	6					E
1853	W. Simpson	IP in IP Tunneling	10/04/1995	8					I
1854	N. Freed, A. Cargille	SMTP Service Extension for Command Pipelining	10/04/1995	7			2197		PS
1855	S. Hambridge	Netiquette Guidelines	10/20/1995	21					FYI 28
1856	H. Clark	The Opstat Client-Server Model for Statistics Retrieval	10/20/1995	21					I
1857	M. Lambert	A Model for Common Operational Statistics	10/20/1995	27				1404	I
1858	P. Ziemba, D. Reed, P. Traina	Security Considerations for IP Fragment Filtering	10/25/1995	10					I
1859	Y. Pouffary	ISO Transport Class 2 Non-use of Explicit Flow Control over TCP RFC1006 extension	10/20/1995	8					I
1860	T. Pummill, B. Manning	Variable Length Subnet Table For IPv4	10/20/1995	3			1878		I
1861	A. Gwinn	Simple Network Paging Protocol - Version 3 - Two-Way Enhanced	10/19/1995	23				1645	I
1862	M. McCahill, M. Schwartz, T. Verschuren, C. Weider, K. Sollins	Report of the IAB Workshop on Internet Information Infrastructure, October 12-14, 1994	11/03/1995	27					I
1863	D. Haskin	A BGP/IDRP Route Server alternative to a full mesh routing	10/20/1995	16				1645	E
1864	J. Myers, M. Rose	The Content-MD5 Header Field	10/24/1995	4				1544	DS
1865	W. Houser, J. Griffin, C. Hage	EDI Meets the Internet: Frequently Asked Questions about Electronic Data Interchange (EDI) on the Internet	01/04/1996	42					I
1866	T. Berners-Lee, D. Connolly	Hypertext Markup Language - 2.0	11/03/1995	77					PS
1867	E. Nebel, L. Masinter	Form-based File Upload in HTML	11/07/1995	13					E
1868	G. Malkin	ARP Extension - UNARP	11/06/1995	4					E
1869	*J. Klensin, N. Freed, M. Rose, E. Stefferud, D. Crocker*	*SMTP Service Extensions*	*11/06/1995*	*11*				*1651*	*STD 10*
1870	*J. Klensin, N. Freed, K. Moore*	*SMTP Service Extension for Message Size Declaration*	*11/06/1995*	*9*				*1653*	*STD 11*
1871	J. Postel	Addendum to RFC 1602 -- Variance Procedure	11/29/1995	4		1602, 1603			BCP2
1872	E. Levinson	The MIME Multipart/Related Content-type	12/26/1995	8			2112		E
1873	E. Levinson, J. Clark	Message/External-Body Content -ID Access Type	12/26/1995	4					E
1874	E. Levinson	SGML Media Types	12/26/1995	6					E
1875	N. Berge	UNINETT PCA Policy Statements	12/26/1995	10					I
1876	C. Davis, P. Vixie, T. Goodwin, I. Dickinson	A Means for Expressing Location Information in the Domain Name System	01/15/1996	18		1034, 1035			E

RFC No.	Authors	Title	Date	Pgs	Updated By	Updates	Obsoleted By	Obsoletes	Notes / Status
1877	S. Cobb	PPP Internet Protocol Control Protocol Extensions for Name Server Addresses	12/26/1995	6					I
1878	T. Pummill, B. Manning	Variable Length Subnet Table For IPv4	12/26/1995	8				1860	I
1879	B. Manning	Class A Subnet Experiment Results and Recommendations	01/15/1996	6					I
1880	J. Postel	Internet official protocol standards	11/29/1995	38			1920	1800	(STD 1)
1881	I. IESG	IPv6 Address Allocation Management	12/26/1995	2					I
1882	B. Hancock	The 12-Days of Technology Before Christmas	12/26/1995	5					I
1883	S. Deering, R. Hinden	Internet Protocol, Version 6 (IPv6) Specification	01/04/1996	37			2460		PS
1884	R. Hinden, S. Deering	IP Version 6 Addressing Architecture	01/04/1996	18			2373		PS
1885	A. Conta, S. Deering	Internet Control Message Protocol (ICMPv6) for the Internet Protocol Version 6 (IPv6)	01/04/1996	20			2463		PS
1886	S. Thomson, C. Huitema	DNS Extensions to support IP version 6	01/04/1996	5					PS
1887	Y. Rekhter, T. Li	An Architecture for IPv6 Unicast Address Allocation	01/04/1996	25					I
1888	J. Bound, A. Lloyd, B. Carpenter, D. Harrington, J. Houldsworth	OSI NSAPs and IPv6	08/16/1996	16					E
1889	H. Schulzrinne, S. Casner, R. Frederick, V. Jacobson	RTP: A Transport Protocol for Real-Time Applications	01/25/1996	75					PS
1890	H. Schulzrinne	RTP Profile for Audio and Video Conferences with Minimal Control	01/25/1996	18					PS
1891	K. Moore	SMTP Service Extension for Delivery Status Notifications	01/15/1996	31					PS
1892	G. Vaudreuil	The Multipart/Report Content Type for the Reporting of Mail System Administrative Messages	01/15/1996	4					PS
1893	G. Vaudreuil	Enhanced Mail System Status Codes	01/15/1996	15					PS
1894	K. Moore, GVaudreuil,	An Extensible Message Format for Delivery Status Notifications	01/15/1996	31					PS
1895	E. Levinson	The Application/CALS-1840 Content-type	02/15/1996	6					I
1896	P. Resnick, A. Walker	The text/enriched MIME Content-type	02/19/1996	21				1523, 1563	I
1897	R. Hinden, J. Postel	IPv6 Testing Address Allocation	01/25/1996	4			2471		E
1898	D. Eastlake, B. Boesch, S. Crocker, M. Yesil	CyberCash Credit Card Protocol Version 0.8	02/19/1996	52					I
1899	J. Elliott	RFC Summary RFC Numbers 1800-1899	1/1997						I
1900	B. Carpenter, Y. Rekhter	Renumbering Needs Work	02/28/1996	4					I
1901	J. Case, M. Rose, K. McCloghrie, S. Waldbusser	Introduction to Community-based SNMP v2	01/22/1996	8					E
1902	J. Case, M. Rose, K. McCloghrie, S. Waldbusser	Structure of Management Information for Version 2 of the Simple Network Management Protocol (SNMPv2)	01/22/1996	40			2578	1442	STD58, H
1903	J. Case, M. Rose, K. McCloghrie, S. Waldbusser	Textual Conventions for Version 2 of the Simple Network Management Protocol (SNMPv2)	01/22/1996	23			2579	1443	STD58, H
1904	J. Case, M. Rose, K. McCloghrie, S. Waldbusser	Conformance Statements for Version 2 of the Simple Network Management Protocol (SNMPv2)	01/22/1996	24			2580	1444	STD58, H
1905	J. Case, M. Rose, K. McCloghrie, S. Waldbusser	Protocol Operations for Version 2 of the Simple Network Management Protocol (SNMPv2)	01/22/1996	24				1448	DS
1906	J. Case, M. Rose, K. McCloghrie, S. Waldbusser	Transport Mappings for Version 2 of the Simple Network Management Protocol (SNMPv2)	01/22/1996	13				1449	DS
1907	J. Case, M. Rose, K. McCloghrie, S. Waldbusser	Management Information Base for Version 2 of the Simple Network Management Protocol (SNMPv2)	01/22/1996	20				1450	DS
1908	J. Case, M. Rose, K. McCloghrie, S. Waldbusser	Coexistence between Version 1 and Version 2 of the Internet-standard Network Management Framework	01/22/1996	10				1452	DS
1909	K. McCloghrie	An Administrative Infrastructure for SNMPv2	02/28/1996	19					E
1910	G. Waters	User-based Security Model for SNMPv2	02/28/1996	44					E
1911	G. Vaudreuil	Voice Profile for Internet Mail	02/19/1996	22			2421, 2422, 2423		E
1912	D. Barr	Common DNS Operational and Configuration Errors	02/28/1996	16				1537	I
1913	C. Weider, J. Fullton, S. Spero	Architecture of the Whois++ Index Service	02/28/1996	16					PS

RFC No.	Authors	Title	Date	Pgs	Updated By	Updates	Obsoleted By	Obsoletes	Notes / Status
1914	P. Faltstrom, R. Schoultz, C. Weider	How to interact with a Whois++ mesh	02/28/1996	10					PS
1915	F. Kastenholz	Variance for The PPP Connection Control Protocol and The PPP Encryption Control Protocol	02/28/1996	7					BCP3
1916	H. Berkowitz, P. Ferguson, W. Leland, P. Nesser	Enterprise Renumbering: Experience and Information Solicitation	02/28/1996	8					I
1917	P. Nesser	An Appeal to the Internet Community to Return Unused IP Networks (Prefixes) to the IANA	02/29/1996	10					BCP4
1918	Y. Rekhter, E. Lear, R. Moskowitz, D. Karrenberg, G. de Groot	Address Allocation for Private Internets	02/29/1996	9				1597, 1627	BCP5
1919	M. Chatel	Classical versus Transparent IP Proxies	03/28/1996	35					I
1920	J. Postel	Internet official protocol standards	03/22/1996	40			2000	1880	(STD 1)
1921	J. Dujonc	TNVIP protocol	03/25/1996	30					I
1922	H. Zhu, D. Hu, Z. Wang, T. Kao, W. Chang, M. Crispin	Chinese Character Encoding for Internet Messages	03/26/1996	27					I
1923	J. Halpern, S. Bradner	RIPv1 Applicability Statement for Historic Status	03/25/1996	3					I
1924	R. Elz	A Compact Representation of IPv6 Addresses	04/01/1996	6					I
1925	R. Callon	The Twelve Networking Truths	04/01/1996	3					I
1926	J. Eriksson	An Experimental Encapsulation of IP Datagrams on Top of ATM	04/01/1996	2					I
1927	C. Rogers	Suggested Additional MIME Types for Associating Documents	04/01/1996	3					I
1928	M. Leech, M. Ganis, Y. Lee, R. Kuris, D. Koblas, L. Jones	SOCKS Protocol Version 5	04/03/1996	9					PS
1929	M. Leech	Username/Password Authentication for SOCKS V5	04/03/1996	2					PS
1930	J. Hawkinson, T. Bates	Guidelines for creation, selection, and registration of an Autonomous System (AS)	04/03/1996	10					BCP6
1931	D. Brownell	Dynamic RARP Extensions and Administrative Support for Automatic Network Address Allocation	04/03/1996	11					I
1932	R. Cole, D. Shur, C. Villamizar	IP over ATM: A Framework Document	04/08/1996	31					I
1933	R. Gilligan, E. Nordmark	Transition Mechanisms for IPv6 Hosts and Routers	04/08/1996	22					PS
1934	K. Smith	Ascend's Multilink Protocol Plus (MP+)	04/08/1996	47					I
1935	J. Quarterman, S. Carl-Mitchell	What is the Internet, Anyway?	04/10/1996	11					I
1936	J. Touch, B. Parham	Implementing the Internet Checksum in Hardware	04/10/1996	21					I
1937	Y. Rekhter, D. Kandlur	"Local/Remote" Forwarding Decision in Switched Data Link Subnetworks	05/08/1996	10					I
1938	N. Haller, C. Metz	A One-Time Password System	05/14/1996	18			2289		PS
1939	*J. Myers, M. Rose*	*Post Office Protocol - Version 3*	*05/14/1996*	*23*	*1957, 2449*			*1725*	*STD53*
1940	D. Estrin, T. Li, Y. Rekhter, K. Varadhan, D. Zappala	Source Demand Routing: Packet Format and Forwarding Specification (Version 1)	05/14/1996	27					I
1941	J. Sellers, J. Robichaux	Frequently Asked Questions for Schools	05/15/1996	70				1578	FYI22
1942	D. Raggett	HTML Tables	05/15/1996	30					E
1943	B. Jennings	Building an X.500 Directory Service in the US	05/15/1996	22					I
1944	S. Bradner, J. McQuaid	Benchmarking Methodology for Network Interconnect Devices	05/17/1996	30			2544		I
1945	T. Berners-Lee, R. Fielding, H. Nielsen	Hypertext Transfer Protocol -- HTTP/1.0	05/17/1996	60					I
1946	S. Jackowski	Native ATM Support for ST2+	05/17/1996	21					I
1947	D. Spinellis	Greek Character Encoding for Electronic Mail Messages	05/17/1996	7					I
1948	S. Bellovin	Defending Against Sequence Number Attacks	05/17/1996	6					I
1949	A. Ballardie	Scalable Multicast Key Distribution	05/17/1996	18					E
1950	L. Deutsch, J. Gailly	ZLIB Compressed Data Format Specification version 3.3	05/23/1996	11					I
1951	L. Deutsch	DEFLATE Compressed Data Format Specification version 1.3	05/23/1996	17					I
1952	L. Deutsch	GZIP file format specification version 4.3	05/23/1996	12					I
1953	P. W. Edwards, R. E. Hoffman, F. Liaw, T. Lyon, G. Minshall	Ipsilon Flow Management Protocol Specification for IPv4 Version 1.0	05/23/1996	20					I

RFC No.	Authors	Title	Date	Pgs	Updated By	Updates	Obsoleted By	Obsoletes	Notes / Status
1954	P. Newman, W. Edwards, R. Hinden, F. Liaw, E. Hoffman, T. G. Minshall	Transmission of Flow Labelled IPv4 on ATM Data Links Ipsilon Version 1.0	05/22/1996	8					I
1955	R. Hinden	New Scheme for Internet Routing and Addressing (ENCAPS) for IPN	06/06/1996	5					I
1956	D. Engebretson, R. Plzak	Registration in the MIL Domain	06/06/1996	2					I
1957	R. Nelson	Some Observations on Implementations of the Post Office Protocol (POP3)	06/06/1996	2		1939			I
1958	B. Carpenter	Architectural Principles of the Internet	06/06/1996	8					I
1959	T. Howes, M. Smith	An LDAP URL Format	06/19/1996	4			2255		PS
1960	T. Howes	A String Representation of LDAP Search Filters	06/19/1996	3			2254	1558	PS
1961	P. McMahon	GSS-API Authentication Method for SOCKS Version 5	06/19/1996	9					PS
1962	D. Rand	The PPP Compression Control Protocol (CCP)	06/19/1996	9	2153				PS
1963	K. Schneider, S. Venters	PPP Serial Data Transport Protocol (SDTP)	08/14/1996	20					I
1964	J. Linn	The Kerberos Version 5 GSS-API Mechanism	06/19/1996	20					PS
1965	P. Traina	Autonomous System Confederations for BGP	06/19/1996	7					E
1966	T. Bates, R. Chandra	BGP Route Reflection An alternative to full mesh IBGP	06/19/1996	7					E
1967	K. Schneider, R. Friend	PPP LZS-DCP Compression Protocol (LZS-DCP)	08/13/1996	18					I
1968	G. Mey,	The PPP Encryption Control Protocol (ECP)	06/19/1996	11					PS
1969	K. Sklower, G. Meyer	The PPP DES Encryption Protocol (DESE)	06/19/1996	10			2419		I
1970	T. Narten, E. Nordmark, W. Simpson	Neighbor Discovery for IP Version 6 (IPv6)	08/16/1996	82			2461		PS
1971	S. Thomson, T. Narten	IPv6 Stateless Address Autoconfiguration	08/16/1996	23			2464		PS
1972	M. Crawford	A Method for the Transmission of IPv6 Packets over Ethernet Networks	08/16/1996	4					PS
1973	W. Simpson	PPP in Frame Relay	06/19/1996	8					PS
1974	R. Friend, W. Simpson	PPP Stac LZS Compression Protocol	08/13/1996	20					I
1975	D. Schremp, J. Black, J. Weiss	PPP Magnalink Variable Resource Compression	08/09/1996	6					I
1976	K. Schneider, S. Venters	PPP for Data Compression in Data Circuit-Terminating Equipment (DCE)	08/14/1996	10					I
1977	V. Schryver	PPP BSD Compression Protocol	08/09/1996	27					I
1978	D. Rand	PPP Predictor Compression Protocol	08/28/1996	9					I
1979	J. Woods	PPP Deflate Protocol	08/09/1996	10					I
1980	J. Seidman	A Proposed Extension to HTML: Client-Side Image Maps	08/14/1996	7					I
1981	J. McCann, S. Deering, J. Mogul	Path MTU Discovery for IP version 6	08/14/1996	15					PS
1982	R. Elz, R. Bush	Serial Number Arithmetic	09/03/1996	7		1034, 1045			PS
1983	G. Malkin	Internet Users' Glossary	08/16/1996	62				1392	FYI18
1984	I. and IESG	IAB and IESG Statement on Cryptographic Technology and the Internet	08/20/1996	5					I
1985	J. De Winter	SMTP Service Extension for Remote Message Queue Starting	08/14/1996	7					PS
1986	W. Polites, W. Wollman, D. Woo, R. Langan	Experiments with a Simple File Transfer Protocol for Radio Links using Enhanced Trivial File Transfer Protocol (ETFTP)	08/16/1996	21					E
1987	P. Newman, W. Edwards, R. Hinden, F. Liaw, E. Hoffman, T. G. Minshall	Ipsilon's General Switch Management Protocol Specification Version 1.1	08/16/1996	44	2297				I
1988	G. McAnally, D. Gilbert, J. Flick	Conditional Grant of Rights to Specific Hewlett-Packard Patents In Conjunction With the Internet Engineering Task Force's Internet-Standard Network Management Framework	08/16/1996	2					I
1989	W. Simpson	PPP Link Quality Monitoring	08/16/1996	16				1333	DS
1990	K. Sklower, B. Lloyd, D. Carr, G. McGregor, T. Coradetti	The PPP Multilink Protocol (MP)	08/16/1996	24				1717	DS
1991	D. Atkins, W. Stallings, P. Zimmermann	PGP Message Exchange Formats	08/16/1996	21					I
1992	I. Castineyra, J. Chiappa, M. Steenstrup	The Nimrod Routing Architecture	08/30/1996	27					I
1993	D. Carr	PPP Gandalf FZA Compression Protocol	08/30/1996	6					I

RFC No.	Authors	Title	Date	Pgs	Updated By	Updates	Obsoleted By	Obsoletes	Notes / Status
1994	W. Simpson	PPP Challenge Handshake Authentication Protocol (CHAP)	08/30/1996	12	2484			1334	DS
1995	M. Ohta	Incremental Zone Transfer in DNS	08/28/1996	8		1035			PS
1996	P. Vixie	A Mechanism for Prompt Notification of Zone Changes (DNS NOTIFY)	08/28/1996	7		1035			PS
1997	R. Chandra, P. Traina, T. Li	BGP Communities Attribute	08/30/1996	5					PS
1998	E. Chen, T. Bates	An Application of the BGP Community Attribute in Multi-home Routing	08/30/1996	9					I
1999	J. Elliott	Summary RFC Numbers 1900-1999	1/97						I
2000	J. Postel	Internet Official Protocol Standard	10/97				2200	1920	(STD1)
2001	W. Stevens	TCP Slow Start, Congestion Avoidance, Fast Retransmit, and Fast Recovery Algorithms	1/97				2581		
2002	C. Perkins	IP Mobility Support	10/22/1996	79	2290				PS
2003	C. Perkins	IP Encapsulation within IP	10/22/1996	14					PS
2004	C. Perkin	Minimal Encapsulation within IP	10/22/1996	5					PS
2005	J. Solomon	Applicability Statement for IP Mobility Support	10/22/1996	5					PS
2006	D. Cong, C. Perkins, M. Hamlen	The Definitions of Managed Objects for IP Mobility Support using SMIv2	10/22/1996	52					PS
2007	J. Foster, M. Isaacs, M. Prior	Catalogue of Network Training Materials	10/14/1996	55					FYI 29
2008	Y. Rekhter, T. Li	Implications of Various Address Allocation Policies for Internet Routing	10/14/1996	13					BCP7
2009	T. Imielinski, J. Navas	GPS-Based Addressing and Routing	11/08/1996	28					E
2010	B. Manning, P. Vixie	Operational Criteria for Root Name Servers	10/14/1996	7					I
2011	K. McCloghrie	SNMPv2 Management Information Base for the Internet Protocol using SMIv2	11/12/1996	18		1213			PS
2012	K. McCloghrie	SNMPv2 Management Information Base for the Transmission Control Protocol	11/12/1996	10		1213			PS
2013	K. McCloghrie	SNMPv2 Management Information Base for the User Datagram Protocol using SMIv2	11/12/1996	6		1213			PS
2014	A. Weinrib, J. Postel	IRTF Research Group Guidelines and Procedures	10/17/1996	13					BCP8
2015	M. Elkins	MIME Security with Pretty Good Privacy (PGP)	10/14/1996	8					PS
2016	L. Daigle, P. Deutsch, B. Heelan, C. Alpaugh, M. Maclachlan	Uniform Resource Agents (URAs)	10/31/1996	21					E
2017	N. Freed, K. Moore, A. Cargille	Definition of the URL MIME External-Body Access-Type	10/14/1996	5					PS
2018	M. Mathis, J. Mahdavi, S. Floyd, A. Romanow	TCP Selective Acknowledgment Options	10/17/1996	12				1072	PS
2019	M. Crawford	Transmission of IPv6 Packets Over FDDI	10/17/1996	6			2467		PS
2020	J. Flick	Definitions of Managed Objects for IEEE 802.12 Interfaces	10/17/1996	31					PS
2021	S. Waldbusser	Remote Network Monitoring Management Information Base Version 2 using SMIv2	03/1997	262223 b					PS
2022	G. Armitage	Support for Multicast over UNI 3.0/3.1 based ATM Networks	11/05/1996	82					PS
2023	D. Haskin, E. Allen	IP Version 6 over PPP	10/22/1996	10			2472		PS
2024	D. Chen, P. Gayek, S. Nix	Definitions of Managed Objects for Data Link Switching using SNMPv2	10/22/1996	90					PS
2025	C. Adams	The Simple Public-Key GSS-API Mechanism (SPKM)	10/22/1996	41					PS
2026	S. Bradner	The Internet Standards Process -- Revision 3	10/29/1996	36				1602	BCP9
2027	J. Galvin	IAB and IESG Selection, Confirmation, and Recall Process: Operation of the Nominating and Recall Committees	10/29/1996	11			2282		BCP10
2028	R. Hovey, S. Bradner	The Organizations Involved in the IETF Standards Process	10/29/1996	7					BCP11
2029	M. Speer, D. Hoffman	RTP Payload Format of Sun's CellB Video Encoding	10/30/1996	6					PS
2030	D. Mills	Simple Network Time Protocol (SNTP) Version 4 for IPv4, IPv6 and OSI	10/31/1996	18				1769	I
2031	E. Huizer	IETF-ISOC relationship	10/29/1996	4					I
2032	T. Turletti, C. Huitema	RTP payload format for H.261 video streams	10/30/1996	11					PS
2033	J. Myers	Local Mail Transfer Protocol	10/30/1996	7					I
2034	N. Freed	SMTP Service Extension for Returning Enhanced Error Codes	10/30/1996	6					PS
2035	L. Berc, W. Fenner, R. Frederick, S. McCanne	RTP Payload Format for JPEG-compressed Video	10/30/1996	16			2435		PS
2036	G. Huston	Observations on the use of Components of the Class A Address Space within the Internet	10/30/1996	9					I

RFC No.	Authors	Title	Date	Pgs	Updated By	Updates	Obsoleted By	Obsoletes	Notes / Status
2037	K. McCloghrie, A. Bierman	Entity MIB	10/30/1996	35			2737		PS
2038	D. Hoffman, G. Fernando, V. Goyal	RTP Payload Format for MPEG1/MPEG2 Video	10/30/1996	11			2250		PS
2039	C. Kalbfleisch	Applicability of Standards Track MIBs to Management of World Wide Web Servers	11/06/1996	14					I
2040	R. Baldwin, R. Rivest	The RC5, RC5-CBC, RC5-CBC-Pad, and RC5-CTS Algorithms	10/30/1996	29					I
2041	B. Noble, R. Katz, G. Nguyen, M. Satyanarayanan	Mobile Network Tracing	10/30/1996	27					I
2042	B. Manning	Registering New BGP Attribute Types	01/1997	4001 b					I
2043	A. Fuqua	The PPP SNA Control Protocol (SNACP)	10/30/1996	7					PS
2044	F. Yergeau	UTF-8, a transformation format of Unicode and ISO 10646	10/30/1996	6			2279		I
2045	N. Freed, N. Borenstein	Multipurpose Internet Mail Extensions (MIME) Part One: Format of Internet Message Bodies	12/02/96	31	2184, 2231			1521, 1522, 1590	DS
2046	N. Freed, N. Borenstein	Multipurpose Internet Mail Extensions (MIME) Part Two: Media Types	12/02/96	44	2646			1521, 1522, 1590	DS
2047	K. Moore	MIME (Multipurpose Internet Mail Extensions) Part Three: Message Header Extensions for Non-ASCII Text	12/02/96	15	2184, 2231			1521, 1522, 1590	DS
2048	N. Freed, J. Klensin J. Postel	Multipurpose Internet Mail Extensions (MIME) Part Four: Registration Procedures	11/1996					1521, 1522, 1590	BCP0013
2049	N. Freed, N. Borenstein	Multipurpose Internet Mail Extensions (MIME) Part Five: Conformance Criteria and Examples	12/02/96	24				1521, 1522, 1590	DS
2050	K. Hubbard, M. Kosters, D. Conrad, J. Postel, D. Karrenberg,	INTERNET REGISTRY IP ALLOCATION GUIDELINES	11/05/1996	13				1466	BCP0012
2051	M. Allen, B. Clouston, Z. Kielczewski, W. Kwan, R. Moore	Definitions of Managed Objects for APPC	10/30/1996	124					PS
2052	A. Gulbrandsen, P. Vixie	A DNS RR for specifying the location of services (DNS SRV)	10/31/1996	10			2782		E
2053	E. Der-Danieliantz	The AM (Armenia) Domain	10/31/1996	3					I
2054	B. Callaghan	WebNFS Client Specification	10/31/1996	16					I
2055	B. Callaghan	WebNFS Server Specification	10/31/1996	10					I
2056	R. Denenberg, J. Kunze, D. Lynch	Uniform Resource Locators for Z39.50	11/05/1996	7					PS
2057	S. Bradner	Source directed access control on the Internet	11/11/1996	20					I
2058	S. Willens, A. Rubens, W. Simpson, C. Rigney	Remote Authentication Dial In User Service (RADIUS)	1/3/1997	64			2138		PS
2059	C. Rigney	RADIUS Accounting	1/3/1997	25			2139		I
2060	M. Crispin	INTERNET MESSAGE ACCESS PROTOCOL - VERSION 4rev1	12/4/1996	82				1730	PS
2061	M. Crispin	IMAP4 COMPATIBILITY WITH IMAP2BIS	12/5/1996	3				1730	I
2062	M. Crispin	Internet Message Access Protocol - Obsolete Syntax	12/4/1996	8					I
2063	N. Brownlee, C. Mills, G. Ruth	Traffic Flow Measurement: Architecture	1/3/1997	37					E
2064	N. Brownlee	Traffic Flow Measurement: Meter MIB	1/3/1997	38			2720		E
2065	D. Eastlake, C. Kaufman	Domain Name System Security Extensions	1/3/1997	41		1034, 1035	2535		PS
2066	R. Gellens	TELNET CHARSET Option	1/3/1997	12					E
2067	J. Renwick	IP over HIPPI	1/3/1997	30				1374	DS
2068	R. Fielding, J. Gettys, J. Mogul, H. Frystyk, T. Berners-Lee	Hypertext Transfer Protocol -- HTTP/1.1	1/3/1997	162			2616		PS
2069	J. Franks, P. Hallam-Baker, J. Hostetler, P. A. Luotonen, E. L. Stewart	An Extension to HTTP: Digest Access Authentication	1/3/1997	18			2617		PS
2070	F. Yergeau, G. Nicol, G. Adams, M. Duerst	Internationalization of the Hypertext Markup Language	1/6/1997	43					PS
2071	P. Ferguson, H. Berkowitz	Network Renumbering Overview: Why would I want it and what is it anyway?	1/8/1997	14					I
2072	H. Berkowitz	Router Renumbering Guide	1/8/1997	48					I
2073	Y. Rekhter, P. Lothberg, R. Hinden, S. Deering, J. Postel	An IPv6 Provider-Based Unicast Address Format	1/8/1997	7			2374		PS

RFC No.	Authors	Title	Date	Pgs	Updated By	Updates	Obsoleted By	Obsoletes	Notes / Status
2074	A. Bierman, R. Iddon	Remote Network Monitoring MIB Protocol Identifiers	1/19/1997	43					PS
2075	C. Partridge	IP Echo Host Service	1/8/1997	5					E
2076	J. Palme	Common Internet Message Headers	2/1997						I
2077	S. Nelson, C. Parks, Mitra	The Model Primary Content Type for Multipurpose Internet Mail Extensions	1/10/1997	13					PS
2078	J. Linn	Generic Security Service Application Program Interface, Version 2	1/10/1997	85			2743	1508	PS
2079	M. Smith	Definition of X.500 Attribute Types and an Object Class to Hold Uniform Resource Identifiers (URIs)	1/10/1997	5					PS
2080	G. Malkin, R. Minnear	RIPng for IPv6	1/10/1997	19					I
2081	G. Malkin	RIPng Protocol Applicability Statement	1/10/1997	4					I
2082	F. Baker, R. Atkinson	RIP-2 MD5 Authentication	1/10/1997	12					PS
2083	T. Boutell	PNG (Portable Network Graphics) Specification Version 1.0	1/16/1997	102					I
2084	G. Bossert, S. Cooper, W. Drummond	Considerations for Web Transaction Security	1/22/1997	6					I
2085	M. Oehler, R. Glenn	HMAC-MD5 IP Authentication with Replay Prevention	2/1997						PS
2086	J. Myers	IMAP4 ACL extension	1/22/1997	8					PS
2087	J. Myers	IMAP4 QUOTA extension	1/22/1997	5					PS
2088	J. Myers	IMAP4 non-synchroniziong literals	1/22/1997	2					PS
2089	B. Wijnen, D. Levi	V2ToV1 Mapping SNMPv2 onto SNMPv1 within a bi-lingual SNMP agent	1/28/1997	12					I
2090	A. Emberson	TFTP Multicast Option	2/97	11857B					E
2091	G. Meyer, S. Sherry	Triggered Extensions to RIP to Support Demand Circuits	1/24/1997	22					PS
2092	S. Sherry, G. Meyer	Protocol Analysis for Triggered RIP	1/24/1997	6					I
2093	H. Harney, C. Muckenhirn	Group Key Management Protocol (GKMP) Specification	7/1997	48678B					E
2094	H. Harney, C. Muckenhirn	Group Key Management Protocol (GKMP) Architecture	7/1997	53097B					E
2095	J. Klensin, R. Catoe, P. Krumviede	IMAP/POP Authorize Extension for Simple Challenge/Response	1/1997	10446B			2195		PS
2096	F. Baker	IP Forwarding Table MIB	1/1997	35930B				1354	PS
2097	G. Pall	The PPP NetBIOS Frames Control Protocol (NBFCP)	1/1997	27104B					PS
2098	Y. Katsube, K. Nagami, H. Esaki	Toshiba's Router Architecture Extensions for ATM: Overview	2/1997	43622B					I
2099	J. Elliott	Request for Comments Summary RFC Numbers 2000-2099	3/1997	40763B					I
2100	J. Ashworth	The Naming of Hosts	4/1997	4077B					I
2101	B. Carpenter, J. Crowcroft, Y. Rekhter	IPv4 Address Behaviour Today	2/1997	31407B					I
2102	R. Ramanathan	Multicast Support for Nimrod: Requirements and Solution Approaches	2/1997	50963B					I
2103	R. Ramanathan	Mobility Support for Nimrod: Challenges and Solution Approaches	2/1997	41352B					I
2104	H. Krawczyk, M. Bellare, R. Canetti	HMAC: Keyed-Hashing for Message Authentication	2/1997	22297B					I
2105	Y. Rekhter, B. Davie, D. Katz, E. Rosen, G. Swallow	Cisco Systems' Tag Switching Architecture Overview	2/1997	33013B					I
2106	S. Chiang, J. Lee, H. Yasuda	Data Link Switching Remote Access Protocol	2/1997	40819B			2114		I
2107	K. Hamzeh	Ascend Tunnel Management Protocol - ATMP	2/1997	44300B					I
2108	K. de Graaf, D. Romascanu, D. McMaster, K. McCloghrie	Definitions of Managed Objects for IEEE 802.3 Repeater Devices using SMIv2	2/1997	166336B				1516	PS
2109	D. Kristol, L. Montulli	HTTP State Management Mechanism	2/1997	43469B					PS
2110	J. Palme, A. Hopmann	MIME E-mail Encapsulation of Aggregate Documents, such as HTML (MHTML)	3/1997	41961B			2557		PS
2111	E. Levinson	Content-ID and Message-ID Uniform Resource Locators	2/1997	9099B			2392		PS
2112	E. Levinson	The MIME Multipart/Related Content-type	2/1997	17052B			2387	1872	PS
2113	D. Katz	IP Router Alert Option	2/1997	7924B					PS

RFC No.	Authors	Title	Date	Pgs	Updated By	Updates	Obsoleted By	Obsoletes	Notes / Status
2114	S. Chiang, J. Lee, H. Yasuda	Data Link Switching Client Access Protocol	2/1997	50872B				2106	I
2115	C. Brown, F. Baker	Management Information Base for Frame Relay DTEs Using SMIv2	September 1997	59950B				1315	DS
2116	C. Apple, K. Rossen	X.500 Implementations Catalog-96	4/1997	243994 B				1632	I FYI0011
2117	D. Estrin, D. Farinacci, A. Helmy, D. Thaler, S. Deering, M. Handley, V. Jacobson, C. Liu, P. Sharma, L. Wei	Protocol Independent Multicast-Sparse Mode (PIM-SM): Protocol Specification	6/1997	151886 B			2362		E
2118	G. Pall	Microsoft Point-To-Point Compression (MPPC) Protocol	3/1997	17443B					I
2119	S. Bradner	Key words for use in RFCs to Indicate Requirement Levels	3/1997	4723B					BCP0014
2120	D. Chadwick	Managing the X.500 Root Naming Context	3/1997	30773B					E
2121	G. Armitage	Issues affecting MARS Cluster Size	3/1997	26781B					I
2122	D. Mavrakis, H. Layec, K. Kartmann	VEMMI URL Specification	3/1997	25043B					PS
2123	N. Brownlee	Traffic Flow Measurement: Experiences with NeTraMet	3/1997	81874B					I
2124	P. Amsden, J. Amweg, P. Calato, S. Bensley, G. Lyons	Cabletron's Light-weight Flow Admission Protocol Specification Version 1.0	3/1997	47912B					I
2125	C. Richards, K. Smith	The PPP Bandwidth Allocation Protocol (BAP) / The PPP Bandwidth Allocation Control Protocol (BACP)	3/1997	49213B					PS
2126	Y. Pouffary, A. Young	ISO Transport Service on top of TCP (ITOT)	3/1997	51032B		1006			PS
2127	G. Roeck	ISDN Management Information Base using SMIv2	3/1997	95994B					PS
2128	G. Roeck	Dial Control Management Information Base using SMIv2	3/1997	66153B					PS
2129	K. Nagami, Y. Katsube, Y. Shobatake, A. Mogi, S. Matsuzawa, T. Jinmei, H. Esaki	Toshiba's Flow Attribute Notification Protocol (FANP) Specification	4/1997	41137B					I
2130	C. Weider, C. Preston, K. Simonsen, H. Alvestrand, R. Atkinson, M. Crispin, P. Svanberg	The Report of the IAB Character Set Workshop held 29 February – 1 March, 1996	4/1997	63443B					I
2131	R. Droms	Dynamic Host Configuration Protocol	3/1997	113738 B				1541	DS
2132	S. Alexander, R. Droms	DHCP Options and BOOTP Vendor Extensions	3/1997	63670B				1533	DS
2133	R. Gilligan, S. Thomson, J. Bound, W. Stevens	Basic Socket Interface Extensions for IPv6	4/1997	69737B			2553		I
2134	ISOC Board of Trustees	Articles of Incorporation of Internet Society	4/1997	9131B					I
2135	ISOC Board of Trustees	Internet Society By-Laws	4/1997	20467B					I
2136	P. Vixie, Ed., S. Thomson, Y. Rekhter, J. Bound	Dynamic Updates in the Domain Name System (DNS UPDATE)	4/1997	56354B		1035			PS
2137	D. Eastlake	Secure Domain Name System Dynamic Update	4/1997	24824B		1035			PS
2138	C. Rigney, A. Rubens, W. Simpson, S. Willens	Remote Authentication Dial In User Service (RADIUS)	4/1997	120407 B				2058	PS
2139	C. Rigney	RADIUS Accounting	4/1997	44919B				2059	I

RFC No.	Authors	Title	Date	Pgs	Updated By	Updates	Obsoleted By	Obsoletes	Notes / Status
2140	J. Touch	TCP Control Block Interdependence	4/1997	26032B					I
2141	R. Moats	URN Syntax	5/1997	14077B					PS
2142	D. Crocker	Mailbox Names for Common Services, Roles and Functions	5/1997	12195B					PS
2143	B. Elliston	Encapsulating IP with the Small Computer System Interface	5/1997	10749B					E
2144	C. Adams	The CAST-128 Encryption Algorithm	5/1997	37532B					I
2145	J. C. Mogul, R. Fielding, J. Gettys, H. Frystyk	Use and Interpretation of HTTP Version Numbers	5/1997	13659B					I
2146	Federal Networking Council	U.S. Government Internet Domain Names	5/1997	26564B				1816	I
2147	D. Borman	TCP and UDP over IPv6 Jumbograms	5/1997	1883B			2675		PS
2148	H. Alvestrand, P. Jurg	Deployment of the Internet White Pages Service	September 1997	31539B					BCP0015
2149	R. Talpade, M. Ammar	Multicast Server Architectures for MARS-based ATM multicasting	5/1997	42007B					I
2150	J. Max, W. Stickle	Humanities and Arts: Sharing Center Stage on the Internet	October 1997	154037 B					I FYI0031
2151	G. Kessler, S. Shepard	A Primer On Internet and TCP/IP Tools and Utilities	6/1997	114130 B				1739	I FYI0030
2152	D. Goldsmith, M. Davis	UTF-7 A Mail-Safe Transformation Format of Unicode	5/1997	28065B				1642	I
2153	W. Simpson	PPP Vendor Extensions	5/1997	10780B		1661, 1962			I
2154	S. Murphy, M. Badger, B. Wellington	OSPF with Digital Signatures	6/1997	72701B					E
2155	B. Clouston, B. Moore	Definitions of Managed Objects for APPN using SMIv2	6/1997	213809 B			2455		PS
2156	S. Kille	MIXER (Mime Internet X.400 Enhanced Relay): Mapping between X.400 and RFC 822/MIME	1/1998	280385 B		0822		0987, 1026, 1138, 1148, 1327, 1495	PS
2157	H. Alvestran	Mapping between X.400 and RFC-822/MIME Message Bodies	1/1997	92554B					PS
2158	H. Alvestrand	X.400 Image Body Parts	1/1998	5547B					PS
2159	H. Alvestrand	A MIME Body Part for FAX	1/1998	11471B					PS
2160	H. Alvestrand	Carrying PostScript in X.400 and MIME	1/1998	7059B					PS
2161	H. Alvestrand	A MIME Body Part for ODA	1/1998	8009B					E
2162	C. Allocchio	MaXIM-11 - Mapping between X.400 / Internet mail and Mail-11 mail	1/1998	58553B				1405	E
2163	C. Allocchio	Using the Internet DNS to Distribute MIXER Conformant Global Address Mapping (MCGAM)	1/1998	58789B				1664	PS
2164	S. Kille	Use of an X.500/LDAP directory to support MIXER address mapping	1/1998	16701B				1838	PS
2165	J. Veizades, E. Guttman, C. Perkins, S. Kaplan	Service Location Protocol	6/1997	169889 B	2608, 2609				PS
2166	D. Bryant, P. Brittain	APPN Implementer's Workshop Closed Pages Document DLSw v2.0 Enhancements	6/1997	75527B					I
2167	S. Williamson, M. Kosters, D. Blacka, J. Singh, K. Zeilstra	Referral Whois (RWhois) Protocol V1.5	6/1997	136355 B				1714	I
2168	R. Daniel, M. Mealling	Resolution of Uniform Resource Identifiers using the Domain Name System	6/1997	46528B					E
2169	R. Daniel	A Trivial Convention for using HTTP in URN Resolution	6/1997	17763B					E
2170	W. Almesberger, J. Le Boudec, P. Oechslin	Application REQuested IP over ATM (AREQUIPA)	6/1997	22874B					I
2171	K. Murakami, M. Maruyama	MAPOS - Multiple Access Protocol over SONET/SDH Version 1	6/1997	17480B					I
2172	M. Maruyama, K. Murakami	MAPOS Version 1 Assigned Numbers	6/1997	4857B					I
2173	K. Murakami, M. Maruyama	A MAPOS version 1 Extension - Node Switch Protocol	6/1997	12251B					I
2174	K. Murakami, M. Maruyama	A MAPOS version 1 Extension - Switch-Switch Protocol	6/1997	47967B					I
2175	K. Murakami, M. Maruyama	MAPOS 16 - Multiple Access Protocol over SONET/SDH with 16 Bit Addressing	6/1997	11677B					I

RFC No.	Authors	Title	Date	Pgs	Updated By	Updates	Obsoleted By	Obsoletes	Notes / Status
2176	K. Murakami, M. Maruyama	IPv4 over MAPOS Version 1	6/1997	12305B					I
2177	B. Leiba	IMAP4 IDLE command	6/1997	6770B					PS
2178	J. Moy	OSPF Version 2	7/1997	495866 B			2328	1583	DS
2179	A. Gwinn	Network Security For Trade Shows	7/1997	20690B					I
2180	M. Gahrns	IMAP4 Multi-Accessed Mailbox Practice	7/1997	24750B					I
2181	R. Elz, R. Bush	Clarifications to the DNS Specification	7/1997	36989B	2535	1034, 1035, 1123			PS
2182	R. Elz, M. Patton R. Bush, S. Bradner,	Selection and Operation of Secondary DNS Servers	7/1997	27456B					BCP
2183	R. Troost, S. Dorner, K. Moore	Communicating Presentation Information in Internet Messages: The Content-Disposition Header Field	8/1997	23150B	2184, 2231	1806			PS
2184	N. Freed, K. Moore	MIME Parameter Value and Encoded Word Extensions: Character Sets, Languages, and Continuations	8/1997	17635B		2045, 2047, 2183	2231		PS
2185	R. Callon, D. Haskin	Routing Aspects of IPv6 Transition	9/1997	31281B					I
2186	D. Wessels, K. Claffy	Internet Cache Protocol (ICP), version 2	9/1997	18808B					I
2187	D. Wessels, K. Claffy	Application of Internet Cache Protocol (ICP), version 2	9/1997	51662B					I
2188	M. Banan, M. Taylor, J. Cheng	AT&T/Neda's Efficient Short Remote Operations (ESRO) Protocol Specification Version 1.2	9/1997	118374 B					I
2189	A. Ballardie	Core Based Trees (CBT version 2) Multicast Routing	9/1997	52043B					E
2190	C. Zhu	RTP Payload Format for H.263 Video Streams	9/1997	26409B					PS
2191	G. Armitage	VENUS - Very Extensive Non-Unicast Service	9/1997	31316B					I
2192	C. Newman	IMAP URL Scheme	9/1997	31426B					PS
2193	M. Gahrns	IMAP4 Mailbox Referrals	9/1997	16248B					PS
2194	B. Aboba, J. Lu, J. Alsop, J. Ding, W. Wang	Review of Roaming Implementations	9/1997	81533B					I
2195	J. Klensin, R. Catoe, P. Krumviede	IMAP/POP AUTHorize Extension for Simple Challenge/Response	9/1997	10468B		2095			PS
2196	B. Fraser	Site Security Handbook	9/1997	191772 B		1244			I FYI0008
2197	N. Freed	SMTP Service Extension for Command Pipelining	9/1997	15003B		1854			DS
2198	C. Perkins, I. Kouvelas, O. Hodson, V. Hardman, M. Handley, J.C. Bolot, A. Vega-Garcia, S. Fosse-Parisis	RTP Payload for Redundant Audio Data	9/1997	25166B					PS
2199	A. Ramos	Request for Comments Summary RFC Numbers 2100-2199	1/1998	47664B					I
2200	J. Postel	Internet Official Protocol Standards	6/1997	94506B			2300	2000	(STD1)
2201	A Ballardie	Core Based Trees (CBT) Multicast Routing Architecture	9/1997	38040B					E
2202	P. Cheng, R. Glenn	Test Cases for HMAC-MD5 and HMAC-SHA-1	9/1997	11945B					I
2203	M. Eisler, A. Chiu, L. Ling	RPCSEC_GSS Protocol Specification	9/1997	50937B					PS
2204	D. Nash	ODETTE File Transfer Protocol	9/1997	151857 B					I
2205	R. Braden, Ed., L. Zhang, S. Berson, S. Herzog, S. Jami	Resource ReSerVation Protocol (RSVP) -- Version 1 Functional Specification	9/1997	223974 B	2750				PS
2206	F. Baker, J. Krawczyk, A. Sastry	RSVP Management Information Base using SMIv2	9/1997	112937 B					PS
2207	L. Berger, T. O'Malley	RSVP Extensions for IPSEC Data Flows	9/1997	30473B					PS
2208	A. Mankin, Ed., F. Baker, B. Braden, S. Bradner, M. O'Dell, A. Romanow, A. Weinrib, L. Zhang	Resource ReSerVation Protocol (RSVP) -- Version 1 Applicability Statement Some Guidelines on Deployment	9/1997	14289B					I

RFC No.	Authors	Title	Date	Pgs	Updated By	Updates	Obsoleted By	Obsoletes	Notes / Status
2209	R. Braden, L. Zhang	Resource ReSerVation Protocol (RSVP) -- Version 1 Message Processing Rules	9/1997	51690B					I
2210	J. Wroclawski	The Use of RSVP with IETF Integrated Services	9/1997	77613B					PS
2211	J. Wroclawski	Specification of the Controlled-Load Network Element Service	9/1997	46523B					PS
2212	S. Shenker, C. Partridge, R. Guerin	Specification of Guaranteed Quality of Service	9/1997	52330B					PS
2213	F. Baker, J. Krawczyk, A. Sastry	Integrated Services Management Information Base using SMIv2	9/1997	36147B					PS
2214	F. Baker, J. Krawczyk, A. Sastry	Integrated Services Management Information Base Guaranteed Service Extensions using SMIv2	9/1997	15971B					PS
2215	S. Shenker, J. Wroclawski	General Characterization Parameters for Integrated Service Network Elements	9/1997	39552B					PS
2216	S. Shenker, J. Wroclawski	Network Element Service Specification Template	9/1997	53655B					I
2217	G. Clark	Telnet Com Port Control Option	10/1997	31664B					E
2218	T. Genovese, B. Jennings	A Common Schema for the Internet White Pages Service	10/1997	16258B					PS
2219	M. Hamilton, R. Wright	Use of DNS Aliases for Network Services	10/1997	17858B					BCP0017
2220	R. Guenther	The Application/MARC Content-type	10/1997	7025B					I
2221	M. Gahrns	IMAP4 Login Referrals	10/1997	9251B					PS
2222	J. Myers	Simple Authentication and Security Layer (SASL)	10/1997	35010B	2444				PS
2223	J. Postel, J. Reynolds	Instructions to RFC Authors	10/1997	37948B				1543, 1111, 825	I
2224	B. Callaghan	NFS URL Scheme	10/1997	22726B					I
2225	M. Laubach, J. Halpern	Classical IP and ARP over ATM	4/1998	65779B				1626, 1577	PS
2226	T. Smith, G. Armitage	IP Broadcast over ATM Networks	10/1997	30661B					PS
2227	J. Mogul, P. Leach	Simple Hit-Metering and Usage-Limiting for HTTP	10/1997	85127B					PS
2228	M. Horowitz, S. Lunt	FTP Security Extensions	10/1997	58733B		0959			PS
2229	R. Faith, B. Martin	A Dictionary Server Protocol	10/1997	59551B					I
2230	R. Atkinson	Key Exchange Delegation Record for the DNS	10/1997	25563B					I
2231	N. Freed, K. Moore	MIME Parameter Value and Encoded Word Extensions: Character Sets, Languages, and Continuations	11/1997	19280B		2045, 2047, 2183		2184	PS
2232	B. Clouston, B. Moore	Definitions of Managed Objects for DLUR using SMIv2	11/1997	37955B					PS
2233	K. McCloghrie, F. Kastenholz	The Interfaces Group MIB using SMIv2	11/1997	148033 B				1573	PS
2234	D. Crocker, Ed., P. Overell	Augmented BNF for Syntax Specifications: ABNF	11/1997	24265B					PS
2235	R. Zakon	Hobbes' Internet Timeline	11/1997	43060B					I FYI0032
2236	W. Fenner	Internet Group Management Protocol, Version 2	11/1997	51048B		1112			PS
2237	K. Tamaru	Japanese Character Encoding for Internet Messages	11/1997	11628B					I
2238	B. Clouston, B. Moore	Definitions of Managed Objects for HPR using SMIv2	11/1997	65498B					PS
2239	K. de Graaf, D. Romascanu, D. McMaster, K. McCloghrie, S. Roberts	Definitions of Managed Objects for IEEE 802.3 Medium Attachment Units (MAUs) using SMIv2	11/1997	80651B			2668		PS
2240	O. Vaughan	A Legal Basis for Domain Name Allocation	11/1997	13602B			2352		I
2241	D. Provan	DHCP Options for Novell Directory Services	11/1997	8419B					PS
2242	R. Droms, K. Fong	NetWare/IP Domain Name and Information	11/1997	10653B					PS
2243	C. Metz	OTP Extended Responses	11/1997	19730B					PS
2244	C. Newman, J. G. Myers	ACAP -- Application Configuration Access Protocol	11/1997	154610 B					PS
2245	C. Newman	Anonymous SASL Mechanism	11/1997	9974B					PS
2246	T. Dierks, C. Allen	The TLS Protocol Version 1.0	1/1999	170401 B					PS

RFC No.	Authors	Title	Date	Pgs	Updated By	Updates	Obsoleted By	Obsoletes	Notes / Status
2247	S. Kille, M. Wahl, A. Grimstad, R. Huber, S. Sataluri	Using Domains in LDAP/X.500 Distinguished Names	1/1998	12411B					PS
2248	N. Freed, S. Kille	Network Services Monitoring MIB	1/1998	34106B				1565	PS
2249	N. Freed, S. Kille	Mail Monitoring MIB	1/1998	52334B				1566	PS
2250	D. Hoffman, G. Fernando, V. Goyal, M. Civanlar	RTP Payload Format for MPEG1/MPEG2 Video	1/1998	34293B				2038	PS
2251	M. Wahl, T. Howes, S. Kille	Lightweight Directory Access Protocol (v3)	12/1997	114488 B					PS
2252	M. Wahl, A. Coulbeck, T. Howes, S. Kille	Lightweight Directory Access Protocol (v3): Attribute Syntax Definitions	12/1997	60204B					PS
2253	M. Wahl, S. Kille, T. Howes	Lightweight Directory Access Protocol (v3): UTF-8 String Representation of Distinguished Names	12/1997	18226B				1779	PS
2254	T. Howes	The String Representation of LDAP Search Filters	12/1997	13511B				1960	PS
2255	T. Howes, M. Smith	The LDAP URL Format	12/1997	20685B				1959	PS
2256	M. Wahl	A Summary of the X.500(96) User Schema for use with LDAPv3	12/1997	32377B					PS
2257	M. Daniele, B. Wijnen, D. Francisco	Agent Extensibility (AgentX) Protocol Version 1	1/1998	177452 B			2741		PS
2258	J. Ordille	Internet Nomenclator Project	1/1998	34871B					I
2259	J. Elliott, J. Ordille	Simple Nomenclator Query Protocol (SNQP)	1/1998	58508B					I
2260	T. Bates, Y. Rekhter	Scalable Support for Multi-homed Multi-provider Connectivity	1/1998	28085B					I
2261	D. Harrington, R. Presuhn, B. Wijnen	An Architecture for Describing SNMP Management Frameworks	1/1998	128036 B			2271		PS
2262	J. Case, D. Harrington, R. Presuhn, B. Wijnen	Message Processing and Dispatching for the Simple Network Management Protocol (SNMP)	1/1998	88254B			2272		PS
2263	D. Levi, P. Meyer, B. Stewart	SNMPv3 Applications	1/1998	143493 B			2273		PS
2264	U. Blumenthal, B. Wijnen	User-based Security Model (USM) for version 3 of the Simple Network Management Protocol (SNMPv3)	1/1998	168759 B			2274		PS
2265	B. Wijnen, R. Presuhn, K. McCloghrie	View-based Access Control Model (VACM) for the Simple Network Management Protocol (SNMP)	1/1998	77807B			2275		PS
2266	J. Flick	Definitions of Managed Objects for IEEE 802.12 Repeater Devices	1/1998	134027 B					PS
2267	P. Ferguson, D. Senie	Network Ingress Filtering: Defeating Denial of Service Attacks which employ IP Source Address Spoofing	1/1998	21032B					I
2268	R. Rivest	A Description of the RC2(r) Encryption Algorithm	1/1998	19048B					I
2269	G. Armitage	Using the MARS Model in non-ATM NBMA Networks	1/1998	12094B					I
2270	J. Stewart, T. Bates, R. Chandra, E. Chen	Using a Dedicated AS for Sites Homed to a Single Provider	1/1998	12063B					I
2271	D. Harrington, R. Presuhn, B. Wijnen	An Architecture for Describing SNMP Management Frameworks	1/1998	128227 B			2571	2261	PS
2272	J. Case, B. Wijnen, D. Harrington, R. Presuhn	Message Processing and Dispatching for the Simple Network Management Protocol (SNMP)	1/1998	88445B			2572	2262	PS
2273	D. Levi, P. Meyer, B. Stewart	SNMPv3 Applications	1/1998	143754 B			2573	2263	PS
2274	U. Blumenthal, B. Wijnen	User-based Security Model (USM) for version 3 of the Simple Network Management Protocol (SNMPv3)	1/1998	168950 B			2574	2264	PS
2275	B. Wijnen, R. Presuhn, K. McCloghrie	View-based Access Control Model (VACM) for the Simple Network Management Protocol (SNMP)	1/1998	77998B			2575	2265	PS

RFC No.	Authors	Title	Date	Pgs	Updated By	Updates	Obsoleted By	Obsoletes	Notes / Status
2276	K. Sollins	Architectural Principles of Uniform Resource Name Resolution	1/1998	64811B					I
2277	H. Alvestrand	IETF Policy on Character Sets and Languages	1/1998	16622B					BCP0018
2278	N. Freed, J. Postel	IANA Charset Registration Procedures	1/1998	18881B					BCP0019
2279	F. Yergeau	UTF-8, a transformation format of ISO 10646	1/1998	21634B				2044	DS
2280	C. Alaettinoglu, T. Bates, E. Gerich, D. Karrenberg, D. Meyer, M. Terpstra, C. Villamizar	Routing Policy Specification Language (RPSL)	1/1998	114985 B			2622		PS
2281	T. Li, B. Cole, P. Morton, D. Li	Cisco Hot Standby Router Protocol (HSRP)	3/1998	35161B					I
2282	J. Galvin	IAB and IESG Selection, Confirmation, and Recall Process: Operation of the Nominating and Recall Committees	2/1998	29852B			2727	2027	I
2283	T. Bates, R. Chandra D. Katz, Y. Rekhter	Multiprotocol Extensions for BGP-4	2/1998	18946B					PS
2284	L. Blunk, J. Vollbrecht	PPP Extensible Authentication Protocol (EAP)	3/1998	29452B	2484				PS
2285	R. Mandeville	Benchmarking Terminology for LAN Switching Devices	2/1998	43130B					I
2286	J. Kapp	Test Cases for HMAC-RIPEMD160 and HMAC-RIPEMD128	2/1998	11849B					I
2287	C. Krupczak, J. Saperia	Definitions of System-Level Managed Objects for Applications	2/1998	98210B					PS
2288	C. Lynch, C. Preston, R. Daniel	Using Existing Bibliographic Identifiers as Uniform Resource Names	2/1998	21628B					I
2289	N. Haller, C. Metz, P. Nesser, M. Straw	A One-Time Password System	2/1998	56495B				1938	DS
2290	J. Solomon, S. Glass	Mobile-IPv4 Configuration Option for PPP IPCP	2/1998	39421B		2002			PS
2291	J. Slein, F. Vitali, E. Whitehead, D. Durand	Requirements for a Distributed Authoring and Versioning Protocol for the World Wide Web	2/1998	44036B					I
2292	W. Stevens, M. Thomas	Advanced Sockets API for IPv6	2/1998	152077 B					I
2293	S. Kille	Representing Tables and Subtrees in the X.500 Directory	3/1998	12539B			1837		PS
2294	S. Kille	Representing the O/R Address hierarchy in the X.500 Directory Information Tree	3/1998	22059B			1836		PS
2295	K. Holtman, A. Mutz	Transparent Content Negotiation in HTTP	3/1998	125130 B					E
2296	K. Holtman, A. Mutz	HTTP Remote Variant Selection Algorithm -- RVSA/1.0	3/1998	26932B					E
2297	P. Newman, W. Edwards, R. Hinden, E. Hoffman, F. Ching Liaw, T. Lyon, G. Minshall	Ipsilon's General Switch Management Protocol Specification Version 2.0	3/1998	280484 B	1987				I
2298	R. Fajman	An Extensible Message Format for Message Disposition Notifications	3/1998	62059B					PS
2299	A. Ramos	Request for Comments Summary	1/1999	51234B					I
2300	J. Postel	Internet Official Protocol Standards	5/1998	128322 B			2400	2200	(STD1)
2301	L. McIntyre, S. Zilles, R. Buckley, D. Venable, G. Parsons, J. Rafferty	File Format for Internet Fax	3/1998	200525 B					PS
2302	G. Parsons, J. Rafferty, S. Zilles	Tag Image File Format (TIFF) - image/tiff MIME Sub-type Registration	3/1998	14375B					PS
2303	C. Allocchio	Minimal PSTN address format in Internet Mail	3/1998	14625B					PS
2304	C. Allocchio	Minimal FAX address format in Internet Mail	3/1998	13236B					PS
2305	K. Toyoda, H. Ohno, J. Murai, D. Wing	A Simple Mode of Facsimile Using Internet Mail	3/1998	24624B					PS
2306	G. Parsons, J. Rafferty	Tag Image File Format (TIFF) - F Profile for Facsimile	3/1998	59358B					I

RFC No.	Authors	Title	Date	Pgs	Updated By	Updates	Obsoleted By	Obsoletes	Notes / Status
2307	L. Howard	An Approach for Using LDAP as a Network Information Service	3/1998	41396B					E
2308	M. Andrews	Negative Caching of DNS Queries (DNS NCACHE)	3/1998	41428B		1034, 1035		1034	PS
2309	B. Braden, D. Clark, J. Crowcroft, B. Davie, S. Deering, D. Estrin, S. Floyd, V. Jacobson, G. Minshall, C. Partridge, L. Peterson, K. Ramakrishnan, S. Shenker, J. Wroclawski, L. Zhang	Recommendations on Queue Management and Congestion Avoidance in the Internet	4/1998	38079B					I
2310	K. Holtman	The Safe Response Header Field	4/1998	8091B					E
2311	S. Dusse, P. Hoffman, B. Ramsdell, L. Lundblade, L. Repka	S/MIME Version 2 Message Specification	3/1998	70901B					I
2312	S. Dusse, P. Hoffman, B. Ramsdell, J. Weinstein	S/MIME Version 2 Certificate Handling	3/1998	39829B					I
2313	B. Kaliski	PKCS 1: RSA Encryption Version 1.5	3/1998	37777B			2437		I
2314	B. Kaliski	PKCS 10: Certification Request Syntax Version 1.5	3/1998	15814B					I
2315	B. Kaliski	PKCS 7: Cryptographic Message Syntax Version 1.5	3/1998	69679B					I
2316	S. Bellovin	Report of the IAB Security Architecture Workshop	4/1998	19733B					I
2317	H. Eidnes, G. de Groot, P. Vixie	Classless IN-ADDR.ARPA delegation	3/1998	17744B					BCP0020
2318	H. Lie, B. Bos, C. Lilley	The text/css Media Type	3/1998	7819B					I
2319	KOI8-U Working Group	Ukrainian Character Set KOI8-U	4/1998	18042B					I
2320	M. Greene, J. Luciani, K. White, T. Kuo	Definitions of Managed Objects for Classical IP and ARP Over ATM Using SMIv2 (IPOA-MIB)	4/1998	102116 B					PS
2321	A. Bressen	RITA -- The Reliable Internetwork Troubleshooting Agent	4/1998	12302B					I
2322	K. van den Hout, A. Koopal, R. van Mook	Management of IP numbers by peg-dhcp	4/1998	12665B					I
2323	A. Ramos	IETF Identification and Security Guidelines	4/1998	9257B					I
2324	L. Masinter	Hyper Text Coffee Pot Control Protocol (HTCPCP/1.0)	4/1998	19610B					I
2325	M. Slavitch	Definitions of Managed Objects for Drip-Type Heated Beverage Hardware Devices using SMIv2	4/1998	12726B					I
2326	H. Schulzrinne, A. Rao, R. Lanphier	Real Time Streaming Protocol (RTSP)	4/1998	195010 B					PS
2327	M. Handley, V. Jacobson	SDP: Session Description Protocol	4/1998	87096B					PS
2328	*J. Moy*	*OSPF Version 2*	*4/1998*	*447367 B*				*2178*	*STD54*
2329	J. Moy	OSPF Standardization Report	4/1998	15130B					I
2330	V. Paxson, G. Almes, J. Mahdavi, M. Mathis	Framework for IP Performance Metrics	5/1998	94387B					I
2331	M. Maher	ATM Signalling Support for IP over ATM - UNI Signalling 4.0 Update	4/1998	61119B					PS
2332	J. Luciani, D. Katz, D. Piscitello, B. Cole, N. Doraswamy	NBMA Next Hop Resolution Protocol (NHRP)	4/1998	126978 B					PS
2333	D. Cansever	NHRP Protocol Applicability Statement	4/1998	20164B					PS
2334	J. Luciani, G. Armitage, J. Halpern, N. Doraswamy	Server Cache Synchronization Protocol (SCSP)	4/1998	98161B					PS
2335	J. Luciani	A Distributed NHRP Service Using SCSP	4/1998	14007B					PS

RFC No.	Authors	Title	Date	Pgs	Updated By	Updates	Obsoleted By	Obsoletes	Notes / Status
2336	J. Luciani	Classical IP to NHRP Transition	7/1998	10500B					I
2337	D. Farinacci, D. Meyer, Y. Rekhter	Intra-LIS IP multicast among routers over ATM using Sparse Mode PIM	4/1998	16357B					E
2338	S. Knight, D. Weaver, D. Whipple, R. Hinden, D. Mitzel, P. Hunt, P. Higginson, M. Shand, A. Lindem	Virtual Router Redundancy Protocol	4/1998	59871B					PS
2339	The Internet Society, Sun Microsystems	An Agreement Between the Internet Society, the IETF, and Sun Microsystems, Inc in the matter of NFS V.4 Protocols	5/1998	5 10745B					I
2340	B. Jamoussi, D. Jamieson, D. Williston, S. Gabe	Nortel's Virtual Network Switching (VNS) Overview	5/1998	30731B					I
2341	A. Valencia, M. Littlewood, T. Kolar	Cisco Layer Two Forwarding (Protocol) "L2F"	5/1998	66592B					H
2342	M. Gahrns, C. Newman	IMAP4 Namespace	5/1998	19489B					PS
2343	M. Civanlar, G. Cash, B. Haskell	RTP Payload Format for Bundled MPEG	5/1998	16557B					E
2344	G. Montenegro	Reverse Tunneling for Mobile IP	5/1998	39468B					PS
2345	J. Klensin, T. Wolf, G. Oglesby	Domain Names and Company Name Retrieval	5/1998	29707B					E
2346	J. Palme	Making Postscript and PDF International	5/1998	12382B					I
2347	G. Malkin, A. Harkin	TFTP Option Extension	5/1998	13060B		1350		1782	DS
2348	G. Malkin, A. Harkin	TFTP Blocksize Option	5/1998	9515B		1350		1783	DS
2349	G. Malkin, A. Harkin	TFTP Timeout Interval and Transfer Size Options	5/1998	7848B		1350		1784	DS
2350	N. Brownlee, E. Guttman	Expectations for Computer Security Incident Response	6/1998	86545B					BCP0021
2351	A. Rober	Mapping of Airline Reservation, Ticketing, and Messaging Traffic over IP	5/1998	43440B					I
2352	O. Vaughan	A Convention For Using Legal Names as Domain Names	5/1998	16354B				2240	I
2353	G. Dudley	APPN/HPR in IP Networks APPN Implementers' Workshop Closed Pages Document	5/1998	116972 B					I
2354	C. Perkins, O. Hodson	Options for Repair of Streaming Media	6/1998	28876B					I
2355	B. Kelly	TN3270 Enhancements	6/1998	89394B				1647	DS
2356	G. Montenegro, V. Gupta	Sun's SKIP Firewall Traversal for Mobile IP	6/1998	53198B					I
2357	A. Mankin, A. Romanow, S. Bradner, V. Paxson	IETF Criteria for Evaluating Reliable Multicast Transport and Application Protocols	6/1998	24130B					I
2358	J. Flick, J. Johnson	Definitions of Managed Objects for the Ethernet-like Interface Types	6/1998	87891B		2665	1650		PS
2359	J. Myers	IMAP4 UIDPLUS extension	6/1998	10862B					PS
2360	G. Scott	Guide for Internet Standards Writers	6/1998	47280B					BCP0022
2361	E. Fleischman	WAVE and AVI Codec Registries	6/1998	97796B					I
2362	D. Estrin, L. Wei, D. Farinacci, A. Helmy, D. Thaler, S. Deering, M. Handley, V. Jacobson, C. Liu, P. Sharma	Protocol Independent Multicast-Sparse Mode (PIM-SM): Protocol Specification	6/1998	159833 B				2117	E
2363	G. Gross, M. Kaycee, A. Li, A. Malis, J. Stephens	PPP Over FUNI	7/1998	22576B					PS
2364	G. Gross, M. Kaycee, A. Li, A. Malis, J. Stephens	PPP Over AAL5	7/1998	23539B					PS
2365	D. Meyer	Administratively Scoped IP Multicast	7/1998	17770B					BCP0023
2366	C. Chung, M. Greene	Definitions of Managed Objects for Multicast over UNI 3.0/3.1 based ATM Networks	7/1998	134312 B		2417			PS
2367	D. McDonald, C. Metz, B. Phan	PF_KEY Key Management API, Version 2	7/1998	146754 B					I

RFC No.	Authors	Title	Date	Pgs	Updated By	Updates	Obsoleted By	Obsoletes	Notes / Status
2368	P. Hoffman, L. Masinter, J. Zawinski	The mail to URL scheme	7/1998	16502B		1738, 1808			PS
2369	G. Neufeld, J. Baer	The Use of URLs as Meta-Syntax for Core Mail List Commands and their Transport through Message Header Fields	7/1998	30853B					PS
2370	R. Coltun	The OSPF Opaque LSA Option	7/1998	15					PS RFC2328
2371	J. Lyon, K. Evans, J. Klein	Transaction Internet Protocol Version 3.0	7/1998	71399B					PS
2372	K. Evans, J. Klein, J. Lyon	Transaction Internet Protocol - Requirements and Supplemental Information	7/1998	53699B					I
2373	R. Hinden, S. Deering	IP Version 6 Addressing Architecture	7/1998	52526B				1884	PS
2374	R. Hinden, M. O'Dell, S. Deering	An IPv6 Aggregatable Global Unicast Address Format	7/1998	25068B				2073	PS
2375	R. Hinden, S. Deering	IPv6 Multicast Address Assignments	7/1998	14356B					I
2376	E. Whitehead, M. Murata	XML Media Types	7/1998	32143B					I
2377	A. Grimstad, R. Huber, S. Sataluri, M. Wahl	Naming Plan for Internet Directory-Enabled Applications	9/1998	38274B					I
2378	R. Hedberg, P. Pomes	The CCSO Nameserver (Ph) Architecture	8/1998	38960B					I
2379	L. Berger	RSVP over ATM Implementation Guidelines	8/1998	15174B					BCP0024
2380	L. Berger	RSVP over ATM Implementation Requirements	8/1998	31234B					PS
2381	M. Garrett, M. Borden	Interoperation of Controlled-Load Service and Guaranteed Service with ATM	8/1998	107299B					PS
2382	E. Crawley, L. Berger, S. Berson, F. Baker, M. Borden, J. Krawczyk	A Framework for Integrated Services and RSVP over ATM	8/1998	73865B					I
2383	M. Suzuki	ST2+ over ATM Protocol Specification - UNI 3.1 Version	8/1998	99889B					I
2384	R. Gellens	POP URL Scheme	8/1998	13649B					PS
2385	A. Heffernan	Protection of BGP Sessions via the TCP MD5 Signature Option	8/1998	12315B					PS
2386	E. Crawley, R. Nair, B. Rajagopalan, H. Sandick	A Framework for QoS-based Routing in the Internet	8/1998	93459B					I
2387	E. Levinson	The MIME Multipart/Related Content-type	8/1998	18864B				2112	P
2388	L. Masinter	Returning Values from Forms: multipart/form-data	8/1998	16531B					PS
2389	P. Hethmon, R. Elz	Feature negotiation mechanism for the File Transfer Protocol	8/1998	18536B					PS RFC0959
2390	T. Bradley, C. Brown, A. Malis	Inverse Address Resolution Protocol	8/1998	20849B				1293	DS
2391	P. Srisuresh, D. Gan	Load Sharing using IP Network Address Translation (LSNAT)	8/1998	44884B					I
2392	E. Levinson	Content-ID and Message-ID Uniform Resource Locators	8/1998	11141B				2111	PS
2393	A. Shacham, R. Monsour, R. Pereira, M. Thomas	IP Payload Compression Protocol (IPComp)	12/1998	20757B					PS
2394	R. Pereira	IP Payload Compression Using DEFLATE	12/1998	11053B					I
2395	R. Friend, R. Monsour	IP Payload Compression Using LZS	12/1998	14882B					I
2396	T. Berners-Lee, R. Fielding, L. Masinter	Uniform Resource Identifiers (URI): Generic Syntax	8/1998	83639B					DS
2397	L. Masinter	The "data" URL scheme	8/1998	9514B					PS
2398	S. Parker, C. Schmechel	Some Testing Tools for TCP Implementors	8/1998	24107B					I FYI0033
2399	A. Ramos	Request for Comments Summary	1/1999	45853B					I
2400	J. Postel, J. Reynolds	Internet Official Protocol Standards	9/1998	110969B			2500	2300	(STD1)
2401	S. Kent, R. Atkinson	Security Architecture for the Internet Protocol	11/1998	168162B				1825	PS

RFC No.	Authors	Title	Date	Pgs	Updated By	Updates	Obsoleted By	Obsoletes	Notes / Status
2402	S. Kent, R. Atkinson	IP Authentication Header	11/1998	52831B				1826	PS
2403	C. Madson, R. Glenn	The Use of HMAC-MD5-96 within ESP and AH	11/1998	13578B					PS
2404	C. Madson, R. Glenn	The Use of HMAC-SHA-1-96 within ESP and AH	11/1998	13089B					PS
2405	C. Madson, N. Doraswamy	The ESP DES-CBC Cipher Algorithm With Explicit IV	11/1998	20208B					PS
2406	S. Kent, R. Atkinson	IP Encapsulating Security Payload (ESP)	11/1998	54202B				1827	PS
2407	D. Piper	The Internet IP Security Domain of Interpretation for ISAKMP	11/1998	67878B					PS
2408	D. Maughan, M. Schertler, M. Schneider, J. Turner	Internet Security Association and Key Management Protocol (ISAKMP)	11/1998	209175 B					PS
2409	D. Harkins, D. Carrel	The Internet Key Exchange (IKE)	11/1998	94949B					PS
2410	R. Glenn, S. Kent	The NULL Encryption Algorithm and Its Use With IPsec	11/1998	11239B					PS
2411	R. Thayer, R. Glenn, N. Doraswamy	IP Security Document Roadmap	11/1998	22983B					I
2412	H. Orman	The OAKLEY Key Determination Protocol	11/1998	118649 B					I
2413	S. Weibel, J. Kunze, C. Lagoze, M. Wolf	Dublin Core Metadata for Resource Discovery	9/1998	15501B					I
2414	M. Allman, S. Floyd, C. Partridge	Increasing TCP's Initial Window	9/1998	32019B					E
2415	K. Poduri, K. Nichols	Simulation Studies of Increased Initial TCP Window Size	9/1998	24205B					I
2416	T. Shepard, C. Partridge	When TCP Starts Up With Four Packets Into Only Three Buffers	9/1998	12663B					I
2417	C. Chung, M. Greene	Definitions of Managed Objects for Multicast over UNI 3.0/3.1 based ATM Networks	9/1998	134862 B				2366	PS
2418	S. Bradner	IETF Working Group Guidelines and Procedures	9/1998	62857B				1603	BCP0025
2419	K. Sklower, G. Meyer	The PPP DES Encryption Protocol, Version 2 (DESE-bis)	9/1998	24414B				1969	PS
2420	H. Kummert	The PPP Triple-DES Encryption Protocol (3DESE)	9/1998	16729B					PS
2421	G. Vaudreuil, G. Parsons	Voice Profile for Internet Mail - version 2	9/1998	123663 B				1911	PS
2422	G. Vaudreuil, G. Parsons	Toll Quality Voice - 32 kbit/s ADPCM MIME Sub-type Registration	9/1998	10157B				1911	PS
2423	G. Vaudreuil, G. Parsons	VPIM Voice Message MIME Sub-type Registration	9/1998	10729B				1911	PS
2424	G. Vaudreuil, G. Parsons	Content Duration MIME Header Definition	9/1998	7116B					PS
2425	T. Howes, M. Smith, F. Dawson	A MIME Content-Type for Directory Information	9/1998	64478B					PS
2426	F. Dawson, T. Howes	vCard MIME Directory Profile	9/1998	74646B					PS
2427	*C. Brown, A. Malis*	*Multiprotocol Interconnect over Frame Relay*	*9/1998*	*74671B*				*1490, 1294*	*STD55*
2428	M. Allman, C. Metz, S. Ostermann	FTP Extensions for IPv6 and NATs	9/1998	16028B					PS
2429	C. Bormann, L. Cline, G. Deisher, T. Gardos, C. Maciocco, D. Newell, J. Ott, G. Sullivan, S. Wenger, C. Zhu	RTP Payload Format for the 1998 Version of ITU-T Rec. H.263 Video (H.263+)	10/1998	43166B					PS
2430	T. Li, Y. Rekhter	A Provider Architecture for Differentiated Services and Traffic Engineering (PASTE)	10/1998	40148B					I
2431	D. Tynan	RTP Payload Format for BT.656 Video Encoding	10/1998	22323B					PS
2432	K. Dubray	Terminology for IP Multicast Benchmarking	10/1998	29758B					I
2433	G. Zorn, S. Cobb	Microsoft PPP CHAP Extensions	10/1998	34502B					I
2434	T. Narten, H. Alvestrand	Guidelines for Writing an IANA Considerations Section in RFCs	10/1998	25092B					BCP026
2435	L. Berc, W. Fenner, R. Frederick, S. McCanne, P. Stewart	RTP Payload Format for JPEG-compressed Video	10/1998	54173B				2035	PS
2436	R. Brett, S. Bradner, G. Parsons	Collaboration between ISOC/IETF and ITU-T	10/1998	31154B					I
2437	B. Kaliski, J. Staddon	PKCS #1: RSA Cryptography Specifications Version 2.0	10/1998	73529B				2313	I

RFC No.	Authors	Title	Date	Pgs	Updated By	Updates	Obsoleted By	Obsoletes	Notes / Status
2438	M. O'Dell, H. Alvestrand, B. Wijnen, S. Bradner	Advancement of MIB specifications on the IETF Standards Track	10/1998	13633B					BCP0027
2439	C. Villamizar, R. Chandra, R. Govindan	BGP Route Flap Damping	11/1998	86376B					PS
2440	J. Callas, L. Donnerhacke, H. Finney, R. Thayer	OpenPGP Message Format	11/1998	141371 B					PS
2441	D. Cohen	Working with Jon, Tribute delivered at UCLA, October 30, 1998	11/1998	11992B					I
2442	N. Freed, D. Newman, J. Belissen, M. Hoy	The Batch SMTP Media Type	11/1998	18384B					I
2443	J. Luciani, A. Gallo	A Distributed MARS Service Using SCSP	11/1998	41451B					E
2444	C. Newman	The One-Time-Password SASL Mechanism	10/1998	13408B		2222			PS
2445	F. Dawson, D. Stenerson	Internet Calendaring and Scheduling Core Object Specification (iCalendar)	11/1998	291838 B					PS
2446	S. Silverberg, S. Mansour, F. Dawson, R. Hopson	iCalendar Transport-Independent Interoperability Protocol (iTIP) Scheduling Events, BusyTime, To-dos and Journal Entries	11/1998	225964 B					PS
2447	F. Dawson, S. Mansour, S. Silverberg	iCalendar Message-Based Interoperability Protocol (iMIP)	11/1998	33480B					PS
2448	M. Civanlar, G. Cash, B. Haskell	AT&T's Error Resilient Video Transmission Technique	11/1998	14655B					I
2449	R. Gellens, C. Newman, L. Lundblade	POP3 Extension Mechanism	11/1998	36017B		1939			PS
2450	R. Hinden	Proposed TLA and NLA Assignment Rule	12/1998	24486B					I
2451	R. Pereira, R. Adams	The ESP CBC-Mode Cipher Algorithms	11/1998	26400B					PS
2452	M. Daniele	IP Version 6 Management Information Base for the Transmission Control Protocol	12/1998	19066B					PS
2453	*G. Malkin*	*RIP Version 2*	*11/1998*	*98462B*				*1388, 1723*	*STD56*
2454	M. Daniele	IP Version 6 Management Information Base for the User Datagram Protocol	12/1998	15862B					PS
2455	B. Clouston, B. Moore	Definitions of Managed Objects for APPN	11/1998	251061 B				2155	PS
2456	B. Clouston, B. Moore	Definitions of Managed Objects for APPN TRAPS	11/1998	44023B					PS
2457	B. Clouston, B. Moore	Definitions of Managed Objects for Extended Border Node	11/1998	54380B					PS
2458	H. Lu, L. Conroy, M. Krishnaswamy, S. Bellovin, F. Burg, A. DeSimone, K. Tewani, P. Davidson, H. Schulzrinne, K. Vishwanathan	Toward the PSTN/Internet Inter-Networking-- Pre-PINT Implementations	11/1998	139100 B					I
2459	R. Housley, W. Ford, W. Polk, D. Solo	Internet X.509 Public Key Infrastructure Certificate and CRL Profile	1/1999	278438 B					PS
2460	S. Deering, R. Hinden	Internet Protocol, Version 6 (IPv6) Specification	12/1998	85490B				1883	DS
2461	T. Narten, E. Nordmark, W. Simpson	Neighbor Discovery for IP Version 6 (IPv6)	12/1998	222516 B				1970	DS
2462	S. Thomson, T. Narten	IPv6 Stateless Address Autoconfiguration	12/1998	61210B				1971	DS
2463	A. Conta, S. Deering	Internet Control Message Protocol (ICMPv6) for the Internet Protocol Version 6 (IPv6) Specification	12/1998	34190B				1885	DS
2464	M. Crawford	Transmission of IPv6 Packets over Ethernet Networks	12/1998	12725B				1972	PS
2465	D. Haskin, S. Onishi	Management Information Base for IP Ver 6: Textual Conventions and General Group	12/1998	77339B					PS

RFC No.	Authors	Title	Date	Pgs	Updated By	Updates	Obsoleted By	Obsoletes	Notes / Status
2466	D. Haskin, S. Onishi	Management Information Base for IP Version 6: ICMPv6 Group	12/1998	27547B					PS
2467	M. Crawford	Transmission of IPv6 Packets over FDDI Networks	12/1998	16028B				2019	PS
2468	V. Cerf	I REMEMBER IANA	10/1998	8543B					I
2469	T. Narten, C. Burton	A Caution On The Canonical Ordering Of Link-Layer Addresses	12/1998	9948B					I
2470	M. Crawford, T. Narten, S. Thomas	Transmission of IPv6 Packets over Token Ring Networks	12/1998	21677B					PS
2471	R. Hinden, R. Fink, J. Postel (dec'd)	IPv6 Testing Address Allocation	12/1998	8031B				1897	E
2472	D. Haskin, E. Allen	IP Version 6 over PPP	12/1998	29696B				2023	PS
2473	A. Conta, S. Deering	Generic Packet Tunneling in IPv6 Specification	12/1998	77956B					PS
2474	K. Nichols, S. Blake, F. Baker, D. Black	Definition of the Differentiated Services Field (DS Field) in the IPv4 and IPv6 Headers	12/1998	50576B				1455, 1349	PS
2475	S. Blake, D. Black, M. Carlson, E. Davies, Z. Wang, W. Weiss	An Architecture for Differentiated Service	12/1998	94786B					I
2476	R. Gellens, J. Klensin	Message Submission	12/1998	30050B					PS
2477	B. Aboba, G. Zorn	Criteria for Evaluating Roaming Protocols	12/1998	23530B					I
2478	E. Baize, D. Pinkas	The Simple and Protected GSS-API Negotiation Mechanism	12/1998	35581B					PS
2479	C. Adams	Independent Data Unit Protection Generic Security Service Application Program Interface (IDUP-GSS-API)	12/1998	156070 B					I
2480	N. Freed	Gateways and MIME Security Multiparts	1/1999	11751B					PS
2481	K. Ramakrishnan, S. Floyd	A Proposal to add Explicit Congestion Notification (ECN) to IP	1/1999	64559B					E
2482	K. Whistler, G. Adams	Language Tagging in Unicode Plain Text	1/1999	27800B					I
2483	M. Mealling, R. Daniel	URI Resolution Services Necessary for URN Resolution	1/1999	30518B					E
2484	G. Zorn	PPP LCP Internationalization Configuration Option	1/1999	8330B		2284, 1994, 1570			PS
2485	S. Drach	DHCP Option for The Open Group's User Authentication Protocol	1/1999	7205B					PS
2486	B. Aboba, M. Beadles	The Network Access Identifier	1/1999	14261B					PS
2487	P. Hoffman	SMTP Service Extension for Secure SMTP over TLS	1/1999	15120B					PS
2488	M. Allman, D. Glover, L. Sanchez	Enhancing TCP Over Satellite Channels using Standard Mechanisms	1/1999	47857B					BCP0028
2489	R. Droms	Procedure for Defining New DHCP Options	1/1999	10484B					BCP0029
2490	M. Pullen, R. Malghan, L. Lavu, G. Duan, J. Ma, H. Nah	A Simulation Model for IP Multicast with RSVP	1/1999	74936B					I
2491	G. Armitage, P. Schulter, M. Jork, G. Harter	IPv6 over Non-Broadcast Multiple Access (NBMA) networks	1/1999	100782 B					PS
2492	G. Armitage, P. Schulter, M. Jork	IPv6 over ATM Networks	1/1999	21199B					PS
2493	K. Tesink, Ed	Textual Conventions for MIB Modules Using Performance History Based on 15 Minute Intervals	1/1999	18749B					PS
2494	D. Fowler, Ed	Definitions of Managed Objects for the DS0 and DS0 Bundle Interface Type	1/1999	47208B					PS
2495	D. Fowler, Ed	Definitions of Managed Objects for the DS1, E1, DS2 and E2 Interface Types	1/1999	155560 B				1406	PS
2496	D. Fowler, Ed	Definitions of Managed Object for the DS3/E3 Interface Type	1/1999	124251 B				1407	PS
2497	I. Souvatzis	Transmission of IPv6 Packets over ARCnet Networks	1/1999	10304B					PS RFC1201

RFC No.	Authors	Title	Date	Pgs	Updated By	Updates	Obsoleted By	Obsoletes	Notes / Status
2498	J. Mahdavi, V. Paxson	IPPM Metrics for Measuring Connectivity	1/1999	17869B			2678		E
2499	A. Ramos	Request for Comments Summary	7/1999	43431B					I
2500	*J. Reynolds, R. Braden*	*Internet Official Protocol Standards*	*6/1999*	*78845B*				*2400*	*STD1*
2501	S. Corson, J. Macker	Mobile Ad hoc Networking (MANET): Routing Protocol Performance Issues and Evaluation Considerations	1/1999	28912B					I
2502	M. Pullen, M. Myjak, C. Bouwens	Limitations of Internet Protocol Suite for Distributed Simulation the Large Multicast Environment	2/1999	28190B					I
2503	R. Moulton, M. Needleman	MIME Types for Use with the ISO ILL Protocol	2/1999	9078B					I
2504	Guttman, L. Leong, G. Malkin	Users' Security Handbook	2/1999	74036B					I FYI0034
2505	G. Lindberg	Anti-Spam Recommendations for SMTP MTAs	2/1999	53597B					BCP0030
2506	K. Holtman, A. Mutz, T. Hardie	Media Feature Tag Registration Procedure	3/1999	24892B					BCP0031
2507	M. Degermark, B. Nordgren, S. Pink	IP Header Compression	2/1999	106292 B					PS
2508	S. Casner, V. Jacobson	Compressing IP/UDP/RTP Headers for Low-Speed Serial Links	2/1999	60474B					PS
2509	M. Engan, S. Casner, C. Bormann	IP Header Compression over PPP	2/1999	18676B					PS
2510	C. Adams, S. Farrell	Internet X.509 Public Key Infrastructure Certificate Management Protocols	3/1999	158178 B					PS
2511	M. Myers, C. Adams, D. Solo, D. Kemp	Internet X.509 Certificate Request Message Format	3/1999	48278B					PS
2512	K. McCloghrie, J. Heinanen, W. Greene, A. Prasad	Accounting Information for ATM Networks	2/1999	29119B					PS
2513	K. McCloghrie, J. Heinanen, W. Greene, A. Prasad	Managed Objects for Controlling the Collection and Storage of Accounting Information for Connection-Oriented Networks	2/1999	60789B					PS
2514	M. Noto, E. Spiegel, K. Tesink	Definitions of Textual Conventions and OBJECT-IDENTITIES for ATM Management	2/1999	37583B					PS
2515	K. Tesink, Ed	Definitions of Managed Objects for ATM Management	2/1999	179993 B				1695	PS
2516	L. Mamakos, K. Lidl, J. Evarts, D. Carrel, D. Simone, R. Wheeler	Method for Transmitting PPP Over Ethernet (PPPoE)	2/1999	32537B					I
2517	R. Moats, R. Huber	Building Directories from DNS: Experiences from WWWSeeker	2/1999	14001B					I
2518	Y. Goland, E. Whitehead, A. Faizi, S. Carter, D. Jensen	HTTP Extensions for Distributed Authoring -- WEBDAV	2/1999	202829 B					PS
2519	E. Chen, J. Stewart	A Framework for Inter-Domain Route Aggregation	2/1999	25394B					I
2520	J. Luciani, H. Suzuki, N. Doraswamy, D. Horton	NHRP with Mobile NHCs	2/1999	16763B					E
2521	P. Karn, W. Simpson	ICMP Security Failures Messages	3/1999	14637B					E
2522	P. Karn, W. Simpson	Photuris: Session-Key Management Protocol	3/1999	157224 B					E
2523	P. Karn, W. Simpson	Photuris: Extended Schemes and Attributes	3/1999	38166B					E
2524	M. Banan	Neda's Efficient Mail Submission and Delivery (EMSD) Protocol Specification Version 1.3	2/1999	153171 B					I
2525	V. Paxson, M Allman, S. Dawson, W. Fenner, J. Griner, I. Heavens, K. Lahey, J. Semke, B. Volz	Known TCP Implementation Problems	3/1999	137201 B					I
2526	D. Johnson, S. Deering	Reserved IPv6 Subnet Anycast Addresses	3/1999	14555B					PS
2527	S. Chokhani, W. Ford	Internet X.509 Public Key Infrastructure Certificate Policy and Certification Practices Framework	3/1999	91860B					I

RFC No.	Authors	Title	Date	Pgs	Updated By	Updates	Obsoleted By	Obsoletes	Notes / Status
2528	R. Housley, W. Polk	Internet X.509 Public Key Infrastructure Representation of Key Exchange Algorithm (KEA) Keys in Internet X.509 Public Key Infrastructure Certificates	3/1999	18273B					I
2529	B. Carpenter, C. Jung	Transmission of IPv6 over IPv4 Domains without Explicit Tunnels	3/1999	21049B					PS
2530	D. Wing	Indicating Supported Media Features Using Extensions to DSN and MDN	3/1999	7824B					PS
2531	G. Klyne, L. McIntyre	Content Feature Schema for Internet Fax	3/1999	88295B					PS
2532	L. Masinter, D. Wing	Extended Facsimile Using Internet Mail	3/1999	22717B					PS
2533	G. Klyne	A Syntax for Describing Media Feature Sets	3/1999	79057B					PS
2534	L. Masinter, A. Mutz, D. Wing, K. Holtman	Media Features for Display, Print, and Fax	3/1999	15466B					PS
2535	D. Eastlake	Domain Name System Security Extensions	3/1999	110958 B		2181, 1035, 1034			PS
2536	D. Eastlake	DSA KEYs and SIGs in the Domain Name System (DNS)	3/1999	11121B					PS
2537	D. Eastlake	RSA/MD5 KEYs and SIGs in the Domain Name System (DNS)	3/1999	10810B					PS
2538	D. Eastlake, O. Gudmundsson	Storing Certificates in the Domain Name System (DNS)	3/1999	19857B					PS
2539	D. Eastlake	Storage of Diffie-Hellman Keys in the Domain Name System (DNS)	3/1999	21049B					PS
2540	D. Eastlake	Detached Domain Name System (DNS) Information	3/1999	12546B					E
2541	D. Eastlake	DNS Security Operational Considerations	3/1999	14498B					I
2542	L. Masinter	Terminology and Goals for Internet Fax	3/1999	46372B					I
2543	M. Handley, H. Schulzrinne, E. Schooler, J. Rosenberg	SIP: Session Initiation Protocol	3/1999	338861 B					PS
2544	S. Bradner, J. McQuaid	Benchmarking Methodology for Network Interconnect Devices	3/1999	66688B				1944	I
2545	P. Marques, F. Dupont	Use of BGP-4 Multiprotocol Extensions for IPv6 Inter-Domain Routing	3/1999	10209B					PS
2546	A. Durand, B. Buclin	6Bone Routing Practice	3/1999	17844B		2772			I
2547	E. Rosen, Y. Rekhter	BGP/MPLS VPNs	3/1999	63270B					I
2548	G. Zorn	Microsoft Vendor-specific RADIUS Attributes	3/1999	80763B					I
2549	D. Waitzman	IP over Avian Carriers with Quality of Service	4/1/1999	9519B		1149			I
2550	S. Glassman, M. Manasse, J. Mogul	Y10K and Beyond	4/1/1999	28011B					I
2551	S. Bradner	The Roman Standards Process --Revision III	4/1/1999	28054B					I
2552	M. Blinov, M. Bessonov, C. Clissmann	Architecture for the Information Brokerage in the ACTS Project GAIA	4/1999	65172B					I
2553	R. Gilligan, S. Thomson, J. Bound, W. Stevens	Basic Socket Interface Extensions for IPv6	3/1999	89215B				2133	I
2554	J. Myers	SMTP Service Extension for Authentication	3/1999	20534B					PS
2555	RFC Editor, et al.	30 Years of RFCs	4/7/1999	42902B					I
2556	S. Bradner	OSI connectionless transport services on top of UDP Applicability Statement for Historic Status	3/1999	5864B					I
2557	J. Palme, A. Hopmann, N. Shelness	MIME Encapsulation of Aggregate Documents, such as HTML (MHTML)	3/1999	61854B				2110	PS
2558	K. Tesink	Definitions of Managed Objects for the SONET/SDH Interface Type	3/1999	138550 B				1595	PS
2559	S. Boeyen, T. Howes, P. Richard	Internet X.509 Public Key Infrastructure Operational Protocols - LDAPv2	4/1999	22889B		1778			PS
2560	M. Myers, C. Adams, R. Ankney, A. Malpani, S. Galperin	X.509 Internet Public Key Infrastructure Online Certificate Status Protocol - OCSP	6/1999	43243B					PS
2561	K. White, R. Moore	Base Definitions of Managed Objects for TN3270E Using SMIv2	4/1999	113705 B					PS
2562	K. White, R. Moore	Definitions of Protocol and Managed Objects for TN3270E Response Time Collection Using SMIv2 (TN3270E-RT-MIB)	4/1999	110633 B					PS

RFC No.	Authors	Title	Date	Pgs	Updated By	Updates	Obsoleted By	Obsoletes	Notes / Status
2563	R. Troll	DHCP Option to Disable Stateless Auto-Configuration in IPv4 Clients	5/1999	17838B					PS
2564	C. Kalbfleisch, C. Krupczak, R. Presuhn, J. Saperia	Application Management MIB	5/1999	183314 B					PS
2565	R. Herriot, Ed., S. Butler, P. Moore, R. Turner	Internet Printing Protocol/1.0: Encoding and Transport	4/1999	80439B					E
2566	R. deBry, T. Hastings, R. Herriot, S. Isaacson, P. Powell	Internet Printing Protocol/1.0: Model and Semantics	4/1999	438887 B					E
2567	F. Wright	Design Goals for an Internet Printing Protocol	4/1999	90260B					E
2568	S. Zilles	Rationale for the Structure of the Model and Protocol for the Internet Printing Protocol	4/1999	23547B					E
2569	R. Herriot, Ed., T. Hastings, N. Jacobs, J. Martin	Mapping between LPD and IPP Protocols	4/1999	57886B					E
2570	J. Case, R. Mundy, D. Partain, B. Stewart	Introduction to Version 3 of the Internet-standard Network Management Framework	4/1999	50381B					I
2571	B. Wijnen, D. Harrington, R. Presuhn	An Architecture for Describing SNMP Management Frameworks	4/1999	139260 B				2271	DS
2572	J. Case, B. Wijnen, D. Harrington, R. Presuhn	Message Processing and Dispatching for the Simple Network Management Protocol (SNMP)	4/1999	96035B				2272	DS
2573	D. Levi, P. Meyer, B. Stewart	SNMP Applications	4/1999	150427 B				2273	DS
2574	U. Blumenthal, B. Wijnen	User-based Security Model (USM) for version 3 of the Simple Network Management Protocol (SNMPv3)	4/1999	190755 B				2274	DS
2575	B. Wijnen, R. Presuhn, K. McCloghrie	View-based Access Control Model (VACM) for the Simple Network Management Protocol (SNMP)	4/1999	79642B				2275	DS
2576	R. Frye, D. Levi, S. Routhier, B. Wijnen	Coexistence between Version 1, Version 2, and Version 3 of the Internet-standard Network Management Framework	03/2000	98589B				1908, 2089	PS
2577	M. Allman, S. Ostermann	FTP Security Considerations	5/1999	17870B					I
2578	*K. McCloghrie, D. Perkins, J. Schoenwaelder*	*Structure of Management Information Version 2 (SMIv2)*	*4/1999*	*89712B*				*1902*	*STD58*
2579	*K. McCloghrie, D. Perkins, J. Schoenwaelder*	*Textual Conventions for SMIv2*	*4/1999*	*59039B*				*1903*	*STD58*
2580	*K. McCloghrie, D. Perkins, J. Schoenwaelder*	*Conformance Statements for SMIv2*	*4/1999*	*54253B*				*1904*	*STD58*
2581	M. Allman, V. Paxson, W. Stevens	TCP Congestion Control	4/1999	31351B				2001	PD
2582	S. Floyd, T. Henderson	The New Reno Modification to TCP's Fast Recovery Algorithm	4/1999	29393B					E
2583	R. Carlson, L. Winkler	Guidelines for Next Hop Client (NHC) Developers	5/1999	21338B					I
2584	B. Clouston, B. Moore	Definitions of Managed Objects for APPN/HPR in IP Networks	5/1999	40187B					PS
2585	R. Housley, P. Hoffman	Internet X.509 Public Key Infrastructure Operational Protocols: FTP and HTTP	5/1999	14813B					PS
2586	J. Salsman, H. Alvestrand	The Audio/L16 MIME content type	5/1999	8694B					I
2587	S. Boeyen, T. Howes, P. Richard	Internet X.509 Public Key Infrastructure LDAPv2 Schema	6/1999	15102B					PS
2588	R. Finlayson	IP Multicast and Firewalls	5/1999	28622B					I
2589	Y. Yaacovi, M. Wahl, T. Genovese	Lightweight Directory Access Protocol (v3): Extensions for Dynamic Directory Services	5/1999	26855B					PS
2590	A. Conta, A. Malis, M. Mueller	Transmission of IPv6 Packets over Frame Relay	5/1999	41817B					PS
2591	D. Levi, J. Schoenwaelder	Definitions of Managed Objects for Scheduling Management Operations	5/1999	52920B					PS

RFC No.	Authors	Title	Date	Pgs	Updated By	Updates	Obsoleted By	Obsoletes	Notes / Status
2592	D. Levi, J. Schoenwaelder	Definitions of Managed Objects for the Delegation of Management Script	5/1999	110629B					PS
2593	J. Schoenwaelder, J. Quittek	Script MIB Extensibility Protocol Version 1.0	5/1999	49663B					E
2594	H. Hazewinkel, C. Kalbfleisch, J. Schoenwaelder	Definitions of Managed Objects for WWW Services	5/1999	88876B					PS
2595	C. Newman	Using TLS with IMAP, POP3 and ACAP	6/1999	32440B					PS
2596	M. Wahl, T. Howes	Use of Language Codes in LDAP	5/1999	17413B					PS
2597	J. Heinanen, F. Baker, W. Weiss, J. Wroclawski	Assured Forwarding PHB Group	6/1999	24068B					PS
2598	V. Jacobson, K. Nichols, K. Poduri	An Expedited Forwarding PHB	6/1999	23656B					PS
2599	A. DeLaCruz	Request for Comments Summary RFC Numbers 2500-2599	3/2000	45845B					I
2600	J. Reynolds, R. Braden	Internet Official Protocol Standards	3/2000	86139B			2700	2500	(STD0001)
2601	M. Davison	ILMI-Based Server Discovery for ATMARP	6/1999	11820B					PS
2602	M. Davison	ILMI-Based Server Discovery for MARS	6/1999	12031B					PS
2603	M. Davison	ILMI-Based Server Discovery for NHRP	6/1999	11865B					PS
2604	R. Gellens	Wireless Device Configuration (OTASP/OTAPA) via ACAP	6/1999	65329B			2636		I
2605	G. Mansfield, S. Kille	Directory Server Monitoring MIB	6/1999	49166B				1567	PS
2606	D. Eastlake, A. Panitz	Reserved Top Level DNS Names	6/1999	8008B					BCP RFC2606
2607	B. Aboba, J. Vollbrecht	Proxy Chaining and Policy Implementation in Roaming	6/1999	33271B					I
2608	E. Guttman, C. Perkins, J. Veizades, M. Day	Service Location Protocol, Version 2	6/1999	129475B		2165			PS
2609	E. Guttman, C. Perkins, J. Kempf	Service Templates and Service: Schemes	6/1999	72842B		2165			PS
2610	C. Perkins, E. Guttman	DHCP Options for Service Location Protocol	6/1999	10859B					PS
2611	L. Daigle, D. van Gulik, R. Iannella, P. Falstrom	URN Namespace Definition Mechanisms	6/1999	26916B					BCP033
2612	C. Adams, J. Gilchrist	The CAST-256 Encryption Algorithm	6/1999	37468B					I
2613	R. Waterman, B. Lahaye, D. Romascanu, S. Waldbusser	Remote Network Monitoring MIB Extensions for Switched Networks Version 1.0	6/1999	88701B					PS
2614	J. Kempf, E. Guttman	An API for Service Location	6/1999	164002B '					I
2615	A. Malis, W. Simpson	PPP over SONET/SDH	6/1999	18708B				1619	PS
2616	R. Fielding, J. Gettys, J. Mogul, H. Frystyk, L. Masinter, P. Leach, T. Berners-Lee	Hypertext Transfer Protocol -- HTTP/1.1	6/1999	422317B				2068	DS
2617	J. Franks, P. Hallam-Baker, J. Hostetler, S. Lawrence, P. Leach, A. Luotonen, L. Stewart	HTTP Authentication: Basic and Digest Access Authentication	6/1999	77638B				2069	DS
2618	B. Aboba, G. Zorn	RADIUS Authentication Client MIB	6/1999	26889B					PS
2619	G. Zorn, B. Aboba	RADIUS Authentication Server MIB	6/1999	30464B					PS
2620	B. Aboba, G. Zorn	RADIUS Accounting Client MIB	6/1999	23960B					I
2621	G. Zorn, B. Aboba	RADIUS Accounting Server MIB	6/1999	27768B					I
2622	C. Alaettinoglu, C. Villamizar, E. Gerich, D. Kessens, D. Meyer, T. Bates, D. Karrenberg, M. Terpstra	Routing Policy Specification Language (RPSL)	6/1999	140811B				2280	PS
2623	M. Eisler	NFS Version 2 and Version 3 Security Issues and the NFS Protocol's Use of RPCSEC_GSS and Kerberos V5	6/1999	42521B					PS

RFC No.	Authors	Title	Date	Pgs	Updated By	Updates	Obsoleted By	Obsoletes	Notes / Status
2624	S. Shepler	NFS Version 4 Design Considerations	6/1999	52891B					I
2625	M. Rajagopal, R. Bhagwat, W. Rickard	IP and ARP over Fibre Channel	6/1999	137741B					PS
2626	P. Nesser II	The Internet and the Millennium Problem (Year 2000)	6/1999	547560B					I
2627	D. Wallner, E. Harder, R. Agee	Key Management for Multicast: Issues and Architectures	6/1999	59263B					I
2628	V. Smyslov	Simple Cryptographic Program Interface (Crypto API)	6/1999	60070B					I
2629	M. Rose	Writing I-Ds and RFCs using XML	6/1999	48677B					I
2630	R. Housley	Cryptographic Message Syntax	6/1999	128599B					PS
2631	E. Rescorla	Diffie-Hellman Key Agreement Method	6/1999	25932B					PS
2632	B. Ramsdell, Ed.	S/MIME Version 3 Certificate Handling	6/1999	27925B					PS
2633	B. Ramsdell, Ed.	S/MIME Version 3 Message Specification	6/1999	67870B					PS
2634	P. Hoffman, Ed.	Enhanced Security Services for S/MIME	6/1999	131153B					PS
2635	S. Hambridge, A. Lunde	DON'T SPEW A Set of Guidelines for Mass Unsolicited Mailings and Postings (spam*)	6/1999	44669B					I FYI0035
2636	R. Gellens	Wireless Device Configuration (OTASP/OTAPA) via ACAP	7/1999	65288,B				2604	I
2637	K. Hamzeh, G. Pall, W. Verthein, J. Taarud, W. Little, G. Zorn	Point-to-Point Tunneling Protocol	7/1999	132565B					I
2638	K. Nichols, V. Jacobson, L. Zhang	A Two-bit Differentiated Services Architecture for the Internet	7/1999	72785B					I
2639	T. Hastings, C. Manros	Internet Printing Protocol/1.0: Implementer's Guide	7/1999	145086B					I
2640	B. Curtin	Internationalization of the File Transfer Protocol	7/1999	57204B		959			PS
2641	D. Hamilton, D. Ruffen	Cabletron's VLAN Hello Protocol Specification Version 4	8/1999	34686B					I
2642	L. Kane	Cabletron's VLS Protocol Specification	8/1999	204347B					I
2643	D. Ruffen, T. Len, J. Yanacek	Cabletron's Secure Fast VLAN Operational Model	8/1999	121786B					I
2644	D. Senie	Changing the Default for Directed Broadcasts in Routers	8/1999	6820B		1812			BCP0034
2645	R. Gellens	ON-DEMAND MAIL RELAY (ODMR) SMTP with Dynamic IP Addresses	8/1999	16302B					PS
2646	R. Gellens	The Text/Plain Format Parameter	8/1999	29175B		2046			PS
2647	D. Newman	Benchmarking Terminology for Firewall Performance	8/1999	45374B					I
2648	R. Moats	A URN Namespace for IETF Documents	8/1999	46826B					I
2649	B. Greenblatt, P. Richard	An LDAP Control and Schema for Holding Operation Signatures	8/1999	20470B					E
2650	D. Meyer, J. Schmitz, C. Orange, M. Prior, C. Alaettinoglu	Using RPSL in Practice	8/1999	55272B					I
2651	J. Allen, M. Mealling	The Architecture of the Common Indexing Protocol (CIP)	8/1999	41933B					PS
2652	J. Allen, M. Mealling	MIME Object Definitions for the Common Indexing Protocol (CIP)	8/1999	42464B					PS
2653	J. Allen, P. Leach, R. Hedberg	CIP Transport Protocols	8/1999	22999B					PS
2654	R. Hedberg, B. Greenblatt, R. Moats, M. Wahl	A Tagged Index Object for use in the Common Indexing Protocol	8/1999	46739B					E
2655	T. Hardie, M. Bowman, D. Hardy, M. Schwartz, D. Wessels	CIP Index Object Format for SOIF Objects	8/1999	34285B					E
2656	T. Hardie	Registration Procedures for SOIF Template Types	8/1999	17409B					E
2657	R. Hedberg	LDAPv2 Client vs the Index Mesh	8/1999	1219251B					E
2658	K. McKay	RTP Payload Format for PureVoice™ Audio	8/1999	21895B					PS
2659	E. Rescorla, A. Schiffman	Security Extensions For HTML	8/1999	8134B					E

RFC No.	Authors	Title	Date	Pgs	Updated By	Updates	Obsoleted By	Obsoletes	Notes / Status
2660	E. Rescorla, A. Schiffman	The Secure HyperText Transfer Protocol	8/1999	95645B					E
2661	W. Townsley, A. Valencia, A. Rubens, G. Pall, G. Zorn, B. Palter	Layer Two Tunneling Protocol "L2TP"	8/1999	168150 B					PS
2662	G. Bathrick, F. Ly	Definitions of Managed Objects for the ADSL Lines	8/1999	247122 B					PS
2663	P. Srisuresh, M. Holdrege	IP Network Address Translator (NAT) Terminology and Considerations	8/1999	72265B					I
2664	R. Plzak, A. Wells, E. Krol	FYI on Questions and Answers - Answers to Commonly Asked "New Internet User" Questions	8/1999	23640B				1594	I FYI0004
2665	J. Flick, J. Johnson	Definitions of Managed Objects for the Ethernet-like Interface Types	8/1999	110038 B				2358	PS
2666	J. Flick	Definitions of Object Identifiers for Identifying Ethernet Chip Sets	8/1999	37699B					I
2667	D. Thaler	IP Tunnel MIB	8/1999	32770B					PS
2668	A. Smith, J. Flick, K. de Graaf, D. Romascanu, D. McMaster, K. McCloghrie, S. Roberts	Definitions of Managed Objects for IEEE 802.3 Medium Attachment Units (MAUs)	8/1999	121843 B				2239	PS
2669	M. St. Johns, Ed.	DOCSIS Cable Device MIB Cable Device Management Information Base for DOCSIS compliant Cable Modems and Cable Modem Termination Systems	8/1999	112880 B					PS
2670	M. St. Johns, Ed.	Radio Frequency (RF) Interface Management Information Base for MCNS/DOCSIS compliant RF interfaces	8/1999	141077 B					PS
2671	P. Vixie	Extension Mechanisms for DNS (EDNS0)	8/1999	15257B					PS
2672	M. Crawford	Non-Terminal DNS Name Redirection	8/1999	18321B					PS
2673	M. Crawford	Binary Labels in the Domain Name System	8/1999	12379B					PS
2674	E. Bell, A. Smith, P. Langille, A. Rijhsinghani, K. McCloghrie	Definitions of Managed Objects for Bridges with Traffic Classes, Multicast Filtering and Virtual LAN Extensions	8/1999	159971 B					PS
2675	D. Borman, S. Deering, R. Hinden	IPv6 Jumbograms	8/1999	17320B				2147	PS
2676	G. Apostolopoulos, S. Kama, D. Williams, R. Guerin, A. Orda, T. Przygienda	QoS Routing Mechanisms and OSPF Extensions	8/1999	124563 B					E
2677	M. Greene, J. Cucchiara, J. Luciani	Definitions of Managed Objects for the NBMA Next Hop Resolution Protocol (NHRP)	8/1999	129699 B					PS
2678	J. Mahdavi, V. Paxson	IPPM Metrics for Measuring Connectivity	9/1999	18087B				2498	PS
2679	G. Almes, S. Kalidindi, M. Zekauskas	A One-way Delay Metric for IPPM	9/1999	43542B					PS
2680	G. Almes, S. Kalidindi, M. Zekauskas	A One-way Packet Loss Metric for IPPM	9/1999	32266B					PS
2681	G. Almes, S. Kalidindi, M. Zekauskas	A Round-trip Delay Metric for IPPM	9/1999	44357B					PS
2682	I. Widjaja, A. Elwalid	Performance Issues in VC-Merge Capable ATM LSRs	9/1999	29491B					I
2683	B. Leiba	IMAP4 Implementation Recommendations	9/1999	56300B					I
2684	D. Grossman, J. Heinanen	Multiprotocol Encapsulation over ATM Adaptation Layer 5	9/1999	51390B				1483	PS
2685	B. Fox, B. Gleeson	Virtual Private Networks Identifier	9/1999	11168B					PS
2686	C. Bormann	The Multi-Class Extension to Multi-Link PPP	9/1999	24192B					PS
2687	C. Bormann	PPP in a Real-time Oriented HDLC-like Framing	9/1999	28699B					PS
2688	S. Jackowski, D. Putzolu, E. Crawley, B. Davie	Integrated Services Mappings for Low Speed Networks	9/1999	36685B					PS

RFC No.	Authors	Title	Date	Pgs	Updated By	Updates	Obsoleted By	Obsoletes	Notes / Status
2689	C. Bormann	Providing Integrated Services over Low-bitrate Links	9/1999	34345B					I
2690	S. Bradner	A Proposal for an MOU-Based ICANN Protocol Support Organization	9/1999	14221B					I
2691	S. Bradner	A Memorandum of Understanding for an ICANN Protocol Support Organization	9/1999	18940B					I
2692	C. Ellison	SPKI Requirements	9/1999	29569B					E
2693	C. Ellison, B. Frantz, B. Lampson, R. Rivest, B. Thomas, T. Ylonen	SPKI Certificate Theory	9/1999	96699B					E
2694	P. Srisuresh, G. Tsirtsis, P. Akkiraju, A. Heffernan	DNS extensions to Network Address Translators (DNS_ALG)	9/1999	67720B					I
2695	A. Chiu	Authentication Mechanisms for ONC RPC	9/1999	39286B					I
2696	C. Weider, A. Herron, A. Anantha, T. Howes	LDAP Control Extension for Simple Paged Results Manipulation	9/1999	12809B					I
2697	J. Heinanen, R. Guerin	A Single Rate Three Color Marker	9/1999	10309B					I
2698	J. Heinanen, R. Guerin	A Two Rate Three Color Marker	9/1999	9368B					I
2699	S. Ginoza	Request for Comments Summary RFC Numbers 2600-2699	5/2000	42462B					I
2700	J. Reynolds, R. Braden	Internet Official Protocol Standards	8/2000	90213B				2600, STD0001	STD0001
2701	G. Malkin	Nortel Networks Multi-link Multi-node PPP Bundle Discovery Protocol	9/1999	17571B					I
2702	D. Awduche, J. Malcolm, J. Agogbua, M. O'Dell, J. McManus	Requirements for Traffic Engineering Over MPLS	9/1999	68386B					I
2703	G. Klyne	Protocol-independent Content Negotiation Framework	9/1999	42071B					I
2704	M. Blaze, J. Feigenbaum, J. Ioannidis, A. Keromytis	The KeyNote Trust-Management System Version 2	9/1999	79998B					I
2705	M. Arango, A. Dugan, I. Elliott, C. Huitema, S. Pickett	Media Gateway Control Protocol (MGCP) Version 1.0	10/1999	304056 B					I
2706	D. Eastlake, T. Goldstein	ECML v1: Field Names for E-Commerce	10/1999	26135B					I
2707	R. Bergman, T. Hastings, S. Isaacson, H. Lewis	Job Monitoring MIB - V1.0	11/1999	255685 B					I
2708	R. Bergman	Job Submission Protocol Mapping Recommendations for the Job Monitoring MIB	11/1999	57489B					I
2709	P. Srisuresh	Security Model with Tunnel-mode IPsec for NAT Domains	10/1999	24552B					I
2710	S. Deering, W. Fenner, B. Haberman	Multicast Listener Discovery (MLD) for IPv6	10/1999	46838B					PS
2711	C. Partridge, A. Jackson	IPv6 Router Alert Option	10/1999	11973B					PS
2712	A. Medvinsky, M. Hur	Addition of Kerberos Cipher Suites to Transport Layer Security (TLS)	10/1999	13763B					PS
2713	V. Ryan, S. Seligman, R. Lee	Schema for Representing Java™ Objects in an LDAP Directory	10/1999	40745B					I
2714	V. Ryan, S. Seligman, R. Lee	Schema for Representing CORBA Object References in an LDAP Directory	10/1999	14709B					I
2715	D. Thaler	Interoperability Rules for Multicast Routing Protocols	10/1999	49638B					I
2716	B. Aboba, D. Simon	PPP EAP TLS Authentication Protocol	10/1999	50108B					E
2717	R. Petke, I. King	Registration Procedures for URL Scheme Names	11/1999	19780B					BCP0035
2718	L. Masinter, H. Alvestrand, D. Zigmond, R. Petke	Guidelines for new URL Schemes	11/1999	19208B					I

RFC No.	Authors	Title	Date	Pgs	Updated By	Updates	Obsoleted By	Obsoletes	Notes / Status
2719	L. Ong, I. Rytina, M. Garcia, H. Schwarzbauer, L. Coene, H. Lin, I. Juhasz, C. Sharp, M. Holdrege	Framework Architecture for Signaling Transport	10/1999	48646B					I
2720	N. Brownlee	Traffic Flow Measurement: Meter MIB	10/1999	103781 B				2064	PS
2721	N. Brownlee	RTFM: Applicability Statement	10/1999	21200B					I
2722	N. Brownlee, C. Mills, G. Ruth	Traffic Flow Measurement: Architecture	10/1999	114064 B					I
2723	N. Brownlee	SRL: A Language for Describing Traffic Flows and Specifying Actions for Flow Groups	10/1999	44406B					I
2724	S. Handelman, S. Stibler, N. Brownlee, G. Ruth	RTFM: New Attributes for Traffic Flow Measurement	10/1999	37951B					E
2725	C. Villamizar, C. Alaettinoglu, D. Meyer, S. Murphy	Routing Policy System Security	12/1999	101649 B					PS
2726	J. Zsako	PGP Authentication for RIPE Database Updates	12/1999	22594B					PS
2727	J. Galvin	IAB and IESG Selection, Confirmation, and Recall Process: Operation of the Nominating and Recall Committees	2/2000	33396B				2282	BCP0010
2728	R. Panabaker, S. Wegerif, D. Zigmond	The Transmission of IP Over the Vertical Blanking Interval of a Television Signal	11/1999	49099B					PS
2729	P. Bagnall, R. Briscoe, A. Poppitt	Taxonomy of Communication Requirements for Large-scale Multicast Applications	12/1999	53322B					I
2730	S. Hanna, B. Patel, M. Shah	Multicast Address Dynamic Client Allocation Protocol (MADCAP)	12/1999	120341 B					PS
2731	J. Kunze	Encoding Dublin Core Metadata in HTML	12/1999	42450B					I
2732	R. Hinden, B. Carpenter, L. Masinter	Format for Literal IPv6 Addresses in URL's	12/1999	7984B					PS
2733	J. Rosenberg, H. Schulzrinne	An RTP Payload Format for Generic Forward Error Correction	12/1999	53120B					PS
2734	P. Johansson	IPv4 over IEEE 1394	12/1999	69314B					PS
2735	B. Fox, B. Petri	NHRP Support for Virtual Private Networks	12/1999	26451B					PS
2736	M. Handley, C. Perkins	Guidelines for Writers of RTP Payload Format Specifications	12/1999	24143B					BCP0036
2737	K. McCloghrie, A. Bierman	Entity MIB (Version 2)	12/1999	125141 B				2037	PS
2738	G. Klyne	Corrections to "A Syntax for Describing Media Feature Sets"	12/1999	8353B					PS
2739	T. Small, D. Hennessy, F. Dawson	Calendar Attributes for vCard and LDAP	January 2000	25892B					PS
2740	R. Coltun, D. Ferguson, J. Moy	OSPF for IPv6.	December 1999	189810 B					PS
2741	M. Daniele, B. Wijnen, M. Ellison, D. Francisco	Agent Extensibility (AgentX) Protocol Version 1	January 2000	199867 B				2257	PS
2742	Agents. L. Heintz, S. Gudur, M. Ellison	Definitions of Managed Objects for Extensible SNMP	1/2000	36644B					PS
2743	J. Linn.	Generic Security Service Application Program Interface Version 2, Update 1	1/2000	229418 B				2078	PS
2744	J. Wray	Generic Security Service API Version 2 : C-bindings	1/2000	218572 B				1509	PS
2745	A. Terzis, B. Braden, S. Vincent, L. Zhang	RSVP Diagnostic Messages	1/2000	52256B					PS
2746	A. Terzis, J. Krawczyk, J. Wroclawski, L. Zhang	RSVP Operation Over IP Tunnels	1/2000	58094B (Status:					PS
2747	F. Baker, B. Lindell, M. Talwar	RSVP Cryptographic Authentication	1/2000	49477B					PS

RFC No.	Authors	Title	Date	Pgs	Updated By	Updates	Obsoleted By	Obsoletes	Notes / Status
2748	J. Boyle, R. Cohen, D. Durham, S. Herzog, R. Rajan, A. Sastry	The COPS (Common Open Policy Service) Protocol	1/2000	90906B					PS
2749	J. Boyle, R. Cohen, D. Durham, S. Herzog, R. Rajan, A. Sastry	COPS usage for RSVP	1/2000	33477B					PS
2750	S. Herzog	RSVP Extensions for Policy Control	1/2000	26379B		2205			PS
2751	S. Herzog	Signaled Preemption Priority Policy Element	1/2000	21451B					PS
2752	S. Yadav, R. Yavatkar, R. Pabbati, P. Ford, T. Moore, S. Herzog	Identity Representation for RSVP	1/2000	33954B					PS
2753	R. Yavatkar, D. Pendarakis, R. Guerin	A Framework for Policy-based Admission Control	1/2000	49763B					I
2754	C. Alaettinoglu, C. Villamizar, R. Govindan	RPS IANA Issues	1/2000	11582B					I
2755	A. Chiu, M. Eisler, B. Callaghan	Security Negotiation for WebNFS	1/2000	23493B					I
2756	P. Vixie, D. Wessels	Hyper Text Caching Protocol (HTCP/0.0)	1/2000	32176B					E
2757	G. Montenegro, S. Dawkins, M. Kojo, V. Magret, N. Vaidya	Long Thin Networks	1/2000	112988B					I
2758	K. White	Definitions of Managed Objects for Service Level Agreements Performance Monitoring	2/2000	145581B					E
2759	G. Zorn	Microsoft PPP CHAP Extensions, Version 2	1/2000	34178B					I
2760	M. Allman, S. Dawkins, D. Glover, J. Griner, D. Tran, T. Henderson, J. Heidemann, J. Touch, H. Kruse, S. Ostermann, K. Scott, J. Semke	Ongoing TCP Research Related to Satellites	2/2000	111141B					I
2761	J. Dunn, C. Martin	Terminology for ATM Benchmarking	2/2000	61219B					I
2762	J. Rosenberg, H. Schulzrinne	Sampling of the Group Membership in RTP	2/2000	25796B					E
2763	N. Shen, H. Smit	Dynamic Hostname Exchange Mechanism for IS-IS	2/2000	8593B					I
2764	B. Gleeson, A. Lin, J. Heinanen, G. Armitage, A. Malis	A Framework for IP Based Virtual Private Networks	2/2000	163215B					I
2765	E. Nordmark	Stateless IP/ICMP Translation Algorithm (SIIT)	2/2000	59465B					PS
2766	G. Tsirtsis, P. Srisuresh	Network Address Translation - Protocol Translation (NAT-PT)	2/2000	49836B					PS
2767	K. Tsuchiya, H. Higuchi, Y. Atarashi	Dual Stack Hosts using the "Bump-In-the-Stack" Technique (BIS)	2/2000	26402B					I
2768	B. Aiken, J. Strassner, B. Carpenter, I. Foster, C. Lynch, J. Mambretti, R. Moore, B. Teitelbaum	Network Policy and Services: A Report of a Workshop on Middleware	2/2000	81034B					I
2769	C. Villamizar, C. Alaettinoglu, R. Govindan, D. Meyer	Routing Policy System Replication	2/2000	95255B					PS
2770	D. Meyer, P. Lothberg	GLOP Addressing in 233/8	2/2000	8988B					E
2771	R. Finlayson	An Abstract API for Multicast Address Allocation	2/2000	20954B					I
2772	R. Rockell, R. Fink	6Bone Backbone Routing Guidelines	2/2000	28565B				2546	I

RFC No.	Authors	Title	Date	Pgs	Updated By	Updates	Obsoleted By	Obsoletes	Notes / Status
2773	R. Housley, P. Yee, W. Nace	Encryption using KEA and SKIPJACK	2/2000	20008B		959			E
2774	H. Nielsen, P. Leach, S. Lawrence	An HTTP Extension Framework	2/2000	39719B					E
2775	B. Carpenter	Internet Transparency	2/2000	42956B					I
2776	M. Handley, D. Thaler, R. Kermode	Multicast-Scope Zone Announcement Protocol (MZAP)	2/2000	61628B					PS
2777	D. Eastlake	Publicly Verifiable Nomcom Random Selection	2/2000	30064B					I
2778	M. Day, J. Rosenberg, H. Sugano	A Model for Presence and Instant Messaging	2/2000	35153B					I
2779	M. Day, S. Aggarwal, G. Mohr, J. Vincent	Instant Messaging / Presence Protocol Requirements	2/2000	47420B					I
2780	S. Bradner, V. Paxson	IANA Allocation Guidelines For Values In the Internet Protocol and Related Headers	2/2000	18954B					BCP0037
2781	P. Hoffman, F. Yergeau	UTF-16, an encoding of ISO 10646	2/2000	29870B					I
2782	A. Gulbrandsen, P. Vixie, L. Esibov	A DNS RR for specifying the location of services (DNS SRV)	2/2000	24013B				2052	PS
2783	J. Mogul, D. Mills, J. Brittenson, J. Stone, U. Windl	Pulse-Per-Second API for UNIX-like Operating Systems, Version 1.0	3/2000	61421B					I
2784	D. Farinacci, T. Li, S. Hanks, D. Meyer, P. Traina	Generic Routing Encapsulation (GRE)	3/2000	16627B					PS
2785	R. Zuccherato	Methods for Avoiding the "Small-Subgroup" Attacks on the Diffie-Hellman Key Agreement Method	3/2000	24415B					I
2786	M. St. Johns	Diffie-Helman USM Key Management Information Base and Textual Convention	3/2000						E
2787									
2788									
2789									
2790	S. Waldbusser, P. Grillo	Host Resources MIB	3/2000					1514	DS
2791									
2792	M. Blaze, J. Ioannidis, A. Keromytis	DSA and RSA Key and Signature Encoding for the KeyNote Trust Management System	3/2000	13461B					I

Notes / Status:

BcPn Best Current Practice number "n"

DS Draft Standard

E Experimental

FYI For Your Information

H Historical Interest

I Informational.

IENn Internet Engineering Note number "n"

PS Provisional Standard.

STDn Official standard, category "n." If in parenthesis the RFC has been obsoleted.

Appendix H
Record of Access Codes

Although everything one is taught about computer and communications systems security is violated by using the following chart, the fact of life is, it is the lesser of various evils in some cases.

Typically in a business environment, only one or two user names and associated passwords must be remembered. In the Internet environment, for example, one may accumulate 25 to 50 (or more) user name/password pairs for as many controlled access sites visited. This leaves the user with several choices:

- Use the same user name and password for all visited sites. Although it may be easy to remember only one set of access codes, it is a very bad idea because a breach of security at any site implies a breach at all sites.
- Use a different user name and/or password at every site. This is the best situation from a security point of view, but the worst from a human's point of view—remembering all of the different combinations can be tough, at best.

This is where the chart is used—record pertinent information about the sites.

REMEMBER, IF YOU CAN FIND AND READ THIS CHART, SO CAN SOMEONE ELSE!

Service Provider or Network Address	User name	Password	Access Number	Voice Number	Cost	Expiration Date	Notes

Index

About the Author

Frank Hargrave has been a self-employed engineer, technical writer, and part-time instructor at ECPI College of Technology since 1991. (ECPI is an abbreviation from Electronics, Computers, Programming, and Information Technologies). Prior to this he worked for several divisions of ITT. At ITT's Advanced Technology Center he developed several subsystems of their central office telephone system called System 12. At ITT Space Communications he was responsible for all digital systems design, for the design of analog line and trunk interface circuits, and was instrumental in developing their computer center. At Dynamics Instrumentation Corporation, he designed analog systems including active filters, precision amplifiers, and high efficiency power supplies. While at Dynamics he designed his first commercial computer—a 1024-bit core memory based satellite communications controller.

He has a Bachelor of Science in mathematics from California State University at Los Angeles and has completed a mini-graduate program in electronics at Stevens Institute in Hoboken, New Jersey. He has taken numerous computer courses at both the IBM and Digital Equipment Corporation schools.

He has written several technical monographs, and has been issued a dozen patents in the field of telecommunications.